Trigonometric Identities

Symmetry	$\sin(-\theta) = -\sin(\theta)$
	$\cos(-\theta) = \cos(\theta)$
Angle addition	$\sin(\theta + \phi) = \sin(\theta)\cos(\phi) + \cos(\theta)\sin(\phi)$
	$\cos(\theta + \phi) = \cos(\theta)\cos(\phi) - \sin(\theta)\sin(\phi)$
Double angle	$\sin(2\theta) = 2\sin(\theta)\cos(\theta)$
	$\cos(2\theta) = \cos^2(\theta) - \sin^2(\theta)$
Basic Identity	$\sin^2(\theta) + \cos^2(\theta) = 1$

Basic Discrete-Time Dynamical Systems

Description	Updating Function	Solution or Behavior
Population of size b_t with constant per capita production r	$b_{t+1} = rb_t$	Exponential growth or decay $b_t = r^t b_0$
Quantity h_t with constant growth by amount c	$h_{t+1} = h_t + c$	Linear growth or decline $h_t = h_0 + ct$
Concentration c_t with lung with fraction $q < 1$ exchanged and ambient concentration γ	$c_{t+1} = (1 - q)c_t + q\gamma$	Approaches equilibrium at $c^* = \gamma$
Selection model: Fraction of mutants p_t where mutant per capita production is s and wild-type per capita production is r	$p_{t+1} = \frac{sp_t}{sp_t + r(1 - p_t)}$	Approaches $p^* = 1$ if $s > r$ and $p^* = 0$ if $s < r$
Logistic model: Population, relative to a maximum of 1, of size x_t with per capita production r	$x_{t+1} = rx_t(1 - x_t)$	Approaches $x^* = 0$ if $r \leq 1$ Approaches positive stable equilibrium if $1 < r \leq 3$ No stable equilibrium if $3 < r \leq 4$

Basic Autonomous Differential Equations

Description	Equation	Solution or Behavior
Population of size $b(t)$ with constant per capita production rate λ	$\frac{db}{dt} = \lambda b$	Exponential growth or decay $b(t) = b(0)e^{\lambda t}$
Quantity $V(t)$ with constant increase or decrease at rate c	$\frac{dV}{dt} = c$	Linear growth or decline $V(t) = V(0) + ct$
Newton's Law of Cooling: The temperature $H(t)$ of an object with cooling rate α and ambient temperature A	$\frac{dH}{dt} = \alpha(A - H)$	Approaches equilibrium at $H^* = A$
Selection model: The proportion of mutants $p(t)$ where mutant per capita production rate is μ and wild-type per capita production rate is λ	$\frac{dp}{dt} = (\mu - \lambda)p(1 - p)$	Approaches $p^* = 1$ if $\mu > \lambda$ $p^* = 0$ if $\mu < \lambda$

Basic Theorems of Calculus

Intermediate Value Theorem: A continuous function f defined for $a \le x \le b$ takes on each value c between $f(a)$ and $f(b)$ at least once.

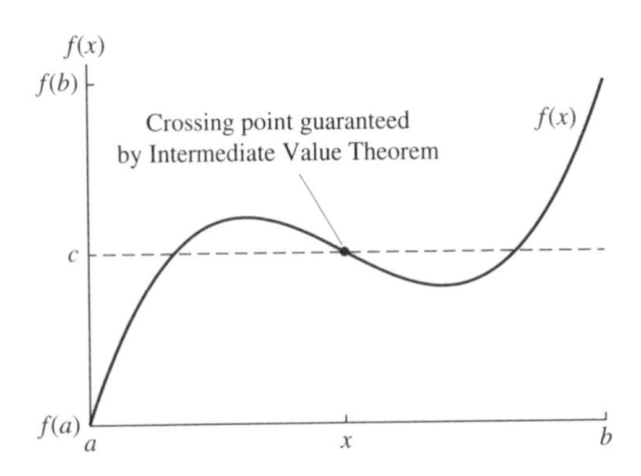

Extreme Value Theorem: A continuous function f defined for $a \le x \le b$ takes on its global maximum and minimum.

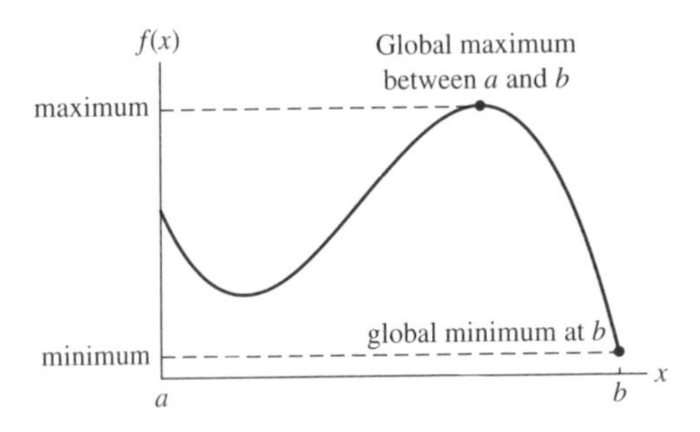

Mean Value Theorem: A differentiable function f defined for $a \le x \le b$ has a point c where $f'(c)$ is equal to the slope of the secant line connecting $(a, f(a))$ and $(b, f(b))$.

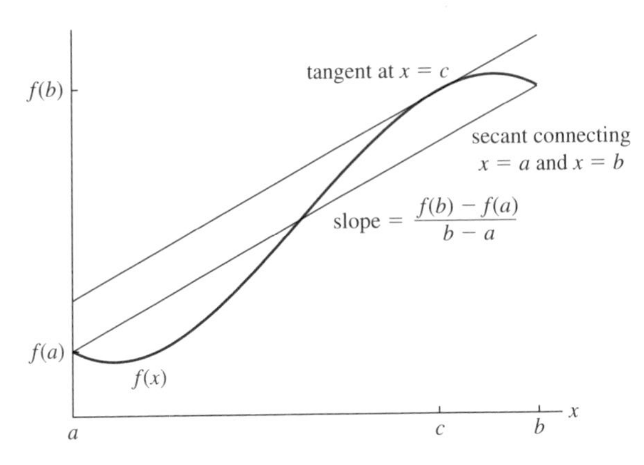

Fundamental Theorem of Calculus: If $F(x) = \int f(x)dx$ then $\int_a^b f(x)dx = F(b) - F(a) = F(x)\big|_a^b$.

Modeling the Dynamics of Life

Calculus and Probability for Life Scientists

Third Edition

This book is dedicated
to Anne, Claire, and Frank
for adding
biological dynamical chaos
to a life of
mathematical stable equilibrium

Modeling the Dynamics of Life

Calculus and Probability for Life Scientists

Third Edition

Frederick R. Adler
University of Utah

Australia • Brazil • Japan • Korea • Mexico • Singapore • Spain • United Kingdom • United States

Modeling the Dynamics of Life: Calculus and Probability for Life Scientists, Third Edition
Frederick R. Adler

Vice President, Editorial Director: P.J. Boardman

Publisher: Richard Stratton

Senior Sponsoring Editor: Molly Taylor

Assistant Editor: Shaylin Walsh Hogan

Editorial Assistant: Alexander Gontar

Associate Media Editor: Andrew Coppola

Senior Marketing Manager: Jennifer Pursley Jones

Marketing Coordinator: Michael Ledesma

Marketing Communications Manager: Mary Anne Payumo

Content Project Manager: Jill Quinn

Art Director: Linda May

Manufacturing Planner: Douglas Bertke

Rights Acquisition Specialist: Shalice Shah-Caldwell

Production Service: Cenveo Publisher Services

Cover Designer: Wing Ngan

Cover Image: "The Complete Encyclopedia of Illustration" by J.G. Heck

Photodisc Royalty Free image; Gettyimages.com

Compositor: Cenveo Publisher Services

Library of Congress Control Number:

ISBN-13: 978-0-8400-6418-9

ISBN-10: 0-8400-6418-7

Brooks/Cole
20 Channel Center Street
Boston, MA 02210
USA

Printed in the United States of America
1 2 3 4 5 6 7 15 14 13 12 11

Contents

Preface

Modeling the Dynamics of Life teaches calculus, probability, and statistics as a way of introducing freshman and sophomore life science majors to the insights mathematics can provide into biology. Why should there be a special book for this audience? Although the importance of quantitative skills in the life sciences is much discussed, current realities tend to conceal their vital role. Too often, biology is the natural science of last resort for students who believe "they aren't cut out for math." Most colleges and universities require little calculus for their biology majors, and those that require a full calculus course doubt its worth when students emerge unable to apply even precalculus mathematics in new contexts. Students are left with similar doubts when the techniques they learned for tests vanish as swiftly from the curriculum as from their memories. Students, biology faculty, and administrators see that biology is burgeoning as a science and as a major, apparently unhindered by pervasive mathematical illiteracy.

In fact, mathematics has played an important role in biology, providing the impetus for breakthroughs in areas including epidemiology, genetics, statistics, and physiology. As a theoretical biologist who uses mathematics to make sense of complex biological systems, I see this role expanding, not contracting. Although a great deal of biology can be done without any mathematics, the powerful new technologies that are transforming fields of biology from genetics and physiology to ecology are increasingly quantitative, as are many of the questions at the frontiers of knowledge. Along with genetics, mathematics is one of two unifying factors in the life sciences. And as biology becomes more important in society, mathematical literacy becomes as necessary for doctors, business people, lawyers, and art historians as for researchers.

Courses on calculus for life scientists serve many communities of students. One of the most important of these are those interested in biomedical sciences, and medical school in particular. Led by Harvard Medical School, medical school requirements are changing and becoming more in line with the content of this book. In particular, Harvard recommends study of calculus as part of a full-year curriculum that integrates statistics with the study of biologically relevant concepts. This book was originally designed to address the key needs of students entering the world of biology in this increasingly quantitative age, and it is heartening that the larger community is coming to appreciate the central role of modeling and interpretation.

Modeling

The central goal of this book is simple: to teach biology majors the mathematical ideas I use every day in my own research and in collaborations with my more empirical colleagues. These ideas are not specific techniques such as differentiation, but concepts of modeling. The skills include describing a system, translating appropriate aspects into equations, and interpreting the results in terms of the original problem. In this process, the science is central and solving the equations is less important.

Even students who will do little modeling on their own will be confronted by the models of others. This book teaches students how to *read* models. Like computers, mathematical models have an aura of authority to those who cannot understand them. As a modeler myself, I want my work to be read, understood, and challenged by knowledgeable scientists. This book aims to give students the skills and confidence to do so.

The Dynamics of Life

Mathematics helps unify biology by identifying dynamical principles that underlie a remarkable diversity of biological processes. This book follows three themes throughout: **growth**, **diffusion**, and **selection**. Each theme is studied in turn with three kinds of models: discrete-time dynamical systems (Chapters 1–3), differential equations (Chapters 4–5), and stochastic processes (Chapters 6–8). The concept of diffusion is treated as a discrete-time dynamical system describing a lung (Section 1.9), as a differential equation describing movement of a chemical across a membrane (Section 5.1), and as a Markov chain describing the random behavior of an individual molecule (Section 6.2). Addressing these themes in different ways shows students how different mathematical ideas describe and explain different facets of a biological process.

Graphical Methods

The focus on reasoning about what models **mean** favors graphical and computer techniques, the methods professionals use when equations cannot be solved. Understanding a discrete-time dynamical system with cobwebbing (Section 1.6) makes visual sense of such advanced topics as stability (Section 3.1) and chaos (Section 3.2). The phase line (Section 5.2) provides a visual presentation that summarizes the meaning of an autonomous differential equation, just as the phase plane (Section 5.6) summarizes a system of coupled differential equations. Approximation methods essential in more advanced applied mathematics are introduced through the method of leading behavior (Section 3.6) and its graphical interpretation.

Problems

More than 3000 problems have been broken into three categories: Mathematical Techniques, Applications, and Computer Exercises. The third edition, in addition to clarifying ambiguous problems and correcting errors in the solutions, includes about 100 new problems to cover new topics. They give students necessary practice in the mechanics of calculations as well as explorations of the art of modeling. Many problems build upon one another and are extended in later sections. To help students with the exercises, I have included many worked examples in the text. Each chapter includes supplementary problems, drawn from previous tests and practice tests. These introduce a variety of new applications and can be used for review and practice sessions.

Complete Solutions Manual

A complete solutions manual containing solutions to all odd- and even-numbered problems and to the supplementary problems (excluding computer problems and projects) is available.

Computer Problems

The book includes well over 100 exercises designed for exploration using a computer or graphing calculator, emphasizing visualization, experimentation, and simulation. Examples include

- Simulation of a simple linear discrete-time dynamical system describing harvesting and observation of threshold behavior to motivate the later discussion of unstable equilibria in Section 3.1 (Section 1.6)

- Exploration of the power of Fourier series through seeing how cosine functions can sum to an approximate square wave (Section 1.8)
- Experimentation with the approximate rate of change to motivate the subsequent derivation of the derivative of cosine (Section 2.1)
- Simulation of a population to show how much individual samples can differ from averages (Section 6.1).

The computer lab is an effective alternative learning environment for students who communicate well with machines, and an excellent place for students to work with one another and with instructors.

In-Depth Explorations of Particular Models

The website associated with the book includes several extended explorations. In association with Chapter 1, the power of cobwebbing is put to work to study the phenomenon of AV block in the heart. The optimization methods in Chapter 3 are used to fully analyze an extension of the basic lung model to study optimal breathing patterns. The phase plane methods of Chapter 5 are used to analyze the Fitzhugh-Nagumo equations describing a neuron. These extended developments show students how to combine a set of processes into a coherent whole, and how to use models to clarify hypotheses and answer specific questions.

Projects

Each chapter includes at least two projects. These are recommended for individual or group exploration. Examples include:

- Study of periodic hematopoiesis with a discrete-time dynamical system (Project 2.1)
- Experimentation with different numerical schemes for solving differential equations (Project 4.1)
- Careful study of models of adaptation by cells (Project 5.1)
- Development of the famous Luria-Delbruck fluctuation test (Project 7.2).

New Topics

Several new topics have been added to the third edition.

- Section 1.7 includes double-log graphs and an introduction to allometry, the study of power function relationships among biological measurements.
- Section 2.9 includes examples of implicit differentiation and related rates with both geometric and dynamical applications.
- Infinite series are introduced first in a new discussion of Taylor series in Section 3.7, and then studied more formally in the context of improper integrals in Section 4.7.
- Integration by partial fractions is introduced in Section 4.3 and used to solve the logistic differential equation in Section 5.4. Trigonometric substitutions are presented in a new series of exercises in Section 4.3.
- Computing volumes of solids of revolution is introduced in Section 4.6.

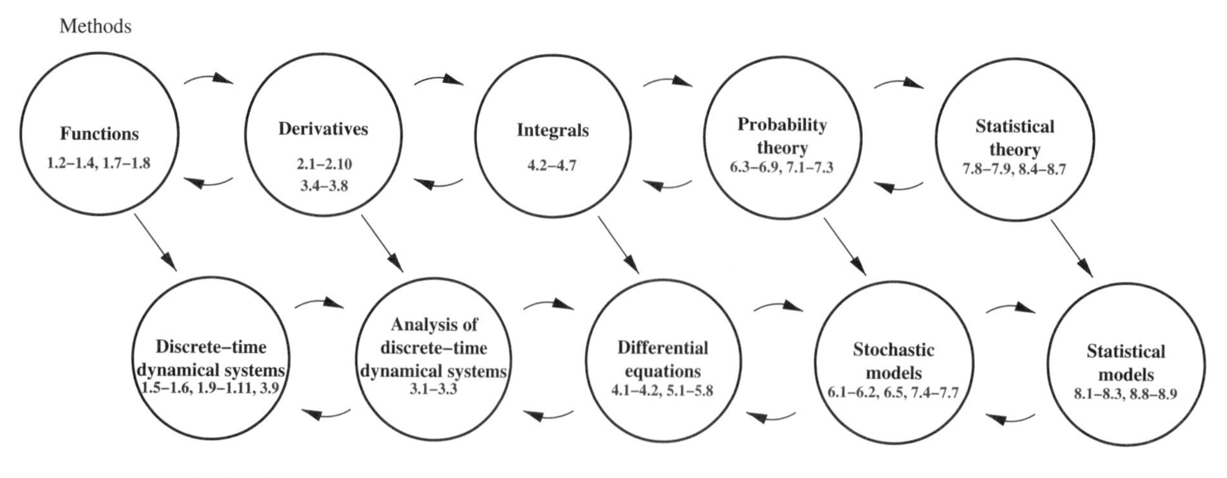

FIGURE

Logical structure of the book

Structure

The book is built upon a backbone of modeling concepts buttressed by a variety of mathematical techniques. At the advanced undergraduate or graduate level, a course can follow the modeling backbone and have students review or learn methods on their own. Such a course can emphasize research projects that combine dynamical and statistical modeling ideas.

The book can also be used for single-semester or single-quarter courses.

- A one-semester calculus course can emphasize less of the modeling material in Chapter 1, and cover Chapters 2–4 and Sections 5.1–5.4 in detail.
- A one-quarter calculus course can emphasize the methods sections in Chapter 1 (1.2–1.4, 1.7–1.8), all of Chapter 2, Sections 3.3–3.4, all of Chapter 4, and Sections 5.1–5.4.
- A one-semester probability and statistics course for students who have had calculus can cover Chapters 6–8, using background on discrete-time dynamical systems from Sections 1.2, 1.5, and 1.10, and a short review of differential equations from Sections 4.1 and 5.1.
- A one-quarter statistics course for students who have had calculus can focus on the more applied sections of Chapter 6 (6.1, 6.2, and 6.5) and Chapter 7 (7.3–7.9), along with all of Chapter 8.

Teaching

When teaching, I attempt to cover one section per 50-minute class period by emphasizing the central concept introduced at the beginning of the section, developing examples that parallel but are not identical to the examples in the book, and choosing exercises that focus on the key points for that section. Ideally, the class will engage in at least one small group exercise to solve a problem that extends the formal example, and back up for some metacognition to reflect upon how that material ties in with the larger themes and techniques of the course.

This approach can differ between the sections in the modeling backbone and those focusing on methods. For example, in the first explicit modeling section (Section 1.5), the class can be based on a model of something in the news, such as the budget deficit, by deriving an equation such as

$$D_{t+1} = 1.1D_t + 200$$

where D is measured in units of billions of dollars. This allows the students to graph the updating function, compose it with itself, work out and graph the inverse, change to units of millions of dollars, meet the concept of the solution, and even challenge themselves to look for a formula for the solution.

The methods sections require extra effort to maintain connection with the biological and dynamical themes. Ideally, as in Section 1.7 on exponential functions, we are motivated to learn these methods in order to solve problems that we care about. When does a population grow? How long does it take to double? When will two populations be equal? The rules of exponents and logs make these hard problems easier to solve.

Some of the methods are a few steps removed from their application. For example, Section 2.6 on the product and quotient rules can be brought to life with an example of where an animal increases in volume, perhaps with a function like

$$V(t) = t^2 + 3t,$$

but decreases in the concentration of some vital nutrient like iron according to

$$I(t) = 4 - t.$$

After using the derivative to understand each of these component functions, the class can work together to understand the total amount of the vital nutrient as the product $T(t) = V(t)I(t)$. Although these functions are somewhat artificial, their qualitative behavior is reasonable, and the consequences can be interpreted.

When possible, it is ideal to combine this in-class study with a computer lab. Given the multiplicity of platforms, the format of these labs is best left up to the instructor, but the most effective labs achieve several goals:

- Reinforce material from class, ideally through creation and interpretation of graphs
- Include a substantial amount of writing
- Encourage improvised collaboration among students, and between students and instructors
- Perhaps most importantly, encourage and reward **experimentation**, the art of using mathematics in combination with the computer as a way to just see what happens.

For example, in the context of the product rule, an instructor can construct complicated functions that capture strange behaviors over time (perhaps by combining polynomials with trigonometric functions) and have students interpret their graphs.

The labs also provide a place where students can work with data before studying the formal sections on statistics. It can be fun to have students collect their own data, such as simple time series on the growth of something they can observe, and come back to these data several times as they learn new methods.

Why Bother?

What are the benefits of this approach? The modeling approach is naturally a **problem-solving** approach. All instructors know that students will not remember every technique they have learned. This book emphasizes understanding what a model is and recognizing what models say. To be able to recognize a differential equation, interpret the terms, and use the solution is far more important than knowing how to find the solution. These reasoning skills, in addition to familiarity with models in general, are what stay with the motivated student and are what matter most in the end.

Most important, the course is fun to teach. Leading students through an integrated course for a full year removes the pressure for instant instructor gratification ("All of my students could take the derivatives of polynomials"). Instead, one can allow understanding to develop as concepts return for the second or third time. Students find this unsettling and yearn for instant gratification, too. But with time, they accept the challenge of thinking. When they begin to apply their new powers to their own problems, when they solve a problem on the computer without being told to, or when they teach me something about biology in the context of a mathematical idea, delayed gratification starts to feel like the best possible kind.

Acknowledgments

This book would never have been written without the support of a Hughes Foundation Grant to the University of Utah, which included as part of its mission an attempt to more effectively teach mathematics to biology majors. The grant was fathered by Gordon Lark, who has been a colleague, a mentor, and a mensch throughout this project. The grant brought together a committee of faculty to guide creation of this book and course: Aaron Fogelson, David Goldenberg, Jim Keener, Mark Lewis, David Mason, Larry Okun, Hans Othmer, Jon Seger, and Ryk Ward. Each added much to this work, generously providing ideas and corrections. Particular thanks to Jon "Preface" Seger and Mark "reasoning about functions" Lewis for discussion and ideas. Frank Wattenberg, Lou Gross, and Simon Levin looked over the book and delivered much-needed advice early in the writing process. Alan Rogers let me use his exercise style and Nelson Beebe helped smooth over many technical problems.

Thanks to Gary Ostedt for getting this project started, Bob Pirtle for carrying on with the Second Edition, Molly Taylor, Daniel Seibert, Jill Quinn, and Shaylin Walsh for their work to make the Third Edition happen, XX for the layout and production, Thanks to XX for cover design, and Anna Campbell Bliss for allowing me to use the beautiful image she created for the building where I am fortunate to work. I am particularly grateful to Professor Jerry Grossman for both teaching and humbling me with his amazing job of fact-checking, with his eagle eye for everything from details of layout and grammar through consistency of terminology to mathematical rigor and precision.

Many reviewers were instrumental in making this book comprehensible. Particular thanks to the class testers: Joe Mahaffy, Kathleen Crowe, Rollie Lamberson, Daniel Bentil, and the student reviewers at San Diego State. The following reviewers were honest in their criticism and generous with their expertise, and constantly provided needed perspective and ideas. I only failed to follow their advice when blinded by over-weaning pride, misguided philosophy, or general sloth. The remaining errors and oversights are entirely my own.

In addition to her extraordinary sartorial advice and culinary support, I thank Anne Collopy for her inspirational example of writing with clear transitions, extended metaphors, and elegant sentence structure. And for filling the work-free interstices of life with the same. Thanks to Claire and Frank for their joy in life and the many much-needed hugs and distractions. All of them bring to life Nietzsche's words: "A musician who *loves* the slow tempo will take the same pieces slower and slower. Thus there is no standstill in any love."

Frederick R. Adler
Salt Lake City, Utah

Chapter 1

Introduction to Discrete-Time Dynamical Systems

This chapter introduces the mathematical tools used to study biology: **functions** and **modeling**. Biological phenomena are described by **measurements**, a set of numerical values with units (like degrees or centimeters). Many **relations** between measurements are described by functions, which take one value as an input and return another as an output. We review the important functions used to describe biological systems: linear, trigonometric, power, logarithmic, and exponential functions.

Modeling is the art of taking a description of a biological phenomenon and converting it into mathematical form. Living things are characterized by change. One goal of modeling is to quantify these **dynamics** with an appropriate function. By using our understanding of the system and carefully following how a set of basic measurements change step by step, we will learn to derive an **updating function** that models the change in a **discrete-time dynamical system**. We will follow this process to derive models of bacterial population growth, gas exchange in the lung, and genetic change in a population of competing bacteria. We will develop a set of algebraic and graphical tools to deduce the dynamics that result from a particular discrete-time dynamical system.

Throughout this chapter, keep the following questions in mind.

- What biological process are we trying to describe?
- What biological questions do we seek to answer?
- What are the basic measurements and their units?
- What are the relationships between the basic measurements?
- What do results mean biologically?

1.1 Biology and Dynamics

Living systems, from cells to organisms to ecosystems, are characterized by change and dynamics. Living things grow, maintain themselves, and reproduce. Even remaining the same requires dynamic responses to a changing environment. Understanding the mechanisms behind these dynamics and deducing their consequences is crucial to understanding biology. This book uses a dynamical approach to address questions about biology.

This dynamical approach is necessarily mathematical because describing dynamics requires quantifying measurements. What is changing? How fast is it changing? What is it changing into?

In this book, we use the language of mathematics to quantitatively describe the working of living systems and develop the mathematical tools needed to compute how they change. From measurements describing the initial state of a system and a set of rules describing how change occurs, we will attempt to predict what will happen to the system. For example, given the position and velocity of a planet (the initial

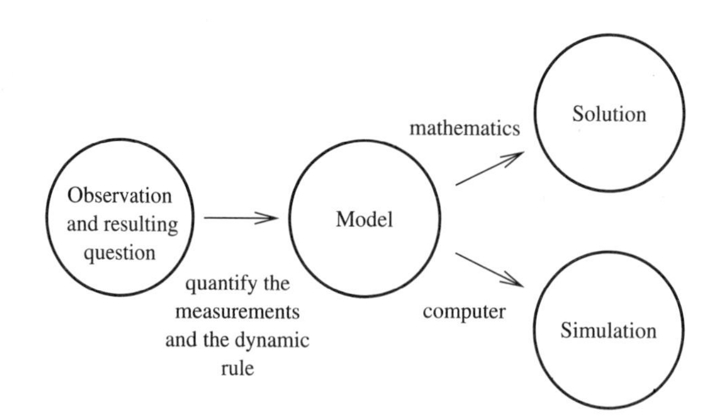

FIGURE 1.1.1

The workings of applied mathematics: the use of mathematics to answer scientific questions

state) and the laws of gravitation and inertia (the set of rules), Isaac Newton invented the mathematical methods of calculus to predict its position at any future time. This example illustrates the approach of **applied mathematics**, *the use of mathematics to answer scientific questions.* Applied mathematics begins with scientific observations and questions, perhaps about the position of a planet, which are then quantified into a **model**. When possible, mathematical methods alone can answer the question. In other cases, computers are used to **simulate** the process and find answers in particular cases, or to complement and check mathematical calculations (Figure 1.1.1).

The Steps in Applied Mathematics

Step	Definition
Quantify the basic measurements	The numerical values that describe the system
Describe a **dynamic rule**	A description of how the basic measurements change
Develop a **model**	A mathematical translation of the observations
Find a **solution**	Use of mathematical methods to predict behavior
Write a **simulation**	Use of a computer to predict behavior

This book is organized around three basic biological processes: **growth**, **maintenance**, and **replication**. Mathematical methods have contributed significantly to the understanding of each of these processes. After briefly describing these contributions, we outline the different types of models and mathematics to be used in this book.

1.1.1 Growth: Models of Malaria

Early in this century, Sir Ronald Ross discovered that malaria is transmitted by certain types of mosquito. Because the disease was (and remains) difficult to treat, one promising strategy for control seemed to be reduction of the number of mosquito. Many people thought that all the mosquitos would have to be killed to eradicate the disease. Because killing every single mosquito was impossible, it was feared that malaria might be impossible to control in this way.

Ross decided to use mathematics to convince people that mosquito control could be effective. The problem can be formulated dynamically as a problem in population growth. His first step was to **quantify the basic measurements**, in this case the numbers of people and mosquitos with and without malaria. The **dynamical rule** describes how these numbers change. Ross knew that an uninfected person can become infected upon being bitten by an infected mosquito, and that an uninfected mosquito can be infected when it bites an infected person (Figure 1.1.2). From these assumptions, he built a **mathematical model** describing the population dynamics of malaria. With this model he proved that the disease *could* be eradicated without killing every single mosquito (a simple version of this model will be studied in Section 5.5). We see evidence of this

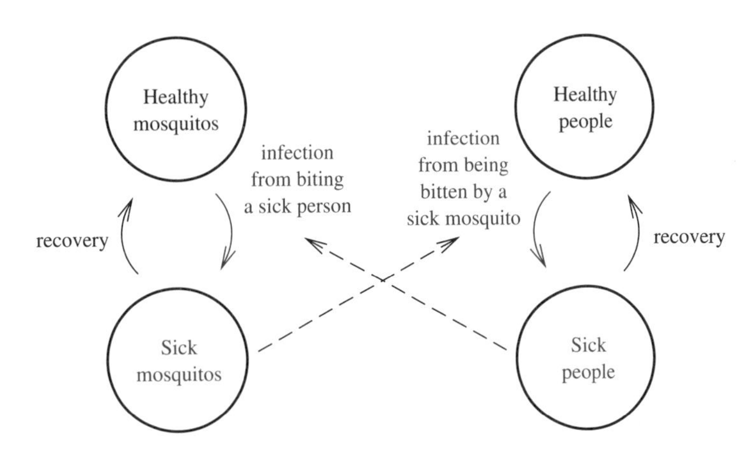

FIGURE 1.1.2
The dynamics of malaria

today in the United States, where malaria has been virtually eliminated although the mosquitos capable of transmitting the disease persist in many regions.

Many dynamic biological processes besides population dynamics are forms of growth. Growth in size is ubiquitous. One might use measurements of size (such as weight, height, or stomach volume) and a rule describing change in size (increase in weight due to large stomach volume) to predict the size of an organism over time. For example, one organism might add a constant amount to its weight every day, and another might add a constant fraction to its weight every day.

Organisms can also grow in complexity. A tree can add branches in addition to adding size. The quantitative description of the system might include the number of branches, their ages, sizes, and pattern. The dynamic rule might give the number of new branches produced each day, the probability that a given branch divides during the next month, or the rate at which new branches are formed. From the description and rule we could compute the number of branches as a function of age.

1.1.2 Maintenance: Models of Neurons

Neurons are cells that transmit information throughout the brain and body. Even the simplest neuron faces a challenging task. It must be able to amplify an appropriate incoming stimulus, transmit it to neighboring neurons, and then turn off and be ready for the next stimulus (Figure 1.1.3). This task is not as simple as it might seem. If we imagine the stimulus to be an input of electrical charge, a plausible sounding rule is "if electrical charge is raised above a certain level, increase it further." Such a rule works well for the first stimulus but provides no way for the cell to turn itself off. How does a neuron maintain functionality?

In the early 1950s, Alan Hodgkin and Andrew Huxley used their own measurements of neurons to develop a mathematical model of dynamics to explain the behavior of neurons. The idea, explained in detail in Section 5.8, is that the neuron has fast and slow mechanisms to open and close specialized ion channels in response to electrical charge. Hodgkin and Huxley measured the dynamic behavior of these channels and showed mathematically that their mechanism explained many aspects of the functioning of

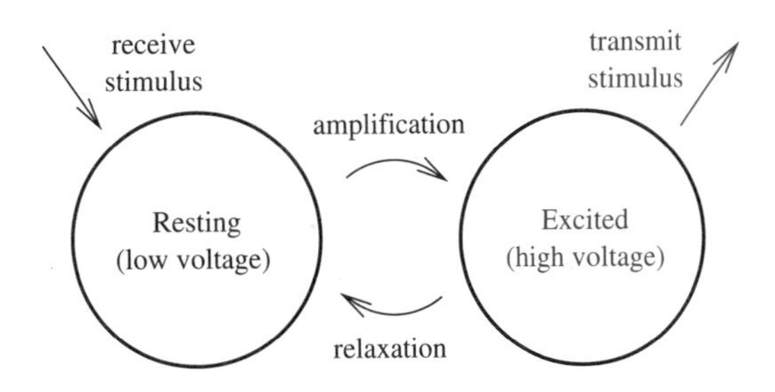

FIGURE 1.1.3
The dynamics of a neuron

neurons. They received the Nobel Prize in Physiology and Medicine for this work in 1963 and, perhaps more impressively, developed a model that is still used today to study neurons and other types of cells.

In general, maintenance of biological systems depends on preserving the distinction between inside and outside while maintaining flows of necessary materials from outside to inside and vice versa. The neuron maintains itself at a different electrical potential from the surrounding tissue in order to be able to respond, while remaining ready to exchange ions with the outside to create the response. As applied mathematicians, we **quantify the basic measurements**, the concentrations of various substances inside and outside the cell. The **dynamic rules** express how concentrations change, generally as a function of properties of the cell membrane. Most commonly, the rule describes the process of **diffusion**, movement of materials from regions of high concentration to regions of low concentration.

1.1.3 Replication: Models of Genetics

Although Mendel's work on genetics from the 1860s had been rediscovered around 1900, many biologists in the following decades remained unconvinced of his proposed mechanism of genetic transmission. In particular, it was unclear whether Darwin's theory of evolution by natural selection was consistent with this, or any other, proposed mechanism.

Working independently, biologists R. A. Fisher, J.B.S. Haldane, and Sewall Wright developed mathematical models of the dynamics of evolution in natural populations. These scientists **quantified the basic measurement**, in this case the number of individuals with a particular allele (a version of a gene). Their **dynamic rules** described how many individuals in a subsequent generation would have a particular allele as a function of numerous factors, including **selection** (differential success of particular types in reproducing), and **drift** (the workings of chance). They showed that Mendel's ideas were indeed consistent with observations of evolution. This work led to the development of methods of genetic analysis used to analyze DNA sequences today. We study a simple model of selection in Section 1.10 and examine some of the consequences of Mendel's laws in Section 6.2.

1.1.4 Types of Dynamical Systems

We will study each of the three processes, growth, maintenance, and replication, with three types of dynamical system, termed **discrete-time**, **continuous time**, and **probabilistic**. The first two types are **deterministic**, meaning that the dynamics includes no chance factors. In this case, the values of the basic measurements can be predicted exactly at all future times. Probabilistic dynamical systems include chance factors and values can be predicted only on average.

Discrete-time dynamical systems describe a sequence of measurements made at equally spaced intervals (Figure 1.1.4). These dynamical systems are described mathematically by a rule that gives the value at one time as a function of the value at the previous time. For example, a discrete-time dynamical system describing population growth is a rule that gives the population in one year as a function of the population in the previous year. A discrete-time dynamical system describing the concentration of oxygen in the lung is a rule that gives the concentration of oxygen in a lung after one breath as a function of the concentration after the previous breath. A discrete-time dynamical system describing the spread of a mutant allele is a rule that gives the number of mutant alleles in one generation as a function of the number in the previous generation. Mathematical analysis of the rule can make scientific predictions, such as the maximum population size, the average concentration of oxygen in the lung, or the final number of mutant alleles. The study of these systems requires the mathematical methods of modeling (Chapter 1) and **differential calculus** (Chapters 2 and 3).

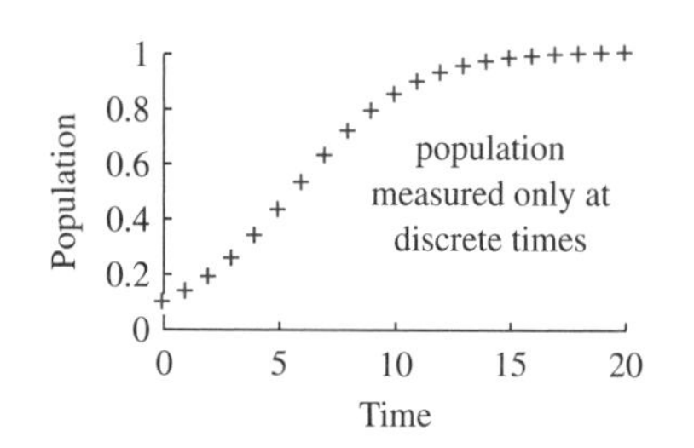

FIGURE 1.1.4

Measurements described by a discrete-time dynamical system

Continuous-Time Dynamical Systems

Continuous-time dynamical systems, usually called **differential equations**, describe measurements that are collected over an entire time interval (Figure 1.1.5). A differential equation consists of a rule that gives the **instantaneous rate of change** of a set of measurements. The miracle of differential equations is that information about a system at one time is sufficient to predict the state of a system at all future times. For example, a continuous-time dynamical system describing the growth of a population is a rule that gives the rate of change of population size as a function of the population size itself. The study of these systems requires the mathematical methods of **integral calculus** (Chapters 4 and 5).

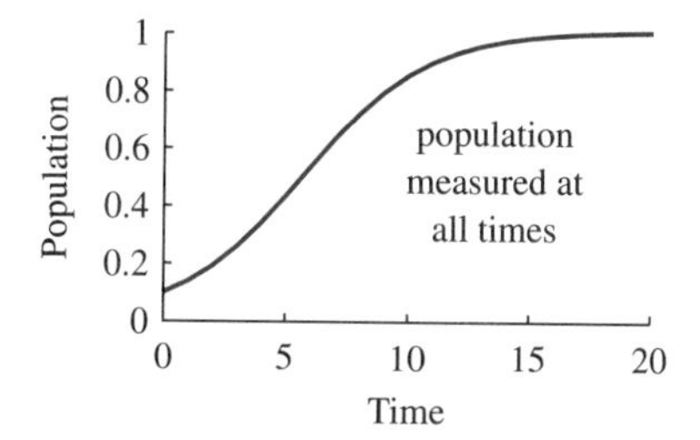

FIGURE 1.1.5

Measurements described by a continuous-time dynamical system

Probabilistic Dynamical Systems

Probabilistic dynamical systems describe measurements, in either discrete or continuous time, that are affected by random factors (Figure 1.1.6). The rule indicating how the measurements at one time depend on measurements at the previous time includes random factors. Rather than knowing with certainty the next measurements, we know only a set of possible outcomes and their associated probabilities and can therefore predict the outcome only in a probabilistic or statistical sense. For example, a probabilistic dynamical system describing population growth is a rule that gives the **probability** that a population has a particular size in one year as a function of the population in the previous year. The study of such systems requires the mathematical methods of **probability theory** (Chapters 6-7).

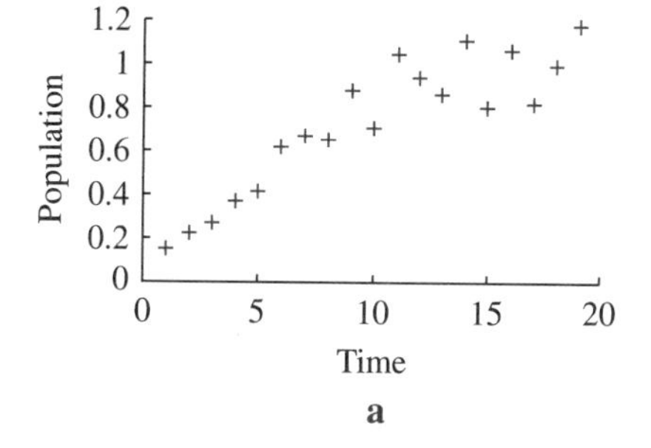

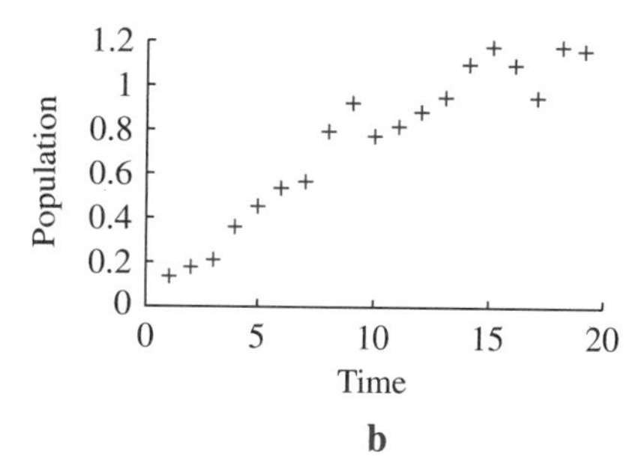

FIGURE 1.1.6

Two sets of measurements described by the same probabilistic dynamical system in discrete time. The two panels show the results of two realizations of the same mathematical model that differ only due to random factors.

1.2 Variables, Parameters, and Functions in Biology

Quantitative science is built upon measurements. Mathematics provides the notation for describing and thinking about measurements and relations between them. In fact, the development of clear notation for measurements and relations was essential for the progress of modern science. In this section, we develop the algebraic notation needed to describe measurements, introducing **variables** to describe measurements that change **during** the course of an experiment and **parameters** that remain constant during an experiment but can change **between** different experiments. The most important types of relations between measurements are described with **functions**, where the value of one can be computed from the value of the other. We will review how to graph functions, how to combine them with **addition**, **multiplication**, and **composition**, and how to recognize which functions have an **inverse** and how to compute it.

1.2.1 Describing Measurements with Variables, Parameters, and Graphs

Algebra uses letters or other symbols to represent numerical quantities.

Definition 1.1 A **variable** is a symbol that represents a measurement that can change during the course of an experiment.

A simple experiment measures how the population of bacteria in a culture changes over time. Because two changing quantities are being measured, time and bacterial population, we need two variables to represent them. In applied mathematics, we choose variables that remind us of the measurement it represents. In this case, we can use t to represent time and b to represent the population of bacteria. Because there are fewer letters than quantities to be measured, the same letter can be used to represent different quantities in different problems. Make sure to explicitly define variables when writing a model and to check their definitions when reading one.

Example 1.2.1 Describing Bacterial Population Growth

t	b
0.0	1.00
1.0	1.24
2.0	1.95
3.0	3.14
4.0	4.81
5.0	6.95
6.0	9.57

The table to the left lists measurements of bacterial population size (in millions), denoted by the variable b, at different times t after the beginning of an experiment.

Thinking about data is often easier with a graph. Graphs are drawn using **Cartesian coordinates**, which use two perpendicular number lines called the **axes** to describe two numbers (Figure 1.2.1). The argument is placed on the horizontal axis (sometimes called the x-axis), and the value on the vertical axis (sometimes called the y-axis). The crossing point of the two axes is the **origin**. The axes are labeled with the variable name, the measurement it represents, and often the units of measurement (Section 1.3). **Never draw a graph without labeling the axes.**

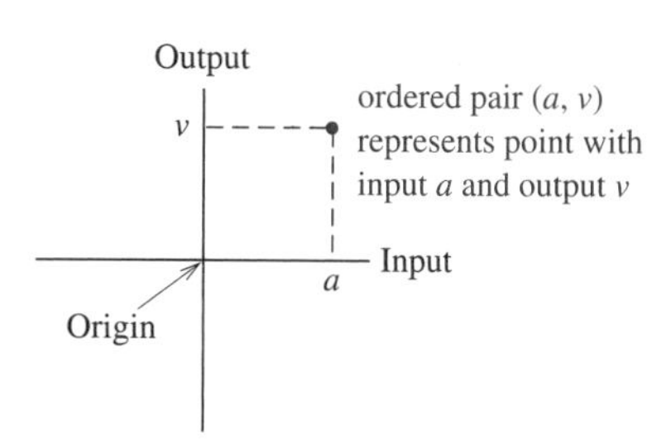

FIGURE 1.2.1 The components of a graph in Cartesian coordinates

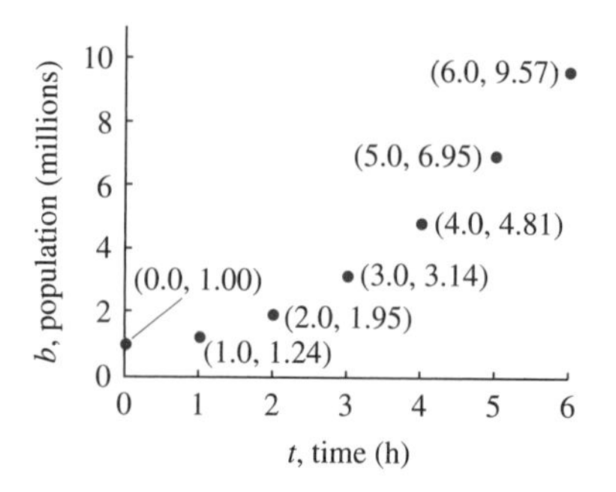

FIGURE 1.2.2 Results of bacterial growth experiment: Cartesian coordinates

Example 1.2.2 Graphing Data Describing Bacterial Population Growth

To graph the six data points in Example 1.2.1, plot each point by moving a distance t to the right of the origin along the horizontal axis and a distance b up from the origin along the vertical axis (Figure 1.2.2). For example, the data point at $t = 4.0$ is graphed by moving a distance 4.0 to the right of the origin on the horizontal axis and a distance 4.81 up from the origin on the vertical axis.

Example 1.2.3 Describing the Dynamics of a Bacterial Population

Suppose several bacterial cultures with different initial population sizes are grown in controlled conditions for one hour and then carefully counted. The population size acts as the basic measurement at both times. We must use different variables to represent these values, and choose to use **subscripts** to distinguish them. In particular, we let b_i (for the **initial** population) represent the population at the beginning of the experiment, and b_f (for the **final** population) represent the population at the end (Figure 1.2.3). The following table presents the results for six colonies.

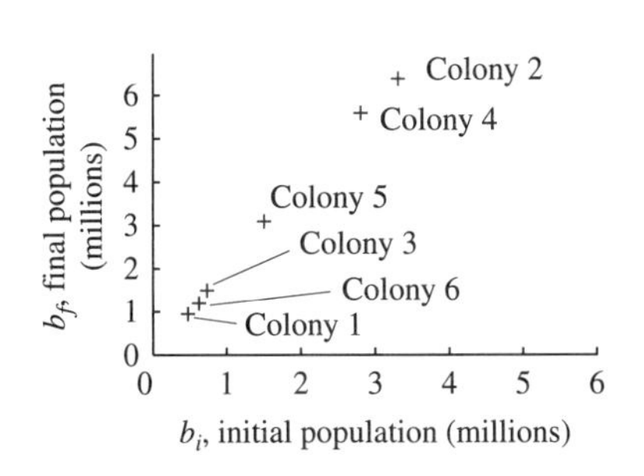

FIGURE 1.2.3

Results of alternative bacterial growth experiment

Colony	Initial Population, b_i	Final Population, b_f
1	0.47	0.94
2	3.3	6.6
3	0.73	1.46
4	2.8	5.6
5	1.5	3.0
6	0.62	1.24

Experiments of this sort form the basis of discrete-time dynamical systems (Section 1.5) and are the central topic of this chapter.

Experiments are done in a particular set of controlled conditions that remain constant during the experiment. However, these conditions might differ between experiments.

Definition 1.2 A **parameter** is a symbol that represents a measurement that does not change during the course of an experiment.

Different experiments tracking the growth of bacterial populations over time might take place at temperatures that are constant during an experiment but that differ between experiments. The temperature, in this case, is represented by a parameter. Parameters are also represented by symbols that recall the measurement. We can use T to represent temperature. In applied mathematics, capital letters (like T) and small letters (like t) are often used in the same problem to represent different quantities.

Example 1.2.4 Variables and Parameters

Suppose a biologist measures growing bacterial populations at three different temperatures. During the course of each experiment, the temperature is held constant, while the population changes.

t	b when $T = 25°$	b when $T = 35°$	b when $T = 45°$
0.0	1.00	1.00	1.00
1.0	1.14	1.45	0.93
2.0	1.30	2.10	0.87
3.0	1.48	3.03	0.81
4.0	1.68	4.39	0.76
5.0	1.92	6.36	0.70
6.0	2.18	9.21	0.66

Figure 1.2.4 compares the sizes of the three populations. The population grows most quickly at the intermediate temperature of 35°C and declines at the high temperature of 45°C.

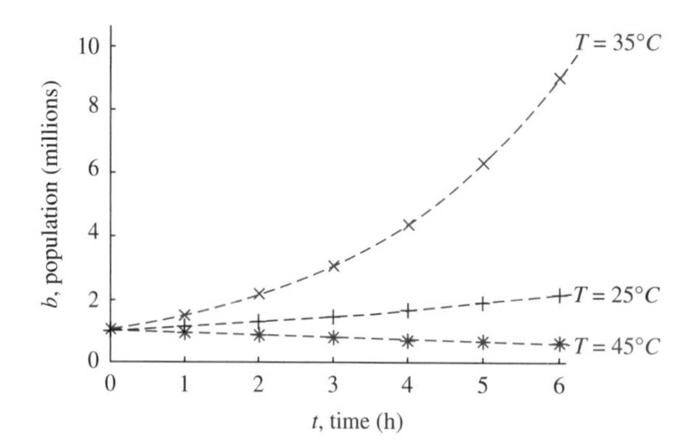

FIGURE 1.2.4

Results of bacterial growth experiment at three temperatures

1.2.2 Describing Relations Between Measurements with Functions

Numbers describe measurements, and **functions** describe **relations** between measurements. For example, bacterial population growth relates two measurements, denoted by the variables t and b. In general, a **relation** between two variables is the set of all pairs of values that are possible.

Example 1.2.5 A Relation Between Temperature and Population Size

T	P
25.0	2.18
25.0	2.45
25.0	2.10
25.0	3.03
35.0	9.21
35.0	7.39
35.0	6.36
45.0	0.66
45.0	0.93

Suppose the temperature T and final population size P are measured for nine populations, with the following results (Figure 1.2.5). These values could result from repeating the experiment in Example 1.2.4 several times and measuring the population at $t = 6.0$.

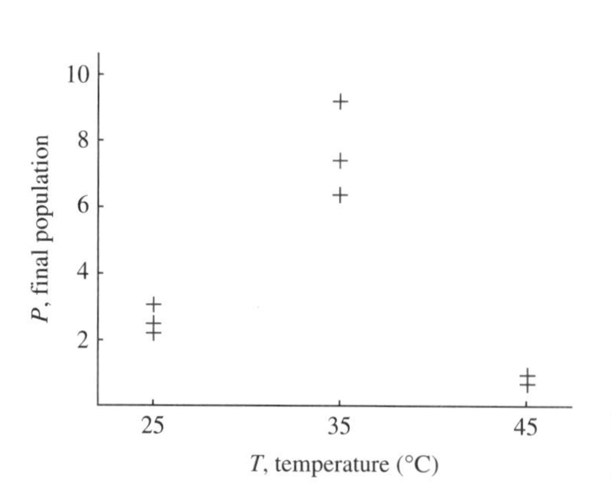

FIGURE 1.2.5

Final population size at three temperatures

Different values of the population P are related to each temperature, perhaps due to differences in experimental conditions.

A **function** describes a specific, and important, type of relation. A function is a mathematical object that takes something (like a number) as input, performs an operation on it, and returns a new object (like another number). The input is called the **argument** (or the **independent variable**), and the output is called the **value** (or the **dependent variable**) (Figure 1.2.6). The set of all possible things that a function can accept as inputs is called the **domain**; the set of all possible things a function **can** return as outputs is called the **co-domain**; and the set of all things the function **does** return as outputs is called the **range**.

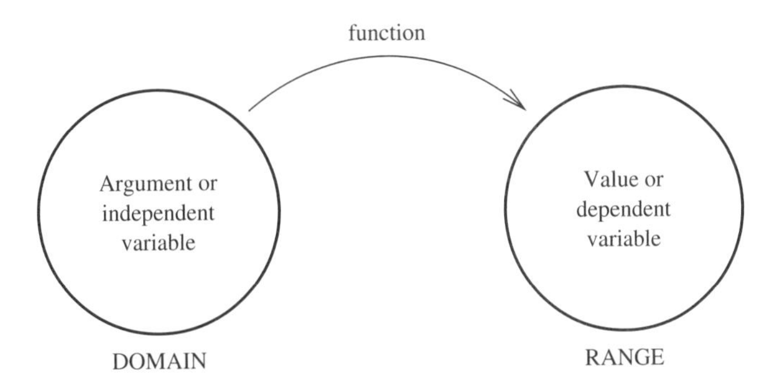

FIGURE 1.2.6
The basic terminology for describing a function

Example 1.2.6 Data That Can Be Described by a Function

The data in Example 1.2.1 can be described by a function. Each value of the input t is associated with only a single value of the output b.

Example 1.2.7 Graphing a Function from Its Formula

To graph a function from a formula, it is easiest to start by plugging in some representative arguments. Suppose we wish to graph the function $f(x) = 4 + x - x^2$ for $x \geq 0$ (a way of restricting the domain to just positive numbers and zero). Evaluating the function at the arguments 0, 1, 2 and 3, we find

$$f(0) = 4 + 0 - 0^2 = 4$$
$$f(1) = 4 + 1 - 1^2 = 4$$
$$f(2) = 4 + 2 - 2^2 = 2$$
$$f(3) = 4 + 3 - 3^2 = -2.$$

We plot the four ordered pairs (0, 4), (1, 4), (2, 2), and (3, −2), and connect them with a smooth curve (Figure 1.2.7). This is precisely the method that calculators and computers use to plot functions, except that they generally use 20 or more points to make a graph. Because the output takes on negative values, the horizontal axis is positioned at a negative value of the output. Rather than drawing axes through the origin (0, 0), graphs of measurements often place the axes to cross at a value that enhances readability.

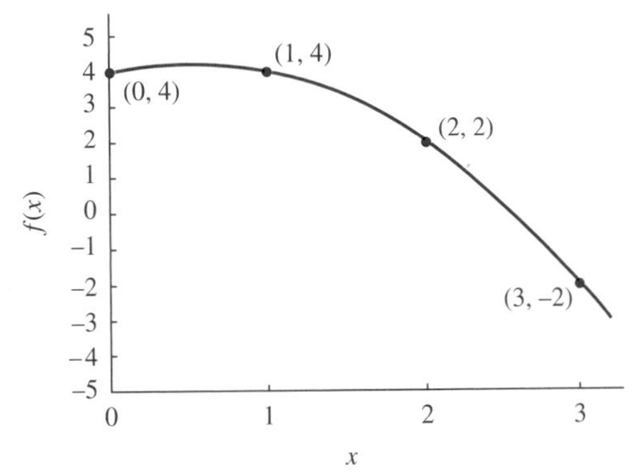

FIGURE 1.2.7
Plotting a function from its formula

One of the great advantages of functional notation is that functions can be evaluated at arguments that consist of parameters and variables (combinations of letters). To do so, replace the basic variable in the formula with the new argument, however complicated.

Example 1.2.8 Evaluating a Function at a Complicated Argument

To evaluate the function $f(x) = 4 + x - x^2$ (Example 1.2.7) at the more complicated argument $2z + 3$, replace all occurrences of x in the formula with the new argument $2z + 3$:

$$f(2z + 3) = 4 + (2z + 3) - (2z + 3)^2.$$

To avoid confusion, place the new argument in parentheses wherever it appears. Although it is not always necessary, this expression can be multiplied out and simplified as

$$
\begin{aligned}
f(2z+3) &= 4 + (2z+3) - (2z+3)^2 && \text{original expression}\\
&= 4 + (2z+3) - (4z^2 + 12z + 9) && \text{expand the square}\\
&= 4 + 2z + 3 - 4z^2 - 12z - 9 && \text{multiply negative sign through}\\
&= 4 + 3 - 9 + 2z - 12z - 4z^2 && \text{group like terms}\\
&= -2 - 10z - 4z^2. && \text{combine like terms}
\end{aligned}
$$

Example 1.2.9 A Function Describing Bacterial Population Growth

The population in Examples 1.2.1 and 1.2.2 obeys the formula

$$b(t) = \frac{t^2}{4.2} + 1.0.$$

The population size b is a function of the time t. The **argument** of the function b is t, the time after the beginning of the experiment. The **value** of the function is the population of bacteria. The formula summarizes the relation between these two measurements: the output is found by squaring the input, dividing by 4.2, and then adding 1.0.

The function b takes time after the beginning of the experiment as its input. Because negative time does not make sense in this case, the **domain** of this function consists of all positive numbers and zero. We write that

$$b \text{ is defined on the domain } t \geq 0.$$

Besides negative numbers, fruits, bacteria, and functions lie outside the domain of b. Because the function b returns population sizes as output, the **range** of b also consists of all positive numbers and zero. We write that

$$b \text{ has range } b \geq 0.$$

Example 1.2.10 A Function with Non-numeric Domain

Animal	Number of Legs
Ant	6
Crab	10
Duck	2
Fish	0
Human being	2
Mouse	4
Spider	8

Consider the following table of data. These data describe a relation between two observations: the identity of the species and the number of legs. We can express this as the function L (to remind us of legs). According to the table,

$$L(\text{Ant}) = 6, \quad L(\text{Crab}) = 10$$

and so forth. The domain of this function is "types of animals," and the range is the non-negative integers (0, 1, 2, 3, ...). We plot the input ("animal") along the horizontal axis and the output ("number of legs") on the vertical axis (Figure 1.2.8).

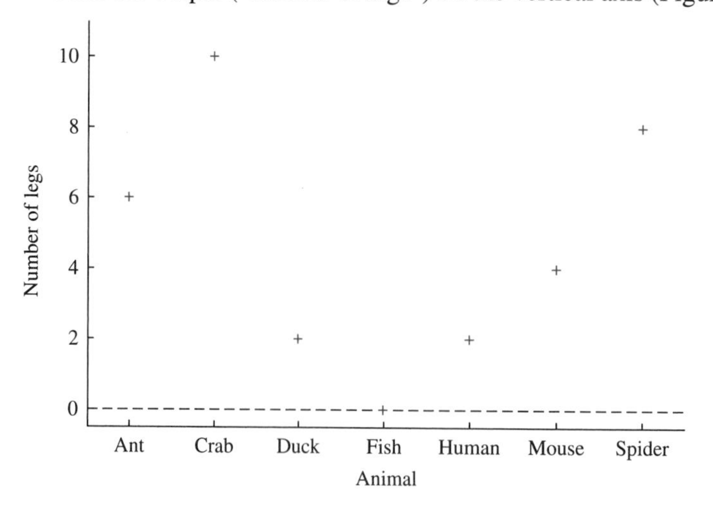

FIGURE 1.2.8

Numbers of legs on various organisms plotted on a graph

It is important to realize that the graph of a function is *not* the function, just as the spot labeled 2 on the number line is not the number 2 and a photograph of a dog is not a dog. The graph is a depiction of the function.

Functions can be described in four ways: (1) numerically (by means of a table), (2) as a formula, (3) as a graph, and (4) verbally. As biologists and applied mathematicians, we need to be able to use all four methods and to translate between them. In particular, we must know how to translate graphical information into words that communicate key observations to colleagues and the public.

Example 1.2.11 Describing Results in Graphs and Words

Time	Population Size
0	0.86
2	1.69
4	2.98
6	4.49
8	5.69
10	6.17
12	5.95
14	5.29
16	4.41
18	3.50
20	2.67
22	1.96
24	1.41

The following table presents a more complicated pattern of population size change.

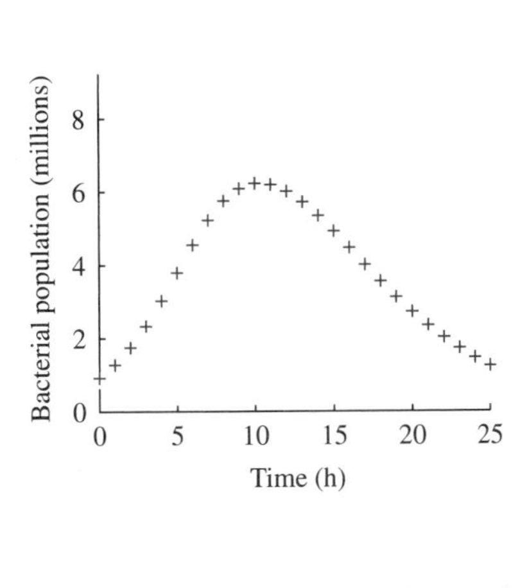

FIGURE 1.2.9
The population of bacteria in a culture

We can see (more easily from Figure 1.2.9 than from the table) that the bacterial population grew during the first ten hours and declined thereafter. The population reached a maximum at time 10. This graph and its description can be used to understand the results even without a mathematical formula.

Example 1.2.12 Sketching a Graph from a Verbal Description

Conversely, it can be useful to sketch a graph of a function from a verbal description. Suppose we are told that a population increases between time 0 and time 5, decreases nearly to 0 by time 12, increases to a higher maximum at time 20, and goes extinct at time 30. A graph translates this information into pictorial form (Figure 1.2.10). Because we were not given exact values, the graph is not exact. It instead gives a **qualitative** picture of the results.

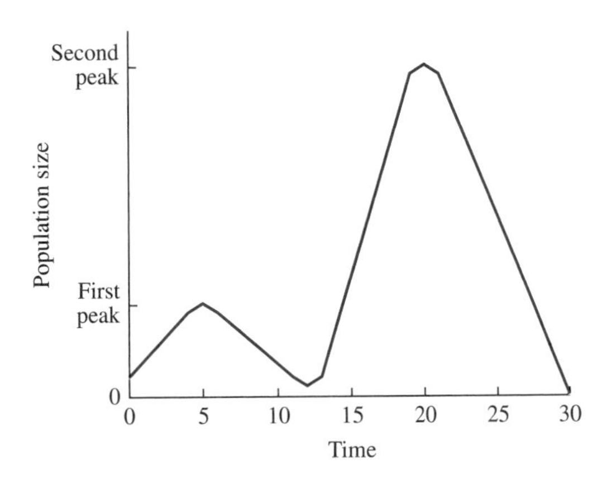

FIGURE 1.2.10
A bacterial population plotted from a verbal description

Not all relations are described by functions. A function must give a unique output for a given input. Relations between measurements can be more complicated. The **vertical line test** provides a graphical method to recognize relations that cannot be described by functions.

The Vertical Line Test A relation is not a function if some vertical line crosses the graph two or more times. In Figure 1.2.11, there are three outputs associated with the input 0.2: 1.12, 1.79, and 3.09. This relationship cannot be described with a function.

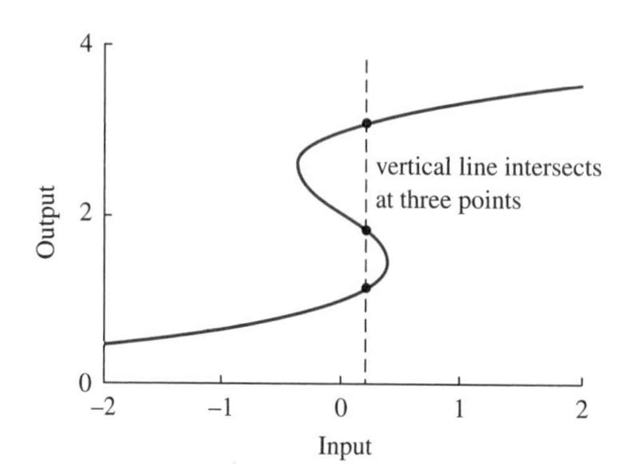

FIGURE 1.2.11
The vertical line test

There is nothing wrong with relations that cannot be described by functions. Experiments, even when performed under apparently identical conditions, rarely produce identical results. As we will see when we study statistics (Chapter 8) functions are a useful mathematical idealization of the expected or average result of an experiment.

Example 1.2.13 A Mathematical Formula That Gives a Relation That Is Not a Function

The set of solutions for x and y from the formula

$$x^2 + y^2 = 1$$

is the circle of radius 1 centered at the origin (Figure 1.2.12). Each value of x between $x = -1$ and $x = 1$ is associated with two different values of y. For example, the value $x = 0.6$ is associated with both $y = 0.8$ and $y = -0.8$.

Example 1.2.14 A Relation That Is Not a Function

Suppose several bacterial cultures with different initial population sizes are grown in controlled conditions for one hour, as in Example 1.2.3, with the following results.

Colony	Initial Population, b_i	Final Population, b_f
1	0.5	0.9
2	0.5	1.0
3	1.0	2.2
4	1.0	1.9
5	1.5	3.0
6	1.5	2.8

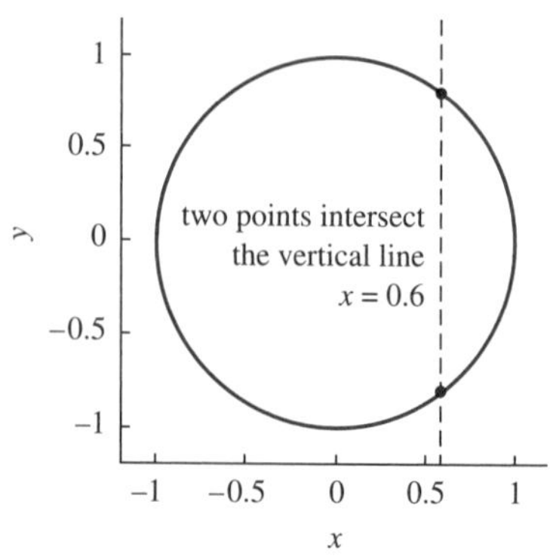

FIGURE 1.2.12
The circle describes a relation that is not a function

Each initial population was used twice, with similar but not identical results (Figure 1.2.13). We cannot treat final population size as a function of initial population size.

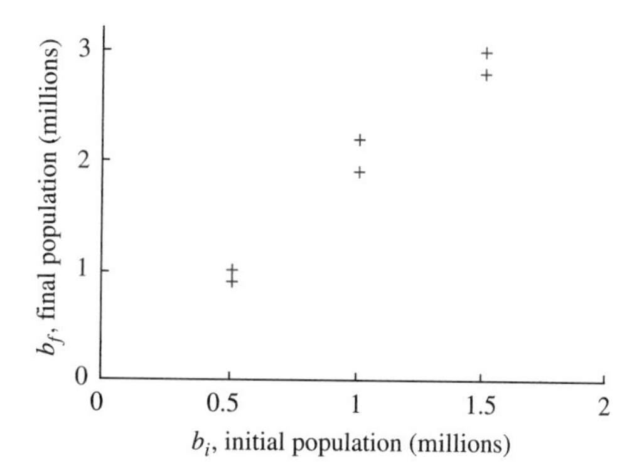

FIGURE 1.2.13
Bacterial growth experiment where results are not a function

1.2.3 Combining Functions

Mathematics makes complicated problems simpler by building complicated structures from simple pieces. By understanding each of the simple pieces and the rules for combining them, a huge array of complicated relations can be analyzed and understood. The most important ways to combine functions are as **sums**, **products**, and **compositions**.

Adding Functions The height of the graph of the sum of two functions is the height of the first plus the height of the second. Geometrically, we can graph each of the pieces and add them together in the same way.

Algebraically, the value of the function $f + g$ is computed as the sum of the values of the functions f and g.

Definition 1.3 The sum $f + g$ of the functions f and g is the function that takes on the value

$$(f + g)(x) = f(x) + g(x).$$

Multiplying Functions The value of the product $f \cdot g$ is computed as the product of the values of the functions f and g.

Definition 1.4 The product $f \cdot g$ of the functions f and g is the function that takes on the value

$$(f \cdot g)(x) = f(x) \cdot g(x).$$

We use the dot $\cdot$ rather than the times sign $\times$ to indicate multiplication to avoid confusion with the variable x.

Example 1.2.15 Adding and Multiplying Functions

Consider the functions $f(x)$ and $g(x)$ with formulas

$$f(x) = 4 + x - x^2$$
$$g(x) = 2x,$$

graphed in Figures 1.2.14 and 1.2.15. The table following the figures computes the values of $f + g$ and $f \cdot g$ at several points.

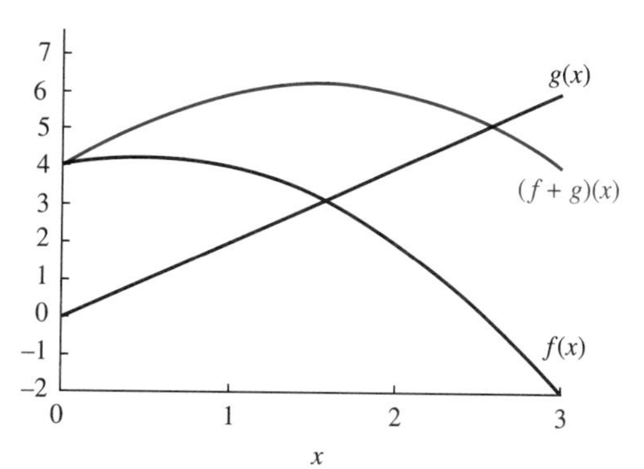

FIGURE 1.2.14
Adding functions

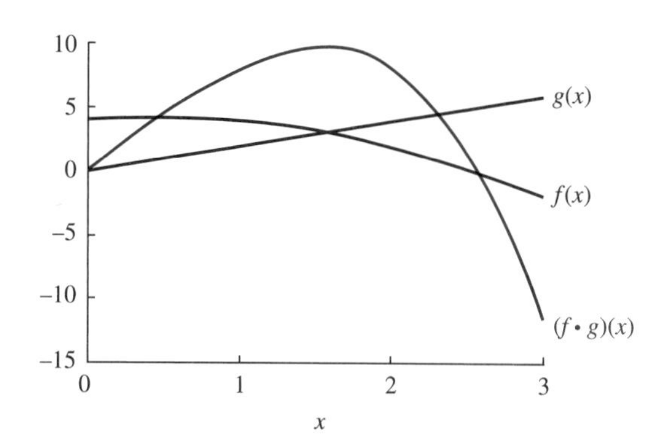

FIGURE 1.2.15
Multiplying functions

x	f(x)	g(x)	(f + g)(x)	(f · g)(x)
0	4	0	4	0
0.5	4.25	1	5.25	4.25
1	4	2	6	8
1.5	3.25	3	6.25	9.75
2	2	4	6	8
2.5	0.25	5	5.25	1.25
3	−2	6	4	−12

Example 1.2.16 Adding Biological Functions

If two bacterial populations are separately counted, the total population is the sum of the two individual populations (Figure 1.2.16). Suppose a growing population is described by the function

$$b_1(t) = t^2 + 1$$

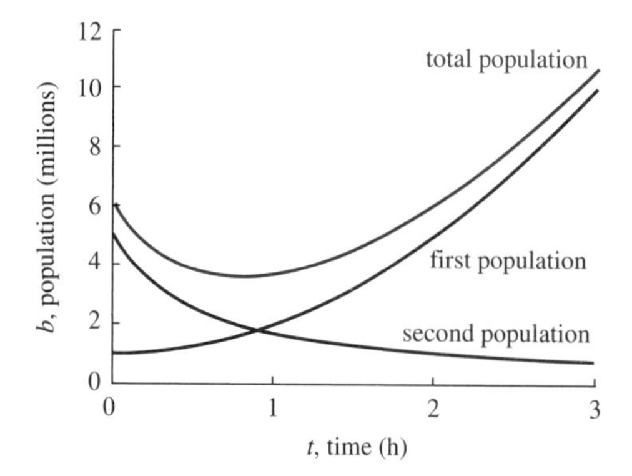

FIGURE 1.2.16
Adding biological functions

and a declining population is described by the function

$$b_1(t) = \frac{5}{1 + 2t}.$$

The individual population sizes and their sum are computed in the following table.

t	$b_1(t)$	$b_2(t)$	$(b_1 + b_2)(t)$
0.00	1.00	5.00	6.00
0.50	1.25	2.50	3.75
1.00	2.00	1.67	3.67
1.50	3.25	1.25	4.50
2.00	5.00	1.00	6.00
2.50	7.25	0.83	8.08
3.00	10.00	0.71	10.71

Example 1.2.17 Multiplying Biological Functions

t	b	μ	$\mu \cdot b$
0.0	1.00	1.00	1.00
1.0	1.24	0.50	0.62
2.0	1.95	0.33	0.65
3.0	3.14	0.25	0.78
4.0	4.81	0.20	0.96
5.0	6.95	0.17	1.16
6.0	9.57	0.14	1.37

Many quantities in science are built as products of simpler quantities. For example, the mass of a population is the product of the mass of each individual and the number of individuals. Consider a population growing according to

$$b(t) = \frac{t^2}{4.2} + 1.0$$

(Example 1.2.9). Suppose that as the population gets larger the individuals become smaller. Let $\mu(t)$ (the Greek letter mu)[1] represent the mass of an individual at time t, and suppose that

$$\mu(t) = \frac{1}{1+t}.$$

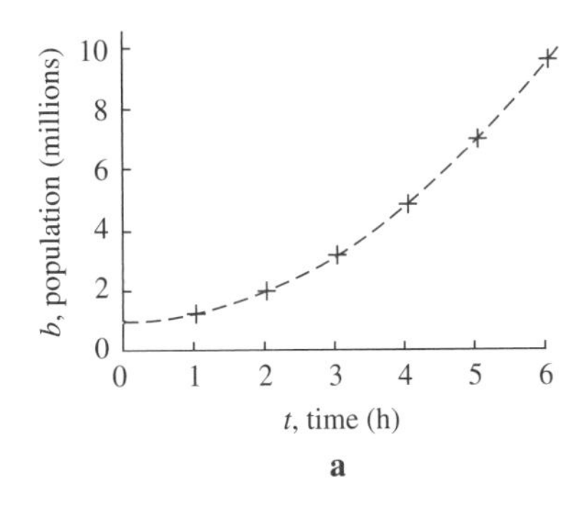

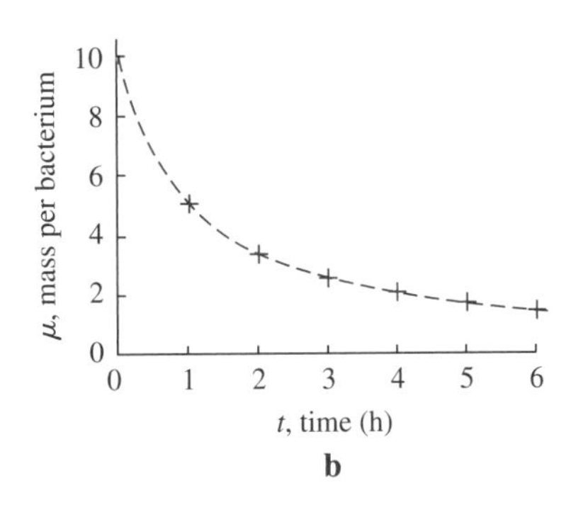

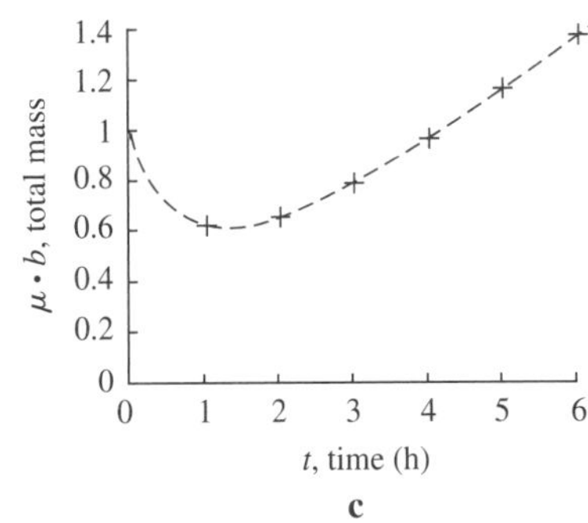

FIGURE 1.2.17
Multiplying biological functions

We can find the total mass by multiplying the mass per individual by the number of individuals, as in the table and Figure 1.2.17. The total mass of this population initially declines and then increases after about 2 hours.

Composition of Functions The most important way to combine functions is through **composition**, where the output of one function acts as the input of another.

Definition 1.5 The composition $f \circ g$ of functions f and g is a function defined by

$$(f \circ g)(x) = f(g(x)). \qquad \textbf{(1.2.1)}$$

We say "f composed with g evaluated at x" or "f of g of x." The function f is called the **outer function**, and the function g is called the **inner function** (Figure 1.2.18).

[1] Applied mathematicians often use Greek letters to represent variables and parameters. The Greek alphabet, along with pronunciations of the letters, is given in Appendix A.

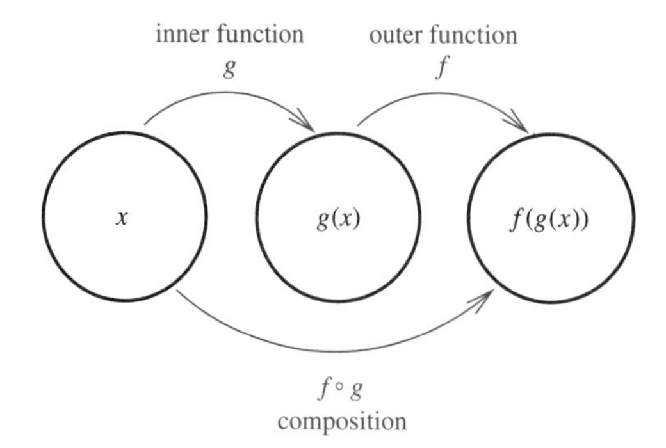

FIGURE 1.2.18
Composition of functions

Example 1.2.18 Computing the Value of a Functional Composition

Consider the functions $f(x)$ and $g(x)$ from Example 1.2.15,

$$f(x) = 4 + x - x^2$$
$$g(x) = 2x.$$

To find the value of the composition $f \circ g$ at $x = 2$, we compute

$$\begin{aligned}(f \circ g)(2) &= f(g(2)) && \text{definition of composition}\\ &= f(2 \cdot 2) && \text{substitute 2 into } g(x)\\ &= f(4) && \text{evaluate } g(2) = 4\\ &= 4 + 4 - 4^2 && \text{substitute 4 into } f(x)\\ &= -8. && \text{evaluate } f(4) = -8\end{aligned}$$

Similarly, to find the value of the composition $g \circ f$ at $x = 2$, we compute

$$\begin{aligned}(g \circ f)(2) &= g(f(2)) && \text{definition of composition}\\ &= g(4 + 2 - 2^2) && \text{substitute 2 into } f(x)\\ &= g(2) && \text{evaluate } f(2) = 2\\ &= 2 \cdot 2 && \text{substitute 2 into } g(x)\\ &= 4. && \text{evaluate } g(2) = 4\end{aligned}$$

Example 1.2.19 Computing the Formula of a Functional Composition

Consider again the functions $f(x)$ and $g(x)$ from Example 1.2.18,

$$f(x) = 4 + x - x^2$$
$$g(x) = 2x.$$

with domains consisting of all numbers. To find the composition $f \circ g$, plug the definition of the **inner function** g into the formula for the **outer function** f, or

$$\begin{aligned}(f \circ g)(x) &= f(g(x)) && \text{the definition}\\ &= f(2x) && \text{write out the formula for the inner function } G\\ &= 4 + (2x) - (2x)^2 && \text{plug the formula for } G \text{ into the outer function } F\\ &= 4 + 2x - 4x^2. && \text{expand the square}\end{aligned}$$

This is the same procedure we used to compute the value of the function $f(x)$ at a complicated argument in Example 1.2.8. In Example 1.2.18 we computed that $(f \circ g)(2) = -8$. If we evaluate by substituting into the formula $(f \circ g)(x) = 4 + 2x - 4x^2$, we find

$$(f \circ g)(2) = 4 + 2 \cdot 2 - 4 \cdot 2^2 = -8,$$

matching our earlier result.

We find the composition $g \circ f$ by following the same steps, or

$$\begin{aligned}(g \circ f)(x) &= g(f(x)) && \text{the definition}\\ &= g(4 + x - x^2) && \text{write out the formula for the inner function } F\\ &= 2(4 + x - x^2) && \text{plug the formula for } F \text{ into the outer function } G\\ &= 8 + 2x - 2x^2. && \text{multiply through}\end{aligned}$$

The key step is substituting the output of the inner function into the outer function; the rest is algebra. In Example 1.2.18 we computed that $(g \circ f)(2) = 4$. If we evaluate by substituting into the formula $g \circ f(x) = 8 + 2x - 2x^2$, we find

$$(g \circ f)(2) = 8 + 2 \cdot 2 - 2 \cdot 2^2 = 4,$$

again matching our earlier result.

Example 1.2.19 illustrates an important point about the composition of functions: the answer is generally different when the functions are composed in a different order. If $f \circ g = g \circ f$, we say that the two functions **commute**. When the two compositions do not match, we say that the two functions do not commute. Without a good reason, never assume that two functions commute. If you think of functions as operations, this should make sense. Sterilizing the scalpel and then making an incision produces a quite different result from making an incision and then sterilizing the scalpel.

Example 1.2.20 Composition of Functions in Biology

Numbers of bacteria are usually measured indirectly, by measuring the optical density of the medium. Water allows less light to come through as the population becomes larger. Suppose that the optical density ρ is a function of the bacterial population size b with formula

$$\rho(b) = \frac{b}{2.0b + 5.0}.$$

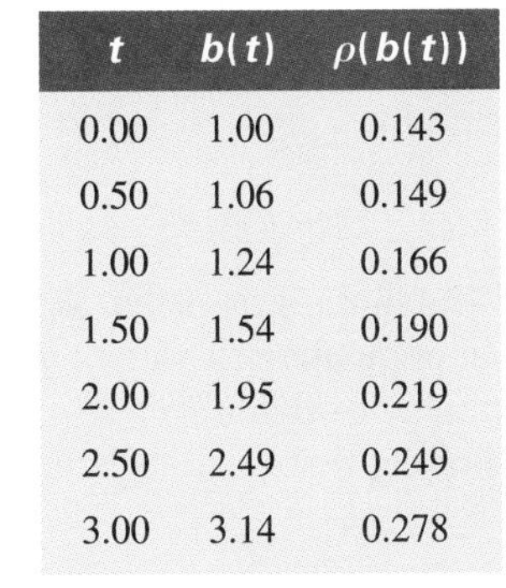

t	$b(t)$	$\rho(b(t))$
0.00	1.00	0.143
0.50	1.06	0.149
1.00	1.24	0.166
1.50	1.54	0.190
2.00	1.95	0.219
2.50	2.49	0.249
3.00	3.14	0.278

Then the optical density as a function of time is the composition of the function $\rho(b)$ with the function $b(t)$ (Figure 1.2.19). Suppose that $b(t) = \frac{t^2}{4.2} + 1.0$ as in Example 1.2.9. Then

$$\rho(b(t)) = \rho\left(\frac{t^2}{4.2} + 1.0\right) = \frac{\frac{t^2}{4.2} + 1.0}{2\left(\frac{t^2}{4.2} + 1.0\right) + 5.0},$$

with values given in the table to the left.

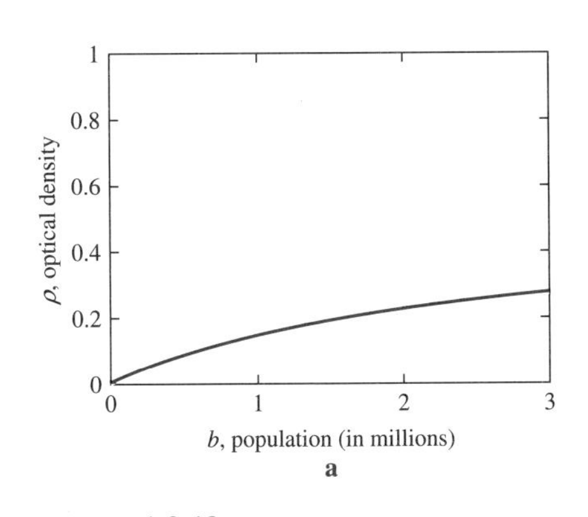

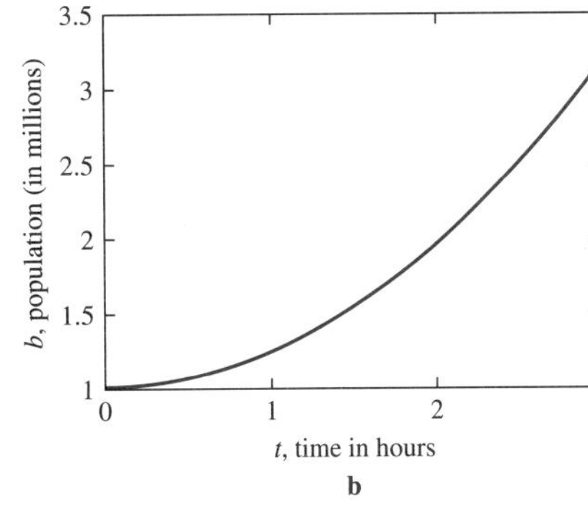

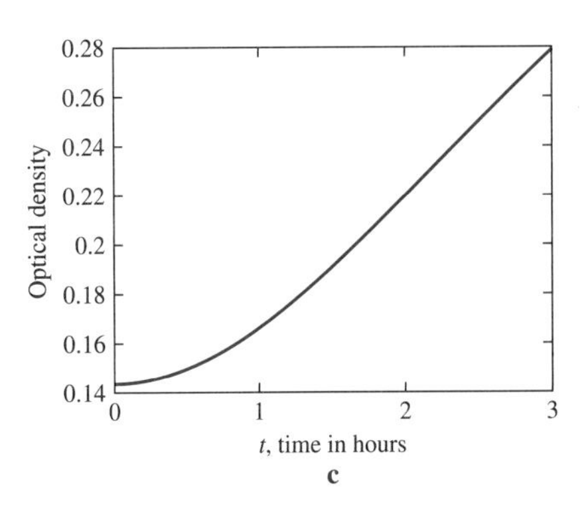

FIGURE 1.2.19
Composing biological functions

The composition $b \circ \rho$ is not merely different from the composition $\rho \circ b$; it does not even make sense. The function b accepts as input only the time t, not the optical density returned as output by the function ρ. We will study this issue more carefully in Section 1.3.

1.2.4 Finding Inverse Functions

A function describes the relation between two measurements and gives a way to compute the output from a given input. Sometimes we wish to reverse the process and figure out which input produced a given output. The **inverse function**, when it exists, provides a way to do this.

Example 1.2.21 A Simple Inverse Operation

What number, when doubled, gives 8? It is not difficult to guess that the answer is 4. However, we can formalize this process using functional notation. Let $f(x) = 2x$ be the function that doubles. Our problem is then solving

$$f(x) = 8.$$

Using the formula for $f(x)$, we find

$$\begin{aligned} 2x &= 8 && \text{the equation to be solved} \\ x &= 4. && \text{divide both sides by 2} \end{aligned}$$

Example 1.2.22 A Simple Inverse Function

Example 1.2.21 undoes the act of multiplying by 2. What function does this in general? If we set $y = f(x)$, we would like to know what value of x produces a given y in general, without picking a particular value such as $y = 8$. We follow the same steps,

$$\begin{aligned} 2x &= y && \text{the equation to be solved} \\ x &= \frac{y}{2}. && \text{divide both sides by 2} \end{aligned}$$

The function f^{-1}, read "f inverse," defined by

$$f^{-1}(y) = \frac{y}{2}$$

is the **inverse** of f; the function that undoes what f did in the first place. Whereas f takes a number as input and returns double that number population as output, f^{-1} takes the doubled number as input and returns the initial number as output.

We can use this inverse like any other function, finding that

$$f^{-1}(8) = \frac{8}{2} = 4,$$

as we found in Example 1.2.21.

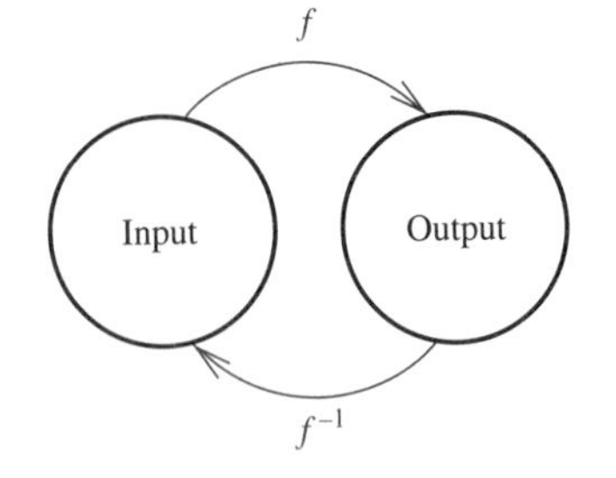

FIGURE 1.2.20
The action of a function and its inverse

The definition of an inverse function in general states precisely that the inverse undoes the action of the original function (Figure 1.2.20).

Definition 1.6 The function f^{-1} is the inverse of f if

$$\begin{aligned} f(f^{-1}(x)) &= x \\ f^{-1}(f(x)) &= x. \end{aligned}$$

Each of f and f^{-1} undoes the action of the other.

The steps for computing the inverse of a function can be summarized in an **algorithm**, which can be thought of as a recipe. This book contains many algorithms for solving particular problems. As with a recipe, following an algorithm without thinking about and checking the steps can lead to disaster. Unlike most algorithms in this book, this one can be impossible to follow because the equation in the second step cannot be solved algebraically.

Algorithm 1.1 Finding the Inverse of a Function

1. Write the equation $y = f(x)$.
2. Solve for x in terms of y.
3. The inverse function is the operation done to y.

It may look odd to have a function defined in terms of y. Do **not** change the letters around to make it look normal. In applied mathematics different letters stand for different things and resent having their names switched as much as we do.

This algorithm may fail in two different ways: a function might not have an inverse, or the inverse might be impossible to compute. There is a useful way to recognize a function that fails to have an inverse. An operation can only be undone if you can deduce the input from the output. If any particular output is associated with more than one input, there is no way to tell where you started based solely on where you ended up.

Example 1.2.23 Finding a More Complicated Inverse

Consider the population following the equation

$$b(t) = \frac{t^2}{4.2} + 1.0$$

(Example 1.2.9, Figure 1.2.21). If we wish to find the time t from the population b, we must solve for t.

$$\begin{aligned} \frac{t^2}{4.2} + 1.0 &= b && \text{the equation to be solved for } t \\ \frac{t^2}{4.2} &= b - 1.0 && \text{subtract 1.0 from both sides} \\ t^2 &= 4.2(b - 1.0) && \text{multiply both sides by 4.2} \\ t &= \sqrt{4.2(b - 1.0)}. && \text{take the square root of both sides} \end{aligned}$$

The last step requires that $b \geq 1.0$ because we cannot take the square root of a negative number.

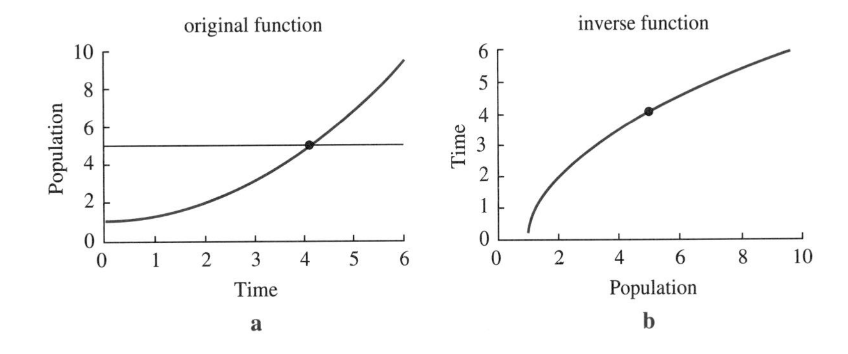

FIGURE 1.2.21
Going backwards with the inverse function

For example, to one decimal place of accuracy, the time associated with a population of 5.0 is

$$t = \sqrt{4.2(5.0 - 1.0)} = 4.099.$$

Example 1.2.24 A Relation That Cannot Be Inverted

Consider the data in the following table (Figure 1.2.22).

Initial Mass	Final Mass	Initial Mass	Final Mass
1.0	7.0	9.0	20.0
2.0	12.0	10.0	18.0
3.0	16.0	11.0	15.0
4.0	19.0	12.0	12.0
5.0	22.0	13.0	9.0
6.0	23.0	14.0	6.0
7.0	23.0	15.0	3.0
8.0	22.0	16.0	1.0

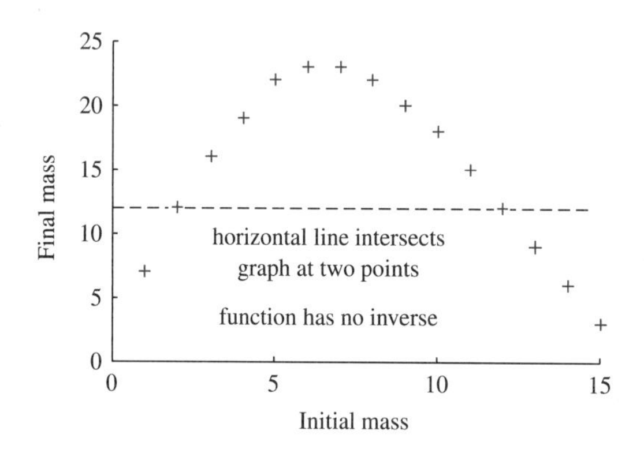

FIGURE 1.2.22

A relation with no inverse

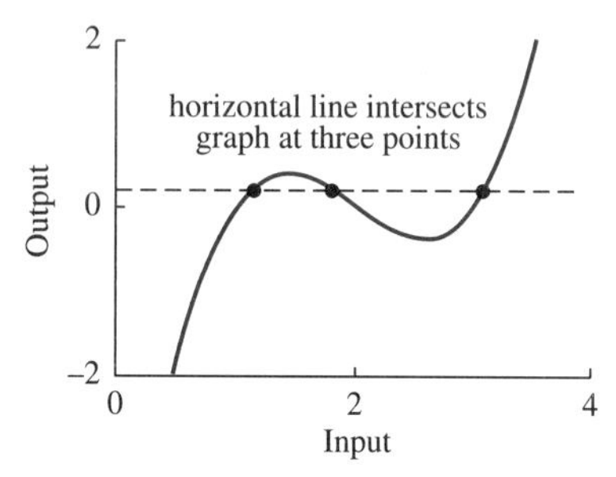

FIGURE 1.2.23

The horizontal line test

Suppose you were told that the mass at the end of experiment was 12.0 grams. Initial masses of 2.0 and 12.0 grams both produce a final mass of 12.0 grams. You cannot tell whether the input was 2.0 or 12.0. This function has no inverse.

This reasoning leads to a useful graphical test for whether a function has an inverse.

The Horizontal Line Test A function has no inverse if it takes on the same value two or more times. This can be established by graphing the function and checking whether the graph crosses some horizontal line two or more times. (Figure 1.2.23)

One can think of functions without inverses as losing information over the course of the experiment: things that started out different ended up the same.

Example 1.2.25 A Function That Has an Inverse on Part of Its Domain

Consider the function $g(x) = x^2$ defined for $x \geq 0$ (Figure 1.2.24). We find the inverse $f^{-1}(y)$ by solving $y = x^2$ for x.

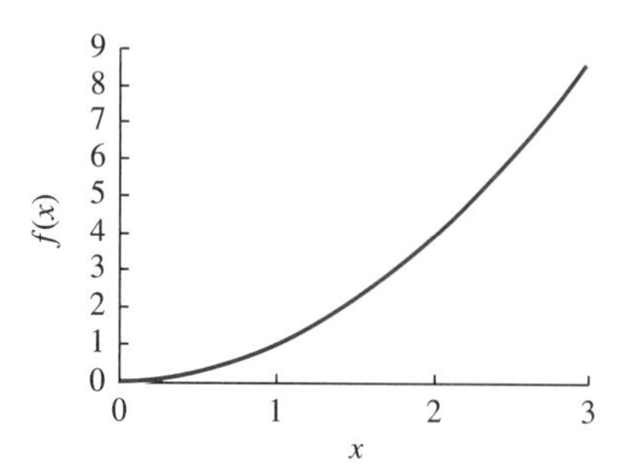

FIGURE 1.2.24

The inverse of x^2 is defined when $x \geq 0$

1. Set $y = x^2$.
2. Then $x = \sqrt{y}$.
3. $f^{-1}(y) = \sqrt{y}$.

Example 1.2.26 A Function Without an Inverse

Consider the function $f(x) = 4 + x - x^2$ (used in Example 1.2.7 and graphed in Figure 1.2.25). We found that the inputs $x = 0$ and $x = 1$ both produce the same output of $f(x) = 4$. If the output is 4, it is impossible to tell which was the input. A graph shows that this function fails the horizontal line test at almost all values in its range.

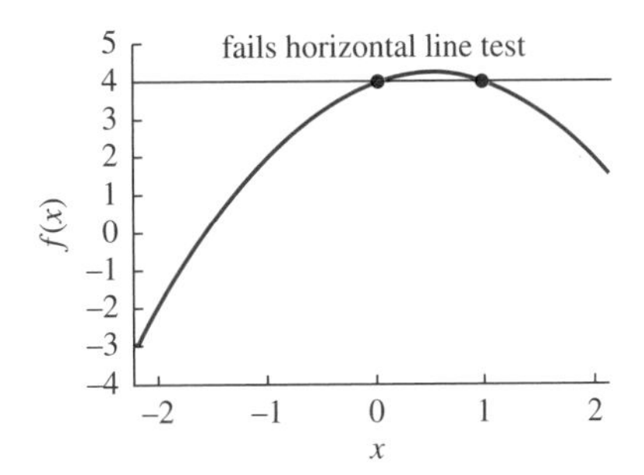

FIGURE 1.2.25
A function with no inverse

In addition, Algorithm 1.1 might fail because the algebra is impossible. Step 2 requires solving an equation. Many equations cannot be solved algebraically.

Example 1.2.27 A Function with an Inverse That Is Impossible to Compute

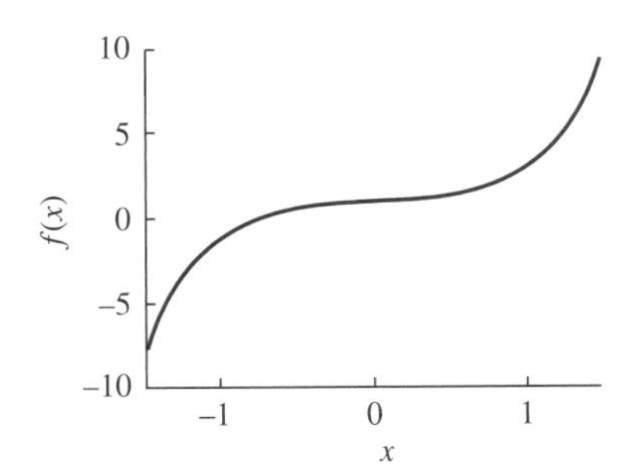

FIGURE 1.2.26
A function with an inverse that is impossible to compute

Consider the function

$$f(x) = x^5 + x + 1.$$

The graph satisfies the horizontal line test (Figure 1.2.26.). We try to find the inverse $f^{-1}(y)$ as follows.

1. Set $y = x^5 + x + 1$.
2. Try to solve for x. Even with the cleverest algebraic tricks, this is impossible (there is a remarkable theorem by the French mathematician Evariste Galois, proven when he was just 20 years old, that there is no formula for the solution of a polynomial with a degree greater than 4).
3. Give up.

In mathematical modeling, however, it is often more important to know that something exists (like the inverse in this case) than to be able to write down a formula. We will later learn a method to compute this inverse numerically with a computer (Section 3.8)

Summary Quantitative science is built upon measurements, and mathematics provides the methods for describing and thinking about measurements and relations between them. **Variables** describe measurements that change during the course of an experiment, and **parameters** describe measurements that remain constant during an experiment but might change between different experiments. **Functions** describe relations between different measurements when a single output is associated with each input, and can be recognized graphically with the **vertical line test**. New functions are built by combining functions through **addition**, **multiplication**, and **composition**. In functional composition, the output of the **inner function** is used as the input of the **outer function**. Many functions do not **commute**, meaning that composing the functions in a different order gives a different result. Finally, we used the **horizontal line test** to check whether a function has an **inverse**. If it does, the inverse can be used to compute the input from the output.

1.2 Exercises

Mathematical Techniques

1–2 ▪ Identify the variables and parameters in the following situations, give the units they might be measured in, and choose an appropriate letter or symbol to represent each.

1. A scientist measures the mass of fish over the course of 100 days, and repeats the experiment at three different levels of salinity: 0%, 2% and 5%.
2. A scientist measures the body temperature of bandicoots every day during the winter, and does so at three different altitudes: 500 m, 750 m, and 1000 m.

3–6 ▪ Compute the values of the following functions at the points indicated and sketch a graph of the function.

3. $f(x) = x + 5$ at $x = 0$, $x = 1$, and $x = 4$.
4. $g(y) = 5y$ at $y = 0$, $y = 1$, and $y = 4$.
5. $h(z) = \frac{1}{5z}$ at $z = 1$, $z = 2$, and $z = 4$.
6. $F(r) = r^2 + 5$ at $r = 0$, $r = 1$, and $r = 4$.

7–10 ▪ Graph the given points and say which point does not seem to fall on the graph of a simple function.

7. $(0, -1)$, $(1, 1)$, $(2, 1)$, $(3, 5)$, $(4, 7)$.
8. $(0, 5)$, $(1, 10)$, $(2, 8)$, $(3, 6)$, $(4, 4)$.
9. $(0, 2)$, $(1, 3)$, $(2, 6)$, $(3, 11)$, $(4, 10)$.
10. $(0, 45)$, $(1, 25)$, $(2, 12)$, $(3, 12.5)$, $(4, 10)$.

11–14 ▪ Evaluate the following functions at the given algebraic arguments.

11. $f(x) = x + 5$ at $x = a$, $x = a + 1$, and $x = 4a$.
12. $g(y) = 5y$ at $y = x^2$, $y = 2x + 1$, and $y = 2 - x$.
13. $h(z) = \frac{1}{5z}$ at $z = \frac{c}{5}$, $z = \frac{5}{c}$, and $z = c + 1$.
14. $F(r) = r^2 + 5$ at $r = x + 1$, $r = 3x$, and $r = \frac{1}{x}$.

15–16 ▪ Sketch graphs of the following relations. Is there a more convenient order for the arguments?

15. A function whose argument is the name of a state and whose value is the highest altitude in that state.

State	Highest Altitude (ft)
California	14,491
Idaho	12,662
Nevada	13,143
Oregon	11,239
Utah	13,528
Washington	14,410

16. A function whose argument is the name of a bird and whose value is the length of that bird.

Bird	Length
Cooper's hawk	50 cm
Goshawk	66 cm
Sharp-shinned hawk	35 cm

17–20 ▪ For each of the following sums of functions, graph each component piece. Compute the values at $x = -2$, $x = -1$, $x = 0$, $x = 1$, and $x = 2$ and plot the sum.

17. $f(x) = 2x + 3$ and $g(x) = 3x - 5$.
18. $f(x) = 2x + 3$ and $h(x) = -3x - 12$.
19. $F(x) = x^2 + 1$ and $G(x) = x + 1$.
20. $F(x) = x^2 + 1$ and $H(x) = -x + 1$.

21–24 ▪ For each of the following products of functions, graph each component piece. Compute the value of the product at $x = -2$, $x = -1$, $x = 0$, $x = 1$, and $x = 2$ and graph the result.

21. $f(x) = 2x + 3$ and $g(x) = 3x - 5$.
22. $f(x) = 2x + 3$ and $h(x) = -3x - 12$.
23. $F(x) = x^2 + 1$ and $G(x) = x + 1$.
24. $F(x) = x^2 + 1$ and $H(x) = -x + 1$.

25–28 ▪ Find the inverses of each of the following functions. In each case, compute the output of the original function at an input of 1.0, and show that the inverse undoes the action of the function.

25. $f(x) = 2x + 3$.
26. $g(x) = 3x - 5$.
27. $G(y) = 1/(2 + y)$ for $y \geq 0$.
28. $F(y) = y^2 + 1$ for $y \geq 0$.

29–32 ▪ Graph each of the following functions and its inverse. Mark the given point on the graph of each function.

29. $f(x) = 2x + 3$. Mark the point $(1, f(1))$ on the graphs of f and f^{-1} (based on Exercise 25).
30. $g(x) = 3x - 5$. Mark the point $(1, g(1))$ on the graphs of g and g^{-1} (based on Exercise 26).
31. $G(y) = 1/(2 + y)$. Mark the point $(1, G(1))$ on the graphs of G and G^{-1} (based on Exercise 27).
32. $F(y) = y^2 + 1$ for $y \geq 0$. Mark the point $(1, F(1))$ on the graphs of F and F^{-1} (based on Exercise 28).

33–36 ▪ Find the compositions of the given functions. Which pairs of functions commute?

33. $f(x) = 2x + 3$ and $g(x) = 3x - 5$.

34. $f(x) = 2x + 3$ and $h(x) = -3x - 12$.

35. $F(x) = x^2 + 1$ and $G(x) = x + 1$.

36. $F(x) = x^2 + 1$ and $H(x) = -x + 1$.

Applications

37–40 ▪ Describe what is happening in the graphs shown.

37. A plot of cell volume against time in days.

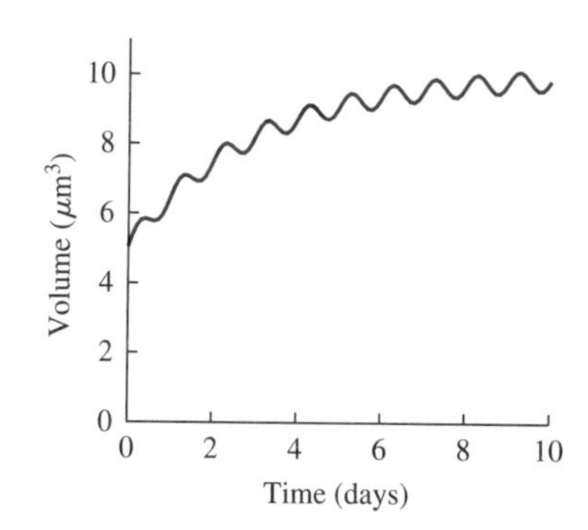

38. A plot of a Pacific salmon population against time in years.

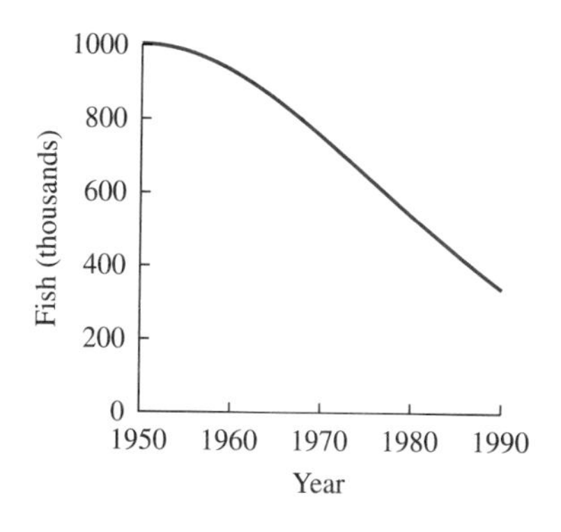

39. A plot of the average height of a population of trees plotted against age in years.

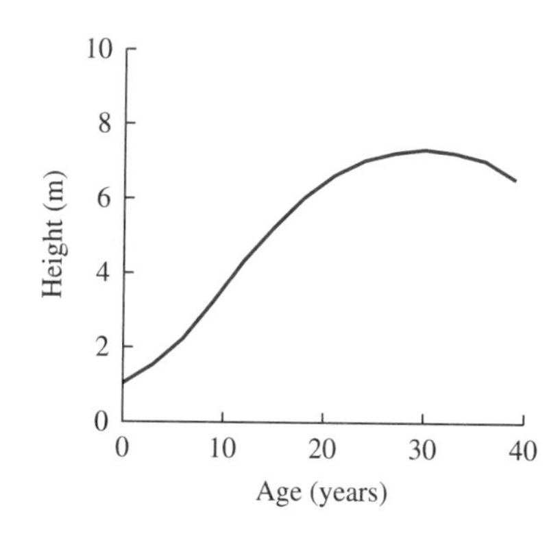

40. A plot of an Internet stock price against time.

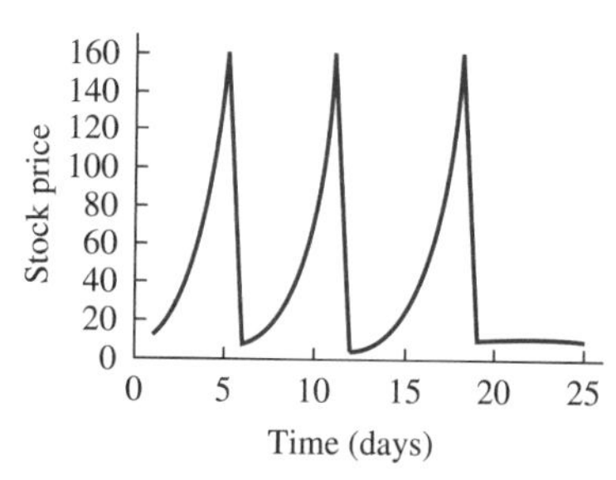

41–44 ▪ Draw graphs based on the following descriptions.

41. A population of birds begins at a large value, decreases to a tiny value, and then increases again to an intermediate value.

42. The amount of DNA in an experiment increases rapidly from a very small value and then levels out at a large value before declining rapidly to 0.

43. Body temperature oscillates between high values during the day and low values at night.

44. Soil is wet at dawn, quickly dries out and stays dry during the day, and then becomes gradually wetter again during the night.

45–48 ▪ Evaluate the following functions over the suggested range, sketch a graph of the function, and answer the biological question.

45. The number of bees b found on a plant is given by $b = 2f + 1$ where f is the number of flowers, ranging from 0 to about 20. Explain what might be happening when $f = 0$.

46. The number of cancerous cells c as a function of radiation dose r (measured in rads) is

$$c = r - 4$$

for r greater than or equal to 5, and is zero for r less than 5. r ranges from 0 to 10. What is happening at $r = 5$ rads?

47. Insect development time A (in days) obeys $A = 40 - \frac{T}{2}$ where T represents temperature in °C for $10 \leq T \leq 40$. Which temperature leads to the more rapid development?

48. Tree height h (in meters) follows the formula

$$h = \frac{100a}{100 + a}$$

where a represents the age of the tree in years for $0 \leq a \leq 1000$. How tall would this tree get if it lived forever?

49–52 ▪ Consider the following data describing the growth of a tadpole.

Age, a (days)	Length, L (cm)	Tail Length, T (cm)	Mass, M (g)
0.5	1.5	1.0	1.5
1.0	3.0	0.9	3.0
1.5	4.5	0.8	6.0
2.0	6.0	0.7	12.0
2.5	7.5	0.6	24.0
3.0	9.0	0.5	48.0

49. Graph length as a function of age.

50. Graph tail length as a function of age.

51. Graph tail length as a function of length.

52. Graph mass as a function of length and then graph length as a function of mass. How do the two graphs compare?

53–56 ▪ The following series of functional compositions describe connections between several measurements.

53. The number of mosquitos (M) that end up in a room is a function of how far the window is open (W, in cm^2) according to $M(W) = 5W + 2$. The number of bites (B) depends on the number of mosquitos according to $B(M) = 0.5M$. Find the number of bites as a function of how far the window is open. How many bites would you get if the window was 10 cm^2 open?

54. The temperature of a room (T) is a function of how far the window is open (W) according to $T(W) = 40 - 0.2W$. How long you sleep (S, measured in hours) is a function of the temperature according to $S(T) = 14 - \frac{T}{5}$. Find how long you sleep as a function of how far the window is open. How long would you sleep if the window was 10 cm^2 open?

55. The number of viruses (V, measured in trillions or 10^{12}) that infect a person is a function of the degree of immunosuppression (I, the fraction of the immune system that is turned off by stress) according to $V(I) = 5I^2$. The fever (F, measured in °C) associated with an infection is a function of the number of viruses according to $F(V) = 37 + 0.4V$. Find fever as a function of immunosuppression. How high will the fever be if immunosuppression is complete ($I = 1$)?

56. The length of an insect (L, in mm) is a function of the temperature during development (T, measured in °C) according to $L(T) = 10 + \frac{T}{10}$. The volume of the insect (V, in cubic mm) is a function of the length according to $V(L) = 2L^3$. The mass (M in milligrams) depends on volume according to $M(V) = 1.3V$. Find mass as a function of temperature. How much would an insect weigh that developed at 25°C?

57–58 ▪ Each of the following measurements is the sum of two components. Find the formula for the sum. Sketch a graph of each component and the total as functions of time for $0 \le t \le 3$. Describe each component and the sum in words.

57. A population of bacteria consists of two types a and b. The first follows $a(t) = 1 + t^2$, and the second follows $b(t) = 1 - 2t + t^2$ where populations are measured in millions and time is measured in hours. The total population is $P(t) = a(t) + b(t)$.

58. The above-ground volume (stem and leaves) of a plant is $V_a(t) = 3.0t + 20.0 + \frac{t^2}{2}$, and the below-ground volume (roots) is $V_b(t) = -1.0t + 40.0$ where t is measured in days and volumes are measured in cm^3. The total volume is $V(t) = V_a(t) + V_b(t)$.

59–62 ▪ Consider the following data describing a plant.

Age, a (days)	Mass, M (g)	Volume, V (cm^3)	Glucose Production, G (mg)
0.5	1.5	5.1	0.0
1.0	3.0	6.2	3.4
1.5	4.3	7.2	6.8
2.0	5.1	8.1	8.2
2.5	5.6	8.9	9.4
3.0	5.6	9.6	8.2

59. Graph M as a function of a. Does this function have an inverse? Could we use mass to figure out the age of the plant?

60. Graph V as a function of a. Does this function have an inverse? Could we use volume to figure out the age of the plant?

61. Graph G as a function of a. Does this function have an inverse? Could we use glucose production to figure out the age of the plant?

62. Graph G as a function of M. Does this function have an inverse? What is strange about it? Could we use glucose production to figure out the mass of the plant?

63–66 ▪ The total mass of a population (in kg) as a function of the time, t, is the product of the number of individuals, $P(t)$, and the mass per person, $W(t)$ (in kg). In each of the following exercises, find the formula for the total mass, sketch graphs of $P(t)$, $W(t)$, and the total mass as functions of time for $0 \le t \le 100$, and describe the results in words.

63. The population of people is $P(t) = 2.0 \times 10^6 + 2.0 \times 10^4 t$, and the mass per person $W(t)$ (in kg) is $W(t) = 80 - 0.5t$.

64. The population is $P(t) = 2.0 \times 10^6 - 2.0 \times 10^4 t$, and the mass per person $W(t)$ is $W(t) = 80 + 0.5t$.

65. The population is $P(t) = 2.0 \times 10^6 + 1000t^2$, and the mass per person $W(t)$ is $W(t) = 80 - 0.5t$.

66. The population is $P(t) = 2.0 \times 10^6 + 2.0 \times 10^4 t$, and the mass per person $W(t)$ is $W(t) = 80 - 0.005t^2$.

Computer Exercises

67–70 ▪ Have your graphics calculator or computer plot the following functions. How would you describe them in words?

67.
a. $f(x) = x^2 e^{-x}$ for $0 \le x \le 20$.

b. $g(x) = 1.5 + e^{-0.1x}\sin(x)$ for $0 \le x \le 20$.

c. $h(x) = \sin(5x) - \cos(7x)$ for $0 \le x \le 5$.

d. $f(x) + h(x)$ for $0 \le x \le 20$ (using the functions in parts **a** and **c**).

e. $g(x) \cdot h(x)$ for $0 \le x \le 20$ (using the functions in parts **b** and **c**).

f. $h(x) \cdot h(x)$ for $0 \le x \le 20$ (using the function in part **c**).

68. Have your computer plot the function

$$h(x) = e^{-x^2} - e^{-1000(x-0.13)^2} - 0.2$$

for values of x between -10 and 10.

a. How would you describe the result in words?

b. Blow up the graph by changing the range to find all points where the value of the function is 0. For example, one such value is between 1 and 2. Plot the function again for x between 1 and 2 to zoom in.

c. If you only found 2 points where $h(x) = 0$, blow up the region between 0 and 1 to try to find two more.

69. Use your computer to find and plot the following functional compositions.

a. $(f \circ g)(x)$ and $(g \circ f)(x)$ if $f(x) = \sin(x)$ and $g(x) = x^2$.

b. $(f \circ g)(x)$ and $(g \circ f)(x)$ if $f(x) = e^x$ and $g(x) = x^2$.

c. $(f \circ g)(x)$ and $(g \circ f)(x)$ if $f(x) = e^x$ and $g(x) = \sin(x)$.

70. Have your graphics calculator or computer plot the following functions for $-2 \le x \le 2$. Do they have inverses?

a. $h_2(x) = x^2 + 2x$.

b. $h_3(x) = x^3 + 2x$.

c. $h_4(x) = x^4 + 2x$.

d. $h_5(x) = x^5 + 2x$.

Have your computer try to find the formula for the inverses of these functions and plot the results. Does the machine always succeed in finding an inverse when there is one? Does it sometimes find an inverse when there is none?

1.3 The Units and Dimensions of Measurements and Functions

Unlike the numbers and functions studied in many mathematics courses, the measurements and relations used by scientists and applied mathematicians have **units** and **dimensions**. Measurements of number, mass, height, and volume are fundamentally different from each other and are said to have different **dimensions**. Measurements of height in inches or in centimeters describe the same quantity but are presented in different **units**. In this section, we learn how to work with the units and dimensions of both measurements and functions. When we wish to express a measurement or relation in different units, we must use appropriate **conversion factors**. Changing the units of a function corresponds to **scaling** or **shifting** the graph of the function. When we wish to express a function in terms of different dimensions, we must **translate** with a **fundamental relation**.

1.3.1 Converting Between Units

The equation

$$2 + 2 = 2$$

looks hopelessly wrong. But

$$2\text{Na}^+ + 2\text{Cl}^- = 2\text{NaCl}$$

is a standard formula from chemistry. The difference is that the terms in the second equation have explicit units: ions of sodium, ions of chlorine, and molecules of salt. Similarly, while it is absurd to write

$$1 = 2.54,$$

it is approximately true that

$$1 \text{ in.} = 2.54 \text{ cm}.$$

(This is the official definition of 1 in.) Numbers with units behave very differently from pure numbers.

In addition to having units, every measurement has an associated number of **significant digits** that describe the precision of that measurement. A length measured as 13.23 cm has four significant digits, meaning that we are highly confident of the first three (13.2 cm) and have only an approximate value of the last. Although measured values are thus only approximate, we will still use the equals sign for any calculations done using all of the significant digits, and reserve the approximation sign ($\approx$) for cases where we are only making rough estimates.

Often, data are presented with more than one unit. To compare measurements, we must be able to convert between different units.

Example 1.3.1 Converting Miles to Centimeters

Suppose we want to know how many centimeters make up a mile. We can do this in steps, first changing miles to feet, then feet to inches, and then inches to centimeters. To convert between units, first **write down the basic identities**

$$5280 \text{ ft} = 1 \text{ mile}$$
$$12 \text{ in.} = 1 \text{ ft}$$
$$2.54 \text{ cm} = 1 \text{ in.}$$

These define how many centimeters are in an inch, how many inches are in a foot, and so forth. We next **divide** to find three **conversion factors**

$$1 = 5280 \frac{\text{ft}}{\text{mile}}$$
$$1 = 12 \frac{\text{in.}}{\text{ft}}$$
$$1 = 2.54 \frac{\text{cm}}{\text{in.}}.$$

Units are manipulated exactly like the numerators and denominators of fractions. We next **multiply** the original measurement by the conversion factors (which are just fancy ways to write the number 1), finding

$$\begin{aligned} 1 \text{ mile} &= 1 \text{ mile} \times 1 \times 1 \times 1 \\ &= 1 \text{ mile} \times 5280 \frac{\text{ft}}{\text{mile}} \times 12 \frac{\text{in.}}{\text{ft}} \times 2.54 \frac{\text{cm}}{\text{in.}} \\ &= 160{,}934 \text{ cm}. \end{aligned}$$

The units cancel like the numerators and denominators of fractions. This method is often called the factor-label method in chemistry.

The steps for converting between units can be summarized in an **algorithm**, which can be thought of as a recipe. This book contains many algorithms for solving particular problems. As with a recipe, following an algorithm without thinking about the steps can lead to disaster.

Algorithm 1.2 The Procedure for Converting Between Units

1. Write down the basic identities that relate the original units to the new units.
2. Divide the basic identities to create conversion factors equal to 1, placing unwanted units where they will cancel.
3. Multiply the original measurement by the appropriate conversion factors.

Example 1.3.2 Using the Algorithm to Change Units of Volume

Suppose a house has an area of 2030 square feet. What is this in square meters? The basic identity relating the new unit to the original unit is

$$0.3048 \text{ m} = 1 \text{ ft}.$$

We want to place feet in the denominator, so we divide to find the conversion factor

$$1 = 0.3048 \frac{\text{m}}{\text{ft}}.$$

Square feet are feet times feet, so we can find

$$\begin{aligned} 2030 \text{ ft}^2 &= 1 \times 1 \times 2030 \text{ ft}^2 \\ &= 0.3048 \frac{\text{m}}{\text{ft}} \times 0.3048 \frac{\text{m}}{\text{ft}} \times 2030 \text{ ft}^2 \\ &= 188.6 \text{ m}^2. \end{aligned}$$

Alternatively, we can create a single conversion factor for changing square feet to square meters

$$1 = 0.3048^2 \frac{\text{m}^2}{\text{ft}^2} = 0.0929 \frac{\text{m}^2}{\text{ft}^2}.$$

Then

$$\begin{aligned} 2030 \text{ ft}^2 &= 1 \times 2030 \text{ ft}^2 \\ &= 0.0929 \frac{\text{m}^2}{\text{ft}^2} \times 2030 \text{ ft}^2 \\ &= 188.6 \text{ m}^2. \end{aligned}$$

Example 1.3.3 Results of Mixing Up Numerator and Denominator

In Example 1.3.1, it would be equally true that

$$\begin{aligned} 1 &= \frac{1}{5280} \frac{\text{mile}}{\text{ft}} \\ 1 &= \frac{1}{12} \frac{\text{ft}}{\text{in.}} \\ 1 &= \frac{1}{2.54} \frac{\text{in.}}{\text{cm}}. \end{aligned}$$

Multiplying by these conversion factors,

$$\begin{aligned} 1\text{mile} &= 1\text{mile} \times 1 \times 1 \times 1 \\ &= 1\text{mile} \times \frac{1}{5280} \frac{\text{mile}}{\text{ft}} \times \frac{1}{12} \frac{\text{ft}}{\text{in.}} \times \frac{1}{2.54} \frac{\text{in.}}{\text{cm}} \\ &= \frac{1}{160934} \frac{\text{mile}^2}{\text{cm}}. \end{aligned}$$

The units did not cancel. Even though the result is **true**, miles and centimeters are left over in a rather inconvenient way. The trick to getting unit conversions to work is making sure that unwanted units in the numerator are canceled by conversion factors with those same units in the denominator, and vice versa.

1.3.2 Translating Between Dimensions

Miles and centimeters measure the same quantity, length, with different rulers. Miles and grams measure completely different quantities, length and mass. **Dimensions** describe the underlying quantities. **Units** are a particular standard for measurement. Measurements in miles and centimeters share the same dimension and can be **converted** into one another. Measurements with different dimensions cannot. The dimensions and units commonly used for biological measurements are listed in Table 1.1.

Suppose we want to measure a bacterial population in grams (mass) rather than numbers or the size of water droplet in cubic centimeters (volume) rather than centimeters (length of radius). We cannot apply a series of identities like 1 in. = 2.54 cm because we are translating between dimensions rather than converting between units. Instead of **identities**, we use **fundamental relations** among measurements with different dimensions (Table 1.2).

Table 1.1 Some Common Quantities, Their Dimensions, and Common Units

Quantity	Dimensions	Sample units
distance	length	meter, micron, inch
duration	time	second, minute, day
mass	mass	gram, kilogram
area	length^2	square meter, acre
volume	length^3	liter, cubic meter, gallon
speed	length/time	meters/second, MPH
acceleration	length/time^2	meters/second^2
force	mass × length/time^2	dynes, pounds
density	mass/length^3	grams/liter

Table 1.2 Important Fundamental Relations in Biology

Relation	Variables	Formula
Geometric Relations		
Volume of a sphere	V = volume r = radius	$V = \frac{4\pi}{3}r^3$
Surface area of a sphere	S = surface area r = radius	$S = 4\pi r^2$
Area of a circle	A = area r = radius	$A = \pi r^2$
Perimeter of a circle	P = perimeter r = radius	$P = 2\pi r$
Volume, area, and thickness	V = volume A = area T = thickness	$V = AT$
Relations Involving Mass		
Total number and mass	m = total mass μ = mass per individual b = number of individuals	$m = \mu b$
Mass, density, and volume	M = total mass ρ = density V = volume	$M = \rho V$

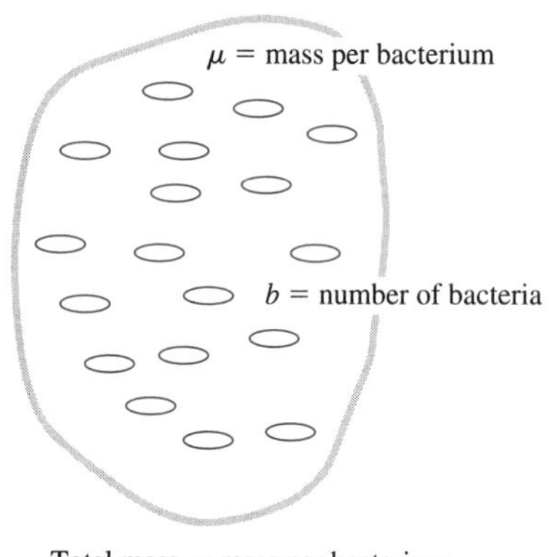

Total mass = mass per bacterium × number of bacteria $m = \mu b$

FIGURE 1.3.1

Fundamental relation between mass and number

Example 1.3.4 Translating Between Number and Total Mass

The fundamental relation between number and total mass is

$$\text{Total mass} = \text{mass per bacterium} \times \text{number of bacteria}.$$

Let m represent the total mass, μ the mass per bacterium, and b the number of bacteria (Figure 1.3.1). The fundamental relation can be written in mathematical symbols as

$$m = \mu b.$$

Like numbers, variables representing measurements have both dimensions and units. The variable m has units of grams, μ has units of grams per bacterium, and b has units of bacteria. If

$$b = 2.0 \times 10^5 \text{ bacteria and } \mu = 3.1 \times 10^{-9}\text{g},$$

then

$$m = (3.1 \times 10^{-9}) \cdot (2.0 \times 10^{5}) = 6.2 \times 10^{-4}\text{g}.$$

Example 1.3.5 Computing the Volume of a Spherical Droplet from Its Radius

Computing the volume of a droplet requires a fundamental relation between radius and volume, which depends on the **shape** of the droplet. Suppose that droplets are perfect spheres. The fundamental relation between radius and volume comes from geometry (Figure 1.3.2). The volume V of a sphere with radius r is

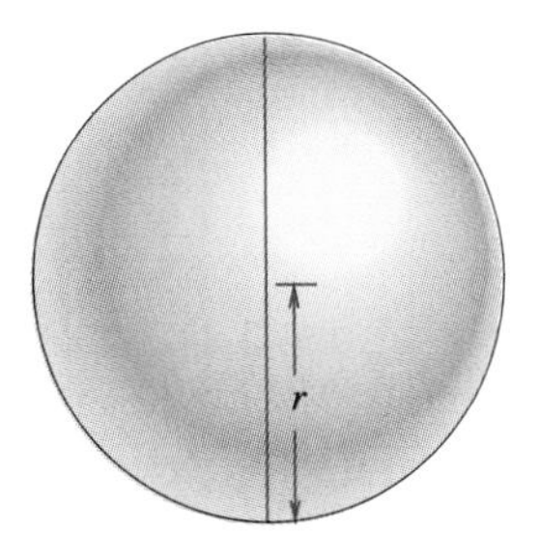

FIGURE 1.3.2
Volume and radius of a spherical droplet

$$V = \frac{4\pi}{3} r^3$$

where r has units of centimeters and V has units of cubic centimeters, or cm^3 (Figure 1.3.2). If a droplet has radius 0.23 cm, the volume is

$$V = \frac{4\pi}{3} 0.23^3 = 0.051.$$

Example 1.3.6 Computing the Mass from the Volume

To compute the mass of the droplet in Example 1.3.5 from its volume we use the fundamental relation

$$\text{mass} = \text{density} \times \text{volume}.$$

If we denote the density by ρ and the mass by M, the fundamental relation can be rewritten in mathematical symbols as

$$M = \rho V.$$

Suppose that the droplet is made of mercury, which has density of $13.58 \frac{\text{g}}{\text{cm}^3}$. The mass M of a droplet with radius 0.23 cm is

$$M = 13.58 \frac{\text{g}}{\text{cm}^3} \cdot 0.051\text{cm}^3 = 0.693 \text{ g}.$$

It is crucial to check the dimensions and units of any unfamiliar equation you encounter. In the equation

$$\text{mass} = \text{density} \times \text{volume},$$

the dimensions of M are mass, the dimensions of V are length^3, and the dimensions of ρ are mass/length^3. Rewriting in dimensions,

$$\text{mass} = \frac{\text{mass}}{\text{length}^3} \times \text{length}^3.$$

The length^3 terms cancel and the dimensions of the two sides match, as they must. This procedure is called **dimensional analysis**. Many errors can be nipped in the bud by checking the dimensions. An equation with inconsistent dimensions is not merely incorrect, it is nonsensical.

1.3.3 Functions and Units: Composition, Scaling, and Shifting

Because functions describe relations between measurements, both their inputs and outputs have units and dimensions. Care must be taken to ensure that functions are composed only when their units and dimensions match.

Example 1.3.7 Composing Functions with Dimensions

Suppose that F takes the radius r of a sphere as input and returns the volume of a sphere as output (as with the spherical droplet in Example 1.3.5). Then F has formula

$$F(r) = \frac{4\pi}{3}r^3.$$

Suppose G takes a volume V as input and returns the mass of an object with that volume as output according to mass = density × volume, or

$$G(V) = 13.58V$$

using the density $\rho = 13.58\frac{\text{g}}{\text{cm}^3}$. The composition $G \circ F$ takes radius as input and returns mass as output in a single step (Figure 1.3.3). The composition is

$$\begin{aligned}(G \circ F)(h) &= G(F(h)) \\ &= G\left(\frac{4\pi}{3}r^3\right) \\ &= 13.58\frac{4\pi}{6}r^3 = 28.44r^3.\end{aligned}$$

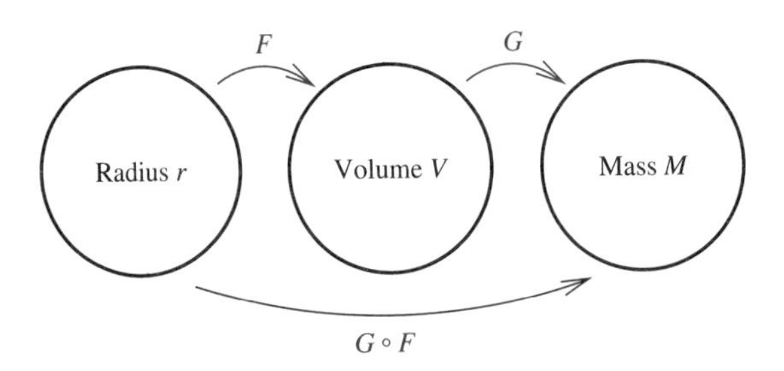

FIGURE 1.3.3
The composition of two functions with units

We could find the mass of a droplet with radius 0.23 cm in two steps by finding the volume and then the mass with the steps

$$\begin{aligned}V &= F(0.2310) = \frac{4\pi}{3}0.23^3 = 0.051 \\ M &= G(0.051) = 13.58 \cdot 0.051 = 0.693.\end{aligned}$$

Alternatively, we could find the mass in a single step by composing the functions $G \circ F$,

$$M = (G \circ F)(0.23) = 13.58\frac{4\pi}{3}0.23^3 = 0.693.$$

F accepts inputs with dimensions of length and returns outputs with dimensions of volume. G accepts inputs with dimensions of volume and returns outputs with dimensions of mass. Because G takes as input precisely what F provides as output, the composition makes sense.

What if we tried to compute $F \circ G$? F cannot accept an input with dimensions of mass, which are the only outputs that G can return. It is impossible to compute the volume of a sphere with a diameter of 4.3 grams. This composition is nonsense. ▲

Changing the units of a measurement that acts as the input or output of a function corresponds to **scaling** or **shifting** the graph of the function. In most cases, the measurement corresponding to a value of zero is the same in different units, and graphs of the function are scaled by changes in units. When units differ in the value corresponding to zero (as with temperature), then the graph of the function is shifted.

Example 1.3.8 Scaling Functions on the Vertical Axis by Changing Units of the Output

Consider the function describing growth of bacterial population

$$b(t) = ct$$

where $c = 2$ million bacteria/hr, t is measured in hours, and $b(t)$ in millions of bacteria.

If bacteria are measured instead in thousands of bacteria, we choose a different variable to represent the new measurement, such as B. The relation becomes

$$\begin{aligned} B(t) &= 2.0 \frac{\text{million bacteria}}{\text{hr}} t \times \frac{1000 \text{ thousand bacteria}}{\text{one million bacteria}} \\ &= 2000.0 \frac{\text{thousand bacteria}}{\text{hr}} t \end{aligned}$$

(Figure 1.3.4). The graph has been changed by **scaling** the vertical axis. It looks the same, except that the numbers labeling the axis are 1000 times larger.

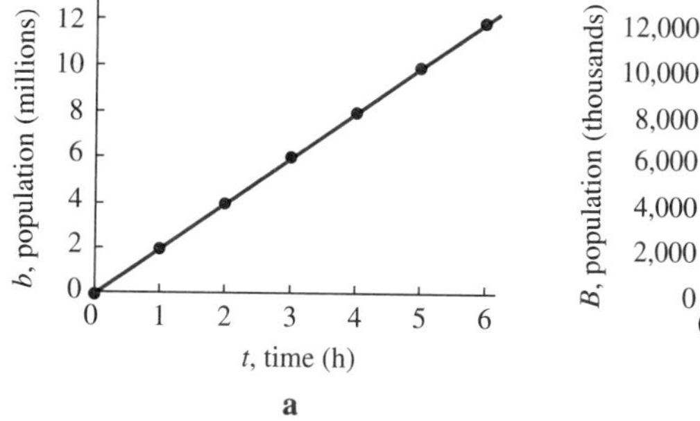

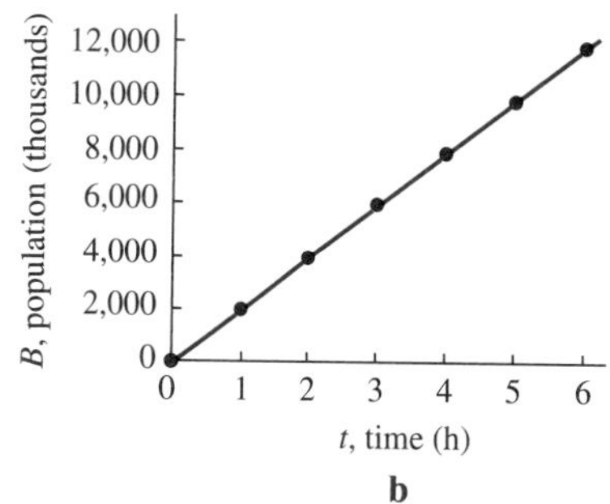

FIGURE 1.3.4
A growing bacterial population: new units on vertical axis

Example 1.3.9 Scaling Functions on the Horizontal Axis by Changing Units of the Input

Suppose again that $b(t) = ct$ with $c = 2$ million bacteria/hr, b is measured in millions, and t in hours. If time were measured in minutes instead of hours, we must define a new variable for time, perhaps m. The relation becomes

$$\begin{aligned} b(m) &= 2.0 \frac{\text{million bacteria}}{\text{hr}} m \times \frac{1 \text{ hour}}{60 \text{ minutes}} \\ &= \frac{1.0}{30.0} \frac{\text{million bacteria}}{\text{min}} m \\ &= 0.0333 \frac{\text{million bacteria}}{\text{min}} m. \end{aligned}$$

The graph has been changed by **scaling** the horizontal axis (Figure 1.3.5). Again, it looks the same, except that the numbers labeling the horizontal axis are 60 times larger.

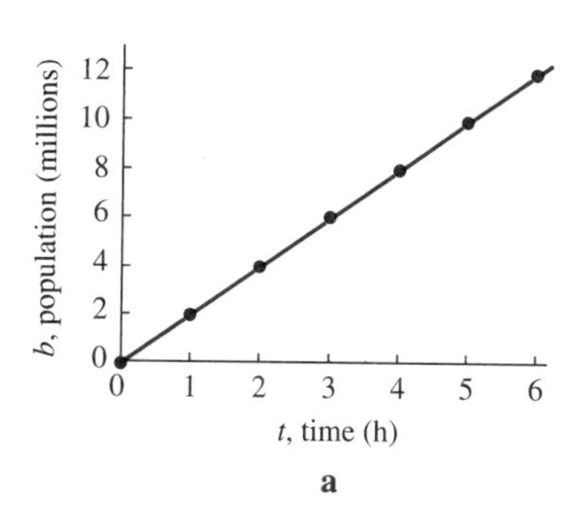

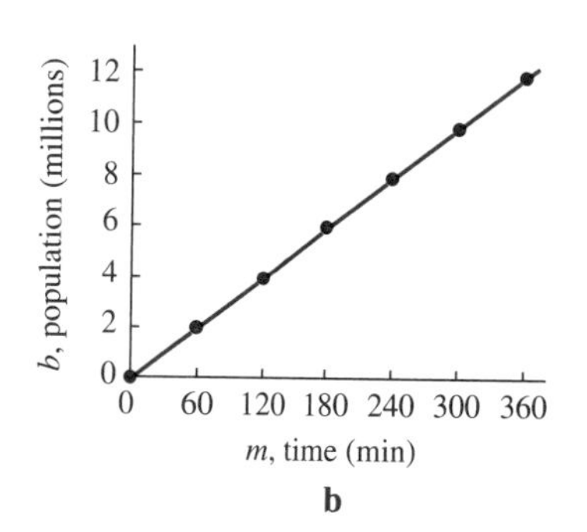

FIGURE 1.3.5
A growing bacterial population: new units on horizontal axis

The data point (1, 2) that indicated that there were 2.0 million bacteria after 1 hour becomes the point (60, 2), indicating 2.0 million bacteria after 60 minutes.

t (h)	m (min)	b (millions)	B (thousands)
0	0	0	0
1	60	2	2,000
2	120	4	4,000
3	180	6	6,000
4	240	8	8,000
5	300	10	10,000
6	360	12	12,000

Example 1.3.10 Shifting Functions Vertically by Changing Units

Temperatures can be measured on scales with different values of zero. For example, 0°C corresponds to 273.15 K. (The units of temperature on the Kelvin scale are referred to as kelvins rather than as degrees, and no degree symbol is used.) Suppose that the temperature of a snake after digesting a mouse with mass m obeys the equation

$$T(m) = 10 + 0.06m$$

where temperature is measured in °C and mouse mass is measured in grams (Figure 1.3.6a). Temperature in K, which we denote by T_K, can be found by adding 273.15 to temperature in °C, or

$$T_K(m) = 273.15 + T(m).$$

In the new units,

$$T_K(m) = 283.15 + 0.06m$$

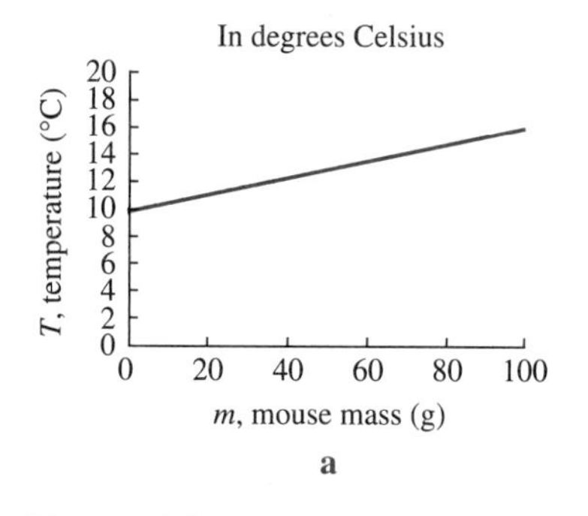

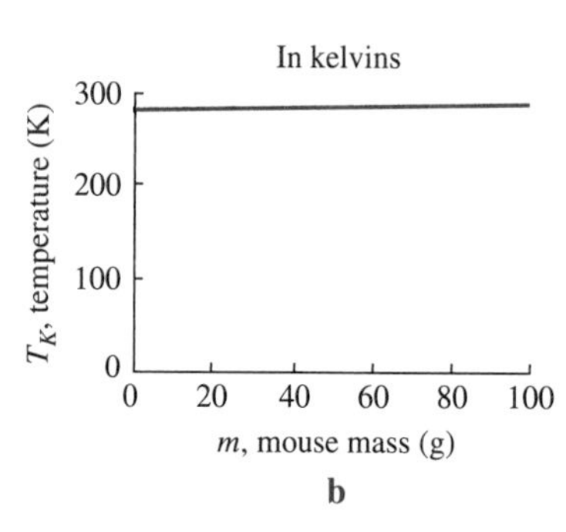

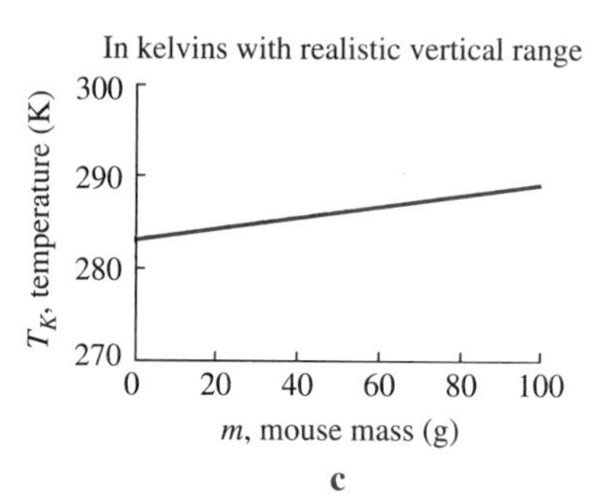

FIGURE 1.3.6
Snake temperature in different units

(Figure 1.3.6b). The graph has been **shifted** vertically. It looks different because 0 K is so far from the temperatures measured. A graph with the horizontal axis set at 270 K is more informative (Figure 1.3.6c).

Example 1.3.11 Shifting Functions Horizontally by Changing Units

Suppose that the sprint speed of a snake is a function of temperature according to the equation

$$s(T) = 4\frac{\text{m}}{\text{s}} + 0.1T$$

where temperature is measured in °C and speed is measured in m/s. Temperature in K can be found by adding 273.15 to temperature in °C, or $T_k = T + 273.15$. To write in the new units, we solve for $T = T_k - 273.15$, giving

$$s(T_K) = 4\frac{\text{m}}{\text{s}} + 0.1(T_K - 273.15)$$

(Figure 1.3.7). The graph has been **shifted** horizontally. However, by including extremely cold temperatures, the function predicts impossible negative speeds. A graph showing only the temperatures between 275 K and 325 K is more informative (Figure 1.3.7c).

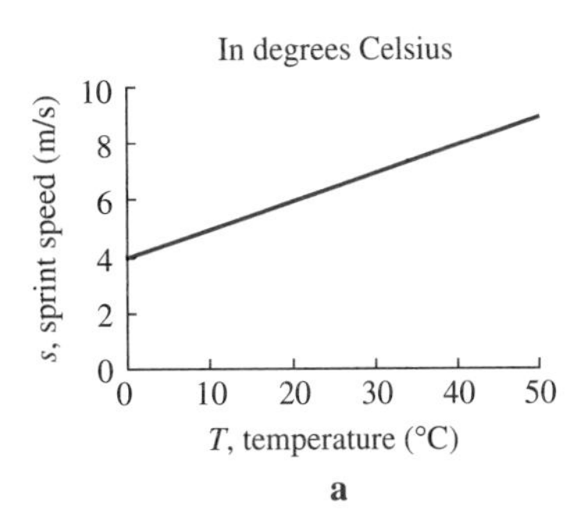

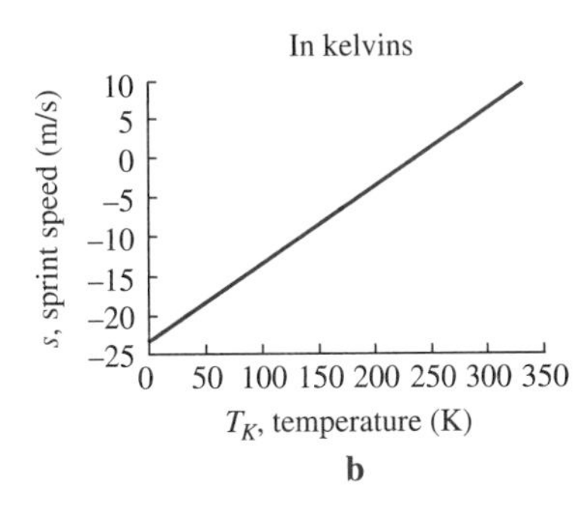

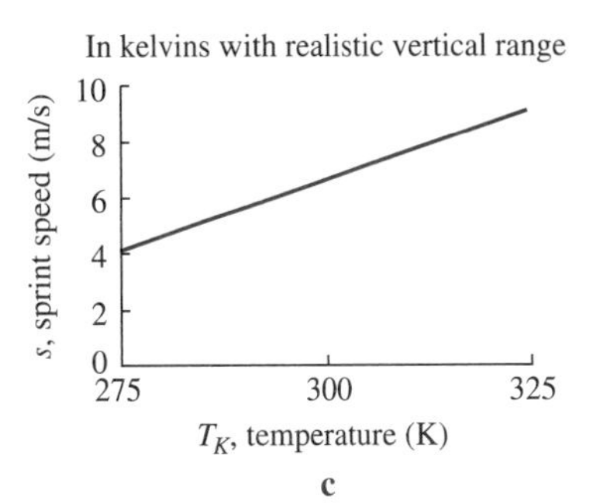

FIGURE 1.3.7
Snake speed in different units

Mathematically, scaling corresponds to multiplying the value or the argument of a function by a constant, while shifting corresponds to adding a constant to the value or argument. In particular, the function $f(x)$ can be scaled or shifted as follows.

- **Vertical Scaling** Multiply the value by the constant a to make the new function $af(x)$ (Figure 1.3.8).

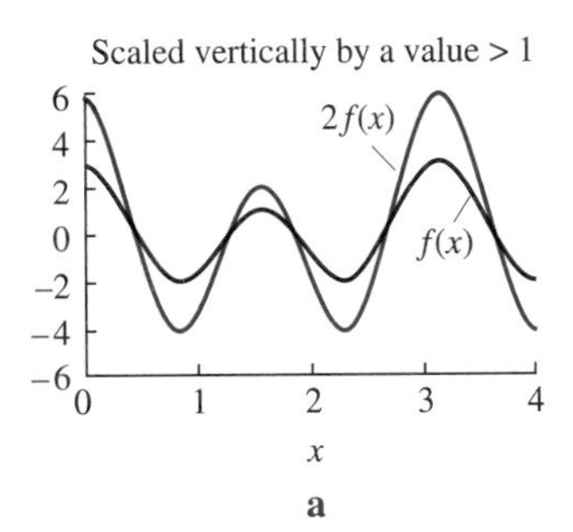

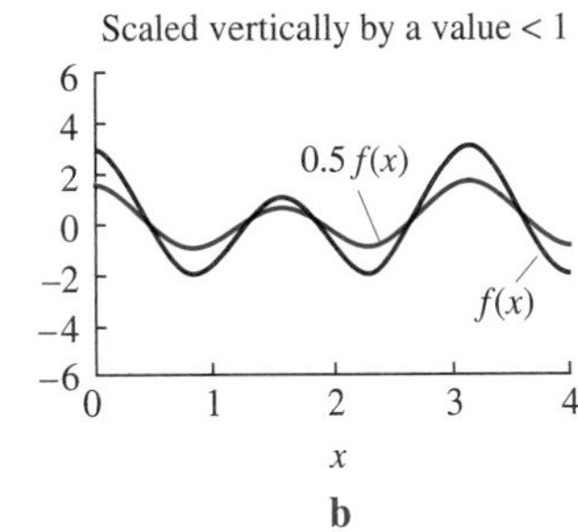

FIGURE 1.3.8
Vertically scaling a function

- **Horizontal Scaling** Multiply the argument by the constant a to make the new function $f(ax)$ (Figure 1.3.9).

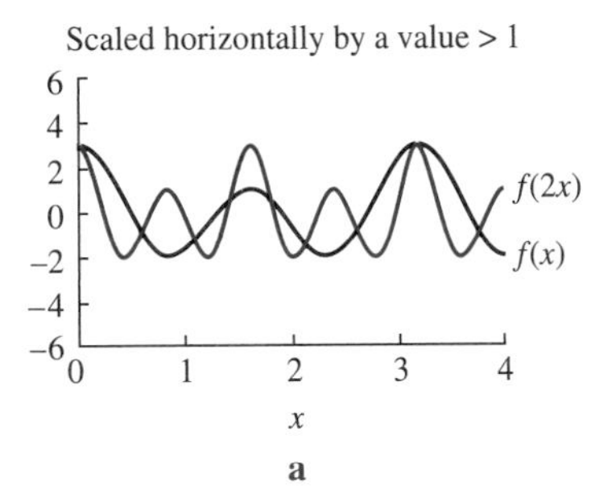

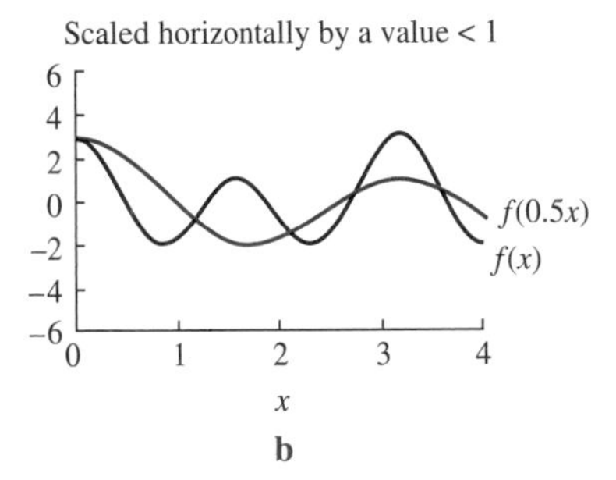

FIGURE 1.3.9
Horizontally scaling a function

- **Vertical Shifting** Add the constant a to the value to make the new function $f(x) + a$. (Figure 1.3.10).

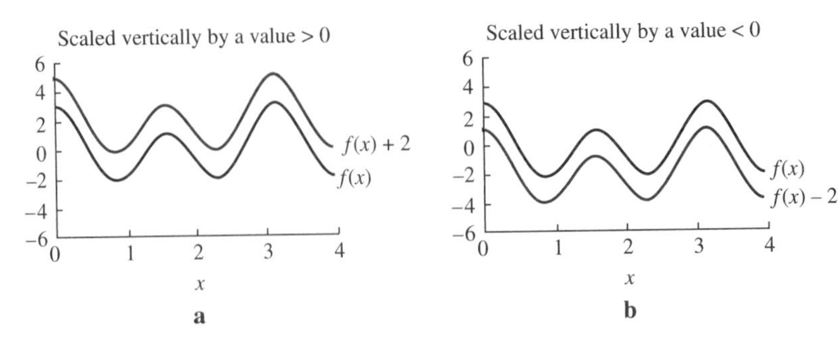

FIGURE 1.3.10
Vertically shifting a function

- **Horizontal Shifting** Add the constant a to the argument to make the new function $f(x + a)$ (Figure 1.3.11).

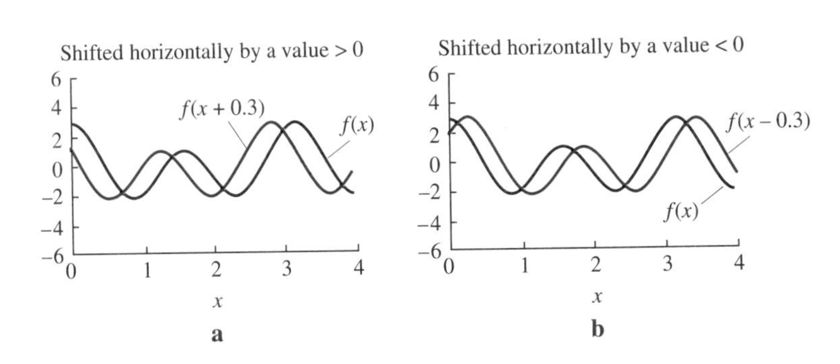

FIGURE 1.3.11
Horizontally shifting a function

Vertical scaling and shifting work as one might expect: multiplying by a value greater than 1 stretches the function (Figure 1.3.8a), and adding a value greater than 0 raises the function (Figure 1.3.10a). Horizontal shifting and scaling, however, might seem to work backwards. Multiplying the argument by a value greater than 1 compresses the function (Figure 1.3.9a), and adding a positive constant to the argument moves the function to the left (Figure 1.3.11a).

Example 1.3.12 Vertically and Horizontally Scaling a Function

Consider the function

$$f(x) = \frac{1}{1 + x^2}$$

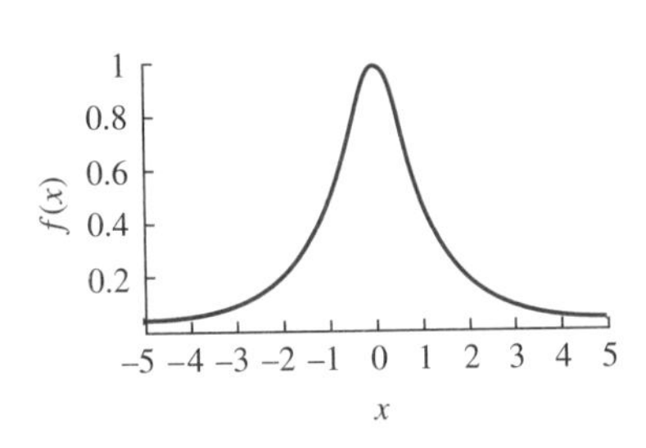

FIGURE 1.3.12
The original function

shown plotted for $-5 \leq x \leq 5$ (Figure 1.3.12). We will scale the value and the argument of this function by values both greater than and less than 1 (Figure 1.3.13).

Consider the input $x = 1.0$. Then

$$2f(1.0) = 2 \cdot \frac{1}{1 + (1.0)^2} = 1.0 \quad \text{scaled vertically by a factor of 2}$$

$$0.5f(1.0) = 0.5 \cdot \frac{1}{1 + (1.0)^2} = 0.25 \quad \text{scaled vertically by a factor of 0.5}$$

$$f(2 \cdot 1.0) = f(2.0) = \frac{1}{1 + (2.0)^2} = 0.2 \quad \text{scaled horizontally by a factor of 2}$$

$$f(0.5 \cdot 1.0) = f(0.5) = \frac{1}{1 + (0.5)^2} = 0.8. \quad \text{scaled horizontally by a factor of 0.5}$$

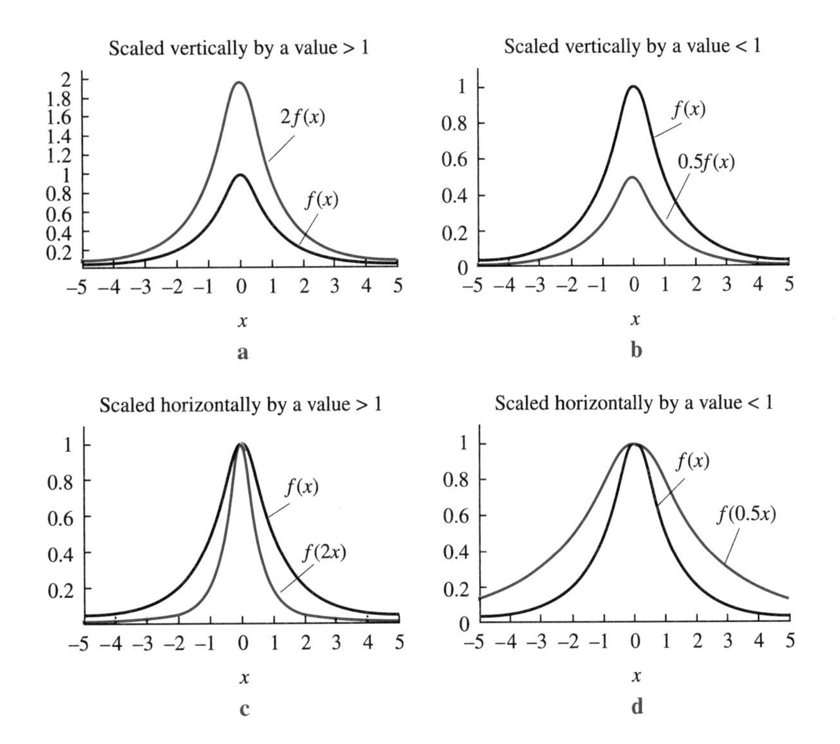

FIGURE 1.3.13
Vertically and horizontally scaling a function

The following table gives several values of the scaled functions.

Argument	Original Function	Vertically Scaled by Value > 1	Vertically Scaled by Value < 1	Horizontally Scaled by Value > 1	Horizontally Scaled by Value < 1
x	$f(x)$	$2f(x)$	$0.5f(x)$	$f(2x)$	$f(0.5x)$
−5.0	0.038	0.077	0.019	0.01	0.14
−4.0	0.059	0.12	0.029	0.015	0.20
−3.0	0.10	0.20	0.05	0.027	0.31
−2.0	0.20	0.40	0.10	0.059	0.50
−1.0	0.50	1.0	0.25	0.20	0.80
0.0	1.0	2.0	0.50	1.0	1.0
1.0	0.50	1.0	0.25	0.20	0.80
2.0	0.20	0.40	0.10	0.059	0.50
3.0	0.10	0.20	0.05	0.027	0.31
4.0	0.059	0.12	0.029	0.015	0.20
5.0	0.038	0.077	0.019	0.01	0.14

Scaling vertically makes the function taller if scaled by a value greater than 1 or shorter if scaled by a value less than 1. Scaling horizontally makes the function thinner if scaled by a value greater than 1 or wider if scaled by a value less than 1.

Example 1.3.13 Vertically and Horizontally Shifting a Function

Consider again the function in Example 1.3.12 (Figure 1.3.12). We will shift the value and the argument of this function by values both greater than and less than 0 (Figure 1.3.14).

Shifting the function vertically corresponds to adding a constant to f.

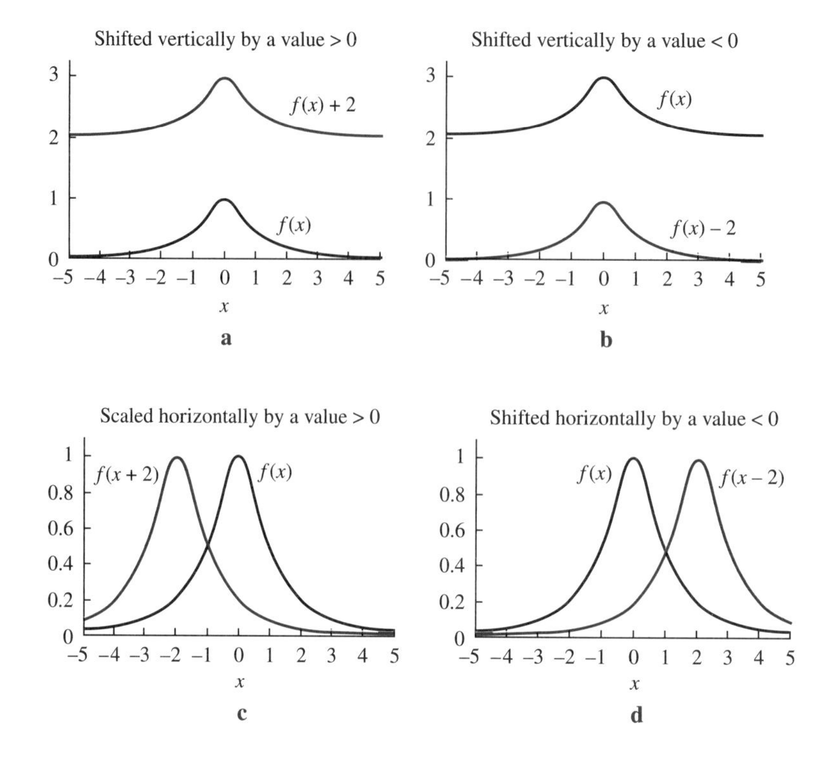

FIGURE 1.3.14
Vertically and horizontally shifting a function

Argument	Original Function	Vertically Shifted by Value > 0	Vertically Shifted by Value < 0	Horizontally Shifted by Value > 0	Horizontally Shifted by Value < 0
x	$f(x)$	$f(x)+2$	$f(x)-2$	$f(x+2)$	$f(x-2)$
−5	0.038	2.04	−1.96	0.1	0.02
−4	0.059	2.06	−1.94	0.2	0.027
−3	0.1	2.1	−1.9	0.5	0.038
−2	0.2	2.2	−1.8	1	0.059
−1	0.5	2.5	−1.5	0.5	0.1
0	1.0	3	−1	0.2	0.2
1	0.5	2.5	−1.5	0.1	0.5
2	0.2	2.2	−1.8	0.059	1
3	0.1	2.1	−1.9	0.038	0.5
4	0.059	2.06	−1.94	0.027	0.2
5	0.038	2.04	−1.96	0.02	0.1

Shifting vertically moves the function up if shifted by a value greater than 0 or down if shifted by a value less than 0. Shifting horizontally moves the function to the right if shifted by a value greater than 0 or to the left if shifted by a value less than 0.

1.3.4 Checking: Dimensions and Estimation

Just as it is essential to check the dimensions and units of equations, it is essential to check the plausibility of the numerical results of calculations. Suppose you wanted to figure out how many tons all the people in the United States weigh. Each person weighs on average of about 100 pounds (counting children) or 1/20 of a ton per person. If there are about 260 million people, they should weigh a net amount of around 13 million tons (using the fundamental relation that total mass is equal to mass per individual times the number of individuals). If you had worked this out with a more complicated set of measurements and found a more precise answer of 14.7 million tons, everything is probably all right. If the complicated method gave an answer of 1.47 million tons, it needs to be checked.

Example 1.3.14 Estimating the Area of a Bacterial Colony

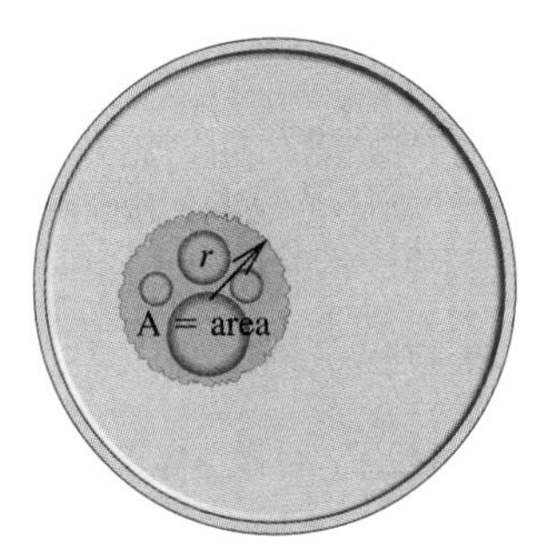

FIGURE 1.3.15
A bacterial colony on a Petri dish

How much area does a colony of 2.0×10^5 bacteria take up on a Petri dish? (Figure 1.3.15) One method is to use our computation of the mass (6.2×10^{-4}g in Example 1.3.4), convert to volume, and find the area by dividing by the thickness. Assuming that bacteria have approximately the density of water, the volume is

$$V = \frac{M}{\rho} = \frac{6.2 \times 10^{-4}\text{gm}}{1.0 \times 10^{-12}\frac{\text{gm}}{\mu\text{m}^3}}$$
$$= 6.2 \times 10^8 \mu\text{m}^3.$$

Here we used the fact that $1.0\ \mu\text{m}^3 = 10^{-12}\text{cm}^3$ to find density in $\frac{\text{gm}}{\mu\text{m}^3}$. The next **fundamental relation** translates between volume and area, and is

$$\text{volume} = \text{area} \times \text{thickness}$$

so that

$$\text{area} = \frac{\text{volume}}{\text{thickness}}.$$

Estimating the thickness of the colony to be about 20 μm (roughly the thickness of a cell),

$$\text{area} \approx \frac{6.2 \times 10^8 \mu\text{m}^3}{20\mu\text{m}}$$
$$\approx 3 \times 10^7 \mu\text{m}^2.$$

This sounds rather large. To convert to centimeters, we use the basic identity

$$1\mu\text{m} = 10^{-4}\text{cm}$$

so that the conversion factor is

$$1 = 10^{-4}\frac{\text{cm}}{\mu\text{m}}.$$

Multiplying,

$$3 \times 10^7 \mu\text{m}^2 = 3 \times 10^7 \mu\text{m}^2 \times 10^{-4}\frac{\text{cm}}{\mu\text{m}} \times 10^{-4}\frac{\text{cm}}{\mu\text{m}}$$
$$= 0.3\ \text{cm}^2.$$

To find the radius, we use the **fundamental geometric relation** between the area A and radius r for a circle

$$A = \pi r^2.$$

The radius r of this colony satisfies

$$\pi r^2 \approx 0.3.$$

Solving for r,

$$r \approx \sqrt{\frac{0.3}{\pi}} \approx 0.3\ \text{cm}.$$

This colony is in fact fairly small, but large enough to be seen.

Example 1.3.15 Fermi's Piano Tuner Problem

The great physicist Enrico Fermi emphasized our ability to combine educated guesses of ordinary quantities to estimate more complicated quantities. For example, we can estimate the number of piano tuners in Salt Lake City and vicinity knowing only that the population is about 1,000,000 people. First, we estimate the number of pianos. If the average family contains four members, the number of families is about 250,000. As a rough guess, suppose that one in five families owns a piano. There will then be 50,000 pianos. If the average piano tuner tunes four pianos every day and works for 250 days per year (50 weeks of 5 days), she would tune 1,000 pianos per year. If each piano is tuned once in two years (another very rough guess), there would be 25,000 pianos tuned per year, requiring 25 piano tuners. A quick check of the phone book indicates that there are in fact about 30 piano tuners. Like all mathematical models, this method requires us to **analyze** the problem by breaking it into component parts. If our estimate proved to be extremely inaccurate, we could check each of our assumptions to find the source of the error.

Summary Understanding scientific equations and formulas requires understanding the **units** and **dimensions** of the measurements and variables. **Dimensions** describe the underlying quantities and tell what sort of thing is being measured. **Units** express numerical values based on a particular scale. Converting between units can be done by starting with **basic identities**, deriving **conversion factors**, and multiplying. Unit conversions correspond to **scaling** or **shifting** the graphs of functions describing the measurements. Translating between measurements with different dimensions requires using **fundamental relations**, such as those between mass and volume or volume and radius. All such relations, and every scientific formula, should be checked for consistency with **dimensional analysis.** Using basic identities and fundamental relations, we can compute useful estimates of quantities, often without using a calculator. Checking results for plausibility can help locate mistakes.

1.3 Exercises

Mathematical Techniques

1–6 ▪ Convert the following into the new units.

1. Find 3.4 pounds in grams (1 ounce is 28.35 grams and 1 pound is 16 ounces).
2. Find 1 yard in mm (1 in. is 25.40 mm and 1 yard is 36 in.).
3. Find 60 years in hours (1.0 year $\approx$ 365.25 days).
4. Find 65 miles per hour in centimeters per second.
5. Find 2.3 grams per cubic centimeter in pounds per cubic foot.
6. Find 9.807 m/sec^2 (the acceleration of gravity) in miles per hour per second.

7–10 ▪ Compute the answers by adding the given quantities.

7. A boy who is 1.34 meters tall grows 2.3 cm.
8. After waiting for 1.2 hours for a plane flight, you are told you will have to wait another 17 minutes. What is the total wait?
9. You purchase 6 apples that weigh 145 g each, and 7 oranges that weigh 123 g each. What is the total weight if you add the apples to the oranges?
10. The density of the apples in the previous problem is 0.8 g/cm^3 and the density of the oranges is 0.95 g/cm^3. What is the total volume if you add the apples to the oranges?

11–14 ▪ Figure out which of the following is larger.

11. The area of a square with side length 1.7 cm or of a disk with radius 1.0 cm.
12. The perimeter of a square with side length 1.7 cm or of a circle with radius 1.0 cm.
13. The volume of a sphere with radius 100 m or of a 50 cm deep lake with an area of 3.0 square km.
14. The surface area of a sphere with radius 100 m or the surface area of a lake with area 3.0 square km.

15–18 ▪ Find the dimensions of the following quantities.

15. Pressure (force per unit area)
16. Energy (force times distance)
17. The rate of change of the area of a colony of bacteria growing on a plate.
18. The force of gravity between two objects is equal to Gm_1m_2/r^2 where m_1 and m_2 are the masses of the two objects, and r is the distance between them. What are the dimensions of the gravitational constant G?

19–22 ▪ Check whether the following formulas are dimensionally consistent.

19. Distance = rate times time.

20. Velocity = acceleration times time.

21. Force = mass times acceleration.

22. Energy = 1/2 mass times the square of velocity (see Exercise 16 for the units of energy).

23–26 ▪ Using the graph of the function $g(x)$, sketch a graph of the shifted or scaled function, say which kind of shift or scale it is, and compare with the original function.

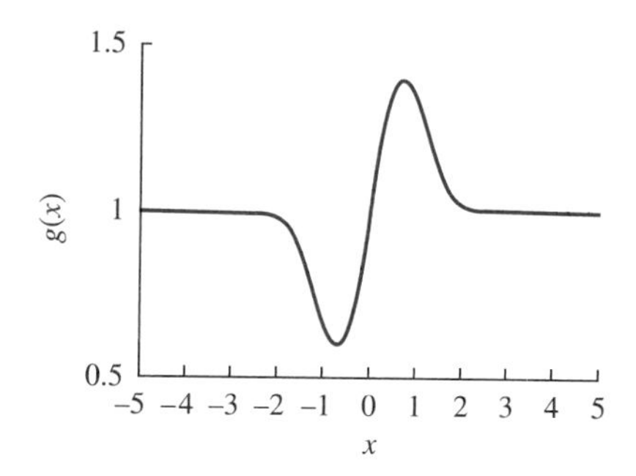

23. $4g(x)$

24. $g(x) - 1$

25. $g(x/3)$

26. $g(x + 1)$

Applications

27–30 ▪ Find the volumes of the following two cartoon trees (drawing a sketch can help) assuming that the height of the first is 23.1 m and that the height of the second is 24.1 m. What is the ratio of the volume of the larger tree to that of the smaller tree?

27. A tree is a perfect cylinder with radius 0.5 m no matter what the height (the volume of a cylinder with height h and radius r is $\pi h r^2$).

28. A tree is a perfect cylinder with radius equal to 0.1 times the height.

29. A tree looks like the tree in Exercise 27, but with half the height in the cylindrical trunk and the other half in a spherical blob on top.

30. A tree looks like the tree in Exercise 27, but with 90% of the height in the cylindrical trunk and the remaining 10% in a spherical blob on top.

31–34 ▪ Find the mass in kilograms of the following objects.

31. A water bed that is 2 m long, 20 cm thick, and 1.5 m wide. The density of water is 1.0 g/cm^3.

32. A spherical cow with diameter 1.3 m and density 1.3 g/cm^3.

33. A coral colony of 3200 individuals each weighing 0.45 g.

34. A circular colony of mold with diameter of 4.8 cm and density of 0.0023 g/cm^2.

35–38 ▪ Change the units in the following functions, and compare a graph in the new units with the original units.

35. (Based on Section 1.2, Exercise 45) The number of bees b on a plant is given by $b = 2f + 1$ where f is the number of flowers. Suppose each flower has 4 petals. Graph the number of bees as a function of the number of petals.

36. The number of cancerous cells c as a function of radiation dose r is

$$c = r - 4$$

for r (measured in rads) greater than or equal to 5, and is zero for r less than 5 (as in Section 1.2, Exercise 46). Suppose that radiation is instead measured in millirads (1 rad = 1000 millirads).

37. Insect development time A (in days) obeys $A = 40 - T/2$ where T represents temperature in °C for °C between 10 and 40 (as in Section 1.2, Exercise 47). Suppose that development time is measured in hours.

38. Tree height h (in meters) follows the formula

$$h = \frac{100a}{100 + a}$$

where a represents the age of the tree in years (as in Section 1.2, Exercise 48). Suppose that tree age is measured instead in decades.

39–44 ▪ Estimate the following.

39. The speed of light in cm per ns (10^{-9} seconds or one nanosecond) (the speed of light is about 186,000 miles/second). A fast computer takes about 0.3 ns per operation. How far does light travel in the time required by one operation?

40. Estimate the speed that your hair grows in miles per hour. (This problem was borrowed from the book *Innumeracy*.)

41. The weight of the earth in kilograms. The earth is approximately a sphere with radius 6500 km and density 5 times that of water.

42. Suppose a person eats 2000 Kcal per day. Using the facts that 1 Kcal is approximately 4.2 Kj (a kilojoule is a unit of energy equal to 1000 joules) and 1 watt is one joule per second (a unit of power), about how many watts does a person use?

43. If a movie is about 2 hours long, how many movies could you watch if you spent half your time watching movies for 60 years?

44. The volume of all the people on earth in cubic kilometers. If a large mine is about 3 km across and 1 km deep, would they all fit?

45–48 ▪ The following problems give several ways to estimate the number of cells in your body. A cell is roughly a sphere 10 μm in radius, where 1 μm is 10^{-6} m.

45. Using the fact that the density of a cell is approximately the density of water and that water weighs 1 g/cm^3, estimate the number of cells in your body.

46. Estimate your volume in cubic meters by pretending you are shaped like a board. Pretending that cells are cubes 20 μm on a side, what do you estimate the number of cells to be by this method?

47. The brain weighs about 1.3 kg and is estimated to have about 100 billion neurons and 10 to 50 times as many other cells (glial cells). Is this consistent with our previous estimates?

48. The nematode *C. elegans* is a cylinder about 1 mm long and 0.1 mm in diameter, consisting of about 1000 cells. Are these cells about the same size as the ones in your body?

49–50 ▪ The following problems regard tying string or gift-wrapping our planet, thought of as a sphere with radius 6500 km.

49. How long a piece of string would be required to go around the equator? If the string were made 1.0 meters longer and stretched out all the way around, how high would it be above the surface? Does the result surprise you?

50. How large a piece of shrink-wrap would be required to cover the entire planet? If the wrap were increased in area by $1.0\ \text{m}^2$ and stretched out all around, how high would it be above the surface? Why do you think the result is so different from the previous problem? (Working this out takes a lot of decimal places.)

1.4 Linear Functions and Their Graphs

Complicated models are built from simple pieces. Throughout the sciences, the simplest building blocks for mathematical models are **linear** functions, functions that have straight lines for their graphs. In this section, we derive formulas for linear functions, including the **point-slope formula** and the **slope-intercept formula**. Because linear functions are simple to work with algebraically, we use them to review methods for solving linear equations to answer scientific questions. In particular, we **interpolate** between known values to make predictions about the results of additional experiments.

1.4.1 Proportional Relations

The simplest relations are **proportional** relations, meaning that the output is **proportional** to the input. Mathematically, this means that the ratio of the output to the input is a constant. The general formula for a proportional relation is

$$f(x) = ax$$

where a is some constant value. The ratio of the output ax to the input x is

$$\frac{\text{output}}{\text{input}} = \frac{ax}{x} = a$$

as long as $x \neq 0$. When the ratio is constant, the value a is called the **constant of proportionality**. Constants of proportionality, like all measurements, have units and dimensions.

Example 1.4.1 A Proportional Relation Between Population Sizes

The function describing the relation

$$b_f = 2.0 b_i$$

(Example 1.2.3) multiplies its input by 2.0 to produce the output (Figure 1.4.1). The ratio of the output population to the input population is

$$\frac{b_f}{b_i} = 2.0,$$

a constant value. The graph of this proportional relation is a line passing through the origin. ▲

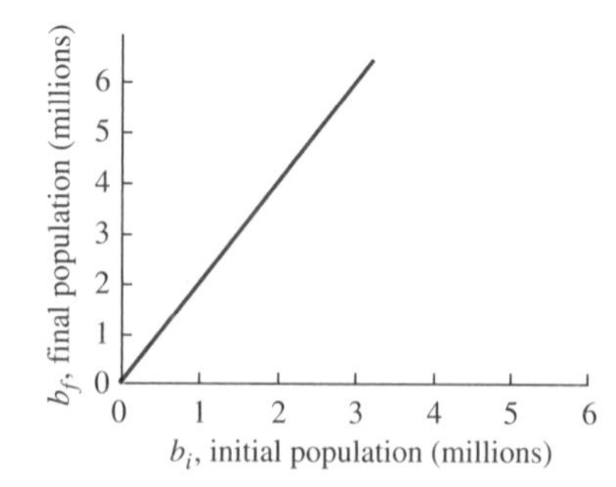

FIGURE 1.4.1

A proportional relation: bacterial populations

Example 1.4.2 A Proportional Relation Between Mass and Volume

In the fundamental relation between mass and volume $M = \rho V$ (Table 1.2), mass is found by multiplying volume by the constant value ρ. The ratio of mass to volume is

$$\frac{\text{mass}}{\text{volume}} = \text{density} = \rho,$$

again a constant. As it must, the constant of proportionality, ρ, has the dimensions of density (mass per unit volume).

The proportion in a proportional relation is the **slope** of the graph. Slope is often defined as "rise over run," but we replace these archaic terms with more scientifically meaningful synonyms.

Definition 1.7 **Slope of a Line**

$$\text{slope} = \frac{\text{change in output}}{\text{change in input}}.$$

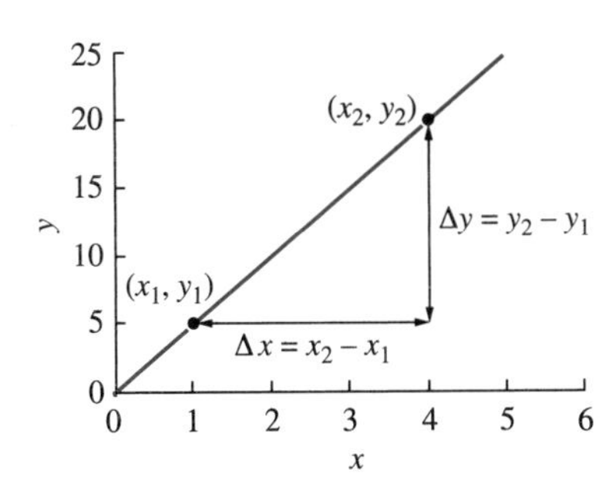

FIGURE 1.4.2
Slope and Δ notation

The "change" is the change between two data points. Suppose we denote two points on the graph by (x_1, y_1) and (x_2, y_2) (Figure 1.4.2). Then

$$\text{slope} = \frac{\text{change in output}}{\text{change in input}} = \frac{y_2 - y_1}{x_2 - x_1}. \tag{1.4.1}$$

Example 1.4.3 Finding the Slope Between Two Data Points

In Figure 1.4.3, the data points are $(x_1, y_1) = (1, 5)$ and $(x_2, y_2) = (4, 20)$. The slope is

$$\text{slope} = \frac{\text{change in output}}{\text{change in input}} = \frac{y_2 - y_1}{x_2 - x_1} = \frac{20 - 5}{4 - 1} = 5.$$

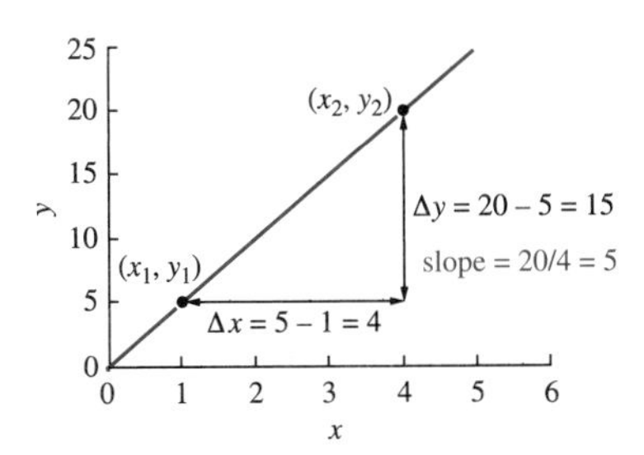

FIGURE 1.4.3
Finding the slope

The changes in the input x and the output y are often written with the shorthand

$$\Delta x = x_2 - x_1$$
$$\Delta y = y_2 - y_1$$

where Δ (the Greek letter "Delta") means "change in." The slope is Δy divided by Δx, or

$$\text{slope} = \frac{\Delta y}{\Delta x}. \tag{1.4.2}$$

This notation will prove very useful when we study derivatives later in this book.

Example 1.4.4 The Slope of a Proportional Relation Between Populations

Recall the data in Example 1.2.3.

Colony	Initial Population, b_i	Final Population, b_f
1	0.47	0.94
2	3.3	6.6
3	0.73	1.46
4	2.8	5.6
5	1.5	3.0
6	0.62	1.24

The first two data points are (0.47, 0.94) and (3.3, 6.6) (Figure 1.4.4). Then

$$\Delta b_i = 3.3 - 0.47 = 2.83$$
$$\Delta b_f = 6.6 - 0.94 = 5.66.$$

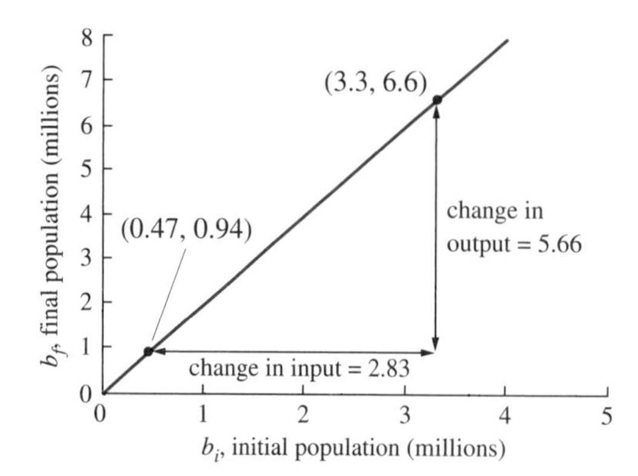

FIGURE 1.4.4
The slope of the proportional relation between bacterial populations

The slope is

$$\text{slope} = \frac{\text{change in output}}{\text{change in input}} = \frac{\Delta b_f}{\Delta b_i} = \frac{6.6 - 0.94}{3.3 - 0.47} = \frac{5.66}{2.83} = 2.0.$$

The slope is equal to the constant of proportionality.

Example 1.4.5 The Slope of a Proportional Relation Between Mass and Volume

Suppose that $\rho = 0.8 \frac{\text{g}}{\text{cm}^3}$ (Figure 1.4.5). A first object with volume $V_1 = 1.0\ \text{cm}^3$ has mass $M_1 = 0.8$ g. A second object with volume $V_2 = 4.0\ \text{cm}^3$ has mass $M_2 = 3.2$ g. We then find

$$\text{change in output} = \Delta M = 3.2 - 0.8 = 2.4\ \text{gm}$$
$$\text{change in input} = \Delta V = 4.0 - 1.0 = 3.0\ \text{cm}^3.$$

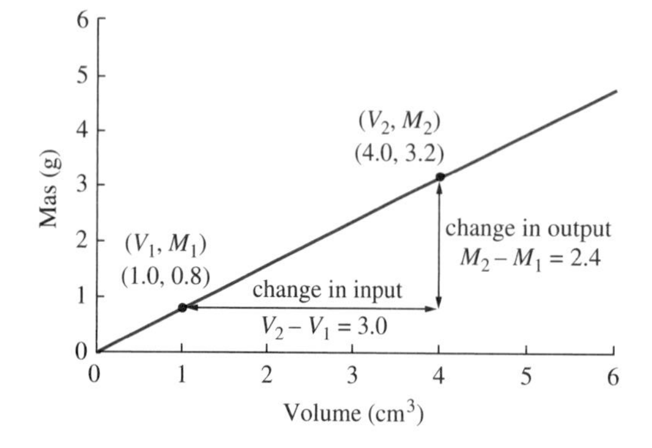

FIGURE 1.4.5
The slope of the proportional relation between mass and volume

The slope is then

$$\text{slope} = \frac{\text{change in output}}{\text{change in input}} = \frac{\Delta M}{\Delta V} = \frac{2.4\ \text{gm}}{3.0\ \text{cm}^3} = 0.8 \frac{\text{g}}{\text{cm}^3}.$$

Again, the slope is equal to the constant of proportionality, here with the units of density.

1.4.2 Linear Functions and the Equation of a Line

Proportional relations are described by functions that perform a single operation on their input: multiplication by a constant. The graphs of such functions are lines with slope equal to the constant of proportionality. Furthermore, these lines pass through the point (0, 0) because an input of 0 produces an output of 0.

Many functions other than those describing proportional relations have linear graphs.

Example 1.4.6 A Linear Function That Is Not a Proportional Relation

The graph of the function

$$y = f(x) = x + 1$$

follows a line (Figure 1.4.6). But the relation between the input x and the output y is **not** a proportional relation. Two points on this line are (1, 2) and (2, 3). At the first, the ratio of output to input is $\frac{2}{1} = 2$. At the second, the ratio of output to input is $\frac{3}{2} = 1.5$. The ratio of output to input is not constant. ▲

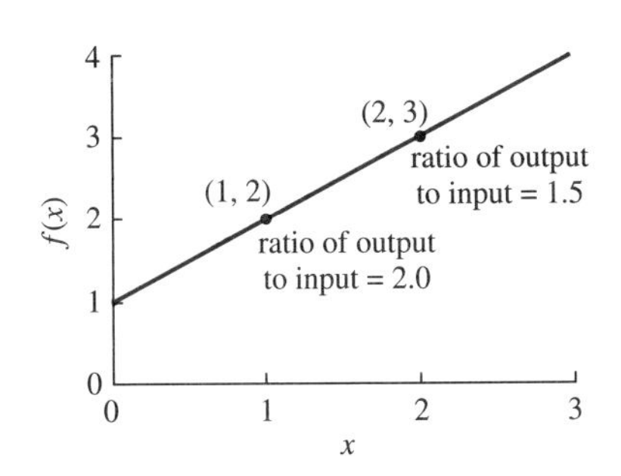

FIGURE 1.4.6

A linear function that is not a proportional relation

For linear functions, it is the ratio of the **change in output** to the **change in input** that is constant. Suppose we start at the point (0, 1) on the graph. The ratio of change in output to change in input between this point and (1, 2) is

$$\frac{\text{change in output}}{\text{change in input}} = \frac{2-1}{1-0} = 1.$$

The ratio of change in output to change in input between this point and (2, 3) is

$$\frac{\text{change in output}}{\text{change in input}} = \frac{3-1}{2-0} = 1.$$

In general, a **line** is characterized by a **constant slope**, like a constant grade on a road. We use this fact to find a formula for a line. First, choose any point that lies on the graph of the function and call it the **base point**. If the base point has coordinates (x_0, y_0), the slope between it and an arbitrary point (x, y) on the line is

$$\text{slope} = \frac{\Delta y}{\Delta x} = \frac{y - y_0}{x - x_0}$$

Because the slope between any two points on the graph is constant,

$$\frac{y - y_0}{x - x_0} = m$$

for some fixed value of m. Multiplying both sides by $(x - x_0)$, we find

$$y - y_0 = m(x - x_0).$$

After solving for y by adding y_0 to both sides, we find the following form.

Definition 1.8 **The Point-Slope Form for a Line**

A line passing through the point (x_0, y_0) with slope m has formula

$$y = m(x - x_0) + y_0.$$

Alternatively, we can multiply out the terms on the right hand side of the point-slope form, finding

$$y = mx + (y_0 - mx_0).$$

We can combine the constants y_0, m, and x_0 into a single new parameter $b = y_0 - mx_0$. The letter b represents the point where the graph crosses the y-axis and is called the ***y*-intercept** (Figure 1.4.7).

Definition 1.9 **The Slope-Intercept Form for a Line**

A line with slope m and y-intercept b has formula

$$y = mx + b.$$

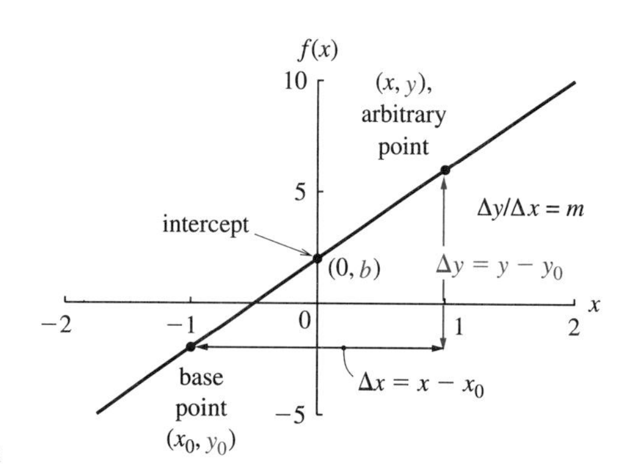

FIGURE 1.4.7
The elements of the general linear graph

In functional notation, if the function $f(x)$ has a linear graph passing through the base point (x_0, y_0) with slope m, then

$$f(x) = m(x - x_0) + y_0$$

in point-slope form. Similarly,

$$f(x) = mx + b$$

in slope-intercept form.

Example 1.4.7 Recognizing the Components of a Linear Function: Slope-Intercept Form

The function $f(x) = -4x + 5$ is a linear function in slope-intercept form. The slope is the factor multiplying the input x, or $m = -4$. The intercept is $b = 5$.

Example 1.4.8 Recognizing the Components of a Linear Function: Point-Slope Form

The function $f(x) = 3(x + 2) + 7$ is a linear function in point-slope form. To find x_0, we must write $x + 2$ as $x - (-2)$. This is in the form $x - x_0$ with $x_0 = -2$. The y coordinate of the point is the added value or $y_0 = 7$. Thus, the base point is $(x_0, y_0) = (-2, 7)$. The slope is the factor multiplying the variable x, or $m = 3$.

Example 1.4.9 Recognizing the Components of a Biological Linear Function

The function describing the relation between initial and final bacterial populations (Example 1.4.1)

$$b_f = f(b_i) = 2.0b_i$$

is a linear function with a slope of 2.0 and a y-intercept of 0 (and is therefore a proportional relation). In applications, inputs and outputs are rarely called x and y. Nonetheless, we recognize linear functions by the operations done to the input variable. If the formula involves only adding, subtracting, and multiplying by constants, the equation describes a linear function.

Example 1.4.10 Recognizing a Nonlinear Function

The function

$$b(t) = \frac{5.0}{1 + 2t}$$

is not a linear function because the input variable t appears in the the denominator. The function

$$b(t) = t^2 + 3t + 2$$

is not linear because the input variable t is squared.

Example 1.4.11 The Linear Relation Between Fahrenheit and Celsius

A once important linear function converts temperature in Fahrenheit into temperature in Celsius (Figure 1.4.8). Recall that

$$F = 1.8C + 32, \tag{1.4.3}$$

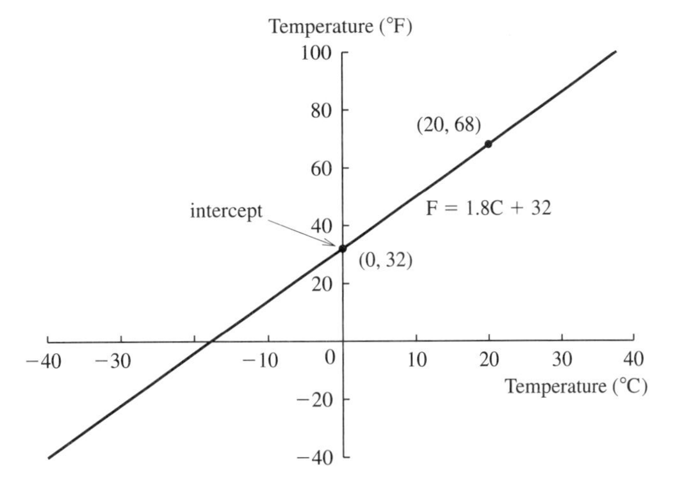

FIGURE 1.4.8
The relation between Fahrenheit and Celsius temperatures

where C represents temperature in Celsius and F represents temperature in Fahrenheit. Unlike almost all unit conversions, this formula does not express a proportional relation. The y-intercept of 32 indicates that 0°C corresponds to 32°F rather than 0°F. The slope, nonetheless, describes the number of degrees Fahrenheit per degree Celsius as in an ordinary conversion. To check the slope, we compute ΔF and ΔC between the points with $C=0$ and $C=20$. Because 20°C corresponds to 68°F, the change in °F (the output) is

$$\Delta F = 68°\text{F} - 32°\text{F} = 36°\text{F}.$$

and the change in °C (the input) is

$$\Delta C = 20°\text{C} - 0°\text{C} = 20°\text{C}.$$

Therefore

$$\text{slope} = \frac{36°\text{F}}{20°\text{C}} = 1.8\frac{°\text{F}}{°\text{C}}.$$

It takes 1.8°F to make up 1.0°C.

1.4.3 Finding Equations and Graphing Lines

Pet store owners can be plagued by parasites. Suppose the employees spend a week observing populations of mites on several lizards and find the following data.

Initial Number, x_i	Final Number, x_f
20	70
30	90
40	110
50	130

Here x_i is the initial number of mites and x_f is the final number. Suppose we wish to estimate the number of mites we would find in a week on a lizard that has 45 mites today.

To do so, we must first find an equation for the function relating x_f and x_i and evaluate it at $x_i = 45$.

We can find the equation with the following steps.

▸▸ Algorithm 1.3 Finding the Equation of a Line from Data

1. Graph the data and check that the points lie on a line.
2. Pick two data points.
3. Find the slope as the change in output divided by the change in input.
4. Find the equation by plugging one point into the point-slope form for a line (Definition 1.8).
5. If needed, convert into slope-intercept form.

▲

Example 1.4.12 Finding the Equation of a Line from Data

We can follow this algorithm to find the equation of the function describing our data.

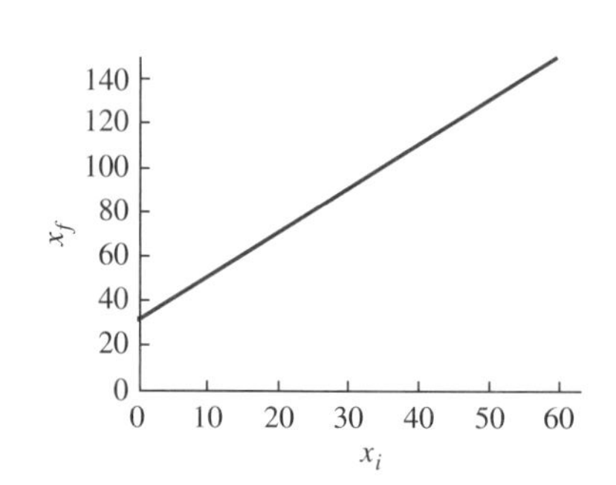

FIGURE 1.4.9

Graph describing a changing mite population

1. The graph in Figure 1.4.9 looks like a line.
2. Pick the first and last data points (any others could be chosen as long as the data lie on a line as shown in Figure 1.4.10.
3. The slope m is
$$m = \frac{\Delta x_f}{\Delta x_i} = \frac{130 - 70}{50 - 20} = \frac{60}{30} = 2.0.$$
4. Using the point (20, 70) as (x_0, y_0) in the point-slope form for a line, the equation is
$$\begin{aligned} x_f &= m(x_i - x_0) + y_0 \\ &= 2.0(x_i - 20) + 70. \end{aligned}$$
5. We can multiply this out to find the slope-intercept form
$$x_f = 2.0x_i + 30.$$
This defines a function $h(x_i)$ with formula
$$h(x_i) = 2.0x_i + 30.$$

This function $h(x_i)$ can be interpreted in biological terms. The y-intercept of 30 is the number of mites we would find on a lizard that started out with no mites. These mites probably arrived from other lizards. The slope of 2.0 is number of additional mites we would find after a week if we added one mite at the beginning. For example, $h(1) = 32$, two more than $h(0) = 30$. The one additional mite left two offspring. ▲

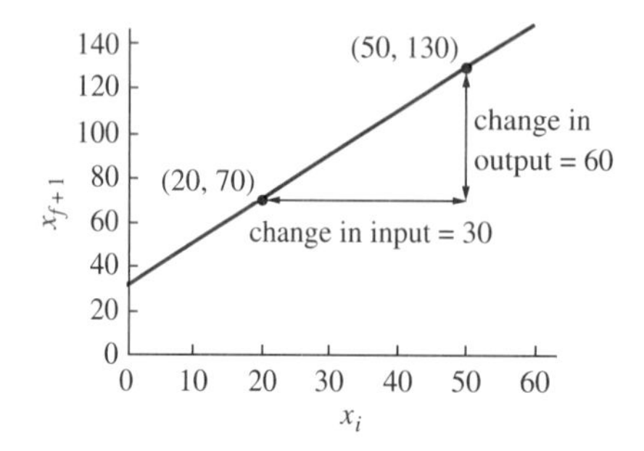

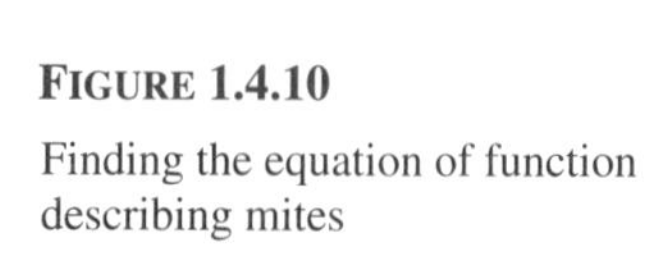

FIGURE 1.4.10

Finding the equation of function describing mites

Example 1.4.13 Using a Linear Function to Interpolate

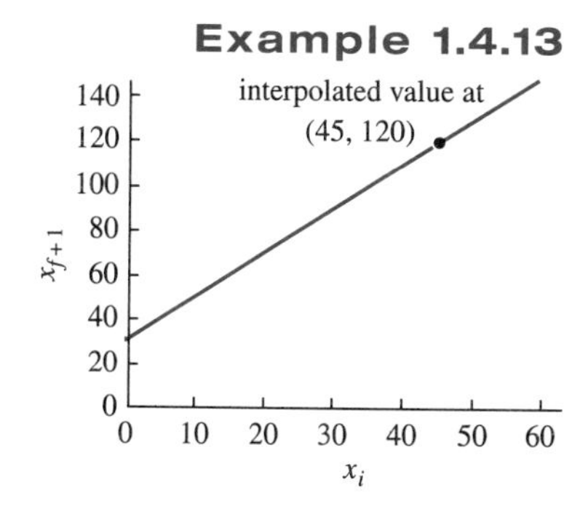

FIGURE 1.4.11
Interpolating a value

To predict the number of mites we will find in a week on a lizard that has 45 mites today (Figure 1.4.11), we substitute $x_i = 45$ into the function $h(x_i) = 2.0x_i + 30$, finding

$$h(x_i) = x_f = 2.0 \cdot 45 + 30 = 120.$$

We have used the formula for the function to **interpolate** a prediction between known values.

In this example, we used a set of numerical data to plot a graph and derive an algebraic formula. In other cases, we are given a formula and need to produce a graph. The easiest way to plot the graph of a linear function from its formula is to plug two reasonable values of the input into the equation, graph the points, and connect them with a line.

Example 1.4.14 Plotting a Line from an Equation

Suppose we wish to plot the linear function $F(x)$ given by

$$F(x) = -2x + 30.$$

Plugging in $x = 0$ gives $F(0) = 30$, and $x = 10$ gives $F(10) = 10$. The graph of the line (Figure 1.4.12) connects the points (0, 30) and (10, 10). This line goes down with a negative slope of -2.

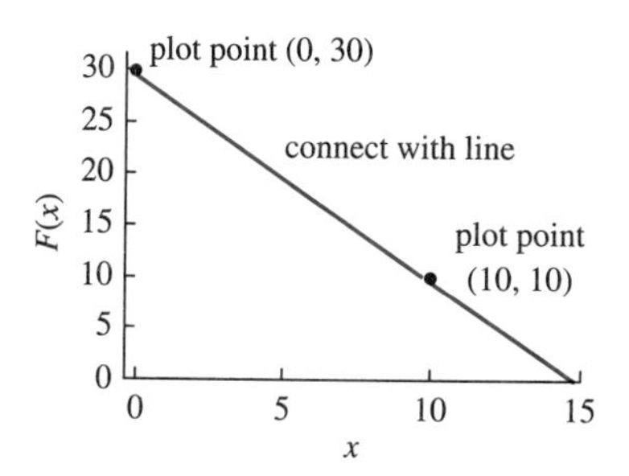

FIGURE 1.4.12
Graphing a line from its equation

When the graph goes down when read from left to right (as in Example 1.4.14), we say that the function is a **decreasing function**. A larger input produces a smaller output. Linear functions with negative slopes are decreasing functions. In contrast, a positive slope corresponds to an **increasing function**. Larger inputs produce larger outputs. A slope of exactly 0 corresponds to a function with equation

$$f(x) = 0 \cdot x + b = b.$$

Such a function always takes on the constant value b, the y-intercept, and has as its graph a horizontal line.

Slope	Graph	Function
positive	goes up	increasing
negative	goes down	decreasing
zero	flat	constant

An example of each type is shown in Figure 1.4.13.

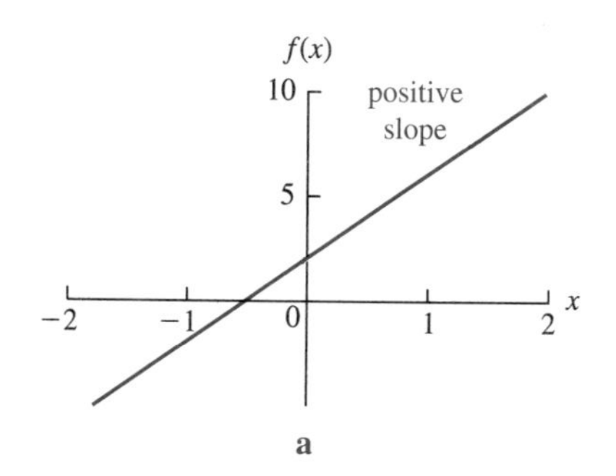

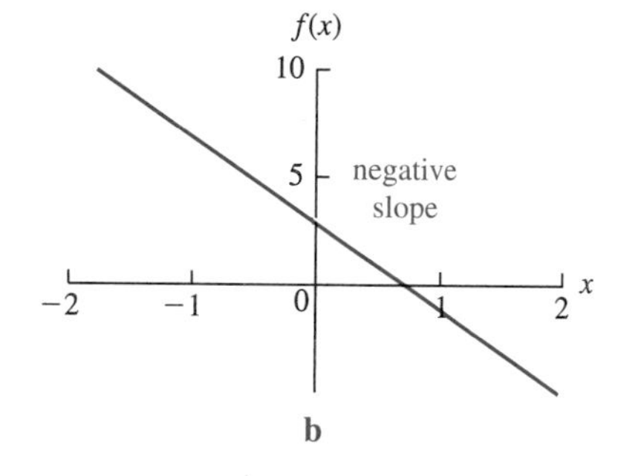

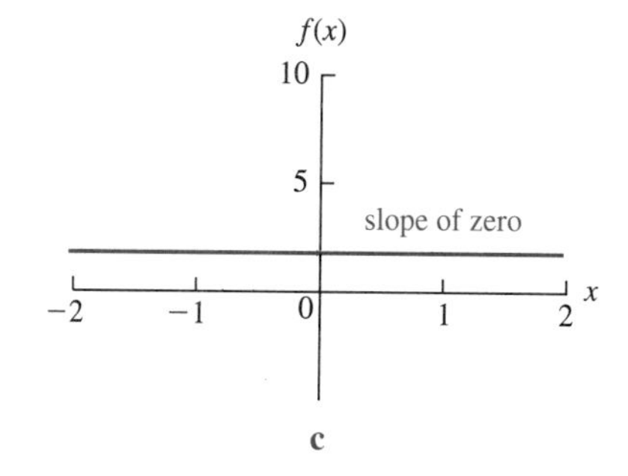

FIGURE 1.4.13
Linear functions with positive, negative, and zero slopes

1.4.4 Solving Equations Involving Lines

Answering questions about linear relations requires solving **linear equations**, which are among the simplest equations to solve.

Example 1.4.15 Solving a Linear Equation

Suppose we wish to find where the line

$$y = 3x + 1$$

takes on the value $y = 7$ (Figure 1.4.14).

$$\begin{aligned} 7 &= 3x + 1 && \text{substitute the value of } y \\ 6 &= 3x && \text{subtract 1 from both sides} \\ 2 &= x. && \text{divide both sides by 3} \end{aligned}$$

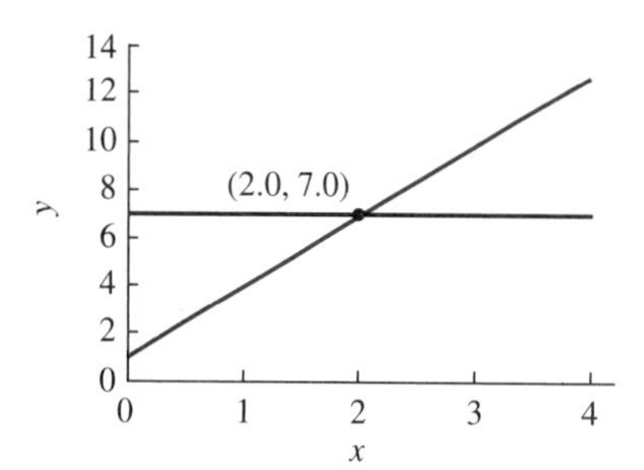

FIGURE 1.4.14
Solving an equation involving a linear function

Graphically, this equation corresponds to finding when the line $y = 3x + 1$ crosses a horizontal line at $y = 7$.

Example 1.4.16 Solving a Linear Equation Involving a Parameter

Suppose we wish to find where the line

$$y = mx + 1$$

takes on the value $y = 7$ for any value of the slope m.

$$\begin{aligned} 7 &= mx + 1 && \text{substitute the value of } y \\ 6 &= mx && \text{subtract 1 from both sides} \\ \frac{6}{m} &= x. && \text{divide both sides by m} \end{aligned}$$

This solution makes sense for any value of $m \neq 0$.

Example 1.4.17 Finding the Intersection of Two Lines

Suppose we wish to find where the lines $f(x) = 3x + 1$ and $g(x) = -4x + 5$ intersect (Figure 1.4.15). We set the two values equal, and solve for x as follows.

$$\begin{aligned} 3x + 1 &= -4x + 5 && \text{set the two formulas equal to each other} \\ 7x + 1 &= 5 && \text{add } 4x \text{ to both sides} \\ 7x &= 4 && \text{subtract 1 from both sides} \\ x &= \frac{4}{7}. && \text{divide both sides by 7} \end{aligned}$$

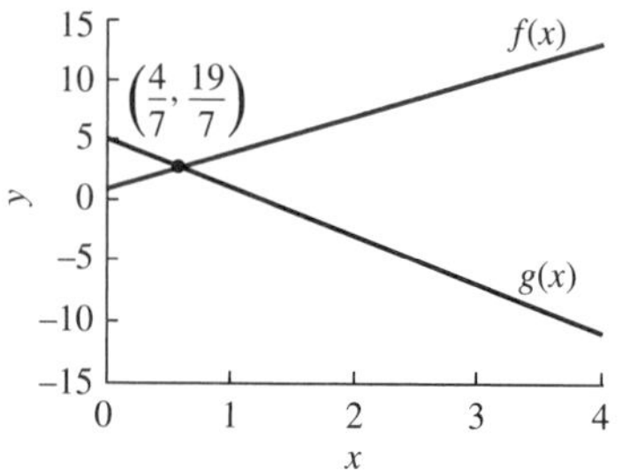

FIGURE 1.4.15
Finding where two lines intersect

This gives the value of x where the two intersect. The value of y can be found by substituting $x = \frac{4}{7}$ into either function, or

$$f\left(\frac{4}{7}\right) = 3 \cdot \frac{4}{7} + 1 = \frac{19}{7}$$
$$g\left(\frac{4}{7}\right) = -4 \cdot \frac{4}{7} + 5 = \frac{19}{7}.$$

Both functions give the same result, as they must.

Example 1.4.18 Solving a Classic Word Problem with Linear Equations

Little Billy's father is three times as old as Billy in 2001. Ten years later, Billy's father will be only twice as old as Billy. What year was Billy born, and how old was his dad? Let B represent Billy's age in 2001, and D his dad's age. Then

$$D = 3B.$$

Ten years later, Billy is $B + 10$ and his dad is $D + 10$. Because dad is then twice as old

$$D + 10 = 2(B + 10).$$

We can rewrite this (in slope-intercept form) to find that

$$D = 2B + 10.$$

This gives two equations for D (Figure 1.4.16). Setting the right-hand sides equal gives the single linear equation

$$3B = 2B + 10.$$

Subtracting $2B$ from both sides gives $B = 10$, Billy's age in 2001. His dad was 3 times as old, or 30. Thus Billy was born in 1991, when his dad was 20. Ten years later, in 2011, Billy will be 20, exactly half as old as his 40-year-old father.

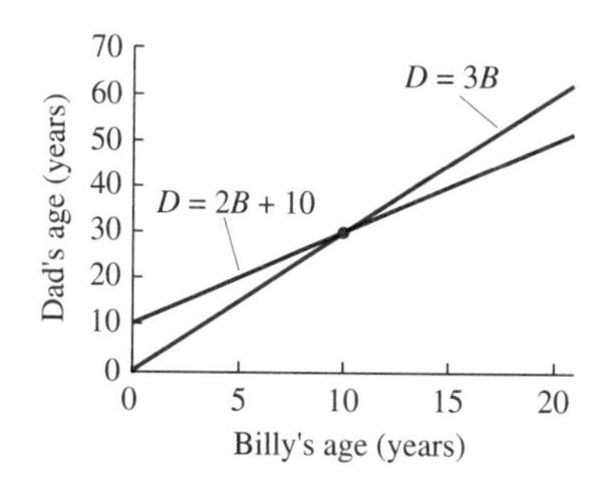

FIGURE 1.4.16
Dad's age as a function of Billy's

Example 1.4.19 A Linear Equation with No Solution

Suppose we are given the following variant of the classic word problem. Little Billy's father is three times as old as Billy in 2001. Ten years later, Billy's father is 10 less than three times Billy's age. What year was Billy born, and how old was his dad? Let B represent Billy's age in 2001, and D his dad's age. Then

$$D = 3B.$$

Ten years later, Billy is $B + 10$ and his dad is $D + 10$. Because dad is 10 less than three times as old

$$D + 10 = 3(B + 10) - 10.$$

Subtracting 10 from both sides and solving for D gives

$$D = 3B + 10.$$

(Figure 1.4.17) Setting these two equations for D equal gives

$$3B = 3B + 10.$$

Subtracting $3B$ from both sides gives $0 = 10$, which is impossible. The original problem has no solution. Graphically, this corresponds to trying to find the intersection of two parallel lines.

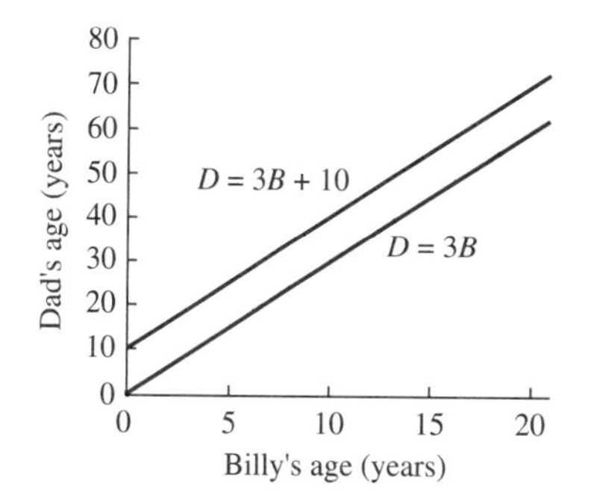

FIGURE 1.4.17
Dad's age as a function of Billy's

Example 1.4.20 A Linear Equation with Many Solutions

Suppose we are given yet another variant of the classic word problem. Little Billy's father is three times as old as Billy in 2001. Ten years later, Billy's father is 20 less than three times Billy's age. What year was Billy born, and how old was his dad? Then

$$D = 3B.$$

Ten years later, Billy is $B + 10$ and his dad is $D + 10$. Because dad is then 20 less than three times as old

$$D + 10 = 3(B + 10) - 20.$$

Solving for D, we find

$$D = 3B.$$

This matches our original equation, and works for any value of B. For example, if Billy was born in 1988 and was 13 in 2001, his dad is three times as old in 2001, or 39. Ten years later, Billy would be 23, and his dad would be 49, exactly 20 less than three times 23. But if Billy had been born in 1992, he would have been 9 in 2001, and his dad would have been 27. Ten years later, Billy would be 19, and dad would be 37, again exactly 20 years less than three times 19. ▲

Example 1.4.21 Solving Another Word Problem with Linear Equations

Recall the lizards with mites obeying the equation

$$x_f = 2.0x_i + 30$$

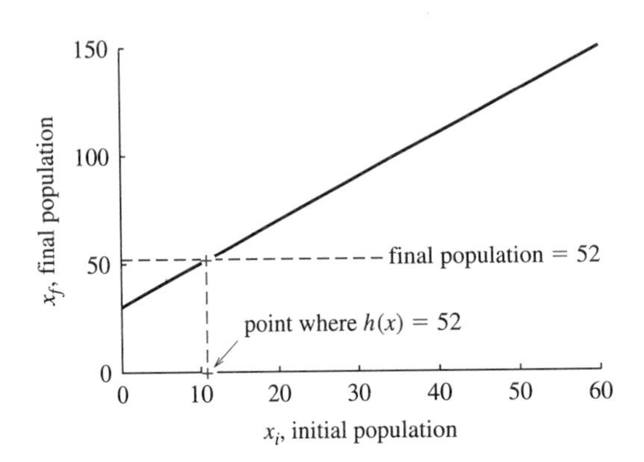

FIGURE 1.4.18
Going backwards with the mite population

(Example 1.4.12). Suppose a lizard ends up with $x_f = 52$ mites (Figure 1.4.18). How many did it have the week before? This is a kind of "inverse" problem; starting from where you ended up, you want to try to end up where you started. Fortunately, these problems can be expressed more clearly in equations than in words. In terms of the variables x_i and x_f, our question can be rephrased. What was x_i if x_f is 52? We want to solve the equation $h(x_i) = 52$ for x_i, or

$$2.0x_i + 30 = 52.$$

The two sides of the equation say the same thing in two ways. The right-hand side gives our measured value 52. The left-hand side gives the measured value as a function of the unknown x_i. We can solve for x_i

$$2.0x_i = 52 - 30 = 22 \quad \text{subtract 30 from both sides}$$

$$x_i = \frac{22}{2.0} = 11. \quad \text{divide both sides by 2}$$

We can check this answer by plugging in, finding

$$h(11) = 2 \cdot 11 + 30 = 52.$$

▲

Summary The graphs of many important functions in biology are lines. We derived the link between lines and **linear functions**. A **proportional relation** is a special type of linear function in which the ratio of the output to the input is always the same. This constant ratio is the **slope** of the graph of the relation. Lines other than proportional relations can be expressed in **point-slope form** or **slope-intercept form**. The slope can be found as the change in output divided by the change in input. Equations of linear functions can be used to **interpolate**, that is, estimate outputs from untested inputs.

1.4 Exercises

Mathematical Techniques

1–4 ▪ For the following lines, find the slopes between the two given points by finding the change in output divided by the change in input. What is the ratio of the output to the input at each of the points? Which are proportional relations? Which are increasing and which are decreasing? Sketch a graph.

1. $y = 2x + 3$, using points with $x = 1$ and $x = 3$.

2. $z = -5w$, using points with $w = 1$ and $w = 3$.

3. $z = 5(w - 2) + 8$, using points with $w = 1$ and $w = 3$.

4. $y - 5 = -3(x + 2) - 6$, using points with $x = 1$ and $x = 3$.

5–6 ▪ Check that the point indicated lies on the line and find the equation of the line in point-slope form using the given point. Multiply out to check that the point-slope form matches the original equation.

5. The line $f(x) = 2x + 3$ and the point $(2, 7)$.

6. The line $g(y) = -2y + 7$ and the point $(3, 1)$.

7–12 ▪ Find equations in slope-intercept form for the following lines. Sketch a graph indicating the original point from point-slope form.

7. The line $f(x) = 2(x - 1) + 3$.

8. The line $g(z) = -3(z + 1) - 3$.

9. A line passing through the point $(1, 6)$ with slope -2.

10. A line passing through the point $(-1, 6)$ with slope 4.

11. A line passing through the points $(1, 6)$ and $(4, 3)$.

12. A line passing through the points $(6, 1)$ and $(3, 4)$.

13–16 ▪ Check whether the following are linear functions.

13. $h(z) = \frac{1}{5z}$.

14. $F(r) = r^2 + 5$.

15. $P(q) = 8(3q + 2) - 6$.

16. $Q(w) = 8(3q + 2) - 6(q + 4)$.

17–18 ▪ Check that the following curves do not have constant slope by computing the slopes between the points indicated. Compare with the graphs in Exercises 5 and 6 in Section 1.2.

17. $h(z) = \frac{1}{5z}$ at $z = 1$, $z = 2$, and $z = 4$, as in Exercise 5. Find the slope between $z = 1$ and $z = 2$, and the slope between $z = 2$ and $z = 4$.

18. $F(r) = r^2 + 5$ at $r = 0$, $r = 1$, and $r = 4$, as in Exercise 6. Find the slope between $r = 0$ and $r = 1$, and the slope between $r = 1$ and $r = 4$.

19–24 ▪ Solve the following equations. Check your answer by plugging in the value you found.

19. $2x + 3 = 7$.

20. $\frac{1}{2}z - 3 = 7$.

21. $2x + 3 = 3x + 7$.

22. $-3y + 5 = 8 + 2y$.

23. $2(5(x - 1) + 3) = 5(2(x - 2) + 5)$.

24. $2(4(x - 1) + 3) = 5(2(x - 2) + 5)$.

25–28 ▪ Solve the following equations for the given variable, treating the other letters as constant parameters.

25. Solve $2x + b = 7$ for x.

26. Solve $mx + 3 = 7$ for x.

27. Solve $2x + b = mx + 7$ for x. Are there any values of b or m for which this has no solution?

28. Solve $mx + b = 3x + 7$ for x. Are there any values of b or m for which this has no solution?

29–32 ▪ Most unit conversions are proportional relations. Find the slope and graph the relations between the following units.

29. Graph the length of a fish in inches on the horizontal axis and centimeters on the vertical axis. Use the fact that 1 in. = 2.54 cm. Mark the point corresponding to a length of 1 in.

30. Graph the length of a fish in centimeters on the horizontal axis and inches on the vertical axis. Use the fact that 1 in. = 2.54 cm. Mark the point corresponding to a length of 1 in.

31. Graph the mass of a fish in grams on the horizontal axis and its weight in pounds on the vertical axis. Use the identity 1 lb = 453.6 g. Mark the point corresponding to a weight of 1 lb.

32. Graph the weight of a fish in pounds on the horizontal axis and its mass in grams on the vertical axis. Use the identity 1 lb = 453.6 g. Mark the point corresponding to a weight of 1 lb.

33–34 ▪ Not very many functions commute with each other. The following problems ask you to find all linear functions that commute with the given linear function.

33. Find all functions of the form $g(x) = mx + b$ that commute with the function $f(x) = x + 1$. Can you explain your answer in words?

34. Find all functions of the form $g(x) = mx + b$ that commute with the function $f(x) = 2x$. Can you explain your answer in words?

Applications

35–38 ▪ Many fundamental relations express a proportional relation between two measurements with different dimensions. Find the slopes and the equations of the relations between the following quantities.

35. Volume = area × thickness. Find the volume V as a function of the area A if the thickness is 1.0 cm.

36. Volume = area × thickness. Find the volume V as a function of thickness T if the area is 7.0 cm^2.

37. Total mass = mass per bacterium × number of bacteria. Find total mass M as a function of the number of bacteria b if the mass per bacterium is 5.0×10^{-9} g.

38. Total mass = mass per bacterium × number of bacteria. Find total mass M as a function of mass per bacterium m if the total number is 10^6.

39–42 ▪ A ski slope has a slope of −0.2. You start at an altitude of 10,000 ft.

39. Write the equation giving altitude a as a function of horizontal distance moved d.

40. Write the equation of the line in meters.

41. What will be your altitude when you have gone 2000 ft horizontally?

42. The ski run ends at an altitude of 8000 ft. How far will you have gone horizontally?

43–46 ▪ The following data give the elevation of the surface of the Great Salt Lake in Utah.

Year, y	Elevation, E (ft)
1965	4,193
1970	4,196
1975	4,199
1980	4,199
1985	4,206
1990	4,203
1995	4,200

43. Graph these data.

44. During which periods is the surface elevation changing linearly?

45. What was the slope between 1965 and 1975? What would the surface elevation have been in 1990 if things had continued as they began? How different is this from the actual depth?

46. What was the slope during between 1985 and 1995? What would the surface elevation have been in 1965 if things had always followed this trend? How different is this from the actual depth?

47–50 ▪ Graph the following relations between measurements of a growing plant, checking that the points lie on a line. Find the equations in both point-slope and slope-intercept form.

Age, a (days)	Mass, M (g)	Volume, V (cm^3)	Glucose production, G (mg)
0.5	2.5	5.1	0.0
1.0	4.0	6.2	3.4
1.5	5.5	7.3	6.8
2.0	7.0	8.4	10.2
2.5	8.5	9.5	13.6
3.0	10.0	10.6	17.0

47. Mass as a function of age. Find the mass on day 1.75.

48. Volume as a function of age. Find the volume on day 2.75.

49. Glucose production as a function of mass. Estimate glucose production when the mass reaches 20.0 g.

50. Volume as a function of mass. Estimate the volume when the mass reaches 30.0 g. How will the density at that time compare with the density when $a = 0.5$?

51–44 ▪ Consider the data in the following table (adapted from *Parasitoids* by H. C. F. Godfray), describing the number of wasps that can develop inside caterpillars of different weights.

Weight of Caterpillar (g)	Number of Wasps
0.5	80
1.0	115
1.5	150
2.0	175

51. Graph these data. Which point does not lie on the line?

52. Find the equation of the line connecting the first two points.

53. How many wasps does the function predict would develop in a caterpillar weighing 0.72 g?

54. How many wasps does the function predict would develop in a caterpillar weighing 0.0 g? Does this make sense? How many would you really expect?

55–58 ▪ The world record times for various races are decreasing at roughly linear rates (adapted from *Guinness Book of Records*, 1990).

55. The men's Olympic record for the 1500 meters was 3:36.8 in 1972 and 3:35.9 in 1988. Find and graph the line connecting these. (Don't forget to convert everything into seconds.)

56. The women's Olympic record for the 1500 meters was 4:01.4 in 1972 and 3:53.9 in 1988. Find and graph the line connecting these.

57. If things continue at this rate, when will women finish the race in exactly no time? What might happen before that date?

58. If things continue at this rate, when will women be running this race faster than men?

Computer Exercises

59. Try Exercise 58 on the computer. Compute the year when the times will reach 0. Give your best guess of the times in the year 1900.

60. Graph the ratio of temperature measured in Fahrenheit to temperature measured in Celsius for $-273 \leq {}^\circ C < 200$. What happens near $^\circ C = 0$? What happens for large and small values of $^\circ C$? How would the results differ if the zero for Fahrenheit were changed to match that of Celsius?

1.5 Discrete-Time Dynamical Systems

Suppose we collect data on how much several bacterial cultures grow in one hour, or how much trees grow in one year. How can we predict what will happen in the long run? In this section, we begin addressing these dynamic problems, which form the theme of this chapter and indeed of much of this book. We follow the basic steps of applied mathematics: **quantifying the basic measurement** and describing the **dynamic rule**. We will learn how to summarize the rule with a **discrete-time dynamical system** or an **updating function** that describes change. From the discrete-time dynamical system and a starting point, called an **initial condition**, we will compute a **solution** that gives the values of the measurement as a function of time.

1.5.1 Discrete-Time Dynamical Systems and Updating Functions

A discrete-time dynamical system describes the relation between a quantity measured at the beginning and the end of an experiment or a time interval. If the measurement is represented by the variable m, we will use the notation m_t to denote the measurement at the beginning of the experiment and m_{t+1} to denote the measurement at the end of the experiment (Figure 1.5.1). Think of t as the current time, and $t+1$ as the time one step into the future. The relation between the initial measurement m_t and the final measurement m_{t+1} is given by the **discrete-time dynamical system**

$$m_{t+1} = f(m_t). \tag{1.5.1}$$

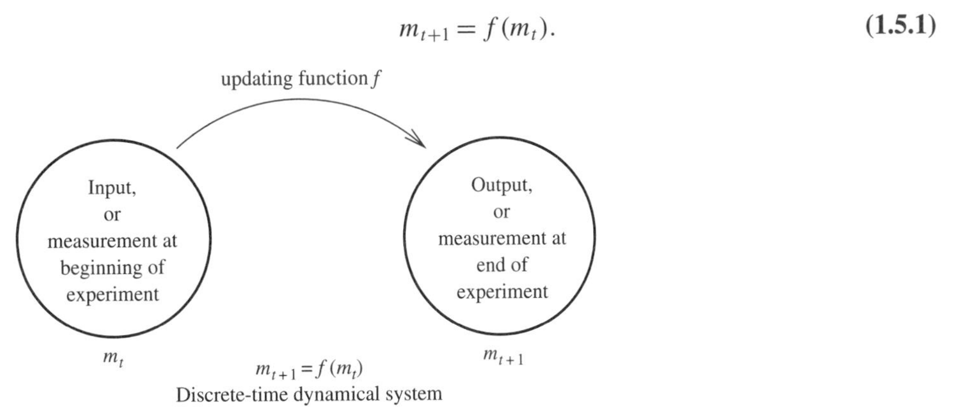

FIGURE 1.5.1
Notation for a discrete-time dynamical system

The **updating function** f accepts the initial value m_t as input and returns the final value m_{t+1} as output.

We will begin by applying this notation to several examples of discrete-time dynamical systems.

Example 1.5.1 A Discrete-Time Dynamical System for a Bacterial Population

Recall the data introduced in Example 1.2.3. Several bacterial cultures with different initial population sizes are grown in controlled conditions for one hour and then carefully recounted.

Colony	Initial Population, b_t	Final Population, b_{t+1}
1	0.47	0.94
2	3.3	6.6
3	0.73	1.46
4	2.8	5.6
5	1.5	3.0
6	0.62	1.24

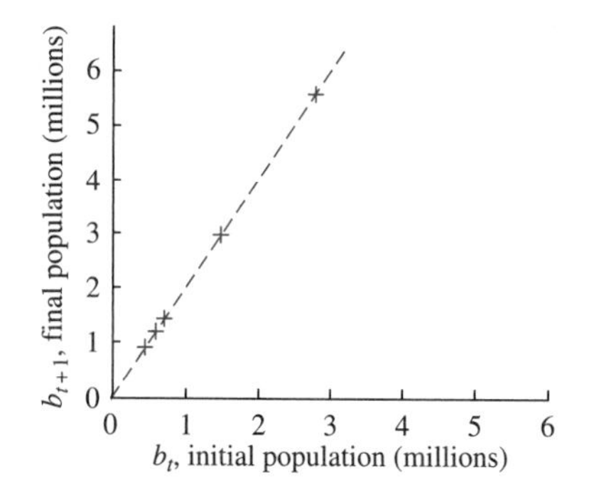

FIGURE 1.5.2
Graph of the updating function for a bacterial population

We have replaced b_i (the initial population) with b_t (the population at time t) and b_f (the final population) with b_{t+1} (the population at time $t+1$).

In each colony, the population doubled in size. We can describe this with the discrete-time dynamical system

$$b_{t+1} = 2.0b_t.$$

The updating function f describes the rule applied to the initial population,

$$f(b_t) = 2.0b_t.$$

The graph of the updating function plots the initial measurement b_t on the horizontal axis and the final measurement b_{t+1} on the vertical axis (Figure 1.5.2).

Example 1.5.2 A Discrete-Time Dynamical System for Tree Growth

Suppose you measure the heights of several trees in one year, and then again the next year (Figure 1.5.3). Denoting the initial height by h_t and the final height by h_{t+1}, we might find the data in the following table (all measured in meters).

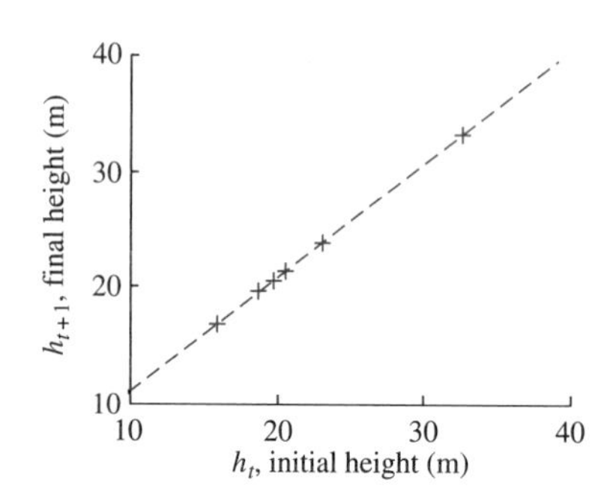

FIGURE 1.5.3
Data describing the growth of six trees

Tree	Initial Height, h_t	Final Height, h_{t+1}	Change in Height
1	23.1	24.1	1.0
2	18.7	19.8	1.1
3	20.6	21.5	0.9
4	16.0	17.0	1.0
5	32.5	33.6	1.1
6	19.8	20.6	0.8

The trees increase in height by about 1.0 m per year.

If we approximate this by assuming that trees grow by exactly 1.0 m per year, the discrete-time dynamical system that expresses this relation is

$$h_{t+1} = h_t + 1.0.$$

The updating function, which we can denote by g, has the formula

$$g(h_t) = h_t + 1.0.$$

For example, for a tree beginning with height 12.2 m, the discrete-time dynamical system predicts a final height of

$$h_{t+1} = g(12.2) = 12.2 + 1.0 = 13.2 \text{ m}.$$

In this example, the data points do not exactly match the discrete-time dynamical system. The updating function captures the major trend in the data while ignoring the noise. Including only the trend corresponds to the use of a **deterministic** dynamical system to describe the behavior. To include the noise, we must use a **probabilistic** dynamical system (Chapter 6). We will specifically address the problem of finding an updating function that captures the major trends in the data when we study the technique of data-fitting called linear regression (Section 8.9).

Example 1.5.3 Discrete-Time Dynamical System for Mites

Recall the lizards infested by mites (Example 1.4.12). The final number of mites x_{t+1} is related to the initial number of mites x_t by the formula

$$x_{t+1} = 2x_t + 30.$$

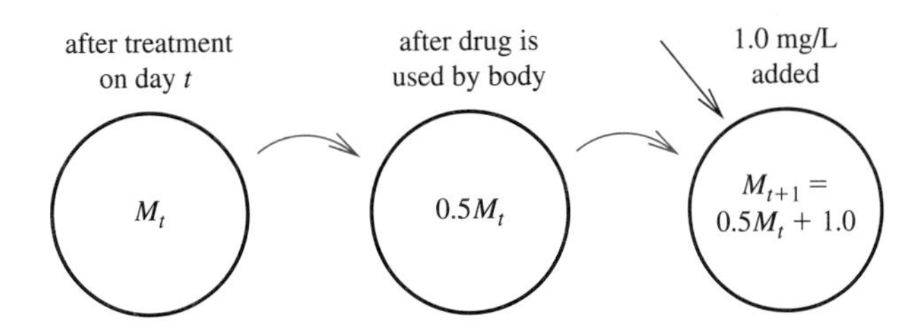

FIGURE 1.5.4
The dynamics of medication concentration in the blood

This discrete-time dynamical system has the associated updating function

$$h(x_t) = 2x_t + 30.$$

The discrete-time dynamical systems for bacterial populations, tree height, and mite number were all derived from data. Often, dynamical rules can instead be derived directly from the principles governing a system.

Example 1.5.4 A Discrete-Time Dynamical System for Medication Concentration

Suppose we know the following facts about the dynamics of medication. Each day, a patient uses up half of the medication in his bloodstream. However, he is given a new dose sufficient to raise the concentration in the bloodstream by 1.0 milligrams per liter (Figure 1.5.4). Let M_t denote the concentration at time t. The discrete-time dynamical system is

$$M_{t+1} = 0.5M_t + 1.0.$$

The term $0.5M_t$ indicates that only half of the initial medication **remains** the next day. The factor 0.5 is the slope of this linear function. The second term, the intercept, indicates that 1.0 milligrams per liter of medication is added each day (Figure 1.5.5). We can graph this linear function by substituting two reasonable values for M_t. If $M_t = 0$, then $M_{t+1} = 1$, the y-intercept of this line. If $M_t = 1$, then $M_{t+1} = 1.5$.

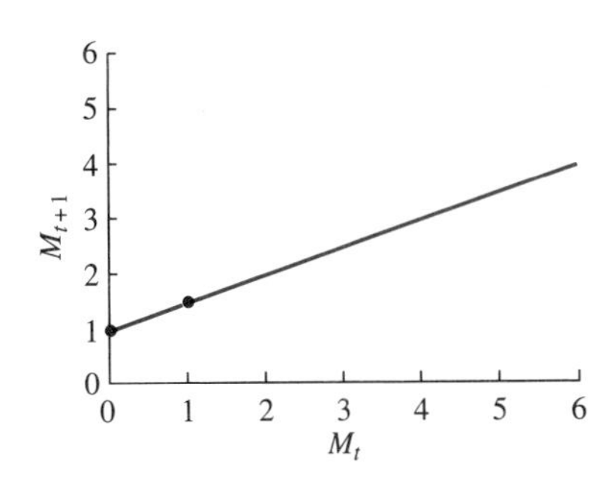

FIGURE 1.5.5
A graph of the updating function for medication concentration

1.5.2 Manipulating Updating Functions

All of the operations that can be applied to ordinary functions can be applied to updating functions, but with special interpretations. We will study **composition** of an updating function with itself, find the **inverse** of an updating function, and **convert the units** or **translate the dimensions** of a discrete-time dynamical system.

Composition Consider the discrete-time dynamical system

$$m_{t+1} = f(m_t)$$

with updating function f. What does the composition $f \circ f$ mean? The updating function **updates** the measurement by one time step. Then

$$\begin{aligned} (f \circ f)(m_t) &= f(f(m_t)) && \text{definition of composition} \\ &= f(m_{t+1}) && \text{definition of updating function} \\ &= m_{t+2}. && \text{updating function applied to } m_{t+1} \end{aligned}$$

Therefore,

$$(f \circ f)(m_t) = m_{t+2}.$$

The composition of an updating function with itself corresponds to a two-step updating function (Figure 1.5.6).

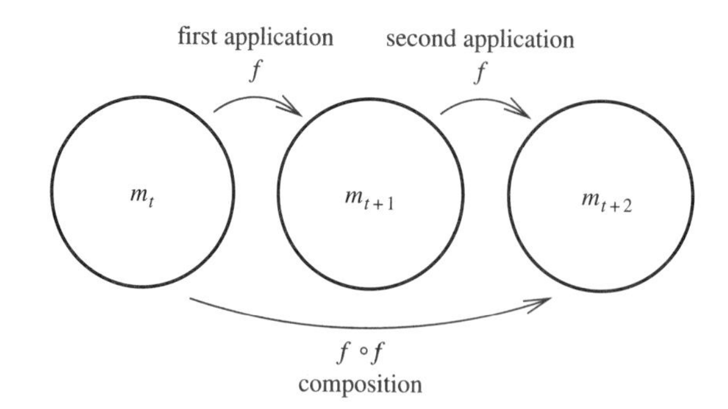

FIGURE 1.5.6
Composition of an updating function with itself

Example 1.5.5 Composition of the Bacterial Population Updating Function with Itself

The bacterial updating function is $f(b_t) = 2b_t$. The function $f \circ f$ takes the population size at time t as input and returns the population size 2 hours later, at time $t = 2$, as output. We can compute $f \circ f$ with the steps

$$\begin{aligned}(f \circ f)(b_t) &= f(f(b_t)) \\ &= f(2.0b_t) \\ &= 4.0b_t.\end{aligned}$$

After two hours, the population is four times larger, having doubled twice. In this case, composition of f with itself looks like multiplication. This simple rules works **only** for an updating function expressing a proportional relation.

Example 1.5.6 Composition of the Mite Population Updating Function with Itself

The composition of the mite updating function $h(x_t) = 2x_t + 30$ with itself gives

$$\begin{aligned}(h \circ h)(x_t) &= h(h(x_t)) \\ &= h(2x_t + 30) \\ &= 2(2x_t + 30) + 30 \\ &= 4x_t + 90.\end{aligned}$$

Suppose we started with $x_t = 10$ mites. After 1 week, we would find $h(10) = 2 \cdot 10 + 30 = 50$ mites. After a second week, we would find $h(50) = 2 \cdot 50 + 30 = 130$ mites. Using the composition of the updating function with itself, we can compute the number of mites after 2 weeks, skipping over the intermediate value of 50 mites after 1 week, finding

$$(h \circ h)(10) = 4 \cdot 10 + 90 = 130.$$

Inverses Consider the general discrete-time dynamical system

$$m_{t+1} = f(m_t)$$

with updating function f. What does the inverse f^{-1} mean? The updating function updates the measurement by one time step, and the inverse function undoes the action of the updating function. Therefore,

$$f^{-1}(m_{t+1}) = m_t.$$

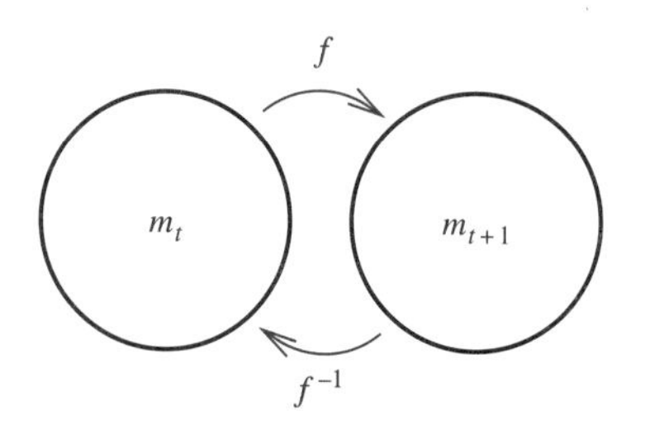

FIGURE 1.5.7
Inverse of an updating function

The inverse of an updating function corresponds to an "updating" function that goes **backwards** in time (Figure 1.5.7).

Example 1.5.7 Inverse of the Bacterial Population Updating Function

The bacterial updating function is $f(b_t) = 2b_t$. We find the inverse by writing the discrete-time dynamical system

$$b_{t+1} = 2.0b_t$$

and solving for the input variable b_t (Algorithm 1.1). In this case, dividing both sides by 2.0 gives

$$b_t = \frac{b_{t+1}}{2.0}.$$

The inverse function is

$$f^{-1}(b_{t+1}) = \frac{b_{t+1}}{2.0}.$$

If **multiplying** by 2.0 describes how the population changes forward in time, **dividing** by 2.0 describes how it changes backward in time.

If $b_t = 2.0$, then $b_{t+1} = 2.0b_t = 2.0 \cdot 2.0 = 4.0$. If we go backwards from $b_{t+1} = 4.0$ using the inverse of the updating function, we find

$$b_t = f^{-1}(4.0) = \frac{4.0}{2.0} = 2.0,$$

exactly where we started.

Example 1.5.8 Inverse of the Mite Updating Function

To find the inverse of the mite updating function $h(x_t) = 2.0x_t + 30$, we use Algorithm 1.1

$$2.0x_t + 30 = x_{t+1} \quad \text{the original equation}$$
$$2.0x_t = x_{t+1} - 30 \quad \text{subtract 30 from both sides}$$
$$x_t = \frac{x_{t+1} - 30}{2.0}. \quad \text{divide both sides by 2.0}$$

Therefore,

$$x_t = h^{-1}(x_{t+1}) = \frac{x_{t+1} - 30}{2.0} = 0.5x_{t+1} - 15.$$

Suppose we started with $x_t = 10$ mites. After 1 week, we would find

$$h(10) = 2 \cdot 10 + 30 = 50.$$

Applying the inverse, we find

$$h^{-1}(50) = \frac{50 - 30}{2.0} = 10.$$

The inverse function takes us back to where we started.

1.5.3 Discrete-Time Dynamical Systems: Units and Dimensions

The updating function $f(b_t) = 2.0b_t$ accepts as input positive numbers with units of bacteria. If we measure this quantity in different units, we must convert the updating function itself into the new units. If we measure a different quantity like total mass or volume, we can translate the updating function into different dimensions.

Example 1.5.9 Describing the Dynamics of Tree Height in Centimeters

Suppose we wish to study tree height (Example 1.5.2) in units of centimeters rather than meters. In meters, the discrete-time dynamical system is

$$g(h_t) = h_t + 1.0 \text{ m}.$$

First, we define a new variable to represent the measurement in the new units. Let H_t be tree height measured in centimeters rather than meters. Then $H_t = 100h_t$, because there are 100 centimeters in a meter. We wish to find a discrete-time dynamical system that gives a formula for H_{t+1} in terms of H_t (Figure 1.5.8).

$$\begin{aligned} H_{t+1} &= 100h_{t+1} && \text{definition of } H_{t+1} \\ &= 100(h_t + 1.0) && \text{discrete-time dynamical system for } h_{t+1} \\ &= 100h_t + 100.0 && \text{multiply through by 100} \\ &= H_t + 100.0. && \text{definition of } H_t \end{aligned}$$

The discrete-time dynamical system in the new units corresponds to adding 100 centimeters to the height, which is equivalent to adding 1 meter. Although the underlying process is the same, the discrete-time dynamical system and the corresponding updating function are different, just as the numerical values of measurements are different in different units.

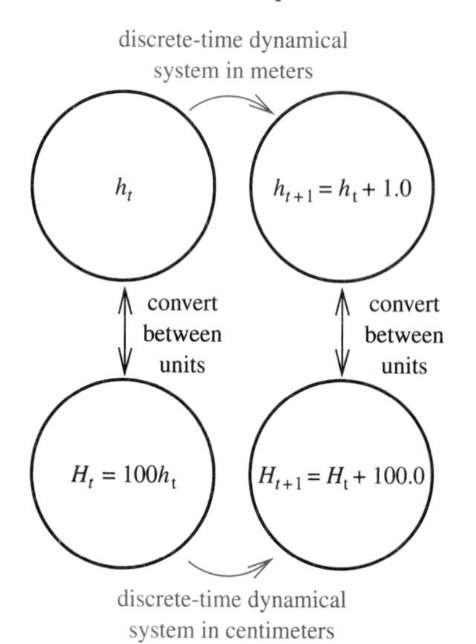

FIGURE 1.5.8
Finding the discrete-time dynamical system for trees in centimeters

Example 1.5.10 Describing the Dynamics of Bacterial Mass

Suppose we wish to study the bacterial population in terms of mass rather than number. At the beginning, the mass, denoted by m_t, is

$$m_t = \mu b_t$$

where μ is the mass per bacterium (as in Example 1.3.4). The updated mass m_{t+1} is

$$\begin{aligned} m_{t+1} &= \mu b_{t+1} && \text{the definition of } m_{t+1} \\ &= \mu \cdot 2.0 b_t && \text{substituting the original updating function} \\ &= 2.0\mu b_t && \text{rearranging the terms} \\ &= 2.0 m_t. && \text{recognize that } m_t = \mu b_t \end{aligned}$$

This new discrete-time dynamical system doubles its input like the original discrete-time dynamical system, but takes mass as its input rather than numbers of bacteria (Figure 1.5.9).

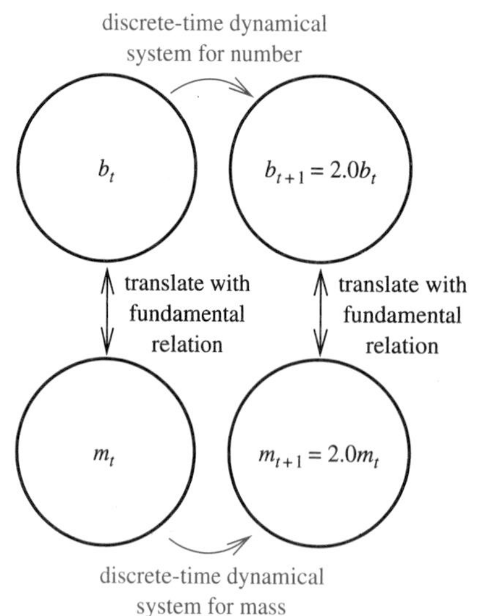

FIGURE 1.5.9
Finding the discrete-time dynamical system for bacteria in terms of mass

1.5.4 Solutions

A discrete-time dynamical system describes some quantity at the end of an experiment as a function of that same quantity at the beginning. What if we were to continue the experiment? A bacterial population growing according to $b_{t+1} = 2.0b_t$ would double again and again. A tree growing according to $h_{t+1} = h_t + 1.0$ would add more and more meters to its height. An infested lizard would become even more infested.

To describe a situation in which a dynamical process is repeated many times, we let m_0 represent the measurement at the beginning, m_1 the measurement after one time step, m_2 the measurement after two time steps, and so forth (Figure 1.5.10). In general, we define

$$m_t = \text{measurement } t \text{ hours after the beginning of the experiment.}$$

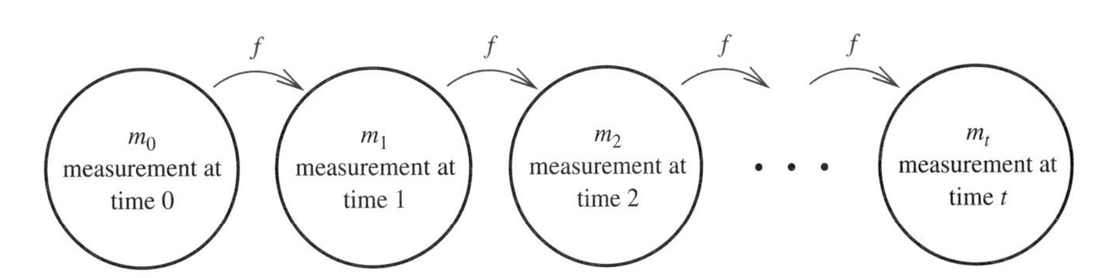

FIGURE 1.5.10
The repeated action of an updating function

Our goal is to find the values of m_t for all values of t. Before we can do so, however, we must know where we **started**. Without knowing where you started, it is impossible to answer a question like "where are you after driving 5 miles south?" The starting value is known as the **initial condition**.

Definition 1.10 The sequence of values of m_t for $t = 0, 1, 2, \ldots$ is the **solution** of the discrete-time dynamical system $m_{t+1} = f(m_t)$ starting from the **initial condition** m_0.

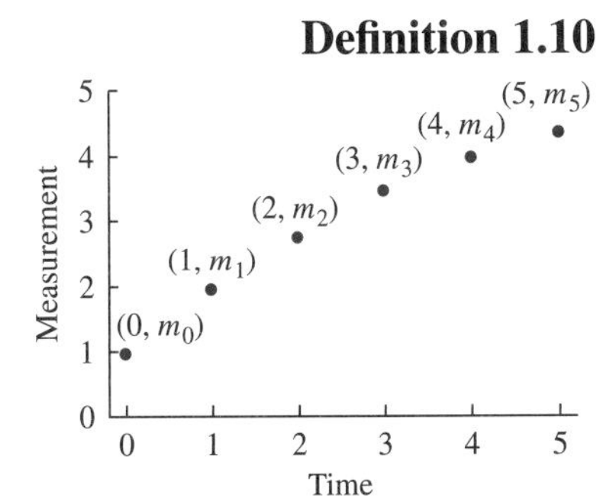

FIGURE 1.5.11
The graph of a solution

There is a mathematical theory devoted to **sequences** like the solutions of discrete-time dynamical system, concerned often with whether the values **converge** to a particular value.

The graph of a solution is a discrete set of points with the time t on the horizontal axis and the measurement m_t on the vertical axis. The initial point has coordinates $(0, m_0)$ to describe the initial condition. The next point, with coordinates $(1, m_1)$, describes the measurement at $t = 1$, and so forth (Figure 1.5.11). It is possible to find a formula for the solution for simple discrete-time dynamical systems, but not in many more complicated cases.

Example 1.5.11 A Solution of the Bacterial Discrete-Time Dynamical System

Suppose we begin with one million bacteria, which corresponds to an initial condition of $b_0 = 1.0$ (with bacterial population measured in millions). If the bacteria obey the discrete-time dynamical system $b_{t+1} = 2.0b_t$,

$$b_1 = 2.0b_0 = 2.0 \cdot 1.0 = 2.0$$
$$b_2 = 2.0b_1 = 2.0 \cdot 2.0 = 4.0$$
$$b_3 = 2.0b_2 = 2.0 \cdot 4.0 = 8.0.$$

Examining these results, we notice that

$$b_1 = 2.0 \cdot 1.0$$
$$b_2 = 2.0^2 \cdot 1.0$$
$$b_3 = 2.0^3 \cdot 1.0.$$

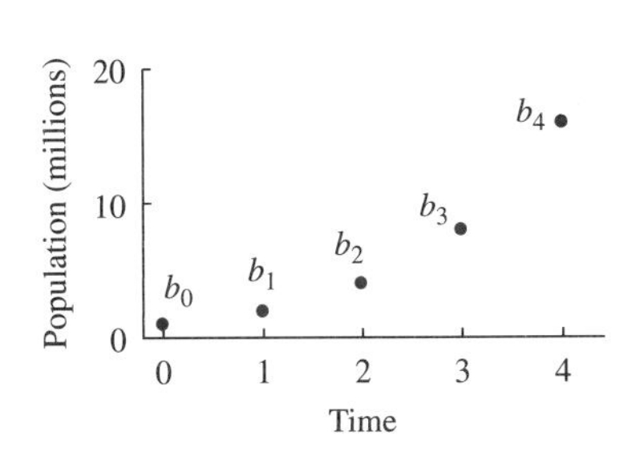

FIGURE 1.5.12
A solution: Bacterial population size as a function of time

After 3 hours, the population has doubled three times, and is $2.0^3 = 8.0$ times the original population. We graph the solution by plotting the time t on the horizontal axis and the number of bacteria after t hours (b_t) on the vertical axis (Figure 1.5.12). The graph consists only of a discrete set of points describing the hourly measurements hence the name "discrete-time dynamical system." Sometimes, we will connect the points in a solution with line segments to make the pattern easier to see.

After t hours, the population has doubled t times and has reached the size

$$b_t = 2.0^t. \tag{1.5.2}$$

This formula describes the solution of the discrete-time dynamical system with initial condition $b_0 = 1.0$. It predicts the population after t hours of reproduction for any value of t. For example, we can compute

$$b_8 = 2.0^8 \cdot 1.0 = 256.0$$

without ever computing b_1, b_2 or other intermediate values.

Example 1.5.12 A Solution with a Different Initial Condition

Suppose we started the system with a different initial condition of $b_0 = 0.3$ million. We can find subsequent values by repeatedly applying the discrete-time dynamical system,

$$b_1 = 2.0 \cdot 0.3 = 0.6$$
$$b_2 = 2.0 \cdot 0.6 = 1.2$$
$$b_3 = 2.0 \cdot 1.2 = 2.4.$$

If we look for the pattern in this case,

$$b_1 = 2.0 \cdot 0.3$$
$$b_2 = 2.0^2 \cdot 0.3$$
$$b_3 = 2.0^3 \cdot 0.3.$$

After t hours, the population has doubled t times as before and reached the size

$$b_t = 2.0^t \cdot 0.3 \text{ million bacteria.}$$

The solution is different from the one found in Example 1.5.11 with a different initial condition (Figure 1.5.13). Although the two solutions get further and further apart, the ratio always remains the same (Exercise 55).

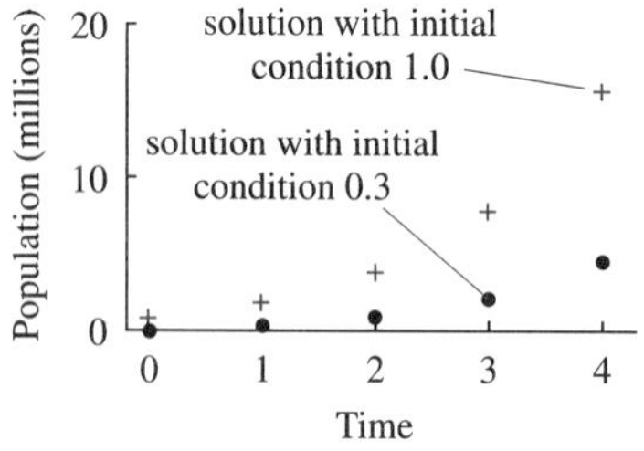

FIGURE 1.5.13

Solutions starting from two different initial conditions

Example 1.5.13 Two Solutions of the Tree Height Discrete-Time Dynamical System

Tree height obeys the discrete-time dynamical system

$$h_{t+1} = h_t + 1.0$$

(Example 1.5.2). Suppose the tree begins with a height of $h_0 = 10.0$ m. Then

$$h_1 = h_0 + 1.0 = 11.0 \text{ m}$$
$$h_2 = h_1 + 1.0 = 12.0 \text{ m}$$
$$h_3 = h_2 + 1.0 = 13.0 \text{ m.}$$

Each year, the height of the tree increases by 1.0 m. After 3 years, the height is 3.0 m greater than the original height. After t years the tree has added 1.0 m to its height t times, meaning that the height will have increased by a total of t m (Figure 1.5.14). Therefore the solution is

$$h_t = 10.0 + t.$$

This formula predicts the height after t years of growth for any t. We can compute

$$h_8 = 10.0 + 8.0 = 18.0 \text{ m}$$

without computing h_1, h_2 or other intermediate values.

If the tree began at a height of 2.0 m, the size for the first few years would be

$$h_1 = h_0 + 1.0 = 3.0 \text{ m}$$
$$h_2 = h_1 + 1.0 = 4.0 \text{ m}$$
$$h_3 = h_2 + 1.0 = 5.0 \text{ m.}$$

Again, the tree adds t meters of height in t years, so the height is

$$h_t = 2.0 + t.$$

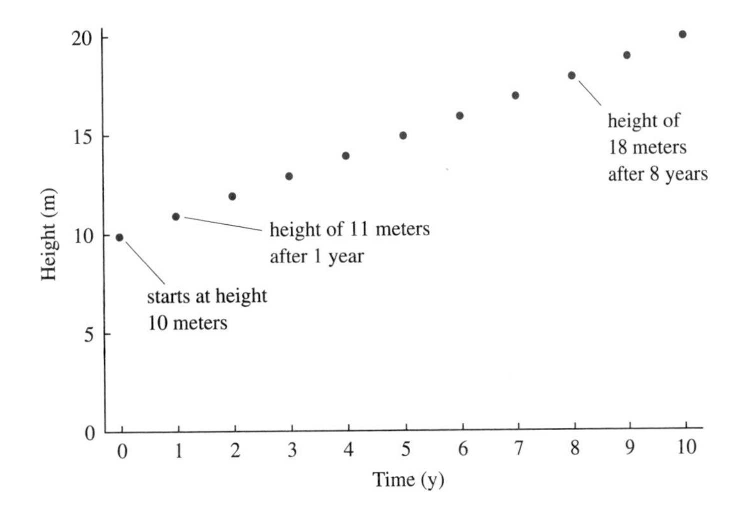

FIGURE 1.5.14
A solution: tree height as a function of time

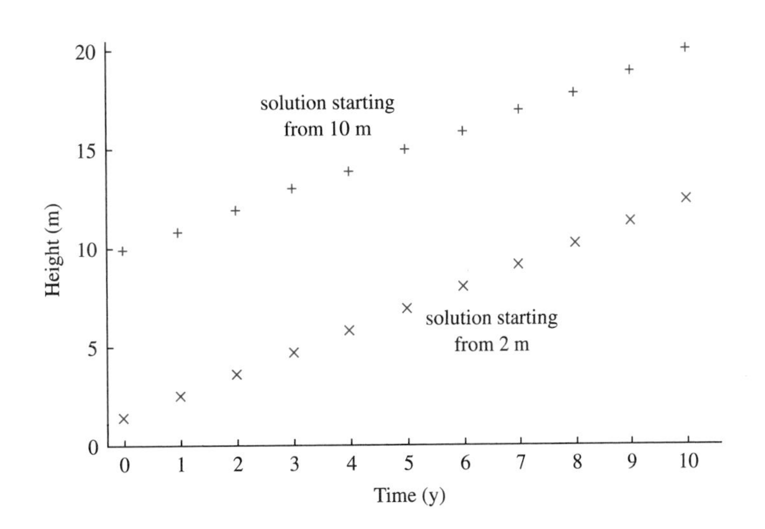

FIGURE 1.5.15
Two solutions for tree height as functions of time

The solution with this smaller initial condition is always exactly 8.0 m less than the solution with the $h_0 = 10$ (Figure 1.5.15).

Is it always possible to guess the formula for a solution in this way? We will next see some cases where computing the solution step by step is straightforward, but finding a solution is tricky. Remarkably, there are simple discrete-time dynamical systems for which it is **impossible** to write a formula for a solution. For example, **chaotic** dynamical systems have solutions that are so unpredictable that no formula can describe them (Subsection 3.2.3).

Example 1.5.14 Finding a Solution of the Medication Discrete-Time Dynamical System

Consider the discrete-time dynamical system for medication (Example 1.5.4), given by

$$M_{t+1} = 0.5M_t + 1.0.$$

Suppose we begin from an initial condition of $M_0 = 5.0$ milligrams per liter. Then

$$M_1 = 0.5 \cdot 5.0 + 1.0 = 3.5$$
$$M_2 = 0.5 \cdot 3.5 + 1.0 = 2.75$$
$$M_3 = 0.5 \cdot 2.75 + 1.0 = 2.375$$
$$M_4 = 0.5 \cdot 2.375 + 1.0 = 2.1875.$$

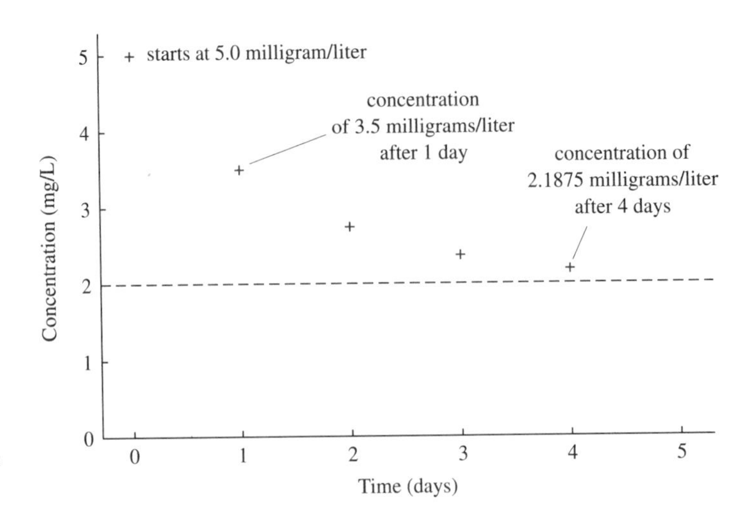

FIGURE 1.5.16
Medication concentration as a function of time

The values are getting closer and closer to 2.0 (Figure 1.5.16). More careful examination indicates that the results move exactly **half way** toward 2.0 each step. In particular,

$$\begin{aligned}
M_0 - 2.0 &= 5.0 - 2.0 = 3.0 \\
M_1 - 2.0 &= 3.5 - 2.0 = 1.5 = 0.5 \cdot 3.0 \\
M_2 - 2.0 &= 2.75 - 2.0 = 0.75 = 0.5 \cdot 1.5 \\
M_3 - 2.0 &= 2.375 - 2.0 = 0.375 = 0.5 \cdot 0.75 \\
M_4 - 2.0 &= 2.1875 - 2.0 = 0.1875 = 0.5 \cdot 0.375.
\end{aligned}$$

Can we convert these observations into the formula for a solution? If we write the concentration as 2.0 plus the difference,

$$\begin{aligned}
M_0 &= 2.0 + 3.0 \\
M_1 &= 2.0 + 0.5 \cdot 3.0 \\
M_2 &= 2.0 + 0.5^2 \cdot 3.0 \\
M_3 &= 2.0 + 0.5^3 \cdot 3.0
\end{aligned}$$

we might see that

$$M_t = 2.0 + 0.5^t \cdot 3.0.$$

Finding patterns in this way and translating them into formulas can be tricky. It is much more important to be able to **describe** the behavior of solutions with a graph or in words. In this case, our calculations quickly revealed that the solution moved closer and closer to 2.0. In Section 1.6, we will develop a powerful graphical method to deduce this pattern with a minimum of calculation. ▲

Example 1.5.15 A Second Solution of the Medication Discrete-Time Dynamical System

If we begin with an initial concentration of $M_0 = 1.0$ milligrams per liter, then

$$\begin{aligned}
M_1 &= 0.5 \cdot 1.0 + 1.0 = 1.5 \\
M_2 &= 0.5 \cdot 1.5 + 1.0 = 1.75 \\
M_3 &= 0.5 \cdot 1.75 + 1.0 = 1.875 \\
M_4 &= 0.5 \cdot 1.875 + 1.0 = 1.9375.
\end{aligned}$$

Unlike bacterial populations (Example 1.5.12) and tree size (Example 1.5.13), the graphs of solutions starting from different initial conditions look completely different (Figure 1.5.17).

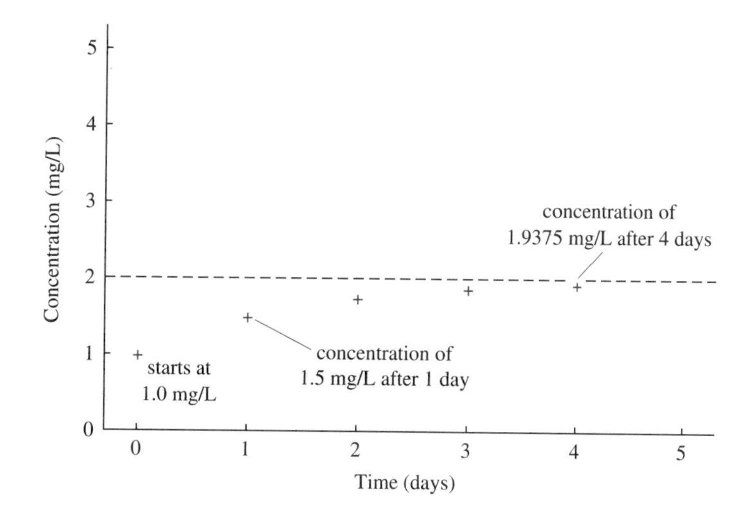

FIGURE 1.5.17
Another solution of medication concentration as a function of time

However, the values still get closer and closer to 2.0, and the difference from 2.0 is reduced by a factor of 0.5 each day,

$$M_0 - 2.0 = 1.0 - 2.0 = -1.0$$
$$M_1 - 2.0 = 1.5 - 2.0 = -0.5$$
$$M_2 - 2.0 = 1.75 - 2.0 = -0.25$$
$$M_3 - 2.0 = 1.875 - 2.0 = -0.125$$
$$M_4 - 2.0 = 1.9375 - 2.0 = -0.0625.$$

We can find the formula using the same idea as before. If we write

$$M_0 = 2.0 - 1.0$$
$$M_1 = 2.0 - 0.5 \cdot 1.0$$
$$M_2 = 2.0 - 0.5^2 \cdot 1.0$$
$$M_3 = 2.0 - 0.5^3 \cdot 1.0$$

we can see that

$$M_t = 2.0 - 0.5^t \cdot 1.0.$$

In Section 2.2, we will use the fundamental idea of the **limit** to study more carefully what it means for the sequence of values that define a solution to get closer and closer to 2.0.

Example 1.5.16 A Solution of the Mite Discrete-Time Dynamical System

Recall the discrete-time dynamical system

$$x_{t+1} = 2x_t + 30$$

for mites. If we started our lizard off with $x_0 = 10$ mites, we compute

$$x_1 = 2.0x_0 + 30 = 50$$
$$x_2 = 2.0x_1 + 30 = 130$$
$$x_3 = 2.0x_2 + 30 = 290.$$

The pattern is not too obvious in this case. There is a pattern, which is a good challenge to find (Exercise 35).

Example 1.5.17 Finding an Updating Function from a Solution

Suppose we have measured a sequence of values over time and wish to find the discrete-time dynamical system that generated them. Consider the following population of birds in a newly founded refuge.

Decade After Founding, t	Population Size, B_t
0	300
1	400
2	700
3	1600

These values correspond to the solution of an unknown discrete-time dynamical system, and we begin, as usual, by graphing them. The solution rises quickly.

To find the discrete-time dynamical system, we can rewrite these data in terms of the current and next value.

Old Population, B_t	New Population, B_{t+1}
300	400
400	700
700	1600

We can seek the equation describing this relationship with Algorithm 1.3.

1. The graph in Figure 1.5.18b looks like a line.

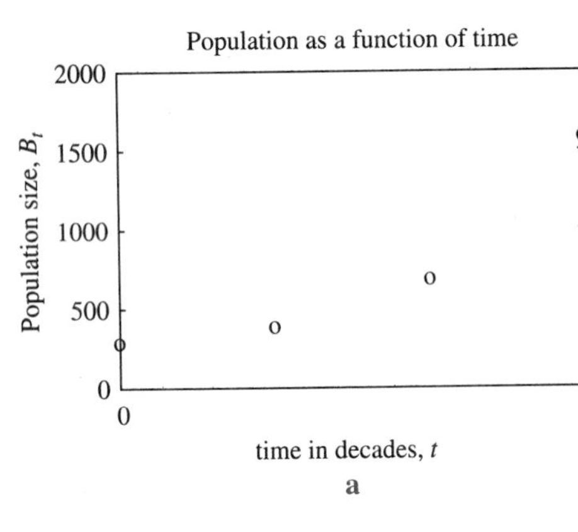

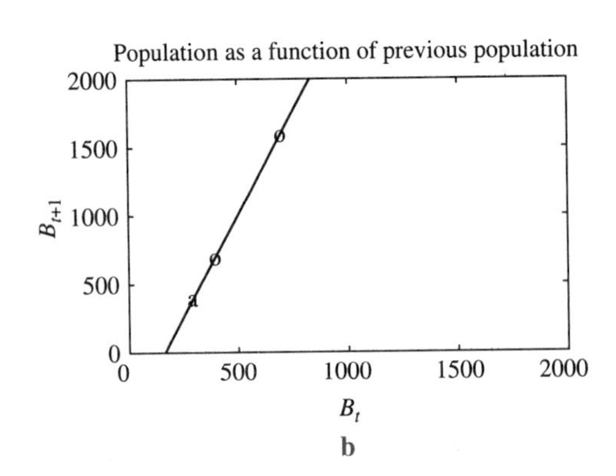

FIGURE 1.5.18
Finding the equation of a discrete-time dynamical system from data

2. Pick the first two data points.
3. The slope m is

$$m = \frac{\Delta B_{t+1}}{\Delta B_t} = \frac{700 - 400}{400 - 300} = \frac{300}{100} = 3.0.$$

4. Using the point (300, 400) as the base point in the point-slope form for a line, the equation is

$$B_{t+1} = 3.0(B_t - 300) + 400.$$

5. We can multiply this out to find the slope-intercept form

$$B_{t+1} = 3.0B_t - 500.$$

This population seems to have the potential to triple every decade, but 500 individuals are removed, perhaps by poachers.

Summary Starting from data or an understanding of a biological process, we can derive a **discrete-time dynamical system**, the **dynamical rule** that tells how a measurement changes from one time step to the next. The **updating function** describes the relation between measurements at times t and $t+1$. The **composition** of the updating function with itself produces a two-step discrete-time dynamical system, while the **inverse** of the updating function produces a backwards discrete-time dynamical system. Like all biological relations, a discrete-time dynamical system can be described in different units and dimensions. Repeated application of a discrete-time dynamical system starting from an **initial condition** generates a **solution**, the value of the measurement as a function of time. With the proper combination of diligence, cleverness, and luck, it is sometimes possible to find a formula for the solution. Given data from a solution, we can sometimes work backwards to find the underlying discrete-time dynamical system.

1.5 Exercises

Mathematical Techniques

1–4 ▪ Write the updating function associated with each of the following discrete-time dynamical systems and evaluate it at the given arguments. Which are linear?

1. $p_{t+1} = p_t - 2$, evaluate at $p_t = 5$, $p_t = 10$, and $p_t = 15$.
2. $\psi_{t+1} = \frac{\psi_t}{2}$, evaluate at $\psi_t = 4$, $\psi_t = 8$, and $\psi_t = 12$.
3. $x_{t+1} = x_t^2 + 2$, evaluate at $x_t = 0$, $x_t = 2$, and $x_t = 4$.
4. $Q_{t+1} = \frac{1}{Q_t + 1}$, evaluate at $Q_t = 0$, $Q_t = 1$, and $Q_t = 2$.

5–8 ▪ Compose the updating function associated with each discrete-time dynamical system with itself. Find the two-step discrete-time dynamical system. Check that the result of applying the original discrete-time dynamical system twice to the given initial condition matches the result of applying the new discrete-time dynamical system to the given initial condition once.

5. Volume follows $v_{t+1} = 1.5v_t$, with $v_0 = 1220\mu\text{m}^3$.
6. Length obeys $l_{t+1} = l_t - 1.7$, with $l_0 = 13.1$ cm.
7. Population size follows $n_{t+1} = 0.5n_t$, with $n_0 = 1200$.
8. Medication concentration obeys $M_{t+1} = 0.75M_t + 2.0$ with $M_0 = 16.0$.

9–12 ▪ Find the backwards discrete-time dynamical system associated with each discrete-time dynamical system. Use it to find the value at the previous time.

9. $v_{t+1} = 1.5v_t$. Find v_0 if $v_1 = 1220\mu\text{m}^3$ (as in Exercise 5).
10. $l_{t+1} = l_t - 1.7$. Find l_0 if $l_1 = 13.1$cm (as in Exercise 6).
11. $n_{t+1} = 0.5n_t$. Find n_0 if $n_1 = 1200$ (as in Exercise 7).
12. $M_{t+1} = 0.75M_t + 2.0$. Find M_0 if $M_1 = 16.0$ (as in Exercise 8).

13–14 ▪ Find the composition of each of the following mathematically elegant updating functions with itself, and find the inverse function.

13. The updating function $f(x) = \frac{x}{1+x}$. Put things over a common denominator to simplify the composition.
14. The updating function $h(x) = \frac{x}{x-1}$. Put things over a common denominator to simplify the composition.

15–18 ▪ Find and graph the solutions of the following discrete-time dynamical systems for five steps with the given initial condition. Compare the graph of the solution with the graph of the updating function.

15. $v_{t+1} = 1.5v_t$, with $v_0 = 1220\mu\text{m}^3$.
16. $l_{t+1} = l_t - 1.7$, with $l_0 = 13.1$ cm.
17. $n_{t+1} = 0.5n_t$, with $n_0 = 1200$.
18. $M_{t+1} = 0.75M_t + 2.0$ with $M_0 = 16.0$.

19–22 ▪ Using a formula for the solution, you can project far into the future without computing all the intermediate values. Find the following, and say whether the results are reasonable.

19. Find a formula for v_t for the discrete-time dynamical system in Exercise 15, and use it to find the volume at $t = 20$.
20. Find a formula for v_t for the discrete-time dynamical system in Exercise 16, and use it to find the length at $t = 20$.
21. Find a formula for v_t for the discrete-time dynamical system in Exercise 17, and use it to find the number at $t = 20$.
22. Find a formula for v_t for the discrete-time dynamical system in Exercise 18, and use it to find the concentration at $t = 20$ (use the method in Example 1.5.14 after finding the value it seems to be approaching).

23–26 ▪ Experiment with the following mathematically elegant updating functions and try to find the solution.

23. Consider the updating function

$$f(x) = \frac{x}{1+x}$$

from Exercise 13. Starting from an initial condition of $x_0 = 1$, compute x_1, x_2, x_3, and x_4, and try to spot the pattern.

24. Use the updating function in Exercise 23 but start from the initial condition $x_0 = 2$.

25. Consider the updating function

$$g(x) = 4 - x.$$

Start from initial condition of $x_0 = 1$, and try to spot the pattern. Experiment with a couple of other initial conditions. How would you describe your results in words?

26. Consider the updating function

$$h(x) = \frac{x}{x-1}$$

from Exercise 14. Start from initial condition of $x_0 = 3$, and try to spot the pattern. Experiment with a couple of other initial conditions. How would you describe your results in words?

Applications

27–30 ▪ Consider the following actions. Which of them commute (produce the same answer when done in either order)?

27. A population doubles in size; 10 individuals are removed from a population. Try starting with 100 individuals, and then try to figure out what happens in general.

28. A population doubles in size; population size is divided by 4. Try starting with 100 individuals, and then try to figure out what happens in general.

29. An organism grows by 2.0 cm; an organism shrinks by 1.0 cm.

30. A person loses half his money. A person gains $10.

31–34 ▪ Use the formula for the solution to find the following, and say whether the results are reasonable.

31. Using the solution for tree height $h_t = 10.0 + t$ m (Example 1.5.13), find the tree height after 20 years.

32. Using the solution for tree height $h_t = 10.0 + t$ m (Example 1.5.13), find the tree height after 100 years.

33. Using the solution for bacterial population number $b_t = 2.0^t \cdot 1.0$ (Equation 1.5.2), find the bacterial population after 20 hours. If an individual bacterium weighs about 10^{-12} grams, how much will the whole population weigh?

34. Using the solution for bacterial population number $b_t = 2.0^t \cdot 1.0$ (Equation 1.5.2), find the bacterial population after 40 hours. How much would this population weigh?

35–36 ▪ Find a formula for the solution of the given discrete-time dynamical system.

35. Find the pattern in the number of mites on a lizard with $x_0 = 10$ and following the discrete-time dynamical system $x_{t+1} = 2x_t + 30$. (*Hint:* Add 30 to the number of mites.)

36. Find the pattern in the number of mites on a lizard with $x_0 = 10$ and following the discrete-time dynamical system $x_{t+1} = 2x_t + 20$.

37–40 ▪ The following tables display data from four experiments:

a. Cell volume after 10 minutes in a watery bath

b. Fish mass after 1 week in a chilly tank

c. Gnat population size after 3 days without food

d. Yield of several varieties of soybean before and after fertilization

For each, graph the new value as a function of the initial value, write the discrete-time dynamical system, and fill in the missing value in the table.

37.

Cell Volume *Parasitoids* (μm^3)	
Initial, v_t	Final, v_t
1220	1830
1860	2790
1080	1620
1640	2460
1540	2310
1420	??

38.

Fish Mass (g)	
Initial, m_t	Final, m_{t+1}
13.1	10.4
18.2	15.5
17.3	14.6
16.0	13.3
20.5	17.8
2.5	??

39.

Gnat Number	
Initial, n_t	Final, n_{t+1}
1.2×10^3	6.0×10^2
2.4×10^3	1.2×10^3
1.6×10^3	8.0×10^2
2.0×10^3	1.0×10^3
1.4×10^3	7.0×10^2
8.0×10^2	??

40.

Soybean Yield per Acre	
Initial, Y_t	Final, Y_{t+1}
100	210
50	110
200	410
75	160
95	200
250	??

41–44 ▪ Recall the data used for Exercises 49–52 in Section 1.2.

Age, a (days)	Length, L (cm)	Tail Length, T (cm)	Mass, M (g)
0.5	1.5	1.0	1.5
1.0	3.0	0.9	3.0
1.5	4.5	0.8	6.0
2.0	6.0	0.7	12.0
2.5	7.5	0.6	24.0
3.0	9.0	0.5	48.0

These data define several discrete-time dynamical systems. For example, between the first measurement (on day 0.5) and the second (on day 1.0), the length increases by 1.5 cm. Between the second measurement (on day 1.0) and the third (on day 1.5), the length again increases by 1.5 cm.

41. Graph the length at the second measurement as a function of length at the first, the length at the third measurement as a function of length at the second, and so forth. Find the discrete-time dynamical system that reproduces the results.

42. Find and graph the discrete-time dynamical system for tail length.

43. Find and graph the discrete-time dynamical system for mass.

44. Find and graph the discrete-time dynamical system for age.

45–48 ▪ Suppose students are permitted to take a test again and again until they get a perfect score of 100. We wish to write a discrete-time dynamical system describing these dynamics.

45. In words, what is the argument of the updating function? What is the value?

46. What are the domain and range of the updating function? What value do you expect if the argument is 100?

47. Sketch a possible graph of the updating function.

48. Based on your graph, how would a student do on her second try if she scored 20 on her first try?

49–50 ▪ Consider the discrete-time dynamical system $b_{t+1} = 2.0b_t$ for a bacterial population (Example 1.5.1).

49. Write a discrete-time dynamical system for the total volume of bacteria (suppose each bacterium takes up $10^4 \mu m^3$).

50. Write a discrete-time dynamical system for the total area taken up by the bacteria (suppose the thickness is 20 μm).

51–52 ▪ Recall the equation $h_{t+1} = h_t + 1.0$ for tree height.

51. Write a discrete-time dynamical system for the total volume of the cylindrical trees in Section 1.3, Exercise 27.

52. Write a discrete-time dynamical system for the total volume of a spherical tree (this is kind of tricky).

53–54 ▪ Consider the following data describing the level of medication in the blood of two patients over the course of several days.

Day	Medication Level in Patient 1	Medication Level in Patient 2
0	20.0	0.0
1	16.0	2.0
2	13.0	3.2
3	10.75	3.92

53. Graph three points on the updating function for the first patient. Find the discrete-time dynamical system for the first patient.

54. Graph three points on the updating function for the second patient and find the discrete-time dynamical system.

55–56 ▪ For the following discrete-time dynamical systems, compute solutions with the given initial condition. Then find the difference between the solutions as a function of time, and the ratio of the solutions as a function of time. In which cases is the difference constant, and in which cases is the ratio constant? Can you explain why?

55. Two bacterial populations follow the discrete-time dynamical system $b_{t+1} = 2.0b_t$, but the first starts with initial condition $b_0 = 1.0 \times 10^6$ and the second starts with initial condition $b_0 = 3.0 \times 10^5$.

56. Two trees follow the discrete-time dynamical system $h_{t+1} = h_t + 1.0$, but the first starts with initial condition $h_0 = 10.0$ m and the second starts with initial condition $h_0 = 2.0$ m.

57–60 ▪ Follow the steps to derive discrete-time dynamical systems describing the following contrasting situations.

57. A population of bacteria doubles every hour, but 1.0×10^6 individuals are removed after reproduction to be converted into valuable biological by-products. The population begins with $b_0 = 3.0 \times 10^6$ bacteria.

a. Find the population after 1, 2, and 3 hours.

b. How many bacteria were harvested?

c. Write the discrete-time dynamical system.

d. Suppose you waited to harvest bacteria until the end of 3 hours. How many could you remove and still match the population b_3 found in part **a**? Where did all the extra bacteria come from?

58. Suppose a population of bacteria doubles every hour, but that 1.0×10^6 individuals are removed **before** reproduction to be converted into valuable biological by-products. Suppose the population begins with $b_0 = 3.0 \times 10^6$ bacteria.

a. Find the population after 1, 2, and 3 hours.

b. Write the discrete-time dynamical system.

c. How does the population compare with that in the previous problem? Why is it doing worse?

59. Suppose the fraction of individuals with some superior gene increases by 10% each generation.

a. Write the discrete-time dynamical system for the fraction of organisms with the gene (denote the fraction at time t by f_t and figure out the formula for f_{t+1}).

b. Write the solution with $f_0 = 0.0001$.

c. Will the fraction reach 1.0? Does the discrete-time dynamical system make sense for all values of f_t?

60. The Weber-Fechner law describes how human beings perceive differences. Suppose, for example, that a person first hears a tone with a frequency of 400 hertz (cycles per second). He is then tested with higher tones until he can hear the difference. The ratio between these values describes how well this person can hear differences.

a. Suppose the next tone he can distinguish has a frequency of 404 hertz. What is the ratio?

b. According to the Weber-Fechner law, the next higher tone will be greater than 404 by the same ratio. Find this tone.

c. Write the discrete-time dynamical system for this person. Find the fifth tone he can distinguish.

d. Suppose the experiment is repeated on a musician, and she manages to distinguish 400.5 hertz from 400 hertz. What is the fifth tone she can distinguish?

61–62 ▪ The total mass of a population of bacteria will change if either the number of bacteria changes, the mass per bacterium changes, or both. The following problems derive discrete-time dynamical systems when both change.

61. The number of bacteria doubles each hour, and the mass of each bacterium triples during the same time.

62. The number of bacteria doubles each hour, and the mass of each bacterium increases by 1.0×10^{-9}g. What seems to go wrong with this calculation? Can you explain why?

1.6 Analysis of Discrete-Time Dynamical Systems

We have defined discrete-time dynamical systems that describe what happens during a single time step, and defined the solution as the sequence of values taken on over many time steps. Often enough, finding a formula for the solution is difficult or impossible. Nonetheless, we can often deduce the behavior of the solution with simpler methods. This section introduces two such methods. **Cobwebbing** is a graphical technique that makes it possible to sketch solutions without computing anything. Algebraically, we will learn how to solve for **equilibria**, points where the discrete-time dynamical system leaves the value unchanged.

1.6.1 Cobwebbing: A Graphical Solution Technique

Suppose we have a general discrete-time dynamical system

$$m_{t+1} = f(m_t)$$

with the updating function graphed in Figure 1.6.1. By adding the diagonal (the line $m_{t+1} = m_t$) to the graph, we can find the behavior of solutions graphically. The technique is called **cobwebbing**.

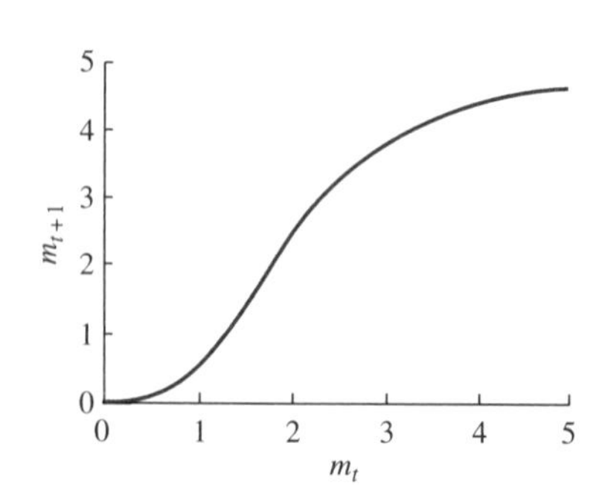

FIGURE 1.6.1
Graph of the updating function

Suppose we are given some initial condition m_0. To find m_1, we evaluate the updating function at m_0, or

$$m_1 = f(m_0).$$

Graphically, m_1 is the point on the graph of the updating function directly above m_0 (Figure 1.6.2a). Similarly, m_2 is the point on the graph of the updating function directly above m_1 and so on.

The missing step is moving m_1 from the vertical axis onto the horizontal axis. The trick is to **reflect** it off the diagonal line that has equation $m_{t+1} = m_t$. Move the point (m_0, m_1) horizontally until it intersects the diagonal. Moving a point horizontally does not change the vertical coordinate. The intersection with the diagonal occurs at the point (m_1, m_1) (Figure 1.6.2b). The point $(m_1, 0)$ lies directly below (Figure 1.6.2c).

What have we done? Starting from the initial value m_0, plotted on the horizontal axis, we used the updating function to find m_1 on the vertical axis and the reflecting trick to project m_1 onto the horizontal axis. We then can find m_2, by moving vertically to the graph of the updating function (Figure 1.6.2d). To find m_3, we move horizontally to the diagonal to reach the point (m_2, m_2), and then vertically to the point (m_2, m_3).

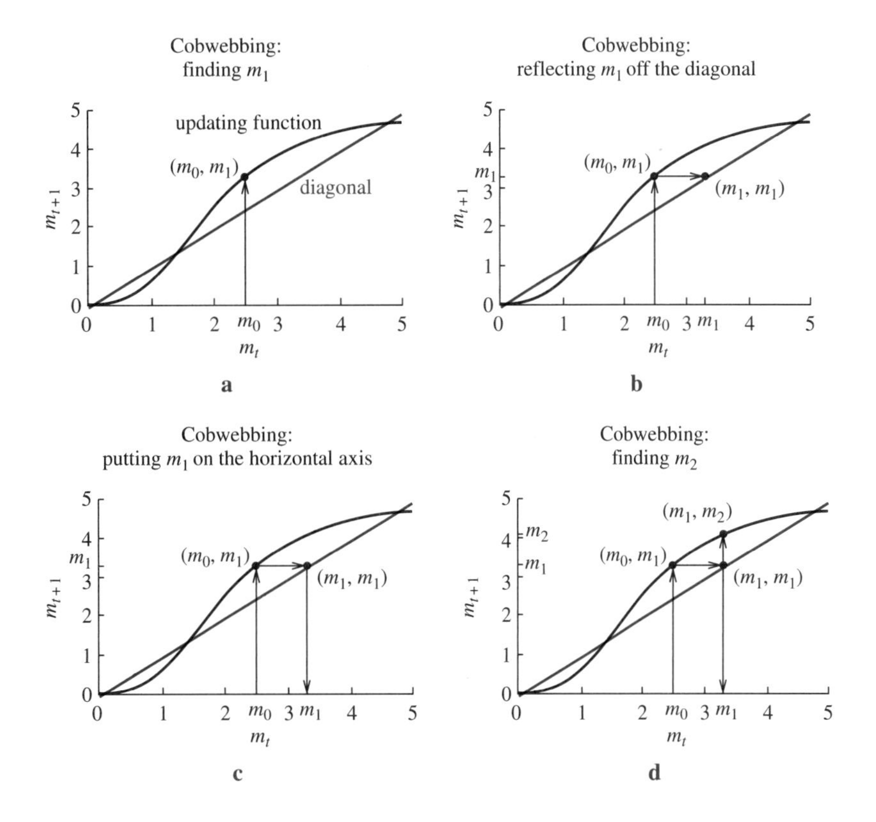

FIGURE 1.6.2
Cobwebbing: The first steps

Because the lines reaching all the way to the horizontal axis are unnecessary, they are generally omitted to make the diagram more readable (Figure 1.6.3).

Having found m_1, m_2, and m_3 on our cobwebbing graph, we can sketch a graph of the solution that shows the measurement as a function of time. In Figure 1.6.2, we began at $m_0 = 2.5$. This is plotted as the point $(0, m_0) = (0, 2.5)$ in the solution (Figure 1.6.4) The value m_1 is approximately 3.2 and is plotted as the point $(1, m_1)$ in the solution. The values of m_2 and m_3 increase more slowly, and are plotted thus on the graph. Without plugging numbers into the formula, we have used the **graph** of the updating function to figure out the behavior of a solution starting from a given initial condition.

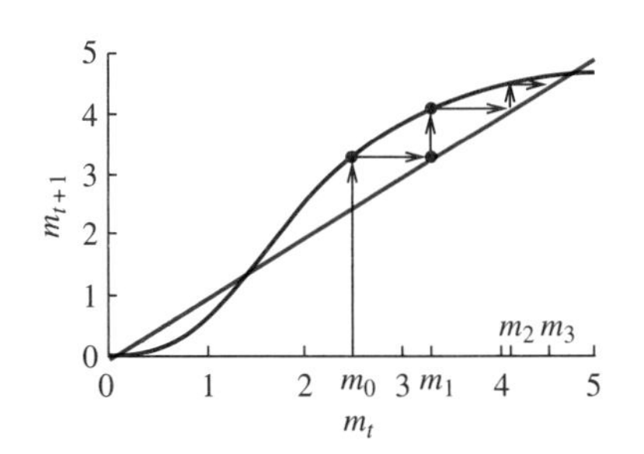

FIGURE 1.6.3
Cobwebbing

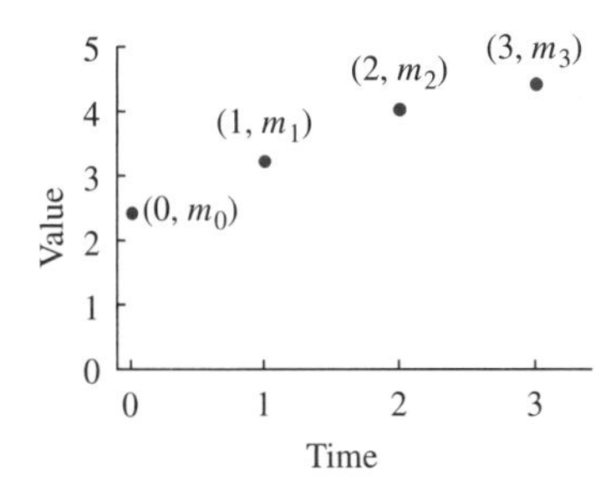

FIGURE 1.6.4
The solution derived from a cobweb diagram

Similarly, we can find how the concentration would behave over time if we started from the different initial condition $m_0 = 1.2$ (Figure 1.6.5). In this case, the diagonal lies below the graph of the updating function, so reflecting off the diagonal moves points to the left. Therefore, the solution decreases.

The steps for cobwebbing are summarized in the following algorithm.

▸▸ Algorithm 1.4 Using Cobwebbing to Find a Solution

1. Graph the updating function and the diagonal.
2. Starting from the initial condition go "up or down to the updating function and over to the diagonal."

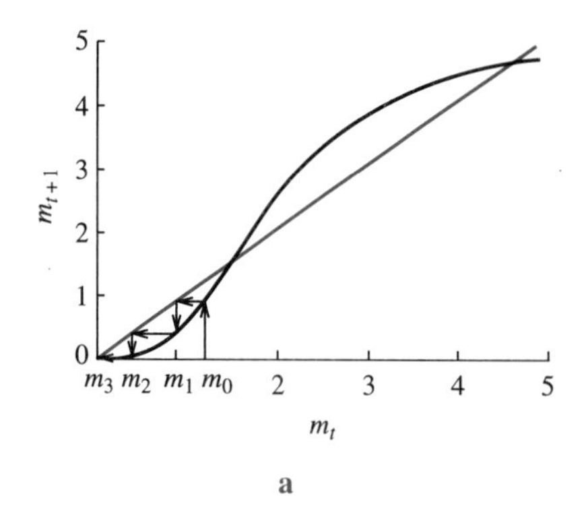

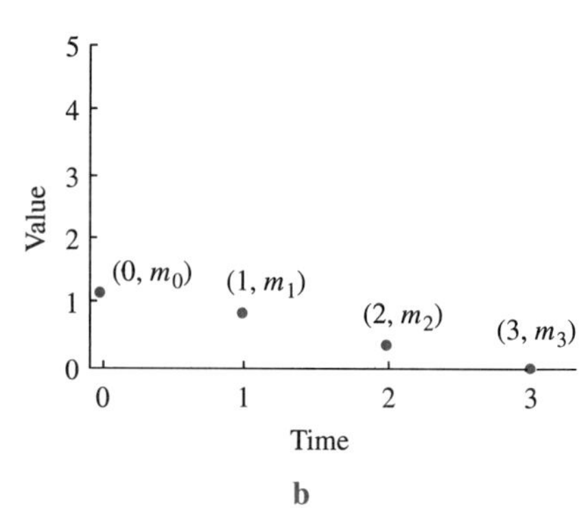

FIGURE 1.6.5
Cobweb and solution with a different initial condition

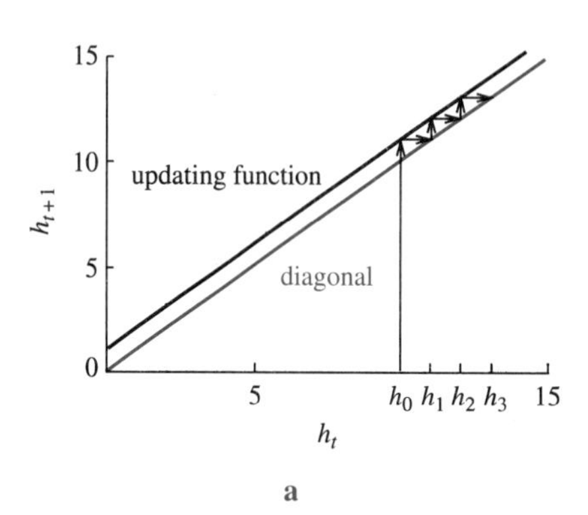

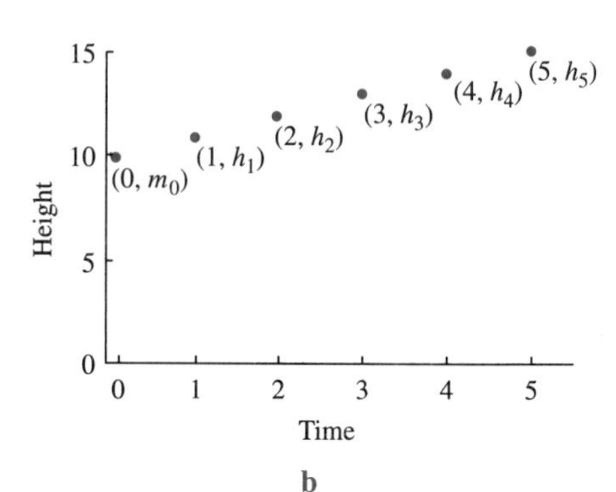

FIGURE 1.6.6
Cobweb and solution of tree growth model

3. Repeat for as many steps as needed to find the pattern.
4. Sketch the solution at times 0, 1, 2, and so forth as the consecutive horizontal coordinates of intersections with the diagonal.

Example 1.6.1 Cobwebbing and Solution of the Tree Growth Model

Consider the discrete-time dynamical system for a growing tree (Example 1.5.2)

$$h_{t+1} = h_t + 1.0.$$

The graph of the updating function $g(h_t) = h_t + 1.0$ is a line with slope 1 and intercept 1.0, and is thus parallel to the diagonal $h_{t+1} = h_t$ (Figure 1.6.6). Starting from an initial condition of 10.0, the cobweb moves up steadily, as does the solution. The graphical solution is consistent with the exact solution $h_t = 10.0 + t$ m (Example 1.5.13), although it does not provide exact **quantitative** predictions.

Example 1.6.2 Cobwebbing and Solution of the Medication Model

Consider the discrete-time dynamical system for medication (Example 1.5.4)

$$M_{t+1} = 0.5M_t + 1.0.$$

The updating function is a line with slope 0.5 and intercept 1, and is thus less steep than the diagonal $M_{t+1} = M_t$. If we begin at $M_0 = 5$, the cobweb and solution decrease more and more slowly over time (Figure 1.6.7). If we begin instead at $M_0 = 1$, the cobweb and solution increase over time (Figure 1.6.8).

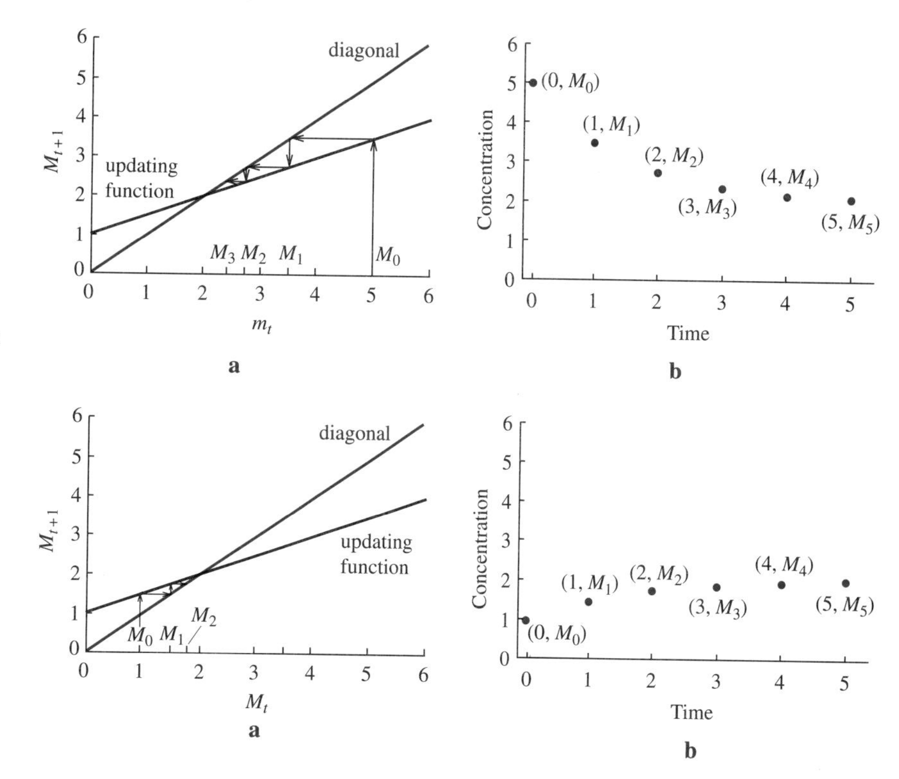

FIGURE 1.6.7
Cobweb and solution of the medication model: $M_0 = 5.0$

FIGURE 1.6.8
Cobweb and solution of the medication model: $M_0 = 1.0$

1.6.2 Equilibria: Graphical Approach

The points where the graph of the updating function intersects the diagonal play a special role in cobweb diagrams. These points also play an essential role in understanding the behavior of discrete-time dynamical systems. Consider the discrete-time dynamical systems plotted in Figure 1.6.9. The first describes a population of plants (denoted by P_t at time t) and the second a population of birds (denoted by B_t at time t). Each graph includes the diagonal line used in cobwebbing.

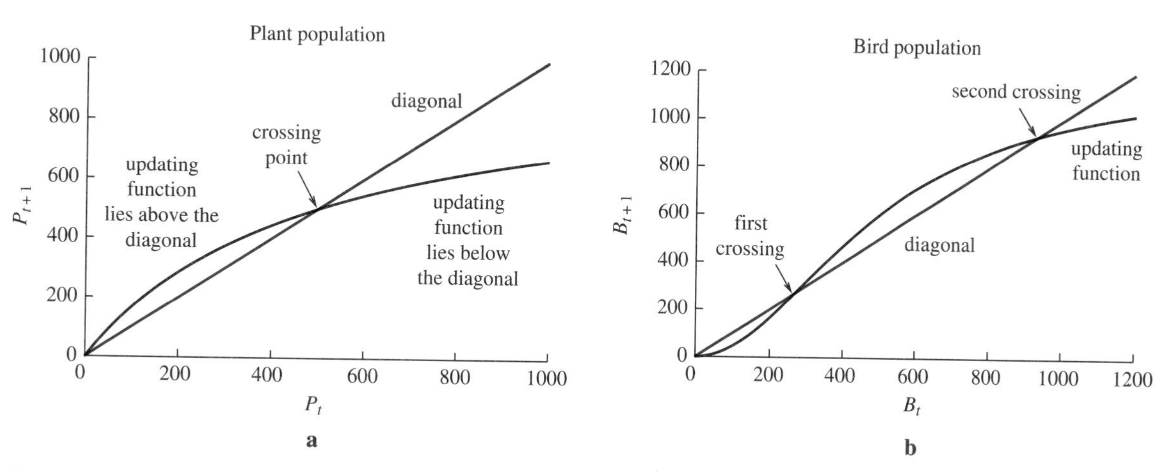

FIGURE 1.6.9
Dynamics of two populations

If we begin cobwebbing from an initial condition where the graph of the updating function lies **above** the diagonal, the population increases (Figure 1.6.10a). In contrast, if we begin cobwebbing from an initial condition where the graph of the updating function lies **below** the diagonal, the population decreases (Figure 1.6.10b). The plant population will thus increase if the initial condition lies to the left of the crossing point, and decrease if it lies to the right of the crossing point.

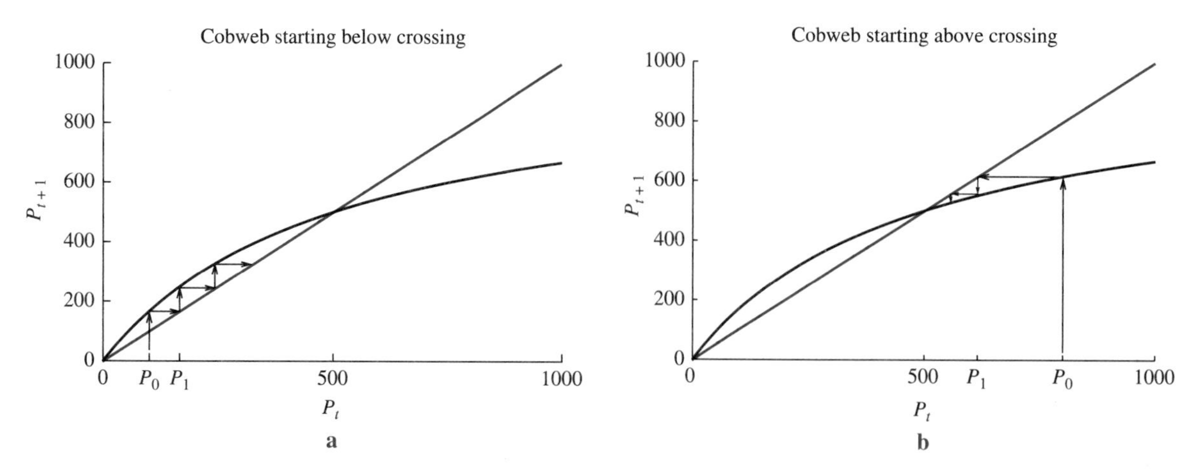

FIGURE 1.6.10
Behavior of plant population starting from two initial conditions

Similarly, the updating function for the bird population lies below the diagonal for initial conditions less than the first crossing and the population decreases (Figure 1.6.11a). The updating function is above the diagonal for initial conditions between the crossings and the population increases (Figure 1.6.11b). Finally, the updating function is again below the diagonal for initial conditions greater than the second crossing and the population decreases (Figure 1.6.11c).

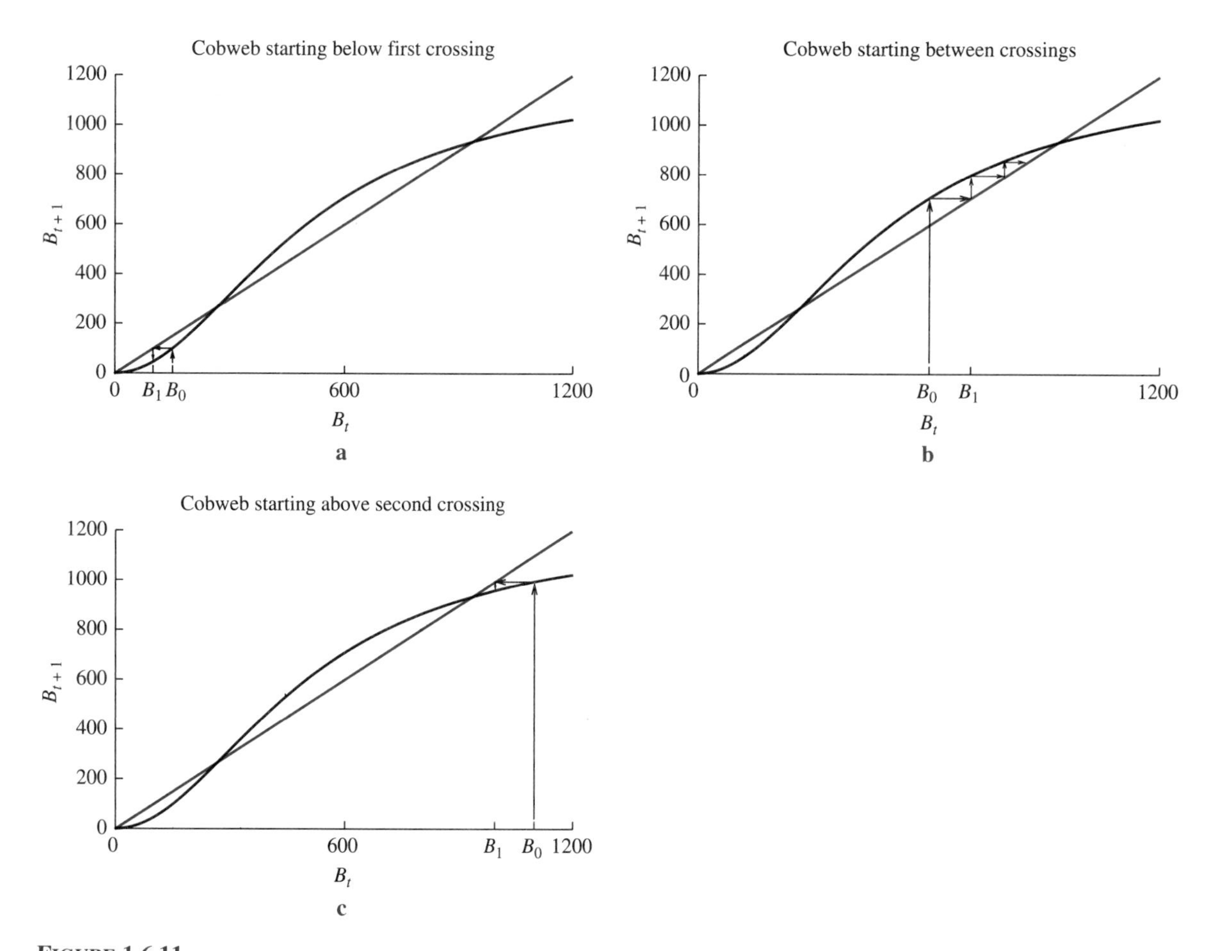

FIGURE 1.6.11
Behavior of bird population starting from three initial conditions

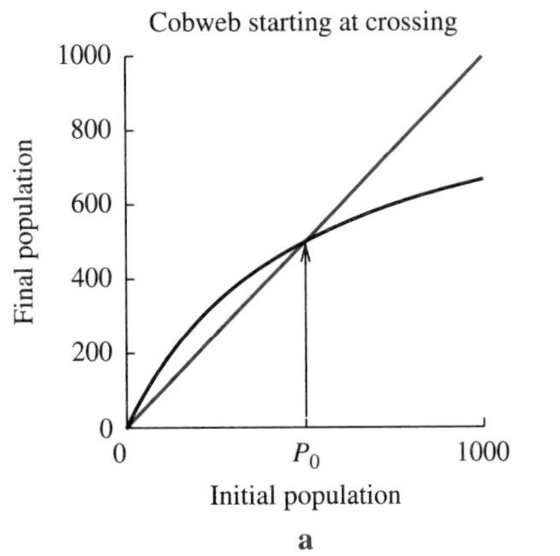

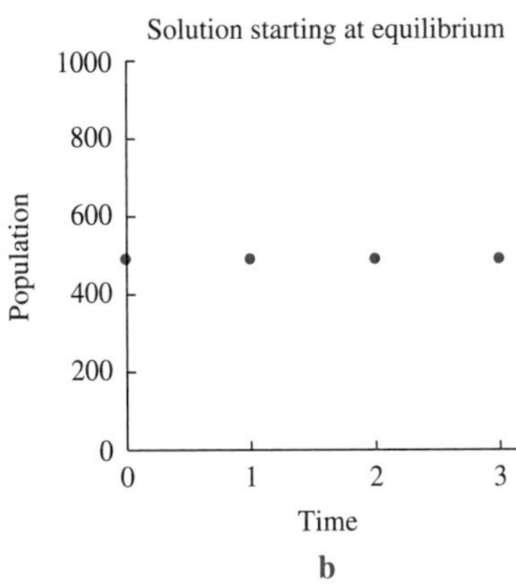

FIGURE 1.6.12
Behavior of plant population starting from an equilibrium

What happens where the updating function crosses the diagonal? At these points, the population neither increases nor decreases, and thus remains the same. Such a point is called an **equilibrium**.

Definition 1.11 A point m^* is called an equilibrium of the discrete-time dynamical system

$$m_{t+1} = f(m_t)$$

if $f(m^*) = m^*$.

This definition says that the discrete-time dynamical system leaves m^* unchanged. These points can be found graphically by looking for intersections of the graph of the updating function with the diagonal line.

When there is more than one equilibrium, they are called **equilibria**. The plant population has two equilibria, one at $P = 0$ and one at $P = 50$. If we start cobwebbing from an initial condition exactly equal to an equilibrium, not much happens. The cobweb goes up to the crossing point and gets stuck there (Figure 1.6.12a). The solution is a horizontal sequence of dots (Figure 1.6.12b).

Why does the graphical method for finding equilibria work? The diagonal has equation

$$m_{t+1} = m_t$$

and can be thought of as a discrete-time dynamical system that leaves **all** inputs unchanged, and always returns an output equal to its input. The intersections of the graph of the updating function with the diagonal are thus points where the updating function leaves its input unchanged. These are the equilibria.

1.6.3 Equilibria: Algebraic Approach

When we know the formula for the discrete-time dynamical system, we can solve for the equilibria algebraically.

Example 1.6.3 The Equilibrium of the Medication Discrete-Time Dynamical System

Recall the discrete-time dynamical system for medication

$$M_{t+1} = 0.5M_t + 1.0$$

(Figure 1.6.13, Example 1.5.4). Let M^* stand for an equilibrium. The equation for equilibrium says that M^* is unchanged by the discrete-time dynamical system, or

$$M^* = 0.5M^* + 1.0.$$

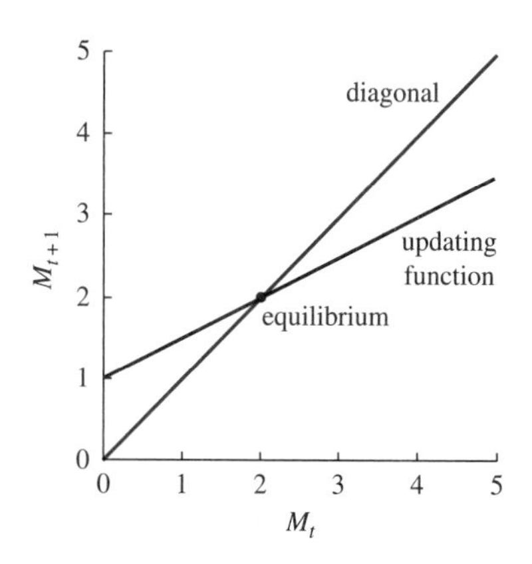

FIGURE 1.6.13

Equilibrium of the medication discrete-time dynamical system

We can solve this linear equation.

$$M^* = 0.5M^* + 1.0 \quad \text{the original equation}$$
$$M^* - 0.5M^* = 1.0 \quad \text{subtract } 0.5M^* \text{ to get unknowns on one side}$$
$$0.5M^* = 1.0 \quad \text{do the subtraction}$$
$$M^* = \frac{1.0}{0.5} = 2.0. \quad \text{divide by 0.5}$$

The equilibrium value is 2.0 milligrams per liter. We can check this by plugging $M_t = 2.0$ into the discrete-time dynamical system, finding that

$$M_{t+1} = 0.5 \cdot 2.0 + 1.0 = 2.0.$$

A concentration of 2.0 is indeed unchanged over the course of a day. Furthermore, we have seen that solutions approach this equilibrium (Examples 1.5.14 and 1.5.15).

Example 1.6.4 The Equilibrium of the Bacterial Discrete-Time Dynamical System

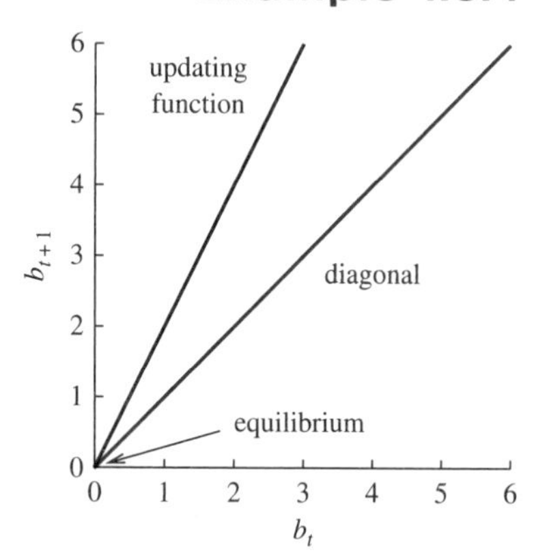

FIGURE 1.6.14

Equilibrium of the bacterial discrete-time dynamical system

To find the equilibria for the bacterial population discrete-time dynamical system

$$b_{t+1} = 2b_t$$

(Figure 1.6.14, Example 1.5.1), we write the equation for equilibria,

$$b^* = 2b^*.$$

We then solve this equation

$$b^* = 2b^* \quad \text{the original equation}$$
$$b^* - b^* = 2b^* - b^* \quad \text{subtract } b^* \text{ from both sides}$$
$$0 = b^*. \quad \text{do the subtraction}$$

Consistent with our picture, the only equilibrium is at $b_t = 0$. The only number that remains the same after doubling is 0.

Example 1.6.5 A Discrete-Time Dynamical System with No Equilibrium

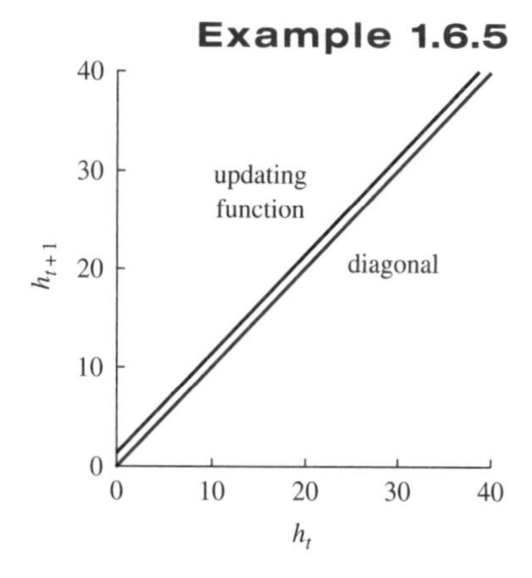

FIGURE 1.6.15

A discrete-time dynamical system with no equilibrium

The updating function for a growing tree following the discrete-time dynamical system

$$h_{t+1} = h_t + 1.0$$

has a graph that is parallel to the diagonal (Figure 1.6.15). To solve for the equilibria, we try

$$h^* = h^* + 1 \quad \text{the equation for the equilibrium}$$
$$h^* - h^* = 1 \quad \text{subtract } h^* \text{ to get unknowns on one side}$$
$$0 = 1. \quad \text{do the subtraction}$$

This looks bad. The graph of the updating function and the graph of the diagonal do not intersect because they are parallel lines. Something that grows 1.0 m per year cannot remain unchanged.

Example 1.6.6 A Biologically Unrealistic Equilibrium

The graph of the updating function associated with a mite population that follows the discrete-time dynamical system

$$x_{t+1} = 2x_t + 30$$

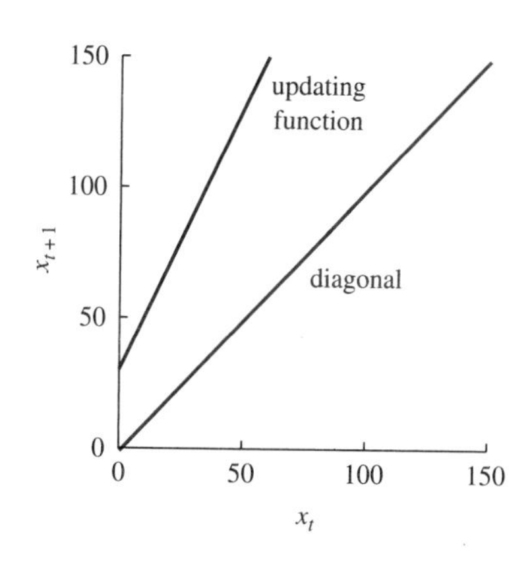

FIGURE 1.6.16
The discrete-time dynamical system for mites

lies above the diagonal for all positive values of x_t (Figure 1.6.16). To solve for the equilibria, try

$$\begin{aligned} x^* &= 2x^* + 30 && \text{the equation for the equilibrium} \\ x^* - 2x^* &= 30 && \text{subtract } 2x^* \text{ to get unknowns on one side} \\ -x^* &= 30 && \text{do the subtraction} \\ x^* &= -30. && \text{divide both sides by } -1 \end{aligned}$$

This looks like nonsense. However, if we check by substituting $x_t = -30$ into the discrete-time dynamical system, we find

$$x_{t+1} = 2 \cdot (-30) + 30 = -30,$$

which is indeed equal to x_t.

Although there is a **mathematical** equilibrium, there is no **biological** equilibrium. If we extend the graph to include biologically meaningless negative values, we see that the graph of the updating function does intersect the diagonal (Figure 1.6.17).

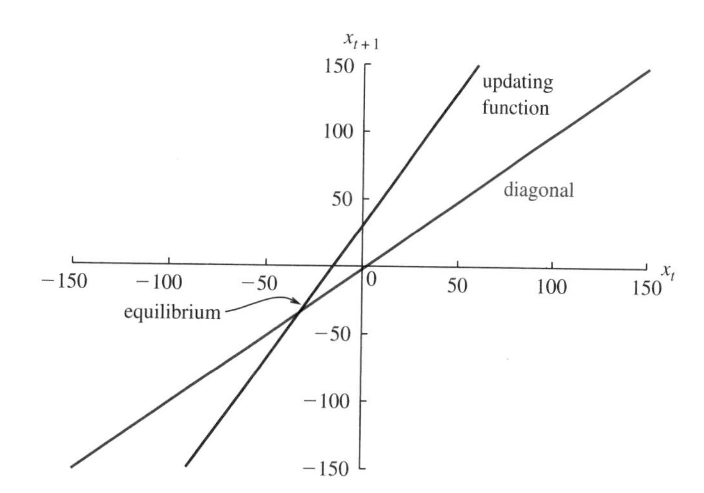

FIGURE 1.6.17
Extending the discrete-time dynamical system for mites to include a negative domain

Example 1.6.7 The Loading Dose for a Medication

In the discrete-time dynamical system for medication (Example 1.6.3) the solution only slowly increases to its equilibrium at $M^* = 2$. To reach this equilibrium immediately, the dose on the first day can be increased. Rather than giving a dose of 1.0 as on later days, suppose a larger dose of 2.0 were given. Then, on this first day,

$$M_1 = 0.5M_0 + 2.0.$$

Because this is the first dose given, $M_0 = 0$, meaning that $M_1 = 2.0$, equal to the equilibrium. Returning thereafter to the normal dose of 1.0, the system

$$M_{t+1} = 0.5M_t + 1.0$$

will remain at equilibrium. This **loading dose** avoids the slow build-up to the desired concentration.

Algebra Involving Parameters Studying the general form of a discrete-time dynamical system, using parameters instead of numbers, can simplify the algebra and make results easier to understand. When we work with parameters, however, we must be more careful with the algebra.

Algorithm 1.5 Solving for Equilibria

1. Write the equation for the equilibrium.
2. Use subtraction to move all the terms to one side, leaving 0 on the other.
3. Factor (if possible).
4. Set each factor equal to 0 and solve for the equilibria (if possible).
5. Think about the results.

As always, we begin by setting up the problem. The next three steps give a safe method to do the algebra (although the algebra may be impossible). The final step is perhaps the most important. A result is worthwhile only if it makes sense.

Example 1.6.8 Finding Equilibria of the Bacterial Model in General

Consider the bacterial discrete-time dynamical system where the factor of 2.0 has been replaced with a general **per capita reproduction** of r,

$$b_{t+1} = rb_t.$$

The factor r describes how much the population grows (or declines) in one hour. Applying Algorithm 1.5 gives

$$\begin{aligned} b^* &= rb^* && \text{the equation for the equilibrium} \\ b^* - rb^* &= 0 && \text{move everything to one side} \\ b^*(1-r) &= 0 && \text{factor out the common factor of } b^* \\ b^* = 0 \text{ or } 1 - r &= 0 && \text{set both factors to 0} \\ b^* = 0 \text{ or } r &= 1. && \text{solve each term} \end{aligned}$$

There are two possibilities. The first matches what we found earlier; a population of 0 is at equilibrium. This makes sense because an extinct population cannot reproduce. The second is new. If the per capita reproduction r is exactly 1, any value of b_t is an equilibrium. In this case, each bacterium exactly replaces itself. The population size will remain the same no matter what its size, even though the individual bacteria are reproducing and dying.

Example 1.6.9 Equilibria of the Medication Model with a Dosage Parameter

Consider the medication discrete-time dynamical system with the parameter S

$$M_{t+1} = 0.5M_t + S,$$

where S represents the daily dosage. The algorithm for finding equilibria gives

$$\begin{aligned} M^* &= 0.5M^* + S && \text{the equation for the equilibrium} \\ M^* - 0.5M^* - S &= 0 && \text{move everything to one side} \\ 0.5M^* - S &= 0 && \text{simplify} \\ M^* &= 2.0S. && \text{nothing to factor, solve for } M^* \end{aligned}$$

The equilibrium value is proportional to S, the daily dosage.

Example 1.6.10 Equilibria of the Medication Model with Absorption

Consider the medication discrete-time dynamical system with parameter α

$$M_{t+1} = (1-\alpha)M_t + 1.0,$$

where the parameter α represents the fraction of existing medication absorbed by the body during a given day. For example, if $\alpha = 0.1$, 10% of the medication is absorbed by the body and 90% remains.

$$
\begin{aligned}
M^* &= (1-\alpha)M^* + 1.0 && \text{the equation for the equilibrium}\\
M^* - (1-\alpha)M^* - 1.0 &= 0 && \text{move everything to one side}\\
M^* - M^* + \alpha M^* - 1.0 &= 0 && \text{distribute negative sign through quantity}\\
\alpha M^* - 1.0 &= 0 && \text{cancel } M^* - M^*\\
M^* &= \frac{1.0}{\alpha}. && \text{solve for } M^*
\end{aligned}
$$

The equilibrium value is proportional to the reciprocal of α and is thus larger when the fraction absorbed is larger. If $\alpha = 0.1$, the equilibrium is

$$M^* = \frac{1.0}{0.1} = 10.0.$$

In contrast, if the body absorbs 50% of the medication each day, leading to a larger value of $\alpha = 0.5$, then

$$M^* = \frac{1.0}{0.5} = 2.0.$$

The body that absorbs more reaches a lower equilibrium.

Example 1.6.11 Equilibria of the Medication Model with Two Parameters

Consider the medication discrete-time dynamical system with two parameters (extending Examples 1.6.9 and 1.6.10),

$$M_{t+1} = (1-\alpha)M_t + S.$$

The algorithm for finding equilibria gives

$$
\begin{aligned}
M^* &= (1-\alpha)M^* + S && \text{the equation for the equilibrium}\\
M^* - (1-\alpha)M^* - S &= 0 && \text{move everything to one side}\\
M^* - M^* + \alpha M^* - S &= 0 && \text{distribute negative sign through quantity}\\
\alpha M^* - S &= 0 && \text{cancel } M^* - M^*\\
M^* &= \frac{S}{\alpha}. && \text{solve for } M^*
\end{aligned}
$$

The equilibrium value is larger if S is larger or if α is smaller. It makes sense because the equilibrium concentration can be increased in two ways: by increasing the dosage or by decreasing the fraction absorbed.

Summary We have developed a graphical technique to estimate solutions called **cobwebbing**. By examining the diagrams used for cobwebbing, we found that intersections of the graph of the updating function with the diagonal line play a special role. These **equilibria** are points that are unchanged by the discrete-time dynamical system. Algebraically, we find equilibria by solving the equation that describes such points. We can often solve for equilibria in general, without substituting numerical values for the parameters. Solving the equations in this way can help clarify the underlying biological process.

1.6 Exercises

Mathematical Techniques

1–2 ▪ The following steps are used to build a cobweb diagram. Follow them for the given discrete-time dynamical system based on bacterial populations.

a. Graph the updating function.

b. Use your graph of the updating function to find the point (b_0, b_1).

c. Reflect it off the diagonal to find the point (b_1, b_1).

d. Use the graph of the updating function to find (b_1, b_2).

e. Reflect off the diagonal to find the point (b_2, b_2).

f. Use the graph of the updating function to find (b_2, b_3).

g. Sketch the solution as a function of time.

1. The discrete-time dynamical system $b_{t+1} = 2.0b_t$ with $b_0 = 1.0$.

2. The discrete-time dynamical system $n_{t+1} = 0.5n_t$ with $n_0 = 1.0$.

3–6 ▪ Cobweb the following discrete-time dynamical systems for three steps starting from the given initial condition. Compare with the solution found earlier.

3. $v_{t+1} = 1.5v_t$, starting from $v_0 = 1220\mu m^3$ (as in Section 1.5, Exercise 5).

4. $l_{t+1} = l_t - 1.7$, starting from $l_0 = 13.1$ cm (as in Section 1.5, Exercise 6).

5. $n_{t+1} = 0.5n_t$, starting from $n_0 = 1200$ (as in Section 1.5, Exercise 7).

6. $M_{t+1} = 0.75M_t + 2.0$ starting from the initial condition $M_0 = 16.0$ (as in Section 1.5, Exercise 8).

7–12 ▪ Graph the updating functions associated with the following discrete-time dynamical systems, and cobweb for five steps starting from the given initial condition.

7. $x_{t+1} = 2x_t - 1$, starting from $x_0 = 2$.

8. $z_{t+1} = 0.9z_t + 1$, starting from $z_0 = 3$.

9. $w_{t+1} = -0.5w_t + 3$, starting from $w_0 = 0$.

10. $x_{t+1} = 4 - x_t$, starting from $x_0 = 1$ (as in Section 1.5, Exercise 25).

11. $x_{t+1} = \dfrac{x_t}{1 + x_t}$, starting from $x_0 = 1$ (as in Section 1.5, Exercise 23).

12. $x_{t+1} = \dfrac{x_t}{x_t - 1}$ for $x_t > 1$, starting from $x_0 = 3$ (as in Section 1.5, Exercise 26).

13–16 ▪ Find the equilibria of the following discrete-time dynamical system from the graphs of their updating functions Label the coordinates of the equilibria.

13.

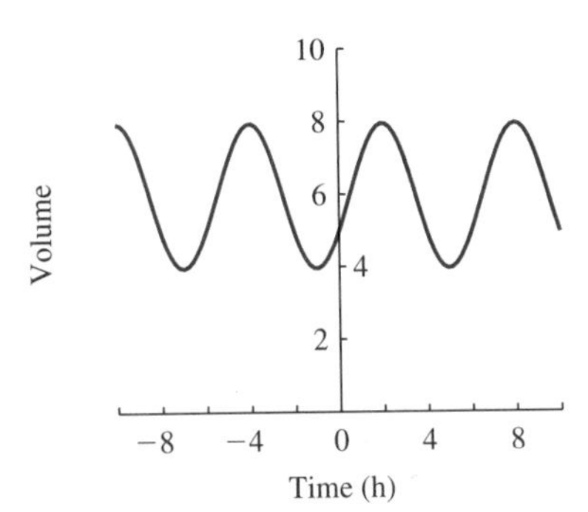

14.

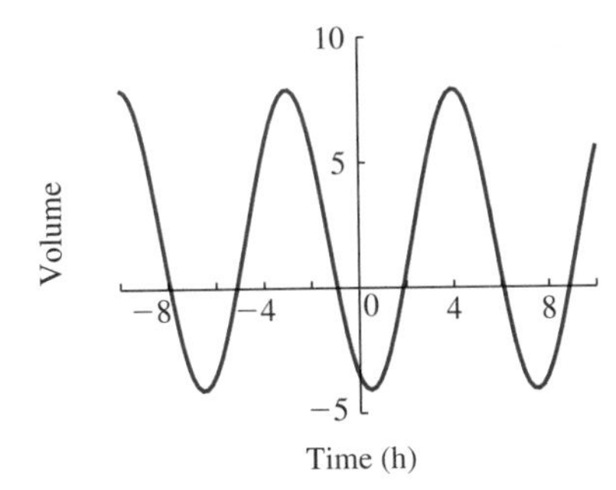

15.

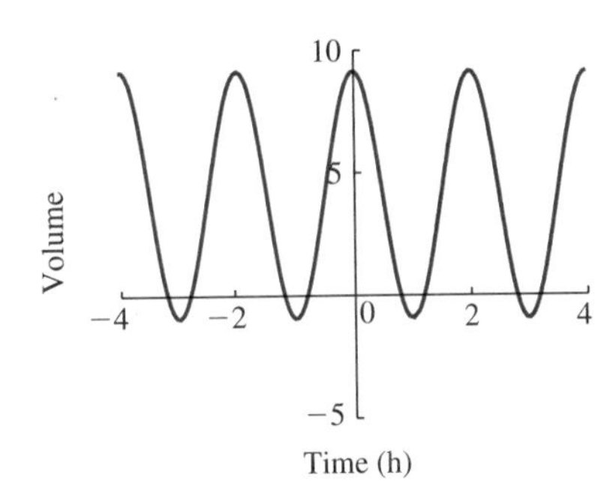

16.

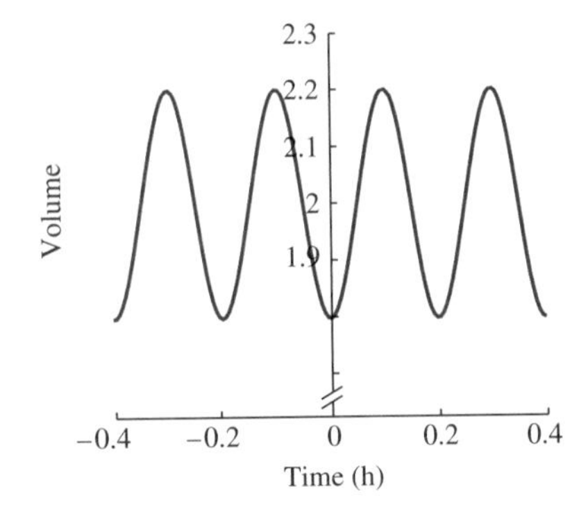

17–22 ▪ Sketch graphs of the following updating functions over the given range and mark the equilibria. Find the equilibria algebraically if possible.

17. $f(x) = x^2$ for $0 \le x \le 2$.

18. $g(y) = y^2 - 1$ for $0 \le y \le 2$.

19–22 ▪ Graph the following discrete-time dynamical systems. Solve for the equilibria algebraically, and identify equilibria and the regions where the updating function lies above the diagonal on your graph.

19. $c_{t+1} = 0.5c_t + 8.0$, for $0 \le c_t \le 30$.

20. $b_{t+1} = 3b_t$, for $0 \le b_t \le 10$.

21. $b_{t+1} = 0.3b_t$, for $0 \le b_t \le 10$.

22. $b_{t+1} = 2.0b_t - 5.0$, for $0 \le b_t \le 10$.

23–30 ▪ Find the equilibria of the following discrete-time dynamical systems. Compare with the results of your cobweb diagram from the earlier problem.

23. $v_{t+1} = 1.5v_t$ (as in Section 1.5, Exercise 5).

24. $l_{t+1} = l_t - 1.7$ (as in Section 1.5, Exercise 6).

25. $x_{t+1} = 2x_t - 1$ (as in Exercise 7).

26. $z_{t+1} = 0.9z_t + 1$ (as in Exercise 8).

27. $w_{t+1} = -0.5w_t + 3$ (as in Exercise 9).

28. $x_{t+1} = 4 - x_t$ (as in Exercise 10).

29. $x_{t+1} = \dfrac{x_t}{1 + x_t}$ (as in Exercise 11).

30. $x_{t+1} = \dfrac{x_t}{x_t - 1}$ for $x_t > 1$ (as in Exercise 12).

31–34 ▪ Find the equilibria of the following discrete-time dynamical systems that include parameters. Identify values of the parameter for which there is no equilibrium, for which the equilibrium is negative, and for which there is more than one equilibrium.

31. $w_{t+1} = aw_t + 3$.

32. $x_{t+1} = b - x_t$.

33. $x_{t+1} = \dfrac{ax_t}{1 + x_t}$.

34. $x_{t+1} = \dfrac{x_t}{x_t - K}$.

Applications

35–40 ▪ Cobweb the following discrete-time dynamical systems for five steps starting from the given initial condition.

35. An alternative tree growth discrete-time dynamical system with form $h_{t+1} = h_t + 5.0$ with initial condition $h_0 = 10$.

36. The lizard-mite system (Example 1.5.3) $x_{t+1} = 2x_t + 30$ with initial condition $x_0 = 0$.

37. The model defined in Section 1.5, Exercise 37 starting from an initial volume of 1420.

38. The model defined in Section 1.5, Exercise 38 starting from an initial mass of 13.1.

39. The model defined in Section 1.5, Exercise 39 starting from an initial population of 800.

40. The model defined in Section 1.5, Exercise 40 starting from an initial yield of 20.

41–42 ▪ Reconsider the data describing the levels of a medication in the blood of two patients over the course of several days (measured in mg per liter), used in Section 1.5, Exercises 53 and 54.

Day	Medication Level in Patient 1	Medication Level in Patient 2
0	20.0	0.0
1	16.0	2.0
2	13.0	3.2
3	10.75	3.92

41. For the first patient, graph the updating function and cobweb starting from the initial condition on day 0. Find the equilibrium.

42. For the second patient, graph the updating function and cobweb starting from the initial condition on day 0. Find the equilibrium.

43–44 ▪ Cobweb and find the equilibrium of the following discrete-time dynamical system.

43. Consider a bacterial population that doubles every hour, but 1.0×10^6 individuals are removed after reproduction (Section 1.5, Exercise 57). Cobweb starting from $b_0 = 3.0 \times 10^6$ bacteria.

44. Consider a bacterial population that doubles every hour, but 1.0×10^6 individuals are removed before reproduction (Section 1.5, Exercise 58). Cobweb starting from $b_0 = 3.0 \times 10^6$ bacteria.

45–46 ▪ Consider the following general models for bacterial populations with harvest.

45. Consider a bacterial population that doubles every hour, but h individuals are removed after reproduction. Find the equilibrium. Does it make sense?

46. Consider a bacterial population that increases by a factor of r every hour, but 1.0×10^6 individuals are removed after reproduction. Find the equilibrium. What values of r produce a positive equilibrium?

47–48 ▪ Consider the general model $M_{t+1} = (1 - \alpha)M_t + S$ for medication (Example 1.6.11). Find the loading dose (Example 1.6.7) in the following cases.

47. $\alpha = 0.2$, $S = 2$.

48. $\alpha = 0.8$, $S = 4$.

Computer Exercises

49. Use your computer (it may have a special feature for this) to find and graph the first 10 points on the solutions of the following discrete-time dynamical systems. The first two describe populations with reproduction and immigration of 100 individuals per generation, and the last two describe populations that have 100 individuals harvested or removed each generation.

a. $b_{t+1} = 0.5b_t + 100$ starting from $b_0 = 100$.

b. $b_{t+1} = 1.5b_t + 100$ starting from $b_0 = 100$.

c. $b_{t+1} = 1.5b_t - 100$ starting from $b_0 = 201$.

d. $b_{t+1} = 1.5b_t - 100$ starting from $b_0 = 199$.

e. What happens if you run the last one for 15 steps? What is wrong with the model?

50. Compose the medication discrete-time dynamical system $M_{t+1} = 0.5M_t + 1.0$ with itself 10 times. Plot the resulting function. Use this composition to find the concentration after 10 days starting from concentrations of 1.0, 5.0, and 18.0 milligrams per liter. If the goal is to reach a stable concentration of 2.0 milligrams per liter, do you think this is a good therapy?

1.7 Expressing Solutions with Exponential Functions

The solution associated with the bacterial discrete-time dynamical system given by $b_{t+1} = 2.0b_t$ is

$$b_t = 2.0^t$$

when $b_0 = 1.0$. As a function of t, the solution is an example of an **exponential function**. To find how long it will take the population to reach a particular target value such as 100 requires solving an equation where the variable t appears in the exponent. Solving for t can be simplified by converting this function into a standard form with the base e, and working with the inverse of the exponential function, the **natural logarithm**. In this section, we will study the **laws of exponents** and the **laws of logarithms** that make this conversion convenient. We will generalize the bacterial population growth discrete-time dynamical system to include the death of some bacteria and show that the solution is again an exponential function, but with **base** equal to the **per capita reproduction** of the bacteria.

1.7.1 Bacterial Population Growth in General

The bacteria studied hitherto have doubled in number each hour. Each bacterium divided once and both "daughter" bacteria survived. Suppose instead that only a fraction σ ("sigma") of the daughters survive. Instead of 2.0 offspring per bacteria, we find an average of 2σ offspring (Figure 1.7.1). For example, if only 75% of offspring survived ($\sigma = 0.75$), there are an average of 1.5 surviving offspring per parent. Let

$$r = 2\sigma.$$

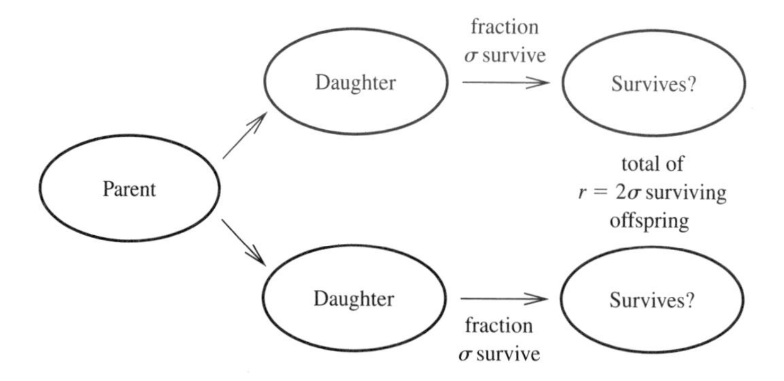

FIGURE 1.7.1
Bacterial population growth with reproduction and mortality

The new parameter r represents the number of new bacteria produced per bacterium and is called the **per capita reproduction**.

In terms of the parameter r, the discrete-time dynamical system is

$$b_{t+1} = rb_t.$$

This fundamental equation of population biology says that the population at time $t + 1$ is equal to the per capita reproduction (the number of new bacteria per old bacterium) times the population at time t (the number of old bacteria), or

$$\text{new population} = \text{per capita reproduction} \times \text{old population.}$$

Example 1.7.1 Discrete-Time Dynamical System If Most Offspring Survive

If $\sigma = 0.75$, then $r = 2 \cdot 0.75 = 1.5$. The discrete-time dynamical system is

$$b_{t+1} = 1.5b_t.$$

If $b_0 = 100$, then $b_1 = 1.5 \cdot 100 = 150$. The population increases by 50% each hour.

Example 1.7.2 Discrete-Time Dynamical System If Few Offspring Survive

If $\sigma = 0.25$, then $r = 2 \cdot 0.25 = 0.5$. The discrete-time dynamical system is

$$b_{t+1} = 0.5b_t.$$

If $b_0 = 100$, then $b_1 = 0.5 \cdot 100 = 50$. Because the value of the survival σ is so small, this population decreases by 50% each hour.

Starting from a population with b_0 bacteria, we can apply the discrete-time dynamical system repeatedly to derive a solution, much as we did in Example 1.5.11 with the particular value $r = 2$ (Figure 1.7.2). We find

$$b_1 = rb_0$$
$$b_2 = rb_1 = r^2 b_0$$
$$b_3 = rb_2 = r^3 b_0.$$

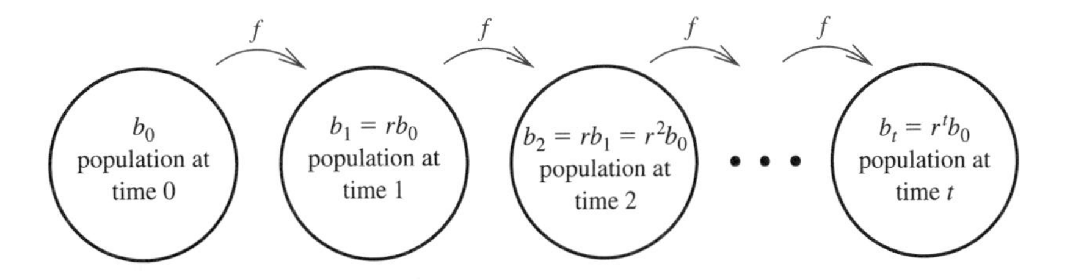

FIGURE 1.7.2
Bacterial population growth

Each hour, the initial population b_0 is multiplied by the per capita reproduction r. After t hours, the initial population b_0 has been multiplied by t factors of r. Therefore

$$b_t = r^t b_0.$$

How do these solutions behave for different values of the per capita reproduction r? Results with four values of r starting from $b_0 = 1.0$ are given in the following table.

t	$r = 2.0$	$r = 1.5$	$r = 1.0$	$r = 0.5$
0	1.0	1.0	1.0	1.0
1	2.0	1.5	1.0	0.5
2	4.0	2.25	1.0	0.25
3	8.0	3.37	1.0	0.125
4	16.0	5.06	1.0	0.0625
5	32.0	7.59	1.0	0.0312
6	64.0	11.4	1.0	0.0156
7	128.0	17.1	1.0	0.00781
8	256.0	25.6	1.0	0.00391

In the first two columns, $r > 1$ and the population increases each hour (Figure 1.7.3a and b). In the third column, $r = 1$ and the population remains the same hour after hour

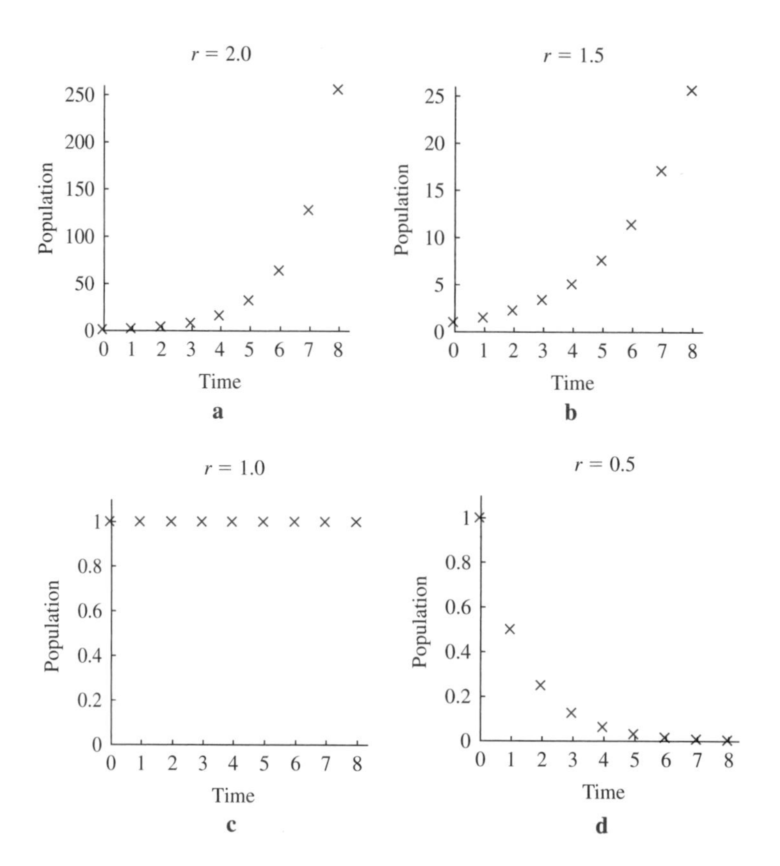

FIGURE 1.7.3
Growing and declining bacterial populations

(Figure 1.7.3c). In the final column, $r < 1$ and the population decreases each hour (Figure 1.7.3d). We summarize these observations in the following table.

Value of r	Behavior of Population
$r > 1$	population increases
$r = 1$	population remains constant
$r < 1$	population decreases

A population with $r = 1$ exactly replaces itself each generation and retains a constant size, even though the individuals in the population change. This is consistent with our finding that any value of b is an equilibrium when $r = 1$ (Example 1.6.8).

1.7.2 Laws of Exponents and Logs

In the solution $b_t = r^t b_0$, the variable t appears in the exponent, in contrast to a function like $f(t) = t^3$ where the variable t is raised to a power. For any positive number a, the **exponential function to the base a** is written

$$f(x) = a^x$$

and said to be "a to the xth power." This function takes x as input and returns as output x factors of a multiplied together. The notation generalizes that used in equations like

$$a^2 = a \cdot a.$$

The key to using exponential functions is knowing the **laws of exponents**, summarized in the table. The table also includes examples using $a = 2$ that can help in remembering when to add and when to multiply.

Laws of Exponents

	General Formula	Example with $a=2$, $x=2$ and $y=3$
Law 1	$a^x \cdot a^y = a^{x+y}$	$2^2 \cdot 2^3 = 2^5 = 32$
Law 2	$(a^x)^y = a^{xy}$	$(2^2)^3 = 2^6 = 64$
Law 3	$a^{-x} = \frac{1}{a^x}$	$2^{-2} = \frac{1}{2^2} = \frac{1}{4}$
Law 4	$\frac{a^y}{a^x} = a^{y-x}$	$\frac{2^3}{2^2} = 2^{3-2} = 2$
Law 5	$a^1 = a$	$2^1 = 2$
Law 6	$a^0 = 1$	$2^0 = 1$

The exponential function is defined for all values of x, including negative numbers and fractions. What does it mean to multiply half an a or -3 a's together? These expressions must be computed with the laws of exponents.

Example 1.7.3 Negative Powers

To compute a^{-3}, apply law 3 to find

$$a^{-3} = \frac{1}{a^3}.$$

For example

$$2^{-3} = \frac{1}{2^3} = \frac{1}{8} = 0.125.$$

Negative powers are in the denominator.

Example 1.7.4 Fractional Powers

To compute $a^{0.5}$, we raise this unknown quantity to the 2nd power (square it), and use law 2 to find

$$(a^{0.5})^2 = a^{0.5 \cdot 2} = a^1 = a.$$

Therefore, a to the 0.5th power is the number that, when squared, gives back a. In other words, a to the 0.5 power is the square root of a. For example

$$2^{0.5} = \sqrt{2} = 1.41421.$$

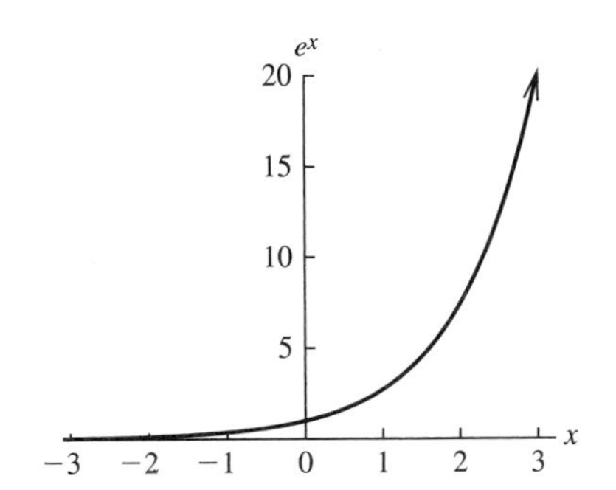

FIGURE 1.7.4
Graph of the exponential function

For reasons that will make sense only with a bit of calculus (Section 2.8), the base most commonly used throughout the sciences is the irrational number

$$e = 2.718281828459\ldots$$

The function

$$f(x) = e^x$$

said "e to the x" is called the **exponential function to the base *e***, or simply the **exponential function** (Figure 1.7.4). Calculators and computers often abbreviate this as exp. The domain of this function consists of all numbers, and the range is all **positive** numbers.

Example 1.7.5 Examples of the Laws of Exponents with the Base e

- $e^3 \cdot e^4 = e^{3+4} = e^7$ (law 1).
- $(e^3)^4 = e^{3 \cdot 4} = e^{12}$ (law 2).
- $e^{-2} = \frac{1}{e^2}$ (law 3).
- $\frac{e^4}{e^3} = e^{4-3} = e^1 = e$ (laws 4 and 5).

- $e^0 = 1$ (law 6).
- $e^3 + e^4$ cannot be simplified with a law of exponents.

The graph of the exponential function crosses every positive horizontal line only once, and thus passes the horizontal line test for having an inverse (see Section 1.2.4). The inverse is the natural log.

Definition 1.12 The inverse function of the exponential function e^x is called the **natural logarithm** (or natural log). The natural log of x is written $\ln(x)$. The natural logarithm has a domain consisting of all positive numbers.

From the definition of the inverse (Definition 1.6),

$$\ln(e^x) = x$$
$$e^{\ln(x)} = x$$

(Figure 1.7.5).

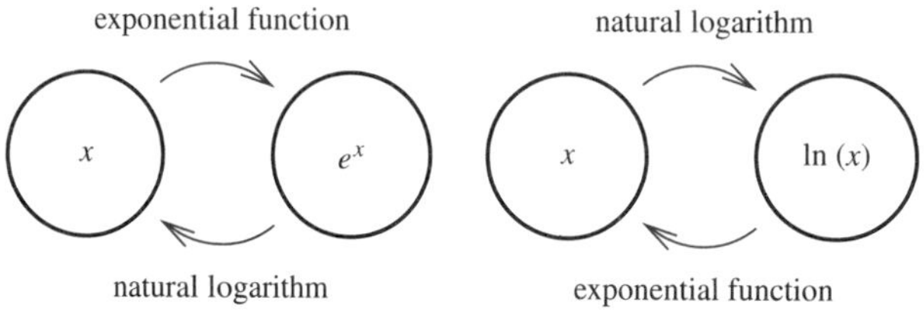

exponential function — natural logarithm — x — e^x — natural logarithm — exponential function — x — $\ln(x)$

FIGURE 1.7.5
The exponential function and natural logarithm are inverses

The graph of the natural logarithm increases from "negative infinity" near $x = 0$, through 0 at $x = 1$, and rises more and more slowly as x becomes larger (Figure 1.7.6). It is impossible to compute the natural log of a negative number.

Example 1.7.6 Exponential and Logarithmic Functions

- If $\ln(100) = 4.605$, then $e^{4.605} = 100$.
- If $e^5 = 148.41$, then $\ln(148.41) = 5$.
- If $\ln(0.1) = -2.302$, then $e^{-2.302} = 0.1$.
- If $e^{-3} = 0.04979$, then $\ln(0.4979) = -3$.

The key to understanding natural logarithms is knowing the laws of logs, which are the laws of exponents in reverse.

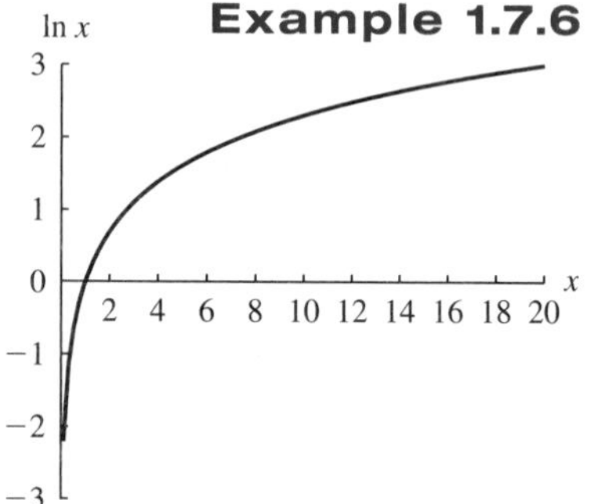

FIGURE 1.7.6
Graph of the natural logarithm

Example 1.7.7 The Laws of Logs in Action

The Laws of Logs

Law 1	$\ln(xy) = \ln(x) + \ln(y)$
Law 2	$\ln(x^y) = y\ln(x)$
Law 3	$\ln(1/x) = -\ln(x)$
Law 4	$\ln(x/y) = \ln(x) - \ln(y)$
Law 5	$\ln(e) = 1$
Law 6	$\ln(1) = 0$

- $\ln(3) + \ln(4) = \ln(3 \cdot 4) = \ln(12)$, using law 1.
- $\ln(3^4) = 4\ln(3)$, using law 2.
- $\ln(1/3) = -\ln(3)$, using law 3.

- $\ln(4/3) = \ln(4) - \ln(3)$, using law 4.
- $\ln(3) \cdot \ln(4)$ cannot be simplified with a law of logs.

In some disciplines, people use the **exponential function with base 10**, or

$$f(x) = 10^x.$$

Its inverse is the **logarithm to the base 10**, written

$$\log_{10} x$$

and said "log base 10 of x." Just as $\ln(x) = y$ implies that $x = e^y$,

$$\log_{10} x = y$$

implies

$$x = 10^y.$$

For example, if $\log_{10} x = 2.3$, $x = 10^{2.3} = 199.5$. In most ways, the exponential function with base 10 and the log base 10 work much like the exponential function with base e and the natural logarithm. All laws of exponents and logs are the same except for law 5, which becomes

$$\text{Law 5 of exponents:} \quad 10^1 = 10$$
$$\text{Law 5 of logs:} \quad \log_{10}(10) = 1.$$

The base e is more convenient for studying dynamics with calculus.

Example 1.7.8 Converting Logarithms in Base 10 to Natural Logs

Suppose $\log_{10}(x) = y$. How can we find $\ln(x)$? By the definition of $\log_{10}$,

$$x = 10^y.$$

Then

$$\begin{aligned} \ln(x) &= \ln(10^y) && \text{taking the natural log of both sides} \\ &= y\ln(10) && \text{law of logs 2} \\ &= 2.303y && \text{because } \ln(10) = 2.303 \\ &= 2.303\log_{10}(x). && \text{definition of } y \end{aligned}$$

For instance, $\log_{10}(100) = 2$, so $\ln(100) = 2.302 \cdot 2 = 4.604$.

1.7.3 Expressing Results with Exponentials

We can use the laws of exponentials and logs to express

$$b_t = r^t b_0$$

in terms of the exponential function with base e. Because the exponential function and the natural logarithm are inverses, we can rewrite r as

$$r = e^{\ln(r)}.$$

Then, using law 2 of exponents,

$$\begin{aligned} r^t &= \left(e^{\ln(r)}\right)^t \\ &= e^{\ln(r)t}. \end{aligned}$$

The **general solution** for the discrete-time dynamical system

$$b_{t+1} = rb_t$$

with initial condition b_0 can be written in exponential notation as

$$b_t = b_0 e^{\ln(r)t}.$$

Example 1.7.9 Expressing a Solution with the Exponential Function

Consider the case $r = 2.0$ and $b_0 = 1.0$. Because $\ln(2.0) = 0.6931$, the solution is

$$b_t = 1.0e^{\ln(2.0)t} = 1.0e^{(0.6931)t}.$$

What is the value of rewriting the solution in this way? Exponential notation makes it easier to solve equations describing the future behavior of a population.

Example 1.7.10 Using a Solution Expressed with the Exponential Function: Increasing Case

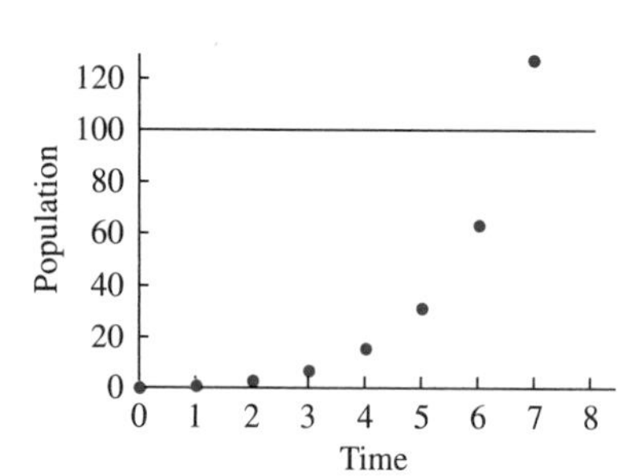

FIGURE 1.7.7
Using solutions to find times

When will the population that obeys $b_{t+1} = 2.0b_t$, with solution

$$b_t = 2.0^t$$

reach 100.0 million? In Example 1.7.9 we wrote this solution in exponential notation. Now we can set $b_t = 100$ and solve for t with the steps

$$\begin{aligned} e^{\ln(2.0)t} &= 100 && \text{equation for } t \\ \ln(2.0)t &= \ln(100) && \text{take the natural log of both sides} \\ t &= \frac{\ln(100)}{\ln(2)} = 6.64. && \text{solve for } t \end{aligned}$$

The population will pass 100.0 million between hours 6 and 7 (Figure 1.7.7). The key step uses the natural log, the inverse of the exponential function, to remove the variable t from the exponent.

Example 1.7.11 Using a Solution Expressed with the Exponential Function: Decreasing Case

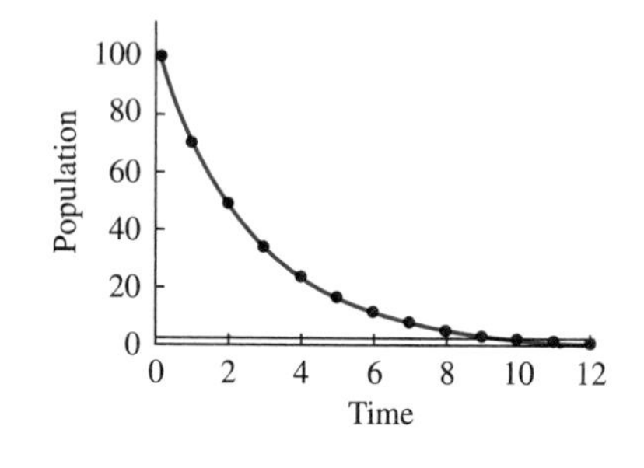

FIGURE 1.7.8
Using solutions to find times

How long it will take a population with $r < 1$ to decrease to some specified value? Suppose $r = 0.7$ and $b_0 = 100.0$. The population decreases because $r < 1$. When will it reach $b_t = 2$ (Figure 1.7.8)? In exponential notation,

$$b_t = 100.0e^{\ln(0.7)t}.$$

Then $b_t = 2.0$ can be solved

$$\begin{aligned} 100.0e^{\ln(0.7)t} &= 2.0 && \text{equation for } t \\ e^{\ln(0.7)t} &= 0.02 && \text{divide both sides by 100} \\ \ln(0.7)t &= \ln(0.02) && \text{take the natural log} \\ t &= \frac{\ln(0.02)}{\ln(0.7)} = 10.96. && \text{solve for } t \end{aligned}$$

All the negative signs cancel, and we see that this population will pass 2.0 just before hour 11.

Throughout the sciences, many measurements other than population sizes are described by exponential functions. In such cases, we write the measurement S as a function of t as

$$S(t) = S(0)e^{\alpha t}.$$

The parameter $S(0)$ represents the value of the measurement at time $t = 0$. The parameter α describes how the measurement changes and has dimensions of 1/time. When $\alpha > 0$ the function is increasing (Figure 1.7.9a and b). When $\alpha < 0$ the function is decreasing (Figure 1.7.9c and d). The function increases most quickly with large positive values of α, and decreases most quickly with large negative values of α.

The **doubling time** is defined as the time it takes the initial value of a growing measurement to double (Figure 1.7.10).

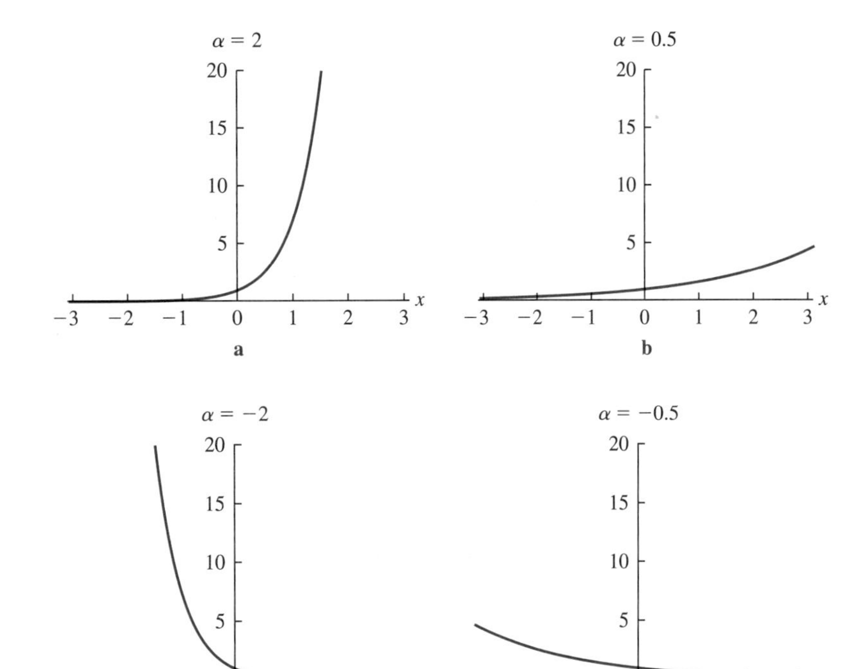

FIGURE 1.7.9
The exponential function with different parameter values in the exponent

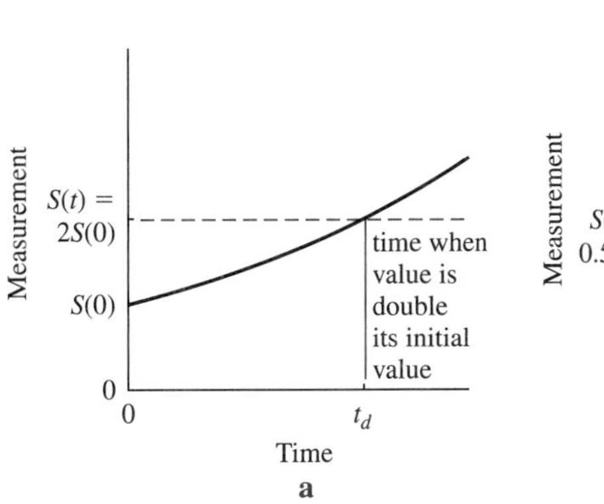

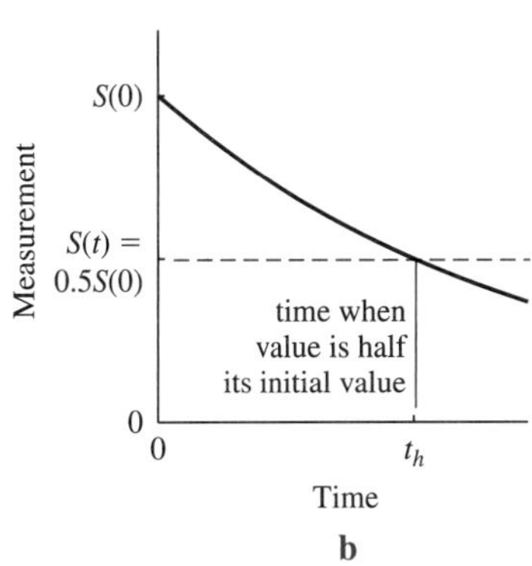

FIGURE 1.7.10
Doubling times and half-lives

Example 1.7.12 Computing a Doubling Time

Suppose

$$S(t) = 150.0e^{1.2t}$$

with t measured in hours. This measurement starts at $S(0) = 150.0$, and doubles when $S(t) = 300.0$, or

$$\begin{aligned} 150.0e^{1.2t} &= 300.0 \\ e^{1.2t} &= 2.0 \\ 1.2t &= \ln(2.0) \\ t &= \frac{\ln(2.0)}{1.2} = 0.5776. \end{aligned}$$

$$S(0.5776) = 150.0e^{1.2 \cdot 0.5776} = 300.0.$$

We can solve for the doubling time in general by finding the time t_d when $S(t_d) = 2S(0)$,

$$\begin{aligned} S(t_d) &= S(0)e^{\alpha t_d} = 2S(0) && \text{equation for } t_d \\ e^{\alpha t_d} &= 2 && \text{divide by } S(0) \\ \alpha t_d &= \ln(2) && \text{take the natural log} \\ t_d &= \frac{\ln(2)}{\alpha} = \frac{0.6931}{\alpha}. && \text{solve for } t_d \end{aligned}$$

The **general formula for the doubling time** is

$$t_d = \frac{0.6931}{\alpha}.$$

The doubling time becomes smaller as α becomes larger, consistent with the fact that measurements with larger values of α increase more quickly. Importantly, the doubling time does not depend on the initial value $S(0)$.

Example 1.7.13 Computing a Doubling Time with the Formula

Suppose $S(t) = 150.0e^{1.2t}$ (Example 1.7.12). Then $\alpha = 1.2$/hour, and the doubling time is

$$t_d = \frac{0.6931}{1.2} = 0.5776 \text{ hours.}$$

When $\alpha < 0$, the measurement is decreasing, and we can ask how long it will take to become half as large. This time, denoted t_h, is called the **half-life**, and can be found with the following steps.

$$\begin{aligned} S(t_h) &= S(0)e^{\alpha t_h} = 0.5S(0) && \text{equation for } t_h \\ e^{\alpha t_h} &= 0.5 && \text{divide by } S(0) \\ \alpha t_h &= \ln(0.5) && \text{take the natural log} \\ t_h &= \frac{\ln(0.5)}{\alpha} = -\frac{0.6931}{\alpha}. && \text{solve for } t_h \end{aligned}$$

The **general formula for the half-life** is

$$t_h = -\frac{0.6931}{\alpha}.$$

The half-life becomes smaller when α grows larger in absolute value. Apply this equation only when $\alpha < 0$.

Example 1.7.14 Computing the Half-Life

If a measurement follows the equation

$$M(t) = 240.0e^{-2.3t},$$

with t measured in seconds, then $\alpha = -2.3$/s and the half-life is

$$t_h = \frac{-0.6931}{-2.3} = 0.3014 \text{ s.}$$

Example 1.7.15 Thinking in Half-Lives

Consider the measurement $M(t)$ given in Example 1.7.14, with a half-life of 0.3014s. To figure out how much the value will have decreased in 2.0s, we could plug into the original formula, finding

$$M(2.0) = 240.0e^{-2.3 \cdot 2.0} = 2.41.$$

The value decreased by a factor of nearly 100. Alternatively, 2.0s is

$$\frac{2.0}{0.3014} = 6.636$$

half-lives. After this many half-lives, the value will have decreased by a factor of $2^{6.636} = 99.45$. We can think of using half-lives as converting the exponential to base 2.

If we are told the initial value and the doubling time or half-life of some measurement, we can find the formula. Instead of solving for the doubling time, we solve for the parameter α.

Example 1.7.16 Finding the Formula from the Doubling Time

Suppose $t_d = 26,200$ years. Because

$$t_d = \frac{0.6931}{\alpha},$$

we can solve for α as

$$\alpha = \frac{0.6931}{t_h} = \frac{0.6931}{26,200} = 2.645 \times 10^{-5}.$$

If $m(0) = 0.031$ then

$$m(t) = 0.031e^{2.645 \times 10^{-5}t}.$$

Example 1.7.17 Finding the Formula from the Half-Life

Suppose $t_h = 6.8$ years. Because

$$t_h = -\frac{0.6931}{\alpha}.$$

we can solve for α as

$$\alpha = -\frac{0.6931}{t_h} = -\frac{0.6931}{6.8} = -0.1019.$$

If $V(0) = 23.1$, then

$$V(t) = 23.1e^{-0.1019t}.$$

When a measurement follows an exponential function, the results are often plotted on a **semilog graph**.

Definition 1.13 A semilog graph plots the logarithm of the output against the input.

Example 1.7.18 A Semilog Graph of a Growing Value

Suppose

$$S(t) = 150.0e^{1.2t}$$

with t measured in hours (Example 1.7.12). To plot a semilog graph of $S(t)$ against t, we find the natural logarithm of $S(t)$.

$$\begin{aligned} \ln(S(t)) &= \ln(150.0e^{1.2t}) && \text{the natural logarithm of } S(t) \\ &= \ln(150.0) + \ln(e^{1.2t}) && \text{break up with law of logs 2} \\ &= 5.01 + 1.2t. && \text{evaluate } \ln(150.0) \text{ and cancel ln and exponential function} \end{aligned}$$

The semilog graph is a line with intercept 5.01 and slope 1.2 (Figure 1.7.11). The semilog graph is useful for exponentially growing measurements because it contracts the large range of values and converts an exponential curve into a straight line.

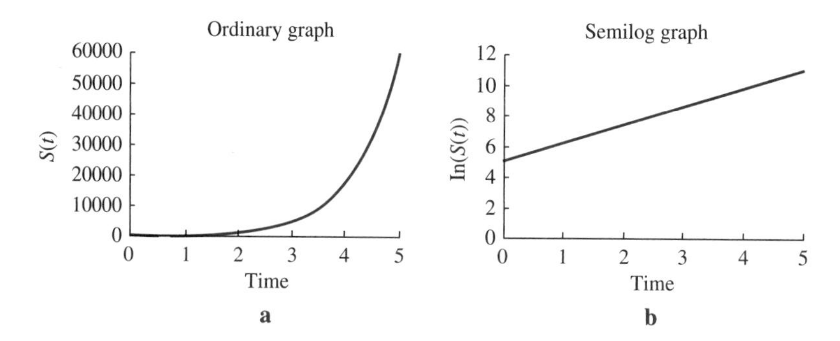

FIGURE 1.7.11
Original graph and semilog graph

Example 1.7.19 A Semilog Graph of Some Data

Suppose we are to graph the following data.

Time	Value
0	120.12
1	24.34
2	2.19
3	0.89
4	0.056
5	0.078
6	0.125
7	0.346
8	1.128

The graph of the original data is difficult to read because the large vertical scale makes the small values almost indistinguishable (Figure 1.7.12a). If we take the logarithm of the data, however, the values are much easier to compare (Figure 1.7.12b).

Time	Value	Logarithm of Value
0	120.12	4.79
1	24.34	3.19
2	2.190	0.78
3	0.89	−0.11
4	0.056	−2.88
5	0.078	−2.55
6	0.125	−2.08
7	0.346	−1.06
8	1.128	0.12

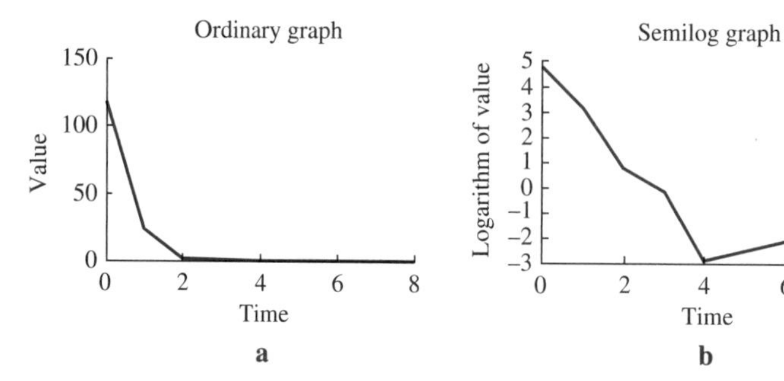

FIGURE 1.7.12
Original graph and semilog graph

The value reaches a minimum at time 4 and increases steadily thereafter. ▲

The semilog graph is particularly useful for graphing outputs that follow an exponential function or have a wide range of values. If both the input and the output satisfy these conditions, taking the logarithm of both values can help illustrate their relationship. Such a graph is called a **double-log plot**.

Definition 1.14 A double-log graph plots the logarithm of the output against the logarithm of the input.

Example 1.7.20 Using Ordinary, Semilog, and Double-Log Graphs to Illustrate Data

Suppose

$$S(t) = 150.0e^{1.2t}$$

(Example 1.7.12) and that

$$M(t) = 13.2e^{2.0t}.$$

Values for times t from 0 up to 5 are given in the following table.

Time	S	M	$\ln(S)$	$\ln(M)$
0	150.0	13.20	5.011	2.58
1	498.0	97.54	6.211	4.58
2	1,653.0	720.70	7.411	6.58
3	5,490.0	5,325.00	8.611	8.58
4	18,230.0	39,350.00	9.811	10.58
5	60,510.0.0	290,700.00	11.010	12.58

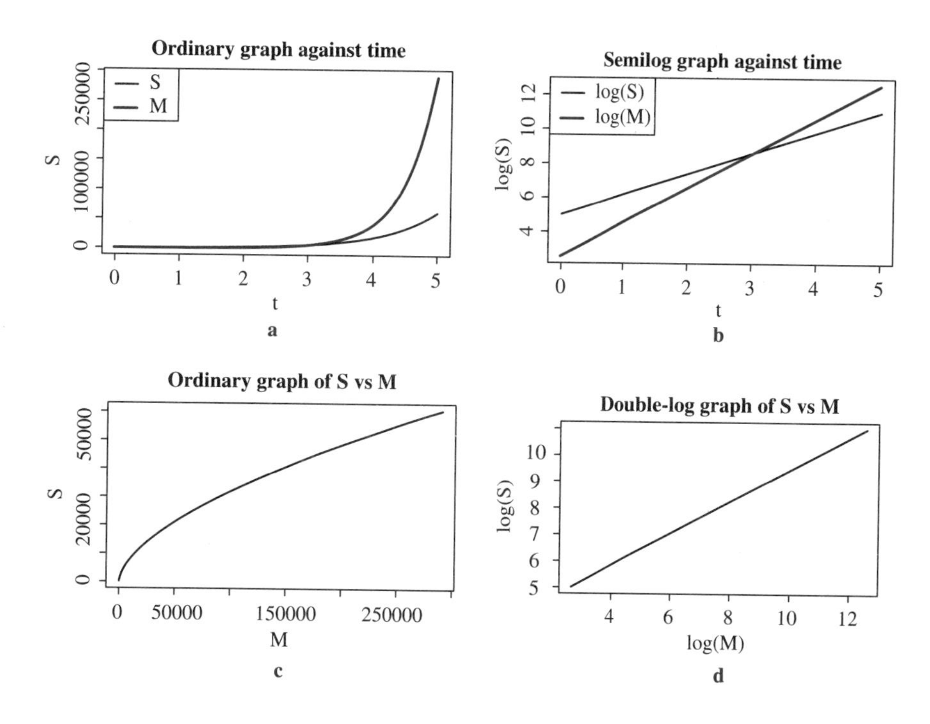

FIGURE 1.7.13
Presenting data with semilog and double-log graphs

To illustrate the behavior of S and M as functions of time, we can use either ordinary or semi-log graphs (Figure 1.7.13a,b). If instead we wish to show the relationship between these two measurements, the double-log plot can be useful (Figure 1.7.13c,d). The curved relationship caused by the differences in the growth rates of the two measurements becomes a linear relationship on a double-log graph.

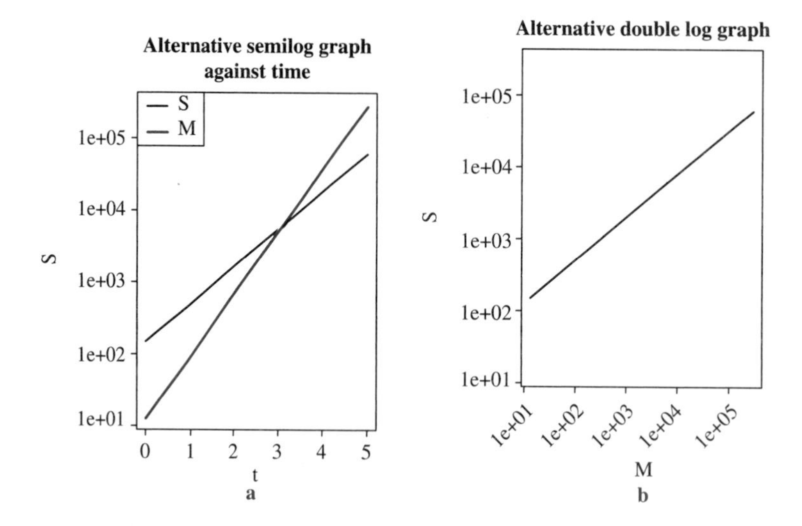

FIGURE 1.7.14
Transforming the axes on semilog and double-log graphs

Example 1.7.21 Alternative Way to Present Semilog and Double-Log Graphs

Instead of transforming the measurements by taking the natural log, we can transform the axes themselves. The values on the axis are not distributed on a linear scale as in ordinary graph where the values 10, 20, 30 and so forth are evenly spaced. Instead, the powers of 10 such as 10, 100, and 1000 are evenly spaced, providing a way to display data that take on a larger range of values without changing the values themselves (Figure 1.7.14).

Many relationships in biology are linear on double-log graphs even though the underlying measurements do not grow exponentially. Such measurements follow a **power function**, of the form

$$y = cx^p$$

and are often said to have an **allometric relationship**.

Using the laws of logs to decompose the product and the power, we find that

$$\ln(y) = \ln(c) + p\ln(x).$$

The natural log of the multiplicative constant c becomes the intercept and the power p becomes the slope of a linear relationship between $\ln(x)$ and $\ln(y)$. When two measurements each grow exponentially, they will always be related by a power function.

Example 1.7.22 Find a Power Function Relationship

In Example 1.7.20, we studied the relationship between the measurements $S(t) = 150.0e^{1.2t}$ and $M(t) = 13.2e^{2.0t}$, finding that S is a linear function of M. The points $(\ln[M(0)], \ln[S(0)]) = (2.58, 5.011)$ and $(\ln[M(1)], \ln[S(1)]) = (4.58, 6.211)$ must lie on this line. To use the equation for the point-slope form of a line, we first find the slope as

$$m = \frac{6.211 - 5.011}{4.58 - 2.58} = 0.6.$$

Using (2.58, 5.011) as the base point,

$$\begin{aligned} \ln[S(t)] &= 0.6(\ln[M(t)] - 2.58) + 5.011 \\ &= 0.6\ln[M(t)] + 3.468. \end{aligned}$$

Example 1.7.23 The Allometric Relationship Between Surface Area and Volume

A sphere with radius r has surface area $S = 4\pi r^2$ and volume $V = \frac{4}{3}\pi r^3$ (Table 1.2). Because the power of the radius in the surface area is 2/3 that in volume, we can write

$$S = cV^{2/3}$$

where the constant for a sphere is $(36\pi)^{1/3}$. This allometric relationship holds for any given shape, but with a different value of the constant c. ▲

Example 1.7.24 The Allometric Relationship of Energy Use and Body Mass

A famous allometric relationship concerns the link between metabolic rate, the amount of energy an organism uses, and body mass. For mammals, this takes roughly the form

$$E = 0.018M^{0.75}$$

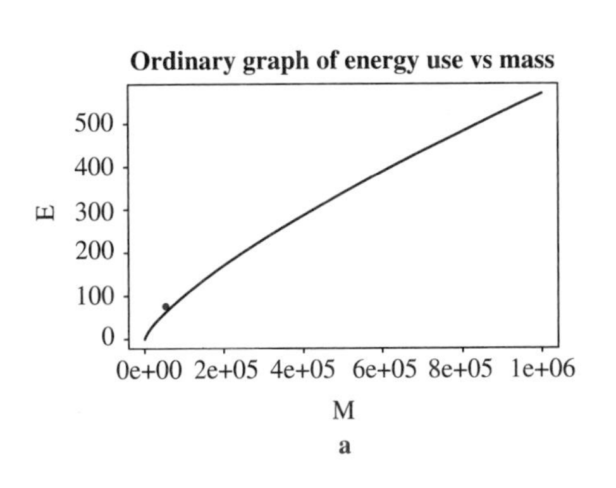

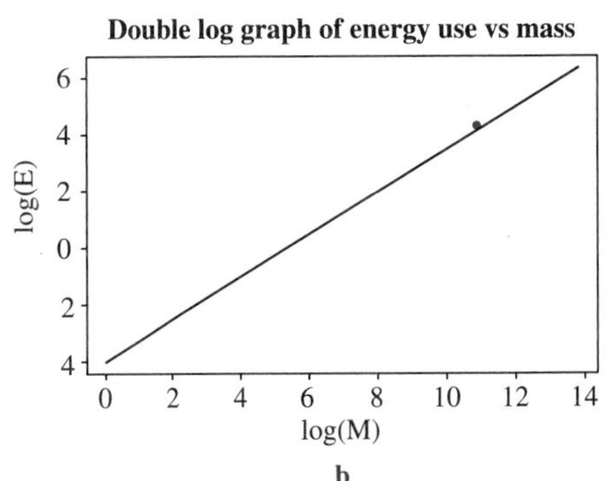

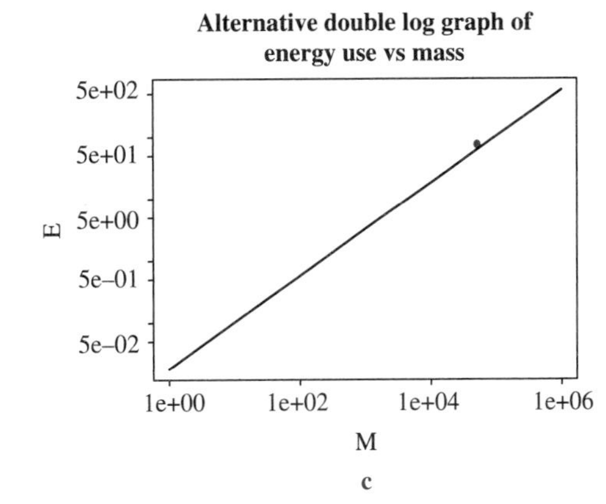

FIGURE 1.7.15
Energy use and body mass

where E is measured in Kcal/hr and M is measured in grams. This relationship is shown in three ways, with each including a point representing the energy use of a 75 kg person who consumes about 2000 calories per day (Figure 1.7.15). ▲

Summary We generalized the discrete-time dynamical system for bacterial population growth to include mortality, writing the discrete-time dynamical system in terms of the **per capita reproduction** r. A population grows if $r > 1$ and declines if $r < 1$. The solution can be expressed as an **exponential function to base r**. For convenience, exponential functions are often expressed to the base e, often the **exponential function**. Using the laws of exponents, any exponential function can be expressed to the base e. The inverse of the exponential function is the **natural logarithm** or natural log. This function can be used to solve equations involving the exponential function, including finding **doubling times** and **half-lives**. Measurements that cover a large range of positive values can be conveniently displayed on a **semilog graph**, which reduces the range, and produces a linear graph if the measurements follow an exponential function. Double-log graphs help to display data where both the input and output cover a large range of values, and are linear when the two measurements are both exponential functions of time.

1.7 Exercises

Mathematical Techniques

1–10 ▪ Use the laws of exponents to rewrite the following (if possible). If no law of exponents applies, say so.

1. 43.2^0
2. 43.2^1
3. 43.2^{-1}
4. $43.2^{-0.5} + 43.2^{0.5}$
5. $43.2^{7.2}/43.2^{6.2}$
6. $43.2^{0.23} \cdot 43.2^{0.77}$
7. $(3^4)^{0.5}$
8. $(43.2^{-1/8})^{16}$
9. $2^{2^3} \cdot 2^{2^2}$
10. $4^2 \cdot 2^4$

11–22 ▪ Use the laws of logs to rewrite the following if possible. If no law of logs applies or the quantity is not defined, say so.

11. $\ln(1)$
12. $\ln(-6.5)$
13. $\log_{43.2} 43.2$
14. $\log_{10}(3.5 + 6.5)$
15. $\log_{10}(5) + \log_{10}(20)$
16. $\log_{10}(0.5) + \log_{10}(0.2)$
17. $\log_{10}(500) - \log_{10}(50)$
18. $\log_{43.2}(5 \cdot 43.2^2) - \log_{43.2}(5)$
19. $\log_{43.2}(43.2^7)$
20. $\log_{43.2}(43.2^7)^4$
21. Using the fact that $\log_7 43.2 = 1.935$, find $\log_7\left(\frac{1}{43.2}\right)$.
22. Using the fact that $\log_7 43.2 = 1.935$, find $\log_7(43.2)^3$.

23–26 ▪ Solve the following equations for x and check your answer.

23. $7e^{3x} = 21$.
24. $4e^{2x+1} = 20$.
25. $4e^{-2x+1} = 7e^{3x}$.
26. $4e^{2x+3} = 7e^{3x-2}$.

27–30 ▪ Sketch graphs of the following exponential functions. For each, find the value of x where it is equal to 7.0. For the increasing functions, find the doubling time, and for the decreasing functions, find the half-life. For what value of x is the value of the function 3.5? For what value of x is the value of the function 14.0?

27. e^{2x}
28. e^{-3x}
29. $5e^{0.2x}$
30. $0.1e^{-0.2x}$

31–32 ▪ Sketch graphs of the following updating functions over the given range and mark the equilibria.

31. $h(z) = e^{-z}$ for $0 \le z \le 2$.
32. $F(x) = \ln(x) + 1$ for $0 \le x \le 2$. (Although this cannot be solved algebraically, you can guess the answer.)

33–36 ▪ Find the equations of the lines after transforming the variables to create semilog or double-log plots.

33. Suppose $M(t) = 43.2e^{5.1t}$. Find the slope and intercept of $\ln(M(t))$.
34. Suppose $L(t) = 0.72e^{-2.34t}$. Find the slope and intercept of $\ln(L(t))$.
35. Suppose $M(t) = 43.2e^{5.1t}$ and $S(t) = 18.2e^{4.3t}$. Find the slope and intercept of $\ln(M(t))$ as a function of $\ln(S(t))$.
36. Suppose $L(t) = 0.72e^{-2.34t}$ and $K(t) = 4.23e^{0.91t}$. Find the slope and intercept of $\ln(L(t))$ as a function of $\ln(K(t))$.

Applications

37–40 ▪ Find the solution of each discrete-time dynamical system, express it in exponential notation, and solve for the time when the value reaches the given target. Sketch a graph of the solution.

37. A population follows the discrete-time dynamical system $b_{t+1} = rb_t$ with $r = 1.5$ and $b_0 = 1.0 \times 10^6$. When will the population reach 1.0×10^7?
38. A population follows the discrete-time dynamical system $b_{t+1} = rb_t$ with $r = 0.7$ and $b_0 = 5.0 \times 10^5$. When will the population reach 1.0×10^5?
39. Cell volume follows the discrete-time dynamical system $v_{t+1} = 1.5v_t$ with initial volume of 1350 μm^3 (as in Exercise 37). When will the volume reach 3250 μm^3?
40. Gnat number follows the discrete-time dynamical system $n_{t+1} = 0.5n_t$ with an initial population of 5.5×10^4. When will the population reach 1.5×10^3?

41–44 ▪ Suppose the size of an organism at time t is given by

$$S(t) = S_0 e^{\alpha t}$$

where S_0 is the initial size. Find the time it takes for the organism to double or quadruple in size in the following circumstances.

41. $S_0 = 1.0$ cm and $\alpha = 1.0$/day.
42. $S_0 = 2.0$ cm and $\alpha = 1.0$/day.
43. $S_0 = 2.0$ cm and $\alpha = 0.1$/hour.
44. $S_0 = 2.0$ cm and $\alpha = 0.0$/hour

45–48 ▪ The amount of carbon-14 (14_c) left t years after the death of an organism is given by

$$Q(t) = Q_0 e^{-0.000122t}$$

where Q_0 is the amount left at the time of death. Suppose $Q_0 = 6.0 \times 10^{10}$ 14_c atoms.

45. How much is left after 50,000 years? What fraction is this of the original amount?

46. How much is left after 100,000 years? What fraction is this of the original amount?

47. Find the half-life of 14_c.

48. About how many half-lives will occur in 50,000 years? Roughly what fraction will be left? How does this compare with the answer of Exercise 45?

49–52 ▪ Suppose a population has a doubling time of 24 years and an initial size of 500.

49. What is the population in 48 years?

50. What is the population in 12 years?

51. Find the equation for population size $P(t)$ as a function of time.

52. Find the one-year discrete-time dynamical system for this population (figure out the factor multiplying the population in one year).

53–56 ▪ Suppose a population is dying with a half-life of 43 years. The initial size is 1600.

53. How long will it take to reach 200?

54. Find the population in 86 years.

55. Find the equation for population size $P(t)$ as a function of time.

56. Find the one year discrete-time dynamical system for this population (figure out the factor multiplying the population in one year).

57–60 ▪ Plot semilog graphs of the values.

57. The growing organism in Exercise 41 for $0 \leq t \leq 10$. Mark where the organism has doubled in size and when it has quadrupled in size.

58. The carbon-14 in Exercise 45 for $0 \leq t \leq 20,000$. Mark where the amount of carbon has gone down by half.

59. The population in Exercise 49 for $0 \leq t \leq 100$. Mark where the population has doubled.

60. The population in Exercise 53 for $0 \leq t \leq 100$. Mark where the population has gone down by half.

61–64 ▪ The following pairs of measurements can be described by ordinary, semilog, and double-log graphs.

a. Graph each measurement as a function of time on both ordinary and semilog graphs.

b. Graph the second measurement as a function of the first on both ordinary and double-log graphs.

61. The antler size $A(t)$ in centimeters of an elk increases with age t in years according to $A(t) = 53.2e^{0.17t}$ and its shoulder height $L(t)$ increases according to $L(t) = 88.5e^{0.1t}$.

62. Suppose a population of viruses in an infected person grows according to $V(t) = 2.0e^{2.0t}$ and that the immune response (described by the number of antibodies) increases according to $I(t) = 0.01e^{3.0t}$ during the first week of an infection. When will the number of antibodies equal the number of viruses?

63. The growth of a fly in an egg can be described allometrically (see H. F. Nijhout and D. E. Wheeler, 1996). During growth, two **imaginal disks** (the first later becomes the wing and the second becomes the haltere) expand according to $S_1(t) = 0.007e^{0.1t}$ and $S_2(t) = 0.007e^{0.4t}$ where size is measured in mm^3 and time is measured in days. Development takes about 5 days.

64. While the imaginal disks are growing (Exercise 63), the yolk of the egg is shrinking according to $Y(t) = 4.0e^{-1.2t}$. Create graphs comparing $S_1(t)$ and $Y(t)$.

65–66 ▪ For each of the given shapes, find the constant c in the power relationship $S = cV^{2/3}$ between the surface area S and volume V. By how much is does c exceed the value $(36\pi)^{1/3} = 4.836$ for the sphere (which is in fact the minimum for any shape).

65. For a cube with side length w.

66. For a cylinder with radius r and height $3r$.

67–68 ▪ Many measurements in biology are related by power functions. For each of the following, graph the second measurement as a function of the first on both ordinary and double-log graphs.

67. The $-3/2$ law of self-thinning in plants argues that the mean weight W of surviving trees in a stand increases while their number N decreases, related by

$$W = cN^{-3/2}.$$

Suppose 10^4 trees start out with mass of 0.001 kg. Graph the relationship, and find how heavy the trees would be when only 100 remain alive, and again when only 1 remains alive. Is the total mass larger or smaller than when it started?

68. Suppose that the population density D of a species of mammal is a decreasing function of its body mass M according to the relationship

$$D = cM^{-3/4}.$$

Suppose that an unlikely 1 g mammal would have a density of 10^4 per hectare. What is the predicted density of species with mass of 1000 g? A species with a mass of 100 kg? According to the metabolic scaling law (Example 1.7.24), which species will use the most energy?

Computer Exercises

69. Use your computer to find the following. Plot the graphs to check.

a. The doubling time of $S_1(t) = 3.4e^{0.2t}$.

b. The doubling time of $S_2(t) = 0.2e^{3.4t}$.

c. The half-life of $H_1(t) = 3.4e^{-0.2t}$.

d. The half-life of $H_2(t) = 0.2e^{-3.4t}$.

70. Solve for the times when the following hold. Plot the graphs to check your answer.

a. $S_1(t) = S_2(t)$ with S_1 and S_2 from the previous problem.

b. $H_1(t) = 2H_2(t)$ with H_1 and H_2 from the previous problem.

c. $H_1(t) = 0.5H_2(t)$ with H_1 and H_2 from the previous problem.

71. Plot the following functions.

a. $\ln(x)$ for $10 \le x \le 100{,}000$.

b. $\ln[\ln(x)]$ for $10 \le x \le 100{,}000$.

c. $\ln[\ln[\ln(x)]]$ for $10 \le x \le 100{,}000$.

d. e^x for $0 \le x \le 2$.

e. e^{e^x} for $0 \le x \le 2$.

f. $e^{e^{e^x}}$ for $0 \le x \le 2$. Will your machine let you do it? Can you compute the value of $e^{e^{e^2}}$?

72. Compute the following. Does this give you any idea why e is special?

a. $2^{0.001}$

b. $10^{0.001}$

c. $0.5^{0.001}$

d. $e^{0.001}$

73. Suppose that the antler size $A(t)$ in centimeters of an elk increases with age in years during the first five years of growth according to the exponential function

$$A(t) = 53.2e^{0.17t}$$

and that the shoulder height $L(t)$ increases according to

$$L(t) = 88.5e^{0.1t}.$$

a. Plot A and L as functions of t on ordinary and on semilog graphs.

b. Plot A as a function of L with an ordinary and with a semilog graph.

c. Find when the antler size would exceed the shoulder height of the elk.

1.8 Oscillations and Trigonometry

Linear and exponential functions describe many important relations between measurements. However, these functions cannot describe **oscillations**, processes that repeat in cycles. Heartbeats and breathing are examples of biological oscillations. In addition, the daily and seasonal cycles imposed by the movements of the earth drive sleep-wake cycles, seasonal population cycles, and the tides. In this section, we will use **trigonometric functions** to describe simple oscillations. Four numbers are needed to describe such oscillations with the **cosine** function: the **average**, the **amplitude**, the **period**, and the **phase**.

1.8.1 Sine and Cosine: A Review

Like many functions, the trigonometric functions have two interpretations: geometric and dynamic. Geometrically, the trigonometric functions are used to compute angles and distances. After briefly reviewing the geometry behind the **sine** and **cosine** functions, we will use them to study the dynamics of biological oscillations.

In applied mathematics, angles are measured in **radians**, defined as the distance along the perimeter of the circle of radius 1. Because the full circumference of a circle with radius 1 is 2π, 2π radians corresponds to $360°$, or one complete revolution (Figure 1.8.1). The **basic identity** between radians and degrees is given by

$$2\pi \text{ rad} = 360°.$$

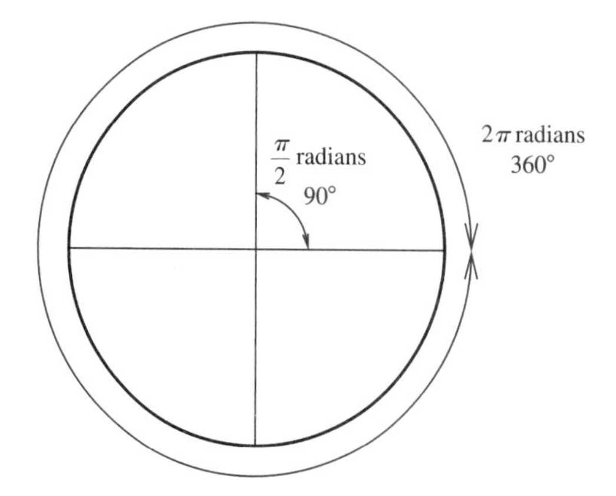

FIGURE 1.8.1
Degrees and radians

From this, we derive the **conversion factors**

$$1 = \frac{2\pi \text{ rad}}{360^\circ} = \frac{\pi \text{ rad}}{180^\circ}$$
$$1 = \frac{360^\circ}{2\pi \text{ rad}} = \frac{180^\circ}{\pi \text{ rad}}.$$

Example 1.8.1 Converting Degrees to Radians

To find 60° in radians, we convert

$$60^\circ = 60^\circ \times \frac{\pi \text{ rad}}{180^\circ} = \frac{\pi}{3}\text{rad}.$$

Example 1.8.2 Converting Radians to Degrees

Similarly, to find 1.0 rad in degrees, we convert

$$1.0 \text{ rad} = 1.0 \text{ rad} \times \frac{180^\circ}{\pi \text{ rad}} = 57.3^\circ.$$

The **sine** and **cosine** functions take angles as inputs and return numbers between -1 and 1 as outputs. We write $\sin(\theta)$ and $\cos(\theta)$ to denote these functions, where the Greek letter θ ("theta") is often used for angles. The sine and cosine give the Cartesian coordinates of points on the circle (Figure 1.8.2).

Definition 1.15 The Cartesian coordinates of the point on the unit circle an angle θ measured counter-clockwise from (1, 0) are $(\cos(\theta), \sin(\theta))$.

Values of these functions for representative inputs are given in the following table.

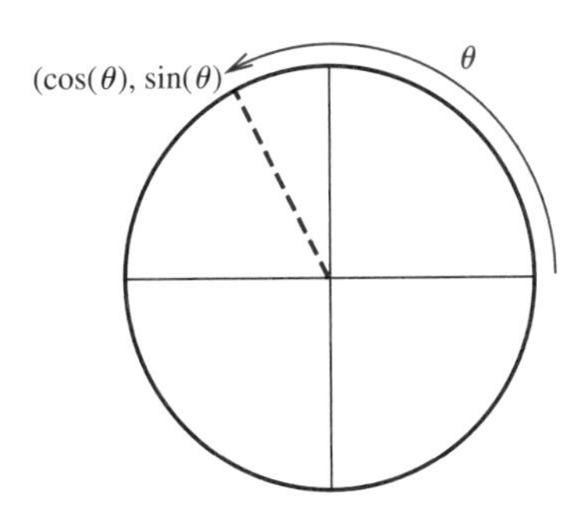

FIGURE 1.8.2
The definition of $\sin(\theta)$ and $\cos(\theta)$

Radians	Degrees	$\cos(\theta)$	$\sin(\theta)$	Radians	Degrees	$\cos(\theta)$	$\sin(\theta)$
0	0°	1	0	π	180°	-1	0
$\frac{\pi}{6}$	30°	$\frac{\sqrt{3}}{2}$	$\frac{1}{2}$	$\frac{7\pi}{6}$	210°	$-\frac{\sqrt{3}}{2}$	$-\frac{1}{2}$
$\frac{\pi}{4}$	45°	$\frac{\sqrt{2}}{2}$	$\frac{\sqrt{2}}{2}$	$\frac{5\pi}{4}$	225°	$-\frac{\sqrt{2}}{2}$	$-\frac{\sqrt{2}}{2}$
$\frac{\pi}{3}$	60°	$\frac{1}{2}$	$\frac{\sqrt{3}}{2}$	$\frac{4\pi}{3}$	240°	$-\frac{1}{2}$	$-\frac{\sqrt{3}}{2}$
$\frac{\pi}{2}$	90°	0	1	$\frac{3\pi}{2}$	270°	0	-1
$\frac{2\pi}{3}$	120°	$-\frac{1}{2}$	$\frac{\sqrt{3}}{2}$	$\frac{5\pi}{3}$	300°	$\frac{1}{2}$	$-\frac{\sqrt{3}}{2}$
$\frac{3\pi}{4}$	135°	$-\frac{\sqrt{2}}{2}$	$\frac{\sqrt{2}}{2}$	$\frac{7\pi}{4}$	315°	$\frac{\sqrt{2}}{2}$	$-\frac{\sqrt{2}}{2}$
$\frac{5\pi}{6}$	150°	$-\frac{\sqrt{3}}{2}$	$\frac{1}{2}$	$\frac{11\pi}{6}$	330°	$\frac{\sqrt{3}}{2}$	$-\frac{1}{2}$
π	180°	-1	0	2π	360°	1	0

Both the sine and cosine functions repeat every 2π radians (Figure 1.8.3). The value 2π is called the **period** of the oscillation. This means that adding or subtracting multiples of 2π from the argument does not change the value, so

$$\cos(\theta) = \cos(\theta + 2\pi) = \cos(\theta + 4\pi) = \cos(\theta + 2n\pi)$$
$$\cos(\theta) = \cos(\theta - 2\pi) = \cos(\theta - 4\pi) = \cos(\theta - 2n\pi)$$

for any value of θ and any integer n.

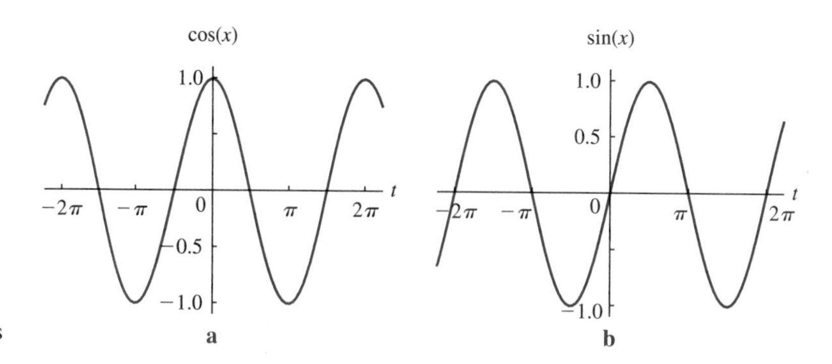

FIGURE 1.8.3
Graphs of the cosine and sine functions

Example 1.8.3 Periodicity of the Cosine Function

$$\cos\left(\frac{\pi}{4}\right) = \cos\left(\frac{\pi}{4} + 2\pi\right) = \cos\left(\frac{\pi}{4} + 4\pi\right) = \frac{\sqrt{2}}{2}$$

$$\cos\left(\frac{\pi}{4}\right) = \cos\left(\frac{\pi}{4} - 2\pi\right) = \cos\left(\frac{\pi}{4} - 4\pi\right) = \frac{\sqrt{2}}{2}.$$

The graphs of sine and cosine have the same shape, but are **shifted** from each other by $\pi/2$ rad. In equations,

$$\sin(\theta) = \cos\left(\theta - \frac{\pi}{2}\right).$$

Example 1.8.4 Relation Between Sine and Cosine

$$\sin\left(\frac{2\pi}{3}\right) = \cos\left(\frac{2\pi}{3} - \frac{\pi}{2}\right) = \cos\left(\frac{\pi}{6}\right) = \frac{\sqrt{3}}{2}.$$

Because we can compute the sine function in terms of the cosine function, we will use cosine to describe oscillations.

1.8.2 Describing Oscillations with the Cosine

Oscillations that are shaped like the graph of the sine or cosine function are called **sinusoidal**. There are four numbers needed to describe an oscillation with the cosine function: the **average**, the **amplitude**, the **period**, and the **phase** (Figure 1.8.4).

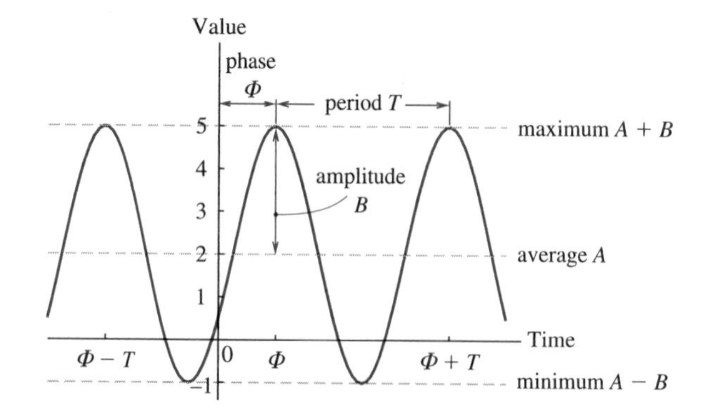

FIGURE 1.8.4
The four numbers that describe a sinusoidal oscillation

- The **amplitude** is the difference between the maximum and the average.
- The **average** is the middle value on the curve.
- The **period** is the time between successive peaks.
- The **phase** is the time of the first peak.

We can **build** the oscillation pictured in Figure 1.8.4 from the cosine function by **shifting** and **scaling** both vertically and horizontally (Section 1.3).

Example 1.8.5 Building an Oscillation by Shifting and Scaling the Cosine Function

Suppose we wish to build a function with an amplitude of 2.0, an average of 3.0, a period of 4.0, and a phase of 1.0. We can construct the formula in steps (Figure 1.8.5).

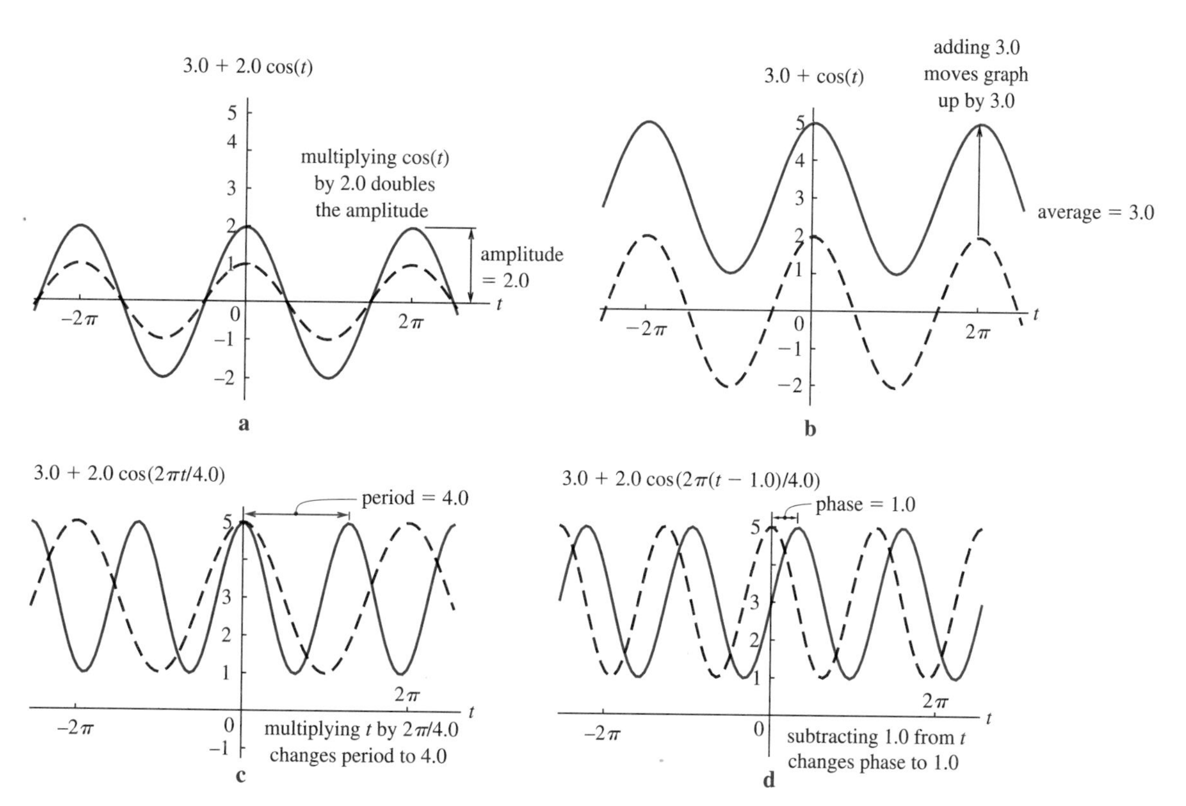

FIGURE 1.8.5
Building a function with different average, amplitude, period, and phase

1. To increase the amplitude by a factor of 2.0, we scale vertically by **multiplying** the cosine by 2.0. The function is now
$$f(t) = 2.0\cos(t).$$
2. To raise the **average** from 0 to 3.0, we vertically shift the function by **adding** 3.0 to the function, making
$$f(t) = 3.0 + 2.0\cos(t).$$
3. Next, we wish to decrease the period from 2π to 4.0. We do this by scaling horizontally by a factor of $\frac{2\pi}{4}$, or by **multiplying** the t inside the cosine by $\frac{2\pi}{4.0}$. Our function is now
$$f(t) = 3.0 + 2.0\cos\left(\frac{2\pi}{4.0}t\right).$$
4. Finally, we shift the curve horizontally so that the first peak is at 1.0 instead of 0.0. We do this by **subtracting** 1.0 from t, arriving at the final answer of
$$f(t) = 3.0 + 2.0\cos\left(\frac{2\pi}{4.0}(t - 1.0)\right).$$

In general, a sinusoidal oscillation with amplitude B, average A, period T, and phase ϕ ("phi") can be described as a function of time t with the formula

$$f(t) = A + B\cos\left(\frac{2\pi}{T}(t-\phi)\right). \tag{1.8.1}$$

This function has a maximum at $t=\phi$, a minimum at $t=\phi+\frac{T}{2}$, and takes on its average value at $t=\phi+\frac{T}{4}$ and $t=\phi+\frac{3T}{4}$. It repeats every T (Figure 1.8.6).

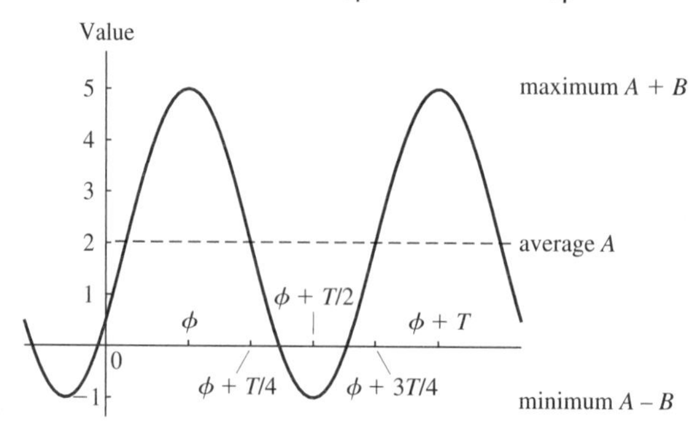

FIGURE 1.8.6
The guideposts for plotting $f(t) = A + B\cos\left(\frac{2\pi}{T}(t-\phi)\right)$

Example 1.8.6 Plotting a Sinusoidal Function from Its Equation

Suppose we wish to plot

$$f(t) = 2.0 + 0.4\cos\left(\frac{2\pi}{10.0}(t-7.0)\right).$$

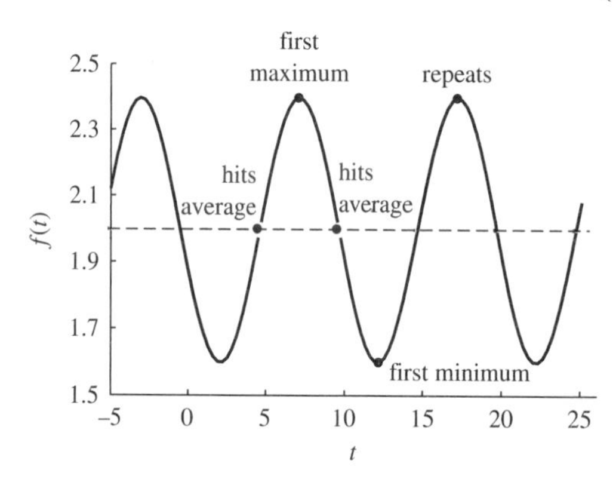

FIGURE 1.8.7
Graphing a sinusoidal oscillation from its equation

The amplitude is 0.4, and the average is 2.0; the period is 10.0 and the phase is 7.0 (Figure 1.8.7). The maximum is the sum of the average and amplitude or $2.0 + 0.4 = 2.4$, and the minimum is the average minus the amplitude or $2.0 - 0.4 = 1.6$. The first maximum occurs at the phase, or $t = 7.0$. The average occurs 1/4 and 3/4 of the way through each cycle, or at $t = 7.0 + \frac{1}{4}(10.0) = 9.5$ and $t = 7.0 + \frac{3}{4}(10.0) = 14.5$. The minimum occurs halfway through the first period at $t = 7.0 + \frac{1}{2}(10.0) = 12.0$. The cycle repeats at $t = 17.0$, 27.0 and so forth.

Example 1.8.7 The Daily and Monthly Temperature Cycles

Women have two cycles affecting body temperature: a daily and a monthly rhythm (Figure 1.8.8). The key facts about these two cycles are given in the following table:

	Minimum	Maximum	Average	Time of Maximum	Period
Daily cycle	36.5	37.1	36.8	2:00 p.m.	24 hours
Monthly cycle	36.6	37.0	36.8	Day 16	28 days

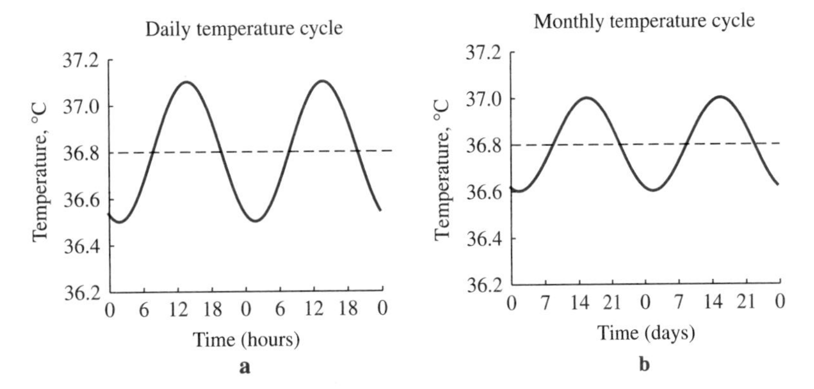

FIGURE 1.8.8

The daily and monthly temperature cycles

Assuming these cycles are sinusoidal, we can use this information to describe these cycles with the cosine function.

The amplitude of a cycle is

$$\text{amplitude} = \text{maximum} - \text{average}.$$

For the daily cycle, the amplitude is

$$\text{daily cycle amplitude} = 37.1 - 36.8 = 0.3.$$

For the monthly cycle, the amplitude is

$$\text{monthly cycle amplitude} = 37.0 - 36.8 = 0.2.$$

The phase depends on the time chosen as the starting time. We define the daily cycle to begin at midnight and the monthly cycle to begin at menstruation. The maximum of the daily cycle occurs 14 hours after the start, and that of the monthly cycle 16 days after the start. The oscillations can be described by the fundamental formula (Equation 1.8.1). For the daily cycle, with t measured in hours, the formula $P_d(t)$ is

$$P_d(t) = 36.8 + 0.3\cos\left(\frac{2\pi}{24}(t-14)\right).$$

For the monthly cycle, with t measured in days, the formula $P_m(t)$ is

$$P_m(t) = 36.8 + 0.2\cos\left(\frac{2\pi}{28}(t-16)\right).$$

▲

1.8.3 More Complicated Shapes

Real oscillations are not perfectly sinusoidal. Nonetheless, the cosine function is useful for describing more complicated oscillations. There is a powerful theory, beyond the scope of this book, called **Fourier series**, that shows how almost any oscillation can be written as the sum of many cosine functions with different amplitudes, periods, and phases (Exercise 49).

As an illustration, we will combine the daily and monthly temperature cycles. To do so, we must write both cycles in the same time units, days. In days, the period of the daily cycle is 1.0 days and the phase (the time of the maximum) is

$$\text{phase in days} = \frac{14 \text{ hours}}{24 \text{ hours}} \cdot \frac{1.0 \text{ days}}{24 \text{ hours}} = 0.583 \text{ days}.$$

The equation for the daily cycle, with t measured in days, is

$$P_d(t) = 36.8 + 0.3\cos[2\pi(t - 0.583)].$$

To figure out how the daily and monthly cycles combine, we cannot simply add them together because

$$P_d(t) + P_m(t) = 36.8 + 0.3\cos[2\pi(t - 0.583)] + 36.8 + 0.2\cos\left(\frac{2\pi(t-16)}{28}\right)$$

$$= 73.6 + 0.3\cos[2\pi(t - 0.583)] + 0.2\cos\left(\frac{2\pi(t-16)}{28}\right)$$

which has an average of 73.6.

To keep the average at the appropriate value of 36.8, we add only the two cosine terms to the average, getting a formula for the combined cycle of

$$P_t(t) = 36.8 + 0.2\cos\left(\frac{2\pi(t-16)}{28}\right) + 0.3\cos(2\pi(t - 0.583))$$

(Figure 1.8.9). In the course of one month, there is a single slow cycle, with 28 daily cycles superimposed. The maximum possible temperature can be found by adding the sum of the amplitudes of the daily and monthly cycles to the overall average, or

$$\text{maximum possible temperature} = 36.8 + (0.2 + 0.3) = 37.3.$$

This maximum occurs only if each cycle takes on its maximum at the same time, which does not happen exactly in this case. It is closest at 2:00 P.M. on the 16th day of the cycle.

The minimum possible temperature can be found by subtracting the sum of the amplitudes of the daily and monthly cycles from the overall average, or

$$\text{minimum possible temperature} = 36.8 - (0.2 + 0.3) = 36.3.$$

This minimum occurs only if each cycle takes on its minimum at the same time, which does not happen exactly in this case. It is closest at 2:00 A.M. on the 2nd day of the cycle.

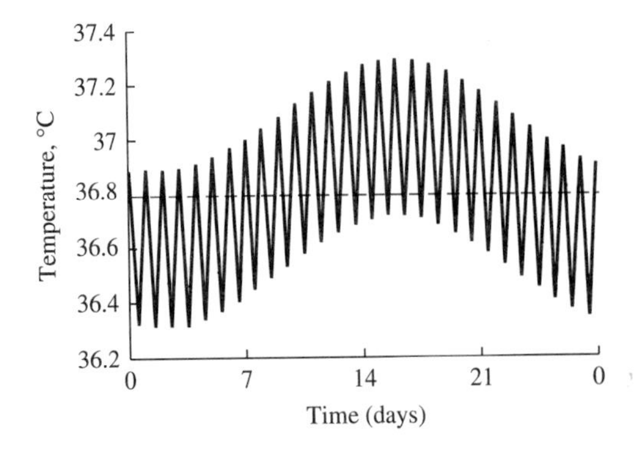

FIGURE 1.8.9
The combined effect of the daily and monthly temperature cycles

Summary **Sinusoidal** oscillations can be described mathematically with the **cosine** function. Four factors change the shape of the graph: the **amplitude** (the distance from the middle to the minimum or maximum), the **average** (the middle value), the **period** (the time between successive maxima), and the **phase** (the time of the first maximum). Functions with these parameters can be created by **shifting** and **scaling** the cosine function vertically and horizontally. Oscillations with more complicated shapes can be described by adding together appropriate cosine functions.

1.8 Exercises

Mathematical Techniques

1–6 ▪ Use the table or a calculator to find the values of sine and cosine for the following inputs (in radians), and plot them on **a.** a graph of $\sin(\theta)$, **b.** a graph of $\cos(\theta)$, **c.** as the coordinates of a point on the circle.

1. $\theta = \pi/2$.

2. $\theta = 3\pi/4$.

3. $\theta = \pi/9$.

4. $\theta = 5.0$.

5. $\theta = -2.0$.

6. $\theta = 3.2$.

7–14 ▪ Convert the following angles from degrees to radians or vice versa.

7. 30°

8. 330°

9. 1°

10. -30°

11. 2.0 rad

12. $\pi/5$ rad

13. $-\pi/5$ rad

14. 30 rad

15–20 ▪ The other trigonometric functions (tangent, cotangent, secant, and cosecant) are defined in terms of sin and cos by

$$\tan(x) = \frac{\sin(x)}{\cos(x)}, \quad \cot(x) = \frac{\cos(x)}{\sin(x)},$$

$$\sec(x) = \frac{1}{\cos(x)}, \quad \csc(x) = \frac{1}{\sin(x)}.$$

Calculate the value of each of these functions at the following angles (all in radians). Plot the points on a graph of each function.

15. $\pi/2$

16. $3\pi/4$

17. $\pi/9$

18. 5.0.

19. -2.0

20. 3.2

21–26 ▪ The following are some of the most important trigonometric identities. Check them at **a.** $\theta = 0$, **b.** $\theta = \pi/4$, **c.** $\theta = \pi/2$, **d.** $\theta = \pi$.

21. $\cos\left(\frac{\theta}{2}\right) = \sqrt{\frac{1+\cos(\theta)}{2}}$. Only check at points **a**, **c**, and **d**.

22. $\sin^2(\theta) + \cos^2(\theta) = 1$.

23. $\cos(\theta - \pi) = -\cos(\theta)$.

24. $\cos\left(\theta - \frac{\pi}{2}\right) = \sin(\theta)$.

25. $\cos(2\theta) = \cos^2(\theta) - \sin^2(\theta)$.

26. $\sin(2\theta) = 2\sin(\theta)\cos(\theta)$.

27–30 ▪ Convert the following sinusoidal oscillations to the standard form and sketch a graph.

27. $r(t) = 5.0[2.0 + 1.0\cos(2\pi t)]$.

28. $g(t) = 2.0 + 1.0\sin(t)$. (*Hint:* Use Exercise 24.)

29. $f(t) = 2.0 - 1.0\cos(t)$. (*Hint:* Use Exercise 23.)

30. $h(t) = 2.0 + 1.0\cos(2\pi t - 3.0)$.

Applications

31–34 ▪ Find the average, minimum, maximum, amplitude, period, and phase from the graphs of the following oscillations.

31.

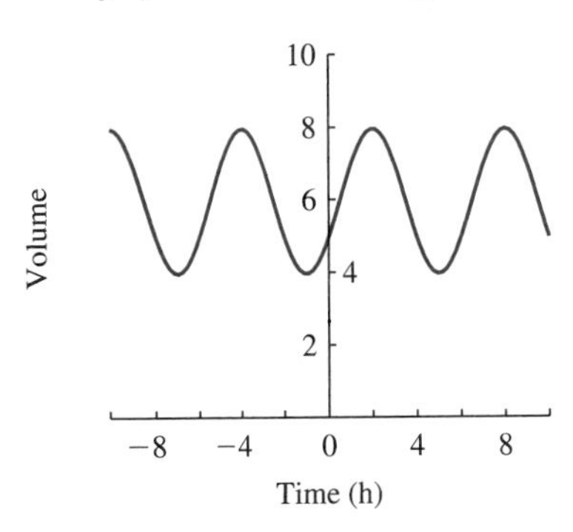

32.

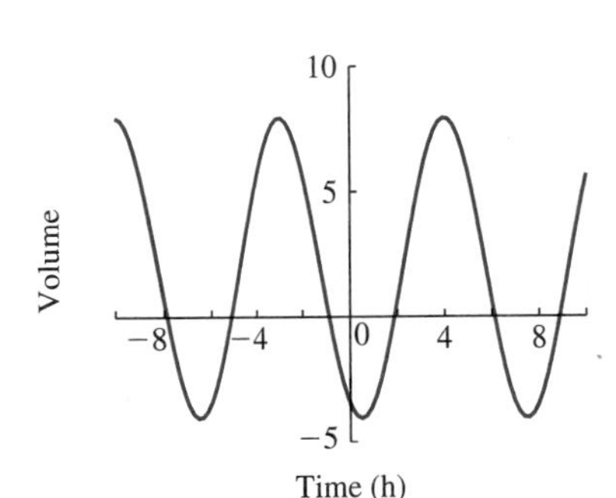

33.

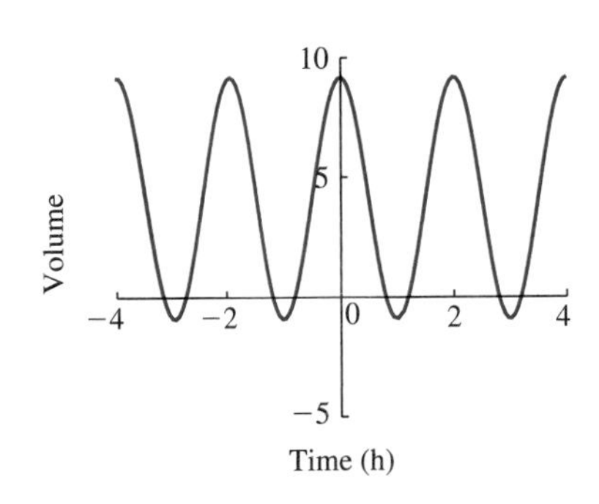

34.

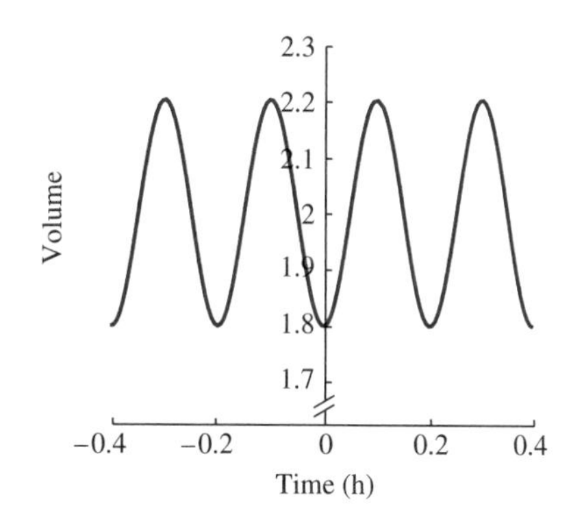

35–38 ▪ Graph the following functions. Give the average, maximum, minimum, amplitude, period, and phase of each and mark them on your graph.

35. $f(x) = 3.0 + 4.0\cos\left(2\pi\frac{x - 1.0}{5.0}\right)$.

36. $g(t) = 4.0 + 3.0\cos[2\pi(t - 5.0)]$.

37. $h(z) = 1.0 + 5.0\cos\left(2\pi\frac{z - 3.0}{4.0}\right)$.

38. $W(y) = -2.0 + 3.0\cos\left(2\pi\frac{y + 0.1}{0.2}\right)$.

39–44 ▪ Oscillations are often combined with growth or decay. Plot graphs of the following functions, and describe in words what you see. Make up a biological process that might have produced the result.

39. $f(t) = 1 + t + \cos(2\pi t)$ for $0 < t < 4$.

40. $h(t) = t + 0.2\sin(2\pi t)$ for $0 < t < 4$.

41. $g(t) = e^t\cos(2\pi t)$ for $0 < t < 3$.

42. $W(t) = e^{-t}\cos(2\pi t)$ for $0 < t < 3$.

43. $H(t) = \cos(e^t)$ for $0 < t < 3$.

44. $b(t) = \cos(e^{-t})$ for $0 < t < 3$.

45–48 ▪ Sleepiness has two cycles, a circadian rhythm with a period of approximately 24 hours and an ultradian rhythm with a period of approximately 4 hours. Both have phase 0 (starting at midnight) and average 0, but the amplitude of the circadian rhythm is 1.0 sleepiness unit and the ultradian is 0.4 sleepiness unit.

45. Find the formula and sketch the graph of sleepiness over the course of a day due to the circadian rhythm.

46. Find the formula and sketch the graph of sleepiness over the course of a day due to the ultradian rhythm.

47. Sketch the graph of the two cycles combined.

48. At what time of day are you sleepiest? At what time of day are you least sleepy?

Computer Exercises

49. Consider the following functions.

$$f_1(x) = \cos\left(x - \frac{\pi}{2}\right)$$
$$f_3(x) = \frac{\cos\left(3x - \frac{\pi}{2}\right)}{3}$$
$$f_5(x) = \frac{\cos\left(5x - \frac{\pi}{2}\right)}{5}$$
$$f_7(x) = \frac{\cos\left(7x - \frac{\pi}{2}\right)}{7}.$$

a. Plot them all on one graph.

b. Plot the sum $f_1(x) + f_3(x)$.

c. Plot the sum $f_1(x) + f_3(x) + f_5(x)$.

d. Plot the sum $f_1(x) + f_3(x) + f_5(x) + f_7(x)$.

e. What does this sum look like?

f. Try to guess the pattern, and add on $f_9(x)$ and $f_{11}(x)$. This is an example of a **Fourier series**, a sum of cosine functions that add up to a **square wave** that jumps between values of -1 and 1.

50. Use a computer to cobweb and graph solutions of the following discrete-time dynamical systems. Try three different initial conditions for each. Can you make any sense of what happens? Why don't the solutions follow a sinusoidal oscillation?

a. $x_{t+1} = \cos(x_t)$.

b. $y_{t+1} = \sin(y_t)$.

c. $z_{t+1} = \sin(z_t) + \cos(z_t)$.

51. Plot the function $f(x) = \cos(2\pi \cdot 440x) + \cos(2\pi \cdot 441x)$. Describe the result. If these were sounds, what might you hear? (This corresponds to playing two notes with the same amplitude and slightly different frequencies.)

1.9 A Model of Gas Exchange in the Lung

The exchange of materials between an organism and its environment is one of the most fundamental biological processes. By following the amount of chemical step by step through the breathing process, we can derive a discrete-time dynamical system that models this process for a simplified pair of lungs. This discrete-time dynamical system describes how the outside air mixes with internal air, and takes the form of a **weighted average**. This model provides a framework we use to study more complicated biological processes such as absorption or release of chemical.

1.9.1 A Model of the Lungs

Consider a simplified breathing process. An adult male lung has a volume of about 6.0 L when full. With each breath, 0.6 L of the air in the lungs are exhaled, and replaced by 0.6 L of outside (or **ambient**) air. After exhaling, the volume of the lungs is 5.4 L, and returns to 6.0 L after inhaling (Figure 1.9.1).

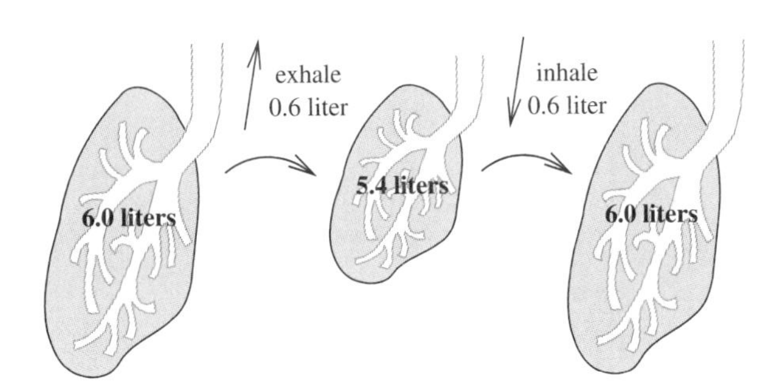

FIGURE 1.9.1
Gas exchange in the lung: the volume

Suppose further that the lung contains a particular chemical with a concentration of 2.0 mmol/L before exhaling that the lungs contain. (A mole is a convenient chemical unit indicating 6.02×10^{23} molecules, and a millimole is 6.02×10^{20} molecules). The ambient air has a chemical concentration of 5.0 mmol/L. What is the chemical concentration after one breath?

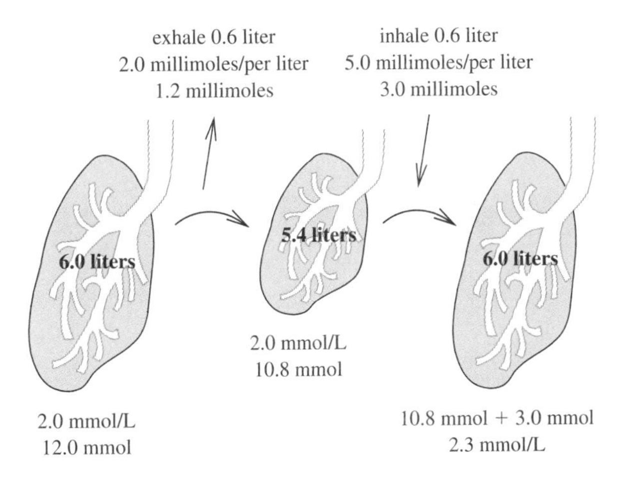

FIGURE 1.9.2
Gas exchange in the lung: the concentration

We must track three quantities through these steps: the volume (Figure 1.9.1), the total amount of chemical, and the chemical concentration (Figure 1.9.2). To find the total amount from the concentration, we use the fundamental relation

$$\text{total amount} = \text{concentration} \times \text{volume}.$$

Conversely, to find the concentration from the total amount, we rearrange the fundamental relation as

$$\text{concentration} = \frac{\text{total amount}}{\text{volume}}.$$

One basic biological assumption underlies our reasoning—that air breathed out has a concentration equal to that of the whole lung. This means that the air in the lungs is completely mixed each breath, which is not exactly true. Assuming that neither air nor chemical is produced or used while breathing, we can track through the process step by step.

Step	Volume (L)	Total Chemical (mmol)	Concentration (mmol/L)	What We Did
Air in lungs before breath	6.0	12.0	2.0	Multiplied volume of lungs (6.0) by concentration (2.0) to get 12.0.
Air exhaled	0.6	1.2	2.0	Multiplied volume exhaled (0.6) by concentration (2.0) to get 1.2.
Air in lungs after exhalation	5.4	10.8	2.0	Multiplied volume remaining (5.4) by concentration (2.0) to get 10.8.
Air inhaled	0.6	3.0	5.0	Multiplied volume inhaled (0.6) by ambient concentration (5.0) to get 3.0.
Air in lungs after breath	6.0	13.8	2.3	Found total by adding $10.8 + 3.0 = 13.8$, and divided by volume (6.0) to get 2.3.

Breathing creates a discrete-time dynamical system. The original concentration of 2.0 mmol/L is updated to 2.3 mmol/L after a breath. To write the discrete-time dynamical system, we must figure out the concentration after a breath, c_{t+1}, as a function of the concentration before the breath, c_t. We follow the same steps, but replace 2.0 with c_t (Figure 1.9.3).

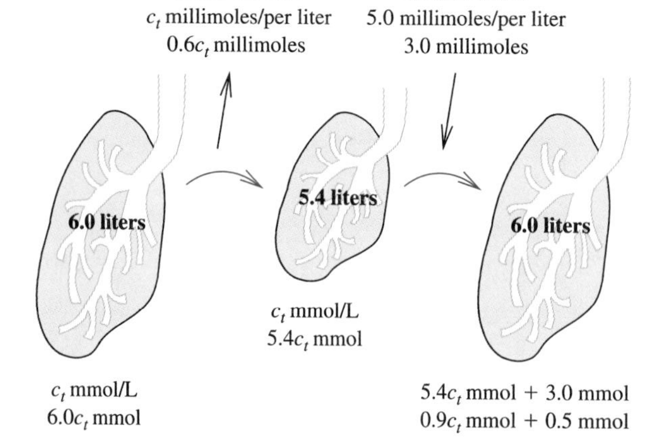

FIGURE 1.9.3
Gas exchange in the lung: finding the discrete-time dynamical system

Step	Volume (L)	Total Chemical (mmol)	Concentration (mmol/L)	What We Did
Air in lungs before breath	6.0	$6.0c_t$	c_t	Multiplied volume of lungs (6.0) by concentration (c_t) to get $6.0c_t$.
Air exhaled	0.6	$0.6c_t$	c_t	Multiplied volume exhaled (0.6) by concentration (c_t) to get $0.6c_t$.
Air in lungs after exhalation	5.4	$5.4c_t$	c_t	Multiplied volume remaining (5.4) by concentration (c_t) to get $5.4c_t$.
Air inhaled	0.6	3.0	5.0	Multiplied volume inhaled (0.6) by ambient concentration (5.0) to get 7.5.
Air in lungs after breath	6.0	$3.0 + 5.4c_t$	$0.5 + 0.9c_t$	Added inhaled chemical (3.0) to remaining chemical ($+5.4c_t$) and divided by volume (6.0) to get $0.5 + 0.9c_t$.

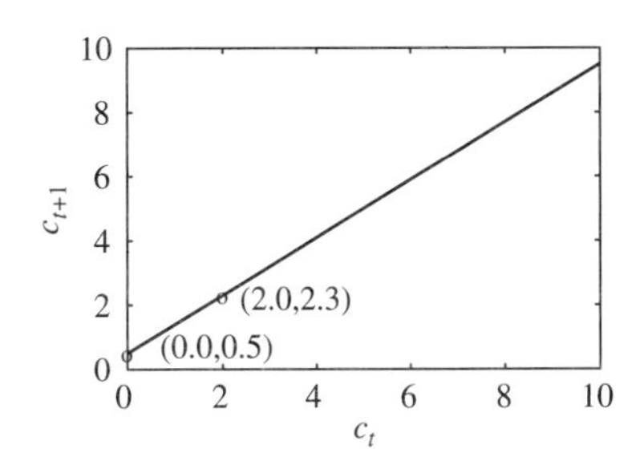

FIGURE 1.9.4

Updating function for the lung model

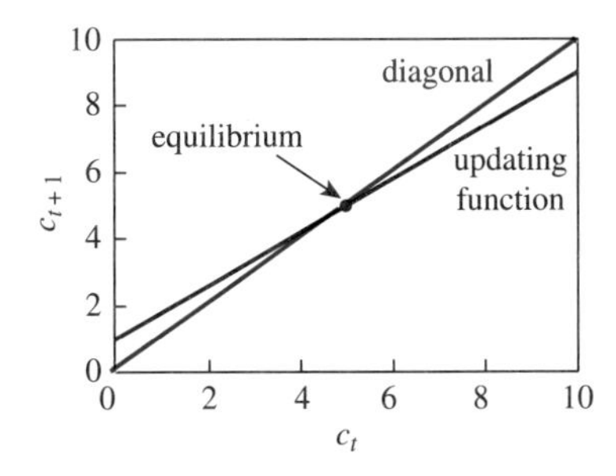

FIGURE 1.9.5

Equilibrium of the lung discrete-time dynamical system

The discrete-time dynamical system is therefore

$$c_{t+1} = 0.5 + 0.9c_t.$$

Checking, an input of $c_t = 2.0$ gives

$$c_{t+1} = 0.5 + 0.9 \cdot 2.0 = 2.3$$

as found above. The graph of the updating function is a line with y-intercept 0.5 and slope 0.9 (Figure 1.9.4). We graph it by connecting the y-intercept (0, 0.5) with another point, such as (2.0, 2.3).

We can solve for equilibria and use cobwebbing to better understand this discrete-time dynamical system. Let c^* stand for an equilibrium. The equation for equilibrium says that an input of c^* is unchanged by the discrete-time dynamical system, or

$$c^* = 0.5 + 0.9c^*.$$

The solutions of this equation are equilibria (Figure 1.9.5). To solve,

$$\begin{aligned} c^* &= 0.5 + 0.9c^* && \text{the original equation} \\ c^* - 0.9c^* &= 0.5 && \text{subtract } 0.9c^* \text{ to get unknowns on one side} \\ 0.1c^* &= 0.5 && \text{do the subtraction} \\ c^* &= \frac{0.5}{0.1} = 5.0. && \text{divide by } 0.1 \end{aligned}$$

The equilibrium value is 5.0 mmol/L. We can **check** this by plugging $c_t = 5.0$ into the discrete-time dynamical system, finding

$$c_{t+1} = 0.5 + 0.9 \cdot 5.0 = 5.0.$$

A concentration of 5.0 is indeed unchanged by the breathing process.

We can use cobwebbing to check whether solutions move toward or away from this equilibrium. Recall that cobwebbing is a graphical procedure for finding approximate solutions (Section 1.6), with steps summarized in the phrase "up or down to the updating function and over to the diagonal." Both the cobweb starting from $c_0 = 10.0$ (Figure 1.9.6) and the one starting from $c_0 = 0.0$ (Figure 1.9.7) produce solutions that approach the equilibrium at $c^* = 5.0$.

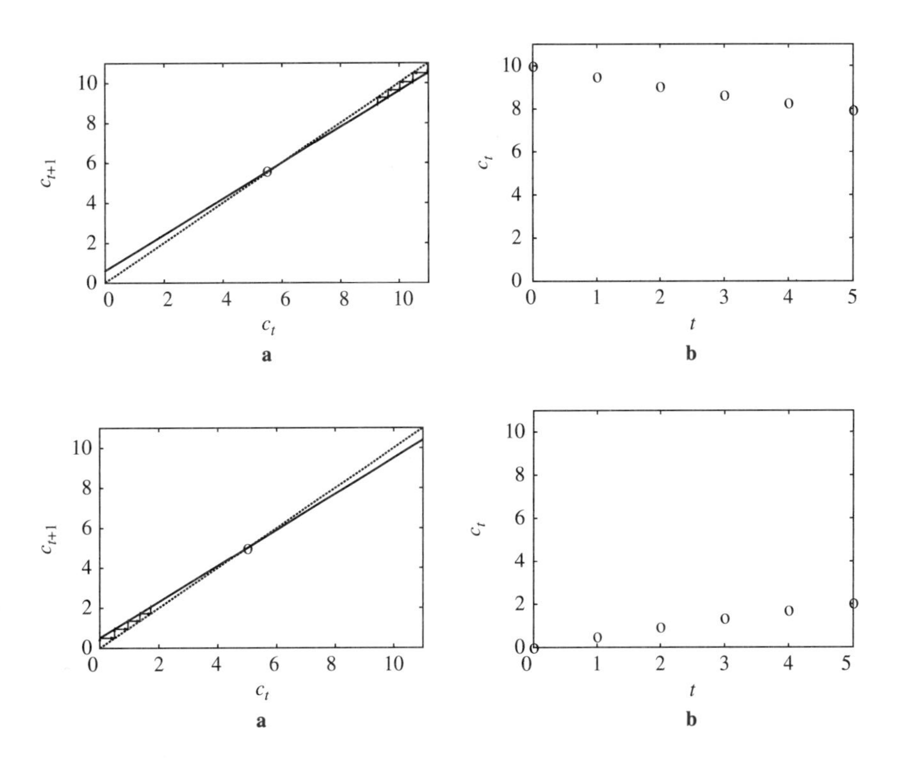

FIGURE 1.9.6

Cobweb and solution of the lung discrete-time dynamical system with $c_0 = 10.0$

FIGURE 1.9.7

Cobweb and solution of the lung discrete-time dynamical system with $c_0 = 0.0$

1.9.2 The Lung System in General

In the previous subsection, we assumed that the lungs had a volume of 6.0 L, that 0.6 L of air were exhaled and inhaled, and that the ambient concentration of chemical was 5.0 mmol/L. Suppose, more generally, that the lungs have a volume of V liters, that W liters of air are exhaled and inhaled each breath, and that the ambient concentration of chemical is γ ("gamma"). We can find the discrete-time dynamical systemgiving c_{t+1} as a function of c_t by again following the breathing process step by step (Figure 1.9.8).

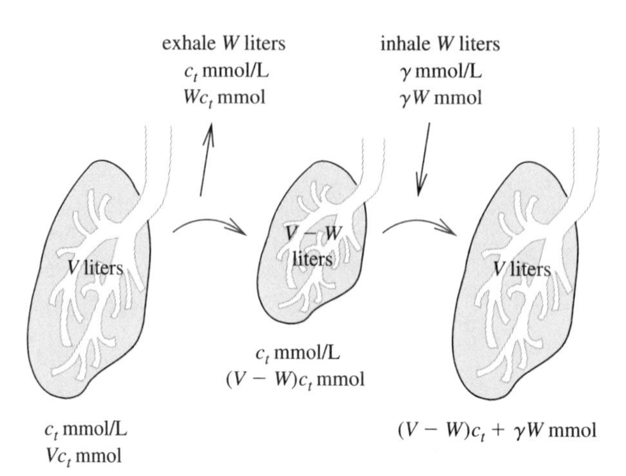

FIGURE 1.9.8
Gas exchange in the lungs: general case

Step	Volume (L)	Total Chemical (mmol)	Concentration (mmol/L)	What We Did
Air in lungs before breath	V	$c_t V$	c_t	Multiplied volume of lungs (V) by concentration (c_t) to get $c_t V$.
Air exhaled	W	$c_t W$	c_t	Multiplied volume exhaled (W) by concentration (c_t) to get $c_t W$.
Air in lungs after exhalation	$V - W$	$c_t(V - W)$	c_t	Multiplied volume remaining ($V - W$) by concentration (c_t) to get $c_t(V - W)$.
Air inhaled	W	γW	γ	Multiplied volume inhaled (W) by ambient concentration (γ) to get γW.
Air in lungs after breath	V	$c_t(V - W) + \gamma W$	$\frac{c_t(V - W) + \gamma W}{V}$	Found total by adding $c_t(V - W)$ to γW and divided by volume (V).

The new concentration appears at the end of the last line of the table, giving the discrete-time dynamical system

$$c_{t+1} = \frac{c_t(V - W) + \gamma W}{V}.$$

This equation can be simplified by multiplying out the first term and dividing out the V,

$$\begin{aligned} c_{t+1} &= \frac{c_t(V - W) + \gamma W}{V} \\ &= \frac{c_t V - c_t W + \gamma W}{V} \\ &= c_t - c_t \frac{W}{V} + \gamma \frac{W}{V}. \end{aligned}$$

The two values W and V appear only as the ratio $\frac{W}{V}$, which is the fraction of the total volume exchanged each breath. For example, when $W = 0.6$ L and $V = 6.0$ L, $\frac{W}{V} = 0.1$, meaning that 10% of air is exhaled each breath. We define a new parameter

$$q = \frac{W}{V} = \text{fraction of air exchanged}$$

to represent this quantity. We can then write the discrete-time dynamical system as

$$c_{t+1} = c_t - c_t q + \gamma q$$

or, after combining terms with c_t, as **the general lung discrete-time dynamical system**,

$$c_{t+1} = (1 - q)c_t + q\gamma. \tag{1.9.1}$$

Example 1.9.1 Finding the Discrete-Time Dynamical System with Specific Parameter Values

In the original example, $W = 0.6$ and $V = 6.0$, giving $q = \frac{W}{V} = 0.1$. Using $\gamma = 5.0$, the general equation matches our original discrete-time dynamical system because

$$c_{t+1} = (1 - 0.1)c_t + 0.1 \cdot 5.0 = 0.9c_t + 0.5.$$

After a breath, the air in the lungs is a mix of old air and ambient air (Figure 1.9.9). The fraction $1 - q$ is old air that remains in the lungs, and the remaining fraction q is ambient air. If $q = 0.5$, half of the air in the lungs after a breath came from outside, and c_{t+1} is the average of the previous concentration and the ambient concentration. If q is small, little of the internal air is replaced with ambient air and c_{t+1} is close to c_t. If q is near 1, most of the internal air is replaced with ambient air. The air in the lungs then resembles ambient air, and c_{t+1} is close to the ambient concentration γ.

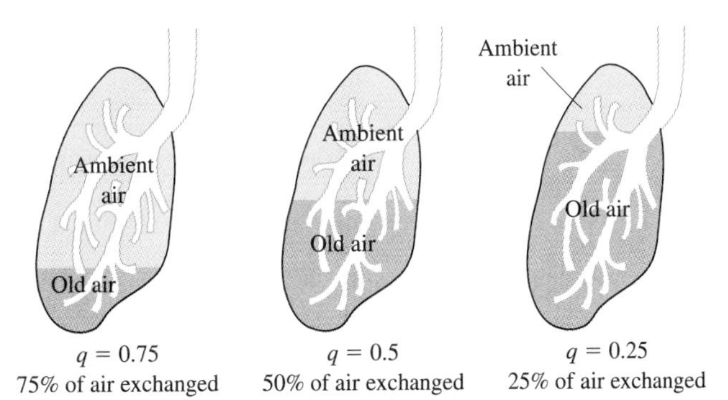

FIGURE 1.9.9
Effects of different values of q

The new concentration c_{t+1} is a **weighted average** of the old concentration c_t and the ambient concentration γ.

Definition 1.16 A weighted average of two values x and y is a sum of the form $qx + (1 - q)y$ for some value of q between 0 and 1.

When $q = 1/2$, the weighted average is the ordinary average. The concentration in the lungs after breathing is a weighted average: a fraction $1 - q$ of air is left over from the previous breath, and a fraction q is ambient air.

Example 1.9.2 A Weighted Average

Suppose $x = 2$ and $y = 5$. Then the weighted average that places a weight $q = 0.8$ on x and $1 - q = 0.2$ on y is

$$qx + (1 - q)y = 0.8 \cdot 2 + 0.2 \cdot 5 = 2.6.$$

Less weight is placed on y, and the weighted average is closer to x.

Example 1.9.3 A Contrasting Weighted Average

Suppose $x = 2$ and $y = 5$, as in Example 1.9.2. The weighted average that places a weight $q = 0.2$ on x and $1 - q = 0.8$ on y is

$$qx + (1 - q)y = 0.2 \cdot 2 + 0.8 \cdot 5 = 4.4.$$

More weight is placed on y, and the weighted average is closer to y.

Example 1.9.4 An Ordinary Average

Suppose $x = 2$ and $y = 5$, as in Examples 1.9.2 and 1.9.3. The ordinary average places equal weight $q = 0.5$ on x and $1 - q = 0.5$ on y, and is equal to

$$qx + (1 - q)y = 0.5 \cdot 2 + 0.5 \cdot 5 = 3.5.$$

This value is exactly in the middle between x and y.

Example 1.9.5 The Weighted Average Applied to Liquids

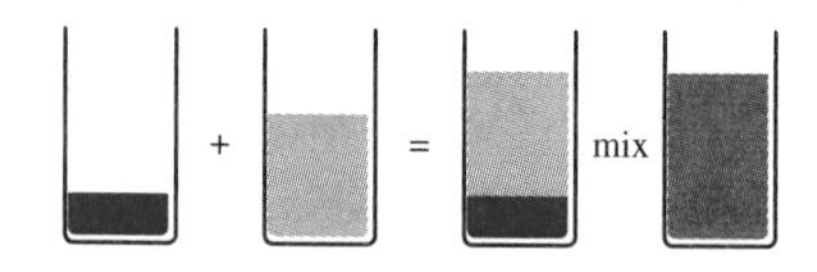

FIGURE 1.9.10
Mixing liquids as a weighted average

Suppose 1.0 L of liquid with a concentration of 10.0 mmol/L of salt are mixed with 3.0 L of liquid with a concentration of 5.0 mmol/L of salt (Figure 1.9.10). What is the concentration of the resulting mixture? We can think of this as a weighted average. The 4.0 L of the mixture contains 1.0 L of the high-salt solution (or a fraction of 0.25) and 3.0 L of the low-salt solution (or a fraction of 0.75). The resulting concentration is the weighted average

$$0.25 \cdot 10.0 \text{ mmol/L} + 0.75 \cdot 5.0 \text{ mmol/L} = 6.25 \text{ mmol/L}.$$

We could work this out more explicitly by computing the total amount of salt and the total volume. There are 10.0 mmol of salt from the first solution and 15.0 mmol from the second (multiplying the concentration of 5.0 mmol/L by the volume of 3.0 L), for a total of 25.0 mmol in 4.0 L. The concentration is

$$\frac{25.0 \text{ mmol}}{4.0 \text{ L}} = 6.25 \text{ mmol/L}.$$

The weighted average provides a simpler way to find this answer.

Example 1.9.6 A Weighted Average with More Than Two Components

Weighted averages also work when more than two solutions are mixed. Suppose 1.0 L of liquid with a concentration of 10.0 mmol/L of salt are mixed with 3.0 L of liquid with a concentration of 5.0 mmol/L of salt and 1.0 L of liquid with a concentration of 2.0 mmol/L of salt. What is the concentration of the resulting mixture? In this case, the 5.0 L of the mixture are composed of 20% (or 0.20) of the high-salt concentration solution, and 60% (or 0.60) of the medium salt concentration solution and 20% (or 0.20) of the low-salt concentration solution. The resulting concentration is the weighted average

$$0.20 \cdot 10.0 \text{ mmol/L} + 0.60 \cdot 5.0 \text{ mmol/L} + 0.20 \cdot 2.0 \text{ mmol/L} = 5.4 \text{ mmol/L}.$$

The Equilibrium of the Lung Discrete-Time Dynamical System The general discrete-time dynamical system for the lung model is

$$c_{t+1} = (1 - q)c_t + q\gamma.$$

Following the steps for finding equilibria gives

$$\begin{aligned}
c^* &= (1-q)c^* + q\gamma && \text{the equation for the equilibrium}\\
c^* - (1-q)c^* - q\gamma &= 0 && \text{move everything to one side}\\
c^* - c^* + qc^* - q\gamma &= 0 && \text{multiply } c^* \text{ through } (1-q)\\
qc^* - q\gamma &= 0 && \text{do the subtraction}\\
q(c^* - \gamma) &= 0 && \text{factor out the } q\\
q = 0 &\text{ or } c^* - \gamma = 0 && \text{set both factors to 0}\\
q = 0 &\text{ or } c^* = \gamma && \text{solve each term}
\end{aligned}$$

The key algebraic step comes after factoring. Remember that the product of two terms (like q and $c^* - \gamma$) can equal 0 only if one of the terms is equal to 0.

What do these results mean? The first case, $q = 0$, occurs when no air is exchanged. Because lungs that are exchanging no air are, there is no expression for c^* in this case, **any** value of c_t is an equilibrium. This make sense because a lung that is exchanging no air is, in fact, at equilibrium. The second case is more interesting. It says that the equilibrium value of the concentration is equal to the ambient concentration. Exchanging air with the outside world has no effect when the inside and the outside match. Doing the calculation in general explains why the equilibrium of 5.0 mmol/L found in Subsection 1.9.1 had to match the ambient concentration of 5.0 mmol/L.

1.9.3 Lung Dynamics with Absorption

Our model of chemical dynamics in the lungs ignored any absorption of the chemical by the body. We can now consider the dynamics of oxygen, which is of course absorbed by blood. How will this change the discrete-time dynamical system and the resulting solution and equilibrium?

We can use the weighted average to derive the discrete-time dynamical system including absorption. Suppose that a fraction q of air is exchanged each breath, that ambient air has a concentration of γ, and that a fraction α of chemical is absorbed before breathing out (Figure 1.9.11). After absorption, the concentration in the lungs is $(1-\alpha)c_t$. Mixing produces a weighted average with a fraction $1-q$ of this old air and a fraction q of ambient air, giving the discrete-time dynamical system

$$c_{t+1} = (1-q)(1-\alpha)c_t + q\gamma.$$

If $\alpha = 0$, this reduces to the original model of a lungs without absorption.

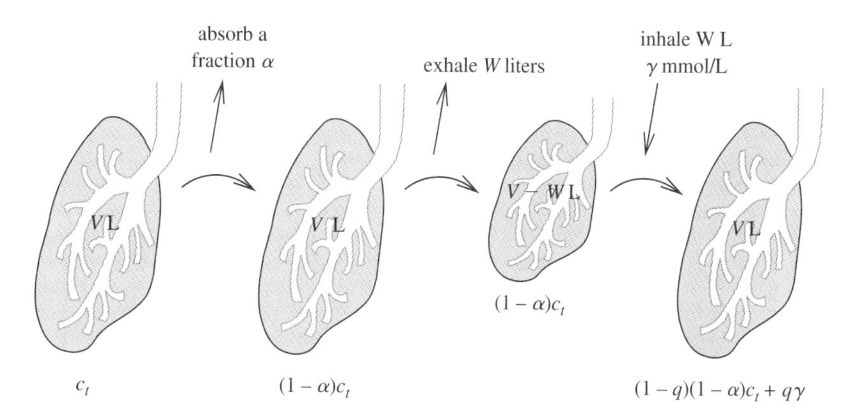

FIGURE 1.9.11
Dynamics of a lungs with absorption

Example 1.9.7 Absorption of Oxygen by the Lung

Consider again a lungs that has a volume of 6.0 L and that replaces 0.6 L each breath with ambient air (as in Figure 1.9.2). Suppose that we are tracking oxygen, with an ambient concentration of 21%, and assume that 30% of the oxygen in the lungs is absorbed each breath. Our parameter values are

$$q = 0.1$$
$$\alpha = 0.3$$
$$\gamma = 0.21.$$

The discrete-time dynamical system is then

$$c_{t+1} = 0.9 \cdot 0.7c_t + 0.1 \cdot 0.21 = 0.63c_t + 0.021.$$

The equilibrium concentration in the lungs then solves

$$\begin{aligned} c^* &= 0.63c^* + 0.021 \\ 0.37c^* &= 0.021 \\ c^* &= 0.057. \end{aligned}$$

The equilibrium concentration of oxygen in the lungs, which is equal to the concentration of oxygen in the air breathed out, would be about 5.7%, or roughly one fourth of the ambient concentration. ▲

As a result of absorption, the equilibrium concentration will be lower than the ambient concentration. By solving for the equilibrium of the system in general, we can investigate how the equilibrium depends on the fraction absorbed. To find the equilibrium, we solve

$$\begin{aligned} c^* &= (1-q)(1-\alpha)c^* + q\gamma \\ c^* - (1-q)(1-\alpha)c^* &= q\gamma \\ c^*(1-(1-q)(1-\alpha)) &= q\gamma \\ c^* &= \frac{q\gamma}{1-(1-q)(1-\alpha)}. \end{aligned}$$

As a check, if we substitute $\alpha = 0$, we find

$$c^* = \frac{q\gamma}{1-(1-q)} = \frac{q\gamma}{q} = \gamma,$$

matching the result without absorption.

The total oxygen absorbed per breath will be the product of the fraction absorbed α, the concentration c^*, and the volume V, or

$$\text{Total absorbed per breath} = \alpha c^* V. \tag{1.9.2}$$

Example 1.9.8 The Equilibrium Concentration of Oxygen as a Function of α

With the parameter values $q = 0.1$ and $\gamma = 0.21$, we find

$$c^* = \frac{0.1 \cdot 0.21}{1-0.9(1-\alpha)} = \frac{0.021}{1-0.9(1-\alpha)}.$$

By substituting values of α ranging from $\alpha = 0$ to $\alpha = 1$, we can plot the equilibrium concentration as a function of absorption (Figure 1.9.12). With $V = 6.0$,

$$\text{Total absorbed per breath} = 6.0\frac{0.021\alpha}{1-0.9(1-\alpha)}.$$

▲

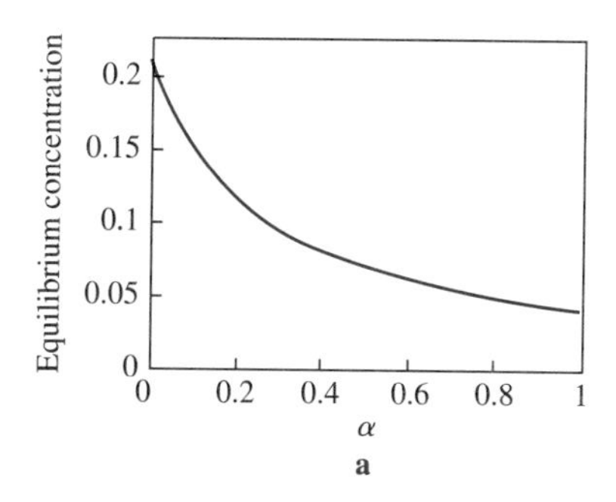

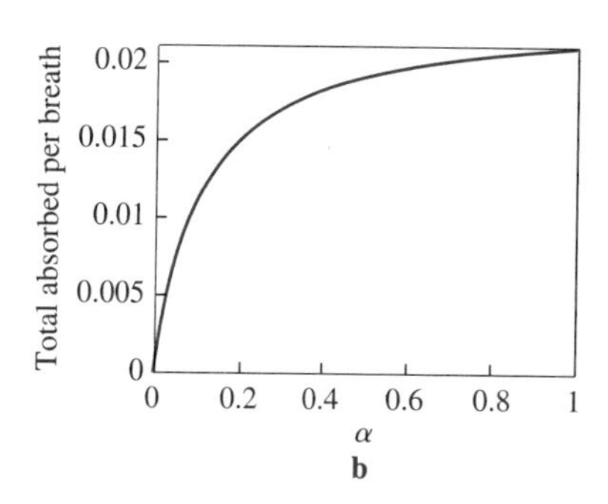

FIGURE 1.9.12
Equilibrium and absorption as a function of α

Example 1.9.9 Finding α from the Equilibrium Concentration of Oxygen

The actual oxygen concentration in exhaled air is approximately 15%, although this values varies depending on activity level. What fraction of oxygen is in fact absorbed? We can find this by solving for the value of α that produces $c^* = 0.15$.

$$
\begin{aligned}
0.15 &= \frac{0.021}{1 - 0.9(1-\alpha)} \\
0.15(1 - 0.9(1-\alpha)) &= 0.021 \\
0.15(0.1 + 0.9\alpha) &= 0.021 \\
0.1 + 0.9\alpha &= \frac{0.021}{0.15} = 0.14 \\
0.9\alpha &= 0.04 \\
\alpha &= 0.044.
\end{aligned}
$$

Rather surprisingly, the lungs absorbs less than 5% of the available oxygen, leading to exhaled air that has nearly 30% less oxygen than ambient air.

Summary Starting from an understanding of how a lungs exchanges air, we derived a discrete-time dynamical system for the concentration of a chemical in the lungs. The discrete-time dynamical system can be described as a **weighted average** of the internal concentration and the **ambient concentration**. The equilibrium is equal to the ambient concentration, and cobwebbing diagrams indicate that solutions approach this equilibrium. Including absorption produces a slightly more complicated model with an equilibrium that is less than the ambient concentration. We used this model to investigate the dynamics of oxygen in the lungs.

1.9 Exercises

Mathematical Techniques

1–4 ▪ Use the idea of the weighted average to find the following.

1. 1.0 L of water at 30°C is mixed with 2.0 L of water at 100°C. What is the temperature of the resulting mixture?

2. 2.0 ml of water with a salt concentration of 0.85 mol/L, is mixed with 5.0 ml of water with a salt concentration of 0.70 mol/L. What is the concentration of the mixture?

3. In a class of 52 students, 20 scored 50 on a test, 18 scored 75, and the rest scored 100. What was the average score?

4. In a class of 100 students, 10 score at 20, 20 score at 40, 30 score at 60, and 40 score at 80. What is the average score in the class?

5–8 ▪ Express the following weighted averages in terms of the given variables.

5. 1.0 L of water at temperature T_1 is mixed with 2.0 L of water at temperature T_2. What is the temperature of the resulting mixture? Set $T_1 = 30$ and $T_2 = 100$ and compare with the result of Exercise 1.

6. V_1 liters of water at 30°C is mixed with V_2 liters of water at 100°C. What is the temperature of the resulting mixture? Set $V_1 = 1.0$ and $V_2 = 2.0$ and compare with the result of Exercise 1.

7. V_1 liters of water at temperature T_1 is mixed with V_2 liters of water at temperature T_2. What is the temperature of the resulting mixture?

8. V_1 liters of water at temperature T_1 is mixed with V_2 liters of water at temperature T_2 and V_3 liters of water at temperature T_3. What is the temperature of the resulting mixture?

9–12 ▪ The following are similar to examples of weighted averages with absorption.

9. 1.0 L of water at 30°C is to be mixed with 2.0 L of water at 100°C, as in Exercise 1. Before mixing, however, the temperature of each moves half-way to 0 °C (so the 30°C water cools to 15°C). What is the temperature of the resulting mixture? Is this half the temperature of the result in Exercise 1?

10. 2.0 ml of water with a salt concentration of 0.85 mol/L, is to be mixed with 5.0 ml of water with a salt concentration of 0.70 mol/L, as in Exercise 2. Before mixing, however, evaporation leads the each concentration of each component to double. What is the concentration of the mixture? Is it exactly twice the concentration found in Exercise 2?

11. In a class of 52 students, 20 scored 50 on a test, 18 scored 75, and the rest scored 100. The professor suspects cheating, however, and deducts 10 from each score. What is the average score after the deduction? Is it exactly 10 less than the average found in Exercise 3?

12. In a class of 100 students, 10 score at 20, 20 score at 40, 30 score at 60, and 40 score at 80 as in Exercise 4. Because students did so poorly, the professor moves each score half way up toward 100 (so the students with 20 are moved up to 60). What is the average score in the class? Is the new average the old average moved half way to 100?

Applications

13–16 ▪ Suppose that the volume of the lungs is V, the amount breathed in and out is W, and the ambient concentration is γ mmol/L. For each of the given sets of parameter values and the given initial condition, find the following:

a. The amount of chemical in the lungs before breathing

b. The amount of chemical breathed out

c. The amount of chemical in the lungs after breathing out

d. The amount of chemical breathed in

e. The amount of chemical in the lungs after breathing in

f. The concentration of chemical in the lungs after breathing in

g. Compare this result with the result of using the general lungs discrete-time dynamical system (equation 1.9.1). Remember that $q = W/V$.

13. $V = 2.0$ L, $W = 0.5$ L, $\gamma = 5.0$ mmol/L, $c_0 = 1.0$ mmol/L.

14. $V = 1.0$ L, $W = 0.1$ L, $\gamma = 8.0$ mmol/L, $c_0 = 4.0$ mmol/L.

15. $V = 1.0$ L, $W = 0.9$ L, $\gamma = 5.0$ mmol/L, $c_0 = 9.0$ mmol/L.

16. $V = 10.0$ L, $W = 0.2$ L, $\gamma = 1.0$ mmol/L, $c_0 = 9.0$ mmol/L.

17–20 ▪ Find and graph the updating function in the following cases. Cobweb for three steps starting from the points indicated in the earlier problems. Sketch the solutions.

17. The situation in Exercise 13.

18. The situation in Exercise 14.

19. The situation in Exercise 15.

20. The situation in Exercise 16.

21–24 ▪ Compute the equilibrium of the lungs discrete-time dynamical system and check that $c^* = \gamma$.

21. $V = 2.0$ L, $W = 0.5$ L, $\gamma = 5.0$ mmol/L, $c_0 = 1.0$ mmol/L (as in Exercise 13).

22. $V = 1.0$ L, $W = 0.1$ L, $\gamma = 8.0$ mmol/L, $c_0 = 4.0$ mmol/L (as in Exercise 14).

23. $V = 1.0$ L, $W = 0.9$ L, $\gamma = 5.0$ mmol/L, $c_0 = 9.0$ mmol/L (as in Exercise 15).

24. $V = 10.0$ L, $W = 0.2$ L, $\gamma = 1.0$ mmol/L, $c_0 = 9.0$ mmol/L (as in Exercise 16).

25–26 ▪ Compare the equilibrium and total amount absorbed per breath for different values of q. Use an ambient concentration of $\gamma = 0.21$ and a volume of $V = 6.0$L.

25. Suppose $q = 0.4$ and $\alpha = 0.1$. Why is the equilibrium concentration higher than with $q = 0.2$ even though the person is breathing more?

26. Suppose $q = 0.1$ and $\alpha = 0.05$. Think of this as a person gasping for breath. Why is the concentration nearly the same as in Example 1.9.9? Does this mean that gasping for breath is OK?

27–30 ▪ The following problems investigate absorption that is not proportional to the concentration in the lungs. Assume $\gamma = 0.21$ and $q = 0.1$, and find the equilibrium concentration.

27. Oxygen concentration is reduced by 2% each breath (that is, if the concentration before absorption were 18%, it would be 16% after absorption). Find the discrete-time dynamical system and the equilibrium. Are there values of c_t for which the system does not make sense?

28. Oxygen concentration is reduced by 1% each breath. Find the discrete-time dynamical system and the equilibrium. Are there values of c_t for which the system does not make sense?

29. The amount absorbed is $0.2(c_t - 0.05)$ if $c_t \geq 0.05$. This models a case where the only oxygen available is that in excess of the concentration in the blood, which is roughly 5%.

30. The amount absorbed is $0.1(c_t - 0.05)$ if $c_t \geq 0.05$.

31–32 ▪ Find the value of the parameter that produces an exhaled concentration of exactly 0.15. Assume $\gamma = 0.21$ and $q = 0.1$.

31. Oxygen concentration is reduced by an amount A (generalizing the case in Exercises 27 and 28). How does the amount absorbed with this value of A compare with the amount of oxygen absorbed in Example 1.9.9?

32. The amount absorbed is $\alpha(c_t - 0.05)$ (generalizing the case where only available oxygen is absorbed in Exercises 29 and 30). How does the amount absorbed with this value of α compare with the amount of oxygen absorbed in Example 1.9.9?

33–34 ▪ The following problems investigate production of carbon dioxide by the lungs. Suppose that the concentration increases by an amount S before the air is exchanged. Assume an ambient concentration of carbon dioxide of $\gamma = 0.0004$ and $q = 0.1$.

33. Suppose $S = 0.001$. Write the discrete-time dynamical system and find its equilibrium. Compare the equilibrium with the ambient concentration.

34. The actual concentration of carbon dioxide in exhaled air is about 0.04, or 100 times the ambient concentration. Find the value of S that gives this as the equilibrium.

35–36 ▪ A bacterial population that has per capita production $r < 1$ but that is supplemented each generation follows a discrete-time dynamical system much like that of the lungs. Use the following steps to build the discrete-time dynamical system.

a. Starting from 3.0×10^6 bacteria, find the number after reproduction.

b. Find the number after the new bacteria are added.

c. Find the discrete-time dynamical system.

35. A population of bacteria has per capita production $r = 0.6$, and 1.0×10^6 bacteria are added each generation.

36. A population of bacteria has per capita production $r = 0.2$, and 5.0×10^6 bacteria are added each generation.

37–40 ▪ Find the equilibrium population of bacteria in the following cases with supplementation.

37. A population of bacteria has per capita production $r = 0.6$, and 1.0×10^6 bacteria are added each generation (as in Exercise 35).

38. A population of bacteria has per capita production $r = 0.2$, and 5.0×10^6 bacteria are added each generation (as in Exercise 36).

39. A population of bacteria has per capita production $r = 0.5$, and S bacteria are added each generation. What happens to the equilibrium when S is large? Does this make biological sense?

40. A population of bacteria has per capita production $r < 1$, and 1.0×10^6 bacteria are added each generation. What happens to the equilibrium if $r = 0$? What happens if r is close to 1? Do these results make biological sense?

41–44 ▪ Lakes receive water from streams each year and lose water to outflowing streams and evaporation. The following values are based on the Great Salt Lake in Utah. The lake receives $3.0 \times 10^6 \text{m}^3$ of water per year with salinity of 1 part per thousand (concentration 0.001). The lake contains $3.3 \times 10^7 \text{m}^3$ of water and starts with no salinity. Assume that the water that flows out has a concentration equal to that of the entire lake. Compute the discrete-time dynamical system by finding (a) the total salt before the inflow, (b) total water, (c) total salt and salt concentration after inflow, and (d) total water, total salt, and salt concentration after outflow or evaporation,

41. There is no evaporation, and $3.0 \times 10^6 \text{m}^3$ of water flows out each year.

42. $1.5 \times 10^6 \text{m}^3$ of water flows out each year, and $1.5 \times 10^6 \text{m}^3$ evaporates. No salt is lost through evaporation.

43. A total of $3.0 \times 10^6 \text{m}^3$ of water evaporates, and there is no outflow.

44. Assume instead that $2.0 \times 10^6 \text{m}^3$ of water evaporates and there is no outflow. The volume of this lake is increasing.

45–48 ▪ Find the equilibrium concentration of salt in a lake in the following cases. Describe the result in words by comparing the equilibrium salt level with the salt level of the water flowing in.

45. The situation described in Exercise 41.

46. The situation described in Exercise 42.

47. The situation described in Exercise 43.

48. The situation described in Exercise 44.

49–50 ▪ A lab is growing and harvesting a culture of valuable bacteria described by the discrete-time dynamical system

$$b_{t+1} = rb_t - h.$$

The bacteria have per capita production r, and h bacteria are harvested each generation.

49. Suppose that $r = 1.5$ and $h = 1.0 \times 10^6$ bacteria. Sketch the updating function, and find the equilibrium both algebraically and graphically.

50. Without setting r and h to particular values, find the equilibrium algebraically. Does the equilibrium get larger when h gets larger? Does it get larger when r gets larger? If the answers seem odd (as they should), look at a cobweb diagram to try to figure out why.

Computer Exercise

51. Investigate which factor is most important in absorbing oxygen at the maximum rate: the volume V of the lungs, the amount exchanged W, or the fraction absorbed α, using Equation 1.9.2. If an athlete could train to increase one of these values, which would be the most effective?

1.10 An Example of Nonlinear Dynamics

The discrete-time dynamical systems we have studied in detail (bacterial populations, tree height, mite populations, and the lung) are said to be **linear** because the updating function is linear. We now derive a model of two competing bacterial populations that leads naturally to a discrete-time dynamical system that is not linear. **Nonlinear dynamical systems** can have much more complicated behavior than a linear system. For example, they may have more than one equilibrium. By comparing the two equilibria in this model of selection, we will catch a glimpse of an important theme of this book, the **stability** of equilibria.

1.10.1 A Model of Selection

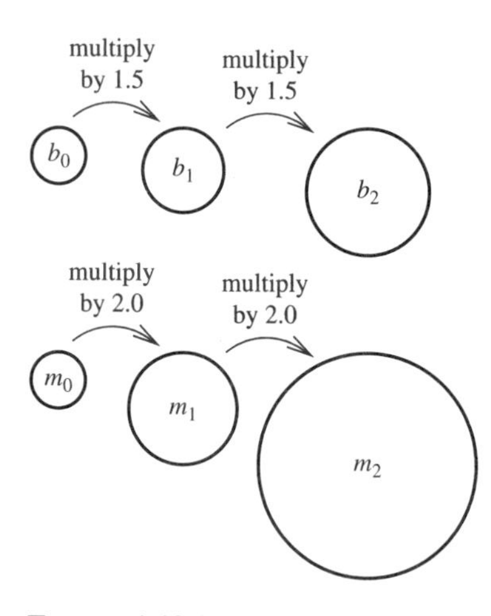

FIGURE 1.10.1
An invasion by mutant bacteria

Our original model of bacterial growth followed the population of a single type of bacteria, denoted by b_t at time t. Suppose that a mutant type with population m_t appears and begins competing. If the original (or **wild**) type has a per capita reproduction of 1.5 and the mutant type has a per capita reproduction of 2.0, the two populations will follow the discrete-time dynamical systems

$$\begin{aligned} b_{t+1} &= 1.5b_t \quad \text{discrete-time dynamical system for wild type} \\ m_{t+1} &= 2.0m_t \quad \text{discrete-time dynamical system for mutants.} \end{aligned} \tag{1.10.1}$$

The per capita reproduction of the mutant type is greater than that of the wild type, perhaps because it is better able to survive in a particular laboratory. Over time, we would expect the population to include a larger and larger proportion of mutant bacteria (Figure 1.10.1). The establishment of this mutant is an example of **selection**. Selection occurs when the frequency of a gene (the mutation) changes over time.

Imagine observing this mixed population for many hours. Counting all of the bacteria each hour would be impossible. Nonetheless, we could track the mutant invasion by taking a sample and measuring the fraction of the mutant type by counting or using a specific stain. If this fraction becomes larger and larger, we would know that the mutant type was taking over.

How can we model the dynamics of the fraction? The vital first step is to **define a new variable**. In this case, we set p_t to be the fraction of mutants at time t. Then

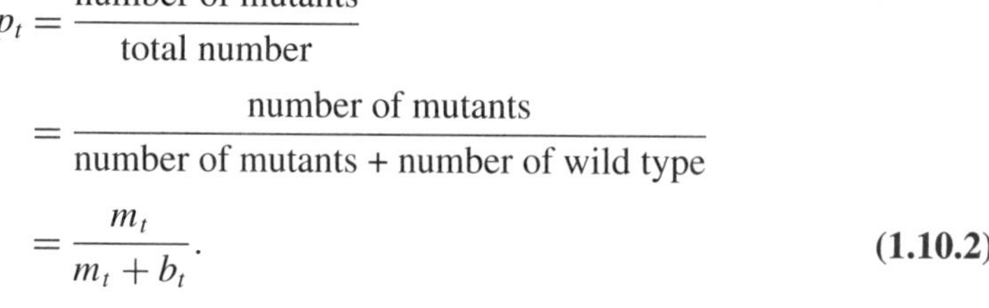

$$\begin{aligned} p_t &= \frac{\text{number of mutants}}{\text{total number}} \\ &= \frac{\text{number of mutants}}{\text{number of mutants} + \text{number of wild type}} \\ &= \frac{m_t}{m_t + b_t}. \end{aligned} \tag{1.10.2}$$

Example 1.10.1 Finding the Fraction of Mutants

If $b_t = 3.0 \times 10^6$ and $m_t = 2.0 \times 10^5$ there are a total of $b_t + m_t = 3.0 \times 10^6 + 2.0 \times 10^5 = 3.2 \times 10^6$ bacteria. The fraction of the mutant type is

$$p_t = \frac{2.0 \times 10^5}{2.0 \times 10^5 + 3.0 \times 10^6} = \frac{2.0 \times 10^5}{3.2 \times 10^6} = 0.0625.$$

What is the fraction of the wild type? It is the number of wild type divided by the total number of bacteria, or

$$\begin{aligned} \text{fraction of wild type} &= \frac{\text{number of wild type}}{\text{total number}} \\ &= \frac{\text{number of wild type}}{\text{number of mutants} + \text{number of wild type}} \\ &= \frac{b_t}{m_t + b_t} \end{aligned}$$

(Figure 1.10.2).

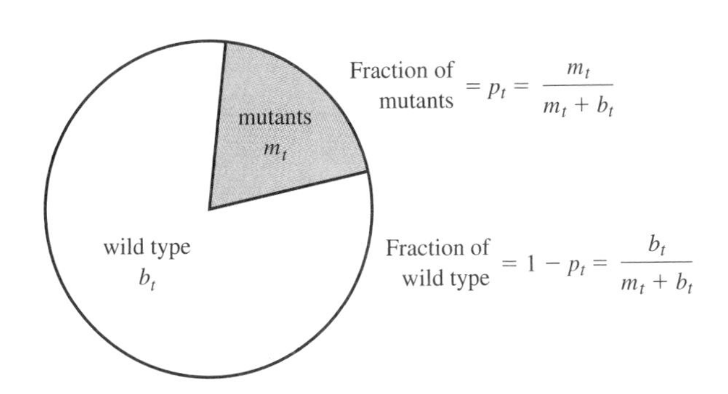

FIGURE 1.10.2
The fraction of mutants and wild type

Example 1.10.2 Finding the Fraction of Wild Type

If $m_t = 2.0 \times 10^5$ and $b_t = 3.0 \times 10^6$ (as in Example 1.10.1), then

$$\begin{aligned}\text{fraction of wild type} &= \frac{3.0 \times 10^6}{2.0 \times 10^5 + 3.0 \times 10^6}\\ &= \frac{3.0 \times 10^6}{3.2 \times 10^6} = 0.9375.\end{aligned}$$

We do not need to give a new name to this fraction because we know that

$$\text{fraction of mutants} + \text{fraction of wild type} = 1.$$

Because all bacteria are of these two types, the fractions must add up to 1.

Example 1.10.3 The Fraction of Mutants and the Fraction of Wild Type

In Examples 1.10.1 and 1.10.2, the fraction of mutants is 0.0625 and that of wild type is 0.9375. These two fractions do indeed add up to $0.0625 + 0.9375 = 1$, as they must.

Solving for the fraction of wild type,

$$\begin{aligned}\text{fraction of wild type} &= 1 - \text{fraction of mutants}\\ &= 1 - p_t.\end{aligned}$$

Putting this together with our original calculation, we find

$$\frac{b_t}{m_t + b_t} = 1 - p_t. \qquad \textbf{(1.10.3)}$$

Our goal is to find p_{t+1}, the fraction of the mutant type after one hour.

Example 1.10.4 Finding the Updated Fraction

If $m_t = 2.0 \times 10^5$ and $b_t = 3.0 \times 10^6$ (Example 1.10.1), the updated populations are

$$\begin{aligned}m_{t+1} &= 2.0 m_t = 4.0 \times 10^5\\ b_{t+1} &= 1.5 b_t = 4.5 \times 10^6.\end{aligned}$$

The updated fraction of the mutant type, p_{t+1}, is

$$p_{t+1} = \frac{4.0 \times 10^5}{4.0 \times 10^5 + 4.5 \times 10^6} = 0.0816.$$

As expected, the fraction has increased. We might expect that the fraction of mutants would increase by a factor equal to the ratio $\frac{2.0}{1.5} = 1.333$ of the per capita reproduction of the two types. In fact,

$$\frac{p_{t+1}}{p_t} = \frac{0.0816}{0.0625} = 1.3056,$$

which is slightly less. We will soon see why the mutant increases more slowly than we might at first expect.

We can follow these same steps to find the discrete-time dynamical system for p_t. By definition

$$p_{t+1} = \frac{m_{t+1}}{m_{t+1} + b_{t+1}}.$$

Using the discrete-time dynamical systems for the two types (Equation 1.10.1), we find

$$p_{t+1} = \frac{2.0m_t}{2.0m_t + 1.5b_t}. \tag{1.10.4}$$

Although mathematically correct, this is not a satisfactory discrete-time dynamical system. We have supposed that the actual values of m_t and b_t are impossible to measure. The discrete-time dynamical system must give the new fraction p_{t+1} in terms of the old fraction p_t, which we can measure by sampling.

Writing in terms of p_t can be done with an algebraic trick: dividing the numerator and denominator by the same thing. Because the definition of p_t has the total population $m_t + b_t$ in the denominator, we divide it into the numerator and denominator, finding

$$p_{t+1} = \frac{2.0\frac{m_t}{m_t+b_t}}{2.0\frac{m_t}{m_t+b_t} + 1.5\frac{b_t}{m_t+b_t}}.$$

We can simplify by substituting

$$p_t = \frac{m_t}{m_t + b_t}$$

(Equation 1.10.2) and

$$1 - p_t = \frac{b_t}{m_t + b_t}$$

(Equation 1.10.3), finding

$$p_{t+1} = \frac{2.0p_t}{2.0p_t + 1.5(1 - p_t)}. \tag{1.10.5}$$

This is the discrete-time dynamical system we sought, giving a formula for the fraction at time $t + 1$ in terms of the fraction at time t.

Example 1.10.5 Using the Discrete-Time Dynamical System to Find the Updated Fraction

If $p_t = 0.0625$, as in Example 1.10.1, the discrete-time dynamical system tells us that

$$p_{t+1} = \frac{2.0 \cdot 0.0625}{2.0 \cdot 0.0625 + 1.5(1 - 0.0625)} = 0.0816.$$

This matches the answer we found before, but is based only on **measurable quantities**.

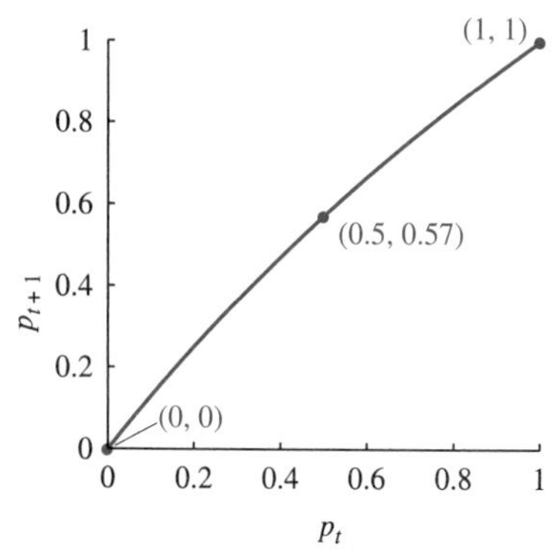

This calculation illustrates one of the great strengths of mathematical modeling. Our **derivation** of this measurable discrete-time dynamical system used the values m_t and b_t, which might not be possible to measure. But just because things cannot be measured in practice does not mean they cannot be measured in principle. These values do exist and can be worked with mathematically. One can think of mathematical models as a way to "see the invisible."

The discrete-time dynamical system for the fraction (Equation 1.10.5) is not linear because it involves division. The graph of the function is curved (Figure 1.10.3). We plotted it by substituting in representative values for a fraction, which must lie between 0 and 1. The points (0, 0), (0.5, 0.57), and (1, 1) lie on the graph, and we connected them with a smooth curve.

FIGURE 1.10.3
Graph of updating function from the selection model

The equilibria of this discrete-time dynamical system are found by solving

$p^* = \dfrac{2.0p^*}{2.0p^* + 1.5(1 - p^*)}$	the equation for the equilibrium
$p^*(2.0p^* + 1.5(1 - p^*)) = 2.0p^*$	multiply through by denominator
$p^*(2.0p^* + 1.5(1 - p^*)) - 2.0p^* = 0$	move everything to one side
$p^*(2.0p^* + 1.5(1 - p^*) - 2.0) = 0$	factor out p^*
$p^*(2.0p^* + 1.5 - 1.5p^* - 2.0) = 0$	multiply out terms in parentheses
$p^*(0.5p^* - 0.5) = 0$	simplify

Therefore $p^* = 0$ or $0.5p^* - 0.5 = 0$, which has solution $p^* = 1$. These equilibria correspond to extinction of the mutant (at $p^* = 0$) and extinction of the wild type (at $p^* = 1$).

1.10.2 The Discrete-Time Dynamical System and Equilibria

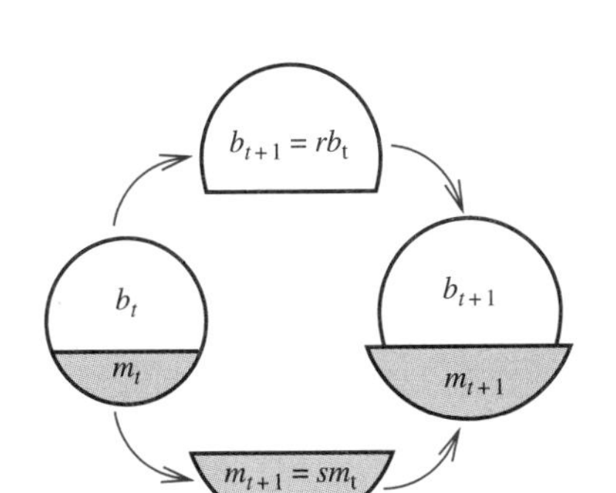

FIGURE 1.10.4
The general case

We can gain a better understanding of this process by studying the general case. Suppose that the mutant type has per capita reproduction s and the wild type has per capita reproduction r (Figure 1.10.4). The populations follow

$$m_{t+1} = sm_t$$
$$b_{t+1} = rb_t. \qquad (1.10.6)$$

We can follow the steps above to derive the discrete-time dynamical system for the fraction.

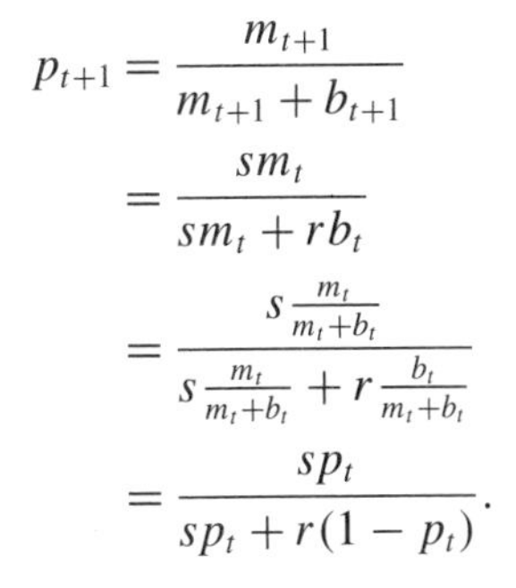

$$p_{t+1} = \frac{m_{t+1}}{m_{t+1} + b_{t+1}} = \frac{sm_t}{sm_t + rb_t} = \frac{s\frac{m_t}{m_t+b_t}}{s\frac{m_t}{m_t+b_t} + r\frac{b_t}{m_t+b_t}} = \frac{sp_t}{sp_t + r(1-p_t)}.$$

This gives the general form

$$p_{t+1} = \frac{sp_t}{sp_t + r(1-p_t)}. \qquad (1.10.7)$$

Example 1.10.6 Substituting Parameters into the General Discrete-Time Dynamical System

The derivation in the previous subsection considered the case $s = 2.0$ and $r = 1.5$. Substituting these parameter values into the general form for bacterial selection gives

$$p_{t+1} = \frac{2.0p_t}{2.0p_t + 1.5(1-p_t)},$$

matching what we found before. ▲

When $s > r$, the graph of the updating function lies above the diagonal except at the intersection points $p_t = 0$ and $p_t = 1$ (Figure 1.10.5a). This means that any value of

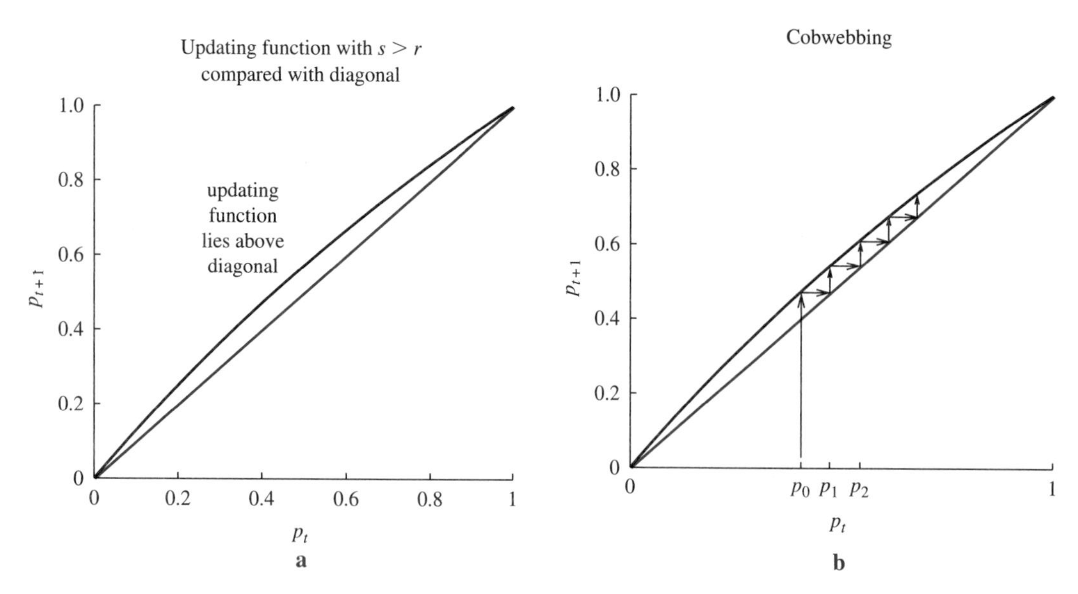

FIGURE 1.10.5
Dynamics when mutants reproduce more quickly

p_t between 0 and 1 will be increased by the discrete-time dynamical system, consistent with the higher per capita reproduction of the mutants. The cobwebbing moves up, indicating this increase (Figure 1.10.5b).

What happens if the per capita reproduction of the wild type exceeds that of the mutants? With $r = 2.0$ and $s = 1.5$, the discrete-time dynamical system is

$$p_{t+1} = \frac{1.5p_t}{1.5p_t + 2.0(1 - p_t)}.$$

The three points (0, 0), (0.5, 0.43), and (1, 1) lie on the graph, which itself lies below the diagonal (Figure 1.10.6a). Values of p_t between 0 and 1 are decreased by the discrete-time dynamical system as shown by the decreasing cobweb (Figure 1.10.6b). This is consistent with the lower reproduction of the mutants.

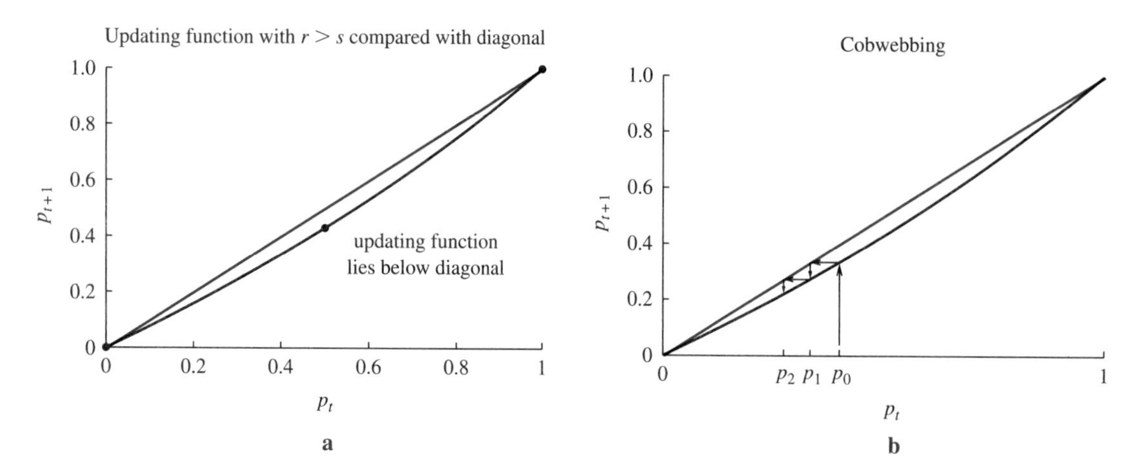

FIGURE 1.10.6
Dynamics when wild type reproduces more quickly

Finally, what happens if the two types grow at exactly the same rate? If $r = s$, the discrete-time dynamical system simplifies

$$\begin{aligned} p_{t+1} &= \frac{sp_t}{sp_t + s(1 - p_t)} \\ &= \frac{sp_t}{sp_t + s - sp_t} \\ &= \frac{sp_t}{s} \\ &= p_t. \end{aligned}$$

In this case, the discrete-time dynamical system leaves all values unchanged. When both types reproduce at the same rate, the fraction of the mutant neither increases nor decreases and every value of p_t is an equilibrium. This makes biological sense; there is no selection when the two types have the same growth rate. This does not say that the **total number** of bacteria is unchanged; if $r = s = 2.0$, the total number will double each hour. Only the **fraction** of mutants remains the same.

We can use the five steps of Algorithm 1.5 to find the equilibria.

$$p^* = \frac{sp^*}{sp^* + r(1 - p^*)} \quad \text{the equation for the equilibrium}$$

$$(sp^* + r(1 - p^*))p^* = sp^* \quad \text{multiply both sides by the denominator}$$

$$(sp^* + r(1 - p^*))p^* - sp^* = 0 \quad \text{move everything to one side}$$

$$(sp^* + r - rp^*)p^* - sp^* = 0 \quad \text{multiply out inner term}$$

$$sp^{*2} + rp^* - rp^{*2} - sp^* = 0 \quad \text{multiply out all terms}$$

$$sp^{*2} - sp^* + rp^* - rp^{*2} = 0 \quad \text{collect like terms}$$

$$(s - r)p^*(p^* - 1) = 0 \quad \text{factor}$$

$$s - r = 0 \text{ or } p^* = 0 \text{ or } p^* = 1 \quad \text{set each factor to 0}$$

$$s = r \text{ or } p^* = 0 \text{ or } p^* = 1. \quad \text{solve each term}$$

Factoring involves some tricky algebra, which is worth checking.

What do these three equilibria mean? If $s = r$, the discrete-time dynamical system leaves all values unchanged, and every value of p_t is an equilibrium. Otherwise, the equilibria are $p_t = 0$ and $p_t = 1$. When $p_t = 0$, the population consists entirely of the wild type. Because our model includes no mutation or immigration, there is nowhere for the mutant type to arise. Similarly, when $p_t = 1$ the population consists entirely of the mutant type and the wild type will never arise. These equilibria correspond to the extinction equilibrium for a population of one type of bacteria: at $p_t = 0$ the mutants are extinct, and at $p_t = 1$ the wild type are extinct.

1.10.3 Stable and Unstable Equilibria

Figure 1.10.7 shows many steps of cobwebbing, with $s = 2.0$ and $r = 1.5$ starting near the equilibrium $p^* = 0$. The solution moves slowly away from 0, more swiftly past the half way point at $p_t = 0.5$, and then slowly approaches the other equilibrium at $p_t = 1$.

If we started **exactly** at $p_0 = 0$, the solution would remain at $p_t = 0$ for all times t. Similarly, if we started **exactly** at $p_t = 1$, the solution would remain at $p_t = 1$ for all times t. The two equilibria behave quite differently, however, if our starting point is nearby. A solution starting **near** $p_0 = 0$ moves steadily **away** from the equilibrium (Figure 1.10.7). A solution starting **near** $p_0 = 1$ moves **toward** the equilibrium (Figure 1.10.8).

This situation is analogous to keeping a ball from rolling around on a surface. If we place it on the bottom of a small depression, it will remain at an equilibrium (Figure 1.10.9a). If it is moved slightly away from this equilibrium, it will come back, much like the solution that starts near the equilibrium $p^* = 1$ (Figure 1.10.8). We call an equilibrium with this property **stable**. Similarly, if we place it exactly on top of a small hill, it will remain there (Figure 1.10.9b). However, if it is moved even slightly

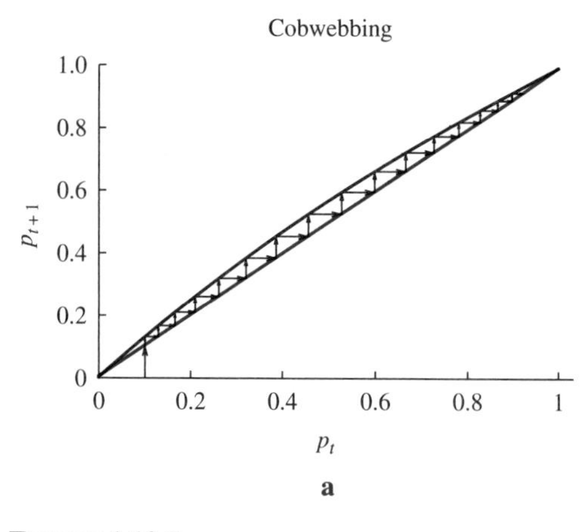

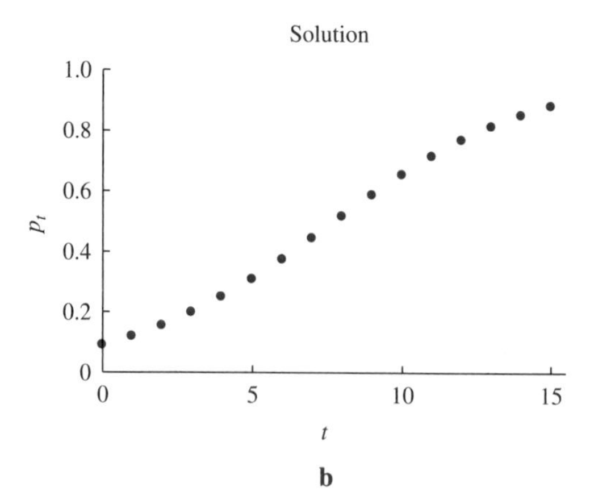

FIGURE 1.10.7
Solution of the selection model starting near $p_0 = 0$

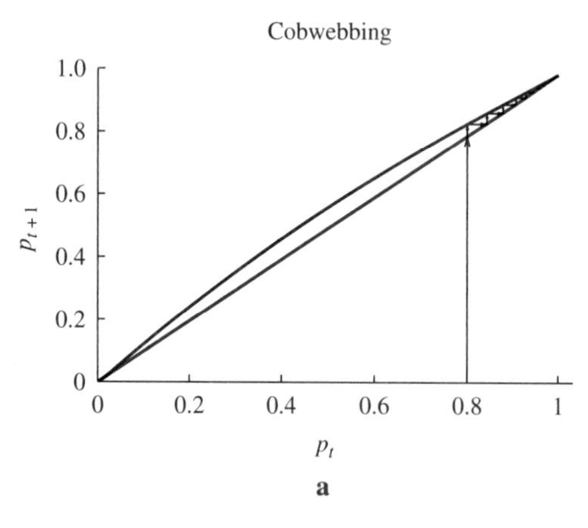

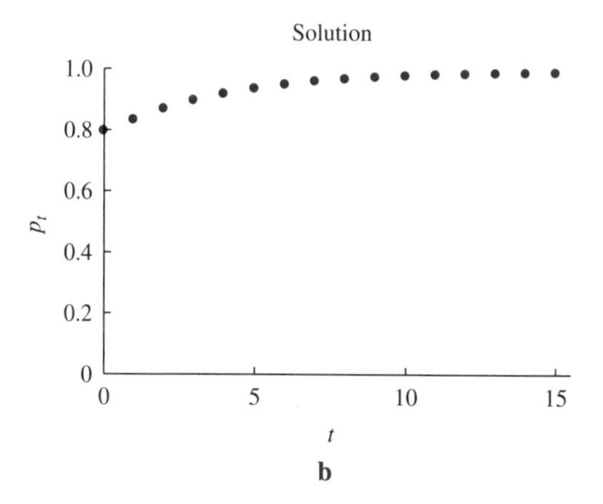

FIGURE 1.10.8
Solution of the selection model starting near $p_0 = 1$

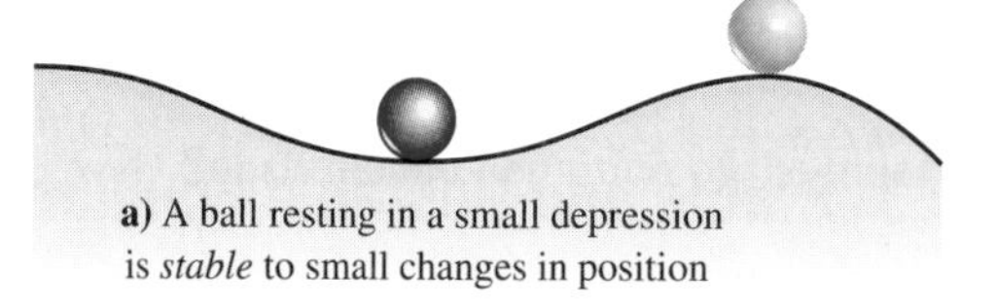

b) A ball resting on a small hill is
unstable to small changes in position
a) A ball resting in a small depression
is stable to small changes in position

FIGURE 1.10.9
Stable and unstable resting points for a ball

away from this equilibrium, it will roll farther and farther away, much like the solution that starts near the equilibrium $p^* = 0$ (Figure 1.10.7). We call an equilibrium with this property **unstable**.

This leads us to an informal definition of stable and unstable equilibria.

Definition 1.17 An equilibrium is **stable** if solutions that begin near that equilibrium stay near, or approach, that equilibrium. An equilibrium is **unstable** if solutions that begin near that equilibrium move away from that equilibrium.

In Section 3.1, we will derive powerful methods to analyze discrete-time dynamical systems and determine whether their equilibria are stable or unstable. Because these techniques require the **derivative**, a central idea from calculus, we must first study the foundational notions of limits and rate of change.

Summary As an example of a **nonlinear dynamical system**, a discrete-time dynamical system with a curved graph, we derived the equation for the fraction of mutants invading a population of wild-type bacteria. This dynamical system, unlike the linear ones studied hitherto, has two equilibria. One of these equilibria is **unstable**; solutions starting nearby move farther and farther away. The other is **stable**; solutions starting nearby move closer and closer to the equilibrium.

1.10 Exercises

Mathematical Techniques

1–4 ▪ A population consists of 200 red birds and 800 blue birds. Find the fraction of red birds and blue birds after the following. Check that the fractions add up to 1.

1. The population of red birds doubles, and the population of blue birds remains the same.
2. The population of blue birds doubles, and the population of red birds remains the same.
3. The population of red birds is multiplied by a factor of r, and the population of blue birds remains the same.
4. The population of blue birds is multiplied by a factor of s, and the population of red birds remains the same.

5–6 ▪ Sketch graphs of the following functions.

5. $f(x) = \frac{x}{x+1}$ for $0 \le x \le 2$ (the updating function in Section 1.5, Exercise 23).
6. $g(x) = \frac{3x}{2x+1}$ for $0 \le x \le 2$.

7–10 ▪ Using the discrete-time dynamical system and the derivation of Equation 1.10.7, find p_t, m_{t+1}, b_{t+1} and p_{t+1} in the following situations.

7. $s = 1.2, r = 2.0, m_t = 1.2 \times 10^5, b_t = 3.5 \times 10^6$.
8. $s = 1.2, r = 2.0, m_t = 1.2 \times 10^5, b_t = 1.5 \times 10^6$.
9. $s = 0.3, r = 0.5, m_t = 1.2 \times 10^5, b_t = 3.5 \times 10^6$.
10. $s = 1.8, r = 1.8, m_t = 1.2 \times 10^5, b_t = 3.5 \times 10^6$.

11–12 ▪ Solve for the equilibria of the following discrete-time dynamical systems.

11. $p_{t+1} = \frac{p_t}{p_t + 2.0(1 - p_t)}$.
12. $p_{t+1} = \frac{4.0p_t}{4.0p_t + 0.5(1 - p_t)}$.

13–14 ▪ Find all non-negative equilibria of the following mathematically elegant discrete-time dynamical systems.

13. $x_{t+1} = \frac{x_t}{1 + ax_t}$ where a is a positive parameter. What happens to this system if $a = 0$?
14. $x_{t+1} = \frac{x_t}{a + x_t}$ where a is a positive parameter. What happens to this system if $a = 0$?

15–18 ▪ Identify stable and unstable equilibria on the following graphs of updating functions.

15.

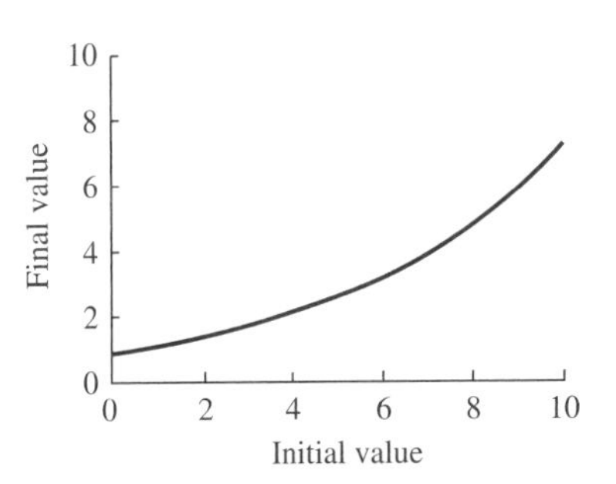

16.

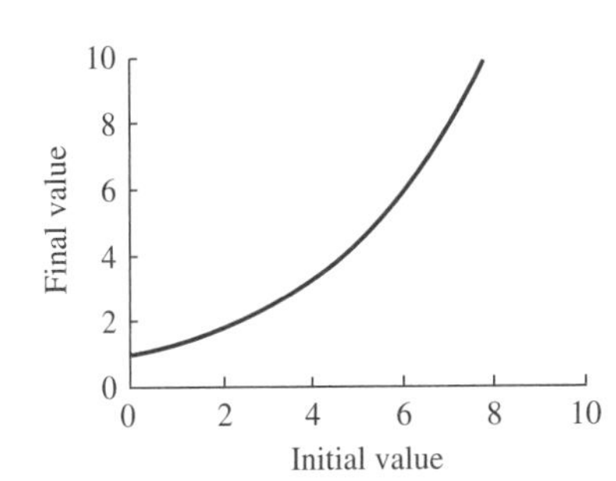

17.

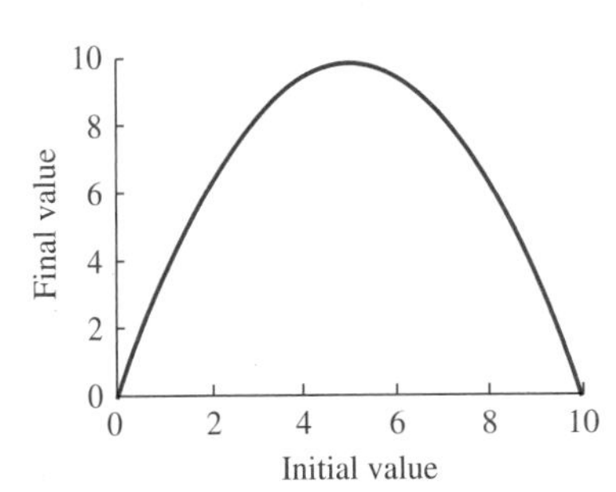

18.

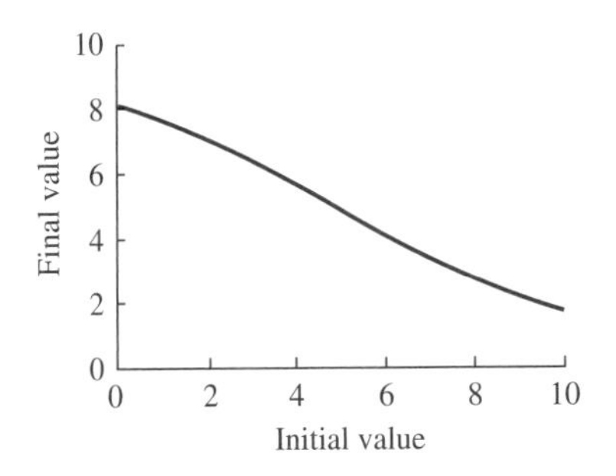

Applications

19–22 ▪ Find and graph the updating functions for the following cases of the selection model (Equation 1.10.7). Cobweb starting from $p_0 = 0.1$ and $p_0 = 0.9$. Which equilibria are stable?

19. $s = 1.2, r = 2.0$.

20. $s = 1.8, r = 0.8$.

21. $s = 0.3, r = 0.5$. Compare with Exercise 19.

22. $s = 1.8, r = 1.8$.

23–24 ▪ For each of the following discrete-time dynamical systems, indicate which of the equilibria are stable and which are unstable.

23.

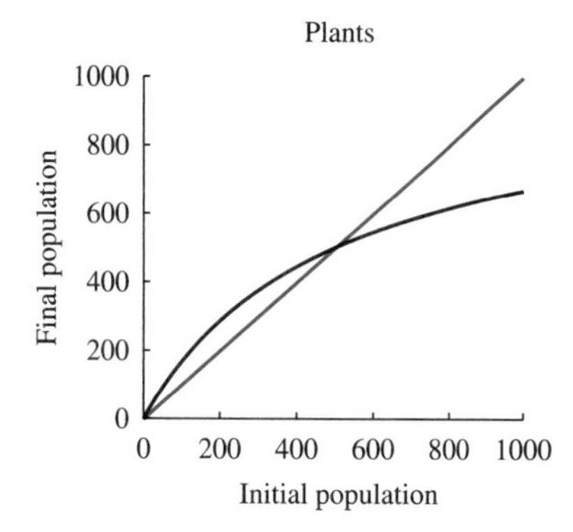

24.

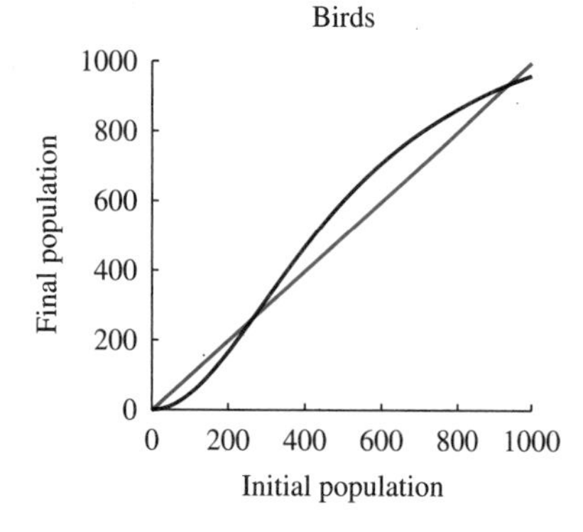

25–28 ▪ This section has ignored the important evolutionary force of mutation. This series of problems builds models that consider mutation without reproduction. Suppose that 20% of wild-type bacteria transform into mutants and that 10% of mutants transform back into wild type ("revert"). In each case, find the following.

a. The number of wild type bacteria that mutate and the number of mutants that revert.

b. The number of wild-type bacteria and the number of mutants after mutation and reversion.

c. The total number of bacteria before and after mutation. Why is it the same?

d. The fraction of mutants before and after mutation.

25. Begin with 1.0×10^6 wild type and 1.0×10^5 mutants.

26. Begin with 1.0×10^5 wild type and 1.0×10^6 mutants.

27. Begin with b_t wild type and m_t mutants. Find the discrete-time dynamical system for the fraction p_t of mutants. Find the equilibrium fraction of mutants. Cobweb starting from the initial condition in Exercise 25. Is the equilibrium stable?

28. Begin with b_t wild type and m_t mutants, but suppose that a fraction 0.1 mutate and a fraction 0.2 revert. Find the discrete-time dynamical system and the equilibrium fraction of mutants.

29–32 ▪ This series of problems combines mutation with selection. In one simple scenario, mutations occur in only one direction (wild type turn into mutants but not vice versa), but wild type and mutants have different levels of per capita production. Suppose that a fraction 0.1 of wild type mutate each generation, but that each wild-type individual produces 2.0 offspring while each mutant produces only 1.5 offspring. In each case, find the following.

a. The number of wild-type bacteria that mutate.

b. The number of wild-type bacteria and the number of mutants after mutation.

c. The number of wild-type bacteria and the number of mutants after reproduction.

d. The total number of bacteria after mutation and reproduction.

e. The fraction of mutants after mutation and reproduction.

29. Begin with 1.0×10^6 wild type and 1.0×10^5 mutants.

30. Begin with 1.0×10^5 wild type and 1.0×10^6 mutants.

31. Begin with b_t wild type and m_t mutants. Find the discrete-time dynamical system for the fraction p_t of mutants. Find the equilibrium fraction of mutants. Cobweb starting from the initial condition in Exercise 29. Is the equilibrium stable?

32. Begin with b_t wild type and m_t mutants, but suppose that a fraction 0.2 mutate and that the per capita production of mutants is 1.0. Find the discrete-time dynamical system and the equilibrium fraction of mutants.

33–36 ▪ The model of selection studied in this section is similar to a model of migration. Suppose two nearby islands have populations of butterflies, with x_t on the first island and y_t on the second. Each year, 20% of the butterflies from the first island fly to the second and 30% of the butterflies from the second fly to the first. Suppose there are no births and deaths.

33. Suppose there are 100 butterflies on each island at time $t = 0$. How many are on each island at $t = 1$? At $t = 2$?

34. Suppose there are 200 butterflies on the first island and none on the second at time $t = 0$. How many are on each island at $t = 1$? At $t = 2$?

35. Find equations for x_{t+1} and y_{t+1} in terms of x_t and y_t.

36. Divide both sides of the discrete-time dynamical system for x_t by $x_{t+1} + y_{t+1}$ to find a discrete-time dynamical system for the fraction p_t on the first island. What is the equilibrium fraction?

37–38 ▪ The following two problems extend the migration models to include some reproduction. Each year, 20% of the butterflies from the first island fly to the second and 30% of the butterflies from the second fly to the first. Again, x_t represents the number of butterflies on the first island, y_t represents the number of butterflies on the second island, and p_t represents the fraction of butterflies on the first island. In each case, find the following:

a. Start with 100 butterflies on each island and find the number after migration and after reproduction.

b. Find equations for x_{t+1} and y_{t+1} in terms of x_t and y_t.

c. Find the discrete-time dynamical system for p_{t+1} in terms of p_t.

d. Find the equilibrium p^*.

e. Sketch a graph and cobweb from a reasonable initial condition.

37. Each butterfly that begins the year on the first island produces one offspring after migration (whether they find themselves on the first or the second island). Those that begin the year on the second island do not reproduce. Assume that no butterflies die.

38. Now suppose that the butterflies that do not migrate reproduce (making one additional butterfly each) and those that do migrate fail to reproduce from exhaustion. No butterflies die.

39–42 ▪ The model describing the dynamics of the concentration of medication in the bloodstream,

$$M_{t+1} = 0.5M_t + 1.0,$$

becomes nonlinear if the fraction of medication used is a function of the concentration. In the basic model, half is used no matter how much there is. More generally,

$$\begin{aligned}\text{new concentration} = \text{old concentration} - &\\ \text{fraction used} \times \text{old} &\\ \text{concentration} + &\\ \text{supplement}&\end{aligned}$$

Suppose that the fraction used is a **decreasing function** of the concentration. The following problems look at two cases. In each, find the equilibrium concentration.

39. Suppose that

$$\text{fraction used} = \frac{0.5}{1.0 + 0.1M_t}.$$

Write the discrete-time dynamical system and solve for the equilibrium and compare with $M^* = 2.0$ for the basic model.

40. Suppose that

$$\text{fraction used} = \frac{0.5}{1.0 + 0.4M_t}.$$

Write the discrete-time dynamical system, solve for the equilibrium, and compare with $M^* = 2.0$ for the basic model.

41. Suppose that

$$\text{fraction used} = \frac{\beta}{1.0 + 0.1M_t}.$$

for some parameter $\beta \le 1$. Write the discrete-time dynamical system and solve for the equilibrium. Sketch a graph of the equilibrium as a function of β. Cobweb starting from $M_0 = 1.0$ in the cases $\beta = 0.05$ and $\beta = 0.5$.

42. Suppose that

$$\text{fraction used} = \frac{0.5}{1.0 + \alpha M_t}$$

for some parameter α. Write the discrete-time dynamical system and solve for the equilibrium. Sketch a graph of the equilibrium as a function of α. What happens when $\alpha > 0.5$? Can you explain this in biological terms? Cobweb starting from $M_0 = 1.0$ in the cases $\alpha = 0.1$ and $\alpha = 1.0$.

43–46 ▪ Our models of bacterial population growth neglect the fact that bacteria produce fewer offspring in large populations. The following problems introduce two important models of this process, having the form

$$b_{t+1} = r(b_t)b_t$$

where the per capita production r is a function of the population size b_t. In each case,

a. Graph the per capita production as a function of population size.

b. Write the discrete-time dynamical system and graph the updating function.

c. Find the equilibria.

d. Cobweb and say whether the equilibrium seems to be stable.

43. One widely used nonlinear model of competition is the "logistic" model, where per capita production is a linearly decreasing function of population size. Suppose that the per capita production is $r(b) = 2\left(1 - \frac{b}{1.0 \times 10^6}\right)$.

44. In an alternative model, the per capita production decreases as the reciprocal of a linear function. Suppose that the per capita production is $r(b) = \frac{2}{1 + \frac{b}{1.0 \times 10^6}}$.

45. In another alternative model, called the Ricker model, the per capita production decreases exponentially. Suppose that per capita production is $r(b) = 2e - \frac{b}{1.0 \times 10^6}$.

46. In a model with an **Allee effect**, organisms reproduce poorly when the population is small. In one case, per capita production follows $r(b) = \frac{4b}{1 + b^2}$.

Computer Exercises

47. Consider the discrete-time dynamical system

$$x_{t+1} = rx_t(1 - x_t)$$

similar to the form in Exercise 43. Plot the updating function and have your computer find solutions for 50 steps starting from $x_0 = 0.3$ for the following values of r:

a. Some value of r between 0 and 1. What is the only equilibrium?

b. Some value of r between 1 and 2. Where are the equilibria? Which one seems to be stable?

c. Some value of r between 2 and 3. Where are the equilibria? Which one seems to be stable?

d. Try several values of r between 3 and 4. What is happening to the solution? Is there any stable equilibrium?

e. The solution is **chaotic** when $r=4$. One property of chaos is **sensitive dependence on initial conditions**. Compare a solution starting from $x_0=0.3$ with one starting at $x_0=0.30001$. Even though they start off very close, they soon separate and become completely different. Why might this be a problem for a scientific experiment?

48. Consider the discrete-time dynamical system

$$x_{t+1}=e^{ax_t}$$

for the following values of the parameter a. Use your computer to graph the function and the diagonal to look for equilibria. Cobweb starting from $x_0=1$ in each case.

a. $a=0.3$.

b. $a=0.4$.

c. $a=1/e$.

49. Consider the equation describing the dynamics of selection

$$p_{t+1}=\frac{sp_t}{sp_t+r(1-p_t)}$$

but with two cultures 1 and 2. In 1, the mutant does better than the wild type, and in 2 the wild type does better. In particular, suppose that $s=2.0$ and $r=0.3$ in culture 1 and that $s=0.6$ and $r=2.0$ in culture 2. Define discrete-time dynamical systems f_1 and f_2 to describe the dynamics in the two cultures.

a. Graph the functions f_1 and f_2 along with the identity function. Find the first five values of solutions starting from $p_0=0.02$ and $p_0=0.98$ in each culture. Explain in words what each solution is doing and why.

b. Suppose you change the experiment. Begin by taking a population with a fraction p_0 of mutants. Split this population in half, and place one half in culture 1 and the other half in culture 2. Let the bacteria reproduce once in each culture, and then mix them together. Split the mix in half and repeat the process. The updating function is

$$f(p)=\frac{f_1(p)+f_2(p)}{2}.$$

Can you derive this? Plot this updating function along with the identity function. Have your computer find the equilibria and label them on your graph. Do they make sense?

c. Use cobwebbing to figure out which equilibria are stable.

d. Find one solution starting from $p_0=0.001$ and another starting from $p_0=0.999$. Are these results consistent with the stability of the equilibria? Explain in words why the solutions do what they do.

1.11 An Excitable Systems: The Heart

We can use cobwebbing and equilibria to study a simplified model of the heart. Our goal is to understand how simple changes in the parameters of a heart can produce heartbeat patterns called **second-degree block**. With these syndromes, people's hearts either beat half as often as they should or beat normally for a while, skip a beat, and return to beating. These conditions are solutions for the same model that describes a normal heartbeat, but with different parameter values.

1.11.1 A Simple Heart

Figure 1.11.1 shows the basic apparatus for beating of the heart. The sinoatrial node (SA node) is the pacemaker, sending regular signals to the atrioventrical node (AV node). The AV node then tells the heart to beat if conditions are suitable.

The AV node can be thought of as keeping track of the condition of the heart with an electrical potential. Denote the potential after responding to a signal from the SA node as V_t. Two processes go into updating this potential. During the time τ between signals from the SA node, the electrical potential of the AV node decays exponentially at rate α. Setting $\hat{V}_t$ to be the potential of the AV node just before receiving the next

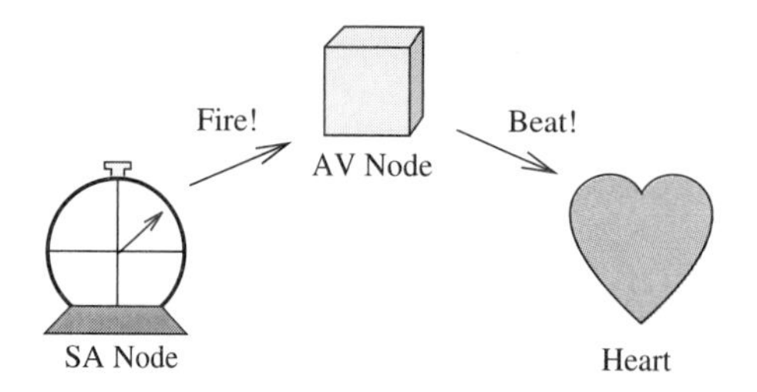

FIGURE 1.11.1
A mathematician's version of the heart

signal from the SA node,

$$\hat{V}_t = e^{-\alpha\tau} V_t$$

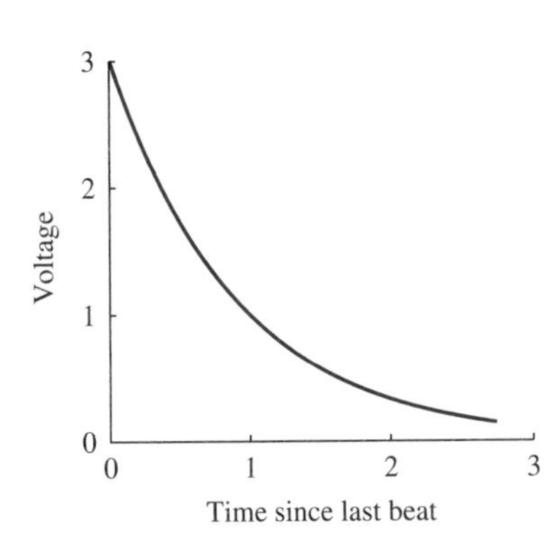

FIGURE 1.11.2

The exponential decay of voltage between beats

(Figure 1.11.2). Whether the heart beats depends on the state of the AV node when a signal arrives. If the potential $\hat{V}_t$ is too high, the heart has not had enough time to recover from the last beat, and the AV node ignores the signal. Otherwise, the AV node accepts the signal, tells the heart to beat, and increases its potential by u.

Let V_c be the threshold potential (Figure 1.11.3). If $\hat{V}_t > V_c$, the heart is not ready to beat and

$$V_{t+1} = \hat{V}_t \quad \text{if } \hat{V}_t > V_c.$$

If $\hat{V}_t \leq V_c$, the AV node responds and tells the heart to beat, and

$$V_{t+1} = \hat{V}_t + u \quad \text{if } \hat{V}_t \leq V_c.$$

To translate this description into a discrete-time dynamical system, we must write V_{t+1} entirely in terms of V_t, eliminating the $\hat{V}_t$ terms. Because $\hat{V}_t = e^{-\alpha\tau} V_t$, the two cases can be summarized as

$$V_{t+1} = \begin{cases} e^{-\alpha\tau} V_t & \text{if } e^{-\alpha\tau} V_t > V_c \\ e^{-\alpha\tau} V_t + u & \text{if } e^{-\alpha\tau} V_t \leq V_c. \end{cases}$$

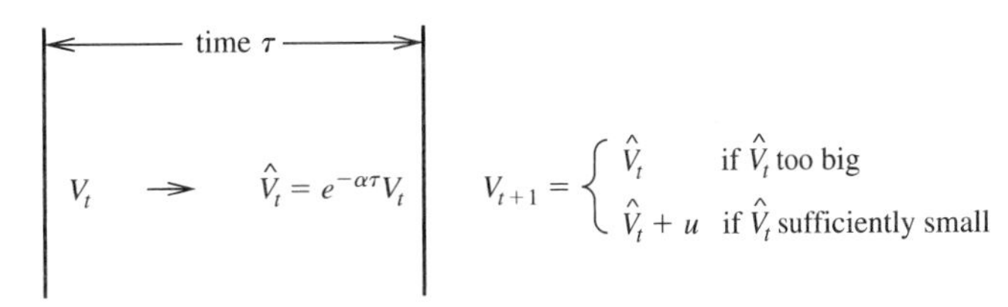

FIGURE 1.11.3

Schematic diagram of the potential of the AV node

For convenience we substitute the new parameter c for $e^{-\alpha\tau}$. A value of c near 1 means that the potential decays very little, and a value of c near 0 means that the potential decays a great deal (Figure 1.11.4).

Example 1.11.1 The Relation Between c, α, and τ

If $\tau = 1$ and $\alpha = \ln(3) = 1.099$, then $c = e^{-\alpha\tau} = 1/3$. The potential decays to 1/3 of its initial value between beats.

Example 1.11.2 The Relation Between c, α, and τ

If $\tau = 1$ and $\alpha = \ln(1.5) = 0.405$, then $c = e^{-\alpha\tau} = 2/3$. The potential decays less, to 2/3 of its initial value between beats, because α is smaller.

Using this new notation,

$$V_{t+1} = \begin{cases} cV_t & \text{if } cV_t > V_c \\ cV_t + u & \text{if } cV_t \leq V_c. \end{cases}$$

This updating function is graphed with $u = 1$, $c = 0.4$ and $V_c = 1$ in Figure 1.11.5.

This function, unlike those we have studied hitherto, has a jump (where $cV_t = V_c$). This jump reflects the sharp response threshold. In the real heart, the threshold is not precise, and the two branches of the updating function are connected (Figure 1.11.6).

We can use the graphical method for finding equilibria as intersections with the diagonal to study this discrete-time dynamical system. Each piece of the updating function is a line with slope $c < 1$. There are two possible pictures. Either the upper branch of the updating function crosses the diagonal at an equilibrium (Figure 1.11.7a),

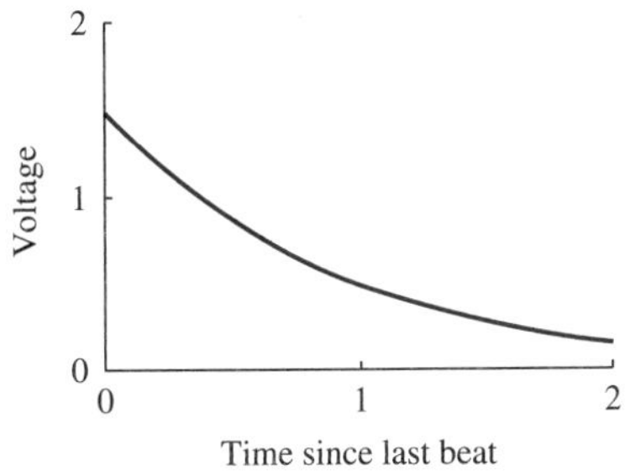

FIGURE 1.11.4

Graph of the potential of the AV node without beating

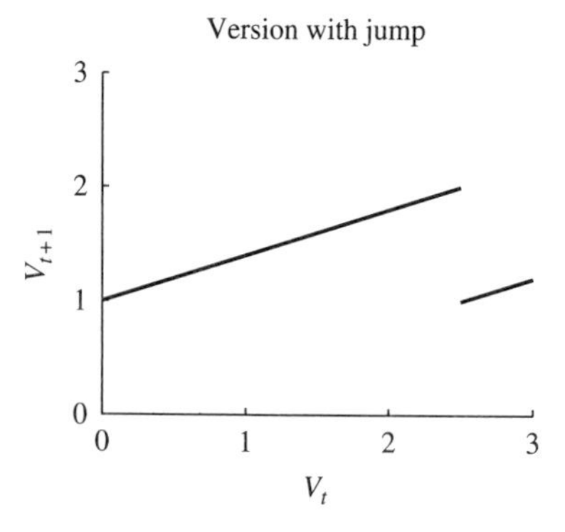

FIGURE 1.11.5
The updating function for the potential of the AV node

FIGURE 1.11.6
A smoothed updating function for the potential of the AV node

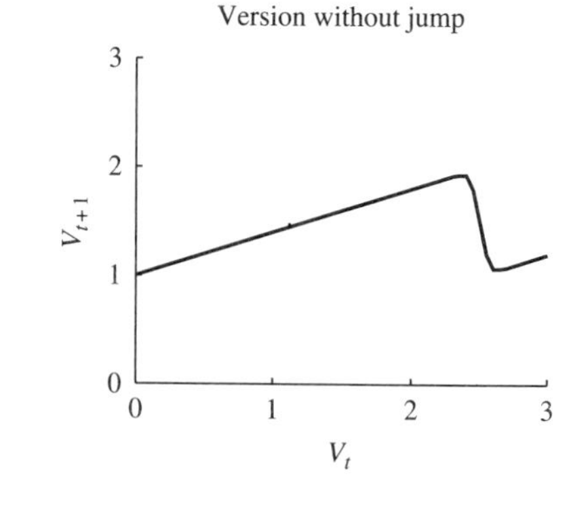

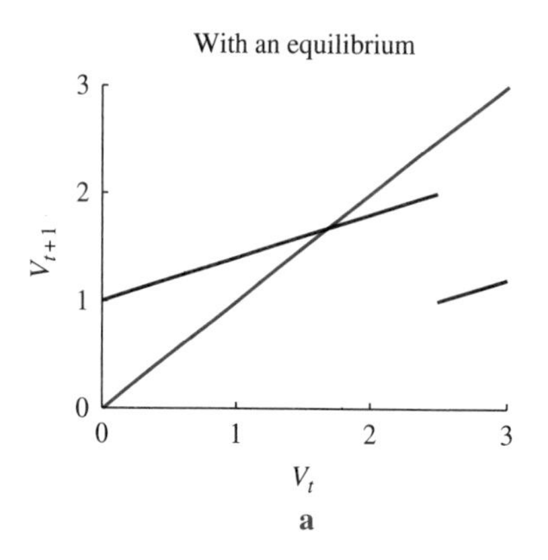

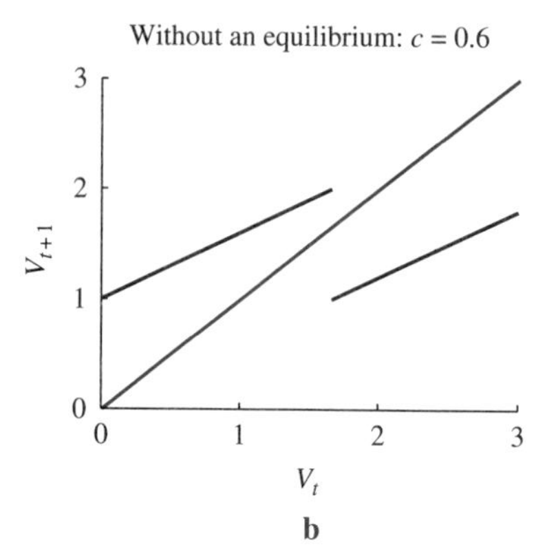

FIGURE 1.11.7
The heart updating function with and without an equilibrium: $u = 1$, $V_c = 1$.

or the diagonal sneaks through the gap between the two branches and there is no equilibrium at all (Figure 1.11.7b).

What does this equilibrium mean? An equilibrium is a point where different processes balance. In the present case, an equilibrium represents a value of the potential where the decay (by a factor of c) is exactly balanced by the response to the signal (an increase of u). This means that the heart will beat steadily.

What are the algebraic conditions for an equilibrium? An equilibrium is a value of V_t that solves $V_{t+1} = V_t$. The heart must proceed through the cycle

$$V_t \overset{\text{decay}}{\longrightarrow} \hat{V}_t = cV_t \overset{\text{decay}}{\longrightarrow} cV_t + u = V_t$$

and end up where it started. Setting V^* to be the equilibrium, we can solve

$$V^* = cV_t + u$$

to find

$$V^* = \frac{u}{1-c}. \tag{1.11.1}$$

This equilibrium exists only if the heart is indeed ready to beat when the next signal comes, or if

$$cV^* = c\frac{u}{1-c} \le V_c. \tag{1.11.2}$$

Example 1.11.3 Case Where Heart Beats with Every Signal

Suppose that $u = 1$, $V_c = 1$, $\tau = 1$, and $\alpha = \ln(3) = 1.099$. We found in Example 1.11.1 that $c = e^{-\alpha\tau} = 1/3$. The equilibrium is

$$V^* = \frac{1}{1-c} = \frac{1}{1-1/3} = 1.5.$$

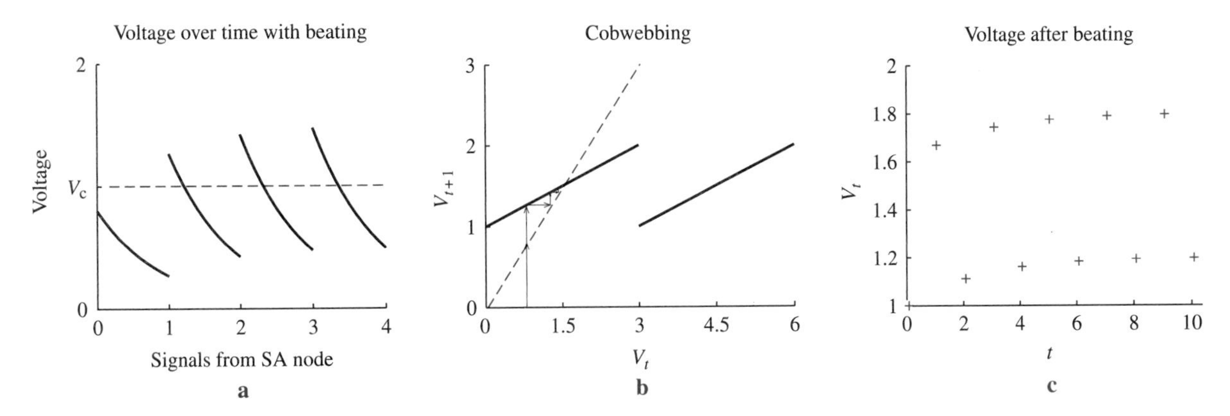

FIGURE 1.11.8
The behavior of a heart with an equilibrium

This equilibrium exists only if $cV^* = 0.5$ is less than $V_c = 1$. Because it is, the heart will beat every time, with voltage decaying from 1.5 to 0.5 between beats and increasing back to 1.5 on the beat (Figure 1.11.8a).

Example 1.11.4 Case Where Heart Fails to Beat with Every Signal

What happens if α, the recovery rate, becomes smaller? We found in Example 1.11.2 that with $\alpha = \ln(1.5) = 0.405$, then $c = e^{-\alpha\tau} = 2/3$. The equilibrium is

$$V^* = \frac{1}{1 - 2/3} = 3.0$$

This equilibrium exists only if $cV^* = 2.0$ is less than $V_c = 1.0$. Because it is not, the heart cannot beat every time. If α is too small, the AV node recovers too slowly from one signal to be ready to respond to the next. Similarly, if the time τ between beats is decreased by too much, the heart might not have time to recover, and there will be no equilibrium. The AV node cannot respond to every signal when signals from the SA node arrive too frequently. The more complicated dynamics that result are our next topic.

1.11.2 Second-Degree Block

When the heart fails to beat in response to every signal from the SA node, the condition is called **second-degree block**. In one type, called **2:1 AV block**, the heart beats only with every other stimulus. In another, called the **Wenckebach phenomenon**, the heart beats normally for a while, skips a beat, and then resumes normal beating and repeats the cycle. Our model of the heart can help us understand these two conditions.

Graphically, 2:1 AV block corresponds to the situation in Figure 1.11.9. There is no equilibrium. The potential of the AV node alternates between a high value and a low value. When high, the potential does not decay sufficiently to respond to the next signal. After another cycle (time τ), however, the potential has reached a low enough value to respond.

To find the conditions for 2:1 AV block, we use techniques similar to those used to find an equilibrium. Suppose the potential is V_t just after beating (Figure 1.11.9). If the node responds to the second signal but not the first

$$V_t \overset{\text{decay}}{\longrightarrow} cV_t \overset{\text{signal ignored}}{\longrightarrow} cV_t \overset{\text{decay}}{\longrightarrow} c^2V_t \overset{\text{signal obeyed}}{\longrightarrow} c^2V_t + u. \tag{1.11.3}$$

If the potential after these two full cycles comes back exactly to where it started, the heart beats with every other signal, producing 2:1 AV block. The updated potential after two cycles matches the original potential if V_t is equal to some value $\bar{V}$ that satisfies the following equations

$$c^2\bar{V} + u = \bar{V}$$
$$c\bar{V} > V_c \tag{1.11.4}$$
$$c^2\bar{V} < V_c$$

which has solution

$$\bar{V} = \frac{u}{1 - c^2}, \tag{1.11.5}$$

if the inequalities are satisfied. Note the similarity to the equation for an equilibrium for this model (Equation 1.11.1).

Example 1.11.5 2:1 AV Block

In Figure 1.11.9, we have set $u = V_c = 1.0$ and $c = 2/3$, corresponding to the second case considered above. We have seen that there is no equilibrium. Equation 1.11.5 implies that $\bar{V} = 1.8$. We can follow the dynamics through a complete cycle, finding

$$\bar{V} = 1.8 \overset{\text{decay}}{\longrightarrow} c\bar{V} = 1.2 \overset{\text{signal ignored}}{\longrightarrow} c\bar{V} = 1.2 \overset{\text{decay}}{\longrightarrow} c^2\bar{V} = 0.8$$
$$\overset{\text{signal obeyed}}{\longrightarrow} c^2\bar{V} + u = 1.8.$$

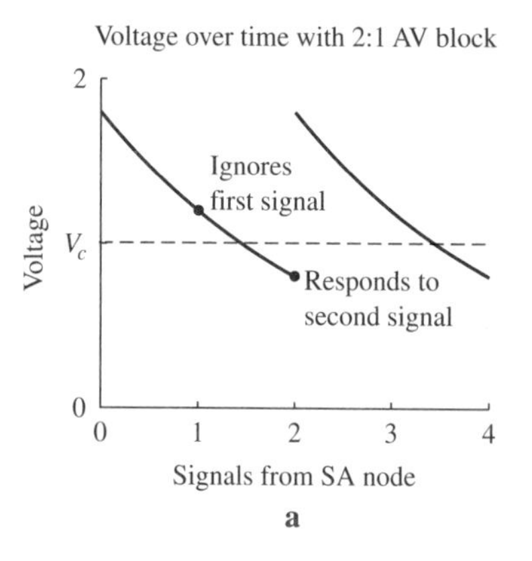

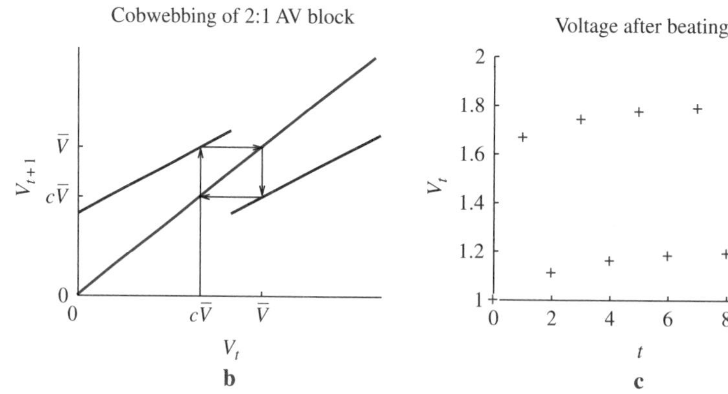

FIGURE 1.11.9
The dynamics of 2:1 AV block

The AV node does not respond to the first signal because $1.2 > V_c = 1.0$, but is ready to respond to the second and return to its original potential. ▲

1.11.3 The Wenckebach Phenomenon

With the parameter values $u = V_c = 1$, we can compute the conditions on c for the existence of an equilibrium. The equation for an equilibrium is

$$V^* = \frac{1}{1 - c}$$

requiring that $cV^* \leq 1$ (Equation 1.11.2). The equilibrium at V^* exists only if

$$\frac{c}{1 - c} \leq 1.$$

We can solve for c, finding

$$c \leq 1 - c$$
$$2c \leq 1$$
$$c \leq 0.5.$$

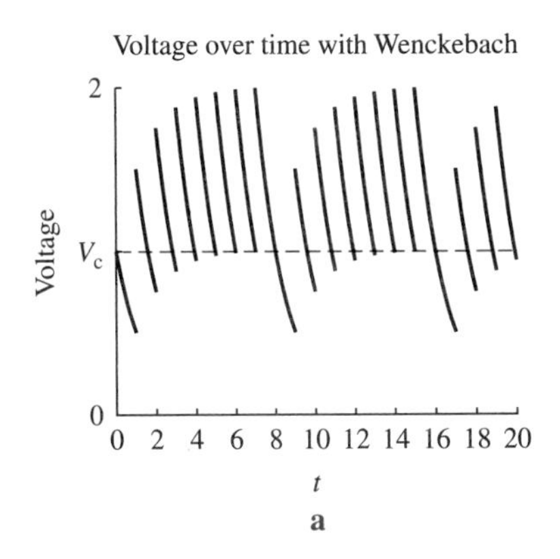

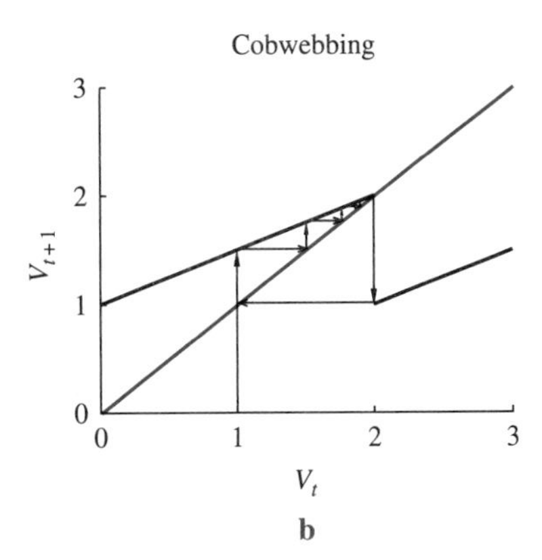

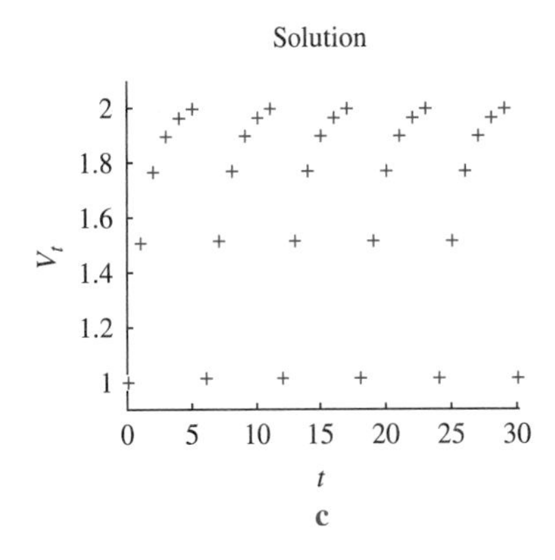

FIGURE 1.11.10
The Wenckebach phenomenon

The value $c = 1/3$ that produced an equilibrium and normal beating (Figure 1.11.8) is well below this value. The value $c = 2/3$ that produced 2:1 AV block (Figure 1.11.9) is well above this value. What happens if c is only slightly above 0.5 and the heart can nearly recover?

Figure 1.11.10 shows the behavior of the system when $c = 0.5001$, a hair above the threshold for existence of an equilibrium. The heart beats 12 times, building up to a higher and higher potential. Eventually, the potential becomes too high, the AV node cannot recover, and the heart fails to beat. After this rest, the potential drops, and the process begins again. This is the Wenckebach phenomenon.

Actual measurements of the Wenckebach phenomenon correspond in part to this model, but show that the heart beats a bit more slowly before missing a beat. Why might this be the case? Our model assumes that the SA node sends out precise pulses at precise times. If the signals from the SA node take a little while to build up, an AV node at low potential will respond right at the beginning of a signal from the SA node. An AV node close to the threshold will be slower and might respond near the end of the signal from the SA node, delaying the heartbeat slightly. This slowing indicates that the AV node will soon exceed the threshold and that the heart will miss a beat.

Summary A simplified model of the heart includes two phases: decay of potential in the AV node (recovery from the last beat) and response to a rhythmic signal from the SA node. We derived conditions for the heart to beat properly with each signal and showed that if the recovery time is not long enough, two types of **second-degree block** can result. In the first, **2:1 AV block**, the heart beats with every other signal. In the second, the **Wenckebach phenomenon**, the heart misses a beat only occasionally.

1.11 Exercises

Applications

1–4 ▪ In the following circumstances, compute $\hat{V}_t$ and V_{t+1} and state whether the heart will beat.

1. $V_c = 20.0$ mV, $u = 10.0$ mV, $c = 0.5$, $V_t = 30.0$ mV.
2. $V_c = 20.0$ mV, $u = 10.0$ mV, $c = 0.6$, $V_t = 30.0$ mV.
3. $V_c = 20.0$ mV, $u = 10.0$ mV, $c = 0.7$, $V_t = 30.0$ mV.
4. $V_c = 20.0$ mV, $u = 10.0$ mV, $c = 0.8$, $V_t = 30.0$ mV.

5–8 ▪ Describe the long-term dynamics in each of the given cases. Find which ones will beat every time, which display 2:1 AV block, and which show some sort of Wenckebach phenomenon.

5. The case in Exercise 1.
6. The case in Exercise 2.
7. The case in Exercise 3.
8. The case in Exercise 4.

9–12 ▪ Use the parameter values in Exercise 1 (except for the values of c), and state whether the heart would beat every time with the given values of α and τ.

9. $\alpha = 1.0, \tau = 1.0$.

10. $\alpha = 1.0, \tau = 0.5$.

11. $\alpha = 2.0, \tau = 0.5$.

12. $\alpha = 0.5, \tau = 0.5$.

13–14 ▪ Consider the following continuous system that approximates the discontinuous model studied in this chapter.

$$V_{t+1} = cV_t + u(V_t)$$

where

$$u(V_t) = \frac{2(1-c)}{1+V_t^n}$$

for the following values of n. Find the equilibria and their stability as a function of c, and describe the dynamics.

13. Suppose $n = 2$. Show that $V_t = 1$ is an equilibrium. Sketch a graph and cobweb with $c = 1/4$. Does the equilibrium seem to be stable?

14. Suppose $n = 4$. Show that $V_t = 1$ is an equilibrium. Sketch a graph and cobweb with $c = 1/4$. Does the equilibrium seem to be stable?

15–18 ▪ Population models with thresholds can also have unusual behavior. Evaluate the following models where individuals emigrate when the population is overly crowded. In particular, suppose h individuals leave if the population is larger than some critical value N_c

$$N_{t+1} = \begin{cases} rN_t - h & \text{if } N_t > N_c \\ rN_t & \text{if } N_t \le N_c. \end{cases}$$

15. Suppose $h = 1000$ and $N_c = 1000$ and $r = 1.5$. Investigate some solutions starting with different values of $N_0 < 1000$. What is happening?

16. Find the equilibrium when $h = 1000$ and $N_c = 1000$ and $r = 1.5$. What would happen to solutions starting with values greater than the equilibrium? Use this information, and that in the previous problem, to sketch a cobweb diagram.

17. Redo Exercise 15 with $r = 1.65$. How do the results differ from those in Exercise 15?

18. Find the equilibrium when $h = 1000$ and $N_c = 1000$ and $r = 1.65$. Can you explain why solutions that start below the equilibrium can shoot off to infinity?

Computer Exercises

19. Study the dynamics of Exercises 1–4 for values of c ranging from 0.4 up to 1.0. Are there any cases where the behavior is neither 2:1 AV block nor the Wenckebach phenomenon? How would you describe these behaviors.

20. What happens to the dynamics of the example illustrated in Figure 1.11.10 if c is made even closer to 0.5? What does it look like on a cobwebbing diagram? If $c = 0.5000000000001$, do you think it would be possible to distinguish the Wenckebach phenomenon from normal beating? Is it?

Supplementary Problems

1. Suppose you have a culture of bacteria, where the density of each bacterium is 2.0 g/cm^3.

a. If each bacterium is $5\mu m \times 5\mu m \times 20\mu m$ in size, find the number of bacteria if their total mass is 30 grams. Recall that $1\mu m = 10^{-6}$ meters.

b. Suppose that you learn that the sizes of bacteria range from $4\mu m \times 5\mu m \times 15\mu m$ to $5\mu m \times 6\mu m \times 25\mu m$. What is the range of the possible number of bacteria making up the total mass of 30 grams?

2. Suppose the number of bacteria in a culture is a linear function of time.

a. If there are 2.0×10^8 bacteria in your lab at 5 P.M. on Tuesday, and 5.0×10^8 bacteria the next morning at 9 A.M., find the equation of the line describing the number of bacteria in your culture as a function of time.

b. At what time will your culture have 1.1×10^9 bacteria?

c. The lab across the hall also has a bacterial culture where the number of bacteria is a linear function of time. If they have 2.0×10^8 bacteria at 5 P.M. on Tuesday, and 3.4×10^8 bacteria the next morning at 9 A.M., when will your culture have twice as many bacteria as theirs?

3. Consider the functions $f(x) = e^{-2x}$ and $g(x) = x^3 + 1$.

a. Find the inverses of f and g, and use these to find when $f(x) = 2$ and when $g(x) = 2$.

b. Find $f \circ g$ and $g \circ f$ and evaluate each at $x = 2$.

c. Find the inverse of $g \circ f$. What is the domain of this function?

4. A lab has a culture of a new kind of bacteria where each individual takes 2 hours to split into three bacteria. Suppose that these bacteria never die and that all offspring are OK.

a. Write an updating function describing this system.

b. Suppose there are 2.0×10^7 bacteria at 9 A.M. How many will there be at 5 P.M.?

c. Write an equation for how many bacteria there are as a function of how long the culture has been running.

d. When will this population reach 10^9?

5. The number of bacteria (in millions) in a lab are as follows

Time, t (h)	Number, b_t
0.0	1.5
1.0	3.0
2.0	4.5
3.0	5.0
4.0	7.5
5.0	9.0

a. Graph these points.

b. Find the line connecting them and the time t at which the value does not lie on the line.

c. Find the equation of the line and use it to find what the value at t would have to be to lie on the line.

d. How many bacteria would you expect at time 7.0 hours?

6. The number of bacteria in another lab follows the discrete-time dynamical system

$$b_{t+1} = \begin{cases} 2.0b_t & b_t \leq 1.0 \\ -0.5(b_t - 1.0) + 2.0 & b_t > 1.0 \end{cases}$$

where t is measured in hours and b_t in millions of bacteria.

a. Graph the updating function. For what values of b_t does it make sense?

b. Find the equilibrium.

c. Cobweb starting from $b_0 = 0.4$ million bacteria. What do you think happens to this population?

7. Convert the following angles from degrees to radians and find the sine and cosine of each. Plot the related point both on a circle and on a graph of the sine or cosine.

a. $\theta = 60°$.

b. $\theta = -60°$.

c. $\theta = 110°$.

d. $\theta = -190°$.

e. $\theta = 1160°$.

8. Suppose the temperature H of a bird follows the equation

$$H = 38.0 + 3.0 \cos\left(\frac{2\pi(t - 0.4)}{1.2}\right)$$

where t is measured in days and H is measured in degrees C.

a. Sketch a graph of the temperature of this bird.

b. Write the equation if the period changes to 1.1 days. Sketch a graph.

c. Write the equation if the amplitude increases to 3.5 degrees. Sketch a graph.

d. Write the equation if the average decreases to 37.5 degrees. Sketch a graph.

9. The butterflies on a particular island are not doing too well. Each autumn, every butterfly produces on average 1.2 eggs and then dies. Half of these eggs survive the winter and produce new butterflies by late summer. At this time, 1000 butterflies arrive from the mainland to escape overcrowding.

a. Write a discrete-time dynamical system for the population on this island.

b. Graph the updating function and cobweb starting from 1000.

c. Find the equilibrium number of butterflies.

10. A culture of bacteria has mass 3.0×10^{-3} grams and consists of spherical cells of mass 2.0×10^{-10} grams and density 1.5 grams/cm^3.

a. How many bacteria are in the culture?

b. What is the radius of each bacterium?

c. If the bacteria were mashed into mush, how much volume would they take up?

11. A person develops a small liver tumor. It grows according to

$$S(t) = S(0)e^{\alpha t}$$

where $S(0) = 1.0$ gram and $\alpha = 0.1$/day. At time $t = 30$ days, the tumor is detected and treatment begins. The size of the tumor then decreases linearly with slope of -0.4 grams/day.

a. Write the equation for tumor size at $t = 30$.

b. Sketch a graph of the size of the tumor over time.

c. When will the tumor disappear completely?

12. Two similar objects are left to cool for one hour. One starts at 80°C and cools to 70°C and the other starts at 60°C and cools to 55°C. Suppose the discrete-time dynamical system for cooling objects is linear.

a. Find the discrete-time dynamical system. Find the temperature of the first object after 2 hours. Find the temperature after 1 hour of an object starting at 20°C.

b. Graph the updating function and cobweb starting from 80°C.

c. Find the equilibrium. Explain in words what the equilibrium means.

13. A culture of bacteria increases in area by 10% each hour. Suppose the area is 2.0 cm^2 at 2:00 P.M.

a. What will the area be at 5:00 P.M.?

b. Write the relevant discrete-time dynamical system and cobweb starting from 2.0.

c. What was the area at 1:00 P.M.?

d. If all bacteria are the same size and each adult produces two offspring each hour, what fraction of offspring must survive?

e. If the culture medium is only 10 cm^2 in size, when will it be full?

14. Candidates Dewey and Howe are competing for fickle voters. A total of 100,000 people are registered to vote in the election, and each will vote for one of these two candidates. Voters often switch their allegiance. 20% of Dewey's supporters switch to Howe. Howe's supporters are more likely to switch when Dewey is doing well: the fraction switching from Howe to Dewey is proportional to Dewey's percentage of the vote—none switch if Dewey commands 0% of the vote, and 50% switch if Dewey commands 100% of the vote. Suppose Howe starts with 90% of the vote.

a. Find the number of votes Dewey and Howe have after a week.

b. Find Dewey's percentage after a week.

c. Find the discrete-time dynamical system describing Dewey's percentage.

d. Graph the updating function and find the equilibrium or equilibria.

e. Who will win the election?

15. A certain bacterial population has the following odd behavior. If the population is less than 1.5×10^8 in a given generation, each bacterium produces two offspring. If the population is greater than or equal to 1.5×10^8 in a given generation, it will be exactly 1.0×10^8 in the next.

a. Cobweb starting from an initial population of 10^7.

b. Graph a solution starting from an initial population of 10^7.

c. Find the equilibrium or equilibria of this population.

16. An organism is breathing a chemical that modifies the depth of its breaths. In particular, suppose that the fraction q of air exchanged is given by

$$q = \frac{c_t}{c_t + \gamma}$$

where γ is the ambient concentration and c_t is the concentration in the lung. After a breath, a fraction q of the air came from outside, and a fraction $1 - q$ remained from inside. Suppose $\gamma = 0.5$ mol/L.

a. Describe in words the breathing of this organism.

b. Find the discrete-time dynamical system for the concentration in the lung.

c. Find the equilibrium or equilibria.

17. Lint is building up in a dryer. With each use, the old amount of lint x_t is divided by $1 + x_t$ and 0.5 lintons (the units of lint) are added.

a. Find the discrete-time dynamical system and graph the updating function.

b. Cobweb starting from $x_0 = 0$. Graph the associated solution.

c. Find the equilibrium or equilibria.

18. Suppose people in a bank are waiting in two separate lines. Each minute several things happen: some people are served, some people join the lines, and some people switch lines. In particular, suppose that 1/10 of the people in the first line are served, and 3/10 of the people in the second line are served. Suppose the number of people who join each line is equal to 1/10 of the total number of people in both lines and that 1/10 of the people in each line switch to the other.

a. Suppose there are 100 people in each line at the beginning of a minute. Find how many people are in each line at the end of the minute.

b. Write a discrete-time dynamical system for the number of people in the first line, and another discrete-time dynamical system for the number of people in the second.

c. Write a discrete-time dynamical system for the fraction of people in the first line.

19. A gambler faces off against a small casino. She begins with \$1000, and the casino with \$11,000. In each round, the gambler loses 10% of her current funds to the casino, and the casino loses 2% of its current funds to the gambler.

a. Find the amount of money each has after one round.

b. Find a discrete-time dynamical system for the amount of money the gambler has and another for the amount of money the casino has.

c. Find the discrete-time dynamical system for the fraction p of money the gambler has.

d. Find the equilibrium fraction of the money held by the gambler.

e. Using the fact that the total amount of money is constant, find the equilibrium amount of money held by the gambler.

20. Let V represent the volume of a lung and c the concentration of some chemical inside. Suppose the internal surface area is proportional to volume and that a lung with volume 400 cm^3 has a surface area of 100 cm^2. The lung absorbs the chemical at a rate per unit surface area of

$$R = \alpha \frac{c}{4.0 \times 10^{-2} + c}.$$

Time is measured in seconds, surface area in cm^2, and volume in cm^3. The parameter α takes on the value 6.0 in the appropriate units.

a. Find surface area as a function of volume. Make sure your dimensions make sense.

b. What are the units of R? What must be the units of α?

c. Suppose that $c = 1.0 \times 10^{-2}$ and $V = 400.0$ cm^3. Find the total amount of chemical absorbed.

d. Suppose that $c = 1.0 \times 10^{-2}$. Find the total chemical absorbed as a function of V.

21. Suppose a person's head diameter D and height H grow according to

$$D(t) = 10.0e^{0.03t}$$

$$H(t) = 50.0e^{0.09t}$$

during the first 15 years of life.

a. Find D and H at $t = 0$, $t = 7.5$, and $t = 15$.

b. Sketch graphs of these two measurements as functions of time.

c. Sketch semilog graphs of these two measurements as functions of time.

d. Find the doubling time of each measurement.

22. On another planet, people have three hands and like to compute tripling times instead of doubling times.

a. Suppose a population follows the equation $b(t) = 3.0 \times 10^3 e^{0.333t}$ where t is measured in hours. Find the tripling time.

b. Suppose a population has a tripling time of 33 hours. Find the equation for population size $b(t)$ if $b(0) = 3.0 \times 10^3$.

23. A Texas millionaire (with $1,000,001 in assets in 2010) got rich by clever investments. She managed to earn 10% interest per year for the last 20 years and plans to do the same in the future.

a. How much did she have in 1990?

b. When will she have $5,000,001?

c. Write the discrete-time dynamical system and graph the updating function.

d. Write and graph the solution.

24. A major university hires a famous Texas millionaire to manage its endowment. The millionaire decides to follow this plan each year:

- Spend 25% of all funds above $100 million on university operations.
- Invest the remainder at 10% interest.
- Collect $50 million in donations from wealthy alumni.

a. Suppose the endowment has $340 million to start. How much will it have after spending on university operations? After collecting interest on the remainder? After the donations roll in?

b. Find the discrete-time dynamical system.

c. Graph the updating function and cobweb starting from $340 million.

25. Another major university hires a different famous Texas millionaire to manage its endowment. This millionaire starts with $340 million, brings back $355 million the next year, and claims to be able to guarantee a linear increase in funds thereafter.

a. How much money will this university have after 8 years?

b. Graph the endowment as a function of time.

c. Write the discrete-time dynamical system, graph, and cobweb starting from $340 million.

d. Which university do you think will do better in the long run? Which Texan would you hire?

26. A heart receives a signal to beat every second. If the voltage when the signal arrives is below 50 microvolts, the heart beats and increases its voltage by 30 microvolts. If the voltage when the signal arrives is greater than 50 microvolts, the heart does not beat and the voltage does not change. The voltage of the heart decreases by 25% between beats in either case.

a. Suppose the voltage of the heart is 40 microvolts right after one signal arrives. What is the voltage before the next signal, and will the heart beat?

b. Graph the updating function for the voltage of this heart.

c. Will this heart exhibit normal beating or some sort of AV block?

27. Suppose vehicles are moving at 72 kph (kilometers per hour). Each car carries an average of 1.5 people, and all are carefully keeping a 2-second following distance (getting no closer than the distance a car travels in 2 seconds) on a three-lane highway.

a. How far between vehicles?

b. How many vehicles per kilometer?

c. How many people will pass a given point in an hour?

d. If commuter number oscillates between this maximum (at 8:00 A.M.) and a minimum that is one third as large (at 8:00 P.M.) on a 24-hour cycle, give a formula for the number of people passing the given point as a function of time of day.

28. Suppose traffic volume on a particular road has been as follows:

Year	Vehicles
1970	40,000
1980	60,000
1990	90,000
2000	135,000

a. Sketch a graph of traffic over time.

b. Find the discrete-time dynamical system that describes this traffic.

c. What was the traffic volume in 1960?

d. Give a formula for the predicted traffic in the year 2050.

e. Find the half-life or doubling-time of traffic.

29. In order to improve both the economy and quality of life, policies are designed to encourage growth and decrease traffic flow. In particular, the number of vehicles is encouraged to increase by a factor of 1.6 over each 10-year period, but the commuters from 10,000 vehicles are to choose to ride comfortable new trains instead of driving.

a. If there were 40,000 vehicles in 1970, how many would there be in 1980?

b. Find the discrete-time dynamical system describing this traffic.

c. Find the equilibrium.

d. Graph the updating function and cobweb starting from an initial number of 40,000.

e. In the long run, will there be more or less traffic with this policy than with the policy that led to the data in the previous problem? Why?

Projects

1. Combine the model of selection from Section 1.10 with the models of mutation and reversion. Assume that wild type have per capita reproduction r, mutants have per capita reproduction s, a fraction μ of the offspring of the wild type mutate into the mutant type, and a fraction of ν of the offspring of the mutant type revert. First set $b_t = 4.0 \times 10^6$, $m_t = 2.0 \times 10^5$, $\mu = 0.2$, $\nu = 0.1$, $r = 1.5$, and $s = 2.0$.

a. How many of the wild-type individuals will there be after reproduction and before mutation? How many of these will mutate? How many will remain the wild type?

b. How many mutant individuals will there be after reproduction and before mutation? How many of these will revert? How many will not?

c. Find the total number of wild type after reproduction, mutation, and reversion.

d. Find the total number of mutants after reproduction, mutation, and reversion.

e. Find the total number of bacteria after reproduction, mutation, and reversion. Why is it different from the initial number?

f. Find the fraction of mutants to begin with, as well as the fraction after reproduction, mutation, and reversion.

Now, treat b_t and m_t as variables. The following steps will help you find the updating function.

a. Use the above steps to find m_{t+1} in terms of b_t and m_t.

b. Find the total number of bacteria, $b_{t+1} + m_{t+1}$ in terms of b_t and m_t.

c. Divide your equation for m_{t+1} by $b_{t+1} + m_{t+1}$ to find an expression for p_{t+1} in terms of b_t and m_t.

d. Divide the numerator and denominator by $m_t + b_t$ as in the derivation of equation 1.10.5 and write the updating function in terms of p_t.

e. Find the equilibrium of this updating function.

Finally, we can do this in general by treating r, s, μ, and ν as parameters. Do exactly the above steps to find the updating function. After doing so, study the following special cases. In each case, explain your answer.

a. $r = s$ (no selection) and mutation in only one direction ($\mu = 0.0$ and $\nu > 0$). Find the updating function and the equilibrium. What does this mean?

b. $r = s$ and $\mu = \nu$. Find the updating function and the equilibrium. What does this mean?

c. $r > s$, $\nu = 0$, and $\mu > 0$. This means that the wild type has a reproductive advantage but keeps mutating. Find the updating function and the equilibrium. The result is called mutation-selection balance. Can you guess why?

2. Suppose a measurement follows

$$f(t) = \cos\left(\frac{2\pi t}{r}\right)$$

where the period r is an unknown number. You guess that the period is 1.0, and you figure you can get a good fix on the behavior of the measurement by checking every 1.0 time steps (once per period). Try the following for these values of r: 1.0, 0.6, 0.601, 0.602, 0.603, and 0.618. Experimenting with other values of r is highly recommended.

a. Graph the actual data as a continuous function of time $0 \le t \le 100$.

b. Graph what you find by plotting data points every 1.0 time unit.

c. Would you be able to figure out what was going on?

Chapter 2

Limits and Derivatives

Discrete-time dynamical systems describe biological change when measurements are made at discrete intervals. We now develop methods to describe a value that changes **continuously**. This description requires the two central ideas of differential calculus, the **limit** and the **instantaneous rate of change** or **derivative**. We will find a **geometric** interpretation of the derivative as the **slope of the tangent line**, which will help us to graph and analyze complicated functions.

Achieving these goals requires both understanding the idea of the derivative and having the tools to compute the derivatives of a wide range of functions. Complicated functions are built by combining simple pieces, and their derivatives are computed with a set of rules for combining the derivatives of sums, powers, products, quotients, and compositions. Starting with the derivatives of linear, exponential, logarithmic, and trigonometric functions, we will be able to differentiate pretty much any function we can write down. The ability to compute the derivative, combined with its dual interpretation as the instantaneous rate of change and as the slope, opens up a dazzling array of applications.

2.1 Introduction to Derivatives

Discrete-time dynamical systems are a powerful tool for describing the dynamics of biological systems when change can be accurately described by measurements made at **discrete times**. To fully understand other systems, however, measurements must be made at all times, or **continuously**. We have only to think of the growth of a plant or the motion of an animal to realize that some change is best described by a continuous set of measurements.

In this chapter, we will develop the tools needed to describe measurements that change continuously. We will switch from thinking about what **happens** to what is **happening** to a measurement. The central idea is the **instantaneous rate of change**, or the **derivative**. Graphically, the derivative is equal to the slope of the **tangent line** to a curve.

These two ideas of the derivative, as the instantaneous rate of change of a measurement and as the slope of a curve, are the keys to appreciating the many applications of **calculus**. Understanding that the derivative must be computed as a **limit** is the key to using calculus correctly.

2.1.1 The Average Rate of Change

Suppose we measure a bacterial population continuously and find that the population size $b(t)$ follows the equation

$$b(t) = 2.0^t$$

(Figure 2.1.1). To describe the growth of this population, we first define the concept of **rate**, the value of one measurement divided by time. Rates thus describe how a

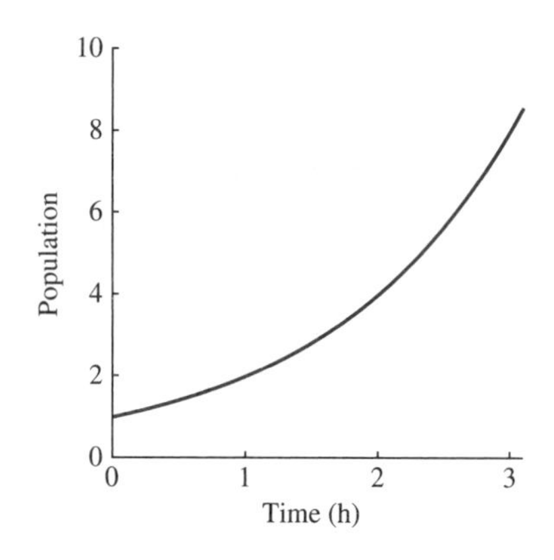

FIGURE 2.1.1

A bacterial population measured continuously

measurement changes over time, and provide the key tool for modeling growth or change in continuous time.

If we checked the population size every hour, we would find the following data (plotted in Figure 2.1.2).

t (h)	$b(t)$ (million)	Change, $\Delta b = b(t) - b(t-1)$
0	1.0	–
1	2.0	1.0
2	4.0	2.0
3	8.0	4.0
4	10.0	8.0
5	16.0	16.0

We can describe these values with the discrete-time dynamical system

$$b_{t+1} = 2.0 b_t.$$

To begin working toward a description of how the population size is **changing** at any given time, we have added a column Δb, denoting the change in the population size between times $t-1$ and t. Recall that Δ (the Greek letter "Delta") means "change in" (Subsection 1.4.1). For example, the bacterial population size increased by 2.0 million between $t = 1.0$ and $t = 2.0$. The change in population size is not defined at $t = 0$ because there was no previous measurement.

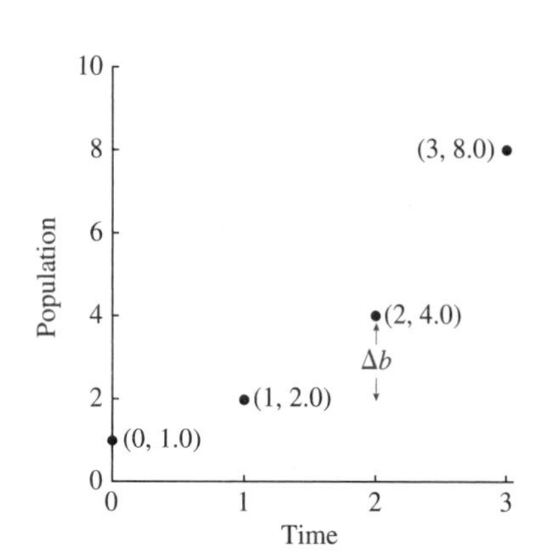

FIGURE 2.1.2

Hourly measurement of a bacterial population

For more accuracy, we could check the population every half hour (Figure 2.1.3). We have added another column to the table, indicating the **average rate of change** of the population, defined as

$$\text{average rate of change} = \frac{\text{change in population}}{\text{change in time}} = \frac{\Delta b}{\Delta t}.$$

The average rate of change is an indication of how fast the population size is changing.

t (h)	$b(t)$ (million)	Change $\Delta b = b(t) - b(t-0.5)$	Average Rate of Change $\frac{\Delta b}{\Delta t} = \frac{b(t) - b(t-0.5)}{0.5}$
0.0000	1.0000	–	–
0.5000	1.4142	0.4142	0.8284
1.0000	2.0000	0.5858	1.1716
1.5000	2.8284	0.8284	1.6568
2.0000	4.0000	1.1716	2.3431
2.5000	5.6569	1.6569	3.3137
3.0000	8.0000	2.3431	4.6863

Example 2.1.1 Computing the Average Rate of Change

Between $t = 1.0$ and $t = 1.5$, the population grew by 0.8284 million. Because this change took only $\Delta t = 0.5$ hour,

$$\text{average rate of change} = \frac{\Delta b}{\Delta t} = \frac{0.8284}{0.5} = 1.6568$$

in units of million bacteria per hour.

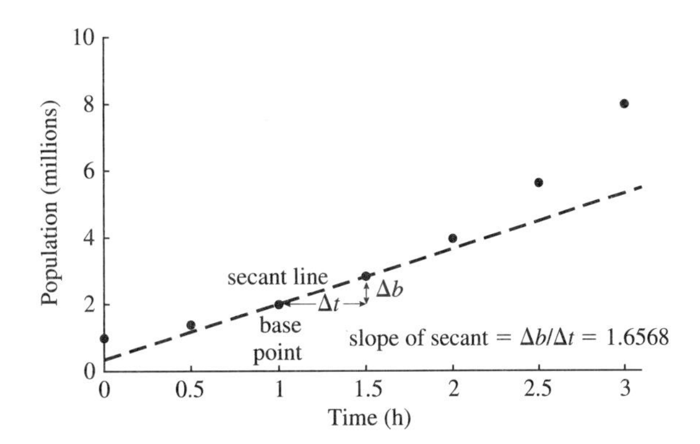

FIGURE 2.1.3
Half-hourly measurement of a bacterial population

There is an important graphical interpretation of the average rate of change, as the slope of the line connecting the two points on a graph.

Definition 2.1 **Secant Line**

A line connecting two points on the graph of a function is called a **secant line**.

We will often single out one of the two points as a **base point** to focus on for more detailed study.

Example 2.1.2 The Slope of a Secant Line

Suppose we choose $(1, 2.0)$ as the base point. The average rate of change between $t = 1.0$ and $t = 1.5$, found to be 1.6568 in Example 2.1.1, is equal to the slope of the secant line connecting the data points at these two times (Figure 2.1.3).

The **equation** of a secant line can be found using the point-slope form for a line (Definition 1.8). Suppose we are given a base point $[x_0, f(x_0)]$ on the graph of a function $f(x)$. If the second point is $[x_1, f(x_1)]$, the slope is

$$m = \frac{\Delta f}{\Delta x} = \frac{f(x_1) - f(x_0)}{x_1 - x_0}.$$

The equation of the secant line, which we will represent by the function $f_s(x)$ is

$$f_s(x) = f(x_0) + m(x - x_0),$$

where we used $[x_0, f(x_0)]$ as the base point and m as the slope (Figure 2.1.4).

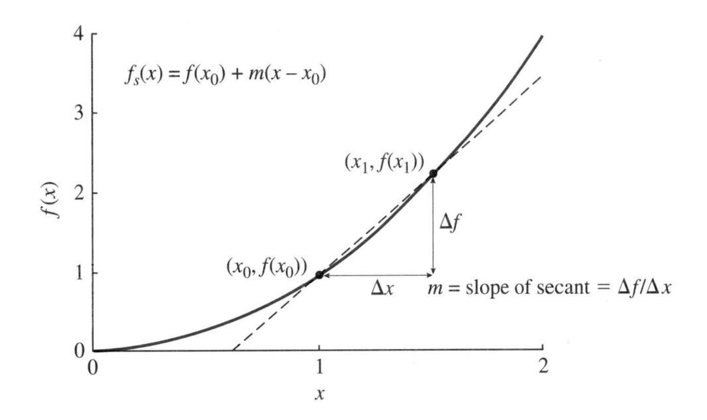

FIGURE 2.1.4
The secant line, its slope, and its equation

Example 2.1.3 The Equation of a Secant Line

In Example 2.1.2, we computed the slope of the line connecting the data points at $t = 1.0$ and $t = 1.5$, to be 1.6568. To use point-slope form, we identify the base point

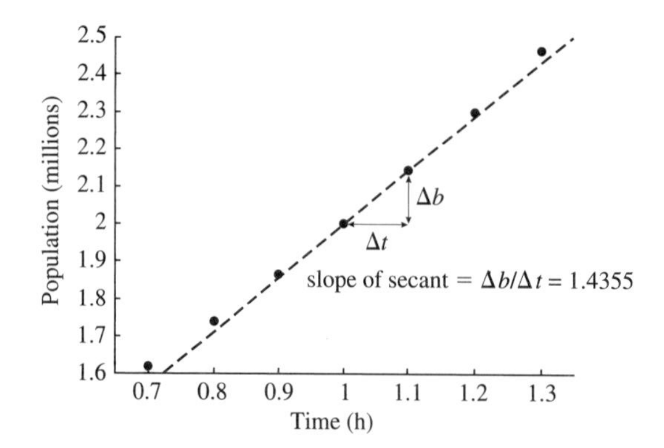

FIGURE 2.1.5
A different secant line to the same data

(x_0, y_0) as $[1.0, b(1.0)] = (1.0, 2.0)$. The equation of the secant line is then

$$b_s(t) = b(1.0) + 1.6568(t - 1.0) = 2.0 + 1.6568(t - 1.0).$$

This could be converted to slope-intercept form, but this is usually not necessary.

Example 2.1.4 A Secant Line with a Different Base Point Has a Different Equation

To find the secant line connecting the data points at $t = 1.5$ and $t = 2.0$, we first find the slope (also computed in Table 2.1.1), as the average rate of change, or

$$\text{slope} = \frac{b(2.0) - b(1.5)}{2.0 - 1.5} = \frac{4.0 - 2.8284}{0.5} = 2.3431.$$

The base point (x_0, y_0) is $[2.0, b(2.0)] = (2.0, 4.0)$. The equation of the secant line is then

$$b_s(t) = b(2.0) + 2.3431(t - 2.0) = 4.0 + 2.3431(t - 2.0).$$

This secant line has a different slope and base point than the secant line (Example 2.1.3) because we are connecting different points on the graph.

To find an accurate estimate of the rate of change right at $t = 1.0$, we should look only at the change near that point (Figure 2.1.5). The following table shows the results when measurements are taken every 0.1 hour.

t	$b(t)$	Change $\Delta b = b(t) - b(t - 0.1)$	Average Rate of Change $\frac{\Delta b}{\Delta t}$
0.7000	1.6245	0.1088	1.0879
0.8000	1.7411	0.1166	1.1660
0.9000	1.8661	0.1250	1.2496
1.0000	2.0000	0.1339	1.3393
1.1000	2.1435	0.1435	1.4355
1.2000	2.2974	0.1538	1.5385
1.3000	2.4623	0.1649	1.6489

The measurements over this shorter interval of time appear to nearly follow a line (Figure 2.1.6). This is the key observation in calculus: over short intervals, curves look like lines. The art of calculus is finding the **slope** of the line that most closely matches the curve.

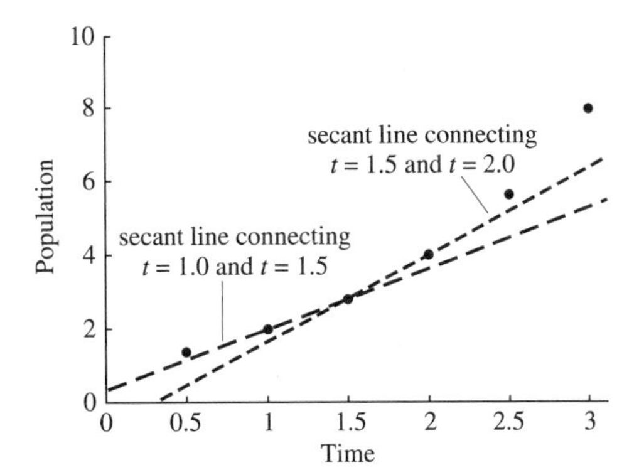

FIGURE 2.1.6
Measurement of a bacterial population every 0.1 hour

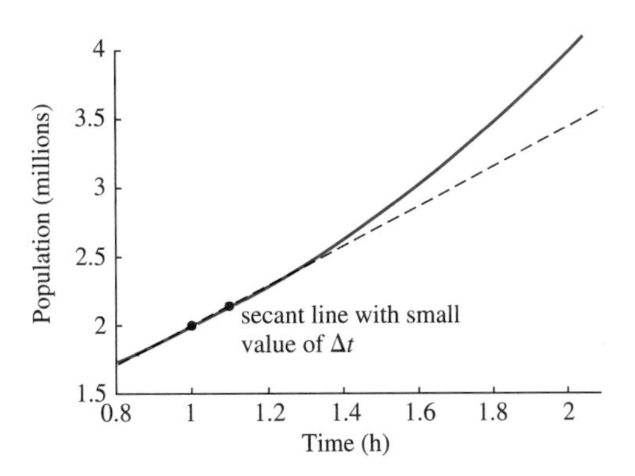

FIGURE 2.1.7
A secant line with a tiny value of Δt

2.1.2 Instantaneous Rates of Change

As the time Δt between measurements becomes smaller, the average rate of change between times $t = 1.0$ and $t = 1.0 + \Delta t$ becomes a better and better description of what is happening exactly at $t = 1.0$. Average rates of change over smaller and smaller intervals are computed in the following table.

Δt	$1.0 + \Delta t$	$b(1.0 + \Delta t)$	Δb	$\frac{\Delta b}{\Delta t}$
1.0	2.0	4.0000	2.0000	2.0000
0.5	1.5	2.8284	0.8284	1.6568
0.1	1.1	2.1435	0.1435	1.4354
0.01	1.01	2.0139	0.0139	1.3911
0.001	1.001	2.00139	0.00139	1.3868
0.0001	1.0001	2.000139	0.000139	1.3863

As Δt becomes smaller, there is less and less time for anything to happen, and the **change** in population size Δb also becomes small (Figure 2.1.7). However, the average **rate** of change does not become tiny because the change takes place over shorter and shorter time intervals. In fact, the average rate of change approaches a value of about 1.386. We would like to define this value as the **instantaneous rate of change**, the rate of change **exactly** at $t = 1.0$.

What do we mean by instantaneous rate of change? The most familiar instantaneous rate of change is speed, the rate of change of position, as measured by a speedometer. But how does a speedometer **measure** this rate? In fact, most speedometers directly measure the speed by converting rotation of the drive shaft into a magnetic field. However, for an organism without wheels, speed must be estimated by measuring position at one time and again at a later time, and then dividing the distance moved by the time elapsed as

$$\text{average rate of change} = \frac{\text{change in position}}{\text{change in time}}.$$

Again, this value only tells what happened on **average** during the interval.

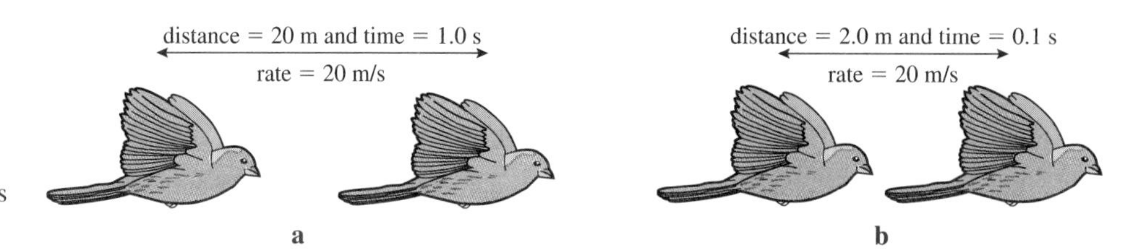

FIGURE 2.1.8
A bird moving at constant speed travels a shorter distance in a shorter time

Suppose the bird is flying at a constant speed of 20.0 m/s. It moves 20.0 m in 1.0 s, for an average rate of change of 20.0 m/s (Figure 2.1.8a). It moves only 2.0 m in 0.1 s, again with an average rate of change of 20.0 m/s (Figure 2.1.8b). The distance moved becomes smaller as the time between checks becomes smaller. The average rate of change of position remains the same. If we continued to make more and more accurate measurements, assessing the position of the bird every Δt seconds, we would find the following.

Δt	Distance Moved	Average Rate of Change
1.0	20.0	20.0
0.1	2.0	20.0
0.01	0.2	20.0
0.001	0.02	20.0

The average rate of change is always 20.0 because the bird is moving at a constant speed. This constant speed must be the instantaneous rate of change.

With the bacterial population, it is not as easy to guess the instantaneous rate of change. Ideally, we would compute the instantaneous rate of change by picking $\Delta t = 0$, the smallest possible value, resulting in

Δt	$1.0 + \Delta t$	$b(1.0 + \Delta t)$	Δb	$\frac{\Delta b}{\Delta t}$
0.0	1.0	2.0	0.0	$\frac{0.0}{0.0}$ = undefined!

Dividing by 0 is a mathematical misdemeanor. Dividing 0 by 0 is a mathematical felony. Not only have we failed to find the answer, but we have violated a major mathematical law.

Gradually choosing a smaller and smaller value of Δt corresponds to picking the second point on the curve closer and closer to the base point (Figure 2.1.9). With $\Delta t = 0$, the second point lies right on top of the base point. It is impossible to draw a unique line through a single point. In fact, there are an infinite number of possible lines through this point (Figure 2.1.10). Only one of these lines is close to the secant lines in Figure 2.1.9, the **tangent line**. This line does touch the curve in only a single point, but does so by just kissing the side of it rather than rudely crossing straight through. In fact, if we zoom in on the tangent line, it looks more and more similar to the curve (Figure 2.1.11).

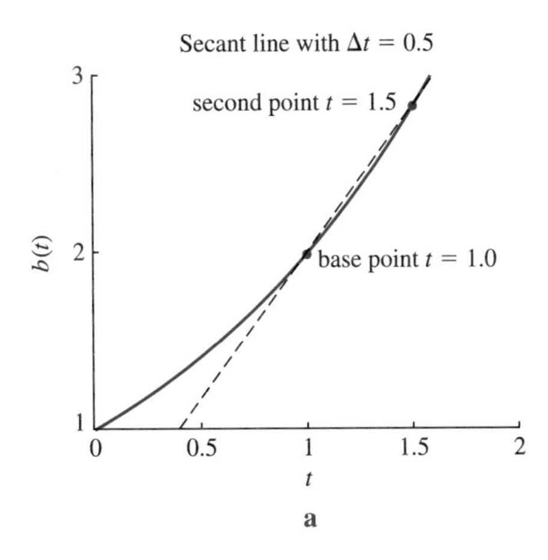

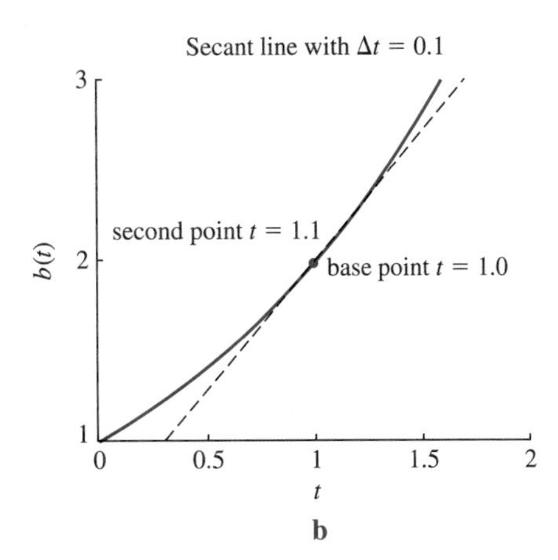

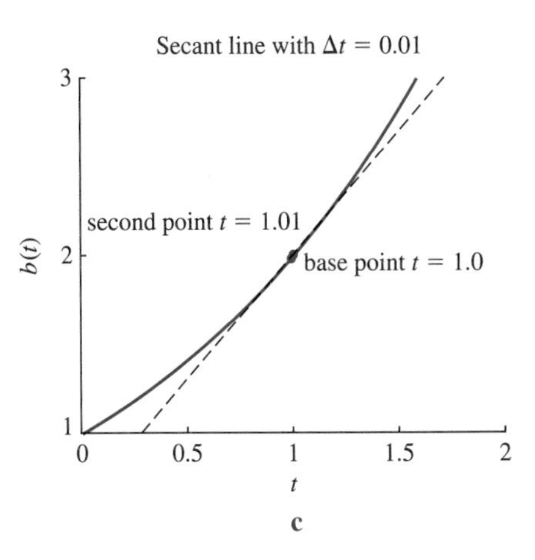

FIGURE 2.1.9
Secant lines with successively closer points

Example 2.1.5 Finding a Tangent Line

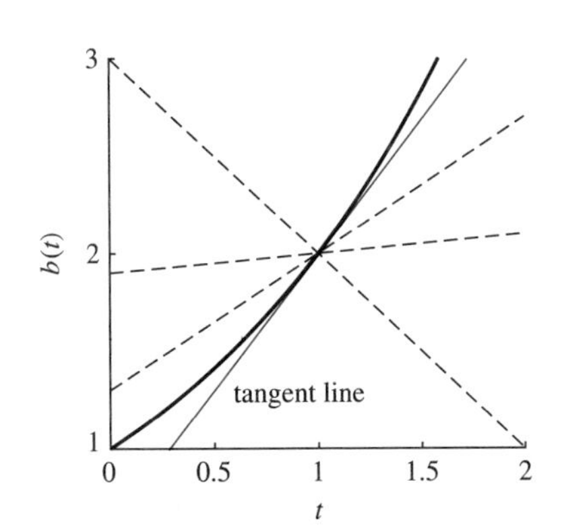

FIGURE 2.1.10
The many lines through a single point

Suppose we wish to find the slope of the tangent line to the function $f(x) = \ln(x)$ at $x = 2$ (Figure 2.1.12). We can evaluate the slope of the tangent as a function of Δx for both positive and negative values of Δx.

Δx	$2.0 + \Delta x$	$f(2.0 + \Delta x)$	Δf	$\frac{\Delta f}{\Delta x}$
1.00000	3.00000	1.09861	0.40547	0.40547
0.50000	2.50000	0.91629	0.22314	0.44629
0.10000	2.10000	0.74194	0.04879	0.48790
0.01000	2.01000	0.69813	0.00499	0.49875
0.00100	2.00100	0.69365	0.00050	0.49988
−1.00000	1.00000	0.00000	−0.69315	0.69315
−0.50000	1.50000	0.40547	−0.28768	0.57536
−0.10000	1.90000	0.64185	−0.05129	0.51293
−0.01000	1.99000	0.68813	−0.00501	0.50125
−0.00100	1.99900	0.69265	−0.00050	0.50013

As Δx becomes small, the slope of the secant gets close to 0.5.

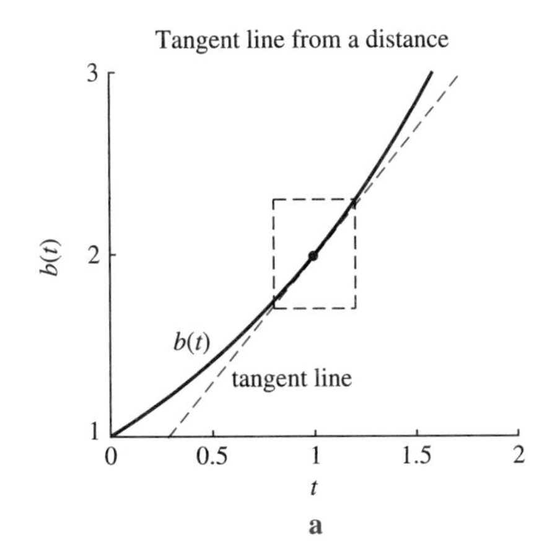

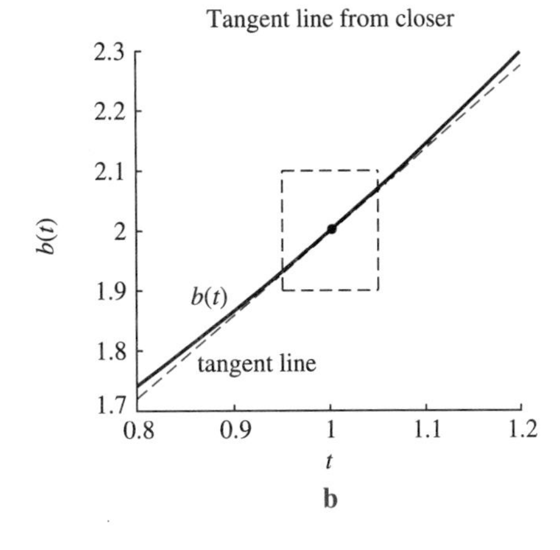

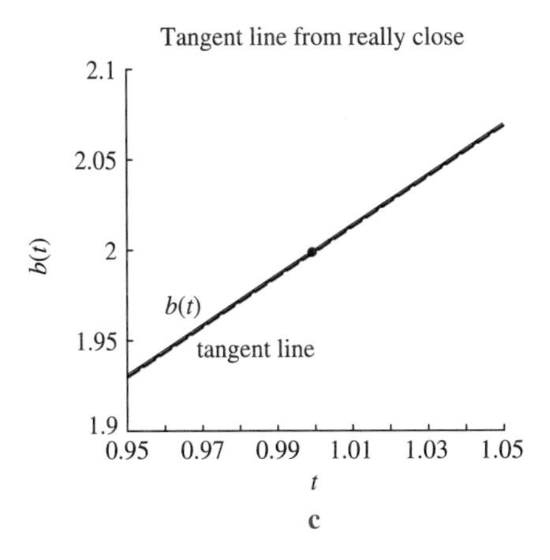

FIGURE 2.1.11
Zooming in on the tangent line

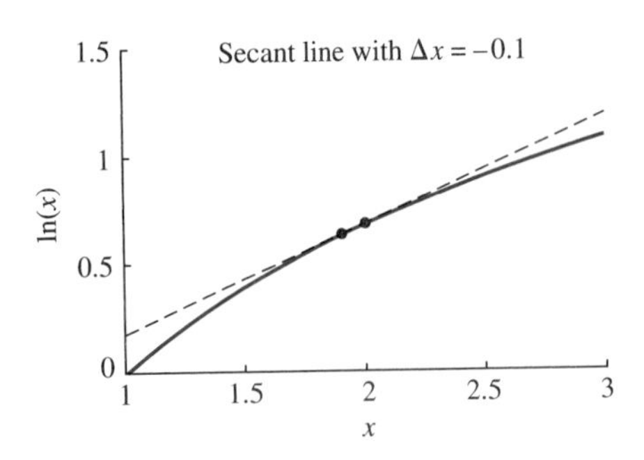

FIGURE 2.1.12
A secant line to $\ln(x)$

We find the instantaneous rate of change by computing the average rate of change with smaller and smaller values of Δt, but without ever reaching $\Delta t = 0$. Graphically, this corresponds to moving the second point closer and closer to the base point, and seeing that the secant line gets closer and closer to the tangent line. Just as the slope of the secant line is equal to the average rate of change, the slope of the tangent line is equal to the instantaneous rate of change.

2.1.3 Limits and Derivatives

We now need a way to work with the idea of "smaller and smaller" or "closer and closer." The mathematical notion is called the **limit**. The expression

$$\lim_{\Delta t \to 0} \frac{\Delta b}{\Delta t},$$

said "the limit as Delta t approaches 0 of Delta b over Delta t," means what we **would like to** get by plugging in $\Delta t = 0$, but **cannot** get exactly because we cannot divide 0 by 0. Near $t = 1.0$, we found that $\Delta b/\Delta t$ has a value of about 1.386, so we define that value to be the limit.

Using the limit, we can define the derivative.

Definition 2.2 The instantaneous rate of change of a function $f(t)$ at $t = t_0$ is called the **derivative of** f and is computed as

$$\lim_{\Delta t \to 0} \frac{\Delta f}{\Delta t}$$

where $\Delta f = f(t_0 + \Delta t) - f(t_0)$.

The derivative measures how fast a measurement is changing at a particular instant. There are two different notations for the derivative:

$$\begin{aligned} \text{the derivative of } f \text{ at } t_0 &= \frac{df}{dt}\bigg|_{t_0} \quad \text{differential notation} \\ &= f'(t_0). \quad \text{prime notation} \end{aligned}$$

In differential notation, the d's represent small versions of the letter Δ. The vertical line with the t_0 indicates the point where we are evaluating the derivative. Prime notation defines a new function $f'(t_0)$ that outputs the rate of change of $f(t_0)$ at any input time t_0. **Differential notation** is convenient for writing differential equations and the complicated expressions needed to calculate derivatives. **Prime notation** is convenient for analyzing discrete-time dynamical systems and for describing rates of change.

Geometrically, the slope of the tangent matches the slope of the curve. We use this observation to define the slope of a curve.

Definition 2.3 The slope of the graph of a function is equal to the slope of the tangent line to the graph, which is itself equal to derivative of the function.

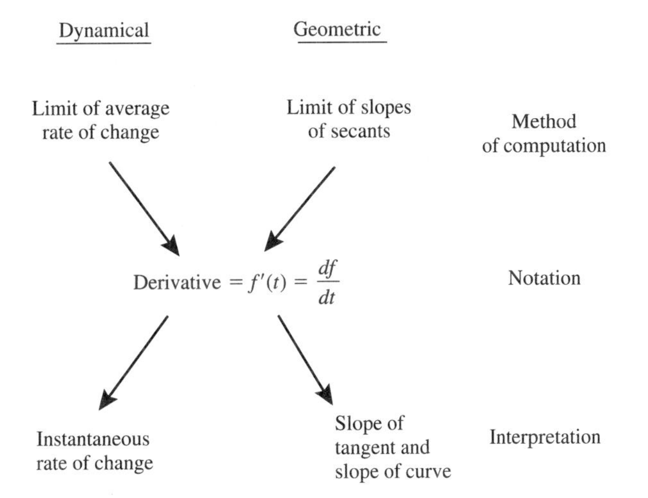

FIGURE 2.1.13
The inter related meanings of the derivative

These two ways of thinking about the derivative are linked in Figure 2.1.13.

The tangent line is a good approximation because it lies so close to the curve (Figure 2.1.11). The central trick in calculus is realizing that from up close, a smooth curve looks like its tangent line, and that the slope of the tangent can be computed with the derivative. There are, however, examples of curves that are not smooth and have no well-defined tangent line (Section 2.4).

2.1.4 Differential Equations: A Preview

Using the derivative, we can find the instantaneous rate of change of the bacterial population size at $t = 1.0$, or indeed at any time. However, this description does not tell us the **rule** being followed by the population. We can use the derivative along with our intuition about the behavior of growing populations to derive such a rule. Instead of finding a discrete-time dynamical system, a formula for the new population size as a function of the old population size, we will derive a **differential equation**, a formula for the **rate of change** of the population size as a function of the current population size.

The following table estimates the instantaneous rate of change with the average rate of change by using a fairly small value of $\Delta t = 0.1$. In other words, we estimate that the derivative $\frac{db}{dt}$ at each time t is approximately

$$\frac{db}{dt} \approx \frac{\Delta b}{\Delta t} = \frac{b(t) - b(t - 0.1)}{0.1}.$$

We have another new column to the table, the **per capita rate of change**, defined as

$$\text{per capita rate of change} = \frac{\text{instantaneous rate of change}}{\text{population}}.$$

When studying the discrete-time dynamical system for a growing bacterial population, we found that the **per capita reproduction** provided a useful description. Similarly, the per capita rate of change, or per capita **rate** of reproduction, provides a useful description of a continuous set of measurements of population size.

Mathematically,

$$\text{per capita rate of reproduction} = \frac{\frac{db}{dt}}{b(t)} = 0.6697.$$

Solving for the derivative $\frac{db}{dt}$,

$$\frac{db}{dt} = 0.6697b(t). \qquad \textbf{(2.1.1)}$$

t	$b(t)$	Change Δb	Average Rate of Change $\frac{\Delta b}{\Delta t} \approx \frac{db}{dt}$	Per Capita Rate of Change $\frac{\frac{db}{dt}}{b(t)}$
0.0000	1.0000	–	–	–
0.1000	1.0718	0.0718	0.7177	0.6697
0.2000	1.1487	0.0769	0.7692	0.6697
0.3000	1.2311	0.0824	0.8245	0.6697
0.4000	1.3195	0.0884	0.8836	0.6697
0.5000	1.4142	0.0947	0.9471	0.6697
0.6000	1.5157	0.1015	1.0150	0.6697
0.7000	1.6245	0.1088	1.0879	0.6697
0.8000	1.7411	0.1166	1.1660	0.6697
0.9000	1.8661	0.1250	1.2496	0.6697
1.0000	2.0000	0.1339	1.3393	0.6697
1.1000	2.1435	0.1435	1.4355	0.6697
1.2000	2.2974	0.1538	1.5385	0.6697
1.3000	2.4623	0.1649	1.6489	0.6697
1.4000	2.6390	0.1767	1.7673	0.6697
1.5000	2.8284	0.1894	1.8941	0.6697
1.6000	3.0314	0.2030	2.0301	0.6697
1.7000	3.2490	0.2176	2.1758	0.6697
1.8000	3.4822	0.2332	2.3319	0.6697
1.9000	3.7321	0.2499	2.4993	0.6697
2.0000	4.0000	0.2679	2.6787	0.6697

This is a **differential equation**. A differential equation sets the derivative of some function (usually written in differential notation) equal to some combination of measurements not involving derivatives. In this case, the rate of change of population size is proportional to the population size itself. Of course, this differential equation is not exactly correct because we estimated the instantaneous rate of change as the average rate of change with $\Delta t = 0.1$. Exercises 31–34 follow the same steps to find a more accurate differential equation by using smaller values of Δt.

Differential equations and discrete-time dynamical systems are two types of rule describing how a measurement changes. How do they differ? A discrete-time dynamical system, like $b_{t+1} = 2.0b_t$, gives the new value of the population size (at time $t + 1$) as a function of the previous value (at time t). A differential equation instead gives the **rate of change** of the population size as a function of the population. Unlike a discrete-time dynamical system, it does not involve the population size at two different times (like b_{t+1} and b_t). The differential equation instead relates two different pieces of information about the population at one time, the population size $b(t)$ and the rate of change $\frac{db}{dt}$ (Figure 2.1.14).

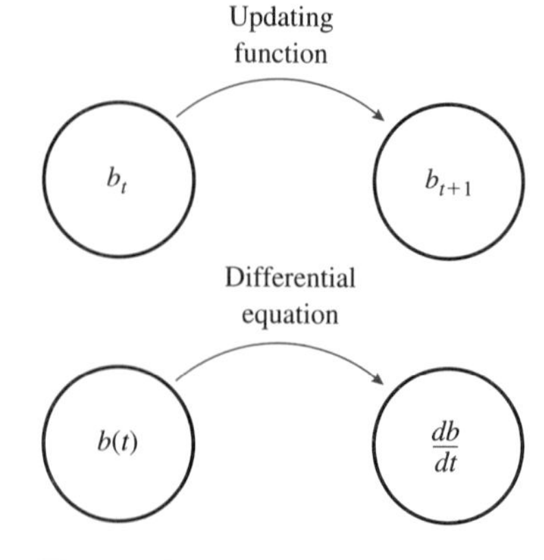

FIGURE 2.1.14
The difference between discrete-time dynamical systems and differential equations

Consider again the bird moving at a constant speed of 20.0 m/s. The instantaneous rate of change of position, or derivative, is equal to 20.0 m/s. If we denote the position at time t by $P(t)$, we can describe the bird with the differential equation

$$\frac{dP}{dt} = 20.0.$$

This differential equation says mathematically that the rate of change of position is exactly 20.0 m/s at all times.

Differential equations are the single most powerful tool in applied mathematics. Their discovery by Sir Isaac Newton, the great English physicist, represents one of the greatest insights in the history of science. If differential equations seem hard to understand, remember that it took hundreds of years of work by the world's leading mathematicians and scientists to realize their importance. The detailed study of differential equations in this book begins in Chapter 4, after we learn how to compute and apply derivatives in other circumstances.

Summary To describe how a measurement changes, we defined the **average rate of change** of the population size, the change in population size divided by the change in time. The average rate of change corresponds graphically to the slope of the **secant line** connecting two data points. To find the **instantaneous rate of change** or **derivative**, we estimated the average rate of change over shorter and shorter intervals. Because it is impossible to use an interval of length zero, we found the **limit** as the interval approaches zero. Graphically, the instantaneous rate of change corresponds to the slope of the **tangent line**, which is defined to be equal to the slope of the curve itself. In some cases, we can write a **differential equation**, a formula for the derivative.

2.1 Exercises

Mathematical Techniques

1–6 ▪ For each of the following functions, find the average rate of change between the given base point t_0 and times $t_0 + \Delta t$ for the four following values of Δt: $\Delta t = 1.0$, $\Delta t = 0.5$, $\Delta t = 0.1$ and $\Delta t = 0.01$.

1. $f(t) = 2 + 3t$ with base point $t_0 = 1.0$.
2. $g(t) = 2 - 3t$ with base point $t_0 = 0.0$.
3. $h(t) = 2t^2$ with base point $t_0 = 1.0$.
4. $h(t) = t^2 + 1$ with base point $t_0 = 0.0$.
5. $G(t) = e^{2t}$ with base point $t_0 = 0.0$.
6. $G(t) = e^{-t}$ with base point $t_0 = 0.0$.

7–12 ▪ For each of the following functions, find the equation of the secant line connecting the given base point t_0 and times $t_0 + \Delta t$ for $\Delta t = 1.0$, $\Delta t = 0.5$, $\Delta t = 0.1$, and $\Delta t = 0.01$. Sketch the function and each of the secant lines.

7. $f(t) = 2 + 3t$ with base point $t_0 = 1.0$ (based on Exercise 1).
8. $g(t) = 2 - 3t$ with base point $t_0 = 0.0$ (based on Exercise 2).
9. $h(t) = 2t^2$ with base point $t_0 = 1.0$ (based on Exercise 3).
10. $h(t) = t^2 + 1$ with base point $t_0 = 0.0$ (based on Exercise 4).
11. $G(t) = e^{2t}$ with base point $t_0 = 0.0$ (based on Exercise 5).
12. $G(t) = e^{-t}$ with base point $t_0 = 0.0$ (based on Exercise 6).

13–18 ▪ Using the results in Exercises 1–6, take a guess at the limit of the slopes of the secants, and find the slope and equation of the tangent line.

13. $f(t) = 2 + 3t$ with base point $t_0 = 1.0$. Call the tangent line function $\hat{f}(t)$ (based on Exercise 1).
14. $g(t) = 2 - 3t$ with base point $t_0 = 0.0$. Call the tangent line function $\hat{g}(t)$ (based on Exercise 2).
15. $h(t) = 2t^2$ with base point $t_0 = 1.0$. Call the tangent line function $\hat{h}(t)$ (based on Exercise 3).
16. $h(t) = t^2 + 1$ with base point $t_0 = 0.0$. Call the tangent line function $\hat{h}(t)$ (based on Exercise 4).
17. $G(t) = e^{2t}$ with base point $t_0 = 0.0$. Call the tangent line function $\hat{G}(t)$ (based on Exercise 5).
18. $G(t) = e^{-t}$ with base point $t_0 = 0.0$. Call the tangent line function $\hat{G}(t)$ (based on Exercise 6).

19–20 ▪ Important concepts have many names and formulas. The following problems ask you to recall them.

19. Give two other names for the instantaneous rate of change of the function $g(t)$.
20. Give three different notations for the instantaneous rate of change of a function $g(t)$.

Applications

21–22 ▪ For each equation for population size, find the following and illustrate on a graph.

a. The population at times 0, 1, and 2.
b. The average rate of change between times 0 and 1.
c. The average rate of change between times 1 and 2.

21. A population of bacteria is described by the formula $b(t) = 1.5^t$ where the time t is measured in hours.
22. A population of bacteria described by the formula $b(t) = 1.2^t$ where the time t is measured in hours.

23–26 ▪ For each equation for population size, find the following.

a. The average rate of change between times 0 and 1.0.
b. The average rate of change between times 0 and 0.1.
c. The average rate of change between times 0 and 0.01.
d. The average rate of change between times 0 and 0.001.
e. What do you think the limit is?
f. Graph the tangent line.

23. A population following $b(t) = 1.5^t$.

24. A population following $b(t) = 2.0^t$.

25. A population following $h(t) = 5.0t^2$.

26. A bacterial population following $b(t) = (1.0 + 2.0t)^3$.

27–30 ▪ For the following bacterial populations, find the average rate of change during the first hour, and during the first and second half hours. Graph the data and the secant lines associated with the average rates of change. Which populations change faster during the first half hour?

27. $b(t) = 3.0(2.0^t)$.

28. $b(t) = e^{0.5t}$.

29. $b(t) = 2.0e^{-0.5^t}$.

30. $b(t) = 3.0(0.5^t)$.

31–34 ▪ Follow the steps in the text used to derive the approximate differential Equation 2.1.1

$$\frac{db}{dt} = 0.6697b(t)$$

with the following values of Δt. This requires computing the value of the function $b(t) = 2.0^t$ at times separated by Δt, and finding the average rate of change between those times.

31. $\Delta t = 1.0$.

32. $\Delta t = 0.5$.

33. $\Delta t = 0.01$.

34. $\Delta t = 0.001$.

35–36 ▪ Consider the following data on a tree.

Age	Height (m)	Mass (metric tons)
0	10.11	30.0
1	11.18	39.1
2	12.40	50.6
3	13.74	65.8
4	15.01	85.9
5	16.61	111.6
6	18.27	144.2
7	20.17	187.7
8	22.01	244.1
9	24.45	319.2
10	26.85	414.2

For each of the measurements,

a. Estimate the rate of change at each age.

b. Graph the rate of change as a function of age.

c. Find and graph the rate of change divided by the value as a function of age.

d. Use these results to describe the growth of this tree with a differential equation.

35. The height

36. The mass

37–43 ▪ The procedure banks use to compute continuously compounded interest is similar to the process we used to derive a differential equation. Suppose several banks claim to be giving 5% annual interest and that you have \$1000 to deposit.

37. How much would you have after a year from a bank that has no compounding?

38. A bank that compounds twice yearly really gives 2.5% interest twice. How much would you have after a year from this bank? How much better is this than a bank with no compounding?

39. A bank that compounds monthly really gives 5/12% interest each month. How much would you have after a year from this bank?

40. How much would you have after a year from a bank that compounded daily?

41. Write down a limit that expresses the amount of money you would get from a bank that compounded continuously, and try to guess the answer.

42. Follow the same steps to compare yearly, monthly, and daily compounding for a bank giving 20% interest.

43. Follow the same steps to compare yearly, monthly, and daily compounding for a bank giving 100% interest (in a time of severe inflation). Why do you think compound interest makes a bigger difference when the interest rate is higher?

Computer Exercises

44. Consider the function

$$f(x) = \sqrt{1 - x^2}$$

defined for $-1 \le x \le 1$. This is the equation for a semi circle. The tangent line at the base point $(\sqrt{2}/2, \sqrt{2}/2)$ has slope -1. Graph this tangent line. Now zoom in on the base point. Does the circle look more and more like the tangent line? How far do you need to go before the circle looks flat? Would a tiny insect be able to tell that his world was curved?

45. Suppose a bacterial population oscillates with the formula

$$b(t) = 2.0 + \cos(t).$$

a. Graph this function

b. Find and graph the function that gives the rate of change between times t and $t + 1$ as a function of t for $0 \le t \le 10$.

c. Find and graph the function that gives the rate of change between times t and $t + 0.1$ as a function of t for $0 \le t \le 10$.

d. Try the same with smaller values of Δt. Do you have any idea what the limit might be?

46. Follow the steps in Exercise 45 for a bacterial population that follows the formula

$$b(t) = e^{-0.1t}[2.0 + \cos(t)].$$

2.2 Limits

The derivative, the mathematical version of the instantaneous rate of change and the slope of a curve, includes a **limit** in its definition. We will now study the mathematical and scientific basis of this fundamental idea. By understanding the useful properties of limits, we will be able to calculate limits of polynomials and rational functions. The basic idea of the limit extends in many ways. We will define **left-hand limits** and **right-hand limits** for functions that make sense on only one side and **infinite limits** for quantities or functions that become extremely large.

2.2.1 Limits of Functions

We begin by formalizing the steps we used to compute the instantaneous rate of change of the population following the law

$$b(t) = 2.0^t$$

at $t = 1.0$. First, we found the change in the population, Δb, between times 1.0 and $1.0 + \Delta t$ as

$$\begin{aligned}\Delta b &= b(1.0 + \Delta t) - b(1.0) \\ &= 2^{1.0+\Delta t} - 2.0.\end{aligned}$$

The average rate of change, the change in the population divided by the change in time, is then

$$\frac{\Delta b}{\Delta t} = \frac{2^{1.0+\Delta t} - 2.0}{\Delta t}.$$

The instantaneous rate of change (or derivative) at $t = 1.0$ is

$$\frac{db}{dt} = b'(1.0) = \lim_{\Delta t \to 0} \frac{2^{1.0+\Delta t} - 2.0}{\Delta t}.$$

As Δt becomes smaller and smaller, the average rate of change gets closer and closer to a value near 1.386. We could not take the smallest possible value, $\Delta t = 0$, because that would have led to division of zero by zero.

We can think of the average rate of change

$$\frac{\Delta b}{\Delta t} = \frac{2^{1.0+\Delta t} - 2.0}{\Delta t}.$$

as a **function** of Δt that is defined at all points except $\Delta t = 0$ (Figure 2.2.1). We guessed that the limit is 1.386 because values get closer and closer to 1.386. But what does it mean for a function to get "closer and closer" to a "limit"? What, indeed, does it mean to be "close"?

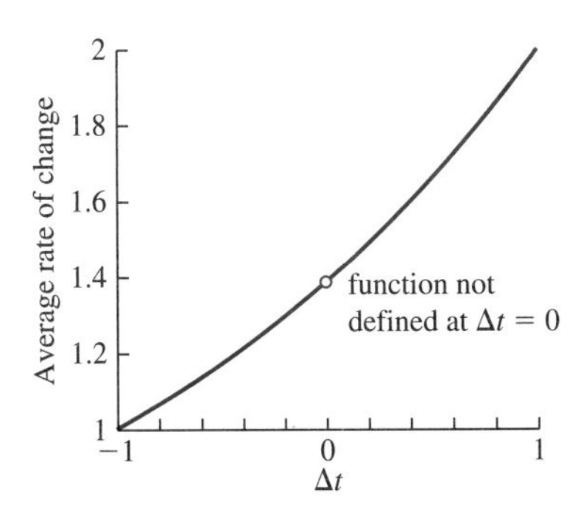

FIGURE 2.2.1
The limit of the average rate of change

As is often the case, this mathematical question can be understood by thinking of it **scientifically**. What does it mean scientifically for two quantities to be close to one another?

Two quantities are close when they are too similar to distinguish with a precise measuring device.

Consider the measurement of temperature. With a crude measuring device, like waving your hand around in the air, it might be difficult to distinguish temperatures of 20°C and 25°C. Without a more precise device, the two temperatures are effectively the same. With an ordinary thermometer, we can distinguish 20°C and 25°C, but may not be able to distinguish 20°C and 20.1°C. Two temperatures are **exactly** equal if no thermometer, no matter how precise, can distinguish them.

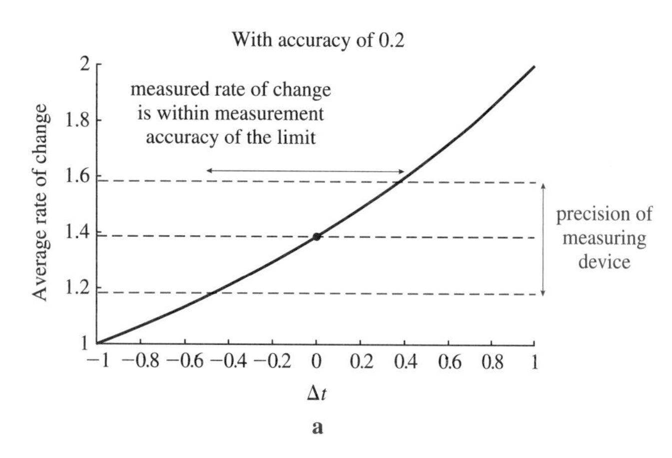

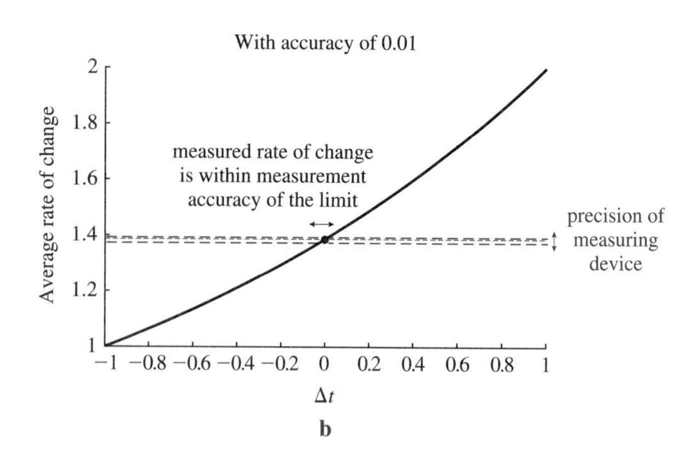

FIGURE 2.2.2

An experimental output approaching a limit

We can translate this scientific idea into the mathematical idea of a limit. The average rate of change is a function of Δt (Figure 2.2.1). If we can measure the rate of change with an accuracy of only 0.1, we do not need a very small value of Δt to be within 0.1 of the limit (Figure 2.2.2a). With a more precise measurement of the average rate of change with an accuracy of 0.01, we need a much smaller Δt to be within measurement accuracy of the limit (Figure 2.2.2b). No matter how precisely we could measure rate of change, however, we can always pick a sufficiently small Δt to be within measurement accuracy of the limit.

Example 2.2.1 How Close Must the Input Be?

Consider the function

$$f(x) = \frac{3x + 2x^2}{x}$$

which is not defined at $x = 0$. For any $x \neq 0$, $f(x) = 3 + 2x$ because we can divide the numerator through by x. We will soon prove that $\lim_{x \to 0} f(x) = f(0) = 3$. The idea of the limit says we can pick values of x close enough to 0 to guarantee getting as close as we might wish to this limit. For example, if we wish to be within 0.1 of the limit, we require

$$\begin{aligned} 2.9 &< f(x) < 3.1 \\ 2.9 &< 3 + 2x < 3.1 \\ -0.1 &< 2x < 0.1 \\ -0.05 &< x < 0.05. \end{aligned}$$

Graphically, the inputs on the horizontal axis must lie within the region that produces outputs within the range on the vertical axis (Figure 2.2.3). If we wish to be within

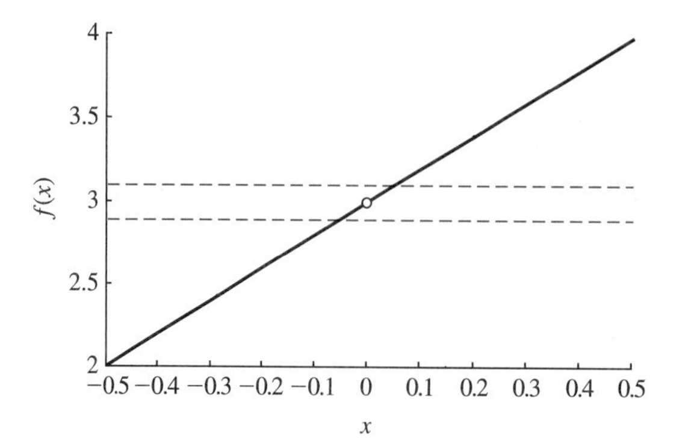

FIGURE 2.2.3

Finding how close the input must be

0.01 of the limit, we require

$$\begin{aligned}2.99 &< f(x) < 3.01\\ 2.99 &< 3 + 2x < 3.01\\ -0.01 &< 2x < 0.01\\ -0.005 &< x < 0.005.\end{aligned}$$

Getting closer to the limit requires inputs that are closer to 0.

Example 2.2.2 Finding the Limit of the Average Rate of Change Algebraically

Consider finding the instantaneous rate of change of the distance traveled by a falling cat that follows approximately the quadratic function (Figure 2.2.4)

$$y(t) = 5.0t^2,$$

at least until the cat reaches terminal velocity, a concept we study in Section 5.2. We wish to find the rate of change at $t = 1.0$. The change Δy between times 1.0 and $1.0 + \Delta t$ is

$$\begin{aligned}\Delta y &= y(1.0 + \Delta t) - y(1.0)\\ &= 5.0(1.0 + \Delta t)^2 - 5.0\\ &= 5.0(1.0 + 2.0\Delta t + \Delta t^2) - 5.0\\ &= 10.0\Delta t + 5.0\Delta t^2.\end{aligned}$$

FIGURE 2.2.4
The distance traveled by a falling cat and the average rate of change

Then

$$\begin{aligned}\text{average rate of change} &= \frac{\Delta y}{\Delta t}\\ &= \frac{10.0\Delta t + 5.0\Delta t^2}{\Delta t}\end{aligned}$$

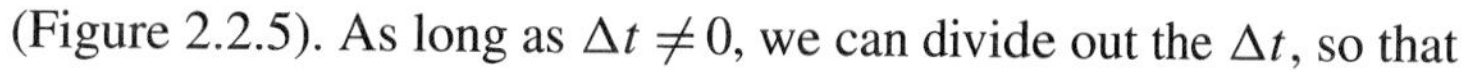

(Figure 2.2.5). As long as $\Delta t \neq 0$, we can divide out the Δt, so that

$$\text{average rate of change} = 10.0 + 5.0\Delta t.$$

To find the exact rate of change, we must take the limit as $\Delta t \to 0$. For small values of Δt, the average rate of change gets as close as we might wish to 10.0. The limit is exactly 10.0.

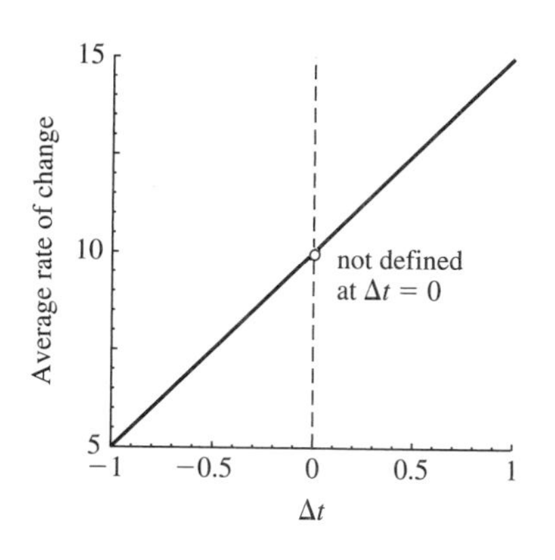

FIGURE 2.2.5
The average rate of change for the distance traveled by a falling cat

2.2.2 Left-Hand and Right-Hand Limits

Sometimes a function is defined only on one side of a point. For example, the value of the function shown in Figure 2.2.6a seems to get closer and closer to 2.0, even though it is defined only for $x < 1$. If we wish to find the limit as x approaches 1, we must plug

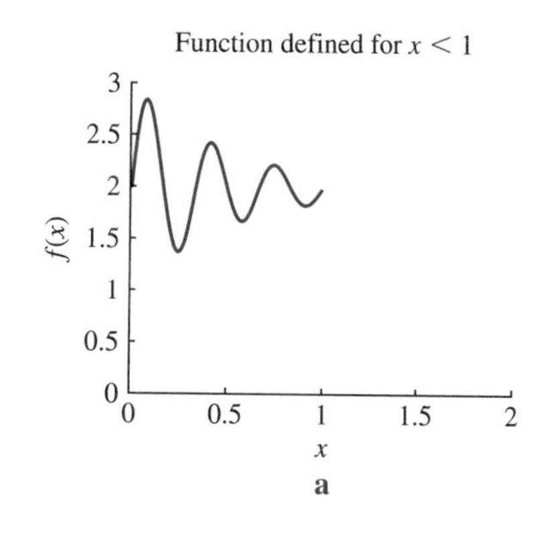

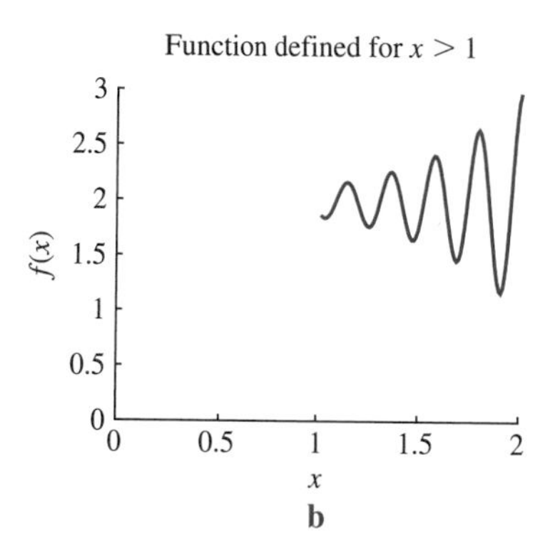

FIGURE 2.2.6
Functions with left-hand and right-hand limits

in values of x that lie in the domain. We write

$$\lim_{x \to 1^-} f(x) = 2,$$

where the minus sign above the 1 indicates that x approaches 1 from **below**. This is called the **left-hand limit**, stated as "the limit of $f(x)$ as x approaches 1 from below."

Similarly, if the domain of f contains only values of $x > 1$ (Figure 2.2.6b), we write

$$\lim_{x \to 1^+} f(x) = 2.$$

The plus sign above the 1 indicates that x approaches 1 from **above**, and this is called the **right-hand limit**. We say "the limit of $f(x)$ as x approaches 1 from above."

Left- and right-hand limits are most often used to describe functions and quantities that have only positive numbers in their domains, like $\sqrt{x}$, $\ln(x)$ and population sizes. Suppose we were asked to find

$$\lim_{x \to 0^+} \sqrt{x}.$$

We must compute the right-hand limit because we cannot take the square root of a negative number. Because the square root of 0 is 0 and the graph gets closer and closer to 0, the limit itself is 0.

Example 2.2.3 A Function with Unequal Left-Hand and Right-Hand Limits

Consider the "signum" function, defined by

$$\begin{cases} \text{signum}(x) = -1 \ \text{ if } x < 0 \\ \text{signum}(x) = 0 \ \text{ if } x = 0 \\ \text{signum}(x) = 1 \ \text{ if } x > 0. \end{cases}$$

This function gives the "sign" of a number, telling whether it is negative, zero, or positive (Figure 2.2.7). The limit as x approaches 0 is not the value $\text{signum}(0) = 0$ because the function does not get close to 0. The left-hand limit is -1, and the right-hand limit is 1. This is an example of a function where the left-hand and right-hand limits match neither each other nor the value of the function.

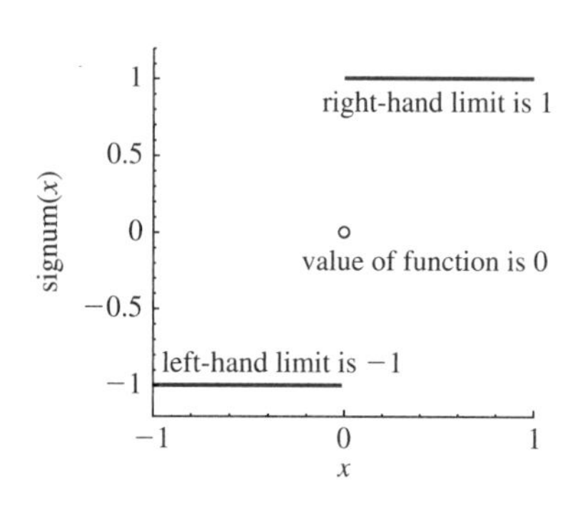

FIGURE 2.2.7
The signum function

2.2.3 Properties of Limits

Limits have many properties that simplify their computation. The basic idea is an important one; we first find the limits of some simple functions, and then we build limits of more complicated functions by combining these simple pieces with rules that tell how limits add, multiply, and divide.

The simple functions we begin with are constant functions, with formula

$$f(x) = c$$

and the **identity function** with formula

$$f(x) = x.$$

Theorem 2.1 **Limits of Basic Functions**

a. Suppose $f(x) = c$ for all x. Then, for any value of a,

$$\lim_{x \to a} f(x) = c.$$

b. Suppose $f(x) = x$ for all x. Then, for any value of a,

$$\lim_{x \to a} f(x) = a.$$

Example 2.2.4 Finding the Limit of a Constant

Suppose the function $h(x) = 5$, a constant (Figure 2.2.8). Then Theorem 2.1 implies that

$$\lim_{x \to 7} h(x) = 5.$$

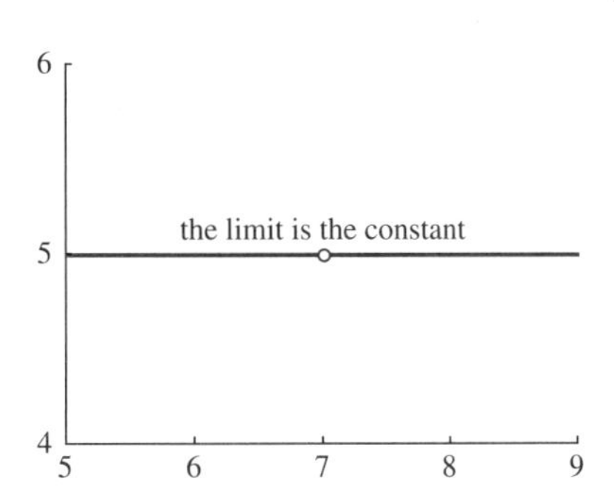

FIGURE 2.2.8
Limit of a constant function

Example 2.2.5 Finding the Limit of the Identify Function

Suppose the function $g(x) = x$, the identity function (Figure 2.2.9). Then Theorem 2.1 implies that

$$\lim_{x \to 7} g(x) = 7.$$

Second, we can combine limits by adding, multiplying, and dividing (as long as we avoid dividing by 0).

Theorem 2.2 **Rules for Combining Limits**

Suppose $f(x)$ and $g(x)$ are functions with well-defined limits at $x = a$.

a. The limit of the sum is the sum of the limits, or

$$\lim_{x \to a}[f(x) + g(x)] = \lim_{x \to a} f(x) + \lim_{x \to a} g(x).$$

b. The limit of the product is the product of the limits, or

$$\lim_{x \to a} f(x)g(x) = [\lim_{x \to a} f(x)] \cdot [\lim_{x \to a} g(x)].$$

c. The limit of the product of a constant and a function is the product of the constant and the limit of the function.

$$\lim_{x \to a}[cf(x)] = c \cdot \lim_{x \to a} f(x).$$

d. Suppose $\lim_{x \to a} g(x) \neq 0$. Then the limit of the quotient is the quotient of the limits, or

$$\lim_{x \to a} \left(\frac{f(x)}{g(x)} \right) = \frac{\lim_{x \to a} f(x)}{\lim_{x \to a} g(x)}.$$

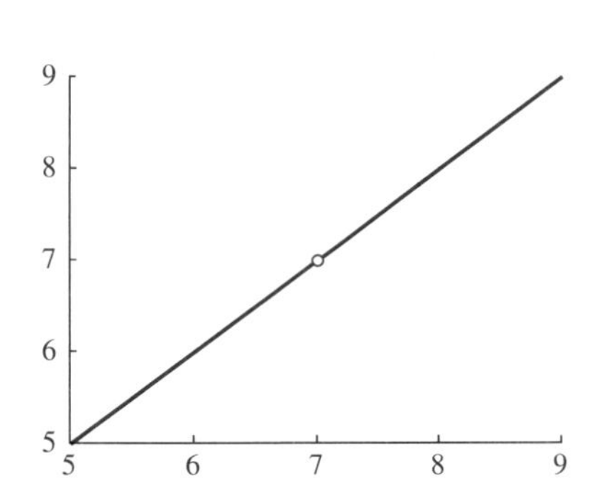

FIGURE 2.2.9
Limit of the identity function

Together, these theorems imply that we can find the limit of any **polynomial** by **plugging in the value**. Remember that a polynomial is a function that can be written as a sum of constants multiplied by powers of the argument.

Example 2.2.6 Finding the Limit of a Polynomial

The function $f(x) = 2x^2 + 3x$ is a polynomial. We can find the limit by substituting the value. For example,

$$\lim_{x \to 5} 2x^2 + 3x = 2 \cdot 5^2 + 3 \cdot 5 = 65.$$

Example 2.2.7 Finding the Instantaneous Rate of Change of Position of a Cat

We found the average rate of change of a cat's position to be $10.0 + 5.0\Delta t$ as a function of the interval Δt (Example 2.2.2). Because this is a polynomial, we can find the limit by substituting the value $\Delta t = 0$, giving

$$\lim_{\Delta t \to 0} 10.0 + 5.0\Delta t = 10.0 + 5.0 \cdot 0 = 10.0.$$

The instantaneous rate of change, or the **velocity**, is indeed 10.0 m/s. We use velocity as a more precise term for speed.

More generally, these theorems make it easy to compute the limit of a **rational function**, defined as the **ratio of polynomials**.

Example 2.2.8 The Limit of a Rational Function

We can compute the limit

$$\lim_{x \to 5} \frac{2x^2 + 3x}{2x + 1} = \frac{2 \cdot 5^2 + 3 \cdot 5}{2 \cdot 5 + 1} = \frac{65}{11} = 5.91$$

by plugging in $x = 5$ because the denominator is not equal to 0 at $x = 5$.

Left- and right-hand limits share the properties of ordinary limits summarized in Theorems 2.1 and 2.2: they add, multiply, and divide.

Example 2.2.9 Combining Right-Hand Limits

We can use the rules for combining limits to find

$$\begin{aligned} \lim_{x \to 0^+} (3\sqrt{x} + 2) &= \lim_{x \to 0^+} 3\sqrt{x} + \lim_{x \to 0^+} 2 && \text{limit of sum} \\ &= 3 \lim_{x \to 0^+} \sqrt{x} + 2 && \text{constant comes out} \\ &= 3 \cdot 0 + 2 = 2 && \text{plug in.} \end{aligned}$$

2.2.4 Infinite Limits

One sometimes hears it said that a function gets "infinitely large" near some input. What might such a statement mean scientifically? As with ordinary limits, we can clarify this idea by thinking of measurements. It is impossible to actually measure (or even imagine) a value of infinity. Instead, we encounter values that exceed the capacity of our measuring device. A thermometer would explode if exposed to a temperature above its capacity. A value can be thought of as **infinitely** large if it exceeds the capacity of **every possible** measuring device. We say "the limit of $f(x)$ as x approaches a is equal to infinity" and write

$$\lim_{x \to a} f(x) = \infty.$$

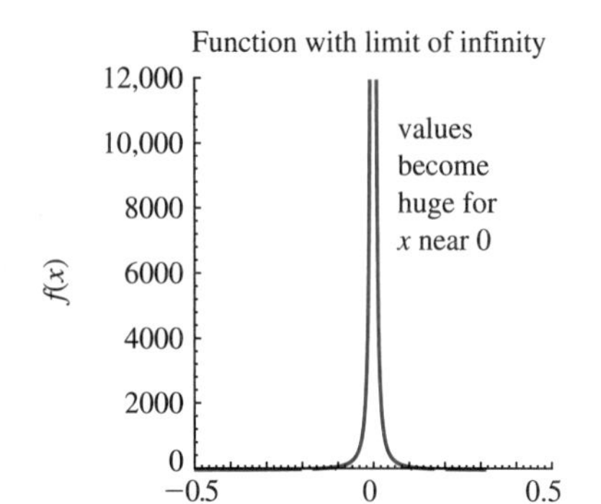

FIGURE 2.2.10
A function with a limit of infinity

Example 2.2.10 A Function with a Limit of Infinity

x	$f(x)$
1.0	1.0
0.1	100.0
0.01	10000.0
0.001	1.0×10^6
−1.0	1.0
−0.1	100.0
−0.01	10000.0
−0.001	1.0×10^6

Consider the function

$$f(x) = \frac{1}{x^2}$$

(Figure 2.2.10). If we plug in values of x closer and closer to 0, $f(x)$ becomes larger and larger.

In fact, almost any calculator will give an "error" if x is too small (although some calculators are smart enough to say "infinity"). In this case,

$$\lim_{x \to 0} f(x) = \infty.$$

Similarly, a limit is equal to **negative infinity**, written $-\infty$, when the output becomes smaller than any given value. We write

$$\lim_{x \to a} f(x) = -\infty.$$

Example 2.2.11 A Function with a Limit of Negative Infinity

The function

$$g(x) = \frac{-1}{x^2}$$

has a limit of negative infinity as x approaches 0 (Figure 2.2.11).

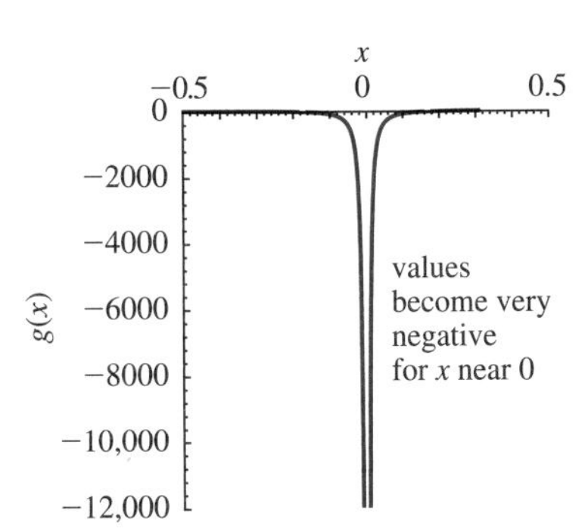

FIGURE 2.2.11
A function with a limit of negative infinity

x	g(x)
1.0	−1.0
0.1	−100.0
0.01	−10000.0
0.001	-1.0×10^6
−1.0	−1.0
−0.1	−100.0
−0.01	−10000.0
−0.001	-1.0×10^6

Example 2.2.12 A Function with Different, but Infinite, Left-Hand and Right-Hand Limits

Consider the function

$$h(x) = \frac{1}{x}.$$

(Figure 2.2.12). Suppose we wish to find the limit as x approaches 0. This function increases to positive infinity for $x > 0$ and to negative infinity for $x < 0$.

FIGURE 2.2.12
A function with a left-hand limit of $-\infty$ and a right-hand limit of ∞

x	h(x)
1.0	1.0
0.1	10.0
0.01	100.0
0.001	1000.0
−1.0	−1.0
−0.1	−10.0
−0.01	−100.0
−0.001	−1000.0

The left-hand and right-hand limits help us to separate out these two types of behavior. As we move toward 0 from the left (negative x), the values of the function get smaller and smaller. As we move toward 0 from the right (positive x), the values of the function get larger and larger.

Example 2.2.13 The Limit of $\ln(x)$ as $x \to 0^+$

During our study of the natural logarithm, we indicated that $\ln(x)$ is not defined for negative values of x, and that the graph of $\ln(x)$ rises from negative infinity near $x = 0$ (Figure 2.2.13).

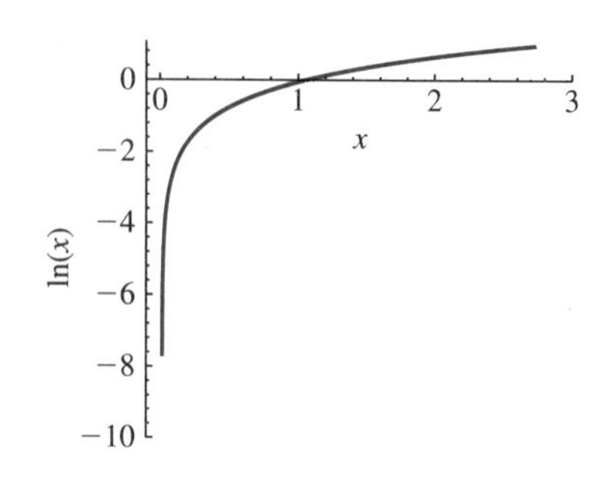

FIGURE 2.2.13
The natural logarithm approaching negative infinity

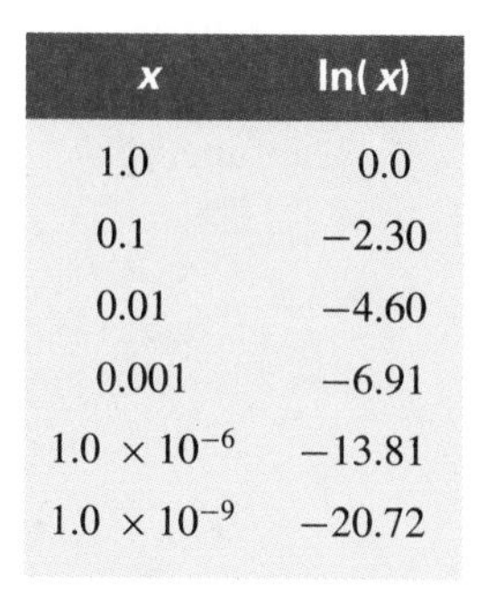

x	$\ln(x)$
1.0	0.0
0.1	−2.30
0.01	−4.60
0.001	−6.91
1.0×10^{-6}	−13.81
1.0×10^{-9}	−20.72

The value of $\ln(x)$ becomes more and more negative, although very slowly. We can now describe this in terms of the limit as

$$\lim_{x \to 0^+} \ln(x) = -\infty.$$

Example 2.2.14 How Close Must the Input Be?

Consider the function $\ln(x)$. If the limit as $x \to 0^+$ is indeed infinity, we should be able to choose inputs x so small that the output is less than -10, or -100, or any large negative number. When is $\ln(x) < -10$? Solving,

$$\ln(x) < -10$$

$$x < e^{-10} = 4.5 \times 10^{-5}.$$

Graphically, the inputs on the horizontal axis must lie close enough to zero for the output to be below the horizontal line (Figure 2.2.14). A tiny input is required to get an output even this small. We will later think of this as meaning that $\ln(x)$ approaches negative infinity "slowly."

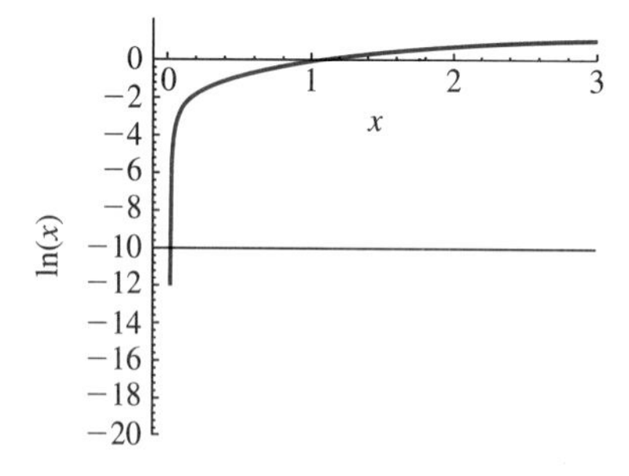

FIGURE 2.2.14
Finding how small the input must be

Summary Based on scientific reasoning, we have developed an intuitive idea of the limit of a function. A function $f(x)$ approaches a particular limit if the values of f become very close to that limit. We defined **left-hand limits** and **right-hand limits** for situations where functions or quantities are meaningful on only one side of a point. Limits can be added, multiplied, and divided. Finally, we defined limits of **infinity** and **negative infinity**, based on the idea that a quantity is effectively infinite when it exceeds the capacity of any measuring device.

2.2 Exercises

Mathematical Techniques

1–8 ▪ Using a computer or calculator, estimate the following limits. Sketch the function.

1. $\lim_{x\to 0}(1+x)^{1/x}$.
2. $\lim_{x\to 0}\dfrac{\sin(x)}{x}$.
3. $\lim_{x\to 0}\dfrac{1-\cos(x)}{x}$.
4. $\lim_{x\to 0^+} x^x$ (the function is defined only for positive values of x).
5. $\lim_{x\to 0^+} x\ln(x)$.
6. $\lim_{x\to 0}\dfrac{e^{2x}-1}{x}$.
7. $\lim_{x\to 1^-}\dfrac{\ln(1-x)}{x}$.
8. $\lim_{x\to 1^+}\sqrt{\ln(x)}$.

9–12 ▪ Using the results from the earlier problems, find the combined limits using Theorem 2.2. Say how the new function was built.

9. $\lim_{x\to 0} 5(1+x)^{1/x}$ (based on Exercise 1).
10. $\lim_{x\to 0} 3\dfrac{\sin(x)}{x}+4$ (based on Exercise 2).
11. $\lim_{x\to 0}(1+x)^{1/x}\dfrac{1-\cos(x)}{x}$ (based on Exercises 1 and 3).
12. $\lim_{x\to 0^+}\dfrac{(1+x)^{1/x}}{x^x}$ (based on Exercises 1 and 4).

13–16 ▪ The given functions all have limits of 0 as $x \to 0+$. For each function, find how close the input must be to 0 for the output to be a) within 0.1 of 0, and b) within 0.01 of the limit. Sketch a graph of each function for $x < 1$, and say which functions approach 0 quickly and which approach 0 slowly.

13. $f_1(x)=\sqrt{x}$.
14. $f_2(x)=x$.
15. $f_3(x)=x^2$.
16. $f_4(x)=x^4$.

17–20 ▪ The given functions all have limits of ∞ as $x \to 0^+$. For each function, find how close the input must be to 0 for the output to be a) greater than 10, and b) greater than 100. Sketch a graph of each function for $x < 1$, and say which functions approach infinity quickly and which approach infinity slowly.

17. $g_1(x)=\dfrac{1}{\sqrt{x}}$.
18. $g_2(x)=\dfrac{1}{x}$.
19. $g_3(x)=\dfrac{1}{x^2}$.
20. $g_4(x)=\dfrac{1}{x^4}$.

21–24 ▪ From the following pictures, find the left-hand and right-hand limits as x approaches 1.

21.

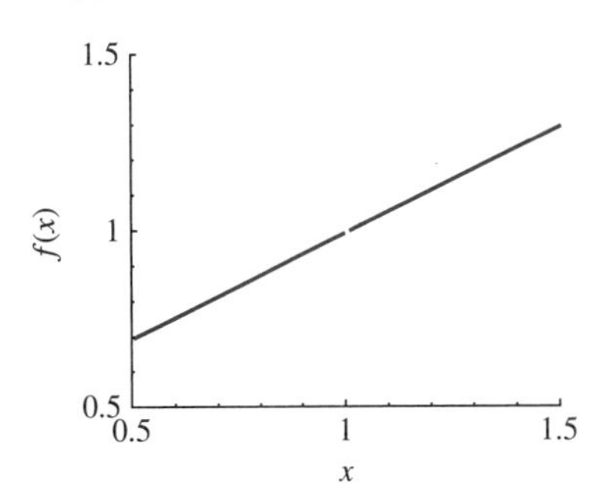

22.

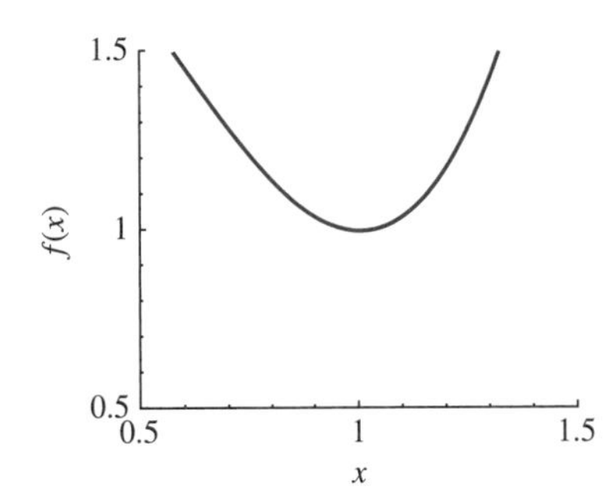

23.

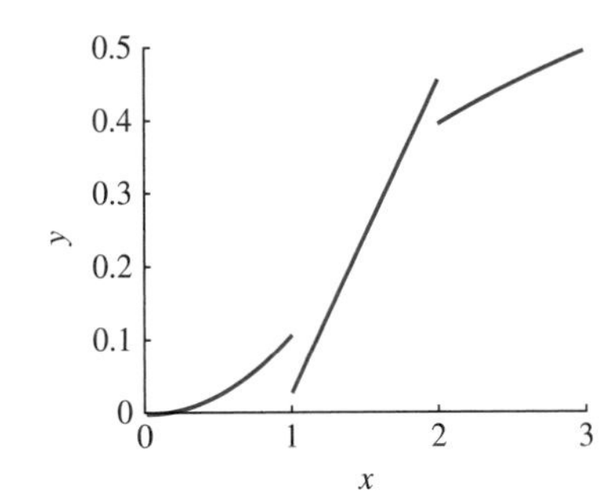

24.

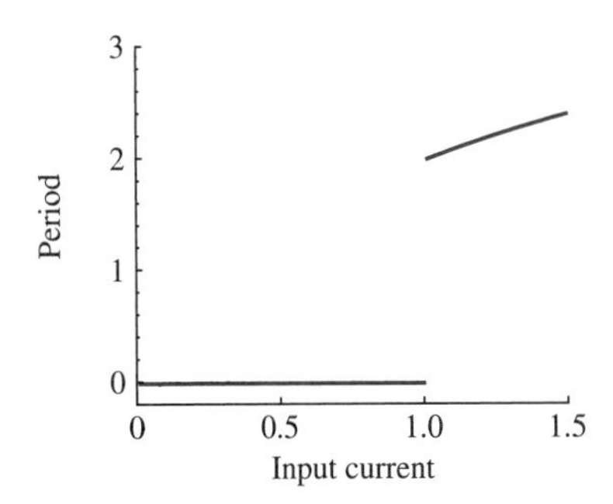

25–30 ▪ Find the average rate of change of the following functions as a function of Δx, and find the limit as $\Delta x \to 0$. Graph the function and indicate the rate of change on your graph.

25. $f(x)=5x+7$ near $x=0$.
26. $f(x)=5x+7$ near $x=1$.
27. $f(x)=5x^2$ near $x=0$.

28. $f(x) = 5x^2$ near $x = 1$.

29. $f(x) = 5x^2 + 7x + 3$ near $x = 1$.

30. $f(x) = 5x^2 + 7x + 3$ near $x = 2$.

Applications

31–32 ▪ Suppose we are interested in measuring the properties of a substance at temperature of absolute zero (which is 0 degrees Kelvin). However, we cannot measure these properties directly because it is impossible to reach absolute 0. Instead, properties are measured for small values of the temperature T, measured in kelvins. For each of the following, find

a. The limit as $T \to 0^+$.

b. How close would we be to the limit if we measured the property at 2K?

c. How close would we be to the limit if we measured the property at 1K?

d. About how cold would the temperature have to be for the property to be within 1% of its limit?

31. The volume $V(T)$ (in cm^3) follows $V(T) = 1 + T^2$.

32. The hardness $H(T)$ follows $H(T) = \dfrac{10.0}{1+T}$.

33–36 ▪ For each of the following populations, the instantaneous rate of change of the population size at $t = 0$ is exactly 1.0 million bacteria per hour. If you computed the average rate of change between $t = 0$ and $t = \Delta t$, how small would Δt have to be before your value was within 1% of the instantaneous rate of change?

33. $b(t) = t + t^2$.

34. $b(t) = t + 0.1t^2$.

35. $b(t) = e^t$ (this cannot be solved algebraically).

36. $b(t) = \sin(t)$ (this cannot be solved algebraically).

37–40 ▪ Scientifically, two quantities are close if it requires an accurate measuring device to detect the difference. In real life, accuracy costs money. How much would it cost to measure the differences in the following circumstances?

37. A piano tuner is trying to get the note A on a piano to have a frequency of exactly 440 hertz (H), or cycles per second. An electronic tuner capable of detecting a difference of x cycles per second costs $\dfrac{5}{x}$ dollars.

a. How much would it cost to make sure the note was within 1.0 H of 440?

b. How much would it cost to make sure the note was within 0.1 H of 440?

c. How much would it cost to make sure the note was within 0.01 H of 440?

38. The army is developing satellite-based targeting systems. A system that can send a missile within y meters of its target costs $\dfrac{1}{y^2}$ million dollars.

a. How much would it cost to hit within 10 m?

b. How much would it cost to hit within 1 m?

c. How much would it cost to hit within 1 cm?

39. Suppose a body has temperature B and is cooling toward room temperature of 20°C according to the function $B(t) = 20 + 17e^{-t}$ where t is measured in hours. A \$10 thermocouple can detect a difference of 0.1°C, a \$100 thermocouple can detect a difference of 0.01°C, and so forth.

a. How much would it cost to detect the difference after 1 h?

b. How much would it cost to detect the difference after 5 h?

c. How much would it cost to detect the difference after 10 h?

40. Some dangerously radioactive and toxic radium was dumped in the desert in 1950. It has a half-life of 50 years, and the initial level of radioactivity was $r = 10.0$ rads. Nobody remembers where it was. How much will it cost to find it in the following years if detecting radioactivity costs $\dfrac{5}{r}$ thousand dollars?

a. How much would it cost to find in the year 2000?

b. How much would it cost to find in the year 2050?

c. How much would it cost to find in the year 2130?

41–42 ▪ Scientifically, a quantity is large if it requires a tough measuring device to assess. In real life, toughness costs money. How much would it cost to measure the value in the following circumstances?

41. We are interested in measuring the pressure at different depths below the surface of the ocean. Pressure increases by approximately 1 atmosphere for every 10 m of depth below the surface (for example, at a depth of 20 m, there are approximately 3 atmospheres of pressure, 2 due the ocean and 1 to the atmosphere itself). Measuring a pressure of x atmospheres without crushing the device costs x^2 dollars.

a. How much would it cost to measure the pressure 100 m down?

b. How much would it cost to measure the pressure 1000 m down?

c. How much would it cost to measure the pressure 5000 m down?

42. Solar scientists want to measure the temperature inside the sun by sending in probes. Imagine that temperature increases by 1 million°C for every 10,000 km below the surface. A probe that can handle a temperature of x million degrees costs x^3 million dollars.

a. How much would it cost to measure the temperature 10,000 km down?

b. How much would it cost to measure the temperature 100,000 km down?

c. How much would it cost to measure the temperature 200,000 km down?

Computer Exercise

43. There are various ways to find the smallest and largest numbers your calculator or computer can handle.

a. Try doubling some number until you get an overflow.

b. Try halving some number until you get an underflow.

c. Compute $1.0 + \epsilon$ for smaller and smaller values of ϵ. At what point does your calculator return 1.0 as the answer?

d. Compare the answers of **b** and **c**.

2.3 Continuity

Even with their useful properties, computing limits can be rather slow. When we are finding the limit of some function $f(x)$ as x approaches a, we would like to simply compute $f(a)$. In this section, we will learn the conditions under which this method works. The chief new tool is the **continuous function**, the mathematically precise definition of a function without "jumps." The definition of the continuous function depends on the limit, and continuous functions share the useful properties of limits. Scientifically, continuous functions represent relations between measurements where a small change in the input produces a small change in the output. Functions that are not continuous do occur in biological systems, and can generate novel phenomena, such as **hysteresis**.

2.3.1 Continuous Functions

We can find

$$\lim_{x \to 0} 2x + 1 = 1$$

in two ways. We can plug in values of x close to 0 and see that the results get close to 1. Alternatively, we could break the function into component parts and use the properties of limits. But why not simply compute $g(0) = 1$ to find the limit?

We can do just this when the function is **continuous**. What is a continuous function? Our intuitive notion of "continuous" is "connected," without jumps. Figure 2.3.1 illustrates functions with and without jumps. The graph of a continuous function can be drawn without lifting one's pencil. The mathematical definition captures this intuitive idea.

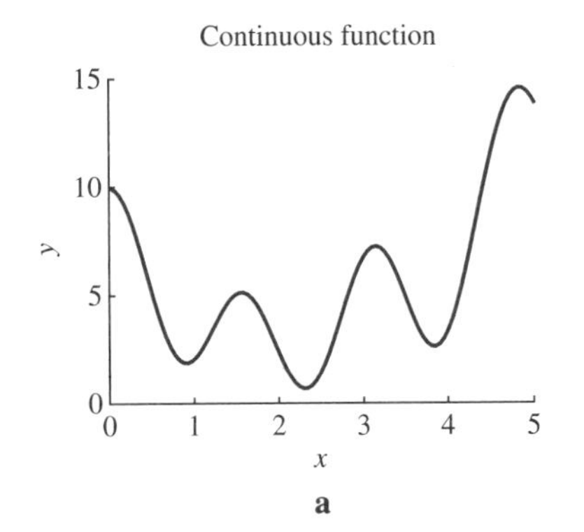

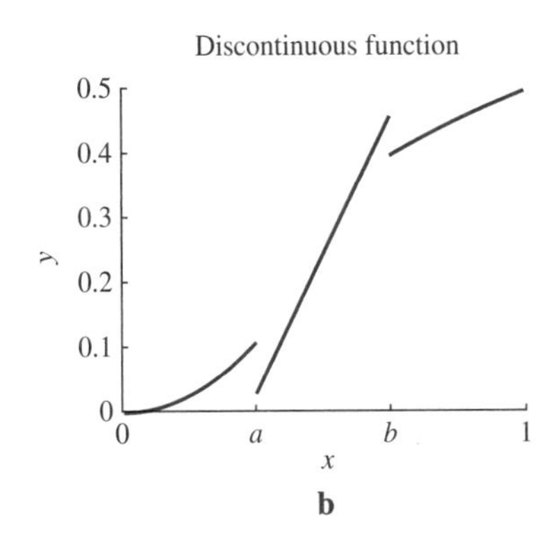

FIGURE 2.3.1
Continuous and discontinuous functions

Definition 2.4 A function f is **continuous** at a point a if

$$\lim_{x \to a} f(x) = f(a).$$

Otherwise, we say the function is **discontinuous** at the point a.

This definition requires that the left- and right-hand limits are equal. As the graph approaches a particular point in the domain, the value of a continuous function gets closer and closer to the value of the function at that point. Otherwise, the graph would have a jump and could not be drawn without lifting a pencil.

Continuity is defined at **points**. A function can be continuous at some points and discontinuous at others. The points a and b are points of discontinuity of the function in Figure 2.3.1b; the function is continuous at all other points. When a function is said to be continuous without specifying a particular point, the function is continuous at **all** points of its domain.

Definition 2.5 A function is **continuous** if it is continuous at every point in its domain.

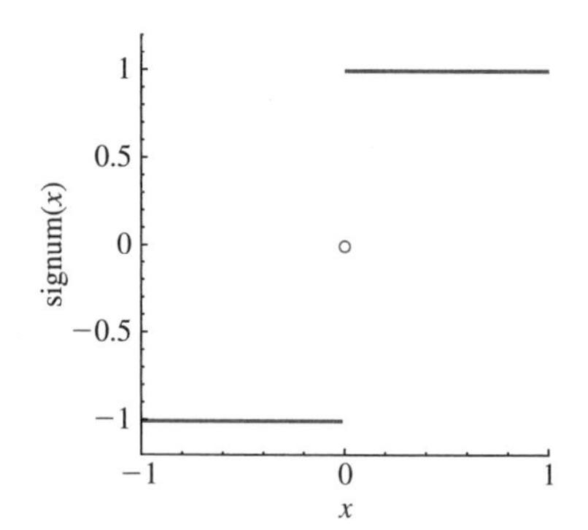

FIGURE 2.3.2
The signum function is not continuous at 0

What sorts of functions are not continuous? We have encountered examples of several types. Some functions have jumps, like the signum function which is equal to −1 for negative arguments, 0 when the argument is 0, and 1 for positive arguments (Figure 2.3.2). At the point $x = 0$, the value of the function is 0, the left-hand limit is −1, and the right-hand limit is 1. Because the left-hand limit and the right-hand limit do not match, the limit does not exist. Furthermore, neither of these limits matches the value of the function.

Second, functions with limits of infinity or negative infinity are not defined at all points and cannot be graphed without a jump (strictly speaking, these functions can be continuous at all points in their domain). In Figure 2.3.3a, the limit is infinity from both below and above. However, the function cannot be evaluated at $x = 0$ and has a jump. In Figure 2.3.3b, the limit is negative infinity from below and positive infinity from above. This function has an infinitely large jump.

Continuous functions are useful because we can find limits by plugging in. How can we recognize a continuous function? Like limits, continuous functions can be combined in many useful ways, including addition, multiplication, division, and composition. Mathematically, we build the edifice of continuous functions on a stable foundation of two theorems: one giving the basic continuous functions and the other saying how they can be combined.

Theorem 2.3 The Basic Continuous Functions

a. The constant function $f(x) = c$ is continuous.

b. The identity function $f(x) = x$ is continuous.

c. The exponential function $f(x) = e^x$ is continuous.

d. The logarithmic function $f(x) = \ln(x)$ is continuous for $x > 0$.

e. The cosine function $f(x) = \cos(x)$ is continuous.

Theorem 2.4 Combining Continuous Functions

Suppose $f(x)$ and $g(x)$ are continuous at $x = a$.

a. The sum $f(x) + g(x)$ is continuous at $x = a$.

b. The product $f(x)g(x)$ is continuous at $x = a$.

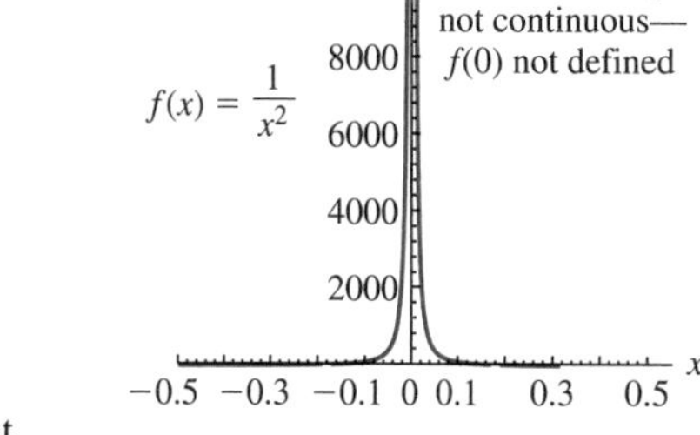

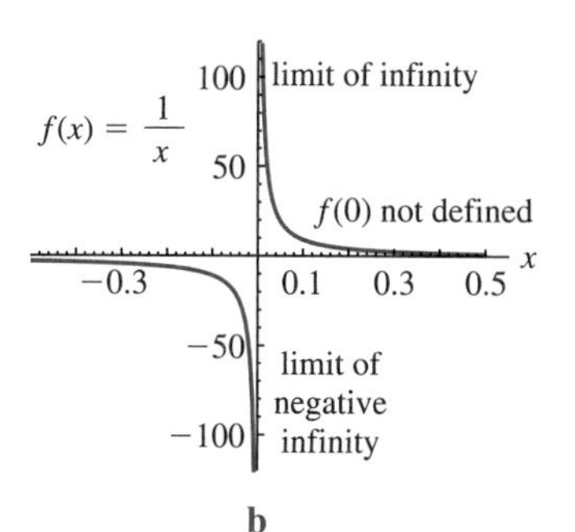

FIGURE 2.3.3
Two functions that are not defined at $x = 0$

c. The quotient $\dfrac{f(x)}{g(x)}$ is continuous at $x = a$ if $g(a) \neq 0$.

d. If $f(x)$ is continuous at $x = g(a)$, then the composition $(f \circ g)(x)$ is continuous at $x = a$.

The rules for recognizing continuous functions can be summarized in three tables: the basic continuous functions (Table 2.3.1), the rules for combining (Table 2.3.2), and the three trouble signs (Table 2.3.3).

Table 2.3.1 Basic continuous functions

Type of Function	Example	Where Continuous
Linear	$f(x) = mx + b$	All x
Polynomial	$f(x) = ax^3 + bx^2 + cx + d$	All x
Exponential	$f(x) = e^{\alpha x}$	All x
Cosine	$f(x) = \cos(x)$	All x
Logarithm	$f(x) = \ln(x)$	$x > 0$

Table 2.3.2 Continuous combinations of continuous functions $f(x)$ and $g(x)$

Type of Combination	Formula	Where Continuous	Example
Sum	$f(x) + g(x)$	where defined	$2x + 1 + e^x$
Product	$f(x)g(x)$	where defined	$(2x + 1)e^x$
Quotient	$\dfrac{f(x)}{g(x)}$	where $g(x) \neq 0$	$\dfrac{e^x}{2x + 1}$ if $2x + 1 \neq 0$
Composition	$(f \circ g)(x)$	where defined	e^{2x+1}

Table 2.3.3 The three trouble signs: points where a function may be undefined or discontinuous

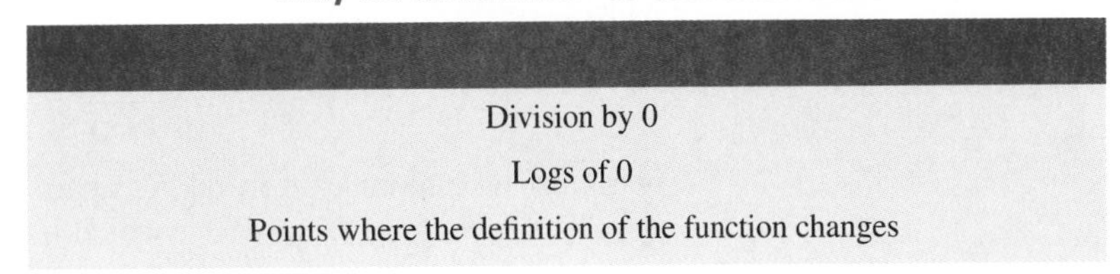

Division by 0
Logs of 0
Points where the definition of the function changes

Example 2.3.1 Recognizing a Discontinuous Function

The function $f(x) = \sin(\frac{1}{x})$ oscillates faster and faster as x approaches zero (Figure 2.3.4). There is no limit as $x \to 0$. We recognize that there might be a point of

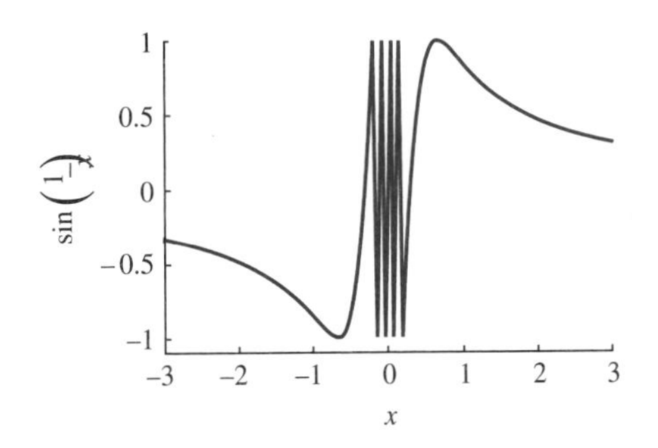

FIGURE 2.3.4
The function $f(x) = \sin(1/x)$

discontinuity at $x=0$ because this composition of functions involves division by 0. ▲

Example 2.3.2 Recognizing a Cryptic Discontinuous Function

The function $f(x)=\tan(x)$ does not have any obvious trouble signs. But if we recall that

$$\tan(x)=\frac{\sin(x)}{\cos(x)},$$

we note that there is division by 0 when $\cos(x)=0$ (Figure 2.3.5). This occurs at $x=\frac{-\pi}{2}$, $x=\frac{\pi}{2}$ and so forth. These are points where the function is not defined. ▲

Example 2.3.3 Evaluating the Limit of a Continuous Function I

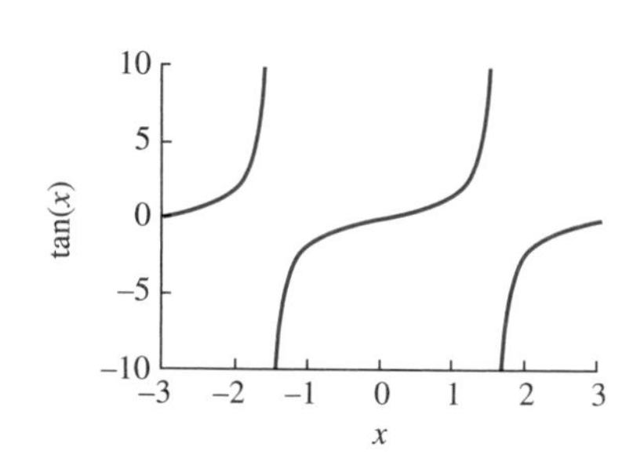

FIGURE 2.3.5
The function $f(x)=\tan(x)$

We can find

$$\lim_{x\to 0} f(x)$$

where

$$f(x)=e^{x^2+3}$$

by evaluating $f(0)$ if f is continuous at $x=0$. This function is built up as a **composition** of the polynomial function x^2+3 and the exponential function. Both of the pieces are continuous (Table 2.3.1). The composition of these pieces is also continuous (Table 2.3.2). Therefore,

$$\lim_{x\to 0} e^{x^2+3}=e^{0^2+3}=e^3=20.08.$$ ▲

Example 2.3.4 Evaluating the Limit of a Continuous Function II

If we wish to find

$$\lim_{x\to 0} g(x)$$

where

$$g(x)=\frac{f(x)}{h(x)}=\frac{e^{x^2+3}}{2+x+\ln(x+1)},$$

we must take more care in taking apart the function. First, it is the quotient of the previous function $f(x)=e^{x^2+3}$, which is continuous, and a new function $h(x)=2+x+\ln(x+1)$. $h(x)$ is the sum of a continuous linear function $2+x$ and the natural log, which is continuous when its argument is positive. The quotient will then be continuous at 0 as long as $h(0)\neq 0$. But

$$h(0)=2+0+\ln(0+1)=2+0+0=2\neq 0.$$

Therefore $h(x)$ is continuous and

$$\lim_{x\to 0}\frac{e^{x^2+3}}{2+x+\ln(x+1)}=\frac{e^{0^2+3}}{2}=\frac{e^3}{2}=10.04.$$ ▲

Example 2.3.5 Evaluating the Limit of a Continuous Function III

The function

$$P_d(t)=36.8+0.3\cos\left(\frac{2\pi(t-14)}{24}\right)$$

is continuous because it can be written as the composition

$$P_d(t)=g(\cos(h(t)))$$

where

$$g(x) = 36.8 + 0.3x$$
$$h(t) = \frac{2\pi(t-14)}{24}.$$

Both g and h are continuous linear functions, while cos is one of the basic continuous functions.

Example 2.3.6 Recognizing Disguised Division by 0

Division by 0 can be disguised. Suppose we are asked to find

$$\lim_{x\to 1^-} F(x) = \lim_{x\to 1^-} (1-x)^{-3}.$$

Negative exponents can be written in the denominator (law of exponents 3), so

$$F(x) = \frac{1}{(1-x)^3}.$$

Because the denominator is equal to 0 at $x = 1$, this function may not be continuous. If we try to evaluate $F(1)$, we find ourselves dividing by 0. The limit is in fact equal to infinity.

Example 2.3.7 A Discontinuous Function That Is Defined in Pieces

Functions defined in pieces look like the signum function (Figure 2.3.2) or the functions used to describe the heart (Section 1.11). An example is the discrete-time dynamical system

$$V_{t+1} = \begin{cases} 3.0V_t & \text{if } V_t > 2.0 \\ 2.0V_t + 1.0 & \text{if } V_t \le 2.0. \end{cases}$$

The only way to check whether the updating function is continuous is to check whether the two parts match up at $V_t = 2$, the point where the definition changes (Figure 2.3.6). The value at $V_t = 2$ jumps from 6.0 to 5.0. These do not match, and the function is not continuous.

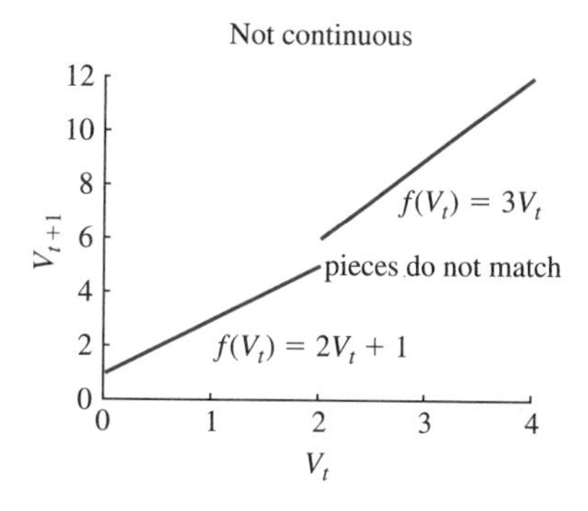

FIGURE 2.3.6
A discontinuous function that is defined in pieces

Example 2.3.8 A Continuous Function That Is Defined in Pieces

In contrast, the function

$$V_{t+1} = \begin{cases} 3.0V_t & \text{if } V_t > 2.0 \\ 2.0V_t + 2.0 & \text{if } V_t \le 2.0 \end{cases}$$

is continuous because the two pieces match up, each taking the value $V_{t+1} = 6.0$ at $V_t = 2.0$ (Figure 2.3.7).

2.3.2 Input and Output Tolerances

In applied mathematics, functions describe relationships between measurements. Continuous functions represent a special sort of relationship between measurements where a small change in input produces only a small change in the output.

Example 2.3.9 Input and Output Tolerances for a Continuous Function I

Recall the discrete-time dynamical system

$$b_{t+1} = 2.0b_t$$

describing a bacterial population (Example 1.5.1). Suppose we want a population of 2.0×10^6 at time $t = 1$. To hit 2.0×10^t exactly, we require $b_0 = 1.0 \times 10^6$. How close must b_0 be to 1.0×10^6 for b_1 to be within 0.1×10^6 of our target (Figure 2.3.8)? We require

$$1.9 \times 10^6 \le b_1 \le 2.1 \times 10^6$$
$$1.9 \times 10^6 \le 2.0b_0 \le 2.1 \times 10^6$$
$$0.95 \times 10^6 \le b_0 \le 1.05 \times 10^6.$$

As long as we can guarantee that b_0 is within 0.05×10^6 of 1.0×10^6, the output is within our tolerance. A small error in the input produces a small (but somewhat larger) error in the output. In other words, the output tolerance of 0.1×10^6 translates into an input tolerance of 0.05×10^6.

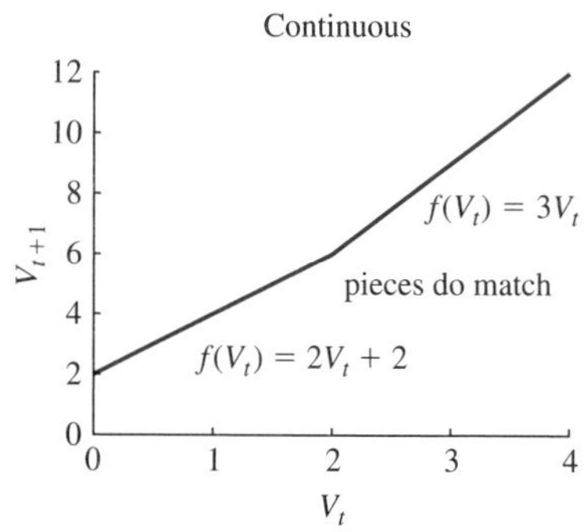

FIGURE 2.3.7
A continuous function that is defined in pieces

Example 2.3.10 Input and Output Tolerances for a Continuous Function II

Consider the discrete-time dynamical system

$$M_{t+1} = 0.5M_t + S$$

for the concentration of medication in the blood (Example 1.5.4), modified so that the dosage is a variable S rather than the constant 1.0. Suppose the concentration on day t is $M_t = 1.0$ and we wish to hit $M_t = 2.0$. What value of S should be used? The value of M_{t+1} is

$$M_{t+1} = 0.5 \cdot 1.0 + S = 0.5 + S.$$

Therefore, to hit $M_{t+1} = 2.0$ exactly, we require $S = 1.5$. Nothing in biology is exact, and perhaps we need only be within 0.2 of the target. Because M_{t+1} is a continuous function of S, there will be a tolerance around $S = 1.5$. For M_{t+1} to be between 1.8 and 2.2,

$$1.8 \le 0.5 + S \le 2.2$$
$$1.3 \le S \le 1.7$$

(Figure 2.3.9). If the input S is within 0.2 of 1.5, the output M_{t+1} will be within the desired tolerance of 2.0.

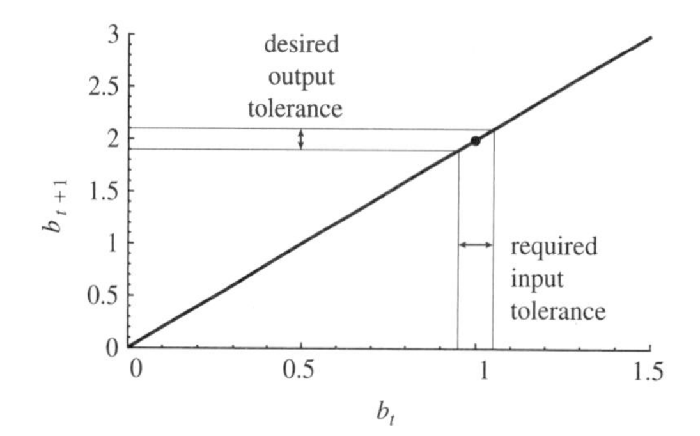

FIGURE 2.3.8
Tolerances and continuous functions

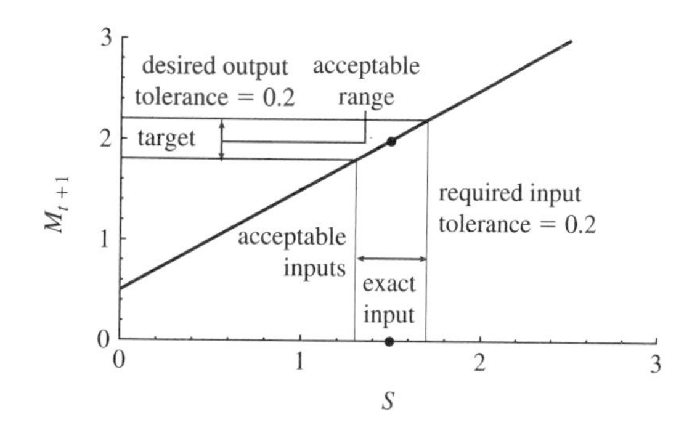

FIGURE 2.3.9
Tolerances and continuous functions

Example 2.3.11 Failure to Find Input Tolerance of a Discontinuous Function

In contrast, if the updating function is discontinuous, a tiny change in the input can produce a huge change in the output. Suppose we wish to have a voltage $V_1 = 5.0$ from the discrete-time dynamical system

$$V_{t+1} = \begin{cases} 3.0V_t & \text{if } V_t > 2.0 \\ 2.0V_t + 1.0 & \text{if } V_t \le 2.0 \end{cases}$$

(Example 2.3.7). We could hit $V_1 = 5.0$ exactly with $V_0 = 2.0$. But if the input is even a tiny bit too high, the output will be different. If $V_0 = 2.001$, then $V_1 = 6.003$ (Figure 2.3.10). If the output tolerance was 0.1, meaning that values between 4.9 and 5.1 were satisfactory, there is no input tolerance at all. ▲

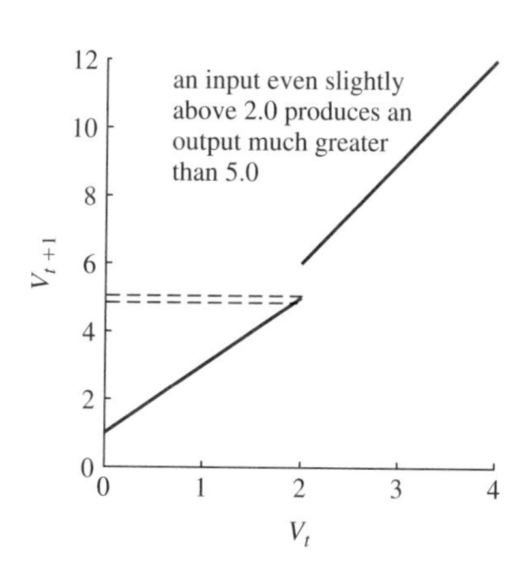

FIGURE 2.3.10
A discontinuous function with no input tolerance

Relations that are described by functions that are not continuous are difficult to deal with experimentally. Because no measurement can be controlled exactly, we hope that small errors in one measurement will not result in large errors in another. Systems with threshold behavior, or systems that display hysteresis (as in the next section), can include discontinuous relations and must be treated with caution.

2.3.3 Hysteresis

Different left-hand and right-hand limits arise in biological systems with **hysteresis** (Figure 2.3.11). In the experiment shown, a neuron is stimulated with different levels of electric current. Low levels of current generally have no effect, while large levels tend to induce an oscillation. If the current is started at a low level and steadily increased, the neuron switches from no oscillation (designated as period 0 on the graph) to an oscillation with period 2.0 when the input current crosses 3.0. If the current is instead started at a high value and steadily decreased, the neuron switches from an oscillation of period 1 to no oscillation when the input current crosses 2.0. In both cases, the jump indicates that the left-hand and right-hand limits do not match.

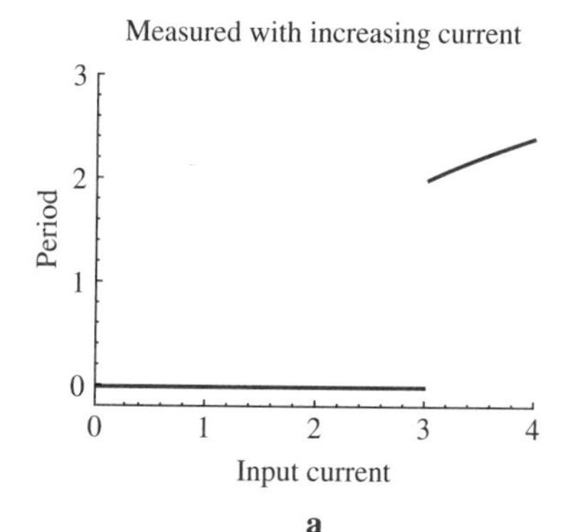

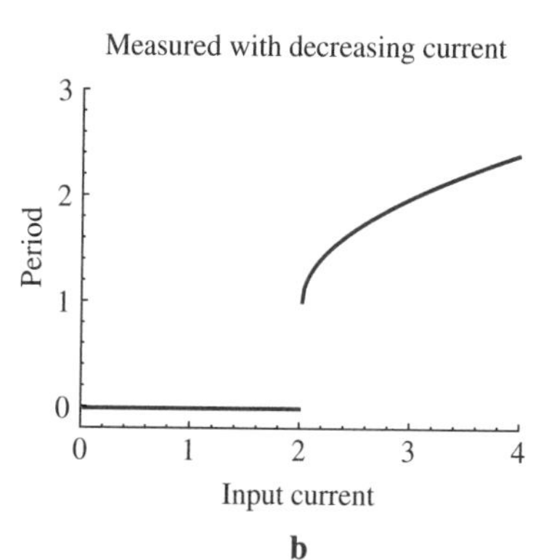

FIGURE 2.3.11
Hysteresis

If the device starts off tilted to the left, the ball is on the left when device is flat

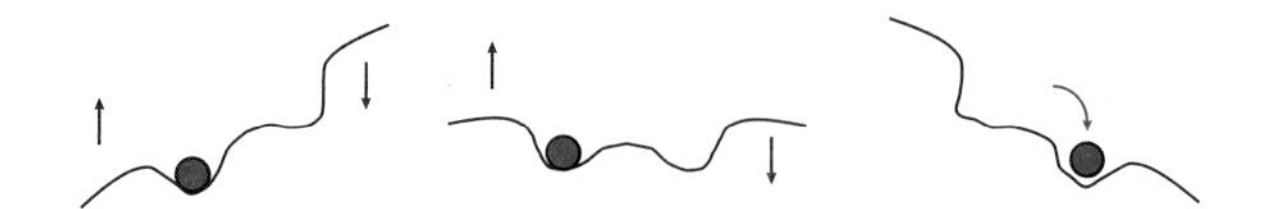

If the device starts off tilted to the right, the ball is on the right when device is flat.

FIGURE 2.3.12
Hysteresis and stability

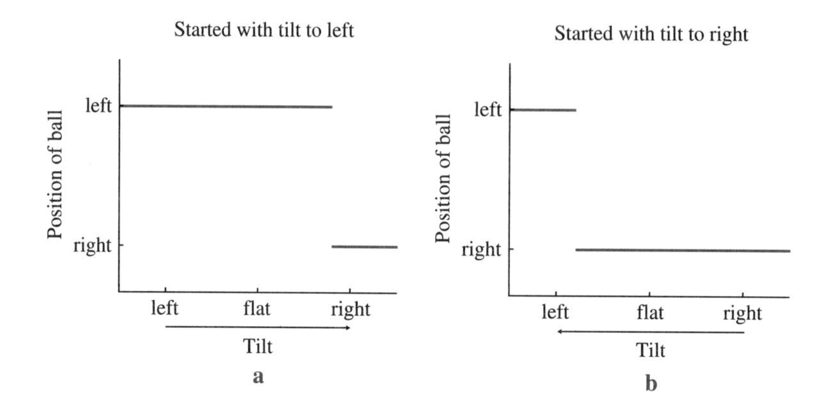

FIGURE 2.3.13
The position of a ball as a function of tilt

More oddly, the two graphs jump at different points. If we wish to **measure** the left-hand limit, we would slowly increase the current toward the jump, finding the results in Figure 2.3.11a. If we wish to measure the right-hand limit, we would slowly decrease the current toward the jump, finding the results in Figure 2.3.11b. However, both the position and the size of the jump depend on which way we change the current.

Why do they fail to match? The idea is related to the idea of equilibria. When the current increases, the equilibrium without an oscillation eventually becomes **unstable** and the cell begins to oscillate. When the current decreases, the oscillation itself becomes unstable and the cell stops oscillating. For input currents between 2.0 and 3.0, **both** the oscillation and the lack of oscillation are stable, and the behavior of the neuron depends on what it was last doing. Hysteresis is a simple form of memory, because knowing the input current at one time is insufficient to describe the behavior of the system.

Hysteresis can be pictured with a physical device. Figure 2.3.12 compares two experiments on a device with two possible stable resting places for a ball. In the top panel, the device begins tilted to the left, with the ball resting in the left depression and is then slowly rotated toward the right. When the device is flat, the ball is still on the left side. As the device is tilted to the right, the ball will roll into the right depression. In the bottom panel, the device begins tilted to the right, with the ball resting in the right depression, and is slowly rotated toward the left. When the device is flat, the ball is still on the right side. The position of the ball is an example of hysteresis (Figure 2.3.13). If you were asked to guess the position of the ball when the device is flat, you could only say that it depended on how it got there.

Summary Many important mathematical and scientific functions are **continuous**, meaning that they have graphs with no jumps. In these cases, limits can be computed by evaluating the function. Many continuous functions are built from four basic pieces: polynomials (including linear functions), exponentials, logarithms, and cosines, combined with addition, multiplication, division (except by zero), and functional composition. Most functions encountered in biology are continuous unless they exhibit one of the three warning flags: division by zero, taking the log of zero, or being defined in pieces. Continuous functions describe scientific measurements when small changes in the input produce only small changes in the output. With a discontinuous function, even a tiny error in input can produce a large jump in the output. We examined discontinuous functions created by **hysteresis**, where the results of an experiment depend on the order in which conditions are tested.

2.3 Exercises

Mathematical Techniques

1–10 ▪ Describe how the following functions are built out of the basic continuous functions. Identify points where they might not be continuous.

1. $l(t) = 5t + 6$.
2. $p(x) = x^5 + 6x^3 + 7$.
3. $f(x) = \frac{e^x}{x+1}$.
4. $h(y) = y^2 \ln(y - 1)$ for $y > 1$.
5. $g(z) = \frac{\ln(z-1)}{z^2}$ for $z > 1$.
6. $F(t) = \cos(e^{t^2})$.
7. $\sin(x)$ (recall that $\sin(x) = \cos(x - \frac{\pi}{2})$).
8. $a(t) = t^2$ if $t > 0$ and 0 if $t = 0$.
9. $r(w) = (1 - w)^{-4}$.
10. $q(z) = (1 + z^2)^{-2}$.

11–20 ▪ Using the given functions, find the limits by plugging in (if possible). Say whether the limit is infinity or negative infinity. Compute the value of the function 0.1 and 0.01 above and below the limiting argument to see if your answer is correct.

11. $\lim_{t\to 5} l(t)$ where $l(t) = 5t + 6$ (based on Exercise 1).
12. $\lim_{x\to 2} p(x)$ where $p(x) = x^5 + 6x^3 + 7$ (based on Exercise 2).
13. $\lim_{x\to 0} f(x)$ where $f(x) = \frac{e^x}{x+1}$ (based on Exercise 3).
14. $\lim_{y\to 1} h(y)$ where $h(y) = y^2 \ln(y-1)$ for $y > 1$ (based on Exercise 4).
15. $\lim_{z\to 2} g(z)$ where $g(z) = \frac{\ln(z-1)}{z^2}$ for $z > 1$ (based on Exercise 5).
16. $\lim_{y\to 2} h(y)$ where $h(y) = y^2 \ln(y-1)$ for $y > 1$ (based on Exercise 4).
17. $\lim_{z\to 0} g(z)$ where $g(z) = \frac{\ln(z-1)}{z^2}$ for $z > 1$ (based on Exercise 5).
18. $\lim_{t\to 2} F(t)$ where $F(t) = \cos(e^{t^2})$ (based on Exercise 6).
19. $\lim_{w\to 1} r(w)$ where $r(w) = (1 - w)^{-4}$ (based on Exercise 9).
20. $\lim_{t\to 0} a(t)$ where $a(t) = t^2$ if $t > 0$ and 0 otherwise (based on Exercise 8).

21–26 ▪ For the following functions, find the input tolerance necessary to achieve the given output tolerance.

21. How close must the input be to $x = 0$ for $f(x) = x + 2$ to be within 0.1 of 2?
22. How close must the input be to $x = 1$ for $f(x) = 2x + 1$ to be within 0.1 of 3?
23. How close must the input be to $x = 1$ for $f(x) = x^2$ to be within 0.1 of 1?
24. How close must the input be to $x = 2$ for $f(x) = 5x^2$ to be within 0.1 of 20?
25. Consider the Heaviside function, defined by

$$\begin{cases} H(x) = 0 & \text{if } x < 0 \\ H(x) = 1 & \text{if } x \geq 0 \end{cases}$$

How close must the input be to $x = 1$ for $H(x)$ to be within 0.1 of 1?

26. How close must the input be to $x = 0$ for $H(x)$ to be within 0.1 of 0?

27–28 ▪ We can build different continuous approximations of signum (the function giving the sign of a number) as follows. For each, case

a. Graph the continuous function.

b. Find the formula.

c. How close would the input have to be to 0 for the output to be within 0.1 of 0?

27. A continuous function that is -1 for $x \leq -0.1$, 1 for $x \geq 0.1$, and is linear for $-0.1 < x < 0.1$.
28. A continuous function that is -1 for $x \leq -0.01$, 1 for $x \geq 0.01$, and is linear for $-0.01 < x < 0.01$.

Applications

29–32 ▪ Find the accuracy of input necessary to achieve the desired output accuracy.

29. Suppose the mass of an object as a function of volume is given by $M = \rho V$. If $\rho = 2.0$ g/cm^3, how close must V be to 2.5 cm^3 for M to be within 0.2 g of 5.0 g?
30. The area of a disk as a function of radius is given by $A = \pi r^2$. How close must r be to 2.0 cm to guarantee an area within 0.5 cm^2 of 4π ?

31. The flow rate F through a vessel is proportional to the fourth power of the radius, or

$$F(r) = ar^4.$$

Suppose $a = 1.0/\text{cm s}$. How close must r be to 1.0 cm to guarantee a flow within 5% of 1 mL/s?

32. Consider an organism growing according to $S(t) = S(0)e^{\alpha t}$. Suppose $\alpha = 0.001/\text{s}$, and $S(0) = 1.0$ mm. At time 1000 s, $S(t) = 2.71828$ mm. How close must t be to 1000 s to guarantee a size within 0.1 mm of 2.71828 mm?

33–36 ▪ Suppose a population of bacteria follows the discrete-time dynamical system

$$b_{t+1} = 2.0b_t$$

and we wish to have a population within 1.0×10^8 of 1.0×10^9 at $t = 10$.

33. What values of b_9 produce a result within the desired tolerance? What is the input tolerance?

34. What values of b_5 produce a result within the desired tolerance? What is the input tolerance? Why is it harder to hit the target from here?

35. What values of b_0 produce a result within the desired tolerance? What is the input tolerance?

36. How would your answers differ if the discrete-time dynamical system were $b_{t+1} = 5b_t$? Would the tolerances be larger or smaller? Why?

37–40 ▪ Suppose the amount of toxin in a culture declines according to $T_{t+1} = 0.5T_t$ and we wish to have a concentration within 0.02 of 0.5 g/L at $t = 10$.

37. What values of T_9 produce a result within the desired tolerance? What is the input tolerance?

38. What values of T_5 produce a result within the desired tolerance? What is the input tolerance?

39. What values of T_0 produce a result within the desired tolerance? What is the input tolerance?

40. How would your answers differ if the discrete-time dynamical system were $T_{t+1} = 0.1T_t$? Would the tolerances be larger or smaller? Why?

41–44 ▪ Suppose a neuron has the following response to inputs. If it receives a voltage input V greater than or equal to a threshold of V_0, it outputs a voltage of kV for some constant k. If it receives an input less than the threshold value of V_0, it outputs a fixed voltage V^*.

41. Suppose that $k = 2.0$, $V_0 = 50$, and $V^* = 80$. Write and graph the function giving output in terms of input as a function defined in pieces.

42. Suppose that $k = 1.5$, $V_0 = 60$, and $V^* = 100$. Write and graph the function giving output in terms of input as a function defined in pieces.

43. If $k = 2.0$ and $V_0 = 50$, what would V^* have to be for the function to be continuous? Graph the resulting function.

44. If $V_0 = 50$ and $V^* = 80$, what would k have to be to make the function continuous? Graph the resulting function.

45–46 ▪ The following questions are based on examples of hysteresis involving children.

45. A child outside is swinging on a swing that makes a horrible screeching noise. Starting from when the swing is furthest back, the pitch of the screeching noise increases as it swings forward and then decreases as it swings back.

a. Draw a graph of the pitch as a function of position without hysteresis.

b. Draw a graph with hysteresis. Which graph seems more likely?

c. Imagine what each sounds like. Which is more irritating?

46. Little Billy walks due east to school, but must cross from the south side to the north side of the street. Because he is a very careful child, he crosses quickly at the first possible opportunity.

a. Graph little Billy's latitude as a function of distance from home on the way to school.

b. Graph little Billy's latitude as a function of distance from home on the way home.

c. Is this an example of hysteresis?

Computer Exercise

47. Graph the function $f(x) = \sin(\frac{1}{x})$. What happens near $x = 0$?

2.4 Computing Derivatives: Linear and Quadratic Functions

We learned about continuous functions and limits as a means to compute a particular, and particularly important, limit: the limit of the average rate of change. When it exists, this limit is the **instantaneous rate of change** or the **derivative**. First, we ask whether every function has a derivative, finding examples of functions that do not. Those that do are called **differentiable functions**. We then use the definition of the derivative to compute the derivatives of two simple functions, a **linear function** and a **quadratic function**. Along the way, we will see that the derivative, thought of as the slope of a graph, can help us reason more effectively about the measurements that graphs depict. We can tell whether a function is increasing or decreasing simply by checking whether the derivative is positive or negative.

2.4.1 Differentiable Functions

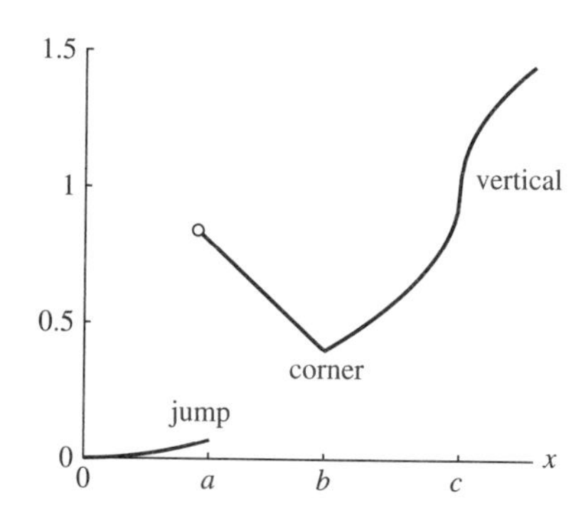

FIGURE 2.4.1

The graph of a function that is not differentiable at $x = a$, $x = b$, and $x = c$

Recall the two interpretations of the derivative: dynamically as the instantaneous rate of change and graphically as the slope of the tangent line. Do all functions (thought of as measured quantities that depend on time in this case) have a well-defined instantaneous rate of change and a well-defined tangent line?

The function shown in Figure 2.4.1 is not differentiable at three points in three different ways. At **a**, the function is not continuous. There is no way to draw a tangent. At **b**, there is a corner. No line can hug the curve at this point. At **c**, the tangent line is vertical. The tangent line, although it can be drawn, is not a function.

If Figure 2.4.1 represents a graph of position against time, the derivative is the velocity. At a point of discontinuity (such as point **a**) the object jumped instantly from one place to another. The idea of speed or instantaneous rate of change makes no sense. At a corner (such as point **b**), the object has instantly changed direction and has no well-defined speed. At a point with a vertical tangent line (such as point **c**), the tangent line has "infinite slope," meaning that the object has "infinite" speed at this time.

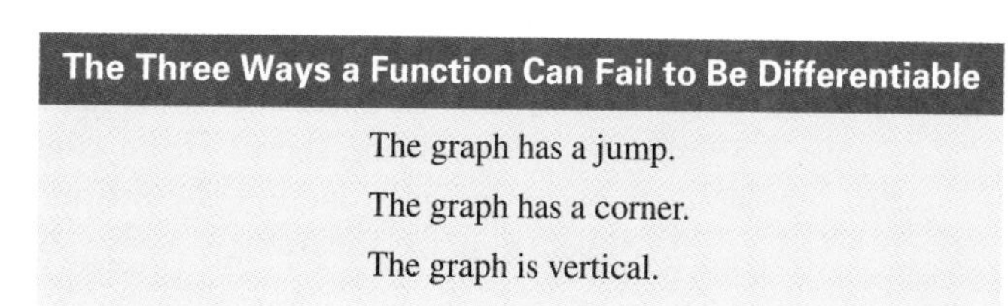

The Three Ways a Function Can Fail to Be Differentiable

The graph has a jump.
The graph has a corner.
The graph is vertical.

Differentiability, like continuity, is a property of a function at a point. If a function has a well-defined tangent line with finite slope at some point, the function is said to be **differentiable** at that point. If we say that a function is differentiable without mentioning a particular point, the function is differentiable on its entire domain. The function shown in Figure 2.4.1 is differentiable everywhere except at **a**, **b**, and **c**. Points where a function fails to have a derivative are one type of **critical point**. We define the second type in the next subsection.

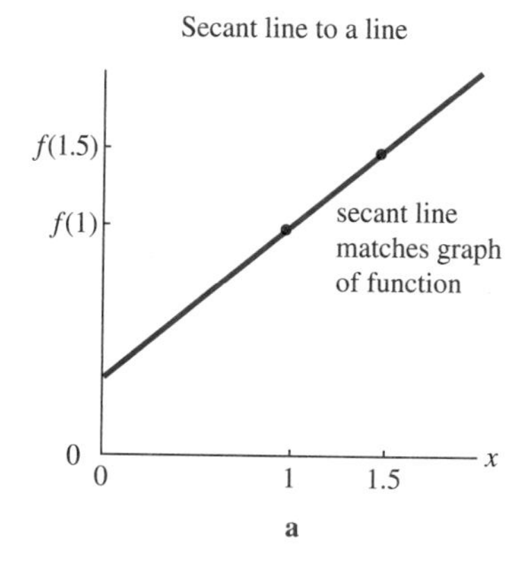

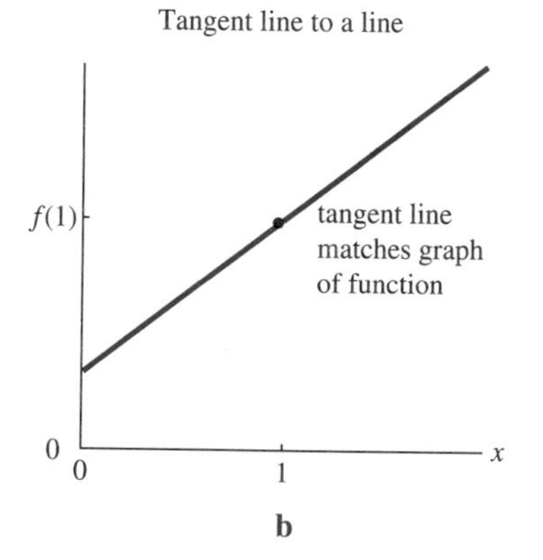

FIGURE 2.4.2

The secant and tangent line for a line

2.4.2 The Derivative of a Linear Function

What is the slope of the graph of the function $f(x) = 2x + 1$ at $x = 1$? We now have two ways to answer this question. Because f is a linear function, we know that the slope is the factor 2 multiplying x. Alternatively, we could compute the slope by computing the derivative, which requires finding the slope of a secant line and taking the limit as Δx approaches 0. These two methods should give the same answer.

The secant line to $f(x) = 2x + 1$ connects two points on the graph of the function. Because the graph of the function is a line, the graph of the secant line matches the graph of the function (Figure 2.4.2). For example, with base point $x_0 = 1.0$ and $\Delta x = 0.5$,

$$\begin{aligned}
\text{slope of secant} &= \frac{\Delta f}{\Delta x} \\
&= \frac{f(x_0 + \Delta x) - f(x_0)}{\Delta x} \\
&= \frac{f(1.0 + 0.5) - f(1.0)}{0.5}
\end{aligned}$$

$$= \frac{f(1.5) - f(1.0)}{0.5}$$
$$= \frac{4.0 - 3.0}{0.5}$$
$$= 2.0.$$

The slope of the secant matches the slope of the function.

In general, with base point x_0 and second point $x_0 + \Delta x$,

$$\begin{aligned}\text{slope of secant} &= \frac{\Delta f}{\Delta x} \\ &= \frac{f(x_0 + \Delta x) - f(x_0)}{\Delta x} \\ &= \frac{2(x_0 + \Delta x) + 1 - (2x_0 + 1)}{\Delta x} \\ &= \frac{2\Delta x}{\Delta x} \\ &= 2\end{aligned}$$

as long as $\Delta x \neq 0$. Every secant line has slope 2. Because it passes through a common base point, and only one line with a given slope passes through that point, the secant coincides exactly with the graph of the original function.

The slope of the tangent is the limit as Δx approaches 0 of the slope of the secant. The slope of the secant is the **constant** function 2 (except that it is not defined for $\Delta x = 0$). This function is **continuous** (Theorem 2.3a), and we can find the limit by evaluating the function at $\Delta x = 0$, so

$$\text{slope of tangent} = \lim_{\Delta x \to 0} 2 = 2.$$

The derivative does match the usual idea of the slope of a line.

In general, we can follow the same steps with the linear function $f(x) = mx + b$. The secant line connecting points x_0 and $x_0 + \Delta x$ has slope

$$\begin{aligned}\text{slope of secant} &= \frac{\Delta f}{\Delta x} \\ &= \frac{f(x_0 + \Delta x) - f(x_0)}{\Delta x} \\ &= \frac{(m(x_0 + \Delta x) + b) - (mx_0 + b)}{\Delta x} \\ &= \frac{m\Delta x}{\Delta x} \\ &= m\end{aligned}$$

as long as $\Delta x \neq 0$. The derivative of a linear function, the limit of m as Δx approaches 0, is equal to the slope m of the line.

In differential and prime notation, if $f(x) = mx + b$,

$$\frac{df}{dx} = m \qquad \text{differential notation}$$
$$f'(x_0) = m. \qquad \text{prime notation}$$

Sometimes we use a different version of differential notation and write

$$\frac{d(mx + b)}{dx} = m$$

or

$$\frac{d}{dx}(mx + b) = m.$$

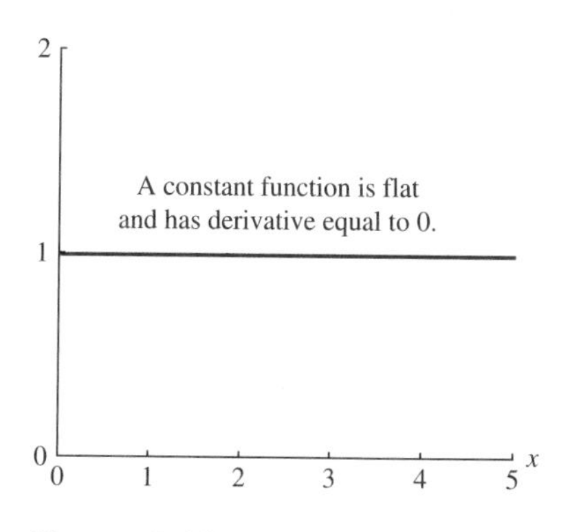

FIGURE 2.4.3
The slope of a constant function is 0

One special case is worthy of note. The function

$$f(x) = b,$$

the constant function, is a linear function with slope 0. According to the general formula for the derivative of a linear function,

$$\frac{df}{dx} = 0$$

when $f(x)$ is constant. The rate of change of something that does not change is 0 (Figure 2.4.3).

Lines can have positive, negative, or zero slope, corresponding to increasing, decreasing, and constant functions (Section 1.4). Because the derivative is equal to the slope, this same correspondence holds in general. If the derivative is positive, the rate of change of the function is positive and the function is **increasing** (Figure 2.4.4a). If the derivative is negative, the function is **decreasing** (Figure 2.4.4b). If the derivative is exactly zero, the function is neither increasing nor decreasing (Figure 2.4.4c) at that point.

Points where the derivative is zero are the other type of **critical point** and will prove extremely useful for finding maxima and minima of functions (Section 3.3). The definition combines the two types of critical point.

Definition 2.6 A function f has a **critical point** at x if $f(x) = 0$ or if the derivative is not defined at x.

The derivative encodes more information than simply whether the function is increasing or decreasing. We know that lines with large slopes are steep. Large values of the derivative are thus associated with points where the graph of the function is steep.

Example 2.4.1 Graphing and Interpreting the Derivative

The derivative of the function plotted in Figure 2.4.5a is shown in Figure 2.4.5b. Steep increasing portions of the graph correspond to large positive values of the derivative.

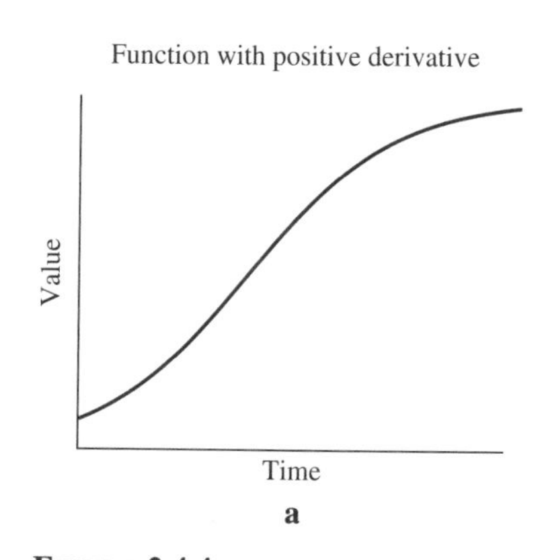

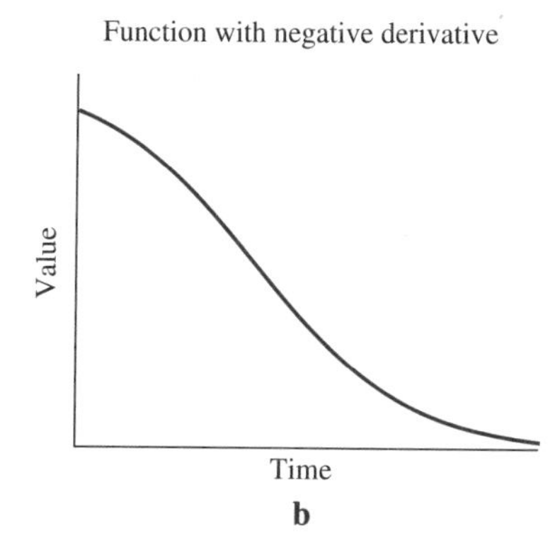

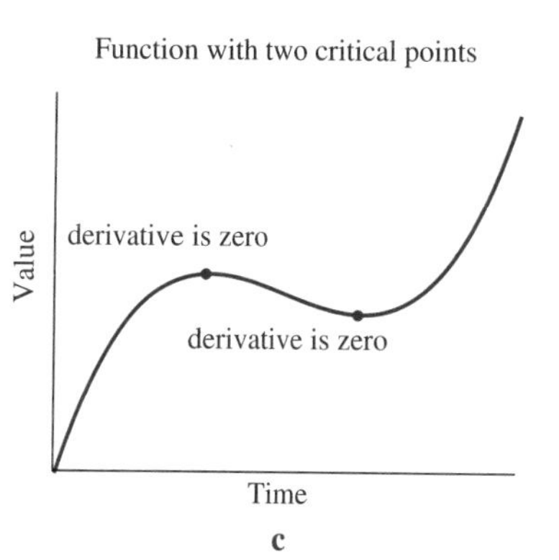

FIGURE 2.4.4
Functions with positive and negative derivatives

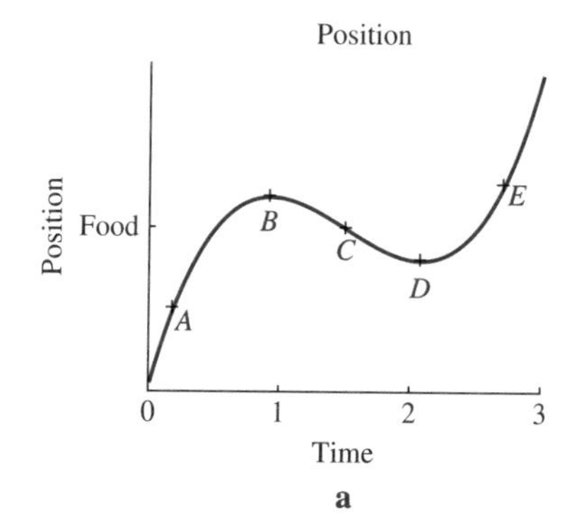

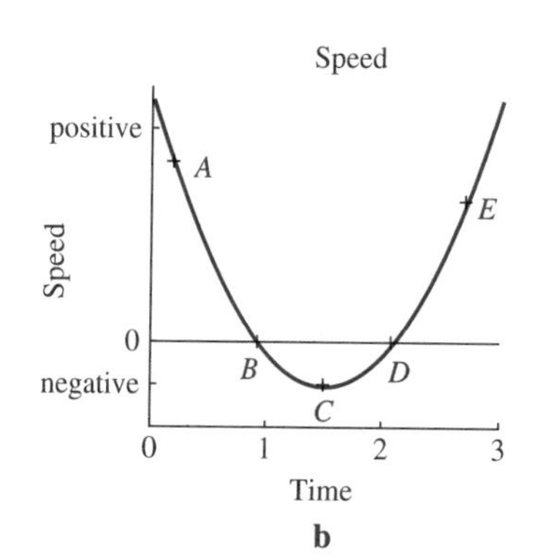

FIGURE 2.4.5
A function and its derivative

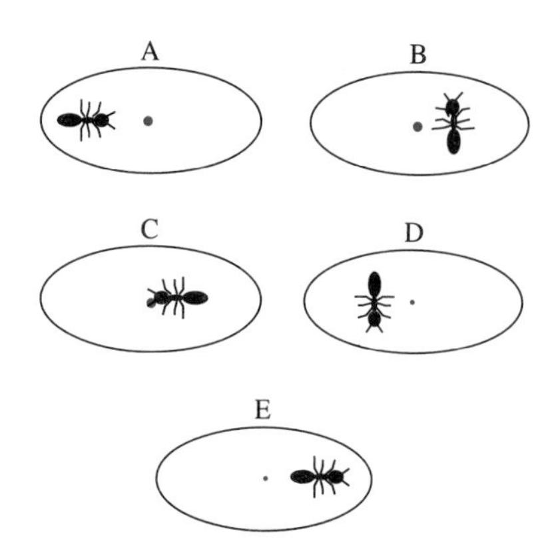

FIGURE 2.4.6
The experiences of an ant

Steep decreasing portions of the graph correspond to large negative values of the derivative. The derivative passes through zero wherever the graph is flat.

Suppose Figure 2.4.5a represents the horizontal position of an ant as a function of time. There is a bit of food at the position marked **C**. At the point marked **A** the ant is moving forward quickly toward the food. At **B**, the ant has overshot the food and stopped. At **C** the ant is walking slowly through the food in a reverse direction. By **D** the ant has passed the food and stopped, preparatory to turning around and walking quickly away at **E** (Figure 2.4.6).

Example 2.4.2 Identifying Regions with Positive and Negative Rates of Change

Suppose the volume of a cell is given graphically (Figure 2.4.7a). The volume is increasing when the graph is increasing, or between times 0 and about 0.6, and again between times 1.5 and about 2.6. It is decreasing between time 0.6 and 1.5, and again after time 2.6. The graph of the derivative is positive when the function is increasing and negative when the function is decreasing. Furthermore, the derivative takes on its largest positive value when the function is increasing most steeply at about time 2, and its largest negative value when the function is decreasing most quickly at about time 1.

2.4.3 A Quadratic Function

Consider a cat that leaps from a building in a constant gravitational field. The cat falls a distance y (measured in meters) in time t (measured in seconds) where $y(t)$ satisfies

$$y(t) = 5t^2.$$

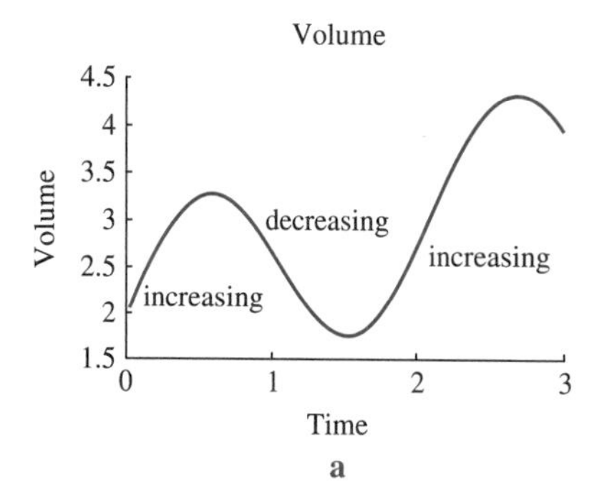

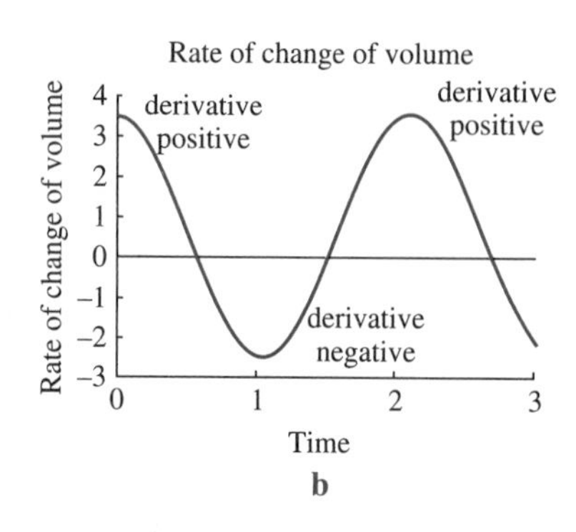

FIGURE 2.4.7
The changing volume of a cell

What is the downward velocity of the cat after 1.0 s? The velocity is the rate of change of position. The average velocity between $t = 1.0$ and $t = 1.0 + \Delta t$ is found by dividing the distance traveled by the elapsed time, or

$$\begin{aligned}\text{average velocity} &= \frac{\text{distance}}{\text{time}}\\ &= \frac{\Delta y}{\Delta t}\\ &= \frac{y(1.0+\Delta t) - y(1.0)}{\Delta t}\\ &= \frac{5(1.0+\Delta t)^2 - 5}{\Delta t}\\ &= \frac{(5 + 10\Delta t + 5\Delta t^2) - 5}{\Delta t}\\ &= \frac{10\Delta t + 5\Delta t^2}{\Delta t}\\ &= 10 + 5\Delta t\end{aligned}$$

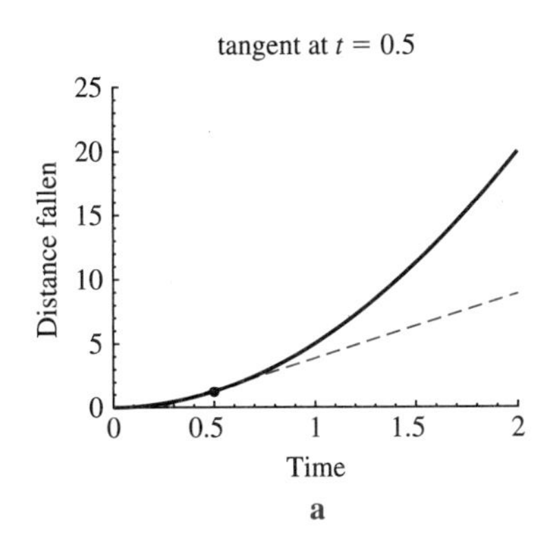

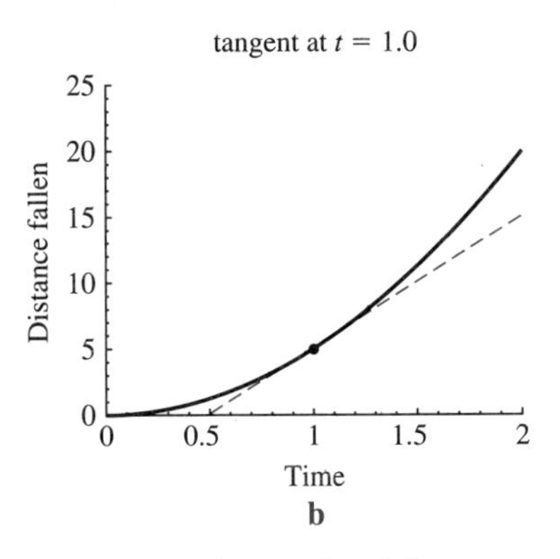

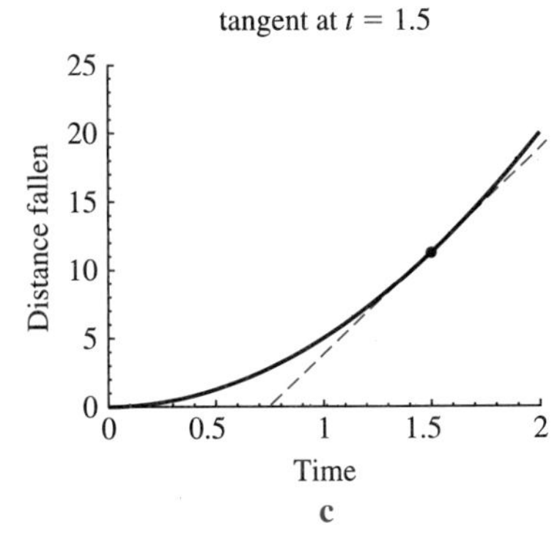

FIGURE 2.4.8

The slope of a quadratic function is constantly changing

as long as $\Delta t \neq 0$. To compute the derivative, we must take the limit of this function as Δt approaches 0. The average downward velocity is a linear function of Δt and is therefore continuous (Table 2.3.1). We can thus find the limit by evaluating at $\Delta t = 0$,

$$\begin{aligned}\text{velocity} &= \lim_{\Delta t \to 0} (10 + 5\Delta t)\\ &= 10 + 5 \cdot 0 = 10.\end{aligned}$$

What are the units of the velocity? The change in distance Δy is measured in meters, and the change in time Δt is measured in seconds. The velocity at $t = 1.0$ is therefore 10 m/s.

We can follow this procedure to find the exact downward velocity at any time t. The velocity is a function of time because it is constantly changing (Figure 2.4.8). The average velocity between times t and $t + \Delta t$ is

$$\begin{aligned}\text{average velocity} &= \frac{\Delta y}{\Delta t}\\ &= \frac{y(t+\Delta t) - y(t)}{\Delta t}\\ &= \frac{5(t+\Delta t)^2 - 5t^2}{\Delta t}\\ &= \frac{5(t^2 + 2t\Delta t + \Delta t^2) - 5t^2}{\Delta t}\\ &= \frac{(5t^2 + 10t\Delta t + 5\Delta t^2) - 5t^2}{\Delta t}\\ &= \frac{10t\Delta t + 5\Delta t^2}{\Delta t}\\ &= 10t + 5\Delta t\end{aligned}$$

as long as $\Delta t \neq 0$. Taking the limit by evaluating this continuous function at $\Delta t = 0$,

$$\begin{aligned}\text{velocity} &= \lim_{\Delta t \to 0} 10t + 5\Delta t\\ &= 10t + 5 \cdot 0 = 10t.\end{aligned}$$

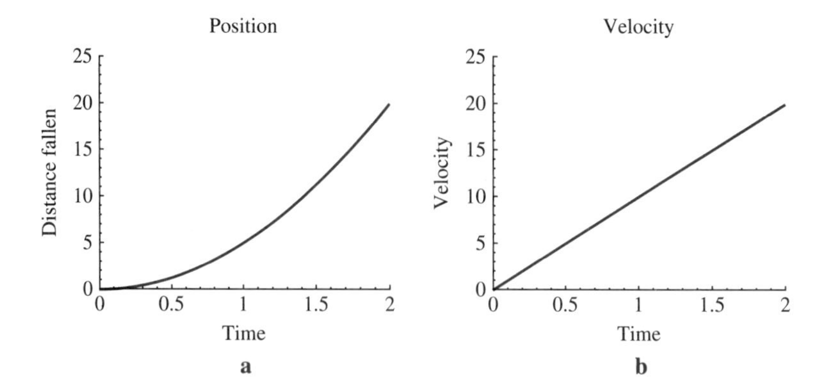

FIGURE 2.4.9
Position and velocity as functions of time

In differential and prime notation, if $y(t) = 5t^2$,

$$\frac{dy}{dt} = 10t \quad \text{differential notation}$$

$$y'(t) = 10t \quad \text{prime notation}$$

$$\frac{d}{dt}(5t^2) = 10t. \quad \text{alternative differential notation}$$

The function $y'(t)$ gives the downward velocity as a function of time (Figure 2.4.9). Does the result make sense? At time $t = 0$, the velocity is 0. As time increases, the velocity increases linearly, consistent with the behavior of a falling object.

When written in differential notation, the equation

$$\frac{dy}{dt} = 10t$$

is a **differential equation**, giving a formula for the rate of change as a function of time. In this case, we know the **solution** $y(t) = 5t^2$ of the differential equation because we began with a formula for $y(t)$ itself.

Example 2.4.3 The Behavior of a Falling Object

An object dropped from a height of 50 m on a planet where gravity has acceleration a has distance above the ground of

$$M(t) = 50.0 - \frac{a}{2}t^2.$$

We can use the derivative to find the velocity of the object as a function of time. To find the velocity, we compute the derivative as

$$\begin{aligned}
\text{average velocity} &= \frac{\Delta M}{\Delta t} \\
&= \frac{M(t + \Delta t) - M(t)}{\Delta t} \\
&= \frac{[50.0 - \frac{a}{2}(t + \Delta t)^2] - (50.0 - \frac{a}{2}t^2)}{\Delta t} \\
&= \frac{-\frac{a}{2}(t + \Delta t)^2 + \frac{a}{2}t^2}{\Delta t} \\
&= \frac{-\frac{a}{2}[(t + \Delta t)^2 - t^2]}{\Delta t} \\
&= \frac{-\frac{a}{2}(t^2 + 2t\Delta t + \Delta t^2 - t^2)}{\Delta t}
\end{aligned}$$

$$= \frac{-\frac{a}{2}(2t\Delta t + \Delta t^2)}{\Delta t}$$
$$= -\frac{a}{2}(2t + \Delta t)$$

as long as $\Delta t \neq 0$. The limit of this continuous function can be found by plugging in $\Delta t = 0$, giving an instantaneous velocity of

$$v(t) = -\frac{a}{2}(2t + 0) = -at.$$

This velocity is negative because the distance above the ground is decreasing. The speed (the absolute value of velocity) increases linearly with time.

Example 2.4.4 How Fast Does a Falling Object Hit the Ground?

On earth, the acceleration of gravity is 9.8 m/s^2. We can use this to find the velocity of this object when it hits the ground. To find the time when it hits the ground, we solve $M(t) = 0$ for t,

$$50.0 - 4.9t^2 = 0$$
$$50.0 = 4.9t^2$$
$$\frac{50.0}{4.9} = t^2$$
$$10.2 = t^2$$
$$t = \sqrt{10.2} = 3.19.$$

The velocity is $v(3.19) = -9.8 \cdot 3.19 = 31.3$ m/s.

Summary A function might fail to have a derivative at a particular point in three ways: if the graph has a jump, has a corner, or is vertical. The graphs of linear and quadratic functions have none of these problems, and we can compute their derivatives directly from the definition. The derivative of a linear function is equal to the usual slope. The derivative is positive when the function is increasing and negative when the function is decreasing. Points where the function is neither increasing nor decreasing are characterized by a derivative of zero. **Critical points** are points where either the derivative is zero or is not defined. Finally, we studied a quadratic function describing the position of an object in a constant gravitational field, finding that the derivative (the **velocity**) is a linear function of time.

2.4 Exercises

Mathematical Techniques

1–4 ▪ Expand the following binomials.

1. $(x + \Delta x)^2$
2. $(x + 2\Delta x)^2$
3. $(3x + 2\Delta x)^2$
4. $\left(2x + \frac{\Delta x}{2}\right)^2$

5–6 ▪ On the following graphs, identify points where

a. the function is not continuous,

b. the function is not differentiable (and say why),

c. the derivative is zero (critical points).

5.

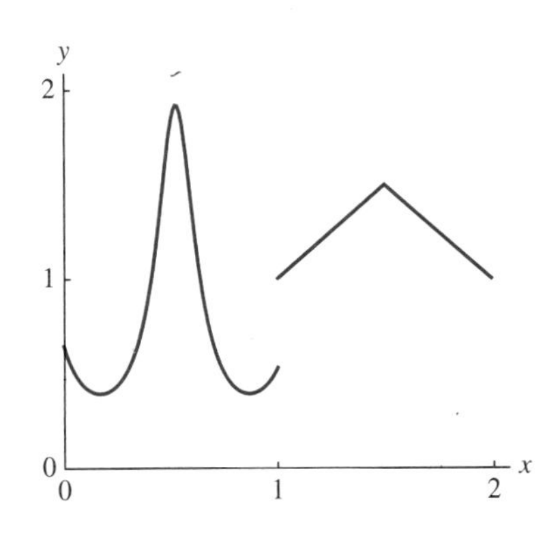

6.

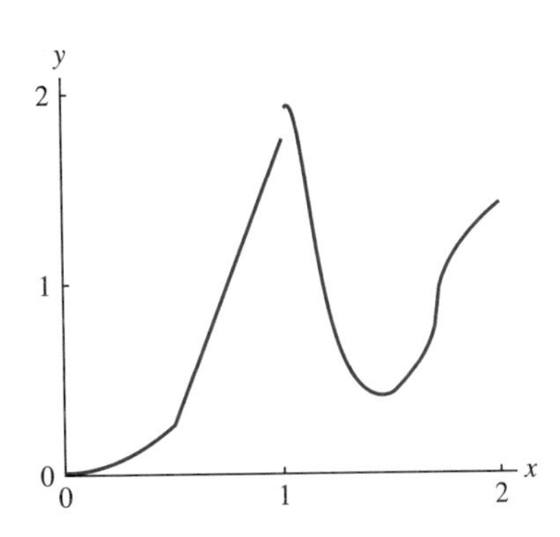

7–10 ▪ Find the derivatives of the following functions. Write your answers in both differential and prime notation. Which functions are increasing and which are decreasing?

7. $M(x) = 0.5x + 2$.

8. $L(t) = 2t + 30$.

9. $g(y) = -3y + 5$.

10. $Q(z) = -3.5 \times 10^8$.

11–12 ▪ For each of the following quadratic functions, find the slope of the secant line connecting $x = 1$ and $x = 1 + \Delta x$, and the slope of the tangent line at $x = 1$ by taking the limit.

11. $f(x) = 4 - x^2$.

12. $g(x) = x + 2x^2$.

13–14 ▪ For each of the following quadratic functions, find the slope of the secant line connecting x and $x + \Delta x$, and the slope of the tangent line as a function of x. Write your result in both differential and prime notation.

13. $f(x) = 4 - x^2$ (based on Exercise 11).

14. $g(x) = x + 2x^2$ (based on Exercise 12).

15–16 ▪ For each of the following quadratic functions, graph the function and the derivative. Identify critical points, points where the function is increasing, and points where the function is decreasing.

15. $f(x) = 4 - x^2$ (based on Exercise 13).

16. $g(x) = x + 2x^2$ (based on Exercise 14).

17–18 ▪ On the figures, label the following points and sketch the derivative.

a. One point where the derivative is positive.

b. One point where the derivative is negative.

c. The point with maximum derivative.

d. The point with minimum (most negative) derivative.

e. Points with derivative of zero (critical points).

17.

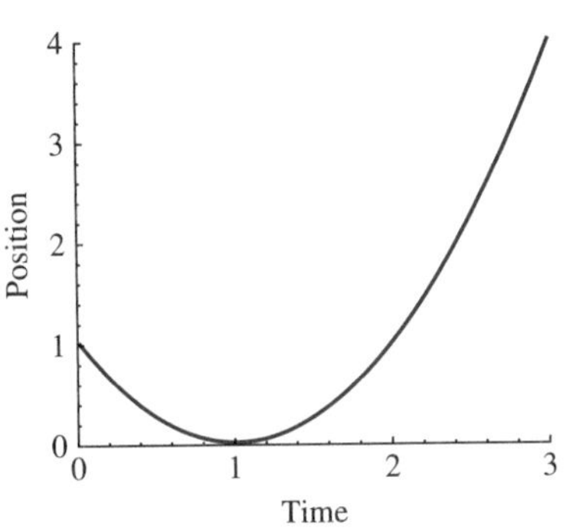

18.

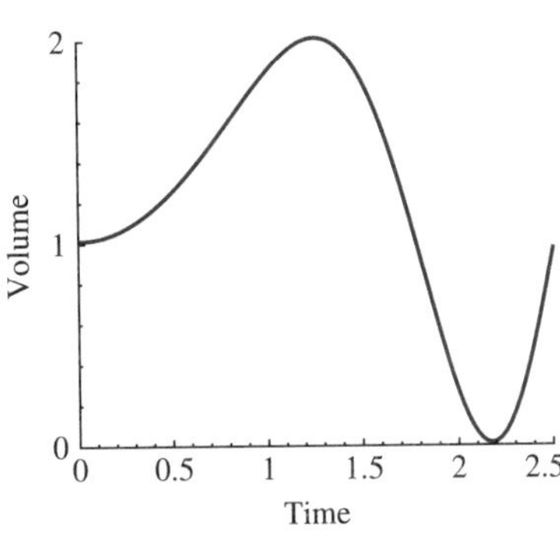

19–22 ▪ On the figures, identify which of the curves is a graph of the derivative of the other.

19.

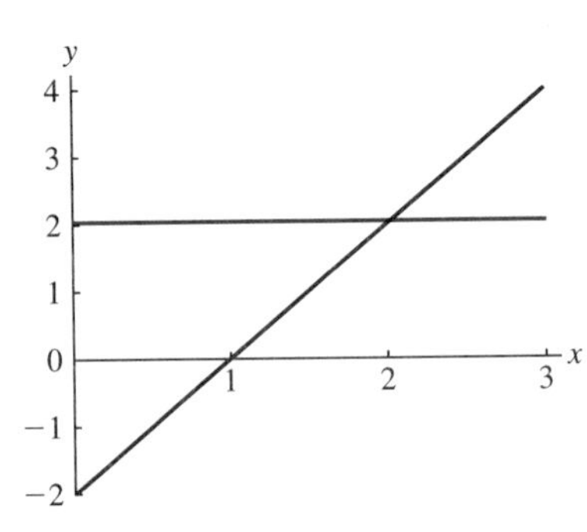

20.

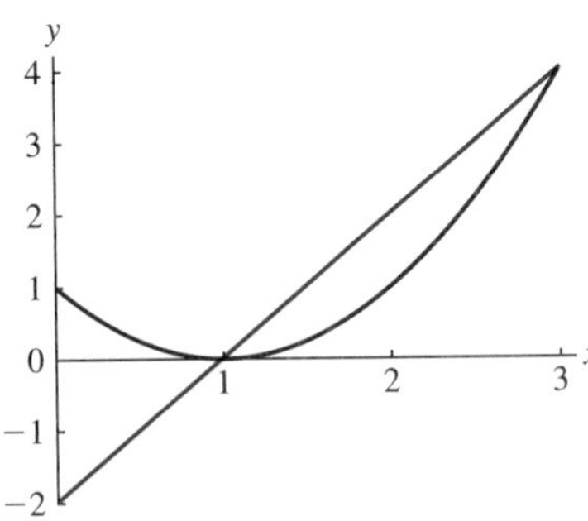

21.

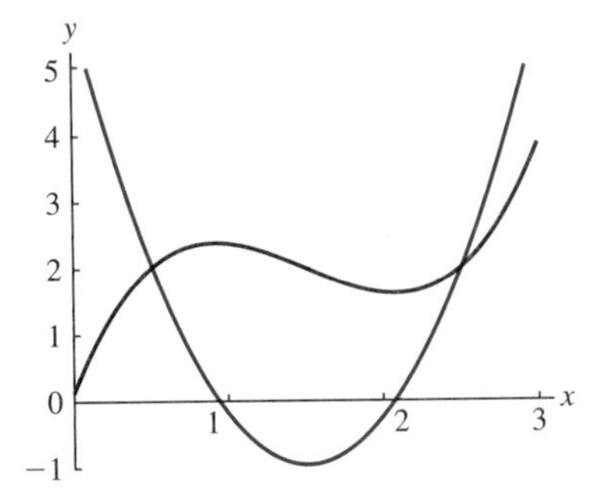

22.

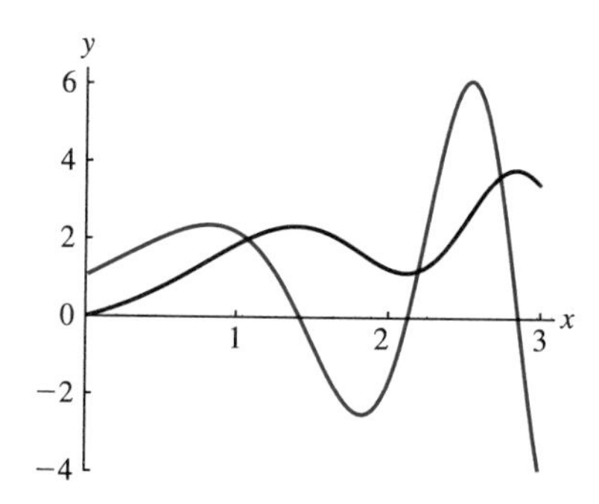

23–26 ▪ The following functions all fail to be differentiable at $x = 0$. In each case, graph the function, see what happens if you try to compute the derivative as the limit of the slopes of secant lines, and say something about the tangent line.

23. The absolute value function $f(x) = |x|$.

24. The square root function $f(x) = \sqrt{x}$. (Because this function is only defined for $x \geq 0$, you can only use $\Delta x > 0$.)

25. The Heaviside function (Section 2.3, Exercise 25), defined by

$$\begin{cases} H(x) = 0 & \text{if } x < 0 \\ H(x) = 1 & \text{if } x \geq 0. \end{cases}$$

26. The signum function defined by

$$\begin{cases} S(x) = -1 & \text{if } x < 0 \\ S(x) = 0 & \text{if } x = 0 \\ S(x) = 1 & \text{if } x \geq 0. \end{cases}$$

Applications

27–30 ▪ The following graphs show the temperature of different solutions with chemical reactions as functions of time. Graph the rate of change of temperature in each case, and indicate when the solution is warming up and when it is cooling down.

27.

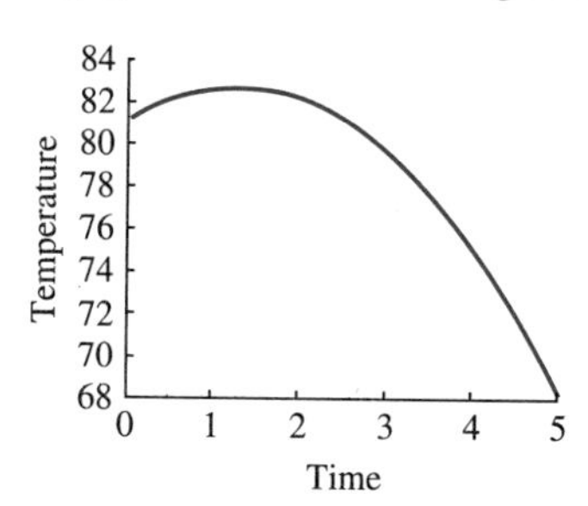

28.

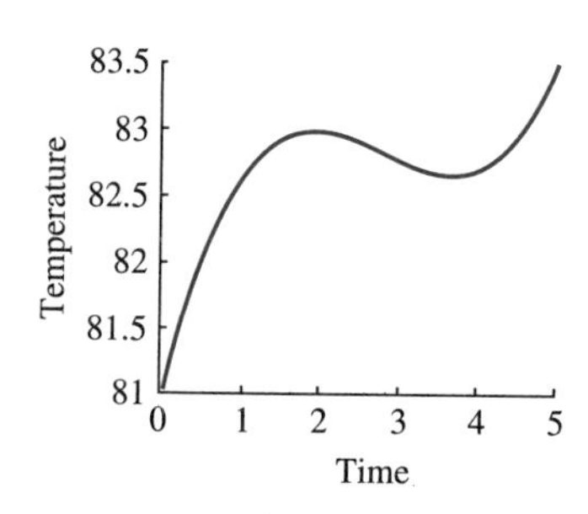

29.

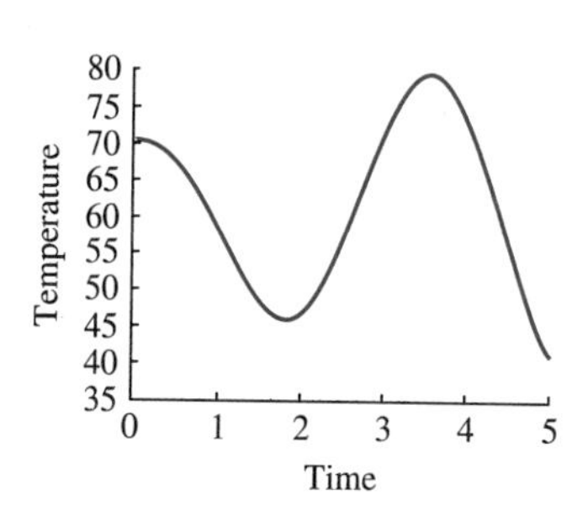

30.

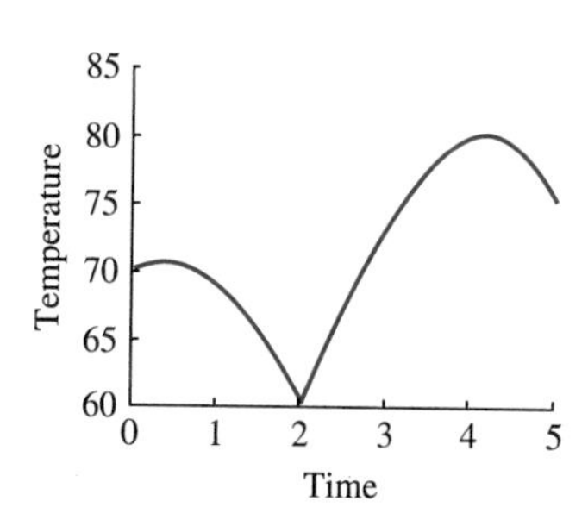

31–34 ▪ A bear sets off in pursuit of a hiker. Graph the position of a bear and hiker as functions of time from the following descriptions.

31. Both move at constant speed, but the bear is faster and eventually catches the hiker.

32. Both increase speed until the bear catches the hiker.

33. The bear increases speed and the hiker steadily slows down until the bear catches the hiker.

34. The bear runs at constant speed, the hiker steadily runs faster until the bear gives up and stops. The hiker slows down and stops soon after that.

35–38 ▪ An object dropped from a height of 100 m has distance above the ground of

$$M(t) = 100 - \frac{a}{2}t^2$$

where a is the acceleration of gravity. For each of the following planets with the given acceleration, find the time when the object hits the ground, and the speed of the object at that time.

35. On Earth, where $a = 9.78\text{m/sec}^2$.

36. On the moon, where $a = 1.62\text{m/sec}^2$.

37. On Jupiter, where $a = 22.88\text{m/sec}^2$.

38. On Mars' moon Deimos, where $a = 2.15 \times 10^{-3}\text{m/sec}^2$.

Computer Exercise

39. Have your computer plot the following for the function

$$y(t) = 10t^{2.5}.$$

Use base point $t = 1$.

a. The secant line with $\Delta t = 1$.

b. The secant line with $\Delta t = -1$.

c. The secant line with $\Delta t = 0.1$.

d. The secant line with $\Delta t = -0.1$.

e. Use smaller and smaller values of Δt and try to estimate the slope of the tangent. Graph it, then zoom in on your graph. Does the tangent look right?

2.5 Derivatives of Sums, Powers, and Polynomials

In the previous section, we used the definition of the derivative and the limit to compute some derivatives. Derivatives of complicated functions can be computed easily when the functions have been built from simpler component parts. If we have one set of rules for computing derivatives of these components, and another for putting these derivatives together, we will be able to find the derivative of almost any function. The next set of sections is dedicated to finding these rules. We will derive ways to compute derivatives of the most important components used to build biological functions: power functions, exponential functions, and trigonometric functions. We will then combine these with the **sum rule**, **product rule**, **quotient rule**, and **chain rule** (for computing derivatives when functions have been combined with functional composition).

This section presents three rules for putting derivatives together. First, **the sum rule** says that the derivative of the sum is the sum of the derivatives. Second, the **power rule** gives a formula for the derivative of a power function. Third, the **constant product rule** states that multiplying a function by a constant multiplies the derivative by that same constant. With these tools, we can compute the derivative of any **polynomial**.

2.5.1 The Sum Rule

Consider the function

$$s(t) = 5t^2 + 15t.$$

This is the sum of two functions we studied in the previous section: the quadratic function $y(t) = 5t^2$ and a linear function $l(t) = 15t$. The derivatives of the pieces are

$$y'(t) = 10t$$
$$l'(t) = 15.$$

How can we use our knowledge of the derivatives of the building blocks $y(t)$ and $l(t)$ to find the derivative of the sum $s(t)$? The technique is provided by the following theorem, which says that **the derivative of the sum** of two functions is **the sum of the derivatives**.

Theorem 2.5 The Sum Rule for Derivatives

Suppose

$$s(x) = f(x) + g(x)$$

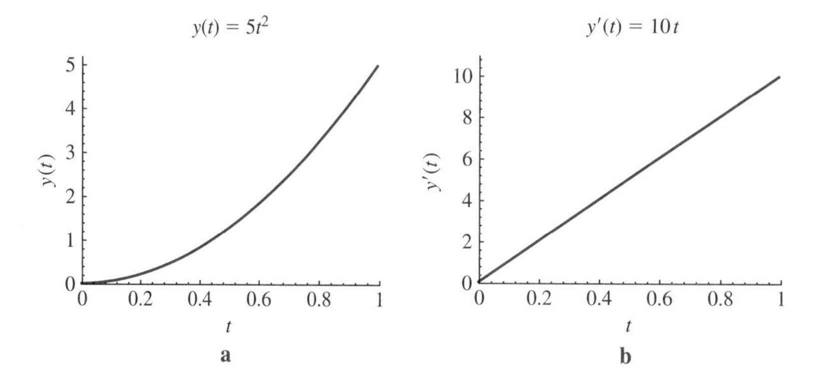

FIGURE 2.5.1
Graph and derivative of $y(t)$

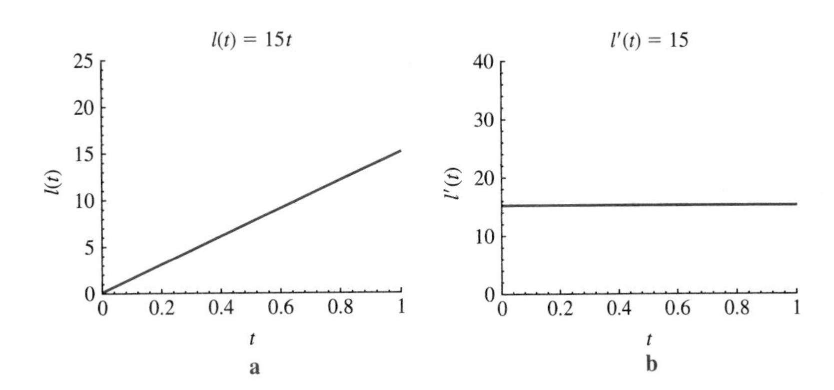

FIGURE 2.5.2
Graph and derivative of $l(t)$

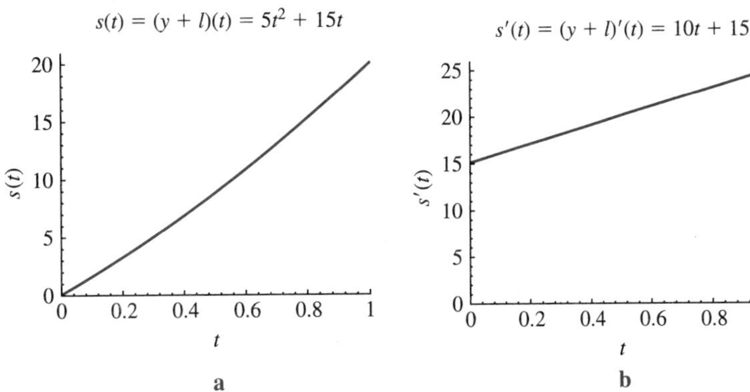

FIGURE 2.5.3
Graph and derivative of $s(t) = y(t) + l(t)$

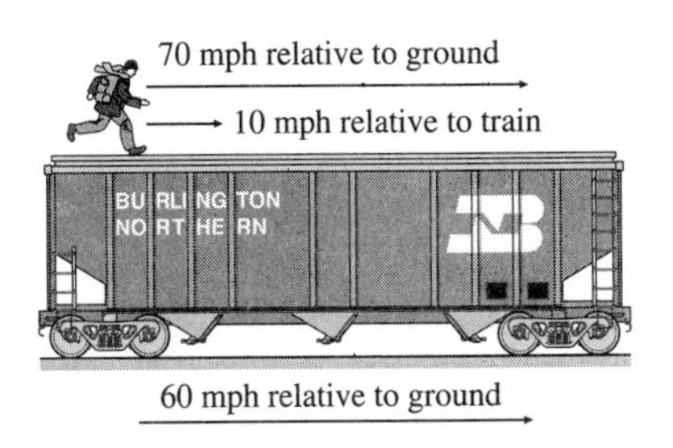

FIGURE 2.5.4
Velocities add

where f and g are both differentiable functions. Then

$$s'(x) = f'(x) + g'(x) \quad \text{prime notation}$$

$$\frac{ds}{dx} = \frac{df}{dx} + \frac{dg}{dx}. \quad \text{differential notation}$$

Applying the sum rule to our example with $y(t) = 5t^2$ and $l(t) = 15t$,

$$s'(t) = y'(t) + l'(t) = 10t + 15.$$

The functions and their derivatives are shown in figures 2.5.1, 2.5.2 and 2.5.3.

Why does the sum rule work? In one interpretation, the derivative of position is the velocity of a moving object. Suppose a train has a speed of 60 mph and a person is running forward on the train at a speed of 10 mph. Relative to the ground, the person is moving at $60 + 10 = 70$ mph, the sum of the two velocities (Figure 2.5.4). Similarly, if the person is running backwards on the train at a velocity of 10 mph, her velocity relative to the ground will be $60 - 10 = 50$ mph. Velocities add (and subtract). Derivatives, the mathematical version of velocity, also add.

There is another geometric way to understand the sum rule. Suppose that we measure the height of a tree as the sum of the trunk length and the crown height (Figure 2.5.5). The increase in trunk height $T(t)$ between times t and $t + \Delta t$ is

$$\Delta T = T(t + \Delta t) - T(t),$$

and the increase in crown height $C(t)$ between times t and $t + \Delta t$ is

$$\Delta C = C(t + \Delta t) - C(t).$$

The total increase in height $H(t)$ is

$$\Delta H = \Delta T + \Delta C,$$

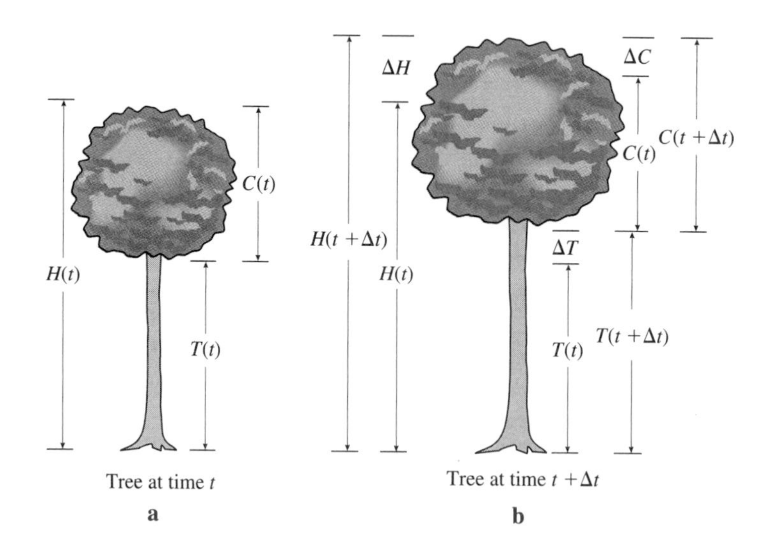

FIGURE 2.5.5
The height of a tree as a sum

the sum of the changes in trunk and crown. Therefore,

$$\begin{aligned}\frac{dH}{dt} &= \lim_{\Delta t \to 0} \frac{\Delta H}{\Delta t} \\ &= \lim_{\Delta t \to 0} \frac{\Delta T + \Delta C}{\Delta t} \\ &= \lim_{\Delta t \to 0} \frac{\Delta T}{\Delta t} + \lim_{\Delta t \to 0} \frac{\Delta C}{\Delta t} \\ &= \frac{dT}{dt} + \frac{dC}{dt}.\end{aligned}$$

The total rate of change of height is the sum of the rate of change of trunk height and the rate of change of crown height.

Example 2.5.1 Applying the Sum Rule

Consider the function

$$p(t) = 5t^2 - 8t$$

In differential notation,

$$\begin{aligned}\frac{dp}{dt} &= \frac{d}{dt}(5t^2 - 8) \\ &= \frac{d(5t^2)}{dt} - \frac{d}{dt}8t \\ &= 10t - 8\end{aligned}$$

because the derivative of the linear function $-8t$ is the constant -8. ▲

Example 2.5.2 The Constant Sum Rule

We measured the position of our falling cat from the point where it jumped, finding

$$y(t) = 5t^2.$$

Suppose we measured the position of the cat from a point 2 m higher instead. The distance fallen would be

$$s(t) = 5t^2 + 2$$

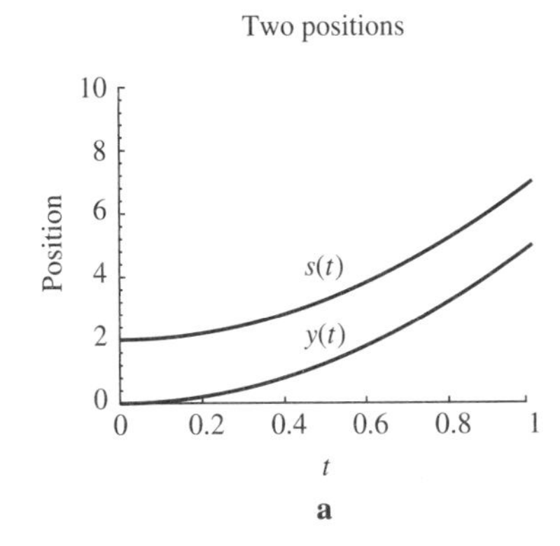

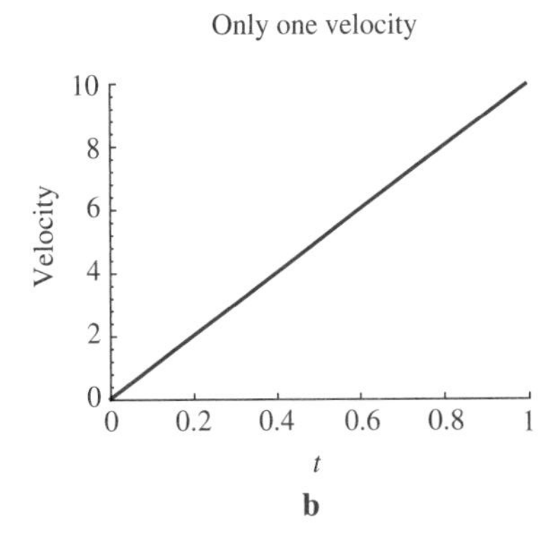

FIGURE 2.5.6
Two functions with the same derivative

(Figure 2.5.6a). In differential notation, the velocity of the cat is

$$\begin{aligned}\frac{ds}{dt} &= \frac{d}{dt}(5t^2+2)\\ &= \frac{d(5t^2)}{dt} + \frac{d}{dt}2\\ &= 10t + 0 = 10t\end{aligned}$$

because the derivative of the constant 2 is equal to 0 (the slope of a horizontal line). The derivative of $s(t)$ exactly matches that of $y(t)$ itself (Figure 2.5.6b). This makes physical sense because velocity does not depend on where we begin measuring distance. Graphically, two curves differing only by a constant have the same slope at each point (Figure 2.5.6). Mathematically, adding a constant to a function does not change the derivative.

We summarize the sum rule, along with the special cases involving constants in the following tables.

Prime Notation

Rule	Function	Derivative
Sum rule	$f(x)+g(x)$	$f'(x)+g'(x)$
Constant sum rule	$f(x)+c$	$f'(x)$

Differential Notation

Rule	Function	Derivative
Sum rule	$\frac{d(f+g)}{dx}$	$\frac{df}{dx}+\frac{dg}{dx}$
Constant sum rule	$\frac{d(f+c)}{dx}$	$\frac{df}{dx}$

2.5.2 Derivatives of Power Functions

Power functions take the form

$$f(x) = x^p$$

for some power p. We begin by finding the derivatives of the simplest power functions

$$f(x) = x^n$$

where n is a positive integer (1, 2, 3, ...).

There are two ways to compute the derivative of $f(x) = x^n$. One uses the **binomial theorem**, and the other uses **mathematical induction**. We use the binomial theorem to derive the result in the text, leaving mathematical induction to the exercises (Exercises 19–22 in section 2.6). The binomial theorem gives a formula for computing powers of sums.

Theorem 2.6 **The Binomial Theorem (simplified version)**

The expansion of $(x+y)^n$ is a polynomial of degree n in x, with expansion

$$(x+y)^n = x^n + nx^{n-1}y + \dots.$$

The dots represent terms with smaller powers of x and larger powers of y. For small values of n,

$$\begin{aligned}(x+y)^2 &= x^2+2xy+y^2\\ (x+y)^3 &= x^3+3x^2y+3xy^2+y^3\\ (x+y)^4 &= x^4+4x^2y+6x^2y^2+4xy^3+y^4.\end{aligned}$$

Example 2.5.3 The Binomial Theorem Applied with $n = 2$

To find $(x+2)^2$, we substitute 2 for y, finding

$$(x+2)^2 = x^2 + 2x \cdot 2 + 2^2 = x^2 + 4x + 4.$$

Example 2.5.4 The Binomial Theorem Applied with $n = 3$

To find $(2z+5)^3$, we substitute $2z$ for x and 5 for y, finding

$$(2z+5)^3 = (2z)^3 + 3(2z)^2 \cdot 5 + 3 \cdot (2z) \cdot 5^2 + 5^3 = 8z^3 + 60z^2 + 150z + 125.$$

Example 2.5.5 Using the Binomial Theorem to Compute the Derivative of x^2

We follow the same steps used to find the derivative of the quadratic function $f(x) = x^2$. We use h instead of Δx to make the calculation more readable.

$$\begin{aligned}
\frac{dx^2}{dx} &= \lim_{h \to 0} \frac{f(x+h) - f(x)}{h} \\
&= \lim_{h \to 0} \frac{(x+h)^2 - x^2}{h} \\
&= \lim_{h \to 0} \frac{(x^2 + 2xh + h^2) - x^2}{h} \\
&= \lim_{h \to 0} \frac{2xh + h^2}{h} \\
&= \lim_{h \to 0} 2x + h \\
&= 2x.
\end{aligned}$$

We computed the limit by evaluating the continuous function $2x + h$ at $h = 0$.

Example 2.5.6 Using the Binomial Theorem to Compute the Derivative of x^3

When $n = 3$, we expand $(x+h)^3$, finding

$$\begin{aligned}
\frac{dx^3}{dx} &= \lim_{h \to 0} \frac{(x+h)^3 - x^3}{h} \\
&= \lim_{h \to 0} \frac{(x^3 + 3x^2h + 3xh^2 + h^3) - x^3}{h} \\
&= \lim_{h \to 0} \frac{3x^2h + 3xh^2 + h^3}{h} \\
&= \lim_{h \to 0} 3x^2 + 3xh + h^2 \\
&= 3x^2.
\end{aligned}$$

We used the expansion

$$(x+h)^3 = x^3 + 3x^2h + 3xh^2 + h^3,$$

finding that the only term that appears in the derivative is the second. The first term canceled when we subtracted $f(x) = x^3$, and the terms with h^2 and h^3 disappear because they approach 0 even after being divided by h.

Now consider the function $f(x) = x^n$. Using the binomial theorem to expand the term $f(x+h)$ in the definition of the derivative gives

$$\begin{aligned}
\frac{df}{dx} &= \lim_{h \to 0} \frac{(x+h)^n - x^n}{h} \\
&= \lim_{h \to 0} \frac{(x^n + nx^{n-1}h + Eh^2) - x^n}{h} \\
&= \lim_{h \to 0} \frac{nx^{n-1}h + Eh^2}{h}
\end{aligned}$$

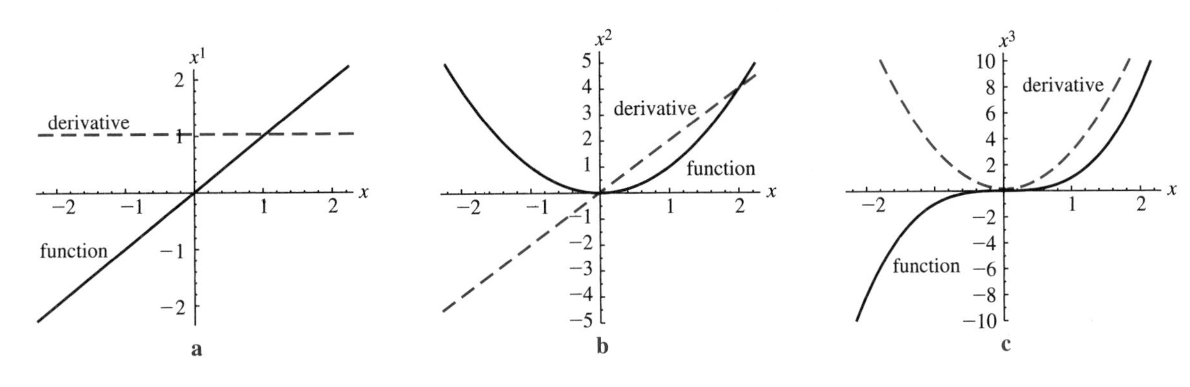

FIGURE 2.5.7
The first few power functions and their derivatives

$$= \lim_{h \to 0} nx^{n-1} + Eh$$
$$= nx^{n-1}$$

where E is the expression (a polynomial in x and h) that multiplies h^2 in the additional terms in the binomial theorem.

We have derived the **power rule**,

Theorem 2.7 The Power Rule for Derivatives

Suppose

$$f(x) = x^n$$

when n is a positive integer. Then

$$f'(x) = nx^{n-1} \quad \text{prime notation}$$
$$\frac{df}{dx} = nx^{n-1}. \quad \text{differential notation}$$

For the first few values of n,

Function	Derivative
x^1	1
x^2	$2x$
x^3	$3x^2$
x^4	$4x^3$

In words, to find the derivative of a power function, multiply by the power out front and then decrease the power by 1.

Example 2.5.7 The Power Rule in Action

With $n = 1$, the power function is a line with constant slope of 1 (Figure 2.5.7a). With $n = 2$, the power function is a parabola. The derivative is a line with slope 2 (Figure 2.5.7b). The function changes from decreasing to increasing at $x = 0$, indicated by the derivative crossing from negative to positive values. With $n = 3$, the power function is always increasing, but has a critical point at $x = 0$ (Figure 2.5.7c).

Negative and Fractional Powers The power rule also works for negative and fractional powers.

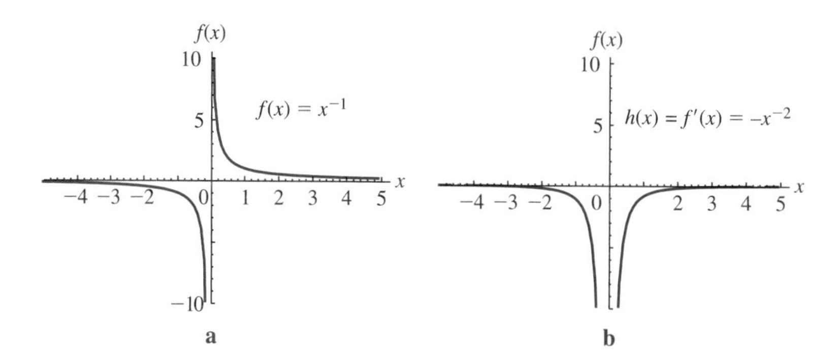

FIGURE 2.5.8

Power function with a negative power and its derivative

Theorem 2.8 **The Power Rule for Derivatives**

Suppose

$$f(x) = x^p$$

defined for $x > 0$, where $p \neq 0$. Then

$$f'(x) = px^{p-1} \quad \text{prime notation}$$

$$\frac{df}{dx} = px^{p-1}. \quad \text{differential notation}$$

Example 2.5.8 The Power Rule Applied to Negative Powers

For the first few negative powers,

Function	Derivative
x^{-1}	$-x^{-2}$
x^{-2}	$-2x^{-3}$
x^{-3}	$-3x^{-4}$

To find the derivative, multiply out front by the power and decrease the power by 1. Be careful to **subtract** 1 from the power even when it is negative. The function and derivative with power equal to -1 are shown in Figure 2.5.8. The derivative is always negative, indicating that the function is decreasing (except at $x = 0$ where it is not defined). Near $x = 0$, the slope approaches negative infinity.

Example 2.5.9 The Power Rule Applied to the Square Root

The power rule can be used to find the derivative of $\sqrt{x} = x^{1/2}$.

$$\frac{d}{dx}x^{\frac{1}{2}} = \frac{1}{2}x^{\frac{1}{2}-1}$$
$$= \frac{1}{2}x^{-\frac{1}{2}}$$
$$= \frac{1}{2\sqrt{x}}.$$

Near $x = 0$, the slope becomes infinite. For large x, the slope becomes small (Figure 2.5.9).

Example 2.5.10 Using the Power Rule Along with the Sum Rule

Consider the function

$$g(z) = z^{-2} + z^4 + 3.$$

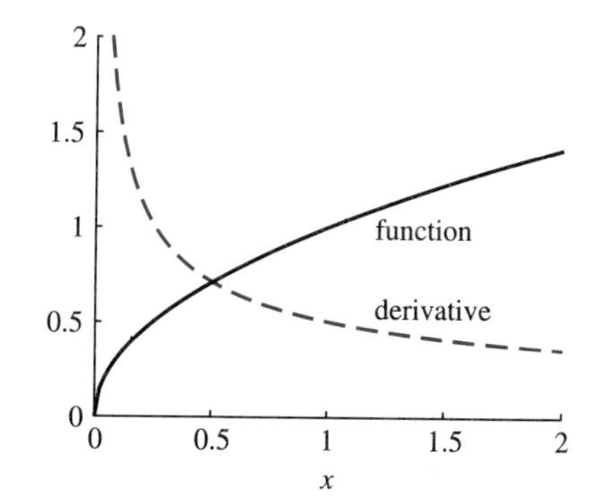

FIGURE 2.5.9
The square root function and its derivative

We find the derivative with the following steps

$$\frac{dg}{dz} = \frac{d(z^{-2})}{dz} + \frac{d(z^4)}{dz} + \frac{d3}{dz} \quad \text{sum rule}$$
$$= \frac{d(z^{-2})}{dz} + \frac{d(z^4)}{dz} \quad \text{constant sum rule}$$
$$= (-2z^{-3}) + 4z^3 \quad \text{power rule}$$
$$= -2z^{-3} + 4z^3. \quad \text{multiply out}$$

2.5.3 Derivatives of Polynomials

A **polynomial** is a function built from power functions with positive integer powers that are multiplied by constants and added up. For example, the function

$$p(x) = 2x^3 - 5x^2 + 7x - 8$$

is a polynomial constructed by multiplying x^3 by 2, subtracting 5 times x^2, adding 7 times x^1, and subtracting 8. The **degree** of a polynomial is the highest power that appears in the formula, in this case, 3. Using the power rule, the sum rule, and a new rule, the **constant product rule**, we can find the derivative of any polynomial.

Theorem 2.9 **The Constant Product Rule for Derivatives**

If c is a constant and $f(x)$ is a differentiable function, then

$$\frac{d}{dx}cf(x) = c\frac{d}{dx}f(x).$$

The constant product rule says constants come "outside" of derivatives. For example, doubling a function doubles the derivative (Figure 2.5.10).

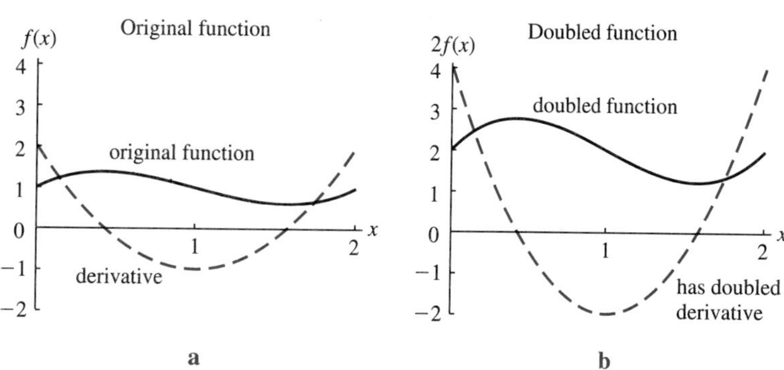

FIGURE 2.5.10
The constant product rule for derivatives

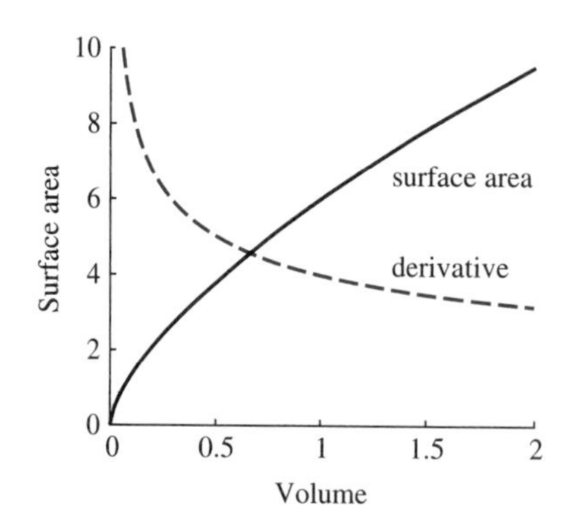

FIGURE 2.5.11
The derivative of the power relations between surface area and volume

Example 2.5.11 Finding the Derivative of a Power Function

The derivative of the function $8t^4$ is

$$\begin{aligned} \frac{d8t^4}{dt} &= 8\frac{dt^4}{dt} && \text{constant product rule} \\ &= 8 \cdot 4t^3 && \text{power rule} \\ &= 32t^3. && \text{multiply constants} \end{aligned}$$

Example 2.5.12 The Power Rule Applied to Comparing Surface Area and Volume

We can use the generalized power rule to interpret power relations between measurements. The surface area S of an object is proportional to the volume V raised to the 2/3 power, or

$$S = cV^{\frac{2}{3}}$$

where the constant c depends on the shape of the object (Example 1.7.23). If we imagine increasing the volume of an object, the surface area will increase, with

$$\frac{dS}{dV} = \frac{2}{3}cV^{-\frac{1}{3}}.$$

Surface area does increase, but at a **decreasing rate** (Figure 2.5.11). In words, as an object gets larger, the surface area increases more slowly than the volume.

Example 2.5.13 Finding the Derivative of a Complicated Polynomial

Using the power rule, the constant product rule, and the sum rule, we can find the derivative of the polynomial $p(x)$

$$p(x) = 2x^3 - 7x^2 + 7x + 1$$

by breaking it into component parts as follows.

$$\begin{aligned} \frac{dp}{dx} &= \frac{d(2x^3)}{dx} + \frac{d(-7x^2)}{dx} + \frac{d(7x)}{dx} + \frac{d(+1)}{dx} && \text{sum rule} \\ &= 2\frac{dx^3}{dx} - 7\frac{d(x^2)}{dx} + 7\frac{dx}{dx} + 1\frac{d1}{dx} && \text{constant product rule} \\ &= 2 \cdot 3x^2 - 7 \cdot 2x + 7 \cdot 1 && \text{power rule} \\ &= 6x^2 - 14x + 7. && \text{the answer} \end{aligned}$$

The derivative of a polynomial is always another polynomial. Because the power rule reduces the power by one, the degree of the derivative is always one less than the degree of the original polynomial. The function and its derivative are plotted in Figure 2.5.12.

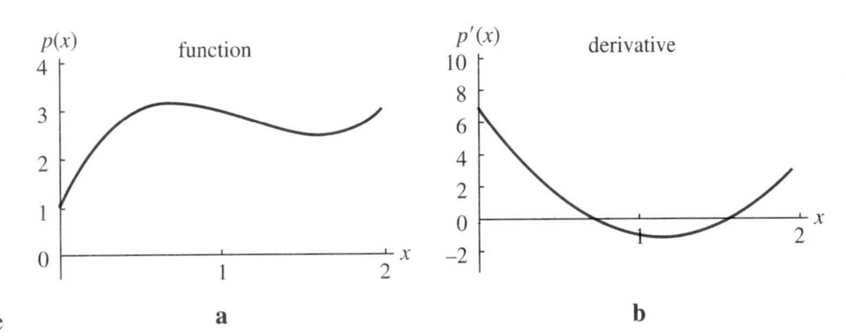

FIGURE 2.5.12
The polynomial $p(x)$ with its derivative

Example 2.5.14 Using Polynomials to Describe Bird Reproduction

Polynomials are convenient to work with mathematically and describe many biological processes. Consider the following model of offspring production. Let N be the number of eggs a bird lays. Suppose that each chick survives with probability

$$P(N) = 1 - 0.1N$$

(Figure 2.5.13a). The more offspring a parent has, the smaller chance each one has to survive. If the bird lays 1 egg, the chick has a 90% chance of surviving. If the bird lays 5 eggs, each chick has only a 50% chance of surviving. If the bird lays 10 eggs, none of the chicks survive. Mathematically, the total number of offspring that survive, $S(N)$, is the product of the number of eggs N and the probability of survival $P(N)$, or

$$S(N) = N \cdot P(N) = N(1 - 0.1N) = N - 0.1N^2$$

(Figure 2.5.13b). For example, when $N = 1$, 0.9 offspring survive on average. When $N = 5$, 50% of the 5 chicks survive, producing 2.5 offspring on average. When $N = 10$, none of the chicks survive.

The formula for $S(N)$ is a polynomial. We use the constant product and power rules to find

$$\begin{aligned}\frac{dS}{dN} &= \frac{d(N - 0.1N^2)}{dN}\\ &= \frac{dN}{dN} - 0.1\frac{dN^2}{dN}\\ &= 1 - 0.1(2N)\\ &= 1 - 0.2N.\end{aligned}$$

At $N = 0$, the derivative is positive, meaning that a higher value of N produces more surviving offspring. At $N = 5$ the derivative is zero. For $N > 5$, the derivative is negative. The negative derivative indicates that the number of surviving offspring **decreases** when the number of eggs becomes too large. ▲

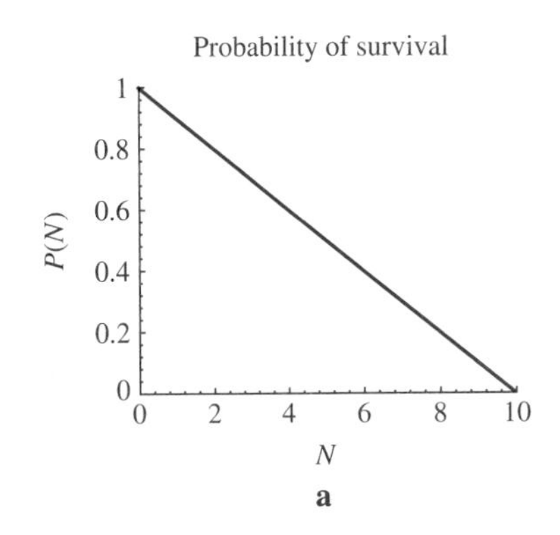

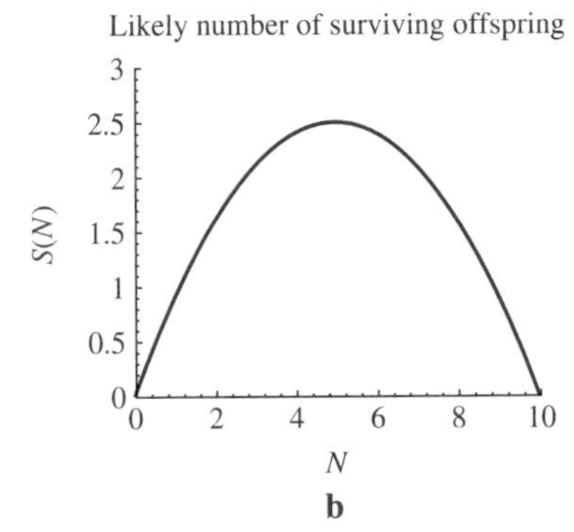

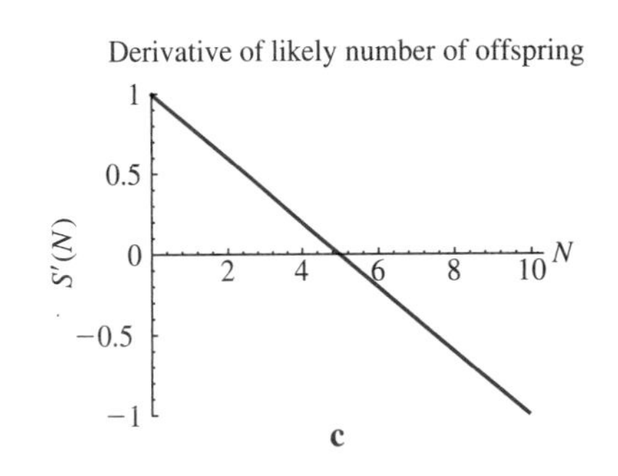

FIGURE 2.5.13
A model of offspring number and survival

Summary By combining three rules, the **sum rule**, the **power rule**, and the **constant product rule**, we can find the derivative of any polynomial. The **sum rule** states that the derivative of the sum of two functions is equal to the sum of the derivatives. This corresponds to the physical fact that velocities add. As a special case, we found the **constant sum rule**, which states that adding a constant to a function does not change the derivative. The **power rule** gives the formula for the derivative of the power function x^p as px^{p-1} for any non-zero power. The **constant product rule** says that constants come "outside" the derivative; multiplying a function by a constant multiplies the derivative by that same constant.

2.5 Exercises

Mathematical Techniques

1–8 ▪ Find the derivatives of the following power functions.

1. x^5
2. x^{-5}
3. $x^{0.2}$
4. $x^{-0.2}$
5. x^e
6. x^{-e}
7. $x^{1/e}$
8. $x^{-1/e}$

9–12 ▪ Find the derivatives of the following polynomial functions. Say where you used the sum, constant product, and power rules.

9. $f(x) = 3x^2 + 3x + 1$.
10. $s(x) = 1 - x + x^2 - x^3 + x^4$.
11. $g(z) = 3z^3 + 2z^2$.
12. $p(x) = 1 + x + \frac{x^2}{2} + \frac{x^3}{6} + \frac{x^4}{24}$.

13–16 ▪ Use the binomial theorem to compute the following.

13. $(x + 1)^3$
14. $(x + 2)^3$
15. $(2x + 1,)^3$
16. $(x + 1)^4$

17–20 ▪ Use the derivative to sketch a graph of each of the following functions.

17. $f(x) = 1 - 2x + x^2$ for $0 \le x \le 2$.
18. $g(x) = 4x - x^2$ for $0 \le x \le 5$.
19. $h(x) = x^3 - 3x$ for $0 \le x \le 2$.
20. $F(x) = x + \frac{1}{x}$ for $0 < x \le 2$.

21–26 ▪ Try to guess functions that have the following as their derivatives.

21. 2
22. $2x$
23. $15x^{14}$
24. x^{14}
25. $-x^{-2}$
26. $3x^{-4}$

Applications

27–30 ▪ Find the derivatives of the following functions.

27. In its early phase, the number of AIDS cases in the United States grew approximately according to a cubic equation, $A(t) = 175t^3$ where t is measured in years since the beginning of the epidemic in 1972. Find and interpret the derivative.
28. $F(r) = 1.5r^4$, where F represents the flow in cubic centimeters per second through a pipe of radius r. If $r = 1$, how much will a small increase in radius change the flow (try it with $\Delta r = 0.1$)? If $r = 2$, how will a small increase in radius change the flow?
29. The area of a circle as a function of radius is $A(r) = \pi r^2$, with area measured in cm^2 and radius measured in centimeters. Find the derivative of area with respect to radius. On a geometric diagram, illustrate the area corresponding to $\Delta A = A(r + \Delta r) - A(r)$. What is a geometric interpretation of the derivative? Do the units make sense?
30. The volume of a sphere as a function of radius is $V(r) = \frac{4}{3}\pi r^3$. Find the derivative of volume with respect to radius. On a geometric diagram, illustrate the volume corresponding to $\Delta V = V(r + \Delta r) - V(r)$. What is a geometric interpretation of this derivative? Do the units make sense?

31–32 ▪ One car is towing another using a rigid 50-ft pole. Sketch the positions and speeds of the two cars as functions of time in the following circumstances.

31. The car starts from a stop, slowly speeds up, cruises for a while, and then abruptly stops.
32. The car starts from a stop, goes slowly in reverse for a short time, stops, goes forward slowly, and then more quickly.

33–36 ▪ A passenger is traveling on a luxury train that is moving west at 80 mph (miles per hour).

33. The passenger starts running east at 10 mph. What is her velocity relative to the ground?

34. While running, the passenger flips a dinner roll over her shoulder (west) at 25 mph. What is the velocity of the roll relative to the train? What is the velocity of the roll relative to the ground?

35. A roll weevil jumps east off the roll at 5 mph. What is the velocity of the roll relative to the passenger? What is the velocity of the roll relative to the train? What is the velocity of the roll relative to the ground?

36. A roll weevil flea jumps west off the roll weevil at 15 mph. What is the velocity of the roll weevil flea relative to the roll? What is the velocity of the roll weevil flea relative to the passenger? What is the velocity of the roll weevil flea relative to the train? What is the velocity of the roll weevil flea relative to the ground?

37–38 ▪ Each of the following measurements is the sum of two components.

a. Find the formula for the sum.

b. Find the derivative of each component. What are the units?

c. Find the derivative of the sum and check that the sum rule worked.

d. Describe in words what is happening.

e. Sketch a graph of each component and the total as functions of time.

37. A population of bacteria consists of two types a and b. The first follows $a(t) = 1 + t^2$, and the second follows $b(t) = 1 - 2t + t^2$ where populations are measured in millions and time is measured in hours. The total population is $P(t) = a(t) + b(t)$.

38. The above-ground volume (stem and leaves) of a plant is $V_a(t) = 3.0t + 20.0 + \frac{t^2}{2}$, and the below-ground volume (roots) is $V_b(t) = -1.0t + 40.0$ where t is measured in days and volumes are measured in cm^3. The total volume is $V(t) = V_a(t) + V_b(t)$.

39–42 ▪ An object tossed upward at 10 m/s from a height of 100 m has distance above the ground of

$$M(t) = 100 + 10t - \frac{a}{2}t^2$$

where a is the acceleration of gravity. For each of the following planets with the given acceleration, find the time when the object reaches a critical point, how high it gets, the time when it hits the ground, and the speed of the object at that time. Sketch the position of the object as a function of time.

39. On Earth, where $a = 9.78\ \text{m/sec}^2$.

40. On the moon, where $a = 1.62\ \text{m/sec}^2$.

41. On Jupiter, where $a = 22.88\ \text{m/sec}^2$.

42. On Mars' moon Deimos, where $a = 2.15 \times 10^{-3}\ \text{m/sec}^2$.

43–46 ▪ Try different power functions of the form t^n to guess solutions of the following differential equations describing the size of an organism. Find the size at $t = 1$ and $t = 2$. Can you explain why some grow so much faster than others?

43. $\frac{dS}{dt} = 6t$.

44. $\frac{dS}{dt} = 6t^2$.

45. $\frac{dS}{dt} = 6\frac{S(t)}{t}$.

46. $\frac{dS}{dt} = 12\frac{S(t)}{t}$.

Computer Exercise

47. Consider the polynomials

$$p_0(x) = 1$$
$$p_1(x) = 1 + x$$
$$p_2(x) = 1 + x + \frac{x^2}{2}$$
$$p_3(x) = 1 + x + \frac{x^2}{2} + \frac{x^3}{6}.$$

To find the fourth polynomial in the series, add a term equal to the last term of $p_3(x)$ multiplied by $\frac{x}{4}$. To find the fifth polynomial in the series, add a term equal to the last term of $p_4(x)$ multiplied by $\frac{x}{5}$ and so forth.

a. Find the first six polynomials in this series.

b. Plot them all for $-3 \le x \le 3$.

c. Take the derivative of $p_6(x)$. What is it equal to? Compare the graph of $p_6(x)$ with the graph of its derivative.

d. Can you guess what function these polynomials are approaching?

2.6 Derivatives of Products and Quotients

The previous section derived the three rules needed to find the derivative of a polynomial. The next two ways to combine functions are multiplication and division. In this section, we learn how to compute the derivatives of products and quotients of known functions by deriving the **product rule** and **quotient rule**. These two rules allow us to study the derivatives of **rational functions**, functions that are the ratios of two polynomials.

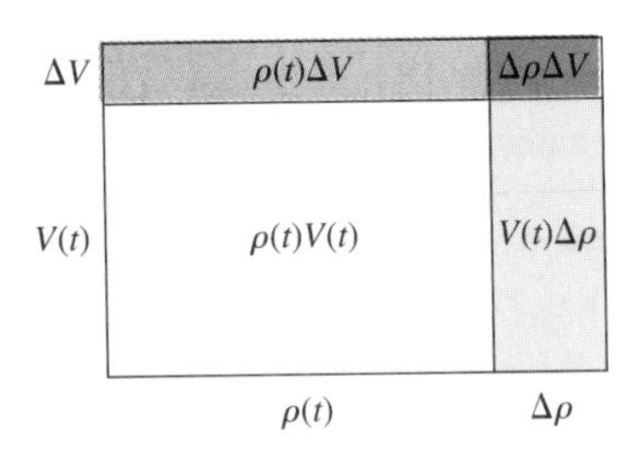

FIGURE 2.6.1
The product rule

2.6.1 The Product Rule

Tempting as it might sound, the derivative of the product is **not** the product of the derivatives. Finding the derivative of a product is a bit more complicated. Like the derivative of the sum, there is a useful geometric argument to help derive and understand the formula.

Suppose the density $\rho(t)$ and the volume $V(t)$ of an object are functions of time. The mass $M(t)$, the product of the density and volume, is also a function of time and has the formula

$$M(t) = \rho(t)V(t).$$

What is the derivative of the mass? For example, if both the density and volume are increasing, the mass is also increasing. But if the density increases while the volume decreases, we cannot tell whether mass increases without more information.

$M(t)$ can be represented geometrically as the area of a rectangle with base $\rho(t)$ and height $V(t)$ (Figure 2.6.1). Between time t and $t + \Delta t$, the density changes from $\rho(t)$ to $\rho(t) + \Delta\rho$ and the volume changes from $V(t)$ to $V(t) + \Delta V$. The mass at time $t + \Delta t$ is then

$$\begin{aligned} M(t + \Delta t) &= (\rho(t) + \Delta\rho)(V(t) + \Delta V) \\ &= \rho(t)V(t) + \rho(t)\Delta V + V(t)\Delta\rho + \Delta\rho\Delta V \\ &= M(t) + \rho(t)\Delta V + V(t)\Delta\rho + \Delta\rho\Delta V. \end{aligned}$$

The change in mass ΔM is

$$\begin{aligned} \Delta M &= M(t + \Delta t) - M(t) \\ &= \rho(t)\Delta V + V(t)\Delta\rho + \Delta\rho\Delta V. \end{aligned}$$

Each component of this algebraic computation corresponds to one of the regions in Figure 2.6.1.

How does this help us compute the derivative of the product? To find the derivative of M with respect to t, we divide ΔM by Δt and take the limit. ΔM divided by Δt is

$$\begin{aligned} \frac{\Delta M}{\Delta t} &= \frac{\rho(t)\Delta V + V(t)\Delta\rho + \Delta\rho\Delta V}{\Delta t} \\ &= \rho(t)\frac{\Delta V}{\Delta t} + V(t)\frac{\Delta\rho}{\Delta t} + \frac{\Delta\rho\Delta V}{\Delta t}. \end{aligned}$$

Taking the limit

$$\lim_{\Delta t \to 0} \frac{\Delta M}{\Delta t} = \rho(t) \lim_{\Delta t \to 0} \frac{\Delta V}{\Delta t} + V(t) \lim_{\Delta t \to 0} \frac{\Delta\rho}{\Delta t} + \lim_{\Delta t \to 0} \frac{\Delta\rho\Delta V}{\Delta t}$$

where we used the properties of limits (Theorems 2.1 and 2.2) to break up the sum and take constants (terms without Δt) outside the limits. The first two pieces are

$$\rho(t) \lim_{\Delta t \to 0} \frac{\Delta V}{\Delta t} = \rho(t)V'(t)$$

$$V(t) \lim_{\Delta t \to 0} \frac{\Delta\rho}{\Delta t} = V(t)\rho'(t)$$

by the definition of the derivative. The final piece can be analyzed as a product,

$$\lim_{\Delta t \to 0} \frac{\Delta\rho\Delta V}{\Delta t} = \lim_{\Delta t \to 0} \Delta\rho \lim_{\Delta t \to 0} \frac{\Delta V}{\Delta t}.$$

However,

$$\lim_{\Delta t \to 0} \Delta\rho = 0$$

because ρ is a continuous function. Therefore,

$$\lim_{\Delta t \to 0} \frac{\Delta\rho\Delta V}{\Delta t} = 0 \cdot V'(t) = 0.$$

Geometrically, the region labeled $\Delta\rho\Delta V$ becomes very small as Δt approaches 0 and disappears in the limit. We summarize this calculation in the following theorem.

Theorem 2.10 **The Product Rule for Derivatives**

Suppose

$$p(x) = f(x)g(x)$$

where f and g are both differentiable. Then

$$p'(x) = f'(x)g(x) + g'(x)f(x) \quad \text{prime notation}$$

$$\frac{dp}{dx} = \frac{df}{dx}g(x) + f(x)\frac{dg}{dx}. \quad \text{differential notation}$$

2.6.2 Special Cases and Examples

The constant product rule is a special case of the product rule. If we measure distance fallen in centimeters rather than meters, the distance fallen will be 100 times the original distance $y(t)$, or

$$p(t) = 100y(t).$$

This function is a product of the constant function 100 and the function $y(t)$. The velocity is then

$$\frac{dp}{dt} = \frac{d(100y)}{dt} \quad \text{definition of } p(t)$$

$$= 100\frac{dy}{dt} + y(t)\frac{d100}{dt} \quad \text{the product rule}$$

$$= 100\frac{dy}{dt} + y(t) \cdot 0 = 100\frac{dy}{dt}. \quad \text{the derivative of the constant 100 is 0}$$

In centimeters per second, the velocity is 100 times the velocity in meters per second. Mathematically, multiplying a function by a constant multiplies the derivative by that same constant.

We summarize these rules in the following tables.

Prime Notation

Rule	Function	Derivative
Product rule	$f(x)g(x)$	$f(x)g'(x) + g(x)f'(x)$
Constant product rule	$cf(x)$	$cf'(x)$

Differential Notation

Rule	Function	Derivative
Product rule	$\frac{d(fg)}{dx}$	$f(x)\frac{dg}{dx} + g(x)\frac{df}{dx}$
Constant product rule	$\frac{d(cf)}{dx}$	$c\frac{df}{dx}$

Example 2.6.1 Using the Product Rule

To find the derivative of

$$p(x) = (x - 1)(x + 1),$$

think of $p(x)$ as the product of $f(x) = x - 1$ and $g(x) = x + 1$. Both $f(x)$ and $g(x)$ are linear functions with slope 1; therefore

$$f'(x) = 1$$
$$g'(x) = 1.$$

Then, by the product rule

$$p'(x) = (x - 1) \cdot 1 + (x + 1) \cdot 1 = 2x.$$

To check, we can multiply the function out, finding

$$p(x) = x^2 - 1.$$

Using the power and constant sum rules,

$$p'(x) = 2x + 0 = 2x,$$

matching the result found with the product rule.

Example 2.6.2 Applying the Product Rule

Suppose the volume V of a plant is increasing according to

$$V(t) = 100.0 + 12.0t$$

where t is measured in hours and V is measured in cm^3, and that the density is decreasing according to

$$\rho(t) = 0.8 - 0.05t$$

where ρ is measured in grams per cm^3. Is the mass increasing or decreasing? We can use the product rule to find out.

$$\begin{aligned} M'(t) &= \rho(t)V'(t) + V(t)\rho'(t) \\ &= (0.8 - 0.05t)12.0 + (100.0 + 12.0t)(-0.05) \\ &= 9.6 - 0.6t + -5.0 - 0.6t = 4.6 - 1.2t. \end{aligned}$$

At $t = 0$, the derivative is positive and the mass is increasing. By $t = 4$ the decrease in density has overwhelmed the increase in volume and the mass is decreasing (Figures 2.6.2 and 2.6.3).

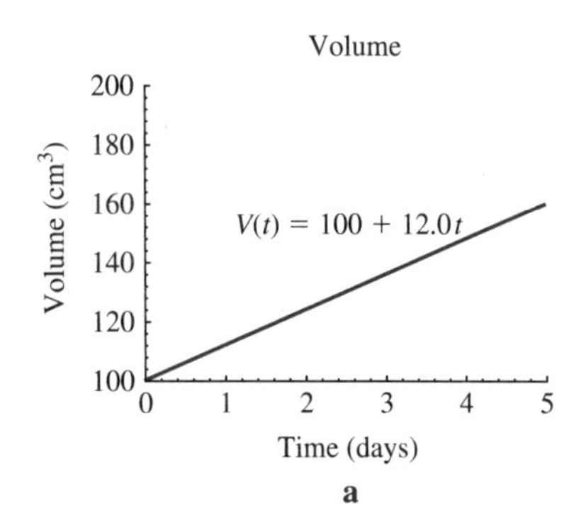

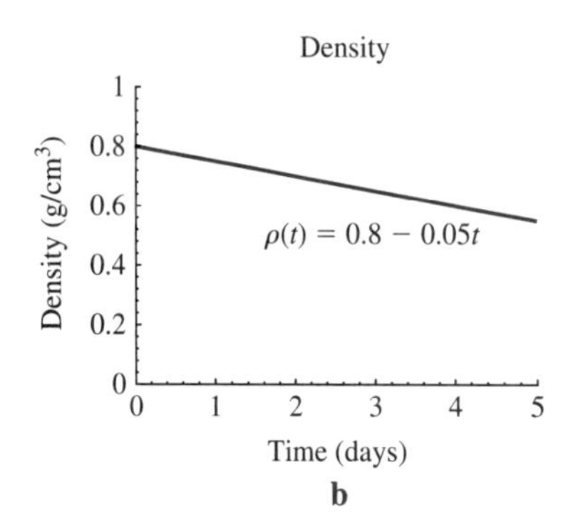

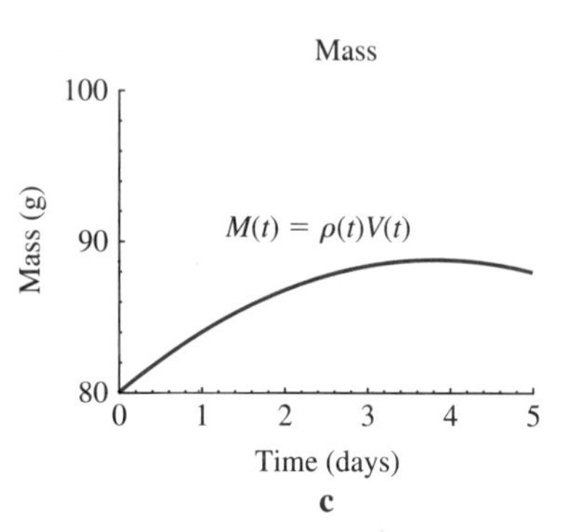

FIGURE 2.6.2
The volume, density, and mass of a plant

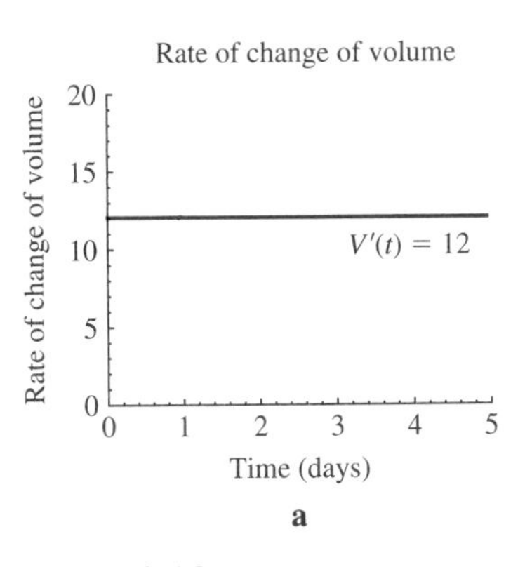

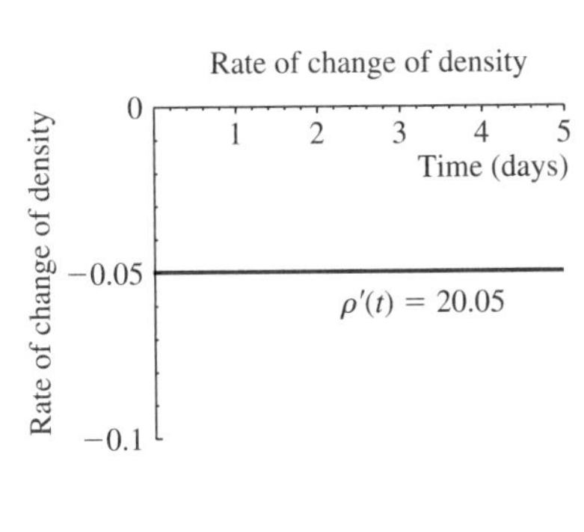

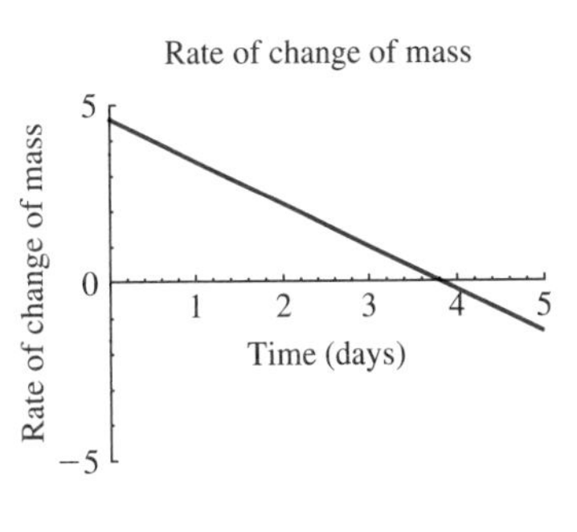

FIGURE 2.6.3
The rates of change of volume, density, and mass of a plant

2.6.3 The Quotient Rule

Suppose we are interested in finding the derivative of a function $w(x)$ defined as the quotient of $u(x)$ and $v(x)$,

$$w(x) = \frac{u(x)}{v(x)}.$$

With a bit of a trick, we can find $w'(x)$ from the product rule. First, multiply both sides by $v(x)$,

$$v(x)w(x) = u(x).$$

Now take the derivative of both sides using the product rule,

$$v(x)w'(x) + v'(x)w(x) = u'(x).$$

This can be thought of as an equation for $w'(x)$, the derivative of the quotient, and can be solved for $w'(x)$ as

$$v(x)w'(x) = u'(x) - v'(x)w(x)$$
$$w'(x) = \frac{u'(x) - v'(x)w(x)}{v(x)}.$$

To write the answer entirely in terms of the component functions $u(x)$ and $v(x)$, substitute in $w(x) = \frac{u(x)}{v(x)}$. Putting the result over a common denominator gives

$$w'(x) = \frac{u'(x) - v'(x)u(x)/v(x)}{v(x)}$$
$$= \frac{u'(x)v(x) - v'(x)u(x)}{(v(x))^2}.$$

We summarize this result in the following theorem.

Theorem 2.11 The Quotient Rule for Derivatives

Suppose

$$w(x) = \frac{u(x)}{v(x)}$$

where $u(x)$ and $v(x)$ are differentiable and $v(x) \neq 0$. Then

$$w'(x) = \frac{u'(x)v(x) - v'(x)u(x)}{(v(x))^2} \quad \text{prime notation}$$

$$\frac{dw}{dx} = \frac{v(x)\frac{du}{dx} - u(x)\frac{dv}{dx}}{(v(x))^2}. \quad \text{differential notation}$$

Example 2.6.3 Applying the Quotient Rule

We can use the derivative to check whether the ratio of polynomials

$$f(x) = \frac{x^3 + 2x}{1 + x^2}$$

is an increasing function (Figure 2.6.4). We write

$$f(x) = \frac{u(x)}{v(x)}$$

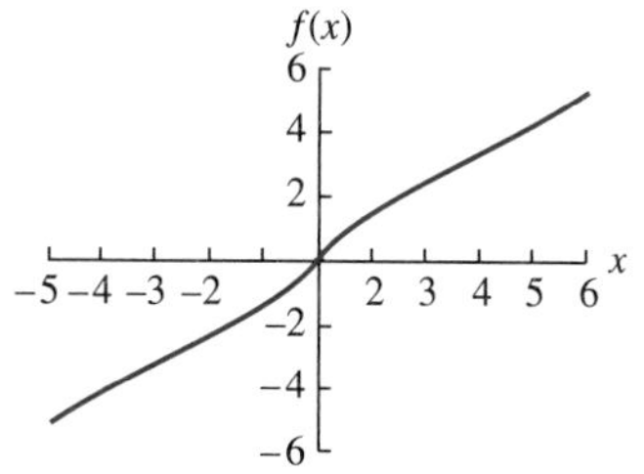

where we define the numerator as $u(x) = x^3 + 2x$ and the denominator as $v(x) = 1 + x^2$. Then

$$u'(x) = 3x^2 + 2 \text{ and } v'(x) = 2x.$$

The quotient rule says that

$$\begin{aligned} f'(x) &= \frac{u'(x)v(x) - v'(x)u(x)}{(v(x))^2} \\ &= \frac{(3x^2 + 2)(1 + x^2) - 2x(x^3 + 2x)}{(1 + x^2)^2}. \end{aligned}$$

FIGURE 2.6.4
An increasing rational function

If needed, this could be simplified as

$$\begin{aligned} f'(x) &= \frac{(3x^2 + 2)(1 + x^2) - 2x(x^3 + 2x)}{(1 + x^2)^2} \\ &= \frac{(3x^4 + 5x^2 + 2) - (2x^4 + 4x^2)}{(1 + x^2)^2} \\ &= \frac{x^4 + x^2 + 2}{(1 + x^2)^2}. \end{aligned}$$

All the components of this expression are positive, meaning that the derivative is positive and this function is always increasing. This would be difficult to show simply by plotting points.

Example 2.6.4 Applying the Quotient Rule

Consider the discrete-time dynamical system for the fraction of invading mutants p_t

$$p_{t+1} = \frac{2.0p_t}{2.0p_t + 1.5(1 - p_t)}$$

(Section 1.10). The updating function is

$$f(p) = \frac{2.0p}{2.0p + 1.5(1 - p)}.$$

The numerator is $u(p) = 2.0p$ and the denominator is $v(p) = 2.0p + 1.5(1 - p)$, with derivatives

$$\frac{du}{dp} = 2.0$$

$$\frac{dv}{dp} = 2.0 - 1.5 = 0.5.$$

Therefore, the derivative is

$$\begin{aligned} \frac{df}{dp} &= \frac{v(p)\frac{du}{dp} - u(p)\frac{dv}{dp}}{v(p)^2} \\ &= \frac{(2.0p + 1.5(1 - p))\frac{d(2.0p)}{dp} - 2.0p\frac{d(2.0p+1.5(1-p))}{dp}}{(2.0p + 1.5(1 - p))^2} \end{aligned}$$

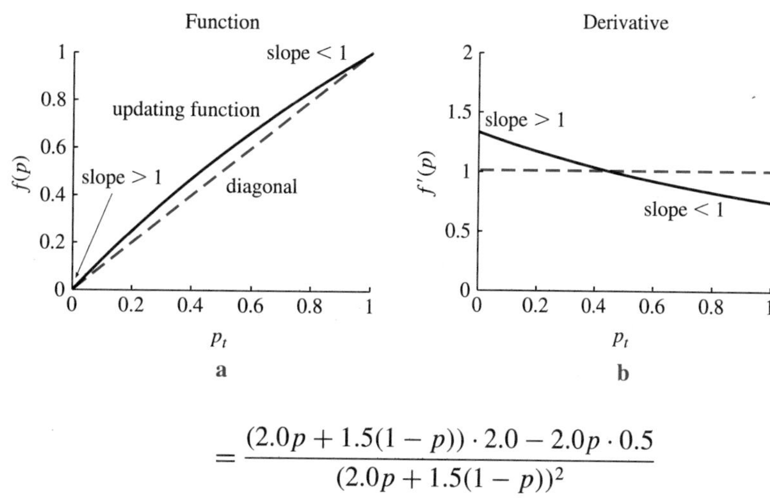

FIGURE 2.6.5
An updating function and its derivative

$$= \frac{(2.0p + 1.5(1-p)) \cdot 2.0 - 2.0p \cdot 0.5}{(2.0p + 1.5(1-p))^2}$$
$$= \frac{3.0}{(2.0p + 1.5(1-p))^2}.$$

At $p = 0$, the derivative is

$$\frac{3.0}{(2.0 \cdot 0 + 1.5(1-0))^2} = 1.333$$

and at $p = 1$, the derivative is

$$\frac{3.0}{(2.0 \cdot 1 + 1.5(1-1))^2} = 0.75.$$

Compared to the diagonal line, which has slope 1, the graph of this function is steep near $p = 0$ and less steep near $p = 1$ (Figure 2.6.5).

The **Hill functions** are an important family of functions used to describe biological processes. In the simplest case, they take the form,

$$h(x) = \frac{x^n}{1 + x^n} \tag{2.6.1}$$

where n can be any positive number. We will compute the derivatives of these functions with $n = 1$ and $n = 2$.

Example 2.6.5 Computing the Derivatives of Hill Functions

With $n = 1$, $u(x) = x$ and $v(x) = 1 + x$, so

$$\frac{dh}{dx} = \frac{(1+x)\frac{dx}{dx} - x\frac{d(1+x)}{dx}}{(1+x)^2}$$
$$= \frac{(1+x) - x}{(1+x)^2}$$
$$= \frac{1}{(1+x)^2}.$$

This derivative is always positive, and takes on the value $h'(0) = 1$ at $x = 0$.

When $n = 2$, $u(x) = x^2$ and $v(x) = 1 + x^2$, so

$$\frac{dh}{dx} = \frac{(1+x^2)\frac{d(x^2)}{dx} - x^2\frac{d(1+x^2)}{dx}}{(1+x^2)^2}$$
$$= \frac{(1+x^2)2x - x^2 \cdot 2x}{(1+x^2)^2}$$
$$= \frac{2x}{(1+x^2)^2}.$$

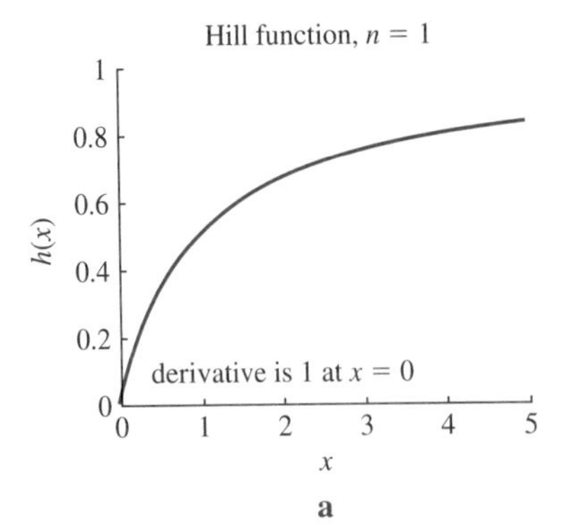

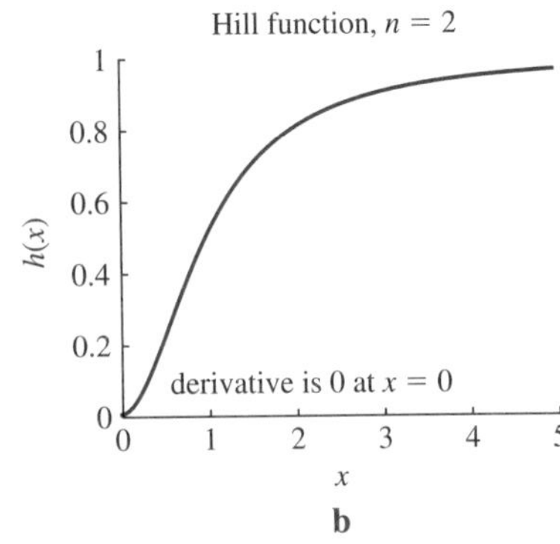

FIGURE 2.6.6
Two Hill functions

This derivative too is always positive, but takes on the value $h'(0) = 0$ at $x = 0$. Both Hill functions are increasing, but with different shapes (Figure 2.6.6). These functions are useful for describing the response to a stimulus.

Summary The **product rule** states that the derivative of the product of two functions is equal to the first times the derivative of the second plus the second times the derivative of the first. The constant product rule is a special case of the product rule. We derived the **quotient rule** for differentiating quotients. With these rules, we can find derivatives of **rational functions**, the ratios of polynomials, such as the **Hill functions**.

2.6 Exercises

Mathematical Techniques

1–6 ▪ Find the derivatives of the following functions using the product rule.

1. $f(x) = (2x + 3)(-3x + 2)$.
2. $g(z) = (5z - 3)(z + 2)$.
3. $r(y) = (5y - 3)(y^2 - 1)$.
4. $s(t) = (t^2 + 2)(3t^2 - 1)$.
5. $h(x) = (x + 2)(2x + 3)(-3x + 2)$ (apply the product rule twice).
6. $F(w) = (w - 1)(2w - 1)(3w - 1)$ (apply the product rule twice).

7–12 ▪ Find the derivatives of the following functions using the quotient rule.

7. $f(x) = \dfrac{1 + x}{2 + x}$.
8. $f(x) = \dfrac{x^2}{1 + 2x^3}$.
9. $g(z) = \dfrac{1 + z^2}{1 + 2z^3}$.
10. $h(z) = \dfrac{1 + 2z^3}{1 + z^2}$.
11. $F(x) = \dfrac{1 + x}{(2 + x)(3 + x)}$.
12. $G(x) = \dfrac{(1 + x)(2 + x)}{3 + x}$.

13–14 ▪ For the following functions, use base point $x_0 = 1.0$ and $\Delta x = 0.1$ to compute Δf and Δg. Find $\Delta(fg)$ (the change in the product) by computing $f(x_0 + \Delta x)g(x_0 + \Delta x) - f(x_0)g(x_0)$. Check that $\Delta(fg) = g(x_0)\Delta f + f(x_0)\Delta g + \Delta f \Delta g$. Try the same with $\Delta x = 0.01$ and see whether the term $\Delta f \Delta g$ becomes very small.

13. $f(x) = 2x + 3$ and $g(x) = -3x + 2$.
14. $f(x) = x^2 + 2$ and $g(x) = 3x^2 - 1$.

15–16 ▪ Suppose $p(x) = f(x)g(x)$. Test out the **incorrect** formula $p'(x) = f'(x)g'(x)$ on the following functions.

15. $f(x) = x$, $g(x) = x^2$.
16. $f(x) = 1$, $g(x) = x^3$.

17–18 ▪ Suppose that $f(x)$ is a positive increasing function defined for all x.

17. Use the product rule to show that $f(x)^2$ is also increasing.
18. Use the quotient rule to show that $\dfrac{1}{f(x)}$ is decreasing.

19–22 ▪ For positive integer powers, it is possible to derive the power rule with **mathematical induction**. The idea is to show that a formula is true for $n = 1$, and then that if it is true for some particular n, it must then also be true for $n + 1$.

19. Check that the power rule is true for $n = 1$.
20. Use the product rule on $x^2 = x \cdot x$ to check the power rule for $n = 2$ using only the power rule with $n = 1$.

21. Use the product rule on $x^3 = x^2 \cdot x$ to check the power rule for $n = 3$ using only the power rule with $n = 1$ and $n = 2$.

22. Assuming that the power rule is true for n, find $\frac{d(x^{n+1})}{dx}$ using the product rule, and check that it too satisfies the power rule.

Applications

23–26 ▪ The total mass of a population is the product of the number of individuals and the mass of each individual. In each case, time is measured in years and mass is measured in kilograms.

a. Find the total mass as a function of time.

b. Compute the derivative.

c. Find the population, the mass of each individual, and the total mass at the time when the derivative is equal to zero.

d. Sketch a graph of the total mass over the next 100 years.

23. The population P is $P(t) = 2.0 \times 10^6 + 2.0 \times 10^4 t$ and the weight per person $W(t)$ is $W(t) = 80 - 0.5t$ (as in Section 1.2, Exercise 63).

24. The population P is $P(t) = 2.0 \times 10^6 - 2.0 \times 10^4 t$ and the weight per person $W(t)$ is $W(t) = 80 + 0.5t$ (as in Section 1.2, Exercise 64).

25. The population P is $P(t) = 2.0 \times 10^6 + 1000t^2$ and the weight per person $W(t)$ is $W(t) = 80 - 0.5t$ (as in Section 1.2, Exercise 65).

26. The population P is $P(t) = 2.0 \times 10^6 + 2.0 \times 10^4 t$ and the weight per person $W(t)$ is $W(t) = 80 - 0.005t^2$ (as in Section 1.2, Exercise 66).

27–28 ▪ In each of the following situations (extending Exercise 38 in Section 2.5), the mass is the product of the density and the volume. In each case, time is measured in days and density is measured in grams per cm^3.

a. Find the mass as a function of time.

b. Compute the derivative.

c. Sketch a graph of the mass over the next 30 days.

27. The above-ground volume is $V_a(t) = 3.0t + 20.0$ and the above-ground density is $\rho_a(t) = 1.2 - 0.01t$.

28. The below-ground volume is $V_b(t) = -1.0t + 40.0$ and the below-ground density is $\rho_b(t) = 1.8 + 0.02t$.

29–32 ▪ Suppose that the fraction of chicks that survive, $P(N)$, as a function of the number N of eggs laid, is given by the following forms (variants of the model studied in Example 2.5.14). The total number of offspring that survive is $S(N) = N \cdot P(N)$. Find the number of surviving offspring when the bird lays 1, 5, or 10 eggs. Find $S'(N)$. Sketch a graph of $S(N)$. What do you think is the best strategy for each bird?

29. $P(N) = 1 - 0.08N$

30. $P(N) = 1 - 0.16N$.

31. $P(N) = \frac{1}{1 + 0.5N}$.

32. $P(N) = \frac{1}{1 + 0.1N^2}$.

33–34 ▪ Find the derivative of the updating function from Equation 1.10.7, $f(p) = \frac{sp}{sp + r(1-p)}$ with the following values of the parameters s and r.

33. $s = 1.2$, $r = 2.0$.

34. $s = 1.8$, $r = 0.8$.

35–36 ▪ Suppose that the mass $M(t)$ of an insect (in grams) and the volume $V(t)$ (in cm^3) are known functions of time (in days).

a. Find the density $\rho(t)$ as a function of time.

b. Find the derivative of the density.

c. At what times is the density increasing?

d. Sketch a graph of the density over the first 5 days.

35. $M(t) = 1 + t^2$ and $V(t) = 1 + t$.

36. $M(t) = 1 + t^2$ and $V(t) = 1 + 2t$.

37–38 ▪ In a discrete-time dynamical system describing the growth of a population in the absence of immigration and emigration, the final population is the product of the initial population and the per capita production. Represent the initial population by b_t. In each case, find the final population as a function $f(b_t)$ of the initial population, find the derivative, and sketch the function.

37. Per capita production is $2.0\left(1 - \frac{b_t}{1000}\right)$. Sketch $f(b_t)$ for $0 \le b_t \le 1000$.

38. Per capita production is $\frac{2.0}{1 + \frac{b_t}{1000}}$. Sketch $f(b_t)$ for $0 \le b_t \le 2000$.

39–40 ▪ The following steps should help you to figure out what happens to the Hill function $h_n(x) = \frac{x^n}{1 + x^n}$ for large values of n.

a. Compute the value of the function at $x = 0$, $x = 1$, and $x = 2$.

b. Compute the derivative and evaluate at $x = 0$, $x = 1$, and $x = 2$.

c. Sketch a graph.

d. $h_n(x)$ can be thought of as representing a response to a stimulus of strength x. Would the response work as a good filter, giving a small output for inputs less than 1 and a large output for inputs greater than 1?

39. With $n = 3$

40. With $n = 10$

Computer Exercises

41. Consider the functions

$$g_n(x) = 1 + x + x^2 + \dots + x^n$$

for various values of n. We will compare these functions with

$$g(x) = \frac{1}{1 - x}.$$

a. Plot $g_1(x)$, $g_3(x)$, $g_5(x)$, and $g(x)$ the intervals $0 \leq x \leq 0.5$ and $0 \leq x \leq 0.9$.

b. Take the derivative of $g(x)$. Can you see how the derivative is related to $g(x)$ itself? In other words, what function could you apply to the formula for $g(x)$ to get the formula for $g'(x)$?

c. Apply this same function to $g_1(x)$, $g_3(x)$, and $g_5(x)$. Can you see why these functions are good approximations to $g(x)$? Can you see why these approximations are best for small values of x? What happens to the approximations for x near 1?

42. Consider the function

$$r(x) = \frac{u(x)}{v(x)} = \frac{1+x}{2+x^2+x^3}.$$

a. Make one graph of $u(x)$ and $v(x)$ for $0 \leq x \leq 1$, and another of $r(x)$. Could you have guessed the shape of $r(x)$ from looking at the graphs of $u(x)$ and $v(x)$?

b. What happens at the critical point?

c. Find the exact location of the critical point $x = x_c$.

d. Compare $\frac{u'(x_c)}{u(x_c)}$ with $\frac{v'(x_c)}{v(x_c)}$. Why are they equal?

2.7 The Second Derivative, Curvature, and Acceleration

A function with a positive derivative is increasing, and one with a negative derivative is decreasing. In this section, we extend the graphical interpretation of the derivative by examining the derivative of the derivative, or the **second derivative**. In particular, we will see that the second derivative tells whether the graph of a function curves upward (**concave up**) or downward (**concave down**). Furthermore, just as the first derivative of position is the **velocity**, the second derivative of position is the **acceleration**.

2.7.1 The Second Derivative

Consider the two graphs in Figure 2.7.1. Both show functions that increase, but in different ways. In the first, the slope becomes steeper and steeper, indicating that the measurement is increasing faster and faster. In the second, the slope becomes smaller and smaller. Although this measurement is increasing, it does so at a decreasing rate. These differences are clear on graphs of the derivatives of the two functions. Each derivative is **positive** because the functions are increasing. The derivative of the first function is **increasing** (Figure 2.7.2a) because the graph gets steeper and steeper. The derivative of the second function is decreasing (Figure 2.7.2b) because the graph becomes less and less steep.

The derivative of any function is positive when the function is increasing, and negative when the function is decreasing. We can apply this observation to the derivative itself. Because the **derivative** of the first function is increasing, the **derivative of the derivative** must be positive. Similarly, because the derivative of the second function is decreasing, the **derivative of the derivative** must be negative (Figure 2.7.3).

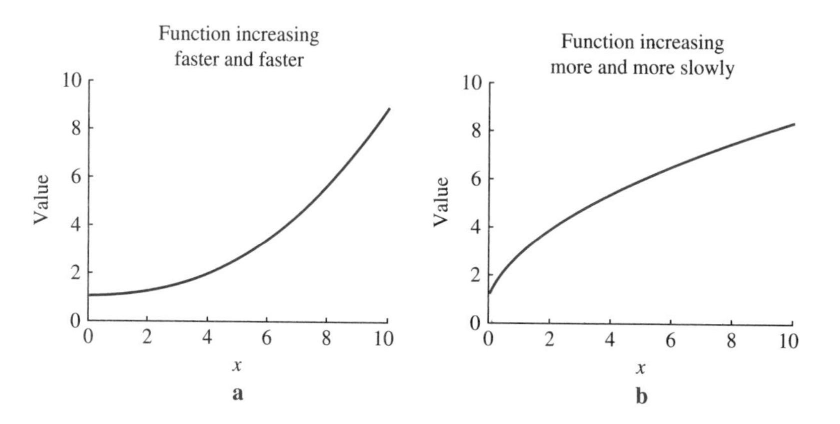

FIGURE 2.7.1
Two different increasing functions

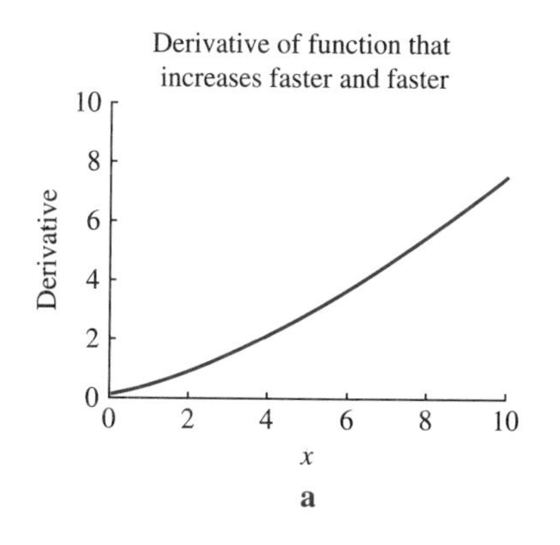

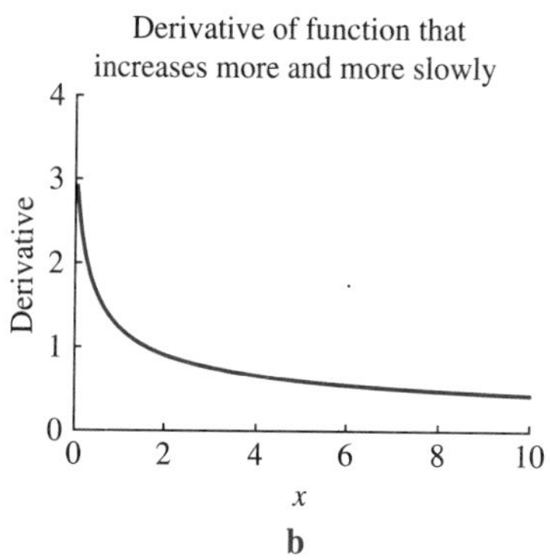

FIGURE 2.7.2

Derivatives of two increasing functions

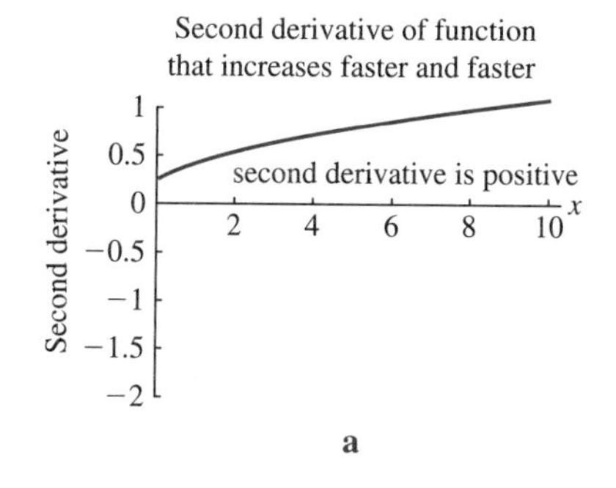

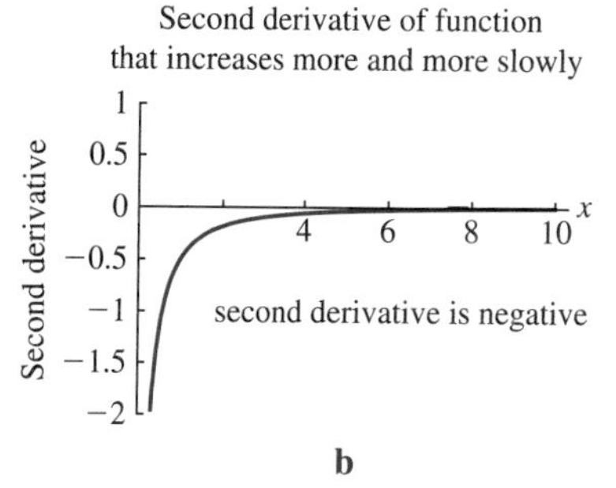

FIGURE 2.7.3

Second derivatives of two increasing functions

Definition 2.7 **The Second Derivative**

The derivative of the derivative is called the **second derivative**. We write the second derivative of f in prime notation as

$$\text{the second derivative of } f = f''(x). \tag{2.7.1}$$

In differential notation,

$$\text{the second derivative of } f = \frac{d^2 f}{dx^2}. \tag{2.7.2}$$

For clarity, the derivative itself is often called the **first derivative**.

The differential notation for the second derivative may look odd. Think of the object $\frac{d}{dx}$ as a sort of function (called an **operator** by mathematicians) that takes one function as input and returns another function as output. This operator returns as output the derivative of its input. To find the second derivative, apply this operator twice. More generally, the result of taking n derivatives is written

$$\begin{aligned} \text{the } n\text{th derivative of } f &= f^{(n)}(x) && \text{prime notation} \\ &= \frac{d^n f}{dx^n}. && \text{differential notation} \end{aligned}$$

The second derivative has a general interpretation in terms of the **curvature** of the graph. Consider the function shown in Figure 2.7.4a. There are many ways to describe this graph. First of all, the graph of the function itself is positive, meaning that the associated measurement takes on only positive values. Furthermore, the function is increasing for $x < 1$, decreasing for $1 < x < 2$, and increasing for $x > 2$. The derivative, therefore, is positive for $x < 1$, negative for $1 < x < 2$, and positive for $x > 2$ (Figure 2.7.4b).

With more careful examination, we can see that the graph breaks into two regions of **curvature**. Between $x = 0$ and $x = 1.5$, the graph curves downward. This portion of the graph is said to be **concave down**. The slope of the curve is a **decreasing** function in this region (Figure 2.7.4b), implying that the **second derivative is negative** (Figure 2.7.4c). Between $x = 1.5$ and $x = 3$, the graph curves upward, like a bowl.

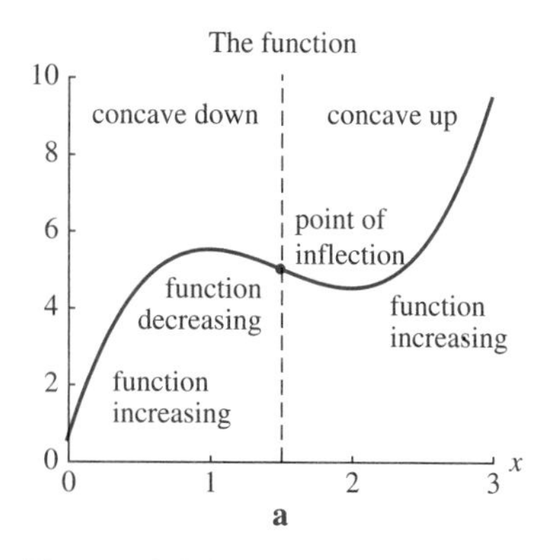

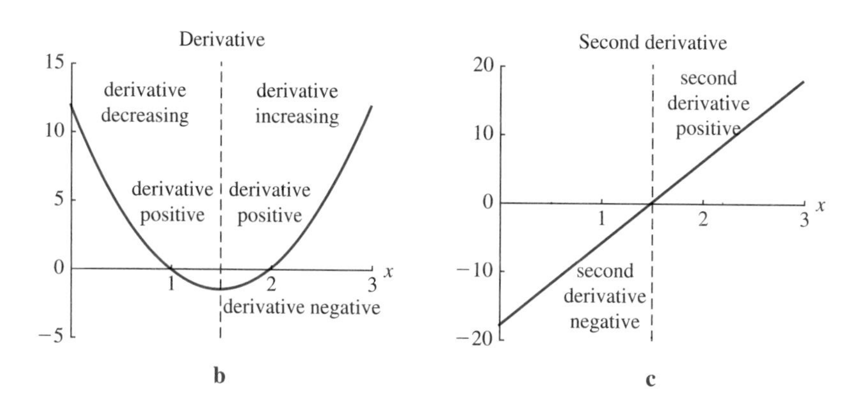

FIGURE 2.7.4

A function, its derivative, and second derivative

This portion of the graph is said to be **concave up**. In this region, the slope of the curve becomes steeper and steeper, the derivative is increasing, and the **second derivative is positive** (Figure 2.7.4c).

At a point where the **second derivative** is zero, the curvature of the graph can change. If the curvature changes sign (from positive to negative or from negative to positive), such a point is called a **point of inflection**. In Figure 2.7.4a, the graph has a point of inflection at $x = 1.5$, where the function switches from concave down to concave up and the second derivative switches from negative to positive. Neither the function itself nor the derivative changes sign at this point. Points of inflection can occur when the derivative is positive, negative, or zero.

Example 2.7.1 Graphing a Power Function with the Second Derivative

Consider the power function $C(x) = x^3$ (Figure 2.7.5). Using the power rule, $C'(x) = 3x^2$ and $C''(x) = 6x$. Therefore, this function has both derivative and second derivative equal to 0 at the same point, $x = 0$. Furthermore, the second derivative changes from negative for $x < 0$ to positive for $x > 0$, meaning that it has both a critical point and a point of inflection at $x = 0$.

Example 2.7.2 A Point with Second Derivative Equal to Zero That Is Not a Point of Inflection

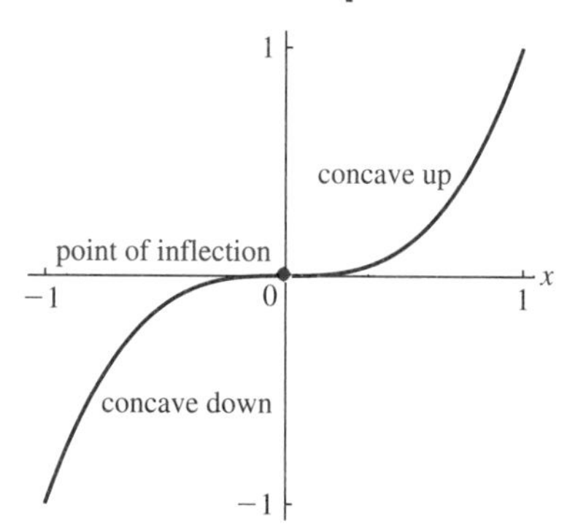

FIGURE 2.7.5

The function $C(x) = x^3$ has a point of inflection at a critical point

Consider the power function $Q(x) = x^4$ (Figure 2.7.6). Using the power rule, $Q'(x) = 4x^3$ and $Q''(x) = 24x^2$. This function has both derivative and second derivative equal to 0 at $x = 0$. However, the second derivative is positive for $x < 0$ and for $x > 0$. Because the function is concave up on both sides of $x = 0$, this is not a point of inflection.

2.7.2 Using the Second Derivative for Graphing

When we can compute the derivative and second derivative of a function, we can use these tools to quickly sketch graphs.

Example 2.7.3 The Second Derivative of a Linear Function

As a test, we check the second derivative of a linear function

$$f(x) = 2x + 1.$$

The first derivative is $f'(x) = 2$, a constant, implying that the second derivative is 0, consistent with the fact that a line is straight and has no curvature (Figure 2.7.7).

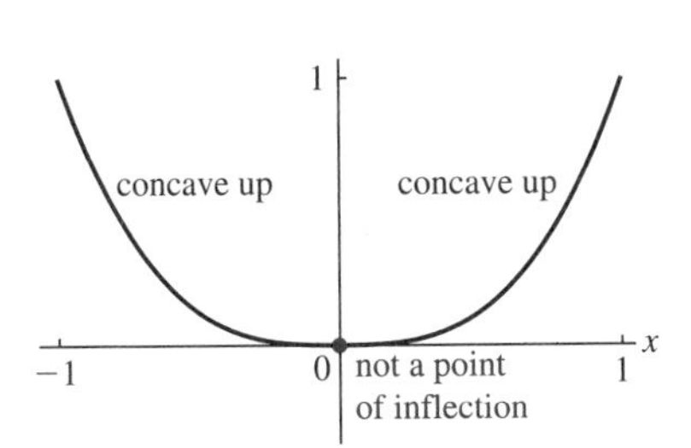

FIGURE 2.7.6
The function $C(x)=x^4$ does not have a point of inflection at $x=0$

FIGURE 2.7.7
A linear function has no curvature

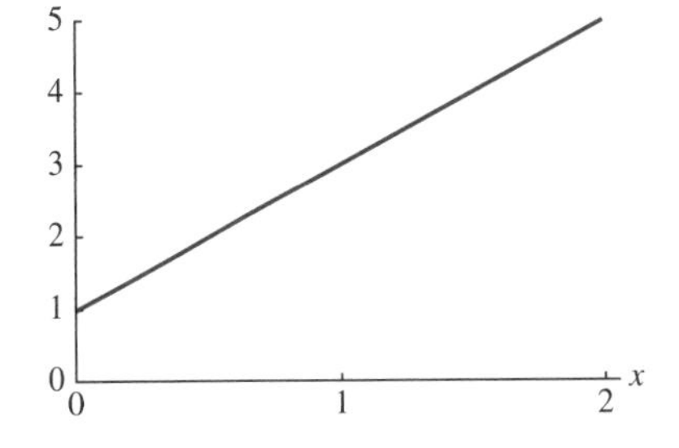

Example 2.7.4 A Quadratic Function That Is Concave Up

Consider now the quadratic function

$$y(t)=5t^2.$$

The first derivative is $y'(t)=10t$, using the constant product and power rules. Because $y'(t)$ is a linear function, the second derivative is $y''(t)=10$, which is a positive constant (Figure 2.7.8).

Example 2.7.5 A Quadratic Function That Is Concave Down

The quadratic function

$$z(t)=5-5t^2.$$

has first derivative $z'(t)=-10t$ and second derivative $z''(t)=-10$, a negative constant (Figure 2.7.9).

A quadratic function with a **positive** coefficient on the quadratic term has a positive second derivative and is concave up at all points. A quadratic function with a **negative**

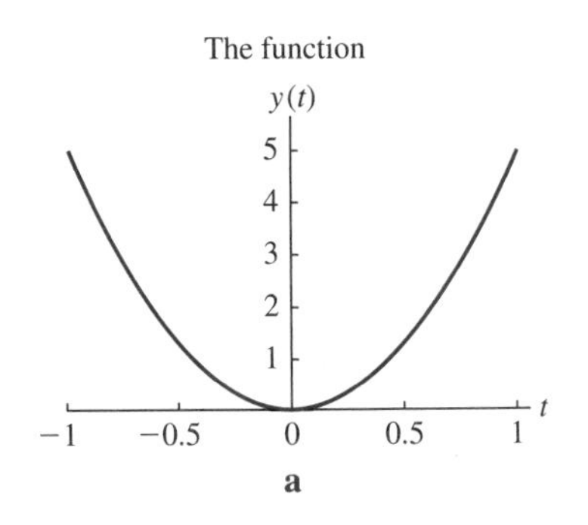

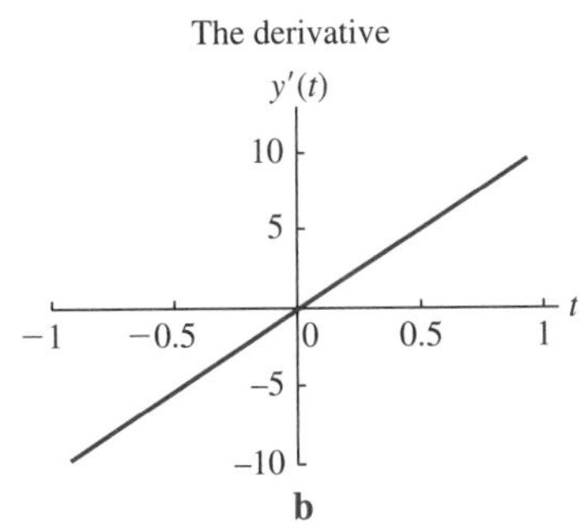

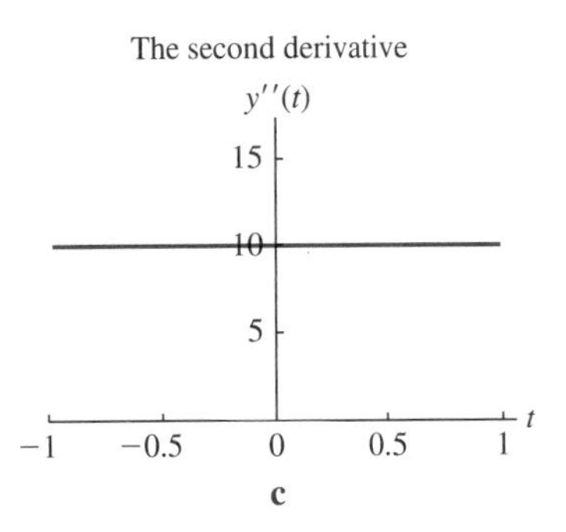

FIGURE 2.7.8
Concave up quadratic function

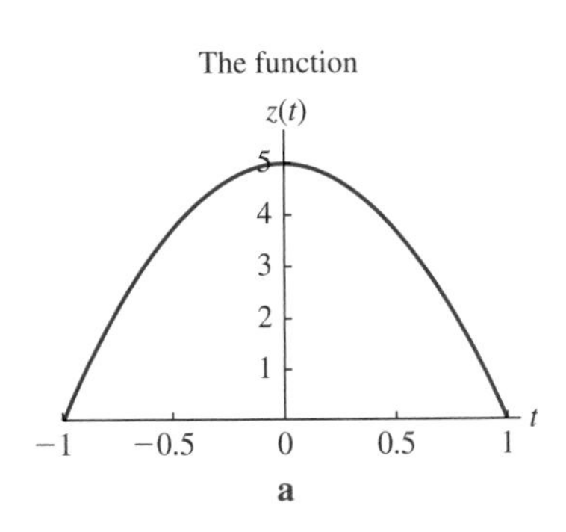

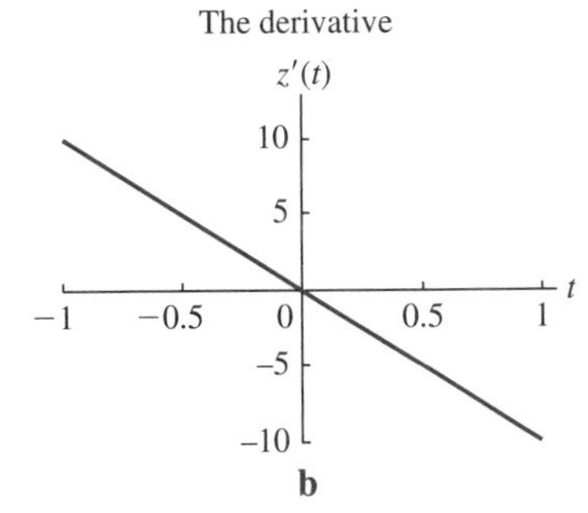

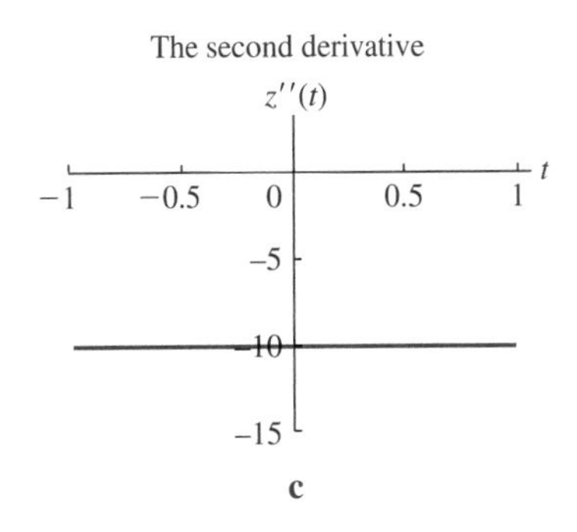

FIGURE 2.7.9
Concave down quadratic function

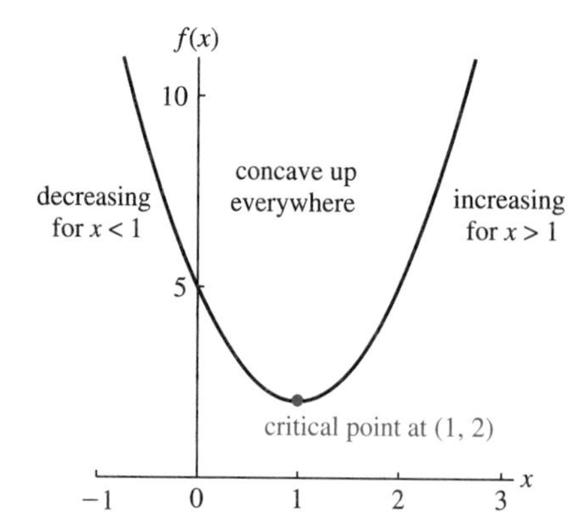

FIGURE 2.7.10

Graphing a quadratic using the first and second derivatives

coefficient on the quadratic term has a negative second derivative and is concave down at all points.

Example 2.7.6 Graphing a Quadratic Function with the Second Derivative

The quadratic function

$$f(x) = 3x^2 - 6x + 5$$

has a positive coefficient 3 in front of x^2, and therefore must be concave up. The derivative is

$$f'(x) = 6x - 6.$$

Solving for $f'(x) = 0$ gives a critical point at $x = 1$. Because the graph of this function is an upward-pointing parabola, this critical point must be the bottom of the bowl. Using the fact that $f(1) = 2$, we can easily sketch a graph of this function (Figure 2.7.10).

The second derivative helps to categorize all **power functions**. Consider the power function

$$g(x) = x^p.$$

According to the power rule, the first and second derivatives are

$$g'(x) = px^{p-1}$$
$$g''(x) = p(p-1)x^{p-2}.$$

By substituting in various values of p, we can create the following table and graph the three possibilities (Figure 2.7.11). This table and the graphs apply only for $x > 0$, the values for which power functions are most often applied.

Behavior of Power Function $g(x) = x^p$ for $x > 0$		
Power	**First Derivative**	**Second Derivative**
$p > 1$	Positive	Positive
$0 < p < 1$	Positive	Negative
$p < 0$	Negative	Positive

Example 2.7.7 The Cubic Function Revisited

The cubic power function $g(x) = x^3$ has first derivative $g'(x) = 3x^2$ and second derivative $g''(x) = 6x$. For $x > 0$, both are positive (Figure 2.7.11a). This function is increasing and concave up on this domain, consistent with the fact that the power is greater than 1.

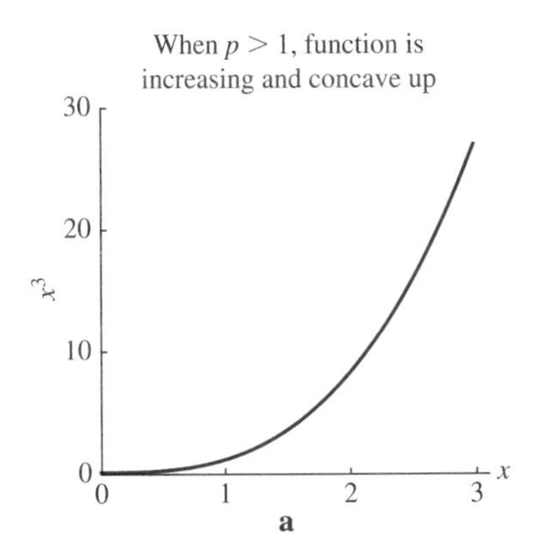

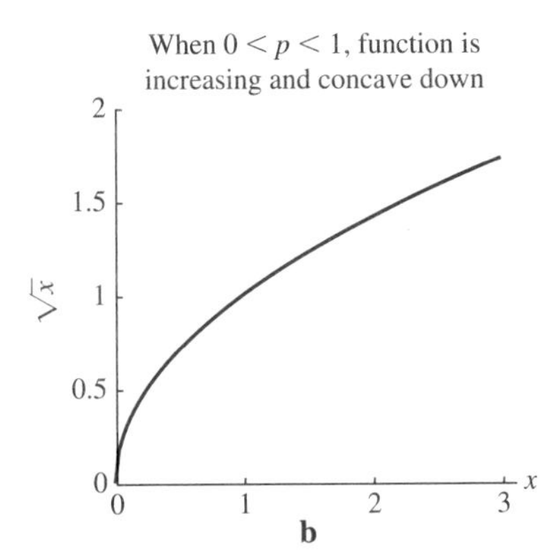

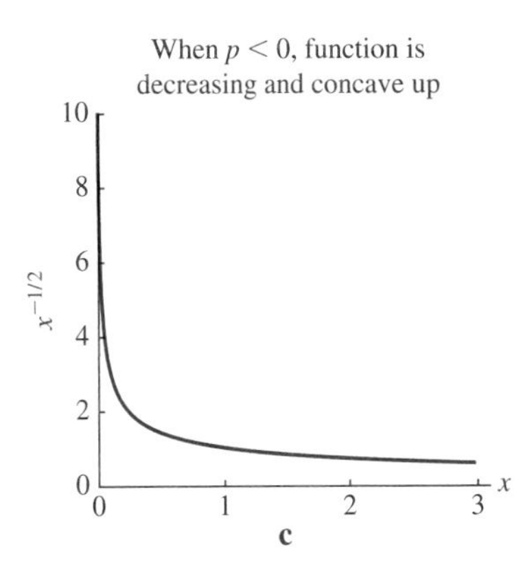

FIGURE 2.7.11

Graphs of the basic power functions

Example 2.7.8 Graphing the Square Root Function

The square root function $g(x) = \sqrt{x}$ is the power function

$$g(x) = x^{\frac{1}{2}}.$$

The first and second derivatives are

$$g'(x) = \frac{1}{2}x^{-\frac{1}{2}}$$
$$g''(x) = -\frac{1}{2} \cdot \frac{1}{2}x^{-\frac{3}{2}} = -\frac{1}{4}x^{-\frac{3}{2}}.$$

In accordance with the table, the second derivative is negative and the function is concave down (Figure 2.7.11b).

Example 2.7.9 Graphing the Reciprocal of the Square Root Function

The function

$$g(x) = x^{-\frac{1}{2}}$$

has first and second derivatives

$$g'(x) = -\frac{1}{2}x^{-\frac{3}{2}}$$
$$g''(x) = \frac{1}{2} \cdot \frac{3}{2}x^{-\frac{5}{2}} = \frac{3}{4}x^{-\frac{5}{2}}.$$

This decreasing function is concave up (Figure 2.7.11c).

Example 2.7.10 Graphing a Cubic Polynomial

Suppose we wish to study the polynomial

$$p(x) = 2x^3 - 7x^2 + 5x + 2.$$

What does the graph look like? Does $p(x)$ take on negative values for $x > 0$? To begin, we find the derivatives

$$p'(x) = 6x^2 - 14x + 5$$
$$p''(x) = 12x - 14.$$

The point of inflection occurs where $12x - 14 = 0$, or at $x = 1.167$. The graph is concave down for $x < 1.167$ and concave up for $x > 1.167$. The critical points occur at solutions of $p'(x) = 0$. Applying the quadratic formula, these occur at

$$\frac{7 - \sqrt{19}}{6} = 0.440, \quad \frac{7 + \sqrt{19}}{6} = 1.893.$$

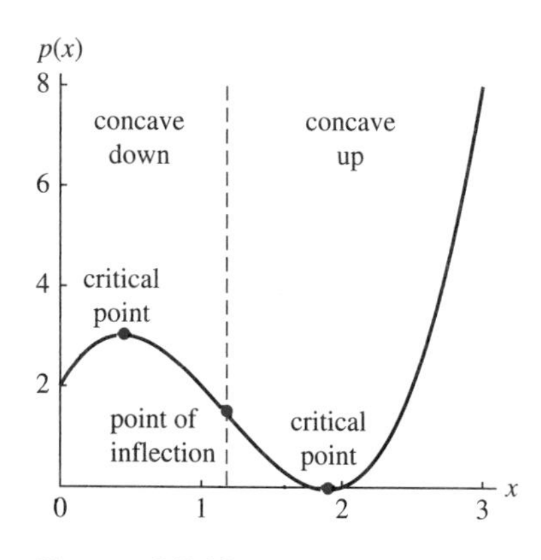

FIGURE 2.7.12

Using the first and second derivatives to graph a polynomial

At these points,

$$p(0.440) = 3.105, \quad p(1.893) = -0.052.$$

Figure 2.7.12 combines this information in a single graph. This function does indeed take on negative values, but only right near the critical point at $x = 1.893$.

The guideposts for reading and interpreting a graph are summarized in the following table.

What the Graph Does	What the Derivative Does
Jump	Function discontinuous, derivative not defined
Corner	Derivative not defined
Graph vertical	Derivative not defined (infinite)
Graph increasing	Derivative positive
Graph decreasing	Derivative negative
Graph horizontal	Derivative equal to zero
Graph concave up	Second derivative positive
Graph concave down	Second derivative negative
Graph switches curvature	Point of inflection, second derivative zero

2.7.3 Acceleration

When considering position as a function of time, the second derivative has a fundamental physical interpretation as the **acceleration**. Suppose the position of an object is $y(t)$. The derivative $\frac{dy}{dt}$ is the velocity, and the second derivative is the rate of change of velocity. Formally,

$$\frac{d^2y}{dt^2} = \text{acceleration}. \tag{2.7.3}$$

A positive acceleration indicates that an object is speeding up, and a negative acceleration that it is slowing down.

Equation 2.7.3 is a fundamental type of differential equation studied in physics, saying that acceleration is proportional to force (here the force of gravity). In fact, we can rewrite Newton's famous law

$$F = ma,$$

where F is the force, m the mass of the object, and a the resulting acceleration, as

$$a = \frac{d^2y}{dt^2} = \frac{F}{m}.$$

If we know the force F and the mass m, we have a differential equation for the position y.

Example 2.7.11 A Falling Cat

A cat that has fallen a distance

$$y(t) = 5.0t^2$$

in time t has second derivative, and acceleration, of

$$\frac{d^2y}{dt^2} = 10.0.$$

This acceleration is positive because we are measuring how far the cat has fallen, and the downward speed is increasing.

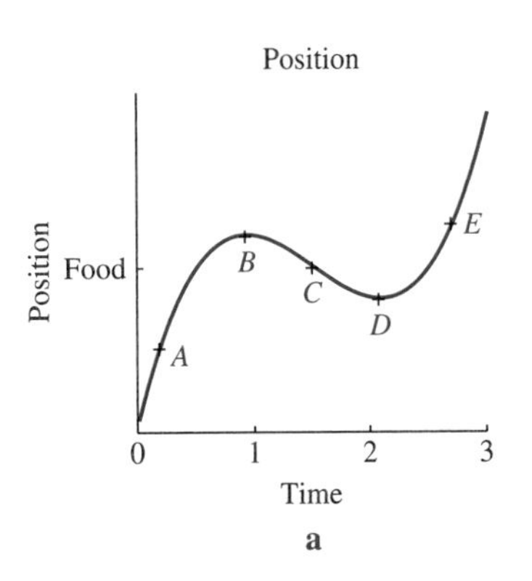

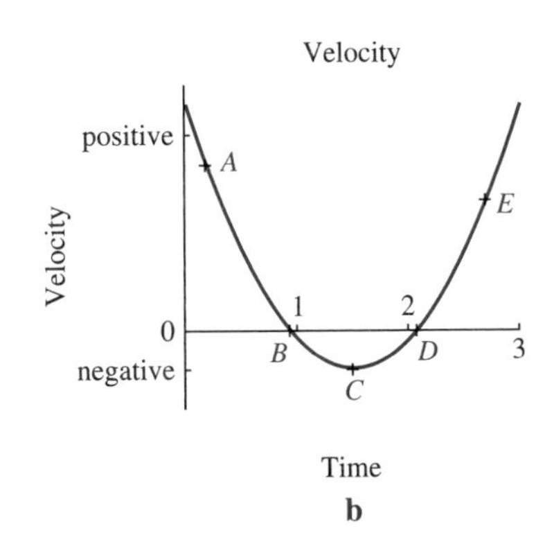

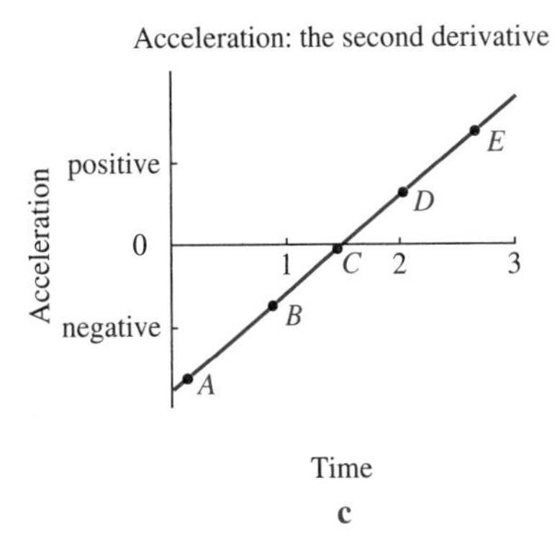

FIGURE 2.7.13
The acceleration of an ant

Example 2.7.12 Acceleration of an Ant

Consider again the ant pictured in Figures 2.4.5 and 2.4.6. The position, velocity, and acceleration are shown in Figure 2.7.13. This ant is slowing down, or decelerating, until point C, and then speeds up afterward. Deceleration includes both slowing down in the forward direction and speeding up in the negative direction. Physically, the ant acts as if there is a force pushing it to the left before it reaches C. After that time, the ant accelerates. Acceleration includes any change tending to move the ant more toward the right.

Summary The **second derivative**, defined as the derivative of the derivative, is positive when the graph of a function is **concave up** and negative when the graph of a function is **concave down**. A point where a function changes curvature is called a **point of inflection**. Using the second derivative, we can graph quadratic functions, power functions, and complicated polynomials. Physically, the second derivative of the position is the **acceleration**, the tool used to write physical laws as differential equations.

2.7 Exercises

Mathematical Techniques

1–4 ▪ On the figures, label

a. One critical point.

b. One point with a positive derivative.

c. One point with a negative derivative.

d. One point with a positive second derivative.

e. One point with a negative second derivative.

f. One point of inflection.

1.

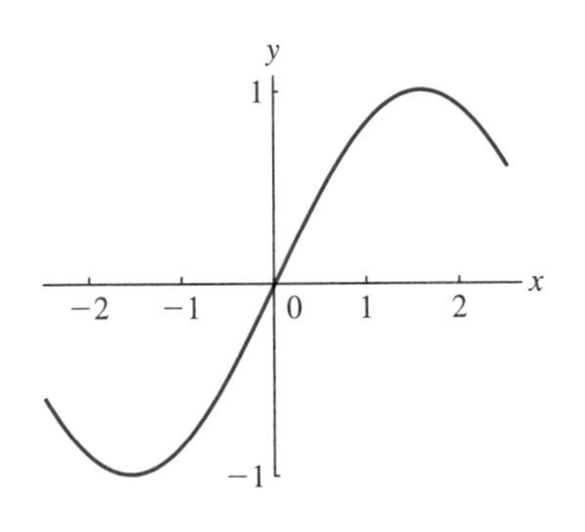

2.

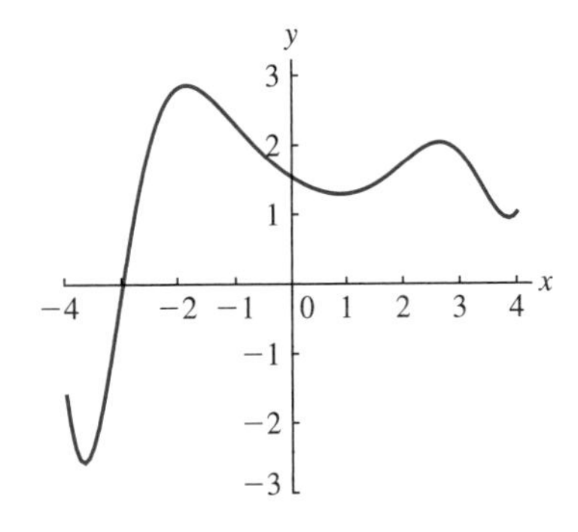

3.

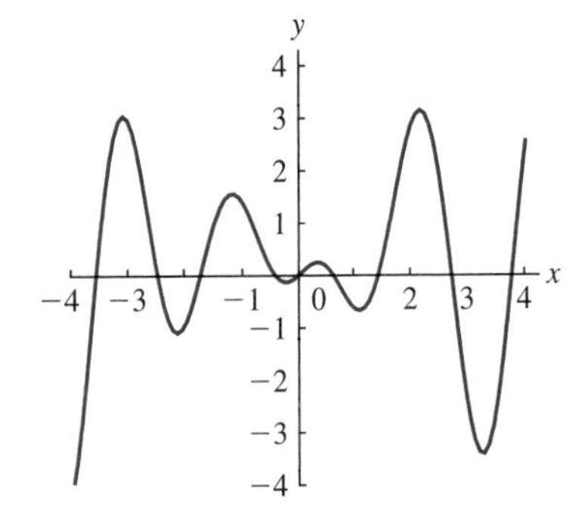

4.

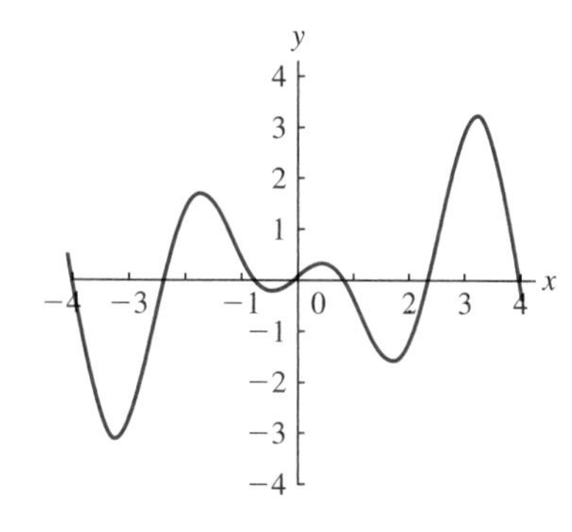

5–8 ▪ Draw graphs of functions with the following properties.

5. A function with a positive, increasing derivative.

6. A function with a positive, decreasing derivative.

7. A function with a negative, increasing (becoming less negative) derivative.

8. A function with a negative, decreasing (becoming more negative) derivative.

9–10 ▪ Although there is no easy way to recognize all points with positive or negative **third** derivative, it is possible for some points (usually points of inflection).

9. On the figure for Exercise 1, find one point with negative **third** derivative.

10. On the figure for Exercise 2, find one point with positive third derivative.

11–18 ▪ Find the first and second derivatives of the following functions.

11. $s(x) = 1 - x + x^2 - x^3 + x^4$.

12. $g(z) = 3z^3 + 2z^2$.

13. $h(y) = y^{10} - y^9$.

14. $p(x) = 1 + x + \frac{x^2}{2} + \frac{x^3}{6} + \frac{x^4}{24}$.

15. $F(z) = z(1+z)(2+z)$.

16. $R(s) = (1+s^2)(2+s)$.

17. $f(x) = \frac{3+x}{2x}$.

18. $G(y) = \frac{2+y}{y^2}$.

19–26 ▪ Find the first and second derivatives of the following functions and use them to sketch a graph.

19. $f(x) = x^{-3}$ for $x > 0$.

20. $g(z) = z + \frac{1}{z}$ for $z > 0$.

21. $h(x) = (1-x)(2-x)(3-x)$.

22. $M(t) = \frac{t}{1+t}$ for $t > 0$.

23. $f(x) = 2x^3 + 1$ for $-5 \le x \le 5$.

24. $f(x) = \frac{1}{x^2}$ for $0 < x \le 2$.

25. $f(x) = 10x^2 - 50x$ for $-5 \le x \le 5$.

26. $f(x) = x - x^2$ for $0 \le x \le 1$.

27–30 ▪ Some higher derivatives can be found without a lot of calculation.

27. Find the 10th derivative of x^9.

28. Describe the graph of the 5th derivative of x^5.

29. Is the eighth derivative of $p(x) = 7x^8 - 8x^7 - 5x^6 + 6x^5 - 4x^3$ positive or negative?

30. Find the fifth derivative of $x(1+x)(2+x)(3+x)(4+x)$.

Applications

31–34 ▪ The following equations give the positions as functions of time of objects tossed from towers in various exotic solar system locations. For each,

a. Find the velocity and the acceleration of this object.

b. Sketch a graph of the position for $0 \le t \le 3$.

c. How high was the tower? Which way was the object thrown? How does the acceleration compare with that on earth (9.8 m/sec^2)?

31. An object on Saturn that follows $p(t) = -5.2t^2 - 2.0t + 50.0$.

32. An object on the Sun that follows $p(t) = -137t^2 + 20.0t + 500.0$.

33. An object on Pluto that follows $p(t) = -0.325t^2 - 20.0t + 500.0$.

34. An object on Mercury that follows $p(t) = -1.85t^2 + 20.0t$.

35–38 ▪ The total mass is the product of the following functions for mass and number as functions of time in years (based on Section 2.6, Exercises 23–26). Find the second derivative of each and check your graph.

35. The population P is $P(t) = 2.0 \times 10^6 + 2.0 \times 10^4 t$ and the weight per person $W(t)$ is $W(t) = 80 - 0.5t$ (based on Section 2.6, Exercise 23).

36. The population P is $P(t) = 2.0 \times 10^6 - 2.0 \times 10^4 t$ and the weight per person $W(t)$ is $W(t) = 80 + 0.5t$ (based on Section 2.6, Exercise 24).

37. The population P is $P(t) = 2.0 \times 10^6 + 1000t^2$ and the weight per person $W(t)$ is $W(t) = 80 - 0.5t$ (based on Section 2.6, Exercise 25).

38. The population P is $P(t) = 2.0 \times 10^6 + 2.0 \times 10^4 t$ and the weight per person $W(t)$ is $W(t) = 80 - 0.005t^2$ (based on Section 2.6, Exercise 26).

39–40 ▪ The following graphs show the position of a roller coaster as a function of time. When is the roller coaster going most quickly? When is it accelerating most quickly? When is it decelerating most quickly?

39.

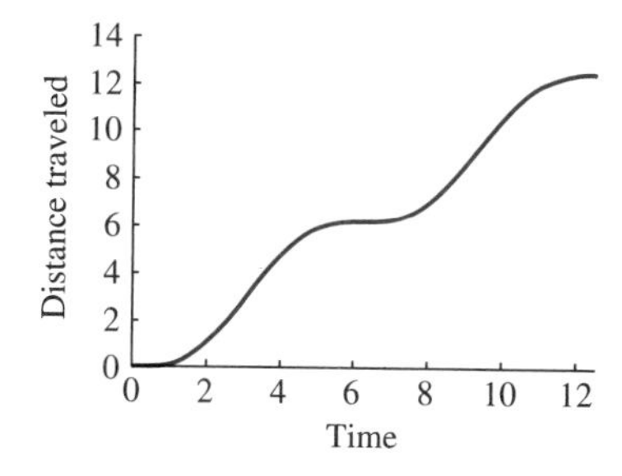

40.

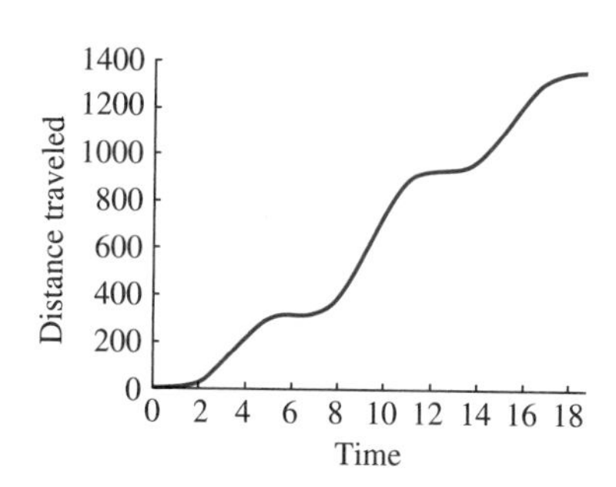

41–42 ▪ In a model of a growing population, we find the new population by multiplying the old population by the per capita production. For each case, find the second derivative of the new population as a function of the old population and sketch a graph.

41. Per capita production is $2.0\left(1 - \frac{b_t}{1000}\right)$. Consider values of b_t less than 1000.

42. Per capita production is $2.0b_t\left(1 - \frac{b_t}{1000}\right)$. Consider values of b_t less than 1000.

43–44 ▪ We can use the second derivative to study Hill functions $h_n(x) = \frac{x^n}{1+x^n}$ for $x > 0$.

43. Find the second derivative of the Hill function with $n = 1$ and describe the curvature of the graph.

44. Find the second derivative of the Hill function with $n = 2$ and describe the curvature of the graph.

Computer Exercises

45. In the current universe, acceleration due to gravity is constant, so that position follows a differential equation rather like

$$\frac{d^2p}{dt^2} = -g$$

when g points in the downward direction. One can imagine a universe where gravity changed over time, making objects accelerate according to

$$\frac{d^2p}{dt^2} = -gt^n$$

for some power n. Set $g = 10$ m/s.

a. Find a solution of the normal differential equation.

b. Find solutions of the modified differential equation for different values of n. Would objects fall faster or slower?

46. Have your computer find all critical points and points of inflection of the function

$$P(x) = 8x^5 - 18x^4 - x^3 + 18x^2 - 7x.$$

Show that these match what you see on a graph.

2.8 Derivatives of Exponential and Logarithmic Functions

With the sum, product, power, and quotient rules, we can differentiate polynomials and ratios of polynomials. To be able to differentiate all biologically important functions, however, we need three more building blocks: exponential, logarithmic, and trigonometric functions. We here find and apply the derivatives of the exponential and logarithmic functions.

2.8.1 The Exponential Function

Suppose we wish to find the derivative of the function $b(t) = 2^t$. None of our rules tell us how to find a formula for the derivative. The power rule looks promising, but only applies to functions of the form t^n, not when t is in the exponent. The best way to find the derivative of an unfamiliar function is to return to the definition of the derivative (Definition 2.2),

$$\frac{db}{dt} = \lim_{h \to 0} \frac{b(t+h) - b(t)}{h}.$$

The derivative of the function 2^t is

$$\begin{aligned}
\frac{d2^t}{dt} &= \lim_{h \to 0} \frac{2^{t+h} - 2^t}{h} && \text{definition of the derivative} \\
&= \lim_{h \to 0} \frac{2^t 2^h - 2^t}{h} && \text{law 1 of exponents} \\
&= \lim_{h \to 0} \frac{2^t (2^h - 1)}{h} && \text{factor out } 2^t \\
&= 2^t \left(\lim_{h \to 0} \frac{2^h - 1}{h} \right). && \text{pull constant out of limit}
\end{aligned}$$

The quantity 2^t acts as a constant because the limit depends on h, not on t. The derivative is the product of two factors, the function 2^t and the *number*

$$\lim_{h \to 0} \frac{2^h - 1}{h}.$$

This is related to a limit we studied in Section 2.1, where we estimated the values of 0.6697 by plugging in $h = 0.1$. Exercise 34 in Section 2.1 checked a small value of h, finding a result of approximately 0.693. We will finally learn how to compute this number exactly in Section 2.9. Based on our best estimate, the derivative is

$$\frac{db}{dt} = (0.693)2^t = 0.693b(t).$$

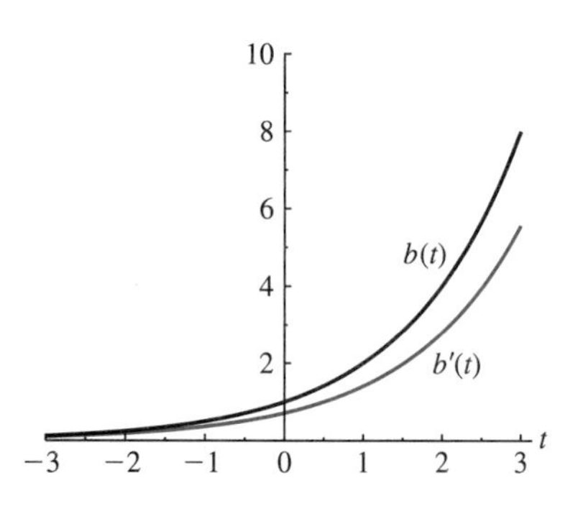

FIGURE 2.8.1
The function 2^t and its derivative

The derivative of this function is 0.693 times the function itself (Figure 2.8.1). The graph of the derivative looks like the graph of the function, but is shifted down by a constant factor.

What happens when we try find the derivative of the exponential function itself? We can follow the same steps to compute

$$\begin{aligned}
\frac{de^x}{dx} &= \lim_{h \to 0} \frac{e^{x+h} - e^x}{h} && \text{definition of the derivative} \\
&= \lim_{h \to 0} \frac{e^x e^h - e^x}{h} && \text{law 1 of exponents} \\
&= \lim_{h \to 0} \frac{e^x (e^h - 1)}{h} && \text{factor out } e^x \\
&= e^x \lim_{h \to 0} \frac{e^h - 1}{h}. && \text{pull constant out of limit}
\end{aligned}$$

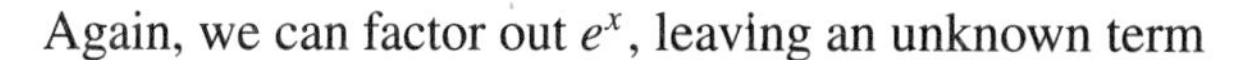

Again, we can factor out e^x, leaving an unknown term

$$\lim_{h \to 0} \frac{e^h - 1}{h}.$$

We can guess the limit by evaluating the function for smaller and smaller values of h.

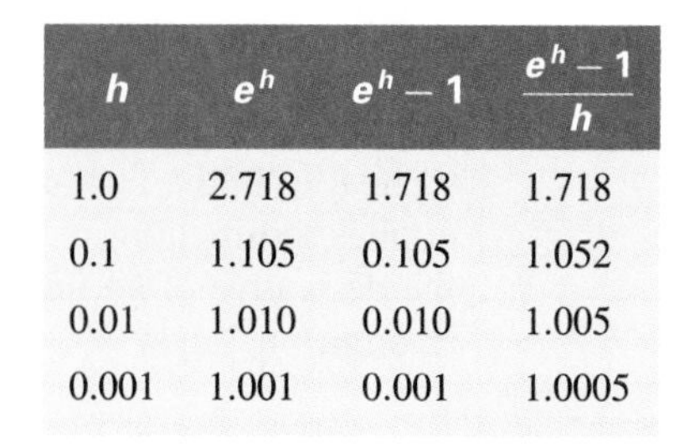

h	e^h	$e^h - 1$	$\frac{e^h - 1}{h}$
1.0	2.718	1.718	1.718
0.1	1.105	0.105	1.052
0.01	1.010	0.010	1.005
0.001	1.001	0.001	1.0005

The limit seems to be 1.0. In fact, the number e is **defined** to be the number for which the limit is 1. Mathematically,

Definition 2.8 The number e is the number for which

$$\lim_{h \to 0} \frac{e^h - 1}{h} = 1.$$

It is a mathematical fact that e is the irrational number 2.7182818284590..., rather than some familiar number like 2 or 3.

Therefore

$$\frac{de^x}{dx} = e^x.$$

The exponential function is its own derivative. At every point on the graph, the slope of the curve is equal to the height of the curve (Figure 2.8.2).

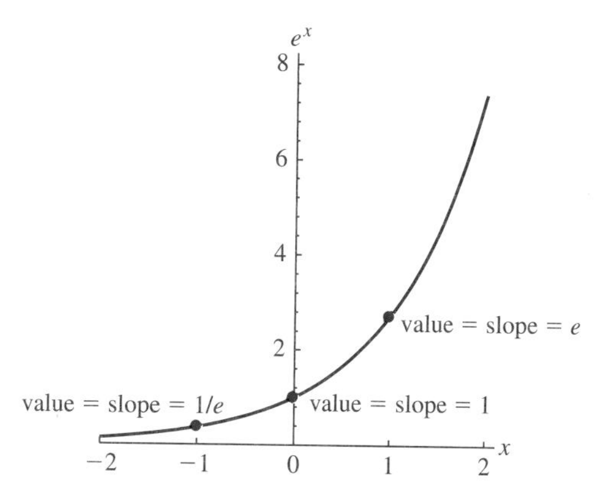

FIGURE 2.8.2
The function e^x and its derivative

Example 2.8.1 A Related Function That Is Its Own Derivative

Are there any other functions with this remarkable property? The function $g(x) = 2e^x$ has the derivative

$$\begin{aligned} \frac{d}{dx} 2e^x &= 2\frac{de^x}{dx} && \text{the constant product rule} \\ &= 2e^x. && \text{derivative of the exponential function} \end{aligned}$$

In general, any constant multiple of the exponential function is its own derivative, but these are the **only** functions with this property. Therefore, functions of the form $b(t) = Ke^t$ are solutions of the differential equation

$$\frac{db}{dt} = b$$

which says that the rate of change of the population is equal to the population size. When we study differential equations in more detail, we will find that the constant K appearing in front of the exponential depends on the **initial condition**, just like the solution of a discrete-time dynamical system.

Example 2.8.2 Computing the Derivative of e^{3t} from the Definition

How does including a constant in the exponent change the derivative?

$$\begin{aligned} \frac{de^{3t}}{dt} &= \lim_{h \to 0} \frac{e^{3(t+h)} - e^{3t}}{h} && \text{definition of the derivative} \\ &= \lim_{h \to 0} \frac{e^{3t}e^{3h} - e^{3t}}{h} && \text{law 1 of exponents} \\ &= \lim_{h \to 0} \frac{e^{3t}(e^{3h} - 1)}{h} && \text{factor out } e^{3t} \\ &= e^{3t} \lim_{h \to 0} \frac{e^{3h} - 1}{h}. && \text{pull constant out of limit} \end{aligned}$$

Factoring out e^{3t} leaves the unknown term

$$\lim_{h\to 0} \frac{e^{3h}-1}{h}.$$

We can find this limit with a clever trick, changing to a new variable $h' = 3h$. If we replace every appearance of h with $\frac{h'}{3}$ and note that $h' \to 0$ if $h \to 0$, we find

$$\begin{aligned}
\lim_{h\to 0} \frac{e^{3h}-1}{h} &= \lim_{h'\to 0} \frac{e^{h'}-1}{h'/3} && \text{substitute } h'/3 \text{ for } h \\
&= \lim_{h'\to 0} 3\frac{e^{h'}-1}{h'} && \text{move the 3 to the numerator} \\
&= 3 \lim_{h'\to 0} \frac{e^{h'}-1}{h'} && \text{pull constant outside limit} \\
&= 3. && \text{recognize that limit is equal to 1 from definition of } e
\end{aligned}$$

Therefore,

$$\frac{de^{3t}}{dt} = 3e^{3t}.$$

The derivative of this function is 3 times the original function, meaning that it is a solution of

$$\frac{db}{dt} = 3b.$$

Example 2.8.3 Finding the Derivative of e^{-x}

We could find the derivative of e^{-x} using the same trick as in Example 2.8.2. Alternatively, we can use the quotient rule to find

$$\begin{aligned}
\frac{de^{-x}}{dx} &= \frac{d}{dx}\frac{1}{e^x} \\
&= \frac{e^x \frac{d1}{dx} - 1\frac{de^x}{dx}}{e^{2x}} \\
&= \frac{-e^x}{e^{2x}} \\
&= \frac{-1}{e^x} = -e^{-x}.
\end{aligned}$$

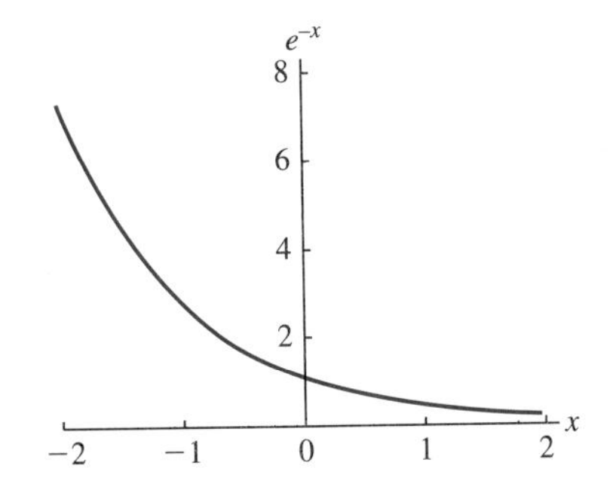

FIGURE 2.8.3
The function $F(x) = e^{-x}$

The derivative of this function is the negative of the original function (Figure 2.8.3). It is a solution of

$$\frac{db}{dt} = -b.$$

Furthermore,

$$\begin{aligned}
\frac{d^2e^{-x}}{dx^2} &= \frac{d}{dx}(-e^{-x}) && \text{the derivative we just found} \\
&= -\frac{d}{dx}e^{-x} && \text{pull out negative sign using the constant product rule} \\
&= -(-e^x) = e^{-x}. && \text{the derivative we just found and canceling negative signs}
\end{aligned}$$

This function is its own second derivative. Because the exponential function takes on only positive values, this function has a negative first derivative and a positive second derivative everywhere.

Example 2.8.4 Graphing a Combined Exponential and Polynomial Function

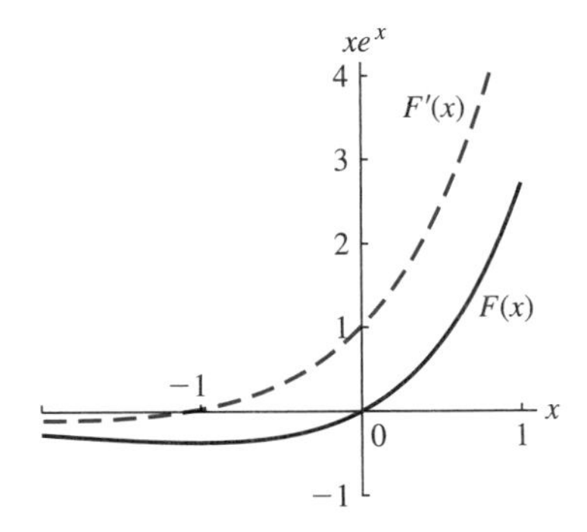

FIGURE 2.8.4

The function $F(x) = xe^x$ and its derivative

Using the derivative of the exponential function and the basic rules for differentiation, we can find the derivatives of more complicated functions. Consider the function $F(x) = xe^x$ (Figure 2.8.4). This is a **product** of a power function and the exponential function. Therefore,

$$\begin{aligned} \frac{dF}{dx} &= x\frac{de^x}{dx} + e^x\frac{dx}{dx} && \text{product rule} \\ &= xe^x + e^x && \text{derivative of exponential and power rule} \\ &= (1+x)e^x. && \text{factoring} \end{aligned}$$

This derivative is negative when $x < -1$, zero at $x = -1$, and positive for $x > -1$.

2.8.2 The Natural Logarithm

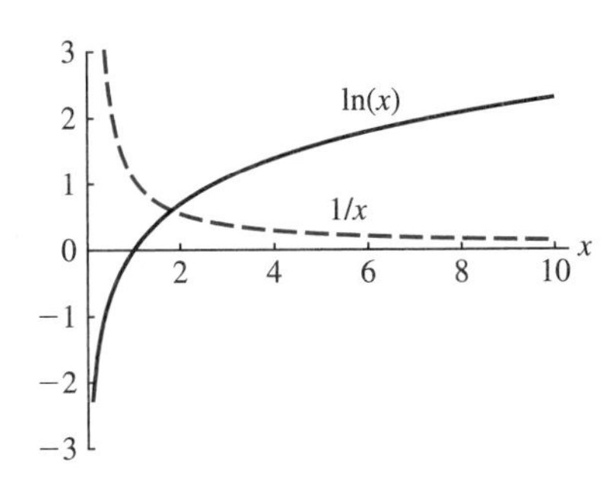

FIGURE 2.8.5

The natural logarithm and its derivative

The formula for the derivative of the inverse of the exponential function, the natural logarithm, is also simple:

$$\frac{d\ln(x)}{dx} = \frac{1}{x}$$

(Figure 2.8.5).

We will use the fact that the natural log and exponential functions are inverses to derive this formula in Example 2.9.5. Because $\ln(x)$ is only defined for $x > 0$, the derivative is also defined only on this domain. The derivative of the natural logarithm is a power function with a power of -1. We can therefore find the second derivative with the power rule, computing

$$\frac{d^2\ln(x)}{dx^2} = -\frac{1}{x^2}.$$

The graph of the natural logarithm is concave down, increasing more and more slowly.

Example 2.8.5 The Derivative of $\ln(x)$

The derivative of the function $g(x) = x\ln(x)$ is

$$\begin{aligned} \frac{dg}{dx} &= x\frac{d\ln(x)}{dx} + \ln(x)\frac{dx}{dx} && \text{product rule} \\ &= x\cdot\frac{1}{x} + \ln(x)\cdot 1 && \text{derivative of log and power rule} \\ &= 1 + \ln(x). && \text{simplifying} \end{aligned}$$

Example 2.8.6 Finding the Derivative Using a Law of Logs

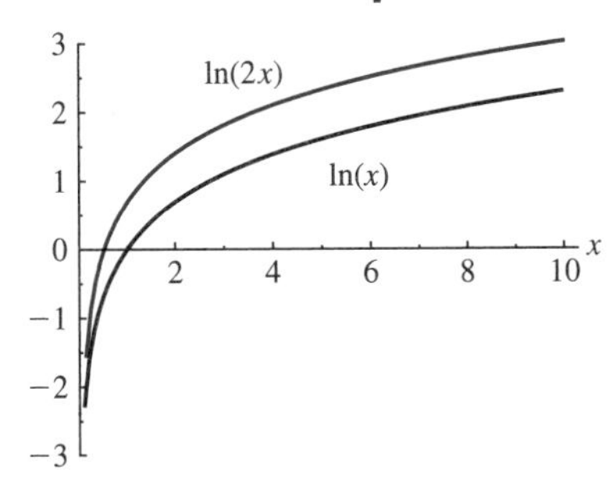

FIGURE 2.8.6

The functions $\ln(x)$ and $\ln(2x)$ have the same derivative

We can find the derivative of the function $\ln(2x)$ using law 1 of logs that states that $\ln(2x) = \ln(2) + \ln(x)$.

$$\begin{aligned} \frac{d\ln(2x)}{dx} &= \frac{d(\ln(2)+\ln(x))}{dx} && \text{law 1 of logs} \\ &= \frac{d\ln(2)}{dx} + \frac{d\ln(x)}{dx} && \text{sum rule} \\ &= 0 + \frac{1}{x} = \frac{1}{x}. && \text{constant sum rule and derivative of log} \end{aligned}$$

The derivative of $\ln(2x)$ matches the derivative of $\ln(x)$ because the graphs of the two functions differ only by a constant (Figure 2.8.6).

2.8.3 Applications

An important family of functions often used in probability theory are the gamma distributions, with formulas

$$g(x) = x^n e^{-x}$$

for various powers n (the real gamma distributions are multiplied by a constant). What do their graphs look like? We can use the result in Example 2.8.3 and the product rule to find the first and second derivatives and deduce the shape of the graph.

Example 2.8.7 The Gamma Distribution with $n = 1$

With $n = 1$, the gamma distribution has derivative

$$\begin{aligned} g'(x) &= \frac{dx}{dx}e^{-x} + x\frac{de^{-x}}{dx} \\ &= e^{-x} - xe^{-x} = (1-x)e^{-x}. \end{aligned}$$

This derivative is positive for $x < 1$ and negative for $x > 1$. Furthermore,

$$\begin{aligned} g''(x) &= \frac{d(1-x)}{dx}e^{-x} + (1-x)\frac{de^{-x}}{dx} \\ &= -e^{-x} - (1-x)e^{-x} = (x-2)e^{-x}. \end{aligned}$$

The second derivative is negative when $x < 2$ and positive for $x > 2$. Using the facts that $g(0) = 0$ and $g(x) > 0$ when $x > 0$, we can draw an accurate graph of this function (Figure 2.8.7).

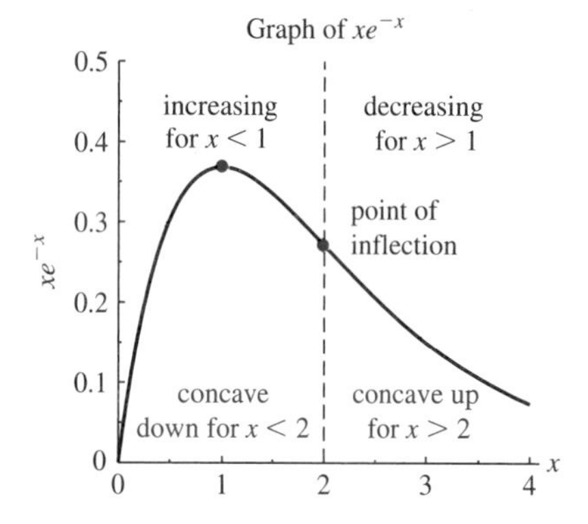

FIGURE 2.8.7
Graph of the gamma distribution with $n = 1$

Example 2.8.8 The Gamma Distribution with $n = 2$

With $n = 2$, the gamma distribution is

$$g(x) = x^2 e^{-x}.$$

The derivative and second derivative are

$$\begin{aligned} g'(x) &= \frac{dx^2}{dx}e^{-x} + x^2\frac{de^{-x}}{dx} \\ &= 2xe^{-x} - x^2e^{-x} = (2x - x^2)e^{-x} \\ g''(x) &= \frac{d(2x-x^2)}{dx}e^{-x} + (2x-x^2)\frac{de^{-x}}{dx} \\ &= (2-2x)e^{-x} - (2x-x^2)e^{-x} = (x^2-4x+2)e^{-x}. \end{aligned}$$

The derivative is positive for $x < 2$ and negative for $x > 2$. There are two points of inflection, found by solving

$$x^2 - 4x + 2 = 0$$

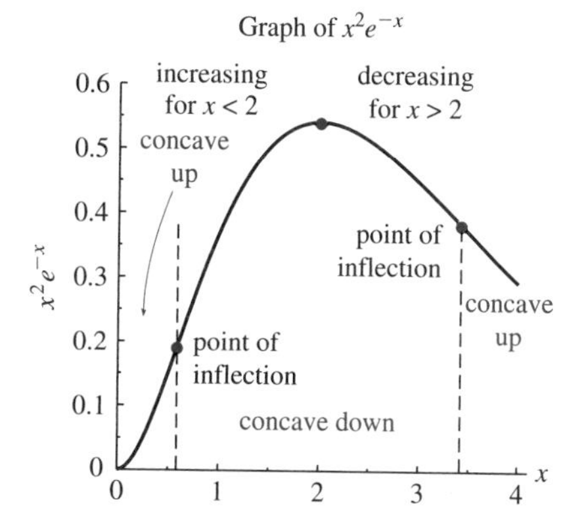

FIGURE 2.8.8
Graph of the gamma distribution with $n=2$

for the roots

$$x=2-\sqrt{2}=0.586, \quad x=2+\sqrt{2}=3.414.$$

The second derivative is negative between these roots (Figure 2.8.8).

Summary We added the derivatives of the exponential function and the natural log to our collection of building blocks. With these pieces and the basic rules for derivatives, we can find derivatives of many more complicated functions, including the gamma distributions.

2.8 Exercises

Mathematical Techniques

1–18 ▪ Compute the first and second derivatives of the following functions.

1. $F(x)=x^2+4e^x$.
2. $V(y)=1+y+2y^2+3y^3-4e^y$.
3. $f(x)=x^2e^x$.
4. $h(x)=(2-x^2)e^x$.
5. $g(x)=\dfrac{e^x}{x}$.
6. $G(z)=\dfrac{e^z}{z^2}$.
7. $f(x)=\dfrac{1+x}{e^x}$.
8. $H(t)=\dfrac{1+t^2}{e^t}$.
9. $f(x)=x+4\ln(x)$.
10. $h(x)=2x^2-2\ln(x)$.
11. $g(z)=(z+4)\ln(z)$.
12. $f(x)=x^2\ln(x)$.
13. $F(w)=e^w\ln(w)$.
14. $F(y)=\dfrac{\ln(y)}{e^y}$.
15. $s(x)=\ln(x^2)$ (use a law of logs).
16. $r(x)=\ln(x^2e^x)$ (use a law of logs).
17. $F(z)=\dfrac{1+e^{-z}}{1+e^z}$.
18. $p(x)=\dfrac{1-e^x}{1+e^x}$.

19–24 ▪ Use the first and second derivatives to sketch graphs of the following functions on the given domains. Identify regions where the function is increasing, and where it is concave up.

19. $f(x)=(1-x)e^x$ for $-2\le x\le 1$.
20. $g(x)=(2-x)e^x$ for $-2\le x\le 1$.
21. $G(z)=\dfrac{e^z}{z^2}$ for $1\le z\le 3$.
22. $F(z)=\dfrac{z^3}{e^z}$ for $0\le z\le 5$.
23. $L(x)=\dfrac{x}{2}-\ln(x)$ for $1\le x\le 3$.
24. $M(x)=(x+2)\ln(x)$ for $1\le x\le 3$.

25–26 ▪ We can return to the definition to find derivatives of other exponential functions. For each of the following

a. Write down the definition of the derivative of this function.

b. Simplify with a law of exponents and factor.

c. Estimate the limit by plugging in small values of h.

d. Exponentiate the limit to figure out what it is.

25. $f(x)=5^x$.
26. $g(x)=e^{2x}$. After following the steps, use the fact that $g(x)=e^x\cdot e^x$ to find the derivative with the product rule.

27–30 ▪ Polynomials form a useful set of functions in part because the derivative of a polynomial is another polynomial. Another set of functions with this useful property is called the **generalized**

polynomials, formed as products of polynomials and exponential functions. One simple group of generalized polynomials are the products of linear functions with the exponential function, taking the form

$$h(x) = (ax + b)e^x$$

for various values of a and b. We will call these **generalized first-order polynomials**.

27. Set $a = 1$ and $b = 1$. Find the first and second derivatives of $h(x)$.

28. Use the results of Exercise 27 to guess the tenth derivative of $h(x)$ when $a = 1$ and $b = 1$.

29. Find a generalized first-order polynomial such that $h(1) = 0$. Where is the critical point? Where is the point of inflection?

30. Let x^* be the solution of the equation $h(x^*) = 0$. Show that the critical point of a generalized first-order polynomial is $x^* - 1$ and the point of inflection is $x^* - 2$.

31–32 ▪ We can return to the definition to figure out the derivative of the natural log. We will first find the derivative at different values of x. For each value of x,

a. Write down the definition of the derivative for $\ln(x)$.

b. Plug in some small values of h to guess the limit.

c. Check that your answer matches the value of the derivative according to the formula.

31. $x = 1$

32. $x = 2$

33–34 ▪ Instead of plugging different values of x into the definition of the derivative for $\ln(x)$, as in Exercises 31 and 32, we can use a law of logs to find the derivative in general.

33. Write down the definition of the derivative at $x = 2$, and use a law of logs to try to convert the limit into something that looks like the limit at $x = 1$, as in Exercise 31. *Hint:* Substitute a new variable for $h/2$.

34. Write down the definition of the derivative for general x, and use a law of logs to try to convert the limit into something that looks like the limit at $x = 1$, as in Exercise 31. *Hint:* Substitute a new variable for h/x.

Applications

35–38 ▪ Suppose a population of bacteria grows according to $P(t) = 10e^t$. Find the first and second derivative to graph the total mass when the mass per individual $m(t)$ has the following forms. Is the total mass ever greater than it is at $t = 0$? When does the total mass reach zero?

35. $m(t) = 1 - \frac{t}{2}$.

36. $m(t) = 1 - t$.

37. $m(t) = 1 - t^2$.

38. $m(t) = 1 - \frac{t^2}{4}$.

39–40 ▪ Find the first and second derivatives of the following functions (related to the gamma distribution) and sketch graphs for $0 \le x \le 2$.

39. $G(x) = \sqrt{x}e^{-x}$.

40. $G(x) = \frac{1}{\sqrt{x}}e^{-x}$.

41–44 ▪ The following are differential equations that could describe a bacterial population. For each, describe in words what the equation says and check that the given solution works. Say whether the solution is an increasing or decreasing function.

41. $\frac{db}{dt} = e^t$ has solution $b(t) = e^t$.

42. $\frac{db}{dt} = b(t)$ has solution $b(t) = e^t$.

43. $\frac{db}{dt} = -b(t)$ has solution $b(t) = e^{-t}$.

44. $\frac{db}{dt} = e^{-b(t)}$ has solution $b(t) = \ln(t)$.

Computer Exercises

45. Consider again the function

$$g(x) = x^n e^{-x}$$

describing the gamma distribution.

a. Find the critical point in general.

b. Find the point or points of inflection. What happens when $n < 1$?

c. Graph this function for $n = 0.5$, $n = 2$, and $n = 5$. What happens for very large values of n?

46. Consider the generalized polynomial

$$R(x) = x^4 + (88x^3 - 76x^2 - 65x + 25)e^x.$$

As defined in Exercise 27, a generalized polynomial is formed by multiplying and adding polynomials and exponential functions.

a. Find all critical points and points of inflection of $R(x)$ for $-1 \le x \le 1$.

b. Find the fifth derivative of $R(x)$. Does it look any simpler than $R(x)$ itself? Compare with what happens when you take many derivatives of a polynomial or of the exponential function.

2.9 The Chain Rule

Composition is the final and perhaps most important way to combine functions. If we know the derivatives of the components parts, the **chain rule** gives the formula for the derivative of the composition. Using the chain rule (along with the sum,

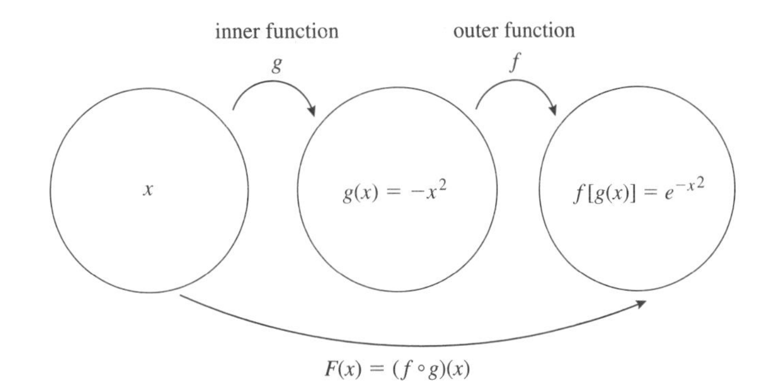

FIGURE 2.9.1
$F(x)$ written as a composition of functions

product, power, and quotient rules and the derivatives of special functions), we can find the derivative of **any** function that can be built from polynomial, exponential, and trigonometric functions. We can apply the chain rule to compute the derivatives of **inverse functions**, and to find derivatives of relations that are not functions using **implicit differentiation**.

2.9.1 The Derivative of a Composite Function

Consider the function

$$F(x) = e^{-x^2}$$

that we will use later to describe the **normal distribution**. This function is the **composition** of the exponential function $f(g) = e^g$ with the quadratic function $g(x) = -x^2$, or

$$F(x) = (f \circ g)(x)$$

(Figure 2.9.1) where $f \circ g$ is the notation for composition introduced in Subsection 1.2.3 (Definition 1.5).

Example 2.9.1 Another Functional Composition

The function

$$H(y) = \frac{1}{1 + y^2}$$

is the composition of the reciprocal function $r(p) = \frac{1}{p}$ and the polynomial $p(y) = 1 + y^2$, or

$$H(y) = (r \circ p)(y)$$

(Figure 2.9.2).

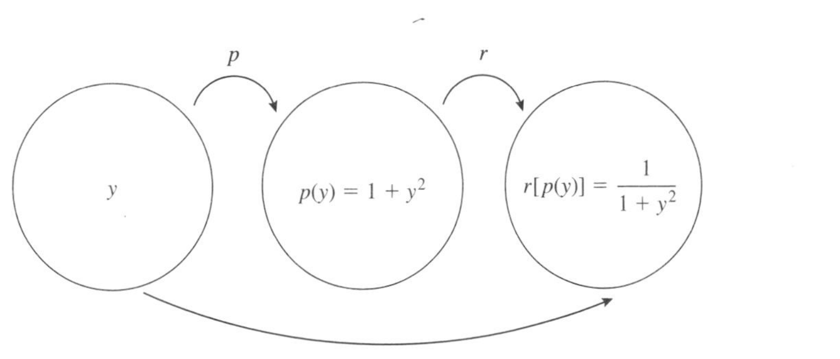

FIGURE 2.9.2
$H(y)$ written as a composition of functions

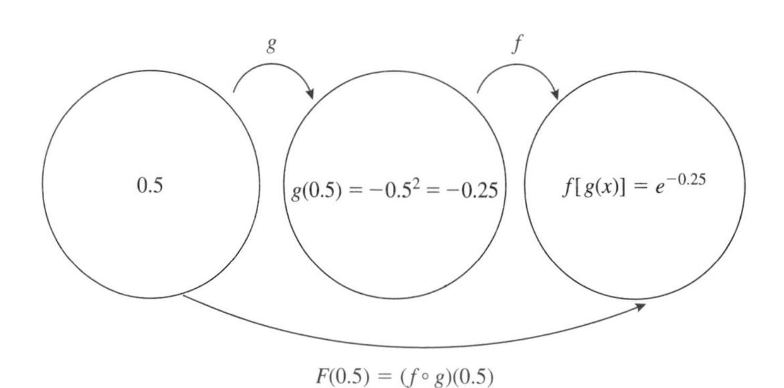

FIGURE 2.9.3
Computing $F(0.5)$

In each case, we know the derivative of the component functions. Can we use this information to find the derivative of the composition?

First, we need to recognize a function as a composition. To figure out how a function was built, think about computing the value on a calculator. To calculate $F(0.5)$, first find the negative of the square of 0.5 (the **inner function** g) as $g(0.5) = -(0.5)^2 = -0.25$, and then exponentiate that (the **outer function** f) to find the result $f(-0.25) = e^{-0.25} = 0.779$ (Figure 2.9.3).

Example 2.9.2 Calculating $H(1)$

To calculate $H(1)$ using the function $H(y) = \frac{1}{1+y^2}$ from Example 2.9.1, first evaluate the polynomial $1 + y^2$ at $y = 1$ (the inner function p) to find $p(1) = 1 + 1^2 = 2$, and then take the reciprocal of that (the outer function r) to find the $r(2) = \frac{1}{2} = 0.5$ (Figure 2.9.4).

The chain rule tells us how to compute the derivative by combining information about the inner and outer functions and their derivatives.

Theorem 2.12 The Chain Rule for Derivatives

Suppose

$$F(x) = (f \circ g)(x)$$

where f and g are both differentiable functions. Then

$$F'(x) = f'(g(x))g'(x) \qquad \text{prime notation}$$

$$\frac{dF}{dx} = \frac{df}{dg}\frac{dg}{dx}. \qquad \text{differential notation}$$

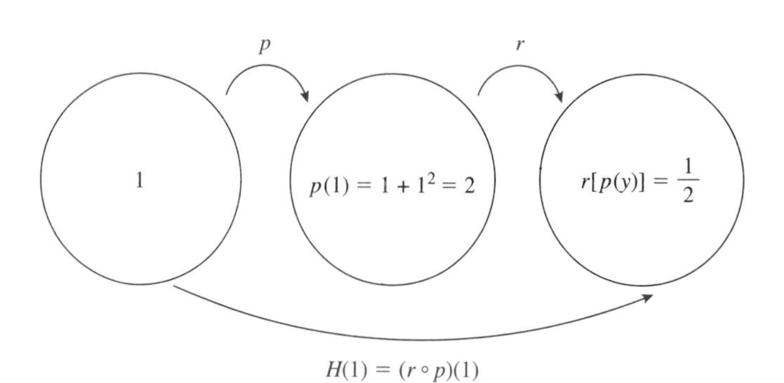

FIGURE 2.9.4
Computing $H(1)$

Proof: We know that

$$f'(y) = \lim_{\Delta y \to 0} \frac{f(y + \Delta y) - f(y)}{\Delta y}$$

at any point y. If we choose $y = g(x)$ and $\Delta y = \Delta g = g(x + \Delta x) - g(x)$, then

$$f'(g(x)) = \lim_{\Delta g \to 0} \frac{f(g(x) + \Delta g) - f(g(x))}{\Delta g} \quad \text{substitute in for } y \text{ and } \Delta y$$

$$= \lim_{\Delta x \to 0} \frac{f(g(x + \Delta x)) - f(g(x))}{g(x + \Delta x) - g(x)} \quad \text{expand } \Delta g \text{ and use fact that } \Delta x \to 0 \text{ if } \Delta g \to 0$$

$$= \lim_{\Delta x \to 0} \frac{\frac{f(g(x+\Delta x)) - f(g(x))}{\Delta x}}{\frac{g(x+\Delta x) - g(x)}{\Delta x}} \quad \text{divide top and bottom by } \Delta x$$

$$= \frac{(f \circ g)'(x)}{g'(x)} \quad \text{apply definition of derivative to top and bottom}$$

$$= \frac{F'(x)}{g'(x)}.$$

Multiplying both sides by $g'(x)$ gives the chain rule.

The expression $f'(g(x))$ means the derivative of the function f evaluated at the point $g(x)$. The expression

$$\frac{df}{dg}$$

refers to the same quantity, but means the derivative of f thought of as a function of g, again evaluated at the point $g(x)$.

How do we *use* the chain rule? The following algorithm gives the necessary steps.

Algorithm 2.1 Using the Chain Rule

1. Write the function as a composition.
2. Take the derivatives of the component pieces.
3. Multiply the derivatives together.
4. Put everything in terms of the original variable.

We apply this method to the two functions F and H defined above. To find the derivative of $F(x)$;

1. Write $F(x) = f(g(x))$ where
$$f(g) = e^g$$
$$g(x) = -x^2.$$
2. Find the derivative of each component function,
$$f'(g) = e^g$$
$$g'(x) = -2x.$$
3. The derivative is the product
$$F'(x) = e^g \cdot (-2x).$$
4. Substitute the definition of the inner function $g(x)$ to write the answer in terms of x.
$$F'(x) = e^g \cdot (-2x) = e^{-x^2} \cdot (-2x) = -2xe^{-x^2}.$$

This derivative is positive for $x < 0$ and negative for $x > 0$ (Figure 2.9.5).

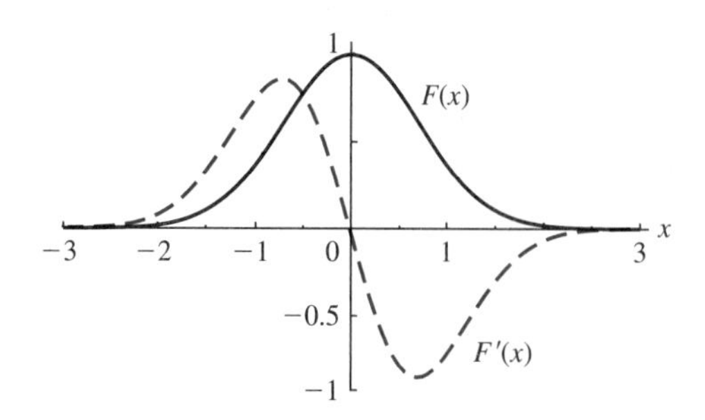

FIGURE 2.9.5
The function $F(x) = e^{-x^2}$ and its derivative

This information is organized in the following table.

The Derivative of $F(x) = (f \circ g)(x) = e^{-x^2}$

Function	Derivative (prime notation)	Derivative (differential notation)
$g(x) = -x^2$	$g'(x) = -2x$	$\frac{dg}{dx} = -2x$
$f(g) = e^g$	$f'(g) = e^g$	$\frac{df}{dg} = e^g$
$F(x) = f(g(x))$	$F'(x) = -2xe^g = -2xe^{-x^2}$	$\frac{dF}{dx} = -2xe^g = -2xe^{-x^2}$

Example 2.9.3 Finding the Derivative of $H(y)$

To find the derivative of the function $H(y)$ defined by

$$H(y) = \frac{1}{1+y^2}$$

in Example 2.9.1, we follow Algorithm 2.1.

1. Write $H(y) = (r \circ p)(y)$ where $p(y) = 1 + y^2$ and $r(p) = \frac{1}{p}$.
2. Find the derivatives of r and p,
$$p'(y) = 2y$$
$$r'(p) = -\frac{1}{p^2}.$$
3. The derivative is the product
$$H'(y) = -\left(\frac{1}{p^2}\right)2y.$$
4. Substitute the definition of the inner function p to write the answer in terms of y,
$$H'(y) = -\frac{2y}{(1+y^2)^2}.$$

This derivative is positive for $y < 0$ and negative for $y > 0$, leading to a graph quite similar to that of $F(x) = e^{-x^2}$ (Figure 2.9.6).

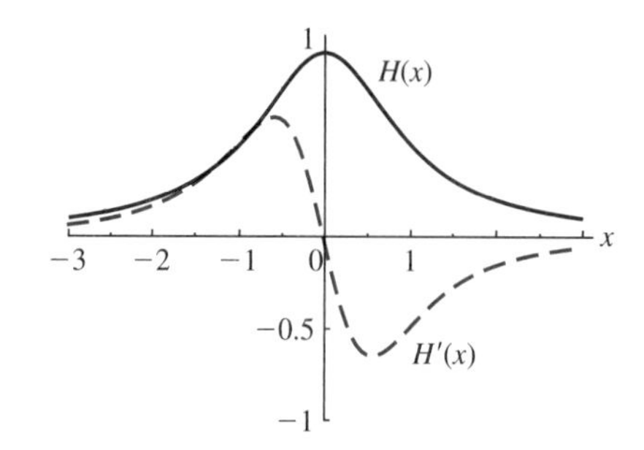

FIGURE 2.9.6
The function $H(y) = \frac{1}{1+y^2}$ and its derivative

As a table, written only in prime notation,

The Derivative of $H(y) = (r \circ p)(y) = \frac{1}{1+y^2}$

Function	Derivative
$p(y) = 1 + y^2$	$p'(y) = 2y$
$r(p) = \frac{1}{p}$	$r'(p) = -\frac{1}{p^2}$
$H(y) = r[p(y)]$	$H'(y) = -\frac{2y}{p^2} = -\frac{2y}{(1+y^2)^2}$

For more complicated functions, the chain rule can be applied several times. Some functions must be broken down into three or more component pieces. The version of the chain rule for triple compositions is

$$(f \circ g \circ h)'(x) = f'\{g[h(x)]\}g'[h(x)]h'(x), \tag{2.9.1}$$

or in differential notation

$$\frac{d(f \circ g \circ h)}{dx} = \frac{df}{dg}\frac{dg}{dh}\frac{dh}{dx}. \tag{2.9.2}$$

As in the case with two functions, the derivative of the composition is found by multiplying the derivatives of the component functions.

Example 2.9.4 Using the Chain Rule on a Triple Composition

Consider the function

$$F(x) = \ln[(1-x)^4 + 1].$$

1. Write as the composition $F(x) = f\{g[h(x)]\}$ where
$$h(x) = 1 - x$$
$$g(h) = h^4 + 1$$
$$f(g) = \ln(g).$$
(Figure 2.9.7).
2. Take the derivatives of the components,
$$h'(x) = -1$$
$$g'(h) = 4h^3$$
$$f'(g) = \frac{1}{g}.$$
3. Multiply the derivatives together,
$$F'(x) = \frac{1}{g} \cdot 4h^3 \cdot (-1).$$

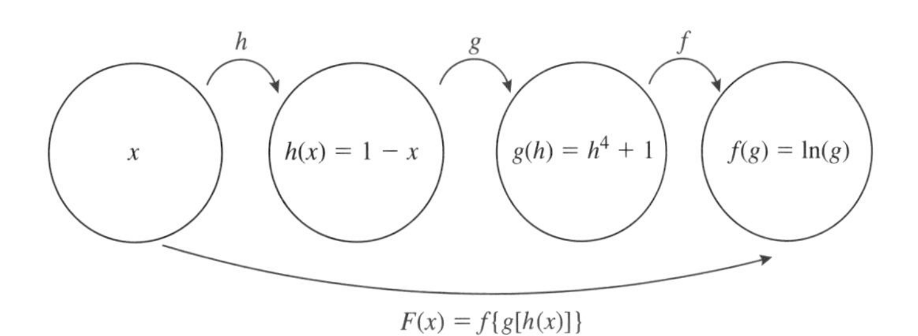

FIGURE 2.9.7
The function $F(x) = \ln[(1-x)^4 + 1]$ written as a composition

4. Substitute the definitions of the functions $h(x)$ and $g(h)$ to write the answer in terms of x,

$$\begin{aligned} F'(x) &= \frac{1}{h^4+1} \cdot 4h^3 \cdot (-1) \\ &= \frac{1}{(1-x)^4+1} \cdot 4(1-x)^3 \cdot (-1) \\ &= -4\frac{1}{(1-x)^4+1}(1-x)^3. \end{aligned}$$

In tabular form,

The Derivative of $F(x) = f\{g[h(x)]\} = \ln[(1-x)^4+1]$

Function	Derivative
$h(x) = 1-x$	$h'(x) = -1$
$g(h) = h^4+1$	$g'(h) = 4h^3$
$f(g) = \ln(g)$	$f'(g) = \frac{1}{g}$
$F(x) = f\{g[h(x)]\}$	$F'(x) = -1 \cdot 4h^3 \frac{1}{g} = -4(1-x)^3 \frac{1}{(1-x)^4+1}$

2.9.2 Derivatives of Inverse Functions

Many important functions, like the natural logarithm, are defined as the **inverses** of other functions. The inverse f^{-1} of the function $f(x)$ is defined in terms of functional composition as

$$(f \circ f^{-1})(x) = x.$$

(Definition 1.6). We can use the chain rule to find the derivative. However, in this case we will use the derivative of the composition (which has the simple formula x) to find the derivative of the component part f^{-1}.

We can take the derivatives of both sides, finding

$$(f \circ f^{-1})'(x) = f'[f^{-1}(x)](f^{-1})'(x) = 1$$

because

$$\frac{d}{dx}x = 1.$$

We can solve this equation for $(f^{-1})'(x)$, the derivative of the inverse, finding

$$(f^{-1})'(x) = \frac{1}{f'[f^{-1}(x)]}.$$

Theorem 2.13 The Derivative of an Inverse Function

Suppose f is a differentiable function with inverse f^{-1} and that

$$f'[f^{-1}(x)] \neq 0.$$

Then

$$(f^{-1})'(x) = \frac{1}{f'[f^{-1}(x)]}.$$

Example 2.9.5 Checking the Derivative of $\ln(x)$

We can use this theorem to check the derivative of $\ln(x)$, which is the inverse of the exponential function $f(x) = e^x$. We can set up the problem as

$$\begin{aligned} f'(x) &= e^x \\ f^{-1}(x) &= \ln(x). \end{aligned}$$

Therefore

$$
\begin{aligned}
(f^{-1})'(x) &= \frac{1}{f'[f^{-1}(x)]} \\
&= \frac{1}{e^{f^{-1}(x)}} \\
&= \frac{1}{e^{\ln(x)}} \\
&= \frac{1}{x}.
\end{aligned}
$$

Example 2.9.6 Checking the Derivative of $\sqrt{x}$

We can also check the derivative of $\sqrt{x}$ by using its definition as the inverse of the quadratic function $S(x) = x^2$, and

$$
\begin{aligned}
S'(x) &= 2x \\
S^{-1}(x) &= \sqrt{x}.
\end{aligned}
$$

Therefore

$$
\begin{aligned}
(f^{-1})'(x) &= \frac{1}{f'[f^{-1}(x)]} \\
&= \frac{1}{2f^{-1}(x)} \\
&= \frac{1}{2\sqrt{x}},
\end{aligned}
$$

matching the power rule for fractional powers (Theorem 2.8).

The formula for the derivative of the inverse has a nice geometric interpretation (Figure 2.9.8). The graph of the inverse function is the mirror image of the graph of the function itself. Therefore, the tangent line to the inverse function is the mirror image of the tangent line to the function. The tangent to the inverse at the point $(f(x_0), x_0)$, the reflection of the point $(x_0, f(x_0))$ will have slope

$$
\text{slope of tangent to inverse at } (f(x_0), x_0) = \frac{1}{\text{slope of tangent to function at } (x_0, f(x_0))}.
$$

Therefore

$$
(f^{-1})'(f(x_0)) = \frac{1}{f'(x_0)}.
$$

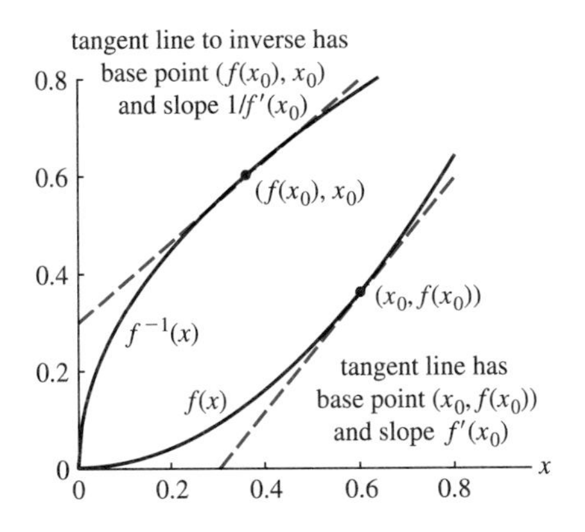

FIGURE 2.9.8
The tangent lines to a function and its inverse

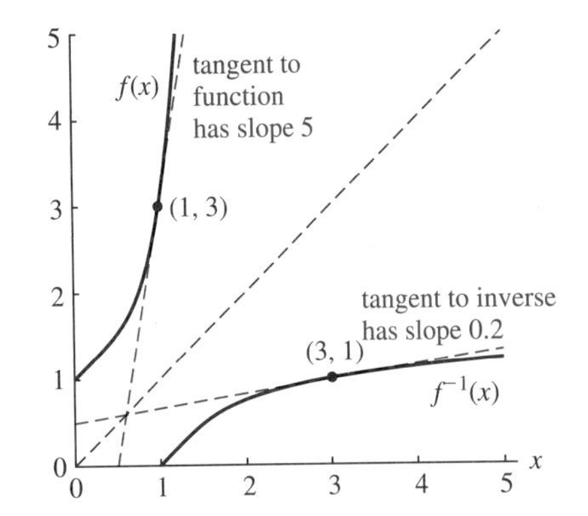

FIGURE 2.9.9
The tangent line to an uncomputable inverse

Example 2.9.7 Sketching the Derivative That Cannot Be Computed

With this geometric interpretation, we can find the slope of a curve for which we have no formula. When studying the inverse in Subsection 1.2.4, we found that the function

$$f(x) = x^5 + x + 1$$

has an inverse, even though we cannot find a formula (Example 1.2.27). Suppose, nonetheless, that we wish to find the slope of the inverse at the point (3, 1). This point lies on the inverse because $f(1) = 3$. Furthermore,

$$f'(x) = 5x^4 + 1$$

so $f'(1) = 6$. (Figure 2.9.9) Therefore, the slope of the inverse is the reciprocal, or

$$(f^{-1})'(3) = \frac{1}{5}.$$

2.9.3 Applications

We can use the chain rule to find the derivative of $b(t) = 2.0^t$ (Section 2.1) and finally explain the factor of 0.693. The trick is to write the function in terms of the exponential function as

$$\begin{aligned} 2.0^t &= \left(e^{\ln(2.0)}\right)^t && \text{exponential and natural log are inverses} \\ &= e^{\ln(2.0)t}. && \text{law 2 of exponents} \end{aligned}$$

The derivative of this composition can be found by following the chain rule, Algorithm 2.1.

1. $b(t)$ is a composition of $f(g(t))$ where
$$f(g) = e^g$$
$$g(t) = \ln(2.0)t.$$
2. The derivatives of the components are
$$f'(g) = e^g$$
$$g'(t) = \ln(2.0).$$
3. Using the chain rule, the derivative of b is
$$b'(t) = e^g \ln(2.0).$$

4. Putting back in terms of t,

$$\begin{aligned} b'(t) &= e^{\ln(2.0)t}\ln(2.0) && \text{plug in definition of } g \\ &= (e^{\ln(2.0)})^t \ln(2.0) && \text{law 2 of exponents} \\ &= 2.0^t \ln(2.0) && \text{exponential and log are inverses} \\ &= \ln(2.0)2.0^t \approx 0.693 \cdot 2.0^t. && \text{reorder and evaluate} \end{aligned}$$

The mysterious factor 0.693 that cropped up in Section 2.1 is the natural logarithm of 2.0. Perhaps better than any other calculation, this shows the pivotal role of the number e and the exponential function. Although the function 2.0^t does not contain any reference to e, a natural log is spontaneously generated by the derivative.

Exponential measurements are generally written in the form

$$M(t) = M(0)e^{\alpha t}$$

(Section 1.7). The constant $M(0)$ represents the value of the measurement at $t = 0$. The parameter α determines whether and how fast the measurement is increasing as a function of time. If α is negative, the measurement is decreasing and decreases faster the more negative α is. If α is positive, the measurement is increasing, and increases faster the larger α is.

We can find the derivative of $M(t)$ with the chain rule.

1. $M(t)$ is the composition of $f(g(t))$ where

$$\begin{aligned} f(g) &= M(0)e^g \\ g(t) &= \alpha t. \end{aligned}$$

2. The derivatives of the components are

$$\begin{aligned} f'(g) &= M(0)e^g \\ g'(t) &= \alpha. \end{aligned}$$

3. Using the chain rule, the derivative of M is

$$M'(t) = M(0)e^g \alpha.$$

4. Putting back in terms of t,

$$\begin{aligned} M'(t) &= M(0)e^{\alpha t}\alpha && \text{plug in definition of } g \\ &= \alpha M(0)e^{\alpha t} && \text{reorder} \\ &= \alpha M(t). && \text{rewrite in terms of } M(t) \end{aligned}$$

The derivative of the general exponential function can be found by multiplying the original function by the parameter α that appears in the exponent.

Example 2.9.8 Finding the Derivative of e^{3t}

In Example 2.8.2, we used the definition of the derivative to find that

$$\frac{de^{3t}}{dt} = 3e^{3t}.$$

We now see that this is one case of the general rule. The derivative of an exponential function is found by multiplying the original function e^{3t} by the parameter 3 that appears in the exponent.

Example 2.9.9 Finding the Solution of a Differential Equation

We can use this rule to find a solution of the differential equation

$$\frac{db}{dt} = 0.5b,$$

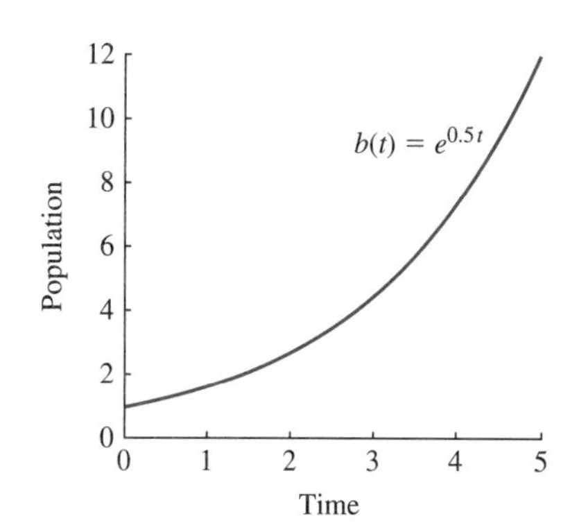

FIGURE 2.9.10

The growth of a population obeying $\frac{db}{dt} = 0.5b$

which says that the rate of change of the population is 0.5 times the population size. One solution is

$$b(t) = e^{0.5t}$$

because this function has parameter $\alpha = 0.5$ in the exponent (Figure 2.9.10).

2.9.4 Implicit Differentiation and Related Rates

We have compiled an arsenal of methods to compute the derivatives of relations that are defined by *functions*. When it is impossible or inconvenient to solve for one variable in terms of the other, we can still compute derivatives with the technique of **implicit differentiation**.

Example 2.9.10 Finding the Slope of a Circle

The set of points that satisfy

$$x^2 + y^2 = 1$$

is the circle of radius 1 centered at the origin. As we saw in Example 1.2.13, this relation is not a function because each value of $-1 < x < 1$ is associated with two values of y. However, every point on the circle does have a well-defined tangent that we can find with implicit differentiation. We begin by thinking of y as a function of x, and taking the derivative of both sides with respect to x.

$$\frac{d}{dx}(x^2 + y^2) = \frac{d}{dx}1 \qquad \text{take derivative of both sides}$$

$$\frac{dx^2}{dx} + \frac{d}{dx}y^2 = 0 \qquad \text{sum rule and constant rule}$$

$$2x + 2y\frac{dy}{dx} = 0. \qquad \text{power rule and chain rule}$$

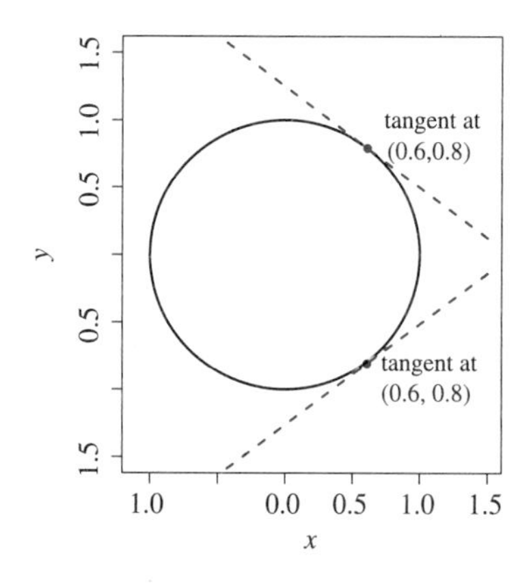

FIGURE 2.9.11

Two tangent lines to the circle $x^2 + y^2 = 1$

We can then solve for dy/dx as

$$\frac{dy}{dx} = -\frac{x}{y}.$$

At the point (0.6, 0.8) the slope of the tangent line is

$$-\frac{x}{y} = -\frac{0.6}{0.8} = -0.75.$$

(Figure 2.9.11) Because the circle fails the vertical line test at $x = 0.6$, there is a second value of y at the point (0.6, −0.8). At this point, the slope of the tangent line is

$$-\frac{x}{y} = -\frac{0.6}{-0.8} = 0.75.$$

Implicit differentiation can be used to solve for the rate of change of one variable in terms of the rate of change of a related variable. Instead of differentiating the relation

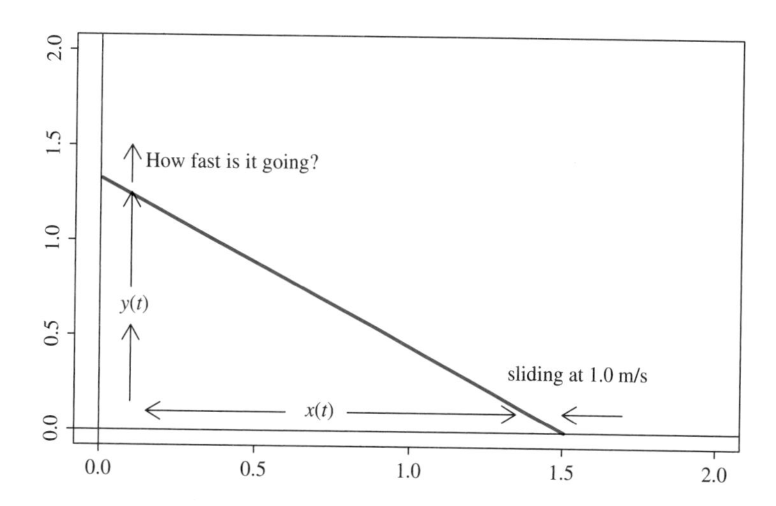

FIGURE 2.9.12
A sliding board

between them with respect to one of the variables, we differentiate the relation with respect to time.

Example 2.9.11 A Classic Related Rates Problem

Suppose a person is pushing a rigid board of length 2.0 m against a wall, with the bottom moving at a rate of 1.0 m/s along the floor (Figure 2.9.12). How fast is the top end moving upward?

To begin, we define $x(t)$ as the distance between the bottom of the board from the wall and $y(t)$ as the distance between the top of the board and the floor. These two values are related by the Pythagorean theorem by

$$x^2(t) + y^2(t) = 4.0.$$

We take the derivative of both sides with respect to t, finding

$$2x(t)\frac{dx}{dt} + 2y(t)\frac{dy}{dt} = 0.$$

Solving for dy/dt gives

$$\frac{dy}{dt} = -\frac{x}{y}\frac{dx}{dt}.$$

We have that $\frac{dx}{dt} = -1.0$ because it is moving to the left, or toward smaller values of x. Evaluating at various points gives

t	$(x(t), y(t))$	$\frac{dy}{dt}$
0.0	(2.0, 0.000)	∞
0.1	(1.9, 0.624)	3.042
0.5	(1.5, 1.322)	1.134
1.0	(1.0, 1.732)	0.577
1.5	(0.5, 1.936)	0.258
1.9	(0.1, 1.997)	0.050
2.0	(0.0, 2.000)	0.000

Mathematically, the vertical position would begin moving infinitely fast. Physically, of course, this is impossible, and in fact pushing on the board starting from a perfectly flat position would not move it at all. Instead, it would need a slight boost to start sliding up the wall.

Example 2.9.12 Related Rates Applied to Metabolism

We saw in Example 1.7.24 that energy use is approximately proportional to the body mass raised to the 3/4 power. However, different body tissues use different amounts of energy. For example, if a body were made of two tissue types, with masses M_1 and M_2, total energy expenditure might be

$$E = 0.01M_1^{0.75} + 0.022M_2^{0.75}.$$

The second tissue uses more than twice as much energy as the first.

In a growing organism, all three of these measurements change over time. At some time, we have that $M_1 = 10.0$g and $M_2 = 20.0$g. Suppose we measure the rate of change of energy consumption to be 0.02 kCal/h^2, and the rate of growth of the first tissue type to be 1.0 g/h, but wish to compute the rate of growth of the second, more difficult to study, tissue type. We take the derivative of both sides, using the power rule and the chain rule on the right-hand side to find

$$\frac{dE}{dt} = 0.01 \cdot 0.75M_1^{-0.25}\frac{dM_1}{dt} + 0.022 \cdot 0.75M_2^{-0.25}\frac{dM_2}{dt}.$$

Substituting the measured values gives

$$0.02 = 0.01 \cdot 0.75 \cdot 10.0^{-0.25} \cdot 1.0 + 0.022 \cdot 0.75 \cdot 20.0^{-0.25}\frac{dM_2}{dt}$$

$$0.02 = 0.00422 + 0.00780\frac{dM_2}{dt}.$$

Solving for $\frac{dM_2}{dt}$ gives

$$\frac{dM_2}{dt} = \frac{0.02 - 0.00422}{0.00780} = 2.023.$$

The second tissue is growing twice as fast as the first.

Summary We derived the **chain rule** for differentiating the composition of two functions. Almost any complicated function can be differentiated in four steps (1) breaking the function into components, (2) taking the derivative of each component, (3) multiplying the derivatives together, and (4) rewriting in terms of the original variable. The same technique works for compositions of more than two functions. The chain rule can be used to derive the formula for the derivatives of inverse functions and provides a useful tool for finding derivatives of general exponential functions. The method of **implicit differentiation** uses the chain rule to compute derivatives based on relations that are not expressed as functions.

2.9 Exercises

Mathematical Techniques

1–16 ▪ Compute the following derivatives using the chain rule.

1. $g(x) = (1 + 3x)^2$.
2. $h(x) = (1 + 2x)^3$.
3. $f_1(t) = (1 + 3t)^{30}$.
4. $f_2(t) = (1 + 2t^2)^{15}$.
5. $r(x) = \frac{(1 + 3x)^2}{(1 + 2x)^3}$.
6. $p(z) = (1 + 3z)^2(1 + 2z)^3$.
7. $F(z) = \left(1 + \frac{2}{1+z}\right)^3$.
8. $G(w) = \left[\left(2 - \frac{3}{1-w}\right)^2 + 1\right]^2$.
9. $f(x) = e^{-3x}$.
10. $h(x) = 2^x 3^x$.
11. $g(y) = \ln(1 + y)$.
12. $A(z) = \ln(1 + e^z)$.
13. $G(x) = 8e^{x^2}$.
14. $s(w) = 4.2\sqrt{1 + e^w}$.
15. $L(x) = \ln(\ln(x))$.
16. $q(y) = y^y$. *Hint:* Rewrite with the exponential function.

17–24 ▪ Compute the derivative of each of the following functions in the two ways given.

17. $F(x) = \frac{1}{1+e^x}$ with a) the quotient rule and b) the chain rule.

18. $H(y) = \frac{1}{1+y^3}$ with a) the quotient rule and b) the chain rule.

19. $g(x) = \ln(3x)$ with a) a law of logs and b) the chain rule.

20. $h(x) = \ln(x^3)$ with a) a law of logs and b) the chain rule.

21. $F(x) = (1+2x)^2$ by a) expanding the binomial and taking the derivative of the polynomial and b) applying the chain rule.

22. $F(x) = (1+2x)^3$ by a) expanding the binomial and taking the derivative of the polynomial and b) applying the chain rule.

23. $F(x) = x^3$ with a) the power rule and b) the chain rule after writing $F(x)$ using the exponential function.

24. $F(x) = x^{-5}$ with a) the power rule and b) the chain rule after writing $F(x)$ using the exponential function.

25–30 ▪ Find the derivatives of the inverses of the following functions in two ways: first by finding the inverse and taking its derivative directly, and then by using the formula for the derivative of the inverse (Theorem 2.13).

25. $f(x) = 3x + 1$.

26. $g(x) = -x + 3$.

27. $h(x) = 2 + x^3$.

28. $F(x) = 1 - e^{-x}$.

29. $q(x) = x + x^2$ for $x \geq 0$.

30. $N(x) = e^{x^2}$ for $x \geq 0$.

31–32 ▪ Check that implicit differentiation gives the same formula for dy/dx as does solving for y in terms of x and then finding the derivative in the ordinary way.

31. $xy = 1$.

32. $e^y = x$.

33–34 ▪ We can use laws of exponents and the chain rule to check the power rule.

33. Write $f(x) = x^n$ using the exponential function.

34. Take the derivative and then rewrite as a power function.

35–36 ▪ The equation for the top half of a circle is $f(x) = \sqrt{1-x^2}$. In addition to using implicit differentiation (Example 2.9.10), we can find the slope of the tangent with the chain rule, or find it geometrically.

35. Find the derivative of $f(x)$ with the chain rule.

36. Find the slope of the ray connecting the center of the circle at $(0, 0)$ to the point $(x, f(x))$ on the circle. Then use the fact that the tangent to a circle is perpendicular to the ray to find the slope of the tangent. Check that it matches the result with the chain rule.

37–38 ▪ Use implicit differentiation to find the slope of the tangent to the following relations at the given points. What do the shapes look like?

37. $y^2 + 4x^4 = 4x^2$ at all points where $x = -0.5$, $x = 0.5$, $x = -1$, and $x = 1$. What happens at $x = 0$?

38. $y^2 + 4xy + 4x^4 = 0$ at all points where $x = -0.5$, $x = 0.5$, $x = -1$, and $x = 1$. What happens at $x = 0$?

Applications

39–42 ▪ The following functional compositions describe connections between measurements (as in Section 1.2, Exercises 53–56). Find the derivative of the composition using the chain rule.

39. The number of mosquitoes (M) that end up in a room is a function of how far the window is open (W, in cm^2) according to $M(W) = 5W + 2$. The number of bites (B) depends on the number of mosquitoes according to $B(M) = 0.5M$. Find the derivative of B as a function of W.

40. The temperature of a room (T) is a function of how far the window is open (W) according to $T(W) = 40 - 0.2W$. How long you sleep (S, measured in hours) is a function of the temperature according to $S(T) = 14 - T/5$. Find the derivative of S as a function of W.

41. The number of viruses (V, measured in trillions) that infect a person is a function of the degree of immunosuppression (I, the fraction of the immune system that is turned off by medication) according to $V(I) = 5I^2$. The fever (F, measured in °C) associated with an infection is a function of the number of viruses according to $F(V) = 37 + 0.4V$. Find the derivative of F as a function of I.

42. The length of an insect (L, in mm) is a function of the temperature during development (T, measured in °C) according to $L(T) = 10 + T/10$. The volume of the bug (V, in cubic mm) is a function of the length according to $V(L) = 2L^3$. The mass (M in milligrams) depends on volume according to $M(V) = 1.3V$. Find the derivative of M as a function of T.

43–44 ▪ The amount of carbon-14 (^{14}C) left t years after the death of an organism is given by

$$Q(t) = Q_0 e^{-0.000122t}$$

where Q_0 is the amount left at the time of death. Suppose $Q_0 = 6.0 \times 10^{10}$ ^{14}C atoms/g.

43. Find the derivative of $Q(t)$.

44. Evaluate the derivative at $t = 0$ and after one and two half-lives. Find and explain their relationship.

45–46 ▪ The method of implicit differentiation is often applied to related rates problems involving distances.

45. Suppose a cheetah is dashing due north at a rate of 30 m/s toward a popular watering hole, and a misguided gazelle is running due east toward the same spot at a rate of 20 m/s. The cheetah begins from a distance of 120 m, while the gazelle begins from a distance of 80 m. Find equations for the position $y(t)$ of the cheetah, $x(t)$ for the position of gazelle, and the distance $r(t)$ between them, and use it to find the rate

of change of the distance between them at $t = 0$, $t = 2$, and $t = 4$.

46. Suppose a cheetah is dashing due south at a rate of 30 m/s toward a popular watering hole, and a misguided gazelle is running due east toward the same spot at a rate of 20 m/s. The cheetah begins from a distance of 100 m, while the gazelle begins from a distance of 80 m. Find equations for the position $y(t)$ of the cheetah, $x(t)$ for the position of the gazelle, and the distance $r(t)$ between them, and use it to find the rate of change of the distance between them at $t = 0$, $t = 2$, $t = 3$, and $t = 4$. What is the minimum distance between them? Does this occur before or after the first of them has reached the watering hole?

47–48 ▪ We can describe the energy use of an organism with two tissue types by looking at the fraction of each type rather than the total mass of each as in Example 2.9.12, by writing

$$E = c_1(pM)^{0.75} + c_2((1 - p)M)^{0.75}.$$

Here p is the fraction of tissue of the first type. Suppose $c_1 = 0.01$, $c_2 = 0.022$.

47. Suppose that at some time $M = 30.0$, $p = 0.3$, $dM/dt = 2.0$, and $dE/dt = 0.02$. Find dp/dt. Why might this be a useful calculation?

48. Suppose that at some time $M = 30.0$, $dM/dt = 2.0$ and $dE/dt = 0.01$ without setting a value for p. Find dp/dt for $p = 0.1$, $p = 0.5$, and $p = 0.9$. Is there are value of p for which $dp/dt = 0$? Can you explain what is going on?

49–52 ▪ Check the given solutions to the following differential equations. Which solutions are increasing and which are decreasing?

49. $b(t) = 100e^{3t}$ is a solution of $\frac{db}{dt} = 3b(t)$.

50. $b(t) = 10e^{-2t}$ is a solution of $\frac{db}{dt} = -2b(t)$.

51. $b(t) = 3e^{2t} - 0.5$ is a solution of $\frac{db}{dt} = 1 + 2b(t)$.

52. $b(t) = 5 + 20e^{-2t}$ is a solution of $\frac{db}{dt} = 10 - 2b(t)$.

Computer Exercise

53. Consider again energy use E by an organism with mass M, a fraction p of which is in a tissue with parameter c_1 and the remaining $1 - p$ in a tissue with parameter c_2,

$$E = c_1(pM)^{0.75} + c_2((1 - p)M)^{0.75}.$$

Think of M, p, and E as functions of time. Suppose that M increases exponentially over time. Experiment with different equations for p as a function of time. When does $E(t)$ grow exponentially?

2.10 Derivatives of Trigonometric Functions

The last group of special functions important in biology are the trigonometric functions. We here compute the derivatives of these special functions. Like the exponential function, the derivatives of sine and cosine have special properties that link them to solutions of important differential equations.

2.10.1 Deriving the Derivatives of Sine and Cosine

Because the functions sine and cosine cannot be built out of polynomials and exponential functions with products, sums, and quotients (except by using the more advanced topic of **complex numbers**), we must compute the derivative from its definition. For the sine and cosine functions,

$$\frac{d\sin(x)}{dx} = \lim_{h \to 0} \frac{\sin(x + h) - \sin(x)}{h}$$
$$\frac{d\cos(x)}{dx} = \lim_{h \to 0} \frac{\cos(x + h) - \cos(x)}{h}.$$

These formulas do us little good unless we can expand the expressions for $\sin(x + h)$ and $\cos(x + h)$. When finding the derivative of the exponential function, we used a law of exponents to rewrite e^{x+h} as $e^x e^h$. Trigonometric functions have similarly useful sum laws, called the **angle addition formulas**.

$$\sin(x + h) = \sin(x)\cos(h) + \cos(x)\sin(h). \quad \textbf{(2.10.1)}$$
$$\cos(x + h) = \cos(x)\cos(h) - \sin(x)\sin(h). \quad \textbf{(2.10.2)}$$

We will use the sum law for sine to try to find the derivative.

$$\lim_{h\to 0} \frac{\sin(x+h)-\sin(x)}{h}$$

$$= \lim_{h\to 0} \frac{\sin(x)\cos(h)+\cos(x)\sin(h)-\sin(x)}{h} \quad \text{apply sum law for sine}$$

$$= \lim_{h\to 0} \frac{\sin(x)(\cos(h)-1)+\cos(x)\sin(h)}{h} \quad \text{combine terms involving } \sin(x)$$

$$= \lim_{h\to 0} \frac{\sin(x)(\cos(h)-1)}{h} + \lim_{h\to 0} \frac{\cos(x)\sin(h)}{h} \quad \text{break up limit}$$

$$= \sin(x) \lim_{h\to 0} \frac{\cos(h)-1}{h} + \cos(x) \lim_{h\to 0} \frac{\sin(h)}{h}. \quad \text{pull terms without } h\text{'s outside}$$

Although this might not look much better than the limit we started with, it has an important simplification, much like that found with the exponential function. All terms involving x have come outside the limits. If we could figure out the numerical values of

$$\lim_{h\to 0} \frac{\cos(h)-1}{h}$$

and

$$\lim_{h\to 0} \frac{\sin(h)}{h}$$

we would be done.

In Subsection 2.10.4 we derive that

$$\lim_{h\to 0} \frac{\cos(h)-1}{h} = 0$$

$$\lim_{h\to 0} \frac{\sin(h)}{h} = 1.$$

Substituting these values into the formulas, we have found that

$$\frac{d\sin(x)}{dx} = \sin(x) \lim_{h\to 0} \frac{\cos(h)-1}{h} + \cos(x) \lim_{h\to 0} \frac{\sin(h)}{h}$$

$$= \sin(x)\cdot 0 + \cos(x)\cdot 1 = \cos(x).$$

Similarly, the derivative of $\cos(x)$ is

$$\lim_{h\to 0} \frac{\cos(x+h)-\cos(x)}{h}$$

$$= \lim_{h\to 0} \frac{\cos(x)\cos(h)-\sin(x)\sin(h)-\cos(x)}{h} \quad \text{apply sum law for cosine}$$

$$= \lim_{h\to 0} \frac{\cos(x)(\cos(h)-1)-\sin(x)\sin(h)}{h} \quad \text{combine terms involving } \cos(x)$$

$$= \lim_{h\to 0} \frac{\cos(x)(\cos(h)-1)}{h} - \lim_{h\to 0} \frac{\sin(x)\sin(h)}{h} \quad \text{break up limit}$$

$$= \cos(x) \lim_{h\to 0} \frac{(\cos(h)-1)}{h} - \sin(x) \lim_{h\to 0} \frac{\sin(h)}{h} \quad \text{pull terms without } h\text{'s out}$$

$$= \cos(x)\cdot 0 - \sin(x)\cdot 1 \quad \text{use the limits found earlier}$$

$$= -\sin(x). \quad \text{multiply out}$$

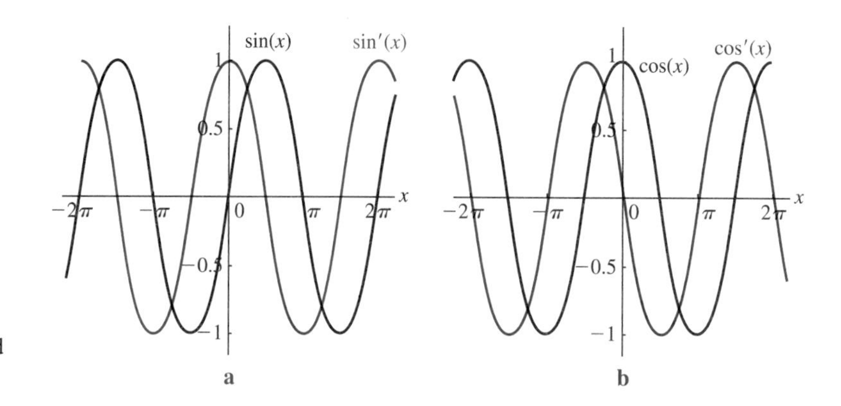

FIGURE 2.10.1
The graphs and derivatives of sine and cosine

In summary,

$$\frac{d\sin(x)}{dx} = \cos(x)$$
$$\frac{d\cos(x)}{dx} = -\sin(x).$$

To recall where the negative sign goes, remember the graphs of sine and cosine. At $x = 0$, the cosine is flat, but beginning to decrease, meaning that the derivative must begin at 0 and become negative. At $x = 0$, sine is increasing, meaning the derivative should take on a positive value at that point (Figure 2.10.1a).

What is the second derivative of each of these functions?

$$\frac{d^2\sin(x)}{dx^2} = \frac{d\cos(x)}{dx} = -\sin(x)$$
$$\frac{d^2\cos(x)}{dx^2} = -\frac{d\sin(x)}{dx} = -\cos(x).$$

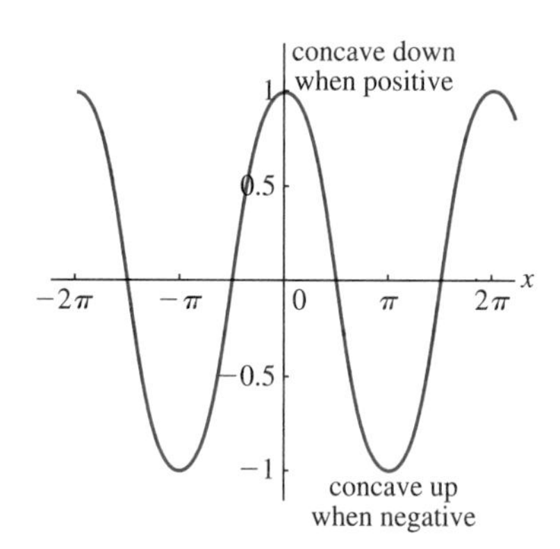

FIGURE 2.10.2
The curvature of cosine

Each of these functions has the remarkable property that it is equal to the **negative** of its second derivative. These functions are concave down when positive and concave up when negative. Furthermore, both have points of inflection every time they cross zero (Figure 2.10.2). This property may remind you of the fact that the exponential function is its own derivative (and that e^{-x} is its own second derivative). There is a deep connection between these two types of functions that requires the more advanced topic of **complex numbers** to understand.

Example 2.10.1 A Derivative Involving a Trigonometric Function

We can use rules for differentiation to find the derivatives of more complicated functions (Figure 2.10.3). For example, is the function

$$F(x) = x + \sin(x)$$

an increasing function? We can check by computing the derivative

$$\frac{d(x+\sin(x))}{dx} = \frac{dx}{dx} + \frac{d\sin(x)}{dx} \quad \text{sum rule}$$
$$= 1 + \cos(x) \quad \text{power rule and derivative of sine}$$

Because $\cos(x)$ is never smaller than -1, the derivative is never negative, and this function is increasing. The function has critical points where $\cos(x) = -1$, such as $x = -\pi$ and $x = \pi$.

We have written sinusoidal oscillations as

$$f(t) = A + B\cos\left(\frac{2\pi}{T}(t - \phi)\right)$$

(Equation 1.8.1 in Subsection 1.8.2), where A is the average, B is the amplitude, T is the period, and ϕ is the phase. We can find the derivative of $f(t)$ with the chain rule.

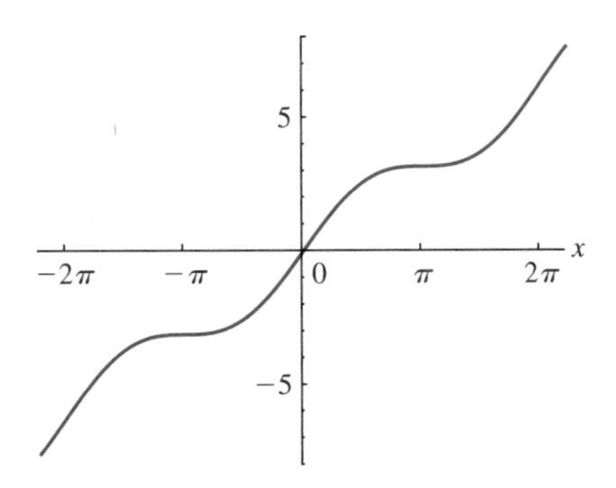

FIGURE 2.10.3
The graph of $x + \sin(x)$

1. $f(t)$ is the composition of $h(g(t))$ where
$$h(g) = A + B\cos(g)$$
$$g(t) = \frac{2\pi}{T}(t - \phi).$$
2. The derivatives of the components are
$$h'(g) = -B\sin(g)$$
$$g'(t) = \frac{2\pi}{T}.$$
3. Using the chain rule, the derivative of f is
$$f'(t) = -\frac{2\pi B}{T}\sin(g).$$
4. Putting back in terms of t,
$$f'(t) = -\frac{2\pi B}{T}\sin(\frac{2\pi}{T}(t - \phi)).$$

This gives the derivative of the general function describing a sinusoidal oscillation.

Example 2.10.2 Derivatives of Oscillations

The chain rule can be applied to find the derivatives of the daily and monthly temperature cycles introduced in Section 1.8 (written in units of days),
$$P_d(t) = 36.8 + 0.3\cos(2\pi(t - 0.583))$$
$$P_m(t) = 36.8 + 0.2\cos\left(\frac{2\pi(t - 16)}{28}\right).$$

Using the rule for finding the derivative of a general trigonometric function, along with the constant sum rule,
$$\frac{dP_d}{dt} = 0.3\frac{d}{dt}\cos(2\pi(t - 0.583))$$
$$= -0.3 \cdot 2\pi\sin(2\pi(t - 0.583))$$
$$\frac{dP_d}{dt} = 0.2\frac{d}{dt}\cos\left(\frac{2\pi(t - 16)}{28}\right)$$
$$= -0.2 \cdot \frac{2\pi}{28}\sin\left(\frac{2\pi(t - 16)}{28}\right).$$

The scales on the graphs of the two derivatives are very different (Figures 2.10.4 and 2.10.5). The daily oscillation has a much greater rate of change than the monthly one, nearly 2 degrees per day, with the other reaching only 0.05 degrees per day. Biologically, this corresponds to the fact that the daily oscillation is much faster. Mathematically,

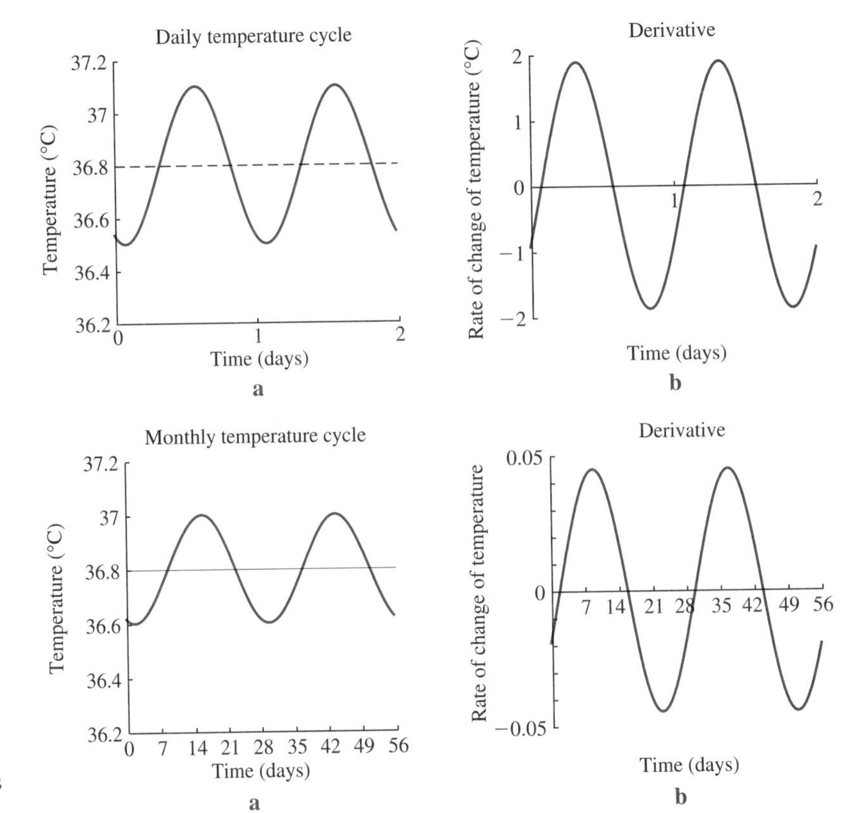

FIGURE 2.10.4

The daily temperature cycle and its derivative

FIGURE 2.10.5

The monthly temperature cycle and its derivative

this results from the factor of 28 dividing the amplitude of the derivative of the monthly rhythm. ▲

2.10.2 Other Trigonometric Functions

We can use the quotient rule to find the derivatives of the other trigonometric functions. For $\tan(\theta)$,

$$\begin{aligned}
\frac{d}{d\theta}\tan(\theta) &= \frac{d}{d\theta}\frac{\sin(\theta)}{\cos(\theta)} \\
&= \frac{\cos(\theta)\frac{d\sin(\theta)}{d\theta} - \sin(\theta)\frac{d\cos(\theta)}{d\theta}}{\cos^2\theta} && \text{quotient rule} \\
&= \frac{\cos(\theta)\cos(\theta) + \sin(\theta)\sin(\theta)}{\cos^2(\theta)} && \text{derivatives of sine and cosine} \\
&= \frac{1}{\cos^2\theta} && \sin^2(\theta) + \cos^2(\theta) = 1 \\
&= \sec^2\theta. && \text{definition of the secant function}
\end{aligned}$$

On a graph of $\tan(\theta)$, the function has slope of 1 at $\theta = 0$ and is always increasing except at points where it is not defined (Figure 2.10.6).

Similarly (Exercises 9–12),

$$\begin{aligned}
\frac{d}{dx}\cot(x) &= -\csc^2(x) \\
\frac{d}{dx}\sec(x) &= \tan(x)\sec(x) \\
\frac{d}{dx}\csc(x) &= -\cot(x)\csc(x).
\end{aligned}$$

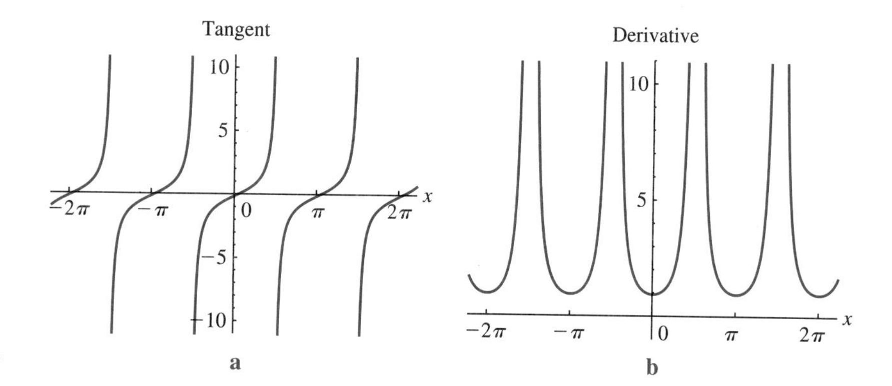

FIGURE 2.10.6
The tangent function and its derivative

Applies no force when not stretched

Applies outward force when compressed

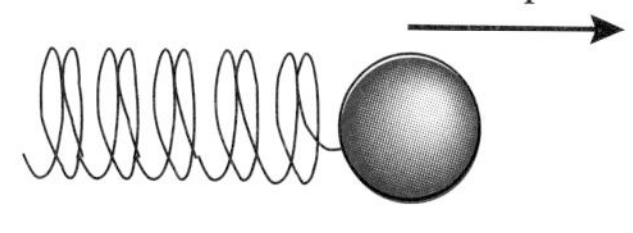

Applies inward force when stretched

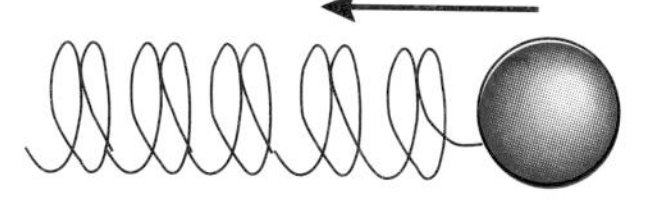

FIGURE 2.10.7
A perfect spring

2.10.3 Applications

We have studied the behavior of a cat falling in a constant gravitational field. Consider now an object attached to a perfect spring (Figure 2.10.7). The spring produces an outward force when compressed and an inward force when stretched. Assume that

$$\text{Force} = -k(\text{amount of stretch}).$$

The negative sign indicates that the force acts opposite the direction of stretch (Figure 2.10.7). If the spring is stretched to the right, it produces a force to the left, and if the spring is stretched to the left, it produces a force to the right. The constant k is the **spring constant**, with larger values indicating a stiffer spring.

Newton's law says that acceleration, the second derivative of position, is proportional to force according to

$$F = ma$$

where m is the mass of the object.

Suppose first that both the mass of the object m and the spring constant k are equal to 1.0. Then, if the position of the object is x,

$$a = \frac{d^2x}{dt^2} = -x.$$

The solution of this differential equation is a function that has second derivative equal to the negative of itself. We have just met two such functions: sine and cosine. The position of an object attached to such an ideal spring (without friction) will follow exactly a sinusoidal oscillation. This spring is called the "simple harmonic oscillator" in physics and is one of the reasons why the sinusoidal functions are so important (in the same way that the simple differential equation $\frac{db}{dt} = b$ is one of the reasons why exponential functions are so important).

Example 2.10.3 A Solution of the Spring Equation: Stretched Initial Condition

Suppose the spring is stretched one unit and then the object is released. The subsequent movement follows the function

$$p(t) = \cos(t).$$

We can check that $p(0) = 1$ and that $\frac{dp}{dt} = -\sin(0) = 0$, meaning that the object does begin at rest at a position of 1 (Figure 2.10.8). Furthermore,

$$\frac{d^2p}{dt^2} = -\cos(t) = -p(t).$$

This function solves the differential equation in the case $k = 1.0$ and $m = 1.0$, and describes an oscillation with period $T = 2\pi$, the period of the cosine function itself. ▲

Example 2.10.4 A Solution of the Spring Equation: Compressed Initial Condition

If the object is released when the spring is *compressed* by one unit, the solution is

$$p(t) = -\cos(t).$$

We can check that $p(0) = -1$ and that $\frac{dp}{dt} = -\sin(0) = 0$, so the object is at rest at time 0 (Figure 2.10.9). Furthermore,

$$\frac{d^2p}{dt^2} = \cos(t) = p(t).$$ ▲

Example 2.10.5 A Solution of the Spring Equation: Slightly Compressed Initial Condition

If the object is released when the spring is stretched outward by 0.2 unit, the solution is

$$p(t) = 0.2\cos(t).$$

This is 0.2 times the solution in Example 2.10.3 (Figure 2.10.10). Therefore, $p(0) = 0.2$ and $\frac{dp}{dt} = -0.2\sin(0) = 0$. ▲

Stretch 1 0.5 0 −0.5 −1 Time π 2π

Solution starting from an outward stretch of 1

FIGURE 2.10.8

Solution of the spring equation with $p(0) = 1$

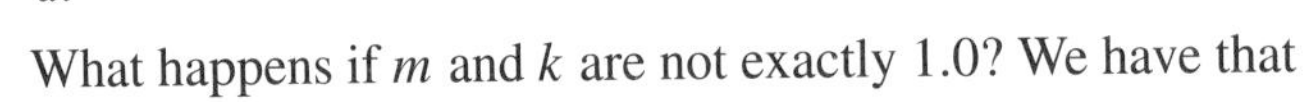

What happens if m and k are not exactly 1.0? We have that

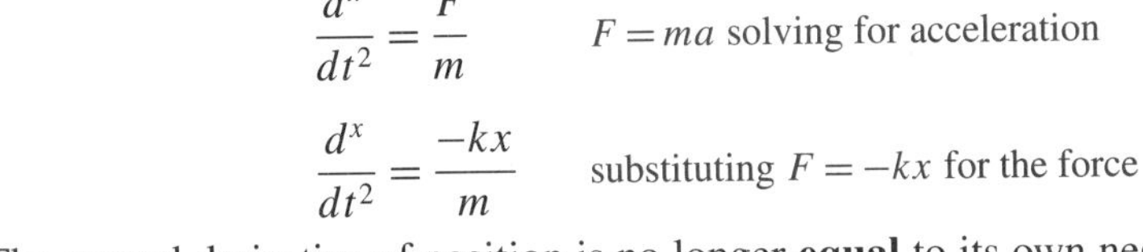

$$\frac{d^x}{dt^2} = \frac{F}{m} \quad F = ma \text{ solving for acceleration}$$

$$\frac{d^x}{dt^2} = \frac{-kx}{m} \quad \text{substituting } F = -kx \text{ for the force}$$

The second derivative of position is no longer **equal** to its own negative, but equal to $-k/m$ times its negative.

Experience with springs shows that stiffer springs oscillate more quickly, with a smaller period. Perhaps a cosine function with a different period T will satisfy this equation. The function

$$p(t) = B\cos\left(\frac{2\pi}{T}t\right)$$

is a special case of the general sinusoidal oscillation (Subsection 1.8.2) with period T and amplitude B, but with average and phase set to 0. Using the chain rule, we find

$$\frac{dp}{dt} = -B\frac{2\pi}{T}\sin\left(\frac{2\pi}{T}t\right)$$

$$\frac{d^2p}{dt^2} = B\left(\frac{2\pi}{T}\right)^2\cos\left(\frac{2\pi}{T}t\right)$$

$$= -\left(\frac{2\pi}{T}\right)^2 p(t).$$

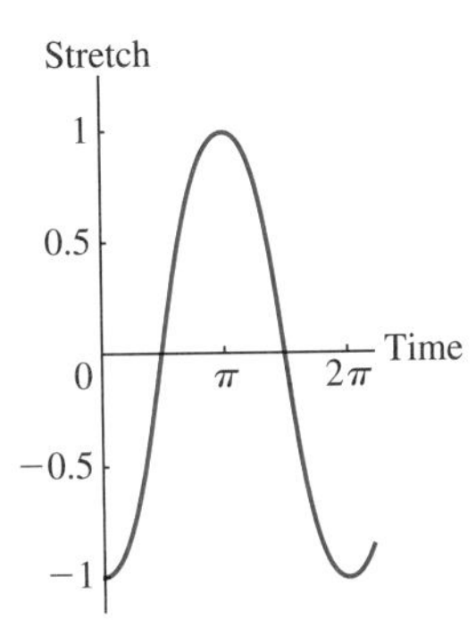

Solution starting from an inward stretch (compression) of 1

FIGURE 2.10.9

Solution of the spring equation with $p(0) = -1$

This function is a solution of

$$\frac{d^x}{dt^2} = \frac{-kx}{m}$$

if

$$\left(\frac{2\pi}{T}\right)^2 = \frac{k}{m} \quad \textbf{(2.10.3)}$$

Example 2.10.6 The Period of Oscillation When the Spring Is Strong

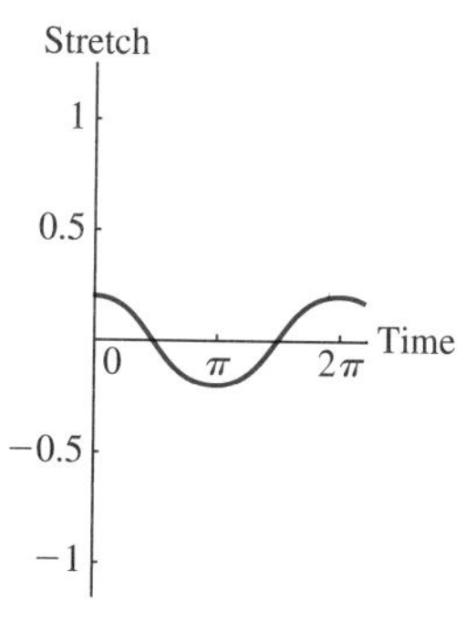

FIGURE 2.10.10
Solution of the spring equation with $p(0) = 0.2$

Suppose that $k = 2.0$, twice as strong as the spring in Example 2.10.3, but that the mass is still $m = 1.0$. According to Equation 2.10.3, the period T must solve

$$\left(\frac{2\pi}{T}\right)^2 = \frac{2.0}{1.0} = 2.0.$$

Isolating T gives

$$(2\pi)^2 = 2.0T^2$$
$$\frac{(2\pi)^2}{2.0} = T^2$$
$$\frac{2\pi}{\sqrt{2.0}} = T.$$

This period is smaller by a factor of $\sqrt{2.0}$ than the period $T = 2\pi$ of the spring with $k = 1.0$ (Figure 2.10.11).

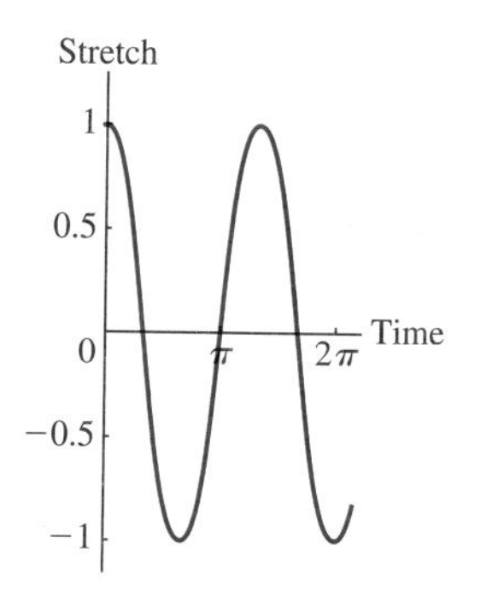

FIGURE 2.10.11
Solution of the spring equation with $p(0) = 1$ and $k = 2.0$

2.10.4 Derivation of the Key Limits

Finding the derivatives of sin and cos requires computing the following limits

$$\lim_{h \to 0} \frac{\cos(h) - 1}{h}$$
$$\lim_{h \to 0} \frac{\sin(h)}{h}.$$

Testing with a calculator gives the following. It seems as though the first limit is 0 and the second is 1.

h	$\frac{\cos(h) - 1}{h}$	$\frac{\sin(h)}{h}$
1.0	−0.45970	0.84147
0.1	−0.04996	0.99833
0.01	−0.00500	0.99998
0.001	−0.00050	1.00000

We can see why this occurs from a geometric diagram of sine and cosine. First, we show that

$$\lim_{h \to 0} \frac{\sin(h)}{h} = 1.$$

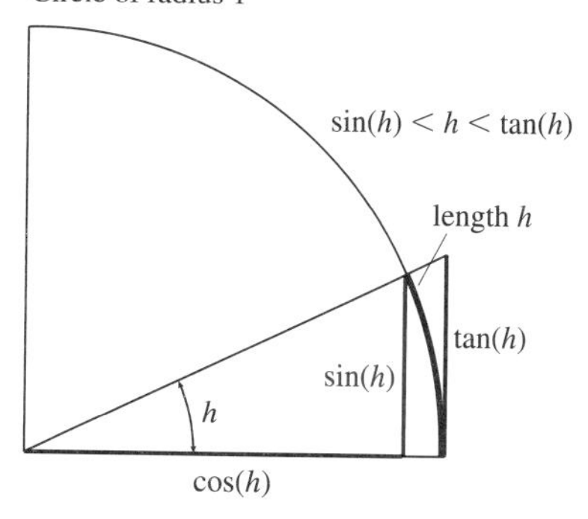

FIGURE 2.10.12
Geometric components of sine

In Figure 2.10.12, the length of the arc is equal to the angle in radians, and is greater than $\sin(h)$, the length of the vertical line segment. Therefore,

$$\frac{\sin(h)}{h} \leq 1.$$

Furthermore, the length of the line segment tangent to the circle is equal to $\tan(h)$ which is greater than the length of the arc, so $\tan(h) > h$ or

$$\frac{\sin(h)}{\cos(h)} > h$$

implying that

$$\frac{\sin(h)}{h} > \cos(h).$$

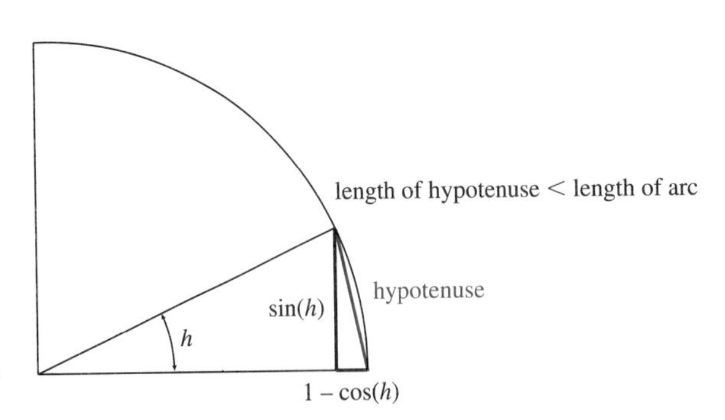

FIGURE 2.10.13
The geometry of the limit of $\frac{\cos(h)-1}{h}$

Therefore,

$$\lim_{h\to 0}\cos(h) \le \lim_{h\to 0}\frac{\sin(h)}{h} \le \lim_{h\to 0} 1.$$

Because $\cos(h)$ is a continuous function, $\lim_{h\to 0}\cos(h) = \cos(0) = 1$. Furthermore, the limit of the constant 1 is also 1. Therefore,

$$1 \le \lim_{h\to 0}\frac{\sin(h)}{h} \le 1$$

implying that

$$\lim_{h\to 0}\frac{\sin(h)}{h} = 1$$

as found in the table in Section 2.10.4.

To find the other limit

$$\lim_{h\to 0}\frac{\cos(h)-1}{h},$$

we can use the fact that $\sin(h)$ and $1-\cos(h)$ are two sides of a right triangle with hypotenuse of length less than the length of the arc (Figure 2.10.13). By the Pythagorean theorem, the square of the length of hypotenuse is the sum of the squares of the sides,

$$\begin{aligned} \sin^2(h) + (1-\cos(h))^2 &< h^2 && \text{hypotenuse is shorter than arc} \\ \sin^2(h) + 1 - 2\cos(h) + \cos^2(h) &< h^2 && \text{expand the quadratic} \\ 2 - 2\cos(h) &< h^2 && \text{use fact that } \sin^2(h)+\cos^2(h)=1 \\ 1-\cos(h) &< \tfrac{h^2}{2}. && \text{divide by 2} \end{aligned}$$

So

$$\lim_{h\to 0}\frac{\cos(h)-1}{h} < \lim_{h\to 0}\frac{h}{2} = 0.$$

This limit, in accord with the calculations in the table in Section 2.10.4, is 0.

Summary To complete our collection of building blocks, we derived the derivatives of the trigonometric functions.

Function	Derivative
$\sin(x)$	$\cos(x)$
$\cos(x)$	$-\sin(x)$

The sinusoidal functions provide the solutions of the differential equation describing a spring, the **simple harmonic oscillator**. They describe the position of an object that moves back to a central state with force proportional to the displacement.

2.10 Exercises

Mathematical Techniques

1–8 ▪ Find the derivatives of the following functions.

1. $f(x) = x^2 \sin(x)$.
2. $g(x) = x^2 \cos(x)$.
3. $h(x) = \sin(\theta)\cos(\theta)$.
4. $q(x) = \dfrac{\sin(\theta)}{1 + \cos(\theta)}$.
5. $F(z) = 3 + \cos(2z - 1)$.
6. $G(t) = 1 + 2\cos\left(\dfrac{2\pi}{5}(t - 3)\right)$.
7. $f(x) = e^{\cos(x)}$.
8. $f(x) = \cos(e^x)$.

9–12 ▪ Use the definitions and the derivatives of $\sin(\theta)$ and $\cos(\theta)$ to check the derivatives of the other trigonometric functions, and to find their second derivatives.

9. $\tan(\theta)$
10. $\cot(\theta)$
11. $\sec(\theta)$
12. $\csc(\theta)$

13–16 ▪ Use the angle addition formulas (Equations 2.10.1 and 2.10.2) to expand the following and use the product and sum rules to compute the derivatives of the following functions. Compare the result with what you get with the chain rule.

13. $\cos(2\theta)$. Simplify the answer in terms of $\sin(2\theta)$.
14. $\sin(2\theta)$. Simplify the answer in terms of $\cos(2\theta)$.
15. Take the derivative of $\cos(\theta + \phi)$ with respect to θ, thinking of ϕ as a constant. Simplify the answer in terms of $\sin(\theta + \phi)$.
16. $\sin(\theta + \phi)$. Simplify the answer in terms of $\cos(\theta + \phi)$.

17–22 ▪ Compute the derivatives of the following functions. Find the value of the function and the slope at 0, $\pi/2$, and π. Sketch a graph of the function on the given domain. How would you describe the behavior in words?

17. $a(x) = 3x + \cos(x)$ for $0 \le x \le 2\pi$.
18. $b(y) = y^2 + 3\cos(y)$ for $0 \le y \le 2\pi$.
19. $c(z) = e^{-z}\sin(z)$ for $0 \le z \le 2\pi$.
20. $r(t) = t\cos(t)$ for $0 \le t \le 4\pi$.
21. $s(t) = e^{0.2t}\cos(t)$ for $0 \le t \le 40$.
22. $p(t) = e^t(1 + 0.2\cos(t))$ for $0 \le t \le 40$.

23–26 ▪ We can use Theorem 2.13 to find the derivatives of the inverse trigonometric functions.

23. Find the derivative of $\sin^{-1}(x)$. If you use the identity $\cos^2(x) + \sin^2(x) = 1$, you can write the answer without any trigonometric functions.
24. Find the derivative of $\cos^{-1}(x)$. Write the answer without any trigonometric functions.
25. Find the derivative of $\tan^{-1}(x)$. If you use the identity $1 + \tan^2(x) = \sec^2(x)$, you can write the answer without any trigonometric functions.
26. Find the derivative of $\sec^{-1}(x)$. Write the answer without any trigonometric functions.

27–30 ▪ Show that the following are solutions of the given differential equation.

27. $s(t) = \sin(t) + 4$ is a solution of $\dfrac{ds}{dt} = \cos(t)$.
28. $y(t) = t^2\sin(t)$ is a solution of $\dfrac{dy}{dt} = 2t\sin(t) + t^2\cos(t)$.
29. $g(t) = \cos(t) + 2\sin(t) - 3e^t$ is a solution of $\dfrac{d^4g}{dt^4} = g(t)$.
30. $h(t) = \cos(t) + 2\sin(t) - 3e^t$ is a solution of $\dfrac{d^{40}h}{dt^{40}} = h(t)$.

Applications

31–34 ▪ Find the derivatives of the following functions (from Section 1.8, Exercises 35-38). Sketch a graph and check that your derivative has the correct sign when the argument is equal to 0.

31. $f(x) = 3.0 + 4.0\cos(\frac{2\pi}{5.0}(x - 1.0))$.
32. $g(t) = 4.0 + 3.0\cos(2\pi(t - 5.0))$.
33. $h(z) = 1.0 + 5.0\cos(\frac{2\pi}{4.0}(z - 3.0))$.
34. $W(y) = -2.0 + 3.0\cos(\frac{2\pi}{0.2}(y + 0.1))$.

35–36 ▪ Consider the function $p(t) = \cos(\frac{2\pi t}{T})$ where T is a constant.

35. Consider a spring with $k = 0.1$ and $m = 1.0$. Find the period T that produces a solution of the spring equation. Is this spring stronger or weaker than one with $k = 1.0$, and does the oscillation have a larger or smaller period?
36. Consider a spring with $k = 1.0$ and $m = 5.0$. Find the period T that produces a solution of the spring equation. Does a heavier object oscillate more slowly than a light one?

37–38 ▪ Consider the combination of the temperature cycles (Subsection 1.8.3)

$$P_d(t) = 36.8 + 0.3\cos(2\pi(t - 0.583))$$
$$P_m(t) = 36.8 + 0.2\cos\left(\frac{2\pi(t - 16)}{28}\right)$$

given by

$$P_t(t) = 36.8 + 0.2\cos\left(\frac{2\pi(t - 16)}{28}\right) + 0.3\cos(2\pi(t - 0.583)).$$

37. Find the derivative of P_t.
38. Sketch a graph of the derivative over one month. If you measured only the derivative, which oscillation would you see?

39–40 ▪ The spring we studied had no friction. Friction acts as a force much like the spring itself, but is proportional to velocity rather than displacement. One possible equation describing this is

$$\frac{d^2x}{dt^2} = -2x - 2\frac{dx}{dt}.$$

39. Explain each term in this equation and show that $x(t) = e^{-t}\cos(t)$ is a solution.

40. Graph the solution and explain what is going on. Is friction strong in this system?

Computer Exercises

41. This problem requires a computer program with a built-in ability to solve differential equations. The spring equation

$$\frac{d^2y}{dt^2} = -y(t)$$

is only an approximation to the behavior of a pendulum, which is in fact better described by the equation

$$\frac{d^2y}{dt^2} = -\sin(y(t)).$$

It is impossible to write down a solution of this equation.

a. Starting from $y(0) = 0.1$ and $\frac{dy}{dt} = 0$ at $t = 0$, the solution of the spring equation is $y(t) = 0.1\cos(t)$. Compare this with the solution of the pendulum equation for one period (from $t = 0$ to $t = 2\pi$). Graph the two solutions.

b. Do the same starting from $y(0) = 0.2$.

c. Do the same starting from $y(0) = 0.5$.

d. Do the same starting from $y(0) = 1.0$.

e. Do the same starting from $y(0) = 1.5$.

f. How long does it take the pendulum to swing all the way back in each case? Does the period of a pendulum depend on the amplitude?

42. Consider the family of functions

$$h(t) = \cos(3.0t) + 1.5\cos(3.6t) + 2.0\cos(vt)$$

for various values of v ranging from 2.0 to 3.0. Graph them. Why do they look so weird? When do they look least weird?

43. Find the derivatives of the following functions. Where are the critical points and points of inflection? What happens as more and more cosines are piled up? Explain this in terms of the updating function

$$x_{t+1} = \cos(x_t).$$

a. $\cos(\cos(x))$.

b. $\cos(\cos(\cos(x)))$.

c. $\cos(\cos(\cos(\cos(x))))$.

d. $\cos(\cos(\cos(\cos(\cos(x)))))$.

Supplementary Problems

1–6 ▪ Find the limits of the following functions or say why you can't.

1. $\lim_{x\to 0} 1 + x^2$.

2. $\lim_{x\to -2} 1 + x^2$.

3. $\lim_{x\to 1} \frac{1}{x^2 - 1}$.

4. $\lim_{x\to 0} \frac{e^x}{1 + e^{2x}}$.

5. $\lim_{x\to 1^+} \frac{1}{x - 1}$.

6. $\lim_{x\to 1^-} 1/(x - 1)$.

7–12 ▪ Find the derivatives of the following functions. Note any points where the derivative does not exist.

7. $F(y) = y^4 + 5y^2 - 1$.

8. $a(x) = 4x^7 + 7x^4 - 28$.

9. $H(c) = \frac{c^2}{1 + 2c}$.

10. $h(z) = \frac{z}{1 + \ln(z^2)}$ for $z \geq 0$.

11. For $y \geq 0$, $b(y) = \frac{1}{y^{0.75}}$.

12. For $z \geq 0$, $c(z) = \frac{z}{(1 + z)(2 + z)}$.

13–18 ▪ Find the derivatives of the following functions

13. $g(x) = (4 + 5x^2)^6$.

14. $c(x) = (1 + \frac{2}{x})^5$.

15. $s(t) = \ln(2t^3)$.

16. $p(t) = t^2 e^{2t}$.

17. $s(x) = e^{-3x+1} + 5\ln(3x)$.

18. $g(y) = e^{3y^3+2y^2+y}$.

19–22 ▪ Find the derivatives and other requested quantities for the following functions.

19. $f(t) = e^t\cos(t)$. Find one critical point.

20. $g(x) = \ln(1 + x^2)$. Find one point where $g(x)$ is decreasing.

21. $h(y) = \frac{1 - y}{(1 + y)^3}$. Find all values where h is increasing.

22. $c(z) = \frac{e^{2z} - 1}{z}$. What is $c(0)$? (You need to take the limit).

23–26 ▪ Find all critical points and points of inflection of the following. Sketch graphs.

23. $f(x) = e^{-x^3}$.

24. $g(x) = e^{-x^4}$.

25. $h(y) = \cos(y) + \frac{y}{2}$.

26. $F(c) = e^{-2c} + e^c$.

27–34 ▪ Solve the following.

27. A population has size

$$N(t) = 1000 + 10t^2.$$

where t is measured in years.

a. What units should follow the 1000 and the 10?

b. Graph the population between $t = 0$ and $t = 10$.

c. Find the population after 7 yr and 8 yr. Find the approximate growth rate between these two times. Where does this approximate growth rate appear on your graph?

d. Find and graph the derivative $N'(t)$.

e. Find the per capita rate of growth. Is it increasing?

28. Suppose the volume of a cell is described by the function $V(t) = 2 - 2t + t^2$ where t is measured in minutes and V is measured in 1000s of μm^3.

a. Graph the secant line from time $t = 2$ to $t = 2.5$.

b. Find the equation of this line.

c. Find the value of the derivative at $t = 2$, and express it in both differential and prime notation. Don't forget the units.

d. What is happening to cell volume at time $t = 2$?

e. Graph the derivative of V as a function of time.

29. Suppose a machine is invented to measure the amount of knowledge in a student's head in units called "factoids." One student is measured at $F(t) = t^3 - 6t^2 + 9t$ factoids at time t, where t is measured in weeks.

a. Find the rate at which the student is gaining (or losing) knowledge as a function of time (be sure to give the units).

b. During what time between $t = 0$ and 11 is the student losing knowledge?

c. Sketch a graph of the function $F(t)$.

30. The following measurements are made of a plant's height in centimeters and its rate of growth.

Day	Height (cm)	Growth Rate
1	8.0	5.2
2	15.0	9.4
3	28.0	17.8
4	53.0	34.6

a. Graph these data. Make sure to give units.

b. Write the equation of a secant line connecting two of these data points and use it to guess the height on day 5. Why did you pick the points you did?

c. Write the equation of a tangent line and use it to guess the height on day 5.

d. Which guess do you think is better?

e. How do you think the "growth rate" might have been measured?

31. Suppose the fraction of mutants in a population follows the discrete-time dynamical system

$$p_{t+1} = \frac{2.0p_t}{2.0p_t + 1.5(1 - p_t)}.$$

Suppose you wish the fraction at time $t = 1$ to be close to $p_1 = 0.4$.

a. What would p_0 have to be to hit 0.4 exactly?

b. How close would p_0 have to be to this value to produce p_1 within 0.1 of the target?

c. What happens to the input tolerance as the output tolerance becomes smaller (as p_1 is required to be closer and closer to 0.4)?

32. Suppose the position of an object attached to a spring is

$$x(t) = 2.0 \cos\left(\frac{2\pi t}{3.2}\right).$$

a. Find the derivative of $x(t)$. What does it mean physically?

b. Find the second derivative of $x(t)$. What does this mean physically?

c. What is the position and velocity at $t = 0$?

d. What is the position and velocity at $t = 1.6$?

e. What is the position and velocity at $t = 3.2$?

f. What differential equation does this object follow?

33. Suppose a system exhibits hysteresis. While increasing the temperature T, the voltage response follows

$$V_i(T) = \begin{cases} T & \text{if } T \le 50 \\ 100 & \text{if } 50 < T \le 100. \end{cases}$$

While decreasing the temperature, however, the voltage response follows

$$V_d(T) = \begin{cases} T & \text{if } T \le 20 \\ 100 & \text{if } 20 < T \le 100. \end{cases}$$

a. Graph these functions.

b. Find the left- and right-hand limits of V_i and V_d as T approaches 50.

c. Find the left and right-hand limits of V_i and V_d as T approaches 20.

34. Suppose the total product generated by a chemical reaction is

$$P(t) = \frac{t}{1+2t}$$

where t is measured in hours.

a. Find the average rate of change between $t = 1$ and $t = 2$.

b. Find the equation of the secant line between these times.

c. Graph the secant line and the function $P(t)$.

d. Sketch a graph of the tangent line at $t = 2$. Based on your graph, is the slope of the secant larger, smaller, or the same as the slope of the tangent?

e. Write the limit you would take to find the instantaneous rate of change at $t = 2$.

35–36 ▪ In each of the following cases, suppose a window is described by the graph. Time is measured in hours after going to sleep at 10:00 P.M., and the area open is in square feet. In each case,

a. Label a point where the first derivative is positive and the second derivative is negative,

b. Label a point of inflection,

c. Sketch a graph of the rate of change,

d. Sketch a graph of the second derivative.

35.

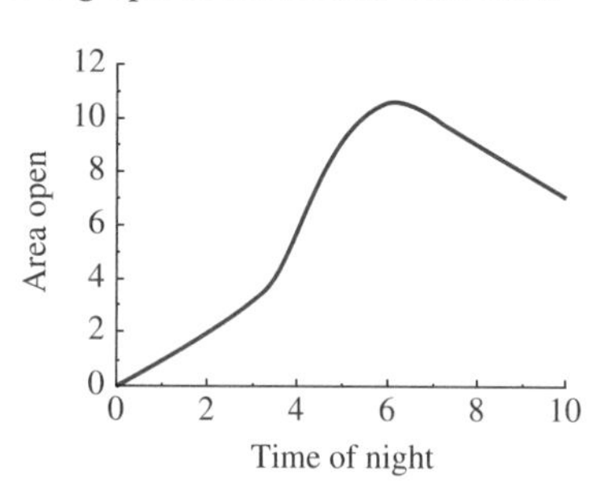

36.

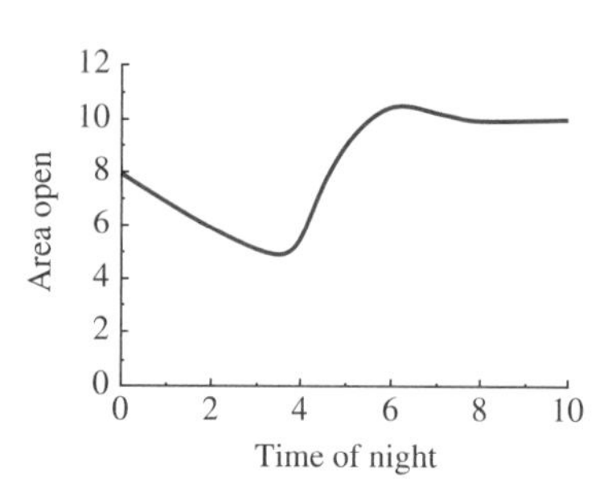

Projects

1. Periodic hematopoiesis is a disease characterized by large oscillations in the red blood cell count. Red blood cells are generated by a feedback mechanism that approximately obeys the following equation.

$$x_{t+1} = \frac{\tau}{1+\gamma\tau} F(x_t) + \frac{1}{1+\gamma\tau} x_t.$$

The function $F(x_t)$ describes production as a function of number of cells and takes the form of a Hill function

$$F(x) = F_0 \frac{\theta^n}{\theta^n + x^n}.$$

The terms are described as follows.

Parameter	Meaning	Normal Value
x_t	Number of cells at time t (billions)	About 330
τ	Time for cell development	5.7 days
γ	Fraction of cells that die each day	0.0231
F_0	Maximum production of cell	76.2 billion
θ	Value where cell production is halved	247 billion
n	Shape parameter of Hill function	7.6

a. Graph and explain $F(x)$ with these parameter values. Does this sort of feedback system make sense?

b. Use a computer to find the equilibrium with healthy parameters.

c. Graph and cobweb the updating function. If you start near the equilibrium, do values remain nearby? What would a solution look like?

d. Certain autoimmune diseases increase γ, the death rate of cells. Study what happens to the equilibrium and the solution as γ increases. Explain your results in biological terms.

e. Explore what would happen if the value of n were decreased. Does the equilibrium become more sensitive to small changes in γ?

M. C. Mackey. Mathematical models of hematopoietic cell replication and control. Pages 149–178 in H. G. Othmer and F. R. Adler and M. A. Lewis and J. D. Dallon, editors, *Case Studies in Mathematical Modeling*, Prentice Hall, Upper Saddle River, N.J., 1997.

2. An object attached to a spring with friction of strength α that oscillates with period T can have solution

$$x(t) = e^{-\alpha t} \cos\left(\frac{2\pi t}{T}\right).$$

a. Take the first and second derivatives.

b. Show that $x(t)$ is a solution of

$$\frac{d^2x}{dt^2} = -\left(\left(\frac{2\pi}{T}\right)^2 + \alpha^2\right)x - 2\alpha\frac{dx}{dt}.$$

c. What is the strength of the spring?

d. Find two different values of spring strength and friction that produce the same period T.

3. The functions that we have seen which are not differentiable only fail at a few points. Surprisingly, it is possible to find continuous curves that have a corners at every point. These curves are called fractals. The following construction creates a fractal called Koch's snowflake.

a. Draw an equilateral triangle.

b. Take the middle third of each side and expand it as shown.

c. Do the same for the middle third of each straight piece.

d. Repeat the process for as long as you can.

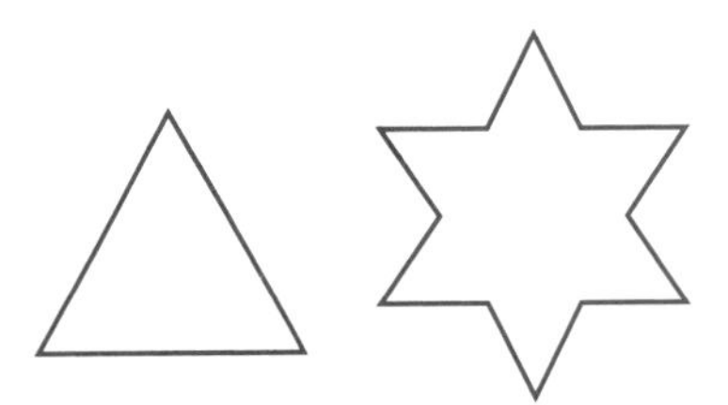

If this process is continued for an infinite number of steps, the resulting curve has a corner at every point. How long is the curve? What is the area it encloses?

Chapter

3 Applications of Derivatives and Dynamical Systems

We have developed techniques for computing derivatives of many functions, and we have interpreted the derivative as both the rate of change of a measurement and the slope of the graph of a function. The derivative has a remarkable range of applications for understanding the behavior of biological measurements, with the general theme of **reasoning about functions**, the mathematical version of reasoning about qualitative relationships between measurements. To begin, we use the slope interpretation to find that the derivative of the updating function determines whether an equilibrium of a discrete-time dynamical system is **stable** or **unstable**. We will then use the derivative to find **maxima** and **minima** of functions. When combined with appropriate models of a biological process, these methods can be used to solve **optimization** problems.

Next, we will develop tools for reasoning about continuous and differentiable functions with a minimum of calculation. We will show that certain equations have solutions or that a maximum exists based on the general properties of functions, and we will deduce more specific conclusions by using a system of simple comparisons called the **method of leading behavior**. We will use the tangent line approximation to solve impossible equations with **Newton's method**, and extend the theme of approximating complicated functions with simple functions to **Taylor polynomials**.

3.1 Stability and the Derivative

In our initial study of discrete-time dynamical systems, we observed that equilibria can be **stable** or **unstable**. By examining the graph and cobweb of a discrete-time dynamical system, we will develop a method for evaluating the stability of an equilibrium by computing the **derivative of the updating function**. This method explains why some solutions, like that for a growing bacterial population, move away from equilibrium, while others, like that of the lung model, move toward their equilibrium.

3.1.1 Motivation

Figures 3.1.1–3.1.5 review the cobwebbing diagrams for five of the discrete-time dynamical systems we have studied, and the following points review the terminology of discrete-time dynamical systems.

- A **discrete-time dynamical system** is a rule (described by its **updating function**) that takes a measurement at one time step as an input and returns the measurement at the next time step as an output (Section 1.5).
- A **solution** gives the values of the measurement as a function of time (Section 1.5).
- **Cobwebbing** is a graphical method for finding solutions of discrete-time dynamical systems (Section 1.6).
- An **equilibrium** is a point where the discrete-time dynamical system leaves the value unchanged (Definition 1.11 in Section 1.6). Graphically, an equilibrium is a point where the graph of the updating function crosses the diagonal line.

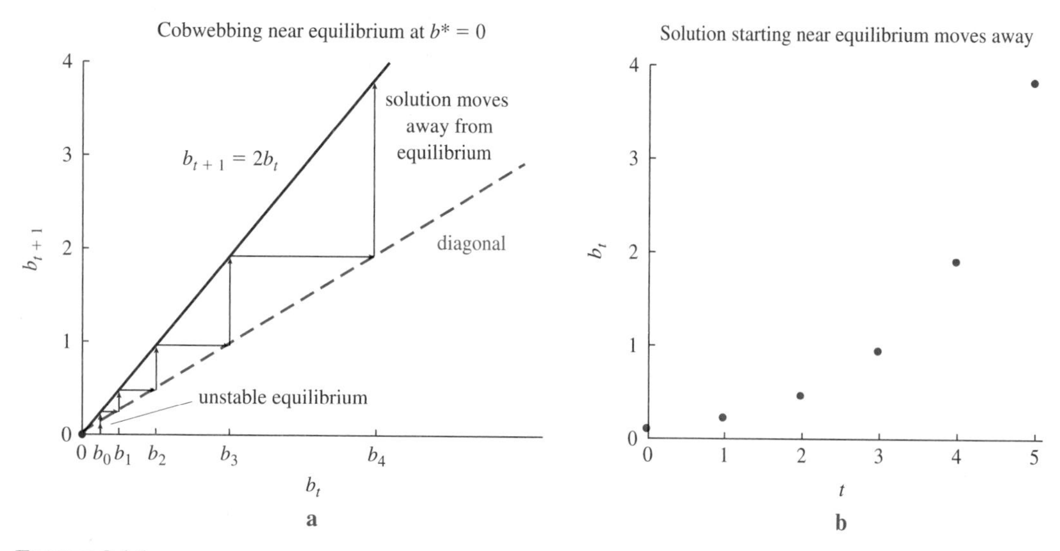

FIGURE 3.1.1

Bacterial population growth discrete-time dynamical system $b_{t+1} = 2b_t$

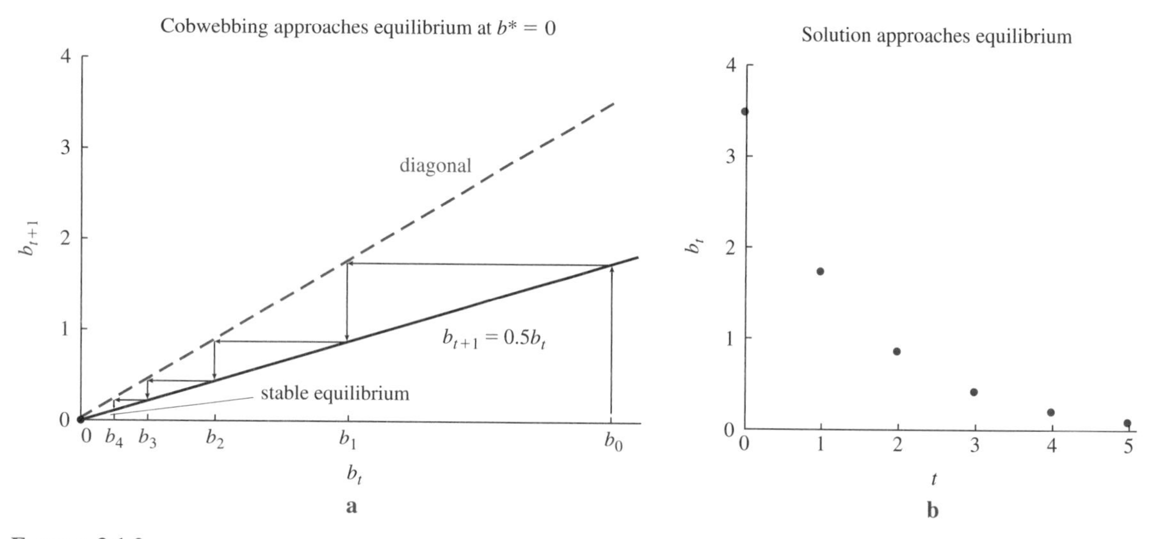

FIGURE 3.1.2

Bacterial population growth discrete-time dynamical system $b_{t+1} = 0.5b_t$

- An equilibrium is **stable** if solutions that start near the equilibrium move closer to the equilibrium, and **unstable** if solutions that start near the equilibrium move away from the equilibrium (Section 1.10).

The figures show four stable equilibria: $b^* = 0$ in Figure 3.1.2, $c^* = 5$ in Figure 3.1.3, $p^* = 1$ in Figure 3.1.4, and $p^* = 0$ in Figure 3.1.5. There are three unstable equilibria: $b^* = 0$ in Figure 3.1.1, $p^* = 0$ in Figure 3.1.4, and $p^* = 1$ in Figure 3.1.5. What is the pattern? How can we recognize which equilibria are stable?

Think about cobwebbing near a stable equilibrium, such as the one in Figure 3.1.3. If the initial condition is slightly less than the equilibrium, the solution **increases** because the graph of the updating function lies above the diagonal. Similarly, if we start slightly above the equilibrium, the solution **decreases** because the graph of the updating function lies below the diagonal. In other words, if the graph of the updating function crosses from above the diagonal to below the diagonal, the equilibrium is stable. Does this work for the other stable equilibria? It does if we imagine extending the graphs beyond the biologically meaningful realm. For example, in Figure 3.1.4,

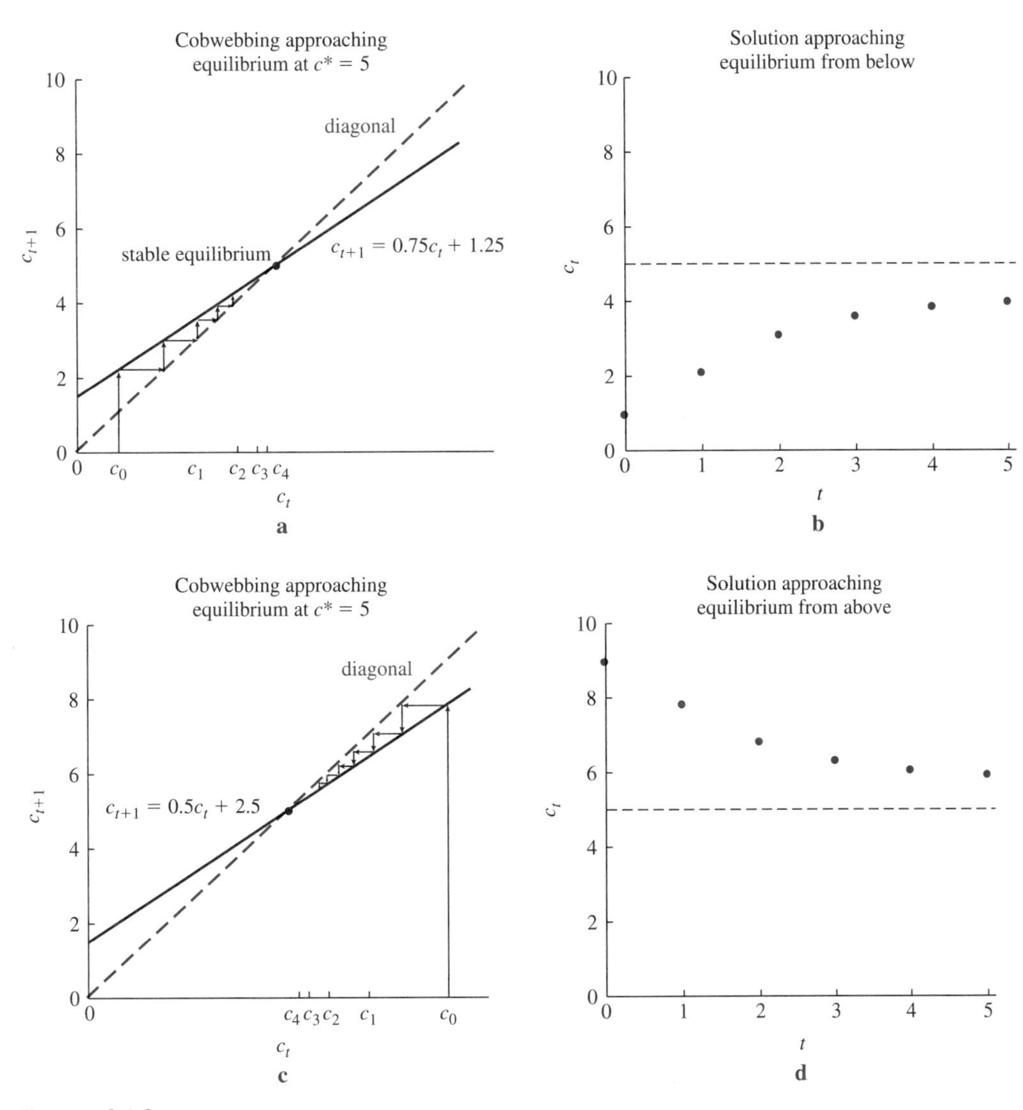

FIGURE 3.1.3

Lung discrete-time dynamical system $c_{t+1} = 0.75c_t + 1.25$

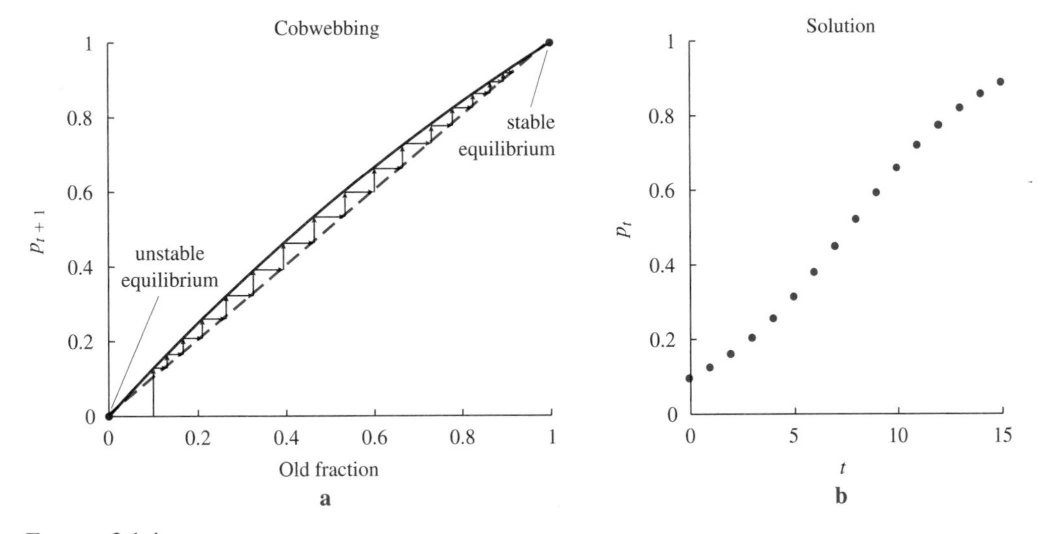

FIGURE 3.1.4

Bacterial selection discrete-time dynamical system $p_{t+1} = \dfrac{2.0p_t}{2.0p_t + 1.5(1 - p_t)}$

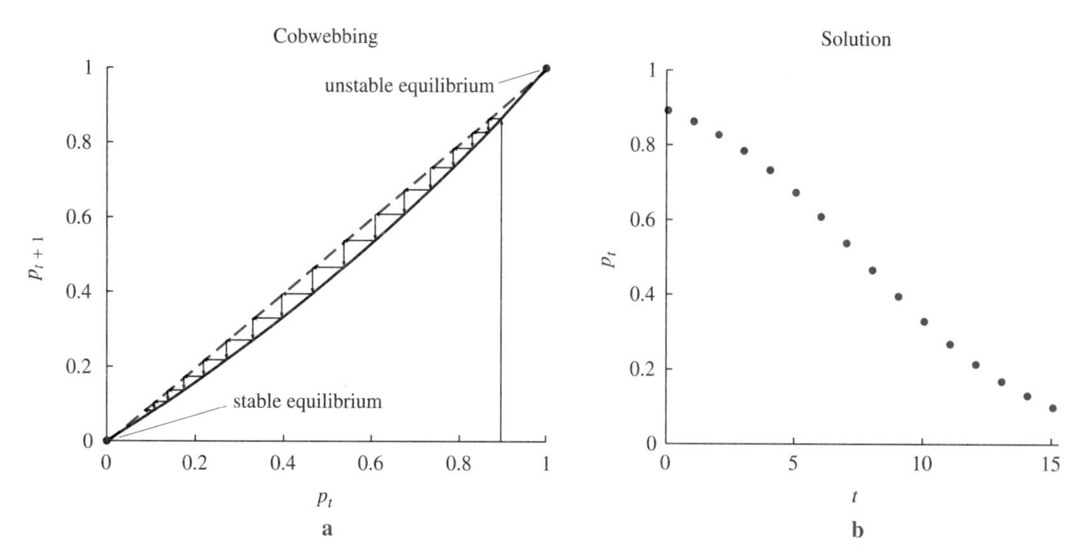

FIGURE 3.1.5
Bacterial selection discrete-time dynamical system $p_{t+1} = \dfrac{1.5p_t}{1.5p_t + 2.0(1 - p_t)}$

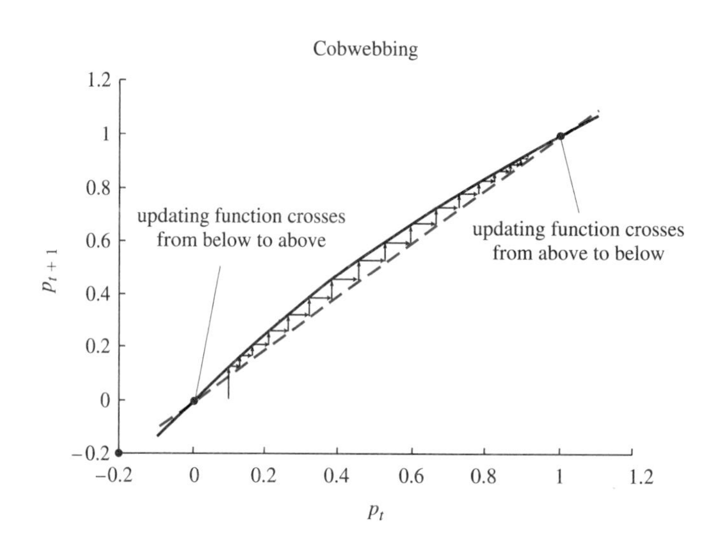

FIGURE 3.1.6
Bacterial selection discrete-time dynamical system with extended graph

the updating function intersects the diagonal from above near $p^* = 1$. If we extend the curve beyond $p^* = 1$, it crosses from above to below (Figure 3.1.6).

Unstable equilibria, in contrast, are points where the updating function crosses the diagonal from below to above. In this case, a solution that starts below the equilibrium will decrease further, moving away from the equilibrium. A solution that starts above the equilibrium will increase further, again moving away from the equilibrium. At the unstable equilibrium at $p^* = 0$ in Figure 3.1.4, the updating function crosses from below to above if we extend it below $p^* = 0$ (Figure 3.1.6).

We can summarize our observations with the following condition (Figure 3.1.7).

Graphical criterion for an equilibrium to be stable or unstable

- An equilibrium is stable if the graph of the updating function crosses the diagonal from above to below.
- An equilibrium is unstable if the graph of the updating function crosses the diagonal from below to above.

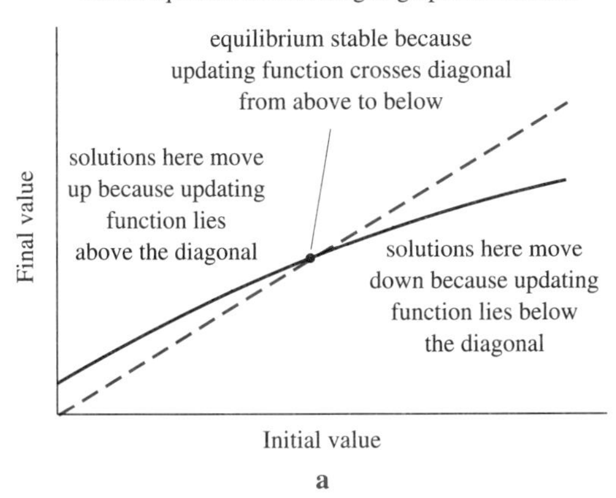
Stable equilibrium according to graphical criterion
equilibrium stable because
updating function crosses diagonal
from above to below
solutions here move
up because updating
function lies
above the diagonal
solutions here move
down because updating
function lies below
the diagonal
Final value
Initial value
a
Unstable equilibrium according to graphical criterion
equilibrium unstable because
updating function crosses diagonal
from below to above
solutions here move
down because
updating function
lies below the
diagonal
solutions
here move up
because updating
function lies
above the diagonal
Final value
Initial value
b

FIGURE 3.1.7
Graphical criterion for stability

3.1.2 Stability and the Slope of the Updating Function

When does the updating function cross the diagonal from below to above? The diagonal is the graph of $y = x$, a line with a slope equal to 1. The graph of the updating function will cross from below to above when its slope is **greater than 1**. Conversely, a curve will cross from above to below when the slope is **less than 1**. We can therefore translate the Graphical Criterion for stability into a condition about the **derivative of the updating function**, because the derivative is equal to the slope (Figure 3.1.8).

Slope criterion for an equilibrium to be stable or unstable

- An equilibrium is stable if the derivative of the updating function is less than 1 at the equilibrium.
- An equilibrium is unstable if the derivative of the updating function is greater than 1 at the equilibrium.

A nonlinear discrete-time dynamical system can have many equilibria. With the aid of the Graphical and Slope Criteria for stability, we can recognize stable and unstable equilibria by examining the slope of the updating function (Figure 3.1.9). We will see in the next section, however, that the dynamics can be much more complicated when the updating function has a negative slope.

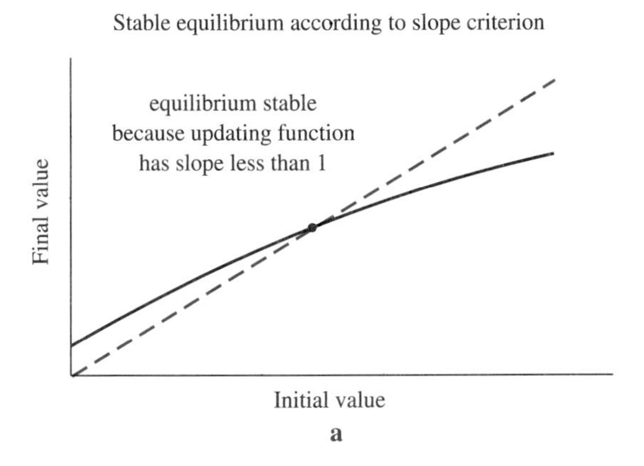
Stable equilibrium according to slope criterion
equilibrium stable
because updating function
has slope less than 1
Final value
Initial value
a

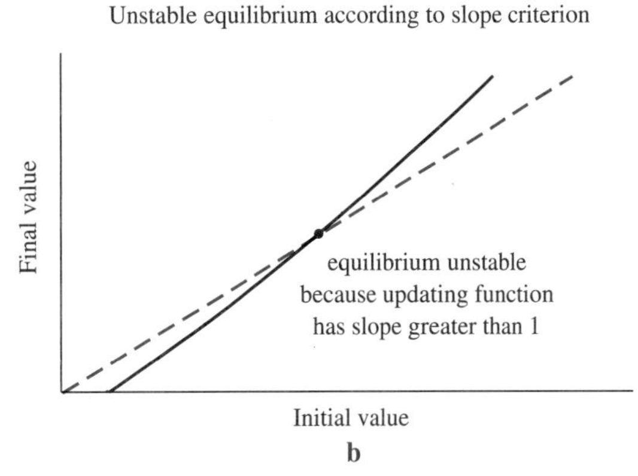
Unstable equilibrium according to slope criterion
equilibrium unstable
because updating function
has slope greater than 1
Final value
Initial value
b

FIGURE 3.1.8
Slope criterion for stability

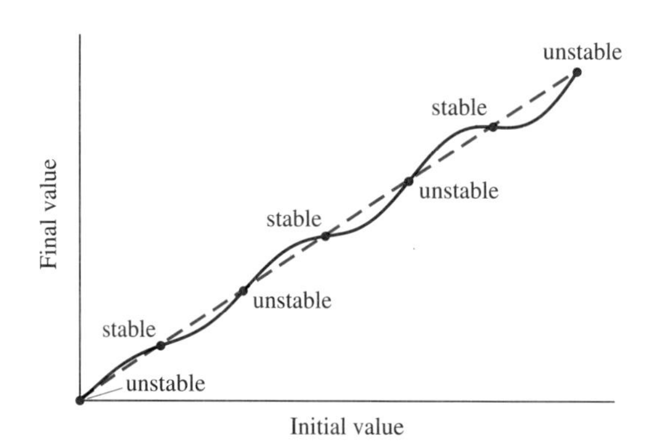

FIGURE 3.1.9

Recognizing the stability of a dynamical system with many equilibria

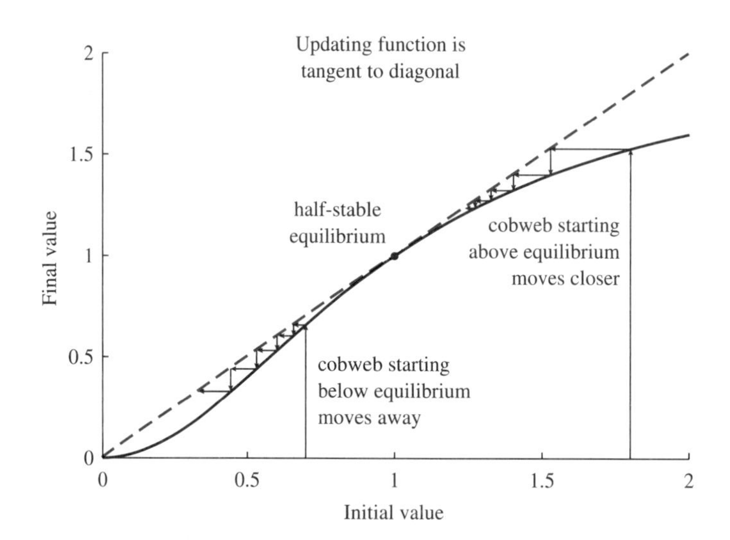

FIGURE 3.1.10

An unusual equilibrium

The Slope Criterion for stability says nothing about what happens when the slope is exactly 1. Figure 3.1.10 shows an equilibrium where the updating function does not cross the diagonal, but remains below the diagonal both to the left and to the right of the equilibrium. A solution that starts from any point below the equilibrium will decrease and move away from the equilibrium, as in the unstable case. A solution that starts from any point above the equilibrium will decrease and move toward the equilibrium, as in the stable case. This sort of half-stable equilibrium is possible because the diagonal is tangent to the updating function. It touches the curve, but does not cross.

3.1.3 Evaluating Stability with the Derivative

We can test the Slope Criterion for stability by computing the derivatives of the discrete-time dynamical systems pictured in figures 3.1.1–3.1.5.

Example 3.1.1 Using the Derivative to Check Stability: Bacterial Population Growth

In Figure 3.1.1, the discrete-time dynamical system is

$$b_{t+1} = 2.0b_t,$$

with the updating function

$$f(b) = 2.0b.$$

The only equilibrium is $b^* = 0$, corresponding to extinction. The updating function is a line with slope of 2.0. More formally,

$$f'(b) = 2.0.$$

Because this slope is greater than 1, the equilibrium is unstable. This makes biological sense because the population is doubling each generation and growing away from a population size of 0.

Example 3.1.2 Using the Derivative to Check Stability: Bacterial Population Decline

In Figure 3.1.2, the discrete-time dynamical system is

$$b_{t+1} = 0.5b_t,$$

with the updating function

$$f(b) = 0.5b.$$

Again, the only equilibrium is $b^* = 0$. This updating function, however, has a slope of 0.5 because

$$f'(b) = 0.5.$$

The slope is less than 1 and this equilibrium is stable. This makes biological sense because this population is being halved each generation, decreasing toward extinction.

Example 3.1.3 Using the Derivative to Check Stability: The Lung Model

The results with the lung discrete-time dynamical system

$$c_{t+1} = 0.75c_t + 1.25,$$

are a little less obvious (Figure 3.1.3). If we designate the updating function $g(c) = 0.75c + 1.25$, we find

$$g'(c) = 0.75$$

because the updating function is linear with slope of 0.75. This slope is less than 1, implying that the single equilibrium is stable. In Section 1.9 we deduced that the equilibrium is always equal to the ambient concentration. Because the equilibrium is stable, we now know that the concentration in the lung (in the absence of absorption) gets closer and closer to the ambient concentration.

Example 3.1.4 Using the Derivative to Check Stability: The Selection Model

What happens when the discrete-time dynamical system is nonlinear? Figure 3.1.4 uses the bacterial selection discrete-time dynamical system

$$p_{t+1} = \frac{2.0p_t}{2.0p_t + 1.5(1 - p_t)},$$

where p_t represents the proportion of mutant bacteria at time t, 2.0 is the per capita production of mutant bacteria, and 1.5 is the per capita production of wild type bacteria. There are two equilibria, at $p^* = 0$ and $p^* = 1$ (Section 1.10). The first corresponds to extinction of the mutant and the second to extinction of the wild type. By writing the updating function as

$$f(p) = \frac{2.0p}{2.0p + 1.5(1 - p)},$$

we computed the derivative in Example 2.6.4, finding

$$f'(p) = \frac{3.0}{[2.0p + 1.5(1 - p)]^2}.$$

At the equilibrium $p = 0$,

$$f'(0) = \frac{3.0}{[2.0 \cdot 0 + 1.5(1 - 0)]^2} = \frac{3.0}{2.25} = 1.333,$$

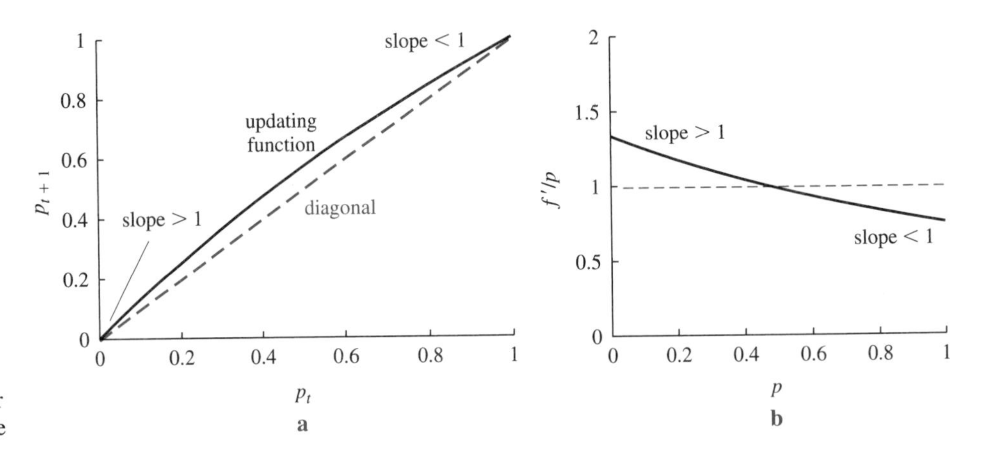

FIGURE 3.1.11
The discrete-time dynamical system for the fraction of mutants and its derivative

which is greater than 1. This equilibrium is unstable. At the equilibrium $p = 1$,

$$f'(1) = \frac{3.0}{[2.0 \cdot 1 + 1.5(1 - 1)]^2} = \frac{3.0}{4.0} = 0.75,$$

which is less than 1. This equilibrium is stable (Figure 3.1.11). Biologically, any population that begins with a positive fraction of mutants will be taken over by mutants.

Example 3.1.5 Using the Derivative to Check Stability: The Selection Model with Inferior Mutants

In the case with $s = 1.5$ and $r = 2.0$, the updating function is

$$f(p) = \frac{1.5p}{1.5p + 2.0(1 - p)}.$$

The derivative (as in Example 2.6.4) is

$$\begin{aligned}\frac{df}{dp} &= \frac{1.5p + 2.0(1-p)\frac{d(1.5p)}{dp} - 1.5p\frac{d[1.5p+2.0(1-p)]}{dp}}{[1.5p + 2.0(1 - p)]^2} \\ &= \frac{1.5p + 2.0(1 - p) \cdot 1.5 + 1.5p \cdot 0.5}{[1.5p + 2.0(1 - p)]^2} \\ &= \frac{3.0}{[1.5p + 2.0(1 - p)]^2}.\end{aligned}$$

With these parameter values, $f'(0) = 0.75 < 1$ and $f'(1) = 1.333 > 1$. The $p = 0$ equilibrium is stable and the $p = 1$ equilibrium is unstable. When the wild type has an advantage ($r > s$), mutants cannot invade.

Example 3.1.6 Stability of the Medication Discrete-Time Dynamical System

Consider the discrete-time dynamical system for medication concentration in the bloodstream,

$$M_{t+1} = 0.5M_t + 1.0,$$

describing a patient who is administered 1.0 unit of medication each day but uses up half of the previous amount (Example 1.5.4). The equilibrium concentration is 2.0. Because the updating function is linear with a slope of $0.5 < 1$, this equilibrium is stable (Figure 3.1.12).

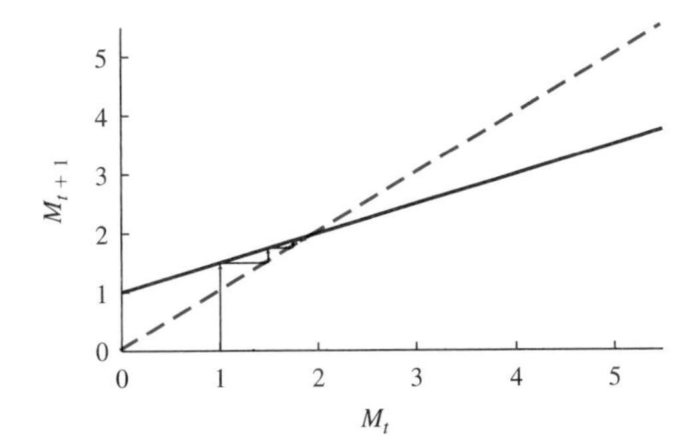

FIGURE 3.1.12
Stability of the medication discrete-time dynamical system

Example 3.1.7 Stability of the Medication Discrete-Time Dynamical System If Run Backwards

What happens if we run this system backwards in time (Subsection 1.2.4)? The backwards discrete-time dynamical system can be found by solving for M_t as

$$M_t = 2.0(M_{t+1} - 1) = 2.0M_{t+1} - 2.0.$$

The backwards updating function is

$$B(M) = 2.0M - 2.0.$$

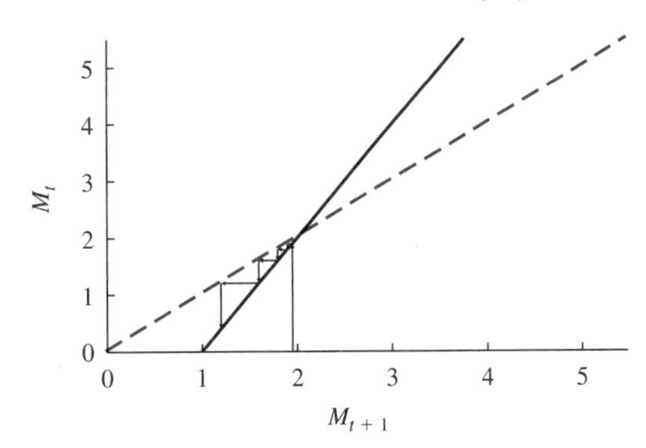

FIGURE 3.1.13
Stability of the medication discrete-time dynamical system if run backwards

It shares the equilibrium at 2.0 (Figure 3.1.13). However, because the backwards updating function is the mirror image, the slope at the equilibrium is the reciprocal of 0.5, or 2.0, because

$$B'(M) = 2.0.$$

An equilibrium that is stable when time runs forward is *unstable* when time runs backwards.

Example 3.1.8 Stability of Limited Population

Consider a population where the per capita production is a decreasing function of the population size x according to

$$\text{per capita production} = \frac{2.0}{1 + 0.001x}.$$

The per capita production is 2.0 when $x = 0$ and decreases as x becomes larger (Figure 3.1.14a). For example, when $x = 500$,

$$\text{per capita production} = \frac{2.0}{1 + 0.001 \cdot 500} = 1.33.$$

Using the fact that the updated population is the per capita production (offspring per individual) times the old population (the number of individuals), the discrete-time dynamical system for this population is

$$x_{t+1} = (\text{per capita production}) \times x_t$$
$$x_{t+1} = \frac{2.0x_t}{1 + 0.001x_t}$$

(Figure 3.1.14b).

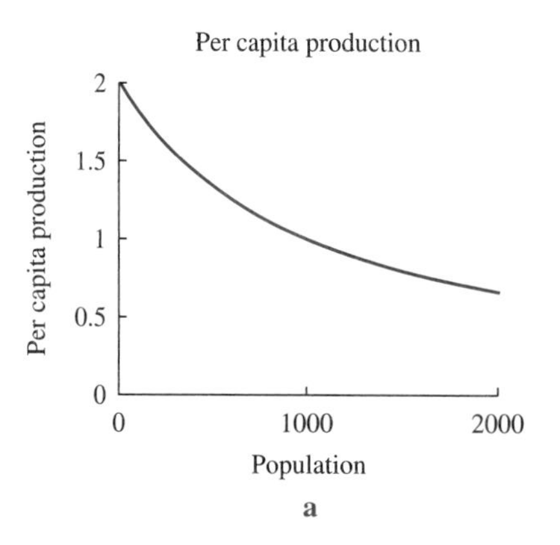

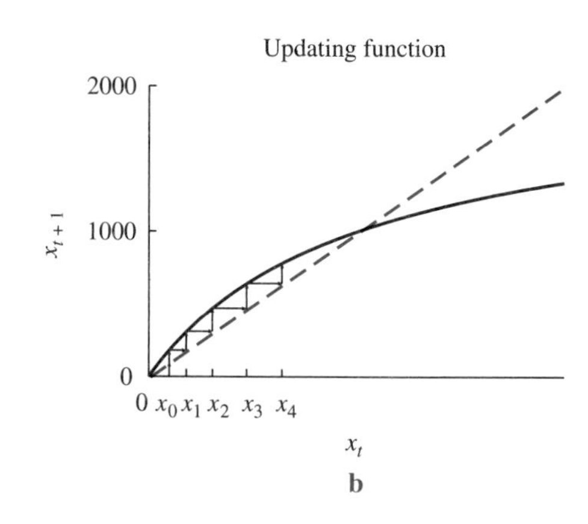

FIGURE 3.1.14
A discrete-time dynamical system describing competition

To find the equilibria, we follow Algorithm 1.5.

$$x^* = \frac{2.0x^*}{1 + 0.001x^*} \qquad \text{the original equation}$$

$$x^* - \frac{2.0x^*}{1+0.001x^*} = 0 \qquad \text{subtract to get unknowns on one side}$$

$$(1+0.001x^*)x^* - 2.0x^* = 0 \qquad \text{multiply both sides by } 1+0.001x^*$$

$$x^*(1+0.0001x^* - 2.0) = 0 \qquad \text{factor}$$

$$x^* = 0 \quad \text{or} \quad 1+0.001x^* - 2.0 = 0 \qquad \text{set both factors to 0}$$

$$x^* = 0 \quad \text{or} \quad 0.001x^* = 1 \qquad \text{begin isolating } x^*$$

$$x^* = 0 \quad \text{or} \quad x^* = 1000. \qquad \text{solve the second equation}$$

The equilibrium $x^* = 0$ corresponds to extinction. The equilibrium $x^* = 1000$ is the point where per capita production has been so reduced by crowding that the population can only break even.

The equilibrium at $x^* = 0$ appears to be unstable and the equilibrium at $x^* = 1000$ appears to be stable (Figure 3.1.14b). In accord with the Graphical Criterion for stability, the graph of the updating function crosses the diagonal from below to above at $x^* = 0$ and from above to below at $x^* = 1000$. We therefore expect the derivative of the updating function to be greater than 1 at $x^* = 0$ and less than 1 at $x^* = 1000$.

In functional notation, the updating function is

$$f(x) = \frac{2.0x}{1+0.001x}.$$

The derivative of the updating function, found with the quotient rule, is

$$\begin{aligned} f'(x) &= \frac{(1+0.001x)\frac{d(2.0x)}{dx} - 2.0x\frac{d(1+0.001x)}{dx}}{(1+0.001x)^2} \\ &= \frac{2.0(1+0.001x) - 2.0x \cdot 0.001}{(1+0.001x)^2} \\ &= \frac{2.0}{(1+0.001x)^2}. \end{aligned}$$

Evaluated at the equilibria,

$$f'(0) = \frac{2.0}{(1+0.001 \cdot 0)^2} = 2.0$$

$$f'(1000) = \frac{2.0}{(1+0.001 \cdot 1000)^2} = 0.5.$$

As indicated by the graph, the equilibrium at $x = 0$ is unstable and the equilibrium at $x = 1000$ is stable. If reduced to a low level, this population would increase back toward the stable equilibrium at $x^* = 1000$. If artificially increased to a high level, the population would decrease toward the stable equilibrium at $x^* = 1000$.

Example 3.1.9 Stability of Limited Population with Lower Production

Consider a variant of the model in Example 3.1.8 where the per capita production is a decreasing function of the population size x according to

$$\text{per capita production} = \frac{1.0}{1 + 0.001x}.$$

The per capita production is 1.0 when $x = 0$ and decreases as x becomes larger (Figure 3.1.15a). The discrete-time dynamical system for this population is

$$x_{t+1} = (\text{per capita production}) \cdot x_t$$
$$x_{t+1} = \frac{x_t}{1 + 0.001x_t}$$

(Figure 3.1.15b).

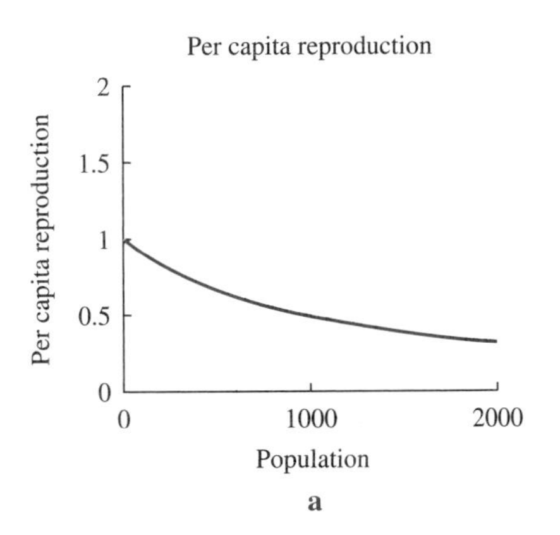

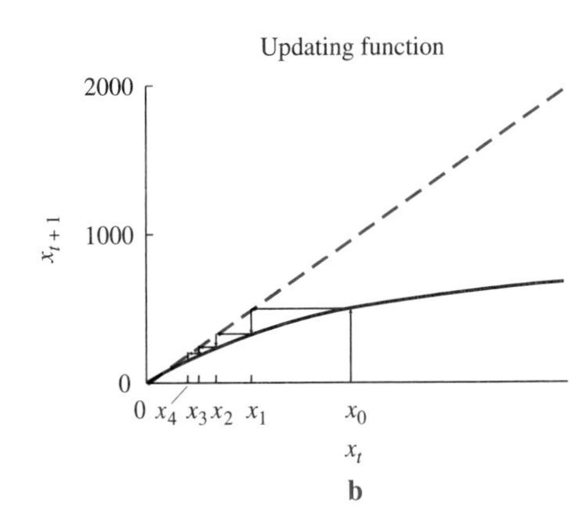

FIGURE 3.1.15
A discrete-time dynamical system with reduced per capita production

To find the equilibria, we follow Algorithm 1.5.

$$x^* = \frac{x^*}{1 + 0.001x^*} \quad \text{the original equation}$$
$$x^* - \frac{x^*}{1 + 0.001x^*} = 0 \quad \text{subtract to get unknowns on one side}$$
$$(1 + 0.001x^*)x^* - x^* = 0 \quad \text{multiply both sides by } 1 + 0.001x^*$$
$$x^*(1 + 0.0001x^* - 1) = 0 \quad \text{factor}$$
$$x^* \cdot 0.0001x^* = 0. \quad \text{simplify}$$

This has only a single equilibrium at $x^* = 0$, corresponding to extinction.

In functional notation, the updating function is

$$f(x) = \frac{x}{1 + 0.001x}.$$

The derivative of the updating function, found with the quotient rule (Subsection 2.6.3) is

$$f'(x) = \frac{(1 + 0.001x)\frac{dx}{dx} - x\frac{d(1+0.001x)}{dx}}{(1 + 0.001x)^2}$$
$$= \frac{(1 + 0.001x) - x \cdot 0.001}{(1 + 0.001x)^2}$$
$$= \frac{1}{(1 + 0.001x)^2}.$$

Evaluated at the equilibria,

$$f'(0) = \frac{1}{(1 + 0.001 \cdot 0)^2} = 1.0.$$

The updating function is tangent to the diagonal at the equilibrium at $x = 0$, meaning that we cannot determine the stability from the slope criterion. By more carefully examining the function, we see that the graph of the updating function lies below the diagonal for $x > 0$. Algebraically,

$$f(x) = \frac{x}{1 + 0.001x} < x$$

because the denominator is greater than 1 for $x > 0$. Therefore, any positive population decreases, and the equilibrium at $x = 0$ is stable.

Summary We used cobwebbing and logic to derive two criteria for the stability of equilibria. When the updating function crosses the diagonal from below to above, initial conditions both above and below the equilibrium are pushed away, making the equilibrium unstable (Graphical Criterion for stability). For the updating function to cross from below to above, the slope at the equilibrium must be greater than 1 (Slope Criterion for stability). The opposite occurs at stable equilibria where the updating function crosses from above to below and has slope less than 1. In the special case that the slope is exactly equal to 1, the equilibrium might be neither stable nor unstable.

3.1 Exercises

Mathematical Techniques

1–4 ▪ Find the equilibria of the following discrete-time dynamical systems from their graphs and apply the Graphical Criterion for stability to find which are stable. Check by cobwebbing.

1.

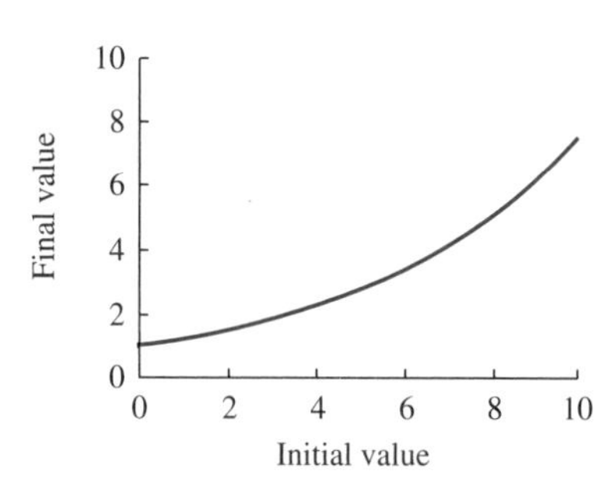

2.

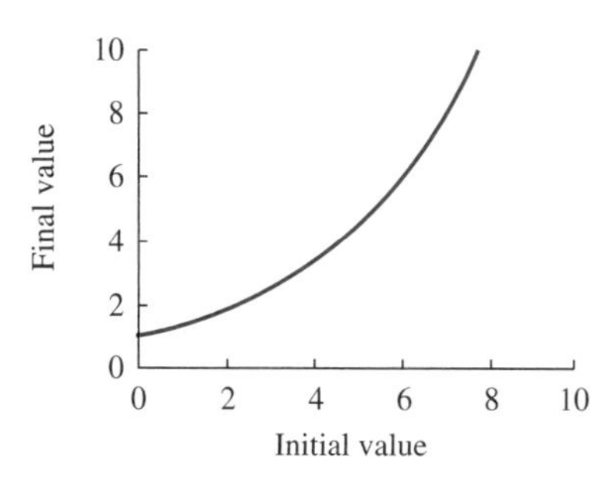

3.

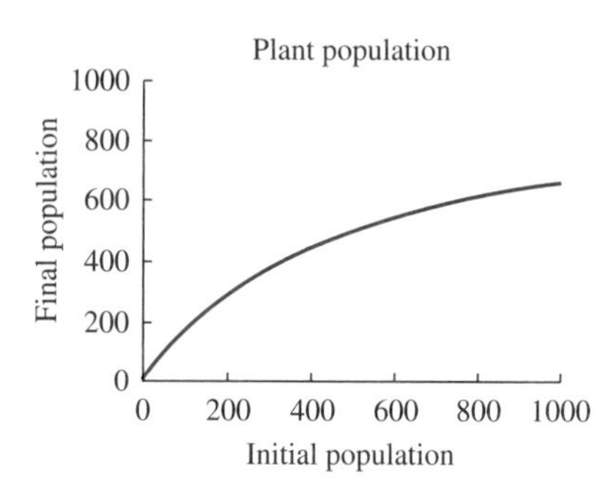

4.

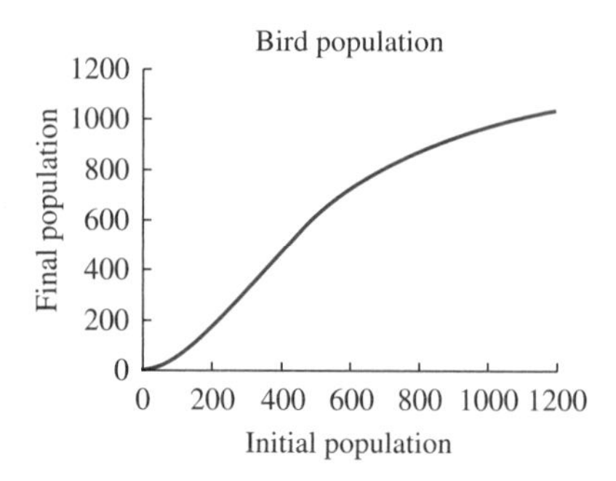

5–12 ▪ Graph the following discrete-time dynamical systems, find the equilibria algebraically, and check whether the stability derived from the Slope Criterion for stability matches that found with cobwebbing.

5. $c_{t+1} = 0.5c_t + 8.0$, for $0 \le c_t \le 30$ (from Section 1.6, Exercise 19).

6. $b_{t+1} = 3b_t$, for $0 \le b_t \le 10$ (from Section 1.6, Exercise 20).

7. $b_{t+1} = 0.3b_t$, for $0 \le b_t \le 10$ (from Section 1.6, Exercise 21).

8. $b_{t+1} = 2.0b_t - 5.0$, for $0 \le b_t \le 10$ (from Section 1.6, Exercise 22).

9. $f(x) = x^2$ for $0 \le x \le 2$ (from Section 1.6, Exercise 17).

10. $g(y) = y^2 - 1$ for $0 \le y \le 2$ (from Section 1.6, Exercise 18).

11. $x_{t+1} = \dfrac{x_t}{1 + x_t}$ (from Section 1.6, Exercise 29).

12. $x_{t+1} = -1 + 4x_t - 3x_t^2 + x_t^3$ (the only equilibrium is $x^* = 1$).

13–18 ▪ The unusual equilibrium in the text has an updating function that lies below the diagonal both to the left and to the right of the equilibrium. There are several other ways that an updating function can be tangent to the diagonal at an equilibrium. In each case, cobweb starting from points to the left and to the right of the equilibrium, and describe the stability.

13. Graph an updating function that lies above the diagonal both to the left and to the right of an equilibrium.

14. Graph an updating function that is tangent to the diagonal at an equilibrium but crosses from below to above. Show by cobwebbing that the equilibrium is unstable. What is the second derivative at the equilibrium?

15. Graph an updating function that is tangent to the diagonal at an equilibrium but crosses from above to below. Show by cobwebbing that the equilibrium is stable. What is the second derivative at the equilibrium?

16. Sketch the graph of an updating function that has a corner at an equilibrium and is stable.

17. Sketch the graph of an updating function that has a corner at an equilibrium and is unstable.

18. Sketch the graph of an updating function that has a corner at an equilibrium and is neither stable nor unstable.

19–20 ▪ Another peculiarity of an updating function that is tangent to the diagonal at an equilibrium is that slight changes in the graph can produce big changes in the number of equilibria. The following are based on Figure 3.1.10.

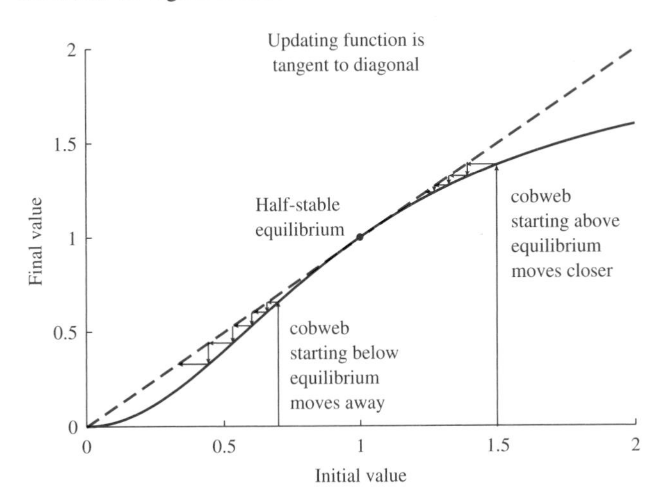

19. Move the curve slightly down (while keeping the diagonal in the same place). How many equilibria are there now? What happens when you cobweb starting from a point at the right-hand edge of the figure?

20. Move the curve slightly up (again keeping the diagonal in the same place). How many equilibria are there? Describe their stability.

21–22 ▪ Find the inverse of each of the following updating functions, and compute the slope of both the original updating function and the derivative at the equilibrium.

21. The updating function $f(x) = \dfrac{x}{1+x}$ (as in Section 3.1, Exercise 21).

22. The updating function $f(x) = \dfrac{x}{x-1}$ (as in Section 3.1, Exercise 22).

Applications

23–26 ▪ Recall the updating function with the fraction p of mutant bacteria given by

$$f(p) = \frac{sp}{sp + r(1-p)},$$

where s is the per capita production of the mutant and r is the per capita production of the wild type. Find the derivative in the following cases, and evaluate at the equilibria $p = 0$ and $p = 1$. Are the equilibria stable?

23. $s = 1.2$, $r = 2.0$

24. $s = 3.0$, $r = 1.2$

25. $s = 1.5$, $r = 1.5$

26. In general (without substituting numerical values for s and r), can both equilibria be stable? What happens if $r = s$?

27–28 ▪ Find the equilibrium population of bacteria in the following cases with supplementation. Graph the updating function for each, and use the Slope Criterion for stability to check the stability.

27. A population of bacteria has per capita production $r = 0.6$ and 1.0×10^6 bacteria are added each generation (as in Section 1.9, Exercise 35).

28. A population of bacteria has per capita production $r = 0.2$ and 5.0×10^6 bacteria are added each generation (as in Section 1.9, Exercise 36).

29–30 ▪ A lab is growing and harvesting a culture of valuable bacteria described by the discrete-time dynamical system

$$b_{t+1} = rb_t - h.$$

The bacteria have per capita production r, and h are harvested each generation (as in Section 1.9, Exercises 49 and 50). Graph the updating function for each, and use the Slope Criterion for stability to check the stability.

29. Suppose that $r = 1.5$ and $h = 1.0 \times 10^6$ bacteria.

30. Without setting r and h to particular values, find the equilibrium algebraically. When is the equilibrium stable?

31–32 ▪ The model describing the dynamics of the concentration of medication in the bloodstream,

$$M_{t+1} = 0.5M_t + 1.0,$$

becomes nonlinear if the fraction of medication used is a function of the concentration (as in Section 1.10, Exercise 39). In each case, use the Slope Criterion for stability to check the stability of the equilibrium.

31. The nonlinear discrete-time dynamical system

$$M_{t+1} = M_t - \frac{0.5}{1.0 + 0.1M_t}M_t + 1.0.$$

32. The nonlinear discrete-time dynamical system

$$M_{t+1} = M_t - \frac{1.0}{1.0 + 0.1M_t}M_t + 1.0.$$

How does this differ from the model in Exercise 31? Why is the equilibrium smaller?

33–36 ▪ An equilibrium that is stable when time goes forward should be unstable when time goes backward. Find the inverses of the updating functions associated with the following discrete-time dynamical systems, and find the derivative at the equilibria.

33. $c_{t+1} = 0.5c_t + 8.0$, for $0 \le c_t \le 30$ (as in Exercise 5).

34. $b_{t+1} = 3b_t$ (as in Exercise 6).

35. $x_{t+1} = \dfrac{2x_t}{1 + 0.001x_t}$ (Example 3.1.8).

36. $p_{t+1} = \dfrac{2p_t}{2p_t + (1 - p_t)}$.

37–38 ▪ Consider a population x_t with per capita production of $r\dfrac{x_t}{1.0 + x_t^2}$. After writing the discrete-time dynamical system, compute the following for the given values of the parameter r.

a. Find the equilibria.

b. Graph the updating function.

c. Indicate which equilibria are stable and which are unstable, and check with the Slope Criterion for stability.

d. Describe in words how the population would behave.

37. $r = 1.0$

38. $r = 2.5$

39–42 ▪ Consider the discrete-time dynamical system for a heart studied in Section 1.11.

$$V_{t+1} = \begin{cases} cV_t & \text{if } cV_t > V_c \\ cV_t + u & \text{if } cV_t \le V_c. \end{cases}$$

In each case, sketch the updating function. Why is the equilibrium stable when it exists?

39. $V_c = 20.0$ millivolts, $u = 10.0$ millivolts, $c = 0.5$.

40. $V_c = 20.0$ millivolts, $u = 10.0$ millivolts, $c = 0.6$.

41. $V_c = 20.0$ millivolts, $u = 10.0$ millivolts, $c = 0.7$.

42. $V_c = 20.0$ millivolts, $u = 10.0$ millivolts, $c = 0.8$.

Computer Exercises

43. Consider the discrete-time dynamical system found in Section 1.10, Exercise 49. In that exercise, there were two cultures, 1 and 2. In culture 1, the mutant does better than the wild type, and in culture 2 the wild type does better. In particular, suppose that $s = 2.0$ and $r = 0.3$ in culture 1, and that $s = 0.6$ and $r = 2.0$ in culture 2. Define updating functions f_1 and f_2 to describe the dynamics in the two cultures. The overall updating function after mixing equal amounts from the two is

$$f(p) = \frac{f_1(p) + f_2(p)}{2}.$$

a. Write the updating function explicitly.

b. Find the equilibria.

c. Find the derivative of the updating function.

d. Evaluate the stability of the equilibrium.

44. Consider the discrete-time dynamical systems with the following discrete-time dynamical systems. Check the stability of the equilibria.

a. $x_{t+1} = \cos(x_t)$

b. $y_{t+1} = \sin(y_t)$

c. $z_{t+1} = \sin(z_t) + \cos(z_t)$

45. Consider the discrete-time dynamical system

$$z_{t+1} = a^{z_t}$$

(thanks to Larry Okun). We will study this for different values of a.

a. Follow the dynamics starting from initial condition $z_0 = 1.0$ for $a = 1.0, 1.1, 1.2, 1.3, 1.4$. Keep running the system until it seems to reach an equilibrium.

b. Do the same, but increase a slowly past a critical value of approximately 1.4446679. Solutions should creep up for a while, and then increase very quickly. Graph the updating function for values above and below the critical value and try to explain why.

c. At the critical value, the slope of the updating function is 1 at the equilibrium. Show that this occurs with $a = e^{1/e}$. What is the equilibrium?

46. Consider a population with

$$\text{per capita production } = r\frac{x_t}{1.0 + x_t^2}.$$

Start with $r = 3.0$ and find the equilibrium. Then try smaller and smaller values of r and track what happens to the equilibrium, using the previous equilibrium as a starting point. What happens when r crosses 2? Follow the equilibrium population down to $r = 1$. If this were a real population and r was a measure of the quality of habitat, how would you interpret this behavior? Then do the same but start with $r = 1$ and increase r up to 3. Can you explain what is going on?

3.2 More Complicated Dynamics

The derivative or slope of an updating function determines whether an equilibrium is stable or unstable. The Slope Criterion for stability tells how to assess stability by checking whether the slope is greater than or less than 1. We have yet, however, to consider cases where the slope of the updating function is **negative** at an equilibrium. By studying several such cases, we will see that rather exotic behaviors are possible. Using the idea of **qualitative dynamical systems**, where we approximate a nonlinear discrete-time dynamical system with its **tangent line approximation** at an equilibrium, we will find a general condition that includes these new cases.

3.2.1 The Logistic Dynamical System

In the previous section, we looked at a model where large population size reduced per capita production according to

$$\text{per capita production} = \frac{2.0}{1+0.001x_t}.$$

(Example 3.1.8). Another widely studied model is the **logistic dynamical system** (introduced in Section 1.10, Exercise 43). In this model, the per capita production of a population decreases linearly with population size according to

$$\text{per capita production} = r\left(1-\frac{N_t}{K}\right),$$

where N_t represents population size. The parameter r is the greatest possible production, and K is the maximum possible population. For populations greater than K, the per capita production is assumed to be 0. Using the fact that the new population is the per capita production times the old population, the discrete-time dynamical system for the logistic dynamical system is

$$N_{t+1} = r\left(1-\frac{N_t}{K}\right)N_t.$$

To make the algebra simpler, we define the new variable

$$x_t = \frac{N_t}{K}$$

to represent the fraction of the maximum possible population. If K is 1000, a population of $N_t = 500$ corresponds to the fraction $x_t = 0.5$. We can write the discrete-time dynamical system for the x_t as

$$\begin{aligned} x_{t+1} &= \frac{N_{t+1}}{K} \\ &= \frac{r\left(1-\frac{N_t}{K}\right)N_t}{K} \\ &= r(1-x_t)x_t. \end{aligned}$$

The factors on the right-hand side are usually written in a different order, giving the logistic dynamical system

$$x_{t+1} = rx_t(1-x_t). \tag{3.2.1}$$

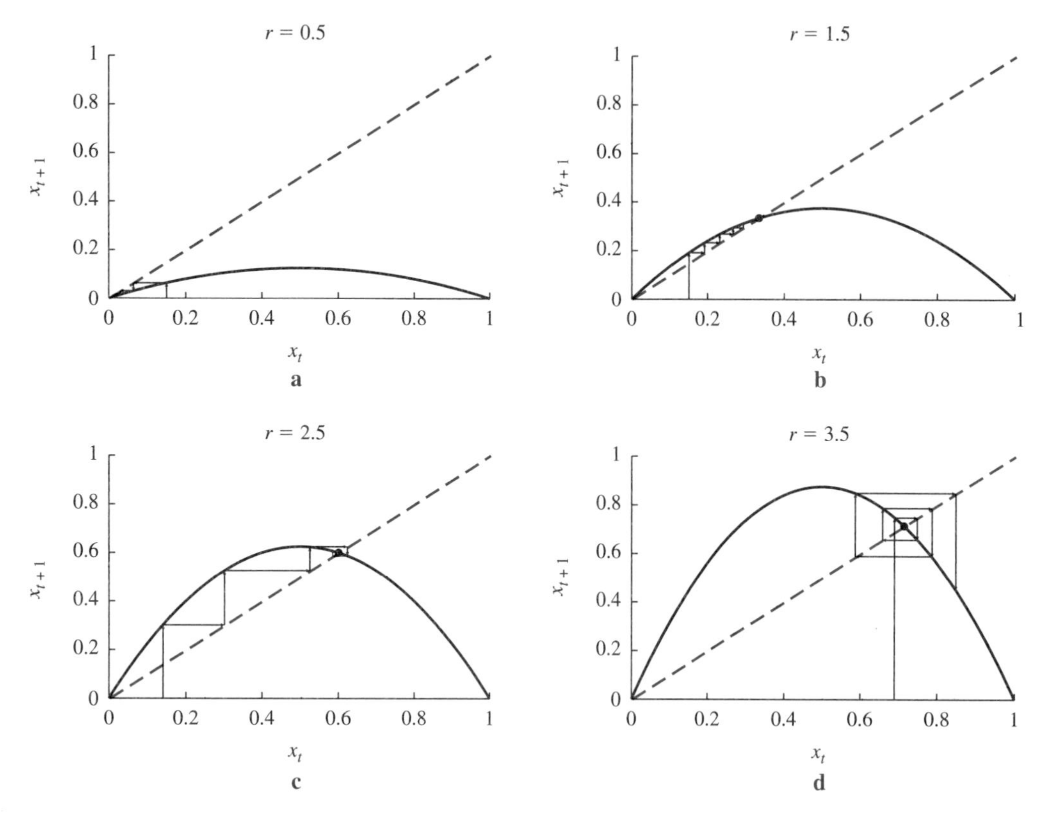

FIGURE 3.2.1
The behavior of the logistic dynamical system from simulations

Graphs and cobwebs of this dynamical system for several values of r are shown in Figure 3.2.1.

We begin by finding the equilibria by setting $x_t = x_{t+1} = x^*$ and solving with Algorithm 1.5.

$x^* = rx^*(1 - x^*)$	the original equation
$x^* - rx^*(1 - x^*) = 0$	place unknowns on one side<
$x^*[1 - r(1 - x^*)] = 0$	factor
$x^* = 0 \quad \text{or} \quad 1 - r(1 - x^*) = 0$	set each factor equal to 0
$x^* = 0 \quad \text{or} \quad x^* = 1 - \frac{1}{r}.$	solve each piece

Do these solutions make sense? The first, $x^* = 0$, is the extinction equilibrium that shows up in all population models without immigration. The second equilibrium depends on the maximum per capita production r. If $r < 1$, this equilibrium is negative and biologically impossible. A population with a **maximum** per capita production less than 1 cannot replace itself and will go extinct. For larger values of r, the equilibrium becomes larger, consistent with the fact that a population with higher potential production grows to a larger value.

We will examine four cases: $r = 0.5$, $r = 1.5$, $r = 2.5$, and $r = 3.5$. The equilibria and their behavior as deduced from cobwebbing are summarized in the following table.

r	x^*	Stability
0.5	0	Stable
1.5	0	Unstable
1.5	$1 - \frac{1}{1.5} \approx 0.333$	Stable
2.5	0	Unstable
2.5	$1 - \frac{1}{2.5} = 0.600$	Stable (oscillates)
3.5	0	Unstable
3.5	$1 - \frac{1}{3.5} \approx 0.714$	Unstable (oscillates)

The results match the Graphical Criterion for stability pretty well until $r = 3.5$ (Figure 3.2.1d). The updating function crosses the diagonal from above to below at the positive equilibrium, so we expect this equilibrium to be stable. However, a solution starting near the positive equilibrium $x^* \approx 0.714$ moves away, and does so by jumping back and forth. Closer examination of the case with $r = 2.5$ (Figure 3.2.1c) reveals that solutions jump back and forth as they approach the equilibrium at $x^* = 0.6$. What is going on?

3.2.2 Qualitative Dynamical Systems

An equilibrium where the updating function has a positive slope is stable precisely when that slope is less than 1. These results match the behavior of the bacterial growth model

$$b_{t+1} = rb_t$$

at the $b^* = 0$ equilibrium, which is stable when $r < 1$.

What happens if $r < 0$? Of course, this case does not make biological sense, but it will provide insight into what happens near equilibria with negative slopes. Why should this work? Remember the central trick in calculus: replacing problems about curves with simpler problems about lines.

Consider a nonlinear discrete-time dynamical system $x_{t+1} = f(x_t)$ and its tangent line $\hat{f}(x_t)$ at an equilibrium x^* (Figure 3.2.2). Because the graph of $\hat{f}(x_t)$ is the tangent line at the equilibrium x^* of the original system, the two systems share the equilibrium at x^* and have the same slope. According to the Slope Criterion for stability, the equilibrium of $\hat{f}$ should be stable precisely when the equilibrium of f is. Furthermore, solutions remain very close to each other (Figure 3.2.2c). Starting from a point near x^*, the solutions are nearly identical because the tangent line lies so close to the curve.

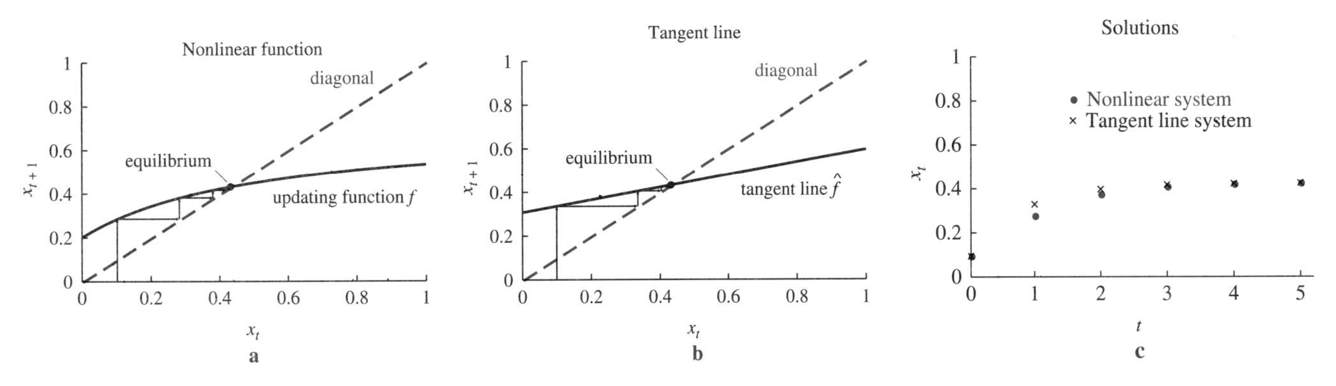

FIGURE 3.2.2
Comparing an updating function with its tangent line

The original updating function and the tangent line are not exactly equal. The exact values in the solution will therefore also be different. When we ask about stability, however, we are concerned with general behavior rather than exact values. Studying general aspects of dynamical systems without requiring exact measurements is the realm of **qualitative dynamical systems**. The approach is essential in biology, where measurements often include a great deal of noise. For example, we neither know nor believe that the updating function for the lung model (Section 1.9) is *exactly* linear. Nonetheless, we would like to use the linear model to make predictions. Some predictions are **qualitative**, verbal descriptions of behavior such as "the concentration of a chemical in the lung will approach an equilibrium." Others are **quantitative**, numerical descriptions of behavior such as "the concentration of a chemical in the lung will reach 5.23 mmol/L after seven breaths."

The qualitative theory of dynamical systems requires comparing the dynamics produced by similar discrete-time dynamical systems. If two nearly indistinguishable discrete-time dynamical systems produce qualitatively different dynamics, the underlying biological system might be highly sensitive to small changes in conditions. Such situations do occur. In this chapter, we restrict attention to cases where similar discrete-time dynamical systems produce qualitatively similar dynamics. Other cases are treated in more advanced texts on **bifurcation theory**. We will meet some of the more important bifurcations in the context of differential equations in Section 5.3, Exercises 17–20.

Using this philosophy, we can try to figure out what happens at an equilibrium where the slope is negative by studying the linear bacterial growth model

$$b_{t+1} = rb_t$$

with negative values for the per capita production r.

Example 3.2.1 The Consequences of "Negative" Per Capita Production: $r = -0.5$

If $r = -0.5$, the discrete-time dynamical system is

$$b_{t+1} = -0.5b_t$$

(Figure 3.2.3). Starting from an initial condition of $b_0 = 0.5$, the solution is

$$b_1 = -0.25$$
$$b_2 = 0.125$$
$$b_3 = -0.0625$$
$$b_4 = 0.03125.$$

The absolute value decreases while the sign switches back and forth.

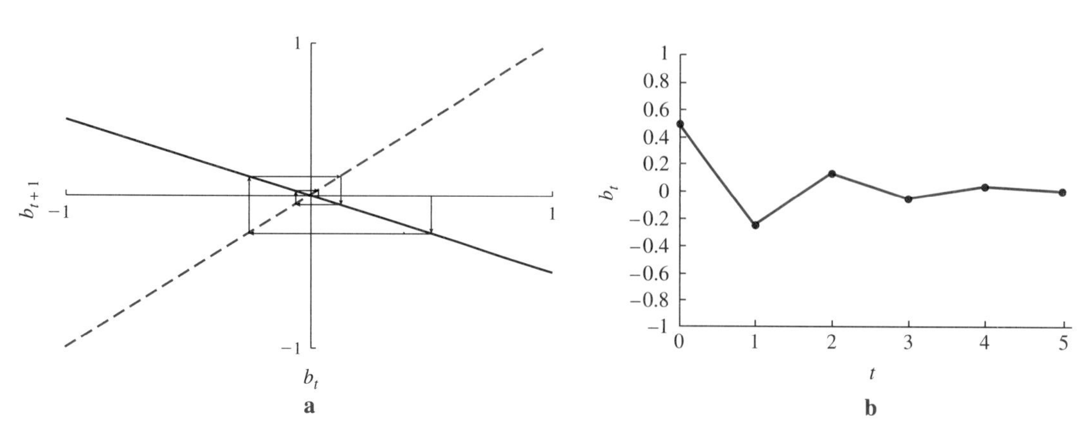

FIGURE 3.2.3

Cobwebbing and solutions with $r = -0.5$

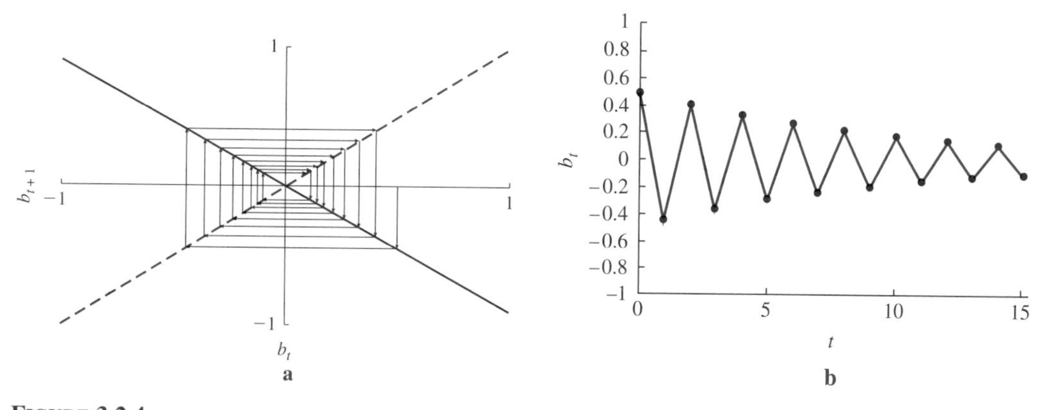

FIGURE 3.2.4
Cobwebbing and solutions with $r = -0.9$

Example 3.2.2 The Consequences of "Negative" Per Capita Production: $r = -0.9$

If $r = -0.9$, the discrete-time dynamical system is

$$b_{t+1} = -0.9 b_t.$$

(Figure 3.2.4). Starting from an initial condition of $b_0 = 0.5$, the solution is

$$b_1 = -0.45$$
$$b_2 = 0.405$$
$$b_3 = -0.3645$$
$$b_4 = 0.32805.$$

The absolute value decreases slowly while the sign switches back and forth.

Example 3.2.3 The Consequences of "Negative" Per Capita Production: $r = -2.0$

With $r = -2.0$ and an initial condition of $b_0 = 0.02$, the solution is

$$b_1 = -0.04$$
$$b_2 = 0.08$$
$$b_3 = -0.16$$
$$b_4 = 0.32.$$

(Figure 3.2.5). Again, the sign switches back and forth, but the absolute value now increases.

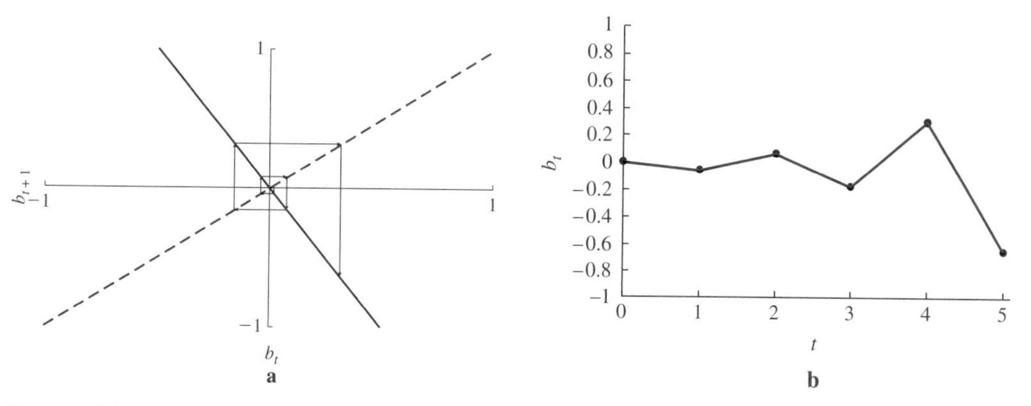

FIGURE 3.2.5
Cobwebbing and solutions with $r = -2.0$

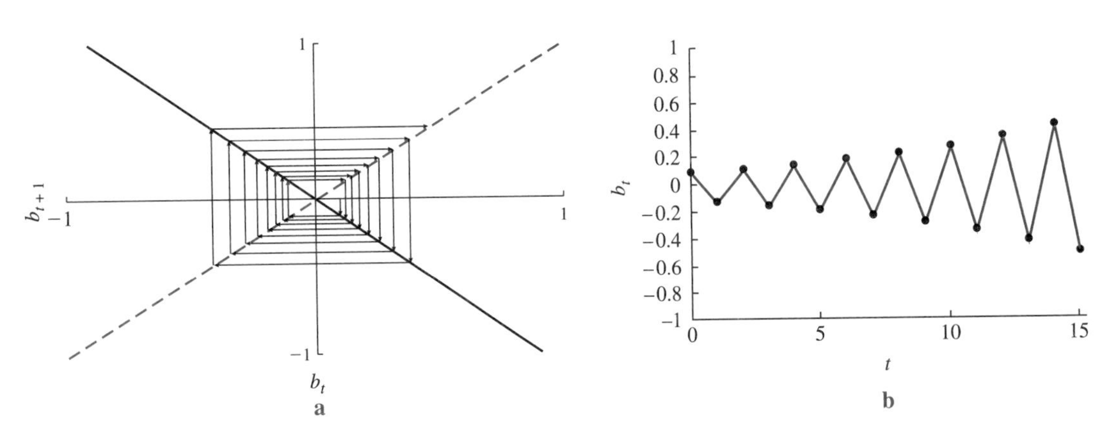

FIGURE 3.2.6
Cobwebbing and solutions with $r = -1.1$

Example 3.2.4 The Consequences of "Negative" Per Capita Production: $r = -1.1$

With $r = -1.1$ and an initial condition of $b_0 = 0.1$, the solution is

$$b_1 = -0.11$$
$$b_2 = 0.121$$
$$b_3 = -0.1331$$
$$b_4 = 0.14641.$$

(Figure 3.2.6). The sign switches back and forth, and the absolute value increases slowly. ▲

With negative r, the system is stable when $r > -1$ and unstable when $r < -1$. Putting this together with the results with positive r gives

Behavior of Solutions of $b_{t+1} = rb_t$

r	Stability	Behavior
$r > 1$	Unstable	Moves away from equilibrium
$0 < r < 1$	Stable	Moves toward equilibrium
$-1 < r < 0$	Stable	Oscillates toward equilibrium
$r < -1$	Unstable	Oscillates away from equilibrium

An equilibrium is stable if $r < 1$ in **absolute value** or $|r| < 1$, and unstable if $r > 1$ in absolute value or $|r| > 1$.

Because the behavior of a nonlinear dynamical system resembles that of its tangent line, we can extend these results to the following theorem.

Theorem 3.1 **Stability Theorem for Discrete-Time Dynamical Systems**

Suppose that the discrete-time dynamical system

$$x_{t+1} = f(x_t)$$

has an equilibrium at x^*. Let $f'(x)$ be the derivative of f with respect to x. The equilibrium at x^* is stable if

$$|f'(x^*)| < 1$$

and unstable if

$$|f'(x^*)| > 1.$$

The proof depends on the Mean Value Theorem (Section 3.4).

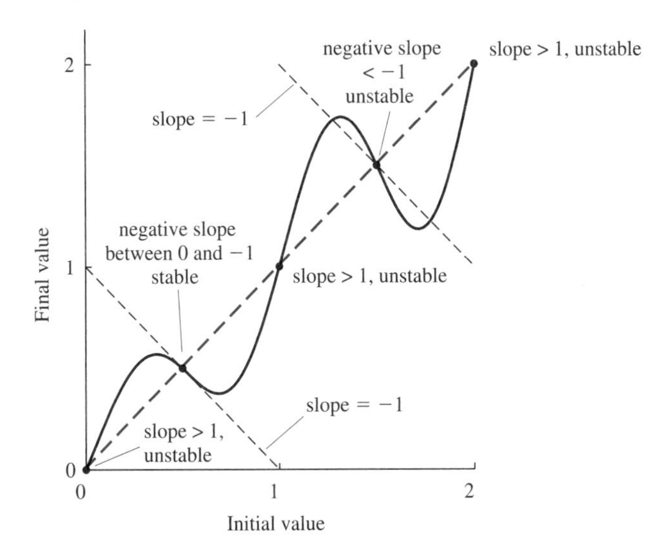

FIGURE 3.2.7
Applying the stability theorem for discrete-time dynamical systems

Using this method, we can read off the stability of equilibria of more complicated discrete-time dynamical systems (Figure 3.2.7). By including lines with slope -1 on our graph in addition to the diagonal, we can see that the equilibrium at $(0.5, 0.5)$ is stable because the slope is between 0 and -1, while the equilibrium at $(1.5, 1.5)$ is stable because the slope is less than -1.

3.2.3 Analysis of the Logistic Dynamical System

Do these results help make sense of the behavior of the logistic dynamical system? The derivative of the updating function is

$$\begin{aligned}
f'(x) &= \frac{d}{dx} rx(1-x) && \text{derivative of updating function} \\
&= r\left(\frac{dx}{dx}(1-x) + \frac{d(1-x)}{dx}x\right) && \text{constant product rule and product rule} \\
&= r\,[(1-x) - x] && \text{evaluate derivatives of linear functions} \\
&= r(1-2x). && \text{combine terms}
\end{aligned}$$

We next evaluate the derivative at the two equilibria. At $x^* = 0$,

$$f'(0) = r(1 - 2 \cdot 0) = r.$$

According to the stability theorem for discrete-time dynamical systems, the equilibrium at 0 is stable if $r < 1$ and unstable if $r > 1$. When $r > 1$, the derivative at the positive equilibrium $x^* = 1 - \frac{1}{r}$ is

$$\begin{aligned}
f'\left(1 - \frac{1}{r}\right) &= r\left[1 - 2\left(1 - \frac{1}{r}\right)\right] \\
&= r\left[1 - 2 + \frac{2}{r}\right)\right] \\
&= r - 2r + \frac{2r}{r} \\
&= r - 2r + 2 \\
&= 2 - r.
\end{aligned}$$

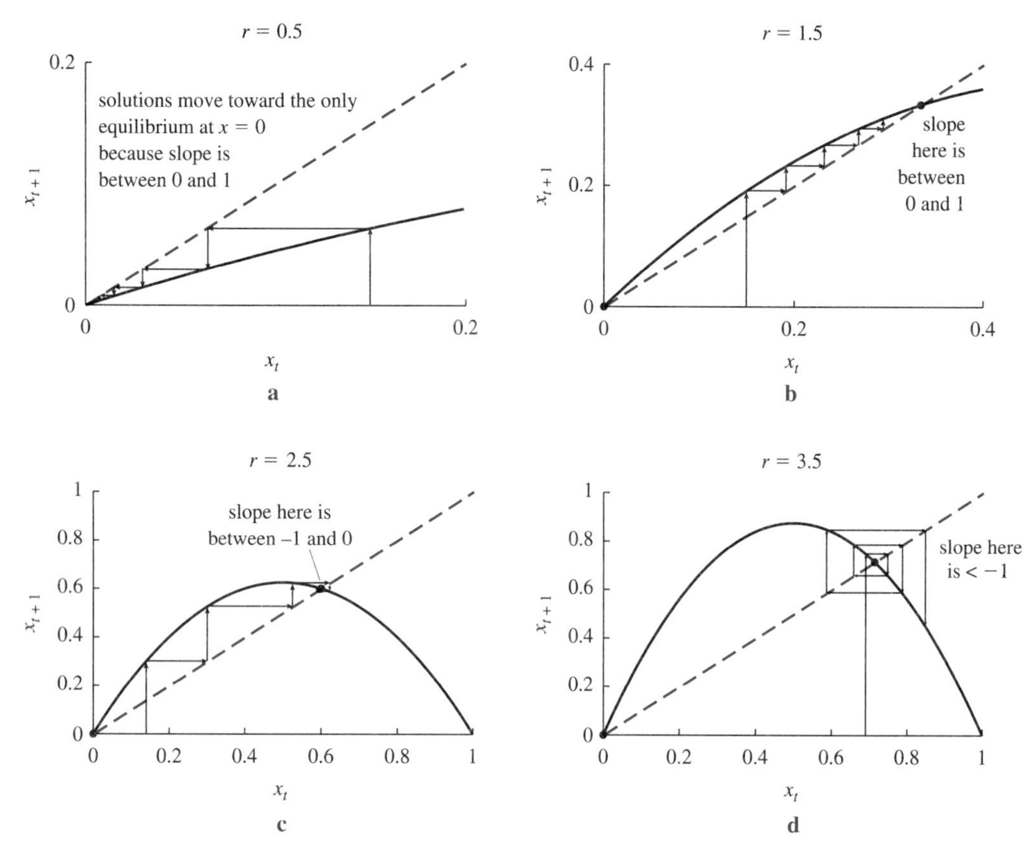

FIGURE 3.2.8
The behavior of the logistic dynamical system explained

The results are summarized in the following table and Figure 3.2.8.

r	x^*	$f'(x^*)$	Stability
0.5	0	0.5	Stable
1.5	0	1.5	Unstable
1.5	0.333	0.5	Stable
2.5	0	2.5	Unstable
2.5	0.600	−0.5	Stable (oscillates)
3.5	0	3.5	Unstable
3.5	0.714	−1.5	Unstable (oscillates)

When $r = 3.5$, neither equilibrium is stable. This population has nowhere to settle down and will continue jumping around indefinitely. Some of the interesting dynamics that can result are shown in Figure 3.2.9. The solutions of this simple discrete-time dynamical system for $3.5 < r \leq 4$ are extremely complicated, including examples of **chaos** (Section 1.10, Exercise 47).

A commonly used model for reproduction in fisheries is the **Ricker model**

$$x_{t+1} = r x_t e^{-x_t}$$

(introduced in Section 1.10, Exercise 45). The per capita production declines exponentially from a maximum of r as the population becomes larger. Like the logistic dynamical system, production decreases rapidly, but this model is a bit more realistic

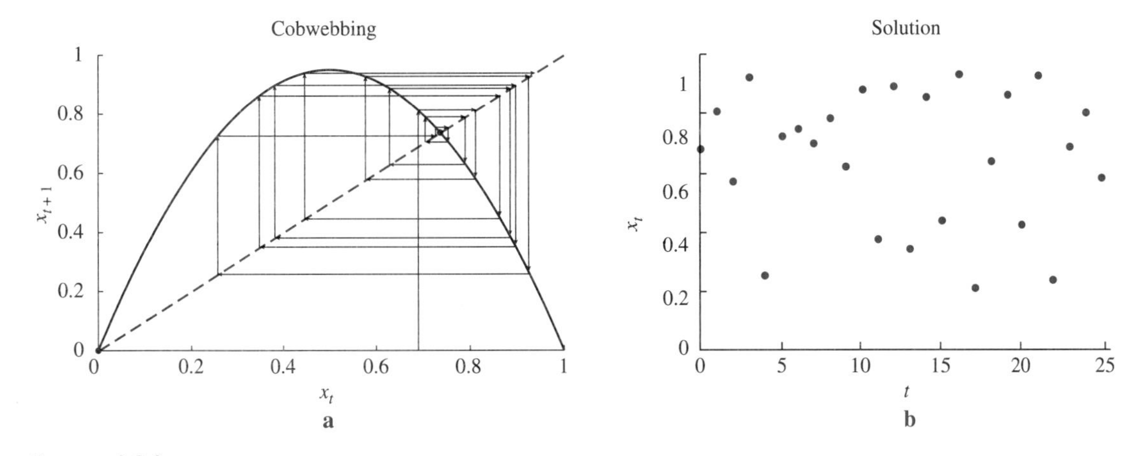

FIGURE 3.2.9

The long-term behavior of the logistic dynamical system with $r = 3.8$

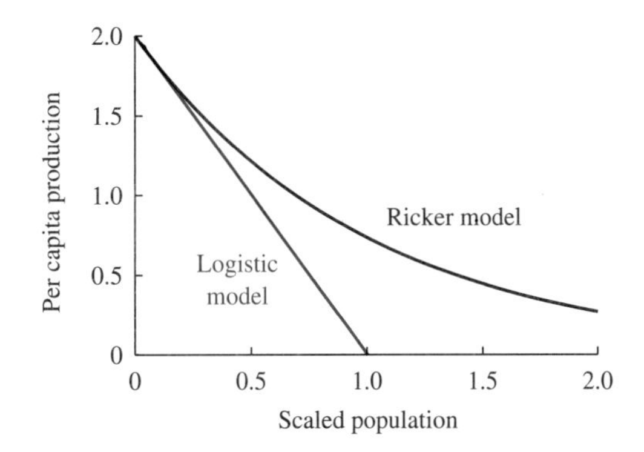

FIGURE 3.2.10

The per capita production in the Ricker and logistic models

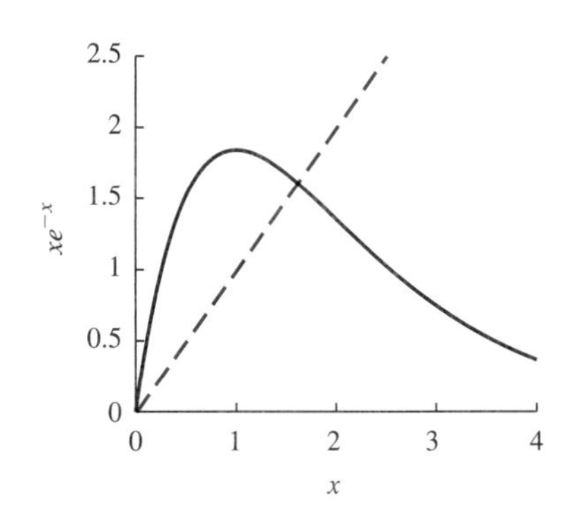

FIGURE 3.2.11

The Ricker discrete-time dynamical system with $r = 5.0$

because per capita production never reaches 0 (Figure 3.2.10). If we write the updating function as Figure 3.2.11

$$R(x) = rxe^{-x},$$

the first derivative is

$$R'(x) = r(1 - x)e^{-x}$$

(as in Section 2.8, Example 2.8.7).

Where are the equilibria of the Ricker model? We solve

$$\begin{aligned} x^* &= rx^*e^{-x^*} \\ x^* - rx^*e^{-x^*} &= 0 \\ x^*(1 - re^{-x^*}) &= 0 \\ x^* = 0 &\text{ or } 1 - re^{-x^*} = 0 \\ x^* = 0 &\text{ or } e^{x^*} = r \\ x^* = 0 &\text{ or } x^* = \ln(r). \end{aligned}$$

At the equilibrium $x = 0$, $R'(0) = r$. This equilibrium is stable when the maximum per capita production $r < 1$ (Figure 3.2.12a) and unstable otherwise. The equilibrium $x^* = \ln(r)$ is positive when $r > 1$. At this point, the derivative of the updating function is

$$\begin{aligned} R'(\ln(r)) &= r[1 - \ln(r)]e^{-\ln(r)} \\ &= r[1 - \ln(r)]\frac{1}{r} \\ &= 1 - \ln(r). \end{aligned}$$

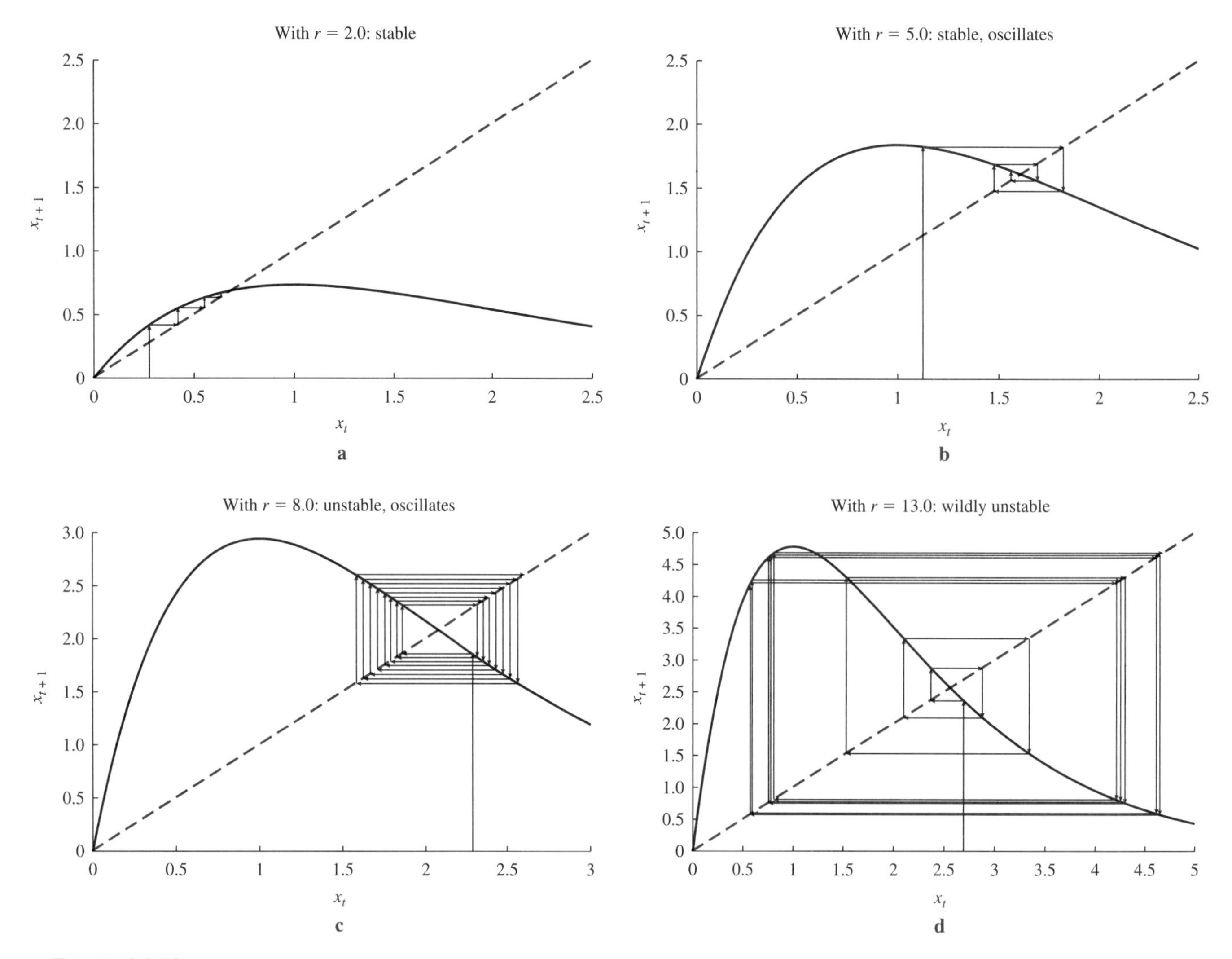

FIGURE 3.2.12
The dynamics of the Ricker model with various values of r

Results for various values of r are summarized in the following table.

r	x^*	$f'(x^*)$	Stability
0.5	0	0.5	Stable
2.0	0	2.0	Unstable
2.0	$\ln(2.0) \approx 0.693$	$1 - \ln(2.0) \approx 0.307$	Stable
4.0	0	4.0	Unstable
4.0	$\ln(4.0) \approx 1.386$	$1 - \ln(4.0) \approx -0.386$	Stable (oscillates)
8.0	0	8.0	Unstable
8.0	$\ln(8.0) \approx 2.079$	$1 - \ln(8.0) \approx -1.079$	Unstable (oscillates)
13.0	0	8.0	Unstable
13.0	$\ln(13.0) \approx 2.565$	$1 - \ln(13.0) \approx -1.565$	Unstable (oscillates)

As long as $\ln(r) < 2$, or $r < e^2 = 7.39$, the positive equilibrium is stable (Figure 3.2.12b). Like the logistic model, this discrete-time dynamical system produces unstable dynamics when the maximum possible per capita production is large (Figure 3.2.12c). In this case, fish can produce so many offspring that they destroy their resource base and induce a population crash in the following year, a phenomenon called **overcompensation**.

Summary The approach of **qualitative dynamical systems** lets us think of behavior in general terms without depending on quantitative details about the discrete-time dynamical system. As an example, we compare the dynamics generated by a nonlinear discrete-time dynamical system with those generated by the tangent line at the equilibrium. Because stability of equilibria can be understood by studying linear discrete-time dynamical systems, we examined linear systems with negative slopes, finding that such systems oscillate. Based on these results, we stated the **stability theorem for discrete-time dynamical systems**: an equilibrium is stable if the **absolute value** of the slope is less than 1. We applied this condition to the **logistic dynamical system** and the **Ricker model** and found that there are parameter values with no stable equilibrium when **overcompensation**, decreased production at high population size, is sufficiently strong.

3.2 Exercises

Mathematical Techniques

1–4 ▪ Draw the tangent line approximating the given system at the specified equilibrium and compare the cobweb diagrams. Use the stability theorem to check whether the equilibrium is stable.

1. The bacterial selection equation $p_{t+1} = \dfrac{1.5p_t}{1.5p_t + 2.0(1 - p_t)}$ at the equilibrium $p^* = 0$.
2. As in Exercise 1, but at the equilibrium $p^* = 1$.
3. $x_{t+1} = 1.5x_t(1 - x_t)$ at the equilibrium $x^* = 0$.
4. $x_{t+1} = 1.5x_t(1 - x_t)$ at the equilibrium $x^* = 1/3$.

5–8 ▪ Starting from the given initial condition, find the solution for five steps of each of the following.

5. $y_{t+1} = 1.2y_t$ with $y_0 = 2.0$. When will the value exceed 100?
6. $y_{t+1} = -1.2y_t$ with $y_0 = 2.0$. When will the value exceed 100?
7. $y_{t+1} = 0.8y_t$ with $y_0 = 2.0$. When will the value be less than 0.2?
8. $y_{t+1} = -0.8y_t$ with $y_0 = 2.0$. When will the value be between 0.0 and 0.2?

9–12 ▪ Consider the linear discrete-time dynamical system $y_{t+1} = 1.0 + m(y_t - 1.0)$. For each of the following values of m,

a. Find the equilibrium.
b. Graph and cobweb.
c. Compare your results with the stability condition.

9. $m = 0.9$.
10. $m = 1.5$.
11. $m = -0.5$.
12. $m = -1.5$.

13–16 ▪ The following discrete-time dynamical systems have slope of exactly -1 at the equilibrium. Check this, and then iterate the function for a few steps starting from near the equilibrium to see whether it is stable, unstable, or neither.

13. $x_{t+1} = 4 - x_t$ (as in Section 1.6, Exercise 10).
14. $x_{t+1} = \dfrac{x_t}{x_t - 1}$ for $x_t > 1$ (as in Section 1.6, Exercise 30).
15. $x_{t+1} = 3x_t(1 - x_t)$, the logistic system with $r = 3$.
16. $x_{t+1} = \dfrac{2}{1 + x_t^2}$ (the equilibrium is $x^* = 1$).

17–18 ▪ Equilibria where the slope of the tangent line is exactly 0 are also special. Show that the following systems satisfy this special relationship. What does this say about the stability of the equilibrium? (Think about a linear dynamical system with a slope of 0. How quickly do solutions approach the equilibrium?) Draw a cobweb diagram to illustrate these results.

17. The logistic dynamical system with $r = 2$.
18. The dynamical system $x_{t+1} = x_t e^{1-x_t}$.

Applications

19–22 ▪ We have studied several systems where the fraction of medication absorbed depends on the concentration of medication in the bloodstream. These take the form

$$M_{t+1} = M_t - f(M_t)M_t + 1.0$$

where $f(M_t)$ is the fraction absorbed and 1.0 is the supplement. If the fraction absorbed increases, it seems possible that the equilibrium level will become unstable (high levels are rapidly reduced). For each of the following forms for $f(M_t)$, the equilibrium level is $M^* = 2$. Find the slope of the updating function at the equilibrium and check if it is stable. In which cases does the solution oscillate?

19. $f(M_t) = \dfrac{M_t}{2 + M_t}$.
20. $f(M_t) = \dfrac{M_t^2}{4 + M_t^2}$.
21. $f(M_t) = \dfrac{M_t^4}{16 + M_t^4}$.
22. $f(M_t) = \dfrac{M_t^8}{256 + M_t^8}$.

23–26 ▪ Find values of r satisfying the following conditions for the Ricker model. In each case, graph and cobweb.

23. The value of r where the positive equilibrium switches from having a positive to a negative slope.

24. One value of r between 1 and the value found in Exercise 47.

25. One value of r between the value found in Exercise 47 and $r = e^2$.

26. One value of r greater than $r = e^2$.

27–28 ▪ The logistic model quantifies a competitive interaction, wherein per capita production is a decreasing function of population size. In some situations, per capita production is enhanced by population size. For each of the following cases where per capita growth increases as a function of population size,

a. Write the updating function.

b. Find the equilibria and their stability.

c. Graph the updating function and cobweb. Which equilibrium is stable?

d. Explain what this population is doing.

27. Per capita production $= 0.5 + 0.5b_t$.

28. Per capita production $= 0.5 + 0.5b_t^2$.

29–30 ▪ Consider a modified version of the logistic dynamical system $x_{t+1} = rx_t(1 - x_t^n)$. For the following values of n,

a. Sketch the updating function with $r = 2$.

b. Find the equilibria.

c. Find the derivative of the updating function at the equilibria.

d. For what values of r is the $x = 0$ equilibrium stable? For what values of r is the positive equilibrium stable?

29. $n = 2$.

30. $n = 3$.

31–32 ▪ Expanding oscillations can result from improperly tuned feedback systems. Suppose that a thermostat is supposed to keep a room at 20°C. Follow the steps to figure out what is happening in each of the given cases.

a. Suppose the temperature produced is a linear function of the temperature on the thermometer. Find this function.

b. You continue to respond to a temperature x°C above 20 by setting the thermostat to $20 - x$°C, and to a temperature x°C below 20° by setting the thermostat to $20 + x$°C. Find the temperature for the next few days.

c. Denote the temperature on day t by T_t. Find a formula for the thermostat setting z_t in response.

d. Use the answer to **a** to find T_{t+1}. Write the updating function.

e. Use the stability condition to describe what will happen in this room.

31. You come in one morning and find that the temperature is 21°C. To correct this, you move the thermostat down by 1°C to 19°C. But the next day the temperature has dropped to 18°C.

32. You come in one morning and find that the temperature is 21°C. To correct this, you move the thermostat down by 1°C to 19°C. But the next day the temperature has dropped to 18.5°C.

33–34 ▪ The model of bacterial selection includes no **frequency-dependence**, meaning that the per capita production of the different types does not depend on the fraction of types in the population. Each of the following discrete-time dynamical systems for the number of mutants a_t and the number of wild type b_t depends on the fraction of mutants p_t. For each, explain in words how each type is affected by the frequency of the mutants, find the discrete-time dynamical system for p_t, find the equilibria, evaluate their stability, and plot a cobweb diagram. Do any of them oscillate?

33. Suppose

$$a_{t+1} = 2(1 - p_t)a_t$$
$$b_{t+1} = (1 + p_t)b_t.$$

34. Suppose

$$a_{t+1} = 2(1 - p_t)^2 a_t$$
$$b_{t+1} = (1 + p_t)b_t.$$

35–38 ▪ Crowded plants grow to smaller size. Smaller plants make fewer seeds. The following describe the dynamics of a population described by n, the total number of seeds, and s, the size of the adult produced. In each case,

a. Start from $n = 20.0$ and find the total number of seeds for the next 2 yr (if the number of seeds per plant is a fraction, don't worry. Just think of it as an average).

b. Write the discrete-time dynamical system for the number of seeds.

c. Find the equilibrium number of seeds.

d. Graph the updating function and cobweb.

e. How does this result relate to the stability condition?

35. If there are n seeds, each sprouts and grows to a size $s = \frac{100.0}{n}$. An adult of size s produces $s - 1.0$ seeds (because it must use 1.0 units of energy to survive).

36. If there are n seeds, each sprouts and grows to a size $s = \frac{100}{n}$. An adult of size s produces $s - 0.5$ seeds.

37. If there are n seeds, each sprouts and grows to a size $s = \frac{100}{n}$. Suppose that an adult of size s produces $s - 2.0$ seeds.

38. If there are n seeds, each sprouts and grows to a size $s = \frac{100}{n+5}$. An adult of size s produces $s - 1$ seeds.

Computer Exercises

39. Study the behavior of the logistic dynamical system first for values of r near 3.0, and then for values between 3.5 and 4.0. Try the following with five values of r near 3.0 (such as 2.9, 2.99, 3.0, 3.01, and 3.1) and ten values of r between 3.5 and 4.0.

a. Use your computer to find solutions for 100 steps.

b. Look at the last 50 or so points on the solution and try to describe what is going on.

c. The case with $r = 4.0$ is famously chaotic. One of the properties of chaotic systems is "sensitivity to initial conditions." Run the system for 100 steps from one initial condition, and then run it again from an initial condition that is very close. If you compare your two solutions, they should be similar for a while, but eventually become completely different. What if a real system had this property?

40. Consider a more general version of the logistic dynamical system

$$x_{t+1} = rx_t(1 - x_t^p).$$

a. Sketch the updating function for several values of $p > 2$ (such as $p = 3$, $p = 4$, and $p = 10$).

b. Find the equilibria.

c. Find the derivative of the updating function at the equilibria.

d. For what values of r is the $x = 0$ equilibrium stable? For what values of r is the positive equilibrium stable?

e. What happens to solutions when r is too large?

41. Consider a population following

$$x_{t+1} = rx_t^2 e^{-x_t}$$

a. Graph the updating function for the following values of r: $r = 1$, $r = 2.6$, $r = e$, $r = 2.8$, $r = 3.6$, $r = \frac{e^2}{2}$, $r = 3.8$, $r = 6.6$, $r = \frac{e^3}{3}$, $r = 6.8$, and $r = 10.0$.

b. Find the equilibria for these values of r (the equation cannot be solved in general, but your computer should have a routine for solving, or just guess). Make sure you find them all.

c. Find the derivative of the updating function at each equilibrium.

d. Find which of the equilibria are stable.

e. When all the equilibria are unstable, how might this model behave differently from the Ricker model with $r > e^2$? Can you explain why?

3.3 Maximization

A bee arrives at a flower. The more time she spends eating nectar, the more slowly the nectar comes out. But she knows that she must fly a long distance to find the next flower. When should she give up and leave? On a larger scale, a fisherman must decide how many fish to catch. The more he catches, the more he gets that year, but the more he depletes the fishery for next year. How many fish should he catch to catch the most fish in the long run? In fisheries jargon, what is the **maximum sustained yield** from the fish population?

Both of these are problems in **optimization**, finding the best solution to a problem. In this section, we will learn how to use the derivative to find **optima**. When the problem has been set up correctly, these best solutions are points where a function takes on a **minimum** or **maximum** value. In particular, **critical points** (Definition 2.6) where the derivative is equal to 0 or the derivative is not defined are candidates for being minima or maxima, which can sometimes be distinguished with the **second derivative**. We will use these techniques to figure out the optimal behavior for the bee and the fisherman.

3.3.1 Minima and Maxima

The graph of the function

$$f(x) = xe^{-x}$$

rises to a peak, known as a **maximum** of the function (Figure 3.3.1). How do we find where the maximum occurs? A function has a maximum where it switches from increasing to decreasing, just as a trail reaches a peak when it switches from ascending to descending. Mathematically, a maximum occurs when the derivative switches from positive to negative.

We have found the derivative of this function before (Section 2.8, Example 2.8.7 and in Subsection 3.2.3),

$$\frac{df}{dx} = \frac{d}{dx}xe^{-x} = (1 - x)e^{-x}$$

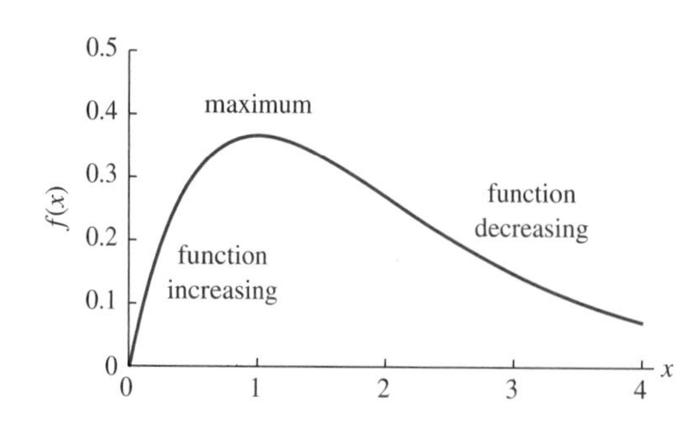

FIGURE 3.3.1
A function with a peak

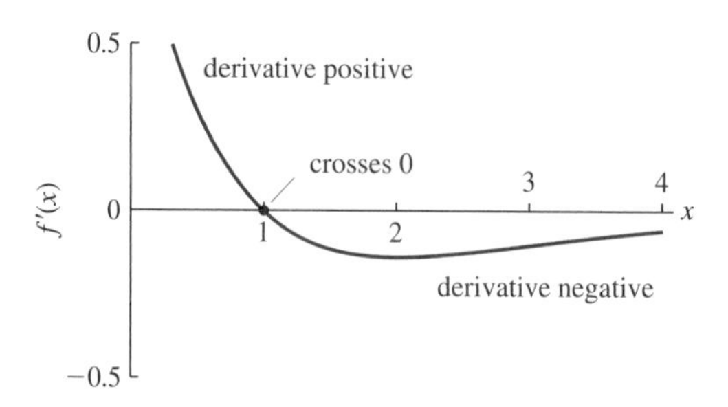

FIGURE 3.3.2
The derivative of a function with a peak

(Figure 3.3.2). The derivative is positive for $x < 1$, negative for $x > 1$, and 0 for $x = 1$. The point where $f'(x) = 0$ is a **critical point** (Definition 2.6). Because the function switches from increasing to decreasing, the critical point at $x = 1$ is a maximum.

Example 3.3.1 A Quadratic Function with a Maximum

Similarly, the graph of

$$g(x) = x(1 - x)$$

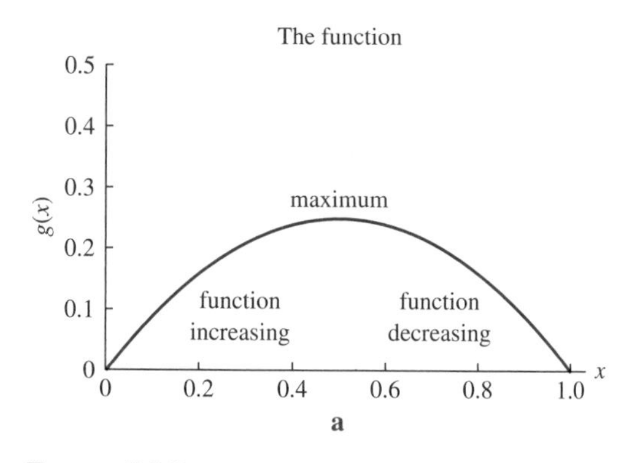

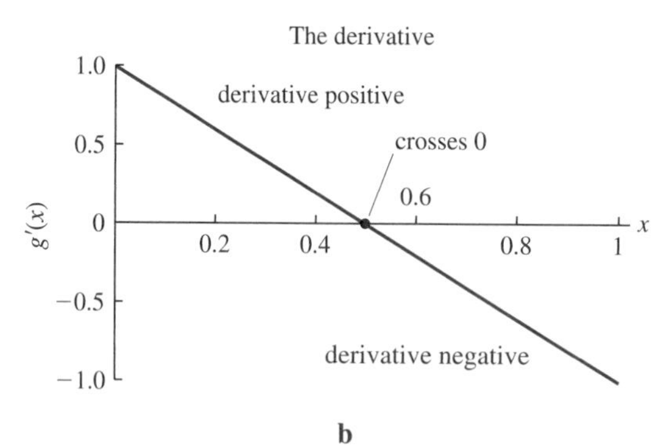

FIGURE 3.3.3
Another function with a peak

for $0 \le x \le 1$ rises to a peak for some value of x in this domain (Figure 3.3.3). The derivative is

$$\frac{dg}{dx} = \frac{d}{dx}(x - x^2) = 1 - 2x.$$

The derivative is exactly 0 when

$$1 - 2x = 0$$
$$x = 0.5.$$

Because $f'(x) > 0$ for $x < 0.5$ and $f'(x) < 0$ for $x > 0.5$, the function switches from increasing to decreasing, which means this critical point is a maximum. ▲

Example 3.3.2 Finding the Maximum of a Cubic Function

Consider the function

$$h(x) = x^3 - x$$

for $-1 < x < 1$. Does this function have a maximum? To begin, take the derivative

$$\frac{dh}{dx} = 3x^2 - 1.$$

Next, set the derivative equal to 0 to find critical points

$$3x^2 - 1 = 0.$$

Therefore,

$$3x^2 = 1$$
$$x^2 = \frac{1}{3}$$
$$x = \pm\sqrt{\frac{1}{3}}.$$

These points are candidates for where the function changes from increasing to decreasing. How can we tell what is happening at these points? One simple method is to evaluate the function, finding

$$h\left(\sqrt{\frac{1}{3}}\right) = -0.385$$
$$h\left(-\sqrt{\frac{1}{3}}\right) = 0.385.$$

The point $x = \sqrt{\frac{1}{3}}$ is a minimum and the point $x = -\sqrt{\frac{1}{3}}$ is a maximum (Figure 3.3.4). ▲

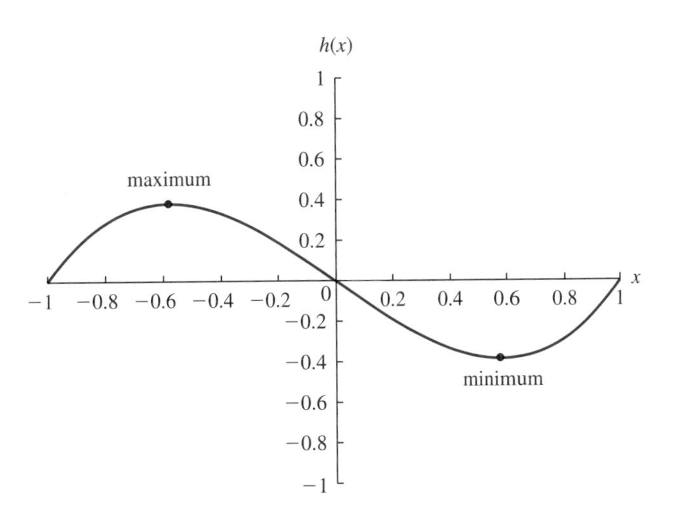

FIGURE 3.3.4
A function with a minimum and a maximum: $h(x) = x^3 - x$

To develop a general method for finding maxima and minima of functions, we must classify the types of maxima and minima. Maxima are classified as **global** or **local**. The global maximum of a function f is the largest value taken on by the function anywhere in its domain. Similarly, the global minimum is the smallest value taken on by the function. A local maximum is a "peak" where the function takes on its largest value in

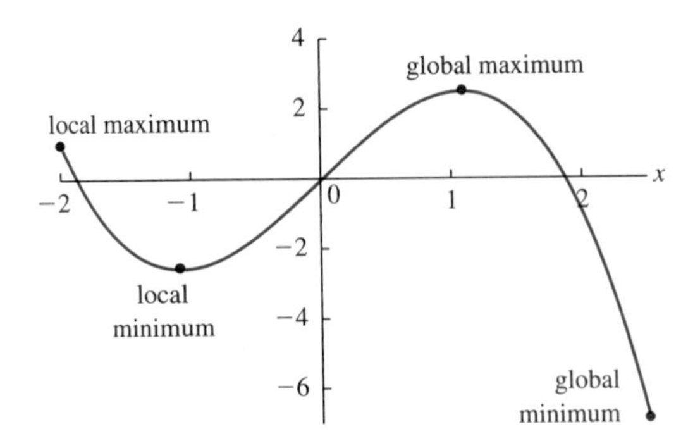

FIGURE 3.3.5
Local and global maxima

a region of the domain (Figure 3.3.5). The peak of the tallest mountain in Utah is a local maximum, but not a global one (Mt. Everest is taller). A function may have many local maxima with different values, just as a mountain range has many peaks with different heights. It may also have more than one input at which a global maximum occurs (if there were other mountains that were exactly as tall as Mt. Everest).

A special kind of maximum occurs at the boundary of the domain. The function in Figure 3.3.5 has a local maximum at the left-hand edge of its domain and a global minimum at the right. Maxima and minima at the boundaries can also be analyzed with the derivative. A function with a negative derivative at the left-hand boundary has a local maximum at that point, while one with a negative derivative at the right-hand boundary has a local minimum at that point. The function in Figure 3.3.5 matches these criteria. A function with positive derivative is increasing and has a local minimum at the left-hand boundary and a local maximum at the right.

Algorithm 3.1 Finding Global Maxima and Minima

1. Compute the value of the function at the endpoints.
2. Find all critical points.
3. Compute the value of the function at all critical points.
4. The largest of the numbers found in steps **1** and **3** is the global maximum. The smallest of the numbers found in steps **1** and **3** is the global minimum.

Remember that critical points include points of two types: those where the derivative is equal to zero, and those where the function is not differentiable.

Example 3.3.3 The Global Maximum and Minimum of a Function with a Corner

Consider the absolute value function $a(x) = |x|$ defined for $-2 \le x \le 3$ (Figure 3.3.6). Following the algorithm,

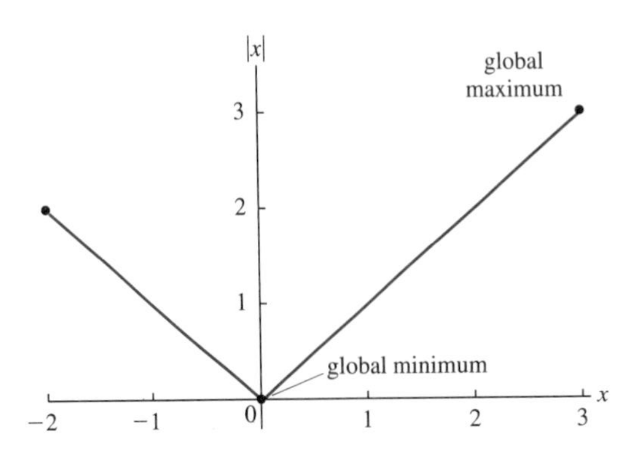

FIGURE 3.3.6
The minimum of the absolute value function

1. The endpoints are -2 and 3, where $a(-2) = |-2| = 2$, $a(3) = |3| = 3$.
2. The function is not differentiable at $x = 0$ (because of the corner). For $x > 0$, we have that $a(x) = x$, so $a'(x) = 1$. For $x = 0$, we have that $a(x) - = x$, so $a'(x) = -1$. Thus $x = 0$ is the only critical point because the derivative is never 0.
3. The value at the critical point is $a(0) = |0| = 0$.
4. The largest of these values is 3, so the global maximum is at the right-hand endpoint. The smallest is 0, so $x = 0$ is the global minimum.

Example 3.3.4 The Global Maximum and Minimum of a Cubic Function

To find the global minima and maxima of the differentiable function $h(x) = x^3 - x$ for $-1 \leq x \leq 1$ (Figure 3.3.4), we follow these steps:

1. $h(-1) = 0$ and $h(1) = 0$.
2. We found the critical points earlier to be at $x = \sqrt{\frac{1}{3}}$ and $x = -\sqrt{\frac{1}{3}}$.
3. $h(\sqrt{\frac{1}{3}}) = -0.385$ and $h(-\sqrt{\frac{1}{3}}) = 0.385$.
4. The largest of these values is 0.385, so the global maximum occurs at $x = -\sqrt{\frac{1}{3}}$. The smallest is -0.385, so the global minimum occurs at $x = \sqrt{\frac{1}{3}}$.

Example 3.3.5 The Global Maximum and Minimum on a Different Domain

If we extended the domain of the function $h(x) = x^3 - x$ to include $-1 \leq x \leq 2$, the global maximum moves to $x = 2$ (Figure 3.3.7). The critical point $x = -\sqrt{\frac{1}{3}}$ is only a **local** maximum.

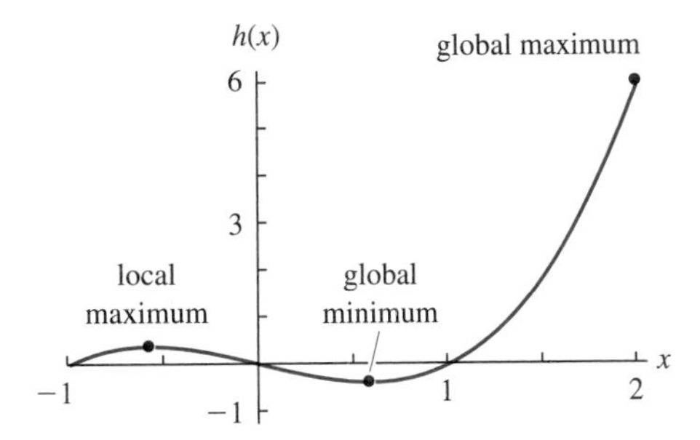

FIGURE 3.3.7
The global maximum of $h(x) = x^3 - x$ changes on an extended domain

The second derivative provides a method to check whether a critical point where the derivative is equal to 0 is a maximum or a minimum. On the graph of $h(x) = x^3 - x$, we can see that a critical point is a local maximum when the graph is **concave down** and is a local minimum when the graph is **concave up**.

We can formalize these observations into another algorithm.

Algorithm 3.2 Identifying Local Maxima and Minima with the Second Derivative

1. Find all critical points where the function is differentiable.
2. Compute the sign of the second derivative at these critical points.
3. Critical points where the second derivative is positive correspond to local minima, and critical points where the second derivative is negative correspond to local maxima.

The second derivative of $h(x) = x^3 - x$ is

$$\frac{d^2}{dx^2}(x^3 - x) = \frac{d}{dx}(3x^2 - 1) = 6x.$$

This function is negative when $x < 0$, so the critical point $x = -\sqrt{\frac{1}{3}}$ is a local maximum, and the critical point $x = \sqrt{\frac{1}{3}}$ is a local minimum (Figure 3.3.8c).

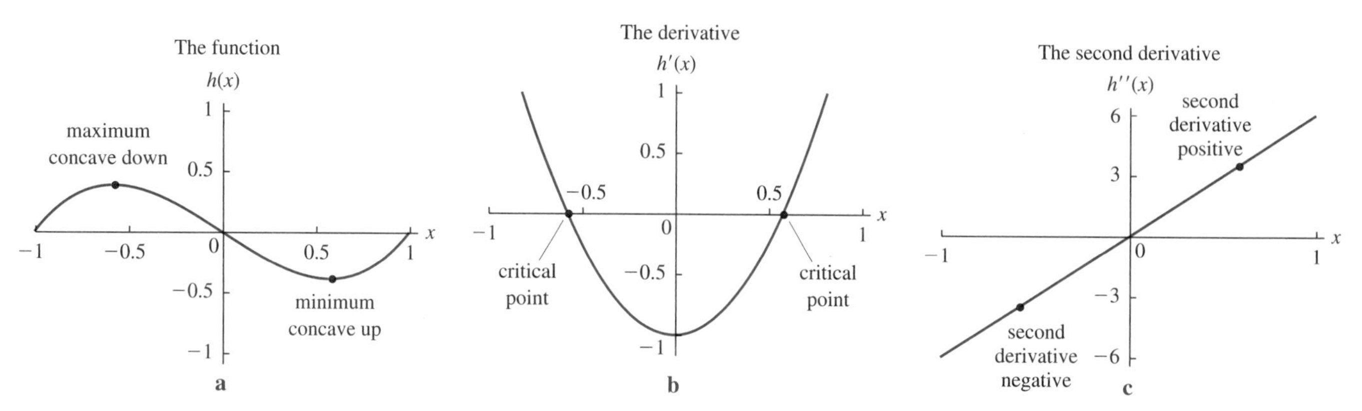

FIGURE 3.3.8
The function $h(x) = x^3 - x$, its derivative, and its second derivative

Example 3.3.6 Using the Second Derivative to Find Maxima

Consider again the function $f(x) = xe^{-x}$ with a critical point at $x = 1$. The second derivative of $f(x) = xe^{-x}$ is

$$\begin{aligned}
\frac{d^2 f}{dx^2} &= \frac{d}{dx}(1 - x)e^{-x} && \text{take derivative of derivative} \\
&= \frac{d(1-x)}{dx}e^{-x} + \frac{d(e^{-x}x)}{dx}(1-x) && \text{product rule} \\
&= -e^{-x} + -e^{-x}(1-x) && \text{linear function and exponential rules} \\
&= (x-2)e^{-x}. && \text{factoring}
\end{aligned}$$

At the critical point $x = 1$,

$$f''(1) = (1-2)e^{-1} = -e^{-1} < 0.$$

Because the second derivative is negative at the critical point, the function switches from increasing to decreasing at $x = 1$. There is a local maximum at $x = 1$.

Example 3.3.7 Using the Second Derivative to Identify Maxima of a Quadratic

The second derivative of $g(x) = x(1 - x)$, studied in Example 3.3.1, is

$$\frac{d^2 g}{dx^2}[x(1-x)] = \frac{d}{dx}(1 - 2x) = -2.$$

Because the second derivative is negative, the critical point at $x = 0.5$ is a local maximum (Figure 3.3.1b).

If the second derivative is 0 at a critical point, we cannot tell whether the critical point is a local minimum, a local maximum, or neither.

Example 3.3.8 A Function with a Point of Inflection at a Critical Point

The function $C(x) = x^3$ (Figure 3.3.9) has both a critical point and a point of inflection at $x = 0$ because

$$C'(x) = 3x^2$$
$$C''(x) = 6x.$$

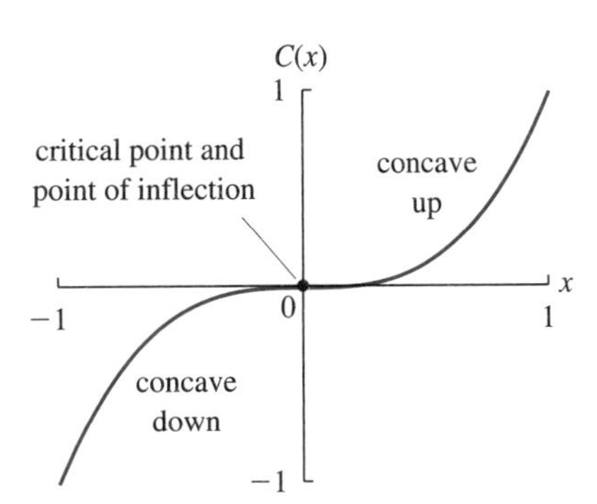

FIGURE 3.3.9
The function $C(x) = x^3$

Because the derivative is positive for all values of $x \neq 0$, $C(x)$ is increasing everywhere except at $x = 0$. The critical point is neither a minimum nor a maximum, but just an instantaneous pause.

Example 3.3.9 A Function with Second Derivative Equal to 0 at a Critical Point

The quartic function $Q(x) = x^4$ (considered in Section 2.7, Example 2.7.2 and Figure 3.3.10) has both a critical point and a point of inflection at $x = 0$ because

$$Q'(x) = 4x^3$$
$$Q''(x) = 12x^2.$$

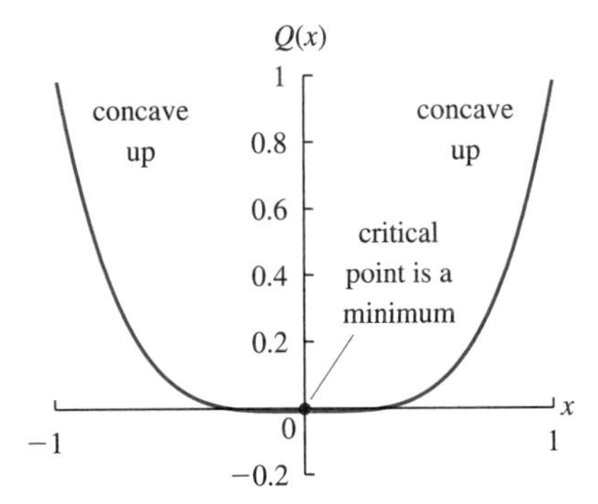

FIGURE 3.3.10
The function $Q(x) = x^4$

In this case, the second derivative is positive for all values of $x \neq 0$. Therefore, $Q(x)$ is concave up everywhere except at $x = 0$. The critical point is a minimum.

3.3.2 Maximizing Food Intake Rate

Consider the bee described in the introduction. After she finds a flower, she sucks up nectar at a slower and slower rate as the flower is depleted. However, she does not want to leave too soon because she must fly some distance to find the next flower. When should she give up and leave? If she stays a long time at each flower, she will get most of the nectar from each flower, but will visit few flowers. If she stays a short time at

FIGURE 3.3.11
The bee's maximization problem

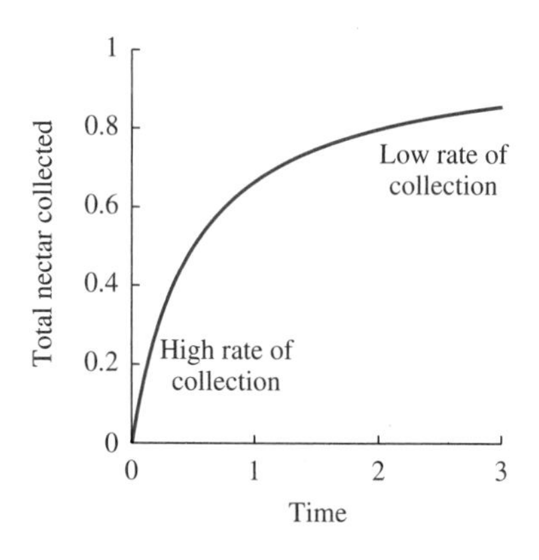

FIGURE 3.3.12
Nectar collected as a function of time

each flower, she gets to skim the best nectar off the top, but spends most of her time flying to new flowers (Figure 3.3.11).

First, we must formulate this as a **maximization problem**. What is the bee trying to achieve? Her job is to bring back as much nectar as possible over the course of a day. To do so, she should maximize the **rate per visit**, which includes the travel time τ between flowers. Suppose $F(t)$ is the amount of nectar collected by a bee that stays on a flower for time t (Figure 3.3.12). The rate at which nectar is collected is

$$\begin{aligned} R(t) &= \frac{\text{food per visit}}{\text{total time per visit}} \\ &= \frac{\text{food per visit}}{\text{time on flower} + \text{travel time}} \\ &= \frac{F(t)}{t+\tau}. \end{aligned}$$

Example 3.3.10 Computing the Maximum with Particular Parameter Values

As a particular case, assume that the travel time τ is equal to 1.0 min, and that

$$F(t) = \frac{t}{t+0.5}.$$

Then

$$R(t) = \frac{F(t)}{t+1.0} = \frac{t}{(t+0.5)(t+1.0)}.$$

The derivative can be found with the quotient rule and a lot of algebra to be

$$\begin{aligned} \frac{dR}{dt} &= \frac{(t+0.5)(t+1.0) - t[(t+0.5)+(t+1.0)]}{[(t+0.5)(t+1.0)]^2} \\ &= \frac{0.5 - t^2}{[(t+0.5)(t+1.0)]^2}. \end{aligned}$$

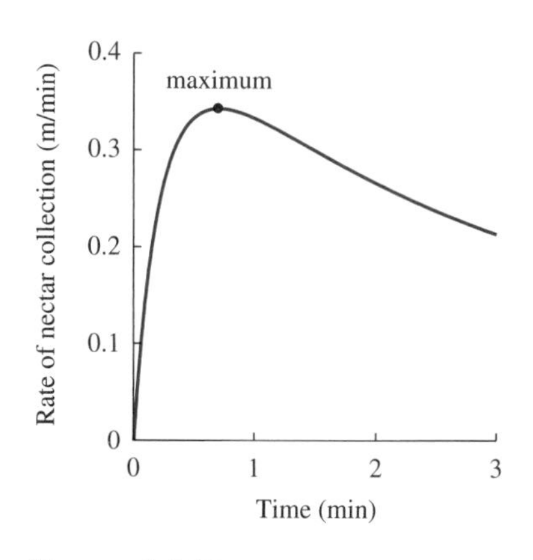

FIGURE 3.3.13
The average rate of return for the bee

This derivative is 0 when the numerator is 0 or when

$$0.5 - t^2 = 0,$$

which has positive solution $t = 0.707$ (Figure 3.3.13). The numerator is positive for $t < 0.707$ and negative for $t > 0.707$, which means that this value is a maximum.

By looking at the problem in general, without substituting in a specific functional form for $F(t)$ or a value of τ, we can find a graphical method to solve the bee's problem.

We differentiate $R(t)$ with the quotient rule, finding

$$\frac{dR}{dt} = \frac{(t+\tau)F'(t) - F(t)}{(t+\tau)^2}.$$

Because the denominator is positive, the critical points occur where the numerator is 0, or when

$$(t+\tau)F'(t) = F(t).$$

Solving for $F'(t)$ gives

$$F'(t) = \frac{F(t)}{t+\tau} = R(t).$$

The solution of this equation is a local maximum as long as $F(t)$ is concave down (Exercise 40). This fundamental equation says that the bee should leave when the derivative of F, the instantaneous rate of food collection, is equal to the average rate. This is called the **Marginal Value Theorem** and is a powerful tool in both ecology and economics. The idea is simple; leave when you can do better elsewhere.

Graphically, the slope of the food collection curve at the critical point is equal to the slope of the line connecting that point with a point at negative τ (Figure 3.3.14a), because that line has a vertical change $F(t)$ and a horizontal change $t+\tau$. The slope of the line is

$$\frac{F(t)}{t+\tau} = R(t).$$

From the graph, we can see that the optimal time to remain becomes shorter when the travel time between flowers is shorter (Figure 3.3.14b).

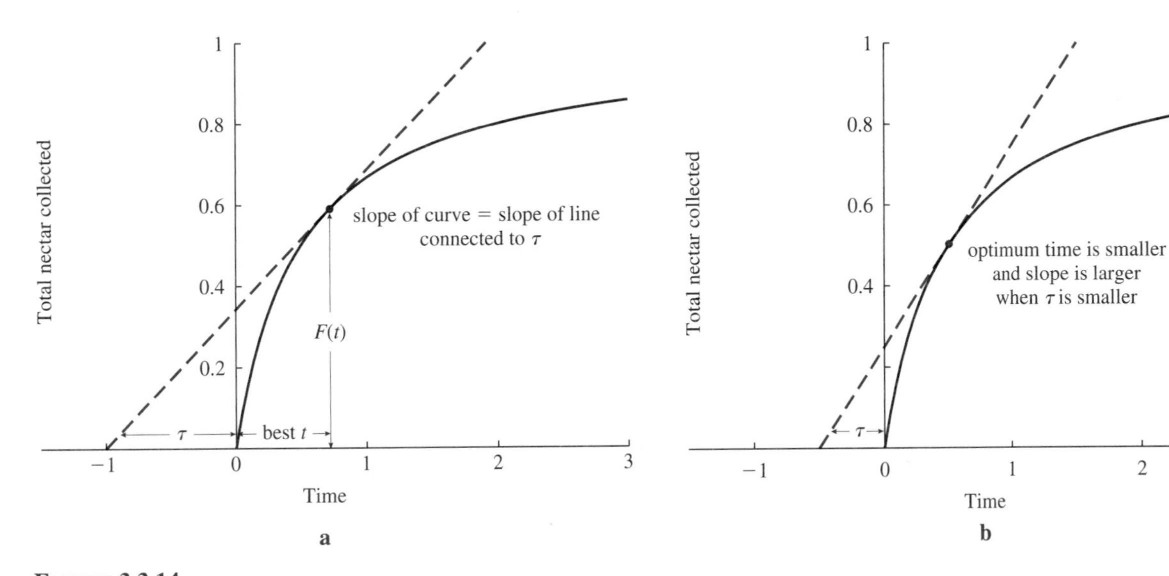

FIGURE 3.3.14
The Marginal Value Theorem: graphical method

Example 3.3.11 Using the Marginal Value Theorem to Solve a Maximization Problem

Consider the situation in Example 3.3.10, with the nectar collected as a function of time equal to

$$F(t) = \frac{t}{t+0.5}$$

and travel time of $\tau = 1.0$. In Example 3.3.10 we found that

$$R(t) = \frac{F(t)}{t+1.0} = \frac{t}{(t+0.5)(t+1.0)}$$

by writing the formula for $R(t)$ and differentiating. Alternatively, we can find the maximum with the Marginal Value Theorem by solving for the time when $F'(t) = R(t)$. The derivative of $F(t)$ can be found with the quotient rule as

$$F'(t) = \frac{(t+0.5)\frac{dt}{dt} - t\frac{d(t+0.5)}{dt}}{(t+0.5)^2} = \frac{(t+0.5)-t}{(t+0.5)^2} = \frac{0.5}{(t+0.5)^2}.$$

To find the maximum, we solve $F'(t) = R(t)$, or

$$\begin{aligned}
\frac{0.5}{(t+0.5)^2} &= \frac{t}{(t+0.5)(t+1.0)} && \text{the original equation}\\
\frac{0.5}{t+0.5} &= \frac{t}{t+1.0} && \text{multiply both sides by } t+0.5\\
0.5(t+1.0) &= t(t+0.5) && \text{cross multiply}\\
0.5t+0.5 &= t^2+0.5t && \text{multiply out}\\
0.5 &= t^2 && \text{subtract } 0.5t\\
t &= 0.707. && \text{solve for } t
\end{aligned}$$

This matches the result found by differentiating $R(t)$ directly. ▲

Example 3.3.12 The Effect of Increasing Travel Time

Suppose that the nectar collected as a function of time follows

$$F(t) = \frac{t}{t+0.5}$$

as in Example 3.3.11, but that the travel time is $\tau = 10.0$. Then

$$R(t) = \frac{F(t)}{t+10.0} = \frac{t}{(t+0.5)(t+10.0)}.$$

The Marginal Value Theorem states that the optimal time for the bee to depart is when $F'(t) = R(t)$. We found the derivative of function $F(t)$ in Example 3.3.11, so we can find the maximum by solving

$$\begin{aligned}
\frac{0.5}{(t+0.5)^2} &= \frac{t}{(t+0.5)(t+10.0)} && \text{the equation } F'(t)=R(t)\\
\frac{0.5}{t+0.5} &= \frac{t}{t+10.0} && \text{multiply both sides by } t+0.5\\
0.5(t+10.0) &= t(t+0.5) && \text{cross multiply}\\
0.5t+5.0 &= t^2+0.5t && \text{multiply out}\\
5.0 &= t^2 && \text{subtract } 0.5t\\
t &= 2.236. && \text{solve for } t
\end{aligned}$$

As expected, the optimal time to remain on the flower becomes longer when travel time becomes longer. Nonetheless, the bee still spends a smaller **fraction** of its time on flowers in this case. With $\tau = 10.0$, the fraction of time on flowers is

$$\text{fraction of time on flower} = \frac{\text{time on flower}}{\text{time on flower} + \text{travel time}} = \frac{2.236}{2.236+10.0} = 0.183.$$

This bee is predicted to spend 18.3% of its time on flowers. The bee in Example 3.3.11, in contrast, is predicted to spend

$$\text{fraction of time on flower} = \frac{0.707}{0.707 + 1.0} = 0.414$$

or 41.4% of its time.

3.3.3 Maximizing Fish Harvest

Consider a variant of the logistic dynamical system (Equation 3.2.1) that includes harvesting,

$$N_{t+1} = 2.5N_t(1 - N_t) - hN_t \tag{3.3.1}$$

(Figure 3.3.15). N_t denotes the population of fish at the beginning of one fishing season, and N_{t+1} represents the population at the beginning of the next season. The population is measured as the fraction of the maximum possible population size. The term $-hN_t$ is the harvest, where h is called the "harvesting effort." Harvesting effort depends on the number of ships, the number of fishing days, the quality of the fishing vessels, and many other factors. Total harvest is the product of harvesting effort and population size.

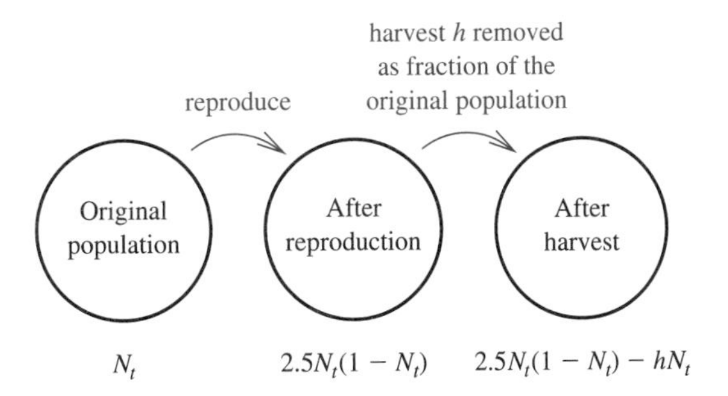

FIGURE 3.3.15
The dynamics of a simple fishery

What harvesting effort brings in the maximum long-term harvest? No harvesting effort ($h = 0$) brings in nothing. An enormous harvesting effort might bring in many fish in the short term but end up depleting the population. We suspect that an intermediate harvesting effort will maximize the long-term harvest.

The long-term behavior of this system is described by a stable equilibrium. What are the equilibria? The equilibrium value N^* can be found with Algorithm 1.5.

$$\begin{aligned} N^* &= 2.5N^*(1 - N^*) - hN^* && \text{the original equation} \\ N^* - 2.5N^*(1 - N^*) + hN^* &= 0 && \text{move everything to one side} \\ N^*[1 - 2.5(1 - N^*) + h] &= 0 && \text{factor} \\ N^* = 0 \text{ or } 1 - 2.5(1 - N^*) + h &= 0 && \text{set each piece equal to 0} \\ N^* = 0 \text{ or } N^* &= 1 - \frac{1 + h}{2.5} && \text{do the algebra} \end{aligned}$$

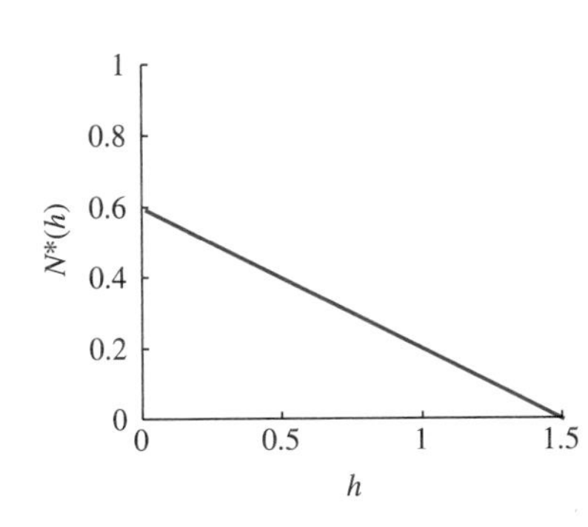

FIGURE 3.3.16
The positive equilibrium population as a function of harvesting effort with $r = 2.5$

(Figure 3.3.16). We have the usual extinction equilibrium, and a second equilibrium that is positive only if $\frac{1+h}{2.5} < 1$. If we choose $h > 1.5$, then the only equilibrium is $N^* = 0$ and the population goes extinct. (The peculiar possibility that $h > 1$ occurs because our model measures the population before reproduction and collects the harvest after reproduction.)

Suppose that we have chosen a harvesting effort h and that the population has reached the positive equilibrium N^* (conditions for stability are derived in Exercise 47). The equilibrium harvest, denoted $P(h)$, is the product of the harvesting effort h and the population size N^*, so

$$P(h) = hN^* = h\left(1 - \frac{1 + h}{2.5}\right). \tag{3.3.2}$$

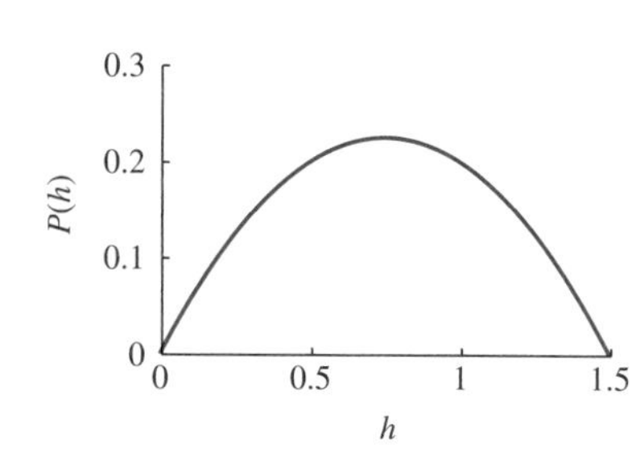

FIGURE 3.3.17

Long-term harvest as a function of harvesting effort with $r = 2.5$

(Figure 3.3.17). With our algorithm, we can find the value of h that maximizes harvest by checking the endpoints and locating critical points. The endpoints are $h = 0$ and $h = 1.5$ (above which the population goes extinct), where $P(0) = 0$ and $P(1.5) = 0$ as we suspected. To find critical points, solve

$$P'(h) = 1 - \frac{1 + 2h}{2.5} = 0$$

for h, finding

$$h = \frac{1.5}{2} = 0.75.$$

This must be a maximum based on our knowledge of the graph of the function (Figure 3.3.17). We would do best by letting the fish reproduce and then collecting a harvest equal to 75% of the population before reproduction. The payoff $P(h)$ is

$$P(0.75) = 0.75\left(1 - \frac{1 + 0.75}{2.5}\right) = 0.225.$$

A larger harvest would deplete the population. With $h = 1.0$,

$$P(1.0) = 1.0\left(1 - \frac{1 + 1.0}{2.5}\right) = 0.2,$$

which is smaller because the equilibrium is only 0.2. In contrast, a smaller harvest is inefficient. With $h = 0.5$,

$$P(0.5) = 0.5\left(1 - \frac{1 + 0.5}{2.5}\right) = 0.2,$$

which is smaller even though the equilibrium is larger.

Summary We have seen how to use the derivative to find **maxima** and **minima** of functions by locating **critical points** where the derivative is either 0 or is undefined. To find **global** maxima and minima, values at critical points must be compared with values at the endpoints, where maxima and minima often occur. When it is defined, the second derivative is positive at a local minimum and negative at a local maximum. We first applied this method to find the optimal length of time a bee should spending harvesting nectar from a flower. The solution is an example of the **Marginal Value Theorem**, which states that the best time to leave occurs when the rate of collecting resources falls below the average rate. We next applied the methods of maximization to a discrete-time dynamical system with harvesting to find the optimal way to harvest a fish population.

3.3 Exercises

Mathematical Techniques

1–6 ▪ Find all critical points of the following functions.

1. $a(x) = \frac{x}{1+x}$.
2. $f(x) = 1 + 2x - 2x^2$.
3. $c(w) = w^3 - 3w$.
4. $g(y) = \dfrac{y}{1 + y^2}$.
5. $h(z) = e^{z^2}$.
6. $c(\theta) = \cos(2\pi\theta)$.

7–12 ▪ Find the global minimum and maximum of the following functions on the interval given. Don't forget to check the endpoints.

7. $a(x) = \dfrac{x}{1 + x}$ for $0 \le x \le 1$.
8. $f(x) = 1 + 2x - 2x^2$ for $0 \le x \le 2$.
9. $c(w) = w^3 - 3w$ for $-2 \le w \le 2$.
10. $g(y) = \dfrac{y}{1 + y^2}$ for $0 \le y \le 2$.
11. $h(z) = e^{z^2}$ for $0 \le z \le 1$.
12. $F(x) = |1 - x|$ for $0 \le x \le 3$. (This function cannot be differentiated at $x = 1$).

13–18 ▪ Find the second derivative at the critical points of the following functions. Classify the critical points as minima and maxima. Use the second derivative to draw an accurate graph of the function for the given range.

13. $a(x) = \dfrac{x}{1 + x}$ (as in Exercise 1) for $0 \le x \le 1$.
14. $f(x) = 1 + 2x - 2x^2$ (as in Exercise 2) for $0 \le x \le 2$.

15. $c(w) = w^3 - 3w$ (as in Exercise 3) for $-2 \le w \le 2$.

16. $g(y) = \frac{y}{1+y^2}$ (as in Exercise 4) for $0 \le y \le 2$.

17. $h(z) = e^{z^2}$ (as in Exercise 5) for $-1 \le z \le 1$.

18. $c(\theta) = \cos(2\pi\theta)$ (as in Exercise 6) for $-1 \le \theta \le 1$.

19–22 ▪ Suppose $f(x)$ is a positive function with a maximum at x^*. We can often find maxima and minima of other functions composed with $f(x)$. For each of the functions $h(x) = g(f(x))$,

a. Show that h has a critical point at x^*.

b. Compute the second derivative at this point.

c. Check whether your function has a minimum or a maximum and explain.

d. Check your result using the function $f(x) = xe^{-x}$ for $x \ge 0$ which has a maximum at $x = 1$. Sketch a graph of $f(x)$ and $h(x)$ in this case.

19. $g(f) = \dfrac{1}{f}$.

20. $g(f) = 1 - f$.

21. $g(f) = \ln(f)$.

22. $g(f) = f - f^2$.

Applications

23–24 ▪ Solve the following optimization problems.

23. Organic waste deposited in a lake at $t = 0$ decreases the oxygen content of the water. Suppose the oxygen content is $C(t) = t^3 - 30t^2 + 6000$ for $0 \le t \le 25$. Find the maximum and minimum oxygen content during this time.

24. The size of a population of bacteria introduced to a nutrient grows according to

$$N(t) = 5000 + \frac{30{,}000t}{100 + t^2}.$$

Find the maximum size of this population for $t \ge 0$.

25–26 ▪ No calculus book is complete without optimization problems involving fences.

25. A farmer owns 1000 m of fence and wants to enclose the largest possible rectangular area. The region to be fenced has a straight canal on one side, and a perpendicular and perfectly straight ancient stone wall on another. The area thus needs to be fenced on only two sides. What is the largest area she can enclose?

26. A farmer owns 1000 m of fence and wants to enclose the largest possible rectangular area. The region to be fenced has a straight canal on one side, and thus needs to be fenced on only three sides. What is the largest area she can enclose?

27–30 ▪ Consider the bee faced by the problem in Subsection 3.3.2. Find the optimal strategy with the following travel times τ and illustrate the graphical method of solution. For each particular value of τ, find the equation of the tangent line at the optimal t and show that it goes through the point $(-\tau, 0)$.

27. $\tau = 2.0$.

28. $\tau = 0.5$.

29. $\tau = 0.1$.

30. Find the solution in general (without substituting a value for τ). What is the limit as τ approaches 0? Does this answer make sense?

31–34 ▪ Suppose that the total food collected by a bee follows

$$F(t) = \frac{t}{c + t}$$

where c is some parameter. If $\tau = 1.0$, find the optimal departure time in the following circumstances. Sketch the plot from the graphical method.

31. $c = 2.0$.

32. $c = 1.0$.

33. $c = 0.1$.

34. Find the solution in general (without substituting a value for c). What does the parameter c mean biologically (think about how long it takes the bee to collect half the nectar)? Explain in words why the bee leaves sooner when c is smaller.

35–38 ▪ Mathematical models can help us estimate values that are difficult to measure. Consider again a bee sucking nectar from a flower, with

$$F(t) = \frac{t}{0.5 + t}.$$

We also measure that the bee remains a length of time t on the flower. Estimate the travel time τ assuming that the bee understands the Marginal Value Theorem for the following values of t.

35. $t = 1.0$.

36. $t = 0.1$.

37. $t = 4.0$.

38. Find the solution in general (without substituting a value for t).

39–40 ▪ We never showed that the value found in computing the optimal t with the Marginal Value Theorem is in fact a maximum. For each of the following forms for the function $F(t)$, find the second derivative of $R(t)$ and the point where $F'(t) = \dfrac{F(t)}{t + \tau}$ and check whether the solution is a maximum.

39. Suppose $F(t) = \dfrac{t}{1 + t}$ and travel time is $\tau = 1$.

40. Suppose $F(t)$ is any function with $F''(t) < 0$ and travel time is $\tau = 1$.

41–44 ▪ Animals must survive predation in addition to maximizing their rate of food intake. One theory assumes that they try to maximize the ratio of food collected to predation risk. Suppose that different flowers with nectar of quality n (the rate of food collection) attract $P(n)$ predators. For example, flowers with higher-quality nectar (large values of n) might attract more predators (large value of $P(n)$). Bees must decide which flowers to select. For each of the following forms of $P(n)$, find the function the bees are trying to maximize, and find the optimal n.

41. Suppose that $P(n) = 1 + n^2$. Find the optimal n for the bees.

42. Suppose that $P(n) = 1 + n$. Find the optimal n for the bees and draw a graph like that for the Marginal Value Theorem. Does this make sense? Why is the result so different?

43. Find the condition for the maximum for a general function $P(n)$ by solving for $P'(n)$. Use this condition to find the optimal n for the cases $P(n) = 1 + n^2$ and $P(n) = 1 + n$.

44. Find a graphical interpretation of the condition in the previous problem and test it on $P(n) = 1 + n^2$ and $P(n) = 1 + n$.

45–46 ▪ Find the maximum harvest from a population following the discrete-time dynamical system

$$N_{t+1} = rN_t(1 - N_t) - hN_t$$

for the given values of r.

a. Find the equilibrium population as a function of h. What is the largest h consistent with a positive equilibrium?

b. Find the equilibrium harvest as a function of h.

c. Find the harvesting effort that maximizes harvest.

d. Find the maximum harvest.

45. $r = 2.0$.

46. $r = 1.5$.

47–48 ▪ Find the conditions for stability of the equilibrium of

$$N_{t+1} = rN_t(1 - N_t) - hN_t$$

for the following values of r. Show that the equilibrium N^* is stable when h is set to the value that maximizes the long-term harvest. Graph the updating function and cobweb.

47. $r = 2.5$, as in the text.

48. $r = 1.5$, as in Exercise 46.

49–50 ▪ Calculate the maximum long-term harvest for an alternative model of competition obeying the discrete-time dynamical system

$$N_{t+1} = \frac{rN_t}{1 + kN_t} - hN_t$$

Try the following steps for the given values of the parameters r and k.

a. Find the equilibrium as a function of h.

b. What is the largest value of h consistent with a positive equilibrium?

c. Find the harvest level that produces the maximum long-term harvest.

d. Sketch a graph of $P(h)$ and compute the value at the maximum.

e. How do the results compare with those using the discrete-time dynamical system in the text?

49. With $r = 2.5$ and $k = 1$.

50. With $r = 1.5$ and $k = 1$.

51–54 ▪ The model of fish harvesting studied in the text includes nothing about harvesting cost. Suppose that the population follows

$$N_{t+1} = 2.5N_t(1 - N_t) - hN_t$$

as in the text, but that the payoff is

$$P(h) = hN^* - ch$$

where c is the cost per unit effort of harvesting. Find the optimal harvest for the following values of c, the associated equilibrium population N^*, and the associated payoff $P(h)$. Do your answers all make sense? What should the fisherman do if c becomes too large?

51. $c = 0.1$.

52. $c = 0.2$.

53. $c = 0.5$.

54. $c = 1.0$.

Computer Exercise

55. Suppose a population follows the updating function

$$N_{t+1} = 2.5N_t(1 - N_t) - hN_t$$

but can be harvested only every second year. This means that harvest alternates between the chosen value h and 0.

a. Find the 2-yr updating function.

b. Find the optimal harvest.

c. Compare with the results in the text. Is it better to harvest less often?

3.4 Reasoning About Functions

Continuous and differentiable functions have many useful properties that can be used to reason about the biological processes they describe without doing a great deal of algebra. In particular, we can make deductions about the solutions of equations, the existence of maxima and minima, or the values of the derivative. We will use the **Intermediate Value Theorem** to show that a discrete-time dynamical system has an equilibrium without solving any equations, the **Extreme Value Theorem** to show that a function has a maximum without computing any derivatives, and the **Mean Value Theorem** to find the value of a derivative without taking any limits.

3.4.1 Continuous Functions: The Intermediate Value Theorem

Consider a model for chemical concentration in the lung which includes absorption,

$$c_{t+1} = (1-q)[1-\alpha(c_t)]c_t + q\gamma. \qquad \textbf{(3.4.1)}$$

In this discrete-time dynamical system, c_t represents the concentration before a breath, c_{t+1} the concentration before the next breath, q the fraction of air exchanged, γ the concentration of chemical in the ambient air, and $\alpha(c_t)$ the fraction of chemical **absorbed** as a function of the chemical concentration in the lung (Subsection 1.9.3). Suppose that $\alpha(c_t)$ is

$$\alpha(c_t) = 0.5(1-e^{-0.5c_t}).$$

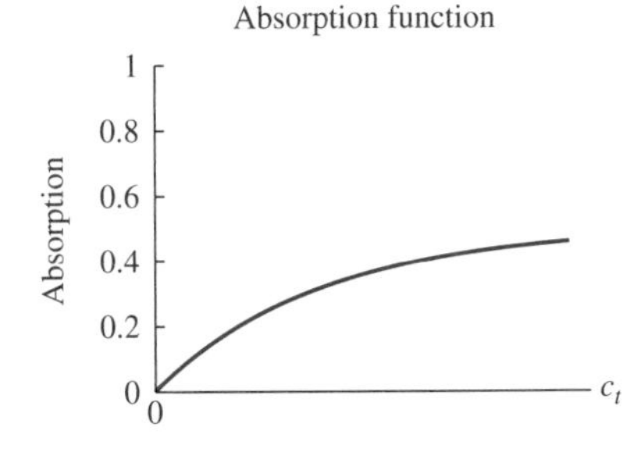

FIGURE 3.4.1
Dynamics of a lung with absorption

(Figure 3.4.1b). Absorption is equal to 0 when $c_t = 0$ because there is nothing to absorb, and the fraction absorbed increases to 0.5 as the concentration becomes larger. What happens to the concentration in the lung? Does it still have an equilibrium?

We begin by trying to finding an equilibrium.

$$c^* = (1-q)\big[1-0.5(1-e^{-5c^*})\big]c^* + q\gamma \qquad \text{the original equation}$$
$$0 = (1-q)\big[1-0.5(1-e^{-5c^*})\big]c^* + q\gamma - c^*. \qquad \text{place unknowns on one side}$$

This equation cannot be factored or solved. Is there any way to establish whether there is an equilibrium and to have some idea where it is?

We can answer these questions without doing any algebra by reasoning about the general process of breathing. One process, breathing in, adds chemical to the lungs, and two processes, breathing out and absorption, remove chemical from the lungs. Without absorption, we found that the equilibrium concentration is γ, the ambient concentration. With absorption, then, we expect the equilibrium to be decreased below γ. We can test this intuition by reasoning about the updating function.

Suppose the lung starts out with $c_t = 0$, the lowest possible concentration. Then

$$c_{t+1} = (1-q)[1-\alpha(0)]\cdot 0 + q\gamma = q\gamma.$$

As long as $q > 0$ (meaning that some air is exchanged) and $\gamma > 0$ (meaning that some chemical is available), $c_{t+1} = q\gamma > 0$. In other words, the amount of chemical has **increased**. Conversely, suppose the lung starts out with a concentration of $c_t = \gamma$. We expect that this concentration will decrease because of absorption.

$$\begin{aligned} c_{t+1} &= (1-q)[1-\alpha(\gamma)]\gamma + q\gamma && \text{discrete-time dynamical system with } c_t = \gamma \\ &= (1-q)\gamma + q\gamma - (1-q)\alpha(\gamma)\gamma && \text{separate out term with } \alpha \\ &= \gamma - (1-q)\alpha(\gamma)\gamma && \text{sum of first two terms is } \gamma \\ &< \gamma. && \text{as long as } \alpha(\gamma) > 0 \text{ and } q < 1 \end{aligned}$$

Absorption reduces the concentration below the ambient concentration, the equilibrium value without absorption.

The graph of the updating function therefore lies above the diagonal at $c = 0$ and below at $c = \gamma$ (Figure 3.4.2). Furthermore, the updating function is **continuous** because

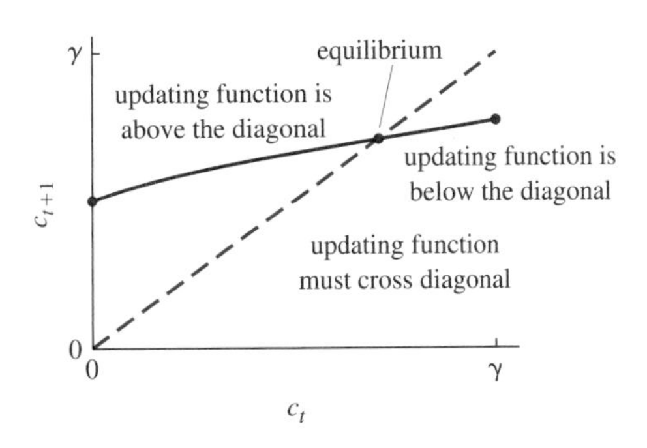

FIGURE 3.4.2
Reasoning about the equilibrium of the lung discrete-time dynamical system with absorption

it is built by combining continuous linear and exponential functions using only multiplication and composition (Section 2.3). The graph of a continuous function has "no jumps," meaning that it is impossible to draw a graph connecting these two points without crossing the diagonal. Such a crossing point is an equilibrium (Section 1.6).

We have found, without solving any equations, that this discrete-time dynamical system *must* have an equilibrium between 0 and γ, in accord with our biological intuition. Furthermore, the updating function must cross the diagonal from above to below. However, we cannot be sure that the equilibrium is stable because the updating function could be decreasing steeply at the equilibrium (Exercise 45).

How do we prove mathematically that the updating function must cross the diagonal? This result follows from the **Intermediate Value Theorem** for continuous functions.

Theorem 3.2 **Intermediate Value Theorem**

If $f(x)$ is continuous for $a \leq x \leq b$ and $f(a) < c < f(b)$, then there is some x between a and b such that $f(x) = c$.

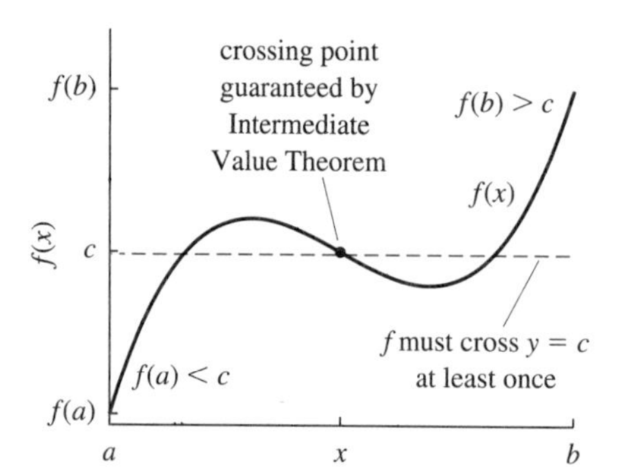

FIGURE 3.4.3
The Intermediate Value Theorem

The proof of this simple theorem is subtle, requiring deep facts about real numbers and continuity. The concept of the Intermediate Value Theorem is shown in Figure 3.4.3. The theorem guarantees that there is **at least** one solution of $f(x) = c$, but there may be more.

Example 3.4.1 The Intermediate Value Theorem Applied to Height

Physically, if you grew from 2 feet to 6 feet in height, you must have been exactly 4 feet tall at some time (Figure 3.4.4).

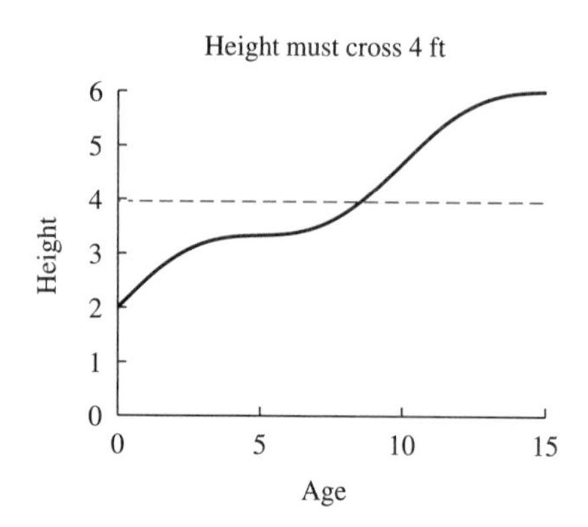

FIGURE 3.4.4
The Intermediate Value Theorem applied to height

Example 3.4.2 The Intermediate Value Theorem Applied to Speed

If you accelerate from 0 to 60 miles per hour (mph), you must have been going exactly 31.4159 mph at some time (Figure 3.4.5).

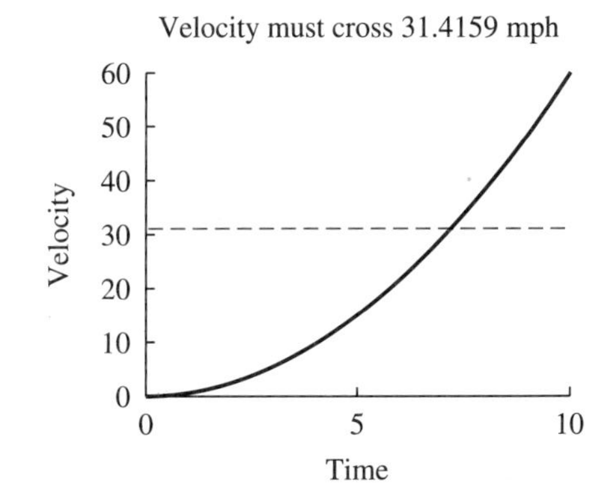

FIGURE 3.4.5
The Intermediate Value Theorem applied to speed

Example 3.4.3 The Intermediate Value Theorem Applied to Solving an Equation

Suppose we wish to show there is a value of $x > 0$ that solves

$$e^x = 5x + 10.$$

Define the function as $f(x) = e^x - 5x - 10$. Then $f(x) = 0$ at any solution of the original equation (Figure 3.4.6). Checking $x = 0$ gives $f(0) = e^0 - 5 \cdot 0 - 10 = -9 < 0$. If we can find a positive value of x such that $f(x) > 0$, we will have found a solution. Trying successive integers gives

x	$f(x)$
1	$e^1 - 5 - 10 = -12.28 < 0$
2	$e^2 - 10 - 10 = -12.61 < 0$
3	$e^3 - 15 - 10 = -4.91 < 0$
4	$e^4 - 20 - 10 = 24.60 > 0$

There must be a solution for $3 < x < 4$.

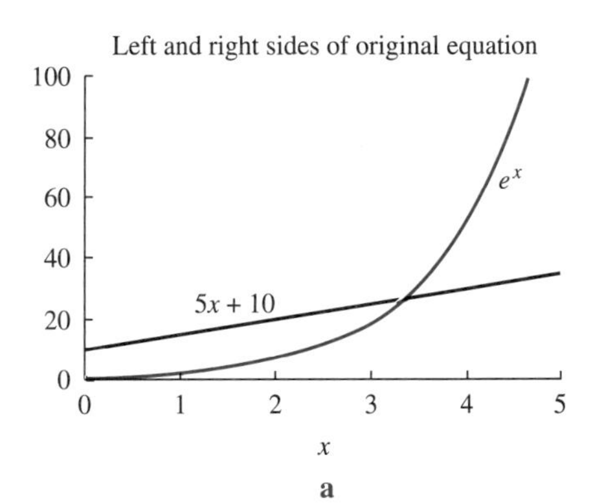

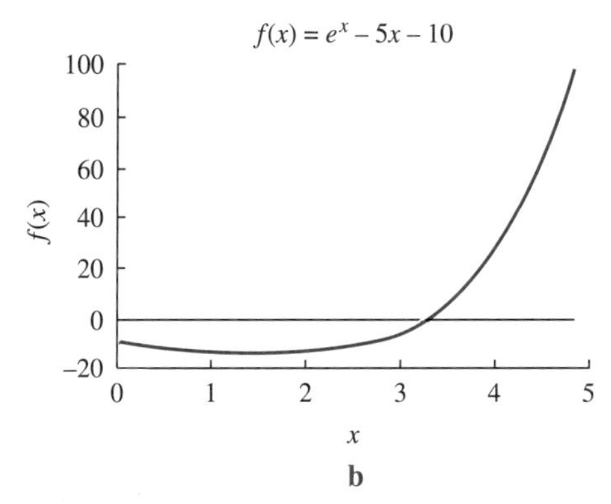

FIGURE 3.4.6
The Intermediate Value Theorem applied to solving an equation

We need to make a transformation to apply the Intermediate Value Theorem to the equilibria of the lung discrete-time dynamical system with absorption. An equilibrium is a point where the **change in concentration** is equal to 0 (Figure 3.4.7). The change in concentration Δc is

$$\Delta c = c_{t+1} - c_t.$$

At $c_t = 0$, $c_{t+1} > c_t$ and $\Delta c > 0$. At $c_t = \gamma$, $c_{t+1} < \gamma$ and $\Delta c < 0$. The Intermediate Value Theorem guarantees that there must be some point where $\Delta c = 0$ (Figure 3.4.7). This point is the equilibrium.

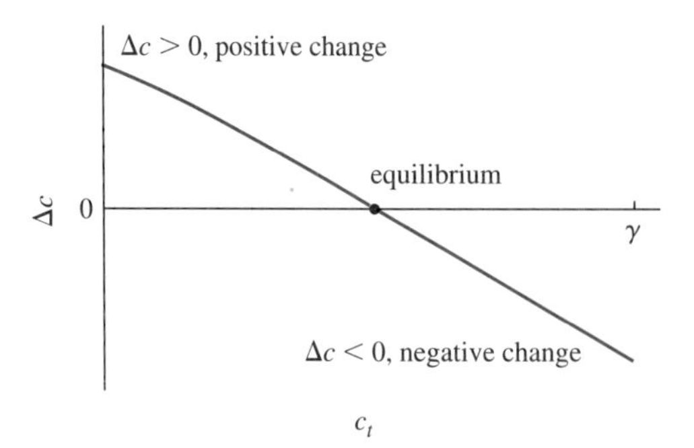

FIGURE 3.4.7
Reasoning about the equilibrium of the lung discrete-time dynamical system using the Intermediate Value Theorem

3.4.2 Maximization: The Extreme Value Theorem

Suppose that the per capita production of a population of fish is

$$\text{per capita production} = 2.5e^{-N_t}$$

and that a factor h times the pre-productive population is harvested each year (modified from equation 3.3.1 and Section 1.9, Exercise 50). The discrete-time dynamical system for the population is

$$N_{t+1} = 2.5N_t e^{-N_t} - hN_t. \tag{3.4.2}$$

We want to find the harvesting effort h that maximizes long-term harvest. Recall the steps used in Subsection 3.3.3: First find the equilibrium N^* as a function of h, then find the fish harvested as $P(h) = hN^*$, and then compute the maximum of $P(h)$ by differentiating.

To find the equilibria, follow the usual steps.

$$\begin{aligned}
N^* &= 2.5N^*e^{-N^*} - hN^* && \text{the original equation} \\
N^* - 2.5N^*e^{-N^*} + hN^* &= 0 && \text{move everything to one side} \\
N^*(1 - 2.5e^{-N^*} + h) &= 0 && \text{factor} \\
N^* = 0 \text{ or } (1 - 2.5e^{-N^*} &+ h) = 0. && \text{set each piece equal to 0}
\end{aligned}$$

Solving the second part requires a bit of algebra.

$$\begin{aligned}
1 - 2.5e^{-N^*} + h &= 0 && \text{original equation} \\
1 + h &= 2.5e^{-N^*} && \text{move unknowns to one side} \\
\frac{1+h}{2.5} &= e^{-N^*} && \text{divide by 2.5} \\
\ln\left(\frac{1+h}{2.5}\right) &= -N^* && \text{take the natural logarithm} \\
N^* &= -\ln\left(\frac{1+h}{2.5}\right) && \text{solve for } N^* \\
N^* &= \ln\left(\frac{2.5}{1+h}\right). && \text{use law 3 of logs}
\end{aligned}$$

Is this value positive? Recall that $\ln(x) > 0$ if $x > 1$. Therefore N^* is positive if

$$\begin{aligned}
\frac{2.5}{1+h} &> 1 \\
2.5 &> 1 + h \\
1.5 &> h.
\end{aligned}$$

If $h = 1.5$, then the equilibrium is 0.

The harvest $P(h)$ is the factor h times the total population N^*, so

$$P(h) = hN^* = h \ln\left(\frac{2.5}{1+h}\right).$$

Does this equation have a maximum? We have two algorithms for finding a maximum, each requiring that we find critical points by computing the derivative and finding where it is equal to 0. In this case,

$$P'(h) = \ln\left(\frac{2.5}{1+h}\right) - \frac{h}{1+h}.$$

Finding critical points requires solving the equation $P'(h) = 0$. This equation cannot be solved algebraically.

Nonetheless, we can still prove that this function has a maximum. We know that $P(0) = 0$ (no harvesting) and that $P(1.5) = 0$ (no fish). Furthermore, $P(h) > 0$ if $0 < h < 1.5$ because both h and N^* are positive (Figure 3.4.8). A function that is 0 at its endpoints and positive in between must have a maximum. Mathematically, this result follows from the **Extreme Value Theorem**.

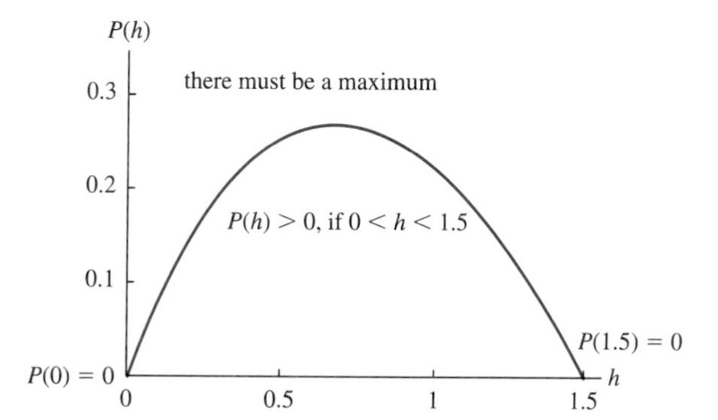

FIGURE 3.4.8
Reasoning about harvest: the Extreme Value Theorem

Theorem 3.3 **Extreme Value Theorem**

If $f(x)$ is continuous for $a \le x \le b$, then there is a point c_h, $a \le c_h \le b$, where $f(x)$ takes on its global maximum and a point c_l, $a \le c_l \le b$, where $f(x)$ takes on its global minimum.

Again, the proof of this theorem in general is quite subtle. The conclusions are illustrated in Figure 3.4.9. The theorem does not guarantee that the maximum and minimum must occur between a and b. Either might lie at one of the endpoints (the minimum in Figure 3.4.9 lies at the endpoint b).

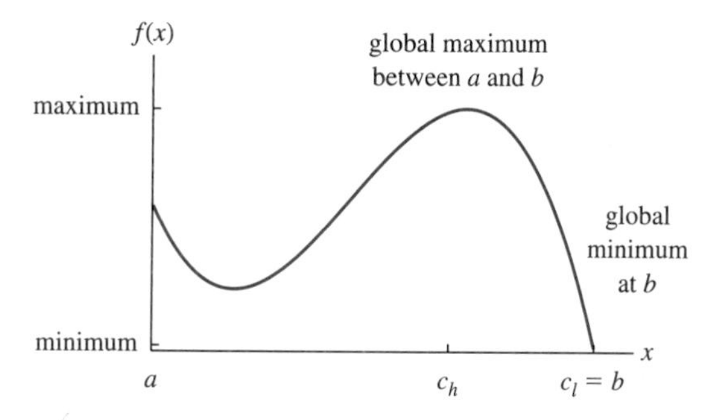

FIGURE 3.4.9
The Extreme Value Theorem

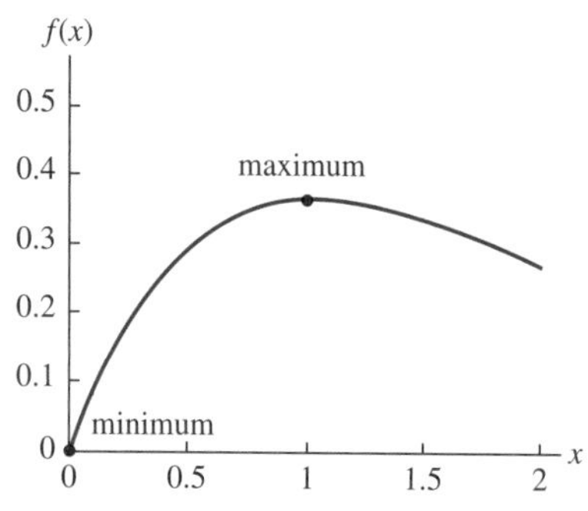

FIGURE 3.4.10
The function $f(x) = xe^{-x}$

Example 3.4.4 Applying the Extreme Value Theorem

Consider the function $f(x) = xe^{-x}$ (the gamma distribution with $n = 1$ studied in Section 2.8, Example 2.8.7). On the interval $0 \le x \le 2$, the Extreme Value Theorem guarantees that the function has a maximum and a minimum on this interval (Figure 3.4.10). In Subsection 3.3.1, we found that this function has a maximum at $x = 1$ and a minimum at $x = 0$.

The function $P(h)$ is continuous between $h = 0$ and $h = 1.5$ because it involves no division or logs of 0. Therefore, the Extreme Value Theorem guarantees that it has a maximum. We know that the maximum does not lie at the endpoints because $P(h)$ takes on positive values, larger than the value at each endpoint, for all values of h between 0 and 1.5. The maximum guaranteed by the Extreme Value Theorem must occur for $0 < h < 1.5$. The minimum guaranteed by the Extreme Value Theorem is shared by the two endpoints.

Example 3.4.5 Failure of Extreme Value Theorem When a Function Is Not Continuous

Why must the function in the Extreme Value Theorem be continuous? Consider the function $f(x) = \frac{1}{x}$ defined for x between -1 and 1, and with $f(0) = 0$ (Figure 3.4.11a). This function has a right limit of infinity and a left limit of negative infinity at $x = 0$, and has neither a maximum nor a minimum value.

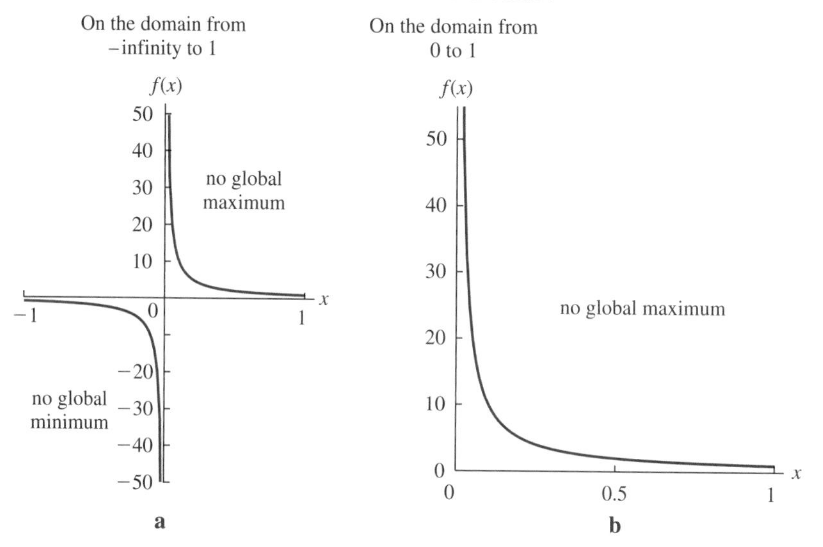

FIGURE 3.4.11
The function $f(x) = \frac{1}{x}$ fails to satisfy the conditions for the Extreme Value Theorem

Even if we quarantine the trouble point $x = 0$ to the end of the interval, the Extreme Value Theorem breaks down. If we define the function only for $0 < x \le 1$, $f(x)$ is perfectly continuous. However, because it is not defined at the endpoint, it fails to have a maximum (Figure 3.4.11b).

3.4.3 Rolle's Theorem and the Mean Value Theorem

The Intermediate Value Theorem and the Extreme Value Theorem guarantee that a continuous function must take on particular values. **Rolle's Theorem** and the **Mean Value Theorem** guarantee that the **derivative** takes on particular values.

Rolle's theorem is closely related to the Extreme Value Theorem. If $P(h)$ is differentiable, the derivative of $P(h)$ will be 0 at its maximum (Figure 3.4.8). Rolle's Theorem states that a differentiable function that takes on equal values at its endpoints, must have a derivative equal to 0 at some point in between.

Theorem 3.4 **Rolle's Theorem**

If $f(x)$ is differentiable for all x, $a \le x \le b$, and $f(a) = f(b)$, then there exists some c with $a < c < b$ and $f'(c) = 0$.

As with the Intermediate Value Theorem, this theorem guarantees only that there is at least one point with derivative 0; there may be more than one (Figure 3.4.12).

Example 3.4.6 Application of Rolle's Theorem

The function $g(x) = x(1-x)$ (studied in Example 3.3.1) takes on the value 0 for both $x = 0$ and $x = 1$ (Figure 3.4.13). Therefore, Rolle's theorem guarantees that there must be at least one point with derivative equal to 0 for $0 \le x \le 1$. By taking the derivative, we found that there is only a single such point, $x = 0.5$.

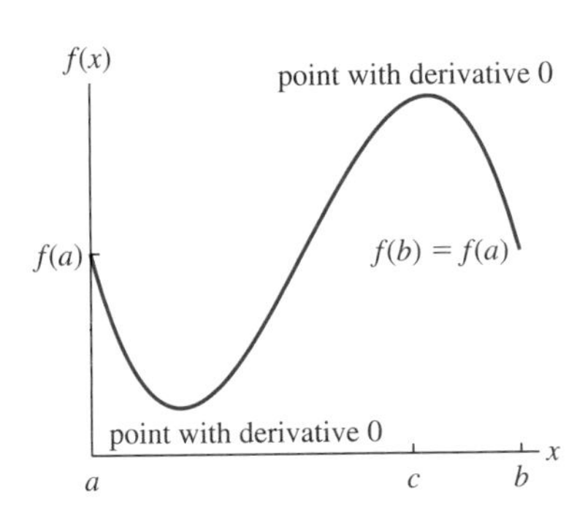

FIGURE 3.4.12
Rolle's Theorem

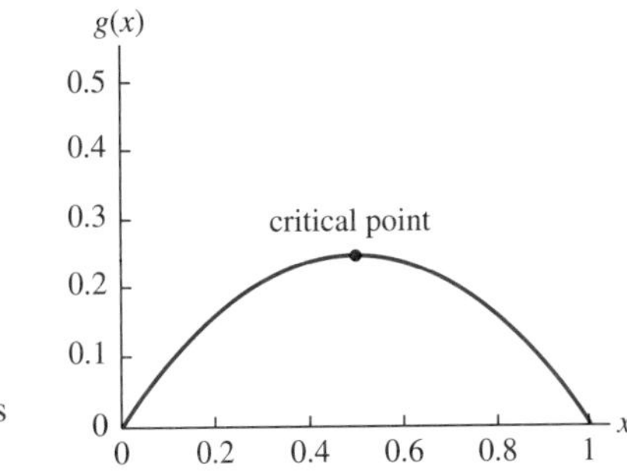

FIGURE 3.4.13
Application of Rolle's Theorem

Example 3.4.7 Application of Rolle's Theorem to $P(h)$

This theorem applies directly to the harvest $P(h)$. Rolle's theorem guarantees the existence of a critical point with derivative equal to 0 for $0 < h < 1.5$. Because the function is positive, the value of the function P at the critical point must be greater than the value at the endpoints.

The proof of Rolle's Theorem uses the Extreme Value Theorem to show that the function must have a minimum or maximum in the interior of the interval (unless the function is constant) and to show that an interior minimum or maximum must occur at a critical point. If the function is differentiable, then the critical point must be one where the derivative is equal to 0.

The Mean Value Theorem is a tilted version of Rolle's theorem.

Theorem 3.5 **Mean Value Theorem:**

If $f(x)$ is differentiable for $a \le x \le b$, then there exists some c such that $a < c < b$ and

$$f'(c) = \frac{f(b) - f(a)}{b - a}.$$

This theorem says that the slope of the tangent line at some point in the interval matches the slope of the secant line spanning that interval (Figure 3.4.14). Alternatively, if the

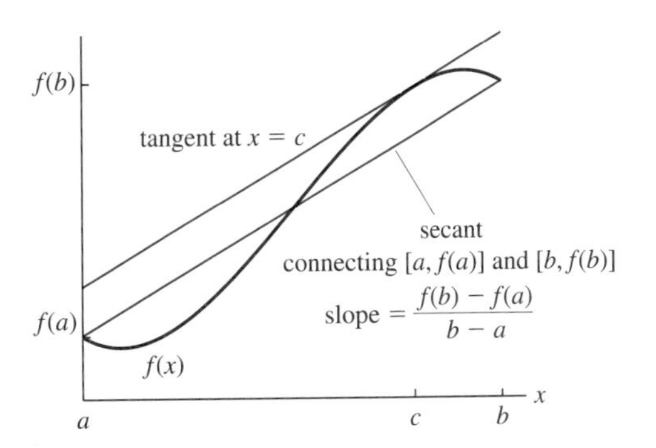

FIGURE 3.4.14
The Mean Value Theorem

average rate of change during the interval from a to b is A, there is a point in the interval when the instantaneous rate of change is equal to the average rate of change.

Example 3.4.8 Application of the Mean Value Theorem to Velocity

The most popular application of this theorem involves velocity. If a car travels 140 miles in two hours, the average rate of change over this time is 70 mph. A graph of position versus time must pass through the two points (0, 0) and (2, 140), and the line connecting these is a secant line. The Mean Value Theorem guarantees that the instantaneous speed (on the speedometer) must have been exactly 70 mph at some time (Figure 3.4.15).

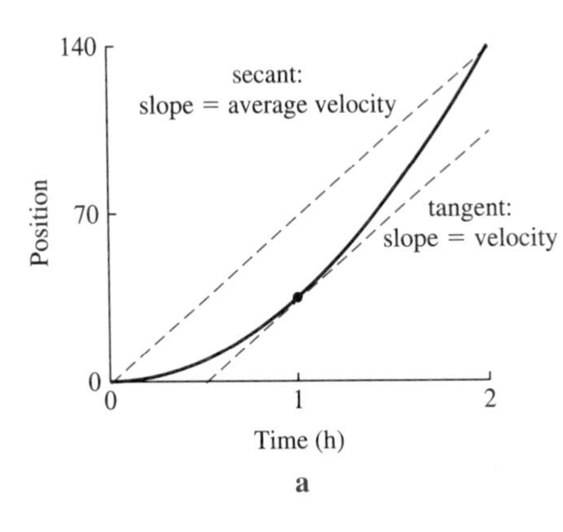

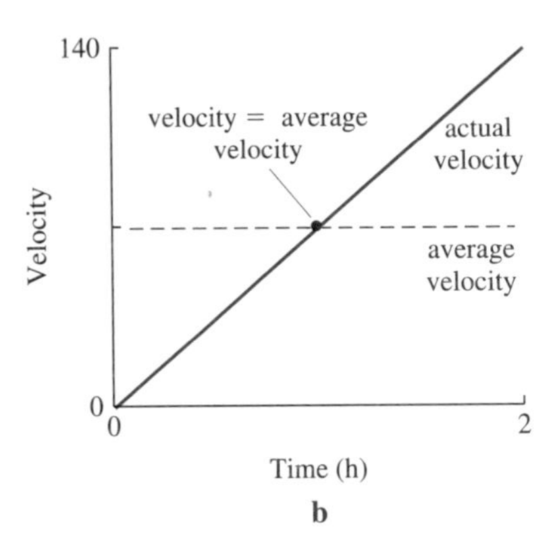

FIGURE 3.4.15
The Mean Value Theorem applied to velocity

Example 3.4.9 Finding the Point Guaranteed by the Mean Value Theorem

The Mean Value Theorem guarantees that, for some $1 \leq x \leq 2$, the slope of $h(x) = \ln(x)$ must exactly match the slope of the secant connecting $x = 1$ and $x = 2$ (Figure 3.4.16).

$$\text{slope of secant} = \frac{h(2) - h(1)}{2 - 1} = \frac{\ln(2) - \ln(1)}{1} = \ln(2).$$

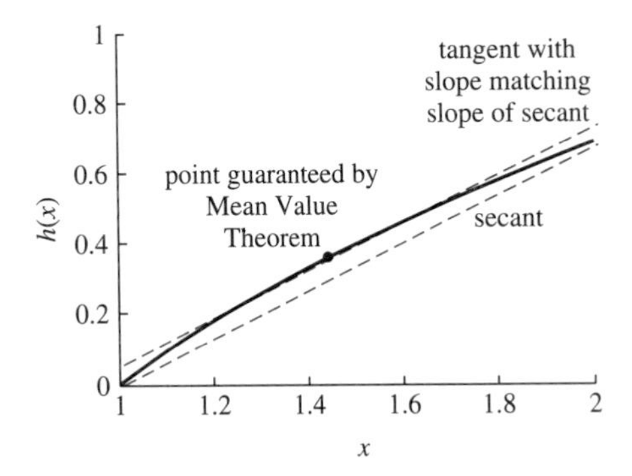

FIGURE 3.4.16
The Mean Value Theorem applied to $\ln(x)$

Where is $h'(x) = \ln(2)$? We recall that

$$\frac{d}{dx}\ln(x) = \frac{1}{x},$$

so the value solves

$$h'(x) = \frac{1}{x} = \ln(2).$$

Therefore, $x = \frac{1}{\ln(2)} = 1.442.$

Summary With a minimum of algebra, we can use theorems about continuous and differentiable functions to deduce mathematical conclusions. We began by showing that a version of the lung model with absorption must have an equilibrium by using the **Intermediate Value Theorem**, which guarantees that a continuous function takes on all values between those at its endpoints. We then argued that a new version of the harvesting model must take on a maximum value with the **Extreme Value Theorem**, which states that a continuous function that is defined on a domain including the endpoints must have a maximum and a minimum. We applied **Rolle's Theorem** to the same problem, arguing that a differentiable function that takes on equal values at its endpoints must have a critical point with a derivative equal to 0 in between. Its generalization, the **Mean Value Theorem**, states that the derivative of a differentiable function must at some point match the slope of the secant.

3.4 Exercises

Mathematical Techniques

1–6 ▪ Use the Intermediate Value Theorem to show that the following equations have solutions for $0 \le x \le 1$.

1. $e^x + x^2 - 2 = 0$.
2. $e^x - 3x^2 = 0$.
3. $e^x + x^2 - 2 = x$.
4. $e^x + x^2 - 2 = \cos(2\pi x) - 1$.
5. $xe^{-3(x-1)} - 2 = 0$ (you will need to check an intermediate point).
6. $x^3 e^{-4(x-1)} - 1.1 = 0$ (you will need to check an intermediate point).

7–10 ▪ Use the Extreme Value Theorem to show that each of the following functions has a positive maximum on the interval $0 \le x \le 1$.

7. $f(x) = x(x-1)e^x$.
8. $f(x) = ex - xe^x$.
9. $f(x) = 5x(1-x)(2-x) - 1$.
10. $f(x) = 8x(1-x)^2 - 1$.

11–14 ▪ Find the points guaranteed by the Mean Value Theorem and sketch the associated graph.

11. The slope of the function $f(x) = x^2$ must match the slope of the secant connecting $x = 0$ and $x = 1$.
12. The slope of the function $f(x) = x^2$ must match the slope of the secant connecting $x = 0$ and $x = 2$.
13. The slope of the function $g(x) = \sqrt{x}$ must match the slope of the secant connecting $x = 0$ and $x = 1$.
14. The slope of the function $g(x) = \sqrt{x}$ must match the slope of the secant connecting $x = 0$ and $x = 2$.

15–20 ▪ Draw functions with the following properties.

15. A function with a global minimum and global maximum between the endpoints.
16. A function with a global maximum at the left endpoint and global minimum between the endpoints.
17. A differentiable function with a global maximum at the left endpoint, a global minimum at the right endpoint, and no critical points.
18. A function with a global maximum at the left endpoint, a global minimum at the right endpoint, and at least one critical point.
19. A function with a global minimum and global maximum between the endpoints, but no critical points.
20. A function that never reaches a global maximum.

21–22 ▪ Check whether the Intermediate Value Theorem and Mean Value Theorem fail in the following cases where the function is not continuous.

21. Consider the Heaviside function (Section 2.3, Exercise 25) defined by

$$\begin{cases} H(x) = 0 & \text{if } x < 0 \\ H(x) = 1 & \text{if } x \ge 0. \end{cases}$$

Show that there is no solution to the equation $H(x) = 0.5$, and that there is no tangent that matches the slope of the secant connecting $x = -1$ and $x = 1$.

22. Consider the absolute value function $g(x) = |x|$. Does this satisfy the conditions for the Intermediate Value Theorem? Show that there is no tangent that matches the slope of the secant connecting $x = -1$ and $x = 2$.

23–24 ▪ There is a clever proof of the Mean Value Theorem from Rolle's Theorem. The idea is to tilt the function f so that it takes on the same values at the endpoints a and b. In particular, we apply Rolle's Theorem to the function

$$g(x) = f(x) - (x-a)\frac{f(b) - f(a)}{b - a}.$$

For the following functions, show that $g(a) = g(b)$, apply Rolle's Theorem to g, and find the derivative of f at a point where $g'(x) = 0$.

23. $f(x) = x^2$, $a = 1$, and $b = 2$.
24. In general, without assuming a particular form for $f(x)$ or values for a and b.

Applications

25–28 ▪ Try to apply the Intermediate Value Theorem to the following problems.

25. The price of gasoline rises from \$1.199 to \$1.279. Why is it not necessarily true that the price was exactly \$1.25 at some time?

26. A pot is dropped from the top of a 500-ft building exactly 200 ft above your office. Must it have fallen right past your office window?

27. A cell takes up 1.5×10^{-9} milliliters of water in the course of an hour. Must the cell have taken up exactly 1.0×10^{-9} milliliters at some time? Is it possible that the cell took up exactly 2.0×10^{-9} milliliters at some time?

28. The population of bears in Yellowstone Park has increased from 100 to 1000. Must it have been exactly 314 at some time? What additional assumption would guarantee this?

29–32 ▪ The Intermediate Value Theorem can often be used to prove that complicated discrete-time dynamical systems have equilibria.

29. Prove that the discrete-time dynamical system $x_{t+1} = \cos(x_t)$ has an equilibrium between 0 and $\frac{\pi}{2}$.

30. A lung follows the discrete-time dynamical system $c_{t+1} = 0.25e^{-3c_t}c_t + 0.75\gamma$ where $\gamma = 5.0$. Show that there is an equilibrium between 0 and γ.

31. A lung follows the discrete-time dynamical system $c_{t+1} = 0.75\alpha(c_t)c_t + 0.25\gamma$ where $\gamma = 5.0$ and the function $\alpha(c_t)$ is positive, decreasing, and $\alpha(0) = 1$. Show that there is an equilibrium between 0 and γ.

32. A lung follows the discrete-time dynamical system $c_{t+1} = f(c_t)$. We know only that neither c_{t+1} nor c_t can exceed 1 mole/liter. Use the Intermediate Value Theorem to show that this discrete-time dynamical system must have an equilibrium.

33–34 ▪ The Intermediate Value Theorem has applications in agricultural transport.

33. A farmer sets off on Saturday morning at 6 a.m. to bring a crop to market, arriving in town at noon. On Sunday she sets off in the opposite direction at 6 a.m. and returns home along the same route, arriving once again at noon. Use the Intermediate Value Theorem to show that at some point along the path, her watch must read exactly the same time on the two days.

34. Suppose instead that the farmer sets off one morning at 6 a.m. to bring a crop to market and arrives in town at noon. Having received a great price for her crop, she buys a new car and drives home the next day along the same route, leaving at 10 a.m. and arriving home at 11 a.m. Is it still true that at some point along the route her perfectly accurate watch must read exactly the same time on the two days? If so, must that time occur between 10 and 11 a.m.?

35–38 ▪ An organism grows from 4 kg to 60 kg in 14 yr. Suppose that mass is a differentiable function of time.

35. Why must the mass have been exactly 10 kg at some time?

36. Why must the rate of increase have been exactly 4.0 kg/yr at some time?

37. Draw a graph of mass against time where the mass is increasing, is equal to 10.0 kg at 13 yr, and has a growth rate of exactly 4.0 kg/yr after 1 yr.

38. Draw a graph of mass against time where the organism reaches 10.0 kg at 1 yr, and has a growth rate of exactly 4.0 kg/yr at 13 yr.

39–42 ▪ Draw the positions of cars from the following descriptions of 1-h trips. What speed must the car achieve according to the Mean Value Theorem? What speeds must the car achieve according to the Intermediate Value Theorem?

39. A car starts at 60 mph and slows down to 0 mph. The average speed is 20 mph after 1 h.

40. A car starts at 60 mph, steadily slows down to 20 mph, and then speeds up to 50 mph by the end of 1 h. The average speed over the whole time is 40 mph.

41. A car drives 60 miles in 1 h, and never varies speed by more than 10 mph.

42. In a test, a car drives zero net distance in 1 h by switching from reverse to forward at some point. The test includes achieving the maximum possible reverse speed (20 mph) and the maximum possible forward speed (120 mph).

43–44 ▪ The Marginal Value Theorem states that the best time t to leave a patch of cabbage is the solution t of the equation

$$F'(t) = \frac{F(t)}{t + \tau}$$

where τ is the travel time to the next cabbage patch and $F(t)$ is the total amount of food gathered in one location in time t. Suppose that $\tau = 1$ and $F(t) = 1 - e^{-t}$.

43. Sketch the associated figure (as in Figure 3.3.14) and estimate the solution.

44. Use the Intermediate Value Theorem to prove that there is a solution.

Computer Exercise

45. Consider the model

$$c_{t+1} = (1 - q)[1 - \alpha(c_t)]c_t + q\gamma$$

where

$$\alpha(c) = \frac{c^n}{1 + c^n}.$$

Study the behavior of the model for $n = 1$, $n = 5$, and $n = 15$, and for values of q between 0 and 1 (you can pick any value of γ). When is the equilibrium stable? Can you explain in biological terms why the equilibrium is stable when $n = 15$ and q is either near 0 or near 1?

3.5 Limits at Infinity

Models of biological systems sometimes lead to complicated functional forms. Reasoning about these functions often requires computing the behavior of the function at the endpoints of its domain. When there is no natural upper bound to the domain, we must figure out what happens to the function as its argument gets very large. To do so, we generalize the limit to include **limits at infinity** and study the behavior of the exponential, power, and logarithmic functions, finding which approach infinity, 0 or other values. Because many biological processes involve more than one basic function, we must be able to **compare** them. The key tool we will use is a way to formalize whether one function approaches infinity or 0 **faster** or **slower** than another. The concept of limits at infinity can be used to study **limits of sequences**, the output of discrete-time dynamical systems with stable equilibria.

3.5.1 The Behavior of Functions at Infinity

Suppose the function

$$\alpha(c) = 0.5(1 - e^{-0.5c})$$

describes the fraction of chemical absorbed by a lung during each breath. The total amount absorbed with each breath is

$$\begin{aligned}\text{amount absorbed } &= \alpha(c)cV \\ &= 0.5(1 - e^{-0.5c})cV,\end{aligned}$$

the product of the fraction absorbed, the concentration c, and the volume V. What does the graph of the function $\alpha(c)$ look like? What does the graph of the total amount absorbed look like?

The amount of a chemical or resource used as a function of the amount available is important throughout biology. For consumers, like predators, this relation is called the **functional response**. In chemical reactions, this relation is often described by **Michaelis-Menten** or **Monod** reaction kinetics.

Several possible absorption functions are given in Table 3.1. Each describes the total amount of chemical absorbed as a function of the concentration c. The parameter A is a measure of efficiency, with small values producing low absorption and large values producing high absorption. The parameters k and β describe the shape of the function.

In each case, absorption is 0 when $c = 0$. The behavior of the absorption function for large values of c describes absorption at high concentrations. How do we compute and describe the functions as their arguments become large?

Table 3.1 Some Different Absorption Functions

Function of c	Description	Figure
Ac	Linear absorption	Figure 3.5.1a
$\dfrac{Ac}{k+c}$	Saturated absorption	Figure 3.5.1b
$\dfrac{Ac^2}{k+c^2}$	Saturated absorption with threshold	Figure 3.5.1c
$Ace^{-\beta c}$	Saturated absorption with overcompensation I	Figure 3.5.1d
$\dfrac{Ac}{k+c^2}$	Saturated absorption with overcompensation II	Figure 3.5.1e
$Ac(1+kc)$	Enhanced absorption	Figure 3.5.1f

Recall that infinity is the mathematician's abstraction of the scientist's idea of "very large." Because infinity is not a number that we can plug into equations, we can only **approach** it. The limit (Section 2.2) tells us how a function behaves as the argument approaches some finite value. We now extend this idea to find limits as the argument becomes very large, or "approaches infinity."

To say that a function $f(x)$ approaches a limit L as x approaches infinity means that the value gets closer and closer to L as x gets huge (Figure 3.5.2a). In other words, a function f approaches the limit L as x approaches infinity if the measurement eventually becomes indistinguishable from the limit.

A function approaches infinity as x approaches infinity if the value gets larger and larger as x gets huge (Figure 3.5.3a). In other words, the function f approaches infinity as x approaches infinity if the output eventually overflows any given measurement device. Similarly, a function approaches negative infinity as x approaches infinity if the value gets smaller and smaller (more and more negative numbers) as x gets huge (Figure 3.5.3b).

Before studying the limits of the absorption functions shown in Figure 3.5.1 as c approaches infinity, we will learn how to compare the limits of power, logarithmic, and exponential functions.

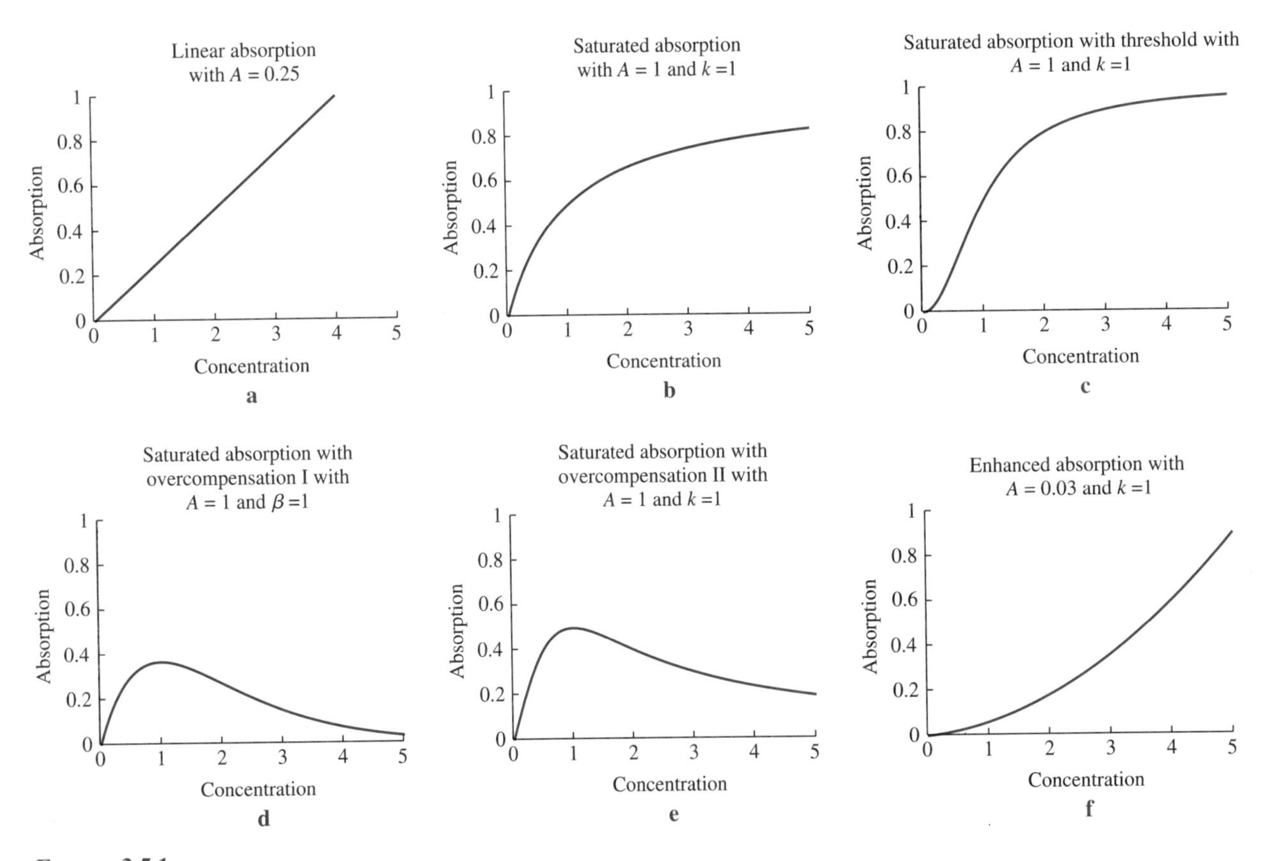

FIGURE 3.5.1
Various absorption functions

Comparing Functions that Approach ∞ at ∞ Many of the fundamental functions of biology approach infinity at infinity, including the natural logarithm $\ln(x)$, the power function x^n with positive n, and the exponential function e^x.

To reason about more complicated functions, like those in Table 3.1, we must **compare** the behavior of these basic functions. The graph of e^x increases very quickly, while that of $\ln(x)$ increases slowly. Is there a precise way in which the exponential function increases "faster" than the logarithmic function? What does it mean to approach infinity "faster" or "slower"? These relations are summarized in the following definition.

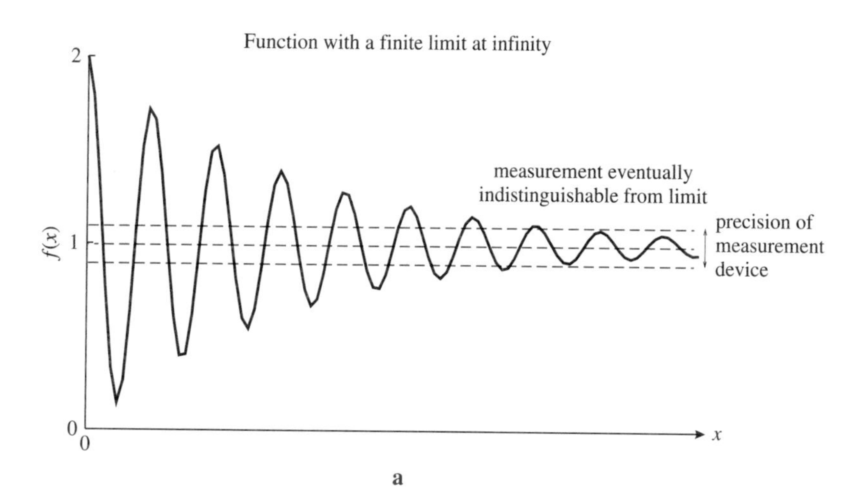

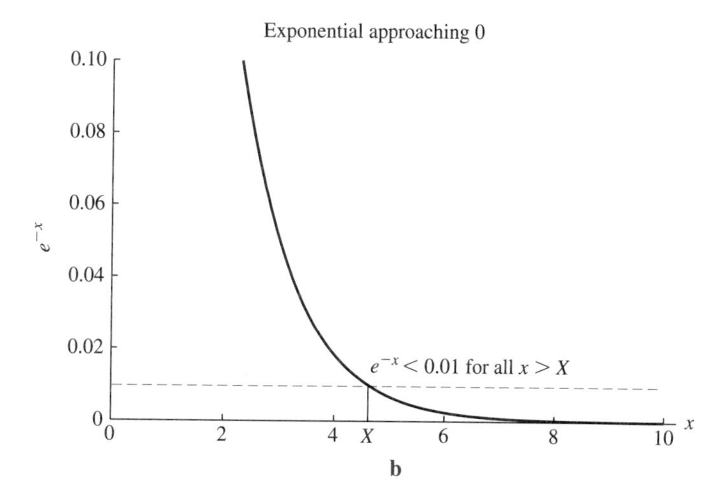

FIGURE 3.5.2
Functions with finite limits at infinity

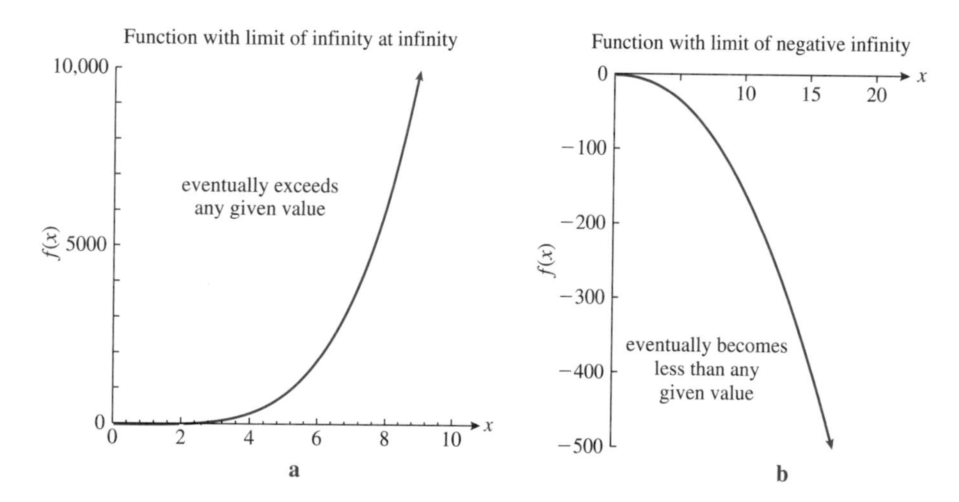

FIGURE 3.5.3
Functions with infinite limits at infinity

Definition 3.1 Suppose

$$\lim_{x \to \infty} f(x) = \infty$$
$$\lim_{x \to \infty} g(x) = \infty.$$

1. The function $f(x)$ approaches infinity **faster** than $g(x)$ if

$$\lim_{x \to \infty} \frac{f(x)}{g(x)} = \infty.$$

2. The function $f(x)$ approaches infinity **slower** than $g(x)$ if

$$\lim_{x \to \infty} \frac{f(x)}{g(x)} = 0.$$

3. $f(x)$ and $g(x)$ approach infinity at the **same rate** if

$$\lim_{x \to \infty} \frac{f(x)}{g(x)} = L$$

where L is any finite number other than 0.

When $f(x)$ approaches infinity faster than $g(x)$, $f(x)$ gets farther and farther ahead (Figure 3.5.4a). When $f(x)$ approaches infinity more slowly than $g(x)$, $f(x)$ falls farther and farther behind $g(x)$ (Figure 3.5.4b). When the two functions approach infinity at the same rate, neither gets ahead or falls behind (Figure 3.5.4c). *Faster*, *slower*, and at *the same rate* act like *greater than, less than*, and *equal* for numbers. They provide a way to compare the "sizes" of functions.

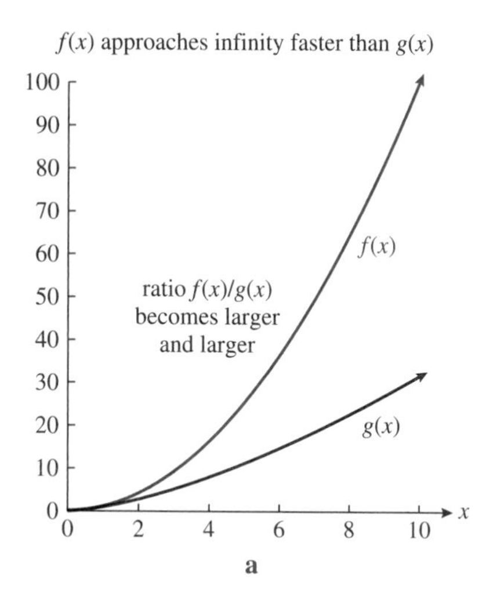

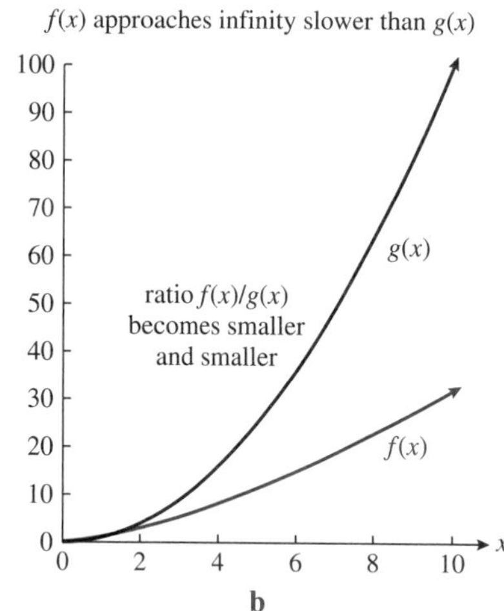

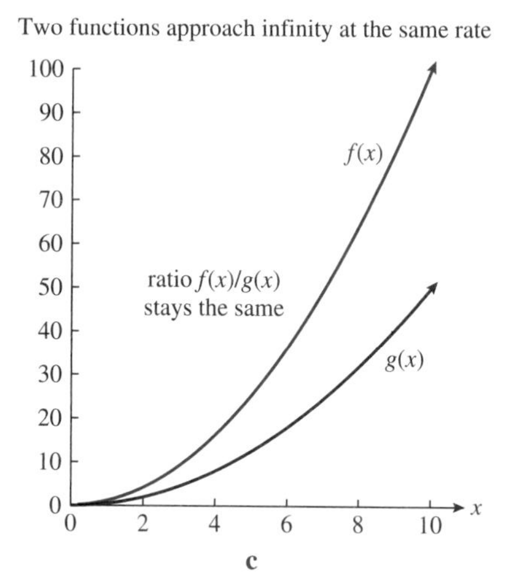

FIGURE 3.5.4

Comparing functions at infinity

The basic functions are shown in increasing order in Table 3.2 and Figure 3.5.5. The constant a in front of each function can be any positive number and does not change

Table 3.2 The Basic Functions in Increasing Order of Speed

Function	Comments
$a \ln(x)$	Goes to infinity slowly
ax^n with $n > 0$	Approaches infinity faster for larger n
$ae^{\beta x}$ with $\beta > 0$	Approaches infinity faster for larger β

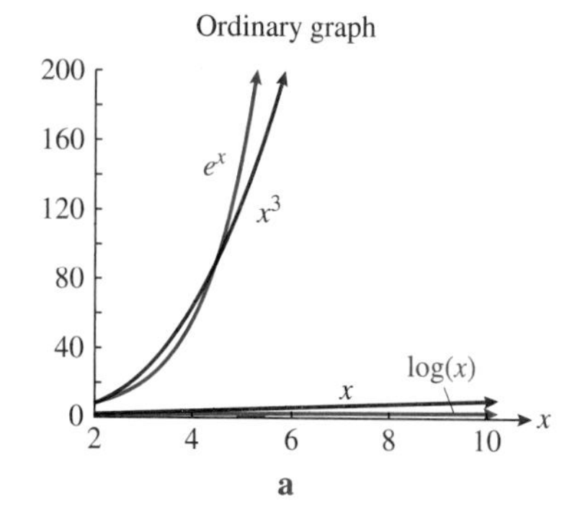

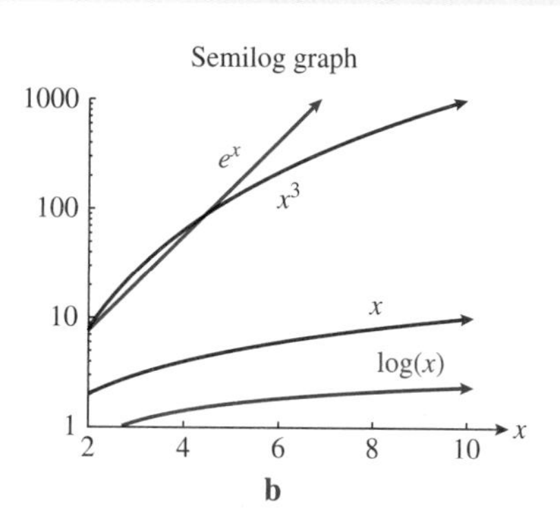

FIGURE 3.5.5

The behavior of the basic functions that approach infinity

the order of the functions. Any power function, however small the power n, beats the logarithm. Any exponential function with a positive parameter β in the exponent beats any power function.

Example 3.5.1 Ordering a Set of Functions

To order the functions

$$0.1e^{2x},\ 4.5\ln(x),\ 23.2x^{0.5},\ 10.1e^{0.2x},\ 0.03x^4$$

in increasing order, first spot functions of the three types: logarithmic, power, and exponential. There is only one logarithmic function, which is therefore the slowest. There are two power functions, with $23.2x^{0.5}$ having the smaller power and $0.03x^4$ the larger power. There are two exponential functions, with $10.1e^{0.2x}$ having the smaller parameter (0.2) inside the exponent, and the exponential function $0.1e^{2x}$ having the larger parameter (2) inside the exponent. In increasing order, these functions are $4.5\ln(x)$, $23.2x^{0.5}$, $0.03x^4$, $10.1e^{0.2x}$, $0.1e^{2x}$. The constants in front do not affect the ordering.

There are two cautions regarding this method. First, comparing functions that are not logarithmic, power, or exponential functions requires different techniques (Section 3.6). Second, these results hold only for very large values of x.

Example 3.5.2 Functions That Take a Long Time to Get into Order

The power function x^2 does eventually grow faster than $1000x$ because it has a larger power (Figure 3.5.6). However, $x^2 < 1000x$ for $x < 1000$. If x cannot realistically take on values greater than 1000, the comparison in Table 3.2 is not relevant. When a comparison includes a large or small parameter (like 1000 in this case), we must first check whether the faster function becomes larger for biologically reasonable values of the argument.

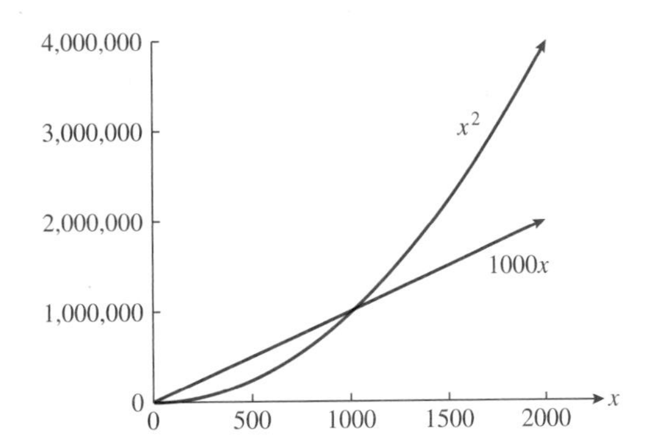

FIGURE 3.5.6
A faster function eventually overtaking a slower function

Functions Approaching 0 at ∞ We use a similar approach to compare the rate at which functions approach a limit of 0.

Definition 3.2 Suppose

$$\lim_{x\to\infty} f(x) = 0$$

$$\lim_{x\to\infty} g(x) = 0.$$

1. The function $f(x)$ approaches 0 **faster** than $g(x)$ as x approaches infinity if

$$\lim_{x\to\infty} \frac{f(x)}{g(x)} = 0.$$

2. The function $f(x)$ approaches 0 **slower** than $g(x)$ as x approaches infinity if

$$\lim_{x\to\infty} \frac{f(x)}{g(x)} = \infty.$$

3. $f(x)$ approaches 0 at the same rate as $g(x)$ if

$$\lim_{x \to \infty} \frac{f(x)}{g(x)} = L$$

where L is any finite number other than 0.

Be careful not to confuse this with the definition for functions approaching infinity (Definition 3.1). The function $f(x)$ approaches 0 faster if it becomes *small* faster than $g(x)$ (Figure 3.5.7).

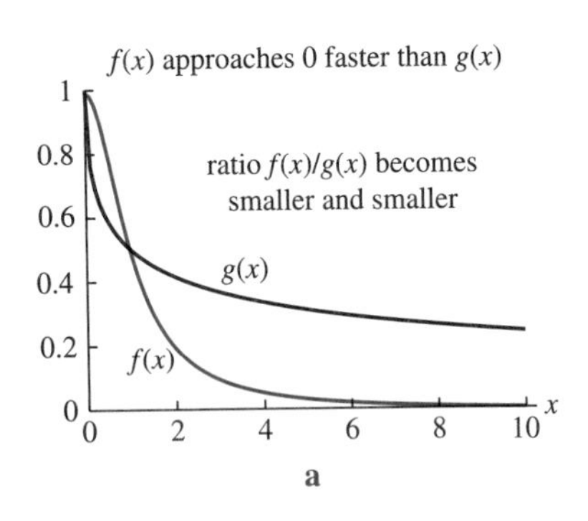

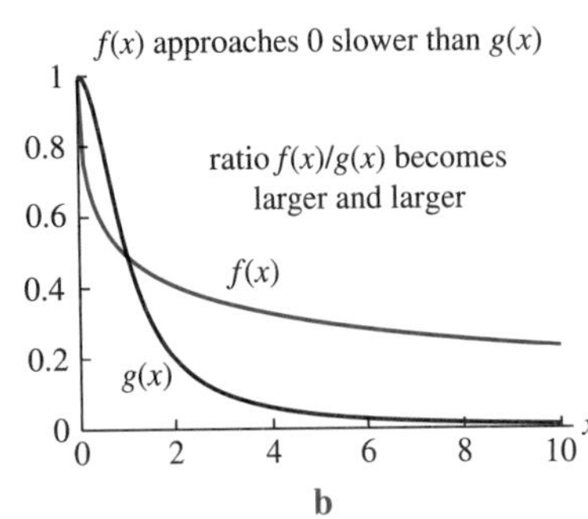

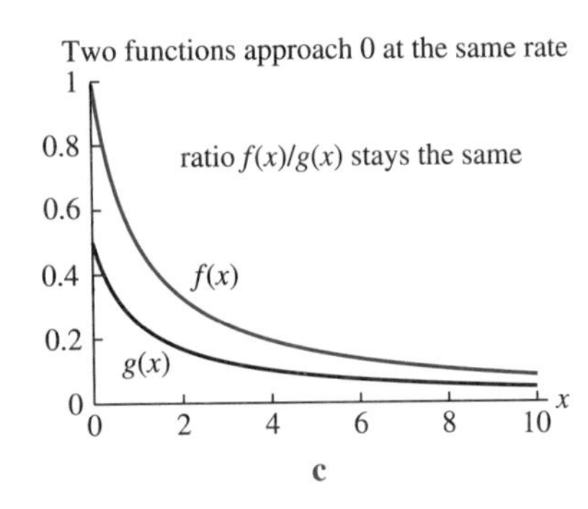

FIGURE 3.5.7
Comparing functions that approach a limit of 0 at infinity

The basic examples are reciprocals of the functions in Table 3.2 (Table 3.3 and Figure 3.5.8). If $f(x)$ approaches **infinity** quickly, the reciprocal $\frac{1}{f(x)}$ approaches **0** quickly. Because these functions decrease so quickly, they are easier to distinguish on a semilog graph (Figure 3.5.8b). Again, the positive constant c does not change the ordering of the functions.

Table 3.3 The Basic Functions Approaching 0

Function	Comments
ax^{-n} with $n > 0$	Approaches 0 faster for larger n
$ae^{-\beta x}$ with $\beta > 0$	Approaches 0 faster for larger β
$ae^{-\beta x^2}$ with $\beta > 0$	Approaches 0 really fast

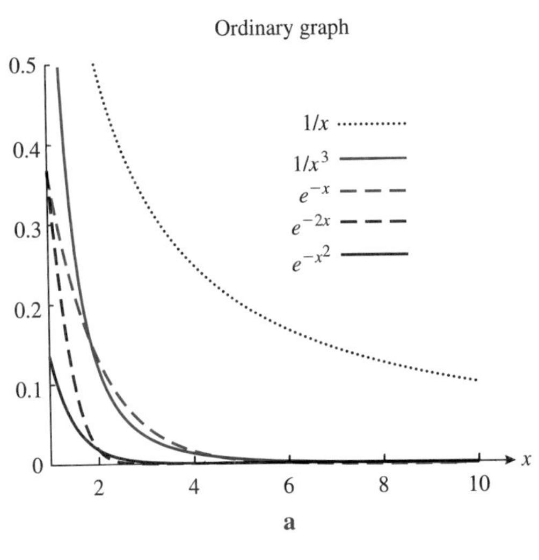

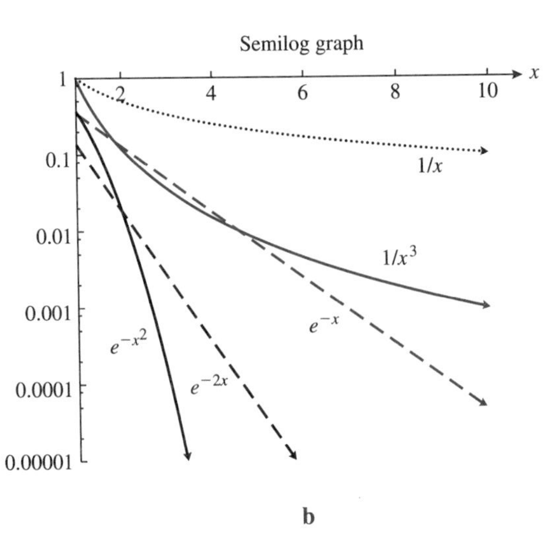

FIGURE 3.5.8
The behavior of the basic functions that approach 0

Example 3.5.3 Ordering Functions That Approach 0

To order the functions

$$0.1e^{-2x},\ 23.2x^{-0.5},\ 10.1e^{-0.2x},\ 0.03x^{-4}$$

from the one that approaches 0 fastest to the one that approaches 0 slowest, first identify the functions as exponential or power functions. The fastest is the exponential function with the most negative parameter $0.1e^{-2x}$, followed by the exponential function with the less negative parameter $10.1e^{-0.2x}$, then the power function $0.03x^{-4}$ with the more negative power, and finally the power function $23.2x^{-0.5}$ with the less negative power. From fastest to slowest, they are $0.1e^{-2x}$, $10.1e^{-0.2x}$, $0.03x^{-4}$, $23.2x^{-0.5}$.

3.5.2 Application to Absorption Functions

How can we use these facts about the basic functions to understand the absorption functions in Table 3.1? The results are summarized in Table 3.4. We compare the numerator and denominator of the absorption function as functions of the concentration c. If the numerator grows faster than the denominator, absorption grows without bound as c gets large. If the numerator and denominator grow at the same rate, absorption approaches a constant as c gets large (Figure 3.5.1b and 3.5.1c). If the denominator grows faster than the numerator, absorption approaches 0 as c gets large (Figure 3.5.1d and 3.5.1e).

Table 3.4 Analyzing Absorption Functions

Number	Numerator	Denominator	Behavior at infinity
22.1a	Linear	None	Approaches infinity
22.1b	Linear	Linear	Approaches constant
22.1c	Quadratic	Quadratic	Approaches constant
22.1d	Linear	Exponential	Approaches 0
22.1e	Linear	Quadratic	Approaches 0

With linear absorption (Figure 3.5.1a), absorption grows without bound (there is no denominator to balance the numerator). Because there are almost always limits to absorption, the saturated absorption functions (Figure 3.5.1b and 3.5.1c) provide more reasonable models.

The equation in Figure 3.5.1b, a ratio of linear functions, is among the most important in biology and is known as the **Michaelis-Menten** or **Monod** equation. In the next section we will deduce the difference in shape between the equations in Figure 3.5.1b and 3.5.1c and formalize the calculation of the behavior at infinity with the **method of leading behavior**.

In the equations Figure 3.5.1d and 3.5.1e, the denominator grows faster than the numerator (both exponential and quadratic functions grow faster than linear functions). These functions begin at 0, increase to a maximum, and eventually decrease again to 0. This behavior is a form of overcompensation (Subsection 3.2.3) because absorption decreases when the concentration is too large.

3.5.3 Limits of Sequences

We use the same idea to define the limit of a solution of a discrete-time dynamical system. The solution of the bacterial discrete-time dynamical system $b_{t+1} = rb_t$ is

$$b_t = b_0 r^t = b_0 e^{\ln(r)t}.$$

This differs from an ordinary function in that populations are defined only at integer values of t. The solution is a list of numbers known as a **sequence**.

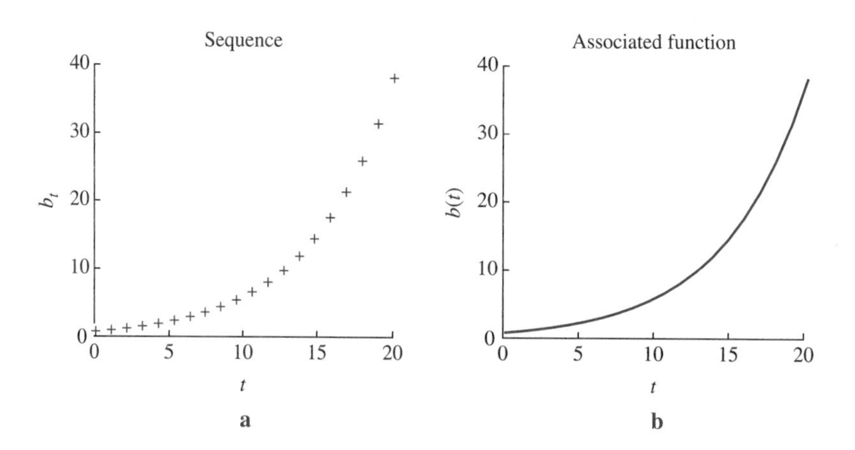

FIGURE 3.5.9
A sequence and its associated function

To find the limit, define the **associated function** $b(t)$ as the exponential function

$$b(t) = b_0 e^{\ln(r)t}$$

defined for all values of t. This function fills in the gaps in the sequence (Figure 3.5.9). If the associated function has a limit, whether 0, infinity, or some other value, the sequence will share that limit. In this case, the associated function is an exponential function. If $r > 1$, the parameter in the exponent is positive and the limit of the function, and therefore of the sequence, is infinity. If $r < 1$, the parameter in the exponent is negative and the limit of the function, and therefore of the sequence, is 0. If $r = 1$, the function has the constant value b_0 and the function and sequence share the limit b_0.

This definition has an important connection with the idea of a **stable equilibrium**. If the sequence of points that represents a solution approaches a particular value as a limit, that limit is a stable equilibrium (Figure 3.5.10a).

Example 3.5.4 The Limit of a Solution of the Medication Discrete-Time Dynamical System

In Section 1.5, Example 1.5.14, we found the solution of the discrete-time dynamical system

$$M_{t+1} = 0.5M_t + 1.0$$

describing the concentration of medication in the bloodstream to be

$$M_t = 2.0 + 0.5^t \cdot 3.0.$$

We now know that the associated function $M(t)$ can be written as

$$M(t) = 2.0 + 3.0 \cdot e^{\ln(0.5)t}.$$

The second term is an exponential function with a negative coefficient in the power ($\ln(0.5) = -0.693 < 0$), implying that this solution approaches 0 as the argument t approaches infinity. The whole associated function then approaches the value 2.0. The sequence of points that represents the solution also approaches 2.0.

Example 3.5.5 A Case Where the Sequence and Associated Function Behave Differently

If the associated function has a limit, the sequence shares that limit. The sequence, however, may have a limit even when the associated function does not. Consider the sequence

$$a_t = \sin(2\pi t).$$

The associated function $a(t) = \sin(2\pi t)$ has no limit because it oscillates forever. The sequence, on the other hand, takes on only the value 0 and has the limit 0 (Figure 3.5.11).

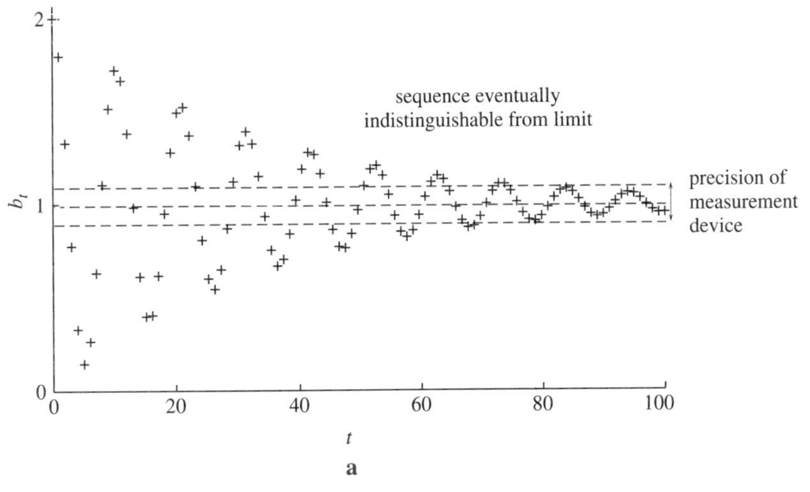

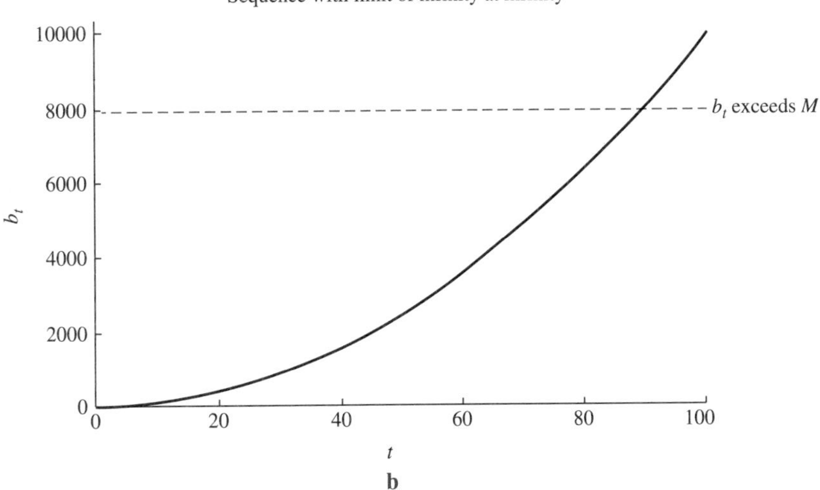

FIGURE 3.5.10
Limits of sequences at infinity

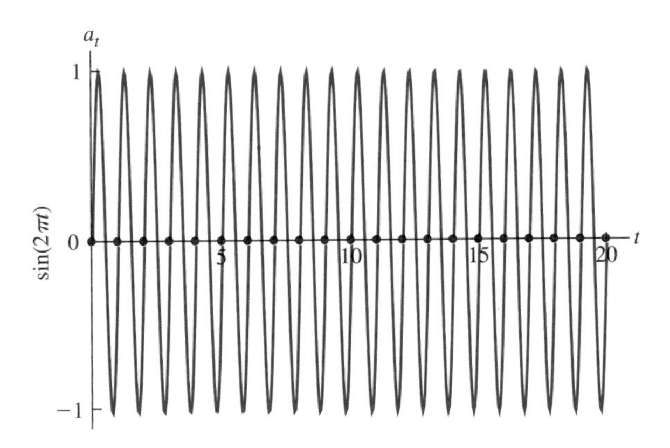

FIGURE 3.5.11
A sequence with a different limit from its associated function

Summary In order to reason about the behavior of functions with large inputs, we defined the **limit** of a function as its argument approaches infinity. As with ordinary limits, the limit formalizes the idea that the value gets closer and closer to some particular number (if the limit is finite) or larger than any given number (if the limit is infinite). One function approaches infinity **faster** than another if the limit of the ratio is infinity. Exponential functions approach infinity faster than power functions, which in turn approach infinity faster than the logarithmic functions. Conversely, one function approaches **0** faster than another if the limit of the ratio is 0. Exponential functions with negative parameters approach 0 faster than power functions with negative powers. Using limits at infinity, we analyzed the limits of **sequences**, lists of numbers generated as solutions of discrete-time dynamical systems by studying the **associated function**.

3.5 Exercises

Mathematical Techniques

1–8 ▪ Find the following limits.

1. $\lim_{x\to\infty} x^{-0.25}$.
2. $\lim_{x\to\infty} \ln(x/5)$.
3. $\lim_{x\to\infty} 0.8^x$.
4. $\lim_{x\to\infty} 1 - e^{-4x}$.
5. $\lim_{x\to\infty} 1 + x^4$.
6. $\lim_{x\to\infty} 1.2^{x+2}$.
7. $\lim_{x\to\infty} e^{-x^2}$.
8. $\lim_{x\to\infty} (x^2+2)^{0.25}$.

9–14 ▪ For each pair of functions, say which approaches infinity faster as x approaches infinity. Explain which rule you used to compare each pair. Compute the value of each function at $x = 1$, $x = 10$, and $x = 100$. How do these compare with the order of the functions in the limit? If they are different, how large would x have to be for the values to match the order in the limit?

9. x^2 and e^{2x}
10. x^3 and $1000x$
11. $x^{3.5}$ and $0.1x^{10}$
12. $5e^x$ and e^{5x}
13. $0.1x^{0.5}$ and $30\ln(x)$
14. $10x^{0.1}$ and $x^{0.5}$

15–20 ▪ For each pair of functions, say which approaches 0 more quickly as x approaches infinity. Explain which rule you used to compare each pair. Compute the value of each function at $x = 1$, $x = 10$, and $x = 100$. How do these compare with the order of the functions in the limit? If they are different, how large would x have to be for the values to match the order in the limit?

15. e^{-2x} and x^{-10}
16. $10e^{-x}$ and $0.1e^{-0.2x}$
17. $1000/x$ and $x^{-3.5}$
18. $x^{-0.1}$ and $25x^{-0.2}$
19. x^{-2} and $30/\ln(x)$
20. $1/\ln(x)$ and $30x^{-0.1}$

21–24 ▪ The following are possible absorption functions. What happens to each as c approaches infinity? Assume that all parameters take on positive values.

21. $\dfrac{\beta c^2}{1+e^c}$
22. $\dfrac{Ac}{\ln(1+c)}$
23. $\dfrac{\gamma(e^c-1)}{e^{2c}}$
24. $\dfrac{c^2}{1+10c}$

Applications

25–30 ▪ Find the derivatives of the following absorption functions (from Table 3.1 with particular values of the parameters). Compute the value at $c = 0$ and the limit of the derivative as c approaches infinity. Are your results consistent with the figures?

25. $\alpha(c) = 5c$.
26. $\alpha(c) = \dfrac{5c}{1+c}$.
27. $\alpha(c) = \dfrac{5c^2}{1+c^2}$.
28. $\alpha(c) = \dfrac{5c}{e^{2c}}$.
29. $\alpha(c) = \dfrac{5c}{1+c^2}$.
30. $\alpha(c) = 5c(1+c)$.

31–34 ▪ A bacterial population that obeys the discrete-time dynamical system $b_{t+1} = rb_t$ with the initial condition b_0 has the solution $b_t = b_0 r^t$. For the following values of r and b_0, state which populations increase to infinity and which decrease to 0. For those increasing to infinity, find the time when the population will reach 10^{10}. For those decreasing to 0, find the time when the population will reach 10^3.

31. $b_0 = 10^8$ and $r = 1.1$.
32. $b_0 = 10^8$ and $r = 1.5$.
33. $b_0 = 10^8$ and $r = 0.5$.
34. $b_0 = 10^8$ and $r = 0.9$.

35–36 ▪ In the polymerase chain reaction used to amplify DNA, some sequences of DNA produced are too long and others are the right length. Denote the number of overly long pieces after t generations of the process by l_t and the number of pieces of the right length by r_t. The dynamics follow approximately

$$l_{t+1} = l_t + 2$$
$$r_{t+1} = 2r_t$$

because two new overly long pieces are produced at each step while the number of good pieces doubles. Suppose that $l_0 = 0$ and $r_0 = 2$.

35. Find expressions for l_t and r_t and compute the fraction of pieces that are too long after 1, 5, 10, and 20 generations of the process.

36. Find the ratio of the number of pieces that are too long to the total number of pieces as a function of time. What is the limit? How long would you have to wait to make sure that less than one in a million pieces are too long? (This can't be solved exactly, just plug in some numbers.)

37–40 ▪ Consider the discrete-time dynamical system for medication given by $M_{t+1} = 0.5M_t + 1.0$ with $M_0 = 3$.

37. Find the equilibrium.

38. The solution is $M_t = 2.0 + 0.5^t \times 3.0$. Find the limit as $t \to \infty$.

39. How long will the solution take to be within 1% of the equilibrium?

40. What are two ways to show that this equilibrium is stable?

41–42 ▪ The amount of food a predator eats as a function of prey density is called the **functional response**. Functional response is often broken into three categories.

- Type I: Linear.
- Type II: Increasing, concave down, finite limit.
- Type III: Increasing with finite limit, concave up for small prey densities, concave down for large prey densities.

41. Sketch a picture of the type II response. Which of the absorption functions does it resemble? What is the optimal prey density for a predator?

42. Sketch a picture of the type III response. Which of the absorption functions does it resemble? What is the optimal prey density for a predator?

43–46 ▪ Now suppose that the number of prey that escape increases linearly with the number of prey (the prey join together and fight back). Let p be the number of prey and $F(p)$ be the functional response. The number of prey captured is then $F(p) - cp$. The constant c represents how effectively the prey can fight.

43. Write the equation for the optimal prey density (the value giving the maximum rate of prey capture) in terms of $F'(p)$. Find the optimal prey density if $F(p) = p$. Make sure to separately consider cases with $c < 1$ and $c > 1$.

44. Draw a picture illustrating the optimal prey density in a case with a type II functional response.

45. Draw a picture illustrating the optimal prey density in a case with a type III functional response.

46. Find the optimal prey density if $F(p) = \frac{p}{1+p}$. Make sure to separately consider cases with $c < 1$ and $c > 1$.

Computer Exercise

47. Use your computer to find out how large x must be before the faster function finally overtakes the slower function.

    a. $e^{0.1x}$ catches up with x^3.

    b. $0.1e^x$ catches up with x^3.

    c. $0.1e^{0.1x}$ catches up with x^3.

    d. $0.1x$ catches up with $\ln(x)$.

    e. $x^{0.1}$ catches up with $\ln(x)$.

3.6 Leading Behavior and L'Hôpital's Rule

Finding limits at infinity gives general information about the behavior of functions. In particular, we learned how to compare the **ratios** of different functions by describing which increased to infinity or decreased to 0 faster. We now study a much larger class of functions, **sums** of functions and the ratios of sums. The technique is called the **method of leading behavior** and consists of focusing on the largest piece of the function. By determining the leading behavior of a function at both infinity and 0, we can deduce a great deal about the **shape** of the function by using the technique of **matched leading behaviors**. **L'Hôpital's** rule provides an alternative way to compare the behavior of functions.

3.6.1 Leading Behavior of Functions at Infinity

Suppose we wish to describe how the sum of several functions, such as

$$f(x) = 5e^{2x} + 34e^x + 45x^5 + 56\ln(x) + 10,$$

behaves for large values of x. We might suspect that this function will be dominated by the fastest term, the one that increases faster than all the others in the sense of Definition 3.1. The **method of leading behavior** is based on this idea.

Definition 3.3 The **leading behavior** of a function at infinity is the term that is largest as the argument approaches infinity. We write f_∞ to represent the leading behavior of the function f at infinity.

The largest term in the function $f(x)$ is $5e^{2x}$ because the exponential term with the largest parameter in the exponent grows to infinity most quickly. Therefore,

$$f_\infty(x) = 5e^{2x}.$$

In what sense does the leading behavior describe a function? The graph of the leading behavior looks indistinguishable from the graph of the original function when x is large, even though the two functions are quite different for small x (Figure 3.6.1). More mathematically, if we divide a function by its leading behavior, the limit is 1 (they approach infinity at the same rate). For example,

$$\begin{aligned}\lim_{x\to\infty} \frac{f(x)}{f_\infty(x)} &= \lim_{x\to\infty} \frac{5e^{2x} + 34e^x + 45x^5 + 56\ln(x) + 10}{5e^{2x}} \\ &= \lim_{x\to\infty} 1 + \frac{34e^x}{5e^{2x}} + \frac{45x^5}{5e^{2x}} + \frac{56\ln(x)}{5e^{2x}} + \frac{10}{5e^{2x}} \\ &= 1 + 0 + 0 + 0 + 0 = 1.\end{aligned}$$

All terms after the first approach 0 because $5e^{2x}$ approaches infinity the fastest.

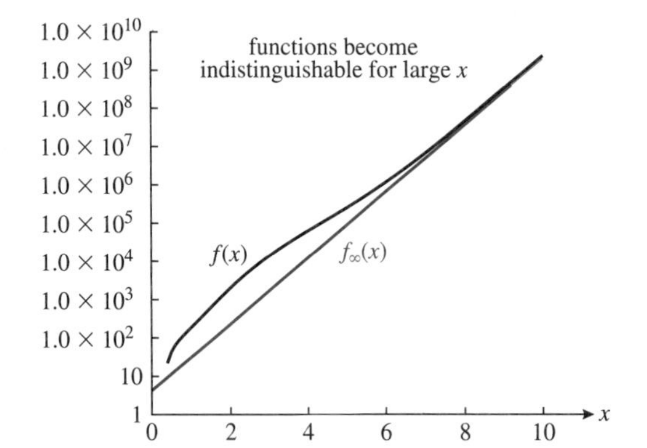

FIGURE 3.6.1
A comparison of a function and its leading behavior in a semi-log plot

Example 3.6.1 Comparing Functions with the Method of Leading Behavior

Complicated functions can be compared by comparing their leading behaviors. To find whether the function $f(x)$ increases to infinity faster than the function

$$g(x) = 23e^{2.5x} + 3e^{2x} + 2x^6,$$

we need only compare the leading behavior of $f(x)$ with that of $g(x)$. The leading behavior of the function $g(x)$ is the first term, as it is the term with the largest parameter in the exponent, so

$$g_\infty(x) = 23e^{2.5x}.$$

To see whether $f(x)$ approaches infinity faster than $g(x)$, we compute the limit of the ratio of the functions as x approaches infinity. Because each function is well represented by its leading behavior,

$$\begin{aligned}\lim_{x\to\infty} \frac{f(x)}{g(x)} &= \lim_{x\to\infty} \frac{f_\infty(x)}{g_\infty(x)} \\ &= \lim_{x\to\infty} \frac{5e^{2x}}{23e^{2.5x}} \\ &= 0\end{aligned}$$

because the exponent in the denominator has a larger parameter. The function $g(x)$ approaches infinity faster than $f(x)$.

However, the leading behavior gives more information than the limit. For large x,

$$\frac{f(x)}{g(x)} \approx \frac{f_\infty(x)}{g_\infty(x)} = \frac{5e^{2x}}{23e^{2.5x}} = \frac{5}{23}e^{-0.5x}.$$

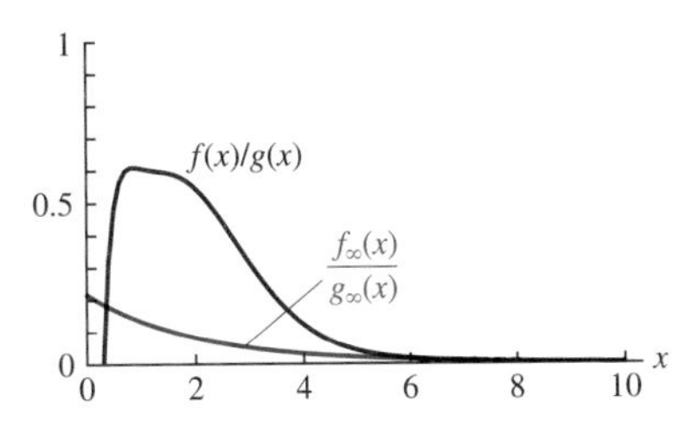

FIGURE 3.6.2

The ratio of two functions compared with the ratio of their leading behaviors

This term not only tells us that $g(x)$ approaches infinity faster than $f(x)$, but it also gives us an idea of how much faster. The ratio of the functions behaves much like the ratio of the leading behaviors for large x (Figure 3.6.2).

The same definition of leading behavior works for sums of functions that approach 0. The largest term is the term approaching 0 the *most slowly* in the sense of Definition 3.2.

Example 3.6.2 Comparing Functions That Approach 0 with the Method of Leading Behavior

The leading behavior of the function

$$h(x) = 5e^{-2x} + 34e^{-x} + 45x^{-5}$$

is the power term $45x^{-5}$ because it approaches 0 the most slowly and is therefore the largest for large x,

$$h_\infty(x) = 45x^{-5}$$

(Figure 3.6.3).

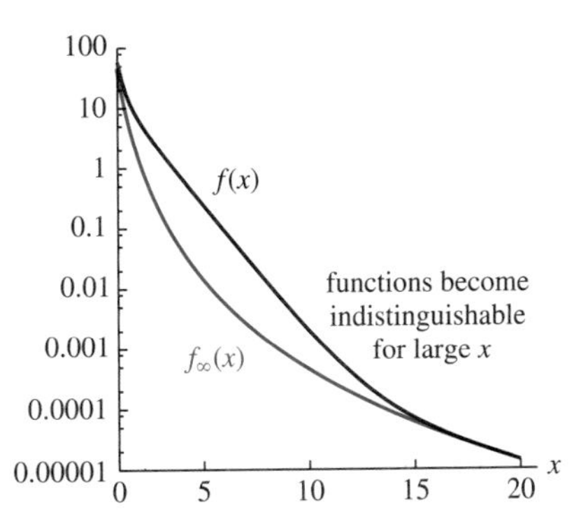

FIGURE 3.6.3

A function that approaches 0 and its leading behavior

Example 3.6.3 The Leading Behavior of a Sum That Includes a Term That Does Not Approach 0

Similarly, the function

$$H(x) = 5e^{-2x} + 34e^{-x} + 45x^{-5} + 5,$$

has leading behavior

$$H_\infty(x) = 5$$

because the constant 5 does not approach 0 and is therefore the largest (Figure 3.6.4).

Example 3.6.4 Application to Absorption Function 3.5.1b

We can apply the idea of leading behavior to supplement our reasoning about the absorption functions in Section 3.5.1. The absorption function with saturation (Equation 3.5.1b) has the formula

$$\alpha(c) = \frac{Ac}{k + c},$$

where c is the concentration and α and k are constant parameters. Both numerator and denominator increase to infinity. The leading behavior of the denominator is c, the larger of the two terms. Replacing the denominator with the leading behavior produces

$$\alpha_\infty(c) = \frac{Ac}{c} = A$$

for large values of c (Figure 3.6.5).

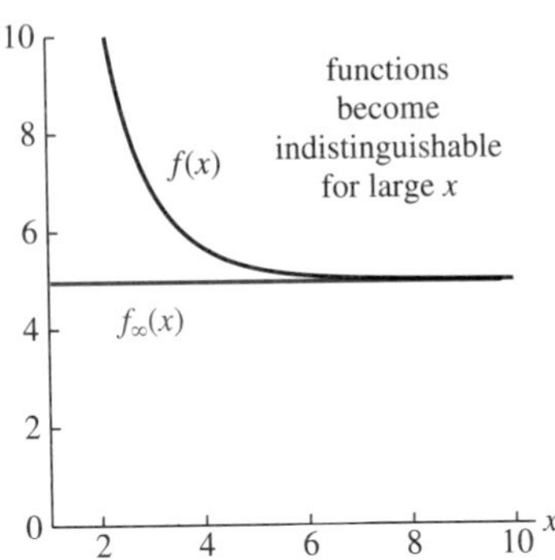

FIGURE 3.6.4
A function that approaches a finite value and its leading behavior

FIGURE 3.6.5
An absorption function with saturation

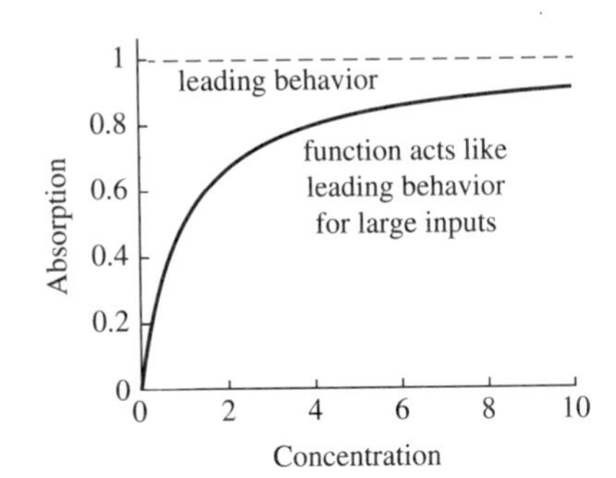

Example 3.6.5 Application to Absorption Function 3.5.1c

Similarly, the absorption function with saturation and a threshold (Equation 3.5.1c) is has the formula

$$\alpha(c) = \frac{Ac^2}{k + c^2}.$$

The numerator is a single term, which is its own leading behavior. The leading behavior of the denominator is c^2 because it increases faster than the constant k. For large values of c,

$$\alpha_\infty(c) = \frac{Ac^2}{c^2} = A.$$

This function also saturates at α (Figure 3.6.6).

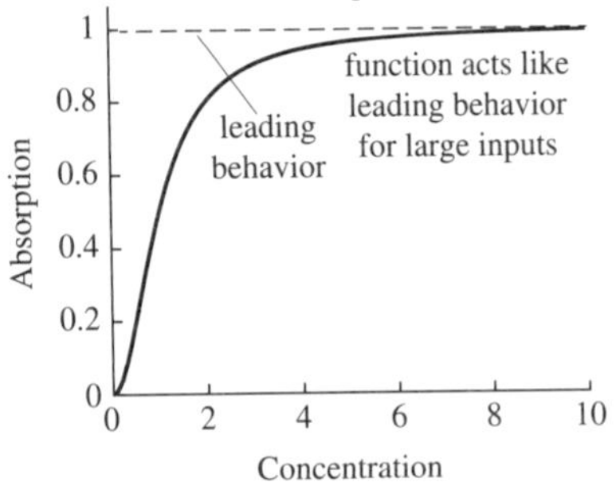

FIGURE 3.6.6
An absorption function with saturation

Example 3.6.6 Application to Absorption Function 3.5.1d

The method of leading behavior tells us nothing new about the absorption function

$$\alpha(c) = Ace^{-\beta c}.$$

Placing the exponential term in the denominator tells us that that the limit as c approaches infinity is 0 because an exponential function grows faster than a linear function. Because the numerator and denominator each have only a single term, we cannot simplify further.

Example 3.6.7 Application to Absorption Function 3.5.1e

The alternative absorption function with overcompensation (Equation 3.5.1e),

$$\alpha(c) = \frac{Ac}{k + c^2},$$

can be simplified with the method of leading behavior. In this case, the leading behavior of the denominator is the quadratic term c^2, so that

$$\alpha_\infty(c) = \frac{Ac}{c^2} = \frac{\alpha}{c}.$$

Absorption decreases to 0, and does so like the function $\frac{\alpha}{c}$ (Figure 3.6.7).

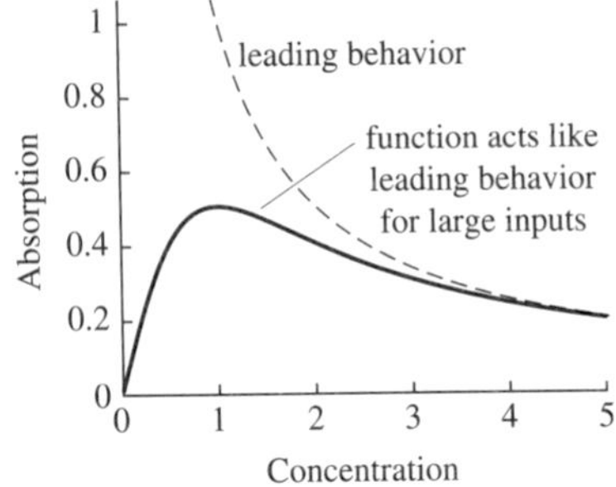

FIGURE 3.6.7
An absorption function with saturation

3.6.2 Leading Behavior of Functions at 0

In each comparison of an absorption function with its leading behavior, we have done well for large values of the concentration and poorly for small values. To complete our analysis of the absorption functions, we need a better sense of what is happening near 0. Once again, the method of leading behavior can be used to identify which small terms can be ignored.

First, we define the leading behavior of a function at 0.

Definition 3.4 The **leading behavior** of a function at 0 is the term that is largest as the argument approaches 0. We write f_0 to represent the leading behavior of the function f at 0.

The idea of largest is the same at 0 and at infinity, and is formalized in the following definition.

Definition 3.5

1. The function $f(x)$ is **larger** than $g(x)$ as x approaches 0 if

$$\lim_{x \to 0} \frac{f(x)}{g(x)} = \infty.$$

2. The function $f(x)$ is **smaller** than $g(x)$ if

$$\lim_{x \to 0} \frac{f(x)}{g(x)} = 0.$$

This definition really includes two cases: $f(x)$ and $g(x)$ approach infinity as x approaches 0, and $f(x)$ and $g(x)$ approach 0 as x approaches 0. In the first case, all power functions of the form

$$f(x) = x^{-n}$$

with positive values of n approach infinity as x approaches 0 (from the right). The larger the value of n, the faster the function approaches infinity and thus the **larger** it is.

Example 3.6.8 The Leading Behavior with Negative Powers at 0

The sum

$$f(x) = x^{-1} + 5x^{-5}$$

has leading behavior

$$f_0(x) = 5x^{-5}$$

because $5x^{-5}$ approaches infinity faster than x^{-1}. The limit of the ratio of the terms is

$$\lim_{x \to 0} \frac{5x^{-5}}{x^{-1}} = \lim_{x \to 0} 5x^{-4} = \infty.$$

Power functions with large negative powers do everything fast, approaching ∞ quickly at 0 and 0 quickly at ∞ and are thus **largest** for small values of x and **smallest** for large values of x (Figure 3.6.8).

If two functions $f(x)$ and $g(x)$ approach 0 as x approaches 0, the **larger** function approaches 0 more **slowly**. All power functions of the form

$$f(x) = x^n$$

for positive values of n approach 0 as x approaches 0. As before, power functions with large powers do everything quickly. The larger the power of n, the **faster** the function approaches 0 for x near 0 and the faster it approaches ∞ as x approaches ∞ (Figure 3.6.9). These functions are small for small x and large for large x.

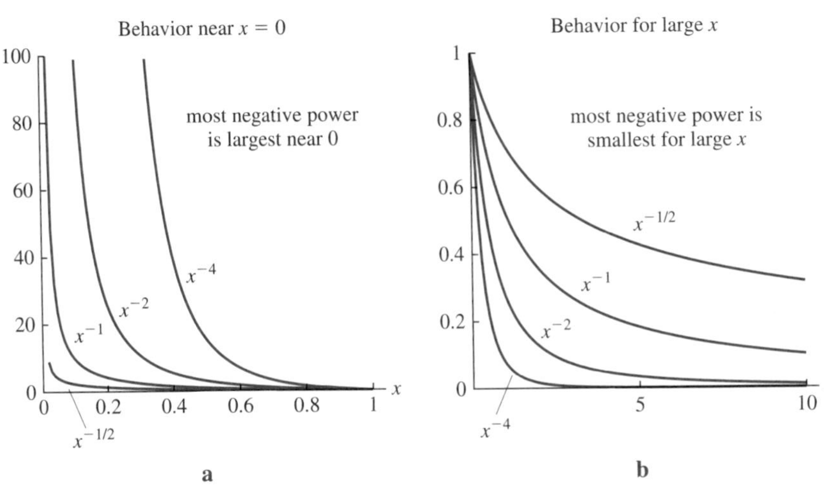

FIGURE 3.6.8
Power functions with negative powers

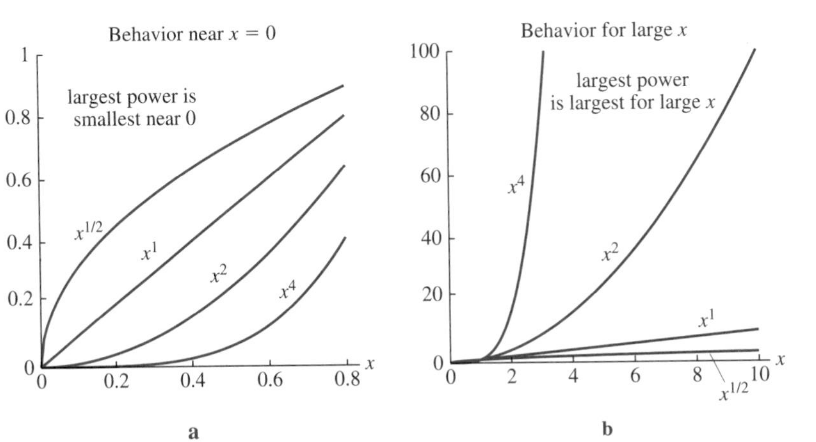

FIGURE 3.6.9
Power functions with positive powers

Example 3.6.9 The Leading Behavior of a Polynomial at 0

Consider the function

$$F(x) = 4x + x^3.$$

The leading behavior near $x = 0$ is given by the first term x because it has the smallest power, or

$$F_0(x) = 4x.$$

The leading behavior for large x, on the other hand, is given by the second term $4x^3$ because it has the larger power, or

$$F_\infty(x) = x^3$$

(Figure 3.6.10).

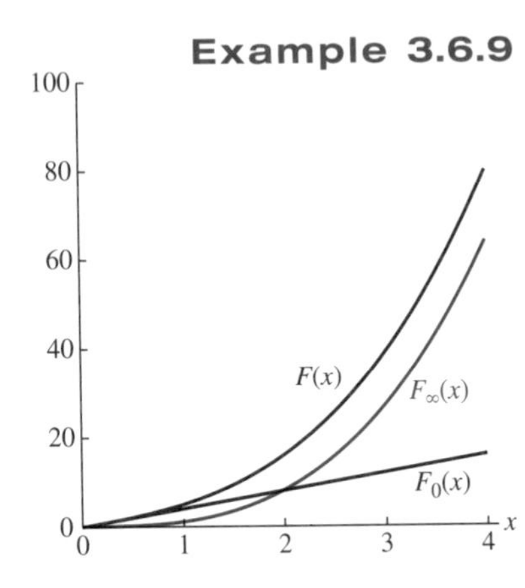

FIGURE 3.6.10
A function compared with the leading behavior at 0 and infinity

3.6.3 The Method of Matched Leading Behaviors

The idea of studying a function for both large and small values of x can be formalized as the **method of matched leading behaviors**.

Algorithm 3.3 The Method of Matched Leading Behaviors

1. Find the leading behavior at 0 and at infinity.
2. Sketch graphs of each leading behavior.
3. Connect the graphs with a smooth curve.

Example 3.6.10 Applying Matched Leading Behaviors to the Absorption Function 3.5.1b

Consider the absorption function with saturation (Equation 3.5.1b),

$$\alpha(c) = \frac{Ac}{k+c}.$$

Near 0, we first simplify by finding the leading behavior of the denominator as

$$(k+c)_0 = k$$

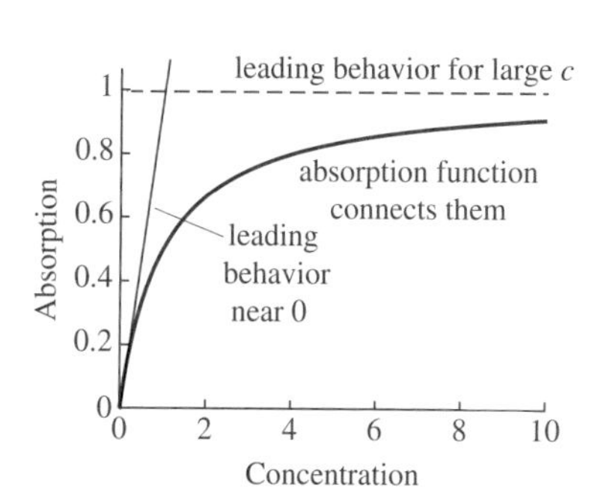

FIGURE 3.6.11
Saturated absorption

because the constant value k is much **larger** than c for small c. Therefore,

$$\alpha_0(c) = \frac{Ac}{k}.$$

For large values of c, the denominator becomes

$$(k+c)_\infty = c$$

because the constant value k is much **smaller** than c for large c. Therefore,

$$\alpha_\infty(c) = \frac{Ac}{c} = A.$$

These linear functions can be easily connected with a smooth curve, as illustrated in Figure 3.6.11 with parameter values $A = 1$ and $k = 1$.

Example 3.6.11 Applying Matched Leading Behaviors to the Absorption Function 3.5.1c

For absorption with saturation and a threshold

$$\alpha(c) = \frac{Ac^2}{k+c^2}$$

(Equation 3.5.1c), we can again use the method of matched leading behaviors to plot an accurate graph. For small values of c, the denominator has leading behavior

$$(k+c^2)_0 = k$$

because the constant value k is much larger than c^2 for small c. Therefore,

$$\alpha(c)_0 = \frac{Ac^2}{k}.$$

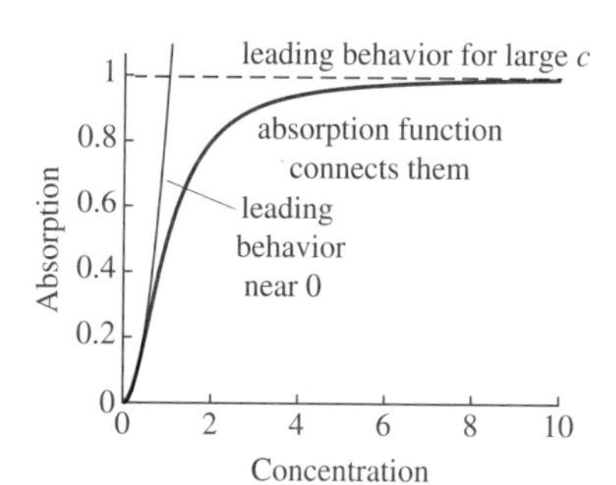

FIGURE 3.6.12
Saturated absorption with threshold

This part of the curve looks like a parabola. For large values of c, the denominator becomes

$$(k+c^2)_\infty = c^2$$

because the constant value k is much smaller than c^2 for large c. Therefore,

$$\alpha_\infty(c) = \frac{Ac^2}{c^2} = A.$$

Connecting these portions of the graph with a smooth curve shows that the graph is concave up for small c and concave down for large c (sketched with parameter values $A = 1$ and $k = 1$ in Figure 3.6.12).

Example 3.6.12 Applying Matched Leading Behaviors to the Absorption Function 3.5.1e

Finally, saturated absorption with overcompensation II (Equation 3.5.1e)

$$\alpha(c) = \frac{Ac}{k+c^2}$$

has the same denominator as in Example 3.6.11, so

$$(k+c^2)_0 = k$$
$$(k+c^2)_\infty = c^2.$$

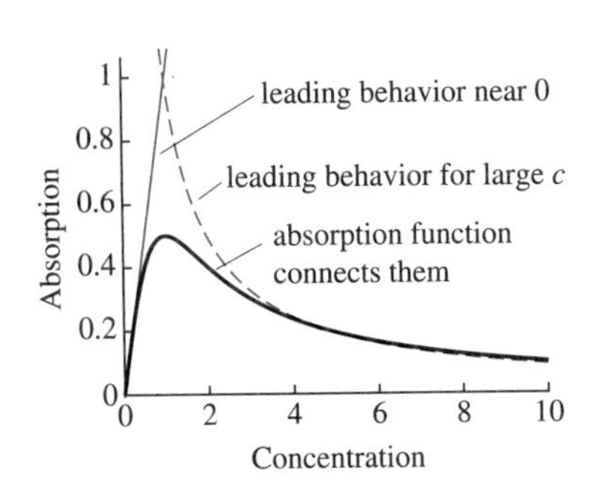

FIGURE 3.6.13
The method of matched leading behaviors

Therefore,

$$\alpha_0(c) = \frac{Ac}{k}$$

$$\alpha_\infty(c) = \frac{Ac}{c^2} = \frac{\alpha}{c}.$$

This curve begins by increasing like a line with slope $\frac{\alpha}{k}$, but it eventually begins decreasing like $\frac{\alpha}{c}$. As a consequence, this curve must have a maximum (Figure 3.6.13).

3.6.4 L'Hôpital's Rule

Suppose we wish to find

$$\lim_{x \to 0} \frac{e^{2x} - 1}{x}. \tag{3.6.1}$$

Both numerator and denominator approach 0. We cannot simplify the numerator with the method of leading behavior because both terms approach a limit of 1 rather than 0 or ∞.

There is a general and powerful rule for dealing with cases like this. Instead of comparing the functions, we compare their derivatives. This is known as L'Hôpital's rule. First, we define the cases where this rule applies, called indeterminate forms.

Definition 3.6 The limit of the ratio

$$\lim_{x \to a} \frac{f(x)}{g(x)}$$

(where a could be ∞) is called an **indeterminate form** if

$$\lim_{x \to a} f(x) = \lim_{x \to a} g(x) = 0$$

or

$$\lim_{x \to a} f(x) = \lim_{x \to a} g(x) = \infty.$$

If we tried to plug $x = a$ into an indeterminate form, we would be committing a mathematical felony. These limits can often be computed with L'Hôpital's rule.

Theorem 3.6 **L'Hôpital's Rule**

Suppose that f and g are differentiable functions and that

$$\lim_{x \to a} \frac{f(x)}{g(x)}$$

is an indeterminate form. If

$$\lim_{x \to a} \frac{f'(x)}{g'(x)} = L$$

then

$$\lim_{x \to a} \frac{f(x)}{g(x)} = L.$$

In words, if the ratio of functions is an indeterminate form, the limit of the ratio of functions is equal to the limit of the ratio of their derivatives.

Example 3.6.13 Comparing Functions Using L'Hôpital's Rule

With this theorem we can prove that exponentials grow faster than power functions and that power functions grow faster than logarithms as x approaches infinity (Table 3.2).

L'Hôpital's rule says that

$$\begin{aligned} \lim_{x\to\infty} \frac{e^x}{x} &= \lim_{x\to\infty} \frac{\frac{de^x}{dx}}{\frac{dx}{dx}} && \text{take derivative of numerator and denominator because this is an indeterminate form} \\ &= \lim_{x\to\infty} \frac{e^x}{1} && \text{compute derivatives} \\ &= \infty. && \text{the exponential function has a limit of infinity} \end{aligned}$$

Example 3.6.14 Comparing the Logarithmic Function with a Linear Function

To compare the behavior of x and $\ln(x)$ at infinity using L'Hôpital's rule, we compute

$$\begin{aligned} \lim_{x\to\infty} \frac{x}{\ln(x)} &= \lim_{x\to\infty} \frac{\frac{dx}{dx}}{\frac{d\ln(x)}{dx}} \\ &= \lim_{x\to\infty} \frac{1}{\frac{1}{x}} \\ &= \lim_{x\to\infty} x \\ &= \infty. \end{aligned}$$

If a single application of L'Hôpital's rule results in an another indeterminate form, a second or third application might make it possible to evaluate the limit.

Example 3.6.15 Case Requiring Multiple Applications of L'Hôpital's Rule

To show that e^x increases faster than x^3, we must take the derivative three times.

$$\begin{aligned} \lim_{x\to\infty} \frac{e^x}{x^3} &= \lim_{x\to\infty} \frac{e^x}{3x^2} && \text{indeterminate form, take derivatives} \\ &= \lim_{x\to\infty} \frac{e^x}{6x} && \text{still indeterminate, take derivatives} \\ &= \lim_{x\to\infty} \frac{e^x}{6} = \infty. && \text{limit of the exponential function is infinity} \end{aligned}$$

Example 3.6.16 L'Hôpital's Rule Can Succeed When the Method of Leading Behaviors Fails

L'Hôpital's rule is indispensable when the method of leading behaviors fails. We have seen that

$$\lim_{x\to 0} \frac{e^{2x}-1}{x}.$$

(Equation 3.6.1) is indeterminate. We apply L'Hôpital's rule, finding that

$$\lim_{x\to 0} \frac{e^{2x}-1}{x} = \lim_{x\to 0} \frac{2e^{2x}}{1} = 2.$$

Example 3.6.17 L'Hôpital's Rule Used Incorrectly

L'Hôpital's rule almost always gives the wrong answer if applied to an expression that is not an indeterminate form. We can compute directly that

$$\lim_{x\to 0} \frac{e^{2x}-1}{x+1} = 0$$

because the numerator approaches 0 and the denominator approaches 1. If we mistakenly apply L'Hôpital's rule, we find

$$\lim_{x\to 0} \frac{e^{2x}-1}{x+1} \text{ “=” } \lim_{x\to 0} \frac{2e^{2x}}{1} = 2.$$

Why does L'Hôpital's rule work? Recall that the derivative can be used to approximate functions with the tangent line. The idea behind L'Hôpital's rule is to replace $f(x)$ with $\hat{f}(x)$ and $g(x)$ with $\hat{g}(x)$ and prove that

$$\lim_{x \to a} \frac{f(x)}{g(x)} = \lim_{x \to a} \frac{\hat{f}(x)}{\hat{g}(x)}.$$

Remember that

$$\hat{f}(x) = f(a) + f'(a)(x - a)$$
$$\hat{g}(x) = g(a) + g'(a)(x - a).$$

If this is an indeterminate form of the $\frac{0}{0}$ type, it must be true that $f(a) = g(a) = 0$. Therefore,

$$\frac{\hat{f}(x)}{\hat{g}(x)} = \frac{f'(a)}{g'(a)}$$

if $x \neq a$.

Summary We have learned two ways to compute the behavior of complicated functions. The **method of leading behavior** is a way to examine sums of functions and determine which piece increases to infinity **fastest** or to 0 **slowest**. For large values of the input, this piece is called the **leading behavior at infinity** and can be used to approximate the behavior of the function. For small values of the input, the largest piece is called the **leading behavior at 0** and can be used to approximate the behavior of the function. By graphing the leading behavior of functions at both 0 and infinity, we can use the **method of matched leading behaviors** to sketch an accurate graph. In other cases, we can evaluate **indeterminate forms** with **L'Hôpital's rule**, which says that a ratio of functions that both approach 0 or both approach infinity has the same limit as the ratio of their derivatives.

3.6 Exercises

Mathematical Techniques

1–6 ▪ Find the leading behavior of the following functions at 0 and ∞.

1. $f(x) = 1 + x$.
2. $g(y) = y + y^3$.
3. $h(z) = z + e^z$.
4. $F(x) = 1 + 2x + 3e^x$.
5. $m(a) = 100a + 30a^2 + \frac{1}{a}$.
6. $G(c) = e^{-4c} + \frac{5}{c^2} + \frac{3}{c^5} + 10e^{-3c}$

7–16 ▪ For each pair of functions, use the basic functions (when possible) to say which approaches its limit more quickly, and then check with L'Hôpital's rule.

7. x^2 and e^{2x} as $x \to \infty$.
8. x^2 and $1000x$ as $x \to \infty$.
9. $0.1x^{0.5}$ and $30\ln(x)$ as $x \to \infty$.
10. x and $\ln(x)^2$ as $x \to \infty$.
11. e^{-2x} and x^{-2} as $x \to \infty$.
12. $1/\ln(x)$ and $30x^{-0.1}$ as $x \to \infty$.
13. x^{-1} and $-\ln(x)$ as $x \to 0$. Use your result to figure out
$$\lim_{x \to 0} x\ln(x).$$
14. x^{-1} and $\frac{1}{e^x - 1}$ as $x \to 0$.
15. x^2 and x^3 as $x \to 0$.
16. x^2 and $e^x - x - 1$ as $x \to 0$.

17–22 ▪ For each of the following functions, find the leading behavior of the numerator, the denominator, and the whole function at both 0 and ∞. Find the limit of the function at 0 and ∞ (and check with L'Hôpital's rule when appropriate). Use the method of matched leading behaviors to sketch a graph.

17. $\alpha(c) = \frac{2c^2}{1 + c}$.
18. $\alpha(c) = \frac{c^2}{1 + 2c}$.
19. $\alpha(c) = \frac{1 + c + c^2}{1 + c}$.
20. $\alpha(c) = \frac{1 + c}{1 + c + c^2}$.
21. $\alpha(c) = \frac{3c}{1 + \ln(1 + c)}$.
22. $\alpha(c) = \frac{e^c + 1}{e^{2c} + 1}$.

Applications

23–28 ▪ Use the method of leading behavior, L'Hôpital's rule, and the method of matched leading behaviors to graph the following absorption functions.

23. $\alpha(c) = \frac{5c}{1+c}$.

24. $\alpha(c) = \frac{c}{5+c}$.

25. $\alpha(c) = \frac{5c^2}{1+c^2}$.

26. $\alpha(c) = \frac{5c}{e^{2c}}$.

27. $\alpha(c) = \frac{5c}{1+c^2}$.

28. $\alpha(c) = 5c(1+c)$.

29–32 ▪ Use the method of matched leading behaviors to graph the following Hill functions (Chapter 2, Equation 2.6.1) and their variants.

29. $h_3(x) = \frac{x^3}{1+x^3}$.

30. $g_3(x) = \frac{x^3}{10+x^3}$.

31. $h_{10}(x) = \frac{x^{10}}{1+x^{10}}$.

32. $g_{10}(x) = \frac{x^{10}}{0.1+x^{10}}$.

33–36 ▪ The following discrete-time dynamical systems describe the populations of two competing strains of bacteria

$$a_{t+1} = sa_t$$
$$b_{t+1} = rb_t.$$

For the following values of the initial conditions a_0 and b_0, and the per capita production s and r,

a. Find the number of each type as a function of time.

b. Find the fraction of type a as a function of time.

c. Use leading behavior or L'Hôpital's rule to find the limit of the fraction as $t \to \infty$.

d. Compute the fraction at $t = 0, 10, 20$, and 50 and compare with your limit.

33. $a_0 = 10^4$, $b_0 = 10^6$, $s = 2.0$, $r = 1.5$.

34. $a_0 = 10^4$, $b_0 = 10^6$, $s = 1.5$, $r = 2.0$.

35. $a_0 = 10^4$, $b_0 = 10^5$, $s = 0.8$, $r = 1.2$.

36. $a_0 = 10^4$, $b_0 = 10^5$, $s = 0.5$, $r = 0.3$.

37–40 ▪ Many of our absorption equations are of the form

$$\alpha(c) = A\frac{r(c)}{k+r(c)}$$

where α and k are positive parameters, and where $r(0) = 0$, $\lim_{c\to\infty} r(c) = \infty$ and $r'(c) > 0$. In each of the following cases, identify $r(c)$ and show that $\alpha(c)$ is increasing. Use L'Hôpital's rule to find the limit as $c \to \infty$, and use the method of leading behavior to describe absorption near $c = 0$ and $c = \infty$.

37. $\alpha(c) = \frac{Ac}{k+c}$ (from Table 3.1).

38. $\alpha(c) = \frac{Ac^2}{k+c^2}$ (from Table 3.1).

39. $\alpha(c) = \frac{Ac^n}{k+c^n}$ for some positive value of n.

40. Try without plugging in a particular form for $r(c)$.

Computer Exercises

41. Consider the following functions.

$$f(c) = \frac{c^2}{1+c^2}$$
$$g(c) = \frac{c}{1+c^2}.$$

Find the leading behavior of each at 0 and infinity. Suppose we approximated each by a function defined in pieces

$$\check{f}(c) = \begin{cases} f_0(c) & \text{if } f_0(c) < f_\infty(c) \\ f_\infty(c) & \text{if } f_0(c) > f_\infty(c) \end{cases}$$

and similarly for $\check{g}(c)$. Plot this approximation in each case. Find and plot the following ratios.

$$\frac{\check{f}(c)}{f(c)} \text{ and } \frac{\check{g}(c)}{g(c)}$$

When is the approximation best? When is it worst?

42. Consider the following function.

$$\check{f}(x) = \frac{1+x+x^2+x^3+x^4}{5+4x+3x^2+2x^3+x^4}$$

Find the leading behavior for large x. Next find a function that keeps both the largest and second largest term from the numerator and denominator. How much better is this new approximation? How much improvement do you get by adding in more and more terms?

3.7 Approximating Functions with Lines and Polynomials

The method of leading behavior provides a way to **approximate** complicated functions with simpler functions. In this section, we extend the related idea of the **tangent line approximation** in several ways. First, we compare the tangent line approximation with the **secant line approximation**, showing that the tangent line is the **best** linear approximation to a curve near the point of tangency, but that the secant line can be more useful over larger ranges. The tangent line matches the value and derivative of a function at a point. More accurate approximations can be found by also matching the second, third, and higher derivatives. If we use **polynomials** to match these higher derivatives, the resulting approximation is called a **Taylor polynomial**.

3.7.1 The Tangent and Secant Lines

Suppose we wish to approximate the exponential function e^x near 0 with a line, perhaps to compare a complicated dynamical system with its linear approximation. The best such approximation is given by the tangent line (Figure 3.7.1).

Definition 3.7 The general formula for the tangent line or tangent line approximation to the function f at base point a is

$$\hat{f}(x) = f'(a)(x-a) + f(a).$$

The function $\hat{f}(x)$ has as its graph the tangent line to $f(x)$ at a. At the point a, $\hat{f}(a) = f(a)$, and $\hat{f}'(a) = f'(a)$, meaning that the linear function $\hat{f}(a)$ matches both the value and the slope of the original nonlinear function at the point of tangency a.

Example 3.7.1 The Tangent Line to the Exponential Function

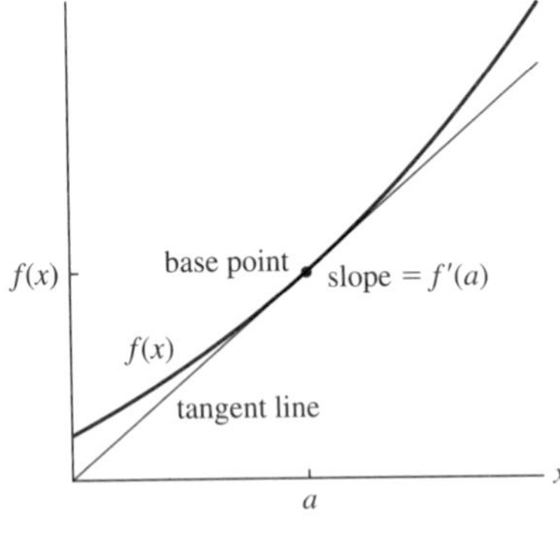

FIGURE 3.7.1
Approximating a function with the tangent line

The tangent line to the exponential function

$$g(x) = e^x$$

at $x = 0$ matches the value $g(0) = e^0 = 1$ and the slope $g'(0) = e^0 = 1$. Hence,

$$\hat{g}(x) = g(a) + g'(0)(x-0) = 1 + x$$

(Figure 3.7.2). The graph of the tangent line hugs the curve near the point of tangency and provides a good way to approximate values. For example,

$$\hat{g}(0.1) = 1.1,$$

close to the exact value $e^{0.1} \approx 1.10507$.

Before the age of computers, this sort of approximation was indispensable. As we will soon see when we study Newton's method for solving equations (Section 3.8) and Euler's method for solving differential equations (Subsection 4.1.3), the tangent line approximation remains necessary for more complicated problems.

Example 3.7.2 The Tangent Line to the Logarithmic Function

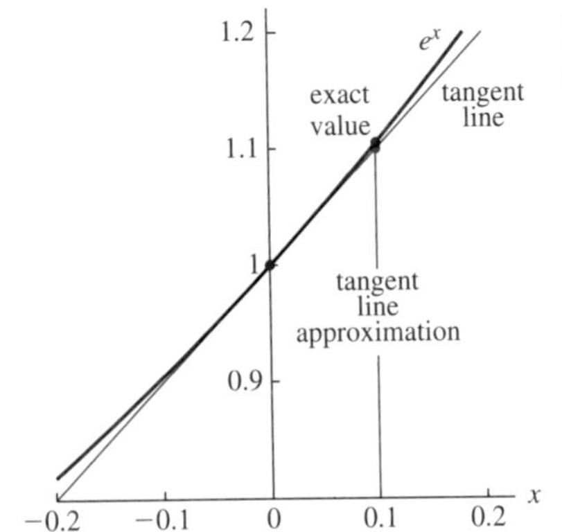

FIGURE 3.7.2
Approximating $e^{0.1}$ with the tangent line

To estimate $\ln(0.9)$ without a calculator, note that the input value 0.9 is near 1.0. Because $\ln(1.0) = 0$, the answer is close to 0. To do better, we match both the value and the slope of the function $\ln(x)$ at $x = 1$. In this case,

$$\frac{d}{dx}\ln(x) = \frac{1}{x}$$

so the slope at $x = 1$ is 1 (Figure 3.7.3). The tangent line is

$$\hat{\ln}(x) = 0 + 1(x-1) = x - 1.$$

Substituting in $x = 0.9$, we find an approximate value of -0.1, close to the exact value of $\ln(0.9) \approx -0.10536$.

Recall the bacterial population growing according to

$$b(t) = 2.0^t$$

(Subsection 2.1.1). Can we find a good linear approximation for this function for times between 0 and 1? One method is to use the tangent line at $t = 0$. First, we find the

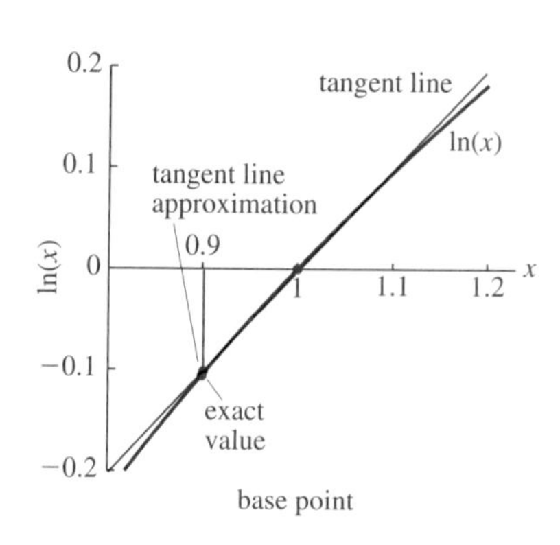

FIGURE 3.7.3

Approximating ln(0.9) with the tangent line

derivative

$$\begin{aligned}\frac{db}{dt} &= \frac{d}{dt}2.0^t \\ &= \frac{d}{dt}e^{\ln(2.0)t} \\ &= \ln(2.0)e^{\ln(2.0)t},\end{aligned}$$

using the rule for finding derivatives of the general exponential function (Subsection 2.9.3). The tangent line at $t = 0$ is therefore

$$\hat{b_0}(t) = b(0) + b'(0)(t - 0) = 1.0 + 0.693t,$$

where the subscript 0 indicates the base point. If we are interested in approximating values near $t = 0$, the tangent line is accurate (Figure 3.7.4a).

If instead we are interested in approximating values near $t = 1$, this tangent line is quite inaccurate. Near $t = 1$, the tangent line is

$$\hat{b_1}(t) = b(1) + b'(1)(t - 1) \approx 2.0 + 1.386(t - 1)$$

because $b(1) = 2.0$ and $b'(1) = 2.0\ln(2.0) = 1.386$. Near the base point $t = 1$, this tangent line provides an accurate approximation (Figure 3.7.4b).

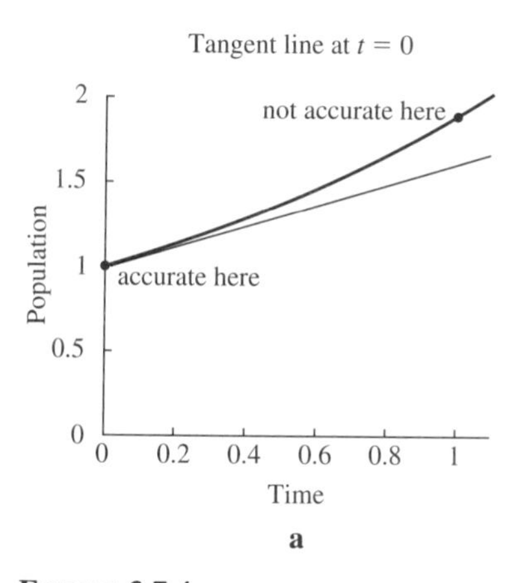

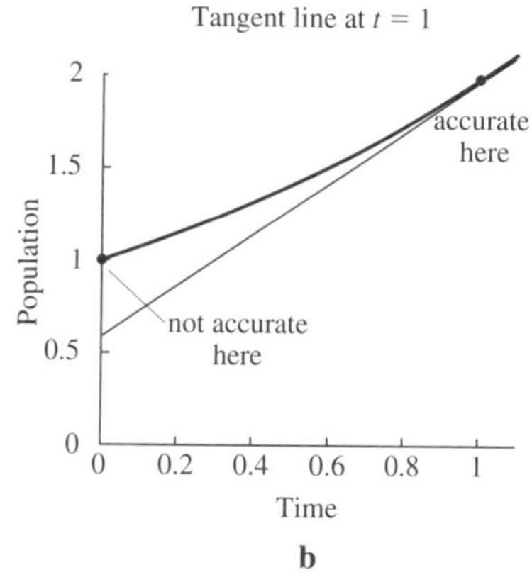

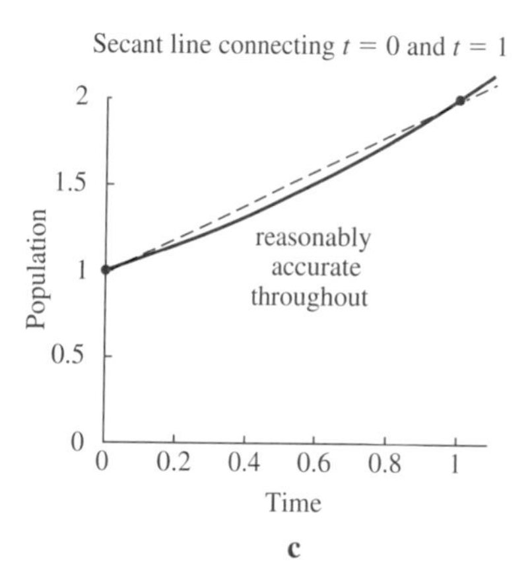

FIGURE 3.7.4

Two tangent lines and a secant line as approximations

Over the whole interval from $t = 0$ to $t = 1$, the secant line is reasonably good everywhere. The secant line has slope

$$\text{slope of secant} = \frac{\Delta b}{\Delta t} = \frac{b(1.0) - b(0.0)}{1.0 - 0.0} = 1.0,$$

which lies between that of the two tangent lines, and has the equation

$$b_s(t) = b(0) + 1.0(t - 0) = 1 + t.$$

How can we quantify the accuracy of these alternative approximations? Results near both endpoints and the middle of the interval are given in the following table.

t	$b(t)$	$b_0(t)$	$b_1(t)$	$b_s(t)$
0.01	1.00695	1.00693	0.62757	1.01
0.10	1.07177	1.06931	0.75244	1.10
0.50	1.41421	1.34657	1.30685	1.50
0.90	1.86607	1.62383	1.86137	1.90
0.99	1.98618	1.68622	1.98614	1.99

Each tangent line is an excellent approximation near its point of tangency. The secant line is always fairly close. This is one of the two primary strengths of the secant line, which is also called **linear interpolation**. The other is that the secant can be directly estimated from data, even when we do not know the underlying equation. We will see how this idea is used when solving equations (Section 3.8) and fitting data with lines (Section 8.9).

Nonetheless, the tangent line is the **best** possible approximation near the point of tangency. No other line is closer. Suppose we compare the tangent and secant lines near the base point 0. To quantify how close the estimates are to the exact values, define errors e_0 and e_s for the two lines as

$$e_0 = \hat{b}_0(t) - b(t)$$
$$e_s = \hat{b}_s(t) - b(t).$$

To check how small these are for t near 0, we divide the errors e_0 and e_s by t.

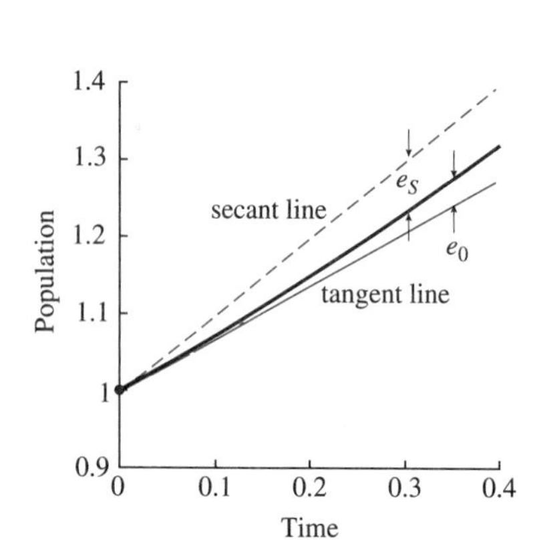

FIGURE 3.7.5
The errors associated with tangent and secant line approximations

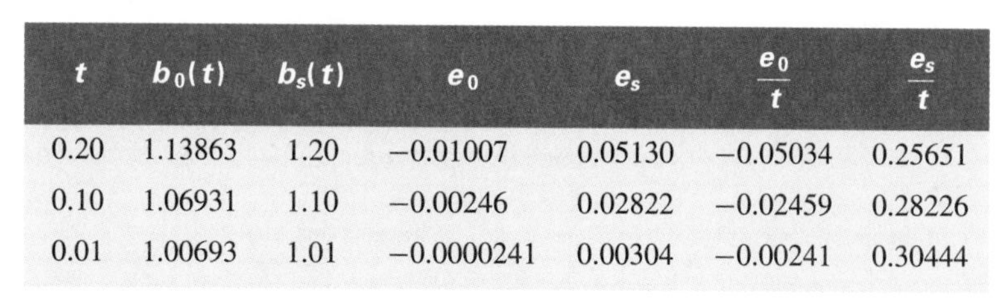

t	$b_0(t)$	$b_s(t)$	e_0	e_s	$\frac{e_0}{t}$	$\frac{e_s}{t}$
0.20	1.13863	1.20	−0.01007	0.05130	−0.05034	0.25651
0.10	1.06931	1.10	−0.00246	0.02822	−0.02459	0.28226
0.01	1.00693	1.01	−0.0000241	0.00304	−0.00241	0.30444

Both errors e_0 and e_s get smaller as t gets smaller (Figure 3.7.5). With the tangent line approximation, the **relative** error $\frac{e_0}{t}$ also gets smaller as t gets smaller. With the secant line approximation, the relative error $\frac{e_s}{t}$ remains roughly constant.

3.7.2 Quadratic Approximation

The tangent and secant lines provide ways to approximate curves with lines. We can do better by approximating curves with simpler curves.

Example 3.7.3 Approximating the Exponential Function with a Quadratic

Suppose that we wish to approximate the exponential function near 0 with a quadratic function. Just as we can match the value of the function and the first derivative with the tangent line, we can match the function, the first derivative, *and* the second derivative with a quadratic. For the exponential function $g(x) = e^x$, $g'(x) = e^x$ and $g''(x) = e^x$. Therefore,

$$g(0) = 1$$
$$g'(0) = 1$$
$$g''(0) = 1.$$

This **does not** mean that the quadratic approximation is $\hat{g}(x) = 1 + x + x^2$. We must be a bit more careful. The derivative is

$$\frac{d}{dx}(1 + x + x^2) = 1 + 2x$$

and the second derivative is

$$\frac{d^2}{dx^2}(1 + x + x^2) = 2.$$

This is double what it should be. Thanks to the constant product rule for the derivatives, we can correct this by dividing the quadratic term by 2. We thus guess that

$$\hat{g}(x) = 1 + x + \frac{x^2}{2}.$$

Checking, we find that

$$\frac{d}{dx}\hat{g}(x) = 1 + x$$

and that $\hat{g}(0) = 1$. Also,

$$\frac{d^2}{dx^2}\hat{g}(x) = 1.$$

Therefore, the value, the first derivative, and the second derivative of this quadratic approximation match those of the exponential function (Figure 3.7.6).

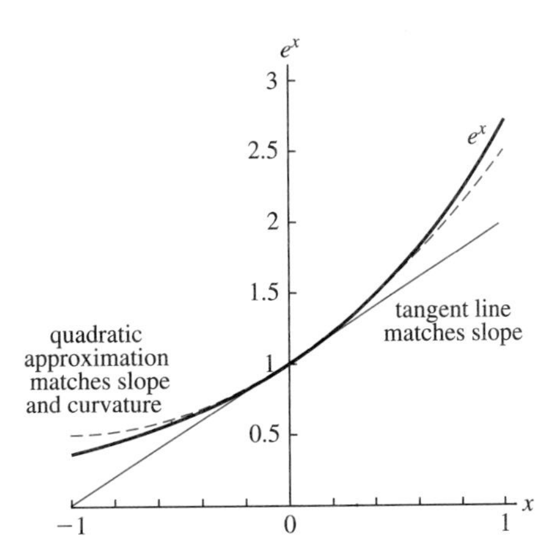

FIGURE 3.7.6
The linear and quadratic approximations of e^x

To find the quadratic approximation to a function $f(x)$ at the base point a in general, we match the value, the derivative, and the second derivative at the point a. To make the calculation easier, we write the quadratic in a version of point-slope form:

$$\hat{f}(x) = c_0 + c_1(x - a) + c_2(x - a)^2.$$

We want to choose the values c_0, c_1, and c_2 so that

$$\hat{f}(a) = f(a)$$
$$\hat{f}'(a) = f'(a)$$
$$\hat{f}''(a) = f''(a).$$

Then

$$\hat{f}'(x) = c_1 + 2c_2(x - a)$$
$$\hat{f}''(x) = 2c_2$$

and

$$\hat{f}(a) = c_0 + c_1(a - a) + c_2(a - a)^2 = c_0$$
$$\hat{f}'(a) = c_1 + 2c_2(a - a) = c_1$$
$$\hat{f}''(a) = 2c_2.$$

Therefore,

$$c_0 = f(a)$$
$$c_1 = f'(a)$$
$$2c_2 = f''(a).$$

The approximating quadratic is therefore

$$\hat{f}(x) = f(a) + f'(a)(x - a) + \frac{f''(a)}{2}(x - a)^2.$$

The first two terms in this approximate function exactly match the tangent line. The last term is an additional correction. Recall from the method of leading behavior that $(x-a)^2$ is smaller than $x-a$ when $x-a$ itself is small. This new term has little effect near a.

Example 3.7.4 Approximating a Cubic with a Quadratic

Suppose we wish to approximate the function $g(x)=1+2x+x^2+3x^3$ with a quadratic function for x near 1. The derivative and second derivative of g are

$$g'(x)=2+2x+9x^2$$
$$g''(x)=2+18x.$$

Therefore,

$$g(1)=1+2\cdot 1+1^2+3\cdot 1^3=7$$
$$g'(1)=2+2\cdot 1+9\cdot 1^2=13$$
$$g''(1)=2+18\cdot 1=20.$$

Then

$$\hat{g}(x)=7+13x+10x^2.$$

The approximating quadratic is **not** found by just ignoring the cubic term (except if the base point is 0; see Exercise 21).

Example 3.7.5 Approximating the Logarithmic Function with a Quadratic

In Example 3.7.2, we used the tangent line to estimate $\ln(0.9)$ by finding the tangent line to $\ln(x)$ at the base point $x=0$. We can get a more accurate estimate by finding the quadratic approximation at this same base point. Denoting the function $\ln(x)$ by $f(x)$, we have

$$f'(x)=\frac{d}{dx}\ln(x)=\frac{1}{x}$$
$$f''(x)=\frac{d^2}{dx^2}\ln(x)=-\frac{1}{x^2}.$$

Evaluating at the point $x=1$,

$$f(1)=\ln(1)=0$$
$$f'(1)=\frac{1}{1}=1$$
$$f''(1)=-\frac{1}{1^2}=-1.$$

Therefore, the approximating quadratic is

$$\hat{f}(x)=0+1(x-1)-\frac{1}{2}(x-1)^2=x-1-\frac{1}{2}(x-1)^2.$$

Substituting in $x=0.9$, we find an approximate value of -0.105, very close to the exact value of $\ln(0.9)\approx -0.10536$. As we can see from Figure 3.7.7a, the approximation is nearly indistinguishable for x near 1. However, over a larger range, the function and the approximation diverge (Figure 3.7.7b).

Example 3.7.6 Finding Approximating Quadratics at Three Points

The function

$$h(x)=xe^{-x}$$

has the derivatives

$$h'(x)=(1-x)e^{-x}$$
$$h''(x)=(x-2)e^{-x}$$

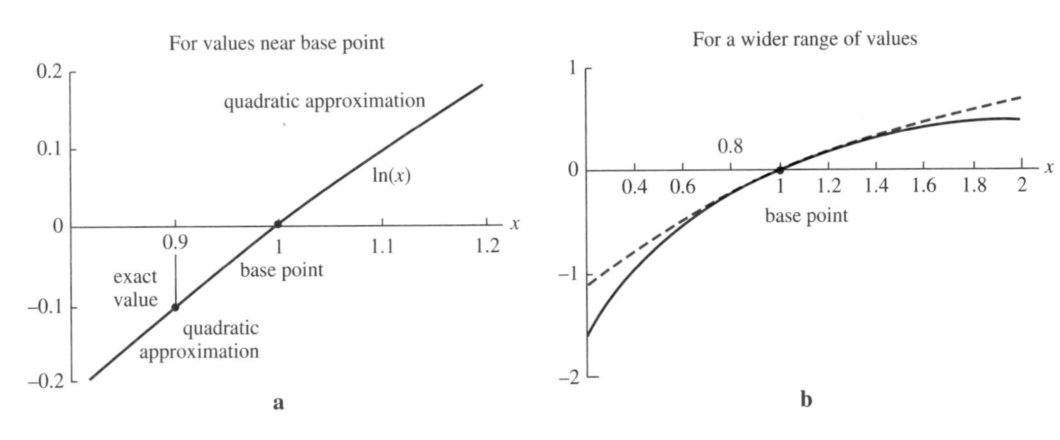

FIGURE 3.7.7
Approximating ln(0.9) with a quadratic

(Example 2.8.7). This function has a critical point at $x = 1$ and a point of inflection at $x = 2$. Using the formula for the approximating quadratic at three points, $x = 0$, $x = 1$, and $x = 2$, to find $h_0(x)$, $h_1(x)$, and $h_2(x)$ gives

$$\hat{h_0}(x) = h(0) + h'(0)x + \frac{h''(0)}{2}x^2 = x - x^2$$

$$\hat{h_1}(x) = h(1) + h'(1)(x-1) + \frac{h''(1)}{2}(x-1)^2 = \frac{1}{e} - \frac{1}{e}(x-1)^2$$

$$\hat{h_2}(x) = h(2) + h'(2)(x-2) + \frac{h''(2)}{2}(x-2)^2 = \frac{1}{e^2} - \frac{1}{e^2}(x-2)$$

(Figure 3.7.8). At the critical point, the approximating quadratic has no $x - 1$ term because $h'(1) = 0$. At the point of inflection, the quadratic term drops out, leaving us with the tangent line.

3.7.3 Taylor Polynomials and Taylor Series

The idea of matching derivatives can be extended to the third, fourth, and higher derivatives. With each added derivative, the approximation becomes more accurate but requires a polynomial of higher **degree** (largest power). The derivation of the following formula is the same as the derivation of the quadratic approximation. The approximating polynomial is called a **Taylor polynomial** of degree n.

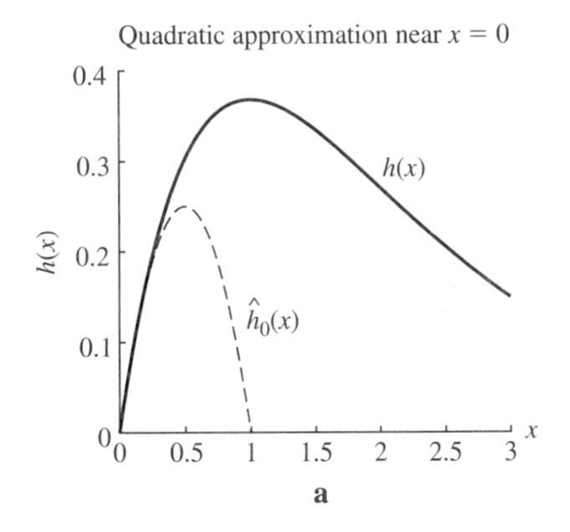

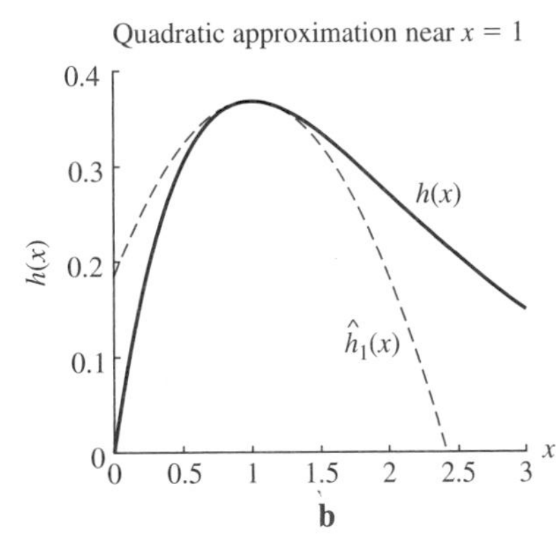

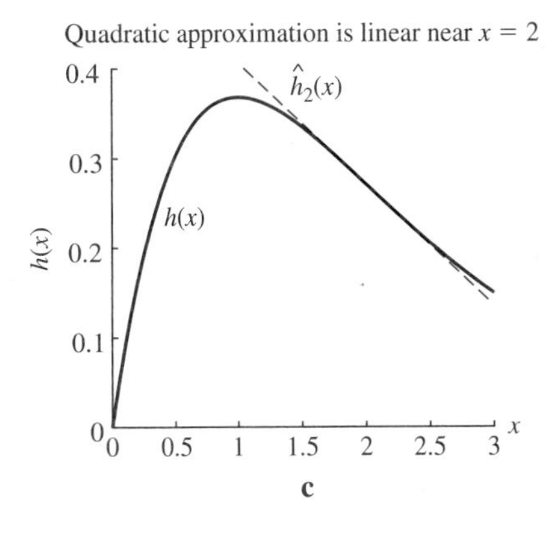

FIGURE 3.7.8
Three quadratic approximations to $h(x) = xe^{-x}$

Definition 3.8 The Taylor Polynomial

Suppose the first n derivatives of the function f are defined at $x = a$. Then the Taylor polynomial of degree n matching the values of the first n derivatives is

$$P_n(x) = f(a) + f'(a)(x-a) + \frac{f''(a)}{2}(x-a)^2 + \cdots + \frac{f^{(i)}(a)}{i!}(x-a)^i + \cdots + \frac{f^{(n)}(a)}{n!}(x-a)^n.$$

We used two pieces of new notation in this definition. First, the notation

$$f^{(i)}(x)$$

indicates the ith derivative of f. For example, we could write

$$f^{(2)}(x) = f''(x)$$

for the second derivative.

Second, the terms with exclamation points are called **factorials**. The value of $i!$ is the product of i and all numbers smaller than i, or

$$i! = i \cdot (i-1) \cdot (i-2) \cdot \ldots \cdot 3 \cdot 2 \cdot 1.$$

For example,

$$2! = 2 \cdot 1 = 2$$
$$3! = 3 \cdot 2 \cdot 1 = 6$$
$$4! = 4 \cdot 3 \cdot 2 \cdot 1 = 24$$
$$5! = 5 \cdot 4 \cdot 3 \cdot 2 \cdot 1 = 120.$$

The values of factorials increase very quickly, meaning that later terms in a Taylor polynomial become ever smaller.

Example 3.7.7 Approximating the Logarithmic Function with a Cubic Function

In Example 3.7.5, we used a quadratic function to estimate $\ln(0.9)$. To find the cubic (third order) approximation, we must also match the third derivative. Denoting the function $\ln(x)$ by $f(x)$, we have

$$f'(x) = \frac{d}{dx}\ln(x) = \frac{1}{x}$$
$$f''(x) = \frac{d^2}{dx^2}\ln(x) = -\frac{1}{x^2}$$
$$f'''(x) = \frac{d^3}{dx^3}\ln(x) = \frac{2}{x^3}.$$

Evaluating at the point $x = 1$,

$$f(1) = \ln(1) = 0$$
$$f'(1) = \frac{1}{1} = 1$$
$$f''(1) = -\frac{1}{1^2} = -1$$
$$f'''(1) = \frac{2}{1^3} = 2.$$

Therefore, the approximating cubic is

$$\hat{f}(x) = 0 + 1(x-1) - \frac{1}{2}(x-1)^2 + \frac{2}{6}(x-1)^3.$$

Substituting in $x = 0.9$, we find an approximate value of -0.10533, even closer to the exact value of $\ln(0.9) \approx -0.10536$.

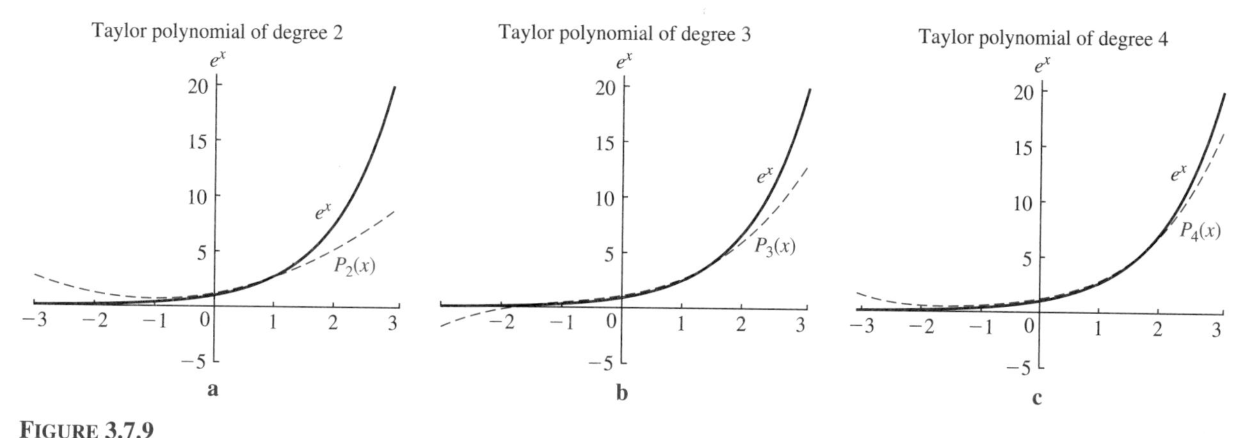

FIGURE 3.7.9

Three Taylor polynomial approximations to $g(x) = e^x$

Example 3.7.8 Taylor Polynomials for the Exponential Function

Consider again the exponential function $g(x) = e^x$. Using our new notation,

$$g^{(i)}(x) = g(x)$$

because each derivative of the exponential function is equal to the function itself. For $a = 0$, then the Taylor polynomial of degree 4 (Figure 3.7.9) is

$$P_4(x) = 1 + x + \frac{1}{2}x^2 + \frac{1}{6}x^3 + \frac{1}{24}x^4.$$

We estimate that

$$e^{0.1} \approx P_4(0.1) = 1 + 0.1 + \frac{1}{2}0.1^2 + \frac{1}{6}0.1^3 + \frac{1}{24}0.1^4 \approx 1.1051708.$$

The exact answer of $e^{0.1} \approx 1.1051709$.

For some functions, we can compute the Taylor polynomial to any degree by finding all of the higher derivatives. The resulting **infinite sum** is called the **Taylor series** for that function.

Definition 3.9 **The Taylor Series**

Suppose all of the derivatives of the function f are defined at $x = a$. Then the Taylor series for f near a is

$$T_f(x) = f(a) + f'(a)(x-a) + \frac{f''(a)}{2!}(x-a)^2 + \frac{f(3)(a)}{3!}(x-a)^3 + \cdots + \frac{f(n)(a)}{n!}(x-a)^n + \ldots$$

The dots indicate that we are adding over all values of n.

For any value of x, this Taylor series is an infinite sum of numbers. Strictly speaking, the meaning of an infinite sum is not defined in elementary arithmetic. Instead, the value must be defined as a limit.

Definition 3.10 **The Sum of an Infinite Series**

For a set of values x_n

$$x_0 + x_1 + x_2 + \ldots x_i + \ldots = \lim_{n \to \infty} x_0 + x_1 + x_2 + \ldots x_n$$

if the limit exists.

This definition converts the infinite sum into the limit of a sequence, a sequence of sums with only a finite number of elements, such as we studied in Subsection 3.5.3. When the sequence has a finite limit, we say that the series **converges**.

Example 3.7.9 Taylor Series for the Exponential Function

As we saw in Example 3.7.8, all derivatives of the exponential function $g(x) = e^x$ are equal to 1 at the base point $a = 0$. The Taylor series is

$$T_g(x) = 1 + x + \frac{x^2}{2} + \frac{x^3}{6} + \frac{x^4}{24} + \dots$$

Evaluating at $x = 1$ gives the remarkable identity

$$e = g(1) = 1 + 1 + \frac{1^2}{2} + \frac{1^3}{6} + \frac{1^4}{24} + \dots$$

The sequence of sums is

$$\begin{aligned} 1 &= 1 \quad (n = 0) \\ 1 + 1 &= 2 \quad (n = 1) \\ 1 + 1 + \tfrac{1}{2} &= 2.5 \quad (n = 2) \\ 1 + 1 + \tfrac{1}{2} + \tfrac{1}{6} &= 2.6667 \; (n = 3) \\ 1 + 1 + \tfrac{1}{2} + \tfrac{1}{6} + \tfrac{1}{24} &= 2.7083 \; (n = 4) \\ 1 + 1 + \tfrac{1}{2} + \tfrac{1}{6} + \tfrac{1}{24} + \tfrac{1}{120} &= 2.7167 \; (n = 5). \end{aligned}$$

These values converge quite quickly to e and provide a way to compute this **transcendental number**, a number that is not the solution of any polynomial equation.

There are some functions for which the Taylor series does not converge, and even some for which the Taylor series does not converge to the value of the function. Here we only examine cases where the Taylor series is well-behaved.

Example 3.7.10 Taylor Series and the Geometric Series

Consider the function

$$f(x) = \frac{1}{1 - x}.$$

Then

$$\begin{aligned} f'(x) &= \frac{1}{(1 - x)^2} \\ f''(x) &= \frac{2}{(1 - x)^3} \\ f'''(x) &= \frac{6}{(1 - x)^4} \\ &\dots \\ f^{(i)}(x) &= \frac{i!}{(1 - x)^{i+1}}. \end{aligned}$$

Evaluating at the base point $a = 0$ gives $f^{(i)}(0) = i!$, and thus the Taylor series

$$T_f(x) = 1 + x + x^2 + x^3 + x^4 + \dots.$$

For $x < 1$, this sum converges to the function $f(x)$, giving the result

$$\frac{1}{1 - x} = 1 + x + x^2 + x^3 + x^4 + \dots.$$

Evaluating at $x = \frac{1}{2}$ gives the identity

$$1 + \frac{1}{2} + \left(\frac{1}{2}\right)^2 + \left(\frac{1}{2}\right)^3 + \left(\frac{1}{2}\right)^4 + \dots = \frac{1}{1 - \frac{1}{2}} = 2.$$

The sequence of sums is

$$\begin{aligned}
1 &= 1 \quad (n=0)\\
1+\tfrac{1}{2} &= 1.5 \quad (n=1)\\
1+\tfrac{1}{2}+\tfrac{1}{4} &= 1.75 \quad (n=2)\\
1+\tfrac{1}{2}+\tfrac{1}{4}+\tfrac{1}{8} &= 1.875 \quad (n=3)\\
1+\tfrac{1}{2}+\tfrac{1}{4}+\tfrac{1}{8}+\tfrac{1}{16} &= 1.9375 \ (n=4).
\end{aligned}$$

This concept of the infinite sum addresses one of the great paradoxes of the ancient world, known as **Zeno's paradox**, which asks how any moving object can arrive at its destination because it must first travel half the distance, and then half of the remaining distance, and so forth.

The **geometric series** is defined as the sum

$$S = 1 + r + r^2 + r^3 + r^4 + \ldots$$

The Taylor series in Example 3.7.10 shows that $S = \frac{1}{1-r}$ if $r < 1$. The geometric series arises in many applications in biology, including population growth and the geometric distribution that plays an important role in probability and statistics (Section 7.6).

Example 3.7.11 Application of the Geometric Series

Suppose a bacterial population starts at population b_0 and declines by a factor of $r = 0.9$ each generation. The population follows the discrete-time dynamical system

$$b_{t+1} = rb_t$$

with the solution

$$b_t = b_0 r^t$$

(Section 1.5, Example 1.5.11). Suppose that each of these bacteria makes a quantity q of some valuable medical product. How much will this population make in the long run? How much more does it make than in the first generation?

The total produced is

$$\begin{aligned}
\text{total product} &= qb_0 + qb_1 + qb_2 + qb_3 + \ldots\\
&= q(b_0 + b_1 + b_2 + b_3 + \ldots)\\
&= q(b_0 + b_0 r + b_0 r^2 + b_0 r^3 + \ldots)\\
&= qb_0(1 + r + r^2 + r^3 + \ldots)\\
&= qb_0 \frac{1}{1-r}\\
&= qb_0 \frac{1}{1-0.9} = 10qb_0.
\end{aligned}$$

The population of b_0 would produce a quantity qb_0 in the first generation, and ten times as much as it slowly dies out.

Summary Using lines to approximate curves is one of the central ideas in calculus. We have compared the tangent line with three other approximations. The secant line, or **linear interpolation**, has the joint virtues of using actual data and remaining fairly accurate over a broad domain. The tangent line is the **best linear approximation** near the base point. To do even better, we can use a quadratic polynomial to match both the first and second derivatives of the original function at the base point. This idea can be expanded to the **Taylor polynomial**, which is a polynomial of degree n that matches the first n derivatives of the function, and to the **Taylor series**, which matches all of the derivatives of the function.

3.7 Exercises

Mathematical Techniques

1–6 ▪ Use the tangent line and secant line to estimate the following values. Make sure to identify the base point a you used for your tangent line approximation and the second point you used for the secant line approximation. Use a calculator to compare the estimates with the exact answer.

1. 2.02^3
2. 3.03^2
3. $\sqrt{4.01}$
4. $\sqrt{6}$
5. $\sin(0.02)$
6. $\cos(-0.02)$

7–12 ▪ Use the quadratic approximation to estimate the following values. Compare the estimates with the exact answer.

7. 2.02^3 (based on Exercise 1).
8. 3.03^2 (based on Exercise 2).
9. $\sqrt{4.01}$ (based on Exercise 3).
10. $\sqrt{6}$ (based on Exercise 4).
11. $\sin(0.02)$ (based on Exercise 5).
12. $\cos(-0.02)$ (based on Exercise 6).

13–16 ▪ Use the tangent line approximation to evaluate the following in two ways. First, find the tangent line to the whole function using the chain rule. Second, break the calculation into two pieces by writing the function as a composition, approximate the inner function with its tangent line, and use this value to plug into the tangent line of the outer function. Do your answers match?

13. $(1 + 3 \times 1.01)^2$
14. $\ln(\sqrt{0.98})$
15. $e^{\sin(0.02)}$
16. $\sin\left(\ln\left[(1+0.1)^3\right]\right)$

17–20 ▪ For the following functions, find the tangent line approximation of the two values and compare with the true value. Indicate which approximations are too high and which are too low. From graphs of the functions, try to explain what it is about the graph that causes this.

17. $e^{0.1}$ and $e^{-0.1}$.
18. $\ln(1.1)$ and $\ln(0.9)$.
19. 1.1^2 and 0.9^2.
20. $\sqrt{1.1}$ and $\sqrt{0.9}$.

21–26 ▪ Find the third order Taylor polynomials for the following functions.

21. $f(x) = x^3 + 4x^2 + 3x + 1$ for x near 0.
22. $g(x) = 4x^4 + x^3 + 4x^2 + 3x + 1$ for x near 0.
23. $f(x) = 4x^2 + 3x + 1$ for x near 1.
24. $g(x) = 4x^4 + x^3 + 4x^2 + 3x + 1$ for x near 1.
25. $h(x) = \ln(x)$ for x near 1.
26. $h(x) = \sin(x)$ for x near 0.

27–30 ▪ Use the Taylor series from the text to evaluate the following sums. Check by adding up through the $n = 5$ term. How close do they get to their limit?

27. $1 + 2 + \frac{2^2}{2} + \frac{2^3}{3!} + \frac{2^4}{4!} + \dots$
28. $1 - 1 + \frac{1}{2} - \frac{1}{3!} + \frac{1}{4!} - \frac{1}{5!} + \dots$
29. $1 + \frac{1}{3} + \frac{1}{9} + \frac{1}{27} + \frac{1}{81} + \dots$
30. $1 + \frac{3}{4} + \frac{9}{16} + \frac{27}{64} + \frac{81}{256} + \dots$

31–32 ▪ Find the Taylor polynomial of degree n and the Taylor series for the following functions. Add up the given series by assuming that the sum of the Taylor series is equal to the function.

31. Find the Taylor polynomial for $h(x) = \cos(x)$ with the base point $a = 0$. Use your result to find $1 - \frac{1}{2!} + \frac{1}{4!} - \frac{1}{6!} + \frac{1}{8!} + \dots$
32. $f(x) = \ln(1 - x)$ with the base point $a = 0$. Use your result to find $1 - \frac{1}{2} + \frac{1}{3} - \frac{1}{4} + \frac{1}{5} + \dots$

33–36 ▪ Write the tangent line approximation for the numerators and denominators of the following functions and show that the result of applying L'Hôpital's rule matches that of comparing the linear approximations.

33. $f(x) = \frac{2x + x^2}{3x + 2x^2}$ at $x = 0$.
34. $f(x) = \frac{\ln(1 + x)}{e^{2x} - 1}$ at $x = 0$.
35. $f(x) = \frac{\ln(x)}{x^2 - 1}$ at $x = 1$.
36. $f(x) = \frac{\cos(x) + 1}{\sin(x)}$ at $x = \pi$.

Applications

37–42 ▪ Compare the tangent line approximation of the following absorption functions with the leading behavior at $c = 0$. If they do not match, can you explain why?

37. $\alpha(c) = \frac{5c}{1 + c}$ (as in Section 3.6, Exercise 23).
38. $\alpha(c) = \frac{c}{5 + c}$ (as in Section 3.6, Exercise 24).
39. $\alpha(c) = \frac{5c^2}{1 + c^2}$ (as in Section 3.6, Exercise 25).
40. $\alpha(c) = \frac{5c}{e^{2c}}$ (as in Section 3.6, Exercise 26).
41. $\alpha(c) = \frac{5c}{1 + c^2}$ (as in Section 3.6, Exercise 27).
42. $\alpha(c) = 5c(1 + c)$ (as in Section 3.6, Exercise 28).

43–46 ▪ Consider a declining population following the formula

$$b(t) = \frac{1}{1+t}$$

(measured in millions). Approximate the population at each of the following times using the tangent with base point $t = 0$, the tangent with base point $t = 1$, and the secant connecting times $t = 0$ and $t = 1$. Graph each of the relevant tangents and secants. Which method is best for what?

43. $t = 0.1$.

44. $t = 0.5$.

45. $t = 0.9$.

46. $t = 1.1$.

47–48 ▪ Consider the following table giving mass as a function of age.

Age, *a* (days)	Mass, *M* (g)
0.5	0.125
1.0	1.000
1.5	3.375
2.0	8.000

The data follow the equation $M(a) = a^3$. Estimate each of the following using the tangent line approximation and the secant line approximation. Which approximation is closer to the exact answer? Which method would be best if you did not know the formula for $M(a)$?

47. $M(1.25)$

48. $M(1.45)$

49–50 ▪ Consider the following table giving temperature as a function of time.

Time, *t*	Temperature, *T* (°C)
0.0	0.172
1.0	1.635
2.0	6.492
3.0	11.95
4.0	20.24

The data follow the equation $T(t) = t + t^2$, but there is some noise in each of the measurements, so that the values are not exactly on the curve. Estimate each of the following using the tangent line approximation and a secant line approximation. Which method do you think deals best with the noise?

49. Estimate $T(1)$ using the values at $t = 0$ and $t = 2$.

50. Estimate $T(3)$ using the values at $t = 2$ and $t = 4$.

51–52 ▪ Even after they have been controlled, diseases can affect many individuals. Suppose a medication has been introduced, and each person infected with a new variety of influenza infects only r people on average, and those people infect r people themselves and so forth. Suppose 1.0×10^5 people are initially infected. Find the total number of people infected by a disease in the following cases.

51. Suppose $r = 0.5$.

52. Suppose $r = 0.95$. How does the total number of people infected compare with the previous problem?

Computer Exercises

53. Find the Taylor polynomials for $\cos(x)$ and $\sin(x)$ with the base point $x = 0$ up to degree 10. Can you see the pattern? Graph the functions, P_2, P_5, and P_{10} on domains around 0 that get larger and larger. What happens to the approximation for values of x far from 0?

54. Find the Taylor polynomial for the function defined by

$$f(x) = \begin{cases} e^{-\frac{1}{x}} & \text{if } x \neq 0 \\ 0 & \text{if } x = 0 \end{cases}$$

with the base point $x = 0$ up to degree 10 (you will have to take the limit as $x \to 0$ to compute the derivatives). Can you see the pattern? Graph the function on the domain $-1 \le x \le 1$. What happens to the approximation for values of x far from 0? Do the Taylor polynomials make sense? What is the Taylor series for this function?

55. A simple equation that is impossible to solve algebraically is

$$e^x = x + 2.$$

a. Graph the two sides and convince yourself there is a solution.

b. Replace e^x with its tangent line at $x = 0$, and try to solve for the point where the tangent line is equal to $x + 2$. This is an approximate solution. What goes wrong in this case?

c. Replace e^x with its quadratic approximation at $x = 0$, and solve for the point where it is equal to $x + 2$.

d. Replace e^x with its tangent line at $x = 1$ and solve.

e. Replace e^x with its quadratic approximation at $x = 1$ and solve.

f. How close are these solutions to the exact answer?

3.8 Newton's Method

With powerful calculators and computers, it might seem unnecessary to approximate a function with a tangent line. On a calculator, exponentiation is no harder than multiplying or adding. While computing specific functional values is easy, **solving equations** for specific values can be difficult. We have seen how to use the Intermediate Value

Theorem to show that an equation **has** a solution. When we cannot solve the equation with algebraic methods, **Newton's method** can be implemented on a computer to find the exact value. The method replaces the original equation with the tangent line approximation and derives a discrete-time dynamical system that converges with remarkable speed to the solution.

3.8.1 Finding the Equilibrium of the Lung Model with Absorption

Suppose a lung is following

$$c_{t+1} = [1 - q)(1 - \alpha(c_t)]c_t + q\gamma$$

where

$$\alpha(c_t) = 0.5(1 - e^{-0.5c_t})$$

(Equation 3.4.1). In this discrete-time dynamical system c_t represents the concentration, q the fraction of air exchanged, γ the concentration of chemical in the ambient air, and $\alpha(c_t)$ the fraction of chemical absorbed as a function of the chemical concentration in the lung. In Section 3.4, we used the Intermediate Value Theorem to show that this function has an equilibrium between 0 and γ. What if we wish to **compute** the exact value of the equilibrium with a particular set of parameter values?

If we set $q = 0.5$ and $\gamma = 5.0$, the discrete-time dynamical system is

$$c_{t+1} = 0.5[1 - 0.5(1 - e^{-0.5c_t})]c_t + 2.5.$$

It is impossible to set $c_{t+1} = c_t = c^*$ and algebraically solve the equation

$$c^* = 0.5[1 - 0.5(1 - e^{-0.5c^*})]c^* + 2.5$$

for the equilibrium value because equations involving both polynomial and exponential function cannot be solved except in unusual circumstances. How can we use a computer or calculator to find the value?

It looks as though the equilibrium we seek is stable (Figure 3.8.1a). A solution will approach the equilibrium (Figure 3.8.1b). This seems like a good way to get the computer to find the answer. The results of solving the discrete-time dynamical system starting from $c_0 = 5.0$ (the equilibrium $c^* = \gamma$ for the case without absorption) are given in the following table. The columns give the concentrations, the difference from the

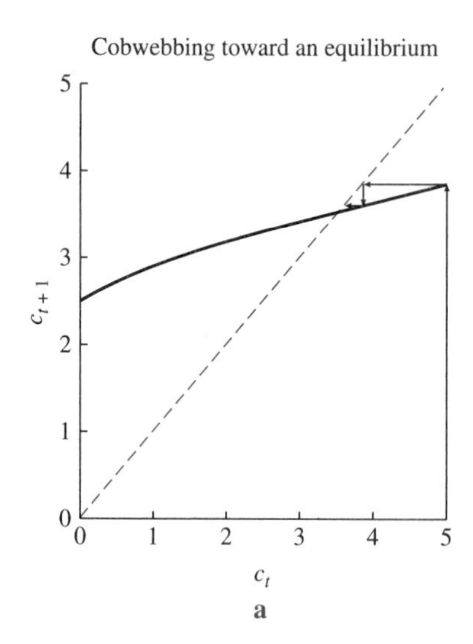

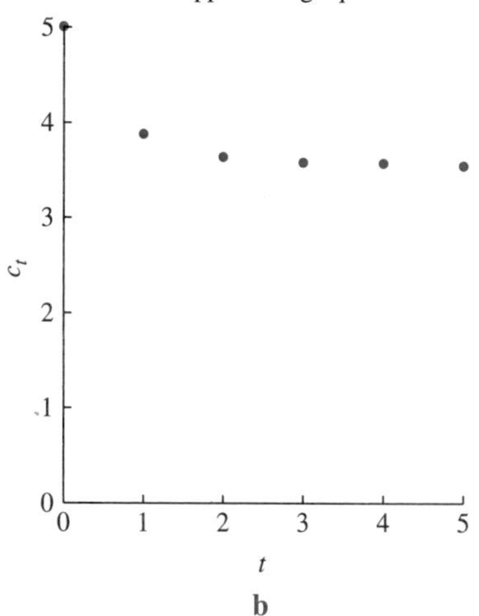

FIGURE 3.8.1
The iterative method of solving an equation

Iteration	Concentration	Distance from Equilibrium	Factor by Which Distance Decreased
0	5.0000000000	1.4654361738	—
1	3.8526062482	0.3180424220	0.2170291874
2	3.6034690548	0.0689052286	0.2166542067
3	3.5495215521	0.0149577259	0.2170767914
4	3.5378126599	0.0032488337	0.2172010455
5	3.5352695692	0.0007057430	0.2172296671
6	3.5347171389	0.0001533127	0.2172359626
7	3.5345971314	0.0000333052	0.2172373338
8	3.5345710613	0.0000072351	0.2172376315
9	3.5345653979	0.0000015717	0.2172376945
10	3.5345641676	0.0000003414	0.2172377007

true solution (which we do not really know yet), and the ratio of the distance on the current step to the distance in the previous step. For example, the factor in the second row is

$$\frac{0.3180424220}{1.4654361738} \approx 0.2170291874.$$

After ten steps, the first five digits have stopped changing, meaning that we have found the equilibrium to about five decimal places of accuracy. The factor in the final column is approximately equal to the slope of the tangent at the unknown equilibrium. Each step gets us almost five times closer to the answer, and we have a highly accurate answer in only ten steps.

Example 3.8.1 Finding the Optimal Behavior for a Mathematically Sophisticated Bee

Suppose now we wish to find the optimal behavior for a bee (as in Subsection 3.3.2), but that the food intake follows the function

$$F(t) = 1 - e^{-t}.$$

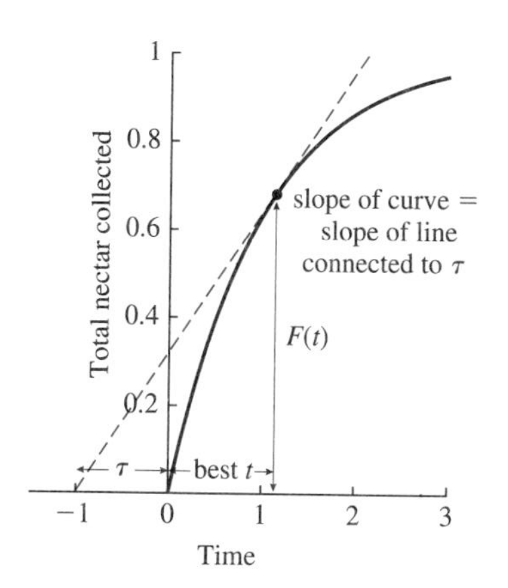

FIGURE 3.8.2

Applying the Marginal Value Theorem when $F(t) = 1 - e^{-t}$

The optimum occurs at the value of t where the tangent line passes through the point $(-\tau, 0)$ where τ is the travel time (Figure 3.8.2). To compute this value, we must solve

$$F'(t) = \frac{F(t)}{t+\tau}.$$

Substituting the equations $F(t) = 1 - e^{-t}$ and $F'(t) = e^{-t}$ and setting $\tau = 1$ gives

$$e^{-t} = \frac{1 - e^{-t}}{t+1}.$$

We do not know how to solve an equation like this. We can rearrange and isolate e^t:

$$\begin{aligned} (t+1)e^{-t} &= 1 - e^{-t} && \text{multiply both sides by } t+1 \\ t+1 &= e^t - 1 && \text{multiply both sides by } e^t \\ 0 &= e^t - t - 2. && \text{move all terms to one side} \end{aligned}$$

At $t = 0$, $e^0 - 0 - 2 < 0$, while at $t = 2$, $e^2 - 2 - 2 > 0$. Because the continuous function $e^t - t - 2$ changes sign between $t = 0$ and $t = 2$, the Intermediate Value Theorem (Theorem 3.2) guarantees at least one solution in between (Figure 3.8.3). However, we cannot solve this algebraically, and it is not written in the form of a discrete-time dynamical system. How can we compute the solution? ▲

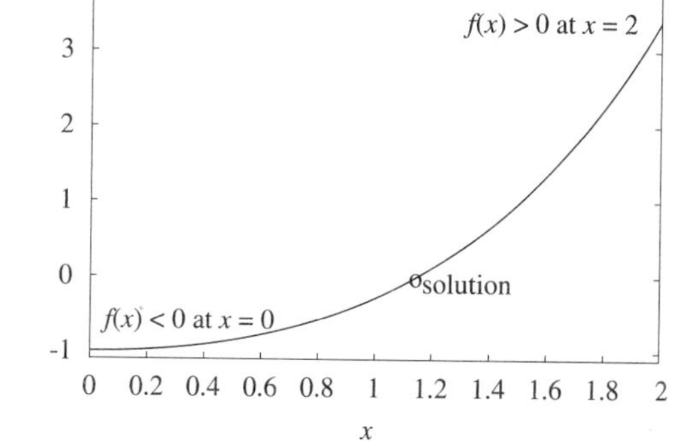

FIGURE 3.8.3

Finding the solution of an algebraically intractable equation

3.8.2 Newton's Method

Newton's method is a method for solving this kind of equation numerically. When it works, the method is incredibly fast, *doubling* the number of digits of accuracy with each step.

Suppose that we want to solve the equation

$$f(x) = 0.$$

If we have some idea that a solution is near the value x_0, we can replace the original equation with the approximate equation

$$\hat{f}(x) = 0,$$

where $\hat{f}(x)$ is the tangent line approximation at x_0 (Figure 3.8.4). The equation for the tangent line $\hat{f}(x)$ is

$$\hat{f}(x) = f(x_0) + f'(x_0)(x - x_0),$$

so our approximate equation $\hat{f}(x) = 0$ is

$$f(x_0) + f'(x_0)(x - x_0) = 0.$$

As long as we can compute the derivative of $f(x)$, we can exactly solve this linear equation for x (Figure 3.8.4). Graphically, the solution of the original equation is the point where the curve intersects the horizontal axis. The solution of the approximate equation is the point where the tangent line intersects the horizontal axis.

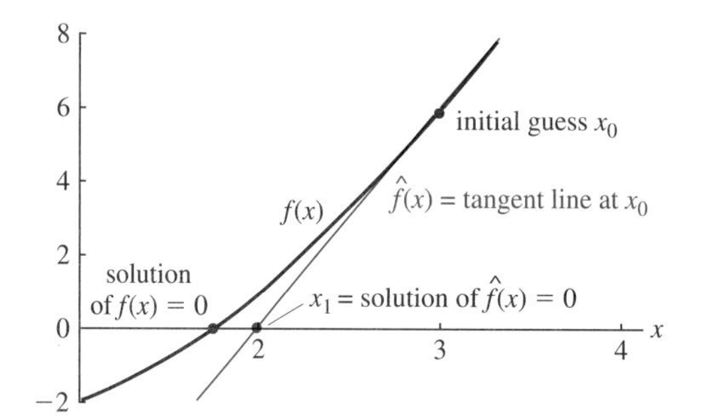

FIGURE 3.8.4

Newton's method: the first step

Example 3.8.2 Computing a Square Root with Newton's Method: First Step

Suppose we wish to solve the equation

$$f(x) = x^2 - 3 = 0.$$

The solution of this equation is $\sqrt{3}$, a numerical value that we may not know and is hard to compute by hand. The first step in Newton's method is to take a guess. We might begin with a rather poor guess of $x_0 = 3$ (Figure 3.8.5). Next, we need to find the tangent line approximation by using the derivative

$$f'(x) = 2x.$$

The tangent line approximation is

$$\begin{aligned}\hat{f}(x) &= f(3) + f'(3)(x - 3)\\ &= (3^2 - 3) + 2 \cdot 3(x - 3)\\ &= 6 + 6(x - 3).\end{aligned}$$

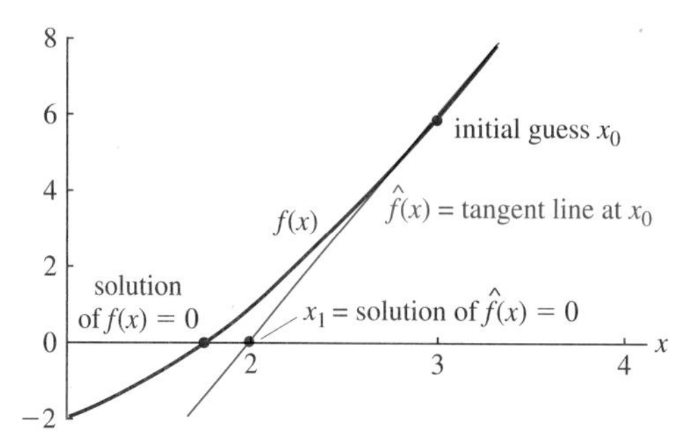

FIGURE 3.8.5

Newton's method for finding a square root: the first step

The approximate equation is

$$\hat{f}(x) = 6 + 6(x - 3) = 0,$$

which has the solution

$$\begin{aligned} 6 + 6(x - 3) &= 0 \\ 6 + 6x - 18 &= 0 \\ 6x &= 12 \\ x &= 2. \end{aligned}$$

This is closer to the right answer because 2^2 is much closer to 3.

We have replaced the difficult equation $f(x) = 0$ with the **linear equation** $\hat{f}(x) = 0$. It is possible to solve this linear equation for x (as long as $f'(x_0) \neq 0$). We find

$$\begin{aligned} f(x) &= 0 && \text{original equation} \\ f(x_0) + f'(x_0)(x - x_0) &= 0 && \text{substitute tangent line approximation} \\ f'(x_0)(x - x_0) &= -f(x_0) && \text{begin solving for } x \\ x - x_0 &= \frac{-f(x_0)}{f'(x_0)} && \text{divide by } f'(x_0) \\ x &= x_0 - \frac{f(x_0)}{f'(x_0)}. && \text{solve for } x \end{aligned}$$

This value of x is the point where the tangent line intersects the horizontal axis. Our hope is that this point is closer to the unknown exact answer than the original guess. If so, we can use x as a new guess, x_1, with the formula

$$x_1 = x_0 - \frac{f(x_0)}{f'(x_0)}. \tag{3.8.1}$$

Starting from the point x_1, we can follow the same steps to find the tangent line and solve for the intersection with the horizontal axis. The new guess, x_2, will have the same formula but with x_1 substituted for x_0, or

$$x_2 = x_1 - \frac{f(x_1)}{f'(x_1)}.$$

Example 3.8.3 Computing a Square Root with Newton's Method: Second Step

In the example with $f(x) = x^2 - 3$, our first guess was $x_0 = 3$. By finding the tangent line and solving the equation, we found $x_1 = 2$. Alternatively, we could use Equation 3.8.1, with $f(3) = 6$ and $f'(3) = 6$, to find

$$\begin{aligned} x_1 &= 3 - \frac{f(3)}{f'(3)} \\ &= 3 - \frac{6}{6} = 2. \end{aligned}$$

Starting from the new guess $x_1 = 2$,

$$\begin{aligned} x_2 &= x_1 - \frac{f(x_1)}{f'(x_1)} \\ &= 2 - \frac{f(2)}{f'(2)} \\ &= 2 - \frac{2^2 - 3}{2 \cdot 2} \\ &= 2 - \frac{1}{4} = 1.75 \end{aligned}$$

(Figure 3.8.6). This is much closer to the exact answer because $1.75^2 = 3.0625$.

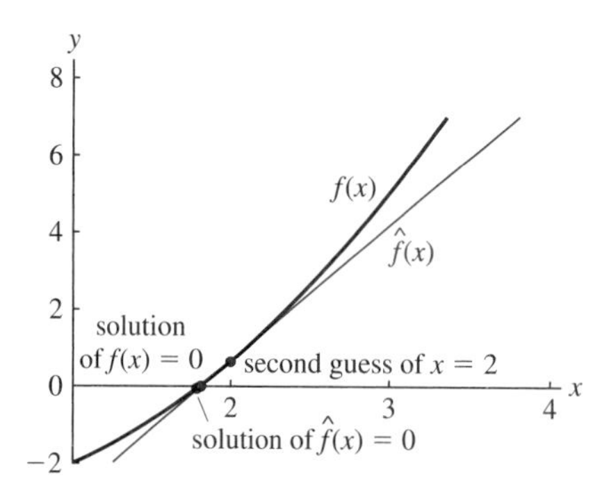

FIGURE 3.8.6
Newton's method: the second step

At each step, we applied the **Newton's method discrete-time dynamical system**

$$x_{t+1} = x_t - \frac{f(x_t)}{f'(x_t)},$$

which leads to the following algorithm.

Algorithm 3.4 Newton's Method for Solving a Nonlinear Equation

To solve the equation $f(x) = 0$,

1. Come up with a first guess called x_0.
2. Use the Newton's method discrete-time dynamical system

$$x_{t+1} = x_t - \frac{f(x_t)}{f'(x_t)}$$

to find x_1, x_2, and so forth until the answer converges. ▲

Why does Newton's method approach a point where $f(x_t) = 0$? If $f(x_t) = 0$, then the Newton's method discrete-time dynamical system gives

$$x_{t+1} = x_t - \frac{f(x_t)}{f'(x_t)} = x_t.$$

Newton's method transforms the problem of solving the equation $f(x) = 0$ into the problem of finding the equilibrium of a discrete-time dynamical system. We can solve this problem by repeatedly applying the discrete-time dynamical system to some initial guess and hoping that it converges.

Example 3.8.4 Computing a Square Root with Newton's Method

With the function $f(x) = x^2 - 3$, the Newton's method discrete-time dynamical system is

$$\begin{aligned} x_{t+1} &= x_t - \frac{f(x_t)}{f'(x_t)} \\ &= x_t - \frac{x_t^2 - 3}{2x_t}. \end{aligned}$$

The results of cobwebbing this equation and finding the solution are shown in Figure 3.8.7 and the following table.

Iteration	Value
0	3.0000000000
1	2.0000000000
2	1.7500000000
3	1.7321428571
4	1.7320508100

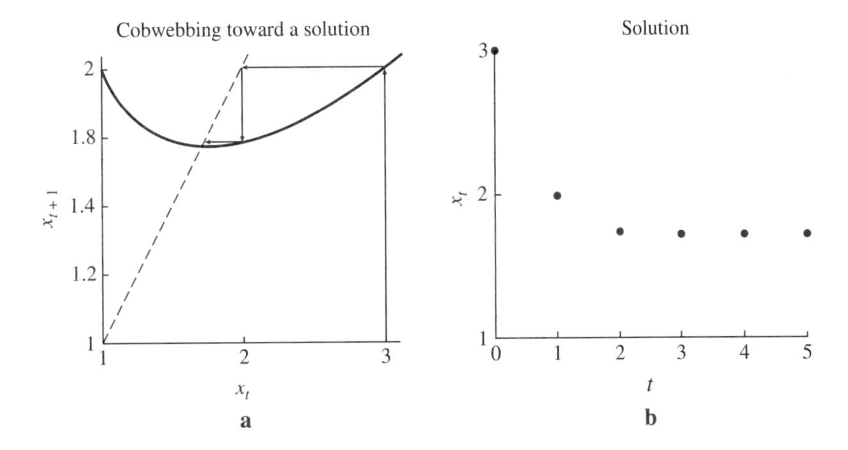

FIGURE 3.8.7

The Newton's method discrete-time dynamical system

Newton's method converges to the answer very quickly, and is correct to eight decimal places after only four steps. And the calculation involved nothing more complicated than multiplication, subtraction, and division (enhanced by the awesome power of the derivative).

Example 3.8.5 Finding the Optimal Behavior of a Bee with Newton's Method

We can apply Newton's method to find the optimal behavior of a bee, requiring the solution of

$$e^t - t - 2 = 0.$$

We define the function $f(x) = e^x - x - 2$ (changing the variable to x to avoid confusion with subscripts). To apply Newton's method, we need to find the derivative of $f(x)$, or $f'(x) = e^x - 1$. The Newton's method discrete-time dynamical system is then

$$x_{t+1} = x_t - \frac{e^{x_t} - x_t - 2}{e^{x_t} - 1}.$$

To start the algorithm, we need a guess. From our graphs (Figures 3.8.2 and 3.8.3), it looks at though the solution is fairly near $x_0 = 1$. The results are

Iteration	Value
0	1.0
1	1.163953414
2	1.146421185
3	1.146193259
4	1.146193221

After only four steps, the values have converged to about seven decimal places, surely good enough for a bee.

Example 3.8.6 Finding an Equilibrium with Newton's Method

In our original problem of finding the equilibrium of a complicated lung discrete-time dynamical system, we already have a discrete-time dynamical system whose solution converges to the equilibrium. Do we gain anything by replacing the original discrete-time dynamical system with the Newton's method discrete-time dynamical system?

The original lung discrete-time dynamical system is

$$c_{t+1} = g(c_t) = 0.5[1 - 0.5(1 - e^{-0.5c_t})]c_t + 2.5.$$

To apply Newton's method, we first replace the equation for equilibrium $g(c) = c$ with the equation $f(c) = g(c) - c = 0$, or

$$f(c) = 0.5[1 - 0.5(1 - e^{-0.5c})]c + 2.5 - c = 0.$$

To find the Newton's method discrete-time dynamical system, we compute the derivative

$$f'(c) = 0.25 + 0.25e^{-0.5c} - 0.125ce^{-0.5c} - 1$$

(this takes some algebra along with the product and chain rules). The Newton's method discrete-time dynamical system is

$$\begin{aligned} c_{t+1} &= c_t - \frac{f(c_t)}{f'(c_t)} \\ &= c_t - \frac{0.5[1 - 0.5(1 - e^{-0.5c_t})]c_t + 2.5 - c_t}{0.25 + 0.25e^{-0.5c_t} - 0.125c_te^{-0.5c_t} - 1}. \end{aligned}$$

We have replaced the original, somewhat messy, discrete-time dynamical system with a truly huge discrete-time dynamical system. Both share the equilibrium point we seek. Do we do better by repeatedly applying the Newton's method discrete-time dynamical system? Starting from the same initial guess $c_0 = 5$, we find the data in the following table.

Iteration	Value	Distance from Equilibrium	Factor by which Distance Decreased
0	5.0000000000	1.4654361738	—
1	3.5304554457	−0.0041083804	−0.0028035205
2	3.5345638801	0.0000000539	−0.0000131336
3	3.5345638261	-2.0872×10^{-14}	−0.0000003868

We have a full **13 decimal places** of accuracy after only three steps. The factor by which the distance decreases gets smaller and smaller, rather than remaining constant (as in Table 3.8.1). With a sufficiently accurate computer, we could **double** this number of digits with only one more step. It takes the ordinary discrete-time dynamical system dozens of steps to achieve this level of accuracy.

3.8.3 Why Newton's Method Works and When It Fails

Figure 3.8.8 indicates why Newton's method is so fast. The two discrete-time dynamical systems share the same equilibrium point, but the slope of the Newton's method discrete-time dynamical system is **0** at the equilibrium. The slope of the curve at an equilibrium determines stability. If the absolute value of the slope is less than 1, the distance from the equilibrium decreases and the solution moves toward the equilibrium. If the slope is very small (near 0), the distance decreases very quickly. If the slope of a discrete-time dynamical system is exactly 0 at the equilibrium, the solution shoots right in toward the equilibrium (Figure 3.8.8b). We call an equilibrium where the slope is 0 **superstable**.

Is the slope of the updating function for Newton's method (Algorithm 3.4) really equal to 0 at the equilibrium? The derivative of the updating function for Newton's method,

$$h(x) = x - \frac{f(x)}{f'(x)},$$

can be computed with the quotient rule as

$$h'(x) = 1 - \frac{f'(x)f'(x) - f(x)f''(x)}{f'(x)^2} = \frac{f(x)f''(x)}{f'(x)^2}.$$

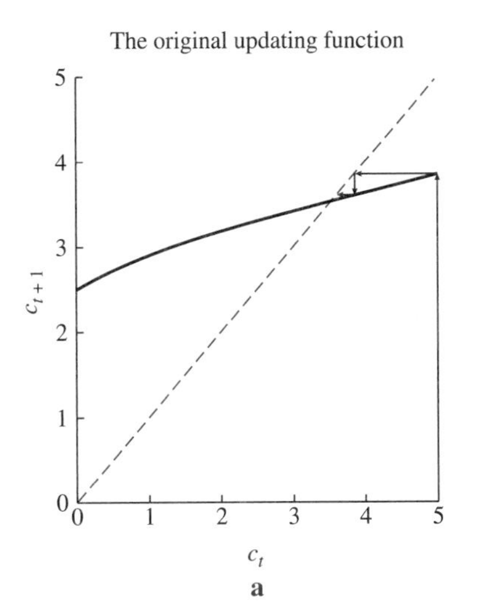

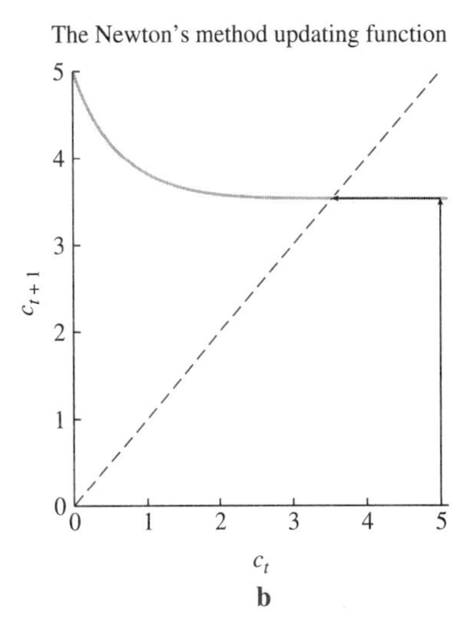

FIGURE 3.8.8

Comparison of discrete-time dynamical systems from original and Newton's methods

The equilibrium of the updating function $h(x)$ is any point x^* where $f(x^*) = 0$. As long as $f'(x^*) \neq 0$,

$$h'(x^*) = \frac{f(x^*)f''(x^*)}{f'(x^*)^2} = 0.$$

The equilibrium of the Newton's method discrete-time dynamical system is superstable.

Example 3.8.7 The Superstable Equilibrium for Newton's Method

When $f(x) = x^2 - 3$, the updating function for Newton's' method is

$$h(x_t) = x_t - \frac{x_t^2 - 3}{2x_t}.$$

The derivative is

$$\begin{aligned} h'(x_t) &= 1 - \frac{2x_t \cdot 2x_t - 2 \cdot (x_t^2 - 3)}{4x_t^2} \\ &= 1 - \left[1 - \frac{2 \cdot (x_t^2 - 3)}{4x_t^2}\right] \\ &= \frac{2 \cdot (x_t^2 - 3)}{4x_t^2}. \end{aligned}$$

At any point where $x_t^2 - 3 = 0$, the derivative is indeed 0 (Figure 3.8.7).

As with many finely-tuned machines, things can go drastically wrong with Newton's method. In Figure 3.8.9, our first guess was not very good and the solution eventually shot off to very large values. This picture shows geometrically why the slope $f'(x)$ must be different from 0. If we start at a point on the original curve where $f'(x) = 0$,

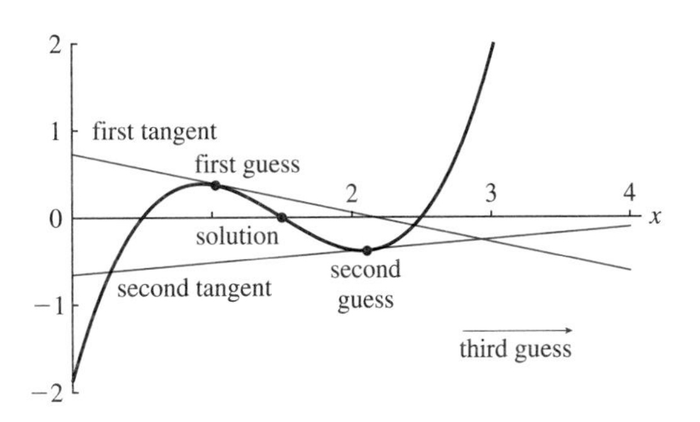

FIGURE 3.8.9

Newton's method failing miserably

the tangent never intersects the horizontal axis. The next step of Newton's method is not defined.

Furthermore, for functions with multiple solutions, Newton's method may not converge to the desired point. In Figure 3.8.9, Newton's method will eventually converge to the solution farthest to the right, which is not close to the first guess. To avoid these problems, computer algorithms usually start with the Intermediate Value Theorem to get close to a particular solution and then capitalize on the speed of Newton's method to gain accuracy.

Summary Solving equations for equilibria or other quantities algebraically is often impossible. Newton's method is a technique that can be implemented on the computer. By replacing the original equation with the tangent line approximation, we derived the **Newton's method discrete-time dynamical system** that converges very quickly to the solution. The awesome speed of Newton's method results from the fact that the solution is a **superstable** equilibrium, where the slope of the graph of the updating function is 0 at the equilibrium.

3.8 Exercises

Mathematical Techniques

1–4 ▪ Try Newton's method graphically for two steps starting from the given points on the figure.

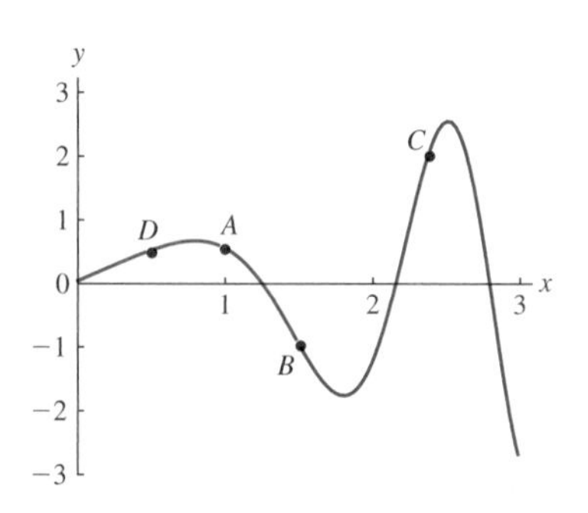

1. The point marked A.
2. The point marked B.
3. The point marked C.
4. The point marked D.

5–8 ▪ Use Newton's method for three steps to solve the following equations. Find and sketch the tangent line for the first step, then find the Newton's method discrete-time dynamical system to check your answer for the first step and compute the next two values. Compare with the result on your calculator.

5. $x^2 - 5x + 1 = 0$ (this can be solved exactly with the quadratic formula).
6. $\sqrt[3]{20}$ (solve $f(x) = x^3 - 20 = 0$)
7. A positive solution of $e^{\frac{x}{2}} = x + 1$ (solve $h(x) = e^{\frac{x}{2}} - x - 1 = 0$).
8. The point where $\cos(x) = x$ (in radians, of course).

9–12 ▪ Many equations can also be solved by repeatedly applying a discrete-time dynamical system. Compare the following discrete-time dynamical systems with the Newton's method discrete-time dynamical system. Show that each has the same equilibrium, and see how close you get in three steps.

9. Use the discrete-time dynamical system $x_{t+1} = e^{x_t} - 2$ to solve $e^x = x + 2$.
10. Use the discrete-time dynamical system $x_{t+1} = x_t^3 + x_t - 20$ to solve $x^3 = 20$ (based on Exercise 6).
11. Use the discrete-time dynamical system $x_{t+1} = e^{\frac{x_t}{2}} - 1$ to solve $e^{\frac{x}{2}} = x + 1$ (based on Exercise 7).
12. Use the discrete-time dynamical system $x_{t+1} = \cos(x_t)$ to solve $\cos(x) = x$ (based on Exercise 8).

13–14 ▪ Find the value of the parameter r for which the given discrete-time dynamical system will converge most rapidly to its positive equilibrium. Follow the system for four steps starting from the given initial condition.

13. The logistic dynamical system $x_{t+1} = rx_t(1 - x_t)$. Start from $x_0 = 0.75$.
14. The Ricker dynamical system $x_{t+1} = rx_t e^{-x_t}$. Start from $x_0 = 0.75$.

15–18 ▪ As mentioned in the text, although Newton's method works incredibly well most of the time, it can perform poorly or even fail in many circumstances. For each of the following, graph the function and illustrate the following problems.

15. Find two initial values from which Newton's method fails to solve $x(x - 1)(x + 1) = 0$. Which starting points converge to a negative solution?
16. Find two initial values from which Newton's method fails to solve $x^3 - 6x^2 + 9x - 1 = 0$. Graphically indicate a third such value.
17. Use Newton's method to solve $x^2 = 0$ (the solution is 0). Why does it approach the solution so slowly?
18. Use Newton's method to solve $\sqrt{|x|} = 0$ (the solution is 0). This is the square root of the absolute value of x. Why does the method fail?

19–20 ▪ Suppose we wish to solve the equation $f(x)=0$ but cannot compute the derivative $f'(x)$ (this kind of problem arises when the function $f(x)$ must be evaluated with a complicated computer program). One method approximates the derivative $f'(x)$ with $f(x+1)-f(x)$ (the secant line). For each of the following cases, write an approximate Newton's method discrete-time dynamical system, illustrate the procedure on a diagram, and try it for five steps to see how quickly it approaches the solution.

19. $f(x)=e^x-x-2$ (from Example 3.8.1).

20. $f(x)=x^3-20$ (from Exercise 6).

21–22 ▪ There is an alternative way to approximate the slope of the function at the value x_t:

$$f'(x_t)\approx\frac{f(x_t)-f(x_{t-1})}{x_t-x_{t-1}}.$$

For each of the following equations, use this estimate to write an approximate Newton's method discrete-time dynamical system and illustrate the idea on a diagram. How is it different from an ordinary discrete-time dynamical system? Run it for a few steps starting from x_0 and x_1 from the earlier problem. Does it converge faster than the earlier approximation? Why? How does it compare with Newton's method itself?

21. $f(x)=e^x-x-2$ (from Exercise 19).

22. $f(x)=x^3-20$ (from Exercise 20).

Applications

23–24 ▪ Suppose the total amount of nectar that comes out of a flower after time t follows

$$F(t)=\frac{t^2}{1+t^2}.$$

After noting how this function differs from the forms studied in the text, write the equation used to find the optimum time to remain, and then solve it with Newton's method (and algebraically if possible) when the travel time τ takes on the following values. Draw the associated Marginal Value Theorem diagram.

23. $\tau=0$.

24. $\tau=1$.

25–26 ▪ Suppose a fish population follows the discrete-time dynamical system

$$N_{t+1}=rN_te^{-N_t}-hN_t.$$

For the following values of r, find the equilibrium N^* as a function of h, write the equation for the critical point of the payoff function $P(h)=hN^*$, and use Newton's method to find the best h.

25. $r=2.5$.

26. $r=1.5$.

27–28 ▪ Consider a variant of the medication discrete-time dynamical system

$$M_t=p(M_t)M_t+1.0,$$

where the function $p(M_t)$ represents the fraction used (see Section 1.10, Exercise 39). Suppose that $p(M_t)=\alpha e^{-0.1M_t}$. For the following values of α, use the Intermediate Value Theorem to show that there is an equilibrium, follow the solution of the discrete-time dynamical system until it gets close to the equilibrium (about three decimal places), and find the equilibrium with Newton's method.

27. $\alpha=0.5$.

28. $\alpha=0.9$.

29–30 ▪ Thomas Malthus predicted doom for the human species when he argued that populations grow exponentially but their resources only grow linearly. Find the time when the population runs out of resources in the following cases.

29. The population grows according to $b(t)=100e^{0.1t}$, and resources grow according to $R(t)=400+100t$. The population starves when $b(t)=R(t)$.

30. The population grows according to $b(t)=100e^{0.1t}$, and resources grow according to $R(t)=4000+500t$. The population starves when $b(t)=R(t)$.

31–32 ▪ A lung follows the discrete-time dynamical system

$$c_{t+1}=0.75\alpha(c_t)c_t+0.25\gamma,$$

where $\gamma=5.0$ and the function $\alpha(c_t)$ is positive, decreasing, and $\alpha(0)=1$. We used the Intermediate Value Theorem (Exercise 31 in Section 3.4) to show that there is an equilibrium for any such function $\alpha(c)$. Use Newton's method to solve for the equilibrium for the following forms of $\alpha(c)$.

31. $\alpha(c)=e^{-c}$.

32. $\alpha(c)=e^{-0.1c}$.

Computer Exercises

33. An alternative method of solution, which is much slower but much safer, is called **bisection**. The method is based on the Intermediate Value Theorem. We will use it to solve $g(x)=e^x-x-2=0$.

a. We know there is a solution between $x=0$ and $x=2$. Show that there is a solution between $x=1$ and $x=2$.

b. By computing $g(1.5)$, show there is a solution between 1.0 and 1.5.

c. Compute $g(1.25)$. There is either a solution between 1.0 and 1.25 or between 1.25 and 1.5. Which is it?

d. Compute the value of g at the midpoint of the previous interval, and find an interval half as big that contains a solution.

e. Continue **bisecting** the interval until your answer is right to three decimal places.

f. About how many more steps would it take to reach six decimal places?

34. Solve the equation $e^x-x-2=0$ by using the quadratic approximation for the function and solving each step by using the quadratic formula. Compare how fast it converges with Newton's method. Which method do you think is better?

3.9 Panting and Deep Breathing

Why do some animals pant and others breathe slowly and deeply? Panting has the advantage of taking more breaths per second, but the breaths are shallower and leave the lung less time to absorb oxygen. Conversely, deep breathing has the advantage of giving the lung more time to absorb oxygen but at the cost of taking fewer breaths per second. We will derive discrete-time dynamical systems describing a lung with absorption in order to generate hypotheses to explain these different breathing strategies. By writing a detailed model of the lung that includes absorption and different breathing rates, we can ask which kind of breathing maximizes the rate of oxygen absorption. We will find that the answer depends on the exact shape of the absorption function.

3.9.1 Breathing at Different Rates

Suppose lung expansion as a function of time behaves as in Figure 3.9.1. The lung receives a signal every T s to switch from inhaling to exhaling. The volume of air in the lung increases at a constant rate until the lung is full and then decreases at a constant rate until the next signal is received. In Figure 3.9.1a, the signal arrives frequently and the animal breathes quickly and shallowly. In Figure 3.9.1b, the signal arrives after a longer delay and the animal breathes slowly and deeply. In Figure 3.9.1c, the signal does not arrive until the lung has emptied and rested.

Suppose an animal is running and needs to gain oxygen as quickly as possible. Should it pant (breathe rapidly and shallowly) or should it breathe more slowly and deeply? The answer depends on how absorption depends on the internal concentration and the rate of breathing.

We can write a discrete-time dynamical system describing the concentration c_t of oxygen in the lung as

$$c_{t+1} = (1 - q)[c_t - c_t A(T)] + q\gamma.$$

As in Section 1.9, q represents the fraction of air exchanged and γ the ambient concentration. The function $A(T)$ denotes the fraction of oxygen absorbed as a function of time T between breaths. This is a different sort of absorption function from $\alpha(c)$ studied in Section 3.5, where absorption depended on the concentration rather than the time. If $A(T) = 0$, there is no absorption and we recover the original lung discrete-time dynamical system (Equation 1.9.1).

In addition to changing the fraction of oxygen absorbed, changing the breathing rate changes the fraction of air exchanged. The fraction of air exchanged is proportional to the amount of time spent inhaling (Figure 3.9.1). We can therefore write

$$q = rT$$

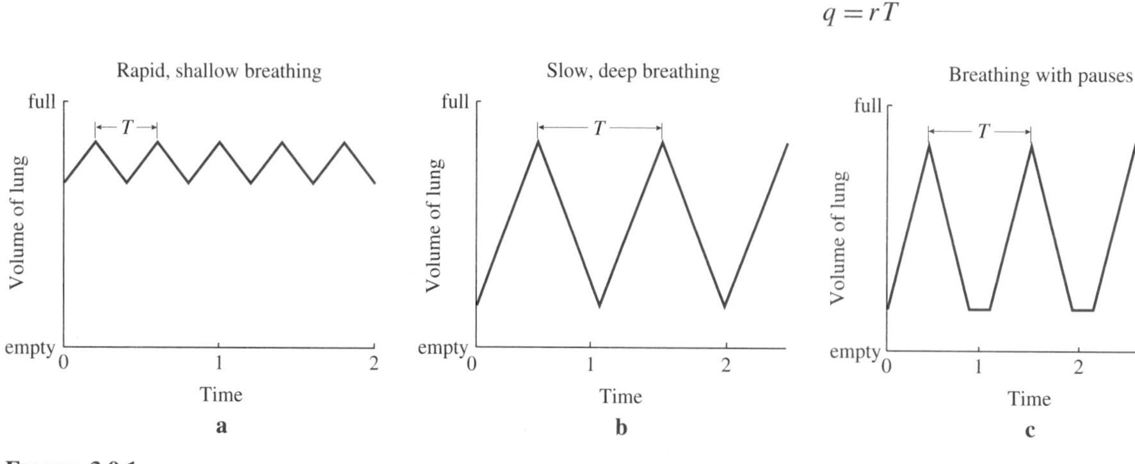

FIGURE 3.9.1

Three types of breathing

for some value of r. Suppose that $T = 1.0$ represents maximum exhalation, or the longest possible time. If $r = 1.0$ and $T = 1.0$, all the air in the lung is exchanged. For most organisms, complete exchange is impossible. The value of r is the fraction of air exchanged if breathing totally exchanges air and must be less than or equal to 1.

The discrete-time dynamical system is then

$$c_{t+1} = (1 - rT)[c_t - c_t A(T)] + rT\gamma. \tag{3.9.1}$$

If we knew the equation for $A(T)$, we could compare total absorption with different values of T.

3.9.2 Deep Breathing

It is simplest to assume that the fraction absorbed is proportional to T. If the lung has twice as much time to absorb oxygen, twice as much oxygen will be absorbed. If $A(T)$ is proportional to T, then

$$A(T) = \alpha T \tag{3.9.2}$$

for some value α (Figure 3.9.2a). If $\alpha = 1.0$ and $T = 1.0$ (maximum exchange), all of the oxygen is absorbed. Because complete absorption is impossible, the value of α must be less than or equal to 1.

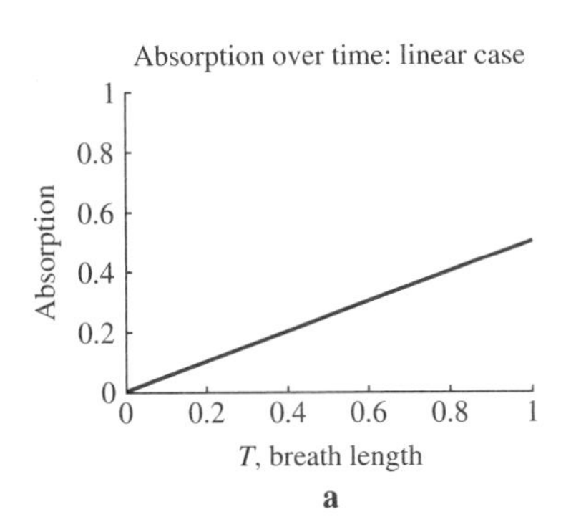

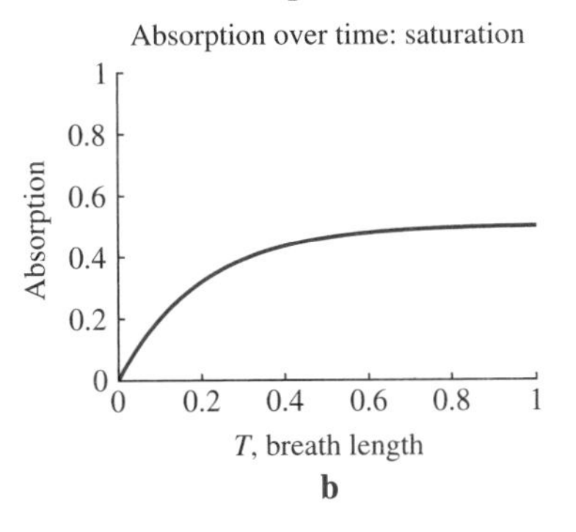

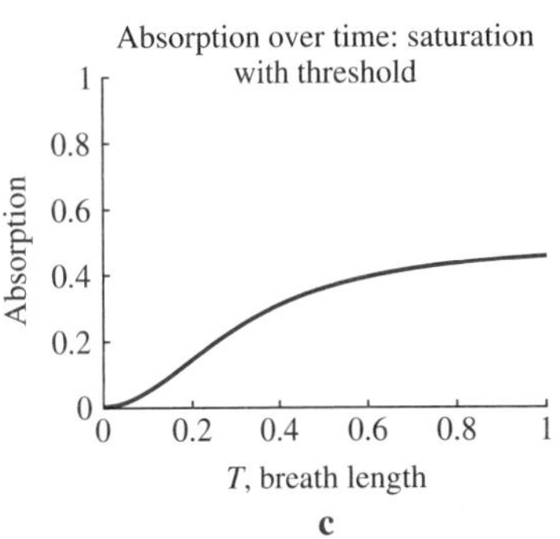

FIGURE 3.9.2
Three different absorption functions

Substituting $A(T) = \alpha T$ into the general discrete-time dynamical system (Equation 3.9.1) gives

$$c_{t+1} = (1 - rT)(c_t - \alpha c_t T) + r\gamma T. \tag{3.9.3}$$

How do we compute how much oxygen the lung absorbs as a function of T? As in the fisheries model (Subsection 3.3.3), there are two steps. First, we find the equilibrium. Then, we compute how much oxygen is absorbed at the equilibrium.

The equilibrium can be found in the usual way (Algorithm 1.5),

$$
\begin{aligned}
c^* &= (1 - rT)(c^* - \alpha c^* T) + r\gamma T && \text{equation for equilibrium} \\
c^* - (1 - rT)(c^* - \alpha c^* T) &= r\gamma T && \text{move all the } c^*\text{'s to one side} \\
c^*[1 - (1 - rT)(1 - \alpha T)] &= r\gamma T && \text{factor out } c^* \\
c^* &= \frac{\gamma r T}{1 - (1 - rT)(1 - \alpha T)}. && \text{solve for } c^*
\end{aligned}
$$

With a bit of algebra, we can cancel a T from the top and bottom, finding

$$
\begin{aligned}
c^* &= \frac{\gamma r T}{1 - (1 - rT - \alpha T + r\alpha T^2)} \\
&= \frac{\gamma r T}{rT + \alpha T - r\alpha T^2} \\
&= \frac{\gamma r}{r + \alpha - r\alpha T}.
\end{aligned} \tag{3.9.4}
$$

At what rate does this lung absorb chemical? The amount absorbed per breath is given by

$$\text{amount absorbed} = c^* A(T),$$

the equilibrium concentration times the fraction absorbed. This does not describe the **rate**, however. A lung that absorbs more oxygen over a longer time might absorb at a lower **rate**. The rate of absorption is

$$\begin{aligned}\text{rate of absorption} &= \frac{\text{amount absorbed}}{\text{time}}\\ &= \frac{c^* A(T)}{T}.\end{aligned}$$

Our goal is to maximize this rate. In this case,

$$\begin{aligned}\frac{c^* A(T)}{T} &= \frac{\alpha c^* T}{T}\\ &= \frac{\alpha\gamma r}{r + \alpha - r\alpha T}.\end{aligned} \tag{3.9.5}$$

The only appearance of T in this formula is in the denominator, and the larger the value of T, the larger the function (Exercise 4). This rate takes on its maximum at $T = 1$, the maximum possible value of T (Figure 3.9.3). In this case, the optimal breathing rate is slow, like a well-conditioned runner.

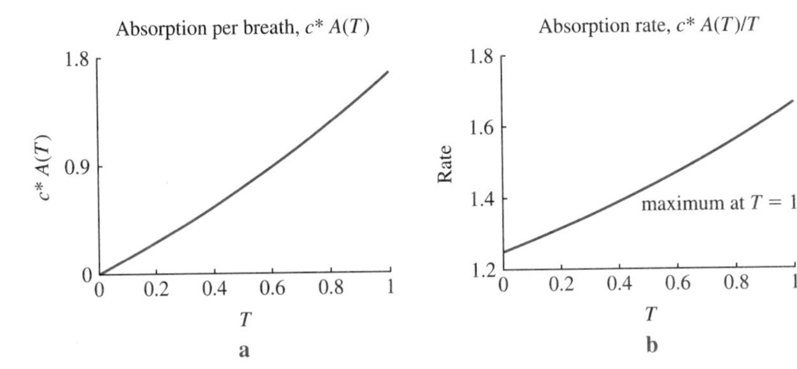

FIGURE 3.9.3
Absorption and rate of absorption at equilibrium as functions of T: linear case

3.9.3 Panting

What if absorption becomes less efficient as the length of the breath increases? One function describing saturation is

$$A(T) = \alpha(1 - e^{-kT}). \tag{3.9.6}$$

In this case, long breaths might not make sense because absorption becomes less and less efficient over time (Figure 3.9.2b).

Substituting into the discrete-time dynamical system (Equation 3.9.1),

$$c_{t+1} = (1 - rT)[c_t - \alpha c_t(1 - e^{-kT})] + rT\gamma. \tag{3.9.7}$$

We can follow the same steps as before to find the equilibrium and maximize the rate of absorption. In this case,

$$c^* = \frac{\gamma r T}{1 - (1 - rT)[1 - \alpha(1 - e^{-kT})]} \tag{3.9.8}$$

(Exercise 5). The absorption per breath is

$$\begin{aligned}c^* A(T) &= \alpha c^*(1 - e^{-kT})\\ &= \frac{\alpha\gamma r T(1 - e^{-kT})}{1 - (1 - rT)[1 - \alpha(1 - e^{-kT})]}\end{aligned}$$

and the rate of absorption is

$$\frac{c^* A(T)}{T} = \frac{\alpha \gamma r(1-e^{-kT})}{1-(1-rT)[1-\alpha(1-e^{-kT})]}.$$

With the parameter values $\alpha = 0.5$, $r = 0.5$, $\gamma = 5$, and $k = 5$, the absorption and absorption rate are shown in Figure 3.9.4. Although the amount of oxygen absorbed **per breath** increases with longer breaths, the absorption **rate** decreases. The optimal value of T is $T = 0$. This organism does best by breathing as fast as it can, like an overheated dog.

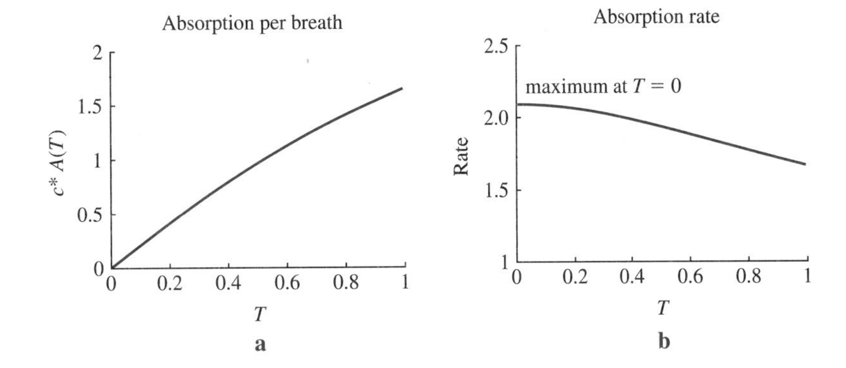

FIGURE 3.9.4

Absorption and rate as functions of T: saturating case

3.9.4 Intermediate Optimum

What if absorption takes a little time to get started? It might have a graph like that in Figure 3.9.2c. One function of T that produces this shape is

$$A(T) = \frac{\alpha T^2}{k + T^2} \tag{3.9.9}$$

(one of the Hill functions studied in Section 2.6, Example 2.6.5).

Two processes are involved: absorption saturates for large T, and it takes some time to get started.

We follow the same steps to find the equilibrium and maximize the rate of absorption. The equilibrium is

$$c^* = \frac{\gamma r T}{1-(1-rT)\left(1-\alpha\frac{T^2}{k+T^2}\right)} \tag{3.9.10}$$

(Exercise 5). The absorption is

$$\begin{aligned} c^* A(T) &= \alpha c^* \frac{T^2}{k+T^2} \\ &= \frac{\alpha \gamma r T \frac{T^2}{k+T^2}}{1-(1-rT)\left(1-\alpha\frac{T^2}{k+T^2}\right)} \end{aligned}$$

and the rate of absorption is

$$\frac{c^* A(T)}{T} = \frac{\alpha \gamma r \frac{T^2}{k+T^2}}{1-(1-rT)\left(1-\alpha\frac{T^2}{k+T^2}\right)}.$$

With the parameter values $\alpha = 0.5$, $r = 0.5$, $\gamma = 5$, and $k = 0.1$, the absorption and rate of absorption are plotted in Figure 3.9.5b. Long breaths are not best because of saturation and panting is not optimal because of the delay. Instead, there is an intermediate maximum. The position of this maximum depends on the parameters (the detailed shape of the graph of absorption as a function of time) and could be longer or shorter depending on the organism (Exercise 8).

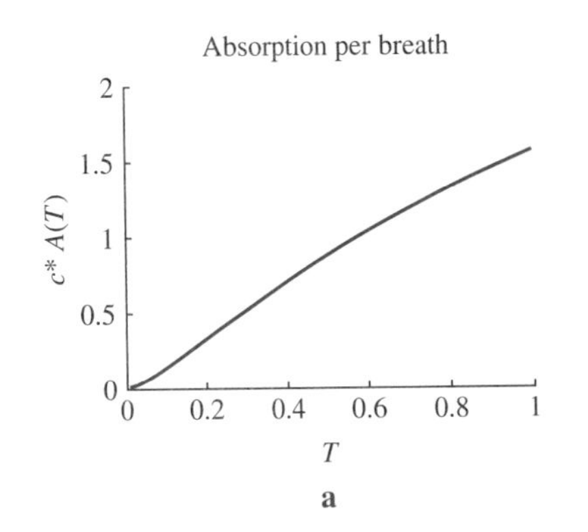

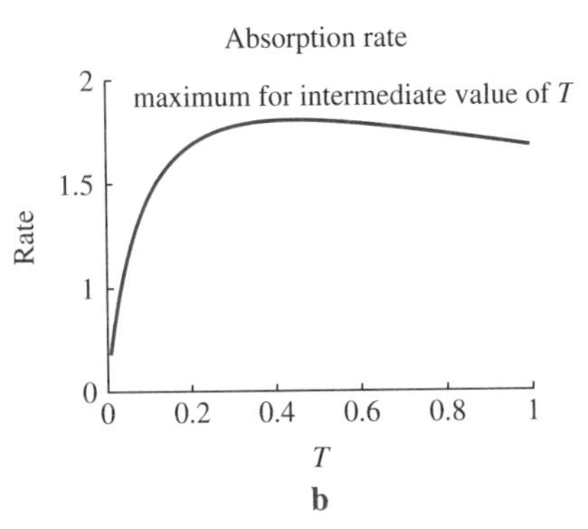

FIGURE 3.9.5

Absorption and rate as a functions of T: case with saturation and threshold

Have we explained why some animals pant and others breath deeply? No, these models provide a hypothesis; animals that take long breaths should have lungs that do not saturate quickly (like $A(T)=\alpha T$), and animals that pant should have lungs that do saturate (like $A(T)=\alpha(1-e^{-kT})$). The model has given us an idea what to measure next; oxygen absorption as a function of breath length. If measurements seem to be consistent with the assumptions of the model and its predictions, we could attempt to discover the physiological basis of the different absorption functions.

Summary By incorporating explicit consideration of breath length into our model for the concentration of a chemical in the lung, we found optimal breathing rates. If absorption is a linear function of time, slow deep breaths maximize the rate of oxygen uptake. If absorption saturates with time, panting can be best. If absorption begins slowly, an intermediate breathing rate is best. This provides a hypothesis to explain the different breathing behavior of organisms.

3.9 Exercises

Applications

1–4 ▪ Show that the rate of absorption is maximized at $T=1$ with the absorption function

$$A(T)=\alpha T$$

(Equation 3.9.2) for the following values of α, r, and γ.

1. With $\alpha=0.5$, $r=1$, and $\gamma=5.0$.
2. With $\alpha=0.5$, $r=0.5$, and $\gamma=5.0$. Why is the optimal absorption lower than in Exercise 1?
3. With $r=1.0$ and $\gamma=5.0$, but without picking a value for α. How does the optimal absorption depend on α?
4. In general, without picking values for any of the parameters.

5–6 ▪ Check the following formulas.

5. The equilibrium given in Equation 3.9.8.
6. The equilibrium given in Equation 3.9.10.

7–10 ▪ Find the value of T that maximizes the rate of absorption with the absorption function

$$A(T)=\alpha\frac{T^2}{k+T^2}$$

(Equation 3.9.9) for the following parameter values.

7. $\alpha=0.5$, $r=0.5$, $\gamma=5.0$, and $k=0.1$. You should find that the best T is $1/\sqrt{5.0}\approx 0.447$.
8. Check that
$$T=\sqrt{\frac{k}{1-\alpha}}$$
in general.
9. How does the optimal T change if r becomes larger?
10. How does the optimal T change if α becomes larger? Does this make sense?

11–12 ▪ Solve for the following without substituting in a particular functional form for $A(T)$.

11. The equilibrium concentration.
12. The equilibrium rate of absorption.

13–16 ▪ Substitute the following forms for $A(T)$ into the expressions found in Exercise 12 and compare with the results in the text.

13. With $A(T)=\alpha T$.
14. With $A(T)=\alpha(1-e^{-kT})$.
15. With $A(T)=\dfrac{\alpha T^2}{k+T^2}$.

16. With $A(T) = \frac{\alpha T}{k+T}$ (a case not considered in the text).

17–21 ▪ Find

$$\lim_{T \to 0} \frac{c^* A(T)}{T}$$

for the following forms of $A(T)$. Do the results make sense?

17. $A(T) = \alpha T$.

18. $A(T) = \alpha(1 - e^{-kT})$.

19. $A(T) = \alpha T^2$.

20. In general, assuming $A(0) = 0$.

21. Do a complete analysis of the case

$$A(T) = \frac{\alpha T}{k+T}.$$

Computer Exercises

22. Use a computer to reproduce all of the figures in this section.

23. Use the computer to experiment with the effects of the parameter α on the absorption function in Equation 3.9.6. Set $r = 0.5$, $\gamma = 5.0$, and $k = 5$. Test values of α ranging from 0.1 to 1.0. Can you explain your results?

Supplementary Problems

1–2 ▪ For the functions shown:

a. Sketch the derivative.

b. Label local and global maxima.

c. Label local and global minima.

d. Find subsets of the domain with positive second derivative.

1.

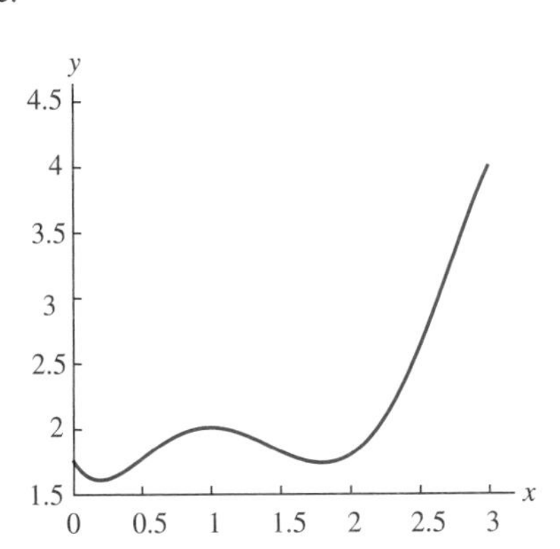

2.

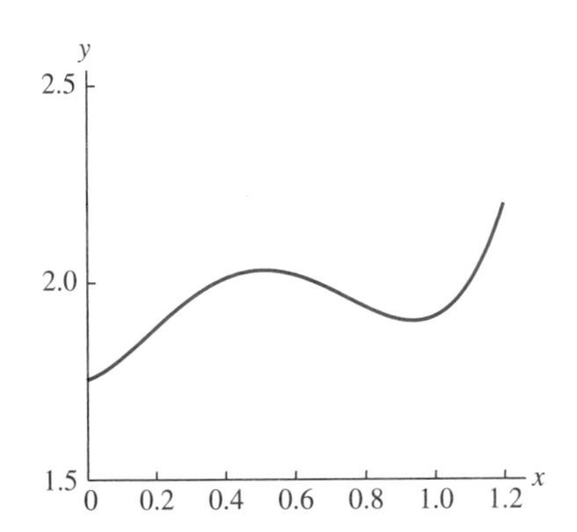

3–4 ▪ Use the tangent line and the quadratic Taylor polynomial to find approximate values of the following. Write down the function or functions you use and the equation of the tangent line. Check with your calculator.

3. $1/(3 + 1.01^2)$

4. $e^{3(1.02)^2 + 2(1.02)}$

5–6 ▪ Write the tangent line approximation for the following functions and estimate the requested values.

5. $f(x) = \frac{1+x}{1+e^{3x}}$. Estimate $f(-0.03)$.

6. $g(y) = (1 + 2y)^4 \ln(y)$. Estimate $g(1.02)$.

7–8 ▪ Sketch graphs of the following functions. Find all critical points, and state whether they are minima or maxima. Find the limit of the function as $x \to \infty$.

7. $(x^2 + 2x)e^{-x}$ for positive x.

8. $\ln(x)/(1+x)$ for positive x. Do not solve for the maximum, just show that there must be one.

9–12 ▪ Find the Taylor polynomial of degree 2 approximating each of the following.

9. $f(x) = \frac{1+x}{1+x^2}$ for x near 0.

10. $f(x) = \frac{1+x}{1+x^2}$ for x near 1.

11. $g(x) = \frac{1+x}{1+e^x}$ for x near 0.

12. $h(x) = \frac{x}{2-e^x}$ for x near 0.

13–14 ▪ Combine the Taylor polynomials from the previous set of problems with the leading behavior of the functions for large x to sketch graphs.

13. $g(x)$

14. $h(x)$

15. Between days 0 and 150 (measured from November 1), the snow at a certain ski resort is given by

$$S(t) = -\frac{1}{4}t^4 + 60t^3 - 4000t^2 + 96000t,$$

where S is measured in microns (1 micron is 10^{-4} cm).

a. A yeti tells you that this function has critical points at $t = 20$, $t = 40$, and $t = 120$. Confirm this assertion.

b. Find the global maximum and global minimum amounts of snow in feet. Remember that 1 in. = 2.54 cm.

c. Use the second derivative test to identify the critical points as local maxima and minima.

d. Sketch the function.

16. Let $r(x)$ be a function giving the per capita production as a function of population size x, with the formula

$$\text{per capita production} = \frac{4x}{1+3x^2}.$$

a. Find the population size that produces the highest per capita production.

b. Find the highest per capita production.

c. Check with the second derivative test.

17. An organism is replacing 25% of the air in its lung each breath, and the external concentration of a chemical is $\gamma = 5.0 \times 10^{-4}$ mol/L. Suppose the body uses a fraction α of the chemical just after breathing. That is, the chemical follows

$$c_t \xrightarrow{\text{absorption}} (1-\alpha)c_t \xrightarrow{\text{breathing}} \text{mix of 75\% used air and 25\% ambient air.}$$

a. Write the discrete-time dynamical system for this process.

b. Find the equilibrium level in the lung as a function of α.

c. Find the amount absorbed by the body with each breath at equilibrium as a function of α.

d. Find the value of α that maximizes the amount of chemical absorbed at equilibrium.

e. Explain your result in words.

18. The formula for an ellipse is

$$\frac{x^2}{a^2} + \frac{y^2}{b^2} = 1.$$

a. Use implicit differentiation to find the slope $\frac{dy}{dx}$.

b. Find all points on the ellipse where the tangent line is horizontal.

c. Find all points on the ellipse where the tangent line is vertical.

19. Suppose the volume of a plant cell follows $V(t) = 1000 \times (1 - e^{-t})\ \mu\text{m}^3$ for t measured in days. Suppose the fraction of cell in a vacuole (a water-filled portion of the cell) is $H(t) = \frac{e^t}{(1+e^t)}$.

a. Sketch a graph of the total size of the cell as a function of time.

b. Find the volume of the cell outside the vacuole.

c. Find and interpret the derivative of this function. (Don't forget the units of measure.)

d. Find when the volume of the cell outside the vacuole reaches a maximum.

20. Consider the function

$$F(t) = \frac{\ln(1+t)}{t+t^2}.$$

a. What is $\lim_{t\to 0} F(t)$?

b. What is $\lim_{t\to 0} F'(t)$?

c. What is $\lim_{t\to 0} F''(t)$?

d. Sketch a graph of this function.

21. During Thanksgiving dinner, the table is replenished with food every 5 min. Let F_t represent the fraction of the table laden with food.

$$F_{t+1} = F_t - \text{ amount eaten } + \text{ amount replenished}.$$

Suppose that

$$\text{amount eaten} = \frac{bF_t}{1+F_t}$$

$$\text{amount replenished} = a(1-F_t)$$

and that $a = 1.0$ and $b = 1.5$.

a. Explain the terms describing the amount eaten and the amount replenished.

b. If the table starts out empty, how much food is there after 5 min? How much is there after 10 min?

c. Use the quadratic formula to find the equilibria.

d. How much food will there be 5 min after the table is 60% full? Sketch the solution.

22. Let N_t represent the difference between the sodium concentration inside and outside a cell at some time. After 1 s, the value of N_{t+1} is

$$\begin{cases} N_{t+1} = 0.5N_t & \text{if } N_t < 2 \\ N_{t+1} = 4.0N_t - 7 & \text{if } 2 < N_t < 4 \\ N_{t+1} = -0.25N_t + 10 & \text{if } 4 < N_t. \end{cases}$$

a. Graph the updating function and show that it is continuous.

b. Find the equilibria and their stability.

c. Find all initial conditions which end up at $N = 0$.

23. Consider looking for a positive solution of the equation

$$e^x = 2x + 1.$$

a. Draw a graph and pick a reasonable starting value.

b. Write down the Newton's method iteration for this equation.

c. Find your next guess.

d. Show explicitly that the slope of the updating function for this iteration is zero at the solution.

24. Consider trying to solve the equation

$$\ln(x) = \frac{x}{3}.$$

a. Convince yourself there is indeed a solution and find a reasonable guess.

b. Use Newton's method to update your guess twice.

c. What would be a bad choice of an initial guess?

25. Suppose a bee gains an amount of energy

$$F(t) = \frac{3t}{1+t}$$

after it has been on a flower for time t, but that it uses $2t$ energy units in that time (it has to struggle with the flower).

a. Find the net energy gain as a function of t.

b. Find when the net energy gain per flower is maximum.

c. Suppose the travel time between flowers is $\tau = 1$. Find the time spent on the flower that maximizes the rate of energy gain.

d. Draw a diagram illustrating the results of parts **b** and **c**. Why is the answer to **c** smaller?

26. Consider the function for net energy gain from the previous problem.

a. Use the Extreme Value Theorem to show that there must be a maximum.

b. Use the Intermediate Value Theorem to show that there must be a residence time t that maximizes the rate of energy gain.

27. A peculiar variety of bacteria enhances its own per capita production. In particular, the number of offspring per bacteria increases according to the function

$$\text{per capita production} = r\left(1 + \frac{b_t}{K}\right).$$

Suppose that $r = 0.5$ and that $K = 10^6$.

a. Graph the per capita production as a function of population size.

b. Find the discrete-time dynamical system for this population and graph the updating function.

c. Find the equilibria.

d. Find their stability.

28. A type of butterfly has two morphs, a and b. Each type reproduces annually after predation. 20% of type a are eaten, and 10% of type b are eaten. Each type doubles its population when it reproduces. However, the types do not breed true. Only 90% of the offspring of type a are of type a, the rest being of type b. Only 80% of the offspring of type b are of type b, the rest being of type a.

a. Suppose there are 10,000 of each type before predation and production. Find the number of each type after predation and production.

b. Find the discrete-time dynamical systems for types a and b.

c. Find the discrete-time dynamical system for the fraction of type a, which you can denote by the variable p.

d. Find the equilibria.

e. Find their stability.

29. A population of size x_t follows the rule

$$\text{per capita production} = \frac{4x_t}{1+3x_t^2}.$$

a. Find the updating function for this population.

b. Find the equilibrium or equilibria.

c. What is the stability of each equilibrium?

d. Find the equation of the tangent line at each equilibrium.

e. What is the behavior of the approximate dynamical system defined by the tangent line at the middle equilibrium?

30. Consider a population following the discrete-time dynamical system

$$N_{t+1} = \frac{rN_t}{1+N_t^2}.$$

a. What is the per capita production?

b. Find the equilibrium as a function of r.

c. Find the stability of the equilibrium as a function of r.

d. Does this positive equilibrium become unstable as r becomes large?

Projects

1. Consider the following alternative version of Ricker model for a fishery.

$$x_{t+1} = rx_te^{-x_{t-1}}$$

The idea is that the per capita production this year (year t) is a decreasing function of the population size in the previous year (year $t-1$).

a. Discuss why this model might make more sense than the basic Ricker model.

b. Find the equilibrium of this equation (set $x_{t-1} = x_t = x_{t+1} = x^*$).

c. Use a computer to study the stability of the equilibrium. Do you have any idea why it might be different from that of the basic Ricker model?

d. Write a model where the per capita production depends on the population size two years ago. Find the equilibrium and use a computer to test stability.

e. Subtract off a harvest. Find the maximum sustained yield for different values of h. Is the equilibrium still stable?

f. Models with an extra delay are supposed to be simplified versions of models with two variables. The per capita production is an increasing function of the amount of food available in that year, while the amount of food available is a decreasing function of the number of fish the previous year. Try to write a pair of equations describing this situation. Try to find equilibria, and discuss whether your results make sense.

2. Many bees collect both pollen and nectar. Pollen is used for protein, and nectar is used for energy. Suppose the amount of nectar harvested during t s on a flower is

$$F(t) = \frac{t}{1+t},$$

and that the amount of pollen harvested during t s on a flower is

$$G(t) = \frac{t}{2+t}.$$

The bee collects pollen and nectar simultaneously. Travel time between flowers is $\tau = 1.0$ s.

a. What is the optimal time to leave to collect nectar at the maximum rate?

b. What is the optimal time to leave to collect pollen at the maximum rate? Why are the two times different?

c. Suppose that the bee values pollen twice as much as nectar. Find a single function $V(t)$ that gives the value of resource collected by time t. What is the optimal time for the bee to leave? (Solving the equation requires Newton's method.)

d. Suppose that the bee values pollen k times as much as nectar. Compute the optimal time to leave and graph it as a function of k. Do the values at $k = 0$ and the limit as k approaches infinity make sense?

e. Suppose that the bee first collects nectar and then switches to pollen. Assume it spends 1.0 s collecting nectar. How long should it spend on pollen? Then suppose it spends 1.0 s collecting pollen. How long should it spend on nectar?

f. Suppose again that the bee values pollen k times as much as nectar. Experiment to try to find a solution that is best when nectar and pollen are harvested sequentially.

3. We know enough to study the solutions of the logistic dynamical system for values of r between up to 3.5. Define the updating function to be

$$f(x) = rx(1 - x).$$

a. Find the two-step updating function $g = f \circ f$.

b. Graph it along with the diagonal for values of r ranging from 2.8 up to 3.6.

c. Write the equation for the equilibria of g.

d. Two of the equilibria, 0 and $x^* = 1 - \frac{1}{r}$, match those for f. Why?

e. The terms x and $x - x^*$ factor out of the equilibrium equation $g(x) - x = 0$. Why?

f. Factor $g(x) - x$ and find the other two equilibria, x_1 and x_2. For what values of r do they make sense? What do they mean? (Think about $f(x_1)$ and $f(x_2)$.)

g. For what values of r are x_1 and x_2 stable?

h. Describe the dynamics of f and g in these cases.

i. What happens for values of r just above the point where x_1 and x_2 become unstable?

Chapter 4

Differential Equations, Integrals, and Their Applications

Biological systems are constantly changing. Describing this change and deducing its consequences constitute the dynamical approach to the understanding of life. In the first three chapters of this book, we used discrete-time dynamical systems to study measurements that change over discrete-time intervals. Given the state of a system (such as a population size) at one time, the discrete-time dynamical system gives the state of the system at a later time. In the course of analyzing the behavior of discrete-time dynamical systems, we developed the idea of the derivative, which describes both the **slope** of a graph and the **instantaneous rate of change** of a measurement.

With the derivative, we can study a different kind of dynamical system, the **differential equation**. The instantaneous rate of change of a measurement takes the place of the updating function and provides the information needed to deduce the future state of a system from its present state. Because the system can be measured at any time, these systems are a type of **continuous-time dynamical system**.

Differential equations were invented by Isaac Newton to study gravitation. They have proven to be the most powerful method for describing dynamics in all of the sciences. Like discrete-time dynamical systems, they can be applied to the three main areas of focus of this book (growth, maintenance, and replication), in addition to a wide range of other problems throughout biology.

4.1 Differential Equations

When we have a measurement, we can differentiate to find the rate of change (Figure 4.1.1). What if we know the rate of change and want to compute the measurement (Figure 4.1.2)? If we know, for example, that the velocity of an object is $v(t)$ as a function of t, then the position $p(t)$ must satisfy the **differential equation**

$$\frac{dp}{dt} = v(t).$$

because the velocity is the derivative of the position. Until now, we have thought of ourselves as knowing the position and taking the derivative to find the velocity. With a differential equation, we know the velocity and wish to find some method to compute the position.

When might we measure the rate of change of some quantity in this way? Two cases, position and cell sodium concentration, illustrate when this might occur. When you are lost, it is much easier to look at the speedometer to determine your speed than to locate landmarks and identify them on a map to determine your position (in the absence of a GPS device). In fact, the speedometer in a car directly measures the rate of rotation of the tires, and the odometer determines distance by adding up the number of rotations. Similarly, it might be easier to measure how many ions enter and leave a cell each second by measuring changes in electrical charge than to track down and count every sodium ion floating around in a cell.

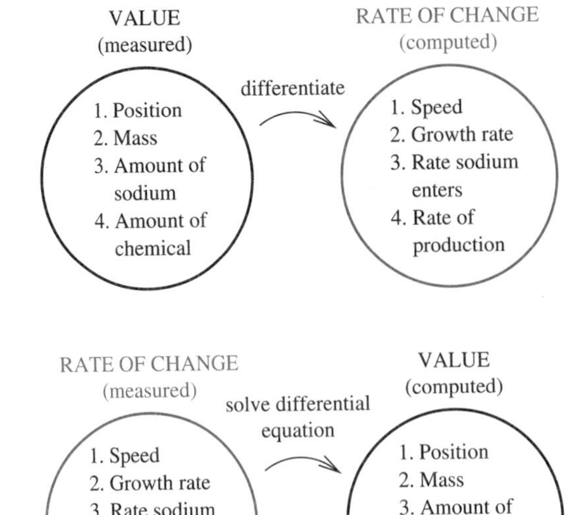

FIGURE 4.1.1
Finding the rate of change with the derivative

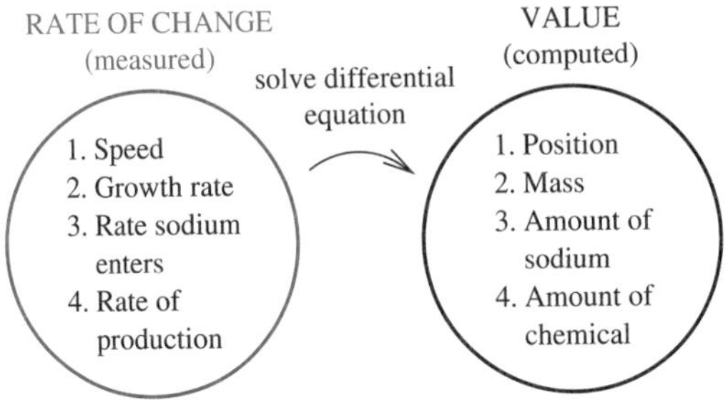

FIGURE 4.1.2
Finding a measurement from the rate of change

In cases when we **measure** the rate of change, we can write down what is called a **pure-time differential equation**,

$$\text{derivative of unknown measurement} = \text{measured rate of change.}$$

There is another way to arrive at a differential equation describing a measurement. Based on biological principles, we might know a **rule** describing how a measurement changes, just as we did when deriving discrete-time dynamical systems. For example, when resources are not limiting, the rate of population growth is proportional to population size. In these cases, we can write down an **autonomous differential equation**,

$$\text{derivative of unknown quantity} = \text{some function of unknown quantity.}$$

Given that the unknown quantity appears on both sides of this equation, it is quite remarkable that it can be solved.

Our plan of attack in this chapter is to begin with a detailed study of pure-time differential equations, developing the method of **integration** to solve them. After studying several other applications of the integral, we return to autonomous differential equations in Chapter 5, and show how to use both integration and graphical methods to analyze these important equations.

4.1.1 Differential Equations: Examples and Terminology

Suppose that 1.5 μm^3 of water enters a cell each second. The volume V of the cell is thus increasing by 1.5 $\mu m^3/s$. Mathematically, the rate of change of V is 1.5 $\mu m^3/s$, or

$$\frac{dV}{dt} = 1.5. \tag{4.1.1}$$

The quantity being differentiated (V in this case) is called the **state variable**. Because we *measured* the rate of change, this is a **pure-time differential equation**.

Our goal is to find the volume V of the cell as a function of time. Such a function is called a **solution**, just as with a discrete-time dynamical system. When it works, the best method for solving a differential equation is *guessing*. What function has a constant rate of change equal to 1.5? Geometrically, what function has a constant slope of 1.5? The answer is a line with slope 1.5. One such line is

$$V(t) = 1.5t$$

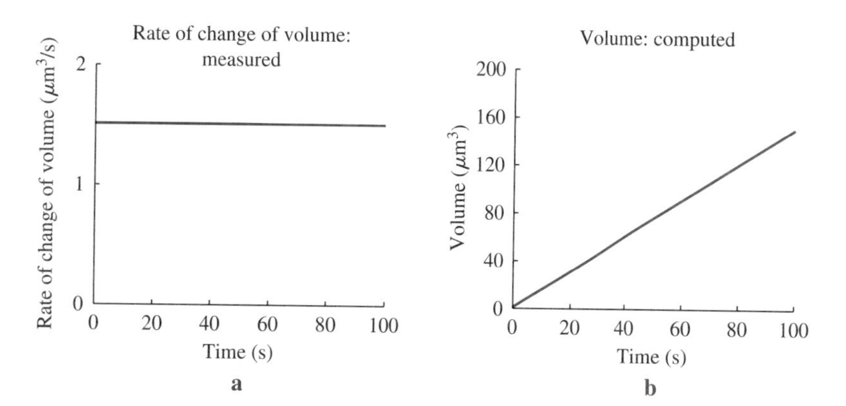

FIGURE 4.1.3

A function that solves the differential equation $\frac{dV}{dt} = 1.5$.

(Figure 4.1.3). We can check whether this guess solves the differential equation by taking the derivative,

$$\frac{dV}{dt} = \frac{d}{dt}1.5t = 1.5.$$

Suppose we know that the volume of the cell at time $t = 0$ is 300.0 μm^3. Although the guess $V(t) = 1.5t$ is one solution of the differential equation, it does not match this **initial condition**. As with discrete-time dynamical systems, a solution also depends on where something starts. From our knowledge of lines, we realize that any equation for $V(t)$ with the form

$$V(t) = 1.5t + c$$

has a constant slope of 1.5 when c is a constant. In terms of the differential equation,

$$\frac{dV}{dt} = \frac{d}{dt}(1.5t + c) = 1.5.$$

The value of c can be chosen to match the initial condition. In this case, because

$$V(0) = 300.0 = 1.5 \cdot 0 + c,$$

we know that $c = 300.0$. The solution of the differential equation

$$\frac{dV}{dt} = 1.5$$

with the initial condition $V(0) = 300.0$ is

$$V(t) = 1.5t + 300.0$$

(Figure 4.1.4). From knowledge of the initial condition and a measurement of the rate of change, we have a formula giving the volume at any time. At $t = 50$, the volume is $1.5 \cdot 50 + 300 = 375.0$ μm^3.

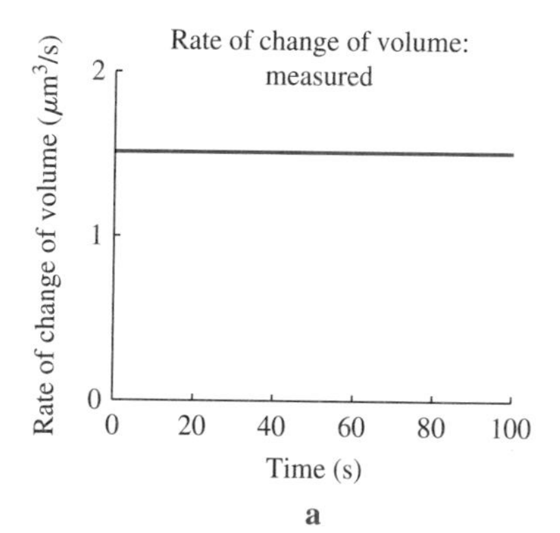

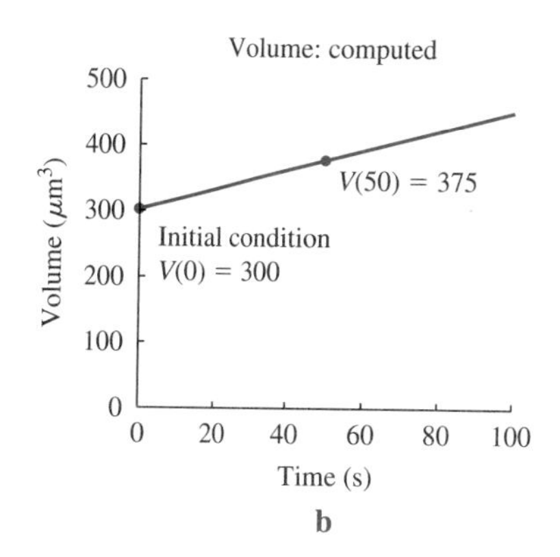

FIGURE 4.1.4

Volume as a function of time

Example 4.1.1 A Pure-Time Differential Equation for Chemical Production

As a more complicated example, suppose we measure that the rate of chemical production is e^{-t}, in units of moles per second (Figure 4.1.5a). The **state variable** P, the total chemical produced, follows the pure-time differential equation

$$\frac{dP}{dt} = e^{-t}. \tag{4.1.2}$$

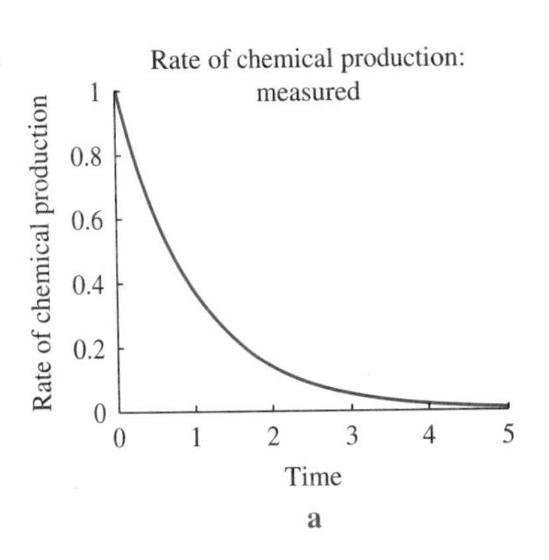

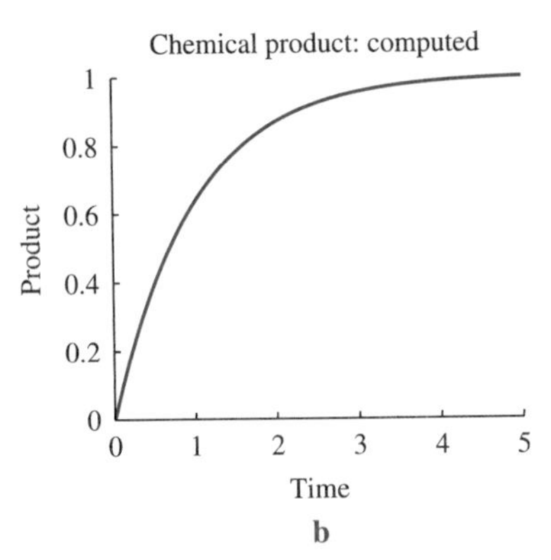

FIGURE 4.1.5
Product as a function of time

The rate at which the chemical is produced becomes slower and slower as time progresses. The solution for $P(t)$ can be sketched graphically. It must be a function that increases more and more slowly (Figure 4.1.5b). The solution sketched has initial condition $P(0) = 0$.

The differential equations for volume and chemical product (Equations 4.1.1 and 4.1.2) are pure-time differential equations because we **measured** the rate of change of the state variable. The equation for chemical production indicates the source for the name. The formula for the rate of change, e^{-t}, depends **purely** on the **time** t.

Example 4.1.2 An Autonomous Differential Equation for Population Growth

Suppose instead we wish to describe a phenomenon with a differential equation derived from biological principles. For a population, the simplest rule is that the rate of production is proportional to the population size. The discrete-time dynamical system describing this process is

$$b_{t+1} = rb_t \tag{4.1.3}$$

where the parameter r is the per capita production (the number of offspring per bacterium). If population size can be measured **continuously** (at any time), the change in population is expressed as the **rate of change**, in this case equal to the rate of production. Therefore,

$$\frac{db}{dt} = \lambda b \tag{4.1.4}$$

where b represents the bacterial population and the constant of proportionality λ is the per capita growth rate. Although b represents a function of time, it is conventional to write just b instead of $b(t)$ on both sides of the equation. The dimensions of λ are 1/time. Because it was derived from a rule, this is an **autonomous differential equation**.

The autonomous differential equation in Example 4.1.2 differs fundamentally from a pure-time differential equation. In the pure-time differential equations, the rate of change is a *measured* function of time. In an autonomous differential equation, the rate of change of population is derived from a *rule* and is a function of the state variable, population size in this example. In Chapter 5 we will study methods to solve autonomous differential equations, along with a variety of algebraic and graphical techniques to analyze the results.

Type of Differential Equation	Example	When Used
Pure-time differential equation	$\frac{dP}{dt} = e^{-t}$	When the rate of change is a measured function of time
Autonomous differential equation	$\frac{db}{dt} = 2b$	When a rule gives the rate of change as a function of the state variable

4.1.2 Graphical Solution of Pure-Time Differential Equations

By checking whether the graph of a function is increasing or decreasing, we can sketch a graph of the derivative (Section 2.7, "The Second Derivative, Curvature, and Accelerations"). Solving a pure-time differential equation reverses this process. From the graph of the derivative, we wish to sketch a graph of the function itself.

The guideposts for this process are the reverse of those for reading and interpreting a graph (Subsection 2.7.2, "Using the Second Derivative for Graphing"), summarized in the following table.

What the Derivatives Does	What the Graph Does
Derivative positive	Graph increasing
Derivative negative	Graph decreasing
Derivative zero	Graph horizontal

Example 4.1.3 Graphical Solution of a Pure-Time Differential Equation I

Suppose we wish to solve the pure-time differential equation

$$\frac{dG}{dt} = 4 - 2t$$

with initial condition $G(0) = 10$. A graph of the rate of change function $4 - 2t$ is a line that is positive for $t < 2$, zero at $t = 2$, and negative for $t > 2$ (Figure 4.1.6a).

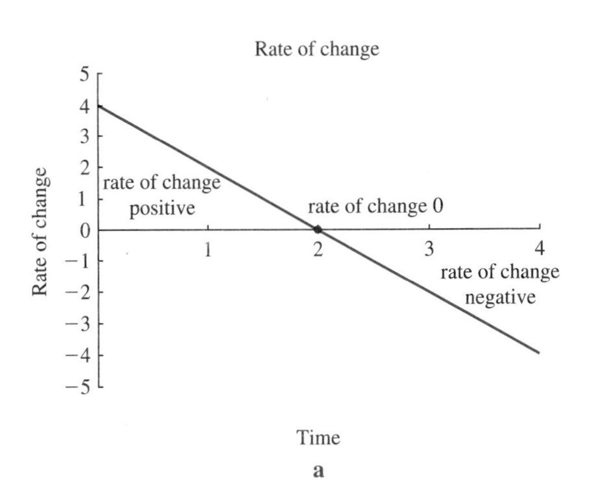

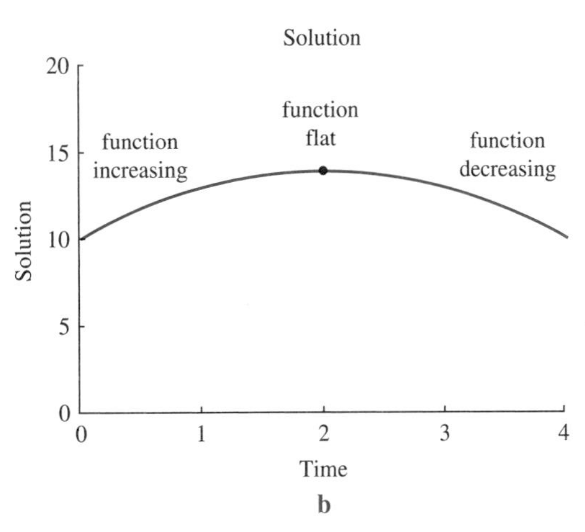

FIGURE 4.1.6
Graphical solution of a pure-time differential equation

The solution begins at $G(0) = 10$ (from the initial condition), increases until $t = 2$, has a critical point at $t = 2$, and decreases thereafter (Figure 4.1.6b). ▲

Example 4.1.4 Graphical Solution of a Pure-Time Differential Equation II

Suppose we wish to find a measurement $M(t)$ with rate of change given in Figure 4.1.7a with initial condition $M(0) = 15.0$. The rate of change is positive for $1 < t < 7$, and negative outside this range.

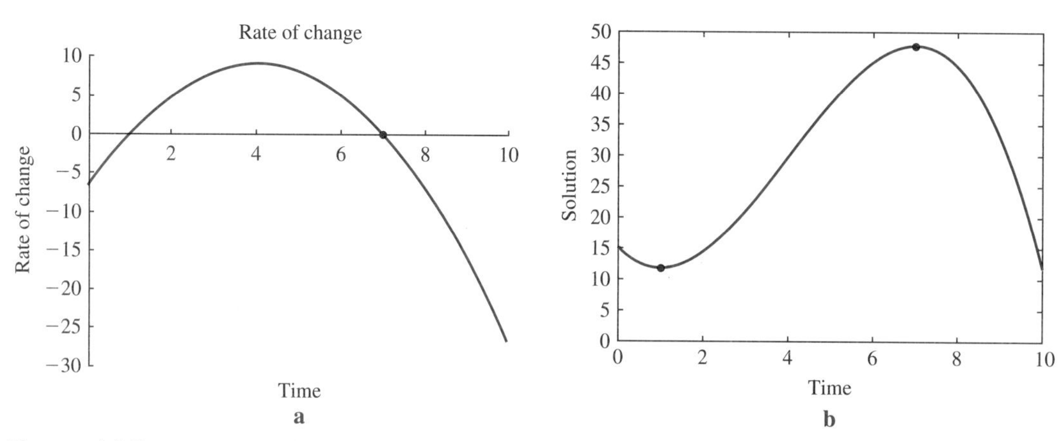

FIGURE 4.1.7

Finding a solution from a graph of the rate of change

The solution begins at $M(0) = 15.0$ (from the initial condition). It is increasing in the range from $t = 1$ to $t = 7$ and decreasing outside that range (Figure 4.1.7). Furthermore, the solution is increasing most rapidly when the rate of change is largest (at around $t = 4$) and decreasing most rapidly when the rate of change is most negative (for t near 10). ▲

4.1.3 Euler's Method for Solving Differential Equations

There are four ways to solve a differential equation. In some simple cases, we can guess the right answer (and then check it, of course). Second, we can sketch solutions of pure-time differential equations using our understanding of the derivative. In the next section, we will learn the third and most important method, **integration**. Even that method, however, does not work for many equations. A fourth approach can be implemented on the computer, called **Euler's method**. Like Newton's method for solving equations (Section 3.8), Euler's method works by converting the original problem into a problem about a discrete-time dynamical system. Also like Newton's method, Euler's method begins with the **tangent-line approximation**, replacing a problem about **curves** with a problem about **lines**.

Example 4.1.5 Applying the Tangent Line Approximation to a Pure-Time Differential Equation

Consider again the pure-time differential describing the volume of a cell,

$$\frac{dV}{dt} = 1.5$$

(Equation 4.1.1) with initial condition $V(0) = 300.0$. How can we use this information to estimate the volume at time $t = 1$ if we fail to guess the solution? We have two pieces of information: the initial condition (a base point) and the differential equation (the

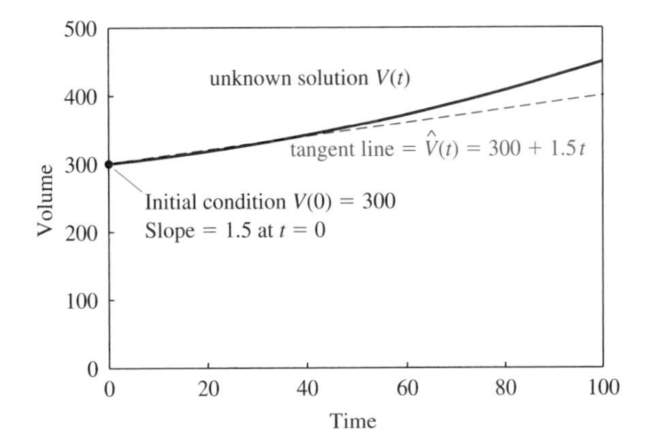

FIGURE 4.1.8

The tangent line approximation of a differential equation

derivative or slope). The differential equation and the initial condition provide exactly the information needed to write down the tangent line approximation for $V(t)$ near the base point $t = 0$ (Figure 4.1.8). In this case, the approximate function $\hat{V}(t)$ with base point 0 is

$$\begin{aligned}\hat{V}(0 + \Delta t) &= V(0) + V'(0)\Delta t && \text{equation for tangent line}\\ &= 300.0 + 1.5 \cdot \Delta t && V(0) = 300.0 \text{ (initial condition) and } V'(0) = 1.5 \text{ (differential equation)}\\ &= 300.0 + \Delta t.\end{aligned}$$

Although we do not know the solution $V(t)$, we do know its tangent line. This is a graphical way of expressing the fact that a pure-time differential equation gives the rate of change of an unknown measurement.

Euler's method begins by using this tangent line to estimate the state variable at some time after $t = 0$. With $\Delta t = 1$,

$$\hat{V}(1) = 300.0 + 1.5 = 301.5.$$

To continue, we use this new information and the differential equation to estimate the volume at $t = 2$. The tangent line with base point $t = 1$ is

$$\begin{aligned}\hat{V}(1 + \Delta t) &= V(1) + V'(1)\Delta t && \text{equation for tangent line}\\ &\approx \hat{V}(1) + V'(1)\Delta t && \text{substitute } \hat{V}(1) \text{ for unknown } V(1)\\ &= 301.5 + 1.5\Delta t.\end{aligned}$$

We estimate that

$$\hat{V}(1 + 1) = 301.5 + 1.5 \cdot 1 = 303.0.$$

Euler's method can be continued step-by-step to find approximate solutions for as long as wanted. The method works as follows.

▸▸ Algorithm 4.1 Euler's Method for Solving a Pure-Time Differential Equation

Suppose a measurement m obeys the pure-time differential equation

$$\frac{dm}{dt} = f(t)$$

with initial condition $m(t_0) = m_0$.

1. Choose a **time step** Δt (the length of time between estimated values).
2. Use the initial condition and the differential equation to find the tangent line $\hat{m}(t)$ with base point $t = t_0$ and use it to estimate $\hat{m}(t_0 + \Delta t)$.

3. Use the estimate $\hat{m}(t_0 + \Delta t)$ and the differential equation to find the tangent line $\hat{m}(t)$ with base point $t_0 + \Delta t$ and estimate $\hat{m}(t_0 + 2\Delta t)$.
4. Repeat the previous step as long as needed.

In Example 4.1.5, the state variable is V, the rate of change obeys the function $f(t) = 1.5$, the initial condition at $t_0 = 0$ is $V(t_0) = V_0 = 300.0$, and we chose a time step of $\Delta t = 1$.

Example 4.1.6 Finding a Discrete-Time Dynamical System to Describe Euler's Method

With the differential equation $\frac{dV}{dt} = 1.5$, we can write a simple discrete-time dynamical system to summarize the steps in Euler's method. Suppose we have an estimate $\hat{V}(t)$ for $V(t)$ at some time t. The tangent line approximation with base point t is

$$\begin{aligned}
\hat{V}(t + \Delta t) &= V(t) + V'(t)\Delta t && \text{equation for tangent line} \\
&\approx \hat{V}(t) + V'(t)\Delta t && \text{substitute } \hat{V}(t) \text{ for unknown } V(t) \\
&= \hat{V}(t) + 1.5 \cdot \Delta t && V'(1) = 1.5 \text{ (from the differential equation)} \\
&= \hat{V}(t) + 1.5\Delta t. && \text{simplify}
\end{aligned}$$

Therefore,

$$\hat{V}(t + 1) = \hat{V}(t) + 1.5.$$

This discrete-time dynamical system says that 1.5 μm^3 is added to the volume each second. We have seen this discrete-time dynamical system as a model for a growing tree (Example 1.5.2) with solution

$$\hat{V}(t) = 300.0 + 1.5t.$$

Because the true solution is in fact a line, the results of Euler's method match the exact solution found by guessing.

Example 4.1.7 Applying Euler's Method to a Differential Equation for Chemical Production

The results are more interesting when we apply Euler's method to the differential equation for chemical production

$$\frac{dP}{dt} = e^{-t}.$$

Suppose the initial condition is $P(0) = 0$. Following Algorithm 4.1,

1. Pick a time step of $\Delta t = 1$ (we will next try a smaller step that should be more accurate).
2. The tangent line with base point $t = 0$ is

$$\begin{aligned}
\hat{P}(0 + \Delta t) &= P(0) + P'(0)\Delta t && \text{equation for tangent line} \\
&= 0 + 1 \cdot \Delta t && P(0) = 0 \text{ (initial condition) and } P'(0) = e^{-0} = 1 \text{ (differential equation)} \\
&= \Delta t. && \text{simplify}
\end{aligned}$$

   We therefore estimate that

$$\hat{P}(1) = 1.$$

3. For the next step,

$$\begin{aligned}
\hat{P}(1 + \Delta t) &= P(1) + P'(1)\Delta t && \text{equation for tangent line} \\
&\approx \hat{P}(1) + P'(1)\Delta t && \text{substitute } \hat{P}(1) \text{ for unknown } P(1) \\
&= 1 + 0.367\Delta t. && \hat{P}(1) = 1 \text{ and } P'(1) = e^{-1} = 0.367
\end{aligned}$$

We therefore estimate that

$$\hat{P}(2) = 1.0 + 0.367 \cdot 1.0 = 1.367.$$

4. To find $\hat{P}(3)$,

$$\begin{aligned}\hat{P}(2+\Delta t) &= P(2) + P'(2)\Delta t \\ &\approx \hat{P}(2) + P'(2)\Delta t \\ &= 1.367 + e^{-2} \cdot \Delta t \\ &= 1.367 + 0.135\Delta t\end{aligned}$$

$$\hat{P}(3) = 1.502.$$

To find $\hat{P}(4)$,

$$\begin{aligned}\hat{P}(3+\Delta t) &= P(3) + P'(3)\Delta t \\ &\approx \hat{P}(3) + P'(3)\Delta t \\ &= 1.502 + e^{-3} \cdot \Delta t \\ &= 1.502 + 0.050\Delta t\end{aligned}$$

$$\hat{P}(4) = 1.552.$$

(Figure 4.1.9).

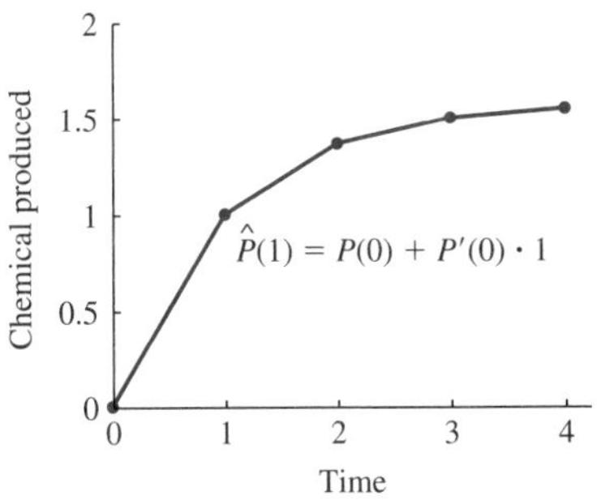

FIGURE 4.1.9

Euler's method applied to a pure-time differential equation $\Delta t = 1$

Example 4.1.8 Euler's Method with a Smaller Time Step

If we try the same method with a smaller time step of $\Delta t = 0.5$, we get

t	$\hat{P}(t+\Delta t)$	$t+\Delta t$	$\hat{P}(t+\Delta t)$
0.0	$P(0) + P'(0)\Delta t = 0.0 + e^{-0}\Delta t$	0.5	0.5
0.5	$\hat{P}(0.5) + P'(0.5)\Delta t = 0.5 + e^{-0.5}\Delta t$	1.0	0.803
1.0	$\hat{P}(1.0) + P'(1.0)\Delta t = 0.803 + e^{-1.0}\Delta t$	1.5	0.987
1.5	$\hat{P}(1.5) + P'(1.5)\Delta t = 0.987 + e^{-1.5}\Delta t$	2.0	1.098
2.0	$\hat{P}(2.0) + P'(2.0)\Delta t = 1.098 + e^{-2.0}\Delta t$	2.5	1.166
2.5	$\hat{P}(2.5) + P'(2.5)\Delta t = 1.166 + e^{-2.5}\Delta t$	3.0	1.207
3.0	$\hat{P}(3.0) + P'(3.0)\Delta t = 1.207 + e^{-3.0}\Delta t$	3.5	1.232
3.5	$\hat{P}(3.5) + P'(3.5)\Delta t = 1.232 + e^{-3.5}\Delta t$	4.0	1.247

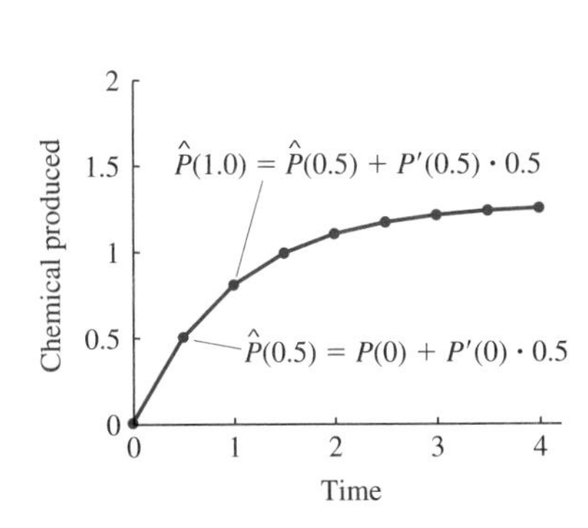

FIGURE 4.1.10

Euler's method applied to a pure-time differential equation $\Delta t = 0.5$

The smaller time step of $\Delta t = 0.5$ gives a more accurate answer, but requires more calculation (Figure 4.1.10).

Euler's method applies equally well to autonomous differential equations, but we save that method for Section 5.1.

Summary A **differential equation** expresses the rate of change of a quantity, the **state variable**, as a function of time or the state variable itself. If the rate of change has been measured as a function of time, the equation is a **pure-time differential equation**. If the rate of change has been derived from a rule, the equation is an **autonomous** differential equation. A **solution** gives the value of the state variable as a function of time. The solution also depends on the **initial condition**, the initial value of the state variable. We have identified four ways to solving differential equations: (1) in simple cases, it is possible to guess the right answer, (2) when we can graph the rate of change, it is possible to graph the solution, (3) integration, the mathematical method we will learn next, (4) When guessing fails, **Euler's method**, based on the tangent line approximation, can convert a differential equation into a discrete-time dynamical system.

4.1 Exercises

Mathematical Techniques

1–4 ▪ Identify the following as pure-time differential equations or autonomous differential equations.

1. $\frac{dx}{dt} = t$.
2. $\frac{dy}{dt} = 2y$.
3. $\frac{dw}{dt} = \frac{2}{1+t}$.
4. $\frac{dz}{dt} = 2\sqrt{z}$.

5–8 ▪ Use the graph of the rate of change to sketch a graph of the function, starting from the given initial condition.

5. Start from $x(0) = 1$.

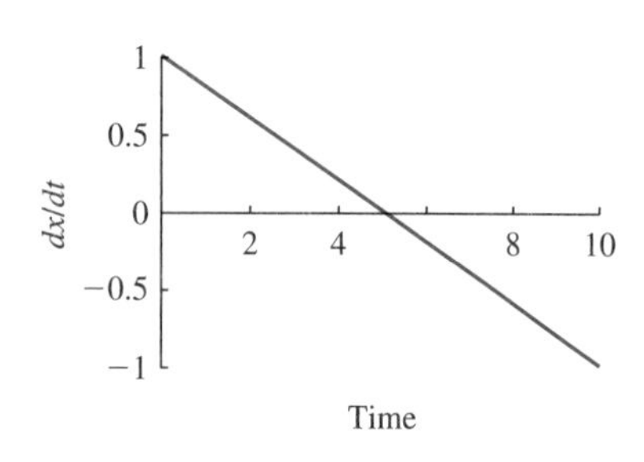

6. Start from $y(0) = -1$.

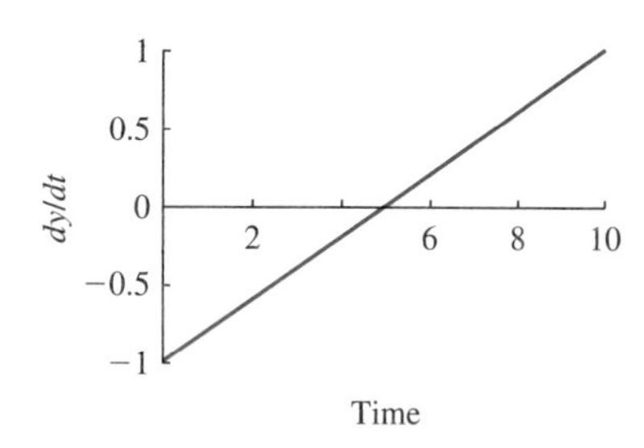

7. Start from $z(0) = 2$.

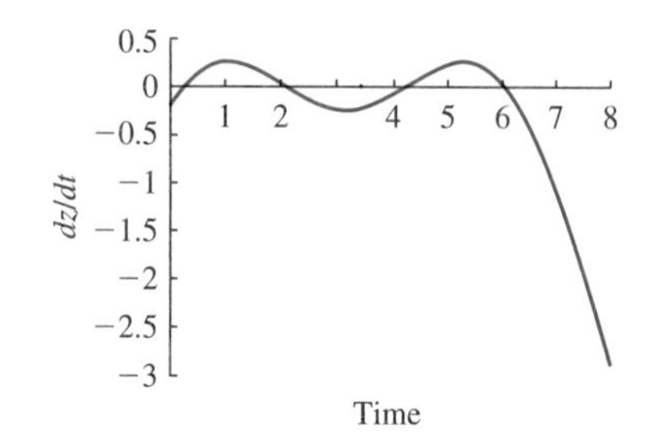

8. Start from $w(0) = 1$.

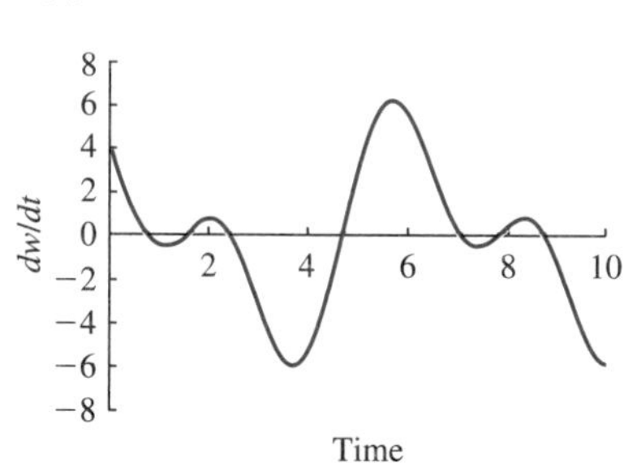

9–10 ▪ Check that the following are solutions of the given differential equation. What was the initial condition (the value of the state variable at $t = 0$)?

9. $\frac{dx}{dt} = t$ has solution $x(t) = 1 + \frac{t^2}{2}$.
10. $\frac{dw}{dt} = \frac{2}{1+t}$ has solution $w(t) = 2\ln(1+t) + 3$.

11–12 ▪ Apply Euler's method to the following differential equations to estimate the solution at $t = 1$ starting from the given initial condition. First, use one step with $\Delta t = 1$, and then use two steps with $\Delta t = 0.5$. Compare with the exact result from the earlier problem.

11. $\frac{dx}{dt} = t$ with initial condition $x(0) = 1$ (as in Exercise 9).
12. $\frac{dw}{dt} = \frac{2}{1+t}$ with initial condition $w(0) = 3$ (as in Exercise 10).

Applications

13–16 ▪ For each of the following descriptions of the volume of a cell, write a differential equation, find and graph the solution, and say whether the solution makes sense for all time.

13. A cell starts at a volume of 600 μm^3 and loses volume at a rate of 2 μm^3 per second.
14. A cell starts at a volume of 400 μm^3 and gains volume at a rate of 3 μm^3 per second.
15. A cell starts at a volume of 900 μm^3 and loses volume at a rate of $2t$ μm^3 per second.
16. A cell starts at a volume of 1000 μm^3 and loses volume at a rate of $3t^2$ μm^3 per second.

17–18 ▪ The following describe the velocities of different animals. In each case,

a. Draw a graph of the velocity as a function of time.

b. Write a differential equation for the position.

c. Guess the solution of this equation.

d. Draw a graph of the position as a function of time.

e. How long will it take to reach its goal?

17. A snail starts crawling across a sidewalk, trying to reach the other side which is 50 cm away. The velocity of the snail t minutes after it starts is t cm/min.

18. A cheetah is standing 1 m from the edge of the jungle. It starts sprinting across the savanna to attack a zebra that is 200 m from the edge of the jungle. After t seconds, the velocity of the cheetah is e^t m/s.

19–24 ▪ Apply Euler's method with the given value of Δt to the differential equation. Compare the approximate result with the exact result from the earlier problem.

19. The cell in Exercise 13. Use a step size of $\Delta t = 10$ to estimate the volume at $t = 40$.

20. The cell in Exercise 14. Use a step size of $\Delta t = 5$ to estimate the volume at $t = 30$.

21. The cell in Exercise 15. Use a step size of $\Delta t = 10$ to estimate the volume at $t = 30$.

22. The cell in Exercise 16. Use a step size of $\Delta t = 2$ to estimate the volume at $t = 10$.

23. The snail in Exercise 17. Use a step size of $\Delta t = 2$ to estimate the position at $t = 10$.

24. The cheetah in Exercise 18. Use a step size of $\Delta t = 1$ to estimate the position at $t = 5$.

25–26 ▪ Use the hints to "guess" the solution of the following differential equations describing the rate of production of some chemical. In each case, check your solution, and graph the rate of change and the solution.

25. $\frac{dP}{dt} = e^{-t+1}$ with initial condition $P(0) = 0$. Start by finding the derivative of e^{-t+1}, correct it by multiplying by some constant, and then add an appropriate value to match the initial condition.

26. $\frac{dP}{dt} = e^{-2t}$ with initial condition $P(0) = 0$. Start by finding the derivative of e^{-2t}, correct it by multiplying by some constant, and then add an appropriate value to match the initial condition.

27–30 ▪ For each of the following measurements, give circumstances under which you could measure the following.

a. The value but not the rate of change.

b. The rate of change but not the value.

27. Position (rate of change is velocity).

28. Mass (rate of change is growth rate).

29. Sodium concentration (with rate of change equal to the rate at which sodium enters and leaves).

30. Total chemical (with rate of change equal to the chemical production rate).

Computer Exercise

31. Apply Euler's method to solve the differential equation

$$\frac{db}{dt} = \frac{e^t}{t+1} + t$$

with the initial condition $b(0) = 1.0$. Compare with the sum of the results of

$$\frac{db_1}{dt} = \frac{e^t}{t+1}$$

and

$$\frac{db_2}{dt} = t.$$

Do you think there is a sum rule for differential equations? If your computer has a method for solving differential equations, find the solution and compare it with your approximate solution from Euler's method.

4.2 Solving Pure-Time Differential Equations

Pure-time differential equations describe situations where we have measured the rate of change of a quantity as a function of time and are interested in finding the quantity itself. We have seen three ways to solve pure-time differential equations: guessing, graphically, and Euler's method. We now develop the most important mathematical method, called the **antiderivative** or the **indefinite integral**.

4.2.1 Pure-Time Differential Equations and Antiderivatives

The general form of a pure-time differential equation is

$$\frac{dF}{dt} = f(t) \tag{4.2.1}$$

where $F(t)$ is the unknown state variable and $f(t)$ is the measured rate of change. The rate of change $f(t)$ depends only on the time t and not on the state variable F. We studied two pure-time differential equations in Section 4.1: an equation for volume with constant influx (Equation 4.1.1) and an equation for chemical product formation (Equation 4.1.2).

Pure-time differential equations are the easiest differential equations to solve—and even they are often impossible. Solving a pure-time differential equation for $F(t)$ requires finding a function that has a derivative equal to $f(t)$. Because this process undoes taking the derivative, F is called an **antiderivative** of f.

Definition 4.1 An antiderivative of the function f is a function F with derivative equal to f, written

$$F(t) = \int f(t)dt.$$

We say "F(t) is equal to the integral of f of t dt." The curvy symbol is an **integral sign** and can be thought of as a kind of function that takes one function f as input and returns another function F as output. $f(t)$ is called the **integrand**. The dt is obligatory and indicates that t is the variable.

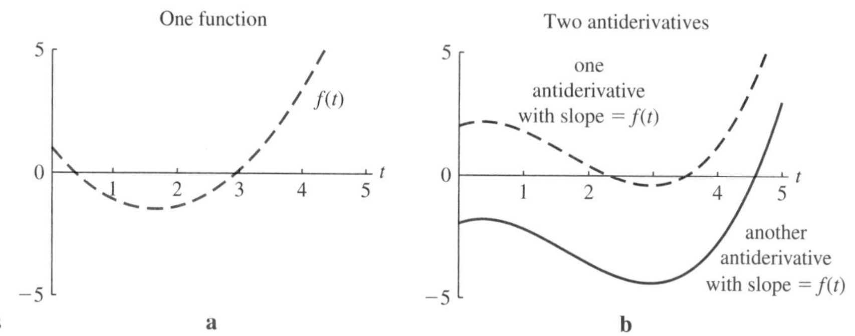

FIGURE 4.2.1
The function $f(t)$ and two antiderivatives

Every differentiable function has only one derivative. But every function has a whole family of antiderivatives. The two functions graphed in Figure 4.2.1 share the same derivative $f(t)$ and are **both** antiderivatives of f. All antiderivatives must share the same slope, however, and can only differ from each other by a constant. The set of all antiderivatives of f is known as the **indefinite integral** of f. We write

$$\int f(t)dt = F(t) + c$$

to indicate we can add any constant c to any particular antiderivative $F(t)$. The constant c is called an **arbitrary constant** because it can take on any value.

It might seem like a nuisance that we must include an arbitrary constant. It is not. Antiderivatives are the way to find the solutions of pure-time differential equations. A differential equation requires an initial condition to get the dynamics started and the arbitrary constant allows us to find a solution that matches the initial condition.

4.2.2 Rules for Antiderivatives

To find the indefinite integrals (antiderivatives) of interesting functions, we begin by using three of the basic rules of differentiation, the power rule: (Theorem 2.7), the constant product rule (Theorem 2.9) and the sum rule (Theorem 2.5).

The Power Rule for Integrals The power rule for derivatives says that

$$\frac{dx^{n+1}}{dx} = (n+1)x^n.$$

Dividing both sides by $n+1$ (using the constant product rule for derivatives),

$$\frac{d}{dx}\left(\frac{x^{n+1}}{n+1}\right) = x^n.$$

We have found a function with derivative equal to x^n, which gives the **power rule** for indefinite integrals.

Theorem 4.1 **The Power Rule for Integrals**

Suppose

$$f(x) = x^n$$

where $n \neq -1$. Then

$$\int x^n dx = \frac{x^{n+1}}{n+1} + c.$$

We will fill in the case $n = -1$ in Subsection 4.3.1.

Example 4.2.1 Applying the Power Rule for Integrals to a Quadratic

Applying the power rule for integrals with $n = 2$, the indefinite integral of x^2 is

$$\int x^2 dx = \frac{x^{2+1}}{2+1} + c = \frac{x^3}{3} + c.$$

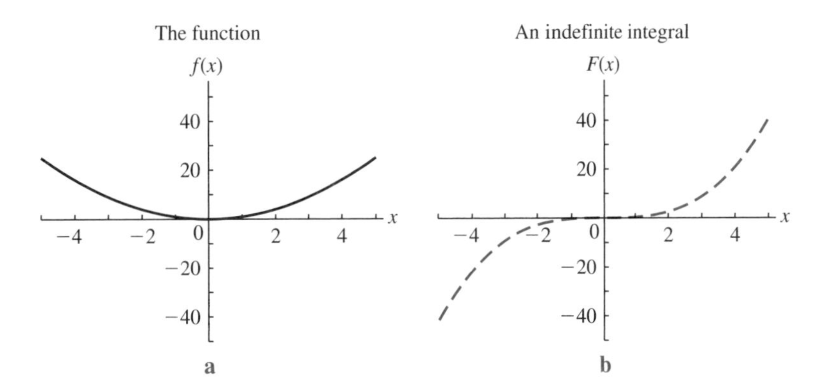

FIGURE 4.2.2
A quadratic function and its integral

Graphically, the slope of the indefinite integral must match the original function. In Figure 4.2.2, the slope of the indefinite integral $F(x) = \frac{x^3}{3}$ is the original function $f(x) = x^2$. The indefinite integral is increasing because the original function is never negative, and increases most steeply where $f(x)$ takes on its largest values.

Example 4.2.2 Applying the Power Rule for Integrals with a Negative Power

With $n = -2$, the indefinite integral of $\frac{1}{x^2}$ is

$$\int \frac{1}{x^2} dx = \int x^{-2} dx = \frac{x^{-2+1}}{-2+1} = -x^{-1} + c = -\frac{1}{x} + c.$$

Example 4.2.3 Applying the Power Rule for Integrals with Fractional Powers

The power rule also works with fractional powers. When $n = 0.5$

$$\int t^{0.5} dt = \frac{t^{0.5+1}}{0.5+1} + c = \frac{t^{1.5}}{1.5} + c.$$

When $n = -0.5$ (Figure 4.2.3)

$$\int t^{-0.5} dt = \frac{t^{-0.5+1}}{-0.5+1} + c = \frac{t^{0.5}}{0.5} + c = 2.0t^{0.5} + c.$$

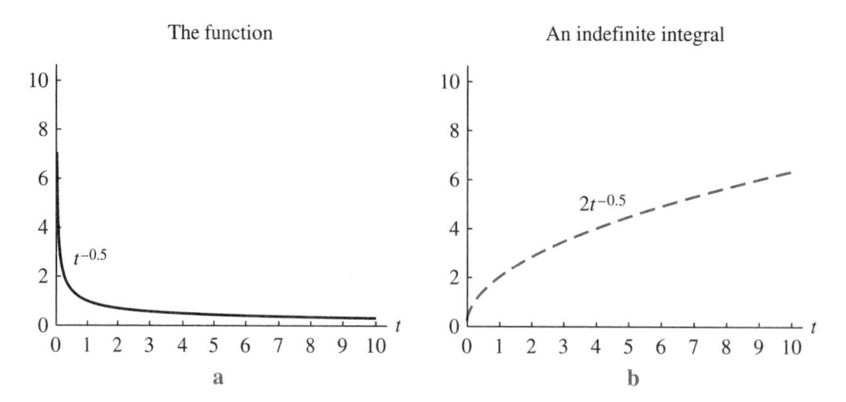

FIGURE 4.2.3
The indefinite integral of $t^{-0.5}$

The Constant Product Rule for Integrals The constant product rule for derivatives states that

$$\frac{d}{dx}[af(x)] = a\frac{df}{dx}.$$

The derivative of a constant times a function is the constant times the derivative. Indefinite integrals work the same way.

Theorem 4.2 **The Constant Product Rule for Integrals**

Suppose

$$\int f(x)dx = F(x) + c.$$

Then

$$\int af(x)dx = a\int f(x)dx = aF(x) + c$$

for any constant a.

We do not need to multiply the arbitrary constant c by the factor a because the product ac is just another way to designate an arbitrary value.

Example 4.2.4 The Constant Product Rule for Integrals I

If we multiply the function $f(x) = x^2$ by 5, we multiply the indefinite integral by 5, so

$$\int 5x^2 dx = 5\int x^2 dx \quad \text{pull constant 5 out front}$$

$$= \frac{5x^3}{3} + c. \quad \text{find integral with power rule, } n = 2$$

Example 4.2.5 The Constant Product Rule for Integrals II

Similarly,

$$\int \frac{-3}{x^2}dx = -3\int x^{-2}dx \quad \text{pull constant } -3 \text{ outside}$$

$$= -3\frac{-1}{x} + c = \frac{3}{x} + c. \quad \text{find integral with power rule, } n = -2$$

The Sum Rule for Integrals The sum rule for derivatives says that the derivative of the sum is the sum of the derivatives, or

$$\frac{d(f+g)}{dx} = \frac{df}{dx} + \frac{dg}{dx}.$$

Again, indefinite integrals work the same way.

Theorem 4.3 **The Sum Rule for Integrals**

Suppose

$$\int f(x)dx = F(x) + c$$

$$\int g(x)dx = G(x) + c$$

then

$$\int f(x) + g(x)dx = \int f(x)dx + \int g(x)dx = F(x) + G(x) + c.$$

There is only a single arbitrary constant c added on at the end.

Example 4.2.6 The Sum Rule Applied to Power Functions

We have found that

$$\int 5x^2 dx = \frac{5x^3}{3} + c$$

$$\int -3x^{-2} dx = \frac{3}{x} + c.$$

Therefore,

$$\int (5x^2 - 3x^{-2})dx = \frac{5x^3}{3} + \frac{3}{x} + c.$$

The power rule, constant product rule, and sum rule are summarized in the following table.

The power, constant product, and sum rules for integrals

Rule	Formula
Power rule	$\int x^n dx = \frac{x^{n+1}}{n+1} + c$ if $n \neq -1$
Constant product rule	$\int af(x)dx = a\int f(x)dx$
Sum rule	$\int[f(x) + g(x)]dx = \int f(x)dx + \int g(x)dx$

4.2.3 Solving Polynomial Differential Equations

With the power, constant product, and sum rules we can find the indefinite integral of any polynomial and can therefore solve any pure-time differential equation where the rate of change is a polynomial function of time.

The simplest polynomial is a constant. We can solve the differential equation

$$\frac{dV}{dt} = 1.5$$

with the power rule and constant product rule because the rate of change is $1.5 = 1.5t^0$. Evaluating the indefinite integral

$$\begin{aligned} V(t) &= \int 1.5\, dt \\ &= 1.5 \int t^0 dt \\ &= 1.5\frac{t^{0+1}}{0+1} + c \\ &= 1.5t + c. \end{aligned}$$

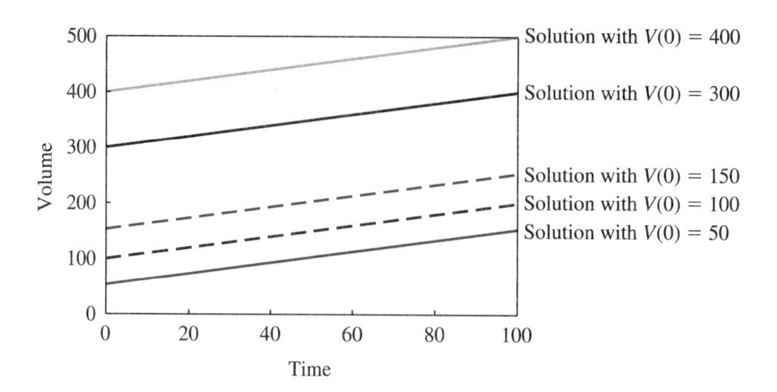

FIGURE 4.2.4
Several solutions of $\frac{dV}{dt} = 1.5$

The constant c is determined by the initial conditions. Suppose the volume is 300.0 at time 0. This gives an equation for the arbitrary constant c,

$$V(0) = 0 + c = 300.0$$

or $c = 300.0$. The solution of the pure-time differential equation

$$\frac{dV}{dt} = 1.5$$

with initial condition $V(0) = 300.0$ is therefore

$$V(t) = 1.5t + 300.0$$

(Figure 4.2.4).

Example 4.2.7 Solving When the Initial Condition Is Not Given at $t = 0$

Suppose we have the polynomial pure-time differential equation

$$\frac{dP}{dt} = 3.0 - 1.5t$$

with initial condition $P(2) = 10.0$. We can use the indefinite integral to find the solution $P(t)$ as

$$\begin{aligned} P(t) &= \int 3.0 - 1.5t\,dt && \text{solution is indefinite integral of rate of change} \\ &= 3.0 \int dt - 1.5 \int t\,dt && \text{sum and constant product rules} \\ &= 3.0t - 1.5\frac{t^2}{2} + c && \text{power rule} \\ &= 3.0t - 0.75t^2 + c. \end{aligned}$$

To find the constant c, we substitute the initial condition, finding

$$P(2) = 10.0 = 3.0 \cdot 2 - 0.75 \cdot 2^2 + c = 3.0 + c$$

from which we see that $c = 7.0$, and the solution is $P(t) = 3.0t - 0.75t^2 + 7.0$.

Problems involving finding the indefinite integrals of polynomials arise when studying the motion of objects with **constant acceleration**. Suppose we are told that a rock is thrown downward from a 100 m tall building with initial downward velocity of 5.0 m/s. The acceleration of gravity is -9.8 m/s^2 (Figure 4.2.5). How long will it take the rock to hit the ground? How fast will it be going?

Let a represent acceleration, v the velocity, and p the position of the rock. The basic differential equations from physics are

$$\frac{dv}{dt} = a \quad \text{acceleration is the rate of change of velocity}$$

$$\frac{dp}{dt} = v \quad \text{velocity is the rate of change of position}$$

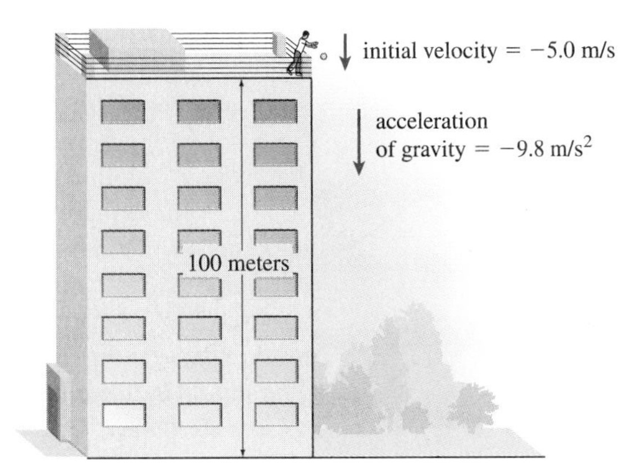

FIGURE 4.2.5
A falling rock

The first says that acceleration is the rate of change of velocity, and the second that velocity is the rate of change of position. We can use these equations to solve for the position of the rock.

Example 4.2.8 The Physics of Constant Acceleration: Finding the Velocity

To find velocity, we have *measured* its rate of change, the acceleration, to be $a = -9.8$ and the initial condition $v(0) = -5.0$. Both values are negative because both are downward. We therefore have enough information to solve the differential equation

$$\frac{dv}{dt} = -9.8$$

with the initial condition $v(0) = -5.0$. The rate of change is constant, so

$$v(t) = -9.8t + c.$$

We find the constant c with the initial condition by solving

$$v(0) = -9.8 \cdot 0 + c = -5.0$$

so $c = -5.0$. Therefore, the equation for the velocity is

$$v(t) = -9.8t - 5.0.$$

As time passes, the velocity becomes more and more negative as the rock falls faster and faster (Figure 4.2.6a).

Example 4.2.9 The Physics of Constant Acceleration: Finding the Position

We now have enough information to solve for the position p; the pure-time differential equation

$$\frac{dp}{dt} = -9.8t - 5.0$$

and the initial condition $p(0) = 100$. We first solve the differential equation with the indefinite integral, finding

$$\begin{aligned} p(t) &= \int (-9.8t - 5.0)dt && p(t) \text{ is the indefinite integral of } v(t) \\ &= \int -9.8t\,dt - \int 5.0dt && \text{the sum rule} \\ &= -9.8 \int t\,dt - \int 5.0dt && \text{the constant product rule} \\ &= -9.8\frac{t^2}{2} - 5.0t + c = -4.9t^2 - 5.0t + c. && \text{the power rule} \end{aligned}$$

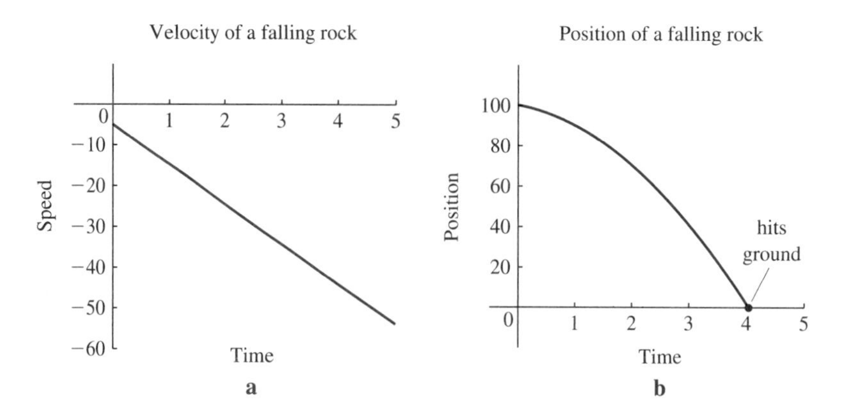

FIGURE 4.2.6
The velocity and position of a falling rock

We use the initial condition to find the constant c by solving

$$p(0) = -4.9 \cdot 0^2 - 5.0 \cdot 0 + c = 100.0$$

so $c = 100.0$. Therefore, the equation for the position is

$$p(t) = -4.9t^2 - 5.0t + 100$$

(Figure 4.2.6b).

The solution tells when the rock will hit the ground and how fast it will be going. It hits the ground at the time when $p(t) = 0$, the solution of

$$-4.9t^2 - 5.0t + 100 = 0.$$

For convenience, we multiply by -1, so

$$4.9t^2 + 5.0t - 100 = 0.$$

With the quadratic formula, we find

$$\begin{aligned} t &= \frac{-5.0 \pm \sqrt{5.0^2 - 4 \cdot 4.9 \cdot (-100)}}{2 \cdot 4.9} \\ &= \frac{-5.0 \pm \sqrt{1985}}{9.8} \\ &= \frac{-5.0 \pm 44.55}{9.8} \\ &= 4.036 \text{ or } -5.506. \end{aligned}$$

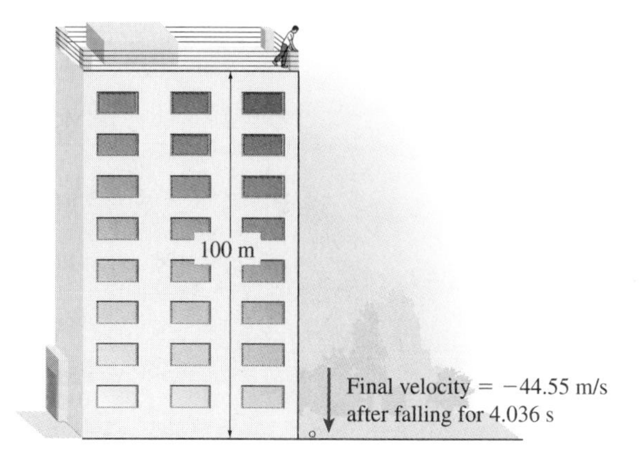

FIGURE 4.2.7
A falling rock hits the ground

The positive solution is the one that makes sense. How fast will the rock be going when it hits the ground? We use the formula for $v(t)$ to find

$$v(4.036) = -9.8 \cdot 4.036 - 5.0 = -44.55\frac{\text{m}}{\text{s}}.$$

(Figure 4.2.7).

Example 4.2.10 A Differential Equation for AIDS

Polynomial pure-time differential equations also arise in public health policy. During the early years of the AIDS epidemic, workers at the Centers for Disease Control found that the number of new AIDS cases followed the formula

$$\text{rate at which new AIDS cases were reported} = 523.8t^2 \qquad \textbf{(4.2.2)}$$

where t is measured in years after the beginning of 1981. The Centers for Disease Control formula can be written as a polynomial pure-time differential equation

$$\frac{dA}{dt} = 523.8t^2.$$

To solve this equation for the total number of AIDS cases, we require an initial condition. Surveys indicated that about 340 people had been infected at the beginning of 1981 ($t = 0$ in the model), so $A(0) = 340$. We can now solve for $A(t)$, finding

$$A(t) = \int 523.8t^2 dt = 523.8\frac{t^3}{3} = 174.6t^3 + c.$$

The constant c is the solution of the equation

$$A(0) = 174.6 \cdot 0^3 + c = 340$$

so $c = 340$. The solution is

$$A(t) = 174.6t^3 + 340$$

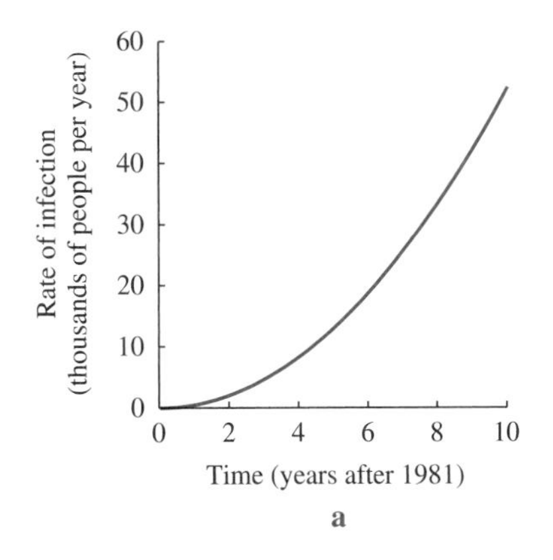

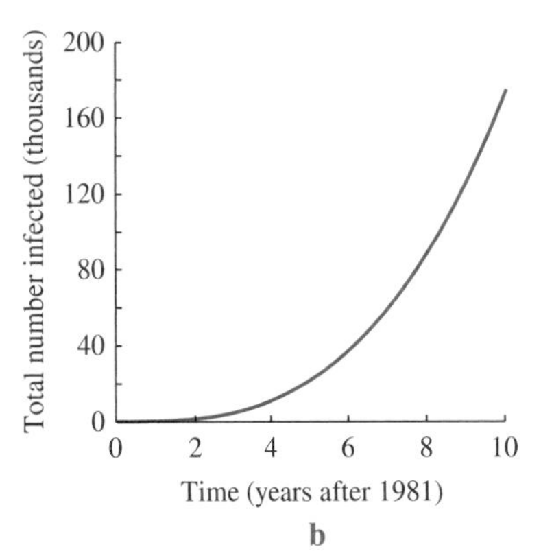

FIGURE 4.2.8

The early course of the AIDS epidemic in the United States

(Figure 4.2.8).

Summary Solving pure-time differential equations requires undoing the derivative. We defined the **antiderivative** of a function f as any function whose derivative is f. Unlike the derivative, the antiderivative includes an **arbitrary constant** to indicate that a whole family of functions share the same slope. The **indefinite integral** is the whole set of antiderivatives. From the rules for computing derivatives, we found the **power, constant product**, and **sum** rules for integrals. With these rules, we can find the indefinite integral of any polynomial. We applied this method to solve pure-time differential equations.

4.2 Exercises

Mathematical Techniques

1–6 ▪ From the graphs, sketch an antiderivative of the function that passes through the given point.

1. An antiderivative that passes through the point (0, 500).

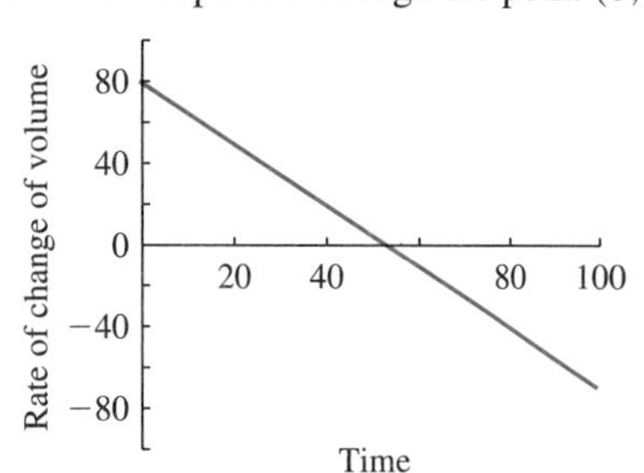

2. An antiderivative that passes through the point (50, 5000).

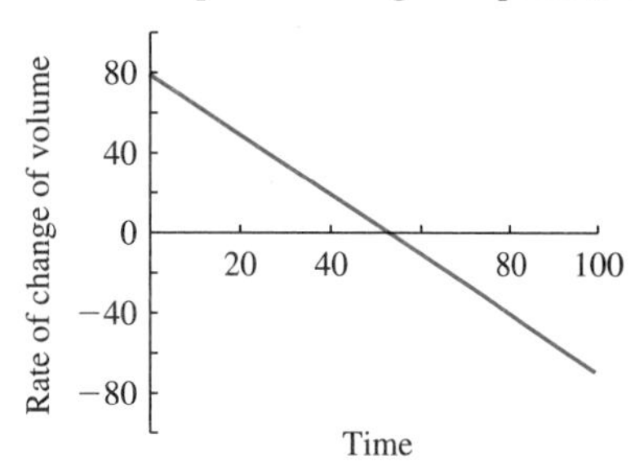

3. An antiderivative that passes through the point (100, 3000).

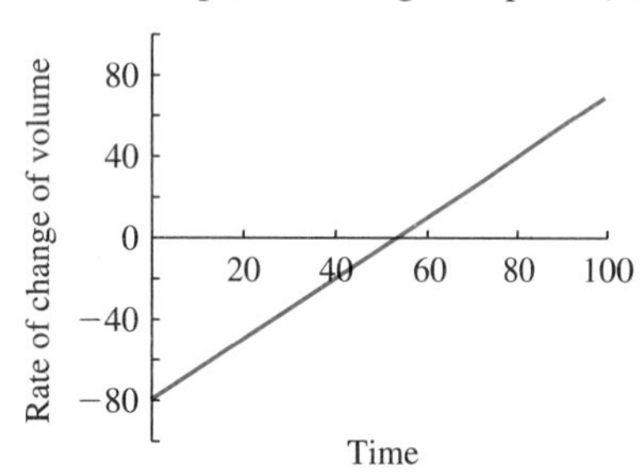

4. An antiderivative that passes through the point (0, 2500).

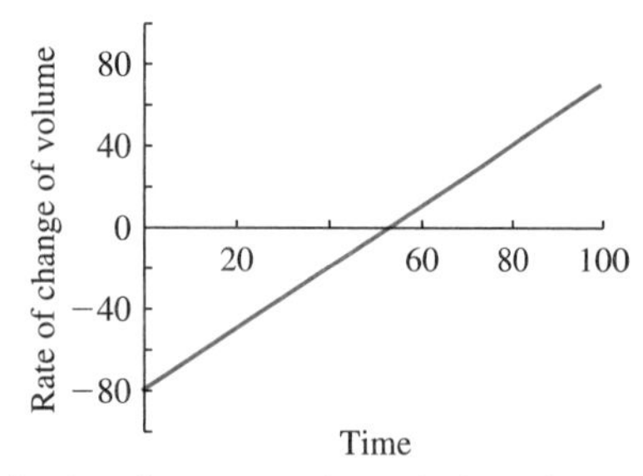

5. An antiderivative that passes through the point (100, 1000).

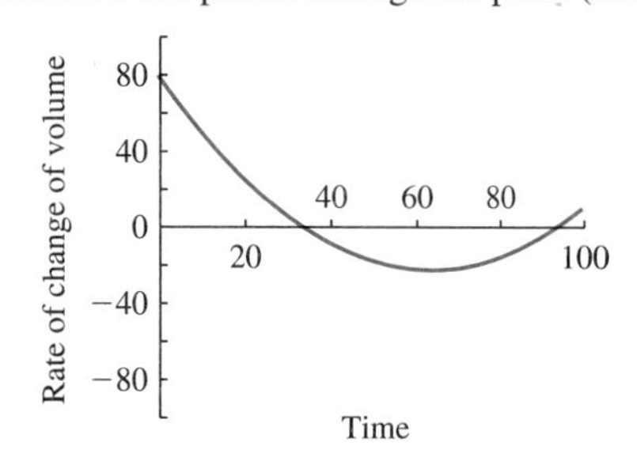

6. An antiderivative that passes through the point (0, 1000).

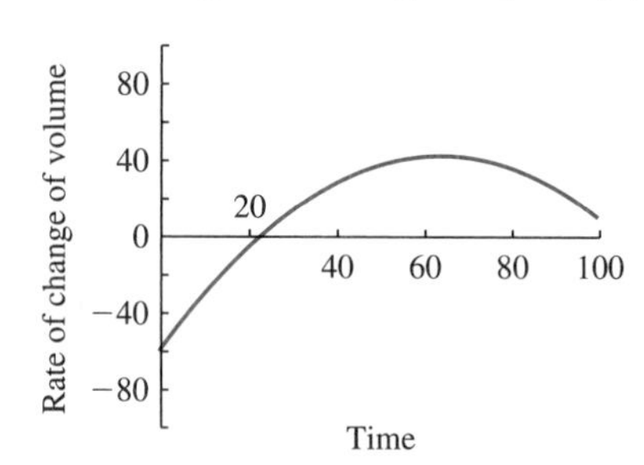

7–14 ▪ Find the indefinite integrals of the following functions.

7. $7x^2$

8. $10t^9 + 6t^5$

9. $72t + 5$

10. $y^4 + 5y^3$

11. $\dfrac{5}{x^3}$

12. $3z^{\frac{3}{7}}$

13. $\dfrac{2}{\sqrt[3]{t}} + 3$

14. $5z^{-1.2} - 1.2$

15–20 ▪ Use indefinite integrals to solve the following differential equations. Sketch a graph of the rate of change and the solution on the given domain.

15. $\dfrac{dV}{dt} = 2t^2 + 5$ with $V(1) = 19.0$. Sketch the rate of change and solution for $0 \le t \le 5$.

16. $\dfrac{dV}{dt} = 2t^2 + 5$ with $V(0) = 19.0$. Sketch the rate of change and solution for $0 \le t \le 5$.

17. $\dfrac{df}{dt} = 5t^3 + 5t$ with $f(0) = -12.0$. Sketch the rate of change and solution for $0 \le t \le 2$.

18. $\dfrac{dg}{dt} = -3t + t^2$ with $g(0) = 10.0$. Sketch the rate of change and solution for $0 \le t \le 5$.

19. $\dfrac{dM}{dt} = t^2 + \dfrac{1}{t^2}$ with $M(3) = 10.0$. Sketch the rate of change and solution for $0 \le t \le 3$.

20. $\dfrac{dp}{dt} = 5t^3 + \dfrac{5}{t^2}$ with $p(1) = 12.0$. Sketch the rate of change and solution for $1 \le t \le 3$.

21–22 ▪ There are no simple integral versions of product and quotient rules for derivatives. Use the given functions to show that proposed rule does not work.

21. Use the functions $f(x) = x^2$ and $g(x) = x^3$ to show that the product of integrals is **not** equal to the integral of the product.

22. Use the functions $f(x) = x^2$ and $g(x) = x^3$ to show that

$$\int f(x)g(x)dx \neq g(x)\int f(x)dx + f(x)\int g(x)dx.$$

Applications

23–24 ▪ Suppose a cell is taking water into two vacuoles. Let V_1 denote the volume of the first vacuole and V_2 the volume of the second. In each of the following cases,

a. Solve the given differential equations for $V_1(t)$ and $V_2(t)$.

b. Write a differential equation for $V = V_1 + V_2$, including the initial condition.

c. Show that the solution of the differential equation for V is the sum of the solutions for V_1 and V_2.

23.

$$\frac{dV_1}{dt} = 2.0t + 5.0$$

$$\frac{dV_2}{dt} = 5.0t + 2.0$$

with initial conditions $V_1(0) = V_2(0) = 10\ \mu\text{m}^3$.

24.

$$\frac{dV_1}{dt} = 3.6t^2 + 5.0t$$

$$\frac{dV_2}{dt} = 5.2t^3 + 2.0$$

with initial conditions $V_1(0) = 5.0\ \mu\text{m}^3$ and $V_2(0) = 10.0\ \mu\text{m}^3$.

25–26 ▪ Suppose organisms grow in mass according to the differential equation

$$\frac{dM}{dt} = \alpha t^n$$

where M is measured in grams and t is measured in days. For each of the following values for n and α, find

a. The units of α.

b. Suppose that $M(0) = 5.0$ gm. Find the solution.

c. Sketch a graph of the rate of change and the solution.

d. Describe your results in words.

25. $n = 1, \alpha = 2.0$

26. $n = -1/2, \alpha = 2.0$

27–30 ▪ Suppose an object is thrown from a height of $h = 100$ m with velocity $v = 5.0$ m/s (upward) to find its trajectory in a local gravitational field of strength a. For the following values of a,

a. Find the velocity and position of the object as functions of time.

b. How high will the object get?

c. How long will it take to pass the initial height of 100 m on the way down? How fast will it be moving?

d. How long will it take to hit the ground? How fast will it be moving? How fast is this in miles per hour?

e. Graph the velocity and position as functions of time.

27. On Earth where $a = -9.8\ \text{m/s}^2$

28. On the moon where $a = -1.62\ \text{m/s}^2$

29. On Jupiter where $a = -22.88\ \text{m/s}^2$

30. On Mars' moon Deimos where $a = -2.15 \times 10^{-3}\ \text{m/s}^2$

31–34 ▪ The velocities of four objects are measured at discrete times.

Time	Velocity of Object 1	Velocity of Object 2	Velocity of Object 3	Velocity of Object 4
0	1.0	9.0	25.0	1.0
1	3.0	7.0	16.0	3.0
2	5.0	5.0	9.0	6.0
3	7.0	3.0	4.0	10.0
4	9.0	1.0	1.0	15.0

Use Euler's method to estimate the position at $t = 4$ starting from the given initial condition. Next, find a formula for the velocities and use it to find the exact position at $t = 4$. Graph your results.

31. Object 1, with initial condition $p(0) = 10.0$. To find the formula for the velocities, note that they increase linearly in time.

32. Object 2, with initial condition $p(0) = 10.0$. To find the formula for the velocities, note that they decrease linearly in time.

33. Object 3, with initial condition $p(0) = 10.0$. To find the formula for the velocities, compare them with the perfect square numbers.

34. Object 4, with initial condition $p(0) = 10.0$. The velocities follow a quadratic equation of the form $v(t) = \frac{t^2}{2} + at + b$ for some value of a.

35–38 ▪ Consider again the velocities of four objects used in the previous set of problems. There is a more accurate variant of Euler's method that approximates the rate of change during a time interval as the average of the rate of change at the beginning and at the end of the interval. For example, if $v(0) = 1.0$ and $v(1) = 3.0$, we approximate the rate of change for $0 \le t \le 1$ as 2.0. Use this variant of Euler's method to estimate the position at $t = 4$ and compare with the exact position at $t = 4$. Graph your results.

35. Object 1, with initial condition $p(0) = 10.0$.

36. Object 2, with initial condition $p(0) = 10.0$.

37. Object 3, with initial condition $p(0) = 10.0$.

38. Object 4, with initial condition $p(0) = 10.0$.

4.3 Integration of Special Functions, Integration by Substitution, by Parts, and by Partial Fractions

We have seen how to solve any pure-time differential equation with the indefinite integral when the rate of change can be expressed as a polynomial or other combination of power functions. There are not too many other functions for which we can find the indefinite integral, and thus not too many pure-time differential equations we can solve algebraically. In this section, we will learn first how to find integrals of exponential, logarithmic, and cosine functions. Using the chain rule for derivatives, we derive the method of **substitution**, which is used to evaluate integrals involving parameters. Using the product rule for derivatives, we derive the method of **integration by parts**, which adds a few more integrals to our tool chest.

4.3.1 Integrals of Special Functions

The derivatives of special functions help us to find a few more indefinite integrals. Recall that

$$\frac{d}{dx}\ln(x) = \frac{1}{x}$$

$$\frac{d}{dx}e^x = e^x.$$

Because the indefinite integral, the antiderivative, undoes the action of the derivative,

$$\int \frac{1}{x}dx = \ln(|x|) + c \tag{4.3.1}$$

$$\int e^x dx = e^x + c. \tag{4.3.2}$$

Try not to be confused by the absolute value bars inside the function $\ln(|x|) + c$ (Figure 4.3.1). The function $\frac{1}{x}$, unlike the natural logarithm, is defined for negative values of x. It must be the slope of some function. The graph of $\frac{1}{x}$ for negative x is the negative of the graph for positive x. The antiderivative for negative x must be the mirror image of the antiderivative for positive x. This integral fills in the $n = -1$ case for the power rule for integrals (Theorem 4.1).

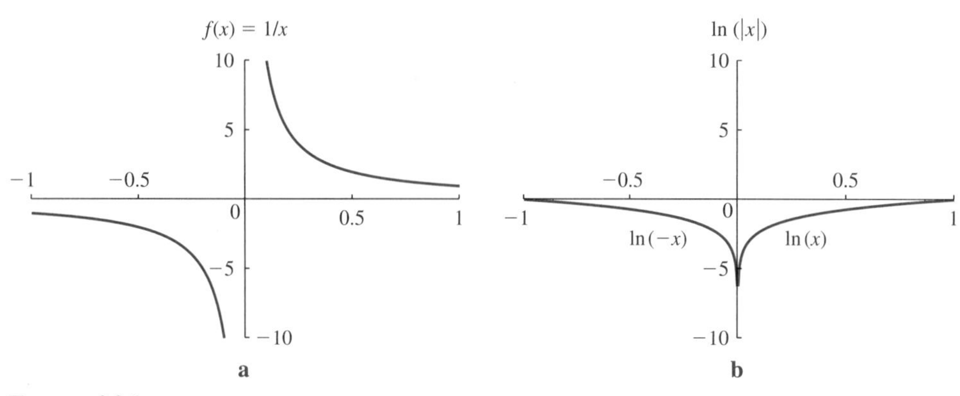

FIGURE 4.3.1
The integral of $1/x$

Using the derivatives of sine and cosine

$$\frac{d}{dx}\sin(x) = \cos(x)$$
$$\frac{d}{dx}\cos(x) = -\sin(x),$$

we see that

$$\int \cos(x)dx = \sin(x) + c \qquad \textbf{(4.3.3)}$$

$$\int \sin(x)dx = -\cos(x) + c. \qquad \textbf{(4.3.4)}$$

Using the constant product and sum rules, we can integrate a few more complicated functions.

Example 4.3.1 Integrating a Complicated Function

We can compute

$$\int \frac{1}{x} + 2e^x + 5\sin(x)dx = \int \frac{1}{x}dx + 2\int e^x dx + 5\int \sin(x)dx$$
$$= \ln(|x|) + 2e^x - 5\cos(x) + c$$

according to the sum rule (Theorem 4.3).

These elements form the basic building blocks of integration. The edifice that can be built from these blocks, even with several additional techniques, is rather paltry. Many perfectly reasonable functions cannot be integrated at all. Only three are needed for the applications in this book: integration by substitution, integration by parts, and integration by partial fractions. Some sophisticated techniques are available to integrate a few additional functions and can be studied in a more advanced text.

4.3.2 The Chain Rule and Integration by Substitution

The chain rule (Theorem 2.12) tells how to differentiate the compositions of functions, stating that

$$\frac{d(f \circ g)}{dx} = \frac{df}{dg}\frac{dg}{dx},$$

as expressed in differential notation. Integration by substitution is easier to understand in differential notation, and we will use it here. Because the indefinite integral, the antiderivative, undoes the action of the derivative,

$$\int \frac{df}{dg}\frac{dg}{dx}dx = (f \circ g)(x) + c.$$

If we were lucky enough to find and recognize an integral of this form, we could use this method to integrate.

Example 4.3.2 Integrating a Complicated Function with a Lucky Guess

Suppose we are asked to find

$$\int 2xe^{x^2}dx.$$

If we were clever or lucky enough, we might notice that

$$2xe^{x^2} = \frac{dg}{dx}\frac{df}{dg}$$

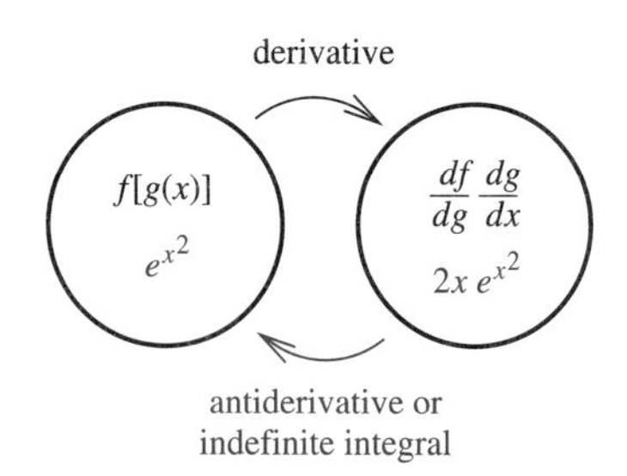

FIGURE 4.3.2
The chain rule and the indefinite integral

if $g(x) = x^2$ and $f(g) = e^g$. In other words,

$$\frac{d(f \circ g)}{dx} = \frac{df}{dg}\frac{dg}{dx}$$
$$= e^g \cdot 2x$$
$$= 2xe^{x^2}.$$

Therefore, because the indefinite integral undoes the action of the derivative,

$$\int 2xe^{x^2}\,dx = e^{x^2} + c$$

(Figure 4.3.2).

Integration by substitution is a way to recognize these patterns, when they exist, without relying entirely on inspired guessing. There is, however, no guarantee that the technique will work.

Algorithm 4.2 Integration by Substitution

1. Define a new variable as some function of the old variable.
2. Take the derivative of the new variable with respect to the old variable.
3. Treat the derivative like a fraction and move the dx to the other side.
4. Put everything in the integral in terms of the new variable.
5. Try to integrate the expression in terms of the new variable.
6. After integrating, try to put everything back in terms of the old variable.

Example 4.3.3 Integrating a Complicated Composition with Substitution

In Example 4.3.1, Algorithm 4.2 proceeds as follows.

1. Define a new variable to be the function of x most tangled up in the expression. In this case, x^2 is inside the exponential, so define y by
$$y = x^2.$$
Our hope is to write everything in terms of the new variable y. The e^{x^2} term is now e^y, which we know how to integrate. However, there remains the piece $2xdx$.
2. Find the derivative of y with respect to x as
$$\frac{dy}{dx} = 2x.$$
3. This step looks weird but is legal. Multiply both sides by dx to find
$$dy = 2xdx.$$
4. By good luck, dy exactly matches the remaining piece $2xdx$. In terms of y, the integral is
$$\int 2xe^{x^2}\,dx = \int e^y\,dy.$$
5. By more good luck, this is an expression we know how to integrate, finding
$$\int e^y\,dy = e^y + c.$$
6. Put everything back in terms of x (using $y = x^2$), so
$$\int 2xe^{x^2}\,dx = e^{x^2} + c.$$

Example 4.3.4 Integrating a Complicated Composition with Clever Substitution

We will follow these steps to evaluate

$$\int \frac{e^{2t}}{(1+e^{2t})^2} dt.$$

1. It is not obvious what to substitute. One good guess is to use the expression that is squared in the denominator,
$$u = 1 + e^{2t}.$$
2. Take the derivative, finding
$$\frac{du}{dt} = 2e^{2t}.$$
3. Solve for du as
$$du = 2e^{2t}dt.$$
4. Put everything in terms of u. Noting that $e^{2t}dt = du/2$, so
$$\int \frac{e^{2t}}{(1+e^{2t})^2} dt = \int \frac{1}{u^2} \frac{du}{2}.$$
5. We can attack the new integral with the constant product and power rules, finding
$$\int \frac{1}{u^2} \frac{du}{2} = \frac{1}{2} \int u^{-2} du$$
$$= \frac{1}{2}(-u^{-1}) + c = -\frac{1}{2u} + c.$$
6. Put everything back in terms of t, arriving at
$$\int \frac{e^{2t}}{(1+e^{2t})^2} dt = -\frac{1}{2u} + c = -\frac{1}{2(1+e^{2t})} + c.$$

Example 4.3.5 The Failure of Integration by Substitution

The problem with integration by substitution is that terms must match up exactly. Suppose we wanted to evaluate

$$\int 2x^2 e^{x^2} dx.$$

1. Try the substitution $y = x^2$ as before.
2. Then
$$\frac{dy}{dx} = 2x.$$
3. $dy = 2xdx$.
4. Putting everything in terms of y is messy:
$$2x^2 dx = x \cdot 2xdx = xdy.$$
To get rid of the remaining x, we have to solve for x in terms of y, finding $x = \sqrt{y}$. Substituting this value gives the new integral
$$\int \sqrt{y} e^y dy.$$
5. We have no idea how to integrate this.
6. Give up.

In fact, there is no way to compute this indefinite integral in terms of basic functions.

4.3.3 Using Substitution to Eliminate Constants

Substitution is guaranteed to work in one case: when the integral would be possible if we could remove some excess constants. For example, the differential equation

$$\frac{dL}{dt} = 6.48e^{-0.09t} \tag{4.3.5}$$

can be used to describe the growth of a fish. If we could just get rid of the -0.09 in the exponent there would be no problem finding the integral and solving the differential equation. Substitution can remove the excess constant.

Alternatively, suppose that the rate of change of energy stores from an organism is proportional to how far its temperature is above 36°C, so that

$$\frac{dE}{dt} = 36.0 - P_d(t) = -0.8 - 0.3\cos[2\pi(t - 0.583)]$$

where we assumed that temperature follows the daily temperature cycle $P_d(t) = 36.8 + 0.3\cos[2\pi(t - 0.583)]$ (studied in Subsection 1.8.2). We would have no problem integrating and finding the total energy loss if we could make that $2\pi(t - 0.583)$ inside the cosine into a simple t. Again, substitution will work.

Example 4.3.6 Applying Substitution to Fish Growth

Suppose fish size begins at $L(0) = 0.0$ (measured from fertilization) and follows the differential equation

$$\frac{dL}{dt} = 6.48e^{-0.09t}.$$

This equation expresses the fact that fish grow more and more slowly, but never stop growing throughout their lives. These match growth data for walleye in North Caribou Lake, Ontario. To solve it, we need to find the indefinite integral of the rate of change $6.48e^{-0.09t}$ and use the initial condition to compute the arbitrary constant. To find the integral, we follow Algorithm 4.2.

1. Define a new variable $z = -0.09t$.
2. $\frac{dz}{dt} = -0.09$.
3. $dz = -0.09dt$.
4. To get dt in terms of dz, solve to find
$$dt = -\frac{dz}{0.09}$$
and write
$$\int 6.48e^{-0.09t}dt = \int -6.48\frac{e^z}{0.09}dz = \int -72.0e^z dz.$$
5. Integrate, finding
$$\int -72.0e^z dz = -72.0\int e^z dz = -72.0e^z + c.$$
6. Put everything back in terms of t, arriving at
$$\int 6.48e^{-0.09t}dt = -72.0e^z + c = -72.0e^{-0.09t} + c. \tag{4.3.6}$$

To finish solving the differential equation, we must find the arbitrary constant. The equation is

$$L(0) = 0.0 = -72.0e^{-0.09\cdot 0} + c = -72.0 + c$$

so $c = 72.0$. The solution is

$$L(t) = 72.0 - 72.0e^{-0.09t} = 72.0(1 - e^{-0.09t}). \tag{4.3.7}$$

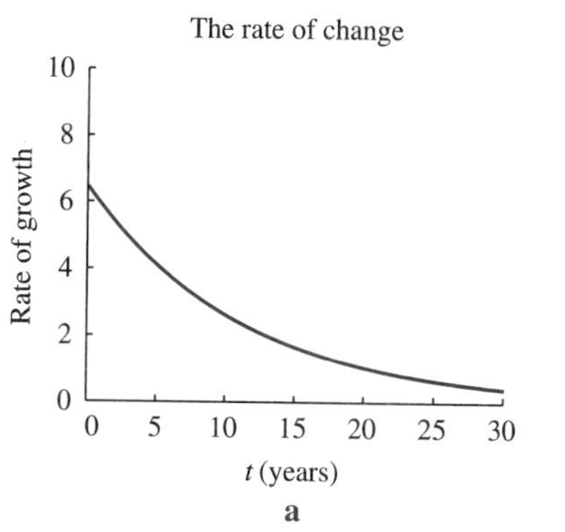

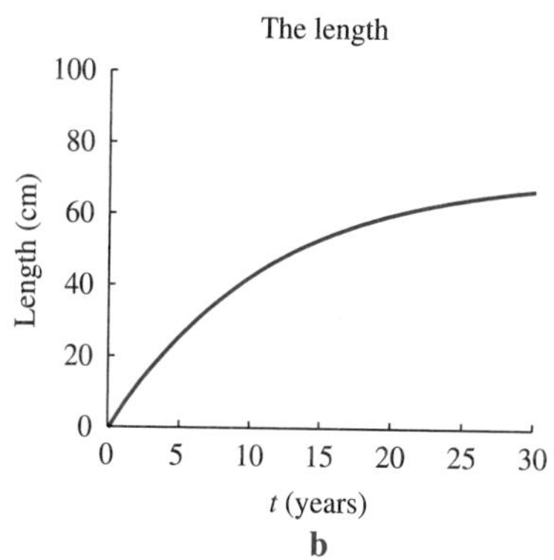

FIGURE 4.3.3
Growth of a fish

This is called the **Bertalanffy growth equation** (Figure 4.3.3). The size approaches a limit of 72.0 cm, although the fish never quite stops growing. ▲

We can use this solution to find the age at which fish mature. Walleye begin to reproduce when they reach about 45 cm in length (but continue to grow after that). How long will it take this fish to mature?

Example 4.3.7 Finding the Age When a Fish Matures

We must solve

$$L(t) = 45$$

or

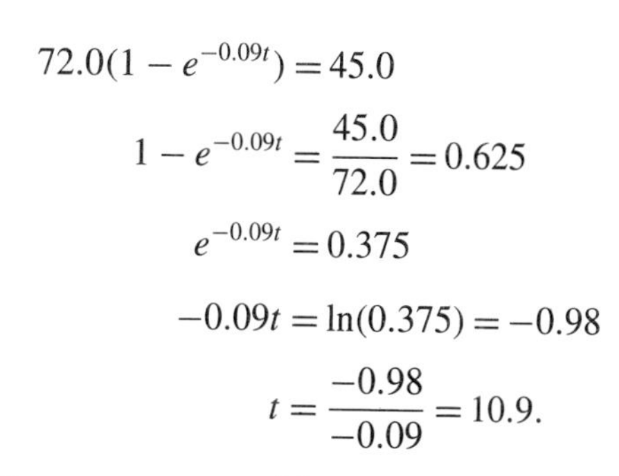

$$72.0(1 - e^{-0.09t}) = 45.0$$

$$1 - e^{-0.09t} = \frac{45.0}{72.0} = 0.625$$

$$e^{-0.09t} = 0.375$$

$$-0.09t = \ln(0.375) = -0.98$$

$$t = \frac{-0.98}{-0.09} = 10.9.$$

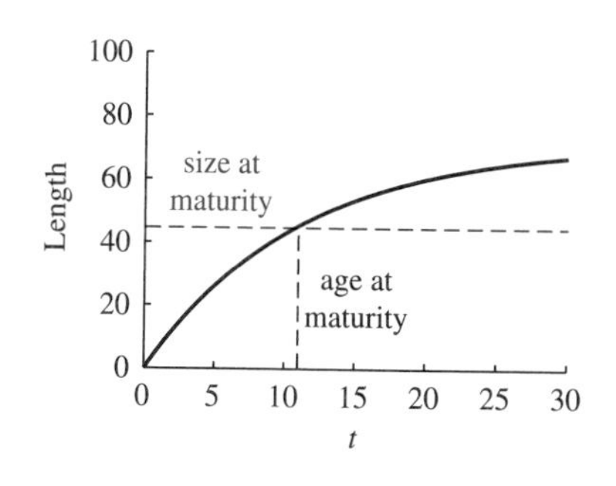

FIGURE 4.3.4
Finding the age of maturity of a walleye

This fish will take 10.9 years to mature (Figure 4.3.4). ▲

We can use substitution to evaluate the indefinite integral of $e^{\alpha x}$ for any value of α.

1. Define a new variable $y = \alpha x$.
2. $\frac{dy}{dx} = \alpha$.
3. $dy = \alpha dx$.
4. To get dx in terms of dy, solve to find $dx = \frac{dy}{\alpha}$ and write
$$\int e^{\alpha x} dx = \int \frac{e^y}{\alpha} dy.$$
5. Integrate, finding
$$\int \frac{e^y}{\alpha} dy = \frac{1}{\alpha} \int e^y dy = \frac{1}{\alpha} e^y + c.$$
6. Put everything back in terms of x, arriving at
$$\int e^{\alpha x} dx = \frac{e^{\alpha x}}{\alpha} + c. \tag{4.3.8}$$

Example 4.3.8 Applying Substitution to Energy Use

We can use substitution to find the total energy used in a day from the differential equation

$$\frac{dE}{dt} = 36.0 - P_d(t) = -0.8 - 0.3\cos[2\pi(t - 0.583)],$$

with initial condition $E(0) = 0$ because no energy has been used by time 0. To solve the equation, we must find the indefinite integral

$$E(t) = \int \{-0.8 - 0.3\cos[2\pi(t - 0.583)]\}\, dt.$$

The hard term to deal with is $\cos[2\pi(t - 0.583)]$. We attack with the method of substitution.

1. Define a new variable $y = 2\pi(t - 0.583)$, the linear function of t inside the cosine.
2. $\frac{dy}{dt} = 2\pi$.
3. $dy = 2\pi\, dt$.
4. To write dt in terms of dy, solve to find
$$dt = \frac{dy}{2\pi}$$
and write
$$\int \cos[2\pi(t - 0.583)] = \int \cos(y)\frac{dy}{2\pi}.$$
5. Integrate, finding
$$\int \cos(y)\frac{dy}{2\pi} = \frac{1}{2\pi}\int \cos(y)dy$$
$$= \frac{1}{2\pi}\sin(y) + c.$$
6. Put everything back in terms of t, arriving at
$$\int \cos[2\pi(t - 0.583)] = \frac{1}{2\pi}\sin[2\pi(t - 0.583)] + c.$$

The solution of the whole equation is

$$E(t) = \int -0.8 - 0.3\cos[2\pi(t - 0.583)]dt$$
$$= -0.8t - \frac{0.3}{2\pi}\sin[2\pi(t - 0.583)] + c$$
$$= -0.8t - 0.048\sin[2\pi(t - 0.583)] + c.$$

To find the initial condition, we must solve

$$E(0) = 0 = -0.8 \cdot 0 - 0.048\sin[2\pi(0 - 0.583)] + c = -0.024 + c,$$

so that $c = 0.024$. The solution is

$$E(t) = -0.8t - 0.048\sin[2\pi(t - 0.583)] + 0.024$$

(Figure 4.3.5).

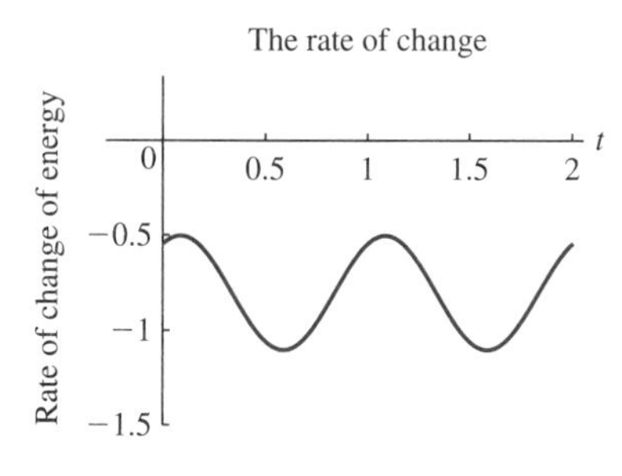

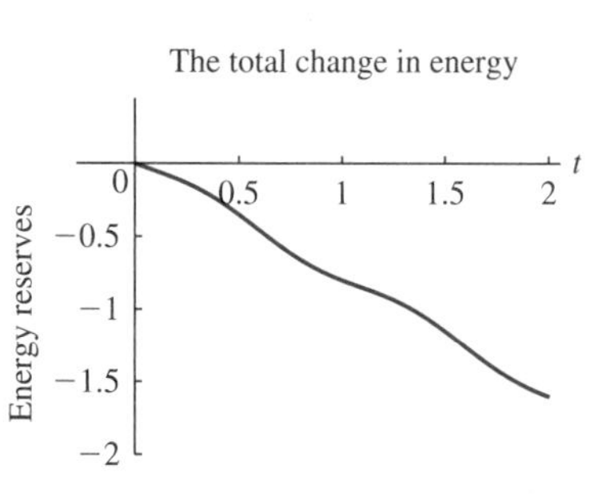

FIGURE 4.3.5
Loss of energy due to temperature

4.3.4 Integration by Parts and by Partial Fractions

The sum rule and constant product rule for derivatives translate directly into similar rules for integrals. The chain rule for derivatives translates into integration by substitution, although finding the correct substitution can be difficult or impossible. The product rule for derivatives also leads to a corresponding method for integrals, called **integration by parts**. Like substitution, it provides a method that can sometimes transform unfamiliar integrals into familiar forms but is not guaranteed to work.

Recall the product rule for derivatives, written for the functions $u(x)$ and $v(x)$,

$$\frac{d[u(x)v(x)]}{dx} = u(x)\frac{dv}{dx} + v(x)\frac{du}{dx}.$$

Solving for $u(x)\frac{dv}{dx}$ gives

$$u(x)\frac{dv}{dx} = \frac{d[u(x)v(x)]}{dx} - v(x)\frac{du}{dx}.$$

Integrating both sides, gives the formula for integration by parts

$$\int u(x)\frac{dv}{dx}dx = \int \frac{d[u(x)v(x)]}{dx}dx - \int v(x)\frac{du}{dx}dx$$

$$\int u(x)\frac{dv}{dx}dx = u(x)v(x) - \int v(x)\frac{du}{dx}dx. \tag{4.3.9}$$

The first term on the right-hand side simplifies because the indefinite integral of the derivative of $u(x)v(x)$ is the function $u(x)v(x)$ itself (from the definition of the antiderivative).

This may not look very useful. If we could **recognize** the integrand $f(x)$ as the product

$$f(x) = u(x)\frac{dv}{dx}$$

and recognize that

$$\int v(x)\frac{du}{dx}dx$$

is a familiar integral, the method would work. Although the cases where these tricks work are not many, several are important.

Example 4.3.9 An Example of Integration by Parts

Suppose we wish to find the indefinite integral

$$\int xe^x dx.$$

This is written as a product, so we can set $u(x) = x$ and $\frac{dv}{dx} = e^x$. Then

$$v(x) = \int e^x dx = e^x$$

(where we chose $c = 0$ for the arbitrary constant) and $\frac{du}{dx} = 1$. Substituting into the formula for integration by parts gives

$$\begin{aligned}\int u(x)\frac{dv}{dx}dx &= u(x)v(x) - \int v(x)\frac{du}{dx}dx \\ &= xe^x - \int e^x dx \\ &= xe^x - e^x + c.\end{aligned}$$

Checking, we use the product rule to find

$$\frac{d}{dx}xe^x - e^x + c = e^x + xe^x - e^x = xe^x.$$

Example 4.3.10 Integration by Parts Applied to $\ln(x)$

We think of the integrand $\ln(x)$ as the product $\ln(x)\cdot 1$. Then let $u(x)=\ln(x)$ and $\frac{dv}{dx}=1$. To use the formula, compute $v(x)=x$ (again picking $c=0$ for the arbitrary constant) and $\frac{du}{dx}=\frac{1}{x}$, giving

$$\begin{aligned}\int u(x)\frac{dv}{dx}dx &= u(x)v(x) - \int v(x)\frac{du}{dx}dx\\ &= x\ln(x) - \int x\frac{1}{x}dx\\ &= x\ln(x) - \int 1dx\\ &= x\ln(x) - x + c.\end{aligned}$$

Checking,

$$\frac{d}{dx}x\ln(x) - x + c = \ln(x) + 1 - 1 = \ln(x).$$

As with substitution, integration by parts can fail.

Example 4.3.11 Failure of Integration by Parts

Suppose we wish to find the indefinite integral

$$\int \frac{e^x}{x}dx.$$

This can be thought of as the product of e^x and $\frac{1}{x}$. We could set $u(x)=\frac{1}{x}$ and $\frac{dv}{dx}=e^x$. Then $v(x)=\int e^x dx = e^x$ (as in Example 4.3.9) and $\frac{du}{dx}=-\frac{1}{x^2}$. Substituting, we get

$$\int \frac{e^x}{x}dx = \frac{e^x}{x} + \int \frac{e^x}{x^2}dx.$$

The new integral is, if anything, worse than the one we started with.

Alternatively, we could set $u(x)=e^x$ and $\frac{dv}{dx}=\frac{1}{x}$. Then $v(x)=\int \frac{1}{x}dx=\ln(x)$ and $\frac{du}{dx}=e^x$. Substituting yields

$$\int \frac{e^x}{x}dx = e^x\ln(x) - \int \ln(|x|)e^x dx.$$

Again, the new integral is no more familiar than the original one.

Like the function e^{x^2}, the function $\frac{e^x}{x}$ cannot be integrated in terms of the basic functions. This does not mean that the integral is not a reasonable function. We could use the Taylor series found in Example 3.7.9, divide each term by x, and integrate each term to find a Taylor polynomial for this integral (Exercise 37).

Unlike the other methods of integration which are based on the rules of differentiation, **integration by partial fractions** is based on an algebraic trick. It works for functions that can be written as ratios of polynomials, **rational functions**, for which the denominator can be factored. The method proves quite useful for solving the autonomous differential equation analog of the discrete-time logistic equation (Section 3.2) that we will study in Section 5.4.

We consider here only functions written as the ratio of a constant over a factored quadratic, leaving a few more complicated cases to the exercises (Exercises 39 and 40).

The algebraic trick breaks a single function with a quadratic in the denominator into a sum of two functions with linear denominators, which can then be directly integrated with substitution.

The function

$$f(x) = \frac{K}{(x + r_1)(x + r_2)}$$

can be expressed as

$$f(x) = \frac{A_1}{x + r_1} + \frac{A_2}{x + r_2}.$$

If we could find the constants A_1 and A_2, we can then compute

$$\begin{aligned}
\int \frac{K}{(x + r_1)(x + r_2)} dx &= \int \frac{A_1}{x + r_1} dx + \int \frac{A_2}{x + r_2} dx && \text{sum rule} \\
&= A_1 \int \frac{1}{x + r_1} dx + A_2 \int \frac{1}{x + r_2} dx && \text{constant product rule} \\
&= A_1 \int \frac{1}{y} dy + A_2 \int \frac{1}{z} dz && \text{substitute } y = x + r_1 \text{ and } z = x + r_2 \\
&= A_1 \ln(|y|) + A_2 \ln(|z|) + c && \text{special function integral} \\
&= A_1 \ln(|x + r_1|) + A_2 \ln(|x + r_2|) + c. && \text{put back in terms of } x
\end{aligned}$$

Although it is possible to find the constants A_1 and A_2 in general, the algebra is simpler in particular cases.

Example 4.3.12 Integration by Partial Fractions

To integrate the function $\frac{5}{(x + 1)(x + 2)}$, we write

$$\frac{5}{(x + 1)(x + 2)} = \frac{A_1}{x + 1} + \frac{A_2}{x + 2}.$$

To solve for A_1 and A_2 we place the right hand sum over a common denominator as

$$\begin{aligned}
\frac{A_1}{x + 1} + \frac{A_2}{x + 2} &= \frac{A_1(x + 2)}{(x + 1)(x + 2)} + \frac{A_2(x + 1)}{(x + 1)(x + 2)} \\
&= \frac{(A_1 + A_2)x + 2A_1 + A_2}{(x + 1)(x + 2)}.
\end{aligned}$$

We then set this numerator equal to the numerator, the constant 5, of the original function

$$5 = (A_1 + A_2)x + 2A_1 + A_2.$$

For the linear function of x on the right hand side to equal the constant on the left, the slope $A_1 + A_2$ must be equal to 0, or $A_2 = -A_1$. The constant $2A_1 + A_2$ is then equal to $2A_1 - A_1 = A_1$, which is in turn equal to 5. Therefore,

$$\frac{5}{(x + 1)(x + 2)} = \frac{5}{x + 1} - \frac{5}{x + 2}.$$

We then have that

$$\begin{aligned}
\int \frac{5}{(x + 1)(x + 2)} dx &= \int \frac{5}{x + 1} dx - \frac{5}{x + 2} dx && \text{split up with partial fractions} \\
&= 5 \ln(x + 1) - 5 \ln(x + 2) + c && \text{integrate each piece with the sum rule} \\
&= 5[\ln(x + 1) - \ln(x + 2)] + c && \text{factor out the 5} \\
&= 5 \ln\left(\frac{x + 1}{x + 2}\right) + c. && \text{combine with a law of logs}
\end{aligned}$$

Summary Based on their derivatives, we found indefinite integrals of exponential, logarithmic, and trigonometric functions. To evaluate complicated integrals, we developed **integration by substitution**, the integral version of the chain rule of differentiation. This technique can be used to integrate a few complicated combinations of functions, and can reliably eliminate excess constants from integrals. **Integration by parts**, the integral version of the product rule, can be used to compute a few additional integrals. Finally, **integration by partial fractions** can be used to integrate many rational functions.

4.3 Exercises

Mathematical Techniques

1–6 ▪ Find the indefinite integrals of the following functions.

1. $\frac{3}{z^2} + \frac{z^2}{3}$

2. $3e^x + 2x^3$

3. $e^x + \frac{1}{x}$

4. $\frac{2}{t} + \frac{t}{2}$

5. $2\sin(x) + 3\cos(x)$

6. $x^2 - 20\sin(x)$

7–12 ▪ Use substitution, if needed, to find the indefinite integrals of the following functions. Check with the chain rule.

7. $3e^{\frac{x}{5}}$

8. $\cos[2\pi(x-2)]$

9. $\left(1 + \frac{t}{2}\right)^4$

10. $(1+2t)^{-4}$

11. $\frac{1}{4+t}$

12. $\frac{1}{1+4t}$

13–20 ▪ Use substitution to find the indefinite integrals of the following functions.

13. $\frac{e^x}{1+e^x}$

14. $e^t(1+e^t)^4$

15. $2y\sqrt{1+y^2}$

16. $\cos(x)e^{\sin(x)}$

17. $\tan(\theta)$ (write it as $\frac{\sin(\theta)}{\cos(\theta)}$ and use a substitution for the denominator.)

18. $\frac{t}{1+t}$

19. $\frac{1}{x\ln(x)}$

20. $\frac{\ln(x)}{x}$

21–24 ▪ Use integration by parts to evaluate the following. Check your answer by taking the derivative.

21. $\int xe^{2x}dx$

22. $\int x\cos(3x)dx$

23. $\int x^2e^xdx$

24. $\int x^2e^{-x}dx$

25–26 ▪ Use integration by partial fractions to compute the following indefinite integrals.

25. $\int \frac{2}{1-x^2}dx$

26. $\int \frac{4}{x(2+3x)}dx$

27–28 ▪ Integration by substitution, in combination with trigonometric identities, can be used to integrate some surprising functions.

27. Substitute $x = \tan(\theta)$ to integrate $\frac{1}{1+x^2}$.

28. Substitute $x = \sin(\theta)$ to integrate $\frac{1}{\sqrt{1-x^2}}$.

29–30 ▪ Integration by parts along with substitution can be used to integrate some of the inverse trigonometric functions.

29. The result of Section 2.10, Exercise 25 gives the derivative of $\tan^{-1}(x)$. Use integration by parts and a substitution to find $\int \tan^{-1}(x)dx$.

30. The result of Section 2.10, Exercise 23 gives the derivative of $\sin^{-1}(x)$. Use integration by parts and a substitution to find $\int \sin^{-1}(x)dx$.

31–32 ▪ In Examples 4.3.9 and 4.3.10, we chose the constant $c = 0$ when finding $v(x)$. Follow the steps for integration by parts, but leave c as an arbitrary constant. Do you get the same answer?

31. Find the indefinite integral $\int xe^xdx$ as in Example 4.3.9.

32. Find the indefinite integral $\int \ln(x)dx$ as in Example 4.3.10.

33–34 ▪ Sometimes integrating by parts seems to lead in a circle, but the answer can still be found. Try the following.

33. Find the indefinite integral $e^x\sin(x)$ using integration by parts twice.

34. Find the indefinite integral $\frac{\ln(x)}{x}$ using integration by parts.

35–38 ▪ When a function has a well-behaved Taylor series, we can find a Taylor series for some integrals by integrating term by term (proving that these integrals converge to the correct answer requires methods from more advanced courses).

35. Integrate the Taylor series for e^x term by term, and find the value of the constant for which the integral matches the function e^x.

36. Integrate the Taylor series for $\frac{1}{1-x}$ term by term, and check if the answer matches the Taylor series for $-\ln(1-x)$ (use the results of Section 3.7, Exercise 32).

37. Integrate the Taylor series for $\frac{e^x}{x}$ term by term. Does this look like a familiar series?

38. Integrate the Taylor series for e^{x^2} term by term. Does this look like a familiar series?

39–40 ▪ Integration by partial fractions works on many more cases than presented in the main text. We here look at functions with a linear function, rather than a constant, as the numerator.

39. Find the integral of $f(x) = \frac{x}{(x+1)(x+2)}$.

40. Find the integral of $g(z) = \frac{z-1}{(z+1)(z+3)}$.

Applications

41–44 ▪ The following differential equations for the production of a chemical P share the properties that the rate of change at $t = 0$ is 5.0, and the limit as $t \to \infty$ is 0. For each, find the solution starting from the initial condition $P(0) = 0$, sketch the solution, and say what happens to $P(t)$ as $t \to \infty$. Compute $P(10)$ and $P(100)$. Why do some increase to infinity while others do not?

41. $\frac{dP}{dt} = \frac{5}{1+2.0t}$.

42. $\frac{dP}{dt} = 5.0e^{-0.2t}$.

43. $\frac{dP}{dt} = 2.5\left(\frac{1}{1+t} + e^{-t}\right)$.

44. $\frac{dP}{dt} = \frac{5.0}{(1+t)^2}$.

45–46 ▪ Use integration by parts by find solutions of the following differential equations.

45. Suppose the mass M of a toad grows according to the differential equation $\frac{dM}{dt} = (t + t^2)e^{-2t}$ with $M(0) = 0$. When does this toad grow fastest? Find $M(1)$. How much larger would the toad be if it always grew at the maximum rate?

46. Suppose the mass W of a worm grows according to the differential equation $\frac{dW}{dt} = (4t - t^2)e^{-3t}$ with $W(0) = 0$. When does this worm grow fastest? Find $W(2)$. How much larger would the worm be if it always grew at the maximum rate?

47–48 ▪ The following problems give the parameters for Walleye in a variety of locations. For each location, the differential equation has the form $\frac{dL}{dt} = \alpha e^{-\beta t}$. Find

a. The solution of the differential equation if $L(0) = 0$.

b. Find the limit of size as t approaches infinity.

c. Assume that all walleye mature at 45 cm in length. How old are these walleye when they mature?

d. Graph the size and compare with Ontario walleye (Figure 4.3.3).

47. In Texas, where $\alpha = 64.3$ and $\beta = 1.19$.

48. In Saskatchewan, where $\alpha = 6.48$ and $\beta = 0.06$.

49–52 ▪ The population of lemmings $L(t)$ at the top of a cliff is increasing with the given formula. However, lemmings leap off the cliff at a rate equal to $0.1L$ and pile up at the bottom.

a. Write a pure-time differential equation for the number of lemmings $B(t)$ piled up at the bottom of the cliff.

b. Solve using the initial condition $B(0) = 0$.

c. Graph the number of lemmings at the top and the number at the bottom of the cliff.

d. Find the limit of the ratio $\frac{B(t)}{L(t)}$ at t approaches infinity.

49. $L(t) = 1000e^{0.2t}$.

50. $L(t) = 1000e^{0.05t}$.

51. $L(t) = 100t$.

52. $L(t) = 100 + 100e^{0.5t}$.

53–56 ▪ Growth rates of insects depend on the temperature T. Suppose that the length of an insect L follows the differential equation

$$\frac{dL}{dt} = 0.001T(t)$$

with t measured in days starting from January 1 and temperature measured in °C. Insects hatch with an initial size of 0.1 cm. For each of the following equations for $T(t)$,

a. Sketch a graph of the temperature over the course of a year.

b. Suppose an insect starts growing on January 1. How big will it be after 30 days?

c. Suppose an insect starts growing on June 1 (day 151). How big will it be after 30 days?

53. $T(t) = 0.001t(365 - t)$.

54. $T(t) = 40.0 - 5.0 \times 10^{-6}t^2(365 - t)$.

55. $T(t) = 20.0 + 10.0\cos\left[\frac{2\pi(t - 190.0)}{365}\right]$.

56. $T(t) = 20.0 + 10.0\cos\left[\frac{2\pi(t - 90.0)}{182.5}\right]$.

Computer Exercises

57. Consider again the fish in Caribou Lake, growing according to

$$\frac{dL}{dt} = \beta e^{-\alpha t}$$

where $\beta = 6.48$ and $\alpha = 0.09$. However, suppose there is some variability among fish in the values of these two parameters.

a. Solve for growth trajectories of 5 fish with values of β evenly spread from 10% below to 10% above 6.48.

b. Solve for growth trajectories of 5 fish with values of α evenly spread from 10% below to 10% above 0.09.

c. Which parameter has a greater effect on the size of the fish?

58. Clever genetic engineers design a set of fish that grow according to the following equations.

$$\frac{dL_1}{dt} = 1, \quad \frac{dL_2}{dt} = \frac{1}{1+\sqrt{t}}, \quad \frac{dL_3}{dt} = \frac{1}{1+t}, \quad \frac{dL_4}{dt} = \frac{1}{1+t^2}$$
$$\frac{dL_5}{dt} = \frac{1}{1+t^3}, \quad \frac{dL_6}{dt} = e^{-t}, \quad \frac{dL_7}{dt} = e^{-t^2}.$$

Suppose they all start at size 0. Use your computer to sketch the growth of these fish for $0 \le t \le 5$. Then zoom in near $t = 0$. Could you tell which fish was which?

4.4 Integrals and Sums

When we cannot guess the answer or use one of the rules of integration to solve a pure-time differential equation, we can use Euler's method to approximate the solution. A graphical analysis of this method shows the link between integrals and sums. Using this link, we define the **definite integral** as the limit of **Riemann sums**. The definite integral represents the total amount of change during some period of time. This interpretation of the integral provides insight into a wide range of applications.

4.4.1 Approximating Integrals with Sums

Consider the pure-time differential equation for the volume V of water in a vessel

$$\frac{dV}{dt} = t^2$$

where t is measured in seconds and V is measured in cm^3. We are asked to find the total quantity of water that entered during the first second. We could solve the differential equation with the indefinite integral to find this quantity. Alternatively, we can take the limit of an approximation in the spirit of Euler's method.

Suppose we had measured the rate at which water was entering the vessel only every 0.2 s. For lack of knowledge of what happens between measurements, we might assume that the rate is *constant* between measurements.

Time (s)	Rate (cm^3/s)
0.0	0.00
0.2	0.04
0.4	0.16
0.6	0.36
0.8	0.64
1.0	1.00

There are two ways to use this information to approximate the total amount of water entering during this second. In one, called the **left-hand estimate**, we approximate the rate at which water enters between measurements as the value at the *beginning* of the interval (Figure 4.4.1a). In the other, the **right-hand estimate**, we approximate the rate at which water enters between measurements as the value at the *end* of the interval (Figure 4.4.1b).

	Left-Hand Estimate			Right-Hand Estimate		
Time Interval	Rate During Interval	Influx During Interval	Net Influx	Rate During Interval	Influx During Interval	Net Influx
0.0 – 0.2	0.00	0.000	0.000	0.04	0.008	0.008
0.2 – 0.4	0.04	0.008	0.008	0.16	0.032	0.040
0.4 – 0.6	0.16	0.032	0.040	0.36	0.072	0.112
0.6 – 0.8	0.36	0.072	0.112	0.64	0.128	0.240
0.8 – 1.0	0.64	0.128	0.240	1.00	0.200	0.440

The estimate replaces the curve with a **step function**, a series of horizontal lines anchored on the actual function. Step functions are easy to deal with because rates are constant during each interval.

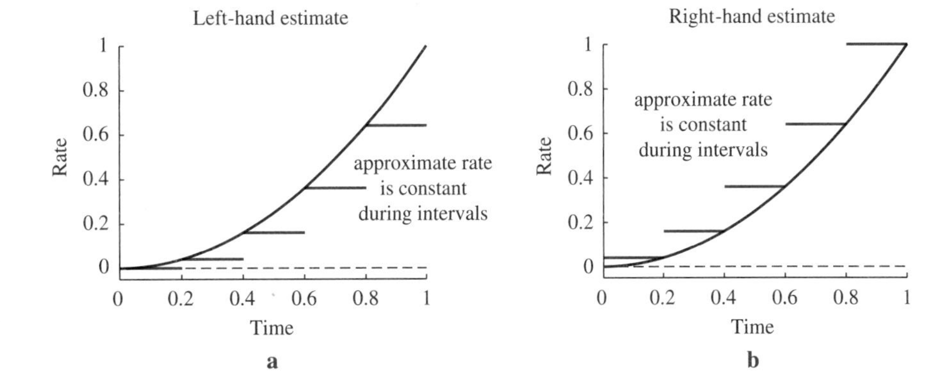

FIGURE 4.4.1
Left-hand and right-hand estimates of the integral

The computation of the net or total influx proceeds as follows. In the first time interval, from $t = 0.0$ to $t = 0.2$, the left-hand estimate approximates the influx with the rate of flow at the beginning of the interval, or 0.0. The influx is thus approximately 0.0 during this first interval. In the second interval, from 0.2 to 0.4, the left-hand estimate approximates the rate of influx with 0.04, producing an influx of 0.008 (multiplying the rate by the time). Adding this to the 0.0 from the first interval gives a net of 0.008. Continuing in this way, we estimate a total influx of water of 0.24 cm^3 during the first second.

The right-hand estimate approximates the rate of influx during the first interval by 0.04, the rate at the end of the interval. The total influx during this interval is estimated as 0.008. During the second interval, the right-hand estimate approximates the rate of influx with 0.16, and the influx by 0.032. Adding this to the 0.008 from the first interval gives a net of 0.040. Continuing in this way, we estimate a total influx of water of 0.44 cm^3.

We have computed two estimates of the total, a left-hand estimate of 0.24 and a right-hand estimate of 0.44. The two differ by quite a bit. The figures tell us why (Figure 4.4.1). The step function associated with the left-hand estimate lies well below the exact measurement, while the step function associated with the right-hand estimate lies well above the exact measurement.

By using more measurements, we expect to get a more accurate answer. We can follow the same steps assuming that we have measured every 0.1 s (Figure 4.4.2).

	Left-Hand Estimate			Right-Hand Estimate		
Time Interval	Rate During Interval	Influx During Interval	Net Influx	Rate During Interval	Influx During Interval	Net Influx
0.0–0.1	0.00	0.000	0.000	0.01	0.001	0.001
0.1–0.2	0.01	0.001	0.001	0.04	0.004	0.005
0.2–0.3	0.04	0.004	0.005	0.09	0.009	0.014
0.3–0.4	0.09	0.009	0.014	0.16	0.016	0.030
0.4–0.5	0.16	0.016	0.030	0.25	0.025	0.055
0.5–0.6	0.25	0.025	0.055	0.36	0.036	0.091
0.6–0.7	0.36	0.036	0.091	0.49	0.049	0.140
0.7–0.8	0.49	0.049	0.140	0.64	0.064	0.204
0.8–0.9	0.64	0.064	0.204	0.81	0.081	0.285
0.9–1.0	0.81	0.081	0.285	1.00	0.100	0.385

The two estimates are closer to each other.

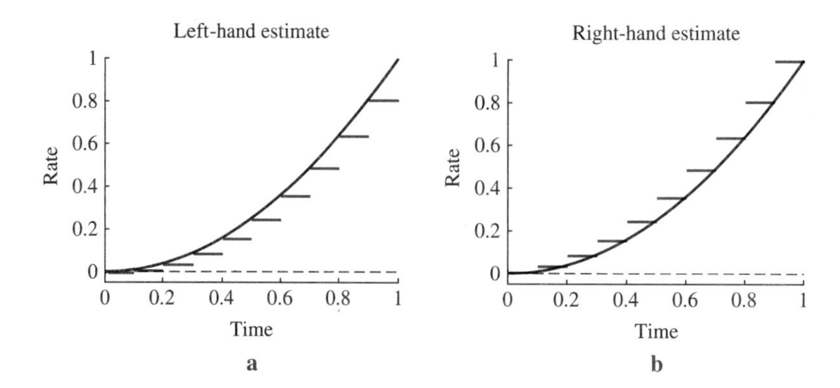

FIGURE 4.4.2
Left-hand and right-hand estimates of the integral: n = 10

4.4.2 Approximating Integrals in General

Our example was rather specific, working on a particular function (t^2) and two particular numbers of intervals (5 and 10). We will write out the estimates first for a general number of intervals n, and then for a general function f.

Generalizing this method of approximating integrals is easier if we use **summation notation** to write the expressions more compactly. As an example of the notation, we write

$$\sum_{i=1}^{3} x_i = x_1 + x_2 + x_3.$$

The character Σ is a capital Greek "sigma," standing for "sum." This expression is read "the sum from i equal 1 to 3 of x sub i." This means that we add up x_i for i taking on values from 1 to 3. The letter i is the **index**. The values x_1, x_2, and x_3 are called **terms**.

Example 4.4.1 Summation Notation

If $x_1 = 5$, $x_2 = 3$, and $x_3 = 1$, then

$$\sum_{i=1}^{3} x_i = x_1 + x_2 + x_3 = 5 + 3 + 1 = 9.$$

In this case,

$$\sum_{i=1}^{3} x_i^2 = x_1^2 + x_2^2 + x_3^2 = 5^2 + 3^2 + 1^2 = 35.$$

To begin generalizing the method for estimating integrals, let n denote the number of intervals to be used. Let t_i be the time at the end of the ith interval. With $n = 5$, we found that $t_1 = 0.2$ and $t_2 = 0.4$ and so on. Set t_0 to be the time at the beginning, so $t_0 = 0.0$ in this case. Finally, set Δt to be the length of the intervals, or 0.2 when $n = 5$ (Figure 4.4.3). We can then write the left-hand and right-hand estimates of volume as sums,

$$\begin{aligned} I_l &= 0.0^2(0.2) + 0.2^2(0.2) + 0.4^2(0.2) + 0.6^2(0.2) + 0.8^2(0.2) \\ &= t_0^2\Delta t + t_1^2\Delta t + t_2^2\Delta t + t_3^2\Delta t + t_4^2\Delta t \end{aligned} \quad (4.4.1)$$

and

$$\begin{aligned} I_r &= 0.2^2(0.2) + 0.4^2(0.2) + 0.6^2(0.2) + 0.8^2(0.2) + 1.0^2(0.2) \\ &= t_1^2\Delta t + t_2^2\Delta t + t_3^2\Delta t + t_4^2\Delta t + t_5^2\Delta t. \end{aligned} \quad (4.4.2)$$

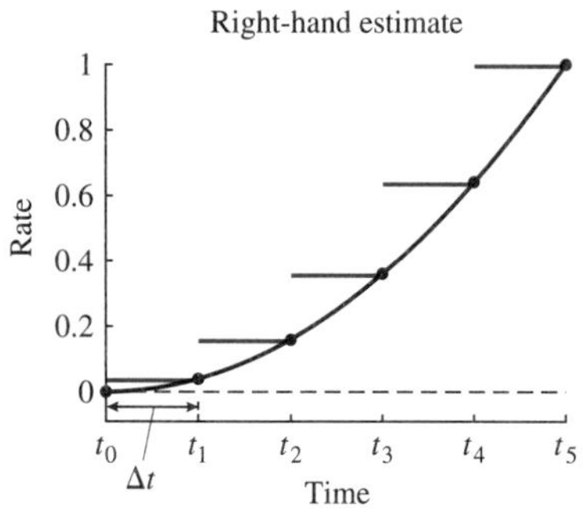

FIGURE 4.4.3
Writing the left-hand and right-hand estimates as sums: n = 5

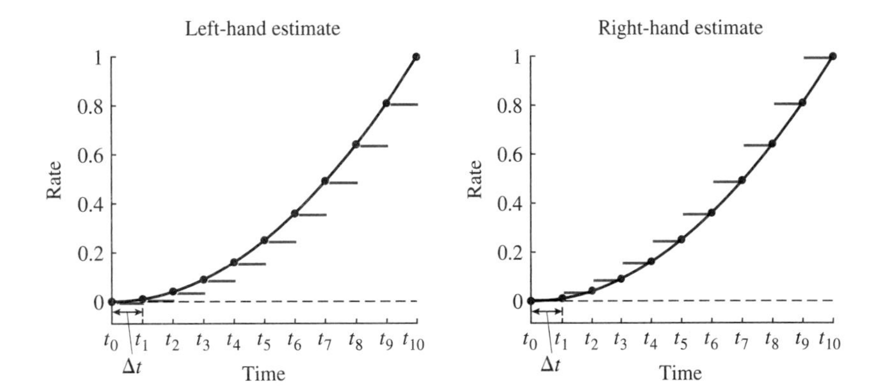

FIGURE 4.4.4
Writing the left-hand and right-hand estimates as sums: n = 10

We can use this notation to rewrite our expressions for I_l and I_r as

$$I_l = \sum_{i=0}^{4} t_i^2 \Delta t$$

$$I_r = \sum_{i=1}^{5} t_i^2 \Delta t.$$

These equations are a convenient shorthand for Equations 4.4.1 and 4.4.2 and are called **Riemann sums**.

With this notation, we can quickly write down what happens if we break the interval into 10 pieces. The length Δt is 0.1, and $t_1 = 0.1$, $t_2 = 0.2$ up to $t_{10} = 1.0$ (Figure 4.4.4). We can write our estimates as

$$I_l = \sum_{i=0}^{9} t_i^2 \Delta t$$

$$I_r = \sum_{i=1}^{10} t_i^2 \Delta t.$$

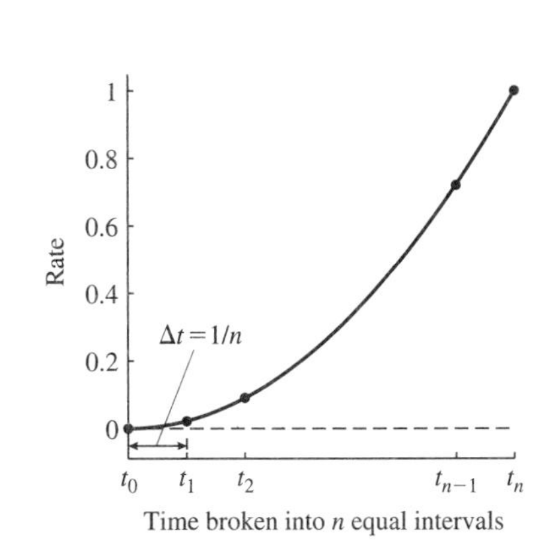

FIGURE 4.4.5
Writing the left-hand and right-hand estimates as sums: general case

The only thing that has changed are the number of terms added up. We are adding up twice as many terms, but each is smaller because Δt is half as big. Again, these Riemann sums should approximate the total change during this time interval.

Each term in this sum corresponds to one piece of the step function. The first term is the approximate total amount that entered during the first time interval of length Δt. In the left-hand estimate, this term is the product of t_0^2, the rate at the beginning of this interval, times the width of the interval Δt.

We are now ready to write down the general formula. Instead of 5 or 10, we plug n into our formulas for I_l and I_r. Breaking 0 to 1 into n intervals produces intervals of length $\Delta t = \frac{1}{n}$. The times are $t_i = i\,\Delta t$. For instance, $t_1 = \Delta t$, $t_2 = 2\Delta t$ up to $t_n = 1.0$ (Figure 4.4.5). We can write

$$I_l = \sum_{i=0}^{n-1} t_i^2 \Delta t$$

$$I_r = \sum_{i=1}^{n} t_i^2 \Delta t$$

for the Riemann sums.

In Table 4.4.1 we show the results of computing these estimates for different values of n. The last column denoted by I_a is the average of I_l and I_r and converges rapidly to 0.333333 (Exercises 17–20). We call this the **averaged estimate**.

Table 4.4.1 Estimates of the integral with different n

n	Δt	I_l	I_r	I_a
5	0.200	0.240000	0.440000	0.340000
10	0.100	0.285000	0.385000	0.335000
20	0.050	0.308750	0.358750	0.333750
30	0.033	0.316852	0.350185	0.333519
40	0.025	0.320938	0.345938	0.333438
50	0.020	0.323400	0.343400	0.333400
100	0.010	0.328350	0.338350	0.333350
500	0.002	0.332334	0.334334	0.333334
1000	0.001	0.332833	0.333833	0.333333

Example 4.4.2 Finding Left-Hand, Right-Hand, and Averaged Estimates for an Exponential Function

Suppose we wish to find the amount of chemical P produced after 2.0 seconds in a reaction that obeys

$$\frac{dP}{dt} = e^{-t}.$$

(Example 4.1.1). Using $n = 10$ steps, the interval from $t = 0$ to $t = 2$ breaks into intervals of width $\Delta t = 0.2$ (Figure 4.4.6).

	Left-Hand Estimate			Right-Hand Estimate		
Time Interval	Rate During Interval	Influx During Interval	Net Influx	Rate During Interval	Influx During Interval	Net Influx
0.0–0.2	1.000	0.200	0.200	0.819	0.164	0.164
0.2–0.4	0.819	0.164	0.364	0.670	0.134	0.298
0.4–0.6	0.670	0.134	0.498	0.549	0.110	0.408
0.6–0.8	0.549	0.110	0.608	0.449	0.090	0.497
0.8–1.0	0.449	0.090	0.697	0.368	0.074	0.571
1.0–1.2	0.368	0.074	0.771	0.301	0.060	0.631
1.2–1.4	0.301	0.060	0.831	0.247	0.049	0.681
1.4–1.6	0.248	0.049	0.881	0.202	0.040	0.721
1.6–1.8	0.202	0.040	0.921	0.165	0.033	0.754
1.8–2.0	0.165	0.033	0.954	0.135	0.027	0.781

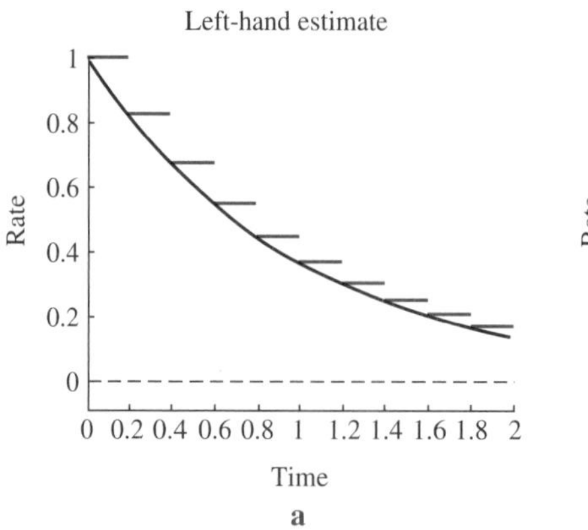

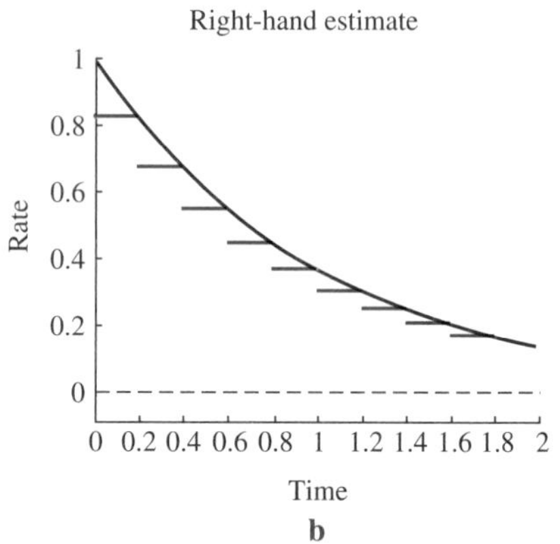

FIGURE 4.4.6
Left-hand and right-hand estimates of the integral

Using the substitution $u = -t$, we can evaluate the exact amount produced as a definite integral, finding

$$\int_0^2 e^{-t} dt = \int_0^{-2} -e^u du$$

$$= \int_{-2}^0 e^u du$$

$$= e^u|_{-2}^0 = 1 - e^{-2} = 0.865.$$

The averaged estimate, found by averaging the left-hand and right-hand estimates, is $\frac{0.954 + 0.781}{2} = 0.868$, quite close to the exact answer.

4.4.3 The Definite Integral

We have used sums to approximate the total volume that has been added according to the differential equation

$$\frac{dV}{dt} = t^2$$

during the time interval from $t = 0$ and $t = 1$. The general problem is to find the total change during the time interval from $t = a$ to $t = b$ when the state variable M obeys the pure-time differential equation

$$\frac{dM}{dt} = f(t).$$

Sums that approximate the integral of the function f are then

$$I_l = \sum_{i=0}^{n-1} f(t_i)\Delta t \quad (4.4.3)$$

and

$$I_r = \sum_{i=1}^{n} f(t_i)\Delta t. \quad (4.4.4)$$

The **Riemann integral** or **definite integral** is defined as the **limit** of I_r or I_l as $n \to \infty$. For example,

$$\int_0^1 t^2 dt = \lim_{n\to\infty} \sum_{i=0}^{n-1} t_i^2 \Delta t \quad (4.4.5)$$

$$= \lim_{n\to\infty} \sum_{i=1}^{n} t_i^2 \Delta t \quad (4.4.6)$$

where $\Delta t = \frac{1}{n}$. The expression

$$\int_0^1 t^2 dt$$

is read "the integral from 0 to 1 of $t^2 dt$." The little numbers on the integral indicate the interval to be integrated over and are called the **limits of integration**. In this case, we are integrating (finding the total amount that entered) from time 0 to time 1. The dt in the integral can be thought of as the width of an infinitesimally small interval Δt. When limits of integration are present, the expression is called a **definite integral**. Remember that the definite integral is a number and the indefinite integral is a function. In Section 4.5 we will see the connection between them.

In general, we substitute in limits of integration from a to b and a function f.

Definition 4.2 The Riemann integral of a function f on the interval from a to b is

$$\int_a^b f(t)dt = \lim_{n\to\infty} \sum_{i=0}^{n-1} f(t_i)\Delta t = \lim_{n\to\infty} \sum_{i=1}^{n} f(t_i)\Delta t$$

where the values $t_0, \ldots, t_n$ break the interval from a to b into n equal pieces of length

$$\Delta t = \frac{b-a}{n}.$$

The elements of this definition are illustrated in Figure 4.4.7.

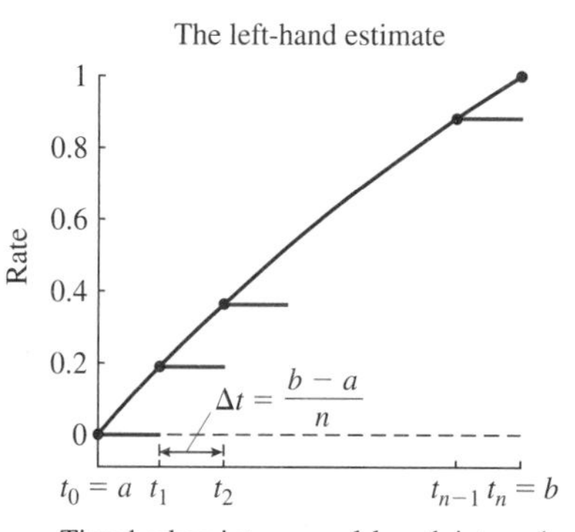

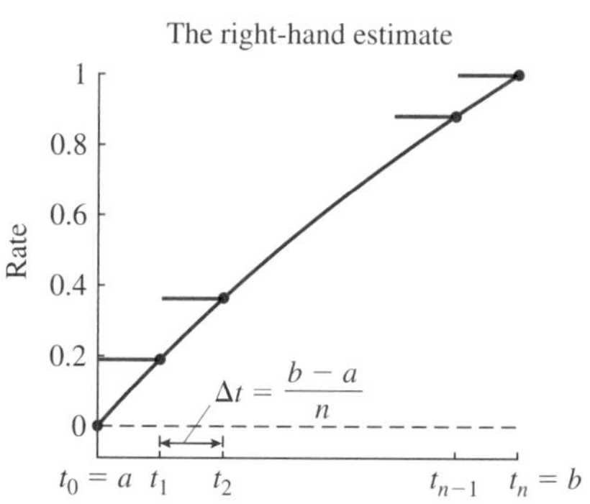

FIGURE 4.4.7
The Riemann integral in general

The steps to compute the Riemann sums that approximate the definite integral are given in the following algorithm.

Algorithm 4.3 Computing Riemann Sums

To evaluate the Riemann sums to approximate the integral $\int_a^b f(t)dt$ using n intervals.

1. Find the step size $\Delta t = \frac{b-a}{n}$.
2. Find the endpoints of the intervals as $t_0 = a$, $t_1 = a + \Delta t$, $t_2 = a + 2\Delta t$ up to $t_n = b$.
3. To evaluate the left-hand estimate, compute
$$I_l = \sum_{i=0}^{n-1} f(t_i)\Delta t.$$
4. To evaluate the right-hand estimate, compute
$$I_r = \sum_{i=1}^{n} f(t_i)\Delta t.$$

There are a couple of things that need to be proved regarding the definition of the Riemann integral. First, we should show that the Riemann sums really have limits as n approaches infinity. Second, we should show that the limits of the left-hand and right-hand Riemann sums are equal. For our purposes, we will believe these statements. The proofs are rather difficult and can be found in more advanced Calculus texts. The Riemann integral works for functions that are continuous at all but a finite number of points. In fact, as long as the function f does not approach infinity at any point, it is difficult to find functions for which either of these statements is false.

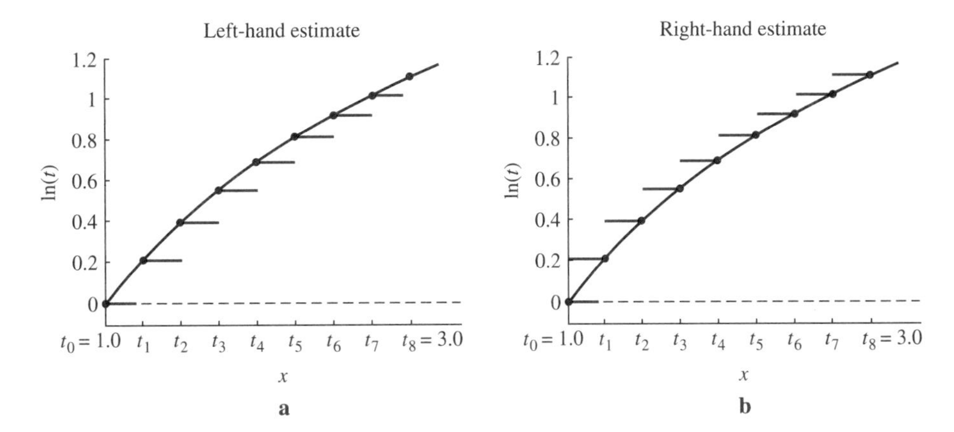

FIGURE 4.4.8
Approximating an integral

Example 4.4.3 Using Riemann Sums to Approximate an Integral

Suppose we wish to evaluate

$$A = \int_1^3 \ln(t)dt$$

by breaking the interval from $t = 1.0$ to $t = 3.0$ into $n = 8$ pieces. (Figure 4.4.8)

We must find the step size (width Δt of each interval), find the endpoints of the eight intervals, and evaluate the sum using the left-hand or right-hand estimate. The length of each piece is

$$\Delta t = \frac{3.0 - 1.0}{8} = 0.25.$$

We next find the values $t_0, \ldots, t_8$ that break the interval into the eight equal pieces as

$$t_0 = 1.0, t_1 = 1.25, t_2 = 1.5, t_3 = 1.75, t_4 = 2.0, t_5 = 2.25, t_6 = 2.5, t_7 = 2.75, t_8 = 3.0.$$

To use the left-hand estimate, we evaluate

$$\begin{aligned} I_l &= \sum_{i=0}^{7} \ln(t_i)\Delta t \\ &= \ln(1.0) \cdot 0.25 + \ln(1.25) \cdot 0.25 + \ln(1.5) \cdot 0.25 + \ln(1.75) \cdot 0.25 + \\ &\quad \ln(2.0) \cdot 0.25 + \ln(2.25) \cdot 0.25 + \ln(2.5) \cdot 0.25 + \ln(2.75) \cdot 0.25 \\ &= 1.155. \end{aligned}$$

To use the right-hand estimate, we evaluate

$$\begin{aligned} I_l &= \sum_{i=1}^{8} \ln(t_i)\Delta t \\ &= \ln(1.25) \cdot 0.25 + \ln(1.5) \cdot 0.25 + \ln(1.75) \cdot 0.25 + \ln(2.0) \cdot 0.25 + \\ &\quad \ln(2.25) \cdot 0.25 + \ln(2.5) \cdot 0.25 + \ln(2.75) \cdot 0.25 + \ln(3.0) \cdot 0.25 + \\ &= 1.4297. \end{aligned}$$

(Figure 4.3.8)

Example 4.4.4 Riemann Sums and Euler's Method

Consider using Euler's method to estimate $V(1)$ using $\Delta t = 0.2$ starting from the initial condition $V(0) = 0$, assuming that V follows the differential equation

$$\frac{dV}{dt} = t^2.$$

We find that

$$\hat{V}(0.2) = V(0) + 0.0^2 \cdot 0.2 = 0.0$$
$$\hat{V}(0.4) = \hat{V}(0.2) + 0.2^2 \cdot 0.2 = 0.008$$
$$\hat{V}(0.6) = \hat{V}(0.4) + 0.4^2 \cdot 0.2 = 0.04$$
$$\hat{V}(0.8) = \hat{V}(0.6) + 0.6^2 \cdot 0.2 = 0.112$$
$$\hat{V}(1.0) = \hat{V}(0.8) + 0.8^2 \cdot 0.2 = 0.240.$$

Euler's method gives exactly the same answer as the left-hand estimate because both methods make the same approximation: they assume that the rate during an interval is equal to the rate at the beginning of the interval.

Summary We have approximated integrals with **Riemann sums**, expressing our results in **summation notation**. By making the sums correspond more and more closely to the function, the estimates get more and more accurate. The **definite integral** is defined as the limit of the Riemann sums between particular **limits of integration**.

4.4 Exercises

Mathematical Techniques

1–4 ▪ Evaluate the following sums.

1. $\sum_{i=1}^{5} x_i$, for $x_i = \frac{1}{i}$.

2. $\sum_{j=1}^{6} y_j$, for $y_j = \frac{1}{2^j}$.

3. $\sum_{i=1}^{5} x_i^2$, for $x_i = \frac{1}{i}$.

4. $\sum_{k=1}^{7} z_k$, for $z_k = 2^k$.

5–8 ▪ Find the value of Δt and the values of $t_0, t_1, \ldots t_n$ when the interval from $t = a$ to $t = b$ is broken into n equal intervals of width Δt.

5. $a = 0, b = 2, n = 5$.

6. $a = 0, b = 2, n = 10$.

7. $a = 2, b = 3, n = 5$.

8. $a = 2, b = 3, n = 100$.

9–12 ▪ Find the left-hand and right-hand estimates for the definite integrals of the following functions.

9. $f(t) = 2t$, limits of integration 0 to 1, $n = 5$.

10. $f(t) = 2t$, limits of integration 0 to 2, $n = 5$.

11. $f(t) = 1 + t^3$, limits of integration 0 to 2, $n = 5$.

12. $f(t) = 1 + t^3$, limits of integration 0 to 1, $n = 5$.

13–16 ▪ Write the left-hand and right-hand Riemann sums for the following cases using summation notation.

13. $f(t) = 2t$, limits of integration 0 to 1, $n = 5$ (as in Exercise 9).

14. $f(t) = 2t$, limits of integration 0 to 2, $n = 5$ (as in Exercise 10).

15. $f(t) = 1 + t^3$, limits of integration 0 to 2, $n = 5$ (as in Exercise 11).

16. $f(t) = 1 + t^3$, limits of integration 0 to 1, $n = 5$ (as in Exercise 12).

17–20 ▪ Another way to think about the column I_a in Table 4.4.1 is to think that the value of the function is approximated by the average of the values at the beginning and the end of the time interval. We pretend that the rate of change is $\frac{f(t_{i+1}) + f(t_i)}{2}$ during the interval from t_i to t_{i+1}. Use this method to estimate the following integrals. In each case,

a. Draw a graph illustrating the estimate.

b. Write an expression for I_a using summation notation.

c. Compute the sum.

17. $f(t) = 2t$, limits of integration 0 to 1, $n = 5$ (as in Exercise 9).

18. $f(t) = 2t$, limits of integration 0 to 2, $n = 5$ (as in Exercise 10).

19. $f(t) = 1 + t^3$, limits of integration 0 to 2, $n = 5$ (as in Exercise 11).

20. $f(t) = 1 + t^3$, limits of integration 0 to 1, $n = 5$ (as in Exercise 12).

21–24 ▪ One other estimate of the integral, called I_m, can be computed by pretending that the value during the interval from t_i to t_{i+1} is the value of the function at the midpoint, or $f[(t_{i+1} + t_i)/2]$. Use this method to estimate the following integrals. In each case,

a. Draw a graph illustrating this estimate for $n = 5$. Make sure you see the difference from I_a.

b. Write an expression for I_m using summation notation.

c. Compute the sum.

21. $f(t) = 2t$, limits of integration 0 to 1, $n = 5$ (as in Exercise 9).

22. $f(t) = 2t$, limits of integration 0 to 2, $n = 5$ (as in Exercise 10).

23. $f(t) = 1 + t^3$, limits of integration 0 to 2, $n = 5$ (as in Exercise 11).

24. $f(t) = 1 + t^3$, limits of integration 0 to 1, $n = 5$ (as in Exercise 12).

Applications

25–28 ▪ Use summation notation and find the total number of offspring for each of the following organisms.

25. The organism has 2 offspring in year 1, 3 offspring in year 2, 5 offspring in year 3, 4 offspring in year 4, and 1 offspring in year 5.

26. The organism has 0 offspring in year 1, 8 offspring in year 2, 15 offspring in year 3, 24 offspring in year 4, 31 offspring in year 5, 11 offspring in year 6, and 3 offspring in year 7.

27. The organism has $B_i = i(6 - i)$ offspring in years 0 through 6.

28. The organism has $B_i = \frac{i(i+1)}{2} + 4$ offspring in years 0 through 7.

29–32 ▪ Use Euler's method to estimate the solutions of the following differential equations with the following parameters. Suppose that $V(0) = 0$ in each case. Your answer should exactly match one of the estimates in Exercise 12.

29. $\frac{dV}{dt} = 2t$, estimate $V(1)$ using $\Delta t = 0.2$ (as in Exercise 9).

30. $\frac{dV}{dt} = 2t$, estimate $V(2)$ using $\Delta t = 0.4$ (as in Exercise 10).

31. $\frac{dV}{dt} = 1 + t^3$, estimate $V(2)$ using $\Delta t = 0.4$ (as in Exercise 11).

32. $\frac{dV}{dt} = 1 + t^3$, estimate $V(1)$ using $\Delta t = 0.2$ (as in Exercise 12).

33–36 ▪ Suppose the speed of a bee is given in the following table.

Time (s)	Speed (cm/s)
0.0	127.0
1.0	122.0
2.0	118.0
3.0	115.0
4.0	113.0
5.0	112.0
6.0	112.0
7.0	113.0
8.0	116.0
9.0	120.0
10.0	125.0

33. Using the measurements on even-numbered seconds, find the left-hand and right-hand estimates for the distance the bee moved during the experiment.

34. Using all the measurements, find the left-hand and right-hand estimates for the distance the bee moved during the experiment.

35. Use the method of the last column of Table 4.4.1 (or Exercises 17–20) to estimate the distance moved (using all measurements).

36. Figure out a way to use the measurements on odd-numbered seconds to estimate the distance the bee moved during the experiment. Think about the method in Exercises 21–24.

37–40 ▪ Biologists measure the number of aspen that germinate in four sites over eight years, but can only measure two sites per year. In the following table, NA indicates that no measurement was made in that year. In each case, compare the estimate with twice the number that germinated in the four years studied. Why are they different?

Year	Site 1	Site 2	Site 3	Site 4
1990	12	23	NA	NA
1991	NA	NA	34	10
1992	16	NA	NA	15
1993	17	21	NA	NA
1994	NA	NA	40	18
1995	NA	23	31	NA
1996	NA	27	NA	8
1997	13	NA	37	NA

The goal is to estimate the total number of aspen that germinated in each of the four sites during all eight years.

37. In site 1, estimate the total number of aspen using a modification of the left-hand estimate.

38. In site 2, estimate the total number of aspen using a modification of the left-hand estimate.

39. In site 3, estimate the total number of aspen using a modification of the right-hand estimate.

40. In site 4, the first and last years are both NA's. Come up with some variation on the left-hand or right-hand estimate to estimate the total number of aspen.

Computer Exercise

41. We will compare various methods used to estimate the solution of

$$\frac{dp}{dt} = \ln(1 + \sqrt{t} - t^3)$$

with $p(0) = 0$. We wish to find $p(1)$.

a. Graph the rate of change as a function of t.

b. Use the right-hand estimate with $\Delta t = 0.2$, 0.1, and 0.02.

c. Use the left-hand estimate with $\Delta t = 0.2$, 0.1, and 0.02.

d. Try to figure out from your graph why the two estimates are the same.

4.5 Definite and Indefinite Integrals

We now have two types of integral. The **indefinite integral** is a *function* that solves the pure-time differential equation

$$\frac{dF}{dt} = f(t).$$

We write

$$\int f(t)dt = F(t) + c$$

to indicate this solution, found by computing an antiderivative. This integral is **indefinite** because it includes an arbitrary constant c that must be computed from the initial conditions of the differential equation.

The **definite integral** is a *number*, the change in value between the two times represented by the limits of integration. If the rate of change is $f(t)$,

$$\text{the total change between } a \text{ and } b = \int_a^b f(t)dt.$$

The definite integral is defined as the limit of **Riemann sums**. What is the relationship between these two mathematical approaches? In this section, we derive the connection, called the **Fundamental Theorem of Calculus**.

4.5.1 The Fundamental Theorem of Calculus: Computing Definite Integrals with Indefinite Integrals

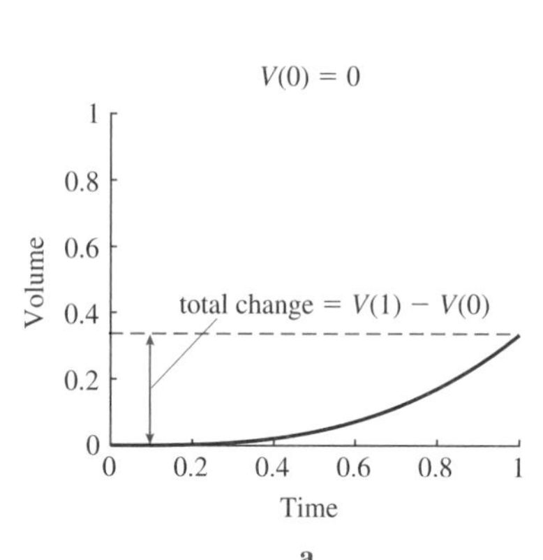

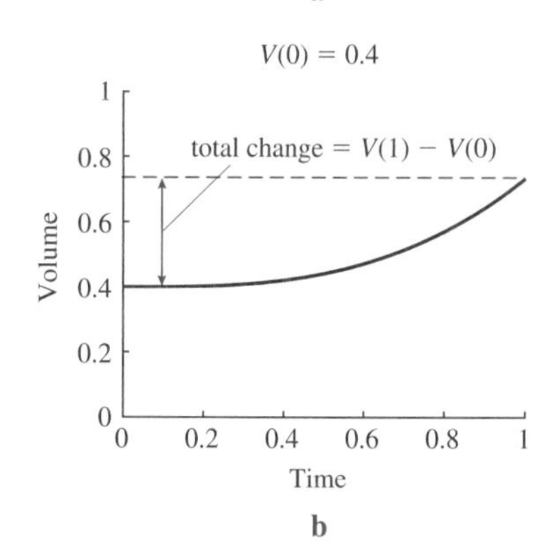

FIGURE 4.5.1
Total change

Consider again the pure-time differential equation for volume

$$\frac{dV}{dt} = t^2.$$

The change between times 0 and 1 is given by the definite integral

$$(\text{change between } t = 0 \text{ and } t = 1) = \int_0^1 t^2 dt.$$

We can use Riemann sums to approximate the definite integral, but the procedure requires a great deal of calculation.

If we think about the meaning of the differential equation, subtraction provides another way to express the change between the two times as

$$(\text{change between } t = 0 \text{ and } t = 1) = V(1) - V(0).$$

The total amount that entered between times 0 and 1 must equal the difference between the volumes at times 0 and 1 (Figure 4.5.1). Therefore, if V is a solution of the differential equation,

$$\int_0^1 t^2 dt = V(1) - V(0).$$

We can thus compute the value of the *definite integral* without any Riemann sums if we can use the *indefinite integral* to compute the solution $V(t)$ and the value $V(1)$.

Example 4.5.1 Using the Fundamental Theorem to Find Total Change

Suppose that $V(0) = 0$. We solve the differential equation $\frac{dV}{dt} = t^2$ with the indefinite integral and the power rule, finding

$$V(t) = \int t^2 dt = \frac{t^3}{3} + c.$$

With the initial condition $V(0) = 0$, $c = 0$, and the solution is

$$V(t) = \frac{t^3}{3}.$$

With this formula, we can find the total change during the first minute by subtracting, or

$$\begin{aligned}\text{total change between times 0 and 1} &= V(1) - V(0) \\ &= \frac{1^3}{3} - \frac{0^3}{3} = \frac{1}{3} = 0.3333.\end{aligned}$$

This matches the estimates we found using Riemann sums (Table 4.4.1), but took a lot less work. ▲

Example 4.5.2 Using the Fundamental Theorem to Find Total Change: Different Initial Conditions

Suppose instead that $V(0) = 0.4$. Then

$$V(t) = \frac{t^3}{3} + c$$

as before. With the initial condition $V(0) = 0.4$, we find $c = 0.4$. The solution is

$$V(t) = \frac{t^3}{3} + 0.4.$$

We can again find the total change during the first minute by subtracting, or

$$\begin{aligned}\text{total change between times 0 and 1} &= V(1) - V(0) \\ &= \left(\frac{1^3}{3} + 0.4\right) - \left(\frac{0^3}{3} + 0.4\right) = 0.3333.\end{aligned}$$

Although the two solutions do not match, they are *parallel* (Figure 4.5.1). Both increase by exactly the same amount. ▲

Example 4.5.3 Using the Fundamental Theorem to Find Total Change: Irrelevance of Initial Conditions

More generally, suppose we denote the volume at time 0 by the unknown value V_0. The solution of the differential equation is still

$$V(t) = \frac{t^3}{3} + c,$$

but now the arbitrary constant c must satisfy

$$V(0) = V_0 = \frac{0^3}{3} + c = c,$$

or $c = V_0$. The solution is

$$V(t) = \frac{t^3}{3} + V_0,$$

and the change in volume is

$$\begin{aligned}V(1) - V(0) &= \left(\frac{1^3}{3} + V_0\right) - \left(\frac{0^3}{3} + V_0\right) \\ &= \frac{1}{3} = 0.3333.\end{aligned}$$

▲

The total change does not depend on the initial condition of the pure-time differential equation. The initial condition is required to answer a question like "where are you after driving 5 miles due north?" But the *change* in position is clear; you are 5 miles north of where you started.

This idea is the essence of the **Fundamental Theorem of Calculus**. A definite integral can be evaluated in two ways:

1. By evaluating Riemann sums and taking the limit
2. By solving the appropriate pure-time differential equation and subtracting.

When the differential equation can be solved with the indefinite integral, the second method is much simpler. However, when the indefinite integral is impossible, Riemann sums provide an guaranteed, if laborious, alternative.

It is easier to write the Fundamental Theorem of Calculus, with one bit of new notation. To represent the change in the value of $F(x)$ between $x=a$ and $x=b$, we use the shorthand

$$F(x)|_a^b = F(b) - F(a).$$

Instead of reading this notation in a new way, we say "$F(b)$ minus $F(a)$."

Theorem 4.4 The Fundamental Theorem of Calculus

For any continuous function $f(x)$ and any indefinite integral

$$F(x) = \int f(x)dx,$$

$$\int_a^b f(x)dx = F(b) - F(a) = F(x)|_a^b.$$

The Fundamental Theorem applies to more than just continuous functions, including any function that you can graph (any function that is continuous at all but a finite number of points). This theorem is **fundamental** because it describes the relation between definite and indefinite integrals. In Section 4.5, we sketch a proof in the case where the function $f(x)$ is continuous. First, we learn how to use this theorem.

Example 4.5.4 Using the Fundamental Theorem of Calculus

To find $\int_1^2 \frac{1}{x}$, we find any indefinite integral of $\frac{1}{x}$ as

$$\int \frac{1}{x}dx = \ln(x).$$

Then

$$\int_1^2 \frac{1}{x} = \ln(x)|_1^2 = \ln(2) - \ln(1) = 0.693.$$

Example 4.5.5 Using the Fundamental Theorem of Calculus to Find How Far a Rock Falls

Suppose that the position of a rock $p(t)$ follows the differential equation

$$\frac{dp}{dt} = v(t) = -9.8t - 5.0$$

(as in "Solving Polynomial Differential Equations," Section 4.2) where t is measured in seconds and p is measured in meters. How far does the rock fall between $t=1$ and $t=3$? The total change in position is given by the definite integral of the rate of change of position

$$\int_1^3 v(t)dt.$$

According to the Fundamental Theorem of Calculus, we can compute this value by finding *any* indefinite integral of $v(t)$ and subtracting the values at $t=1$ and $t=3$.

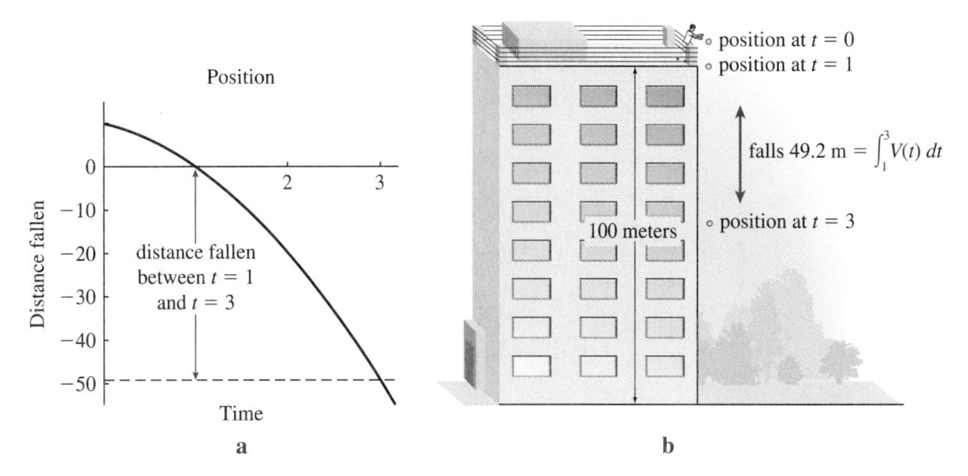

FIGURE 4.5.2
The position of a falling rock

One indefinite integral is

$$\int v(t)dt = \int (-9.8t - 5.0)dt$$
$$= -9.8\frac{t^2}{2} - 5.0t$$
$$= -4.9t^2 - 5.0t,$$

where we set the arbitrary constant to $c = 0$ for convenience. Then

$$\text{total change in position between } t = 1 \text{ and } t = 3 = -4.9t^2 - 5.0t|_1^3$$
$$= (-4.9 \cdot 3^2 - 5.0 \cdot 3) - (-4.9 \cdot 1^2 - 5.0 \cdot 1)$$
$$= -49.2.$$

The rock will have fallen 49.2 m during this time (Figure 4.5.2).

Example 4.5.6 Using the Fundamental Theorem of Calculus to Find How Much a Fish Grows

Suppose the change in length of a fish follows the equation

$$\frac{dL}{dt} = 6.48e^{-0.09t}$$

(Example 4.3.6) with t measured as age in years and L measured in centimeters. How much does the fish grow between ages 2 and 5? The total change is

$$\int_2^5 6.48e^{-0.09t}dt.$$

To evaluate, we find the indefinite integral

$$\int 6.48e^{-0.09t}dt = -72.0e^{-0.09t}$$

(the answer we found using substitution in Equation 4.3.6), but with the arbitrary constant set to 0 for convenience. The *change* in length is then

$$-72.0e^{-0.09t}|_2^5 = -72.0e^{-0.09 \cdot 5} - (-72.0e^{-0.09 \cdot 2}) = -45.9 - (-60.1) = 14.2.$$

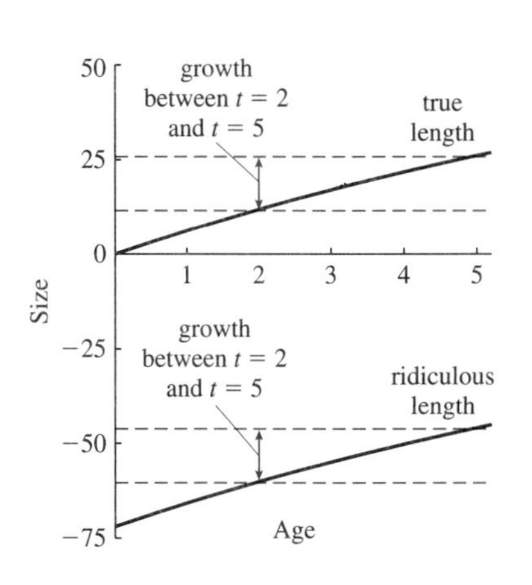

FIGURE 4.5.3
The growth of a walleye

Neither of the component terms (−45.9 and −60.1) makes biological sense. Their difference, however, gives the correct answer (Figure 4.5.3).

The solution using the realistic initial condition $L(0) = 0$ is

$$L(t) = 72.0 - 72.0e^{-0.09t} = 72.0(1 - e^{-0.09t})$$

(Equation 4.3.7). The change is

$$\begin{aligned} L(t)|_2^5 &= L(5) - L(2) \\ &= 72.0(1 - e^{-0.09\cdot 5}) - 72.0(1 - e^{-0.09\cdot 2}) \\ &= 26.1 - 11.9 = 14.2. \end{aligned}$$

The difference 14.2 is now found by subtracting the actual lengths at ages 5 and 2 (Figure 4.5.3). ▲

Integration by parts (Section 4.3) also works on definite integrals. In addition to carrying along the limits of integration, we must use the Fundamental Theorem of Calculus to include those limits in the term $u(x)v(x)$. The formula for definite integration by parts is

$$\int_a^b u(x)\frac{dv}{dx}dx = u(x)v(x)|_a^b - \int_a^b v(x)\frac{du}{dx}dx.$$

Example 4.5.7 An Example of Definite Integration by Parts

Suppose the amount of chemical produced in a reaction obeys

$$\frac{dC}{dt} = te^{-t}.$$

If there is no product at time 0, or $C(0) = 0$, how much will there be at time $t = 2$? The amount produced is the integral of the rate of production, or

$$C(2) = \int_0^2 te^{-t}dt.$$

To integrate by parts, we set $u(t) = t$ and $\frac{dv}{dt} = e^{-t}$. Then $\frac{du}{dt} = 1$ and $v(t) = -e^{-t}$. Therefore,

$$\begin{aligned} \int_0^2 te^{-t}dt &= -te^{-t}|_0^2 - \int_0^2 (-e^{-t})dt \\ &= -te^{-t}|_0^2 - e^{-t}|_0^2 \\ &= -(2e^{-2} - 0e^0) - (e^{-2} - e^0) = 1 - 3e^{-2} = 0.594. \end{aligned}$$ ▲

4.5.2 The Summation Property of the Definite Integral

What happens if the function we want to integrate takes on both positive and negative values? If our function represents the rate at which water enters a vessel, a positive value means that water is entering and a negative value means that water is leaving. Suppose

$$\frac{dV}{dt} = t^2 - t$$

where t is measured in seconds and V is measured in cm^3. Water flows out during the first second when the rate is negative, and it flows in during the next second when the rate is positive (Figure 4.5.4a).

To find the total change in volume from time 0 until time 1, we use the indefinite integral and the Fundamental Theorem of Calculus to compute

$$\begin{aligned} \int_0^1 t^2 - t\,dt &= \left(\frac{t^3}{3} - \frac{t^2}{2}\right)|_0^1 \\ &= \left(\frac{1}{3} - \frac{1}{2}\right) - (0 - 0) \\ &= -0.167\text{ cm}^3. \end{aligned}$$

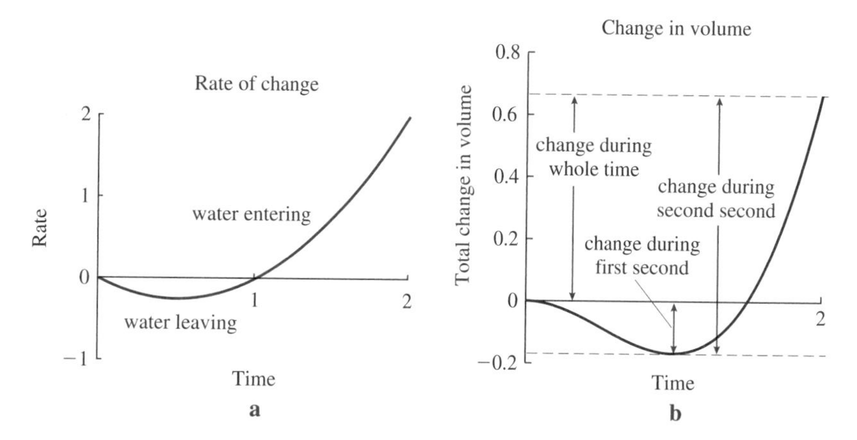

FIGURE 4.5.4
Positive and negative rates

This means that 0.167 cm^3 of water *left* the vessel during the first second. During the next second, the change in volume is

$$\int_1^2 (t^2 - t)\, dt = \left(\frac{t^3}{3} - \frac{t^2}{2}\right)\Big|_1^2$$
$$= \left(\frac{8}{3} - 2\right) - \left(\frac{1}{3} - \frac{1}{2}\right)$$
$$= 0.833 \text{ cm}^3.$$

This means that 0.833 cm^3 *entered* the vessel during the second second.

What happens during the whole period between $t = 0$ and $t = 2$? Because 0.167 cm^3 left during the first second and 0.833 cm^3 entered during the second,

$$\begin{aligned}(\text{change between } t = 0 \text{ and } t = 2) &= (\text{change between } t = 0 \text{ and } t = 1)\\ &\quad + (\text{change between } t = 1 \text{ and } t = 2)\\ &= -0.167 + 0.833 = 0.666.\end{aligned}$$

Alternatively, evaluating the definite integral from $t = 0$ to $t = 2$,

$$\int_0^2 (t^2 - t)\, dt = \left(\frac{t^3}{3} - \frac{t^2}{2}\right)\Big|_0^2$$
$$= \left(\frac{8}{3} - 2\right) - (0 - 0) = 0.667.$$

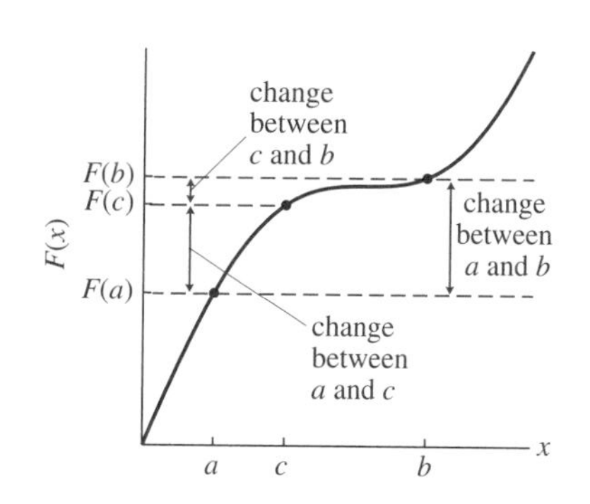

FIGURE 4.5.5
The summation property of the definite integral

This calculation points out an important property of the definite integral. The total change between times a and b is the change between a and some intermediate time c plus the change between time c and the final time b (Figure 4.5.5). This important and useful fact can be stated formally as a theorem.

Theorem 4.5 Summation Property of the Definite Integral

Suppose $f(x)$ is continuous on the interval from a to b, and that $a \le c \le b$. Then

$$\int_a^b f(x)\, dx = \int_a^c f(x)\, dx + \int_c^b f(x)\, dx.$$

The proof depends on a slightly more general definition of the Riemann integral based on breaking the interval from a to b into n intervals that need not be equal in length.

Example 4.5.8 The Summation Property Applied to a Falling Rock

If we wish to find how far the rock studied in Example 4.5.5 falls between $t = 1$ and $t = 3$, we can add the change of position between $t = 1$ and $t = 2$ to the change of

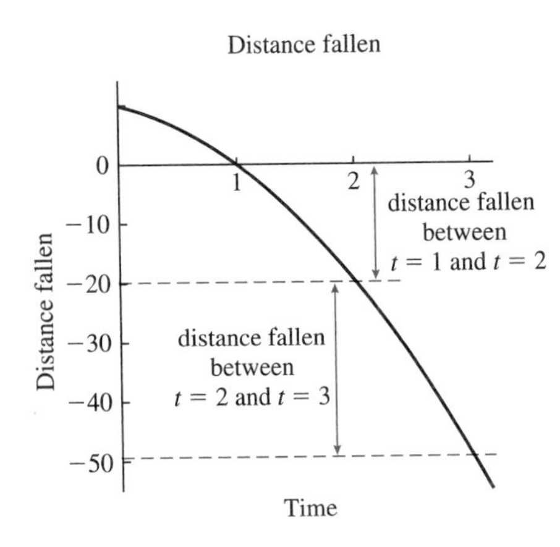

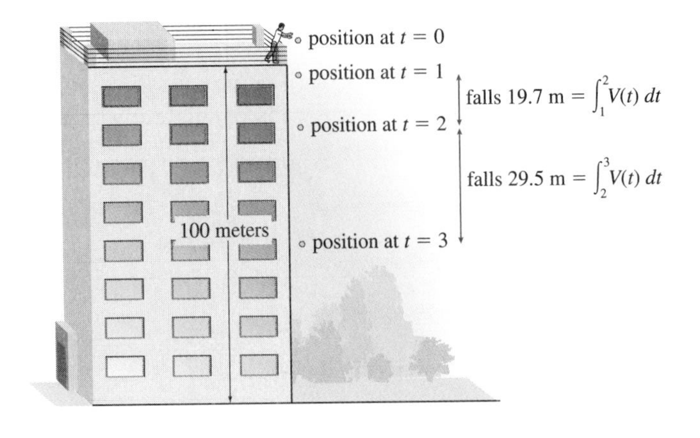

FIGURE 4.5.6
The summation property of distance traveled

position between $t = 2$ and $t = 3$. Between $t = 1$ and $t = 2$, the change of position is

$$\int_1^2 v(t)dt = \int_1^2 (-9.8t - 5.0)dt$$
$$= -4.9t^2 - 5.0t|_1^2$$
$$= (-4.9 \cdot 2^2 - 5.0 \cdot 2) - (-4.9 \cdot 1^2 - 5.0 \cdot 1) = -19.7$$

and between $t = 2$ and $t = 3$, the change of position is

$$\int_2^3 v(t)dt = \int_2^3 -9.8t - 5.0dt$$
$$= -4.9t^2 - 5.0t|_2^3$$
$$= (-4.9 \cdot 3^2 - 5.0 \cdot 3) - (-4.9 \cdot 2^2 - 5.0 \cdot 2) = -29.5.$$

The total change of position, found earlier to be −49.2 m, is the sum of −19.7 m and −29.5 m (Figure 4.5.6).

4.5.3 Euler's Method and the Fundamental Theorem of Calculus

Riemann sums provide a way to estimate a definite integral, while Euler's method provides a way to estimate the numerical solution of a differential equation. Because Euler's method corresponds to the left-hand Riemann sum (Example 4.4.4), we can use the Fundamental Theorem to **prove** that using Euler's method with smaller and smaller step sizes gives an accurate approximation of the solution of a differential equation.

Example 4.5.9 Riemann Sums and Euler's Method

Consider again finding a numerical solution to the differential equation

$$\frac{dP}{dt} = e^{-t}$$

with initial condition $P(0) = 0$ (from Example 4.1.7). To find $P(1)$, we can follow Algorithm 4.1. If we choose a step size of $\Delta t = 0.2$, then

t	pt$\hat{P}(t + \Delta t)$	$t + \Delta t$	pt$\hat{P}(t + \Delta t)$
0.0	$P(0) + P'(0)\Delta t = 0.0 + e^{-0} \cdot 0.2$	0.2	0.2
0.2	$\hat{P}(0.2) + P'(0.2)\Delta t = 0.2 + e^{-0.2} \cdot 0.2$	0.4	0.364
0.4	$\hat{P}(0.4) + P'(0.4)\Delta t = 0.364 + e^{-0.4} \cdot 0.2$	0.6	0.498
0.6	$\hat{P}(0.6) + P'(0.6)\Delta t = 0.498 + e^{-0.6} \cdot 0.2$	0.8	0.608
0.8	$\hat{P}(0.8) + P'(0.8)\Delta t = 0.608 + e^{-0.8} \cdot 0.2$	1.0	0.697

Euler's method corresponds to assuming that the rate of change is constant during each interval. As shown in the Figure 4.5.7, this corresponds exactly to the approximation used in the left-hand Riemann sum.

In this case, we can find the exact answer as

$$P(1) = \int_0^1 e^{-t} dt = -e^{-t}|_0^1 = 1 - e^{-1} = 0.63212.$$

Example 4.5.10 Euler's Method Converges on the Correct Answer

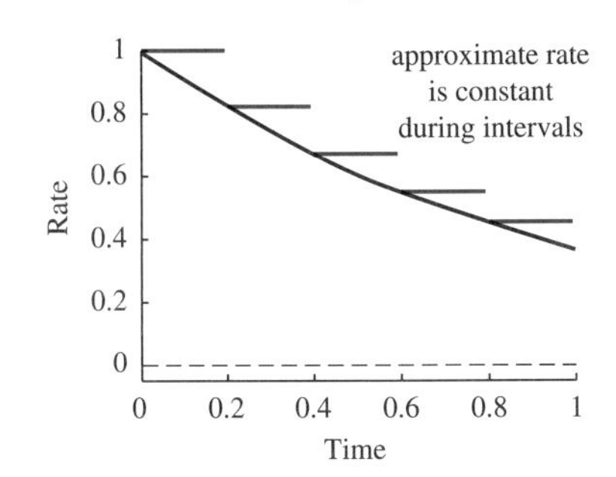

FIGURE 4.5.7
Euler's method applied to a pure-time differential equation matches the left-hand estimate

For the case in Example 4.5.9, choosing smaller and smaller step sizes gives the following results.

Δt	$\hat{P}(1)$
0.1	0.66425
0.01	0.63529
0.001	0.63244
0.0001	0.63215

As they must, the values converge to the exact answer of $1 - e^{-1} = 0.63212$ as the step size becomes smaller.

Example 4.5.11 Solving the Impossible

Consider finding a numerical solution to the differential equation

$$\frac{dP}{dt} = e^{-t^2}$$

with initial condition $P(0) = 0$. Although similar to the differential equation in Example 4.5.9, it is impossible to evaluate the integral in terms of elementary functions and compute the exact answer. If we choose a step size of $\Delta t = 0.2$, then we get

t	$\hat{P}(t+\Delta t)$	$t+\Delta t$	$\hat{P}(t+\Delta t)$
0.0	$P(0) + P'(0)\Delta t = 0.0 + e^{-0^2} \cdot 0.2$	0.2	0.2
0.2	$\hat{P}(0.2) + P'(0.2)\Delta t = 0.2 + e^{-0.2^2} \cdot 0.2$	0.4	0.392
0.4	$\hat{P}(0.4) + P'(0.4)\Delta t = 0.392 + e^{-0.4^2} \cdot 0.2$	0.6	0.563
0.6	$\hat{P}(0.6) + P'(0.6)\Delta t = 0.563 + e^{-0.6^2} \cdot 0.2$	0.8	0.702
0.8	$\hat{P}(0.8) + P'(0.8)\Delta t = 0.702 + e^{-0.8^2} \cdot 0.2$	1.0	0.808

Choosing smaller and smaller step sizes gives the following results.

Δt	$\hat{P}(1)$
0.1	0.77782
0.01	0.74998
0.001	0.74714
0.0001	0.74686

Evaluation on an accurate computer gives an answer of 0.74682 to five decimal places.

Although the Fundamental Theorem of Calculus guarantees that Euler's method will converge to the correct answer, it tends to converge rather slowly. A project at the end of this chapter gives a chance to explore some more advanced methods that combine the proven accuracy of Euler's method with much more rapid convergence.

4.5.4 Proof of the Fundamental Theorem of Calculus

Theorem 4.4 **The Fundamental Theorem of Calculus**

For any continuous function $f(x)$ and any indefinite integral

$$F(x) = \int f(x)dx,$$

$$\int_a^b f(x)dx = F(b) - F(a) = F(x)|_a^b.$$

Proof: First we show that the function

$$G(x) = \int_a^x f(s)ds$$

is an antiderivative of $f(x)$. The derivative of $G(x)$ is

$$\lim_{h \to 0} \frac{G(x+h) - G(x)}{h}.$$

The quantity

$$\frac{G(x+h) - G(x)}{h} = \frac{1}{h}\left[\int_a^{x+h} f(s)ds - \int_a^x f(s)ds\right]$$

$$= \frac{1}{h}\int_x^{x+h} f(s)ds,$$

where we used the summation property of the definite integral (Theorem 4.5). Because f is continuous, it takes on a minimum value (which we call f_{min}) and a maximum value (which we call f_{max}) on the interval between x and $x+h$. Then

$$f_{min} \le \frac{1}{h}\int_x^{x+h} f(s)ds \le f_{max}.$$

because $f_{min} \le f(s)$ and $f(s) \le f_{max}$. As $h \to 0$, we know that $f_{min} \to f(x)$ and $f_{max} \to f(x)$ because $f(x)$ is continuous. Therefore,

$$\lim_{h \to 0} f_{min} \le \lim_{h \to 0} \frac{G(x+h) - G(x)}{h} \le \lim_{h \to 0} f_{max}$$

$$f(x) \le G'(x) \le f(x),$$

implying that $G'(x) = f(x)$. We have that $F'(x) = f(x)$ because $F(x)$ is an indefinite integral of $f(x)$. Therefore $F'(x) = G'(x)$, and the function $F(x) - G(x)$ has derivative equal to zero. The only functions with derivative equal to zero are constants, so $F(x) = G(x) + c$ for some constant c. Therefore,

$$F(b) - F(a) = G(b) - G(a) = G(b) = \int_a^b f(x)dx$$

where we used the fact that $G(a) = 0$ (an integral over a domain of length zero is zero). This proves the theorem.

Summary The **Fundamental Theorem of Calculus** describes the connection between definite and indefinite integrals; the definite integral is equal to the difference between the values of the indefinite integral at the limits of integration. The Fundamental Theorem simplifies calculations of total change from pure-time differential equations by eliminating the need to solve for the arbitrary constant. The **Summation Property of Definite Integrals** states that the total change over two time intervals is the sum of the changes in each. The equivalence between Euler's method and the left-hand Riemann sum thus guarantees that it converges to the correct solution of a pure-time differential equation as the step size approaches 0.

4.5 Exercises

Mathematical Techniques

1–4 ▪ Compute the following definite integrals and compare with your answer from the earlier problem.

1. $\int_0^1 2t\,dt$. Compare with Section 4.4, Exercises 9 and 17.
2. $\int_0^2 2t\,dt$. Compare with Section 4.4, Exercises 10 and 18.
3. $\int_0^2 1+t^3\,dt$. Compare with Section 4.4, Exercises 11 and 19.
4. $\int_0^1 1+t^3\,dt$. Compare with Section 4.4, Exercises 12 and 20.

5–18 ▪ Compute the following definite integrals.

5. $\int_0^1 7x^2\,dx$
6. $\int_0^1 10t^9+6t^5\,dt$
7. $\int_{-1}^2 72t+5\,dt$
8. $\int_{-2}^2 y^4+5y^3\,dy$
9. $\int_1^2 \frac{5}{x^3}\,dx$
10. $\int_1^4 3z^{\frac{3}{7}}\,dz$
11. $\int_1^8 \frac{2}{\sqrt[3]{t}}+3\,dt$
12. $\int_{1.2}^{2.4} 5z^{-1.2}-1.2\,dz$
13. $\int_2^3 \frac{3}{z^2}+\frac{z^2}{3}\,dz$
14. $\int_0^1 3e^x+2x^3\,dx$
15. $\int_1^4 e^x+\frac{1}{x}\,dx$
16. $\int_{-3}^{-1} \frac{2}{t}+\frac{t}{2}\,dt$
17. $\int_0^{\pi} 2\sin(x)+3\cos(x)\,dx$
18. $\int_{-\pi/2}^{\pi/2} x^2-20\sin(x)\,dx$

19–24 ▪ Compute the definite integrals of the following functions from $t=1$ to $t=2$, from $t=2$ to $t=3$, and finally from $t=1$ to $t=3$ to check the summation property of definite integrals.

19. $g(t)=t^2$.
20. $h(t)=1+t^3$.
21. $L(t)=\frac{5}{t^3}$.
22. $B(t)=3t^{\frac{3}{7}}$.
23. $F(t)=e^t+\frac{1}{t}$.
24. $G(t)=\frac{2}{t}+\frac{t}{2}$.

25–28 ▪ Another way to write the Fundamental Theorem of Calculus (sometimes called the First Fundamental Theorem of Calculus) relates the definite integral and the derivative. It states that if we treat the definite integral $\int_a^x f(s)ds$ as a function of x, then

$$\frac{d}{dx}\int_a^x f(s)\,ds = f(x)$$

for any value of a. Check this in the following cases by computing the definite integral and then taking its derivative.

25. $f(x)=x^2$ with $a=0$.
26. $f(x)=1+x^3$ with $a=1$.
27. $f(x)=\left(1+\frac{x}{2}\right)^4$ with $a=-1$.
28. $f(x)=(1+2x)^{-4}$ with $a=0$. Taking the derivative with the chain rule returns the original function.

Applications

29–34 ▪ Find the change in the state variable between the given times first by solving the differential equation with the given initial conditions and then by using the definite integral.

29. The change of position by a rock between times $t=1$ and $t=5$ with position following the differential equation $\frac{dp}{dt}=-9.8t-5.0$ and initial condition $p(0)=200$.
30. The amount a fish grows between ages $t=1$ and $t=5$ if it follows the differential equation $\frac{dL}{dt}=6.48e^{-0.09t}$ with initial condition $L(0)=5.0$.
31. The amount a fish grows between ages $t=0.5$ and $t=1.5$ if it follows the differential equation $\frac{dL}{dt}=64.3e^{-1.19t}$ with initial condition $L(0)=5.0$.
32. The number of new AIDS cases between 1985 and 1987 if the number of AIDS cases follows $\frac{dA}{dt}=523.8t^2$ with initial condition $A(0)=13,400$ and t measured in years since 1981.
33. The amount of chemical produced between times $t=5$ and $t=10$ if the amount P follows $\frac{dP}{dt}=\frac{5}{1+2.0t}$ with initial condition $P(0.0)=2.0$, and t is measured in minutes and P in moles.
34. The amount of chemical produced between times $t=5$ and $t=10$ if the amount P follows $\frac{dP}{dt}=5.0e^{-2.0t}$ with initial condition $P(0.0)=2.0$, and t is measured in minutes and P in moles.

35–38 ▪ Check the summation property for the solutions of the differential equations by showing that change in value of the whole interval is equal to the change during the first half of the interval plus the change during the second half of the interval.

35. The position of a rock obeys the differential equation $\frac{dp}{dt} = -9.8t - 5.0$ with initial condition $p(0) = 200$ (as in Exercise 29.) Show that the distance moved between times $t = 1$ and $t = 5$ is equal to the sum of the distance moved between $t = 1$ and $t = 3$ and the distance moved between $t = 3$ and $t = 5$.

36. The growth of a fish obeys the differential equation $\frac{dL}{dt} = 6.48e^{-0.09t}$ with initial condition $L(0) = 5.0$ (as in Exercise 30.) Show that the growth between times $t = 1$ and $t = 5$ is equal to the sum of the growth between $t = 1$ and $t = 3$ and the growth between $t = 3$ and $t = 5$.

37. The growth of a fish obeys the differential equation $\frac{dL}{dt} = 64.3e^{-1.19t}$ with initial condition $L(0) = 5.0$ (as in Exercise 31.) Show that the growth between times $t = 0.5$ and $t = 1.5$ is equal to the sum of the growth between $t = 0.5$ and $t = 1.0$ and the growth between $t = 1.0$ and $t = 1.5$.

38. The number of AIDS cases obeys $\frac{dA}{dt} = 523.8t^2$ with initial condition $A(0) = 13,400$ (as in Exercise 32.) Show that the number of new cases between times $t = 4$ and $t = 6$ is equal to the sum of the number of new cases between $t = 4$ and $t = 5$ and the number of new cases between $t = 5$ and $t = 6$.

39–40 ▪ Two rockets are shot from the ground. Each has a different upward acceleration, and a different amount of fuel. After the fuel runs out, each rocket falls with an acceleration of -9.8 m/s^2. For each rocket,

a. Write down and solve differential equations describing the velocity and position of the rocket while it still has fuel.

b. Find the velocity and height of the rocket when it runs out of fuel.

c. Write down and solve differential equations describing the velocity and position of the rocket after it has run out of fuel. What is the initial condition for each?

d. Find the maximum height reached by the rocket. Does it rise more with or without fuel. Why?

e. Find the velocity when it hits the ground.

39. The upward acceleration is 12.0 m/s^2 and it has 10 seconds worth of fuel.

40. The upward acceleration is 2.0 m/s^2 and it has 60 seconds worth of fuel.

Computer Exercise

41. Toward the end of the universe, acceleration due to gravity begins to break down. Suppose that

$$a = -9.8\frac{1}{1+t}$$

where time is measured in seconds after the beginning of the end. An object begins falling from 10.0 m above the ground.

a. Find the velocity at time t.

b. Find the position at time t.

c. Graph acceleration, velocity, and position on the same graph. Which of these measurements are integrals of each other?

d. When will this object hit the ground?

4.6 Applications of Integrals

In this section, we introduce several of the remarkable applications of the definite integral. First, we notice that the graph describing the Riemann sum can be interpreted geometrically as a way to approximate the **area under a curve**. In fact, geometric problems of this sort provided the motivation for the near discovery of the integral by Archimedes about 2000 years ago, long before Newton introduced the study of differential equations. The idea of chopping quantities into small bits, adding them up with Riemann sums, and computing exact answers with the definite integral has a remarkably broad range of other applications, including finding the **average value** of a function and finding the total mass from a **density**.

4.6.1 Integrals and Areas

Suppose we want to find the area under the curve $f(x) = x^2$ between $x = 0$ to $x = 1$ (Figure 4.6.1). We can approximate the area with little rectangles, producing a picture identical to those used to find the left-hand Riemann sum (Figure 4.4.1). In particular, when we break the area into five rectangles as shown, their areas match the values found in computing the left-hand estimate.

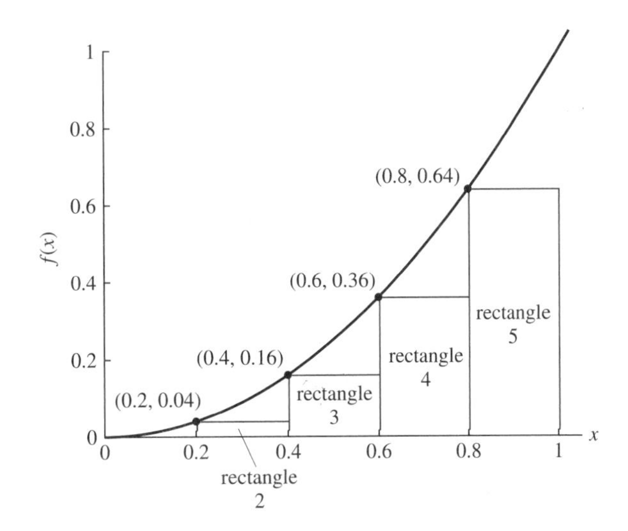

FIGURE 4.6.1
The area under $f(x) = x^2$

Rectangle	Base	Height	Area	Total
1	0.2	0.0	0.0	0.0
2	0.2	0.04	0.008	0.008
3	0.2	0.16	0.032	0.040
4	0.2	0.36	0.072	0.112
5	0.2	0.64	0.128	0.240

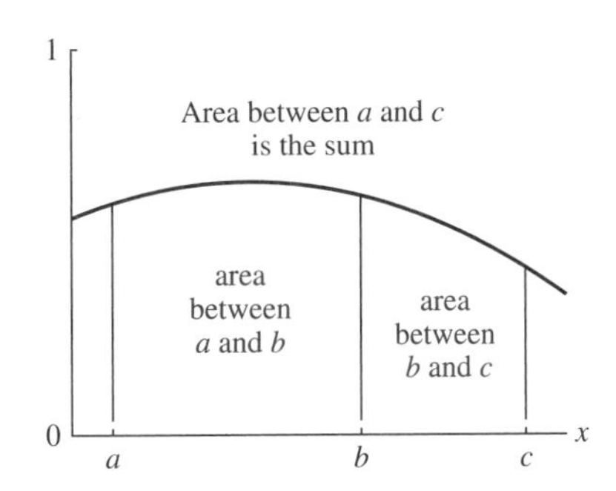

FIGURE 4.6.2
Areas and the summation property of definite integrals

Because the Riemann sums converge both to the area and to the definite integral,

$$(\text{area under } f(x) = x^2 \text{ between 0 and 1}) = \int_0^1 x^2 dx.$$

We have already computed this integral several times, finding the pleasingly simple answer of 1/3.

In general, we can find the area under the positive curve $f(x)$ from a to b by computing the definite integral between the limits of integration

$$(\text{area under } f(x) \text{ between } a \text{ and } b) = \int_a^b f(x)\, dx. \quad \textbf{(4.6.1)}$$

The summation property of the definite integral (Theorem 4.5) also has a convenient geometric interpretation (Figure 4.6.2). The area between a and c is the sum of the area between a and b and the area between b and c.

Areas are positive, but definite integrals can give negative values. How are these interpreted?

Example 4.6.1 What to Do with a Negative Area

The integral

$$\int_0^2 x^2 - x\, dx$$

does not give the sum of the two shaded areas in Figure 4.6.3, but *subtracts* the area below the x-axis from the area above the axis.

To find the total shaded area, it is necessary to integrate the **absolute value** of the function, or

$$\int_0^2 |x^2 - x|\, dx.$$

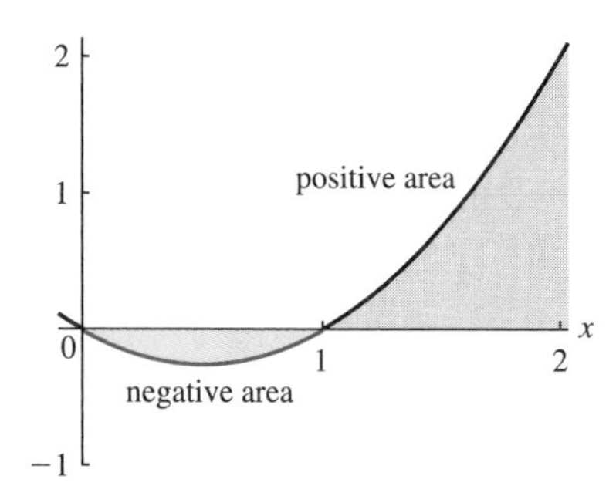

FIGURE 4.6.3
Positive and negative area

The only way to evaluate this is to find where the integrand is positive and negative. In this case, the function is negative between 0 and 1 and positive between 1 and 2. Therefore, $|x^2 - x| = x - x^2$ for $0 \le x \le 1$ and $|x^2 - x| = x^2 - x$ for $1 \le x \le 2$,

$$\int_0^2 |x^2 - x|\,dx = \int_0^1 x - x^2\,dx + \int_1^2 x^2 - x\,dx$$
$$= \left(\frac{x^2}{2} - \frac{x^3}{3}\right)\Big|_0^1 + \left(\frac{x^3}{3} - \frac{x^2}{2}\right)\Big|_1^2$$
$$= \left(\frac{1}{2} - \frac{1}{3}\right) - (0 - 0) + \left(\frac{8}{3} - 2\right) - \left(\frac{1}{3} - \frac{1}{2}\right) = 1.$$

Another quirk of using definite integrals to find areas arises when the limits of integration are in the "wrong" order: when the "lower" limit is a larger number than the "upper" limit.

Example 4.6.2 What to Do When the Limits of Integration Are in the Wrong Order

For example,

$$\int_1^0 t^2 dt = \frac{t^3}{3}\Big|_1^0 = 0 - \frac{1}{3} = -\frac{1}{3}.$$

The answer is negative, because definite integrals treat left to right as the positive direction and models areas measured from right to left as being taken away. When computing areas, make sure that the limits of integration are in the right order.

When finding the indefinite integrals of complicated functions, we often use substitution. Substitution works for definite integrals but requires an additional step. Suppose we wish to find the area under the curve $f(t) = (1 + 2t)^2$ from $t = 0$ to $t = 1$ (Figure 4.6.4a). One way is to multiply out the function and integrate,

$$\int_0^1 (1 + 2t)^2 dt = \int_0^1 1 + 4t + 4t^2 dt$$
$$= \left(t + 2t^2 + \frac{4t^3}{3}\right)\Big|_0^1$$
$$= 1 + 2 + \frac{4}{3} = 4.333.$$

To use substitution, we use a modified version of Algorithm 4.2.

Algorithm 4.4 Computing a Definite Integral with Substitution

1. Define a new variable as some function of the old variable.
2. Take the derivative of the new variable with respect to the old variable.
3. Treat the derivative like a fraction and move the dx to the other side.

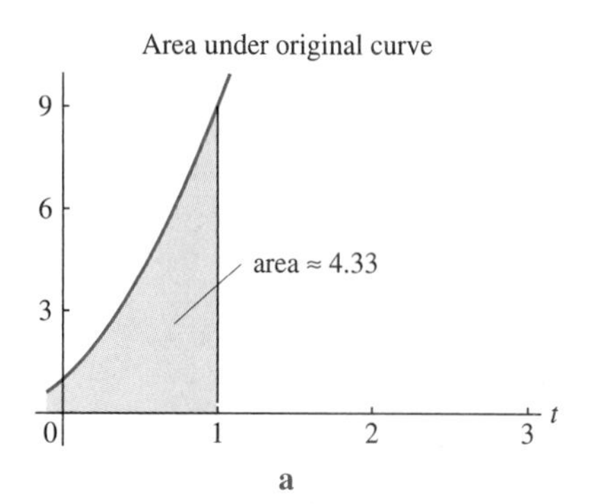

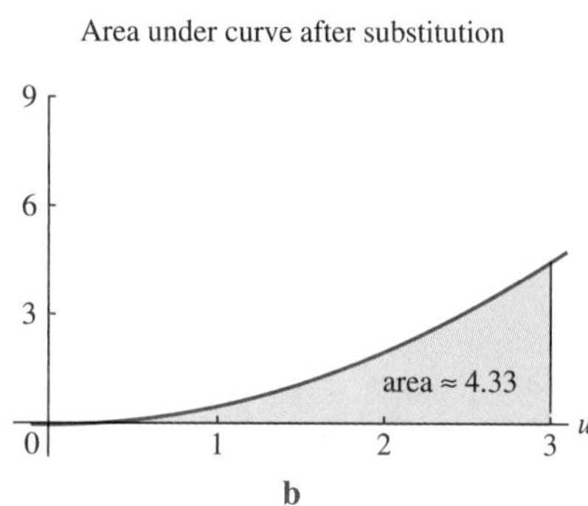

FIGURE 4.6.4
Computing an area with substitution

4. Put everything in the integral in terms of the new variable.
5. Change the limits of integration into the new variable.
6. Try to integrate.

Example 4.6.3 Using Integration by Substitution to Compute an Area

To find the area under the curve $(1+2t)^2$ between $t=0$ and $t=1$,

1. Set $u = 1+2t$.
2. $\frac{du}{dt} = 2$.
3. $du = 2dt$ or $dt = \frac{du}{2}$.
4. Put the integrand in terms of u, or
$$(1+2t)^2 dt = u^2 \frac{du}{2}.$$
5. The new step is to **change the limits of integration**. The original limits are from $t=0$ to $t=1$. When $t=0$, $u=1+2\cdot 0=1$, and when $t=1$, $u=1+2\cdot 1=3$. The new integral, after substituting for every t, is
$$\int_1^3 \frac{u^2}{2} du.$$
6. Work this out to find
$$\int_1^3 \frac{u^2}{2} du = \frac{u^3}{6}\Big|_1^3 = \frac{27}{6} - \frac{1}{6} = 4.333.$$

We do not have to convert back to the old variable at the end because the answer is a number rather than a function. The graph of the area under this function is shown in Figure 4.6.4b. The height of the u curve is half that of the t curve, but its width has been doubled, thus preserving the area.

The Riemann sum concept, that of breaking an area or volume into small pieces and then integrating to find the total, can be used to do much more than finding areas under curves. We here show how to use integration to find the area inside a circle and the volume of a cone. Project 2 looks in more detail at how Archimedes first developed this method to estimate the value of π.

Example 4.6.4 Finding the Area Inside a Circle

The number π is defined as the ratio of the perimeter to the diameter of a circle, giving the formula
$$\text{perimeter of circle} = 2\pi r.$$

The appearance of the number π in the formula for the area inside the circle is not a matter of definition. To compute that area, we can picture the area inside a circle as broken into little rings extending from radius x to radius $x+\Delta x$ (Figure 4.6.5). Such

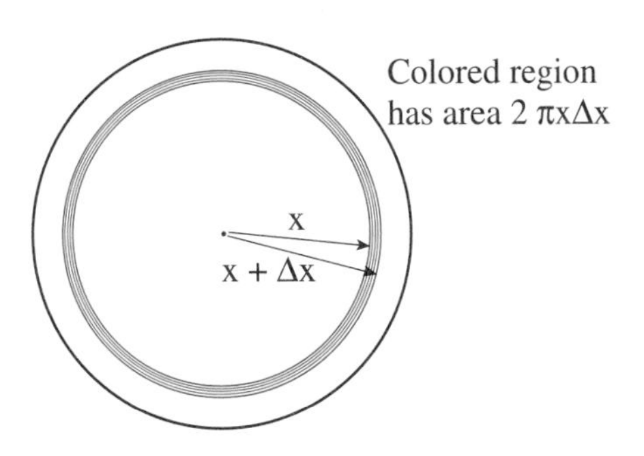

FIGURE 4.6.5
Breaking a circle into small rings

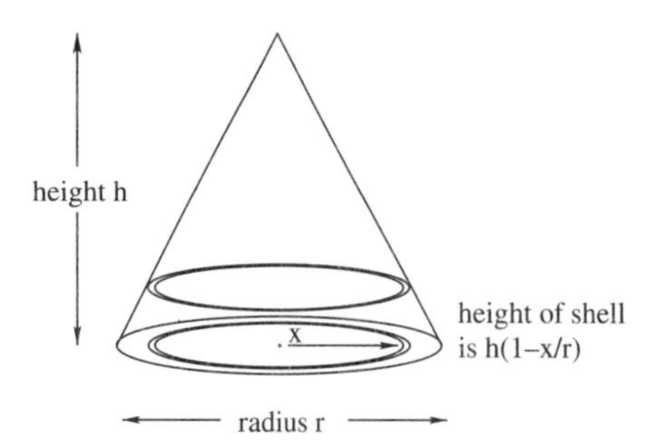

FIGURE 4.6.6
Breaking a cone into small shells

a ring has area $2\pi x\Delta x$ because it is approximately a rectangle with length $2\pi x$ and width Δx. The area inside a circle is then

$$\int_0^r 2\pi x dx = \pi x^2|_0^r = \pi r^2.$$

Example 4.6.5 Finding the Volume of a Cone

We can use the definition of the perimeter of a circle to compute the volume of a cone by breaking it into a set of cylindrical shells (Figure 4.6.6). Consider a circle drawn at a radius x inside the base of a cone of radius r and height h. If we stretch that circle upward it will hit the cone at a height $(1 - \frac{x}{r})h$. A shell of thickness Δx will have volume $2\pi x(1 - \frac{x}{r})h\Delta x$. The entire volume can be found by adding up the volumes of these shells over all x, or

$$\int_0^r 2\pi x(1 - \frac{x}{r})hdx = \pi x^2 h|_0^r - \frac{2\pi}{3r}x^3 h|_0^r = \pi r^2 h - \frac{2\pi}{3}r^2 h = \frac{\pi}{3}r^2 h.$$

This is one-third of the volume of the cylinder with the same radius and height.

4.6.2 Integrals and Averages

Example 4.6.6 Finding an Average Rate

Suppose that water is flowing into a vessel at a rate of $1 - e^{-t}$ cm^3 per second for the 2 seconds between $t = 0.0$ and $t = 2.0$. What is the *average* rate at which water enters during this time? The average is the total amount of water that entered divided by the time, or

$$\text{average rate} = \frac{\text{total water entering}}{\text{total time}}.$$

The total amount of water that entered is

$$\text{total water entering} = \int_{0.0}^{2.0} 1 - e^{-t} dt$$
$$= t + e^{-t}\,|_{0.0}^{2.0}$$
$$= 2.0 + e^{-2.0} - (0.0 + e^{-0.0}) = 1.135.$$

The average rate is

$$\text{average rate} = \frac{\text{total water entering}}{\text{total time}} = \frac{1.135}{2.0} = 0.568$$

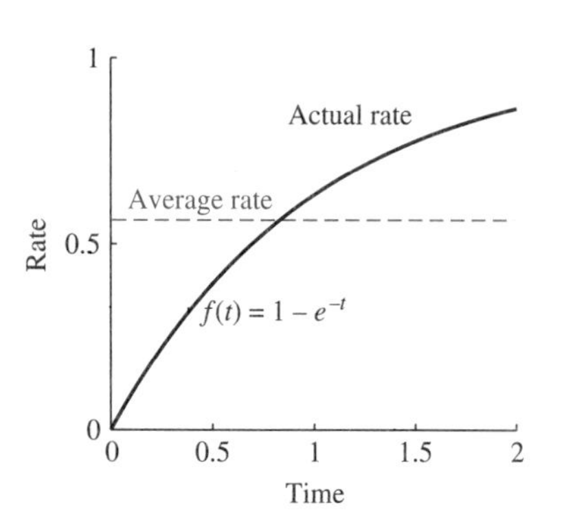

FIGURE 4.6.7
The average value of a rate

(Figure 4.6.7). If water entered at the constant rate of 0.568 cm^3/s for 2.0 s, then 1.135 cm^3 would enter, equal to the amount of water that entered at the variable rate $1 - e^{-t}$ during those same 2.0 s. Geometrically, the area under the horizontal line at 0.568 is equal to the area under the curve $1 - e^{-t}$.

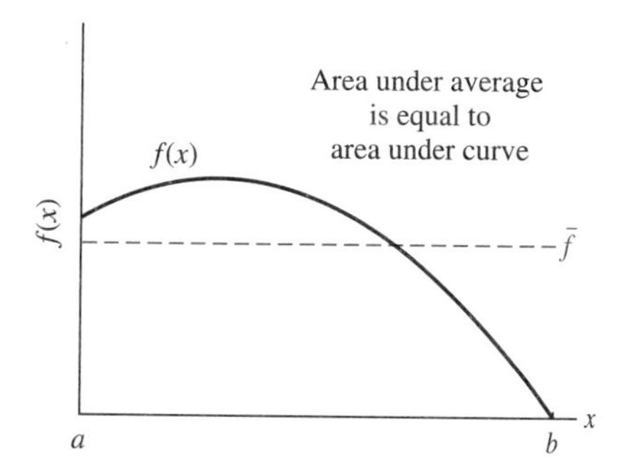

FIGURE 4.6.8
The average value in general

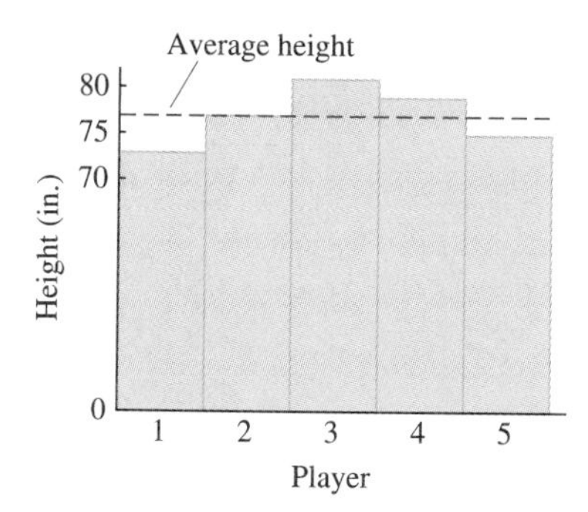

FIGURE 4.6.9
The average height of a basketball team

The general formula for the average value of a function f, often denoted by $\bar{f}$, on the interval from a to b is

$$\text{average value of } f = \frac{1}{b-a}\int_a^b f(x)dx. \tag{4.6.2}$$

The area under the horizontal line representing the average is equal to the area under the curve (Figure 4.6.8).

How does this compare with the usual meaning of average? Suppose you wanted to find the average height of a starting basketball team. If h_i denotes the height of player i, the average $\bar{h}$ is ordinarily found by adding up the heights and dividing by 5, or

$$\bar{h} = \frac{1}{5}\sum_{i=1}^{5} h_i.$$

Figure 4.6.9 shows a graphical representation of this team with $h_1 = 73$, $h_2 = 77$, $h_3 = 81$, $h_4 = 79$, and $h_5 = 75$. Each player is represented by a bar with height equal to a player's height. The bars define a function f as follows:

$$f(x) = \begin{cases} h_1 & \text{for } 0 \le x < 1 \\ h_2 & \text{for } 1 \le x < 2 \\ h_3 & \text{for } 2 \le x < 3 \\ h_4 & \text{for } 3 \le x < 4 \\ h_5 & \text{for } 4 \le x < 5. \end{cases}$$

From this graph, we can find the total height of the team as the integral

$$\begin{aligned}\text{total height} &= \int_0^5 f(x)dx \\ &= \int_0^1 h_1 dx + \int_1^2 h_2 dx + \int_2^3 h_3 dx + \int_3^4 h_4 dx + \int_4^5 h_5 dx \\ &= h_1 + h_2 + h_3 + h_4 + h_5 = 385.\end{aligned}$$

To integrate a function that is defined in pieces, integrate each piece separately and add up the results. Substituting the definition of the average (Equation 4.6.2),

$$\text{average height} = \frac{\text{total height}}{\text{total number}} = \frac{1}{5}(h_1 + h_2 + h_3 + h_4 + h_5) = \frac{385}{5} = 77,$$

matching the standard result.

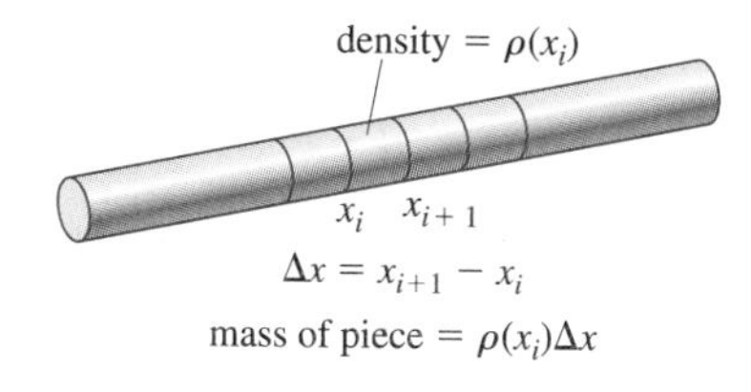

FIGURE 4.6.10
The mass of a bar

4.6.3 Integrals and Mass

Integration can be used to find the mass of an object with known **density**. Here we consider only the one-dimensional case, such as a thin bar. Suppose the density of the bar, measured in g/cm, is $\rho(x)$ at x (Figure 4.6.10). To estimate the mass of the bar, we break it into n small pieces of length Δx. The mass of the bit between x_i and $x_i + \Delta x$ is approximately $\rho(x_i)\Delta x$, the density at the left end of the bit times the length. Adding up all the little pieces,

$$\text{mass of bar} \approx \sum_{i=1}^{n} \rho(x_i)\Delta x.$$

This has the exact form of a Riemann sum. The limit as Δx approaches zero and n approaches infinity is equal to the definite integral (Definition 4.2), so

$$\text{mass of bar} = \int_a^b \rho(x)dx. \tag{4.6.3}$$

Example 4.6.7 Finding the Mass of a Bar

Consider a 100 cm vertical bar composed of a substance that has settled and become denser near the ground. Let z denote the height above ground and suppose the density is given by

$$\rho(z) = e^{-0.01z}$$

in g/cm. The mass is

$$\begin{aligned}\int_0^{100} e^{-0.01z}dz &= \frac{1}{-0.01}(e^{-0.01z})|_0^{100}\\ &= -100(e^{-0.01z})|_0^{100}\\ &= -100(e^{-0.01\cdot 100}) - [-100(e^{-0.01\cdot 0})]\\ &= 100(1 - e^{-1.0}) = 63.21.\end{aligned}$$

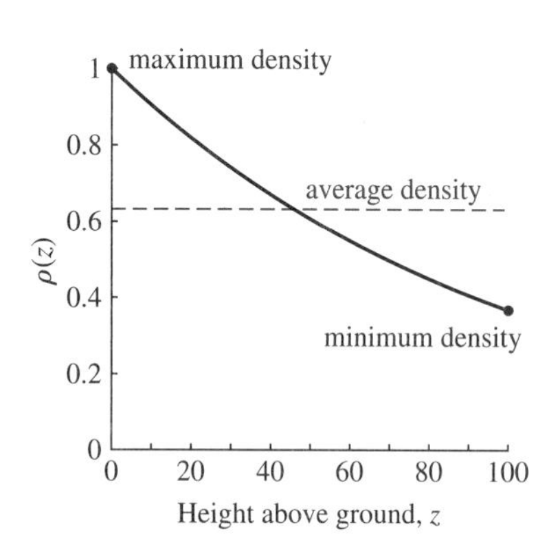

FIGURE 4.6.11

The average density of a bar

We used the general formula for the integral of $e^{\alpha x}$ from "Using Substitution to Eliminate Constants," Subsection 4.3.3, to find the indefinite integral in the first step. Does this result make sense? The density at the bottom of the bar is $\rho(0) = 1.0$ g/cm. If the entire bar had this maximum density, the mass would be 100 g. The density at the top of the bar is $\rho(100) = e^{-1} = 0.368$ g/cm, so the mass of the bar would be 36.8 g if the entire bar had this minimum density. Our result lies between these extremes. Furthermore,

$$\text{average density} = \frac{\text{total mass}}{\text{total length}} = \frac{\int_0^{100} \rho(x)\,dx}{100} = \frac{63.2}{100.0} = 0.632.$$

The result lies between the minimum density of 0.368 and the maximum of 1.0 (Figure 4.6.11).

Example 4.6.8 Counting Otters with Integration

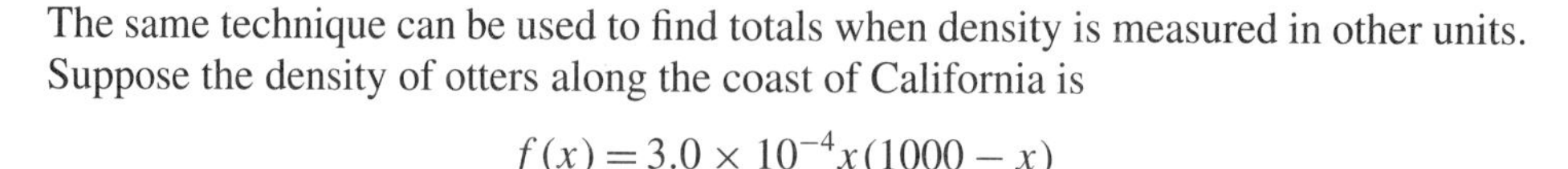

The same technique can be used to find totals when density is measured in other units. Suppose the density of otters along the coast of California is

$$f(x) = 3.0 \times 10^{-4}x(1000 - x)$$

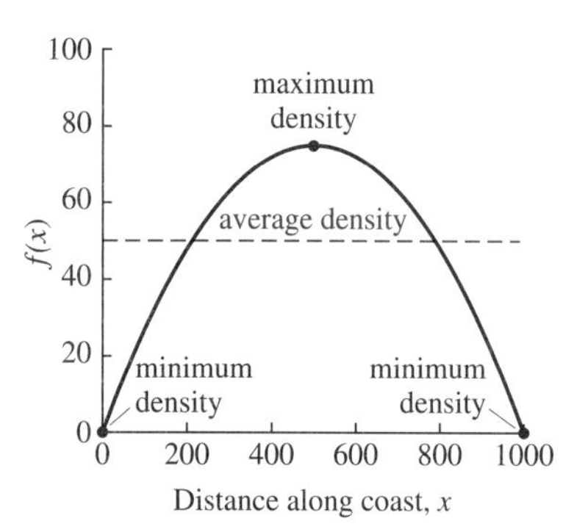

FIGURE 4.6.12

The average density of otters

in otters per mile, where x is measured in miles from the Mexican border and can take on values between 0 and 1000. The population density takes on a maximum value of 75 otters/mile halfway up the coast at $x = 500$ and a minimum value of 0 at $x = 0$ and $x = 1000$ (Figure 4.6.12).

The total number T is the definite integral of the density, or

$$\begin{aligned}T &= \int_0^{1000} 3.0 \times 10^{-4}x(1000 - x)\,dx\\ &= \int_0^{1000} 0.3x - 3.0 \times 10^{-4}x^2\,dx\\ &= 0.15x^2|_0^{1000} - 1.0 \times 10^{-4}x^3|_0^{1000}\\ &= 1.5 \times 10^5 - 1.0 \times 10^5\\ &= 0.5 \times 10^5 = 50{,}000 \text{ otters.}\end{aligned}$$

The average density is

$$\text{average density} = \frac{\text{total number}}{\text{total distance}}$$
$$= \frac{50{,}000 \text{ otters}}{1000 \text{ miles}}$$
$$= 50.0\frac{\text{otters}}{\text{mile}}.$$

This value lies between the maximum density of 75 otters/mile and the minimum density of 0 otters/mile. The average value is not the average of these minimum and maximum values.

Summary The definite integral can be used to find the **area under a curve**. The area computed is negative if the value of the function itself is negative, or if the limits of integration are in decreasing order. Integration by substitution, introduced for indefinite integrals, can be extended to definite integrals but requires the additional step of expressing the limits of integration in terms of the new variable. We used the definite integral to calculate areas and volumes by breaking shapes into small regions, and the **average value of a function** by dividing the integral (the total amount) by the length of the interval. Similarly, we computed masses or total numbers from **densities**. In each case, the underlying idea is that of the Riemann sum; chopping things up into small pieces and adding up the results.

4.6 Exercises

Mathematical Techniques

1–6 ▪ Use substitution to evaluate the following definite integrals.

1. $\int_0^5 3e^{\frac{x}{5}}\,dx$
2. $\int_2^5 \cos[2\pi(x-2)]\,dx$
3. $\int_0^4 \left(1+\frac{t}{2}\right)^4 dt$
4. $\int_1^{10}(1+2t)^{-4}\,dt$
5. $\int_{-3}^0 \frac{1}{4+t}\,dt$
6. $\int_0^2 \frac{1}{1+4t}\,dt$

7–12 ▪ Find the areas under the following curves. If you use substitution, draw a graph to compare the original area with that in transformed variables.

7. Area under $f(x)=3x^3$ from $x=0$ to $x=3$.
8. Area under $g(x)=e^x$ from $x=0$ to $x=\ln(2)$.
9. Area under $h(x)=e^{x/2}$ from $x=0$ to $x=\ln(2)$.
10. Area under $f(t)=(1+3t)^3$ from $t=0$ to $t=2$.
11. Area under $G(y)=(3+4y)^{-2}$ from $y=0$ to $y=2$.
12. Area under $s(z)=\sin(z+\pi)$ from $z=0$ to $z=\pi$.

13–18 ▪ The definite integral can be used to find the area between two curves. In each case,

a. Sketch the graphs of the two functions over the given range, and shade the area between the curves.

b. Sketch the graph of the difference between the two curves. The area **under** this curve matches the area **between** the original curves.

c. Find the area under the difference curve (remembering to use absolute value).

13. $f(x)=2x$ and $g(x)=x^2$ for $0\le x\le 2$.
14. $f(x)=e^x$ and $g(x)=x+1$ for $-1\le x\le 1$.
15. $f(x)=2x$ and $g(x)=x^2$ for $0\le x\le 4$.
16. $f(x)=x^2$ and $g(x)=x^3$ for $0\le x\le 2$.
17. $f(x)=e^x$ and $g(x)=\frac{e^{2x}}{2}$ for $0\le x\le 1$.
18. $f(x)=\sin(2x)$ and $g(x)=\cos(2x)$ for $0\le x\le \pi$.

19–22 ▪ Find the average value of the following functions over the given range. Sketch a graph of the function along with a horizontal line at the average to make sure that your answer makes sense.

19. x^2 for $0<x<3$.
20. $\frac{1}{x}$ for $0.5<x<2.0$.
21. $x-x^3$ for $-1<x<1$.
22. $\sin(2x)$ for $0\le x\le \pi/2$.

23–24 ▪ Use integration by parts to evaluate the following as definite integrals.

23. Find the area under the curve $g(x)=x\ln(x)$ for $1\le x\le 2$. Sketch a graph to see if your answer makes sense.
24. Find the area under the curve $g(x)=x\sin(2\pi x)$ for $0\le x\le 2$. Sketch a graph to see if your answer makes sense.

25–26 ▪ We have used little vertically oriented rectangles to compute areas. There is no reason why little horizontal rectangles cannot be used. Here are the steps to find the area under the curve $y = f(x)$ from $x = 0$ to $x = 1$ by using such horizontal rectangles.

a. Draw a picture with five horizontal rectangles, each of height 0.2, approximately filling the region to the right of the curve.

b. Calculate an upper and lower estimate of the length of each rectangle based on the length of the upper and lower boundaries.

c. Add these up to find upper and lower estimates of the area.

d. Think now of a very thin rectangle at height y. How long is the rectangle?

e. Write down a definite integral expression for the area.

f. Evaluate the integral and check that the answer is correct.

25. With $f(x) = x^2$.

26. With $f(x) = \sqrt{x}$.

27–28 ▪ There is often more than one way to divide up a region to find an area or volume.

27. Use the fact that the area of a circle of radius r is πr^2 to find the volume of a cone of height 1 that has radius r at a height r. Think of the cone as being built of a stack of little circular disks with some small thickness Δr.

28. Break a sphere of radius r into horizontal discs to find the volume. The trick is to figure out the area of each disc at height z where z ranges from $-r$ to r.

29–32 ▪ Some books *define* the natural log function with the definite integral as the function $l(a)$ for which

$$l(a) = \int_1^a \frac{1}{x} dx.$$

Using this definition, we can prove the laws of logarithms (page 84).

29. Show that $l(6) - l(3) = l(2)$. (Use the summation property of the definite integral to write the difference as an integral, and then use the substitution $y = \frac{x}{3}$.)

30. Find the integral from a to $2a$ by following the same steps. (Make the substitution $y = \frac{x}{a}$.)

31. Show that $l(10^2) = 2 \cdot l(10)$. (Try the substitution $y = \sqrt{x}$ in $\int_1^{10^2} \frac{1}{x}\, dx$.)

32. Show that $l(a^b) = b \cdot l(a)$. (Try the substitution $y = \sqrt[b]{x}$ in $\int_1^{a^b} \frac{1}{x}\, dx$.)

Applications

33–34 ▪ The average of a step function computed with the definite integral matches the average computed in the usual way. Test this in the following situations by finding the average of the values directly, and then as the integral of a step function.

33. Suppose a math class has four equally weighted tests. A student gets 60 on the first test, 70 on the second, 80 on the third, and 90 on the last.

34. A math class has 20 students. In a quiz worth 10 points, 4 students get 6, 7 students get 7, 5 students get 8, 3 students get 9, and 1 student gets 10.

35–38 ▪ Suppose water is entering a tank at a rate of $g(t) = 360t - 39t^2 + t^3$ where g is measured in liters per hour and t is measured in hours. The rate is 0 at times 0, 15, and 24.

35. Find the total amount of water entering during the first 15 hr, from $t = 0$ to $t = 15$. Find the average rate at which water entered during this time.

36. Find the total amount and average rate from $t = 15$ to $t = 24$.

37. Find the total amount and average rate from $t = 0$ to $t = 24$.

38. Suppose that energy is produced at a rate of

$$E(t) = |g(t)|$$

in J/h (Joules per hour). Find the total energy generated from $t = 0$ to $t = 24$. Find the average rate of energy production. Check that $g(t)$ changes sign from positive to negative at $t = 15$.

39-40 ▪ Several very skinny 2.0 m long snakes are collected in the Amazon. Each has density of $\rho(x)$ given by the following formulas, where ρ is measured in g/cm and x is measured in centimeters from the tip of the tail. For each snake,

a. Find the minimum and maximum density of the snake. Where does the maximum occur?

b. Find the total mass of the snake.

c. Find the average density of the snake. How does this compare with the minimum and maximum?

d. Graph the density and average.

39. $\rho(x) = 1.0 + 2.0 \times 10^{-8}x^2(300 - x)$.

40. $\rho(x) = 1.0 + 2.0 \times 10^{-8}x^2(240 - x)$.

41–42 ▪ A piece of *E. coli* DNA has about 4.7×10^6 nucleotides, and is about 1.6×10^6 nm long. The genetic code consists of four possible nucleotides, called A, C, G, and T. For each of the following cases,

a. Use the given information to find the formula for the number of A's, C's, G's and T's per thousand as a function of distance along the DNA strand.

b. Find the total number of A's, C's, G's, and T's in the DNA.

c. Find the mean number of A's, C's, G's, and T's in the DNA per thousand.

41. Suppose that the number of A's per thousand increases linearly from 150 at one end of the DNA strand to 300 at the other. The number of C's per thousand decreases linearly from 350 at one end to 200 at the other, and the number of G's per thousand increases linearly from 220 at one end to 320 at the other. The remainder is made up of T's.

42. Suppose that the number of A's per thousand increases linearly from 200 at one end of the DNA strand to 250 at the other. The number of C's per thousand increases linearly from 250 at one end to 300 at the other, and the number of G's per thousand decreases linearly from 300 at one end to 200 at the other. The remainder is made up of T's.

43–46 ▪ Suppose water is entering a series of vessels at the given rate. In each case, find the total amount of water entering during the first second, and the average rate during that time. Compare the average rate with rate at the "average time," at $t = 0.5$ halfway through the time period from 0 to 1. In which case is the average rate greater than the rate at the average time? Graph the flow rate function, and mark the flow rate at the average time. Can you guess what it is about the shape of the graph that determines how the average rate compares with the rate at the average time?

43. Water is entering at a rate of $t^3 \text{cm}^3/\text{s}$.

44. Water is entering at a rate of $\sqrt{t}$ cm^3/s.

45. Water is entering at a rate of t cm^3/s.

46. Water is entering at a rate of $4t(1-t)$ cm^3/s.

Computer Exercises

47. Find the area between the two curves $f(x) = \cos(x)$ and $g(x) = 0.1x$ for $0 \leq x \leq 10$.

a. Graph the two functions and make the three regions.

b. Have your computer find where each region begins and ends.

c. Integrate to find the area of each region.

d. Add them up.

48. Suppose the volume of water in a vessel obeys the differential equation

$$\frac{dV}{dt} = f(t) = 1 + 3t + 3t^2$$

with $V(0) = 0$.

a. Graph the functions f and V for $t = 0$ to $t = 2$ and label the curves.

b. Find the volume at time $t = 10$. What is the definite integral that has the same answer? Shade the area on your graph from part **a** and write the associated integral.

c. Define a function $A(T)$ that gives the **average** rate of change of volume as a function of time (the total volume added between times $t = 0$ and $t = T$ divided by the elapsed time). Graph this on the same graph as $f(t)$. Label the curves (and write the formula for $A(t)$ as a definite integral). Why is the average rate A greater than the instantaneous rate f?

d. Graph $f(t)$ between $t = 0$ and $t = 10$ and the constant function with a rate equal to the average at time $T = 10$. What is the area under the line? Does it match what you found in part **b**? Why should it? Mark the point where the average and instantaneous rates are equal. Use your computer to solve for this point.

4.7 Improper Integrals

So far, we have considered only definite integrals of functions that do not approach infinity between finite limits of integration. Integrals of this sort are called **proper integrals**. **Improper integrals** are of two types: integrals with infinite limits of integration and integrals of functions that approach infinity somewhere between the limits of integration. We now learn how to compute and apply improper integrals.

4.7.1 Infinite Limits of Integration

"Infinite" measurements cannot crop up in biological experiments. Nonetheless, infinity is a useful mathematical abstraction of "very long" or "very far." Consider again the equation for chemical production with exponentially declining rate (Equation 4.1.2),

$$\frac{dP}{dt} = e^{-t}$$

in moles per second. The amount of product produced between $t = 0$ and $t = T$ is given by the definite integral

$$\text{production between 0 and } T = \int_0^T e^{-t}\,dt.$$

The longer we wait, the more product has been produced. Let P_∞ denote the amount that would be produced if the experiment ran forever. We would like to write

$$P_\infty = \int_0^\infty e^{-t}\,dt.$$

To be honest, however, we have never defined what this expression means. We *defined* the definite integral to be the limit of Riemann sums. Computing Riemann sums requires breaking the region between the limits of integration into n equally sized regions. But the infinite region from zero to infinity cannot be broken into n finite regions of equal size.

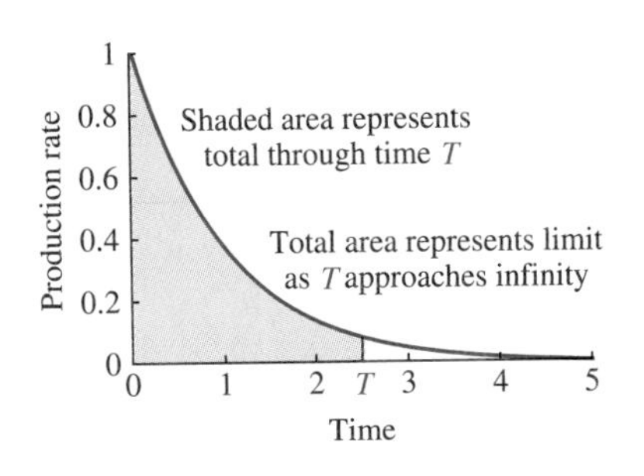

FIGURE 4.7.1
Computing an integral with infinite limits of integration

Instead, we can think of this **improper integral** as the *limit* of **proper integrals** with formula

$$\int_0^\infty e^{-t}dt = \lim_{T\to\infty}\int_0^T e^{-t}dt.$$

This formula captures the spirit of what we want. The proper integral gives the amount of production until time T and the limit allows us to make T "very large" (Figure 4.7.1). In this case,

$$\int_0^T e^{-t}dt = -e^{-t}|_0^T = -e^{-T} + 1.$$

Therefore

$$\begin{aligned}\int_0^\infty e^{-t}dt &= \lim_{T\to\infty} 1 - e^{-T}\\ &= \lim_{T\to\infty} 1 - \lim_{T\to\infty} e^{-T}\\ &= 1 - 0 = 1.0.\end{aligned}$$

Exactly 1.0 mol of the product would be created after an infinite amount of time. If we wait for a "long time," such as 10 s, the exact amount is

$$1 - e^{-10.0} = 0.99995,$$

quite close to the limit.

Formally, we define

Definition 4.3 The improper integral of the function f from a to ∞ is

$$\int_a^\infty f(t)dt = \lim_{T\to\infty}\int_a^T f(t)dt.$$

4.7.2 Improper Integrals: Examples

Our definition does not guarantee that the limit is finite. Nothing mathematical prevents the definite integral from 0 to T from getting larger and larger as T approaches infinity. Consider a different chemical produced at the diminishing rate

$$\frac{dQ}{dt} = \frac{1}{1+t},$$

again in moles per second. This rate decreases to 0 more slowly than the exponential function (Section 3.6). How much product would this reaction produce after a long time?

The total product ever produced, Q_∞, is computed with the improper integral

$$\begin{aligned}Q_\infty &= \int_0^\infty \frac{1}{1+t}dt\\ &= \lim_{T\to\infty}\int_0^T \frac{1}{1+t}dt.\end{aligned}$$

To compute this integral, we must use the substitution $u = 1 + t$. Then $dt = du$ and the limits of integration of $t = 0$ to $t = T$ become $u = 1$ and $u = 1 + T$. Then

$$\begin{aligned}Q_\infty &= \lim_{T\to\infty}\int_0^T \frac{1}{1+t}dt\\ &= \lim_{T\to\infty}\int_1^{T+1} \frac{1}{u}du\\ &= \lim_{T\to\infty} \ln(u)|_1^{T+1}\\ &= \lim_{T\to\infty} \ln(T+1) - 0\\ &= \infty.\end{aligned}$$

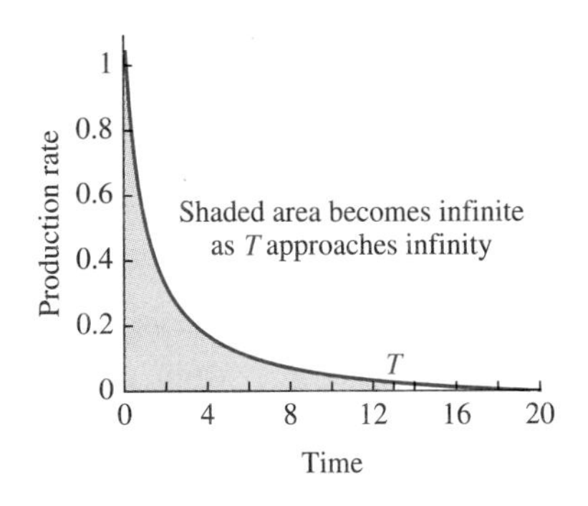

FIGURE 4.7.2
A divergent integral

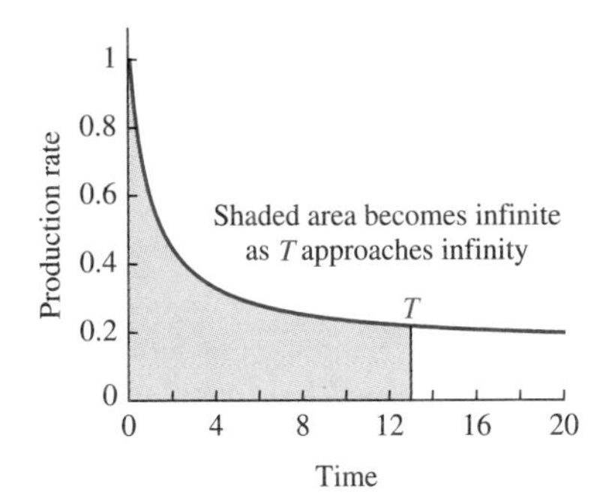

FIGURE 4.7.3
A divergent integral where the integrand does not decrease to 0

If this rule were followed forever, the amount of product would be infinite (Figure 4.7.2). In this case, we say that the integral **diverges**. When the limit exists, we say that the integral **converges**. Such a result might seem absurd and irrelevant. But this very absurdity provides a valuable negative result; no real system could follow this law indefinitely. Even though the rate gets smaller and smaller, the total production increases without bound.

What laws can be maintained indefinitely without producing an infinite amount of product? Any function that does not decrease to zero has an infinite improper integral (Figure 4.7.3) because the total production gets larger and larger without bound. In terms of area, the region under the curve can be thought of as including a rectangle with positive height and an infinite length.

Of functions that decrease to zero as their arguments approach infinity, we have learned to integrate only those of the form $1/t^p$ for $p > 0$ and $e^{-\alpha t}$ for $\alpha > 0$. We can use the power rule to compute the integral of $1/t^p$ when $p \neq 1$. (We set the lower limit of integration to 1 to avoid the point $t = 0$ where the integrand approaches infinity). When $p > 1$,

$$\begin{aligned}\int_1^{\infty} \frac{1}{t^p}\, dt &= \lim_{T \to \infty} \int_1^{T} \frac{1}{t^p}\, dt \\ &= \lim_{T \to \infty} \frac{t^{1-p}}{1-p} \Big|_1^T \\ &= \lim_{T \to \infty} \frac{T^{1-p}}{1-p} - \frac{1}{1-p}.\end{aligned}$$

When $p > 1$, the power $1 - p$ is negative. Therefore T^{1-p} approaches 0 as T approaches infinity. The improper integral is

$$\begin{aligned}\int_1^{\infty} \frac{1}{t^p}\, dt &= \lim_{T \to \infty} \frac{T^{1-p}}{1-p} - \frac{1}{1-p} \\ &= 0 - \frac{1}{1-p} \\ &= \frac{1}{p-1}.\end{aligned}$$

This integral converges.

Example 4.7.1 Improper Integrals of Power Functions with $p > 1$

The integral of the function $\frac{1}{t^p}$ for $p = 1.2$ is

$$\int_1^{\infty} \frac{1}{t^{1.2}} dt = \frac{1}{1.2 - 1} = 5.0$$

(Figure 4.7.4a).

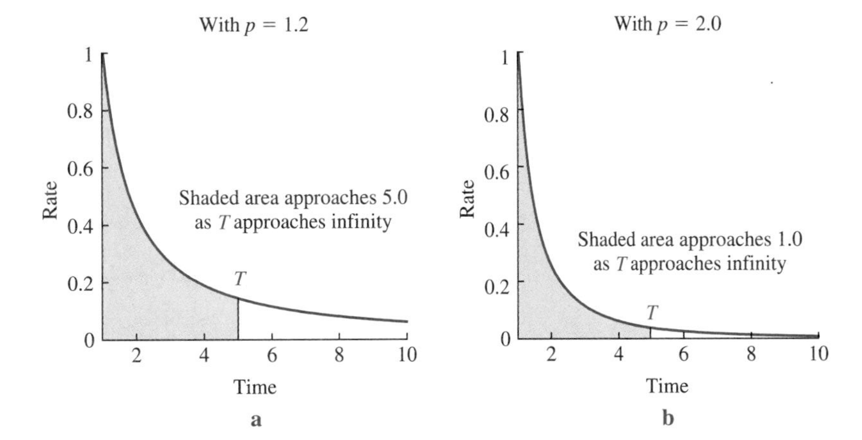

FIGURE 4.7.4
Power functions with $p > 1$

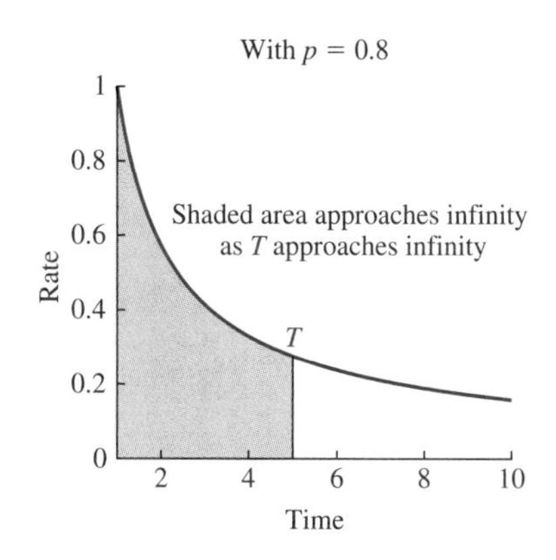

FIGURE 4.7.5
A power function with $p < 1$

With the larger value $p = 2.0$,

$$\int_1^{\infty} \frac{1}{t^{2.0}} dt = \frac{1}{2.0 - 1} = 1.0.$$

(Figure 4.7.4b) The value of the improper integral (the total area under the curve or the limiting amount of product produced) becomes smaller as the value of p becomes larger. ▲

With $p < 1$, the integration is the same. But when we take the limit, we find

$$\int_1^{\infty} \frac{1}{t^p} dt = \lim_{T\to\infty} \frac{T^{1-p}}{1-p} - \frac{1}{1-p} = \infty$$

because T is taken to the *positive* power $1 - p$ when $p < 1$. This integral diverges. Recall that $\frac{1}{t^p}$ decreases to 0 faster for larger values of p. If the value of p is sufficiently large (greater than 1), the area under the curve is finite. When p is too small (less than 1), the function decreases to 0 more slowly and the area is infinite (Figure 4.7.5).

When the rate decreases exponentially with formula $e^{-\alpha t}$, the improper integral is

$$\begin{aligned}\int_0^{\infty} e^{-\alpha t}\, dt &= \lim_{T\to\infty} \int_0^T e^{-\alpha t}\, dt \\ &= \lim_{T\to\infty} -\frac{e^{-\alpha t}}{\alpha}\Big|_0^T \\ &= \lim_{T\to\infty} -\frac{e^{-\alpha T}}{\alpha} + \frac{1}{\alpha} \\ &= \frac{1}{\alpha}\end{aligned}$$

where we used the substitution $y = \alpha t$. This exponential integral converges for every positive value of α, but the integral takes on a larger value for smaller α, consistent with the fact that $e^{-\alpha t}$ decreases to 0 more slowly when α is small.

Example 4.7.2 Improper Integrals of Exponential Functions

The integral of the function $e^{-0.2t}$ (Figure 4.7.6a) is

$$\int_0^{\infty} e^{-0.2t} dt = \frac{1}{0.2} = 5.0.$$

With the larger value $\alpha = 1.0$,

$$\int_0^{\infty} e^{-1.0t} dt = \frac{1}{1.0} = 1.0.$$

The value of the improper integral (the total area under the curve or the limiting amount of product produced) becomes smaller as the value of α becomes larger (Figure 4.7.6). ▲

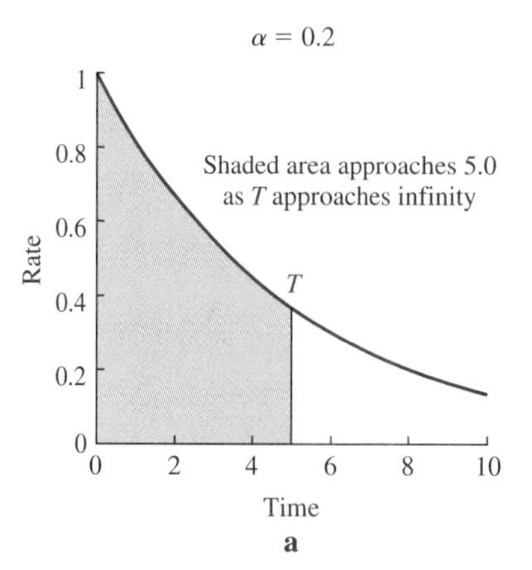

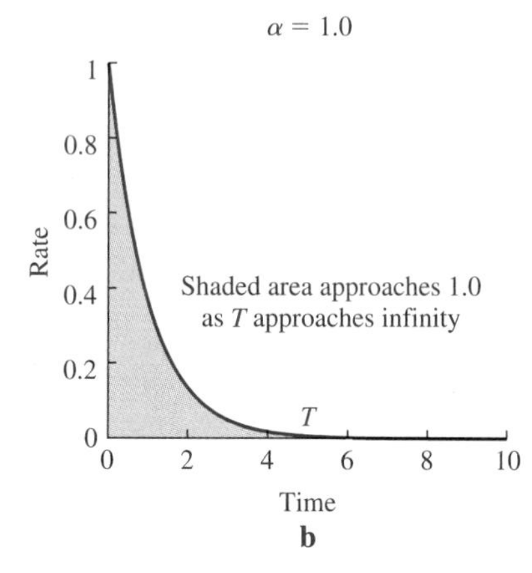

FIGURE 4.7.6
Improper integrals with different parameters in the exponent

Example 4.7.3 Checking Whether a Quantity Will Increase Without Bound

Suppose we wish to know whether a tree growing in height H according to the pure-time differential equation

$$\frac{dH}{dt} = \frac{1}{t^{3/4}}$$

with initial condition $H(1) = 5.0$ will reach arbitrarily large size. The solution is

$$H(t) = 5.0 + \int_1^t \frac{1}{s^{3/4}} ds,$$

adding the initial condition $H(0) = 5.0$ to the subsequent change. Because the power 3/4 is less than 1, the integral

$$\int_1^\infty \frac{1}{s^{3/4}} ds$$

diverges. Therefore $H(t)$ approaches a limit of infinity as t becomes large, and this tree would exceed any given height (if it lived long enough). We suspect, however, that it would be knocked over by a hurricane or just collapse under its own weight if it became too tall.

The **comparison test** provides a way to establish whether an integral converges or diverges. If we can compare our function with a simple function, we can often establish that the integral converges or diverges.

The Comparison Test

1. **Proving convergence:** Suppose $0 < f(x) < g(x)$ for all $x > a$. Then $\int_a^\infty f(x)$ converges if $\int_a^\infty g(x)$ converges.
2. **Proving divergence:** Suppose $f(x) > g(x) > 0$ for all $x > a$. Then $\int_a^\infty f(x)$ diverges if $\int_a^\infty g(x)$ diverges.

As shown in Figure 4.7.7a, if the function $f(x)$ lies below a function $g(x)$ that converges, then $f(x)$ also converges. Conversely, as shown in Figure 4.7.7b, if the function $f(x)$ lies above a function $g(x)$ that diverges, then $f(x)$ also diverges.

Example 4.7.4 Using the Comparison Test to Establish That an Integral Converges

Consider the integral

$$\int_1^\infty \frac{1}{t + e^t} dt$$

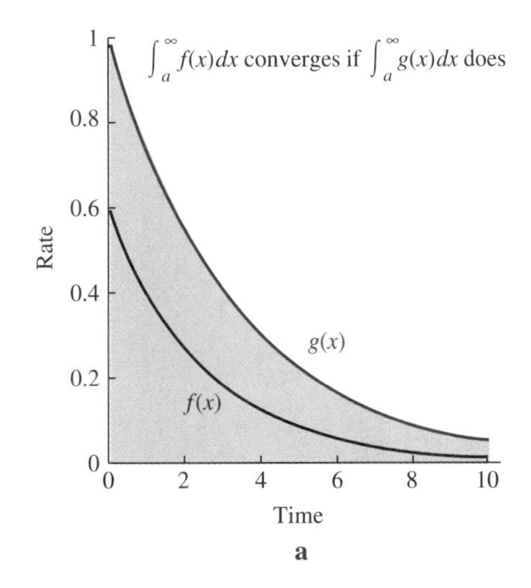

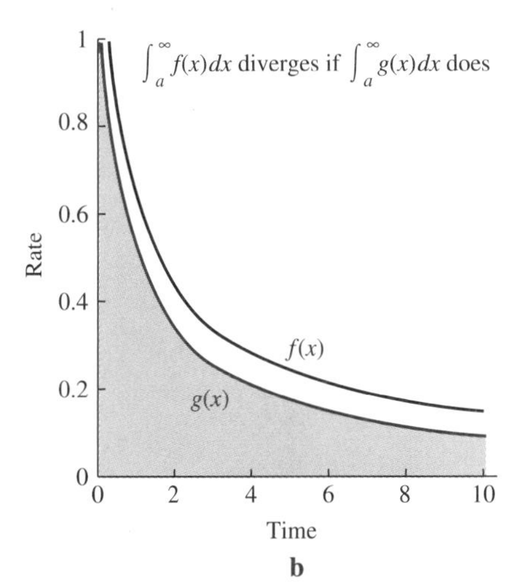

FIGURE 4.7.7
The comparison test with infinite limits of integration

Because $t > 0$ for all values of t,

$$\frac{1}{t + e^t} < \frac{1}{e^t} = e^{-t}.$$

Therefore

$$\int_1^\infty \frac{1}{t + e^t}\,dt < \int_1^\infty e^{-t} = 1.$$

Example 4.7.5 Using the Comparison Test to Establish That an Integral Diverges

Consider evaluating

$$\int_1^\infty \frac{1}{t + \sqrt{t}}.$$

For $t > 1$, $\sqrt{t} < t$, so

$$\frac{1}{t + \sqrt{t}} > \frac{1}{t + t} = \frac{1}{2t}.$$

$\int_1^\infty \frac{1}{2t}$, diverges, implying that the original integral does also.

4.7.3 Application to Infinite Series

As with average, we can think of the sum of an infinite series as the area under a set of rectangles. By comparing the rectangles with known functions, we can sometimes determine whether a series converges (adds up to a finite value) or diverges (approaches infinity) (Definition 3.10). In particular, if the rectangles lie entirely below a curve known to have a finite area, then their total area too must be finite (Figure 4.7.8a). On the other hand, if the tops of the rectangles lie above a curve known to have infinite area, then their total area must be infinite (Figure 4.7.8b).

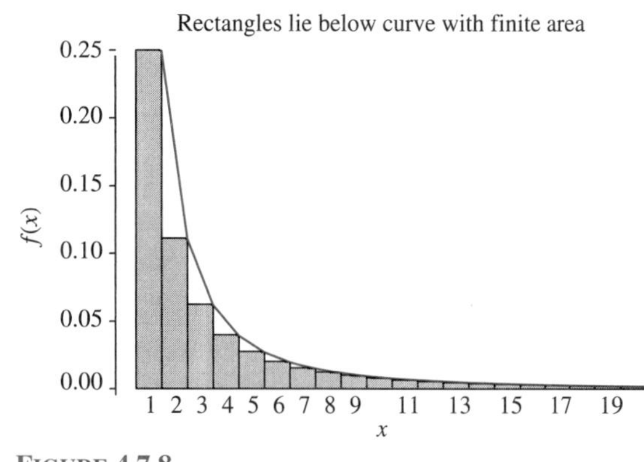

FIGURE 4.7.8
Using the comparison test to check the sum of an infinite series

Example 4.7.6 The Harmonic Series

The harmonic series is defined as the sum

$$1 + \frac{1}{2} + \frac{1}{3} + \frac{1}{4} + \frac{1}{5} + \ldots.$$

The values of this sum increase quite slowly as shown in the following table. But do they increase to infinity?

Sums of the harmonic series

1	1.000000
2	1.500000
3	1.833333
4	2.083333
5	2.283333
6	2.450000
7	2.592857
8	2.717857
9	2.828968
10	2.928968

We can graph this as a set of rectangles, the first of height 1, the second of height $\frac{1}{2}$, and so forth (Figure 4.7.9). These rectangles lie above the graph of the function $h(x) = \frac{1}{1+x}$, and therefore the area of the rectangles exceeds the area under the curve. But

$$\int_0^{\infty} \frac{1}{1+x} = \lim_{X\to\infty} \ln(1+X) = \infty.$$

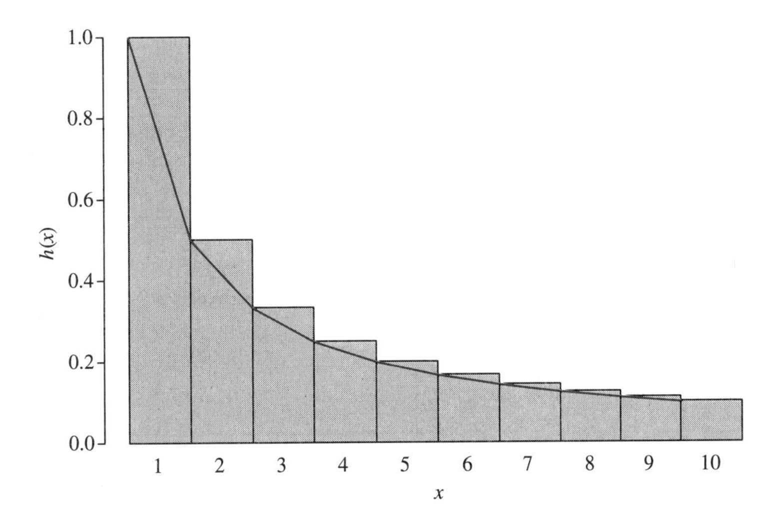

FIGURE 4.7.9
The harmonic series compared with $\frac{1}{1+x}$

Therefore the sum does approach infinity, but as slowly as the natural log function.

Example 4.7.7 The Geometric Series

The geometric series is defined as the sum

$$1 + r + r^2 + r^3 + r^4 + r^5 + \ldots.$$

We have used that Taylor series to compute the sum of this series, and we know that it is finite when $r < 1$. We can also show that the sum is finite by comparison. Each term of the series, r^i, matches the value of the function $e^{-\alpha x}$ if $r = e^{-\alpha}$ or $\alpha = -\ln(r)$. By shifting the graph over by one, each bar lies below the graph of the function $e^{-\alpha(x-1)}$. The integral of this exponential function is finite. (Figure 4.7.10)

For the specific choice of $r = 0.8$, $\alpha = -\ln(0.8) = 0.2231436$. From the sum of the geometric series, we know that

$$1 + 0.8 + 0.8^2 + 0.8^3 + 0.8^4 + 0.8^5 + \ldots = \frac{1}{1-0.8} = 5.$$

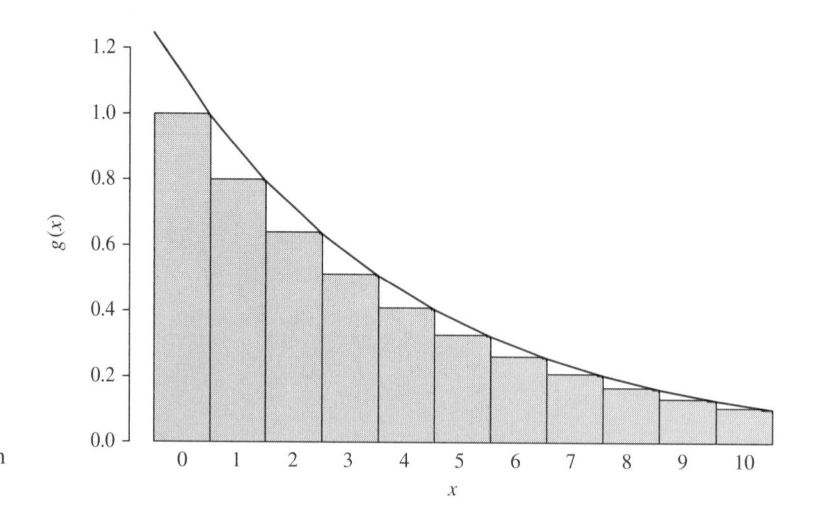

FIGURE 4.7.10
The geometric series compared with an exponential function

With the comparison test,

$$\begin{aligned}\int_0^{\infty} e^{-\alpha(x-1)} &= \int_{-1}^{\infty} e^{-\alpha y} dy \\ &= -\frac{1}{\alpha} e^{-\alpha y}\Big|_{-1}^{\infty} \\ &= \frac{e^{\alpha}}{\alpha}.\end{aligned}$$

Evaluating at $\alpha = 0.2231436$, we find $\frac{e^{\alpha}}{\alpha} = 5.601775$. This is indeed greater than the area under the rectangles, and the sum of the geometric series.

4.7.4 Infinite Integrands

Although functions with infinite integrands crop up less frequently in biological problems, it is useful to be familiar with their behavior. Consider trying to find the area under the curve $f(x) = \frac{1}{\sqrt{x}}$ between $x = 0$ and $x = 1$ (Figure 4.7.11). This function approaches infinity as x approaches 0.

Although one can, with some care, define this integral as a limit of right-hand Riemann sums, we instead use the limit to define and compute this second type of improper integral.

Definition 4.4 If the function f approaches infinity as x approaches 0 but nowhere else between $x = 0$ and $x = a$, the improper integral of the function f from 0 to a is defined by

$$\int_0^a f(x)dx = \lim_{\epsilon \to 0^+} \int_{\epsilon}^a f(x)dx.$$

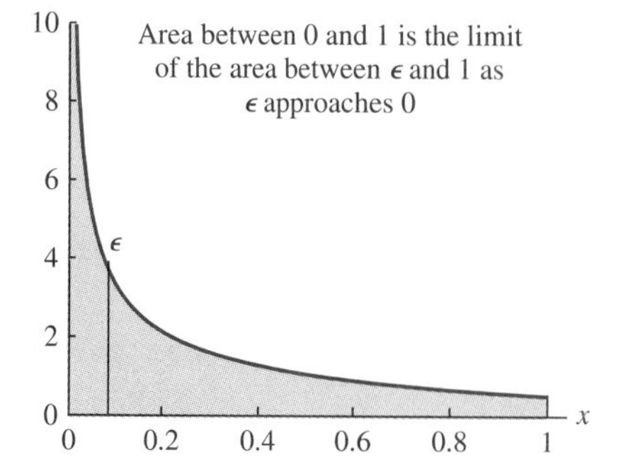

FIGURE 4.7.11
Computing an integral with an infinite integrand

The limit with the 0^+ indicates that ϵ approaches 0 from the right ("Limits of Functions," Section 2.2).

Example 4.7.8 A Finite Area Even When a Function Approaches Infinity

With this definition,

$$\begin{aligned}\int_0^1 \frac{1}{\sqrt{x}}dx &= \lim_{\epsilon\to 0^+}\int_\epsilon^1 \frac{1}{\sqrt{x}}dx\\ &= \lim_{\epsilon\to 0^+} 2\sqrt{x}|_\epsilon^1\\ &= \lim_{\epsilon\to 0^+} 2-2\sqrt{\epsilon}\\ &=2.\end{aligned}$$

Although the function is ill-behaved, the area is perfectly well defined. ▲

Example 4.7.9 An Infinite Area Under a Function That Approaches Infinity

If $g(x)=\frac{1}{x^2}$,

$$\begin{aligned}\int_0^1 \frac{1}{x^2}dx &= \lim_{\epsilon\to 0^+}\int_\epsilon^1 \frac{1}{x^2}dx\\ &= \lim_{\epsilon\to 0^+} -\frac{1}{x}|_\epsilon^1\\ &= \lim_{\epsilon\to 0^+} -1+\frac{1}{\epsilon}\\ &=\infty.\end{aligned}$$

▲

For power functions of the form $\frac{1}{t^p}$ and $p\neq 1$, we can derive a general result for convergence over the interval from 0 to a.

$$\begin{aligned}\int_0^a \frac{1}{t^p}dt &= \lim_{\epsilon\to 0^+}\int_\epsilon^a \frac{1}{t^p}dt\\ &= \lim_{\epsilon\to 0^+} \frac{t^{1-p}}{1-p}|_\epsilon^a\\ &= \lim_{\epsilon\to 0^+} \frac{a^{1-p}}{1-p}-\frac{\epsilon^{1-p}}{1-p}.\end{aligned}$$

If $p<1$, the integral converges, while if $p>1$, the limit is infinite and the integral diverges.

We can use the comparison test to check whether more complicated functions have finite integrals. The idea is the same as before: an integral is finite if the function is less than a function with a finite integral, and it is infinite if the function is greater than a function with an infinite integral.

Example 4.7.10 Using the Comparison Test When the Integrand Approaches Infinity

Consider the function

$$\int_0^1 \frac{1}{\sqrt{t}+t^2}dt.$$

Because $t^2<\sqrt{t}$ for $t<1$, we have that (Figure 4.7.12)

$$\frac{1}{\sqrt{t}+t^2}<\frac{1}{\sqrt{t}+\sqrt{t}}=\frac{1}{2\sqrt{t}}$$

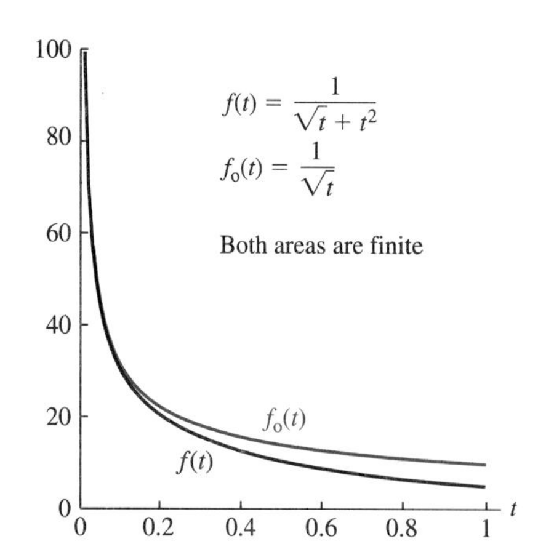

FIGURE 4.7.12
A convergent improper integral checked with the comparison test

Therefore,

$$\int_0^1 \frac{1}{\sqrt{t}+t^2}\,dt < \int_0^1 \frac{1}{2\sqrt{t}}\,dt = \sqrt{t}\,|_0^1 = 1.$$

Summary We have introduced **improper integrals** of two types: integrals with infinite limits of integration and integrals with infinite integrands. The first are defined as the limit when one limit of integration increases to infinity and the second as the limit when one limit of integration approaches a point where the function itself approaches infinity. When the limit exists, we say that the integral **converges**; if not, we say the integral **diverges**. We found conditions for convergence or divergence of integrals of power and exponential functions and extended these results with the **comparison test**. This test can be used to decide whether an infinite series converges or diverges.

4.7 Exercises

Mathematical Techniques

1–4 ▪ Many improper integrals can be evaluated by comparing functions with the method of leading behavior. State which of the given pair of functions approaches its limit more quickly, and demonstrate the result with L'Hôpital's rule when needed.

1. Which function approaches 0 faster as x approaches infinity: e^{-x} or $\frac{1}{x}$?
2. Which function approaches 0 faster as x approaches infinity: $\frac{1}{1+x^2}$ or $\frac{1}{1+x}$.
3. Which function approaches infinity faster as x approaches 0: $\frac{1}{x^2}$ or $\frac{1}{x}$?
4. Which function approaches infinity faster as x approaches 0: $\frac{1}{x}$ or $\frac{1}{\sqrt{x}}$?

5–10 ▪ Evaluate the following improper integrals or say why they don't converge.

5. $\int_0^\infty e^{-3t}\,dt$
6. $\int_0^\infty e^t\,dt$
7. $\int_1^\infty \frac{1}{\sqrt{x}}dx$
8. $\int_5^\infty \frac{1}{x^2}\,dx$
9. $\int_0^\infty \frac{1}{(1+3x)^{3/2}}\,dx$
10. $\int_0^\infty \frac{1}{(2+5x)^4}\,dx$

11–14 ▪ Evaluate the following improper integrals or say why they don't converge.

11. $\int_0^1 \frac{1}{x}\,dx$
12. $\int_0^{.001} \frac{1}{x^2}\,dx$
13. $\int_0^{.001} \frac{1}{\sqrt[3]{x}}\,dx$
14. $\int_0^\infty \frac{1}{\sqrt[3]{x}}\,dx$

15–18 ▪ Use the comparison test to deduce whether the following improper integrals converge. If they do, find an upper bound on the value.

15. $\int_0^1 \frac{1}{\sqrt[3]{x}+x^3}\,dx$
16. $\int_0^1 \frac{1}{x^2+e^x}\,dx$

17. $\int_1^\infty \frac{1}{\sqrt[3]{x}+x^3}\,dx$

18. $\int_1^\infty \frac{1}{\sqrt{x}+e^x}\,dx$

19–22 ▪ The method of leading behavior can be used to deduce whether some improper integrals converge. Choose the leading behavior of the denominator of each function and compare with the results using the comparison test.

19. $\int_0^1 \frac{1}{\sqrt[3]{x}+x^3}\,dx$

20. $\int_0^1 \frac{1}{x^2+e^x}\,dx$

21. $\int_1^\infty \frac{1}{\sqrt[3]{x}+x^3}\,dx$

22. $\int_1^\infty \frac{1}{\sqrt{x}+e^x}\,dx$

23–26 ▪ Compare the following series with the given integral to determine whether the sum approaches infinity.

23. Compare $\sum_{i=0}^{\infty} \frac{1}{2^i}$ with $\int_0^\infty \frac{2}{2^x}dx$.

24. Compare $\sum_{i=0}^{\infty} \frac{1}{3^i}$ with $\int_0^\infty \frac{1}{3^x}dx$.

25. Compare $\sum_{i=1}^{\infty} \frac{1}{i^2}$ with $\int_1^\infty \frac{1}{x^2}dx$.

26. Compare $\sum_{i=0}^{\infty} \frac{i}{2^i}$ with $\int_0^\infty \frac{2x}{2^x}dx$.

Applications

27–30 ▪ Write pure-time differential equations to describe the following situations, find out what happens over the long term, and state whether the rule could be followed indefinitely.

27. The volume of a cell is increasing at a rate of $\frac{100}{(1+t)^2}$ μm^3/s, starting from a size of 500 μm^3.

28. The concentration of a toxin in a cell is increasing at a rate of $50e^{-2t}$ μmol/L/s, starting from a concentration of 10 μmol/L. If the cell is poisoned when the concentration exceeds 30 μmol/L, could this cell survive?

29. A population of bacteria is increasing at a rate of $\frac{1000}{(2+3t)^{0.75}}$ bacteria per hour, starting from a population of 10^6. Could this sort of growth be maintained indefinitely? When would the population reach 2.0×10^6? Would you say that this population is growing quickly?

30. A population of bacteria is increasing at a rate of $\frac{1000}{(2+3t)^{1.5}}$ bacteria per hour, starting from a population of 1000. Could this sort of growth be maintained indefinitely? Would the population reach 2000?

Supplementary Problems

1. The voltage v of a neuron follows the differential equation

$$\frac{dv}{dt} = 1.0 + \frac{1}{1+0.02t} - e^{0.01t}$$

over the course of 100 ms, where t is measured in ms and v in millivolts. We start at $v(0) = -70$.

a. Sketch a graph of the rate of change. Indicate on your graph the times when the voltage reaches minima and maxima (you don't need to solve for the numerical values).

b. Sketch a graph of the voltage as a function of time.

c. What is the voltage after 100 ms?

2. Consider again the differential equation in the previous problem,

$$\frac{dv}{dt} = 1.0 + \frac{1}{1+0.02t} - e^{0.01t}$$

with $v(0) = -70$.

a. Use Euler's method to estimate the voltage after 1 ms, and again 1 ms after that.

b. Estimate the voltage after 2 ms using left-hand and right-hand Riemann sums.

c. Which of your estimates matches Euler's method and why?

3. A neuron in your brain sends a charge down an 80 cm long axon (a long skinny thing) toward your hand at a speed of 10 m per second. At the time when the charge reaches your elbow, the voltage in the axon is −70 mV except on the 6 cm long piece between 47 and 53 cm from your brain. On this piece, the voltage is

$$v(x) = -70.0 + 10.0[9.0 - (x - 50.0)^2]$$

where $v(x)$ is the voltage at a distance of x centimeters from the brain.

a. How long will it take the information to get to your hand? How long did it take to reach your elbow?

b. Sketch a graph of the voltage along the whole axon.

c. Find the average voltage of the 6 cm piece.

d. Find the average voltage of the whole axon.

4. The charge in a dead neuron decays according to

$$\frac{dv}{dt} = \frac{1}{\sqrt{1+4t}} - \frac{2}{(1+4t)^{\frac{3}{2}}}$$

starting again from $v(0) = -70$ at $t = 0$.

a. Is the voltage approaching 0 as $t \to \infty$? How do you know that it will eventually reach 0?

b. Write an equation (but don't solve it) for the time when the voltage reaches 0.

c. What is wrong with this model?

5. Consider the differential equations

$$\frac{db}{dt} = 2b$$

and

$$\frac{dB}{dt} = 1 + 2t.$$

a. Which of these is a pure-time differential equation? Describe circumstances when you might find each of these equations.

b. Suppose $b(0) = B(0) = 1$. Use Euler's method to find estimates for $b(0.1)$ and $B(0.1)$.

c. Use Euler's method again to find estimates for $b(0.2)$ and $B(0.2)$.

6. Consider the differential equation

$$\frac{dp}{dt} = e^{-4t}$$

where $p(t)$ is product in moles at time t and t is measured in seconds.

a. Explain in words what is going on.

b. Suppose $p(0) = 1$. Find $p(1)$.

7. Consider the differential equation

$$\frac{dV}{dt} = 4 - t^2$$

where $V(t)$ is volume in liters at time t and t is measured in minutes.

a. Explain in words what is going on.

b. At what time is the volume a maximum?

c. Break the interval from $t = 0$ to $t = 3$ into three parts and find the left-hand and right-hand estimates of the volume at $t = 3$ (assume $V(0) = 0$).

d. Write down the definite integral expressing volume at $t = 3$ and evaluate.

8. Find the area under the curve $f(x) = 3 + (1 + x/3)^2$ between $x = 0$ and $x = 3$.

9. The population density of trout in a stream is

$$\rho(x) = |-x^2 + 5x + 50|$$

where ρ is measured in trout per mile and x is measured in miles. x runs from 0 to 20.

a. Graph $\rho(x)$ and find the minimum and maximum.

b. Find the total number of trout in the stream.

c. Find the average density of trout in the stream.

d. Indicate on your graph how you would find where the actual density is equal to the average density.

10. The amount of product is described by the differential equation

$$\frac{dp}{dt} = \frac{1}{\sqrt{1 + 3t}}$$

starting at time $t = 0$. Suppose p is measured in moles, t in hours, and that $p(0) = 0$.

a. Find the limiting amount of product.

b. Find the average rate at which product is produced as a function of time, and compute the limit as $t \to \infty$.

c. Find the limit as $t \to 0$ of the average rate at which product is produced.

11. A student is hooked up to an EEG during a test, and her α brain wave power follows

$$A(t) = \frac{50}{2.0 + 0.3t} + 10e^{0.0125t}$$

where t runs from 0 to 120 minutes.

a. Convince yourself that brain wave activity has a minimum value some time during the test. Sketch a graph of the function. Find the maximum.

b. Find the total brain wave power during the test.

c. Find the average brain wave power during the test. Sketch the corresponding line on your graph.

d. Estimate the minimum value from your average value.

e. Draw a graph showing how you would estimate the total using the right-hand approximation with $n = 6$. Write down the associated sum. Do you think your estimate is high or low?

12. The density of sugar in a hummingbird's 20 mm long tongue is

$$s(x) = \frac{1.2}{1.0 + 0.2x}$$

where x is measured in mm from the end of the tongue and s is measured in mol/m.

a. Find the total amount of sugar in the hummingbird's tongue.

b. Find the average density of sugar in the tongue.

c. Compare the average with the minimum and maximum density. Does your answer make sense?

13. Consider the function $G(h)$ giving the density of nutrients in a plant stem as a function of the height h

$$G(h) = 5 + 3e^{-2h}$$

where G is measured in mol/m and h is measured in m.

a. Find the total amount of nutrient if the stem is 2.0 m tall.

b. Find the average density in the stem.

c. Find the exact and approximate amount between 1.0 and 1.01 m.

Projects

1. As noted in the text, Euler's method is not a very good way to numerically solve differential equations. In this project, you will compare Euler's method with the **midpoint** or **second-order Runge-Kutta** method and the **implicit Euler** method.

Suppose we know (or estimate) that the solution of the general differential equation

$$\frac{dy}{dt} = f(t, y)$$

is some value $y = y_0$ at time $t = t_0$. Our goal is to estimate the value of $y_1 = y(t_0 + h)$ with **step size** h. We will use y_1, y_2, and so forth to represent these approximate values.

We have seen that Euler's method uses the tangent line to estimate

$$y_1 = y(t_0) + hf(t_0, y_0).$$

This method has the problem that it only uses the derivative at the beginning of the interval between times t_0 and $t_0 + h$. If the derivative changes significantly during this interval, the method can be very inaccurate.

An alternative called the midpoint method uses an estimate of the derivative at time $t_0 + \frac{h}{2}$ instead. The next step is

$$k = hf(t_0, y_0)$$

$$y_1 = y(t_0) + hf\left(t_0 + \frac{h}{2}, y_0 + \frac{k}{2}\right).$$

k is the estimated of change from Euler's method, and we use it to guess the value of y half way through the interval. This method is more accurate than Euler's method.

Both of these methods have the problem that they are **unstable**, meaning that the approximate solutions they produce can fail to approach a stable equilibrium if the step size h is too large. The **implicit Euler** scheme is stable for linear equations. The idea is to use the derivative at the end of the interval instead of the beginning, or

$$y_1 = y(t_0) + hf(t_0 + h, y_1).$$

Because we do not know the value of y_1, we have to solve for it.

This project involves comparing these three methods on four differential equations

$$\frac{db}{dt} = b$$

$$\frac{dx}{dt} = -x$$

$$\frac{dV}{dt} = -e^{-t}$$

$$\frac{dy}{dt} = -e^{-t} + e^{-y}.$$

Suppose the state variable in each equation has the initial value of 1.0 at $t = 0$.

a. First, find the solution of each at $t = 1$. Try each method (if you can figure out how to get the implicit method to work on the last equation) with values of h ranging from 1.0 down to 0.001.

b. The equations for x and V should both approach 0 exponentially as t becomes large. Use large values of h (such as 10 or 100) in each of the three methods. How well do they do for large t?

c. The midpoint method is related to approximating the solution with a quadratic Taylor polynomial (Section 3.7). Develop an extension of Euler's method based on the quadratic approximation and compare with the midpoint method.

W. H. Press, S. A. Teukolsky, W. T. Vettering, and B. P. Flannery, *Numerical Recipes in FORTRAN: The Art of Scientific Computing*. Cambridge University Press, Cambridge, 1992.

2. Back in the days of the Greeks, people were fascinated by geometric problems such as finding the area of a circle. Archimedes, perhaps the greatest mathematician in the ancient world, came up with an idea closely related to the Riemann sum to solve this problem. Using some modern tools, we can use his ideas to compute the value of π.

a. Break a circle of radius 1 into n wedges, each with angle $\theta = \frac{2\pi}{n}$. Show that the area of a right triangle inside each wedge is $\sin(\theta)/2$. Make sure to draw a picture.

b. Approximate the area of a circle using $n = 4$. Look up (or remember) the half-angle formula giving $\sin\left(\frac{\theta}{2}\right)$ in terms of $\cos(\theta)$ to approximate the area with $n = 8$.

c. Continue in this way to approximate the area with $n = 16$ and so forth.

d. Use the same approach but approximate each wedge with a right triangle that lies outside the circle. Show that the area of the triangle is $\tan(\theta)/2$.

e. Approximate the area of a circle using $n = 8$ (what happens when $n = 4$?). Look up (or remember) the half-angle formula giving $\tan\left(\frac{\theta}{2}\right)$ in terms of $\tan(\theta)$ to approximate the area with $n = 8$.

f. Continue in this way to approximate the area with $n = 16$ and so forth.

g. Can you think of a better way Archimedes could have used to compute the value of π?

P. Beckmann, *A History of π*. Golem Press, Boulder, Colo. 1982.

Chapter 5

Analysis of Autonomous Differential Equations

We have developed the indefinite integral and definite integral as tools to solve **pure-time differential equations**, and linked them with the Fundamental Theorem of Calculus. We will now use integrals and graphical tools to study **autonomous differential equations**, when the rate of change is a **rule** relating change to the current state. In particular, we will use the **phase-line diagram** to find **equilibria** and their **stability**, much as we used cobwebbing to find equilibria and stability of discrete-time dynamical systems.

Because most biological systems involve several interacting measurements, we will extend our methods to address **systems of autonomous differential equations**, where the rate of change of each measurement depends on the value of the others. Generalization of the phase line to the **phase plane** allows study of two-dimensional systems.

5.1 Basic Differential Equations

Temperature change, chemical exchange in the lung, and selection can be simply and accurately modeled with **autonomous differential equations**. In each case, the model describes the same biological situation as a related discrete-time dynamical system, but the equation usually looks completely different. After reviewing the terminology for autonomous differential equations, we will derive these fundamental equations from biological or physical laws.

5.1.1 Review of Autonomous Differential Equations

We have developed techniques to solve pure-time differential equations with the general form

$$\frac{dV}{dt} = f(t). \tag{5.1.1}$$

In these equations the rate of change is a function of time and not of the state variable. Change is imposed from outside the system, as in a lake level controlled by the weather or an engineer. In most biological systems, however, the rate of change depends on the current state of the system as well as external factors. A lake that experienced increased evaporation at high lake levels would change as a function of the level itself and not just as a function of the weather or decisions of engineers.

The class of differential equations where the rate of change depends only on the state of the system and not on external circumstances is particularly important. These equations are called **autonomous differential equations** (Section 4.1). For example, an autonomous differential equation for the growth of a bacterial population $b(t)$ is

$$\frac{db}{dt} = 2.0b.$$

The rate of change of the population size b depends only on the population size b and not on the time t.

The general autonomous differential equation for a measurement m is

$$\frac{dm}{dt} = g(m). \tag{5.1.2}$$

It differs from the general pure-time differential equation in that the rate of change $g(m)$ depends on m rather than t. Autonomous differential equations are slightly harder to solve than pure-time differential equations, but the techniques of solution are different and the solutions are often more interesting.

When the rate of change depends on both the state variable and the time, the equation is called a **nonautonomous differential equation**. The general form is

$$\frac{dm}{dt} = g(m, t). \tag{5.1.3}$$

These equations are generally much more difficult to solve than either pure-time or autonomous differential equations.

Although the *rate-of-change* function in an autonomous differential equation does not depend explicitly on time, the *solution* does.

Example 5.1.1 The Autonomous Differential Equation for Population Growth

Suppose that a bacterial culture with 1.0×10^6 bacteria is refrigerated to prevent production, and is warmed to room temperature at 9:00 a.m. and allowed to reproduce. The population $b(t)$ might follow the autonomous differential equation

$$\frac{db}{dt} = 2.0b.$$

Before we develop more formal mathematical methods, we can find the solution by guessing. What function has a derivative equal to double itself? Remember that the derivative of the exponential function $b(t) = b_0 e^{\lambda t}$ is

$$\frac{db}{dt} = \lambda b_0 e^{\lambda t} = \lambda b(t)$$

(Subsection 2.9.3). The guess $b(t) = 1.0 \times 10^6 e^{2.0t}$ is a solution of the differential equation because

$$\begin{aligned}
\frac{db}{dt} &= \frac{d}{dt}(1.0 \times 10^6 e^{2.0t}) && \text{plugging in the formula} \\
&= 1.0 \times 10^6 \frac{d}{dt} e^{2.0t} && \text{constant product rule} \\
&= 1.0 \times 10^6 \cdot 2.0 e^{2.0t} && \text{derivative of exponential} \\
&= 2.0 \cdot 1.0 \times 10^6 e^{2.0t} && \text{reorganizing} \\
&= 2.0 b(t). && \text{recognizing the formula for } b(t)
\end{aligned}$$

Furthermore, this formula matches the initial condition because

$$b(0) = 1.0 \times 10^6 e^{2.0 \cdot 0} = 1.0 \times 10^6.$$

Like a population described by the discrete-time dynamical system $b_{t+1} = rb_t$ (equation 4.1.3 on page 352), this population grows exponentially. Although the **formula** for the rate of change always looks the same, the **value** of the rate of change (and the population size itself) becomes larger over time (Figure 5.1.1).

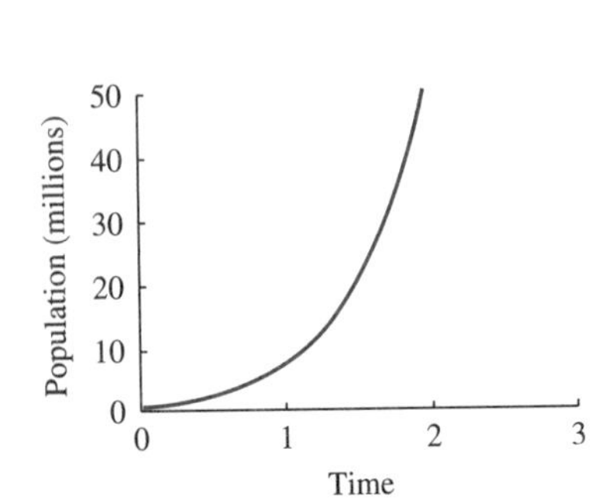

FIGURE 5.1.1
An autonomous bacterial population

Euler's method (described for pure-time differential equations in Algorithm 4.1) extends directly to autonomous differential equations.

Algorithm 5.1 Euler's Method for Solving an Autonomous Differential Equation

Suppose a measurement m obeys the autonomous differential equation

$$\frac{dm}{dt} = f(m)$$

with initial condition $m(t_0) = m_0$.

1. Choose a **time step** Δt (the length of time between estimated values).
2. Use the initial condition and the differential equation to find the tangent line $\hat{m}(t)$ with base point $t = t_0$ and slope $f(m_0)$. Use it to estimate $\hat{m}(t_0 + \Delta t)$.
3. Use the estimate $\hat{m}(t_0 + \Delta t)$ and the differential equation to find the tangent line $\hat{m}(t)$ with base point $t_0 + \Delta t$ and estimate $\hat{m}(t_0 + 2\Delta t)$.
4. Repeat the previous step to find $\hat{m}(t_0 + 3\Delta t)$ and so forth.

Example 5.1.2 Applying Euler's Method to a Differential Equation for Population Growth

Suppose we wish to apply Euler's method to the autonomous differential equation for population growth

$$\frac{db}{dt} = 2.0b$$

with initial condition $b(0) = 1.0$ to estimate $b(1)$.

1. Pick a time step of $\Delta t = 0.25$ (a smaller value should be more accurate).
2. The tangent line with base point $t = 0$ is

$$\begin{aligned} \hat{b}(0 + \Delta t) &= b(0) + b'(0)\Delta t && \text{equation for tangent line} \\ &= 1.0 + 2.0 \cdot \Delta t. && b(0) = 1.0 \text{ (initial condition) and } b'(0) = 2.0b(0) = 2.0 \text{ (differential equation)} \end{aligned}$$

   We therefore estimate that

$$\hat{b}(0.25) = 1.0 + 2.0 \cdot 0.25 = 1.5.$$

3. For the next step, to find $\hat{b}(0.5)$, we compute

$$\begin{aligned} \hat{b}(0.25 + \Delta t) &= b(0.25) + b'(0.25)\Delta t && \text{equation for tangent line} \\ &\approx \hat{b}(0.25) + b'(0.25)\Delta t && \text{substitute } \hat{b}(0.25) \text{ for unknown } b(1) \\ &\approx \hat{b}(0.25) + 2.0\hat{b}(0.25)\Delta t && \text{substitute } 2.0\hat{b}(0.25) \text{ for unknown } b'(1) \\ &= 1.5 + 3.0\Delta t. && \hat{b}(0.25) = 1.5 \end{aligned}$$

   We therefore estimate that

$$\hat{b}(0.5) = 1.5 + 3.0 \cdot 0.25 = 2.25.$$

4. To find $\hat{b}(0.75)$,

$$\begin{aligned} \hat{b}(0.5 + \Delta t) &= \hat{b}(0.5) + b'(0.5)\Delta t \\ &\approx \hat{b}(0.5) + 2.0\hat{b}(0.5)\Delta t \\ &= 2.25 + 4.5 \cdot \Delta t \\ \hat{b}(0.75) &= 2.25 + 4.5 \cdot 0.25 = 3.375. \end{aligned}$$

   Finally, to find $\hat{b}(1.0)$,

$$\begin{aligned} \hat{b}(0.75 + \Delta t) &= \hat{b}(0.75) + b'(0.75)\Delta t \\ &\approx \hat{b}(0.75) + 2.0\hat{b}(0.75)\Delta t \\ &= 3.375 + 6.75 \cdot \Delta t \\ \hat{b}(0.75) &= 3.375 + 6.75 \cdot 0.25 = 5.0625. \end{aligned}$$

These values, along with a comparison with the exact solution $b(t) = e^{(2.0t)}$, are plotted in Figure 5.1.2.

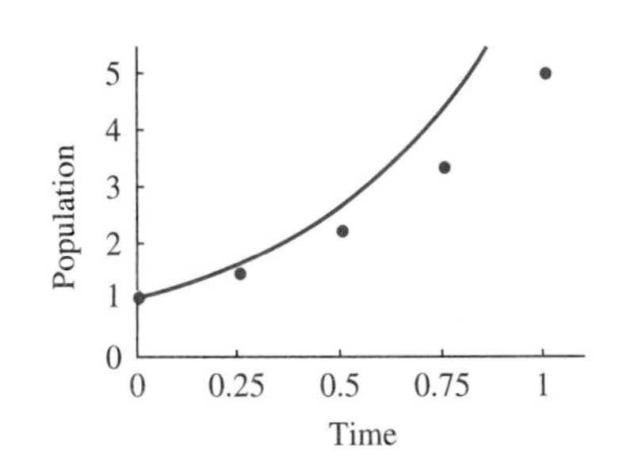

FIGURE 5.1.2
Euler's method applied to an autonomous differential equation

In this example, the approximation given by Euler's method falls further and further behind. Unlike applying Euler's method to pure-time differential equations, where the rate of change is exact, both the value of the function and the rate of change are approximate when applied to an autonomous differential equation.

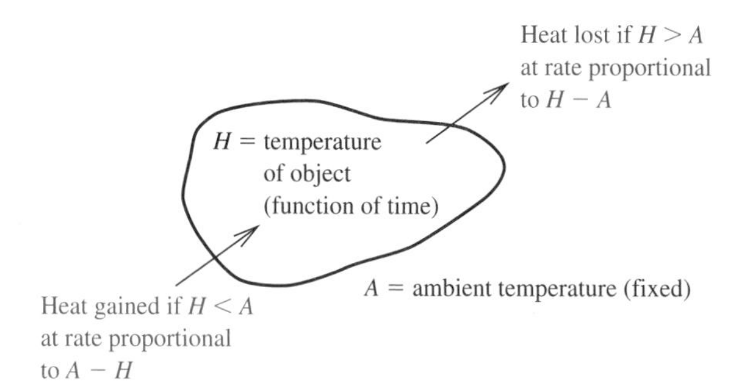

FIGURE 5.1.3
Newton's law of cooling

5.1.2 Newton's Law of Cooling

Both the concentration of a substance in a fluid and its temperature change continuously in time, making these appropriate quantities to model with differential equations. As we saw in our study of pure-time differential equations, it can be easier to measure the amount of fluid crossing a membrane than to measure the entire quantity of fluid inside. More importantly, the rules describing the movement of fluids and heat can be described compactly and intuitively with autonomous differential equations.

The basic idea is expressed by **Newton's law of cooling**, which states that

> the rate at which heat is lost from an object is proportional to the difference between the temperature of the object and the ambient temperature.

Denote the temperature of an object by H, and the fixed ambient temperature by A (Figure 5.1.3). Newton's law of cooling can be written as the differential equation

$$\frac{dH}{dt} = \alpha(A - H). \tag{5.1.4}$$

where α is a positive constant with dimensions of 1/time. This is an **autonomous** differential equation because it describes a **rule**, and because the rate of change is a function of the state variable H and not the time t. Remember that H is a function of time but that α and A are constants.

If the temperature of the object is higher than the ambient temperature ($H > A$), the rate of change of temperature is negative and the object cools. If the temperature of the object is lower than the ambient temperature ($H < A$), the rate of change of temperature is positive and the object warms up. If the temperature of the object is equal to the ambient temperature, the rate of change of temperature is zero, and the temperature of the object does not change.

The parameter α, with dimensions of 1/time, depends on the **specific heat** of the material and on the shape. Materials with high specific heat, like water, retain heat for a long time, have low values of α, and experience small rates of temperature change for a given temperature difference. Materials with low specific heat, like metals, lose heat rapidly, have large values of α, and experience high rates of temperature change for a given temperature difference. The parameter α also depends on the surface area to volume ratio of the object. An object with a large exposed surface relative to its volume will heat or cool rapidly, as a shallow puddle freezes quickly on a cold evening. An object with a small surface area heats or cools more slowly.

Although it is possible to guess the solution or solve the equation mathematically with the method of separation of variables (Section 5.4), we will first study Newton's law of cooling with Euler's method.

Example 5.1.3 Applying Euler's Method to Newton's Law of Cooling

Suppose that α has the relatively small value of 0.1 per minute, that the ambient temperature A is 20°C, and that the initial condition is $H(0) = 40$°C. To study how the

temperature changes over the first 10 minutes, we choose a time step of $\Delta t = 1$. The results for Euler's method are shown in the following table.

t	$\hat{H}(t+\Delta t)$	$t+\Delta t$	$\hat{H}(t+\Delta t)$
0	$H(0) + H'(0)\Delta t = 40.0 + 0.1(20 - 40)\Delta t$	1	38.0
1	$\hat{H}(1) + H'(1)\Delta t = 38.0 + 0.1(20 - 38.0)\Delta t$	2	36.2
2	$\hat{H}(2) + H'(2)\Delta t = 36.2 + 0.1(20 - 36.2)\Delta t$	3	34.6
3	$\hat{H}(3) + H'(3)\Delta t = 34.6 + 0.1(20 - 34.6)\Delta t$	4	33.1
4	$\hat{H}(4) + H'(4)\Delta t = 33.1 + 0.1(20 - 33.1)\Delta t$	5	31.8
5	$\hat{H}(5) + H'(5)\Delta t = 31.8 + 0.1(20 - 31.8)\Delta t$	6	30.6
6	$\hat{H}(6) + H'(6)\Delta t = 30.6 + 0.1(20 - 30.6)\Delta t$	7	29.6
7	$\hat{H}(7) + H'(7)\Delta t = 29.6 + 0.1(20 - 29.6)\Delta t$	8	28.6
8	$\hat{H}(8) + H'(8)\Delta t = 28.6 + 0.1(20 - 28.6)\Delta t$	9	27.7
9	$\hat{H}(9) + H'(9)\Delta t = 27.7 + 0.1(20 - 27.7)\Delta t$	10	27.0

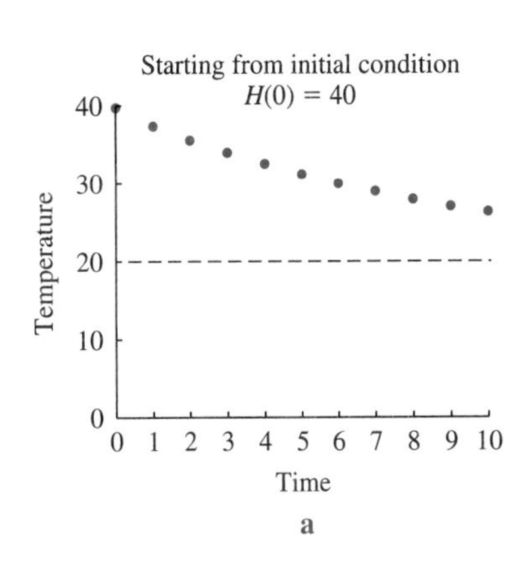

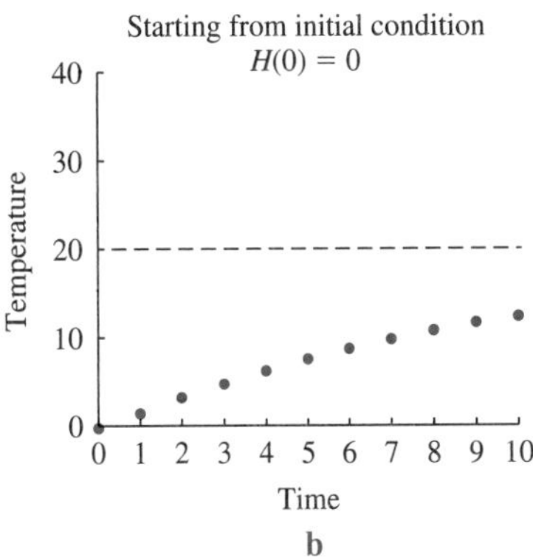

FIGURE 5.1.4
Newton's law of cooling approximated with Euler's method: small α

If we start instead at $H(0) = 0°C$,

t	$\hat{H}(t+\Delta t)$	$t+\Delta t$	$\hat{H}(t+\Delta t)$
0	$H(0) + H'(0)\Delta t = 0.0 + 0.1(20 - 0.0)\Delta t$	1	2.0
1	$\hat{H}(1) + H'(1)\Delta t = 2.0 + 0.1(20 - 2.0)\Delta t$	2	3.8
2	$\hat{H}(2) + H'(2)\Delta t = 3.8 + 0.1(20 - 3.8)\Delta t$	3	5.4
3	$\hat{H}(3) + H'(3)\Delta t = 5.4 + 0.1(20 - 5.4)\Delta t$	4	6.9
4	$\hat{H}(4) + H'(4)\Delta t = 6.9 + 0.1(20 - 6.9)\Delta t$	5	8.2
5	$\hat{H}(5) + H'(5)\Delta t = 8.2 + 0.1(20 - 8.2)\Delta t$	6	9.4
6	$\hat{H}(6) + H'(6)\Delta t = 9.4 + 0.1(20 - 9.4)\Delta t$	7	10.4
7	$\hat{H}(7) + H'(7)\Delta t = 10.4 + 0.1(20 - 10.4)\Delta t$	8	11.4
8	$\hat{H}(8) + H'(8)\Delta t = 11.4 + 0.1(20 - 11.4)\Delta t$	9	12.3
9	$\hat{H}(9) + H'(9)\Delta t = 12.3 + 0.1(20 - 12.3)\Delta t$	10	13.0

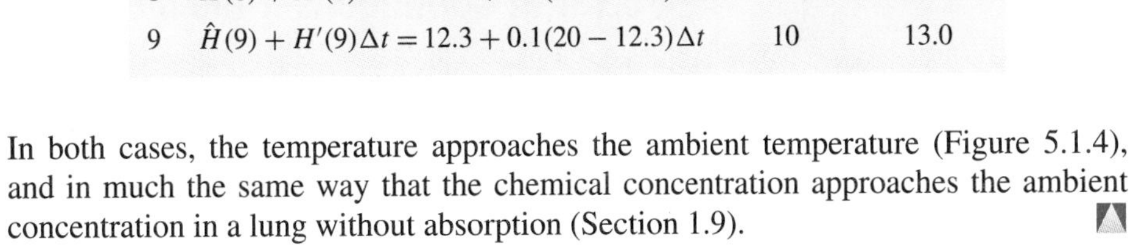

In both cases, the temperature approaches the ambient temperature (Figure 5.1.4), and in much the same way that the chemical concentration approaches the ambient concentration in a lung without absorption (Section 1.9).

5.1.3 Diffusion Across a Membrane

The processes of heat exchange and chemical exchange have many parallels (which is remarkable given that heat is **not** a substance, as was once thought). The underlying mechanism is that of **diffusion**. Substances leave the cell at a rate proportional to their concentration inside and enter the cell at a rate proportional to their concentration outside. The constants of proportionality depend on the properties of the substance and the properties of the membrane separating the two regions (Figure 5.1.5).

Denote the concentration inside the cell by C, the concentration outside by Γ (the Greek capital letter gamma), and the constant of proportionality by β. The rate at which the chemical leaves the cell is βC, and the rate at which it enters is $\beta\Gamma$. Therefore,

$$\begin{aligned}\frac{dC}{dt} &= \text{rate at which chemical enters} - \text{rate at which chemical leaves}\\ &= \beta\Gamma - \beta C\\ &= \beta(\Gamma - C).\end{aligned} \tag{5.1.5}$$

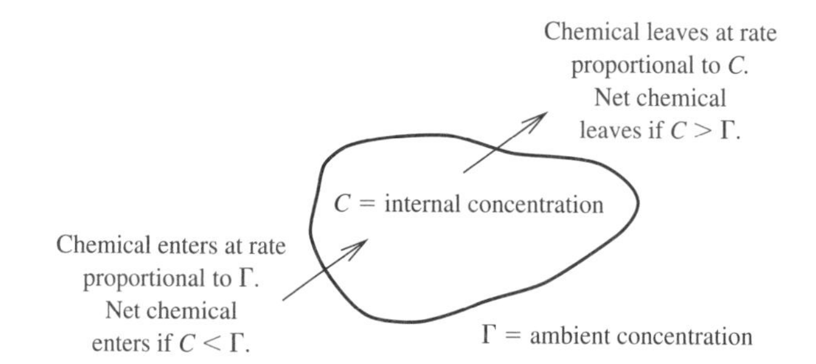

FIGURE 5.1.5
Passive diffusion between two regions

Except for the letters, this is exactly the same as Newton's law of cooling! The rate of change of concentration is proportional to the difference in concentration inside and outside the cell.

This model looks completely different from the discrete-time dynamical system for the lung,

$$c_{t+1} = (1 - q)c_t + q\gamma$$

(Equation 1.9.1), even though it describes a very similar situation. To be fair, the differential equation gives the rate of change of C, not the new value. The *change* in concentration for the lung described by the discrete-time dynamical system is

$$\Delta c = c_{t+1} - c_t = -qc_t + q\gamma = q(\gamma - c_t).$$

The two models now look similar, with the fraction exchanged q playing the role of the rate β.

Because this model is identical to Newton's law of cooling, their solutions must match. The concentration of chemical will get closer and closer to the ambient concentration, just as it did with the discrete-time dynamical system.

5.1.4 A Continuous-Time Model of Selection

Suppose we have two strains of bacteria, with population sizes denoted a and b, with rate of growth proportional to population size (Equation 4.1.4). If the per capita growth rate of strain a is μ and that of b is λ, they follow the equations

$$\frac{da}{dt} = \mu a$$

$$\frac{db}{dt} = \lambda b.$$

If $\mu > \lambda$, type a has a higher per capita growth than b and we expect it to take over the population. However, we can often measure only the *fraction* of bacteria of type a and not the total number. We would like to write a differential equation for the fraction, and can do so by following steps much like those used to find the discrete-time dynamical system for competing bacteria (Section 1.10).

The fraction p of type a is

$$p = \frac{a}{a + b}.$$

Also,

$$1 - p = \frac{b}{a + b}$$

because $1 - p$ is the fraction of bacteria of type b.

We can use the quotient rule (Theorem 2.11) to compute the derivative of p, finding

$$
\begin{aligned}
\frac{dp}{dt} &= \frac{d}{dt}\left(\frac{a}{a+b}\right) && \text{the derivative of } p \\
&= \frac{(a+b)\frac{da}{dt} - a\frac{d(a+b)}{dt}}{(a+b)^2} && \text{the quotient rule} \\
&= \frac{a\frac{da}{dt} + b\frac{da}{dt} - a\frac{da}{dt} - a\frac{db}{dt}}{(a+b)^2} && \text{the sum rule} \\
&= \frac{b\frac{da}{dt} - a\frac{db}{dt}}{(a+b)^2} && \text{cancel the } a\tfrac{da}{dt} \text{ terms} \\
&= \frac{\mu ab - \lambda ab}{(a+b)^2} && \text{plug in differential equations for } a \text{ and } b \\
&= \frac{(\mu - \lambda)ab}{(a+b)^2}. && \text{factor}
\end{aligned}
$$

We are not yet done, however. To write this as an autonomous differential equation, we must express the derivative as a function of the state variable p, and not the unmeasurable total populations a and b. Then

$$
\begin{aligned}
\frac{dp}{dt} &= \frac{(\mu - \lambda)ab}{(a+b)^2}. && \text{previous equation} \\
\frac{dp}{dt} &= (\mu - \lambda)\left(\frac{a}{a+b}\right)\left(\frac{b}{a+b}\right) && \text{factor} \\
\frac{dp}{dt} &= (\mu - \lambda)p(1-p). && \text{write in terms of } p \text{ and } 1-p
\end{aligned}
$$

Although our derivation was similar, this differential equation looks nothing like the discrete-time dynamical system for this system,

$$p_{t+1} = \frac{sp_t}{sp_t + r(1-p_t)}$$

(Equation 1.10.7). This is a key point. Never try to make a discrete-time dynamical system into a differential equation by cheerfully changing the letters. It is safer to go back to the underlying mechanism. In addition, the parameters in the two equations have different meanings. The parameters s and r in the discrete-time dynamical system represent the per capita production, the number of bacteria per bacterium in one generation, while the parameters μ and λ in the differential equation represent the per capita growth **rates**, the rates at which one bacterium produces new bacteria.

Like the discrete-time dynamical system describing selection, this differential equation is **nonlinear** because the state variable p appears inside a nonlinear function, in this case a quadratic. As before, nonlinear equations are generally much more difficult to solve. In this case, however, the structure of the system makes it possible to find a solution. We begin by solving each of the differential equations for a and b. Suppose the initial number of type a bacteria is a_0 and the initial number of type b is b_0 (even though these values cannot be measured). The solutions are

$$
\begin{aligned}
a(t) &= a_0 e^{\mu t} \\
b(t) &= b_0 e^{\lambda t}
\end{aligned}
$$

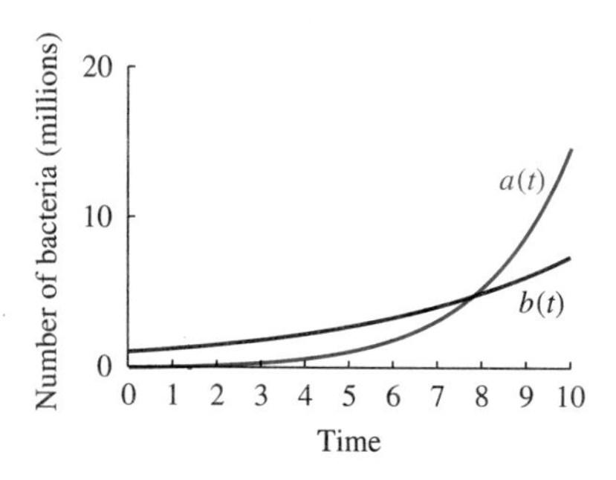

FIGURE 5.1.6

Exponential growth of two bacterial populations: $a_0 = 0.1$, $b_0 = 1.0$, $\mu = 0.5$, $\lambda = 0.2$

(Figure 5.1.6). We can check these directly,

$$\begin{aligned}\frac{da}{dt} &= \frac{d}{dt} a_0 e^{\mu t} && \text{taking the derivative}\\ &= a_0 \frac{d}{dt} e^{\mu t} && \text{the constant product rule}\\ &= a_0 \mu e^{\mu t} && \text{the derivative of an exponential function}\\ &= \mu a(t). && \text{recognize the formula for } a\end{aligned}$$

Checking the solution for $b(t)$ is similar.

We can now find the equation for the fraction p as a function of t as

$$\begin{aligned}p(t) &= \frac{a(t)}{a(t) + b(t)}\\ &= \frac{a_0 e^{\mu t}}{a_0 e^{\mu t} + b_0 e^{\lambda t}}.\end{aligned}$$

Again, we are not quite done. The initial conditions a_0 and b_0 cannot be measured, only the initial fraction $p(0) = p_0$. However, we do know that

$$p_0 = \frac{a_0}{a_0 + b_0}.$$

Dividing the numerator and denominator of our expression for $p(t)$ by $a_0 + b_0$,

$$\begin{aligned}p(t) &= \frac{\frac{a_0}{a_0+b_0} e^{\mu t}}{\frac{a_0}{a_0+b_0} e^{\mu t} + \frac{b_0}{a_0+b_0} e^{\lambda t}}\\ &= \frac{p_0 e^{\mu t}}{p_0 e^{\mu t} + (1 - p_0) e^{\lambda t}}.\end{aligned} \quad \textbf{(5.1.6)}$$

Checking this solution requires some algebraic gymnastics.

Example 5.1.4 Behavior of the Solution of the Selection Model

We can use the methods of reasoning about functions (Section 3.4) to study this solution. Suppose we use the parameter values $\mu = 0.5$ and $\lambda = 0.2$ and the initial condition

$$p_0 = \frac{a_0}{a_0 + b_0} = \frac{0.1}{0.1 + 1.0} = 0.091$$

(as in Figure 5.1.6). The solution is

$$\begin{aligned}p(t) &= \frac{0.091 e^{0.5t}}{0.091 e^{0.5t} + (1 - 0.091) e^{0.2t}}\\ &= \frac{0.091 e^{0.5t}}{0.091 e^{0.5t} + 0.909 e^{0.2t}}.\end{aligned}$$

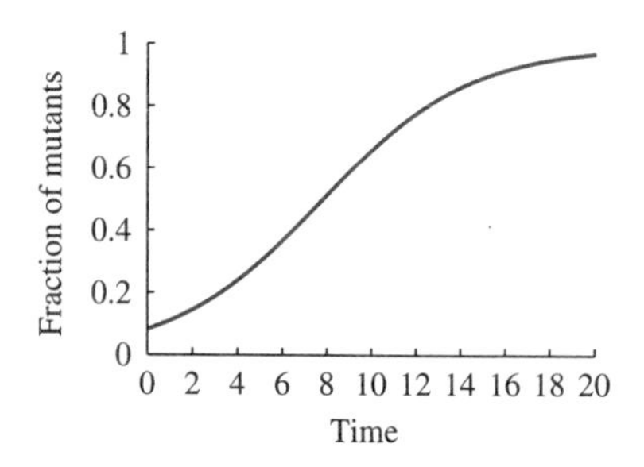

FIGURE 5.1.7

Solution for the fraction of mutant bacteria

Because $0.5 > 0.2$, the leading behavior of the denominator for large values of t is the term $0.091e^{0.5t}$ (Section 3.6). Therefore, for large t

$$p(t) = \frac{0.091 e^{0.5t}}{0.091 e^{0.5t}} = 1.$$

The fraction of mutants does approach 1 as t approaches infinity (Figure 5.1.7).

Summary The rate of change of an **autonomous differential equation** depends on the state variable and not on the time. The rate of change of a **nonautonomous differential equation** can depend on both the state variable and the time. We have introduced autonomous differential equations describing three processes: cooling, diffusion, and selection, finding that the first two processes are governed by the same equation. The equation for selection, derived by returning to the underlying model, is a **nonlinear** equation, but can be solved by combining the solutions for each separate population.

5.1 Exercises

Mathematical Techniques

1–4 ▪ Identify the following as pure-time, autonomous, or nonautonomous differential equations. In each case, identify the state variable.

1. $\frac{dF}{dt} = F^2 + kt$.
2. $\frac{dx}{dt} = \frac{x^2}{(x-\lambda)}$.
3. $\frac{dy}{dt} = \mu e^{-t} - 1$.
4. $\frac{dm}{dt} = \frac{e^{\alpha m} m^2}{(mt - \lambda)}$.

5–8 ▪ For the given time, value of the state variable, and values of the parameters, say whether the state variable is increasing, decreasing, or remaining unchanged.

5. $t=0$, $F=1$, and $k=1$ in the differential equation in Exercise 1.
6. $t=0$, $x=1$, and $\lambda=2$ in the differential equation in Exercise 2.
7. $t=1$, $y=1$, and $\mu=2$ in the differential equation in Exercise 3.
8. $t=2$, $m=0$, $\alpha=2$, and $\lambda=1$ in the differential equation in Exercise 4.

9–14 ▪ The following exercises compare the behavior of two similar-looking differential equations, the pure-time differential equation $\frac{dp}{dt} = t$ and the autonomous differential equation $\frac{db}{dt} = b$.

9. Use integration to solve the pure-time differential equation starting from the initial condition $p(0)=1$, find $p(1)$, and sketch the solution.
10. Solve the pure-time differential equation starting from the initial condition $p(1)=1$, find $p(2)$, and add the curve to your graph.
11. Check that the solution of the autonomous differential equation starting from the initial condition $b(0)=1$ is $b(t)=e^t$. Find $b(1)$ and sketch the solution.
12. Check that the solution of the autonomous differential equation starting from the initial condition $b(1)=1$ is $b(t)=e^{t-1}$. Find $b(2)$ and add to the sketch of the solution.
13. Exercises 9 and 10 give the value of p one time unit after it took on the value 1. Why don't the two answers match? (This behavior is typical of pure-time differential equations.)
14. Exercises 11 and 12 give the value of b one time unit after it took on the value 1. Why do the two answers match? (This behavior is typical of autonomous differential equations.)

15–18 ▪ Check the solution of the autonomous differential equation, making sure that it also matches the initial condition.

15. Check that $x(t) = -\frac{1}{2} + \frac{3}{2}e^{2t}$ is a solution of the differential equation $\frac{dx}{dt} = 1 + 2x$ with initial condition $x(0)=1$.
16. Check that $b(t) = 10e^{3t}$ is a solution of the differential equation $\frac{db}{dt} = 3b$ with initial condition $b(0)=10$.
17. Check that $G(t) = 1 + e^t$ is a solution of the differential equation $\frac{dG}{dt} = G - 1$ with initial condition $G(0)=2$.
18. Check that $z(t) = 1 + \sqrt{1+2t}$ is a solution of the differential equation $\frac{dz}{dt} = \frac{1}{z-1}$ with initial condition $z(0)=2$.

19–22 ▪ Use Euler's method to estimate the solution of the differential equation at the given time, and compare with the value given by the exact solution. Sketch a graph of the solution along with the lines predicted by Euler's method.

19. Estimate $x(2)$ if x obeys the differential equation $\frac{dx}{dt} = 1 + 2x$ with initial condition $x(0)=1$. Use Euler's method with $\Delta t = 1$ for two steps. Compare with the exact answer in Exercise 15.
20. Estimate $b(1.0)$ if b obeys the differential equation $\frac{db}{dt} = 3b$ with initial condition $b(0)=10$. Use Euler's method with $\Delta t = 0.5$ for two steps. Compare with the exact answer in Exercise 16.
21. Estimate $G(1.0)$ if G obeys the differential equation $\frac{dG}{dt} = G - 1$ with initial condition $G(0)=2$. Use Euler's method with $\Delta t = 0.2$ for five steps. Compare with the exact answer in Exercise 17.
22. Estimate $z(4.0)$ if z obeys the differential equation $\frac{dz}{dt} = \frac{1}{z-1}$ with initial condition $z(0)=2$. Use Euler's method with $\Delta t = 1.0$ for four steps. Compare with the exact answer in Exercise 18.

23–26 ▪ The derivation of the differential equation for p in the text requires combining two differential equations for a and b. Often, one can find a differential equation for a new variable derived from a single equation. In the following cases, use the chain rule to derive a new differential equation.

23. Suppose $\frac{dx}{dt} = 2x - 1$. Set $y = 2x - 1$ and find a differential equation for y. The end result should be simpler than the original equation.
24. Suppose $\frac{db}{dt} = 4b + 2$. Set $z = 4b + 2$ and find a differential equation for z. The end result should be simpler than the original equation.
25. Suppose $\frac{dx}{dt} = x + x^2$. Set $y = \frac{1}{x}$ and find a differential equation for y. This transformation changes a nonlinear differential equation for x into a linear differential equation for y (this is called a Bernoulli differential equation).
26. Suppose $\frac{dx}{dt} = 2x + \frac{1}{x}$. Set $y = x^2$ and find a differential equation for y. This is another example of a Bernoulli differential equation.

Applications

27–30 ▪ The simple model of bacterial growth assumes that per capita production does not depend on population size.

The following problems help you derive models of the form

$$\frac{db}{dt} = \lambda(b)b$$

where the per capita production λ is a function of the population size b.

27. One widely used nonlinear model of competition is the logistic model, where per capita production is a linearly decreasing function of population size. Suppose that per capita production has a maximum at $\lambda(0) = 1$ and that it decreases with a slope of -0.002. Find $\lambda(b)$ and the differential equation for b. Is $b(t)$ increasing when $b = 10$? Is $b(t)$ increasing when $b = 1000$?

28. Suppose that per capita production decreases linearly from a maximum of $\lambda(0) = 4$ with slope -0.001. Find $\lambda(b)$ and the differential equation for b. Is $b(t)$ increasing when $b = 1000$? Is $b(t)$ increasing when $b = 5000$?

29. In some circumstances, individuals reproduce better when the population size is large, and fail to reproduce when the population size is small (the Allee effect introduced in Exercise 46). Suppose that per capita production is an increasing linear function with $\lambda(0) = -2$ and a slope of 0.01. Find $\lambda(b)$ and the differential equation for b. Is $b(t)$ increasing when $b = 100$? Is $b(t)$ increasing when $b = 300$?

30. Suppose that per capita production increases linearly with $\lambda(0) = -5$ and a slope of 0.001. Find $\lambda(b)$ and the differential equation for b. Is $b(t)$ increasing when $b = 1000$? Is $b(t)$ increasing when $b = 3000$?

31–36 ▪ The derivation of the movement of chemical assumed that chemical moved as easily into the cell as out of it. If the membrane can act as a filter, the rates at which chemical enters and leaves might differ, or might depend on the concentration itself. In each of the following cases, draw a diagram illustrating the situation and write the associated differential equation. Let C be the concentration inside the cell, Γ the concentration outside, and β the constant of proportionality relating the concentration and the rate.

31. Suppose that no chemical re-enters the cell. This should look like the differential equation for a population. What would be happening to the concentration?

32. Suppose that no chemical leaves the cell. What would happen to the concentration?

33. Suppose that the constant of proportionality governing the rate at which chemical enters the cell is three times as large as the constant governing the rate at which it leaves. Would the concentration inside the cell be increasing or decreasing if $C = \Gamma$? What would this mean for the cell?

34. Suppose that the constant of proportionality governing the rate at which chemical enters the cell is half as large as the constant governing the rate at which it leaves. Would the concentration inside the cell be increasing or decreasing if $C = \Gamma$? What would this mean for the cell?

35. Suppose that the constant of proportionality governing the rate at which chemical enters the cell is proportional to $1 + C$ (because the chemical helps to open special channels). Would the concentration inside the cell be increasing or decreasing if $C = \Gamma$? What would this mean for the cell?

36. Suppose that the constant of proportionality governing the rate at which chemical enters the cell is proportional to $1 - C$ (because the chemical helps to close special channels). Would the concentration inside the cell be increasing or decreasing if $C = \Gamma$? What would this mean for the cell?

37–38 ▪ The model of selection includes no interaction between bacterial types a and b (the per capita production of each type is a constant). Write a pair of differential equations for a and b with the following forms for the per capita production, and derive an equation for the fraction p of type a. Assume that the basic per capita production for type a is $\mu = 2$, and that for type b is $\lambda = 1.5$.

37. The per capita production of each type is reduced by a factor of $1 - p$ by a factor of $1 - p$, so that the per capita production of type a is $2(1 - p)$. This is a case where a large proportion of type a reduces the production of both types. Will type a take over?

38. The per capita production of type a is reduced by a factor of $1 - p$ and the per capita production of type b is reduced by a factor of p. This is a case where a large proportion of type a reduces the production of type a, and a large proportion of type b reduces the production of type b. Do you think that type a will still take over?

39–40 ▪ We will find later (with separation of variables) that the solution for Newton's law of cooling with initial condition $H(0)$ is

$$H(t) = A + [H(0) - A]e^{-\alpha t}.$$

For each set of given parameter values,

a. Write and check the solution.

b. Find the temperature at $t = 1$ and $t = 2$.

c. Sketch of graph of your solution. What happens as t approaches infinity?

39. Set $\alpha = 0.2$/min, $A = 10$°C, and $H(0) = 40$.

40. Set $\alpha = 0.02$/min, $A = 30$°C, and $H(0) = 40$.

41–42 ▪ Use Euler's method to estimate the temperature for the following cases of Newton's law of cooling. Compare with the exact answer.

41. $\alpha = 0.2$/min and $A = 10$°C and $H(0) = 40$. Estimate $H(1)$ and $H(2)$ using $\Delta t = 1$. Compare with Exercise 39.

42. $\alpha = 0.02$/min and $A = 30$°C and $H(0) = 40$. Estimate $H(1)$ and $H(2)$ using $\Delta t = 1$. Compare with Exercise 40. Why is the result so close?

43–44 ▪ Use the solution for Newton's law of cooling (Exercises 39 and 40) to find the solution expressing the concentration of chemical inside a cell as a function of time in the following examples. Find the concentration after 10 seconds, 20 seconds, and 60 seconds. Sketch your solutions for the first minute.

43. $\beta = 0.01$/s, $C(0) = 5.0$ mmol/cm^3, and $\Gamma = 2.0$ mmol/cm^3.

44. $\beta = 0.1$/s, $C(0) = 5.0$ mmol/cm^3, and $\Gamma = 2.0$ mmol/cm^3.

45–48 ▪ Recall that the solution of the discrete-time dynamical system $b_{t+1} = rb_t$ is $b_t = r^t b_0$. This is closely related to the differential equation $\frac{db}{dt} = \lambda b$.

45. For what values of b_0 and r does this solution match $b(t) = 1.0 \times 10^6 e^{2t}$ (the solution of the differential equation with $\lambda = 2$ and $b(0) = 1.0 \times 10^6$) for all values of t?

46. For what values of b_0 and r does this solution match $b(t) = 100e^{-3t}$ (the solution of the differential equation with $\lambda = -3$ and $b(0) = 100$) for all values of t?

47. For what values of λ do solutions of the differential equation grow? For what values of r do solutions of the discrete-time dynamical system grow?

48. What is the relation between r and λ? That is, what value of r produces the same growth as a given value of λ?

49–50 ▪ Use Euler's method to estimate the value of $p(t)$ from the selection differential equation (Equation 5.1.4) for the given parameter values. Compare with the exact answer using the equation for the solution. Graph the solution, including the estimates from Euler's method.

49. Suppose $\mu = 2.0$/h, $\lambda = 1.0$/h, and $p(0) = 0.1$. Estimate the proportion after 2 hours using a time step of $\Delta t = 0.5$.

50. Suppose $\mu = 2.5$/h, $\lambda = 3.0$/h, and $p(0) = 0.6$. Estimate the proportion after 1 hour using a time step of $\Delta t = 0.25$.

51–52 ▪ Suppose that an endothermic (warm-blooded) animal generates heat at a rate proportional to its metabolic rate with constant c_1, and loses heat at a rate proportional to its surface area with constant c_2. Heat loss is also proportional to the difference between its temperature and the ambient temperature A as in Newton's Law of Cooling. The temperature T of a conveniently spherical creature of radius r would then follow

$$\frac{dT}{dt} = c_1(r^3)^{\frac{3}{4}} - c_2 r^2 (T - A)$$

51. Find the equilibrium temperature as a function of r. For the same values of c_1 and c_2, which animals stay warmest?

52. Suppose that $c_2 = 1.0$ and that $A = -20.0$°C. Find the value of c_1 required to maintain an equilibrium temperature of 40.0°C when $r = 1.0$ and when $r = 2.0$. Which organism needs to generate less heat to maintain its temperature?

Computer Exercises

53. Suppose that the ambient temperature oscillates with period T according to

$$A(t) = 20.0 + \cos\left(\frac{2\pi t}{T}\right).$$

The differential equation is

$$\frac{dH}{dt} = \alpha[A(t) - H].$$

which has a solution with $A(0) = 20.0$ of

$$H(t) = 20.0 + \frac{\alpha^2 T^2 \cos\left(\frac{2\pi t}{T}\right) + 2\alpha\pi T \sin\left(\frac{2\pi t}{T}\right) - \alpha^2 T^2 e^{-\alpha t}}{\alpha^2 T^2 + 4\pi^2}$$

a. Use a computer algebra system to check that this answer works.

b. Plot $H(t)$ for five periods using values of T ranging from 0.1 to 10.0 when $\alpha = 1.0$. Compare $H(t)$ with $A(t)$. When does the temperature of the object most closely track the ambient temperature?

54. Apply Euler's method to solve the differential equation

$$\frac{db}{dt} = b + t$$

with the initial condition $b(0) = 1.0$. Compare with the sum of the solutions of $\frac{db_1}{dt} = b_1$ and $\frac{db_2}{dt} = t$. Do you think there is a sum rule for differential equations? If your computer has a method for solving differential equations, find the solution and compare with your approximate solution from Euler's method.

5.2 Equilibria and Display of Autonomous Differential Equations

Euler's method provides an algorithm for computing approximate solutions of autonomous differential equations. As with discrete-time dynamical systems, much can be deduced about the dynamics by algebraically solving for **equilibria** and plotting them on a graphical summary of the dynamics, here called a **phase-line diagram**.

5.2.1 Equilibria

One of the most important tools for analyzing discrete-time dynamical systems is the equilibrium, a state of the system that remains unchanged by the dynamics. Solutions starting from an equilibrium do not change over time. For example, solutions of the bacterial selection model (Equation 1.10.5) approach an equilibrium at $p = 1$ when the mutants are superior to the wild type whether the model is a discrete-time dynamical system (Figure 5.2.1a) or an autonomous differential equation (Figure 5.2.1b). In either case, if we started with all mutant bacteria, the population would never change.

Equilibria are as useful in analyzing autonomous differential equations as in analyzing discrete-time dynamical systems. As before, an equilibrium is a value of the state variable that remains unchanged by the dynamical system.

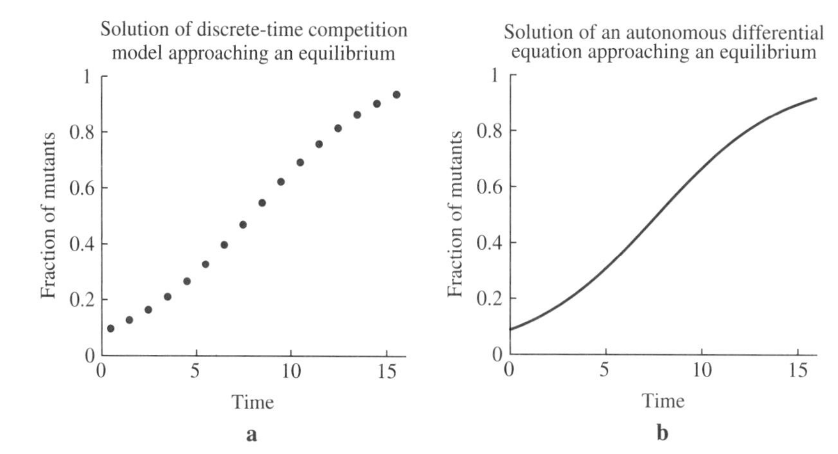

FIGURE 5.2.1
Solutions and equilibria: discrete-time dynamical systems and differential equations

Definition 5.1 A value m^* of the state variable is called an equilibrium of the autonomous differential equation

$$\frac{dm}{dt} = f(m)$$

if

$$f(m^*) = 0.$$

At a point where $f(m^*) = 0$, the rate of change is zero. A rate of change of zero indicates precisely that the value of the measurement remains the same.

The steps for finding equilibria closely resemble those for finding equilibria of a discrete-time dynamical system.

Algorithm 5.2 Finding Equilibria of an Autonomous Differential Equation

1. Make sure that the differential equation is autonomous.
2. Write the equation for the equilibria.
3. Factor.
4. Set each factor equal to 0 and solve for the equilibria.
5. Meditate upon the results.

Example 5.2.1 The Equilibria for the Bacterial Population Growth Model

The simplest autonomous differential equation we have considered is the equation for bacterial growth

$$\frac{db}{dt} = \lambda b.$$

What are the equilibria of this differential equation? The rate of change depends only on b and not on the time t, so this equation is autonomous. The algebraic steps are

$$\lambda b^* = 0 \quad \text{the equation for the equilibrium}$$

$$\lambda = 0 \text{ or } b^* = 0. \quad \text{set both factors to 0}$$

If $\lambda = 0$, every value of b is an equilibrium because the population does not change. This equilibrium corresponds to the case $r = 1$ of the related discrete-time dynamical system, $b_{t+1} = rb_t$. The other equilibrium at $b^* = 0$ indicates extinction, again matching the results for the discrete-time dynamical system. A solution starting from $b = 0$ remains there forever (Figure 5.2.2).

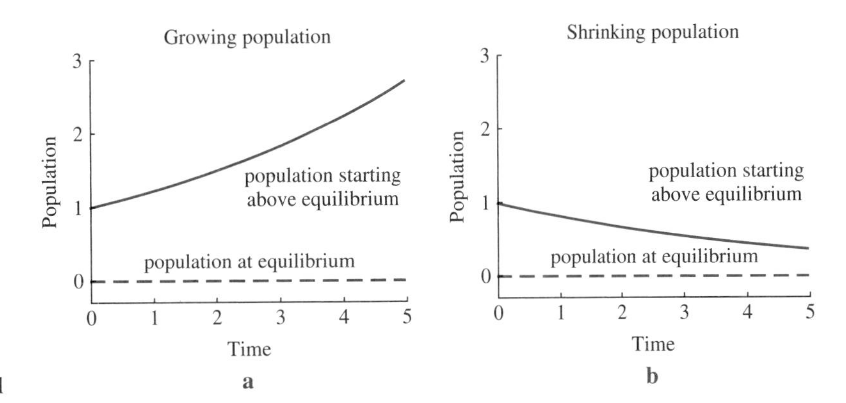

FIGURE 5.2.2
Solutions of the bacterial growth model

Example 5.2.2 The Equilibria for Newton's Law of Cooling

For Newton's law of cooling, the differential equation is

$$\frac{dH}{dt} = \alpha(A - H).$$

The rate of change depends only on H and not on the time t, so this equation is autonomous. To find the equilibria, we follow Algorithm 5.2.

$$\alpha(A - H^*) = 0 \quad \text{the equation for the equilibrium}$$

$$\alpha = 0 \text{ or } (A - H^*) = 0 \quad \text{set both factors to 0}$$

$$\alpha = 0 \text{ or } H^* = A. \quad \text{solve each term}$$

The solutions are $\alpha = 0$ and $H^* = A$. If $\alpha = 0$, no heat is exchanged with the outside world. Such an object is always at equilibrium. This corresponds to the case $q = 0$ of the lung model; a lung that exchanges no air with the outside world is at equilibrium. The other equilibrium is $H^* = A$. The object is in equilibrium when its temperature is equal to the ambient temperature. A solution with initial condition $H(0) = A$ remains there forever. This corresponds to the equilibrium $c^* = \gamma$ for the lung model (Equation 1.9.1); a lung that exchanges some air eventually comes to match the external environment.

Example 5.2.3 The Equilibria for Bacterial Selection Model

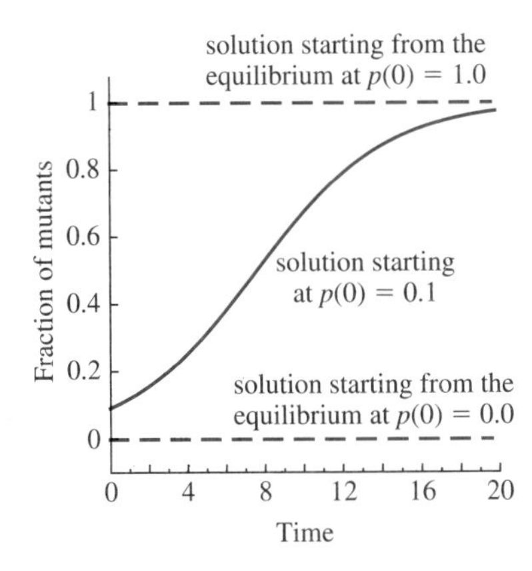

FIGURE 5.2.3
Three solutions of the bacterial selection equation if $\mu > \lambda$

We follow the same algorithm to find the equilibria of the selection model

$$\frac{dp}{dt} = (\mu - \lambda)p(1 - p).$$

The equation is autonomous because the rate of change is not a function of t. The equation for the equilibria is

$$(\mu - \lambda)p^*(1 - p^*) = 0 \quad \text{the equation for the equilibrium}$$

$$\mu - \lambda = 0 \text{ or } p^* = 0 \text{ or } 1 - p^* = 0 \quad \text{set each factor to 0}$$

$$\mu = \lambda \text{ or } p^* = 0 \text{ or } p^* = 1. \quad \text{solve each term}$$

If $\mu = \lambda$, the reproductive rates of the two types match and the fraction of mutants does not change. The other two equilibria correspond to extinction. If $p^* = 0$, there are no mutants, and if $p^* = 1$, there are no wild types. Solutions starting from either of these points remain there (Figure 5.2.3). Although the differential equation describing these dynamics looks very different from the discrete-time dynamical system model describing the same process (Equation 1.10.5), the equilibria match.

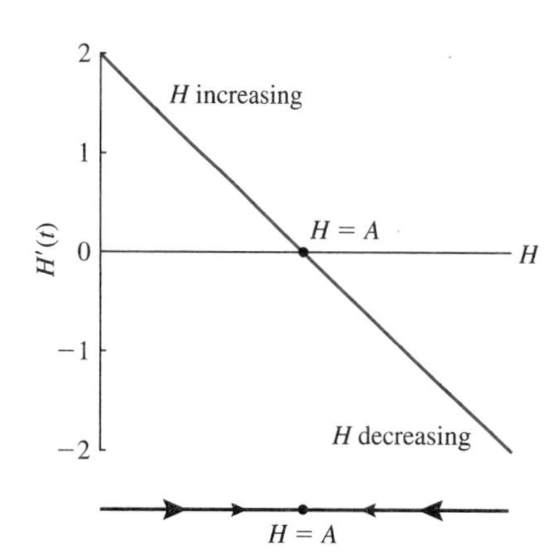

FIGURE 5.2.4
The rate of change and phase-line diagram for Newton's law of cooling

5.2.2 Graphical Display of Autonomous Differential Equations

The graphical technique of cobwebbing helped us sketch solutions of discrete-time dynamical systems with a minimum of algebra. There is a similarly useful graphical way to display autonomous differential equations, called the **phase-line diagram**. The diagram summarizes where the state variable is increasing, decreasing, and unchanged. According to the interpretation of the derivative, the state variable is increasing when the rate of change is positive, decreasing when the rate of change is negative, and unchanging when the rate of change is zero. By figuring out where the rate of change is positive, negative, and zero we can get a good idea of how solutions behave.

The rate of change for Newton's law of cooling is graphed as a function of the state variable H (top of Figure 5.2.4). The rate of change is positive when $H < A$, zero at $H = A$, and negative when $H > A$. The temperature is therefore increasing when $H < A$, constant at the equilibrium $H = A$, and decreasing when $H > A$.

This description can be translated into a one-dimensional drawing of the dynamics, called a **phase-line diagram**, shown at the bottom of Figure 5.2.4. The line represents the state variable, in this case the temperature, sometimes referred to as the **phase** of the system. To construct the diagram, draw rightward pointing arrows at points where the temperature is increasing, leftward pointing arrows at points where temperature is decreasing, and big dots where the temperature is fixed. The directions of the arrows come from the values of the rate of change. When the rate of change is positive, the temperature is increasing and the arrow points to the right.

Solutions follow the arrows. Starting below the equilibrium, the arrows push the temperature up toward the equilibrium. Starting above the equilibrium, the arrows push the temperature down toward the equilibrium. Unlike the solutions of discrete-time dynamical systems, the temperature cannot overshoot the equilibrium.

More information can be encoded on the diagram by making the size of the arrows correspond to the magnitude of the rate of change. In Figure 5.2.4, the arrows are larger when H is farther from the equilibrium where the rate of change of temperature is larger. The solutions (Figure 5.2.5) change rapidly far from the equilibrium A and more slowly near A.

Example 5.2.4 Phase-Line Diagram for the Selection Model: $\mu > \lambda$

When $\mu > \lambda$, the rate of change of the selection model

$$\frac{dp}{dt} = (\mu - \lambda)p(1 - p)$$

is positive except at the equilibria (Figure 5.2.6). All the arrows on the phase-line diagram point to the right, pushing solutions toward the equilibrium at $p = 1$. The rate

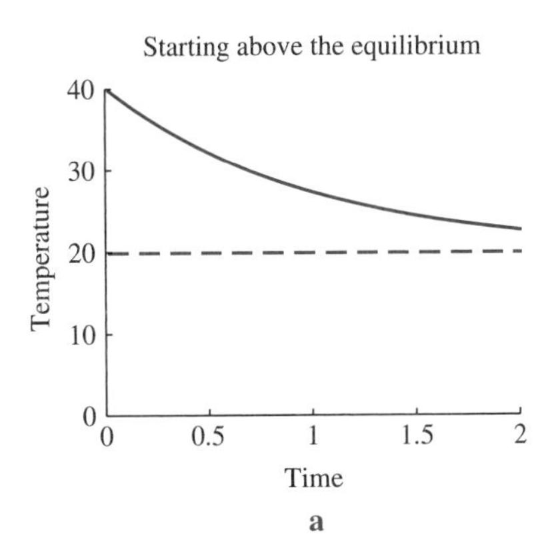

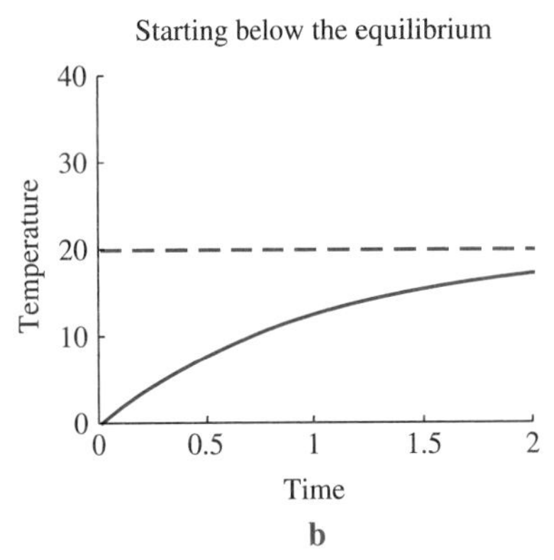

FIGURE 5.2.5

Two solutions of Newton's law of cooling

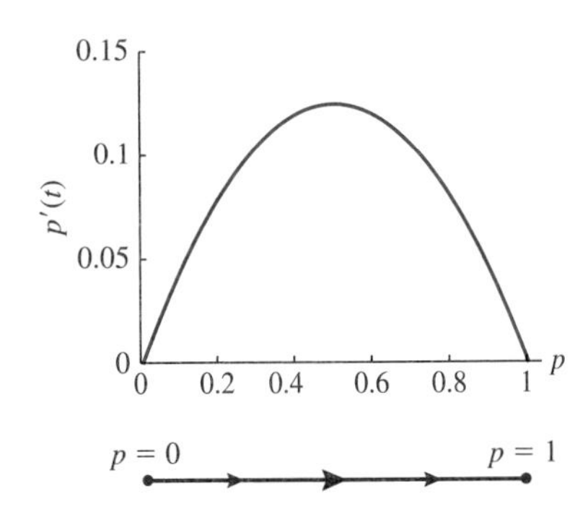

FIGURE 5.2.6

The rate of change and phase-line diagram for the selection model

of change takes on its maximum value at $p = 0.5$, meaning that the largest arrow is at $p = 0.5$. A solution starting near $p = 0$ begins by increasing slowly, increases faster as it passes $p = 0.5$, and then slows down as it approaches $p = 1$ (Figure 5.2.3). Compared to the complicated algebra required to find a formula for this solution (Subsection 5.1.4), this method requires only the ability to graph and interpret a quadratic.

The technique of phase-line diagrams is appropriate **only** for autonomous differential equations. Were we to try drawing arrows for a nonautonomous differential equation, the arrows would change with time. This is difficult to achieve in a drawing (although computers can do it through animation).

The phase-line diagram for the selection model (Figure 5.2.6) has two different kinds of equilibria. The arrows push the solution away from the equilibrium at $p = 0$ and toward the equilibrium at $p = 1$. We propose an informal definition of stable and unstable equilibria similar to that for discrete-time dynamical systems (Definition 1.17).

Definition 5.2 An equilibrium of an autonomous differential equation is **stable** if solutions that begin near the equilibrium approach the equilibrium. An equilibrium of an autonomous differential equation is **unstable** if solutions that begin near the equilibrium move away from the equilibrium.

Summary **Equilibria** of an autonomous differential equation are found by setting the rate of change equal to zero and solving for the state variable. By graphing the rate of change as a function of the state variable, we can identify values where the state variable is increasing or decreasing. This information can be translated into a **phase-line diagram** on which arrows pointing to the right indicate increasing solutions and arrows pointing to the left indicate decreasing solutions. This diagram can be used to sketch solutions by following the arrows.

5.2 Exercises

Mathematical Techniques

1–4 ▪ Find the equilibria of the following autonomous differential equations.

1. $\dfrac{dx}{dt} = 1 - x^2$.
2. $\dfrac{dx}{dt} = 1 - e^x$.
3. $\dfrac{dy}{dt} = y\cos(y)$.
4. $\dfrac{dz}{dt} = \dfrac{1}{z} - 3$.

5–8 ▪ Find the equilibria of the following autonomous differential equations that include parameters.

5. $\dfrac{dx}{dt} = 1 - ax$.
6. $\dfrac{dx}{dt} = cx + x^2$.
7. $\dfrac{dW}{dt} = \alpha e^{\beta W} - 1$.
8. $\dfrac{dy}{dt} = ye^{-\beta y} - ay$.

9–10 ▪ From the following graphs of the rate of change as a function of the state variable, draw the phase-line diagram.

9.

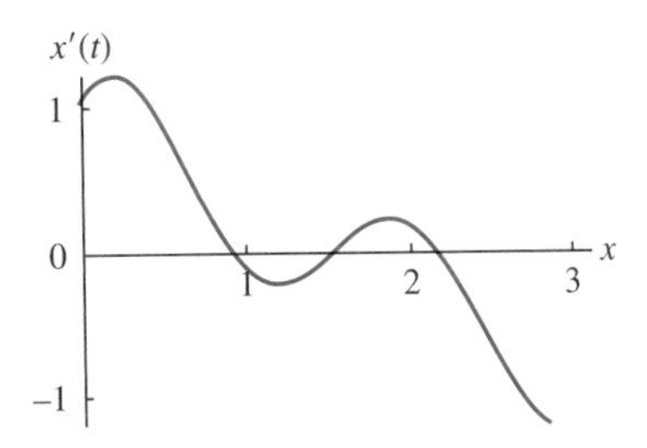

10.

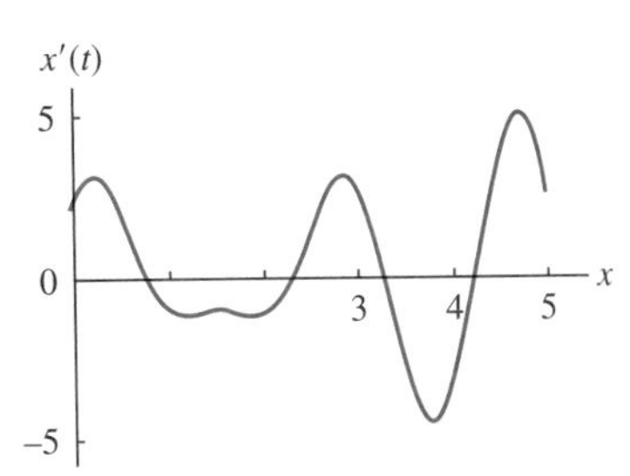

11–12 ▪ From the following phase-line diagrams, sketch a solution starting from the specified initial condition.

11.

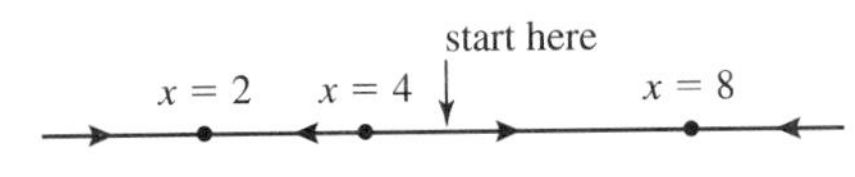

12.

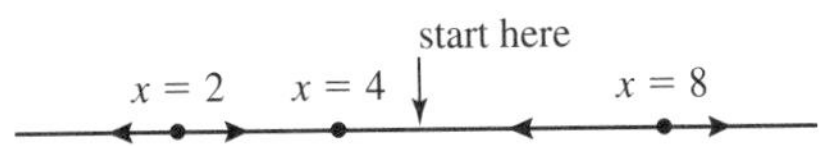

13–14 ▪ From the given phase-line diagram, sketch a possible graph of the rate of change of x as a function of x.

13. The phase line in Exercise 11.

14. The phase line in Exercise 12.

15–18 ▪ Graph the rate of change as a function of the state variable and draw the phase-line diagram for the following differential equations.

15. $\frac{dx}{dt} = 1 - x^2$ (as in Exercise 1). Graph for $-2 \leq x \leq 2$.

16. $\frac{dx}{dt} = 1 - e^x$ (as in Exercise 2). Graph for $-2 \leq x \leq 2$.

17. $\frac{dy}{dt} = y\cos(y)$ (as in Exercise 3). Graph for $-2 \leq y \leq 2$.

18. $\frac{dz}{dt} = \frac{1}{z} - 3$ (as in Exercise 4). Graph for $0 < z \leq 1$.

Applications

19–20 ▪ Suppose a population is growing at constant rate λ, but that individuals are harvested at a rate of h. The differential equation describing such a population is

$$\frac{db}{dt} = \lambda b - h.$$

For each of the following values of λ and h, find the equilibrium, draw the phase-line diagram, and sketch one solution with initial condition below the equilibrium and another with initial condition above the equilibrium. Explain your result in words.

19. $\lambda = 2.0$, $h = 1000$.

20. $\lambda = 0.5$, $h = 1000$.

21–24 ▪ Find the equilibria, graph the rate of change $\frac{db}{dt}$ as a function of b, and draw a phase-line diagram for the following models describing bacterial population growth.

21. The model in Section 5.1, Exercise 27. Check that your arrows are consistent with the behavior of $b(t)$ at $b = 10$ and $b = 1000$.

22. The model in Section 5.1, Exercise 28. Check that your arrows are consistent with the behavior of $b(t)$ at $b = 1000$ and $b = 5000$.

23. The model in Section 5.1, Exercise 29. Check that your arrows are consistent with the behavior of $b(t)$ at $b = 100$ and $b = 300$.

24. The model in Section 5.1, Exercise 30. Check that your arrows are consistent with the behavior of $b(t)$ at $b = 1000$ and $b = 3000$.

25–26 ▪ Find the equilibria, graph the rate of change $\frac{dC}{dt}$ as a function of C, and draw a phase-line diagram for the following models describing chemical diffusion.

25. The model in Section 5.1, Exercise 33. Check that the direction arrow is consistent with the behavior of $C(t)$ at $C = \Gamma$.

26. The model in Section 5.1, Exercise 34. Check that the direction arrow is consistent with the behavior of $C(t)$ at $C = \Gamma$.

27–28 ▪ Find the equilibria, graph the rate of change $\frac{dp}{dt}$ as a function of p, and draw a phase-line diagram for the following models describing selection.

27. The model in Section 5.1, Exercise 37. What happens to a solution starting from a small, but positive, value of p?

28. The model in Section 5.1, Exercise 38. What happens to a solution starting from a small, but positive, value of p?

29–36 ▪ Find the equilibria and draw the phase-line diagram for the following differential equations, in addition to answering the questions.

29. Suppose the population size of some species of organism follows the model

$$\frac{dN}{dt} = \frac{3N^2}{2 + N^2} - N$$

where N is measured in hundreds. Why might this population behave as it does at small values? This is another example of the Allee effect discussed in Section 5.1, Exercise 29.

30. Suppose the population size of some species of organism follows the model

$$\frac{dN}{dt} = \frac{5N^2}{1 + N^2} - 2N$$

where N is measured in hundreds. What is the critical value below which this population is doomed to extinction (as in Exercise 29)?

31. The drag on a falling object is proportional to the square of its speed. In a differential equation

$$\frac{dv}{dt} = a - Dv^2$$

where v is speed, a is acceleration, and D is drag. Suppose that $a = 9.8$ m/s^2 and that $D = 0.0032$ per meter (values for a skydiver). Check that the units in the differential equation are consistent. What does the equilibrium speed mean?

32. Consider the same situation as in Exercise 31 but for a skydiver diving head down with her arms against her sides and her toes pointed, thus minimizing drag. The drag D is reduced to $D = 0.00048$ per meter. Find the equilibrium speed. How does it compare to the ordinary skydiver?

33. According to Torricelli's law of draining, the rate that a fluid flows out of a cylinder through a hole at the bottom is proportional to the square root of the depth of the water. Let y represent the depth of water in centimeters. The differential equation is

$$\frac{dy}{dt} = -c\sqrt{y}$$

where $c = 2.0\sqrt{\text{cm}}/\text{s}$. Show that the units are consistent. Use your phase-line diagram to sketch solutions starting from $y = 10.0$ and $y = 1.0$.

34. Write a differential equation describing the depth of water in a cylinder where water enters at a rate of 4.0 cm/s but drains out as in Exercise 33. Use your phase-line diagram to sketch solutions starting from $y = 10.0$ and $y = 1.0$.

35. One of the most important differential equations in chemistry uses the **Michaelis-Menton** or **Monod** equation. Suppose S is the concentration of a substrate that is being converted into a product. Then

$$\frac{dS}{dt} = -k_1 \frac{S}{k_2 + S}$$

describes how substrate is used. Set $k_1 = k_2 = 1$. How does this equation differ from Torricelli's law of draining (Exercise 33)?

36. Write a differential equation describing the amount of substrate if substrate is added at rate R but is converted into product as in Exercise 35. Find the equilibrium, and draw the phase-plane diagram and a representative solution with $R = 0.5$ and $R = 1.5$. Can you explain your results?

37–38 ▪ Small organisms like bacteria take in food at rates proportional to their surface area but use energy at higher rates.

37. Suppose that energy is used at a rate proportional to the mass. In this case,

$$\frac{dV}{dt} = a_1 V^{2/3} - a_2 V$$

where V represents the volume in cubic centimeters and t is time measured in days. The first term says that surface area is proportional to volume to the 2/3 power. The constant a_1 gives the rate at which energy is taken in and has units of centimeters per day. a_2 is rate at which energy is used and has units of per day. Check the units. Find the equilibrium. What happens to the equilibrium as a_1 becomes smaller? Does this make sense? What happens to the equilibrium as a_2 becomes smaller? Does this make sense?

38. Suppose that energy is used at a rate proportional to the mass to the 3/4 power (as in Example 1.7.24). In this case,

$$\frac{dV}{dt} = a_1 V^{2/3} - a_2 V^{3/4}.$$

Find the units of a_2 if V is measured in cm^3 and t is measured in days (they should look rather strange). Find the equilibrium. What happens to the equilibrium as a_1 becomes smaller? Does this make sense? What happens to the equilibrium as a_2 becomes smaller? Does this make sense?

Computer Exercises

39. Consider the differential equation

$$\frac{db}{dt} = -b^{-p}$$

for various positive values of the parameter p starting from $b(0) = 1.0$. For which values of p does the solution approach the equilibrium at $b = 0$ most quickly? Plot the solution on a semilog graph. When does the solution approach zero faster than an exponential function?

40. In Exercise 53 we considered the equation

$$\frac{dH}{dt} = \alpha[A(t) - H].$$

where

$$A(t) = 20.0 + \cos\left(\frac{2\pi t}{T}\right).$$

Using either the solution given in that problem or a computer system that can solve the equation, show that the solution always approaches the "equilibrium" at $H = A(t)$. Use the parameter values $\alpha = 1.0$ and values of T ranging from 0.1 to 10.0. For which value of T does the solution get closest to $A(t)$? Can you explain why?

5.3 Stable and Unstable Equilibria

Phase-line diagrams for autonomous differential equations have two types of equilibria: stable and unstable. Solutions starting near a stable equilibrium move closer to the equilibrium. The equilibrium of Newton's law of cooling (Figure 5.2.4) and the equilibrium at $p = 1$ of the selection equation with superior mutants (Figure 5.2.6) are stable in this sense. Solutions starting near an unstable equilibrium move farther from the equilibrium. The equilibrium at $p = 0$ of the selection equation (Figure 5.2.6) is

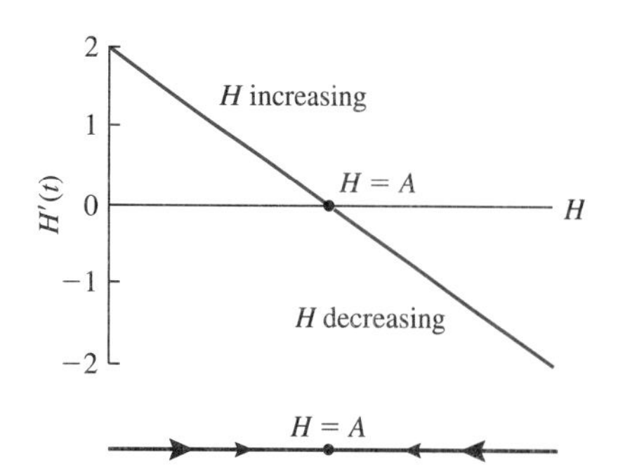

FIGURE 5.3.1
The rate of change and phase-line diagram for Newton's law of cooling revisited

unstable. By reasoning about the graph of the rate-of-change function, we can use the derivative to recognize stable and unstable equilibria.

5.3.1 Recognizing Stable and Unstable Equilibria

Consider again Newton's law of cooling

$$\frac{dH}{dt} = \alpha(A - H)$$

(Figure 5.3.1). Our phase-line diagram and our intuition suggest that the equilibrium $H = A$ is stable. Arrows to the left of the equilibrium point right, pushing solutions up toward the equilibrium (a cool object warms up). Arrows to the right of the equilibrium point left, pushing solutions down toward the equilibrium (a hot object cools off). The equilibrium must be stable.

The rate-of-change function is positive to the left of the equilibrium and negative to the right of a stable equilibrium. Therefore, the rate-of-change function must be **decreasing** at the equilibrium, implying that the **derivative** of the rate-of-change function is negative. An equilibrium is stable if the derivative of the rate-of-change function **with respect to the state variable** is negative at the equilibrium. The right-hand equilibrium in Figure 5.3.2 satisfies this criterion.

Example 5.3.1 The Derivative of the Rate-of-Change Function for Newton's Law of Cooling

The function $\alpha(A - H)$ gives the rate of change of H in Newton's law of cooling. If H is slightly larger than the equilibrium value A, the rate of change is negative. If H is slightly smaller than A, the rate of change is positive. To check stability with the derivative criterion, we compute the derivative of the rate-of-change function with respect to the state variable H,

$$\frac{d}{dH}\alpha(A - H) = -\alpha < 0.$$

The rate of change is a decreasing function at the equilibrium and the equilibrium is therefore stable (Figure 5.3.1).

At an unstable equilibrium, the arrows push solutions away. Solutions starting above the equilibrium increase, and those starting below decrease. In terms of the rate of change, an equilibrium is unstable if the rate of change is negative below the equilibrium and positive above it. This implies that the rate-of-change function is **increasing** and that the **derivative** of the rate-of-change function is positive. An equilibrium is unstable if the derivative of the rate-of-change function with respect to the state variable is positive at the equilibrium. The left-hand equilibrium in Figure 5.3.2 satisfies this criterion.

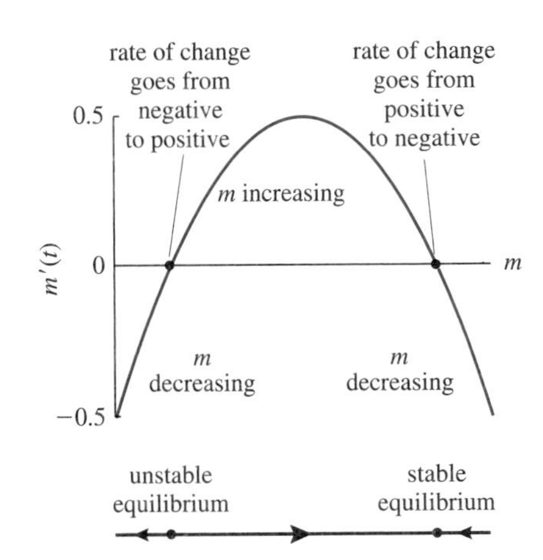

FIGURE 5.3.2
Behavior of functions with positive and negative derivatives

Example 5.3.2 The Rate of Change for a Perverse Version of Newton's Law of Cooling

Imagine a perverse version of Newton's law of cooling given by

$$\frac{dH}{dt} = \alpha(H - A)$$

with $\alpha > 0$. Objects cooler than A decrease in temperature, and objects warmer than A increase in temperature. The rate of change and the phase-line diagram are shown in Figure 5.3.3. The derivative of the rate-of-change function with respect to H at $H = A$ is

$$\frac{d}{dH}\alpha(H - A) = \alpha > 0.$$

The rate of change is an increasing function, and the equilibrium is unstable.

We summarize these results in the following simple and powerful theorem.

Theorem 5.1 **Stability Theorem for Autonomous Differential Equations**

Suppose

$$\frac{dm}{dt} = f(m)$$

is an autonomous differential equation with an equilibrium at m^*. Let $f'(m)$ represent the derivative of f with respect to m. The equilibrium at m^* is stable if

$$f'(m^*) < 0$$

and unstable if

$$f'(m^*) > 0.$$

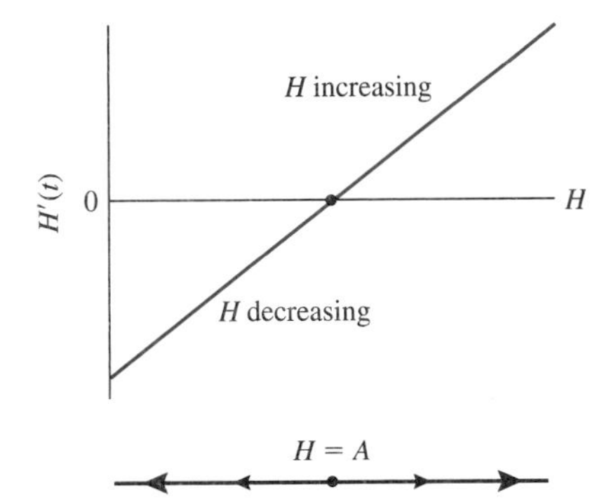

FIGURE 5.3.3
A perverse version of Newton's law of cooling

5.3.2 Applications of the Stability Theorem

We now apply the stability theorem for autonomous differential equations to our models of population growth and selection to check whether the phase-line diagrams are correct.

Example 5.3.3 Applying the Stability Theorem to Bacterial Population Dynamics

The equation for population growth is

$$\frac{db}{dt} = \lambda b.$$

If $\lambda \neq 0$, the only equilibrium is $b^* = 0$. The derivative of the rate of change λb with respect to b is

$$\frac{d}{db}\lambda b = \lambda.$$

According to the stability theorem for autonomous differential equations, the equilibrium is stable if the derivative is negative. If $\lambda > 0$, the equilibrium is unstable (Figure 5.3.4a), and if $\lambda < 0$, the equilibrium is stable (Figure 5.3.4b).

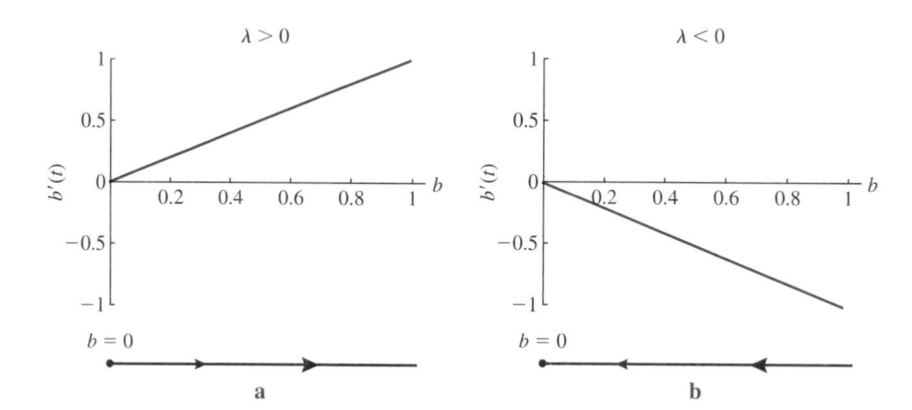

FIGURE 5.3.4
Phase-line diagrams for the bacterial growth model

The solution of the bacterial growth model with initial condition $b(0) = b_0$ is

$$b(t) = b_0 e^{\lambda t}$$

(Example 5.1.1). If $\lambda > 0$, the solution increases exponentially without bound from any positive initial condition, consistent with the instability of the equilibrium in this case (Figure 5.3.4a). If $\lambda > 0$, the solution decreases exponentially toward zero, consistent with the stability of the equilibrium in this case (Figure 5.3.4b).

There are two important differences between using the stability theorem for autonomous differential equations and solving the equation to figure out whether the equilibrium is stable. The stability theorem helps us recognize stable and unstable equilibria with a minimum of work; we solve for the equilibria and compute the derivative of the rate-of-change function with respect to b. Finding what happens near an equilibrium by computing the solution requires more work; finding the solution and then evaluating its limit as t approaches infinity. We seem to have gotten something for nothing. In mathematics, as in life, this is rarely the case. The other difference between the two methods is that the simpler calculation based on the stability theorem gives less information about the solutions, telling us only what happens to solutions that start **near** the equilibrium.

Example 5.3.4 Applying the Stability Theorem to the Selection Equation

In cases where finding the exact solution is difficult or impossible, this **local information** about solutions near an equilibrium may be all we can find. Computing the solution of the selection equation

$$\frac{dp}{dt} = f(p) = (\mu - \lambda)p(1-p)$$

is possible, but tricky (Subsection 5.1.4). Finding the stability of the equilibria with the stability theorem is straightforward. The differential equation has equilibria at $p=0$ and $p=1$ (Subsection 5.2.1). The derivative of the rate-of-change function $f(p)$ with respect to p is

$$f'(p) = \frac{d}{dp}[(\mu - \lambda)p(1-p)] = (\mu - \lambda)(1-2p).$$

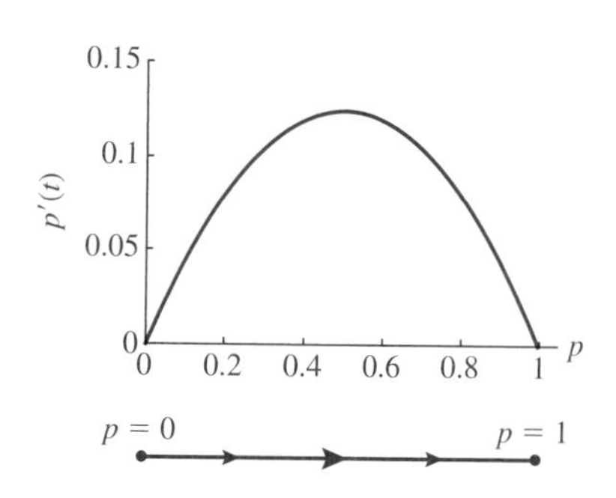

FIGURE 5.3.5

The rate of change and phase-line diagram for the selection model revisited

Then

$$f'(0) = \mu - \lambda$$

$$f'(1) = \lambda - \mu.$$

If $\mu > \lambda$, the equilibrium $p=0$ has a positive derivative and is unstable, and the equilibrium $p=1$ has a negative derivative and is stable, matching Figure 5.3.5. These results make biological sense; the mutant is superior in this case. If $\mu < \lambda$, the mutant is at a disadvantage, and the $p=0$ equilibrium is stable and the $p=1$ equilibrium is unstable. ▲

5.3.3 A Model of a Disease

Suppose a disease is circulating in a population. Individuals recover from this disease unharmed but are susceptible to re-infection. How many people will be sick at any given time? Is there any way that such a disease will die out? The factors affecting the dynamics are sketched in Figure 5.3.6.

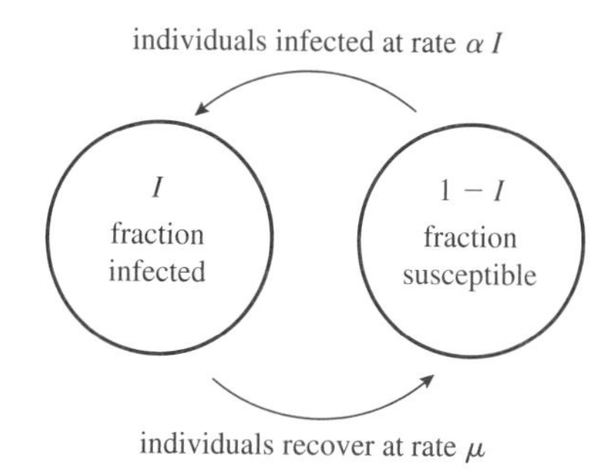

FIGURE 5.3.6

Factors involved in disease dynamics

Let I denote the fraction of infected individuals in the population. Each uninfected, or susceptible, individual has a chance of getting infected when she encounters an infectious individual (depending, perhaps, on whether she gets sneezed on). It seems plausible that a susceptible individual will run into infectious individuals at a rate proportional to the fraction of infectious individuals. Then

$$\text{per capita rate at which a susceptible individual is infected} = \alpha I.$$

The parameter α combines the rate at which people are encountered and the probability that an encounter produces an infection.

What fraction of individuals are susceptible? Every individual is either infected or not, so a fraction $1 - I$ are susceptible. The overall rate at which new individuals are infected is the per capita rate times the fraction or

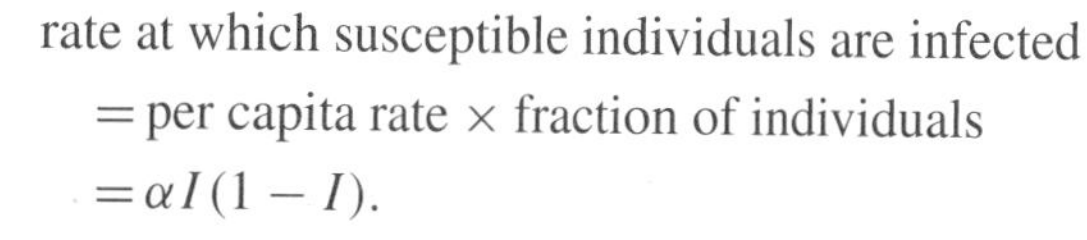

$$\begin{aligned} &\text{rate at which susceptible individuals are infected} \\ &\quad = \text{per capita rate} \times \text{fraction of individuals} \\ &\quad = \alpha I(1-I). \end{aligned}$$

Individuals recover from the disease at a rate proportional to the fraction of infected individuals,

$$\text{rate at which infected individuals recover} = \mu I.$$

Putting the two processes together

$$\frac{dI}{dt} = \alpha I(1-I) - \mu I. \tag{5.3.1}$$

Suppose we start with a few infected individuals, a small value of I. Will the disease persist? The equilibria and their stability can provide the answer. To find the equilibria, we use Algorithm 5.2. Equation 5.3.1 is autonomous because the rate of

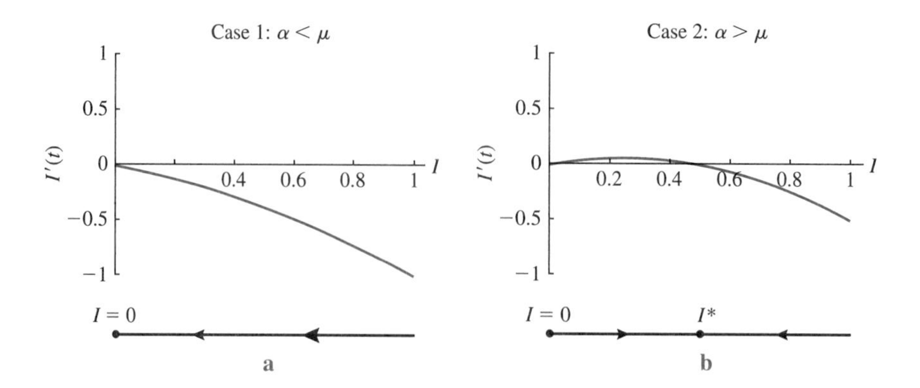

FIGURE 5.3.7
Phase-line diagrams for the disease model

change depends only on the state variable I and not on time.

$$\begin{aligned} \alpha I^*(1 - I^*) - \mu I^* &= 0 && \text{the equation for the equilibrium} \\ I^*[\alpha(1 - I^*) - \mu] &= 0 && \text{factor out } I^* \\ I^* = 0 \text{ or } \alpha(1 - I^*) - \mu &= 0 && \text{set each factor to 0} \\ I^* = 0 \text{ or } I^* &= 1 - \frac{\mu}{\alpha}. && \text{solve each term} \end{aligned}$$

Do these make sense? There are no sick people if $I = 0$. In the absence of some process bringing in infected people (such as migration or mutation), a population without illness will remain so. The other equilibrium depends on the parameters α and μ. There are two cases, depending upon which parameter is larger.

Case 1: If $\alpha < \mu$, $1 - \frac{\mu}{\alpha} < 0$, which is nonsense. When the recovery rate is large, the only biologically plausible equilibrium is $I = 0$.

Case 2: If $\alpha > \mu$, $1 > 1 - \frac{\mu}{\alpha} > 0$, which is biologically possible. There are two equilibria if the infection rate is larger than the recovery rate.

To draw the phase-line diagrams for the two cases, we must compute the stability of the equilibria. The derivative of the rate-of-change function with respect to the state variable I is

$$\frac{d}{dI}[\alpha I(1 - I) - \mu I] = \alpha - 2\alpha I - \mu. \tag{5.3.2}$$

Case 1: If $\alpha < \mu$, the derivative at $I = 0$ is

$$\alpha - 2\alpha \cdot 0 - \mu = \alpha - \mu < 0.$$

The single equilibrium is stable.

Case 2: If $\alpha > \mu$, the derivative at $I = 0$ is

$$\alpha - 2\alpha \cdot 0 - \mu = \alpha - \mu > 0$$

and this equilibrium is unstable. At $I^* = 1 - \frac{\mu}{\alpha}$, the derivative of the rate-of-change function is

$$\alpha - 2\alpha \cdot \left(1 - \frac{\mu}{\alpha}\right) - \mu = \mu - \alpha < 0.$$

The positive equilibrium is stable. The phase-line diagrams for the two cases are drawn in Figure 5.3.7.

From our phase-line diagram, we can deduce the behavior of solutions. In case 1, with $\alpha < \mu$, all solutions converge to the equilibrium at $I = 0$ (Figure 5.3.8a). If control measures could be implemented to increase the recovery rate μ or reduce the transmission α, the disease could be completely eliminated from a population. The transmission α need not be reduced to zero to eliminate the disease. This is an example of the **threshold theorem** referred to in Section 1.1. In case 2, with $\mu < \alpha$, the equilibrium I^* is positive and stable. Solutions starting near 0 increase to I^*, and the disease remains present in the population (Figure 5.3.8b). Such a disease is called **endemic**.

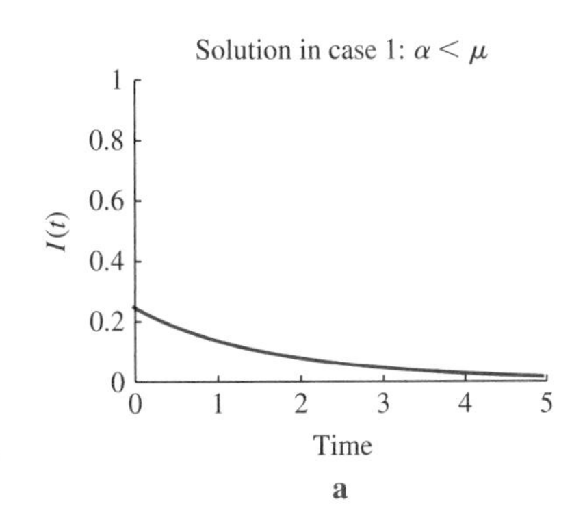

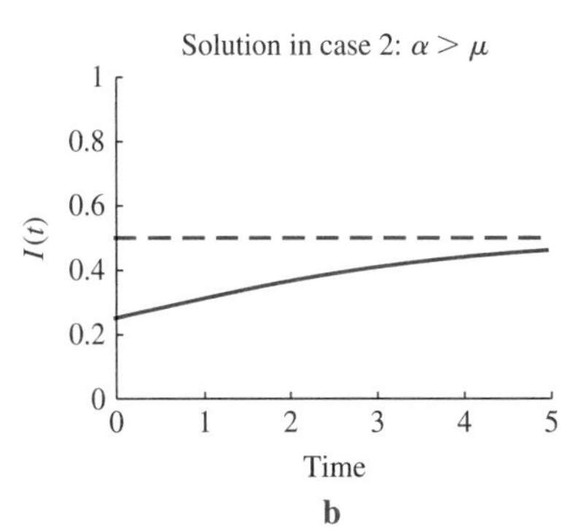

FIGURE 5.3.8
Solutions of the disease equation in two cases

Summary By examining the phase-line diagram near an equilibrium, we found the stability theorem for autonomous differential equations. An equilibrium of an autonomous differential equation is stable if the derivative of the rate-of-change function with respect to the state variable is negative and unstable if this derivative is positive. Calculating stability in this way is easier than solving the equation, but does not provide exact information about the behavior of solutions far from the equilibrium. We applied this method to several familiar models and to a model of a disease, showing that a disease could be eradicated without entirely stopping transmission.

5.3 Exercises

Mathematical Techniques

1–2 ▪ From the following graphs of the rate of change as a function of the state variable, identify stable and unstable equilibria by checking whether the rate of change is an increasing or decreasing function of the state variable.

1.

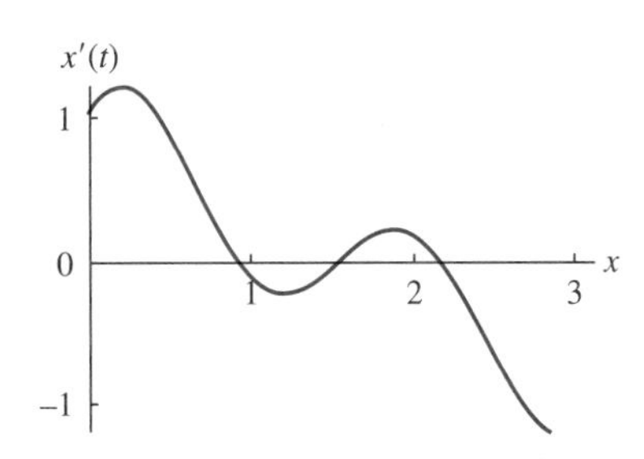

2.

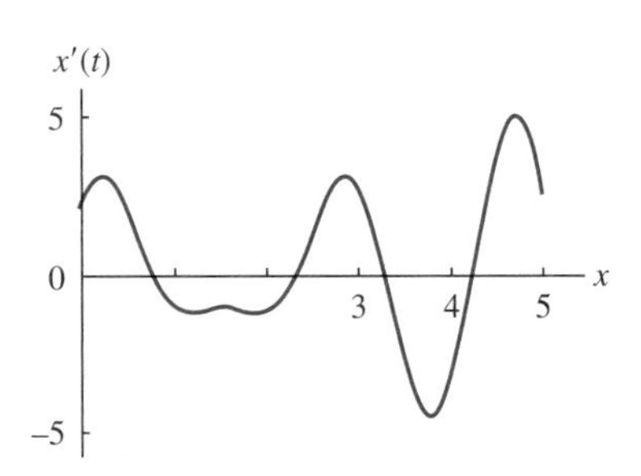

3–6 ▪ Use the stability theorem to evaluate the stability of the equilibria of the following autonomous differential equations.

3. $\frac{dx}{dt} = 1 - x^2$ (as in Section 5.2, Exercise 1). Compare your results with the phase line in Section 5.2, Exercise 15.

4. $\frac{dx}{dt} = 1 - e^x$ (as in Section 5.2, Exercise 2). Compare your results with the phase line in Section 5.2, Exercise 16.

5. $\frac{dy}{dt} = y\cos(y)$ (as in Section 5.2, Exercise 3). Compare your results with the phase line in Section 5.2, Exercise 17.

6. $\frac{dz}{dt} = \frac{1}{z} - 3$ (as in Section 5.2, Exercise 4). Compare your results with the phase line in Section 5.2, Exercise 18.

7–10 ▪ Find the stability of the equilibria of the following autonomous differential equations that include parameters.

7. $\frac{dx}{dt} = 1 - ax$ (as in Section 5.2, Exercise 5). Suppose that $a > 0$.

8. $\frac{dx}{dt} = cx + x^2$ (as in Section 5.2, Exercise 6). Suppose that $c > 0$.

9. $\frac{dW}{dt} = \alpha e^{\beta W} - 1$ (as in Section 5.2, Exercise 7). Suppose that $\alpha > 0$ and $\beta < 0$.

10. $\frac{dy}{dt} = ye^{-\beta y} - ay$ (as in Section 5.2, Exercise 8). Suppose that $\beta < 0$ and $a > 1$.

11–14 ▪ As with discrete-time dynamical systems, equilibria can act strange when the slope of the rate-of-change function is exactly equal to the critical value of zero.

11. Consider the differential equation $\frac{dx}{dt} = x^2$. Find the equilibrium, graph the rate of change as a function of x, and draw the phase-line diagram. Would you consider the equilibrium to be stable or unstable?

12. Consider the differential equation $\frac{dx}{dt} = -(1 - x)^4$. Find the equilibrium, graph the rate of change as a function of x, and draw the phase-line diagram. Would you consider the equilibrium to be stable or unstable?

13. Graph a rate-of-change function that has a slope of 0 at the equilibrium but the equilibrium is stable. What is the sign of the second derivative at the equilibrium? What is the sign of the third derivative at the equilibrium?

14. Graph a rate-of-change function that has a slope of 0 at the equilibrium but the equilibrium is unstable. What is the sign of the second derivative at the equilibrium? What is the sign of the third derivative at the equilibrium?

15–16 ▪ The fact that the rate-of-change function is continuous means that many behaviors are impossible for an autonomous differential equation.

15. Try to draw a phase-line diagram with two stable equilibria in a row. Use the Intermediate Value Theorem to sketch a proof of why this is impossible.

16. Why is it impossible for a solution of an autonomous differential equation to oscillate?

17–20 ▪ When parameter values change, the number and stability of equilibria sometimes change. Such changes are called bifurcations and play a central role in the study of differential equations. The following illustrate several of the more important bifurcations. In each case, graph the value of equilibria as functions of the parameter value, using a solid line when an equilibrium is stable and a dashed line when an equilibrium is unstable. This picture is called a **bifurcation diagram**.

17. Consider the equation

$$\frac{dx}{dt} = ax - x^2$$

for both positive and negative values of x. Find the equilibria as functions of a for values of a between -1 and 1. Draw a bifurcation diagram and describe in words what happens at $a = 0$. The change that occurs at $a = 0$ is called a **transcritical bifurcation**.

18. Consider the equation

$$\frac{dx}{dt} = a - x^2$$

for both positive and negative values of x. Find the equilibria as functions of a for values of a between -1 and 1. Draw a bifurcation diagram and describe in words what happens at $a = 0$. The change that occurs at $a = 0$ is called a **saddle-node bifurcation**.

19. Consider the equation

$$\frac{dx}{dt} = ax - x^3$$

for both positive and negative values of x. Find the equilibria as functions of a for values of a between -1 and 1. Draw a bifurcation diagram and describe in words what happens at $a = 0$. The change that occurs at $a = 0$ is called a **pitchfork bifurcation**.

20. Consider the equation

$$\frac{dx}{dt} = ax + x^3$$

for both positive and negative values of x. Find the equilibria as functions of a for values of a between -1 and 1. Draw a bifurcation diagram and describe in words what happens at $a = 0$. The change that occurs at $a = 0$ is a slightly different type of **pitchfork bifurcation** (Exercise 19) called a **subcritical bifurcation** (Exercise 19 is **supercritical**). How does your picture differ from the simple mirror image of that in Exercise 19?

Applications

21–24 ▪ Use the stability theorem to check the phase-line diagrams for the following models of bacterial population growth.

21. The model in Section 5.1, Exercise 27 and Section 5.2, Exercise 21.

22. The model in Section 5.1, Exercise 28 and Section 5.2, Exercise 22.

23. The model in Section 5.1, Exercise 29 and Section 5.2, Exercise 23.

24. The model in Section 5.1, Exercise 30 and Section 5.2, Exercise 24.

25–26 ▪ Use the stability theorem to check the phase-line diagrams for the following models of bacterial population growth.

25. The model in Section 5.1, Exercise 37 and Section 5.2, Exercise 27.

26. The model in Section 5.1, Exercise 38 and Section 5.2, Exercise 28.

27–30 ▪ A **reaction-diffusion equation** describes how chemical concentration changes due to two factors simultaneously, reaction and movement. A simple model has the form

$$\frac{dC}{dt} = \beta(\Gamma - C) + R(C).$$

The first term describes diffusion, and the second term $R(C)$ is the reaction, which could have a positive or negative sign (depending on whether chemical is being created or destroyed). Suppose that $\beta = 1.0$/min, and $\Gamma = 5.0$ mol/L. For each of the following forms of $R(C)$,

a. Describe how the reaction rate depends on the concentration.

b. Find the equilibria and their stability.

c. Describe how absorption changes the results.

27. Suppose that $R(C) = -C$.

28. Suppose that $R(C) = 0.5C$.

29. Suppose that $R(C) = \dfrac{C}{2+C}$.

30. Suppose that $R(C) = -\dfrac{C}{2+C}$.

31–40 ▪ Apply the stability theorem for autonomous differential equations to the following equations. Show that your results match what you found in your phase-line diagrams, and give a biological interpretation.

31. The equation $\frac{dN}{dt} = \frac{3N^2}{2+N^2} - N$ (Section 5.2, Exercise 29).

32. The equation $\frac{dN}{dt} = \frac{5N^2}{1+N^2} - 2N$ (Section 5.2, Exercise 30).

33. The equation $\frac{dv}{dt} = a - Dv^2$ with $a = 9.8$ m/s^2 and $D = 0.0032$ per meter (Section 5.2, Exercise 31).

34. The equation $\frac{dv}{dt} = a - Dv^2$ with $a = 9.8$ m/s^2 and $D = 0.00048$ per meter (Section 5.2, Exercise 32).

35. The equation $\frac{dy}{dt} = -c\sqrt{y}$ with $c = 2.0\sqrt{\text{cm}}/\text{s}$ (Section 5.2, Exercise 33).

36. The equation from the previous problem but with water entering at a rate of 4.0 cm/s (Section 5.2, Exercise 34).

37. The equation $\frac{dS}{dt} = -\frac{S}{1+S}$ (Section 5.2, Exercise 35).

38. The equation from the previous problem but with substrate added at rate $R = 0.5$ (Section 5.2, Exercise 36).

39. The equation $\frac{dV}{dt} = a_1 V^{2/3} - a_2 V$ (Section 5.2, Exercise 37).

40. The equation $\frac{dV}{dt} = a_1 V^{2/3} - a_2 V^{3/4}$ (Section 5.2, Exercise 38).

41–44 ▪ Exercises 17–20 show how the number and stability of equilibria can change when a parameter changes. Often, bifurcations have important biological applications, and bifurcation diagrams help in explaining how the dynamics of a system can suddenly change when a parameter changes only slightly. In each case, graph the equilibria against the parameter value, using a solid line when an equilibrium is stable and a dashed line when an equilibrium is unstable to draw the **bifurcation diagram**.

41. Consider the logistic differential equation (Section 5.1, Exercise 27) with harvesting proportional to population size, or $\frac{db}{dt} = b(1 - b - h)$ where h represents the fraction harvested. Graph the equilibria as functions of h for values of h between 0 and 2, using a solid line when an equilibrium is stable and a dashed line when an equilibrium is unstable. Even though they do not make biological sense, include negative values of the equilibria on your graph. You should find a **transcritical bifurcation** (Exercise 17) at $h = 1$.

42. Suppose $\mu = 1$ in the basic disease model $\frac{dI}{dt} = \alpha I(1 - I) - \mu I$. Graph the two equilibria as functions of α for values of α between 0 and 2, using a solid line when an equilibrium is stable and a dashed line when an equilibrium is unstable. Even though they do not make biological sense, include negative values of the equilibria on your graph. You should find a **transcritical bifurcation** (Exercise 17) at $\alpha = 1$.

43. Consider a version of the equation in Section 5.2, Exercise 29 that includes the parameter r,

$$\frac{dN}{dt} = \frac{rN^2}{1+N^2} - N.$$

Graph the equilibria as functions of r for values of r between 0 and 3, using a solid line when an equilibrium is stable and a dashed line when an equilibrium is unstable. The algebra for checking stability is messy, so it is only necessary to check stability at $r = 3$. You should find a **saddle-node bifurcation** (Exercise 18) at $r = 2$.

44. Consider a variant of the basic disease model given by

$$\frac{dI}{dt} = \alpha I^2(1 - I) - I.$$

Graph the equilibria as functions of α for values of α between 0 and 5, using a solid line when an equilibrium is stable and a dashed line when an equilibrium is unstable. The algebra for checking stability is messy, so it is only necessary to check stability at $\alpha = 5$. You should find a **saddle-node bifurcation** (Exercise 18) at $\alpha = 4$.

45–46 ▪ Right at a bifurcation point, the stability theorem fails because the slope of the rate-of-change function at the equilibrium is exactly zero. In each of the following cases, check that the stability theorem fails, and then draw a phase-line diagram to find the stability.

45. Analyze the stability of the positive equilibrium in the model from Exercise 44 when $\alpha = 4$, the point where the bifurcation occurs.

46. Analyze the stability of the disease model when $\alpha = \mu = 1$, the point where the bifurcation occurs in Exercise 42.

5.4 Solving Autonomous Differential Equations

The method of phase-line diagrams allows us to sketch solutions with a minimum of algebra. What if we want an exact formula for the solution? Guessing sometimes works, as with pure-time differential equations. For a more dependable method, we would like to transform the problem of solving an autonomous differential equation into an integration problem. The technique for doing this is called **separation of variables** and is among the most powerful techniques in applied mathematics.

5.4.1 Separation of Variables

Consider the autonomous differential equation for bacterial growth

$$\frac{db}{dt} = \lambda b. \tag{5.4.1}$$

We cannot integrate to solve this as we could with a pure-time differential equation, because integrating the function λb requires knowing the solution $b(t)$.

The trick of separation of variables, as the name implies, is to separate the two variables: the b's and the t's. To separate variables, we divide both sides by b to get all the b's on the left hand side and multiply by dt to get all the t's on the right hand side, arriving at

$$\frac{db}{b} = \lambda dt.$$

As with substitution, we can treat dt like the denominator of a fraction. The next step is to compute the indefinite integral of each side. The left hand side has a db and can be integrated using b as the variable, or

$$\int \frac{1}{b}\, db = \ln(|b|) + c_1.$$

The right hand side has a dt and can be integrated using t as the variable, or

$$\int \lambda\, dt = \lambda t + c_2.$$

The absolute value bars around b (Equation 4.3.1) are unnecessary because b represents a population and must be positive.

The idea of separation of variables is that if $\frac{db}{b} = \lambda dt$, the integrals must also be equal. Therefore,

$$\ln(b) + c_1 = \lambda t + c_2.$$

Because c_1 and c_2 are arbitrary constants, we can combine the two constants into one by setting $c = c_2 - c_1$,

$$\ln(b) = \lambda t + c.$$

Finally, solve for b in terms of t. In this case, we can do so by exponentiating both sides, finding

$$b = e^{\lambda t + c}.$$

This solution might look strange because the arbitrary constant is inside the exponential function. It is often more convenient to move the constant outside. In particular, if we define the new constant

$$K = e^c,$$

we find that

$$e^{\lambda t + c} = e^{\lambda t} e^c = K e^{\lambda t}.$$

We find the constant K by plugging in the initial conditions.

Example 5.4.1 An Increasing Solution of the Bacterial Growth Equation

If $b(0) = 1000$, then

$$1000 = b(0) = K e^{\lambda \cdot 0} = K.$$

The solution with $\lambda = 2$ and $b(0) = 1000$ is

$$b(t) = 1000 e^{2t}.$$

(Figure 5.4.1).

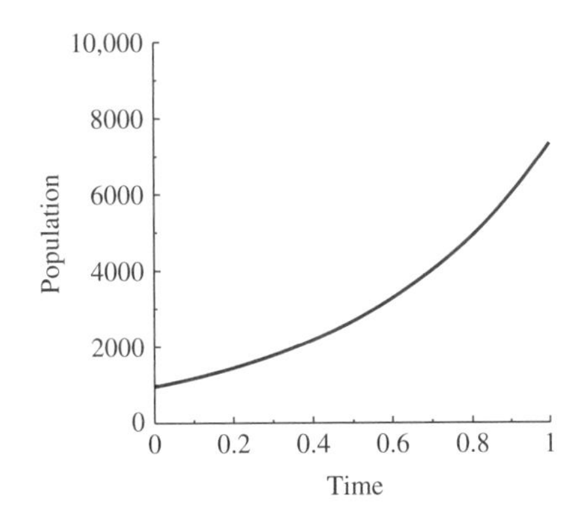

FIGURE 5.4.1
Solutions of the bacterial growth equation found with separation of variables

Example 5.4.2 A Decreasing Solution of the Bacterial Growth Equation

In contrast, if $b(0) = 9000$ and $\lambda = -1.5$, then

$$9000 = b(0) = Ke^{\lambda \cdot 0} = K.$$

The solution is

$$b(t) = 9000e^{-1.5t}$$

(Figure 5.4.2).

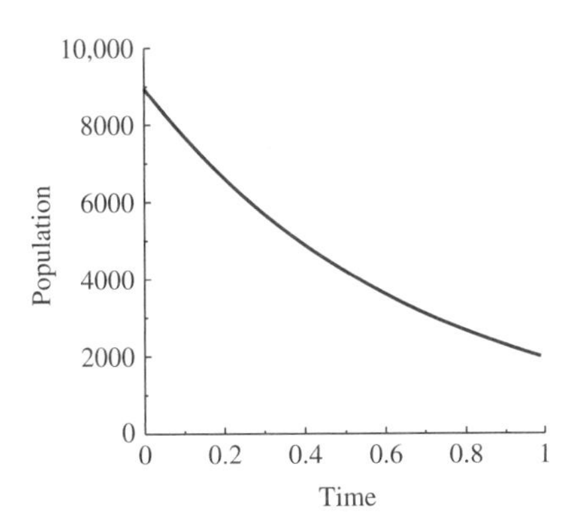

FIGURE 5.4.2
Solutions of the bacterial growth equation found with separation of variables

The following are the steps for **separation of variables.**

Algorithm 5.3 Separation of variables

1. Move all instances of the state variable to the left-hand side and all instances of the time t to the right-hand side.
2. Integrate both sides.
3. Set the two integrals equal to each other.
4. Combine the two arbitrary constants into one.
5. Solve for the state variable in terms of the time t, possibly rewriting the constant in a more convenient form.
6. Solve for the constant with the initial conditions.

In our example, the first step of separating the state variable b from the time t, could have been done differently had we chosen to move λ to the left-hand side:

$$\frac{db}{\lambda b} = dt.$$

Following Algorithm 5.3,

$$\frac{1}{\lambda}\ln(b) + c_1 = t + c_2 \quad \text{integrating}$$

$$\frac{1}{\lambda}\ln(b) = t + c \quad \text{combining constants}$$

$$b = e^{\lambda t}e^{\lambda c}. \quad \text{solving for } b$$

This has the same form we found before, but the constant looks different. If we set $K = e^{\lambda c}$, we again find the solution $b(t) = Ke^{\lambda t}$.

5.4.2 Solving Pure-Time Differential Equations with Separation of Variables

The technique of separation of variables is essentially the same method we used to solve pure-time differential equations.

Example 5.4.3 Separation of Variables Applied to a Pure-Time Differential Equation

Consider the pure-time differential equation for volume

$$\frac{dV}{dt} = t^2$$

with initial condition $V(0) = 1.0$ cm^3. We first solved this equation by finding the indefinite integral of t^2. The method of separation of variables requires computing the same integral. To begin, we multiply both sides by dt to move all instances of the time t to the right-hand side. All instances of the state variable V are already on the left-hand side because this is a pure-time differential equation. Following Algorithm 5.3,

$$dV = t^2 dt \quad \text{separating variables}$$

$$V + c_1 = \frac{t^3}{3} + c_2 \quad \text{integrating}$$

$$V = \frac{t^3}{3} + c. \quad \text{combining constants}$$

The last step, solving for V in terms of t, is already done. Furthermore, the constant c appears in its traditional spot, added to an antiderivative. We find the constant c with the initial condition,

$$1.0 = V(0) = \frac{0^3}{3} + c = c.$$

The solution of the equation is therefore $V = \frac{t^3}{3} + 1.0$. ◭

Because the only V in this pure-time equation appears as dV, the left-hand side is easy to integrate. The function of t, however, might be difficult or impossible to integrate. In contrast, the only t in an autonomous differential equation appears as dt, making the t integral simple. The function of V might be difficult or impossible to integrate. With an autonomous differential equation, there is the additional difficult or impossible step of solving for the state variable (b in the example above). Exercises 15 and 16 give examples of autonomous differential equations for which this last step is impossible.

5.4.3 Applications of Separation of Variables

We can use separation of variables to solve Newton's law of cooling,

$$\frac{dH}{dt} = \alpha(A - H).$$

Following the algorithm,

$$\frac{dH}{A-H} = \alpha dt \qquad \text{separating variables}$$

$$-\ln(|A-H|) + c_1 = \alpha t + c_2 \qquad \text{integrating}$$

$$-\ln(|A-H|) = \alpha t + c \qquad \text{combining constants}$$

$$|A-H| = e^{-\alpha t - c} \qquad \text{solving for } |A-H|$$

$$|A-H| = Ke^{-\alpha t}. \qquad \text{rewriting the constant as } K = e^{-c}$$

To get rid of the absolute value bars, we consider two cases. If $H < A$, $|A - H| = A - H$ so

$$|A-H| = A - H = Ke^{-\alpha t}.$$

We can now solve for H, finding

$$H(t) = A - Ke^{-\alpha t}.$$

Example 5.4.4 Separation of Variables Applied to Newton's Law of Cooling

If $\alpha = 0.1$, $A = 40°\text{C}$, and $H(0) = 10°\text{C}$, the solution is

$$H(t) = 40 - Ke^{-0.1t}.$$

We find the arbitrary constant K by plugging in $t = 0$,

$$10 = H(0) = 40 - Ke^{-0.1 \cdot 0} = 40 - K.$$

This has solution $K = 30$, so that

$$H(t) = 40 - 30e^{-0.1t}$$

(Figure 5.4.3a). The solution approaches the ambient temperature 40°C because

$$\lim_{t \to \infty} \left(40 - 30e^{-0.1t}\right) = 40,$$

consistent with our finding from the phase-line diagram that the equilibrium is stable. The solution gives more information by showing that the solution approaches the equilibrium exponentially.

If $H > A$, $|A - H| = H - A$ and

$$|A-H| = H - A = Ke^{-\alpha t}.$$

Next, we solve for H as

$$H(t) = A + Ke^{-\alpha t}.$$

Example 5.4.5 The Solution with a Different Initial Condition

If $\alpha = 0.1$, $A = 40°\text{C}$, and $H(0) = 60°\text{C}$, the solution is

$$H(t) = 40 + Ke^{-0.1t}.$$

We find the arbitrary constant K by plugging in $t = 0$,

$$60 = H(0) = 40 + Ke^{-0.1 \cdot 0} = 40 + K.$$

This has solution $K = 20$, so that

$$H(t) = 40 + 20e^{-0.1t}$$

(Figure 5.4.3b). Again, the solution approaches the ambient temperature 40°C because

$$\lim_{t \to \infty} \left(40 + 20e^{-0.1t}\right) = 40.$$

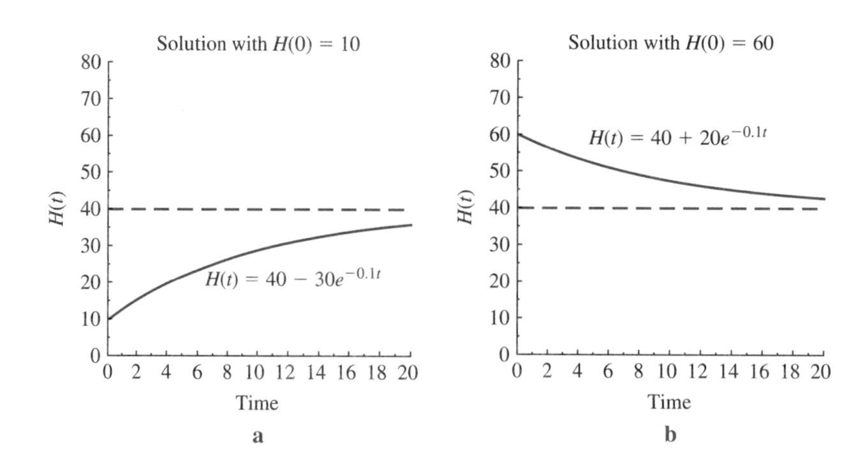

FIGURE 5.4.3

Solutions of Newton's law of cooling found with separation of variables

Example 5.4.6 A Population Explosion

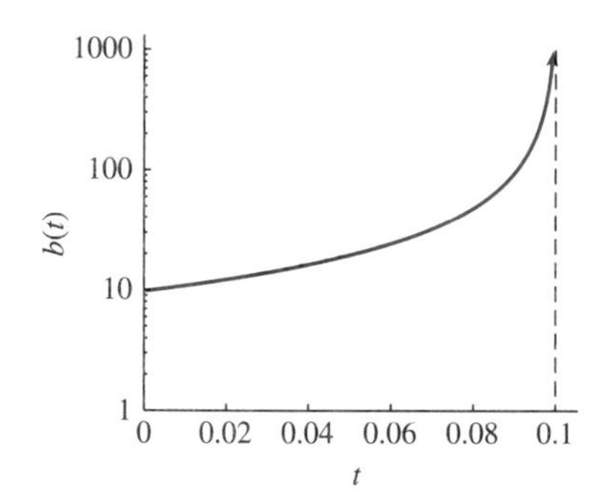

FIGURE 5.4.4

Semilog plot of a population with self-enhancing growth

As a more unusual example, suppose that the per capita production of some population with size b is

$$\text{per capita production} = b.$$

This means that individuals reproduce faster and faster the more of them there are. We expect that this population will grow very quickly, and can use separation of variables to compute how quickly. The differential equation for growth is

$$\begin{aligned} \frac{db}{dt} &= \text{per capita production} \cdot b \\ &= b \cdot b = b^2. \end{aligned}$$

Separation of variables proceeds as follows:

$$\begin{aligned} \frac{db}{b^2} &= dt && \text{separating variables} \\ -\frac{1}{b} + c_1 &= t + c_2 && \text{integrating} \\ -\frac{1}{b} &= t + c && \text{combining constants} \\ b &= \frac{-1}{t+c}. && \text{solving for } b \end{aligned}$$

The result looks dangerously negative. Proceeding anyway, suppose that $b(0) = 10$. We solve for c by plugging in the initial conditions, finding

$$10 = b(0) = \frac{-1}{c}$$

so that $c = -0.1$. Therefore,

$$b(t) = \frac{-1}{t - 0.1} = \frac{1}{0.1 - t}$$

(Figure 5.4.4). The population blasts off to infinity at time $t = 0.1$. What happens at $t = 0.11$? The solution does not exist. As far as this differential equation is concerned, the world comes to an end at $t = 0.1$. Models including self-enhancing growth tend to have this sort of "chain reaction" property.

Example 5.4.7 Using Integration by Partial Fractions to Solve the Logistic Equation

The **logistic differential equation**, like the logistic discrete-time dynamical system, gives a simple model of competition within a species. Consider a population of size N, and assume that per capita production rate decreases linearly from a maximum of r when $N = 0$ and reaches 0 when $N = K$. The value K is called the **carrying capacity**, and can be thought of as the maximum population for which the birth rate exceeds the death rate. Then

$$\text{per capita production rate} = r\left(1 - \frac{N}{K}\right),$$

just as in the discrete-time dynamical system derived in Subsection 3.2.1. In that case, however, we had to assume that the population went extinct if $N > K$ because negative values of per capita production lead to negative population sizes. With a differential equation, a negative per capita production rate is perfectly reasonable, meaning only that the population size is decreasing.

The logistic differential equation is

$$\frac{dN}{dt} = rN\left(1 - \frac{N}{K}\right). \tag{5.4.2}$$

Separation of variables begins by moving all N's to the left-hand side, or

$$\frac{dN}{N\left(1 - \frac{N}{K}\right)} = r\,dt.$$

The left-hand side is now the rational function

$$\frac{1}{N\left(1 - \frac{N}{K}\right)}$$

which we can integrate with integration by partial fractions (Subsection 4.3.4). This method works by separating the quotient into a sum of simpler quotients. In this case, we solve

$$\begin{aligned}
\frac{1}{N\left(1 - \frac{N}{K}\right)} &= \frac{A}{N} + \frac{B}{1 - \frac{N}{K}} \\
&= \frac{A\left(1 - \frac{N}{K}\right) + BN}{1 - \frac{N}{K}} \\
&= \frac{A - A\frac{N}{K} + BN}{1 - \frac{N}{K}}.
\end{aligned}$$

For the two sides to be equal, we require the numerators to match, or

$$1 = A - A\frac{N}{K} + BN.$$

Matching terms with and without N implies that $A = 1$ and $B = \frac{1}{K}$.

We can now apply separation of variables

$$\begin{aligned}
\frac{dN}{N\left(1 - \frac{N}{K}\right)} &= r\,dt && \text{separating variables} \\
\frac{dN}{N} + \frac{dN}{K - N} &= r\,dt && \text{breaking into partial fractions} \\
\ln(N) - \ln(K - N) &= rt + c && \text{integrating and combining constants} \\
\ln\left(\frac{N}{K - N}\right) &= rt + c && \text{applying a law of logs} \\
\frac{N}{K - N} &= Ce^{rt} && \text{exponentiating both sides} \\
N &= Ce^{rt}(K - N) && \text{multiplying out} \\
(1 + Ce^{rt})N &= CKe^{rt} && \text{isolating } N \\
N &= \frac{CKe^{rt}}{1 + Ce^{rt}}. && \text{solving for } N
\end{aligned}$$

In this calculation, we assumed that $N < K$, and could thus leave the absolute value bars out of the $\ln(K - N)$. As with Newton's law of cooling, the result comes out the same if $N > K$ (Exercise 17). The constant C can be found with the initial condition $N(0) = N_0$.

$$\begin{aligned} N_0 &= \frac{CK}{1+C} \\ N_0(1+C) &= CK \\ C(K - N_0) &= N_0 \\ C &= \frac{N_0}{K - N_0}. \end{aligned}$$

Substituting in and placing over a common denominator gives

$$N = \frac{N_0 K e^{rt}}{K - N_0 + N_0 e^{rt}}.$$

Summary We developed the method of **separation of variables** to solve autonomous differential equations. The method works by isolating the state variable on the left-hand side and time on the right-hand side, and then integrating to find a solution. Autonomous differential equations describing population growth and Newton's law of cooling can be solved with this technique. Solving pure-time differential equations with integration is a special case of separation of variables. In combination with integration by partial fractions, we can use separation of variables to solve the logistic differential equation for a population growing with competition.

5.4 Exercises

Mathematical Techniques

1–6 ▪ Use separation of variables to solve the following autonomous differential equations. Check your answers by differentiating.

1. $\frac{db}{dt} = 0.01b$, $b(0) = 1000$.
2. $\frac{db}{dt} = -3b$, $b(0) = 1.0 \times 10^6$.
3. $\frac{dN}{dt} = 1 + N$, $N(0) = 1$.
4. $\frac{dN}{dt} = 4 + 2N$, $N(0) = 1$.
5. $\frac{db}{dt} = 1000 - b$, $b(0) = 500$.
6. $\frac{db}{dt} = 1000 - b$, $b(0) = 1000$.

7–10 ▪ Use separation of variables to solve the following pure-time differential equations. Check your answers by differentiating.

7. $\frac{dP}{dt} = \frac{5}{1+2t}$, with $P(0) = 0$.
8. $\frac{dP}{dt} = 5e^{-2t}$, with $P(0) = 0$.
9. $\frac{dL}{dt} = 1000e^{0.2t}$, with $L(0) = 1000$.
10. $\frac{dL}{dt} = 64.3e^{-1.19t}$, with $L(0) = 0$.

11–14 ▪ Find the solution of the differential equation $\frac{db}{dt} = b^p$ in the following cases. At what time does it approach infinity? Sketch a graph.

11. $p = 2$ (as in the text) and $b(0) = 100$.
12. $p = 2$ and $b(0) = 0.1$.
13. $p = 1.1$ and $b(0) = 100$.
14. $p = 1.1$ and $b(0) = 0.1$.

15–16 ▪ The following autonomous differential equations can be solved except for the step of finding x in terms of t. Use the steps to figure out how the solution behaves.

a. Solve the equation with separation of variables.

b. It is impossible to algebraically solve for x in terms of t. However, you can still find the arbitrary constant. Find it.

c. Although you cannot find x as a function of t, you can find t as a function of x. Graph this function for $1 \le x \le 10$.

d. Sketch the solution for x as a function of t.

15. The autonomous differential equation $\frac{dx}{dt} = \frac{x}{1+x}$ with $x(0) = 1$. This describes a population with per capita production that decreases like $\frac{1}{1+x}$.
16. The autonomous differential equation $\frac{dx}{dt} = \frac{x}{1+x^2}$ with $x(0) = 1$. This describes a population with per capita production that decreases like $\frac{1}{1+x^2}$. Describe in words how the solution differs from that in Exercise 15.

17–18 ▪ Check the following results about the solution of the logistic equation from Example 5.4.7.

17. Show that the solution comes out the same if $N > K$.

18. Differentiate to check that the solution given in Example 5.4.7 solves the differential equation.

Applications

19–22 ▪ Using the method used to find the solution derived for Newton's law of cooling, find the solution of the chemical diffusion equation $\frac{dC}{dt} = \beta(\Gamma - C)$ with the following parameter values and initial conditions. Find the concentration after 10 seconds. How long would it take for the concentration to get halfway to the equilibrium value?

19. $\beta = 0.01/\text{s}$, $C(0) = 5.0\ \text{mmol/cm}^3$, and $\Gamma = 2.0\ \text{mmol/cm}^3$.

20. $\beta = 0.01/\text{s}$, $C(0) = 1.0\ \text{mmol/cm}^3$, and $\Gamma = 2.0\ \text{mmol/cm}^3$. Why do you think the time matches that in Exercise 19?

21. $\beta = 0.1/\text{s}$, $C(0) = 5.0\ \text{mmol/cm}^3$, and $\Gamma = 2.0\ \text{mmol/cm}^3$.

22. $\beta = 0.1/\text{s}$, $C(0) = 1.0\ \text{mmol/cm}^3$, and $\Gamma = 2.0\ \text{mmol/cm}^3$.

23–24 ▪ Consider Torricelli's law of draining $\frac{dy}{dt} = -2\sqrt{y}$ (Section 5.2, Exercise 33) with the constant set to 2.

23. Suppose the initial condition is $y(0) = 4$. Find the solution with separation of variables and graph the result. What really happens at time $t = 2$? And what happens after this time? How does this differ from the solution of the equation $\frac{dy}{dt} = -2y$?

24. Suppose the initial condition is $y(0) = 16$. Find the solution with separation of variables and graph the result. When does the solution reach 0? What would the depth be at this time if draining followed the equation $\frac{dy}{dt} = -2y$?

25–26 ▪ Suppose a population is growing at constant rate λ, but that individuals are harvested at a rate of h, following the differential equation $\frac{db}{dt} = \lambda b - h$. For each of the following values of λ and h, use separation of variables to find the solution, and compare graphs of the solution with those found earlier from a phase-line diagram.

25. $\lambda = 2.0$, $h = 1000$ (as in Section 5.2, Exercise 19).

26. $\lambda = 0.5$, $h = 1000$ (as in Section 5.2, Exercise 20).

27–28 ▪ Use separation of variables to solve for C in the following models describing chemical diffusion, and find the solution starting from the initial condition $C = \Gamma$.

27. The model in Section 5.1, Exercise 33.

28. The model in Section 5.1, Exercise 34.

29–32 ▪ Separation of variables can help to solve some nonautonomous differential equations. For example, suppose that the per capita production rate is $\lambda(t)$, a function of time, so that

$$\frac{db}{dt} = \lambda(t)b.$$

This equation can be separated into parts depending only on b and on t by dividing both sides of the equation by b and multiplying by dt. For the following functions $\lambda(t)$, give an interpretation of the equation, and solve the equation with the initial condition $b(0) = 10^6$. Sketch each solution. Check your answer by substituting into the differential equation.

29. $\lambda(t) = t$.

30. $\lambda(t) = \frac{1}{1+t}$.

31. $\lambda(t) = \cos(t)$.

32. $\lambda(t) = e^{-t}$.

33–34 ▪ Use integration by partial fractions to find solutions of the following models. Describe the solution in words.

33. The model from Section 5.1, Exercise 29 with initial condition $b(0) = 100$.

34. The model from Section 5.1, Exercise 30 with initial condition $b(0) = 10{,}000$.

35–36 ▪ Our model of selection (Equation 5.1.4) looks much like the logistic equation studied in Example 5.4.7, but can be extended to be **frequency-dependent** when the fitness of one or both types depends on how common it is.

35. Find the solution of Equation 5.1.4 as a special case of the logistic equation and compare with the solution in Equation 5.1.6.

36. Suppose that $\mu = 2 - 2p$ and $\lambda = 1$. Find the equilibria and their stability. What method would you use to try to find the solution after separating variables?

37–40 ▪ There are many important differential equations for which separation of variables fails, but which can be solved with other techniques. An important category involves Newton's law of cooling when the ambient temperature is changing. Consider, in particular, the case where $A(t) = e^{\beta t}$. Assume that the constant α is 1.0, so that the differential equation is $\frac{dH}{dt} = -H + A(t)$. It is impossible to separate variables in this equation.

37. Create the new variable $y = e^t H$ and find a differential equation for y.

38. Identify the type of differential equation, and solve it with the initial condition $H(0) = 0$.

39. Graph your solution and the ambient temperature when β is small, say $\beta = 0.1$. Describe the result.

40. Graph your solution and the ambient temperature when β is large, say $\beta = 1.0$. Why are the two curves so much farther apart?

Computer Exercise

41. We have seen that solutions of the differential equation

$$\frac{db}{dt} = b^2$$

approach infinity in a finite amount of time. We will try to stop it by multiplying the rate of change by a decreasing function of t in the nonautonomous equation

$$\frac{db}{dt} = g(t)b^2.$$

Try the following three functions for $g(t)$,

$$g_1(t) = e^{-t}$$
$$g_2(t) = \frac{1}{1+t^2}$$
$$g_3(t) = \frac{1}{1+t}.$$

a. Which of these functions decreases fastest and should best be able to stop $b(t)$ from approaching infinity?

b. Have your computer solve the equation in each case with initial conditions ranging from $b(0) = 0.1$ to $b(0) = 5.0$. Which solutions approach infinity?

c. Use separation of variables to try to find the solution in each of these cases (have your computer help with the integral of $g_2(t)$). Can you figure out when the solutions approach infinity? How well does this match your results from part **b**?

5.5 Two-Dimensional Differential Equations

We now begin the study of problems in multiple dimensions. The discrete-time dynamical systems and differential equations we have considered hitherto have described the dynamics of a single state variable: a population, a concentration, a fraction, and so forth. Most biological systems cannot be fully described without multiple measurements and multiple state variables. The tools for studying these systems, differential equations and discrete-time dynamical systems, remain the same, as does the goal of figuring out what will happen. The methods for getting from the problem to the answer, however, are more complicated. In this section, we introduce two important equations for population dynamics and an extension of Newton's law of cooling that takes into account the changing temperature of the room. These are called **systems of autonomous differential equations** or **coupled autonomous differential equations**. Euler's method can be applied to find approximate solutions of these systems of equations.

5.5.1 Predator-Prey Dynamics

Our basic model of population growth followed a single species, existing undisturbed in isolation. Life is rarely so peaceful. Imagine a bacterial population disrupted by the arrival of a predator, perhaps some sort of amoeba. We expect the bacteria to do worse when more predatory amoebas are around. Conversely, we expect the amoebas to do better when more of their bacterial prey are around. We now translate these intuitions into differential equations.

Denote the population of bacteria at time t by $b(t)$ and the population of amoebas by $p(t)$ to represent predation. We can build the equations by considering the factors affecting the population of each type. Suppose that the bacteria would grow exponentially in the absence of predators according to

$$\frac{db}{dt} = \lambda b.$$

How might predation affect the bacterial population? One ecologically naive but mathematically convenient approach is to assume that the organisms obey

The Principle of Mass Action: Individual bacteria encounter amoebas at a rate proportional to the number of amoebas.

Doubling the number of predators therefore doubles the rate at which bacteria run into predators. If running into an amoeba spells doom, this doubles the rate at which each bacterium risks being eaten.

In equations,

$$\text{rate at which an individual bacterium is eaten} = \epsilon p$$

where ϵ is the constant of proportionality. Therefore

$$\text{per capita growth rate of bacteria} = \lambda - \epsilon p.$$

Because the growth rate of a population is the product of the per capita growth rate and the population size,

$$\begin{aligned}\frac{db}{dt} &= \text{per capita growth rate} \times \text{population size}\\ &= (\lambda - \epsilon p)b.\end{aligned}$$

Suppose that the predators have a negative growth rate of δ in the absence of their prey, so

$$\frac{dp}{dt} = -\delta p.$$

Solutions of this equation converge to zero. These predators must eat to live. How might predation affect the amoeba population? As above, we assume mass action: doubling the number of bacteria doubles the rate at which predators run into bacteria, and thus doubles the rate at which they eat:

$$\text{rate at which an amoeba eats bacteria} = \eta b$$

(η is the Greek letter "eta" used to remind us of eating). If the per capita reproduction is increased by the eating rate,

$$\text{per capita growth rate of predators} = -\delta + \eta b.$$

Multiplying the per capita growth rate of the amoebas by their population size gives

$$\frac{dp}{dt} = \text{per capita growth rate} \times \text{population size}$$
$$= (-\delta + \eta b)p.$$

Example 5.5.1 Per Capita Growth Rates with Particular Parameter Values

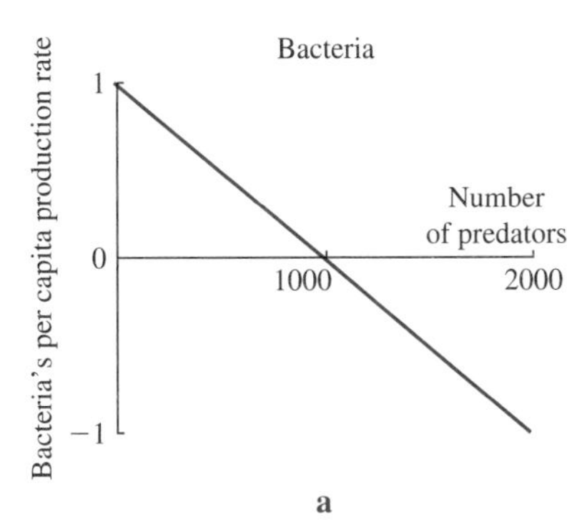

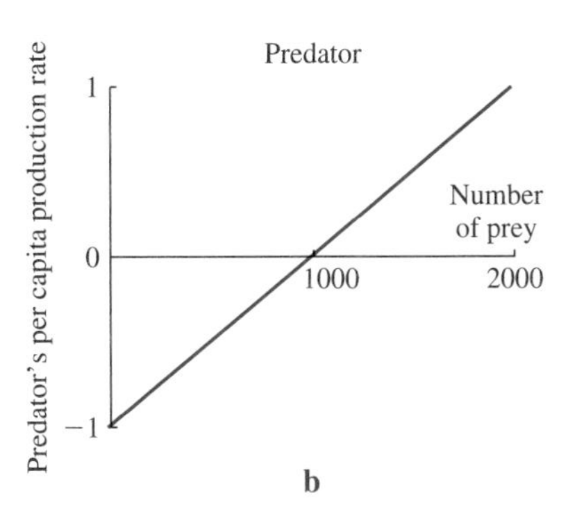

FIGURE 5.5.1
Per capita growth rates of prey and predator species

With $\lambda = 1.0$ and $\epsilon = 0.001$, the bacteria have a per capita growth rate of 1.0 when predators are absent, which decreases to 0 when the predator population is 1000 (Figure 5.5.1a). Similarly, with $\delta = 1.0$ and $\eta = 0.001$, the predator population declines at a per capita rate of -1.0 when prey are absent, but begins to grow when the prey population exceeds 1000 (Figure 5.5.1b).

We combine the differential equations into a **system of autonomous differential equations**

$$\frac{db}{dt} = (\lambda - \epsilon p)b$$
$$\frac{dp}{dt} = (-\delta + \eta b)p. \tag{5.5.1}$$

These are also called **coupled autonomous differential equations** because the rate of change of the bacterial population depends both on their own population size and that of the amoebas and the rate of change of the amoebas depends on both their own population and that of the bacteria. Two separate measurements are required to find the rate of change of either population. These equations are **autonomous** because neither rate of change depends explicitly on time.

The dynamics of a disease and a host, studied as a one-dimensional model in Subsection 5.3.3, have many similarities to those of a predator and its prey. The simplest two-dimensional model involves the same assumptions as the basic model, but explicitly tracks the number of susceptible individuals (S) and the number of infected individuals (I) rather than the fraction. Individuals become infected, and thus move from the S class into the I class, at a rate proportional to the product of the number of infected individuals with the number of susceptible people. Individuals recover, and move from the I class into the S class, at a rate proportional to the number of infected individuals (Figure 5.5.2a). Equations modeling this situation are

$$\frac{dI}{dt} = \alpha I S - \mu I$$
$$\frac{dS}{dt} = -\alpha I S + \mu I. \tag{5.5.2}$$

Although the I equation is essentially the same as the predator equation, with a constant per capita death and a mass action growth term, the S equation differs from the prey

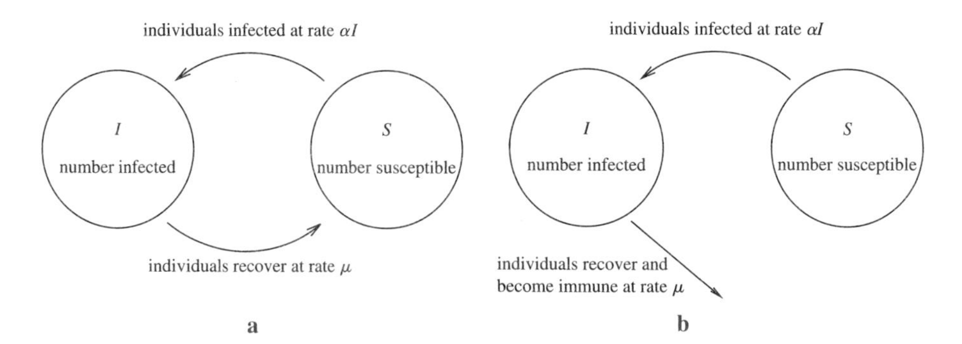

FIGURE 5.5.2
Per capita growth rates of prey and predator species

equation in that growth comes not from reproduction, but instead from recovery of sick individuals.

In an important and realistic extension, individuals who leave the infected class through recovery become permanently immune rather than becoming susceptible again (Figure 5.5.2b). This is true for many viruses, such as influenza. The equations differ from Equation 5.5.2 in that recovery does not add to the susceptible class. We do not need a separate equation for the recovered individuals because they have no further effect on the dynamics of the disease. The equations are

$$\frac{dI}{dt} = \alpha I S - \mu I$$

$$\frac{dS}{dt} = -\alpha I S. \tag{5.5.3}$$

In both equations 5.5.2 and 5.5.3, we ignore births and deaths, assuming that the disease itself is not deadly and that we are tracking it over a sufficiently short time that other births and deaths are few. Extensions that include births, deaths, and partial immunity are introduced in Exercises 35–40.

5.5.2 Dynamics of Competition

We can use similar reasoning to describe the competitive interaction between two populations. Our basic model of selection is based on the **uncoupled** pair of differential equations

$$\frac{da}{dt} = \mu a$$

$$\frac{db}{dt} = \lambda b.$$

Although there are two measurements, the rate of change of each type does not depend on the population size of the other. This lack of interaction is an idealization. Two populations in a single vessel will probably interact. Suppose that the per capita growth rate of each type declines as a linear function of the total number $a + b$ according to

$$\text{per capita growth rate of type } a = \mu \left(1 - \frac{a+b}{K_a}\right)$$

$$\text{per capita growth rate of type } b = \lambda \left(1 - \frac{a+b}{K_b}\right).$$

We have written the growth rates in this form to separate out the maximum per capita production μ of type a and λ of type b. Growth of type a becomes negative when the total population exceeds K_a, and growth of type b becomes negative when the total population exceeds K_b. The values K_a and K_b are sometimes called the **carrying capacities** of types a and b, respectively.

Example 5.5.2 Per Capita Growth Rates with Particular Parameter Values

With $\mu = 2.0$, $\lambda = 2.0$, $K_a = 1000$, and $K_b = 500$,

$$\text{per capita growth rate of type } a = 2.0\left(1 - \frac{a+b}{1000}\right)$$

$$\text{per capita growth rate of type } b = 2.0\left(1 - \frac{a+b}{500}\right)$$

(Figure 5.5.3).

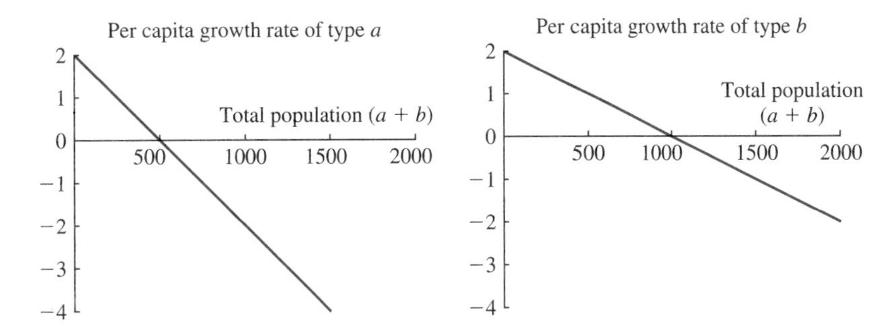

FIGURE 5.5.3
Per capita growth rates of competing species

We find the rates of change of the populations by multiplying the per capita growth rates by the population sizes, finding the coupled system of differential equations

$$\frac{da}{dt} = \mu\left(1 - \frac{a+b}{K_a}\right)a$$

$$\frac{db}{dt} = \lambda\left(1 - \frac{a+b}{K_b}\right)b. \quad (5.5.4)$$

The behavior of each population depends both on its own size and that of the other species. Neither depends explicitly on time, so this system of equations is autonomous.

Example 5.5.3 Competition Model with Parameter Values

With $\mu = 2.0$, $\lambda = 2.0$, $K_a = 1000$, and $K_b = 500$, the model becomes

$$\frac{da}{dt} = 2.0a\left(1 - \frac{a+b}{1000}\right)$$

$$\frac{db}{dt} = 2.0b\left(1 - \frac{a+b}{500}\right).$$

5.5.3 Newton's Law of Cooling

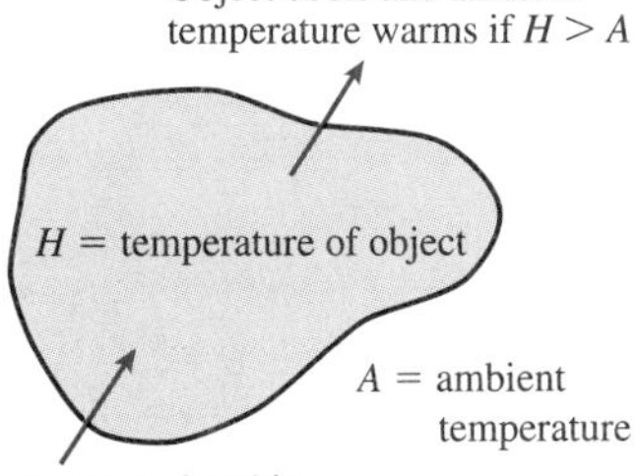

FIGURE 5.5.4
Newton's law of cooling revisited

Both Newton's law of cooling and the model of chemical diffusion across a membrane ignored the fact that the ambient temperature or concentration might also change. A small hot object placed in a large room will have little effect on the room temperature, but a large hot object will not only cool off itself but will also heat up the room. We require a system of autonomous differential equations to describe both temperatures simultaneously (Figure 5.5.4). We will now derive the coupled differential equations that describe this situation.

Newton's law of cooling expresses the rate of change of the temperature H of an object as a function of the ambient temperature A as

$$\frac{dH}{dt} = \alpha(A - H).$$

This equation is valid even if the ambient temperature A is itself changing. If $A < H$, heat is leaving the object and warming the room. The room follows the same law as the object itself, but we expect that the factor α, which depends on the size, shape, and material of the object, will be different,

$$\frac{dA}{dt} = \alpha_2(H - A).$$

The rate of change of temperature of each object depends on the temperature of the other. Together, these equations give the system a couple of autonomous differential equations

$$\frac{dH}{dt} = \alpha(A - H)$$

$$\frac{dA}{dt} = \alpha_2(H - A).$$

What is the relation between α and α_2? In general, α_2 will be smaller as the room becomes larger. If the "object" is made of the same stuff as the "room," the ratio of α to α_2 is equal to the ratio of their volumes.

Example 5.5.4 Newton's Law of Cooling When Object Is Smaller Than Room

If the room is three times bigger than the object but is made of the same substance (such as a balloon of air), then

$$\alpha_2 = \frac{\alpha}{3}$$

(Figure 5.5.5).

Example 5.5.5 Newton's Law of Cooling When Object Cools Faster Than Room

Suppose that the specific heat of the room in Example 5.5.4 is 10.0 times smaller than that of the object (meaning that a small amount of heat warms the room a great deal). Then α_2 is smaller by a factor of 3.0 because the room is larger, but larger by a factor of 10.0 because the specific heat is smaller, giving

$$\alpha_2 = 10.0\frac{\alpha}{3} = 3.33\alpha.$$

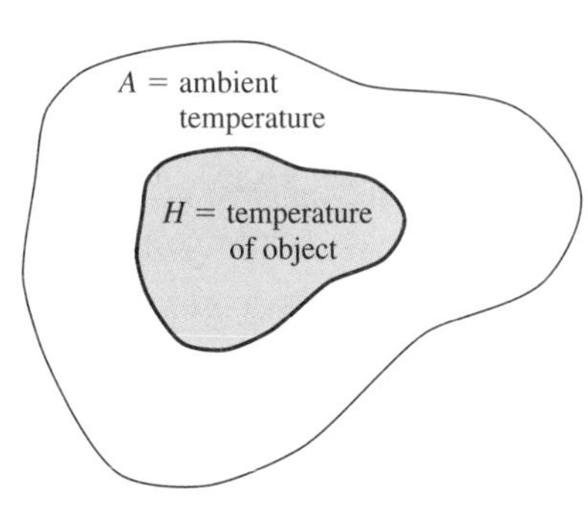

If the ambient space is 3 times as large as the object, the rate of change of the ambient temperature is one third smaller than that of the object

FIGURE 5.5.5

Newton's law of cooling applied to objects of different sizes

5.5.4 Applying Euler's Method to Systems of Autonomous Differential Equations

How do we find solutions of systems of autonomous differential equations? In general, this is a very difficult problem. In the next section, we will extend the method of equilibria and phase-line diagrams to sketch solutions. Surprisingly, perhaps, Euler's method for finding approximate solutions with the tangent line approximation works exactly the same way for systems as for single equations. We will apply the method to figure out how solutions behave for the systems describing predator-prey dynamics and competition.

Example 5.5.6 Applying Euler's Method to the Predator-Prey Equations

Consider the predator-prey equations (Equation 5.5.1) with $\lambda = 1$, $\delta = 1$, $\epsilon = 0.001$, and $\eta = 0.001$,

$$\frac{db}{dt} = (1.0 - 0.001p)b$$

$$\frac{dp}{dt} = (-1.0 + 0.001b)p.$$

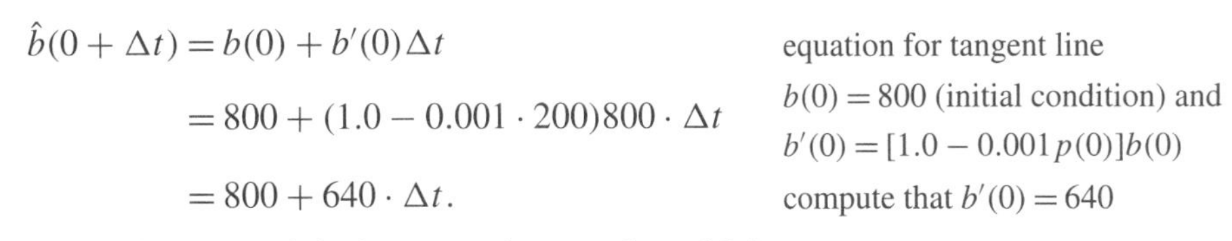

Suppose the initial condition is $b(0) = 800$ and $p(0) = 200$. Initially, there are 800 prey and 200 predators. What will happen to these populations after 2.0 time units? We break this interval into 10 units of length $\Delta t = 0.2$. After one of these intervals, we can approximate each population with the tangent line as

$$\begin{aligned}\hat{b}(0+\Delta t) &= b(0) + b'(0)\Delta t && \text{equation for tangent line}\\ &= 800 + (1.0 - 0.001\cdot 200)800\cdot\Delta t && b(0)=800 \text{ (initial condition) and } b'(0)=[1.0-0.001p(0)]b(0)\\ &= 800 + 640\cdot\Delta t. && \text{compute that } b'(0)=640\end{aligned}$$

After a time $\Delta t = 0.2$, the approximate value of b is

$$\hat{b}(0.2) = 800 + 640\cdot 0.2 = 928.$$

At the same time, the value for p can be found with the tangent line approximation to be

$$\begin{aligned}\hat{p}(0+\Delta t) &= p(0) + p'(0)\Delta t && \text{equation for tangent line}\\ &= 200 + (-1.0 + 0.001\cdot 800)200\cdot\Delta t && p(0)=800 \text{ (initial condition) and } p'(0)=[-1.0+0.001b(0)]p(0)\\ &= 200 - 40\cdot\Delta t. && \text{compute that } p'(0)=-40\end{aligned}$$

After a time $\Delta t = 0.2$, the approximate value of p is

$$\hat{p}(0.2) = 200 - 40\cdot 0.2 = 192$$

(Figure 5.5.6).

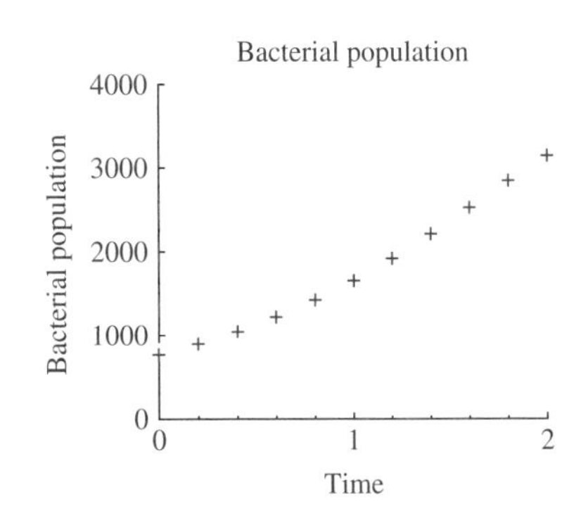

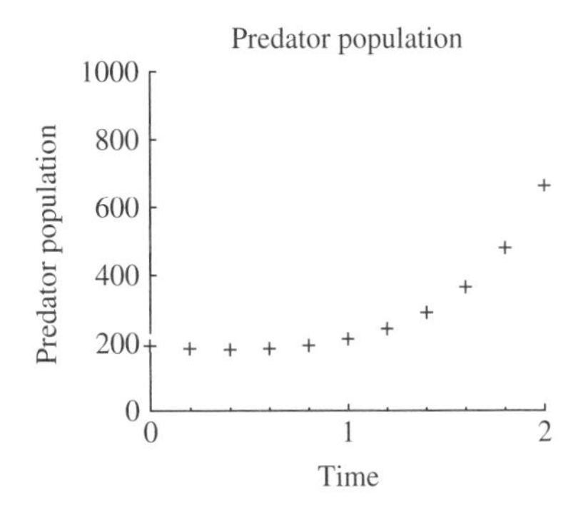

FIGURE 5.5.6
Euler's method applied to the predator-prey equations

The next step uses the same idea, but requires the approximate values $\hat{b}(0.2)$ and $\hat{p}(0.2)$ found in the first step.

$$\begin{aligned}\hat{b}(0.2+\Delta t) &= \hat{b}(0.2) + b'(0.2)\Delta t && \text{equation for tangent line}\\ &= 928 + (1.0 - 0.001\cdot 192)928\cdot\Delta t && b(0.2)=928 \text{ (initial condition) and } b'(0.2)=[1.0-0.001\hat{p}(0.2)]\hat{b}(0.2)\\ &= 928 + 749.8\cdot\Delta t. && \text{compute that } b'(0.2)=749.8\end{aligned}$$

Therefore,

$$\hat{b}(0.4) = 928 + 749.8\cdot 0.2 = 1078.$$

Things seem to be going pretty well for the prey. For the predators,

$$\begin{aligned}\hat{p}(0.2+\Delta t) &= \hat{p}(0.2) + p'(0.2)\Delta t && \text{equation for tangent line}\\ &= 192 + (-1.0 + 0.001\cdot 928)192\cdot\Delta t && p(0.2)=192 \text{ (initial condition) and } p'(0.2)=[-1.0+0.001\hat{b}(0.2)]\hat{p}(0.2)\\ &= 192 - 13.8\cdot\Delta t. && \text{compute that } p'(0.2)=-13.8\end{aligned}$$

Therefore,

$$\hat{p}(0.4) = 192 - 13.8\cdot 0.2 = 189.2.$$

t	$\hat{b}$	$\hat{p}$
0.0	800.0	200.0
0.2	928.0	192.0
0.4	1078.0	189.2
0.6	1252.8	192.2
0.8	1455.2	201.9
1.0	1687.4	220.3
1.2	1950.6	250.6
1.4	2242.9	298.2
1.6	2557.8	372.3
1.8	2878.8	488.3
2.0	3173.5	671.8

The predator numbers are fading.

We can continue following these steps to find the approximate solution at time $t = 2.0$. By following along step by step, we would see that the bacterial population eventually begins to decline.

Euler's method (described for a single autonomous differential equation in Algorithm 5.1) extends directly to coupled autonomous differential equations.

▶▶ Algorithm 5.4 Euler's Method for Solving Coupled Autonomous Differential Equations

Suppose a pair of measurements m and n obey the autonomous coupled differential equations

$$\frac{dm}{dt} = f(m,n)$$

$$\frac{dn}{dt} = g(m,n).$$

with initial condition $m(t_0) = m_0$, $n(t_0) = n_0$.

1. Choose a **time step** Δt (the length of time between estimated values).
2. Use the initial condition and the differential equation to find the tangent line $\hat{m}(t)$ with base point $t = t_0$ and slope $f(m_0, n_0)$. Use it to estimate $\hat{m}(t_0 + \Delta t)$. Similarly find the tangent line $\hat{n}(t)$ with base point $t = t_0$ and slope $g(m_0, n_0)$. Use it to estimate $\hat{g}(t_0 + \Delta t)$.
3. Use the estimate $\hat{m}(t_0 + \Delta t)$ and the differential equation to find the tangent line $\hat{m}(t)$ with base point $t_0 + \Delta t$ to estimate $\hat{m}(t_0 + 2\Delta t)$. Use the estimate $\hat{n}(t_0 + \Delta t)$ and the differential equation to find the tangent line $\hat{n}(t)$ with base point $t_0 + \Delta t$ to estimate $\hat{n}(t_0 + 2\Delta t)$.
4. Repeat the previous step to find $\hat{m}(t_0 + 3\Delta t)$ and $\hat{n}(t_0 + 3\Delta t)$ and so forth.

Example 5.5.7 Applying Euler's Method to the Coupled Differential Equations Describing Competition

Suppose we wish to apply Euler's method to the coupled differential equations for competition using the parameter values in Example 5.5.2

$$\frac{da}{dt} = 2.0a\left(1 - \frac{a+b}{1000}\right)$$

$$\frac{db}{dt} = 2.0b\left(1 - \frac{a+b}{500}\right).$$

with initial condition $a(0) = 750$ and $b(0) = 750$ to estimate $a(0.3)$ and $b(0.3)$.

1. Pick a time step of $\Delta t = 0.1$ (a smaller value should be more accurate).
2. The tangent line with base point $t = 0$ is

$$\hat{b}(0 + \Delta t) = b(0) + b'(0)\Delta t \quad \text{equation for tangent line}$$

$$= 750 + 2.0 \cdot 750\left(1 - \frac{750 + 750}{1000}\right)\Delta t \quad b(0) = 750 \text{ (initial condition) and } b'(0) = 2.0a(1 - \tfrac{a+b}{1000}) \text{ (differential equation for } b)$$

$$= 750 - 750\Delta t \quad \text{plug in values}$$

$$\hat{a}(0 + \Delta t) = a(0) + a'(0)\Delta t \quad \text{equation for tangent line}$$

$$= 750 + 2.0 \cdot 750\left(1 - \frac{750 + 750}{500}\right)\Delta t \quad a(0) = 750 \text{ (initial condition) and } a'(0) = 2.0a(1 - \tfrac{a+b}{500}) \text{ (differential equation for } a)$$

$$= 750 - 3000\Delta t. \quad \text{plug in values}$$

We therefore estimate that

$$\hat{b}(0.1) = 750 - 750 \cdot 0.1 = 675$$

$$\hat{a}(0.1) = 750 - 3000 \cdot 0.1 = 450.$$

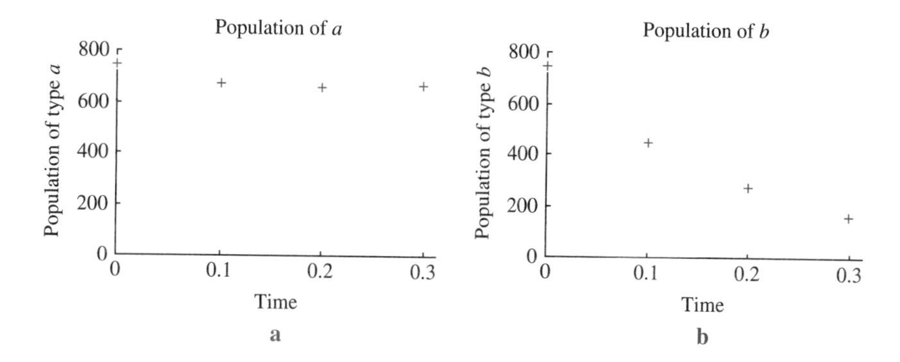

FIGURE 5.5.7
Euler's method applied to the competition equations

3. For the next step, to find $\hat{b}(0.2)$, we compute

$$\begin{aligned}
\hat{b}(0.1+\Delta t) &= \hat{b}(0.1) + b'(0.1)\Delta t && \text{equation for tangent line} \\
&= 675 + 2.0 \cdot 675\left(1 - \frac{675+450}{1000}\right)\Delta t && \hat{b}(0.1)=675 \text{ and } b'(0.1) = 2.0a(1-\tfrac{a+b}{1000}) \text{ (differential equation for } b) \\
&= 675 - 168.75\Delta t && \text{plug in values} \\
\hat{a}(0.1+\Delta t) &= \hat{a}(0.1) + a'(0.1)\Delta t && \text{equation for tangent line} \\
&= 450 + 2.0 \cdot 450\left(1 - \frac{675+450}{500}\right)\Delta t && \hat{a}(0.1)=450 \text{ and } a'(0.1) = 2.0a(1-\tfrac{a+b}{500}) \text{ (differential equation for } a) \\
&= 450 - 1687.5\Delta t. && \text{plug in values}
\end{aligned}$$

We therefore estimate that

$$\hat{b}(0.2) = 675 - 168.75 \cdot 0.1 = 658.125$$
$$\hat{a}(0.2) = 450 - 1687.5 \cdot 0.1 = 281.25.$$

4. For the next step, to find $\hat{b}(0.3)$, we compute

$$\begin{aligned}
\hat{b}(0.2+\Delta t) &= \hat{b}(0.2) + b'(0.2)\Delta t && \text{equation for tangent line} \\
&= 658.125 + 2.0 \cdot 658.125 \\
&\quad \times \left(1 - \frac{658.125+281.25}{1000}\right)\Delta t && \hat{b}(0.2)=658.125 \text{ and } b'(0.2) = 2.0a(1-\tfrac{a+b}{1000}) \\
&= 658.125 + 79.797\Delta t && \text{plug in values} \\
\hat{a}(0.2+\Delta t) &= \hat{a}(0.2) + a'(0.2)\Delta t && \text{equation for tangent line} \\
&= 281.25 + 2.0 \cdot 281.25 \\
&\quad \times \left(1 - \frac{658.125+281.25}{500}\right)\Delta t && \hat{a}(0.2)=281.25 \text{ and } a'(0.2) = 2.0a(1-\tfrac{a+b}{500}) \\
&= 281.25 - 1156.6\Delta t. && \text{plug in values}
\end{aligned}$$

We therefore estimate that

$$\hat{b}(0.3) = 658.125 + 79.797 \cdot 0.1 = 666.1$$
$$\hat{a}(0.3) = 281.25 - 1156.6 \cdot 0.1 = 165.6.$$

These values are plotted in Figure 5.5.7.

Euler's method is an effective way to compute solutions, but requires a great deal of numerical calculation. In the next section, we will begin to develop methods to predict how solutions will behave using graphical techniques.

Summary We have introduced three **coupled autonomous differential equations**, pairs of differential equations in which the rate of change of each state variable depends on its own value and on the value of the other state variable. We derived models of a predator and its prey, a competitive analog of the system studied to describe selection, and a version of Newton's law of cooling that keeps track of the change in room temperature. Each system is **autonomous** because the rates of change depend only on the state variables and not on time. **Euler's method** can be used to compute approximate solutions of these equations using the tangent line approximation.

5.5 Exercises

Mathematical Techniques

1–4 ▪ Consider the following special cases of the predator-prey equations

$$\frac{db}{dt} = (\lambda - \epsilon p)b$$
$$\frac{dp}{dt} = (-\delta + \eta b)p.$$

Write the differential equations and say what they mean.

1. $\epsilon = \eta = 0$.
2. $\delta = 0$.
3. $\eta = 0$.
4. $\epsilon = 0$.

5–6 ▪ If each state variable in a system of autonomous differential equations does not respond to changes in the value of the other, but depends only on a constant value, the two equations can be considered separately. For example, in the competition equation,

$$\frac{da}{dt} = \mu\left(1 - \frac{a+b}{K_a}\right)a$$

we could treat b as a constant value. In each case, find the equilibrium and draw a phase-line diagrams for a. Set the parameters to $\mu = 2$ and $K_a = 1000$.

5. Suppose that types a and b do not interact (equivalent to setting $b = 0$ in the differential equation for a).
6. Suppose that types a and b interact with a fixed population of 500 of the other (set $b = 500$ in the differential equation for a).

7–10 ▪ Apply Euler's method to the competition equations

$$\frac{da}{dt} = \mu\left(1 - \frac{a+b}{K_a}\right)a$$
$$\frac{db}{dt} = \lambda\left(1 - \frac{a+b}{K_b}\right)b$$

starting from the given initial conditions. Assume that $\mu = 2.0$, $\lambda = 2.0$, $K_a = 1000$, and $K_b = 500$.

7. Start from $a = 750$ and $b = 500$. Take two steps, with a step length of $\Delta t = 0.1$.
8. Start from $a = 250$ and $b = 500$. Take two steps, with a step length of $\Delta t = 0.1$.
9. Start from $a = 750$ and $b = 500$. Take four steps, with a step length of $\Delta t = 0.05$. How do your results compare with those in Exercise 7?
10. Start from $a = 250$ and $b = 500$. Take four steps, with a step length of $\Delta t = 0.05$. How do your results compare with those in Exercise 8?

11–14 ▪ Apply Euler's method to Newton's law of cooling

$$\frac{dH}{dt} = \alpha(A - H)$$
$$\frac{dA}{dt} = \alpha_2(H - A)$$

with the given parameter values and starting from the given initial conditions.

11. Suppose $\alpha = 0.3$ and $\alpha_2 = 0.1$. Start from $H = 60$ and $A = 20$. Take two steps, with a step length of $\Delta t = 0.1$.
12. Suppose $\alpha = 0.3$ and $\alpha_2 = 0.1$. Start from $H = 0$ and $A = 20$. Take two steps, with a step length of $\Delta t = 0.1$.
13. Suppose $\alpha = 3.0$ and $\alpha_2 = 1.0$. Start from $H = 60$ and $A = 20$. Take two steps, with a step length of $\Delta t = 0.25$. Do the results look reasonable?
14. Suppose $\alpha = 3.0$ and $\alpha_2 = 1.0$. Start from $H = 0$ and $A = 20$. Take two steps, with a step length of $\Delta t = 0.5$. Do the results look reasonable?

15–20 ▪ The spring equation or simple harmonic oscillator

$$\frac{d^2x}{dt^2} = -x$$

studied in Subsection 2.10.3 describes how acceleration (the second derivative of the position x) is equal to the negative of the position. The spring constant k has been set to 1 for simplicity. This one equation for the second derivative can be written as a system of two autonomous differential equations.

15. Write the velocity v in terms of the derivative of the position x, and the acceleration in terms of the derivative of the velocity v. Use the spring equation to write the derivative of the velocity in terms of the position x. Write the spring equation as a pair of equations for position and velocity.
16. Friction also creates acceleration proportional to the negative of the velocity (see Section 2.10, Exercise 39). A simple case obeys the equation

$$\frac{d^2x}{dt^2} = -2x - 2v.$$

    Write this as a pair of coupled differential equations for x and v.
17. We know that one solution of the basic spring equation in Exercise 15 is $x(t) = \cos(t)$. Find $v(t)$ and check that the

solution matches the system of equations. What are the initial position and velocity?

18. One solution of the spring equation with friction in Exercise 16 is $x(t) = e^{-t}\cos(t)$. Find $v(t)$ and check that the solution matches the system of equations. What are the initial position and velocity?

19. Use Euler's method with $\Delta t = 0.2$ for five steps to estimate $x(1)$ and $v(1)$ for the model in Exercise 15 with initial position $x(0) = 1$ and initial velocity $v(0) = 0$. Compare with the results in Exercise 17.

20. Use Euler's method with $\Delta t = 0.2$ for five steps to estimate $x(1)$ and $v(1)$ for the model in Exercise 16 with initial position $x(0) = 1$ and initial velocity $v(0) = -1$. Compare with the results in Exercise 18.

21–22 ▪ Nonautonomous differential equations can be written as a system of coupled autonomous differential equations by writing a separate differential equation for the variable t.

21. Write the differential equation $\frac{dF}{dt} = F^2 + kt$ (from Section 5.1, Exercise 1) as two equations for F and t.

22. Write the differential equation $\frac{dm}{dt} = \frac{e^{\alpha m} m^2}{(mt - \lambda)}$ (from Section 5.1, Exercise 4) as two equations for m and t.

Applications

23–26 ▪ Consider the following types of predator-prey interactions. Graph the per capita rates of change and write the associated system of autonomous differential equations.

23.
$$\text{per capita growth of prey} = 1.0 - 0.05p$$
$$\text{per capita growth of predators} = -1.0 + 0.02b.$$

24.
$$\text{per capita growth of prey} = 2.0 - 0.01p$$
$$\text{per capita growth of predators} = 1.0 + 0.01b.$$

How does this differ from the basic predator-prey system (Equation 5.5.1)?

25.
$$\text{per capita growth of prey} = 2.0 - 0.0001p^2$$
$$\text{per capita growth of predators} = -1.0 + 0.01b$$

26.
$$\text{per capita growth of prey} = 2.0 - 0.01p$$
$$\text{per capita growth of predators} = -1.0 + 0.0001b^2$$

27–30 ▪ Write systems of differential equations describing the following situations. Feel free to make up parameter values as needed.

27. Two predators that must eat each other to survive.

28. Two predators that must eat each other to survive, but with per capita production of each reduced by competition with its own species.

29. Two competitors where the per capita production of a is decreased by the total population, and the total population of b is decreased by the population of b.

30. Two competitors where the per capita production of each type is affected only by the population size of the other type.

31–32 ▪ Follow these steps to derive the equations for chemical exchange between two adjacent cells of different size. Suppose the concentration in the first cell is designated by the variable C_1, and the concentration in the second cell is designated by the variable C_2. In each case,

a. Write an expression for the total amount of chemical A_1 in the first cell and A_2 in the second.

b. Suppose that the amount of chemical moving from the first cell to the second cell is β times C_1, and the amount of chemical moving from the second cell to the first cell is β times C_2. Write equations for the rates of change of A_1 and A_2.

c. Divide by the volumes to find differential equations for C_1 and C_2.

d. In which cell is the concentration changing faster?

31. Suppose that the size of the first cell is 2.0 μL and the second is 5.0 μL.

32. Suppose that the size of the first cell is 5.0 μL and the second is 12.0 μL.

33–34 ▪ Write systems of autonomous differential equations describing the temperature of an object and the temperature of the room in the following cases.

33. The size of the room is 10.0 times that of the object, but the specific heat of the room is 0.2 times that of the object (meaning that a small amount of heat can produce a large change in the temperature of the room).

34. The size of the room is 5.0 times that of the object, but the specific heat of the room is 2.0 times that of the object (meaning that a large amount of heat produces only a small change in the temperature of the room).

35–40 ▪ There are many important extensions of the two-dimensional disease model (Equation 5.5.2) that include more categories of people, and that model processes of birth, death, and immunity.

35. Suppose that individuals who leave the infected class through recovery (at per capita rate μ) become susceptible again, but that individuals in the infected class also die at additional per capita rate k.

36. Suppose that one-third of the individuals who leave the infected class through recovery become permanently immune, and the other two-thirds become susceptible again.

37. Suppose that all individuals become susceptible upon recovery (as in the basic model) but that there is a source of mortality, so that both infected and susceptible individuals die at per capita rate k.

38. Suppose that all individuals become susceptible upon recovery (as in the basic model) but that there is a source of mortality, where susceptible individuals die at per capita rate k, but infected individuals die at a per capita rate that is twice as large.

39. Suppose that all individuals become susceptible upon recovery (as in the basic model) but that all individuals give birth at rate b. The offspring of susceptible individuals are susceptible and the offspring of infected individuals are infected.

40. Suppose that all individuals become susceptible upon recovery (as in the basic model) but that all individuals give birth at rate b, and that all offspring are susceptible.

41–42 ▪ Write the following models of frequency dependence as systems of autonomous differential equations for a and b.

41. The situation in Section 5.1, Exercise 37.

42. The situation in Section 5.1, Exercise 38.

Computer Exercises

43. Euler's method for systems of differential equations can be implemented as an updating system, a coupled pair of discrete-time dynamical systems. For example, with the predator-prey equations

$$\frac{db}{dt} = (1.0 - 0.001p)b$$
$$\frac{dp}{dt} = (-1.0 + 0.001b)p,$$

Euler's method is

$$\hat{b}(t + \Delta t) = \hat{b}(t) + b'(t)\Delta t$$
$$= \hat{b}(t) + [1.0 - 0.001\hat{p}(t)]\hat{b}(t)\Delta t$$

$$\hat{p}(t + \Delta t) = \hat{p}(t) + p'(t)\Delta t$$
$$= \hat{p}(t) + (-1.0 + 0.001\hat{b}(t))\hat{p}(t)\Delta t.$$

Starting from the initial condition $(b(0), p(0)) = (800, 200)$, follow this system until it loops around near its initial condition. Use the following values of Δt.

a. $\Delta t = 1.0$

b. $\Delta t = 0.2$

c. $\Delta t = 0.1$

d. $\Delta t = 0.01$

44. Follow the same steps as in the previous problem for the spring equations

$$\frac{dx}{dt} = v$$
$$\frac{dv}{dt} = -x.$$

derived in Exercise 16. How close is your estimated solution to the exact solution? What happens if you keep running for many cycles?

5.6 The Phase Plane

Systems of autonomous differential equations are generally impossible to solve exactly. We have seen how to use Euler's method to find approximate solutions. As with autonomous differential equations and discrete-time dynamical systems, we can deduce a great deal about the behavior of solutions from an appropriate graphical display. For systems of autonomous differential equations, the tool is the **phase-plane** diagram, an extension of the phase-line diagram. Our goal is again to find **equilibria**, points where each of the state variables remain unchanged. Finding these points on the phase plane requires a new tool, the **nullcline**, a graph of the set of points where each state variable separately remains unchanged.

5.6.1 Equilibria and Nullclines: Predator-Prey Equations

A single autonomous differential equation has an equilibrium where the rate of change of the state variable is zero. An **equilibrium** of a two-dimensional system of autonomous differential equations is a point where the rate of change of **each** state variable is zero.

Consider again a predator and its prey described by the system of autonomous differential equations

$$\frac{db}{dt} = (1.0 - 0.001p)b$$
$$\frac{dp}{dt} = (-1.0 + 0.001b)p.$$

with $\lambda = 1$, $\delta = 1$, $\epsilon = 0.001$, and $\eta = 0.001$ (Example 5.5.6). The rates of change of both b and p are equal to zero when the following hold simultaneously,

$$\frac{db}{dt} = (1.0 - 0.001p)b = 0$$
$$\frac{dp}{dt} = (-1.0 + 0.001b)p = 0.$$

Solving **simultaneous equations** can be much harder than solving a single equation. Our technique is to break the problem down into pieces and use a graph to combine the results.

To begin, we review solving simpler linear simultaneous equations.

Example 5.6.1 Solving Linear Simultaneous Equations

To solve the equations

$$y = 3x - 1$$
$$y = 2x + 3$$

for x and y, we set the two expressions for y equal to each other. This gives the equation

$$3x - 1 = 2x + 3$$

which can be solved to give $x = 4$. Substituting into either equation then gives $y = 11$ (Figure 5.6.1).

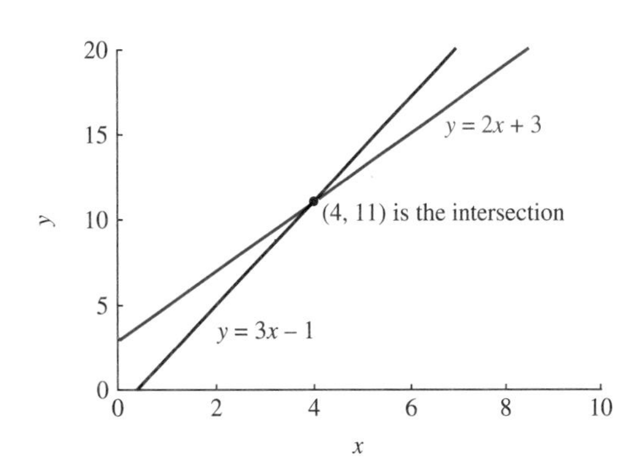

FIGURE 5.6.1
Two linear simultaneous equations

How do we graph the solutions of an equation? We are used to graphing functions written in the form $y = f(x)$, placing x on the horizontal axis and y on the vertical axis. It is easy to graph because we have **solved** for y. To graph the solutions of

$$\frac{db}{dt} = (1.0 - 0.001p)b = 0,$$

we must pick one of the variables b or p to place on the vertical axis like y. We can pick either one, and here choose p. Our next step is to plot an empty graph with the vertical axis labeled p and the horizontal axis labeled b (Figure 5.6.2). This is a graph of the **phase plane**. Like the phase line, it is a picture of all possible values the state variables can take. For example, the point (1200, 1500) represents the system with 1200 prey and 1500 predators.

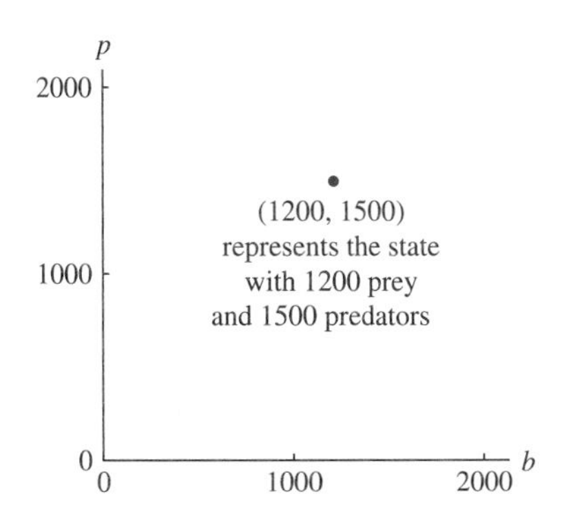

FIGURE 5.6.2
The phase plane with p on the vertical axis and b on the horizontal axis

Now, we can solve the equation

$$\frac{db}{dt} = (1.0 - 0.001p)b = 0$$

to find all values of b and p where the state variable b does not change. The first step in solving almost any equation is to **factor** and set each term equal to zero. This equation, conveniently enough, is already factored as $1.0 - 0.001p$ times b. The two equations to solve are therefore

$$1.0 - 0.001p = 0 \quad \text{or} \quad b = 0.$$

Solving means isolating p, the variable chosen as the vertical axis. On the first piece,

$$p = 1000.$$

In the phase plane, the graph of this function is a horizontal line at $p = 1000$ (Figure 5.6.3a). The second piece, $b = 0$, does not include the variable p. This means that **any** value of p where $b = 0$ is a solution. The graph of this piece is a vertical line at $b = 0$ (Figure 5.6.3b).

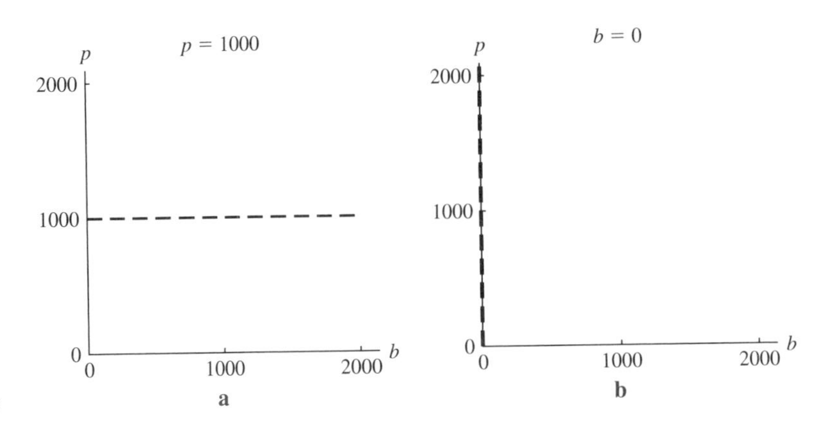

FIGURE 5.6.3
The two components of the b-nullcline

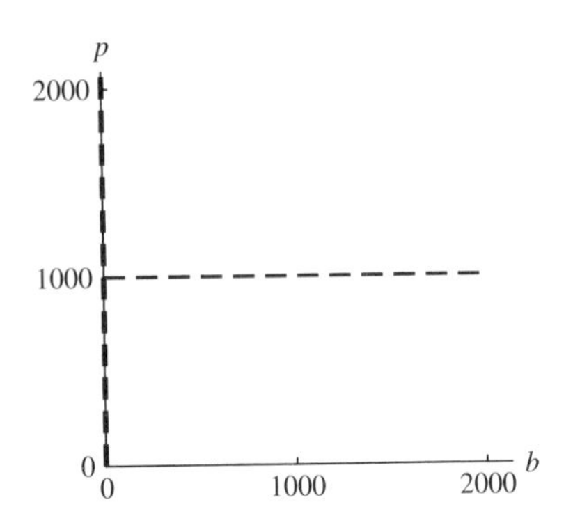

FIGURE 5.6.4
The b-nullcline: points in the phase plane where b does not change

These solutions represent the set of values where the rate of change of b is zero. This entire set is called the ***b*-nullcline** and is graphed in the phase plane in Figure 5.6.4. This might look like a strange graph, consisting as it does of two distinct pieces. Although no **function** could have a graph like this, it is typical of the behavior of nullclines.

We use the same method to find the p-nullcline, the set of points where the rate of change of p is zero. The equation

$$\frac{dp}{dt} = (-1.0 + 0.001b)p = 0$$

has a solution where either of the factors is equal to zero. Because the equation is in factored form with factors $-1.0 + 0.001b$ and p, solutions occur where

$$-1.0 + 0.001b = 0 \quad \text{and} \quad p = 0.$$

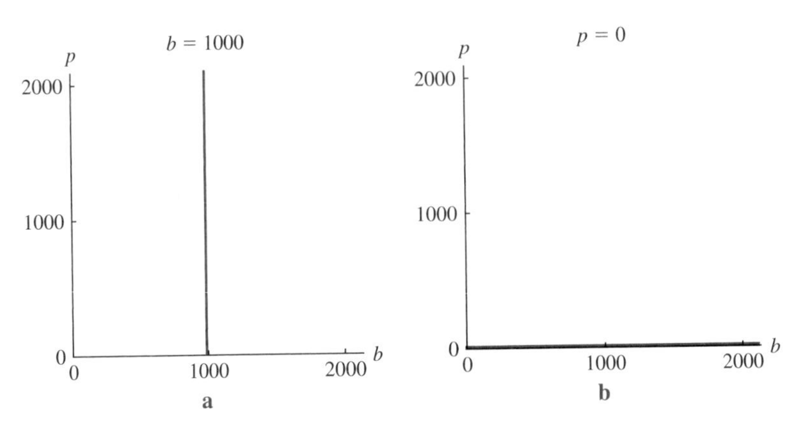

FIGURE 5.6.5
The two components of the p-nullcline

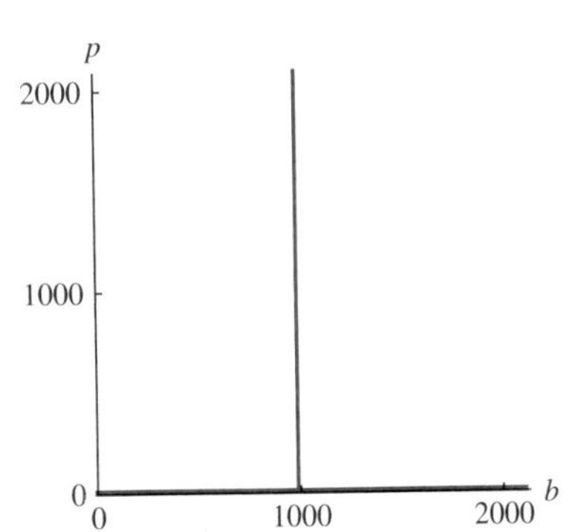

FIGURE 5.6.6
The p-nullcline: points in the phase plane where p does not change

To graph them, we again solve for the vertical variable p. Remember to use the **same** vertical variable for both nullclines. The first piece lacks any occurrence of the vertical variable p to solve for. This indicates that the solution is a vertical line, found by solving for the horizontal variable. The solution of $-1.0 + 0.001b = 0$ is $b = 1000$, so the first piece of the p-nullcline is a vertical line at $b = 1000$ (Figure 5.6.5a). The second piece is the horizontal line $p = 0$. The entire p-nullcline is the combination of these two pieces (Figure 5.6.6).

The state variable b does not change on the b-nullcline. The state variable p does not change on the p-nullcline. Therefore, neither b nor p changes at any point where the two nullclines intersect. These intersections, therefore, are the equilibria. To find equilibria, plot both nullclines on the same graph, being careful to distinguish which piece belongs to which nullcline (Figure 5.6.7). There are two intersections of the nullclines, at (0, 0) and (1000, 1000). At the first equilibrium, both populations are extinct. At the second, both populations are positive, and the system is balanced.

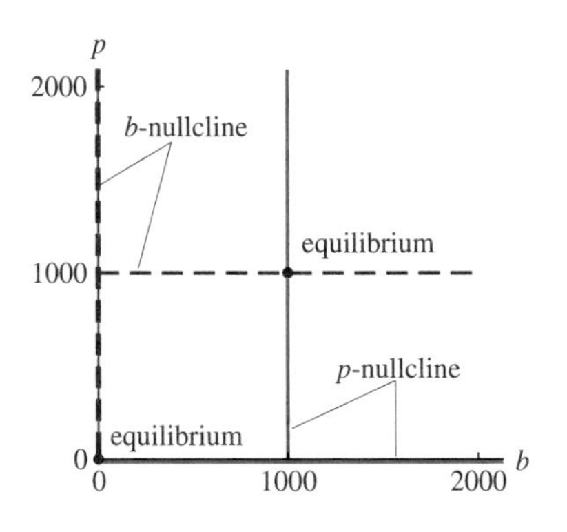

FIGURE 5.6.7
The nullclines of the predator-prey system

The points (0, 1000) and (1000, 0) are not equilibria because they do not lie at the intersection of the two nullclines. The point (0, 1000) lies on both pieces of the b-nullcline but not on the p-nullcline. We can check whether a point is an equilibrium by plugging into the system of differential equations. At $b = 0$ and $p = 1000$,

$$\frac{db}{dt} = (1.0 - 0.001p)b = (1.0 - 0.001 \cdot 1000) \cdot 0 = 0$$
$$\frac{dp}{dt} = (-1.0 + 0.001b)p = (-1.0 + 0.001 \cdot 0) \cdot 1000 = -1000.$$

Because the rate of change of the predator population p is not equal to zero, this is not an equilibrium. At this point, there are no prey and the predator population declines.

The following algorithm gives the steps to find the nullclines and equilibria of a system of autonomous differential equations.

Algorithm 5.5 Finding the Nullclines and Equilibria of Coupled Autonomous Differential Equations

1. Pick one of the variables to act as the vertical variable in the phase plane.
2. Write the equation for the first nullcline by setting the rate of change of the first state variable equal to zero.
   (a) Factor.
   (b) Solve each piece for the vertical variable. If there is no vertical variable in the piece, solve for the horizontal variable.
   (c) Graph each piece in the phase plane. If the equation contains no vertical variable, the graph is a vertical line.
3. Follow the same steps to find the second nullcline using the equation for the second state variable. Graph in the phase plane using a different color or style.
4. Find the intersections, which are the equilibria of the system.

Solutions that begin at an equilibrium remain there. To figure out what other solutions do, we would need to assess the stability of our equilibria. The techniques to do this, extensions of the methods used in one dimension, involve **linear algebra**, a subject that lies just beyond the scope of this book. In the next section, however, we will learn to analyze at least some situations by drawing **direction arrows** on our phase-plane diagram.

Example 5.6.2 The Model of a Disease with Immunity

As we have seen, models of diseases are similar in many ways to predator-prey models. Consider the case where individuals who recover are immune and thus removed from the population as far as the disease is concerned (Equation 5.5.3). With parameters $\alpha = 2.0$ and $\mu = 1.0$, the equations are

$$\frac{dI}{dt} = \alpha IS - \mu I = 2.0IS - 1.0I$$
$$\frac{dS}{dt} = -\alpha IS = -2.0IS.$$

1. We choose to place S on the vertical axis.
2. The I-nullcline satisfies the equation $2.0IS - 1.0I = 0$.
   (a) We factor as $(2.0S - 1.0)I = 0$.
   (b) Solve the first piece to find $S = 0.5$ and the second as $I = 0$ because it lacks the vertical variable S.
   (c) The graph consists of two pieces (Figure 5.6.8).

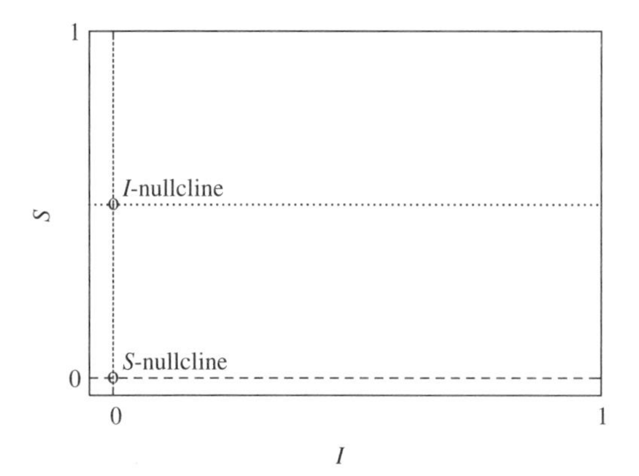

FIGURE 5.6.8

The nullclines of the disease model with immunity

3. The S-nullcline satisfies the equation $-2.0IS = 0$.
   (a) We factor as $-2.0SI = 0$.
   (b) Solve the first piece to find $S = 0$ and the second as $I = 0$.
   (c) The graph consists of two pieces (Figure 5.6.8).
4. These interact along the entire line $I = 0$, so this population will be in equilibrium as long as there is no disease.

5.6.2 Equilibria and Nullclines: Competition Equations

We can use Algorithm 5.5 to find the equilibria of competing bacterial types that follow the equations

$$\frac{da}{dt} = 2.0\left(1 - \frac{a+b}{1000}\right)a$$
$$\frac{db}{dt} = 2.0\left(1 - \frac{a+b}{500}\right)b$$

with $\mu = 2.0$, $\lambda = 2.0$, $K_a = 1000$, and $K_b = 500$ (Example 5.5.2).

1. We begin by choosing the vertical variable, picking b because it comes later in the alphabet.
2. To find the a-nullcline, we solve
$$\frac{da}{dt} = 2.0\left(1 - \frac{a+b}{1000}\right)a = 0.$$
   (a) The factors are $2.0\left(1 - \frac{a+b}{1000}\right)$ and a.
   (b) Solutions occur where
$$2.0\left(1 - \frac{a+b}{1000}\right) = 0 \text{ or } a = 0.$$
   We solve the first piece for b, the vertical variable, finding
$$2.0\left(1 - \frac{a+b}{1000}\right) = 0$$
$$\left(1 - \frac{a+b}{1000}\right) = 0$$
$$\frac{a+b}{1000} = 1$$
$$a + b = 1000$$
$$b = 1000 - a.$$

The second piece has no vertical variable b and is therefore the vertical line at $a = 0$.

(c) The graph is shown in Figure 5.6.9a.

3. To find the b-nullcline, we solve

$$\frac{db}{dt} = 2.0\left(1 - \frac{a+b}{500}\right)b.$$

(a) The factors are $2.0\left(1 - \frac{a+b}{500}\right)$ and b.

(b) Solutions occur where

$$2.0\left(1 - \frac{a+b}{500}\right) = 0 \text{ or } b = 0.$$

We solve the first piece for the vertical variable, finding

$$\begin{aligned} 2.0\left(1 - \frac{a+b}{500}\right) &= 0 \\ \left(1 - \frac{a+b}{500}\right) &= 0 \\ \frac{a+b}{500} &= 1 \\ a + b &= 500 \\ b &= 500 - a. \end{aligned}$$

The second piece is a horizontal line at $b = 0$.

(c) The graph is shown in Figure 5.6.9b.

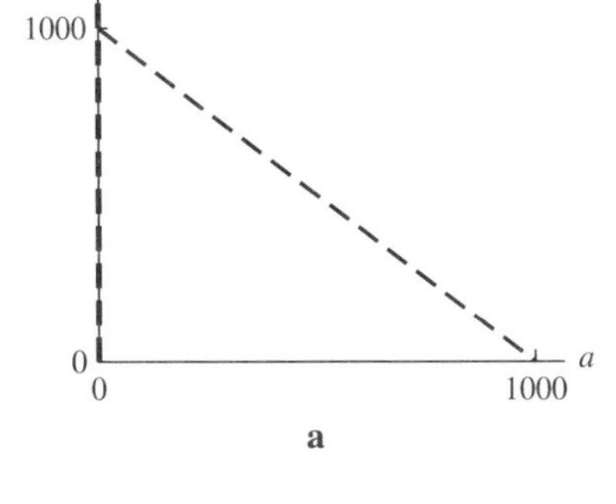

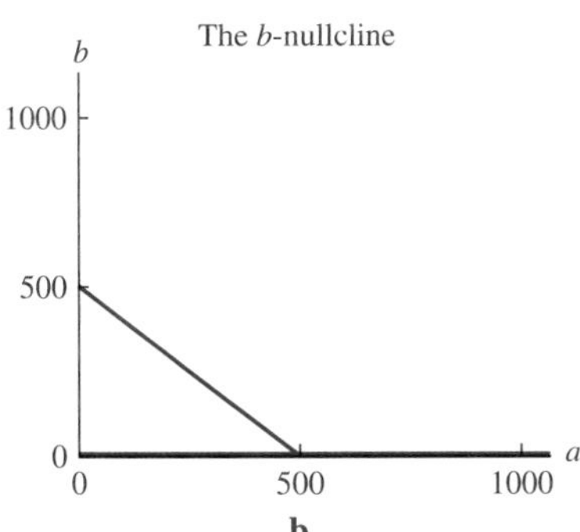

FIGURE 5.6.9

The nullclines of the competition system

4. To find the equilibria, we plot both nullclines on the same graph of the phase plane (Figure 5.6.10). There are three intersections and thus three equilibria: at (0, 0), (1000, 0), and (0, 500). Both types are extinct at the first, type b dominates the population at (0, 500), and type a dominates the population at (0, 1000).

The points (0, 1000) and (500, 0) are not equilibria. At (0, 1000),

$$\frac{db}{dt} = 2.0\left(1 - \frac{1000}{500}\right) = -2.0 < 0.$$

The point lies on both pieces of the a-nullcline but not on the b-nullcline. Similarly, at (500, 0)

$$\frac{da}{dt} = 2.0\left(1 - \frac{500}{1000}\right) = 1.0 > 0.$$

This point lies on both branches of the b-nullcline but not on the a-nullcline.

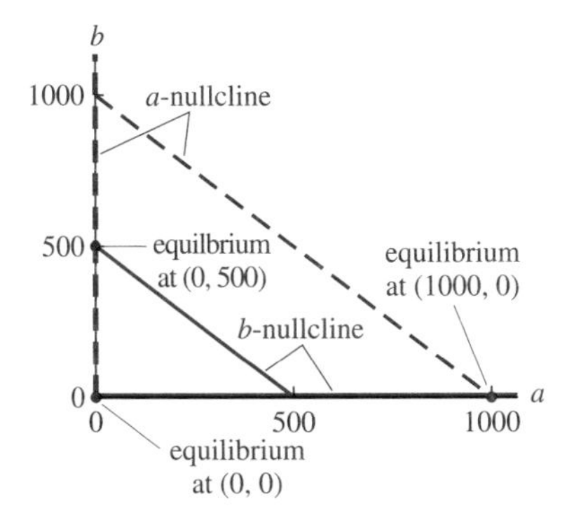

FIGURE 5.6.10

The nullclines and equilibria of the competition system

5.6.3 Equilibria and Nullclines: Newton's Law of Cooling

We use the same steps to find the equilibria for Newton's law of cooling

$$\frac{dH}{dt} = \alpha(A - H)$$

$$\frac{dA}{dt} = \alpha_2(H - A)$$

assuming that both α and α_2 are greater than 0.

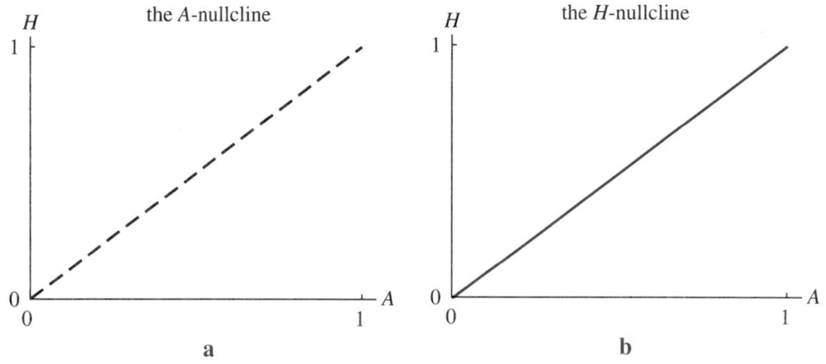

the A-nullcline

the H-nullcline

a

b

FIGURE 5.6.11
The nullclines of Newton's law of cooling

1. To begin, we choose the temperature of the object H as the vertical variable.
2. The A-nullcline is the set of points where
$$\frac{dA}{dt} = \alpha_2(H - A) = 0.$$
   (a) This has only one component and does not need to be factored (unless $\alpha_2 = 0$).
   (b) If $\alpha > 0$, we solve for the vertical variable H,
$$H = A.$$
   (c) The graph is shown in Figure 5.6.11a.
3. The H-nullcline is the set of points where
$$\frac{dH}{dt} = \alpha(A - H) = 0.$$
   (a) This has only one component and does not need to be factored
   (b) Solving for the vertical variable H, we again find
$$H = A$$
   (c) The graph is shown in Figure 5.6.11b.
4. The two nullclines exactly overlap. Because all points where the two nullclines intersect are equilibria, every point on the line $H = A$ is an equilibrium (Figure 5.6.12).

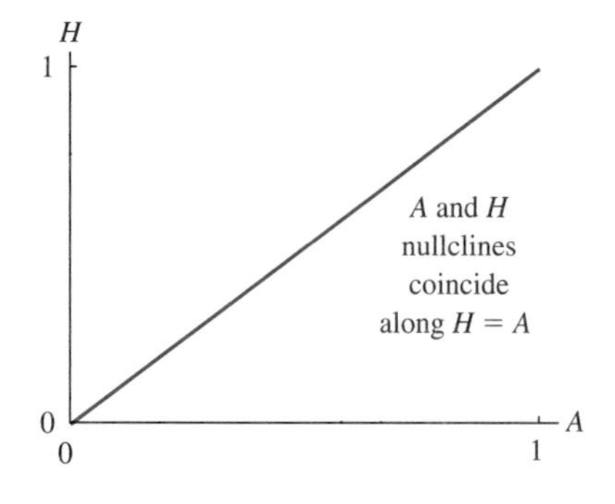

FIGURE 5.6.12
The nullclines and equilibria of Newton's law of cooling

This peculiar result makes sense. When the two temperatures are the same, no matter what they are, there is no further change in temperature by either the object or the room.

Summary We have introduced the **phase plane**, **nullclines**, and **equilibria** as tools to study autonomous systems of differential equations. Equilibria occur where the rate of change of each state variable is 0. They can be found on the phase plane (the Cartesian plane with axes labeled by the state variables) by graphing the two **nullclines**. The nullcline associated with a state variable is the set of values where it remains unchanged.

5.6 Exercises

Mathematical Techniques

1–4 ▪ Finding equilibria of coupled differential equations requires solving **simultaneous equations**. The following are linear equations, where the only possibilities are no solutions, one solution, or a whole line of solutions. For each of the following pairs, solve each equation for y in terms of x, set the two equations for x equal, and solve for x. Check that both equations give the same value for y. Sketch a graph with y on the vertical axis.

1.
$$-3 - y + 3x = 0$$
$$-2 + 2y - 4x = 0$$

2.
$$-6+3y+3x=0$$
$$-2-2y-6x=0$$

3.
$$3-3y+3x=0$$
$$-2+2y-2x=0$$

What goes wrong? Use your graph with y on the vertical axis to explain the problem.

4.
$$8-2y+4x=0$$
$$-2+y-2x=0$$

What goes wrong? Use your graph with y on the vertical axis to explain the problem.

5–10 ▪ Finding equilibria of nonlinear coupled differential equations requires solving nonlinear simultaneous equations that can have any number of solutions. For each of the following pairs, solve each equation for y in terms of x, set the two equations for x equal, and solve for x. Check that both equations give the same value for y. Sketch a graph with y on the vertical axis.

5.
$$-5-y+3x^2+2x=0$$
$$-2+2y-4x=0$$

6.
$$-3-y+3x^2+2x=0$$
$$-2+2y-4x+2x^2=0$$

7.
$$-y^2+x^2=0$$
$$-2+2y-4x=0$$

Solving the first equation for y in terms of x does not give a function. Graph the relation and find the solutions.

8.
$$y(y-x^2)=0$$
$$-6+2y-4x=0$$

Solving the first equation for y in terms of x does not give a function. Graph the relation and find the solutions.

9.
$$x(y-x)=0$$
$$-6+2y-4x=0$$

Solving the first equation for y in terms of x does not give a function (and includes a vertical section). Graph the relation and find the solutions.

10.
$$(x-1)(y^2-x^2)=0$$
$$-2+2y-6x=0$$

Solving the first equation for y in terms of x does not give a function and includes a vertical section. Graph the relation and find the solutions.

11–16 ▪ Graph the nullclines in the phase plane and find the equilibria of the following.

11. Predator-prey model
$$\frac{db}{dt}=(\lambda-\epsilon p)b$$
$$\frac{dp}{dt}=(-\delta+\eta b)p$$
(Equation 5.5.1) with $\lambda=1.0$, $\delta=3.0$, $\epsilon=0.002$, $\eta=0.005$.

12. Predator-prey model with $\lambda=1.0$, $\delta=3.0$, $\epsilon=0.005$, $\eta=0.002$.

13. Newton's law of cooling
$$\frac{dH}{dt}=\alpha(A-H)$$
$$\frac{dA}{dt}=\alpha_2(H-A)$$
with $\alpha=0.01$ and $\alpha_2=0.1$.

14. Newton's law of cooling with $\alpha=0.1$ and $\alpha_2=0.5$.

15. Competition model
$$\frac{da}{dt}=\mu\left(1-\frac{a+b}{K_a}\right)a$$
$$\frac{db}{dt}=\lambda\left(1-\frac{a+b}{K_b}\right)b$$
(Equation 5.5.4) with $\lambda=2.0$, $\mu=1.0$, $K_a=10^6$, $K_b=10^7$.

16. Competition model with $\lambda=1.0$, $\mu=2.0$, $K_a=10^6$, $K_b=10^7$. How do the results compare with those in Exercise 15? Why?

17–20 ▪ Redraw the phase planes for the following problems, but make the other choice for the vertical variable. Check that you get the same equilibrium.

17. The equations in Exercise 11.
18. The equations in Exercise 12.
19. The equations in Exercise 15.
20. The equations in Exercise 16.

21–22 ▪ If each state variable in a system of autonomous differential equations does not respond to changes in the value of the other, but depends only on a constant value, the two equations can be considered separately. In this case, the phase plane is particularly simple. Find the nullclines and equilibria in the following cases.

21. The situation in Section 5.5, Exercise 5.
22. The situation in Section 5.5, Exercise 6.

Applications

23–26 ▪ Find the nullclines and equilibria for the following predator-prey models.

23. The model in Section 5.5, Exercise 23.
24. The model in Section 5.5, Exercise 24.
25. The model in Section 5.5, Exercise 25.
26. The model in Section 5.5, Exercise 26.

27–30 ▪ Find and graph the nullclines, and find the equilibria for the following models.

27. The model found in Section 5.5, Exercise 27.
28. The model found in Section 5.5, Exercise 28.

29. The model found in Section 5.5, Exercise 29.

30. The model found in Section 5.5, Exercise 30.

31–32 ▪ The models of diffusion derived in Section 5.5, Exercises 31 and 32 assume that the membrane between the vessels is equally permeable in both directions. Suppose instead that the constant of proportionality governing the rate at which chemical moves differs in the two directions. In each of the following cases,

a. Find the rate at which chemical moves from the smaller to the larger vessel.

b. Find the rate at which chemical moves from the larger to the smaller vessel.

c. Find the rate of change of the amount of chemical in each vessel.

d. Divide by the volumes V_1 and V_2 to find the rate of change of concentration.

e. Find and graph the nullclines.

f. What are the equilibria? Do they make sense?

31. The constant of proportionality governing the rate at which chemical enters the cell is three times as large as the constant governing the rate at which it leaves (as in Section 5.1, Exercise 33).

32. The constant of proportionality governing the rate at which chemical enters the cell is half as large as the constant governing the rate at which it leaves (as in Section 5.1, Exercise 34).

33–34 ▪ In our model of competition, the per capita growth rate of types a and b are functions only of the total population size. This means that reproduction is reduced just as much by an individual of type a as by an individual of type b. In many systems, each type interferes differently with type a than with type b. Check that the given set of equations matches the assumptions in each of the following cases, and find and graph the equilibria and nullclines.

33. Suppose that individuals of type b reduce the per capita growth rate of type a by half as much as individuals of type a, and that individuals of type a reduce the per capita type b growth rate by twice as much as individuals of type b. The equations are

$$\frac{da}{dt} = \left(1 - \frac{a + b/2}{1000}\right) a$$

$$\frac{db}{dt} = \left(1 - \frac{2a + b}{1000}\right) b.$$

34. Suppose that individuals of type b reduce the per capita type a growth rate by half as much as individuals of type a, and that individuals of type a reduce the per capita type b growth rate by half as much as individuals of type b. The equations are

$$\frac{da}{dt} = \left(1 - \frac{a + b/2}{1000}\right) a$$

$$\frac{db}{dt} = \left(1 - \frac{a/2 + b}{1000}\right) b.$$

(There should be four equilibria.)

35–40 ▪ Draw the nullclines and find equilibria of the following extensions of the basic disease model.

35. The model in Section 5.5, Exercise 35. Find the nullclines and equilibria of this model when $\alpha = 2.0$, $\mu = 1.0$, and $k = 1.0$.

36. The model in Section 5.5, Exercise 36. Find the nullclines and equilibria of this model when $\alpha = 2.0$ and $\mu = 1.0$.

37. The model in Section 5.5, Exercise 37. Find the nullclines and equilibria of this model when $\alpha = 2.0$, $\mu = 1.0$, and $k = 0.5$.

38. The model in Section 5.5, Exercise 38. Find the nullclines and equilibria of this model when $\alpha = 2.0$, $\mu = 1.0$, and $k = 4.0$.

39. The model in Section 5.5, Exercise 39. Find the nullclines and equilibria of this model when $\alpha = 2.0$, $\mu = 1.0$, and $b = 2.0$.

40. The model in Section 5.5, Exercise 40. Find the nullclines and equilibria of this model when $\alpha = 2.0$, $\mu = 1.0$, and $b = 1.0$.

Computer Exercise

41. One complicated equation for chemical kinetics is the Schnakenberg reaction. Let A and B denote the concentrations of two chemicals A and B. A is added at constant rate k_1, B is added at constant rate k_4, A breaks down at rate k_2, and B is converted into A with an **autocatalytic reaction**. The equations are

$$\frac{dA}{dt} = k_1 - k_2 A + k_3 A^2 B$$

$$\frac{dB}{dt} = k_4 - k_3 A^2 B.$$

The final term is somewhat like the term αIS in the epidemic equation studied in the previous three problems, but differs in that the rate of the reaction becomes faster the larger the concentration of A. Suppose that $k_2 = k_3 = 1.0$.

a. Have your computer draw the nullclines and find the equilibria in the case $k_1 = 0.2$ and $k_4 = 2.0$.

b. Do the same with $k_1 = -0.2$ and $k_4 = 2.0$.

c. Try to explain your results.

5.7 Solutions in the Phase Plane

We have seen how to find nullclines and equilibria in the phase plane. Our real goal is to find **solutions**, descriptions of how the state variables change over time. We begin by graphing the results from Euler's method in the phase plane. To deduce the behavior of solutions without all the calculations necessary for Euler's method, we will add

direction arrows to the phase-plane diagram. Like the arrows that appear in phase-line diagrams, they indicate where solutions are increasing or decreasing and can be used to sketch **phase-plane trajectories** or **solutions in the phase plane**.

5.7.1 Euler's Method in the Phase Plane

We applied Euler's method to the predator-prey equations

$$\frac{db}{dt} = (1.0 - 0.001p)b$$
$$\frac{dp}{dt} = (-1.0 + 0.001b)p$$

in Subsection 5.5.4, and computed the values for ten steps of length $\Delta t = 0.2$ starting from the initial condition $b(0) = 800$ and $p(0) = 200$.

t	$\hat{b}$	$\hat{p}$
0.0	800.0	200.0
0.2	928.0	192.0
0.4	1078.0	189.2
0.6	1252.8	192.2
0.8	1455.2	201.9
1.0	1687.4	220.3
1.2	1950.6	250.6
1.4	2242.9	298.2
1.6	2557.8	372.3
1.8	2878.8	488.3
2.0	3173.5	671.8

We have plotted these values in the phase plane (Figure 5.7.1). The value at $t = 0$ is plotted as the point (800, 200) in the phase plane, the value at $t = 0.2$ is plotted as (928, 192), and so forth. Following the points through time, we see that the prey population increases steadily, and the predator population begins by decreasing and then increases.

Euler's method does not provide an exact solution, however, and takes a lot of calculation. More precise techniques correct the errors produced by using the tangent line approximation and can be programmed on the computer to generate solutions accurate to any desired level. A solution generated with one such method (called the Runge-Kutta method) is plotted in Figure 5.7.2. The initial condition in this graph is $b(0) = 800$ and $p(0) = 200$. The solution consists of two curves, one for each of the state variables. At any time t, the values of b and p can be read from the graphs of b and p.

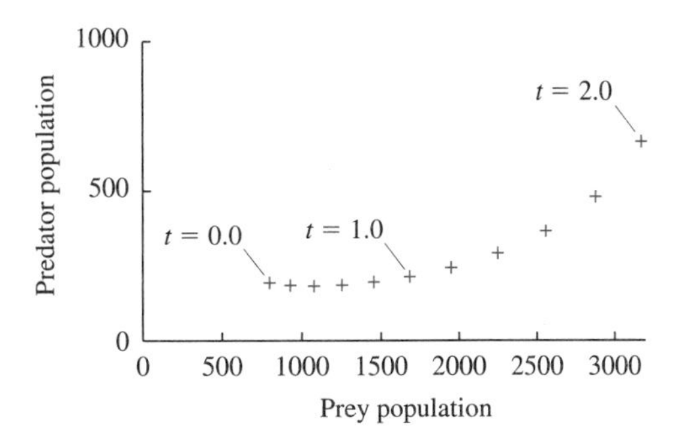

FIGURE 5.7.1
Results from Euler's method plotted in the phase plane

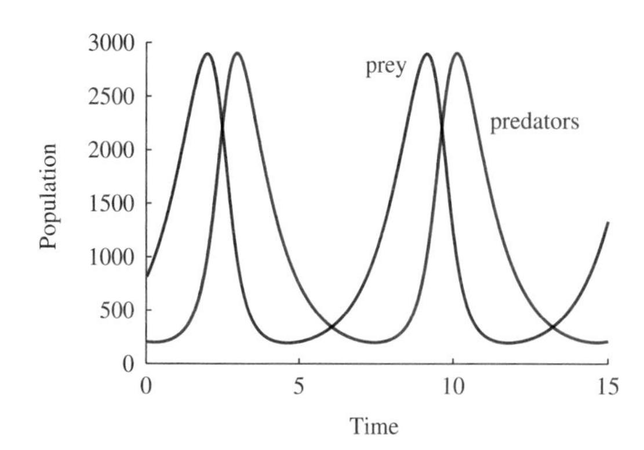

FIGURE 5.7.2
Solutions of the predator-prey system

The populations oscillate, with the peak predator population occurring after the peak prey population. When prey are plentiful, the predators increase in number. They eat the prey, eventually decreasing the prey population. This reduces the predator food supply, which leads to an eventual reduction in the predator population. The prey can then increase, beginning the cycle again.

As with the results from Euler's method, we can graph these solutions in the phase plane as a **phase-plane trajectory** (Figure 5.7.3). The initial condition (800, 200) is plotted at the point (800, 200) (labeled $t = 0$). At $t = 2$, the population of prey is 2894 and the population of predators is 1051, so the point (2894, 1051) is plotted in the phase plane. Because time does not appear explicitly as part of a phase-plane trajectory, we have labeled several points with the time. We have graphed three things on one graph: the time, the prey population, and the predator population. This is called a **parametric graph**.

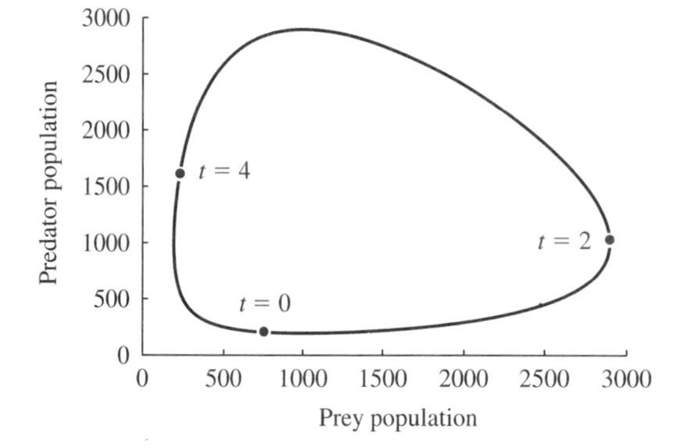

FIGURE 5.7.3
Solution of the predator-prey equations in the phase plane

How can we translate back and forth between the solutions plotted as functions of time and the phase-plane trajectory? Starting from the solution, you can plot the number of predators against the number of prey at several times (such as $t = 0$, $t = 1$, and so forth), and connect the dots. Starting from the phase-plane trajectory, you can sketch the solutions by tracing along the graph at a constant speed. The horizontal location gives the prey population, or the height of the graph of $b(t)$. The vertical location gives the predator population, or the height of the graph of $p(t)$. In our example, we see that the prey population begins by increasing, reaches a maximum value of nearly 3000, then decreases to a value of about 200 before beginning to increase again. The predator population decreases slightly below 200 before beginning an increase to nearly 3000. The predators reach their maximum after the prey, and then decrease again.

Example 5.7.1 Graphing a Solution and a Phase-Plane Trajectory

Suppose that $x(t) = 1 - t$ and $y(t) = 1 + 2t$ are the solutions of two coupled differential equations for $0 \le t \le 1$. Both $x(t)$ and $y(t)$ are linear functions of time. We can plot $x(t)$ by computing $x(0) = 1$ and $x(1) = 0$ and connecting the points (0, 1) and (1, 0) with a

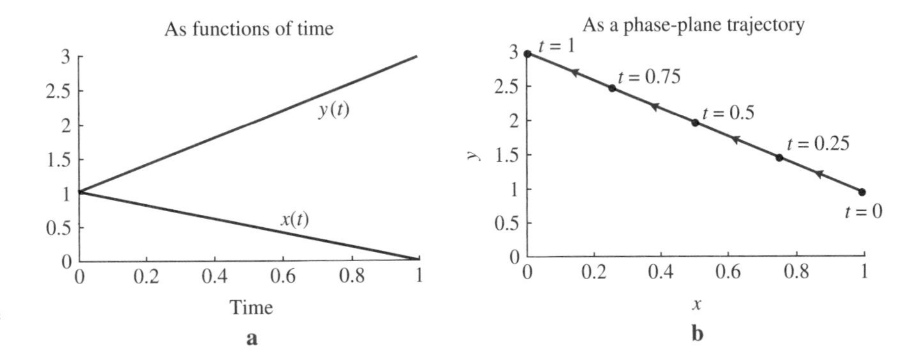

FIGURE 5.7.4
Graphing a solution and a phase-plane trajectory

line. Similarly, we can plot $y(t)$ by computing $y(0) = 1$ and $y(1) = 3$ and connecting the points (0, 1) and (1, 3) with a line (Figure 5.7.4a). To plot the phase-plane trajectory, we can make a chart of values.

t	$x(t)$	$y(t)$
0.0	1.0	1.0
0.25	0.75	1.5
0.5	0.5	2.0
0.75	0.25	2.5
1.0	0.0	3.0

The values $x(t)$ and $y(t)$ are plotted in Figure 5.7.4, labeled by the corresponding value of t.

Example 5.7.2 Translating a Graph of a Solution into a Phase-Plane Trajectory

Suppose instead that we are given only the graph of a solution in Figure 5.7.5a.

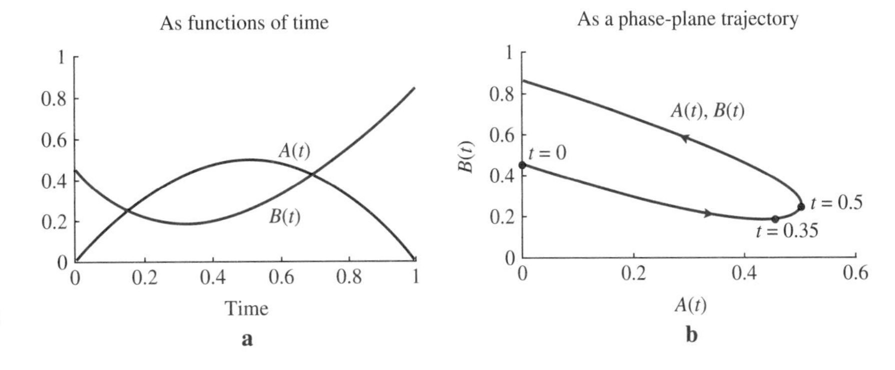

FIGURE 5.7.5
Translating a graph of a solution into a phase-plane trajectory

To plot the phase-plane trajectory, we identify the values of t for which the state variables A and B are increasing and decreasing. From $t = 0$ until about $t = 0.35$, $A(t)$ is increasing and $B(t)$ is decreasing. From $t = 0.35$ until $t = 0.5$, $A(t)$ is increasing and $B(t)$ is increasing. From $t = 0.5$ on, $A(t)$ is decreasing and $B(t)$ is increasing.

t	Behavior of $A(t)$	Behavior of $B(t)$
0.0–0.35	Increasing	Decreasing
0.35–0.5	Increasing	Increasing
0.5–1.0	Decreasing	Increasing

Example 5.7.3 Translating a Phase-Plane Trajectory into a Graph of a Solution

Suppose instead that we are given only the graph of the phase-plane trajectory in Figure 5.7.6a. To plot the solution, we use the phase-plane trajectory to identify values

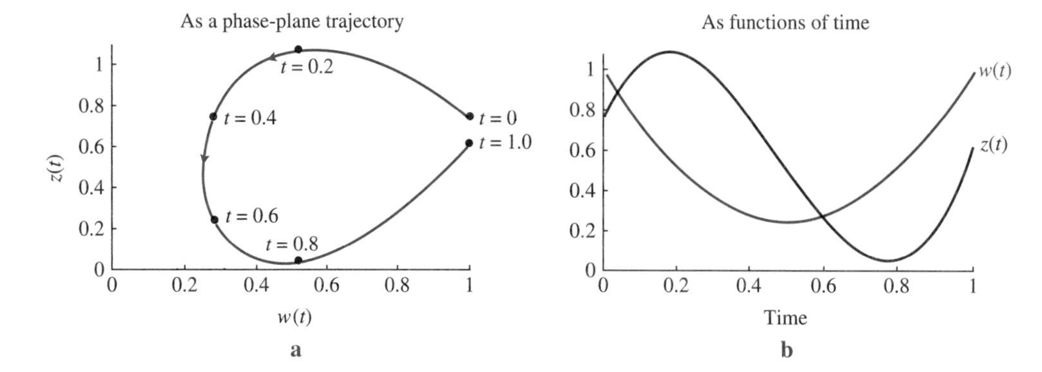

FIGURE 5.7.6
Translating a phase-plane trajectory into a graph of a solution

of t for which the state variables w and z are increasing and decreasing. From $t = 0$ until about $t = 0.2$, $w(t)$ is decreasing and $z(t)$ is increasing. From $t = 0.2$ until $t = 0.5$, $w(t)$ is decreasing and $z(t)$ is decreasing. From $t = 0.5$ until $t = 0.75$, $w(t)$ is increasing and $z(t)$ is decreasing. From $t = 0.75$ on, both $w(t)$ and $z(t)$ are increasing.

t	Behavior of $w(t)$	Behavior of $z(t)$
0.0–0.2	Decreasing	Increasing
0.2–0.5	Decreasing	Decreasing
0.5–0.75	Increasing	Decreasing
0.75–1.0	Increasing	Increasing

5.7.2 Direction Arrows: Predator-Prey Equations

The numerical solutions (Figures 5.7.2 and 5.7.3) give an accurate description of the dynamics. Our goal, however, is to **understand** the behavior of the populations without solving the equations. For one-dimensional systems, we sketched solutions from the direction arrows and equilibria on a phase-line diagram. We know how to draw nullclines and find equilibria on a phase-plane diagram. The next step is to figure out how to draw the direction arrows.

We have redrawn the nullclines and equilibria in the phase plane for the predator-prey system (Figure 5.7.7). The nullclines break the phase plane into four regions,

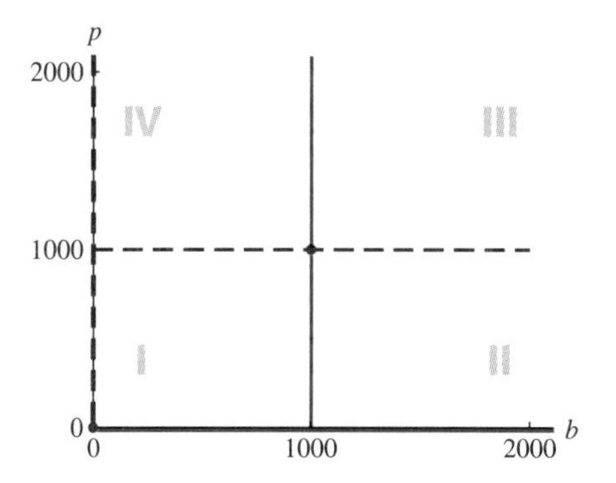

FIGURE 5.7.7
Regions of the phase plane for the predator-prey equations

labeled **I**, **II**, **III**, and **IV**. In each, we wish to determine whether the populations of predators and prey are increasing or decreasing.

There are three different approaches to figuring this out,

- Method I: Pick a pair of values (b, p) in the region and plug into the differential equation. Check whether $\frac{db}{dt}$ and $\frac{dp}{dt}$ are positive or negative.
- Method II: Algebraic manipulation of inequalities.
- Method III: Reasoning about the equations.

Which method is best depends on the equations. We will apply all three to the predator-prey phase plane.

In region **I**, the predator population is below the b-nullcline $p = 1000$, and the prey population is below the p-nullcline $b = 1000$. Method I requires picking a pair of values in this region. One point in this region is (500, 500). We can substitute this into the differential equation to check whether the populations are increasing or decreasing. With these values,

$$\frac{db}{dt} = (1.0 - 0.001 \cdot 500) \cdot 500 = 250 > 0$$

$$\frac{dp}{dt} = (-1.0 + 0.001 \cdot 500) \cdot 500 = -250 < 0.$$

The prey population is increasing because $\frac{db}{dt} > 0$, and the predator population is decreasing because $\frac{dp}{dt} < 0$. We indicate this by an arrow pointing toward larger values of the prey (to the right) and toward smaller values of the predator (down) (Figure 5.7.8). This is a **direction arrow**.

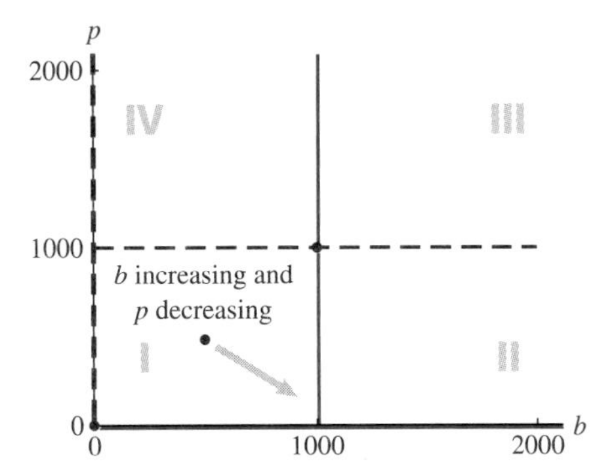

FIGURE 5.7.8
Direction arrow in region **I** of the predator-prey phase plane

Method II is algebraic. If we are in region **I**,

$$0 < b < 1000$$

$$0 < p < 1000.$$

Then

$$\frac{db}{dt} = (1.0 - 0.001p)b > 0$$

because both $1.0 - 0.001p > 0$ and $b > 0$. The prey population increases in this region. Similarly,

$$\frac{dp}{dt} = (-1.0 + 0.001b)p < 0$$

because $-1.0 + 0.001b < 0$ and $p > 0$. The predator population decreases. Again, the direction arrow points to the right and down.

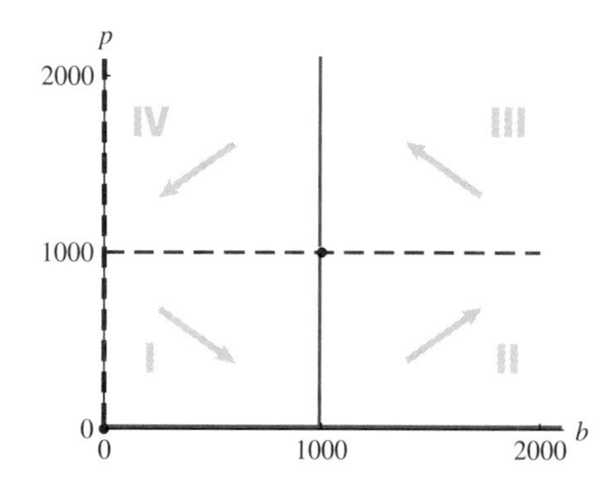

FIGURE 5.7.9

Direction arrows for the predator-prey equations

Method III uses reasoning about the equations. In region **I**, the prey and predator populations are both low. This means that the prey are happy (few predators to eat them) and the predators are sad (too few prey to eat). The prey population will increase and the predator population will decrease.

We can use any of the three methods to find the direction arrows in the three remaining regions (Figure 5.7.9). Using the reasoning method, in region **II**, the prey population is large and the predator population is small. Both species should be happy and increase, generating a direction arrow that points up and to the right. In region **III**, both populations are large. This is bad for the prey and good for the predators. The direction arrow therefore points toward lower values of prey (to the left) and larger values of predators (up). Finally, in region **IV**, the prey population is small and the predator population is large. Neither species does well under these circumstances, and the direction arrow points left and down.

As on a phase-line diagram, solutions or phase-plane trajectories follow the arrows (Figure 5.7.10). Starting in region **I**, where predator and prey populations are small, the arrow points down and to the right, pushing the solution into region **II**. Both populations then increase, following the arrows up and to the right into region **III**. The predators then increase while the prey decrease, moving the population into region **IV**, from which both populations decrease into region **I**. We cannot tell from this description whether the phase-plane trajectory circles around, spirals toward, or spirals away from the equilibrium.

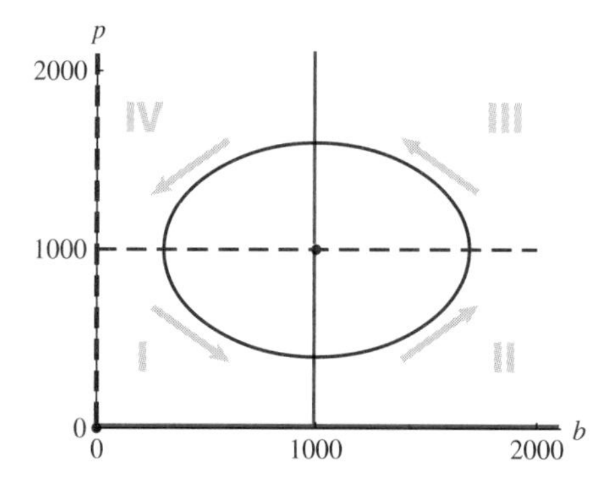

FIGURE 5.7.10

Direction arrows and solution for the predator-prey equations

What happens to direction arrows right on the nullclines? For example, what happens to a population with $b = 1000$ and $p < 1000$? This point lies on the p-nullcline, meaning that the population of predators remains unchanged. If an increasing predator population is associated with an upward pointing arrow and a decreasing predator

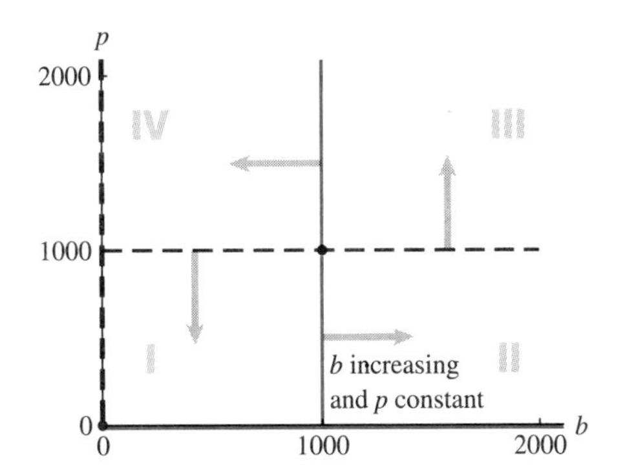

FIGURE 5.7.11
Direction arrows on nullclines

population is associated with a downward pointing arrow, an unchanging predator population must be associated with an arrow that points neither up nor down. Such an arrow is horizontal. The population of prey, however, is changing. Because the number of predators is small, the prey population will increase and the arrow will point to the right (Figure 5.7.11). The other four direction arrows on nullclines can be found in a similar way.

Finding the direction arrows on the nullclines is useful for two reasons. First, it provides a useful check on the rest of the direction arrows. The directions can only change one at a time; when we move from region **I** to region **II**, the arrow switches from pointing down and to the right to pointing up and to the right. The only change was the vertical direction, and this change happens right at the nullcline where the arrow is horizontal. Second, these direction arrows can help in sketching more accurate phase-plane trajectories. Because solutions must follow the arrows, solutions must be horizontal when they cross the p-nullcline and vertical when they cross the b-nullcline. The solution sketched in Figure 5.7.10 satisfies these criteria.

Example 5.7.4 Nullclines and Direction Arrows for the Disease Model with Immunity

The nullclines for the disease model with immunity break the phase plane into two regions, points with $S < 0.5$ and points with $S > 0.5$.

$$\frac{dI}{dt} = 2.0IS - 1.0I$$

$$\frac{dS}{dt} = -2.0IS.$$

The number of infected people is increasing when $S > 0.5$, and decreasing when $S < 0.5$, while the number of susceptible people is always decreasing. On the I-nullcline where $S = 0.5$, the direction arrow points straight down because I is not changing, while S is decreasing (Figure 5.7.12). Solutions must approach some point on the line of equilibria where $I = 0$. The point the solution approaches depends on the initial condition, and more advanced methods are needed to find this value.

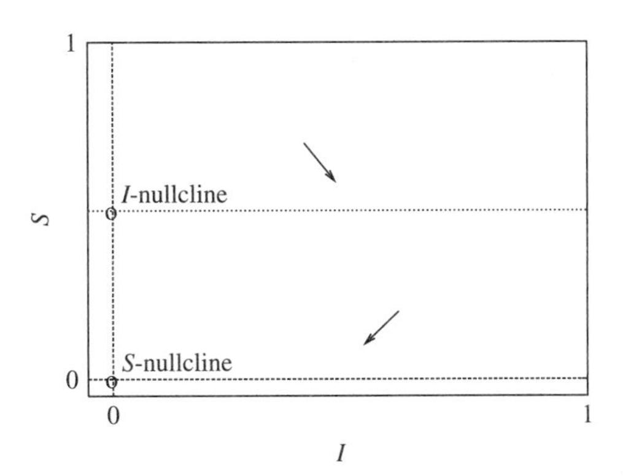

FIGURE 5.7.12
Direction arrows and nullclines for disease model

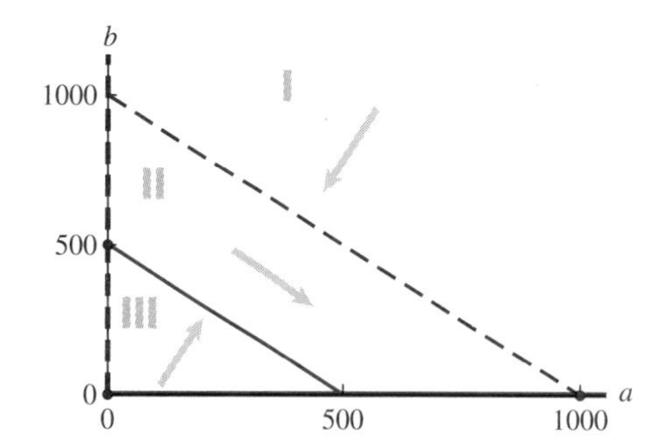

FIGURE 5.7.13
Direction arrows for the competition system

5.7.3 Direction Arrows for the Competition Equations

For the competition model, the nullclines break the phase plane into three regions (Figure 5.7.13). We have again used the parameter values $K_a = 1000$ and $K_b = 500$, and we assume that μ and λ are positive. The a-nullcline lies above the b-nullcline. We can determine the direction arrows by using each of the three methods: plugging in values, algebra, and reasoning. With the first method, a point in region **I** is (1000, 1000). Then

$$\frac{da}{dt} = \mu\left(1 - \frac{1000 + 1000}{1000}\right) \cdot 1000 = -1000\mu < 0$$

$$\frac{db}{dt} = \lambda\left(1 - \frac{1000 + 1000}{500}\right) \cdot 1000 = -3000\lambda < 0.$$

Because both derivatives are negative, both a and b are decreasing, meaning that the direction arrow points down and to the left. It is a bit harder to find a point in region **II**. This region is defined by

$$1000 > a + b > 500,$$

so one point in the region is (400, 400). Then

$$\frac{da}{dt} = \mu\left(1 - \frac{400 + 400}{1000}\right) \cdot 400 = 80\mu > 0$$

$$\frac{db}{dt} = \lambda\left(1 - \frac{400 + 400}{500}\right) \cdot 400 = -240\lambda < 0.$$

The direction arrow points down and to the right. The point (100, 100) lies in region **III** where

$$\frac{da}{dt} = \mu\left(1 - \frac{100 + 100}{1000}\right) \cdot 100 = 80\mu > 0$$

$$\frac{db}{dt} = \lambda\left(1 - \frac{100 + 100}{500}\right) \cdot 100 = 60\lambda > 0.$$

The arrow points up and to the right, because the rate of change of each state variable is positive.

Algebraically, in region **I**, $a + b > 1000 > 500$, so

$$1 - \frac{a+b}{1000} < 0$$

$$1 - \frac{a+b}{500} < 0.$$

Therefore,

$$\frac{da}{dt} = \mu\left(1 - \frac{a+b}{1000}\right)a < 0$$

$$\frac{db}{dt} = \lambda\left(1 - \frac{a+b}{500}\right)b < 0.$$

Both populations decrease, and the direction arrow points down and to the left. In region **II**, $500 < a + b < 1000$ and

$$\frac{da}{dt} = \mu\left(1 - \frac{a+b}{1000}\right)a > 0$$

$$\frac{db}{dt} = \lambda\left(1 - \frac{a+b}{500}\right)b < 0.$$

The population of a increases, the population of b decreases, and the direction arrow points down and to the right. In region **III**, $a + b < 500 < 1000$, so that

$$\frac{da}{dt} = \mu\left(1 - \frac{a+b}{1000}\right)a > 0$$

$$\frac{db}{dt} = \lambda\left(1 - \frac{a+b}{500}\right)b > 0.$$

Both populations increase, and the direction arrow points up and to the right.

With the reasoning method, think of $K_a = 1000$ and $K_b = 500$ as the largest total populations that types a and b can tolerate. In region **I**, the total population exceeds both K_a and K_b. Both types suffer from overpopulation and have shrinking populations, generating a direction arrow that points down and to the left. In region **II**, the total population lies between 500 and 1000. Type b cannot withstand the competition that type a can tolerate, producing a direction arrow that points down and to the right. In region **III**, the total population is less than both K_a and K_b. Both types can grow, generating a direction arrow pointing up and to the right.

The two phase-plane trajectories plotted in Figure 5.7.14a follow the arrows and are forced toward the equilibrium at (1000, 0) where type b has been driven extinct. This figure includes the direction arrows on the nullclines. On the a-nullcline, the population of b is decreasing because the total population is 1000, exceeding the tolerance of type b. The direction arrow therefore points straight down. On the b-nullcline, the population of a is increasing because the total population is 500, less than K_a for type a. The direction arrow points straight to the right. The solution starting in region **III** is plotted as a function of time in Figure 5.7.14b. Even though type b is doomed to extinction, it has an initial period of growth. The solution for b reaches a maximum when the trajectory crosses the b-nullcline.

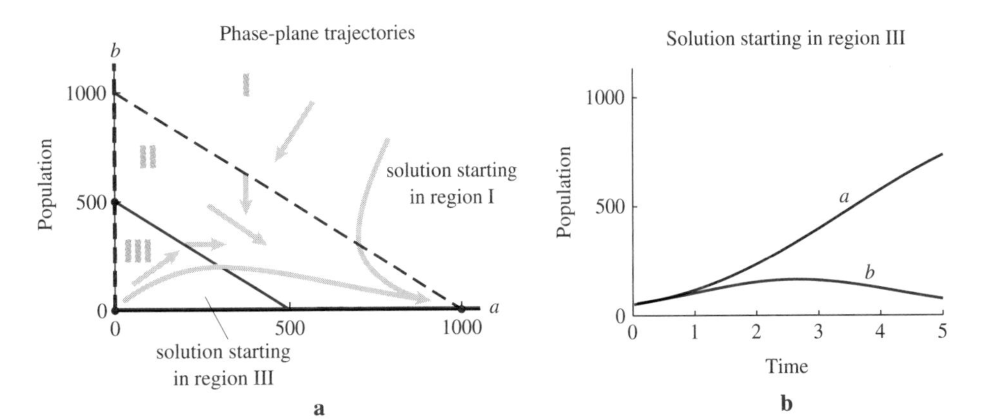

FIGURE 5.7.14
Solutions of the competition system

This model differs in several ways from the differential equations

$$\frac{da}{dt} = \mu a$$
$$\frac{db}{dt} = \lambda b$$

(which we reduced to a single equation for the fraction p of type a in Subsection 5.1.4). When $\lambda < \mu$, type b grows more slowly and goes extinct. In the competition model, the type better able to withstand competition (the one with the larger value of K) eventually wins out, even if its growth rate is lower.

5.7.4 Direction Arrows for Newton's Law of Cooling

The direction arrows are simpler for Newton's law of cooling. The nullclines coincide and break the plane into only two regions (Figure 5.7.15). In region **I**, $H > A$. Therefore,

$$\frac{dH}{dt} = \alpha(A - H) < 0$$
$$\frac{dA}{dt} = \alpha_2(H - A) > 0,$$

meaning that H decreases and A increases. The direction arrow points down and to the right. In region **II**, $H < A$. Therefore,

$$\frac{dH}{dt} = \alpha(A - H) > 0$$
$$\frac{dA}{dt} = \alpha_2(H - A) < 0,$$

meaning that H increases and A decreases. The direction arrow points up and to the left. Solutions are pushed toward the line of equilibria. Physically, this means that the object and the room will tend to approach the same temperature.

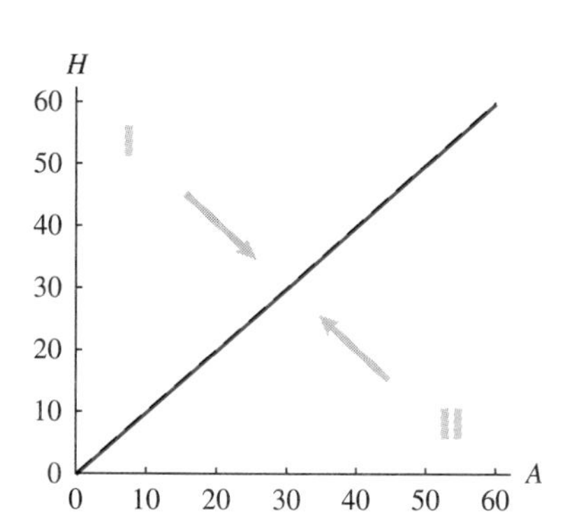

FIGURE 5.7.15

Direction arrows for Newton's law of cooling

Summary We have seen how to plot solutions of two-dimensional differential equations as functions of time and as **phase-plane trajectories**. The nullclines break the phase-plane into regions, in each of which we can find a **direction arrow** indicating whether the state variables are increasing or decreasing. For additional information on the behavior of the phase-plane trajectories, we can draw vertical or horizontal direction arrows on the nullclines. Phase-plane trajectories follow the direction arrows.

5.7 Exercises

Mathematical Techniques

1–4 ▪ Suppose the following functions are solutions of some differential equation. Graph these as functions of time and as phase-plane trajectories for $0 \le t \le 2$. Mark the position at $t = 0$, $t = 1$, and $t = 2$.

1. $x(t) = t$, $y(t) = 3t$.

2. $a(t) = 2e^{-t}$, $b(t) = e^{-2t}$.

3. $f(t) = 1 + t$, $g(t) = e^{-t}$.

4. $x(t) = 1 + t(t - 2)$, $y(t) = t(3 - t)$.

5–6 ▪ From the following graphs of solutions of differential equations as functions of time, graph the matching phase-plane trajectory.

5.

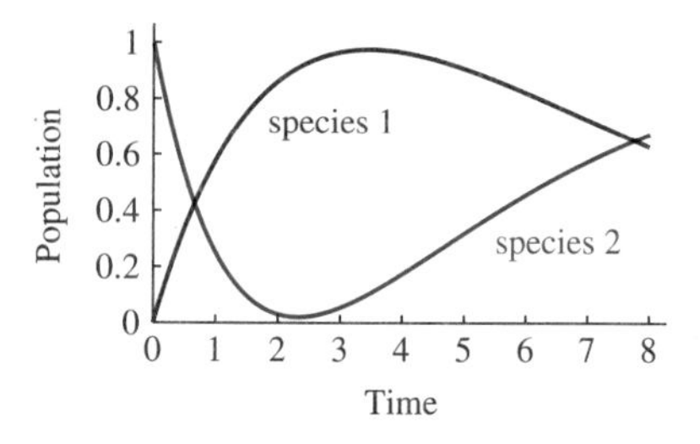

6.

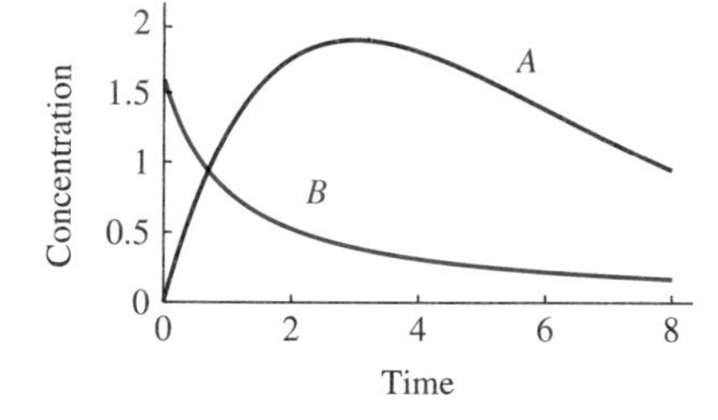

7–8 ▪ From the following graphs of phase-plane trajectories, graph the matching solutions of differential equations as functions of time.

7.

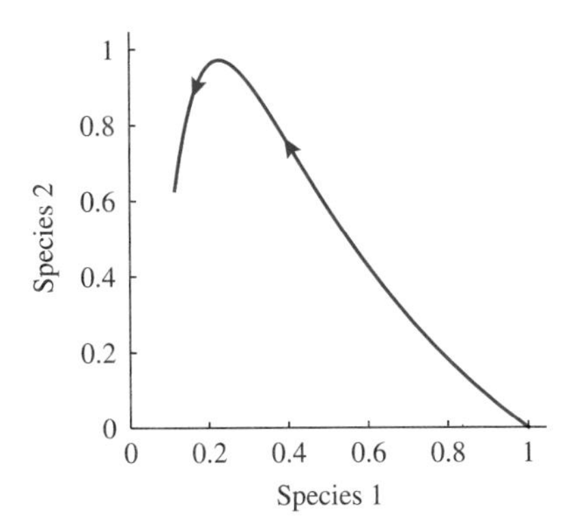

8.

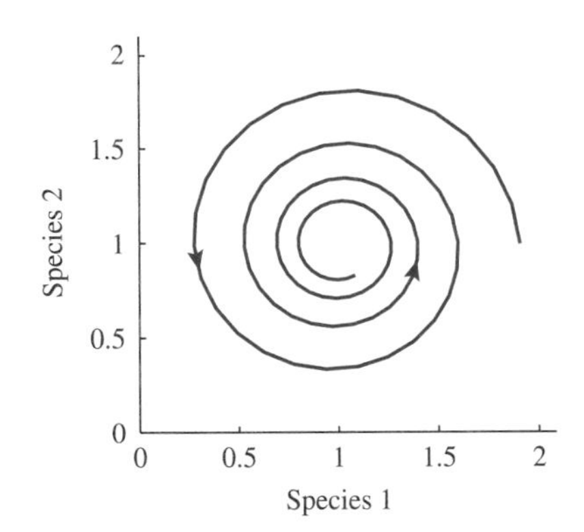

9–10 ▪ On the following phase-plane diagrams, use the direction arrows to sketch phase-plane trajectories starting from two different initial conditions.

9.

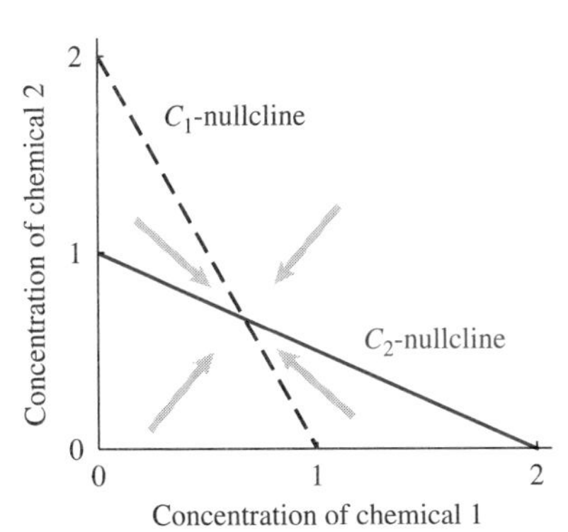

10.

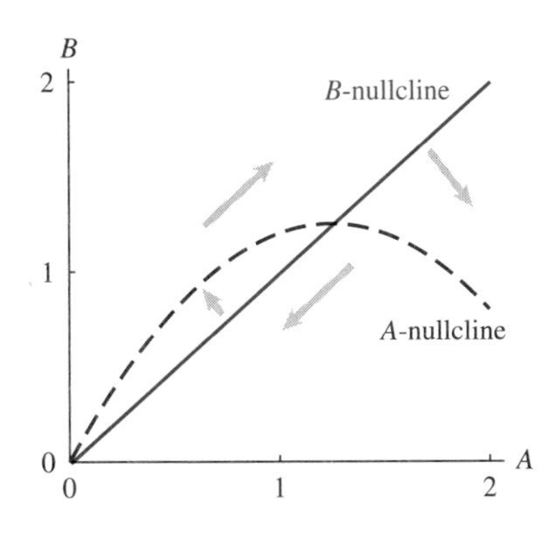

11–12 ▪ Use the information in the phase-plane diagram to draw direction arrows on the nullclines.

11. The diagram in Exercise 9.

12. The diagram in Exercise 10.

13–14 ▪ Compare solutions estimated with Euler's method with the phase-plane diagram and direction arrows found in the text for the competition equations (Figure 5.7.13)

$$\frac{da}{dt} = \mu\left(1 - \frac{a+b}{K_a}\right)a$$

$$\frac{db}{dt} = \lambda\left(1 - \frac{a+b}{K_b}\right)b$$

starting from the given initial conditions. Assume that $\mu = 2.0$, $\lambda = 2.0$, $K_a = 1000$, and $K_b = 500$.

13. Start from $a = 750$ and $b = 500$. Take two steps, with a step length of $\Delta t = 0.1$, as in Section 5.5, Exercise 7.

14. Start from $a = 250$ and $b = 500$. Take two steps, with a step length of $\Delta t = 0.1$, as in Section 5.5, Exercise 8.

15–18 ▪ Compare solutions estimated with Euler's method with the phase-plane diagram and direction arrows found in the text for Newton's law of cooling (Figure 5.7.15)

$$\frac{dH}{dt} = \alpha(A - H)$$

$$\frac{dA}{dt} = \alpha_2(H - A)$$

with the given parameter values and starting from the given initial conditions.

15. Suppose $\alpha = 0.3$ and $\alpha_2 = 0.1$. Start from $H = 60$ and $A = 20$. Take two steps, with a step length of $\Delta t = 0.1$, as in Section 5.5, Exercise 11.

16. Suppose $\alpha = 0.3$ and $\alpha_2 = 0.1$. Start from $H = 0$ and $A = 20$. Take two steps, with a step length of $\Delta t = 0.1$, as in Section 5.5, Exercise 12.

17. Suppose $\alpha = 3.0$ and $\alpha_2 = 1.0$. Start from $H = 60$ and $A = 20$. Take two steps, with a step length of $\Delta t = 0.25$, as in Section 5.5, Exercise 13. Does this diagram help explain what went wrong?

18. Suppose $\alpha = 3.0$ and $\alpha_2 = 1.0$. Start from $H = 0$ and $A = 20$. Take two steps, with a step length of $\Delta t = 0.5$, as in Section 5.5, Exercise 14. Does this diagram help explain what went wrong?

19–20 ▪ Draw the nullclines and direction arrows for the following models of springs. Make sure to include positive and negative values for the position x and the velocity v.

19. The model in Section 5.5, Exercise 15.

20. The model in Section 5.5, Exercise 16.

21–22 ▪ Sketch the given solution of the following models of springs first as a pair of functions of time, and then in the phase-plane. Check that the solution follows the arrows.

21. The solution $x(t) = \cos(t)$ (Section 5.5, Exercise 17) of the spring equation in Section 5.5, Exercise 15.

22. The solution $x(t) = e^{-t}\cos(t)$ (Section 5.5, Exercise 18) of the spring equation with friction in Section 5.5, Exercise 16.

Applications

23–36 ▪ For the following problems, add direction arrows to the phase-plane.

23. The model in Section 5.5, Exercise 23.
24. The model in Section 5.5, Exercise 24.
25. The model in Section 5.5, Exercise 27.
26. The model in Section 5.5, Exercise 28.
27. The model in Section 5.5, Exercise 29.
28. The model in Section 5.5, Exercise 30.
29. The model in Section 5.5, Exercise 33.
30. The model in Section 5.5, Exercise 34.
31. The model in Section 5.5, Exercise 35.
32. The model in Section 5.5, Exercise 36.
33. The model in Section 5.5, Exercise 37.
34. The model in Section 5.5, Exercise 38.
35. The model in Section 5.5, Exercise 39.
36. The model in Section 5.5, Exercise 40.

37–44 ▪ For the following problems, use the direction arrows on your phase-plane to sketch a solution starting from the given initial condition.

37. The model in Exercise 25 starting from (1500, 200). Is there another path for the solution that is consistent with the direction arrows?
38. The model in Exercise 26 starting from (1500, 200).
39. The model in Exercise 27 starting from (200, 300).
40. The model in Exercise 28 starting from (200, 300).
41. The model in Exercise 31 starting from (0.5, 1).
42. The model in Exercise 32 starting from (0.5, 1).
43. The model in Exercise 35 starting from (0.5, 1).
44. The model in Exercise 36 starting from (0.5, 0,5). Can you be sure that the solution behaves exactly like your picture?

Computer Exercises

45. Consider the following differential equations describing diffusion and utilization of a chemical.

$$\frac{dC}{dt} = \alpha(\Gamma - C) - \frac{\delta C}{1 + C}$$

$$\frac{d\Gamma}{dt} = \frac{\alpha}{K}(C - \Gamma) + S$$

The parameters have the following meanings.

Name	Meaning	Values to Use
α	Diffusion rate	1.0
δ	Use efficiency	4.0 and 1.0
K	Ratio of volumes	2.0
S	Supplementation rate	1.0

a. Set $\delta = 4$ and the rest of the parameters to their designated values. Plot the nullclines and find the equilibrium.

b. Follow the same steps with $\delta = 1$. Is there an equilibrium? Can you say why not? (No math jargon allowed.) Sketch C and Γ as functions of time.

c. Try to figure out the critical value of δ where the behavior changes.

46. Many biological systems need to be able to respond to **changes** in the level of some signal (like a hormone) without responding to the actual level. For example, a cell might have no response to a low level of hormone. If the hormone level rapidly increases, the cell responds. But if the hormone level then remains constant at the higher level, the cell again stops responding. This process is sometimes called **adaptation**.

One mechanism for this process is summarized in the following model. Internal response is a function of the fraction p of cell surface receptors that are bound by the hormone. This fraction increases when the hormone level H is high. However, hormone also dissociates from bound receptors. Assume this happens at a rate A, but that this rate is controlled by the cell. One possible set of equations is

$$\frac{dp}{dt} = k_1 H(1 - p) - Ap$$

$$\frac{dA}{dt} = \epsilon(H - A).$$

Suppose that $k_1 = 0.5$ and that ϵ is a small value (like 0.1 or 0.01). The value of H is determined by conditions external to the cell and does not have its own differential equation.

a. Find the nullclines and equilibria of this model assuming that H is a constant. Does H appear in your final results? Explain why the cell might want to respond in the same way to any constant level of H.

b. Use your computer to simulate the response when the level of H jumps quickly from $H = 1$ to $H = 10$. One way to do this is to solve the equations with $H = 10$ using as initial conditions the equilibrium values of p and A when $H = 1$. Draw graphs of p and A in the phase-plane and as functions of time. Explain what is happening.

c. Do the same when the level of H drops rapidly from $H = 10$ to $H = 1$.

5.8 The Dynamics of a Neuron

In Section 1.11, we studied a discrete-time dynamical system describing the heart, an important excitable system that responds to a periodic stimulus. To follow more precisely the response to a single stimulus, or to figure out mechanisms for **creating** a periodic stimulus, we need to use differential equations. In this section, we present a simplified (but venerable and valuable) model of a neuron.

5.8.1 A Mathematician's View of a Neuron

Some basic properties of a neuron are illustrated in Figure 5.8.1. The key measurements are of the concentrations of two ions, sodium and potassium. Like most cells, a resting neuron maintains an excess of potassium and deficit of sodium. A neuron maintains a negative **resting potential**, meaning that there is an overall excess of negative charge inside the cell. Sodium and potassium ions are both positively charged. Because sodium concentrations are much higher than potassium concentrations, the deficit of these positive charges inside the cell contributes most of the negative charge of the cell. Cells use a significant amount of energy to run pumps to maintain this distribution of ions.

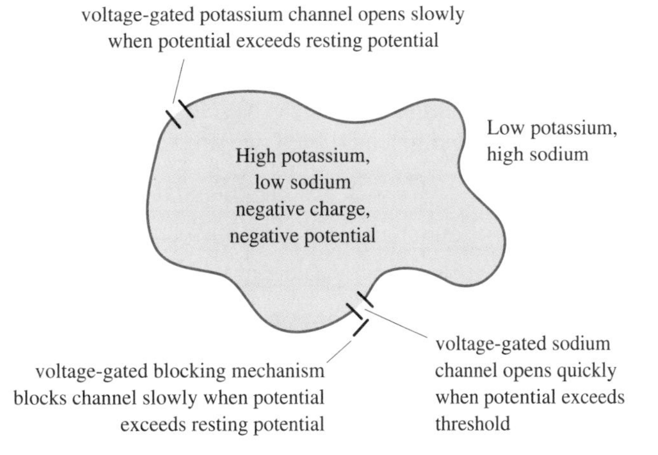

FIGURE 5.8.1
The basic elements of neural dynamics

These gradients of sodium, potassium, and charge across the cell membrane are used by the neuron to amplify and transmit information. The process depends on a set of **voltage-gated channels** for each ion. Such channels open and close in response to voltage (another name for potential) differences and are closed when the cell is at rest.

When a burst of positive charges enters the cell (and makes the potential of the cell less negative), voltage-gated sodium channels open (Figure 5.8.2). Because there is an excess of sodium outside the cell, more sodium ions enter, further increasing the potential of the cell until it actually becomes positive.

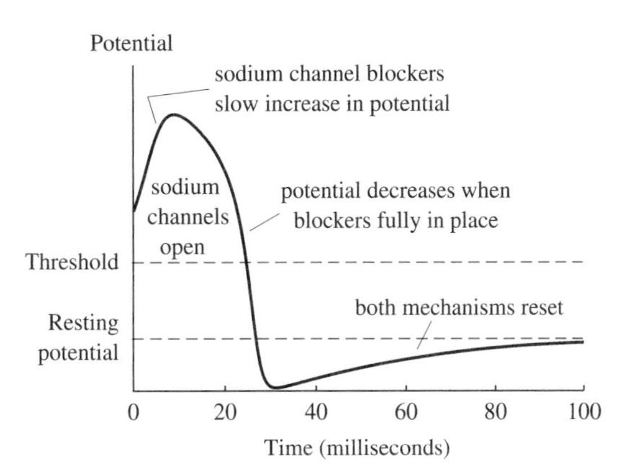

FIGURE 5.8.2
An action potential

Two things then occur. One slower mechanism acts to block the voltage-gated sodium channels, and another slow mechanism begins to open voltage-gated potassium channels. Both of these processes act to diminish the build-up of positive charge in the cell. Blocking the sodium channels halts the entry of sodium, and opening potassium channels allows the exit of positively charged potassium. Neither process happens as fast as the opening of the sodium channels, and cannot halt the increase in potential immediately, just as a weak backward force applied to a heavy moving object only gradually slows it down and reverses the direction. When the potential of the cell has again decreased to the resting potential or below, these mechanisms slowly turn off. Soon, the cell is ready to begin the cycle again.

How does a neuron use this ability? This sharp peak of electrical excitation, called an **action potential**, can be transmitted precisely to other neurons. Action potentials are part of the language of the brain, like the 0's and 1's used by computers.

Our model is designed to study the structure of the interaction of two processes: a fast process with **positive feedback** (a slight increase in cell potential quickly induces sodium channels to open and increase the potential further) and a slow process with **negative feedback** (the increase in cell potential induces two mechanisms that halt the increase). Because the system has two partially redundant mechanisms for negative feedback, we focus on one, the slow opening of potassium channels.

5.8.2 The Mathematics of Sodium Channels

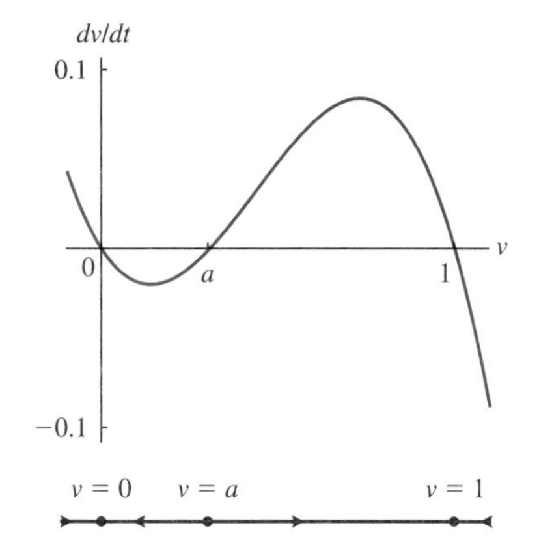

FIGURE 5.8.3
The phase-line diagram without the slow mechanism for opening potassium channels

The potential in a cell can be scaled to more convenient mathematical values. Denoting the scaled potential by v, we set

$$\begin{cases} v = 0 & \text{resting potential} \\ v = a & \text{the threshold above which the neuron fires} \\ v = 1 & \text{potential with sodium channels completely open.} \end{cases} \qquad \textbf{(5.8.1)}$$

Small deviations above resting potential are not amplified, and the potential returns to rest. However, if the cell potential is raised above the threshold a, the cell moves toward the higher potential found when the sodium channels are open. A phase-line diagram for cell potential following this description is shown in Figure 5.8.3.

One convenient equation consistent with this diagram is

$$\frac{dv}{dt} = -v(v-a)(v-1) = f(v). \qquad \textbf{(5.8.2)}$$

The right-hand side is in factored form, and the values $v = 0$, $v = a$, and $v = 1$ are all equilibria. We can take the derivative of the rate-of-change function to check the stability of the equilibria (Theorem 5.1), finding

$$f'(v) = -(v-a)(v-1) - v(v-1) - v(v-a).$$

Then

$$f'(0) = -a < 0$$
$$f'(a) = a(1-a) > 0$$
$$f'(1) = -(1-a) < 0.$$

The equilibrium at $v = 0$ is stable, the one at $v = a$ is unstable, and the one at $v = 1$ is stable. The graph of f and the associated phase-line diagram are consistent with our biological assumptions (Figure 5.8.3). This cubic (Equation 5.8.2) is the simplest equation describing the basic rules of sodium dynamics, and we use it to illustrate the behavior of the system.

Suppose a cell at resting potential receives an influx of positive charges from another neuron that raises its potential slightly above the threshold a. The solution moves upward, amplifying the signal, the first task of a functioning neuron. However,

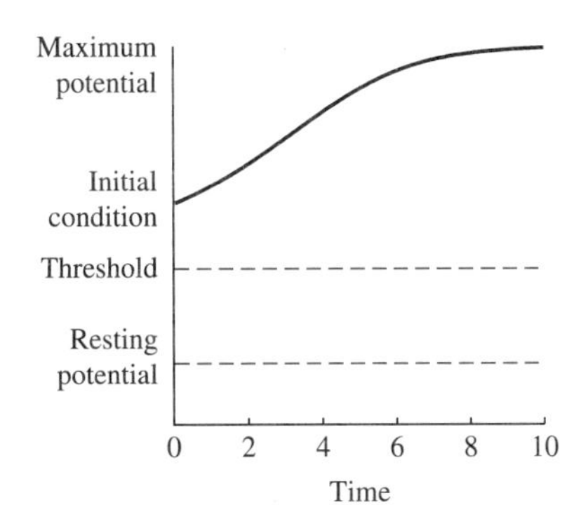

FIGURE 5.8.4
Solution without the slow mechanism for opening potassium channels

this neuron gets stuck at the higher equilibrium $v = 1$ (Figure 5.8.4). The neuron cannot be restimulated.

5.8.3 The Mathematics of Slow Potassium Channels

Without the slow mechanism for opening potassium channels, a neuron could respond only to a single stimulus. How does opening the potassium channels help the neuron function?

Let w represent the degree to which potassium channels are open. At $v = 0$, this mechanism is turned off, so $w = 0$. As v gets closer to 1, the mechanism becomes stronger and stronger, so w takes on larger and larger values. A simple equation that seems to model this behavior is

$$\frac{dw}{dt} = \epsilon v$$

for positive ϵ. However, if v remains positive (above resting potential) for a long time, w might approach infinity. There must be an upper bound on how much potassium these channels can let in.

To incorporate the requirement that the potassium channels have a maximum possible capacity, we use an equation resembling Newton's law of cooling,

$$\frac{dw}{dt} = \epsilon(v - \gamma w). \tag{5.8.3}$$

For every fixed value of v, this equation has an equilibrium at

$$w = \frac{v}{\gamma}.$$

If $v = 0$, w approaches 0 and the mechanism does not operate. If, however, $v = 1$, w increases to an equilibrium value of $\frac{1}{\gamma}$. This represents the maximum possible degree of potassium channel opening. A smaller γ produces a larger equilibrium value. If $\gamma = 1.0$ and $v = 1$, the equilibrium is $w = 1.0$. If $\gamma = 10.0$ and $v = 1$, the equilibrium is $w = 0.1$ (Figure 5.8.5).

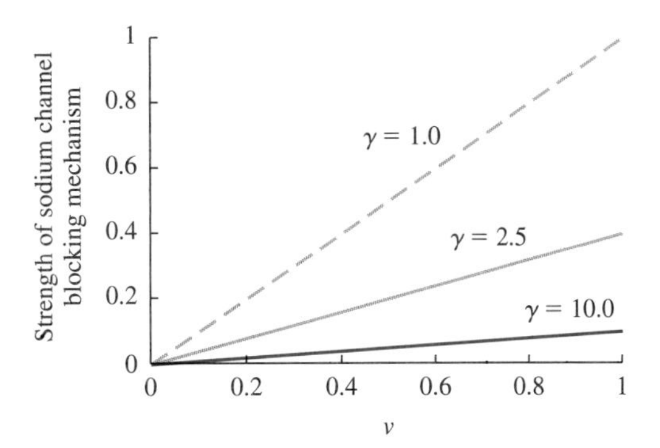

FIGURE 5.8.5
Equilibrium level of potassium channel opening for fixed voltage

The parameter ϵ does not affect the equilibrium level. It instead changes the rate at which the equilibrium is approached. A small value of ϵ produces a small rate of change and a slow response of w. Because the opening mechanism is slow, ϵ has a fairly small value.

5.8.4 The Fitzhugh-Nagumo Equations

How do potassium channels modify the rate of change of the potential v? In other words, how do we couple the dynamics of v to the value of w? When w is large, the potassium channels are wide open, allowing the exit of positively charged ions that rapidly reduce the potential. For convenience, we assume that the rate at which the potential decreases is proportional to w. This gives the coupled system of equations

$$\frac{dv}{dt} = -v(v-a)(v-1) - w \quad \textbf{(5.8.4)}$$

$$\frac{dw}{dt} = \epsilon(v - \gamma w). \quad \textbf{(5.8.5)}$$

These are called the **Fitzhugh-Nagumo** equations. Can they reproduce the firing behavior of the neuron?

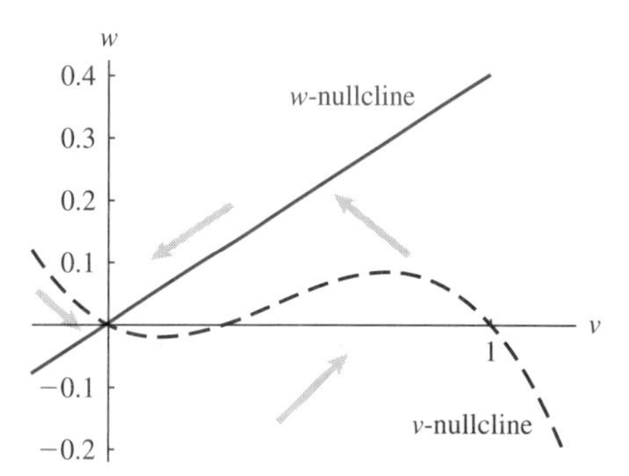

FIGURE 5.8.6
The nullclines and direction arrows for the Fitzhugh-Nagumo equations

Figure 5.8.6 shows the nullclines and direction arrows for the Fitzhugh-Nagumo equation with $\gamma = 2.5$ and $a = 0.3$, where we have chosen w as the vertical variable. The v-nullcline can be found by solving

$$\frac{dv}{dt} = -v(v-a)(v-1) - w = 0$$

for w as

$$w = -v(v-a)(v-1).$$

The w-nullcline is found by solving

$$\frac{dw}{dt} = \epsilon(v - \gamma w)$$

for w as

$$w = \frac{v}{\gamma}.$$

In the case shown, the nullclines intersect only at the point (0, 0). At this equilibrium, the cell is at resting potential ($v = 0$) and the potassium channels are closed ($w = 0$). In other words, the only equilibrium describes a cell completely at rest. Although we do not have the techniques to prove it (linear algebra is needed), this equilibrium is stable, meaning that solutions that start nearby return to the equilibrium. This hardly seems to be the recipe for useful dynamics.

We can draw direction arrows by reasoning about the equations. w is subtracted from the rate of change of v, so large values of w (above the v-nullcline) correspond

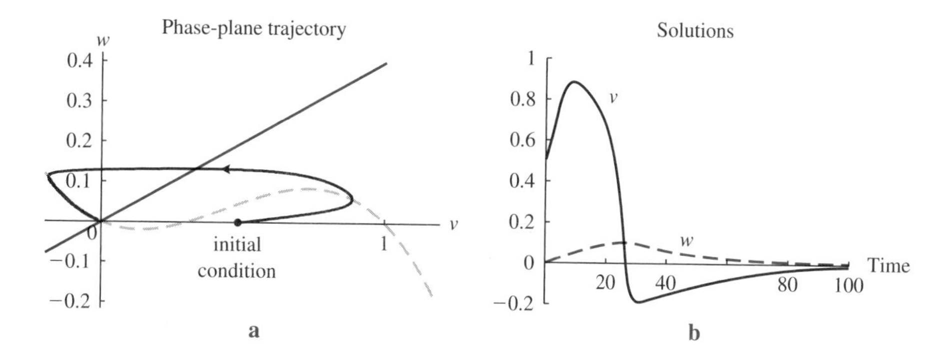

FIGURE 5.8.7

Results of perturbing above resting potential

to decreasing v and arrows that point to the left. Small values of w (below the v-nullcline) correspond to increasing v, and arrows that point to the right. Similarly, $\epsilon\gamma w$ is subtracted from the rate of change of w, so large values of w (above the w-nullcline) correspond to decreasing w and arrows that point down. Small values of w (below the w-nullcline) correspond to increasing w, and arrows that point up (Figure 5.8.6).

What happens when the potential of the cell is raised above the threshold a? Mathematically, this corresponds to initial conditions $(v_0, 0)$ with $v_0 > a$. With the equation for sodium channels opening alone (Equation 5.8.2), the potential moves up to the equilibrium at $v = 1$ and remains there, leaving the cell paralyzed and useless.

The results of this same experiment with the slow mechanism for opening potassium channels are quite different (Figure 5.8.7). Initially, the potential of the cell increases (the phase-plane trajectory moves to the right) due to the rapid opening of the sodium channels (Figure 5.8.7). Slowly, however, the potassium channels open, and the trajectory moves toward larger values of w. When the trajectory crosses the v-nullcline, the potential begins to decrease (the trajectory moves to the left), and when it crosses the w-nullcline, the potassium channels begin to close (the trajectory heads down). The trajectory undershoots $v = 0$ before the slow potassium channels finish closing. This system has the basic properties of a neuron: the ability to quickly amplify a signal and return to a state of readiness for the next signal. In addition, the mathematical analysis predicts an unexpected (but real) phenomenon, the undershoot of the potential. This undershoot results from the slowness of the mechanism for opening the potassium channels.

If the perturbation of the cell is insufficient to open the sodium channels ($v_0 < a$), the cell does not amplify the signal. In this way, the cell acts as a filter that can ignore small stimuli (Exercise 6).

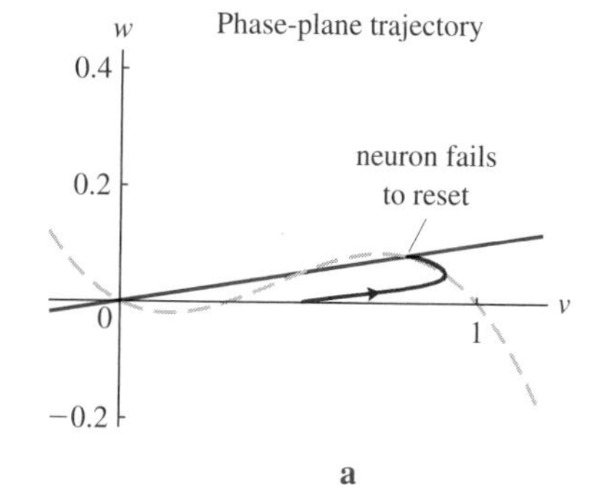

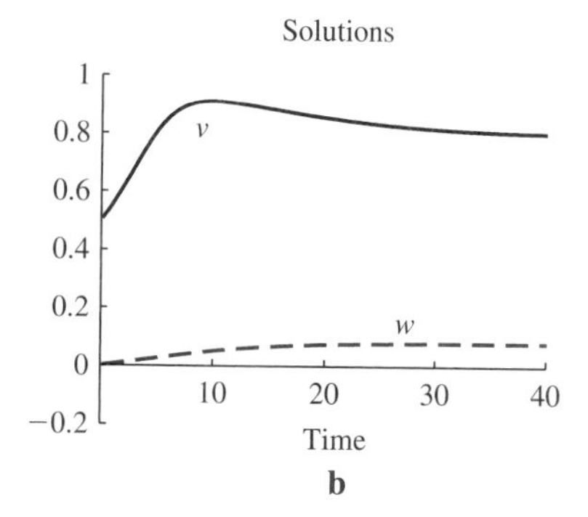

FIGURE 5.8.8

Dynamics with weak potassium channel opening response

5.8.5 Weak Potassium Channel Opening Mechanism

The larger the value of γ, the smaller the response of the potassium channel opening mechanism. A cell without this mechanism (Equation 5.8.2) could respond only to a single stimulus. What happens if this mechanism is weak?

When the value of γ is large, the w-nullcline swings down to intersect the v-nullclines in three places (Figure 5.8.8). Although we cannot prove it without linear algebra, the central equilibrium is unstable, and the outer two are stable. If we stimulate the cell by raising the potential of the cell above the threshold a, the potential rises further as the sodium channels open. However, the potassium channels cannot overcome the opening, and the trajectory is trapped at the upper equilibrium (Figure 5.8.8).

Once the w-nullcline lowers enough to cross the v-nullcline, the neuron stops firing. Further increases in γ change where the potential gets stuck but not the *qualitative* behavior of the cell. In effect, a cell with weak potassium channels acts like a cell with no potassium channels at all.

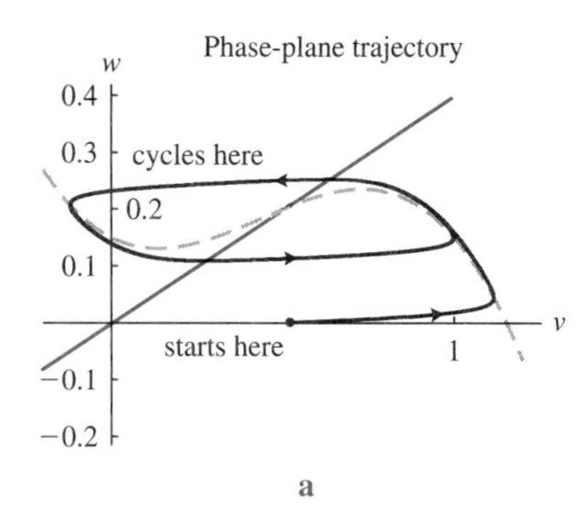

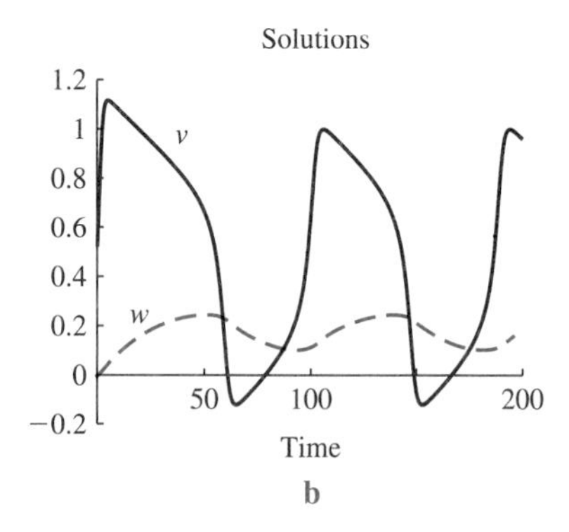

FIGURE 5.8.9

Dynamics with a constant applied current

5.8.6 The Effects of Constant Applied Current

We have imagined a resting cell receiving a single pulse of positively charged ions. One interesting experiment alters this situation by giving the cell a constant input of positive ions. Can the cell convert a constant input into a usable output?

The modification of the Fitzhugh-Nagumo equations is not complicated. The current describes the rate at which ions enter the cell and is therefore proportional to the derivative of the potential. If the current applied is I_a, we add I_a to the rate of change of potential, finding

$$\frac{dv}{dt} = -v(v-a)(v-1) - w + I_a. \tag{5.8.6}$$

Because we are not studying actual numerical values, we have set the constant of proportionality in front of I_a to 1, assuming that one unit of current raises the potential by one unit in one unit of time. The constant is positive because current increases the potential.

The v-nullcline is

$$w = -v(v-a)(v-1) + I_a$$

and the w-nullcline is unchanged (Figure 5.8.9a). As in the original case (Figure 5.8.6), the nullclines intersect at only a single point. In this case, however, the equilibrium is unstable (we would again need linear algebra to prove it). A phase-plane trajectory and the corresponding values of v and w as functions of time are shown in Figure 5.8.9.

The neuron shows a periodic "bursting" behavior. Real neurons also produce periodic spikes when subjected to a constant current. This periodic output translates a current, an analog input, into periodic bursts, a digital output, and can be used by the body as a pacemaker for a periodic process.

A cautionary note is in order. The equations we have been studying are a simplified version of the *four-dimensional* Hodgkin-Huxley equations (a system of four coupled autonomous differential equations). The value of studying this simplified model lies in its ability to qualitatively mimic more complicated and accurate models while remaining easy to understand with phase-plane techniques.

Summary We used some basic facts about sodium and potassium channels to derive equations describing **action potentials**, the firing of neurons. In particular, we have seen how one mechanism that opens **voltage-gated sodium channels** can amplify an incoming stimulus and how slower potassium channels can counteract that effect. We combined these two processes into the **Fitzhugh-Nagumo** equations and used phase-plane analysis to study these equations in several circumstances. The basic equations display excitability, the ability to temporarily amplify a signal and reset. Reducing the strength of the potassium channels can make the cell unable to respond to more than a single stimulus. An excitable cell responds to a constant stimulus by producing a periodic sequence of action potentials.

5.8 Exercises

Applications

1. Give an equation like Equation 5.8.2 describing a system with two thresholds. The system has a stable equilibrium at a resting potential at 0, but will be pushed to a higher positive equilibrium if the potential is raised above a particular positive threshold, and to a negative equilibrium if the potential is dropped below a particular negative threshold. Draw a phase-line diagram and check that your equations match it.

2. Draw phase-line diagrams for Equation 5.8.2 in the following cases.

   a. $a = 0.01$. How might this neuron malfunction?

   b. $a = 0.5$.

   c. $a = 0.99$. Why might this neuron work poorly?

3. Use the method of separation of variables to solve

$$\frac{dw}{dt} = \epsilon(v - \gamma w)$$

(Equation 5.8.3) assuming that v is constant. Show how the dynamics are slowed when ϵ is small, but that ϵ does not affect the equilibrium.

4. Assuming that w is constant (perhaps the potassium channels are jammed in a particular state) and $a = 0.3$, figure out the dynamics of

$$\frac{dv}{dt} = -v(v-a)(v-1) - w$$

(Equation 5.8.4) thought of as a one-dimensional differential equation. Try it with $w = 0.01$, $w = 0.05$, and $w = 0.1$.

5. Draw direction arrows on the nullclines for the Fitzhugh-Nagumo equations.

6. Sketch the phase-plane trajectory and solutions of the Fitzhugh-Nagumo equations when the initial stimulus is less than the threshold.

7. Sketch the phase-plane diagram and solution for the Fitzhugh-Nagumo equations for the following values of ϵ. (Think of changing ϵ as changing the direction arrows: the arrows are nearly horizontal when ϵ is small because w, the vertical variable, changes slowly.)
   a. ϵ very small.
   b. ϵ rather large.

8. With positive applied current, there could be three intersections of the nullclines (as in Figure 5.8.8). Draw a phase-plane diagram illustrating this scenario, and take a guess at the dynamics. Try to make sense of the results biologically.

9. There could also be a single intersection with positive applied current, but on the rightmost decreasing part of the v-nullcline. Draw such a phase plane. Assuming that this equilibrium is stable, sketch the dynamics. Why might this cell also be thought of as excitable?

10. What happens to the phase plane and the cell if the applied current is negative? Can the cell lose its ability to respond if the applied current is negative and large?

11. Suppose that a higher applied current I_a in Equation 5.8.6 produces faster bursting. How might the body use this ability to translate signal strength into response speed?

Computer Exercises

12. Use a computer to study a cell that is forced by an external current I_a that oscillates. Try different periods and amplitudes of the oscillation. What happens? Do you see any strange behaviors?

13. The behaviors observed in the previous problem can occur when an object that naturally oscillates at one frequency is **forced** at a different frequency. For example, a spring following the equation

$$\frac{d^2x}{dt^2} = -x$$

naturally oscillates with period 2π. Forcing can be added with the modified equation

$$\frac{d^2x}{dt^2} = -x + A\cos\left(\frac{2\pi t}{T}\right).$$

Study this equation with the following values of T, trying a range of values of A from 0.1 to 10.0.
   a. $T = 2\pi$.
   b. $T = \frac{\pi}{2}$.
   c. $T = 4\pi$.
   d. $T = 3.0$.
   e. $T = 4.0$.
   f. $T = 3.14$.

Supplementary Problems

1. Consider the differential equation

$$\frac{dC}{dt} = 3(\Gamma - C) + 1$$

where C is the concentration of some chemical in a cell, measured in moles per liter, and Γ is a constant.
   a. What kind of differential equation is this? Explain the terms in the equation.
   b. Draw the phase-line diagram.
   c. Verify the stability of the equilibrium by using the derivative.
   d. Sketch solutions as functions of time starting from two initial conditions: $C(0) = 0$ and $C(0) = \Gamma + 1$.

2. Consider again the differential equation

$$\frac{dC}{dt} = 3(\Gamma - C) + 1$$

where Γ is a constant.
   a. Solve the equation when $C(0) = 0$.
   b. Check your answer.
   c. Find $C(0.4)$.
   d. After what time will the solution be within 5% of its limit?

3. Consider the system of equations

$$\frac{d\Gamma}{dt} = (C - \Gamma) - \frac{\Gamma^2}{3}$$
$$\frac{dC}{dt} = 3(\Gamma - C) + 1.$$

C is the internal concentration of a chemical and Γ the external concentration.

a. Describe in words the two processes affecting concentration in the external environment. How big is the external environment relative to the cell?

b. Draw a phase plane replete with nullclines, equilibria, and direction arrows. *Hint*: It is easier to put C on the vertical axis.

4. Describe conditions when you might observe population growth described by the following.

a. One-dimensional autonomous differential equation.

b. One-dimensional nonautonomous differential equation.

c. One-dimensional pure-time differential equation.

d. Two-dimensional autonomous differential equation.

5. Consider the differential equation

$$\frac{dV}{dt} = 12 - t^2$$

where $V(t)$ is volume in liters at time t and t is measured in seconds.

a. What kind of differential equation is this?

b. Graph the rate-of-change function and use it to sketch a graph of the solution.

c. At what time does V take on its maximum?

d. Suppose $V(0) = 0$. Use Euler's method to estimate $V(0.1)$.

e. Suppose $V(0) = 0$. Find the time T when $V(t)$ is 0 again.

f. What is the average volume between 0 and T.

6. Consider the differential equation

$$\frac{dx}{dt} = 3x(x-1)^2$$

a. Draw the phase-line diagram of this equation.

b. Find the stability of the equilibria using the derivative.

c. Sketch trajectories of x as a function of time for initial conditions $x(0) = -0.5$, $x(0) = 0.5$, $x(0) = 1.5$.

7. Suppose the per capita production rate of a bacterial population is given by

$$\text{per capita production} = \frac{1}{\sqrt{b}}$$

where $b(t)$ is the population size at time t and t is measured in hours.

a. Find the differential equation describing this population.

b. Solve the equation and check your answer.

c. What is the population after 2 hours if the population starts at $b(0) = 10000$?

d. Does this population grow faster or slower than one growing exponentially? Why?

8. Differential equations to describe an epidemic are sometimes given as

$$\frac{dS}{dt} = \beta(S+I) - cSI$$

$$\frac{dI}{dt} = cSI - \delta I.$$

Here S measures the number of susceptible people and I the number of infected people.

a. Compare with the predator-prey equations. What is different in these equations? What biological process does each term on the right-hand side describe?

b. Sketch the nullclines and find the equilibria if $\beta = 1$ and $c = \delta = 2$. (Draw only the parts where S and I are positive.)

c. Sketch direction arrows on your phase-plane diagram.

9. Consider the following differential equation describing the concentration of sodium ions in a cell following consumption of a bag of Doritos at time $t = 0$ seconds.

$$\frac{dN}{dt} = 2 - 10t.$$

Suppose $N(0) = 50$ mm/cm^3.

a. What kind of differential equation is this?

b. Sketch a graph of the rate of change as a function of time.

c. Sketch a graph of the concentration as a function of time.

d. Use Euler's method to estimate $N(0.1)$.

e. Find $N(1)$ exactly.

f. At what times is $N(t) = 50$?

10. Consider the following differential equation describing the concentration of sodium ions in a cell.

$$\frac{dN}{dt} = 2(N-50) - (N-50)^2.$$

a. What kind of differential equation is this? What does each term mean?

b. Sketch a graph of the rate of change as a function of concentration.

c. Draw the phase-line diagram.

d. Sketch solutions starting from $N(0) = 48$, $N(0) = 51$, and $N(0) = 54$.

e. Suppose $N(0) = 51$. Estimate $N(0.1)$.

f. What method would you use to solve this equation?

11. Two types of bacteria with populations a and b are living in a culture. Suppose

$$\frac{da}{dt} = 2a\left(1 - \frac{a}{500} - \frac{b}{200}\right)$$

$$\frac{db}{dt} = 3b\left(1 + \frac{a}{1000} - \frac{b}{100}\right).$$

a. What kind of differential equation is this? Explain the terms.

b. Draw the phase plane, including nullclines, equilibria, and direction arrows.

12. Suppose the scaled potential v and level of potassium channel opening w in a neuron were described by

$$\frac{dv}{dt} = v(1 - v) - w$$
$$\frac{dw}{dt} = v - 2w.$$

a. Draw a phase line for v assuming that w is fixed at 0.

b. Draw the phase plane for the full system including equilibria, nullclines, and direction arrows.

13. The length L of a microtubule is found to follow the differential equation

$$\frac{dL}{dt} = -L(2 - L)(1 - L)$$

where L is measured in microns.

a. Draw the phase-line diagram.

b. Check the stability of the equilibria using the derivative.

c. Sketch trajectories starting from $L(0) = 0.5$, $L(0) = 1.5$, and $L(0) = 2.5$.

14. A different lab finds that

$$\frac{dL}{dt} = -2.0L + 5.4LE$$

$$\frac{dE}{dt} = 3.5 - LE$$

where E is the level of some component of the microtubules and L is the length of the microtubule.

a. Explain the terms in these equations.

b. Draw the phase-plane diagram, including nullclines, equilibria, and direction arrows.

15. Consider the differential equation describing a population of mathematically sophisticated bacteria,

$$\frac{db}{dt} = b \ln\left(\frac{2b + 1}{2 + b}\right).$$

a. What kind of differential equation is this?

b. Draw the derivative as a function of the state variable and draw the phase-line diagram.

c. Sketch trajectories starting from $b(0) = 0.8$ and $b(0) = 1.2$.

d. Check the stability of the equilibrium at $b = 0$ by taking the derivative of the rate-of-change function.

e. Use Euler's method to estimate $b(0.01)$ if $b(0) = 0.5$.

16. Suppose a population of size N has per capita production equal to $1 - 2e^{-0.01N}$.

a. Write the differential equation for population size.

b. Find the equilibrium or equilibria.

c. Use the stability theorem to determine stability.

d. Based on these calculations, draw a phase-line diagram.

Projects

1. The Computer Exercise 46 in Section 5.7 presents one possible model of adaptation. This project studies a simpler alternative model (proposed by H. G. Othmer). Because this model is so simple, we can compare the results of phase-plane analysis with actual solutions of the equations.
Consider the equations

$$\frac{dp}{dt} = k(H - A - p)$$

$$\frac{dA}{dt} = \epsilon(H - A)$$

where p represents the response of the cell and H the external condition driving the response. The variable A describes some internal state of the cell. First, suppose that H is constant.

a. Find the nullclines and equilibria and draw the phase plane, including direction arrows.

b. The equation for A is the same as Newton's law of cooling. Write down the solution for A with an arbitrary initial condition.

c. Substitute this solution into the equation for p. There is a clever way to solve the resulting nonautonomous equation. Make up a new variable $q(t) = e^{kt} p(t)$. With a bit of manipulation, you can write a pure-time differential equation for q. Solve this equation and find the solution for $p(t)$.

d. Plot this solution as a phase-plane trajectory for several different initial conditions. Are the results consistent with the phase plane?

Using these results, we can study what happens when H changes. Suppose first that the cell is at equilibrium with $H = 1$, and then H jumps up to 10. Set $k = 1$ and $\epsilon = 0.1$.

a. How long will it take before A has roughly reached its new equilibrium value?

b. What is the value of p at this time? How long will it take p to nearly reach its equilibrium value again?

c. Suppose now that $H = 1$ for a time T, jumps to $H = 10$ for a time T, jumps back to 1, and so forth. Experiment with different values of T. How large must T be before the cell produces a healthy response to each change? What happens when T is much smaller than this value? What might the cell do to be able to respond more quickly?

2. (Based on the research of J. Cherry.) Consider an interaction between two mutually inhibiting proteins with concentrations x and y, given by the differential equations

$$\frac{dx}{dt} = f(y) - x$$
$$\frac{dy}{dt} = g(x) - y.$$

Both $f(y)$ and $g(x)$ are decreasing functions.

a. Explain each of the terms in these equations.

b. Try to imagine a biological situation they might describe.

c. Sketch the nullclines (remember that both f and g are decreasing).

d. Show that equilibria occur where $f[g(x)] = x$. What discrete-time dynamical system shares the equilibria of the system of differential equations?

Next try the following steps for functions of three different forms:

$$\text{Case 1:}\quad f(y) = \frac{1}{1+\alpha y},\quad g(x) = \frac{1}{1+\alpha x}$$
$$\text{Case 2:}\quad f(y) = e^{-\alpha y},\quad g(x) = e^{-\alpha x}$$
$$\text{Case 3:}\quad f(y) = \frac{1}{1+\alpha y^2},\quad g(x) = \frac{1}{1+\alpha x^2}.$$

Experiment with different values of α.

a. Find the function $f[g(x)]$ and graph it on a cobwebbing diagram. Does the equilibrium look stable?

b. Draw the nullclines and direction arrows for the differential equation. Is the equilibrium stable?

c. Why does the stability you "found" in **a** match that in **b**?

d. Try to figure out whether it is possible to have three equilibria (it is possible in Cases 2 and 3).

Why might it be important for a biological system to have three equilibria? How could it operate as a switch?

Chapter

6

Probability Theory and Descriptive Statistics

The dynamical models studied so far in this book, discrete-time dynamical systems and differential equations, are **deterministic**. The state at one time exactly **determines** the state at every future time. When population size is accurately described by a discrete-time dynamical system, the population size at any future time can be computed exactly by repeatedly applying the discrete-time dynamical system starting from a known initial condition. When population size is accurately described by a differential equation, the population size at any future time can be computed exactly by **integrating** the differential equation starting from a known initial condition.

These perfectly predictable results are a mathematical idealization. If we track several experimental populations, each experiencing apparently identical conditions, the results will not be identical. Various more or less "random" events can affect the outcome. Lower temperatures might decrease reproduction. An infection might diminish one population. All the offspring in one experiment might be males. Describing these chance events and deducing their consequences is the realm of **probability theory**.

By studying idealized deterministic models, we have learned how to describe and analyze important biological processes. Our goal now is to extend those models and insights to begin the analysis of data. Real data include so-called noise, and deducing the underlying process can be difficult. How do we distinguish the signal from the noise? This is the realm of **statistics**, the art of describing and reasoning about real data. Our approach to statistics is motivated by a single principle:

Correctly understanding and applying statistics requires understanding the underlying probabilistic model.

This chapter develops the language and techniques of **probability theory** and **descriptive statistics**. Probability theory provides a precise mathematical description of chance, and descriptive statistics (like the average) usefully summarize results.

6.1 Introduction to Probabilistic Models

We begin by introducing the basic terminology of **probability theory** and the distinction and relation between **probability** and **statistics**. Using examples from population growth, we illustrate three types of stochastic models and the sorts of questions probabilists and statisticians might ask about them.

6.1.1 Probability and Statistics

Describing chance events and deducing their consequences is the realm of **probability theory**. To avoid the ambiguous colloquial connotations of the words "random" and "chance," we use the term **stochastic** to describe unpredictable effects. A stochastic model describes a biological process that includes chance events. The goals of a stochastic model are the same as the goals of a deterministic model: to produce an accurate description of a biological process and to deduce its consequences.

Conceptually, using probability theory to analyze a stochastic model is similar to using discrete-time dynamical systems and differential equations to analyze a deterministic model. First, one must come up with a sufficiently detailed description of a biological system to write down a model. Then, one uses appropriate mathematical tools to figure out what the model does. In the probabilistic case, however, **describing** the results is more difficult because a single number will not suffice.

Because of this difficulty with description, probabilistic systems are inextricably linked with **statistics**. Suppose we ran an experiment 1000 times, perhaps allowing a bacterial population to grow for 24 hours. We would get 1000 different, though related, numbers. Listing them all would convey little information. It would be better to state one or two numbers, perhaps the average or the minimum and maximum, to summarize the data in a meaningful and useful way. The choice and computation of these numbers is the realm of **descriptive statistics**, and the numbers chosen are called **statistics**. More precisely,

Definition 6.1 A **statistic** is a numerical value that summarizes the results of one or more experiments.

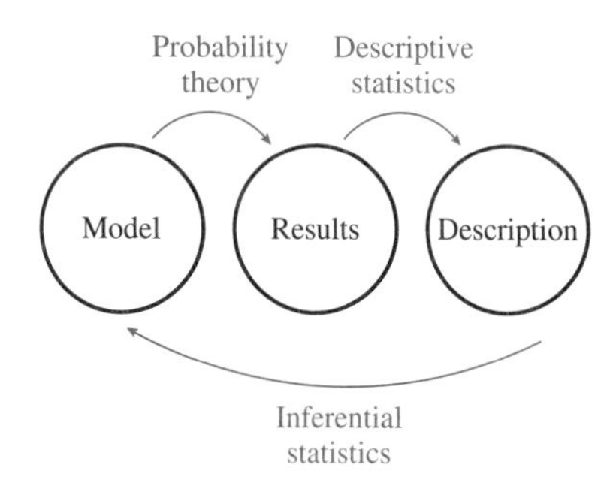

FIGURE 6.1.1
The relations among probability theory, descriptive statistics, and inferential statistics

Even if one had a computer-like mind and could contemplate enormous data sets in a single thought, statistics are necessary to **compare** data sets. Two sets of 1000 measurements might have no exact values in common, but could be essentially identical when considered in their entirety. This similarity can be revealed by comparing appropriate statistics. The average, for example, might be nearly the same for each data set.

Often, we wish to do more with our results than to summarize or describe them. Instead, we seek to use them to make **deductions** or **inferences** about the underlying biological process. If the world were deterministic, this task would be relatively easy. In a stochastic world, this task is more difficult and requires the techniques of **inferential statistics**. These techniques use statistical descriptions of results to make inferences about what happened. For example, we might wish to infer a numerical estimate (like the per capita production) or the answer to a biological question (is one population growing more rapidly than another?). The relations among probability theory, descriptive statistics, and inferential statistics are illustrated in Figure 6.1.1.

In this section and the next, we develop stochastic models of the three fundamental processes studied in this book: population growth, diffusion, and selection. In each case, we pose questions that we will later answer with probabilistic and statistical methods.

6.1.2 Stochastic Population Growth: Stochastic Production

Recall the discrete-time dynamical system describing population growth

$$b_{t+1} = rb_t.$$

The new population can be computed as the old population multiplied by the per capita production r. In a real-world system, the per capita production may vary stochastically. We indicate this by rewriting the equation as

$$b_{t+1} = r_t b_t. \quad \textbf{(6.1.1)}$$

The dependence of the per capita production r on the time t indicates that the production varies over time, due perhaps to such factors as temperature, food availability, and predation.

Example 6.1.1 Deterministic Production

In a deterministic model, the per capita production is a fixed constant, perhaps $r_t = 1.1$ for all t. The population size increases by 10% each year. If the initial population $b_0 = 1$, the population b_t after t years can be found with the exact formula

$$b_t = 1.1^t b_0.$$

The first ten values of population size are given in the following table.

Time	Per Capita Production	Population Size	Log Population Size
0	1.1	1.00	0.000
1	1.1	1.10	0.095
2	1.1	1.21	0.191
3	1.1	1.33	0.286
4	1.1	1.46	0.381
5	1.1	1.61	0.477
6	1.1	1.77	0.572
7	1.1	1.95	0.667
8	1.1	2.14	0.762
9	1.1	2.36	0.858
10	1.1	2.59	0.953

When plotted on a semilog graph (Definition 1.13), the logarithm of population size follows a line with equation

$$\ln(b_t) = \ln(1.1^t b_0) = \ln(b_0) + t\ln(1.1) = 0.0953t.$$

(Figure 6.1.2).

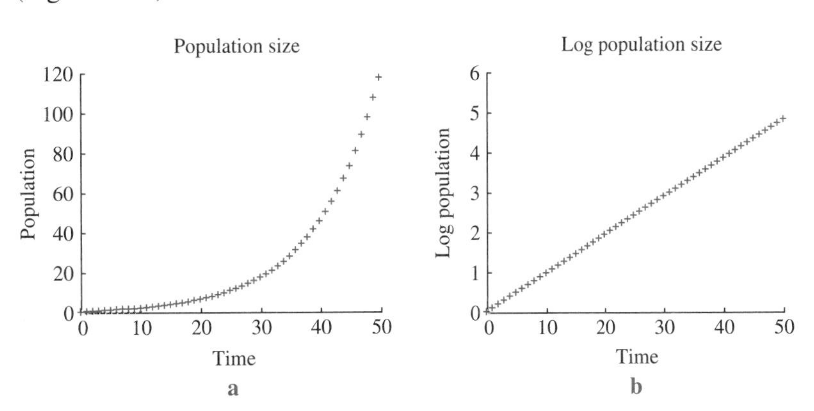

FIGURE 6.1.2
Deterministic population growth

Example 6.1.2 Population Growth with Stochastic Production

Suppose that the per capita production r only *averages* 1.1. If it is chosen from the range 1.0 to 1.2, then there are some bad years when the population does not grow at all ($r = 1.0$) and some good years when the population grows by 20% ($r = 1.2$). The results are similar to those without stochasticity, but the population wiggles around the line (Figure 6.1.3). If we run the same experiment twice, we get different results

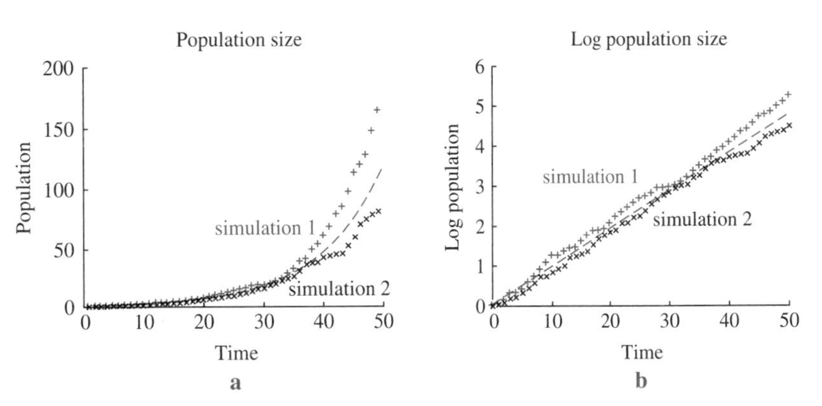

FIGURE 6.1.3
Stochastic population growth

because different stochastic factors affect the output. The first ten values from the two simulations are given in the following table.

	Simulation 1			Simulation 2		
Time	Per Capita Production	Population Size	Log Size	Per Capita Production	Population Size	Log Size
0	1.04	1.00	0.00	1.00	1.00	0.00
1	1.11	1.04	0.04	1.08	1.00	0.00
2	1.18	1.15	0.14	1.12	1.08	0.08
3	1.01	1.36	0.30	1.03	1.22	0.20
4	1.09	1.37	0.31	1.12	1.25	0.22
5	1.19	1.49	0.40	1.14	1.40	0.33
6	1.15	1.77	0.57	1.12	1.59	0.46
7	1.19	2.05	0.72	1.17	1.77	0.57
8	1.19	2.44	0.89	1.01	2.06	0.72
9	1.19	2.90	1.07	1.10	2.08	0.73
10	1.00	3.46	1.24	1.10	2.30	0.83

Example 6.1.3 Population Growth with Highly Stochastic Production II

The per capita production could still average 10% per year but be even more variable and range from 0.7 (population shrinks by 30%) up to 1.5 (population grows by 50%). Again, the population growth roughly follows the line, but it jumps around even more (Figure 6.1.4). The first ten values from the two simulations are given in the following table.

	Simulation 1			Simulation 2		
Time	Per Capita Production	Population Size	Log Size	Per Capita Production	Population Size	Log Size
0	0.86	1.00	0.00	0.70	1.00	0.00
1	1.12	0.86	−0.15	1.03	0.70	−0.35
2	1.42	0.96	−0.04	1.19	0.72	−0.33
3	0.73	1.37	0.31	0.80	0.86	−0.15
4	1.06	1.01	0.01	1.18	0.69	−0.37
5	1.46	1.07	0.06	1.24	0.82	−0.20
6	1.31	1.56	0.44	1.16	1.01	0.01
7	1.46	2.04	0.71	1.36	1.18	0.16
8	1.47	2.98	1.09	0.74	1.60	0.47
9	1.46	4.39	1.48	1.11	1.18	0.16
10	0.70	6.41	1.86	1.09	1.31	0.27

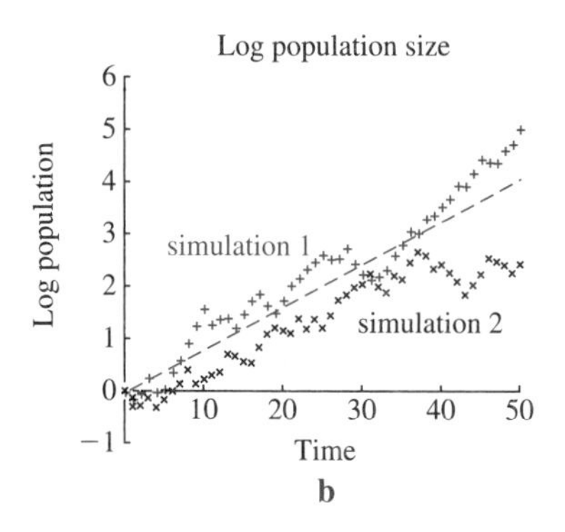

FIGURE 6.1.4
Highly stochastic population growth

As probabilists, we wish to ask and answer questions about these results. In the deterministic case (Example 6.1.1), we can ask what the population is after 50 years or how long it would take the population to reach exactly 100 (Exercises 21–24). In the stochastic cases (Examples 6.1.2 and 6.1.3), we cannot exactly predict the population in 50 years. Instead, we might predict the **average** population after 50 years, or the

probability that the population is greater than 125 after 50 years. Alternatively, we could ask how long it would take, on average, for the population to reach 100. As we can see in Example 6.1.3, the population actually declines for the first few years. What is the probability that the population is less than the original population after 10 years?

As statisticians, we are faced with different problems. Instead of a model, we have only the data. Our goal is to describe and answer questions about what happened. For example, we might want to know the average per capita production over this time. This is straightforward in the deterministic case (Example 6.1.1) because the values never change. In the stochastic cases (Examples 6.1.2 and 6.1.3), however, our task is not so simple. How do we go about drawing a line that seems to fit the data? Is the slope of that line a good estimate of the average logarithm of the per capita production? This is a problem in **parameter estimation**, the subject of Section 8.1.

Alternatively, a scientist might want to answer specific questions about her data, an area of inquiry called **hypothesis testing** (Section 8.4). From looking at the two populations in Example 6.1.3, it might seem that one is growing faster than the other. Can we use the data to tell whether it really is? More generally, the data could be used to ask whether the simple growth model (Equation 6.1.1) is an adequate description. Does the per capita production become smaller for larger populations? These sorts of questions lie at the heart of the scientific process: using data to test hypotheses about biological mechanisms.

6.1.3 Stochastic Population Growth: Stochastic Immigration

Populations can also grow by immigration. In real-world situations, immigration, like production, is stochastic.

Example 6.1.4 Population Growth with Deterministic Immigration

The simplest model of deterministic immigration into a population of size N_t at time t is

$$N_{t+1} = N_t + 1,$$

meaning that exactly one immigrant arrives per day. This population grows linearly with slope 1 (Figure 6.1.5).

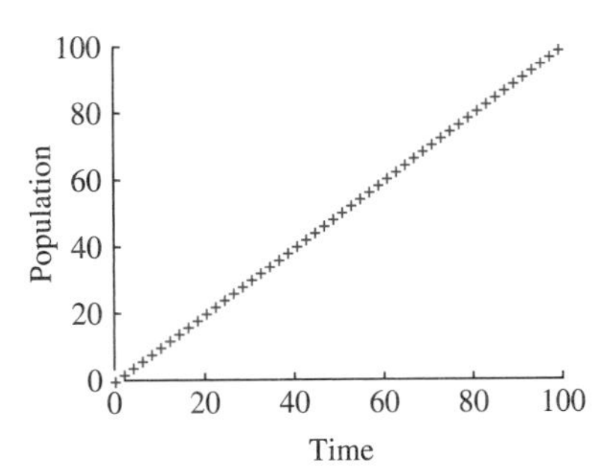

FIGURE 6.1.5
Deterministic immigration

Example 6.1.5 Population Growth with Stochastic Immigration

Suppose instead that these animals travel in pairs, and that one pair, or two individuals, arrives each day with probability 0.5. This population, denoted S_t at time t, obeys the rule

$$S_{t+1} = \begin{cases} S_t + 2 & \text{with probability } 0.5 \\ S_t & \text{with probability } 0.5. \end{cases}$$

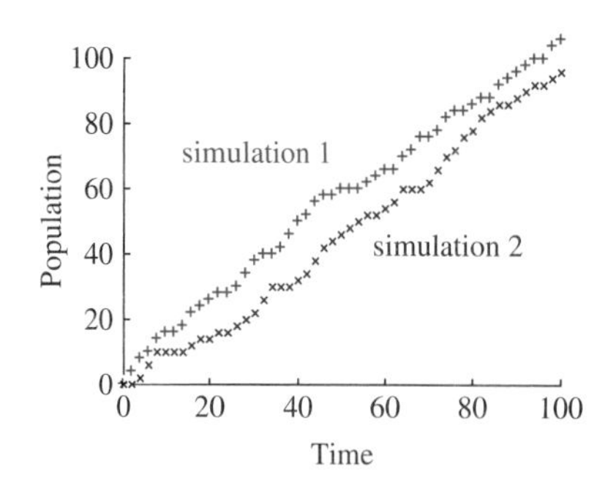

FIGURE 6.1.6
Stochastic immigration

This can be thought of as a coin-tossing experiment where "heads" means that a pair of immigrants arrives and "tails" mean that no pair arrives. This stochastic dynamical system adds up twice the number of "heads." Two simulations are shown. Although we expect 100 immigrants after 100 days, there are 96 in one case and 106 in the other due to the vagaries of chance. The results for the first 10 years are given in the following table, and results for a simulation of 100 years are given in Figure 6.1.6.

	Simulation 1		Simulation 2	
Time	Immigrants	Population	Immigrants	Population
0	2	0	0	0
1	2	2	0	0
2	2	4	2	0
3	2	6	0	2
4	2	8	2	2
5	0	10	2	4
6	2	10	2	6
7	2	12	2	8
8	2	14	0	10
9	0	16	0	10
10	0	16	0	10

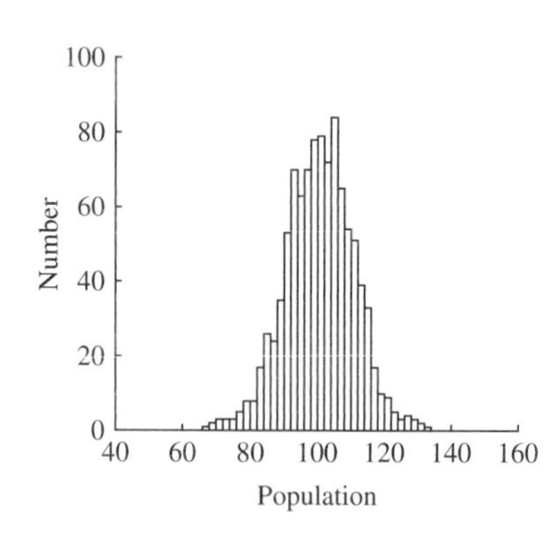

FIGURE 6.1.7
A summary of the results of 1000 simulations of the immigration model

As probabilists, we use the model to predict what happens to this population after a given number of days. We might want to know how likely it is that exactly 86 immigrants arrived during the first 100 days, or that more than 120 immigrants arrived in this same time. A useful tool for describing the results of many stochastic simulations is the **histogram**, to be studied in detail in Section 6.6 (Figure 6.1.7). The height of each bar indicates how many simulations out of 1000 ended up exactly at the population shown on the horizontal axis. For example, 24 populations ended up with exactly 86 immigrants.

As statisticians, our task is different. Suppose we have observed one or more populations over a period of time (Example 6.1.5). We might then try to estimate the average rate at which immigrants arrive, another problem in **parameter estimation**. This is simple in the deterministic case; the line with slope 1 indicates that exactly 1 immigrant arrived per year (Figure 6.1.5). The data in Figure 6.1.6 roughly follow a line, but which line?

Alternatively, we might wish to ask particular scientific questions about the data, using the method of **hypothesis testing**. Is the immigration rate higher in one population than the other? Is it more likely that an immigrant arrives in the second population in years when an immigrant arrives in the first? Is the immigration rate becoming larger over time?

6.1.4 Markov Chains

In our stochastic equations for population growth by production and immigration, population growth does not depend on the population size. As in realistic deterministic models of population growth, however, the growth of a population often depends on the size of a population. One tool for describing cases where the current state affects the future state in the presence of stochastic effects is called a **Markov chain**.

Definition 6.2 A discrete-time Markov chain is a stochastic dynamical system where the probability of arriving in a particular state at a particular time depends on the state at the previous time.

Example 6.1.6 A Markov Chain for Presence and Absence

A simple and widely applicable Markov chain follows a population on an island, noting each year only whether the island is occupied or not. This description requires only two values: 0 (to represent absence) and 1 (to represent presence). The dynamics depend on the probabilities of transitions between these two states. Suppose that an empty island is settled in a given year with probability 0.2 and remains empty otherwise, while the population on an occupied island goes extinct with probability 0.1 in a given year and remains occupied otherwise (Figure 6.1.8).

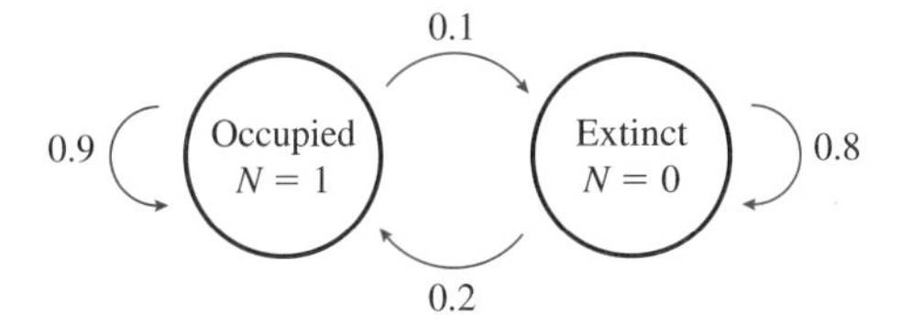

FIGURE 6.1.8
A Markov chain for presence and absence

We will soon develop a convenient probabilistic language for writing and describing this model (Subsection 6.5.3). An island following this model might start out occupied, become extinct for 9 years, be occupied for 5 years, and so forth (Figure 6.1.9).

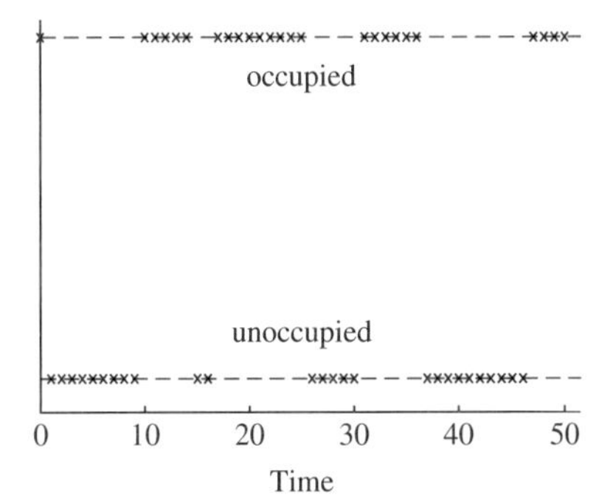

FIGURE 6.1.9
An island switching between occupied and extinct according to a Markov chain

Faced by this model, a probabilist might ask how often the island will be occupied, or for how long. Faced by the data in Figure 6.1.9, a statistician might attempt to estimate the probability that an occupied island becomes extinct. Scientific questions about the data might include asking whether occupied islands are more likely to go extinct after they have been occupied for a while, or whether conditions on the island are getting better or worse over time.

Summary We compared and contrasted **deterministic** models, in which the future is exactly determined, with **stochastic** models, in which "chance" events make the future unpredictable. It is the realm of **probability theory** to write and mathematically analyze models that describe these situations. Description and comparison of large quantities of data require **statistics**, single numbers that summarize important properties of the data. **Inferential statistics** are used to figure out what sort of process might have generated that data. We examined three generalizations of deterministic models of population growth: a model where per capita production changes stochastically over time, a model where immigration is a random event, and a **Markov chain** describing how an island might alternate stochastically between being occupied and empty.

6.1 Exercises

Mathematical Techniques

1–4 ▪ In each of the following populations, per capita production changes over time according to a fixed pattern (rather than a random pattern). For each population, find the population during the first 6 years and sketch a graph. Do you think it will grow or shrink in the long run? Find the average per capita production in each case. Does an average greater than 1.0 always mean that the population is growing?

1. A population starts at size 1000 and grows with per capita production r_t that alternates between 0.6 and 1.5.
2. A population starts at size 1000 and grows with per capita production r_t that alternates between 0.7 and 1.5.
3. A population starts at size 1000 and grows with per capita production r_t that has a 3 year cycle, first 0.7, then 0.9, and then 1.6.
4. A population starts at size 1000 and grows with per capita production r_t that has a 3 year cycle, first 0.7, then 0.9, and then 1.5.

5–8 ▪ In each of the following populations, immigration changes over time according to a fixed pattern (rather than a random pattern). For each population, find the population in each of the first 6 years and sketch a graph. Do you think it will grow or shrink in the long run? Find the average immigration in each case. Does an average greater than 0 mean that the population is growing?

5. A population starts at size 100, and receives 10 immigrants in the first year, loses 5 emigrants in the second year, receives 10 immigrants in the third year, and so forth.
6. A population starts at size 100, and receives 10 immigrants in the first year, loses 12 emigrants in the second year, receives 10 immigrants in the third year, and so forth.
7. A population starts at size 50, and receives 10 immigrants in the first year, loses 12 emigrants in the second year, and gains 5 immigrants in the third year. It then repeats this 3 year cycle.
8. A population starts at size 20, and has a 6 year cycle: gain 1, lose 2, gain 3, lose 4, gain 3, lose 2.

9–12 ▪ The following table gives the per capita production for four populations over a period of 10 years. Find the population over the 10 years starting from the given initial population, sketch a graph, and check whether the population has increased or decreased. Find the average per capita production (add up the 10 values for per capita production and divide by 10). Does an average greater than 1 always mean that the population is growing?

Year	Per Capita Production of Population 1	Per Capita Production of Population 2	Per Capita Production of Population 3	Per Capita Production of Population 4
1	1.030	0.670	0.960	0.997
2	0.886	0.870	0.841	1.030
3	0.564	1.020	1.450	1.100
4	1.050	1.480	0.966	1.040
5	0.507	0.602	1.260	1.110
6	0.919	0.941	0.769	0.991
7	0.632	0.911	1.270	1.020
8	0.712	1.350	0.967	1.110
9	1.360	0.883	1.180	0.935
10	1.250	1.420	0.883	0.958

9. Population 1 starting from 100 individuals.
10. Population 2 starting from 50 individuals.
11. Population 3 starting from 500 individuals.
12. Population 4 starting from 200 individuals.

13–16 ▪ The following table gives the immigration (positive values) or emigration (negative values) for four populations over a period of 10 years. Find the population after 10 years starting from the given initial population, and check whether the population has increased or decreased. Find the average immigration (add up the 10 values and divide by 10). Does an average greater than 0 mean that the population is growing?

Year	Change in Population 1	Change in Population 2	Change in Population 3	Change in Population 4
1	2	1	1	0
2	−1	3	−1	3
3	1	−1	0	2
4	−2	3	1	−2
5	−3	0	−3	3
6	−2	0	−1	1
7	−2	4	−3	−3
8	3	−3	1	0
9	0	−1	0	1
10	4	3	3	−1

13. Population 1 starting from 10 individuals.

14. Population 2 starting from 20 individuals.

15. Population 3 starting from 10 individuals.

16. Population 4 starting from 50 individuals.

Applications

17–20 ▪ The following table describes populations that are growing through production. For each, sketch a graph, find the per capita production in each year, and describe the growth of the population in words. When will the population reach 500?

Year	Population 1	Population 2	Population 3	Population 4
1	100	100	100	100
2	110	110	98	84
3	121	124	103	66
4	133	134	118	102
5	146	150	135	151
6	161	167	161	144
7	177	181	153	201
8	195	200	166	174
9	214	222	183	278
10	236	249	230	160
11	259	277	235	112

17. What is the per capita production of population 1 during these years?

18. What is the per capita production of population 2 during these years?

19. What is the per capita production of population 3 during these years?

20. What is the per capita production of population 4 during these years?

21–24 ▪ The following table describes populations that are growing through immigration or emigration. For each, sketch a graph, find the number of individuals that arrived or left in each year, and describe the growth of the population in words. When will the population reach 100?

Year	Population 1	Population 2	Population 3	Population 4
1	50	50	50	50
2	52	51	54	54
3	54	52	56	59
4	56	54	60	59
5	58	57	62	63
6	60	59	64	65
7	62	62	68	72
8	64	65	73	69
9	66	66	75	74
10	68	67	77	70
11	70	69	79	70

21. What is the change in population 1 during these years?

22. What is the change in population 2 during these years?

23. What is the change in population 3 during these years?

24. What is the change in population 4 during these years?

25–26 ▪ Describe what would happen to a population following the Markov chain for occupation and extinction (Figure 6.1.8) in the following special cases.

25. The probability of an empty island being occupied and the probability of an occupied island becoming empty are both equal to 1.

26. The probability of an empty island being occupied is 0 and the probability of an occupied island becoming empty is 0.5.

27–30 ▪ Suppose the states of populations on four islands are described in the following table.

Year	Island 1	Island 2	Island 3	Island 4
0	Occupied	Occupied	Occupied	Occupied
1	Extinct	Occupied	Occupied	Extinct
2	Occupied	Extinct	Occupied	Occupied
3	Extinct	Occupied	Extinct	Occupied
4	Occupied	Occupied	Occupied	Extinct
5	Occupied	Extinct	Occupied	Occupied
6	Occupied	Occupied	Extinct	Extinct
7	Extinct	Occupied	Occupied	Occupied
8	Occupied	Extinct	Occupied	Extinct

For each, illustrate what is happening with a graph, and describe it in words. Does any of the islands have a pattern that can be described deterministically?

27. On island 1.
28. On island 2.
29. On island 3.
30. On island 4.

31–34 ▪ The simple models of stochasticity described in the book leave out a lot of biological detail. Use your imagination to add some of that detail back.

31. Think of two biological factors that are neglected in the stochastic model $b_t = r_t b_t$, Equation 6.1.1.
32. Think of three factors neglected in the stochastic model of immigration (Example 6.1.5).
33. Think of two factors neglected in the Markov chain model of presence and absence on a island (Figure 6.1.8).
34. Write a simple description of a population with stochastic production and stochastic immigration.

Computer Exercises

35. Your computer should have several types of **random number generators**. One type chooses a random number between an upper and a lower limit. Consider a population growing according to the rule
$$b_{t+1} = r b_t$$
where r is a "random number" chosen from the range 0.5 to 1.5. Start two simulations with $b_0 = 100$ and run them for 50 generations. Try this a few times until you get a plot that looks interesting. Why are the two solutions so different? If someone showed you these data without telling you they were generated on a computer, how would you describe and interpret the results?

36. Consider a population growing due to immigration
$$N_{t+1} = \begin{cases} N_t + 1 & \text{with probability } 0.5 \\ N_t & \text{with probability } 0.5. \end{cases}$$
We can define an updating function
$$g(N) = N + p$$
where p is a random number chosen by the computer to take on the value 0 with probability 0.5 and 1 with probability 0.5. Generate two 50-generation solutions starting from populations of 0. If someone showed you these data, how would you describe and interpret the results?

37. Consider a population on an island described by $M = 1$ if the island is occupied and $M = 0$ if the island is unoccupied. Suppose this population follows the rule
$$M_{t+1} = 1 \begin{cases} \text{with probability } 0.3 \text{ if } M_t = 0 \\ \text{with probability } 0.9 \text{ if } M_t = 1 \end{cases}$$
$$M_{t+1} = 0 \begin{cases} \text{with probability } 0.7 \text{ if } M_t = 0 \\ \text{with probability } 0.1 \text{ if } M_t = 1. \end{cases}$$
where $q_{0.3}$ is a random number that is equal to 1 with probability 0.3 and 0 with probability 0.7 and $q_{0.9}$ is a random number that is equal to 1 with probability 0.9 and 0 with probability 0.1. Generate a solution for 100 generations of this population starting from the occupied state. When does the population first go extinct? For how long? What is the final state of the population? Is the island occupied more often than unoccupied? Does this make sense?

6.2 Stochastic Models of Diffusion and Genetics

Chance plays a fundamental role in the other two biological themes of this book: diffusion and genetics. At the molecular level, the process of diffusion describes how individual molecules randomly enter or leave a particular cell or compartment. The deterministic models describing this process (a discrete-time dynamical system given by the lung model of Section 1.9 and an autonomous differential equation given by the model of diffusion across a membrane in Subsection 5.1.3) are only appropriate for large numbers of molecules. We develop two stochastic models for a single molecule in this section: a discrete-time dynamical system for the probability that a molecule remains inside a cell after a certain amount of time, and a Markov chain to model a molecule that can re-enter the cell.

Genetic systems are unavoidably probabilistic. No two offspring, even from the same parents, are exactly the same. Models of genetics (along with models of gambling) provided much of the motivation for the development of probability theory. This section introduces three basic models of genetics, all addressing how variation is lost or maintained in populations. The first deduces some of the consequences of inbreeding, the second the consequences of outcrossing, and the third the consequences of "blending" inheritance.

6.2.1 Stochastic Diffusion: Discrete-Time Model

At a macroscopic level, molecules come in huge numbers like Avogadro's number, 6.02×10^{23}. Within a single cell, however, important molecules can come in much smaller numbers. When there are very few molecules, chance effects can be important. For example, if only ten molecules of some enzyme exist in a cell, it is quite possible that all of them will wander simultaneously into one half of the cell. It is effectively impossible that all of the many billions of water molecules in the cell would do the same.

To study enzymes or toxins that can have important effects on cell function even in small numbers, we must use stochastic models of diffusion. These models take into account the fact that molecules bounce around unpredictably. Stochastic models are also appropriate for common molecules that take on functional importance on the rare occasions when they bind with particular proteins.

Suppose a single molecule of some toxin is drifting around in a cell. It leaves during any 1 minute interval with probability 0.1 and cannot re-enter once it leaves (Figure 6.2.1).

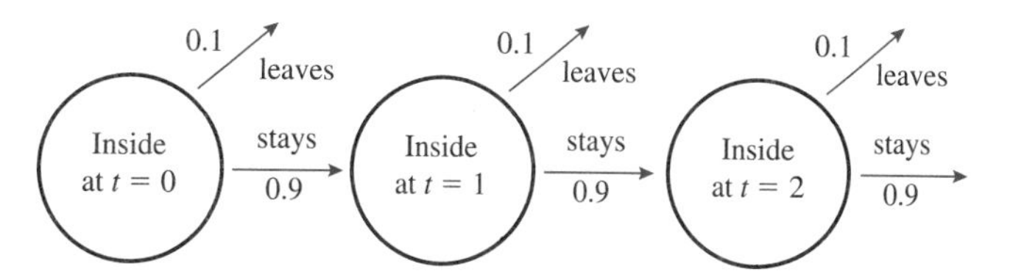

FIGURE 6.2.1
A stochastic, toxic molecule

The fates of several such molecules are given in the following table.

	Molecule									
Time	1	2	3	4	5	6	7	8	9	10
1	In	In	In	In	In	In	In	In	In	In
2	In	Out	In	In	In	In	Out	In	In	In
3	In	Out	In	In	In	In	Out	In	In	In
4	In	Out	Out	In	In	In	Out	In	In	In
5	In	Out	Out	In	In	In	Out	In	In	In
6	In	Out	Out	Out	In	In	Out	Out	In	In
7	In	Out	Out	Out	In	Out	Out	Out	In	In
8	In	Out	Out	Out	In	Out	Out	Out	In	Out
9	In	Out	Out	Out	In	Out	Out	Out	In	Out
10	Out	Out	Out	Out	In	Out	Out	Out	In	Out

The first molecule left at time 10, the second at time 2, and so forth. Molecules 5 and 9 remained inside at time 10. But what is the **probability** that a molecule is still in the cell after 10 minutes?

The first step in writing a model is to define the **state variable**, the quantity we wish to follow. Let p_t be the probability that the molecule is still in the cell after t minutes. We can derive a discrete-time dynamical system describing this new kind of state variable.

The molecule begins inside the cell, so

$$p_0 = 1$$

(a probability of 1 corresponds to certainty). After 1 minute, the molecule has left with probability 0.1 and remains with probability 0.9, so

$$p_1 = 0.9 \cdot p_0 = 0.9.$$

After the second minute, the molecule is inside only if it was inside at time 1 (probability $p_1 = 0.9$) and remained there during the next minute (probability 0.9). Therefore,

$$p_2 = 0.9 \cdot p_1 = 0.9 \cdot 0.9 = 0.81.$$

In general,

$$p_{t+1} = 0.9p_t. \tag{6.2.1}$$

The probability that the molecule is inside at the beginning of 1 minute is 90% of the probability that it was inside at the beginning of the previous minute.

This discrete-time dynamical system has the same form as the equation $b_{t+1} = rb_t$ describing a bacterial population with $r = 0.9$. With the initial condition $p_0 = 1$, the solution is

$$p_t = 0.9^t. \tag{6.2.2}$$

The probability that a molecule is inside decreases exponentially. As time becomes large, this means that the molecule has almost certainly left.

What does this probability mean? If many molecules follow this process, a **fraction** of approximately p_t would be inside the cell after t minutes.

Example 6.2.1 Predicted and Actual Numbers of Molecules

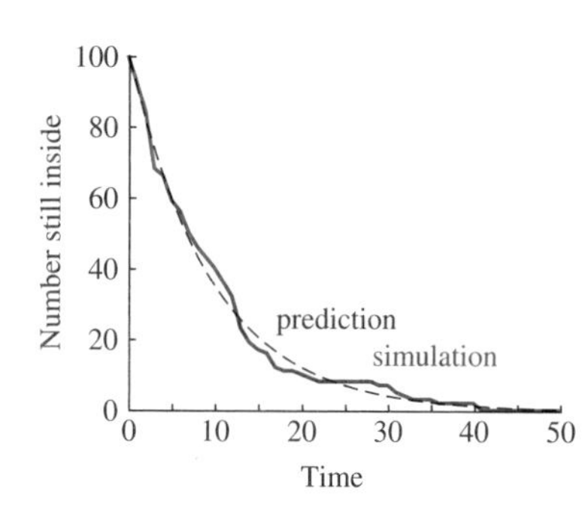

FIGURE 6.2.2
The fate of many toxic molecules

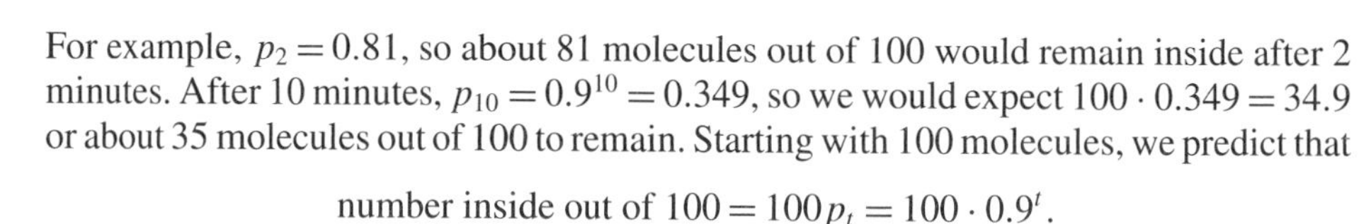

For example, $p_2 = 0.81$, so about 81 molecules out of 100 would remain inside after 2 minutes. After 10 minutes, $p_{10} = 0.9^{10} = 0.349$, so we would expect $100 \cdot 0.349 = 34.9$ or about 35 molecules out of 100 to remain. Starting with 100 molecules, we predict that

$$\text{number inside out of } 100 = 100p_t = 100 \cdot 0.9^t.$$

A computer simulation of 100 molecules compares well with the predicted number, although the actual number of molecules inside is not exactly equal to the predicted number (Figure 6.2.2). The data for the first 10 minutes are given in the following table.

Time	Actual Number	Predicted Number
0	100	100.0
1	92	90.0
2	84	81.0
3	68	72.9
4	66	65.6
5	59	59.0
6	56	53.1
7	50	47.8
8	46	43.0
9	43	38.7
10	40	34.9

Example 6.2.2 Predicted and Actual Numbers of Molecules II

With a larger number of molecules, the simulation and prediction are much closer. Starting with 10^4 molecules, we predict that

$$\text{number inside out of } 10^4 = 10^4 p_t = 10^4 \cdot 0.9^t.$$

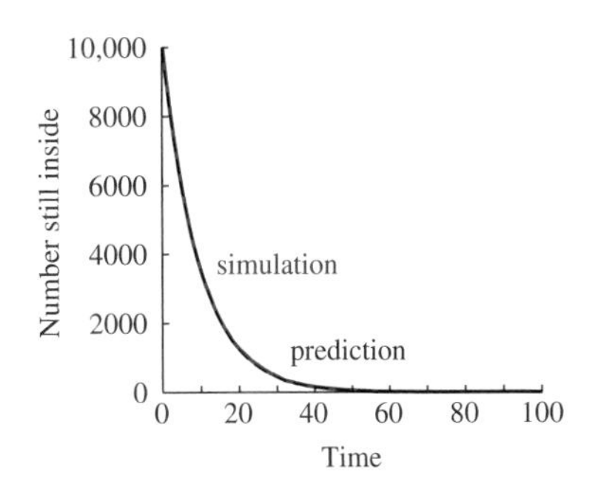

FIGURE 6.2.3
The fate of many toxic molecules

After 2 minutes, we predict that $10^4 \cdot 0.9^2 = 8100$ molecules will remain. Figure 6.2.3 compares a computer simulation of 10^4 molecules with the predicted number. The prediction is too close to distinguish on the graph, but is still not exact. The data for the first 10 minutes are given in the following table.

Time	Actual Number	Predicted Number
0	10,000	10,000.0
1	9004	9000.0
2	8100	8100.0
3	7303	7290.0
4	6552	6561.0
5	5875	5904.9
6	5276	5314.4
7	4699	4783.0
8	4237	4304.7
9	3818	3874.2
10	3458	3486.8

Our result does not tell us what a particular molecule will do because the behavior of a particular molecule depends on stochastic factors. Each molecule in the table behaves quite differently from the others. Molecule 2 left after the first minute, while molecules 5 and 9 did not leave during the first 10 minutes. When probabilities are very small, however, certain possibilities can be ruled out. After 100 minutes, the probability of finding a particular molecule inside is $p_{100} = 0.9^{100} = 2.656 \times 10^{-5}$. Only about 26 out of 1,000,000 molecules would remain. If toxic effects are not observed in 100 minutes, we can be pretty sure they will never occur.

6.2.2 Stochastic Diffusion: Markov Chain Model

Suppose now that the molecule can re-enter the cell. Possible data might appear as in the following table.

	Time									
Molecule	1	2	3	4	5	6	7	8	9	10
1	In	In	In	In	In	In	In	In	In	Out
2	In	Out	Out	Out	Out	Out	Out	Out	Out	Out
3	In	In	In	In	In	Out	Out	Out	Out	Out
4	In	In	Out	Out	Out	Out	Out	Out	Out	Out
5	In	In	In	In	In	In	In	In	In	In
6	In	In	In	In	In	In	In	In	Out	Out
7	In	In	Out	Out	Out	Out	Out	Out	Out	Out
8	In	In	In	In	Out	Out	Out	Out	Out	Out
9	In	In	In	In	In	In	In	In	In	In
10	Out	Out	Out	Out	Out	Out	Out	Out	Out	Out

We want to know how likely we are to find such a molecule inside. Suppose that if the molecule is inside, it leaves during the next minute with probability 0.2 and otherwise remains inside. If it is outside, it enters during the next minute with probability 0.1 (Figure 6.2.4). This molecule, perhaps a toxin, is more likely to leave when inside than

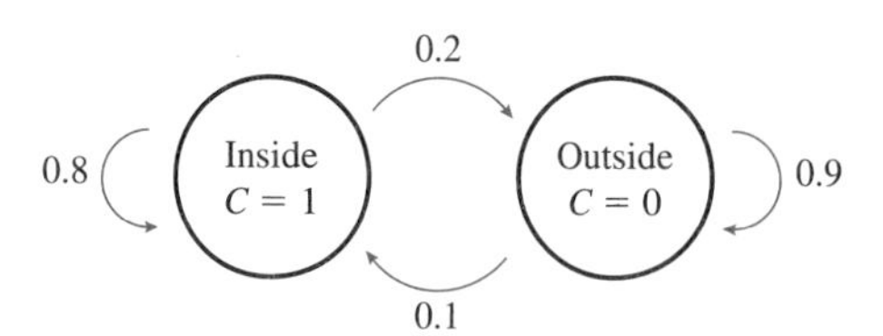

FIGURE 6.2.4
A Markov chain model of a molecule

to enter when outside. This could be a consequence of filtering by the cell membrane or of a larger volume outside than inside the cell.

The molecule follows a rule much like the occupied or empty island (Example 6.1.6). What is the long-term probability of finding the molecule inside the cell?

Set p_t to be the probability that the molecule is inside at time t. The probability that it is outside is then $1 - p_t$. There are two ways the molecule can be inside at time $t + 1$: it was inside at time t and remained inside or was outside at time t and entered. If it was inside at time t (probability p_t), it remains with probability 0.8. If it was outside at t (probability $1 - p_t$), it enters with probability 0.1. The updated probability is

$$\begin{aligned}(\text{probability inside at } t+1) &= (\text{probability inside at } t) \times (\text{probability did not leave}) \\ &\quad + (\text{probability outside at } t) \times (\text{probability entered}) \\ p_{t+1} &= 0.8p_t + 0.1(1 - p_t). \qquad (6.2.3)\end{aligned}$$

We can rewrite this more simply as

$$p_{t+1} = 0.8p_t + 0.1 - 0.1p_t = 0.7p_t + 0.1.$$

We can find the equilibria and apply the method of cobwebbing to describe the behavior of this discrete-time dynamical system (Figure 6.2.5). To find the equilibrium, we solve for the value p^* which the discrete-time dynamical system leaves unchanged. In this case, p^* satisfies the equation

$$p^* = 0.7p^* + 0.1.$$

To solve,

$$\begin{aligned} p^* &= 0.7p^* + 0.1 \\ p^* - 0.7p^* &= 0.1 \\ 0.3p^* &= 0.1 \\ p^* &= \frac{0.1}{0.3} = 1/3. \end{aligned}$$

Furthermore, the updating function $f(p) = 0.7p + 0.1$ is a line with slope 0.7, which is less than 1.0, meaning this equilibrium is stable (Theorem 3.1).

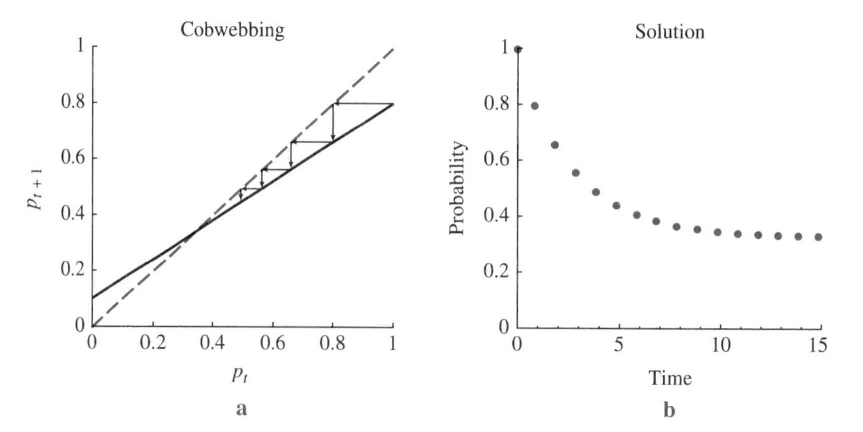

FIGURE 6.2.5
Cobwebbing and solution for probability discrete-time dynamical system

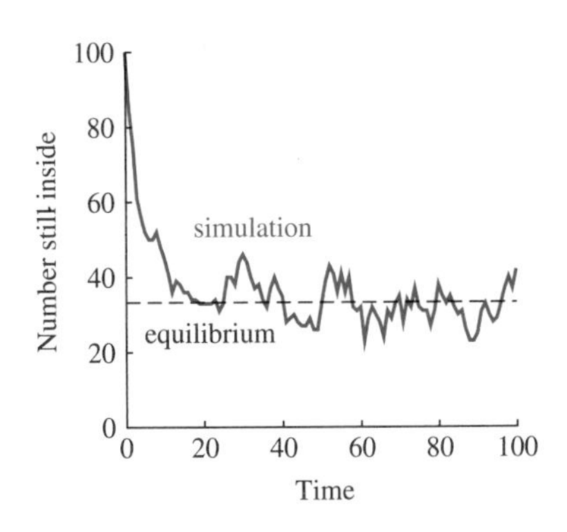

FIGURE 6.2.6
The fates of 100 Markov molecules

What does this tell us? The probability of finding a particular molecule inside the cell after a long time is 1/3. If many molecules were observed, about 1/3 of them would be inside the cell after a long time. The fates of 100 computer molecules that began inside the cell are shown in Figure 6.2.6. The fraction inside approaches 1/3, but there is much variation around this average. Without this variation, the model matches our model of deterministic chemical exchange.

6.2.3 The Genetics of Inbreeding

One of the consequences of inbreeding (breeding with close relatives) is the loss of genetic variability. The simplest form of inbreeding is **selfing** (or self-pollination) in plants. Rather than wait for pollen to arrive from other plants in the wind or on the bodies of bees, some plants use their own. Many important crop plants use this dependable system of fertilization.

Suppose a plant is **diploid**, meaning that it has two copies of each gene (one from the ovule or egg and one from the pollen). Plants with two different **alleles** (variants) of this gene are called **heterozygous**, while those with two copies of the same allele are called **homozygous**. What is the probability that an offspring of a heterozygous plant is heterozygous?

Let **a** and **A** denote the two different alleles. A heterozygous plant has genotype **Aa** (plants of type **aA** are the same). There are four possible ways to combine genes in the offspring: an offspring could get **a** from the ovule and **a** from the pollen, **a** from the ovule and **A** from the pollen, **A** from the ovule and **a** from the pollen, or **A** from the ovule and **A** from the pollen (Figure 6.2.7). In the absence of other factors, each of these possibilities is equally likely and has probability 0.25 (Figure 6.2.8). Two of these four possibilities produce indistinguishable heterozygous offspring, so that the probability an offspring is heterozygous is 0.5.

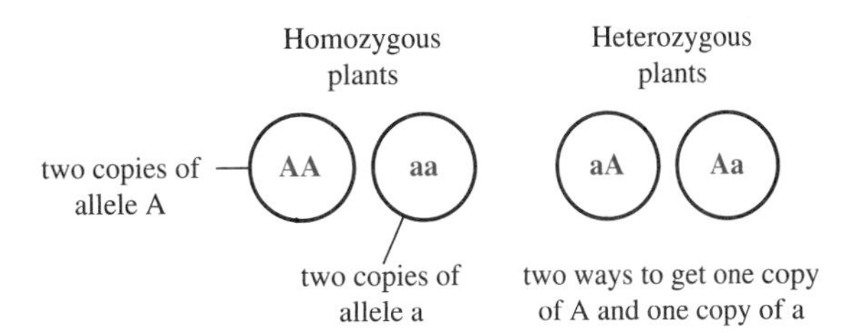

FIGURE 6.2.7
The four possible genotypes

We can set up a discrete-time dynamical system to compute the probability that a plant will be heterozygous after t generations of selfing. Let h_t represent the probability a plant is heterozygous in generation t. Numbering the parent plant as generation 0, we have assumed that $h_0 = 1$. Half of the offspring are heterozygous after one generation, so $h_1 = 0.5$. What is the discrete-time dynamical system? In the absence of mutation, a homozygous parent in generation t will produce only homozygous offspring. If the parent is heterozygous, its offspring will be heterozygous with probability 0.5 (Figure 6.2.9). The discrete-time dynamical system is

$$h_{t+1} = 0.5h_t.$$

Half of the offspring of heterozygous plants are heterozygous.

This discrete-time dynamical system has the form of the bacterial growth equation $b_{t+1} = rb_t$ with $r = 0.5$, and has solution

$$h_t = 0.5^t.$$

After one generation, the probability that a plant is heterozygous is 0.5, and after two generations, the probability is $h_2 = 0.5^2 = 0.25$ (Figure 6.2.9). After ten generations, the probability that a plant is heterozygous is

$$h_{10} = 0.5^{10} = 0.001.$$

Only about 1 out of a 1000 progeny of this plant would be heterozygous after ten generations.

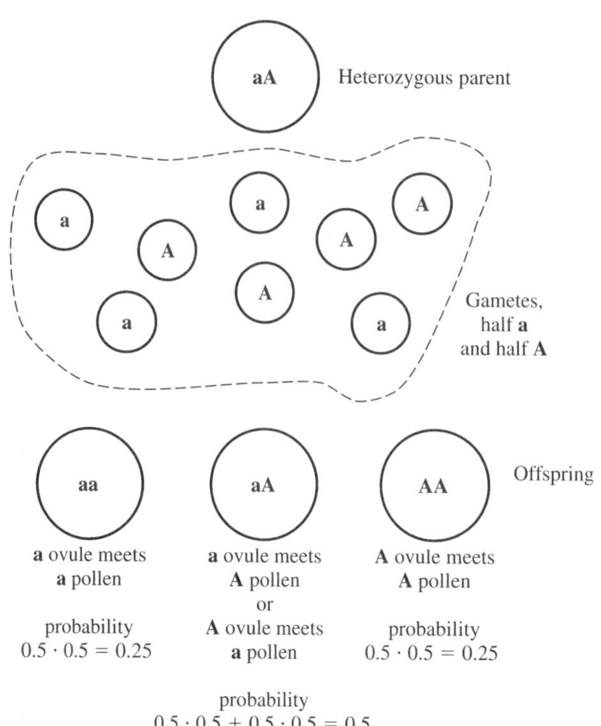

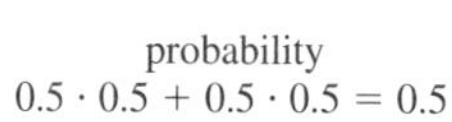

FIGURE 6.2.8
The probabilities of the four genotypes

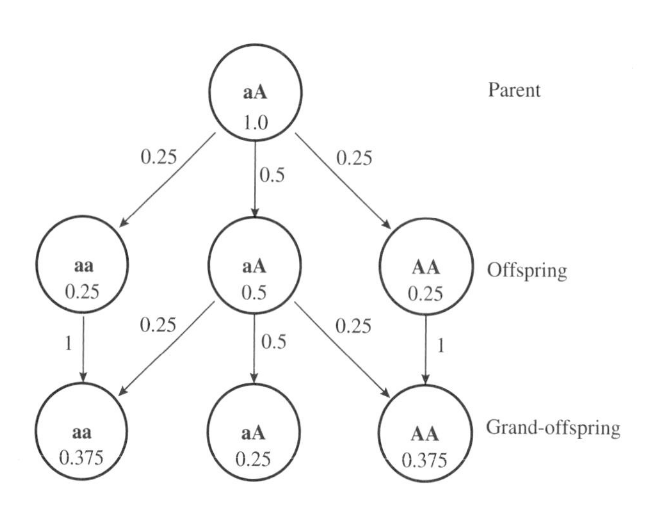

FIGURE 6.2.9
Selfing and heterozygosity

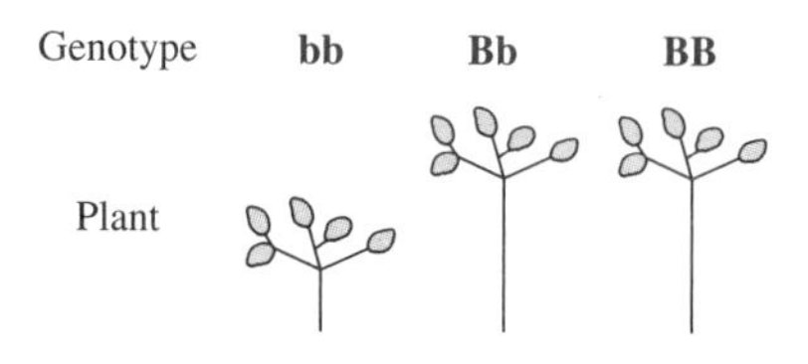

FIGURE 6.2.10
The action of a dominant gene

If fewer and fewer plants are heterozygous, more and more are homozygous. In each generation, the number of **aa** homozygotes is equal to the number of **AA** homozygotes. In the long run, half the plants will be of each homozygous type. Thus, the population started off with exactly half **a** alleles (1 out of 2) and ends up the same, even though they are now packaged in homozygotes rather than heterozygotes.

6.2.4 The Dynamics of Height

Even with modern techniques, it is difficult to identify the genetic basis of complex traits like plant height. Based on a general knowledge of genetics, we can propose various hypotheses about the underlying genetics and use probability theory to deduce the resulting dynamics. As statisticians, we can compare data with the results of probabilistic models to check whether the original hypotheses make sense.

The basic model of underlying genetics dates back to Mendel. Consider a population of plants that *cannot* self-pollinate. Assume first that height is determined by a single gene. Plants with at least one copy of the **B** allele are tall, while those with two

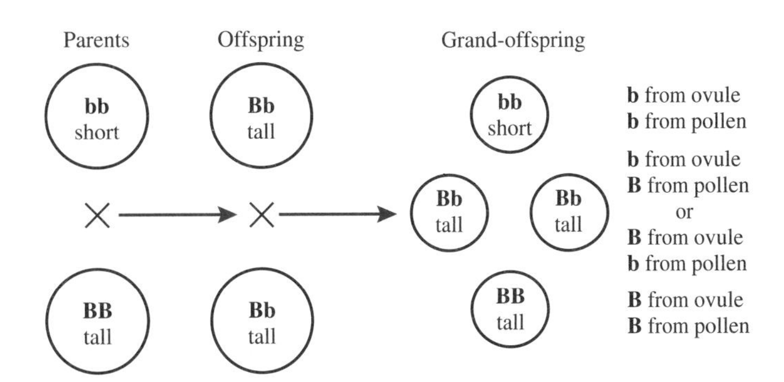

FIGURE 6.2.11
An experiment in outcrossing

copies of the **b** allele are short (the **B** allele is said to be **dominant**) (Figure 6.2.10). A breeding experiment is diagrammed in Figure 6.2.11. A short plant with genotype **bb** is crossed with a tall **BB** parent. All the offspring are of type **Bb** and are tall. These offspring are then crossed with each other.

What fraction of the next generation will be short? Short plants must have received a **b** allele from each parent. The probability of getting the **b** allele from the ovule is 0.5, equal to that of getting the **b** allele from the pollen. The probability of getting both is the product $0.5 \cdot 0.5 = 0.25$. This probability describes the fraction of short plants in an infinitely large number of offspring. With 100 offspring, we expect to find 25 short plants and 75 tall plants, but might find 28 short plants and 72 tall plants.

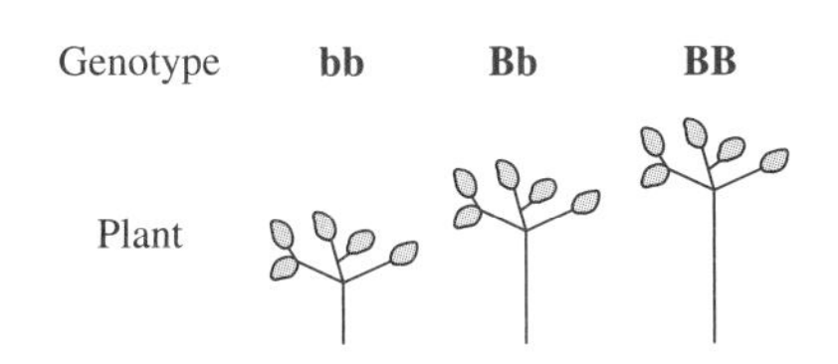

FIGURE 6.2.12
The action of an additive gene

Suppose instead that the effects of the genes are **additive**, meaning that a plant with genotype **bb** is short, a plant with genotype **BB** is tall, and a plant with genotype **Bb** is intermediate (Figure 6.2.12). Computing as above, we see that 25% of second-generation plants will have genotype **bb**, 25% will have genotype **BB**, and the remaining 50% will have genotype **Bb**. The heights of 100 simulated plants are shown in Figure 6.2.13. Twenty-one turned out short, 48 medium, and 31 tall. These fractions do not exactly match probabilities because we measured only 100 plants.

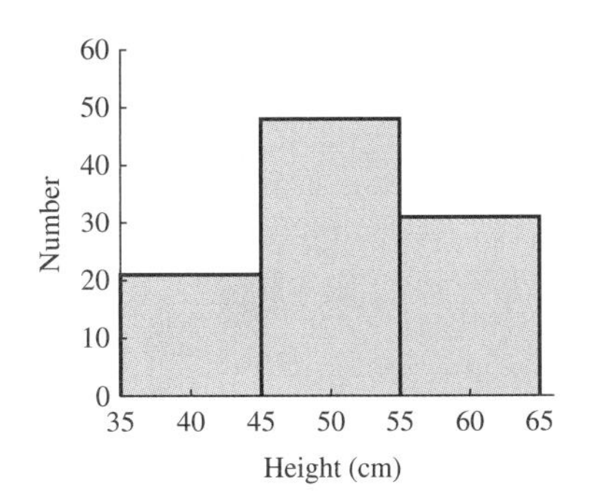

FIGURE 6.2.13
Heights of 100 plants produced by a single additive gene

This figure illustrates two widely observed properties: there is a single most common observation, and this most common observation is in the middle. These facets become more clear when height is determined by many genes. Suppose a plant has ten different genes that affect height. Let $\mathbf{b_i}$ be the "short" allele of gene i and $\mathbf{B_i}$ be the "tall" allele. Each tall allele makes the plant 1 cm taller. A plant with no tall alleles has a baseline height of 40 cm. A plant with two copies of the tall allele of each gene will have a height of 60 cm. In general,

$$\text{height of plant} = 40 + \text{ total number of } \mathbf{B} \text{ alleles} \quad \textbf{(6.2.4)}$$

(Figure 6.2.14). We can use these plants to conduct the same experiment shown in Figure 6.2.11, crossing a very short plant with a very tall plant for two generations. After the first generation, all plants have a height of exactly 50 cm. What happens in the next generation? The result of a simulation of 200 plants is shown in Figure 6.2.15. The distribution of heights is approximately bell-shaped and is an instance of the **binomial distribution**, which we study in detail in Section 7.4.

Plant 1

$\mathbf{B_1\, b_2\, b_3\, B_4\, b_5\, b_6\, B_7\, b_8\, b_9\, b_{10}}$
$\mathbf{b_1\, b_2\, b_3\, b_4\, b_5\, b_6\, b_7\, b_8\, b_9\, b_{10}}$ 3 **B**'s, height 43

Plant 2

$\mathbf{B_1\, B_2\, b_3\, B_4\, B_5\, b_6\, B_7\, B_8\, b_9\, B_{10}}$
$\mathbf{B_1\, B_2\, B_3\, B_4\, b_5\, B_6\, b_7\, b_8\, B_9\, B_{10}}$ 14 **B**'s, height 54

FIGURE 6.2.14
How height might be produced by 10 additive genes

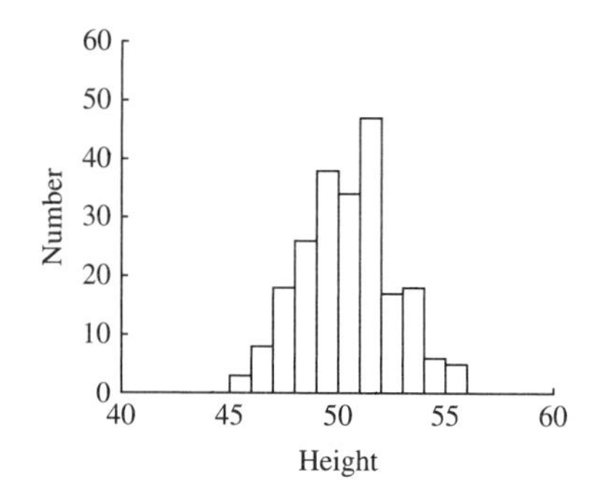

FIGURE 6.2.15

Heights of 200 plants produced by ten additive genes

Although the sizes of these second-generation plants cluster around the average of 50 cm, there is still significant variability. If these plants continued to breed among themselves, little further variability would be lost. This variability provides the material used by breeders and natural selection.

6.2.5 The Dynamics of Blending Inheritance

Consider an alternative situation where offspring height is **exactly** intermediate between the heights of their parents, called **blending inheritance**. If two parents with heights 50 were crossed, all offspring would have height 50. Suppose we raised a populations of plants, half with height 40 and half with height 60, and crossed them. The offspring would all have height 50. Unlike the cases we have studied hitherto, however, the grand-offspring produced by crossing the offspring would also have height 50 and so forth (Figure 6.2.16). All variability disappears permanently. Interestingly, this results from using a deterministic rather than a stochastic model of inheritance.

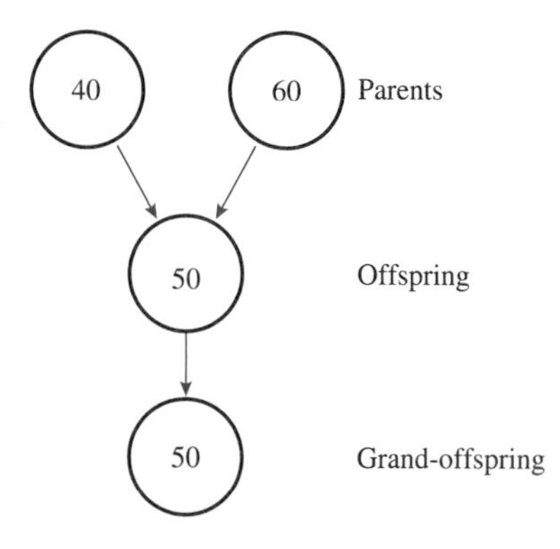

FIGURE 6.2.16

The action of blending inheritance

When Darwin was developing the theory of evolution by natural selection, little was known about the mechanisms of inheritance. It was thought that offspring traits resulted from a process like blending some "tall substance" from the mother with some "short substance" from the father. When a mathematician demonstrated these results, Darwin was concerned that his theory would not work because natural selection, like plant breeding, depends on the persistence of extreme types. Darwin modified later editions of his book *The Origin of the Species* (1859) to include the generally discredited idea of inheritance of acquired characteristics (resemblance of offspring to traits acquired by parents during their lifetimes) because he could see no other way to address the problem. The resolution of this difficulty did not come for many decades, until Mendel's work was rediscovered around 1900 and mechanisms of inheritance began to be understood.

Summary We introduced and analyzed two models describing the behavior of a molecule diffusing out of a cell. First, we assumed that the molecule has a fixed probability of leaving during each minute. In this case, the probability that the molecule is inside follows a simple discrete-time dynamical system and decreases exponentially. Second, we assumed that the molecule could re-enter the cell and found a Markov chain describing the probability that a molecule is inside the cell. In this case, the probability approaches an intermediate equilibrium.

We also examined three different genetic systems, selfing, outcrossing, and blending inheritance, to study how variability within a population is maintained between generations. Selfing reduces **heterozygosity** by 50% each generation. Crossing plants produces a wide range of different offspring types, with the exact pattern depending on whether alleles are **dominant**, **recessive**, or **additive**, and on how many genes are involved in the trait. Population variability in not lost. With **blending inheritance**, wherein offspring have height equal to the average of the heights of their parents, all plants converge to the same height.

6.2 Exercises

Mathematical Techniques

1–2 ▪ For the given probability that a molecule leaves a cell, write the discrete-time dynamical system for the probability that it remains inside (assuming it can never return) and find the solution. Compute the probability that the molecule remains inside after 10 seconds, and the time before it will have left with probability 0.9.

1. The probability it leaves is 0.3 each second.

2. The probability it leaves is 0.03 each second.

3–4 ▪ The following probabilities describe molecules that can hop into and out of a cell. For each, find a discrete-time dynamical system for the probability that the molecule is inside. Find the probability that a molecule that begins inside is inside at $t = 2$, and the probability that a molecule that begins outside is inside at $t = 2$. Compute the equilibrium, and use it to estimate how many out of 100 molecules would be inside after a long time.

3. The probability it leaves is 0.3 each second, and the probability it returns is 0.2 each second.

4. The probability it leaves is 0.03 each second, and the probability it returns is 0.1 each second.

5–8 ▪ Draw cobweb diagrams based on the discrete-time dynamical systems in the earlier exercise.

5. The molecule in Exercise 1.

6. The molecule in Exercise 2.

7. The molecule in Exercise 3.

8. The molecule in Exercise 4.

9–10 ▪ In many ways, probabilities act like fluids. For each of the following models of chemical exchange, let c_t represent the amount in container 1 and d_t the amount in container 2 at time t. Write a discrete-time dynamical system for the amount of chemical in each container. Define p_t to be the fraction of chemical in container 1, and write a discrete-time dynamical system giving p_{t+1} as a function of p_t. Find the equilibrium fraction of chemical in the first container.

9. Each second, 30% of the chemical in container 1 enters container 2, and 20% of the chemical in container 2 returns to container 1 (compare with Exercise 3).

10. Each second, 3% of the chemical in container 1 enters container 2, and 10% of the chemical in container 2 returns to container 1 (compare with Exercise 4).

11–14 ▪ Compute the following probabilities for a selfing plant using Figure 6.2.9.

11. The fraction of grand-offspring (second generation) with genotype **AA**.

12. The fraction of third-generation offspring with genotype **AA**.

13. The fraction of fourth-generation offspring with genotype **AA**.

14. The fraction of offspring with genotype **AA** in generation t.

15–16 ▪ Suppose that the fraction of homozygous and heterozygous offspring that survive self-fertilization by a heterozygote is measured. Find the fraction of surviving offspring that are heterozygous in the following cases.

15. All of the homozygous offspring survive, and half of the heterozygous offspring survive.

16. Half of the homozygous offspring survive, and one third of the heterozygous offspring survive.

17–22 ▪ Consider the following case of blending inheritance. A population of plants starts out with an equal number of individuals of heights 40 and 60 cm. The parents mate randomly, and the height of an offspring is exactly equal to the average height of its parents.

17. What are all the possible matings? What would be the heights of the offspring? What is the probability that a 40-cm tall plant mates with a 40-cm tall plant?

18. What is the probability that a 60-cm tall plant mates with a 60-cm tall plant, and the probability that a 40-cm tall plant mates with a 60-cm tall plant?

19. Suppose that these offspring now mate with each other. Find all possible matings and the resulting offspring heights.

20. Find the probability of each of the possible matings of the offspring. Out of 100 plants, about how many would have height 50 cm?

21. With blending inheritance, the height of the offspring is equal to the average height of the four grandparents. Find the probability that all four grandparents have height 40 cm and thus the probability that a plant in the second generation has height 40 cm.

22. Find all the ways that the heights of the grandparents average to exactly 50 cm.

Applications

23–26 ▪ In each of the following circumstances, find the discrete-time dynamical system describing the probability, find the solution, and use it to answer the question.

23. A certain highly mutable gene has a 1.0% chance of mutating each time a cell divides. Suppose that there are 15 cell divisions between each pair of generations. What is the chance that the gene mutates in one generation, during the course of those 15 cell divisions? If there were 100 such genes, about how many would have mutated in one generation?

24. A herd of lemmings is standing at the top of a cliff. Each jumps off with probability 0.2 each hour. What is the probability that a particular lemming remains on top of the cliff after 3 hours? If 5000 lemmings are standing around on top of the cliff, about how many will remain after 3 hours?

25. A molecule has a 5.0% chance of binding to an enzyme each second and remains permanently attached thereafter. If the molecule starts out unbound, find the probability that it is bound after 10 seconds. How long would it take for the molecule to have bound with probability 0.95?

26. In tropical regions, caterpillars suffer extremely high parasitism, sometimes as high as 15% per day. In other words, a caterpillar is attacked by a parasitoid with probability 0.15 each day. If a caterpillar takes 25 days to develop, what is the probability it survives? If a female lays 50 eggs, about how many would survive? How much lower would the parasitism rate have to be for 2 out of the 50 caterpillars to survive?

27–30 ▪ In each of the following circumstances, find the discrete-time dynamical system describing the probability, find the solution, and use it to answer the question.

27. Suppose that each mutant gene in Exercise 23 has a 1.0% chance of mutating back to the original type each cell division. Use the Markov chain approach to find the fraction of mutant genes after 15 cell divisions. How much difference does the correction mechanism make?

28. Suppose that the lemmings in Exercise 24 can sometimes crawl back up the cliff. In particular, suppose that a lemming at the bottom of the cliff climbs back up with probability 0.1

each hour. What is the probability that a particular lemming is on top of the cliff after 3 hours? If 5000 lemmings are standing around on top of the cliff to begin with, about how many will be there after 3 hours? How much difference does crawling back up make?

29. Suppose that bound molecules in Exercise 25 have a 2.0% chance of unbinding from the enzyme each second. Find the fraction of molecules that are bound in the long run. What is the probability that a molecule is bound after 10 seconds?

30. Suppose that the caterpillars in Exercise 26 have some chance of eliminating their attacker, thus becoming a caterpillar again. In particular, suppose that a caterpillar has a 0.03 chance of eliminating a parasitoid each day. Find the probability that a caterpillar is a caterpillar after 25 days. Is a 3% recovery rate enough for about 2 out of 50 eggs to end up as caterpillars?

31–34 ▪ In normal plants, the probability that an offspring of a heterozygous parent is heterozygous is 0.5. If the survival of heterozygous offspring differs from that of homozygous offspring, the probability that a surviving offspring is heterozygous may not be equal to 0.5. For the following values of the probability, write a discrete-time dynamical system for the fraction of heterozygous offspring over time, find the solution, and compute the fraction that will be heterozygous after ten generations. How does this compare with the fraction for a normal plant?

31. The probability that an offspring is heterozygous is 0.6.

32. The probability that an offspring is heterozygous is 0.4.

33. The probability that an offspring is heterozygous is 0.2.

34. The probability that an offspring is heterozygous is 0.9.

35–38 ▪ A heterozygous plant with genotype **Aa** self-pollinates. Find the probability that an offspring is tall for the following genetic systems.

35. Only plants that have two **A** alleles are tall (the allele **A** is recessive).

36. Plants that have either one or two **A** alleles are tall (the allele **A** is dominant).

37. Heterozygous plants are tall, and homozygous plants are short.

38. Half of heterozygous plants and one fourth of homozygous plants are tall (it depends on their position in the greenhouse).

39–42 ▪ A heterozygous plant with genotype **Aa** self-pollinates, and then its offspring also self-pollinate. Find the probability that the offspring of the offspring are tall for the following genetics systems.

39. Only plants that have two **A** alleles are tall.

40. Plants that has either one or two **A** alleles are tall.

41. Heterozygous plants are tall, and homozygous plants are short.

42. Half of heterozygous plants and one fourth of homozygous plants are tall (it depends on their position in the greenhouse).

43–48 ▪ Often geneticists want to change one allele in an outcrossing organism while keeping the rest of the genome the same. For example, they might wish to take a specially designed stock of flies and alter the eye color from red to white. Suppose that the white-eye allele is dominant, meaning that flies with one or two white-eye alleles will have white eyes. One procedure used is to take a white-eyed fly and cross it with the red-eyed stock. The white-eyed offspring are then considered to be the first generation, and are crossed with the red-eyed stock. Their white-eyed offspring are considered to be the second generation, and are again crossed with the red-eyed stock, and so forth. The special red-eyed stock is homozygous for the desirable allele **A** at some other locus, but the white-eyed fly is homozygous for the inferior **a** allele at that locus.

43. What is the genotype at the eye color locus in the first generation?

44. What is the genotype at the eye color locus in the second and subsequent generations?

45. What fraction of flies will have the **a** allele (at the second locus) after one generation?

46. What fraction of flies will have the **a** allele (at the second locus) after two generations?

47. What fraction of flies will have the **a** allele (at the second locus) after t generations?

48. How many back-crosses would be necessary to purge 99.9999% of the inferior genes from the white-eyed fly?

49–52 ▪ One force that can alter the ratio of heterozygotes produced by a selfing heterozygote is **meiotic drive**. This means that one allele, say **A**, pushes its way into more than half of the gametes (ovules or pollen).

49. Suppose meiotic drive affects the pollen only and that 80% of the pollen grains from a heterozygote carry the **A** allele. Ovules are normal, and 50% of them carry the **A** allele. What fraction of offspring from a selfing heterozygote will be heterozygous?

50. Suppose meiotic drive affects both pollen and ovules and that 80% of the pollen grains and ovules from a heterozygote carry the **A** allele. What fraction of offspring from a selfing heterozygote will be heterozygous?

51. Suppose meiotic drive affects both pollen and ovules but that 80% of the pollen grains carry the **A** allele while 80% of ovules carry the **a** allele. What fraction of offspring from a selfing heterozygote will be heterozygous?

52. How many generations would it take before the probability of a descendent of a plant described in Exercise 51 would have less than a 0.01 chance of being a heterozygote? Compare this with the number of generations required in the absence of meiotic drive.

53–56 ▪ Heterozygosity in inbreeding organisms can be restored by mutation. Suppose that mutations always create brand new alleles. Suppose that each parental allele has a probability 0.01 of mutating.

53. Suppose first that the parent has genotype **AA**. What is the probability that the allele that came from the pollen is type

A? What is the probability that the allele that came from the ovule is type **A**?

54. The probability that both alleles in the offspring are type **A** is the **product** of the probability that the allele from the pollen is **A** and the probability that the allele from the ovule is **A** (we will derive this in Section 6.5). What is the probability that the offspring of a homozygous parent is homozygous? What is the probability that the offspring of a homozygous parent is heterozygous?

55. Suppose a plant is heterozygous with genotype **Aa**. What is the probability that the allele that came from the pollen is type **A**? What is the probability that the allele that came from the ovule is type **A**?

56. Find the probability that the offspring is **AA**. Find the probability that the offspring is **aa**. What is the probability that the offspring of a heterozygous parent is homozygous? What is the probability that the offspring of a heterozygous parent is heterozygous? How does this compare with the result in the absence of mutation?

Computer Exercises

57. Consider a molecule that leaves a cell each minute with probability 0.1. If $M_t = 1$ when the molecule is inside at time t and $M_t = 0$ if the molecule is outside,

$$M_{t+1} = q_{0.9} M_t$$

where $q_{0.9}$ takes on the value 1 with probability 0.9 and 0 with probability 0.1. Define an updating function F and replicate the computer experiment 10 times. Count up how many times the molecule is inside at $t = 5$.

The probability m_t that the molecule is inside at time t has updating function

$$f(m) = 0.9m.$$

Plot the solution of this deterministic discrete-time dynamical system with $m_0 = 1$. Compare the mathematically expected fraction with the fraction you counted in the stochastic version. What could you do to make the fraction end up closer to the probability?

58. We can simulate a Markovian molecule that has a 20% chance of jumping back in with the updating function

$$G(M) = Mq_{0.9} + (1 - M)q_{0.2}.$$

The **probability** m that the molecule is inside follows the related updating function

$$g(m) = 0.9m + 0.2(1 - m).$$

Plot a solution of the probability equation and a simulation of a single molecule starting from $M = m = 1$ (use enough steps to see what is going on). Solve for the equilibrium probability. Why doesn't the simulation seem to approach an equilibrium? What do the two curves have to do with each other?

59. Plants with many genes affecting a trait can be simulated by adding up the effects of each gene. For example, suppose that each of 20 genes adds 1 to the height with probability 0.5 and 0 with probability 0.5. The total height is the sum of 20 such numbers. Find a way to create 100 such plants. Which height is most common? What is the largest plant? What is the smallest plant?

6.3 Probability Theory

In presenting stochastic models of populations, diffusion, and genetics, we have been vague about the mathematical underpinnings of stochastic events and their probabilities. We now begin the study of **probability theory** to make precise these concepts and help us to analyze models. The basic language of probability theory comes from **set theory**, but is rephrased to describe **events**, or possible occurrences. Probability theory requires assigning numbers—probabilities—to events in mathematically consistent and scientifically useful ways.

6.3.1 Sample Spaces, Experiments, and Events

Imagine doing idealized experiments that have a specific set of possible outcomes.

Definition 6.3 The **sample space** is the set of all possible results of an experiment.

Example 6.3.1 The Sample Space for Stochastic Production

The value taken by r_t in the growth model (Subsection 6.1.2) can be thought of as the result of an experiment. In a given year, the result is a particular positive number equal to the per capita production. The sample space is the set of all positive numbers (Figure 6.3.1).

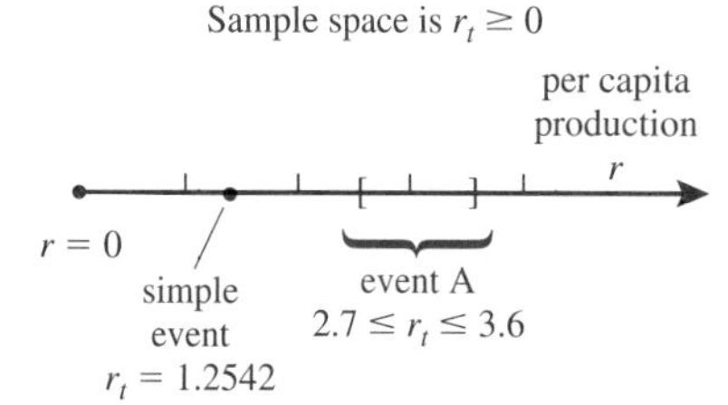

FIGURE 6.3.1
Sample space, simple event, and event for per capita production

Example 6.3.2 The Sample Space for Stochastic Immigration

The random component of the immigration model (Subsection 6.1.3) is whether or not a pair of immigrants arrive. The sample space consists of the two events, "pair" or "no pair."

Probability theory helps us compute the probabilities of the different outcomes of an experiment. First, we need a bit of technical language to describe outcomes.

Definition 6.4 Any particular outcome is known as a **simple event**. An **event** is a set of simple events or, equivalently, a subset of the sample space.

Example 6.3.3 Simple Events for Stochastic Production

The particular outcome $r_t = 1.22475$ is one simple event in the bacterial growth model (Figure 6.3.1). Another simple event is $r_t = 0.1$.

Example 6.3.4 Events for Stochastic Production

Events for stochastic population growth include $r_t > 1$ and $1.0 \leq r_t \leq 3.0$ (Figure 6.3.1). Each of these possible outcomes includes many particular single outcomes. Events are sometimes referred to with a capital letter. The letter A in Figure 6.3.1 refers to the event $2.7 \leq r_t \leq 3.6$.

Example 6.3.5 Sample Space and Events for Diffusion

Suppose ten molecules begin inside a cell and each leaves with probability 0.1 each minute for 10 minutes (Subsection 6.2.1). Many different measurements could describe the results of this experiment. A scientist might count the number of molecules left in the cell, denoting it by N (Figure 6.3.2). The sample space is the set of integers from 0 to 100. The simple events are $N = 0$, $N = 1, \ldots, N = 10$, the specific outcomes of the experiment. Events include $4 \leq N \leq 6$, $N < 6$, and "N is odd." Each event is a set of possible specific outcomes.

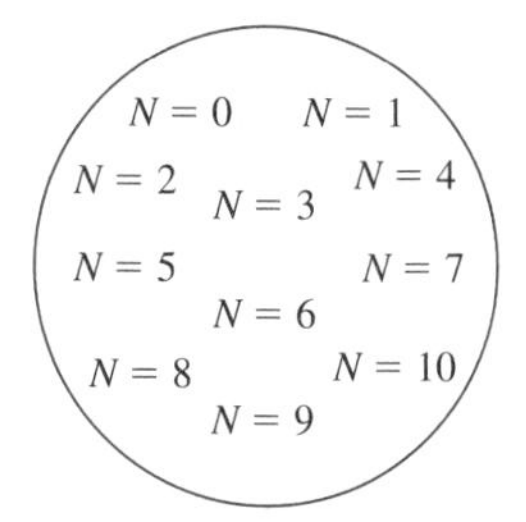

FIGURE 6.3.2
One set of simple events for ten diffusing molecules

Example 6.3.6 Alternative Sample Space and Events for Diffusion

Alternatively, the scientist could measure exactly when each molecule left the cell. The sample space in this case is much larger, all sets of ten positive integers. A simple event is one precise listing of when the various molecules left, such as "molecule 1 left at $t = 47$, molecule 2 left at $t = 13$, and so on."

First Simple Event		Second Simple Event	
Molecule	Time it Left	Molecule	Time it Left
1	47	1	3
2	13	2	2
3	11	3	7
4	1	4	8
5	9	5	3
6	91	6	5
7	23	7	2
8	3	8	10
9	13	9	9
10	5	10	4

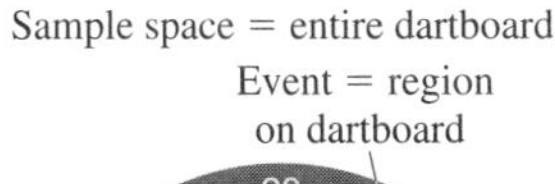

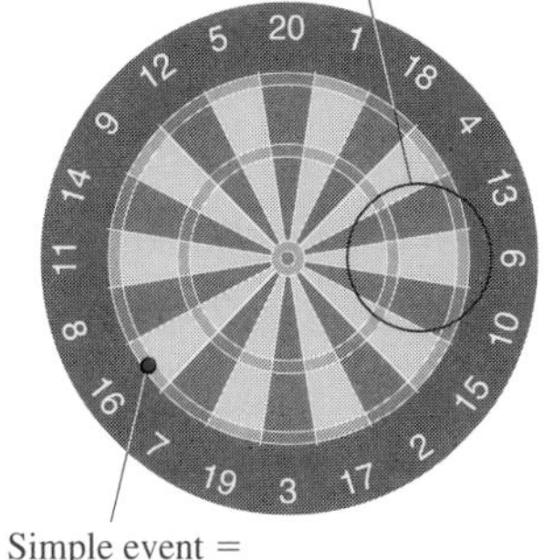

FIGURE 6.3.3
A dartboard

Events include "no molecules left at exactly time 10," "all the molecules left at odd times," "all the molecules left before time 20," and "half the molecules left between times 3 and 7, inclusive." The first simple event in the table is an element of the first two events (all left at odd times, none at time 10), while the second simple event in the table is an element of the last two (all left before time 20, and exactly five out of ten left between times 3 and 7 inclusive).

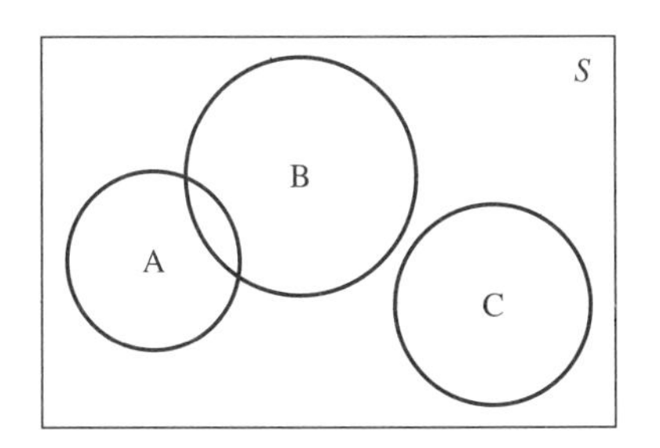

FIGURE 6.3.4
The Venn diagram

One can think of the sample space as a dartboard, and an experiment as throwing a dart (Figure 6.3.3). A simple event corresponds to a single point on the dartboard. Events correspond to regions of the board. One simple event is hitting exactly in the middle of the top half of the dartboard. Events include hitting the bull's-eye (hitting any of the points in this region), scoring 20, and hitting the top half of the board.

6.3.2 Set Theory

A probability model assigns a probability to each event. Doing this requires knowledge of what the events are (the experiment and its possible outcomes) and how "likely" they are to occur. For a dartboard, one probability model is that the chance of hitting a region is proportional to the size of that region. This means that the dart thrower has terrible aim and is just as likely to hit near the edge as near the bull's-eye. A quite different probability model would be associated with a skillful British thrower who gets mainly bull's-eyes.

To explain how one assigns probabilities to all possible events (there can be quite a few), we need to review some **set theory**. Sets can be visually portrayed on **Venn diagrams**, which are no more than abstracted drawings of dartboards (Figure 6.3.4). The region inside the rectangle represents the whole sample space S, all the possible things that can happen. The circular regions A, B, and C represent events, particular things that can happen, to which we would like to assign probabilities.

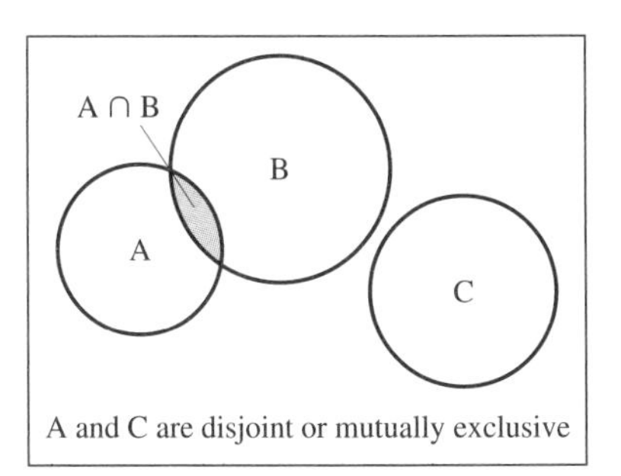

FIGURE 6.3.5
Venn diagram for the intersection of A and B

Set theory has three basic operations: intersection, union, and complementation. The **intersection** of two sets is the region contained in both (this region may be the empty or "null" set), corresponding to the English word "and." It means that both events occurred. In Figure 6.3.5, the shaded region indicates where both events A **and** B occurred. This is written $A \cap B$, and read "A intersection B." On a dartboard, if the event A is "I hit the bull's-eye" and event B is "I hit the top half of the dartboard," the intersection is "I hit the top half of the bull's-eye." Intersection is commutative, meaning that $A \cap B = B \cap A$. If the intersection of two sets is the null set (as with sets A and C in Figure 6.3.5), the sets are said to be **disjoint**, and their associated events are said to be **mutually exclusive**. Mutually exclusive events cannot occur simultaneously.

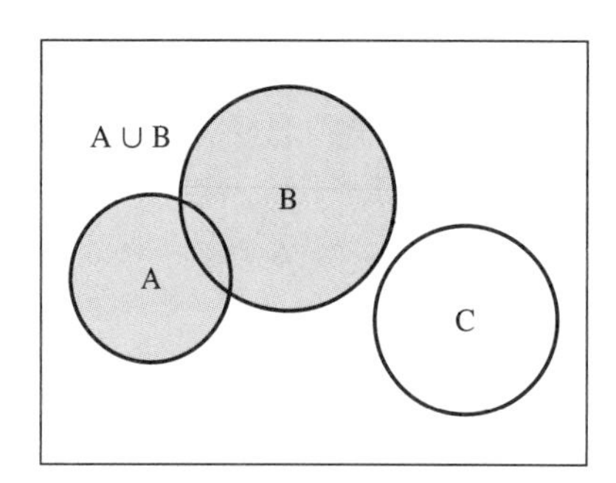

FIGURE 6.3.6
Venn diagram for the union of A and B

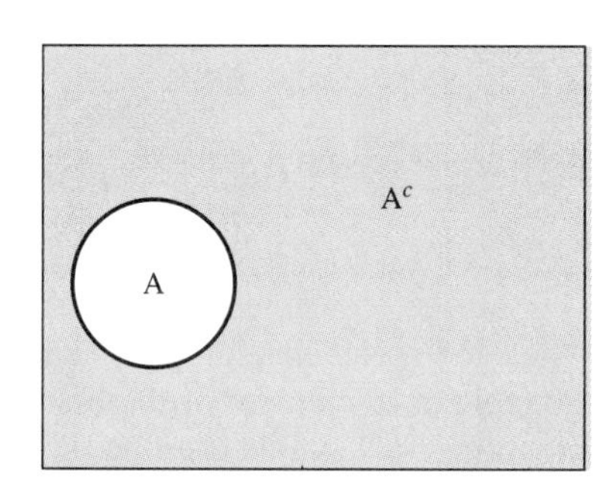

FIGURE 6.3.7
Venn diagram for the complement of A

On a dartboard, hitting the bull's-eye and scoring a 3 are mutually exclusive events. Any pair of distinct simple events are mutually exclusive because it is impossible for a dart to hit two different points.

The **union** of two sets is the region contained in either one, corresponding to the English word "or." It means that one or the other event occurred (Figure 6.3.6) and includes the possibility that both occurred. This is written A ∪ B and read "A union B." On a dartboard, if the events are "I hit the bull's-eye" and "I hit in the top half of the dartboard," the union is hitting either the bull's-eye or the top half of the board.

The final operation of set theory is taking the **complement**. The complement of a set is everything outside the set (Figure 6.3.7). This corresponds to the English word "not." If the event is "I hit the bull's-eye," the complement is "I did not hit the bull's-eye." We denote the complement of the set A by A^c (it is sometimes denoted A′).

Example 6.3.7 Intersection, Union, and Complement of Events

Consider the simple case where the sample space consists of the numbers 1, 2, 3, 4, 5, and 6, or $S = \{1, 2, 3, 4, 5, 6\}$. If $A = \{1, 3\}$ and $B = \{1, 4, 5\}$, then $A \cap B = \{1\}$, the only number in both sets. The union is $A \cup B = \{1, 3, 4, 5\}$, the numbers that appear in one or the other set. The complement of the event A is $A' = \{2, 4, 5, 6\}$, the numbers that do not appear in the event A.

Example 6.3.8 Intersection, Union, and Complement of Events Describing Per Capita Production

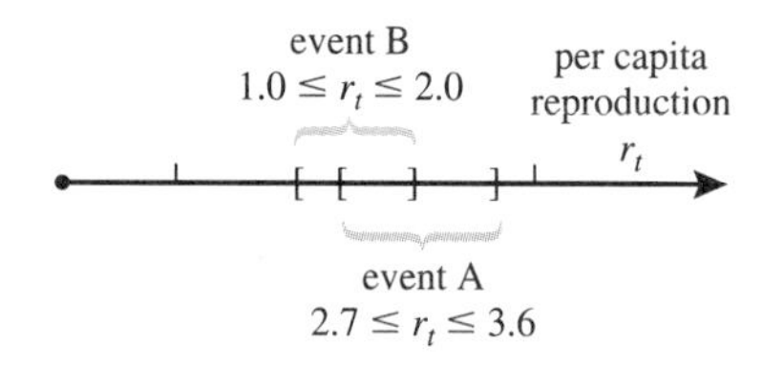

FIGURE 6.3.8
Two events in the model of stochastic production

Events for stochastic population growth include the event A given by $2.7 \le r_t \le 3.6$ (Example 6.3.4) and the event B given by $1.0 \le r_t \le 3.0$ (Figure 6.3.8). The intersection of these two events is all values of r that lie in both regions, so $A \cap B = 2.7 \le r_t \le 3.0$. The union of the two events is all values of r that lie in one or the other region, so $A \cup B = 1.0 \le r \le 3.6$. Finally, the complement of event A is all values outside the region, so $A' = r < 2.7$ or $r > 3.6$.

6.3.3 Assigning Probabilities to Events

A **probability model** assigns probabilities to all events. To make sense, these assigned probabilities must satisfy several requirements or axioms. We write the probability of event A by Pr(A), said "the probability of A."

The Four Requirements for a Probabilistic Model

1. If $\mathcal{S}$ is the sample space, $\Pr(\mathcal{S}) = 1$ (Figure 6.3.9a).
2. $0 \le \Pr(A) \le 1$ for any event A (Figure 6.3.9b).
3. If A and B are mutually exclusive events, $\Pr(A \cup B) = \Pr(A) + \Pr(B)$ (Figure 6.3.9c).
4. If A^c is the complement of A, $\Pr(A^c) = 1 - \Pr(A)$ (Figure 6.3.9d).

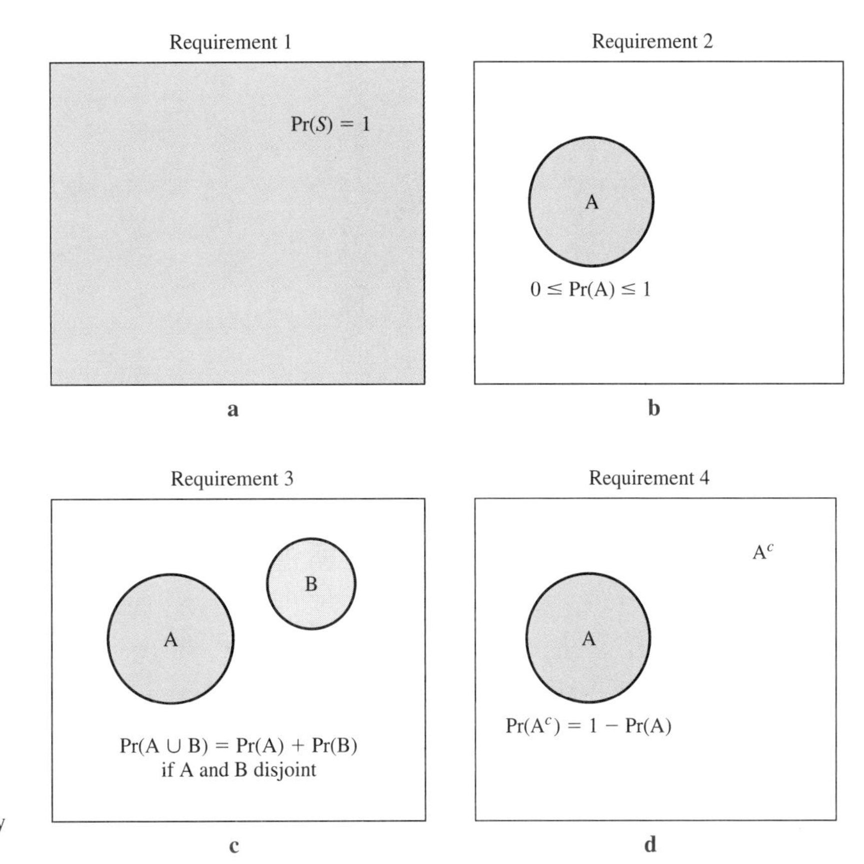

FIGURE 6.3.9

The four requirements of a probability model

What do these requirements mean? A probability of 1 means certainty. **Requirement 1** says that if you do the experiment, something must happen. Every dart that hits the dartboard must hit the dartboard somewhere. **Requirement 2** says that the probability of any event lies between 0 (never happens) and 1 (always happens). **Requirement 3** is mathematically the most important, abstracting the intuitive idea of probability. The probability of hitting one of two nonintersecting regions on a dartboard is the sum of the probabilities of hitting either one of them. If you have a 0.1 probability of hitting the bull's-eye and a 0.2 probability of scoring a 3, the probability of hitting the bull's-eye **or** getting a 3 is $0.1 + 0.2 = 0.3$. **Requirement 4** is a special case of requirement 3, because a set and its complement do not intersect and have a union equal to the whole space $\mathcal{S}$. If the probability of hitting the bull's-eye is 0.1, the probability of missing it, the complement, is $1 - 0.1 = 0.9$.

A valuable way to think about probabilities on a Venn diagram is as **areas**. The area of the whole universe (the sample space $\mathcal{S}$) is 1. The area of any subset, like the event A, must be between 0 and 1 (requirement 2). The total area of any two subsets that do not overlap is the sum of their areas (requirement 3). The area outside a subset can be found by subtracting that area from the total area of 1 (requirement 4).

How do we go about assigning probabilities to all events in ways consistent with the four requirements? There are two mathematically distinct cases: a finite number of simple events and an infinite number. In the model of immigration, there are only two simple events ("pair" and "no pair"), while the model of per capita production for one generation has an infinite number of possible outcomes (r could take on any positive value). We address the finite case in this section, saving the more difficult infinite case for Subsection 6.6.2.

With a finite number of possible outcomes (a finite number of simple events), we can construct consistent assignments of probabilities directly. Denote the simple events by $E_1, \ldots, E_n$ and associate each with a probability satisfying requirement 2,

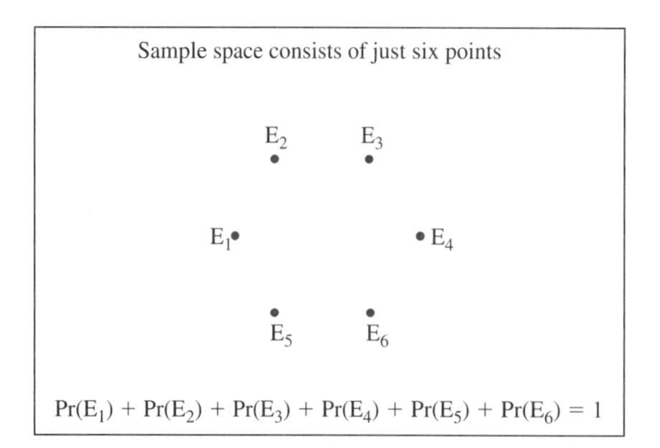

FIGURE 6.3.10
A finite set of simple events

or $0 \le \Pr(E_i) \le 1$ for each i (illustrated with $n = 6$ in Figure 6.3.10). To guarantee that the rest of the requirements are met, we need only make sure that the sum of the probabilities of all simple events is equal to 1, or that

$$\sum_{i=1}^{n} \Pr(E_i) = 1.$$

Example 6.3.9 One Way of Assigning Probabilities in the Immigration Model

Suppose that the probability that a pair of immigrants arrives is 0.5. Because the complement of this event is "no pair," the probability of the complement must be $1 - 0.5 = 0.5$.

Example 6.3.10 All Ways of Assigning Probabilities in the Immigration Model

More generally, if we pick any probability p of there being a pair of immigrants (with $0 \le p \le 1$), the probability of "no pair" must be $1 - p$. These are *all* of the possible consistent assignments of probabilities to the events in this case.

Example 6.3.11 Assigning Probabilities with Four Simple Events

Consider three molecules leaving a cell. We measure the number, N, left in the cell after 10 minutes. The simple events are $N = 0$, $N = 1$, $N = 2$, and $N = 3$ (Figure 6.3.11). One assignment of probabilities is $\Pr(N = 0) = 0.4$, $\Pr(N = 1) = 0.3$, $\Pr(N = 2) = 0.2$, and $\Pr(N = 3) = 0.1$. Because these add to 1, the assignment is mathematically consistent.

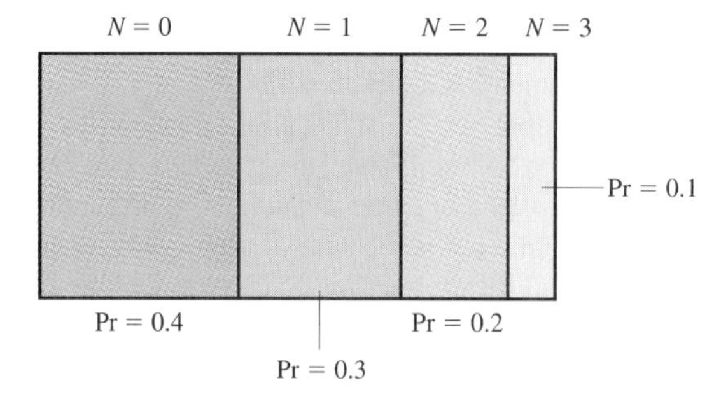

FIGURE 6.3.11
A Venn diagram for number of molecules left in a cell

Knowing the probability of each simple event is sufficient information to find the probability of **any** event by writing that event as the union of simple events (which are necessarily mutually exclusive).

Example 6.3.12 Computing the Probability of an Event

Using the probabilities found in Example 6.3.11, we can find the probability of the event "N is odd." This event is the union of two simple events $N = 1$ and $N = 3$.

We can use requirement 3 to compute

$$\Pr(N \text{ is odd}) = \Pr(N=1) + \Pr(N=3) = 0.3 + 0.1 = 0.4.$$

Example 6.3.13 Computing the Probability of Another Event

The probability of the event "$N \neq 1$" can be computed in two ways. First, this event is the complement of the event $N = 1$, and can be found with requirement 4 as

$$\Pr(N \neq 1) = 1 - \Pr(N = 1) = 1 - 0.3 = 0.7.$$

Alternatively, "$N \neq 1$" is the union of the disjoint events $N = 0$, $N = 2$, and $N = 3$, so

$$\Pr(N \neq 1) = \Pr(N = 0) + \Pr(N = 2) + \Pr(N = 3) = 0.4 + 0.2 + 0.1 = 0.7.$$

There are many consistent ways to assign probabilities to the simple events. Each assignment represents a different model of the underlying process. We can determine which is right from the biological mechanisms driving the underlying model. It is *mathematically* consistent to set the probability of a heterozygous offspring produced by selfing from a heterozygous parent to be 0.57721, but not *biologically* consistent with our understanding of genetics. The mathematical requirements define the universe of *possible* assignments. The biological mechanisms restrict our attention to *reasonable* assignments. Much of the work for the rest of the book will be to derive these biologically reasonable assignments.

With our new terminology, we can restate the connection between probability and statistics. Suppose we are given the results of an experiment and want to understand the process that produced those results. First, we use our understanding of biology to develop a stochastic model of the system. Next, we use techniques from probability theory to assign probabilities to every possible outcome. As statisticians, we then compare the measured results with these mathematical probabilities. If the results match, the data are consistent with the model and we can estimate parameters and make predictions about other experiments. If the results do not match, we must reject or revise our model.

Summary This section introduced and defined the basic language of probability theory. Probabilistic models are mathematical experiments with possible outcomes known as **simple events**. The set of all possible simple events is called the **sample space**, subsets of which are called **events**. We portrayed the sample space on a **Venn diagram**, illustrating the **intersection** of events A and B (both events happened), the **union** of events A and B (one or both events happened), and the **complement** of A (event A did not happen). Sets with null intersections are called **disjoint**, and their associated events are called **mutually exclusive**. We presented the consistency requirements for assigning probabilities to different events and showed how to make a mathematically consistent assignment when the number of simple events is finite.

6.3 Exercises

Mathematical Techniques

1–4 ▪ For the given sets A and B, find $A \cap B$, $A \cup B$, and A^c (the complement of A).

1. A and B are subsets of the set S={0, 1, 2, 3, 4}. A={0, 1, 2} and B={0, 2, 4}.
2. A and B are subsets of the set S={0, 1, 2, 3, 4, 5}. A={0, 1, 2} and B={0, 2, 4, 5}.
3. A and B are subsets of the set of all positive integers, {1, 2, 3, ...}, with A={1, 2, 6, 10} and B={2, 4, 5}.
4. A and B are subsets of the set of all positive integers, {1, 2, 3, ...}, with set A being all even numbers and set B being all multiples of 3.

5–8 ▪ For the given sets and sample spaces, show that the assignment of probabilities is mathematically consistent and use them to compute the requested probability.

5. The sample space is S={0, 1, 2, 3, 4}. Suppose that

$$\Pr(\{0\}) = 0.2$$
$$\Pr(\{1\}) = 0.3$$
$$\Pr(\{2\}) = 0.4$$
$$\Pr(\{3\}) = 0.1$$
$$\Pr(\{4\}) = 0.0.$$

Find Pr(A) and Pr(A^c) if A={0, 1, 2} and Pr(B) if B={0, 2, 4}. Is $\Pr(A \cup B) = \Pr(A) + \Pr(B)$? Why or why not?

6. The sample space is S={0, 1, 2, 3, 4}. Suppose that

$$\Pr(\{0\}) = 0.1$$
$$\Pr(\{1\}) = 0.3$$
$$\Pr(\{2\}) = 0.4$$
$$\Pr(\{3\}) = 0.1$$
$$\Pr(\{4\}) = 0.1.$$

Find Pr(A) and Pr(A^c) if A={0, 2} and Pr(B) if B={3, 4}. Is $\Pr(A \cup B) = \Pr(A) + \Pr(B)$? Why or why not?

7. The sample space is S={0, 1, 2, 3, 4}. Suppose that $\Pr(\{0\}) = 0.2$, $\Pr(\{1\}) = 0.1$, $\Pr(\{2\}) = 0.4$, and $\Pr(\{3\}) = 0.1$. Find Pr(A) and Pr(A^c) if A={4} and Pr(B) if B={3, 4}.

8. The sample space is S={0, 1, 2, 3, 4}. Suppose that $\Pr(\{0\}) = 0.3$, $\Pr(\{1\}) = 0.2$, $\Pr(\{2\}) = 0.4$, and $\Pr(\{4\}) = 0.1$. Find Pr({3}), Pr({1,2,3}), and Pr({2,4}).

9–12 ▪ Draw Venn diagrams with sets A, B, and C satisfying the following requirements.

9. A and B disjoint, B and C disjoint, A and C not disjoint.

10. A and B disjoint, B and C not disjoint, A and C not disjoint.

11. No two sets disjoint, but A∩B∩C empty.

12. No two sets disjoint, and A∩B∩C nonempty.

13–16 ▪ The following formula gives the probability of the union of any two events, whether or not they are disjoint,

$$\Pr(A \cup B) = \Pr(A) + \Pr(B) - \Pr(A \cap B).$$

As indicated in the figure, adding the area in A and B counts the area in the intersection $A \cap B$ twice. Subtracting the area of the intersection corrects for this double counting.

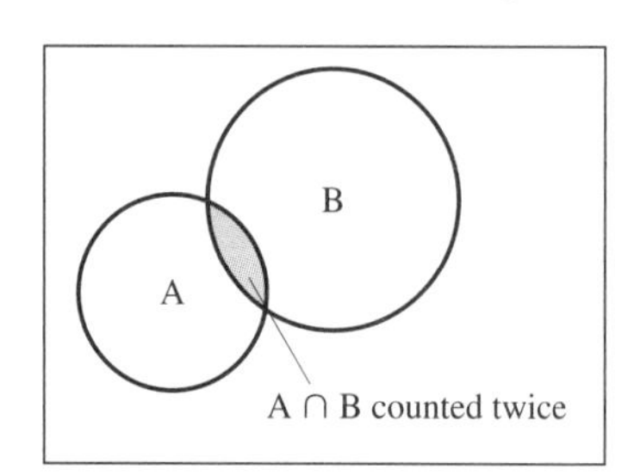

Test this formula on the following examples.

13. The sets A and B in Exercise 5.

14. The sets A and B in Exercise 6.

15. Using the probabilities in Exercise 5, check the formula on the sets C = {1, 2, 3} and D = {0, 1, 2}.

16. Using the probabilities in Exercise 6, check the formula on the sets C = {2, 3, 4} and D = {0, 1, 2}.

Applications

17–24 ▪ Give the sample spaces associated with the following experiments. Say how many simple events there are and list them if there are fewer than ten. If there are more than ten, list three simple events.

17. We cross two plants with genotype **bB** and check the genotype of one offspring.

18. We cross two plants with genotype **bB** and check the genotype of two offspring.

19. A molecule jumps in and out of a cell. We record whether the molecule is inside or outside the cell at times 1 and 5.

20. A molecule jumps in and out of a cell. We record whether the molecule is inside or outside the cell at times 2, 5, and 10.

21. Two molecules jump in and out of a cell. We record how many molecules are inside at times 1 and 5.

22. Three molecules jump in and out of a cell. We record how many molecules are inside at times 3 and 5.

23. We count how many out of 16 plants are taller than 50 cm.

24. We measure the heights of two plants.

25–28 ▪ We start 100 molecules in a cell and count the number, N, that remain after 10 minutes. Give five simple events that are included in the following events.

25. $N < 10$.

26. $N > 90$.

27. N is odd.

28. $30 \leq N \leq 32$ or $68 \leq N \leq 70$.

29–32 ▪ We start 100 molecules in a cell and count the number, N, that remain after 10 minutes. Find the union and intersection of the following events.

29. Event A is $N < 10$, and event B is $N > 5$.

30. Event A is $N > 10$, and event B is $N < 5$.

31. Event A is $N > 10$, and event B is $N > 5$.

32. Event A is $20 > N > 10$, and event B is $15 > N > 5$.

33–36 ▪ We follow four individually labeled molecules and record the minute t_i when molecule i leaves the cell. For example, if $t_1 = 1$, $t_2 = 3$, $t_3 = 6$, and $t_4 = 2$, the first molecule left during minute 1, the second left during minute 3, the third left during minute 6, and the fourth left during minute 2. Give three simple events that are included in the following events.

33. All molecules left before minute 5.

34. Molecules 1, 2, and 4 left before minute 5, and molecule 3 left after minute 7.

35. All odd-numbered molecules left at odd times.

36. All odd-numbered molecules left at odd times, and all even-numbered molecules left at even times.

37–38 ▪ Give two mathematically consistent ways of assigning probabilities to the results of the following experiments. Try to make one of your assignments biologically reasonable.

37. The situation in Exercise 17.

38. The situation in Exercise 18.

39–42 ▪ Give two assignments, different from those in the text, of probabilities when counting the number of molecules inside a cell starting from an initial number of 3. Compute $\Pr(N \text{ is odd})$ and $\Pr(N \neq 1)$ in each case.

39. Create an assignment where $\Pr(N = 1)$ is larger than the probability of any other simple event (but none is zero).

40. Create an assignment where $\Pr(N = 1)$ is equal to the probability of each other simple event.

41. Create an assignment where $\Pr(N = 1)$ is smaller than the probability of any other simple event, but not equal to 0.

42. Create an assignment where the probabilities get larger as the number of molecules gets larger.

6.4 Conditional Probability

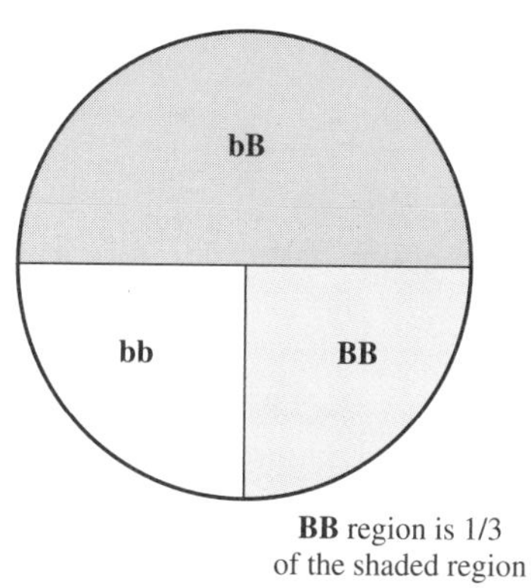

FIGURE 6.4.1
A Venn diagram for genetics

Probability theory can be thought of as the study of scientific problems with limited information. For example, we sometimes wish to make deductions about a process that we cannot measure directly. **Conditional probability** is the tool needed to compute the probability of one event conditional on knowing that another occurred. We will use conditional probability to break up the sample space with the **law of total probability**, and to reason about complicated systems with **Bayes' theorem**.

6.4.1 Conditional Probability

Suppose plant height is governed by a single dominant allele **B**. Two plants of genotype **Bb** (created by crossing tall and short inbred plants) are crossed (Figure 6.4.1). We are given a tall offspring from this cross. What is the probability that its genotype is **BB**? It could be either **Bb** or **BB**. It might seem tempting to guess that these genotypes are equally probable, meaning that the probability of **BB** is 1/2. As is so often the case, temptation must be resisted. We must use **conditional probability** to figure out the probability correctly.

Example 6.4.1 Computing a Conditional Probability by Counting

We can compute these probabilities by thinking of a large number of offspring, perhaps 100. If everything behaved exactly according to theory, there would be 25 of type **bb**, 50 of type **Bb**, and 25 of type **BB**. Knowing that our plant is tall eliminates the 25 short **bb** plants. Our plant is one of the 75 tall offspring, 1/3 of which are of type **BB**. This is the probability of genotype **BB conditional** on the plant being tall. This probability exceeds the probability 1/4 of genotype **BB** in the absence of information about height. ▲

To express this more compactly, we need some new notation. Let G represent the event "genotype **BB**" and T represent the event "tall." We know that $\Pr(G) = 1/4$ and $\Pr(T) = 3/4$. To denote the probability of G conditional on T, we write

$$\Pr(G \mid T) = \text{ the probability of G conditional on T.} \qquad \textbf{(6.4.1)}$$

The expression $\Pr(G \mid T)$ means the probability of G given that T is true.

The mathematical definition of conditional probability looks rather different.

Definition 6.5 The probability of event A conditional on event B is

$$\Pr(A \mid B) = \frac{\Pr(A \cap B)}{\Pr(B)}$$

if $\Pr(B) \neq 0$.

Example 6.4.2 Computing a Conditional Probability from the Definition

In Example 6.4.1, $\Pr(G \cap T) = 1/4$ (the probability that an offspring has genotype **BB** and is tall) and $\Pr(T) = 3/4$, so

$$\Pr(G \mid T) = \frac{\Pr(G \cap T)}{\Pr(T)} = \frac{1/4}{3/4} = 1/3,$$

as we found before.

Why does this rule work? The probability of G conditional on T can be thought of as the probability of G after restricting our attention to cases where T is true. Using the area interpretation of probabilities, the probability of G when restricted to the smaller universe T is the fraction of area of T [Pr(T)] that is also included in G [$\Pr(G \cap T)$].

Example 6.4.3 Computing a Different Conditional Probability

Using the same diagram (Figure 6.4.1), we can find the probability that a plant is homozygous (event H) conditional on it being tall (event T).

$$\Pr(H \mid T) = \frac{\Pr(H \cap T)}{\Pr(T)}.$$

The region $H \cap T$ is precisely the event G, that a plant has genotype **BB**. Therefore,

$$\Pr(H \mid T) = \frac{\Pr(G)}{\Pr(T)} = \frac{1/4}{3/4} = 1/3.$$

The conditional probability is the same because the only way that a homozygous plant could be tall is by having genotype **BB**.

Example 6.4.4 Computing the Reverse Conditional Probability

What is the probability that a plant is tall (event T) conditional on it being homozygous (event H)? From the definition,

$$\Pr(T \mid H) = \frac{\Pr(T \cap H)}{\Pr(H)}.$$

The region $T \cap H$ is the event G, that a plant has genotype **BB**. Furthermore, half the offspring are homozygous, so $\Pr(H) = 1/2$. Therefore,

$$\Pr(T \mid H) = \frac{\Pr(G)}{\Pr(H)} = \frac{1/4}{1/2} = 1/2.$$

This conditional probability is different from $\Pr(H \mid T)$, as is generally the case.

Example 6.4.5 Working Out Conditional Probabilities with Numbers

We can also find the conditional probabilities in Examples 6.4.3 and 6.4.4 by thinking of a large number of plants. Out of 100 plants, we expect 25 to be tall homozygotes with genotype **BB**, 50 to be tall heterozygotes with genotype **Bb**, and 25 to be short homozygotes with genotype **bb**. Thus, 25 of the 75 tall plants are homozygotes, giving $\Pr(H \mid T) = 1/3$ (as in Example 6.4.3). Similarly, 25 of the 50 homozygous plants are tall, giving $\Pr(T \mid H) = 1/2$ (as in Example 6.4.4).

The definition of conditional probability (Definition 6.5) can be rewritten in multiplicative form as

$$\Pr(A \cap B) = \Pr(A \mid B)\Pr(B). \quad \textbf{(6.4.2)}$$

The probability of the intersection is the probability that both A and B occurred. This requires that B occurred [probability Pr(B)] and that A occurred **conditional on B** [probability Pr(A | B)]. The probability that both A and B occurred is the product.

Example 6.4.6 Checking the Multiplicative Form

The probability that a plant is both tall and homozygous can be expressed as two different products. First, the plant could be homozygous and also tall conditional on being homozygous, or

$$\Pr(T \cap H) = \Pr(T \mid H)\Pr(H) = 1/2 \cdot 1/2 = 1/4.$$

Alternatively, the plant could be tall and also homozygous conditional on being tall, or

$$\Pr(T \cap H) = \Pr(H \mid T)\Pr(T) = 1/3 \cdot 3/4 = 1/4.$$

Example 6.4.7 Using Conditional Probability to Study Color-Blindness

The probability of color-blindness depends on a person's sex. A person can be male (event M) or female (event F), and we suppose that Pr(M) = Pr(F) = 0.5. Let the event C indicate that a person is color-blind, and let the event N indicate that he or she is not. Suppose that

$$\Pr(M \cap C) = 0.025, \Pr(F \cap C) = 0.005,$$

meaning that 2.5% of people are color-blind males and 0.5% of people are color-blind females. We can use this information to find the probability that a male is color-blind, and that a color-blind person is male. To find the first, we use the definition of conditional probability

$$\Pr(C \mid M) = \frac{\Pr(C \cap M)}{\Pr(M)} = \frac{0.025}{0.5} = 0.05.$$

To find the second, we again use the definition, finding

$$\Pr(M \mid C) = \frac{\Pr(M \cap C)}{\Pr(C)}.$$

We have not been given the probability of color-blindness, so we must add together the component probabilities to find it. A color-blind person is either male or female, so

$$\Pr(C) = \Pr(M \cap C) + \Pr(F \cap C) = 0.025 + 0.005 = 0.03.$$

Therefore,

$$\Pr(M \mid C) = \frac{0.025}{0.03} = 0.833.$$

About 83% of color-blind people are male.

Example 6.4.8 Checking Color-Blindness Calculation with Large Numbers

We can check the probabilities in Example 6.4.7 by counting how many people out of 10,000 are of each type. There would be 5000 males and 5000 females. A fraction 0.025 of the people, or 250, are color-blind males. A fraction 0.005 of the people, or 50, are color-blind females. Thus, 250 out of 5000 males are color-blind, meaning that the probability a male is color-blind is $\frac{250}{5000} = 0.05$. Conversely, a total of 300 people are color-blind, 250 of whom are male. The probability that a color-blind person is male is then $\frac{250}{300} = 0.833$.

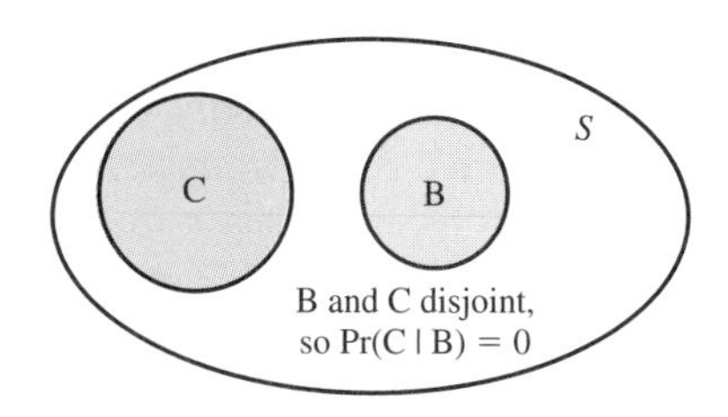

FIGURE 6.4.2
Conditional probability: disjoint events

Two special cases can help clarify the meaning of conditional probability. Consider first two disjoint events B and C (Figure 6.4.2). B∩C is empty, so $\Pr(B \cap C) = 0$ and

$$\Pr(B \mid C) = \frac{\Pr(B \cap C)}{\Pr(C)} = 0.$$

This means that once we know that C has occurred, we can be sure that B did **not** occur. The event "the dart scored less than 10" is disjoint from the event "the dart hit the bull's-eye." Once you know that your shot hit the bull's-eye, you can be sure that your dart did not score less than 10. Knowledge about one event provides complete information about the other.

Example 6.4.9 Conditional Probabilities of Disjoint Events

Consider again the plants in Subsection 6.4.1, where the allele **B** is dominant. Plants with genotypes **BB** and **Bb** are tall, while those with genotype **bb** are short. Let T represent the event tall, and bb the event that a plant has genotype **bb**. These two events are disjoint because no plant with genotype **bb** is tall, or $bb \cap T$ is empty. Therefore $\Pr(bb \mid T) = 0$. If a plant is tall, we can be certain that it does not have the genotype **bb**. ▲

A is a subset of C,
so Pr(C | A) = 1
and Pr(A | C) ≥ Pr(A)

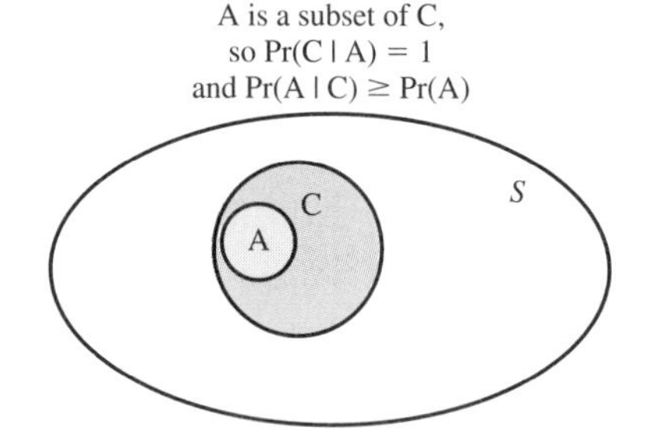

FIGURE 6.4.3
Conditional probability: subsets

A second special case involves two events A and C where A is a subset of C (Figure 6.4.3). Therefore, A∩C = A and $\Pr(A \cap C) = \Pr(A)$. From the definition of conditional probability,

$$\Pr(C \mid A) = \frac{\Pr(C \cap A)}{\Pr(A)} = \frac{\Pr(A)}{\Pr(A)} = 1$$

[as long as $\Pr(A) \neq 0$]. Knowing that A occurred guarantees that C occurred.

For example, consider the events C "the dart scored more than 15 points with that shot" and A "the dart hit the bull's-eye." The event A is a subset of the event C. Knowing that the shot was a bull's-eye guarantees that the shot scored more than 15. Knowing that a shot scored more than 15 points makes it more probable, although not certain, that the shot was a bull's-eye.

Example 6.4.10 Conditional Probabilities of Subsets

In the case described in Example 6.4.9, the event BB that a plant has genotype **BB** is a subset of the event T (that a plant is tall). Thus, $\Pr(T \mid BB) = 1$. ▲

Example 6.4.11 Conditional Probabilities in a Disease Test

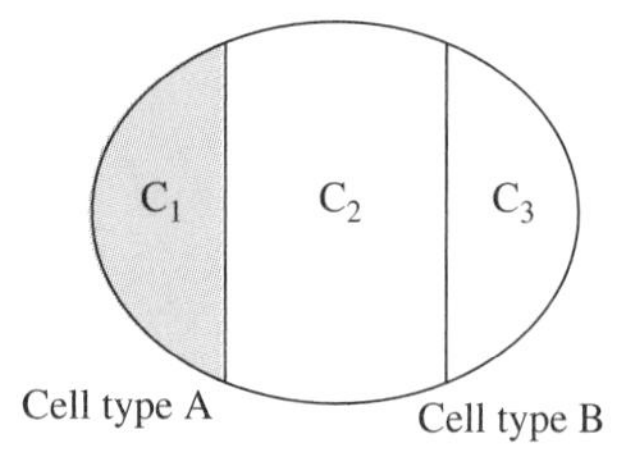

FIGURE 6.4.4
Conditional probability: disease test

A type of cancer can occur as three different conditions, but a screen can distinguish only two of them by examining cell types. In particular, patients with cell type A always have condition 1 (event C_1), while patients with cell type B can have either condition 2 (event C_2) or condition 3 (event C_3) (Figure 6.4.4). The events C_1 and B are disjoint, thus $\Pr(C_1 \mid B) = 0$. Furthermore, the event C_2 is a subset of the event B, meaning that $\Pr(B \mid C_2) = 1$. However, we cannot be sure that a person with cell type B has condition 2. We do know, however, that $\Pr(C_2 \mid B) > \Pr(C_2)$ because C_2 takes up a larger fraction of the area in event B than it does of the entire sample space. ▲

Example 6.4.12 Conditional Probabilities on Tropical Islands

To clarify the area interpretation of conditional probability, consider the scenario illustrated in Figure 6.4.5. A shipwrecked traveler finds herself stranded on a pair of islands. After constructing a rude raft, she sets off to catch fish, but is engulfed in fog. Fortunately, she finds her way back to land in a wooded region. Because the wooded area W is a subset of island B, $\Pr(B \mid W) = 1$. Furthermore, because the wooded area W is disjoint from island A, $\Pr(A \mid W) = 0$. She thus knows precisely which island she is on.

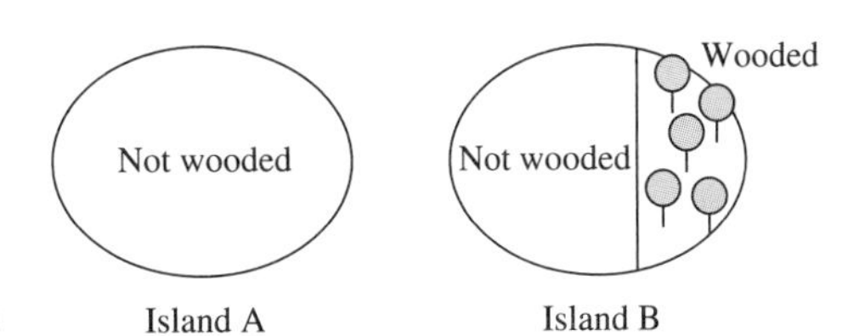

FIGURE 6.4.5
Conditional probability: tropical islands

Alternatively, suppose that she can use the ocean currents to land on island B. The probability she lands in sheltered wooded area is Pr(W | B), which is not equal to 1 because only part of island B is wooded. However, Pr(W | B) > Pr(W) because the fraction of wooded area on island B is greater than the fraction of wooded area on both islands combined.

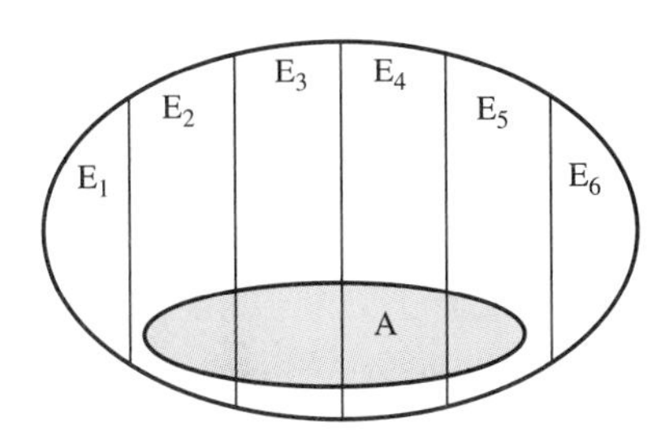

FIGURE 6.4.6
$E_1, \ldots, E_6$ form a mutually exclusive and collectively exhaustive set of events

6.4.2 The Law of Total Probability

Conditional probability allows us to break events into their component parts, which we can then reassemble with **the law of total probability**. Let $E_1, E_2, \ldots, E_n$ form a set of mutually exclusive (no pair intersect) and collectively exhaustive (the union is the full set) events (Figure 6.4.6). The sets E_i break the whole sample space into disjoint pieces, so the sum of their probabilities must be 1, or

$$\sum_{i=1}^{n} \Pr(E_i) = 1. \tag{6.4.3}$$

Example 6.4.13 Mutually Exclusive and Collectively Exhaustive Events: Genetics

In the model of genetics, the events T that the plant is tall and S that the plant is short are mutually exclusive and collectively exhaustive. No plant is both tall and short, but every plant must be one or the other.

Example 6.4.14 Events That Are Not Mutually Exclusive: Genetics

In the model of genetics, the events that a plant has at least one **B** allele and that a plant has at least one **b** allele are not mutually exclusive, because heterozygous plants have one allele of each type. These events are, however, collectively exhaustive.

Example 6.4.15 Mutually Exclusive and Collectively Exhaustive Events: Population Growth

The three events $A = \{0.0 \le r_t < 2.0\}$, $B = \{2.0 \le r_t \le 3.6\}$, $C = \{3.6 < r_t\}$ break possible growth rates r_t into three mutually exclusive and collectively exhaustive sets. (Figure 6.4.7) Exactly one of these events must occur.

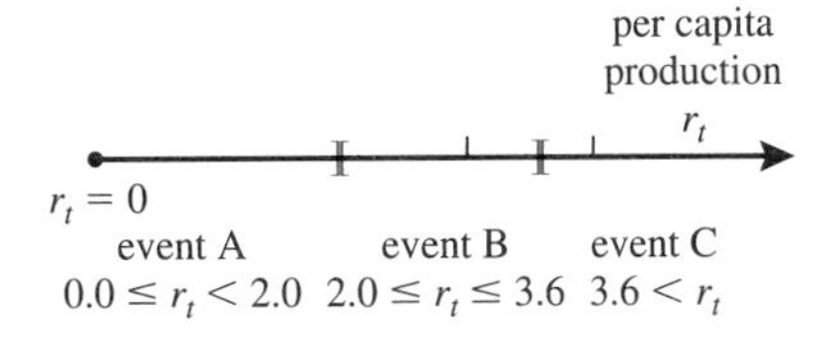

FIGURE 6.4.7
Three mutually exclusive and collectively exhaustive events

We can use mutually exclusive and collectively exhaustive events to decompose other events and their probabilities.

Theorem 6.1 **The Law of Total Probability**

Suppose that $E_1, E_2, \ldots, E_n$ form a set of mutually exclusive and collectively exhaustive events. Then for any event A,

$$\Pr(A) = \sum_{i=1}^{n} \Pr(A|E_i)\Pr(E_i).$$

Proof: This law follows from the requirements on probabilities. The set A can be written as a union of the disjoint sets $A \cap E_i$, so that, by requirement 3,

$$\Pr(A) = \sum_{i=1}^{n} \Pr(A \cap E_i).$$

But $\Pr(A \cap E_i) = \Pr(A|E_i)\Pr(E_i)$, implying the law.

Many useful instances require breaking the sample space into only two disjoint sets E and its complement E^c.

Example 6.4.16 Law of Total Probability Applied to Color-Blindness

Suppose 1% of females and 5% of males are color-blind. What is the probability that a person chosen at random is color-blind? Let C represent the event "color-blind," F the event "female," and M the event "male." If the number of males and females is equal, $\Pr(F) = \Pr(M) = 0.5$. By the law of total probability,

$$\Pr(C) = \Pr(C \mid F)\Pr(F) + \Pr(C|F^c)\Pr(F^c)$$
$$= (0.01)(0.5) + (0.05)(0.5) = 0.03.$$

Example 6.4.17 Law of Total Probability Applied to Cancer

Individuals are diagnosed with a particular type of cancer that can take on three different forms, D_1, D_2, and D_3. It is known that 20% of people will eventually be diagnosed with disease 1, 30% with disease 2, and 50% with disease 3. The probability of requiring chemotherapy differs among the three forms of disease: 80% with disease 1, 30% with disease 2, and 10% with disease 3. Based solely on the preliminary test, what is the probability of requiring chemotherapy, the event C?

We can use the law of total probability to find Pr(C) as

$$\Pr(C) = \Pr(C \mid D_1)\Pr D_1 + \Pr(C \mid D_2)\Pr D_2 + \Pr(C \mid D_3)\Pr D_3$$
$$= 0.8 \cdot 0.2 + 0.3 \cdot 0.3 + 0.1 \cdot 0.5 = 0.3.$$

Based on information at the time of the initial diagnosis, 30% of individuals will eventually require chemotherapy.

6.4.3 Bayes' Theorem and the Rare Disease Example

If we know Pr(A), Pr(B), and Pr(A | B), we can find the probability of B conditional on A with **Bayes' theorem**, which forms the basis for **Bayesian statistics** and **Bayesian decision theory**.

Theorem 6.2 **Bayes' Theorem**

For any events A and B where $\Pr(A) \neq 0$,

$$\Pr(B \mid A) = \frac{\Pr(A \mid B)\Pr(B)}{\Pr(A)}.$$

Proof: The theorem follows from the multiplicative formula (Definition 6.5),

$$\Pr(A \cap B) = \Pr(A \mid B)\Pr(B).$$

Symmetrically,

$$\Pr(A \cap B) = \Pr(B \mid A)\Pr(A).$$

Equating these two expressions for $\Pr(A \cap B)$ and solving for $\Pr(B \mid A)$ gives Bayes' theorem.

Example 6.4.18 Applying Bayes' Theorem to Color-Blindness

Recall the situation in Example 6.4.7, where a person is male with probability 0.5, males are color-blind with probability 0.05, and all people are color-blind with probability 0.03. We can use Bayes' theorem to find the probability that a color-blind person is male,

$$\Pr(\mathrm{M} \mid \mathrm{C}) = \frac{\Pr(\mathrm{C} \mid \mathrm{M})\Pr(\mathrm{M})}{\Pr(\mathrm{C})}$$

$$= \frac{0.05 \cdot 0.5}{0.03} = 0.833.$$

A particularly important application of Bayes' theorem arises when testing for a rare disease. Suppose a disease infects only 1% of people. A diagnostic test always picks up the disease, but generates 5% false positives. A patient tests positive. What is the probability that she has the disease? This is a conditional probability problem; how likely is it that she has the disease **conditional** on her testing positive? We solve this problem first by working out the component probabilities explicitly and then with Bayes' theorem.

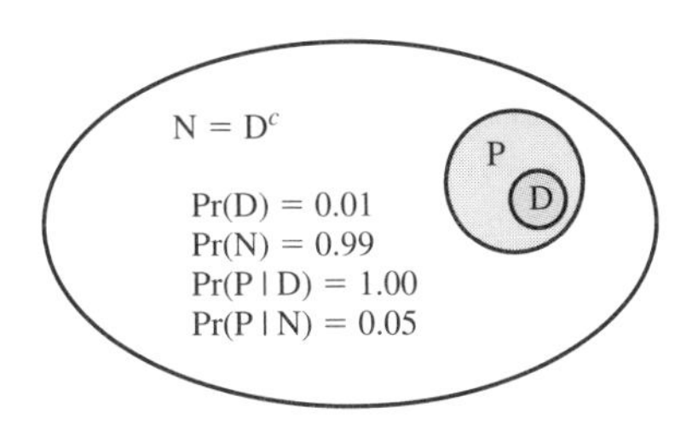

FIGURE 6.4.8

The probabilities in the rare disease model

Before applying the formal rules of conditional probability, we can do a more intuitive calculation. Out of 10,000 people, 1% (100) would have the disease. All 100 of these would test positive. Of the 9900 who did not have the disease, 5% (495) would test positive. There would then be a total of 595 positive tests, only 100 of whom actually have the disease, corresponding to a fraction of $\frac{100}{595} = 0.168$.

This calculation can be formalized with the laws of conditional probability (Figure 6.4.8). Let D denote the event of an individual having the disease, N the event of not having the disease, and P the event of a positive result on the test. We can translate our assumptions as

$$\Pr(\mathrm{D}) = 0.01$$

$$\Pr(\mathrm{N}) = 0.99$$

$$\Pr(\mathrm{P} \mid \mathrm{D}) = 1.00$$

$$\Pr(\mathrm{P} \mid \mathrm{N}) = 0.05.$$

We are interested in computing $\Pr(\mathrm{D} \mid \mathrm{P})$.

Bayes' theorem says that

$$\Pr(\mathrm{D} \mid \mathrm{P}) = \frac{\Pr(\mathrm{P} \mid \mathrm{D})\Pr(\mathrm{D})}{\Pr(\mathrm{P})}.$$

To apply this, we must first compute $\Pr(\mathrm{P})$ with the law of total probability.

$$\Pr(\mathrm{P}) = \Pr(\mathrm{P} \mid \mathrm{D})\Pr(\mathrm{D}) + \Pr(\mathrm{P} \mid \mathrm{N})\Pr(\mathrm{N})$$

$$= (1.0)(0.01) + (0.05)(0.99) = 0.0595.$$

We now have all the needed information, finding

$$\Pr(\mathrm{D} \mid \mathrm{P}) = \frac{\Pr(\mathrm{P} \mid \mathrm{D})\Pr(\mathrm{D})}{\Pr(\mathrm{P})}$$

$$= \frac{1.0 \cdot 0.01}{0.0595} = 0.168.$$

Even though this diagnostic test is fairly good, an individual with a positive result is still unlikely to have the disease. For this reason, many screenings for disease do not test the entire population but focus on risk groups where the disease is more common. Within a risk group, the test can be more useful. Suppose instead that $\Pr(\mathrm{D}) = 0.4$ in some group. Then

$$\begin{aligned}\Pr(P) &= \Pr(P \mid D)\Pr(D) + \Pr(P \mid N)\Pr(N)\\ &= (1.0)(0.4) + (0.05)(0.6) = 0.43.\end{aligned}$$

Furthermore,

$$\begin{aligned}\Pr(D \mid P) &= \frac{\Pr(P \mid D)\Pr(D)}{\Pr(P)}\\ &= \frac{1.0 \cdot 0.4}{0.43} = 0.93.\end{aligned}$$

Within this risk group, a positive test is a strong indicator of disease.

Summary Understanding probabilistic systems requires understanding relations between different events. The key tool is **conditional probability**, the probability that one event occurs **conditional** on some second event having occurred, found as the probability that both occurred divided by the probability that the second event occurred. By breaking the sample space into components, probabilities can be computed using the **law of total probability**. The order of the conditional probabilities can be reversed with **Bayes' theorem**. We applied these rules to find the probability of a patient having a rare disease conditional on a positive test.

6.4 Exercises

Mathematical Techniques

1–4 ▪ For the given sample spaces, find a set of mutually exclusive and collective exhaustive events with the given number of elements.

1. $S = \{0, 1, 2, 3, 4\}$. Find a set of two mutually exclusive and collective exhaustive events.
2. $S = \{0, 1, 2, 3, 4\}$. Find a set of three mutually exclusive and collective exhaustive events.
3. $S = \{1, 2, 3, 4, \ldots\}$, the set of all positive integers. Find a set of two mutually exclusive and collective exhaustive events.
4. $S = \{1, 2, 3, 4, \ldots\}$, the set of all positive integers. Find a set of three mutually exclusive and collective exhaustive events.

5–8 ▪ In each of the following cases, find $\Pr(A \cap B)$, $\Pr(A \mid B)$, and $\Pr(B \mid A)$.

5. As in Section 6.3, Exercise 5, the sample space is $S=\{0, 1, 2, 3, 4\}$, $\Pr(\{0\}) = 0.2$, $\Pr(\{1\}) = 0.3$, $\Pr(\{2\}) = 0.4$, $\Pr(\{3\}) = 0.1$, $\Pr(\{4\}) = 0.0$, $A=\{0, 1, 2\}$, and $B=\{0, 2, 4\}$.
6. As in Section 6.3, Exercise 5, the sample space is $S=\{0, 1, 2, 3, 4\}$, $\Pr(\{0\}) = 0.2$, $\Pr(\{1\}) = 0.3$, $\Pr(\{2\}) = 0.4$, $\Pr(\{3\}) = 0.1$, and $\Pr(\{4\}) = 0.0$. Suppose now that $A=\{1, 2, 3\}$ and $B=\{2, 3, 4\}$.
7. As in Section 6.3, Exercise 6, the sample space is $S=\{0, 1, 2, 3, 4\}$, $\Pr(\{0\}) = 0.1$, $\Pr(\{1\}) = 0.3$, $\Pr(\{2\}) = 0.4$, $\Pr(\{3\}) = 0.1$, $\Pr(\{4\}) = 0.1$, $A=\{0, 2\}$, and $B=\{3, 4\}$.
8. As in Section 6.3, Exercise 6, the sample space is $S=\{0, 1, 2, 3, 4\}$, $\Pr(\{0\}) = 0.1$, $\Pr(\{1\}) = 0.3$, $\Pr(\{2\}) = 0.4$, $\Pr(\{3\}) = 0.1$, and $\Pr(\{4\}) = 0.1$. Suppose now that $A=\{1, 2\}$ and $B=\{1, 2, 3, 4\}$.

9–12 ▪ Somebody invents a three-sided die that gives scores of 1, 2, or 3, each with probability 1/3. Two such die are rolled. Use the law of total probability to find the following probabilities, and then find them directly by counting.

9. The probability that the total on the two die is 4 or more. (To use the law of total probability, find the probability that the score is 4 or more if the first die gives a 1, if the first die gives a 2, and if the first die gives a 3.)
10. The probability that the total on the two die is 5 or more.
11. The probability that the total on the two die is odd.
12. The probability that the second roll was larger than the first.

13–16 ▪ Consider again the three-sided die that gives a score of 1 with probability 1/3, a score of 2 with probability 1/3, and a score of 3 with probability 1/3. Two such die are rolled. Use Bayes' theorem to find the following conditional probabilities. Check your result by direct counting.

13. Find the probability that the first roll is a 3 if the total of the two rolls is greater than or equal to 4 (based on Exercise 9).
14. Find the probability that the first roll is a 3 if the total of the two rolls is greater than or equal to 5 (based on Exercise 10).
15. Find the probability that the first roll is a 3 if the total of the two rolls is odd (based on Exercise 11).
16. Find the probability that the first roll is a 1 if the second roll is greater than the first (based on Exercise 12).

17–20 ▪ Four balls are placed in a jar, two red, one blue, and one yellow. Two are removed at random.

17. You are told that the first ball removed was red. What is the probability that the second is red?

18. You are told that at least one of the two removed is red. What is the probability that both are red?

19. As in Exercise 17, but the first ball is replaced (but remembered) before the second ball is drawn. What is the probability that the second is red?

20. As in Exercise 18, but the first ball is replaced (but remembered) before the second ball is drawn. What is the probability that both are red?

Applications

21–26 ▪ Give a set of three mutually exclusive and collectively exhaustive sets for each of the following sample spaces.

21. The situation in Exercise 17.

22. The situation in Exercise 18.

23. The situation in Exercise 19.

24. The situation in Exercise 20.

25. The situation in Exercise 23.

26. The situation in Exercise 24.

27–28 ▪ An ecologist is looking for the effects of eagle predation on the behavior of jackrabbits. In each of the following cases,

a. Draw a Venn diagram to illustrate the situation.

b. Find the probability that she saw a jackrabbit conditional on her seeing an eagle. How might you interpret this result? Compare with the overall probability of seeing a jackrabbit.

c. Find the probability that she saw an eagle conditional on her seeing a jackrabbit. How might you interpret this result?

27. She sees an eagle with probability 0.2 during an hour of observation, a jackrabbit with probability 0.5, and both with probability 0.05.

28. She sees an eagle with probability 0.2 during an hour of observation, a jackrabbit with probability 0.5, and both with probability 0.15.

29–30 ▪ A lab is attempting to stain many cells. Young cells stain properly 90% of the time and old cells stain properly 70% of the time.

29. If 30% of the cells are young, what is the probability that a cell stains properly?

30. If 70% of the cells are young, what is the probability that a cell stains properly?

31–32 ▪ Further study of the cell-staining problem (Exercises 29 and 30) reveals that new cells stain properly with probability 0.95, 1-day-old cells stain properly with probability 0.9, 2-day-old cells stain properly with probability 0.8, and 3-day-old cells stain properly with probability 0.5. Suppose

$$\Pr(\text{cell is 0 day old}) = 0.4$$
$$\Pr(\text{cell is 1 day old}) = 0.3$$
$$\Pr(\text{cell is 2 days old}) = 0.2$$
$$\Pr(\text{cell is 3 days old}) = 0.1.$$

31. Find the probability that a cell stains properly.

32. The lab finds a way to eliminate the oldest cells (more than 3 days old) from its stock. What is the probability of proper staining? Write this as a conditional probability.

33–36 ▪ Use Bayes' theorem to compute the following. Say whether the stain is a good indicator of the age of the cell.

33. For the cells in Exercise 29, what is the probability that a cell that stains properly is young? How does this compare with the unconditional probability of 0.3?

34. For the cells in Exercise 30, what is the probability that a cell that stains properly is young? How does this compare with the unconditional probability of 0.7?

35. For the cells in Exercise 31, what is the probability that a cell that stains properly is less than 1 day old? How does this compare with the unconditional probability of 0.4?

36. For the cells in Exercise 32, what is the probability that a cell that stains properly is less than 1 day old? How does this compare with the unconditional probability?

37–38 ▪ Consider a disease with an imperfect test. Let D denote the event of an individual having the disease, N the event of not having the disease, and P the event of a positive result on the test. In each of the following cases, find $\Pr(D \mid P)$.

37. $\Pr(D) = 0.2$, $\Pr(N) = 0.8$, $\Pr(P \mid D) = 1.00$, and $\Pr(P \mid N) = 0.05$. Compare with the results in the text, when the disease was much less common.

38. $\Pr(D) = 0.8$, $\Pr(P \mid D) = 1.00$, and $\Pr(P \mid N) = 0.1$. Compare with the results in the text and Exercise 37.

39–40 ▪ In the following cases, the test does not catch every sick person. Let D denote the event of an individual having the disease, N the event of not having the disease, and P the event of a positive result on the test. Find $\Pr(D \mid P)$ and $\Pr(D|P^c)$ (the probability that a person who did not test positive has the disease).

39. $\Pr(D) = 0.2$, $\Pr(N) = 0.8$, $\Pr(P \mid D) = 0.95$, and $\Pr(P \mid N) = 0.05$. Compare your results with Exercise 37.

40. $\Pr(D) = 0.8$, $\Pr(P \mid D) = 0.95$, and $\Pr(P \mid N) = 0.1$. Compare your results with Exercise 38.

41–44 ▪ Consider a dominant gene where plants with genotype **BB** or **Bb** are tall, while plants with genotype **bb** are short. Find the probability that a tall plant has genotype **Bb** when it results from the following crosses.

41. A plant with genotype **Bb** is crossed with the offspring from a cross between a **BB** plant and a **Bb** plant.

42. A plant with genotype **Bb** is crossed with the offspring from a cross between a **Bb** plant and a **Bb** plant.

43. Two offspring from the cross between a **BB** plant and a **Bb** plant are crossed with each other.

44. Two tall offspring from the cross between a **Bb** plant and a **Bb** plant are crossed with each other.

45–50 ▪ A popular probability problem refers to a once popular game show called "Let's Make a Deal." In this game, the host (named Monty Hall) hands out large prizes to contestants for no reason at all. In one situation, Monty would show the contestant three doors, named door 1, door 2, and door 3. One would hide a new car, one $500 worth of false eyelashes, and the other a goat (deemed worthless by the purveyors of the show). The contestant picks door 1. But instead of showing her the prize, Monty opens door 3 to reveal the goat.

45. If the car is really behind door 1, what happens if she switches?

46. If the car is really behind door 2, what happens if she switches?

47. Should the contestant switch her guess to door 2?

48. If she uses the right strategy, what is her probability of getting the new car?

49. It is later revealed that Monty does not always show what is behind one of the other doors, but does so only when the contestant guessed right in the first place (the so-called "Machiavellian Monty"). How often would a contestant who used the strategy in Exercise 47 get the new car?

50. What is the right strategy to use for dealing with the Machiavellian Monty? How well would the contestant do?

Computer Exercises

51. Use the command q_p (Section 6.1, Exercise 37) that returns 1 with probability p and 0 with probability $1 - p$ to simulate the rare disease example.

a. Simulate 100 people who have the disease with probability 0.05 and count up the number with the disease.

b. For each of the remaining people, assume that the probability of a false positive is 0.1. Simulate them and count up the number of positives.

c. What fraction of positive tests identify people who are sick? How does this compare with the mathematical expectation?

d. Try the same experiment where the probability that each person has the disease is 0.4.

52. Suppose cells fall into three categories: those that are dead, those that are alive but do not stain properly, and those that are alive and do stain properly. Let D_t denotes the number of cells that are dead, N_t the number that are alive but do not stain properly, and S_t those that are alive and do stain properly. Each day, there are two possible transitions.

- Cells that are alive die with probability 0.9.
- Cells that stain properly cease to stain properly with probability 0.8.

a. Start with $S_0 = 100$, $D_0 = 0$, and $N_0 = 0$. Use your computer to simulate the numbers in the next generation.

b. Follow these cells until there are no more cells that stain properly. How long did it take?

c. At each time, what is the fraction of cells that stain properly? What is the fraction of *living* cells that stain properly? Estimate the probability that a cell stains properly and the probability it stains properly conditional on being alive.

6.5 Independence and Markov Chains

Conditional probability can be used to describe whether or not two events are related. If knowing about one gives no additional information about the other, we say that the events are **independent**. Markov chains, probabilistic models that evolve over time, describe events that are related over time, and can be conveniently and powerfully formulated in terms of conditional probability.

6.5.1 Independence

Suppose two parents each with genotype **Bb** are crossed. It is generally thought that the allele contributed by the mother is **independent** of the allele contributed by the father. Scientifically, this means that knowing that the mother provided allele **b** tells us nothing about the allele provided by the father. In other words, knowledge of one allele in the offspring gives no new information about the other.

How do we describe independence mathematically? Let B_m designate the event that the allele from the mother is **B**, b_m be the event that the allele from the mother is **b**, and B_f and b_f describe the allele from the father in the same way (Figure 6.5.1). Suppose we learn that the mother provided allele **b**, so that event b_m occurred. What is the probability that the father provided allele **b** also? Biologically, this probability is the same as what it was without the additional information. In the mathematical language of conditional probability,

$$\Pr(b_f | b_m) = \Pr(b_f) = 0.5.$$

The father, at least under ordinary circumstances, provides the **b** allele with probability 0.5 **independent** of what the mother does.

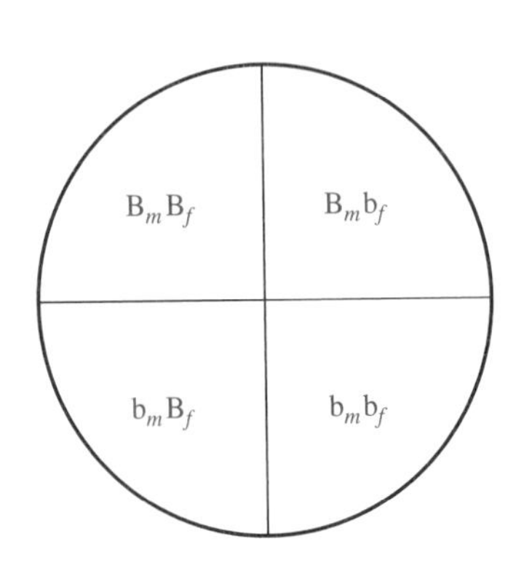

FIGURE 6.5.1
Genetic probabilities: independent case

We formalize this idea in the definition of independence.

Definition 6.6 Event A is **independent** of event B if

$$\Pr(A \mid B) = \Pr(A).$$

This compact definition matches our finding in the genetic case. The additional information (event B in the definition or b_m in the example) does not change our calculation of the probability of the other event. On a dartboard, suppose B is "I hit the right side of the dartboard" and A is "I hit the bull's-eye." For most players, these events are independent. If someone tells you that he hit the right side of the board, you are unlikely to change your guess of whether he hit the bull's-eye.

The definition of conditional probability does not require that we know the probabilities of the events. Suppose, for example, that genotype has no effect on the height of a plant. If T represents the event that the plant is tall, then

$$\Pr(T \mid BB) = \Pr(T \mid Bb) = \Pr(T \mid bb) = \Pr(T).$$

Example 6.5.1 The Distinction Between a Fair Coin and an Independent Coin

Consider two different coins. The first gives heads independently with probability 0.6 on every toss. This coin is unfair, but the probability on each toss does not depend on the previous toss. Another coin might give heads with probability 0.5, but exactly repeat the value of the previous toss. This coin is fair, because it gives heads just as often as tails, but is not independent. It will produce strings of heads or tails.

Independence is **reciprocal**. If A is independent of B, B is independent of A. Suppose we know that A is independent of B, but we are told that event A occurred. Then

$$\begin{aligned}
\Pr(B \mid A) &= \frac{\Pr(A \mid B)\Pr(B)}{\Pr(A)} && \text{Bayes' theorem} \\
&= \frac{\Pr(A)\Pr(B)}{\Pr(A)} && \text{A is independent of B} \\
&= \Pr(B). && \text{divide out factors of } \Pr(A)
\end{aligned}$$

Because the relation is reciprocal, we can simply say "A and B are independent."

Example 6.5.2 The Reciprocity of Independence: Genetic Example

In the genetic example, we assumed that the allele contributed by the father is independent of the allele contributed by the mother. The reciprocal property guarantees that the converse is also true; the allele contributed by the mother is independent of the allele contributed by the father. In terms of conditional probability,

$$\Pr(b_m \mid b_f) = \Pr(b_m) = 0.5.$$

If information about a first event (the allele contributed by the mother) gives no information about a second (the allele contributed by the father), then the converse must also be true; information about the second event (the allele contributed by the father) gives no information about the first (the allele contributed by the mother).

6.5.2 The Multiplication Rule for Independent Events

An extremely useful property of independent events follows from the multiplication rule for conditional probabilities

$$\Pr(A \cap B) = \Pr(A \mid B)\Pr(B)$$

(Equation 6.4.2). When events A and B are independent, $\Pr(A \mid B) = \Pr(A)$. This gives the following theorem.

Theorem 6.3 Multiplication Rule for Independent Events

Suppose A and B are any two independent events. Then

$$\Pr(A \cap B) = \Pr(A)\Pr(B).$$

In words, the probability that two independent events both occur is equal to the product of the probabilities for each.

Example 6.5.3 Application of the Multiplication Rule to Genetics

We already used the multiplication rule to derive the probabilities of genotypes **bb** and **BB** (Subsection 6.2.4). When meiosis is fair, each allele has a 0.5 chance of making it into the offspring. Therefore, the probability of a **b** from the mother is 0.5, as is the probability of a **b** from the father. An offspring has genotype **bb** if it gets a **b** from *both* parents. If the alleles from the parents arrive *independently*, then

$$\begin{aligned}\Pr(\text{bb}) &= \Pr(\mathbf{b}\text{ from mother}) \times \Pr(\mathbf{b}\text{ from father})\\ &= 0.5 \cdot 0.5 = 0.25.\end{aligned}$$

Example 6.5.4 Failure of the Multiplication Rule When Events Are Not Independent

We had to make *three* assumptions in the calculation in Example 6.5.3: fair meiosis in the mother and the father (a probability 0.5 of each allele from each), and independence. Even with fair meiosis, independence might fail. For example, it is possible that both parents contribute the **B** allele with probability 0.5, but do not contribute them independently. Figure 6.5.2 shows a case where the probability that the mother contributes the **B** allele is much greater when the father contributes the **B** allele than when he contributes the **b** allele. In this case,

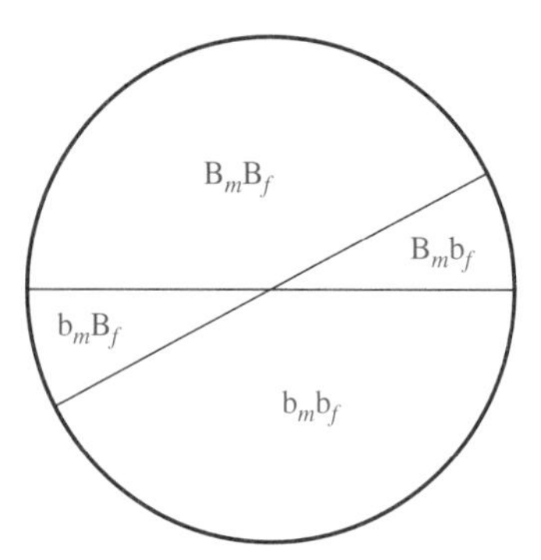

Allele contributed by father tends to match the allele contributed by mother

FIGURE 6.5.2
Genetic probabilities: dependent case

$$\Pr(B_f | B_m) = 0.8 \text{ and } \Pr(b_f | b_m) = 0.8.$$

Therefore,

$$\Pr(B_f \cap B_m) = \Pr(B_f | B_m)\Pr(B_m) = 0.8 \cdot 0.5 = 0.4.$$

Because the alleles contributed by the two parents tend to match, the offspring is more likely to be homozygous.

The multiplication rule can be used to define the idea of **mutual independence**. We think of events $A_1, \ldots, A_n$ as **mutually independent** if knowledge of all but one gives no information about the other. The events are mutually independent if probabilities can be computed with multiplication. For example, three events A, B, and C are mutually independent if

$$\Pr(A \cap B \cap C) = \Pr(A)\Pr(B)\Pr(C)$$

$$\Pr(A \cap B) = \Pr(A)\Pr(B)$$

$$\Pr(A \cap C) = \Pr(A)\Pr(C)$$

$$\Pr(B \cap C) = \Pr(B)\Pr(C).$$

Example 6.5.5 Success of the Multiplication Rule for Independent Events

As an example of the need to establish independence before applying the multiplication rule for many events, consider the following situation. A class consists of ten students, each of whom skips class with probability 0.2. What is the probability no students come to class on a particular day? Suppose first that students behave independently; each makes the decision about whether or not to come without regard for what her classmates do. We find the probability that nobody comes to class by multiplying the probabilities for each student

$$\text{Pr(Nobody came to class)} = 0.2^{10} = 0.0000001$$

or 1 day in 10 million.

Example 6.5.6 Failure of the Multiplication Rule for Non-Independent Events I

If students do pay attention to each other's behavior, the results can be quite different. First, suppose two of the students detest each other. Each immediately goes to class when he hears that the other has decided not to. The behaviors of these two students are not independent, and the multiplication rule will not work. At least one of these surly students is sure to attend, meaning that the probability that nobody comes is 0.

Example 6.5.7 Failure of the Multiplication Rule for Non-Independent Events II

Alternatively, suppose that the students have a leader who makes the other students do exactly as she does. Again, the multiplication rule fails. In this case, the professor would find herself teaching to an empty room on 1 day out of 5 and a full room on the other 4 days.

Whether events are independent is generally a matter of scientific judgment. Do we really know that alleles behave independently? Are we sure that hitting the right side of the bull's-eye is exactly as likely as hitting the left side? Testing these apparently obvious assumptions can provide novel insights into the way things work. Often, we explicitly build the assumption of independence into a mathematical model. If the assumption is later shown to be false or the results of the model do not match those of the experiment, the model must be modified.

Example 6.5.8 Pseudo-Replication: A Failure of Independence

Scientific results depend on replication, the ability to repeat the result. Suppose a new treatment cures ten out of ten patients, where only two out of ten got better with an older treatment. Is this convincing evidence that the new treatment is better? When we study the statistical methods of hypothesis testing (Section 8.4), we will see, as intuition suggests, that such a difference is indeed convincing. However, the argument, and our intuition, depend on the patients being independent. Suppose the ten patients who received the new treatment all live in one city with a cold dry climate, and the ten who received the older treatment live in another city with a hot humid climate. The patients in each group are not independent, and it is impossible to know whether the results are due to the new medication or to the climate. Careful clinical trials require randomizing patients to avoid this confusion.

6.5.3 Markov Chains and Conditional Probability

Much philosophical energy has been expended regarding the distinction between related events and "cause and effect." We will avoid this issue and think of cause and effect as a lack of temporal independence; knowing that one event occurred makes it more or less likely that some other event will occur later. Our study of discrete-time dynamical systems and autonomous differential equations was essentially the study of cause and effect in deterministic systems. We defined a Markov chain (Definition 6.2 in Subsection 6.2.2) as a stochastic dynamical system in which the probability of arriving in a particular state at a particular time depends on the state at the previous time. Markov chains are one way to formalize cause and effect for stochastic systems.

We have considered two Markov chains: one for presence and absence (Subsection 6.1.4) and one for molecular position (Subsection 6.2.2). For comparison, we will also consider a sequence of independent coin tosses and the behavior of a shy rabbit jumping in and out of its hole. In each case, the system has two possible states ("in" and "out," or "heads" and "tails"), and the state at one time gives information about the state at the next.

Example 6.5.9 A Markov Chain for Molecular Position

Recall the model for molecular position (Subsection 6.2.2). Let I_t denote the event that the molecule was inside during minute t and O_t denote the event that the molecule was outside during minute t. We can translate the assumptions into conditional probability notation as

$$\Pr(I_{t+1}|I_t) = 0.8$$
$$\Pr(I_{t+1}|O_t) = 0.1$$
$$\Pr(O_{t+1}|I_t) = 0.2$$
$$\Pr(O_{t+1}|O_t) = 0.9.$$

(Figure 6.5.3).

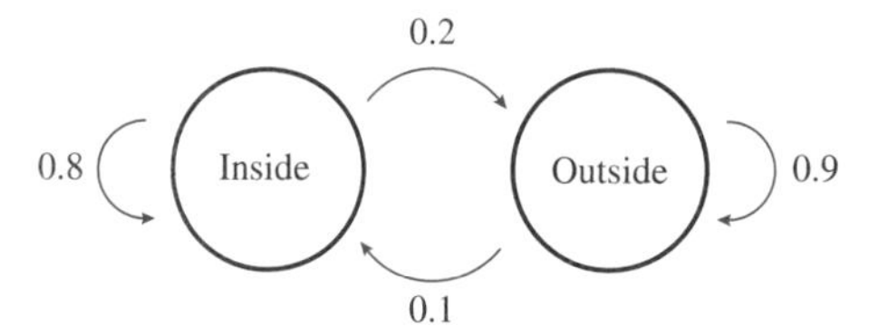

FIGURE 6.5.3
A Markov chain for a molecule

The position at time $t + 1$ is not independent of the position at time t because

$$\Pr(I_{t+1}|I_t) = 0.8 \neq \Pr(I_{t+1}|O_t) = 0.1.$$

Because it is much more likely that the molecule remains inside than that it enters, the positions at different times are not independent.

Example 6.5.10 The Markov Chain for a Fair Coin

How about a "fair" coin? Let H_t denote the event "heads" and T_t denote the event "tails" on toss t. We expect

$$\Pr(H_{t+1}|H_t) = 0.5$$
$$\Pr(H_{t+1}|T_t) = 0.5$$
$$\Pr(T_{t+1}|H_t) = 0.5$$
$$\Pr(T_{t+1}|T_t) = 0.5$$

(Figure 6.5.4). The probability of heads on toss $t + 1$ is the same whether toss t was heads or tails. These conditional probabilities encode the assumption of independence.

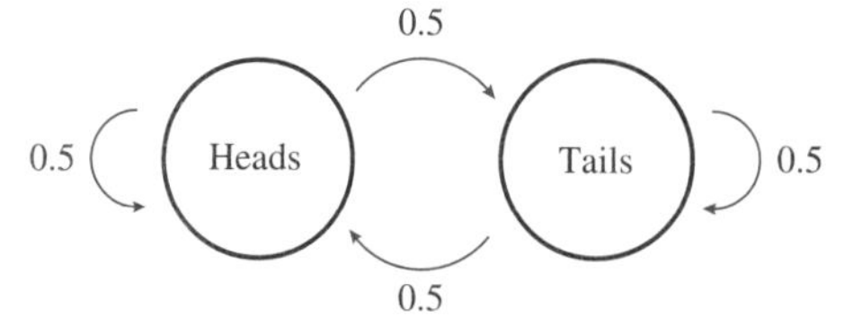

FIGURE 6.5.4
A Markov chain for a coin

Example 6.5.11 The Markov Chain for a Nervous Rabbit

Finally, consider a rabbit jumping in and out of its burrow with position denoted by the same notation as the molecule (Figure 6.5.5). The rabbit, however, is very nervous and

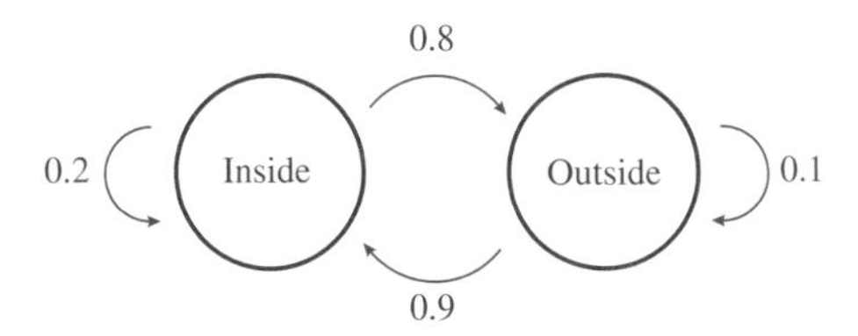

FIGURE 6.5.5
A Markov chain for a rabbit

tends to switch positions. In fact,

$$\Pr(I_{t+1}|I_t) = 0.2$$
$$\Pr(I_{t+1}|O_t) = 0.9$$
$$\Pr(O_{t+1}|I_t) = 0.8$$
$$\Pr(O_{t+1}|O_t) = 0.1.$$

The behavior during one minute is not independent of the behavior in the previous minute because, for instance, the rabbit is much more likely to be inside after it was outside (probability 0.9) than after it was inside (probability 0.1).

Example 6.5.12 The Long-Term Consequences for the Molecule

In a computer simulation, the molecule maintains position for long times with infrequent switches (Figure 6.5.6).

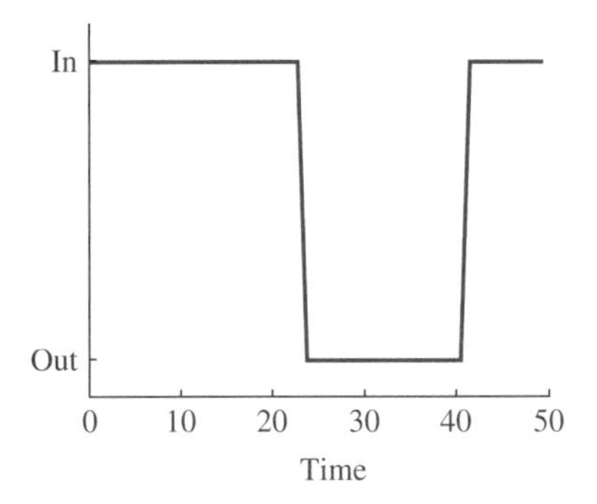

FIGURE 6.5.6
Long-term results for the molecule

Example 6.5.13 The Long-Term Consequences for the Coin

The coin switches between heads and tails "randomly" (colloquial usage tends to confound "random" and "independent") (Figure 6.5.7).

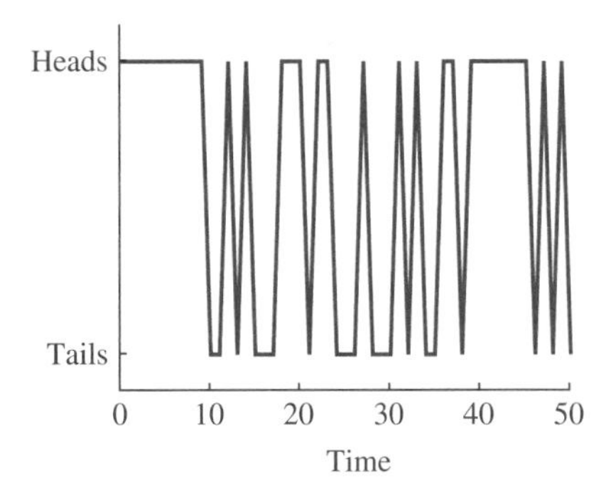

FIGURE 6.5.7
Long-term results for the coin

Example 6.5.14 The Long-Term Consequences for the Rabbit

The rabbit jumps back and forth with no long periods in the same position (Figure 6.5.8).

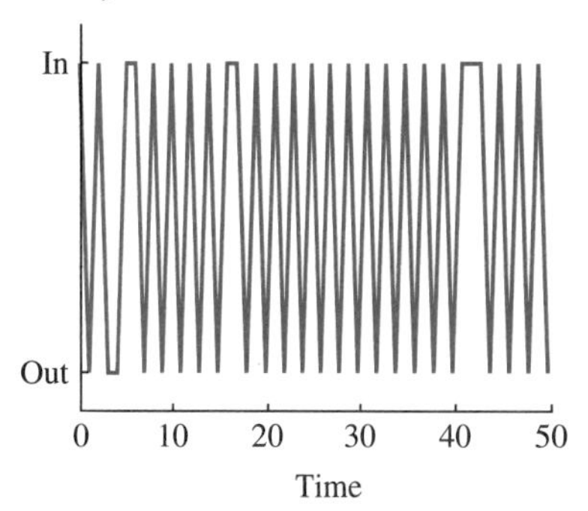

FIGURE 6.5.8
Long-term results for the rabbit

Summary Two events are **independent** if the probability of one conditional on the other is equal to the unconditional probability, meaning that information about one event conveys nothing about the other. The probability that two independent events happen simultaneously can be found with the **multiplication rule**. The multiplication rule generalizes to sets of **mutually independent** events. Markov chains provide an example of temporal dependence, where the state at one time depends probabilistically on the state at the previous time.

6.5 Exercises

Mathematical Techniques

1–4 ▪ Check whether the following events are independent by checking three equations:

$$\Pr(A) = \Pr(A \mid B) \quad \text{A is independent of B}$$
$$\Pr(B) = \Pr(B \mid A) \quad \text{B is independent of A}$$
$$\Pr(A \cap B) = \Pr(A)\Pr(B). \quad \text{the multiplication rule}$$

Do you ever find a case where only one or two of these equations is satisfied?

1. As in Section 6.3, Exercise 5 and Section 6.4, Exercise 5, the sample space is S={0, 1, 2, 3, 4}, $\Pr(\{0\}) = 0.2$, $\Pr(\{1\}) = 0.3$, $\Pr(\{2\}) = 0.4$, $\Pr(\{3\}) = 0.1$, $\Pr(\{4\}) = 0.0$, A={0, 1, 2}, and B={0, 2, 4}.
2. As in Section 6.3, Exercise 5 and Section 6.4, Exercise 6, the sample space is S={0, 1, 2, 3, 4}, $\Pr(\{0\}) = 0.2$, $\Pr(\{1\}) = 0.3$, $\Pr(\{2\}) = 0.4$, $\Pr(\{3\}) = 0.1$, $\Pr(\{4\}) = 0.0$, A={1, 2, 3}, and B={2, 3, 4}.
3. The sample space is S={1, 2, 3, 4}, $\Pr(\{1\}) = 0.48$, $\Pr(\{2\}) = 0.12$, $\Pr(\{3\}) = 0.32$, $\Pr(\{4\}) = 0.08$, A={3, 4}, and B={1, 3}.
4. The sample space is S={1, 2, 3, 4}, $\Pr(\{1\}) = 0.4$, $\Pr(\{2\}) = 0.4$, $\Pr(\{3\}) = 0.1$, $\Pr(\{4\}) = 0.1$, A={1, 2}, and B={1, 3}.

5–10 ▪ Consider again the three-sided die that gives scores of 1, 2, or 3, each with probability 1/3 (Section 6.4, Exercises 9–12). Suppose that the results of rolls are independent. Use the multiplication rule to find the following probabilities.

5. The probability of rolling a 1 followed by a 3.
6. The probability of rolling three 1's in a row.
7. The probability of rolling two odd values in a row.
8. The probability of rolling an odd value followed by an even value.
9. The probability of rolling six 3's in a row.
10. The probability of rolling exactly the sequence 1, 2, 3, 3, 2, 1.

11–12 ▪ In each of the following problems, the sample space is S = {1, 2, 3, 4}. From the probabilities of the given events A and B, and the assumption that A and B are independent, find $\Pr(\{1\})$, $\Pr(\{2\})$, $\Pr(\{3\})$, and $\Pr(\{4\})$.

11. A={1, 2}, B={1, 3}, $\Pr(A) = 0.4$, $\Pr(B) = 0.6$.
12. A={1, 4}, B={1, 3}, $\Pr(A) = 0.8$, $\Pr(B) = 0.3$.

13–14 ▪ Show that the multiplication rule (Theorem 6.3) does not work in the following cases.

13. For two events A and B that are disjoint, as long as $\Pr(A) > 0$ and $\Pr(B) > 0$.
14. For two events A and B where A is a subset of B, as long as $\Pr(A) > 0$ and $\Pr(B) < 1$.

15–18 ▪ Write the information from each of the following two-state Markov chains in terms of conditional probability. Write a discrete-time dynamical system for the probability, and find the long-term probability.

15. The mutants described in Section 6.2, Exercise 27, where a gene has a 1.0% chance of mutating each time a cell divides, and a mutant gene has a 1.0% chance of reverting to wild type.
16. The lemmings described in Section 6.2, Exercise 28, where a lemming has a probability 0.2 of jumping off the cliff each hour and a probability 0.1 of crawling back up.

17. The molecules described in Section 6.2, Exercise 29, where a molecule has a probability 0.05 of binding and a probability of 0.02 of unbinding each second.

18. The caterpillars described in Section 6.2, Exercise 30, where a caterpillar has a probability 0.15 of being taken over by a parasitoid each day and a probability 0.03 of recovering.

19–20 ▪ The formula for the probability of the union of two events,

$$\Pr(A \cup B) = \Pr(A) + \Pr(B) - \Pr(A \cap B),$$

(from Section 6.3, Exercises 13–16) is simpler when events are independent.

$$\Pr(A \cup B) = \Pr(A) + \Pr(B) - \Pr(A)\Pr(B).$$

Test the formula on the following independent events.

19. Using the probabilities found in Exercise 11, where A={1, 2}, B={1, 3}, Pr(A) = 0.4, Pr(B) = 0.6.

20. Using the probabilities found in Exercise 12, where A={1, 4}, B={1, 3}, Pr(A) = 0.8, Pr(B) = 0.3.

Applications

21–22 ▪ An ecologist is looking for the effects of eagle predation on the behavior of jackrabbit (as in Section 6.4, Exercises 27 and 28). Assuming that the jackrabbits and eagles behave independently,

a. Find the probability that the ecologist sees both a jackrabbit and an eagle during a particular hour of observation.

b. Draw a Venn diagram to illustrate the situation.

c. Find the probability that she saw a jackrabbit conditional on her seeing an eagle. How might you interpret this result? Compare with the overall probability of seeing a jackrabbit.

d. Find the probability that she saw an eagle conditional on her seeing a jackrabbit. How might you interpret this result?

21. She sees an eagle with probability 0.2 during an hour of observation and a jackrabbit with probability 0.5.

22. She sees an eagle with probability 0.4 during an hour of observation and a jackrabbit with probability 0.8.

23–24 ▪ Someone comes up with a cut-rate "test" for a disease. This test gives a positive result with probability 0.5 whether or not the patient has the disease. In each of the following cases, find the probability of having the disease conditional on a positive test in two ways.

a. Work it out directly as on page 539.

b. Use independence.

23. 1% of people have the disease.

24. 10% of people have the disease.

25–28 ▪ Consider again the molecules in Section 6.2, Exercises 1–4. Suppose that we wish to consider two molecules instead of one molecule, both starting inside the cell. Find the following probabilities.

25. What is the probability that both of the molecules remain inside after 1 second (using parameters from Section 6.2, Exercise 1)?

26. What is the probability that both of the molecules have moved outside after 1 second (using parameters from Section 6.2, Exercise 2)?

27. What is the probability that both of the molecules remain inside after 2 seconds (using parameters from Section 6.2, Exercise 3)?

28. What is the probability that both of the molecules have moved outside after 2 seconds (using parameters from Section 6.2, Exercise 4)?

29–30 ▪ One force that can alter the ratio of heterozygotes produced by a selfing heterozygote is **meiotic drive** (Section 6.2, Exercises 49–52), where one allele, say **A**, pushes its way into more than half of the gametes. Another possibility is that the alleles in surviving offspring are not independent. Compare the fraction of heterozygotes produced in the following cases.

29. Compare a case of meiotic drive where 60% of both pollen and ovules carry the **A** allele independently, with a case of nonindependent assortment where an offspring gets an **A** allele from the pollen with probability 0.6 when the ovule provides an **A** and gets an **A** allele from the pollen with probability 0.4 when the ovule provides an **a**. The ovule provides **A** with probability 0.5.

30. Compare a case of meiotic drive where 70% of the pollen and 40% of the ovules carry the **A** allele independently, with a case of nonindependent assortment where an offspring gets an **A** allele from the pollen with probability 0.7 when the ovule provides an **A** and gets an **A** allele from the pollen with probability 0.3 when the ovule provides an **a**. The ovule provides **A** with probability 0.5.

31–32 ▪ A species of bird comes in three colors: red, blue, and green. Twenty percent are red, 30% are blue, and 50% are green. Females prefer red to blue and blue to green and mate with the best male they find.

31. Females pick the better of the first two males they meet. What is the probability a female mates with a green bird? What did you have to assume about independence?

32. Females pick the better of the first two males they meet. What is the probability a female mates with a blue bird and the probability a female mates with a red bird?

33–36 ▪ A small class has only three students. Each comes to class with probability 0.9. Find the probability that all the students come to class and the probability that no students come to class in the following circumstances.

33. The students act independently.

34. Student 2 comes to class with probability 1.0 if student 1 does. Student 3 ignores them.

35. Student 2 comes to class with probability 8/9 if student 1 does. Student 3 ignores them.

36. Student 3 comes to class with probability 1.0 if both the others come and with probability 1/2 if only one comes. Students 1 and 2 ignore each other.

37–38 ▪ In DNA, there are four nucleotides: A, T, C, and G. A pairs with T, and C pairs with G. In many organisms, mutations that change an AT pair into a GC pair are more common than those that change a GC pair into an AT pair. Write down Markov chains describing the following situations.

37. The probability of a switch from AT to GC is 0.002, while a switch from GC to AT occurs with probability 0.001.

38. The probability of a switch from AT to GC is 0.004, while a switch from GC to AT occurs with probability 0.003.

39–42 ▪ Suppose a molecule is transferred among three cells according to a Markov chain. Write down conditional probabilities to describe the following situations. It can help to draw a picture.

39. The position of the molecule in one minute is independent of the position in the previous minute.

40. The molecule leaves a cell with probability 0.1. When it does so, it enters each of the other cells with equal probability. Are the positions independent over time?

41. Imagine the three cells arranged in a ring. The molecule leaves a cell with probability 0.1, and when it does so, it always moves clockwise.

42. Imagine the three cells arranged in a line. The molecule makes a given move with probability 0.1. If it is at the end, it moves to the middle. If it is in the middle, it enters the end cells with equal probability.

43–44 ▪ From each of the following sets of data, estimate the probability that a molecule that is inside a cell leaves during a given second. Write a discrete-time dynamical system for the probability that the molecule is inside and find the probability it is inside after 3 seconds. How does this probability compare with the fraction of molecules that actually were inside at $t = 3$?

43. Ten molecules start inside a cell. They are first observed outside the cell in the given second.

Molecule	Time First Observed Outside
1	11
2	1
3	2
4	3
5	1
6	2
7	4
8	7
9	1
10	4

44. Ten molecules start inside a cell. They are first observed outside the cell in the given second.

Molecule	Time First Observed Outside
1	4
2	16
3	14
4	10
5	4
6	1
7	1
8	11
9	2
10	12

45–46 ▪ From each of the following sets of data, estimate the probability that a molecule that is inside a cell leaves during a given second, and the probability that it returns. Write a discrete-time dynamical system for the probability that the molecule is inside and find the equilibrium probability. How does the equilibrium probability compare with the fraction of times the molecule actually was inside?

45. One molecule is observed for 20 seconds, and follows

Time	Location	Time	Location
1	In	11	Out
2	In	12	In
3	Out	13	Out
4	Out	14	Out
5	In	15	In
6	In	16	In
7	Out	17	In
8	In	18	Out
9	In	19	In
10	Out	20	Out

46. One molecule is observed for 20 seconds, and follows

Time	Location	Time	Location
1	In	11	Out
2	In	12	Out
3	In	13	Out
4	In	14	Out
5	Out	15	In
6	Out	16	In
7	Out	17	Out
8	In	18	Out
9	In	19	Out
10	In	20	In

Computer Exercises

47. Figure out a way to simulate 100 offspring from the mechanisms in Section 6.2, Exercises 49 and 50. How close are your results to the mathematically expected results?

48. The updating function for the position of a molecule is given by
$$h(x) = q_{0.7}x + q_{0.3}(1-x)$$
where $x = 1$ represents inside, $x = 0$ represents outside, and q is defined in Section 6.1, Exercise 37. Run this system for 50 steps. Based on your data, estimate $\Pr(x_{t+1} = 1)$, $\Pr(x_{t+1} = 1 | x_t = 1)$, and $\Pr(x_{t+1} = 1 | x_t = 0)$. Compare these results with what you would expect based on the discrete-time dynamical system.

6.6 Displaying Probabilities

In order to fully understand the assignment of probabilities in a model, we need an efficient and readable way to display them. For a small number of simple events, we can use **pie charts**. In general, however, the best and most widely used method is the **histogram** or bar graph, which plots the probability of each event as the height of the bar. A related graph is the **cumulative distribution**, which adds up probabilities in order before graphing them. Specifying and characterizing the probabilities of simple events when the sample space is infinite is more difficult. By a limiting process, we develop two mathematical and graphical tools, the **probability density function**, or **p.d.f.**, and the **cumulative distribution function**, or **c.d.f.**

6.6.1 Probability and Cumulative Distributions

Suppose we cross two heterozygous plants of intermediate height, each with one short gene and one tall gene. If a single offspring were produced, there are three simple events: the offspring could be short, tall, or intermediate. Under the assumptions that the alleles each have a fair chance of 0.5 and come from the two parents independently, the offspring will be short (height 40) with probability 0.25, tall (height 60) with probability 0.25, and intermediate (height 50) with probability 0.5, or in tabular form

Height	Probability
40	0.25
50	0.5
60	0.25

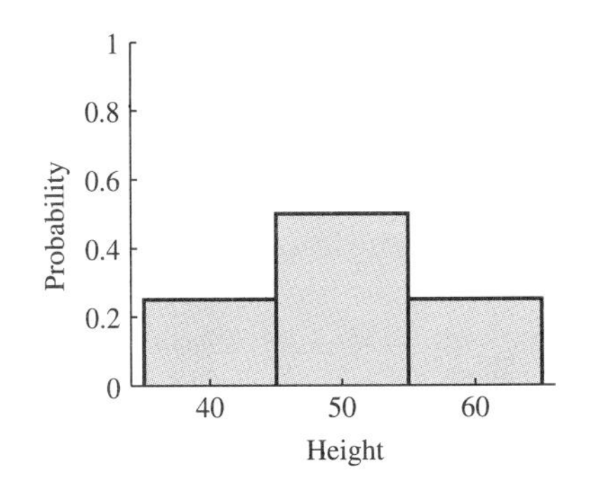

FIGURE 6.6.1
A histogram with three simple events

The biological assumptions guide us to an assignment of probabilities to all of the simple events. These probabilities add to 1, so our biologically derived assignment is indeed mathematically consistent.

These probabilities can be displayed as a **probability distribution** or **histogram** (Figure 6.6.1). The vertical axis is the probability. The height of the bar above the event that the height is 40 is the probability 0.25 and so forth. Histograms are convenient because we can immediately see which event is most probable (height is 50). Furthermore, the distribution is **symmetric** (would look the same in a mirror), meaning that a plant is just as likely to be short as to be tall.

Example 6.6.1 Histogram for Molecules at Three Times

Consider ten toxic molecules diffusing out of a cell. Suppose that the underlying mathematical model is that each molecule has a 10% chance of leaving the cell each minute. We will later learn how to compute the probability that N molecules remain after different lengths of time (Section 7.6). Results after 1, 4, and 8 minutes are displayed in Figure 6.6.2 and the following table.

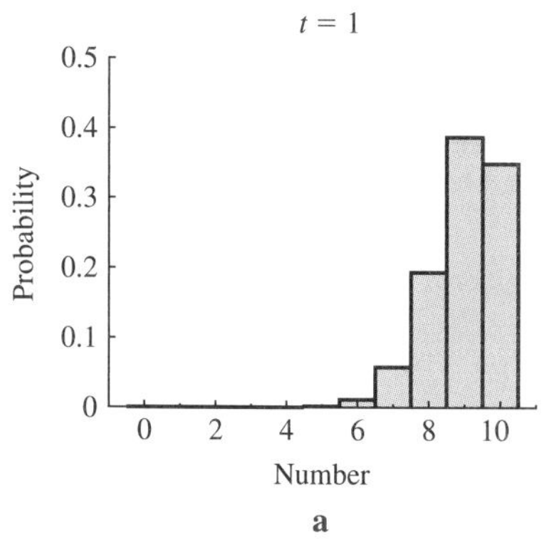

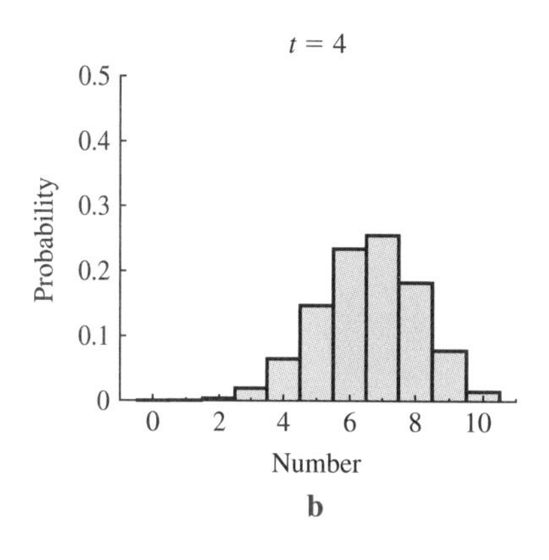

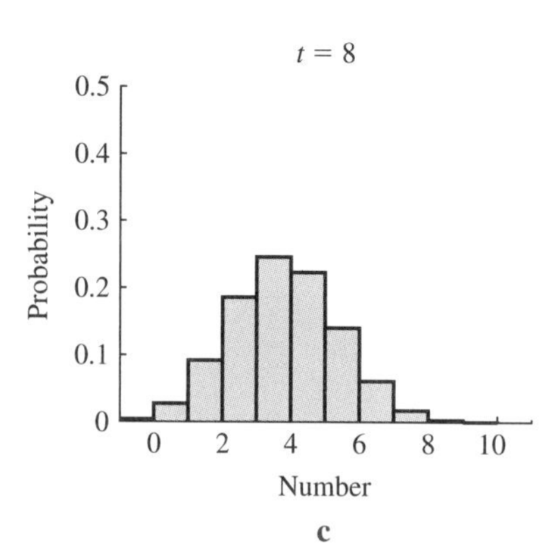

FIGURE 6.6.2

Diffusion of ten molecules

Number Left	Probability at $t=1$	Probability at $t=4$	Probability at $t=8$
0	0.000	0.000	0.004
1	0.000	0.000	0.027
2	0.000	0.004	0.092
3	0.000	0.019	0.186
4	0.000	0.064	0.246
5	0.001	0.147	0.223
6	0.011	0.234	0.141
7	0.057	0.255	0.061
8	0.194	0.183	0.017
9	0.387	0.077	0.003
10	0.349	0.015	0.000

The simple events are arrayed along the horizontal axis, and above each rises a bar with height equal to the probability of that simple event. ▲

When the simple events have a natural numerical order, the **cumulative distribution** provides another useful way to plot results. The cumulative distribution plots the probability not of a single result (a simple event), but of all results less than or equal to a single result. The probability that the plants shown in Figure 6.6.1 have heights less than or equal to a given value are given in the following table.

Height	Probability Height Is Less Than or Equal to This Value
40	0.25
50	0.75
60	1.00

FIGURE 6.6.3

A cumulative distribution with three simple events

The heights of bars increase up to 1.0, as they must, because all plants must have height less than or equal to 60 in this model (Figure 6.6.3).

Example 6.6.2 Cumulative Distributions for Molecules at Three Times

Instead of plotting the probability that N molecules remain, the cumulative distribution plots the probability that N or fewer molecules remain (Figure 6.6.4). For example, the probability that five or fewer molecules remain after 8 minutes is

$$\begin{aligned}\Pr(N \le 5) &= \Pr(N=0) + \Pr(N=1) + \Pr(N=2) \\ &\quad + \Pr(N=3) + \Pr(N=4) + \Pr(N=5) \\ &= 0.004 + 0.027 + 0.092 + 0.186 + 0.246 + 0.223 = 0.778.\end{aligned}$$

The complete results are compiled in the following table.

Number Left	Probability at $t=1$	Probability at $t=4$	Probability at $t=8$
0	0.000	0.000	0.004
1	0.000	0.000	0.031
2	0.000	0.004	0.123
3	0.000	0.024	0.309
4	0.000	0.088	0.555
5	0.002	0.235	0.778
6	0.013	0.470	0.919
7	0.070	0.725	0.980
8	0.264	0.908	0.997
9	0.651	0.985	1.000
10	1.000	1.000	1.000

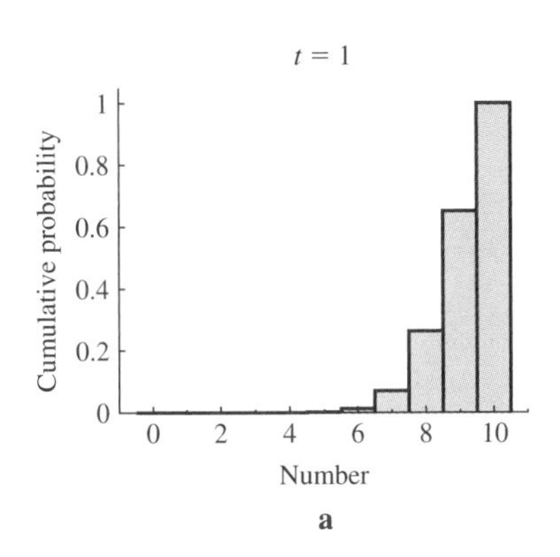

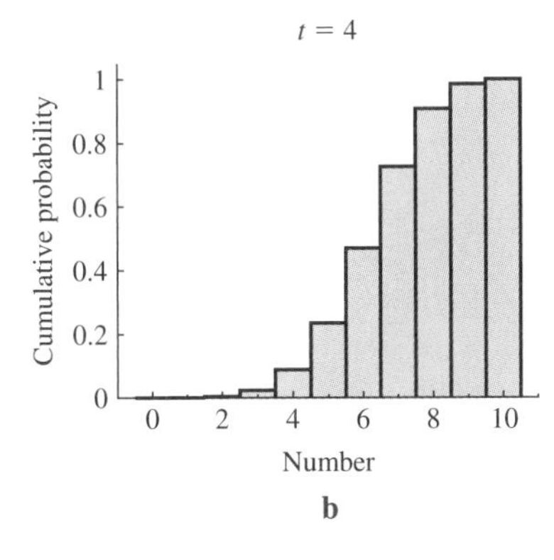

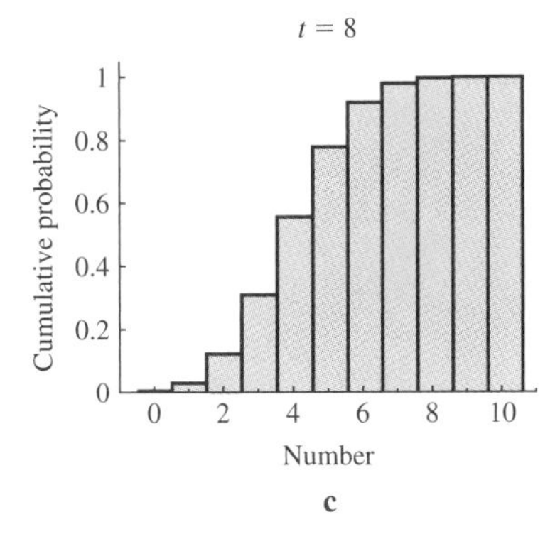

FIGURE 6.6.4
Diffusion of ten molecules: cumulative distributions

The probabilities must increase to 1.0 because there are always ten or fewer molecules remaining. Graphically, the heights of the bars in a cumulative distribution plot always increase to 1. The fact that the bars increase more quickly when t is larger means that there are *fewer* molecules left at the later times.

6.6.2 The Probability Density Function: Derivation

So far, we have computed and displayed probabilities only when the number of simple events is finite. We now address the case when the simple events form a continuum. As with the derivative and the integral, the key is to think of the continuum as the limit of smaller and smaller units.

A perfect "random number generator" picks a number between 0 and 1 (but not the value 1.0 exactly) with any possible value being equally probable. It is just as likely to pick the value 0.5 as it to pick 0.2345 or any particular irrational number like $\sqrt{0.5}$. Because there are an infinite number of possible events (numbers chosen), we cannot describe the results with a histogram, because the histogram would require an infinite number of bars. Furthermore, as we will see, the "probability" of any particular simple event (such as choosing *exactly* 0.2345) is zero. How can we describe an infinite number of probabilities, each equal to zero, which must add up to a total probability of 1.0?

Scientifically, we break out of this apparent paradox by thinking of a more realistic case, where the random number generator produces only a finite number of significant digits. Alternatively, we could imagine that the random number generator itself produces an infinite number of digits, but we might observe only the first one, two, four, or eight places.

Example 6.6.3 Perfect Random Numbers and Imperfect Measurements

The following table gives the exact values generated by an idealized random number generator and the observed values out to the given number of decimal places.

Exact Value	Value to 1 Decimal Place	Value to 2 Decimal Places	Value to 4 Decimal Places	Value to 8 Decimal Places
0.5	0.5	0.50	0.5000	0.50000000
0.2345	0.2	0.23	0.2345	0.23450000
$\sqrt{0.5}$	0.7	0.70	0.7071	0.70710678
$1/\pi$	0.3	0.31	0.3183	0.31830989

These values are not **rounded** to the nearest value, but have been **truncated** by removing later digits. With two decimal places of accuracy, we observe the value 0.70 for $\sqrt{0.5}$ simply by discarding later digits.

These imperfect measurements produce only a finite number of possible results. With a single decimal place of accuracy, there are 10 simple events (the values 0.0, 0.1, 0.2, 0.3, 0.4, 0.5, 0.6, 0.7, 0.8, 0.9), while with 2 decimal places there are 100 simple events.

Example 6.6.4 Histogram for Random Numbers with One Decimal Place

Because the random number generator is equally likely to produce any given value, the first decimal place is just as likely to be 0 as it is to be 1, 2, or any value up to 9. These 10 probabilities are thus equal to each other. In order to add up to 1.0, each must be exactly 0.1 (Figure 6.6.5).

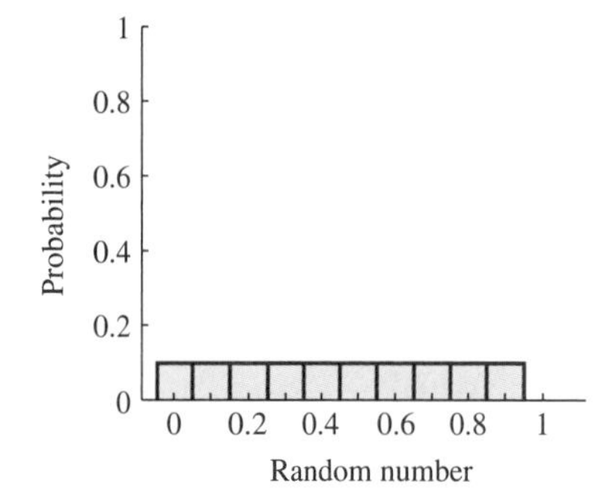

FIGURE 6.6.5
Histogram with a single decimal place

Example 6.6.5 Histogram for Random Numbers with Two Decimal Places

Because the random number generator is equally likely to produce any given value, the first two decimal places are just as likely to be 00 as they are to be 19, 82, or any pair of numbers up to 99. These 100 probabilities are equal to each other, and must each equal 0.01 in order to add up to 1.0 (Figure 6.6.6). There are now a lot of skinny, short bars in the histogram.

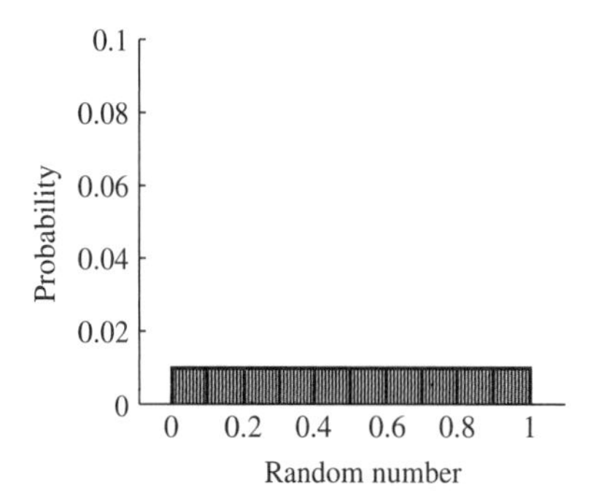

FIGURE 6.6.6
Histogram with two decimal places

To make the histograms more readable, we will draw a new type of histogram where the **area** (rather than the height) represents the probability. This new drawing keeps the bars from getting unreadably short. More importantly, these histograms look like the p.d.f. we seek.

Example 6.6.6 New Type of Histogram for Random Numbers with One Decimal Place

With one decimal place of accuracy, each probability is 0.1, and the histogram in Example 6.6.4 showed this with bars of *height* 0.1. However, each bar is also 0.1 wide. For the bars to have **area** equal to 0.1, they must have height 1.0 (Figure 6.6.7).

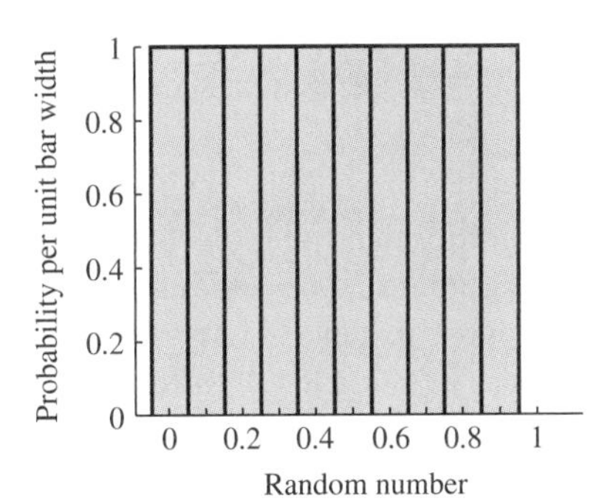

FIGURE 6.6.7

New histogram with a single decimal place

Example 6.6.7 New Type of Histogram for Random Numbers with Two Decimal Places

With two decimal places of accuracy, each probability is 0.01, and the histogram in Example 6.6.5 showed this with tiny bars of height 0.01. Each bar is also 0.01 wide, and for the bars to have area equal to 0.01, they must have height 1.0 (Figure 6.6.8).

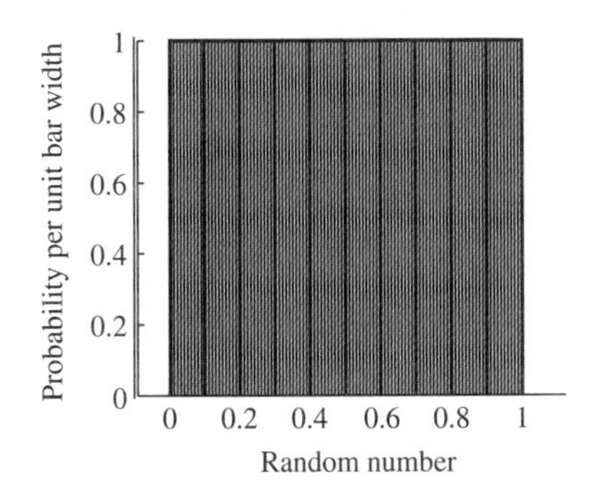

FIGURE 6.6.8

New histogram with two decimal places

The new type of histogram has bars of height exactly 1.0 for any number of decimal places. The bars lie under a graph of the function $f(x) = 1.0$ for $0 \le x < 1$. This function is the **p.d.f.** associated with the ideal random number generator that produces an infinite number of decimal places of accuracy. In Subsection 6.6.3, we will see how to use this function to compute probabilities. First, however, we consider another example of a p.d.f.

Consider again a single molecule diffusing out of a cell and think of measuring the time of departure. The simple events are "the molecule left exactly at time t" for any positive value of t. What is the probability that the molecule left after exactly $t = \pi$ seconds? Strictly speaking, this probability is 0. A difference even in the millionth decimal place would mean a completely different simple event. But this logic holds for any time t. The probability that it left *exactly* at any particular time is 0.

We break out of this apparent paradox by thinking of imperfect measuring devices. If our device has precision of only 10 seconds, we cannot distinguish measurements in the interval from 0 to 10, 10 to 20, and so forth. A true value of 3.14159 cannot be told from a true value of 2.71828, for example. We can break the possibilities from 0 to 50 into five blocks of width 10. The results (which we will derive in Section 7.6) are

Time Range	Probability	Height of Bar
0–10	0.632	0.063
10–20	0.233	0.023
20–30	0.086	0.009
30–40	0.031	0.003
40–50	0.012	0.001

Again, we draw the new type of histogram where the area rather than height of the bar represents the probability that the result lies in that interval (Figure 6.6.9a). For

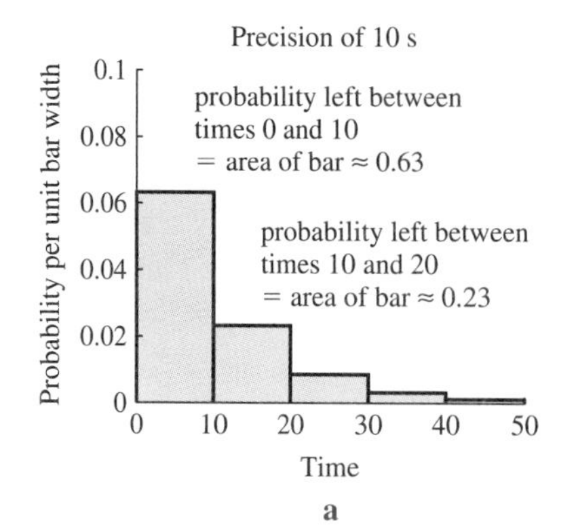

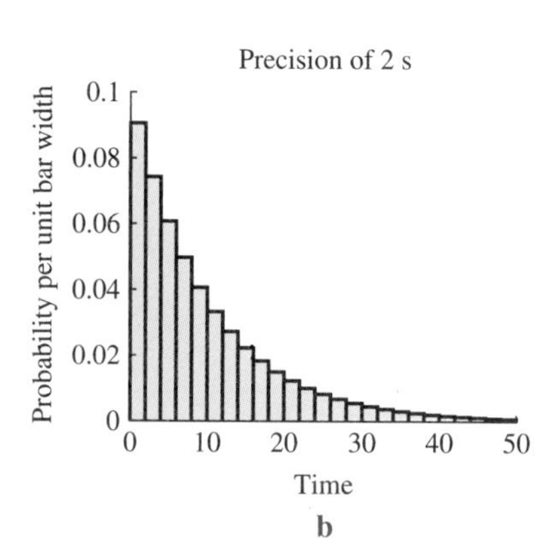

FIGURE 6.6.9

Probability that molecule left during particular interval with two measuring devices

example, the probability that the molecule left between times 0 and 10 is 0.63, so we draw a bar with height 0.063 and width 10.

Using a more precise device with accuracy 2.0, we break the possibilities into intervals from 0 to 2, 2 to 4, and so forth, finding

Time Range	Probability	Height of Bar	Time Range	Probability	Height of Bar
0–2	0.181	0.091	26–28	0.013	0.007
2–4	0.148	0.074	28–30	0.011	0.006
4–6	0.122	0.061	30–32	0.009	0.005
6–8	0.099	0.050	32–34	0.007	0.004
8–10	0.081	0.041	34–36	0.006	0.003
10–12	0.067	0.033	36–38	0.005	0.002
12–14	0.055	0.027	38–40	0.004	0.002
14–16	0.045	0.022	40–42	0.003	0.002
16–18	0.037	0.018	42–44	0.003	0.001
18–20	0.030	0.015	44–46	0.002	0.001
20–22	0.025	0.012	46–48	0.002	0.001
22–24	0.020	0.010	48–50	0.001	0.001
24–26	0.016	0.008			

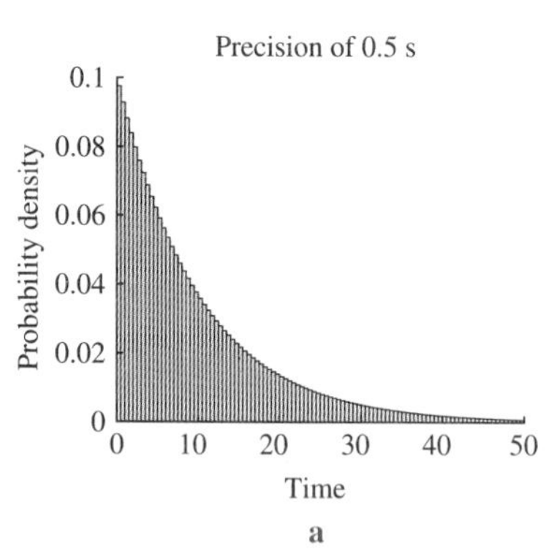

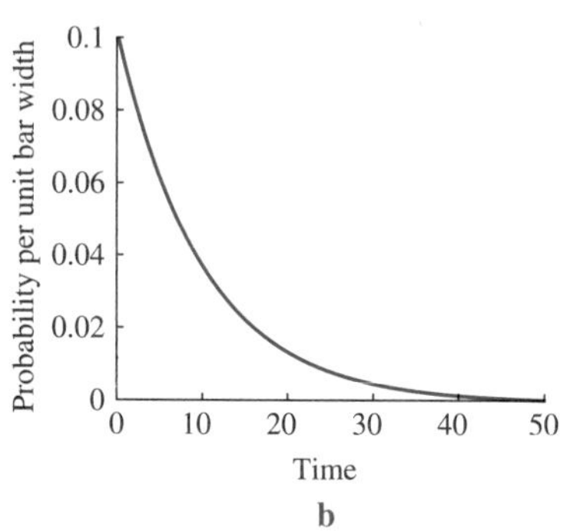

FIGURE 6.6.10

Probability density dunction for a diffusing molecule

We again draw bars with area equal to the probability (Figure 6.6.9b). The probability that the molecule left between times 2 and 4 is 0.148. Because the width of the bar is 2.0, the bar must have a height of 0.074 to have an area of 0.148.

With an accuracy of 0.5 (Figure 6.6.10a), the histogram looks a lot like the curve sketched in Figure 6.6.10b. What does this curve mean? Suppose we wanted to find the probability that the molecule left between times 0 and 10. With a precision of 10 (Figure 6.6.9a), this is the area under the first bar. With a precision of 2 (Figure 6.6.9b), this is the sum of the areas under the first five bars. These areas are approximately the area under the curve. With a mathematically exact device, the probability is equal to the area under the curve between 0 and 10. If we knew the formula for the curve, we could find probabilities by integrating.

6.6.3 Using the Probability Density Function

Before learning how to use the p.d.f., what curves can act as a p.d.f.? The requirements are spelled out in the following definition.

Definition 6.7 Suppose simple events can be indexed by the real number x. Let a be the smallest possible simple event and b the largest. A function $f(x)$ can be a p.d.f. if

1. $f(x) \geq 0$ for all x with $a \leq x \leq b$.
2. $\int_a^b f(x)\,dx = 1$.

The first requirement says that probabilities must be positive, and the second says that they must add up to 1. The definition does *not* require the value of the function to be less than 1 as in a probability distribution.

How do we use a p.d.f.? The probability is the area, and the area is the integral. Therefore, if X denotes the outcome of the experiment and f is the p.d.f., then

$$\Pr(a \leq X \leq b) = \int_a^b f(x)\,dx. \tag{6.6.1}$$

A probability density acts like an ordinary density (like grams per centimeter). Even though there is no mass exactly at a particular point, the total mass can be found by integrating.

Example 6.6.8 Using the p.d.f. to Compute Probabilities

The p.d.f. for the random number generator is the function $f(x) = 1.0$ for $0 \leq x < 1$. To find the probability that a random number is between 0.2 and 0.44, we find

$$\begin{aligned}\Pr(0.2 \leq \text{random number} \leq 0.44) &= \int_{0.2}^{0.44} f(x)\,dx \\ &= \int_{0.2}^{0.44} 1.0\,dx \\ &= x\Big|_{0.2}^{0.44} \\ &= 0.44 - 0.2 = 0.24.\end{aligned}$$

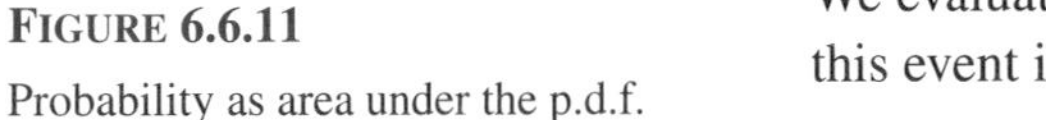

FIGURE 6.6.11
Probability as area under the p.d.f.

We evaluated this definite integral using the fact that $\int 1.0\,dx = x$. The probability of this event is the **area under the p.d.f. (Figure 6.6.11).**

The exact p.d.f. for the time when a diffusing toxic molecule leaves the cell is

$$f(t) = 0.1e^{-0.1t}$$

when the rate of departure is 0.1.

Example 6.6.9 Finding a Probability for a Molecule

To find the probability that the molecule left between time 0 and time 10, integrate the p.d.f. from $t = 0$ to $t = 10$,

$$\Pr(0 \leq \text{departure time} \leq 10) = \int_0^{10} 0.1e^{-0.1t}dt.$$

Evaluating this integral requires integration by substitution (Subsection 4.3.3), which gives

$$\int_0^{10} 0.1e^{-0.1t}dt = e^{-0.1t}dt + c.$$

Then

$$\int_0^{10} 0.1e^{-0.1t}dt = -e^{-0.1t}|_0^{10} = -e^{-1.0} + e^{-0.0} = 0.632.$$

If we ran this experiment 1000 times, about 632 molecules would leave before time 10 (Figure 6.6.12).

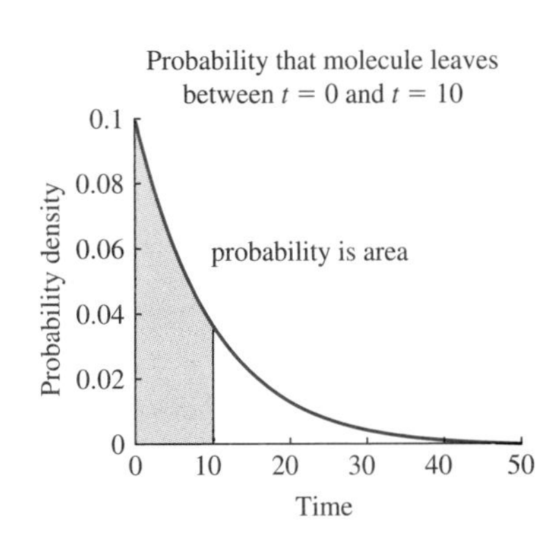

FIGURE 6.6.12
Finding the probability as area under the p.d.f.

Example 6.6.10 Finding Another Probability for a Molecule

The probability a molecule leaves between time 5 and 15 is

$$\Pr(5 \leq \text{departure time} \leq 15) = \int_{5}^{15} 0.1e^{-0.1t}dt$$

$$= -e^{-0.1t}\Big|_{5}^{15}$$

$$= -e^{-1.5} + e^{-0.5} = 0.383$$

(Figure 6.6.13). ▲

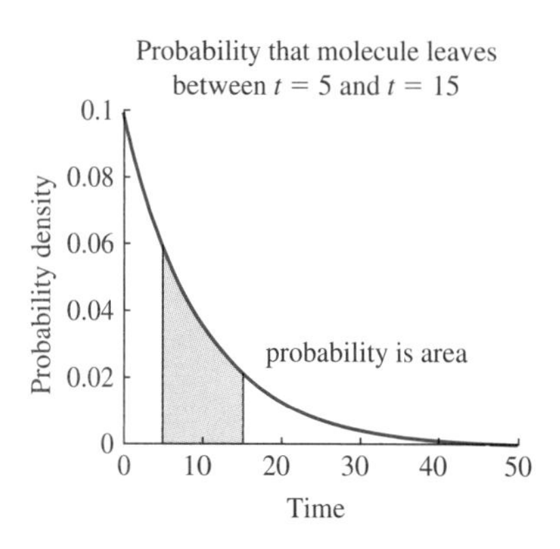

FIGURE 6.6.13
Finding probabilities as areas under a p.d.f.

Example 6.6.11 Finding a Probability for Two Times That Are Close

To better understand the definition, it helps to think of finding the probability that the molecule leaves between two times that are close together, like $t = 2.0$ and $t = 2.01$. The exact probability is

$$\Pr(2.0 \leq \text{departure time} \leq 2.1) = \int_{2.0}^{2.1} 0.1e^{-0.1t}dt$$

$$= -e^{-0.1t}\Big|_{2.0}^{2.1}$$

$$= -e^{-0.21} + e^{-0.2} = 0.00815.$$

Because the interval is narrow, we can approximate this region as a rectangle with base 0.1 and height $f(2.0)$

$$\Pr(2.0 \leq \text{departure time} \leq 2.1) \approx \text{base} \times \text{height}$$

$$= 0.1 f(2.0)$$

$$= 0.1 \cdot 0.1e^{-0.2} = 0.00819$$

(Figure 6.6.14). The probability that it leaves near time 2.0 is **proportional** to the height of the p.d.f. at that point. ▲

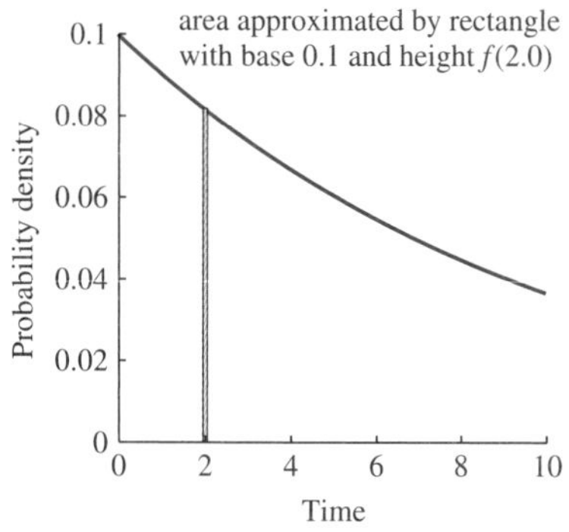

FIGURE 6.6.14
Approximating area with a rectangle

Example 6.6.12 A p.d.f. That Takes on Values Greater Than 1

The p.d.f. for a molecule that leaves a cell with rate $\lambda = 10.0$ has formula

$$f(t) = 10.0e^{-10.0t}$$

(Figure 6.6.15). The vertical range rises above 1. This is mathematically consistent because the height is not *equal* to the probability, but only *proportional* to it. The large value of the p.d.f. at $t = 0$ indicates that the molecule is very likely to leave near time $t = 0$, consistent with its high rate of departure. The probability that a molecule leaves

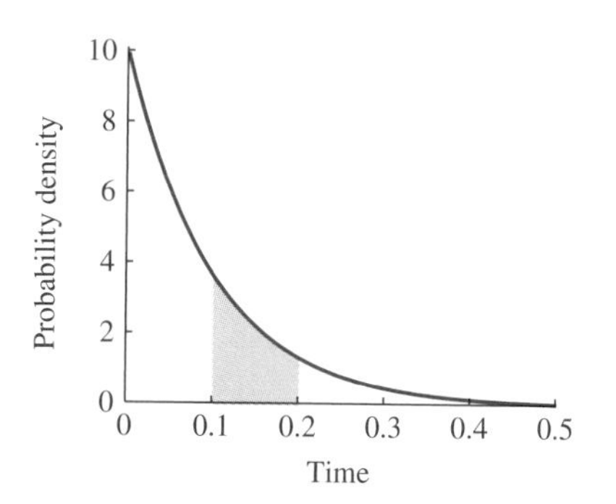

FIGURE 6.6.15
A p.d.f. that takes on values greater than 1

between times 0.1 and 0.2 is

$$\begin{aligned}\Pr(0.1 \leq \text{departure time} \leq 0.2) &= \int_{0.1}^{0.2} 10.0e^{-10.0t}dt \\ &= -e^{-10.0t}\Big|_{0.1}^{0.2} \\ &= -e^{-2.0} + e^{-1.0} = 0.232.\end{aligned}$$

Although the values of the p.d.f. are greater than 1, the area is less than 1.

Example 6.6.13 The Normal p.d.f. for Height

The bell curve or normal p.d.f., which we will study in detail in Section 7.8, characterizes many measurements and is the most important p.d.f. in statistics (Figure 6.6.16). In the case shown, the p.d.f. takes on its maximum where the height is equal to 60, which is right in the middle of this symmetric picture. The domain of this distribution includes all values from $-\infty$ to ∞. However, because the value of the p.d.f. becomes very small, the probabilities of absurd results (like negative height) are negligible.

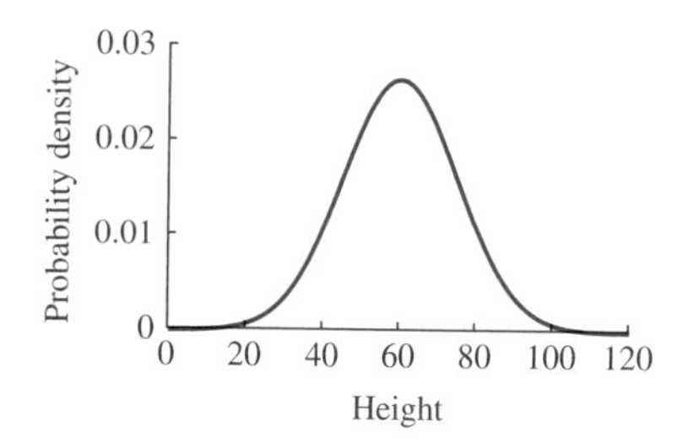

FIGURE 6.6.16
Another p.d.f. that takes on values greater than 1

6.6.4 The Cumulative Distribution Function

What is the probability that our original molecule with p.d.f.

$$f(t) = 0.1e^{-0.1t}$$

left before some given time t? To find the probability it left before time $t = 10$, we integrate

$$\Pr(\text{molecule left before time } 10) = \int_0^{10} f(t)dt.$$

The probability it left before a general time t is

$$\Pr(\text{molecule left before time } t) = \int_0^{t} f(s)ds.$$

(We changed t to s inside the integral to avoid using the same letter to mean two different things.)

Thought of as a function of t, this probability defines the c.d.f. Formally,

Definition 6.8 Suppose $f(x)$ is a p.d.f. with smallest simple event a (which could be $-\infty$) and largest simple event b (which could be ∞). Then the function $F(x)$ defined by

$$F(x) = \int_a^x f(y)dy$$

is the **cumulative distribution function**, or **c.d.f.**

In general, we use a capital letter, such as F, to denote the c.d.f. associated with a p.d.f. with the related small letter, such as f.

Example 6.6.14 The c.d.f. for a Molecule

For the molecule, the c.d.f. is

$$\begin{aligned} F(t) &= \int_0^t f(s)ds \\ &= \int_0^t 0.1e^{-0.1s}ds \\ &= -e^{-0.1s}\Big|_0^t \\ &= 1 - e^{-0.1t}. \end{aligned}$$

$F(t)$ gives the probability that the molecule left at or before time t (Figure 6.6.17). The probability that the molecule left before time 10 is

$$\Pr(\text{molecule left before time 10}) = F(10.0) = 1 - e^{-0.2} = 0.632$$

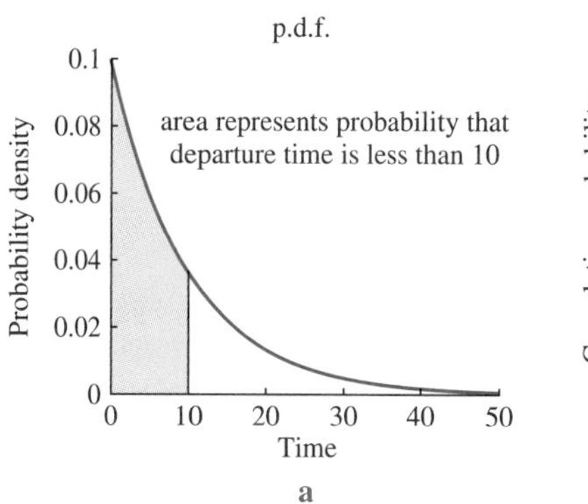

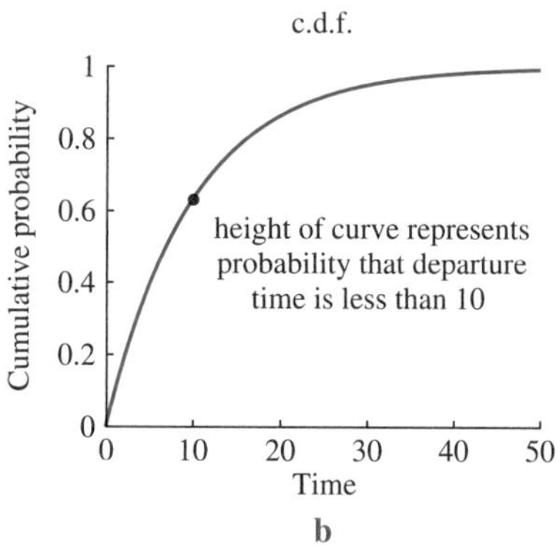

FIGURE 6.6.17
Two ways to describe a molecule

as we found before. Like a cumulative distribution, the c.d.f. must increase to 1. In this case

$$\lim_{t\to\infty} F(t) = \lim_{t\to\infty} 1 - e^{-0.1t} = 1.$$

Example 6.6.15 The c.d.f. for a Random Number Generator

The p.d.f. for a random number generator is $f(x) = 1.0$ for $0 \le x < 1$. Therefore, the c.d.f. is

$$\begin{aligned} F(x) &= \int_0^x f(y)\,dy \\ &= \int_0^x 1.0\,dy \\ &= y\Big|_0^x = x. \end{aligned}$$

Thus, the probability that a random number is less than 0.4 is $F(0.4) = 0.4$ (Figure 6.6.18).

The Fundamental Theorem of Calculus (Theorem 4.4) makes working with the c.d.f. convenient. First, because the c.d.f. is the integral of the p.d.f., the p.d.f. must be the derivative of the c.d.f., or

$$\frac{dF}{dt} = f(t).$$

Second, we can use the c.d.f. to find probabilities by subtracting. The Fundamental Theorem says that

$$\int_a^b f(x)\,dx = F(b) - F(a).$$

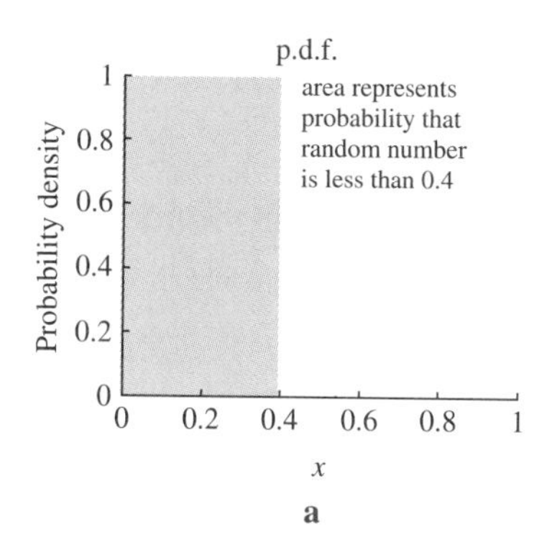

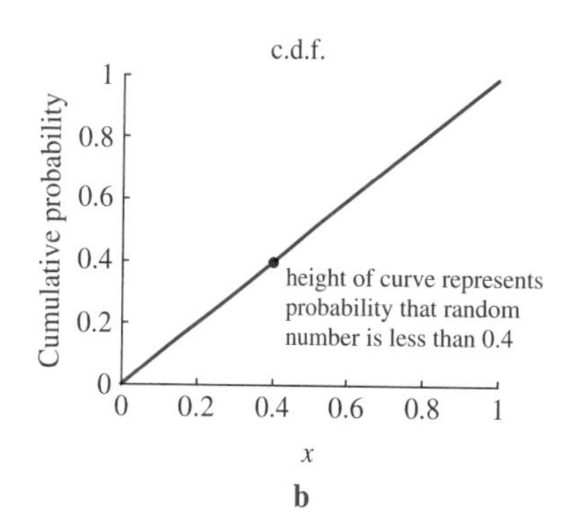

FIGURE 6.6.18
The p.d.f. and the c.d.f. for a random number generator

Example 6.6.16 Applying the Fundamental Theorem of Calculus to the Molecule

Using the c.d.f.

$$F(t) = 1 - e^{-0.1t},$$

the probability that the departure time lies between $t = 5$ and $t = 15$ is

$$F(15) - F(5) = (1 - e^{-0.1 \cdot 15}) - (1 - e^{-0.1 \cdot 5}) = 0.776 - 0.393 = 0.383$$

as we found in Example 6.6.10.

Example 6.6.17 A Linear p.d.f. and Its Associated c.d.f.

Consider the p.d.f.

$$f(x) = 2 - 2x$$

for $0 \le x \le 1$ (Figure 6.6.19). First, we check that this is indeed a p.d.f. because

$$\int_0^1 f(x) = \int_0^1 2 - 2x\,dx = 2x - x^2|_0^1 = (2 - 1) - (0 - 0) = 1.$$

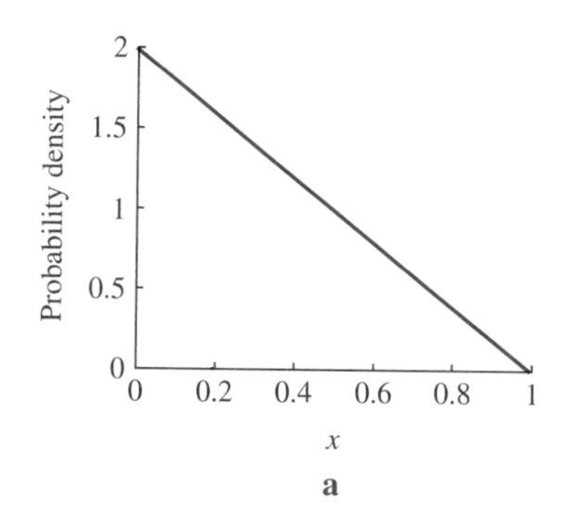

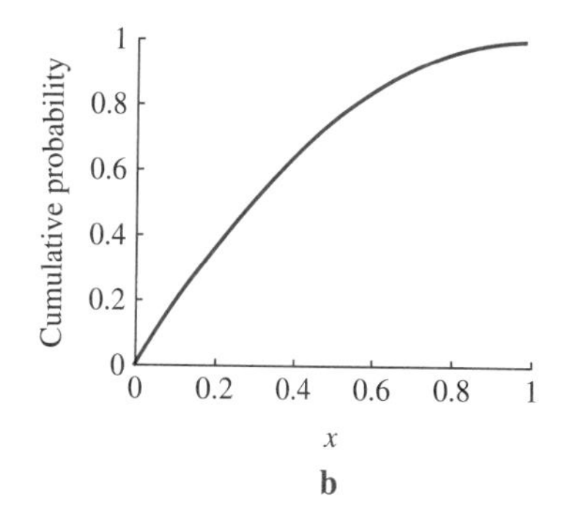

FIGURE 6.6.19
A p.d.f. and its associated c.d.f.

Again, although the function itself takes on values greater than 1, the total area is equal to 1. The c.d.f. is

$$\begin{aligned} F(x) &= \int_0^x f(y)\,dy \\ &= \int_0^x 2 - 2y\,dy \\ &= 2y - y^2|_0^x \\ &= 2x - x^2. \end{aligned}$$

Example 6.6.18 Computing a Probability with the Linear p.d.f. and Its Associated c.d.f.

With the p.d.f. and c.d.f. in Example 6.6.17, we can find the probability that x is between 0.3 and 0.6 either by integrating the p.d.f. or by evaluating the c.d.f. at the endpoints and subtracting (Figure 6.6.20). In the first case, we find

$$\begin{aligned}\Pr(0.3 \le x \le 0.6) &= \int_{0.3}^{0.6} 2 - 2x\,dx \\ &= 2x - x^2\Big|_{0.3}^{0.6} \\ &= (2\cdot 0.6 - 0.6^2) - (2\cdot 0.3 - 0.3^2) = 0.33.\end{aligned}$$

Alternatively,

$$\begin{aligned}\Pr(0.3 \le x \le 0.6) &= F(0.6) - F(0.3) \\ &= (2\cdot 0.6 - 0.6^2) - (2\cdot 0.3 - 0.3^2) = 0.33.\end{aligned}$$

Graphically, the first calculation corresponds to the area between the two values, whereas the second corresponds to the change in height of the c.d.f.

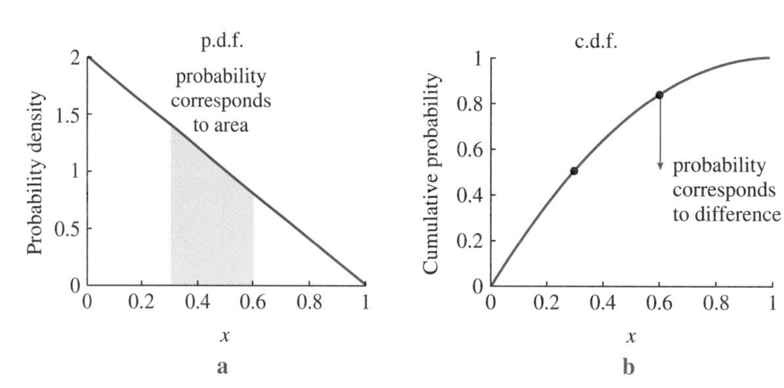

FIGURE 6.6.20
Computing probabilities with a p.d.f. and its associated c.d.f.

Example 6.6.19 The c.d.f. Associated with a p.d.f. That Takes on Values Greater Than 1

The c.d.f. for the p.d.f. for a fast molecule (Example 6.6.12) is very steep near 0, where the probability density is high (Figure 6.6.21). The cumulative probability increases from 0 to 1 as it must.

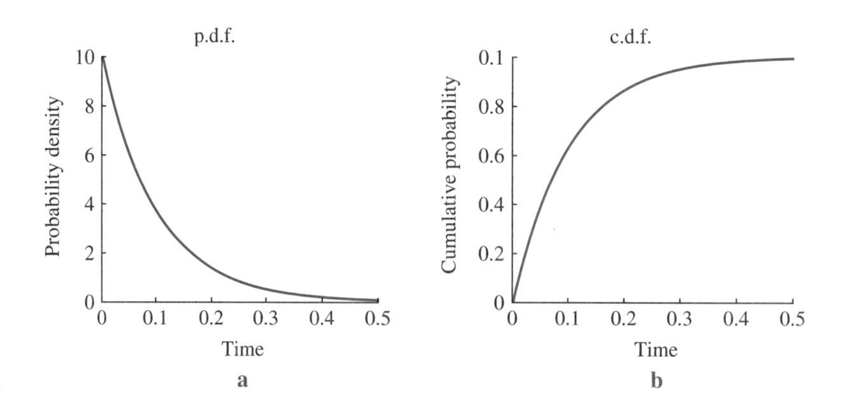

FIGURE 6.6.21
The p.d.f. and c.d.f. for a fast molecule

Example 6.6.20 The c.d.f. Associated with the Normal Distribution

The c.d.f. for the p.d.f. for the normal distribution (Example 6.6.13) swoops up from 0 to 1, increasing most steeply where the p.d.f. takes on its largest value and increasing slowly where the p.d.f. takes on small values (Figure 6.6.22).

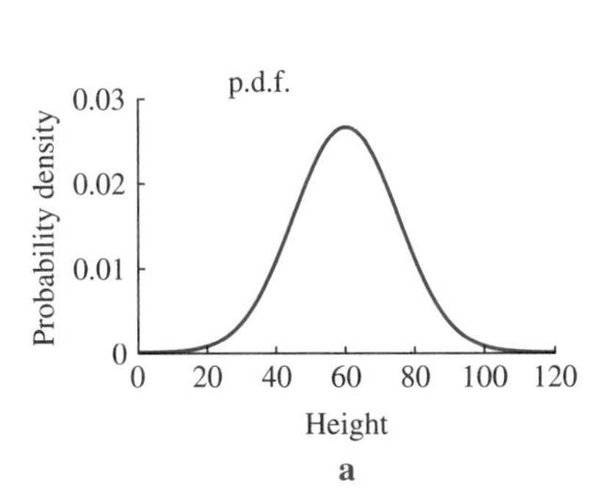

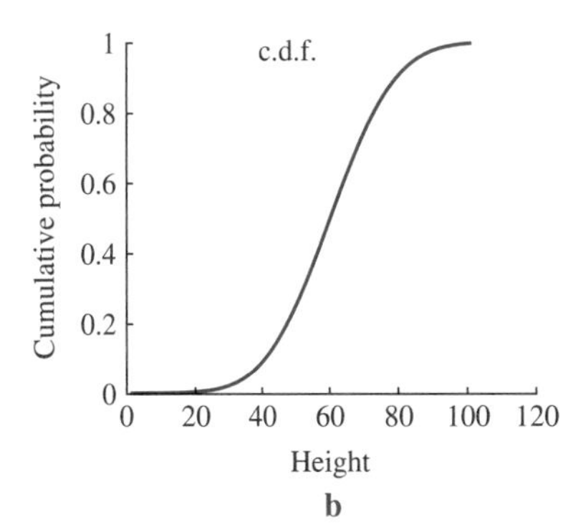

FIGURE 6.6.22
The p.d.f. and c.d.f. for the normal distribution

Summary We have seen two ways to graph the probabilities of discrete and continuous sets of simple events. When the number of events is finite, the probability distribution can be displayed with the **histogram** or the **cumulative distribution**. Histograms show which simple events are most likely and whether the distribution is **symmetric**. The cumulative distribution plots the probability that the result is less than or equal to a given value. Because simple events have probability 0 when the sample space is a continuum, the histogram must be replaced with a **probability density function (p.d.f.).** Definite integrals of this function give the probability that the result lies between the limits of integration. The **cumulative distribution function** (**c.d.f.**) gives the probability that the result is less than or equal to a particular value.

6.6 Exercises

Mathematical Techniques

1–4 ▪ Draw histograms describing the probabilities of the outcomes of four experiments to count the number of mutants in a bacterial cultural.

	Probability			
Number of Mutants	Experiment a	Experiment b	Experiment c	Experiment d
0	0.1	0.6	0.3	0.1
1	0.2	0.3	0.2	0.3
2	0.3	0.1	0.2	0.1
3	0.3	0.0	0.2	0.4
4	0.1	0.0	0.1	0.1

1. Experiment a.

2. Experiment b.

3. Experiment c.

4. Experiment d.

5–8 ▪ Find and sketch the cumulative distribution associated with the histogram from the earlier problem.

5. The histogram in Exercise 1.

6. The histogram in Exercise 2.

7. The histogram in Exercise 3.

8. The histogram in Exercise 4.

9–12 ▪ On each histogram, find the most and least likely simple events. Is the histogram symmetric?

9.

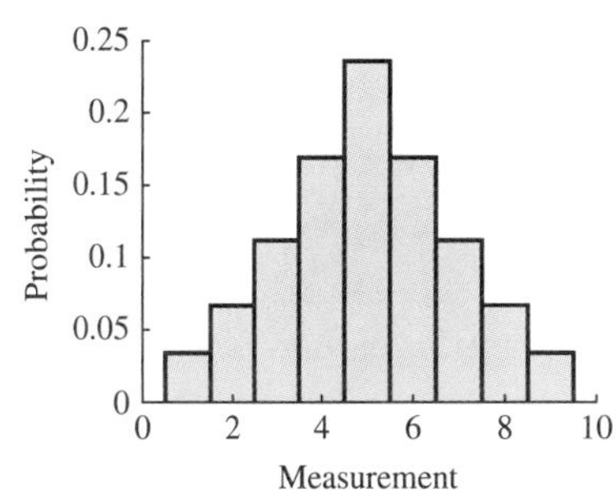

10.

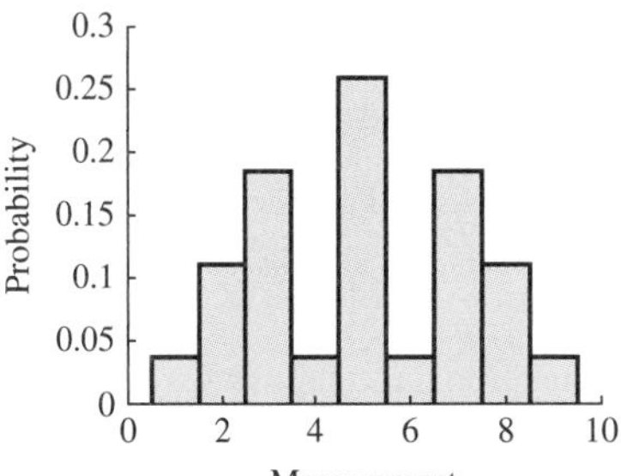

11.

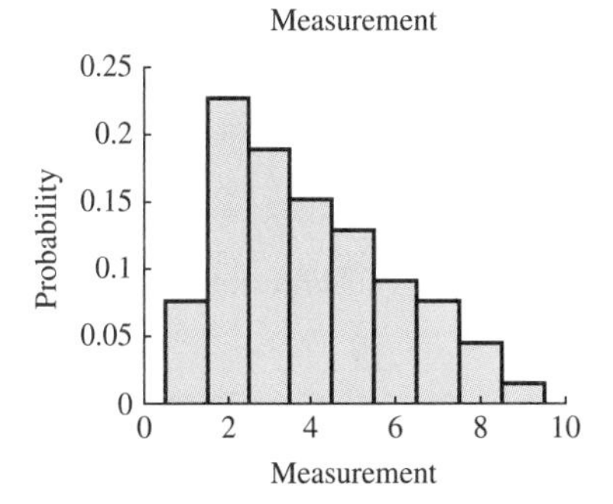

12.

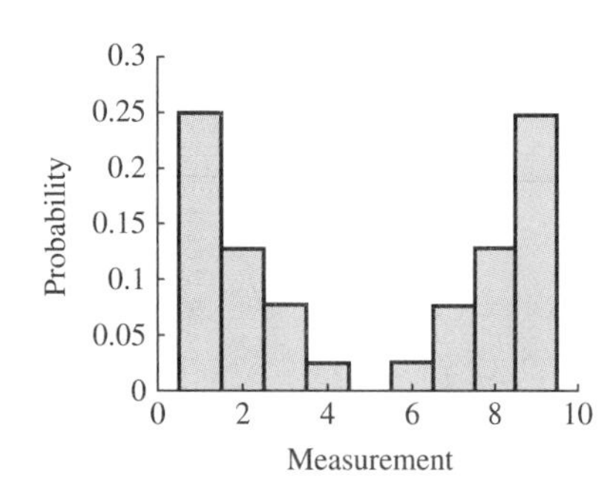

13–16 ▪ Using the histogram indicated, estimate the probabilities of the following events.

a. The measurement is equal to 7.

b. The measurement is less than or equal to 4.

c. The measurement is greater than 4.

13. The histogram in Exercise 9.

14. The histogram in Exercise 10.

15. The histogram in Exercise 11.

16. The histogram in Exercise 12.

17–20 ▪ For each of the following p.d.f.'s,

a. Check that the area under the curve is exactly 1.

b. Sketch a graph.

c. Indicate the maximum of the p.d.f., and explain why you are not worried that it is sometimes greater than 1.

17. The p.d.f. is $f(x) = 2x$ for $0 \le x \le 1$.

18. The p.d.f. is $f(x) = 1 - \frac{x}{2}$ for $0 \le x \le 2$.

19. The p.d.f. is $h(t) = \frac{1}{t}$ for $1 \le t \le e$.

20. The p.d.f. is $g(t) = 6t(1-t)$ for $0 \le t \le 1$.

21–24 ▪ Find and sketch the c.d.f. associated with the given p.d.f. and check that it increases to a value of 1.

21. The p.d.f. is $f(x) = 2x$ for $0 \le x \le 1$ (as in Exercise 17).

22. The p.d.f. is $f(x) = 1 - \frac{x}{2}$ for $0 \le x \le 2$ (as in Exercise 18).

23. The p.d.f. is $h(t) = \frac{1}{t}$ for $1 \le t \le e$ (as in Exercise 19).

24. The p.d.f. is $g(t) = 6t(1-t)$ for $0 \le t \le 1$ (as in Exercise 20).

25–28 ▪ Find the probability in two ways:

a. By integrating the given p.d.f.

b. By using the c.d.f.

Make sure that your answers match. Shade the given areas on a graph of the p.d.f.

25. The p.d.f. is $f(x) = 2x$ for $0 \le x \le 1$ (as in Exercises 17 and 21). Find the probability that the measurement is between 0.2 and 0.6.

26. The p.d.f. is $f(x) = 1 - \frac{x}{2}$ for $0 \le x \le 2$ (as in Exercises 18 and 22). Find the probability that the measurement is between 1.0 and 1.5.

27. The p.d.f. is $h(t) = \frac{1}{t}$ for $1 \le t \le e$ (as in Exercises 19 and 23). Find the probability that the measurement is between 2.0 and 2.5.

28. The p.d.f. is $g(t) = 6t(1-t)$ for $0 \le t \le 1$ (as in Exercises 20 and 24). Find the probability that the measurement is between 0.5 and 0.8.

29–30 ▪ Sketch the c.d.f. associated with each of the following p.d.f.'s.

29.

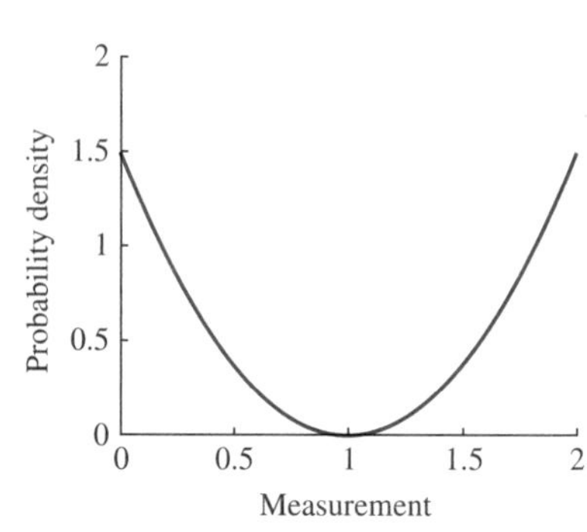

30.

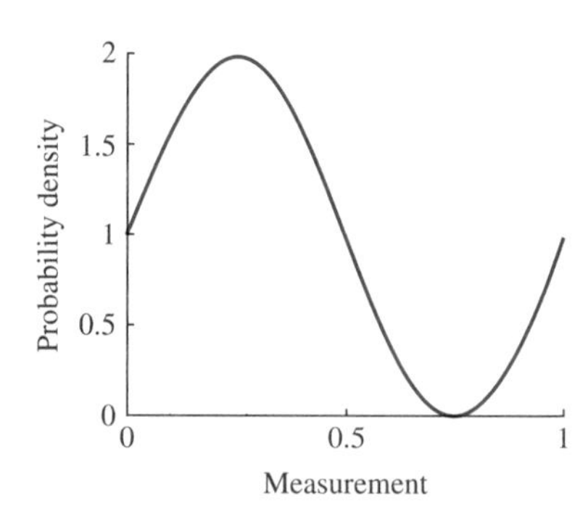

31–34 ▪ Sketch the p.d.f. associated with each of the following c.d.f.'s.

31.

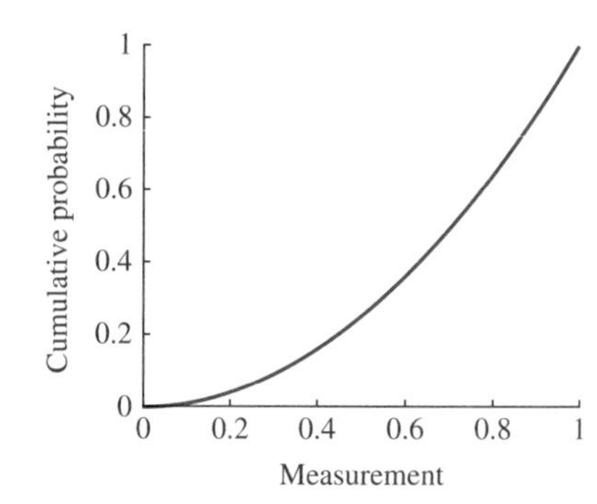

32.

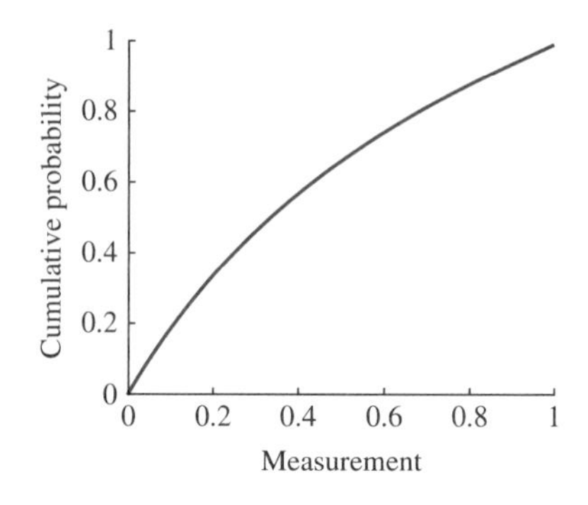

33.

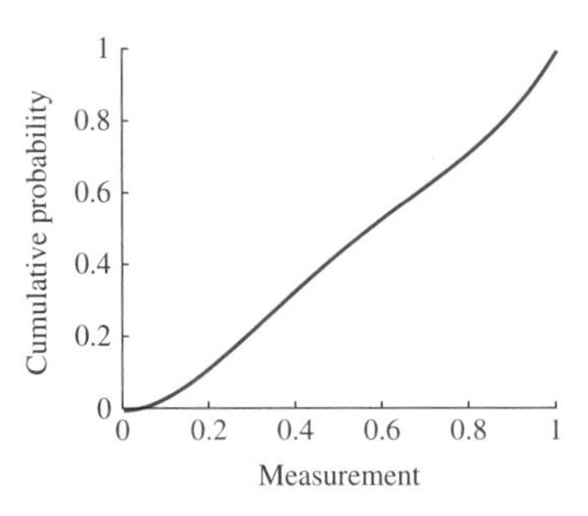

34.

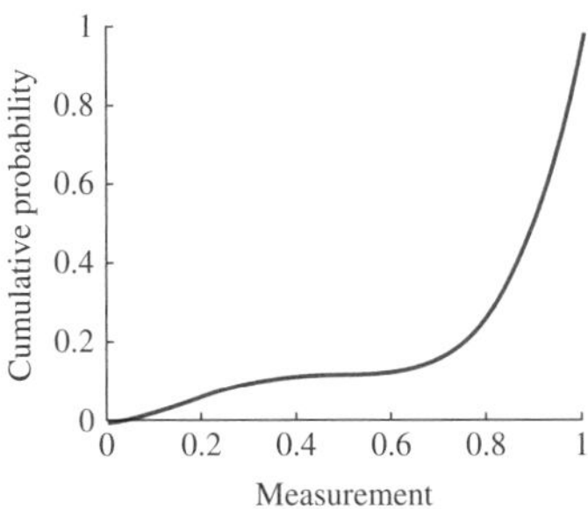

Applications

35–36 ▪ Draw histograms of the distributions of cell age from the assumptions in the earlier problem. Find and graph the cumulative distribution.

35. The cells in Section 6.4, Exercise 31, where

$$\Pr(\text{cell is 0 day old}) = 0.4$$
$$\Pr(\text{cell is 1 day old}) = 0.3$$
$$\Pr(\text{cell is 2 days old}) = 0.2$$
$$\Pr(\text{cell is 3 days old}) = 0.1.$$

36. The cells in Section 6.4, Exercise 32, where the cells with age greater than or equal to 3 days have been eliminated from the culture.

37–40 ▪ One hundred pairs of plants are crossed, and each pair produces ten offspring. The number of tall offspring is then counted. For the given experiment, draw a histogram of the probability of each result, and find the requested probability.

	Frequency in 100 Experiments			
Number of Tall Offspring	Experiment a	Experiment b	Experiment c	Experiment d
0	0	0	0	5
1	0	0	2	11
2	0	1	5	35
3	0	1	10	27
4	2	6	21	13
5	8	19	27	8
6	9	20	19	1
7	31	27	8	0
8	22	16	5	0
9	22	9	3	0
10	6	1	0	0

37. Draw the histogram for experiment a, and find the probability that between 4 and 6 plants (inclusive) are tall.

38. Draw the histogram for experiment b, and find the probability that between 4 and 6 plants (inclusive) are tall.

39. Draw the histogram for experiment c, and find the probability that between 4 and 6 plants (inclusive) are tall.

40. Draw the histogram for experiment d, and find the probability that between 4 and 6 plants (inclusive) are tall.

41–44 ▪ An experiment to see which color of male birds female birds prefer is repeated two times. The first time, females mate with red males with probability 0.5, with blue males with probability 0.3, and with green males with probability 0.2. The second time, females mate with red males with probability 0.4, with blue males with probability 0.35, and with green males with probability 0.25. At the end, the results of the two experiments are combined.

41. Suppose that 100 female birds were tested in each experiment. Find the number out of 200 that mated with each type of male, and convert the results into a probability distribution.

42. Suppose that 100 female birds were tested in the first experiment and 200 females in the second. Find the number out of 300 that mated with each type of male, and convert the results into a probability distribution.

43. Suppose that an equal number of female birds were used in each experiment. Use the law of total probability to find the probability distribution in the combined experiment.

44. Suppose that three times as many females birds were used in the first experiment. Use the law of total probability to find the probability distribution in the combined experiment.

45–46 ▪ The p.d.f. for the waiting time X until an event occurs often follows the **exponential distribution** (to be studied in Section 7.6), with the form $g(x) = \alpha e^{-\alpha x}$ for some positive value of α, defined for $x \geq 0$. For each of the following values of α,

a. Find the c.d.f.

b. Plot the p.d.f. and c.d.f.

c. Check that the p.d.f. is the derivative of the c.d.f.

d. Find $\Pr(X \leq 1)$. Indicate this on both of your graphs.

e. Find $\Pr(1 \leq X \leq 3)$.

f. Find $\Pr(1 \leq X \leq 1.01)$ and show that it is approximately $g(1) \cdot 0.01$.

45. $\alpha = 0.5$.

46. $\alpha = 2.0$.

Computer Exercise

47. The p.d.f. for the normal distribution shown in Example 6.6.13 has the rather unlikely looking formula

$$f(x) = \frac{1}{30\pi} e^{-\frac{(x-60.0)^2}{450}}.$$

Graph this function. It is impossible to integrate this function, but figure out how to get your computer to graph the associated c.d.f. Use this to compute the probability that the value lies between 45.0 and 75.0.

6.7 Random Variables

FIGURE 6.7.1
The dart score is a random variable

Many measurements can be made from a single experiment. Rather than setting up a whole new sample space and set of simple events for each measurement, we exploit the common probabilistic structure by using **random variables**. We here introduce the two main types of random variables, **discrete** and **continuous**. The probabilities associated with different values of random variables are displayed in the same way as simple events, with **probability distributions** and p.d.f.'s. We define the mathematical version of the average of a measurement, the **expectation** of a random variable.

6.7.1 Types of Random Variables

Throwing a dart at a dartboard is an experiment of sorts and can be thought of as a metaphor for any experiment. There is a continuum of simple events, one for each point on the board. Often enough, we do not care *exactly* where the dart hit, but only what it scored (Figure 6.7.1). The score is an example of a **random variable**. Because we can figure out the score from the exact position where the dart hit, it is a **function** of the exact location. This function has domain equal to the sample space (all the points on the board) and range equal to the set of possible scores. Formally,

Definition 6.9 A **random variable** is a function from the sample space to some subset of the real numbers. When the number of values of the function is finite, it is called a **discrete random variable.** When the number of values of the function is infinite, it is called a **continuous random variable**.

More scientifically, a random variable is a measurement that depends on the result of an experiment. When the measurement can take on only a finite number of values, the associated random variable is a **discrete random variable**. When the measurement can take on an infinite range of values, the associated random variable is a **continuous random variable**.

We usually denote random variables by capital letters, such as S for score. For example, we denote the event that a dart scored 17 as $S = 17$. There are many random variables associated with a given experiment, just as many different quantities can be measured during that experiment. We distinguish them by representing them with different letters, just as we do with any measured quantity.

Random variables are displayed in the same way as probabilities. Discrete random variables are shown with histograms and cumulative distributions. Continuous random variables are plotted with probability density functions (p.d.f.'s) and cumulative distribution functions (c.d.f.'s)

Example 6.7.1 A Simple Random Variable Associated with Darts

Suppose a player wins if he scores 10 or more, and loses if he scores 9 or less. A win might be worth 1 point in a tournament, and a loss is worth 0. The random variable W that describes whether the player wins takes on the value 1 if the dart hits anywhere that scores 10 or more, and takes on the value 0 if the dart scores less than 10. In other words, we let W represent the number of wins produced by that dart.

A random variable taking on only the values 0 and 1 is called a **Bernoulli random variable**.

Definition 6.10 A Bernoulli random variable is a discrete random variable that takes on the value 1 with probability p and the value 0 with probability $1 - p$.

Example 6.7.2 Histogram for a Bernoulli Random Variable Associated with Darts

If the probability that the dart in Example 6.7.1 scores 10 or more is 0.8, the random variable takes the value 1 with probability 0.8 and 0 with probability 0.2, with the histogram shown in Figure 6.7.2. As before, the height of the bar gives the probability of a particular event.

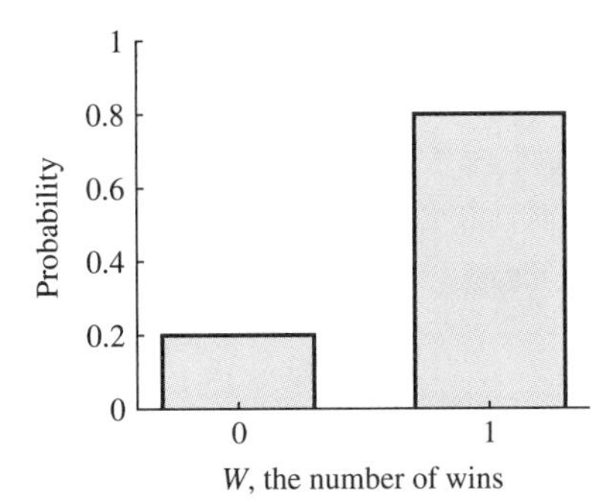

FIGURE 6.7.2
Histogram for a Bernoulli random variables

Example 6.7.3 A Coin Flip as a Bernoulli Random Variable

Suppose we count the number of heads produced by a single flip of a fair coin, representing this by the random variable H (Figure 6.7.3). Flipping heads is the event $H=1$, and $\Pr(H=1)=0.5$. Flipping tails is the event $H=0$, and $\Pr(H=0)=0.5$.

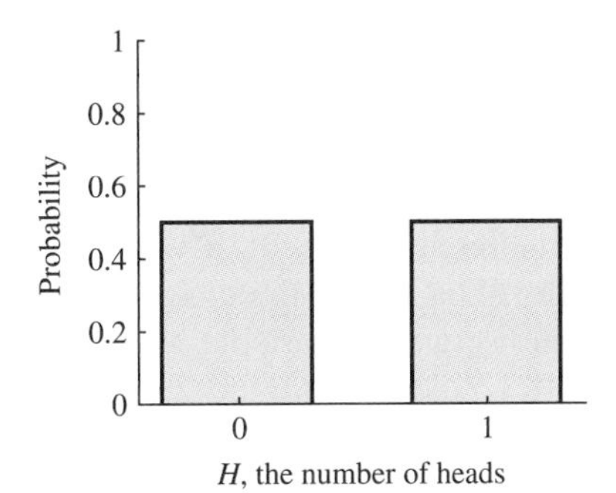

FIGURE 6.7.3
Histogram for coin

Example 6.7.4 Stochastic Immigration as a Bernoulli Random Variable

Recall the stochastic immigration described in Example 6.1.5, where a pair of immigrants arrives with probability 0.5 and no immigrants arrive with probability 0.5. We can describe this process with a Bernoulli random variable, denoted perhaps by the letter I. This random variable takes on the value $I=1$ if one pair arrives, and the value $I=0$ if none arrives. We have assumed that $\Pr(I=1)=0.5$ and $\Pr(I=0)=0.5$, exactly like the fair coin in Example 6.7.3.

The dart score is an intermediate case. The number of possible scores is finite, but greater than two. Random variables that take on a finite number of values are called **discrete random variables**. A Bernoulli random variable is a special case of a discrete random variable.

Example 6.7.5 The Score: A Discrete Random Variable for a Dart

Let the letter S represent that score. To indicate that a dart scored 20, we write $S=20$. The score can take on 44 possible scores. To plot the histogram for the discrete random variable S representing the score, we array the different possible scores along the horizontal axis and indicate their probabilities with the bars (Figure 6.7.4). The scores have been resolved into the 44 possible scores from a dart throw, each with its own probability. For example, the probability that $S=0$ is 0.2 and the probability that $S=20$ is 0.06.

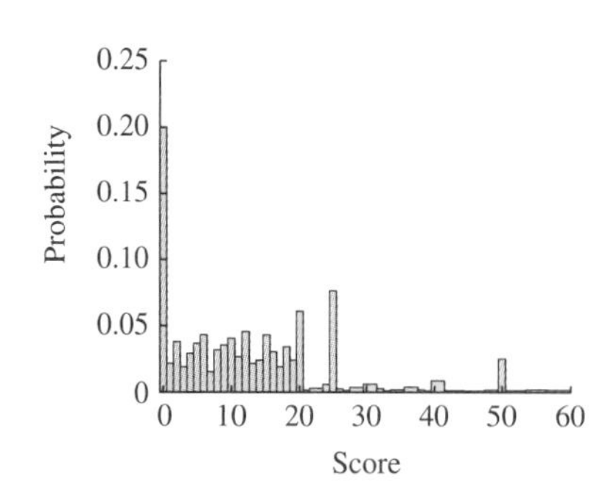

FIGURE 6.7.4
Probability distribution for dart score

Example 6.7.6 A Random Variable with Four Possible Values

Recall the three molecules leaving a cell described in Example 6.3.11. We measured the number, N, left in the cell after 10 minutes. This random variable could take on four values, represented by the events $N = 0$, $N = 1$, $N = 2$, and $N = 3$. Suppose the probabilities are

$$\Pr(N = 0) = 0.4$$
$$\Pr(N = 1) = 0.3$$
$$\Pr(N = 2) = 0.2$$
$$\Pr(N = 3) = 0.1.$$

The histogram describing molecule number is shown in Figure 6.7.5.

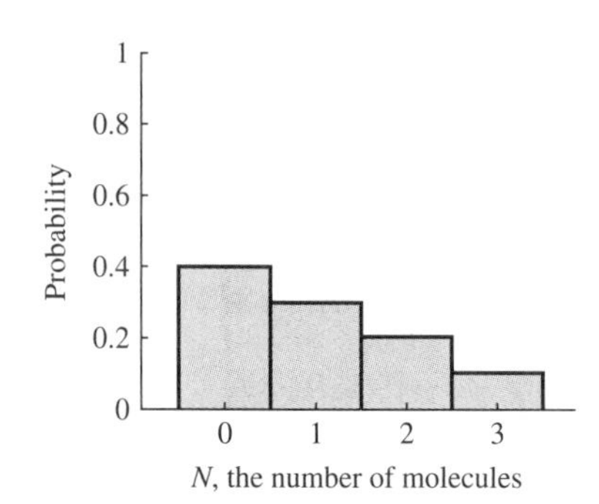

FIGURE 6.7.5
Probability distribution for molecule number

Random variables that take on a whole range of values are called **continuous random variables**. For a dart, the exact distance of the dart from the center of the board is a continuous random variable, which we might denote by R. We indicate the event "the dart hit 2.5 inches from the center" by $R = 2.5$.

Example 6.7.7 The Time a Molecule Leaves a Cell as a Continuous Random Variable

The molecule described in Subsection 6.6.2 could leave the cell at any positive time. We can denote the measured time with the random variable T. This random variable can take on any value. The event that the molecule left at exactly time 1.43 is represented by $T = 1.43$.

Example 6.7.8 A Random Number as a Continuous Random Variable

The random number generator described in Subsection 6.6.2 also can be thought of as a continuous random variable. If we denote the number produced by the letter X, the event $X = \sqrt{0.5}$ indicates that the number chosen was exactly $\sqrt{0.5}$.

Example 6.7.9 A Truncated Random Number is a Discrete Random Variable

When a random number is "measured" differently, by saving only the first digit, the measurement is a discrete random variable that takes on ten values. We might represent this measurement with the letter X_1. The event $X_1 = 7$ indicates that the first digit of the random number was 7.

The probability that a continuous random variable takes on any particular value is often 0. To find actual probabilities, we must integrate its p.d.f., as in Subsection 6.6.3. If the random variable X has p.d.f. $f(x)$, then the probability that the value of X lies between a and b is

$$\Pr(a \leq X \leq b) = \int_a^b f(x)\,dx.$$

Example 6.7.10 The p.d.f. for a Random Number Generator

We found that the p.d.f. for a random number generator is $f(x) = 1.0$ for $0 \le x < 1$. The event that a random number lies between 0.2 and 0.44, as in Example 6.6.8, is then

$$\begin{aligned}\Pr(0.2 \le X \le 0.44) &= \int_{0.2}^{0.44} f(x)\,dx \\ &= \int_{0.2}^{0.44} 1.0\,dx \\ &= x\big|_{0.2}^{0.44} \\ &= 0.44 - 0.2 = 0.24.\end{aligned}$$

Example 6.7.11 The p.d.f. for the Time a Molecule Leaves a Cell

The p.d.f., for the molecule,

$$f(t) = 0.1e^{-0.1t}$$

describes the probabilities associated with the random variable T defined in Example 6.7.7. The event that the molecule left between times 0 and 10 is represented as $0 \le T \le 10$. The probability is then

$$\begin{aligned}\Pr(0 \le T \le 10) &= \int_0^{10} 0.1e^{-0.1t}dt \\ &= -e^{-0.1t}\big|_0^{10} \\ &= -e^{-1.0} + e^{-0.0} = 0.632,\end{aligned}$$

exactly as in Example 6.6.9.

Example 6.7.12 A Possible p.d.f. for a Random Variable Describing Per Capita Production

Recall the per capita production described in Subsection 6.1.2, where the growth of a population depends stochastically on unpredictable factors like the weather. If we suppose that per capita reproduction can never be greater than 4.0, one possible p.d.f. for the random variable R is

$$g(r) = \frac{1}{8}(4 - r)$$

(Figure 6.7.6). This is a mathematically consistent p.d.f. because it takes on only positive values, and because

$$\begin{aligned}\int_0^4 \frac{1}{8}(4 - r)dr &= \frac{r}{2} - \frac{r^2}{16}\Big|_0^4 \\ &= \frac{4}{2} - \frac{4^2}{16} = 1.\end{aligned}$$

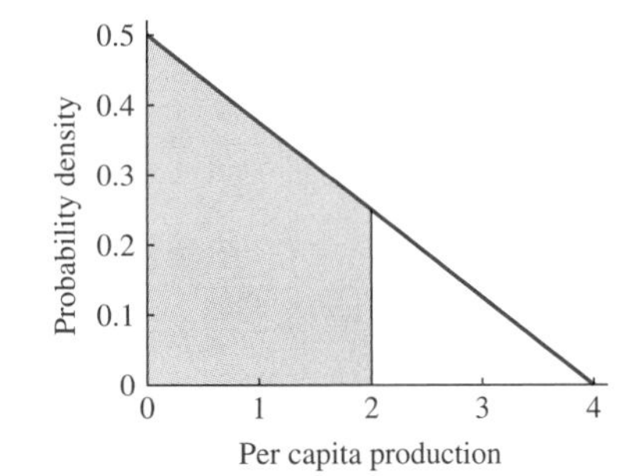

FIGURE 6.7.6

The p.d.f. for a continuous random variable

As before, $g(r)$ is proportional to the probability that R takes on a value near r. Because the p.d.f. is decreasing, this population is more likely to have a small value near its minimum at $R = 0$ than a large value near its maximum at $R = 4$.

Example 6.7.13 Applying the p.d.f. to Find a Probability

The probability that the per capita production is less than 2.0 is found by integrating the p.d.f. over the interval from 0.0 to 2.0, or

$$\begin{aligned}\Pr(0 \le R \le 2) &= \int_0^2 \frac{1}{8}(4 - r)dr \\ &= \frac{r}{2} - \frac{r^2}{16}\Big|_0^2 \\ &= \frac{2}{2} - \frac{2^2}{16} = 0.75.\end{aligned}$$

Example 6.7.14 Finding the c.d.f. and Computing a Probability

The c.d.f. $G(r)$ gives the probability that $R \leq r$. The c.d.f. associated with per capita production is

$$\begin{aligned} G(r) &= \Pr(R \leq r) \\ &= \int_0^r g(s)ds \\ &= \int_0^r \frac{1}{8}(4-s)ds \\ &= \frac{r}{2} - \frac{r^2}{16} \end{aligned}$$

(Figure 6.7.7). We could find the probability that the per capita production is less than 2.0 with the c.d.f. as

$$G(2) = \frac{2}{2} - \frac{2^2}{16} = 0.75,$$

matching the answer in Example 6.7.13.

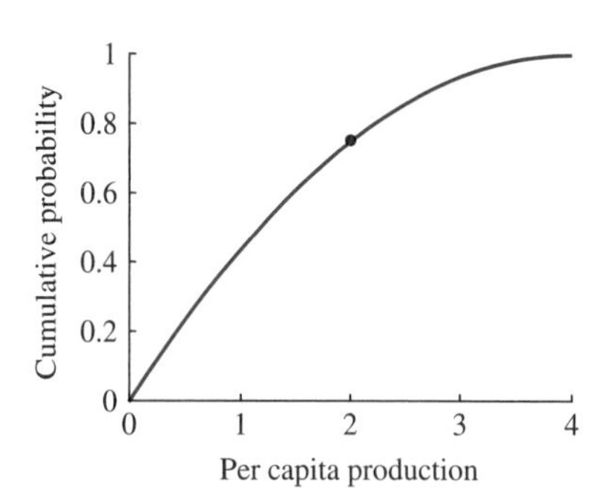

FIGURE 6.7.7
The c.d.f. for a continuous random variable

Example 6.7.15 An Alternative p.d.f. for Per Capita Production

A different population might have a different p.d.f. One possibility is

$$h(r) = \frac{r}{8}$$

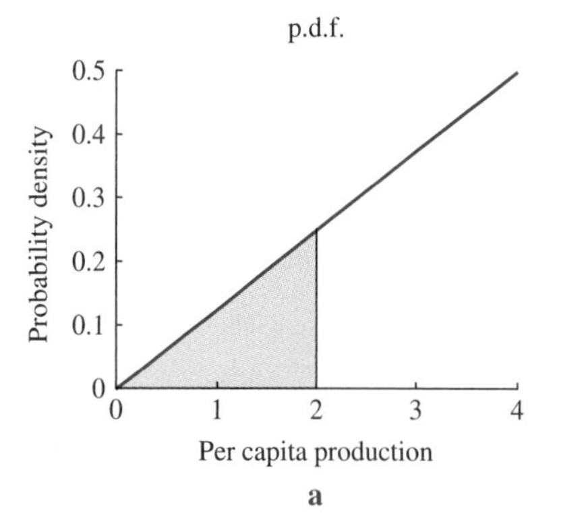

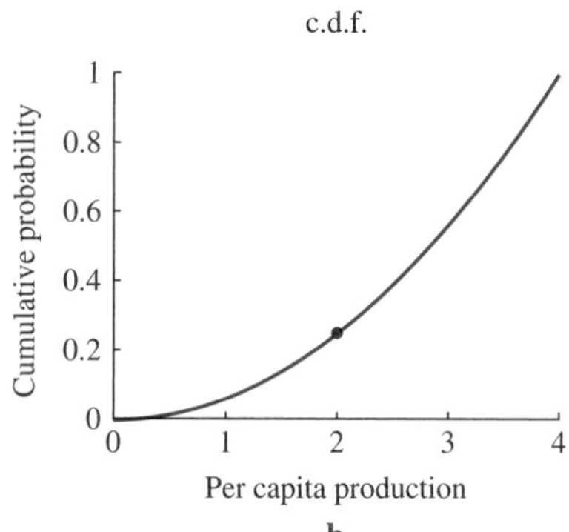

FIGURE 6.7.8
The p.d.f. and c.d.f. for a second population

for $0.0 \leq r \leq 4.0$ (Figure 6.7.8). This population is more likely to have a large per capita production. In fact, the probability that it is less than 2.0 is only

$$\int_0^2 \frac{r}{8}dr = \frac{r^2}{16}\Big|_0^2 = \frac{2^2}{16} = 0.25.$$

The c.d.f. for this population is

$$\begin{aligned} H(r) &= \int_0^r h(s)ds \\ &= \int_0^r \frac{s}{8} ds \\ &= \frac{s^2}{16}\Big|_0^r \\ &= \frac{r^2}{16}. \end{aligned}$$

6.7.2 Expectation of a Discrete Random Variable

Expectation is the mathematician's word for average. If ten students show up for class 75% of the time and six show up 25% of the time (Figure 6.7.9), the **average** number who show up is

$$\text{average number of students in class } = 10 \cdot 0.75 + 6 \cdot 0.25 = 9.$$

The average (also known as the **mean** or **arithmetic mean**) number of students is nine, even though the exact number never takes on this value. We multiplied each value of the measurement (the number of students) by the probability of that measurement and added up the resulting products.

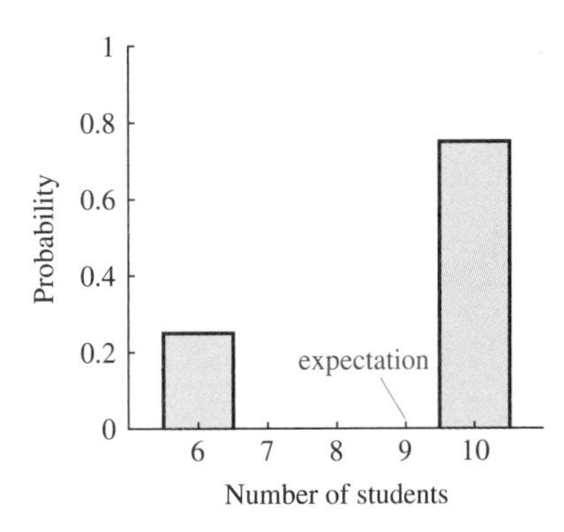

FIGURE 6.7.9
A simple discrete random variable

The definition of the **expectation** of a random variable formalizes this approach. Let $v_1, \ldots, v_n$ denote the n possible values of a random variable V. Suppose that $V = v_i$ with probability p_i. The expectation of V is often denoted by $E(V)$ or $\bar{V}$, and is defined as

Definition 6.11 The **expectation** $E(V)$ or $\bar{V}$ of a discrete random variable V taking on the value v_i with probability p_i is

$$E(V) = \bar{V} = \sum_{i=1}^{n} v_i p_i.$$

To find the expectation, multiply the values by the probabilities and add up the resulting products.

Example 6.7.16 Using the Definition to Find the Expectation of a Discrete Random Variable

In the class, there are two possible values, which we can denote as $v_1 = 10$ and $v_2 = 6$. The associated probabilities are $p_1 = 0.75$ and $p_2 = 0.25$. If we represent the number of students with the random variable N, then

$$EN = v_1 p_1 + v_2 p_2 = 10 \cdot 0.75 + 6 \cdot 0.25 = 9.0.$$

Example 6.7.17 The Expectation of a Bernoulli Random Variable for Immigration

Recall the random variable describing immigration in Example 6.7.4, where a pair of immigrants arrives with probability 0.5 and no immigrants arrive with probability 0.5. This random variable takes on the value $I = 1$ if one pair arrives, and the value $I = 0$ if none arrives. We have assumed that $\Pr(I = 1) = 0.5$ and $\Pr(I = 0) = 0.5$. Then

$$EI = 1 \cdot 0.5 + 0 \cdot 0.5 = 0.5.$$

On average, one half a pair arrives each year.

Example 6.7.18 The Expectation of a Discrete Random Variable for Molecule Number

The number of molecules remaining in a cell (from Example 6.7.6) are described by the random variable N with four values,

$$\Pr(N = 0) = 0.4, \ \Pr(N = 1) = 0.3, \ \Pr(N = 2) = 0.2, \ \Pr(N = 3) = 0.1.$$

Then

$$EN = 0 \cdot 0.4 + 1 \cdot 0.3 + 2 \cdot 0.2 + 3 \cdot 0.1 = 1.0.$$

On average, one molecule remains.

Example 6.7.19 Expected Number of Molecules Revisited

Consider again the number of molecules left in a cell at three different times (Example 6.6.1).

Number Left	Probability at $t = 1$	Probability at $t = 4$	Probability at $t = 8$
0	0.000	0.000	0.004
1	0.000	0.000	0.027
2	0.000	0.004	0.092
3	0.000	0.019	0.186
4	0.000	0.064	0.246
5	0.001	0.147	0.223
6	0.011	0.234	0.141
7	0.057	0.255	0.061
8	0.194	0.183	0.017
9	0.387	0.077	0.003
10	0.349	0.015	0.000

We can think of the measurements at times 1, 4, and 8 as three random variables, which we denote by N_1, N_4, and N_8, respectively. We can use our formula to find the expectation of each. At $t = 1$,

$$E(N_1) = \bar{N}_1 = 0 \cdot 0.0 + 1 \cdot 0.0 + 2 \cdot 0.0 + 3 \cdot 0.0 + 4 \cdot 0.0 + 5 \cdot 0.001 + 6 \cdot 0.011 + 7 \cdot 0.057 + 8 \cdot 0.194 + 9 \cdot 0.387 + 10 \cdot 0.349 = 9.0.$$

On average, there are exactly nine molecules in the cell after 1 minute. Similarly, we can use the formula for expectation to compute

$$E(N_4) = \bar{N}_4 = 6.55$$

$$E(N_8) = \bar{N}_8 = 4.30$$

(Exercises 5 and 6).

The expectation is probably the most widely used **statistic**. Recall that a statistic is one number that summarizes one or more experiments (Definition 6.1). The expectation summarizes the probability distribution in one way. The information in this statistic is not complete; we cannot figure out the values and probabilities from the expectation. Knowing that nine students attend class on average does not tell us whether exactly nine ever do, or with what probability.

6.7.3 The Expectation of a Continuous Random Variable

The idea of adding the value times the probability to find the expectation does not quite work for continuous random variables because each simple event has probability 0. By returning to the Riemann sum approach, we will find the expectation of a continuous random variable with an **integral**.

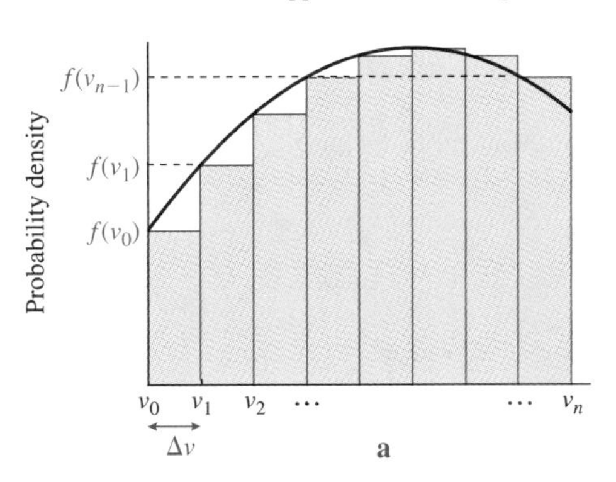

FIGURE 6.7.10
Approximating a continuous random variable

By breaking the set of all possible values of the random variable into pieces (Figure 6.7.10), we can approximate a continuous random variable with a discrete random variable. As in the left-hand Riemann approximation to a function (Section 4.4), we approximate the values of the continuous random variable with a discrete random variable that takes on the values $v_0, \ldots, v_{n-1}$ with probabilities $f(v_0)\Delta v, \ldots, f(v_{n-1})\Delta v$. Scientifically, this corresponds to being unable to distinguish the values in the interval v_i to $v_i + \Delta v$ from v_i itself. The expectation of the approximate discrete random variable is

$$\sum_{i=0}^{n-1} v_i f(v_i)\Delta v.$$

This has precisely the form of a Riemann sum. Taking the limit as $n \to \infty$ and $\Delta v \to 0$ gives a definite integral, leading us to the definition of the expectation of a continuous random variable.

Definition 6.12 The **expectation** $E(V)$ or $\bar{V}$ of a continuous random variable V taking on values between a and b with p.d.f. $f(v)$ is

$$E(V) = \bar{V} = \int_a^b vf(v)dv.$$

Again, we can think of the expectation as the "sum" of the value times the probability, where here the sum is an integral, and the probability is the p.d.f.

Example 6.7.20 Finding the Expectation of the Per Capita Production

Consider the p.d.f. describing the per capita production R

$$g(r) = \frac{1}{8}(4 - r)$$

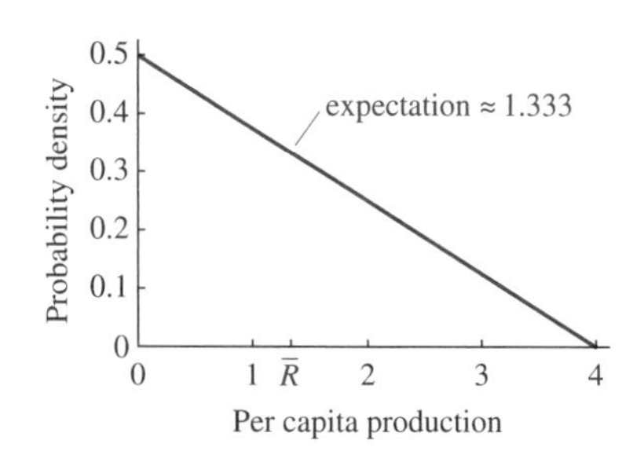

FIGURE 6.7.11

The p.d.f. and expectation for per capita production

defined for $0 \le r \le 4$ (Figure 6.7.11, Example 6.7.12). To find the expectation $\bar{R}$, we integrate

$$\begin{aligned}\bar{R} &= \int_0^4 rg(r)dr \\ &= \int_0^4 r\frac{1}{8}(4-r)dr \\ &= \int_0^4 \frac{1}{8}(4r - r^2)dr \\ &= \frac{r^2}{4} - \frac{r^3}{24}\Big|_0^4 = \frac{4}{3}.\end{aligned}$$

The mean production is less than 2.0 (the midpoint of the domain) because the p.d.f. is larger for values near 0.

Example 6.7.21 Finding the Expectation of the Per Capita Production II

For the population in Example 6.7.15, the p.d.f. for the per capita production R is

$$h(r) = \frac{r}{8}$$

(Figure 6.7.12). The expectation is

$$\bar{R} = \int_0^4 rh(r)dr = \int_0^4 r\frac{r}{8}dr = \int_0^4 \frac{r^2}{8}dr = \frac{r^3}{24}\Big|_0^4 = \frac{8}{3}.$$

The mean production is greater than 2.0 because the p.d.f. is larger for values near 4.0.

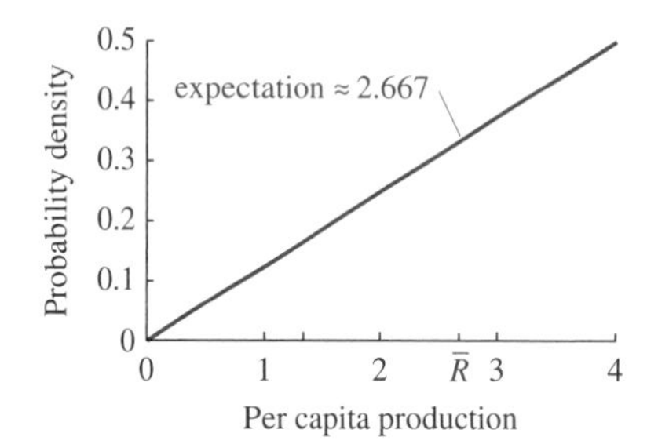

FIGURE 6.7.12

The p.d.f. and expectation for per capita production in the second population

Example 6.7.22 The Expectation of a Random Number

The p.d.f. for the random variable X describing a random number is $f(x) = 1.0$ for $0 \le x < 1$ (Example 6.7.10). The expectation is then

$$E(X) = \int_0^1 xf(x)\,dx = \int_0^1 x\,dx = \frac{x^2}{2}\Big|_0^1 = \frac{1^2}{2} - \frac{0^2}{2} = \frac{1}{2}.$$

Summary A measurement made from an experiment is thought of as a **random variable**, a function from the sample space (all possible experimental outcomes) to the real numbers. Important examples include **Bernoulli random variables** that take on the two values 0 and 1, **discrete random variables** that take on a finite number of values, and **continuous random variables** that take on a continuum of values. Random variables are described graphically with the same tools as probabilities: histograms and cumulative distributions in the discrete case and probability density functions and cumulative distribution functions in the continuous case. The **expectation** of a random variable, defined to match the intuitive idea of average, is found by multiplying the value by the probability and summing (in the discrete case) or integrating (in the continuous case).

6.7 Exercises

Mathematical Techniques

1–4 ▪ For the data presented in Section 6.6, Exercises 1–4, write the results in terms of a random variable and find the expectation. What fraction of experiments have a result less than the expectation?

1. Experiment a.
2. Experiment b.
3. Experiment c.
4. Experiment d.

5–6 ▪ Find the expected number of molecules in the cell using the data in Example 6.7.19 at the following times.

5. At time 4.
6. At time 8.

7–10 ▪ Find the expectation of the continuous random variables with the given p.d.f. Find the probability that the random variable has a value less than the expectation.

7. The p.d.f. of a random variable X is $f(x) = 2x$ for $0 \le x \le 1$ (as in Section 6.6, Exercise 17).
8. The p.d.f. of a random variable X is $f(x) = 1 - \frac{x}{2}$ for $0 \le x \le 2$ (as in Section 6.6, Exercise 18).
9. The p.d.f. of a random variable T is $h(t) = \frac{1}{t}$ for $1 \le t \le e$ (as in Section 6.6, Exercise 19).
10. The p.d.f. of a random variable T is $g(t) = 6t(1-t)$ for $0 \le t \le 1$ (as in Section 6.6, Exercise 20).

11–14 ▪ For each continuous random variables with the given p.d.f., find a discrete random variable that approximates it. Graph the histogram for this discrete random variable and compare it with the p.d.f. for the continuous random variable. Find the expectation of the discrete random variable, and check whether it is equal to the expectation of the continuous random variable.

11. The p.d.f. of a random variable X is $f(x) = 2x$ for $0 \le x \le 1$ (as in Section 6.6, Exercises 17 and 7). Suppose that measurements are very imprecise, and that all values of $X \le 0.5$ are recorded as 0.25 and all values of $X > 0.5$ are recorded as 0.75. Write a random variable describing these imprecise measurements, find the associated probabilities, and compute the expectation.
12. The p.d.f. of a random variable X is $f(x) = 1 - \frac{x}{2}$ for $0 \le x \le 2$ (as in Section 6.6, Exercises 18 and 8). Suppose that measurements are very imprecise, and that all values of $X \le 1$ are recorded as 0.5 and all values of $X > 1$ are recorded as 1.5. Write a random variable describing these imprecise measurements, find the associated probabilities, and compute the expectation.
13. The p.d.f. of a random variable X is $f(x) = 2x$ for $0 \le x \le 1$ (as in Section 6.6, Exercises 17 and 7). Suppose that measurements are imprecise, and that all values of $X \le 0.25$ are recorded as 0.25, all values of $0.25 < X \le 0.5$ are recorded as 0.5, all values of $0.5 < X \le 0.75$ are recorded as 0.75, and all values of $0.75 < X$ are recorded as 1.0.
14. The p.d.f. of a random variable X is $f(x) = 1 - \frac{x}{2}$ for $0 \le x \le 2$ (as in Section 6.6, Exercises 18 and 8). Suppose that measurements are imprecise, and that all values of $X \le 0.5$ are recorded as 0.25, all values of $0.5 < X \le 1.0$ are recorded as 0.75, all values of $1.0 < X \le 1.5$ are recorded as 1.25, and all values of $1.5 < X$ are recorded as 1.75.

15–16 ▪ Check that the following could be p.d.f.'s and compute their expectations. Does anything seem odd about them?

15. $f(x) = \frac{1}{2\sqrt{x}}$ for $0 < x \le 1$.
16. $g(t) = \frac{1}{t^2}$ for $1 \le t < \infty$.

Applications

17–22 ▪ Think about one or more molecules independently leaving a cell, each with probability 0.9 in a given second. Find the random variable describing the following events, find the probabilities of the outcomes, and find the expectation.

17. A Bernoulli random variable describing whether a molecule is in or out at time 1.
18. A Bernoulli random variable describing whether a molecule is in or out at time 2.
19. A Bernoulli random variable describing whether two molecules are together (both in or both out) or separate at time 1.
20. A Bernoulli random variable describing whether three out of three molecules remain inside at time 1.
21. A discrete random variable that counts the number out of two molecules that are in at time 1.
22. A discrete random variable that counts the number out of two molecules that are in at time 2.

23–26 ▪ For the data presented in Section 6.6, Exercises 37–40, write the results in terms of a random variable and find the expectation. What fraction of experiments have a result less than the expectation?

23. Experiment a
24. Experiment b
25. Experiment c
26. Experiment d

27–28 ▪ Consider again the cells in Section 6.4, Exercises 29 and 30, but suppose that older cells, instead of not staining as often, do not stain as well. In particular, they produce a brightness of 7, while young cells have a brightness of 9. Write a random variable describing the brightness, and find its probability distribution and its expectation.

27. Suppose 30% of the cells are young.
28. Suppose 70% of the cells are young.

29–30 ▪ Consider again the cells in Section 6.4, Exercises 31 and 32, but suppose that older cells, instead of not staining as often, do not stain as well. New cells have a brightness of 9.5, 1-day-old cells have a brightness of 9.0, 2-day-old cells have a brightness of 8.0, and 3-day-old cells have a brightness of 5.0. Write a random variable describing the brightness, and find its probability distribution and its expectation.

29. Suppose

$$\Pr(\text{cell is 0 day old}) = 0.4$$
$$\Pr(\text{cell is 1 day old}) = 0.3$$
$$\Pr(\text{cell is 2 days old}) = 0.2$$
$$\Pr(\text{cell is 3 days old}) = 0.1.$$

30. The lab finds a way to eliminate the oldest cells (more than 3 days old) from its stock.

31–34 ▪ Suppose immigration and emigration change the sizes of four populations with the following probabilities.

Population a		Population b		Population c		Population d	
Number	Probability	Number	Probability	Number	Probability	Number	Probability
−1	0.4	−1	0.1	−1	0.4	−10	0.4
0	0.2	0	0.3	0	0.2	0	0.2
1	0.3	1	0.2	1	0.3	1	0.3
2	0.1	2	0.4	100	0.1	2	0.1

Write the result as a random variable and find the expectation.

31. Population a. How many immigrants do you think would arrive (or depart) in 10 years? Will the population grow?

32. Population b. How many immigrants do you think would arrive (or depart) in 10 years? Will the population grow?

33. Population c. How many immigrants would arrive (or depart) in 10 years? Will the population grow? Does the expectation seem close to the "middle" of the distribution?

34. Population d. About how many immigrants would arrive (or depart) in 10 years? Will the population grow? Does the expectation seem close to the "middle" of the distribution?

35–36 ▪ As in Section 6.6, Exercises 45 and 46, the p.d.f. for the waiting time X until an event occurs often follows the **exponential distribution**, with the form $g(x) = \lambda e^{-\lambda x}$ for some positive value of λ, defined for $x \geq 0$. Use the following indefinite integral fact to find the expectation for the following values of λ.

$$\int \lambda x e^{-\lambda x}\, dx = \frac{-\lambda x e^{-\lambda x} - e^{-\lambda x}}{\lambda}.$$

35. $\lambda = 0.5$.

36. $\lambda = 2.0$.

37–40 ▪ The formula for the expectation in mathematics is identical to the formula for the **center of mass** in physics. Mass acts like probability, and distance acts like the measured random variable. For example, suppose two people are on a see saw, and one is 2 m to the right of center and weighs 60 kg, and the other is 3 m to the right of center and weighs 50 kg. The proportion of mass in the first person is 60/110, and the proportion in the second is 50/110. The center of mass is $2 \cdot \frac{60}{110} + 3 \cdot \frac{50}{110} = \frac{270}{110} = 2.455$. In terms of balancing, these two people act like a single 110 kg person at a position 2.455 meters to the right of center.

37. A 20 kg child is the end of a long see saw, 3 m to the right of center. An 80 kg adult is 1 m to the left of center. Find the center of mass. Who will go up?

38. A 20 kg child is 3 m to the right of center, her 30 kg older brother is 1 m to the right of center, and an 80 kg adult is 1 m to the left of center. Find the center of mass. Who will go up?

39. A 20 kg child is 3 m to the right of center. An 80 kg adult wishes to balance the see saw by sitting a distance x to the left of center. They will balance if the center of mass is exactly 0. Solve for x.

40. A 20 kg child is 3 m to the right of center, her 30 kg older brother is 1 m to the right of center, and an 80-kg adult wishes to balance the see saw by sitting a distance x to the left of center. Solve for x.

41–44 ▪ The relation between the mathematical expectation and the center of mass in physics also holds for continuous distributions. Mass density acts like probability density (after the mass density has been divided by the total density). For example, suppose the density of a 1 m long bar is $\rho(x) = 4x$ kg/m. Then the total mass is $\int_0^1 \rho(x)\,dx = \int_0^1 4x\,dx = 2x^2|_0^1 = 2$. Dividing $\rho(x)$ by this total mass of 2 gives a density $f(x) = 2x$, which has integral 1, like a p.d.f. The expectation of a random variable X with this p.d.f. is

$$E(X) = \int_0^1 x f(x)\,dx = \int_0^1 2x^2\,dx = \frac{2x^3}{3}\Big|_0^1 = \frac{2}{3}.$$

41. Find the center of mass of a bar with mass density $\rho(x) = x(1 - x^2)$ for $0 \leq x \leq 1$. Is it to the left or the right of the center of the bar at $x = 1/2$? Is it at the point where $\rho(x)$ takes on its maximum? Sketch a graph.

42. Find the center of mass of a bar with mass density $\rho(x) = x^2(2 - x)$ for $0 \leq x \leq 2$. Is it to the left or the right of the center of the bar at $x = 1$? Is it at the point where $\rho(x)$ takes on its maximum?

43. As in Exercise 37, a 4 m long wooden board has a fulcrum placed 1 m from the left end. A 20 kg child sits at the right end of the board (at position $x = 3$) and an 80 kg adult sits at the left end (at position $x = -1$). Suppose that the board has density of 5 kg/m. Find the center of mass. Who will go up?

44. As in Exercise 37, a 4 m long wooden board has a fulcrum placed 1 m from the left end. A 20 kg child sits at the right end of the board (at position $x = 3$), and an 80 kg adult sits at the left end (at position $x = -1$). Suppose that the board has density of y kg/m. Find the value of y for which the board will balance.

Computer Exercises

45. Your computer should have a way to roll a random die (giving results 1 through 6 each with equal probability). Roll such a die 5, 10, 20, 50, and 100 times and find the average score. How close is each to the expectation?

46. Consider again the population growing by stochastic immigration as in Section 6.1, Exercise 36, where an immigrant arrives each year with probability 0.5.

a. Generate two 50-generation solutions starting from populations of 0.

b. What is the expected population as a function of time?

c. Does the solution get closer or farther from the expectation?

d. Find the average number of immigrants that have arrived by each time (if 13 immigrants arrived in the first 25 time steps, the average is 13/25 = 0.52 per year).

e. Does the average get closer or farther from the expectation?

47. Your computer should have a way to choose a random number between 0 and 1 (using the uniform p.d.f.). Pick 5, 10, 20, 50, and 100 such numbers and find the average. How close is each to the expectation?

6.8 Descriptive Statistics

The expectation is an important statistic that gives one idea of the "average" of a random variable. Many other statistics usefully summarize random variables. Like the expectation, they give an idea of what to "expect" from an experiment, but have different applications. The **median** lies right in the middle of the probabilities, the **mode** is the most probable single value, and the **geometric mean** is the appropriate average for random variables that are multiplied rather than added.

6.8.1 The Median

The expectation, as a generalization of the average, might seem like the most natural measure of the central value of a random variable. The expectation, however, can be rather far from the "middle" of a distribution. A few extreme values can pull the expectation far from the bulk of the values. Two alternative statistics, the **median** and the **mode**, capture different intuitive ideas of the middle.

The median is defined to be right in the middle of the probabilities. Picking a value less than the median is exactly as likely as picking a value greater than the median. Consider the probability distribution of dart scores (Figure 6.8.1a). The median is the score that gets beaten exactly half the time. The best way to find the median is to plot the cumulative distribution and find where it crosses 0.5 (Figure 6.8.1b). In this case, the median is near 10. It is just as likely to get a score less than 10 as a score greater than 10.

When is the median a good measure of the middle? Unlike the expectation, the median is not sensitive to a few large values. In the figure, the probability of getting a bull's-eye is 0.025. If the score for a bull's-eye were increased to 6.02×10^{23}, the

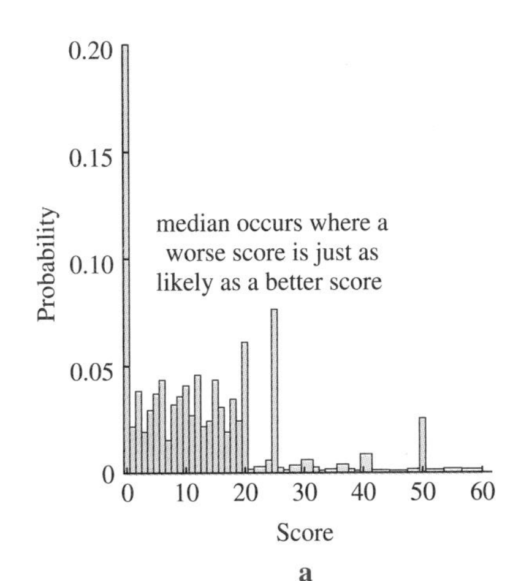

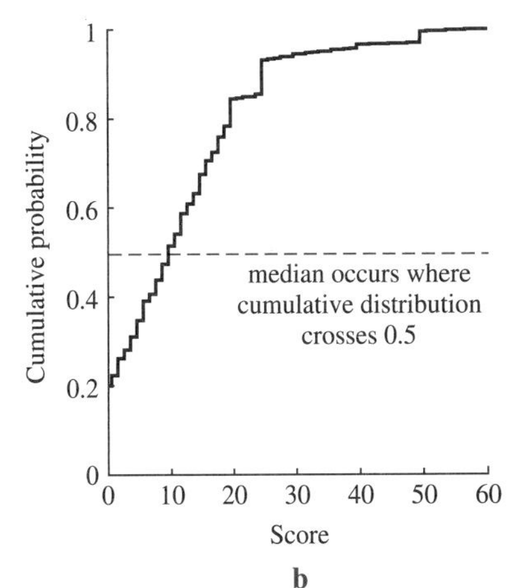

FIGURE 6.8.1
The median of the dart scores

expectation would increase to 1.5×10^{22}. The median would remain the same. When we are interested in a "typical" score, the median can be a better measure of the middle.

Algorithm 6.1 Finding the Median of a Discrete Random Variable

1. Compute the cumulative distribution.
2. Find the point where the cumulative distribution crosses 0.5. If it hits exactly 0.5 at some value v_i, this is the median. If it hits between values v_i and v_{i+1}, the median lies between v_i and v_{i+1}.

The median is generally harder to compute than the mean. To find the median, we have to find the cumulative distribution, and solve for when it is equal to 0.5. The expectation can be computed by multiplying and adding, without any sorting or solving.

Example 6.8.1 Finding the Median of a Discrete Random Variable

Recall the probability distribution for the number of molecules out of ten left in a cell after 8 minutes,

Number Left	Probability at $t=8$	Cumulative Probability
0	0.004	0.004
1	0.027	0.031
2	0.092	0.123
3	0.186	0.309
4	0.246	0.555
5	0.223	0.778
6	0.141	0.919
7	0.061	0.980
8	0.017	0.997
9	0.003	1.000
10	0.000	1.000

From the cumulative distribution, we can see that the probabilities cross 0.5 between 3 and 4 (Figure 6.8.2). This is the median number of molecules.

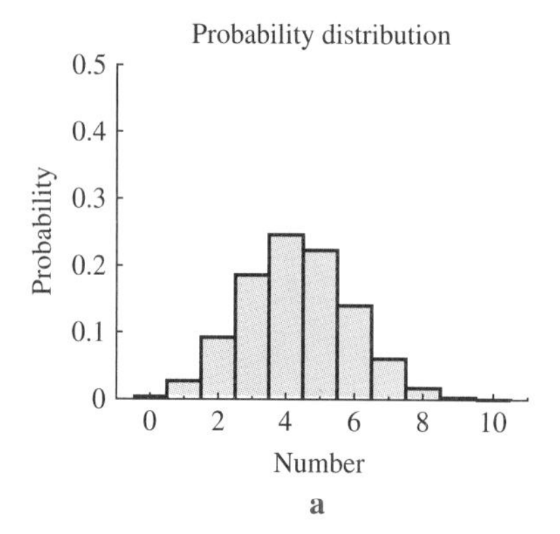

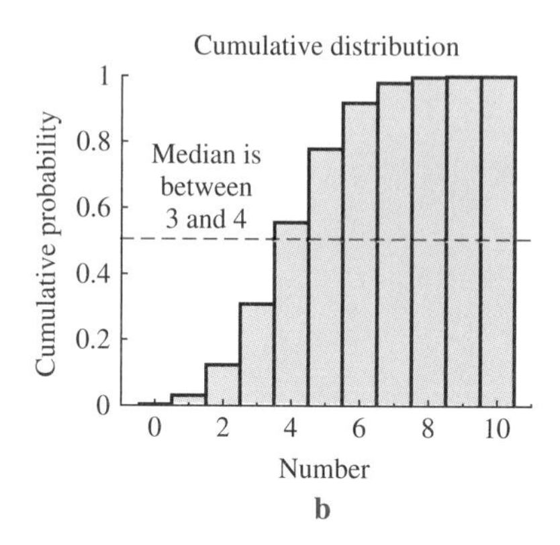

FIGURE 6.8.2
The median number of molecules

Example 6.8.2 The Median of a Random Variable with Four Possible Values

The random variable N describing the number of molecules left in a cell after 10 minutes (Example 6.7.6) takes on only four values, with probabilities

$$\Pr(N=0)=0.4$$
$$\Pr(N=1)=0.3$$
$$\Pr(N=2)=0.2$$
$$\Pr(N=3)=0.1.$$

The cumulative distribution is

$$\Pr(N<=0)=0.4$$
$$\Pr(N<=1)=0.7$$
$$\Pr(N<=2)=0.9$$
$$\Pr(N<=3)=1.0.$$

The median, which lies between 0 and 1, is not a particularly useful statistic for a random variable that takes on only a few values.

The median is most useful for continuous random variables, which take on an infinite number of values. As before, the median is defined as the point where the cumulative probability is 0.5. With a continuous random variable, the c.d.f. is a continuous function that increases from 0 to 1. The Intermediate Value Theorem (Subsection 3.4.1) guarantees that the cumulative probability is exactly equal to the intermediate value 0.5 at some point.

More formally, the median $\tilde{X}$ of the random variable X with p.d.f. $f(x)$ and c.d.f. $F(x)$ is the solution of the equation

$$\Pr(X \le \tilde{X}) = F(\tilde{X}) = 0.5$$

(Figure 6.8.3).

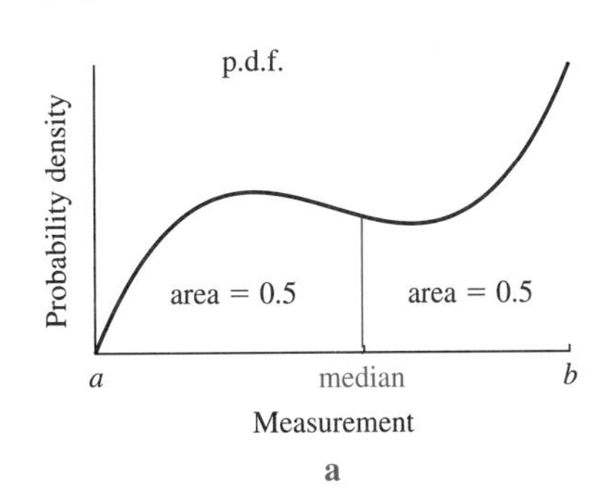

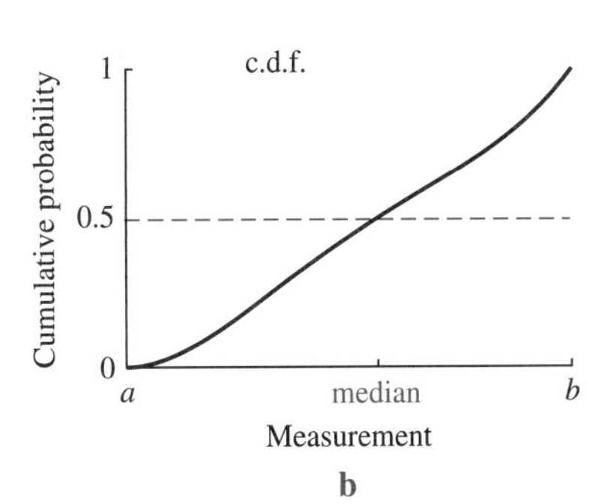

FIGURE 6.8.3
The median of a continuous random variable

Algorithm 6.2 Finding the Median of a Continuous Random Variable

1. Compute the cumulative distribution.
2. Solve for the point where the cumulative distribution crosses 0.5.

Example 6.8.3 The Median for a Random Number Generator

The c.d.f. for a random number generator (Example 6.6.15) with p.d.f. $f(x)=1$ for $0 \le x < 1$ is

$$F(x) = \int_0^x f(y)dy = \int_0^x 1dy = y|_0^x = x - 0 = x.$$

The median is the solution of the equation $F(x)=0.5$. But because $F(x)=x$, this solution is $x=0.5$, matching the mean found in Example 6.7.22. As we might expect, a random number between 0 and 1 is just as likely to be less than 0.5 as is it to be greater.

Example 6.8.4 The Median for a Departing Molecule

The time T at which a molecule leaves a cell has p.d.f.

$$f(t) = 0.1e^{-0.1t}$$

if the departure rate is 0.1/s (Figure 6.8.4, Example 6.7.11) and c.d.f.

$$F(t) = 1 - e^{-0.1t}$$

(Example 6.6.14). The median occurs where $F(t) = 0.5$. Solving

$$\begin{aligned} 1 - e^{-0.1t} &= 0.5 \\ e^{-0.1t} &= 0.5 \\ -0.1t &= \ln(0.5) \\ t = -\frac{\ln(0.5)}{0.1} &= 6.93. \end{aligned}$$

Finding the expectation requires integration by parts and is presented in Section 7.6.

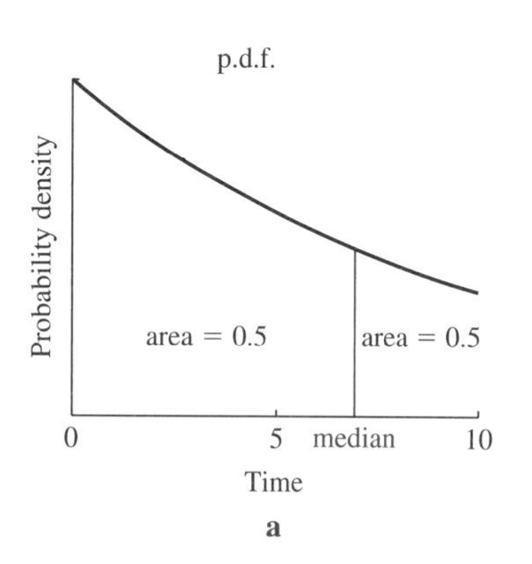

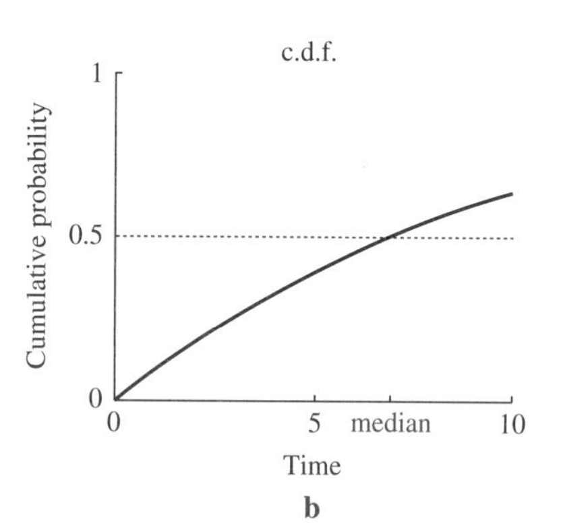

FIGURE 6.8.4
The median for a departing molecule

Example 6.8.5 The Median of the Per Capita Production

Consider the p.d.f. for per capita production given by

$$g(r) = \frac{1}{8}(4 - r)$$

for $0 \le r \le 4$ (Example 6.7.12). We found the c.d.f.

$$G(r) = \frac{r}{2} - \frac{r^2}{16}$$

(Example 6.7.14). The median $\tilde{R}$ is the solution of

$$G(\tilde{R}) = \frac{\tilde{R}}{2} - \frac{\tilde{R}^2}{16} = 0.5.$$

First, we simplify this quadratic.

$$\frac{\tilde{R}}{2} - \frac{\tilde{R}^2}{16} = 0.5 \quad \text{original equation}$$

$$\frac{\tilde{R}^2}{16} - \frac{\tilde{R}}{2} + 0.5 = 0 \quad \text{move everything to left-hand side and multiply by } -1$$

$$\tilde{R}^2 - 8\tilde{R} + 8 = 0. \quad \text{multiply by 16}$$

Next, we use the quadratic formula to solve, finding

$$\begin{aligned}\tilde{R} &= \frac{8 \pm \sqrt{8^2 - 4 \cdot 8}}{2} \\ &= \frac{8 \pm \sqrt{32}}{2} \\ &= 4 \pm 2\sqrt{2} \\ &= 6.82 \text{ or } 1.18.\end{aligned}$$

The first solution lies outside the range of this random variable and cannot be the median. Thus, the second must be the answer, giving $\tilde{R} = 1.18$. Half of the time, the per capita production will be less than 1.18, and the other half of the time, it will be greater. On the graph of the p.d.f., half of the area lies below the median (Figure 6.8.5a), corresponding to the point where the c.d.f. is equal to 0.5 (Figure 6.8.5b). The median is close to the mean of 1.33, but not exactly equal.

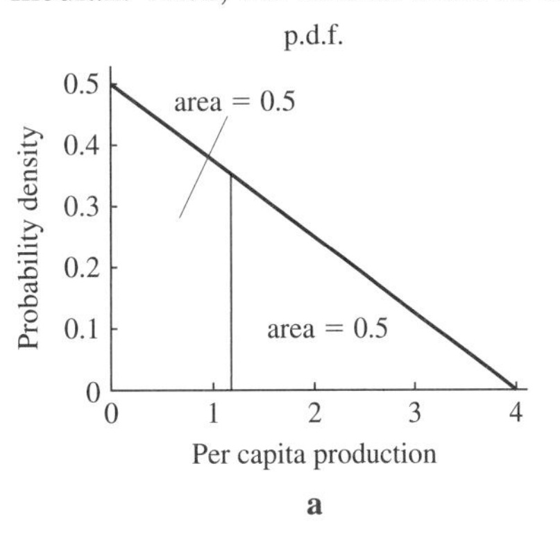

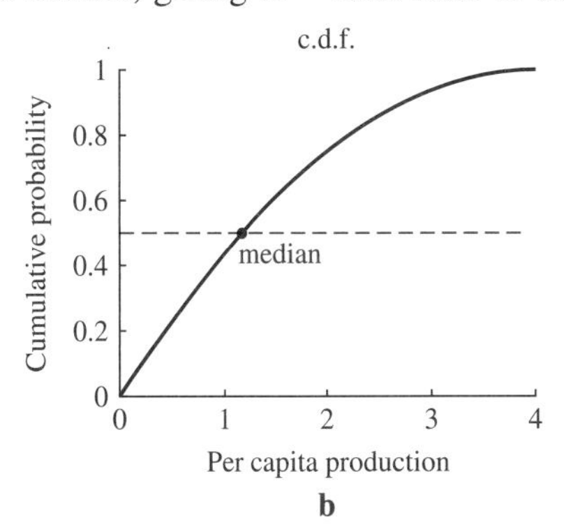

FIGURE 6.8.5
Median for the stochastic production

Example 6.8.6 The Median of the Per Capita Production in a Second Population

For the second population (Example 6.7.15), the p.d.f. is

$$h(r) = \frac{r}{8}$$

for $0 \le r \le 4$. The c.d.f. is

$$H(r) = \frac{r^2}{16}$$

(Example 6.7.15). The median is the solution of

$$H(\tilde{R}) = \frac{\tilde{R}^2}{16} = 0.5.$$

In this case, then $\tilde{R} = \sqrt{8} = 2.83$ (Figure 6.8.6). The median of this random variable is close to, but not equal to, its mean of 2.67.

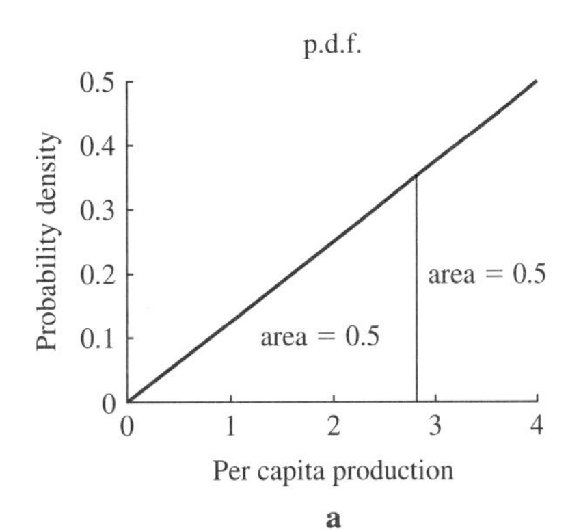

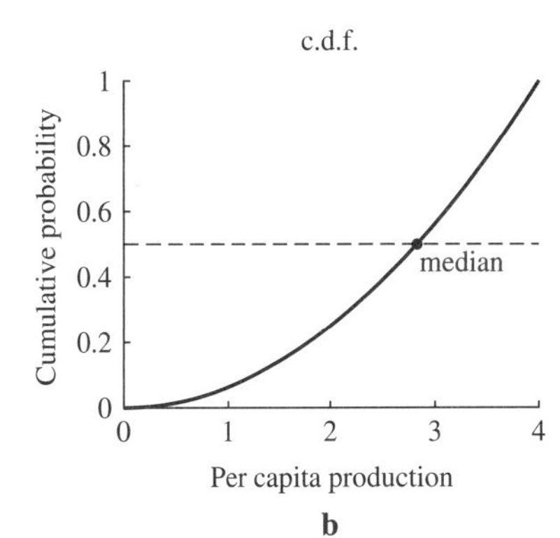

FIGURE 6.8.6
Median for the population

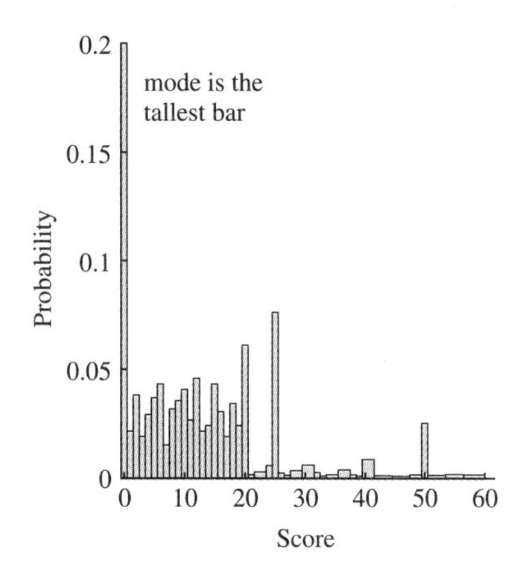

FIGURE 6.8.7
The mode of the dart score

Like most concepts in probability theory, the median can be interpreted in terms of gambling. Because the median score is 10, a single dart is just as likely to score more than 10 as less than 10. Suppose two people had to agree on a number x for the following bet: if the score is less than x, the first person wins, if greater, the second person wins. What value of x produces a fair bet? Two gamblers trained in probability theory would know to choose at the median.

6.8.2 The Mode

For a random variable with many values, the **mode**, defined as the most common measurement, is a useful measure of the middle. The mode of the dart scores is 0, the single most likely score (the tallest bar in Figure 6.8.7). If you threw only one dart, your best guess of the result would be a score of 0. In gambling terms, if you had to bet on any exact score for the dart, you would be wise to bet on 0. Sometimes, we say that the "modal score" is 0.

Example 6.8.7 Modal Number of Molecules at a Given Time

Consider again the number out of ten molecules left in a cell at time 8 (Example 6.7.19). The modal number of molecules left at $t = 8$ is 4, found as the tallest bar in the histogram or the largest value in the probability distribution.

In the same way, the **mode** of a continuous random variable occurs where the p.d.f. takes on its maximum.

Example 6.8.8 The Mode of the Departure Time for a Molecule

The p.d.f.

$$f(t) = 0.1e^{-0.1t}$$

describing the time T when a molecule leaves a cell (Example 6.8.4), has a mode of 0 (Figure 6.8.8). .

Example 6.8.9 The Mode of a More Complicated Continuous Random Variable

Consider the p.d.f. $f(x) = xe^{-x}$ (Figure 6.8.9). This is a simple example of the gamma distribution mentioned in Section 2.8.3. To find the mode, we find where the p.d.f. takes on its maximum value. In this case,

$$f'(x) = e^{-x} - xe^{-x}$$

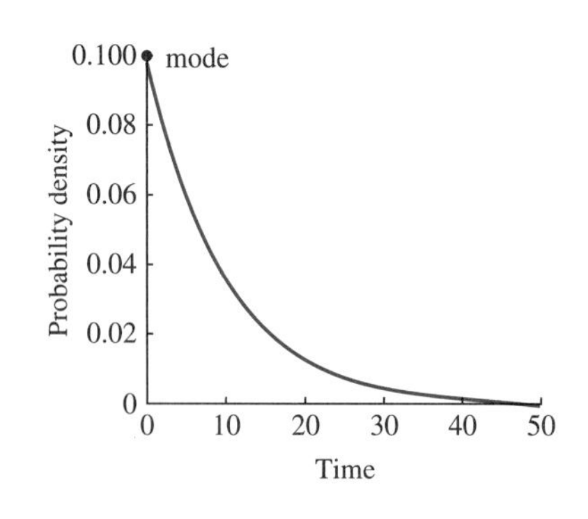

FIGURE 6.8.8
The mode of molecule departure time is at $t = 0$

FIGURE 6.8.9
The mode of a more complicated departure time distribution

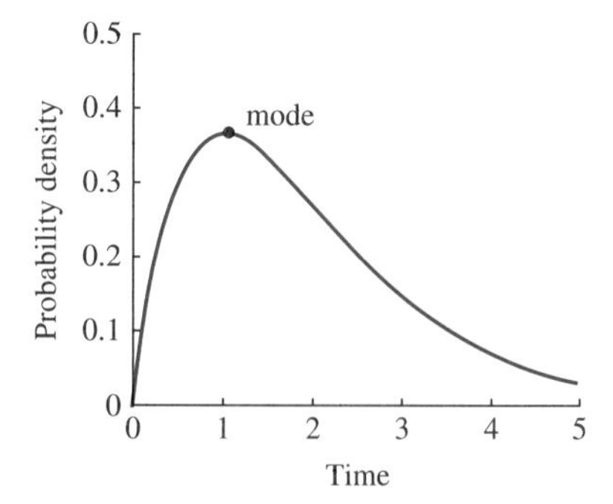

(using the product rule and the chain rule). Solving

$$f'(x) = e^{-x}(1 - x) = 0$$

gives $x = 1$. Because $f(0) = 0$ and $\lim_{x \to \infty} f(x) = 0$, this point is a maximum, and the mode occurs at $x = 1$.

Example 6.8.10 The Modes for Per Capita Production

The mode of the per capita production with p.d.f.

$$g(r) = \frac{1}{8}(4 - r)$$

(Example 6.8.5) is 0. The mode of the per capita production with p.d.f.

$$g(r) = \frac{r}{8}$$

(Example 6.8.6) is 4. In cases like this, the mode does not give a great deal of information.

In summary,

Algorithm 6.3 Finding the Mode of a Random Variable

1. For a discrete random variable, find the largest value in the probability distribution.
2. For a continuous random variable, find the maximum of the probability density function.

Unlike the mean, the mode need not be unique if the maximum of the probability distribution or p.d.f. is not unique. The mode is often the most easily computed measure of the middle, requiring only finding the maximum.

The mean, median, and mode are equal for symmetric distributions and symmetric p.d.f.'s, which have only a single peak in the center (Figure 6.8.10). In this case, which comes up often, we do not have to worry about which statistic is most appropriate.

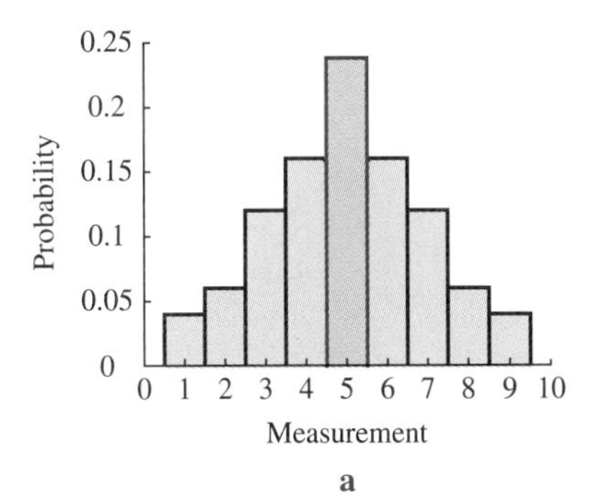

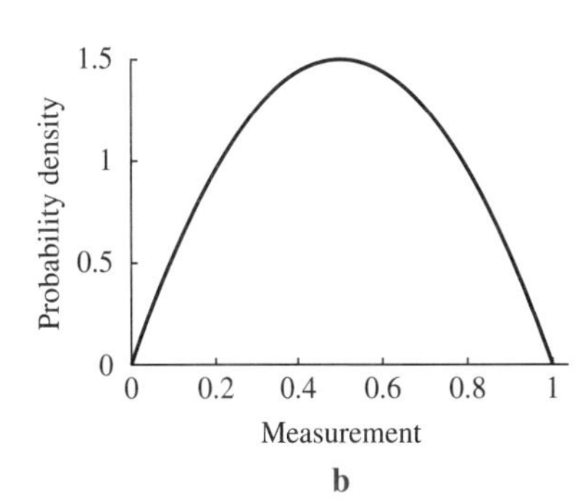

FIGURE 6.8.10
Symmetric distributions

6.8.3 The Geometric Mean

The expectation is computed with addition. Traditionally, mathematical objects involving addition are called **arithmetic**, hence the name arithmetic mean. This mean is appropriate for use with additive biological processes, like the arrival of immigrants. Some biological processes, such as population growth through production, are multiplicative. In this case, the expectation or arithmetic mean can give the wrong idea about the behavior.

Example 6.8.11 The Arithmetic Mean Fails to Identify Whether a Population Will Grow

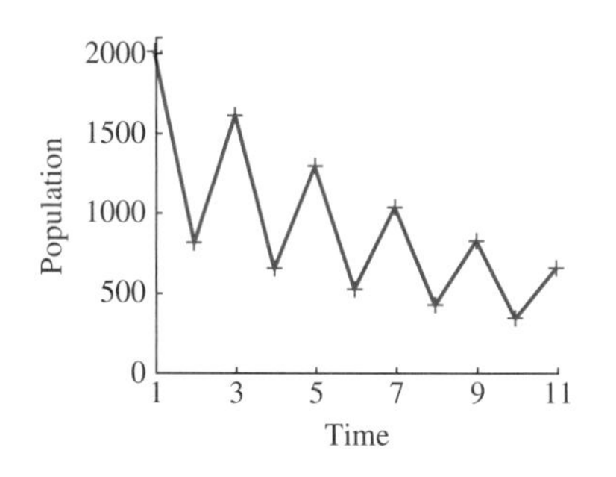

FIGURE 6.8.11

The decline of a population with mean per capita production greater than 1.0

Suppose a population doubles (has per capita production 2.0) in even-numbered years and decreases by 60% (has per capita production 0.4) in odd-numbered years (Figure 6.8.11). Will this population grow? The average per capita production is just the average of 2.0 and 0.4, because these occur equally frequently. Because

$$\frac{2.0+0.4}{2} = 1.2,$$

which is greater than 1.0, we might suspect that this population will grow by about 20% per year. We can check this by working out some actual population sizes. Suppose the population begins at $N_0 = 1000$ in an even-numbered year. Then

$$N_1 = 2.0 \cdot 1000 = 2000$$
$$N_2 = 0.4 \cdot 2000 = 800$$
$$N_3 = 2.0 \cdot 800 = 1600$$
$$N_4 = 0.4 \cdot 1600 = 640.$$

This population is declining. But why? And how fast?

This reasoning is inappropriate because growth is a multiplicative process. Let R_t be a random variable representing the per capita growth in year t. Suppose the population starts at a value of N_0. The population after 1 year is $R_1 N_0$, after 2 years is $R_2 R_1 N_0$, and after t years is

$$N_t = R_t R_{t-1} \ldots R_2 R_1 N_0 \tag{6.8.1}$$

(Figure 6.8.12). Do we expect N_t to be larger or smaller than N_0?

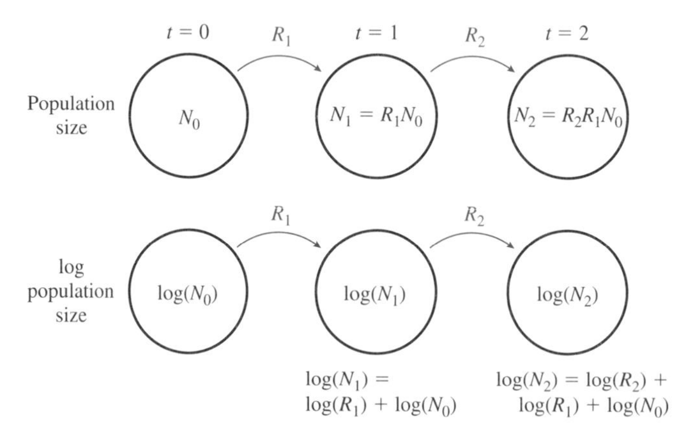

FIGURE 6.8.12

A stochastic production model

The trick is to convert the multiplication into addition with the logarithm. Taking logarithms of both sides of Equation 6.8.1,

$$\ln(N_t) = \ln(R_t) + \ln(R_{t-1}) + \cdots + \ln(R_2) + \ln(R_1) + \ln(N_0).$$

Every year, we **add** the **logarithm** of the per capita production to the log of the population. If the expectation of the terms we are adding is positive, the sum will increase. If the expectation is negative, the sum will decrease. The key quantity is therefore the **expectation of the logarithm** of the per capita production.

Example 6.8.12 The Mean of the Logarithm Shows That Population Declines

The population in Example 6.8.11 can be thought of as having per capita production R, which takes on the value 2.0 with probability 0.5 and the value 0.4 with probability 0.5. We found the arithmetic mean using the usual formula for the expectation, finding

$$E(R) = 2.0 \cdot 0.5 + 0.4 \cdot 0.5 = 1.2.$$

However, the expectation of the logarithm is

$$E[\ln(R)] = \ln(2.0) \cdot 0.5 + \ln(0.4) \cdot 0.5 = -0.11.$$

This value is less than 0, and implies that the population will decline, as we found.

Example 6.8.13 Expectation of the Per Capita Production and Its Natural Logarithm

Suppose a population increases by 50% (per capita production of 1.5) with probability 0.6 and decreases by 50% (per capita production of 0.5) with probability 0.4. Will this population grow? The **expectation** of the per capita production is

$$1.5 \cdot 0.6 + 0.5 \cdot 0.4 = 1.1.$$

Because this "average" growth is greater than 1, we might guess that the population grows.

What is the expectation of the logarithm of the per capita production? We use the same formula for the expectation: multiply the values times the probabilities and add them up. In this case, the values are $\ln(1.5) = 0.405$ with probability 0.6 and $\ln(0.5) = -0.693$ with probability 0.4, so

$$E[\ln(R)] = \ln(1.5) \cdot 0.6 + \ln(0.5) \cdot 0.4 = -0.034.$$

This is negative, meaning that, on average, the logarithm of the population size becomes smaller the more terms we add (Figure 6.8.13).

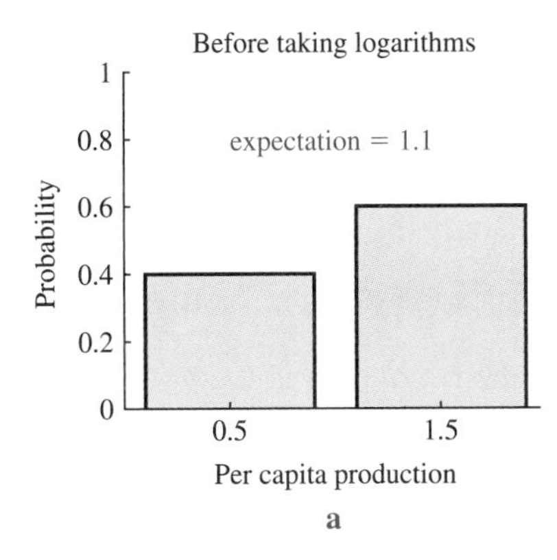

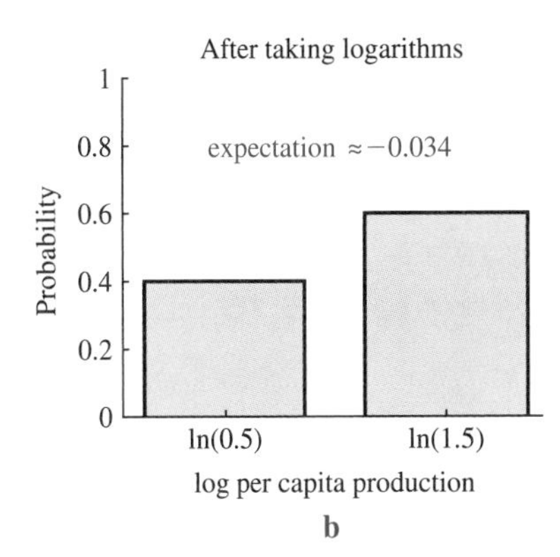

FIGURE 6.8.13
Expectations before and after taking the logarithm

Example 6.8.14 Expected Behavior of a Population

What happens to the population in Example 6.8.13? The logarithm of the population size decreases by −0.034 per generation on average. Therefore,

$$E[\ln(N_{t+1})] = E[\ln(N_t)] - 0.034.$$

The expectation of the population size itself is more difficult to find. We can find this value approximately by exponentiating, giving

$$\bar{N}_{t+1} = e^{-0.034}\bar{N}_t = 0.967\bar{N}_t.$$

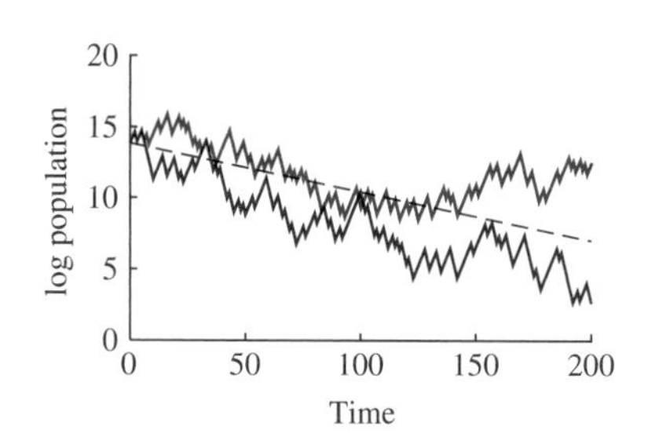

FIGURE 6.8.14

Two simulations of a stochastic population

This population, on average, decreases by more than 3% per generation. The expected population after t generations, starting from an initial size of 1.0×10^6, is approximately

$$\bar{N}_t = 1.0 \times 10^6 \cdot 0.967^t.$$

A simulation of the logarithm of two such populations is shown in Figure 6.8.14. Even though the populations jump up and down quite a bit, both decrease from an initial size of one million down to essentially zero in 200 generations. The line is the logarithm of the expected population, or

$$\begin{aligned}\ln(\bar{N}_t) &= \ln(1.0 \times 10^6 \cdot 0.967^t) \\ &= \ln(1.0 \times 10^6) + t\ln(0.967) \\ &= 13.81 - 0.0336t.\end{aligned}$$

Again, although neither population exactly tracks the expected line, each follows the trend rather well.

This average per capita production is called the geometric mean, with the following definition.

Definition 6.13 Suppose R is a random variable that takes on only positive values. Then

$$\text{the geometric mean of } R = e^{E[\ln(R)]}.$$

The geometric mean can be computed only for random variables that take on strictly positive values because we cannot take the logarithm of a negative number.

▸▸ Algorithm 6.4 Finding the Geometric Mean of a Discrete Random Variable

If R is a discrete random variable that takes on the positive values r_i with probability p_i, then

$$\text{the geometric mean of } R = e^{\sum \ln(r_i) p_i}$$

If the geometric mean is less than 1, the population shrinks. If the geometric mean is greater than 1, the population grows. The key to computing the geometric mean correctly is remembering **not to change the probabilities**. Think of $\ln(R)$ as a new random variable, a new way to measure per capita production. The underlying probabilities remain the same, only the values change (Figure 6.8.13).

Example 6.8.15 Describing a Declining Population with the Geometric Mean

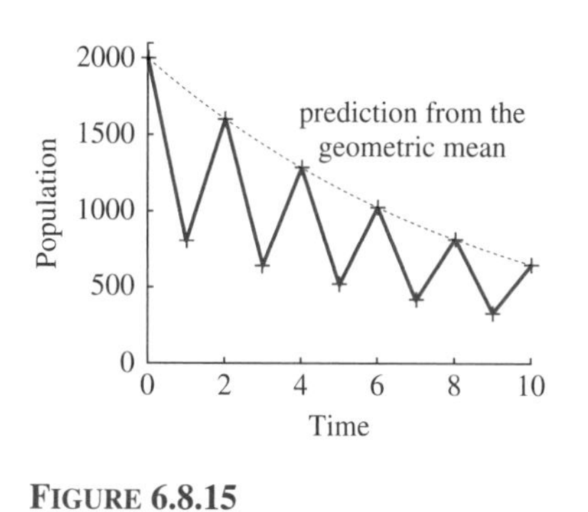

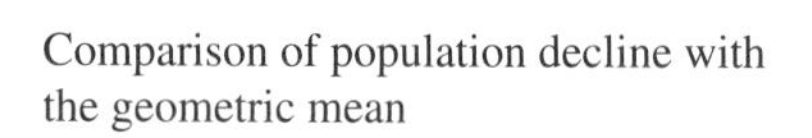

FIGURE 6.8.15

Comparison of population decline with the geometric mean

Consider again the population in Example 6.8.11 that alternately grows by a factor of 2.0 or declines by a factor of 0.4. We found that

$$E[\ln(R)] = -0.111.$$

The geometric mean, then is

$$e^{E(\ln(R))} = e^{-0.111} = 0.894.$$

The geometric mean predicts that this population will decline by a little over 10% per year. A population starting at 2000 with per capita production of 0.894 exactly tracks the true population in the even-numbered years (Figure 6.8.15).

Unlike the arithmetic mean, which emphasizes large values, the geometric mean emphasizes small values. If per capita production near zero is possible, it can swamp out the effects of many good years.

Example 6.8.16 A Few Bad Years Can Outweigh Many Good Years

If a population grows by 22% (per capita production of 1.22) in 90% of years and crashes by 98% (per capita production of 0.02) in 10% of years, the arithmetic mean of the per capita production is $1.22 \cdot 0.90 + 0.02 \cdot 0.1 = 1.10$ as before. The geometric mean, however, is

$$e^{\ln(1.22)0.9+\ln(0.02)0.1} = 0.809.$$

The rare bad years mean that this population will, on average, decline very rapidly (Figure 6.8.16).

The geometric mean of a continuous random variable is also found by exponentiating the expectation of the logarithm of the random variable.

Algorithm 6.5 Finding the Geometric Mean of a Continuous Random Variable

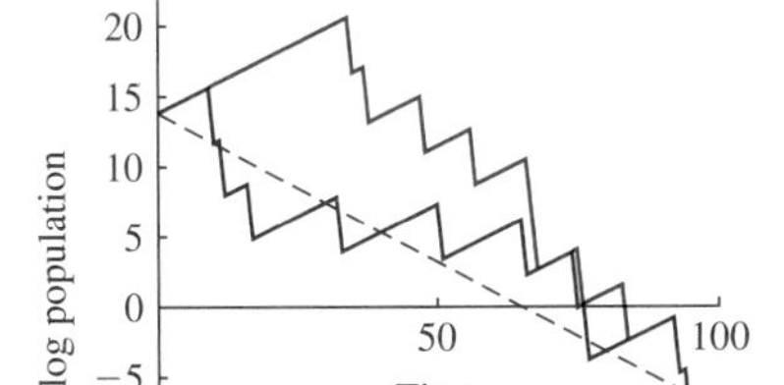

FIGURE 6.8.16
Two simulations of a stochastic population with occasional crashes

Suppose R is a continuous random variable that takes on the positive values with p.d.f. $f(r)$ with range $a > 0$ to b (which could be infinite). Then

$$\text{the geometric mean of } R = e^{\int_a^b \ln(r) f(r) dr}.$$

Unfortunately, computing the integral needed to find the geometric mean is often difficult or impossible. However, the following two indefinite integrals can be found with integration by parts from Subsection 4.3.4. The two simplest forms for $f(r)$ are the constant function $f(r) = k$ or the linear function $f(r) = kr$. In Example 4.3.10, we addressed the constant case, finding

$$\int \ln(r)dr = r\ln(r) - r + c. \tag{6.8.2}$$

Therefore, the indefinite integral with $f(r) = k$ is

$$\int k\ln(r)dr = kr\ln(r) - kr + c.$$

In Exercise 4.6.23, we studied the case $f(r) = r$ by finding

$$\int r\ln(r)dr = \frac{r^2}{2}\ln(r) - \frac{r^2}{4} + c. \tag{6.8.3}$$

Therefore, the indefinite integral with $f(r) = kr$ is

$$\int kr\ln(r)dr = k\frac{r^2}{2}\ln(r) - k\frac{r^2}{4} + c.$$

Example 6.8.17 The Geometric Mean of a Continuous Random Variable

Suppose the per capita production R has p.d.f. $g(r) = 1.0$ for $0.5 \le r \le 1.5$. The per capita production is equally likely to take on any value between 0.5 and 1.5. The arithmetic mean of this symmetric distribution of the per capita production is

$$\int_{0.5}^{1.5} rg(r)dr = \int_{0.5}^{1.5} rdr = \frac{r^2}{2}\Big|_{0.5}^{1.5} = 1.0.$$

The geometric mean is found by computing the expectation of the logarithm,

$$\begin{aligned} E[\ln(R)] &= \int_{0.5}^{1.5} \ln(r)g(r)dr \\ &= \int_{0.5}^{1.5} \ln(r)dr \\ &= \left[r\ln(r) - r)\Big|_{0.5}\right]^{1.5} \\ &= -0.045. \end{aligned}$$

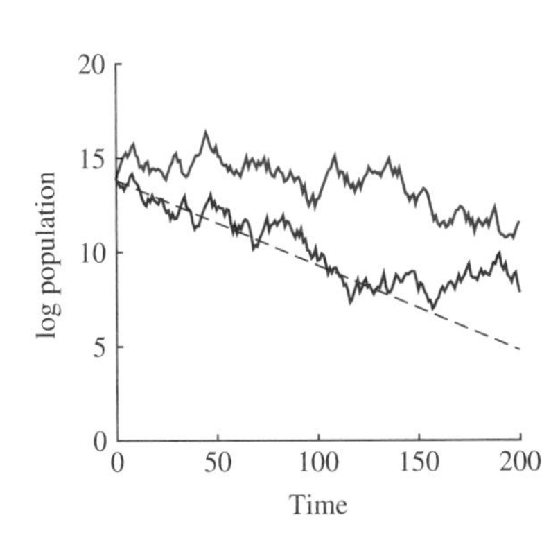

FIGURE 6.8.17
Two simulations of a stochastic population with per capita growth rates between 0.5 and 1.5

Therefore, the geometric mean is $e^{-0.045} = 0.955 < 1.0$. This population will decline (Figure 6.8.17). Once again, the small values pull the geometric mean below the arithmetic mean.

There is a general inequality relating the arithmetic and geometric means of random variables.

Inequality 6.1 Arithmetic-Geometric Inequality
The geometric mean is less than or equal to the arithmetic mean.

Using the arithmetic mean when the geometric mean is appropriate always overestimates the value. We have seen this inequality at work in each of our examples.

Example 6.8.18 The Geometric Mean for Per Capita Production

The geometric mean of the per capita production with p.d.f.

$$g(r) = \frac{1}{8}(4 - r)$$

(Example 6.8.5) is found as

$$\begin{aligned}
E(R) &= \int_0^4 \frac{1}{8} \ln(r)(4 - r)dr \\
&= \frac{1}{2}\int_0^4 \ln(r)dr - \frac{1}{8}\int_0^4 r \ln(r)rdr \\
&= \frac{1}{2}[r \ln(r) - r]\Big|_0^4 - \frac{1}{8}\left[\frac{r^2}{2}\ln(r) - \frac{r^2}{4}\right]\Bigg|_0^4 \\
&= -0.114.
\end{aligned}$$

The geometric mean is $e^{-0.114} = 0.892$, far less than the arithmetic mean of $4/3 = 1.333$ found in Example 6.7.20.

Example 6.8.19 The Geometric Mean for Per Capita Production in a Second Population

The geometric mean of the per capita production with p.d.f.

$$g(r) = \frac{r}{8}$$

(Example 6.8.6) is found as

$$\begin{aligned}
E(R) &= \int_0^4 \frac{r}{8} \ln(r)dr \\
&= \frac{1}{8}\left[\frac{r^2}{2}\ln(r) - \frac{r^2}{4}\right]\Bigg|_0^4 \\
&= 0.886.
\end{aligned}$$

The geometric mean is $e^{0.886} = 2.426$, less than the arithmetic mean of $8/3 = 2.667$ found in Example 6.7.21.

Summary We introduced three statistics to describe the central value of a random variable: the **median**, the **mode**, and the **geometric mean**. The value of a random variable is just as likely to be less than the **median** as greater. The **mode** is the single most likely value of the random variable. The mean, median, and mode are all equal for a **symmetric distribution** with the mode in the center. The **geometric mean**, defined as the exponential of the expectation of the logarithm of the random variable, is the appropriate mean for multiplicative processes.

6.8 Exercises

Mathematical Techniques

1–4 ▪ Consider again the data presented in Section 6.6, Exercises 1–4,

	Probability			
Number of Mutants	Experiment a	Experiment b	Experiment c	Experiment d
0	0.1	0.6	0.3	0.1
1	0.2	0.3	0.2	0.3
2	0.3	0.1	0.2	0.1
3	0.3	0.0	0.2	0.4
4	0.1	0.0	0.1	0.1

Find the median and the mode, and compare with the expectation. When is the median greater than the expectation?

1. Experiment a.
2. Experiment b.
3. Experiment c.
4. Experiment d.

5–8 ▪ Using the histograms (from Section 6.6, Exercises 9–12), estimate the median, the mode, and the expectation.

5.

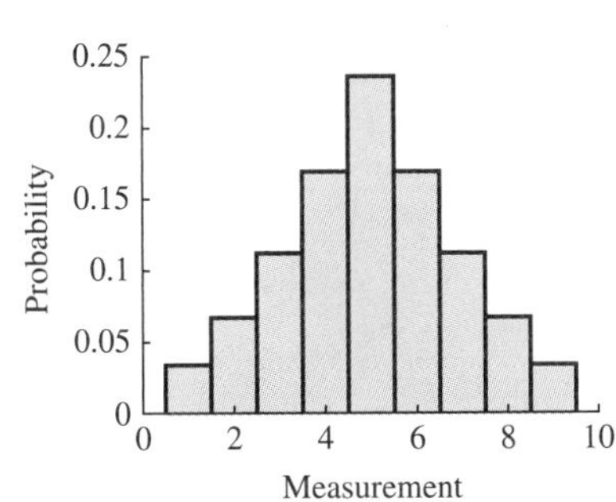

6.

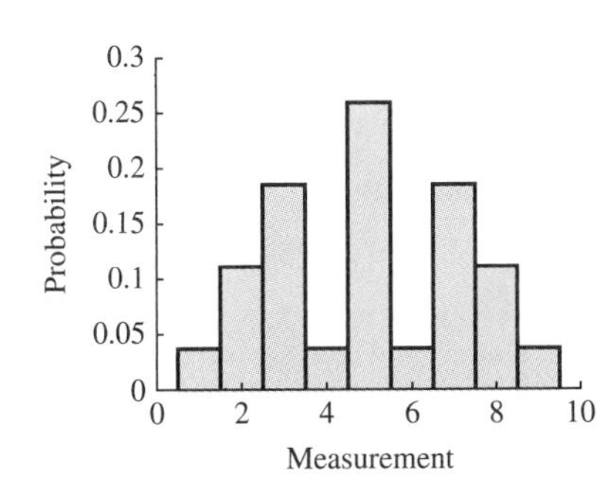

7.

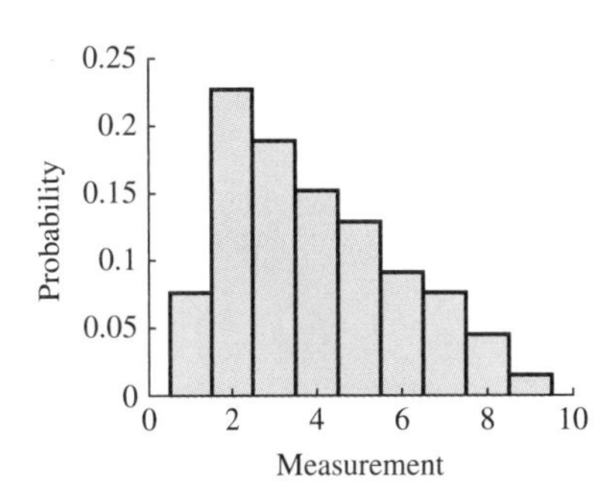

8.

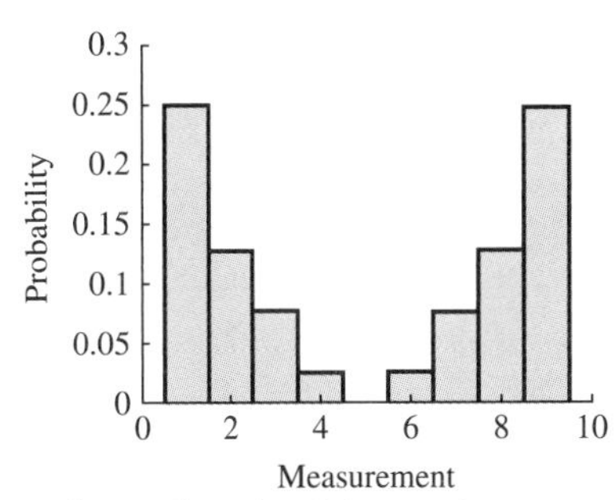

9–12 ▪ Find the median and mode of the continuous random variables with the given p.d.f., and compare with the expectation as found earlier.

9. The p.d.f. of a random variable X is $f(x) = 2x$ for $0 \le x \le 1$ (as in Section 6.6, Exercises 17 and 21, and Section 6.7, Exercise 7).
10. The p.d.f. of a random variable X is $f(x) = 1 - \frac{x}{2}$ for $0 \le x \le 2$ (as in Section 6.6, Exercises 18 and 22, and Section 6.7, Exercise 8).
11. The p.d.f. of a random variable T is $h(t) = \frac{1}{t}$ for $1 \le t \le e$ (as in Section 6.6, Exercises 19 and 23, and Section 6.7, Exercise 9).
12. The p.d.f. of a random variable T is $g(t) = 6t(1 - t)$ for $0 \le t \le 1$ (as in Section 6.6, Exercises 20 and 24, and Section 6.7, Exercise 10).

13–14 ▪ Recall the following slightly peculiar p.d.f.'s from Section 6.7, Exercises 15 and 16. Find the median of each. How does it compare with the expectation?

13. $f(x) = \frac{1}{2\sqrt{x}}$ for $0 < x \le 1$ (from Section 6.7, Exercise 15).
14. $g(t) = \frac{1}{t^2}$ for $1 \le t < \infty$ (from Section 6.7, Exercise 16).

15–18 ▪ Find the arithmetic and geometric means of the following random variables. Check that the arithmetic-geometric inequality holds.

15. The random variable X where $\Pr(X = 1) = 0.3$ and $\Pr(X = 2) = 0.7$.

16. The random variable X where $\Pr(X=1)=0.4$ and $\Pr(X=3)=0.6$.

17. The random variable X where $\Pr(X=1)=0.3$, $\Pr(X=2)=0.3$, and $\Pr(X=3)=0.4$.

18. The random variable X where $\Pr(X=2)=0.1$, $\Pr(X=3)=0.2$, and $\Pr(X=5)=0.7$.

19–22 ▪ Find the geometric mean of the continuous random variables with the given p.d.f. Compare with the expectation.

19. The p.d.f. of a random variable X is $f(x)=2$ for $0.75 \le x \le 1.25$. Use Equation 6.8.2.

20. The p.d.f. of a random variable X is $f(x)=5$ for $1 \le x \le 1.2$.

21. The p.d.f. of a random variable X is $f(x)=2x$ for $0 \le x \le 1$ (as in Section 6.6, Exercise 17 and Section 6.7, Exercise 7). Use Equation 6.8.3, and you will have to use L'Hôpital's rule to evaluate the integral.

22. The p.d.f. of a random variable T is $h(t)=\frac{1}{t}$ for $1 \le t \le e$ (as in Section 6.6, Exercise 19 and Section 6.7, Exercise 9). *HINT:* Use the substitution $u=\ln(t)$ to do the integral.

23–26 ▪ The geometry behind the geometric mean is based on the following argument. If a random variable R takes on each of the values r_1 and r_2 with probability 0.5, a rectangle with sides of length r_1 and r_2 has area equal to that of a square with sides with length equal to the geometric mean.

23. Check this in the case that $r_1=1$ and $r_2=2$.

24. Check this in general, without picking values for r_1 and r_2.

25. Fix $r_1=1$. Find the value of r_2 that maximizes the ratio of the geometric mean to the arithmetic mean.

26. Prove that the geometric mean is always less than or equal to the arithmetic mean (the arithmetic-geometric inequality) when $r_1=1$ (the case described in Exercise 25).

Applications

27–28 ▪ Suppose that incomes in a company have the following probabilities.

Income	Probability
20,000	0.48
30,000	0.04
35,000	0.16
50,000	0.12
57,000	0.04
100,000	0.08
150,000	0.04
top salary	0.04

For the given values of the top salary, find the mean, median, and mode, and say which statistic is most informative about the distribution of salaries. (This is based on an example in the book *How to Lie with Statistics*.)

27. The top salary is \$450,000.

28. The top salary is \$4,500,000.

29–32 ▪ For the data (from Section 6.6, Exercises 37–40), find the median and the mode.

	Frequency in 100 Experiments			
Number of Tall Offspring	Experiment a	Experiment b	Experiment c	Experiment d
0	0	0	0	5
1	0	0	2	11
2	0	1	5	35
3	0	1	10	27
4	2	6	21	13
5	8	19	27	8
6	9	20	19	1
7	31	27	8	0
8	22	16	5	0
9	22	9	3	0
10	6	1	0	0

29. Experiment a

30. Experiment b

31. Experiment c

32. Experiment d

33–34 ▪ As in Section 6.6, Exercises 45 and 46, the p.d.f. for the waiting time X until an event occurs often follows the **exponential distribution**, with the form $g(x)=\lambda e^{-\lambda x}$ for some positive value of λ, defined for $x \ge 0$. Find the median waiting time for the following values of λ and compare with the expectation (found in Section 6.7, Exercises 35 and 36).

33. $\lambda=0.5$.

34. $\lambda=2.0$.

35–36 ▪ The following problems show that most people live in places that are more crowded than average. In each case, find the average size of a city, the average crowding a person experiences, and the fraction of people who live in places more crowded than average.

35. There are two cities, one with 100,000 people and the other with 1,000,000 people.

36. There are three cities, one with 100,000 people, one with 400,000 people, and the other with 1,000,000 people.

37–40 ▪ Find the arithmetic and geometric means of the random variables R for per capita production in the following cases. Check that the arithmetic-geometric inequality holds in each case. Which describes a growing population?

37. $R=4$ with probability 0.5, $R=0.25$ with probability 0.5.

38. $R=4$ with probability 0.25, $R=0.25$ with probability 0.75.

39. $R=4$ with probability 0.75, $R=0.25$ with probability 0.25.

40. $R=5$ with probability 0.25, $R=0.25$ with probability 0.25, $R=1$ with probability 0.5.

41–44 ▪ Suppose populations start at 100. Estimate the population size after 50 generations in the following cases.

41. The situation in Exercise 37.

42. The situation in Exercise 38.

43. The situation in Exercise 39.

44. The situation in Exercise 40.

45–46 ▪ A store has two managers, one who believes that high profits come from lowering prices and getting more customers and another who believes that high profits come from raising prices and making more profit per customer. These managers get to choose prices in alternate weeks. For each case,

a. Find the price after 1, 2, 3, and 4 weeks.

b. Find a formula for the price after t weeks (break into the two cases, t even and t odd).

c. Which of the managers wins?

d. What does this have to do with the geometric mean?

45. In week 1, the "low price" manager cuts prices by 50%. In week 2, the "high price" manager raises prices by 50%, and so forth. Suppose an item started out at $100.

46. In week 1, the "low price" manager cuts prices by 20%. In week 2, the "high price" manager raises prices by 30%, and so forth. Suppose an item started out at $100.

Computer Exercises

47. The p.d.f. for a random variable taking on values between 0.8 and 1.1 with equal probability is $f(x) = 10/3$ for $0.8 \leq x \leq 1.1$. Find the geometric mean r of this random variable. Define two updating functions, a deterministic g for a population that has per capita production of exactly r and a stochastic G. Compare the dynamics of the two populations for 100 steps starting from an initial condition of 100. How similar do they look?

48. Suppose that the per capita production of a population is a random variable with uniform (flat) p.d.f. on the interval from 0.8 to y, where y is an unknown value. To guarantee that the integral is equal to 1, the height of the p.d.f. must be $\frac{1}{y - 0.8}$ rather than 3.333. Why is this?

a. Compute a function $H(y)$, which gives the geometric mean as a function of y.

b. Solve for the value $y = y_{\max}$ for which the geometric mean is 1.0.

c. Define a random variable R to produce random numbers in the interval from 0.8 to $y_{\max}$ and an associated updating function G to describe a population with per capita production equal to R.

d. Generate two trajectories of 100 generations using the updating function G starting from $N = 100$.

e. Where do you expect the trajectories to end up? How close are they?

6.9 Descriptive Statistics for Spread

A measure of the center can be thought of as a "best guess" of the value of a random variable. Our *confidence* in that guess depends on how spread out the probability distribution or p.d.f. is. Knowing that a score has a mean of 12.5 (or a median of 10) does not tell us whether likely scores range from 5 to 15 or from 0 to 60. We introduce the most widely used statistics for describing spread: **range**, **percentiles**, the **variance**, and the **standard deviation**. As always, different statistics are appropriate in different circumstances.

6.9.1 Range and Percentiles

The simplest measure of spread is the **range**, a description of all the values taken on by a random variable, usually summarized by the lowest and highest values. The range of dart scores is from 0 to 60 (Figure 6.9.1a). The range of the number out of 10 molecules remaining is from 0 to 10 (Figure 6.9.1b). Many continuous random variables have a mathematically possible range from 0 (or negative infinity) to infinity. In these cases, the range gives no useful information. Thinking of random variables as functions, the range coincides with the **range** of the function. Perhaps confusingly, however, the range of the random variable is the **domain** of the p.d.f.

Percentiles generalize the median. Exactly 50% of the values of the random variable lie below the median. We define the pth percentile as the value of the random variable that exceeds exactly p% of the values. The median is the 50th percentile. Like the median, percentiles must be computed from the cumulative distribution.

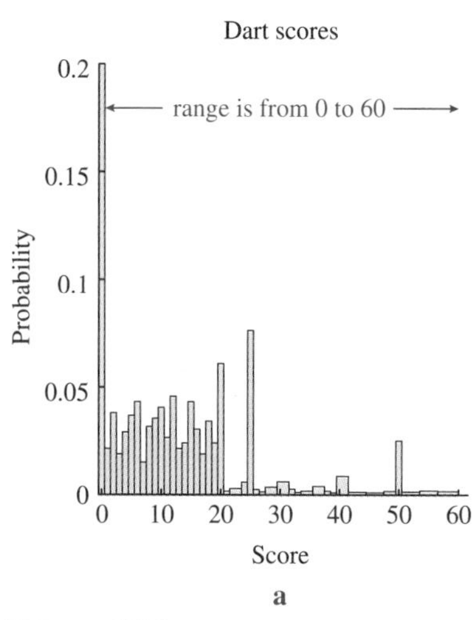

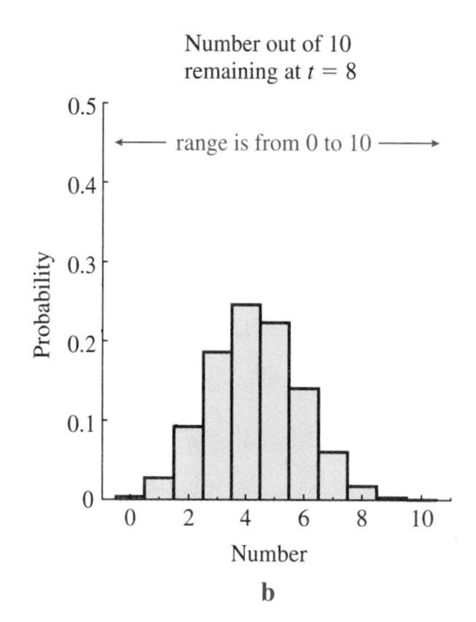

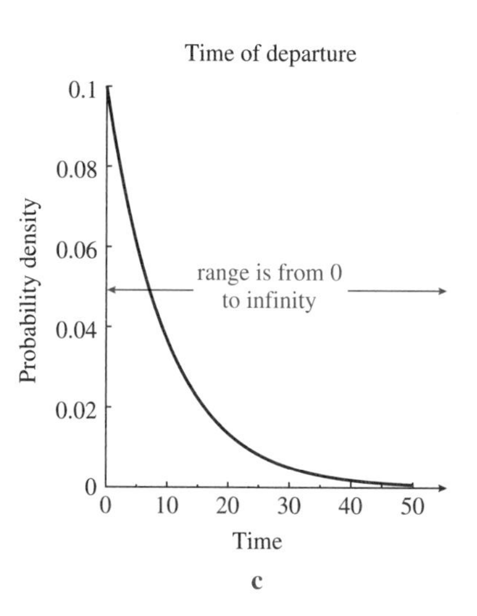

FIGURE 6.9.1
The ranges of three random variables

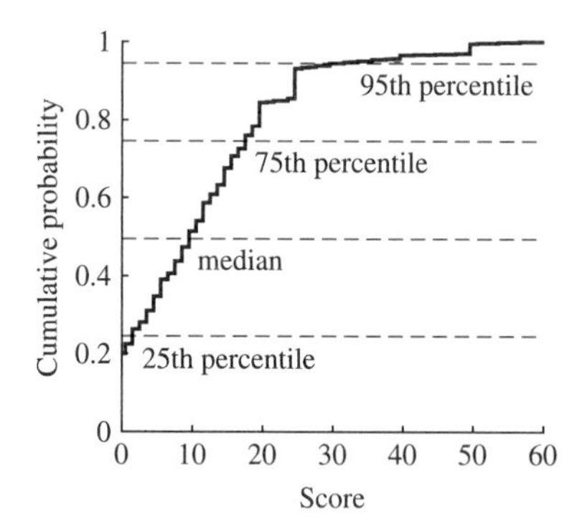

FIGURE 6.9.2
Finding percentiles from a cumulative distribution

Consider again the cumulative distribution for dart scores (Figure 6.9.2). We found the median by solving for where the cumulative distribution crossed the horizontal line at 0.5. We find the 95th percentile by finding where the cumulative distribution crosses the horizontal line at 0.95, in this case at a score of 34.

The lower and upper **quartiles**, defined as the 25th and 75th percentiles, are often used to describe random variables. The 25th percentile of the dart scores is 2, and the 75th percentile is 18. The quartiles and the median divide the values of the random variable into four equally probable sets. One quarter of dart throws score less than 2, another quarter between 2 and 10, another quarter between 10 and 18, and a quarter more than 18. Quartiles can also be thought of as dividing the data into two equal pieces, the central half between the quartiles (between 2 and 18) and the outlying half outside the quartiles (less than 2 or greater than 18).

Example 6.9.1 Percentiles and Test Scores

Percentiles are commonly used in describing test results. The following are data for the quantitative scores on the SAT test (Figure 6.9.3). We see that only 15% of students score lower than 400, while only about 6% score higher than 700.

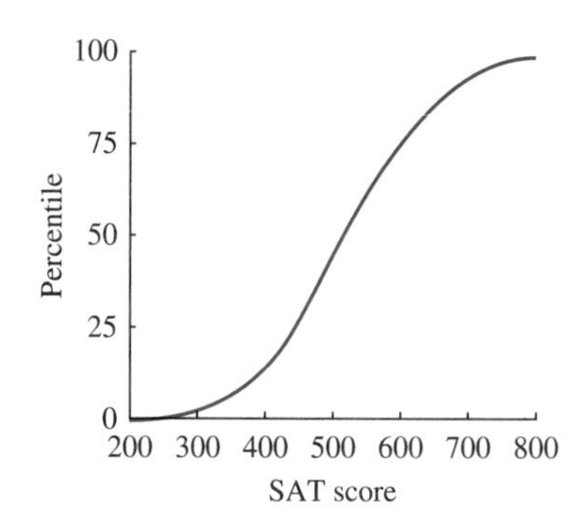

FIGURE 6.9.3
Percentiles from the SAT test

Score	Percentile
200	–
250	1
300	2
350	7
400	15
450	28
500	45
550	61
600	76
650	87
700	94
750	98
800	99+

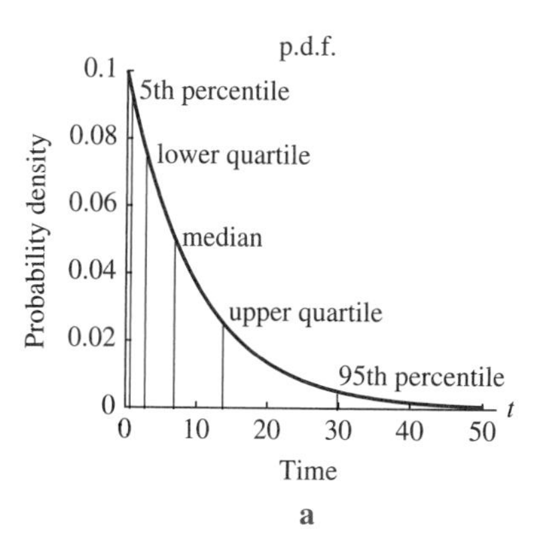

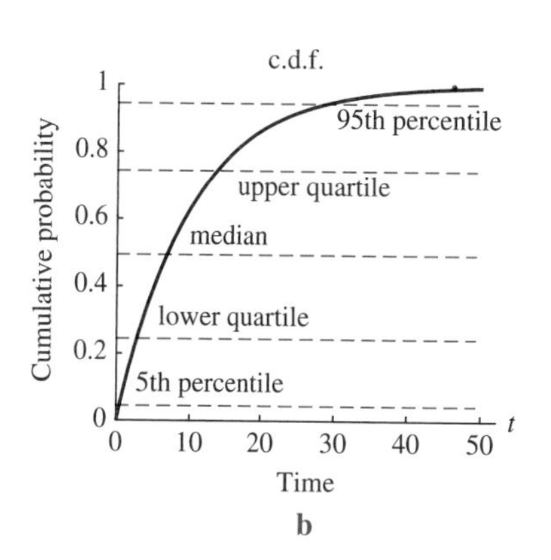

FIGURE 6.9.4
Percentiles for a diffusing molecule

Percentiles of continuous random variables are computed by finding where the c.d.f. crosses the appropriate horizontal line.

Example 6.9.2 Percentiles for Departure Time of a Molecule

The time T when a diffusing molecule leaves a cell is described by the p.d.f.

$$f(t) = 0.1e^{-0.1t}$$

and the c.d.f.

$$F(t) = 1 - e^{-0.1t}$$

(Figure 6.9.4). We found the median value of 6.93 by solving the equation

$$F(t) = 0.5.$$

We find the percentiles in the same way. To find the 95th percentile, solve

$$F(t) = 1 - e^{-0.1t} = 0.95$$
$$e^{-0.1t} = 0.05$$
$$t = -10\ln(0.05) = 29.96.$$

Only 5% of measurements exceed 29.96. The 5th percentile is

$$F(t) = 1 - e^{-0.1t} = 0.05$$
$$e^{-0.1t} = 0.95$$
$$t = -10\ln(0.95) = 0.51.$$

Only 5% of measurements are less than 0.51. On the p.d.f., the area above the 95th percentile is 0.05, equal to the area below the 5th percentile.

The lower quartile solves $F(t) = 0.25$, and the upper quartile solves $F(t) = 0.75$. Using the same procedure, we can solve to find values of 2.88 and 13.86. These quartiles, along with the median, break the random variable into four equally likely intervals (each with area 0.25). The results of an experiment are as likely to lie between 0 and 2.88 as between 6.93 and 13.86.

6.9.2 Mean Absolute Deviation

Percentiles generalize the median. Another group of statistics generalize the mean, measuring the "average distance" from the mean. Their calculation proceeds in two steps: define a new random variable measuring distance from the mean, and compute its expectation.

Example 6.9.3 Mean Distance from the Mean for Molecule Number

Consider again the number of molecules left in a cell after 8 minutes. In addition to tabulating the values and the probabilities, we have also tabulated how far the values are from the mean.

Number Left	Probability	Mean	Difference from Mean
0	0.004	4.3	−4.3
1	0.027	4.3	−3.3
2	0.092	4.3	−2.3
3	0.186	4.3	−1.3
4	0.246	4.3	−0.3
5	0.223	4.3	0.7
6	0.141	4.3	1.7
7	0.061	4.3	2.7
8	0.017	4.3	3.7
9	0.003	4.3	4.7
10	0.000	4.3	5.7

As a measure of spread, we would like to know how far the values are from the mean *on average*. We can compute the mean distance from the mean by multiplying the values (the difference from the mean) by the probabilities,

$$\begin{aligned}\text{mean distance from mean} = &-4.3 \cdot 0.004 - 3.3 \cdot 0.027 - 2.3 \cdot 0.092 \\ &- 1.3 \cdot 0.186 - 0.3 \cdot 0.246 + 0.7 \cdot 0.223 \\ &+ 1.7 \cdot 0.141 + 2.7 \cdot 0.061 + 3.7 \cdot 0.017 \\ &+ 4.7 \cdot 0.003 + 5.7 \cdot 0.000 = 0.0.\end{aligned}$$

Perhaps disappointingly, the mean distance from the mean is 0. This is always the case because the negative values exactly balance out the positive ones. ▲

The simplest way to make the differences from the mean into a positive number is to take their absolute values.

Example 6.9.4 Mean Absolute Deviation for Molecule Number

Number Left	Probability	Mean	Difference from Mean	Absolute Difference from Mean
0	0.004	4.3	−4.3	4.3
1	0.027	4.3	−3.3	3.3
2	0.092	4.3	−2.3	2.3
3	0.186	4.3	−1.3	1.3
4	0.246	4.3	−0.3	0.3
5	0.223	4.3	0.7	0.7
6	0.141	4.3	1.7	1.7
7	0.061	4.3	2.7	2.7
8	0.017	4.3	3.7	3.7
9	0.003	4.3	4.7	4.7
10	0.000	4.3	5.7	5.7

If there are 0 molecules left in the cell, this is 4.3 below the mean. The absolute value of the difference is 4.3. If there are 6 molecules left, this is 1.7 above the mean, and the absolute value of the difference is 1.7. We find the expectation of these distances in the usual way, by taking the values times the probabilities and adding them up,

$$\begin{aligned}\text{mean absolute distance from mean} = 4.3\cdot 0.004 &+ 3.3\cdot 0.027 + 2.3\cdot 0.092\\ &+ 1.3\cdot 0.186 + 0.3\cdot 0.246 + 0.7\cdot 0.223\\ &+ 1.7\cdot 0.141 + 2.7\cdot 0.061 + 3.7\cdot 0.017\\ &+ 4.7\cdot 0.003 + 5.7\cdot 0.000 = 1.27.\end{aligned}$$

This means that the measurement is, on average, 1.27 away from the mean. This gives a rough idea of how "spread out" the distribution is (Figure 6.9.5).

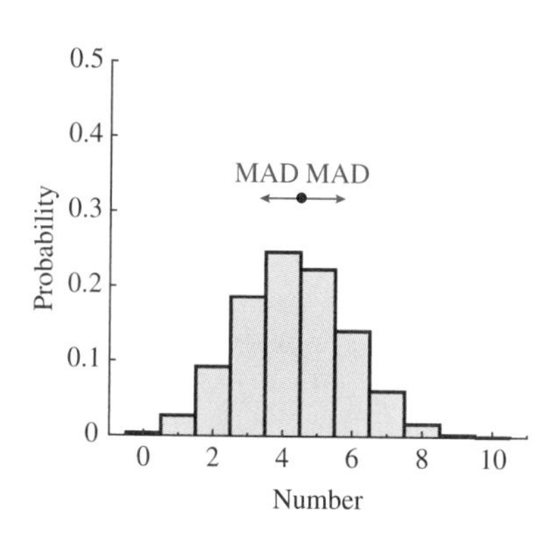

FIGURE 6.9.5
Mean absolute deviation of molecule number

This measure is called the **mean absolute deviation**, or **MAD** for short. The formal definition of the mean absolute deviation is as follows.

Definition 6.14 Suppose that the expectation of the random variable X is $\bar{X}$. Consider the new random variable $Y = |X - \bar{X}|$. Then

$$\text{mean absolute deviation} = E(Y) = E(|X - \bar{X}|).$$

More computationally, the mean absolute deviation is found with the following algorithm.

Algorithm 6.6 Finding MAD for a Random Variable

1. If X is a discrete random variable that takes on the value x_i with probability p_i, then

$$\text{MAD} = \sum_{i=1}^{n} |x_i - \bar{X}| p_i.$$

2. If X is a continuous random variable with p.d.f. $f(x)$ and range from a to b, then

$$\text{MAD} = \int_a^b |x - \bar{X}| f(x) dx.$$

Example 6.9.5 Finding MAD for a Continuous Random Variable

Suppose we wish to find the mean absolute deviation of the continuous random variable R with p.d.f. describing per capita production (Example 6.7.12) given by

$$g(r) = \frac{1}{8}(4 - r),$$

for $0 \le r \le 4$. The mean of this random variable is 4/3. The mean absolute deviation is found by integrating

$$\text{MAD} = \int_0^4 \left| r - \frac{4}{3} \right| g(r) dr.$$

Evaluating an integral with an absolute value requires breaking it up into two pieces, the first where $r \ge 4/3$ and $\left| r - \frac{4}{3} \right| = r - \frac{4}{3}$, and the second where $r < 4/3$ and

$\left|r - \frac{4}{3}\right| = \frac{4}{3} - r$. We find

$$\begin{aligned}
\text{MAD} &= \int_0^4 \left|r - \frac{4}{3}\right| g(r)dr \\
&= \int_0^4 \left|r - \frac{4}{3}\right| \frac{1}{8}(4 - r)dr \\
&= \int_0^{\frac{4}{3}} \left(\frac{4}{3} - r\right) \frac{1}{8}(4 - r)dr + \int_{\frac{4}{3}}^4 \left(r - \frac{4}{3}\right) \frac{1}{8}(4 - r)dr \\
&= \left(\frac{1}{24}r^3 - \frac{1}{3}r^2 + \frac{2}{3}r\right)\Bigg|_0^{\frac{4}{3}} - \left(\frac{1}{24}r^3 - \frac{1}{3}r^2 + \frac{2}{3}r\right)\Bigg|_{\frac{4}{3}}^4 \\
&= 0.79.
\end{aligned}$$

Doing this integral requires the extra step of breaking up the absolute value, which is one reason why the mean absolute deviation is rarely used. ▲

Although the variance, which we introduce next, has nicer mathematical and computational properties than the mean absolute deviation, the mean absolute deviation is a perfectly reasonable measure of spread.

6.9.3 Variance, Standard Deviation, and Coefficient of Variation

A more widely used statistic, based on the same sort of reasoning, is the **variance**. When we measured the distance of our various measurements from the mean, we used the absolute value. An alternative way to make sure that all the distances are positive is to **square** them.

Example 6.9.6 Finding the Variance of Molecule Number

To find the variance, we build a new table that includes the square of the difference from the mean.

Number Left	Probability	Mean	Difference from Mean	Squared Difference from Mean
0	0.004	4.3	−4.3	18.49
1	0.027	4.3	−3.3	10.89
2	0.092	4.3	−2.3	5.29
3	0.186	4.3	−1.3	1.69
4	0.246	4.3	−0.3	0.09
5	0.223	4.3	0.7	0.49
6	0.141	4.3	1.7	2.89
7	0.061	4.3	2.7	7.29
8	0.017	4.3	3.7	13.69
9	0.003	4.3	4.7	22.09
10	0.000	4.3	5.7	32.49

To find the mean squared deviation from the mean, we take these new values, multiply by the probabilities, and add them up. In this case,

$$\begin{aligned}\text{mean squared distance from mean} &= 18.49 \cdot 0.004 + 10.89 \cdot 0.027 + 5.29 \cdot 0.092 \\ &+ 1.69 \cdot 0.186 + 0.09 \cdot 0.246 + 0.49 \cdot 0.223 \\ &+ 2.89 \cdot 0.141 + 7.29 \cdot 0.061 + 13.69 \cdot 0.017 \\ &+ 22.09 \cdot 0.003 + 32.49 \cdot 0.000 = 2.45.\end{aligned}$$

This new statistic is called the **variance**, with the following formal definition.

Definition 6.15 The variance of a random variable X, denoted by σ^2 or $\mathrm{Var}(X)$, is equal to

$$\sigma^2 = E[(X - \bar{X})^2].$$

The variance can be found with the following algorithm, although we will soon develop a computational algorithm that is easier to use.

Algorithm 6.7

1. If X is a discrete random variable that takes on the value x_i with probability p_i, then
$$\sigma^2 = \sum_{i=1}^{n} (x_i - \bar{X})^2 p_i.$$
2. If X is a continuous random variable with p.d.f. $f(x)$ and range from a to b, then
$$\sigma^2 = \int_a^b (x - \bar{X})^2 f(x) dx.$$

Example 6.9.7 Finding the Variance of Per Capita Production

Consider again the random variable with p.d.f.

$$g(r) = \frac{1}{8}(4 - r),$$

$0 \le r \le 4$ describing per capita production (Example 6.9.5). Using the mean of 4/3, the variance is

$$\begin{aligned}\sigma^2 &= \int_0^4 \left(r - \frac{4}{3}\right)^2 g(r) dr \\ &= \int_0^4 \left(r - \frac{4}{3}\right)^2 \frac{1}{8}(4 - r) dr \\ &= -\frac{1}{32} r^4 + \frac{5}{18} r^3 - \frac{7}{9} r^2 + \frac{8}{9} r \Big|_0^4 \\ &= \frac{8}{9} = 0.889.\end{aligned}$$

This still requires some algebra, but at least it is not necessary to split the integral into two pieces.

An important advantage of the variance over the mean absolute deviation is the existence of the following computational formula.

Theorem 6.4 **Computational Formula for the Variance**

If X is a random variable with mean $\bar{X}$, the variance can be written

$$\sigma^2 = E(X^2) - E(X)^2 = E(X^2) - \bar{X}^2.$$

The proof of this theorem is outlined in Exercises 23 and 24. In words, the variance is the expectation of the square minus the square of the expectation.

The theorem is used with the following algorithm.

Algorithm 6.8

1. If X is a discrete random variable that takes on the value x_i with probability p_i, then

$$\sigma^2 = \sum_{i=1}^{n} x_i^2 p_i - \bar{X}^2.$$

2. If X is a continuous random variable with p.d.f. $f(x)$ and range from a to b, then

$$\sigma^2 = \int_a^b x^2 f(x)dx - \bar{X}^2.$$

Example 6.9.8 Using the Computational Formula to Find the Variance of Molecule Number

Why are these formulas easier to use? Consider first the discrete random variable for molecule number. Instead of finding how far each possible measurement is from the mean, we have only to find the square of each measurement, as in the following table.

Number Left	Probability	Square of Number Left
0	0.004	0
1	0.027	1
2	0.092	4
3	0.186	9
4	0.246	16
5	0.223	25
6	0.141	36
7	0.061	49
8	0.017	64
9	0.003	81
10	0.000	100

The variance is the mean of the squares minus the square of the mean. We find the mean of the squares as usual, the sum of the values times the probabilities, or

$$\begin{aligned}\text{mean of squared number} &= 0 \cdot 0.004 + 1 \cdot 0.027 + 4 \cdot 0.092 + 9 \cdot 0.186 \\ &+ 16 \cdot 0.246 + 25 \cdot 0.223 + 36 \cdot 0.141 \\ &+ 49 \cdot 0.061 + 64 \cdot 0.017 + 81 \cdot 0.003 \\ &+ 100 \cdot 0.000 = 20.94.\end{aligned}$$

Using the mean of 4.3 (Example 6.7.19), the variance is then

$$\text{mean of squared number} - \text{square of mean} = 20.94 - 4.3^2 = 2.45,$$

as in Example 6.9.6.

Example 6.9.9 Using the Computational Formula to Find the Variance of Per Capita Production

For the random variable R with p.d.f.

$$g(r) = \frac{1}{8}(4 - r),$$

for $0 \leq r \leq 4$, the computational formula for the variance is

$$\begin{aligned}\sigma^2 &= \int_0^4 r^2 \frac{1}{8}(4 - r)dr - \left(\frac{4}{3}\right)^2 \\ &= \int_0^4 \frac{r^2}{2} - \frac{r^3}{8} dr - \left(\frac{4^2}{3}\right) \\ &= \frac{r^3}{6} - \frac{r^4}{32}\bigg|_0^4 - \left(\frac{4^2}{3}\right) \\ &= \frac{8}{9} = 0.889\end{aligned}$$

again matching the result found earlier, in Example 6.9.7.

Example 6.9.10 The Variance of a Bernoulli Random Variable

An important variance to know and remember is the variance of a **Bernoulli random variable** (Definition 6.10), a random variable B that takes on the value 1 with probability p and 0 with probability $1 - p$. The expectation of B is

$$E(B) = 0 \cdot (1 - p) + 1 \cdot p = p.$$

The variance is

$$\text{Var}(B) = 0^2 \cdot (1 - p) + 1^2 \cdot p - p^2 = p - p^2 = p(1 - p). \qquad (6.9.1)$$

What does the variance mean? A first step in interpreting a new quantity is to figure out its units. The variance of the number of molecules has units of molecules^2, because we squared the difference from the mean. This quantity is rather hard to interpret. To get back to units of molecules, we can take the square root. The square root of the variance is the **standard deviation**, denoted σ.

Definition 6.16 The standard deviation σ is the square root of the variance.

Example 6.9.11 The Standard Deviation of Molecule Number

The standard deviation of molecule number is the square root of the variance of 2.45 (Example 6.9.6). We find that $\sigma = \sqrt{2.45} = 1.56$, which is close to, but not equal to, the mean absolute deviation of 1.27 (found in Example 6.9.4).

Example 6.9.12 The Standard Deviation of Per Capita Production

The standard deviation of per capita production is the square root of the variance of 0.889 found in Example 6.9.7, giving $\sigma = \sqrt{0.889} = 0.94$.

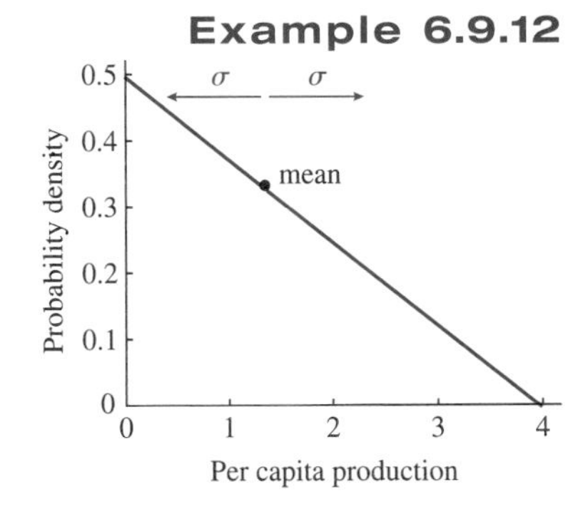

FIGURE 6.9.6

Standard deviation of per capita production

What does the standard deviation tell us? Like the mean absolute deviation, it is a measure of how spread out the distribution is. Roughly speaking, most values of per capita production lie within two standard deviations of the mean, or between $1.33 - 2 \cdot 0.94 = -0.55$ and $1.33 + 2 \cdot 0.94 = 3.21$ (Figure 6.9.6). In this case, the standard deviation does not tell us very much.

The standard deviation is most useful for describing random variables with a "bell-shaped" distribution or p.d.f. (Figure 6.9.7). In this case, the standard deviation is the basis for two rules of thumb.

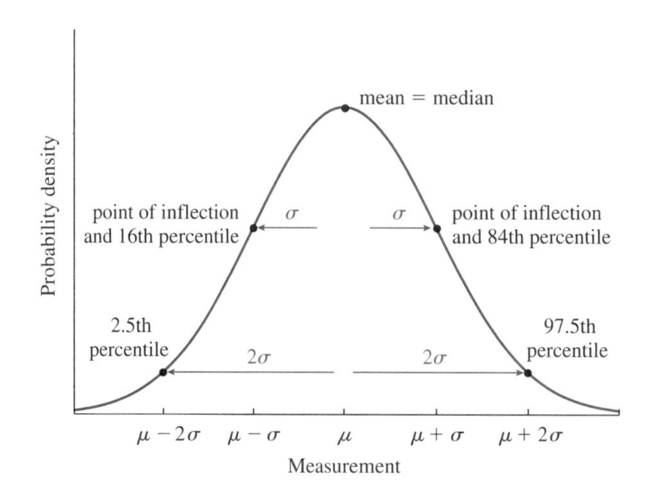

FIGURE 6.9.7
A bell-shaped distribution

Rules of Thumb for Bell-Shaped Distributions

1. The points of inflection are 1 standard deviation from the mean.
2. Approximately 95% of the probability lies within two standard deviations of the mean.

Recall that the points of inflection are where the curvature changes from concave up (curved upward) to concave down (curved downward) (Subsection 2.7.1).

In terms of percentiles, the point of inflection one standard deviation below the mean is the 16th percentile, and the point of inflection one standard deviation above the mean is the 84th percentile. The second rule says that the 2.5th percentile is approximately two standard deviations below the mean and the 97.5th percentile is approximately two standard deviations above the mean. If the mean is designated by μ (the convention for distributions of this shape), these rules are summarized in the following table.

Point	Shape	Percentile
$\mu - 2\sigma$		2.5th
$\mu - \sigma$	point of inflection	16th
μ	maximum	50th
$\mu + \sigma$	point of inflection	84th
$\mu + 2\sigma$		97.5th

Example 6.9.13 Applying the Rules of Thumb for a Bell-Shaped Distribution

Suppose a random variable has mean equal to 50 and standard deviation of 10 (Figure 6.9.8). The mean is equal to the mode, and looks to be about 50. The points of inflection are 40 and 60, about 10 away from the mean. According to the first rule of thumb, we estimate that the standard deviation is about 10. According to the second rule of thumb, the 2.5th percentile will be about 30 and the 97.5th percentile about 70. Therefore, 95% of measurements lie between 30 and 70.

These rules work *only* for symmetric bell-shaped distributions. Because the variance and standard deviation weight outlying points heavily, these statistics are very sensitive to values far from the mean (Exercises 29 and 30).

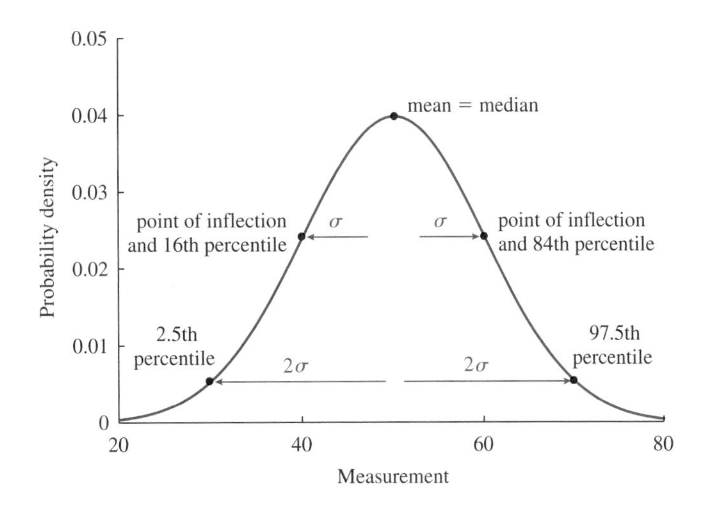

FIGURE 6.9.8

A bell-shaped distribution with mean 50 and standard deviation 10

6.9.4 The Coefficient of Variation

Sometimes we wish to have a number with no units that can be used to compare different distributions. One example is the coefficient of variation, or CV.

Definition 6.17 Suppose X is a random variable that takes on only positive values. The **coefficient of variation** or CV is the ratio of the standard deviation to the mean, or

$$\mathrm{CV} = \frac{\sigma}{\bar{X}}.$$

Because the standard deviation has the same units as the measurement itself, the coefficient of variation has no units and is a dimensionless measure of spread.

Example 6.9.14 The Coefficient of Variation for a Bell-Shaped Distribution

For the random variable in Example 6.9.13, the mean is 50 and the standard deviation is 10, giving a coefficient of variation of

$$\mathrm{CV} = \frac{\sigma}{\bar{X}} = \frac{10.0}{50.0} = 0.2.$$

Example 6.9.15 The Coefficient of Variation for Molecule Number

For the molecules in Example 6.9.11, the coefficient of variation is

$$\mathrm{CV} = \frac{\sigma}{\bar{N}_8} = \frac{1.56}{4.3} = 0.36.$$

The larger the coefficient of variation, the more spread out the distribution. It can be thought of as a measure of uncertainty. If the coefficient of variation is close to or greater than 1, values of the random variable have a large range and are likely to be far from the mean.

Summary We have introduced several statistics that describe the spread of a random variable. The **range** gives the minimum and maximum values. The **percentiles** generalize the median to give the values where the random variable exceeds a given fraction of the distribution. The lower and upper **quartiles** indicate the values lying above 25% and 75% of the distribution, respectively. Statistics of spread that generalize the mean include the **mean absolute deviation**, the **variance** (the mean squared deviation from the mean), the **standard deviation** (the square root of the variance), and the **coefficient of variation** (the standard deviation divided by the mean).

6.9 Exercises

Mathematical Techniques

1–4 ▪ For the given data (first presented in Section 6.6, Exercises 1–4), find the range, MAD, the variance both the direct way and with the computational formula, the standard deviation, and the coefficient of variation.

	Probability			
Number of Mutants	Experiment a	Experiment b	Experiment c	Experiment d
0	0.1	0.6	0.3	0.1
1	0.2	0.3	0.2	0.3
2	0.3	0.1	0.2	0.1
3	0.3	0.0	0.2	0.4
4	0.1	0.0	0.1	0.1

1. Experiment a.
2. Experiment b.
3. Experiment c.
4. Experiment d.

5–8 ▪ Consider the following random variables that take only two values with the given probabilities. For each, find MAD, the variance, the standard deviation, and the coefficient of variation.

5. A Bernoulli random variable with $p = 1/3$.
6. A Bernoulli random variable with $p = 0.9$.
7. A random variable that takes the value 10 with probability 1/3 and the value of 0 with probability 2/3. Compare your answers with the answer to Exercise 5.
8. A random variable that takes the value 11 with probability 0.9 and the value of 10 with probability 0.1. Compare your answers with the answer to Exercise 6.

9–12 ▪ Find the quartiles of a random variable with the given p.d.f. Illustrate the areas on a graph of the p.d.f.

9. The p.d.f. of a random variable X is $f(x) = 2x$ for $0 \le x \le 1$ (as in Section 6.6, Exercise 17 and Section 6.7, Exercise 7).
10. The p.d.f. of a random variable X is $f(x) = 1 - \frac{x}{2}$ for $0 \le x \le 2$ (as in Section 6.6, Exercise 18 and Section 6.7, Exercise 8).
11. The p.d.f. of a random variable T is $h(t) = \frac{1}{t}$ for $1 \le t \le e$ (as in Section 6.6, Exercise 19 and Section 6.7, Exercise 9).
12. The p.d.f. of a random variable T is $g(t) = 6t(1-t)$ for $0 \le t \le 1$ (as in Section 6.6, Exercise 20 and Section 6.7, Exercise 10). This requires a computer (or Newton's method) to solve the equations.

13–16 ▪ Find the variance and standard deviation of a continuous random variable with the given p.d.f.

13. The p.d.f. of a random variable X is $f(x) = 2x$ for $0 \le x \le 1$ (as in Section 6.6, Exercise 17 and Section 6.7, Exercise 7).
14. The p.d.f. of a random variable X is $f(x) = 1 - \frac{x}{2}$ for $0 \le x \le 2$ (as in Section 6.6, Exercise 18 and Section 6.7, Exercise 8).
15. The p.d.f. of a random variable T is $h(t) = \frac{1}{t}$ for $1 \le t \le e$ (as in Section 6.6, Exercise 19 and Section 6.7, Exercise 9).
16. The p.d.f. of a random variable T is $g(t) = 6t(1-t)$ for $0 \le t \le 1$ (as in Section 6.6, Exercise 20 and Section 6.7, Exercise 10).

17–18 ▪ Find MAD for a continuous random variable with the given p.d.f. How does it compare with the standard deviation found in the earlier problem?

17. The p.d.f. of a random variable X is $f(x) = 2x$ for $0 \le x \le 1$ (as in Exercise 13).
18. The p.d.f. of a random variable X is $f(x) = 1 - \frac{x}{2}$ for $0 \le x \le 2$ (as in Exercise 14).

19–22 ▪ Find the probability that the random variable has a value less than one standard deviation below the expectation and less than two standard deviations below the expectation. How do the results compare with the rules of thumb for a bell-shaped distribution?

19. The p.d.f. of a random variable X is $f(x) = 2x$ for $0 \le x \le 1$ (as in Exercise 13).
20. The p.d.f. of a random variable X is $f(x) = 1 - \frac{x}{2}$ for $0 \le x \le 2$ (as in Exercise 14).
21. The p.d.f. of a random variable T is $h(t) = \frac{1}{t}$ for $1 \le t \le e$ (as in Exercise 15).
22. The p.d.f. of a random variable T is $g(t) = 6t(1-t)$ for $0 \le t \le 1$ (as in Exercise 16).

23–24 ▪ The following steps outline the proof of the computational formula for the variance.

23. Multiply out the squared term into three terms and break the sum into three sums.
24. To further simplify, try the following steps.
    a. Factor constants out of the sums.
    b. Remember that $\bar{X}$ is a constant.
    c. Recognize certain sums to be equal to the mean.
    d. Write in terms of the mean and group together like terms.

25–28 ▪ There is a general inequality about any random variable X, called **Chebyshev's inequality**. Suppose X has mean μ and standard deviation σ. Then

$$\Pr(|X - \mu| \ge k\sigma) \le \frac{1}{k^2}$$

for any value of k.

25. What does this mean for $k = 1$? Does this tell us anything?
26. What does this mean for $k = 2$? How much of the probability must lie within two standard deviations of the mean?

27. How much of the probability must lie within three standard deviations of the mean?

28. Compare the result of Exercise 26 with the second rule of thumb for bell-shaped distributions. Which gives more precise information?

Applications

29-30 ▪ Consider again the salaries presented in Section 6.8, Exercises 27 and 28.

Income	Probability
20,000	0.48
30,000	0.04
35,000	0.16
50,000	0.12
57,000	0.04
100,000	0.08
150,000	0.04
top salary	0.04

For each, find MAD, the variance, and coefficient of variation. Which statistics are most sensitive to large values? What are the units of each statistic?

29. With the top salary of $450,000.

30. With the top salary of $4,500,000.

31-34 ▪ For the data first presented in Section 6.6, Exercises 37–40, find the variance and standard deviation. How many of the values lie within two standard deviations of the mean?

	Frequency in 100 Experiments			
Number of Tall Offspring	Experiment a	Experiment b	Experiment c	Experiment d
0	0	0	0	5
1	0	0	2	11
2	0	1	5	35
3	0	1	10	27
4	2	6	21	13
5	8	19	27	8
6	9	20	19	1
7	31	27	8	0
8	22	16	5	0
9	22	9	3	0
10	6	1	0	0

31. Experiment a (Section 6.7, Exercise 23 computes the expectation).

32. Experiment b (Section 6.7, Exercise 24 computes the expectation).

33. Experiment c (Section 6.7, Exercise 25 computes the expectation).

34. Experiment d (Section 6.7, Exercise 26 computes the expectation).

35-38 ▪ Estimate the standard deviation, coefficient of variation, 2.5th percentile, and 97.5th percentile from the following figures.

35.

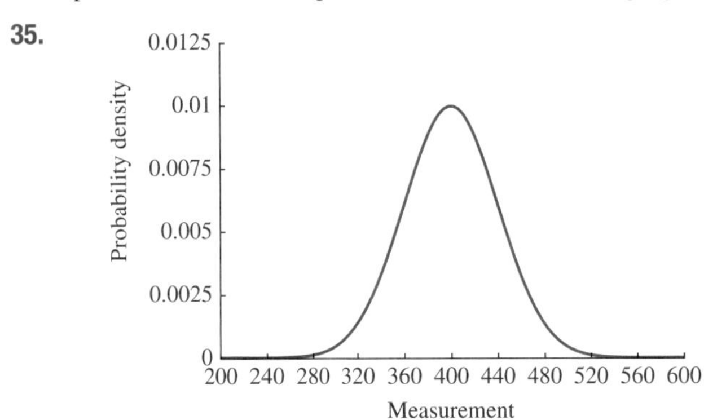

36.

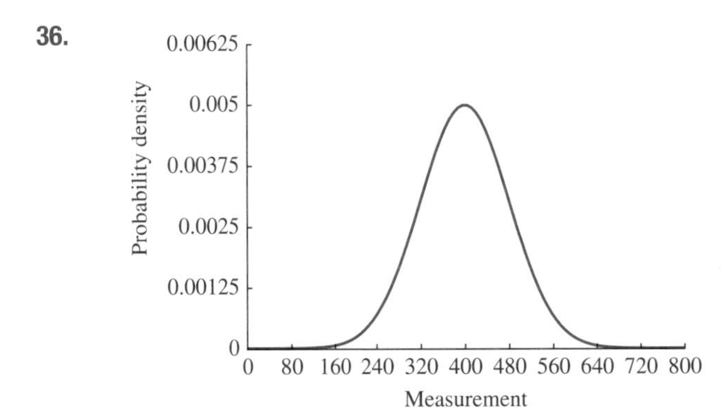

37.

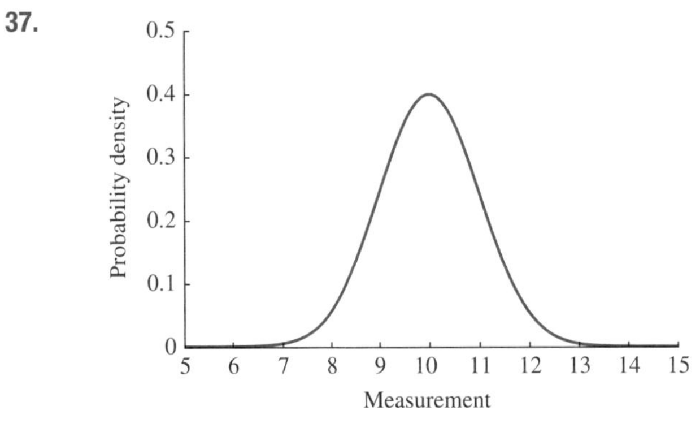

38.

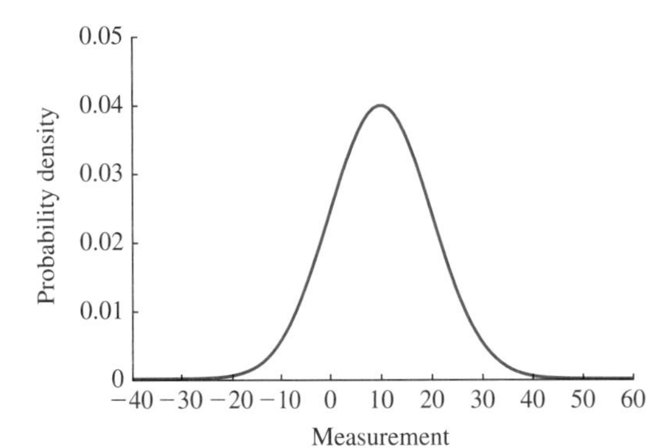

39-42 ▪ Draw bell-shaped p.d.f.'s with the following properties.

39. A p.d.f. with the same standard deviation of 10, but with a mean of 500. Calculate the coefficient of variation.

40. A p.d.f. with the same standard deviation of 10, but with a mean of 5. Calculate the coefficient of variation.

41. A p.d.f. with mean of 50 and coefficient of variation of 0.4. Calculate the standard deviation.

42. A p.d.f. with mean of 50 and coefficient of variation of 0.1. Calculate the standard deviation.

43-44 ▪ Suppose a population obeys

$$N_1 = R_1 N_0$$

where R_1 is a random variable that takes on the value 1.5 with probability 0.6 and 0.5 with probability 0.4. Suppose $N_0 = 1$.

43. Find the variance of the random variable N_1.

44. Find the variance of the random variable $\ln(N_1)$.

Computer Exercises

45. The bell-shaped or normal p.d.f. has the equation

$$f(x) = \frac{1}{\sqrt{2\pi}\sigma} e^{-\frac{(x-\mu)^2}{2\sigma^2}}.$$

μ is the mean and σ the standard deviation. Set $\mu = 50$ and $\sigma = 20$.

a. Plot this function between some reasonable limits.

b. Use integration to compute the mean and standard deviation. Remember that the limits of integration are from $-\infty$ to ∞.

c. Find the coefficient of variation.

d. We can check the rules of thumb regarding the shape of the curve and the standard deviation. Find the first derivative and solve for the maximum. Does it match the mean?

e. Find the second derivative and solve for the points of inflection. Do they match the rules of thumb?

46. Find and graph the cumulative distribution F associated with f from the previous problem. Compute the percentiles associated with points one and two standard deviations above and below the mean. Mark these on your graph. How well do they match the rules of thumb? Find the 5th and 95th percentile and the lower and upper quartiles.

Supplementary Problems

1. An intriguing species of worm spends its time wandering around in search of food. Two percent of the plate is covered with food. A point containing food contains worms with probability 0.1. A point without food contains worms with probability 0.04.

a. Draw a diagram illustrating the events and probabilities.

b. Find the probability that a point containing a worm also contains food.

c. Are the worms and food independent? What does this mean?

2. An experiment involves placing a single worm on a plate. After 3 days, there are no worms with probability 0.1, one worm with probability 0.35, two worms with probability 0.15, three worms with probability 0.3, and four worms with probability 0.1.

a. Sketch the probability distribution.

b. Sketch the cumulative distribution.

c. Find the expected number of worms.

d. What does this mean?

3. Baby worms grow from a length of 0.5 mm to a length of 1.0 mm. After a day of growth, the measurement L representing length has p.d.f.

$$f(l) = 1.5(l + l^2)$$

for $0.5 \le l \le 1.0$.

a. Graph the p.d.f. and check that it is consistent.

b. Find the probability that a worm is less than 0.75 mm long.

c. Find the approximate probability that a worm is between 0.75 and 0.76 mm long.

d. How many worms out of 1000 would you expect to be longer than 0.75 mm?

4. During a 10-minute interval, adult worms switch from eating to egg-laying with probability 0.1 and from egg-laying to eating with probability 0.15.

a. Draw a diagram illustrating this process.

b. Derive a discrete-time dynamical system for the probability the worm is eating.

c. Find the long-term fraction of time spent eating.

d. If egg-laying takes place at a rate of 1.5 per 10 minutes, about how many eggs would this worm lay during its 15-day life (worms never sleep)?

5. Two codons of DNA are being compared (each consists of three nucleotides). The experiment is to check whether the two match. For example, if the first is AAG and the second ATG, the result would be "match, no match, match."

a. List all of the simple events.

b. What simple events are included in the event "they match at the second nucleotide."

c. Find the union and intersection of the events "they match at the second nucleotide" and "they do not match at the first."

d. Suppose the probability of a change is 0.1 at the first site, 0.2 at the second, and 0.4 at the third. What is the probability of no change at any site if the events are independent?

6. A bacterium switches between moving and stopped. It switches from moving to stopped during a given second with probability 0.2 and from stopped to moving with probability 0.1. The bacterium moves at 10 μm/s.

a. How would you simulate this system?

b. What is a discrete-time dynamical system for the probability the bacterium is moving?

c. Find the average speed of the bacterium after a long time.

7. Suppose the per capita production of a population takes on the three values 0.5, 1.0, and 1.5 with the following probabilities

$$\Pr(\text{per capita production} = 0.5) = 0.3$$

$$\Pr(\text{per capita production} = 1.0) = 0.3$$

$$\Pr(\text{per capita production} = 1.5) = 0.4.$$

a. Draw the histogram.

b. Find the expectation.

8. Suppose a measurement X has p.d.f.

$$f(x) = 0.5 + x.$$

for $0 \le x \le 1$.

a. Find $\Pr(0.3 \le X \le 0.7)$. Sketch the region corresponding to this probability on a graph of the p.d.f.

b. Find and graph the c.d.f.

c. Find the expectation.

9. A laboratory is testing for a rare bacterial mutant, which appears in 1% of cells. A test finds the mutant with probability 0.9, and gives a false positive with probability 0.02.

a. Find the probability of a positive test.

b. Find the probability that a cell that tests positive has the mutation.

10. The per capita production R of a population is a function of the annual mean temperature T, satisfying

$$R = 1.5 - \frac{(T-50)^2}{100}.$$

Suppose the annual mean temperature T follows

$$\Pr(T = 40) = \frac{1}{3},\ \Pr(T = 50) = \frac{1}{3},\ \Pr(T = 60) = \frac{1}{3}.$$

a. Draw histograms for the temperature and the per capita production.

b. Find the expectation of T and of R.

11. Suppose a measurement Z has p.d.f.

$$f(z) = \frac{2}{z^2}$$

for $1 \le z \le 2$.

a. Sketch this p.d.f. and show it satisfies the necessary requirements.

b. Find $\Pr(1.2 \le Z \le 1.8)$ and indicate it on your graph.

c. Find the expectation and mark it on your graph.

12. After studying probability theory in the third grade, a boy fancies himself a meteorologist. He advises his family to cancel a picnic with probability 0.8 if it is cloudy, because that is the probability of rain if cloudy. He advises his family to cancel a picnic with probability 0.2 if it is sunny, because that is the probability of rain if sunny. Thirty percent of days are cloudy.

a. Find the probability that a picnic is canceled.

b. Find the probability that it rains on the picnic.

c. Is this strategy better than picnicking only on sunny days?

13. Birds switch between being infested with parasites and being parasite-free. A bird with parasites has a 10% probability of getting rid of them over the winter, and a bird without parasites has a 5% chance of getting infested.

a. Use a diagram or mathematical notation to describe this process.

b. Write a discrete-time dynamical system for the probability a bird is infested.

c. Find the equilibrium fraction of infested birds.

d. Suppose only 80% of birds survive the winter (independent of their being infested) and are replaced by parasite-free baby birds. Find the equilibrium fraction of infested birds.

14. Careful observation reveals that your probability of receiving a phone call during any 20-min interval while you are awake is 0.1. You spend 20 minutes per day in the shower (which is 2% of your waking hours) and during which time the phone rings with probability 0.2.

a. Are the events of showering and receiving phone calls independent? Prove it.

b. Find the probability of receiving a phone call during a 20-min period when you are not in the shower.

c. Find the probability that you are in the shower if the phone rings.

15. A certain basketball player makes a shot with probability 0.3 after making a shot and with probability 0.6 after missing. She complains to the officials about having been fouled with probability 0.2 after making a shot and with probability 0.7 after missing.

a. What is her long-term probability of making a shot?

b. What fraction of times does she complain about a foul?

c. You turn on the TV just in time to see her complaining to the officials. What is the probability she hit her shot?

16. Researchers in the Department of Human Genetics claim to have discovered a gene for mathematical ability. They have identified three alleles of this gene, creatively named A, B, and C. Let G represent the trait "good at math." Testing has shown that $\Pr(G \mid A) = 0.2$, $\Pr(G \mid B) = 0.5$, and $\Pr(G \mid C) = 0.1$. Furthermore, $\Pr(A) = 0.6$ and $\Pr(B) = 0.3$.

a. Draw a picture illustrating this situation.

b. Find Pr(C), Pr(G), and Pr(G and C).

c. Find $\Pr(A \mid G)$ and $\Pr(A \mid G^c)$, where G^c means "not good at math." Do your answers match? Why or why not?

d. If someone walked into your office and said, "I have allele C and should not take math," how would you respond?

17. Researchers in the Department of Horticulture are developing data on the last frost. They measure the time t continuously in months after April 1 (for example, $t = 0.5$ corresponds to the exact middle of April, or midnight on the 15th). Let T be a random variable representing the time of the last frost. They have found that

$$\Pr(T \le t) = 2t - t^2$$

for $0 \le t \le 1$.

a. Check that this c.d.f. makes sense. When is the last possible day for the last frost?

b. Find the median date of the last frost.

c. Find the probability that plants will get damaged by frost if planted at midnight on April 20.

d. Find the p.d.f. of T.

e. Find the mean date of the last frost. Compare with the median and explain why one is greater than the other.

f. (For gardeners only.) When would you plant, and why?

18. Researchers in the Department of Ecology and Evolution have discovered a population of tropical birds whose numbers jump between 5 and 10 and take on no other values. Let H represent the event that the population is "high" (10), and L represent the event that it is "low" (5). Suppose that $\Pr(H \mid L) = 0.3$ and $\Pr(L \mid H) = 0.2$.

a. Give a complete description of the dynamics of this population.

b. Write down a discrete-time dynamical system for the probability that the population is high.

c. Find the long-term probability.

d. The population can double, halve, or stay the same. Find the probability of each of these events after a long time, and compute the geometric mean growth rate. Why does your answer make sense?

Projects

1. We have used the geometric mean to study the growth of populations that follow the updating function

$$N_{t+1} = RN_t$$

where R is some random variable. This analysis was possible because the updating function is **linear**, meaning that nothing complicated happened to the population N_t. Consider the following stochastic variant of the Ricker equation (Section 3.2)

$$N_{t+1} = RN_t e^{-\frac{N_t}{K}}$$

where the per capita production R and the carrying capacity K might both be random variables.

a. Suppose first that $K = 100$ is a fixed value. Choose different random variables for R (with positive values only, of course) and simulate to see when the population survives and when it goes extinct. Does the geometric mean still identify those populations that will survive? Why or why not?

b. Now suppose that $R = 1.1$ is a fixed value. Choose different random variables for K (again with positive values only) and simulate to see when the population survives and when it goes extinct. Does it ever go extinct? What is the average population? Let $\bar{K}$ be the expectation of K. Compare the behavior of your stochastic population with one following the updating function

$$N_{t+1} = RN_t e^{-\frac{N_t}{\bar{K}}}.$$

Is the average of the stochastic population higher or lower than the equilibrium of this deterministic population?

c. Try cases where both R and K are random variables. When the values of R and K are independent, do we need to know anything about K to figure out whether the population will go extinct?

d. Can you find a case where the geometric mean of R is greater than 1 but the population goes extinct? Why or why not?

2. Research the following for at least five different diseases.

- Fraction of people with the disease
- Fraction of people tested
- Fraction of false positives
- Fraction of false negatives
- Cost of the test
- Availability of alternative tests
- Treatment options and effectiveness

Our analysis predicts that for rare diseases, the fraction of people tested should be small, particularly when the probability of a false positive is large. Is this what you find? How might the last three factors affect doctor's decisions?

Chapter

7

Probability Models

The language of probability theory provides a set of raw materials. To succeed in building useful structures, these raw materials must first be combined into a set of building blocks that can be used over and over again. The probability models introduced in this chapter provide such a set of building blocks. In probability theory, as in all of applied mathematics, the key is to describe the biological system and translate the description into mathematical form.

Each probability model describes a different way to combine simple measurements, such as counts (with the **binomial** and **Poisson distributions**), waiting times (with the **geometric** and **exponential distributions**), and sums (with the **normal distribution**). These basic distributions, like polynomials and exponential functions, have a remarkable range of applications and form the basis for the most important methods in statistics.

7.1 Joint Distributions

We have described the behavior of a single measurement with one random variable and its associated probability distribution or probability density function. The chief tool for studying the behavior of *two* measurements simultaneously is the **joint distribution**, which gives the probability of each possible *pair* of values for the random variables. From the joint distribution, we will derive the **marginal distributions**, which give the probabilities for each random variable separately, and the **conditional distributions**, which give the probabilities for one random variable when the other takes on a particular value.

7.1.1 Joint Distributions

Two dart players, each of whom scores according to the probability distribution in Figure 7.1.1, face off in a high-stakes match at their local pub. Do we have enough information to guess what will happen? Suppose one player always throws first. The results depend on how the second player responds to good and bad throws by the first. If the second player ignores what the first does, the throws will be **independent**; the probability that the second player scores a bull's-eye is the same whether the first got a bull's-eye or missed the board. Knowing what the first player did gives no information about the second. Most players, however, are incapable of ignoring success or failure by their opponent. Some might do better under pressure and be more likely to hit the bull's-eye after their opponent did than after their opponent totally missed. Other players tighten up and are less likely to hit the bull's-eye after a bull's-eye than after a miss. Each of these types of second player might have the same histogram. But no sports commentator worth her salary would think these players equivalent.

How can we mathematically summarize this information and distinguish these different types of player?

Example 7.1.1 A Joint Distribution for Independent Dart Throws

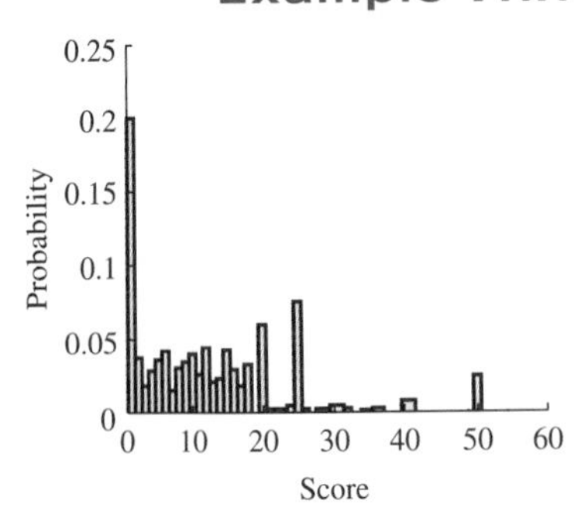

FIGURE 7.1.1
Probability distribution for dart score

Consider a simplified situation: a dart game where hitting the board scores 1 and missing the board scores 0. Suppose that each player scores 1 with probability 0.3. What is the probability that both players score 1? The answer depends on whether the player who throws second responds to the success of the first. Let S_1 be the random variable describing the score of the first player, which can take on the two values 0 and 1, and is thus a Bernoulli random variable. Similarly, S_2 is the random variable describing the score of the second player. We have assumed that $\Pr(S_1=1)=0.3$ and that $\Pr(S_2=1)=0.3$.

If the second player is unaffected by the first, the events will be independent and

$$\Pr(S_1=1 \text{ and } S_2=1)=\Pr(S_1=1)\Pr(S_2=1)=0.3\cdot 0.3=0.09$$
$$\Pr(S_1=1 \text{ and } S_2=0)=\Pr(S_1=1)\Pr(S_2=0)=0.3\cdot 0.7=0.21$$
$$\Pr(S_1=0 \text{ and } S_2=1)=\Pr(S_1=0)\Pr(S_2=1)=0.7\cdot 0.3=0.21$$
$$\Pr(S_1=0 \text{ and } S_2=0)=\Pr(S_1=0)\Pr(S_2=0)=0.7\cdot 0.7=0.49,$$

using the multiplication rule for the probabilities of independent events (Theorem 6.3). A Venn diagram of this situation is shown in Figure 7.1.2.

Player 1 hits and player 2 misses	Both hit
Both miss	Player 2 hits and player 1 misses

FIGURE 7.1.2
Venn diagram for independent dart players

We can summarize these probabilities in a table as

	$S_1=1$	$S_1=0$
$S_2=1$	0.09	0.21
$S_2=0$	0.21	0.49

Each value in the table is the probability that the events in both its column and row occurred. For example, the 0.49 appears in the column $S_1=0$ and the row $S_2=0$, meaning that

$$\Pr(S_1=0 \text{ and } S_2=0)=0.49.$$

Example 7.1.2 A Joint Distribution for Competitive Dart Players

Suppose that the second player becomes more competitive and throws better after the first player scores. Then the probability that $S_2=1$ *conditional* on $S_1=1$ will be greater than 0.3. Assume that this conditional probability is 0.8, or $\Pr(S_2=1 \mid S_1=1)=0.8$. Applying the multiplication rule for conditional probabilities (Equation 6.4.2),

$$\Pr(S_1=1 \text{ and } S_2=1)=\Pr(S_2=1 \mid S_1=1)\Pr(S_1=1)=0.8\cdot 0.3=0.24.$$

The probability that both score is elevated. We can fill in one term in the table.

	$S_1=1$	$S_1=0$
$S_2=1$	0.24	?
$S_2=0$	?	?

How do we find the rest of the probabilities? First, we know that the probability that the first player scores is 0.3. This is the combination of two mutually exclusive events, either both players scored (probability 0.24) or the first scored and the second missed (probability unknown). In equations,

$$\Pr(S_1=1)=\Pr(S_1=1 \text{ and } S_2=1)+\Pr(S_1=1 \text{ and } S_2=0)$$
$$0.3=0.24+\Pr(S_1=1 \text{ and } S_2=0).$$

Solving,

$$\Pr(S_1 = 1 \text{ and } S_2 = 0) = 0.06.$$

We can use the same reasoning to compute that

$$\Pr(S_1 = 0 \text{ and } S_2 = 1) = 0.06.$$

To find the probability that both missed, we can use the fact that all the probabilities must add to 1.

$$\begin{aligned} 1 &= \Pr(S_1 = 1 \text{ and } S_2 = 1) + \Pr(S_1 = 1 \text{ and } S_2 = 0) + \Pr(S_1 = 0 \text{ and } S_2 = 1) \\ &\quad + \Pr(S_1 = 0 \text{ and } S_2 = 0) \\ 1 &= 0.24 + 0.06 + 0.06 + \Pr(S_1 = 0 \text{ and } S_2 = 0), \end{aligned}$$

so that

$$\Pr(S_1 = 0 \text{ and } S_2 = 0) = 0.64.$$

Our complete table is

	$S_1 = 1$	$S_1 = 0$
$S_2 = 1$	0.24	0.06
$S_2 = 0$	0.06	0.64

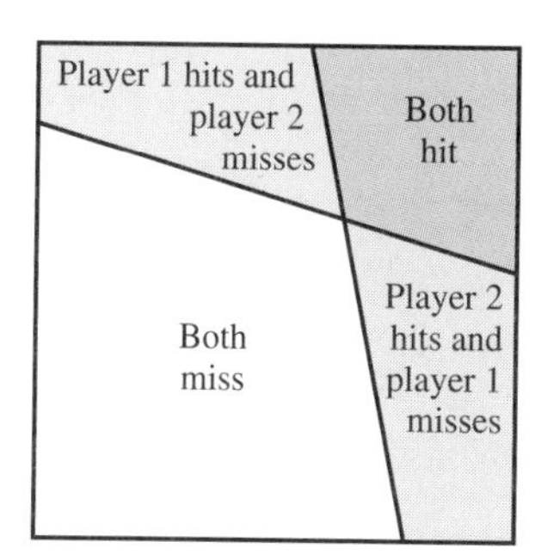

FIGURE 7.1.3

Venn diagram for competitive dart players

It is more likely that both players score or that both miss than in the independent case. This situation is shown as a Venn diagram in Figure 7.1.3.

Example 7.1.3 A Joint Distribution for Intimidated Dart Players

If the second player is intimidated when the first scores, the probability of $S_2 = 1$ conditional on $S_1 = 1$ will be diminished, perhaps to 0.1. In this case,

$$\Pr(S_1 = 1 \text{ and } S_2 = 1) = \Pr(S_1 = 1)\Pr(S_2 = 1 \mid S_1 = 1) = 0.3 \cdot 0.1 = 0.03.$$

Our table begins

	$S_1 = 1$	$S_1 = 0$
$S_2 = 1$	0.03	?
$S_2 = 0$	?	?

Again, we can use the fact that $\Pr(S_1 = 1) = 0.3$ to find that $\Pr(S_1 = 1 \text{ and } S_2 = 0) = 0.27$, and the fact that $\Pr(S_2 = 1) = 0.3$ to find that $\Pr(S_2 = 1 \text{ and } S_1 = 0) = 0.27$. The complete table in this case is

	$S_1 = 1$	$S_1 = 0$
$S_2 = 1$	0.03	0.27
$S_2 = 0$	0.27	0.43

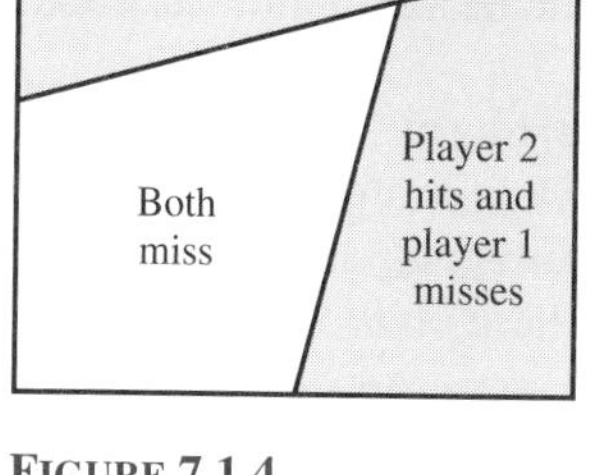

FIGURE 7.1.4

Venn diagram for intimidated dart players

The probability that both players score or both miss is less than in the independent case. This situation is shown as a Venn diagram in Figure 7.1.4.

Each of these tables gives the probability of each possible pair of events and constitutes the **joint probability distribution**. Like an ordinary probability distribution, the joint probability distribution associates a probability with every possible measured result. The set of these probabilities must add to 1. A joint distribution, however, describes two measurements rather than one.

Definition 7.1 Suppose X and Y are discrete random variables taking on the values $x_1 \ldots x_n$ and $y_1 \ldots y_m$, respectively. The joint distribution consists of the probabilities p_{ij} of each composite event $X = x_i$ and $Y = y_j$, or

$$p_{ij} = \Pr(X = x_i \text{ and } Y = y_j).$$

A table of this general joint distribution is

	$Y = y_1$	$Y = y_2$	...	$Y = y_j$	...	$Y = y_m$
$X = x_1$	p_{11}	p_{12}	...	p_{1j}	...	p_{1m}
$X = x_2$	p_{21}	p_{22}	...	p_{2j}	...	p_{2m}
...						
$X = x_i$	p_{i1}	p_{i2}	...	p_{ij}	...	p_{im}
...						
$X = x_n$	p_{n1}	p_{n2}	...	p_{nj}	...	p_{nm}

All the probabilities in a joint distribution must add to 1.

Example 7.1.4 A Joint Distribution for Random Variables Taking on Three Values

Suppose the random variable X can take on the three values 0, 1, and 2, while the random variable Y can take on the three values 2, 4, and 6. One possible joint distribution for these random variables is

	$Y = 2$	$Y = 4$	$Y = 6$
$X = 0$	0.1	0.2	0.0
$X = 1$	0.05	0.1	0.2
$X = 2$	0.2	0.0	0.15

The probability that $X = 0$ and $Y = 2$ is 0.1. The two probabilities of 0.0 indicate that it is impossible to have $X = 2$ and $Y = 4$, or to have $X = 0$ and $Y = 6$.

The joint distribution for more than two random variables is based on the same idea. The joint distribution for three discrete random variables X, Y, and Z is the set of probabilities

$$p_{ijk} = \Pr(X = x_i \text{ and } Y = y_j \text{ and } Z = z_k).$$

Again, we assign a probability to every possible combination of measurements. This joint distribution gives the probability that the three measurements simultaneously take a particular set of values.

For continuous random variables, one can define a **joint probability density function** analogous to the ordinary probability density function (Definition 6.7). The statement and application of this definition requires the **double integral**, a generalization of the integral to functions of two variables. We concentrate here on the discrete case, which is the most important for statistical applications.

7.1.2 Marginal Probability Distributions

Suppose we know the joint distribution of two measurements. We can use this information and the law of total probability to compute the probability distribution for each measurement by itself, called the **marginal probability distribution**.

Example 7.1.5 Finding the Marginal Probability Distributions for Dart Players

To compute the probability that the first player scored from the joint distribution, we return to the argument behind the law of total probability (Theorem 6.1). The event $S_1 = 1$ consists of two disjoint events: $S_1 = 1$ and $S_2 = 1$, and $S_1 = 1$ and $S_2 = 0$.

The probabilities of mutually exclusive events add (Requirement 3 for probabilities in Subsection 6.3.3), so

$$\Pr(S_1 = 1) = \Pr(S_1 = 1 \text{ and } S_2 = 1) + \Pr(S_1 = 1 \text{ and } S_2 = 0).$$

We see that $\Pr(S_1 = 1) = 0.24 + 0.06 = 0.3$, which is consistent with our assumptions. We used this reasoning already to fill in the table of probabilities.

The probabilities of measurements taken singly can be computed conveniently from the tables. $\Pr(S_1 = 1)$ is the sum of the values in the first column, the probabilities associated with events involving a score by the first player. The probability that the second player scores is the sum of the values in the first row, and so forth. In our table showing a joint distribution, we can display this with arrows as follows.

	$S_1 = 1$	$S_1 = 0$		
$S_2 = 1$	0.24	0.06	→	$\Pr(S_2 = 1) = 0.3$
$S_2 = 0$	0.06	0.64	→	$\Pr(S_2 = 0) = 0.7$
	↓	↓		
	$\Pr(S_1 = 1) = 0.3$	$\Pr(S_1 = 0) = 0.7$		

In general,

$$\Pr(X = x_i) = \sum_{j=1}^{m} \Pr(X = x_i \text{ and } Y = y_j).$$

added over all possible values y_j. We call this the marginal probability distribution, defined as follows.

Definition 7.2 Suppose discrete random variables X and Y have the joint distribution

$$p_{ij} = \Pr(X = x_i \text{ and } Y = y_j)$$

where X takes on the values $x_1, x_2, \ldots x_n$ and Y takes on the values $y_1, y_2, \ldots y_m$. The **marginal probability distributions** for the random variables X and Y are

$$p_{X_i} = \Pr(X = x_i) = \sum_{j=1}^{m} p_{ij}$$

$$p_{Y_j} = \Pr(Y = y_j) = \sum_{i=1}^{n} p_{ij}.$$

Example 7.1.6 The Marginal Distribution for Two Random Variables

Consider the random variables in Example 7.1.4. We find the marginal distributions for X and Y by adding the probabilities in the rows and columns.

	$Y = 2$	$Y = 4$	$Y = 6$		
$X = 0$	0.1	0.2	0.0	→	$\Pr(X = 0) = 0.3$
$X = 1$	0.05	0.1	0.2	→	$\Pr(X = 1) = 0.35$
$X = 2$	0.2	0.0	0.15	→	$\Pr(X = 2) = 0.35$
	↓	↓	↓		
	$\Pr(Y = 2) = 0.35$	$\Pr(Y = 4) = 0.3$	$\Pr(Y = 6) = 0.35$		

The new column gives the marginal probabilities for the random variable X, while the new row gives the marginal probabilities for the random variable Y.

The marginal distributions have a simple relationship to the joint distribution when the random variables are independent. The multiplicative rule for independent events (Theorem 6.3) says that

$$\Pr(X = x_i \text{ and } Y = y_j) = \Pr(X = x_i) \cdot \Pr(Y = y_j).$$

This is used as the definition of independent random variables.

Definition 7.3 Two discrete random variables X and Y are independent if the joint probability distribution is equal to the product of the marginal distributions.

Example 7.1.7 Two Random Variables That Are Independent

The random variables S_1 and S_2 in Example 7.1.1 are independent:

	$S_1 = 1$	$S_1 = 0$		
$S_2 = 1$	0.09	0.21	→	$\Pr(S_2 = 1) = 0.3$
$S_2 = 0$	0.21	0.49	→	$\Pr(S_2 = 0) = 0.7$
	↓	↓		
	$\Pr(S_1 = 1) = 0.3$	$\Pr(S_1 = 0) = 0.7$		

Each term in the joint distribution is the product of the corresponding marginal distributions.

$$\Pr(S_1 = 1 \text{ and } S_2 = 1) = 0.09 = \Pr(S_1 = 1)\Pr(S_2 = 1) = 0.3 \cdot 0.3$$

$$\Pr(S_1 = 1 \text{ and } S_2 = 0) = 0.21 = \Pr(S_1 = 1)\Pr(S_2 = 0) = 0.3 \cdot 0.7$$

$$\Pr(S_1 = 0 \text{ and } S_2 = 1) = 0.21 = \Pr(S_1 = 0)\Pr(S_2 = 1) = 0.7 \cdot 0.3$$

$$\Pr(S_1 = 0 \text{ and } S_2 = 0) = 0.49 = \Pr(S_1 = 0)\Pr(S_2 = 0) = 0.7 \cdot 0.7.$$

Example 7.1.8 Two Random Variables That Are Not Independent

We can tell that the random variables X and Y in Example 7.1.6 are not independent because, for example,

$$\Pr(X = 0 \text{ and } Y = 6) = 0.0 \neq \Pr(X = 0)\Pr(Y = 6) = 0.3 \cdot 0.35 = 0.105.$$

7.1.3 Joint Distributions and Conditional Distributions

A more complicated example is given in Table 7.1. Two types of bird are infected by two parasites, a louse and a mite. Because these parasites are rather large, no bird has more than two of either one. Data are collected on the joint probabilities of parasite numbers for each type of bird. For example, the probability that an individual of bird A has one mite and two lice is 0.09, and the probability that it has two mites and one louse is 0.06.

Table 7.1 Lice and mites on two types of birds

		Bird A Lice			Bird B Lice		
		0	1	2	0	1	2
	0	0.20	0.15	0.15	0.33	0.08	0.09
Mites	1	0.12	0.09	0.09	0.05	0.13	0.12
	2	0.08	0.06	0.06	0.02	0.09	0.09

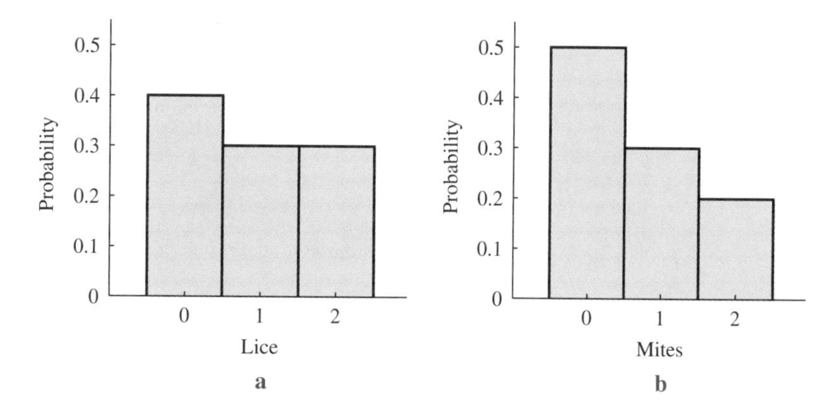

FIGURE 7.1.5
Marginal distributions of lice and mites for bird A

Example 7.1.9 Marginal Distributions for Bird A

To find the marginal distribution describing the number of lice, add up all the numbers in each column. For example, the probability that the first bird has one louse is the sum $0.15 + 0.09 + 0.06 = 0.3$, and so forth (Figure 7.1.5). The marginal distribution of the number of mites is found by adding up the rows. The probability that the bird A has 1 mite is $0.12 + 0.9 + 0.9 = 0.3$. The marginal distributions are computed in the following table.

	$L=0$	$L=1$	$L=2$		
$M=0$	0.20	0.15	0.15	→	$\Pr(M=0)=0.5$
$M=1$	0.12	0.09	0.09	→	$\Pr(M=1)=0.3$
$M=2$	0.08	0.06	0.06	→	$\Pr(M=2)=0.2$
	↓	↓	↓		
	$\Pr(L=0)=0.4$	$\Pr(L=1)=0.3$	$\Pr(L=2)=0.3$		

Example 7.1.10 Marginal Distributions for Bird B

The marginal distributions for both birds have been chosen to match. Checking bird B, we find

	$L=0$	$L=1$	$L=2$		
$M=0$	0.33	0.08	0.09	→	$\Pr(M=0)=0.5$
$M=1$	0.05	0.13	0.12	→	$\Pr(M=1)=0.3$
$M=2$	0.02	0.09	0.09	→	$\Pr(M=2)=0.2$
	↓	↓	↓		
	$\Pr(L=0)=0.4$	$\Pr(L=1)=0.3$	$\Pr(L=2)=0.3$		

Although none of the joint probabilities match, the marginal distributions are identical.

Example 7.1.11 Mites and Lice Infest Bird A Independently

For bird A, the product of these marginal distributions matches the probabilities. For example, the probability of one mite and two lice is 0.06, equal to the product of the marginal probability of one mite (0.3) with the marginal probability of two lice (0.2). We can check the full set of probabilities by reversing the process used to find the marginal distribution, as illustrated in the following table.

	$L=0$	$L=1$	$L=2$		
$M=0$	$0.5\cdot 0.4=0.20$	$0.5\cdot 0.3=0.15$	$0.5\cdot 0.3=0.15$	$\leftarrow$	$\Pr(M=0)=0.5$
$M=1$	$0.3\cdot 0.4=0.12$	$0.3\cdot 0.3=0.09$	$0.3\cdot 0.3=0.09$	$\leftarrow$	$\Pr(M=1)=0.3$
$M=2$	$0.2\cdot 0.4=0.08$	$0.2\cdot 0.3=0.06$	$0.2\cdot 0.3=0.06$	$\leftarrow$	$\Pr(M=2)=0.2$
	$\uparrow$	$\uparrow$	$\uparrow$		
	$\Pr(L=0)=0.4$	$\Pr(L=1)=0.3$	$\Pr(L=2)=0.3$		

The two parasites are indeed distributed independently in this type of bird.

Example 7.1.12 Mites and Lice Do Not Infest Bird B Independently

The marginal distributions for bird B match those for bird A, but the joint probabilities do not. Therefore, the joint probabilities cannot equal the product of the marginal probabilities, and the two parasites do not infest this species of bird independently.

Showing that two random variables are not independent gives little indication of how they are in fact related. When we analyze joint distributions, a new tool called the **conditional distribution** provides a clear way to depict the relation between random variables.

Definition 7.4 Suppose random variables X and Y have joint distribution p_{ij} and marginal distributions $p_{X,i}$ and $p_{Y,j}$. The **conditional distribution** of X conditional on $Y=y_j$ is

$$\Pr(X=x_i|Y=y_j)=\frac{\Pr(X=x_i \text{ and } Y=y_j)}{\Pr(Y=y_j)}$$

$$=\frac{p_{ij}}{p_{Y,j}}.$$

Similarly, the **conditional distribution** of Y conditional on $X=x_i$ is

$$\Pr(Y=y_j|X=x_i)=\frac{\Pr(Y=y_j \text{ and } X=x_i)}{\Pr(X=x_i)}$$

$$=\frac{p_{ij}}{p_{X,i}}.$$

Like any other probability distribution, the probabilities in a conditional distribution must add to 1. This makes comparing these distributions easy. A good way to understand how a joint distribution deviates from independence is to compare the conditional distributions with the marginal distribution.

Example 7.1.13 The Conditional Distributions of Y for Two Random Variables

Consider the random variables in Examples 7.1.4 and 7.1.6, with the joint and marginal distributions given in the following table.

	$Y=2$	$Y=4$	$Y=6$		
$X=0$	0.1	0.2	0.0	$\rightarrow$	$\Pr(X=0)=0.3$
$X=1$	0.05	0.1	0.2	$\rightarrow$	$\Pr(X=1)=0.35$
$X=2$	0.2	0.0	0.15	$\rightarrow$	$\Pr(X=2)=0.35$
	$\downarrow$	$\downarrow$	$\downarrow$		
	$\Pr(Y=2)=0.35$	$\Pr(Y=4)=0.3$	$\Pr(Y=6)=0.35$		

Each row and column has an associated conditional distribution. For example, the first row describes the distribution of values of Y when $X = 0$. The conditional distribution as follows.

$$\Pr(Y=2|X=0) = \frac{\Pr(Y=2 \text{ and } X=0)}{\Pr(X=0)} = \frac{0.1}{0.3} = 0.333$$

$$\Pr(Y=4|X=0) = \frac{\Pr(Y=4 \text{ and } X=0)}{\Pr(X=0)} = \frac{0.2}{0.3} = 0.667$$

$$\Pr(Y=6|X=0) = \frac{\Pr(Y=6 \text{ and } X=0)}{\Pr(X=0)} = \frac{0.0}{0.3} = 0.$$

The conditional distribution when $X = 1$ is quite different.

$$\Pr(Y=2|X=1) = \frac{\Pr(Y=2 \text{ and } X=1)}{\Pr(X=1)} = \frac{0.05}{0.35} = 0.143$$

$$\Pr(Y=4|X=1) = \frac{\Pr(Y=4 \text{ and } X=1)}{\Pr(X=1)} = \frac{0.1}{0.35} = 0.286$$

$$\Pr(Y=6|X=1) = \frac{\Pr(Y=6 \text{ and } X=1)}{\Pr(X=1)} = \frac{0.2}{0.35} = 0.571.$$

Again, the conditional distribution when $X = 2$ is completely different.

$$\Pr(Y=2|X=2) = \frac{\Pr(Y=2 \text{ and } X=2)}{\Pr(X=2)} = \frac{0.2}{0.35} = 0.571$$

$$\Pr(Y=4|X=2) = \frac{\Pr(Y=4 \text{ and } X=2)}{\Pr(X=2)} = \frac{0.0}{0.35} = 0$$

$$\Pr(Y=6|X=2) = \frac{\Pr(Y=6 \text{ and } X=2)}{\Pr(X=2)} = \frac{0.15}{0.35} = 0.429.$$

Histograms of these conditional distributions are indeed completely different (Figure 7.1.6).

Knowing the value of X gives a great deal of information about Y. For example, if $X = 0$, Y is never equal to 6, but if $X = 1$, 6 is the most probable value for Y.

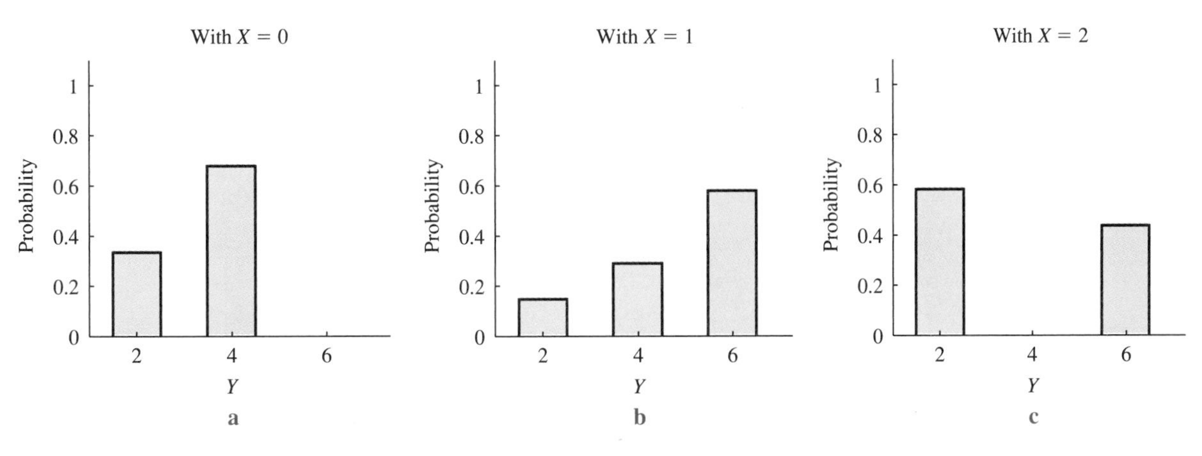

FIGURE 7.1.6
Conditional probabilities of random variable Y for three values of the random variable X

Example 7.1.14 The Conditional Distributions of X for Two Random Variables

For the random variables in Example 7.1.13, we find the conditional distribution of X by fixing each value for Y. With $Y=2$,

$$\Pr(X=0|Y=2) = \frac{\Pr(X=0 \text{ and } Y=2)}{\Pr(Y=2)} = \frac{0.1}{0.35} = 0.286$$

$$\Pr(X=1|Y=2) = \frac{\Pr(X=1 \text{ and } Y=2)}{\Pr(Y=2)} = \frac{0.05}{0.35} = 0.143$$

$$\Pr(X=2|Y=2) = \frac{\Pr(X=2 \text{ and } Y=2)}{\Pr(Y=2)} = \frac{0.2}{0.35} = 0.571.$$

With $Y=4$,

$$\Pr(X=0|Y=4) = \frac{\Pr(X=0 \text{ and } Y=4)}{\Pr(Y=4)} = \frac{0.2}{0.3} = 0.667$$

$$\Pr(X=1|Y=4) = \frac{\Pr(X=1 \text{ and } Y=4)}{\Pr(Y=4)} = \frac{0.1}{0.3} = 0.333$$

$$\Pr(X=2|Y=4) = \frac{\Pr(X=2 \text{ and } Y=4)}{\Pr(Y=4)} = \frac{0.0}{0.3} = 0.0.$$

With $Y=6$,

$$\Pr(X=0|Y=6) = \frac{\Pr(X=0 \text{ and } Y=6)}{\Pr(Y=6)} = \frac{0.0}{0.35} = 0$$

$$\Pr(X=1|Y=6) = \frac{\Pr(X=1 \text{ and } Y=6)}{\Pr(Y=6)} = \frac{0.2}{0.35} = 0.571$$

$$\Pr(X=2|Y=6) = \frac{\Pr(X=2 \text{ and } Y=6)}{\Pr(Y=6)} = \frac{0.15}{0.35} = 0.429.$$

Again, the graphs of these conditional distributions are quite different (Figure 7.1.7), and describe the relationship between these two random variables.

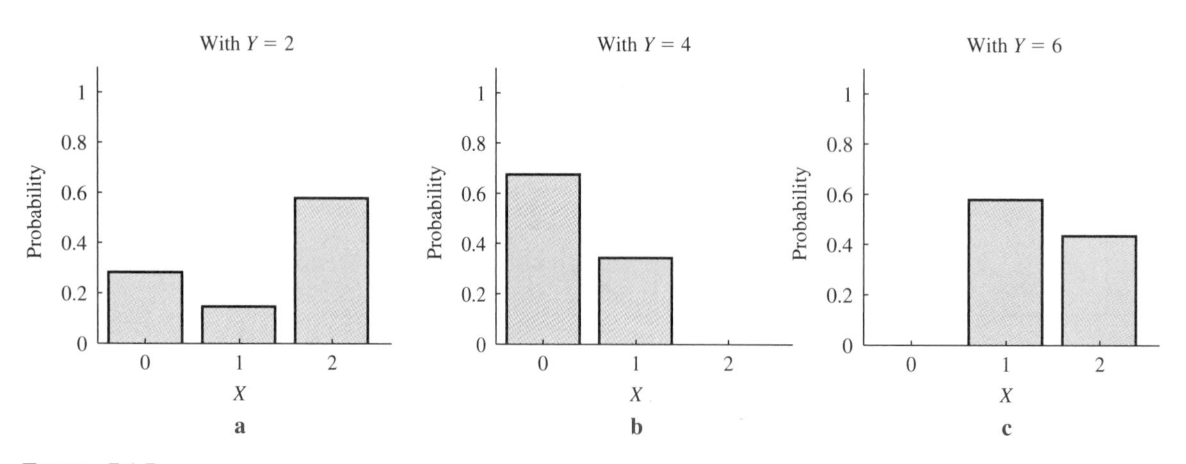

FIGURE 7.1.7
Conditional probabilities of random variable X for three values of the random variable Y

In more applied examples, the conditional probability distributions can be the most useful way to understand and describe relationship between random variables.

Example 7.1.15 Conditional Probabilities Match Marginal Probabilities for Bird A

For bird A, the number of lice conditional on $M = 0$ is

$$\Pr(L=0 \mid M=0) = \frac{\Pr(L=0 \text{ and } M=0)}{\Pr(M=0)} = \frac{0.20}{0.5} = 0.4$$

$$\Pr(L=1 \mid M=0) = \frac{\Pr(L=1 \text{ and } M=0)}{\Pr(M=0)} = \frac{0.15}{0.5} = 0.3$$

$$\Pr(L=2 \mid M=0) = \frac{\Pr(L=2 \text{ and } M=0)}{\Pr(M=0)} = \frac{0.15}{0.5} = 0.3.$$

These values exactly match the marginal distribution for L. Furthermore, the rest of the conditional distributions are identical. With $M = 1$,

$$\Pr(L=0 \mid M=1) = \frac{\Pr(L=0 \text{ and } M=1)}{\Pr(M=1)} = \frac{0.12}{0.3} = 0.4$$

$$\Pr(L=1 \mid M=1) = \frac{\Pr(L=1 \text{ and } M=1)}{\Pr(M=1)} = \frac{0.09}{0.3} = 0.3$$

$$\Pr(L=2 \mid M=1) = \frac{\Pr(L=2 \text{ and } M=1)}{\Pr(M=1)} = \frac{0.09}{0.3} = 0.3.$$

With $M = 2$,

$$\Pr(L=0 \mid M=2) = \frac{\Pr(L=0 \text{ and } M=2)}{\Pr(M=2)} = \frac{0.08}{0.2} = 0.4$$

$$\Pr(L=1 \mid M=2) = \frac{\Pr(L=1 \text{ and } M=2)}{\Pr(M=2)} = \frac{0.06}{0.2} = 0.3$$

$$\Pr(L=2 \mid M=2) = \frac{\Pr(L=2 \text{ and } M=2)}{\Pr(M=2)} = \frac{0.06}{0.2} = 0.3.$$

These identical distributions (Figure 7.1.8) tell us that the number of lice is unaffected by the number of mites, which is another way to say that the two random variables are independent.

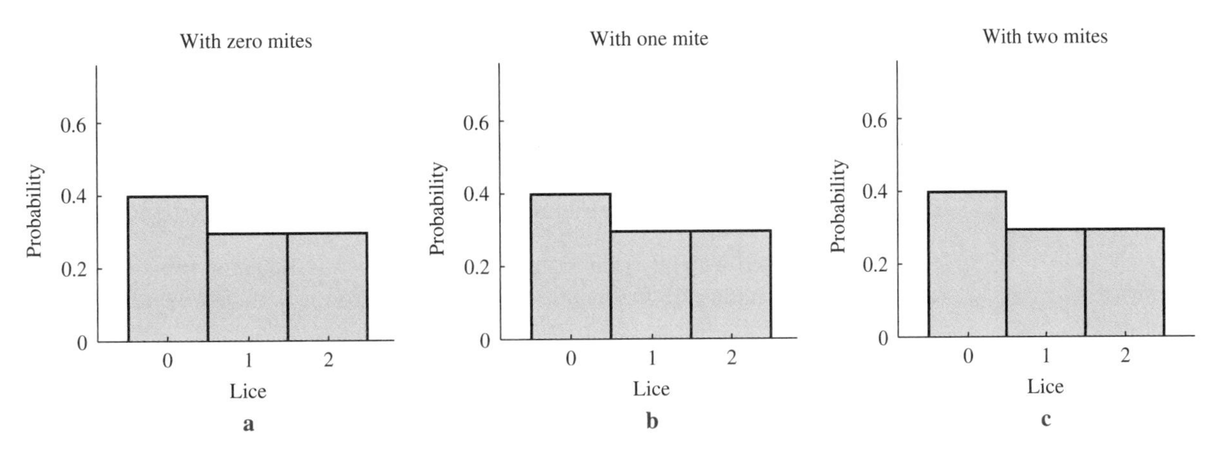

FIGURE 7.1.8
Conditional probabilities of lice on bird A

Example 7.1.16 Conditional Probabilities of Lice on Bird B

For the bird B, the number of lice conditional on $M=0$ is

$$\Pr(L=0 \mid M=0)=\frac{\Pr(L=0 \text{ and } M=0)}{\Pr(M=0)}=\frac{0.33}{0.5}=0.66$$

$$\Pr(L=1 \mid M=0)=\frac{\Pr(L=1 \text{ and } M=0)}{\Pr(M=0)}=\frac{0.08}{0.5}=0.16$$

$$\Pr(L=2 \mid M=0)=\frac{\Pr(L=2 \text{ and } M=0)}{\Pr(M=0)}=\frac{0.09}{0.5}=0.18.$$

With one mite,

$$\Pr(L=0 \mid M=1)=\frac{\Pr(L=0 \text{ and } M=1)}{\Pr(M=1)}=\frac{0.05}{0.3}=0.167$$

$$\Pr(L=1 \mid M=1)=\frac{\Pr(L=1 \text{ and } M=1)}{\Pr(M=1)}=\frac{0.13}{0.3}=0.433$$

$$\Pr(L=2 \mid M=1)=\frac{\Pr(L=2 \text{ and } M=1)}{\Pr(M=1)}=\frac{0.12}{0.3}=0.4.$$

With two mites,

$$\Pr(L=0 \mid M=2)=\frac{\Pr(L=0 \text{ and } M=2)}{\Pr(M=2)}=\frac{0.02}{0.2}=0.10$$

$$\Pr(L=1 \mid M=2)=\frac{\Pr(L=1 \text{ and } M=2)}{\Pr(M=2)}=\frac{0.09}{0.2}=0.45$$

$$\Pr(L=2 \mid M=2)=\frac{\Pr(L=2 \text{ and } M=2)}{\Pr(M=2)}=\frac{0.09}{0.2}=0.45.$$

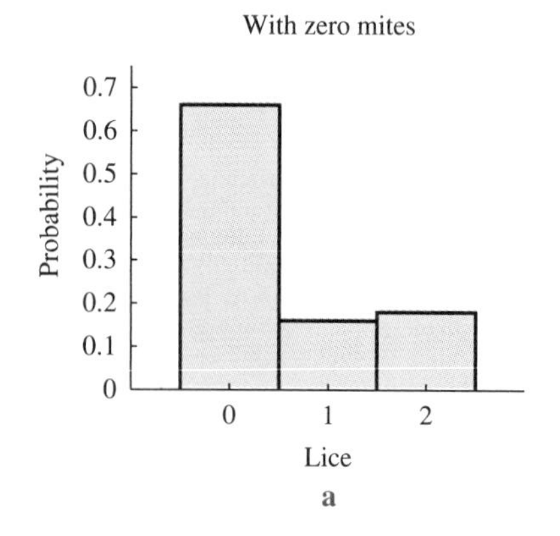

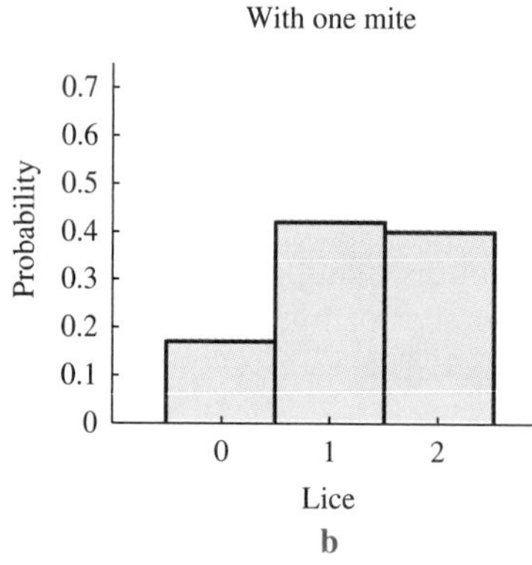

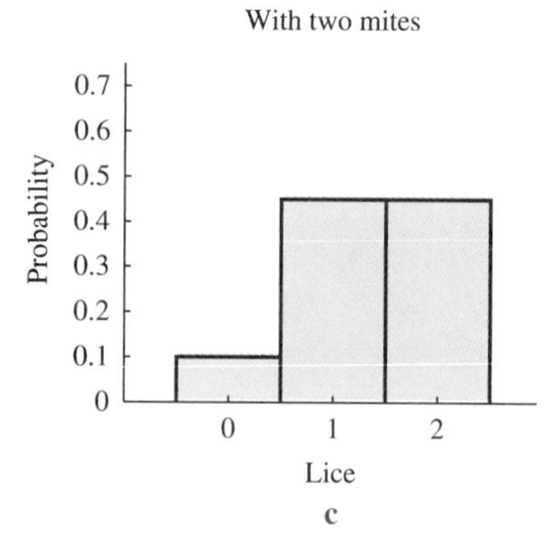

FIGURE 7.1.9

Conditional probabilities of lice on bird B

Birds with no mites are more likely to have no lice, while those with the two mites are likely to have at least one louse (Figure 7.1.9).

As we observed with bird A in Example 7.1.11, the conditional distributions match when random variables are independent. This observation is demonstrated by the following theorem.

Theorem 7.1 Suppose X and Y are random variables. Then their conditional distributions are equal to the marginal distributions if and only if X and Y are independent.

Proof: If X and Y are independent, then

$$p_{ij}=p_{X,i}\,p_{Y,j}$$

(Definition 7.3). Therefore,

$$\Pr(X = x_i | Y = y_j) = \frac{p_{ij}}{p_{Y,j}} = \frac{p_{X,i} p_{Y,j}}{p_{Y,j}} = p_{X,i}.$$

By the same reasoning,

$$\Pr(Y = y_j | X = x_i) = p_{Y,j}$$

so the conditional distributions are both equal to the marginal distributions.

Conversely, suppose that the conditional distributions are equal to the marginal distributions. Then

$$\Pr(X = x_i | Y = y_j) = \frac{\Pr(X = x_i \text{ and } Y = y_j)}{\Pr(Y = y_j)} = \Pr(X = x_i).$$

Therefore,

$$\Pr(X = x_i \text{ and } Y = y_j) = \Pr(Y = y_j) \Pr(X = x_i),$$

and the random variables are independent (Definition 7.3).

If we find that the conditional distributions of one measurement are the same for every value of the other measurement, then the two measurements are independent. This is consistent with our interpretation of independence as a lack of information. Knowing about one measurement gives no new information about another independent measurement.

Summary To describe two discrete random variables simultaneously, we use the **joint distribution**, the set of probabilities describing all possible pairs of measurements. Summing over all possible values of one discrete random variable gives the **marginal probability distribution** for the other. Two random variables are independent when their joint distribution is equal to the product of their marginal distributions. By fixing the value of one random variable, we can find the **conditional distributions**. The conditional distributions are the same when the two random variables are independent. Differences between the conditional distributions are useful for reasoning about the relationship between two measurements.

7.1 Exercises

Mathematical Techniques

1–4 ▪ For the following joint distributions describing the values of the random variables X and Y, find both marginal distributions and the conditional distribution requested. Are the two random variables independent?

1. Find the marginal distributions and the distribution of Y conditional on $X = 0$.

	$X=0$	$X=1$
$Y=0$	0.14	0.26
$Y=1$	0.21	0.39

2. Find the marginal distributions and the distribution of X conditional on $Y = 1$.

	$X=0$	$X=1$
$Y=1$	0.45	0.25
$Y=3$	0.05	0.25

3. Find the marginal distributions and the distribution of Y conditional on $X = 1$.

	$X=0$	$X=1$	$X=2$
$Y=1$	0.1	0.2	0.3
$Y=2$	0.05	0.15	0.2

4. Find the marginal distributions and the distribution of Y conditional on $X = 2$.

	$X=0$	$X=1$	$X=2$
$Y=1$	0.05	0.04	0.01
$Y=2$	0.1	0.08	0.02
$Y=3$	0.35	0.28	0.07

5–8 ▪ Find the expectations of the random variables from their marginal distributions.

5. The random variables X and Y in Exercise 1.

6. The random variables X and Y in Exercise 2.

7. The random variables X and Y in Exercise 3.

8. The random variables X and Y in Exercise 4.

9–12 ▪ Suppose the following random variables are independent. Find the joint distribution and the requested conditional distribution from the given marginal distributions.

9. The random variable X has probability distribution $\Pr(X=0)=0.7$ and $\Pr(X=1)=0.3$. The random variable Y has probability distribution $\Pr(Y=0)=0.3$ and $\Pr(Y=1)=0.7$. Find the distribution of X conditional on $Y=0$.

10. The random variable X has probability distribution $\Pr(X=0)=0.4$ and $\Pr(X=1)=0.6$. The random variable Y has probability distribution $\Pr(Y=1)=0.2$ and $\Pr(Y=3)=0.8$. Find the distribution of X conditional on $Y=3$.

11. The random variable X has probability distribution $\Pr(X=0)=0.8$ and $\Pr(X=1)=0.2$. The random variable Y has probability distribution $\Pr(Y=1)=0.3$, $\Pr(Y=2)=0.5$, and $\Pr(Y=3)=0.2$. Find the distribution of Y conditional on $X=0$.

12. The random variable X has probability distribution $\Pr(X=0)=0.3$, $\Pr(X=1)=0.4$, and $\Pr(X=2)=0.3$. The random variable Y has probability distribution $\Pr(Y=1)=0.6$, $\Pr(Y=2)=0.1$, and $\Pr(Y=3)=0.3$. Find the distribution of Y conditional on $X=2$.

13–16 ▪ Use the given information to construct the entire joint distribution for the following pairs of random variables.

13. Suppose that the random variables X and Y are each Bernoulli random variables (and thus take on only the values 0 and 1). We know that $\Pr(X=0)=0.2$, $\Pr(Y=0)=0.4$, and $\Pr(X=0 \text{ and } Y=0)=0.1$.

14. Suppose that the random variables X and Y are each Bernoulli random variables and that $\Pr(X=0)=0.3$, $\Pr(Y=1)=0.5$, and $\Pr(X=1 \text{ and } Y=0)=0.4$.

15. Suppose that the random variables X and Y are each Bernoulli random variables, and that $\Pr(X=0)=0.3$, $\Pr(Y=0)=0.6$, and $\Pr(X=0|Y=0)=0.5$.

16. Suppose that the random variables X and Y are each Bernoulli random variables, and that $\Pr(X=1)=0.8$, $\Pr(Y=0)=0.4$, and $\Pr(X=0|Y=1)=0.1$.

17–18 ▪ Consider the following joint distribution for the random variables T and N.

	$T=-1$	$T=0$	$T=1$	$T=2$
$N=0$	0.02	0.10	0.03	0.06
$N=1$	0.04	0.02	0.09	0.10
$N=2$	0.09	0.06	0.10	0.11
$N=3$	0.06	0.02	0.09	0.01

17. Suppose measurements can only distinguish two values of T, $T>0$ and $T\leq 0$, and two values of N, $N=0$ and $N>0$. Find the joint distribution for these events.

18. Suppose measurements can only distinguish two values of T, $T>0$ and $T\leq 0$, and two values of N, $N\leq 1$ and $N>1$. Find the joint distribution for these events.

19–20 ▪ When two baseball players bat in the same inning, the first gets a hit 25% of the time and the second gets a hit 35% of the time. In each of the following cases, find the joint distribution, the conditional distribution for the second player conditional on the first getting a hit, and the conditional distribution for the first player conditional on the second getting a hit.

19. The case where the players have the highest possible probability of each getting a hit.

20. The case where the players have the lowest possible probability of each getting a hit.

21–22 ▪ Write the joint distribution describing the states of the following Markov chains at times t and $t+1$. Assume that the marginal distributions at both time t and $t+1$ match the long-term probability.

21. The mutants described in Section 6.2, Exercise 27, where a gene has a 1.0% chance of mutating each time a cell divides, and a mutant gene has a 1.0% chance of reverting to the wild type.

22. The lemmings described in Section 6.2, Exercise 28 and Section 6.5, Exercise 16, where a lemming has a probability 0.2 of jumping off the cliff each hour and a probability 0.1 of crawling back up.

Applications

23–24 ▪ Find the conditional distributions for the number of lice on birds with zero, one, and two mites for the following birds. Describe how the conditional distributions differ from each other.

		Bird C Lice			Bird D Lice		
		0	1	2	0	1	2
	0	0.21	0.13	0.16	0.06	0.21	0.23
Mites	1	0.13	0.07	0.10	0.21	0.03	0.06
	2	0.06	0.10	0.04	0.13	0.06	0.01

23. Bird C

24. Bird D

25–26 ▪ Draw the conditional distribution for the number of mites on birds with zero, one, and two lice for the bird given in the earlier problem.

25. Bird C, from Exercise 23.

26. Bird D, from Exercise 24.

27–28 ▪ Recall the ecologist observing eagles and rabbits (Section 6.4, Exercises 27 and 28). In each of the following cases, find the joint distribution, the marginal distributions, and the conditional distributions. Use the random variables E and J, where $E=0$ represents seeing no eagle, $E=1$ seeing an eagle, $J=0$ seeing no jackrabbit, and $J=1$ seeing a jackrabbit. Use graphs of the conditional distributions to help interpret the results.

27. The ecologist sees an eagle with probability 0.2 during an hour of observation, a jackrabbit with probability 0.5, and both with probability 0.05 (as in Section 6.4, Exercise 27).

28. The ecologist sees an eagle with probability 0.2 during an hour of observation, a jackrabbit with probability 0.5, and both with probability 0.15 (as in Section 6.4, Exercise 28).

29–32 ▪ Find the joint distribution of the two events in the rare disease model (Section 6.4, page 523) where a person either has the disease (event D) or does not (event N) and either tests positive (event P) or does not (event P^c) in the following cases. Use the joint distribution to find $Pr(D \mid P)$.

29. $Pr(D) = 0.2$, $Pr(N) = 0.8$, $Pr(P \mid D) = 1.00$ and $Pr(P \mid N) = 0.05$ (as in Section 6.4, Exercise 37).

30. $Pr(D) = 0.8$, $Pr(P \mid D) = 1.0$, and $Pr(P \mid N) = 0.1$ (as in Section 6.4, Exercise 38).

31. $Pr(D) = 0.2$, $Pr(N) = 0.8$, $Pr(P \mid D) = 0.95$, and $Pr(P \mid N) = 0.05$ (as in Section 6.4, Exercise 39).

32. $Pr(D) = 0.8$, $Pr(P \mid D) = 0.95$, and $Pr(P \mid N) = 0.1$ (as in Section 6.4, Exercise 40).

33–34 ▪ Recall the cells in Section 6.4, Exercises 31 and 32. New cells stain properly with probability 0.95, 1-day-old cells stain properly with probability 0.9, 2-day-old cells stain properly with probability 0.8, and 3-day-old cells stain properly with probability 0.5. Suppose

$$Pr(\text{cell is 0 days old}) = 0.4$$
$$Pr(\text{cell is 1 day old}) = 0.3$$
$$Pr(\text{cell is 2 days old}) = 0.2$$
$$Pr(\text{cell is 3 days old}) = 0.1.$$

In each of the following cases, define two random variables, draw the joint distribution, find the marginal probability distributions, and compute and graph the conditional probability distributions for cell age. Compare the conditional distributions with the marginal distribution.

33. Compare the conditional distributions with the marginal distribution with the probabilities as given.

34. Compare the conditional distributions with the marginal distribution if the lab finds a way to eliminate the oldest cells (those >3 days old) from its stock.

35–36 ▪ Suppose immigration and emigration change the sizes of two populations with the following probabilities (as in Section 6.7, Exercises 31 and 32).

Population a		Population b	
Number	Probability	Number	Probability
−1	0.4	−1	0.1
0	0.2	0	0.3
1	0.3	1	0.2
2	0.1	2	0.4

Let I_a represent the change in population a and I_b the change in population b.

35. Find the joint distribution if immigrants enter the two populations independently.

36. Fill in the rest of the joint distribution.

	$I_a = -1$	$I_a = 0$	$I_a = 1$	$I_a = 2$
$I_b = -1$	0.05	0.03	0.01	?
$I_b = 0$	?	0.05	?	0.02
$I_b = 1$	0.12	?	0.05	?
$I_b = 2$	?	?	0.18	0.05

37–38 ▪ We have seen that meiotic drive and lack of independence can create unusual distributions of genotypes for offspring (Section 6.5, Exercises 29 and 30). In the following cases, find the joint distribution of genotypes in the offspring. Use your joint distribution to find the probability that an offspring is a heterozygote.

37. Compare a case of meiotic drive where 60% of both pollen and ovules carry the **A** allele independently with a case of non-independent assortment where an offspring gets an **A** allele from the pollen with probability 0.6 when the ovule provides an **A** and gets an **A** allele from the pollen with probability 0.4 when the ovule provides an **a**. The ovule provides **A** with probability 0.5 (from Section 6.5, Exercise 29).

38. Compare a case of meiotic drive where 70% of the pollen and 40% of the ovules carry the **A** allele independently with a case of non-independent assortment where an offspring gets an **A** allele from the pollen with probability 0.7 when the ovule provides an **A** and gets an **A** allele from the pollen with probability 0.3 when the ovule provides an **a**. The ovule provides **A** with probability 0.5 (from Section 6.5, Exercise 30).

39–40 ▪ Many matings are observed in a species of bird. Both female and male birds come in three colors: red, blue, and green. For each experiment, find the marginal distributions for both sexes and the conditional distributions of male color for red, blue, and green females, respectively. What might be going on with these birds?

39.

		Male		
		R	B	G
Female	R	0.125	0.195	0.180
	B	0.225	0.027	0.048
	G	0.090	0.102	0.008

40.

		Male		
		R	B	G
Female	R	0.19	0.06	0.00
	B	0.05	0.18	0.02
	G	0.13	0.08	0.29

Computer Exercise

41. Suppose two die are rolled, but the second die must be rerolled until the score is less than or equal to that on the first. For example, if the first die rolls a 3, then the second must be rolled again and again until its value is 3 or less.

a. Roll 100 computer die with these rules and record your results.

b. Try to figure out the mathematical joint distribution for this process.

7.2 Covariance and Correlation

The joint distribution or set of conditional distributions completely describes the simultaneous behavior of two random variables. In some situations, one random variable tends to be large when the other is large; in other situations, one random variable tends to be large when the other is small. These relationships can be summarized by two statistics, the **covariance** and a scaled version called the **correlation**. A **positive covariance** indicates that two random variables tend to be large or small simultaneously, whereas a **negative covariance** indicates that one tends to be large when the other is small.

7.2.1 Covariance

Intuitively, two measurements are "correlated" if they take on high values or low values simultaneously. More precisely, X and Y are positively correlated if X tends to exceed its mean $\bar{X}$ when Y exceeds its mean $\bar{Y}$. We quantify this relation first with the **covariance**.

Definition 7.5 Suppose X and Y are two random variables with expectations $\bar{X}$ and $\bar{Y}$, respectively. The **covariance**, denoted by $\mathrm{Cov}(X, Y)$, of X and Y is

$$\mathrm{Cov}(X, Y) = \mathrm{E}[(X - \bar{X})(Y - \bar{Y})].$$

More particularly, if X and Y are discrete random variables where X takes on the values $x_1, \ldots, x_n$ and Y takes on the values $y_1, \ldots, y_m$, and the variables have joint distribution p_{ij}, then

$$\mathrm{Cov}(X, Y) = \sum_{j=1}^{m} \sum_{i=1}^{n} (x_i - \bar{X})(y_j - \bar{Y}) p_{ij}.$$

Like every other expectation, we take the value (in this case, the product of the deviations from the means) times the probability (the joint probability p_{ij}) and add the products. The values involved are listed in the following table.

X	Y	Difference of X from Mean	Difference of Y from Mean	Product of Differences	Joint Probability
x_i	y_j	$x_i - \bar{X}$	$y_j - \bar{Y}$	$(x_i - \bar{X})(y_j - \bar{Y})$	p_{ij}

The product of the differences is positive in two cases: if both X and Y are larger than their means, or if both X and Y are smaller than their means. The product of the differences is negative if exactly one of X and Y is larger than its mean.

X	Y	Product of Differences
$x_i > \bar{X}$	$y_j > \bar{Y}$	Positive
$x_i < \bar{X}$	$y_j < \bar{Y}$	Positive
$x_i > \bar{X}$	$y_j < \bar{Y}$	Negative
$x_i < \bar{X}$	$y_j > \bar{Y}$	Negative

If there are many terms where the product of the differences is positive, the covariance will be positive.

Example 7.2.1 Computing the Covariance for Two Competitive Dart Players

Recall the joint distribution for the two competitive dart players (Example 7.1.2). To find the covariance, we must find the marginal distributions and expectations for each random variable.

	$S_1 = 1$	$S_1 = 0$		
$S_2 = 1$	0.24	0.06	→	0.3
$S_2 = 0$	0.06	0.64	→	0.7
	↓	↓		
	0.3	0.7		

Then $E(S_1) = 0.3 \cdot 1 + 0.7 \cdot 0 = 0.3$, and $E(S_2) = 0.3 \cdot 1 + 0.7 \cdot 0 = 0.3$.

To find the covariance from the definition, we first build the following table.

S_1	S_2	Difference of S_1 from Mean	Difference of S_2 from Mean	Product of Differences	Joint Probability
1	1	0.7	0.7	0.49	0.24
1	0	0.7	−0.3	−0.21	0.06
0	1	−0.3	0.7	−0.21	0.06
0	0	−0.3	−0.3	0.09	0.64

We then multiply the value (the product of the differences in the second-to-last column) by the probability (in the last column) and add them up. In this case,

$$\text{Cov}(S_1, S_2) = 0.49 \cdot 0.24 - 0.21 \cdot 0.06 - 0.21 \cdot 0.06 + 0.09 \cdot 0.64 = 0.15.$$

This positive covariance indicates that these two random variables tend to take on large values at the same time. In this case, this means that the two players are more likely to both score than they would if scores were independent.

As with the variance, there is a useful computational formula for the covariance.

Theorem 7.2 **Computational Formula for the Covariance**

Suppose X and Y are two random variables with expectations $\bar{X}$ and $\bar{Y}$, respectively. The covariance can be written

$$\text{Cov}(X, Y) = E(XY) - \bar{X}\bar{Y}.$$

The proof is almost identical to the proof for the computation formula for the variance (Theorem 6.4) sketched in Section 6.9, Exercises 23 and 24. In words, the covariance is the expectation of the product minus the product of the expectations.

This theorem gives a useful algorithm for computing the covariance.

Algorithm 7.1 Computing the Covariance with the Computational Formula

If X and Y are discrete random variables where X takes on the values $x_1, \ldots, x_n$ and Y takes on the values $y_1, \ldots, y_m$, and the variables have joint distribution p_{ij}, then

$$\text{Cov}(X, Y) = \sum_{j=1}^{m} \sum_{i=1}^{n} x_i y_j p_{ij} - \bar{X}\bar{Y}.$$

Example 7.2.2 Using the Computational Formula to Compute the Covariance for Competitive Dart Players

Consider again the dart players in Example 7.2.1. We found that $E(S_1) = 0.3$ and $E(S_2) = 0.3$. To find the covariance with the computational formula, we build the following table.

S_1	S_2	Product	Joint Probability
1	1	1	0.24
0	1	0	0.06
1	0	0	0.06
0	0	0	0.64

The covariance is

$$\text{Cov}(S_1, S_2) = 1 \cdot 0.24 + 0 \cdot 0.06 + 0 \cdot 0.06 + 0 \cdot 0.64 - 0.3 \cdot 0.3 = 0.15.$$

This positive value indicates that these two players tend to do well together. This matches the result found with the more complicated direct method in Example 7.2.1.

Example 7.2.3 Finding the Covariance of Independent Random Variables

Consider dart players with the following joint distribution.

	$S_1 = 1$	$S_1 = 0$
$S_2 = 1$	0.09	0.21
$S_2 = 0$	0.21	0.49

With the definition, we can tabulate the data as

S_1	S_2	Difference of S_1 from Mean	Difference of S_2 from Mean	Product of Differences	Joint Probability
1	1	$1 - 0.3 = 0.7$	$1 - 0.3 = 0.7$	0.49	0.09
0	1	$0 - 0.3 = -0.3$	$1 - 0.3 = 0.7$	−0.21	0.21
1	0	$1 - 0.3 = 0.7$	$0 - 0.3 = -0.3$	−0.21	0.21
0	0	$0 - 0.3 = -0.3$	$0 - 0.3 = -0.3$	0.09	0.49

The covariance is

$$\text{Cov}(S_1, S_2) = 0.09 \cdot 0.49 - 0.21 \cdot 0.21 - 0.21 \cdot 0.21 + 0.49 \cdot 0.09 = 0.0.$$

The covariance is 0, as we might expect with *independent* random variables. With the computational formula, we have the following table.

S_1	S_2	Product	Joint Probability
1	1	1	0.09
0	1	0	0.21
1	0	0	0.21
0	0	0	0.49

The covariance is

$$\text{Cov}(S_1, S_2) = 1 \cdot 0.09 + 0 \cdot 0.21 + 0 \cdot 0.21 + 0 \cdot 0.49 - 0.09 = 0.0.$$

Again, the results match, but this method of calculation is easier.

Example 7.2.4 An Example with Negative Covariance

When the second player throws better after the first player misses, we expect the covariance to be negative. The joint distribution is

	$S_1 = 1$	$S_1 = 0$
$S_2 = 1$	0.03	0.27
$S_2 = 0$	0.27	0.43

Using the computational formula, our table is

S_1	S_2	Product	Joint Probability
1	1	1	0.03
0	1	0	0.27
1	0	0	0.27
0	0	0	0.43

and the covariance is

$$\text{Cov}(S_1, S_2) = 1 \cdot 0.03 + 0 \cdot 0.27 + 0 \cdot 0.27 + 0 \cdot 0.43 - 0.09 = -0.06.$$

Example 7.2.5 Finding the Covariance of Mites and Lice

Covariances of joint distributions with more than two values are computed in the same way. Consider again bird B from Table 7.1.

		Bird B Lice		
		0	1	2
	0	0.33	0.08	0.09
Mites	1	0.05	0.13	0.12
	2	0.02	0.09	0.09

Using the conditional distributions, we observed that birds that have no mites also tend to have no lice. We expect, therefore, that the covariance should be positive. To get started, we compute the expected number of mites $\bar{M}$ and lice $\bar{L}$ from the marginal distribution.

$$\bar{M} = 0 \cdot 0.5 + 1 \cdot 0.3 + 2 \cdot 0.2 = 0.7$$
$$\bar{L} = 0 \cdot 0.4 + 1 \cdot 0.3 + 2 \cdot 0.3 = 0.9.$$

Using the computational formula, we can tabulate the values as follows.

M	L	Product	Joint Probability
0	0	0	0.33
0	1	0	0.08
0	2	0	0.09
1	0	0	0.05
1	1	1	0.13
1	2	2	0.12
2	0	0	0.02
2	1	2	0.09
2	2	4	0.09

The covariance is therefore

$$\begin{aligned}\text{Cov}(M, L) &= 0 \cdot 0.33 + 0 \cdot 0.08 + 0 \cdot 0.09 + 0 \cdot 0.05 + 1 \cdot 0.13 + \\ &\quad 2 \cdot 0.12 + 0 \cdot 0.02 + 2 \cdot 0.09 + 4 \cdot 0.09 - 0.7 \cdot 0.9 \\ &= 0.28.\end{aligned}$$

The covariance successfully captures the positive relation between these two measurements.

7.2.2 Correlation

Although the sign of the covariance tells us whether two random variables have a positive or negative relationship, its magnitude and units are difficult to interpret. The covariance we just computed is 0.28 mite-lice. To get rid of the units and make the values easier to use, we transform the covariance into the **correlation**, denoted by ρ, by dividing by the product of the standard deviations.

Definition 7.6 Suppose X and Y are two random variables with standard deviations σ_X and σ_Y, respectively (computed from the marginal distributions). The correlation $\rho_{X,Y}$ of X and Y is

$$\rho_{X,Y} = \frac{\text{Cov}(X, Y)}{\sigma_X \sigma_Y}.$$

The units cancel, giving a pure number mathematically guaranteed to lie between -1 and 1 (for a proof, see a more advanced text on probability theory).

A correlation of 1 means that Y is large exactly when X is. A positive correlation gives a measure of the strength of the tendency of one measurement to be large when the other is. Roughly speaking, if the correlation is 0.7, knowing that the value of X is large makes us 70% certain that Y is also large. If the correlation is -0.7, knowing that the value of X is large makes us 70% certain that Y is small. The magnitude of the correlation is a measure of information, and its sign indicates how the two random variables covary. A correlation of 0 might seem to mean that knowing the value of X gives no information about Y, but we will see in the following section that this interpretation can be misleading.

Example 7.2.6 Computing the Correlation of Two Dart Scores

In the dart score example (Example 7.2.2), S_1 and S_2 are Bernoulli random variables with $p = 0.3$. Therefore,

$$\text{Var}(S_1) = p(1 - p) = (0.3)(0.7) = 0.21$$

$$\text{Var}(S_2) = p(1 - p) = (0.3)(0.7) = 0.21$$

from the formula for the variance of a Bernoulli random variable (Equation 6.9.1). The standard deviations are the square roots of the variances, or

$$\sigma_{S_1} = \sqrt{0.21} = 0.458$$

$$\sigma_{S_2} = \sqrt{0.21} = 0.458.$$

When the two players tend to hit together, we found that the covariance has the positive value 0.15 (Example 7.2.2). This is hard to interpret, but if we normalize this into the correlation, we find

$$\rho_{S_1,S_2} = \frac{0.15}{0.458 \cdot 0.458} = 0.715.$$

The correlation is quite high, indicating a strong relationship.

Example 7.2.7 Computing the Correlation of Two Negatively Correlated Dart Scores

When the second player tends to score after the first misses (Example 7.2.4), the covariance takes on the negative value −0.06. As a correlation,

$$\rho_{S_1,S_2} = \frac{-0.06}{0.458 \cdot 0.458} = -0.286.$$

Because computing the correlation takes many steps, it helps to organize them into an algorithm.

Algorithm 7.2 Finding the Correlation of Two Discrete Random Variables

1. Find the marginal distribution of each.
2. Compute the expectation of each from the marginal distribution.
3. Compute the variance of each from the marginal distribution, and take the square root to find the standard deviation.
4. Compute the covariance with the computational formula.
5. Divide the covariance by the product of the standard deviations to find the correlation.

Example 7.2.8 Computing the Correlation of Two Random Variables

Consider again the random variables X and Y described in Examples 7.1.4 and 7.1.6. To find the correlation, we can follow the algorithm step by step.

1. Find the marginals.

	$Y=2$	$Y=4$	$Y=6$		
$X=0$	0.1	0.2	0.0	→	$\Pr(X=0)=0.3$
$X=1$	0.05	0.1	0.2	→	$\Pr(X=1)=0.35$
$X=2$	0.2	0.0	0.15	→	$\Pr(X=2)=0.35$
	↓	↓	↓		
	$\Pr(Y=2)=0.35$	$\Pr(Y=4)=0.3$	$\Pr(Y=6)=0.35$		

2. Find the expectations of X and Y.

$$\mathrm{E}(X) = 0 \cdot 0.3 + 1 \cdot 0.35 + 2 \cdot 0.35 = 1.05$$

$$\mathrm{E}(Y) = 2 \cdot 0.35 + 4 \cdot 0.3 + 6 \cdot 0.35 = 4.0.$$

3. Find the variances and standard deviations of X and Y.

$$\mathrm{Var}(X) = 0^2 \cdot 0.3 + 1^2 \cdot 0.35 + 2^2 \cdot 0.35 - 1.05^2 = 0.6475$$

$$\mathrm{Var}(Y) = 2^2 \cdot 0.35 + 4^2 \cdot 0.3 + 6^2 \cdot 0.35 - 4.0^2 = 2.8.$$

The standard deviations are then

$$\sigma_X = \sqrt{0.6475} = 0.805$$

$$\sigma_Y = \sqrt{2.8} = 1.673.$$

4. Find the covariance of X and Y with the computational formula.

X	Y	Product	Joint Probability
0	2	0	0.1
1	2	2	0.05
2	2	4	0.2
0	4	0	0.2
1	4	4	0.1
2	4	8	0.0
0	6	0	0.0
1	6	6	0.2
2	6	12	0.15

The covariance is then

$$\begin{aligned}\mathrm{Cov}(X, Y) &= 0 \cdot 0.1 + 2 \cdot 0.05 + 4 \cdot 0.2 + 0 \cdot 0.2 + 4 \cdot 0.1 + 8 \cdot 0.0 \\ &\quad + 0 \cdot 0.0 + 6 \cdot 0.2 + 12 \cdot 0.15 - 1.05 \cdot 4.0 = -0.1.\end{aligned}$$

5. Divide by the product of the standard deviations.

$$\rho_{X,Y} = \frac{-0.1}{0.805 \cdot 1.673} = -0.074.$$

Example 7.2.9 Finding the Correlation of Lice and Mites on Bird B

In Example 7.2.5, we found the expectation and covariance for the numbers of mites and lice on bird B. To find the correlation, we must compute the variance of each marginal distribution. For the mites, we use the computational formula (Theorem 6.4) to find

$$\begin{aligned}\sigma_M^2 &= 0^2 \cdot 0.5 + 1^2 \cdot 0.3 + 2^2 \cdot 0.2 - 0.7^2 = 0.61 \\ \sigma_M &= \sqrt{0.61} = 0.781.\end{aligned}$$

For the lice,

$$\begin{aligned}\sigma_L^2 &= 0^2 \cdot 0.4 + 1^2 \cdot 0.3 + 2^2 \cdot 0.3 - 0.9^2 = 0.69 \\ \sigma_L &= \sqrt{0.69} = 0.831.\end{aligned}$$

The correlation for bird B is therefore

$$\begin{aligned}\rho_{M,L} &= \frac{\mathrm{Cov}(M, L)}{\sigma_M \sigma_L} \\ &= \frac{0.28}{0.781 \cdot 0.831} = 0.43.\end{aligned}$$

Knowing the number of lice on a bird gives a significant amount of information about the number of mites.

7.2.3 Perfect Correlation

What are the covariance and correlation of a random variable with itself? From Definition 7.5,

$$\mathrm{Cov}(X, X) = \mathrm{E}[(X - \bar{X})(X - \bar{X})] = \mathrm{E}[(X - \bar{X})^2] = \mathrm{Var}(X) = \sigma_X^2,$$

because it matches the definition of the variance (Definition 6.15). The correlation is

$$\rho_{X,X} = \frac{\mathrm{Cov}(X, X)}{\sigma_X \cdot \sigma_X} = \frac{\sigma_X^2}{\sigma_X^2} = 1.$$

A random variable is perfectly correlated with itself. Thinking of correlation as a measure of how well the value of one random variable predicts another, it is not surprising that X is a perfect predictor of itself.

Are there situations where two **different** random variables X and Y are perfectly correlated ($\rho_{X,Y} = 1$)? Knowledge of X provides perfect knowledge of Y when Y is a **function** of X. For example, suppose V measures the volume of some rocks and M their mass. If the density of rocks is exactly 5.5 gm/cm^3, knowing that $V = 1.0$ cm^3 ensures that $M = 5.5$ gm. The distribution of rock sizes could be described equally well by either measurement. We expect that this situation guarantees a perfect correlation.

Example 7.2.10 Two Perfectly Correlated Random Variables

Suppose that rocks have volume 1.0 cm^3 with probability 0.5, volume 8.0 cm^3 with probability 0.3, and volume 27.0 cm^3 with probability 0.2. The joint distribution is

	$V = 1.0$	$V = 8.0$	$V = 27.0$
$M = 5.5$	0.5	0.0	0.0
$M = 44.0$	0.0	0.3	0.0
$M = 148.5$	0.0	0.0	0.2

We follow the steps in Algorithm 7.2 to compute the correlation. The marginal distribution for V is

V	Probability
1.0	0.5
8.0	0.3
27.0	0.2

Therefore,

$$\begin{aligned} \mathrm{E}(V) &= 1.0 \cdot 0.5 + 8.0 \cdot 0.3 + 27.0 \cdot 0.2 = 8.3 \\ \mathrm{Var}(V) &= 1.0^2 \cdot 0.5 + 8.0^2 \cdot 0.3 + 27.0^2 \cdot 0.2 - 8.3^2 = 96.61 \\ \sigma_M &= 9.83. \end{aligned}$$

The marginal distribution for M is

M	Probability
5.5	0.5
44.0	0.3
148.5	0.2

Therefore,

$$\begin{aligned} \mathrm{E}(M) &= 5.5 \cdot 0.5 + 44.0 \cdot 0.3 + 148.5 \cdot 0.2 = 45.65 \\ \mathrm{Var}(M) &= 5.5^2 \cdot 0.5 + 44.0^2 \cdot 0.3 + 148.5^2 \cdot 0.2 - 45.65^2 = 2922.45 \\ \sigma_M &= 54.06. \end{aligned}$$

Next, we find the covariance (simplifying our work by ignoring lines with probability 0) as

$$\begin{aligned} \mathrm{Cov}(V, M) &= \mathrm{E}(VM) - \mathrm{E}(V)\mathrm{E}(M) \\ &= 1.0 \cdot 5.5 \cdot 0.5 + 8.0 \cdot 44.0 \cdot 0.2 + 27.0 \cdot 148.5 \cdot 0.2 \\ &\quad -8.3 \cdot 45.65 = 531.4. \end{aligned}$$

The correlation is

$$\rho_{V,M} = \frac{\text{Cov}(V, M)}{\sigma_V \sigma_M} = \frac{531.4}{9.83 \cdot 54.06} = 1.0$$

The correlation between these two measurements is indeed perfect.

Interestingly, however, even when measurement Y is a function of measurement X, we find perfect correlation only when Y is a **linear** function of X. The results are summarized in the following theorem.

Theorem 7.3 Let X and Y be two random variables.

- If $Y = aX + b$ for $a > 0$, $\rho_{X,Y} = 1$.
- If $Y = aX + b$ for $a < 0$, $\rho_{X,Y} = -1$.
- If $Y = aX + b$ for $a = 0$, $\rho_{X,Y}$ is undefined.

In Example 7.2.10, $M = 5.5V$, which corresponds to $a = 5.5$ and $b = 0$ in Theorem 7.3. The case with $a = 0$ occurs when the random variable Y takes on only the single value b. Both Cov(X, Y) and σ_Y are 0 in this case, leading to legal trouble in computing the correlation.

Perfect negative correlation of -1 occurs when the random variables behave in exactly opposite ways.

Example 7.2.11 An Example of Perfect Negative Correlation

Suppose a vain bird spends all its time either sleeping (a time S) or preening (a time P). If the bird sleeps for 10 h, it must preen for 14 h. One set of probabilities is

		Sleeping		
		14	15	16
	10	0.0	0.0	0.4
	9	0.0	0.3	0.0
Preening	8	0.3	0.0	0.0

In general, $P = 24 - S$, fitting Theorem 7.3 with $a = -1$ and $b = 24$. These random variables are perfectly negatively correlated because P is small whenever S is large and vice versa.

However, the correlation of two random variables will not be perfect if the second is a **nonlinear** function of the first.

Example 7.2.12 Imperfect Correlation of Random Variables Related by a Nonlinear Function

Suppose that rocks are perfect cubes with sides of length L. They come in three sizes, $L = 1.0$, $L = 2.0$, and $L = 3.0$. The volume V is the function $V = L^3$ of the side length, and can take on values $V = 1.0$, $V = 8.0$, and $V = 27.0$. The joint probability distribution is

	$L = 1.0$	$L = 2.0$	$L = 3.0$
$V = 1.0$	0.5	0.0	0.0
$V = 8.0$	0.0	0.3	0.0
$V = 27.0$	0.0	0.0	0.2

It seems reasonable to expect that side length and volume will be perfectly correlated because we can compute the exact value of V when we know the exact value of L.

To check this intuition, we compute the correlation. The mean and standard deviation of the marginal distributions of L and V are $\bar{L}=1.7$, $\bar{V}=8.3$, $\sigma_L=0.781$, and $\sigma_V=9.83$. Furthermore, $\mathrm{Cov}(L, V)=7.39$. The correlation is

$$\rho_{L,V}=\frac{\mathrm{Cov}(L, V)}{\sigma_L\sigma_V}=\frac{7.39}{(0.781\cdot 9.83)}=0.963.$$

Even though the volume is a **function** of the length, the correlation is not perfect because the function $V=L^3$ is nonlinear.

The correlation thus measures the strength of the **linear** relation between two random variables. In Section 8.9, we will apply these concepts to data with the statistical technique of linear regression.

Summary We have seen how to compute the **covariance** and **correlation**, two statistics describing the relationship between two random variables. A positive covariance indicates that the two random variables tend to be large at the same time or small at the same time. The correlation is a normalized version of the covariance, guaranteed to take on values between -1 and 1. Values near 1 indicate a strong positive relationship, values near 0 indicate no relationship, and values near -1 indicate a strong negative relationship. The correlation measures the strength of the **linear** relationship between two random variables.

7.2 Exercises

Mathematical Techniques

1–4 ▪ For the following joint distributions, find the covariance of X and Y using the direct method, $\mathrm{Cov}(X, Y)=\mathrm{E}[(X-\bar{X})(Y-\bar{Y})]$. If the covariance is zero, are the random variables independent?

1. (from Section 7.1, Exercises 1 and 5)

	$X=0$	$X=1$
$Y=0$	0.14	0.26
$Y=1$	0.21	0.39

2. (from Section 7.1, Exercises 2 and 6)

	$X=0$	$X=1$
$Y=1$	0.45	0.25
$Y=3$	0.05	0.25

3. (from Section 7.1, Exercises 3 and 7)

	$X=0$	$X=1$	$X=2$
$Y=1$	0.1	0.2	0.3
$Y=2$	0.05	0.15	0.2

4. (from Section 7.1, Exercises 4 and 8)

	$X=0$	$X=1$	$X=2$
$Y=1$	0.05	0.04	0.01
$Y=2$	0.1	0.08	0.02
$Y=3$	0.35	0.28	0.07

5–8 ▪ For the following joint distributions, find the covariance of X and Y using the computational method, $\mathrm{Cov}(X, Y)=\mathrm{E}(XY)-\bar{X}\bar{Y}$.

5. The case in Exercise 1.

6. The case in Exercise 2.

7. The case in Exercise 3.

8. The case in Exercise 4.

9–12 ▪ For the following joint distributions, find the correlation of X and Y.

9. The case in Exercise 1.

10. The case in Exercise 2.

11. The case in Exercise 3.

12. The case in Exercise 4.

13–14 ▪ When two baseball players bat in the same inning, the first gets a hit 25% of the time and the second gets a hit 35% of the time. In each of the following cases, find the covariance, assuming that a hit is worth 1 point and that a miss is worth 0. (Let P_1 be a random variable for the first player and P_2 be a random variable for the second player; let $P_1=1$ mean that player 1 got a hit, and so forth.) Say why the covariance is positive or negative.

13. The case where the players have the highest possible probability of both getting a hit (as in Section 7.1, Exercise 19).

14. The case where the players have the lowest possible probability of both getting a hit (as in Section 7.1, Exercise 20).

15–16 ▪ In most Markov chains, the state of the system is correlated from step to step. Using the joint distribution of the states of the following Markov chains at times t and $t+1$, find the correlation (set the value of one state to be 1 and of the other state to be 0). Assume that the marginal distributions at both time t and $t+1$ match the long-term probability.

15. The mutants described in Section 7.1, Exercise 21, where a gene has a 1.0% chance of mutating each time a cell divides and a 1.0% chance of correcting the mutation.

16. The lemmings described in Section 7.1, Exercise 22, where a lemming has a probability 0.2 of jumping off the cliff each hour and a probability 0.1 of crawling back up.

17–20 ▪ Suppose the random variable X takes on the values 0, 1, and 2 with probabilities 0.2, 0.3, and 0.5, respectively. For each of the following random variables Y, find the joint distribution of X and Y, compute the covariance and the correlation, and sketch a graph showing the relationship, between X and Y that includes the range of values $X=0$, $X=1$, and $X=2$.

17. $Y=3X+1$

18. $Y=X^2$

19. $Y=(X-1)^2$

20. Y takes on the value 2 with probability 1. Show that this is the case with $a=0$ as in Theorem 7.3. What happens when you try to compute the correlation? Explain why (think of correlation as information).

Applications

21–22 ▪ Find the covariance of the number of lice and mites for the following birds from the earlier problem.

21. Bird C from Section 7.1, Exercise 23.

22. Bird D from Section 7.1, Exercise 23.

23–24 ▪ Find the correlation of the number of lice and mites for the following birds.

23. Bird C from Exercise 21.

24. Bird D from Exercise 22.

25–26 ▪ Recall the ecologist observing eagles and rabbits in Exercises 27 and 28. Find the correlation between the random variables E and J where $E=0$ represents seeing no eagle, $E=1$ seeing an eagle, $J=0$ seeing no jackrabbit, and $J=1$ seeing a jackrabbit. What does the result mean?

25. She sees an eagle with probability 0.2 during an hour of observation, a jack-rabbit with probability 0.5, and both with probability 0.05 (as in Section 7.1, Exercise 27).

26. She sees an eagle with probability 0.2 during an hour of observation, a jack-rabbit with probability 0.5, and both with probability 0.15 (as in Section 7.1, Exercise 28).

27–28 ▪ Recall the cells in Section 6.4, Exercises 31 and 32. New cells stain properly with probability 0.95, 1-day-old cells stain properly with probability 0.9, 2-day-old cells stain properly with probability 0.8, and 3-day-old cells stain properly with probability 0.5. Suppose

$$\Pr(\text{cell is 0 days old}) = 0.4$$
$$\Pr(\text{cell is 1 day old}) = 0.3$$
$$\Pr(\text{cell is 2 days old}) = 0.2$$
$$\Pr(\text{cell is 3 days old}) = 0.1.$$

Using the random variables A for age and S for staining, find the covariance in the following cases. Explain why the covariance is positive or negative.

27. With the probabilities as given (as in Section 7.1, Exercise 33).

28. If the lab finds a way to eliminate the oldest cells (>3 days old) from its stock (as in Exercise 34).

29–32 ▪ Consider birds that spend all of their time sleeping, eating, or preening. Let S be the time spent sleeping, E the time spent eating, and P be the time spent preening. How much time is spent on each activity depends on the weather, which is nice with probability 0.7, OK with probability 0.2, and terrible with probability 0.1.

29. Suppose that the bird gives up eating (as in the text) and spends 10 h sleeping when the weather is nice, 14 h sleeping when the weather is OK, and 18 h sleeping when the weather is bad. Find $\rho_{S,P}$.

30. The bird always eats for 4 h per day and spends 10 h sleeping when the weather is nice, 14 h sleeping when the weather is OK, and 18 h sleeping when the weather is bad. Find $\rho_{S,P}$. Why are $\rho_{E,P}$ and $\rho_{E,S}$ not worth finding?

31. When the weather is nice, the bird eats for 1 h and sleeps for 10 h. When the weather is OK, the bird eats for 1 h and sleeps for 14 h. When the weather is bad, the bird eats for 6 h and sleeps for 18 h. Find $\rho_{S,P}$. Why is the correlation not perfect?

32. When the weather is nice, the bird eats for 6 h and sleeps for 10 h. When the weather is OK, the bird eats for 1 h and sleeps for 14 h. When the weather is bad, the bird eats for 1 h and sleeps for 18 h. Find $\rho_{S,P}$. Why is the correlation not perfect?

33–34 ▪ Suppose immigration and emigration change the sizes of two populations with the following probabilities (as in Section 6.7, Exercises 31 and 32).

Population a		Population b	
Number	Probability	Number	Probability
−1	0.4	−1	0.1
0	0.2	0	0.3
1	0.3	1	0.2
2	0.1	2	0.4

Let I_a represent the change in population a and I_b the change in population b.

33. Explicitly compute the covariance if immigrants enter the two populations independently (as in Section 7.1, Exercise 35).

34. Compute the covariance in the case from Section 7.1, Exercise 36.

35–36 ▪ Suppose the following are measurements of the temperature T and insect size S.

	$T=10$	$T=20$	$T=30$
$S=5$	0.2	0.1	0.0
$S=20$	0.1	0.2	0.1
$S=80$	0.0	0.1	0.2

35. Find the correlation of S with T.

36. Find the correlation of $\ln(S)$ with T.

37–38 ▪ Consider the following data for cell age A and the number of toxic molecules N inside.

	$A=0$	$A=1$	$A=2$	$A=3$	$A=4$
$N=0$	0.03	0.00	0.09	0.06	0.09
$N=1$	0.01	0.05	0.06	0.12	0.15
$N=2$	0.02	0.03	0.06	0.08	0.15

37. Find the correlation of A with N.

38. Suppose that the damage done by the toxic molecules is $D=\ln(1+N)$. Find the correlation of A with D.

7.3 Sums and Products of Random Variables

Often we wish to combine measurements by adding or multiplying random variables that may or may not be independent. Many scientific measurements can best be thought of as combinations of simpler measurements in this way. Mathematically, understanding these combined measurements is greatly simplified by three powerful theorems. The first states that the expectation of the sum is the sum of the expectations. The second states that the expectation of the product is the product of the expectations when the two random variables are **independent**. The third states that the variance of a sum of independent random variables is the sum of the variances. These mathematical theorems form the basis of a great deal of statistical reasoning.

7.3.1 Expectation of a Sum

You throw a dart two times. Each time, the score is drawn from the same distribution (such as Figure 7.1.1) with expectation 12.55. What is the expectation of the sum of the two scores? The answer, regardless of whether the throws are independent, is twice the expectation of each score, or 25.1. This is formalized in the following theorem.

Theorem 7.4 If X and Y are any two random variables with finite expectation,

$$E(X+Y)=E(X)+E(Y).$$

The proof is presented at the end of this section.

This theorem extends to any sum of random variables. For example,

$$E(X+Y+Z)=E(X+Y)+E(Z)=E(X)+E(Y)+E(Z)$$

by breaking up the sum one term at a time.

Example 7.3.1 Probability Distribution and Expectation for Total Number of Parasites on Bird A

Consider the lice and mites on bird A, with the joint distribution given in Table 7.1.

		Bird A Lice		
		0	1	2
	0	0.20	0.15	0.15
Mites	1	0.12	0.09	0.09
	2	0.08	0.06	0.06

Suppose we are interested in the total number of parasites found by adding the number of mites and the number of lice. How can we find the expectation of this combined random variable? The straightforward method is to use the joint distribution and compute the probability of zero, one, two, three, or four total parasites. There is only one way for a bird to have no parasites, so

$$\Pr(0 \text{ parasites}) = \Pr(0 \text{ lice and } 0 \text{ mites}) = 0.20.$$

There are two ways to have one parasite (one louse and no mites or one mite and no lice. Therefore,

$$\begin{aligned}\Pr(1 \text{ parasite}) &= \Pr(0 \text{ lice and } 1 \text{ mite}) + \Pr(1 \text{ louse and } 0 \text{ mites})\\ &= 0.12 + 0.15 = 0.27.\end{aligned}$$

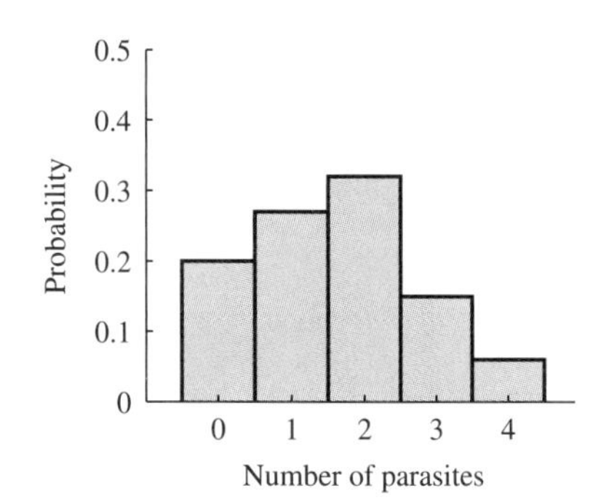

FIGURE 7.3.1
The total number of parasites on bird A

Continuing in this way,

$$\begin{aligned}\Pr(2 \text{ parasites}) &= \Pr(0 \text{ lice and } 2 \text{ mites}) + \Pr(1 \text{ louse and } 1 \text{ mite})\\ &\quad + \Pr(2 \text{ lice and } 0 \text{ mites})\\ &= 0.08 + 0.09 + 0.15 = 0.32\\ \Pr(3 \text{ parasites}) &= \Pr(1 \text{ louse and } 2 \text{ mites}) + \Pr(2 \text{ lice and } 1 \text{ mite})\\ &= 0.06 + 0.09 = 0.15\\ \Pr(4 \text{ parasites}) &= \Pr(2 \text{ lice and } 2 \text{ mites}) = 0.06.\end{aligned}$$

We have found the probability distribution for the total number of parasites in bird A, a random variable we denote by Probability, P_A (Figure 7.3.1). This new measurement is the sum of L and M, or $P_A = L + M$. Remember to think of this as adding *measurements*, not adding *probability distributions*.

Armed with its probability distribution, the expectation is

$$E(P_A) = 0 \cdot 0.20 + 1 \cdot 0.27 + 2 \cdot 0.32 + 3 \cdot 0.15 + 4 \cdot 0.06 = 1.6.$$

Example 7.3.2 The Expected Total Number of Parasites on Bird B

The joint distribution of parasite numbers on bird B is

		Bird B Lice		
		0	1	2
	0	0.33	0.08	0.09
Mites	1	0.05	0.13	0.12
	2	0.02	0.09	0.09

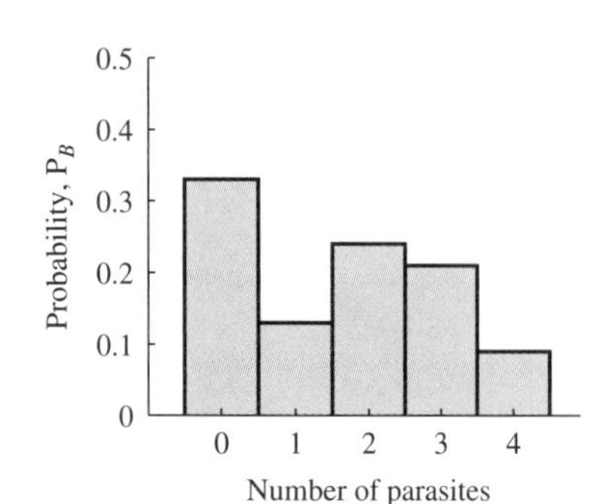

FIGURE 7.3.2
The total number of parasites on bird B

The probability distribution for the total number of parasites P_B on bird B is

$$\begin{aligned}\Pr(0 \text{ parasites}) &= 0.33\\ \Pr(1 \text{ parasite}) &= 0.05 + 0.08 = 0.13\\ \Pr(2 \text{ parasites}) &= 0.02 + 0.13 + 0.09 = 0.24\\ \Pr(3 \text{ parasites}) &= 0.09 + 0.12 = 0.21\\ \Pr(4 \text{ parasites}) &= 0.09.\end{aligned}$$

(Figure 7.3.2). This distribution looks quite different from that of P_A (Figure 7.3.1). However, the expected total number,

$$E(P_B) = 0 \cdot 0.33 + 1 \cdot 0.13 + 2 \cdot 0.24 + 3 \cdot 0.21 + 4 \cdot 0.09 = 1.6,$$

is exactly the same.

Theorem 7.4 shows that these expected numbers **must** be the same for each type of bird and gives a simpler way to compute them: the expectation of the sum is the sum of the expectations. Using the fact that $\mathrm{E}(L)=0.9$ and $\mathrm{E}(M)=0.7$, (found in Example 7.2.5) then

$$\mathrm{E}(P_B)=\mathrm{E}(L+M)=\mathrm{E}(L)+\mathrm{E}(M)=0.9+0.7=1.6,$$

no matter what the joint distribution is.

This calculation would work even if we had never found the probability distribution P_B for the total number of parasites, and even if we did not know the joint distribution of the two random variables L and M. This is a remarkable, and convenient, achievement.

Another useful property of the expectation is that constants come outside.

Theorem 7.5 If X is a random variable with finite expectation and a is a constant,

$$\mathrm{E}(aX)=a\mathrm{E}(X).$$

Example 7.3.3 Multiplying a Random Variable by a Constant I

Suppose a player throws a dart and scores one point with probability $p=0.3$ and zero points with probability $1-p=0.7$. The expectation of the random variable S for the score can be found with the formula for the expectation for a Bernoulli random variable,

$$\mathrm{E}(S)=p=0.3.$$

What happens to the expectation if a new inflated scoring system is used where a player scores ten points with probability $p=0.3$ and zero points with probability $1-p=0.7$? This new random variable, S_I, is related to S by $S_I=10S$. According to Theorem 7.5,

$$\mathrm{E}(S_I)=10\cdot\mathrm{E}(S)=10\cdot 0.3=3.0.$$

We can check this directly by computing

$$\mathrm{E}(S_I)=10\cdot 0.3+0\cdot 0.7=3.0.$$

Example 7.3.4 Multiplying a Random Variable by a Constant II

Suppose each louse weighs exactly 0.05 g. What is the mean weight of lice on a bird? Let the new random variable W_L represent the total weight of lice. Then $W_L=0.05L$. Theorem 7.5 says

$$\mathrm{E}(W_L)=\mathrm{E}(0.05L)=0.05\mathrm{E}(L)=0.05\cdot 0.9=0.045g.$$

We can combine Theorem 7.4 and Theorem 7.5 into a single general theorem that says that the expectation of a new random variable constructed by multiplying simple random variables by constants and adding can be found by multiplying the expectations of the simple random variables by constants and adding.

Theorem 7.6 If $X_1, X_2, \ldots, X_n$ are random variables with finite expectation and $a_1, a_2, \ldots, a_n$ are constants, then

$$\mathrm{E}\left(\sum_{i=1}^{n} a_i X_i\right)=\sum_{i=1}^{n} a_i \mathrm{E}(X_i).$$

Example 7.3.5 Expectation of Sums of Random Variables Multiplied by Constants

Suppose that each louse weighs 0.05 g and that each mite weighs 0.02 g. Consider again Bird A from Table 7.1. The total weight of parasites on the bird is

$$W=0.05L+0.02M.$$

The mean weight of parasites on a bird is

$$\begin{aligned}\mathrm{E}(W)&=\mathrm{E}(0.05L+0.02M)\\&=0.05\mathrm{E}(L)+0.02\mathrm{E}(M)\\&=0.05\cdot 0.9+0.02\cdot 0.7=0.059.\end{aligned}$$

This expectation could be computed by finding the probability distribution of the random variable W (Exercises 29 and 30), but this apparently "straightforward" method requires much more work.

7.3.2 Expectation of a Product

If the expectation of a sum is the sum of the expectations, is the expectation of a product the product of the expectations?

Example 7.3.6 The Expectation of the Product Is Not Equal to the Product of the Expectations

Consider a species of bird that can raise either one or two chicks. In some nests, the chicks are smaller (mass of 40 g), while in others the chicks are larger (mass of 50 g). Let the random variable S denote the mass of each chick, and let C be the number of chicks. The total weight of chicks in the nest is a new random variable $T = S \cdot C$. The following table gives the four possible types of nests and the total weight T of chicks.

		Weight of Each Chick	
		$S = 40$	$S = 50$
Number of chicks	$C = 1$	40	50
	$C = 2$	80	100

Suppose that the joint probability distribution is

	$S = 40$	$S = 50$
$C = 1$	0.6	0.1
$C = 2$	0.0	0.3

This means that 60% of nests have one small chick, 10% have one large chick, 30% have two large chicks, and none have two small chicks.

The marginal distributions are

$$\Pr(S = 40) = 0.6$$
$$\Pr(S = 50) = 0.4$$

and

$$\Pr(C = 1) = 0.7$$
$$\Pr(C = 2) = 0.3.$$

Therefore,

$$E(S) = 40 \cdot 0.6 + 50 \cdot 0.4 = 44$$
$$E(C) = 1 \cdot 0.7 + 2 \cdot 0.3 = 1.3.$$

The total weight T of chicks in the nest is the **product** of the weight of each chick S and the number of chicks C. Can we compute the expectation of the product as the product of the expectations? Is the following true?

$$E(T) = E(S)E(C) = 44 \cdot 1.3 = 57.2$$

From our tables, the probability distribution for T is

$$\Pr(T = 40) = 0.6$$
$$\Pr(T = 50) = 0.1$$
$$\Pr(T = 80) = 0.0$$
$$\Pr(T = 100) = 0.3$$

(Figure 7.3.3).

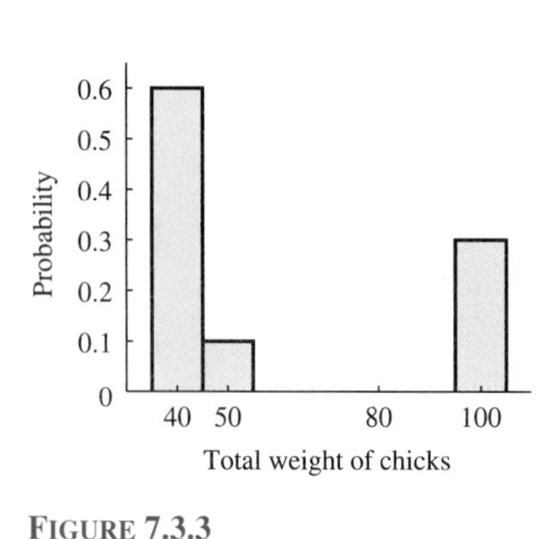

FIGURE 7.3.3

The total weight of chicks in the nest

The expectation is

$$E(T) = 40 \cdot 0.6 + 50 \cdot 0.1 + 80 \cdot 0.0 + 100 \cdot 0.3 = 59.0,$$

which does **not** match the product of the expectations.

Example 7.3.7 The Expectation of the Product Can Equal the Product of the Expectations

Suppose we try the alternative probability distribution

	$S=40$	$S=50$
$C=1$	0.42	0.28
$C=2$	0.18	0.12

The marginal distributions for S and C match those in Example 7.3.6, but the probability distribution for the product T is

$$\Pr(T = 40) = 0.42$$
$$\Pr(T = 50) = 0.28$$
$$\Pr(T = 80) = 0.18$$
$$\Pr(T = 100) = 0.12.$$

Therefore,

$$E(T) = 40 \cdot 0.42 + 50 \cdot 0.28 + 80 \cdot 0.18 + 100 \cdot 0.12 = 57.2.$$

In this case, the expectation of the product T is equal to the product of the expectations of S and C.

What is the difference between these two distributions? A little examination reveals that the first is **independent** because the joint probability distribution is equal to the product of the marginal distributions. The expectation of the product is equal to the product of the expectations only in this case.

Theorem 7.7 If X and Y are independent random variables with finite expectation,

$$E(XY) = E(X)E(Y).$$

Although useful in itself, Theorem 7.7 is more often used to prove the following theorem that verifies our intuition about the covariance.

Theorem 7.8 If X and Y are independent random variables, then

$$\text{Cov}(X, Y) = 0.$$

This theorem follows from the computational formula for the covariance.

$$\begin{aligned} \text{Cov}(X, Y) &= E(XY) - E(X)E(Y) && \text{computation formula for covariance (Theorem 7.2)} \\ &= E(X)E(Y) - E(X)E(Y) && \text{Theorem 7.7} \\ &= 0. \end{aligned}$$

7.3.3 Variance of a Sum

Expectations add. Is this true of any other statistic? It is quite difficult to concoct random variables for which the median, mode, or mean absolute deviation add. The variance is a widely used statistic because the variance of the sum of **independent** random variables is the sum of the variances.

Theorem 7.9 If X and Y are two independent random variables with finite variance,

$$\text{Var}(X + Y) = \text{Var}(X) + \text{Var}(Y).$$

Example 7.3.8 Variance of the Sum Is Equal to the Sum of the Variances If Independent

To check this theorem, we find the variance of the total number of parasites, P_A on bird A, using the probability distributions found earlier. When finding the correlation (Example 7.2.9), we computed that

$$\begin{aligned}\text{Var}(M) &= 0.61\\ \text{Var}(L) &= 0.69.\end{aligned}$$

Because M and L are independent for bird A, we expect that

$$\begin{aligned}\text{Var}(P_A) &= \text{Var}(M) + \text{Var}(L)\\ &= 0.61 + 0.69 = 1.3.\end{aligned}$$

Using the probability distribution found in Example 7.3.1,

$$\text{Var}(P_A) = 0^2 \cdot 0.20 + 1^2 \cdot 0.27 + 2^2 \cdot 0.32 + 3^2 \cdot 0.15 + 4^2 \cdot 0.06 - 1.6^2 = 1.3.$$

Example 7.3.9 Variance of the Sum Is Not Equal to the Sum of the Variances in General

For bird B, where the distributions are not independent, we found the distribution for P_B in Example 7.3.2. The variance is

$$\text{Var}(P_B) = 0^2 \cdot 0.33 + 1^2 \cdot 0.13 + 2^2 \cdot 0.24 + 3^2 \cdot 0.21 + 4^2 \cdot 0.09 - 1.6^2 = 1.86.$$

Because the marginal distributions for these two birds are identical (Examples 7.3.1 and 7.3.2), this variance is not equal to the sum of the variances.

Why is the variance of parasite numbers on bird B larger? There are two ways of thinking about this. The probability distribution for P_B is more spread out, with shorter bars in the middle (compare Figure 7.3.2 with Figure 7.3.1).

Alternatively, we can look to the proof of Theorem 7.9. Before the last step, we found the **general addition formula for variances**.

Theorem 7.10 **General Addition Formula for Variances**

For any two random variables X and Y,

$$\text{Var}(X + Y) = \text{Var}(X) + \text{Var}(Y) + 2\text{Cov}(X, Y).$$

Example 7.3.10 Applying the General Addition Formula for Variances

In Example 7.2.5, we computed that $\text{Cov}(M, L) = 0.28$ for bird B. Therefore,

$$\begin{aligned}\text{Var}(P_B) &= \text{Var}(M) + \text{Var}(L) + 2\text{Cov}(M, L)\\ &= 0.61 + 0.69 + 2 \cdot 0.28 = 1.86.\end{aligned}$$

The positive covariance spreads out the distribution and increases the variance because there are more birds with zero or four parasites.

Multiplying a random variable by a constant multiplies the variance by the **square** of that constant.

Theorem 7.11 If X is a random variable with finite variance and a is a constant,

$$\text{Var}(aX) = a^2\text{Var}(X).$$

Example 7.3.11 Variance of a Random Variable When Multiplied by a Constant

Recall the ordinary and inflated scoring systems in Example 7.3.3, where the inflated score S_I is related to the ordinary score S by $S_I = 10S$. If S is a Bernoulli random variable with $p = 0.3$, then

$$\text{Var}(S) = p(1 - p) = 0.3 \cdot 0.7 = 0.21$$

from the formula for the variance of a Bernoulli random variable (Equation 6.9.1). According to Theorem 7.11,

$$\text{Var}(S_I) = 10^2 \cdot \text{Var}(S) = 100 \cdot 0.21 = 21.0.$$

We can check this variance directly using the computational formula. We found $E(S_I) = 3.0$ in Example 7.3.3, so

$$\begin{aligned}\text{Var}(S_I) &= E(S_I^2) - E(S_I)^2 \\ &= 10^2 \cdot 0.3 + 0^2 \cdot 0.7 - 3.0^2 = 21.0.\end{aligned}$$

Combining Theorem 7.9 with Theorem 7.11 results in the following theorem.

Theorem 7.12 If $X_1, X_2, \ldots, X_n$ are independent random variables with finite variance and $a_1, a_2, \ldots, a_n$ are any constants, then

$$\text{Var}\left(\sum_{i=1}^{n} a_i X_i\right) = \sum_{i=1}^{n} a_i^2 \text{Var}(X_i).$$

Example 7.3.12 Variance of Total Parasite Weight

We defined the random variable W in Example 7.3.5 as

$$W = 0.05L + 0.02M$$

to represent the total weight of parasites. For bird A, L and M are independent. Therefore,

$$\begin{aligned}\text{Var}(W) &= 0.05^2 \text{Var}(L) + 0.02^2 \text{Var}(M) \\ &= 0.05^2 \cdot 0.69 + 0.02^2 \cdot 0.61 = 0.00197.\end{aligned}$$

Computing the variance of this random variable directly, by finding all of the probabilities, would be much more difficult.

The only tricky thing about this theorem is that negative constants are added, not subtracted.

Example 7.3.13 Variance of a Difference

Suppose we are interested in the excess of L over M, or the random variable

$$D = L - M.$$

From Theorem 7.6,

$$E(D) = E(L) - E(M) = 0.9 - 0.7 = 0.2,$$

meaning that these birds have an average of 0.2 more lice than mites. To use Theorem 7.12, the random variables must be independent. For bird A,

$$\text{Var}(D) = \text{Var}(L - M) = \text{Var}(L) + (-1)^2 \text{Var}(M) = 0.69 + 0.61 = 1.3.$$

Squaring the coefficient of M changes the difference into a sum.

Summary We have seen that the expectation of a sum of random variables is equal to the sum of the expectations, regardless of whether the random variables are independent. Only if the random variables are independent, however, is the expectation of the product is equal to the product of the expectations and the variance of the sum equal to the sum of the variances. Multiplying a random variable by a constant multiplies the expectation by the constant and multiplies the variance by the square of the constant. These theorems provide the foundation for building the important probability distributions we study next.

Proofs of theorems

Theorem 7.4: If X and Y are any two random variables with finite expectation,

$$\mathrm{E}(X+Y)=\mathrm{E}(X)+\mathrm{E}(Y).$$

Proof: We present this proof for discrete random variables only. Suppose X takes on the values $x_1, \ldots, x_n$ and Y takes on values $y_1, \ldots, y_m$. Suppose they have marginal distributions $p_{X,i}$ and $p_{Y,j}$ and joint distribution p_{ij}. Then

$\mathrm{E}(X+Y)=\sum_{i=0}^{n}\sum_{j=0}^{m}(x_i+y_j)p_{ij}$	definition of expectation
$=\sum_{i=0}^{n}\sum_{j=0}^{m}x_i p_{ij}+\sum_{i=0}^{n}\sum_{j=0}^{m}y_j p_{ij}$	breaking up the sum
$=\sum_{i=0}^{n}x_i\sum_{j=0}^{m}p_{ij}+\sum_{j=0}^{m}y_j\sum_{i=0}^{n}p_{ij}$	switching sums and removing constants
$=\sum_{i=0}^{n}x_i p_{X,i}+\sum_{j=0}^{m}y_j p_{Y,j}$	definition of marginal distributions
$=\mathrm{E}(X)+\mathrm{E}(Y).$	definition of expectation

Theorem 7.5: If X is a random variable with finite expectation and a is a constant,

$$\mathrm{E}(aX)=a\mathrm{E}(X).$$

Proof: Suppose X takes on values $x_1, \ldots, x_n$ with probability distribution p_i. Then

$\mathrm{E}(aX)=\sum_{i=0}^{n}ax_i p_i$	definition of expectation
$=a\sum_{i=0}^{n}x_i p_i$	pull constant outside sum
$=a\mathrm{E}(X).$	definition of expectation

The proof for continuous random variables is similar, but removes the constant a from an integral at the second step.

Theorem 7.6: If $X_1, X_2, \ldots, X_n$ are random variables with finite expectation and $a_1, a_2, \ldots, a_n$ are constants, then

$$\mathrm{E}\left(\sum_{i=1}^{n}a_i X_i\right)=\sum_{i=1}^{n}a_i\mathrm{E}(X_i).$$

Proof:

$\mathrm{E}\left(\sum_{i=1}^{n}a_i X_i\right)=\sum_{i=1}^{n}\mathrm{E}(a_i X_i)$	Theorem 7.4
$=\sum_{i=1}^{n}a_i\mathrm{E}(X_i).$	Theorem 7.5

Theorem 7.7: If X and Y are independent random variables with finite expectation,

$$\mathrm{E}(XY)=\mathrm{E}(X)\mathrm{E}(Y).$$

Proof: Suppose X takes on values $x_1, \ldots, x_n$ and Y takes on values $y_1, \ldots, y_m$, with marginal distributions $p_{X,i}$ and $p_{Y,j}$ and joint distribution p_{ij}. Then

$\mathrm{E}(XY)=\sum_{i=0}^{n}\sum_{j=0}^{m}x_i y_j p_{ij}$	definition of expectation
$=\sum_{i=0}^{n}\sum_{j=0}^{m}x_i y_j p_{X,i}p_{Y,j}$	independence
$=\sum_{i=0}^{n}x_i p_{X,i}\sum_{j=0}^{m}y_j p_{Y,j}$	breaking up sums
$=\mathrm{E}(X)\mathrm{E}(Y).$	definition of expectation

The proof for continuous random variables is similar but requires manipulating double integrals.

Theorem 7.9: If X and Y are two independent random variables with finite variance,

$$\mathrm{Var}(X+Y)=\mathrm{Var}(X)+\mathrm{Var}(Y).$$

Proof: Suppose X and Y are two independent random variables. Then

$\mathrm{Var}(X+Y)$

$= \mathrm{E}[(X+Y)^2] - [\mathrm{E}(X+Y)]^2$	computational formula for variance
$= \mathrm{E}(X^2+2XY+Y^2) - \mathrm{E}(X)^2 - 2\mathrm{E}(X)\mathrm{E}(Y) - \mathrm{E}(Y)^2$	expand out the squares
$= \mathrm{E}(X^2) + 2\mathrm{E}(XY) + \mathrm{E}(Y^2) - \mathrm{E}(X)^2 - 2\mathrm{E}(X)\mathrm{E}(Y) - \mathrm{E}(Y)^2$	Theorem 7.6
$= \mathrm{E}(X^2) - \mathrm{E}(X)^2 + 2\mathrm{E}(XY) - 2\mathrm{E}(X)\mathrm{E}(Y) + \mathrm{E}(Y^2) - \mathrm{E}(Y)^2$	rearrange the terms
$= \mathrm{Var}(X) + \mathrm{Var}(Y) + 2\mathrm{Cov}(X, Y)$	recognize the pieces
$= \mathrm{Var}(X) + \mathrm{Var}(Y).$	Theorem 7.8

Theorem 7.11: If X is a random variable with finite variance and a is a constant,

$$\mathrm{Var}(aX) = a^2\mathrm{Var}(X).$$

Proof:

$\mathrm{Var}(aX) = \mathrm{E}(a^2X^2) - \mathrm{E}(aX)^2$	computational formula for variance
$= a^2\mathrm{E}(X^2) - [a\mathrm{E}(X)]^2$	Theorem 7.5
$= a^2\mathrm{E}(X^2) - a^2\mathrm{E}(X)^2$	algebra
$= a^2\mathrm{Var}(X).$	recognize variance

Theorem 7.12: If $X_1, X_2, \ldots, X_n$ are independent random variables with finite variance and $a_1, a_2, \ldots, a_n$ are any constants, then

$$\mathrm{Var}\left(\sum_{i=1}^{n} a_i X_i\right) = \sum_{i=1}^{n} a_i^2 \mathrm{Var}(X_i).$$

Proof:

$\mathrm{Var}\left(\sum_{i=1}^{n} a_i X_i\right) = \sum_{i=1}^{n} \mathrm{Var}(a_i X_i)$	Theorem 7.9
$= \sum_{i=1}^{n} a_i^2 \mathrm{Var}(X_i).$	Theorem 7.11

7.3 Exercises

Mathematical Techniques

1–4 ▪ For the following joint distributions, find the probabilities for the random variable $X+Y$ and check that $\mathrm{E}(X+Y) = \mathrm{E}(X) + \mathrm{E}(Y)$.

1. (from Section 7.1, Exercises 1 and 5)

	$X=0$	$X=1$
$Y=0$	0.14	0.26
$Y=1$	0.21	0.39

2. (from Section 7.1, Exercises 2 and 6)

	$X=0$	$X=1$
$Y=1$	0.45	0.25
$Y=3$	0.05	0.25

3. (from Section 7.1, Exercises 3 and 7)

	$X=0$	$X=1$	$X=2$
$Y=1$	0.1	0.2	0.3
$Y=2$	0.05	0.15	0.2

4. (from Section 7.1, Exercises 4 and 8)

	$X=0$	$X=1$	$X=2$
$Y=1$	0.05	0.04	0.01
$Y=2$	0.1	0.08	0.02
$Y=3$	0.35	0.28	0.07

5–8 ▪ For the following joint distributions, find the probabilities for the random variable XY (the product) and check that $\mathrm{E}(XY) = \mathrm{E}(X)\mathrm{E}(Y)$ only if $\mathrm{Cov}(X, Y) = 0$.

5. The random variables X and Y in Exercise 1, with covariance found in Section 7.2, Exercise 1.

6. The random variables X and Y in Exercise 2, with covariance found in Section 7.2, Exercise 2.

7. The random variables X and Y in Exercise 3, with covariance found in Section 7.2, Exercise 3.

8. The random variables X and Y in Exercise 4, with covariance found in Section 7.2, Exercise 4.

9–12 ▪ For the following joint distributions, find the probabilities for the random variable $X - Y$ (the difference), and check that $\text{E}(X - Y) = \text{E}(X) - \text{E}(Y)$ and that $\text{Var}(X - Y) = \text{Var}(X) + \text{Var}(Y)$ if $\text{Cov}(X, Y) = 0$.

9. The random variables X and Y in Exercise 1.

10. The random variables X and Y in Exercise 2, with variances found in Section 7.2, Exercise 10.

11. The random variables X and Y in Exercise 3, with variances found in Section 7.2, Exercise 11.

12. The random variables X and Y in Exercise 4.

13–14 ▪ Consider any random variable X that has a finite expectation and variance.

13. Find $\text{Cov}(X, X)$. How does it compare with $\text{Var}(X)$?

14. Consider the new random variable $Y = -X$. Find $\text{Cov}(X, Y)$. How does it compare with $\text{Var}(X)$?

15–16 ▪ Consider independent random variables X and Y with identical probability distributions as given. Let $Z = X + X$ and $S = X + Y$.

a. Compute the mean and variance of Z directly from its probability distribution.

b. Compute the mean and variance of S directly from its probability distribution.

c. Find $\text{E}(Z)$ by using Theorem 7.4 and $\text{Var}(Z)$ by using the general addition rule for covariances and the covariance of X with itself (Exercise 13).

d. Find $\text{E}(Z)$ by using the fact that $Z = 2X$ and Theorem 7.5. and $\text{Var}(Z)$ by using the fact that $Z = 2X$ and Theorem 7.11.

e. Find $\text{E}(S)$ and $\text{Var}(S)$ from Theorem 7.4 and Theorem 7.9. Why is $\text{Var}(S) < \text{Var}(Z)$?

15. X and Y take the value 0 with probability 0.5 and the value 1 with probability 0.5.

16. X and Y take the value 0 with probability 0.25, the value 1 with probability 0.5, and the value 2 with probability 0.25.

17–20 ▪ Use the following steps to prove that the geometric mean of the product of two random variables X and Y is equal to the product of the geometric means.

17. Write the definition of the geometric mean of the product XY.

18. Use a law of logs to expand the result.

19. Use Theorem 7.4 to break up the expectations.

20. Use a law of exponents and the definition of the geometric mean to prove the result.

21–22 ▪ The harmonic mean is another kind of average (like the geometric mean), defined by $H(X) = \frac{1}{\text{E}(1/X)}$ (the reciprocal of the expectation of the reciprocal). Like the geometric mean, the harmonic mean is only defined for random variables that take on positive values, and it is always less than the expectation. Compute the harmonic mean of the following random variables and check that it is indeed less than the expectation.

21. X takes the value 1 with probability 0.5 and the value 2 with probability 0.5.

22. Suppose X takes the value 1 with probability 0.1 and the value 10 with probability 0.9.

23–24 ▪ Use the following random variables to show that $\text{E}(X/Y) \neq \text{E}(X)/\text{E}(Y)$ (the expectation of the quotient is not equal to the quotient of the expectations) even when two random variables are independent. Use the harmonic mean (Exercises 21 and 22) to guess why the expectation of the quotient is bigger than the quotient of the expectations.

23. Suppose X and Y are independent random variables that each take the value 1 with probability 0.5 and the value 2 with probability 0.5.

24. Suppose X and Y are independent random variables where X takes the value 0 with probability 0.5 and the value 1 with probability 0.5, and Y takes the value 1 with probability 0.1 and the value 10 with probability 0.9.

Applications

25–26 ▪ Consider again the birds suffering from mites and lice in Section 7.1, Exercises 23 and 24. Find the probability distribution of $P = L + M$ and use it to compute $\text{E}(P)$ directly. Compare the result with Theorem 7.4.

		Bird C Lice			Bird D Lice		
		0	1	2	0	1	2
	0	0.21	0.13	0.16	0.06	0.21	0.23
Mites	1	0.13	0.07	0.10	0.21	0.03	0.06
	2	0.06	0.10	0.04	0.13	0.06	0.01

25. Bird C.

26. Bird D.

27–28 ▪ Consider yet again the birds suffering from mites and lice. For each bird, find $\text{Var}(P)$ directly from the probability distribution of $P = L + M$. Show how you could have found the variance with the general addition formula for variances.

27. Bird C.

28. Bird D.

29–30 ▪ Find the probability distribution of $W = 0.05L + 0.02M$ for the following birds. Check that the expectation is 0.059 in both cases.

29. Bird A (from Example 7.3.5).

30. Bird B (from Table 7.1).

31–32 ■ Suppose immigration and emigration change the sizes of four populations with the following probabilities (Section 6.7, Exercises 31–34).

Population a		Population b	
Number	Probability	Number	Probability
−1	0.4	−1	0.1
0	0.2	0	0.3
1	0.3	1	0.2
2	0.1	2	0.4
Population c		**Population d**	
−1	0.4	−10	0.4
0	0.2	0	0.2
1	0.3	1	0.3
100	0.1	2	0.1

31. Find the expectation of the population change in populations a and c summed.

32. Find the expectation of the total population change in b and d summed.

33–38 ■ Suppose annual immigration into a park by three species follows the probabilities in the table.

Number of Immigrants	Species 1	Species 2	Species 3
0	0.3	0.6	0.1
1	0.6	0.2	0.3
2	0.1	0.2	0.6

33. Find the expected number of each type of immigrant. Find the expected total number of immigrants from all species combined.

34. Suppose species 1 in has mass 10 kg, species 2 has mass 5 kg, and species 3 has mass 15 kg. Find the expected mass of the immigrants of each species that arrive. Find the expected total mass of all immigrants.

35. Ignore the third immigrant, species and suppose that species 1 and species 2 arrive independently.

a. Give the joint probability distribution for species 1 and species 2.

b. Find the probability of each possible number of immigrants.

c. Find the expected number of immigrants of each of these two species and compare with the sum of the expected numbers.

d. Find the variance in the total number of immigrants and compare with the sum of the variances.

36. The first two immigrant species are appealing to eco-tourists, and each additional individual brings in $1000. The third species is repellent to eco-tourists and reduces revenue by $500.

a. Find the expected revenue from each species separately and for all species together.

b. Find the variance in revenue due to each species separately.

c. Find the variance in the total revenue if the species immigrate independently.

d. Interpret these results.

37. Consider the situation described in Exercise 35 but suppose that the two species do not arrive independently. Find a set of probabilities for outcomes consistent with the probabilities that you think will have higher variance than the independent case. Compute the expectation and variance.

38. Consider the situation described in Exercise 35 but suppose that the two species do not arrive independently. Find a set of probabilities for outcomes consistent with the probabilities that you think will have lower variance than the independent case. Compute the expectation and variance.

39–42 ■ Consider the following table giving the probability that three types of birds consume two different species of caterpillars that come in two different sizes. The first has food quality (in $kCal/cm^3$) of 1.0, and the second has food quality of 2.0. The two volumes are 0.5 and 1.5 cm^3.

	Volume Eaten by Bird 1		Volume Eaten by Bird 2		Volume Eaten by Bird 3	
Quality	0.5	1.5	0.5	1.5	0.5	1.5
1.0	0.35	0.25	0.24	0.36	0.10	0.50
2.0	0.05	0.35	0.16	0.24	0.30	0.10

39. Which of the joint distributions is independent?

40. Find the mean quality and volume of caterpillars eaten by each bird.

41. Find the expected total calories per caterpillar for each bird.

42. From the expected total calories for each bird, and find the covariance of quality and volume.

43–44 ■ Consider the following data for cell age A and the number of toxic molecules N inside (as in Section 7.2, Exercises 37 and 38).

	$A=0$	$A=1$	$A=2$	$A=3$	$A=4$
$N=0$	0.03	0.00	0.09	0.06	0.09
$N=1$	0.01	0.05	0.06	0.12	0.15
$N=2$	0.02	0.03	0.06	0.08	0.15

Suppose that the probability that a cell is cancerous depends on both the age and the number of toxic molecules. Use the law of total probability to find the overall probability that a cell is cancerous, and show how it is really a version of the sum rule for expectations.

43. Suppose that the probability that a cell is cancerous is $\frac{AN}{10}$.

44. Suppose that the probability that a cell is cancerous is $\frac{AN^2}{20}$.

Computer Exercises

45. Consider Markov chains where I_t represents the event that a molecule is inside a cell at time t and O_t represents the event that a molecule is outside a cell at time t, obeying

$$\Pr(I_{t+1}|I_t) = p$$
$$\Pr(I_{t+1}|O_t) = 1 - p$$
$$\Pr(O_{t+1}|I_t) = 1 - p$$
$$\Pr(O_{t+1}|O_t) = p$$

for various values of p. In each case, the long-term probability that the molecule is inside is 0.5. Suppose you get one point when the molecule is inside and zero points when it is outside.

a. Add up the random variables I_1 through I_{100}, starting from $I_0 = 1$, when $p = 0.01$, $p = 0.1$, $p = 0.25$, $p = 0.5$, $p = 0.75$, $p = 0.9$, and $p = 0.99$.

b. What should the value be? Which value is closest?

c. Try again with the initial value I_0 set to 1 with probability 0.5 and to 0 with probability 0.5. Does this change your results?

46. Recall the dice rolling rules in Section 7.1, Exercise 41. Redo the experiment, and find the average value for each die. Is the sum of the averages equal to the average of the sums?

7.4 The Binomial Distribution

We have developed tools to display and combine probability distributions and probability density functions. We now begin the study of special distributions generated by particular biological models. These distributions arise repeatedly in remarkably diverse contexts. By understanding these distributions and the underlying models, we have a baseline against which to compare and reason about data. It is a good idea to start a table of these special distributions, including the underlying model, the range, the formula, the mean, and the variance for each distribution. In this section we analyze the **binomial distribution**, the distribution constructed from sums of independent Bernoulli random variables.

7.4.1 Definition of the Binomial Distribution

Suppose that each year a population receives an immigrant with probability $p = 0.7$ and no immigrant with probability $1 - p = 0.3$. The arrivals of immigrants are independent; knowing that an immigrant arrived one year says nothing about whether an immigrant will arrive in future years. What is the probability that exactly two immigrants arrive in three years? What is the probability that exactly six immigrants arrive in ten years? We can answer these questions with the **binomial distribution**.

Let the random variable N represent the number of immigrants that arrived. After n years, this random variable could take on any value between 0 and n. To completely understand this process, we must compute the probability that $N = k$ for every value of $0 \leq k \leq n$. We define the binomial probability distribution b as

$$b(k; n, p) = \Pr(N = k). \quad (7.4.1)$$

This gives the probability of exactly k immigrants after n years, where p is the probability that an immigrant arrives in any particular year.

Example 7.4.1 Using the Notation for the Binomial Distribution

The probability that six immigrants arrive in 10 yr if the probability of an immigrant in each year is $p = 0.7$ can be written as

$$\Pr(6 \text{ immigrants in } 10 \text{ yr}) = b(6; 10, 0.7).$$

Example 7.4.2 Using the Binomial Distribution to Describe Coin Flips

The binomial distribution can describe the probability of flipping exactly 50 heads in 100 tosses of a fair coin. The probability on each try is $p = 0.5$, so this probability can be written with the binomial distribution as

$$\Pr(50 \text{ heads in } 100 \text{ flips}) = b(50; 100, 0.5).$$

At the moment, we have no formula for this quantity.

The total number of immigrants in n years is the **sum** of the number of immigrants in each year. The number of immigrants in each year is a Bernoulli random variable, a random variable that takes on only the values 0 and 1 (Definition 6.10). Suppose the number of immigrants over 10 years is

	Year									
	1	2	3	4	5	6	7	8	9	10
Immigrants	0	1	1	0	1	0	1	1	1	0
Total	0	1	2	2	3	3	4	5	6	6

The total number after ten years is $0+1+1+0+1+0+1+1+1+0=6$. The compound random variable N is a **sum** of simpler random variables.

The number of heads or immigrants (often called "successes") has a binomial distribution only if the conditions in the definition are satisfied.

Definition 7.7 A random variable N has a binomial distribution with parameters n and p if it is the sum of n independent Bernoulli random variables, each of which has a probability p of being equal to 1. The range of N is the set of integers from 0 to n, and the probability that $N=k$ is indicated by $b(k; n, p)$.

This definition gives the two key conditions for a random variable to have a binomial distribution:

- The random variable is the sum of n **independent** Bernoulli random variables.
- Each Bernoulli random variable takes on the value 1 with the same probability p.

"All" that remains to be done is to compute $b(k; n, p)$. Before attacking this problem, we will find the mean and variance of the binomial distribution.

7.4.2 The Mean and Variance of the Binomial Distribution

We can use the theorem guaranteeing that the expectation of the sum is the sum of the expectations (Theorem 7.6) to find $E(N)$, the mean number of immigrants after n years. Let I_i represent the number of immigrants in year i, so

$$N = \sum_{i=1}^{n} I_i$$

(Figure 7.4.1). The expectation of N is the sum of the expectations of the I_i. The expectation of a Bernoulli random variable is

$$E(I_i) = 0 \cdot (1-p) + 1 \cdot p = p.$$

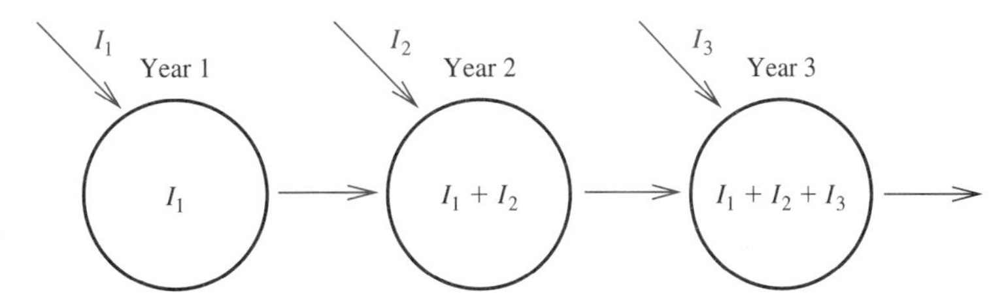

FIGURE 7.4.1

A binomial random variable as a sum of Bernoulli random variables

Therefore,

$$E(N) = \sum_{i=1}^{n} E(I_i) = \sum_{i=1}^{n} p = np. \tag{7.4.2}$$

The mean number of successes is the number of trials times the probability of success per trial.

Example 7.4.3 The Expected Number of Immigrants

If the probability of an immigrant is 0.7 in each year for 10 yr, then $p = 0.7$ and $n = 10$. The expectation is then $E(N) = np = 0.7 \cdot 10 = 7$ immigrants in 10 yr.

Example 7.4.4 The Expected Number of Heads

If the probability of flipping heads is 0.5 for each of 100 flips, then $p = 0.5$ and $n = 100$. The expectation of the number of heads, which we can represent with the random variable H, is $E(H) = np = 0.5 \cdot 100 = 50$ heads in 100 flips.

Similarly, because the Bernoulli random variables are independent (first condition), we can find the variance by adding (Theorem 7.9). The variance of each Bernoulli random variable is

$$\text{Var}(I_i) = p(1 - p)$$

(Section 6.9, Equation 6.9.1). The variance of N is therefore

$$\text{Var}(N) = \sum_{i=1}^{n} \text{Var}(I_i) = \sum_{i=1}^{n} p(1 - p) = np(1 - p). \tag{7.4.3}$$

Example 7.4.5 The Variance of the Number of Immigrants

When immigrants arrive with probability $p = 0.7$ for $n = 10$ yr, as in Example 7.4.3, the variance is

$$\text{Var}(N) = 10 \cdot 0.7 \cdot 0.3 = 2.1.$$

Example 7.4.6 The Variance of the Number of Heads

When flipping for heads with probability $p = 0.5$ for $n = 10$ flips, the variance is

$$\text{Var}(H) = 100 \cdot 0.5 \cdot 0.5 = 25.$$

Using the rule of thumb that most of the distribution is within two standard deviations of the mean (Subsection 6.9.3), the number of heads will usually fall between $50 - 2\sqrt{25} = 40$ and $50 + 2\sqrt{25} = 60$.

7.4.3 Computing the Binomial: $n = 2$ and $n = 3$

We begin finding the binomial probabilities $b(k; n, p)$ with small values of n. In these cases, we can list all the possibilities. After 1 yr,

Year 1	Total Number of Immigrants, k	Probability	$b(k; n, p)$
0	0	$1 - p$	$b(0; 1, p) = 1 - p$
1	1	p	$b(1; 1, p) = p$

Either no immigrant arrived ($k = 0$ with probability $1 - p$) or one immigrant arrived ($k = 1$ with probability p). Therefore, $b(0; 1, p) = 1 - p$ and $b(1; 1, p) = p$. This table merely repeats our assumptions.

After 2 yr, $n = 2$, there are four possibilities (Figure 7.4.2).

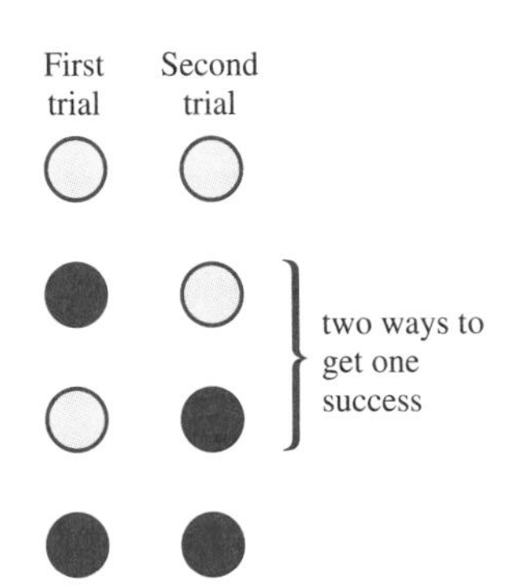

FIGURE 7.4.2
The four possible outcomes with $n = 2$

Year 1	Year 2	Total Number of Immigrants k	Probability	$b(k; n, p)$
0	0	0	$(1-p)^2$	$b(0; 2, p) = (1-p)^2$
1	0	1	$p(1-p)$	$b(1; 2, p) = 2p(1-p)$
0	1	1	$(1-p)p$	
1	1	2	p^2	$b(2; 2, p) = p^2$

The probabilities in the fourth column are computed using the assumption of independence and the multiplicative law for independent probabilities. For example, the second line multiplies the probability of an immigrant in the first year, p, by the probability of no immigrant in the second year, $1 - p$.

The difficult part arises in computing the last column, the probability distribution itself. There is one way to receive zero immigrants (none in either year) and one way to receive two immigrants (one in each year). There are, however, two ways to receive one immigrant, or have $k = 1$. The probabilities of the second and third lines must be added to find $b(1; 2, p) = 2p(1 - p)$. Addition of probabilities is justified because the events in these two lines are mutually exclusive. This is the same logic we used to calculate the probability distribution for the total number of parasites on a bird in Section 7.3.

Example 7.4.7 Applying the Formula for the Binomial Distribution with $n = 2$

Suppose that $p = 0.5$, like a fair coin. The number of heads after 2 flips, denoted by the random variable H, has the following probability distribution (Figure 7.4.3)

$$\Pr(H = 0) = b(0; 2, 0.5) = (1 - p)^2 = (1 - 0.5)^2 = 0.25$$

$$\Pr(H = 1) = b(1; 2, 0.5) = 2p(1 - p) = 2(0.5)(1 - 0.5) = 0.5$$

$$\Pr(H = 2) = b(2; 2, 0.5) = p^2 = 0.5^2 = 0.25.$$

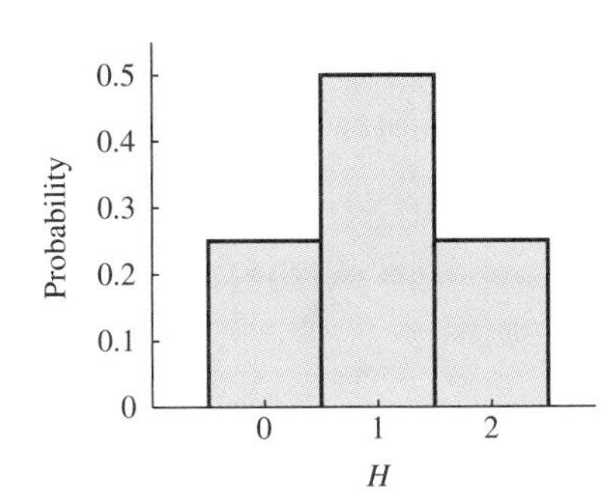

FIGURE 7.4.3
The binomial distribution with $p = 0.5$ and $n = 2$

These probabilities match those of getting the three different types of offspring from a cross of two heterozygous parents (Subsection 6.2.4). $H = 0$ corresponds to inheriting the **b** allele from each parent, $H = 1$ corresponds to inheriting the **b** allele from one parent and the **B** allele from the other, and $H = 2$ corresponds to inheriting the **B** allele from each parent. The probabilities are not all equal because there are two ways to inherit different alleles from the parents.

Example 7.4.8 Applying the Formula for the Binomial Distribution to Immigration

What is the distribution of the number of immigrants after 2 yr? We assumed that an immigrant arrives each year with probability $p = 0.7$. Then

$$\Pr(N = 0) = b(0; 2, 0.7) = (1 - 0.7)^2 = 0.09$$

$$\Pr(N = 1) = b(1; 2, 0.7) = 2(0.7)(1 - 0.7) = 0.42$$

$$\Pr(N = 2) = b(2; 2, 0.7) = 0.7^2 = 0.49.$$

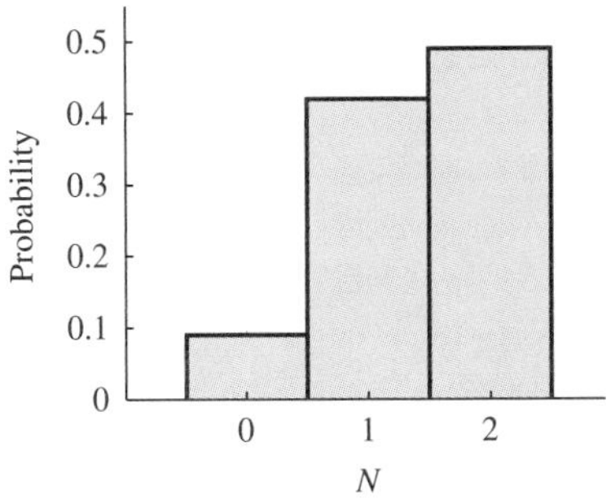

(Figure 7.4.4).

The probability of two immigrants arriving is much higher than the probability of getting two heads with a fair coin or two A alleles in Example 7.4.7.

FIGURE 7.4.4
The binomial distribution with $p = 0.7$ and $n = 2$

We follow the same method to work out the probabilities after 3 yr (Figure 7.4.5).

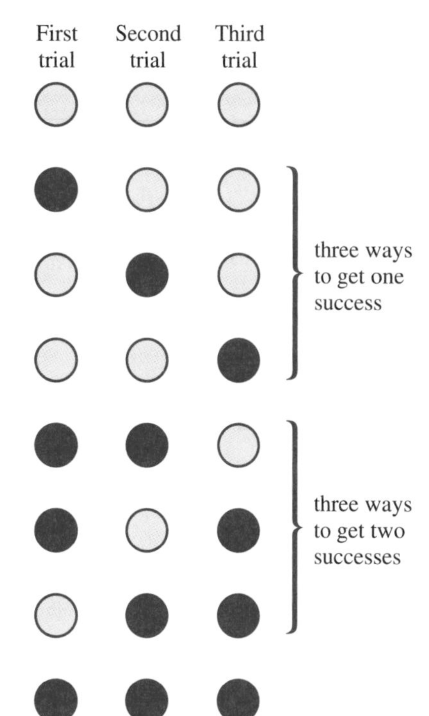

FIGURE 7.4.5
The eight possible outcomes with $n = 3$

Year 1	Year 2	Year 3	Total Number of Immigrants k	Probability	$b(k; n, p)$
0	0	0	0	$(1-p)^3$	$b(0; 3, p) = (1-p)^3$
1	0	0	1	$p(1-p)^2$	
0	1	0	1	$(1-p)p(1-p)$	$b(1; 3, p) = 3p(1-p)^2$
0	0	1	1	$(1-p)^2p$	
1	1	0	2	$p^2(1-p)$	
1	0	1	2	$p(1-p)p$	$b(2; 3, p) = 3p^2(1-p)$
0	1	1	2	$(1-p)p^2$	
1	1	1	3	p^3	$b(3; 3, p) = p^3$

Each probability was again computed with the multiplication rule for independent events. For example, in the second row, the probability that there is an immigrant in the first year, p, is multiplied by the probability of no immigrant in the second year, $1 - p$, and the probability that there is no immigrant in the third year, another $1 - p$, giving $p(10 - p)^2$. Because there are three different ways to have exactly one immigrant in 3 yr, the probability of $k = 1$ is $3p(1-p)^2$.

Example 7.4.9 Applying the Formula for the Binomial Distribution with $n = 3$

We can use this table to find the probability of flipping zero, one, two, or three heads in three tosses of a fair coin.

$$\Pr(H=0) = b(0; 3, 0.5) = (1-p)^3 = (1-0.5)^3 = 0.125$$

$$\Pr(H=1) = b(1; 3, 0.5) = 3p(1-p)^2 = 3(0.5)(1-0.5)^2 = 0.375$$

$$\Pr(H=2) = b(2; 3, 0.5) = 3p^2(1-p) = 3 \cdot 0.5^2(1-0.5) = 0.375$$

$$\Pr(H=3) = b(3; 3, 0.5) = p^3 = 0.5^3 = 0.125.$$

(Figure 7.4.6).

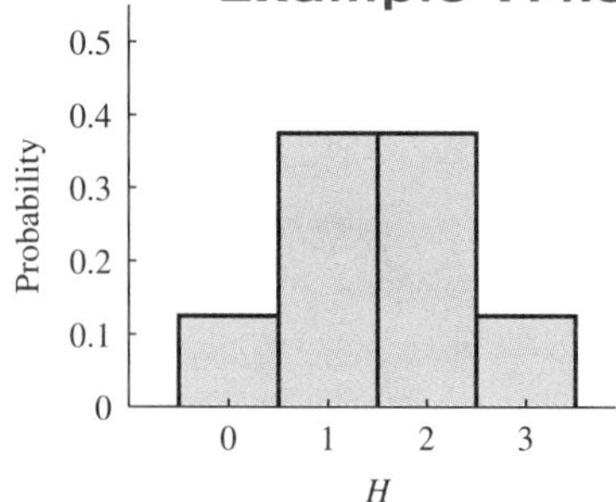

FIGURE 7.4.6
The binomial distribution with $p = 0.5$ and $n = 3$

Example 7.4.10 Applying the Formula for the Binomial Distribution to Immigration

After 3 yr of immigrants arriving with $p = 0.7$,

$$\Pr(N = 0) = b(0; 3, 0.7) = (1 - p)^3 = (1 - 0.7)^3 = 0.027$$
$$\Pr(N = 1) = b(1; 3, 0.7) = 3p(1 - p)^2 = 3(0.7)(1 - 0.7)^2 = 0.189$$
$$\Pr(N = 2) = b(2; 3, 0.7) = 3p^2(1 - p) = 3 \cdot 0.7^2(1 - 0.7) = 0.441$$
$$\Pr(N = 3) = b(3; 3, 0.7) = p^3 = 0.7^3 = 0.343$$

(Figure 7.4.7).

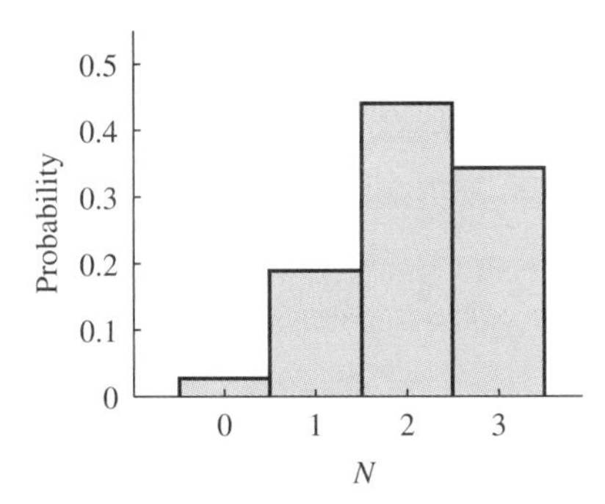

FIGURE 7.4.7
The binomial distribution with $p = 0.7$ and $n = 3$

Example 7.4.11 Checking the Formulas for the Expectation and Variance

We can use these probability distributions to check our calculation of the expectation. With $p = 0.5$ and $n = 3$, the expectation can be found directly as

$$E(H) = 0 \cdot 0.125 + 1 \cdot 0.375 + 2 \cdot 0.375 + 3 \cdot 0.125 = 1.5.$$

This matches the formula (Equation 7.4.2), which states that

$$E(H) = np = 3 \cdot 0.5 = 1.5.$$

Similarly, the variance can be found explicitly with the computation formula as

$$\begin{aligned}\mathrm{Var}(H) &= E(H^2) - E(H)^2 \\ &= 0^2 \cdot 0.125 + 1^2 \cdot 0.375 + 2^2 \cdot 0.375 + 3^2 \cdot 0.125 - 1.5^2 = 0.75.\end{aligned}$$

Again, this matches the formula (Equation 7.4.3), which states that

$$E(H) = np(1 - p) = 3 \cdot 0.5(1 - 0.5) = 0.75.$$

7.4.4 Binomial Distribution: The General Case

Each value of $b(k; n, p)$ is a product of two terms: a constant and a polynomial in p. We can figure out the general formula by working out each term separately, beginning with the polynomial. Each of the three ways of getting one immigrant in 3 yr includes 1 yr with an immigrant and 2 yr with none, although in different orders. In each order, the single immigrant arrives in a particular year with probability p, and no immigrant arrives in each of the other years with probability $1 - p$. Multiplying these three terms together gives the polynomial term $p(1 - p)^2$.

How does the probability of three immigrants in 5 yr depend on the probability p? The following table lists two ways this could happen.

Year 1	Year 2	Year 3	Year 4	Year 5	Probability
1	0	1	1	0	$p^3(1 - p)^2$
0	1	1	0	1	$p^3(1 - p)^2$

The three immigrants correspond to three factors of p, or p^3. The 2 yr without immigrants correspond to two factors of $1 - p$, or $(1 - p)^2$.

In general, to get k immigrants in n years, there must be k successes (a factor of p^k) and $n-k$ failures (a factor of $(1-p)^{n-k}$). Therefore,

$$\text{polynomial term of } b(k; n, p) = p^k(1-p)^{n-k}. \tag{7.4.4}$$

Finding the constant term requires some "combinatoric reasoning," which is essentially just advanced counting. Exactly how many ways can k immigrants arrive in n years? To begin, we give the unknown quantity a new notation and a name.

$$\begin{matrix}\text{the number of ways to have}\\ k \text{ successes in } n \text{ tries}\end{matrix} = \binom{n}{k} = \text{“}n \text{ choose } k\text{”}. \tag{7.4.5}$$

This object is called "n choose k" because it represents the number of ways to choose k things from a set of n.

Example 7.4.12 Using "n choose k" with $n = 2$

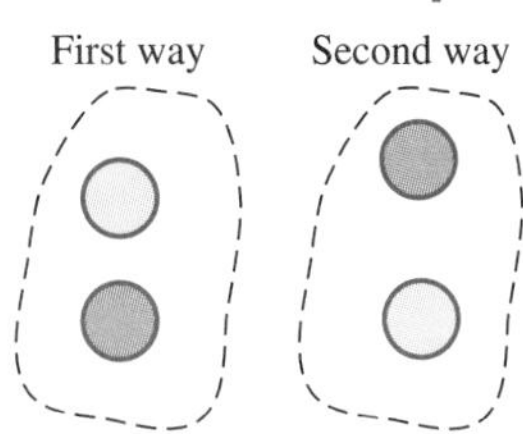

FIGURE 7.4.8
Choosing 1 out of 2

There are two ways to choose 1 yr out of 2 (Figure 7.4.8). Therefore, we can write

$$\binom{2}{1} = 2.$$

This produces the factor 2 in the formula for the binomial distribution with $n = 2$.

$$b(1; 2, p) = 2p(1-p).$$

Example 7.4.13 Using "n choose k" with $k = 1$ and $n = 3$

There are three ways to choose 1 yr out of 3: the first year, the second year, or the third year. Therefore, we write

$$\binom{3}{1} = 3.$$

This produces the factor 3 in the formula for the binomial distribution with $n = 3$ and $k = 1$.

$$b(1; 3, p) = 3p(1-p)^2.$$

How do we go about computing these in general? The trick is thinking of doing the choosing sequentially.

Example 7.4.14 Find $\binom{4}{2}$ Step by Step

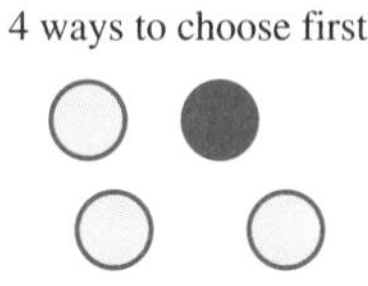

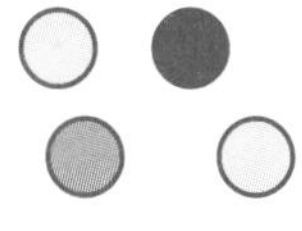

FIGURE 7.4.9
Choosing 2 things out of 4

Consider computing $\binom{4}{2}$, the number of ways of choosing 2 things out of 4 (Figure 7.4.9). There are 4 ways to choose the first element and only 3 ways to choose the second, because the second must be different from the first. We might guess that $\binom{4}{2}$ is given by $4 \cdot 3 = 12$ because we multiply the number of ways to choose the first by the number of ways to choose the second. But this plausible argument is **not** correct.

We can check by listing all possibilities:

$$(1, 2), (1, 3), (1, 4), (2, 1), (2, 3), (2, 4), (3, 1), (3, 2), (3, 4), (4, 1), (4, 2), (4, 3).$$

We found 12, but each is listed twice (Figure 7.4.10). We could get the pair (2, 3) in two ways, either by picking element 2 and then element 3, or by picking element 3 and then element 2. The correct value is

$$\binom{4}{2} = 6.$$

Choosing black then gray

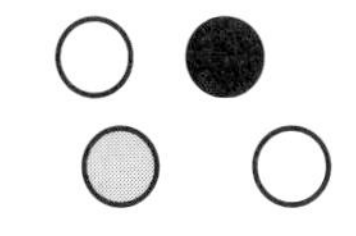

or gray then black

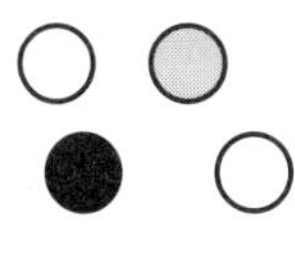

same result
in two ways

FIGURE 7.4.10

Correcting for counting the same possibility more than once

Table 7.2 Values of $k!$ for $k \le 10$

k	$k!$
0	1
1	1
2	2
3	6
4	24
5	120
6	720
7	5040
8	40,320
9	362,880
10	3,628,800

The steps in Example 7.4.14 can be followed in general, giving

$$\binom{n}{k} = \frac{n(n-1)\cdots(n-k+2)(n-k+1)}{\text{number of times each set was counted}}.$$

The numerator is the product of the number of possible first choices (n) by the number of second choices ($n-1$) down to the number of kth choices ($n-k$). The denominator is the number of different ways to order a set of k things:

There are two ways to order two things, (1, 2) and (2, 1). This is why we counted each of our sets of size two twice. There are six ways to order three things:

$$(1, 2, 3), (1, 3, 2), (2, 1, 3), (2, 3, 1), (3, 1, 2), (3, 2, 1).$$

We can use the same sort of reasoning to figure out the pattern. There are three possible first choices, leaving two second choices and one third choice. We multiply these together to find the total number.

In general, with k elements, there are k possible first choices, $k-1$ second choices, and so on. Multiplying these together, we find the factorials introduced in Section 3.7. We formally make the following definition.

Definition 7.8 The number of ways to order k things is called k factorial, with formula

$$k! = k(k-1)\cdots 2\cdot 1$$

For convenience, we define $0! = 1$.

Example 7.4.15 Computing Factorials for Small Values of k

There are two possible orderings for two items, or

$$2! = 2\cdot 1 = 2.$$

There are six possible orderings for three items, or

$$3! = 3\cdot 2\cdot 1 = 6.$$

(Figure 7.4.11).

The two orderings for two items
AB, BA

The six orderings for three items
ABC, ACB, BAC, BCA, CAB, CBA

FIGURE 7.4.11

The smaller factorials

The values of $k!$ become large very quickly.

Putting this together,

$$\binom{n}{k} = \frac{n(n-1)\cdots(n-k+2)(n-k+1)}{k!}.$$

To write this more simply, we can write the number in terms of factorials as

$$n(n-1)\cdots(n-k+2)(n-k+1) = \frac{n(n-1)\cdots 2\cdot 1}{(n-k)(n-k-1)\cdots 2\cdot 1} = \frac{n!}{(n-k)!}.$$

We have finally found the number in front of the polynomial.

Theorem 7.13 The number of ways to choose k things out of n has formula

$$\binom{n}{k} = \frac{n!}{(n-k)!k!}.$$

Example 7.4.16 Using the Formula to Compute $\binom{4}{2}$

With $n=4$ and $k=2$,

$$\binom{4}{2} = \frac{4!}{(4-2)!2!} = \frac{4\cdot 3\cdot 2\cdot 1}{(2\cdot 1)(2\cdot 1)} = \frac{24}{2\cdot 2} = 6,$$

matching the result found by direct counting in Example 7.4.14.

Example 7.4.17 Computing $\binom{n}{k}$ with Larger Values of n and k

Larger values are much harder to compute. For instance,

$$\binom{10}{4} = \frac{10!}{6!4!} = \frac{3628800}{720\cdot 24} = 210.$$

There are 210 different ways to choose 4 things out of 10, or 210 ways that exactly 4 immigrants arrive in 10 yr.

Combining our computation of the constant term "n choose k" with the polynomial term (Equation 7.4.4), we have proven the following.

Theorem 7.14 Under the conditions for the binomial distribution, the probability $b(k; n, p)$ of k successes in n independent trials is

$$b(k; n, p) = \begin{cases} \binom{n}{k} p^k (1-p)^{n-k} = \frac{n!}{(n-k)!k!} p^k (1-p)^{n-k} & \text{if } 0 \le k \le n \\ 0 & \text{otherwise.} \end{cases}$$

The probabilities of zero correspond to impossible events, like 11 immigrants in 10 yr or -3 heads in 50 flips. Table 7.3 gives values for $\binom{n}{k}$ for $n \le 10$.

Example 7.4.18 Using the Formula for the Binomial Distribution: Coin Tosses

What is the probability of exactly 6 heads in 10 flips? With the binomial distribution,

$$b(6; 10, 0.5) = 210(0.5)^6(1-0.5)^4 = 0.205.$$

The entire probability distribution with $n=10$ and $p=0.5$ can be found with this formula (Figure 7.4.12).

Example 7.4.19 Using the Formula for the Binomial Distribution: Immigration

The probability of exactly 6 immigrants in 10 yr with $p=0.7$ is

$$b(6; 10, 0.7) = 210(0.7)^6(1-0.7)^4 = 0.200.$$

Table 7.3 Table of values of "n choose k" for $n \leq 10$

	$k=0$	$k=1$	$k=2$	$k=3$	$k=4$	$k=5$	$k=6$	$k=7$	$k=8$	$k=9$	$k=10$
$n=1$	1										
$n=2$	1	2	1								
$n=3$	1	3	3	1							
$n=4$	1	4	6	4	1						
$n=5$	1	5	10	10	5	1					
$n=6$	1	6	15	20	15	6	1				
$n=7$	1	7	21	35	35	21	7	1			
$n=8$	1	8	28	56	70	56	28	8	1		
$n=9$	1	9	36	84	126	126	84	36	9	1	
$n=10$	1	10	45	120	210	252	210	120	45	10	1

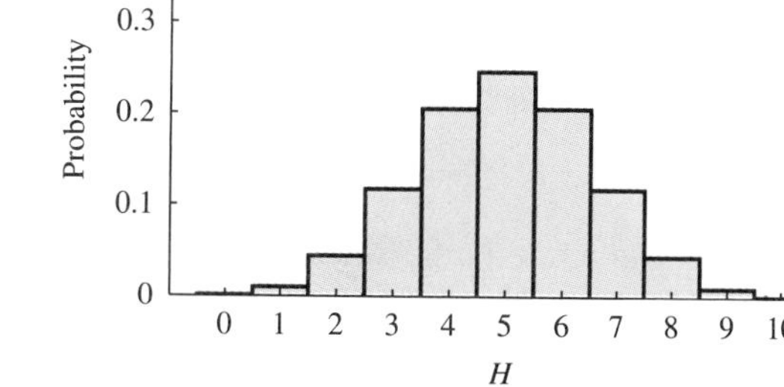

FIGURE 7.4.12

Binomial distribution: $p=0.5$ and $n=10$

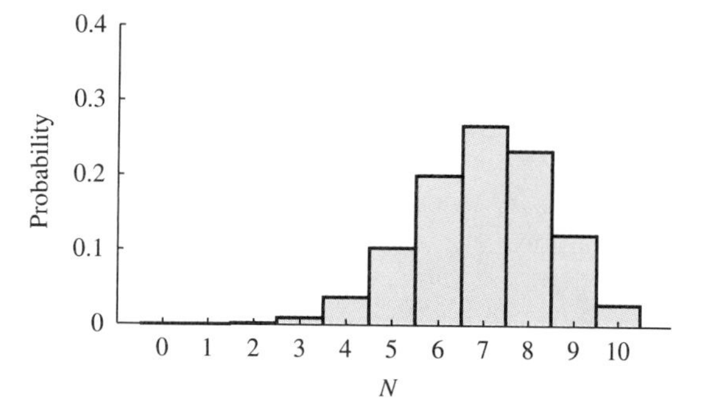

FIGURE 7.4.13

Binomial distribution: $p=0.7$ and $n=10$

The entire distribution has a very different shape, being shifted up toward the mean of 7 (Figure 7.4.13).

Example 7.4.20 The Binomial Distribution with Large Values of n

Computing the probability of 50 heads in 100 flips requires a computer because the factorials are enormous. With a computer, we find that

$$b(50, 100, 0.5) = 0.0796.$$

A fair coin will produce *exactly* 50 heads (the expected number) less than 8% of the time.

7.4.5 Finding the Mode of the Binomial Distribution

The mode of a binomial random variable is the single most probable outcome, and must be an integer. Even though the expectation may not be an integer, the mode is always within 1 of the expectation.

Theorem 7.15 The mode of the binomial probability distribution with parameters n and p is the smallest integer greater than $np - 1 + p$. If $np - 1 + p$ is the integer k, the mode is not unique, with the probability distribution taking on its maximum value at both k and $k + 1$.

Proof: Because the probabilities increase up to some point and then begin decreasing, the mode is where the probabilities start decreasing. This occurs at the smallest value of k for which $b(k + 1; n, p) < b(k; n, p)$. The trick is to find a formula for $b(k + 1; n, p)$ in terms of $b(k; n, p)$.

$$b(k+1; n, p) = \frac{n!}{(k+1)!(n-k-1)!} p^{k+1}(1-p)^{n-k-1}$$

$$= \left(\frac{n-k}{k+1}\right)\left(\frac{p}{1-p}\right) b(k; n, p).$$

Therefore, $b(k + 1; n, p) < b(k; n, p)$ if

$$\left(\frac{n-k}{k+1}\right)\left(\frac{p}{1-p}\right) < 1.$$

Solving for k,

$$k > np - 1 + p.$$

Therefore, the mode is the smallest value of k for which $k > np - 1 + p$. Note that $b(k + 1; n, p) = b(k; n, p)$ if $k = np - 1 + p$, giving a non-unique mode at k and $k + 1$.

Example 7.4.21 Finding the Mode of a Binomial Distribution

To compute the mode, find $np - 1 + p$ and round **up** to the nearest integer. With $n = 10$ and $p = 0.7$ (as in Example 7.4.19),

$$np - 1 + p = 10 \cdot 0.7 - 1 + 0.7 = 6.7.$$

Rounding up to the nearest integer, we find that the mode is $B = 7$, matching the expectation.

Example 7.4.22 Case with Mode Not Equal to Expectation

If $n = 10$ and $p = 0.25$, the expectation is $np = 2.5$. The mode is the smallest integer greater than

$$np - 1 + p = 10 \cdot 0.25 - 1 + 0.25 = 1.75,$$

or 2.0. The mode cannot be *equal* to the mean in this case because the random variable cannot take on the value 2.5.

Summary The number of successes (immigrants or "heads") in n independent trials of a Bernoulli random variable is described by the **binomial distribution**. The quantity $b(k; n, p)$ gives the probability of exactly k successes in n trials when the probability of success in a single trial is p. The expectation and variance of the binomial distribution can be computed using the theorems about the expectation and variance of sums of random variables. With some combinatoric reasoning, we found a formula for $b(k; n, p)$. The **mode** of the binomial distribution is always within 1 of the mean and can be found as the smallest integer above $np - 1 + p$. Along the way, we defined and computed "n choose k," the number of different ways to pick k things out of n in terms of the **factorial**, the number of ways to order k things.

7.4 Exercises

Mathematical Techniques

1–6 ▪ Evaluate the following. Find all the factorials explicitly.

1. $\binom{4}{1}$

2. $\binom{4}{3}$

3. $\binom{5}{2}$

4. $\binom{6}{2}$

5. $\binom{7}{1}$

6. $\binom{7}{2}$

7–12 ▪ Calculate the given value.

7. $b(1; 4, 0.3)$ (Use the value computed in Exercise 1.)

8. $b(3; 4, 0.3)$ (Use the value computed in Exercise 2.)

9. $b(2; 5, 0.4)$ (Use the value computed in Exercise 3.)

10. $b(2; 6, 0.4)$ (Use the value computed in Exercise 4.)

11. $b(1; 7, 0.2)$ (Use the value computed in Exercise 5.)

12. $b(2; 7, 0.6)$ (Use the value computed in Exercise 6.)

13–16 ▪ Find the expectation, variance, and mode for binomial random variables with the following parameters.

13. $n=4$, $p=0.3$.

14. $n=6$, $p=0.4$.

15. $n=7$, $p=0.7$.

16. $n=17$, $p=0.6$.

17–20 ▪ Use the formula for the binomial probability distribution to find and graph the probability distribution in the following cases.

17. A binomial random variable B with $p=0.7$ and $n=2$.

18. A binomial random variable B with $p=0.4$ and $n=2$.

19. A binomial random variable B with $p=0.7$ and $n=3$.

20. A binomial random variable B with $p=0.4$ and $n=3$.

21–24 ▪ Compute the mean and the variance from the probability distribution and make sure that your answers match the formulas in Equations 7.4.2 and 7.4.3.

21. A binomial random variable B with $p=0.7$ and $n=2$ (as in Exercise 17).

22. A binomial random variable B with $p=0.4$ and $n=2$ (as in Exercise 18).

23. A binomial random variable B with $p=0.7$ and $n=3$ (as in Exercise 19).

24. A binomial random variable B with $p=0.4$ and $n=3$ (as in Exercise 20).

25–26 ▪ Suppose the probability of a success is p. Find the probability of each of the following events and compare with the formula for the binomial distribution.

25. List all ways to get three successes out of four trials, and find the probability of each outcome.

26. List all ways to get two successes out of four trials, and find the probability of each outcome.

27–30 ▪ The binomial distribution depends on two key assumptions: the trials must each have the same probability of success, and the trials must be independent. Show that the distribution of outcomes does not match the binomial distribution in the following cases.

27. Suppose two trials are independent, but the first has a probability 0.3 of success, and the second a probability 0.7 of success. Find the probabilities of zero, one, and two successes, and compare with the binomial distribution with $n=2$ and $p=0.5$ (the average of the two probabilities). Does the expected number of successes match the binomial distribution? Does the variance?

28. Suppose two trials are independent, but the first has a probability 0.3 of success, and the second a probability 0.1 of success. Find the probabilities of zero, one, and two successes, and compare with the binomial distribution with $n=2$ and $p=0.2$. Does the expected number of successes match the binomial distribution? Does the variance?

29. Suppose the first trial has probability of success 0.5, and the second is successful with probability 0.8 if the first is and succeeds with probability 0.2 if the first fails. Show that the second trial has a probability 0.5 of success. Find the probabilities of zero, one, and two successes, and compare with the binomial distribution with $n=2$ and $p=0.5$. Does the expected number of successes match the binomial distribution? Does the variance?

30. Suppose the first trial has probability of success 0.2, and the second is successful with probability 0 if the first is and succeeds with probability 0.25 if the first fails. Show that the second trial has a probability 0.2 of success. Find the probabilities of zero, one, and two successes, and compare with the binomial distribution with $n=2$ and $p=0.2$.

31–34 ▪ The values $\binom{n}{k}$ have many beautiful mathematical properties. Here are just a few.

31. Show that $\binom{n}{k}=\binom{n}{n-k}$. Explain why this must be true.

32. Figure out why the following induction should hold, and show that it does.

$$\binom{n}{k}=\binom{n-1}{k}+\binom{n-1}{k-1}$$

33. The values of $\binom{n}{k}$ are also called binomial coefficients. Expand $(x+1)^3$ and $(x+1)^4$ and check that the coefficient of the kth power of x is $\binom{3}{k}$ in the first case and $\binom{4}{k}$ in the second.

34. Explain why the coefficients of the powers of x in the expansion of $(x + 1)^n$ are the binomial coefficients. What is the connection with Exercise 32?

Applications

35–36 ▪ Suppose that the probability that a baby is a boy is 0.5 and that a baby is a girl is also 0.5. Find the probabilities of each of the following families. Assume that the sexes of babies are independent of each other.

35. Family C has eight children, seven of whom are girls. Family D also has eight children, four of whom are girls. Which type of family is more probable?

36. Family C has eight children: three girls, one boy, and then four more girls. Family D also has eight children: two girls, one boy, one girl, three boys, and then a girl. Which type of family is more probable?

37–40 ▪ A group of identical quintuplets named Aaron, Bill, Carl, Dave, and Ed enjoy confusing the teachers at their school. List the number of different possibilities in each case, and then count them up. Make sure your counts match the appropriate value of "n choose k."

37. Only one goes to school.

38. Two go to school.

39. Three go to school.

40. Four go to school.

41–44 ▪ Find the number of ways the following can be ordered. List three of the possible orderings.

41. The order of finishing by three horses (named Speedy, Blinky, and Sparky) in a race.

42. The items in a four-course meal (soup, salad, main dish, and dessert).

43. A five-card poker hand (ace, 2, 5, 10, king).

44. Six occupants of offices along a hall (Al, Brenda, Carla, Dan, Esther, and Frank).

45–52 ▪ Each of five patients has been prescribed a different medication, but the prescriptions were accidentally shuffled. Compute the following probabilities.

45. What is the probability that every one of the patients gets the correct medication?

46. What is the probability that the first patient gets the right medication?

47. What is the probability that the second patient gets the right medication conditional on the first getting the right medication? Are the two events independent?

48. What is the probability that the first two patients get the right medication?

49. What is the probability that the third patient gets the right medication conditional on the first two getting the right medication? What is the probability that the first three patients get the right medication?

50. What is the probability that the second patient gets the wrong medication conditional on the first getting the wrong medication? Are the two events independent?

51. Suppose that four of the patients were prescribed one medication and the other one was prescribed a different one. What is the probability that all five get the right medication?

52. Suppose that three of the patients were prescribed one medication and the other two were prescribed a different one. What is the probability that all five get the right medication?

53–54 ▪ Suppose a heterozygous plant self-pollinates and produces five offspring with independent genotypes.

53. Find the probability distribution for the number of heterozygous offspring. Find the expectation and variance. Sketch the distribution.

54. Suppose that one allele is dominant and produces tall plants. Find and sketch the probability distribution for the number of tall offspring. Find the expectation and variance of the number of tall offspring.

55–56 ▪ We have seen that nonindependence of alleles (possibly caused by differential mortality of genotypes) can lead to deviations from normal proportions of offspring genotypes. Find the probability that a surviving offspring from a selfing heterozygote with genotype **Aa** has zero, one, or two **A** alleles. Which cases follow a binomial distribution?

55. Suppose that all of the homozygous offspring survive and half of the heterozygous offspring survive (as in Section 6.2, Exercise 15).

56. Suppose that all offspring with genotype **AA** survive, half of the offspring with genotype **Aa** survive, and one fourth of the offspring with genotype **aa** survive.

57–58 ▪ We have seen that meiotic drive (where one allele pushes its way into more than half of the gametes) can lead to deviations from normal proportions of offspring genotypes. Find the probability that an offspring from a selfing heterozygote with genotype **Aa** has zero, one, or two **A** alleles. Which cases follow a binomial distribution?

57. Suppose meiotic drive affects the pollen only and that 80% of the pollen grains from a heterozygote carry the **A** allele. Ovules are normal and 50% of them carry the **A** allele (as in Section 6.2, Exercise 49).

58. Suppose meiotic drive affects both pollen and ovules and that 80% of the pollen grains and ovules from a heterozygote carry the **A** allele (as in Section 6.2, Exercise 50).

Computer Exercises

59. There is a remarkable approximation for $n!$ called the **Stirling approximation** (derived by methods related to the method of leading behavior).

$$(n-1)! \approx \sqrt{\frac{2\pi}{n}} \left(\frac{n}{e}\right)^n$$

a. Compare the values for n ranging from 1 to 10.

b. Plot the values of $n!$ and the Stirling approximation for n ranging from 1 to 100 on a semi-log graph. How close are they?

c. Stirling's formula can be made more accurate by multiplying it by $1 + \frac{1}{12n}$. How much better is it for $n = 10$? For $n = 100$?

60. The following values of n and p all give an expectation of 10. Use your computer to plot the binomial distribution in each case. Describe how they are different.

a. $n = 10,\ p = 1$.

b. $n = 15,\ p = \frac{2}{3}$.

c. $n = 20,\ p = 0.5$.

d. $n = 30,\ p = \frac{1}{3}$.

e. $n = 50,\ p = 0.2$.

f. $n = 100,\ p = 0.1$.

g. $n = 1000,\ p = 0.01$.

61. If the probability that a team wins any particular game is 0.6, then the probability that it wins a five-game series is $b(3, 5, 0.6) + b(4, 5, 0.6) + b(5, 5, 0.6)$. (Think about who would win if teams kept playing even after one team had won three games.) Figure out how to write this in terms of the cumulative distribution B, and then how to write it in general for a series of n games with probability q of winning any particular game (assume n is odd). On a single graph, plot the probabilities that a team wins a series of 1 game, 5 games, and 101 games as functions of q. Why is each curve increasing? Explain the shape of the curve with $n = 1$. Explain the shape of the curve with $n = 101$. What is the value of each curve at $q = 0.5$ and why?

7.5 Applications of the Binomial Distribution

Many processes in biology and throughout the sciences fit (or nearly fit) the conditions for the binomial distribution. Using the mean, variance, and formula for the binomial probability distribution, we explore a few of these many applications.

7.5.1 Application to Genetics

Suppose that plants with genotype **BB** and **Bb** are tall and plants with genotype **bb** are short (**B** is a dominant allele). We cross two heterozygous plants with genotype **Bb**. According to our basic probability models (Sections 6.4 and 6.5), an offspring is short with probability 1/4. This does not mean that exactly one out of four offspring will be short. Suppose that a plant produces eight offspring. What is the probability that exactly two are short?

Example 7.5.1 Probability of Exactly Two Out of Eight Short Plants

Let S be the random variable giving the number of short offspring. S will follow the binomial distribution if it satisfies the assumptions that each offspring must have the same probability of being short and that they must be independent. The second assumption would fail if some of the offspring were genetically identical, like identical twins. If we believe that these assumptions hold, S follows the binomial distribution with $n = 8$ and $p = 0.25$ (Figure 7.5.1). The probability that two offspring are short is

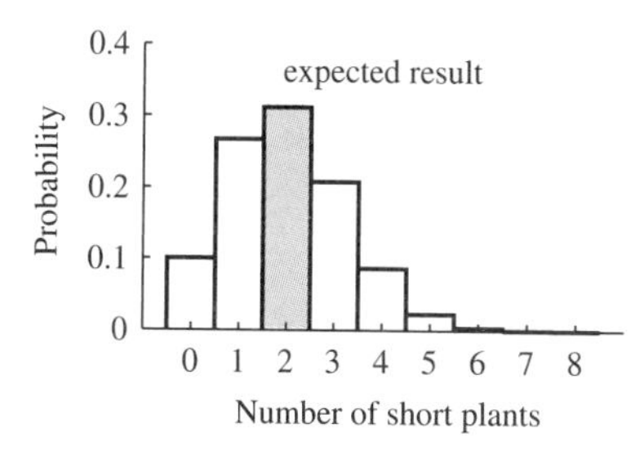

FIGURE 7.5.1

Probability distribution for plant height: Eight offspring

$$\begin{aligned}\Pr(S = 2) &= b(2; 8, 0.25)\\ &= \binom{8}{2} 0.25^2 \times 0.75^6\\ &= 28 \cdot 0.25^2 \times 0.75^6 = 0.311\end{aligned}$$

where we used Table 7.3 to find $\binom{8}{2} = 28$.

Example 7.5.2 Probability of Exactly Four Out of Eight Short Plants

A more confusing result would be that half the plants were short. The probability that four out of eight are short is

$$\begin{aligned}\Pr(S = 4) &= b(4; 8, 0.25)\\ &= \binom{8}{4} 0.25^4 \times 0.75^4\\ &= 70 \cdot 0.25^4 \times 0.75^4 = 0.087.\end{aligned}$$

using the value $\binom{8}{4} = 70$ from Table 7.3. The deceptive result that half the offspring are short would occur fully one fourth as often as the "expected" result (Figure 7.5.2). It might be difficult to guess the underlying genetics from eight offspring.

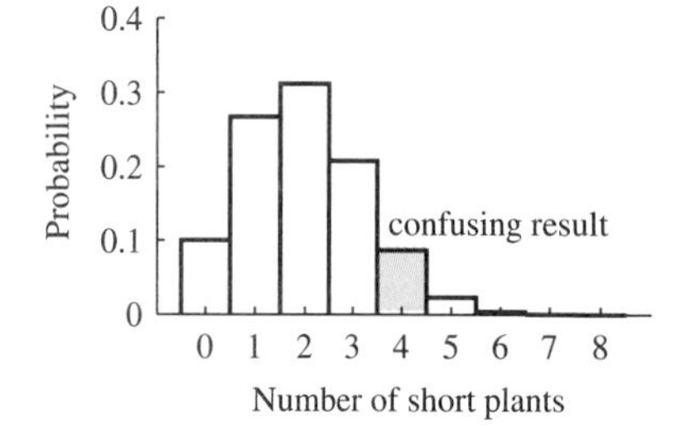

FIGURE 7.5.2

Probability distribution for plant height: Eight offspring

Example 7.5.3 Probability of Exactly Six Out of Eight Short Plants

An even more confusing result would be to get 3/4 short plants, the expected result for a dominant trait. The probability that six out of eight offspring are short is

$$\begin{aligned}\Pr(S=6) &= b(6; 8, 0.25)\\ &= \binom{8}{6} 0.25^6 \times 0.75^2\\ &= 28 \cdot 0.25^6 \times 0.75^2 = 0.0038.\end{aligned}$$

This extremely unlikely result would occasionally occur, but less than 1% of the time.

Example 7.5.4 Probabilities of Expected and Confusing Results with 20 Offspring

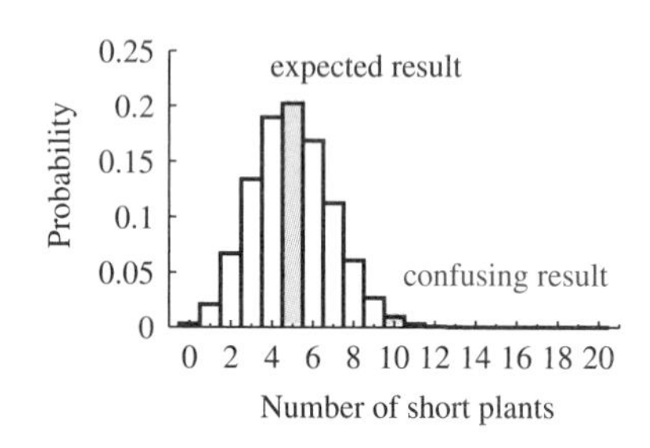

FIGURE 7.5.3

Probability distribution for plant height: 20 offspring

Using a computer, we can find the probability that exactly five out of 20 offspring are short in a cross of two heterozygous plants (the expected result):

$$b(5; 20, 0.25) = 0.202.$$

This probability is less than that of getting two out of eight (Example 7.5.1) because there are more possible outcomes. The deceptive result (Example 7.5.2) of ten out of 20 short plants occurs with probability

$$b(10; 20, 0.25) = 0.01.$$

There is only a 1% chance of being thoroughly misled (Figure 7.5.3). The probability that 3/4 of the offspring are short (the expected result if the **b** allele were dominant, as in Example 7.5.3), is the tiny value

$$b(15; 20, 0.25) = 3.43 \times 10^{-6}.$$

With 20 plants, it is virtually impossible that our plants would so thoroughly deceive us by chance.

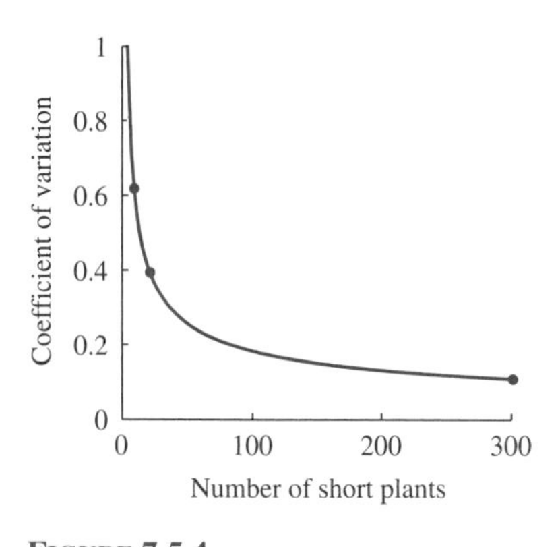

FIGURE 7.5.4

Coefficient of variation as a function of number of offspring

The coefficient of variation of the number of short plants N out of n offspring is

$$\begin{aligned}CV &= \frac{\sigma_N}{\bar{N}}\\ &= \frac{\sqrt{0.25 \cdot 0.75n}}{0.25n}\\ &= \sqrt{\frac{3}{n}}\end{aligned}$$

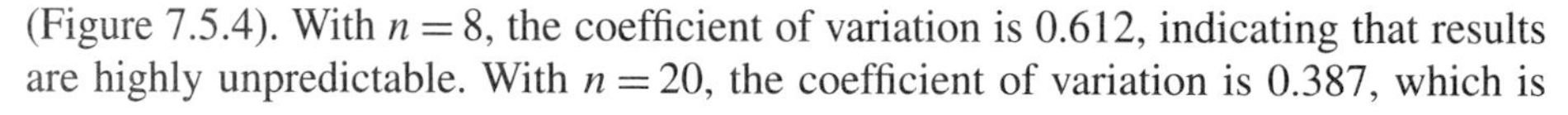

(Figure 7.5.4). With $n = 8$, the coefficient of variation is 0.612, indicating that results are highly unpredictable. With $n = 20$, the coefficient of variation is 0.387, which is smaller but still relatively large. The coefficient of variation does not reach the relatively small value of 0.1 until $n = 300$.

7.5.2 Application to Markov Chains

The binomial probability distribution can be applied to the Markov chain model of a molecule hopping in and out of a cell (Section 6.2.2). If 50 molecules inside a cell leave independently and with same probability, then the number left inside fulfills the conditions for the binomial model.

Example 7.5.5 Applying the Binomial Distribution to a Set of Molecules

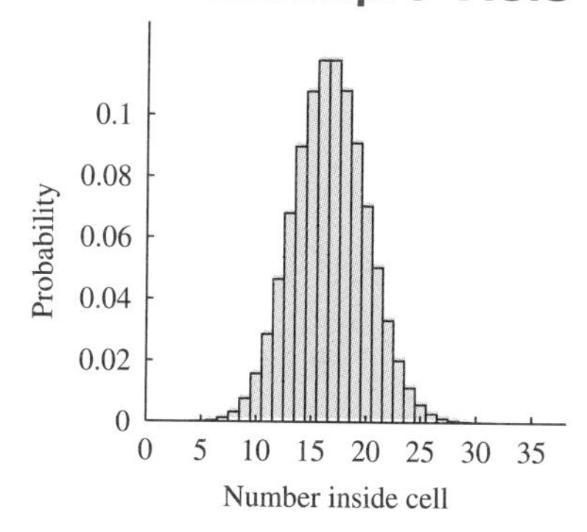

FIGURE 7.5.5

Number of molecules inside a cell

After a long time, we found that a given molecule is inside the cell with probability 1/3 (Figure 7.5.5). Let the random variable M denote the number of molecules in the cell. Then

$$E(M) = np = 50 \cdot \frac{1}{3} = 16.67.$$

The mode of this random variable is the smallest integer greater than $np - 1 + p = 17.0$ or $M = 17$ (Theorem 7.15). The probability that $M = 17$ is

$$\Pr(M = 17) = b\left(17; 50, \frac{1}{3}\right) = 0.1178,$$

which was evaluated with the aid of a computer.

Example 7.5.6 Applying the Binomial Distribution to Replicated Experiments

Suppose we replicate the experiment in Example 7.5.5 20 times and count up how many cells have exactly 17 molecules inside. As long as the experimental conditions for the replicates match and the replicates are independent, the number R matching the mode will have a binomial distribution with $n = 20$ and $p = 0.1178$. The distribution is plotted in Figure 7.5.6. We expect rather few replicates to exactly match the most common result.

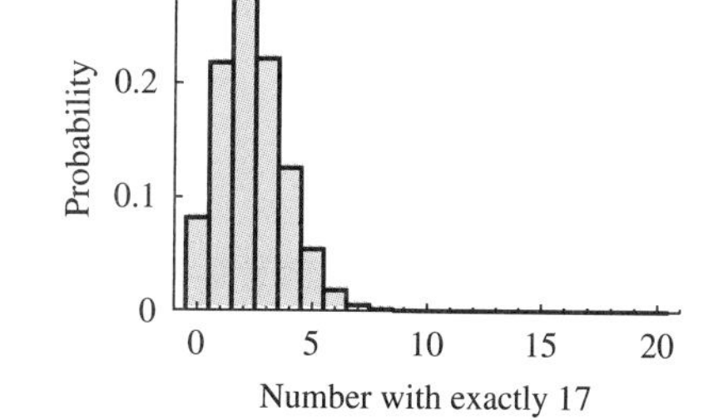

FIGURE 7.5.6

Number of experiments out of 20 with exactly 17 molecules inside a cell

Example 7.5.7 Finding a Cumulative Probability

In the situation in Example 7.5.5, what is the probability that fewer than half the expectation, 8 or fewer, are inside in a given experiment? This is the **sum** of the probabilities for each value less than or equal to 8, or

$$\Pr(M \leq 8) = \sum_{i=0}^{8} b\left(i; 50, \frac{1}{3}\right).$$

The sum can be evaluated term by term on a computer for a total of 0.0051. Only about one in 200 experiments would give a result this extreme.

This calculation would be easy if we had a formula for the **cumulative distribution**.

Definition 7.9 The cumulative distribution for the binomial distribution, written $B(k; n, p)$, is equal to

$$B(k; n, p) = \sum_{i=0}^{k} b(i; n, p).$$

Unfortunately, there is no simple formula for the cumulative distribution associated with the binomial probability distribution. Computers or tables must be used to compute its value. There is thus no convenient way to find the median or other percentiles for a binomial distribution (except when we can use the normal approximation derived in Subsection 7.9.2).

Example 7.5.8 Finding the Cumulative Probability of Exceeding a Given Value

The number inside would be less than half the mean in only one in 200 experiments. How likely is it that more than twice the mean number are inside? We wish to compute $\Pr(M \geq 34)$, because the mean is 16.7. To convert into the form of a cumulative distribution, we write

$$\Pr(M \geq 34) = 1 - \Pr(M \leq 33).$$

In terms of the cumulative distribution, then,

$$\begin{aligned} \Pr(M \geq 34) &= 1 - \Pr(M \leq 33) \\ &= 1 - B\left(33; 50, \frac{1}{3}\right) \\ &= 1.0 - 0.999999422 = 5.78 \times 10^{-7}. \end{aligned}$$

The probability that the molecules will distribute themselves this differently from the expectation (and the mode) is extremely small.

These applications give an indication of how the binomial distribution is used in statistics. If we observe an outcome that is astronomically unlikely for a given model (such as 15 out of 20 offspring being short or 34 or more molecules being inside the cell), it is a good bet that something is wrong with our model. When probabilities are small but not astronomical, knowing when to throw out the model requires more delicate judgment. Furthermore, if we look at many experiments and find that the results differ from the binomial, we question the scientific assumptions of the model.

7.5.3 Applications to Diffusion

Suppose 100 molecules begin inside the cell. Each has a 10% chance of leaving each minute and cannot return. What is the probability that exactly 90 molecules remain after 1 min or that fewer than ten remain after 20 min? What is the mean number inside as a function of t? How spread out is the distribution at different times?

Identical molecules diffusing out of a cell satisfy the conditions for the binomial model as long as they behave independently. Both properties, that the molecules are identical and that they are independent, encode different biological assumptions. If the molecules had different configurations when the experiment began, the assumption that molecules are identical would not be valid. Similarly, if the molecules interacted, the assumption of independence would not be valid.

Example 7.5.9 Finding the Parameters of a Binomial Distribution Describing Diffusing Molecules

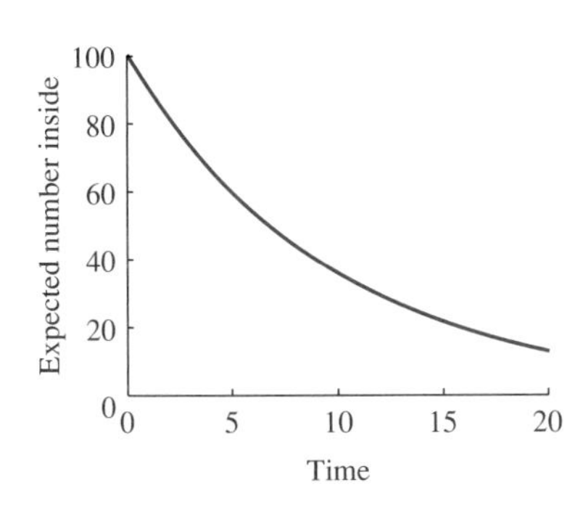

FIGURE 7.5.7

Expected number inside as a function of time

Suppose that 100 molecules follow the assumptions required for a binomial distribution. We must find the probability p_t that a given molecule is inside at time t. This probability follows the discrete-time dynamical system

$$p_{t+1} = 0.9p_t,$$

with solution

$$p_t = 0.9^t$$

(Equation 6.2.2). Let N_t be the random variable measuring the number of molecules inside at time t. Then N_t has a binomial distribution with $n = 100$ and $p = p_t$, so

$$\Pr(N_t = k) = b(k;\ 100,\ p_t).$$

Example 7.5.10 The Expected Number of Molecules Inside

The expected number of molecules in the cell at time t is

$$E(N_t) = np_t = 100(0.9^t)$$

(Figure 7.5.7).

Starting with 100 molecules, a mean of 90 are inside after the first minute, 81 after the second minute, and so on.

Example 7.5.11 The Probability That the Actual Number Matches the Expected Number

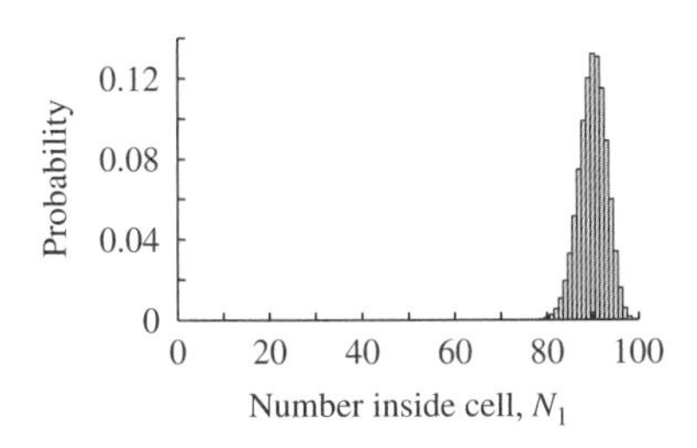

FIGURE 7.5.8

Probability distribution for 100 molecules at $t = 1$

The probability that exactly 90 molecules remain after 1 min is

$$\begin{aligned}\Pr(N_1 = 90) &= b(90;\ 100,\ 0.9)\\ &= \binom{100}{90}(0.9)^{90}(0.1)^{10}\\ &= 0.1318.\end{aligned}$$

The measured number will match the mean number only approximately 13% of the time. If we ran this experiment 100 times, we would observe exactly 90 molecules about 13 times. The average number of molecules, however, would be close to 90 (Figure 7.5.8).

Example 7.5.12 Probabilities After a Long Time

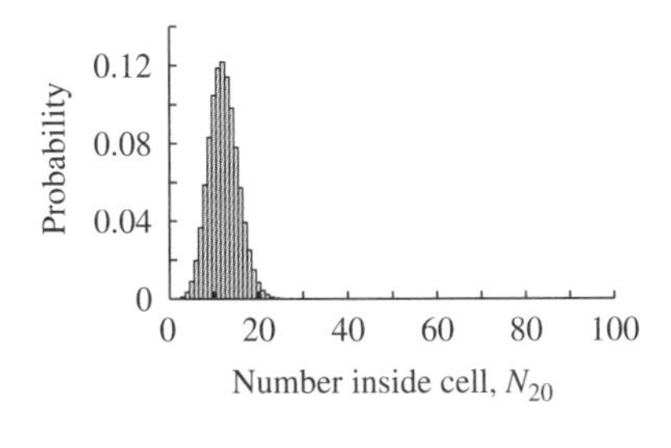

FIGURE 7.5.9

Probability distribution for 100 molecules at $t = 20$

The probability that 10 or fewer molecules remain after 20 min (Figure 7.5.9) must be found with the cumulative distribution as

$$\begin{aligned}\Pr(N_{20} \le 10) &= B(10;\ 100,\ p_{20})\\ &= \sum_{i=0}^{10} b(i;\ 100,\ 0.1216)\\ &= \sum_{i=0}^{10} \binom{100}{i} 0.1216^i (1 - 0.1216)^{100-i}\\ &= 0.3165.\end{aligned}$$

Again, calculations of this sort would be impossible without a computer.

Example 7.5.13 The Variance of the Number of Molecules as a Function of Time

The variance at time t is

$$\mathrm{Var}(N_t) = np_t(1 - p_t) = 100(0.9)^t(1 - 0.9^t)$$

(Figure 7.5.10).

When does the variance take on its maximum? Let

$$v(t) = 100(0.9)^t(1 - 0.9^t).$$

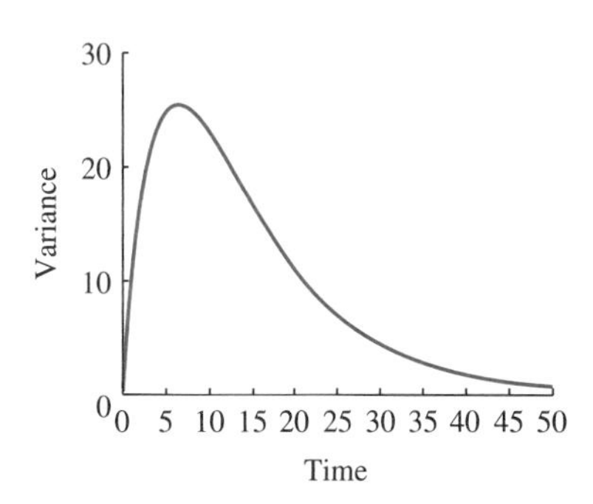

FIGURE 7.5.10

Variance of molecule number as a function of time

We can see that $v(0) = 0$, and $\lim_{t\to\infty} v(t) = 0$ because the function involves raising the number 0.9, which is less than 1.0, to the power t. Then

$$\begin{aligned}
\frac{dv}{dt} &= 100\left[\frac{d0.9^t}{dt}(1 - 0.9^t) - 0.9^t\frac{d0.9^t}{dt}\right] && \text{the constant product and product rules} \\
&= 100\left[\ln(0.9)0.9^t(1 - 0.9^t) - \ln(0.9)0.9^t \cdot 0.9^t\right] && \text{the derivative of an exponential function (Section 2.8)} \\
&= 100\ln(0.9)0.9^t\left(1 - 2 \cdot 0.9^t\right). && \text{factor}
\end{aligned}$$

This derivative is equal to zero when

$$\begin{aligned}
2 \cdot 0.9^t &= 1 \\
0.9^t &= \frac{1}{2} \\
t\ln(0.9) &= \ln\left(\frac{1}{2}\right) \\
t &= \frac{\ln(\frac{1}{2})}{\ln(0.9)} = 6.57.
\end{aligned}$$

The variance takes on a maximum at around $t = 6.6$ min.

Example 7.5.14 The Coefficient of Variation as a Function of Time

How predictable is this process? Does the decreasing variance for larger values of t mean that the number inside becomes more predictable after $t = 6.5$? The coefficient of variation provides a better measure of predictability (Figure 7.5.10). In this case, the coefficient of variation is

$$\begin{aligned}
\mathrm{CV}(N_t) &= \frac{\sqrt{\mathrm{Var}(N_t)}}{\mathrm{E}(N_t)} \\
&= \frac{\sqrt{100(0.9)^t(1 - 0.9^t)}}{100(0.9^t)}
\end{aligned}$$

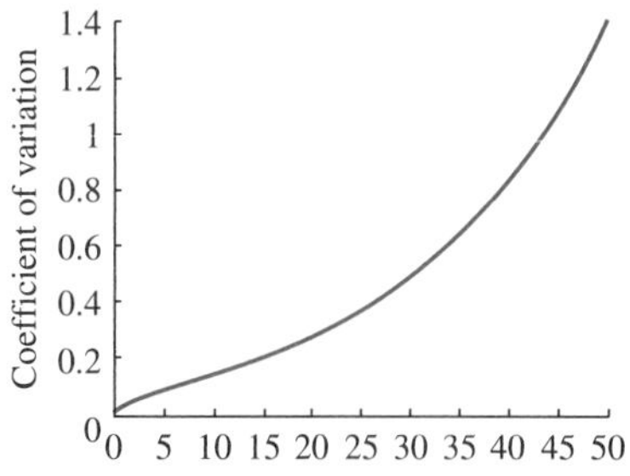

(Figure 7.5.11). Although the variance increases and then decreases, the coefficient of variation steadily increases. As fewer and fewer molecules remain in the cell, the **relative** uncertainty about the number of molecules increases. The fewer molecules that remain, the less predictable the process.

FIGURE 7.5.11

The coefficient of variation of molecule number as a function of time

Summary We have used the binomial distribution to study problems in genetics and diffusion, computing the probability that a particular number of offspring are short or that a particular number of molecules remain inside a cell. The binomial distribution has **cumulative distribution** $B(k; n, p)$, but there is no convenient formula for this quantity. For a process describing the number of molecules remaining in a cell, the expected number of molecules in a cell decreases, the variance first increases and then decreases, and the coefficient of variation increases steadily.

7.5 Exercises

Mathematical Techniques

1–4 ▪ Suppose that the allele **A** for height is dominant, meaning that plants with genotypes **AA** and **Aa** are tall, while those with genotype **aa** are short. If an **Aa** plant is crossed with another **Aa** plant, 3/4 of the offspring should be tall. Assuming that the other conditions for the binomial distribution are met, find the probabilities of the following.

1. Exactly three out of four offspring are tall.
2. Exactly six out of eight offspring are tall.
3. One or fewer out of four offspring are tall.
4. Two or fewer out of eight offspring are tall.

5–8 ▪ Compute the coefficient of variation of the binomial distribution in the following cases.

5. $p = 0.5$, $n = 100$.
6. $p = 0.5$, $n = 25$. Why is the coefficient of variation larger than in Exercise 5?
7. $p = 0.9$, $n = 25$.
8. $p = 0.1$, $n = 25$. Why is the coefficient of variation larger than Exercise 7?

9–12 ▪ When there are more than two outcomes of a trial, the distribution of all possibilities is described by the **multinomial distribution.** Consider an additive pair of alleles **A** and **a**, where an offspring of a cross between two **Aa** individuals is tall (with genotype **AA**) with probability 0.25, is intermediate (with genotype **Aa**) with probability 0.5, and is short (with genotype **aa**) with probability 0.25. Count up all possible ways for the following to happen and find the associated probabilities.

9. Out of three offspring, one is tall, one is intermediate, and one is short.
10. Out of three offspring, one is tall, and two are intermediate.
11. Out of four offspring, two are tall, and two are intermediate.
12. Out of four offspring, one is tall, two are intermediate, and one is short.

13–16 ▪ There is a formula for the multinomial distribution (Exercises 9–12) describing probabilities when there are more than two outcomes of a trial. Suppose there are three possible outcomes of each trial, numbered 1 through 3, with probabilities p_1, p_2, and p_3. If there are n trials, the probability that there are exactly k_1 outcomes of the first type, k_2 outcomes of the second type, and k_3 outcomes of the third type is

$$M(k_1, k_2, k_3; n, p_1, p_2, p_3) = \frac{n!}{k_1!k_2!k_3!} p_1^{k_1} p_2^{k_2} p_3^{k_3}.$$

Consider the genetics described in Exercises 9–12 involving an additive pair of alleles **A** and **a**. Use the formula for the multinomial distribution to compute the following probabilities and compare with the earlier result.

13. Out of three offspring, one is tall, one is intermediate, and one is short (as in Exercise 9).
14. Out of three offspring, one is tall, and two are intermediate (as in Exercise 10).
15. Out of four offspring, two are tall, and two are intermediate (as in Exercise 11).
16. Out of four offspring, one is tall, two are intermediate, and one is short (as in Exercise 12).

Applications

17–20 ▪ Suppose that the alleles **A** and **a** for height are additive, meaning that plants with genotype **AA** are tall, plants with genotype **Aa** are intermediate, and those with genotype **aa** are short. If an **Aa** plant is crossed with another **Aa** plant, 1/4 of the offspring should be tall, 1/2 should be intermediate, and 1/4 should be short. Assuming that the other conditions for the binomial distribution are met, find the probabilities of the following if a cross produces six offspring.

17. Find the expectation and the mode of the number of tall offspring.
18. Find the expectation and the mode of the number of intermediate offspring.
19. Find the probability that the number of tall offspring is less than or equal to the mode.
20. Find the probability that the number of intermediate offspring is less than or equal to the mode.

21–26 ▪ Consider two sets of ten islands. In the first set, each island has a 0.2 chance of switching from empty to occupied, and a 0.1 chance of switching from occupied to empty. The equilibrium fraction occupied is 2/3. In the second set of ten islands, all are occupied if the weather is good (probability 2/3) and all empty if the weather is bad (probability 1/3). Compute the following for both sets of islands.

21. The probability that exactly seven out of ten are occupied.
22. The probability that all ten are occupied.
23. Find the mean number of islands occupied.
24. Find the variance of the number of islands occupied.
25. Find the mode of the number of islands occupied.
26. Sketch the probability distribution for each set of islands. Why does the second set fail to follow the binomial distribution?

27–28 ▪ Find the probabilities of the following events.

27. Consider the mutant genes described in Section 6.2, Exercise 27, where a wild type gene has a 1.0% chance of mutating each time a cell divides and a mutant gene has a 1.0% chance of reverting to wild type. Suppose that four genes start out normal. Find the probability that there are two or more mutants after one division, after two divisions, and after a long time.
28. Consider the lemmings described in Section 6.2, Exercise 28, where a lemming has a probability 0.2 of jumping off the cliff

each hour and a probability 0.1 of crawling back up. Suppose that five lemmings start at top. Find the probability that more than half of the lemmings are at the bottom after 1 h, after 2 h, and after a long time.

29–32 ▪ Starting with five molecules, each leaving with probability 0.2 per minute, compute and graph the probability distribution describing the number remaining at the following times.

29. 1 min.

30. 2 min.

31. 5 min.

32. 10 min.

33–34 ▪ Starting with five molecules, each leaving with probability 0.2/min never to return, find and graph the following probabilities as functions of time.

33. Exactly one remains. At what time is this probability a maximum?

34. Exactly two remain. At what time is this probability a maximum?

35–36 ▪ Suppose that ten independent experiments are run, in which five molecules begin inside a cell and leave with probability 0.2/min and never return.

35. Using the results in Exercise 29, find the probability that exactly three out of ten such experiments have exactly four molecules after 1 min.

36. Using the results in Exercise 31, find the probability that exactly five out of ten such experiments have exactly one molecule after 5 min.

37–40 ▪ 40 molecules begin inside a cell. Each leaves independently with probability 0.2/min.

37. Find the expected number remaining inside as a function of time.

38. Find the mode at $t = 1$, $t = 2$, and $t = 3$. Indicate whether the mode is equal to, greater than, or less than the mean.

39. Compute the variance. At what time is it a maximum?

40. Find the coefficient of variation of the number remaining inside as a function of time. Is it an increasing or a decreasing function?

41–44 ▪ Example 6.1.5 from Section 6.1, which illustrates stochastic immigration, was generated by adding two individuals with probability 0.5 and zero individuals with probability 0.5 for 100 generations. The results in the figure show final populations of 106 and 96.

41. How can these results be described in terms of the binomial distribution?

42. What is the expected number after 100 generations?

43. What is the variance after 100 generations?

44. Write the probability of exactly 106 in terms of the binomial distribution.

45–48 ▪ Unbeknownst to the experimenter, a cell contains two different types of molecule, one which is inside with probability p_1 and the other which is inside with probability p_2. Suppose there are two of each type of molecule.

45. Suppose $p_1 = 0$ and $p_2 = 1$. Find and graph the probability distribution for the total number inside. Find the expectation and the variance.

46. Suppose $p_1 = 0.6$ and $p_2 = 0.8$. Find and graph the probability distribution for the total number inside. Find the expectation and the variance (this can be written as the sum of two binomial random variables).

47. Suppose $p_1 = 0.25$ and $p_2 = 0.75$. Find and graph the probability distribution for the total number inside. Find the expectation and the variance (this can be written as the sum of two binomial random variables).

48. It turns out that all four molecules are different and that $p_1 = 0$, $p_2 = 0.25$, $p_3 = 0.75$, and $p_4 = 1$. Find and graph the probability distribution for the total number inside. Find the expectation and the variance.

Computer Exercises

49. Simulate groups of 8 plants, each of which has a 0.25 chance of being tall. How long does it take until you get half tall plants? If you can automate the process, try the same with groups of 20 plants.

50. The probability p_t that a molecule is inside a cell after t time steps is

$$p_t = (0.9)^t.$$

Suppose 100 molecules start out in a cell. Plot the probabilities that 50 molecules are inside at time t and that 10 are inside at time t as functions of t on a single graph with reasonable axes. Explain the shape of the curves. Is it more likely there are exactly 50 or exactly 10 molecules at time 12? What is a more likely number of molecules at this time? Compute the probability that exactly 10 molecules are inside at time 20 and indicate this point on your graph. Do the same for the probability that exactly 50 molecules are inside at time 8. Is the area under each curve equal to 1? Why or why not?

51. Consider a situation where two types of molecule leave a cell with different probabilities during each minute. The first leaves with probability q_1 and the second leaves with probability q_2. Suppose we begin with n_1 of the first type, n_2 of the second type, and a total of $n = n_1 + n_2$. If we did not know the difference, the average probability is

$$q = q_1\frac{n_1}{n} + q_2\frac{n_2}{n}.$$

Simulate the number of molecules of each type that remain inside after 5 min, and then compare with the number out of n that would remain if they left with the average probability q in each of the following cases. How important is it to distinguish among types of molecule?

a. $n_1 = n_2 = 50$, $q_1 = 0.1$, $q_2 = 0.9$.

b. $n_1 = 90$, $n_2 = 10$, $q_1 = 0.1$, $q_2 = 0.9$.

c. $n_1 = 10$, $n_2 = 90$, $q_1 = 0.1$, $q_2 = 0.9$.

d. $n_1 = n_2 = 50$, $q_1 = 0.4$, $q_2 = 0.6$.

e. $n_1 = 90$, $n_2 = 10$, $q_1 = 0.4$, $q_2 = 0.6$.

f. $n_1 = 10$, $n_2 = 90$, $q_1 = 0.4$, $q_2 = 0.6$.

7.6 Waiting Times: Geometric and Exponential Distributions

The binomial distribution gives the probability that a particular **number** of molecules remain inside a cell at a given time. The **time** that a molecule leaves the cell is a different random variable, described by the **geometric distribution** in the discrete case and the **exponential probability density function** in the continuous case. We derive the formulas and properties of these two important distributions.

7.6.1 The Geometric Distribution

Suppose the position of a molecule is checked once per minute. It leaves a cell with probability q each minute. What is the probability that a molecule leaves at time t? The probability p_t that the molecule is still inside the cell obeys the discrete-time dynamical system

$$p_{t+1} = (1-q)p_t$$

(Equation 6.2.2). For the molecule to be inside at time $t+1$, it must be inside at time t (probability p_t) and remain inside for one time step (probability q). This discrete-time dynamical system has the solution

$$p_t = (1-q)^t. \quad \textbf{(7.6.1)}$$

This matches the solution describing a bacterial population, but with the per capita production r replaced by $1-q$ (Subsection 1.7.1).

Let T be the random variable representing the time at which the molecule leaves, and let

$$g_t = \Pr(T = t)$$

represent the probability that the molecule left during minute t (Figure 7.6.1). To leave at time $T = t$, the molecule must have remained inside for the first $t-1$ minutes and then left during the next minute, or

$$\begin{aligned}
g_t &= \Pr(\text{molecule left at time } t) && \text{definition of } g_t \\
&= \Pr(\text{inside at time } t-1)\times && \text{break up with definition} \\
&\quad\ \Pr(\text{left at } t \text{ conditional on in at } t-1) && \text{of conditional probability} \\
&= q p_{t-1} && \text{probability inside at } t-1 \text{ is } p_{t-1} \\
&= q(1-q)^{t-1}. && p_{t-1} = (1-q)^{t-1} \text{ from Equation 7.6.1}
\end{aligned}$$

The random variable T is said to follow a **geometric distribution** with parameter q. The range of the random variable T is all positive integers (not including 0).

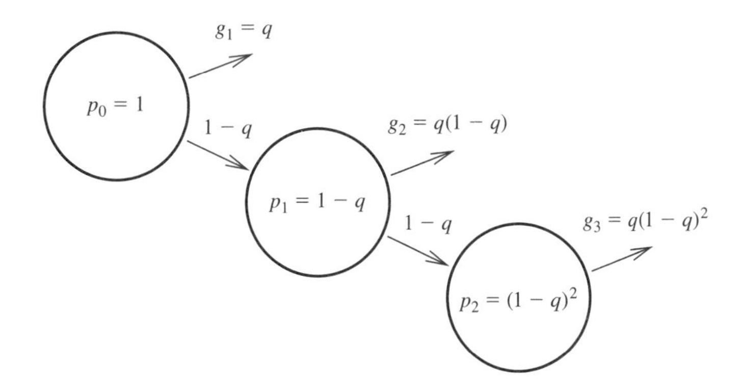

FIGURE 7.6.1
The process underlying the geometric distribution

Example 7.6.1 The Geometric Distribution with $q = 0.1$

If $q = 0.1$, the probability of remaining inside for one step is $1 - q = 0.9$. The probability the molecule is inside at time t is given by $p_t = 0.9^t$. The first few values are

$$p_0 = 1$$
$$p_1 = 0.9$$
$$p_2 = 0.9^2 = 0.81$$
$$p_3 = 0.9^3 = 0.729$$

and so forth. These probabilities decrease toward 0 at t increases (Figure 7.6.2). The probability that the molecule leaves exactly at time t is given by $g_t = q(1 - q)^{t-1}$. The first few values are

$$g_1 = q = 0.1$$
$$g_2 = 0.1 \cdot 0.9 = 0.09$$
$$g_3 = 0.1 \cdot 0.9^2 = 0.081$$
$$g_4 = 0.1 \cdot 0.9^3 = 0.0729.$$

These values also decrease toward 0 as t gets larger. The histograms show only the beginning of the distribution because larger values of T are so improbable that the bars do not show up on the graph. Because g_t describes a probability distribution, the values add up to 1.

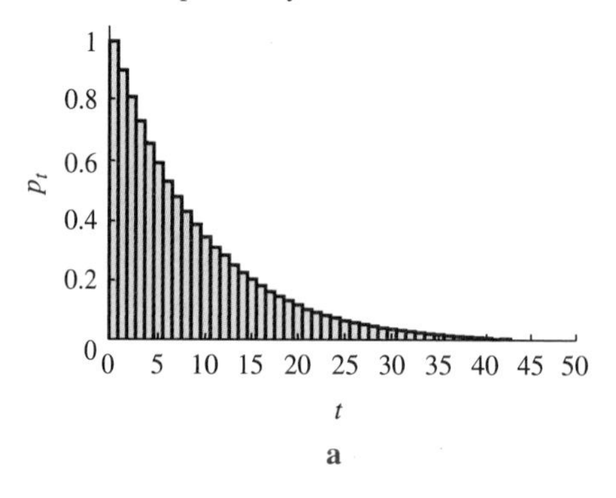

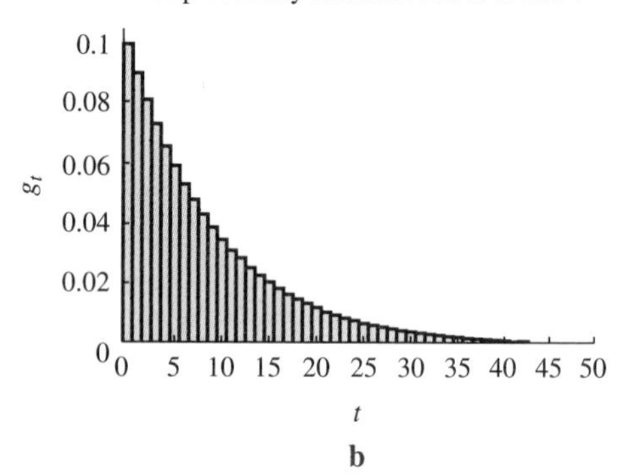

FIGURE 7.6.2
The geometric probability distribution with $q = 0.1$

Example 7.6.2 The Geometric Distribution with $q = 0.5$

If $q = 0.5$, the probability of remaining inside for one step is $1 - q = 0.5$. The probability the molecule is inside at time t is given by $p_t = 0.5^t$. The first few values are

$$p_0 = 1$$
$$p_1 = 0.5$$
$$p_2 = 0.5^2 = 0.25$$
$$p_3 = 0.5^3 = 0.125$$

and so forth (Figure 7.6.3). The probability that the molecule leaves exactly at time t is given by $g_t = q(1 - q)^{t-1}$. The first few values are

$$g_1 = q = 0.5$$
$$g_2 = 0.5 \cdot 0.5 = 0.25$$
$$g_3 = 0.5 \cdot 0.5^2 = 0.125$$
$$g_4 = 0.5 \cdot 0.5^3 = 0.0625.$$

The cumulative distribution G_t gives the probability that the molecule left at or before time t. It can be found from p_t, the probability that the molecule remains inside

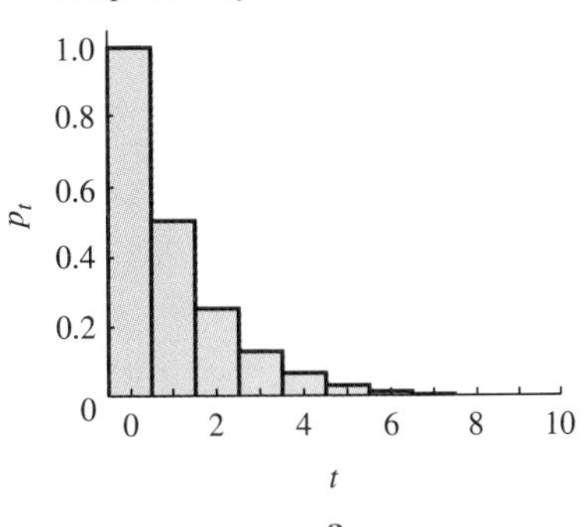

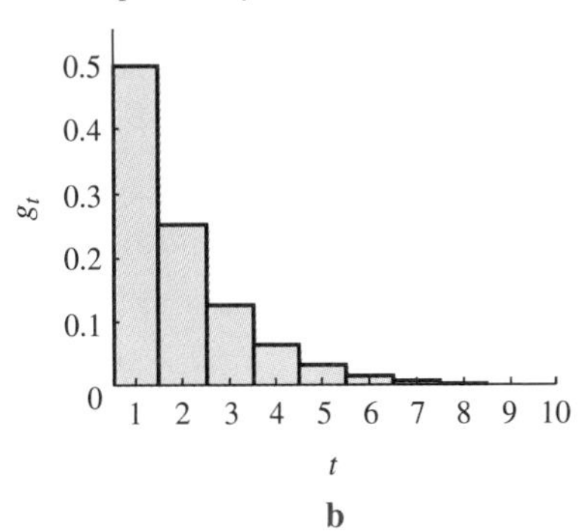

FIGURE 7.6.3
The geometric probability distribution with $q = 0.5$

at time t. The molecule left at or before time t precisely if it is not inside, or

$$\begin{aligned} G_t &= \Pr(\text{left at or before time } t) \\ &= 1 - \Pr(\text{still in at time } t) \\ &= 1 - p_t \\ &= 1 - (1-q)^t. \end{aligned}$$

The cumulative distribution increases to 1, as it must, because

$$\lim_{t\to\infty} G_t = \lim_{t\to\infty} 1 - (1-q)^t = 1$$

(unless $q = 0$, in which case the molecule never leaves).

Example 7.6.3 The Cumulative Distribution If $q = 0.1$

If $q = 0.1$, the cumulative distribution is $G_t = 1 - 0.9^t$ (Figure 7.6.4). The first few values are

$$\begin{aligned} G_1 &= 1 - 0.9 = 0.1 \\ G_2 &= 1 - 0.9^2 = 0.19 \\ G_3 &= 1 - 0.9^3 = 0.271 \\ G_4 &= 1 - 0.9^3 = 0.3439. \end{aligned}$$

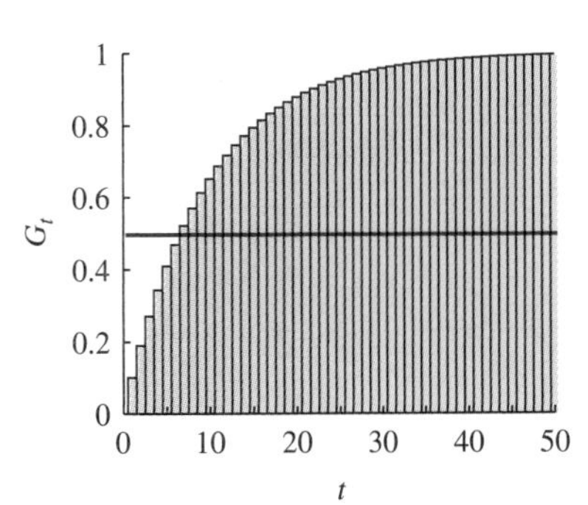

FIGURE 7.6.4
The cumulative distribution for the geometric distribution with $q = 0.1$

From the probability distribution g_t, we can compute the descriptive statistics that summarize the center of this distribution: the mode, the expectation, and the median. The **mode** of the geometric distribution is $T = 1$ even when q is small. The most likely single time for the molecule to leave is immediately.

The **expectation** of T is

$$E(T) = \frac{1}{q}.$$

This computation requires a difficult infinite sum or a clever trick (Exercises 27 and 30). If the value of q is small, meaning that the molecule leaves with low probability each minute, the mean time until departure is large. If the value of q is near 1, meaning that the molecule leaves with high probability each minute, the mean is near 1.

The **median** $\tilde{T}$ occurs where the cumulative distribution is equal to 0.5. Algebraically,

$$\begin{aligned} G_t &= 0.5 \\ 1 - (1-q)^{\tilde{T}} &= 0.5 \\ (1-q)^{\tilde{T}} &= 0.5 \\ \tilde{T}\ln(1-q) &= \ln(0.5) \\ \tilde{T} &= \frac{\ln(0.5)}{\ln(1-q)}. \end{aligned}$$

Example 7.6.4 Mean, Median, and Mode for the Geometric Distribution: $q = 0.1$

When $q = 0.1$, the mode occurs at $T = 1$, as always. The expectation is

$$E(T) = \frac{1}{0.1} = 10.0.$$

On average, we would wait ten time steps for this molecule to leave (Figure 7.6.5). The formula for the median gives

$$\tilde{T} = \frac{\ln(0.5)}{\ln(1 - 0.1)} = 6.57.$$

Because T takes on only integer values, the median, strictly speaking, lies between 6 and 7. The value is well below the expectation. In fact the probability that the value is less than the mean of $\bar{T} = 10$ is G_{10}, given by

$$G_{10} = 1 - 0.9^{10} = 0.65.$$

Almost 2/3 of observations are less than the mean. This is a case where only 1/3 of the measurements are "above average."

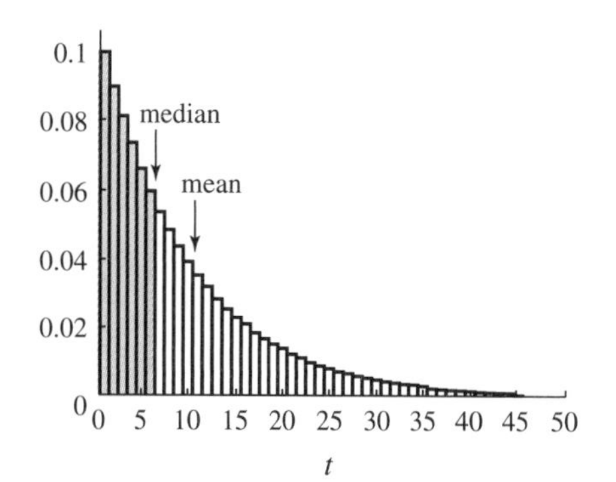

FIGURE 7.6.5

The mean and median of the geometric probability distribution with $q = 0.1$

The variance of a geometric random variable T is

$$\text{Var}(T) = \sigma_T^2 = \frac{1 - q}{q^2},$$

with the computation again requiring an infinite sum or a very clever trick (Exercises 31–34). The standard deviation is the square root of the variance, or

$$\sigma_T = \frac{\sqrt{1 - q}}{q}.$$

The coefficient of variation is the ratio of the standard deviation to the mean, or

$$\text{CV} = \frac{\sigma_T}{E(T)} = \frac{\frac{\sqrt{1-q}}{q}}{\frac{1}{q}} = \sqrt{1 - q}.$$

The coefficient of variation is small only when q is near 1 and all the molecules leave almost immediately.

Example 7.6.5 Statistics of Spread for the Geometric Distribution with $q = 0.5$

When $q = 0.5$,

$$\text{Var}(T) = \frac{1 - 0.5}{0.5^2} = 2.0$$

$$\sigma_T = \sqrt{2.0} = 1.414$$

$$\text{CV} = \frac{\sigma_T}{E(T)} = 0.707.$$

Example 7.6.6 Statistics of Spread for the Geometric Distribution with $q = 0.1$

When $q = 0.1$,

$$\mathrm{Var}(T) = \frac{1-0.1}{0.1^2} = 90$$
$$\sigma_T = \sqrt{90} = 9.49$$
$$\mathrm{CV} = \frac{\sigma_T}{\mathrm{E}(T)} = 0.949.$$

The large value of the coefficient of variation indicates that waiting is highly unpredictable, as experience with infrequent phone calls confirms.

We summarize these results in the following theorem.

Theorem 7.16 A random variable T follows the geometric distribution when it measures the time until the first success, where the probability of a success on each step is independent and equal to q on each time step. The probability distribution g_t and the cumulative distribution G_t are

$$g_t = q(1-q)^{t-1}$$
$$G_t = 1 - (1-q)^t.$$

The mean $\mathrm{E}(T)$, median $\tilde{T}$, and variance σ_T^2 are

$$\mathrm{E}(T) = \frac{1}{q}$$
$$\tilde{T} = \frac{\ln(0.5)}{\ln(1-q)}$$
$$\sigma_T^2 = \frac{1-q}{q^2}.$$

The geometric distribution can be applied in the same wide array of cases as the binomial distribution because the two result from the same process.

Example 7.6.7 Waiting Time Until an Experiment Exactly Matches a Desired Result

In Example 7.5.6, we studied a case where an experiment exactly matches the most probable result (in this case, of finding exactly 17 out of 50 molecules inside a cell) with a probability of $p = 0.1178$. If we repeated this experiment until we achieved that result, the number of replicates T is governed by a geometric distribution with $q = 0.1178$. Then

$$\mathrm{E}(T) = \frac{1}{q} = 8.489$$
$$\tilde{T} = \frac{\ln(0.5)}{\ln(1-q)} = 5.53$$
$$\sigma_T^2 = \frac{1-q}{q^2} = 63.57$$
$$\sigma_T = 7.973.$$

7.6.2 The Exponential Distribution

The exponential distribution is the continuous time version of the geometric distribution. This distribution describes a random variable measuring the **exact** time when a molecule leaves a cell. Finding the distribution requires writing and solving a differential equation.

Differential equations are described by **rates**, and the rate in this problem is the rate at which the molecule leaves. This **probabilistic rate** differs slightly from an ordinary rate. A molecule, after all, is either inside or outside at any given time and does not leak out the way a volume of fluid does. We define a probabilistic rate as a limit in a similar way, however.

Definition 7.10 If an event occurs at a probabilistic rate λ, the probability it occurs in the short time Δt is approximately $\lambda \Delta t$. More precisely, if $p(\Delta t)$ is the probability that an event occurs in time Δt, then $\lim_{\Delta t \to 0} \frac{p(\Delta t)}{\Delta t} = \lambda$.

Because this rate describes probability, which has no units or dimensions, a probabilistic rate has dimensions of 1/time.

Example 7.6.8 Application of the Definition of a Probabilistic Rate

Suppose an event occurs at a probabilistic rate of $\lambda = 10.0$/s. In the short time $\Delta t = 0.01$ s, the event occurs with probability $10 \cdot 0.01 = 0.1$.

Example 7.6.9 The Definition of a Probabilistic Rate Applies Only for Small Values of Δt

If an event occurs at the probabilistic rate of $\lambda = 10$/s, and we wish to know the probability that an event occurs during a time $\Delta t = 0.5$s, the definition gives a probability of $\lambda \Delta t = 10 \cdot 0.5 = 5.0$. This is larger than 1 and cannot represent a real probability.

Let $P(t)$ be the probability that the molecule is inside at time t. To write a differential equation, we must compute the probability that it is inside a short time Δt later. Much as in the derivation of the geometric distribution, a molecule is inside at $t + \Delta t$ if was inside at time t and did not leave in the next Δt, or

$$\begin{aligned}(\text{probability inside at } t + \Delta t) = (&\text{probability inside at } t) \times \\ &(\text{probability it left leave in time } \Delta t).\end{aligned}$$

If the molecule leaves at probabilistic rate λ, the definition tells us that

$$\text{probability did leave in time } \Delta t = \lambda \Delta t$$

for small values of Δt. Therefore,

$$\text{probability it did not leave in time } \Delta t = 1 - \lambda \Delta t.$$

Therefore, for small values of Δt,

$$P(t + \Delta t) = P(t)(1 - \lambda \Delta t).$$

This equation becomes exact as $\Delta t \to 0$. Directly taking the limit of both sides gives

$$\begin{aligned}
\lim_{\Delta t \to 0} P(t + \Delta t) &= \lim_{\Delta t \to 0} P(t)(1 - \lambda \Delta t) && \text{take limit of both sides} \\
P(t + 0) &= P(t)(1 - \lambda \cdot 0) && \text{both sides are continuous (Section 2.3), substitute } \Delta t = 0 \\
P(t) &= P(t). && \text{evaluate both sides}
\end{aligned}$$

This is true, but tells us nothing about the dynamics of $P(t)$.

Instead, we rearrange the equation by placing all the Δt's on the left-hand side.

$$\begin{aligned}
P(t + \Delta t) &= P(t)(1 - \lambda \Delta t) && \text{original equation} \\
P(t + \Delta t) &= P(t) - \lambda \Delta t P(t) && \text{multiply out right-hand side} \\
P(t + \Delta t) - P(t) &= -\lambda \Delta t P(t) && \text{subtract } P(t) \text{ from both sides} \\
\frac{P(t + \Delta t) - P(t)}{\Delta t} &= -\lambda P(t). && \text{divide both sides by } \Delta t
\end{aligned}$$

Upon rearranging, the left-hand side now matches the definition of the derivative (Subsection 2.1.3). Therefore,

$$\lim_{\Delta t \to 0} \frac{P(t+\Delta t) - P(t)}{\Delta t} = \frac{dP}{dt}.$$

Thus $P(t)$ obeys the differential equation

$$\frac{dP}{dt} = -\lambda P(t).$$

We have seen this differential equation before, when describing the decline of a population (Section 5.4). We found the solution of this differential equation with the method of separation of variables (Section 5.4) to be

$$P(t) = Ke^{-\lambda t}.$$

The constant K must be deduced from the initial condition. In this case, we know for certain that the molecule is inside at $t = 0$, so $P(0) = 1$. Therefore $P(0) = 1 = Ke^{-\lambda \cdot 0} = K \cdot 1 = K$. The constant $K = 1$; therefore,

$$P(t) = e^{-\lambda t}.$$

We defined $P(t)$ as the probability that the molecule has **not** left by time t. Finding the cumulative distribution function (c.d.f.) and the probability density function (p.d.f.) describing the time T when it leaves requires a few more steps. If the molecule has not left at some time t, then the time T at which it does leave must be greater than t, or $T > t$. Therefore,

$$P(t) = \Pr(T > t).$$

The c.d.f. $F(t)$, is defined by

$$F(t) = \Pr(T \leq t),$$

the probability that $T < t$. Therefore,

$$F(t) = 1 - P(t)k = 1 - e^{-\lambda t}.$$

The p.d.f. is the **derivative** of the c.d.f. (Subsection 6.6.4), so

$$f(t) = F'(t) = \frac{d}{dt}(1 - e^{-\lambda t}) = \lambda e^{-\lambda t}.$$

We summarize these results in a theorem.

Theorem 7.17 Suppose an event happens at a constant rate λ. The random variable T measuring the time until the first event has probability density function $f(t)$ and cumulative distribution function $F(t)$ with formulas

$$f(t) = \lambda e^{-\lambda t}$$
$$F(t) = 1 - e^{-\lambda t}.$$

Example 7.6.10 The Behavior of the Exponential Distribution with $\lambda = 0.1$

Suppose that a molecule leaves at rate $\lambda = 0.1$ (Figure 7.6.6). The p.d.f. would be

$$f(t) = 0.1e^{-0.1t}.$$

This is the p.d.f. we studied in Examples 6.7.7 and 6.7.11. The c.d.f. is

$$F(t) = 1 - e^{-0.1t},$$

which rises slowly from 0 to 1. Finally, $P(t)$, the probability the molecule is still inside, is

$$P(t) = e^{-0.1t},$$

which decreases slowly from 1.

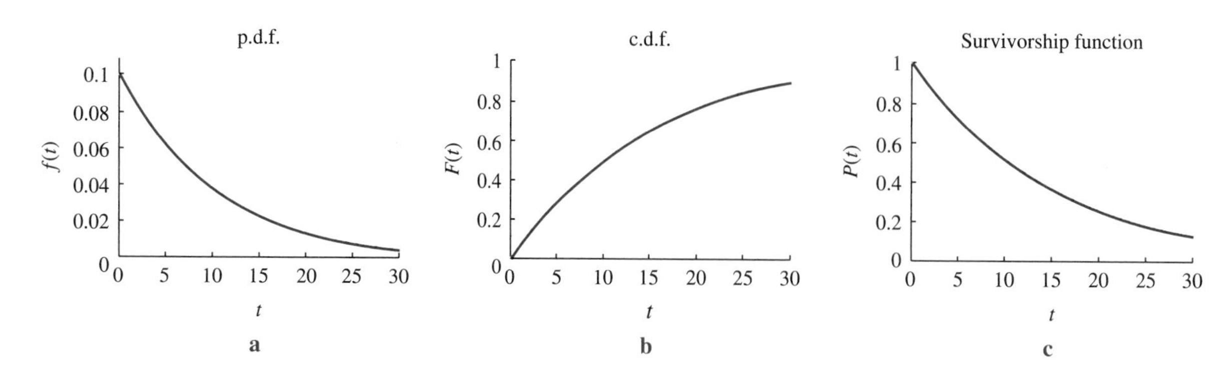

FIGURE 7.6.6
The functions describing the exponential distribution with $\lambda = 0.1$.

Example 7.6.11 Computing Probabilities Associated with the Exponential Distribution with $\lambda = 0.1$

To find probabilities using the p.d.f., we must integrate. For example, the probability that the molecule leaves between times 0 and 10 is

$$\begin{aligned}\Pr(0 \le T \le 10) &= \int_0^{10} 0.1e^{-0.1t}\,dt \\ &= -e^{-0.1t}\Big|_0^{10} \\ &= -e^{-1.0} + e^{-0.0} = 0.632\end{aligned}$$

(Figure 7.6.7). This matches the value we find with the cumulative distribution function, because

$$\begin{aligned}\Pr(0 \le T \le 10) &= \Pr(T \le 10) \\ &= F(10) \\ &= 1 - e^{-1.0} = 0.632.\end{aligned}$$

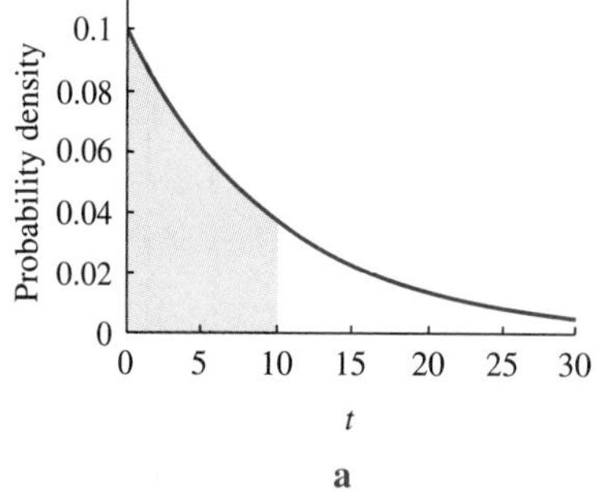

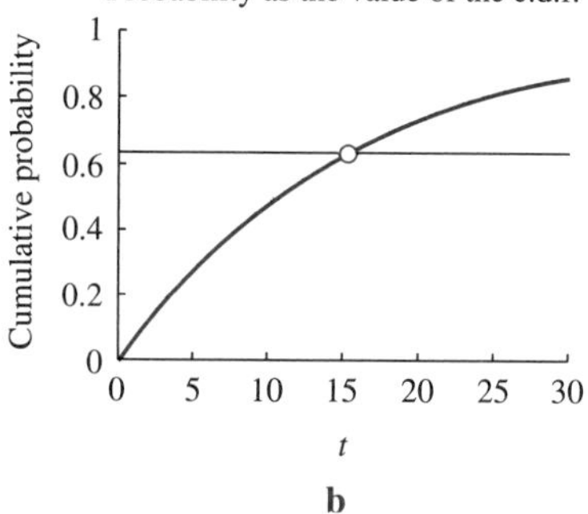

FIGURE 7.6.7
Computing probabilities with the p.d.f. and the c.d.f.

Example 7.6.12 The Behavior of the Exponential Distribution with $\lambda = 10.0$

With the faster rate $\lambda = 10.0$, the p.d.f. begins at a high value and drops off rapidly (Figure 7.6.8). The c.d.f. rapidly increases to 1, and $P(t)$ decreases rapidly from 1.

Using the p.d.f., we can find the basic statistics that describe this distribution.

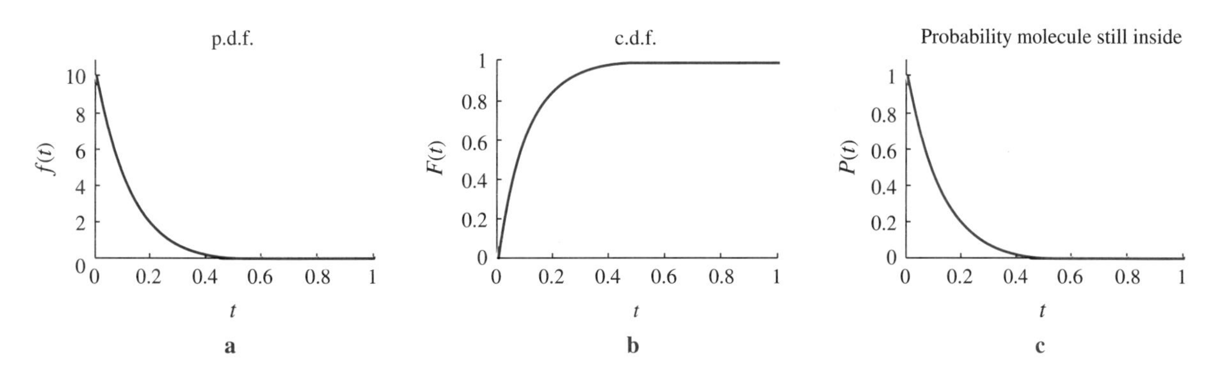

FIGURE 7.6.8
Probability density function, cumulative distribution function, and survivorship function with $\lambda = 10.0$.

Theorem 7.18 Suppose the random variable T follows an exponential distribution with parameter λ. Then

$$\mathrm{E}(T) = \frac{1}{\lambda}$$

$$\mathrm{Var}(T) = \frac{1}{\lambda^2}$$

$$\tilde{T} = \frac{\ln(2.0)}{\lambda}.$$

Proof: The expectation of T is

$$\mathrm{E}(T) = \int_0^\infty t f(t) dt = \int_0^\infty t\lambda e^{-\lambda t} dt = -\frac{\lambda t e^{-\lambda t} + e^{-\lambda t}}{\lambda}\Big|_0^\infty$$

$$= \lim_{s\to\infty} -\frac{\lambda s e^{-\lambda s} + e^{-\lambda s}}{+} \frac{1}{\lambda} = \frac{1}{\lambda}$$

where the integral was evaluated using integration by parts (Subsection 4.3.4, checked in Exercise 35). Integration by parts can again to be used to compute the variance of T (Exercise 36) as follows:

$$\mathrm{Var}(T) = \int_0^\infty t^2 f(t) dt - \mathrm{E}(T)^2$$

$$= -\frac{\lambda^2 t^2 e^{-\lambda t} + 2\lambda t e^{-\lambda t} + 2e^{-\lambda t}}{\lambda^2}\Big|_0^\infty - \frac{1}{\lambda^2}$$

$$= \frac{1}{\lambda^2}.$$

The median is found by solving

$F(t) = 1 - e^{-\lambda t} = 0.5$	definition of median
$e^{-\lambda t} = 0.5$	isolate t
$-\lambda t = \ln(0.5)$	take natural log of both sides
$t = -\dfrac{\ln(0.5)}{\lambda}$	solve for t
$t = \dfrac{\ln(2.0)}{\lambda}.$	use law 3 of logs to write $\ln(2.0) = -\ln(0.5)$

The standard deviation is

$$\sigma = \sqrt{\mathrm{Var}(T)} = \frac{1}{\lambda},$$

so,

$$\mathrm{CV} = \frac{\sigma_T}{\mathrm{E}(T)} = 1 \quad \mathbf{(7.6.2)}$$

for any value of the rate parameter λ.

Example 7.6.13 Descriptive Statistics of the Exponential Distribution with $\lambda = 0.1$

With the rate $\lambda = 0.1$,

$$\mathrm{E}(T) = \frac{1}{0.1} = 10.0$$
$$\mathrm{Var}(T) = \frac{1}{0.1^2} = 100.0$$
$$\tilde{T} = \frac{\ln(2.0)}{\lambda} = 6.93.$$

Because the standard deviation is equal to the mean, all we can say with confidence is that the molecule is sure to leave before too long. After 50 time units (five times larger than the mean), the value of the c.d.f. is

$$F(50) = 1 - e^{-0.1 \cdot 50.0} = 0.993.$$

The probability that the molecule has left is over 99%.

Example 7.6.14 Descriptive Statistics of the Exponential Distribution with $\lambda = 10.0$

With the high rate $\lambda = 10$,

$$\mathrm{E}(T) = \frac{1}{10.0} = 0.1$$
$$\mathrm{Var}(T) = \frac{1}{10.0^2} = 0.01$$
$$\tilde{T} = \frac{\ln(2.0)}{\lambda} = 0.0693.$$

The mean and median are small because the process is rapid.

For some species, the probability that an individual dies during a short interval of time is roughly independent of age. When the rate of death is constant, the lifespan, given by the time of death T, follows an exponential distribution. The probability of being alive, also known as the **survivorship**, is $P(t) = e^{-\lambda t}$. Such an animal is said to have **exponential survivorship**. Exponential survivorship occurs when death results not from aging but from a random event that could happen at any time.

Example 7.6.15 An Insect with Exponential Survivorship

Suppose the survivorship curve for some insect exactly follows the exponential function $S(t) = e^{-2.2t}$ where t is measured in years (Figure 7.6.9). This organism has exponential survivorship with mean $\frac{1}{2.2} = 0.454$ years. Only $S(1.0) = 0.111$, or about 11% of insects, would survive for 1 yr, and only $S(2.0) \approx 0.012$, or about 1.2%, would survive 2 yr.

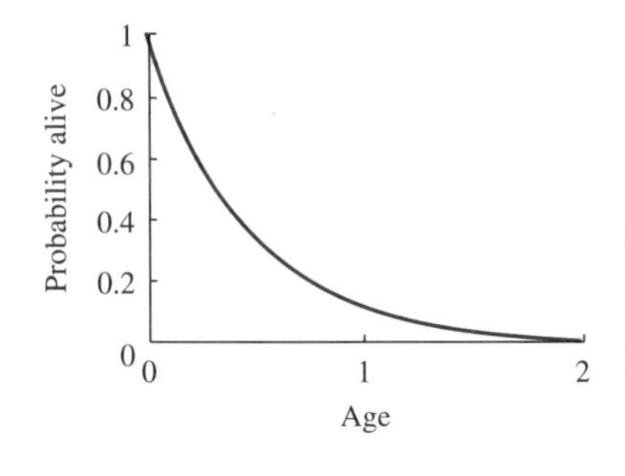

FIGURE 7.6.9 Survivorship function for insect with exponential survivorship

FIGURE 7.6.10 Survivorship function for an organism without exponential survivorship

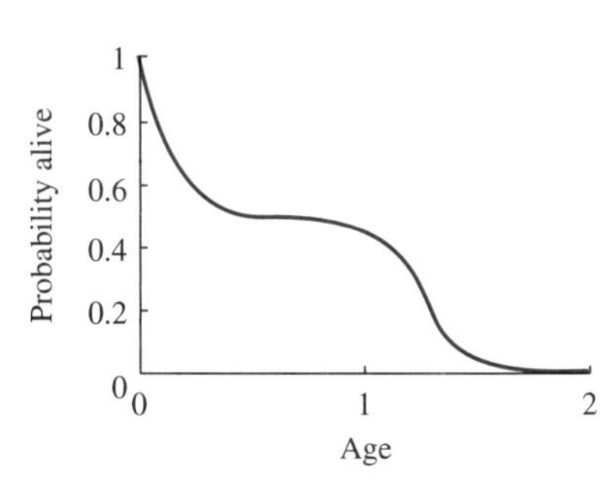

Example 7.6.16 An Organism Without Exponential Survivorship

Many organisms are more likely to die either when very young or when they are old. A graph of the survivorship function dips down quickly for young organisms, indicating many deaths, levels out, and then dips down quickly again (Figure 7.6.10).

Summary The time when a molecule leaves a cell is described by the **geometric distribution** in discrete time and the **exponential distribution** in continuous time. We computed the probabilities, means, and variances associated with these distributions. Each has a large variance, with the standard deviation being equal to the mean for the exponential distribution. The exponential distribution appears throughout biology for any random event that occurs at a constant rate, including mortality for many organisms, which are then said to have **exponential survivorship**.

7.6 Exercises

Mathematical Techniques

1–6 ▪ Compute the following probabilities. In each case, sketch the probability distribution and shade the associated area.

1. The probability that the first success is on the third trial if each trial has a probability 0.2 of success.
2. The probability that the first success is on the fifth trial if each trial has a probability 0.3 of success.
3. The probability that the first success occurs on or before the fourth trial if each has a probability 0.2 of success.
4. The probability that the first success occurs on or before the third trial if each has a probability 0.3 of success.
5. The probability that the first success occurs on or after the third trial if each has a probability 0.2 of success.
6. The probability that the first success occurs on or after the sixth trial if each trial has a probability 0.3 of success.

7–10 ▪ Compute the following probabilities. In each case, sketch the probability distribution function and shade the area associated with the question.

7. Events occur at a rate of 0.5/s. Find the probability that the first event occurs between times 1.0 and 2.0.
8. Events occur at a rate of 1.5/s. Find the probability that the first event occurs between times 0.2 and 1.0.
9. Events occur at rate 0.2/s. Find the probability that the first event occurs before $t = 1$ or after $t = 3$.
10. Events occur at rate 5.0/s. Find the probability that the first event occurs before $t = 0.1$ or after $t = 0.5$.

11–14 ▪ Find the mean, variance, standard deviation, coefficient of variation, and mode for a random variable T that follows a geometric distribution with the given probability of success.

11. The probability of success is $q = 0.2$.
12. The probability of success is $q = 0.3$.
13. The probability of success is $q = 0.7$.
14. The probability of success is $q = 0.9$.

15–18 ▪ Find the mean, variance, standard deviation, coefficient of variation, median, and mode for a random variable T that follows an exponential distribution with the given rate λ.

15. The rate is $\lambda = 0.2$.
16. The rate is $\lambda = 0.5$.
17. The rate is $\lambda = 1.5$.
18. The rate is $\lambda = 5.0$.

19–20 ▪ We can verify that the cumulative distribution for the geometric distribution is $G_t = 1 - (1 - q)^t$ with a mathematical trick.

19. Show that
$$\sum_{i=0}^{n} x^i = \frac{1 - x^{n+1}}{1 - x}$$
for any x by multiplying both sides by $1 - x$ and working out the algebra.
20. Find $\sum_{i=1}^{t} q(1 - q)^{t-1}$ (use $x = 1 - q$).

21–22 ▪ We found that the function $f(x) = \frac{1}{1 - x}$ is equal to the Taylor series
$$T_f(x) = 1 + x + x^2 + x^3 + x^4 + \ldots$$
when $0 < x < 1$ (Example 3.7.10).

21. Use this Taylor series to show that $\sum_{t=1}^{\infty} q(1 - q)^{t-1} = 1$ for any $0 < q < 1$.
22. Differentiate the Taylor series term by term and use it to derive the expectation of a geometric random variable.

23–26 ▪ We can find the exponential p.d.f. as the limit of the geometric distribution. Think of dividing time up into small units of length Δt.

23. If the probability of leaving during 1 min is q, what is the approximate probability of leaving during a short interval of time Δt?
24. Find the number of steps of duration Δt during an interval of length t.

25. Use the geometric distribution with the probability in Exercise 23 and the number of steps in Exercise 24 to find the approximate probability that a molecule is still inside at time t.

26. Find the limit of the result of Exercise 25, using the definition of the number e (Definition 2.8 in Subsection 2.8.1). What exponential distribution has this expression for its survivorship function?

27–30 ▪ The expectation of the geometric distribution can be found using a clever trick.

27. A molecule either leaves immediately or not. Find the probability that it leaves immediately and the time at which it leaves.

28. Find the probability that a molecule does not leave immediately. Show that the expected time to leave in this case is $1 + \mathrm{E}(T)$.

29. Multiply the values by the probabilities in Exercises 27 and 28 to find an expression for $\mathrm{E}(T)$.

30. Solve for $\mathrm{E}(T)$.

31–34 ▪ The variance of the geometric distribution can be found with a clever trick, much like that in Exercises 27–30. Define the random variable R, which is 1 with probability q and $1 + T$ with probability $1 - q$.

31. Why does R have the same probability distribution as T?

32. Show that $\mathrm{E}(R^2) = q + (1 - q)\mathrm{E}[(1 + T)^2]$.

33. Why is $\mathrm{E}(T^2) = q + (1 - q)\mathrm{E}[(1 + T)^2]$.

34. Solve for $\mathrm{E}(T^2)$ and use the computational formula to find $\mathrm{Var}(T)$.

35–36 ▪ Use integration by parts to compute the following statistics for an exponentially distributed random variable with parameter λ.

35. The expectation.

36. The variance.

Applications

37–42 ▪ Find the probabilities of the following events.

37. A certain highly mutable gene has a 1.0% chance of mutating each time a cell divides (as in Section 6.2, Exercise 25). Suppose that there are 15 cell divisions between each pair of generations. What is the chance that the gene first mutates during the last division? What is the chance that the gene mutates at some point during the first 15 divisions?

38. A herd of lemmings is standing at the top of a cliff, and each jumps off with probability 0.2 each hour (as in Section 6.2, Exercise 24). What is the probability that a lemming first jumps on the fifth hour? What is the probability that a particular lemming has not jumped by the end of third hour?

39. A molecule has a 5.0% chance of binding to an enzyme each second and remains permanently attached thereafter (as in Section 6.2, Exercise 25). Find the probability that it binds during the tenth second, and the probability that it binds on the tenth second or before.

40. In tropical regions, growing caterpillars can be eaten with probability 0.15 each day (as in Section 6.2, Exercise 26). What is the probability that a caterpillar is eaten on the fourth day? If a caterpillar takes 25 days to develop, what is the probability it survives?

41. A light bulb blows out with probability 0.01 each day. What is the probability that it blows out on the 50th day? What is the probability it blows out after more than 200 days?

42. 10% of some type of item are defective. Find the probability that the first defective item found is the fifth one inspected. What is the probability that a defective one is found if five are inspected?

43–46 ▪ A lab is screening to find mutant flies. 10% of the offspring of mutant flies have purple eyes, while none of the offspring of wild type flies have purple eyes. Suppose that purple-eyed flies are just as easy to catch as ordinary flies.

43. Use the binomial distribution to find the expected number of purple-eyed flies (from a mutant parent) after n have been checked. How many must be checked for the expectation to equal 1?

44. What is the expected number of mutant flies that must be inspected to find the first one with purple eyes? What is the expected number of purple-eyed flies that will be found if this many are checked? What is the probability that exactly one purple-eyed fly is found if this many are checked?

45. Suppose that half of the parent flies are known to be mutants. Of 20 offspring flies inspected, none have purple eyes. What is the conditional probability that the parents are mutants? If parents are discarded if all 20 offspring have normal eyes, what fraction of mutant parents will be discarded?

46. Suppose that only 5% of the parent flies are known to be mutants. Of 20 flies inspected, none are found to be have purple eyes. What is the conditional probability that the parents are mutants? If parents are discarded if all 20 offspring have normal eyes, what fraction of mutant parents will be discarded?

47–48 ▪ Inspecting for defective items until the first is found only works when all the items to be checked have the same probability of being defective. Suppose that items to be inspected had been deceptively sorted in the factory in order from best to worst. The first item is defective with probability 0.1, the second with probability 0.2, and so forth.

47. What is the probability that the third item inspected is the first defective one?

48. Find the whole probability distribution.

49–52 ▪ Find the probabilities of the following events. Find the associated p.d.f., c.d.f., and survivorship function, and give the expectation and variance.

49. A molecule leaves a cell at rate $\lambda = 0.3$/s. What is the probability it has left by the end of the third second? By the end of the first millisecond? Compare your last answer with the definition of rate.

50. A light bulb blows out at a rate of $\lambda = 0.001$/h. What is the probability that it blows out in less than 500 h? In less than 1 h? Compare your last answer with the definition of rate.

51. Phone calls arrive at a rate of $\lambda = 0.2$/h. What is the probability that there are no calls in 10 h? What is the probability that a call arrives during the 45 s you spend in the bathroom?

52. Raindrops hit a leaf at rate 7.3/min. What is the probability that the first one hits in less than 0.5 min?

53–54 ▪ A population of 100 bacteria are dying independently at rate $\lambda = 2.0$.

53. Find the probability that a given bacterium is alive at time t.

54. What distribution describes the population at time t? Find the expectation and variance of the number alive as functions of time.

55–58 ▪ Compute the following probabilities.

55. The gene in Exercise 37 has not mutated after ten divisions. What is the probability that it mutates by the 15th division?

56. The caterpillar in Exercise 40 is still alive after 10 days. What is the probability that it is eaten by day 25?

57. The molecule in Exercise 49 is still in the cell after 10 s. What is the probability that it leaves before the 13th second?

58. The light bulb in Exercise 50 is OK after 1000 h. What is the probability that it blows out before hour 1500?

59–60 ▪ You are trapped behind an annoyingly slow driver (a.s.d.) in a long no-passing zone. A second a.s.d. merges in front of the first, leaving you twice as trapped. To calm yourself, you attempt to guess which driver will exit first.

59. If the length of time that cars remain on the freeway is an exponential distribution, what is the probability that the new a.s.d. exits first? Under what conditions might this occur?

60. Under what assumptions would you expect the new a.s.d. to exit first? Under what assumptions would you expect the new a.s.d. to exit second?

61–66 ▪ Suppose that some species of insect dies faster as it gets older according to

$$\text{rate of death} = t,$$

where t is the age in years.

61. Write a differential equation describing the probability $P(t)$ the animal is still alive.

62. Solve the equation with separation of variables (Algorithm 5.2).

63. Graph the survivorship function.

64. Find the c.d.f. and p.d.f.

65. Find the probability it survives to age 1.

66. Find the probability it survives to age 2 conditional on surviving to age 1. Why might this value be different from that in Exercise 65?

Computer Exercises

67. The probability p_t that a molecule is inside a cell after t time steps is

$$p_t = (0.9)^t.$$

a. Simulate 100 such molecules. How many are inside at time 3?

b. Of the molecules that are inside at time 3, how many are still inside at time 6?

c. Do the two values match? Should they?

68. Use your computer to find the mean and variance of the exponential distribution with parameter λ. What fraction of measurements are greater than the mean? What fraction of measurements are greater than the mean plus one standard deviation? The mean plus two standard deviations? Why do these results seem so different from those in Subsection 6.9.3?

69. There is a generalization of the geometric distribution called, confusingly enough, the **negative binomial distribution**. Under the same assumptions as the geometric distribution, the random variable T_r is the number of failures that precede the rth success. The case $r = 1$ is the geometric distribution. In general, if the probability of a success on each trial is q, then

$$\Pr(T_r = t) = \binom{t+r-1}{r-1} q^r (1-q)^t.$$

Set $q = 0.1$.

a. Plot the distribution with $r = 1$. Find the expectation.

b. Plot the distribution with $r = 2$. Find the expectation. Does this result make sense?

c. How can you think of the random variable T_2 as the sum of two other random variables?

d. Plot the distribution with $r = 10$. Find the expectation. Does this result make sense? Why is the shape of the distribution so different?

7.7 The Poisson Distribution

The binomial and geometric distributions describe two random variables associated with the same process. The binomial distribution describes the **number** of "successes" in a given number of trials, and the geometric distribution describes the **time** until the first success. The exponential probability density function describes the **time** of the first event when events occur at a constant rate. We now study the **Poisson distribution**, the probability distribution for the **number** of events that occur during a fixed time interval when events occur at a constant rate, as in the exponential distribution. The underlying process generating events is called the **Poisson process**. The following chart relates these four fundamental probability distributions.

	Time	
Measurement	**Discrete**	**Continuous**
count	binomial	Poisson
waiting time	geometric	exponential

7.7.1 The Poisson Process

Consider a cell floating around in a medium with many molecules that bind permanently to receptors on the cell. Suppose that binding events occur at rate λ. We wish to find the probability distribution for the random variable N that gives the number of molecules binding to the cell during a given interval of time.

The **Poisson process** describes events that obey the following assumptions.

- Events occur at a constant probabilistic rate λ.
- Events are independent.

These conditions are similar to those for the process underlying the binomial distribution, where a success occurs independently and with equal probability in each trial, but here events can occur at any time.

Example 7.7.1 A Simulation of the Poisson Process with $\lambda = 1.0$

A simulation of this process for 12 s with $\lambda = 1.0$/s produced the following results, measured to five decimal places.

Event	Time	Event	Time
1	0.219032	6	4.84409
2	0.929085	7	5.54173
3	1.50534	8	6.45588
4	3.79879	9	9.47364
5	4.66062	10	10.6027

Each spike in Figure 7.7.1 indicates exactly when a molecule bound to the cell.

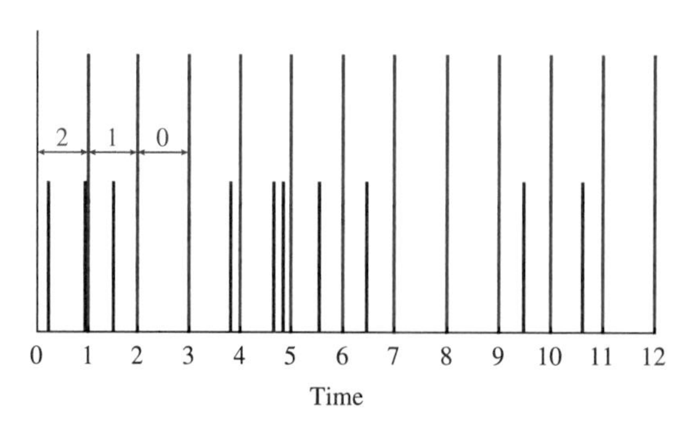

FIGURE 7.7.1
The Poisson process

The waiting times between events in the Poisson distribution follow the exponential distribution with parameter λ. The Poisson process can be simulated efficiently by choosing a series of random numbers that obey the appropriate exponential distribution as the waiting times and then adding them to find the times of the events.

Example 7.7.2 Waiting Times Between Events in the Poisson Process

The data in Example 7.7.1 obey the Poisson process with $\lambda = 1.0$. The waiting times between events (after starting the clock at $t = 0$) are found as the differences between the times when the events occurred.

Event	Waiting Time	Event	Time
1	0.219032	6	0.18347
2	0.71005	7	0.69764
3	0.57625	8	0.91415
4	2.29345	9	3.01776
5	0.86183	10	1.12906

These waiting times come from an exponential distribution with rate parameter $\lambda = 1.0$.

Example 7.7.3 Summarizing a Simulation of the Poisson Process

We can summarize the results in Example 7.7.1 by asking how many molecules bound during each of the 12 1-s intervals: (Figure 7.7.2): two bound during the first second (between times $t = 0.0$ and $t = 1.0$), one bound during the next second (between $t = 1.0$ and $t = 2.0$), none bound during the third second (between times $t = 2.0$ and $t = 3.0$), and so on. A total of four 1-s intervals had zero events, six intervals had one event, and two intervals had two events.

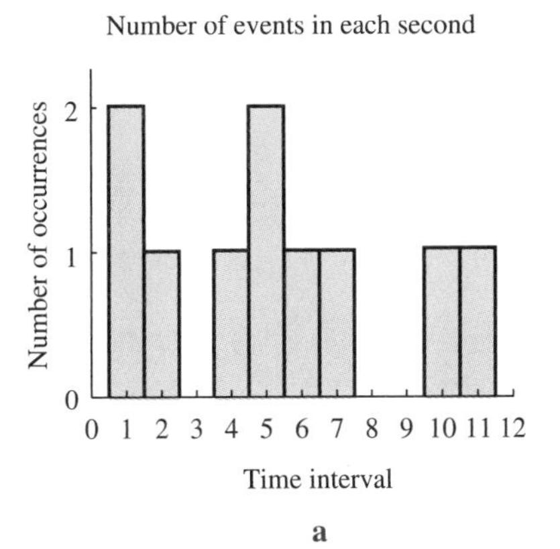

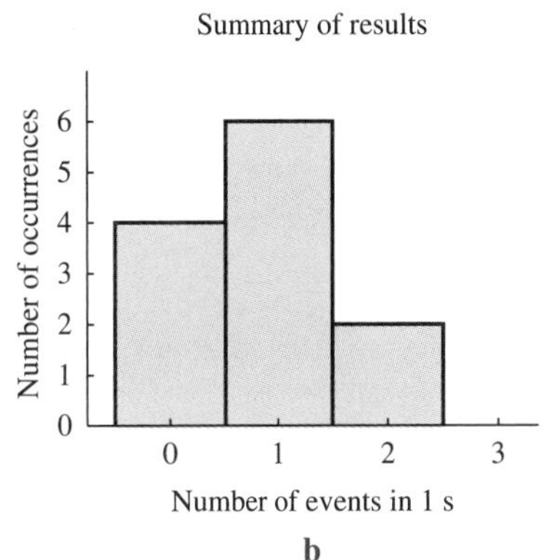

FIGURE 7.7.2
Summaries of a simulation of the Poisson process

Example 7.7.4 A Simulation of the Poisson Process with $\lambda = 4.5$

A simulation with the higher rate $\lambda = 4.5$ produced many more events and much larger numbers in the histogram of results (Figure 7.7.3).

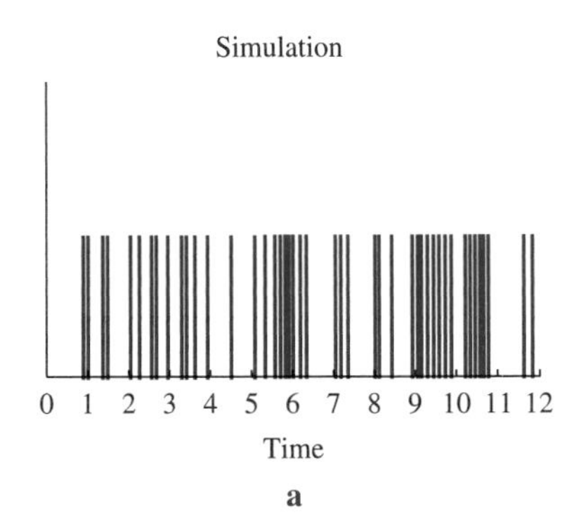

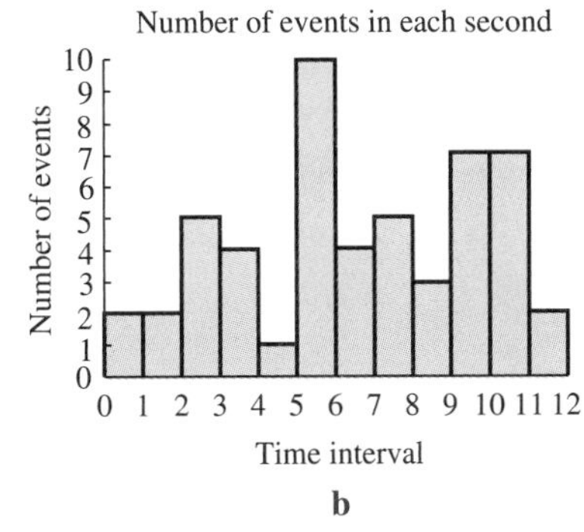

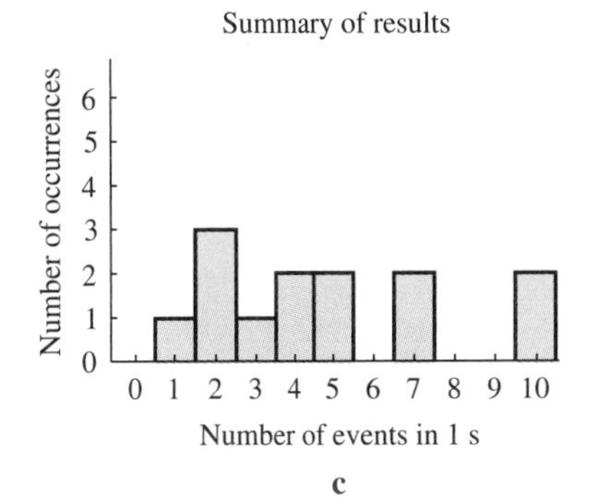

FIGURE 7.7.3
The Poisson process with $\lambda = 4.5$

These two simulations provide a sample of the number of events that occur in 1 s. We would like to compute the mathematically exact probability that exactly k events occurred during an interval of length t when events occur according to a Poisson process at rate λ. Let N be the random variable counting the number of events in an interval of length t. The Poisson distribution gives formulas for $\Pr(N=0)$, $\Pr(N=1)$, $\Pr(N=2)$, and so forth. Although the underlying process occurs in continuous time, the Poisson distribution itself describes a **discrete random variable**, because it describes a count. The Poisson distribution, like the binomial distribution, describes a count of independent events. Because events occur continuously in the Poisson process, there is no upper limit on the number of events possible, in contrast to the upper limit of n successes in n trials with the binomial distribution.

We have already computed the first term in the Poisson distribution, $\Pr(N=0)$. This term describes the probability that no events occurred, which is given by the **exponential survivorship function**

$$\Pr(N=0) = e^{-\lambda t}.$$

Example 7.7.5 Comparing $\Pr(N=0)$ with a Simulation Using $\lambda = 1.0$

With $\lambda = 1.0$ and $t = 1.0$, the probability of zero events is

$$\Pr(N=0) = e^{-1.0 \cdot 1.0} = 0.368.$$

The simulation in Example 7.7.1 has rate $\lambda = 1.0$, and we counted the number of events in 1 s in Example 7.7.3. We found that four out of 12 1-s intervals, or 0.333, experienced zero events, quite close to the probability of 0.368.

Example 7.7.6 Comparing $\Pr(N=0)$ with a Simulation Using $\lambda = 4.5$

With $\lambda = 4.5$ and $t = 1.0$, the probability of zero events is

$$\Pr(N=0) = e^{-4.5 \cdot 1.0} = 0.011.$$

It is quite unlikely that nothing happens in 1 s, consistent with the results in Example 7.7.4, where none of the 12 1-s intervals observed had zero events.

The Poisson distribution fills in the rest of the probabilities, giving $\Pr(N=k)$ for every value of k. We denote $\Pr(N=k)$ by $p(k;\lambda t)$.

Theorem 7.19 Suppose events obey a Poisson process and occur at constant rate λ. The probability that exactly k events occur in time t is

$$\Pr(N=k) = p(k;\lambda t) = \frac{e^{-\lambda t}(\lambda t)^k}{k!}.$$

The range of the random variable N is all non-negative integers, $0, 1, 2, \ldots$.

A hint of one way to derive these probabilities is given in Exercises 35–38.

Example 7.7.7 Checking the Formula for $\Pr(N=0)$

In Examples 7.7.5 and 7.7.6, we used the exponential survivorship function to deduce $\Pr(N=0) = e^{-\lambda t}$. The formula for the Poisson distribution with $k=0$ gives

$$\begin{aligned}\Pr(N=0) &= p(0;\lambda t) \\ &= \frac{e^{-\lambda t}(\lambda t)^0}{0!} \\ &= e^{-\lambda t}\end{aligned}$$

(recall from Table 7.2 that $0! = 1$), matching the earlier formula.

Because the Poisson distribution depends on λ and t only through their **product** λt, we often write $\Lambda = \lambda t$ and rewrite the formula for the Poisson distribution.

Simplified formula for the Poisson distribution:

$$\Pr(N = k) = p(k;\, \Lambda) = \frac{e^{-\Lambda}(\Lambda)^k}{k!}$$

Example 7.7.8 The Poisson Distribution with $\lambda = 1.0$ and $t = 1.0$

If we wish to find the probability distribution for the number of events that occur in time interval $t = 1.0$ at rate $\lambda = 1.0$/min (as in Example 7.7.3), we use the formula for the Poisson distribution with the parameter $\Lambda = \lambda t = 1.0 \cdot 1.0 = 1.0$:

$$\Pr(N = 0) = p(0;\, 1) = \frac{e^{-1.0}(1.0^0)}{0!} = 0.3679$$

$$\Pr(N = 1) = p(1;\, 1) = \frac{e^{-1.0}(1.0^1)}{1!} = 0.3679$$

$$\Pr(N = 2) = p(2;\, 1) = \frac{e^{-1.0}(1.0^2)}{2!} = 0.1839$$

$$\Pr(N = 3) = p(3;\, 1) = \frac{e^{-1.0}(1.0^3)}{3!} = 0.0613.$$

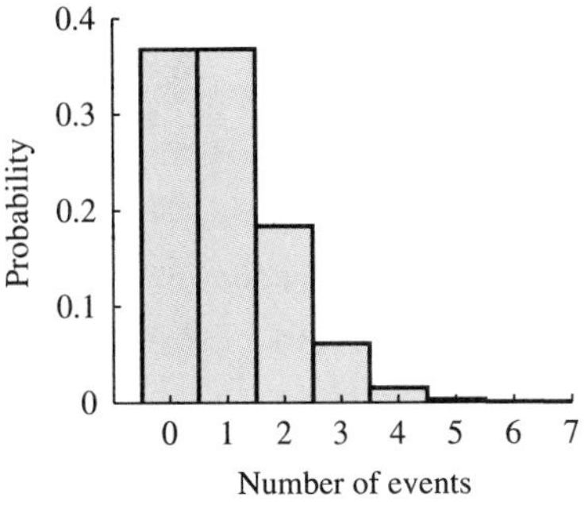

FIGURE 7.7.4
The Poisson distribution with $\Lambda = 1.0$

and so on (Figure 7.7.4). The probability distribution matches the simulated results in Example 7.7.3 reasonably well.

Example 7.7.9 The Poisson Distribution with $\lambda = 4.5$ and $t = 1.0$

With $\lambda = 4.5$, many more events occur per minute (Example 7.7.4). In this case, $\Lambda = \lambda t = 4.5 \cdot 1.0 = 4.5$, and the probabilities are more spread out (Figure 7.7.5).

$$\Pr(N = 0) = p(0;\, 4.5) = \frac{e^{-4.5}(4.5^0)}{0!} = 0.01111$$

$$\Pr(N = 1) = p(1;\, 4.5) = \frac{e^{-4.5}(4.5^1)}{1!} = 0.04999$$

$$\Pr(N = 2) = p(2;\, 4.5) = \frac{e^{-4.5}(4.5^2)}{2!} = 0.11248$$

$$\Pr(N = 3) = p(3;\, 4.5) = \frac{e^{-4.5}(4.5^3)}{3!} = 0.16872$$

$$\Pr(N = 4) = p(4;\, 4.5) = \frac{e^{-4.5}(4.5^4)}{4!} = 0.18981$$

$$\Pr(N = 5) = p(5;\, 4.5) = \frac{e^{-4.5}(4.5^5)}{5!} = 0.17083$$

$$\Pr(N = 6) = p(6;\, 4.5) = \frac{e^{-4.5}(4.5^6)}{6!} = 0.12812$$

$$\Pr(N = 7) = p(7;\, 4.5) = \frac{e^{-4.5}(4.5^7)}{7!} = 0.08236$$

$$\Pr(N = 8) = p(8;\, 4.5) = \frac{e^{-4.5}(4.5^8)}{8!} = 0.04633$$

$$\Pr(N = 9) = p(9;\, 4.5) = \frac{e^{-4.5}(4.5^9)}{9!} = 0.02316$$

$$\Pr(N = 10) = p(10;\, 4.5) = \frac{e^{-4.5}(4.5^{10})}{10!} = 0.01042.$$

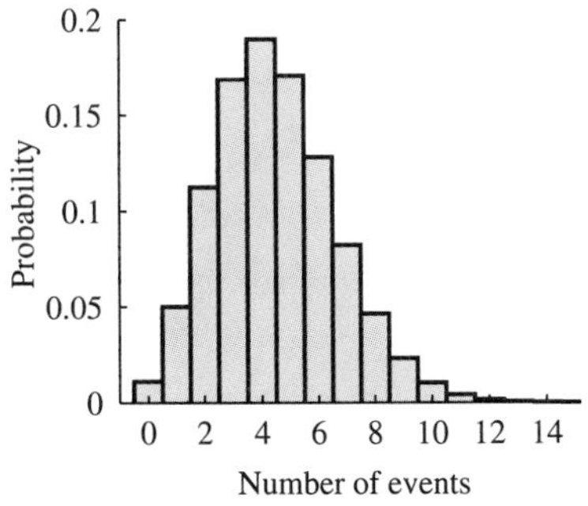

FIGURE 7.7.5
The Poisson distribution with $\Lambda = 4.5$

Example 7.7.10 The Poisson Distribution with $\lambda = 1.0$ and $t = 4.5$

Suppose we count the number of events that occur in intervals of length $t = 4.5$ min when the underlying rate is $\lambda = 1.0$. The parameter Λ is $\Lambda = \lambda t = 1.0 \cdot 4.5 = 4.5$. Because the probability distribution depends only on Λ, the probabilities exactly match those in Example 7.7.9. The number of events that occur with a fast process over a short time interval exactly matches the number with a slow process over a long time interval.

Although they are a bit tricky to derive (Exercises 39 and 40), the formulas for the mean and variance of the Poisson distribution are simple.

Theorem 7.20 Suppose the random variable N has a Poisson distribution with parameter Λ. Then

$$\mathrm{E}(N) = \Lambda$$
$$\mathrm{Var}(N) = \Lambda.$$

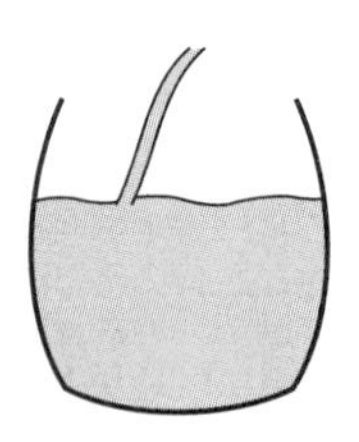

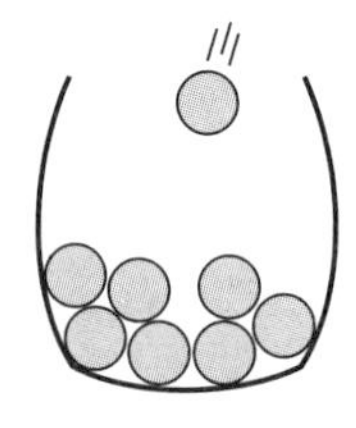

FIGURE 7.7.6
The difference between the Poisson process and an ordinary rate

Why are these formulas so simple? Because λ is a rate, the expectation Λ is the rate times the time. If water is entering a vessel at rate λ for time t, the total amount is exactly λt. The difference between the Poisson process and constant flux is the *variance* (Figure 7.7.6). When the flow rate is 1.0 L/min, exactly 4.5 L enter in 4.5 min. When molecules enter at an average rate of 1.0 each minute, the *expected number* of molecules entering in 4.5 min is 4.5, but with a variance of 4.5 molecules^2 and a standard deviation of $\sqrt{4.5} = 2.12$ molecules.

Be careful to distinguish these results from the exponential distribution, which has *standard deviation* equal to the mean. With the Poisson distribution, the coefficient of variation gets *smaller* as the mean gets larger. By sampling more events, we get a more consistent count. The exponential distribution is always highly variable because it concerns only counting a *single* event.

Example 7.7.11 The Behavior of the Poisson Distribution as a Function of Time

Suppose events occur according to a Poisson process at rate $\lambda = 4.5$. Let $N(t)$ be a random variable representing the number of events that have occurred by time t. Then

$$\mathrm{E}[N(t)] = \Lambda = \lambda t = 4.5t$$
$$\mathrm{Var}[N(t)] = \Lambda = \lambda t = 4.5t$$
$$\sigma_{N(t)} = \sqrt{4.5t}$$
$$\mathrm{CV}_{N(t)} = \frac{\sigma_{N(t)}}{\mathrm{E}[N(t)]} = \frac{\sqrt{4.5t}}{4.5t} = \frac{1}{\sqrt{4.5t}}$$

(Figure 7.7.7).

At $t = 1$,

$$\mathrm{E}[N(t)] = 4.5$$
$$\mathrm{Var}[N(t)] = 4.5$$
$$\sigma_{N(t)} \approx \sqrt{4.5} = 2.121$$
$$\mathrm{CV}_{N(t)} = \frac{1}{\sqrt{4.5}} = 0.471.$$

At the later time $t = 20$,

$$\mathrm{E}[N(t)] = 4.5 \cdot 20 = 90.0$$
$$\mathrm{Var}[N(t)] = 4.5 \cdot 20 = 90.0$$
$$\sigma_{N(t)} = \sqrt{4.5 \cdot 20} = 9.487$$
$$\mathrm{CV}_{N(t)} = \frac{1}{\sqrt{4.5 \cdot 20}} = 0.105.$$

The coefficient of variation decreases over time because there has been time to average out any streaks of unusually dense or unusually sparse events.

The mode of the Poisson distribution, like that for the binomial, is always close to the mean.

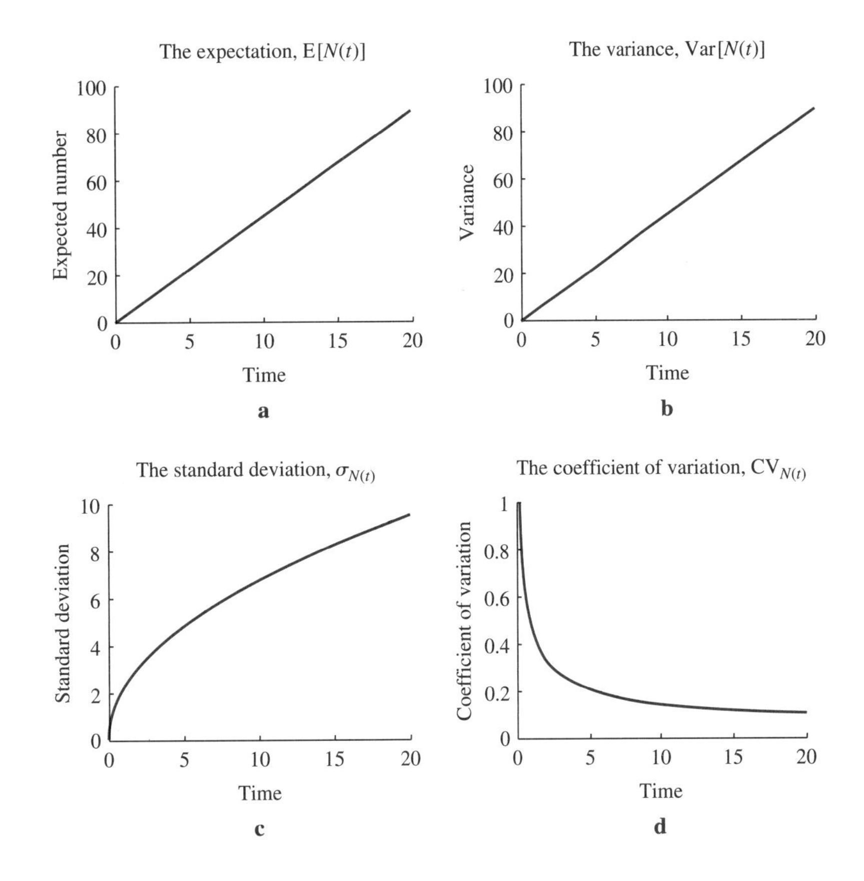

FIGURE 7.7.7

The statistics describing the Poisson distribution as functions of time

Theorem 7.21 Suppose the random variable N has a Poisson distribution with parameter Λ. The mode of N is the largest integer less than Λ.

Proof: The mode can be found with an iterative formula like that for the binomial distribution (Theorem 7.15).

$$p(k+1;\,\Lambda) = \frac{e^{-\Lambda}(\Lambda)^{k+1}}{(k+1)!}$$
$$= p(k;\,\Lambda)\frac{\Lambda}{k+1}.$$

The mode is the smallest value of k for which $p(k+1;\,\Lambda) < p(k;\,\Lambda)$. This occurs when $k+1 > \Lambda$. Therefore, the mode is the largest integer less than Λ.

Example 7.7.12 The Mode with $\Lambda = 4.5$

With $\Lambda = 4.5$, the largest integer less than 4.5 is 4, which matches the mode in Figure 7.7.5.

7.7.2 The Poisson Distribution in Space

Like the other important probability distributions in biology, the Poisson distribution has a remarkable array of applications. One of the most useful is the distribution of objects or events in space rather than time. The parallel with time is strongest in one dimension. Suppose two bacterial clones have been accumulating mutations since they last had a common ancestor, and that an average of 1.3 mutations have become fixed per million nucleotides. We have a piece of DNA 4.7 million nucleotides long from each organism (Figure 7.7.8). In how many sites will the two differ?

The picture of the mutations, which occur over the *length* of the DNA, looks exactly like the simulation of events occurring in *time* (Figure 7.7.1). What assumptions must we

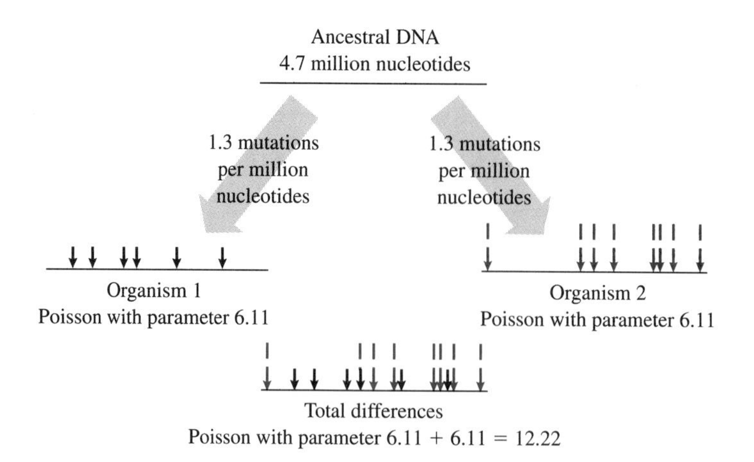

FIGURE 7.7.8
The evolution of two organisms

make to use the Poisson distribution? We must assume that the rate at which mutations occur along the piece of DNA is constant and that mutations occur independently. Both of these assumptions have been the source of important debates among geneticists. Assuming that they are true, the number of mutations M_1 in the first organism has a Poisson distribution with rate $\lambda = 1.3$ per million and "t" equal to 4.7 million. Therefore, $\Lambda = 1.3 \cdot 4.7 = 6.11$ mutations. The same argument holds for organism 2, so M_2 has a Poisson distribution with $\Lambda = 6.11$.

However, both M_1 and M_2 measure the number of differences from the common ancestor, which is extinct and inaccessible. What is the distribution of the number of **differences** between the two organisms? If we superimpose the two pieces of DNA, the number of differences is the **sum** of the number of mutations in each (except for the unlikely event that two mutations occurred in exactly the same spot). In mathematical notation, the total number of differences D is

$$D = M_1 + M_2.$$

The following theorem tells us everything we need to know about D.

Theorem 7.22 Suppose X and Y are independent Poisson distributed random variables with parameters Λ_X and Λ_Y. Then the sum $Z = X + Y$ has a Poisson distribution with parameter $\Lambda_X + \Lambda_Y$.

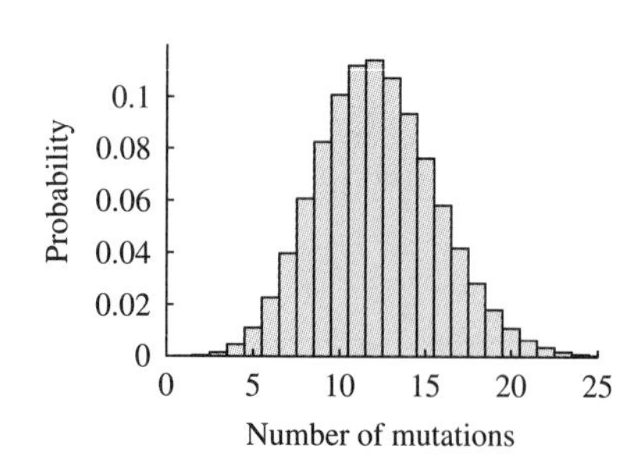

FIGURE 7.7.9
The distribution of the number of mutations separating two organisms

Therefore, D has a Poisson distribution with mean $6.11 + 6.11 = 12.22$ and variance 12.22. Our theorems about expectation and variance of the sum of independent random variables (Theorem 7.4 and Theorem 7.9) guarantee that these must be the mean and variance of D:

$$\mathrm{E}(D) = \mathrm{E}(M_1) + \mathrm{E}(M_2) = 6.11 + 6.11 = 12.22$$
$$\mathrm{Var}(D) = \mathrm{Var}(M_1) + \mathrm{Var}(M_2) = 6.11 + 6.11 = 12.22.$$

Theorem 7.22 tells us more—the entire probability distribution of D. By computing the probability distribution for D, we can compare our result (14 mutations) with the expectation (about 12 mutations). Figure 7.7.9 plots the distribution, from which we can see that the probability of 14 mutations (0.094) is very close to the probability of the mode (12 mutations with probability 0.114).

The Poisson distribution also applies to processes in two or more dimensions. Suppose seeds fall independently in a region at a rate of 0.0023 seeds per cm^2. A simulation of this process in a 9 m^2 region is shown in Figure 7.7.10, with the number of seeds in each block indicated. The number of seeds S per square meter follows a Poisson distribution. In this case, $\lambda = 0.0023$ seeds per cm^2 and "t" is 10,000 cm^2. Therefore,

$$\Lambda = 0.0023 \cdot 10000 = 23.0,$$

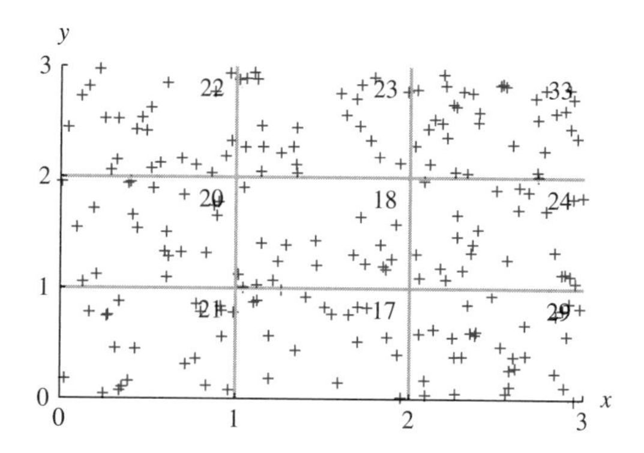

FIGURE 7.7.10
Scatter of seeds in two dimensions

the mean number of seeds per square meter. The probability of k seeds in a given square meter is $p(k;\, 23.0)$ (Figure 7.7.11). The distribution is less spread out and has more of a "bell" shape than with small Λ. The probability that S is within five of the expectation is

$$\Pr(18 \le M \le 28) = \sum_{k=18}^{28} p(k;\, 23.0) = 0.750,$$

where we evaluated the various values $p(k;\, 23.0)$ on a computer (like the binomial, there is no convenient formula for the cumulative distribution). About 75% of square meters surveyed should have within five seeds of the expectation, consistent with the six out of nine in the simulation.

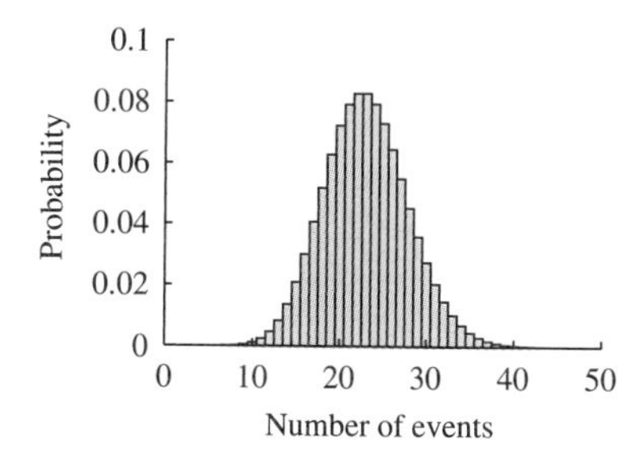

FIGURE 7.7.11
The Poisson distribution with $\Lambda = 23$

7.7.3 The Poisson and the Binomial

The seed example highlights the link between the Poisson and the binomial. The number of seeds in 1 m^2 is a count of the number in each of 10,000 cm^2, each with the tiny probability 0.0023 of containing a seed. If we ignore the probability that some square centimeter has two seeds, the counts follows a binomial distribution with $n = 10000$ and $p = 0.0023$.

Example 7.7.13 Comparison of Probabilities with the Poisson and Binomial Distributions

The probability distributions for the Poisson distribution with $\Lambda = 23$ is nearly identical to the binomial with $n = 10000$ and $p = 0.0023$.

$$p(23;\, 23.0) = 0.08288438$$
$$b(23;\, 10000, 0.0023) = 0.08297986.$$

On a histogram, the two distributions appear identical. ▲

In general, we have the following rule.

Rule for Approximating the Binomial Distribution with the Poisson Distribution
If p is small (less than about 0.01) and n is large (greater than about 100), then

$$b(k;\, n, p) \approx p(k;\, np). \tag{7.7.1}$$

The Poisson distribution can be thought of as the limit of infinitely many trials with infinitesimally unlikely Bernoulli random variables.

Let B be a random variable following a binomial distribution with parameters n and p, and let P be a random variable following a Poisson distribution with parameter $\Lambda = np$. Then

$$\mathrm{E}(B) = \mathrm{E}(P) = np$$
$$\mathrm{Var}(B) = np(1 - p)$$
$$\mathrm{Var}(P) = np.$$

The expectations match, and the variances are very close when p is small. How do B and P differ? The random variable P can take on any non-negative integer value, whereas B can take on only values from 0 to n. However, the probability that $P > n$ (10,000 in the seed example) is astronomically small.

Example 7.7.14 Comparison of the Mean and Variance for Seed Number

If we model seeds as a random variable S that follows a Poisson distribution with parameter $\Lambda = 23.0$, then $E(S) = 23.0$ and $Var(S) = 23.0$. If we think of seed number instead as a random variable S_b that follows a binomial distribution with $n = 10000$ and $p = 0.0023$, then $E(S_b) = 10000 \cdot 0.0023 = 23.0$ and $Var(S_b) = 10000 \cdot 0.0023 \cdot (1 - 0.0023) = 22.9471$. The binomial has the same mean, and a slightly smaller variance.

Summary Events that occur independently at a constant rate λ follow a **Poisson process**. A random variable that counts the number of events that occur in time t follows a **Poisson distribution** with parameter $\Lambda = \lambda t$, the product of the rate and the time. The mean and variance of the Poisson distribution each equal Λ. The Poisson distribution also describes counts of events that occur independently in space. The sum of independent random variables that all follow Poisson distributions also follows a Poisson distribution. A binomial distribution with small p and large n can be approximated by a Poisson distribution with $\Lambda = np$.

7.7 Exercises

Mathematical Techniques

1–4 ▪ The following figures show the results of simulations of the Poisson process with the given value of the rate λ. Find the number of hits per second in each of the simulated seconds. What is the average number of hits per second? How closely does this match what you would expect from the Poisson distribution?

1. $\lambda = 1.5$.

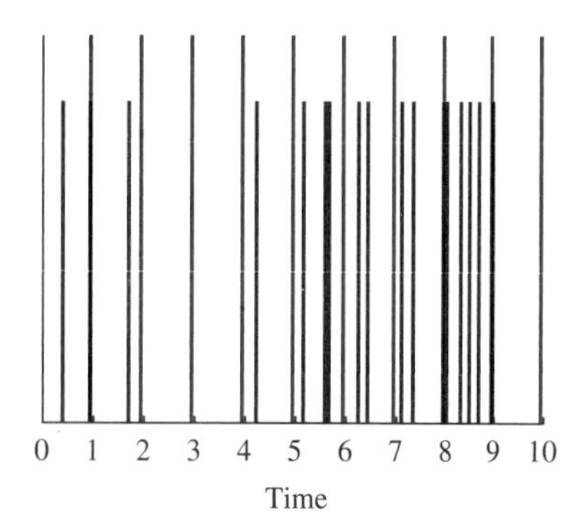

2. $\lambda = 2.0$.

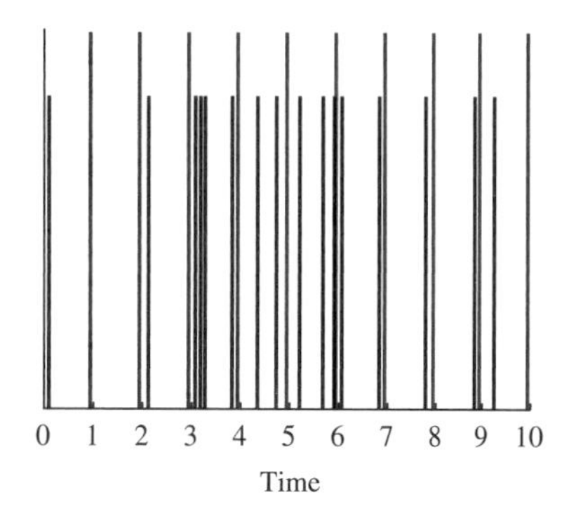

3. $\lambda = 0.5$.

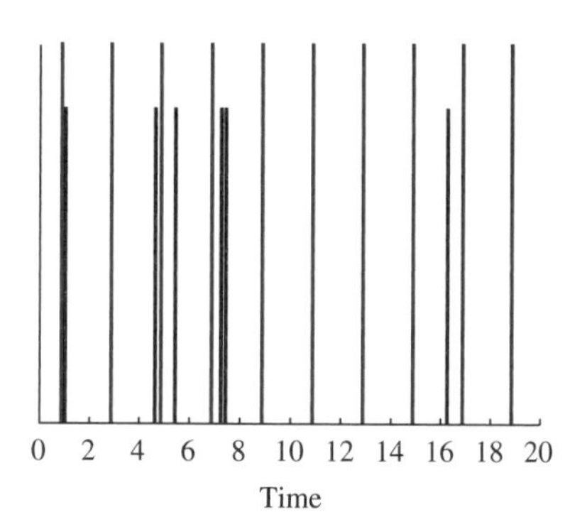

4. $\lambda = 0.8$.

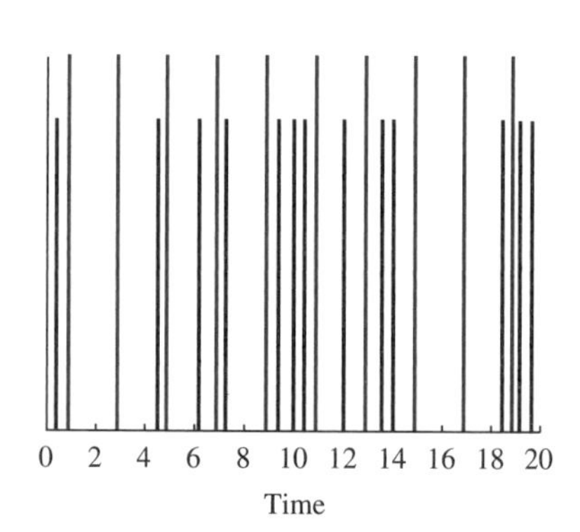

5–10 ▪ Using the simulations from the previous set of problems, compare the observed distribution of the number of hits per second with the Poisson distribution with the given value of λ and $t = 1$ by plotting both distributions.

5. The simulation in Exercise 1 with $\lambda = 1.5$.

6. The simulation in Exercise 2 with $\lambda = 2.0$.

7. The simulation in Exercise 3 with $\lambda = 0.5$.

8. The simulation in Exercise 4 with $\lambda = 0.8$.

9. The simulation in Exercise 3 with $\lambda = 0.5$. Regroup the results into ten intervals with length 2 s, and compare with the Poisson distribution with $\lambda = 0.5$ and $t = 2$.

10. The simulation in Exercise 4 with $\lambda = 0.8$. Regroup the results into ten intervals with length 2 s, and compare with the Poisson distribution with $\lambda = 0.8$ and $t = 2$.

11–14 ▪ The following figures show the results of simulations that do not follow the assumptions of the Poisson process. Can you identify how they differ?

11.

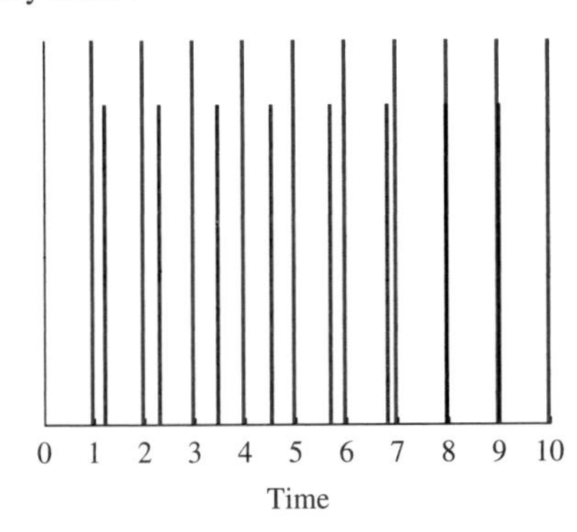

12.

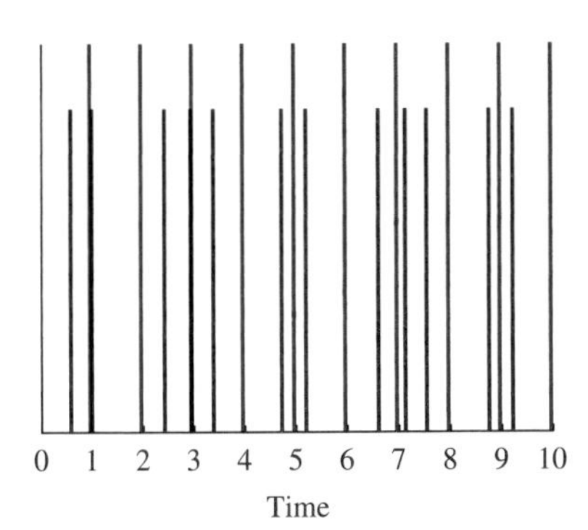

13.

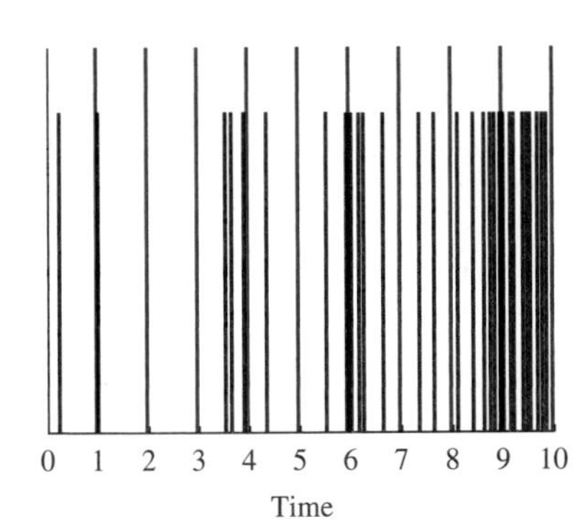

14.

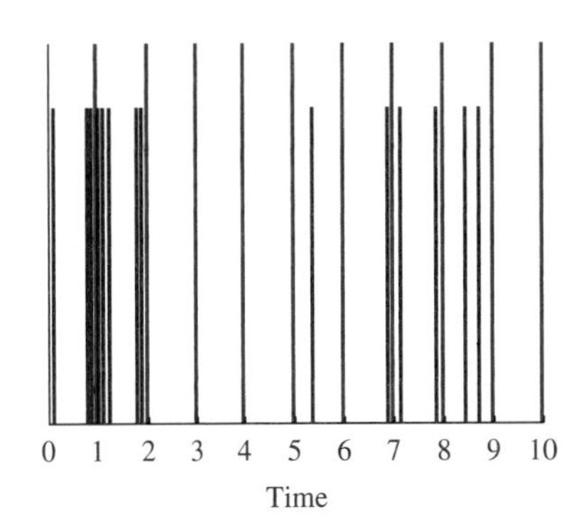

15–18 ▪ Find Λ, the expectation, the variance, the standard deviation, the coefficient of variation, and the mode of Poisson distributions with the following values of λ and t.

15. Molecules leave a cell at rate $\lambda = 0.3$/s and are observed for $t = 3$ s.

16. Phone calls arrive at a rate of $\lambda = 0.2$/h and are monitored for $t = 9$ h.

17. Cosmic rays hit an organism at a rate of 1.2/day and are monitored for one week.

18. Dandelion seeds fall into a garden with area 4.0 m^2 with an average density of 0.9/m^2.

19–22 ▪ Find the probabilities of the following events.

19. Molecules leave a cell at rate $\lambda = 0.3$/s. What is the probability that exactly two have left by the end of the third second?

20. Phone calls arrive at a rate of $\lambda = 0.2$/h. What is the probability that there are exactly five calls in 9 h?

21. Cosmic rays hit an organism at a rate of 1.2/day. What is the probability of being hit ten times in a week?

22. Dandelion seeds fall into a garden with an average density of 0.9/m^2. What is the probability that three fall into a 4.0-m^2 vegetable garden?

23–26 ▪ Find and sketch the Poisson distribution associated with the given rate λ and duration t, and use it to compute the requested probability.

23. Molecules leave a cell at rate $\lambda = 0.3$/s. What is the probability that four or fewer have left by the end of the third second?

24. Phone calls arrive at a rate of $\lambda = 0.2$/h. What is the probability that there are five or more calls in 9 h?

25. Cosmic rays hit an organism at a rate of 1.2/day. What is the probability of being hit between five and ten times (inclusive) in a week?

26. Dandelion seeds fall into a garden with an average density of 0.9/m^2. What is the probability that between two and five seeds (inclusive) fall into a 4.0 m^2 vegetable garden?

27–30 ▪ Molecules leave a cell at rate $\lambda = 0.3$/s. Let $N(t)$ be the random variable measuring the number of cells that have left as a function of t. Consider times t between 0 and 10.

27. Compute and graph $E[N(t)]$ as a function of time.

28. Compute and graph $\mathrm{CV}[N(t)]$ as a function of time.

29. Compute and graph $\Pr[N(t)=1]$ as a function of time. Find the maximum. Why does this graph increase and then decrease?

30. Compute and graph $\Pr[N(t)=2]$ as a function of time. Find the maximum. Why does the maximum occur later than the maximum of $\Pr[N(t)=1]$?

31–32 ▪ Find the parameter Λ for the Poisson distribution that describes the given process.

31. Cells will mutate when hit independently by cosmic rays (at rate 0.3/day) or by X-rays (at rate 0.2/day). Cells are hit by rays for 1 week. Use random variables M_c, M_x, and N to describe the number of mutations caused by cosmic rays, X-rays, and both types of rays together. What are their distributions? Check that $E(N)=E(M_c)+E(M_x)$ and $\mathrm{Var}(N)=\mathrm{Var}(M_c)+\mathrm{Var}(M_x)$.

32. A professor is interrupted independently by phone calls (at rate 1.3/h), by students with questions (at rate 0.6/h), and by colleagues (at rate 0.3/h). How many interruptions might she expect during an 8-h day? What is the expected time between phone calls, between students, between colleagues, and between interruptions of all kinds?

33–34 ▪ In each of the following cases, find the probability exactly with the binomial distribution, and compare your result with what you find with the Poisson approximation.

33. Each cell in a culture of 16 cells has a probability of 0.1 of dying. Find and approximate the probability that exactly one cell dies.

34. Assume you get one call in a given hour with probability 0.2 and zero calls with probability 0.8. Find and approximate the probability of exactly two calls in 10 h.

35–38 ▪ The probabilities for the Poisson distribution can be derived by solving differential equations. Let $P_i(t)$ be the probability of exactly i events by time t, assuming an underlying rate of λ.

35. Write $P_1(t+\Delta t)$ in terms of $P_0(t)$ and $P_1(t)$ (think of the two ways there could have been one molecule at time $t+\Delta t$).

36. Move stuff around so your formula looks like the derivative of $P_1(t)$ and take the limit as $\Delta t \to 0$.

37. Recalling that $P_0(t)=e^{-\lambda t}$, check that $P_1(t)=\lambda t e^{-\lambda t}$ is a solution of your equation with initial condition $P_1(0)=0$.

38. Use the same method to find an equation for $P_2(t)$ and check that $P_2(t)=\dfrac{(\lambda t)^2 e^{-\lambda t}}{2}$ is a solution.

39–40 ▪ We can also use differential equations to derive the formulas for the expectation and variance of the Poisson distribution.

39. Let the expected number of events that have occurred in a Poisson process be $E(t)$. Using the definition of a probabilistic rate to find a formula for $E(t+\Delta t)$, write and solve a differential equation for $E(t)$.

40. Let the variance in the number of events that have occurred in a Poisson process be $V(t)$. Using the definition of a probabilistic rate to find a formula for $V(t+\Delta t)$, write and solve a differential equation for $V(t)$.

Applications

41–44 ▪ Genes in different organisms have different rates of mutation. Compute the following values and probabilities.

41. A gene has a mutation rate of 0.002 mutations per generation. Find the expected number of mutations, the variance, the probability of zero mutations, and the probability of exactly one mutation in a period of 2000 generations.

42. An important gene has a mutation rate of 0.0004 mutations per generation. Find the expected number of mutations, the variance, the probability of zero mutations, and the probability of exactly one mutation in a period of 2000 generations.

43. A gene has a mutation rate of 0.002 mutations per generation. How many generations would it take for the expected number of mutations to be greater than 1? How many generations would it take before the probability of zero mutations is less than 0.001?

44. An important gene has a mutation rate of 0.0004 mutations per generation. How many generations would it take for the expected number of mutations to be greater than 1? How many generations would it take before the probability of zero mutations is less than 0.0001?

45–46 ▪ When two species of organisms have genetically diverged, the number of mutations distinguishing particular genes is the sum of the number in one organism and the number in the other. For the following cases, find the expected number of mutations distinguishing the species.

45. Two populations of fruit flies diverged 10,000 years ago. In the first populations, a gene has a mutation rate of 0.002 mutations per generation. In the second population, a gene has a mutation rate of 0.0004 mutations per generation. Generations in each population are 1 yr.

46. Two populations of flies diverged 1 million years ago. In the first population, a gene has a mutation rate of 1.5×10^{-5} mutations per generation. In the second population, a gene has a mutation rate of 3.0×10^{-6} mutations per generation. Generations in each population are 0.5 yr.

47–48 ▪ In most genes, mutations in some sites ("synonymous sites") have no effect on the protein produced, while those in other sites ("nonsynonymous sites") do affect the protein. In general, mutation rates are higher in synonymous sites because changes are not removed by natural selection.

47. A gene has 200 nonsynonymous sites and 100 synonymous sites. The synonymous sites have mutation rate 6.0×10^{-7}/yr, while nonsynonymous sites have mutation rate 3.0×10^{-9}/yr. What is the expected number of mutations of each type after 1 million years? What is the probability of no mutations of either type?

48. A gene has 300 nonsynonymous sites and 150 synonymous sites. The synonymous sites have mutation rate 6.0×10^{-7}/yr, while nonsynonymous sites have mutation rate 3.0×10^{-9}/yr. What is the expected number of mutations of each type after 2 million years? What is the probability of no mutations of either type?

49–50 ▪ Different environments can lead to different mutation rates. Use the sum rule for the Poisson distribution to find the total expected number of mutations during an experiment by finding the expected number over the entire experiment.

49. Mutations accumulate at a rate of 1.3 per million nucleotides during the first year of a study, and at a rate of 2.2 per million nucleotides during the second year. The DNA is 4.7 million nucleotides long.

50. Mutations accumulate at a rate of 0.3 per million nucleotides during the first 0.5 yr of a study, and at a rate of 3.2 per million nucleotides during the second 1.5 yr. The DNA is 4.7 million nucleotides long.

51–52 ▪ The previous problem shows that the number of mutations follows a Poisson distribution even when the mutation rate changes. In fact, if the rate changes continuously, following some function $\lambda(t)$, the number of mutations between time 0 and time t follows a Poisson distribution with mean

$$\Lambda = \int_0^t \lambda(s) ds.$$

Use this formula to analyze the following circumstances.

51. Due to an increase in radiation levels, the mutation rate increases linearly from 1.0×10^{-6}/yr at $t = 0$ to 3.0×10^{-6}/yr at $t = 2 \times 10^6$. Find the expected number of mutations over the course of 2 million years. What is the average mutation rate over this time?

52. Due to a decrease in radiation, the mutation rate decreases linearly from 4.0×10^{-6}/yr at $t = 0$ to 1.0×10^{-6}/yr at $t = 2 \times 10^6$. Find the expected number of mutations over the course of 2 million years. What is the average mutation rate over this time?

53–54 ▪ Suppose many gnats are flying around in a room. Each leaves independently with a probability that depends on the insect repellent tested. In each case:

a. Describe the random variable with a binomial distribution giving the probability that exactly k have left.

b. Find the Poisson distribution approximating the probability that exactly k have left and the expected number to leave.

c. Find the variance of the binomial random variable and the Poisson approximation.

53. There are 100,000 gnats, and each leaves with probability 0.0067.

54. There are 50,000 gnats, and each leaves with probability 0.037.

55–60 ▪ Suppose an organism would live forever if it weren't for predators that attack at rate λ per year. Fortunately, only a fraction q of attacks are successful. Use the following steps to compute how long the organism will live.

55. What is the probability that the organism has not been attacked at time t?

56. What is the probability that the organism has been attacked once but survived?

57. What is the probability that the organism has been attacked twice but survived?

58. Write down a sum giving the probability that the organism is alive at time t.

59. Write down the first few probabilities in a Poisson distribution with parameter $(1 - q)\lambda$. What do they add up to?

60. Use this last fact to come up with the probability that the organism is alive at time t. What is the average lifetime? Could you have guessed it without doing any calculations?

Computer Exercises

61. Suppose T is an exponentially distributed random variable with parameter $\lambda = 0.3$ that describes the waiting time between events. If the nth event occurs at time t, the $n + 1$st event occurs at time $t + T$. The updating function describing this process is

$$g(t) = t + T,$$

giving the time of the next event as a stochastic function of the time of the previous event.

a. Starting from an initial condition of 0, simulate this updating function for 20 steps. How many events occurred by time 20?

b. Mark the two longest waits and their durations.

c. What is the expected waiting time between events? Sketch the line that should lie close to the graph of your solution.

d. The number of events at time 20 should be described by a Poisson distribution with parameter $\Lambda = 20\lambda$. Plot the histogram of this distribution.

e. Using the number of events that occurred by time 20 in part **a** as k, compute the probability of exactly k events, fewer than k events, and more than k events. Indicate the areas associated with each of these events on your histogram.

62. In the previous exercise, we simulated a Poisson process as a series of exponentially distributed waiting times. An alternative method uses the relation between the Poisson and binomial distributions. Suppose we wish to simulate a process with $\lambda = 1.5$/min for 10 min.

a. Break each minute into n intervals for some large value of n, such as 100. What value of p will produce an average of 1.5 successes per minute?

b. Choose a series of $10n$ independent Bernoulli random variables that equal 1 with probability p. Each value of 1 corresponds to an event in the Poisson process.

c. Produce a graph showing when the events occur.

d. Why does this method fail to exactly reproduce the Poisson process?

7.8 The Normal Distribution

We have studied two distributions associated with counting: the binomial and the Poisson. Counting is the simplest form of adding, adding one for each event that occurred. A more general sum adds random variables that can take on values other than 0 and 1, such as dart scores. The exact distribution resulting from a more general sum is often impossible to compute. Under a wide range of assumptions, however, the **normal distribution** provides a remarkably accurate approximation of the exact distribution. In this section, we sketch the theory behind this approximation.

7.8.1 The Normal Distribution: An Example

The development of statistics was largely inspired by problems in breeding. Here is one sort of problem breeders face. A type of plant has ten different genes affecting height, with an equally common short and tall allele at each locus.

1. Plants with two copies of the short allele gain no height from that locus,
2. Plants with one copy of the short allele and one copy of the tall allele gain 1.0 cm,
3. Plants with two copies of the tall allele gain 2.5 cm.

The complicating fact (see Figure 7.8.1a) is that two copies of the tall allele gives more than double the height change from one copy (contrast with Equation 6.2.4). Suppose the total height gain is the sum of those from each locus. What is the distribution of heights of these plants?

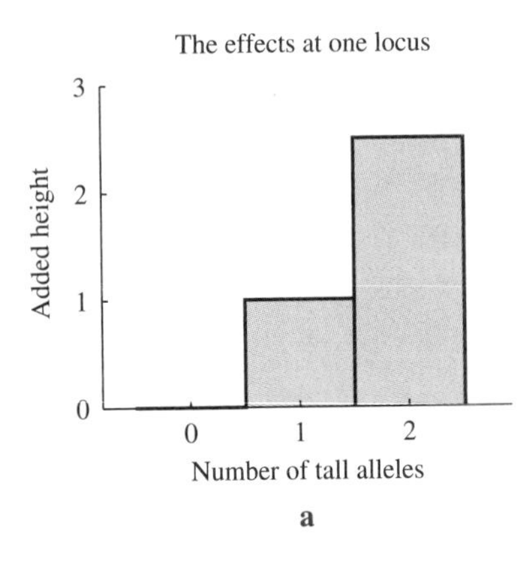

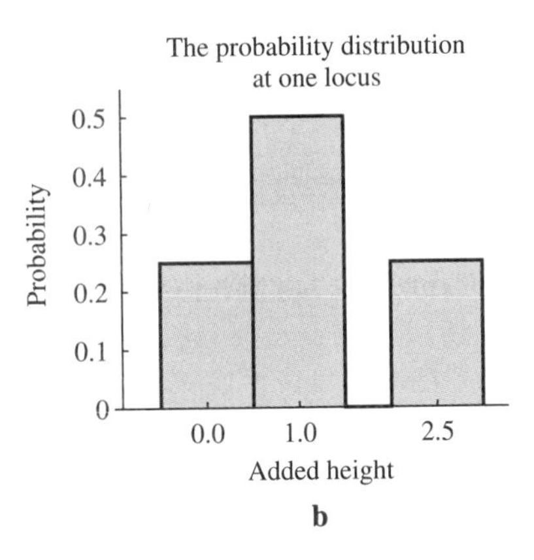

FIGURE 7.8.1
The effects of alleles at one locus

We computed the binomial distribution in a similar case by enumerating all of the possibilities. Enumeration is virtually impossible in this case because there are many different ways of ending up with the same height (Exercise 26).

Let H_i be a random variable measuring the height gain from locus i. Because the short and tall alleles are equally common,

$$\Pr(H_i = 0.0) = 0.25$$
$$\Pr(H_i = 1.0) = 0.5$$
$$\Pr(H_i = 2.5) = 0.25$$

for each i (Figure 7.8.1). Therefore

$$\mathrm{E}(H_i) = 0.0 \cdot 0.25 + 1.0 \cdot 0.5 + 2.5 \cdot 0.25 = 1.125,$$

and

$$\mathrm{Var}(H_i) = 0.0^2(0.25) + 1.0^2(0.5) + 2.5^2(0.25) - 1.125^2 = 0.797.$$

Let H represent the total added height. Because heights have been assumed to add,

$$H = \sum_{i=1}^{10} H_i.$$

The expectation of the sum is the sum of the expectations (Theorem 7.6), so

$$\mathrm{E}(H) = \sum_{i=1}^{10} \mathrm{E}(H_i) = 11.25.$$

Similarly, assuming that the different loci are independent (the alleles at one locus are independent of those at every other), the variance of the sum is the sum of the variances (Theorem 7.9), and we have

$$\mathrm{Var}(H) = \sum_{i=1}^{10} \mathrm{Var}(H_i) = 7.97.$$

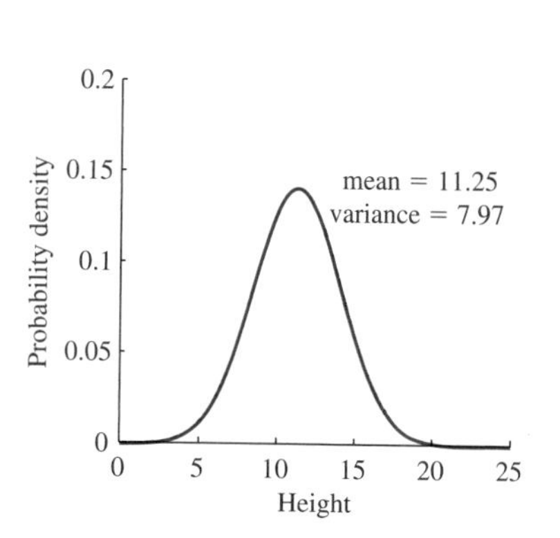

FIGURE 7.8.2
What the mean and variance tell us

We have used the underlying structure of the random variable H to find its mean and variance, which give some information about the whole distribution (Figure 7.8.2). What if we want to answer more detailed questions, like the probability that the height of the plant is increased by less than 6 cm, by between 10 and 15 cm, or by more than 19 cm? The mean and variance alone cannot help us.

There is a powerful and deep theorem that does. The **Central Limit Theorem** says that the sum of independent random variables with the same distribution can be approximated by a **normal distribution**.

Definition 7.11 A continuous random variable X has a normal distribution with mean μ and variance σ^2 if it has probability density function

$$f(x; \mu, \sigma^2) = \frac{1}{\sqrt{2\pi}\,\sigma} e^{-\frac{(x-\mu)^2}{2\sigma^2}}.$$

The range of this random variable is from $-\infty$ and ∞. We write

$$X \sim N(\mu, \sigma^2).$$

The normal distribution has a symmetric bell-shaped p.d.f. that takes on its maximum value at $x = \mu$ and approaches 0 at both ∞ and $-\infty$. The points of inflection are $x = \mu + \sigma$ and $x = \mu - \sigma$, one standard deviation above and below the mean (Figure 7.8.3), which is the first rule of thumb discussed in Subsection 6.9.3 (Exercise 16).

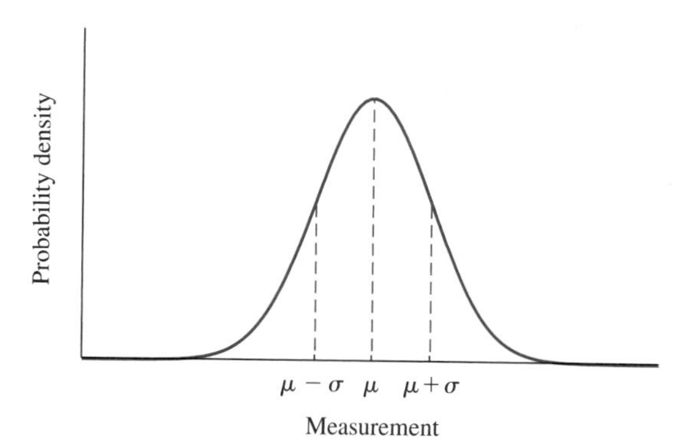

FIGURE 7.8.3
The normal distribution

Example 7.8.1 The First Rule of Thumb Applied to a Normal Distribution

Suppose a normal distribution has mean $\mu = 40.0$ and standard deviation $\sigma = 10.0$. The graph of the probability density function $f(x)$ has its peak at $x = 40.0$ and points of inflection at $x = 40.0 + 10.0 = 50.0$ and $x = 40.0 - 10.0 = 30.0$ (Figure 7.8.4).

The normal distribution is completely described by two numbers: the mean and variance. The normal distribution $X \sim N(11.25, 7.97)$, with mean $\mu = 11.25$ and variance

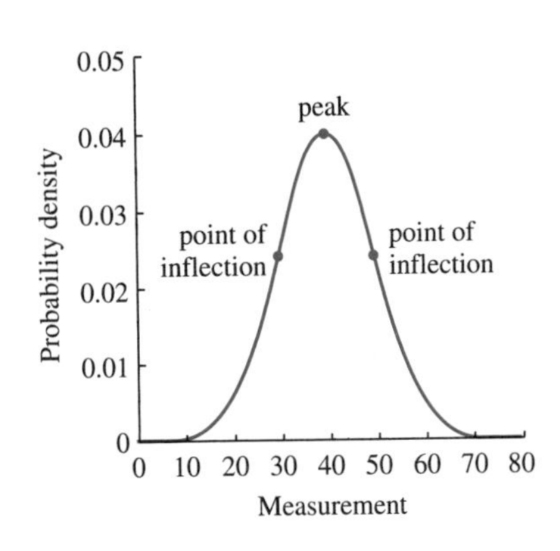

FIGURE 7.8.4
The normal distribution with $\mu = 40.0$ and $\sigma = 10.0$

$\sigma^2 = 7.97$ equal to the mean and variance of H, closely approximated the histogram of results for 200 simulated plants (Figure 7.8.5).

Because this density approximates the true distribution, we can use it to estimate probabilities. In particular, we have

$$\Pr(H \le 6) = \Pr(X \le 6) = \int_{-\infty}^{6} f(x; 11.25, 7.97)dx$$

$$\Pr(10 \le H \le 15) = \Pr(10 \le X \le 15) = \int_{10}^{15} f(x; 11.25, 7.97)dx$$

$$\Pr(H \ge 19) = \Pr(X \ge 19) = \int_{19}^{\infty} f(x; 11.25, 7.97)dx.$$

Some of the limits of integration might seem absurd. Plant heights cannot be negative or greater than 25 in our model, but the limits of integration extend to $-\infty$ and ∞. The normal approximation is just that, an *approximation*. It is a good approximation because the values of the p.d.f. are extremely small in the absurd region. For example, $f(0; 11.25, 7.97) = 5.0 \times 10^{-5}$ and $f(25; 11.25, 7.97) = 1.0 \times 10^{-6}$.

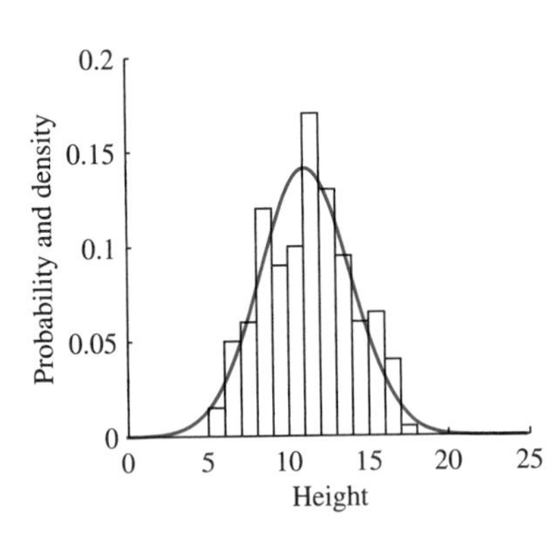

FIGURE 7.8.5
The normal distribution compared with histogram of 200 results from H

It is impossible to write a formula for the integral of the normal distribution, and we must use a table or computer to help with the calculation of probabilities. In the next section, we will find a convenient way to make these calculations.

7.8.2 The Central Limit Theorem for Sums

Why does the normal distribution approximate H so well? We now present a formal statement of the theorem behind the method. One bit of shorthand is often used in this context. Random variables with the same probability distribution or probability density function are said to be **identically distributed**. The random variables H_i in the previous section, or different trials of an experiment, are identically distributed and independent. Sets of random variables that are both **independent and identically distributed** are called **i.i.d.**

Theorem 7.23 The Central Limit Theorem for Sums

Suppose $X_1, \ldots, X_i, \ldots$ are a set of i.i.d. random variables, each with finite expectation μ and finite variance σ^2. Let

$$S_n = \sum_{i=1}^{n} X_i$$

be a random variable representing the sum of the first n. Then the probability density function for S_n is approximately $N(n\mu, n\sigma^2)$ for sufficiently large n.

The proof of this theorem requires technical tools beyond the scope of this book. However, as with the calculation of E(H) and Var(H), we know that

$$\begin{aligned} E(S_n) &= \sum_{i=1}^{n} E(X_i) && \text{expectation of sum is sum of expectations} \\ &= \sum_{i=1}^{n} \mu && E(X_i) = \mu \text{ for all } i \\ &= n\mu && \text{adding up } n\ \mu\text{'s gives } n\mu \end{aligned}$$

and

$$\begin{aligned} \mathrm{Var}(S_n) &= \sum_{i=1}^{n} \mathrm{Var}(X_i) && \text{expectation of variance is sum of variances if independent} \\ &= \sum_{i=1}^{n} \sigma^2 && \mathrm{Var}(X_i) = \sigma^2 \text{ for all } i \\ &= n\sigma^2. && \text{adding up } n\ \sigma^2\text{'s gives } n\sigma^2 \end{aligned}$$

A few of the hypotheses of the theorem might look a bit odd. We have not met any random variables with infinite expectation or infinite variance (although the issue came up in Section 7.3). Such distributions do occasionally arise in practice, but they do not cause much trouble because the attempt to compute the approximate normal distribution falls apart. More important are the assumptions that the random variables are independent and identically distributed. Generalizations of the Central Limit Theorem deal with the case where the random variables are not quite independent (in a suitably precise sense) or have distributions that are not quite the same. For scientific applications, as long as there is good reason to think that the random variables in question are close to satisfying the i.i.d. requirement, application of the Central Limit Theorem is reasonable.

Example 7.8.2 Applying the Central Limit Theorem for Sums

Suppose another set of 20 loci for height are found. Tall alleles at locus i are denoted by A_i, and short alleles by a_i. Assume that the contribution of each locus is described by the following:

1. $a_i a_i$ gain no additional height.
2. $A_i a_i$ gain 0.3 cm.
3. $A_i A_i$ gain 0.4 cm.

In a cross between two parents that are heterozygous at all 20 loci, offspring have genotype $a_i a_i$ with probability 0.25, $A_i a_i$ with probability 0.50, and $A_i A_i$ with probability 0.25. Let G_i be the additional height added by this locus. Then

$$\begin{aligned} E(G_i) &= 0 \cdot 0.25 + 0.3 \cdot 0.5 + 0.4 \cdot 0.25 = 0.25 \\ \mathrm{Var}(G_i) &= 0^2 \cdot 0.25 + 0.3^2 \cdot 0.5 + 0.4^2 \cdot 0.25 - 0.25^2 = 0.0225. \end{aligned}$$

If the total height T added by these loci is the sum $T = \sum_{i=1}^{20} G_i$, then

$$\begin{aligned} E(T) &= 20 \cdot 0.25 = 5.0 \\ \mathrm{Var}(T) &= 20 \cdot 0.0225 = 0.45. \end{aligned}$$

The Central Limit Theorem for sums says that T can be approximated by a random variable Y that follows a normal distribution with $Y \approx N(5.0, 0.45)$.

Why the normal distribution? A hint is given by the **additive property** of the normal distribution. If X and Y are independent random variables with

$$\begin{aligned} X &\approx N(\mu_1, \sigma_1^2) \\ Y &\approx N(\mu_2, \sigma_2^2), \end{aligned}$$

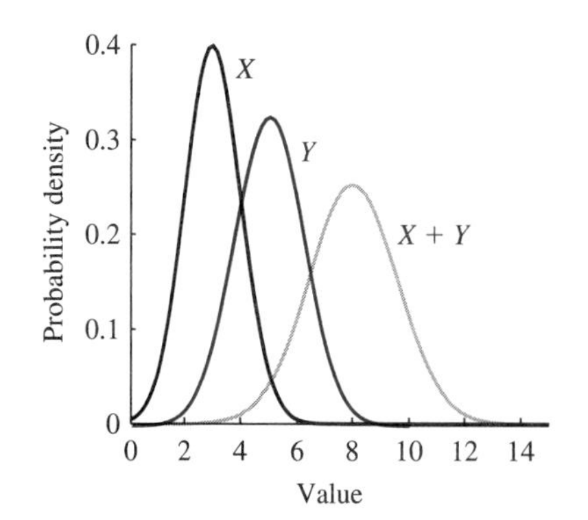

FIGURE 7.8.6
The sum of two independent normally distributed random variables

then

$$X + Y \approx N(\mu_1 + \mu_2, \sigma_1^2 + \sigma_2^2)$$

(Figure 7.8.6). The mean must be $\mu_1 + \mu_2$ because the expectation of the sum is the sum of the expectations (Theorem 7.4). Because the random variables are independent, the variance must be $\sigma_1^2 + \sigma_2^2$ (Theorem 7.9). The normal distribution is special because the sum of independent normally distributed random variables has the same shape, a normal distribution. (This special property also holds for the Poisson distribution, Theorem 7.22).

Example 7.8.3 Height as the Sum of Two Normally Distributed Random Variables

The first set of ten genes studied in Subsection 7.8.1 added height that could be approximated by a normal distribution $X \approx N(11.25, 7.97)$. The second set studied in Example 7.8.2 could be approximated by a normal distribution $Y \approx N(5.0, 0.45)$. The total added height by both sets is then approximated by

$$X + Y \approx N(11.25 + 5.0, 7.97 + 0.45) = N(16.25, 8.42).$$

The Central Limit Theorem states that the sum is approximately normal for "sufficiently large n." How large is large?

Rule of Thumb for Applying the Central Limit Theorem The normal distribution given by the Central Limit Theorem provides a good approximation when $n \geq 30$.

Example 7.8.4 Applying the Central Limit Theorem to a Sum of Exponentially Distributed Random Variables

Suppose T_i describes the waiting time between events $i - 1$ and event i in a Poisson process with rate $\lambda = 1.0$. Then T_i follows an exponential distribution with $\lambda = 1.0$ (as in Example 7.7.2). Therefore,

$$S_n = \sum_{i=1}^{n} T_i$$

is the waiting time until event number n (Figure 7.8.7). The probability density function for S_n is called a **Gamma distribution** (computed in Exercises 17–20) and is compared

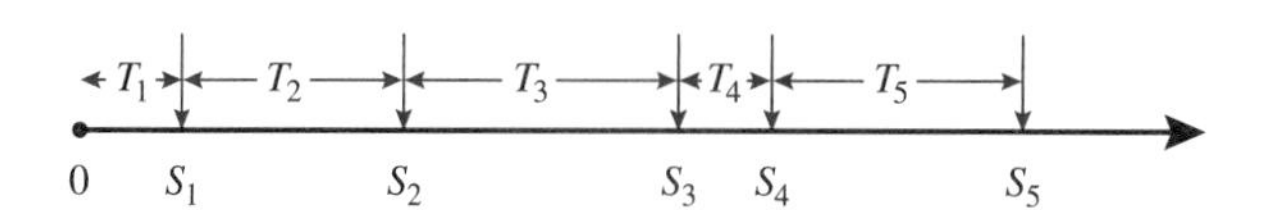

FIGURE 7.8.7
The sum of many waiting times

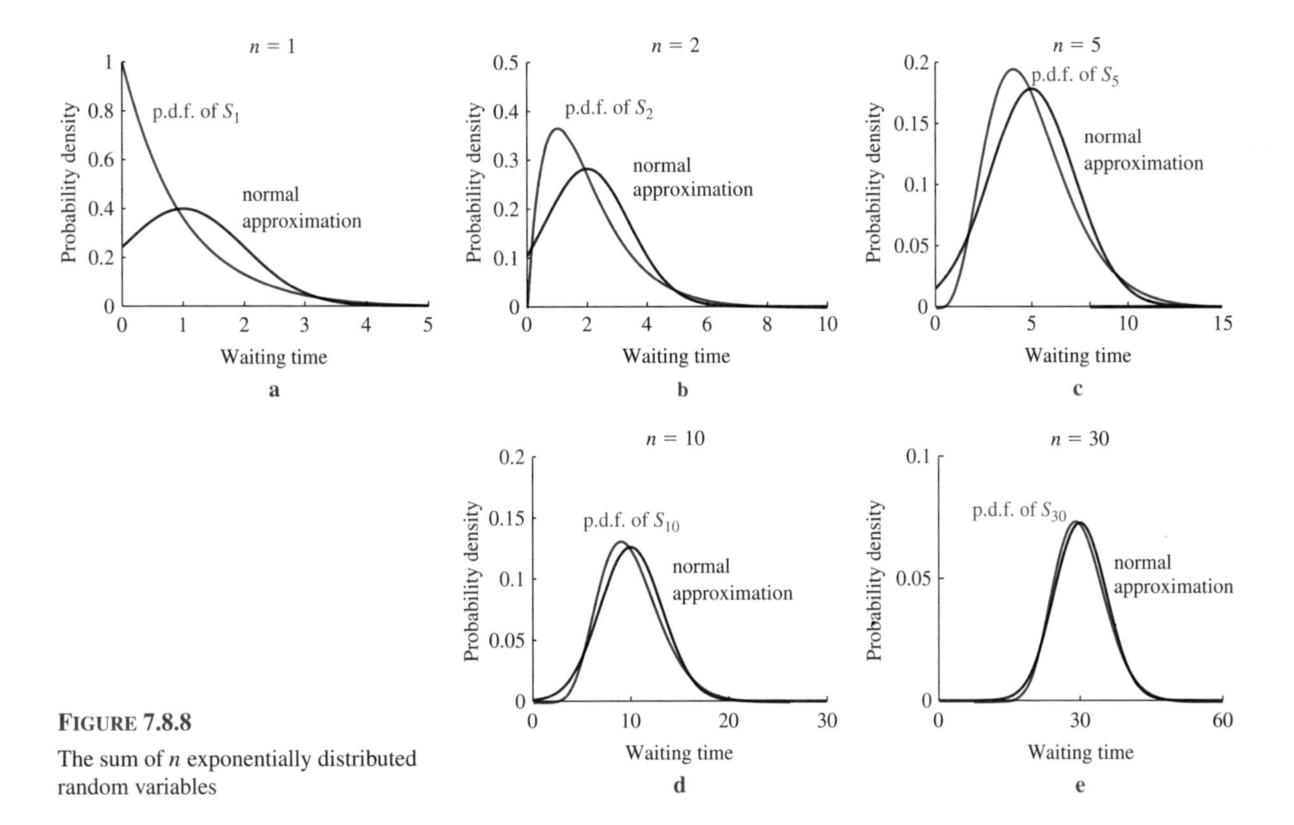

FIGURE 7.8.8
The sum of n exponentially distributed random variables

with the normal approximation for five different values of n (Figure 7.8.8). Although the probability density function for each individual waiting time follows an exponential density function that looks nothing like a normal distribution, the shape becomes more and more "normal" as we add up more and more independent terms. ▲

7.8.3 The Central Limit Theorem for Averages

The most important statistical application of the normal distribution follows from the close connection between averages and sums.

Theorem 7.24 **The Central Limit Theorem for Averages**

Suppose $X_1, \ldots, X_i, \ldots$ are a set of i.i.d. random variables, each with finite expectation μ and finite variance σ^2. Let

$$A_n = \frac{1}{n}\sum_{i=1}^{n} X_i$$

be a random variable equal to the average of the first n. Then the probability density function for A_n is approximately $N(\mu, \frac{\sigma^2}{n})$ for sufficiently large n.

Proof: This theorem follows from Theorem 7.23. By definition,

$$A_n = \frac{S_n}{n}.$$

Dividing a random variable by a constant divides the expectation of that random variable by the constant and the variance by the square of the constant. Therefore,

$$E(A_n) = \frac{E(S_n)}{n} = \mu$$
$$\mathrm{Var}(A_n) = \frac{\mathrm{Var}(S_n)}{n^2} = \frac{n\sigma^2}{n^2} = \frac{\sigma^2}{n}.$$

Example 7.8.5 Applying the Central Limit Theorem for Averages to Heights

Consider again the 20 loci for height in Example 7.8.2, where the expected mean and variance from each locus is

$$E(G_i) = 0.25$$
$$\mathrm{Var}(G_i) = 0.0225.$$

Let

$$A_{20} = \frac{1}{n} \sum_{i=1}^{20} G_i$$

be the average effect of these 20 loci. The random variable A_{20} is approximately normally distributed, and

$$A_{20} \approx N\left(0.25, \frac{0.0225}{20}\right) = N(0.25, 0.001125).$$

In general, the Central Limit Theorem for Averages is applied to study the means of many replicates of an experiment. In statistics, we frequently wish to check whether two populations (perhaps those who have been treated with a new medication and those who have not) differ on average. The Central Limit Theorem for Averages makes the comparison more simple, because the **average** of many samples from even a highly non-normal distribution closely follows the convenient and well-understood normal distribution.

Example 7.8.6 Applying the Central Limit Theorem to the Average of Waiting Times

Suppose that each of ten waiting times follows an exponential distribution with $\lambda = 1.0$. What can we say about the average waiting time? Let T_i be a random variable indicating the waiting time for event i. Each of these random variables satisfies

$$E(T_i) = \frac{1}{\lambda} = 1.0$$
$$\mathrm{Var}(T_i) = \frac{1}{\lambda^2} = 1.0.$$

The average is the sum of these values divided by 10, or

$$A = \frac{1}{10} \sum_{i=1}^{10} T_i.$$

According to the Central Limit Theorem for Averages, the new random variable A will have approximately a normal distribution with mean μ equal to the mean of each of T_i, and variance equal to the variance divided by 10. Mathematically,

$$A \approx N(1.0, 0.1).$$

Example 7.8.7 Comparing the Central Limit Theorem with Simulated Data

Suppose we measure and average the waiting times for ten events, as in Example 7.8.6, but repeat the experiment five times.

Molecule	Experiment 1	Experiment 2	Experiment 3	Experiment 4	Experiment 5
1	4.046	0.437	1.737	0.380	2.603
2	0.849	1.498	0.341	1.202	1.709
3	0.356	0.909	1.455	2.683	2.414
4	1.120	0.976	0.211	0.070	0.476
5	0.467	0.427	0.094	0.298	0.837
6	0.194	0.365	0.316	3.132	0.479
7	0.851	3.026	0.043	0.441	1.374
8	2.230	0.727	0.302	0.533	0.543
9	0.870	0.532	0.595	1.725	0.481
10	0.641	0.295	0.133	0.458	0.351
Average	1.162	0.919	0.523	1.092	1.127

If the waiting times were chosen from an exponential distribution with $\lambda = 1.0$, we found that the average A has mean 1.0 and variance 0.1. According to our rules of thumb (Subsection 6.9.3), most averages should lie within two standard deviations of the mean (Figure 7.8.9). In this case,

$$\sigma_A = \sqrt{0.1} = 0.316,$$

meaning that most results should be between $1.0 - 2 \cdot 0.316 = 0.368$ and $1.0 + 2 \cdot 0.316 = 1.632$. This accords quite well with our results, even though the number of terms averaged is less than the 30 recommended by the rule of thumb. Many of the *individual measurements* lie far outside these bounds, ranging from 0.043 to 4.046. The averages have less variance because these extreme values tend to be averaged out by values that deviate in the other direction.

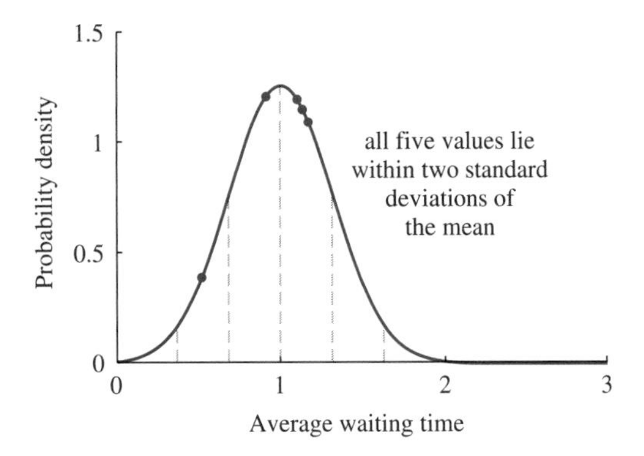

FIGURE 7.8.9
Applying the Central Limit Theorem for Averages

The Central Limit Theorem for Averages is fundamental to estimation techniques in statistics. If we try to estimate the expectation of a distribution by taking independent samples X_i and computing their average, as we do when estimating a value from a series of experimental replicates, we end up with estimates of the true expectation which have a normal distribution centered at the true expectation *even if we do not know the shape of the underlying distribution.*

Summary We have approximated the sum of independent, identically distributed (**i.i.d.**) random variables with a **normally distributed** random variable by matching the mean and variance. This approximation is justified by the **Central Limit Theorem for Sums**.

This theorem can be reinterpreted as the **Central Limit Theorem for Averages**, showing that the averages of independent samples from one distribution are approximately normally distributed with mean equal to the true mean and variance decreasing with the size of the sample.

7.8 Exercises

Mathematical Techniques

1–2 ▪ Suppose the random variables $X_1, X_2 \ldots X_n$ are i.i.d. (independent and identically distributed), and consider the sum $S_n = \sum_{i=1}^{n} X_i$. Find the mean and variance of S_n, and write and sketch the p.d.f. of the approximate normal distribution.

1. Suppose that $n = 16$, and that X_i takes the value 0 with probability 0.5 and the value 1 with probability 0.5.

2. Suppose that $n = 50$, and that X_i takes the value 0 with probability 0.25, the value 1 with probability 0.5, and the value 2 with probability 0.25.

3–4 ▪ Suppose the random variables $X_1, X_2 \ldots X_n$ are i.i.d. and consider the average $A_n = \frac{1}{n} \sum_{i=1}^{n} X_i$. Find the mean and variance of A_n, and write and sketch the p.d.f. of the approximate normal distribution.

3. Suppose that $n = 16$, and that X_i takes the value 0 with probability 0.5 and the value 1 with probability 0.5 (as in Exercise 1).

4. Suppose that $n = 50$, and that X_i takes the value 0 with probability 0.25, the value 1 with probability 0.5, and the value 2 with probability 0.25(as in Exercise 2).

5–8 ▪ Recall the data describing the probabilities of the outcomes of four experiments counting the number of mutants in a bacterial culture (Section 6.6, Exercises 1–4).

	Probability			
Number of Mutants	Experiment a	Experiment b	Experiment c	Experiment d
0	0.1	0.6	0.3	0.1
1	0.2	0.3	0.2	0.3
2	0.3	0.1	0.2	0.1
3	0.3	0.0	0.2	0.4
4	0.1	0.0	0.1	0.1

Suppose that each experiment is repeated (independently) and the total number of mutants is counted.

5. Experiment **a** is repeated 20 times. Write an integral that estimates the probability that there are between 30 and 40 mutants, and shade the corresponding area on a sketch of the approximate normal p.d.f. Take a guess at the probability based on your sketch. (We found the expectation in Section 6.7, Exercise 1 and the variance in Section 6.9, Exercise 1.)

6. Experiment **b** is repeated 80 times. Write an integral that estimates the probability that there are between 30 and 40 mutants, and shade the corresponding area on a sketch of the approximate normal p.d.f. Take a guess at the probability based on your sketch. (We found the expectation in Section 6.7, Exercise 2 and the variance in Section 6.9, Exercise 2.)

7. Experiment **c** is repeated 100 times. Write an integral that estimates the probability that there is an average of less than 1.8 mutants per experiment and shade the corresponding area on a sketch of the approximate normal p.d.f. Take a guess at the probability based on your sketch. (We found the expectation in Section 6.7, Exercise 3 and the variance in Section 6.9, Exercise 3.)

8. Experiment **d** is repeated 35 times. Write an integral that estimates the probability that there is an average of less than 1.8 mutants per experiment and shade the corresponding area on a sketch of the approximate normal p.d.f. Take a guess at the probability based on your sketch. (We found the expectation in Section 6.7, Exercise 4 and the variance in Section 6.7, Exercise 4.)

9–12 ▪ Suppose that X and Y are independent normally distributed random variables with $X \approx N(5.0, 16.0)$ and $Y \approx N(10.0, 9.0)$.

9. Find and sketch the p.d.f. of $3X$.

10. Find and sketch the p.d.f. of $X + Y$.

11. Find and sketch the p.d.f. of the sum of nine independent samples from X.

12. Find and sketch the p.d.f. of the mean of nine independent samples from Y.

13–16 ▪ Show the following facts about the normal distribution.

13. The normal p.d.f. with $\mu = 0$ and $\sigma^2 = 1$ takes on its maximum at $x = 0$.

14. The normal p.d.f. takes on its maximum at $x = \mu$ for any values of μ and σ.

15. The normal p.d.f. with $\mu = 0$ and $\sigma^2 = 1$ has points of inflection at $x = -1$ and $x = 1$.

16. The normal p.d.f. has points of inflection at $x = \mu + \sigma$ and $x = \mu - \sigma$ for any values of μ and σ.

17–20 ▪ Suppose T_1, T_2, $T_3 \ldots$ are i.i.d. exponentially distributed random variables with $\lambda = 1.0$ and that $S_n = \sum_{i=1}^{n} T_i$ (Figure 7.8.7). Let f_n be the p.d.f. for S_n.

17. Use the law of total probability to show that

$$f_2(t) = \int_{s=0}^{t} f_1(s) f_1(t-s) ds.$$

18. Evaluate the integral to find f_2, the p.d.f. for S_2.

19. Use the same trick to find f_3, the p.d.f. for S_3.

20. Can you guess the pattern? Why does the answer look so much like the Poisson distribution?

21–22 ▪ The Central Limit Theorem does not work if random variables are not independent. The simplest case to compute is where they are perfectly correlated with each other. In particular, suppose the random variables $X_1, X_2 \ldots X_n$ are all equal, and consider the sum $S_n = \sum_{i=1}^{n} X_i$. Find the mean and variance of S_n, and sketch its probability distribution.

21. Suppose that $n = 16$ and that X_i takes the value 0 with probability 0.5 and the value 1 with probability 0.5. Compare with the results in Exercise 1.

22. Suppose that $n = 50$ and that X_i takes the value 0 with probability 0.25, the value 1 with probability 0.5, and the value 2 with probability 0.25. Compare with the results in Exercise 2.

Applications

23–28 ▪ Based on the probabilities in Figure 7.8.1b, we can find the probability that a plant gains 5 cm in height from ten genes, $\Pr(H = 5)$.

23. Find the two ways to add up 0's, 1's, and 2.5's to get 5 cm.

24. Find the number of ways each could occur (use binomial coefficients).

25. Find the probability associated with each way.

26. Add them up to find the total probability.

27. Find the normal distribution approximating added height, and find the value of the normal p.d.f. at 5.0 cm.

28. Use the result of the previous problem to approximate the probability that the height is in the interval between 4.5 and 5.5 cm. Compare with the result in Exercise 26.

29–32 ▪ Suppose that scientists develop a new model of human IQ that includes three independent factors: genes of large effect, genes of small effect, and environmental effects. All genes are assumed to be dominant. There are ten smart genes of large effect, each of which adds 2.5 IQ points. There are 20 smart genes of small effect, each of which adds 0.6 IQ points. There are 30 environmental factors, with favorable ones adding 0.9 IQ points. People have a baseline IQ of 80 if they have no favorable effects. Finally, suppose that the probability of getting each smart gene is 0.75, and the probability of getting each favorable environmental effect is 0.5.

29. Find the normal approximation for IQ based only on genes of large effect.

30. Find the normal approximation for IQ based only on genes of small effect.

31. Find the normal approximation for IQ based only on environmental effects.

32. Find the normal approximation for IQ with both genetic and environmental effects. What is the maximum possible IQ with the model?

33–36 ▪ Suppose immigration and emigration change the sizes of four populations with the following probabilities (Section 6.7, Exercises 31–34). Find the p.d.f. of the normal approximation for the average number of immigrants, sketch a graph, and shade and estimate the area corresponding to an increase in the population.

Population a		Population b	
Number	Probability	Number	Probability
−1	0.4	−1	0.1
0	0.2	0	0.3
1	0.3	1	0.2
2	0.1	2	0.4
Population c		**Population d**	
−1	0.4	−10	0.4
0	0.2	0	0.2
1	0.3	1	0.3
100	0.1	2	0.1

33. Suppose immigrants arrive into and emigrate from population **a** for 20 yr.

34. Suppose immigrants arrive into and emigrate from population **b** for 20 yr.

35. Suppose immigrants arrive into and emigrate from population **c** for 10 yr. How accurate do you think the normal approximation is?

36. Suppose immigrants arrive into and emigrate from population **d** for 10 yr. How accurate do you think the normal approximation is?

37–38 ▪ Although the Central Limit Theorem applies to the *sums* of independent and identically distributed random variables, we can use logarithms to analyze *products* of independent and identically distributed random variables. Consider populations growing for the given number of years with the given distribution for the random variable R giving per capita production. Find the normal distribution that approximates the logarithm of the population size P_t assuming that $P_0 = 100$.

37. $R_i = 4$ with probability 0.5, $R_i = 0.25$ with probability 0.5 (as in Section 6.8, Exercise 37). Find the approximate normal distribution for $\ln(P_{50})$, the log of the population size after 50 time steps.

38. $R_i = 4$ with probability 0.25, $R_i = 0.25$ with probability 0.75 (as in Section 6.8, Exercise 38). Find the approximate normal distribution for $\ln(P_{25})$.

39–40 ▪ As in Exercises 37 and 38, the Central Limit Theorem for sums can be used to approximate the logarithm of a product even when the random variables multiplied together are continuous random variables. Find the normal distribution that approximates the logarithm of the population size P_{50} assuming that $P_0 = 1$. You will need the indefinite integrals

$$\int \ln(x)dx = x\ln(x) - x$$

$$\int \ln(x)^2 dx = x\ln(x)^2 - 2x\ln(x) + 2x$$

to evaluate the expectation and variance of $\ln(R)$. Use the rule of thumb that most populations end up within two standard deviations from the mean to give a range of probable population sizes.

39. Let R be a random variable giving the per capita production in a population with p.d.f. $g(x) = 5.0$ for $1.0 \le x \le 1.2$ (the values used in Example 6.1.2). Are the simulations in Example 6.1.2 within this range?

40. Let R be a random variable giving the per capita production in a population with p.d.f. $g(x) = 1.25$ for $0.7 \le x \le 1.5$ (the values used in Example 6.1.3). Are the simulations in Example 6.1.3 within this range?

Computer Exercises

41. Two other values often used to describe distributions are the **skewness** and the **kurtosis**. Suppose a random variable has p.d.f. $f(x)$ for $-\infty < x < \infty$ and mean μ. The skewness k_3 describes symmetry, and has a formula like the variance:

$$k_3 = E[(X-\mu)^3] = \int_{-\infty}^{\infty} (x-\mu)^3 f(x)dx.$$

Instead of taking the difference from the mean and squaring it, we take the difference from the mean and **cube** it. This is sometimes called the *third moment around the mean*. Symmetric distributions have a skewness of 0.

The kurtosis k_4 is based on the *fourth moment around the mean*. However, it includes a correction factor involving the variance σ^2,

$$k_4 = E[(X-\mu)^4] - 3\sigma^4 = \int_{-\infty}^{\infty} (x-\mu)^4 f(x)dx - 3\sigma^4.$$

The correction was added to make the kurtosis of a normal distribution equal to 0.

a. Find the skewness and kurtosis of a normal distribution with mean 0 and variance 1.

b. Find the skewness and kurtosis of a normal distribution with mean 1 and variance 2.

c. Find the skewness and kurtosis of an exponential distribution with mean 1.0. (Remember that the random variable in this case must be positive.) The skewness is positive because the distribution is stretched out to the right.

d. Find the skewness and kurtosis of a uniform distribution with p.d.f. $f(x) = 1$ for $0 \le x \le 1$. The kurtosis is positive because the distribution has "broad shoulders."

42. Program your computer to choose random numbers from an exponential distribution with mean 1.

a. Pick 100 such numbers, and put them into 15 bins (the number between 0 and 0.2, the number between 0.2 and 0.4, etc). Plot a histogram of the number in each bin.

b. Pick 100 pairs of random numbers and take the average of each pair. Put these values into bins and plot a histogram.

c. Pick 100 sets of ten random numbers and take the average of each set. Put these values into bins and plot a histogram.

d. Pick 100 sets of 30 random numbers and take the average of each set. Put these values into bins and plot a histogram. Do the results look more and more like a normal distribution?

7.9 Applying the Normal Approximation

In this section, we put the Central Limit Theorem and normal distribution to work to estimate some actual probabilities. Because the normal distribution is impossible to integrate, we first convert questions about normally distributed random variables into questions about the **standard normal distribution**, amenable to evaluation with the help of a computer or table. We then find the normal distribution approximating a given binomial or Poisson random variable, adjusting for the fact that the normal distribution is continuous, whereas the binomial and Poisson are discrete with the **continuity correction**.

7.9.1 The Standard Normal Distribution

The **standard normal distribution** is a normal distribution with mean 0 and variance 1, or $N(0, 1)$. The p.d.f. is

$$f(x; 0, 1) = \frac{1}{\sqrt{2\pi}} e^{-x^2/2}, \tag{7.9.1}$$

found by substituting $\mu = 0$ and $\sigma^2 = 1$ into the definition of the normal p.d.f. (Definition 7.11). The standard normal p.d.f. is plotted in Figure 7.9.1.

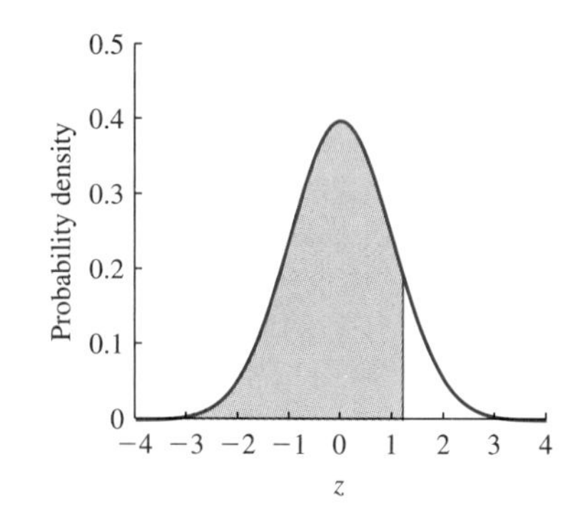

FIGURE 7.9.1
The standard normal curve: shaded area is equal to $\Phi(z)$

The standard normal distribution is useful because tables and computers are set up to compute probabilities on the basis of this distribution. Let Z be a random variable with a standard normal distribution. Although a formula for the c.d.f. of the standard normal cannot be written using elementary functions like the exponential or logarithm, the values are widely available in tables and on computers. This cumulative distribution function, denoted by $\Phi(z)$, is defined by

$$\Phi(z) = \Pr(Z \le z) = \int_{-\infty}^{z} \frac{1}{\sqrt{2\pi}} e^{\frac{-x^2}{2}}\, dx. \tag{7.9.2}$$

The area associated with this definite integral is shown in Figure 7.9.1. A few values of this function are given in the following Table 7.4.

Example 7.9.1 Computing a Probability with $\Phi(z)$

The probability that Z is less than -1.0 is

$$\Pr(Z \le -1.0) = \Phi(-1.0) = 0.1587.$$

Only about 16% of values are less than one standard deviation below the mean of the standard normal.

The c.d.f. $\Phi(z)$ gives the probability that the random variable takes on a value less than z. To find the probability that $Z > z$, we subtract the cumulative probability from 1.

Example 7.9.2 Computing the Probability That $Z > 2.0$

The probability that Z is greater than 2.0 is

$$\Pr(Z > 2.0) = 1 - \Pr(Z \le 2.0) = 1 - \Phi(2.0) = 0.9772 = 0.0228.$$

Very few values are more than two standard deviations above the mean.

Table 7.4 A few values of $\Phi(z)$

z	$\Phi(z)$
−4.0	0.00003
−3.0	0.00135
−2.0	0.02275
−1.0	0.15866
0.0	0.50000
1.0	0.84134
2.0	0.97725
3.0	0.99865
4.0	0.99997

We find the probability that Z lies between two values by subtracting. From the Fundamental Theorem of Calculus (Subsection 6.6.4),

$$\Pr(z_1 \le Z \le z_2) = \int_{z_1}^{z_2} \frac{1}{\sqrt{2\pi}} e^{\frac{-x^2}{2}}\, dx = \Phi(z_2) - \Phi(z_1). \tag{7.9.3}$$

The associated area is illustrated in Figure 7.9.2.

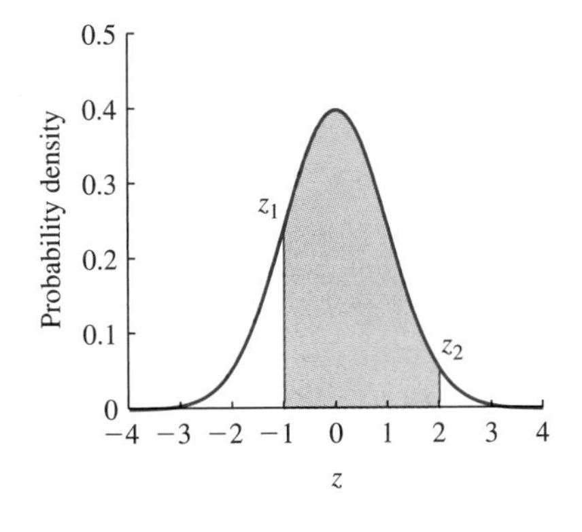

FIGURE 7.9.2
The standard normal curve: shaded area is $\Phi(z_2) - \Phi(z_1)$

Example 7.9.3 Computing the Probability That Z Lies Between Two Values

The probability that a standard normal measurement lies between -1.0 and 2.0 is

$$\begin{aligned}\Pr(-1.0 \le Z \le 2.0) &= \int_{-1.0}^{2.0} \frac{1}{\sqrt{2\pi}} e^{\frac{-x^2}{2}}\, dx \\ &= \Phi(2.0) - \Phi(-1.0) \\ &= 0.9772 - 0.1587 = 0.8185.\end{aligned}$$

Most random variables with normal distributions have neither mean 0 nor variance 1. How can we transform a problem about a random variable X with mean μ and variance σ^2 into a problem about the standard normal?

Theorem 7.25 Suppose $X \approx N(\mu, \sigma^2)$. Then the random variable

$$Z = \frac{X - \mu}{\sigma}$$

has a standard normal distribution.

Proof: We know that

$$\Pr(X \le x) = \int_{-\infty}^{x} \frac{1}{\sqrt{2\pi}\sigma} e^{-\frac{(y-\mu)^2}{2\sigma^2}} dy.$$

By the definition of Z,

$$\Pr(X \le x) = \Pr\left(\frac{X - \mu}{\sigma} \le \frac{x - \mu}{\sigma}\right) = \Pr\left(Z \le \frac{x - \mu}{\sigma}\right).$$

We need to show that

$$\Pr\left(Z \le \frac{x - \mu}{\sigma}\right) = \Phi\left(\frac{x - \mu}{\sigma}\right).$$

We apply the change of variables:

$$z = \frac{x - \mu}{\sigma}.$$

Then $dx = \sigma dz$, and the limits of integration become $z = -\infty$ to $z = \frac{x-\mu}{\sigma}$. Therefore,

$$\Pr(X \le x) = \int_{-\infty}^{x} \frac{1}{\sqrt{2\pi}\sigma} e^{-\frac{(y-\mu)^2}{2\sigma^2}} dy$$

$$= \int_{-\infty}^{\frac{x-\mu}{\sigma}} \frac{1}{\sqrt{2\pi}\sigma} e^{-\frac{z^2}{2}} \sigma dz$$

$$= \int_{-\infty}^{\frac{x-\mu}{\sigma}} \frac{1}{\sqrt{2\pi}} e^{-\frac{z^2}{2}} dz$$

$$= \Phi\left(\frac{x - \mu}{\sigma}\right).$$

Simply put, all normal distributions have the same shape (Figure 7.9.3). The theorem tells us that we can translate a value into a z-**score** by subtracting the mean and dividing by the standard deviation. Be careful to divide by the standard deviation σ, not the variance σ^2. The z-score should be thought of as the difference from the mean measured in units of standard deviations. A value 2.5 standard deviations above the mean is equally unlikely for any normal distribution.

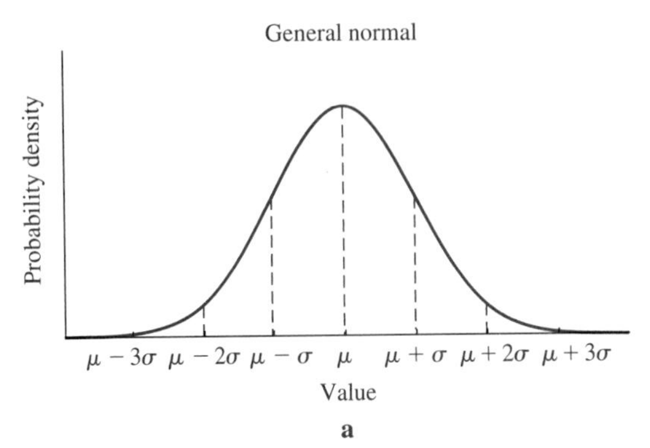

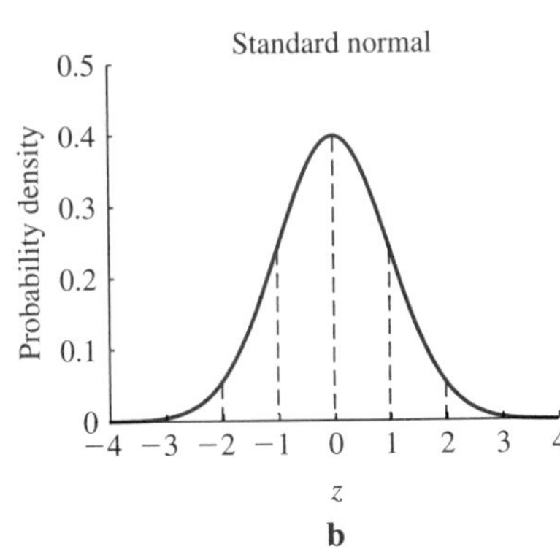

FIGURE 7.9.3

General normal distribution and the standard normal

Example 7.9.4 Computing the z-Score

Suppose the value $x = 33.0$ is measured from a normal distribution with mean $\mu = 50.0$ and variance $\sigma^2 = 100.0$. Then $\sigma = \sqrt{100.0} = 10.0$ and the z-score is

$$z = \frac{x - \mu}{\sigma} = \frac{33.0 - 50.0}{10.0} = -1.7.$$

This value of x lies 1.7 standard deviations below the mean.

Example 7.9.5 Computing a Probability from the z-Score

To find the probability that $X < 33.0$ if $X \approx N(50.0, 100.0)$, we find the z-score to be $z = -1.7$ (Figure 7.9.4, Example 7.9.4) and find

$$\Pr(X < 33.0) = \Pr(Z < -1.7) = \Phi(-1.7) = 0.0446$$

on a table (like that on the inside back cover of this book).

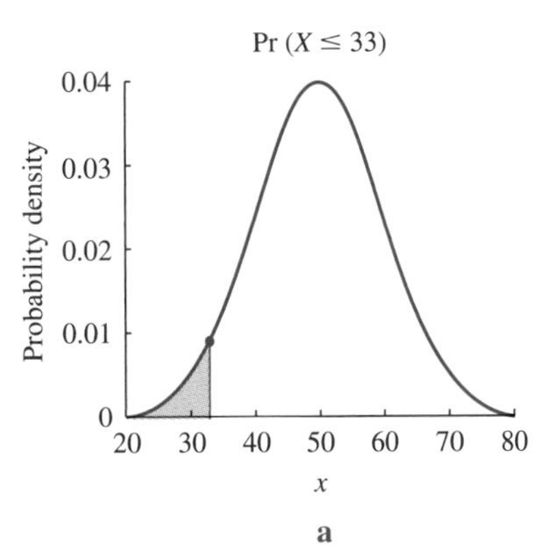

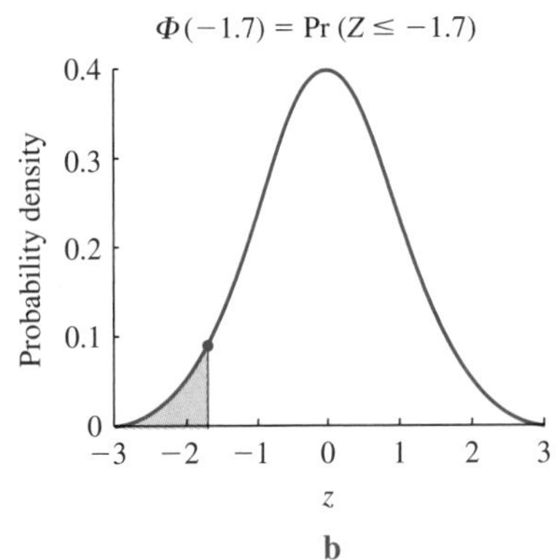

FIGURE 7.9.4
The value and the z-score

Example 7.9.6 Another Probability

To find the probability that $33.0 < X < 55.0$ if $X \approx N(50.0, 100.0)$, we find the z-scores associated with the two values $x = 33.0$ and $x = 55.0$ as

$$z_1 = \frac{33.0 - 50.0}{10.0} = -1.7$$
$$z_2 = \frac{55.0 - 50.0}{10.0} = 0.5.$$

The probability is then

$$\Pr(33.0 < X < 55.0) = \Pr(-1.7 < Z < 0.5) = \Phi(0.5) - \Phi(-1.7)$$
$$= 0.6915 - 0.0446 = 0.6469$$

(Figure 7.9.5).

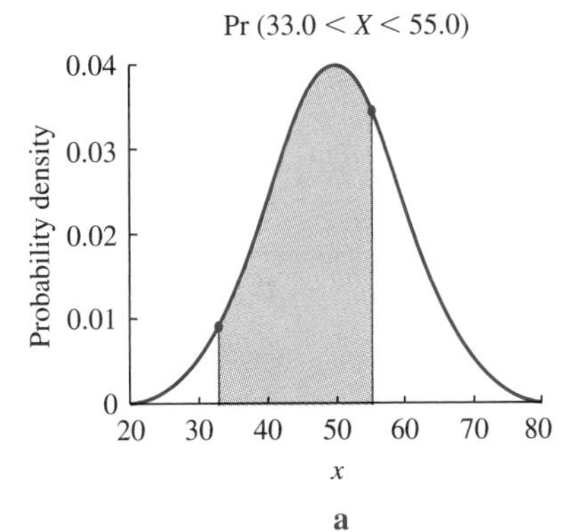

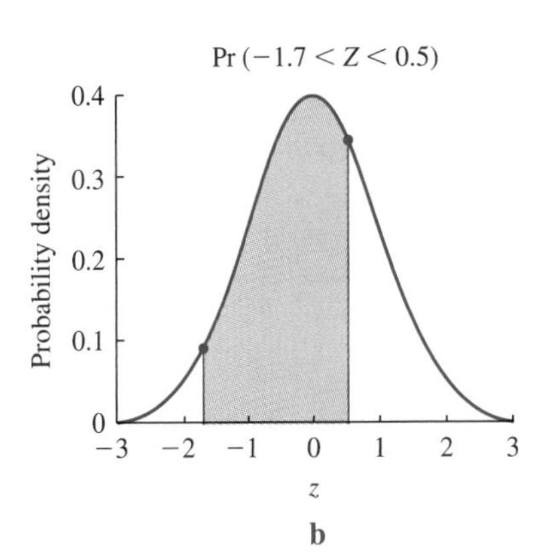

FIGURE 7.9.5
The value and the z-score

Example 7.9.7 Applying the Standard Normal to Plant Heights

We found the added height H of plants to be approximately normally distributed with mean 11.25 and variance 7.97 (Subsection 7.8.1). The standard deviation σ is $\sqrt{7.97} = 2.823$. The probability that a plant gains less than 6.0 cm in height is

$$\begin{aligned}\Pr(H \leq 6.0) &= \Pr\left(\frac{H - 11.25}{2.823} \leq \frac{6.0 - 11.25}{2.823}\right) \\ &= \Pr\left(Z \leq \frac{6 - 11.25}{2.823}\right) \\ &= \Pr(Z \leq -1.86) = \Phi(-1.86) = 0.031.\end{aligned}$$

Only about 3% of plants will gain less than 6.0 centimeters from these ten genes (Figure 7.9.6). Alternatively, we could first compute the z-score

$$z = \frac{6.0 - 11.25}{2.823} = -1.86$$

and then find

$$\Pr(H \leq 6.0) = \Pr(Z \leq -1.86) = \Phi(-1.86) = 0.031.$$

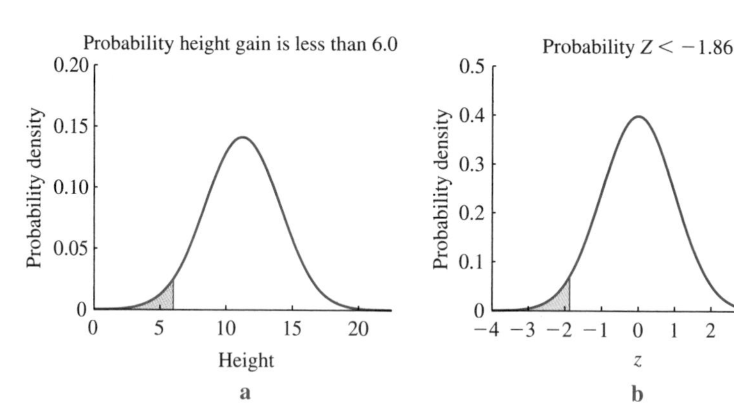

FIGURE 7.9.6
Computing the probability that height gain is less than 6.0

Example 7.9.8 Finding the Probability a Plant Height Gain Is Between Two Values

To find the probability that a plant gains between 10 and 15 cms (Figure 7.9.7), we can first compute the two z-scores, finding

$$\begin{aligned}z_1 &= \frac{10.0 - 11.25}{2.823} = -0.443 \\ z_2 &= \frac{15.0 - 11.25}{2.823} = 1.328.\end{aligned}$$

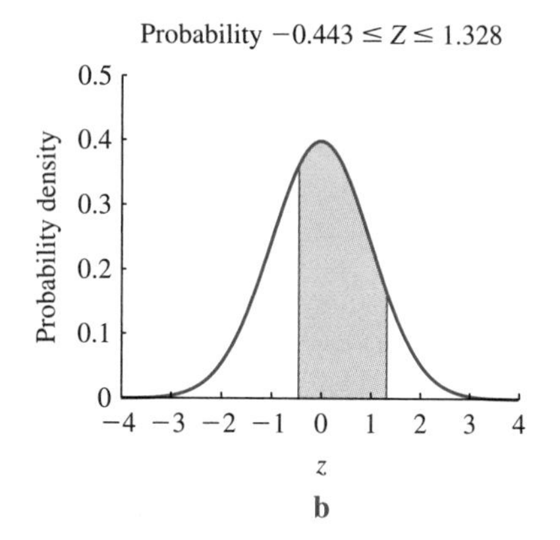

FIGURE 7.9.7
Computing the probability that height gain is between 10 and 15 cm

The probability is then

$$\begin{aligned}\Pr(10 \leq H \leq 15) &= \Pr(-0.443 \leq Z \leq 1.328)\\ &= \Phi(1.328) - \Phi(-0.443)\\ &= 0.908 - 0.329 = 0.579.\end{aligned}$$

Example 7.9.9 Finding the Probability a Plant Height Gain Exceeds Some Value

To find the probability that a plant gains more than 19 cm, we find the z-score

$$z = \frac{19.0 - 11.25}{2.823} = 2.745.$$

The probability is then

$$\Pr(H \geq 19.0) = \Pr(Z \geq 2.745) = 1 - \Phi(2.745) = 0.003.$$

Very few plants will gain this much height.

7.9.2 Normal Approximation of the Binomial

Suppose a heterozygous plant self-pollinates and produces 20 offspring. What is the probability that between 8 and 12 offspring are heterozygous? The number follows a binomial distribution with $n = 20$ and $p = 0.5$. Finding the probability directly requires computing a lot of factorials. Because the number of heterozygous offspring is a *sum*, we can use the Central Limit Theorem for Sums and the normal approximation. Each individual offspring can be described by a Bernoulli random variable with $p = 0.5$:

$$I_i = \begin{cases} 1 & \text{if heterozygous (probability 0.5)} \\ 0 & \text{if homozygous (probability 0.5).} \end{cases}$$

The total number of heterozygotes H is the sum

$$H = \sum_{i=1}^{20} I_i$$

and is therefore a binomial random variable with

$$\begin{aligned}E(H) &= np = 10.0\\ \mathrm{Var}(H) &= np(1 - p) = 5.0.\end{aligned}$$

As long as the I_i are **independent** (which will be the case if no offspring are twins), we can apply the Central Limit Theorem for Sums. The probability distribution for H approximately matches that for a normally distributed random variable X with the same mean and variance, or

$$X \approx N(10.0, 5.0)$$

(Figure 7.9.8).

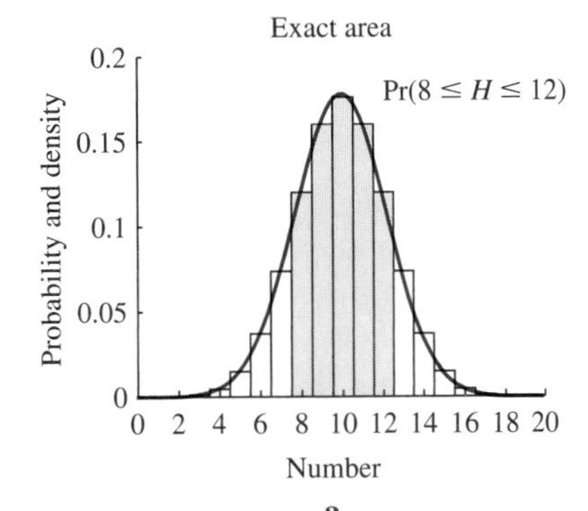

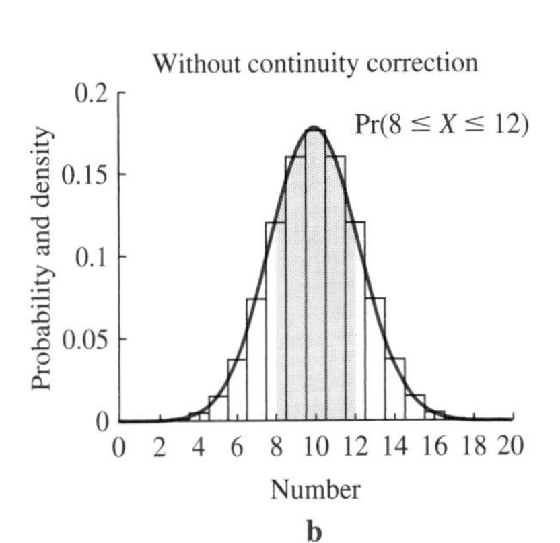

FIGURE 7.9.8
Approximating a binomial distribution with a normal

Although H and X are quite close together, H is a discrete random variable and X is a continuous random variable. We can use this fact to find an estimate better than

$$\Pr(8 \le H \le 12) \approx \Pr(8 \le X \le 12).$$

How can we correct for this difference? The probability that between 8 and 12 offspring are heterozygous is the sum of the areas of the bars centered at 8 through 12 (Figure 7.9.9a). These bars run from 7.5 to 12.5. To find the area under the normal distribution that approximates the area under the bars, we integrate from 7.5 to 12.5. The extra 0.5's are called the **continuity correction**. With this correction, we find $\Pr(7.5 \le X \le 12.5)$ by computing the z-scores:

$$z_1 = \frac{7.5 - 10}{\sqrt{5}} = -1.118$$
$$z_2 = \frac{12.5 - 10}{\sqrt{5}} = 1.118.$$

Then

$$\Pr(7.5 \le X \le 12.5) = \Phi(1.118) - \Phi(-1.118) = 0.8682 - 0.1318 = 0.7364.$$

(Figure 7.9.9b). The exact answer, found on the computer by adding up the binomial probabilities, is 0.7368. The approximation is extremely close. Without the continuity correction, we find $\Pr(8 \le X \le 12)$ by computing the z-scores:

$$z_1 = \frac{8 - 10}{\sqrt{5}} = -0.894$$
$$z_2 = \frac{12 - 10}{\sqrt{5}} = 0.894.$$

Then

$$\Pr(8 \le X \le 12) = \Phi(0.894) - \Phi(-0.894) = 0.8144 - 0.1855 = 0.7289,$$

which is quite a bit low (Figure 7.9.8b).

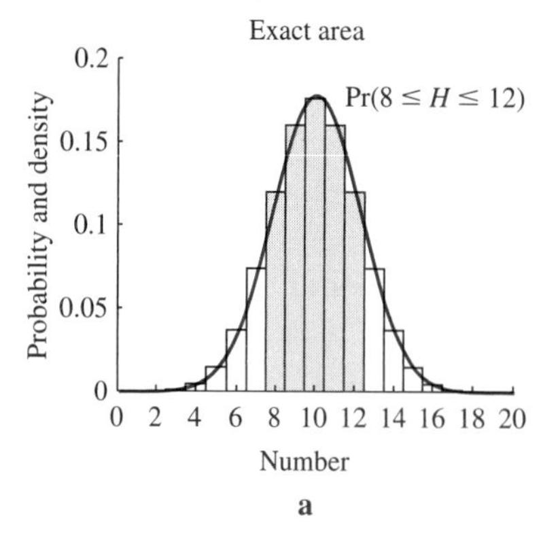

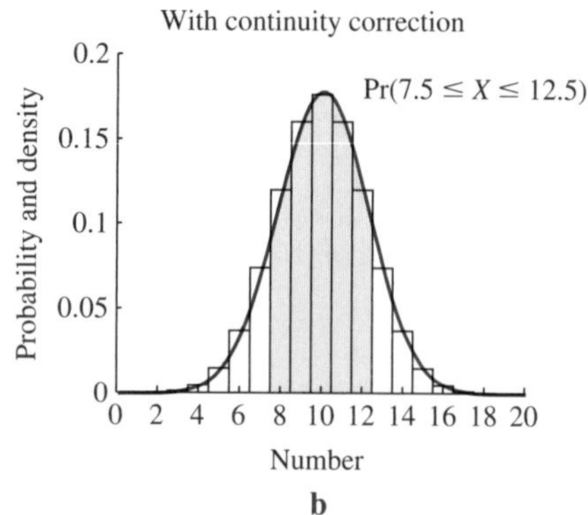

FIGURE 7.9.9

Approximating a binomial distribution with a normal: the continuity correction

Like the Central Limit Theorem, the normal approximation to the binomial works only when the number of terms is sufficiently large. In this case, we must have an expectation of at least five successes **and** five failures. We use the following condition.

Condition for Applying the Normal Approximation to the Binomial Use the normal distribution to a binomial with parameters n and p only when $np \ge 5$ and $n(1 - p) \ge 5$.

The probability of a failure is $1 - p$, so $n(1 - p)$ is the expected number of failures.

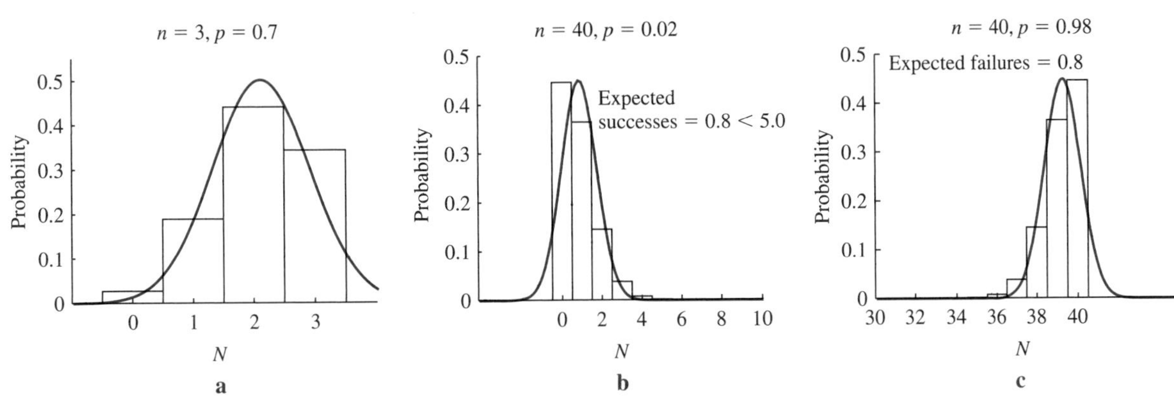

FIGURE 7.9.10
Three cases where the normal approximation is not very good

Example 7.9.10 Three Cases Where the Normal Approximation to the Binomial Does Not Apply

The binomial distribution with $n = 3$ and $p = 0.7$ does not have much of a bell shape (Figure 7.9.10a). The normal approximation is not accurate. With $p = 0.02$ and $n = 40$, the expected number of successes is $np = 0.8$, which is far less than 5 (Figure 7.9.10b). Conversely, if $p = 0.98$ and $n = 40$, the expected number of successes is $np = 39.2$, but the expected number of failures is only $n(1 - p) = 0.8$, far less than 5 (Figure 7.9.10c). Once again, we are not counting enough events for the Central Limit Theorem to apply.

Example 7.9.11 Applying the Normal Approximation to the Genetic Example I

Suppose we wish to figure out the probability of finding exactly eight heterozygous offspring out of 20 offspring from a heterozygous plant. A random variable X that follows a normal distribution with mean $\mu = np = 10$ and variance $\sigma^2 = np(1 - p) = 5$ approximates H. With the continuity correction,

$$\Pr(H = 8) \approx \Pr(7.5 \leq X \leq 8.5).$$

The z-scores are

$$z_1 = \frac{7.5 - 10}{\sqrt{5}} = -1.118$$

$$z_2 = \frac{8.5 - 10}{\sqrt{5}} = -0.671.$$

Therefore,

$$\begin{aligned}\Pr(H = 8) &\approx \Pr(7.5 \leq X \leq 8.5)\\ &= \Phi(-0.671) - \Phi(-1.118)\\ &= 0.2511 - 0.1318 = 0.1199.\end{aligned}$$

The exact answer to four decimal places is $b(8;\, 12,\, 0.5) = 0.1201$.

Example 7.9.12 Applying the Normal Approximation to the Genetic Example II

To find the probability of 15 or more heterozygous offspring,

$$\Pr(H \geq 15) \approx \Pr(X \geq 14.5).$$

The z-score is

$$z_1 = \frac{14.5 - 10}{\sqrt{5}} = 2.012.$$

Therefore,

$$\begin{aligned}\Pr(H \geq 15) &= \Pr(X \geq 14.5)\\ &= 1 - \Phi(2.012)\\ &= 1 - 0.9779 = 0.0221.\end{aligned}$$

It is very unlikely that 15 or more plants will be heterozygous.

7.9.3 Normal Approximation of the Poisson Distribution

Suppose seeds land in a region according to a Poisson process with a mean number of $\Lambda = 23$ per m^2 (as in Figure 7.7.10). What is the probability that between 18 and 28 seeds fall in a given square meter? Finding the exact probability with the Poisson distribution requires a lot of calculation. Again, we can simplify this calculation with the normal approximation to the Poisson distribution.

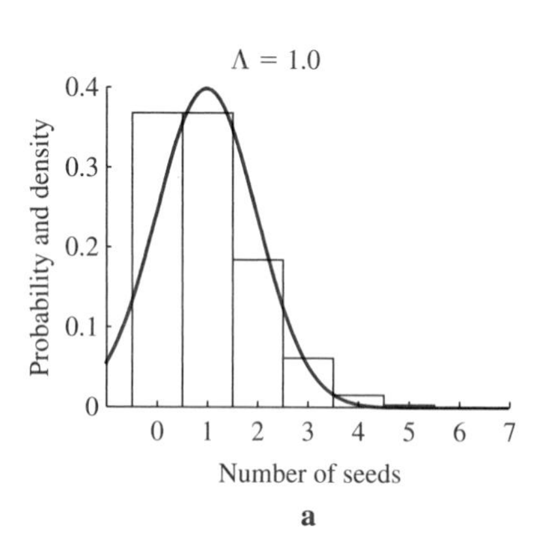

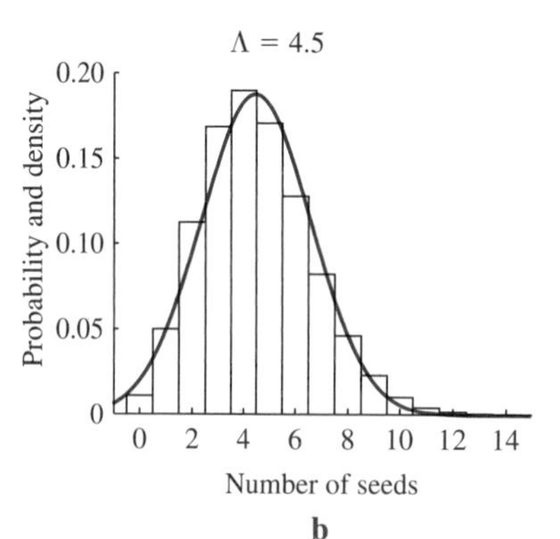

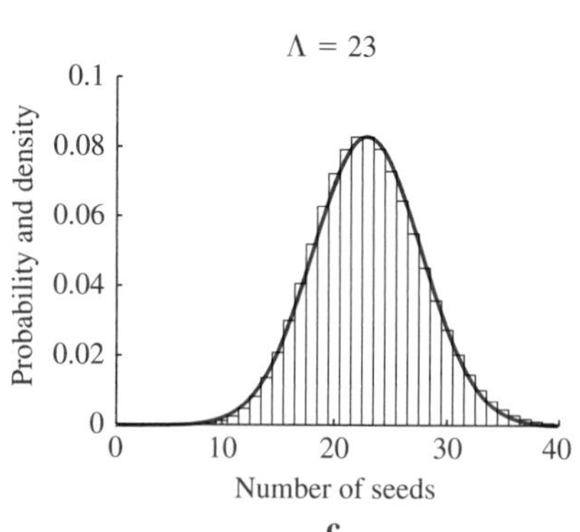

FIGURE 7.9.11
Poisson distribution and normal approximation for small values of Λ

The Poisson distribution with $\Lambda = 1.0$ (Figure 7.9.11a) is not at all bell-shaped; it becomes reasonably bell-shaped when $\Lambda = 4.5$ (Figure 7.9.11b), and quite so when $\Lambda = 23$ (Figure 7.9.11c). As with the binomial distribution, it is accurate to use the normal approximation to the Poisson distribution when there are at least five expected events.

Condition for Applying the Normal Approximation to the Poisson Distribution Use the normal distribution to a Poisson distribution with parameter Λ only when Λ is >5.

We compute probabilities just as with the binomial: find a normally distributed random variable X with matching mean and variance, and compute probabilities using the continuity correction.

To estimate the probability that there are between 18 and 28 seeds in a given square meter, we find the normal approximation

$$X \approx N(23, 23)$$

by matching the mean and variance (Figure 7.9.12). To find the probability, we use the continuity correction and integrate the normal p.d.f. from 17.5 to 28.5. Let S be a random variable representing the number of seeds.

Then

$$\begin{aligned}\Pr(18 \le S \le 28) &= \Pr(17.5 \le X \le 28.5)\\ &= \Phi\left(\frac{17.5 - 23}{\sqrt{23}}\right) - \Phi\left(\frac{28.5 - 23}{\sqrt{23}}\right)\\ &= \Phi(1.147) - \Phi(-1.147)\\ &= 0.8743 - 0.1257 = 0.7486.\end{aligned}$$

The exact answer is 0.7498.

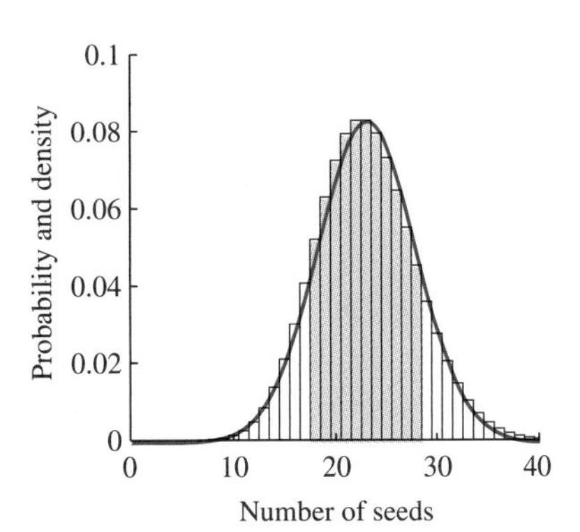

FIGURE 7.9.12
The Poisson distribution and the normal approximation for different values of *Lambda*

Summary We transformed a general normally distributed random variable into a **standard normal distribution** by subtracting the mean and dividing by the standard deviation. The corresponding value is known as the **z-score**. Because the probability density function for the standard normal is impossible to integrate, we use the cumulative distribution function $\Phi(z)$, which is widely available in tables and computers. To approximate discrete random variables described by the binomial and Poisson distributions, we find a normal distribution with matching mean and variance and compute probabilities accurately by including the **continuity correction**, extending the limits of integration by 0.5 on each side.

7.9 Exercises

Mathematical Techniques

1–4 ▪ Find the z-score (the number of standard deviations from the mean) for the following measurements.

1. A value of 11.0 drawn from a normal distribution with mean 13.0 and standard deviation 1.2.
2. A value of 0.9 drawn from a normal distribution with mean 0.5 and standard deviation 0.3.
3. A value of 12.0 drawn from a normal distribution with mean 10.0 and variance 25.0.
4. A value of 7.0 drawn from a normal distribution with mean 10.0 and variance 4.0.

5–10 ▪ Use the cumulative distribution function for the standard normal, $\Phi(z)$, to find the following probabilities. Shade the associated area on two graphs: the given normal distribution, and the standard normal distribution.

5. The probability of a value less than 0.7 drawn from a normal distribution with mean 0 and variance 1.
6. The probability of a value greater than −0.1 drawn from a normal distribution with mean 0 and variance 1.
7. The probability of a value greater than 11.0 drawn from a normal distribution with mean 13.0 and standard deviation 1.2 (as in Exercise 1).
8. The probability of a value less than 0.9 drawn from a normal distribution with mean 0.5 and standard deviation 0.3 (as in Exercise 2).
9. The probability of a value between 10.0 and 12.0 drawn from a normal distribution with mean 10.0 and variance 25.0 (as in Exercise 3).
10. The probability of a value between 7.0 and 13.0 drawn from a normal distribution with mean 10.0 and variance 4.0 (as in Exercise 4).

11–14 ▪ Using a table or computer program that can calculate the cumulative distribution function for the standard normal, find the following probabilities.

11. The masses of a type of insect are normally distributed with a mean of 0.38 g and a standard deviation of 0.09 g. What is the probability that a given insect has mass less than 0.40 g?

12. Scores on a test are normally distributed with mean 70 and standard deviation 10. What is the probability that a student scores more than 85?

13. Measurement errors are normally distributed with a mean of 0 mm and a standard deviation of 0.01 mm. Find the probability that a given measurement is within 0.012 mm of the true value.

14. The number of insects captured in a trap on different nights is normally distributed with mean 2950 and standard deviation 550. What is the probability of capturing between 2500 and 3500 insects?

15–18 ▪ Use the normal approximation to the binomial with and without the continuity correction to find the following probabilities. How much difference does the continuity correction make? Exact answers are given in parentheses.

15. 43% of trees are infested by a certain insect. What is the chance of randomly choosing 40 trees, fewer than 10 of which are infested? (0.0144)

16. 30% of the cells in a small organism are not functioning. What is the probability that an organ consisting of 250 cells is functioning if it requires 170 cells to work? (0.777)

17. In a certain species of wasp, 75% of individuals are female. Find the probability that between 70 and 80 wasps (inclusive) are female in a sample of 100 wasps. (0.7967)

18. 10% of people are known to carry a certain gene. In a sample of 200, what is the probability that between 5 and 15 people carry the gene? (0.1431)

19–22 ▪ Use the normal approximation to the Poisson distribution with and without the continuity correction to find the following probabilities. Compare with the exact answers given in parentheses.

19. Seeds have fallen into a region with an average density of 20 seeds per square meter. What is the probability that a particular square meter contains fewer than 15 seeds? (0.1565)

20. Mutations occur along a piece of DNA at a rate of 0.023 per thousand codons. What is the probability of 30 or more mutations in a piece of DNA 1 million codons long? (0.0915)

21. Insects are caught in a trap at a rate of 0.21 per minute. What is the probability of catching between 10 and 15 insects in an hour? (0.6040)

22. Molecules enter a cell at a rate of 0.045 per second. What is the probability that between 150 and 200 molecules enter during an hour? (0.8352)

23–26 ▪ The normal approximation cannot be used in the following cases. Why not? Find another way to compute the probabilities and compare with the normal approximation.

23. A fair coin is flipped eight times. Find the probability of no more than one head.

24. 5% of people are infected with a disease. Find the probability of choosing 50 people and finding none who are infected.

25. Under the circumstances in Exercise 21, find the probability of catching more than one insect in a minute.

26. Under the circumstances in Exercise 20, find the probability of no mutation in a piece of DNA ten thousand codons long.

27–28 ▪ Suppose that if $X \approx N(\mu, \sigma^2)$ and $Z = \dfrac{X - \mu}{\sigma}$.

27. Show that $E(Z) = 0$ without writing down any integrals.

28. Show that $\mathrm{Var}(Z) = 1$ without writing down any integrals.

29–32 ▪ Recall the data describing the probabilities of the outcomes of four experiments counting the number of mutants in a bacterial culture (Section 7.8, Exercises 5–8). Each experiment is repeated (independently), and the total number of mutants is counted.

29. Experiment **a** is repeated 20 times (as in Section 7.8, Exercise 5). Find the probability that there is a total of between 30 and 40 mutants, and compare with the sketch in that Exercise 5.

30. Experiment **b** is repeated 80 times (as in Section 7.8, Exercise 6). Find the probability that there are between 30 and 40 mutants, and compare with the sketch in that Exercise 6.

31. Experiment **c** is repeated 100 times (as in Section 7.8, Exercise 7). Find the probability that there is an average of less than 1.8 mutants, and compare with the sketch in that Exercise 7.

32. Experiment **d** is repeated 25 times (as in Section 7.8, Exercise 8). Find the probability that there is an average of less than 1.8 mutants, and compare with the sketch in that Exercise 8.

Applications

33–34 ▪ Consider again the populations in Section 7.8, Exercises 33–36. For each population, use the normal approximation to estimate the probability that the population is larger after the given periods of time.

33. Immigrants arrive into and emigrate from population **a** for 20 years (as in Section 7.8, Exercise 33). What is the probability that the population grows if immigrants arrive over a period of 40 years? Why does it become larger?

34. Immigrants arrive into and emigrate from population **b** for 20 years (as in Section 7.8, Exercise 34). What is the probability that the population grows if immigrants arrive over a period of 10 years? Why does it become smaller?

35–38 ▪ Consider again the situation in Section 7.8, Exercises 29–32, describing the putative genetic basis of IQ.

35. What is the probability of an IQ above 100 with just the genes of large effect (as in Section 7.8, Exercise 29)?

36. What is the probability of an IQ above 100 with just the genes of small effect (as in Section 7.8, Exercise 30)?

37. What is the probability of an IQ above 100 with just environmental factors (as in Section 7.8, Exercise 31)?

38. What is the probability of an IQ above 100 with both genetics and environmental factors (as in Section 7.8, Exercise 32)?

39–42 ▪ In a population growing by reproduction, the logarithm of population size can be approximated using the normal distribution and the Central Limit Theorem (Section 7.8, Exercises 37–40). Consider populations growing for the given number of years with the given distribution for the random variable R giving per capita production. Find the probability that the population size P_t lies in the given range assuming that $P_0 = 100$.

39. $R_i = 4$ with probability 0.5, $R_i = 0.25$ with probability 0.5 (as in Section 7.8, Exercise 37). Find the probability that P_{50} lies between 1 and 10000.

40. $R_i = 4$ with probability 0.25, $R_i = 0.25$ with probability 0.75 (as in Section 7.8, Exercise 38). Find the probability that P_{25} lies between 1 and 3.

41. R has p.d.f. $g(x) = 5.0$ for $1.0 \leq x \leq 1.2$ (as in Section 7.8, Exercise 39). Find the probability that P_{50} lies between 50 and 150.

42. R has p.d.f. $g(x) = 1.25$ for $0.7 \leq x \leq 1.5$ (as in Section 7.8, Exercise 40). Find the probability that P_{50} lies between 50 and 150.

43–44 ▪ Suppose that the alleles **A** and **a** for height are additive, meaning that plants with genotype **AA** are tall, plants with genotype **Aa** are intermediate, and those with genotype **aa** are short. If an **Aa** plant is crossed with another **Aa** plant, 1/4 of the offspring should be tall, 1/2 should be intermediate, and 1/4 should be short. Assuming that the other conditions for the binomial distribution are met, find the probabilities of the following. Suppose such a cross produces 36 offspring.

43. Find the binomial distribution and its normal approximation for the number of tall offspring. What is the probability of getting ten or fewer?

44. Find the binomial distribution and its normal approximation for the number of intermediate offspring. What is the probability of getting eight or fewer?

45–46 ▪ Starting with 50 molecules, each leaving with probability 0.2/min, find the normal probability distribution approximating the number remaining at the following times. Use it to estimate the probability that more than 40 molecules remain.

45. 1 min

46. 2 min

47–48 ▪ Figure 6 in Section 6.1, as part of Example 5, illustrating stochastic immigration, was generated by adding two individuals with probability 0.5 and zero with probability 0.5 for 100 generations. The results in the figure show final populations of 106 and 96 people.

47. Use the normal approximation to find the probability that the results lie between 96 and 106 people (inclusive).

48. The population size can be thought of as twice a binomial random variable, because we are counting pairs. Use this idea, and the continuity correction, to estimate the probability that the results lie between 96 and 106 people (inclusive). How do the results compare?

49–50 ▪ Genes in different organisms have different rates of mutation. Use the normal approximation to compute the following probabilities.

49. A gene has a mutation rate of 0.02 mutations per generation. Estimate the probability of more than 50 mutations in a period of 2000 generations.

50. A gene has a mutation rate of 0.002 mutations per generation. Estimate the probability of exactly one mutation in a period of 2000 generations. How does this compare with the exact answer in Section 7.7, Exercise 41?

Computer Exercises

51. Generate 100 numbers from a binomial distribution with $n = 20$ and $p = 0.3$.

   a. Plot the theoretical histogram for this binomial distribution and overlay it with the normal approximation.

   b. Plot a histogram of your results.

   c. Overlay this with a graph of the normal distribution with matching mean and variance. Did you have to modify the formula to make the areas match?

   d. Compare the fraction of times your random number was equal to 4 with the fraction based on the binomial distribution. Shade the region on your graph of the theoretical histogram from part **a**, and compare with the area on the normal curve with the continuity correction.

   e. Do the same for numbers between 7 and 10 (inclusive).

52. Let R_1 be a random variable giving the per capita reproduction in one population with p.d.f. $g(x) = 5.0$ for $1.0 \leq x \leq 1.2$, and let R_2 be a random variable giving the per capita reproduction in a population with p.d.f. $g(x) = 1.25$ for $0.7 \leq x \leq 1.5$ (Section 7.7, Exercise 40). We found that $E[\ln(R_1)] = 0.0939$ and $E[\ln(R_2)] = 0.0723$.

   a. Simulate each population for 50 steps starting from an initial condition of 100.

   b. How do the sizes compare with the expectation?

   c. Simulate each population 100 times, and record the population at $t = 50$. Break them into 15 bins and plot a histogram of the results. Do the results seem to have a normal distribution?

   d. Now take the natural log of each final population size and plot a histogram. Why do these results have a more nearly normal distribution? The actual population sizes are said to have a **lognormal distribution**.

Supplementary Problems

1. A certain student always arrives the day before the test clutching the practice problems, and asks about each with probability 0.6.
   a. Under what conditions is the binomial distribution appropriate?
   b. Under these conditions, what is the probability that the first problem the student asks about is problem 4?
   c. What is the probability that this student asks about exactly four out of the first six problems?
   d. What are the expectation and variance of the number asked about if there are eight practice problems?

2. A second student frequently appears after the first has left. This student also asks about 60% of the problems, but asks about a particular problem with probability 0.8 if the first student does not.
   a. Is the behavior of these students independent? How could you tell?
   b. Give the entire joint distribution. What is the probability that neither asks about problem 5?
   c. Draw the marginal distribution for the second student.
   d. Draw the conditional distributions for the second student.
   e. Find the covariance if asking is worth -1 and not asking is worth 2.

3. During a drizzle, raindrops hit a given square meter at a rate of 12.5 per minute.
   a. What is the expected number of raindrops after 30 s?
   b. What is the probability that exactly five raindrops hit in 30 s?
   c. What is the probability that no raindrops hit in 10 s?
   d. If the drizzle lasts 1 h, would you be surprised to find a square meter with exactly 700 wet spots?

4. During a particularly wet spring, a rainy day follows a rainy day with probability 0.8, and a rainy day follows a dry day with probability 0.5.
   a. Write a discrete-time dynamical system describing this model.
   b. Draw the distributions that describe the weather on one day conditional on the weather of the previous day.
   c. Find the fraction of rainy days after a long time.
   d. Draw the joint distribution of weather on consecutive days after a long time.

5. A band is composed of four guitarists and a drummer. The drummer only knows how to keep a steady beat, and each guitarist knows only the A and D chords. Each time the drummer hits the beat, each guitarist plays an A chord with probability 0.75 and a D chord with probability 0.25.
   a. Under what conditions is the binomial distribution appropriate for describing the number of A chords played on a given beat?
   b. What is the expectation and variance of the number of guitarists playing that A chord on a given beat?
   c. If the binomial distribution is appropriate, find the probability that exactly two guitarists hit the A chord.
   d. Find the probability that all four guitarists play the same chord.
   e. Find the probability that all four guitarists play the same chord on exactly two out of three consecutive beats.

6. People attending a concert by the band in Exercise 5 are observed to be leaving at a rate of three people per minute.
   a. Under what conditions is the Poisson distribution appropriate for describing the number of people who left during a given minute?
   b. Under these conditions, find the probability that nobody leaves during a period of length t and sketch the function.
   c. Find the expected number and the variance of people leaving during a given minute.
   d. Find the probability that exactly three people leave during a given minute.
   e. Find the probability that exactly three people leave during a 2-min interval.

7. After the departure of people in Exercise 6, the remaining throngs show wild enthusiasm after a song with probability 0.3, mild enthusiasm with probability 0.2, and hurl insults with probability 0.5. The band reduces the volume by 20% in response to wild enthusiasm, by 10% in response to mild enthusiasm, and increases volume by 20% in response to insults. Suppose the volume begins at 100.
   a. Find the expected volume after one song.
   b. Would you expect the band to be louder or quieter at the end of a long concert?

8. After a particularly demoralizing performance, two guitarists in the band from Exercise 5 decide to start working together. Recall that they know only two chords, A and D.
   a. If they begin by playing chords independently, give the joint distribution.
   b. If the second player plays an A with probability 90% when the first player does, give the joint distribution.
   c. If playing an A is worth 1 point and a D is worth 0, find the correlation.
   d. Draw the marginal and conditional distributions in this case.

9. An aspiring worm biologist decides to use a random procedure to stock his lab. He chooses among three catalogues: Baxter, Fisher, and Sigma. He orders from Baxter with

probability 0.2, from Fisher with probability 0.7, and from Sigma with probability 0.1. Products are delivered with probability 0.9 from Baxter, 0.95 from Fisher, and 0.8 from Sigma.

a. Find the probability that a product is delivered.

b. Find the probability that a product that was not delivered was ordered from Fisher.

c. Find the joint distribution.

d. What strategy would maximize the probability of delivery?

10. A population switches between phases of growth (G) and decline (D) in subsequent years according to $\Pr(G \mid D) = 0.15$ and $\Pr(D \mid G) = 0.05$. The per capita growth rate is 1.2 during growth and 0.7 during decline.

a. If the population grows in the first year, find the expected size after 2 yr.

b. If the population grows the first year, what is the expected time until the first decline? Sketch the distribution.

c. Will this population grow or decline in the long run?

d. Find the joint distribution and covariance between years after a long time.

11. A staining technique has a 20% chance of success. Let S_n be the random variable representing the number of successful stainings in n attempts.

a. What are the conditions for S_n to have a binomial distribution? What are the parameters?

b. Find $E(S_{10})$ and $\Pr(S_{10} = 2)$.

12. Under the same conditions as in Exercise 11, let T be a random variable representing the number of the first successful attempt. For example, $T = 2$ means that the first try was a failure and the second a success.

a. Graph the probability distribution of T.

b. Find the expectation and variance of T.

13. The density of bacterial colonies on a circular dish of radius 3 cm is 0.12 colonies per square centimeter. Let B be a random variable describing the number of colonies in the whole dish.

a. Find the approximate probability of there being a colony in 1 mm^2.

b. Find the expectation and variance of B.

c. Find the probability of there being no colonies on the whole dish.

14. The woodpecker *Picoides vilosus* enjoys eating both the wood-boring tipulid (*Ctenophora vittata*) and the hickory bark beetle (*Scolytus quadrispinosus*). During any given minute, the woodpecker eats one tipulid with probability 0.5 and none with probability 0.5. Independently, during each minute, it eats one beetle with probability 0.25 and none with probability 0.75.

a. Sketch the distribution describing the time when the first beetle is eaten.

b. Find the expected time, $\bar{B}$, when the first beetle is eaten.

c. Find the probability that no tipulids had been eaten at or before $\bar{B}$, and the expected number of tipulids eaten by time $\bar{B}$.

15. A measurement X has normal distribution with mean 10 and variance 9, and an independent measurement Y has normal distribution with mean 20 and variance 4.

a. Find $\Pr(X \leq 11)$.

b. Find $E(X + Y)$ and $\mathrm{Var}(X + Y)$.

c. Find $\Pr(X + Y \leq 13)$.

16. Bubbles are produced by a chemical reaction at a rate of 1.8 each per minute. Two assiduous students observe two bubbles during the first minute, one during the second minute, and fail to pay attention during the third minute because they are studying math. They then knuckle down and pay attention for the next 30 minutes.

a. What is the probability that they didn't miss anything during that third minute? What is the probability that they missed two or more bubbles?

b. What are the expectation and variance of the number of bubbles seen during the 30 min of paying attention?

c. Write the formula for the probability of exactly 50 bubbles in 30 min.

d. Two other students pay attention for only 15 min and multiply their results by 2. What is the expectation and variance for them? Explain the difference.

17. Because students have proven unable to pay close attention to chemical reactions, the professor designs a new version of the experiment in Problem 16 where bubbles do not pop. Students check for bubbles once per minute, record whether there are any (but don't count them), and pop all the ones that are there. The probability of at least one bubble during a minute is 0.83.

a. What is the probability that the first bubble is seen in the fifth minute?

b. Find the expected number of students in a class of 30 who would see no bubble during the first 4 min. What is the variance? What does it mean?

c. What is the probability that a particular team of students will record bubbles in exactly 25 min out of 30?

d. Estimate the probability of 25 or more minutes with bubbles during 30 min (no factorials allowed). Sketch the probability distribution and the indicate the probability you estimated.

18. Students from Exercise 17 are asked to measure two numbers each minute, the number of bubbles B and the temperature T. B can take on only two values 0 and 1, T can take on the values 20 and 30. The joint distribution is

	$B = 0$	$B = 1$
$T = 20$	0.1	0.1
$T = 30$	0.07	0.73

a. Find the marginal distributions.

b. Find and sketch the conditional distribution of B when $T = 20$ and when $T = 30$.

c. Find the covariance, and explain what you would do to make this a more useful statistic.

19. Two mathematicians propose models for the growth of an elk population. The first says that the number increases by B, where B is a random variable with a Poisson distribution and mean 5. The second says that the population increases by exactly 4% each year. Suppose the actual population begins at 100 and is followed for 10 years (see data in the table below).

Year t	Number of Elk	Year t	Number of Elk
0	100	6	134
1	103	7	153
2	104	8	155
3	115	9	151
4	111	10	156
5	118		

a. Sketch data that could result from each of the models.

b. What is the expected number of elk and the variance after 10 yr according to the models?

c. Which model predicts that the population will grow faster in the long run?

d. What are the strengths and weaknesses of the two models in making sense of the data?

e. If you had to construct a model of this population, what factors would you like to include?

Projects

1. An important class of stochastic processes that describe many types of populations are called **birth-death** processes. In particular, suppose that a population has size $N(t)$ at some time t. The probability that a new individual is born during the short interval of time between t and $t + \Delta t$ is assumed to be $\lambda N(t)\Delta t$. In other words, births occur at a **rate** proportional to the population size. Similarly, the probability that an individual dies during the short interval of time between t and $t + \Delta t$ is assumed to be $\mu N(t)\Delta t$. Deaths also occur at a **rate** proportional to the population size.

We begin by assuming that $\mu = 0$ (no deaths). This model is called a **birth process**.

a. What is the probability that $N(t + \Delta t) = N(t) + 1$?

b. What is the probability that $N(t + \Delta t) = N(t)$?

c. Let $E(t)$ be the expected number of individuals at time t. What is $E(t + \Delta t)$ (use the fact that the expectation of the sum is the sum of the expectations)? Write and solve a differential equation for $E(t)$. Use the initial condition $E(0) = N_0$.

d. Let $V(t)$ be the variance at time t. What is $V(t + \Delta t)$ (use the fact that the variance of the sum of independent random variables is the sum of the variances)? Write and solve a differential equation for $V(t)$. Use the initial condition $V(0) = 0$.

e. Figure out a way to simulate this birth process (use $\lambda = 0.1$).

f. Starting from an initial condition of $N(0) = 30$, compare the values of three simulations of $N(t)$ with the values of $E(t)$ for up to $t = 20$.

g. Using the calculation of $V(t)$, draw curves 2 standard deviations above and below $E(t)$. Do your simulations always remain within these bounds?

Now assume that $\mu \neq 0$.

a. What is the probability that $N(t + \Delta t) = N(t) + 1$?

b. What is the probability that $N(t + \Delta t) = N(t) - 1$?

c. What is the probability that $N(t + \Delta t) = N(t)$?

d. Write and solve a differential equation for the expected population size $E(t)$ with initial condition $E(0) = N_0$. When do you expect this population to grow?

e. Write and solve a differential equation for the variance $V(t)$ with initial condition $V(0) = 0$. Does the death process increase or decrease the variance?

f. Figure out a way to simulate this birth-death process (use $\lambda = 0.1$ and $\mu = 0.08$).

g. Starting from an initial condition of $N(0) = 30$, compare the values of three simulations of $N(t)$ with the values of $E(t)$ for up to $t = 20$. Discuss your results.

h. Starting from an initial condition of $N(0) = 1$, compare the values of three simulations of $N(t)$ with the values of $E(t)$ for up to $t = 20$. Discuss your results. How does the actual population differ from $E(t)$? Draw curves 2 standard deviations above and below $E(t)$. Do your simulation results remain within these bounds?

i. Try the same exercise with $\lambda = \mu = 0.1$. Do you expect this population to go extinct? Can you guess about how long it will take?

2. One of the classic experiments in genetics was designed to determine whether mutations occur in response to selection, or whether they occur randomly before selection. Following an inspirational moment watching slot machines, Salvador Luria realized that random mutations would produce bacterial cultures with either very few or very many mutants, while mutations produced in response to selection would produce a more even distribution of results. This project leads you through the reasoning behind this classic work.

The bacteria in the experiment were subjected to selection by phage, a type of virus that kills any bacterium that is not resistant. The bacteria from a culture were placed on a plate imbued with high concentrations of phage. Any resistant bacterium would survive and grow, producing a colony that could be seen and counted after a day or two. Luria and Delbrück used this count to figure out the exact number of resistant bacteria out of the hundreds of millions in the culture.

The basic population dynamics are assumed to obey the following rules.

- A culture with $N(t)$ individuals at time t reproduces at per capita rate r.
- The population begins with N_0 individuals at time $t = 0$.
- The experiment runs until time $t = T$.

To begin, use these assumptions to find the number of bacteria at time T.

We can now analyze what would happen if mutations occurred in response to selection. If the $N(T)$ bacteria at the end of the experiment had never seen phage, all would be susceptible to attack and none would survive. However, if they had a chance to catch a whiff of phage and mutate before attack, they could survive and produce colonies.

a. If the chance a bacterium smelled phage and quickly mutated was ν and if each bacterium responded independently, what is the distribution of the number of surviving colonies?

b. What is the expected number of surviving colonies?

c. What is the variance?

d. What is the probability of 0 surviving colonies?

The results are quite different if any mutation making bacteria resistant must have already occurred in culture. Any mutant that arose early in the growth phase would produce a large number of progeny, all of which would be resistant. Thus, a few experiments would generate huge numbers of resistant colonies, while most would generate rather few. Denote the number of mutants at time t by $M(t)$ and suppose that bacteria mutate from being susceptible to resistant at a per capita rate μ.

e. Explain why the expected number of mutants $\bar{M}(t)$ follows the differential equation

$$\frac{d\bar{M}}{dt} = r[\mu N(t) + \bar{M}].$$

f. Show that

$$\bar{M}(t) = N(t)(1 - e^{-\mu r t})$$

is a solution (remember to use the formula you found for $N(t)$).

g. Suppose that $\mu r T$ is very small (which is generally the case). Use the tangent line approximation to show that

$$\bar{M}(T) \approx \mu r T N(t).$$

h. Compare this with your result when mutations occur in response to selection. Could you use these results to distinguish the two cases?

i. Let $P(t)$ denote the probability of zero mutants at time t. Write and solve a differential equation for $P(t)$. Could this value be used to distinguish the two cases?

The most important difference between the two cases is that the second has higher variance. The following steps will help you compute $V(t)$, the variance of $M(t)$.

j. What is the probability that a mutation occurs between times t and $t + \Delta t$?

k. If a mutation occurs at time t, how many offspring will it have at time T?

l. Multiply the value times the probability and integrate to check your result for the expected number of mutants.

m. What is the variance of the number produced between t and $t + \Delta t$? What is this approximately equal to if Δt is small?

n. The total number of mutants is the sum of **independent** events occurring during all time intervals from t to $t + \Delta t$. The variance of $M(t)$ is therefore the sum (or the integral) of the variances over each interval. Compute $V(T)$.

o. What is the ratio of $V(t)$ to $\bar{M}(t)$? Will it be large or small when $N(t)$ is large? How does this ratio compare to the case when bacteria respond to selection? Could you use this to distinguish the two cases?

p. What is the coefficient of variation of $M(t)$? How does this compare to the case when bacteria respond to selection?

Reference

S. E. Luria and M. Delbrück. Mutations of bacteria from virus sensitivity to virus resistance. *Genetics*, 28:491–511, 1943.

3. (Based on the paper "Departure time versus departure rate: How to forage optimally when you are stupid" by F. R. Adler and M. Kotar, *Evolutionary Ecology Research*, 11:411–421, 1999.) The Marginal Value Theorem (Subsection 3.3.2) gives the optimal time for a forager to leave a patch. However, some organisms might be unable to leave at the right moment, perhaps because they are too stupid to remember how long they have been there, or have no way to assess the current resource level. Worse yet, some animals might leave a patch simply because they had become lost and could not find the way back. How might such a cognitively challenged organism forage efficiently?

Suppose that the travel time T between patches is equal to 1.0 and that the amount of food gathered by time t is $F(t) = 1 - e^{-\alpha t}$. The parameter α describes how quickly the resource is depleted by the organism.

a. Find the optimal moment for this organism to leave the patch for different values of α (the equation cannot be solved algebraically).

b. Suppose that the organism gets lost at rate λ. What distribution describes the time when this organism leaves?

c. What is the average amount of food found during this time?

d. What is the average time spent in a patch?

e. Write a formula for the average amount of food during the total time for a visit (time in patch plus travel time).

f. Solve for the value of λ that maximizes this (you should be able to do this without picking a particular value of α).

g. Compare the average time spent by this stupid organism with the optimal time computed in **a** for small and large values of α. Can you make sense of the result?

h. Think of another model of departure times (such as a normal distribution) and repeat the analysis.

Chapter 8

Introduction to Statistical Reasoning

Probability theory and the basic models of probability are part of **mathematical modeling**. A model summarizes a biological process, and mathematical tools are used to predict the outcome of that process. If the biological process includes stochastic elements, we predict the **distribution** of results rather than a single specific outcome. When the measured quantity is a count, for example, the binomial distribution or the Poisson distribution might describe the results.

The methods of **statistics** relate real measurements to the distribution predicted by a probabilistic model. Often, we summarize model results and experimental data with descriptive statistics, such as the mean, to make comparison possible. We can thus use measurements to *infer* something about the underlying biological process. If the observed data are a highly unlikely outcome of the model, something must be wrong with or missing from the model. This branch of science is called **inferential statistics** (Figure 8.0.1). The following is our guiding principle for using inferential statistics.

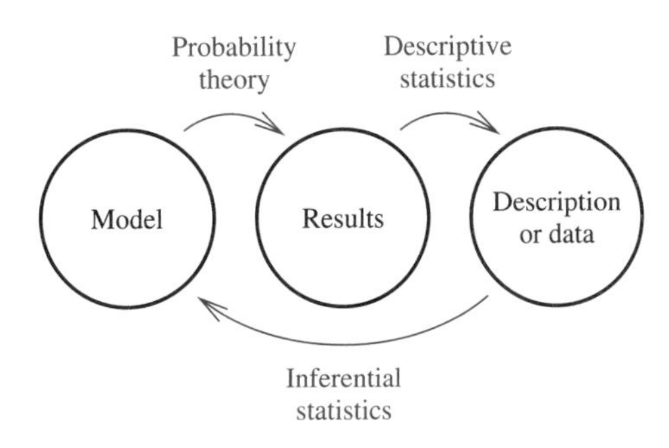

FIGURE 8.0.1
The relations among probability theory, descriptive statistics, and inferential statistics

The key to understanding and correctly applying statistics is understanding the underlying probabilistic model.

8.1 Statistics: Estimating Parameters

The set of measurements used to make statistical inferences is called a **sample**, thought of as part of a **population**. The population is the set of all possible measurements from which the sample is drawn (Figure 8.1.1). A common problem in statistics is to estimate characteristics of a population from a sample. The population can be a real entity (all the people in Utah) or a mathematical abstraction (all possible results of flipping a coin 1 million times).

Statistics (numbers computed from measurements of a sample) are used to estimate the parameters (numerical characteristics) of a population. For example, we might estimate the average height of people in Utah by choosing a sample of 100 people from the phone book and computing the average of their heights. This is an example of **parameter estimation**. In this section, we use the binomial distribution to help us

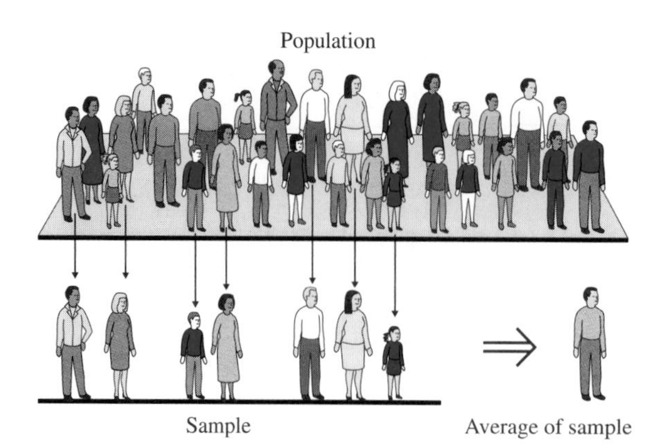

FIGURE 8.1.1
A sample from a larger population

estimate the fraction p of a population with a particular property, and we use exponential distributions to estimate the rate λ at which events occur. This technique is called **maximum likelihood**.

8.1.1 Estimating the Binomial Proportion

One hundred people are tested for a particular allele and twenty test positive. What proportion p of the total population has the allele? The common sense guess is $p = \frac{20}{100} = 0.2$. We have estimated a property of a population from a statistic describing a sample.

To work with this estimate, we must know something about the testing and sampling procedures. First, we must know how dependable the test is. Does it always give the right answer? Estimating results from an undependable test can be quite risky. Second, we would like to assume that the individuals chosen for the sample are independent and have the same value of p, thus satisfying the assumptions of the binomial distribution.

When might our sample fail to satisfy the assumptions? If the allele is most common in people with Japanese ancestry and our sample came from an area with many such people, we would tend to overestimate the true proportion. This sample does not accurately reflect the population as a whole. A similar problem arises if we choose many related individuals. In either case, the sample is not composed of *independent* individuals. Knowing that one brother has the allele makes it much more likely that another does. Choosing a good sample is difficult (techniques are covered in texts on **statistical design**). In general, samples should be chosen "randomly," on the basis of criteria unrelated to any factor influencing the measurement.

Our guess $p = 0.2$ is an **estimator** of the proportion in the whole population.

Definition 8.1 An **estimator** is a value of some parameter describing a population computed from a sample.

Estimated values are generally denoted with a "hat." For example, the estimator $\hat{p}$ is defined as

$$\hat{p} = \frac{\text{number of people with allele}}{\text{number of people tested}}.$$

The estimator $\hat{p}$ is a guess of the true proportion p in the whole population. It would be rather naive to believe that p is *exactly* equal to $\hat{p}$. If the total population consists of 1,782,260 people, there would have to be 356,452 people with the allele for p to be equal to $\hat{p}$. If there were 355,987 people with the allele, the true p would be 0.1997, more than close enough to the estimator for any practical purpose. Part of the process of finding an estimator is to derive a range around $\hat{p}$ that we can be "confident" contains the true p, as we will study in the next section.

8.1.2 Unbiased Estimators

There are many ways to evaluate the quality of different estimators. As an illustration, we will consider one criterion that is important in some contexts. An estimator is a random variable—a number that derives from an experiment just like any other measurement. We can thus evaluate its **expectation**. An estimator is said to be **unbiased** if its expectation is equal to the true value.

Example 8.1.1 Is the Common Sense Estimator of the Proportion Unbiased? A Numerical Experiment

Suppose that the true proportion of individuals carrying a particular allele is $p = 0.18$. The estimator $\hat{p}$ is found by drawing a sample of 100 individuals and computing the proportion of people with that allele. The data analyzed in the previous subsection were precisely of this sort. We found 20 people out of 100 with the allele, and estimated the proportion in the population as $\hat{p} = \frac{20}{100} = 0.2$.

The expectation of $\hat{p}$ is the value we would find on average, if we repeated the experiment many times. The accompanying table gives the results of 10 replicate of the experiment (including the original example).

Sample	Number with Allele	$\hat{p}$
1	20	0.20
2	14	0.14
3	17	0.17
4	18	0.18
5	16	0.16
6	15	0.15
7	22	0.22
8	14	0.14
9	20	0.20
10	16	0.16

The value of the estimator is different for each sample, like any other random variable. The average of the 10 estimators is 0.172, which is close to but not exactly equal to the true value of $p = 0.18$.

To demonstrate that an estimator is unbiased we cannot simply do a small number of simulations and check whether the average value of the estimator is equal to the true value of the parameter. Instead, we use mathematical techniques for finding the expectation to mimic the results of repeating this experiment an *infinite* number of times.

Definition 8.2 An estimator $\hat{p}$ of a parameter with true value p is **unbiased** if

$$\mathrm{E}(\hat{p}) = p$$

and is **biased** if

$$\mathrm{E}(\hat{p}) \neq p.$$

Example 8.1.2 The Common Sense Estimator of the Binomial Proportion Is Unbiased: Proof

To show that the common sense estimator $\hat{p}$ is unbiased, we must describe its probability distribution. The number of people G with the allele is a random variable that follows the binomial distribution with $n = 100$ and $p = 0.18$. Our estimator is another random variable, found by dividing G by 100, or

$$\hat{p} = \frac{G}{100}.$$

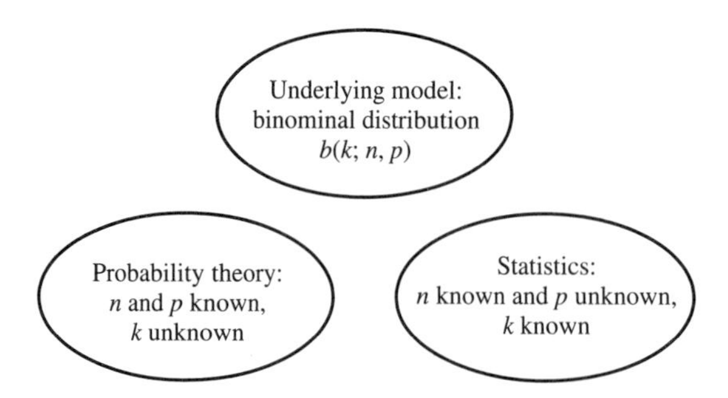

FIGURE 8.1.2

The different known and unknown values in probability and statistics

Using the expectation of the binomial distribution (Subsection 7.4.2), we find

$$\begin{aligned} E(\hat{p}) &= \frac{E(G)}{100} && \text{divide out the constant with Theorem 7.6} \\ &= \frac{0.18 \cdot 100}{100} && E(G) = np \text{ because } G \text{ has a binomial distribution} \\ &= 0.18 = p. && \text{the expectation of the estimator is the true value} \end{aligned}$$

The estimator $\hat{p}$ gives the right answer on average and is indeed unbiased.

By thinking of an estimator as a random variable, we can compute other statistics. An important property, for example, is that an estimator have low variance. In some circumstances, statisticians can show that one estimator has the minimum variance among all unbiased estimators.

8.1.3 Maximum Likelihood

We found the estimator $\hat{p}$ by using "common sense." The method of **maximum likelihood** uses "uncommon sense" derived from understanding the underlying probabilistic model. Unlike common sense, this method works as well on complex models as it does on the stuff of everyday life.

The method proceeds in two steps. First, we compute the **likelihood function**, which gives the probability (or likelihood) of the data as a function of the unknown parameter value. Second, we find the value of the parameter that **maximizes** the likelihood.

Definition 8.3 Suppose a population is described by an unknown parameter p, and that data have been collected from a particular sample. Then

$$L(p) = \Pr(\text{Data if parameter is equal to } p).$$

A likelihood is a probability, but it is given a different name to emphasize the difference between statistics and probability theory (Figure 8.1.2). Both concern the probability of getting a particular set of data from a given underlying model. In the language of conditional probability, this is the probability of the data **conditional** on a given model, or Pr(Data|Model). In probability theory, however, the model is known, and we seek to find the probability of different outcomes. In statistics, the data are known, and we seek to infer something about the model.

Example 8.1.3 The Difference Between Probability and Likelihood

In the genetics example (Subsection 8.1.1), the number of people G with the allele is a random variable that follows the binomial distribution with $n = 100$. In probability theory, we usually think of the value of the other parameter, p, as known. If $p = 0.18$, we find the probability distribution of G as

$$\Pr(G = k) = b(k; 100, 0.18).$$

The measured value k acts as the variable.

In statistics, however, we have measured that $G = 20$ and do not know the underlying probability p. The likelihood function is

$$L(p) = b(20; 100, p).$$

We use the binomial distribution because the underlying model is the same. However, the variable is the unknown parameter p.

We can think of the method of maximum likelihood as a series of thought experiments. How **likely** is it that we would have gotten exactly the same data we did (20 out of 100 in the examples), with different possible values of the unknown parameter p?

Example 8.1.4 Likelihood of Data if $p = 0.18$

Suppose we guessed that the true proportion in the population is $p = 0.18$. Is this plausible? We can quantify this by computing the probability of finding 20 out of 100 with the allele. The number G with the allele has a binomial distribution with $n = 100$ and $p = 0.18$, so

$$\Pr(20 \text{ out of } 100 \text{ with } p = 0.18) = b(20; 100, 0.18) = 0.0870$$

(Figure 8.1.3). This does not seem very likely. Does this mean that the proportion $p = 0.18$ is not plausible?

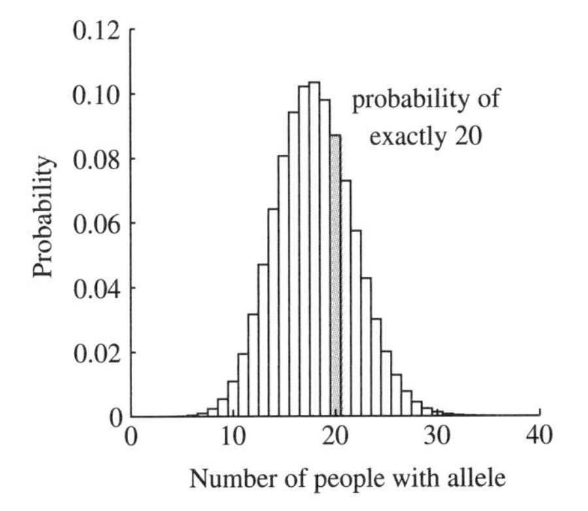

FIGURE 8.1.3
The likelihood of 20 out of 100 if $p = 0.18$

Example 8.1.5 The Likelihood Function for the Genetic Data Set

Guess of p	$L(p) = b(20; 100, p)$
0.10	0.0012
0.12	0.0074
0.14	0.0258
0.16	0.0567
0.18	0.0870
0.20	0.0993
0.22	0.0881
0.24	0.0629
0.26	0.0369
0.28	0.0182
0.30	0.0078

Using the binomial distribution, we can find the likelihood of getting exactly 20 out of 100 people with a given allele for any value of the unknown underlying parameter p. The likelihood function is

$$L(p) = b(20; 100, p) = \binom{100}{20} p^{20}(1-p)^{80},$$

giving the probability of the data (20 out of 100) as a function of the unknown parameter p. Some values are given in the accompanying table. If p were equal to 0.1, we would find exactly 20 out of 100 with the small likelihood $L(0.1) = 0.0012$. Such a result would occur only about once in 1000 trials. If p were 0.2, we would find exactly 20 out of 100 with the rather larger likelihood $L(0.2) = 0.0993$. However, even this probability is rather small. Does this mean that the common-sense estimator of the proportion $\hat{p} = 0.2$ is not plausible either?

Likelihoods tend to take on small values. The method of maximum likelihood gives one way to interpret these probabilities.

Definition 8.4 Suppose $L(p)$ is a likelihood function. The **maximum likelihood estimator** of the parameter p is the point where $L(p)$ takes on its maximum value.

Our estimator is the value of the parameter *most likely* to have produced the data.

Example 8.1.6 Finding the Maximum Likelihood Estimator

The likelihood function for the genetics data is

$$L(p) = b(20;\, 100,\, p) = \binom{100}{20} p^{20}(1-p)^{80}$$

(Figure 8.1.4). The maximum likelihood estimator of p is the point where $L(p)$ takes on its maximum value.

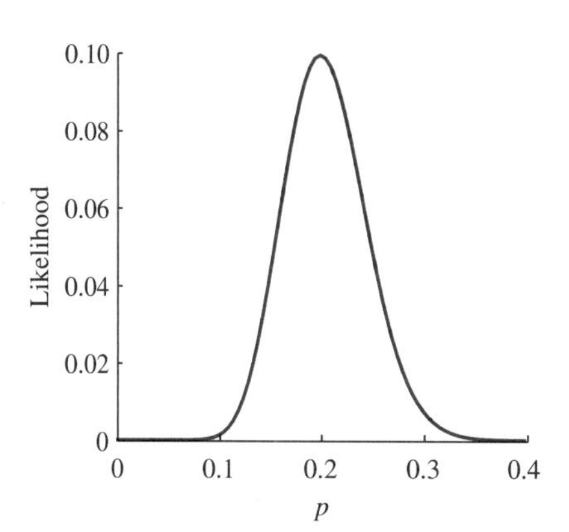

FIGURE 8.1.4

The likelihood of 20 out of 100 as a function of p

We can take the derivative with respect to p by using the product rule (Theorem 2.10), finding

$$\frac{dL}{dp} = \binom{100}{20} \frac{d}{dp}\left[p^{20}(1-p)^{80}\right]$$

$$= \binom{100}{20} \left[20p^{19}(1-p)^{80} - 80p^{20}(1-p)^{79}\right].$$

Setting the result equal to 0 and factoring,

$$0 = \binom{100}{20} \left[20p^{19}(1-p)^{80} - 80p^{20}(1-p)^{79}\right]$$

$$= \binom{100}{20} p^{19}(1-p)^{79}\left[20(1-p) - 80p\right].$$

Because this is in factored form, critical points occur where one of the factors is 0, or

$$\binom{100}{20} p^{19}(1-p)^{79}\left[20(1-p) - 80p\right] = 0$$

$$p^{19} = 0 \quad \text{or} \quad (1-p)^{79} = 0 \quad \text{or} \quad 20(1-p) - 80p = 0$$

$$p = 0 \;\text{ or }\; p = 1 \;\text{ or }\; p = 0.2.$$

To find the maximum, we compute the likelihood at the critical points.

$$L(0) = 0$$

$$L(1) = 0$$

$$L(0.2) = 0.0993.$$

The maximum likelihood estimator is therefore $p = 0.2$, exactly matching the common-sense estimator. The minimum likelihood of 0 at $p = 0$ and at $p = 1$ means that the measured result of 20 out of 100 is *impossible* in these two cases (Figure 8.1.4).

These calculations do not change the fact that the likelihood of the data with $p = 0.2$ is less than 10%. How can we put our trust in this unlikely result? The question results from thinking about the technique backwards. The data may or may not be unlikely, but they are exactly what we observed. No particular result may be likely for an experiment with many possible outcomes.

Example 8.1.7 Interpretations of Very Small Likelihoods

Suppose that an animal is homozygous with the genotype **BB** at each of 100 loci (Figure 8.1.5). The mother is known to be heterozygous with the genotype **Bb** at every locus. There are two possible fathers, one with the genotype **BB** at every locus, and

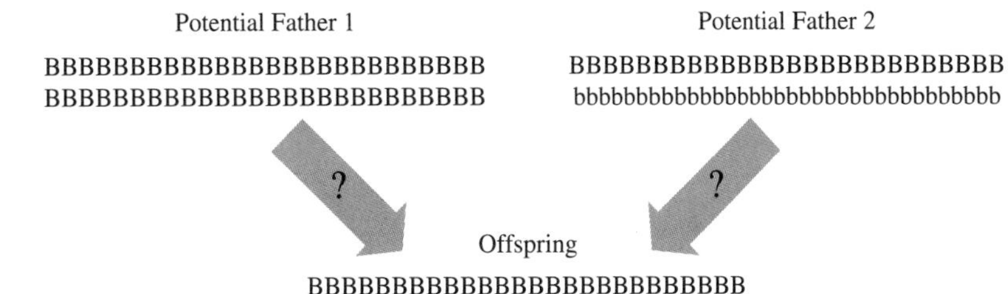

FIGURE 8.1.5
The likelihood of two fathers and their unlikely offspring

one, like the mother, heterozygous with the genotype **Bb** at every locus. Which is more likely to be the real father? With father 1, the offspring has genotype **BB** at a particular locus with probability 0.5 (because it had to get the **B** allele from the mother), and a probability of **BB** at *every* locus of

$$\begin{aligned} L(\text{father 1}) &= \Pr(\mathbf{BB} \text{ at every locus}) \\ &= \Pr(\mathbf{BB} \text{ at locus 1}) \times \Pr(\mathbf{BB} \text{ at locus 2}) \times \ldots \times \Pr(\mathbf{BB} \text{ at locus 100}) \\ &= 0.5^{100} = 7.89 \times 10^{-31}, \end{aligned}$$

which is astronomically small. With father 2, the offspring would have genotype **BB** at a particular locus with probability 0.25 (it had to get the **B** allele from both the mother and the father), and a probability of **BB** at *every* locus of

$$\begin{aligned} L(\text{father 2}) &= \Pr(\mathbf{BB} \text{ at every locus}) \\ &= \Pr(\mathbf{BB} \text{ at locus 1}) \times \Pr(\mathbf{BB} \text{ at locus 2}) \times \ldots \times \Pr(\mathbf{BB} \text{ at locus 100}) \\ &= 0.25^{100} = 6.22 \times 10^{-61}, \end{aligned}$$

which is astronomically smaller. It is virtually impossible that either father would produce such an offspring. But *given that the offspring exists*, the first father is much more likely to have produced it. It is far more reasonable to deduce that the first male fathered the offspring than to deduce that the offspring does not exist.

In the genetics example, the value of the likelihood function is particularly small when $p < 0.1$ or $p > 0.35$. If the true proportion p were smaller than 0.1, we would be quite surprised to see 20 out of 100. The method of maximum likelihood finds the single value of p that is *most likely* to have generated the actual observations. Because $L(0.18)$ is not much lower than $L(0.2)$, that guess is nearly as consistent with the data. We will later develop a rule of thumb to indicate when we can exclude a particular guess (like $p = 0.1$) with confidence (although not with certainty).

8.1.4 Estimating a Rate

Suppose we observe 10 molecules leaving a cell and record the waiting time for each.

Molecule	Waiting Time	Estimated Rate = Reciprocal of Waiting Time
1	$t_1 = 0.311$	3.208
2	$t_2 = 0.791$	1.263
3	$t_3 = 1.196$	0.836
4	$t_4 = 0.539$	1.854
5	$t_5 = 0.575$	1.738
6	$t_6 = 0.908$	1.100
7	$t_7 = 0.088$	11.315
8	$t_8 = 1.764$	0.566
9	$t_9 = 0.619$	1.614
10	$t_{10} = 0.038$	25.949
Average	0.683	4.945

If we suspect that the results follow an exponential distribution, we would like to use the data to estimate the parameter λ, the rate at which molecules leave.

One approach, indicated in the final column, is to estimate the rate for each molecule as the reciprocal of the waiting time, because the expectation of the waiting time with parameter λ is $\frac{1}{\lambda}$ (Theorem 7.18). For example, the first molecule left after a wait of 0.311, which is the mean time if the true rate λ were

$$\lambda = \frac{1}{0.311} = 3.215.$$

The average of these estimated rates is 4.945. Alternatively, we could find the average wait by averaging the waiting times (0.683 in this case) and estimate λ as the reciprocal (1.464). These estimates are quite different. Is either one right?

With an understanding of the underlying exponential model, we can use the method of maximum likelihood. How do we compute a likelihood when the underlying model has a probability density function (p.d.f.) rather than a probability distribution? Because the value of the p.d.f. is proportional to the probability, we find the likelihood function, $L(\lambda)$, in the same way, but we keep in mind that the likelihood cannot be interpreted as a probability.

The probability that the waiting time is near a particular value t is given by the p.d.f.

$$f(t) = \lambda e^{-\lambda t}.$$

For example, the probability density at $t_1 = 0.311$ is

$$f(t_1) = \lambda e^{-0.311\lambda}.$$

We find the likelihood of the entire data set by assuming that all the measurements are independent and multiplying the value of the p.d.f. at each point. Therefore,

$L(\lambda) = f(t_1) \times f(t_2) \times \cdots \times f(t_{10})$	probability of data is product of probabilities
$= \lambda e^{-\lambda t_1} \times \lambda e^{-\lambda t_2} \times \cdots \times \lambda e^{-\lambda t_{10}}$	plug in p.d.f.
$= \lambda^{10} e^{-\lambda t_1} \times e^{-\lambda t_2} \times \cdots \times e^{-\lambda t_{10}}$	substitute t_i into exponential p.d.f.
$= \lambda^{10} e^{-\lambda \sum_{i=1}^{10} t_i}$	use law 1 of exponents to change product into sum
$= \lambda^{10} e^{-6.833\lambda}$	use the fact that the sum of the t_i's is 6.83

(Figure 8.1.6).

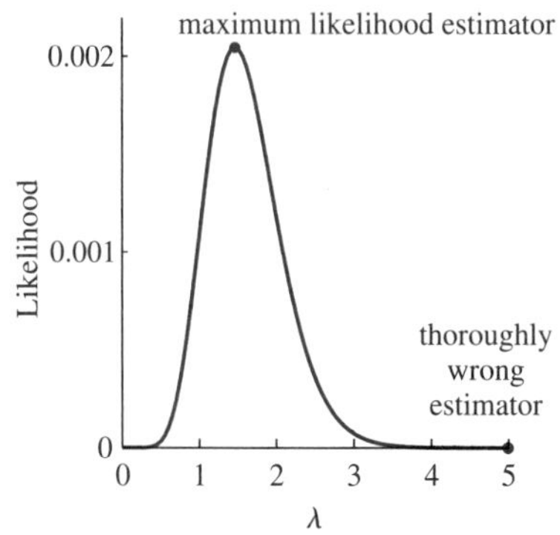

FIGURE 8.1.6
The likelihood function for waiting times

To compute the maximum likelihood estimator, we differentiate the likelihood function. Let $\bar{T}$ denote the average of the waiting times (here 0.6833). We can rewrite the likelihood function as

$$L(\lambda) = \lambda^{10} e^{-10\lambda\bar{T}} \gg$$

because $\sum t_i = 10\bar{T}$ by the definition of the average. Taking the derivative, we get

$$\frac{dL}{d\lambda} = 10\lambda^9 e^{-10\lambda\bar{T}} - 10\bar{T}\lambda^{10} e^{-10\lambda\bar{T}}$$
$$= 10\lambda^9 e^{-10\lambda\bar{T}}(1 - \lambda\bar{T}).$$

This is 0 at $\lambda = 0$ and at $\lambda = \frac{1}{\bar{T}}$. At these points,

$$L(1.464) = 2.05 \times 10^{-3}$$
$$L(0) = 0,$$

so the maximum is 1.464. The maximum likelihood estimator of the rate is the reciprocal of the average waiting time. The likelihood derived by averaging the rates (the value 4.945) is $L(4.945) = 1.895 \times 10^{-8}$, far smaller than the maximum.

Summary We have begun the study of **inferential statistics**, the attempt to infer properties of whole **populations** based on the properties of **samples**. Choosing a sample appropriately is the domain of **statistical design**. Assuming the sample was appropriately chosen, we estimated the proportion of a population with a particular trait first with common sense and then with **maximum likelihood**. The **likelihood function** expresses the probability that the observed data would have occurred for different possible parameter values. We applied the method of maximum likelihood to estimate the rate of an exponential process.

8.1 Exercises

Mathematical Techniques

1–4 ▪ Suppose we wish to calculate the proportion of days that the temperature rises above 20°C. Evaluate the following sampling schemes.

1. Sample 100 consecutive days beginning on January 1.
2. Sample 100 consecutive days beginning on June 1.
3. Sample the temperature on March 15 for 100 years.
4. What might be a good method if only 100 days could be sampled?

5–6 ▪ A clever pollster decides to find the average income of people by calling random individuals on the phone. It turns out, however, that people with cellular phones make $40,000 and people with regular phones make $20,000. Furthermore, 20% of people have cellular phones.

5. What is the true average income in this population?
6. What income would be estimated if people with cell phones were easier to reach, and 50% of the people called had cell phones?

7–8 ▪ Consider again the situation in Exercises 5 and 6, but suppose that incomes for people with cell phones are normally distributed according to $N(40000, 4.0 \times 10^6)$ (measured in 1999 U.S. dollars). The distribution for people without cellular phones is $N(20000, 1.0 \times 10^6)$.

7. What distribution describes the result of sampling 20 people with cellular phones? Use the rule of thumb that 95% of the distribution lies within two standard deviations of the mean to give a probable range. Compare this with the true average of the population.
8. What would be the results of sampling 20 people without cellular phones? Use the rule of thumb that 95% of the distribution lies within two standard deviations of the mean to give a probable range. Compare this with the true average of the population.

9–12 ▪ Find the likelihood as a function of the binomial proportion p for each of the following.

9. Flipping 2 out of 4 heads with a fair coin. Evaluate at $p = 0.5$, the value for a fair coin.
10. Rolling 2 out of 4 sixes with a fair die. Evaluate at $p = 1/6$, the value for a fair die.
11. Flipping 2 out of 12 heads with a fair coin.
12. Rolling 2 out of 12 sixes with a fair die.

13–14 ▪ Find the probability distribution associated with the following random variables, and identify which part corresponds to the likelihood found in the earlier problem.

13. Let H represent the number of heads in four flips of a fair coin. Compare with Exercise 9.
14. The number N of sixes rolled in four rolls of a fair die. Compare with Exercise 10.

15–16 ▪ Find the likelihood as a function of the binomial proportion p in each of the following cases, and find the maximum likelihood.

15. Team A wins five out of six games in a series against team B. Find the maximum likelihood estimator of the probability that team A wins a game against team B. If you were willing to gamble, would it make sense to enter a bet about the next game in the series where you win \$1 if team A wins, but lose \$6 if team A loses?

16. One out of 150 people you know wins \$500 in a raffle that costs \$5 to enter. Find the maximum likelihood estimator of the probability of winning the raffle. What is your best guess of the average payoff?

17–20 ▪ Find the likelihood as a function of the Poisson parameter Λ, find the maximum likelihood estimator, and evaluate the likelihood at the maximum and at the other given value of Λ.

17. Twenty events occur in 1 min. Compare the likelihood with the maximum likelihood estimator of Λ with the likelihood if $\Lambda = 10.0$.

18. Ten high energy cosmic rays hit detector over the course of 1 yr. Compare the likelihood with the maximum likelihood estimator of Λ with the likelihood if $\Lambda = 8.0$.

19. The number of events that occur are counted for 3 min. Twenty events occur the first minute, 16 events occur the second minute, and 21 events occur the third minute. Compare the likelihood with the maximum likelihood estimator of Λ with the likelihood if $\Lambda = 20.0$.

20. Ten high energy cosmic rays hit detector in its first year, 7 in the second year, 11 in the third year, and 8 in the fourth and final year. Compare the likelihood with the maximum likelihood estimator of Λ with the likelihood if $\Lambda = 10.0$.

21–24 ▪ Find the likelihood as a function of the parameter q of a geometric distribution, find the maximum likelihood estimator, and evaluate the likelihood at the maximum and at the other given value of q.

21. Flies are tested for the ability to learn to fly toward the smell of potato, and the first to succeed is the 13th. Compare with the likelihood of $q = 0.1$.

22. Random compounds are tested for the ability to suppress a particular type of tumor, and the first to succeed is the 94th. Compare with the likelihood of $q = 0.005$.

23. The experiment testing flies for the ability to learn to fly toward the smell of potato is repeated three times. In the first experiment the first fly to succeed is the 13th, in the second experiment it is the 8th, and in the third experiment it is the 12th. Compare with the likelihood of $q = 0.1$.

24. The experiment testing compounds for the ability to suppress tumors is repeated twice. In the first experiment the first compound to succeed is the 94th, and in the second experiment it is the 406th. Compare with the likelihood of $q = 0.005$.

25–26 ▪ Write down the equations that would express the fact that the following estimators are unbiased.

25. The estimator of q in Exercise 21.

26. The estimator of Λ in Exercise 17. If you think about the definition of the expectation, you might be able to demonstrate that this estimator is unbiased.

Applications

27–28 ▪ In each of the following cases where a very small number of individuals is tested for an allele, find and graph the likelihood function for the proportion p of individuals in the whole population with this allele, find the maximum likelihood estimator, and make sense of the likelihood at $p = 0$ and $p = 1$.

27. Two individuals are tested for a particular allele and one has it.

28. Three individuals are tested for a particular allele, and all three have it.

29–32 ▪ Thirty out of 100 individuals are found to be infected with a disease. Estimate the proportion of infected women and infected men in the following circumstances. Assuming that the whole population is composed of 50% women, estimate the infected proportion in the whole population.

29. The sample consists of 20 out of 50 infected women and 10 out of 50 infected men.

30. The sample consists of 20 out of 40 infected women and 10 out of 60 infected men.

31. The sample consists of 20 out of 20 infected women and 10 out of 80 infected men.

32. The sample consists of 0 out of 50 infected women and 30 out of 50 infected men.

33–34 ▪ Two couples are trying to have more girl babies. For each, find the likelihood function for the fraction q of female sperm and the maximum likelihood estimator, and compare with the likelihood of $q = 0.5$.

33. The first couple has seven boys before having a girl. Use the geometric distribution to build the likelihood as a function of q.

34. Another couple has four boys, then one girl, then three more boys. Find the likelihood as a function of q.

35–36 ▪ Use the method of maximum likelihood to estimate the rate λ from the accompanying table of data drawn from an exponential distribution.

Event	Waiting Time 1	Waiting Time 2
1	1.565	0.47279
2	0.888	1.69516
3	0.874	3.67104
4	5.156	0.97018
5	0.018	0.37539
6	0.048	0.44228
7	1.496	3.79148
8	0.422	1.19057
9	0.721	0.14163
10	1.119	0.19354

In each case, find the likelihood function, find the maximum likelihood, and say whether it seems probable that the true rate is 1.0.

35. For waiting time 1.

36. For waiting time 2.

37–40 ▪ Mutations are counted in four pieces of DNA that are 1 million base pairs long. There are 14 mutations in the first piece, 17 in the second piece, 8 in the third piece, and 5 in the fourth piece.

37. Write the likelihood function for the expected number of mutations per million bases in the first piece and find the maximum likelihood estimator.

38. Write the likelihood function for the expected number of mutations per million bases in the second piece and find the maximum likelihood estimator.

39. Write the likelihood function for the expected number of mutations per million bases in the first two pieces and find the maximum likelihood estimator. Compare this with the estimated expected number for each of the two pieces separately.

40. Write the likelihood function for the expected number of mutations per million bases in the first four pieces and find the maximum likelihood estimator. Compare this with the estimated expected number for each of the four pieces separately.

41–44 ▪ Mutation rates differ depending on whether changing the nucleotide base changes the amino acid (nonsynonymous sites) or not (synonymous sites). A piece of DNA has 200 nonsynonymous sites with 12 mutations and 100 synonymous sites with 15 mutations. Our goal is to estimate the mutation rate.

41. Estimate λ_s, the mutation rate for synonymous sites, from the given data.

42. Estimate λ_n, the mutation rate for nonsynonymous sites, from the given data.

43. Suppose we assume that the synonymous and nonsynonymous rates are both equal to the same value λ. Estimate λ, and compare with the values of λ_s and λ_n found in Exercises 41 and 42.

44. Suppose instead that the synonymous rate is three times that of the nonsynonymous rate. Formally, $\lambda_n = \lambda$ and $\lambda_s = 3\lambda$. Estimate λ, and compare with the values of λ found in Exercises 41 and 42.

45–48 ▪ Color blindness is due to a recessive allele that appears on the X chromosome. If the color blindness allele has frequency p, a fraction p of males will show the phenotype, and a fraction p^2 of females will show the phenotype (because they require two copies). We wish to estimate the fraction p from a sample of 1000 males, 90 of whom are color-blind, and 1000 females, 13 of whom are color-blind.

45. Estimate p using just the males.

46. Estimate p using just the females.

47. Write the likelihood function for males and females together.

48. Evaluate the likelihood function in Exercise 47 at $p = 0.09$, $p = 0.114$, and $p = 0.1$. Where do you think the maximum might be? If you are very determined, it is possible to solve for the maximum.

Computer Exercises

49. Simulate the following experiments.

a. People have a 0.4 chance of having black hair. Simulate 50 groups of five people. How many groups have exactly two out of five people with black hair?

b. Suppose instead that two out of five people are found with black hair. Simulate 50 groups of five people with different unknown probabilities p that a person has black hair. Use values of p ranging from 0 to 1.0. Which value produces the most groups that exactly match the data?

c. Find and plot the likelihood function describing this case. Where is the maximum? What does it correspond to in your simulation?

d. Explain how the two simulations differ. In each case, indicate what is known and what is unknown.

50. Cells are placed for 1 min in an environment where they are hit by X-rays, some of which are damaging. Cells not hit by the damaging rays are healthy, those hit exactly once are damaged, and those hit more than once are dead. By measuring the states of a number of cells, we wish to infer the rate at which cells are hit by damaging rays. Let x denote the unknown parameter of the Poisson distribution. Use the formula for the Poisson distribution to compute the probabilities p_0 of no hits, p_1 of one hit, and p_m of more than one hit in 1 min as *functions* of x.

a. Suppose the true value of x is 3.0. Plot the histogram.

b. Simulate 50 cells, and count how many you have of each type. (To keep things interesting, keep sampling until you get at least one cell of each type.)

c. Compare the results of your simulation with the idealized histogram.

d. Now pretend that x is unknown. We can use the method of maximum likelihood to analyze our data. Find the likelihood function L of these data (it is the product of the likelihoods for each of the 50 cells) as a function of the unknown parameter y, and let S be the natural log of L. Plot $S(y)$ over a reasonable range.

e. Find the maximum of S and mark it on your graph.

f. Find the $S(x)$ for $x = x_{max}$, $x = 2$, $x = 4$, and the "truth" $x = 3$, and indicate each on your graph.

8.2 Confidence Limits

The method of maximum likelihood gives a way to estimate parameter values from data. We know, however, that our estimator is not *exactly* right. We want a **range** of values that we can be confident contains the true value. The idea is formulated with

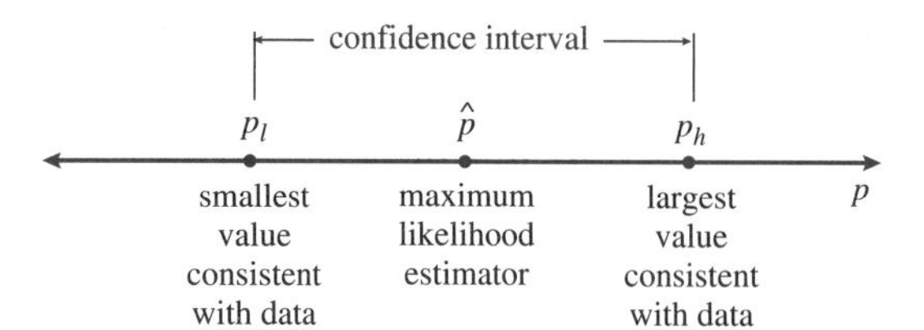

FIGURE 8.2.1
Confidence limits

confidence limits, the range of parameter values that might plausibly have produced our results, and we can be "confident" include the true value. We compute confidence limits exactly in the binomial case and then estimate them with computer simulation and likelihood.

8.2.1 Exact Confidence Limits

When we count 20 out of 100 people who have an allele, our best guess of the population proportion p is $\hat{p} = 0.2$. How can we find an **interval** that contains the population proportion 95% of the time? We attack this problem with the exact probabilities described in this section and apply the normal approximation described in the next section.

A **confidence interval** is described by two numbers, p_l and p_h, between which the true value should lie (Figure 8.2.1). The lower confidence limit is the smallest value $p_l < \hat{p}$ that is consistent with the observed data. To find it, we perform a thought experiment, much like the derivation of the likelihood function. Imagine that the true p were smaller than $\hat{p}$, perhaps 0.15. It is *possible* that our sample would include 20 people with the allele, but not too likely. How do we quantify this?

Computing confidence limits differs in one major way from finding the likelihood. Instead of finding the probability of *exactly* 20 out of 100 with a particular guess of p, we find the probability of 20 *or more* out of 100. We think of this data as extreme as or more extreme than what was actually observed. Why this change? Particular results are usually highly unlikely, whereas the composite result of a *more extreme* result is not.

To find the lower confidence limit, we test possible values of p_l, finding one for which the probability of getting 20 or more is only 0.025. To be 95% confident that our interval captures the true value, there should be only a 2.5% chance that our data were produced by a value as small as the lower confidence limit (saving 2.5% for the upper confidence limit).

Example 8.2.1 Checking Whether $p = 0.15$ Lies Within the Confidence Limits

Suppose the true p were 0.15. The probability of getting 20 or more out of 100 is

$$\Pr(20 \text{ or more out of } 100 \text{ with } p = 0.15) = \sum_{k=20}^{100} b(k; 100, 0.15) = 0.1065$$

(Figure 8.2.2). This value had to be found on a computer. A true value of $p = 0.15$ would produce a result at least as extreme as our actual measurement 10.65% of the time.

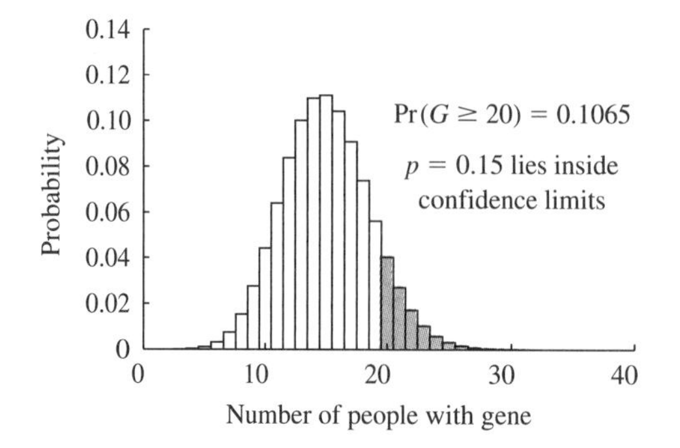

FIGURE 8.2.2
Testing whether $p = 0.15$ is consistent with the data

Because this probability is fairly large, our data might well have been produced with this value of p, and it lies within our confidence limits.

Example 8.2.2 Testing Whether $p = 0.1$ Lies Within the Confidence Limits

Because $p = 0.15$ is consistent with our data, we try smaller values of p that are farther from the maximum likelihood estimator $\hat{p} = 0.2$. If the true value were $p = 0.1$,

$$\Pr(20 \text{ or more out of } 100 \text{ with } p = 0.1) = \sum_{k=20}^{100} b(k; 100, 0.1) = 0.0020$$

(Figure 8.2.3). This small probability, much less than 1%, indicates that our results were very unlikely to have been generated by this value of p. We can be pretty confident (although not certain) that a true value of p this small did not produce our observation. The value $p = 0.1$ lies outside our confidence limits, implying that $p_l > 0.1$.

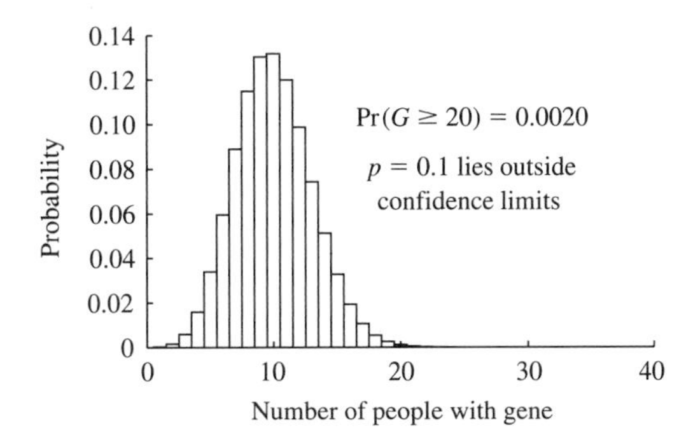

FIGURE 8.2.3

Testing whether $p = 0.1$ is consistent with the data

Example 8.2.3 Testing Whether $p = 0.125$ Lies Within the Confidence Limits

If we guess $p = 0.125$, the probability of data as extreme as or more extreme than the actual results is

$$\Pr(20 \text{ or more out of } 100 \text{ with } p = 0.125) = \sum_{k=20}^{100} b(k; 100, 0.125) = 0.0220$$

(Figure 8.2.4). With this true value of p, approximately 2.2% of samples would exceed our measured value. Because this probability is close to 0.025, the lower confidence limit p_l is close to 0.125.

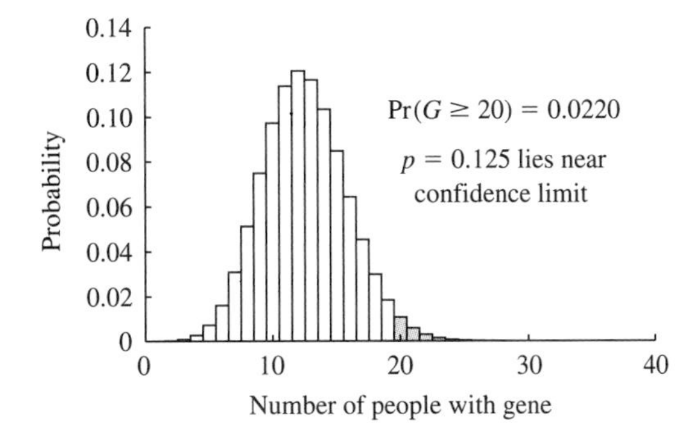

FIGURE 8.2.4

Testing whether $p = 0.125$ is consistent with the data

The precise definition of the lower confidence limit follows

Definition 8.5 The lower 95% confidence limit is the largest value of the parameter that produces a result as large or larger than the measured result 2.5% of the time. Formally, p_l solves the equation

$$\Pr(\text{result greater than or equal to measured data if } p = p_l) = 0.025.$$

Finding the Lower Confidence Limit

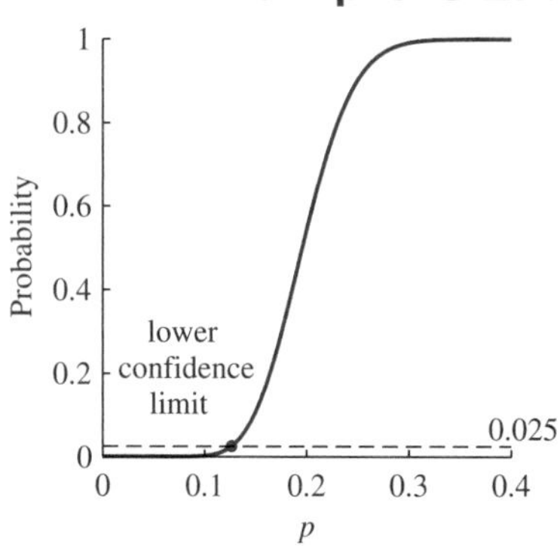

FIGURE 8.2.5
The probability of 20 or more out of 100 as a function of p

In the example where 20 out of 100 people have an allele, the lower confidence limit p_l is the solution of

$$\sum_{k=20}^{100} b(k;\, 100,\, p_l) = 0.025.$$

The graph of the probability of 20 or more as a function of p_l is shown in Figure 8.2.5. With a computer, we can solve for $p_l = 0.127$. If p were 0.127, we would get 20 or more out of 100 in exactly 2.5% of experiments. Values of p less than $p_l = 0.127$ are unlikely to have produced the observed data.

We find the upper confidence limit in the same way, as a value p_h that we can be confident (with 0.025 probability of error) *exceeds* the true p.

Definition 8.6 The upper confidence limit is the smallest value of the parameter that produces a result as small or smaller than the measured result 2.5% of the time. Formally, p_h solves the equation

$$\Pr(\text{result less than or equal to measured data if } p = p_h) = 0.025.$$

Example 8.2.5 Testing Values for the Upper Confidence Limit

We test values p_h in quest of one for which the probability of getting *20 or fewer* out of 100 is less than 0.025. For instance, with $p_h = 0.25$,

$$\Pr(20 \text{ or fewer out of 100 with } p = 0.25) = \sum_{k=0}^{20} b(k;\, 100,\, 0.25) = 0.1488$$

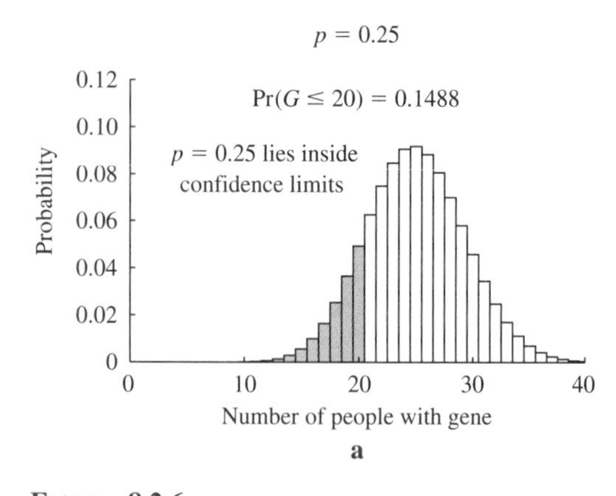

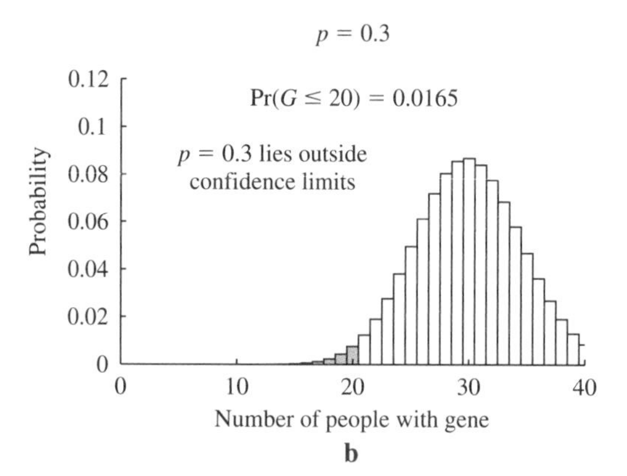

FIGURE 8.2.6
Finding the upper confidence limit: two thought experiments

(Figure 8.2.6a). A true value of $p = 0.25$ would produce a result as extreme as our observations nearly one time out of six, which is not particularly unlikely. If $p = 0.3$,

$$\Pr(20 \text{ or fewer out of 100 with } p = 0.3) = \sum_{k=0}^{20} b(k;\, 100,\, 0.3) = 0.0165$$

(Figure 8.2.6b). This value lies *outside* our confidence limits because a true p of 0.3 would generate 20 or fewer positives out of 100 *less* than 2.5% of the time.

Example 8.2.6 Finding the Upper Confidence Limit

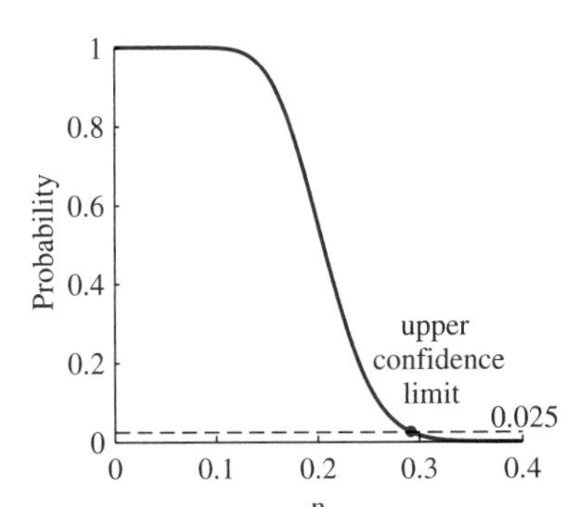

FIGURE 8.2.7

The probability of 20 or fewer out of 100 as a function of p

The exact p_h solves

$$\sum_{k=0}^{20} b(k;\, 100,\, p_h) = 0.025$$

with solution $p_h = 0.292$, again found with the help of a computer (Figure 8.2.7). Values of p greater than p_h are unlikely to have generated the observed data.

The interval between p_l and p_h is called the **95% confidence interval**. If we followed this procedure on many data sets, we would capture the true population proportion 95% of the time. The confidence limits are not a range of values likely to have captured the true proportion in our sample, because that is known to be exactly 0.2. Rather, the confidence limits are a range of values quite likely to contain the true proportion for the whole population.

The equations for p_l and p_h are impossible to solve unless n is quite small (Exercises 7 and 8). When n is large enough for the conditions in Subsection 7.9.2 to apply, we will use the normal approximation to the binomial (Subsection 8.3.5). With small data sets, however, the binomial probability distribution *must* be used.

Example 8.2.7 Finding Confidence Limits with a Small Data Set

Suppose we want to estimate the proportion of bears infested with a parasite. At great hazard, ten bears are captured of which three are infested. We find the 95% confidence interval around the estimate $\hat{p} = 0.3$ by solving for p_l and p_h (Definition 8.5 and Definition 8.6),

$$\sum_{k=3}^{10} b(k;\, 10,\, p_l) = 0.025$$

$$\sum_{k=0}^{3} b(k;\, 10,\, p_h) = 0.025.$$

With a computer, we find $p_l = 0.0667$ and $p_h = 0.6524$. The 95% confidence interval is quite wide, the price paid for a small sample size. Based on our data (and the assumption that our sampling methods are flawless), it is quite possible that as few as 7% or as many as 65% of bears are actually infested. More bears would have to be captured to make the confidence limits narrower.

There is nothing magical about the traditional 95%. With this value, we say that the degree of confidence is 0.95. In statistics, the parameter α represents the probability that the true proportion lies *outside* the confidence limits, and is found by subtracting the degree of confidence from 1. With 95% confidence limits, $\alpha = 1 - 0.95 = 0.05$. The value 0.025 is half the acceptable uncertainty.

Definition 8.7 The lower confidence limit with degree of confidence $1 - \alpha$ is the largest value of the parameter that produces a result as large or larger than the measured result a fraction $\frac{\alpha}{2}$ of the time. Formally, p_l solves the equation

$$\Pr(\text{result greater than or equal to measured data if } p = p_l) = \frac{\alpha}{2}.$$

The upper confidence limit with degree of confidence $1 - \alpha$ is the smallest value of the parameter that produces a result as small or smaller than the measured result a fraction $\frac{\alpha}{2}$ of the time. Formally, p_h solves the equation

$$\Pr(\text{result less than or equal to measured data if } p = p_h) = \frac{\alpha}{2}.$$

To find confidence limits with degree of confidence $1 - \alpha$ when we count 20 out of 100, we solve

$$\sum_{k=20}^{100} b(k;\, 100,\, p_l) = \frac{\alpha}{2}$$

$$\sum_{k=0}^{20} b(k;\, 100,\, p_h) = \frac{\alpha}{2}.$$

Example 8.2.8 Finding 99% Confidence Limits with the Genetic Example

If we want 99% confidence in our genetics example, $\alpha = 0.01$ and

$$\sum_{k=20}^{100} b(k;\, 100,\, p_l) = \frac{0.01}{2} = 0.005$$

$$\sum_{k=0}^{20} b(k;\, 100,\, p_h) = \frac{0.01}{2} = 0.005.$$

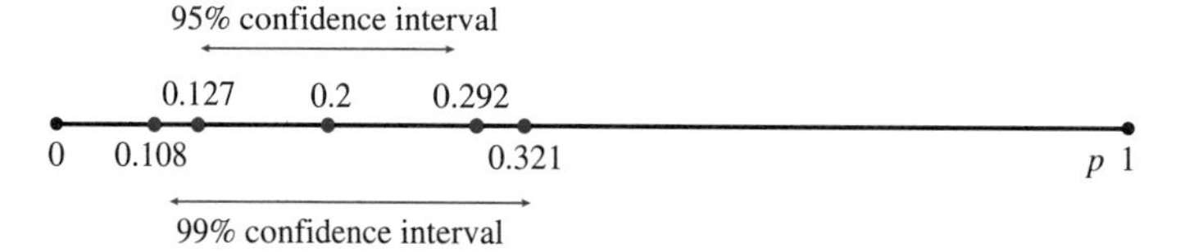

FIGURE 8.2.8
99% confidence limits are wider than 95% confidence limits

Using a computer, we solve for $p_l = 0.1048$ and $p_h = 0.3212$. More confidence requires a wider interval (Figure 8.2.8). The appropriate level of uncertainty depends on the application and the tradeoff between greater certainty and wider confidence limits.

Example 8.2.9 Finding Confidence Limits with the Poisson Model

Suppose we count that 23 seeds have fallen in 1.0 m^2. If the number of seeds obeys the assumptions of the Poisson process (that seeds fall independently and are equally likely to hit any point), we can both estimate the underlying parameter of the Poisson distribution and find confidence limits. To estimate the parameter, we use the method of maximum likelihood. The probability of the data, 23 seeds, conditional on the value of the parameter Λ is

$$L(\Lambda) = \frac{\Lambda^{23} e^{-\Lambda}}{23!}$$

(Figure 8.2.9).

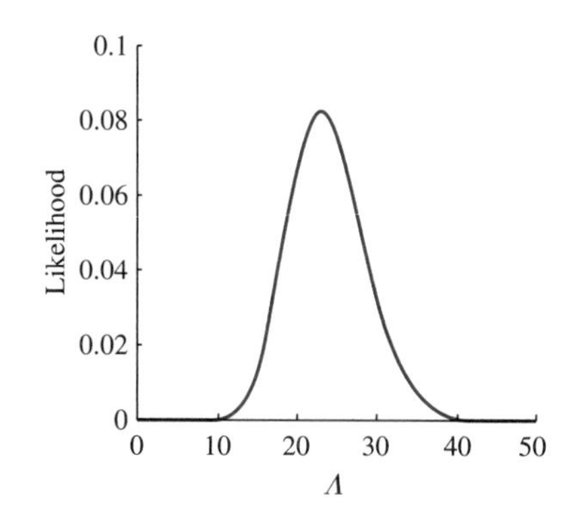

FIGURE 8.2.9
Likelihood function with the Poisson model

To find the maximum we take the derivative to find the critical point.

$$\begin{aligned} L'(\Lambda) &= \frac{1}{23!}\frac{d}{d\Lambda}(\Lambda^{23} e^{-\Lambda}) && \text{constant product rule} \\ &= \frac{1}{23!}\left(23\Lambda^{22} e^{-\Lambda} - \Lambda^{23} e^{-\Lambda}\right) && \text{product rule, power rule, chain rule} \\ &= \frac{\Lambda^{22} e^{-\Lambda}}{23!}(23 - \Lambda). && \text{factor} \end{aligned}$$

Setting this derivative equal to 0, we find

$$L'(\Lambda) = 0 \text{ if } \Lambda = 0 \text{ or } \Lambda = 23.$$

Because $L(0) = 0$ and $\lim_{\Lambda \to \infty} L(\Lambda) = 0$, the maximum occurs at $\Lambda = 23$.

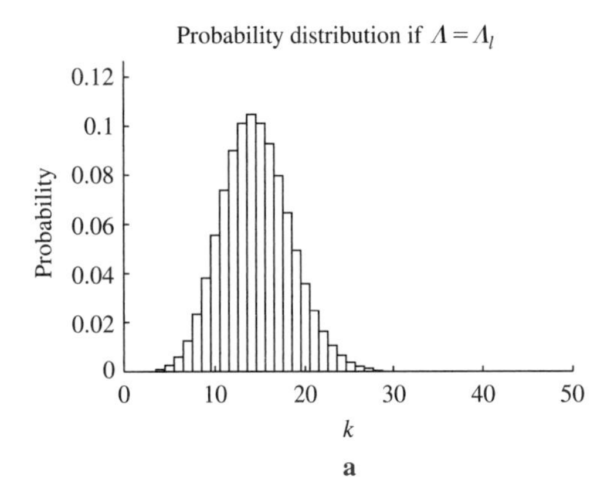

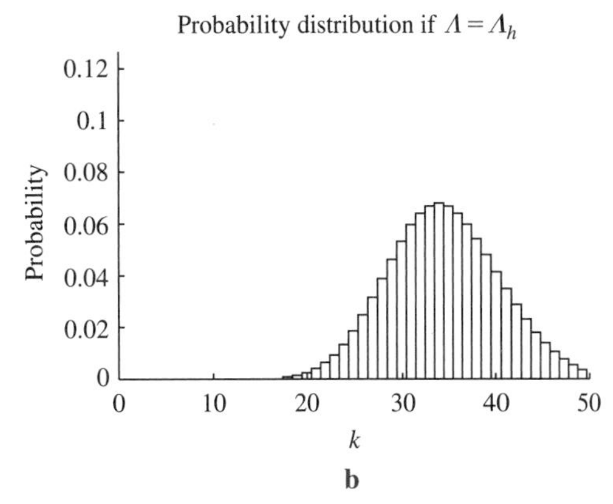

FIGURE 8.2.10
Probability distributions associated with the confidence limits

The lower confidence limit Λ_l solves the equation

$$\Pr(\text{result greater than or equal to measured data if } \Lambda = \Lambda_l) = 0.025$$

or

$$\sum_{k=23}^{\infty} \frac{\Lambda_l^k e^{-\Lambda_l}}{k!} = 0.025,$$

which has solution $\Lambda_l = 14.58$. A result of 23 or more is unlikely with this small a value of Λ (Figure 8.2.10a).

The upper confidence limit Λ_h solves the equation

$$\Pr(\text{result less than or equal to measured data if } \Lambda = \Lambda_h) = 0.025$$

or

$$\sum_{k=0}^{23} \frac{\Lambda_h^k e^{-\Lambda_h}}{k!} = 0.025,$$

which has solution $\Lambda_h = 34.51$. A result of 23 or less is unlikely with this large a value of Λ (Figure 8.2.10b).

8.2.2 Monte Carlo Method

Even with small values of n, the equations for p_l and p_h are virtually impossible to solve. For models more complicated than the binomial, even writing down the equations for confidence limits can be difficult. Fast computers have made an alternative approach, computer simulation, feasible. Any experiment that can be precisely described can be simulated on a computer. Because this computer experiment includes random inputs (as in gambling) the technique is called the **Monte Carlo method**. The logic behind confidence limits, however, remains the same.

Our genetics experiment involves picking 100 individuals, each with an unknown probability p of having the allele. The computer mimics this, but with different guesses for the value of p. In particular, we pick a test value of p, choose 100 Bernoulli random variables each with probability p, and count the number of "successes." To get a clear idea of the implications of the test value p, we run the computer experiment many times. Our goal is to establish whether p is "consistent" with the observed data.

Suppose first we wish to find p_l, the lower confidence limit or the smallest value of p consistent with the actual results (Definition 8.5). If a particular test value produces 20 or more out of 100 often, it is perfectly plausible that it might have generated our experimental results. There is no reason to think that it is too small. If we rarely see 20 or more out of 100, that value is unlikely to have produced the data. Similarly, the upper confidence limit can be found by checking test values larger than $\hat{p}$ and finding the ones that produce 20 or fewer out of 100 infrequently.

▸▸ Algorithm 8.1 Monte Carlo Method for Finding Confidence Limits

1. Compute the maximum likelihood estimator of the parameter.
2. To find the lower confidence limit, simulate the experiment many times on the computer with values of the parameter smaller than the maximum likelihood estimator and find the fraction of times the result is greater than or equal to the measured result.
3. With degree of confidence $1 - \alpha$, the lower confidence limit is the point where this fraction is equal to $\frac{\alpha}{2}$.
4. To find the upper confidence limit, simulate the experiment many times on the computer with a value of the parameter larger than the maximum likelihood estimator and find the fraction of times results are less than or equal to the measured result.
5. The upper confidence limit is the point where this fraction is equal to $\frac{\alpha}{2}$. ▲

Example 8.2.10 Monte Carlo Method Applied to Genetics Example

We can use the Monte Carlo method to estimate the confidence limits associated with detecting 20 out of 100 people with an allele. The results of 1000 repetitions of the experiment of choosing 100 Bernoulli random variables with various values of the probability p are given in Table 8.1. For example, with $p = 0.1$, only 1 out of 1000 simulations produced more than 20 successes. Because this fraction, 0.001, is less than 0.025, this value is less than the lower 95% confidence limit. The 95% confidence limits require that no more than 0.025, or 25 out of 1000, experiments produce 20 or more by chance. This occurs between $p = 0.12$ and $p = 0.14$, bracketing the exact result of

Table 8.1 Monte Carlo simulation results for finding confidence limits

p	Simulations Producing 20 or More	Fraction	Simulations Producing 20 or Fewer	Fraction
0.10	1	0.001	999	0.999
0.12	8	0.008	997	0.997
0.14	80	0.080	954	0.954
0.16	168	0.168	893	0.893
0.18	312	0.312	780	0.780
0.20	561	0.561	553	0.553
0.22	722	0.722	368	0.368
0.24	852	0.852	219	0.219
0.26	927	0.927	107	0.107
0.28	974	0.974	46	0.046
0.30	990	0.990	22	0.022
0.32	995	0.995	6	0.006
0.34	999	0.999	1	0.001

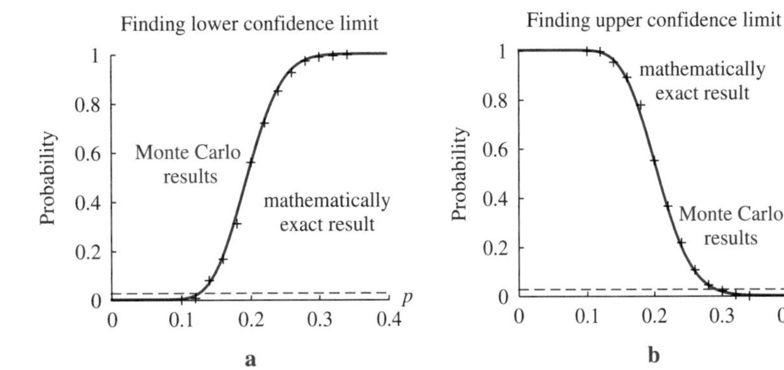

FIGURE 8.2.11

Using the Monte Carlo method to estimate confidence limits

0.127 (Figure 8.2.11a). With $p = 0.3$, 22 out of 1000 simulations produced 20 or fewer successes. This is quite close to 0.025, meaning that the upper confidence limit is near 0.3, close to the exact result of 0.292 (Figure 8.2.11b). We would have to test more values of p and run more simulations to get more precise estimates.

8.2.3 Likelihood, Support, and Confidence Limits

Our intuition about likelihood suggests that values of p with likelihood not too far below the maximum should be part of our confidence interval. For example, $p = 0.18$ has likelihood 0.0870, not far below the maximum of 0.0993 associated with $p = 0.2$ but well above the likelihood of 0.0012 associated with $p = 0.1$ (Example 8.1.5). There is a powerful rule of thumb that uses likelihood to estimate the 95% confidence interval. (Loyal adherents of likelihood do not use confidence intervals and recommend the method of this section as an alternative).

As we have seen, values of the likelihood function can be tiny. The most convenient way to compare likelihoods is to take the natural logarithm. This value is called either the log-likelihood or the support.

Definition 8.8 The **support** for a particular value of an unknown parameter p is

$$S(p) = \ln[L(p)].$$

The numerical value of the support can be thought of as the support that the data provides for a particular hypothesis (a value of the unknown parameter p in this case).

Example 8.2.11 Three Values of the Support in the Genetic Example

We can take the logarithm of the likelihoods for the genetics data set to find the following values of the support.

$$S(0.18) = \ln[L(0.18)] = \ln(0.0870) = -2.441$$

$$S(0.2) = \ln[L(0.2)] = \ln(0.0993) = -2.310$$

$$S(0.1) = \ln[L(0.1)] = \ln(0.0012) = -6.725.$$

Values of the support are negative when the underlying model is discrete because the likelihood is a probability that must be less than 1. The support can take on any value when the underlying model is discrete and the likelihood describes probability density rather than probability. In either case, the support takes on its maximum precisely where the likelihood does (Exercises 27–28).

We use the difference between the values of the support to compare different values of the unknown parameter p.

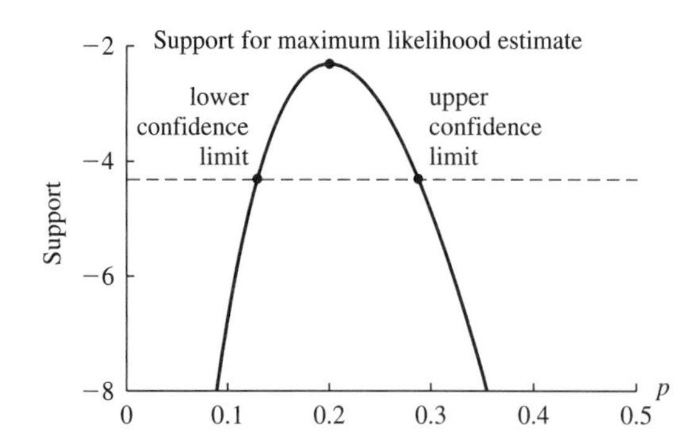

FIGURE 8.2.12
Estimating confidence limits with the support

Rule for Estimating Confidence Limits with the Support Suppose $\hat{p}$ is the maximum likelihood estimate of the unknown parameter p. The approximate upper and lower confidence limits are solutions of

$$S(p) = S(\hat{p}) - 2$$

(Figure 8.2.12).

Example 8.2.12 Testing Whether Two Values Lie Within Confidence Limits

We can use this method to check whether $p = 0.18$ and $p = 0.1$ lie within the confidence limits around the maximum likelihood estimate $p = 0.2$. Subtracting gives

$$S(0.2) - S(0.18) = -2.310 - (-2.441) = 0.131$$
$$S(0.2) - S(0.1) = -2.310 - (-6.725) = 4.415.$$

The support for $p = 0.18$ lies well within 2.0 of the maximum, and $p = 0.18$ lies well within the confidence limits. In contrast, the support for $p = 0.1$ is much more that 2.0 below the maximum, so $p = 0.1$ lies outside the confidence limits.

Example 8.2.13 Solving for Confidence Limits with the Method of Support

In the genetics example, the support $S(p)$ for obtaining exactly 20 out of 100 with true parameter p is

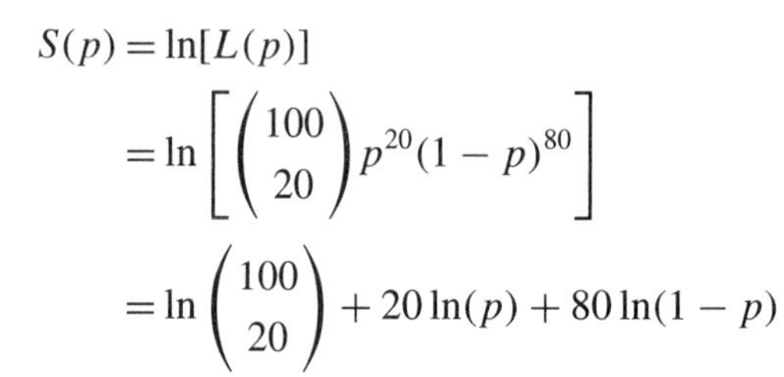

$$\begin{aligned} S(p) &= \ln[L(p)] \\ &= \ln\left[\binom{100}{20} p^{20}(1-p)^{80}\right] \\ &= \ln\binom{100}{20} + 20\ln(p) + 80\ln(1-p) \end{aligned}$$

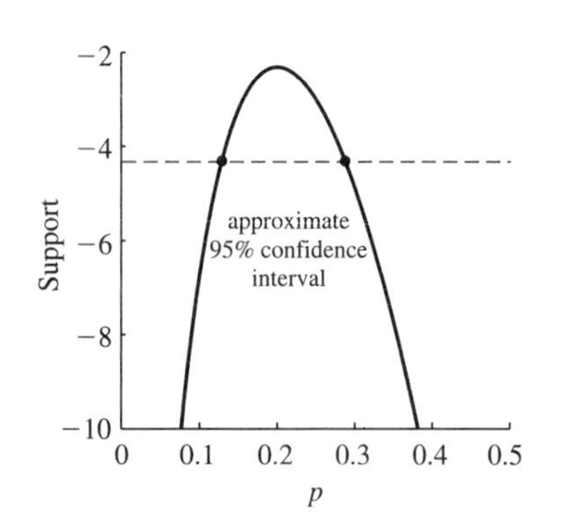

FIGURE 8.2.13
Support given 20 out of 100 as a function of p

(Figure 8.2.13). We used laws 1 and 2 of logs to write the log of a product as the sum of the logs and to pull the powers 20 and 80 out front.

The approximate upper and lower confidence limits are solutions of

$$S(p) = S(0.2) - 2.0.$$

Formally,

$$\left[-\ln\binom{100}{20} + 20\ln(p) + 80\ln(1-p)\right] = \ln\binom{100}{20} + 20\ln(0.2) + 80\ln(1-0.2) - 2.0$$

$$20\ln(p) + 80\ln(1-p) = 20\ln(0.2) + 80\ln(1-0.2) - 2.0.$$

The constants cancel, saving much computation. These equations cannot be solved algebraically, but are much simpler than the equations for the exact confidence limits. By experimentation or with Newton's method (Exercise 25) we find two solutions

$p = 0.128$ and $p = 0.287$ (Figure 8.2.13). These values are quite close to the exact confidence limits of $p_l = 0.127$ and $p_h = 0.292$ (Examples 8.2.4 and 8.2.6).

Example 8.2.14 Solving for Confidence Limits in a Poisson Example with the Method of Support

Consider again finding 23 seeds in 1 m^2. In Example 8.2.9 we found that $\hat{\Lambda} = 23$ is the maximum likelihood estimator for the parameter Λ in the Poisson distribution. The support $S(\Lambda)$ for obtaining exactly 23 seeds as a function of Λ is

$$\begin{aligned} S(\Lambda) &= \ln[L(\Lambda)] \\ &= \ln\left(\frac{\Lambda^{23}e^{-\Lambda}}{23!}\right) \\ &= -\ln(23!) + 23\ln(\Lambda) - \Lambda \end{aligned}$$

using laws 1 and 2 of logs.

The approximate upper and lower confidence limits are solutions of

$$S(\Lambda) = S(23) - 2.0.$$

As before, the constant, here $\ln(23!)$, cancels, giving the equation

$$23\ln(\Lambda) - \Lambda = 23\ln(23) - 25.0.$$

These equations cannot be solved algebraically, but with a computer or Newton's method (Exercise 26), we find two solutions $\Lambda_l = 14.69$ and $\Lambda_h = 33.97$ (Figure 8.2.14). These values are quite close to the exact confidence limits of $\Lambda_l = 14.58$ and $\Lambda_h = 34.51$ (Example 8.2.9).

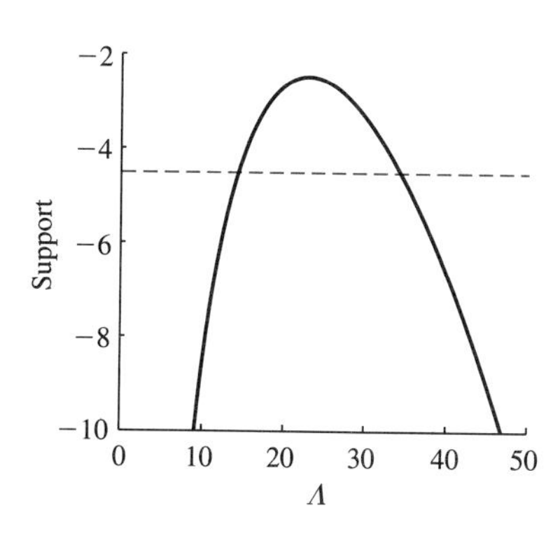

FIGURE 8.2.14
Support for finding 23 as a function of Λ

Summary We have developed the idea of **confidence limits**. True values less than the lower confidence limit are unlikely to have produced results exceeding what we measured and true values greater than the upper confidence limit are unlikely to have produced results less than our measurement. When the equations for the confidence limits are hopelessly intractable, we can conduct computer experiments with the **Monte Carlo method**. Finally, we have seen that the log of the likelihood, the **support**, provides a convenient way to estimate confidence limits.

8.2 Exercises

Mathematical Techniques

1–6 ▪ In the following situations, find the probability of a result as extreme as or more extreme than the actual result for the given value of the parameter.

1. A coin is flipped five times and comes up heads every time. Does the value $p = 0.5$ for the probability of heads lie within the 95% confidence limits?
2. A coin is flipped seven times and comes up heads six out of seven times. Does the value $p = 0.5$ for the probability of heads lie within the 95% confidence limits?
3. A person wins the lottery the second time he plays. Does the value $q = 0.001$ for the probability of success lie within the 99% confidence limits?
4. A person wins the lottery the fifth time he plays. Does the value $q = 0.001$ for the probability of success lie within the 99% confidence limits?
5. Three cosmic rays hit a detector in 1 yr. Does the value $\lambda = 10.0$ for the rate at which rays hit lie within the 98% confidence limits?
6. One cosmic ray hits a detector in 1 yr. Does the value $\lambda = 5.0$ for the rate at which rays hit lie within the 98% confidence limits?

7–10 ▪ Find the exact confidence limits and check if the value for the earlier problem lies within them.

7. Find the 95% confidence limits if a coin is flipped five times and comes up heads every time (Exercise 1).
8. Find the 95% confidence limits if a coin is flipped seven times and comes up heads six out of seven times (Exercise 2).
9. Find the 99% confidence limits if a person wins the lottery the second time he plays (Exercise 3).
10. Find the 98% confidence limits if a one cosmic ray hits a detector in 1 yr (Exercise 6).

11–14 ▪ Find the approximate 95% confidence limits using the method of support and compare with earlier exercises.

11. A coin is flipped five times and comes up heads every time (Exercise 7).

12. A coin is flipped seven times and comes up heads six out of seven times (Exercise 8).

13. A person wins the lottery the second time he plays (Exercise 9, but recall that the earlier exercise found 99% confidence limits).

14. One cosmic ray hits a detector in 1 yr (Exercise 6, but recall that the earlier exercise found 99% confidence limits).

15–18 ▪ Explain how you would use the Monte Carlo method to estimate confidence limits in the following cases.

15. A coin is flipped five times and comes up heads every time, and we wish to find the upper and lower 95% confidence limits.

16. A coin is flipped seven times and comes up heads six out of seven times, and we wish to find upper and lower 95% confidence limits.

17. A person wins the lottery the second time he plays, and we wish to find upper and lower 99% confidence limits.

18. One cosmic ray hits a detector in 1 yr, and we wish to find upper and lower 98% confidence limits.

19–22 ▪ Check whether the expected value of p ($p = 1/2$ for a fair coin and $p = 1/6$ for a fair die) lies within the approximate 95% confidence limits given by the method of support.

19. Flipping 2 out of 4 heads with a fair coin (as in Section 8.1, Exercise 9).

20. Rolling 2 out of 4 sixes with a fair die (as in Section 8.1, Exercise 10).

21. Flipping 2 out of 12 heads with a fair coin (as in Section 8.1, Exercise 11).

22. Rolling 2 out of 12 sixes with a fair die (as in Section 8.1, Exercise 12).

23–24 ▪ Check whether the given value of Λ lies within the approximate 95% confidence limits given by the method of support.

23. Twenty events occur in 1 min with a given value of $\Lambda = 10.0$ (as in Section 8.1, Exercise 17).

24. Ten high energy cosmic rays hit detector over the course of 1 yr, with a given value of $\Lambda = 8.0$ (as in Section 8.1, Exercise 18).

25–26 ▪ Use experimentation and Newton's method to solve the equations for the approximate confidence limits with the method of support.

25. The confidence limits if 20 out of 100 individuals are measured with a particular allele (Example 8.2.13).

26. The confidence limits if 23 seeds are found in 1 m^2 (Example 8.2.14).

27–30 ▪ Prove that the support takes on its maximum where the likelihood does.

27. Using the likelihood function in Section 8.1, Exercise 9.

28. Using the likelihood function in Section 8.1, Exercise 17.

29. Using the likelihood function in Section 8.1, Exercise 19.

30. For a general likelihood function.

Applications

31–34 ▪ Consider a tiny data set where one out of two people is found with an allele.

31. Find the exact 95% confidence limits.

32. Find the exact 99% confidence limits.

33. How would you use the Monte Carlo method to estimate the 99% confidence limits?

34. Use the method of support to estimate 95% confidence limits and compare your results with Exercise 31.

35–38 ▪ Consider a data set where three out of three people are found with an allele.

35. Find the exact 95% confidence limits.

36. Find the exact 99% confidence limits.

37. Why is the upper confidence limit strange?

38. Use the method of support to estimate 95% confidence limits and compare your results with Exercise 35.

39–42 ▪ A person places an advertisement to sell his car in the newspaper and settles down to await calls, which he expects will arrive with an exponential distribution. The first call arrives in 20 min.

39. Find the maximum likelihood estimator of the rate.

40. Find the 95% confidence limits.

41. Find the 98% confidence limits.

42. How many calls might he expect to miss if he went out for a 2-h hike?

43–44 ▪ 14 mutations are counted in one million base pairs.

43. Write the equations for the 95% confidence limits, and solve them numerically if you have a computer.

44. Write the equations for the approximate 95% confidence limits using the method of support, and solve them numerically if you have a computer.

45–46 ▪ Recall the couple that has seven boys before having a girl (Section 8.1, Exercise 33).

45. Find 95% confidence limits around the maximum likelihood estimate of q. How do you interpret these results?

46. Find 99% confidence limits around the maximum likelihood estimate of q. How do you interpret these results?

Computer Exercises

47. Do a simulation to reproduce the results in Table 8.1. Why are your results different from the table in the text? Do you estimate the same confidence limits? Do more simulations to get a more accurate answer. How close can you get to the exact answer?

48. Consider again the data in Section 8.1, Exercise 50. Use the method of support to find approximate 95% confidence limits around your estimated value. Do they include the true value $x = 3.0$?

49. Use the Monte Carlo method to estimate the 95% confidence limits in Exercise 18. How close are your results to those obtained with the other methods?

8.3 Estimating the Mean

For many quantitative measurements, the mean is the most useful statistic for describing a population. This is particularly true when measurements follow a roughly normal distribution. After introducing three methods for estimating the mean, we discuss confidence limits around the **sample mean**, the average of the measured values. When the variance is unknown, we can estimate it as the **sample variance**. The **standard error**, which combines the sample variance and the number of samples, makes it easy to compute confidence limits.

8.3.1 Estimating the Mean

There are many ways to estimate the population mean from a sample. Commonly used methods include

$$\overline{X} = \text{the sample mean} \tag{8.3.1}$$

$$\tilde{X} = \text{the sample median} \tag{8.3.2}$$

$$\overline{X}_{tr(10)} = \text{the trimmed mean.} \tag{8.3.3}$$

The **sample mean** $\overline{X}$ is the average of the sample, the "common sense" estimate of the population mean. In many realistic circumstances, this estimator is undependable because it can be drastically changed by a few extreme values, called **outliers**.

Example 8.3.1 Computing the Sample Mean with an Outlier

Consider the following measurements.

Measurement	Value	Measurement	Value
x_1	43.2	x_{11}	58.3
x_2	51.7	x_{12}	76.1
x_3	81.2	x_{13}	54.8
x_4	67.9	x_{14}	37.2
x_5	48.5	x_{15}	71.4
x_6	39.2	x_{16}	52.1
x_7	361.1	x_{17}	47.5
x_8	62.2	x_{18}	65.7
x_9	39.9	x_{19}	40.3
x_{10}	50.8	x_{20}	48.3

The values seem to be concentrated around 50, except for the extreme measurement x_7. The average of the whole data set is found by adding up all the values and dividing by 20 (the number of samples). In this case,

$$\text{average} = \frac{\sum_{i=1}^{20} x_i}{20} = \frac{1397.4}{20} = 69.87,$$

which is larger than all but four of the measurements. Throwing out the large measurement x_7 reduces the average by quite a bit. In fact,

$$\text{average with } x_7 \text{ removed} = \frac{\sum_{i \neq 7} x_i}{19} = \frac{1036.3}{19} = 54.5,$$

which is much closer to the center of the distribution.

There are several ways to deal with outliers. If there is good reason to believe that those measurements are incorrect, outliers can be removed (but it is essential to record this fact and the reasons in notebooks and papers). Other methods include computing the **sample median** and the **trimmed mean**.

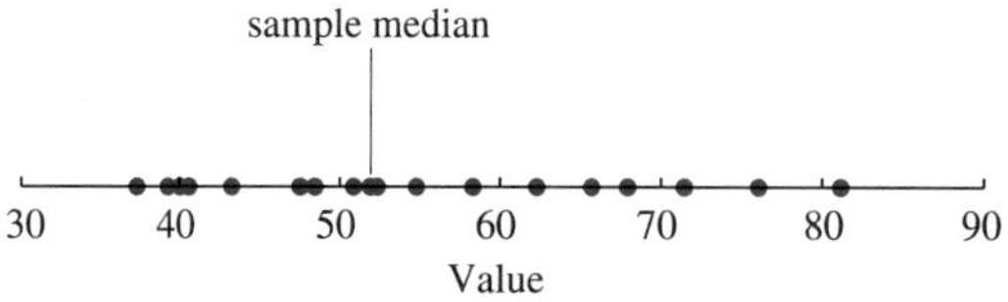

sample median

30 40 50 60 70 80 90

Value

FIGURE 8.3.1

A data set, and the sample median

The **sample median** $\tilde{X}$ is the central value measured (Figure 8.3.1).

$$\text{sample median} = \begin{cases} \text{the middle measurement if } n \text{ is odd} \\ \text{average of two middle measurements if } n \text{ is even} \end{cases}$$

Example 8.3.2 Finding the Sample Median with an Outlier

To find the sample median, we must sort values in increasing order.

Measurement	Value	Measurement	Value
x_{14}	37.2	x_{16}	52.1
x_6	39.2	x_{13}	54.8
x_9	39.9	x_{11}	58.3
x_{19}	40.3	x_8	62.2
x_1	43.2	x_{18}	65.7
x_{17}	47.5	x_4	67.9
x_{20}	48.3	x_{15}	71.4
x_5	48.5	x_{12}	76.1
x_{10}	50.8	x_3	81.2
x_2	51.7	x_7	361.1

We see that x_2 and x_{16} come out in the 10th and 11th positions, so the sample median is the average, or

$$\begin{aligned}\text{sample median} &= \text{average of two middle measurements} \\ &= \frac{51.7 + 52.1}{2} = 51.9.\end{aligned}$$

Because the median uses only relative size, it is insensitive to large values. If x_7 had been recorded wrong and was really 61.1, the median would be unaffected.

The **trimmed mean** $\overline{X}_{tr(10)}$ lops off the largest and smallest 10% of the measurements, whether or not those values might be due to experimental error. The subscript indicates the **percentage** of the sample discarded on each end.

Example 8.3.3 Computing the Trimmed Mean

There are 20 values in the data presented in Example 8.3.1, so the trimmed mean removes the largest two values (10% of 20) and the smallest two values. The sorted values in Example 8.3.2 show that the largest two values are x_3 and x_7 and the smallest two values are x_6 and x_{14}. Removing these, we find the average of the remaining 16 measurements as

$$\overline{X}_{tr(10)} = \frac{\text{sum of middle 16 measurements}}{16} = \frac{878.7}{16} = 54.9.$$

Like the median, the trimmed mean removes the effects of large values.

All of these statistics perform badly at finding the mean of numbers from a distribution that looks nothing like the normal distribution, such as the exponential. Before taking the mean of data like these, it is a good idea to ask whether the mean should be estimated at all. It might make more sense to estimate parameters associated with an explicit underlying model (such as the rate constant for the exponential). When the underlying model resembles the normal distribution, the mean is, as it were, meaningful. With other models, other statistics can be more appropriate for answering scientific questions.

8.3.2 Confidence Limits for the Sample Mean

Suppose our sample data seem to be clustered around an average value that we wish to estimate. If none of the data points are suspect, the sample mean is a good estimator of the population mean. In particular, we use the estimator

$$\hat{\mu} = \frac{\sum_{i=1}^{n} x_i}{n}$$

where $x_1, x_2, \ldots, x_n$ are the n measured values. How can we go about computing confidence limits around this estimator?

The confidence limits depend on the true, unknown, underlying distribution and on the size of our sample. Suppose the true distribution has mean μ and standard deviation σ. Each measurement i from the sample is a random variable X_i. The sample mean is a new random variable defined by

$$\overline{X} = \frac{\sum_{i=1}^{n} X_i}{n}.$$

By the Central Limit Theorem for Averages (Theorem 7.24), the sample mean $\overline{X}$ of a sample of size n is approximately normal with mean μ and variance $\frac{\sigma^2}{n}$. Formally,

$$\overline{X} \sim N\left(\mu, \frac{\sigma^2}{n}\right).$$

Example 8.3.4 Using the Normal Distribution to Describe the Distribution of the Sample Mean

Suppose that each of the random variables X_i has mean $\mu = 50$ and standard deviation $\sigma = 20$. Then $\text{Var}(X_i) = \sigma^2 = 400$. If we make $n = 10$ measurements, the sample mean follows

$$\overline{X} \sim N\left(\mu, \frac{\sigma^2}{n}\right) = N\left(50, \frac{400}{10}\right) = N(50, 40).$$

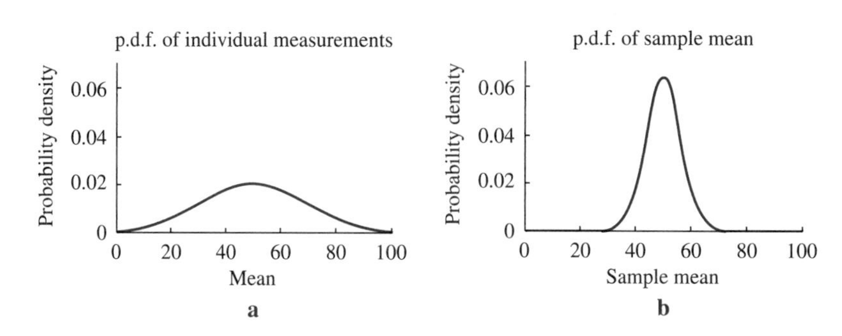

FIGURE 8.3.2
Distribution of the sample mean

The sample mean has variance 40, and thus standard deviation of $\sqrt{40} = 6.32$. The p.d.f. of the sample mean is much narrower than that of the individual measurements (Figure 8.3.2).

The narrower this distribution, the more accurately we can estimate the true mean from the sample mean. There are two ways that the variance of $\overline{X}$ can be made smaller: decreasing the variance σ^2 (more accurate measurement or better control of experimental conditions) and increasing n (making more measurements).

Example 8.3.5 The Effects of Smaller Underlying Variance

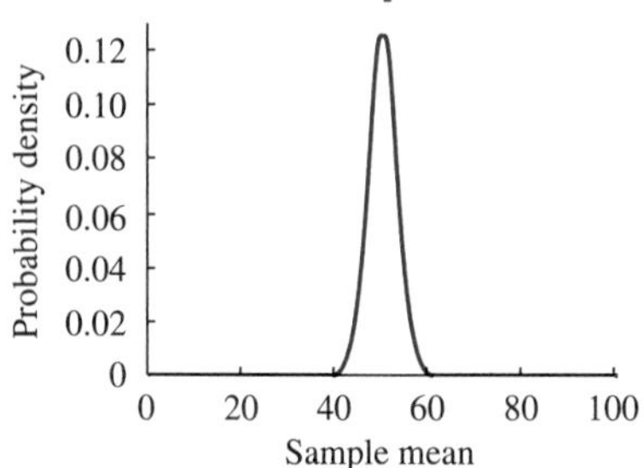

FIGURE 8.3.3
Distribution of the sample mean with lower standard deviation

Suppose that more careful techniques in Example 8.3.4 reduce the standard deviation of the individual measurements X_i from 20 to 10. Then $\text{Var}(X_i) = \sigma^2 = 100$. If we make $n = 10$ measurements, the sample mean follows

$$\overline{X} \sim N\left(\mu, \frac{\sigma^2}{n}\right) = N\left(50, \frac{100}{10}\right) = N(50, 10).$$

The sample mean has variance 10, and thus standard deviation of $\sqrt{10} = 3.16$. The p.d.f. of the sample mean is much narrower than in Example 8.3.4 (Figure 8.3.3).

Example 8.3.6 The Effects of Larger Sample Size

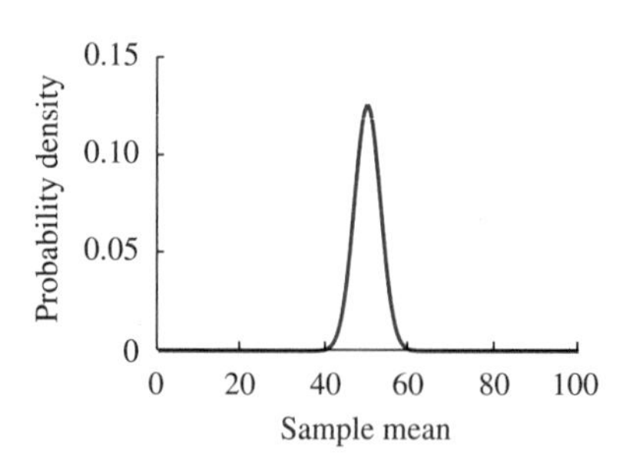

FIGURE 8.3.4
Distribution of the sample mean with larger sample size

Suppose that we increase the number of samples in Example 8.3.4 from 10 to 50. The sample mean then follows

$$\overline{X} \sim N\left(\mu, \frac{\sigma^2}{n}\right) = N\left(50, \frac{400}{50}\right) = N(50, 8).$$

The sample mean has variance 8, and thus standard deviation of $\sqrt{8} = 2.82$. The p.d.f. of the sample mean is again narrower than in Example 8.3.4 (Figure 8.3.4).

If we know the standard deviation σ of the population (we estimate it in Subsection 8.3.3), we can derive confidence limits around our estimator $\hat{\mu}$. As always, the lower confidence limit is the largest value μ_l that would produce a sample mean as large or larger than what we measured 2.5% of the time (Definition 8.5). We find it by a series of thought experiments, computing the probability of results as extreme as or more extreme than our observation as a function of the true mean μ_l.

If the true mean were μ_l, the sample mean X_l is a random variable with

$$X_l \sim N\left[\mu_l, \frac{\sigma^2}{n}\right].$$

Distributions with different possible values of μ_l are shown in Figure 8.3.5. The lower 95% confidence limit is the value μ_l that solves the equation

$$\Pr(X_l \geq \hat{\mu}) = 0.025,$$

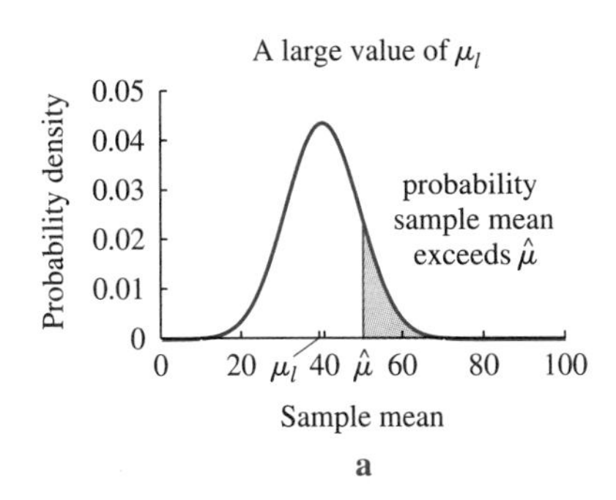

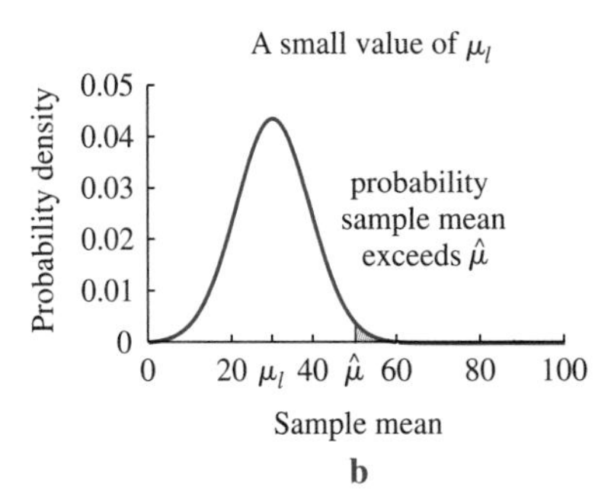

FIGURE 8.3.5
Probability of mean greater than $\hat{\mu}$ with different μ_l

meaning that a sample mean greater than or equal to the observed value of $\hat{\mu}$ is unlikely. Larger values of μ_l are more likely to generate a sample mean that exceeds $\hat{\mu}$ (Figure 8.3.5a).

In our examples with the binomial and Poisson distribution, the equation for the lower confidence limit could not be solved algebraically. Because this case involves the normal distribution, we can transform it into a question about the standard normal. We subtract the mean and divide by the standard deviation to convert to z-scores, giving

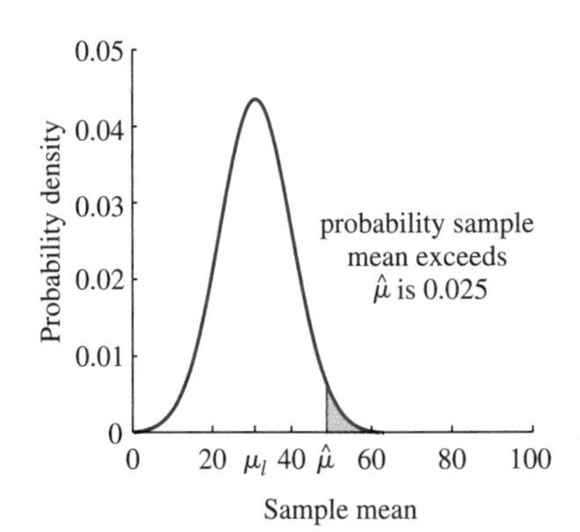

FIGURE 8.3.6
The lower confidence limit

$$\begin{aligned}\Pr(X_l \geq \hat{\mu}) &= \Pr\left(\frac{X_l - \mu_l}{\sigma/\sqrt{n}} \geq \frac{\hat{\mu} - \mu_l}{\sigma/\sqrt{n}}\right) \\ &= \Pr\left(Z \geq \frac{\hat{\mu} - \mu_l}{\sigma/\sqrt{n}}\right) \\ &= 0.025.\end{aligned}$$

This equation has solution where

$$\Phi\left[\frac{\hat{\mu} - \mu_l}{\sigma/\sqrt{n}}\right] = 0.975$$

(Figure 8.3.6).

From detailed tables of Φ, the cumulative distribution function (c.d.f.) for the standard normal, we find that $\Phi(1.96) = 0.975$. The lower confidence limit μ_l satisfies

$$\frac{\hat{\mu} - \mu_l}{\sigma/\sqrt{n}} = 1.96.$$

Solving for μ_l, we get

$$\mu_l = \hat{\mu} - 1.96\frac{\sigma}{\sqrt{n}}.$$

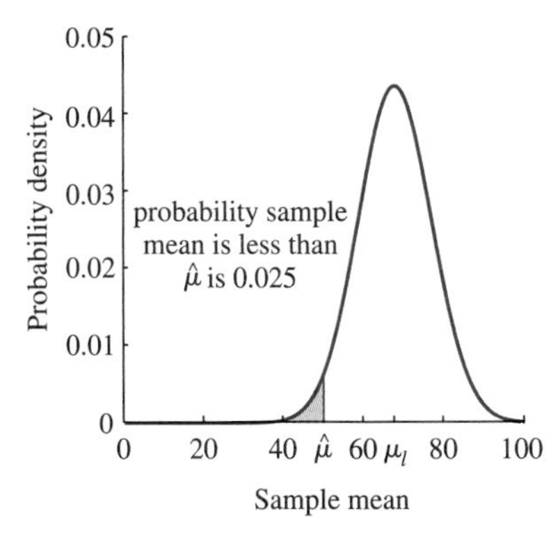

FIGURE 8.3.7
The upper confidence limit

The upper confidence limit μ_h is found in a similar way. Because the normal distribution is symmetric, it lies as far above the sample mean as the lower confidence lies below; that is,

$$\mu_h = \hat{\mu} + 1.96\frac{\sigma}{\sqrt{n}}$$

(Figure 8.3.7).

We summarize these results in the following theorem.

Theorem 8.1 If $\hat{\mu}$ is the sample mean of n measurements from a normal distribution with unknown mean μ and known variance σ^2, then the 95% lower and upper confidence limits are

$$\mu_l = \hat{\mu} - 1.96\frac{\sigma}{\sqrt{n}}$$

$$\mu_h = \hat{\mu} + 1.96\frac{\sigma}{\sqrt{n}}.$$

Example 8.3.7 Finding Confidence Limits for the Sample Mean

Suppose we measure a sample mean of $\hat{\mu} = 47.3$ from a sample of $n = 10$ measurements with known standard deviation $\sigma = 20$. From Theorem 8.1,

$$\mu_l = 47.3 - 1.96\frac{20}{\sqrt{10}} = 34.9$$

$$\mu_h = 47.3 + 1.96\frac{20}{\sqrt{10}} = 59.7.$$

This situation differs importantly from that in Example 8.3.4, where we knew the true mean of the underlying measurement, and used probability theory to find the distribution of the sample mean. This example describes a statistical question. We use data and a probabilistic model to infer something about the underlying process—in this case, confidence limits around the unknown true mean of the distribution.

Example 8.3.8 Finding Confidence Limits with Smaller Standard Deviation

Suppose we measure a sample mean of $\hat{\mu} = 47.3$ from a sample of $n = 10$ measurements with known standard deviation $\sigma = 10$. We expect that this smaller standard deviation, as in Example 8.3.5, will lead to narrower confidence limits. From Theorem 8.1,

$$\mu_l = 47.3 - 1.96\frac{10}{\sqrt{10}} = 41.1$$

$$\mu_h = 47.3 + 1.96\frac{10}{\sqrt{10}} = 53.5.$$

The confidence limits are half as wide as in Example 8.3.7, meaning that our uncertainty about the true mean is lower.

Example 8.3.9 Finding Confidence Limits with Larger Sample Size

Suppose we measure a sample mean of $\hat{\mu} = 47.3$ from a sample of $n = 50$ measurements with known standard deviation $\sigma = 20$. We expect that this larger sample size, as in Example 8.3.6, will lead to narrower confidence limits. From Theorem 8.1,

$$\mu_l = 47.3 - 1.96\frac{20}{\sqrt{50}} = 41.8$$

$$\mu_h = 47.3 + 1.96\frac{20}{\sqrt{50}} = 52.8.$$

The confidence limits are indeed narrower, spanning an interval of 11.0 rather than 24.8 (as in Example 8.3.7).

We could choose a different desired confidence level. To find the 95% confidence limits, we had to find where $\Phi(z)$ is equal 0.025 and 0.975. To find the 99% confidence limits, we solve for where $\Phi(z)$ is equal to 0.005. This occurs at 2.576, so

$$\mu_l = \hat{\mu} - 2.576\frac{\sigma}{\sqrt{n}}$$

$$\mu_h = \hat{\mu} + 2.576\frac{\sigma}{\sqrt{n}}.$$

Wider limits, as always, are required for more confidence.

Example 8.3.10 Finding 99% Confidence Limits

Suppose we measure a sample mean of $\hat{\mu} = 47.3$ from a sample of $n = 10$ measurements with known standard deviation $\sigma = 20$, exactly as in Example 8.3.7. But now we wish to be 99% confident that the true mean lies within our confidence limits. To find 99% confidence limits, we compute

$$\mu_l = 47.3 - 2.576\frac{20}{\sqrt{10}} = 31.0$$

$$\mu_h = 47.3 + 2.576\frac{20}{\sqrt{10}} = 63.6.$$

These confidence limits are *wider* than the 95% confidence limits found in Example 8.3.7. To be more confident that the true value lies in our interval, we must choose a wider interval.

The confidence interval depends on the sample size n, the sample mean $\hat{\mu}$, the standard deviation σ, and the level of confidence. Confidence limits are different for each experiment, because they are arranged around the measured sample mean rather than the unknown true mean. We know only that the true mean μ lies between the 95% confidence limits μ_l and μ_h with probability 0.95.

Example 8.3.11 Confidence Limits Are Different When an Experiment Is Replicated

Suppose the standard deviation of a plant height distribution is known to be 3.2 cm. If we measure 100 plants and find a sample mean of 40.2 cm, the 95% confidence limits are

$$\mu_l = 40.2 - 1.96\frac{3.2}{\sqrt{100}} = 40.2 + 0.63 = 39.57$$

$$\mu_h = 40.2 + 1.96\frac{3.2}{\sqrt{100}} = 40.2 + 0.63 = 40.83$$

(Figure 8.3.8). If we repeat the experiment, the sample mean will be different, as will the confidence limits. If the sample mean were 39.7 in a second replicate with 100 plants, then

$$\mu_l = 39.7 - 1.96\frac{3.2}{\sqrt{100}} = 39.7 + 0.63 = 39.07$$

$$\mu_h = 39.7 + 1.96\frac{3.2}{\sqrt{100}} = 39.7 + 0.63 = 40.33$$

(Figure 8.3.8). If the sample mean were 40.5 with 50 plants, then

$$\mu_l = 40.5 - 1.96\frac{3.2}{\sqrt{50}} = 40.5 + 0.89 = 39.61$$

$$\mu_h = 40.5 + 1.96\frac{3.2}{\sqrt{50}} = 40.5 + 0.89 = 41.39$$

First experiment

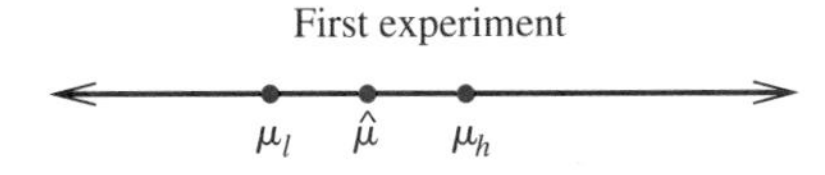

Second experiment

μ_l $\hat{\mu}$ μ_h

Third experiment

μ_l $\hat{\mu}$ μ_h

FIGURE 8.3.8

Confidence limits for the sample mean with known variance: three replicate

(Figure 8.3.8). In this case, these values were chosen on a computer with a true mean of 40.0. This value does lie within the confidence limits for our three experiments.

8.3.3 Sample Variance and Standard Error

If we do not know the true mean μ it seems rather unlikely we would have access to the true variance σ^2. Like the mean, σ^2 can be estimated from the data. Recall that the variance is the mean squared deviation from the mean. If we know μ, the logical estimator is

$$\hat{V} = \frac{\sum_{i=1}^{n}(x_i - \mu)^2}{n}$$

where $x_1, \ldots, x_n$ are again the measured values.

If we do not know μ, we would like to use the sample mean in place of μ. Replacing μ by $\hat{\mu}$ produces an estimate of the variance that tends to be small (the estimator is **biased**). To correct, we use the following definition.

Definition 8.9 The **sample variance** s^2 when the mean is unknown is

$$s^2 = \frac{\sum_{i=1}^{n}(x_i - \hat{\mu})^2}{n-1}.$$

The $n-1$ in the denominator corrects for the fact that we used the estimated mean $\hat{\mu}$ rather than the true mean μ (Exercises 33–36).

Example 8.3.12 Computing the Sample Variance

Consider the data in the accompanying table. The sample mean is

Measurement	Value
x_1	43.2
x_2	51.7
x_3	81.2
x_4	67.9
x_5	48.5

$$\hat{\mu} = \frac{43.2 + 51.7 + 81.2 + 67.9 + 48.5}{5} = 58.5.$$

The sample variance is

$$s^2 = \frac{(43.2 - 58.5)^2 + (51.7 - 58.5)^2 + (81.2 - 58.5)^2 + (67.9 - 58.5)^2 + (48.5 - 58.5)^2}{4}$$
$$= 246.0.$$

Just as for the variance, there is a simpler computational formula for the sample variance s^2.

Theorem 8.2 **Computational Formulas for the Sample Variance**

If x_i are n independent measurements, the sample variance can be written

$$s^2 = \frac{\sum_{i=1}^{n} x_i^2 - n\hat{\mu}^2}{n-1}.$$

The proof uses the same algebra as the computational formula for the variance (Theorem 6.4).

The **sample standard deviation** is s, the square root of the sample variance. Somewhat strangely, even though s^2 is an unbiased estimator of σ^2, s is **not** an unbiased estimator of σ.

Example 8.3.13 Using the Computational Formula to Find the Sample Variance and Standard Deviation

Consider again the data in Example 8.3.12. With the computational formula, the sample variance is

$$s^2 = \frac{43.2^2 + 51.7^2 + 81.2^2 + 67.9^2 + 48.5^2 - 5 \cdot 58.5^2}{4} = 246.0.$$

The sample standard deviation, our estimate of the standard deviation of each measurement in the sample, is

$$s = \sqrt{246.0} = 15.68.$$

The sample variance is even more sensitive to extreme values than the mean. Be careful about computing the sample variance when measurements include extreme values.

Example 8.3.14 The Sample Variance in a Data Set with an Outlier

Consider the data in Example 8.3.12, but suppose that $x_5 = 148.5$ instead of 48.5. The sample mean is now

$$\hat{\mu} = \frac{43.2 + 51.7 + 81.2 + 67.9 + 148.5}{5} = 78.5,$$

an increase of 20.0. The sample variance is

$$s^2 = \frac{43.2^2 + 51.7^2 + 81.2^2 + 67.9^2 + 148.5^2 - 5 \cdot 78.5^2}{4} = 1746.0,$$

and the sample standard deviation is

$$s = \sqrt{1746.0} = 41.78.$$

The sample standard deviation nearly tripled.

We would like to use s in place of σ in our calculation of confidence limits. When is it justified to do so? Two factors are important. When the underlying distribution is not normal, we need to invoke the Central Limit Theorem for Averages, which requires a reasonably large sample. Furthermore, there is additional uncertainty due to the estimated, rather than exact, sample variance. As a rule of thumb, **substitute s for σ only if $n \geq 30$**. Otherwise, an alternative distribution, called **Student's t distribution** must be used, as illustrated in the following subsection.

Theorem 8.3 Suppose $\hat{\mu}$ is the sample mean and s the sample standard deviation of $n \geq 30$ measurements from a normal distribution with unknown mean and variance. The 95% lower and upper confidence limits are

$$\mu_l = \hat{\mu} - 1.96\frac{s}{\sqrt{n}}$$

$$\mu_h = \hat{\mu} + 1.96\frac{s}{\sqrt{n}}.$$

The quantity $s/\sqrt{n}$ is an estimate of the standard deviation of the sample mean, and is called the **standard error of the mean** or the **standard error**. Because it is the standard deviation of the random variable $\overline{X}$, we write

$$\text{standard error} = \sigma_{\overline{X}} = \frac{s}{\sqrt{n}}$$

Example 8.3.15 Finding the Standard Error

Consider again the data in Example 8.3.12. We found the sample standard deviation $s = 15.68$ in Example 8.3.13. The standard error also depends on the number of samples, $n = 5$ in this case, and is equal to

$$\text{standard error} = \frac{s}{\sqrt{n}} = \frac{15.68}{\sqrt{5}} = 7.01.$$

This value is an estimate of our uncertainty about the mean of these five samples.

Example 8.3.16 Computing Confidence Limits with the Standard Error

If the underlying distribution in Example 8.3.15 is normal, the 95% confidence limits lie 1.96 standard errors above and below the sample mean. We found a sample mean of $\hat{\mu} = 58.5$ (Example 8.3.12), and a standard error of 7.01 (Example 8.3.15). The confidence limits are

$$\mu_l = 58.5 - 1.96 \cdot 7.01 = 72.2$$
$$\mu_h = 58.5 + 1.96 \cdot 7.01 = 44.8.$$

8.3.4 The t Distribution

When the variance is unknown and the number of samples is less than about 30, it is inaccurate to use the sample variance as an estimate of the true variance. In this case, we must use the **t distribution**, a generalization of the normal distribution, to estimate confidence limits. The t distribution describes a case where measurements come from a normal distribution with unknown mean and variance.

Critical values of the t distribution are given in Table 8.2. The number of **degrees of freedom** is one less than the number of measurements, or $\nu = n - 1$. If there are 25 measurements, the number of degrees of freedom is $\nu = 24$. In this case, the 95% confidence limit lies 2.064 standard errors from the mean, rather than 1.96 with the normal distribution. The confidence limits are wider because they incorporate the additional uncertainty about the variance (Figure 8.3.9).

Example 8.3.17 Finding Confidence Limits with the t Distribution

Consider again the data in Example 8.3.12. We found the sample standard deviation in Example 8.3.13 to be $s = 15.68$. Using the fact that there are $n = 5$ samples, the standard error is

$$\text{standard error} = \frac{s}{\sqrt{n}} = \frac{15.68}{\sqrt{5}} = 7.01$$

(Example 8.3.15). The t distribution takes into account the uncertainty in the estimate of s^2. With $n = 5$, the number of degrees of freedom is $\nu = n - 1 = 4$. From Table 8.2, the 95% confidence limits lie 2.776 standard errors above and below the sample mean. Using the sample mean of $\hat{\mu} = 58.5$ (Example 8.3.12), the confidence limits are

$$\mu_l = 58.5 - 2.776 \cdot 7.01 = 77.9$$
$$\mu_h = 58.5 + 2.776 \cdot 7.01 = 39.0.$$

These confidence limits are much wider than those found in Example 8.3.16.

Example 8.3.18 Finding 99% Confidence Limits with the t Distribution

Consider again the case in Example 8.3.10, with a sample mean of $\hat{\mu} = 47.3$ from a sample of $n = 10$ measurements. Assume now, however, that the sample variance is estimated as $s^2 = 400$. The standard error is

$$\text{standard error} = \frac{s}{\sqrt{n}} = \frac{\sqrt{400}}{\sqrt{10}} = 6.32.$$

The 99% confidence limits, from the line in the table with $n - 1 = 9$ degrees of freedom, lie 3.250 standard errors above and below the mean. Then

$$\mu_l = 47.3 - 3.250 \cdot 6.32 = 26.7$$
$$\mu_h = 47.3 + 3.250 \cdot 6.32 = 67.8.$$

These confidence intervals are quite wide because we demand a high degree of confidence from a relatively small sample.

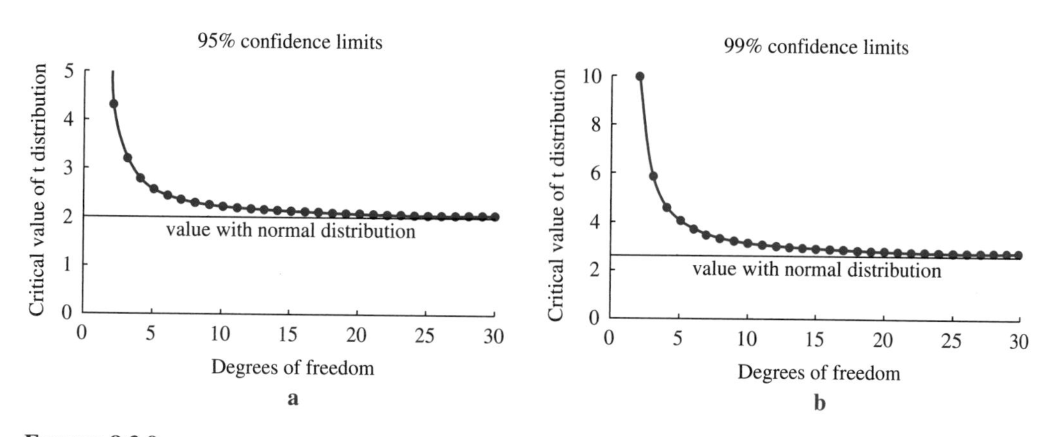

FIGURE 8.3.9
The t distribution: width of confidence limits in standard errors

Table 8.2 Critical values of the t distribution

Degrees of Freedom	95% Confidence Limit	99% Confidence Limit
1	12.71	63.66
2	4.303	9.925
3	3.182	5.841
4	2.776	4.604
5	2.571	4.032
6	2.447	3.707
7	2.365	3.499
8	2.306	3.355
9	2.262	3.250
10	2.228	3.169
11	2.201	3.106
12	2.179	3.055
13	2.160	3.012
14	2.145	2.977
15	2.131	2.947
16	2.120	2.921
17	2.110	2.898
18	2.101	2.878
19	2.093	2.861
20	2.086	2.845
21	2.080	2.831
22	2.074	2.819
23	2.069	2.807
24	2.064	2.797
25	2.060	2.787
26	2.056	2.779
27	2.052	2.771
28	2.048	2.763
29	2.045	2.756
30	2.042	2.750

8.3.5 Using the Normal Approximation to Find Confidence Limits for the Binomial

The binomial distribution is an important special case of a sum of random variables. Suppose we wish to use the normal approximation to compute confidence limits around the estimator $\hat{p}$, for the fraction of individuals in a sample with a particular property. How do we estimate the variance? For the underlying probability model, if we know the value of p we know that the variance is $p(1-p)$. We can use this idea to quickly estimate the variance from the data and apply the normal approximation.

Algorithm 8.2 Finding Confidence Limits with the Normal Approximation: Binomial Case

Suppose we measure m successes out of n trials, where $m \geq 5$ and $n - m \geq 5$ (to guarantee that the normal approximation is appropriate).

1. Estimate the probability of success as
$$\hat{p} = \frac{m}{n}.$$
2. Estimate the sample variance as
$$s^2 = \hat{p}(1-\hat{p}).$$
3. Estimate the confidence limits with Theorem 8.3 as
$$p_l = \hat{p} - 1.96\frac{s}{\sqrt{n}}$$
$$p_h = \hat{p} + 1.96\frac{s}{\sqrt{n}}.$$

If samples include fewer than 30 measurements (or fail to satisfy the condition in Subsection 7.9.2), the exact method should be used. Were we to calculate s^2 from Definition 8.9, we would see that Algorithm 8.2 uses n rather than $n-1$ in the denominator. In practice, this makes little difference because n must be at least 30.

Example 8.3.19 Finding Confidence Limits for the Binomial Proportion

Consider again the data consisting of 20 out of 100 people with a particular allele. Then

1. $\hat{p} = 0.2$.
2. $s^2 = 0.2 \cdot 0.8 = 0.16$.
3. The confidence limits are
$$p_l = 0.2 - 1.96\frac{\sqrt{0.16}}{10} = 0.122$$
$$p_h = 0.2 + 1.96\frac{\sqrt{0.16}}{10} = 0.278.$$

These answers compare well with the exact results of $p_l = 0.127$ (Example 8.2.4) and $p_h = 0.292$ (Example 8.2.6), and with the answers using the method of support of $p_l = 0.128$ and $p_h = 0.287$ (Example 8.2.13).

Example 8.3.20 Finding Confidence Limits from a Poll

Suppose 800 people are polled, and 40% support Donald Duck for governor.

1. $\hat{p} = 0.4$.
2. $s^2 = 0.4 \cdot 0.6 = 0.24$.
3. The confidence limits are

$$p_l = 0.4 - 1.96\frac{\sqrt{0.24}}{\sqrt{800}} = 0.366$$

$$p_h = 0.4 + 1.96\frac{\sqrt{0.24}}{\sqrt{800}} = 0.434.$$

Even with a fairly large sample, there is an uncertainty of more than 3% in the result.

Summary We have studied the problem of estimating the mean and computing confidence limits for the mean of a normal population. Estimates of the mean include the **sample mean**, the **sample median**, and the **trimmed mean**. The statistical properties of the sample mean can be analyzed with the Central Limit Theorem for Averages when the variance is known. When the variance is unknown, it can be estimated with the **sample variance** if sample size is at least 30. The 95% confidence limits are 1.96 **standard errors** above and below the sample mean. If the sample size is smaller, the **t distribution** takes into account the error in estimating the standard deviation, and can be used to find confidence limits. The normal approximation can be used to simplify estimation of confidence limits for the binomial distribution.

8.3 Exercises

Mathematical Techniques

1–4 ▪ Consider the following data on 20 plants.

Plant	Weight	Height	Yield	Number of seeds
1	7.3	2.1	0.045	7
2	10.8	3.3	0.132	18
3	12.1	3.5	0.187	19
4	18.6	4.2	0.129	16
5	20.1	5.0	0.201	17
6	32.3	6.1	0.184	14
7	38.4	5.5	0.231	22
8	40.2	7.2	0.298	21
9	42.1	7.4	0.287	20
10	43.8	6.9	0.310	25
11	43.8	6.7	0.276	24
12	44.9	7.6	0.353	15
13	45.8	5.6	0.462	12
14	46.4	8.0	0.561	23
15	52.7	7.8	0.435	23
16	60.3	9.8	0.598	12
17	98.9	11.2	0.723	25
18	178.3	9.2	0.668	17
19	213.1	13.5	1.022	25
20	298.8	17.3	1.745	21

(*Continued*)

Find the following for the given measurement.

a. The sample mean.

b. The sample median.

c. The trimmed means $\overline{X}_{tr(5)}$, $\overline{X}_{tr(10)}$, and $\overline{X}_{tr(20)}$.

1. The weight W

2. The height H

3. The yield Y

4. The seed number S

5–8 ▪ Find the sample variance, the sample standard deviation, and the standard error for the given measurement.

5. The weight W in Exercise 1.

6. The height H in Exercise 2.

7. The yield Y in Exercise 3.

8. The seed number S in Exercise 4.

9–12 ▪ Find how many measurements lie (a) less than one sample standard deviation from the sample mean and (b) more than two sample standard deviations from the sample mean for the given measurement. Which behave more or less like the normal distribution?

9. The weight W in Exercise 1.

10. The height H in Exercise 2.

11. The yield Y in Exercise 3.

12. The seed number S in Exercise 4.

13–16 ▪ Find the 95% confidence limits around the sample mean for the given measurement, assuming that the sample variance s is a good estimate of the true variance σ.

13. The weight W in Exercises 1 and 5.

14. The height H in Exercises 2 and 6.

15. The yield Y in Exercises 3 and 7.

16. The seed number S in Exercises 4 and 8.

17–20 ▪ Find the 95% confidence limits around the sample mean for the given measurement, without assuming that the sample variance s is a good estimate of the true variance σ (thus using the t distribution).

17. The weight W in Exercises 1 and 5.

18. The height H in Exercises 2 and 6.

19. The yield Y in Exercises 3 and 7.

20. The seed number S in Exercises 4 and 8.

21–24 ▪ Find the given confidence limits around the sample mean for the given measurement, assuming that the sample variance s is a good estimate of the true variance σ.

21. 98% confidence limits around the weight W in Exercise 1.

22. 90% confidence limits around the height H in Exercise 2.

23. 99.8% confidence limits around the yield Y in Exercise 3.

24. 99.9% confidence limits around the seed number S in Exercise 4.

25–26 ▪ Find the normal approximation to the following.

25. The average of 30 numbers chosen from the exponential p.d.f. $g(x) = 2e^{-2x}$.

26. The average of 30 numbers chosen from the p.d.f. $f(x) = 1$ for $0 \le x \le 1$.

27–28 ▪ Find 95% confidence intervals in the following cases, assuming that the standard deviations are known to match those in the earlier problem. Does the confidence interval include the true mean?

27. A sample mean of 0.4 is found in Exercise 25.

28. A sample mean of 0.7 is found in Exercise 26.

29–30 ▪ Use the normal approximation to find 95% confidence limits around the estimated proportion $\hat{p}$ in the following cases.

29. A coin is flipped 100 times and comes out heads 44 times.

30. Of 1000 people polled, 320 favor the use of mathematics in biology.

31–32 ▪ The 95% confidence limits with the t distribution are wider than those with the normal distribution. For approximately how many degrees of freedom do they match the given confidence limits with the normal distribution? How large a sample does this correspond to?

31. The 99% confidence limits with the normal distribution.

32. The 98% confidence limits with the normal distribution.

33–36 ▪ In simple cases, we can see why using a denominator of n in the equation for the sample variance produces a biased estimate. Suppose a population consists of half 0 s and half 1 s.

33. Use the variance of a Bernoulli distribution with $p = 0.5$ to find the exact variance of each measurement.

34. Find all possible samples of size 2 and their associated probabilities.

35. Find the mean squared deviation from the mean for each and average them to find the expected sample variance.

36. Compare with the true answer and show that using a denominator of $n - 1$ rather than n would give the right answer. Try to explain the bias.

Applications

37–40 ▪ Consider the following data on immigration into four populations over 20 yr (based on the probabilities in Section 7.8, Exercises 33–36.) For each, find the sample mean and the sample standard deviation. Compare them with the mathematical mean and standard deviation found in the earlier problem.

Year	Population a	Population b	Population c	Population d
1	0	−1	1	−10
2	−1	0	1	1
3	1	0	1	0
4	1	0	1	1
5	−1	0	−1	−10
6	−1	0	−1	0
7	1	0	0	1
8	2	−1	−1	−10
9	0	2	−1	1

(Continued)

Year	Population a	Population b	Population c	Population d
10	0	−1	−1	−10
11	−1	1	0	1
12	−1	0	100	1
13	−1	2	1	−10
14	−1	2	1	1
15	−1	0	0	1
16	0	0	−1	−10
17	−1	0	−1	−10
18	1	0	1	1
19	−1	0	1	−10
20	0	2	1	−10

37. Population a (see Section 7.8, Exercise 33).

38. Population b (see Section 7.8, Exercise 34).

39. Population c (see Section 7.8, Exercise 35). Why are the mean and variance so different from the mathematical expectations?

40. Population d (see Section 7.8, Exercise 36).

41–44 ▪ Find the 95% confidence intervals around the mean number of immigrants using both the true variance and the sample variance. Does the true mean lie within the confidence limits?

41. Population a

42. Population b

43. Population c

44. Population d

45–48 ▪ Several plants are crossed, producing the following proportions. Find 99% confidence limits around the fraction of tall plants in each case.

45. Of 50 offspring, 35 are tall.

46. Of 500 offspring, 350 are tall.

47. Of 100 offspring, 52 are tall.

48. Of 200 offspring, 13 are tall.

49–50 ▪ Mutations are counted in a large section of the genome. This count finds 14 mutations in one set of 1 million base pairs. Then 30 different sets of 1 million base pairs are measured, and the average number of mutations per million is found to be 13.5.

49. Use the normal approximation to the Poisson distribution to estimate 95% confidence limits around the true mean number of mutations per million base pairs and compare with Section 8.2, Exercise 43.

50. Thirty different sets of 1 million base pairs are measured, and the average number of mutations per million is found to be 13.5. Estimate the standard deviation, and find the standard error of the mean and the 99% confidence limits.

Computer Exercise

51. Use the Monte Carlo method to estimate the confidence limits in Exercises 41 and 42 . How close are your results to the exact answer?

8.4 Hypothesis Testing

Estimators and confidence limits help us quantify what happened in an experiment– to **estimate parameters**. Much of science is built upon a related but fundamentally different task: stating and testing a **hypothesis**. We usually begin by testing whether an experimental result is consistent with the **null hypothesis** which states that nothing interesting happened. For example, the null hypothesis in a trial of new medication would be that the new medication has no effect. The **significance level**, the probability of rejecting a true null hypothesis, and the **power**, the probability of rejecting a false null hypothesis, quantify the effectiveness of a statistical test. We present an approach to **hypothesis testing** based on the logic of confidence limits.

8.4.1 Hypothesis Testing: Terminology

Suppose that extensive sampling of an entire population has shown that exactly 13% of people have a particular allele. We select 100 individuals from this population who have been diagnosed with a particular disease and find that 20 of them have that allele. This is more than the 13 we would expect from the binomial distribution. Would we be justified in reporting that the allele is connected to the disease?

Hypothesis testing begins by setting up a **null hypothesis**, referred to as H_0, and checking whether the data are "consistent" with it. The null hypothesis is a mathematical way of saying that nothing interesting happened. In our example, the null hypothesis is that diseased individuals have the allele with the *same* probability as everybody else.

Example 8.4.1 A Null Hypothesis

Let q be the true unknown proportion of diseased individuals in the population who have the allele. The null hypothesis is $q = 0.13$, which says precisely that the proportion of diseased individuals who have the allele matches the proportion in the population as a whole.

We compare this null hypothesis with an **alternative hypothesis**, referred to as H_a. The alternative hypothesis is that diseased individuals have a **different** probability from everyone else.

Example 8.4.2 Two Alternative Hypotheses

There are at least two possible alternative hypotheses. If we suspect that the allele is positively related to the disease, the alternative hypothesis is $q > 0.13$, meaning that the allele is more common among people with the disease. If we simply suspect a relationship of some sort, the alternative hypothesis is $q \neq 0.13$. This says that the proportion among diseased individuals differs from that in the population as a whole, but does not specify in advance whether the proportion is larger or smaller.

When there is more than one alternative hypothesis, as is usually the case, a scientist should state which of these alternatives will be tested before examining the data. The null hypothesis and alternative hypothesis must be mutually exclusive.

A statistical **test** is a way to **accept** or **reject** the null hypothesis. Rejecting means that we are confident that the null hypothesis is not true. Accepting the null hypothesis means that we cannot prove it wrong. Accepting does *not* mean that the null hypothesis is correct, just that we do not have sufficient information to reject it. There are four possible results.

	H_0 Accepted	H_0 Rejected
H_0 true	correct	type I error
H_0 false	type II error	correct

There are two correct results: accepting a true null hypothesis and rejecting a false null hypothesis. We could also make errors of two types. We make a **type I error** if we reject a true null hypothesis. The probability of committing a type I error is called the **significance level** and is denoted by α. The probability of a type I error depends on our statistical test.

Example 8.4.3 A Type I Error

If individuals with the disease were really the same as everybody else ($q = 0.13$) but we decided otherwise and rejected the null hypothesis based on our sample and statistical test, we would commit a type I error. This type of error might lead to further research about an allele that has no relationship to the disease.

We make a **type II error** if we accept a false null hypothesis. The probability of a type II error, denoted by β, depends on both the statistical test and on the true value of the parameter.

Example 8.4.4 A Type II Error

If the proportion q among diseased individuals were really $q = 0.18$ and our statistical test fails to reject the null hypothesis of $q = 0.13$, we have committed a type II error. The probability of committing a type II error depends on the true value of q. If $q = 0.5$

among diseased individuals, we are quite likely to detect a difference from the null hypothesis and are thus unlikely to commit a type II error. If $q = 0.131$ among diseased individuals, we are unlikely to detect a difference from the very similar null hypothesis and are likely to commit a type II error.

The probability of rejecting a false null hypothesis as a function of the true value of the parameter q is called the **power** of a test.

Definition 8.10 The **power** of a statistical test is the probability that it correctly rejects a false null hypothesis.

$$\text{Power of test} = 1 - \text{ probability of a type II error} = 1 - \beta.$$

Example 8.4.5 Power with Two Values of the True Parameter

For the cases in Example 8.4.4, our power to reject the null hypothesis $q = 0.13$ if the true value is $q = 0.5$ should be high. Our power to reject the null hypothesis if $q = 0.131$ will be low.

Ideally, we could design tests which make neither type I nor type II errors. Because chance plays many tricks, this is impossible. There is a tradeoff between type I and type II errors. A very powerful test that always rejects a false null hypothesis is also more likely to reject a true null hypothesis. A weaker test that never rejects a true null hypothesis will often fail to reject a false null hypothesis. Think of a statistical test as a sort of signal detector (Figure 8.4.1). A detector that is not very powerful will detect only genuine signals (no type I errors) but will miss weaker signals (many type II errors). A very powerful detector will not miss any real signals (no type II errors) but will misidentify noise as genuine alien communication (many type I errors).

Small detector captures few signals but all are from aliens

Few type I errors
Many type II errors

a

Powerful detector captures many signals, many not from aliens

Few type II errors
Many type I errors

b

FIGURE 8.4.1

Detecting aliens: type I and type II errors

8.4.2 Design of Statistical Tests and Computation of p-Values

How do we design a test? The most widely used method is based on **p-values**.

Definition 8.11 The **p-value** is the probability of observing a result at least as extreme as the measured result if the null hypothesis is true.

Chance alone can produce strange results. The p-value quantifies the probability of seeing extreme results just by chance.

Example 8.4.6 Finding a p-Value in the Genetic Example: One-Tailed Test

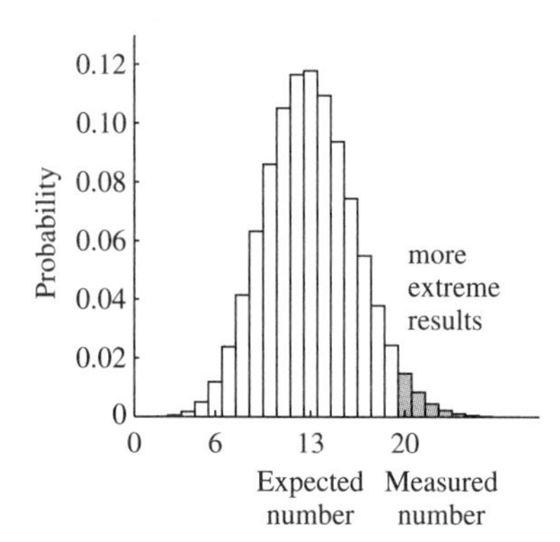

FIGURE 8.4.2

Probability of result as extreme as or more extreme than 20 if $q = 0.13$: one-tailed test

Consider the null hypothesis $q = 0.13$ and the alternative hypothesis $q > 0.13$. Finding 20 out of 100 people with the allele is more than the 13 expected if the null hypothesis were true. To find the p-value we add the probability of all results that are as extreme as or more extreme than the measured value. Any values larger than 20 are more extreme, in the sense of more different from the null hypothesis, than our measurement.

The p-value is thus the probability of finding 20 or more people with the allele if the null hypothesis $q = 0.13$ is true, or

$$\text{P} = \sum_{k=20}^{100} b(k;\, 100, 0.13) = 0.0319$$

(Figure 8.4.2). Because this region includes only one side, this is called a **one-tailed test**. The probability of a result at least as extreme as what we observed is 3.2% if the null hypothesis is true. If we reject the null hypothesis, we have a 3.2% chance of making a type I error, the probability that our extreme result occurred by chance. If we decided to treat diseased individuals with an expensive new gene therapy, there is a 3.2% chance we would be throwing our money away. A p-value less than 0.05 is generally considered to be **significant**.

Example 8.4.7 The p-Value with More Extreme Data

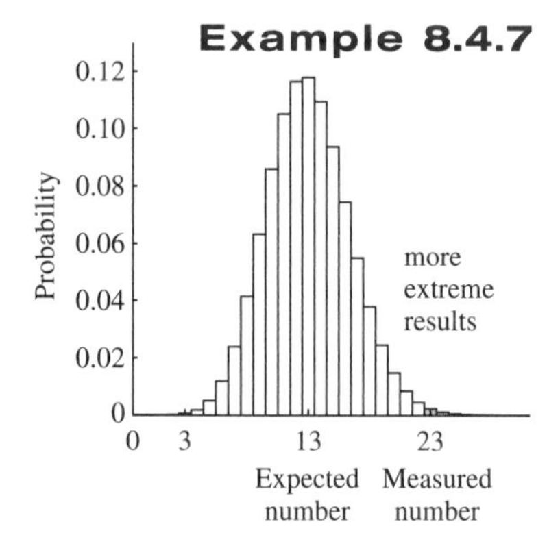

FIGURE 8.4.3

Probability of result as extreme as or more extreme than 23 if $q = 0.13$: one-tailed test

If we had found 23 people out of 100 rather than 20 with the allele, our p-value would be

$$\text{P} = \sum_{k=23}^{100} b(k;\, 100, 0.13) = 0.00424.$$

(Figure 8.4.3).

When the chance of a type I error is less than 0.01, the result is considered to be **highly significant**. Finding 23 out of 100 people is convincing evidence that the true q is greater than 0.13. When the chance of a type I error is less than 0.001 the result is considered to be **very highly significant**. One would be very unlucky indeed to get such an unlikely result by chance.

The terminology used to describe p-values is given in the following table. The cutoffs between the different terms are chosen for convenience. Statistics like the p-value are guides for thinking about and presenting data, not hard and fast rules. An experiment with a p-value of 0.051 should not be interpreted very differently from one with a p-value of 0.049.

p-Value	Significance
$p > 0.1$	not significant
$0.1 > p > 0.05$	trend toward significance
$0.05 > p > 0.01$	significant
$0.01 > p > 0.001$	highly significant
$p < 0.001$	very highly significant

Example 8.4.8 Finding a p-Value in the Genetic Example: Two-Tailed Test

The p-value depends on the alternative hypothesis. Suppose we observe 20 people out of 100 with a particular allele as in Example 8.4.6, but have the alternative hypothesis $q \neq 0.13$. The identification of the set of results as extreme as or more extreme than the observed results is more complicated. In this case, "more extreme" means further from the result expected if the null hypothesis were true (13 out of 100). The measured value is $20 - 13 = 7$ away from this expected result. A result of 6 or fewer is considered as extreme as a result of 20 or more because both lie at least 7 away from 13 (the shaded

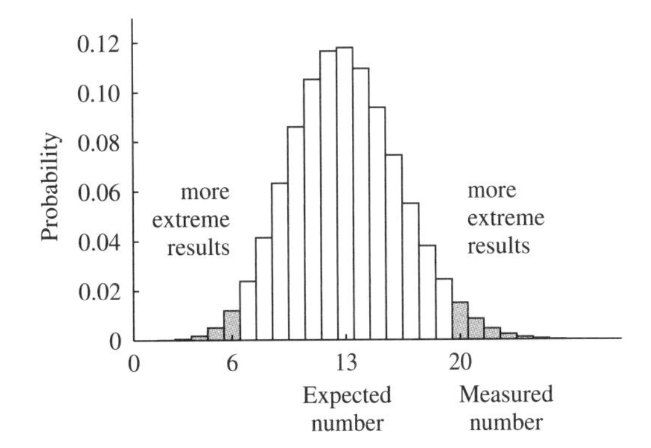

FIGURE 8.4.4
Probability of result more extreme than 20 if $q = 0.13$: two-tailed test

region in Figure 8.4.4). Because the shaded region includes both tails of the distribution, this is called a **two-tailed test**. The p-value is

$$P = \sum_{k=0}^{6} b(k;\, 100, 0.13) + \sum_{k=20}^{100} b(k;\, 100, 0.13) = 0.0512.$$

The result does not quite reach the conventional threshold of 0.05 for significance, although it is close.

Testing hypotheses with p-values is closely related to the calculation of confidence intervals. Both methods involve similar thought experiments, finding the probability of data as extreme as or more extreme than observed. To find confidence limits, we experiment with the underlying parameter to find the smallest and largest values consistent with the data. To test a hypothesis, we use the same logic, but check just the one parameter value characterizing the null hypothesis. A null hypothesis that lies within the 95% confidence limits cannot be rejected, and has a p-value greater than 0.05.

Example 8.4.9 Hypothesis Testing and Confidence Limits

If we counted 20 people out of 100 with the allele, we found that the 95% confidence interval runs from 0.127 to 0.292 (Examples 8.2.4 and 8.2.6). The two-tailed test asks precisely whether the null hypothesis lies between these confidence limits. The null hypothesis $q = 0.13$ lies just within these limits, corresponding to a p-value slightly greater than 0.05.

Why is the significance level different for the one-tailed and two-tailed tests? Both use the same data, the same null hypothesis, and the same underlying model. The two-tailed test, however, uses a less informative alternative hypothesis. It is appropriate to use the one-tailed test only if

1. You have reason to believe that the results will fall on a particular side of the null hypothesis.
2. You make a public announcement of this expectation *before* looking at the data.

If there is laboratory evidence indicating that a particular allele is related to a disease, then a public prediction that the fraction of diseased individuals with the allele will be greater than the 0.13 makes the one-tailed test appropriate.

When the sums required to compute p-values cannot be evaluated or written down, a version of the Monte Carlo method can be used. To test whether data are consistent with the null hypothesis, assume the null hypothesis and simulate many times on a computer. The proportion of times producing a result as extreme as or more extreme than the observation is an estimate of the p-value.

Table 8.3 Testing a hypothesis with the Monte Carlo method

Number Observed	Number of Times Observed Out of 100 Replications
6	1
7	5
8	7
9	7
10	9
11	10
12	9
13	10
14	12
15	6
16	6
17	10
18	2
19	3
20	0
21	2
22	1

Example 8.4.10 Testing a Hypothesis with the Monte Carlo Method

The results of 100 simulations of choosing 100 people with $q = 0.13$ are shown in Table 8.3. Only 3 out of 100 results equal or exceed the measured result of 20 out of 100 (2 with 21 and 1 with 22). If we are using a one-tailed test, this indicates that the significance of the result is 0.03, quite close to the exact significance level of 0.0319 (Example 8.4.6). The Monte Carlo simulation indicates that the result is significant.

The two-tailed test must count all simulations that produce results more than 7 away from the expectation under the null hypothesis and thus includes any where 6 or fewer individuals are observed (Example 8.4.8). There is one additional simulation in this category, for a total of 4 out of 100. The estimated p-value is then 4/100=0.04. Because we ran only 100 simulations, this value differs from the exact significance level of 0.0512 (Example 8.4.8), and in fact lies below the conventional threshold for significance.

8.4.3 Computing the Power of a Statistical Test

The power of a statistical test is the probability of rejecting a false null hypothesis. The power depends on how much the true parameter value differs from the value under the null hypothesis. Because we can never know the true parameter value, all calculations of power involve thought experiments.

Computing the power takes two steps. First, we find the critical value of the measurement required to give significance. Then, we find the probability of exceeding that value for different values of the underlying true parameter.

Example 8.4.11 Finding the Critical Value to Give Significance

How many more than 13 people out of 100 would we have to find before we could reject the null hypothesis at the 0.05 significance level? Suppose that only 19 people out of 100 were found with the allele. The significance level with a one-tailed test is

$$P = \sum_{k=19}^{100} b(k;\,100, 0.13) = 0.056.$$

This result is not strong enough to reject the null hypothesis. We thus require 20 or more people with the allele in a sample of 100 to reject the null hypothesis with a one-tailed test. Think of a test as a buzzer that goes off when the data are extreme enough to reject the null hypothesis. The buzzer goes off if 20 or more people are found with the allele.

Example 8.4.12 Using the Statistical Test to Compute Power If $q = 0.15$

Suppose the "true" q for diseased individuals were 0.15, only a bit larger than 0.13. How likely are we to miss this (that is, to commit a type II error) with the one-tailed test (Example 8.4.11)? The null hypothesis is rejected if we find 20 or more people with the allele.

$$\text{power} = \sum_{k=20}^{100} b(k;\,100, 0.15) = 0.107.$$

This is the power to detect the difference. The probability of missing the difference and committing a type II error is

$$\beta = 1 - \text{power} = \sum_{k=0}^{19} b(k;\,100, 0.15) = 0.893$$

(Figure 8.4.5). A sample of 100 is too small to effectively distinguish 0.13 from 0.15.

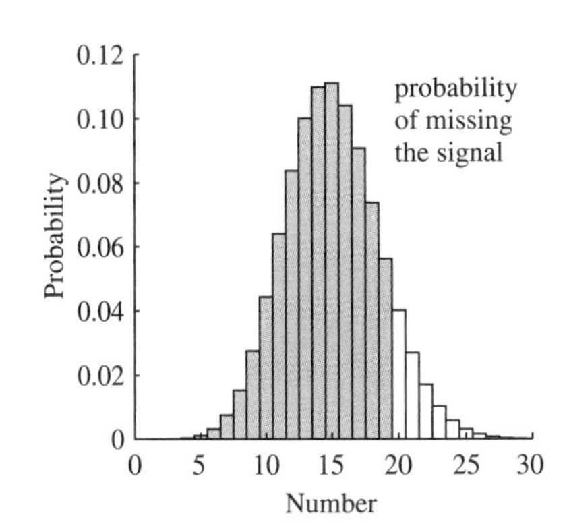

FIGURE 8.4.5
Probability of type II error if $q = 0.15$

Example 8.4.13 Using the Statistical Test to Compute Power If $q = 0.25$

If the true q is 0.25, the probability of finding 20 or more, and thus rejecting the null hypothesis, is

$$\text{power} = \sum_{k=20}^{100} b(k;\,100, 0.25) = 0.900.$$

The probability of a type II error is

$$\beta = 1 - \text{power} = \sum_{k=0}^{19} b(k;\,100, 0.25) = 0.100.$$

We have less than a 10% chance of missing and a 90% chance of recognizing this much stronger signal (Figure 8.4.6).

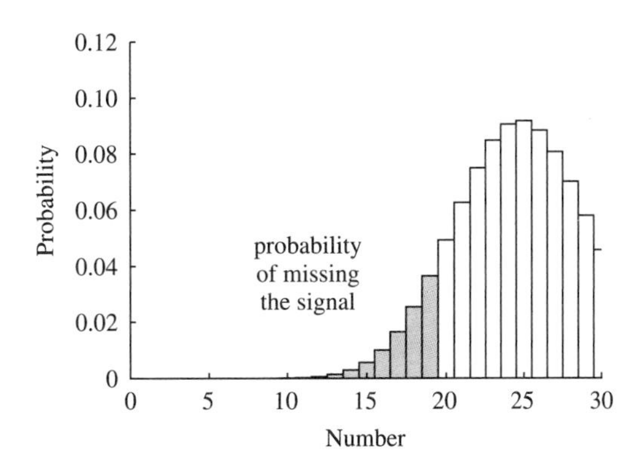

FIGURE 8.4.6
Probability of type II error if $q = 0.25$

The power, unlike the significance level, cannot be calculated without knowing the true value of the parameter. Because the true value is unknown, the best we can do is to understand the power as a function of the true parameter value (Figure 8.4.7).

Example 8.4.14 Graph of Power as a Function of Underlying True Parameter Value

By repeating the calculations in Examples 8.4.12 and 8.4.13 for a whole range of values of q, we can get a sense of how large a difference could be detected with an experiment of this size (Figure 8.4.7). The farther the true parameter is from 0.13, the better we

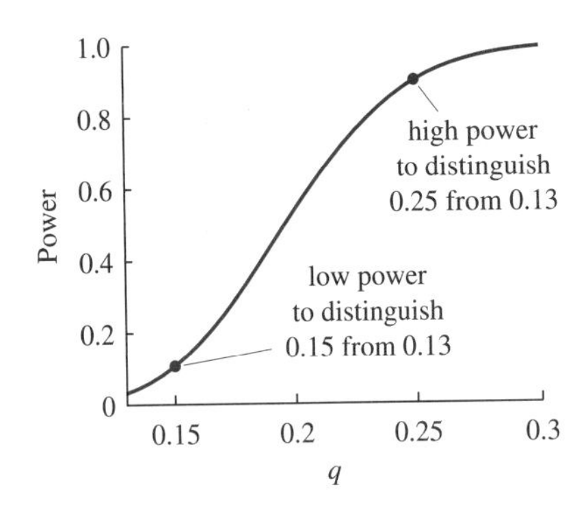

FIGURE 8.4.7
Power as a function of the true parameter value

can distinguish it from 0.13. The power reaches 0.5 for q near 0.2 (consistent with the threshold for our test). Values of q less than this cannot be dependably distinguished from $q = 0.13$. If we had reason to believe that the true q was only about 0.15, we would need to measure many more patients to detect it.

Summary To test whether an observation is consistent with a **null hypothesis** or an **alternative hypothesis**, we have introduced **p-values**, the probability of measuring a result at least as extreme as the actual measurement when the null hypothesis is true. The test is called a **one-tailed test** when extreme results are included on one side, and a **two-tailed test** if extreme results are included on both sides. The **significance level** is the probability of a **type I error**, the rejection of a true null hypothesis. Accepting a false null hypothesis is called a **type II error**. The efficiency with which a test rejects a false null hypothesis is called the **power** of a test and depends on the true value of the parameter, the size of the sample, and the type of statistical test used.

8.4 Exercises

Mathematical Techniques

1–4 ▪ Explain in words what the following statements mean, and restate the result in terms of type I or type II errors as appropriate.

1. The p-value associated with the null hypothesis is 0.2.
2. Out of 1000 simulations of the null hypothesis, 70 produce a result as extreme as or more extreme than the actual observation.
3. The null hypothesis is false, and a test (at significance level 0.05) rejects the null hypothesis with probability 0.6.
4. The null hypothesis is false, and a test (at significance level 0.01) rejects the null hypothesis with probability 0.2.

5–8 ▪ Using the following data, find the p-value associated with the null hypothesis that the true probability of heads is 0.5. Is the result significant?

5. A coin is flipped 5 times and comes up heads every time (as in Section 8.2, Exercise 1).
6. A coin is flipped 7 times and comes up heads 6 out of 7 times (as in Section 8.2, Exercise 2).
7. A coin is flipped 10 times and comes up heads 9 times. You have reason to suspect however, that the coin produces an excess of heads (and have told this to a friend in advance).
8. A coin is flipped 20 times and comes up heads 3 times. You have reason to suspect however, that the coin produces an excess of tails (and have told this to a friend in advance).

9–12 ▪ A coin is flipped 10 times and comes up heads 9 times. You have reason to suspect however, that the coin produces an excess of heads (and have told this to a friend in advance), as in Exercise 7.

9. What is the cutoff number of heads above which you reject the null hypothesis at the $\alpha = 0.05$ significance level?

10. What is the cutoff number of heads above which you reject the null hypothesis at the $\alpha = 0.1$ significance level?

11. Find the power of your test with $\alpha = 0.05$ if the true probability of a head is 0.6.

12. Find the power of your test with $\alpha = 0.05$ if the true probability of a head is 0.9.

13–14 ▪ Find the p-value with both one-tailed and a two-tailed tests.

13. One cosmic ray hits a detector in 1 yr. The null hypothesis is that the rate at which rays hit is $\lambda = 5$/yr (as in Section 8.2, Exercise 6.)

14. Three cosmic rays hit a larger detector in 1 yr. The null hypothesis is that the rate at which rays hit is $\lambda = 10$/yr (as in Section 8.2, Exercise 5.)

15–18 ▪ Find the p-value associated with the following hypotheses.

15. You wait 4000 h for an exponentially distributed event to occur. The null hypothesis is that the mean wait is 1000 h with the alternative hypothesis that the mean wait is greater than 1000 h.

16. You wait 40 h for an exponentially distributed event to occur. The null hypothesis is that the mean wait is 1000 h with the alternative hypothesis that the mean wait is less than 1000 h.

17. The first defective gasket is the 25th. The null hypothesis is that the first defect follows a geometric distribution with mean 10, and the alternative hypothesis is that the mean is greater than 10.

18. The first defective gasket is the 50th. The null hypothesis is that the first defect follows a geometric distribution with mean 1000, and the alternative hypothesis is that the mean is less than 1000.

Applications

19–20 ▪ Think about how thresholds are set in the following situations.

19. Certain screens for cancer work by examining many cells under a microscope and looking for abnormalities. Discuss how setting the threshold for cell abnormality can affect the number of type I and type II errors. What factors would go into deciding where to set the threshold?

20. Certain screens for cancer drugs work by examining many drugs and looking for those that suppress tumor growth. Discuss how setting the threshold for tumor growth reduction can affect the number of type I and type II errors. What factors would go into deciding where to set the threshold?

21–22 ▪ Consider a couple that has seven boys and one girl in a family of eight.

21. Test the hypothesis that boys and girls are equally likely.

22. At birth, boys are slightly more common than girls. Test the hypothesis that 55% of births are boys.

23–24 ▪ Phone calls used to arrive at an average rate of 3.5/h, but after posting your number on your Web page, you receive more calls on subsequent days. For each day,

a. State null and alternative hypotheses.

b. Use the Poisson distribution to compute the probability of this event if the null hypothesis is true.

c. Compute the probability of an event at least this extreme if the null hypothesis is true.

d. Is this result significant? How would you interpret it?

23. You receive 7 calls in 1 h on the first day.

24. You receive 8 calls in 1 hr on the second day.

25–26 ▪ Consider the data in Exercise 23, where calls arrive at a rate of 3.5/h before posting your phone number on your Web page, but 7 arrive in 1 h on the next day.

25. Find the cutoff value for a test with $\alpha = 0.05$. Find the power with $\lambda = 4.0$, $\lambda = 7.0$, and $\lambda = 10.0$. Explain why the power is higher for larger values of Λ.

26. Find the cutoff value for a test with $\alpha = 0.01$. Find the power with $\lambda = 4.0$, $\lambda = 7.0$, and $\lambda = 10.0$. Why does a higher significance level reduce the power?

27–34 ▪ The survival time for one type of cell in culture is exponentially distributed with a mean of 200 h. After applying a new treatment, one cell lasts 800 h. For a second type of cell, the survival time is exponentially distributed with a mean of 100 h. After applying the same new treatment, one cell lasts 30 h.

27. State null and alternative hypotheses for the first type of cell. At what level can you reject the null hypothesis? Is it significant?

28. State null and alternative hypotheses for the second type of cell. At what level can you reject the null hypothesis? Is it significant?

29. What is the shortest survival over 200 h for which you might claim a significant result for the first cell type (at the 0.05 level)?

30. What is the longest survival under 30 h for which you might claim a significant result for the second cell type (at the 0.05 level)?

31. Suppose you adopt the cutoff from Exercise 29. Find and graph the power as a function of the true mean. What is the power of the test if the true mean is 500? What is the power of the test if the true mean is 1000?

32. Suppose you adopt the cutoff from Exercise 30. Find and graph the power as a function of the true mean. What is the power of the test if the true mean is 50? What is the power of the test if the true mean is 10?

33. Solve for the smallest value of the mean for which the power to detect an improvement in cells of the first type is equal to 0.95. Interpret your answer.

34. Solve for the largest value of the mean for which the power to detect harm to cells of the second type is equal to 0.95. Interpret your answer.

35–38 ▪ Consider measuring n plants with a known standard deviation of 3.2 cm. How many plants would have to be measured to achieve the following?

35. 95% confidence limits where the upper and lower confidence limits are 1.0 cm from the sample mean.

36. 99% confidence limits where the upper and lower confidence limits are 1.0 cm from the sample mean.

37. 95% confidence limits where the upper and lower confidence limits are 0.25 cm from the sample mean.

38. 99% confidence limits where the upper and lower confidence limits are 0.25 cm from the sample mean.

Computer Exercises

39. Repeat the calculations in Table 8.3. Is the hypothesis $q = 0.13$ rejected? Try with different numbers of simulations and compare with the theoretical p-values.

40. Generate nine independent random numbers from a normal distribution with mean 10 and variance of 9 and find their average $\overline{X}$. If your value of $\overline{X}$ is within 0.5 of 10.0, try again to guarantee nice pictures and interesting results (real scientists are not allowed to do this sort of thing, of course). The average of your nine measurements comes from a normal distribution with mean and some standard error. Find the standard error and the 95% confidence interval around $\overline{X}$, calling the lower limit μ_l and the upper limit μ_h.

Simulate the following four experiments 100 times: (1) Average nine values from a normal distribution with true mean 10. (2) Average nine values from a normal distribution with true mean μ_l. (3) Average nine values from a normal distribution with true mean μ_h. (4) Average nine values from a normal distribution with true mean $\overline{X}$. Count how many values in each experiment lie below μ_l, between μ_l and 10, between 10 and μ_h, and above μ_h.

Define functions f_1, f_2, f_3, and f_4 to be the p.d.f.'s describing the distribution of elements in the four experiments. Print a graph of each function, and list below it the number of elements from the appropriate experiment lying in the various intervals. Indicate which ones have values predicted by the theory of confidence intervals and what those values should be. Are you bothered by the fact that more than 5% of the elements of experiment 1 lie outside your confidence interval? Why or why not?

8.5 Hypothesis Testing: Normal Theory

In principle, it is possible to compute a p-value when we understand the probability distribution that describes the null hypothesis. Often, however, the computation can be computer-intensive. The Central Limit Theorem shows that the normal distribution often provides a convenient and accurate approximation of more complicated models. In this section, we apply the methods of hypothesis testing when the underlying distributions are normal or approximately so and compute the power of the tests.

8.5.1 Computing p-Values with the Normal Approximation

Suppose we try out a new fertilizer on 25 plants which previously had a mean height of 39.0 cm. The fertilized plants have a mean height of 40.2 cm. Are we convinced that the fertilizer works?

The p-value is the probability of a result as extreme as or more extreme than the observation. Finding the p-value, like finding the confidence limits around the mean, depends on the sample size n and on the standard deviation of the underlying measurements.

Suppose a set of n measurements has sample mean $\overline{X}$, and the null hypothesis is that the true mean is some specified value μ_0. If the variance of the underlying distribution is the known value σ^2, then the distribution X of sample means under the null hypothesis is

$$X \sim N\left(\mu_0, \frac{\sigma^2}{n}\right).$$

Example 8.5.1 Finding the Distribution of Means Under the Null Hypothesis

Suppose that the underlying standard deviation is known to be $\sigma = 3.2$ cm. Then the variance is $\sigma^2 = 3.2^2 = 10.24$ cm^2. With a null hypothesis that $\mu_0 = 39.0$ cm and a sample size $n = 25$, the distribution of the sample mean follows

$$X \sim N\left(39.0, \frac{10.24}{25}\right) = N(39.0, 0.4096)$$

(Figure 8.5.1). The standard error $\sigma_{\overline{X}} = \sqrt{0.4096} = 0.64$.

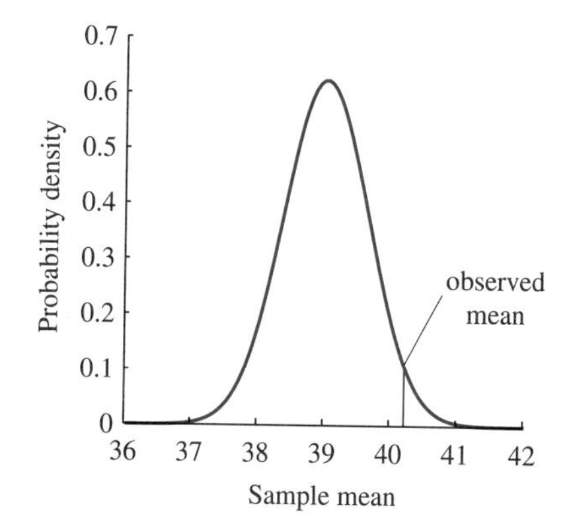

FIGURE 8.5.1

Distribution of the sample mean under the null hypothesis

Example 8.5.2 The Distribution of Means Under the Null Hypothesis with a Larger Sample Size

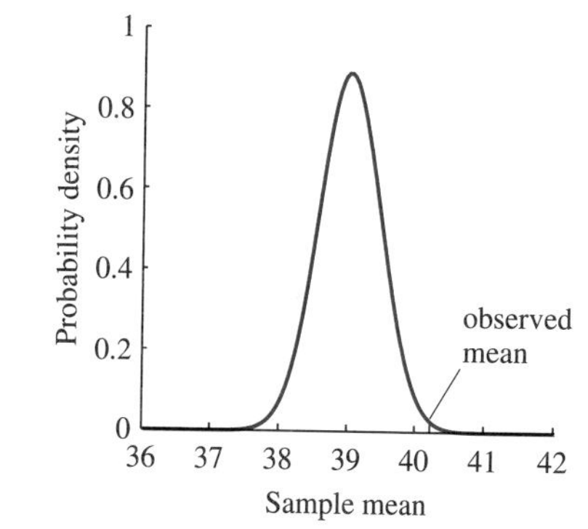

FIGURE 8.5.2

Distribution of means under the null hypothesis with $n = 50$

Suppose that the underlying standard deviation is known to be $\sigma = 3.2$ cm as in Example 8.5.1, but we measure $n = 50$ plants.

$$X \sim N\left(39.0, \frac{10.24}{50}\right) = N(39.0, 0.2048)$$

(Figure 8.5.2). The standard error $\sigma_{\overline{X}} = \sqrt{0.2048} = 0.4526$. Although the underlying measurements are drawn from the same distribution, the distribution of the sample means is narrower, and the observed value $\overline{X} = 40.2$ lies farther into the tails.

The p-value associated with a measured value $\overline{X}$ (the value 40.2 in this example), depends on our alternative hypothesis. Let μ represent the true, unknown mean of the sample. If we had stated in advance that the fertilizer would increase the mean, our alternative would be $\mu > \mu_0$.

Example 8.5.3 Finding the p-Value with a One-Tailed Test

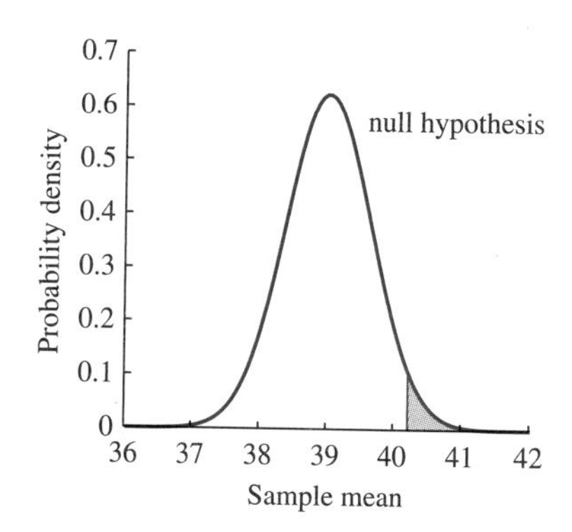

FIGURE 8.5.3

The one-tailed test

To find the p-value, we compute $\Pr(X \geq \overline{X})$. We subtract the mean and divide by the standard error to convert to z-scores and find the probability with the standard normal.

$$\begin{aligned}\Pr(X \geq 40.2) &= \Pr\left(\frac{X - 39.0}{0.64} \geq \frac{40.2 - 39.0}{0.64}\right)\\ &= \Pr(Z \geq 1.875)\\ &= 1 - \Phi(1.875) = 0.030\end{aligned}$$

(Figure 8.5.3). With the one-tailed test, the result is significant at the 0.03 level. There is, however, a 3.0% chance of accepting a fertilizer that does not work (a type I error).

If we have reason to suspect that fertilized plants may be shorter, we use the alternative hypothesis that the true mean $\mu \neq \mu_0$. Remember that one must decide on the appropriate test before looking at the data. Every data set looks odd in one way or another, and snooping around to find these oddities generally leads to mistakes. If the sample mean of fertilized plants had turned out to be 37.6, we cannot pretend we expected a decrease unless we made a public announcement to that effect before examining the data. However, we could repeat the experiment with a new null hypothesis to test whether this reduction in height is significant.

Example 8.5.4 Finding the p-Value with a Two-Tailed Test

Our observed mean $\overline{X} = 40.2$ deviates from the expected mean of $\mu_0 = 39.0$ by 1.2 cm. With a two-tailed test, the set of results more extreme than our observation includes all

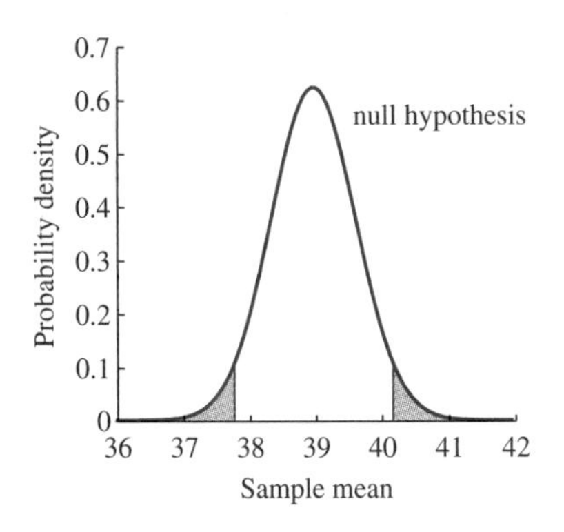

FIGURE 8.5.4

The two-tailed test

results more than 1.2 away from 39.0: results greater than 40.2 and results less than 37.8 (Figure 8.5.4).

$$
\begin{aligned}
\Pr(X \geq 40.2 \text{ or } X \leq 37.8) &= \Pr\left(\frac{X - 39.0}{0.64} \geq \frac{40.2.0 - 39.0}{0.64}\right) \\
&+ \Pr\left(\frac{X - 39.0}{0.64} \leq \frac{37.8.0 - 39.0}{0.64}\right) \\
&= \Pr(Z \geq 1.875) + \Pr(Z \leq -1.875) \\
&= 1 - \Phi(1.875) + \Phi(-1.875) \\
&= 0.061.
\end{aligned}
$$

With this alternative hypothesis, the result is not significant. The loss of significance is a consequence of our less informative alternative hypothesis. In general, the significance level of the two-tailed test is double that of the one-tailed test when the distribution is symmetric.

Example 8.5.5 Finding the p-Value with a Two-Tailed Test: Larger Sample Size

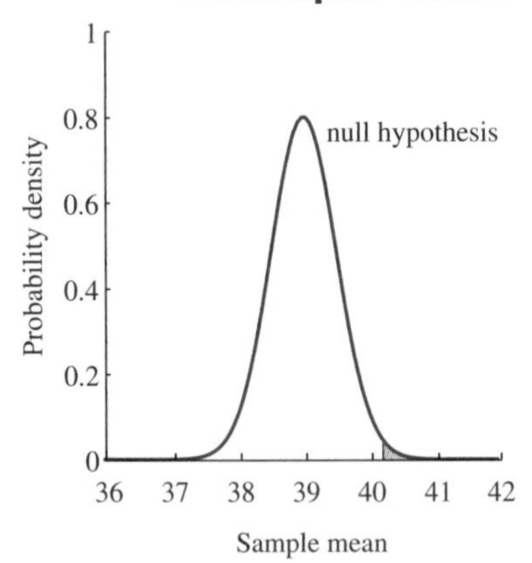

FIGURE 8.5.5

The two-tailed test with a larger sample size

Suppose that we observed $\overline{X} = 40.2$ exactly as in Example 8.5.4, but had measured 50 plants (Figure 8.5.5). The calculations are the same, but we use a standard error of 0.4526 (Example 8.5.2).

$$
\begin{aligned}
\Pr(X \geq 40.2 \text{ or } X \leq 37.8) &= \Pr\left(\frac{X - 39.0}{0.4526} \geq \frac{40.2.0 - 39.0}{0.4526}\right) \\
&+ \Pr\left(\frac{X - 39.0}{0.4526} \leq \frac{37.8 - 39.0}{0.4526}\right) \\
&= \Pr(Z \geq 2.651) + \Pr(Z \leq -2.651) \\
&= 1 - \Phi(2.651) + \Phi(-2.651) \\
&= 0.008.
\end{aligned}
$$

With a larger sample size, the result is highly significant (Figure 8.5.5).

If we have a sufficiently large sample ($n \geq 30$), we can use the sample variance s^2 in place of σ^2 and follow the same steps. Suppose we do not know the true variance σ^2 but have measured $n = 50$ plants.

Example 8.5.6 Using the Sample Variance

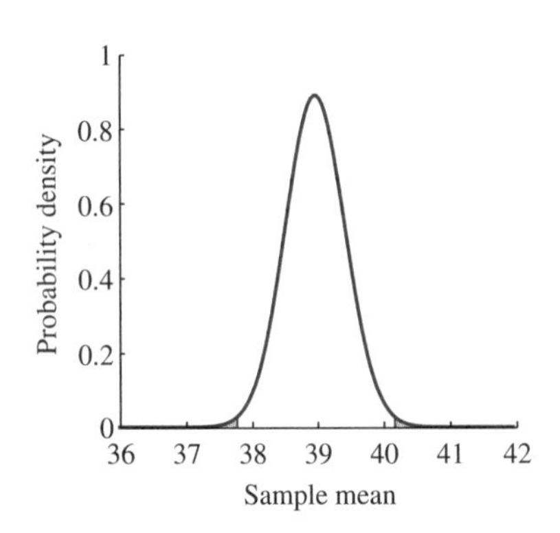

FIGURE 8.5.6

One-tailed test using the sample variance

Suppose we measure 50 plants, and the sample variance is $s^2 = 12.45$. The sample standard deviation is $\sqrt{12.45} = 3.528$. The standard error is

$$
\text{standard error} = \sqrt{\frac{12.45}{50}} = \sqrt{0.249} = 0.499.
$$

Applying the one-tailed test (Figure 8.5.6),

$$
\begin{aligned}
\Pr(X \geq 40.2) &= \Pr\left(\frac{X - 39.0}{0.499} \geq \frac{40.2 - 39.0}{0.499}\right) \\
&= \Pr(Z \geq 2.405) \\
&= 1 - \Phi(2.405) = 0.008.
\end{aligned}
$$

The result is highly significant, the benefit of using a larger sample size (Figure 8.5.6). Although the difference in the means is the same and the variance is slightly larger, the larger sample size makes our result more significant.

We can summarize these methods in the following algorithm.

Algorithm 8.3 Hypothesis Testing with the Normal Distribution

Suppose a set of n measurements have sample mean μ, and the null hypothesis is that the true mean is μ_0. If the variance of the underlying distribution is the known value σ^2, then the distribution of the sample mean $\overline{X}$ under the null hypothesis is

$$\overline{X} \sim N\left(\mu_0, \frac{\sigma^2}{n}\right).$$

Suppose that our measured mean μ differs from μ_0 by $|m - \mu_0| = \delta$.

1. The p-value for the two-tailed test is

$$\mathrm{P} = \Pr(\overline{X} \geq \mu_0 + \delta) + \Pr(X \geq \mu_0 - \delta)$$

2. If $\mu > \mu_0$, the one-tailed test computes

$$\mathrm{P} = \Pr(\overline{X} \geq \mu_0 + \delta).$$

3. If $\mu < \mu_0$, the one-tailed test computes

$$\mathrm{P} = \Pr(\overline{X} \leq \mu_0 - \delta).$$

If the variance of the underlying distribution is unknown but $n \geq 30$, the underlying variance can be estimated as the sample variance s^2 and used in place of σ^2.

Example 8.5.7 Using the Normal Approximation to Test a Hypothesis About a Proportion

Suppose, as in Subsection 8.4.1, that 20 out of 100 diseased people carry a particular allele compared with a population where 13% have the allele. If the null hypothesis is true, then the number G of people with the allele has a binomial distribution with $n = 100$ and $p = 0.13$. Then

$$\mathrm{E}(G) = np = 100 \cdot 0.13 = 13.0$$
$$\mathrm{Var}(G) = np(1 - p) = 100 \cdot 0.13 \cdot 0.87 = 11.31.$$

The normal approximation is therefore

$$X \sim N(13, 11.31).$$

The p-value for a one-tailed test is (Figure 8.5.7a)

$$\begin{aligned}
\Pr(G \geq 20) &= \Pr(X \geq 19.5) && \text{continuity correction} \\
&= 1 - \Phi\left(\frac{19.5 - 13.0}{\sqrt{11.31}}\right) && \text{subtract mean and divide by standard deviation} \\
&= 1 - \Phi(1.933) && \text{compute the } z\text{-score} \\
&= 0.027. && \text{compute value with computer or table}
\end{aligned}$$

The result is significant, and quite close to the significance level of 0.0319 found with the exact method (Figure 8.5.7a, Example 8.4.6). Because of the symmetry of the normal distribution, the p-value associated with the two-tailed test is exactly twice that for the one-tailed test (Figure 8.5.7b), or 0.054, close to the value of 0.0512 found with the exact method (Figure 8.5.7b, Example 8.4.8).

8.5.2 Testing with the t Distribution

If we measured fewer than 30 plants, we could estimate the sample variance in the usual way, but use the t distribution introduced in Subsection 8.3.4. The t distribution takes into account the uncertainty about the sample variance s^2.

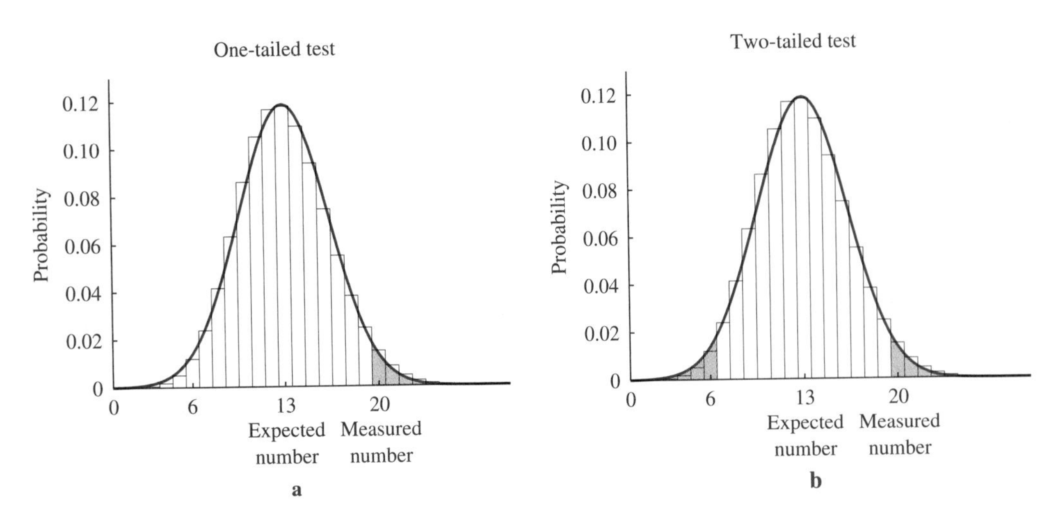

FIGURE 8.5.7
One- and two-tailed tests with the normal approximation to the binomial

Suppose a sample mean of $\overline{X}$ is measured from a sample of size n and sample variance s^2. The standard error of the mean is

$$s_{\overline{X}} = \frac{s}{\sqrt{n}}.$$

The statistic

$$t = \frac{\overline{X} - \mu_0}{s_{\overline{X}}} \tag{8.5.1}$$

measures how far the observed mean lies from the null hypothesis in units of the standard error. To evaluate the significance requires a set of tables for each value of $\nu = n - 1$, the degrees of freedom. The simplified Table 8.2 (reproduced inside the back cover), can be used to determine whether results are significant with a two-tailed test, but more extensive tables are needed to find the exact p-values or deal with one-tailed tests.

Example 8.5.8 Applying the t Distribution

Suppose we measure 20 plants and find a sample mean of $\overline{X} = 40.2$ and a sample variance of $s^2 = 12.45$, (as in Example 8.5.6). The sample standard deviation is $\sqrt{12.45} = 3.528$, and the standard error is

$$s_{\overline{X}} = \frac{3.528}{\sqrt{20}} = 0.789.$$

The t statistic is then

$$t = \frac{40.2 - 39.0}{0.789} = 1.52.$$

To use the table, we compare this value with 2.093, the critical value for 95% confidence limits in Table 8.2 with $\nu = 19$ degrees of freedom. Because our value is much smaller, the null hypothesis cannot be rejected at the 0.05 level.

Example 8.5.9 Applying the t Distribution II

Suppose we measure 20 plants and find a sample variance of $s^2 = 12.45$ (as in Example 8.5.8), but a sample mean of 37.0. The standard error is again 0.789. The t statistic is

$$t = \frac{37.0 - 39.0}{0.789} = -2.535.$$

The negative value indicates that our sample mean was less than the value expected under the null hypothesis. Because we are using a two-tailed test, the sign does not matter (we would obtain the same significance if the mean exceeded 39.0 by the same amount). This value exceeds 2.093, the critical value for 95% confidence limits with 19 degrees of freedom, meaning that the null hypothesis can be rejected at the $P = 0.05$ level. However, it is less than 2.861, the critical value for 99% confidence limits. The null hypothesis cannot be rejected at the $P = 0.01$ level.

8.5.3 The Power of Normal Tests

As we have seen, the significance depends on the size of the deviation between that observed and expected mean, and on the sample size. If we seek to detect a fairly small effect, a large sample is required. But how large a sample do we need to have a good chance to detect that small effect? In statistical language, how large a sample is required to obtain high power?

Computing the power, unlike computing the significance level, depends on the underlying true value of the mean. Because we can never know this value, power is computed as a function of the true mean. This function allows us to determine the **effect size** (the difference between the values divided by the standard error) detectable with a given sample size.

We find power in two steps. First, we must specify our statistical test, the cutoff value of the measured mean that we deem significant. Second, we find the probability of obtaining results as extreme as or more extreme than the cutoff as a function of the underlying true parameter.

Example 8.5.10 Finding the Cutoff Value for a One-Tailed Test

Suppose the heights of 25 plants are measured, the underlying standard deviation is known to be $\sigma = 3.2$ cm, and the null hypothesis is $\mu_0 = 39.0$ (as in Example 8.5.1). The distribution of the sample mean X has mean 39.0 and standard error 0.64. How much larger than 39.0 must the sample mean be to reject the null hypothesis at the 0.05 level? We are looking for the value $\overline{X}$ such that

$$\Pr(X \geq \overline{X}) = 0.05.$$

Transforming this data into a standard normal distribution, this occurs when

$$\begin{aligned} 0.05 &= \Pr\left(\frac{X - 39.0}{0.64} \geq \frac{\overline{X} - 39.0}{0.64}\right) \\ &= \Pr\left(Z \geq \frac{\overline{X} - 39.0}{0.64}\right). \end{aligned}$$

The value that lies above all but 5% of the standard normal is 1.645 [$\Phi(1.645) = 0.95$]. Therefore, the cutoff is where

$$\frac{\overline{X} - 39.0}{0.64} = 1.645$$

which has solution

$$\overline{X} = 39.0 + 1.645 \cdot 0.64 = 40.05.$$

A result is significant with a one-tailed test if it lies 1.645 standard errors above the mean.

Example 8.5.11 Using the Cutoff to Find the Power If the True Mean is 40.0

Suppose the true mean is 40.0. The probability of making a type II error and failing to reject the null hypothesis is the probability of picking 25 plants from an $N(40.0, 10.24)$ distribution and getting an average height of less than 40.05. Let the random variable $\overline{X}$ represent the mean of these 25 plants. Then

$$\overline{X} \sim N\left(40.0, \frac{10.24}{25}\right) = N(40.0, 0.4096)$$

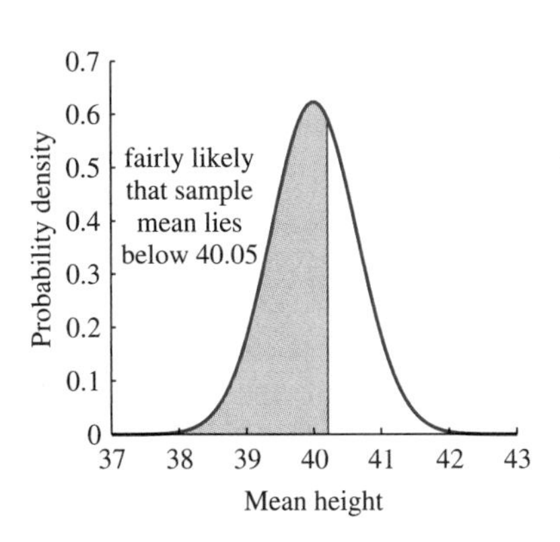

FIGURE 8.5.8
The power if the true mean is 40.0

from the Central Limit Theorem for Averages, and

$$\begin{aligned}\Pr(\overline{X} < 40.05) &= \Pr\left(\frac{\overline{X} - 40.0}{0.64} < \frac{40.05 - 40.0}{0.64}\right)\\ &= \Pr(Z < 0.078)\\ &= \Phi(0.078) = 0.531.\end{aligned}$$

The power of the test is $1 - 0.531 = 0.469$. Our test has a more than 50% chance of missing an effect this small. The area associated with this calculation is shown in Figure 8.5.8.

Example 8.5.12 Finding the Power If the True Mean is 42.0

If the true mean of fertilized plants were 42.0,

$$\overline{X} \sim N\left(42.0, \frac{10.24}{25}\right) = N(42.0, 0.4096).$$

The probability of missing a result this large is

$$\begin{aligned}\Pr(\overline{X} < 40.05) &= \Pr\left(\frac{\overline{X} - 42.0}{0.64} < \frac{40.05 - 42.0}{0.64}\right)\\ &= \Pr(Z < -3.047)\\ &= \Phi(-3.047) = 0.0012\end{aligned}$$

(Figure 8.5.9). The power is 0.9988, and we are very unlikely to miss an effect this large.

These calculations of power depend on the true mean, which we do not know. What good are they? They have shown that our test is unlikely to detect a height increase of 1.0 cm (true mean of 40.0), but quite likely to detect a difference of 3.0 cm (true mean of 42.0). If we are not concerned with small effects like a 1.0 cm change in height, this test is appropriate.

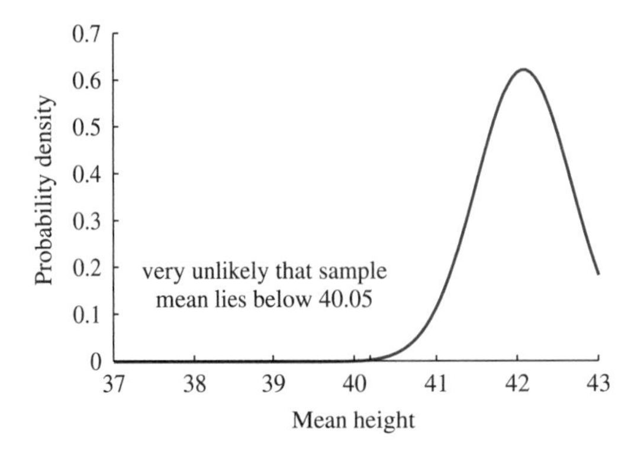

FIGURE 8.5.9
The power if the true mean is 42.0

Example 8.5.13 Finding the Power as a Function of the True Mean

We can compute the power as a function of the true mean μ by following the same steps. If the true mean of fertilized plants is μ, then

$$\overline{X} \sim N\left(\mu, \frac{10.24}{25}\right) = N(\mu, 0.4096).$$

The probability of missing a result this large is

$$\begin{aligned}
\Pr(\overline{X} < 40.05) &= \Pr\left(\frac{\overline{X}-\mu}{0.64} < \frac{40.05-\mu}{0.64}\right) \\
&= \Pr\left(Z < \frac{40.05-\mu}{0.64}\right) \\
&= \Phi\left(\frac{40.05-\mu}{0.64}\right).
\end{aligned}$$

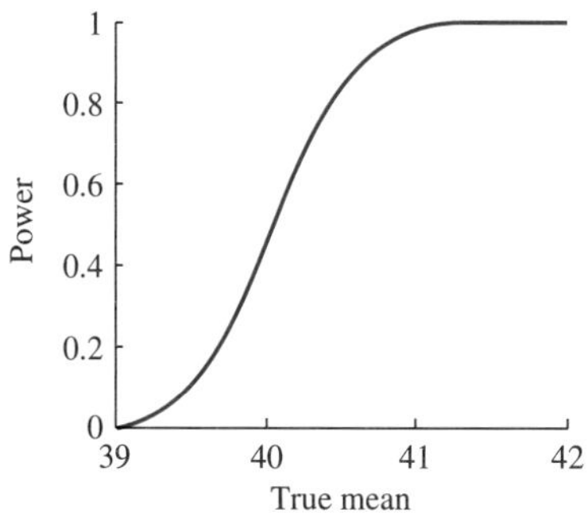

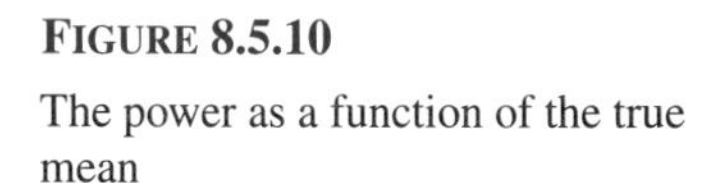

FIGURE 8.5.10
The power as a function of the true mean

(Figure 8.5.10). This function can be graphed accurately by any computer program that has a built-in version of the cumulative distribution for the standard normal.

If we wish to detect a small effect, we must come up with a more precise test. Generally, this requires increasing the sample size to decrease the standard error and increase our resolution. For example, suppose we wish our test to have 90% power to detect a difference of 1.0 cm. How large a sample size do we need?

Example 8.5.14 Finding the Sample Size Needed to Detect a Small Effect

Let the random variable $\overline{X}$ represent the mean of these n plants. Then

$$\overline{X} \sim N\left(40.0, \frac{10.24}{n}\right).$$

We have found that results are significant with a one-tailed test if they lie 1.645 standard errors above the null hypothesis, or

$$\overline{X} \geq 39.0 + 1.645\sqrt{\frac{10.24}{n}} = 39.0 + \frac{5.264}{\sqrt{n}}.$$

The probability of missing the signal is the probability that the sample mean is less than this value, or

$$\begin{aligned}
\Pr\left(\overline{X} < 39.0 + \frac{5.264}{\sqrt{n}}\right) &= \Pr\left(\frac{\overline{X}-40.0}{3.2/\sqrt{n}} < \frac{39.0 + \frac{5.264}{\sqrt{n}} - 40.0}{3.2/\sqrt{n}}\right) \\
&= \Pr\left(Z < 1.645 - \frac{\sqrt{n}}{3.2}\right) = 0.1.
\end{aligned}$$

This has a solution where $\Phi(z) = 0.1$, or at -1.28. Solving for n, we get

$$\begin{aligned}
1.645 - \frac{\sqrt{n}}{3.2} &= -1.28 \\
\sqrt{n} &= 3.2 \cdot (1.645 + 1.28) = 9.36 \\
n &= 88.
\end{aligned}$$

It would take 88 plants to detect a 1.0 cm change in height with 90% probability.

Example 8.5.15 Power with a Binomial

In Example 8.4.14 we plotted the power to detect deviations from $q = 0.13$ using the exact p-values computed from the binomial distribution. We can compare those results with the results obtained with normal approximation.

First, we must find the cutoff value that we deem significant. In Example 8.5.7, we found the normal approximation

$$X \sim N(13, 11.31)$$

by matching the mean and variance associated with the null hypothesis, that 13% out of our sample of 100 people carry a particular allele. The p-value if we measured 20 out of 100 was 0.027. If we measured 19 out of 100, the p-value is

$$\begin{aligned}\Pr(G \geq 19) &= \Pr(X \geq 18.5)\\ &= 1 - \Phi\left(\frac{18.5 - 13.0}{\sqrt{11.31}}\right)\\ &= 1 - \Phi(1.635) = 0.051.\end{aligned}$$

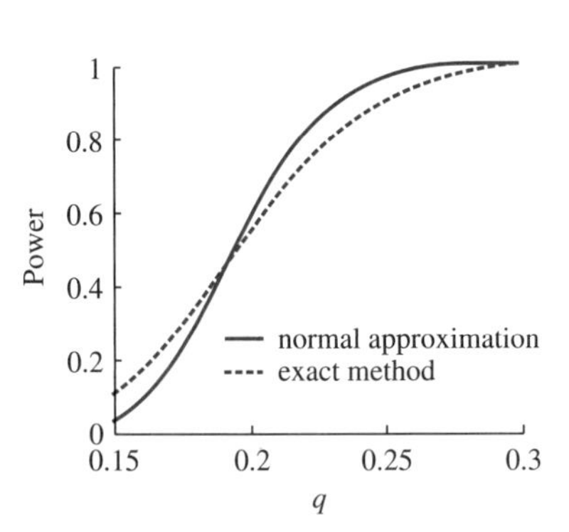

FIGURE 8.5.11

The power as a function of the true proportion

This barely fails to be significant (and compares well with the exact result of 0.056 in Example 8.4.11). Our cutoff value is therefore 20 out of 100.

The power is then the probability of 20 or more out of 100 as a function of the unknown underlying parameter q. The normal approximation to the number G of people with the allele is

$$X \sim N[100q, 100q(1-q)].$$

The probability of a result exceeding our cutoff is

$$\Pr(G \geq 20) = \Pr(X \geq 19.5) = 1 - \Phi\left[\frac{19.5 - 100q}{\sqrt{100q(1-q)}}\right].$$

This function gives the power under the normal approximation. Figure 8.5.11 includes the power using the exact method for comparison. ▲

Decisions regarding the level of power desired, the degree of significance, and the smallest difference that can be detected are part of the discipline of **statistical design**. The underlying principles, however, are those of statistics itself: understanding the connection between measurements and the underlying probabilistic model.

Summary We have computed p-values by using the standard normal distribution when the underlying distribution is normal with known variance or large enough to accurately estimate the sample variance. When the sample is too small to completely trust the sample variance, the t distribution is used in place of the standard normal. In each case, we transform the measured sample mean by subtracting the mean expected under the null hypothesis and dividing by the standard error. By computing the **power** of a test, we can estimate the sample size needed to identify a particular difference.

8.5 Exercises

Mathematical Techniques

1–4 ▪ The weights, heights, yields, and seed number for 10 plants grown in an experimental plot are given in the table. Each measurement is approximately normally distributed. We wish to determine whether plants in the plot differ from those outside. In each case, state one- and two-tailed alternative hypotheses and find their significance.

Plant	Weight	Height	Yield	Number of seeds
1	11.83	46.2	10.11	7
2	7.65	39.4	9.29	18
3	14.57	41.4	12.26	19
4	14.97	40.8	6.61	16
5	14.26	38.5	9.51	17
6	11.26	40.7	8.34	14
7	11.08	39.8	13.35	22
8	7.04	37.9	12.90	21
9	5.44	34.8	10.35	20
10	11.82	43.1	9.84	25

1. The variance for weight is 9.0, and plants outside the plot have mean weight 10.0.
2. The variance for height is 16.0, and plants outside the plot have mean height 38.0.
3. The variance for yield is 6.25, and plants outside the plot have mean yield 9.0.

4. The variance for seed number is 25.0, and plants outside the plot have mean seed number 15.0.

5–8 ▪ Consider again the weights, heights, yields, and seed number for 10 plants given in Exercises 1–4. Find the sample variance for each and use the t distribution to perform a two-tailed test of the hypothesis.

5. Plants outside the plot have mean weight 10.0.
6. Plants outside the plot have mean height 38.0.
7. Plants outside the plot have mean yield 9.0.
8. Plants outside the plot have mean seed number 15.0.

9–12 ▪ Find the smallest values of the sample mean for which the given hypothesis is rejected.

9. Under the conditions in Exercise 1, find the smallest weight that can reject the null hypothesis that the mean weight is 10.0 with a one-tailed test at the 0.01 level.
10. Under the conditions in Exercise 2, find the smallest height that can reject the null hypothesis that the mean height is 38.0 with a two-tailed test at the 0.01 level.
11. Under the conditions in Exercise 3, find the smallest yield that can reject the null hypothesis that the mean yield is 9.0 with a two-tailed test at the 0.001 level.
12. Under the conditions in Exercise 4, find the smallest seed number that can reject the null hypothesis that the mean seed number is 15.0 with a one-tailed test at the 0.001 level.

13–16 ▪ Find the power of the test assuming the given true mean.

13. The true mean weight is 13.0 in Exercise 9.
14. The true mean height is 43.0 in Exercise 10.
15. The true mean yield is 11.0 in Exercise 11.
16. The true mean seed height is 18.0 in Exercise 12.

17–18 ▪ Consider again plants with the null hypothesis that mean height is 39.0. Assume that the standard deviation is known to be 3.2 cm.

17. Show that a measured sample mean of 40.0 is highly significant if the sample size is $n = 88$.
18. Why is the power with this sample size only 90% (as found in the text), rather than more than 99%?

19–20 ▪ Use the normal approximation to test the given hypothesis.

19. A coin is flipped 100 times and comes out heads 44 times. It is thought that the coin is fair (has probability of heads is equal to 0.5). Do the data provide evidence that the coin is unfair?
20. Of 1000 people polled in one state, 320 favor the use of mathematics in biology. The legislature has passed a bill mandating that at least 36% of people must be in favor. Does the poll provide evidence that the proportion is smaller than 0.36? Is the state in violation of the law?

Applications

21–22 ▪ Consider the following data on 30 waiting times for 2 types of events.

Sample	Type a	Type b
1	0.41	0.69
2	0.08	4.83
3	0.52	0.01
4	0.22	1.49
5	0.03	2.91
6	0.11	0.09
7	1.36	1.07
8	0.51	0.17
9	0.29	0.19
10	0.16	0.7
11	0.78	0.04
12	0.41	1.86
13	1.19	1.23
14	0.89	0.50
15	1.87	4.16
16	6.33	0.02
17	1.23	0.02
18	0.18	0.09
19	0.97	0.34
20	0.48	1.63
21	0.27	1.22
22	1.20	0.60
23	0.38	0.83
24	1.35	1.58
25	1.53	2.13
26	1.72	0.01
27	0.19	0.17
28	0.94	0.75
29	1.60	1.79
30	0.87	0.41

21. Find the p-value associated with the null hypothesis that the mean of type **a** is 1.0.
22. Find the p-value associated with the null hypothesis that the mean of type **b** is 1.0.

23–24 ▪ Consider again the data in Exercises 21 and 22. Each type has one or more outliers that strongly affect the mean and standard deviation. Exclude the outlier or outliers and recompute the p-value associated with the null hypothesis that the mean is 1.0. What do you think of this procedure if you were told that the data were generated from an exponential distribution with mean 1.0?

23. The outlier is extreme value 6.33 at time 16.

24. The outliers are the extreme values 4.16 and 4.83.

25–28 ▪ A chronic condition improves spontaneously in 45% of people. A new medication is being tested to try to increase this percentage.

25. Of 25 of 50 patients tested with the new medication, 30 improve. Is this significant?

26. Of 26 of 100 patients tested with the new medication, 60 improve. Is this significantly better?

27. Suppose that the true fraction that improves with the medication is 0.6. What is the power to detect this at the 0.05 level with a sample of 50 patients?

28. Suppose that the true fraction that improves with the medication is 0.6. What is the power to detect this at the 0.05 level with a sample of 100 patients? How much greater is the power?

29–34 ▪ A company develops a new method to reduce error rates in the polymerase chain reaction (PCR). With the old method, the number of errors in a well-studied piece of DNA is known to have a Poisson distribution with mean 35.0. Use a one-tailed test in each case.

29. Find the significance level if the DNA with the new method has only 27 errors. Make sure to start by finding the normal approximation to the null hypothesis.

30. Find the significance level if the DNA with the new method has only 23 errors.

31. What is the largest number of errors that would reject the null hypothesis at the 0.05 level?

32. What is the largest number of errors that would reject the null hypothesis at the 0.01 level?

33. Instead of measuring only a single piece of DNA with the new method, 10 pieces are measured and 300 errors are found. Does the new method reduce the number of errors?

34. Instead of measuring only a single piece of DNA with the new method, 20 pieces are measured and 650 errors are found. Does the new method reduce the number of errors?

35–36 ▪ Use the normal approximation to test the following hypotheses about growing populations. In each case, habitat improvements are tried and the population grows from 1 to 250 individuals in 50 years. Is there reason to think that the habitat improvements helped?

35. The population in Section 7.8, Exercise 39, where per capita production is a random variable with p.d.f. $g(x) = 5.0$ for $1.0 \le x \le 1.2$.

36. The population in Section 7.8, Exercise 40, where per capita production is a random variable with p.d.f. $g(x) = 1.25$ for $0.7 \le x \le 1.5$. Can you explain the difference from the result in the previous problem?

8.6 Comparing Experiments: Normal Theory

In the last two sections, we tested whether the results of an experiment are consistent with a previously known baseline. Our null hypothesis is that the treatment has no effect. More generally, the baseline conditions are not known in advance, and must also be estimated with a **control**. If we are trying to detect how individuals respond to some medication, for example, some are given a control (or placebo) treatment while others are given the actual medication. In cases like this, when a different sample is measured under the two conditions, the data are said to be **unpaired**. In other circumstances, we might compare the state of a single group of patients before and after being given some medication. In this case, the measurement made before treatment acts as the control. Because we have made two measurements on each patient, the data are said to be **paired**. In both cases, the general technique is the same: set up null and alternative hypotheses and check whether the data are consistent with the null hypothesis. We develop methods to test hypotheses for both paired and unpaired data, considering cases with known variance or large sample size (where the standard normal distribution is used) and those with small sample size (where the t distribution is used).

8.6.1 Unpaired Normal Distributions

A widely used and important hypothesis test uses p-values to compare the means of samples from two normally distributed populations. We consider first cases with known variance or large sample size (when the sample variance is a dependable estimate) and then the case of unknown variance.

Let $\overline{X}_1$ and $\overline{X}_2$ be random variables representing the means of samples of size n_1 and n_2 with known variances σ_1^2 and σ_2^2 and unknown means μ_1 and μ_2. Think of the

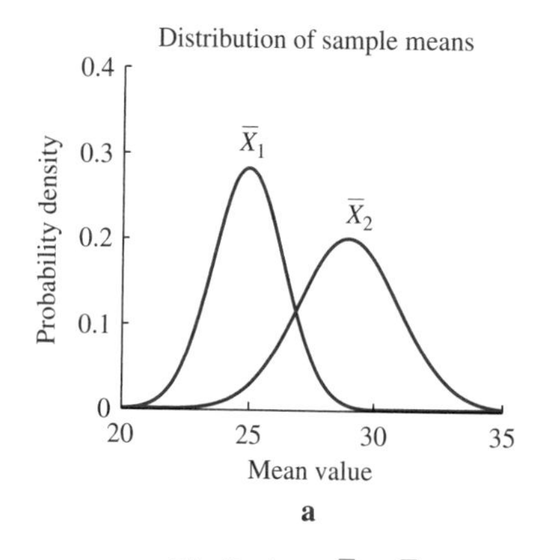

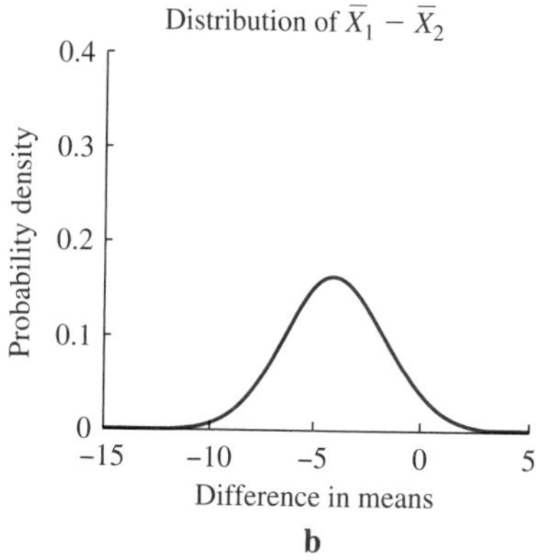

FIGURE 8.6.1

The distribution of sample means and their difference

first population as the control and the second as the experimental population. By the Central Limit Theorem for Averages,

$$\overline{X}_1 \sim N\left(\mu_1, \frac{\sigma_1^2}{n_1}\right)$$

$$\overline{X}_2 \sim N\left(\mu_2, \frac{\sigma_2^2}{n_2}\right).$$

Let the new random variable D,

$$D = \overline{X}_1 - \overline{X}_2,$$

represent the difference of the two means. Then

$$\mathrm{E}(D) = \mathrm{E}(\overline{X}_1) - \mathrm{E}(\overline{X}_2) = \mu_1 - \mu_2$$

$$\mathrm{Var}(D) = \mathrm{Var}(\overline{X}_1) + \mathrm{Var}(\overline{X}_2) = \frac{\sigma_1^2}{n_1} + \frac{\sigma_2^2}{n_2}$$

because the expectation of the difference is the difference of the expectations (Theorem 7.6) and the variance of the difference is the **sum** of the variances (Theorem 7.12). Furthermore, D will have a normal distribution because of the **additive property of the normal distribution** (Subsection 7.8.2). Therefore,

$$D \sim N\left(\mu_1 - \mu_2, \frac{\sigma_1^2}{n_1} + \frac{\sigma_2^2}{n_2}\right).$$

The variance of D is higher because it includes the variance of both $\overline{X}_1$ and $\overline{X}_2$ (Figure 8.6.1).

The calculation of the distribution of D assumes that we know the true means μ_1 and μ_2. Because these values are in fact unknown, in practice we use the random variable D to express the null hypothesis that the treatment has no effect. Mathematically, the treatment has no effect if the true means are equal, or $\mu_1 = \mu_2$. In this case, $\mu_1 - \mu_2 = 0$ and the distribution of D is

$$D \sim N\left(0, \frac{\sigma_1^2}{n_1} + \frac{\sigma_2^2}{n_2}\right). \qquad \textbf{(8.6.1)}$$

This reduces a null hypothesis about *two* unknown values (μ_1 and μ_2) into a null hypothesis about *one* number (their difference $\mu_1 - \mu_2$). This mathematical version of the null hypothesis makes it possible to test whether an observed difference between sample means exceeds what we would expect by chance.

Example 8.6.1 Testing a Fertilizer: Two-Tailed Test

Suppose we treat 25 plants with a new fertilizer and keep 50 untreated plants as controls. Assume that the variance with or without fertilizer has the known value of 10.24 cm^2. The known parameters describing the control population are

$$n_1 = 50 \text{ and } \sigma_1^2 = 10.24$$

and the parameters describing the treatment population are

$$n_2 = 25 \text{ and } \sigma_2^2 = 10.24.$$

If we measure sample means of $\hat{\mu}_1 = 39.0$ cm in the control and $\hat{\mu}_2 = 40.2$ cm in the treated population, do we have reason to suspect that the underlying populations are really different?

Under the null hypothesis, the difference of the sample means has distribution

$$D \sim N\left(0, \frac{10.24}{50} + \frac{10.24}{25}\right) = N(0, 0.6144).$$

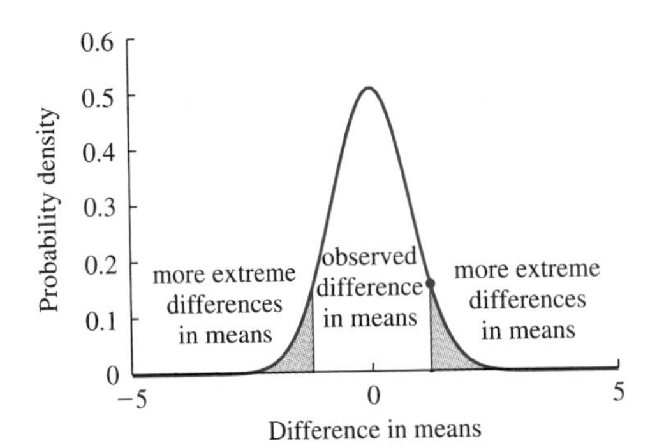

FIGURE 8.6.2
Probability of a more extreme result under the null hypothesis: two-tailed test

What is the probability that the difference exceeds the measured difference of 1.2, the difference between the measured sample means $\hat{\mu}_2$ and $\hat{\mu}_1$? If we had no reason prior reason to expect the fertilizer to work and produce taller plants, we use a two-tailed test and compute

$$\Pr(D \geq 1.2) + \Pr(D \leq -1.2) = 1 - \Phi\left(\frac{1.2}{\sqrt{0.6144}}\right) + \Phi\left(\frac{-1.2}{\sqrt{0.6144}}\right)$$

$$= 1 - \Phi(1.53) + \Phi(-1.53) = 0.126$$

(Figure 8.6.2). The result is not significant.

Example 8.6.2 Testing a Fertilizer: One-Tailed Test

If we had predicted that the treatment would have a larger mean than the control, we could use a one-tailed test. In this case,

$$\Pr(D \geq 1.2) = 1 - \Phi(1.53) = 0.063$$

(Figure 8.6.3). There is a trend toward significance, but not a convincing result. This loss of certainty (compared to the significant finding in the last section) results from the uncertainly in the control population.

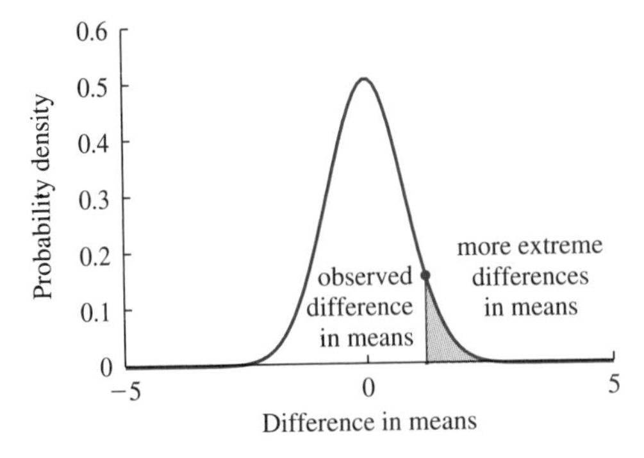

FIGURE 8.6.3
Probability of a more extreme result under the null hypothesis: one-tailed test

When both samples are sufficiently large (greater than 30), the sample variances s_1^2 and s_2^2 can be used in place of σ_1^2 and σ_2^2 in each of the foregoing calculations.

8.6.2 Comparing Population Proportions

The normal approximation provides a convenient way to compare population proportions for large samples. An alternative approach is described in Section 8.8.

Example 8.6.3 Comparing Population Proportions with the Normal Approximation

We measure two populations of cells for expression of a gene and find 96 in a control group of $n_1 = 200$ cells and 54 in a treatment group of $n_2 = 100$ cells. If the cells act independently, the number of cells expressing the gene will follow binomial distribution in each treatment. We estimate the proportions p_1 and p_2 as

$$\hat{p}_1 = \frac{96}{200} = 0.48$$
$$\hat{p}_2 = \frac{54}{100} = 0.54.$$

We can only apply the normal approximation if there are more than five successes and five failures in each sample, as there are in this case. The two proportions differ by 0.06. Is the difference significant?

Let p_1 and p_2 be the unknown true proportion for cells of each type, and let P_1 and P_2 be random variables describing the distribution of sample proportions. The normal approximations are

$$P_1 \sim N\left(p_1, \frac{p_1(1-p_1)}{200}\right)$$
$$P_2 \sim N\left(p_2, \frac{p_2(1-p_2)}{100}\right).$$

Let D be a random variable representing the difference in the two measured proportions. Using the additive relations for the mean and variance of independent normals,

$$D \sim N\left(p_1 - p_2, \frac{p_1(1-p_1)}{200} + \frac{p_2(1-p_2)}{100}\right).$$

Our null hypothesis is that $p_1 = p_2$, so that D has mean 0.

The variance of this distribution depends on the true unknown proportions p_1 and p_2. As in our computation of confidence limits for the binomial (Subsection 8.3.3), we use the estimated proportions $\hat{p}_1$ and $\hat{p}_2$. Our null hypothesis then is that the distribution of the difference D between the sample proportions is

$$D \sim N\left(0.0, \frac{\hat{p}_1(1-\hat{p}_1)}{n_1} + \frac{\hat{p}_2(1-\hat{p}_2)}{n_2}\right).$$

Example 8.6.4 The Distribution of the Difference Under the Null Hypothesis

In Example 8.6.3,

$$N\left(0.0, \frac{0.48 \cdot 0.52}{200} + \frac{0.54 \cdot 0.46}{100}\right) = N(0.0, 0.00373)$$

(Figure 8.6.4).

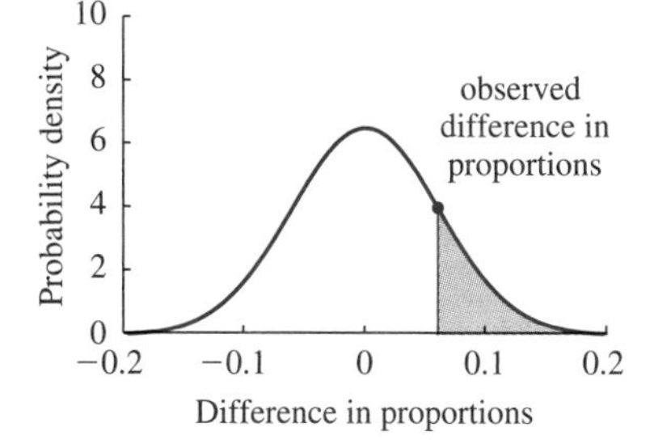

FIGURE 8.6.4

The null hypothesis describing the difference in proportions

How does our measured difference of 0.06 compare? If we had prior reason to believe that the second group of cells would have a higher proportion, we use a one-tailed test, and check $\Pr(D \geq 0.06)$,

$$\Pr(D \geq 0.06) = \Pr\left(\frac{D}{\sqrt{0.00373}} \geq \frac{0.06}{\sqrt{0.00373}}\right)$$
$$= \Pr(Z \geq 0.982) = 1 - \Phi(0.982) = 0.163.$$

The difference in proportions of 0.06 differs from 0 by less than one standard error, and the p-value of 0.163 is nowhere close to significant with the one-tailed test. There is not sufficient evidence to reject the null hypothesis.

Algorithm 8.4 Comparing Proportions from Two Experiments

A control population and treatment population are tested for a trait. A proportion $\hat{p}_1$ out of n_1 individuals from the control and a proportion $\hat{p}_2$ out of n_2 individuals from the treatment are identified with the trait. The null hypothesis is that $p_1 = p_2$, or that the difference of the sample means has distribution

$$D \sim N\left(0.0, \frac{\hat{p}_1(1-\hat{p}_1)}{n_1} + \frac{\hat{p}_2(1-\hat{p}_2)}{n_2}\right).$$

We test whether the measured difference is consistent with the null hypothesis by computing the probability of a result more extreme than $\hat{p}_2 - \hat{p}_1$ using a one-tailed or two-tailed test as appropriate.

Example 8.6.5 Testing for the Effects of an Allele on Disease

We found that a sample of 20 out 100 diseased individuals with a particular allele is nearly significantly different from a population with a known proportion of 13% with that allele (Example 8.4.8). If the population proportion were unknown, we would compare the sample of individuals with the disease to a sample without the disease. Suppose we sample 200 healthy people and find that 26 carry the allele. Then

$$\hat{p}_1 = 0.2, \; n_1 = 100, \; \hat{p}_2 = 0.13, \; n_2 = 200.$$

According to Algorithm 8.4, the difference in proportions under the null hypothesis that $p_1 = p_2$ follows

$$\begin{aligned} D &\sim N\left(0.0, \frac{\hat{p}_1(1-\hat{p}_1)}{n_1} + \frac{\hat{p}_2(1-\hat{p}_2)}{n_2}\right) \\ &= N\left(0.0, \frac{0.2 \cdot 0.8}{100} + \frac{0.13 \cdot 0.87}{200}\right) \\ &= N(0, 0, 0.0022). \end{aligned}$$

Our observed difference of $\hat{p}_1 - \hat{p}_2 = 0.2 - 0.13 = 0.07$ is then $\frac{0.07}{\sqrt{0.0022}} = 1.50$ standard errors from the mean. With a one-tailed test, the significance is

$$p = 1 - \Phi(1.50) = 0.067.$$

The result is not significant, in contrast to the finding in Example 8.4.6, because of our uncertainty about the proportion in the control population. With a two-tailed test, the p-value is doubled, and is equal to 0.134.

8.6.3 Applying the t Distribution with Smaller Samples

As in Subsection 8.5.2, we can use the t distribution to correct for the additional uncertainty created when the underlying variance must be estimated from a small sample.

When the underlying variances of the sample means $\overline{X}_1$ and $\overline{X}_2$ are known or can be estimated accurately with a large sample size, we found the variance of the difference D as the sum of the variances. The variance of D is slightly more complicated to compute from the sample variances.

Suppose that we measure n_1 individuals in treatment 1 and n_2 individuals in treatment 2. Let s_1^2 and s_2^2 be the sample variances. We need to combine these into a single value to estimate the standard error. First, we find an estimate of the **pooled variance** s_p^2, the variance of the population taken as a whole.

$$s_p^2 = \frac{(n_1-1)s_1^2 + (n_2-1)s_2^2}{n_1+n_2-2}. \quad \textbf{(8.6.2)}$$

This equation gives a weighted average (Definition 1.16) of the two sample variances, weighted by fraction of the total degrees of freedom.

Example 8.6.6 Finding the Pooled Sample Variance

Treatment	Control
1.449	2.427
6.299	3.445
7.793	4.424
7.548	1.770
7.553	6.257
4.470	2.956
7.342	2.279
	3.578

Consider the data in the accompanying table describing treatment and control populations. We find the sample variances $s_1^2 = 5.489$ for the treatment and $s_2^2 = 2.043$ for the control with Theorem 8.2. To find the pooled variance, we use the sample sizes $n_1 = 7$ for the treatment and $n_2 = 8$ for the control. Then

$$\begin{aligned} s_p^2 &= \frac{(n_1 - 1)s_1^2 + (n_2 - 1)s_2^2}{n_1 + n_2 - 2} \\ &= \frac{6 \cdot 5.489 + 7 \cdot 2.043}{7 + 8 - 2} \\ &= 3.633. \end{aligned}$$

This value lies between the sample variances for the two populations.

The standard error for the difference in means between the two populations is

$$s_{\overline{X}_1 - \overline{X}_2} = \sqrt{\frac{s_p^2}{n_1} + \frac{s_p^2}{n_2}}. \quad \textbf{(8.6.3)}$$

The subscript on s indicates that this is the standard deviation of the difference between the random variables describing the two sample means. This is identical to the standard deviation of the distribution D found in Equation 8.6.1 except that we have substituted the pooled variance s_p^2 as an estimate for both true variances σ_1^2 and σ_2^2.

Example 8.6.7 Finding the Standard Error for the Difference

With the data in Example 8.6.6, we found $s_p^2 = 3.633$. Then

$$s_{\overline{X}_1 - \overline{X}_2} = \sqrt{\frac{3.633}{7} + \frac{3.633}{8}} = 0.986.$$

We use this standard error to compute the statistic t almost exactly as in Equation 8.5.1 in Subsection 8.5.2. We find the difference in the observed means and divide by the standard error to find

$$t = \frac{\overline{X}_1 - \overline{X}_2}{s_{\overline{X}_1 - \overline{X}_2}}.$$

The number of degrees of freedom is $\nu = n_1 + n_2 - 2$, the sum of the degrees of freedom for each measurement separately. To evaluate the significance, we compare the value of t with those in Table 8.2.

Example 8.6.8 Using the t Test to Compare Two Populations

Evaluating t requires computing both the standard error and the difference in the means. For the data in Example 8.6.7, we found a standard error of 0.986 (Example 8.6.6). The means are

$$\overline{X}_1 = 6.065$$

$$\overline{X}_2 = 3.392.$$

Therefore,

$$t = \frac{6.065 - 3.392}{0.986} = 2.711.$$

With $n_1 + n_2 - 2 = 7 + 8 - 2 = 13$ degrees of freedom, this value of t lies between the values for the 95% confidence limit and the 99% confidence limit, implying that the difference is significant at between the 0.05 and 0.01 levels.

Technically, the t test assumes that the underlying distributions of the measurements are normally distributed. Like all statistical tests, the accuracy of the t test depends on the accuracy of its assumptions. Unlike many other tests, however, the t test is highly **robust**, meaning that it gives meaningful answers even when the distributions are far from normal. More precisely, the test remains powerful at detecting true differences without producing biased p-values that give spurious significant results.

The t test is also the simplest example of one of the most important techniques in statistics, the **analysis of variance**, or ANOVA. This method was developed by R. A. Fisher, whom we have already met as one of the founders of modern population genetics. The idea is to analyze variance within population components (the values s_1^2 and s_2^2) and between population components. If most of the variance occurs within populations, any observed difference in means is swamped. This method can be extended to apply to many more complicated experiments, such as those involving more than two populations or several different factors (such as testing two different medications on both males and females).

8.6.4 Analysis of Paired Data with the t Test

Data are paired when each individual is measured in both treatments. Analysis of paired experiments can be more powerful than unpaired experiments because variation among individuals is better controlled. Rather than analyzing the properties of the populations of individuals with and without the treatment, we analyze how much individuals change under the two conditions.

Example 8.6.9 Paired Data: Analyzing with an Unpaired Test

Suppose that eight patients are evaluated before treatment and, after 1 week, are evaluated again after being given a new medication with the following results.

Patient	Before Treatment	After Treatment
T1	2.929	5.284
T2	3.755	4.384
T3	4.039	4.904
T4	4.237	5.420
T5	8.443	8.356
T6	8.305	9.547
T7	6.291	8.874
T8	3.038	2.689

We can try a t test to evaluate whether the two groups of patients differ. We find that $s_1^2 = 5.07$ and $s_2^2 = 5.96$. Because the two sample sizes are equal, the pooled variance is the ordinary average of these values, so

$$s_p^2 = \frac{s_1^2 + s_2^2}{2} = 5.518.$$

The standard error of the difference is

$$s_{\overline{X}_1} - \overline{X}_2 = \sqrt{\frac{5.518}{8} + \frac{5.518}{8}} = 1.175.$$

The sample means are $\overline{X}_1 = 5.130$ and $\overline{X}_2 = 6.182$. Therefore,

$$t = \frac{\overline{X}_1 - \overline{X}_2}{s_{\overline{X}_1 - \overline{X}_2}} = \frac{5.130 - 6.182}{1.175} = -0.895.$$

The number of degrees of freedom is $\nu = n_1 + n_2 - 2 = 14$. The value of t is not close to the critical value 2.145 from Table 8.2 (and inside back cover). Apparently, this medication has no effect.

This method finds no effect of medication. However, there is reason to think that we can do better, by examining each patient more carefully.

Example 8.6.10 Paired Data: Graphical Analysis

Consider again the patients in Example 8.6.9. If we graph the value after treatment against the value before treatment for each patient (Figure 8.6.5), we see that values for six out of the eight patients increased, and the other two lie close to the diagonal (where treatment is equal to control). Perhaps the **change** in value of each patient is significant.

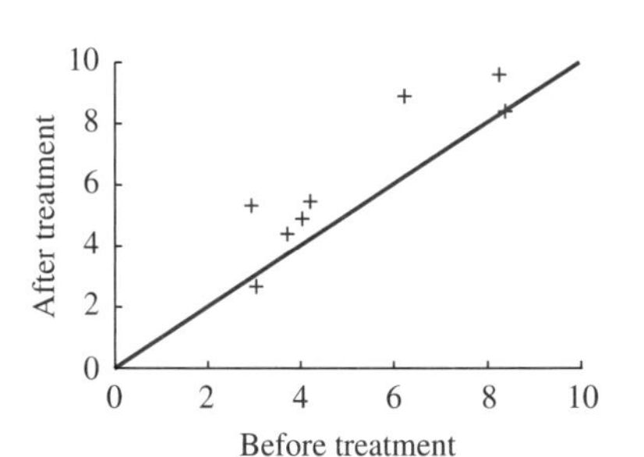

FIGURE 8.6.5

Control versus treatment values

Analysis of paired data as unpaired data loses power because the variability within the population might mask a small change due to the treatment. To avoid this, we create the new random variable

$$Y = X_2 - X_1$$

as the difference between the treatment and control values for each patient. We can then conduct an ordinary one-sample t test, as in Subsection 8.6.3, to check the null hypothesis that the true mean of Y is equal to 0.

Example 8.6.11 Paired Data: Analyzing with a Paired Test

Consider again the data in Example 8.6.9, but with a new column indicating the change in value for each patient.

Patient	Before Treatment	After Treatment	Change
T1	2.929	5.284	2.355
T2	3.755	4.384	0.629
T3	4.039	4.904	0.865
T4	4.237	5.420	1.183
T5	8.443	8.356	−0.087
T6	8.305	9.547	1.242
T7	6.291	8.874	2.583
T8	3.038	2.689	−0.349

The new random variable, the change as denoted by Y, has sample variance equal to $s_Y^2 = 1.081$, and thus has standard error

$$s_{\overline{Y}} = \sqrt{\frac{s_Y^2}{8}} = 0.368.$$

Our null hypothesis is that the mean of Y is equal to 0, while the actual sample mean of Y is 1.053. The t statistic is

$$t = \frac{1.053}{0.368} = 2.86.$$

This value exceeds the critical value for a p-value of 0.05 (and is close to that for a p-value of 0.01). The medication does produce a significant change. ▲

Experiments must be interpreted with care. For example, if patients tend to improve spontaneously, the treatment might in fact have no effect. To address this one could include a placebo control, where some patients are given a sugar pill instead of the proposed medication. This controls for any spontaneous improvement and for the possible psychological benefits of being involved in a clinical trial. These experiments can be analyzed with a t test comparing the changes in the control and treatment populations. In practice, researchers usually use the ANOVA method mentioned earlier, which gives the same results as the t test in this case.

Example 8.6.12 Paired Data: Analyzing a Placebo Control

Suppose that we are comparing patients treated with a medication with another population treated with a placebo.

Patient	Medication	Placebo	Change
P1	7.443	7.069	−0.375
P2	5.798	8.109	2.312
P3	10.361	10.598	0.237
P4	6.584	5.977	−0.607
P5	3.619	3.104	−0.515
P6	5.298	5.781	0.483
P7	6.917	7.528	0.61
P8	6.359	7.245	0.886

We test whether the changes are different for the population given the medication and for the population given the placebo by conducting an **unpaired** t test on the changes. Let P denote the change in the placebo population, and recall that Y represents the change in the population given the medication. The sample means are $\overline{Y} = 1.053$ and $\bar{P} = 0.379$, meaning that patients with medication improve on average by a bit more than patients taking the placebo. But is this greater improvement significant?

The pooled variance of P and Y is the average of the sample variance of the two populations because the sample sizes are equal. The sample variance of the placebo population is $s_P^2 = 0.916$, so the pooled variance is

$$s_p^2 = \frac{s_Y^2 + s_P^2}{2} = 0.999.$$

The standard error of the difference is

$$s_{\overline{Y}-\bar{P}} = \sqrt{\frac{0.999}{8} + \frac{0.999}{8}} = 0.500.$$

Therefore,

$$t = \frac{\overline{Y} - \bar{P}}{s_{\overline{Y}-\bar{P}}} = \frac{1.053 - 0.379}{0.500} = 1.348.$$

The number of degrees of freedom is $\nu = n_1 + n_2 - 2 = 14$. The value of t is less than the critical value 2.145 (Table 8.2, and inside back cover). The difference is not significant. Thus, although we found that patients taking the medication do, indeed, significantly improve, we cannot attribute this improvement to the medication, because patients also improve with the placebo.

Summary We have extended the methods for testing one sample against a known baseline to methods that compare two samples with each other. When the underlying distribution is known to be normal, the null hypothesis of no difference can be phrased as a statement about the distribution of the difference. If the measurements are **unpaired**, meaning a single measurement was made of each individual, we examine the difference between the sample means of the treatment and control. For sufficiently large samples, the normal approximation to the binomial can be used to compare population proportions. When samples are smaller, we use the sample variance and the t test. If measurements are **paired**, meaning that each individual was measured under both treatment and control conditions, we examine the difference between these two measurements for each individual.

8.6 Exercises

Mathematical Techniques

1–4 ▪ Use p-values to test the null hypothesis of equal means against an alternative that $\mu_2 > \mu_1$ when sample means of $\hat{\mu}_1$ and $\hat{\mu}_2$ are found from samples of size n_1 and n_2 with sample variances s_1^2 and s_2^2. Use a one-tailed test. State the significance level of the test.

1. $\hat{\mu}_1 = 70.0, \hat{\mu}_2 = 74.0, n_1 = 50, n_2 = 50, s_1^2 = 25.0, s_2^2 = 25.0$.

2. $\hat{\mu}_1 = 70.0$, $\hat{\mu}_2 = 70.4$, $n_1 = 500$, $n_2 = 500$, $s_1^2 = 25.0$, $s_2^2 = 25.0$. The difference between means is 10 times smaller, but the populations are 10 times larger than in Exercise 1. Why is the significance level different?

3. $\hat{\mu}_1 = 70.0$, $\hat{\mu}_2 = 74.0$, $n_1 = 500$, $n_2 = 500$, $s_1^2 = 250.0$, $s_2^2 = 250.0$. Compare with the answer to Exercise 1.

4. $\hat{\mu}_1 = 70.0, \hat{\mu}_2 = 70.4, n_1 = 500, n_2 = 500, s_1^2 = 2.5, s_2^2 = 2.5$. Compare with the answer to Exercise 1.

5–8 ▪ Find the pooled variance for two populations with the following sample sizes and sample variances.

5. $n_1 = 15, n_2 = 15, s_1^2 = 2.5, s_2^2 = 3.5$.

6. $n_1 = 10, n_2 = 10, s_1^2 = 2.7, s_2^2 = 3.9$.

7. $n_1 = 9, n_2 = 15, s_1^2 = 2.5, s_2^2 = 3.5$. Compare with the value in Exercise 5.

8. $n_1 = 16, n_2 = 10, s_1^2 = 2.7, s_2^2 = 3.9$. Compare with the value in Exercise 6.

9–12 ▪ Find the standard error of the difference of the means in each case.

9. The situation in Exercise 5.

10. The situation in Exercise 6.

11. The situation in Exercise 7.

12. The situation in Exercise 8.

13–16 ▪ Apply a two-tailed t test in the following cases.

13. The situation in Exercise 9 with $\mu_1 = 3.2$ and $\mu_2 = 2.1$.

14. The situation in Exercise 10 with $\mu_1 = 3.8$ and $\mu_2 = 2.1$.

15. The situation in Exercise 11 with $\mu_1 = 3.8$ and $\mu_2 = 2.1$.

16. The situation in Exercise 12 with $\mu_1 = 3.2$ and $\mu_2 = 2.1$.

17–20 ▪ Recall the data in Section 8.5, Exercises 1–4 describing 10 plants in an experimental plot.

Plant	Weight	Height	Yield	Number of seeds
1	11.83	46.2	10.11	7
2	7.65	39.4	9.29	18
3	14.57	41.4	12.26	19
4	14.97	40.8	6.61	16
5	14.26	38.5	9.51	17
6	11.26	40.7	8.34	14
7	11.08	39.8	13.35	22
8	7.04	37.9	12.90	21
9	5.44	34.8	10.35	20
10	11.82	43.1	9.84	25

Suppose that these plants are being compared with populations in a control plot. Use a two-tailed test, and compare the p-values with those found in the earlier problem.

17. There are ten plants in the control plot with mean weight 10.0, and the variance for weight in both populations is known to be 9.0. Compare with the results in Section 8.5, Exercise 1.

18. There are ten plants in the control plot with mean height 36.5, and the variance for height in both populations is known to be 16.0. Compare with the results in Section 8.5, Exercise 2.

19. There are 15 plants in the control plot with mean yield 8.2, and the variance for yield in both populations is known to be 6.25. Compare with the results in Section 8.5, Exercise 3.

20. There are 20 plants in the control plot with mean seed number 15.0, and the variance for seed number in both populations is known to be 25.0. Compare with the results in Section 8.5, Exercise 4.

21–24 ▪ Consider again the data in Exercises 17–20, but suppose that variances are unknown. Use the sample variance for the experimental population found in the earlier problem and the given sample variance for the control population to perform a t test on these unpaired populations.

21. The ten plants in the control plot have mean weight 10.0 and sample variance 8.80. Compare with the results in Exercise 17.

22. The ten plants in the control plot have mean height 36.5 and sample variance of 17.2. Compare with the results in Exercise 18.

23. The 15 plants in the control plot have mean yield 8.2 and sample variance of 8.2. Compare with the results in Exercise 19.

24. The 20 plants in the control plot have mean seed number 15.0 and sample variance of 14.2. Compare with the results in Exercise 20.

25–28 ▪ Test the null hypothesis that the means from two populations are equal in the following cases. $\hat{\mu}_1$ and $\hat{\mu}_2$ are sample means found from samples with size n_1 and n_2 drawn from normal distributions with known variances σ_1^2 and σ_2^2. State the significance level of the test. Use a two-tailed test.

25. $\hat{\mu}_1=20.0, \hat{\mu}_2=25.0, n_1=25, n_2=25, \sigma_1^2=25.0, \sigma_2^2=25.0$.

26. $\hat{\mu}_1=20.0, \hat{\mu}_2=21.0, n_1=25, n_2=50, \sigma_1^2=25.0, \sigma_2^2=25.0$.

27. $\hat{\mu}_1=20.0, \hat{\mu}_2=21.0, n_1=50, n_2=100, \sigma_1^2=16.0, \sigma_2^2=16.0$.

28. $\hat{\mu}_1=20.0, \hat{\mu}_2=21.0, n_1=200, n_2=100, \sigma_1^2=16.0, \sigma_2^2=9.0$.

29–32 ▪ Use the normal approximation to test the null hypothesis that men and women have the same opinions in the following cases. State the significance level of a two-tailed test.

29. Thirty-five out of 50 men believe that if dolphins were so smart they could find their way out of nets, whereas 40 out of 50 women believe this.

30. Three hundred fifty out of 500 men and 400 out of 500 women.

31. Thirty-five out of 50 men and 400 out of 500 women.

32. Seventy out of 1000 men and 40 out of 500 women. Why do you think the difference is not significant even though the samples are very large?

33–34 ▪ Algorithm 8.4 uses $\hat{p}_1$ and $\hat{p}_2$ to estimate the variance under the null hypothesis.

33. Why might it make more sense to use $\hat{p}$, the proportion in the pooled sample? What is the pooled proportion if 96 out of 200 events occur in the control and 54 out of 100 events occur in the treatment?

34. Redo the test using $\hat{p}$. How different are the results? Under what circumstances might it make a larger difference which proportion was used?

35–36 ▪ Show that the two-sample test turns into the one-sample test as n_1 approaches infinity.

35. What is the null hypothesis about the difference between means?

36. What is the distribution of sample means in the treatment population under the null hypothesis?

Applications

37–38 ▪ A cell is placed in a medium with volume equal to that of the cell. Then 100 marked molecules are placed inside, and after 1 h, 40 are found inside and 60 are found outside. In a control, protein in the membrane thought to be involved in transporting the molecule has been removed and 50 out of 100 of the molecules are found inside after the same amount of time.

37. What is the null hypothesis if the cell with the transporter in place is compared with the control? What is the null hypothesis if the treatment is compared with the expectation that molecules end up inside and outside with equal probability?

38. Find the p-value associated with the comparison of the treatment with the control, and the comparison of the treatment with the expectation that molecules end up inside and outside with equal probability. Why do the p-values differ as they do?

39–40 ▪ One organism has 8 mutations in 1 million base pairs, a second has 18 in 1 million, and a third has 28 in 1 million. Use the normal approximation to test whether the following differences are significant.

39. The difference between the first and second organisms.

40. The difference between the second and third organisms. Why is the significance level different from that in Exercise 39 even though the observed difference of 10 mutations is the same in each case?.

41–44 ▪ Consider the following data on ten patients with viral loads measured under control conditions, after treatment A, and then again after treatment B. Use the given test to check whether the treatment has an effect.

Patient	Control	Treatment A	Treatment B
1	6.517	7.142	8.830
2	5.619	6.484	7.097
3	6.707	7.049	8.183
4	4.427	4.514	5.554
5	6.109	8.860	6.092
6	4.828	5.140	5.974
7	4.455	4.612	4.291
8	7.660	9.912	9.941
9	4.399	6.756	5.956
10	4.155	5.670	3.997

41. Use an unpaired test to look for an effect from treatment A.

42. Use an unpaired test to look for an effect from treatment B.

43. Use a paired test to look for an effect from treatment A.

44. Use a paired test to look for an effect from treatment B.

8.7 Analysis of Contingency Tables and Goodness of Fit

Hypothesis tests check how closely data fit a particular model. The idea of **goodness of fit** quantifies this process more generally. Rather than comparing the measured and predicted value of a single statistic, such as the mean, these tests compare the entire distribution of outcomes. For example, in an experiment with four outcomes, we could compare the number of results in each category. The differences between observed and predicted can be combined into a single statistic, which follows the **χ^2 distribution**. Like the t distribution, the χ^2 distributions form a family indexed by the number of **degrees of freedom**. We will use this distribution to examine tests for the binomial proportion, for independence, and for fit to models more generally.

8.7.1 Testing Against a Known Baseline

In Subsections 8.4.1 and 8.5.1, we considered a finding that 20 out of 100 diseased people carry a particular allele in a population where only 13% of people carry that allele. We used both the exact method and the normal approximation to find the probability of a result as extreme as or more extreme if the null hypothesis, that 13% of diseased people also carry the allele, were true. Using the exact method (Example 8.4.8) we found a p-value of 0.0512, and with the normal approximation (Example 8.5.7) we found a similar p-value of 0.054. We now develop a new statistic that summarizes the deviation between the null hypothesis and the observed data.

In our example, there are two possible outcomes: an individual either has the particular allele or does not. More generally, an experiment could have m different outcomes, as in the case where individuals might carry any one of m alleles. If we measure n individuals, a model might predict E_i individuals in category i. Our observations are summarized as O_i individuals in category i. A statistic that summarizes the deviation is called the **chi-squared statistic**, indicated by the symbol χ^2.

Definition 8.12 The **chi-squared statistic** is computed as

$$\chi^2 = \sum_{i=1}^{m} \frac{(O_i - E_i)^2}{E_i}.$$

In words, this statistic is the sum of the squared differences between the observed and expected numbers, divided by the expected number.

Example 8.7.1 Computing the Chi-Squared Statistic

In a table, the result of finding 20 out of 100 people with the particular allele is

	Number with Allele	Number with Other Allele
Observed	20	80
Expected	13	87

The chi-squared statistic is

$$\chi^2 = \frac{(20-13)^2}{13} + \frac{(80-87)^2}{87} = 4.332.$$

Example 8.7.2 Computing the Chi-Squared Statistic with More Than Two Outcomes

In the United States, the frequencies of the four major blood types are **O** at 0.45, **A** at 0.41, **B** at 0.1, and **AB** at 0.04. Although these blood types themselves have not been shown to be associated with any diseases, some diseases are **linked**, meaning that the genes causing those diseases are close to the easily assessed blood type locus. Suppose that we measure 200 individuals with a disease, and find the following.

	Type O	Type A	Type B	Type AB
Observed	80	72	32	16
Expected	90	82	20	8

The expected values are found by multiplying the known frequencies, such as 0.45 for type **O**, by the population of 200. There seems to be an excess of people with the **B** allele. The chi-squared statistic is

$$\chi^2 = \frac{(80-90)^2}{90} + \frac{(72-82)^2}{82} + \frac{(32-20)^2}{20} + \frac{(16-8)^2}{8} = 17.53.$$

This statistic gives a measure of how different the observations are from the expectation. We now need a method for determining whether that deviation is larger than what might be expected by chance. The χ^2 distribution describes the distribution of the statistic χ^2 for large samples, and provides a good approximation for smaller samples. Like the t distribution, the χ^2 distribution is indexed by the number of degrees of freedom, the number of measurements required to completely characterize the results. In cases where we have n individuals in our sample and m possible outcomes, knowing how many individuals fall into the first $m-1$ categories (the values of $O_1, \ldots, O_{m-1}$) is enough to compute O_m. The number of degrees of freedom is thus $\nu = m - 1$.

Example 8.7.3 Computing the Degrees of Freedom with Two Possible Outcomes

If we measure 100 individuals and find 13 with a particular allele, we need make no additional measurements to compute that 87 carry the other allele. There are $m = 2$ outcomes, so there are $\nu = m - 1 = 1$ degrees of freedom. Alternatively, we can find that $\nu = 1$ by noting that a single measurement is sufficient to characterize the data.

Example 8.7.4 Computing the Degrees of Freedom with Four Possible Outcomes

If we measure 200 individuals and find 80 with blood type **O**, 72 with blood type **A**, and 32 with blood type **B**, we can compute that 16 must have blood type **AB**. With $m = 4$ outcomes, there are $\nu = m - 1 = 3$ degrees of freedom.

Graphs of the χ^2 distribution with 1, 2, and 3 degrees of freedom have quite different shapes (Figure 8.7.1). With 1 degree of freedom, the probability density approaches

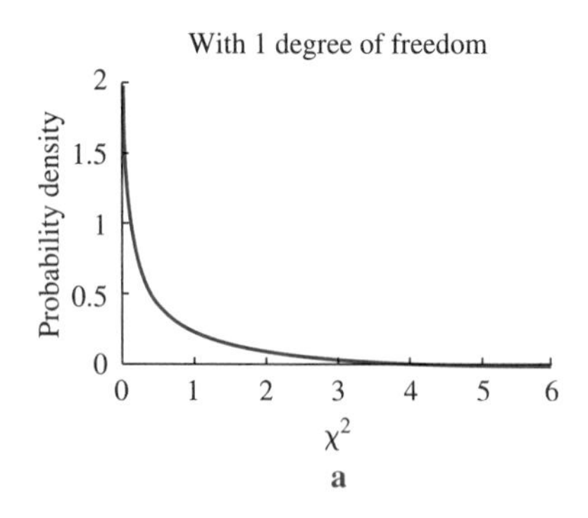

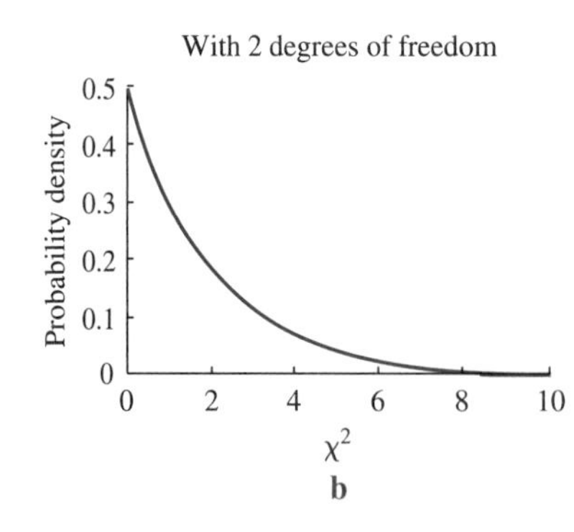

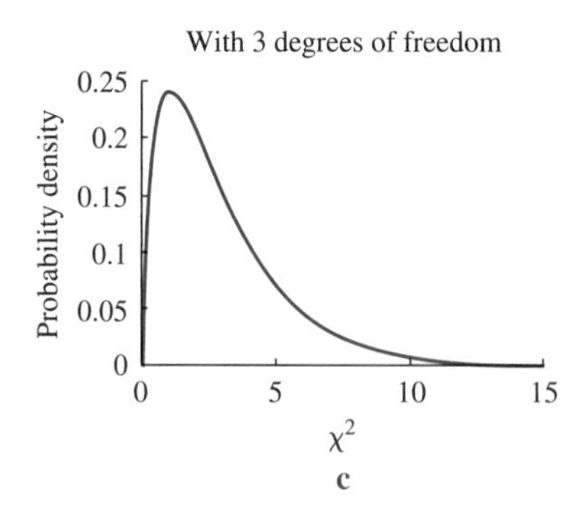

FIGURE 8.7.1
Probability density functions for the χ^2 distribution

infinity as χ^2 approaches 0, meaning that results with very small values are highly probable. The χ^2 distribution with 1 degree of freedom describes the distribution of the square of a random variable that obeys the standard normal. With 2 degrees of freedom,

Table 8.4 Critical values of the χ^2 distribution

Degrees of Freedom	Critical Value for p = 0.05	Critical Value for p = 0.01
1	3.841	6.635
2	5.991	9.210
3	7.815	11.345
4	9.488	13.277
5	11.070	15.086
6	12.592	16.812
7	14.067	18.475
8	15.507	20.090
9	16.919	21.666
10	18.307	23.209

the density function is in fact an exponential distribution with mean 2, but still with a mode at 0. With 3 degrees of freedom, the mode has moved to $\chi^2 = 1$, meaning that results with very small values are now less likely.

To use this distribution, we need critical values for a range of degrees of freedom. If the computed value of the statistic χ^2 exceeds the critical value for p = 0.05 (see the second column in Table 8.4), the result is significant at the 0.05 level, and similarly for p = 0.01. The χ^2 test is always a two-tailed test, because it measures how far the data lie from the model without paying attention to the direction (the difference $O_i - E_i$ is squared).

Example 8.7.5 Computing a p-Value with Chi-Squared Statistic

In Example 8.7.1, we found that $\chi^2 = 4.332$. This exceeds the critical value 3.841 with $\nu = 1$ degrees of freedom, thus the result is significant at the 0.05 level. This result is a bit different from that found in Examples 8.4.8 and 8.5.7 where the p-value slightly exceeded 0.05. As presented here, the χ^2 statistic tends to be a bit too large, because we have not included the continuity correction.

Example 8.7.6 Computing a p-Value with More Than Two Outcomes

In Example 8.7.2, we found that $\chi^2 = 17.53$. This greatly exceeds the critical value 11.345 for p = 0.01 with $\nu = 3$ degrees of freedom. We have strong evidence that the sample differs from the whole population.

As seen in Example 8.7.5, the χ^2 statistical test can produce p-values that are slightly smaller than the exact value. As we found when developing the normal approximation to the binomial (Subsection 7.9.2), this deviation is due in part to the discrete nature of the binomial distribution. We can partially correct using a form of the **continuity correction**. However, there are a series of other corrections that provide further minor improvements, and using the continuity correction alone is considered rather conservative in practice.

Example 8.7.7 Using the Continuity Correction

To include the continuity correction, move each observed value by 0.5 toward the expected.

	Number with Allele	Number with Other Allele
Observed	19.5	80.5
Expected	13	87

The chi-squared statistic is then

$$\chi^2 = \frac{(19.5 - 13)^2}{13} + \frac{(80.5 - 87)^2}{87} = 3.736.$$

This value is slightly less than the critical value for p = 0.05, meaning that the significance level is slightly greater than 0.05, consistent with our results in Examples 8.4.8 and 8.5.7.

In addition to the continuity correction, the χ^2 test must be modified when the expected number in any particular box is less than 5. Usually, this is done by pooling together categories with small numbers of expected outcomes.

Example 8.7.8 Pooling Outcomes to Test for a Poisson Distribution

Suppose we count the number of mutations in 100 pieces of DNA. The expected number of mutations is thought to follow a Poisson distribution with a mean of $\Lambda = 1.5$. With this distribution, the probability of 0 mutations is $e^{-1.5} = 0.2231$, giving an expected number with 0 mutations of $0.2231 \cdot 100 = 22.31$. The entire expected distribution is compared with the data as follows.

	Number of Mutations							
	0	1	2	3	4	5	6	7
Observed	25	27	21	15	5	5	1	1
Expected	22.31	33.47	25.10	12.55	4.71	1.41	0.35	0.076
χ^2	0.324	1.250	0.670	0.478	0.0183	9.120	1.190	11.30

There are fewer observed values close to the expectation (one or two mutations) and slightly more far from the expectation (Figure 8.7.2). Is this difference significant?

The table includes the value of $\chi^2 = \frac{(O_i - E_i)^2}{E_i}$ for each term. We can compute χ^2 as the sum

$$\chi^2 = \sum_{i=0}^{7} \frac{(O_i - E_i)^2}{E_i} = 24.34.$$

If we compare this value with the critical value of the χ^2 distribution with 7 degrees of freedom, the result seems highly significant. If we consider the terms in the sum producing χ^2, approximately half of the value is generated by the last, where

$$\frac{(1 - 0.076)^2}{0.076} = 11.30.$$

This large value is produced not by a large deviation between observed and expected, but by dividing by a small expected value.

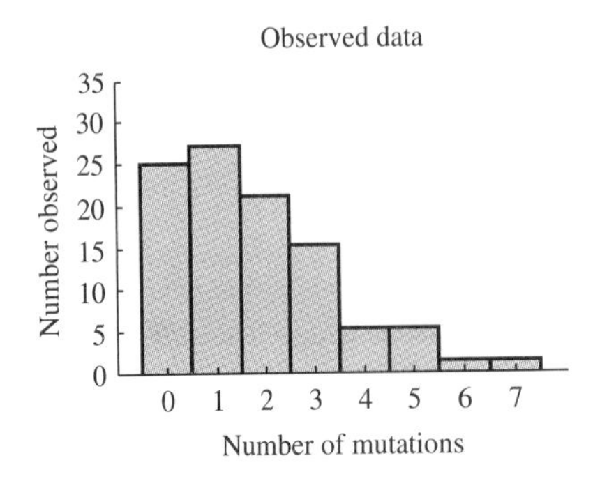

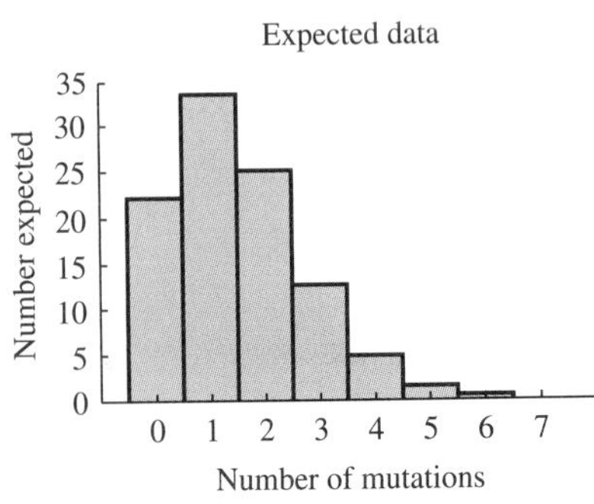

FIGURE 8.7.2
Observed and expected data on mutation number

To avoid this problem, we pool the outcomes with four or more mutations where the expected value is less than 5. Pooling gives a new table of data. The statistic χ^2

	Number of Mutations				
	0	1	2	3	≥ 4
Observed	25	27	21	15	12
Expected	22.31	33.47	25.10	12.55	6.55
χ^2	0.324	1.250	0.670	0.478	4.540

is now

$$\chi^2 = \frac{(25-22.31)^2}{22.31} + \frac{(27-33.47)^2}{33.47} + \frac{(21-25.10)^2}{25.10} + \frac{(15-12.55)^2}{12.55} + \frac{(12-6.55)^2}{6.55} = 7.26.$$

There are now 4 degrees of freedom, because we reduced our data set to $m = 5$ measurements. However, even with fewer degrees of freedom, this much smaller value of χ^2 fails to exceed the critical value for p = 0.05. There is little reason to think that our data fail to follow the Poisson distribution.

8.7.2 Testing with Contingency Tables

The tests against a known baseline in Subsection 8.7.1 are like the one-sample tests in Sections 8.4 and 8.5. These test an **extrinsic** hypothesis, which was developed without any reference to the data. When we instead compare the results of control and treatment populations, as in Section 8.6, our tests must take into account the uncertainty in measurements of the control population. Such tests are of an **intrinsic** hypothesis, because they include estimation of at least one unknown parameter from the data. The approach with the χ^2 test follows the same logic as in Section 8.6: build a null hypothesis based on pooling the control and treatment populations, then quantify the deviation from the null hypothesis by computing the statistic χ^2, and then test with the appropriate number of degrees of freedom.

Example 8.7.9 Building a Contingency Table and the Null Hypothesis

In Example 8.6.5, we compared the number of individuals carrying a given allele in a diseased population (20 out of 100) and a healthy population (26 out of 200). We can arrange these results in a table.

	Number with Allele	Number with Other Allele
Diseased	20	80
Healthy	26	174

Under the null hypothesis that the diseased and healthy individuals have the same probability of carrying the allele, we can create a pooled population by adding the columns, finding

	Number with Allele	Number with Other Allele
Pooled	46	254

In the pooled population, the frequency of the allele is $q = \frac{46}{300} = 0.1533$. To find the expected number with the allele under the null hypothesis, we multiply q by the number of individuals of each type. We thus expect $0.1533 \cdot 100 = 15.33$ diseased individuals with the allele and $0.1533 \cdot 200 = 30.66$ healthy individuals. The expected number with other alleles can be determined by subtracting from the known total of each row.

	Expected Number with Allele	Expected Number with Other Allele
Diseased	15.33	84.67
Healthy	30.66	169.34

We compute the statistic χ^2 in the usual way, by summing the squared differences between observed and expected values divided by expected values for all values in the table.

Example 8.7.10 Computing the Statistic χ^2 from a Table

For the case in Example 8.7.9,

$$\chi^2 = \frac{(20-15.33)^2}{15.33} + \frac{(80-84.67)^2}{84.67} + \frac{(26-30.66)^2}{30.66} + \frac{(174-169.34)^2}{169.34} = 2.52.$$

To evaluate the significance, we require the number of degrees of freedom. In Subsection 8.7.1, the number of degrees of freedom is the number of measurements required to fully specify the data. In this section, we are testing an intrinsic hypothesis where parameters must be estimated. The general rule for finding the number of degrees of freedom is

$$\begin{aligned}\text{degrees of freedom} = &\text{ number of measurements required to fully specify data}\\ &-\text{number of free parameters estimated.}\end{aligned}$$

After finding the degrees of freedom, the value of χ^2 is compared with the critical values of the χ^2 distribution.

Example 8.7.11 Finding the p-Value for the Genetic Example

In Example 8.7.10 we found that $\chi^2 = 2.52$. Two measurements are required to specify the data (the number of diseased individuals with the allele and the number of healthy individuals with the allele), and we estimated the single parameter q from the data. Therefore,

$$\text{degrees of freedom} = 2 - 1 = 1.$$

We thus compare the value $\chi^2 = 2.52$ with the first row of critical values of the χ^2 distribution. Because this value is well below the critical value for p = 0.05, the result is not significant. This is consistent with our finding of $p = 0.134$ using the normal approximation in Example 8.6.5.

As in Subsection 8.7.1, there are various corrections that can be applied, including the continuity correction. In practice, with reasonably large sample sizes, these corrections are usually not that important. We use statistics, and p-values in particular, as a guide to reasoning and further experimentation. The difference between a p-value of 0.04 and 0.06 should not be the difference between joy and despair, but a different shading of confidence in one's results.

One can think of the null hypothesis regarding contingency tables as one of independence. Knowledge of the phenotype of an individual, for example, conveys no information about the genotype when the allele is independent of the disease. Our null hypothesis is that the conditional distributions match.

Example 8.7.12 Using the χ^2 Test to Evaluate Independence

In Section 7.1 we used joint distributions to study independent random variables. One simple example described two dart players, where $S_i = 1$ indicates that player i scored, and $S_i = 0$ indicates that he did not. From the joint distribution, we can recognize independence when the joint probabilities are equal to the product of the marginal probabilities. In Examples 7.1.1 and 7.1.7, we described the mathematical description of probabilities with the joint distribution

	$S_1 = 1$	$S_1 = 0$		
$S_2 = 1$	0.09	0.21	→	$\Pr(S_2 = 1) = 0.3$
$S_2 = 0$	0.21	0.49	→	$\Pr(S_2 = 0) = 0.7$
	↓	↓		
	$\Pr(S_1 = 1) = 0.3$	$\Pr(S_1 = 0) = 0.7$		

As statisticians, we are instead confronted with data, which are counts of different events. For example, we might observe 100 throws, finding

	$S_1 = 1$	$S_1 = 0$
$S_2 = 1$	6	29
$S_2 = 0$	27	38

Are these values consistent with independence? To check, we find the expected numbers if the two were independent, and then use the χ^2 to compare. Our observed marginal probabilities are

	$S_1 = 1$	$S_1 = 0$		
$S_2 = 1$	6	29	→	$\Pr(S_2 = 1) = \frac{35}{100} = 0.35$
$S_2 = 0$	27	38	→	$\Pr(S_2 = 0) = \frac{65}{100} = 0.65$
	↓	↓		
	$\Pr(S_1 = 1) = \frac{33}{100} = 0.33$	$\Pr(S_1 = 0) = \frac{67}{100} = 0.67$		

If S_1 and S_2 were independent, we would expect to see a fraction

$$\Pr(S_1 = 1 \text{ and } S_2 = 1) = \Pr(S_1)\Pr(S_2) = 0.33 \cdot 0.35 = 0.1155.$$

The expected number is this fraction multiplied by 100, the total number of observations. The full table of expected numbers is

	$S_1 = 1$	$S_1 = 0$
$S_2 = 1$	$0.33 \cdot 0.35 \cdot 100 = 11.55$	$0.67 \cdot 0.35 \cdot 100 = 23.45$
$S_2 = 0$	$0.33 \cdot 0.65 \cdot 100 = 21.45$	$0.67 \cdot 0.65 \cdot 100 = 43.55$

We compute the χ^2 statistic as usual as

$$\chi^2 = \frac{(6 - 11.55)^2}{11.55} + \frac{(29 - 23.45)^2}{23.45} + \frac{(27 - 21.45)^2}{21.45} + \frac{(38 - 43.55)^2}{43.55} = 6.12.$$

As in Example 8.7.11, we have 1 degree of freedom. This value exceeds the critical value for p = 0.05, and is close to the value for p = 0.01, giving convincing evidence for a lack of independence of these two players.

The same approach works for testing larger tables. Suppose a table has R rows and C columns, with data

	Column 1	Column 2	...	Column C
Row 1	o_{11}	o_{12}	...	o_{1C}
Row 2	o_{21}	o_{22}	...	o_{2C}
⋮	⋮	⋮	⋮	⋮
Row R	o_{R1}	o_{R2}	...	o_{RC}

There are a total of $R \cdot C$ elements in this table. If the totals along the rows are known, then we need only the first $C - 1$ columns to fully describe the data, for a total of $R \cdot (C - 1)$ elements. To create the table of expected values, we must estimate the proportion in each column. This requires estimating $C - 1$ parameters, because the proportion in the last column is determined by the others. Thus,

$$\text{degrees of freedom} = R(C-1) - (C-1) = (R-1)(C-1). \quad \textbf{(8.7.1)}$$

Example 8.7.13 Applying the Formula for Degrees of Freedom

With a 2×2 table, $R = 2$ and $C = 2$, so

$$\text{degrees of freedom} = (2-1)(2-1) = 1,$$

as we found before.

Example 8.7.14 Testing for Independence in a Larger Table

In Subsection 7.1.3 we studied probabilities describing the numbers of lice and mites on birds. As scientists, we know only a set of measurements describing a population of birds. Suppose we measure 100 birds and find the following results.

	0 Lice	1 Louse	2 Lice
0 Mites	23	11	10
1 Mite	11	14	8
2 Mites	7	5	11

To compute the expected numbers under the null model, we find the fractions in each row and column.

	0 Lice	1 Louse	2 Lice	
0 Mites	23	11	10	$\rightarrow \Pr(M=0) = \frac{44}{100} = 0.44$
1 Mite	11	14	8	$\rightarrow \Pr(M=1) = \frac{33}{100} = 0.33$
2 Mites	7	5	11	$\rightarrow \Pr(M=2) = \frac{23}{100} = 0.23$
	↓	↓	↓	
	$\Pr(L=0) = \frac{41}{100} = 0.41$	$\Pr(L=1) = \frac{30}{100} = 0.30$	$\Pr(L=2) = \frac{29}{100} = 0.29$	

Under the assumption of independence, the expected table is

	0 Lice	1 Louse	2 Lice
0 Mites	$0.41 \cdot 0.44 \cdot 100 = 18.04$	$0.30 \cdot 0.44 \cdot 100 = 13.20$	$0.29 \cdot 0.44 \cdot 100 = 12.76$
1 Mite	$0.41 \cdot 0.33 \cdot 100 = 13.53$	$0.30 \cdot 0.33 \cdot 100 = 9.90$	$0.29 \cdot 0.33 \cdot 100 = 9.57$
2 Mites	$0.41 \cdot 0.23 \cdot 100 = 9.43$	$0.30 \cdot 0.23 \cdot 100 = 6.90$	$0.29 \cdot 0.23 \cdot 100 = 6.67$

We compute the statistic χ^2 by summing the squared deviations from the expected, divided by the expected, finding

$$\chi^2 = \frac{(23-18.04)^2}{18.04} + \frac{(11-13.20)^2}{13.20} + \frac{(10-12.76)^2}{12.76} + \frac{(11-13.53)^2}{13.53} + \frac{(14-9.90)^2}{9.90} + \frac{(8-9.57)^2}{9.57} + \frac{(7-9.43)^2}{9.43} + \frac{(5-6.90)^2}{6.90} + \frac{(11-6.67)^2}{6.67} = 8.72.$$

This table has $R = 3$ rows and $C = 3$ columns, so there are $(R-1)(C-1) = 2 \cdot 2 = 4$ degrees of freedom. This value is slightly less than the critical value of 9.488 with 4 degrees of freedom, so these data do not provide significant evidence that the two parasites deviate significantly from independence.

8.7.3 Testing for Goodness of Fit

In Subsection 8.7.1, we compared data with a predefined set of extrinsic probabilities, perhaps generated by some model. Often we only wish to know whether our data fits a particular model, without knowing the parameter values in advance. This case has an **intrinsic** hypothesis, because the parameter values (generally the maximum likelihood estimates) are derived from the data itself.

Example 8.7.15 Testing for a Binomial Distribution with an Intrinsic Hypothesis

Suppose 44 families of falcons with two offspring are studied to estimate both the fraction of male offspring and whether the sexes of the offspring are independent and follow a binomial distribution. The following results are found.

0 Males	1 Male	2 Males
14	14	16

To find the parameter for the binomial distribution describing the intrinsic hypothesis, we compute that there are $2 \cdot 16 + 14 = 46$ males out of 88 birds. Thus, $\hat{p} = \frac{46}{88} = 0.523$. The probabilities for the binomial distribution with $p = 0.523$ and $n = 2$ are

$$\begin{aligned} \Pr(0 \text{ males}) &= b(0; 2, 0.523) = 0.523^2 = 0.273 \\ \Pr(1 \text{ male}) &= b(1; 2, 0.523) = 2 \cdot 0.523 \cdot (1-0.523) = 0.499 \\ \Pr(2 \text{ males}) &= b(2; 2, 0.523) = (1-0.523)^2 = 0.228. \end{aligned}$$

The expected numbers are these probabilities multiplied by the number of families, 44.

0 Males	1 Male	2 Males
12.03	21.95	10.02

The statistic χ^2 is

$$\chi^2 = \frac{(14-12.03)^2}{12.03} + \frac{(14-21.95)^2}{21.95} + \frac{(16-10.02)^2}{10.02} = 6.77.$$

There are two measurements required to specify the data, and one intrinsic parameter, giving 1 degree of freedom. This value of χ^2 exceeds the critical value for p = 0.01, so the result is highly significant. There are far fewer families with exactly one male than we expect.

Example 8.7.16 Testing for a Poisson Distribution with an Intrinsic Hypothesis

Consider again the situation in Example 8.7.8, where we count the number of mutations in 100 pieces of DNA, but assume that the expected number of mutations is thought to follow a Poisson distribution with an unknown mean Λ.

	Number of Mutations							
	0	1	2	3	4	5	6	7
Observed	25	27	21	15	5	5	1	1

We can use maximum likelihood to estimate the parameter Λ. There are 25 observations with 0 observed mutations, each with probability

$$p(0;\,\Lambda) = e^{-\Lambda}$$

and a total probability, under the assumption of independence, of $p(0;\,\Lambda)^{25}$. The probability of the whole data set, the likelihood, is

$$= p(0;\,\Lambda)^{25}p(1;\,\Lambda)^{27}p(2;\,\Lambda)^{21}p(3;\,\Lambda)^{15}p(4;\,\Lambda)^{5}p(5;\,\Lambda)^{5}p(6;\,\Lambda)^{1}p(7;\,\Lambda)^{1}$$

$$L(\Lambda) = e^{-25\Lambda}\Lambda^{27}e^{-27\Lambda}\frac{\Lambda^{2\cdot 21}e^{-21\Lambda}}{2!^{21}}\frac{\Lambda^{3\cdot 15}e^{-15\Lambda}}{3!^{15}}\frac{\Lambda^{4\cdot 5}e^{-5\Lambda}}{4!^{5}}\frac{\Lambda^{5\cdot 5}e^{-5\Lambda}}{5!^{5}}\frac{\Lambda^{6\cdot 1}e^{-1\Lambda}}{6!}\frac{\Lambda^{7\cdot 1}e^{-1\Lambda}}{7!}$$

$$= \frac{\Lambda^{172}e^{-100\Lambda}}{H}$$

where H is a complicated constant that does not depend on Λ. The derivative is

$$L'(\Lambda) = \frac{172\Lambda^{171}e^{-100\Lambda} - 100\Lambda^{172}e^{-100\Lambda}}{H} \tag{8.7.2}$$

$$= (172 - 100\Lambda)\frac{\Lambda^{171}e^{-100\Lambda}}{H}. \tag{8.7.3}$$

This has a critical point, and maximum, at

$$\hat{\Lambda} = 1.72.$$

The expected numbers in each category are then given by the Poisson distribution with this parameter multiplied by 100, or

	Number of Mutations							
	0	1	2	3	4	5	6	7
Observed	25	27	21	15	5	5	1	1
Expected	17.91	30.80	26.49	15.19	6.53	2.25	0.64	0.16

As in Example 8.7.8, we pool the cases with four or more mutations to avoid expected values less than 5.0, giving the pooled table

	Number of Mutations				
	0	1	2	3	≥ 4
Observed	25	27	21	15	12
Expected	17.91	30.80	26.49	15.19	9.58

The value of the statistic χ^2 is

$$\chi^2 = \frac{(25-17.91)^2}{17.91} + \frac{(27-30.80)^2}{30.80} + \frac{(21-26.49)^2}{26.49} + \frac{(15-15.19)^2}{15.19} + \frac{(12-9.58)^2}{9.58} = 5.03.$$

There are four measurements required to fully specify our data, and we have estimated the single intrinsic parameter $\hat{\Lambda} = 1.72$, so there are now 3 degrees of freedom. As in Example 8.7.8, this does not exceed the critical value for the appropriate χ^2 distribution, and we do not have evidence to reject the hypothesis that these mutations follow a Poisson distribution. ▲

Summary The χ^2 test provides a flexible way to test whether data fit a given prior hypothesis or null hypothesis. The test requires computing the statistic χ^2 as the sum of the squared deviations between the observed and expected values divided by the expected values, and comparing with critical values of the χ^2 distribution for the appropriate number of degrees of freedom. If the prior hypothesis is a set of given probabilities, called an **extrinsic hypothesis**, the number of degrees of freedom is the number of measurements required to fully specify the data. If the null hypothesis requires estimating parameters from the data, an **intrinsic hypothesis**, the number of degrees of freedom is reduced by the number of parameters estimated. If the expected number in any given cell is less than 5, values need to be **pooled** to avoid dividing by small numbers. The comparison of observed and expected values with this test can be thought of more generally as a test of the **goodness of fit**.

8.7 Exercises

Mathematical Techniques

1–4 ▪ Suppose that the number N of molecules of toxin left in a cell after 10.0 min is thought to follow the probability distribution with $\Pr(N=0)=0.4$, $\Pr(N=1)=0.3$, $\Pr(N=2)=0.2$, and $\Pr(N=3)=0.1$ (as in Example 6.3.11). Test whether the following data fit the expectation from this extrinsic hypothesis.

1. There are 35 cells with no molecules, 25 with one molecule, 25 with two molecules, and 15 with three molecules.

2. There are 25 cells with no molecules, 21 with one molecule, 19 with two molecules, and 15 with three molecules.

3. Consider again the data in Exercise 1, but suppose that we can only distinguish cells with two or more molecules from those with one or fewer. Find how many cells are in each of these two categories and compare with the appropriate extrinsic hypothesis. Why might the test give a different result than with the unpooled data?

4. Consider again the data in Exercise 2, but suppose that we can only distinguish cells with no molecules from those with at least one. Find how many cells are in each of these two categories and compare with the appropriate extrinsic hypothesis. Why might the test give a different result than with the unpooled data?

5–6 ▪ The number of molecules remaining in a cell is thought to follow a binomial distribution with the given parameter. In each case, find whether there is reason to reject this model.

5. Suppose there are three molecules, and that the probability of remaining is thought to be $p=0.6$. In a sample of 80 cells, we find 10 with 0 molecules, 20 with 1 molecule, 30 with 2 molecules, and 20 with 3 molecules.

6. Suppose there are 4 molecules, and that the probability of a molecule's remaining is thought to be $p=0.6$. In a sample of 80 cells, we find 5 with no molecules, 20 with one molecule, 20 with two molecules, 20 with three molecules, and 15 with four molecules.

7–8 ▪ Compute the statistic χ^2 in the earlier exercise using the continuity correction. Does it alter the conclusions?

7. The situation in Exercise 5.

8. The situation in Exercise 6.

9–10 ▪ Suppose that the data in Exercises 5 and 6 are thought to follow a binomial distribution with an unknown parameter. Estimate this parameter and test whether the data fit the resulting model.

9. The situation in Exercise 5.

10. The situation in Exercise 6.

11–14 ▪ Consider the following data, which were supposedly generated from 200 replicates of a Poisson process.

Number of Events	Number Observed in Experiment 1	Number Observed in Experiment 2
0	5	25
1	15	33
2	32	35
3	34	23
4	41	15
5	26	21
6	25	19
7	12	9
8	7	3
9	2	5
10	0	2
11	1	4
12	0	1
13	0	2
14	0	1
15	0	0
16	0	1
17	0	0
18	0	1

11. Test the extrinsic hypothesis that $\Lambda=4.5$ for experiment 1.

12. Test the extrinsic hypothesis that $\Lambda=4.0$ for experiment 2.

13. Test the intrinsic hypothesis that the data follow a Poisson distribution for experiment 1.

14. Test the intrinsic hypothesis that the data follow a Poisson distribution for experiment 2.

15–16 ▪ Consider again the data on mites and lice from Example 8.7.14.

	0 Lice	1 Louse	2 Lice
0 Mites	23	11	10
1 Mite	11	14	8
2 Mites	7	5	11

Find the probabilities for each term in the table, and find the conditional distributions.

15. The distribution of the number of lice conditional on 0, 1, and 2 mites. How different are the conditional distributions, and would they lead you to suspect that the two pests do not act independently?

16. The distribution of the number of mites conditional on 0, 1, and 2 lice. How different are the conditional distributions, and would they lead you to suspect that the two pests do not act independently?

17–18 ▪ Test the following tables for independence.

17. Consider the following data on mating in birds.

	Red Male	Blue Male	Green Male
Red female	25	6	2
Blue female	10	14	9
Green female	12	5	15

Do matings deviate from independence, and what might it mean?

18. Consider the following data on student class attendance.

	Student 1 Attends	Student 1 Does Not Attend
Student 2 attends	93	52
Student 2 does not attend	32	5

Is attendance independent, and if not, what might it mean?

19–22 ▪ The significance of deviations from the null hypothesis depends on the sample size. Conduct a χ^2 test for the following samples based on Example 8.7.1. Suppose that 20% of diseased people tested have a particular allele (this would, of course, vary in a series of real experiments), and that 13% of healthy people are known to have the allele.

19. Suppose we tested only 50 diseased people. Find the significance of the result and compare to the results with a sample size of 100.

20. Suppose we tested 200 diseased people. Find the significance and compare to the results with a sample size of 100.

21. Suppose we tested n diseased people. Compute χ^2 as a function of n. Does it increase proportionally to the sample size?

22. Suppose we tested n diseased people. How many people would we need to test to find a result significant at the 0.01 level?

23–26 ▪ A random variable C_ν follows a χ^2 distribution with ν degrees of freedom if

$$C_\nu = X_1^2 + X_2^2 + \ldots + X_\nu^2$$

where $X_1, X_2, \ldots, X_\nu$ follow the standard normal distribution.

23. Find the expectation of C_1.

24. Find the expectation of C_ν.

25. Compute the critical value for p = 0.05 with 1 degree of freedom.

26. Remarkably enough, C_2 is an exponential distribution. Using the mean found in Exercise 24, find the parameter of this distribution, and compute the critical value for p = 0.05.

Applications

27–30 ▪ Use the χ^2 test to check whether the control and treatment differ in the following contingency tables.

27. Consider the following data on the behavior of 50 wild type and 100 mutant worms.

	Produce Eggs	Sterile
Wild type	45	5
Mutant	80	20

28. Consider the following data on the behavior of 100 wild type and 150 mutant worms.

	Crawl Normally	Uncoordinated
Wild type	85	15
Mutant	105	45

29. Consider the following data on the behavior of 80 wild type and 120 mutant worms.

	Produce Eggs and Sperm	Produce Eggs Only	Sterile
Wild type	65	10	5
Mutant	78	22	20

30. Consider the following data on the behavior of 100 wild type and 125 mutant worms.

	Crawl Normally	Spastic	Uncoordinated	Paralyzed
Wild type	75	8	8	9
Mutant	80	17	5	23

31–32 ▪ A recessive allele is expected to be expressed in 25% of offspring from a cross of heterozygous plants. Check whether the following data are consistent with this hypothesis.

31. Ten out of 60 plants are homozygous for the recessive allele.

32. Twenty-one out of 120 plants are homozygous for the recessive allele.

33–34 ▪ Suppose that plants with genotype **WW** have white flowers, those with genotype **WR** or **RW** have pink flowers, and those with genotype **RR** have red flowers. Two **RW** plants are crossed. Check whether the following data are consistent with the expected ratios. If not, try to explain why.

33. Out of 90 offspring, there are 18 white, 40 pink, and 32 red.

34. Suppose 10 additional plants had been measured in Exercise 33, and there were 3 pink ones and 7 red ones.

35–36 ▪ Suppose two traits are controlled by two unlinked loci (so the phenotypes are independent), one for flower color and one for height. Check whether the following data are consistent with the expected numbers in the following scenarios.

35. Suppose that both yellow flower color and shortness are recessive, with white flower color and tallness expressed in the dominant plants. Two parents that are heterozygous for these two traits are crossed, and 80 offspring are checked. Of these, 3 have yellow flowers and are short, 12 have yellow flowers and are tall, 17 have white flowers and are short, and 48 have white flowers and are tall.

36. Suppose that both yellow flower color and shortness are recessive, with white flower color and tallness expressed in the dominant plants. Two parents that are heterozygous for these two traits are crossed, and 87 offspring are checked. Of these, 11 have yellow flowers and are short, 8 have yellow flowers and are tall, 13 have white flowers and are short, and 55 have white flowers and are tall.

37–38 ▪ An ecologist counts the numbers of jack rabbits and eagles observed, and wishes to know whether they are independent (as in Section 6.4, Exercises 27 and 28). E represents the number of eagles seen, and J the number of jackrabbits. Use the χ^2 test to check.

37. Eighty counts are made, with the following results.

	$E=0$	$E=1$
$J=0$	35	15
$J=1$	45	5

Are the results significant? Compare with Section 7.1,

38. Eighty counts are made, with the following results.

	$E=0$	$E=1$
$J=0$	36	4
$J=1$	28	12

Are the results significant? Compare with Section 7.1, Exercise 28.

39–40 ▪ Recall the falcon data studied in Example 8.7.15, where 44 families of two birds were studied, and 14 had no males, 14 had one male, and 16 had 2 males. However, now assume that the order of birth is taken into account, so that there are four possible families (the first offspring could be male or female as could the second). Write a table and evaluate for lack of independence in the following cases, and compare with the results in Example 8.7.15.

39. Of the 14 females with one male, 7 had a male first.

40. Of the 14 females with one male, 3 had a male first.

8.8 Hypothesis Testing with the Method of Support

The p-value is the probability of data as extreme as or *more extreme than* observed if the null hypothesis is true. Although it provides an easily interpreted probability, the inclusion of more extreme data in the calculation cannot be strictly justified on logical grounds. For example, suppose we find that 20 out of 100 diseased individuals carry a particular allele in a population with known allele frequency of 0.13 (as in Subsection 8.4.1). The calculation of the p-value for a one-tailed test sums the probabilities of 20, 21, 22, or more people. But we did not observe 21, 22, or more people with the allele. A. W. F. Edwards has said of p-values that "a hypothesis that may be true may be rejected because it has not predicted observable results that have not occurred." The characterization of possible results as "more extreme" is also problematic. Even in this simple example, we must decide whether 6 out of 100 is really exactly as extreme as 20 out of 100.

Neither of these difficulties in any way invalidates the proven usefulness of p-values. Nonetheless, the **method of support** uses the likelihood function to provide a flexible alternative for hypothesis testing, just as it provides an alternative method for estimation of confidence limits (Subsection 8.2.3). The method applies to comparisons of data with a known baseline, to comparisons of two populations, and to the analysis of contingency tables. The **G test** relates the method of support to the χ^2 distribution, and is an example of the important class of **likelihood ratio tests** that provide a way to compare models with different levels of complexity.

8.8.1 The Method of Support: Comparison of Data with a Known Baseline

As with other statistical techniques, using the method of support tests whether data are consistent with a null hypothesis. With this method, we do so by comparing the support for the null hypothesis with the support for the alternative. If the support for the alternative hypothesis exceeds that for the null by more than 2, we place the null under doubt (devotees of likelihood do not believe in rejecting hypotheses).

Example 8.8.1 Using the Method of Support with the Binomial Distribution

Consider the data set consisting of 20 out of 100 diseased individuals with an allele in a population where 13% of people are known to carry the allele. Recall that the likelihood gives the probability of the data as a function of the unknown parameter q describing the proportion carrying the allele. In this case, the likelihood function (Example 8.1.5) and support function (Example 8.2.13) are

$$L(q) = b(20;\ 100, q)$$

$$S(q) = \ln[b(20;\ 100, q)].$$

The null hypothesis is that diseased individuals do not differ from the general population, and share the known value $q = 0.13$. The support for this null hypothesis is

$$S(0.13) = \ln[b(20;\ 100, 0.13)] = -4.215.$$

The support of the alternative hypothesis is the *maximum* over all possible values in the alternative. Both alternatives $q > 0.13$ and $q \neq 0.13$ include the maximum likelihood estimator $q = 0.2$, for which

$$S(0.2) = -2.310$$

(Figure 8.8.1). The difference is

$$S(0.2) - S(0.13) = -2.310 - (-4.215) = 1.905,$$

which is slightly less than 2. The null hypothesis might well explain the data. This is consistent with the finding that the result is not quite significant using a two-tailed test (Example 8.4.8).

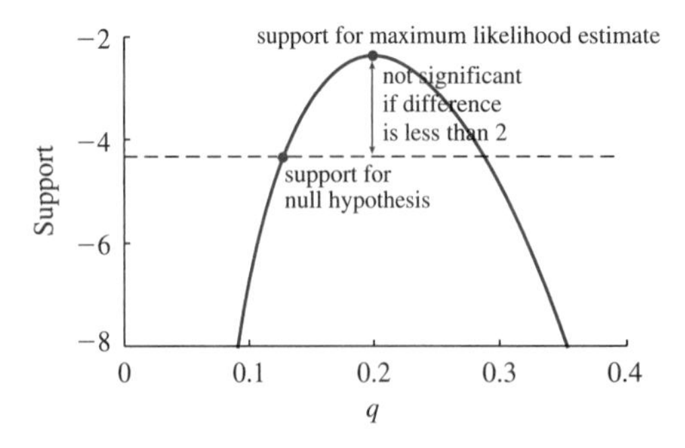

FIGURE 8.8.1
Support function for the genetics data

Example 8.8.2 Testing a Hypothesis About Waiting Times

Molecule	Waiting Time
1	$t_1 = 0.311$
2	$t_2 = 0.791$
3	$t_3 = 1.196$
4	$t_4 = 0.539$
5	$t_5 = 0.575$
6	$t_6 = 0.908$
7	$t_7 = 0.088$
8	$t_8 = 1.764$
9	$t_9 = 0.619$
10	$t_{10} = 0.038$

Consider again the data in Subsection 8.1.4, giving the waiting times for 10 molecules. Suppose the null hypothesis is that these data came from an exponential distribution with $\lambda = 2.0$.

We found the likelihood function

$$L(\lambda) = \lambda^{10} e^{-\lambda \sum_{i=1}^{10} t_i}$$
$$= \lambda^{10} e^{-6.83\lambda}$$

for the unknown rate λ as the product of the values of the p.d.f. $f(t) = \lambda e^{-\lambda t}$ at each of the data points. The support function is the natural logarithm, or

$$S(\lambda) = \ln(\lambda^{10} e^{-6.83\lambda}) \quad \text{support is the log of the likelihood}$$
$$= \ln(\lambda^{10}) + \ln(e^{-6.83\lambda}) \quad \text{log of the product is the sum of the logs (law 1 of logs)}$$
$$= 10\ln(\lambda) - 6.83\lambda. \quad \text{powers come out front (law 2 of logs)}$$

We found the maximum likelihood estimator $\hat{\lambda} = 1.464$ by maximizing this function (Figure 8.8.2).

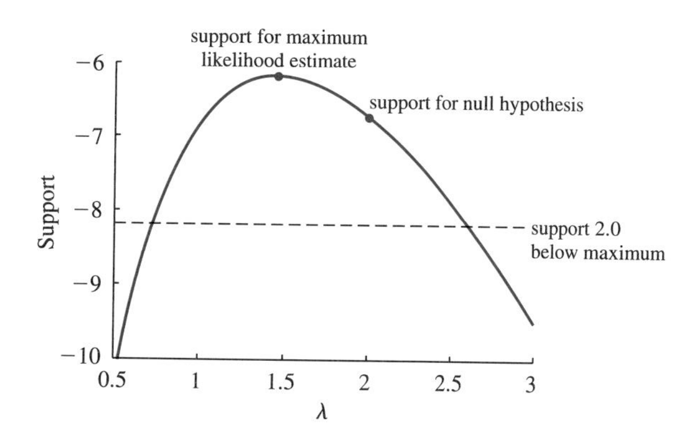

FIGURE 8.8.2
Support function for the waiting time data

To compare the hypotheses, we compute the difference between the maximum support and the support of the null hypothesis as

$$S(1.464) - S(2.0) = -6.187 - (-6.728) = 0.541.$$

Because this value is much less than 2.0, there is no reason to doubt that the data were generated by an exponential process with rate $\lambda = 2.0$.

8.8.2 Comparing Two Populations

Suppose we test two populations of ferocious bears for a parasite. At great risk, we capture ten bears from each population and find two infested bears in the first population and eight in the second. Do we have reason to think that the second population is in worse shape than the first? Our samples are a bit small for the normal approximation or the χ^2 test. The method of support provides a convenient way to compare hypotheses. The method follows the same steps as a one-sample test.

1. Decide on two hypotheses to compare.
2. Write down the support function for each.
3. Find the parameter values for each that are best supported by the data.
4. Compare the results. If the support for the alternative hypothesis exceeds that of the null hypothesis by more than 2, we have evidence that the alternative hypothesis is better supported by the data.

Example 8.8.3 Comparing Two Populations with the Method of Support

The null hypothesis is that the two populations have the same proportion. If so, there is no difference between them, and we treat our data in one unit, as **pooled** (Figure 8.8.3). The maximum likelihood estimator of the pooled proportion p is

$$\hat{p} = \frac{10}{20} = 0.5$$

because 10 out of the 20 bears sampled are infested. Using the fact that the support for the whole data set is the sum of the supports for each component (Theorem 8.4), the support for p with the data as we observed it is

$$\begin{aligned} S_n(p) &= \text{support for } p \text{ with 2 out of 10} + \text{support for } p \text{ with 8 out of 10} \\ &= \ln\left[\binom{10}{2} p^2(1-p)^8\right] + \ln\left[\binom{10}{8} p^8(1-p)^2\right] \\ &= \ln\binom{10}{2} + \ln\binom{10}{8} + 10\ln(p) + 10\ln(1-p). \end{aligned}$$

The maximum occurs at $p = \hat{p} = 0.5$.

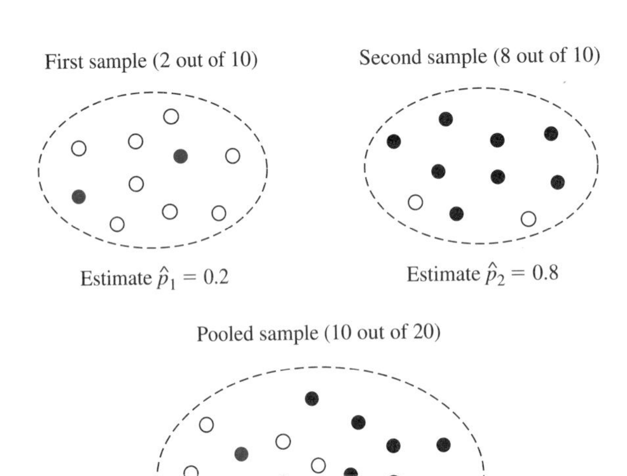

FIGURE 8.8.3
Unpooled and pooled comparison of bear data

The alternative is that the two populations are different. Let the parameters p_1 describe the first population and p_2 describe the second population. The maximum likelihood estimator of p_1 is the proportion in the first sample, or

$$\hat{p}_1 = \frac{2}{10} = 0.2.$$

The maximum likelihood estimator of p_2 is the proportion in the second sample, or

$$\hat{p}_2 = \frac{8}{10} = 0.8.$$

The support for the hypothesis that each sample behaves separately is again the sum of the supports for each, or

$$\begin{aligned}S_a(p_1, p_2) &= \text{support for } p_1 \text{ with 2 out of 10} + \text{support for } p_2 \text{ with 8 out of 10}\\ &= \ln\left[\binom{10}{2} p_1^2(1-p_1)^8\right] + \ln\left[\binom{10}{8} p_2^8(1-p_2)^2\right]\\ &= \ln\binom{10}{2} + \ln\binom{10}{8} + 2\ln(p_1) + 8\ln(1-p_1) + 8\ln(p_2) + 2\ln(1-p_2).\end{aligned}$$

The maximum occurs at $p_1 = \hat{p}_1 = 0.2$ and $p_2 = \hat{p}_2 = 0.8$.

We compare the support of the two hypotheses by computing the difference. The constants involving binomial coefficients cancel, so

$$\begin{aligned}S_a(0.2, 0.8) - S_n(0.5) &= 2\ln(0.2) + 8\ln(1-0.2) + 8\ln(0.8) + 2\ln(1-0.8)\\ &\quad - 10\ln(0.5) + 10\ln(1-0.5)\\ &= 3.85.\end{aligned}$$

We use the rule of thumb that a difference of more than 2 indicates that the alternative hypothesis is better supported by the data. There is good reason to believe that the two samples come from populations that are different.

Example 8.8.4 Testing for a Difference in Mutation Rates

Suppose one organism has 8 mutations and another has 18 in the same set of 1 million base pairs. Is there evidence that the mutation rates differ? We could use the normal approximation to find a p-value (Section 8.6, Exercise 39). Alternatively, we could attack this problem with the method of support.

The maximum likelihood estimator of the parameter Λ is $\hat{\Lambda}_1 = 8$ in the first organism and $\hat{\Lambda}_2 = 18$ in the second organism. The pooled sample consists of 26 mutations in 2 million base pairs, with maximum likelihood estimator of $\hat{\Lambda} = 13$ per million.

The support for the null hypothesis $\hat{\Lambda} = 13$ is

$$\begin{aligned}S_n(\hat{\Lambda}) &= S_1(\hat{\Lambda}) + S_2(\hat{\Lambda}) && \text{support is sum of supports}\\ &= \ln[p(8;13)] + \ln[p(18;13)] && \text{Poisson distribution}\\ &= \ln\left(\frac{e^{-13}13^8}{8!}\right) + \ln\left(\frac{e^{-13}13^{18}}{18!}\right) && \text{formula for Poisson distribution}\\ &= -13 + 8\ln(13) - \ln(8!) - 13 && \text{laws 1 and 2 of logs}\\ &\quad + 18\ln(13) - \ln(18!).\end{aligned}$$

The support for the alternative hypothesis $\Lambda_1 = 8$ and $\Lambda_2 = 18$ is

$$\begin{aligned}S_a(\Lambda_1, \Lambda_2) &= S_1(\Lambda_1) + S_2(\Lambda_2) && \text{support is sum of supports}\\ &= \ln[p(8;8)] + \ln[p(18;18)] && \text{Poisson distribution}\\ &= \ln\left(\frac{e^{-8}8^8}{8!}\right) + \ln\left(\frac{e^{-18}18^{18}}{18!}\right) && \text{formula for Poisson distribution}\\ &= -8 + 8\ln(8) - \ln(8!) - 18 && \text{laws 1 and 2 of logs}\\ &\quad + 18\ln(18) - \ln(18!).\end{aligned}$$

We do not need to compute these values yet because the constants cancel when we do the subtraction. The difference is

$$S_a(8, 18) - S_n(13) = 8\ln(8) + 18\ln(18) - 26\ln(13) = 1.97.$$

The result is near the threshold. It might be worth doing more experiments to demonstrate whether these two organisms are really different.

8.8.3 Support and the Normal Distribution

The method in the previous subsections applies in only limited sets of circumstances. By examining the relationship between the method of support and the normal distribution, we get an indication of the more general method, the likelihood ratio test.

To do so, we use a fundamental property of the support function. If we draw independent random variables with values $X_1, \ldots, X_n$ from a p.d.f. $f(x)$, the likelihood L of this result can be found by multiplying the likelihood of each individual measurement (Theorem 6.3), so

$$L = f(X_1) \cdots f(X_n).$$

Letting $S = \ln(L)$, and using the fact that the logarithm converts multiplication into addition, we get

$$S = \sum_{i=1}^{n} \ln[f(X_i)].$$

Therefore,

Theorem 8.4 The support of a model from a set of independent observations is the sum of the supports from each observation.

Suppose we have measured values $X_1, \ldots, X_n$ from a normal distribution with unknown mean μ and known variance σ^2. The support from observation X_i alone is

$$\begin{aligned} \ln[f(X_i)] &= \ln\left[\frac{1}{\sqrt{2\pi}\sigma} e^{-(X_i-\mu)^2/2\sigma^2}\right] \\ &= \frac{(X_i-\mu)^2}{2\sigma^2} - \ln\left(\sqrt{2\pi}\sigma\right). \end{aligned}$$

The support using the entire data set is then

$$S(\mu) = \sum_{i=1}^{n}\left[\frac{-(X_i-\mu)^2}{2\sigma^2} - \ln\left(\sqrt{2\pi}\sigma\right)\right].$$

After some algebra (Exercises 19 and 20), we find

$$S(\mu) = -\frac{n(\overline{X}-\mu)^2}{2\sigma^2} + c$$

where $\overline{X}$ is the sample mean and c is a constant that depends on σ. To find the maximum, we compute the derivative

$$S'(\mu) = \frac{n(\overline{X}-\mu)}{\sigma^2}.$$

Because $S''(\mu) = -\frac{n}{\sigma^2} < 0$, the critical point at $\mu = \overline{X}$ is a maximum.

If we have a null hypothesis that $\mu = \mu_0$, the difference in support for the null hypothesis and the maximum likelihood alternative is

$$S(\overline{X}) - S(\mu_0).$$

But $S(\overline{X}) = c$, so the difference exceeds 2.0 if

$$\begin{aligned} \frac{n(\overline{X}-\mu_0)^2}{2\sigma^2} &\geq 2.0 \\ (\overline{X}-\mu_0)^2 &\geq \frac{4.0\sigma^2}{n} \\ |\overline{X}-\mu_0| &\geq \frac{2.0\sigma}{\sqrt{n}}. \end{aligned}$$

The alternative hypothesis is better supported if the sample mean differs from the mean under the null hypothesis by more than 2.0 standard errors of the mean. This nearly matches the test for a significance level of 0.05 using p-values with a large sample, where we look for a difference between the sample mean and null hypothesis of 1.96 standard errors.

Example 8.8.5 Using Support to Compare a Single Sample with a Known Baseline

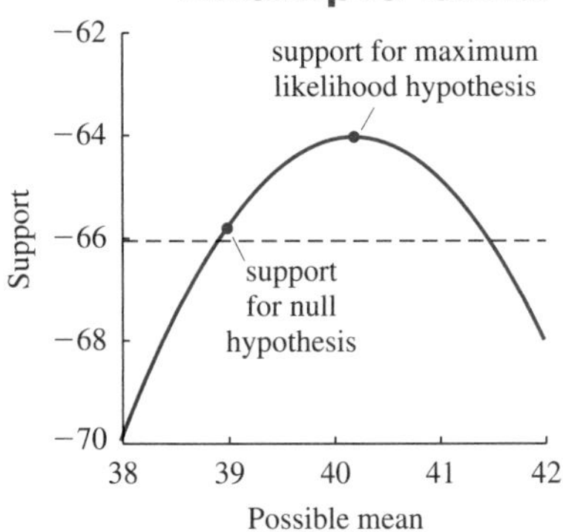

FIGURE 8.8.4
The support as a function of the mean

Suppose we are testing whether the heights of a set of plants differ from a known baseline of 39.0 cm. As in Example 8.5.4, we have a sample of size $n = 25$ with a known variance of $\sigma^2 = 10.24$. The support function is

$$S(\mu) = -\frac{25(40.2 - \mu)^2}{2 \cdot 10.24} + c.$$

We find that

$$S(40.2) - S(39.0) = 1.758$$

(Figure 8.8.4). This value is less than 2.0, matching the non-significant result of the two-tailed test in Example 8.5.4.

Example 8.8.6 Using Support to Compare Two Samples

Consider again the situation in Example 8.6.1, where we treat 25 plants with a new fertilizer and keep 50 untreated plants as controls. The variance with or without fertilizer has the known value of 10.24 cm^2. If we measure sample means of $\hat{\mu}_1 = 39.0$ in the control and $\hat{\mu}_2 = 40.2$ in the treated population, do we have reason to suspect that the underlying populations are really different?

First, we create a pooled population of 75 plants. The mean is a weighted average of the two samples, weighted by their sizes, or

$$\begin{aligned}\hat{\mu}_p &= \frac{50}{75} \cdot \hat{\mu}_1 + \frac{25}{75} \cdot \hat{\mu}_2 \\ &= \frac{50}{75} \cdot 39.0 + \frac{25}{75} \cdot 40.2 = 39.4.\end{aligned}$$

The support function for the whole population is the sum of the supports for the control and the treatment. For the control, the support is

$$S_1(\hat{\mu}_1) = -\frac{50(39.0 - \hat{\mu}_1)^2}{2 \cdot 10.24} + c.$$

For the treatment, the support is

$$S_2(\hat{\mu}_2) = -\frac{25(40.2 - \hat{\mu}_2)^2}{2 \cdot 10.24} + c$$

as in Example 8.8.5. The support for the null hypothesis is

$$S_1(39.4) + S_2(39.4)$$

whereas the best-supported alternative gives

$$S_1(39.0) + S_2(40.2).$$

The difference in these values is

$$[S_1(39.0) + S_2(40.2)] - [S_1(39.4) + S_2(39.4)] = 1.17.$$

This does not exceed 2, and we have no compelling reason to prefer the alternative hypothesis, a result that matches the p-value of 0.126 found in Example 8.6.1.

8.8.4 The G Test: Support and Contingency Tables

The method of support outlined above is a simple special case of a much broader test, the **likelihood ratio test**, which in the special case of contingency table analysis is often called the **G test**. We can think of the G test as an alternative to the χ^2 test, to which it has a beautiful relationship. We begin by again working out the difference in support for the null model and alternative model with the genetics data in Example 8.8.1, finding a general and useful formula. We then state how this value can be translated into a p-value through comparison to the χ^2 distribution with the appropriate number of degrees of freedom.

Example 8.8.7 Recalculating the Difference in Support with the Genetics Data

Consider again the data set consisting of 20 out of 100 diseased individuals with an allele in a population where 13% of people are known to carry the allele. The support, as a function of the parameter q, can be simplified using laws of logs as follows.

$$\begin{aligned} S(q) &= \ln[b(20; 100, q)] \\ &= \ln\left[\binom{80}{20} q^{20}(1-q)^{80}\right] \\ &= \ln\left[\binom{80}{20}\right] + 20\ln(q) + 80\ln(1-q). \end{aligned}$$

The difference in support between the alternative hypothesis, defined by using the maximum likelihood estimator $\hat{q} = 0.2$ and the null hypothesis with $q = 0.13$ is

$$\begin{aligned} S(0.2) - S(0.13) = \ln\left[\binom{80}{20}\right] &+ 20\ln(0.2) \\ &+ 80\ln(1-0.2) \\ &- \ln\left[\binom{80}{20}\right] - 20\ln(0.13) \\ &- 80\ln(1-0.13) && \text{substitute into support function} \\ &= 20\ln(0.2) - 20\ln(0.13) \\ &\quad + 80\ln(0.8) - 80\ln(0.87) && \text{cancel and combine terms} \\ &= 20\ln\left(\frac{0.2}{0.13}\right) + 80\ln\left(\frac{0.8}{0.87}\right) && \text{convert subtraction to division} \\ &= 20\ln\left(\frac{20}{13}\right) + 80\ln\left(\frac{80}{87}\right). && \text{multiply top and bottom by 100} \end{aligned}$$

The values that appear in this calculation are precisely those given in Example 8.7.1:

	Number with Allele	Number with Other Allele
Observed	20	80
Expected	13	87

Let E_1 and O_1 be the expected number and the observed number with the allele, and E_2 and O_2 be the expected number and the observed number without the allele (as in the calculation of the χ^2 statistic). The difference in supports can be written in general as

$$\text{support for alternative} - \text{support for null} = 0_1\ln\left(\frac{0_1}{E_1}\right) + O_2\ln\left(\frac{O_2}{E_2}\right).$$

Definition 8.13 Suppose we measure n individuals, and a model predicts E_i individuals in category i. Our observations are summarized as O_i individuals in category i. The G statistic is

$$G = 2\sum_{i=1}^{m} E_i \ln\left(\frac{O_i}{E_i}\right).$$

The factor of 2 is included because it gives the following relationship.

Theorem 8.5 For suitably large sample sizes, the G statistic follows approximately a χ^2 distribution with number of degrees of freedom equal to the difference in the number of free parameters required to describe the alternative and null hypotheses.

Example 8.8.8 Calculating and Testing G in the Genetics Example

The G statistic is

$$\begin{aligned} G &= 2\sum_{i=1}^{m} O_i \ln\left(\frac{O_i}{E_i}\right) \\ &= 2\left[20\ln\left(\frac{20}{13}\right) + 80\ln\left(\frac{80}{87}\right)\right] \\ &= 3.810. \end{aligned}$$

The alternative model has one free parameter (the value of q) and the null model has none. Therefore, the number of degrees of freedom is

$$\begin{aligned} \text{degrees of freedom} &= \text{free parameters in alternative} - \text{free parameters in null} \\ &= 1 - 0 = 1. \end{aligned}$$

To test the G statistic, we compare with the critical value in Table 8.4. This value is slightly less than the critical value 3.841 for $p = 0.05$ with 1 degree of freedom. Again, this result is consistent with our earlier findings of a result slightly below the threshold for significance (Examples 8.7.7 and 8.8.1). In this case, the G statistic is exactly double the difference in support found earlier.

Example 8.8.9 Comparing Two Populations of Bears with the G Test

In Example 8.8.3, we tested whether two populations of bears, one with two out of ten infested and the other with eight out of ten uninfested, differ. We set up a contingency table just as in Section 8.7.

	Number Infested	Number Uninfested
Population 1	2	8
Population 2	8	2

To compute the G statistic, we build the table of expected values under the null hypothesis using the pooled proportion $p = 0.5$ found in Example 8.8.3.

	Expected Number Infested	Expected Number Uninfested
Population 1	5	5
Population 2	5	5

Then

$$\begin{aligned} G &= 2\sum_{i=1}^{m} O_i \ln\left(\frac{O_i}{E_i}\right) \\ &= 2\left[2\ln\left(\frac{2}{5}\right) + 8\ln\left(\frac{8}{5}\right) + 8\ln\left(\frac{8}{5}\right) + 2\ln\left(\frac{2}{5}\right)\right] \\ &= 7.710. \end{aligned}$$

Again, this statistic is double the value we found with the method of support in Example 8.8.3.

The number of degrees of freedom matches that for the chi-squared test, although found in a different way. In the case of the chi-squared test, we compute the degrees of freedom for a table with two rows ($R = 2$) and two columns ($C = 2$) as $\nu = (R - 1)(C - 1) = 1$. For the G test, we instead find the degrees of freedom by comparing the number of parameters in the two models. The null model has just one, the pooled proportion, while the alternative model has two, the proportions for each population separately. The difference between these two numbers, $\nu = 2 - 1 = 1$, gives the number of degrees of freedom. Our value of 7.710 exceeds the cutoff for p = 0.01 with 1 degree of freedom, giving us good reason to suspect that these populations differ.

Example 8.8.10 Checking for Independence with the G Test

In Example 8.7.12 we checked whether the throws of two dart players seemed to be independent. Our data are

	$S_1 = 1$	$S_1 = 0$
$S_2 = 1$	6	29
$S_2 = 0$	27	38

Under the assumption that the two players were independent, we found an expected table of

	$S_1 = 1$	$S_1 = 0$
$S_2 = 1$	11.55	23.45
$S_2 = 0$	21.45	43.55

The G statistic is then

$$G = 2\left[6\ln\left(\frac{6}{11.55}\right) + 29\ln\left(\frac{29}{23.45}\right) + 27\ln\left(\frac{27}{21.45}\right) + 38\ln\left(\frac{38}{43.55}\right)\right]$$
$$= 6.527.$$

This is slightly larger than the value $\chi^2 = 6.12$ found in Example 8.7.12. With 1 degree of freedom, we again find convincing evidence that these two players do not behave independently.

Summary The **method of support** provides a flexible alternative to p-values in many circumstances, and can be used to test hypotheses regarding a single sample and those comparing samples. In both cases, we compare the support for the null hypothesis with the support for the maximum likelihood alternative. When comparing samples, the null hypothesis is created by **pooling** the samples. If the difference in support between the alternative hypothesis and the null hypothesis is greater than 2, there is reason to doubt the null hypothesis. The value of 2 can be thought of as an approximation of the critical value of 1.96 standard errors in tests using the normal distribution. More generally, the **G statistic**, which is twice the difference in the supports, follows a χ^2 distribution with number of degrees of freedom equal to the difference in the number of parameters in the two models. The **G test** applies to contingency tables and can be used in place of the χ^2 test.

8.8 Exercises

Mathematical Techniques

1–4 ▪ Using the following data, use the method of support to evaluate the null hypothesis that the true probability of heads is 0.5.

1. A coin is flipped 5 times and comes up heads every time (as in Section 8.4, Exercise 5).
2. A coin is flipped 7 times and comes up heads 6 out of 7 times (as in Section 8.4, Exercise 6).
3. A coin is flipped 10 times and comes up heads 9 times (as in Section 8.4, Exercise 7).
4. A coin is flipped 20 times and comes up heads 3 times (as in Section 8.4, Exercise 8).

5–6 ▪ Use the method of support to evaluate the following null hypotheses.

5. One cosmic ray hits a detector in 1 yr. The null hypothesis is that the rate at which rays hit is $\lambda = 5/\text{yr}$ (as in Section 8.4, Exercise 13.)
6. Three cosmic rays hit a larger detector in 1 yr. The null hypothesis is that the rate at which rays hit is $\lambda = 10/\text{yr}$ (as in Section 8.4, Exercise 14.)

7–10 ▪ Find the difference in support of the following hypotheses. Compare with the p-value in the earlier problem.

7. You wait 4000 h for an exponentially distributed event to occur. The null hypothesis is that the mean wait is 1000 h with alternative that the mean wait is greater than 1000 h (as in Section 8.4, Exercise 15).
8. You wait 40 h for an exponentially distributed event to occur. The null hypothesis is that the mean wait is 1000 h with alternative that the mean wait is less than 1000 h (as in Section 8.4, Exercise 16).
9. The first defective gasket is the 25th. The null hypothesis follows a geometric distribution with mean wait 10, and the alternative is that the mean wait is greater than 10 (as in Section 8.4, Exercise 17).
10. The first defective gasket is the 50th. The null hypothesis follows a geometric distribution with mean wait 1000, and the alternative is that the mean wait is less than 1000 (as in Section 8.4, Exercise 18).

11–14 ▪ Use the method of support to check the following hypotheses.

11. The hypothesis in Section 8.5, Exercise 1.
12. The hypothesis in Section 8.5, Exercise 2.
13. The hypothesis in Section 8.5, Exercise 3.
14. The hypothesis in Section 8.5, Exercise 4.

15–18 ▪ How many standard errors from the mean are the following? What are the corresponding p-values for a two-tailed test?

15. The support for the null hypothesis is less than the maximum by 2.
16. The support for the null hypothesis is less than the maximum by 3.
17. The support for the null hypothesis is less than the maximum by 3.5.
18. The support for the null hypothesis is less than the maximum by 4.

19–20 ▪ Follow the steps to show that the support has the simple quadratic form given in the text.

19. Show that

$$S(\mu) = -\frac{1}{2\sigma^2}\left[(n-1)\sigma^2 + n(\overline{X}-\mu)^2\right] - n\left(\sqrt{2\pi}\sigma\right)$$

(expand the quadratic and plug in definitions of $\overline{X}$ and σ^2).

20. Remove the terms that do not depend on μ and show that the maximum occurs at $\mu = \overline{X}$.

Applications

21–22 ▪ Consider the data in Section 8.4, Exercises 23 and 24. Find the difference in support of the null and alternative hypotheses.

21. Day 1, when 7 calls arrive in 1 h while only 3.5 were expected (Section 8.4, Exercise 23).
22. Day 2, when 8 calls arrive in 1 h while only 3.5 were expected (Section 8.4, Exercise 24).

23–24 ▪ Consider again the data on 30 waiting times for 2 types of events used in Section 8.5, Exercises 21 and 22.

Sample	Type a	Type b
1	0.41	0.69
2	0.08	4.83
3	0.52	0.01
4	0.22	1.49
5	0.03	2.91
6	0.11	0.09
7	1.36	1.07
8	0.51	0.17
9	0.29	0.19
10	0.16	0.7
11	0.78	0.04
12	0.41	1.86
13	1.19	1.23
14	0.89	0.50
15	1.87	4.16
16	6.33	0.02

(*Continued*)

Sample	Type a	Type b
17	1.23	0.02
18	0.18	0.09
19	0.97	0.34
20	0.48	1.63
21	0.27	1.22
22	1.20	0.60
23	0.38	0.83
24	1.35	1.58
25	1.53	2.13
26	1.72	0.01
27	0.19	0.17
28	0.94	0.75
29	1.60	1.79
30	0.87	0.41

23. Use maximum likelihood to estimate the rate λ from the waiting times for type a. Compare the support for the null hypothesis that $\lambda = 1.0$ with the support for the maximum likelihood estimate.

24. Use maximum likelihood to estimate the rate λ from the waiting times for type b. Compare the support for the null hypothesis that $\lambda = 1.0$ with the support for the maximum likelihood estimate.

25–26 ▪ In Exercises 23 and 24, the mean and standard deviation are strongly affected by extreme values. Exclude the outlier or outliers and recompute the maximum likelihood estimator of λ. Compare the support for the null hypothesis that $\lambda = 1.0$ with the support for the maximum likelihood estimate. Does the estimator change a great deal? Why does the support become so much larger?

25. For type a, exclude the extreme value 6.33 at time 16.

26. For type b, exclude the extreme values 4.16 and 4.83.

27–30 ▪ Use the method of support to test whether the following samples differ.

27. One player makes 5 out of 10 shots, another makes 9 out of 10.

28. One player makes 5 out of 10 shots, another makes 16 out of 20.

29. A 1 m^2 region in Utah is hit by 4 cosmic rays in 1 yr, and a 1 m^2 region at the North Pole is hit by 10 cosmic rays in 1 yr.

30. Two 1 m^2 regions in Utah are hit by 3 and 5 cosmic rays in 1 yr, and a 1 m^2 region at the North Pole is hit by 10 cosmic rays in 1 yr.

31–32 ▪ Use the G test to test the following by building a table complete with observed and expected values. Compare the G statistic with the difference in support found in the earlier problem.

31. As in Exercise 27, one player makes 5 out of 10 shots, another makes 9 out of 10.

32. As in Exercise 28, one player makes 5 out of 10 shots, another makes 16 out of 20.

33–34 ▪ One organism has 8 mutations in 1 million base pairs, a second has 18 in 1 million, and a third has 28 in 1 million. Use the method of support to test the following differences.

33. Check whether organisms 1 and 2 differ and compare with Section 8.6, Exercise 39.

34. Check whether organisms 2 and 3 differ and compare with Section 8.6, Exercise 40.

35–38 ▪ It has been proposed that a particular salubrious bath extends cell lifespan. Suppose that cell mortality follows an exponential model. Use the method of support to evaluate the following cases.

35. A cell in the salubrious bath survives 30 min, and a cell in standard culture survives only 5 min. Is there reason to think that the salubrious bath lengthens cell life?

36. In a repeated experiment, the cell in the salubrious bath survives 60 min, and a cell in standard culture survives only 3 min. Is there reason to think that the salubrious bath lengthens cell life?

37. Combine the data from Exercises 35 and 36, and evaluate the difference in support between the null and alternative hypotheses.

38. What would happen to the result in Exercise 37 if a third experiment were done and both cells survived 10 min? Why the change?

8.9 Regression

A paired t test (Subsection 8.6.4) can be used to check whether two measurements of a single individual differ. This test, however, does not quantify the **relationship** between the two measurements. We now introduce the method of **regression** to test specific relations between random variables. We examine **linear regression**, designed to find the slope and intercept of the line that lies closest to the data, even if it does not go directly through all points. The method can be derived by minimizing the squared deviation between the linear model and the data, called the **method of least squares**, which is equivalent to finding the maximum likelihood estimators of the slope and intercept.

8.9.1 Comparing a Linear Model with Data

Back in Example 1.5.2, we examined the following data on a set of growing trees.

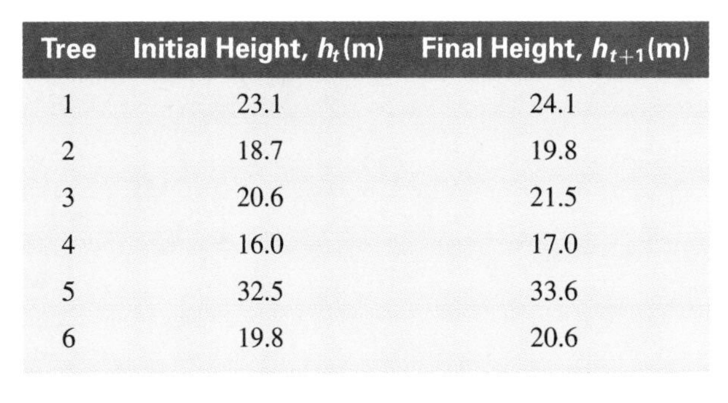

Tree	Initial Height, h_t (m)	Final Height, h_{t+1} (m)
1	23.1	24.1
2	18.7	19.8
3	20.6	21.5
4	16.0	17.0
5	32.5	33.6
6	19.8	20.6

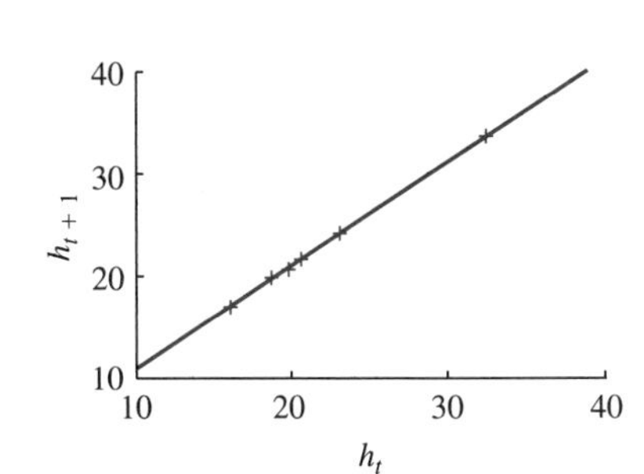

FIGURE 8.9.1
A growing tree

Each tree grows by about 1.0 mm, but the data do not exactly follow the linear relationship

$$h_{t+1} = h_t + 1.0$$

(Figure 8.9.1). In this subsection, we quantify how this idealized model deviates from the data. In the next subsection, we develop a method to find the best linear model.

Linear regression, like linear relationships, studies how two measurements are related to each other. For example, suppose we measure n values of x and y. With the trees, $n = 6$, x corresponds to the initial height, and y corresponds to the final height. If y and x exactly obeyed a linear relationship, we would find

Index	x	y
1	x_1	y_1
2	x_2	y_2
$\vdots$	$\vdots$	$\vdots$
n	x_n	y_n

$$y_i = ax_i + b$$

for each value of i. In reality, the value $ax_i + b$ predicted by the model will differ from the measured value y_i. We thus define

$$\hat{y}_i = ax_i + b.$$

The goal of linear regression is to find the values of a and b that make $\hat{y}_i$ as close to y_i as possible.

Example 8.9.1 Predicted and Observed Values for Growing Trees

For the trees, we can include a column of predicted values for the final height by adding 1.0 to the initial height.

Tree	Initial Height h_t	Observed Final Height h_{t+1}	Predicted Final Height $\hat{h}_{t+1}$
1	23.1	24.1	24.1
2	18.7	19.8	19.7
3	20.6	21.5	21.6
4	16.0	17.0	17.0
5	32.5	33.6	33.5
6	19.8	20.6	20.8

The predicted and observed heights are similar, but not identical.

Example 8.9.2 Predicted and Observed Values of Mass and Toxin Tolerance

Suppose we measure both mass (in μg) and toxin tolerance (the concentration causing cell death) for ten cells, finding

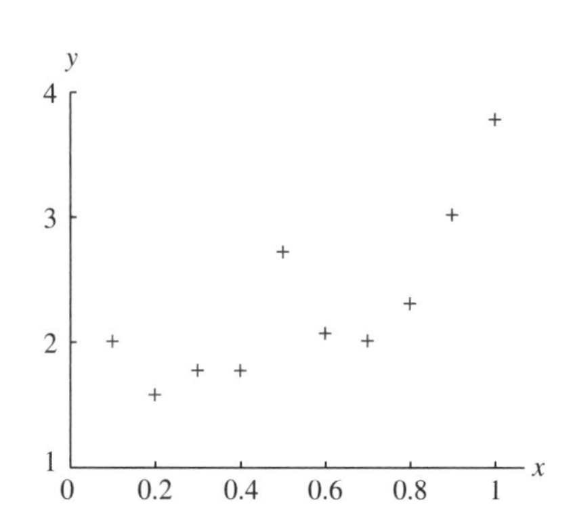

FIGURE 8.9.2
Data that might indicate a relationship between two measurements

Mass	Toxin Tolerance
$x_1 = 0.1$	$y_1 = 2.015$
$x_2 = 0.2$	$y_2 = 1.589$
$x_3 = 0.3$	$y_3 = 1.784$
$x_4 = 0.4$	$y_4 = 1.782$
$x_5 = 0.5$	$y_5 = 2.732$
$x_6 = 0.6$	$y_6 = 2.079$
$x_7 = 0.7$	$y_7 = 2.019$
$x_8 = 0.8$	$y_8 = 2.317$
$x_9 = 0.9$	$y_9 = 3.028$
$x_{10} = 1.0$	$y_{10} = 3.786$

(Figure 8.9.2). It looks as though larger cells are more able to tolerate toxin. Our goal is to quantify this. Suppose an existing model claims that

$$\hat{y}_i = 1.3x_i + 1.7.$$

For example, with $x_1 = 0.1$,

$$\hat{y}_1 = 1.3 \cdot 0.1 + 1.7 = 1.83.$$

We can then add a column of predicted values to the table.

Mass	y_i	$\hat{y}_i$
0.1	2.015	1.83
0.2	1.589	1.96
0.3	1.784	2.09
0.4	1.782	2.22
0.5	2.732	2.35
0.6	2.079	2.48
0.7	2.019	2.61
0.8	2.317	2.74
0.9	3.028	2.87
1.0	3.786	3.00

FIGURE 8.9.3
Testing a proposed linear relationship between two random variables

This line passes pretty close to the data points (Figure 8.9.3).

If the predication were exactly true, random variables X and Y describing the two measurements would exactly obey the rule

$$Y = aX + b.$$

All data, however, have some noise. Mathematically,

$$y_i = ax_i + b + \epsilon_i$$

where ϵ_i is the correction needed for the model to match the data exactly. If the random variable ϵ takes on small values, X is a good predictor of Y. If ϵ takes on large values, the error obscures the signal and X is a poor predictor of Y. The value of ϵ_i is called the **residual**, which can be thought of as the portion of the data that is not explained by the model. It is computed as

$$\epsilon_i = y_i - \hat{y}_i.$$

Example 8.9.3 Finding Residuals for the Tree Data

By subtracting the predicted from the observed final heights of trees, we can add a new column of residuals to our table.

Tree	h_t	h_{t+1}	$\hat{h}_{t+1}$	Residual
1	23.1	24.1	24.1	0.0
2	18.7	19.8	19.7	0.1
3	20.6	21.5	21.6	−0.1
4	16.0	17.0	17.0	0.0
5	32.5	33.6	33.5	0.1
6	19.8	20.6	20.8	−0.2

The residuals can be summarized by the sum of their squares. Just as the variance measures the average distance between data points and their mean, this sum of squares measures the average distance between data and a simple model. This value is called the **sum of squares of errors (SSE)**.

Example 8.9.4 Finding the Sum of the Squared Residuals for the Tree Data

By summing the squares of the residuals in Example 8.9.3, we find

$$\text{SSE} = 0.0^2 + 0.1^2 + (-0.1)^2 + 0.0^2 + 0.1^2 + (-0.2)^2 = 0.07.$$

However, like the variance, SSE can be difficult to interpret. The fundamental idea of statistics is to compare a measure indicating an effect with a null hypothesis. Because the hypothesis is that two measurements have some relationship with each other, the null hypothesis is that they have *no relationship*. Graphically, no relationship is indicated by a horizontal prediction, with all the variation in the vertical variable y_i due to experimental error. The horizontal line that passes closest to the center of the data is the sample mean of the y values, $\overline{Y}$ (Exercise 19). The sum of the squares of the residuals between the horizontal line and the data is called the **sum of squares total (SST)**.

Example 8.9.5 Finding the SST for the Tree Data

The best null model describing the tree data follows a horizontal line at the mean of the h_{t+1} values. In this case,

$$\bar{h}_{t+1} = \frac{24.1 + 19.8 + 21.5 + 17.0 + 33.6 + 20.6}{6} = 22.77.$$

The residuals for the model $h_{t+1} = \bar{h}_{t+1}$ are

Tree	h_t	h_{t+1}	$\overline{h}_{t+1}$	ϵ
1	23.1	24.1	22.77	1.33
2	18.7	19.8	22.77	−2.97
3	20.6	21.5	22.77	−1.27
4	16.0	17.0	22.77	−5.77
5	32.5	33.6	22.77	10.83
6	19.8	20.6	22.77	−2.17

By summing the squares of these residuals, we find

$$\text{SST} = 1.33^2 + (-2.97)^2 + (-1.27)^2 + (-5.77)^2 + 10.83^2 + (-2.17) = 167.49.$$

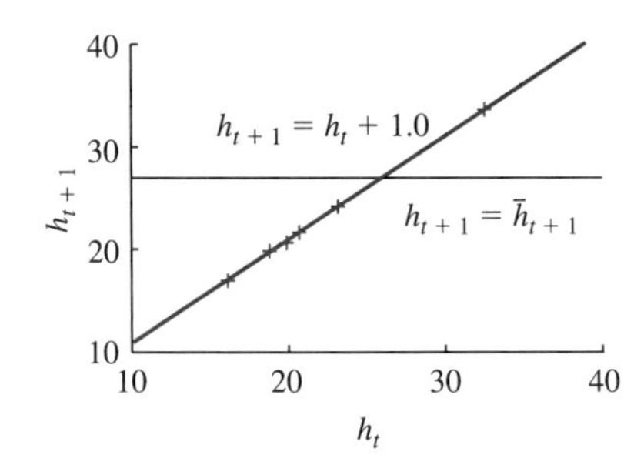

FIGURE 8.9.4
Two models of a growing tree

This value is far larger than SSE = 0.07 found in Example 8.9.4, because the line through the model $h_{t+1} = h_t + 1.0$ lies much closer to the data than the horizontal line at $h_{t+1} = \bar{h}_{t+1}$ (Figure 8.9.4).

We compare the values of SSE and SST by computing the **coefficient of determination**, r^2, as

$$r^2 = 1 - \frac{\text{SSE}}{\text{SST}}.$$

r^2 can take on values between 0 (the horizontal line is as good as the proposed model) and 1 (a perfect linear relation with SSE = 0). If the linear model describes the data well, SSE will be much smaller than SST and r^2 will be near 1. If not, SSE will be nearly as large as SST and r^2 will be near 0. We sometimes think of r^2 as the fraction of the variability in Y "explained" by X.

Example 8.9.6 Finding r^2 for the Tree Data

With the tree data, we found that SSE = 0.07 (Example 8.9.4) and SST = 167.49 (Example 8.9.5). Therefore,

$$r^2 = 1 - \frac{0.07}{167.49} = 0.9996.$$

This value is extremely close to 1.0, quantifying the fact that these data lie almost exactly on the predicted line.

Example 8.9.7 Finding r^2 for the Toxin Data

With the toxin data presented in Example 8.9.2, we find the residuals by subtracting the value predicted by the model $\hat{y}_i = 1.3x_i + 1.7$ from the observed value (Figure 8.9.5).

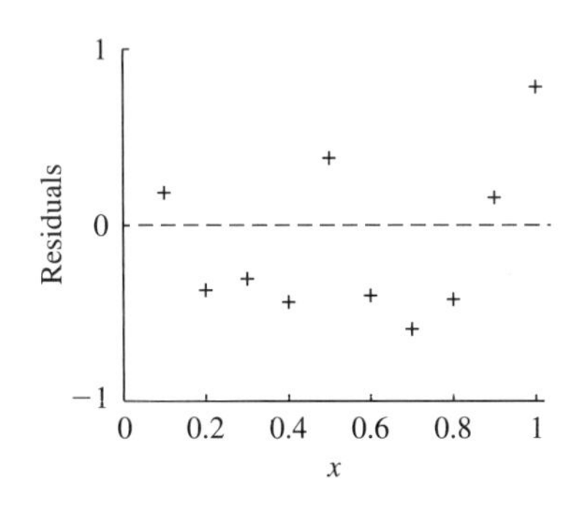

FIGURE 8.9.5
Testing a proposed linear relationship between two random variables

Mass	Toxin Tolerance	Prediction $\hat{y}_i = 1.3x_i + 1.7$	Residual
0.1	2.015	1.83	0.185
0.2	1.589	1.96	−0.371
0.3	1.784	2.09	−0.306
0.4	1.782	2.22	−0.438
0.5	2.732	2.35	0.382
0.6	2.079	2.48	−0.401
0.7	2.019	2.61	−0.591
0.8	2.317	2.74	−0.423
0.9	3.028	2.87	0.158
1.0	3.786	3.00	0.786

SSE is found by adding up the squares of the residuals.

$$\text{SSE} = \sum_{i=1}^{10} (y_i - \hat{y}_i)^2 = 1.936.$$

To find SST, we find residuals from the horizontal line at the mean $\overline{Y}$. In this case,

$$\overline{Y} = 2.313$$

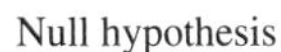

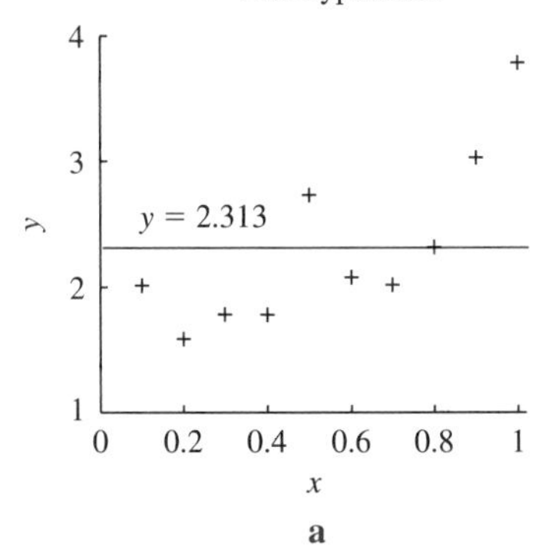

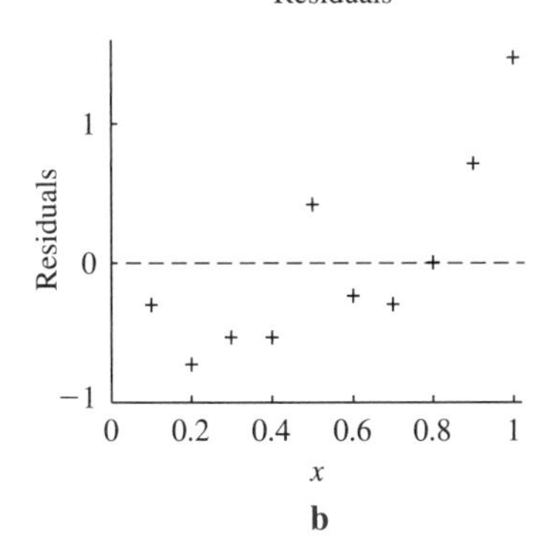

FIGURE 8.9.6
The null model and its residuals

(Figure 8.9.6). The residuals are then

Mass	Toxin Tolerance	Prediction $\hat{y}_i = \overline{y}$	Residual
0.1	2.015	2.313	−0.298
0.2	1.589	2.313	−0.724
0.3	1.784	2.313	−0.529
0.4	1.782	2.313	−0.531
0.5	2.732	2.313	0.419
0.6	2.079	2.313	−0.234
0.7	2.019	2.313	−0.294
0.8	2.317	2.313	0.004
0.9	3.028	2.313	0.715
1.0	3.786	2.313	1.473

The total sum of squares, SST, is

$$\text{SST} = \sum_{i=1}^{10} (y_i - 2.313)^2 = 4.172.$$

Then

$$\begin{aligned} r^2 &= 1 - \frac{\text{SSE}}{\text{SST}} \\ &= 1 - \frac{1.936}{4.172} = 0.535. \end{aligned}$$

About 53% of the variation in tolerance can be attributed to a linear relationship with mass. The remainder could be due to unmeasured factors (age or other cell characteristics) or experimental error.

8.9.2 Finding the Best Line

In most cases, we want to to find the best model rather than comparing a given model with data. The **method of least squares** minimizes SSE, the sum of the squared residuals. To find the best line, we must write SSE as a function of the slope and intercept, and find the minimum.

Suppose

$$\hat{y}_i = ax_i + b$$

for n data points. The sum of the squared residuals, the differences between the measured y_i from the predicted $\hat{y}_i$, is a function of the unknown slope a and the intercept b with formula

$$\text{SSE}(a, b) = \sum_{i=1}^{n} (y_i - \hat{y}_i)^2 = \sum_{i=1}^{n} [y_i - (ax_i + b)]^2.$$

Finding the minimum value of S requires use of **partial derivatives**, a method just beyond the level of this book. The resulting formulas can be written in terms of the sample mean, sample variance, and sample covariance. Recall that the sample variance of the x_i is given by the computational formula

$$s_X^2 = \frac{\sum_{i=1}^{n} x_i^2 - n\overline{X}^2}{n-1}$$

where $\overline{X}$ is the sample mean of the x values (Theorem 8.2). The **sample covariance** is the average of the products minus the product of the averages, again divided by $n - 1$

rather than n, with computational formula

$$\widehat{\text{Cov}}(X, Y) = \frac{\sum_{i=1}^{n} x_i y_i - n\overline{X}\overline{Y}}{n-1}.$$

Recall that the covariance measures the strength of the relation between two sets of measurements.

Theorem 8.6 Suppose two measurements X and Y have sample means $\overline{X}$ and $\overline{Y}$, sample covariance $\widehat{\text{Cov}}(X, Y)$, and sample variance s_X^2. The slope a and intercept b of the line that minimizes SSE are

$$\hat{a} = \frac{\widehat{\text{Cov}}(X, Y)}{s_X^2}$$

$$\hat{b} = \overline{Y} - \hat{a}\overline{X}.$$

We place hats over a and b to indicate that these are **estimators** of the true relation. The estimated slope of the regression is positive if the covariance is positive, negative if the covariance is negative, and 0 if the covariance is 0. Although it is similar to the correlation coefficient, the slope of the regression can take on any value.

Example 8.9.8 Finding the Best Fit Line

To find the best fitting line for the data on size and toxin tolerance presented in Example 8.9.2, we must compute the following four values.

$$\overline{X} = \frac{\sum_{i=1}^{10} x_i}{10} = 0.55$$

$$\overline{Y} = \frac{\sum_{i=1}^{10} y_i}{10} = 2.313$$

$$s_X^2 = \frac{\sum_{i=1}^{10} x_i^2 - 10\overline{X}^2}{9} = 0.0917$$

$$\widehat{\text{Cov}}(X, Y) = \frac{\sum_{i=1}^{10} x_i y_i - 10\overline{X}\,\overline{Y}}{9} = 0.1596.$$

From Theorem 8.6,

$$\hat{a} = \frac{0.1596}{0.0917} = 1.740$$

$$\hat{b} = 2.313 - 1.740 \cdot 0.55 = 1.356.$$

With this line,

Mass	Tolerance	Best Prediction	Residuals
0.1	2.015	1.530	0.485
0.2	1.589	1.704	−0.115
0.3	1.784	1.878	−0.094
0.4	1.782	2.052	−0.270
0.5	2.732	2.226	0.506
0.6	2.079	2.400	−0.321
0.7	2.019	2.574	−0.555
0.8	2.317	2.748	−0.431
0.9	3.028	2.922	0.106
1.0	3.786	3.096	0.690

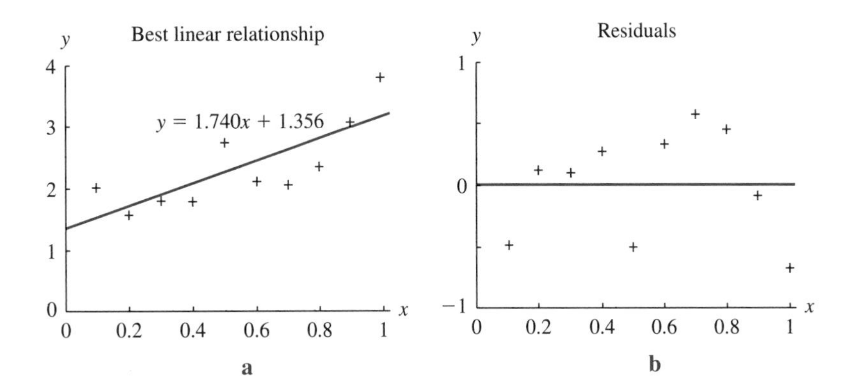

FIGURE 8.9.7
The best linear fit to a data set

By adding up the squares of the residuals, we find SSE = 1.670, smaller than the 1.936 found with our initial guess (Example 8.9.2). This line shoots right through the middle of the points, rather than being a shade high (Figure 8.9.7). Furthermore,

$$r^2 = 1 - \frac{\text{SSE}}{\text{SST}} = 1 - \frac{1.670}{4.172} = 0.600.$$

The coefficient of determination indicates that the effect of cell mass accounts for about 60% of the variation in the data. The remaining deviations from the line must be accounted for by other factors.

Example 8.9.9 Comparison of Regression with Correlation

The correlation between two measurements is the covariance standardized by dividing by the product of the standard deviations. Finding the correlation requires estimating s_Y, the sample standard deviation of Y, in addition to $\widehat{\text{Cov}}(X, Y)$ and s_X. We find that

$$s_Y^2 = \frac{\sum_{i=1}^{10} Y_i^2 - 10\overline{Y}^2}{9} = 0.464.$$

Therefore, the estimated correlation coefficient is

$$\hat{\rho}_{X,Y} = \frac{\widehat{\text{Cov}}(X, Y)}{s_X s_Y} = \frac{0.1596}{\sqrt{0.0917}\sqrt{0.464}} = 0.774.$$

The square of the correlation coefficient is

$$\hat{\rho}_{X,Y}^2 = 0.774^2 = 0.600.$$

The coefficient of determination r^2 is equal to the square of the estimated correlation.

Computing SSE and r^2 for the best fitting line is simplified by the following computational formula.

Theorem 8.7 If the slope $\hat{a}$ and intercept $\hat{b}$ are chosen to minimize SSE as in Theorem 8.6, then

$$\text{SSE} = \sum_{i=1}^{n} y_i^2 - \hat{b}\sum_{i=1}^{n} y_i - \hat{a}\sum_{i=1}^{n} x_i y_i.$$

This formula is simpler because it does not require computing the residuals.

Example 8.9.10 Using the Computational Formula to Find r^2

With the toxin data, we found that $\hat{a} = 1.740$ and $\hat{b} = 1.356$. Furthermore,

$$\sum_{i=1}^{10} y_i^2 = 57.68$$

$$\sum_{i=1}^{10} y_i = 10\overline{Y} = 23.13$$

$$\sum_{i=1}^{10} x_i y_i = 10\overline{Y} = 14.16.$$

Then

$$\text{SSE} = 57.68 - 1.356 \cdot 23.13 - 1.740 \cdot 14.16 = 1.677,$$

matching our earlier value. ▲

Minimizing SSE is justified statistically by the method of maximum likelihood in an important special case. Linear regression describes a model relating the two random variables X and Y, where the additional random variable ϵ describes the error according to

$$Y = aX + b + \epsilon.$$

Suppose that ϵ follows a normal distribution with mean 0 and known variance σ^2. We assume that the variance of the error is independent of X and that there is no error in the measurement of X. If these assumptions are not true, more complicated forms of regression must be used.

With this model, we can compute maximum likelihood estimators of a and b by finding the values most likely to have produced the measured values. For a given guess of the slope a and the intercept b, the residual is

$$\epsilon_i = y_i - (ax_i + b),$$

which must be explained by the error term ϵ. If a and b produce large values of ϵ_i, the model is unlikely to have produced the data, because ϵ has a mean 0.

The support for a particular pair of values a and b from data point i is

$$\ln\left(\frac{1}{\sqrt{2\pi}\sigma} e^{\frac{-[y_i - (ax_i + b)]^2}{2\sigma^2}}\right) = \frac{-[y_i - (ax_i + b)]^2}{2\sigma^2} - \ln(\sqrt{2\pi}\sigma)$$

The support for the model by the whole data set is equal to the sum of the supports from each independent data point (Theorem 8.4) and is therefore

$$S(a, b) = \sum_{i=1}^{n} \frac{-[y_i - (ax_i + b)]^2}{2\sigma^2} - n\ln(\sqrt{2\pi}\sigma).$$

The log likelihood is maximized precisely where the numerator

$$\sum_{i=1}^{n} [y_i - (ax_i + b)]^2$$

is minimized, conveniently independent of the value of σ. The slope and intercept found in Theorem 8.6 are thus the maximum likelihood estimates of these parameters under a specific set of assumptions.

When the distribution of residuals cannot be approximated with a normal distribution, this approach can be modified. **Generalized linear models** can be used when residuals come from other distributions. For example, **logistic regression** deals with the case where the dependent variable y takes on only the values 0 and 1, and is used to study problems in **survivorship analysis**.

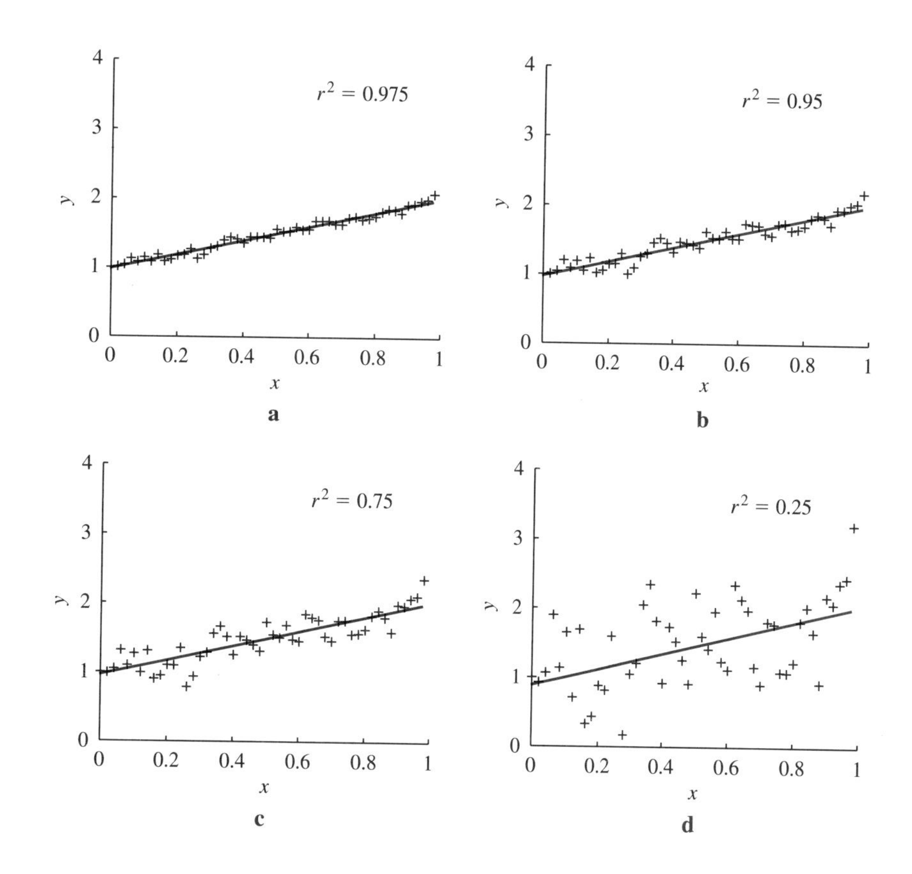

FIGURE 8.9.8
Regressions with different values of r^2

8.9.3 Using Linear Regression

We introduce three issues: interpretation of linear regression, the statistical testing of a regression line, and some situations where linear regression should not be used.

First, what do different values of r^2 mean in practice? When r^2 is near 1 (as in Figure 8.9.8a and b), the data fall nearly along the predicted line. As r^2 becomes smaller, the data look more and more like a "cloud" (Figure 8.9.8c and d).

However, r^2 alone cannot be used to tell whether the relationship between random variables X and Y is **significant**. Determining significance requires computing the slope of the best line through the data and checking whether that slope is significantly different from 0. The method of support can be used to compare the best fit line with the horizontal line through the mean of the y values, but is complicated when the variance of the error distribution ϵ is unknown. Alternatively, the difference between the estimated slope $\hat{a}$ and a slope of 0 obeys at t distribution, but computing the standard error is beyond the scope of this text. Consult a more advanced statistics text for these methods.

Two graphs should be examined in the process of making a regression: a plot of Y against X and a plot of the residuals. The data in Figure 8.9.9 fit the model well; the noise is distributed evenly around the prediction. In Figure 8.9.10, the noise is distributed evenly, but increases in magnitude for larger values of X. The r^2 statistic weights all errors equally, which can be inappropriate in cases like this that are described by the fancy term **heteroscedasticity**.

Figure 8.9.11 shows a curved pattern. The residuals are positive for large and small values of X and negative for medium values, rather than being evenly spread. A linear model is not appropriate. In such cases, we might want to use a different type of model, perhaps

$$Y = aX^2 + b + \epsilon.$$

Again, a statistics text should be consulted for more details about **nonlinear regression** techniques. However, the values of SSE, SST, and r^2 computed have the same meaning as in the linear case.

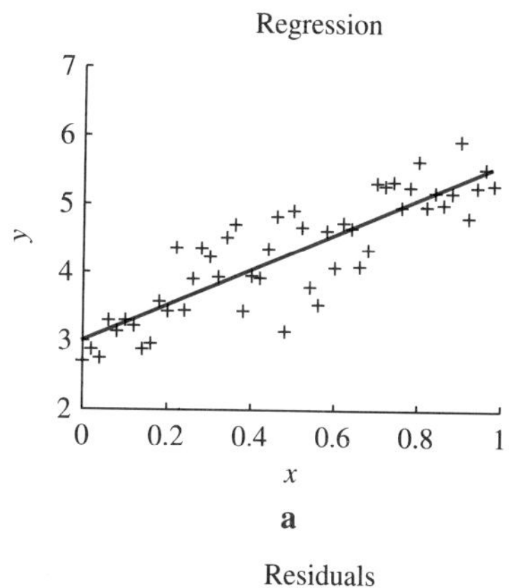

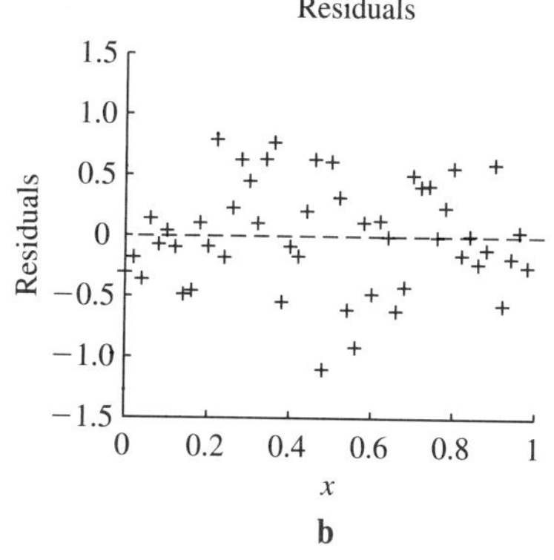

FIGURE 8.9.9
Data where the regression line is an appropriate model

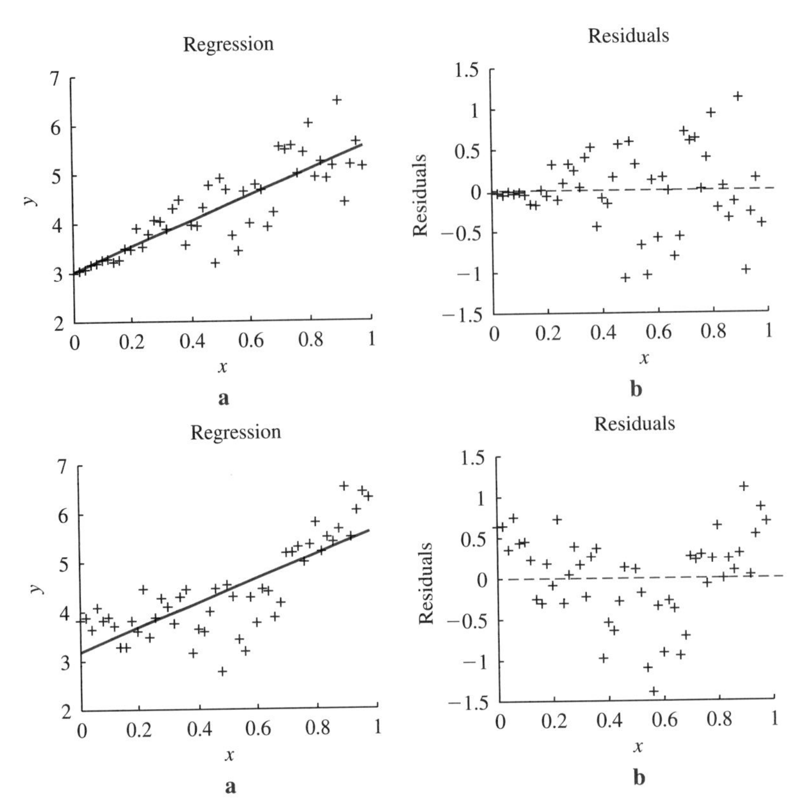

FIGURE 8.9.10

Data which become more spread out for large values of x

FIGURE 8.9.11

Data which seem to have a curved relation

Summary When we believe that two sets of measurements are related by some function, we can quantify the fit by computing the **residuals**, the differences between the data and prediction of a model, and the **sum of the squared residuals (SSE)**. By comparing this with the null model described by a horizontal line, we compute the **coefficient of determination**, r^2, which can be thought of as the fraction of variability in the data explained by the model. Examination of a graph of the residuals can reveal problems with the method, such as changes in the variance and nonlinearity of the data. By applying the **principle of least squares**, we found the line that minimizes SSE and passes most closely through a set of data points. This line is called the **linear regression**. This linear regression line can be found with the method of maximum likelihood when errors are normally distributed.

8.9 Exercises

Mathematical Techniques

1–14 ▪ Consider the data in the following table.

Weight	Yield	Height
0.20	0.599	10.17
0.40	0.909	12.83
0.60	1.220	13.47
0.80	1.363	14.60
1.0	1.523	13.55
1.2	1.995	15.25
1.4	2.518	14.47
1.6	2.330	17.74
1.8	2.963	20.46
2.0	3.628	21.12

(Continued)

1. Plot yield (Y) against weight (W). Suppose we think that the line $Y = W + 0.6$ describes these data. Plot the line on your graph of yield against weight. Find and plot the residuals.
2. Plot height (H) against weight (W). Suppose we think that the line $H = 8W + 5$ describes the data. Plot the line on your graph of height against weight. Find and plot the residuals.
3. Find SSE for the model used in Exercise 1

4. Find SSE for the model used in Exercise 2

5. Find the null model that best fits Y as a function of W and find SST.

6. Find the null model that best fits H as a function of W and find SST.

7. Find r^2 for the model in Exercise 1.

8. Find r^2 for the model in Exercise 2.

9. Compute the best fitting line for yield as a function of weight. Graph the line.

10. Compute the best fitting line for height as a function of weight. Graph the line.

11. Find SSE for the line in Exercise 9, find r^2, and compare with the model in Exercise 1.

12. Find SSE for the line in Exercise 10, find r^2, and compare with the model in Exercise 2.

13. Find the correlation between weight and yield. Check that its square is equal to the value of r^2 found in Exercise 11.

14. Find the correlation between weight and height. Check that its square is equal to the value of r^2 found in Exercise 12.

15–16 ▪ Linear regression has important connections with other techniques in statistics, such as testing whether two populations differ. In the following data set, the independent variable takes on only two values. Find the best fitting line and r^2, and then test whether the two sets of points differ in their mean distribution using the techniques in Section 8.6. Assume that the data are normally distributed with known variance of 25. In each case, graph the regression line and the data.

Diet	Size in Replicate 1	Size in Replicate 2
1	9.51	25.95
1	20.40	19.26
1	14.50	20.05
1	20.76	27.31
1	17.42	21.84
1	23.34	19.45
2	28.20	22.52
2	30.12	25.32
2	24.12	24.04
2	31.43	23.53
2	33.46	28.39
2	29.70	25.17

15. Find the best fitting line and r^2 for replicate 1 and then test whether the diet has a significant effect.

16. Find the best fitting line and r^2 for replicate 2 and then test whether the diet has a significant effect. Compare with the previous problem.

17–20 ▪ Best fit regression lines have many nice properties.

17. Show that the best fitting horizontal line (used to compute SST) passes through the center of the data in the sense that the sum of the residuals is 0.

18. Show that the best linear fit from Theorem 8.6 passes through the center of the data in the sense that the sum of the residuals is 0.

19. Consider models of the form $Y = b$. Show that the sum of the squares of the residuals is minimized when $b = \overline{Y}$, the sample mean of the y_i.

20. Consider models of the form $Y = aX$. Find the slope that minimizes the sum of the squares of the residuals.

21–22 ▪ Consider the following measurements.

x	y
1.0	1.1
2.0	3.9
3.0	8.8
4.0	16.5

21. Find the best linear fit. Plot the line and find r^2. How good is the model?

22. Use the principle of least squares to write the expression you would use to fit a curve of the form $Y = aX^2 + b$. One easy way to solve this is to think of a new measurement $Z = X^2$ and find the linear regression of Y on Z. Plot the linear regression of Y against Z and the curved regression of Y against X^2. Compute r^2. Which model does better?

23–24 ▪ Check that the dimensions for each term of the regression equation for toxin tolerance as a function of mass (Example 8.9.2) are consistent.

23. Find the dimensions of the slope $\hat{a}$.

24. Find the dimensions of the intercept $\hat{b}$ and check that all the parts of the equation match.

Applications

25–26 ▪ Consider the following data describing change in a bacterial population.

Colony	Old Population (b_t)	New Population (b_{t+1})
1	0.47	0.95
2	3.3	6.4
3	0.73	1.5
4	2.8	5.6
5	1.5	3.1
6	0.62	1.2

25. Find r^2 for the line $b_{t+1} = 2b_t$. Graph the data and the line.

26. Find the best fitting line, and compare with a mathematically idealized model. Which makes more sense?

27–30 ▪ Consider the following data on the growth of two bacterial populations.

Year	Population 1	Population 2
0	100	50
1	119	68
2	168	82
3	198	141
4	259	212
5	306	399
6	421	552

27. Find the best fitting line for population 1 as a function of time and compute r^2.

28. Find the best fitting line for population 2 as a function of time and compute r^2.

29. Find the best fitting line for the logarithm of population 1 as a function of time and compute r^2. Is this a better fit?

30. Find the best fitting line for the logarithm of population 2 as a function of time and compute r^2. Is this a better fit?

31–32 ▪ Consider the following data which include one outlying point. Find the best fitting line with and without that point. How much difference does that point make? The idea of removing one point and testing how much the fit changes is an important tool in regression, and is sometimes called the **leverage** of that point.

Feeding Rate	Size in Replicate 1	Size in Replicate 2
1	11.2	9.2
1	12.1	10.2
2	19.2	21.5
2	44.2	17.0
3	31.5	33.6
3	33.4	30.6
4	38.2	43.6
4	44.3	10.6

31. Find the best fitting line and r^2 for replicate 1 with and without the fourth point. Graph the two regression lines and the data.

32. Find the best fitting line and r^2 for replicate 2 with and without the last point. Graph the two regression lines and the data. Why do you think the outlier affects this regression line more?

Computer Exercise

33. The following table gives the winning Olympic times for men and women in the 400 m race.

Year	Men's Record	Women's Record
1896	54.2	
1900	49.4	
1904	49.2	
1906	53.2	
1908	50.0	
1912	48.2	
1920	50.0	
1924	47.6	
1928	47.8	
1932	46.2	
1936	46.5	
1948	46.2	
1952	45.9	
1956	46.7	
1960	44.9	
1964	45.1	52.0
1968	43.86	52.0
1972	44.66	51.08
1976	44.26	49.29
1980	44.60	48.88
1984	44.27	48.83
1988	43.87	48.65

a. Use your computer to find the best linear regression for men and for women.

b. Plot the residuals for each. Does the linear model fit well?

c. Predict the times in the 2000 Olympics for women and men. How well did it actually work?

d. Predict when women will outrun men. Do you believe this?

Supplementary Problems

1. Bacteria can be in one of two states: chemical producing (with probability 0.2) or chemical absorbing (with probability 0.8). Bacteria produce chemical at a rate of 2.0 femtomoles/second or absorb at a rate of 4.0 femtomoles/second. Suppose there are 100 bacteria in a culture, acting independently.

a. Find the mean and variance of the number in the producing and absorbing states.

b. Find the mean and variance of the rate of change of total chemical.

c. Write an exact expression for the probability that the amount of chemical is decreasing at any particular time. How would you evaluate it?

d. If instead of acting independently, all bacteria respond to the same external cue (but with the same probabilities), find the answers to parts **b** and **c**.

2. It is observed that a population grows by a factor R determined by the temperature T as follows:

$$R = \begin{cases} 1.5 \text{ with probability } 0.2 \text{ if T=10} \\ 0.5 \text{ with probability } 0.8 \text{ if T=10} \\ 1.5 \text{ with probability } 0.7 \text{ if T=20} \\ 0.5 \text{ with probability } 0.3 \text{ if T=20.} \end{cases}$$

Furthermore $T = 10$ with probability 0.4 and $T = 20$ with probability 0.6.

a. Will this population grow?

b. Find the correlation of temperature and growth rate.

3. One team hits 30 out of 100 shots, and another hits 40 out of 100. Use the normal approximation to decide whether the second team really shoots better.

4. Cosmic rays are thought to hit an object according to a Poisson process. 19 hit during the first minute and 25 during the second. Use the normal approximation to find 98% confidence limits around the maximum likelihood estimate of the true rate.

5. Molecules bind to a certain type of receptor at a rate of λ per second and never unbind.

a. Suppose the receptor begins in an unbound state. Write and solve a differential equation for the probability the receptor is unbound at time t.

b. A cell has two of these receptors. One binds at time $t = 1.5$ and another at $t = 2.5$. Find the maximum likelihood estimate of λ.

6. Suppose T measures temperature above 37°C in °C, and A measures an activity level. For three cells, $t_1 = 1$, $t_2 = 2$, $t_3 = 3$, $a_1 = 2$, $a_2 = 3$, and $a_3 = 3$

a. Plot A against T.

b. Find the linear regression of A on T.

c. Find the residuals.

7. A model indicates that the probability a cell is healthy after a treatment is $1 - q$, that it is damaged is $2q/3$ and that it is moribund is $q/3$ for some unknown parameter q. 20 cells are tested: 10 are found to be healthy, 6 found to be damaged, and 4 found to be moribund.

a. Find the likelihood function and the maximum likelihood estimate of q.

b. Without the model, the probabilities are $1 - q_1 - q_2$ (healthy), q_1 (damaged), and q_2 (moribund), with maximum likelihood estimates $q_1 = 0.3$ and $q_2 = 0.2$. Find the likelihood in this case. Do you think this model is better supported by the data than the one in **a**?

8. Although students struggled heroically to correctly solve the 8 problems on their final, their professor assigned random grades independently for each problem and student, giving 10 with probability 0.2, 15 with probability 0.3, 20 with probability 0.4, and 25 with probability 0.1.

a. Find the mean and variance of the total score.

b. Find the probability that a student gets a perfect score.

c. Use the normal approximation to find the probability that a student scores above 140. What is the lowest score that scores in the top 25%?

d. Sketch the distribution of scores on this test.

9. A new drug produces measurable reduction in a symptom in 75 out of 100 patients tested. An older drug produces measurable reduction in the symptom 65% of the time.

a. Use the normal approximation to find the 99% confidence interval for the fraction of patients aided by the new drug.

b. Test the hypothesis that the new drug is no better than the old drug. What is the significance level? Make sure to say whether you used a one- or two-tailed test and why.

10. The following data for temperature T and height H are measured:

T	H
10.0	12.0
12.0	13.0
14.0	15.0
16.0	17.0
18.0	20.0

A proposed regression line is $H = T + 1.0$.

a. Graph the data and proposed regression line.

b. Find the residuals.

c. Find SSE (with this line), SST and r^2.

11. The average density of trout along a stream is 0.2 per meter.

a. Find the probability of 3 or more trout in 5 meters.

b. Find the expected number and coefficient of variation of the number of trout in 1 kilometer.

c. If the stream flows at 2 meters per second (and the trout don't swim), find the rate at which trout pass a given point.

12. 50 students from a standard calculus class have a normally distributed scores on a standardized test with mean 60, and 30 students from an innovative new calculus class score a mean of 63. The standard deviation of each and every student's score is known to be 5.0.

a. In mathematical language, give the null hypothesis that students from the reformed class did no better than those from the standard class.

b. What is the significance level of the test?

13. The following data describe 10 measurements taken of height from control and treatment populations known to have normal distributions with standard deviation of 4.0. Test whether the treatment mean is different from the control mean. What is the significance level? What would you do if you did not know the standard deviation?

Control	Treatment
11.0	8.86
10.8	16.7
13.3	12.8
3.03	12.9
14.5	15.2
9.36	21.1
3.77	7.84
6.88	9.36
7.92	7.89
8.97	11.2

14. A lazy scientist wants to estimate the rate at which wolves leave Yellowstone Park by waiting for the first one to leave. This happens after 3.0 months.

a. Find the maximum likelihood estimate of the rate.

b. Find 95% confidence limits around this estimate.

c. How would you use the method of support to approximate these confidence limits? Write the equation you would solve.

d. When is the most likely time for the second wolf to leave? The mean time?

15. A company is testing a new insecticide. It runs three experiments, measuring the number out of 100 that survive at three dosages.

Experiment Number	Dosage (x)	Number Surviving (y)
1	18	3
2	15	4
3	9	5

A highly paid statistician finds the linear regression $y = 7 - 0.2x$.

a. Graph the data and the proposed regression line.

b. Find SSE, SST, and r^2.

c. Write the equation (using the data in the table) that the statistician solved to find this line.

d. Do you think that the pesticide works better at the higher dose? How would you check this statistically?

16. After application of a mutagen, five 100,000 base sequences of mitochondrial DNA have 24, 44, 29, 33, and 30 mutations respectively. Without mutagen, it is known that the mutation rate is 0.0002 per base.

a. Use the method of support to check whether the data are consistent with the null hypothesis.

b. Use the normal approximation to test the hypothesis that the mutagen increases the mutation rate. What does your significance level mean?

c. How would you test the null hypothesis with the Monte Carlo method?

17. The National Park Service has begun an intensive study of the elk in Yellowstone Park. As a first step, they wish to estimate elk density. A preliminary survey locates 36 elk droppings in one square kilometer.

a. What is the sample and what is the population in this case?

b. What assumptions must you make to estimate the true density of elk droppings in the park?

c. What equations would you solve to find exact 99% confidence limits around an estimate of the density?

d. Use the normal distribution to find approximate 99% confidence limits.

e. How might you use these data to estimate the actual number of elk?

18. To corroborate the intensive survey, elk dropping counters walk in straight lines at exactly 2 kilometers per hour, recording the first 10 times when they spot a dropping. The data from one surveyor are

Dropping Number i	Time Since Last Dropping (t_i)
1	0.124
2	0.056
3	0.244
4	0.227
5	0.001
6	0.118
7	0.220
8	0.014
9	0.015
10	0.016

An untrained but trustworthy assistant computes that

$$\sum_{i=1}^{10} t_i = 1.035$$
$$\sum_{i=1}^{10} t_i^2 = 0.193.$$

a. Sketch a graph of where the droppings are.

b. What is your best guess of the rate at which droppings are encountered?

c. Write and sketch the support function.

d. Sketch how you would use this graph to find approximate confidence limits and write the equation you would solve to find them.

e. How might you relate these data to those in the previous problem?

19. In a second stage in the project, managers decide to count both elk and their droppings. In 5 different square kilometer regions, they find

Sample Number i	Number of Droppings d_i	Number of Elk e_i
1	39	85
2	49	101
3	49	104
4	36	78
5	51	111

a. Graph these data.

b. A mathematician proposes two lines, $e_i = 2d_i + 5$ and $e_i = 5d_i + 2$. Sketch them. Which line gives a better fit to the data?

c. Find r^2 for the better of the two lines.

d. Interpret the relation described by this line. Does it make sense?

20. In an even more advanced study, the managers use portable scales to weigh elk inside and outside the park to check whether they are different. They weigh 30 adult males inside the park (weights denoted by $x_1 \ldots x_{30}$) and 40 adult males outside (weights denoted by $y_1 \ldots y_{40}$), all in kilograms. They compute

$$\sum_{i=1}^{30} x_i = 1.829 \times 10^4$$
$$\sum_{i=1}^{40} y_i = 2.316 \times 10^4$$
$$\sum_{i=1}^{30} x_i^2 = 1.146 \times 10^7$$
$$\sum_{i=1}^{40} y_i^2 = 1.368 \times 10^7.$$

a. Find the sample mean and sample variance for each group of elk.

b. Find the standard error for each.

c. State the null hypothesis mathematically.

d. Test the null hypothesis statistically.

21. Having estimated numbers and weights, the managers begin working on health. They capture 40 elk and check them for parasites. They had reported to Congress that no more than 20% of the elk in the park were infested, but find that 13 out of their 40 are.

a. What is the probability that their report to Congress was true but they got unlucky? What does this have to do with significance levels?

b. What would a member of Congress compute if she were using the popular method of support instead?

c. Give one clever excuse the managers could give for their result. How might they argue for an expensive follow-up study?

Projects

1. The statistical tests studied in the text are called **parametric** because they assume a particular form for the underlying distribution. For example, when we compared whether the means of two populations differed in Section 8.6, we assumed that the measurements came from a normal distribution. When there is no reason to believe that the distribution is normal, it would be nice to have **distribution-free** tests that work no matter where the measurements came from. This project studies some of the properties of one such test, the **Wilcoxon rank-sum test**.

Suppose that a set of 20 measurements in the control is

0.46894, 0.35273, 0.92801, 0.50284, 0.94539,
0.29496, 0.15446, 0.99030, 0.10896, 0.99643,
0.00097, 0.92011, 0.04307, 0.00001, 0.00024,
0.39957, 0.00077, 0.35379, 0.00004, 0.05706

and that the measurements in the treatment are

0.48770, 1.82495, 0.90703, 1.37176, 0.90001,
0.93359, 1.38502, 1.56536, 1.71434, 1.28598,
0.89246, 1.74330, 1.47014, 0.90556, 0.90001,
1.82305, 0.90394, 1.80227, 0.90044, 1.83284.

Do these data look normally distributed? What result do you get if you assume they are normally distributed and test for a difference in the means?

An alternative test for weird distributions instead uses the **ranks** of the two sets of data. Pool the data into a single data set, and sort it in order of increasing size. Give each value a rank, ranging from a rank of 1 for the smallest measurement 0.00001 from the control to a rank of 40 for the measurement 1.83284 from the treatment. If you add up the ranks of all measurements that come from the control, the sum is much smaller than the sum of the ranks of all measurements from the treatment.

To determine whether this result is significant, try the following experiment. Suppose that the control and treatment values come from the same distribution. Use a computer to sample 20 values for the control and 20 values for the treatment, and compute the rank sums. Do this many times. How often are the results more extreme than the values found?

Estimate the mean and variance of your simulations. The values should be roughly normally distributed with a mean of 410 and a variance of 1366.7. Use this result to estimate the significance. How well does it compare with the results from your Monte Carlo simulations?

2. Suppose the data in the table are supposed to fit the model

$$Y = \frac{1}{1 + aX}$$

for an unknown value of the parameter a.

X	Y
1	0.365
2	0.192
3	0.124
4	0.129
5	0.0829
6	0.0755
7	0.0738
8	0.0490
9	0.0548
10	0.0486

We wish to use the method of least squares to estimate the parameter a. This problem requires the method of **nonlinear** least squares because a does not simply multiply the value of X. The idea is to set

$$\hat{y}_i = \frac{1}{1 + ax_i}$$

for some value of a. Our goal is to minimize the least squares measure

$$S(a) = \sum_{i=1}^{n} (y_i - \hat{y}_i)^2.$$

Find a way to do this, and plot a graph comparing the actual data with the value predicted by the model.

An alternative method transforms the data to make this into a linear least squares problem. Create a new measurement

$$Z = \frac{1}{Y}.$$

Find Z as a function of X according to the model. Use the method of linear least squares to estimate a. Graph the transformed data along with the value predicted by the model.

Which estimate do you think is more accurate? What would happen to the two estimates if the last value in the data set became $y_{10} = 0.02$?

Answers to Selected Odd Exercises

Answers to all odd exercises can be found at
http://www.brookscole.com/cgi-wadsworth/course_products_wp.pl?fid=M20b&product_isbn_issn=0534404863&discipline_number=1

Chapter 1

Section 1.2, page 22

1. The variables are time and fish mass, which we can call t and m, and the parameter is the salinity level, which we could call s. The time t could be measured in days, the mass m in grams, and the salinity s in percent salt.

3. $f(0)=5$, $f(1)=6$, $f(4)=9$.

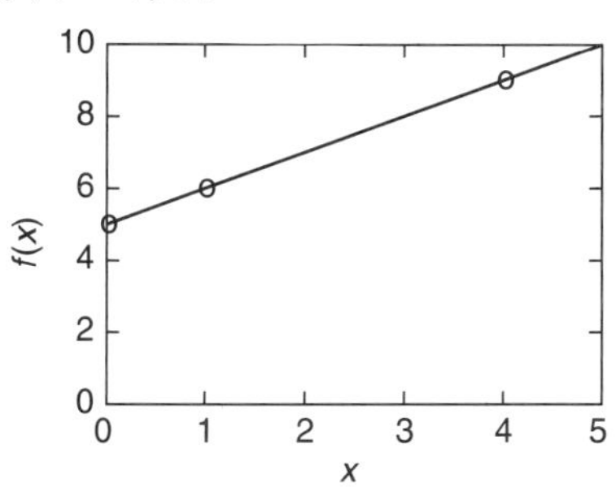

5. $h(1)=\frac{1}{5}$, $h(2)=\frac{1}{10}$, $h(4)=\frac{1}{20}$.

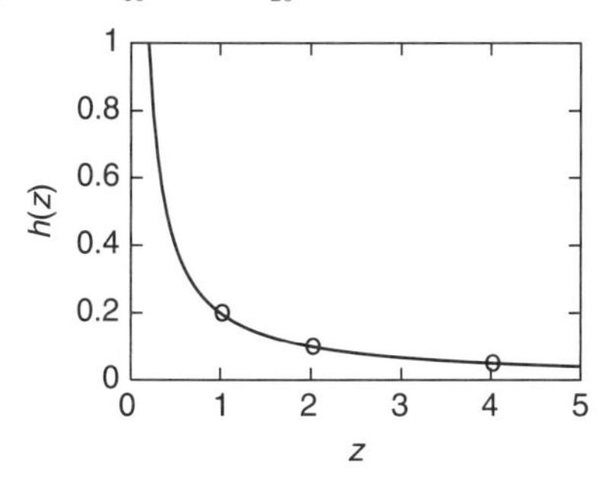

7. (2, 1).

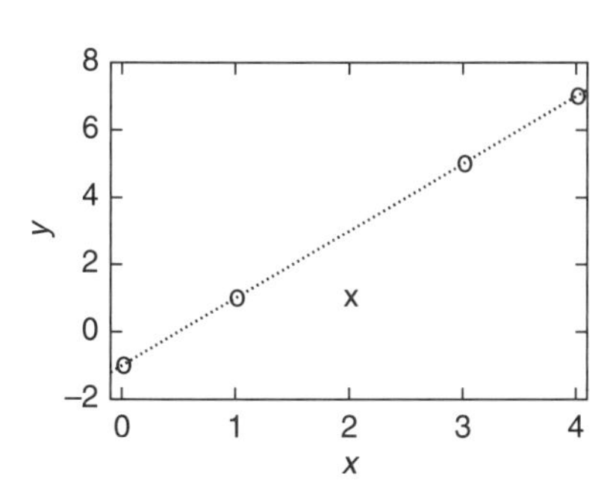

9. (4, 10).

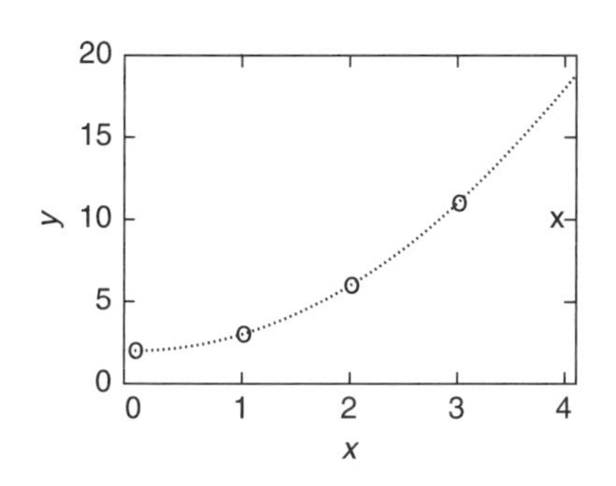

11. $f(a)=a+5$, $f(a+1)=a+6$, $f(4a)=4a+5$.

13. $h(\frac{c}{5})=\frac{1}{c}$, $h(\frac{5}{c})=\frac{c}{25}$, $h(c+1)=\frac{1}{5c+5}$.

15. I put them in increasing order to look nice.

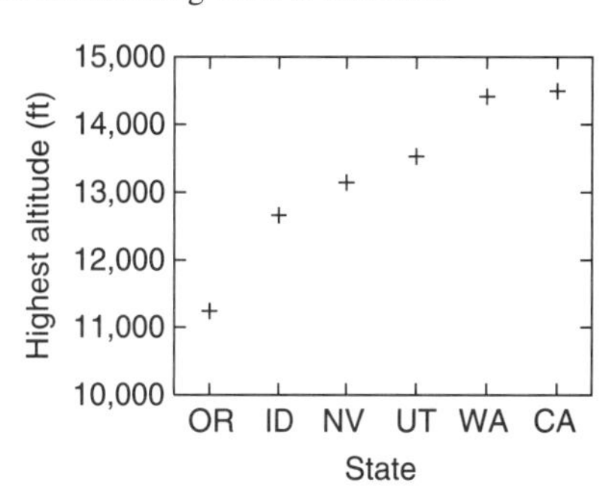

17.

x	$f(x)$	$g(x)$	$(f+g)(x)$
−2	−1	−11	−12
−1	1	−8	−7
0	3	−5	−2
1	5	−2	3
2	7	1	8

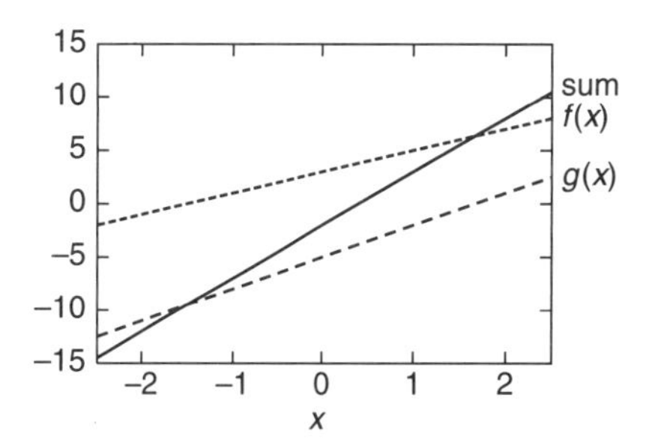

19.

x	$F(x)$	$G(x)$	$(F+G)(x)$
−2	5	−1	4
−1	2	0	2
0	1	1	2
1	2	2	4
2	5	3	8

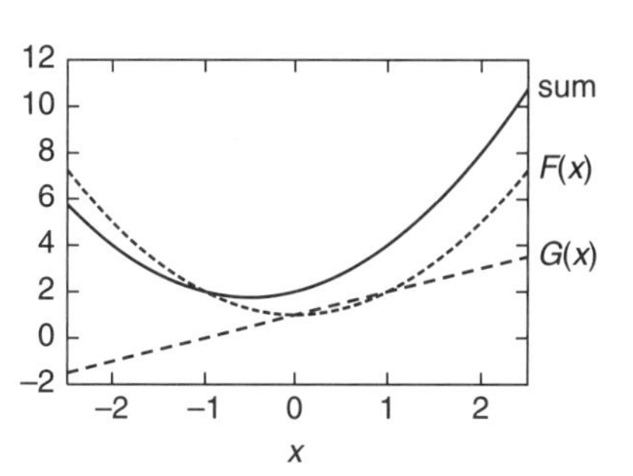

21.

x	$f(x)$	$g(x)$	$(f\cdot g)(x)$
−2	−1	−11	11
−1	1	−8	−8
0	3	−5	−15
1	5	−2	−10
2	7	1	7

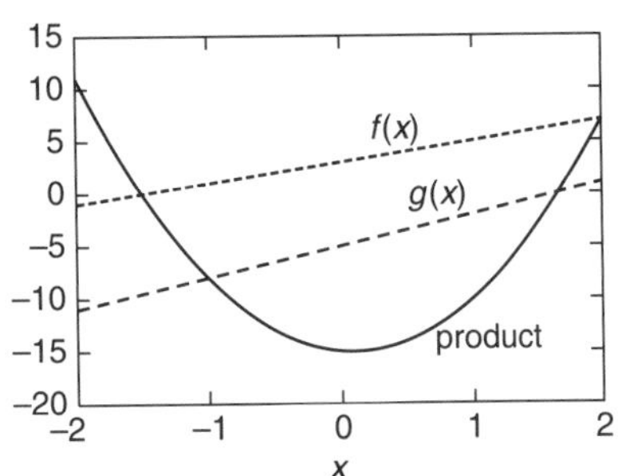

23.

x	$F(x)$	$G(x)$	$(F \cdot G)(x)$
−2	5	−1	−5
−1	2	0	0
0	1	1	1
1	2	2	4
2	5	3	15

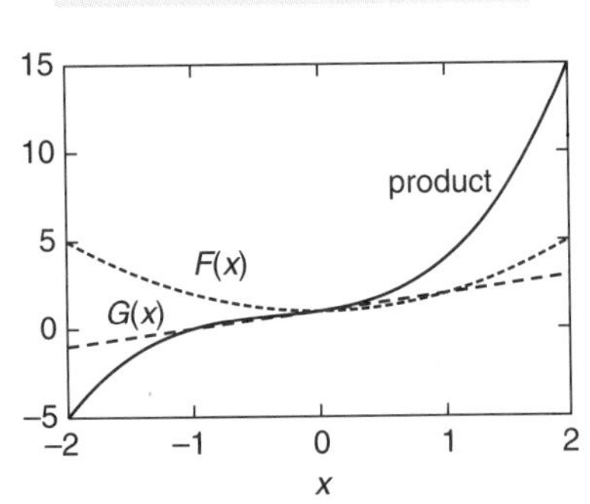

25. If we write $y = 2x + 3$, we can solve for x with the steps

$$y - 3 = 2x \quad \text{subtract 3 from both sides}$$

$$\frac{y-3}{2} = x. \quad \text{divide both sides by 2}$$

Therefore $f^{-1}(y) = \frac{y-3}{2}$. Also, $f(1) = 5$, and $f^{-1}(5) = \frac{5-3}{2} = 1$.

27. Set $z = \frac{1}{2+y}$. We can solve for $y = \frac{1}{z} - 2$ so $G^{-1}(z) = \frac{1}{z} - 2$. Checking, $G(1) = 1/3$, and $G^{-1}(1/3) = 1$.

29.

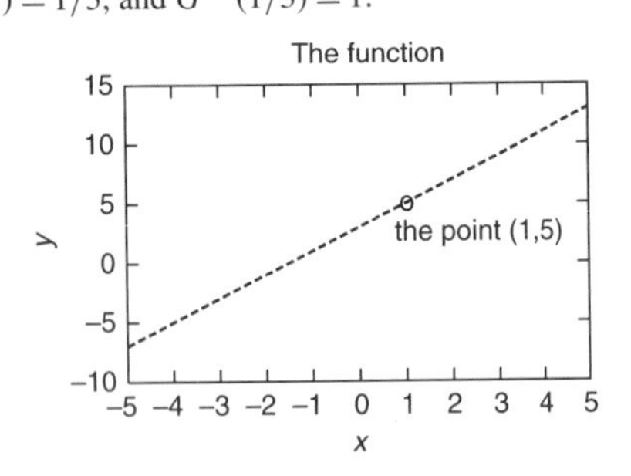

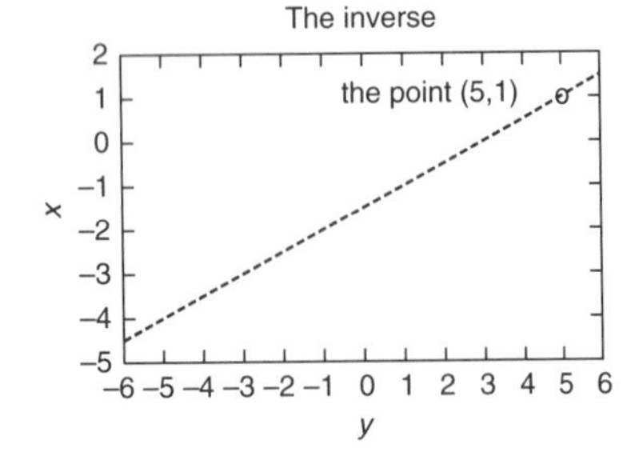

31.

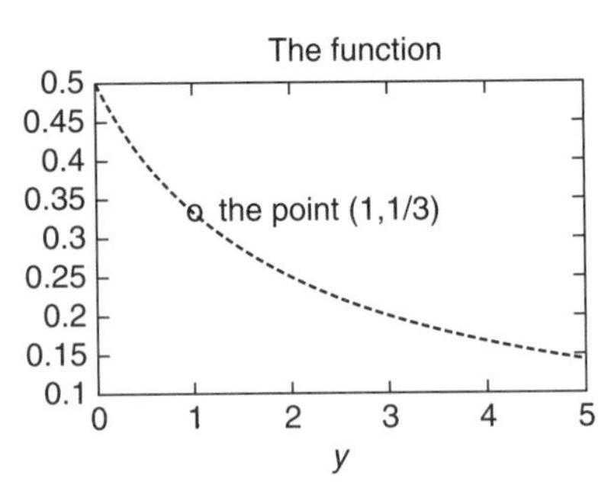

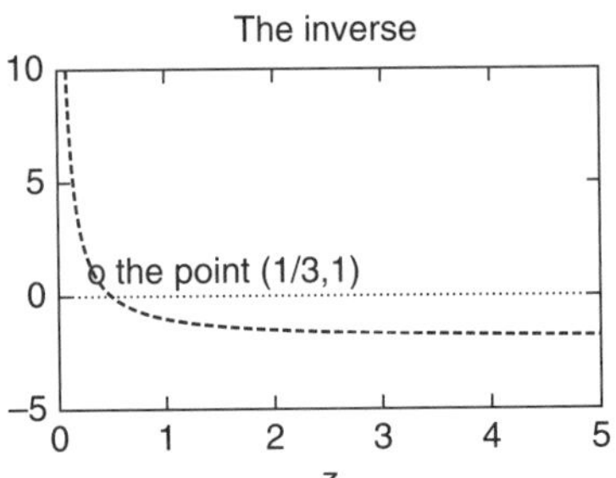

33. $(f \circ g)(x) = f(3x - 5) = 2 \cdot (3x - 5) + 3 = 6x - 7$ and $(g \circ f)(x) = g(2x + 3) = 3 \cdot (2x + 3) - 5 = 6x + 4$. These don't match, so the functions do not commute.

35. $(F \circ G)(x) = F(x+1) = (x+1)^2+1 = x^2+2x+2$ and $(G \circ F)(x) = G(x^2 + 1) = x^2 + 2$. These do not match, so the functions do not commute.

37. The cell volume is generally increasing but decreases during part of its cycle. The cell might get smaller when it gets ready to divide or during the night.

39. The height increases up until about age 30, and then decreases.

41.

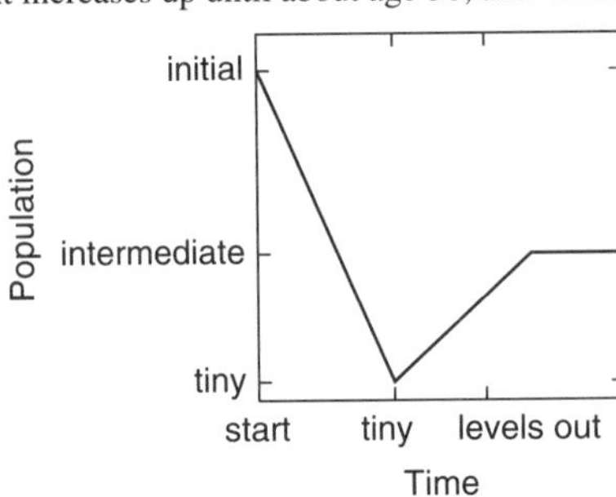

43.

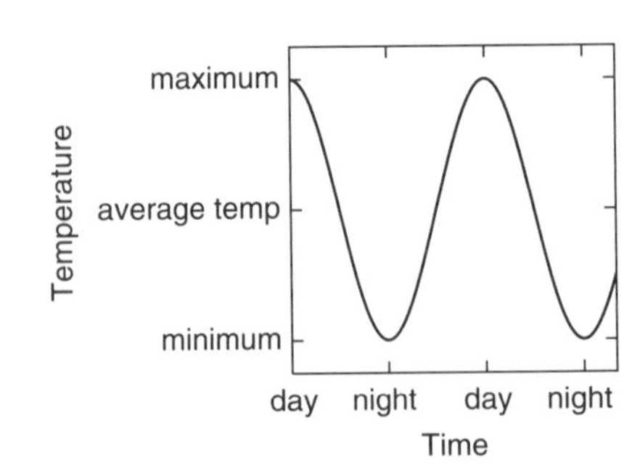

45. When $f = 0$; $b = 1$; when $f = 10$, $b = 21$; when $f = 20$, $b = 41$. Perhaps one bee will check out the plant even if there are no flowers.

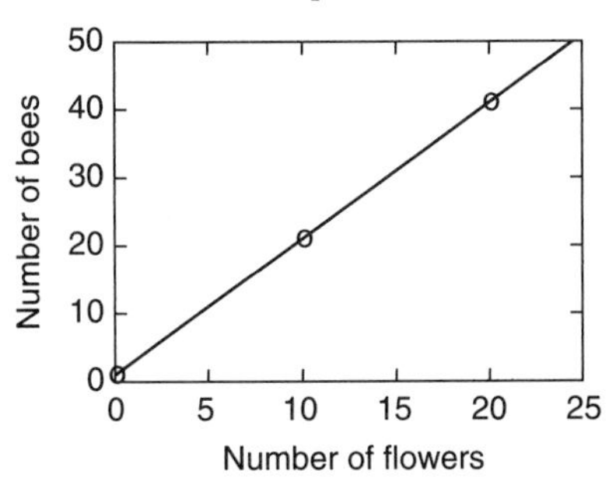

47. When $T = 10$, $A = 35$; when $T = 20$, $A = 30$; when $T = 30$, $A = 25$; when $T = 40$, $A = 20$. The insect develops most quickly (in the shortest time) at the highest temperatures.

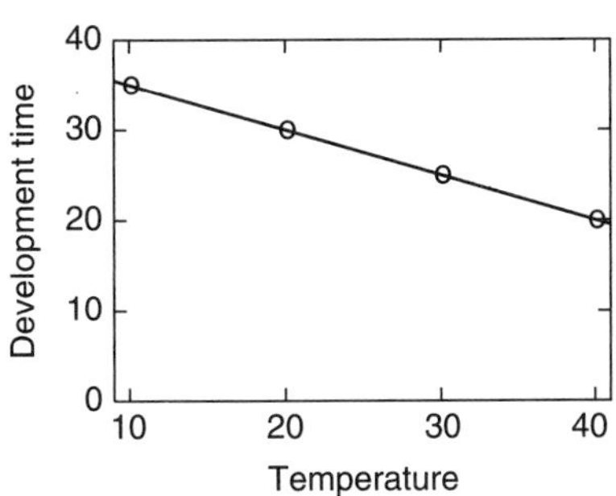

49. Length as a function of age

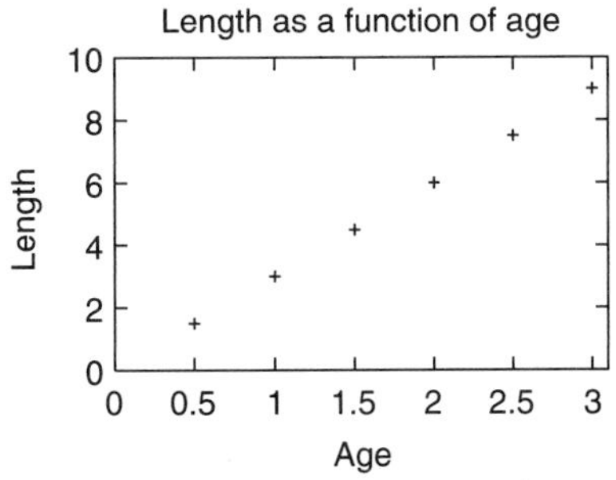

51.

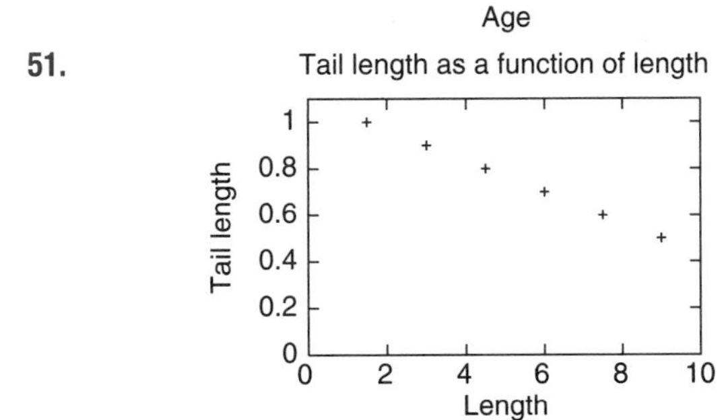

53. $B(M) = B(5W + 2) = 2.5W + 1$. Plugging in $W = 10$ gives 26 bites.

55. $F[V(I)] = F(5I^2) = 37 + 2I^2$. The fever is 39°C if $I = 1$.

57. The formula is $P(t) = (1 + t^2) + (1 - 2t + t^2) = 2 - 2t + 2t^2$.

t	$a(t)$	$b(t)$	$P(t)$
0.00	1.00	1.00	2.00
0.50	1.25	0.25	1.50
1.00	2.00	0.00	2.00
1.50	3.25	0.25	3.50
2.00	5.00	1.00	6.00
2.50	7.25	2.25	9.50
3.00	10.00	4.00	14.00

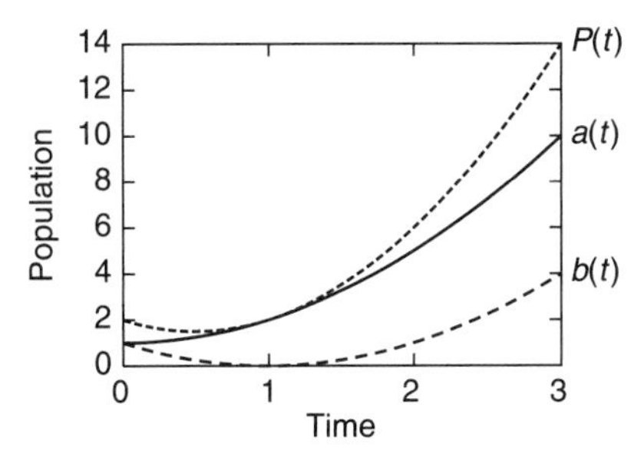

a is increasing, b decreases to 0 at time 1.0 and then increases, and the sum P decreases slightly and then increases.

59. Because the mass is the same at ages 2.5 days and 3.0 days, the function relating a and M has no inverse. Knowing the mass does not give enough information to estimate the age.

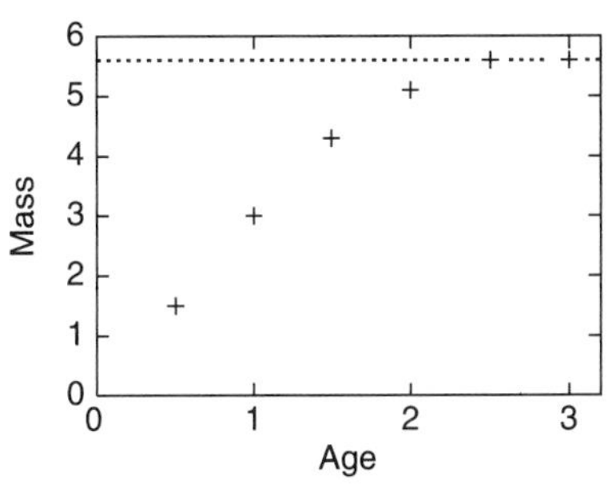

61. Glucose production is 8.2 mg at ages 2.0 days and 3.0 days. We cannot figure out the age from this measurement.

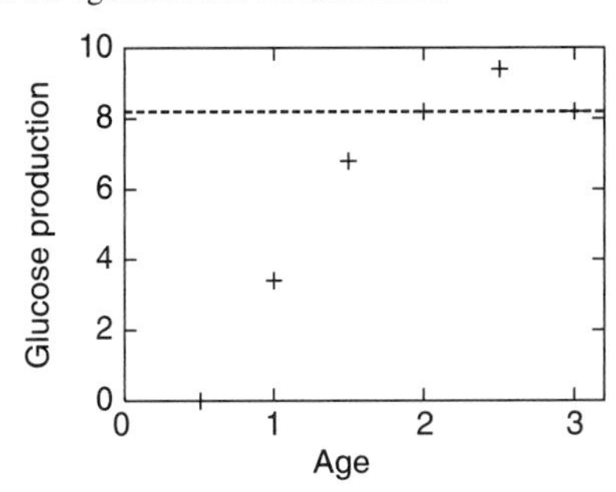

63. Denote the total mass by $T(t)$. Then $T(t) = P(t)W(T) = (2.0 \times 10^6 + 2.0 \times 10^4 t)(80 - 0.5t)$.

t	$P(t)$ (millions)	$W(t)$	$T(t)$ (millions, kg)
0.0	2.0	80.0	160.0
20.0	2.4	70.0	168.0
40.0	2.8	60.0	168.0
60.0	3.2	50.0	160.0
80.0	3.6	40.0	144.0
100.0	4.0	30.0	120.0

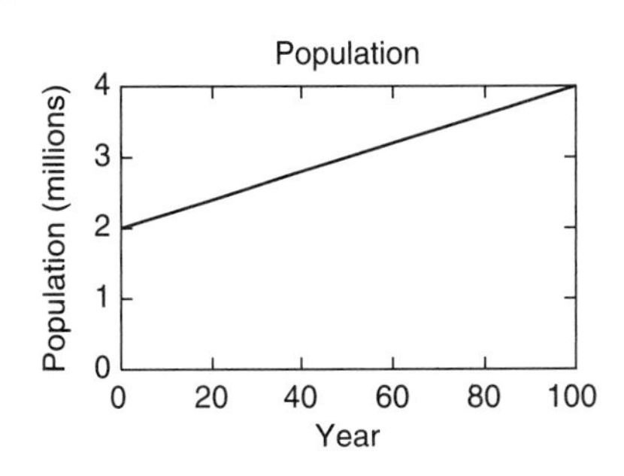

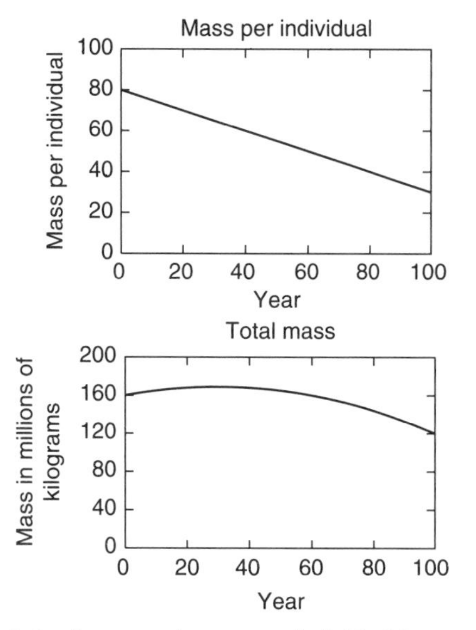

The population increases, the mass per individual decreases, and the total mass increases and then decreases.

65. We denote the total mass by $T(t)$. Then $T(t) = P(t)W(T) = (2.0 \times 10^6 + 1000t^2)(80 - 0.5t)$.

t	$P(t)$ (millions)	$W(t)$	$T(t)$ (millions, kg)
0.0	2.0	80.0	160.0
20.0	2.4	70.0	168.0
40.0	3.6	60.0	216.0
60.0	5.6	50.0	280.0
80.0	8.4	40.0	336.0
100.0	12.0	30.0	360.0

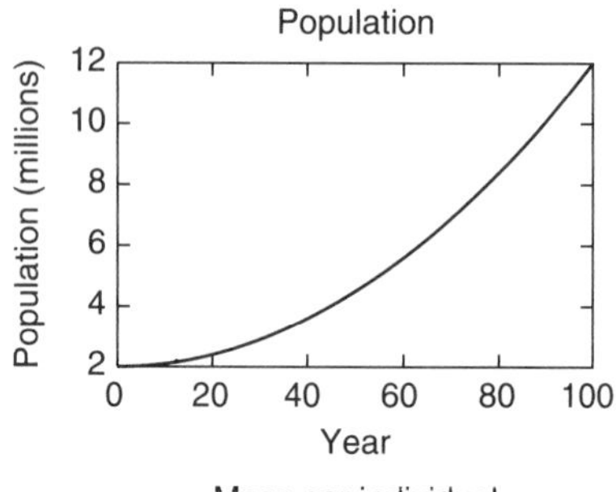

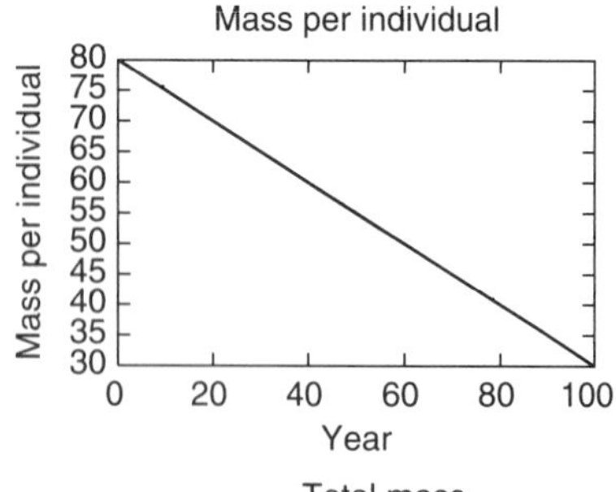

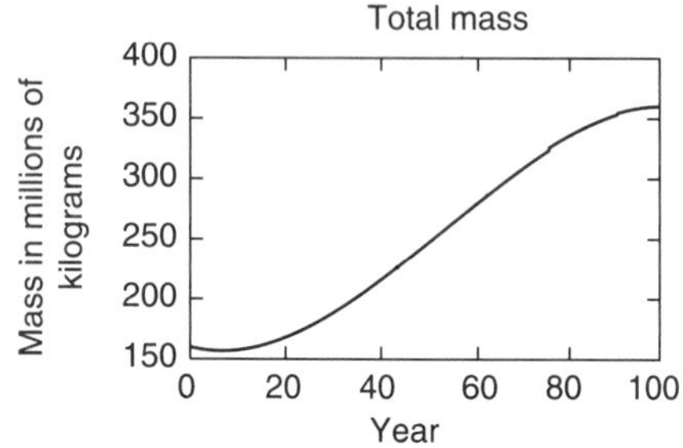

The population decreases, the mass per individual increases, and the total mass increases.

Section 1.3, page 38

1. $3.4 \text{ lb} \times 16\frac{\text{oz}}{\text{lb}} \times 28.35\frac{\text{g}}{\text{oz}} = 1542.24\text{g}$.

3. $60 \text{ yr} \times 365.25\frac{\text{days}}{\text{yr}} \times 24\frac{\text{h}}{\text{day}} = 525960 \text{ h}$.

5. $2.3\frac{\text{g}}{\text{cm}^3} \times \frac{1}{28.35}\frac{\text{oz}}{\text{g}} \times \frac{1}{16}\frac{\text{lb}}{\text{oz}} \times 2.54^3\frac{\text{cm}^3}{\text{in.}^3} \times 12^3\frac{\text{in.}^3}{\text{ft}^3} = 141.58 \text{ lb/ft}^3$.

7. 2.3 cm is 0.023 m, so the final height is $1.34 + 0.023 = 1.363$ m.

9. The total weight of apples is $6 \cdot 145 = 870$ g and the total weight of oranges is $7 \cdot 123 = 861$ g, for a total of $870 + 861 = 1731$ g.

11. The area of the square is $1.7^2 = 2.89 \text{ cm}^2$, but the area of the disk is $\pi r^2 = \pi \cdot 1^2 \text{ cm}^2 = 3.1415 \text{ cm}^2$. The disk has a larger area.

13. The volume of the sphere is $\frac{4}{3}\pi 100^3 = 4.189 \times 10^6\text{m}^3$. The lake has area $3.0 \times 10^6\text{m}^2$ and a depth of 0.5 m, giving a volume of merely $1.5 \times 10^6\text{m}^3$. The sphere is much larger.

15. Pressure is force per unit area, or

$$\frac{\text{force}}{\text{area}} = \frac{\frac{ML}{T^2}}{L^2} = \frac{M}{LT^2}.$$

17. The rate of spread of bacteria on a plate has dimensions of L^2/T, or area per time.

19. This checks, because length $= \frac{\text{length}}{\text{time}} \times$ time.

21. This checks, because $\frac{\text{mass length}}{\text{time}^2} = \text{mass}\frac{\text{length}}{\text{time}^2}$.

23. The vertical axis is scaled by a value greater than 1.

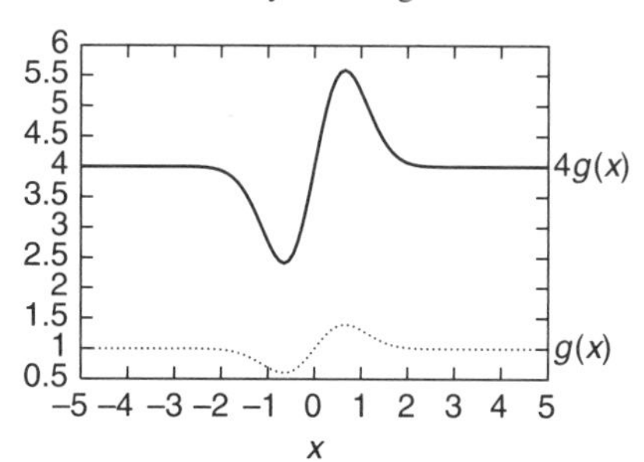

25. The horizontal axis is scaled by a value less than 1.

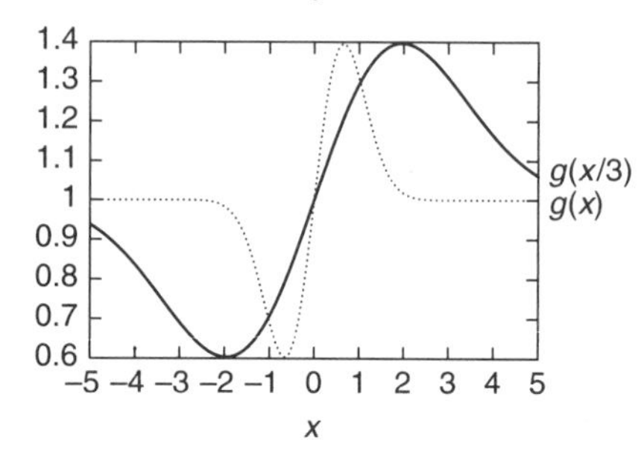

27. The tree with a height of 23.1 m will have a volume of $\pi \cdot 23.1 \cdot 0.5^2 = 18.14 \text{ m}^3$. When the height is 24.1 m, the volume is 18.93 m^3. The ratio is $\frac{18.93}{18.14} = 1.043$.

29. The bottom portion of the 23.1 m-tall tree will have half the volume found earlier, or 9.07 m^3. The top part is 23.1/2 = 11.55 m high, and thus is a sphere with a radius of 11.55/2 = 5.78 m. Substituting into the formula for the volume of a sphere, we find 806.76 m^3. The total volume is 815.83 m^3. The 24.1 m tree has a total volume of $9.46 + 916.13 = 925.59\text{m}^3$. The ratio of the volumes is $\frac{925.59}{815.83} = 1.134$.

31. The volume is $2.0 \cdot 0.2 \cdot 1.5 = 0.6\text{m}^3 = 6.0 \times 10^5\text{cm}^3$. This gives a mass of $6.0 \times 10^5\text{gm} = 600$ kg.

33. $3200 \cdot 0.45\frac{\text{g}}{\text{individual}} = 1440 \text{ g} = 1.44 \text{ kg}$.

35. Let p be the number of petals. Then $f = p/4$, so $b = \frac{p}{2} + 1$. When $p = 0, b = 1$; when $p = 10, b = 6$; when $p = 20, b = 11$. The number of bees increases more slowly as a function of the number of petals.

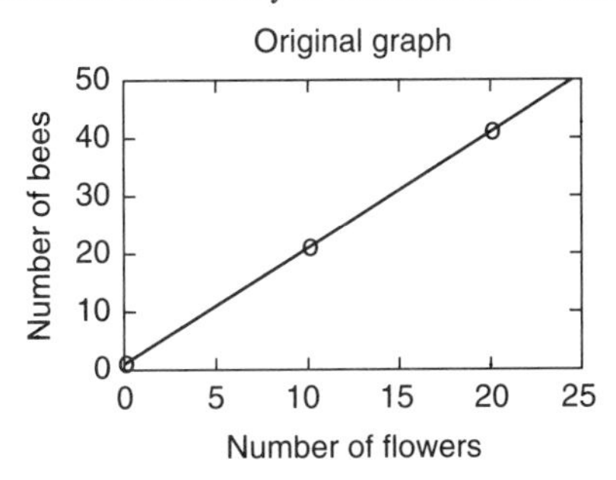

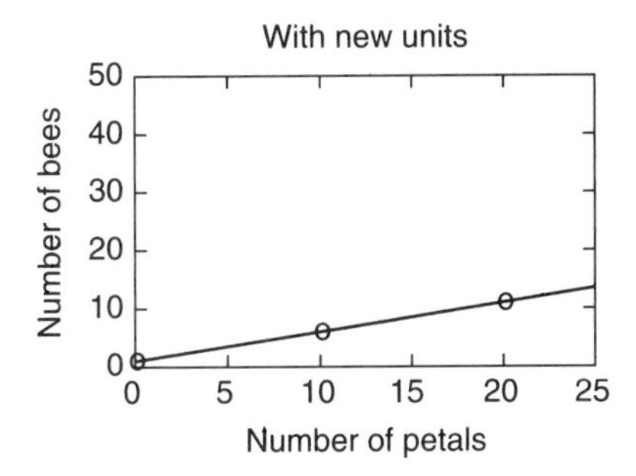

37. Let H be the development time in hours. Then $H = 24A$ so $H = 24(40 - T/2) = 960 - 12T$. If $T = 10$, $H = 840$; when $T = 20$, $A = 720$; when $T = 30$, $A = 600$; when $T = 40$, $H = 480$.

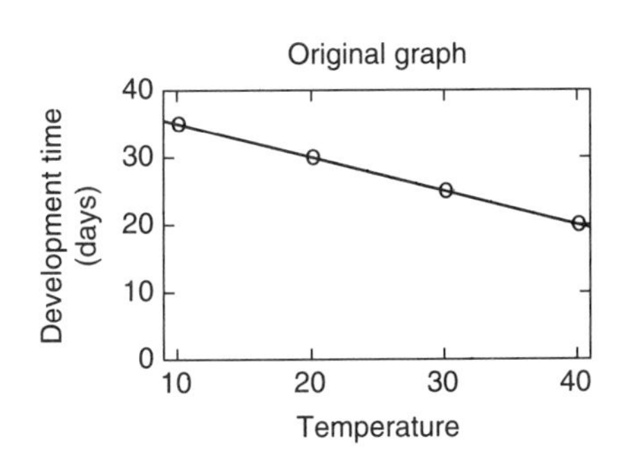

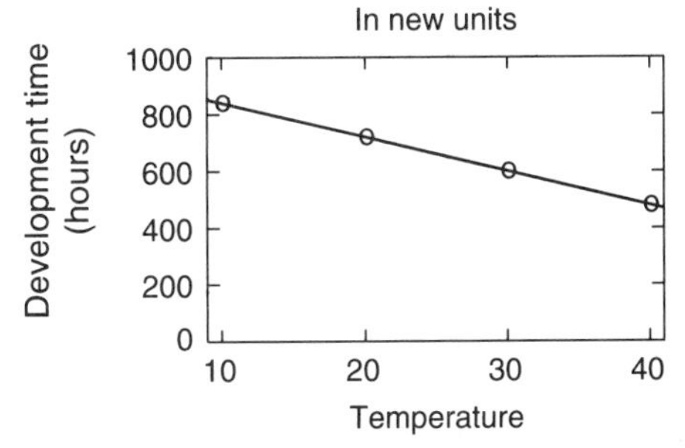

39.

$$186{,}000\frac{\text{mile}}{\text{s}} \approx 200{,}000\frac{\text{mile}}{\text{s}} \approx 200{,}000\frac{\text{mile}}{\text{s}} \times 60{,}000\frac{\text{in.}}{\text{mile}}$$
$$= 1.2 \times 10^{10}\frac{\text{in.}}{\text{s}} \approx 1.2 \times 10^{10}\frac{\text{in.}}{\text{s}} \times 2.5\frac{\text{cm}}{\text{in.}}$$
$$= 3.0 \times 10^{10}\frac{\text{cm}}{\text{s}} = 30\frac{\text{cm}}{\text{ns}}.$$

If a computer is supposed to do an operation in 0.3 ns, it had better not need to move information for more than the distance light can travel in that time, or about 9 cm.

41. The volume is

$$\frac{4}{3}\pi r^3 \approx 3 \cdot 6.5^3 \times 10^9\text{km}^3$$
$$\approx 3 \cdot 300 \times 10^9\text{km}^3$$
$$\approx 1 \times 10^{12}\text{km}^3.$$

One kilometer is 10^5 cm, so the mass of a cubic kilometer of water is 10^{15}gm $= 10^{12}$ kg. Multiplying this by the volume and the density gives a total of 5×10^{24} kg.

43. You would catch 6 movies per day, or about 2000 per year, for a total of 120,000 in your life.

45. The volume of a sphere of radius r is $4\pi r^3/3 \approx 4r^3$. The radius of the cell is 10^{-3} cm, so the volume is about 4×10^{-9} cm^3. The mass of a cell is therefore around 4×10^{-9} g. I weigh about 60 kg, which is 6×10^4 g. The number of cells is then

$$\frac{6 \times 10^4\text{gm}}{4 \times 10^{-9}\text{gm/cell}} = 1.5 \times 10^{13}\text{cells}.$$

47. The brain is about 2 percent of my weight, and should have about 2 percent of my cells, or 3×10^{11}. The number of neurons is 1×10^{11}, but the total number of cells in the brain is between 1×10^{12} and 5×10^{12}, a bit higher than the previous estimates.

49. The length of the string would be $2\pi r = 40840.704$ km. Adding one meter would make it 40,840.705 km. The radius corresponding to this is $r = \frac{40{,}840.705}{2\pi} = 6500.0002$ km. The string would be 0.0002 km, or 0.2 m above the earth. It is amazing that such a relatively tiny change in the length of the string would produce such a big effect.

Section 1.4, page 51

1. The points are (1, 5) and (3, 9). The change in input is 2, the change in output is 4, and the slope is 2. This is not a proportional relation because the ratio of output to input changes from 5 at the first point to 3 at the second point. This relation is increasing because larger values of x lead to larger values of y (and the slope is positive).

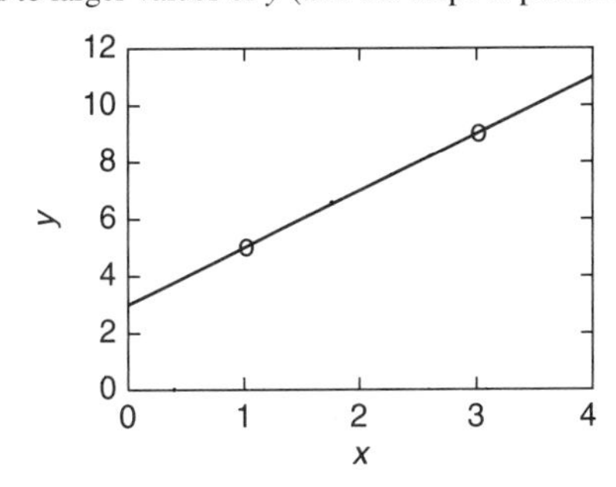

3. The points are (1, 3) and (3, 13). The change in input is 2, the change in output is 10, and the slope is 5. This is a not a proportional relation because the ratio of output to input changes from 3 at the first point to 4.333 at the second point. This relation is increasing because larger values of w lead to larger values of z (and the slope is positive).

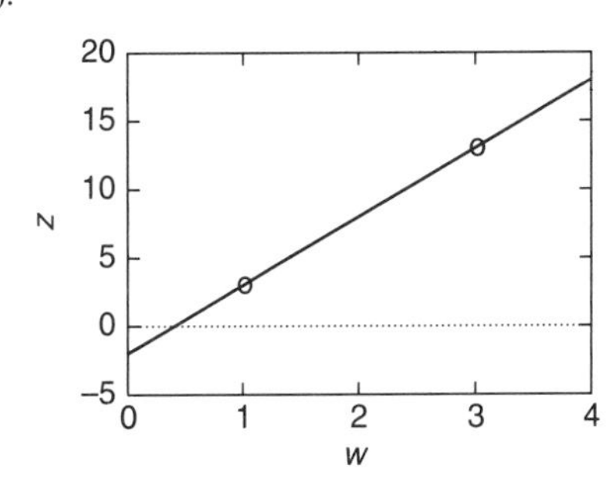

5. The point lies on the line because $f(2)=2\cdot 2+3=7$. The point-slope form is $f(x)=2(x-2)+7$. Multiplying out gives $f(x)=2x-4+7=2x+3$ as it should.

7. Multiplying out, we find that $f(x)=2x+1$. The slope is 2 and the y-intercept is 1. The original point is $(1, 2)$.

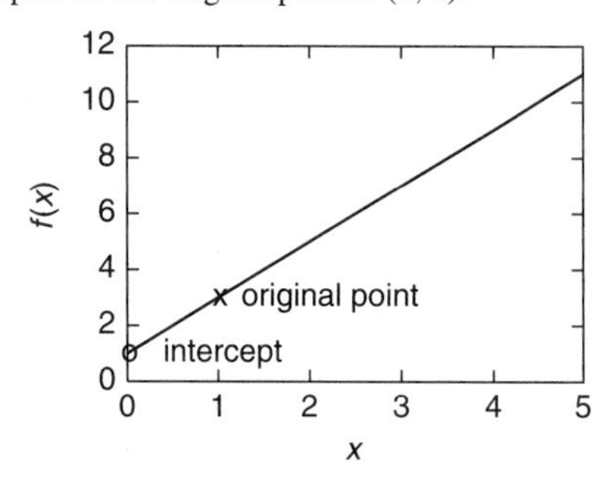

9. In point-slope form, this line has equation $f(x)=-2(x-1)+6$. Multiplying out, we find that $f(x)=-2x+8$. The slope is -2 and the y-intercept is 8.

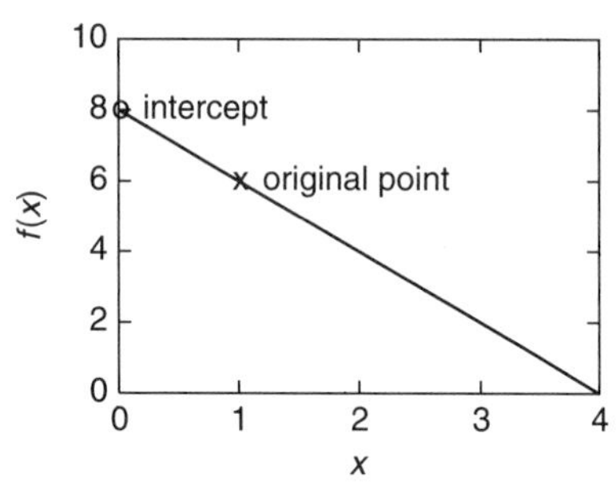

11. The slope between the two points is

$$\text{slope} = \frac{\text{change in output}}{\text{change in input}} = \frac{3-6}{4-1} = -1.$$

In point-slope form, the line has equation $f(x)=-1\cdot(x-1)+6$. In slope-intercept form, it is $f(x)=-x+7$. The slope is -1 and the y-intercept is 7.

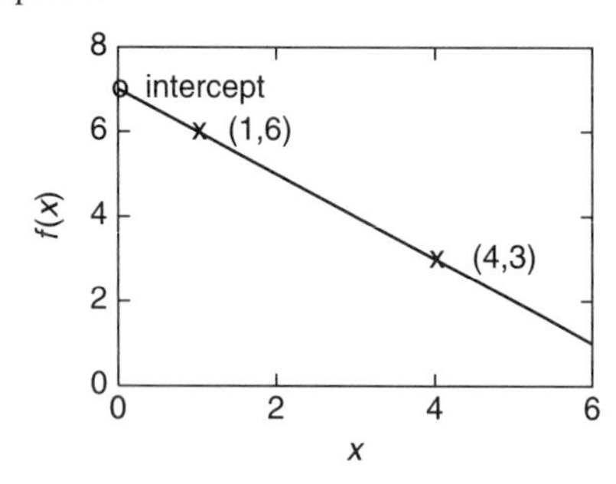

13. This is not linear because the input z appears in the denominator.

15. This is linear because the input q is only multiplied by constants and has constants added to it.

17. $h(1)=\frac{1}{5}$, $h(2)=\frac{1}{10}$, $h(4)=\frac{1}{20}$. The slope between $z=1$ and $z=2$ is -1/10, and between $z=2$ and $z=4$ is -1/40.

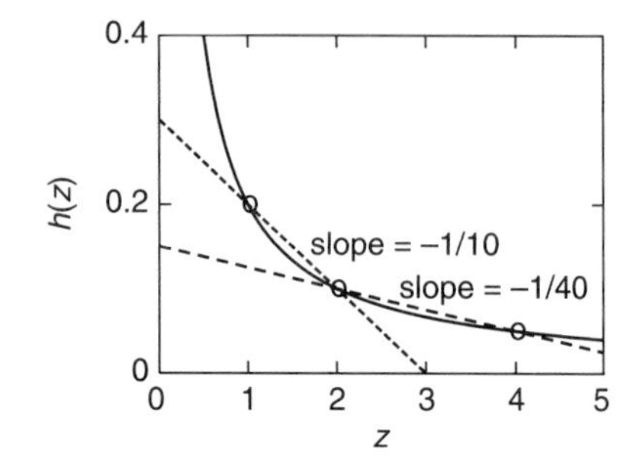

19. $2x=7-3=4$, so $x=4/2=2$. Plugging in, $2\cdot 2+3=7$.

21. $2x-3x=7-3=4$, so $-x=4$ or $x=-4$. Plugging in, $2\cdot(-4)+3=-5=3\cdot(-4)+7$.

23. Multiplying out, we get $10x-4=10x+5$. This has no solution.

25. $2x=7-b$, so $x=\frac{7-b}{2}$.

27. $(2-m)x=7-b$, so $x=\frac{7-b}{2-m}$. There is no solution if $m=2$. However, if $m=2$ and $b=7$, both sides are identical and any value of x works.

29. 1 in. = 2.54 cm. The slope is 2.54 $\frac{\text{cm}}{\text{inch}}$.

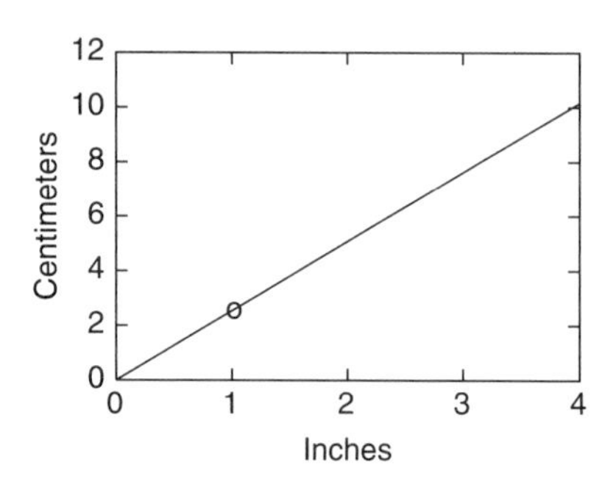

31. $1\text{ g} = \frac{1}{453.6}\text{lb} = 0.0022$ lb. The slope is 0.0022 $\frac{\text{lb}}{\text{g}}$.

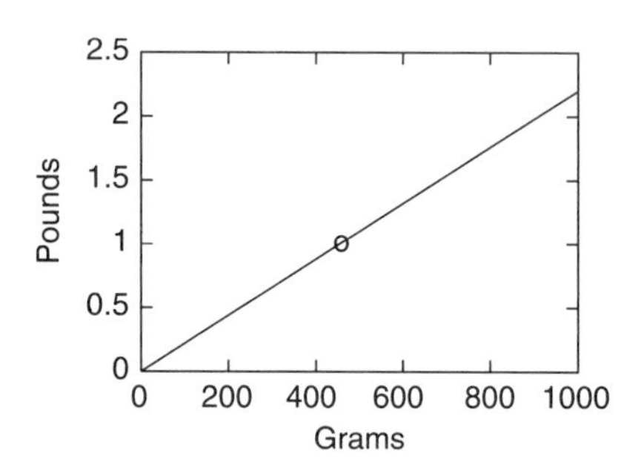

33. $(f\circ g)(x)=(mx+b)+1$ and $(g\circ f)(x)=m(x+1)+b=mx+m+b$. These match only if the intercepts are equal, or $b+1=m+b$. This is true for any b as long as $m=1$. In this case, both f and g have slope 1, meaning that each just adds a constant to its input. The order cannot matter because addition is commutative.

35. The slope is 1.0 cm, and the equation is $V=1.0A$.

37. The slope is 5.0×10^{-9} g, and the equation is $M=5.0\times 10^{-9}b$.

39. The line has slope -0.2 and intercept 10,000. The equation is thus $a=-0.2d+10,000$.

41. $a=-0.2\cdot 2000+10000=9600$ ft.

43.

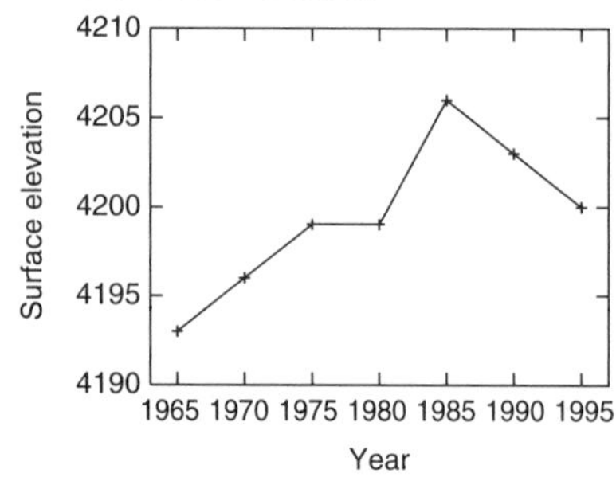

45. The slope is 3 ft every 5 years. It would have been up to 4208 by 1990, 5 ft higher than the actual level.

47. Using the first two rows for mass, we find a slope of

$$\text{slope} = \frac{\text{change in mass}}{\text{change in age}} = \frac{4.0-2.5}{1.0-0.5} = 3.0.$$

Using (1.0, 4.0) as the base point,

$$M = 3.0(a-1.0)+4.0 = 3.0a+1.0.$$

The y-intercept is 1.0, meaning that the mass was 1.0 g at age 0. This might be the mass of a new seedling. Interpolating at $a = 1.75$, $M = 3.0 \cdot 1.75 + 1.0 = 6.25$.

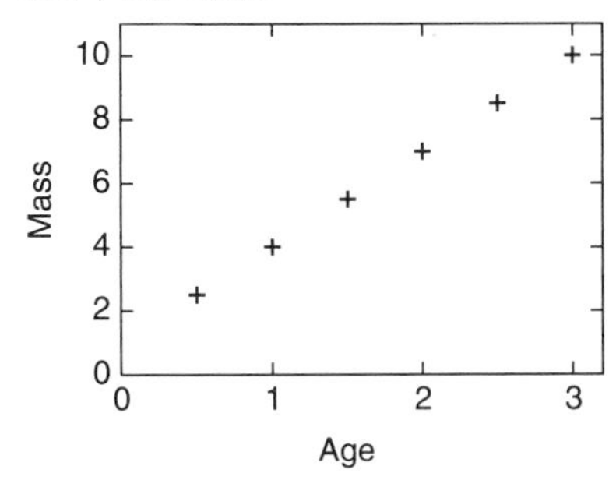

49. The glucose production changes by 3.4 when the mass changes by 1.5, giving a slope of 2.27. The point-slope form of the line, using the last data point, is $G = 2.27(m - 10.0) + 17.0$, which simplifies to $G = 2.27m - 5.7$ in slope-intercept form. The negative intercept must mean that this plant would not start making glucose until the mass got beyond a certain value (around 2). When $m = 20.0$, $G = 39.7$.

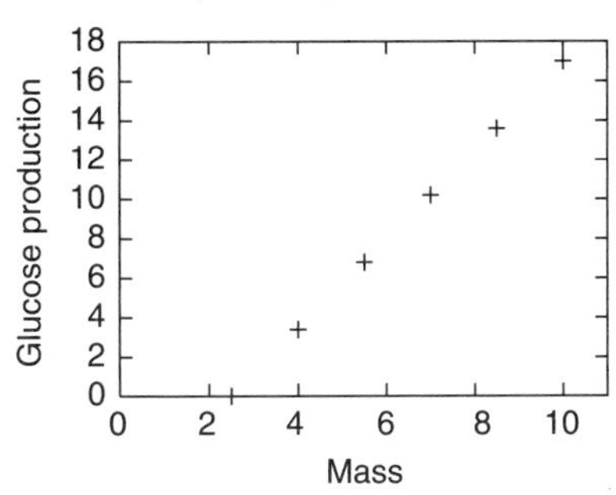

51. The point (2.0, 175) lies below the line.

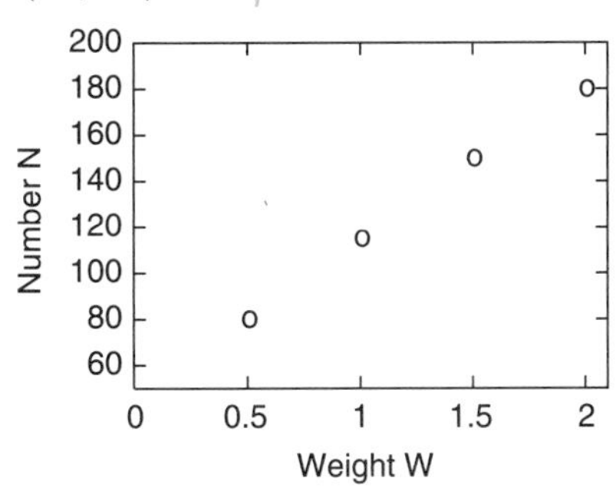

53. $N = 70(0.72 - 0.5) + 80 = 95.4$.

55. The time decreases from 216.8 s to 215.9 s, giving a change of -0.9 s in 16 years or a slope of $-0.9/16 = -0.0563$ s/y. In point-slope form, the men's time m as a function of the year y is

$$m = -0.0563(y - 1972) + 216.8.$$

57. Setting equal to 0 and solving for y,

$$\begin{aligned}-0.469(y - 1972) + 241.4 &= 0\\ 0.469(y - 1972) &= 241.4\\ (y - 1972) &= \frac{241.4}{0.469}\\ y &= \frac{241.4}{0.469} + 1972 = 2486.\end{aligned}$$

It seems likely that this will never happen, so we can assume that women will not improve this quickly forever.

Section 1.5, page 65

1. The updating function is $f(p_t) = p_t - 2$, and $f(5) = 3$, $f(10) = 8$, $f(15) = 13$. This is a linear function.

3. The updating function is $f(x_t) = x_t^2 + 2$, and $f(0) = 2$, $f(2) = 6$, $f(4) = 18$. This is not a linear function because the input x_t is squared.

5. Denote the updating function by $f(v) = 1.5v$. Then $(f \circ f)(v) = f(1.5v) = 1.5(1.5v) = 2.25v$, so $v_{t+2} = 2.25v_t$. Applying f to the initial condition twice gives $f(1220) = 1830$ and $f(1830) = 2745$, which is equal to $2.25 \cdot 1220$.

7. Denote the updating function by $h(n) = 0.5n$, then $(h \circ h)(n) = h(0.5n) = 0.5(0.5n) = 0.25n$ and $n_{t+2} = 0.25n_t$. Applying the updating function to the initial condition twice gives $h(1200) = 600$ and $h(600) = 300$, matching $(h \circ h)(1200) = 0.25 \cdot 1200 = 300$.

9. Solving for v_t gives $v_t = \frac{v_{t+1}}{1.5}$. Then $v_0 = \frac{1220}{1.5} = 813.3$.

11. Solving for n_t gives $n_t = 2n_{t+1}$. Then $n_0 = 2 \cdot 1200 = 2400$.

13.
$$(f \circ f)(x) = f[f(x)] = f\left(\frac{x}{1+x}\right) = \frac{\frac{x}{1+x}}{1 + \frac{x}{1+x}} = \frac{\frac{x}{1+x}}{\frac{1+2x}{1+x}} = \frac{x}{1+2x}.$$

To find the inverse, set $y = f(x)$ and solve

$$\begin{aligned}y &= \frac{x}{1+x}\\ (1+x)y &= x\\ y + xy &= x\\ y &= x - xy\\ \frac{y}{1-y} &= x.\end{aligned}$$

Therefore, $f^{-1}(y) = \frac{y}{1-y}$.

15. $v_t = 1.5^t \cdot 1220\mu\text{m}^3$.

$$\begin{aligned}v_1 &= 1.5 \cdot 1220 = 1830\\ v_2 &= 1.5 \cdot 1830 = 2745\\ v_3 &= 1.5 \cdot 2745 = 4117.5\\ v_4 &= 1.5 \cdot 4117.5 = 6176.25\\ v_5 &= 1.5 \cdot 6176.25 = 9264.375.\end{aligned}$$

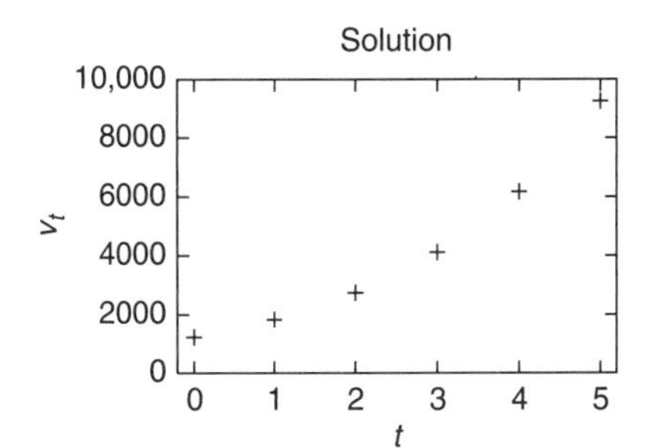

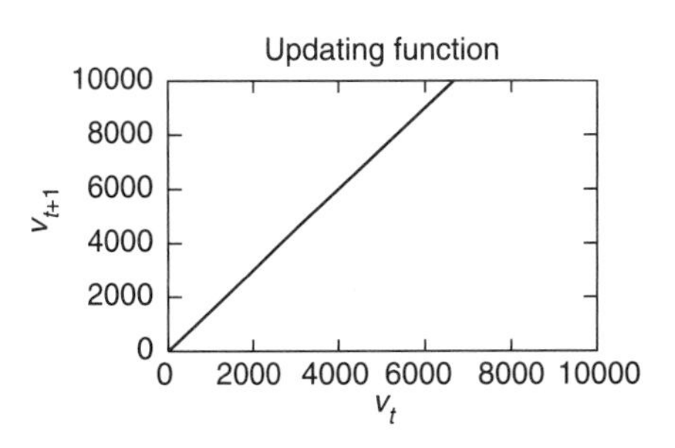

17. $n_t = 0.5^t \cdot 1200$.

$$\begin{aligned}n_1 &= 0.5 \cdot 1200 = 600\\ n_2 &= 0.5 \cdot 600 = 300\\ n_3 &= 0.5 \cdot 300 = 150\\ n_4 &= 0.5 \cdot 150 = 75\\ n_5 &= 0.5 \cdot 75 = 37.5.\end{aligned}$$

Solution

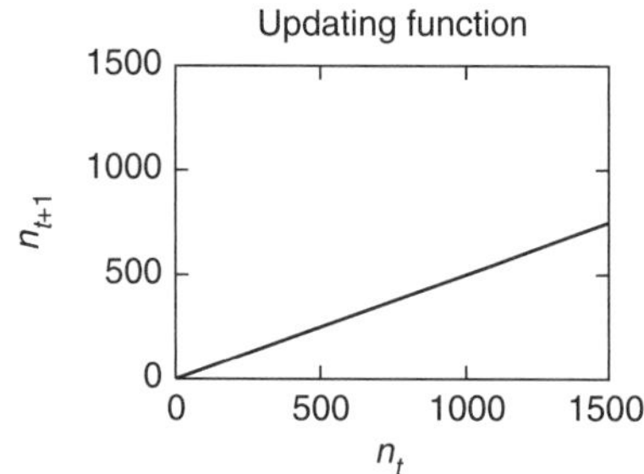

19. Plugging $t = 20$ into the solution $v_t = 1.5^t \cdot 1220\mu\text{m}^3$, we get $v_{20} = 4.05 \times 10^6 \mu\text{m}^3$. This might be reasonable.

21. Plugging $t = 20$ into the solution $n_t = 0.5^t \cdot 1200$, we get $n_{20} = 0.0011$. This is an unreasonably small population.

23. $x_1 = 1/2$, $x_2 = 1/3$, $x_3 = 1/4$, $x_4 = 1/5$. It looks like $x_t = \frac{1}{1+t}$.

25. $x_1 = 3$, $x_2 = 1$, $x_3 = 3$, $x_4 = 1$. It seems to be jumping back and forth between 1 and 3. If I start at $x_0 = 0$, the results jump back and forth between 0 and 4.

27. These do not commute. If you started with 100, doubled (giving 200), and then removed 10, you would end up with 190. If you started with 100, removed 10 (leaving 90) and then doubled, you'd have only 180. In general, if we call the starting population P_t, if we double first and then remove 10, we end up with $P_{t+1} = 2P_t - 10$. If we first remove 10 and then double, we end up with $P_{t+1} = 2(P_t - 10) = 2P_t - 20$, which never matches the result in the other order.

29. These do commute. Either way, it ends up 1.0 cm taller.

31. $h_{20} = 10.0 + 20 = 30.0$ m, a reasonable height for a tree.

33. 1.05×10^6 million bacteria. These will weigh about 10^{-6} grams, which sounds reasonable.

35. The solution is $x_1 = 50$, $x_2 = 130$, $x_3 = 290$. Adding 30, we see that $x_0 + 30 = 40$, $x_1 + 30 = 80 = 40 \cdot 2$, $x_2 + 30 = 160 = 40 \cdot 2^2$, and $x_3 + 30 = 320 = 40 \cdot 2^3$. It looks like $x_t + 30 = 40 \cdot 2^t$, so $x_t = 40 \cdot 2^t - 30$.

37.

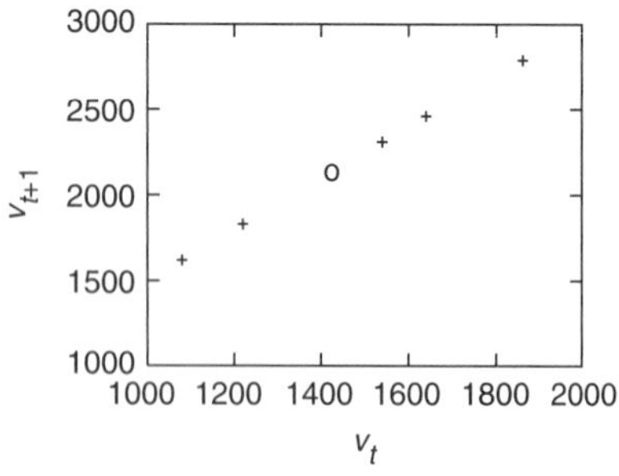

The discrete-time dynamical system is $v_{t+1} = 1.5v_t$ and the missing value is 2130.

39.

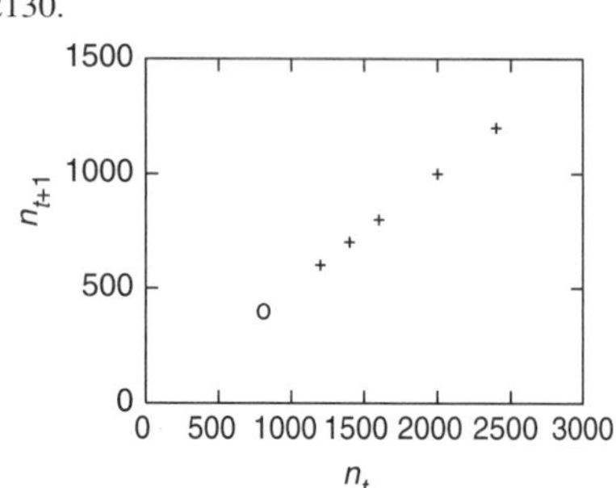

The discrete-time dynamical system is $n_{t+1} = 0.5n_t$ and the missing value is 4.0×10^2.

41. The length increases by 1.5 cm each half day, so $l_{t+1} = l_t + 1.5$cm.

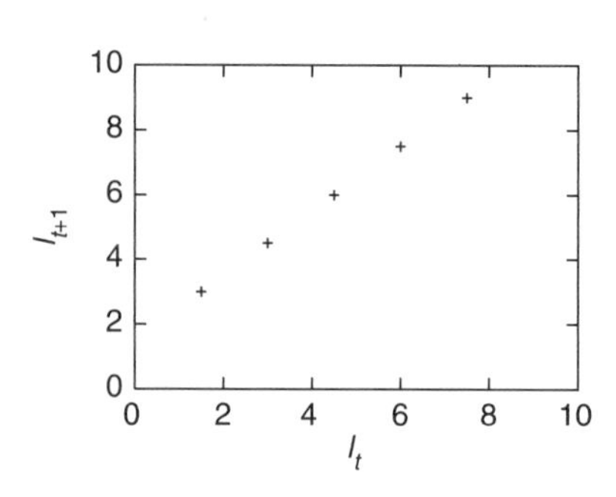

43. The mass doubles each half day, so $m_{t+1} = 2m_t$.

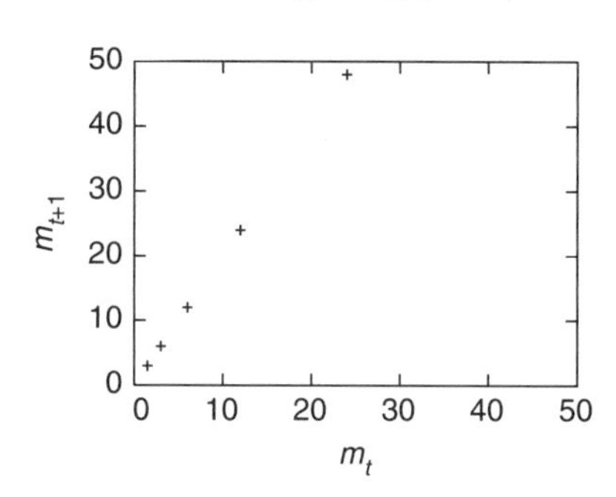

45. The argument is the initial score. The value is the final score.

47.

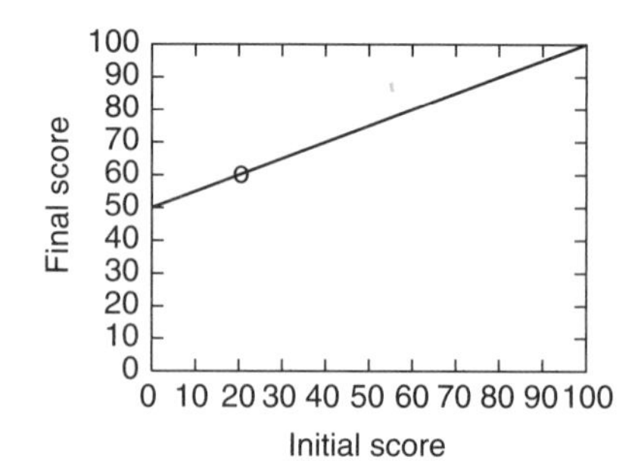

49. Let v_{t+1} and v_t be the total volume before and after the experiment. Then

$$v_t = 10^4 b_t \text{ and } v_{t+1} = 10^4 b_{t+1}.$$

The original discrete-time dynamical system is

$$b_{t+1} = 2.0b_t.$$

Therefore,

$$v_{t+1} = 10^4 b_{t+1} = 10^4(2.0b_t) = 2.0 \cdot 10^4 b_t = 2.0v_t.$$

51. $V_t = \pi h_t 0.5^2$, and $V_{t+1} = \pi h_{t+1} 0.5^2$. Therefore,

$$V_{t+1} = \pi(h_t + 1)0.5^2 = \pi h_t \cdot 0.5^2 + \pi \cdot 0.5^2 = V_t + \pi \cdot 0.5^2.$$

53. The points for the first patient are (20.0, 16.0), (16.0, 13.0), and (13.0, 10.75).

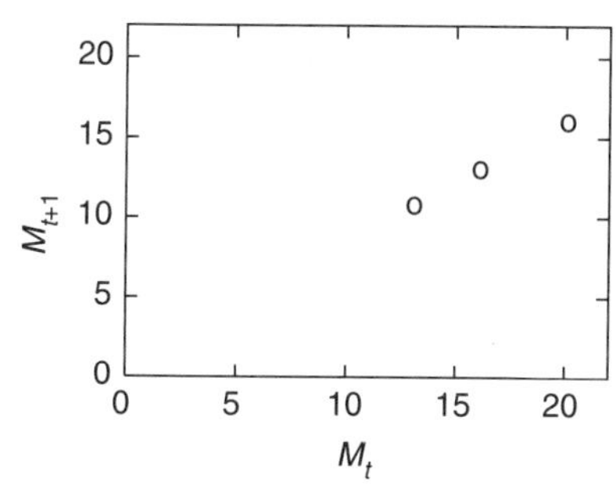

Let the level be M_t at the beginning of the day and M_{t+1} at the end of the day. Using the first two lines in the table, we find the two points to be (20.0, 16.0) and (16.0, 13.0). The slope is

$$\text{slope} = \frac{\text{change in output}}{\text{change in input}} = \frac{13.0 - 16.0}{16.0 - 20.0} = 0.75.$$

In point-slope form, the discrete-time dynamical system is

$$M_{t+1} = 0.75(M_t - 20.0) + 16.0 = 0.75M_t + 1.0.$$

55. The solution for the first is $b_t = 2.0^t \cdot 1.0 \times 10^6$, and the solution for the second is $b_t = 2.0^t \cdot 3.0 \times 10^5$. The difference is $2.0^t \cdot 0.7 \times 10^6$, but the ratio is always approximately 3.33. Both populations are growing at the same rate, but the first has a head start. It is always 3.33 times bigger, which becomes a larger difference as the populations become larger.

57. a. $b_1 = 2b_0 - 1.0 \times 10^6 = 2 \cdot 3.0 \times 10^6 - 1.0 \times 10^6 = 5.0 \times 10^6$, $b_2 = 9.0 \times 10^6$, $b_3 = 17.0 \times 10^6$.
b. There were three harvests of 1.0×10^6 bacteria, for a total of 3.0×10^6.
c. $b_{t+1} = 2.0b_t - 1.0 \times 10^6$.
d. The population would have doubled three times, so there would be 24.0×10^6 bacteria. You could harvest 7.0×10^6 bacteria and still have a population of 17.0×10^6, which means harvesting 4.0×10^6 more than in part **b**. The bacteria removed early never had a chance to reproduce.

59. a. If the old fraction is f_t, then $f_{t+1} = f_t + 0.1f_t$, or $f_{t+1} = 1.1f_t$.
b. $f_t = 0.001 \cdot 1.1^t$.
c. The fraction gets larger and larger and will eventually exceed 1. The discrete-time dynamical system doesn't make sense when $f_{t+1} > 1$ because a fraction can't be bigger than 1.

61. We have that $b_{t+1} = 2b_t$ and $m_{t+1} = 3m_t$. Then the total mass $M_{t+1} = m_{t+1}b_{t+1} = 3m_t \cdot 2b_t = 6m_tb_t = 6M_t$.

Section 1.6, page 78

1.

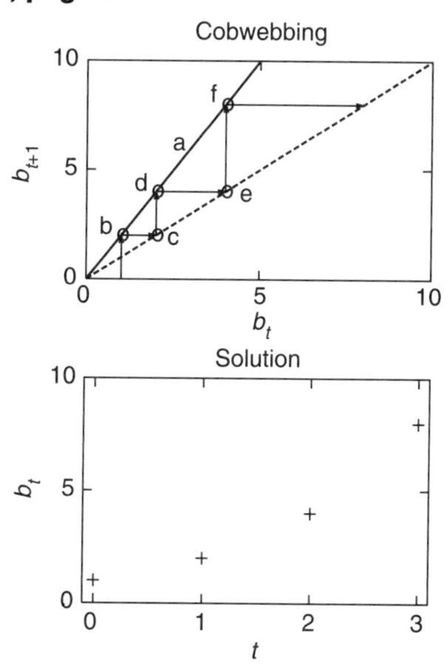

3. The solution was $v_t = 1.5^t \cdot 1220\mu\text{m}^3$, consistent with a cobweb diagram that predicts a solution that increases faster and faster.

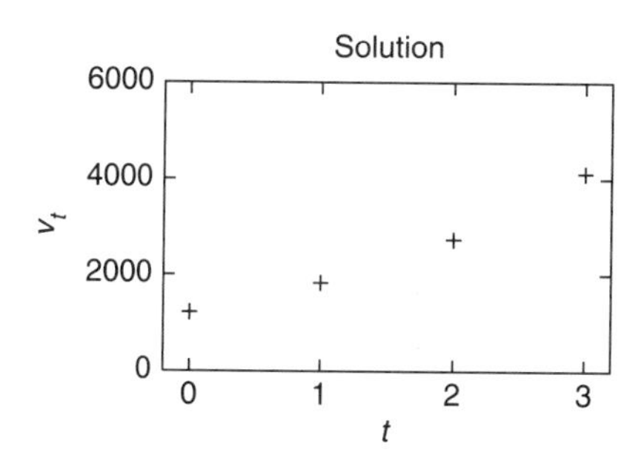

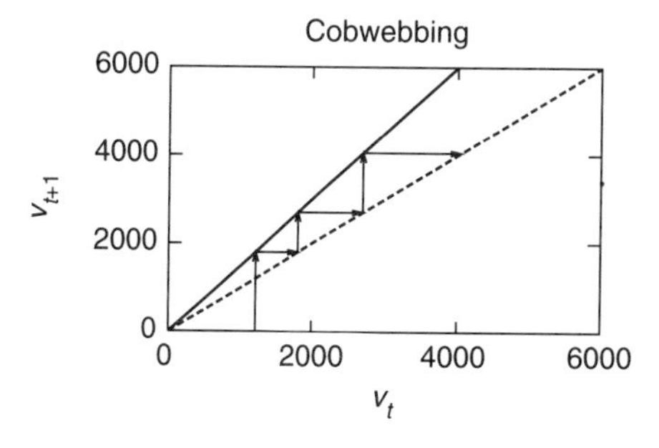

5. The solution was $n_t = 0.5^t \cdot 1200$, consistent with a cobweb diagram that decays toward 0.

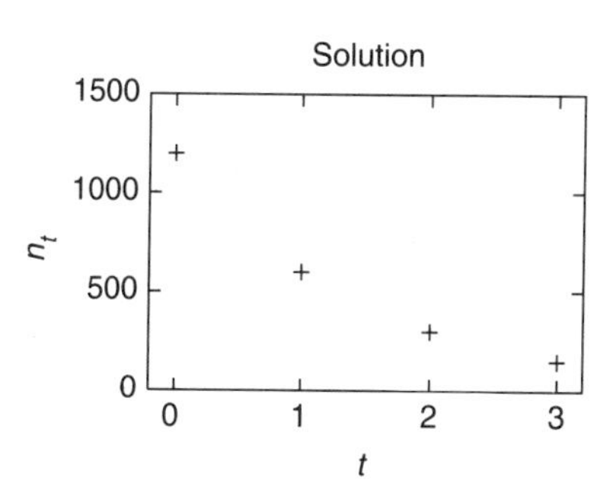

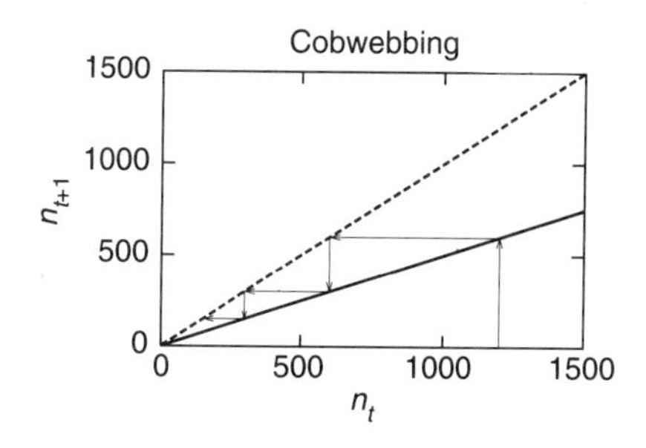

7.

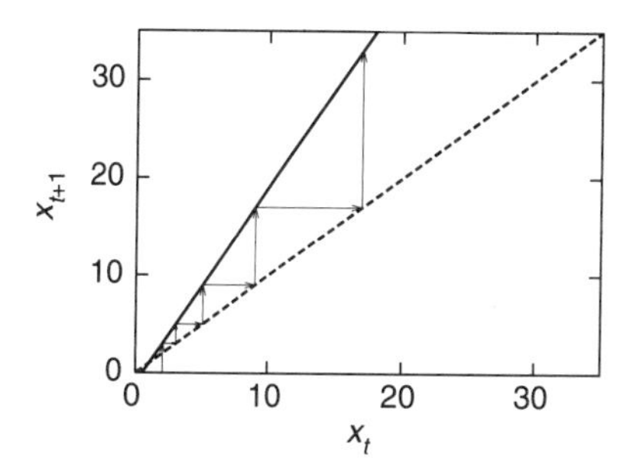

9.

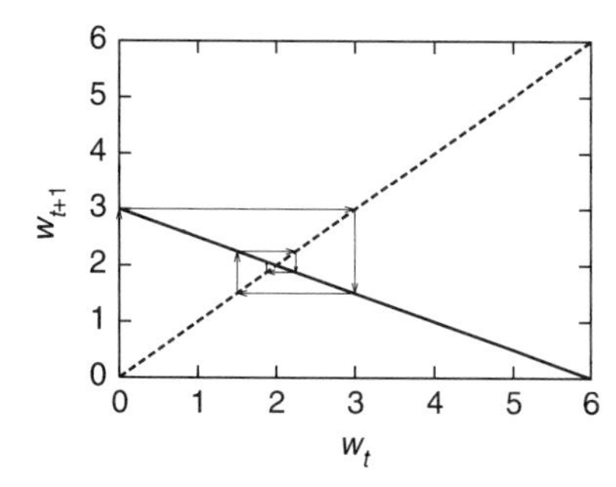

11.

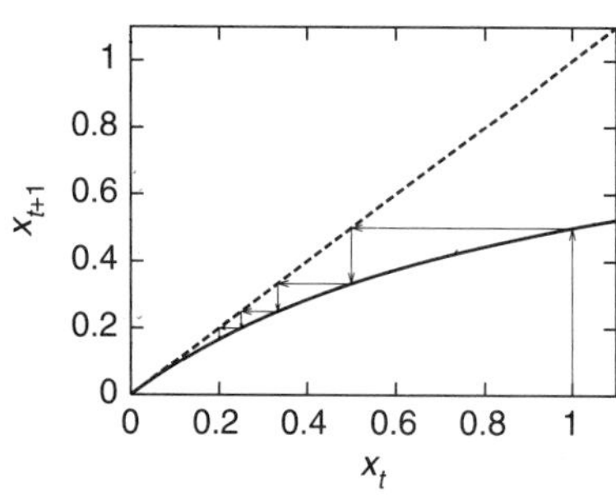

13. The equilibrium seems to be at about 1.3.

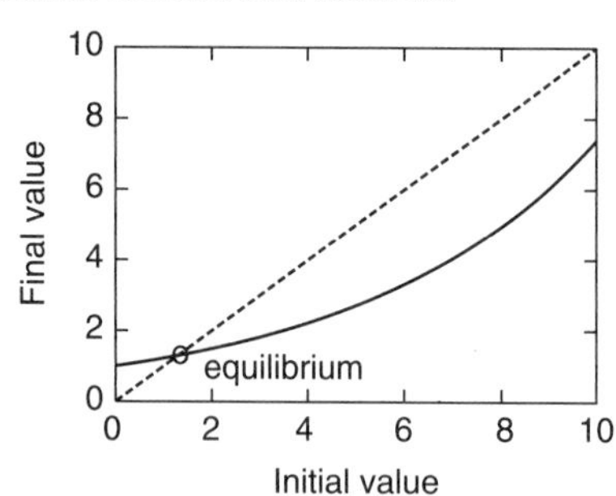

15. The equilibria seem to be at about 0.0 and 7.5.

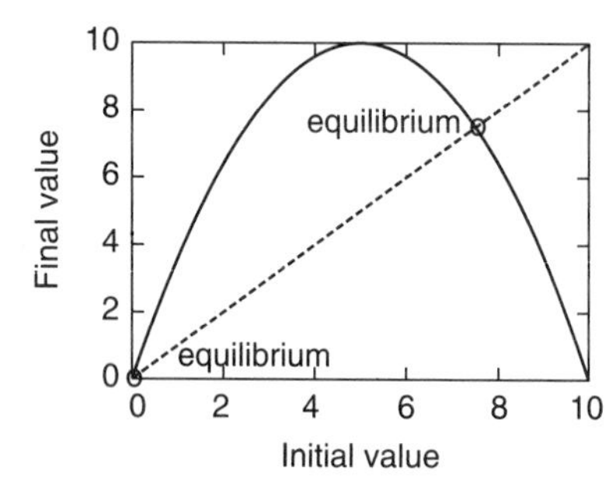

17. $f(x) = x$ when $x^2 = x$, or $x^2 - x = 0$, or $x(x-1) = 0$ which has solutions at $x = 0$ and $x = 1$.

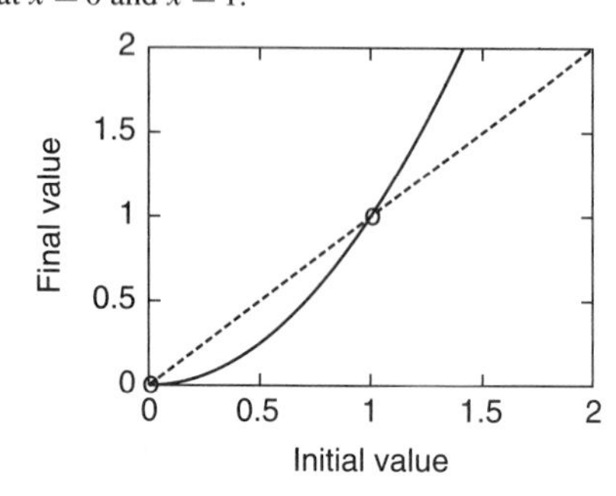

19. The equilibrium is where $c^* = 0.5c^* + 8.0$ or $c^* = 16.0$.

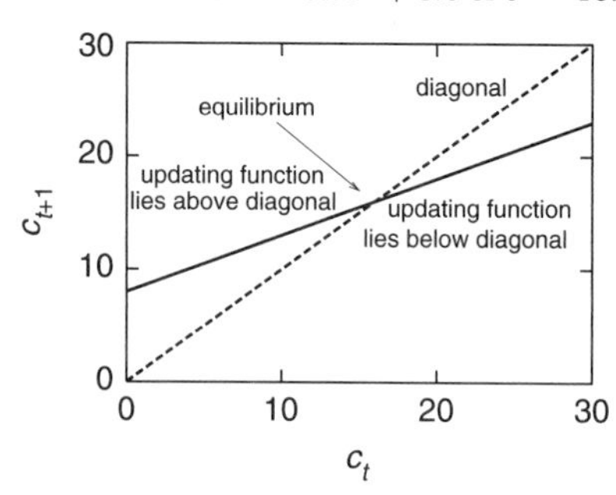

21. The equilibrium is $b^* = 0$.

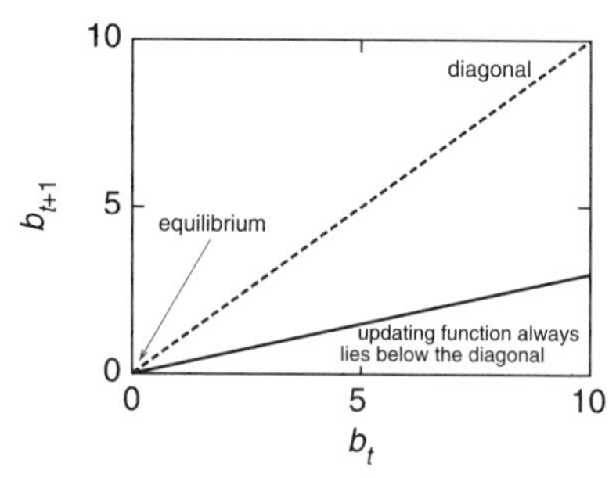

23. $v^* = 1.5v^*$ if $v^* = 0$.

25. $x^* = 2x^* - 1$ has solution $x^* = 1$.

27. $w^* = -0.5w^* + 3$ has solution $w^* = 2$.

29. $x^* = \frac{x^*}{1+x^*}$ has solution $x^* = 0$.

31.

$$\begin{aligned} w^* &= aw^* + 3 \\ w^* - aw^* &= 3 \\ w^* &= \frac{3}{1-a}. \end{aligned}$$

This solution does not exist if $a = 1$, and is negative if $a > 1$.

33.

$$\begin{aligned} x^* &= \frac{ax^*}{1+x^*} \\ x^*(1+x^*) &= ax^* \\ x^*(1+x^*) - ax^* &= 0 \\ x^*(1+x^*-a) &= 0. \end{aligned}$$

There are two solutions, $x^* = 0$ and $x* = a - 1$. The second is negative if $a < 1$. The two solutions are equal (leaving only one) when $a = 1$.

35.

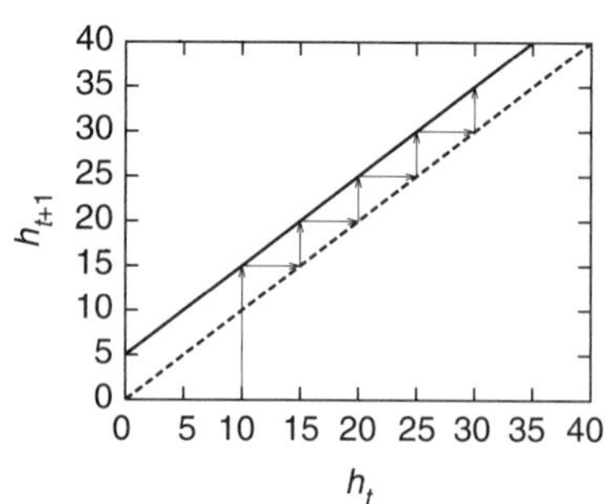

37.

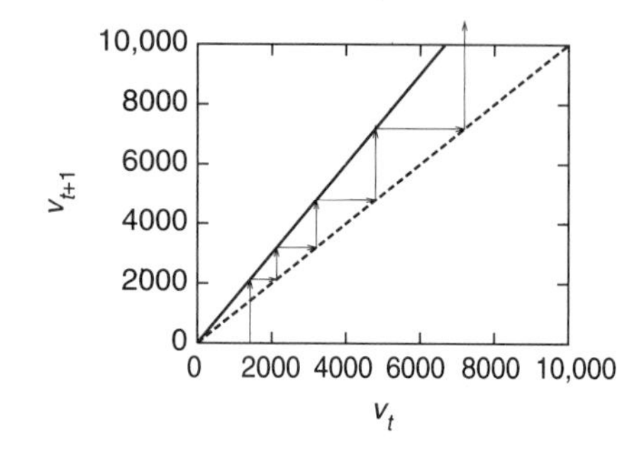

39.

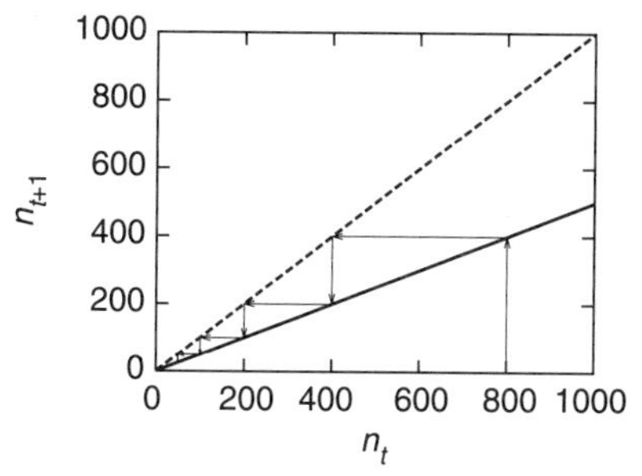

41.

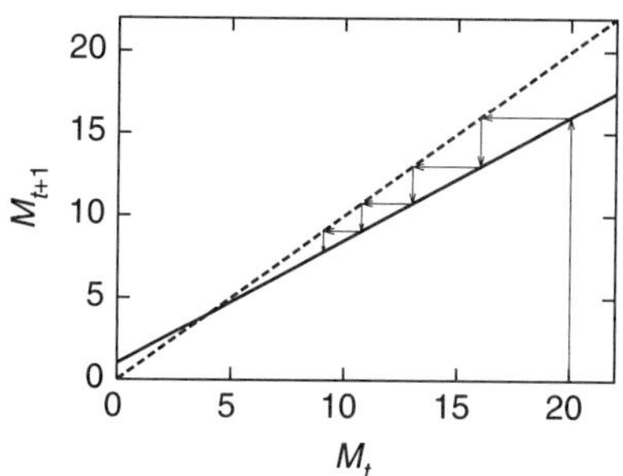

The equilibrium is

$$\begin{aligned} M^* &= 0.75M^* + 1 \\ M^* - 0.75M^* &= 1 \\ 0.25M^* &= 1 \\ M^* &= 4. \end{aligned}$$

43.

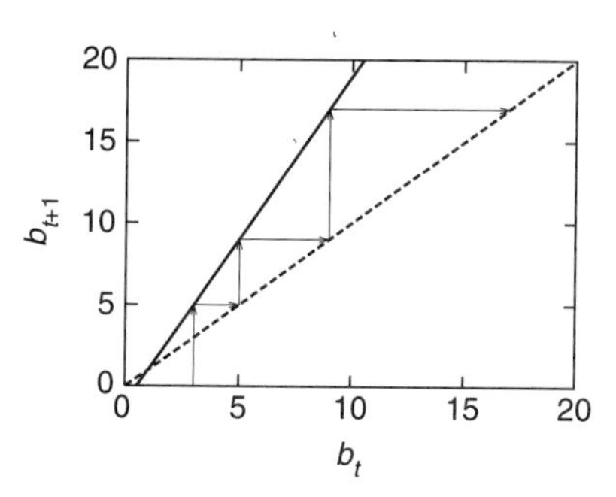

The equilibrium is

$$\begin{aligned} b^* &= 2.0b^* - 1.0 \times 10^6 \\ b^* - 2.0b^* &= -1.0 \times 10^6 \\ -b^* &= -1.0 \times 10^6 \\ b^* &= 1.0 \times 10^6. \end{aligned}$$

The population grows, as we found in Section 1.5, Exercise 57, and seems to be moving away from the equilibrium.

45. The equilibrium is

$$\begin{aligned} b^* &= 2.0b^* - h \\ b^* - 2.0b^* &= -h \\ -b^* &= -h \\ b^* &= h. \end{aligned}$$

It is strange that the equilibrium gets larger as the harvest gets larger. However, the cobwebbing diagram indicates that only populations above the equilibrium will grow, and those below it will shrink. The equilibrium in this case is the minimum population required for the population to survive.

47. The equilibrium is

$$M^* = \frac{S}{\alpha} = \frac{2}{0.2} = 10.0.$$

To achieve this level in one step, we give a loading dose of 10.0 on the first day.

Section 1.7, page 94

1. Law 6: $43.2^0 = 1$.

3. Law 3: $43.2^{-1} = 1/43.2 = 0.023$.

5. Law 4: $43.2^{7.2}/43.2^{6.2} = 43.2^{7.2-6.2} = 43.2^1 = 43.2$.

7. Law 2: $(3^4)^{0.5} = 3^{4\cdot 0.5} = 3^2 = 9$.

9. To use law 1, we need to first multiply out the exponents, finding $2^{2^3} \cdot 2^{2^2} = 2^8 \cdot 2^4 = 2^{12} = 4096$.

11. Law 6: $\ln(1) = 0$.

13. Law 5: $\log_{43.2} 43.2 = 1$.

15. Law 1: $\log_{10}(5) + \log_{10}(20) = \log_{10}(5 \cdot 20) = \log_{10}(100) = \log_{10}(10^2) = 2$.

17. Law 4: $\log_{10}(500) - \log_{10}(50) = \log_{10}(500/50) = \log_{10}(10) = 1$.

19. Law 2: $\log_{43.2}(43.2^7) = 7$.

21. Law 3: $\log_7(\frac{1}{43.2}) = -\log_7 43.2 = -1.935$.

23. $e^{3x} = \frac{21}{7} = 3$. Taking logs of both sides, $3x = \ln(3) = 1.099$, and $x = 0.366$. Checking, $7e^{3\cdot 0.366} = 21.0$.

25. Taking logs of both sides gives $\ln(4) - 2x + 1 = \ln(7) + 3x$. Moving the x's to one side gives $5x = \ln(4) + 1 - \ln(7) = 1 + \ln(\frac{4}{7}) = 0.440$, and $x = 0.088$. Checking, $4e^{-2\cdot 0.088+1} = 7e^{3\cdot 0.088} = 9.11$.

27. $e^{2x} = 7$ at $2x = \ln(7.0) = 1.946$ or $x = 0.973$. This function is increasing, and doubles after a "time" of $x = \frac{\ln(2)}{2} = 0.346$.

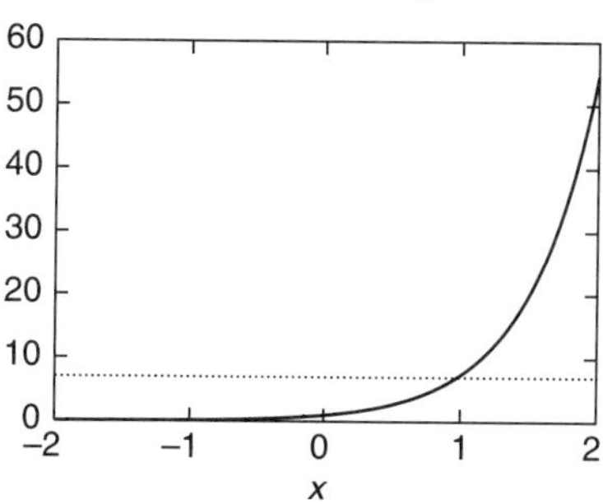

Therefore, $e^{2x} = 14.0$ when $x = 0.973 + 0.346 = 1.319$, and $e^{2x} = 3.5$ when $x = 0.973 - 0.346 = 0.627$.

29. $5e^{0.2x} = 7$ when $e^{0.2x} = 1.4$, or $0.2x = \ln(1.4) = 0.336$ or $x = 1.68$. This function is increasing, and doubles after a "time" of $x = \frac{\ln(2)}{0.2} = 3.46$.

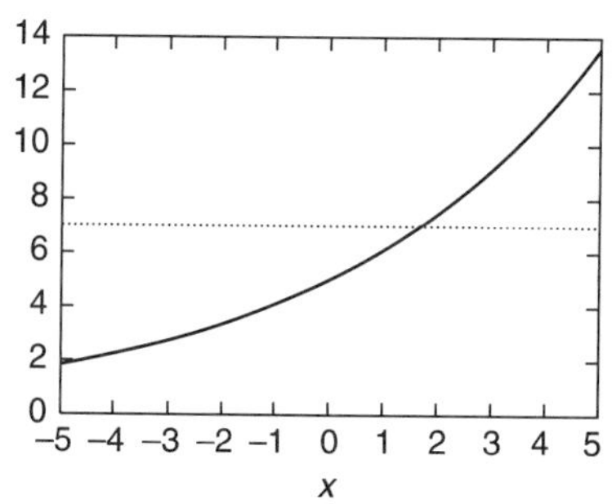

Therefore, $5e^{0.2x} = 14.0$ when $x = 1.68 + 3.46 = 5.14$, and $5e^{0.2x} = 3.5$ when $x = 1.68 - 3.46 = -1.78$.

31.

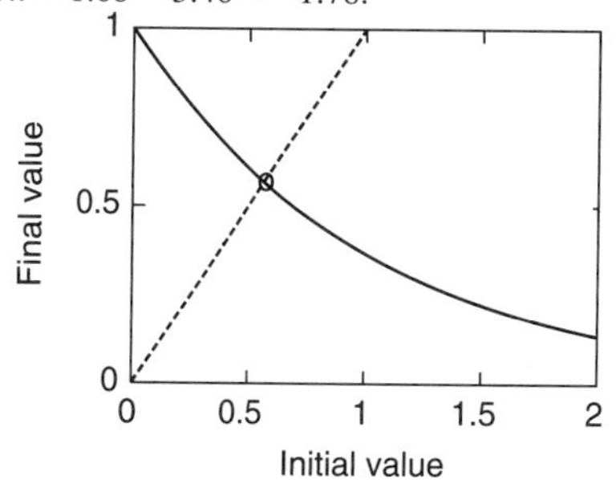

33. $\ln[M(t)] = \ln(43.2e^{5.1t}) = \ln(43.2) + 5.1t = 3.77 + 5.1t$. The slope is 5.1 and the intercept is 3.77.

35. $\ln(M(t)) = \ln(43.2) + 5.1t$ and $\ln(S(t)) = \ln(18.2) + 4.3t$. We can solve for t in terms of $\ln(S(t))$ as $t = \frac{\ln(S(t))-\ln(18.2)}{4.3}$, and then

$$\ln(M(t)) = \ln(43.2) + 5.1\frac{\ln(S(t)) - \ln(18.2)}{4.3} = 0.32 + 1.19\ln(S(t)).$$

The slope is 1.19 and the intercept is 0.32.

37. The solution is $b_t = 1.5^t \cdot 1.0 \times 10^6$. In exponential notation, $b_t = 1.0 \times 10^6 e^{\ln(1.5)t} = 1.0 \times 10^6 e^{0.405t}$. The population reaches 1.0×10^7 when $e^{0.405t} = 10$, or $0.405t = \ln(10) = 2.302$, or $t = \frac{2.302}{0.405} = 5.865$.

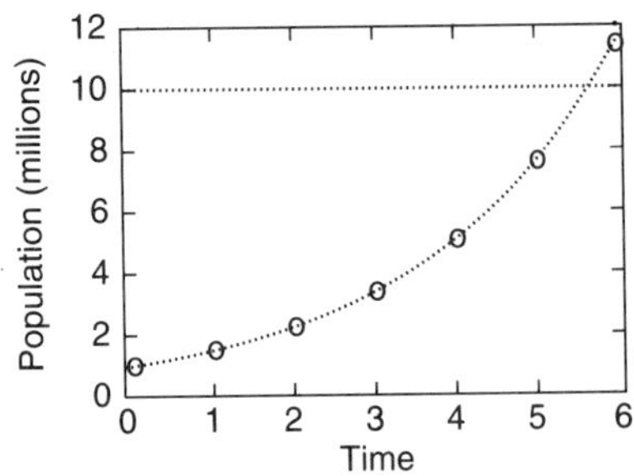

39. The solution is $v_t = 1.5^t \cdot 1350$. In exponential notation, $v_t = 1350e^{\ln(1.5)t} = 1350e^{0.405t}$. The volume reaches 3250 when $e^{0.405t} = \frac{3250}{1350} = 2.407$, or $0.405t = \ln(2.407) = 0.878$, or $t = \frac{0.878}{0.405} = 2.17$.

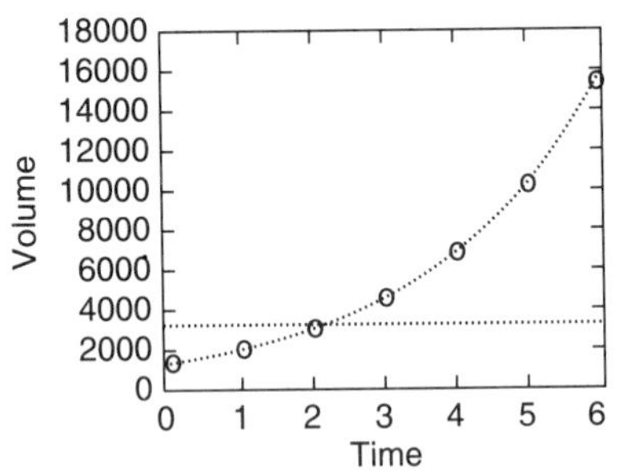

41. $t_d = 0.6931/1.0 = 0.6931$ d. It will take twice this long to quadruple, or 1.386 d.

43. $t_d = 0.6931/0.1 = 6.931$ h. It will quadruple in 13.86 h.

45. $Q(50,000) = 6.0 \times 10^{10}\, e^{-0.000122 \cdot 50,000} = 1.34 \times 10^8$. This is $\frac{1.34\times10^8}{6.0\times10^{10}} = 0.00223$ of the original amount.

47. $t_h = \frac{0.693}{0.000122} = 5680$ years.

49. It will have doubled twice and will be 2000.

51. If the doubling time is 24 yr, the parameter α is $\frac{0.693}{24} = 0.0289$. Therefore, $P(t) = 500e^{0.0289t}$.

53. It will have halved 3 times, which takes 129 years.

55. If the half-life is 43 years, the parameter α is $-\frac{0.693}{43} = -0.0161$, and the equation is $P(t) = 1600e^{-0.0161t}$.

57. We plot the line $\ln[S(t)] = \ln(2.0e^t) = 0.693 + t$.

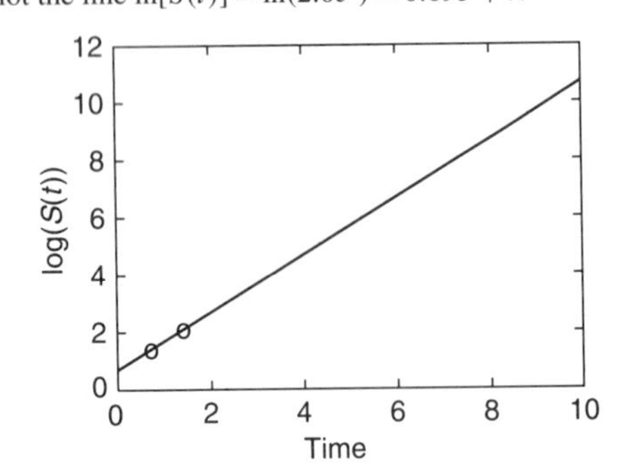

59. $\ln[P(t)] = \ln(500e^{0.0289t}) = 6.21 + 0.0289t$.

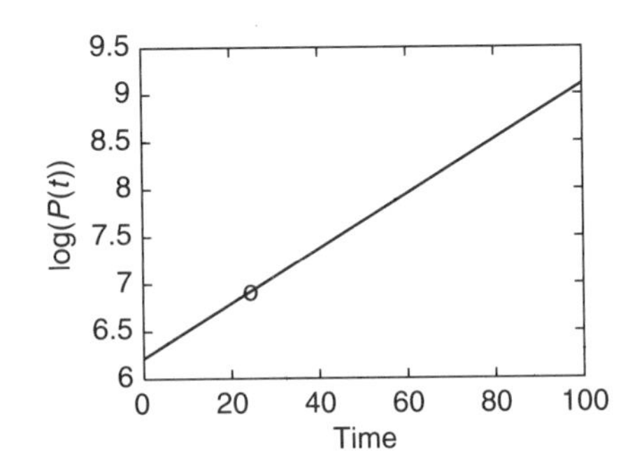

61.

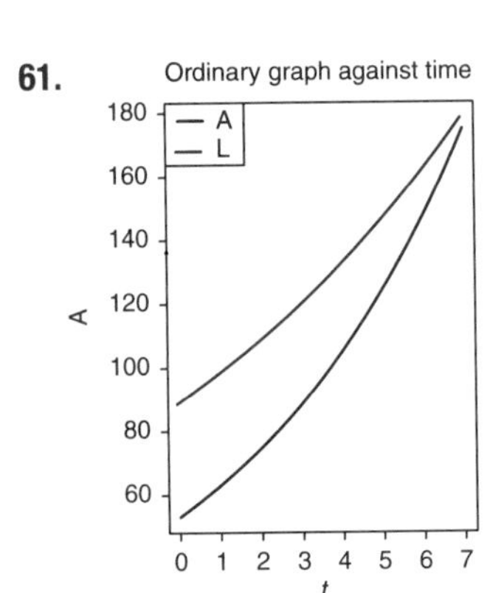

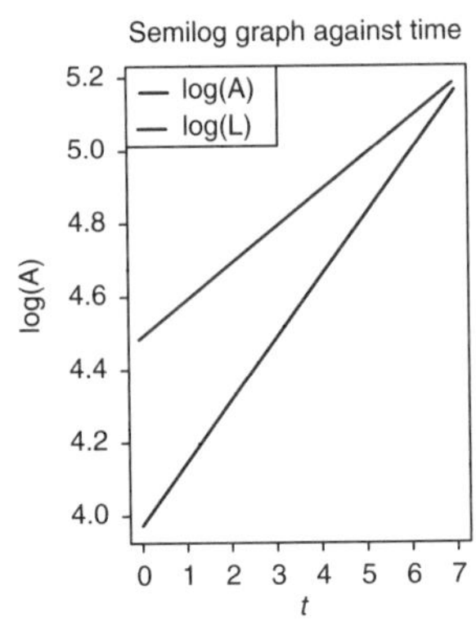

a

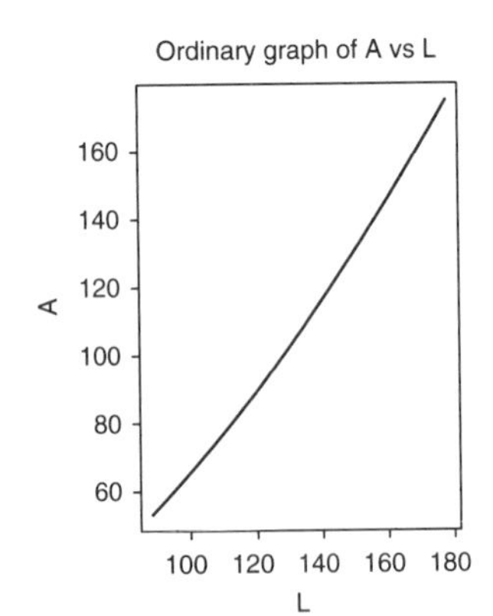

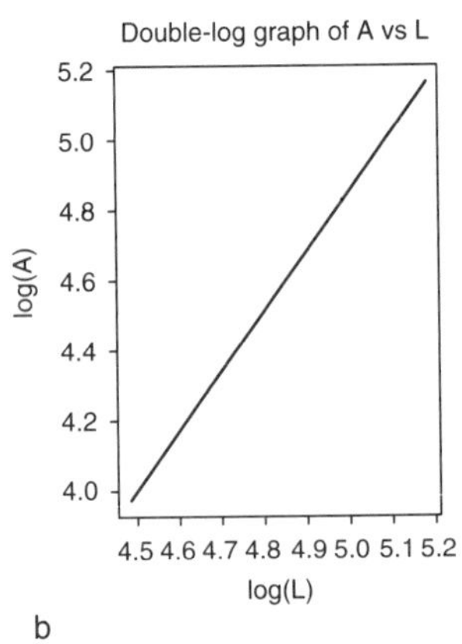

b

63.

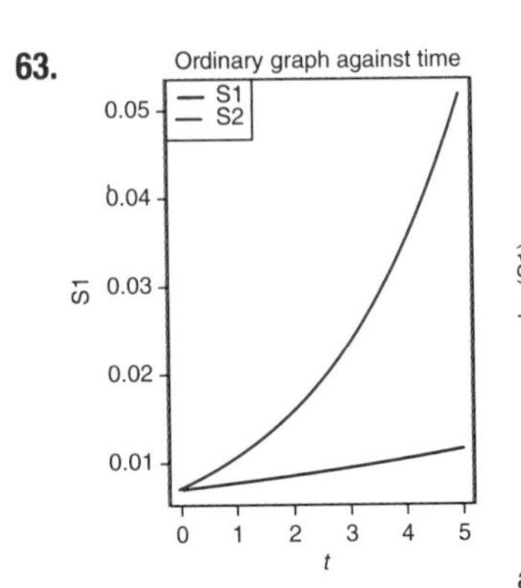

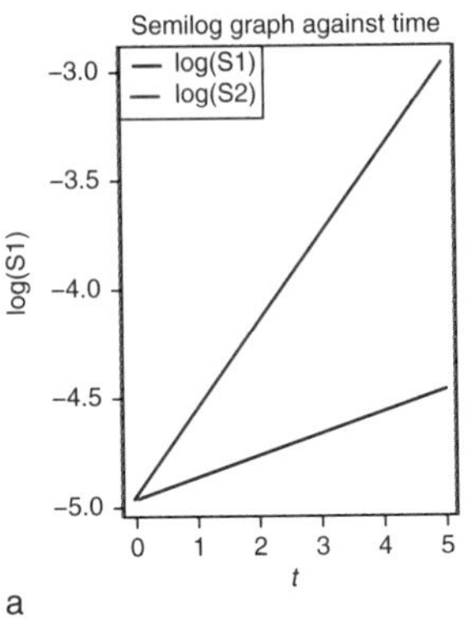

a

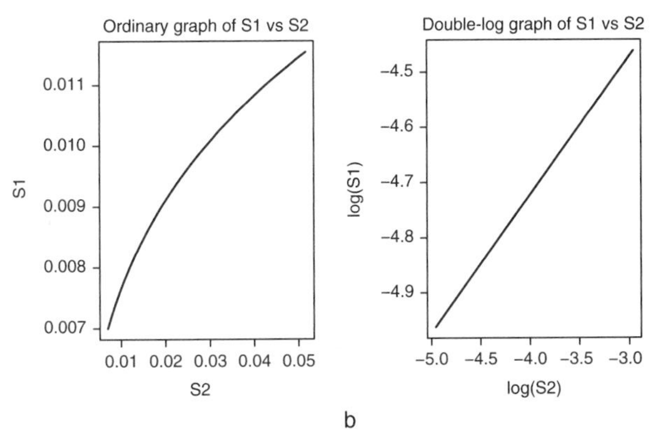

b

65. The cube has a surface area of $S = 6w^2$ and a volume of $V = w^3$. We can solve the second equation for w in terms of V, getting $w = V^{1/3}$, and then substitute that into the first equation to get $S = 6V^{2/3}$. The constant c is 6, which is 24% larger than the value for a sphere.

67. Using our initial condition, we find that $0.001 = c \cdot 10^{-6}$, so $c = 10^3$. The total mass is 10 kg. When only 100 trees remain, $W = 1.0$ kg for a total mass of 100 kg, larger than the initial value. When only 1 tree remains, $W = 1000$ kg for a total mass of 1000 kg, again larger than before.

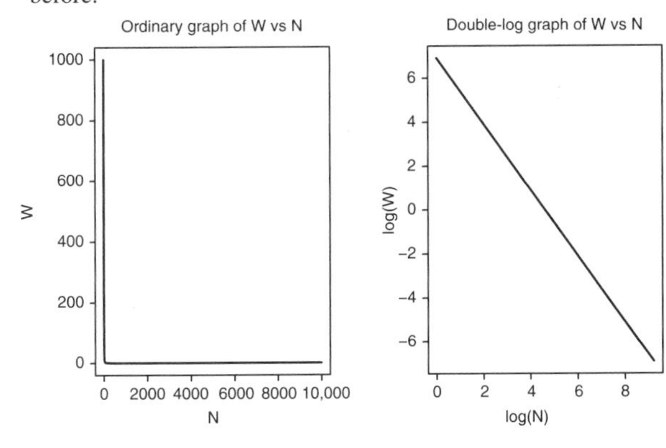

Section 1.8, page 103

1. $\sin(\pi/2) = 1$, $\cos(\pi/2) = 0$.

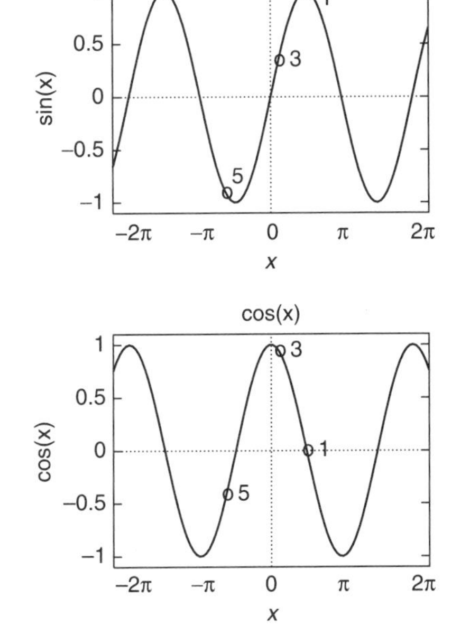

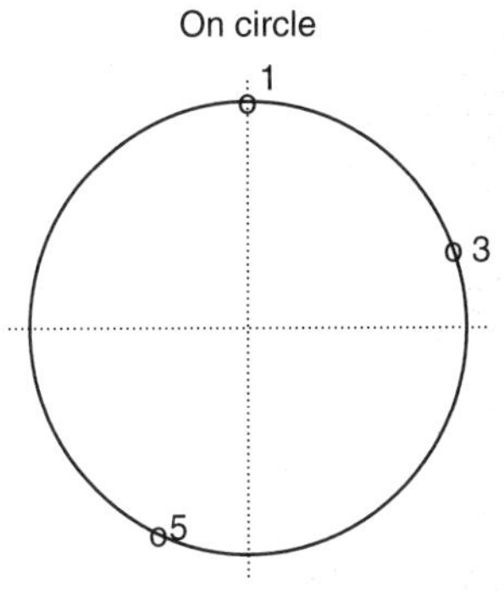

3. $\sin(\pi/9) = 0.342$, $\cos(\pi/9) = 0.940$.

5. $\sin(-2.0) = -0.909$, $\cos(-2.0) = -0.416$.

7. $\pi/6$ rad

9. 0.017 rad

11. 114.6°

13. $-36° = 324°$.

15. No answer, 0, no answer, 1.

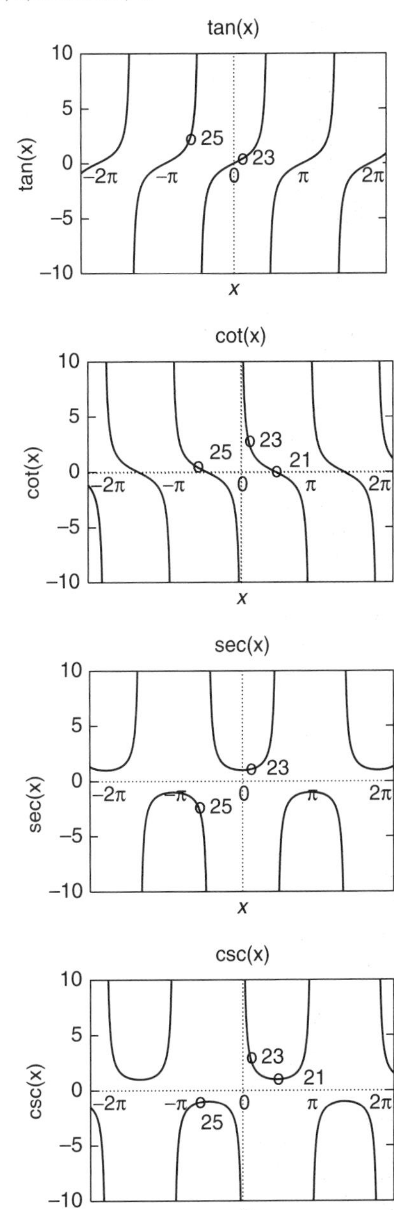

17. $\tan(\pi/9) = 0.36397$, $\cot(\pi/9) = 2.74748$, $\sec(\pi/9) = 1.06418$, $\csc(\pi/9) = 2.92380$.

19. $\tan(-2.0) = 2.18504$, $\cot(-2.0) = 0.45766$, $\sec(-2.0) = -2.40300$, $\csc(-2.0) = -1.09975$.

21. $\cos(0) = \sqrt{\frac{1+1}{2}} = 1$, $\cos(\frac{\pi}{8}) = \sqrt{\frac{1+\sqrt{2}/2}{2}} = 0.924$, $\cos(\frac{\pi}{4}) = \frac{\sqrt{2}}{2} = \sqrt{\frac{1}{2}}$, $\cos(\frac{\pi}{2}) = 0 = \sqrt{\frac{0}{2}}$.

23. $\cos(0 - \pi) = \cos(-\pi) = -1 = -\cos(0)$, $\cos(\pi/4 - \pi) = \cos(-3\pi/4) = -\sqrt{2}/2 = -\cos(\pi/4)$, $\cos(\pi/2 - \pi) = \cos(-\pi/2) = 0 = -\cos(\pi/2)$, $\cos(\pi - \pi) = \cos(0) = 1 = -\cos(\pi)$.

25. $\cos(2 \cdot 0) = \cos^2(0) - \sin^2(0) = 1 - 0 = 1 = \cos(0)$, $\cos(2 \cdot \pi/4) = \cos^2(\pi/4) - \sin^2(\pi/4) = 1/2 - 1/2 = 0 = \cos(\pi/2)$, $\cos(2 \cdot \pi/2) = \cos^2(\pi/2) - \sin^2(\pi/2) = 0 - 1 = -1 = \cos(\pi)$, $\cos(2 \cdot \pi) = \cos^2(\pi) - \sin^2(\pi) = 1 - 0 = 1 = \cos(2\pi)$.

27. Multiplying the factor of 5.0 through gives $r(t) = 10.0 + 5.0\cos(2\pi t)$ with average 10.0, amplitude 5.0, period 1.0, and phase 0.

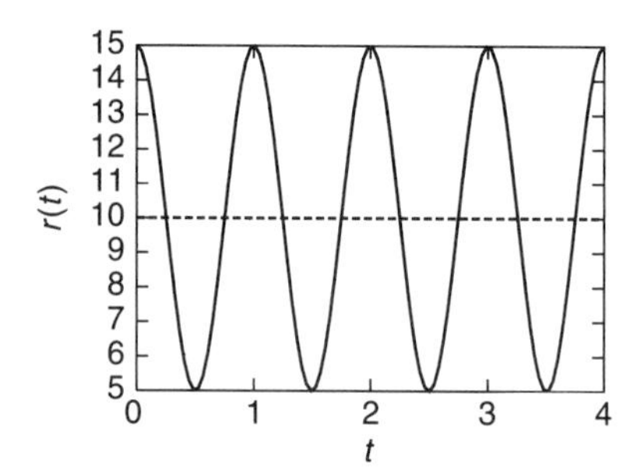

29. Use the fact that $\cos(t - \pi) = -\cos(t)$. Then $f(t) = 2.0 + 1.0\cos(t - \pi)$ with average 2.0, amplitude 1.0, period 2π, and phase π.

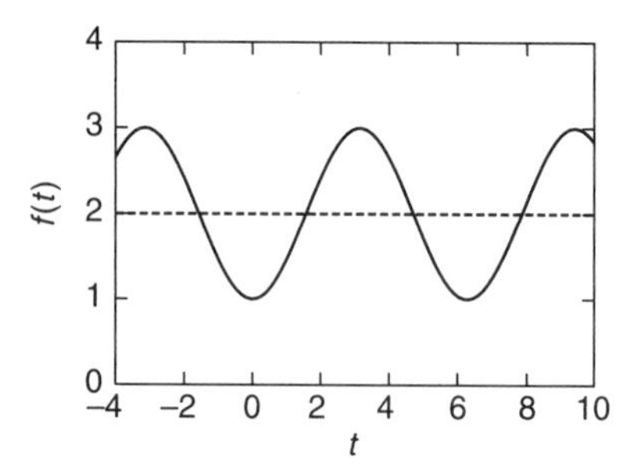

31. Average is 6.0, minimum is 4.0, maximum is 8.0, amplitude is 2.0, period is 6.0, and phase is 2.0.

33. Average is 4.0, minimum is -1.0, maximum is 9.0, amplitude is 5.0, period is 2.0, and phase is 0.0.

35. Average is 3.0, amplitude is 4.0, maximum is 7.0, minimum is -1.0, period is 5.0, and phase is 1.0.

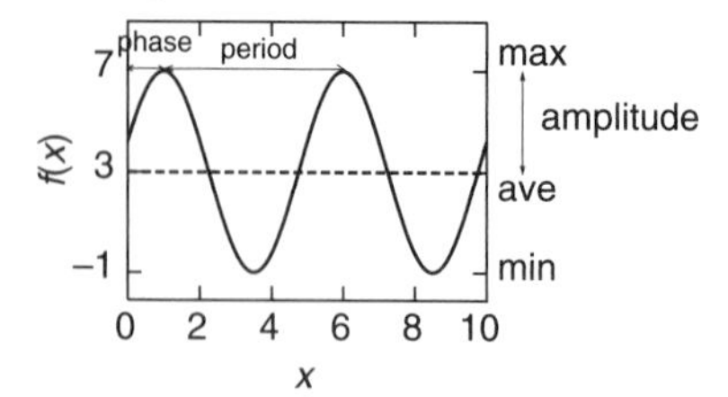

37. Average is 1.0, amplitude is 5.0, maximum is 6.0, minimum is -5.0, period is 4.0, and phase is 3.0.

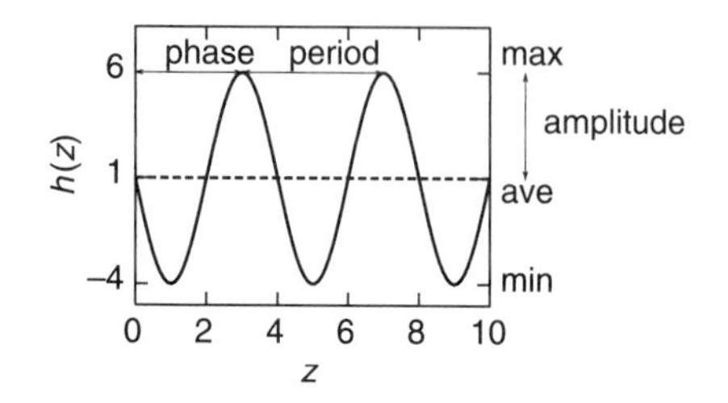

39. This function increases overall, but wiggles around quite a bit. It might describe the size of an organism that grows on average, but grows quickly during the day and shrinks down a bit at night.

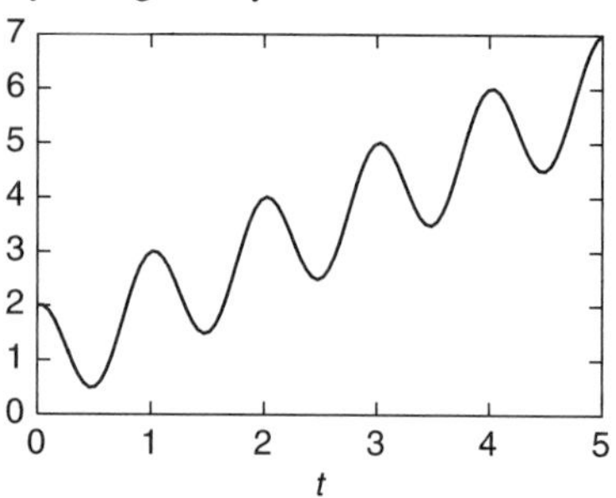

41. This function wiggles up and down with increasing amplitude. Perhaps it describes the insane temperature oscillations that will precede the next ice age.

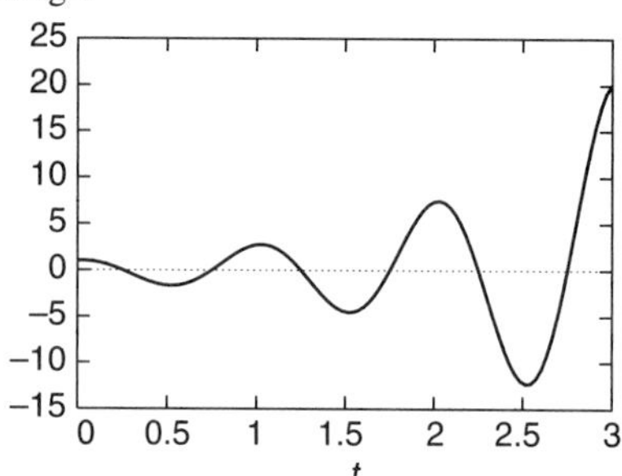

43. This function wiggles up and down with increasing period, but with constant amplitude. Perhaps it describes the position of a child jumping up and down on a mattress in expectation of receiving dessert.

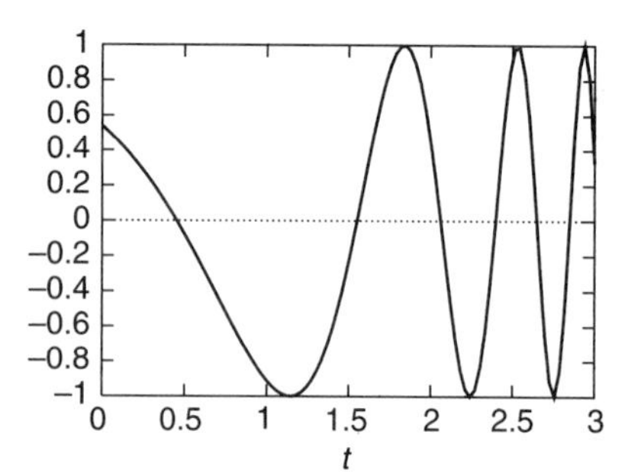

45. Let $S_c(t)$ represent the circadian rhythm. Then $S_c(t) = \cos\left(\frac{2\pi t}{24}\right)$.

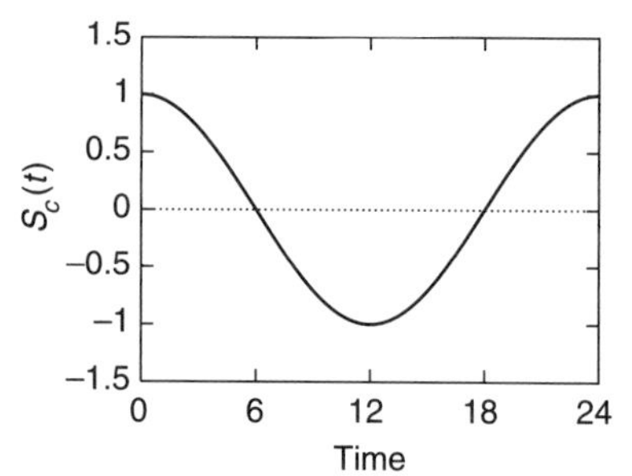

47. To plot, compute the value of the combined function $S_c(t) + S_u(t)$ every hour and smoothly connect the dots.

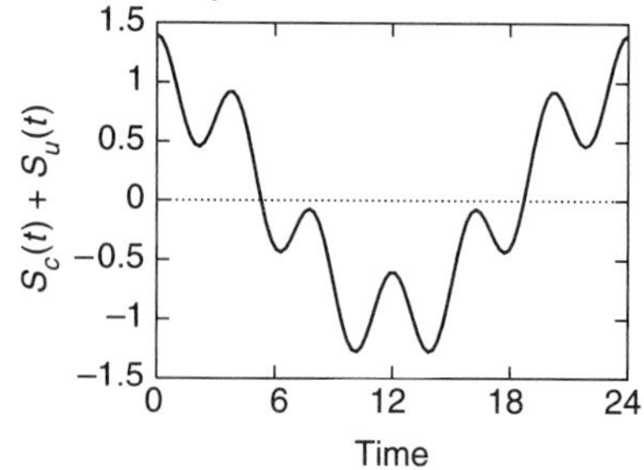

Section 1.9, page 113

1. 2/3 of the water is at 100°C, and 1/3 is at 30°C. The final temperature is then the weighted average $T = \frac{1}{3} \cdot 30°C + \frac{2}{3} \cdot 100°C = 76.7°C$.

3. A fraction $\frac{20}{52}$ got 50, $\frac{18}{52}$ got 75, and $\frac{14}{52}$ got 100, for an average of $\frac{20}{52} \cdot 50 + \frac{18}{52} \cdot 75 + \frac{14}{52} \cdot 100 = 72.1$.

5. 1/3 of the water is at T_1, and 2/3 is at T_2. The final temperature is then the weighted average $T = \frac{1}{3} \cdot T_1 + \frac{2}{3} \cdot T_2$. If $T_1 = 30$ and $T_2 = 100$, we get $T = \frac{1}{3} \cdot 30 + \frac{2}{3} \cdot 100 = 76.7$ as before.

7. A fraction $V_1/(V_1 + V_2)$ is at T_1, and a fraction $V_2/(V_1 + V_2)$ is at T_2. The final temperature is the weighted average

$$T = T_1 \frac{V_1}{V_1 + V_2} + T_2 \frac{V_2}{V_1 + V_2}.$$

9. The 100°C water cools to 50°C, so 2/3 of the water is at 50°C, and 1/3 is at 15°C. The final temperature is then the weighted average $T = \frac{1}{3} \cdot 15°C + \frac{2}{3} \cdot 50°C = 38.3°C$. This is indeed half the value in Exercise 1.

11. After the deduction, a fraction $\frac{20}{52}$ got 40, $\frac{18}{52}$ got 65, and $\frac{14}{52}$ got 90, for an average of $\frac{20}{52} \cdot 40 + \frac{18}{52} \cdot 65 + \frac{14}{52} \cdot 90 = 62.1$. This is indeed 10 less than the average before the deduction.

13. a. amount = volume times concentration, or $V c_0 = 2.0 \cdot 1.0 = 2.0$ mmol.
b. 0.5 L at 1.0 mmol/L = 0.5 mmol.
c. 1.5 L at 1.0 mmol/L = 1.5 mmol.
d. 0.5 L at 5.0 mmol/L = 2.5 mmol.
e. 1.5 + 2.5 = 4.0 mmol.
f. 4.0 mmol/2.0 L = 2.0 mmol/L.
g. $q = 0.5/2.0 = 0.25$. Then $c_{t+1} = (1 - q)c_t + q\gamma = 0.75 \cdot c_t + 0.25 \cdot 5.0$. When $c_0 = 1.0$, $c_1 = 0.75 \cdot 1.0 + 0.25 \cdot 5.0 = 2.0$ mmol/L.

15. Start with 9.0 mmol, breathe out 8.1 mmol, leaving 0.9 mmol, breathe in 4.5 mmol, ending with 5.4 mmol, and a concentration of 5.4 mmol/L. This checks with the discrete-time dynamical system. In this case, $q = 0.9$ and $\gamma = 5.0$, so $c_{t+1} = 0.1c_t + 0.9 \cdot 5.0$. Substituting $c_0 = 9.0$, we find $c_1 = 5.4$.

17. The discrete-time dynamical system has $q = 0.5/2.0 = 0.25$ and $\gamma = 5.0$, and thus has formula $c_{t+1} = (1 - 0.25)c_t + 0.25 \cdot 5.0 = 0.75c_t + 1.25$. We want to start from 1.0 mmol/L.

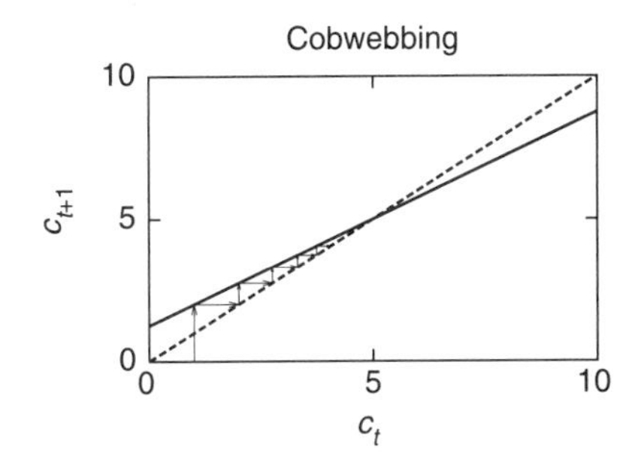

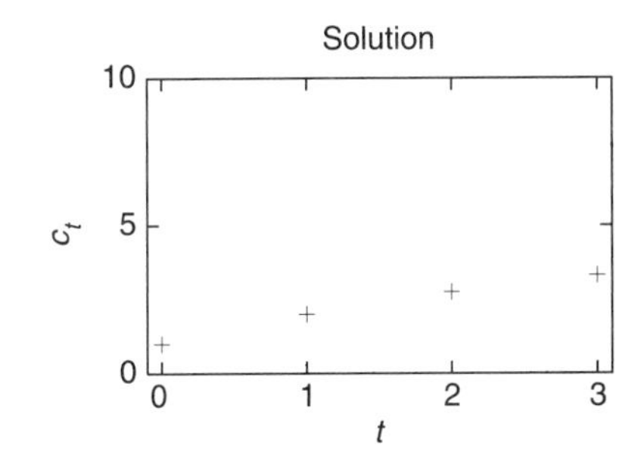

19. The discrete-time dynamical system is $c_{t+1} = 0.1c_t + 4.5$, starting from $c_0 = 9.0$.

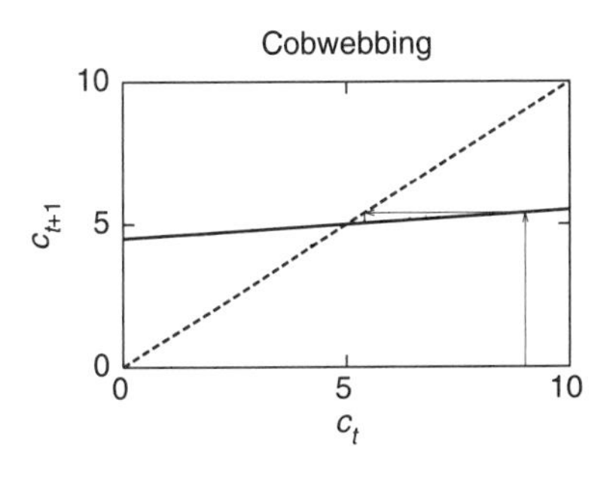

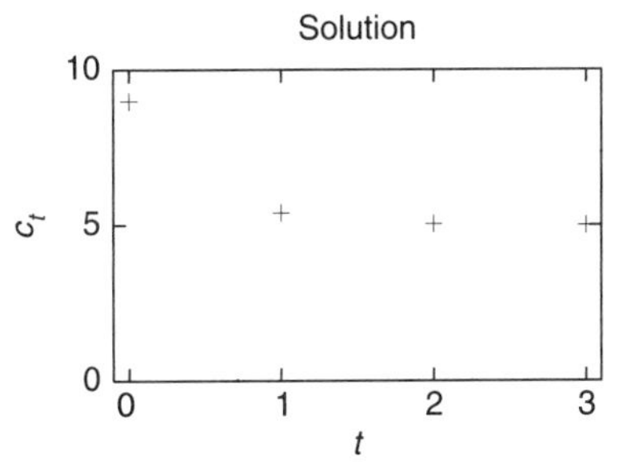

21. The discrete-time dynamical system is $c_{t+1} = 0.75 \cdot c_t + 1.25$. Solving for the equilibrium, we find $c^* = 0.75 \cdot c^* + 1.25$, or $0.25c^* = 1.25$ or $c^* = 5.0$. This matches the value of γ.

23. The discrete-time dynamical system is $c_{t+1} = 0.1 \cdot c_t + 4.5$. Solving for the equilibrium, we find $c^* = 0.1 \cdot c^* + 4.5$, or $0.9c^* = 4.5$ or $c^* = 5.0$. This matches the value of γ.

25. Using the equation $c^* = \frac{q\gamma}{1-(1-q)(1-\alpha)}$ for the equilibrium, we find that $c^* = \frac{0.4 \cdot 0.21}{1-0.6 \cdot 0.9} = 0.183$. The concentration is higher because more of the air in the lung at any one time comes from outside. The total absorbed is $0.1 \cdot 0.183 \cdot 6.0 = 0.11$, also higher than with $q = 0.2$.

27. The concentration after absorption is $c_t - 0.02$. Using the weighted average idea,

$$\begin{aligned} c_{t+1} &= (1 - q)(c_t - 0.02) + q\gamma = 0.9(c_t - 0.02) + 0.021 \\ &= 0.9c_t + 0.003. \end{aligned}$$

The equilibrium solves

$$\begin{aligned} c^* &= 0.9c^* + 0.003 \\ 0.1c^* &= 0.003 \\ c^* &= 0.03. \end{aligned}$$

This function does not really make sense if $c_t < 0.02$ because there not would be enough there to absorb.

29. The concentration after absorption is $c_t - 0.2(c_t - 0.05) = 0.8c_t + 0.01$. Then

$$\begin{aligned} c_{t+1} &= (1 - q)(0.8c_t + 0.01) + q\gamma = 0.9(0.8c_t + 0.01) + 0.021 \\ &= 0.72c_t + 0.03. \end{aligned}$$

The equilibrium is then

$$\begin{aligned} c^* &= 0.72c^* + 0.03 \\ 0.28c^* &= 0.03 \\ c^* &= 0.11. \end{aligned}$$

31. The concentration after absorption is $c_t - A$. Then

$$c_{t+1} = (1 - q)(c_t - A) + q\gamma = 0.9(c_t - A) + 0.021.$$

The equilibrium solves

$$\begin{aligned} c^* &= 0.9c^* + 0.021 - 0.9A \\ 0.1c^* &= 0.021 - 0.9A \\ c^* &= 0.21 - 9A. \end{aligned}$$

Then $c^* = 0.15$ if $0.21 - 9A = 0.15$ or $A = 0.0067$. The lung must reduce the oxygen concentration by 0.67 percent. In Example 1.9.9, we found that the lung absorbs 10 percent of the equilibrium concentration of 0.15, which is equivalent to $A = 0.015$.

33. The concentration before exchanging air is $c_t + 0.001$, so the discrete-time dynamical system is the weighted average

$$c_{t+1} = (1-q)(c_t + 0.001) + q\gamma$$
$$c_{t+1} = 0.9(c_t + 0.001) + 0.1 \cdot 0.0004 = 0.9c_t + 0.00094.$$

The equilibrium is

$$c^* = 0.9c^* + 0.00094$$
$$0.1c^* = 0.00094$$
$$c^* = 0.0094.$$

This is about 11 times higher than the ambient concentration.

35. a. Population after reproduction is $0.6 \cdot 3.0 \times 10^6 = 1.8 \times 10^6$.
b. Population after supplementation is $1.8 \times 10^6 + 1.0 \times 10^6 = 2.8 \times 10^6$.
c. $b_{t+1} = 0.6b_t + 1.0 \times 10^6$.

37. The discrete-time dynamical system is $b_{t+1} = 0.6b_t + 1.0 \times 10^6$. The equilibrium satisfies $b^* = 0.6b^* + 1.0 \times 10^6$ or $0.4b^* = 1.0 \times 10^6$ or $b^* = 2.5 \times 10^6$.

39. The discrete-time dynamical system is $b_{t+1} = 0.5b_t + S$. The equilibrium satisfies $b^* = 0.5b^* + S$, or $0.5b^* = S$ or $b^* = 2S$. The equilibrium becomes larger when S is large. This makes sense because the population will be larger when more bacteria are added.

41. Let s_t be the concentration of salt before inflow and loss through outflow. There is then $3.3 \times 10^7 s_t$ salt in the lake, which receives 3.0×10^3 of salt from inflow, for a total of $3.3 \times 10^7 s_t + 3.0 \times 10^3$ of salt in 3.6×10^7 m^3 of water. The concentration is

$$s_{t+1} = \frac{3.3 \times 10^7 s_t + 3.0 \times 10^3}{3.6 \times 10^7} = 0.917s_t + 8.33 \times 10^{-5}.$$

Because water that flows out is well mixed, this gives the discrete-time dynamical system.

43. There are $3.3 \times 10^7 s_t + 3.0 \times 10^3$ m^3 of salt in 3.6×10^7 m^3 of water. After evaporation, there are 3.3×10^7 m^3 of water, in which the concentration is

$$s_{t+1} = \frac{3.3 \times 10^7 s_t + 3.0 \times 10^3}{3.3 \times 10^7} = s_t + 9.1 \times 10^{-5}.$$

45. The discrete-time dynamical system is $s_{t+1} = 0.917s_t + 8.33 \times 10^{-5}$. The equilibrium solves

$$s^* = 0.917s^* + 8.33 \times 10^{-5}$$
$$0.083s^* = 8.33 \times 10^{-5}$$
$$s^* = 0.001.$$

The water in the lake ends up like the water that flows in.

47. The discrete-time dynamical system is $s_{t+1} = s_t + 9.1 \times 10^{-5}$. The equilibrium equation is $s^* = s^* + 9.1 \times 10^{-5}$ which has no solution. This lake has no equilibrium and will get saltier and saltier.

49. The discrete-time dynamical system is $b_{t+1} = 1.5b_t - 1.0 \times 10^6$, and the equilibrium is $b^* = 2.0 \times 10^6$.

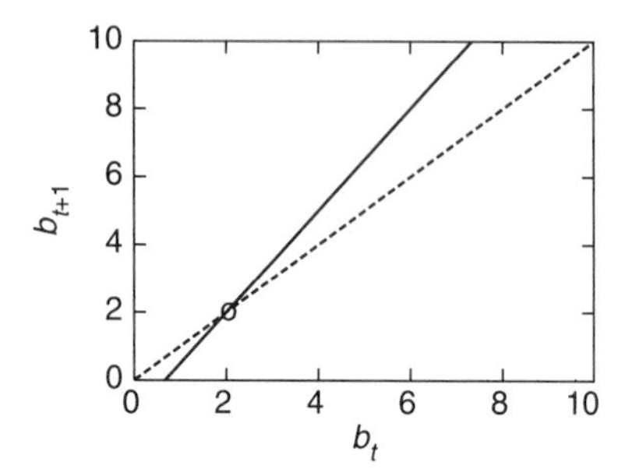

Section 1.10, page 123

1. There are now 400 red birds and 800 blue birds, or 1/3 red birds and 2/3 blue birds. These fractions add to 1.

3. There are now $200r$ red birds and 800 blue birds. The fraction of red birds is $200r$ out of $200r + 800$ or a fraction $\frac{r}{r+4}$. There are 800 blue birds out of $200r + 800$ or a fraction $\frac{4}{r+4}$. These fractions add to 1 no matter what r is.

5.

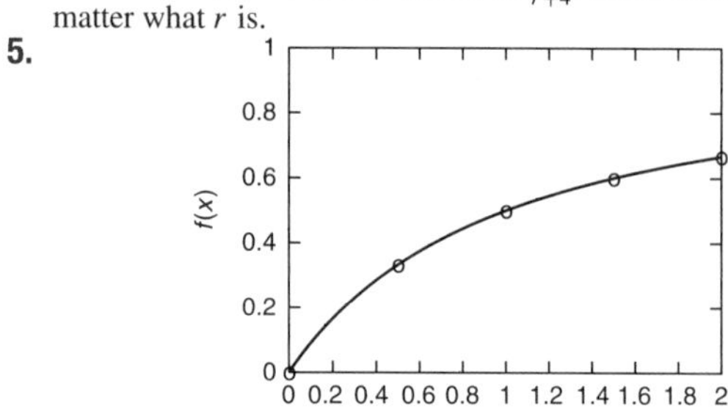

7.

$$p_t = \frac{m_t}{m_t + b_t} = \frac{1.2 \times 10^5}{1.2 \times 10^5 + 3.5 \times 10^6} = 0.033$$

$$m_{t+1} = 1.2m_t = 1.44 \times 10^5$$

$$b_{t+1} = 2.0b_t = 7.0 \times 10^6$$

$$p_{t+1} = \frac{m_{t+1}}{m_{t+1} + b_{t+1}} = \frac{1.44 \times 10^5}{1.44 \times 10^5 + 7.0 \times 10^6} = 0.020.$$

9. $p_t = 0.033$, $m_{t+1} = 0.36 \times 10^5$, $b_{t+1} = 1.75 \times 10^6$, $p_{t+1} = 0.020$.

11. The equation for the equilibria is $p^* = \frac{p^*}{p^* + 2.0(1-p^*)}$. Then

$p^*[p^* + 2.0(1-p^*)] = p^*$	multiply both sides by denominator
$p^*[p^* + 2.0(1-p^*)] - p^* = 0$	subtract p^* from both sides
$p^*[p^* + 2.0(1-p^*) - 1] = 0$	factor out p^*
$p^*(1.0 - p^*) = 0$	simplify
$p^* = 0$ or $p^* = 1.0.$	solve each piece

13. The equation for equilibria is $x^* = \frac{x^*}{1+ax^*}$. Solving,

$x^*(1 + ax^*) = x^*$	multiply both sides by denominator
$x^*(1 + ax^*) - x^* = 0$	subtract x^* from both sides
$x^*(1 + ax^* - 1) = 0$	factor out x^*
$x^*(ax^*) = 0.$	simplify

The only equilibrium is at $x^* = 0$, as long as $a \neq 0$. If $a = 0$, the system is $x_{t+1} = x_t$, which has every value of x^* as an equilibrium.

15. The equilibrium seems to be at about 1.3.

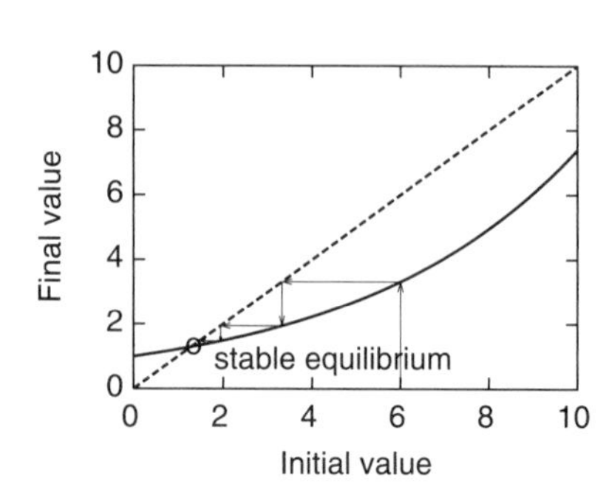

17. The equilibria seem to be at about 0.0 and 7.5.

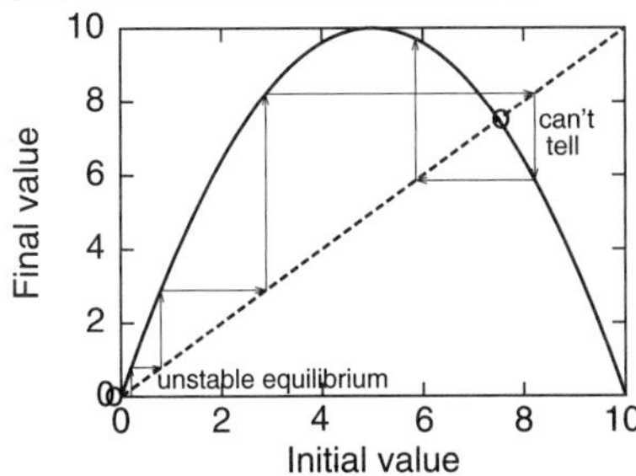

19. The discrete-time dynamical system is

$$p_{t+1} = \frac{1.2p_t}{1.2p_t + 2.0(1 - p_t)}.$$

The equilibrium at $p^* = 0$ is stable and the one at $p^* = 1$ is unstable. This makes sense because the wild type are reproducing faster than the mutants ($r > s$) and should dominate the population.

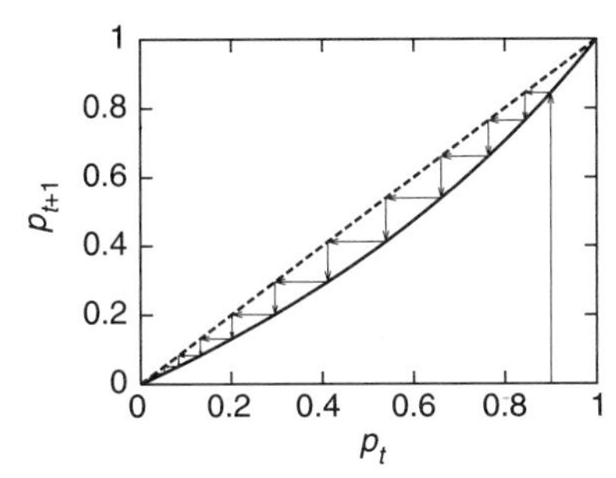

21. The discrete-time dynamical system is

$$p_{t+1} = \frac{0.3p_t}{0.3p_t + 0.5(1 - p_t)}.$$

The equilibrium at $p^* = 0$ is stable and the one at $p^* = 1$ is unstable. The picture looks the same as in **a** because the ratio of r to s is the same. But in this case, both populations are decreasing.

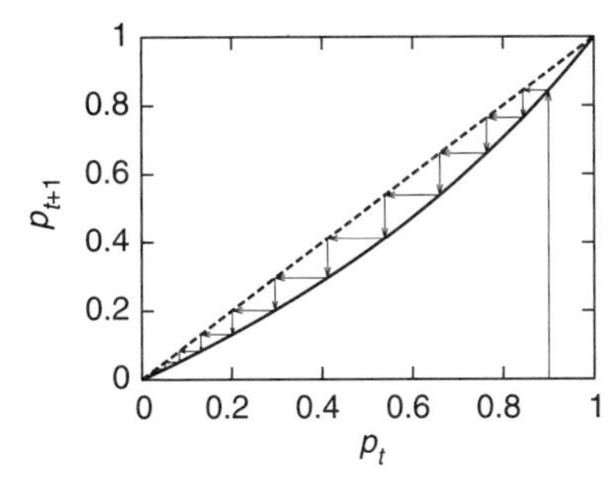

23.

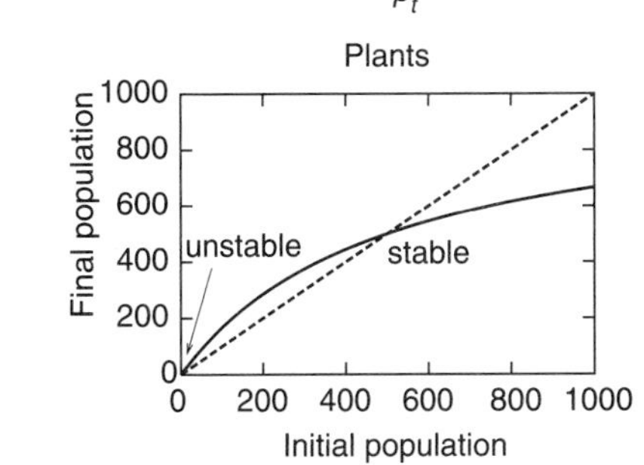

25. a. 2.0×10^5 mutate and 1.0×10^4 revert.
b. There are $1.0 \times 10^6 - 2.0 \times 10^5 + 1.0 \times 10^4 = 8.1 \times 10^5$ wild type, and $1.0 \times 10^5 - 1.0 \times 10^4 + 2.0 \times 10^5 = 2.9 \times 10^5$ mutants.
c. The total number before and after is 1.1×10^6. It does not change because the bacteria are not reproducing or dying, just changing their type.
d. The fraction before is $1.0 \times 10^5 / 1.1 \times 10^6 = 0.091$. The fraction after is $2.9 \times 10^5 / 1.1 \times 10^6 = 0.264$.

27. a. $0.2b_t$ mutate and $0.1m_t$ revert.
b. $b_{t+1} = b_t - 0.2b_t + 0.1m_t = 0.8b_t + 0.1m_t$.
$m_{t+1} = m_t - 0.1m_t + 0.2b_t = 0.9m_t + 0.2b_t$.
c. The total number before is $b_t + m_t$. The total number after is

$$\begin{aligned} b_{t+1} + m_{t+1} &= (0.8b_t + 0.1m_t) + (0.9m_t + 0.2b_t) \\ &= 0.8b_t + 0.2b_t + 0.1m_t + 0.9m_t \\ &= b_t + m_t. \end{aligned}$$

d. Divide the discrete-time dynamical system for m_{t+1} by $b_{t+1} + m_{t+1}$,

$$\begin{aligned} p_{t+1} &= \frac{m_{t+1}}{b_{t+1} + m_{t+1}} = \frac{0.9m_t + 0.2b_t}{b_{t+1} + m_{t+1}} = \frac{0.9m_t + 0.2b_t}{b_t + m_t} \\ &= \frac{0.9m_t}{b_t + m_t} + \frac{0.2b_t}{b_t + m_t} = 0.9p_t + 0.2(1 - p_t) \\ &= 0.2 + 0.7p_t. \end{aligned}$$

e. The equilibrium solves $p^* = 0.2 + 0.7p^*$, or $p^* = 0.667$.

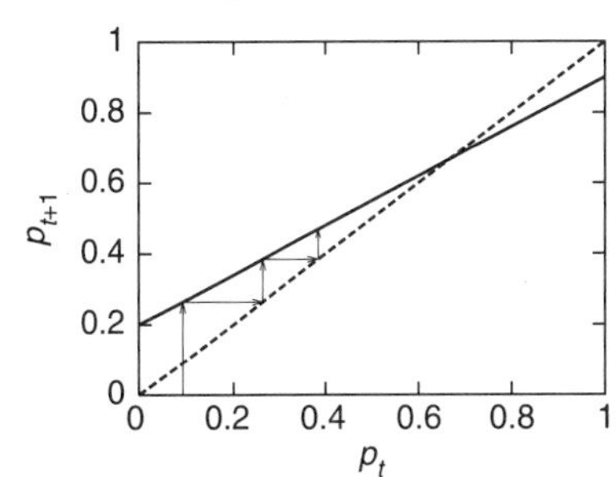

f. The equilibrium seems to be stable. The fraction of mutants will increase until it reaches 66.7 percent.

29. a. 1.0×10^5 mutate.
b. There are $1.0 \times 10^6 - 1.0 \times 10^5 = 9.0 \times 10^5$ wild type, and $1.0 \times 10^5 + 1.0 \times 10^5 = 2.0 \times 10^5$ mutants.
c. There are 1.8×10^6 wild type and 3.0×10^5 mutants.
d. The total number after is 2.1×10^6.
e. The fraction before is $1.0 \times 10^5 / 1.1 \times 10^6 = 0.091$. The fraction after is $3.0 \times 10^5 / 2.1 \times 10^6 = 0.143$.

31. a. $0.1b_t$ mutate.
b. There are $0.9b_t$ wild type and $m_t + 0.1b_t$ mutants after mutation.
c. There are $b_{t+1} = 1.8b_t$ wild type and $m_{t+1} = 1.5m_t + 0.15b_t$ mutants after reproduction.
d. The total number after is $1.95b_t + 1.5m_t$.
e. Divide the discrete-time dynamical system for m_{t+1} by $b_{t+1} + m_{t+1}$,

$$\begin{aligned} p_{t+1} &= \frac{m_{t+1}}{b_{t+1} + m_{t+1}} = \frac{1.5m_t + 0.15b_t}{1.95b_t + 1.5m_t} = \frac{\frac{1.5m_t}{m_t+b_t} + \frac{0.15b_t}{m_t+b_t}}{\frac{1.95b_t}{m_t+b_t} + \frac{1.5m_t}{m_t+b_t}} \\ &= \frac{1.5p_t + 0.15(1 - p_t)}{1.5p_t + 1.95(1 - p_t)}. \end{aligned}$$

f. The equilibrium solves

$$p^* = \frac{1.5p^* + 0.15(1 - p^*)}{1.5p^* + 1.95(1 - p^*)}.$$

Following the algebra gives

$$\begin{aligned} &p^*[1.5p^* + 1.95(1 - p^*)] = 1.5p^* + 0.15(1 - p^*) \\ &p^*[1.5p^* + 1.95(1 - p^*)] \\ &\quad -1.5p^* + 0.15(1 - p^*) = 0 \\ &0.15 - 0.6p^* + 0.45(p^*)^2 = 0 \\ &0.45(p^* - 1/3)(p^* - 1.0) = 0. \end{aligned}$$

Therefore, $p^* = 1/3$ or $p^* = 1.0$.

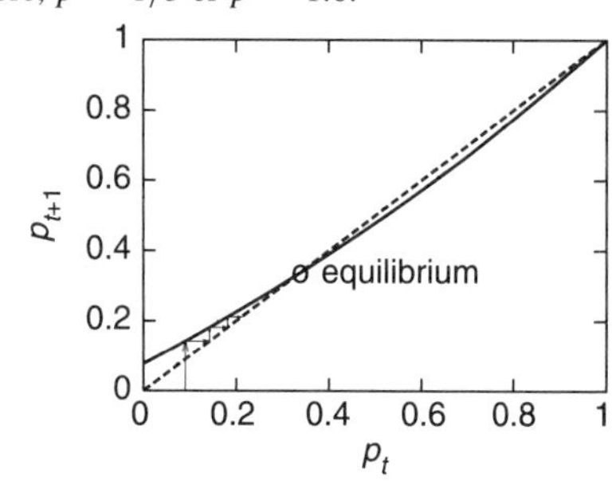

g. The equilibrium at 1/3 seems to be stable. The fraction of mutants will end up at about 33.3 percent.

33. $x_1 = 100 - 20 + 30 = 110$. $y_1 = 100 - 30 + 20 = 90$. $x_2 = 115$ and $y_2 = 85$.

35. $x_{t+1} = x_t - 0.2x_t + 0.3y_t = 0.8x_t + 0.3y_t . y_{t+1} = y_t - 0.3y_t + 0.2x_t = 0.7y_t + 0.2x_t$.

37. a. Consider the first island. After migration, there are 80 butterflies that started on the first island and 30 that started on the second. The 80 reproduce, making a total of 190. On the second island, there are 20 from the first and 70 from the second after migration. The 20 reproduce, making a total of 110. Following the same reasoning, $x_2 = 337$ and $y_2 = 153$.

b. $x_{t+1} = 2(x_t - 0.2x_t) + 0.3y_t = 1.6x_t + 0.3y_t . y_{t+1} = y_t - 0.3y_t + 2 \cdot 0.2x_t = 0.7y_t + 0.4x_t$.

c. $x_{t+1} + y_{t+1} = 2x_t + y_t$. Then

$$p_{t+1} = \frac{x_{t+1}}{x_{t+1} + y_{t+1}} = \frac{1.6x_t + 0.3y_t}{x_{t+1} + y_{t+1}} = \frac{1.6x_t + 0.3y_t}{2x_t + y_t} = \frac{\frac{1.6x_t}{x_t+y_t} + \frac{0.3y_t}{x_t+y_t}}{\frac{2x_t}{x_t+y_t} + \frac{y_t}{x_t+y_t}} = \frac{1.6p_t + 0.3(1 - p_t)}{2p_t + 1 - p_t}.$$

d. We must solve the equation

$$p^* = \frac{1.6p^* + 0.3(1 - p^*)}{2p^* + 1 - p^*} = \frac{0.3 + 1.3p^*}{1 + p^*}.$$

Multiplying out and using the quadratic formula, we find $p^* = 0.718$. This is larger than the result in Exercise 36 because the butterflies from the first island reproduce.

e.

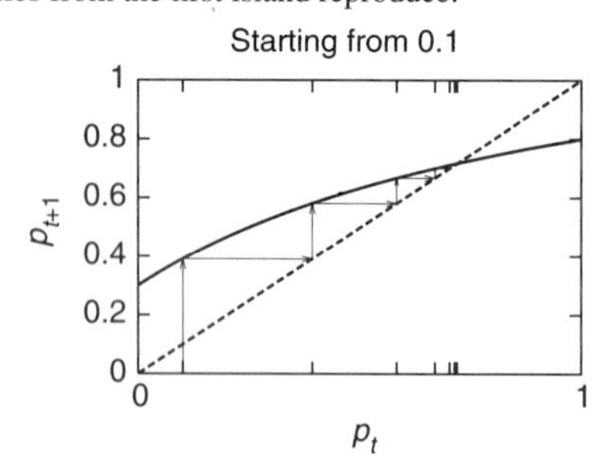

39. The discrete-time dynamical system is

$$M_{t+1} = M_t - \frac{0.5}{1.0 + 0.1M_t} M_t + 1.0.$$

To find the equilibrium,

$$M^* = M^* - \frac{0.5}{1.0 + 0.1M^*} M^* + 1.0$$

$$\frac{0.5}{1.0 + 0.1M^*} M^* = 1.0$$

$$0.5M^* = 1.0 + 0.1M^*$$

$$0.4M^* = 1.0.$$

The equilibrium is therefore 2.5. It is larger because the fraction absorbed is always less than 0.5.

41. The discrete-time dynamical system is

$$M_{t+1} = M_t - \frac{\beta}{1.0 + 0.1M_t} M_t + 1.0.$$

To find the equilibrium,

$$M^* = M^* - \frac{\beta}{1.0 + 0.1M^*} M^* + 1.0$$

$$\frac{\beta}{1.0 + 0.1M^*} M^* = 1.0$$

$$\beta M^* = 1.0 + 0.1M^*$$

$$(\beta - 0.1)M^* = 1.0.$$

The equilibrium is therefore

$$M^* = \frac{1.0}{1.0\beta - 0.1}.$$

This becomes smaller as β becomes larger. Large values of β indicate that the fraction absorbed is large, which makes sense. The fact that the equilibrium is negative when $\beta \leq 0.1$ indicates that there is no equilibrium. The body cannot use up the dose of 1.0 each day, and the level just keeps building up.

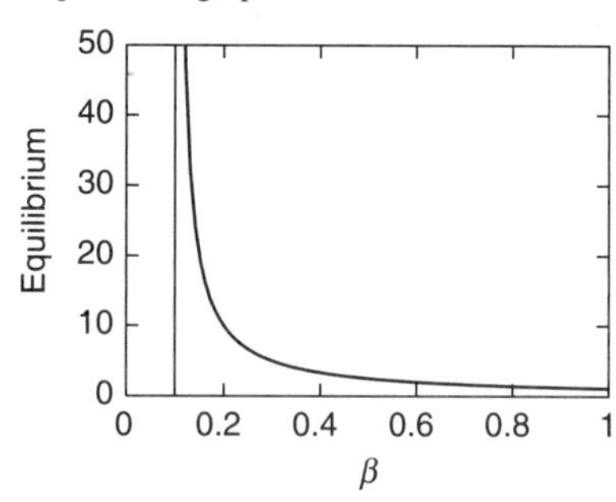

The second diagram looks a lot like the tree growth model. Once the concentration becomes large, it simply increases by 1.0 per day.

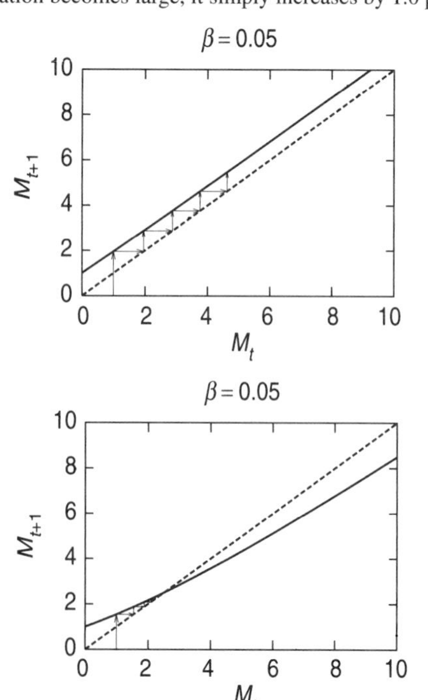

43. a.

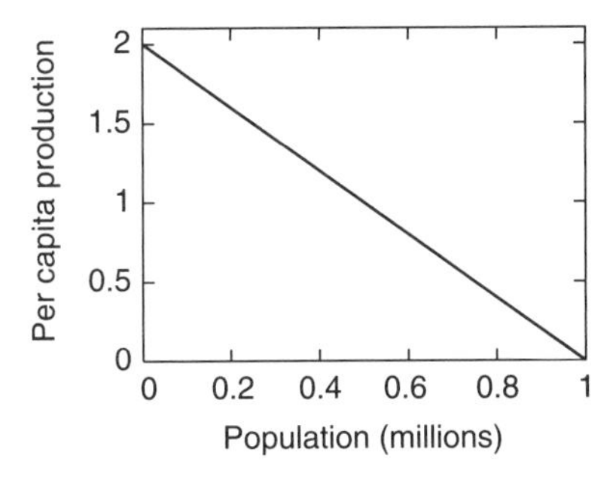

b. $b_{t+1} = 2.0b_t(1 - \frac{b_t}{1.0\times10^6})$.

c. The equilibria are $b^* = 0$ and $b^* = 5.0 \times 10^5$.

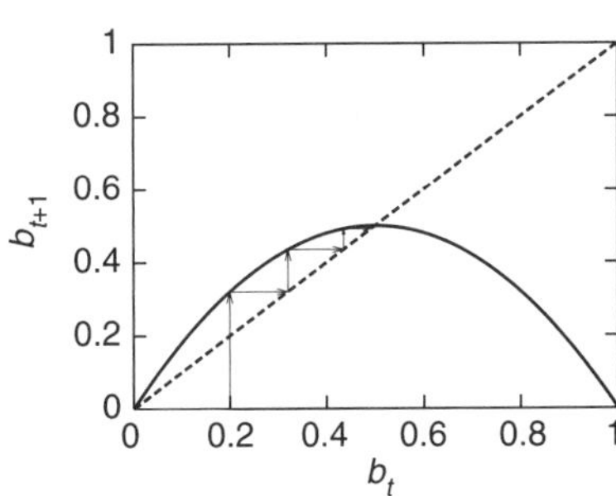

45. a.

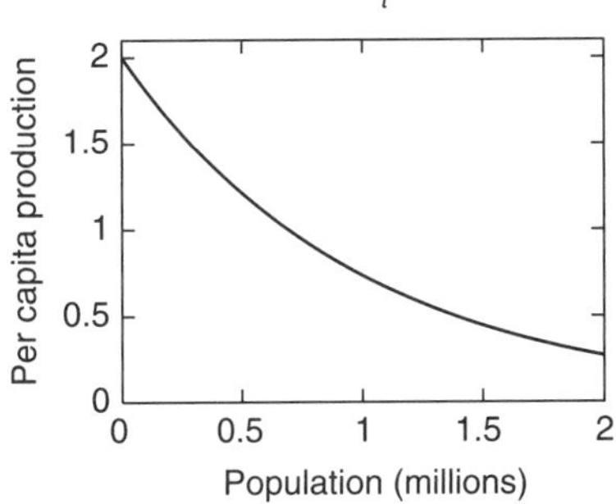

b. $b_{t+1} = 2.0b_t e^{\frac{-b_t}{1.0\times10^6}}$.

c. The equilibria are $b^* = 0$ and $b^* = \ln(2)1.0 \times 10^6 = 6.93 \times 10^5$.

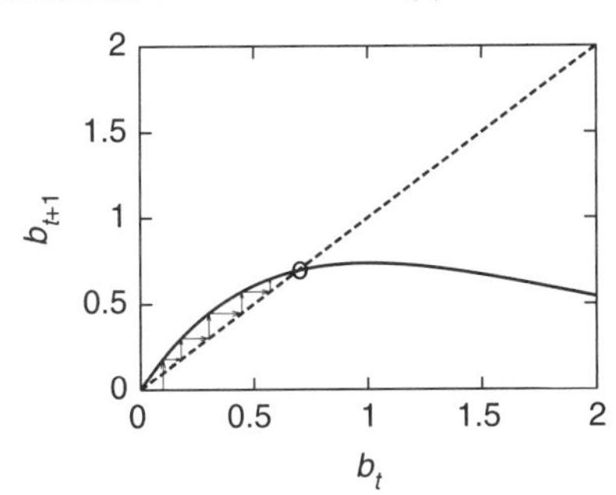

Section 1.11, page 131

1. $\hat{V}_t = 0.5 \cdot 30.0 = 15.0 < V_c$. The heart will beat, and $V_{t+1} = 25.0$ mV.

3. $\hat{V}_t = 0.7 \cdot 30.0 = 21.0 > V_c$. The heart will not beat, and $V_{t+1} = 21.0$ mV.

5. $V^* = u/(1-c) = 20.0$. Because $cV^* = 10.0 < V_c = 20.0$, the inequality in Equation 1.11.2 is satisfied and the equilibrium makes sense. This heart will beat every time.

7. $V^* = 33.3$. Because $cV^* = 23.1 > V_c = 20.0$, the inequality in Equation 1.11.2 is not satisfied. The equilibrium does not make sense. Is this a case of 2:1 AV block? We find that $\bar{V} = 19.61$ from Equation 1.11.5. However, $c\bar{V} < V_c$, so that the second inequality in Equation 1.11.4 is not satisfied. This is not a case of 2:1 AV block. In fact, this turns out to be an example of the Wenckebach phenomenon where the heart skips every fourth beat.

9. $c = e^{-\alpha\tau} = 0.3678$. The heart beats every time.

11. $c = e^{-\alpha\tau} = 0.3678$. The heart beats every time, twice as fast as in part **a**, but with recovery also twice as fast.

13. The dynamical system is

$$V_{t+1} = cV_t + \frac{2(1-c)}{1+V_t^2}.$$

Substituting $V_t = 1$ gives $V_{t+1} = 1$, so this is an equilibrium.

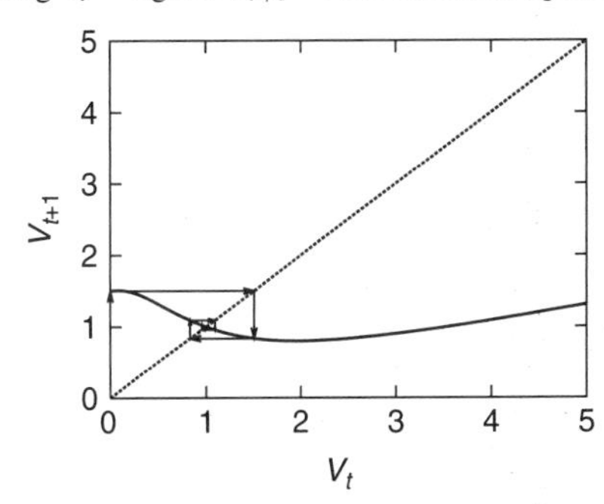

From the cobweb, it seems to be stable.

15. I got the following strange results:

Time	Solution 1	Solution 2	Solution 3
0	500	600.0	800.0
1	750.0	900.0	1200.0
2	1125.0	1350.0	800.0
3	687.50	1025.0	1200.0
4	1031.25	537.5	800.0
5	546.88	806.25	1200.0
6	820.31	1209.37	800.0
7	1230.47	814.062	1200.0
8	845.70	1221.10	800.0
9	1268.55	831.64	1200.0
10	902.83	1247.46	800.0

These are jumping all over the place.

17. I got the following strange results:

Time	Solution 1	Solution 2	Solution 3
0	500.0	600.0	800.0
1	825.0	990.00	1320.0
2	1361.25	1633.50	1178.0
3	1246.06	1695.27	943.70
4	1056.00	1797.20	1557.10
5	742.40	1965.39	1569.22
6	1224.97	2242.89	1589.22
7	1021.20	2700.76	1622.21
8	684.98	3456.26	1676.65
9	1130.21	4702.83	1766.47
10	864.85	6759.67	1914.67

The first jumps around like the earlier ones, but the second seems to be shooting off to infinity.

Chapter 2

Section 2.1, page 137

1. With $\Delta t = 1.0$, $\Delta f = f(2.0) - f(1.0) = 3.0$, so $\frac{\Delta f}{\Delta t} = 3.0$. With $\Delta t = 0.5$, $\Delta f = f(1.5) - f(1.0) = 1.5$, so $\frac{\Delta f}{\Delta t} = 3.0$. With $\Delta t = 0.1$, $\Delta f = f(1.1) - f(1.0) = 0.3$, so $\frac{\Delta f}{\Delta t} = 3.0$. With $\Delta t = 0.01$, $\Delta f = f(1.01) - f(1.0) = 0.03$, so $\frac{\Delta f}{\Delta t} = 3.0$.

3. With $\Delta t = 1.0$, $\Delta h = h(2.0) - h(1.0) = 6.0$, so $\frac{\Delta h}{\Delta t} = 6.0$. With $\Delta t = 0.5$, $\Delta h = h(1.5) - h(1.0) = 2.5$, so $\frac{\Delta h}{\Delta t} = 5.0$. With $\Delta t = 0.1$, $\Delta h = h(1.1) - h(1.0) = 0.42$, so $\frac{\Delta h}{\Delta t} = 4.2$. With $\Delta t = 0.01$, $\Delta h = h(1.01) - h(1.0) = 0.0402$, so $\frac{\Delta h}{\Delta t} = 4.02$.

5. With $\Delta t = 1.0$, $\Delta G = G(1.0) - G(0.0) = 6.389$, so $\frac{\Delta G}{\Delta t} = 6.389$. With $\Delta t = 0.5$, $\Delta G = G(0.5) - G(0.0) = 1.718$, so $\frac{\Delta G}{\Delta t} = 3.436$. With $\Delta t = 0.1$, $\Delta G = G(0.1) - G(0.0) = 0.221$, so $\frac{\Delta G}{\Delta t} = 2.21$. With $\Delta t = 0.01$, $\Delta G = G(0.01) - G(0.0) = 0.0202$, so $\frac{\Delta G}{\Delta t} = 2.02$.

7. Each secant line is $f_s(t) = 2 + 3t$.

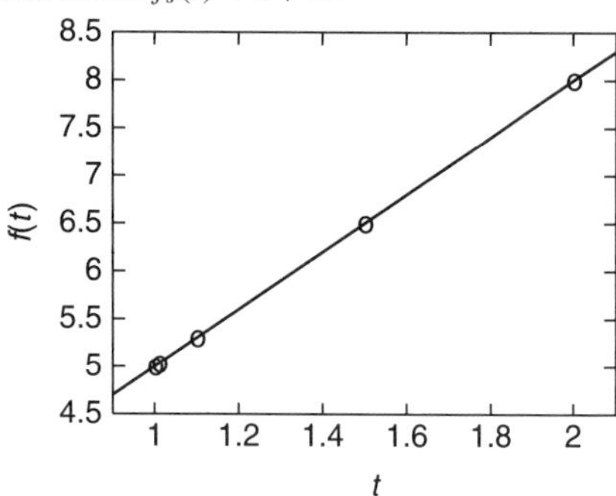

9. The coordinates of the base point are $(1, 2)$, so the secant lines are: with $\Delta t = 1.0$, $h_s(t) = 2 + 6(t - 1)$, with $\Delta t = 0.5$, $h_s(t) = 2 + 5(t - 1)$, with $\Delta t = 0.1$, $h_s(t) = 2 + 4.2(t - 1)$, with $\Delta t = 0.01$, $h_s(t) = 2 + 4.02(t - 1)$.

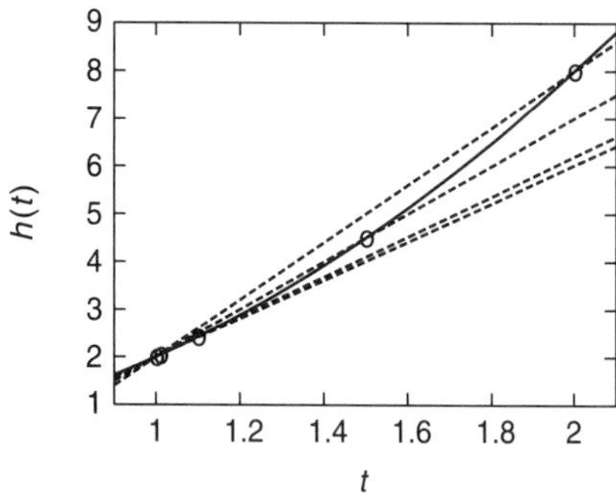

11. The coordinates of the base point are $(0, 1)$, so the secant lines are: with $\Delta t = 1.0$, $G_s(t) = 1 + 6.389t$, with $\Delta t = 0.5$, $G_s(t) = 1 + 3.436t$, with $\Delta t = 0.1$, $G_s(t) = 1 + 2.21t$, with $\Delta t = 0.01$, $G_s(t) = 1 + 2.02t$.

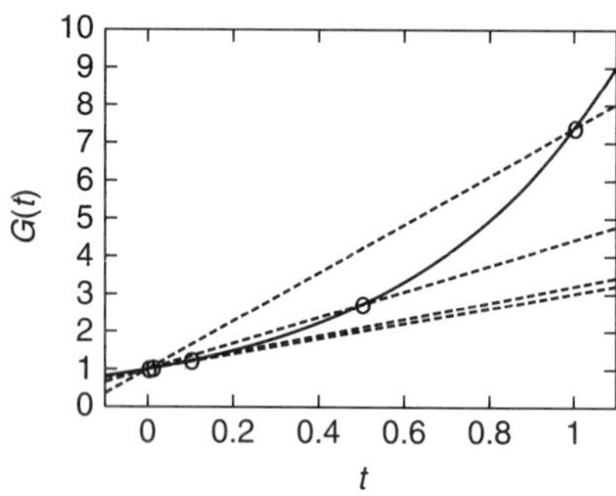

13. The slope is 3, so the tangent line is $\hat{f}(t) = 2 + 3t$.

15. It looks like the slopes are getting close to 4.0, so the tangent line is $\hat{h}(t) = 2 + 4(t - 1)$.

17. It looks like the slopes are getting close to 2.0, so the tangent line is $\hat{G}(t) = 1 + 2t$.

19. The derivative of $g(t)$, the slope of the tangent line.

21.

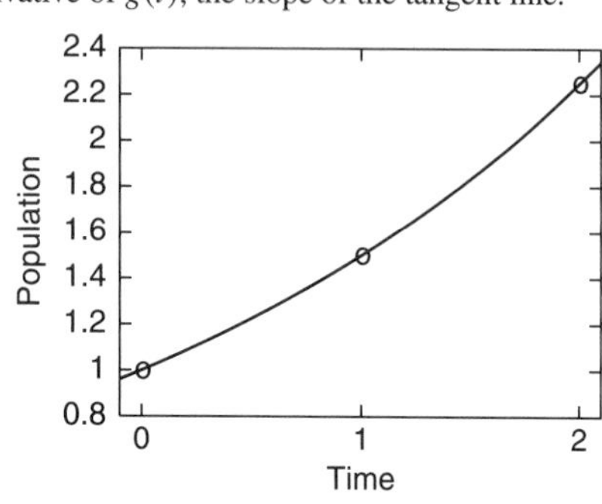

a. $b(0) = 1.0$, $b(1.0) = 1.5$, $b(2.0) = 2.25$.
b. $\Delta b = 1.5 - 1.0 = 0.5$, so $\Delta b/\Delta t = 0.5$.
c. $\Delta b = 2.25 - 1.5 = 0.75$, so $\Delta b/\Delta t = 0.75$.

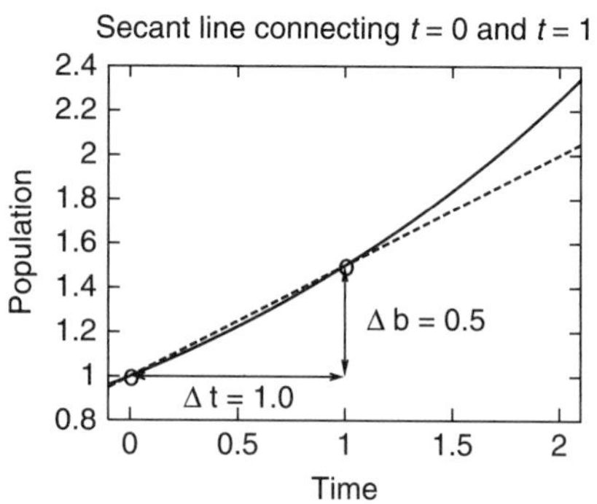

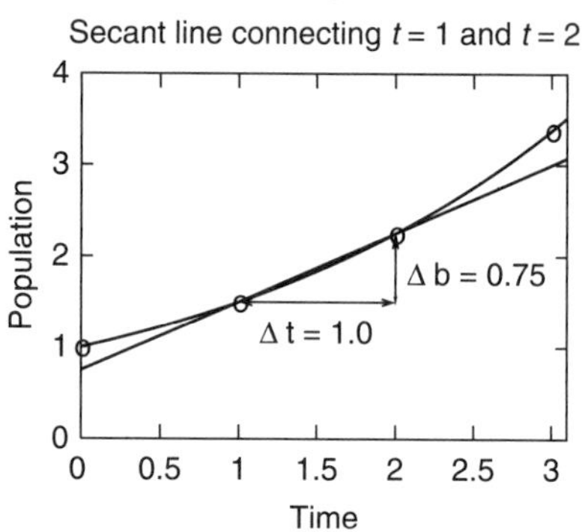

23. a. $\Delta b = 1.5^{1.0} - 1.0 = 0.5$, and $\Delta b/\Delta t = 0.5$.
b. $\Delta b = 1.5^{0.1} - 1.0 = 0.0413$, and $\Delta b/\Delta t = 0.414$.
c. $\Delta b = 1.5^{0.01} - 1.0 = 0.00406$, and $\Delta b/\Delta t = 0.406$.
d. $\Delta b = 1.5^{0.001} - 1.0 = 0.000405$, and $\Delta b/\Delta t = 0.405$.
e. The limit looks like 0.405.
f.

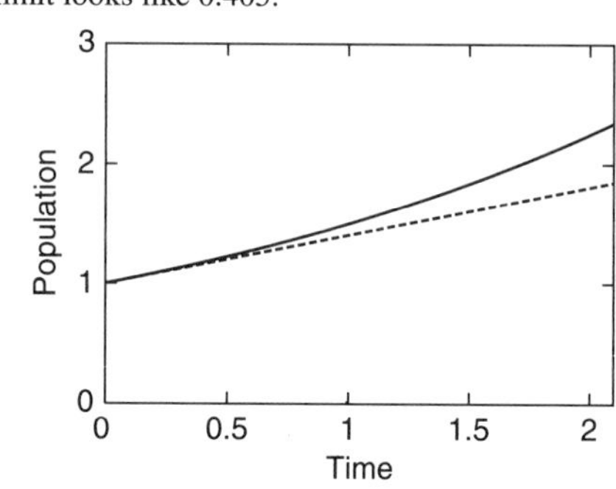

25. a. The slope is $(5 \cdot 1.0^2 - 0.0)/1.0 = 5.0$.
b. The slope is $(5 \cdot 0.1^2 - 0.0)/0.1 = 0.5$.
c. The slope is $(5 \cdot 0.01^2 - 0.0)/0.01 = 0.05$.
d. The slope is $(5 \cdot 0.001^2 - 0.0)/0.001 = 0.005$.
e. The slope gets close to 0.
f.

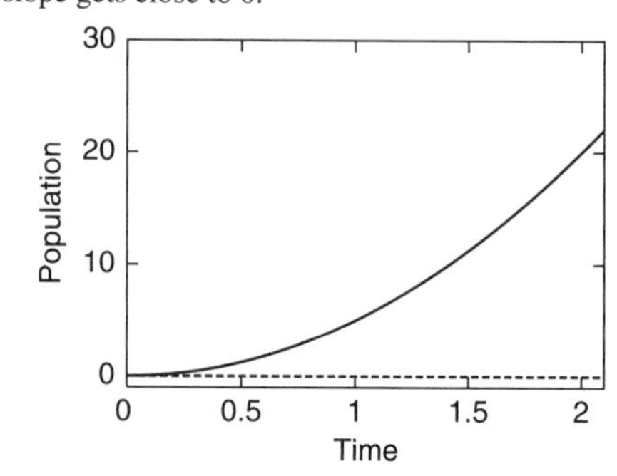

27. During the first hour, 3.0 bacteria/h. During the first half hour, 2.485 bacteria/h. During the second half hour, 3.515 bacteria/h. The population changes faster during the second half hour.

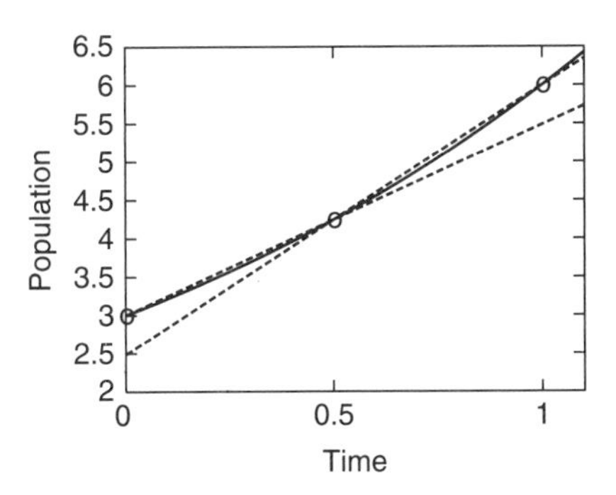

29. During the first hour, −0.79 bacteria/h. During the first half hour, −0.88 bacteria/h. During the second half hour, −0.69 bacteria/h. The population changes faster during the first half hour.

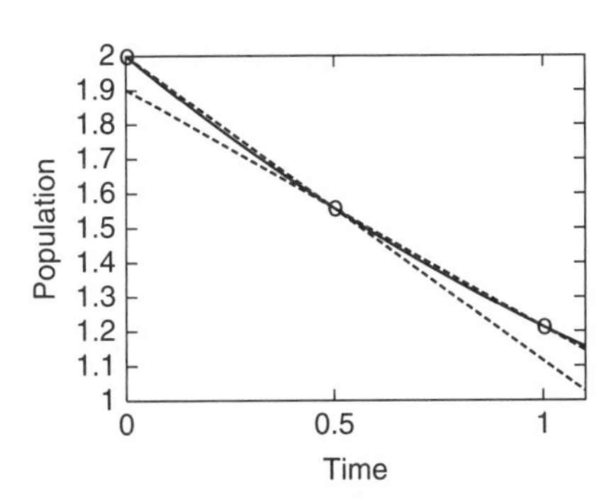

31.

t	$b(t)$	Change Δb	Average Rate of Change $\frac{\Delta b}{\Delta t} \approx \frac{db}{dt}$	Per Capita Rate of Change $\frac{\frac{db}{dt}}{b(t)}$
0.0	1.0	–	–	–
1.0	2.0	1.0	1.0	0.5
2.0	4.0	2.0	2.0	0.5
3.0	8.0	4.0	4.0	0.5

$\frac{db}{dt} = 0.5b(t)$.

33.

t	$b(t)$	Change Δb	Average Rate of Change $\frac{\Delta b}{\Delta t} \approx \frac{db}{dt}$	Per Capita Rate of Change $\frac{\frac{db}{dt}}{b(t)}$
0.0000	1.0000	–	–	–
0.0100	1.0070	0.00696	0.69556	0.69075
0.0200	1.0140	0.00700	0.70039	0.69075
0.0300	1.0210	0.00705	0.70526	0.69075
0.0400	1.0281	0.00710	0.71017	0.69075
0.0500	1.0353	0.00715	0.71511	0.69075

$\frac{db}{dt} = 0.6907b(t)$.

35. a. Using the previous measurement as the starting base point, we can estimate the following:

Age	Rate of Change (m/year)	Rate Divided by Height
1	1.07	0.0957
2	1.22	0.0984
3	1.34	0.0975
4	1.27	0.0846
5	1.60	0.0963
6	1.66	0.0909
7	1.90	0.0942
8	1.84	0.0836
9	2.44	0.0998
10	2.40	0.0894

b.

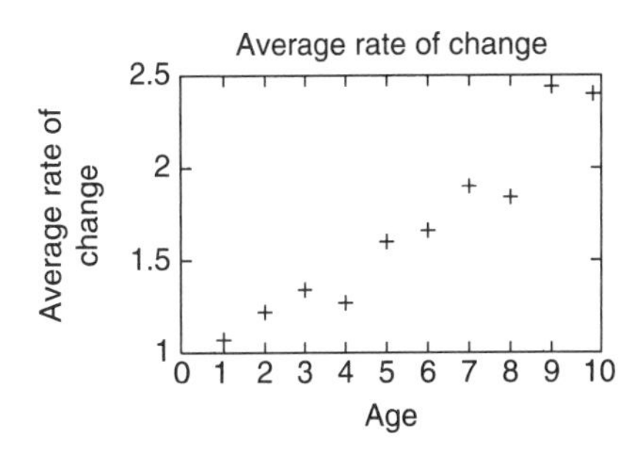

c.

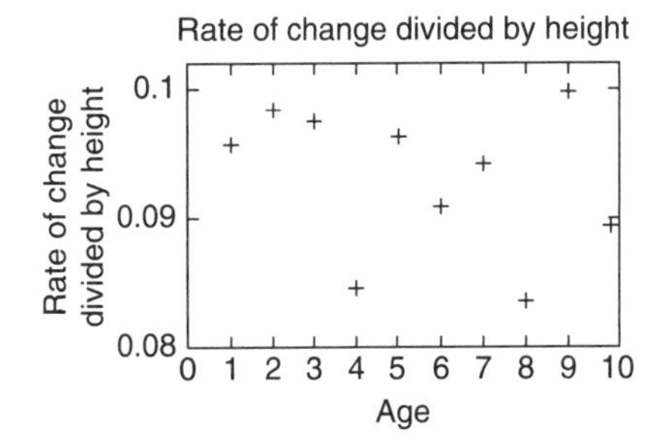

d. Both seem to jump around a bit, but the tree increases its height by about 9 percent per year. The approximate differential equation is

$$\frac{dh}{dt} = 0.09h.$$

37. Getting 5 percent interest corresponds to multiplying the principal by 1.05, giving $1.05 \cdot \$1000 = \$1050.00$.

39. $\frac{5}{12} = 0.417$. We must multiply the principal by 1.00417 twelve times, giving $(1.00417)^{12} \cdot \$1000 = \$1051.16$, giving an extra \$1.16.

41. Breaking into intervals of length Δt, we get

$$\lim_{\Delta t \to 0} (1 + 0.05\Delta t)^{\frac{1}{\Delta t}} \$1000.$$

It looks like the limit is \$1051.27.

43. With yearly compounding, you'd get $2.0 \cdot \$1000 = \$2000.00$. With monthly compounding, you'd get $\frac{100}{12} = 8.33$ for each of 12 months, giving $1.0833^{12} \cdot \$1000 = \$2613.00$. With daily compounding, you'd get $\frac{100}{365} = 0.274$ for each of 365 days, giving $1.00274^{365} \cdot \$1000 = \$2714.60$. The compound interest is more important because the principal changes much more during the year.

Section 2.2, page 157

1. At $x = 0.1$, value is 2.594; at $x = 0.01$ it is 2.705; at $x = 0.001$ it is 2.717; and at $x = 0.0001$ it is 2.718. The limit seems to be e.

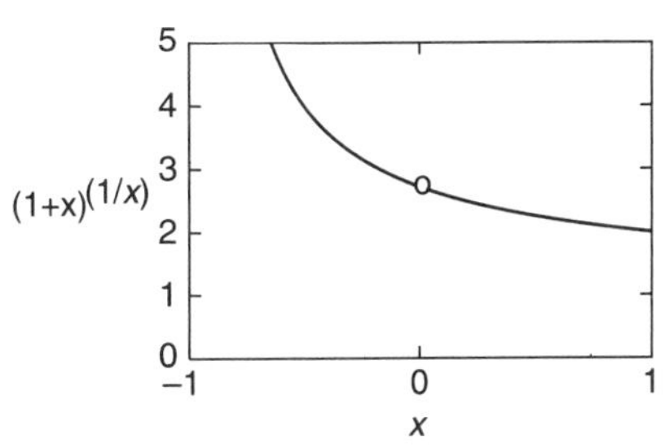

3. At $x = 0.1$, value is 0.499; at $x = 0.01$ it is 0.005; at $x = 0.001$ it is 0.0005. The limit seems to be 0.

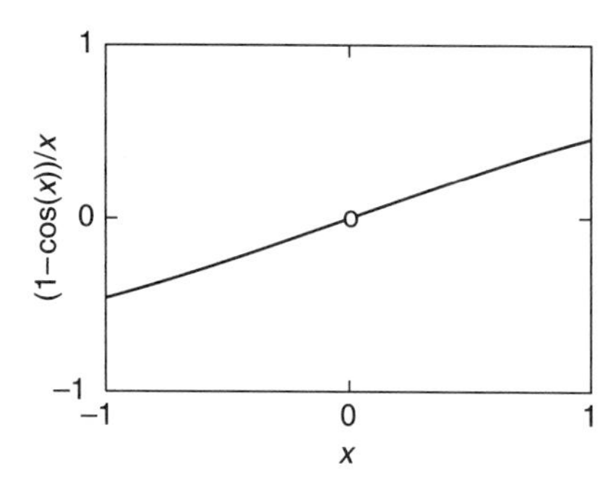

5. $0.1 \ln(0.1) = -0.230$, $0.01 \ln(0.01) = -0.046$, $0.001 \ln(0.001) = -0.006$. It seems to approach 0.

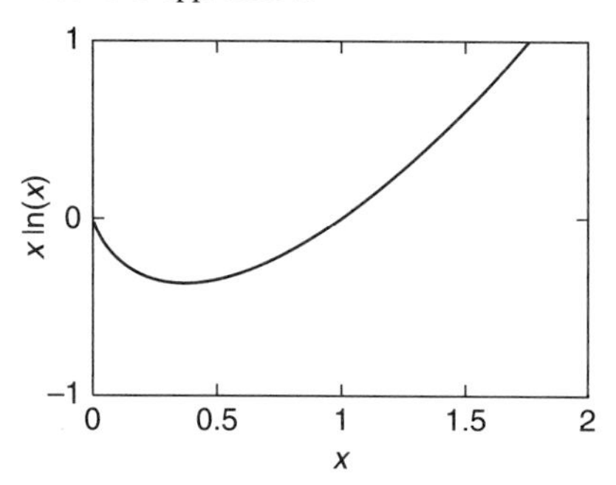

7. Denoting the function by $h(x)$, we find that $h(0.9) = -2.558$, $h(0.99) = -4.652$, $h(0.999) = -6.915$, $h(0.9999) = -9.211$. This seems to go to negative infinity.

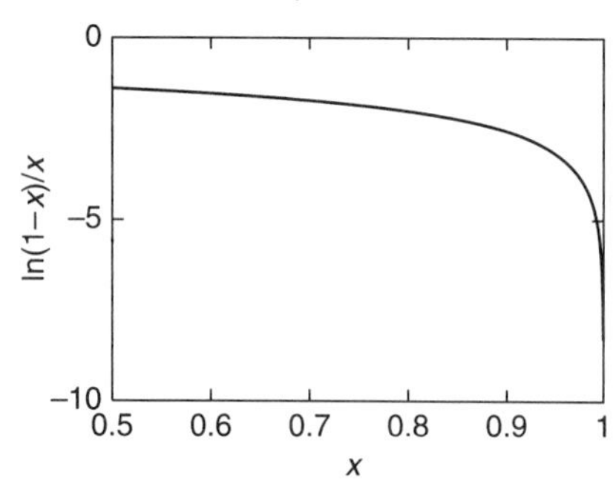

9. This will be 5 times the previous limit, or $5e = 13.59$, using Theorem 2.2c.
11. This will be the product of the two earlier limits, or $e \cdot 0 = 0$, using Theorem 2.2b.
13. $f_1(x) < 0.1$ if $x < 0.01$, $f_1(x) < 0.01$ if $x < 0.0001$. This approaches 0 slowly because tiny values of the input are required to produce small values of the output. Graphically, the value of the function does not become small until x is very small.

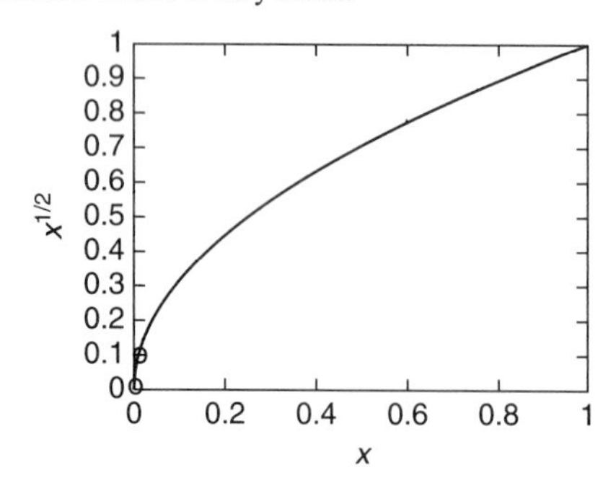

15. $f_3(x) < 0.1$ if $x < 0.316$, $f_3(x) < 0.01$ if $x < 0.1$. This approaches 0 quickly because small values of the input produce tiny values of the output. Graphically, the value of the function becomes small even when x is not that small.

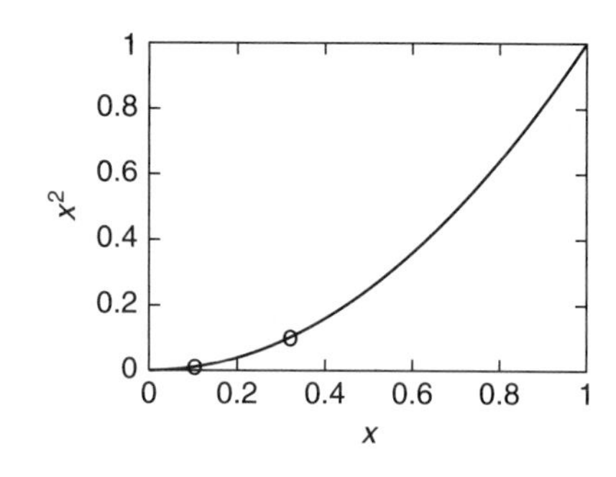

17. $g_1(x) > 10$ if $x < 0.01$. $g_1(x) > 100$ if $x < 0.0001$. This approaches infinity slowly because tiny values of the input are required to produce large values of the output. Graphically, the value of the function only becomes large when x is very small.

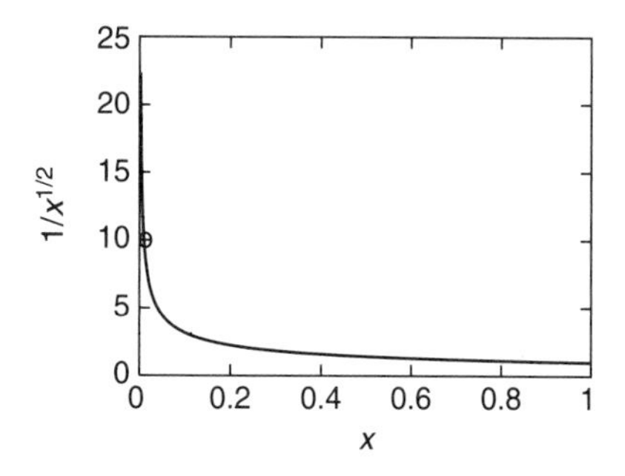

19. $g_3(x) > 10$ if $x < 0.316$. $g_3(x) > 100$ if $x < 0.1$. This approaches infinity quickly because small values of the input produce huge values of the output. Graphically, the value of the function becomes large even when x is not too small.

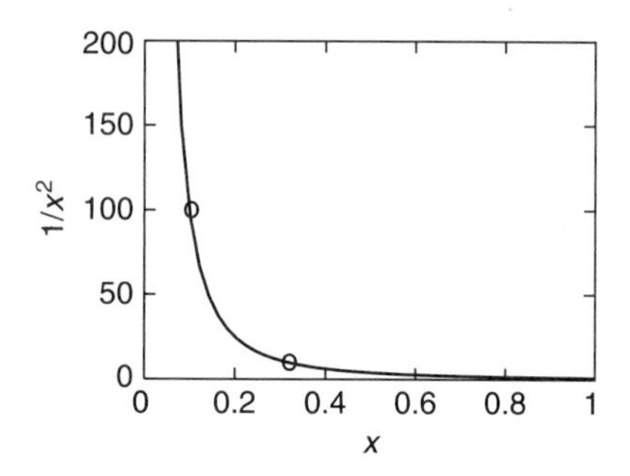

21. Both the left- and the right-hand limits are 1.
23. The left-hand limit is about 0.1, and the right-hand limit is about 0.02.
25. $\Delta f = 5\Delta x$, so the average rate of change is 5 (unless $\Delta x = 0$). The limit is 5.

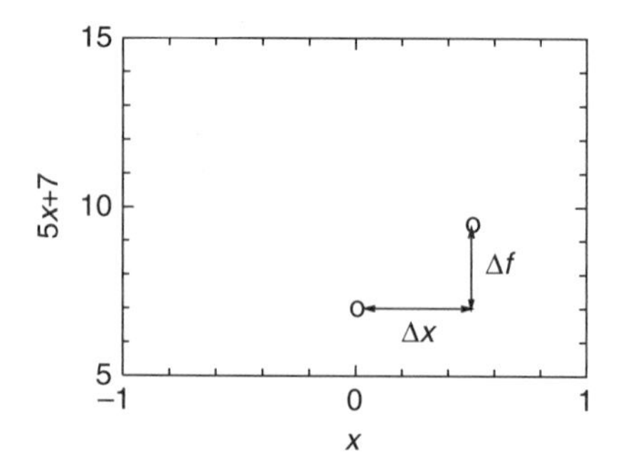

27. $\Delta f = 5\Delta x^2$, so the average rate of change is $5\Delta x$ (unless $\Delta x = 0$). The secant becomes flatter and flatter as Δx becomes smaller, so the limit must be 0.

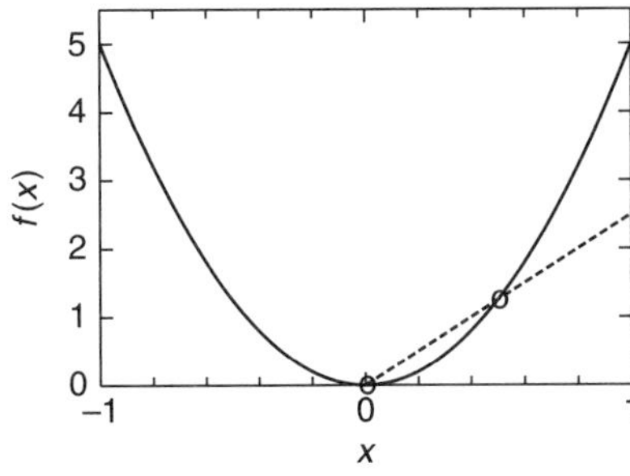

29. $\Delta f = 5(1+\Delta x)^2 + 7(1+\Delta x) - 12$, so the average rate of change is $17 + 5\Delta x$ (unless $\Delta x = 0$). The value gets closer and closer to 17 for small values of Δx, so the limit must be 17.

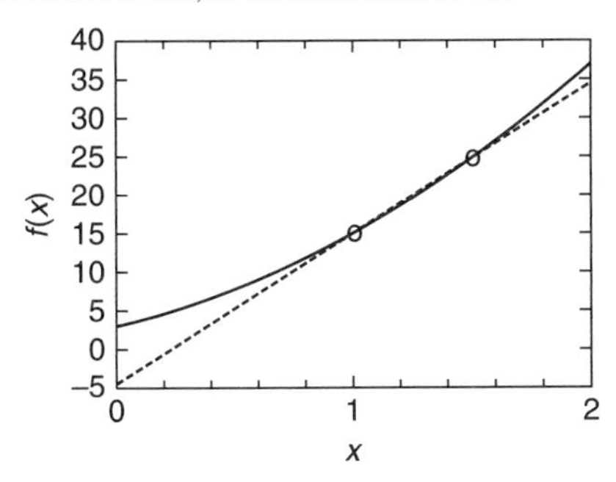

31. a. $\lim_{T \to 0} V(T) = 1.0\ \text{cm}^3$.
b. $V(2.0) = 5.0$.
c. $V(1.0) = 2.0$.
d. We need to solve $V(T) = 1.01$. But $1 + T^2 = 1.01$ if $T^2 = 0.01$ or $T = 0.1$K.

33. The average rate of change between $t = 0$ and $t = \Delta t$ is

$$\frac{b(\Delta t) - b(0)}{\Delta t} = \frac{\Delta t + \Delta t^2 - 0}{\Delta t} = 1 + \Delta t.$$

So Δt would have to be less than 0.01 for average rate of change to be within 1% of the instantaneous rate of change.

35. The average rate of change between $t = 0$ and $t = \Delta t$ is

$$\frac{b(\Delta t) - b(0)}{\Delta t} = \frac{e^{\Delta t} - 1}{\Delta t}.$$

Plugging in small values of Δt, I found that the average rate of change is 1.01 when $\Delta t = 0.02$.

37. a. This would cost only \$5.
b. This would cost \$50.
c. This would cost \$500.

39. a. The difference in temperature is 6.25°C. It looks like a \$1 device would be good enough.
b. The difference in temperature is 0.11°C. This would cost about \$10 to detect.
c. The difference in temperature is 0.00077°C. This would cost over \$1000 to detect.

41. a. The pressure is 11 atmospheres, and would cost \$121.
b. The pressure is 101 atmospheres, and would cost \$10,201.
c. The pressure is 501 atmospheres, and would cost \$251,001.

Section 2.3, page 167

1. This is a **linear** function and is continuous everywhere.

3. This is constructed as the quotient of the continuous exponential function and a continuous linear function. It is guaranteed to be continuous everywhere that the denominator is not equal to 0. The only potential trouble point is $x = -1$.

5. This is a composition of the natural log with a linear function divided by a polynomial. The theorems guarantee continuity except at $z = 1$, where we are taking the natural log of 0. At $z = 0$, where the denominator is 0, the logarithm is not defined.

7. This is a composition of the continuous cosine function with a continuous linear function $\left(x - \frac{\pi}{2}\right)$ and is continuous everywhere.

9. This is the quotient of the constant 1 by the function $(1-w)^4$, which is a polynomial. This is guaranteed to be continuous except where the denominator is 0, or when $w = 1$.

11. $l(5) = 31$, $l(5.1) = 31.5$, $l(5.01) = 31.05$, $l(4.9) = 30.5$, $l(4.99) = 30.95$.

13. $f(0) = e^0/(0+1) = 1$. $f(0.1) = 1.0047$, $f(0.01) = 1.00005$, $f(-0.1) = 1.0054$, and $f(-0.01) = 1.00005$.

15. $g(2) = 0$. $g(2.1) = 0.022$, $g(2.01) = 0.0024$, $g(1.9) = -0.029$, $g(1.99) = -0.0025$.

17. $g(0)$ cannot be computed. In fact, $g(z)$ does not make sense for $z \le 1$.

19. $r(1)$ cannot be computed. $r(1.1) = 1.0 \times 10^4$, $r(1.01) = 1.0 \times 10^8$, $r(0.9) = 1.0 \times 10^4$, $r(0.99) = 1.0 \times 10^8$. This limit is infinity.

21. $f(x) = 2.1$ if $x = 0.1$ and $f(x) = 1.9$ if $x = -0.1$, so $-0.1 \le x \le 0.1$.

23. $f(x) = 1.1$ if $x = \sqrt{1.1} = 1.049$ and $f(x) = 0.9$ if $x = 0.949$, so $0.949 \le x \le 1.049$.

25. Any value of $x \ge 0$ works, so x must be within 1 away from 1.

27. a.

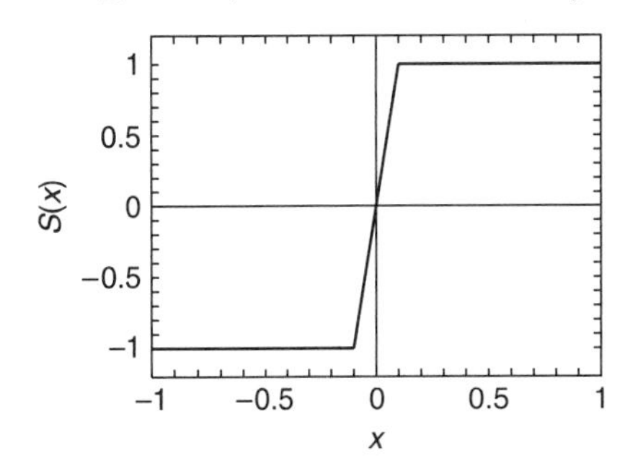

b. The slope on the central part is 10.

$$S(x) = \begin{cases} -1 & \text{if } x \le -0.1 \\ -1 + 10(x + 0.1) & \text{if } 0.1 < x < 0.1 \\ 1 & \text{if } x \ge 0.1. \end{cases}$$

c. The input would have to be between -0.01 and 0.01.

29.
$$\begin{aligned} 4.8 &< M < 5.2 \\ 4.8 &< 2.0V < 5.2 \\ 2.4 &< V < 2.6. \end{aligned}$$

V must be within 0.1 cm³ of 2.5 cm³ for M to be within 0.2 g of 5.0 g.

31.
$$\begin{aligned} 0.95 &< F < 1.05 \\ 0.95 &< r^4 < 1.05 \\ 0.987 &< r < 1.012. \end{aligned}$$

The radius must be within about 1 percent of 1.0 to guarantee a flow within 5 percent.

33. We need to solve

$$\begin{aligned} 9.0 \times 10^8 &\le b_{10} \le 1.1 \times 10^9 \\ 9.0 \times 10^8 &\le 2b_9 \le 1.1 \times 10^9 \\ 4.5 \times 10^8 &\le b_9 \le 5.5 \times 10^8. \end{aligned}$$

The tolerance is 5.0×10^7, or half the width of this interval.

35.

$$\begin{aligned} 9.0 \times 10^8 &\le b_0 \le 1.1 \times 10^9 \\ 9.0 \times 10^8 &\le 1024b_0 \le 1.1 \times 10^9 \\ 8.789 \times 10^5 &\le b_5 \le 10.742 \times 10^5. \end{aligned}$$

The tolerance is 9.76×10^4.

37. To get $T_{10} = 0.52$ requires $T_9 = 1.04$, while getting $T_{10} = 0.48$ requires $T_9 = 0.96$. The tolerance is 0.04 g/L.

39. Between 491.52 and 532.48 g/L. Tolerance is 20.48 g/L.

41. Denote the function by f. Then

$$f(V) = \begin{cases} 80 & \text{if } V < 50 \\ 2V & \text{if } V \ge 50. \end{cases}$$

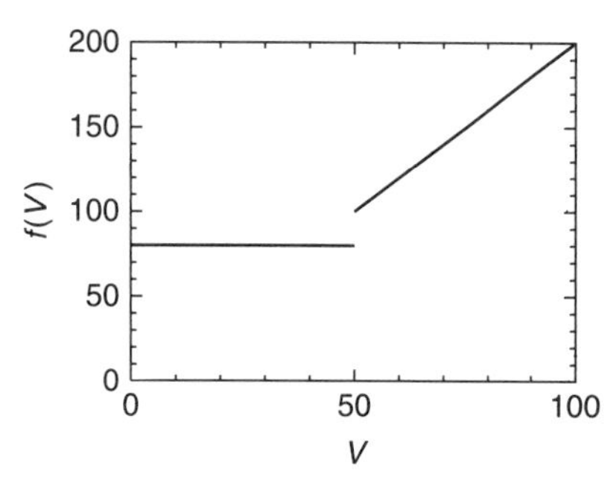

43. We need that $2V = V^*$ at $V = V_0 = 50$. In this case, $V^* = 100$.

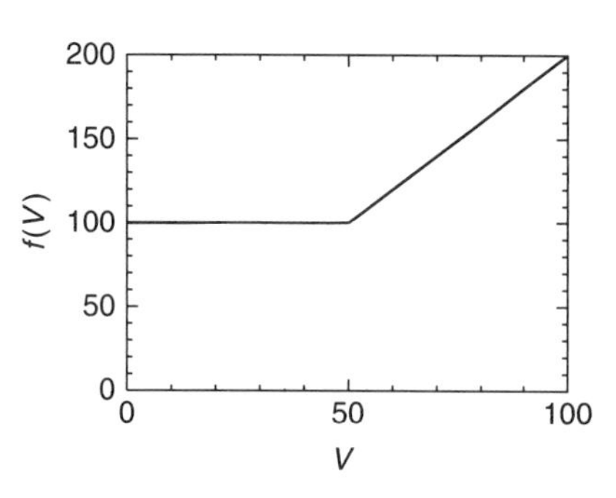

45. a.

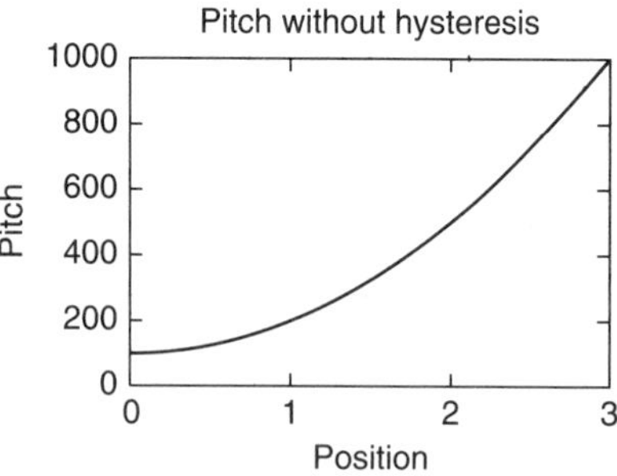

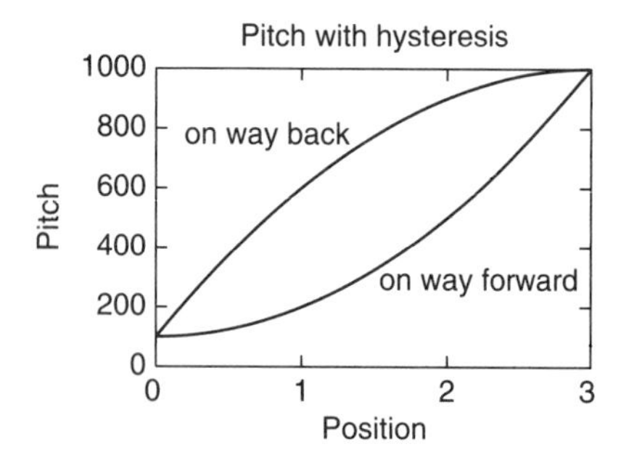

b. The second seems more likely. The pitch is usually different in different directions.
c. Both sound awful.

Section 2.4, page 175

1. $x^2 + 2x\Delta x + \Delta x^2$.
3. $(3x + 2\Delta x)^2 = (3x)^2 + 2 \cdot 3x \cdot 2\Delta x + (2\Delta x)^2 = 9x^2 + 12x\Delta x + 4\Delta x^2$.

5.

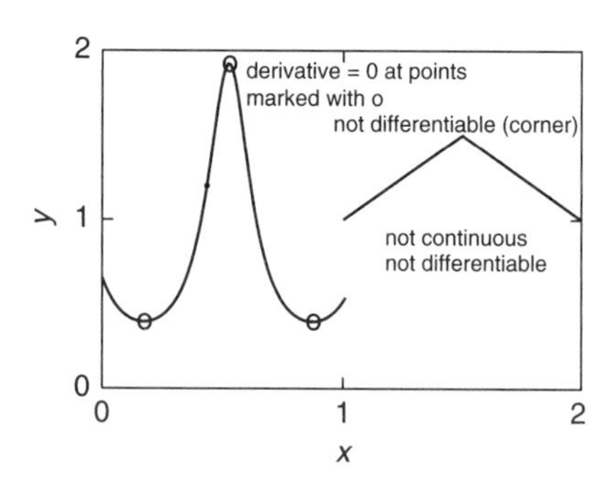

7. $M'(x) = 0.5$, $\frac{dM}{dx} = 0.5$. This function is increasing.
9. $g'(y) = -3$, $\frac{dg}{dy} = -3$. This function is decreasing.
11. $f(1) = 3$, $f(1 + \Delta x) = 3 - 2\Delta x - \Delta x^2$, $\Delta f = -2\Delta x - \Delta x^2$, so the slope of the secant is $\frac{\Delta f}{\Delta x} = -2 - \Delta x$ if $\Delta x \neq 0$. Taking the limit of this constant function by plugging in $\Delta x = 0$ we find that the slope of the tangent is $f'(1) = -2$.
13. $f(x + \Delta x) = 4 - x^2 - 2x\Delta x - \Delta x^2$, $\Delta f = -2x\Delta x - \Delta x^2$, so the slope of the secant is $\frac{\Delta f}{\Delta x} = -2x - \Delta x$ if $\Delta x \neq 0$. Taking the limit, the derivative is $f'(x) = -2x$, or $\frac{df}{dx} = -2x$.
15. The derivative is $f'(x) = -2x$, so the critical point occurs at $x = 0$. For $x < 0$, the derivative is positive while for $x > 0$ it is negative. The function thus switches from increasing to decreasing at $x = 0$.

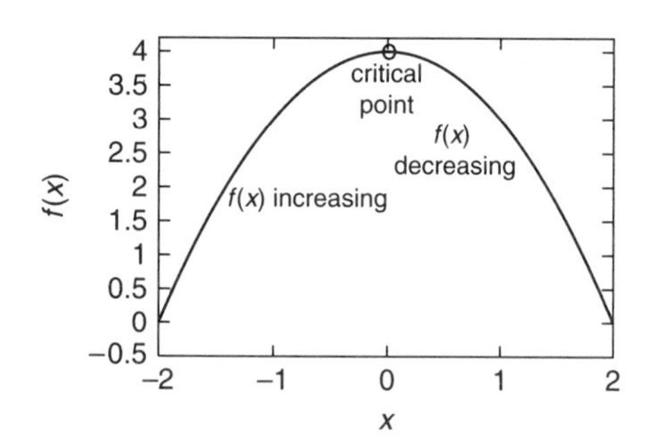

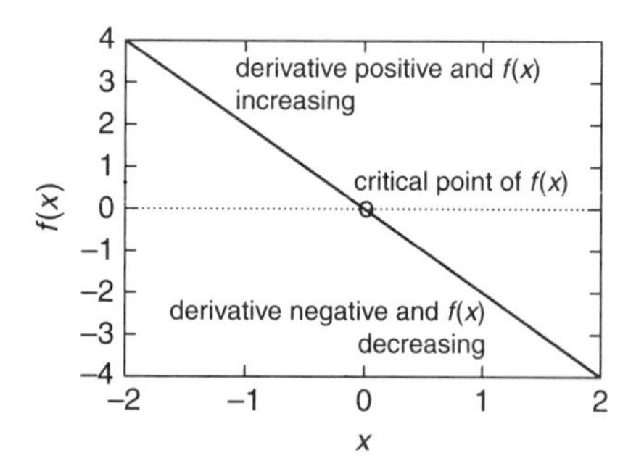

17.

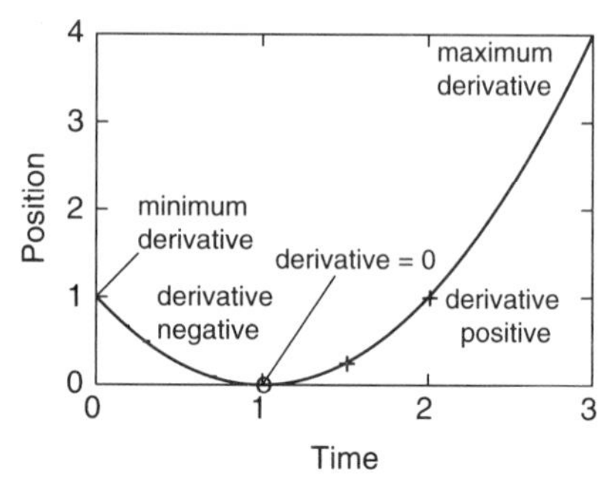

19.

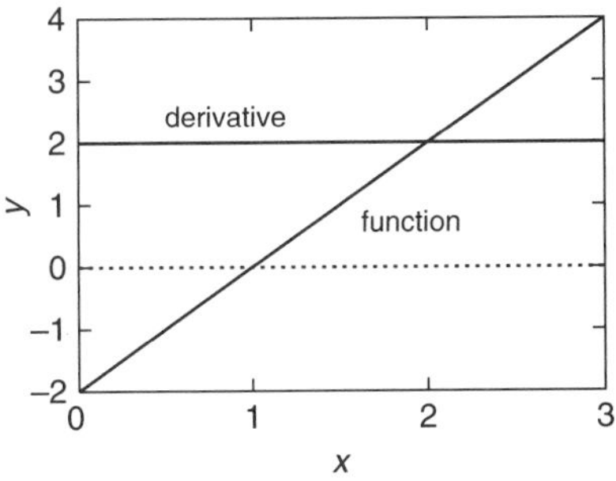

21.

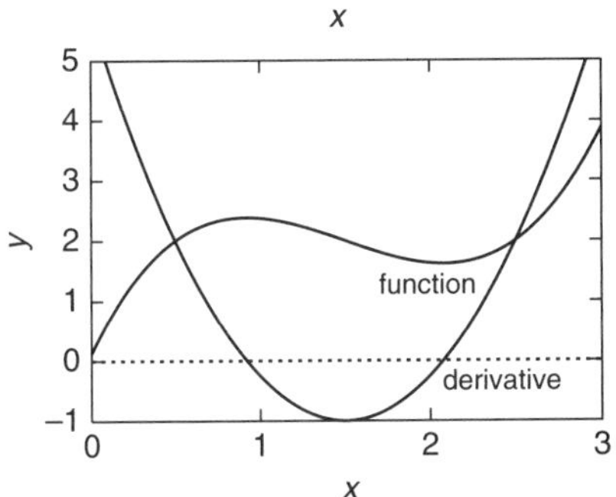

23. At $x = 0$, the slope of any secant with $\Delta x > 0$ is 1, while the slope of any secant with $\Delta x < 0$ is -1. The derivative wants to be both 1 and -1, which is impossible. From the graph, these slopes correspond to two possible tangent lines.

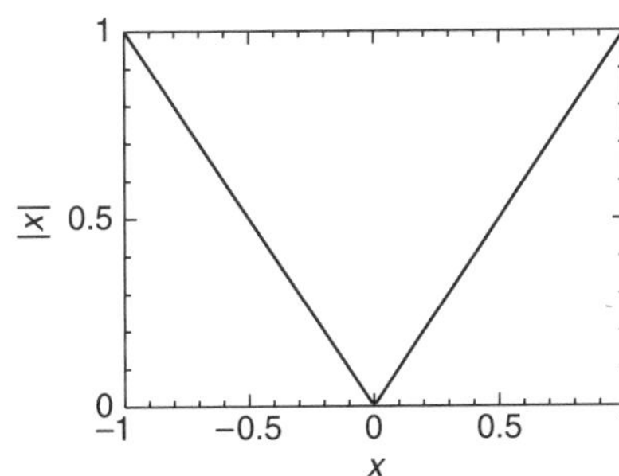

25. The slope of any secant with $\Delta x > 0$ is 0. With $\Delta x < 0$, the slope of the secant is $-1/\Delta x$, which has a limit of negative infinity. There is a sort of half-tangent line with equation $\hat{H}(x) = 1$ for positive Δx, but no candidate tangent line at all for negative Δx.

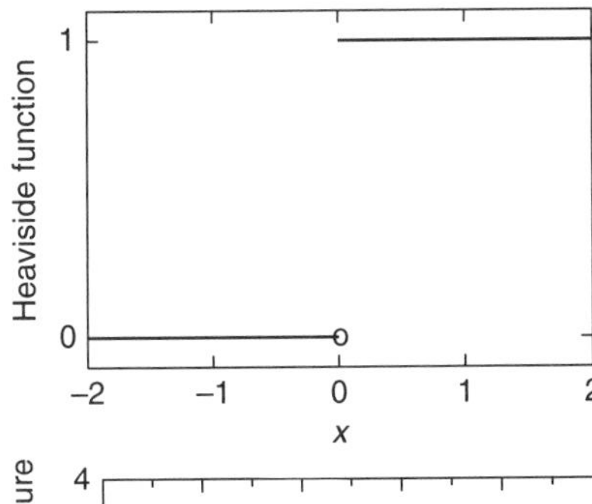

27.

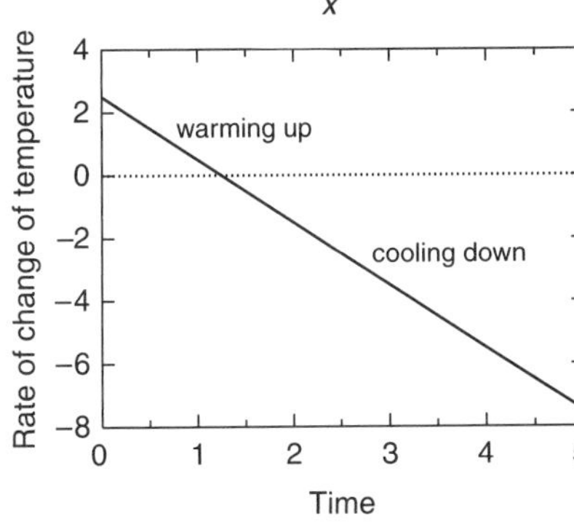

29.

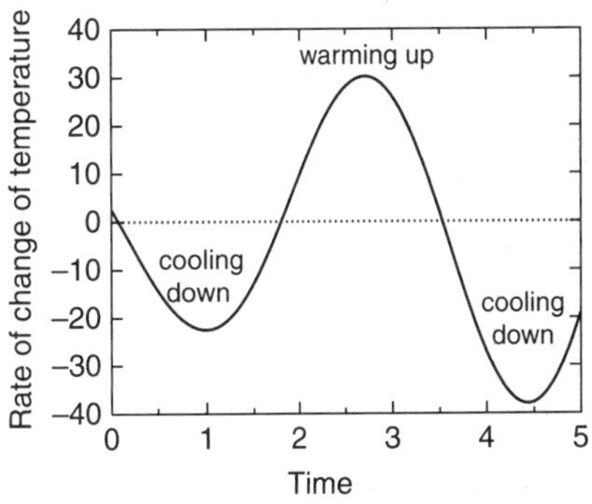

31.

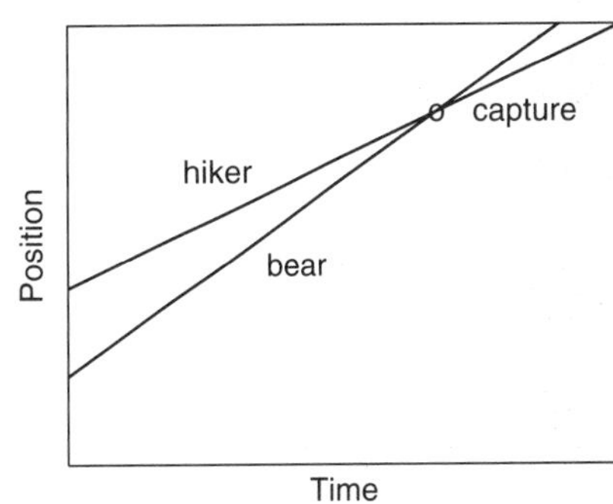

33.

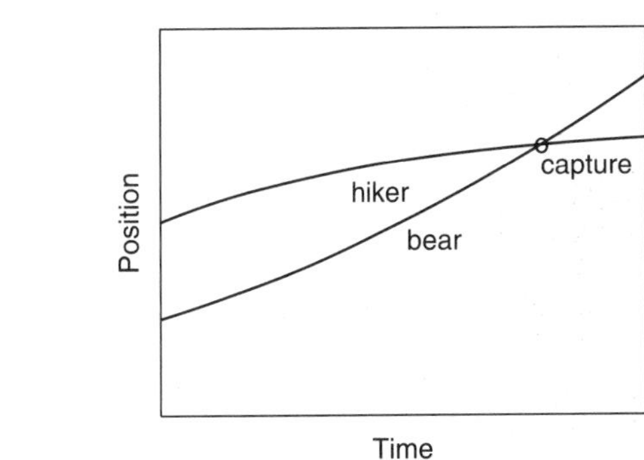

35. The time solves $M(t) = 0$, or $100 = \frac{9.78}{2}t^2$, or $t^2 = 20.44$ or $t = 4.52$ s. The speed is the derivative of the position, or $M'(t) = at = 9.78t$. The speed when it hits the ground is 44.2 m/s.

37. The time solves $M(t) = 0$, or $100 = \frac{22.88}{2}t^2$, or $t = 2.96$ s. The speed is the derivative of the position, or $M'(t) = at = 22.88t$. The speed when it hits the ground is 67.64 m/s.

Section 2.5, page 188

1. Using the power rule with $p = 5$ gives $5x^4$.

3. Using the power rule with $p = 0.2$ gives $0.2x^{-0.8}$.

5. Using the power rule with $p = e$ gives ex^{e-1}.

7. Using the power rule with $p = 1/e$ gives $\frac{1}{e}x^{\frac{1}{e}-1}$.

9.

$$\begin{aligned} f'(x) &= \frac{d(3x^2)}{dx} + \frac{d(3x)}{dx} + \frac{d1}{dx} && \text{sum rule} \\ &= 3\frac{d(x^2)}{dx} + 3\frac{dx}{dx} && \text{constant product rule} \\ &= 3 \cdot 2x + 3 \cdot 1 = 6x + 3. && \text{power rule} \end{aligned}$$

11.

$$\begin{aligned} g'(z) &= \frac{d(3z^3)}{dz} + \frac{d(2z^2)}{dz} && \text{sum rule} \\ &= 3\frac{d(z^3)}{dz} + 2\frac{dz^2}{dz} && \text{constant product rule} \\ &= 3 \cdot 3z^2 + 2 \cdot 2z = 9z^2 + 4z. && \text{power rule} \end{aligned}$$

13. $x^3 + 3x^2 + 3x + 1$.

15. $= (2x+1)^3 = (2x)^3 + 3 \cdot (2x)^2 \cdot 1 + 3 \cdot 2x \cdot 1^2 + 1^3 = 8x^3 + 12x^2 + 6x + 1$.

17. $f'(x) = -2 + 2x$, which is negative for $x < 1$, zero at $x = 1$, and positive for $x > 1$.

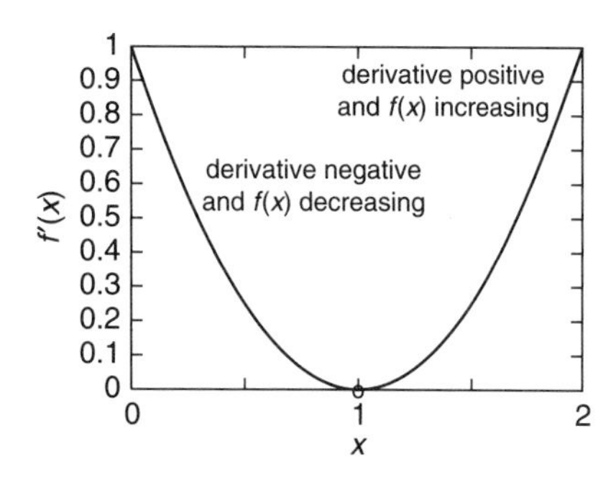

19. $h'(x) = 3x^2 - 3$, which is negative for $x < 1$, zero at $x = 1$, and positive for $x > 1$.

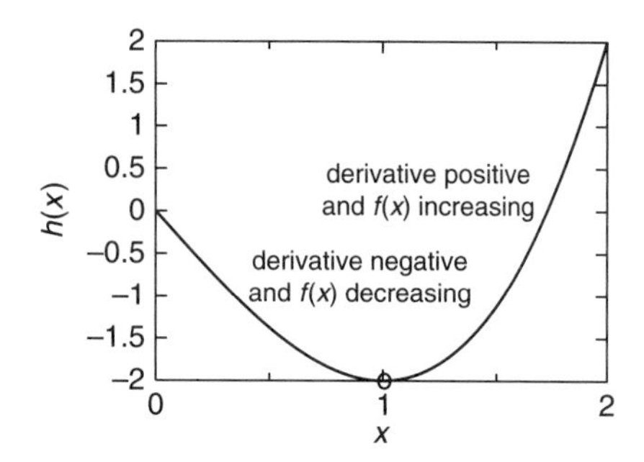

21. The constant 2 is the derivative of a linear function with slope 2. One such function is $2x$.

23. This looks just like the derivative of x^{15}.

25. From the table, this is x^{-1}.

27. $A'(t) = 525t^2$, with units of people per year. New cases went up quickly.

29. $A'(r) = 2\pi r$. The units are centimeters. The derivative is equal to the perimeter of the circle, corresponding to the area of the little ring around the circle.

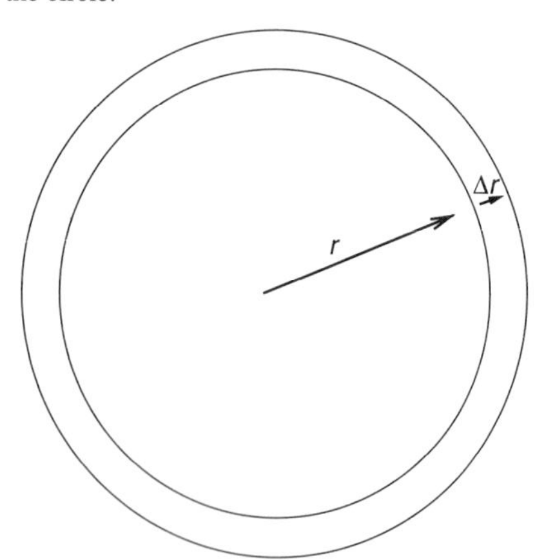

31.

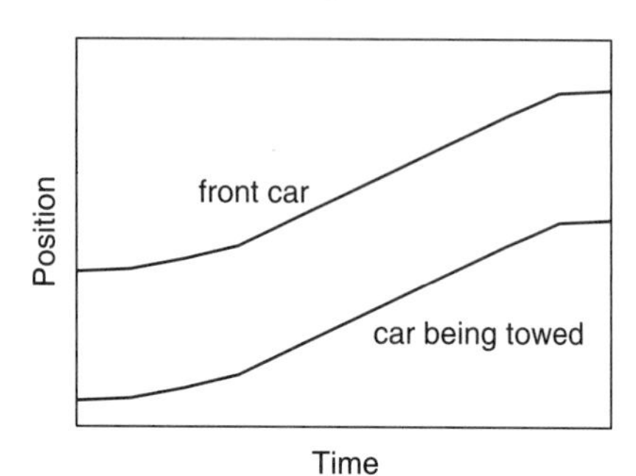

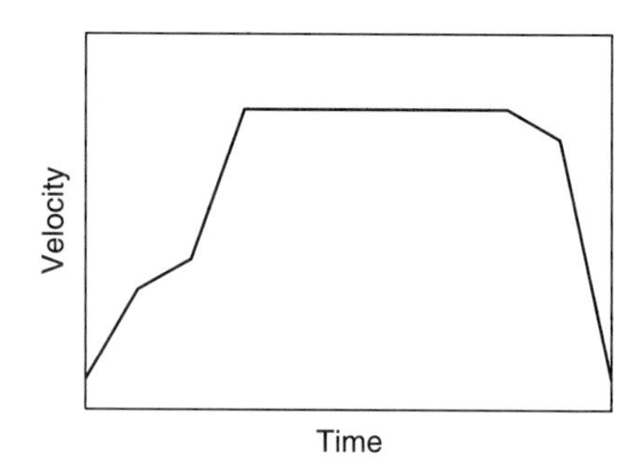

33. Subtract the 10 mph for the passenger from the 80 mph for the train to get 70 mph.

35. Relative to passenger: 20 mph, relative to train: 10 mph, relative to ground: 90 mph.

37. a. $P(t) = 2 - 2t + 2t^2$.

b. $a'(t) = 2t$ and $b'(t) = -2 + 2t$. The units are millions of bacteria per hour.

c. $P'(t) = -2 + 4t$, which is equal to $a'(t) + b'(t) = 2t - 2 + 2t$.

d. The population of a is always growing, the population of b shrinks before it grows, and the whole population also shrinks before it grows.

e.

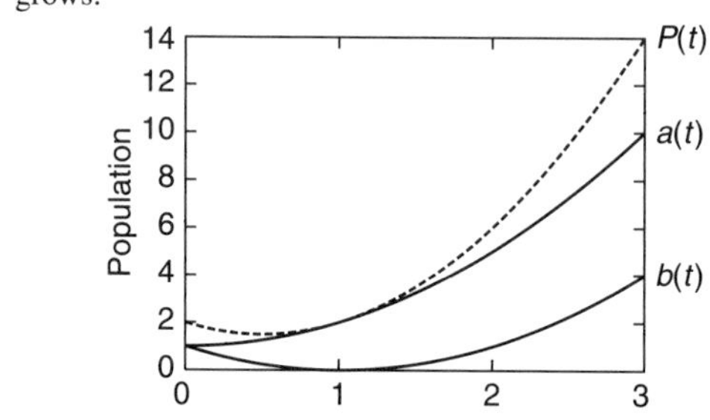

39. The speed is $M'(t) = 10 - at = 10 - 9.78t$. The critical point is then when $t = 1.02$ s. Plugging in $M(1.02) = 105.11$. It hits the ground when $M(t) = 0$, or at $t = 5.66$. The speed is $M'(5.66) = 45.34$ m/s.

41. The speed is $M'(t) = 10 - at = 10 - 22.88t$. The critical point is then when $t = 0.44$ s. Plugging in $M(0.44) = 102.18$. It hits the ground when $M(t) = 0$, or at $t = 3.42$. The speed is $M'(3.42) = 68.38$ m/s.

43. $S(t) = 3t^2$ works. $S(1) = 3$ and $S(2) = 12$. That seems pretty fast.

45. $S(t) = t^6$ because $\frac{dS}{dt} = 6t^5 = 6\frac{S(t)}{t}$. $S(1) = 1$ and $S(2) = 64$. That is quite fast, because the derivative gets bigger the larger the organism is.

Section 2.6, page 196

1. We write $f(x)$ as the product of the functions $g(x) = 2x + 3$ and $h(x) = -3x + 2$. Then $g'(x) = 2$ and $h'(x) = -3$, so

$$f'(x) = g'(x)h(x) + h'(x)g(x) = 2 \cdot (-3x + 2) + (-3) \cdot (2x + 3)$$
$$= -12x - 5.$$

3. We write $r(y)$ as the product of the functions $f(y) = 5y - 3$ and $g(y) = y^2 - 1$. Then $f'(y) = 5$ and $g'(y) = 2y$, so

$$r'(y) = f'(y)g(y) + g'(y)f(y) = 5(y^2 - 1) + 2y(5y - 3).$$

5.
$$\frac{dh}{dx} = \frac{d(x+2)}{dx}(2x+3)(-3x+2) + (x+2)\frac{d(2x+3)(-3x+2)}{dx}$$
$$= 1 \cdot (2x+3)(-3x+2) + (x+2) \cdot (-12x - 5)$$
$$= (2x+3)(-3x+2) - (x+2)(12x+5).$$

7. Set $u(x) = 1 + x$ and $v(x) = 2 + x$. Then $u'(x) = 1$ and $v'(x) = 2$. By the quotient rule,

$$f'(x) = \frac{(2+x) \cdot 1 - (1+x) \cdot 2}{(2+x)^2} = \frac{1}{(2+x)^2}.$$

9. This is the quotient $u(z)/v(z)$ with $u(z)=1+z^2$ and $v(z)=1+2z^3$. Then $u'(z)=2z$ and $v'(z)=6z^2$. Therefore,

$$g'(z)=\frac{(1+2z^3)2z-(1+z^2)6z^2}{(1+2z^3)^2}=\frac{-2z(-1+3z+z^3)}{(1+2z^3)^2}.$$

11. Set $u(x)=1+x$ and $v(x)=(2+x)(3+x)$. Then $u'(x)=1$ and $v'(x)=(2+x)+(3+x)=5+2x$, by the product rule. Applying the quotient rule gives

$$F'(x)=\frac{(2+x)(3+x)\cdot 1-(1+x)(5+2x)}{(2+x)^2(3+x)^2}=\frac{1-2x-x^2}{(2+x)^2(3+x)^2}.$$

13. $\Delta f=0.2$, $\Delta g=-0.3$, and $\Delta(fg)=-1.76$. This is equal to $g(x_0)\Delta f+f(x_0)\Delta g+\Delta f\Delta g=-0.2-1.5-0.06$. With $\Delta x=0.01$, $\Delta f=0.02$, $\Delta g=-0.03$, and $\Delta(fg)=-0.1706$. This is equal to $g(x_0)\Delta f+f(x_0)\Delta g+\Delta f\Delta g=-0.02-0.15-0.0006$. The last term is now much smaller than the rest.

15. With this incorrect rule, we would get that

$$\frac{dx\cdot x^2}{dx}=\frac{dx}{dx}\frac{dx^2}{dx}=1\cdot 2x.$$

However, the product $f(x)g(x)=x\cdot x^2=x^3$ really has derivative $3x^2$ by the power rule.

17.
$$\frac{df(x)\cdot f(x)}{dx}=f'(x)f(x)+f'(x)f(x)=2f'(x)f(x)>0$$

because $f'(x)>0$ and $f(x)>0$. Therefore $f(x)^2$ is increasing.

19. The power rule works for $n=1$ because $x^1=x$ is a linear function with derivative $1=1x^0$.

21. $\frac{dx^3}{dx}=x^2\cdot\frac{dx}{dx}+\frac{dx^2}{dx}\cdot x=x^2+2x^2=3x^2$. It worked.

23. We denote the total mass by $T(t)$.

a. $T(t)=P(t)W(T)=(2.0\times 10^6+2.0\times 10^4t)(80-0.5t)$.
b. $T'(t)=6.0\times 10^5-2.0\times 10^4t$.
c. $T'(t)=0$ when $6.0\times 10^5-2.0\times 10^4t=0$ or when $t=30$ yr. At time 30, the population is 2.6×10^6, the weight per person is 65 kg, and the total weight is 1.69×10^8 kg.
d.

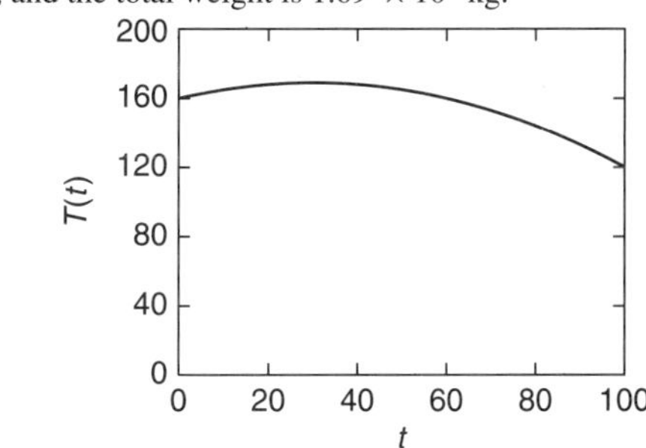

25. We denote the total mass by $T(t)$.

a. $T(t)=P(t)W(T)=(2.0\times 10^6+1.0\times 10^3t^2)(80-0.5t)$.
b. $T'(t)=1.0\times 10^6+1.6\times 10^5t-1.5\times 10^3t^2$.
c. $T'(t)=0$ at $t=6.67$ and $t=100.0$. At time 6.67, the population is 2.04×10^6, the weight per person is 76.67 kg, and the total weight is 1.56×10^8 kg. At time 100, the population is 1.2×10^7, the weight per person is 30.0 kg, and the total weight is 3.6×10^8 kg.
d.

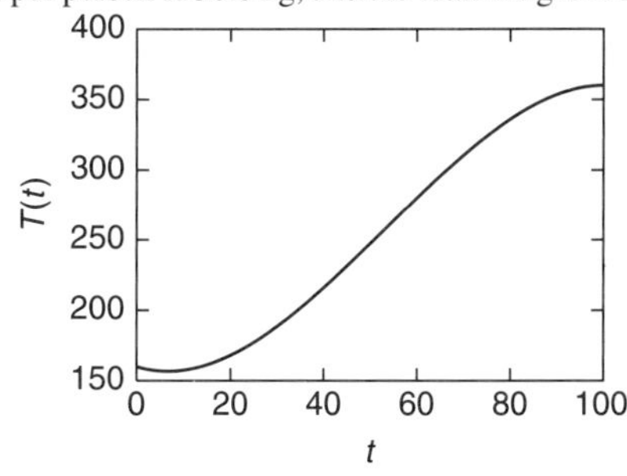

27. Total mass above ground is $(3.0t+20)(1.2-0.01t)$. The derivative is $-0.06t+3.4$. This switches from positive to negative at $t=56.7$, and thus the above-ground mass increases during the entire first 30 days.

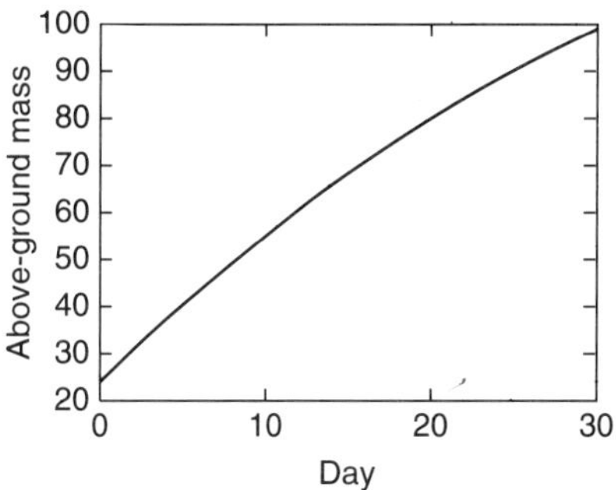

29. $S(1)=0.92$, $S(5)=3.0$, $S(10)=2.0$. $S'(N)=1-0.16N$. This bird does best by laying about 6 eggs.

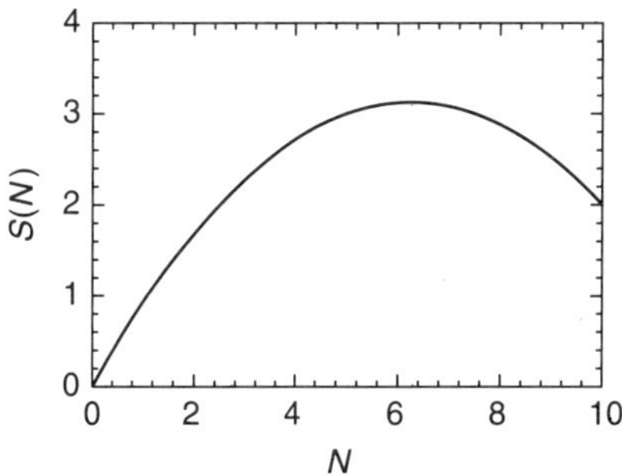

31. $S(1)=0.67$, $S(5)=1.43$, $S(10)=1.67$. $S'(N)=\frac{1}{(1+0.5N)^2}$ which is always positive. This bird does best by laying as many eggs as it can.

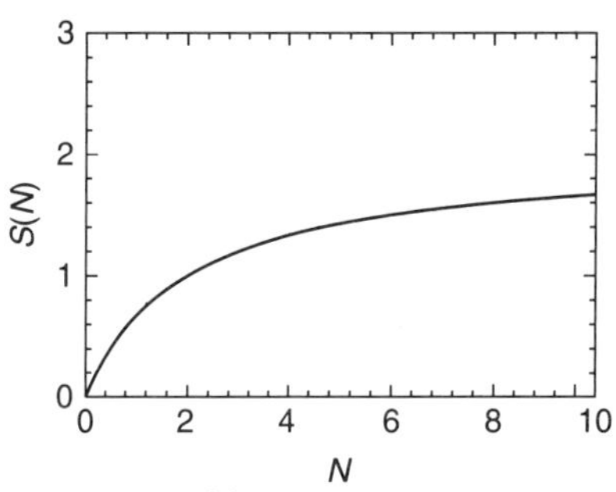

33. This simplifies to $\frac{2.4}{[1.2p+2.0(1-p)]^2}$.

35. a. $\rho(t)=\frac{1+t^2}{1+t}$.
b. Using the quotient rule with $u(t)=1+t^2$ and $v(t)=1+t$ gives $\rho'(t)=\frac{t^2+2t-1}{1+t^2}$.
c. This is positive when $t^2+2t-1>0$. This occurs for t larger than the solution of $t^2+2t-1=0$, which can be found with the quadratic formula to be 0.414. After that, the density increases.

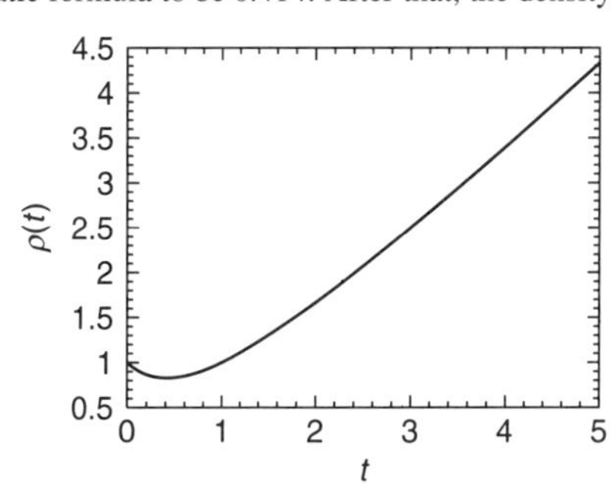

37. The final population is the product of the per capita production and the initial population b_t, or $b_{t+1}=2.0\left(1-\frac{b_t}{1000}\right)b_t$. The derivative of $f(b)=2.0b\left(1-\frac{b}{1000}\right)$ is $f'(b)=2.0-\frac{4.0b}{1000}$, which is positive for $b<500$ and negative for $b>500$.

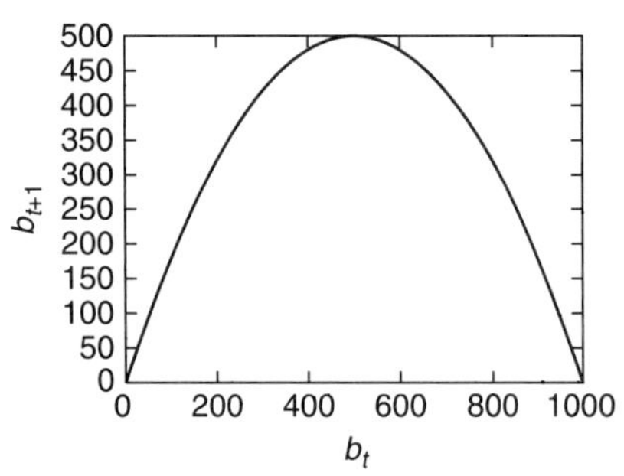

39. a. $h_3(0)=0$, $h_3(1)=0.5$, $h_3(2)=0.89$.

b. Using the quotient rule with $u(x)=x^3$ and $v(x)=1+x^3$ gives $h_3'(x)=\frac{3x^2}{(1+x^3)^2}$. Evaluating, we find $h_3'(0)=0$ $h_3'(1)=0.75$, $h_3'(2)=0.148$.

c.

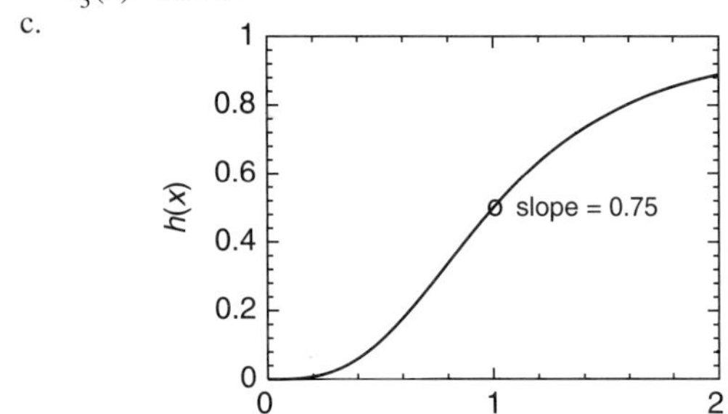

d. This is kind of a mushy response.

Section 2.7, page 205

1.

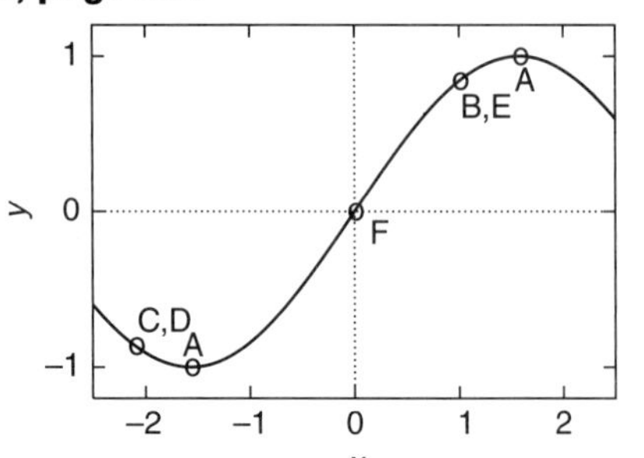

3.

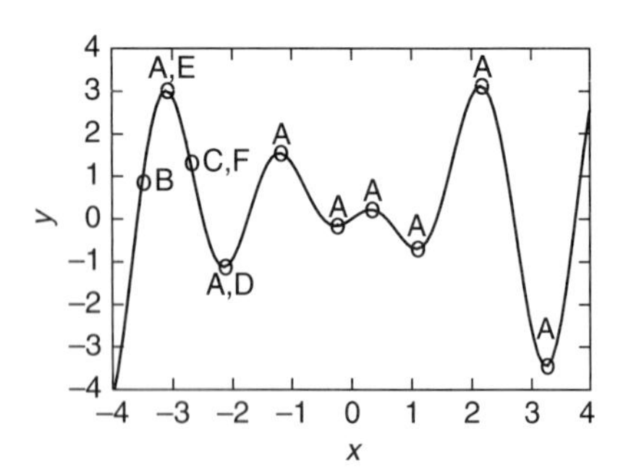

5.

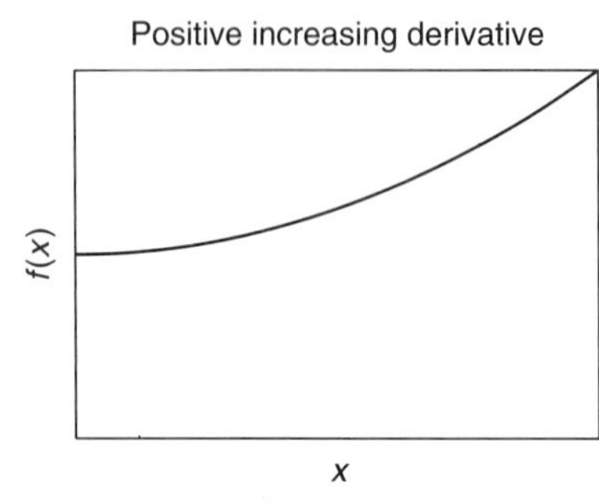

7.

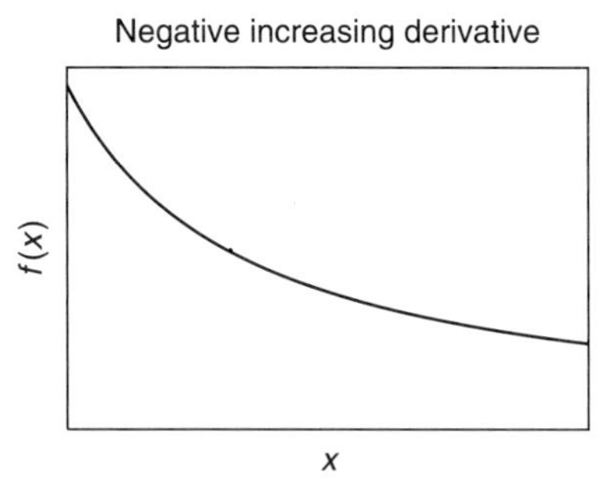

9. The point of inflection at $x=0$ has a negative third derivative because the second derivative changes from positive to negative values and is therefore decreasing.

11. Using the rules for evaluating the derivatives of polynomials, we find

$$s'(x)=\frac{d1}{dx}-\frac{dx}{dx}+\frac{dx^2}{dx}-\frac{dx^3}{dx}+\frac{dx^4}{dx}=0-1+2x-3x^2+4x^3.$$

Similarly,

$$s''(x)=\frac{ds'(x)}{dx}=-\frac{d1}{dx}+2\frac{dx}{dx}-3\frac{dx^2}{dx}+4\frac{dx^3}{dx}$$
$$=0+2-6x+12x^2.$$

13. Using the rules for differentiating polynomials gives $h'(y)=10y^9-9y^8$. $h''(y)=90y^8-72y^7$.

15. This is a triple product, so if $F(z)=f(z)g(z)h(z)$, then $F'(z)=f'(z)g(z)h(z)+f(z)g'(z)h(z)+f(z)g(z)h'(z)$. With $f(z)=z$, $g(z)=1+z$ and $h(z)=2+z$, we have that $f'(z)=g'(z)=h'(z)=1$. Therefore, $F'(z)=(1+z)(2+z)+z(2+z)+z(1+z)$. Each of these terms is a product, and the derivatives can be found with the product rule and added to give $F''(z)=(1+z)+(2+z)+z+(2+z)+z+(1+z)=6z+6$.

17. $f(x)$ is the quotient $u(x)/v(x)$ with $u(x)=3+x$ and $v(x)=2x$. Then $u'(x)=1$ and $v'(x)=2$, so by the quotient rule, $f'(x)=\frac{2x-2(3+x)}{4x^2}=\frac{-3}{2x^2}$. This is the constant -3/2 times a power function with power $p=-2$, so $f''(x)=\frac{-3}{x^3}$.

19. $f'(x)=-3x^{-4}<0$ so the function is decreasing, $f''(x)=12x^{-5}>0$ so the function is always concave up.

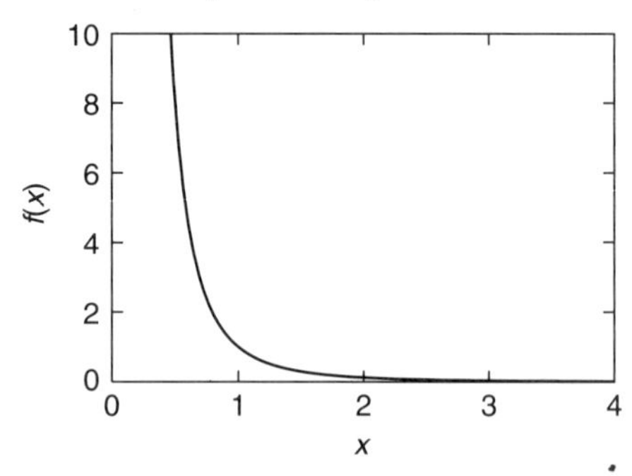

21. $h'(x)=-3x^2+12x-11$, which has solutions at 2.577 and 1.422. $h''(x)=-6x+12$, so the function is concave up for $x<2$ and concave down for $x>2$.

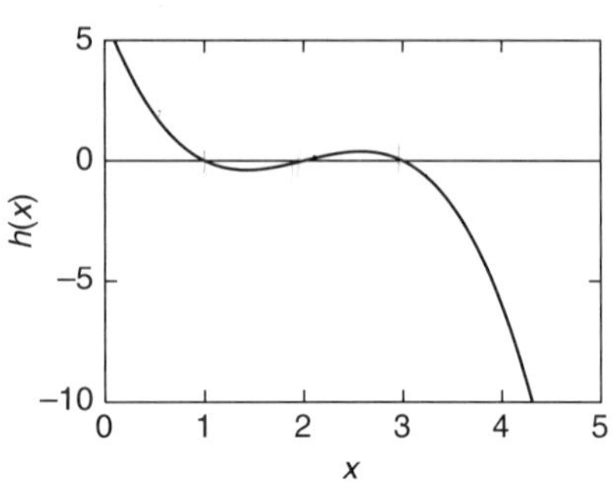

23. This function is a polynomial, so we can find $f'(x) = 6x^2$ by applying the power, constant product, and constant sum rules. The first derivative is always positive. The second derivative is $f''(x) = 12x$, which is positive when $x > 0$ and negative when $x < 0$, producing a point of inflection at $x = 0$.

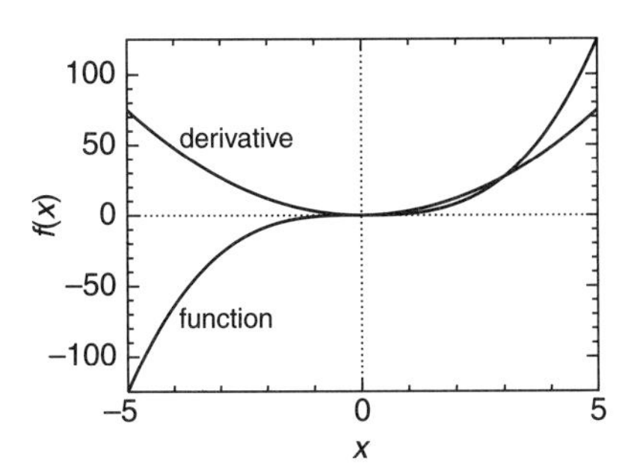

25. The derivative of this quadratic function is $f'(x) = 20x - 50$. The first derivative is positive when $x > 2.5$ and negative when $x < 2.5$. The second derivative is $f''(x) = 20$, which is always positive.

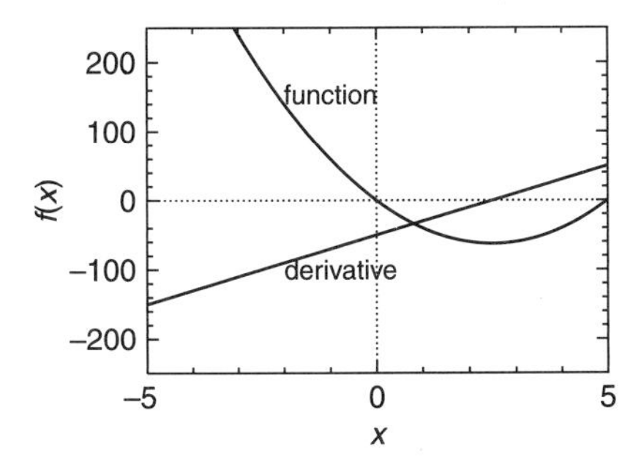

27. Each time we take the derivative of a power function, the power (or the degree) is reduced by 1. After taking 9 derivatives, then, the power will be 0, producing a constant function. The tenth derivative is the derivative of a constant, which is 0.

29. The eighth derivative of any power function with a positive power integer power less than 8 will be zero. Thus the only term with a nonzero derivative is the first. Its derivative will be a constant, found by multiplying a series of positive constant together, and will thus be positive.

31. a. Velocity is $p'(t) = -10.4t - 2.0$. Acceleration is $p''(t) = -10.4$.
b. The position at $t = 3$ is less than 0, so the object has already hit the ground.

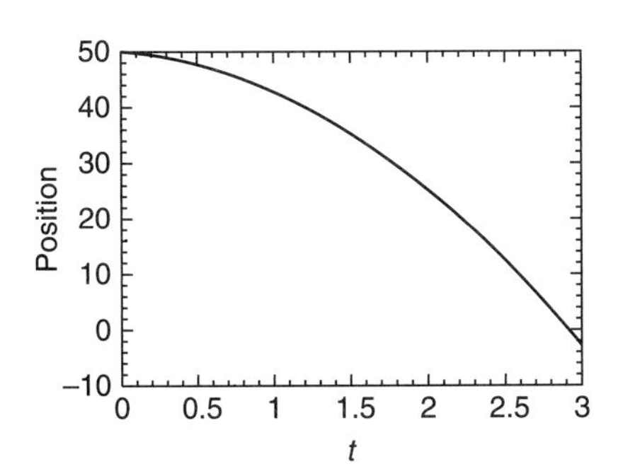

c. The tower was 50.0 m high, and the object was thrown downward at 2.0 m/s. The acceleration of gravity on Saturn is only slightly greater than that on Earth, probably due to Saturn's low density.

33. a. Velocity is $p'(t) = -0.65t - 20.0$. Acceleration is $p''(t) = -0.65$.

b.

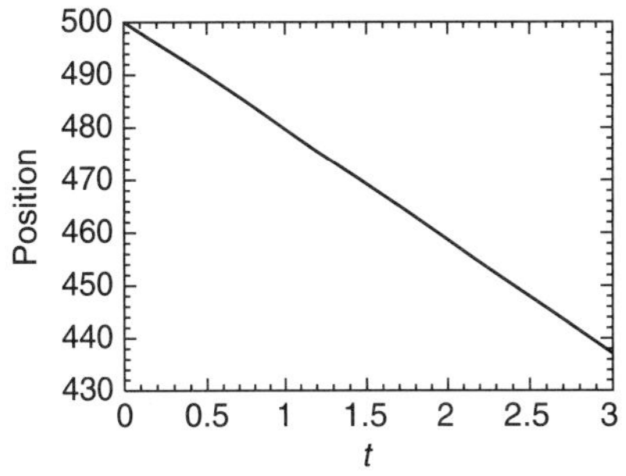

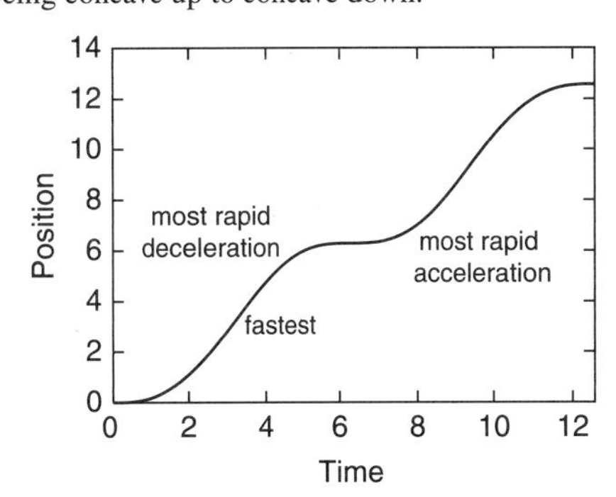

c. The tower was 500.0 m high, and the object was thrown downward at 20.0 m/s. The acceleration of gravity on Pluto is tiny, so the object is falling only slightly faster after 3 seconds than it was when it was thrown.

35. The second derivative is $-20{,}000$, matching the graph, which is always concave down.

37. The second derivative is $160{,}000 - 3000t$, which is positive for $t \leq 53.33$ and negative thereafter. This matches a graph that switches from being concave up to concave down.

39.

41. The derivative of $f(b) = 2.0b\left(1 - \frac{b}{1000}\right)$ is $f'(b) = 2.0 - \frac{4.0b}{1000}$, and the second derivative is $-\frac{4.0}{1000}$ which is always negative. The graph is always concave down.

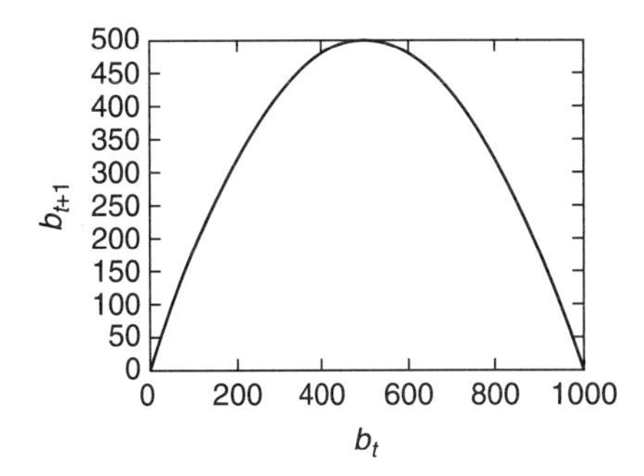

43. Using the quotient rule, we find that the first derivative is

$$h'(x) = \frac{1}{(1+x)^2}.$$

The second derivative can also be found with the quotient rule and product rule as

$$h''(x) = \frac{(1+x)^2 \cdot 0 - 1 \cdot [(1+x) + (1+x)]}{(1+x)^4} = \frac{-2}{(1+x)^3}.$$

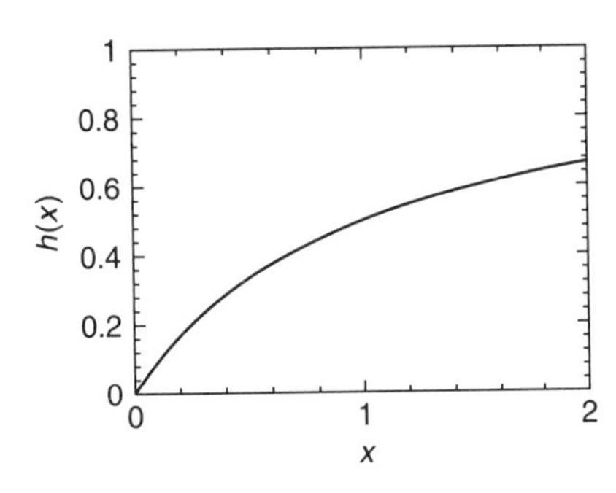

This is always negative, and thus the function is concave down.

Section 2.8, page 213

1.
$$\begin{aligned}\frac{dF}{dx} &= \frac{dx^2}{dx} + \frac{d(4e^x)}{dx} && \text{sum rule}\\ &= 2x + 4\frac{e^x}{dx} && \text{power and constant product rules}\\ &= 2x + 4e^x. && \text{derivative of exponential}\end{aligned}$$

Similarly,

$$\frac{dF'(x)}{dx} = \frac{d(2x)}{dx} + \frac{d(4e^x)}{dx} = 2 + 4e^x.$$

3. This is a product of the functions $u(x) = x^2$ and $v(x) = e^x$ with derivatives $u'(x) = 2x$ and $v'(x) = e^x$, so

$$f'(x) = u'(x)v(x) + v'(x)u(x) = 2xe^x + x^2e^x = (x^2 + 2x)e^x.$$

Similarly, this is the product of $u(x) = x^2 + 2x$ and $v(x) = e^x$ with derivatives $u'(x) = 2x + 2$ and $v'(x) = e^x$, so

$$f''(x) = (2x + 2)e^x + (x^2 + 2x)e^x = (x^2 + 4x + 2)e^x.$$

5. This is a quotient of the functions $u(x) = e^x$ and $v(x) = x$ with derivatives $u'(x) = e^x$ and $v'(x) = 1$, so

$$g'(x) = \frac{u'(x)v(x) - v'(x)u(x)}{v(x)^2} = \frac{xe^x - e^x}{x^2} = \frac{(x-1)e^x}{x^2}.$$

This is the quotient of $u(x) = (x-1)e^x$ and $v(x) = x^2$. The derivative of $v(x)$ is $v'(x) = 2x$, and we can find the derivative of $(x-1)e^x$ with the product rule to be $u'(x) = xe^x$. Therefore

$$g''(x) = \frac{x^3e^x - 2x(x-1)e^x}{x^4} = \frac{e^x(x^2 - 2x + 2)}{x^3}.$$

7. This is a quotient of the functions $u(x) = 1 + x$ and $v(x) = e^x$ with derivatives $u'(x) = 1$ and $v'(x) = e^x$, so

$$f'(x) = \frac{e^x - (1+x)e^x}{e^{2x}} = \frac{-x}{e^x}.$$

This is the quotient of $u(x) = -x$ and $v(x) = e^x$. Then $u'(x) = -1$, and $v'(x) = e^x$. Therefore

$$f''(x) = \frac{-e^x + xe^x}{e^{2x}} = \frac{x-1}{e^x}.$$

9. This is a sum, so

$$f'(x) = \frac{dx}{dx} + \frac{d[4\ln(x)]}{dx} = 1 + 4\frac{d\ln(x)}{dx} = 1 + \frac{4}{x}.$$

By the constant sum rule, the first term adds nothing to the second derivative, so

$$f''(x) = \frac{d1}{dx} + \frac{d(\frac{4}{x})}{dx} = 4\frac{dx^{-1}}{dx} = \frac{-4}{x^2}.$$

11. This is the product of $f(z) = z + 4$ and $g(z) = \ln(z)$ with derivatives $f'(z) = 1$ and $g'(z) = \frac{1}{z}$. By the product rule,

$$g'(z) = \ln(z) + \frac{z+4}{z} = \ln(z) + 1 + \frac{4}{z}.$$

The second derivative can be found with the sum and the power rules to be $g''(z) = \frac{1}{z} - \frac{4}{z^2}$.

13. This is the product of $f(w) = e^w$ and $g(w) = \ln(w)$ with derivatives $f'(w) = e^w$ and $g'(w) = \frac{1}{w}$. By the product rule,

$$F'(w) = e^w\ln(w) + \frac{e^w}{w}.$$

The first term is the original function, with the derivative we just found. The derivative of the second term can be found with the quotient rule (Section 2.8, Exercise 5) as $F''(w) = e^w\ln(w) + 2\frac{e^w}{w} - \frac{e^w}{w^2}$.

15. We can simplify this to $s(x) = 2\ln(x)$ and then use the constant product rule to find $s'(x) = \frac{2}{x}$. We can find the next derivative with the power rule as $s''(x) = \frac{-2}{x^2}$.

17. This is the quotient of $u(z) = 1 + e^{-z}$ and $v(z) = 1 + e^z$ with derivatives $u'(z) = -e^{-z}$ and $v'(z) = e^z$. By the quotient rule,

$$\begin{aligned}F'(z) &= \frac{-e^{-z}(1+e^z) - e^z(1+e^{-z})}{(1+e^z)^2}\\ &= \frac{-e^{-z} - e^z - 2}{(1+e^z)^2}\\ &= \frac{-(1+e^{-z})(1+e^z)}{(1+e^z)^2}\\ &= \frac{-(1+e^{-z})}{1+e^z}.\end{aligned}$$

This is just the negative of the original function, so the second derivative is just the original function again.

19. $f'(x) = -xe^x$, which is positive if $x < 0$ and negative if $x > 0$. $f''(x) = -(1+x)e^x$, which is positive if $x < -1$ and negative if $x > -1$. This function is increasing when $x < 0$ and is concave up when $x < -1$.

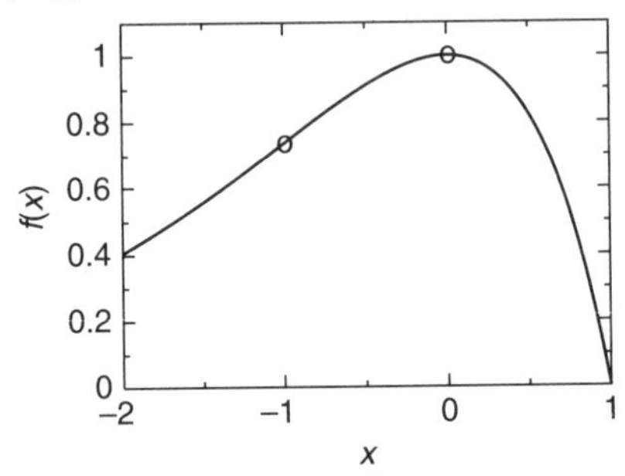

21. $G'(z) = \frac{e^z(z-2)}{z^3}$, which is negative if $z < 2$ and positive if $z > 2$. $G''(z) = \frac{e^z(z^2-4z+6)}{z^4}$ which is always positive. This function is increasing when $z > 2$ and is always concave up.

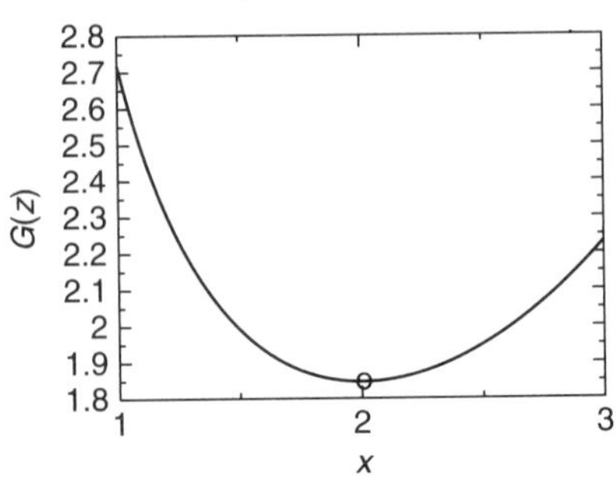

23. $L'(x) = \frac{1}{2} - \frac{1}{x}$ which is negative if $x < 2$ and positive if $x > 2$. $L''(x) = \frac{1}{x^2}$ which is always positive. This function is increasing when $x > 2$ and is always concave up.

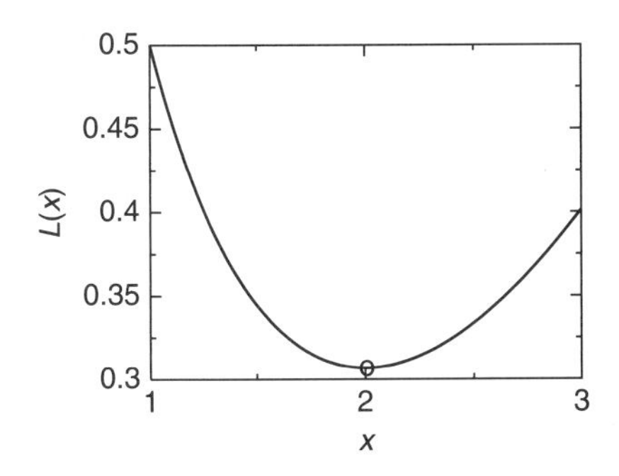

25. a.

$$\frac{df}{dx} = \lim_{h\to 0} \frac{5^{x+h} - 5^x}{h}.$$

b.

$$\frac{df}{dx} = \lim_{h\to 0} \frac{5^x 5^h - 5^x}{h}$$
$$= \lim_{h\to 0} \frac{5^x(5^h - 1)}{h}$$
$$= 5^x \left(\lim_{h\to 0} \frac{5^h - 1}{h} \right).$$

c. For $h = 1$, $(5^1 - 5^0)/1 = 4.0$. For $h = 0.1$, $(5^{0.1} - 5^0)/0.1 = 1.746$. For $h = 0.01$, $(5^{0.01} - 5^0)/0.01 = 1.622$. For $h = 0.001$, $(5^{0.001} - 5^0)/0.001 = 1.611$. For $h = 0.0001$, $(5^{0.0001} - 5^0)/0.0001 = 1.609$.

d. $e^{1.609} = 4.998$, which is suspiciously close to 5.

27. $h'(x) = (x+2)e^x$, $h''(x) = (x+3)e^x$.

29. $h(x) = (x-1)e^x$ seems to work. $h'(x) = xe^x$, so the critical point is at $x = 0$. $h''(x) = (x+1)e^x$, so the point of inflection is at $x = -1$.

31. a. The definition is

$$\frac{d\ln(x)}{dx} = \lim_{h\to 0} \frac{\ln(1+h) - \ln(1)}{h} = \lim_{h\to 0} \frac{\ln(1+h)}{h}.$$

b. The limit seems to be 1.0, meaning that the derivative is 1.0.

c. The derivative of $\ln(x)$ is $\frac{1}{x}$, which is 1.0 at $x = 1$.

33. The definition is

$$\frac{d\ln(x)}{dx} = \lim_{h\to 0} \frac{\ln(2+h) - \ln(2)}{h} = \lim_{h\to 0} \frac{\ln(1+h/2)}{h}.$$

If we write $\Delta x = h/2$, then $h = 2\Delta x$ and the limit can be written

$$\lim_{h\to 0} \frac{\ln(1+h/2)}{h} = \lim_{\Delta x\to 0} \frac{\ln(1+\Delta x)}{2\Delta x}$$
$$= \frac{1}{2} \lim_{\Delta x\to 0} \frac{\ln(1+\Delta x)}{\Delta x} = \frac{1}{2}$$

where we used the limit in Section 2.8, Exercise 31.

35. The total mass is $M(t) = 10(1 - \frac{t}{2})e^t$. $M'(t) = \frac{1-t}{2} 10e^t = 5(1-t)e^t$, which is negative for $t > 1$, and $M''(t) = -5te^t$ which is always negative. The total mass is greater than the initial value when $t = 1$. It reaches zero when $t = 2$.

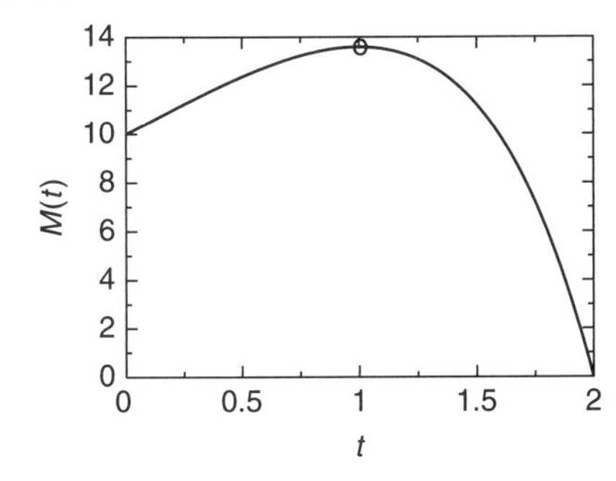

37. The total mass is $M(t) = 10(1 - t^2)e^t$. $M'(t) = 10(1 - 2t - t^2)e^t$, which is negative for $t > \sqrt{2} - 1$, and $M''(t) = 10(-1 - 4t - t^2)e^t$ which is negative for all t. The total mass is greater than the initial value at $t = \sqrt{2} - 1$. It reaches zero when $t = 1$.

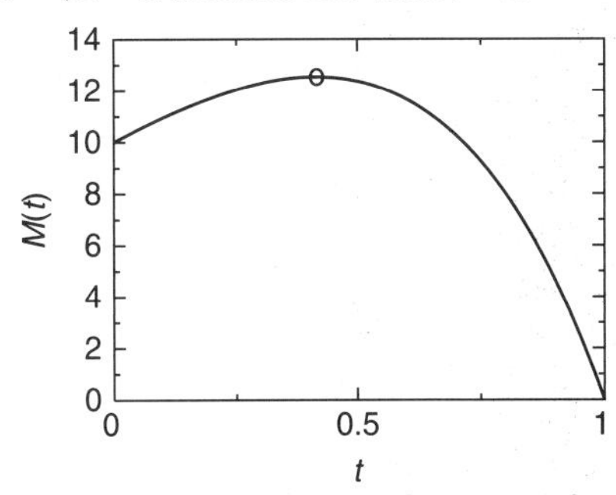

39. $G'(x) = \frac{1-2x}{2\sqrt{x}} e^{-x}$ which is positive for $x < 1/2$ and negative for $x > 1/2$. It also says that the curve is infinitely steep at $x = 0$. $G''(x) = \frac{4x^2 - 4x - 1}{4x^{3/2}} e^{-x}$. Solving $4x^2 - 4x - 1 = 0$ gives points of inflection at $\frac{1 \pm \sqrt{2}}{2}$. Only the larger one is positive. Below this, the function is concave down, and above it, it is concave up.

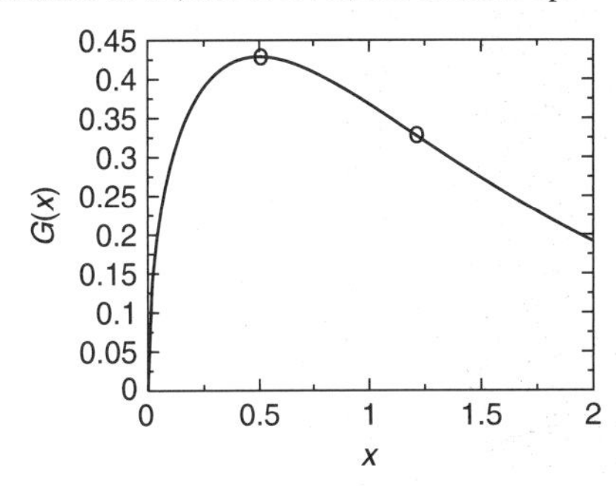

41. This equation says that the rate of change of the population increases exponentially over time. The derivative of $b(t) = e^t$ is e^t, so it works. This solution is increasing.

43. This equation says that the rate of change of the population is equal to the negative of population size, meaning it shrinks faster the larger it is. The derivative of $b(t) = e^{-t}$ is $-e^{-t}$, which is indeed the negative of $b(t)$. This solution is decreasing.

Section 2.9, page 226

1. $g(x) = f[h(x)]$ where $h(x) = 1 + 3x$ and $f(h) = h^2$. Then

$$g'(x) = f'[h(x)]h'(x) = 2h \cdot 3 = 6(1+3x).$$

3. $f_1(t) = h[g(t)]$ where $g(t) = 1 + 3t$ and $h(g) = g^{30}$. Then $g'(t) = 3$ and $h'(g) = 30g^{29}$. Therefore $f_1'(t) = 3 \cdot 30g^{29} = 90(1+t)^{29}$.

5. This is a quotient of two compositions. The numerator is $(1+3x)^2$ with derivative $6(1+3x)$. The denominator is $(1+2x)^3$ with derivative $6(1+2x)^2$. By the quotient rule, the derivative is

$$r'(x) = \frac{6(1+2x)^3(1+3x) - 6(1+3x)^2(1+2x)^2}{(1+2x)^6}$$
$$= \frac{-6x(1+3x)}{(1+2x)^4}.$$

7. $F(z) = g[h(z)]$ where $h(z) = 1 + \frac{2}{1+z}$ and $g(h) = h^3$. $h'(z) = -2(1+z)^{-2}$, found using the chain rule. Then

$$F'(z) = g'[h(z)]h'(z) = 3h^2 \cdot [-2(1+z)^{-2}] = -6\frac{\left(1 + \frac{2}{1+z}\right)^2}{(1+z)^2}.$$

9. Set $f(x) = g[h(x)]$ where $g(h) = e^h$ and $h(x) = -3x$. Then

$$f'(x) = g'[h(x)]h'(x) = e^{h(x)} \cdot (-3) = -3e^{-3x}.$$

11. We can write $g(y) = f[h(y)]$ where $f(h) = \ln(h)$ and $h(y) = 1 + y$. Then $f'(h) = 1/h$ and $h'(y) = 1$. Then $g'(y) = f'(h)h'(y) = \frac{1}{h} = \frac{1}{1+y}$.

13. We can write $G(x) = f[g(x)]$ where $f(g) = 8e^g$ and $g(x) = x^2$. Then $f'(g) = 8e^g$ and $g'(x) = 2x$. Then $G'(x) = 8e^g \cdot 2x = 16xe^{x^2}$.

15. Set $L(x) = f[g(x)]$ where $f(g) = \ln(g)$ and $g(x) = \ln(x)$. Then

$$L'(x) = f'[g(x)]g'(x) = \frac{1}{g(x)}\frac{1}{x} = \frac{1}{x\ln(x)}.$$

17. With the quotient rule, set $u(x) = 1$ and $v(x) = 1 + e^x$. Then

$$H'(x) = \left(\frac{u}{v}\right)'(x) = \frac{v(x)u'(x) - u(x)v'(x)}{v^2(x)}$$
$$= \frac{(1+e^x)\cdot 0 - 1\cdot e^x}{(1+e^x)^2} = -\frac{e^x}{(1+e^x)^2}.$$

With the chain rule, set $H(x) = r[p(x)]$ where $p(x) = 1 + e^x$ and $r(p) = 1/p$. Then $p'(x) = e^x$ and $r'(p) = \frac{-1}{p^2}$. By the chain rule,

$$H'(x) = r'(p)p'(x) = \frac{-1}{p^2}\cdot e^x = \frac{-e^x}{(1+e^x)^2}.$$

19. With law 1 of logs, $\ln(3x) = \ln(3) + \ln(x)$. Then

$$\frac{d\ln(3x)}{dx} = \frac{d\ln(3)}{dx} + \frac{d\ln(x)}{dx} = 0 + \frac{1}{x} = \frac{1}{x}.$$

With the chain rule, set $f(x) = 3x$ and $l(f) = \ln(f)$. Then

$$(l \circ f)'(x) = l'[f(x)]f'(x) = \frac{1}{f}\cdot 3 = \frac{1}{3x}\cdot 3 = \frac{1}{x}.$$

The two answers check.

21. $F(x) = 1 + 4x + 4x^2$ and $F'(x) = 4 + 8x$. With the chain rule, we think of F as the composition $F(x) = g[h(x)]$ where $h(x) = 1 + 2x$ and $g(h) = h^2$. Then $h'(x) = 2$ and $g'(h) = 2h$, so $F'(x) = g'(h)h'(x) = 2h \cdot 2 = 4(1+2x)$. Multiplying out, this matches the result found directly.

23. $F'(x) = 3x^2$. We can write $F(x) = e^{3\ln(x)}$, a composition $f[g(x)]$ where $f(g) = e^g$ and $g(x) = 3\ln(x)$. Then $f'(g) = e^g$ and $g'(x) = 3/x$, so

$$F'(x) = f'(g)g'(x) = e^g \cdot \frac{3}{x} = e^{3\ln(x)}\frac{3}{x} = x^3\frac{3}{x} = 3x^2.$$

25. Solving $f(x) = 3x + 1 = y$ for y gives $f^{-1}(y) = \frac{y-1}{3}$ with derivative $(f^{-1})'(y) = 1/3$. Using Theorem 2.13, $f'(x) = 3$ for any x, so $f'[f^{-1}(y)] = 3$ and $(f^{-1})'(y) = 1/3$.

27. Solving $h(x) = 2 + x^3 = y$ for y gives $h^{-1}(y) = (y-2)^{1/3}$ with derivative $(h^{-1})'(y) = \frac{1}{3}(y-2)^{-2/3}$ using the power rule and the chain rule. Using Theorem 2.13, $h'(x) = 3x^2$, so $h'[h^{-1}(y)] = 3(y-2)^{2/3}$ and $(h^{-1})'(y) = \frac{1}{3}(y-2)^{-2/3}$.

29. Solving $q(x) = y$ for y gives $q^{-1}(y) = \frac{-1\pm\sqrt{4+y}}{2}$. The only possible positive root has the positive sign, so $q^{-1}(y) = \frac{-1+\sqrt{4+y}}{2}$. Taking the derivative with the power rule and chain rules gives $(q^{-1})'(y) = \frac{1}{\sqrt{4+y}}$. Using Theorem 2.13, $q'(x) = 1 + 2x$. so $q'[q^{-1}(y)] = \sqrt{4+y}$ and $(q^{-1})'(y) = \frac{1}{\sqrt{4+y}}$.

31. Taking the derivative of both sides with respect to x,

$$\frac{d}{dx}(xy) = x\frac{dy}{dx} + y = 0$$

from which we find that

$$\frac{dy}{dx} = -\frac{y}{x} = -\frac{1}{x^2}.$$

Solving for y gives $y = 1/x$ which has derivative $-1/x^2$ by the power rule.

33. Because $x = e^{\ln(x)}$, we can write $x^n = \left[e^{\ln(x)}\right]^n$. Using law 2 of logs gives $x^n = e^{n\ln(x)}$.

35. We can write $f(x) = g[h(x)]$ where $h(x) = 1 - x^2$ and $g(h) = \sqrt{h}$. Then $h'(x) = -2x$ and $g'(h) = \frac{1}{2\sqrt{h}}$. Therefore, $f'(x) = -2x \cdot \frac{1}{2\sqrt{h(x)}} = \frac{-x}{\sqrt{1-x^2}}$.

37. Taking derivatives of both sides with respect to x, we find

$$2y\frac{dy}{dx} + 16x^3 = 8x,$$

so we can solve for dy/dx as

$$\frac{dy}{dx} = \frac{8x - 16x^3}{2y}.$$

At $x = 0.5$, we have that $y^2 = 3/4$, or $y = \pm\sqrt{0.75} = \pm 0.866$. At $x = -0.5$, we also have that $y^2 = 3/4$, or $y = \pm\sqrt{0.75} = \pm 0.866$. At $x = 1$ and $x = -1$, we have that $y = 0$. Evaluating the derivatives gives

x	y	dy/dx
0.5	0.866	1.155
0.5	−0.866	−1.155
−0.5	0.866	−1.155
−0.5	−0.866	1.155
1.0	0.0	∞
−1.0	0.0	∞

At $x = 0$, we also have that $y = 0$, but cannot compute the derivative (there are really two tangent lines, one with slope 2 and one with slope −2. This graph looks like the symbol for infinity (a sideways figure 8) with a crossing point at the origin.

39. $B(M) = 0.5M$, so $B'(M) = 0.5$. $M(W) = 5W + 2$ so $M'(W) = 5$. Therefore, the derivative of the composition is $B'(M)M'(W) = 2.5$.

41. $F'(V) = 0.4$ and $V'(I) = 10I$, so the derivative of the composition is $F'(V)V'(I) = 4I$.

43. $Q'(t) = -0.000122 \cdot 6.0 \times 10^{10}e^{-0.000122t} = -7.32 \times 10^6 e^{-0.000122t}$.

45. $y(t) = -120 + 30t$, $x(t) = -80 + 20t$, and

$$r(t) = \sqrt{y(t)^2 + x(t)^2}$$
$$= \sqrt{(-120+30t)^2 + (-80+20t)^2}.$$

Taking the derivative of both sides with respect to t gives

$$\frac{dr}{dt} = \frac{y(t)\frac{dy}{dt} + x(t)\frac{dx}{dt}}{\sqrt{y(t)^2 + x(t)^2}}.$$

t	$x(t)$	$y(t)$	$r(t)$	dr/dt
0	−80	−120	144.22	−36.06
2	−40	−60	72.11	−36.06
4	0.0	0.0	0.0	??

Our formula for dr/dt breaks down at $t = 4$ when the distance between the cheetah and gazelle hits zero. Mathematically, this occurs because we are dividing by zero. However, the distance between the animals is decreasing at 36.06 m/s throughout, and this is the appropriate limit at $t = 4$.

47. Taking the derivative of both sides, we find that

$$\frac{dE}{dt} = c_1\left(0.75p^{-0.25}M^{0.75}\frac{dp}{dt} + 0.75p^{0.75}M^{-0.25}\frac{dM}{dt}\right)$$
$$+ c_2\left(-0.75(1-p)^{-0.25}M^{0.75}\frac{dp}{dt}\right.$$
$$\left. + 0.75(1-p)^{0.75}M^{-0.25}\frac{dM}{dt}\right).$$

Substituting in the given values gives

$$0.02 = 0.01(0.75(0.3)^{-0.25}30.0^{0.75}\frac{dp}{dt}$$
$$+ 0.75(0.3)^{0.75}30.0^{-0.25}\cdot 2.0$$
$$+ 0.022(-0.75(0.7)^{-0.25}30.0^{0.75}\frac{dp}{dt}$$
$$+ 0.75(0.7)^{0.75}30.0^{-0.25}\cdot 2.0,$$

which we can solve for $dp/dt = -0.065$. The fraction of the first type of metabolic tissue is decreasing. This could be useful because the rate of change of p could be more difficult to measure directly than the rate of change of M or E.

49. The derivative is $\frac{db}{dt} = 300e^{3t} = 3 \cdot 100e^{3t} = 3b(t)$. This solution is increasing.

51. The derivative is $\frac{db}{dt} = 6e^{2t} = 1 + 2(3e^{2t} - 0.5) = 1 + 2b(t)$. This solution is increasing.

Section 2.10, page 237

1. This is the product of $g(x) = x^2$ and $h(x) = \sin(x)$ with derivatives $g'(x) = 2x$ and $h'(x) = \cos(x)$. Therefore, by the product rule, $f'(x) = 2x\sin(x) + x^2\cos(x)$.

3. This is the product of $f(\theta) = \sin(\theta)$ and $g(\theta) = \cos(\theta)$ with derivatives $f'(\theta) = \cos(theta)$ and $g'(\theta) = -\sin(\theta)$. Therefore, $h'(x) = \cos^2(\theta) - \sin^2(\theta)$.

5. This is a composition $F(z) = f[g(z)]$ where $g(z) = 2z - 1$ and $f(g) = 3 + \cos(g)$. Then $g'(z) = 2$ and $f'(g) = -\sin(g)$, so by the chain rule, $F'(z) = f'(g)g'(z) = -2\sin(2z - 1)$.

7. This is a composition $f(x) = g[h(x)]$ where $h(x) = \cos(x)$ and $g(h) = e^h$. Then $g'(h) = e^h$ and $h'(x) = -\sin(x)$, so by the chain rule $f'(x) = g'(h)h'(x) = -\sin(x)e^{\cos(x)}$.

9. As in the text, we can use the fact that $\tan(\theta) = \frac{\sin(\theta)}{\cos(\theta)}$ to find that the derivative is $\sec^2(\theta) = \cos(\theta)^{-2}$. With the chain rule, the second derivative is $\frac{2\sin(\theta)}{\cos(\theta)^3}$, which can be written $2\tan(\theta)\sec(\theta)^2$.

11. The first derivative is found with the chain rule

$$\frac{d\sec(\theta)}{d\theta} = \frac{d\cos(\theta)^{-1}}{d\theta} = \sin(\theta)\cos(\theta)^{-2}.$$

This can be rewritten as $\sec(\theta)\tan(\theta)$. We can find the second derivative with the product rule as $\sec(\theta)\tan(\theta)^2 + \sec(\theta)^3$.

13. The double angle formula is $\cos(2\theta) = \cos(\theta)\cos(\theta) - \sin(\theta)\sin(\theta)$. Taking the derivative,

$$\begin{aligned}\frac{d}{d\theta}\cos(2\theta) &= -\sin(\theta)\cos(\theta) - \sin(\theta)\cos(\theta)\\ &\quad - \sin(\theta)\cos(\theta) - \sin(\theta)\cos(\theta)\\ &= -4\sin(\theta)\cos(\theta) = -2\sin(2\theta).\end{aligned}$$

This is just what the chain rule gives.

15. The angle addition formula says that $\cos(\theta + \phi) = \cos(\theta)\cos(\phi) - \sin(\theta)\sin(\phi)$. Taking the derivative,

$$\frac{d}{d\theta}\cos(\theta + \phi) = -\sin(\theta)\cos(\phi) - \cos(\theta)\sin(\phi) = -\sin(\theta + \phi).$$

This matches the result with the chain rule.

17. $a'(x) = 3 - \sin(x)$. $a(0) = 1, a'(0) = 3, a(\pi/2) = 3\pi/2, a'(\pi/2) = 2, a(\pi) = 3\pi - 1, a'(\pi) = 3$. The function is increasing, but with a slight slowing at around $x = \pi$.

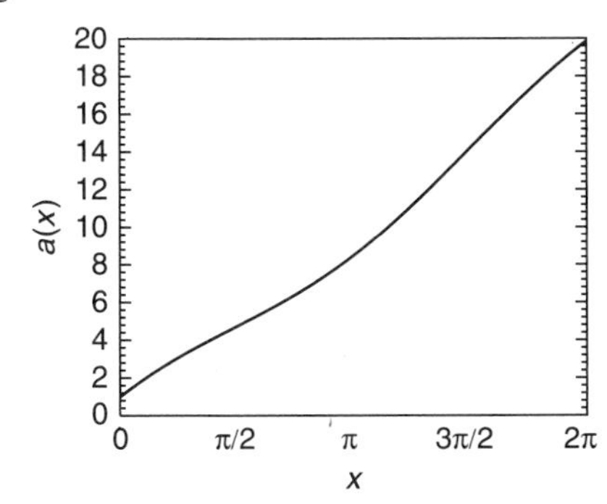

19. $c'(z) = [\cos(z) - \sin(z)]/e^z$. $c(0) = 0$, $c'(0) = 1$, $c(\pi/2) = e^{-\pi/2}$, $c'(\pi/2) = -e^{-\pi/2}$, $c(\pi) = 0$, $c'(\pi) = -e^{-\pi}$. The function zips up to a maximum at around $x = \pi/2$, dips down to zero at $x = \pi$, becomes negative, and then returns to zero at $x = 2\pi$.

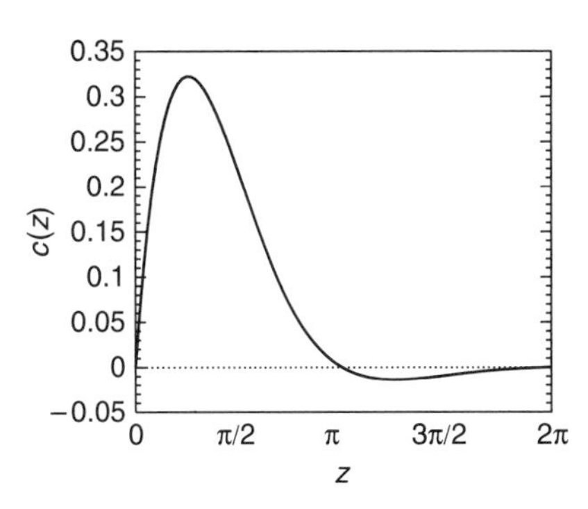

21. $s'(t) = 0.2e^{0.2t}\cos(t) - e^{0.2t}\sin(t)$. $s(0) = 1$, $s'(0) = 0.2$, $s(\pi/2) = 0$, $s'(\pi/2) = -1.369$, $s(\pi) = -1.874$, $s'(\pi) = -0.375$. This oscillation expands exponentially.

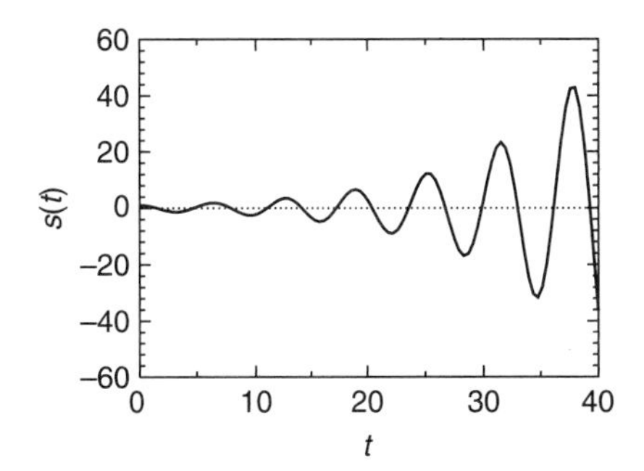

23. The derivative of $\sin(x)$ is $\cos(x)$. Therefore

$$(\sin^{-1})'(x) = \frac{1}{\cos[\sin^{-1}(x)]}.$$

Using the identity, $\cos(x) = \sqrt{1 - \sin^2(x)}$, so

$$\cos[\sin^{-1}(x)] = \sqrt{1 - \sin^2[\sin^{-1}(x)]} = \sqrt{1 - x^2}.$$

Therefore the derivative is $(\sin^{-1})'(x) = \frac{1}{\sqrt{1-x^2}}$.

25. The derivative of $\tan(x)$ is $\sec^2(x)$. Therefore

$$(\tan^{-1})'(x) = \frac{1}{\sec^2[\tan^{-1}(x)]}.$$

Using the identity,

$$\sec^2[\tan^{-1}(x)] = 1 + \tan^2[\tan^{-1}(x)] = 1 + x^2.$$

Therefore the derivative is $(\tan^{-1})'(x) = \frac{1}{1+x^2}$.

27. By the constant sum rule, the derivative of $s(t)$ is $s'(t) = \cos(t)$. This function is a solution of the differential equation.

29. The fourth derivative of each function $\cos(t)$, $\sin(t)$, and e^t is equal to itself. By the constant product rule, then

$$\frac{d^4}{dt^4}\cos(t) = \cos(t)$$

$$\frac{d^4}{dt^4}2\sin(t) = 2\sin(t)$$

$$\frac{d^4}{dt^4}(-3e^t) = -3e^t.$$

By the sum rule, then,

$$\frac{d^4}{dt^4}\cos(t) + 2\sin(t) - 3e^t = \cos(t) + 2\sin(t) - 3e^t,$$

making this function a solution of the differential equation.

31. $f'(x) = -\frac{8.0\pi}{5.0}\sin\left(2\pi\frac{x-1.0}{5.0}\right)$. $f'(0) = -\frac{8.0\pi}{5.0}\sin(-2\pi/5) > 0$, consistent with the fact that this oscillation is increasing at $x = 0$.

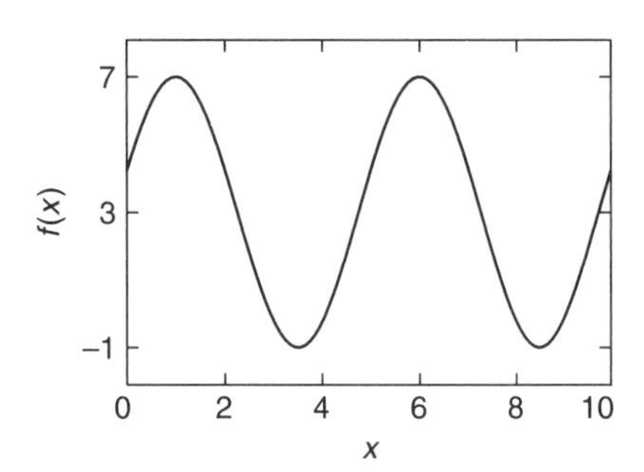

33. $h'(z) = -\frac{5.0\pi}{2.0}\sin\left(2\pi\frac{z-3.0}{4.0}\right)$. $h'(0) = -\frac{5.0\pi}{2.0}\sin(-3\pi/2) < 0$, consistent with the fact that the curve begins by decreasing.

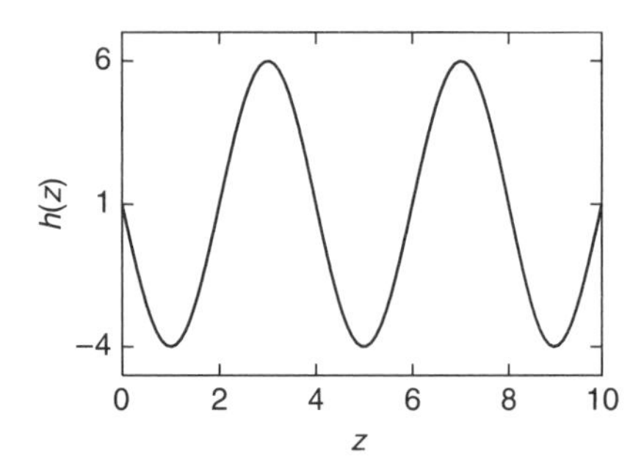

35. Using the chain rule, we find that

$$p'(t) = -\frac{2\pi}{T}\sin\left(\frac{2\pi t}{T}\right), \quad p''(t) = -\left(\frac{2\pi}{T}\right)^2\cos\left(\frac{2\pi t}{T}\right).$$

We thus need

$$\left(\frac{2\pi}{T}\right)^2 = \frac{0.1}{1.0} = 0.1.$$

This has solution $T = \sqrt{0.1}2\pi$. This weaker spring oscillates more slowly with a larger period.

37. The derivative of the sum is

$$-0.2\cdot\frac{2\pi}{28}\sin\left[\frac{2\pi(t-16)}{28}\right] - 0.3\cdot 2\pi\sin[2\pi(t-0.583)].$$

39. This says that acceleration has two pieces: one proportional to the negative of displacement (the $-2x$ term) and one proportional to the negative of velocity (the $2\frac{dx}{dt}$ term).

$$\frac{dx}{dt} = -e^{-t}\cos(t) - e^{-t}\sin(t), \quad \frac{d^2x}{dt^2} = 2e^{-t}\sin(t).$$

But

$$-2x - 2\frac{dx}{dt} = -2e{-}t\cos(t) - 2[-e^{-t}\cos(t) - e^{-t}\sin(t)]$$
$$= 2e^{-t}\sin(t) = \frac{d^2x}{dt^2}.$$

This is indeed a solution.

Chapter 3

Section 3.1, page 254

1. The updating function crosses from above to below and is stable.

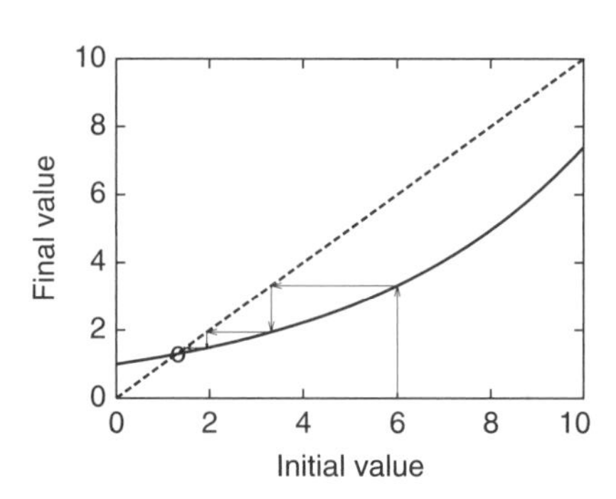

3. At the lower equilibrium the updating function crosses the diagonal from below to above and is therefore unstable. At the upper equilibrium, the updating function crosses from above to below and is stable.

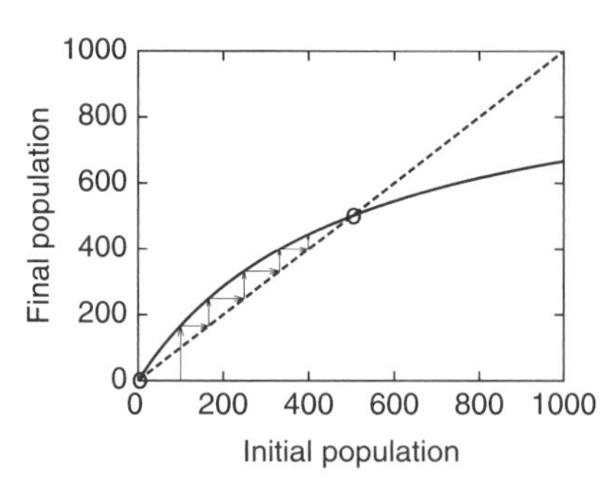

5. The equilibrium satisfies $c^* = 0.5c^* + 8.0$ or $c^* = 16.0$. The slope of the updating function $f(c) = 0.5c + 8.0$ is $f'(c) = 0.5 < 1$. The equilibrium is stable.

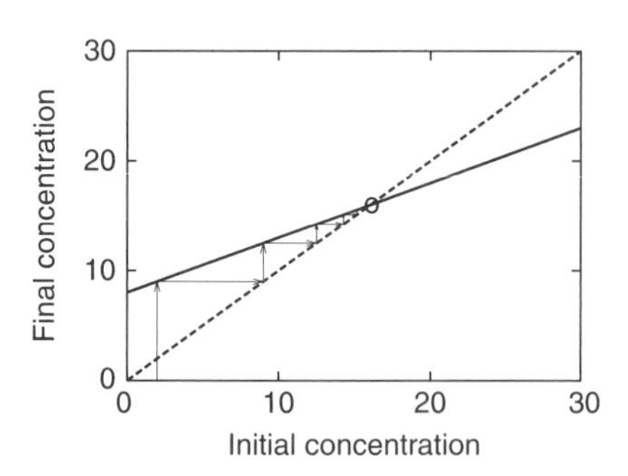

7. The equilibrium is $b^* = 0$. The slope of the updating function $f(b) = 0.3b$ is $f'(b) = 0.3 < 1$. The equilibrium is stable.

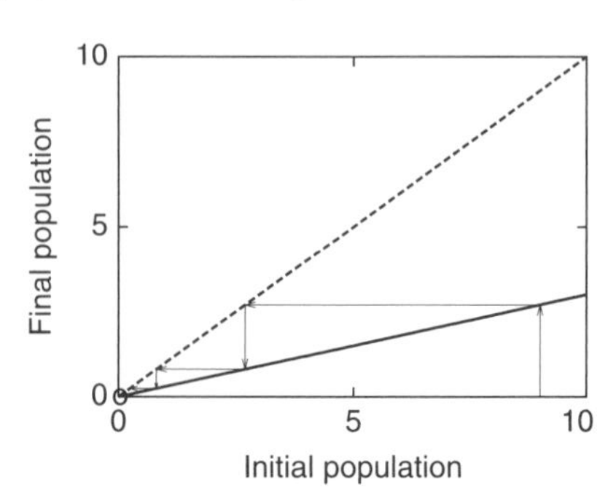

9. $f(x) = x$ when $x^2 = x$, or $x^2 - x = 0$, or $x(x-1) = 0$, which has solutions at $x = 0$ and $x = 1$. The derivative is $f'(x) = 2x$, so $f'(0) = 0 < 1$ (stable) and $f'(1) = 2 > 1$ (unstable).

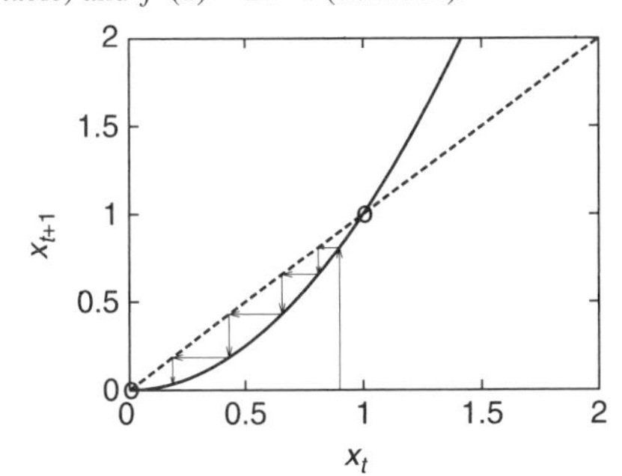

11. $x^* = \frac{x^*}{1+x^*}$ has the solution $x^* = 0$. Also, if $f(x) = \frac{x}{1+x}$, then $f'(x) = \frac{1}{(1+x)^2}$, so $f'(0) = 1$. We can't tell whether this is stable using the Slope Criterion for stability. However, as in Example 3.1.9, the graph of the updating function lies below the diagonal for all $x > 0$, meaning that the equilibrium is stable.

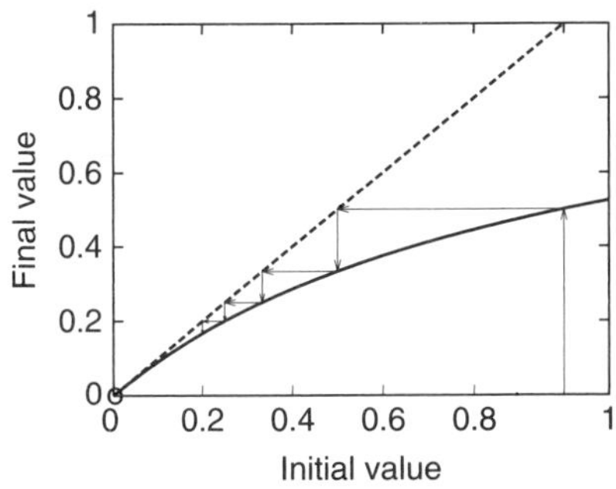

13.

Final value
unstable from above
stable from below
Initial value

15. The second derivative is 0 at the equilibrium.

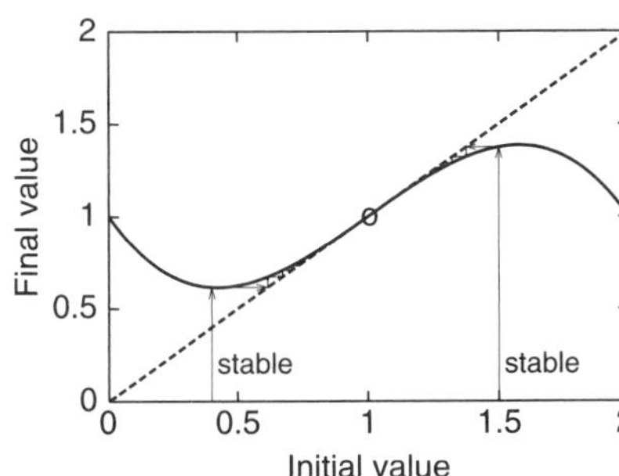

17.

Unstable equilibrium with corner

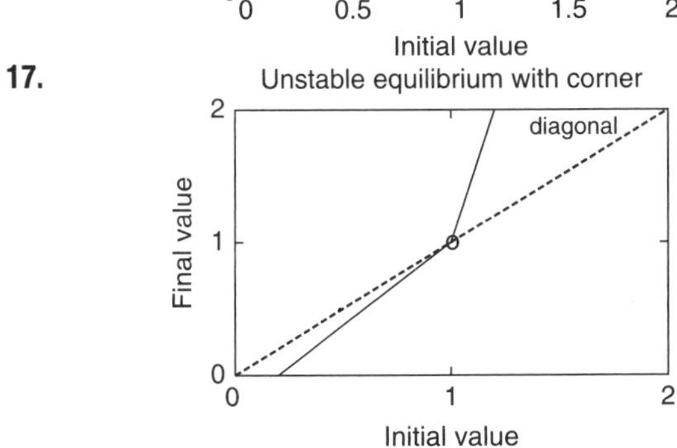

19. There is no equilibrium, and the cobwebbing creeps slowly past the point where the equilibrium used to be.

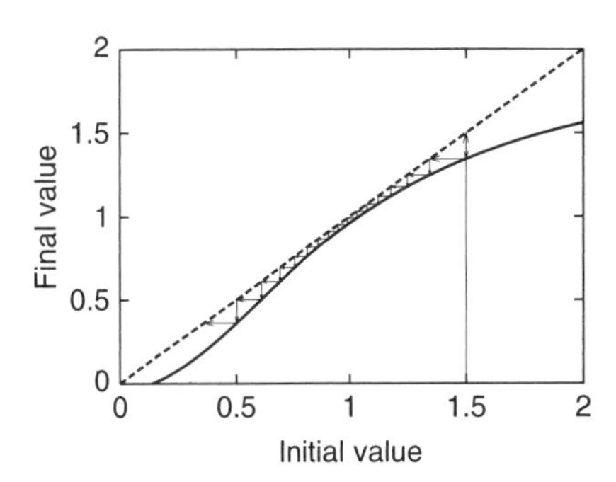

21. The inverse is $f^{-1}(x) = \frac{x}{1-x}$. The only equilibrium is $x = 0$. The slopes of both the original updating function and the inverse are 1 at this point.

23. $f'(p) = \frac{2.4}{[1.2p + 2.0(1-p)]^2}$. $f'(0) = 0.6$, $f'(1) = 1.667$. The equilibrium at $p = 0$ is stable; the equilibrium at $p = 1$ is unstable.

25. $f'(p) = \frac{2.25}{[1.5p + 1.5(1-p)]^2} = 1$. $f'(0) = f'(1) = 1$. Both derivatives are exactly 1, so we cannot tell. This updating function exactly matches the diagonal, meaning that solutions move neither toward nor away from equilibria.

27. The discrete-time dynamical system is $b_{t+1} = 0.6b_t + 1.0 \times 10^6$. The equilibrium satisfies $b^* = 0.6b^* + 1.0 \times 10^6$ or $0.4b^* = 1.0 \times 10^6$ or $b^* = 2.5 \times 10^6$. The derivative of the updating function $f(b) = 0.6b + 1.0 \times 10^6$ is $f'(b) = 0.6$, which is always less than 1. The equilibrium is stable.

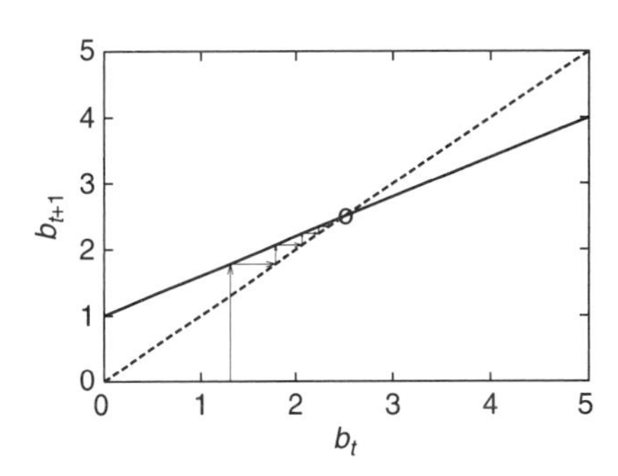

29. The updating function is $b_{t+1} = 1.5b_t - 1.0 \times 10^6$, and the equilibrium is $b^* = 2.0 \times 10^6$. The derivative of the updating function is $f'(b) = 1.5 > 1$, so the equilibrium is unstable.

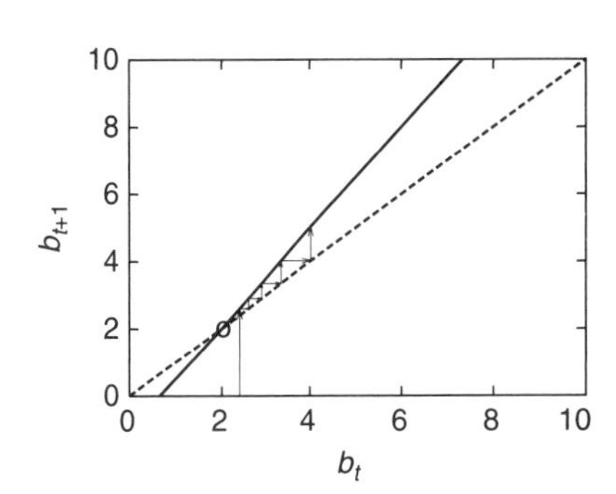

31. We found an equilibrium concentration of $M^* = 2.5$ in Section 1.10, Exercise 39. The derivative of the updating function is

$$f'(M) = 1 - \frac{0.5}{(1 + 0.1M)^2}$$

so $f'(2.5) = 0.68$. The equilibrium is stable.

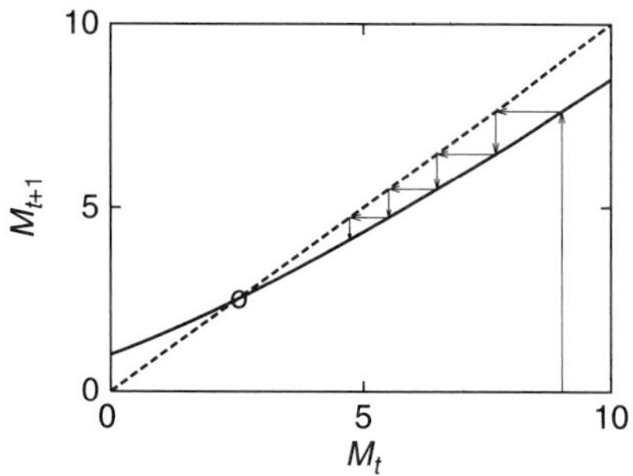

33. Solving for c_t gives an inverse of $c_t = 2.0(c_{t+1} - 8.0) = f^{-1}(c_{t+1})$. The equilibrium is $c^* = 16.0$. The derivative of the backwards updating function is $(f^{-1})'(c) = 2.0$, so the equilibrium is unstable. The same equilibrium was stable in the forward direction.

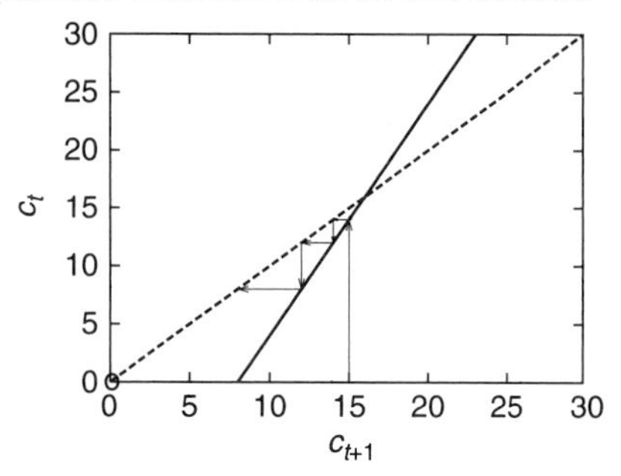

35. Solving for x_t gives an inverse of

$$x_t = \frac{x_{t+1}}{2 - 0.001x_{t+1}} = f^{-1}(x_{t+1}).$$

The equilibria are $x^* = 0$ and $x^* = 1000$. The derivative of the backwards updating function is $(f^{-1})'(x) = \frac{2}{(2-0.001x_{t+1})^2}$. Therefore, $(f^{-1})'(0) = 0.5$ and $(f^{-1})'(1000) = 2$. The equilibrium at $x = 0$ is stable, and the equilibrium at $x = 1000$ is unstable.

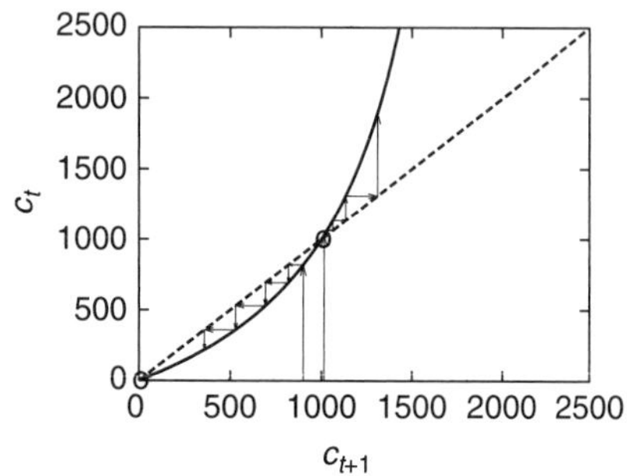

37. $x_{t+1} = \frac{x_t^2}{1.0+x_t^2}$. The equilibria satisfy

$$x = \frac{x^2}{1+x^2}$$
$$x(1+x^2) = x^2$$
$$x(1+x^2-x) = 0$$
$$x = 0 \text{ or } (1+x^2-x) = 0.$$

Because $1 + x^2 - x = 0$ has no real solution (the quadratic formula gives the square root of a negative number), the only equilibrium is $x = 0$.

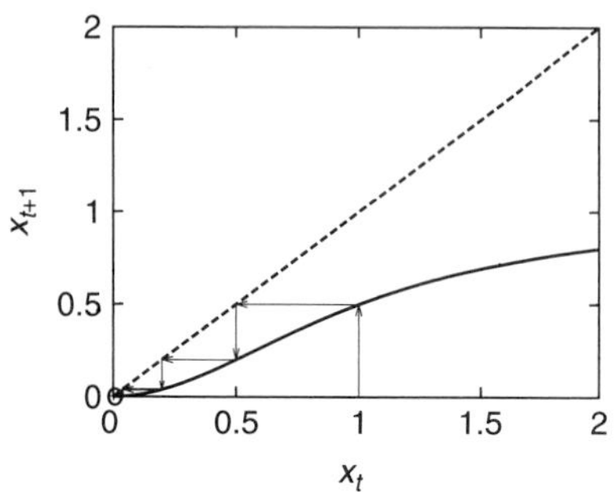

The derivative of the updating function is

$$f'(x) = \frac{2x}{(1+x^2)^2}$$

so $f'(0) = 0$, and the equilibrium is stable. This population is going extinct.

39. Assuming that the heart beats, the equilibrium is $V^* = \frac{u}{1-c} = \frac{10}{1-0.5} = 20.0$. This equilibrium is consistent with beating because $cV^* = 0.5 \times 20.0 = 10.0 < V_c$. The equilibrium is stable because the updating function is a line with slope of $c = 0.5$, which is less than 1.

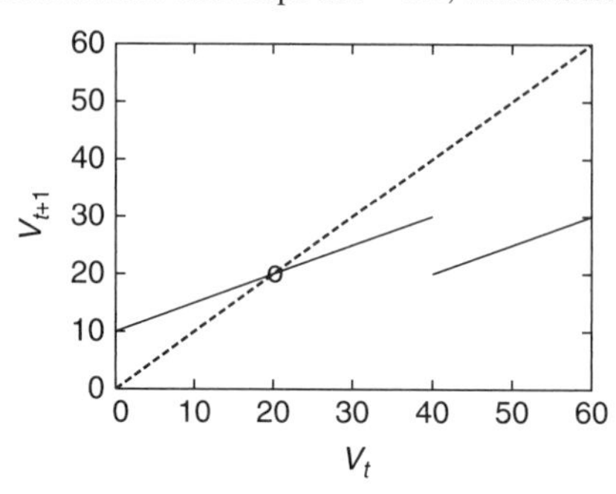

41. Assuming that the heart beats, the equilibrium is $V^* = \frac{u}{1-c} = \frac{10}{1-0.7} = 33.3$. This equilibrium is not consistent with beating because $cV^* = 0.7 \times 33.3 = 23.3 > V_c$. Therefore, the system must oscillate.

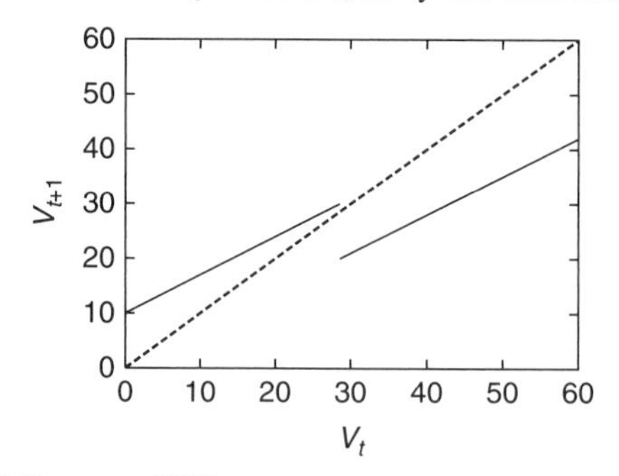

Section 3.2, page 267

1. The derivative of the updating function $f(p) = \frac{1.5p}{1.5p+2.0(1-p)}$ is $f'(p) = \frac{3.0}{[1.5p+2.0(1-p)]^2}$, so $f'(0) = 0.75$. Because $f(0) = 0$, the tangent line is $\hat{f}(p) = f(0) + f'(0)p = 0.75p$, and the equilibrium is stable.

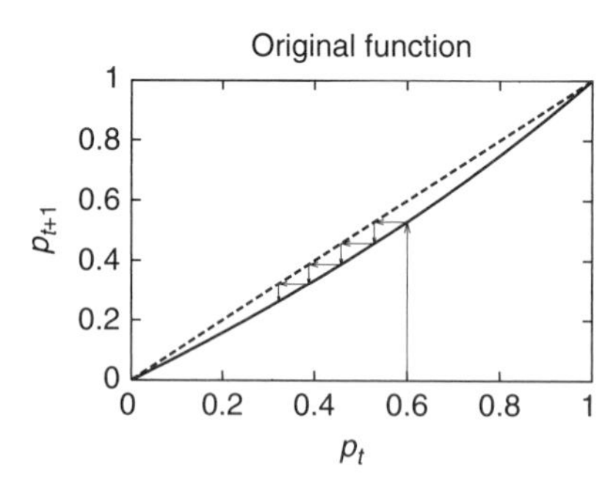

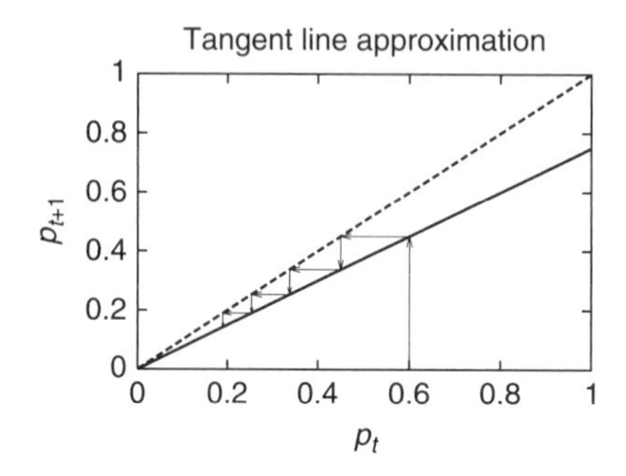

3. The derivative of the updating function $f(x) = 1.5x(1-x)$ is $f'(x) = 1.5(1-2x)$, so $f'(0) = 1.5$. The tangent line is $\hat{f}(x) = f(0) + f'(0)x = 1.5x$, and the equilibrium is unstable.

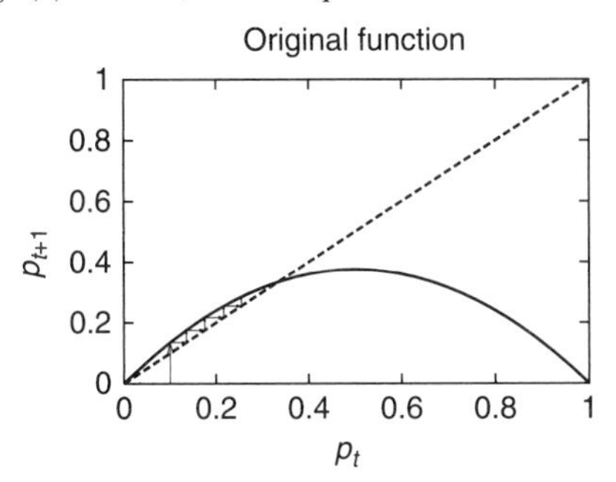

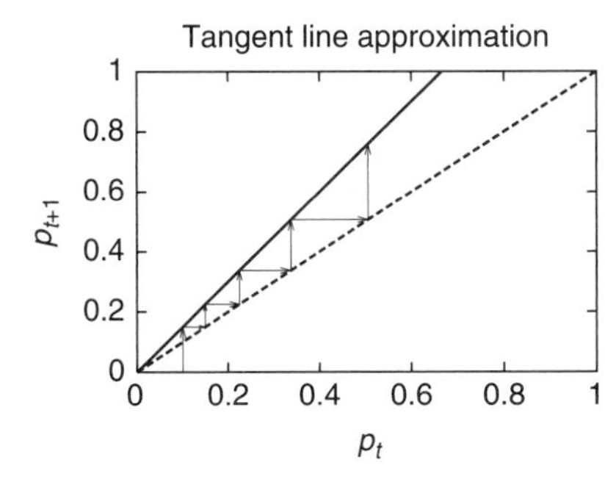

5. The solution is $y_t = 2.0 \times (1.2)^t$ with values 2.0, 2.4, 2.88, 3.456, 4.147, 4.977. Solving $y_t = 100$ gives $1.2^t = 50$ or $t = \ln(50)/\ln(1.2) = 21.45$. It would cross 100 at time step 22.

7. The solution is $y_t = 2.0 \times (0.8)^t$ with values 2.0, 1.6, 1.28, 1.024, 0.819, 0.655. Solving $y_t = 0.2$ gives $0.8^t = 0.1$ or $t = \ln(0.1)/\ln(0.8) = 10.31$. It would cross 0.2 at time step 11.

9. First solve for the equilibrium.

$$\begin{aligned} y^* &= 1.0 + m(y^* - 1.0) \\ 0 &= 1.0 - y^* + m(y^* - 1.0) \\ 0 &= (1-m)(1.0 - y^*). \end{aligned}$$

The equilibrium is $y^* = 1.0$.

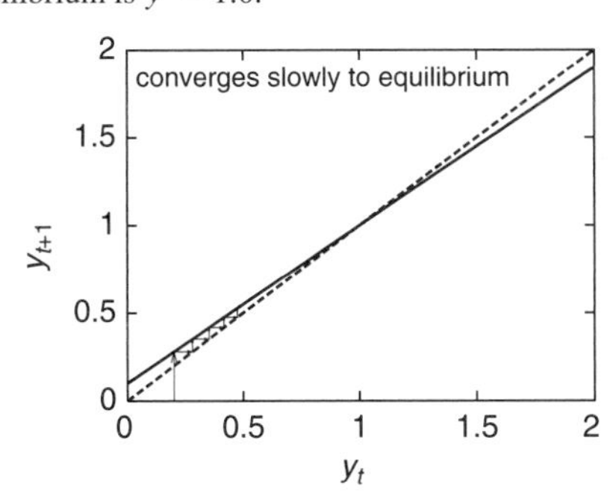

It is stable because the slope m is less than 1.

11. The equilibrium is again $y^* = 1.0$.

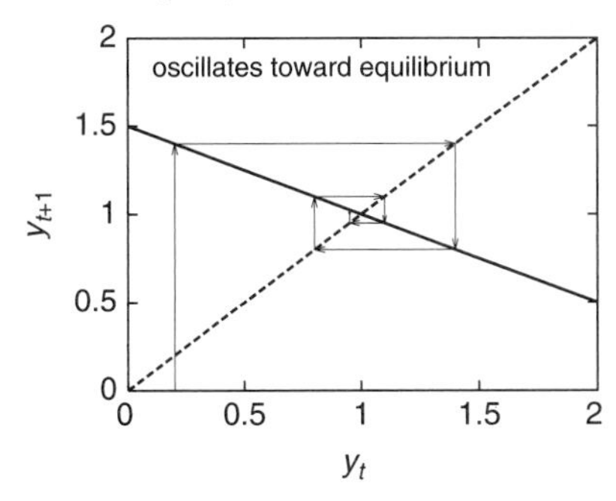

It is stable but oscillatory because $-1 < m < 0$.

13. The slope of the updating function is exactly -1 everywhere. Solutions jump back and forth and are neither stable nor unstable.

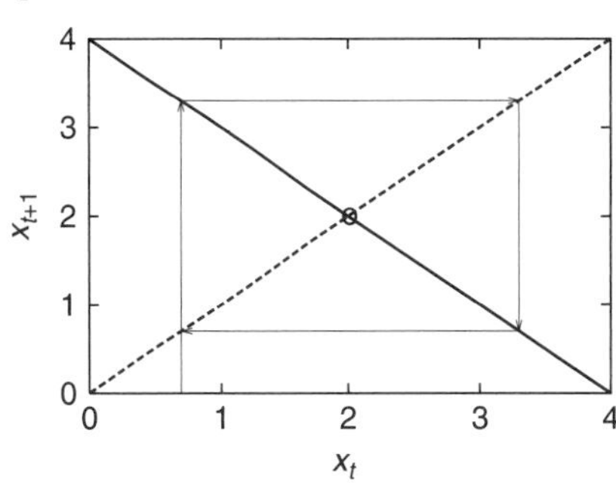

15. $x^* = 3x^*(1-x^*)$ has solutions $x^* = 0$ and $x^* = \frac{2}{3}$. The derivative of the updating function $f(x) = 3x(1-x)$ is $f'(x) = 3(1-2x)$, so $f'(0) = 3$ and $f'\left(\frac{2}{3}\right) = -1$. The zero equilibrium is unstable, but a solution starting at $x_0 = 0.6$ gets slowly closer to x^*, with $x_2 = 0.6048$ and $x_4 = 0.6087$. The positive equilibrium is stable.

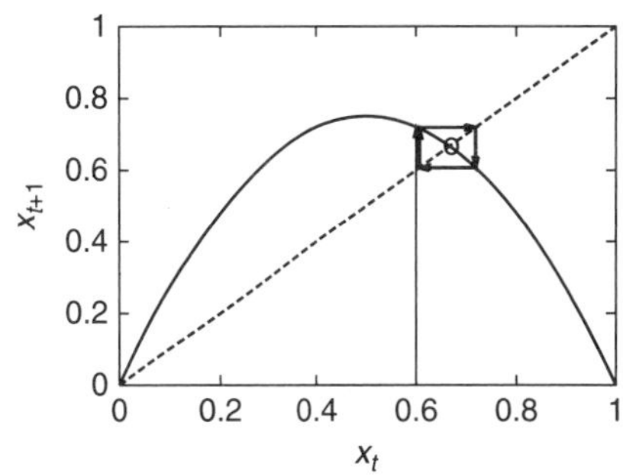

17. $x^* = 0.5$, $f'(x) = 2 - 4x$, and $f'(0.5) = 0$. The equilibrium is highly stable. The solutions of a linear system with slope 0 hit the equilibrium in one step. The solutions in this case move toward the equilibrium very quickly.

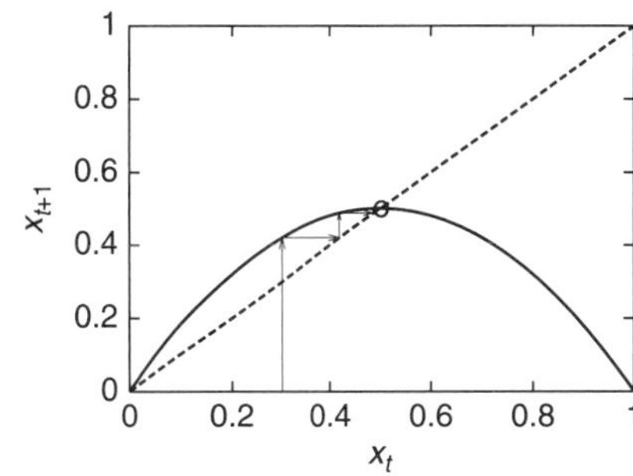

19. The derivative of the updating function $g(M) = M - \frac{M^2}{2+M} + 1$ is $g'(M) = \frac{4}{(2+M)^2}$, so $g'(2) = \frac{1}{4}$. The equilibrium is stable and does not oscillate.

21. The derivative of the updating function $g(M) = M - \frac{M^5}{16+M^4} + 1$ is $g'(M) = \frac{16(3M^4-16)}{(16+M^4)^2}$ and $g'(2) = -\frac{1}{2}$. The equilibrium is stable, but solutions oscillate.

23. The slope at the equilibrium is $1 - \ln(r)$, so it switches sign when $\ln(r) = 1$, or $r = e$.

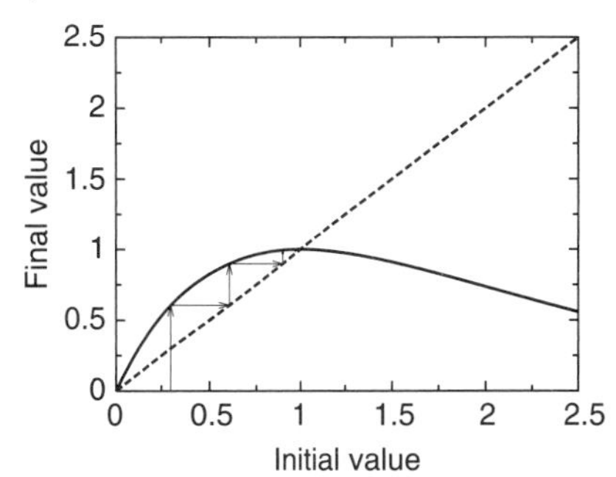

25. I chose the value $r = e^{1.5} = 4.482$.

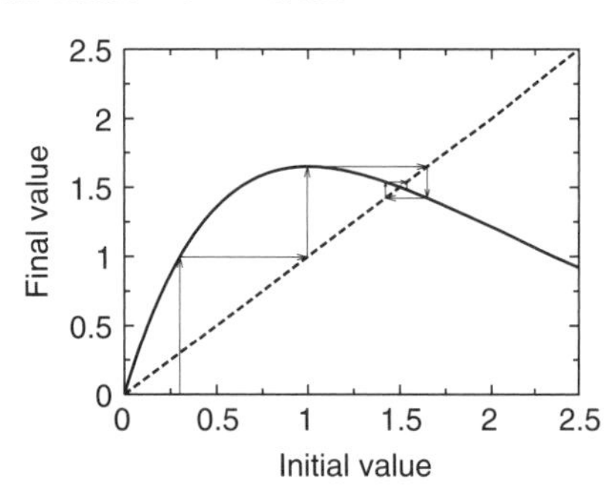

27. a. The discrete-time dynamical system is $b_{t+1} = b_t(0.5 + 0.5b_t)$.
b. The equilibria are $b^* = 0$ and $b^* = 1$. The derivative of the updating function $f(b) = (0.5 + 0.5b)b$ is $f'(b) = 0.5 + b$, so $f'(0) = 0.5$ and $f'(1) = 1.5$. The equilibrium $b = 0$ is stable and $b = 1$ is unstable.
c.

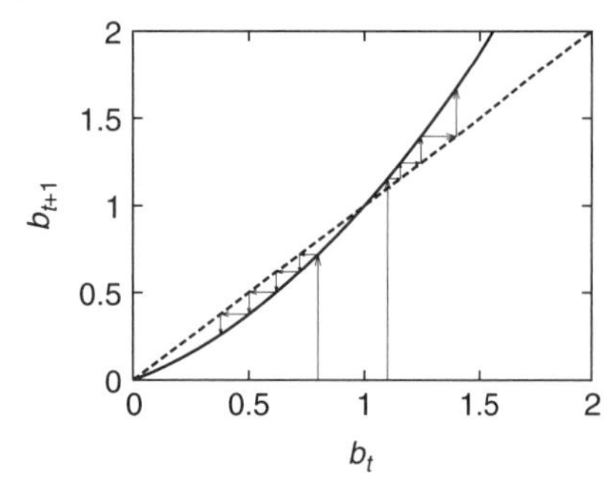

d. Populations starting below 1 die out, while those starting above 1 blast off to infinity. Members of this species do well with a little help from their friends.

29. a.

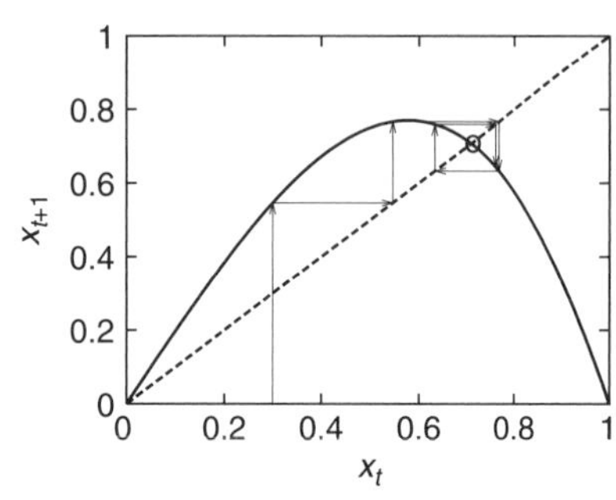

b. The equilibria are $x = 0$ and $x^* = \sqrt{1 - \frac{1}{r}}$ (when $r \geq 1$).
c. $f'(x) = r - 3rx^2$. $f'(0) = r$ and $f'(x^*) = 3 - 2r$.
d. The zero equilibrium is stable when $r < 1$ as usual, and the positive equilibrium is stable when $r < 2$.

31. a. This line passes through the points (20, 21) and (19, 18). Let z be the setting on the thermostat and T be the temperature. This line has slope

$$m = \frac{21 - 18}{20 - 19} = 3.$$

Because it passes through the point (20, 21), we can write the equation in point-slope form as $T = 3(z - 20) + 21$.
b. When it is 18°C, you set the thermometer to 22°C. This results in a temperature of 27°C. Setting the thermometer to 13°C then results in a temperature of 0°C. Things are getting pretty chilly.
c. $z_t = 40 - T_t$
d. $T_{t+1} = 3(z_t - 20) + 21 = 3(40 - T_t - 20) + 21 = 81 - 3T_t$
e. The slope of -3 means that the temperatures will oscillate more and more widely. The system needs a better correction mechanism. I would have it estimate T as a function of z and correct on that basis.

33. The mutants do worse when mutants are common, and the wild type do better. The discrete-time dynamical system is

$$\begin{aligned} p_{t+1} &= \frac{a_{t+1}}{a_{t+1} + b_{t+1}} = \frac{2(1 - p_t)a_t}{2(1 - p_t)a_t + (1 + p_t)b_t} \\ &= \frac{2(1 - p_t)p_t}{2(1 - p_t)p_t + (1 + p_t)(1 - p_t)} = \frac{2p_t}{2p_t + (1 + p_t)}. \end{aligned}$$

We solve for the equilibria by setting $p_{t+1} = p_t = p^*$ or

$$\begin{aligned} p^* &= \frac{2p^*}{2p^* + (1 + p^*)} \\ p^*[2p^* + (1 + p^*)] &= 2p^* \\ p^*(3p^* + 1) - 2p^* &= 0 \\ p^*(3p^* - 1) &= 0. \end{aligned}$$

Therefore, the equilibria are $p = 0$ and $p = \frac{1}{3}$. The derivative of the updating function is $f'(p) = \frac{2}{(1+3p)^2}$, so $f'(0) = 2 > 1$ and $f'(\frac{1}{3}) = \frac{1}{2} < 1$.

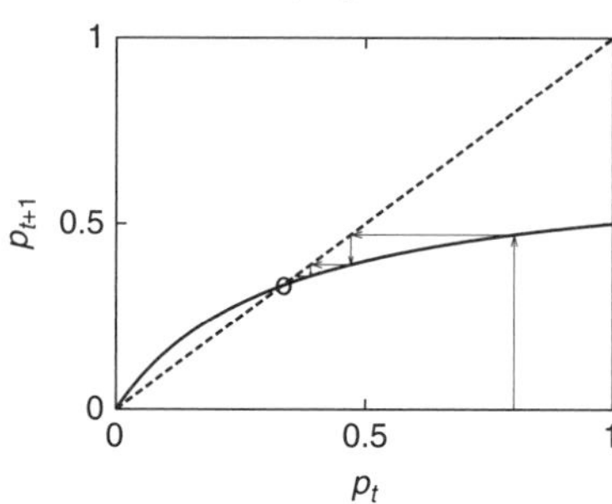

35. a. There are 20 plants, and each grows to size 100.0/20.0 = 5.0 and makes 4.0 seeds. This gives a total of 80.0 plants. These 80.0 plants grow to size $\frac{100.0}{80.0} = 1.25$, so each makes 0.25 seeds (or one in four plants makes a seed). The total number of seeds is then 20.0 in the next generation. The values just keep jumping back and forth.
b. There are n_t plants of size $\frac{100.0}{n_t}$. Each of the n_t plants makes $\frac{100.0}{n_t} - 1.0$ seeds, for a total of $n_{t+1} = 100.0 - n_t$.
c. Solving $n^* = 100.0 - n^*$ gives an equilibrium of $n^* = 50.0$.
d.

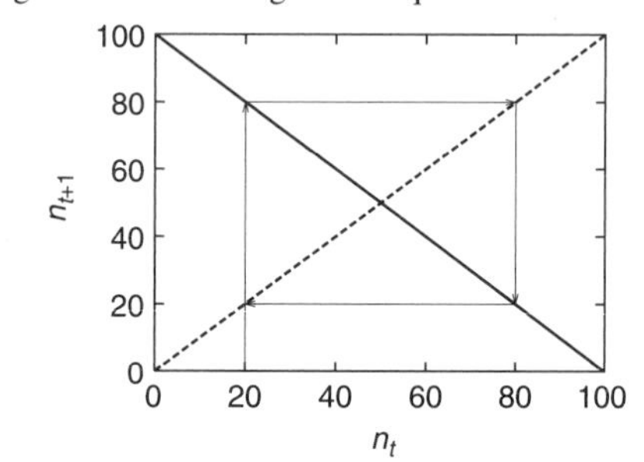

e. The slope is -1, and the equilibrium is neither stable nor unstable.

37. a. There are 20 plants, and each makes 3.0 seeds, or a total of 60 plants. These 60 plants grow to size 1.667, each making a negative number of seeds. This population just went extinct.
b. The discrete-time dynamical system is $n_{t+1} = 100 - 2.0n_t$.
c. Solving $n^* = 100 - 2.0n^*$ gives an equilibrium of $n^* = 33.3$.

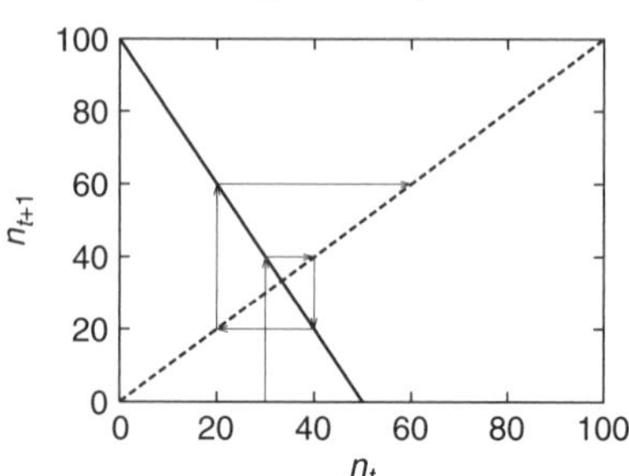

d. The slope is -2.0, and the equilibrium is unstable and oscillatory.

Section 3.3, page 280

1. This is a quotient of $u(x)=x$ and $v(x)=1+x$ with derivatives $u'(x)=1$ and $v'(x)=1$. Therefore $a'(x)=\frac{1+x-x}{(1+x)^2}=\frac{1}{(1+x)^2}$. This function is always positive. There are no critical points.
3. The derivative of this polynomial is $c'(w)=3w^2-3$. Therefore, $c'(w)=0$ if $3w^2=3$, which has solutions at $w=-1$ and $w=1$.
5. This is a composition $f[g(z)]$ with $g(z)=z^2$ and $f(g)=e^g$. These components have derivatives $g'(z)=2z$ and $f'(g)=e^g$. By the chain rule, $h'(z)=f'(g)g'(z)=2ze^{z^2}$. Because the exponential term e^{z^2} is always positive, $h'(z)=0$ only at $z=0$.
7. There are no critical points. The global maximum is at the endpoint $x=1$ where $a(1)=\frac{1}{2}$. The global minimum is at the other endpoint $x=0$ where $a(0)=0$.
9. The critical points are $x=-1$ and $x=1$. $c(-1)=2$ and $c(1)=-2$. At the endpoints, $c(-2)=-2$ and $c(2)=2$. We have two ties. The global maximum value is 2 taken on at $w=-1$ and $w=2$. The global minimum value is -2 taken on at $w=1$ and $w=-2$.
11. The value at the critical point is $h(0)=1$, and the value at the other endpoint is e. The global maximum is thus at $z=1$, and the global minimum is at $z=0$.
13. $a''(x)=\frac{-2}{(1+x)^3}$, negative for $0\le x\le 1$. The graph is increasing, concave down.

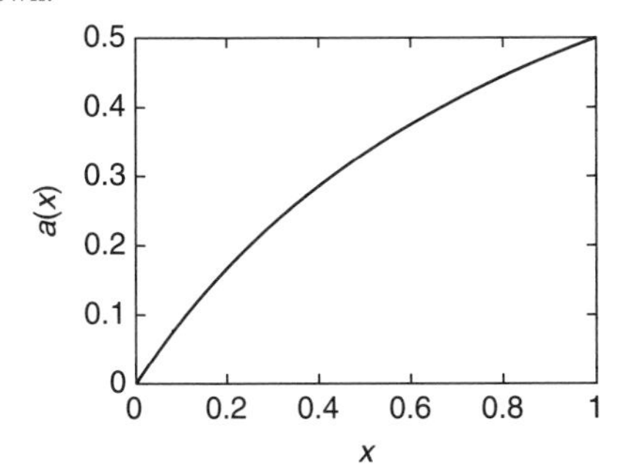

15. $c''(w)=6w$. Therefore, $c''(1)=6$ and $c''(-1)=-6$. This is consistent with the fact that this function has a minimum at 1 and a maximum at -1.

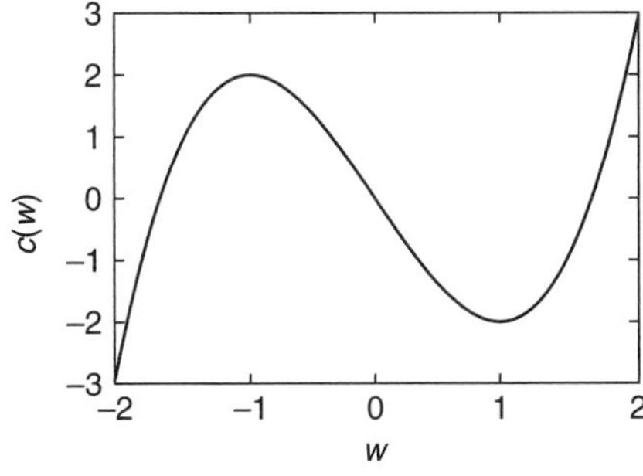

17. $h''(z)=(4z^2+2)e^{z^2}$, which is always positive. The critical point at $z=0$ must be a minimum.

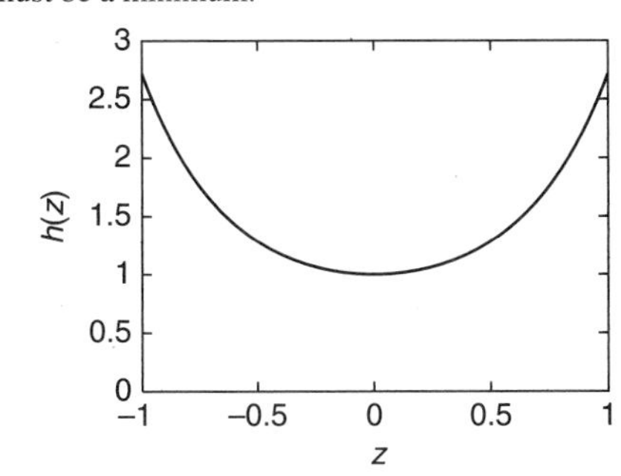

19. a. $h(x)=\frac{1}{f(x)}$, so $h'(x)=\frac{-f'(x)}{f(x)^2}$ by the quotient rule. If $f'(x)=0$, then $h'(x)=0$.

 b.
$$h''(x)=-\frac{f(x)^2f''(x)-2f'(x)^2f(x)}{f(x)^4}$$
by the quotient rule. If $f'(x)=0$, then $h''(x)=\frac{-f''(x)}{f(x)^2}$, which has the opposite sign of the original.

 c. If $f(x)$ has a local maximum, $h(x)$ has a local minimum. This makes sense; a large value of $f(x)$ means a small value of $h(x)$ and vice versa.

 d. The function is $h(x)=\frac{e^x}{x}$. The derivative is $h'(x)=\frac{e^x(x-1)}{x^2}$, which has a critical point at $x=1$. The second derivative is $h''(x)=\frac{e^x(x^2-2x+2)}{x^3}$, which is positive at $x=1$. This is indeed a minimum.

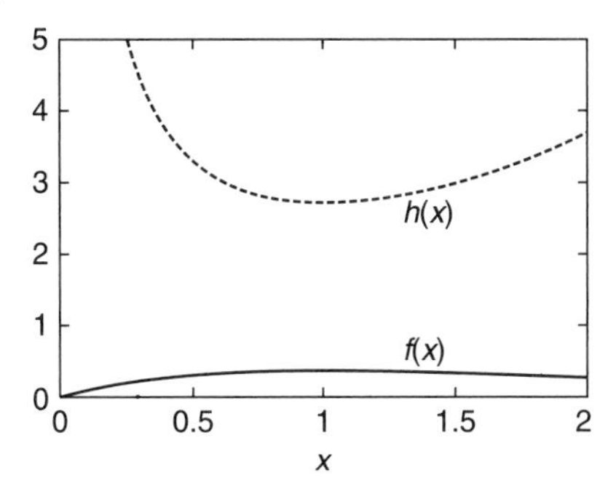

21. a. $h(x)=\ln[f(x)]$, so $h'(x)=\frac{f'(x)}{f(x)}$ by the chain rule. If $f'(x)=0$ then $h'(x)=0$.

 b.
$$h''(x)=\frac{f(x)f''(x)-f'(x)^2}{f(x)^2}$$
by the quotient rule. If $f'(x)=0$ then $h''(x)=\frac{f''(x)}{f(x)}$ which has the same sign as $f''(x)$.

 c. If $f(x)$ has a local maximum, $h(x)$ does also. This makes sense; a large value of $f(x)$ means a large value of $h(x)$ and vice versa.

 d. The function is $h(x)=\ln(xe^{-x})=\ln(x)-x$. The derivative is $h'(x)=\frac{1}{x}-1$, which has a critical point at $x=1$. The second derivative is $h''(x)=\frac{-1}{x^2}$, which is negative at $x=1$. This is indeed a maximum.

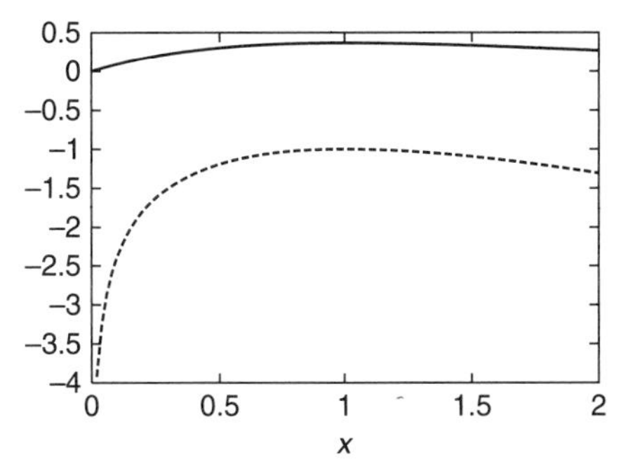

23. We find that $C'(t)=3t^2-60t$, with critical points at $t=0$ and $t=20$. Evaluating at the critical points and endpoints, we find $C(0)=6000$, $C(20)=2000$, and $C(25)=2875$, giving a maximum at $t=0$ and a minimum at $t=20$.
25. Consider the following diagram.

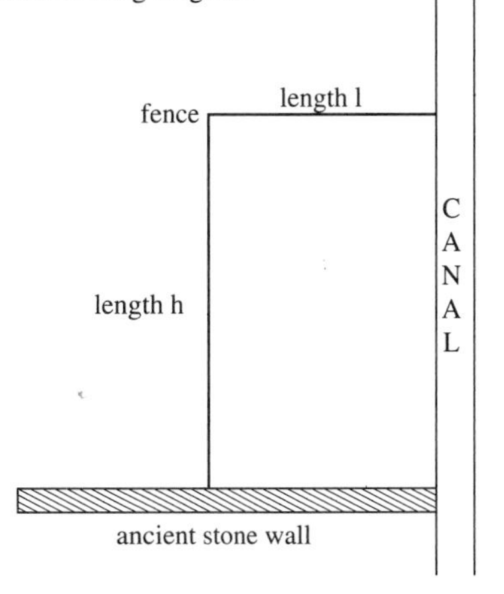

The total length of the fence is $l + h = 1000$m, and the area enclosed is lh. We can solve for h as $h = 1000 - l$, so the area is $A(l) = l(1000 - l)$. Then $A'(l) = 1000 - 2l$, which has a critical point at $l = 500$ (meaning that $h = 500$m). This is a global maximum because the area at the endpoints $l = 0$ and $l = 500$ is 0. The maximum area is then $500 \times 500 = 25000$ m^2.

27. Following the steps in Example 3.3.11 gives an optimum of $t = 1.0$. At this point, the derivative is

$$F'(1.0) = \frac{0.5}{(1.0 + 0.5)^2} = 0.222.$$

The tangent line is $\hat{F}(t) = F(1) + F'(1)(t - 1) = 0.667 + 0.222$ $(t - 1)$. We can check directly that $\hat{F}(-2.0) = 0$.

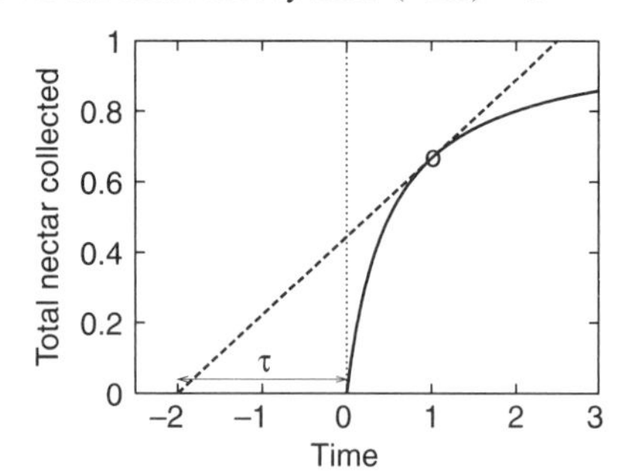

29. $t = \sqrt{0.05} = 0.223$. The tangent line is $\hat{F}(t) = 0.309 + 0.955(t - 0.223)$. It is true that $\hat{F}(-0.1) = 0$.

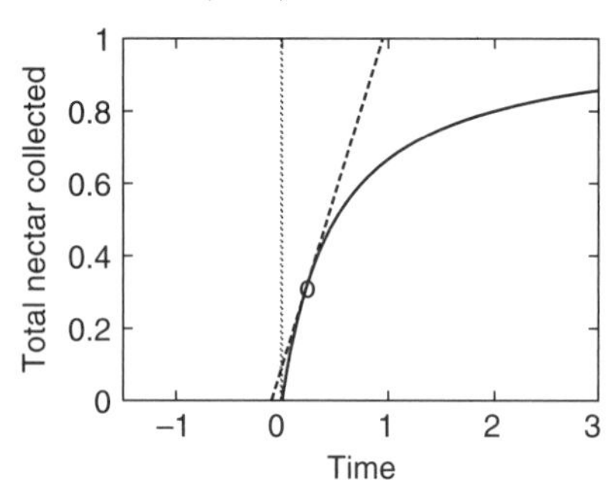

31. $t = \sqrt{2} = 1.414$.

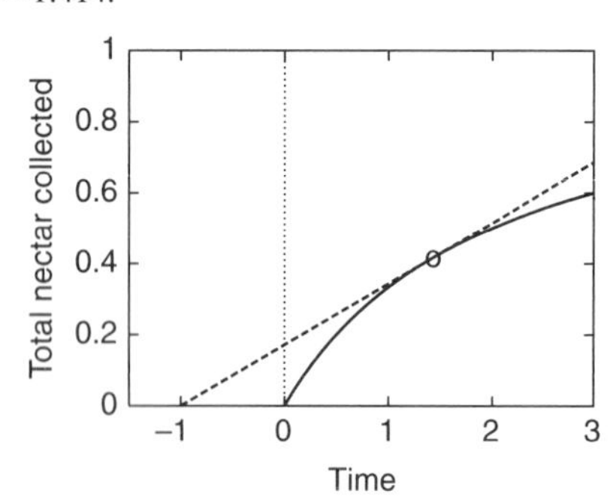

33. $t = \sqrt{0.1} = 0.316$.

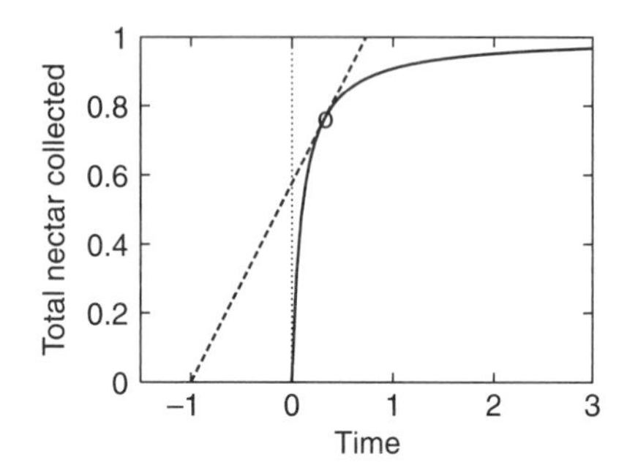

35. The maximum occurs where

$$\frac{0.5}{(t + 0.5)^2} = \frac{t}{(t + 0.5)(t + \tau)}$$

(following the steps in Example 3.3.11). When $t = 1.0$, we find that

$$\begin{aligned} \frac{0.5}{1.5^2} &= \frac{2}{9} \\ \frac{2}{9} &= \frac{1.0}{1.5(1.0 + \tau)} \\ 1.5(1.0 + \tau) &= 4.5 \\ 1.0 + \tau &= 3.0 \\ \tau &= 2.0. \end{aligned}$$

The travel time must be 2.0 min.

37. The maximum occurs where

$$\frac{0.5}{(t + 0.5)^2} = \frac{t}{(t + 0.5)(t + \tau)}.$$

When $t = 4.0$, we find that

$$\begin{aligned} \frac{0.5}{4.5^2} &= \frac{2}{81} \\ \frac{2}{81} &= \frac{4.0}{4.5(4.0 + \tau)} \\ 4.5(4.0 + \tau) &= 162 \\ 4.0 + \tau &= 36 \\ \tau &= 32. \end{aligned}$$

The travel time must be $\tau = 32$ min.

39. $R(t) = \frac{t}{(1+t)^2}$, so

$$R''(t) = \frac{2(t - 2)}{(t + 1)^4},$$

which is negative at $t = 1$, the optimal solution. Therefore this is at least a local maximum.

41. The bees are trying to maximize $R(n) = \frac{n}{P(n)} = \frac{n}{1+n^2}$. Taking the derivative, this has a maximum at $n = 1$.

43. $R(n) = \frac{n}{P(n)}$, so $R'(n) = \frac{P(n) - nP'(n)}{P(n)^2}$. The critical point is where $P(n) = nP'(n)$ or $P'(n) = \frac{P(n)}{n}$. If $P(n) = 1 + n^2$, then $P'(n) = 2n$, so the condition is $2n = \frac{1+n^2}{n}$ or $2n^2 = 1 + n^2$ or $n = 1$. If $P(n) = 1 + n$, then $P'(n) = 1$, so the condition is $1 = \frac{1+n}{n}$, which has no solution.

45. a. $N^* = 0$ or $N^* = 1 - \frac{1+h}{2.0}$. The largest possible h is 1.
b. $P(h) = h\left(1 - \frac{1+h}{2.0}\right)$.
c. $P'(h) = 0$ at $h = 0.5$. This is a maximum because $P''(h) = -1$.
d. $P(0.5) = 0.125$.

47. The derivative of the updating function is $2.5(1 - 2N_t) - h$. The equilibrium is $N^* = 1 - \frac{1+h}{2.5}$. Plugging into the derivative gives $h - 0.5$ (after some algebra). The equilibrium is stable as long as $h < 1.5$. At $h = 0.75$, the slope is 0.25, indicating stability.

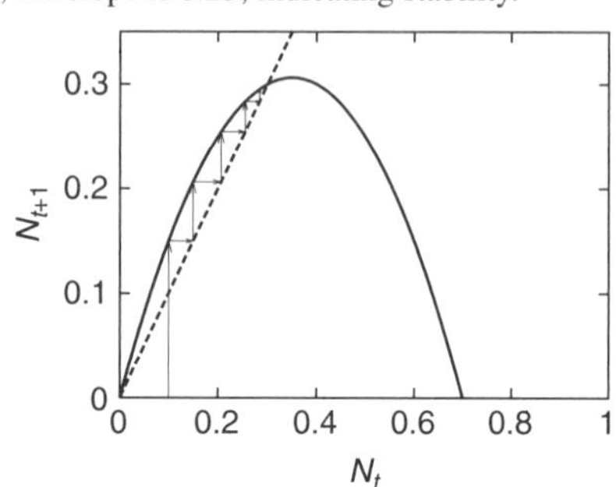

49. a. $N^* = \frac{2.5}{1+h} - 1$.
b. $h = 1.5$.

c. $P(h) = hN^*$. The maximum is at $h = \sqrt{2.5} - 1 = 0.58$. This takes on the value of approximately 0.58 for $r = 2.5$.

d. With $r = 2.5$, $P(0.58) = 0.338$.

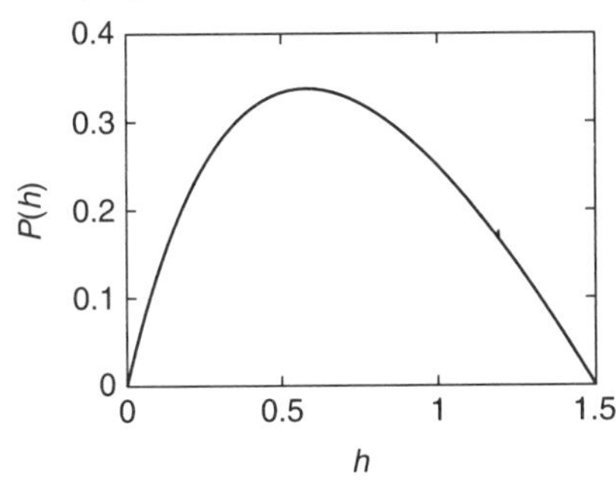

e. The harvest strategy h is lower (0.58 here where it was 0.75 in the text), but the payoff is higher.

51. The equilibrium is $N^* = 1 - \frac{1+h}{2.5}$, and the payoff is $P(h) = h\left(1 - \frac{1+h}{2.5}\right) - 0.1h$, with derivative $P'(h) = 1 - \frac{1+2h}{2.5} - 0.1$. The critical point is $h = 0.625$, the equilibrium population is $N^* = 0.35$, and the payoff is $P(0.625) = 0.156$.

53. The equilibrium is $N^* = 1 - \frac{1+h}{2.5}$, and the payoff is $P(h) = h\left(1 - \frac{1+h}{2.5}\right) - 0.5h$, with derivative $P'(h) = 1 - \frac{1+2h}{2.5} - 0.5$. The critical point is $h = 0.125$, the equilibrium population is $N^* = 0.55$, and the payoff is $P(0.125) = 0.062$.

Section 3.4, page 291

1. Let $f(x) = e^x + x^2 - 2$. Then $f(0) = -1 < 0$ and $f(1) = e - 1 > 0$. By the Intermediate Value Theorem, there must be a solution in between.

3. To get this into the right form, subtract x from both sides to give the equation $e^x + x^2 - 2 - x = 0$. Let $f(x) = e^x + x^2 - 2 - x$. Then $f(0) = -1 < 0$ and $f(1) = e - 2 > 0$. By the Intermediate Value Theorem, there must be a solution where $f(x) = 0$ or where $e^x + x^2 - 2 - x = 0$. This point must also solve the original equation.

5. Let $f(x) = xe^{-3(x-1)} - 2$. Then $f(0) = -2 < 0$ and $f(1) = -1 < 0$. The Intermediate Value Theorem tells us nothing. However, $f\left(\frac{1}{2}\right) = 0.24 > 0$. The Intermediate Value Theorem guarantees solutions between 0 and $\frac{1}{2}$ and also between $\frac{1}{2}$ and 1.

7. $f(0) = f(1) = 0$, and $f(x) > 0$ for $0 \le x \le 1$. Therefore, there must be a positive maximum in that range.

9. $f(0) = f(1) = -1$. There must be a maximum in between. Also, $f(0.5) = 0.875$, which is positive; therefore, there must be a positive maximum.

11. $f'(x) = 2x$. The slope of the secant is $\frac{f(1)-f(0)}{1-0} = 1$. $f'(x) = 1$ at $x = 0.5$.

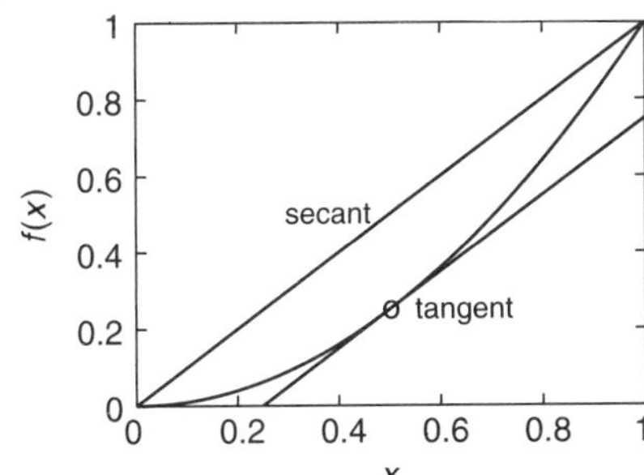

13. $g'(x) = \frac{1}{2\sqrt{x}}$. The slope of the secant is 1, and $g'(x) = 1$ at $x = \frac{1}{4}$.

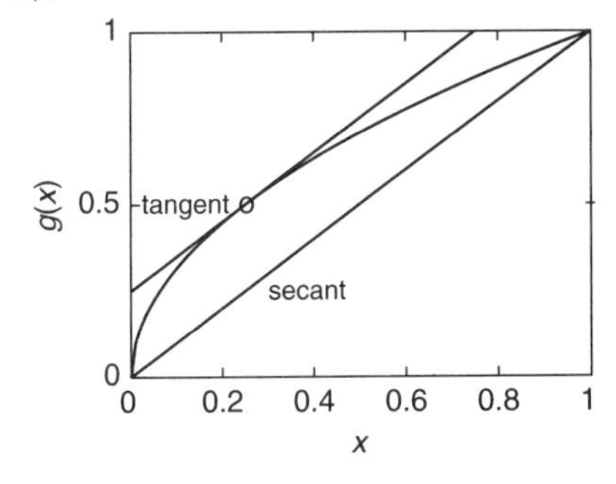

15.

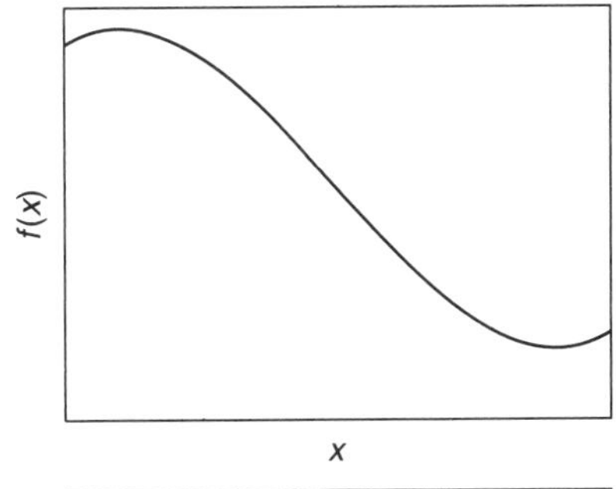

17.

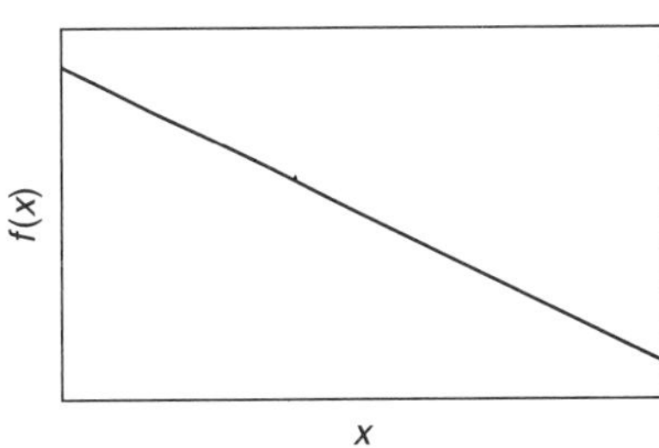

19.

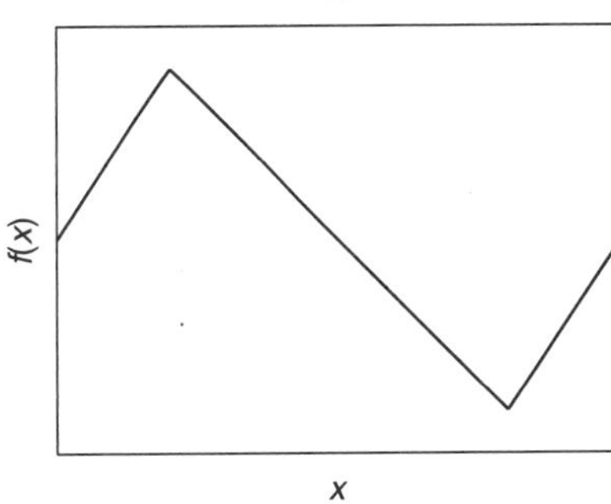

21. The function never takes on any values other than 0 and 1, so can never equal $\frac{1}{2}$. Also, the slope of the secant connecting $x = -1$ and $x = 1$ is $\frac{1}{2}$, but the tangent at every point (except at the point of discontinuity $x = 0$) has slope 0.

23. With these values, $f(b) = f(2) = 4$ and $f(a) = f(1) = 1$, so

$$g(x) = x^2 - 3(x - 1).$$

Then $g(1) = g(2) = 1$. Therefore, there must be some value c between $x = 1$ and $x = 2$ where $g'(c) = 0$. But $g'(x) = 2x - 3 = f'(x) - 3$. The point where $g'(x) = 0$ is then a point where $f'(x) = 3$. This is the point guaranteed by the Mean Value Theorem, because the slope of the secant connecting $x = 1$ and $x = 2$ is 3.

25. The price of gasoline does not change continuously and therefore need not take on all intermediate values.

27. The Intermediate Value Theorem guarantees this crossing. It is possible that it crosses the larger value, but it need not.

29. $x_{t+1} > x_t$ when $x_t = 0$, and $x_{t+1} < x_t$ when $x_t = \frac{\pi}{2}$. Because this discrete-time dynamical system is continuous, there must be a crossing in between.

31. $c_{t+1} > c_t$ when $c_t = 0$, and $c_{t+1} < c_t$ when $c_t = \gamma$. Because this discrete-time dynamical system is continuous, there must be a crossing in between.

33. The times on her watch during the two trips must look something like the following.

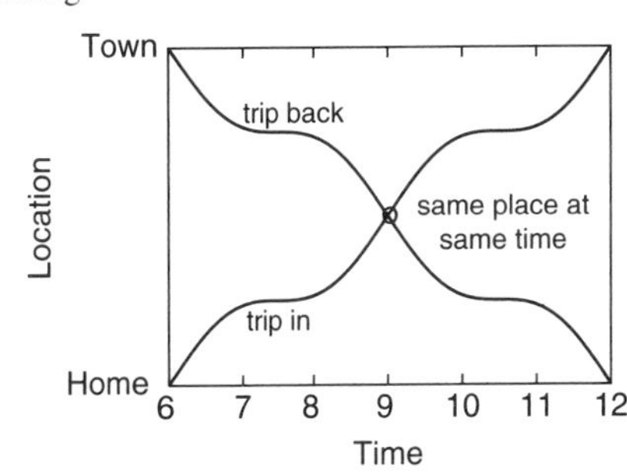

The difference in times on her watch is positive at $t = 6$ and negative at $t = 12$, and must therefore be 0 at some time in between.

35. The Intermediate Value Theorem.

37.

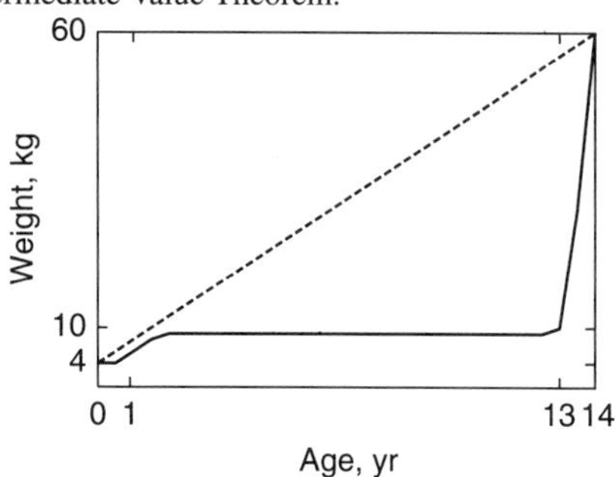

39. The Mean Value Theorem guarantees that the speed at some instant is equal to the average speed of 20 mph. The Intermediate Value Theorem guarantees that the speed must hit **every** value between 0 and 60 mph.

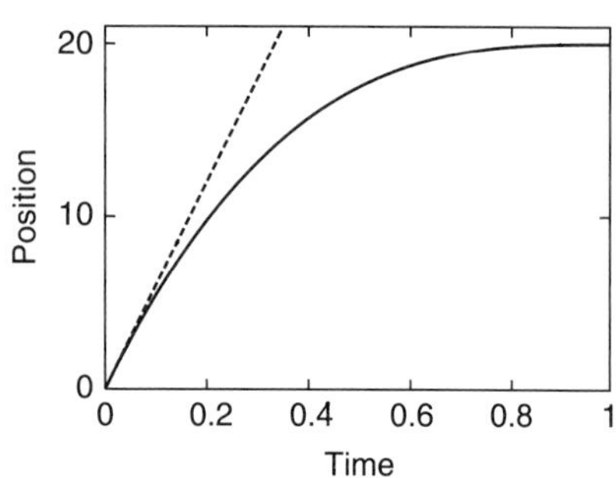

41. The Mean Value Theorem guarantees that the speed at some instant is equal to the average speed of 60 mph. The Intermediate Value Theorem guarantees that the speed must hit **every** value between the minimum and maximum speed, but we do not know what those are.

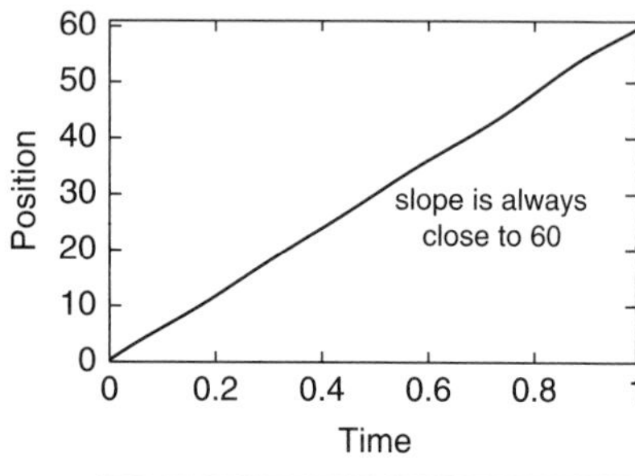

43.

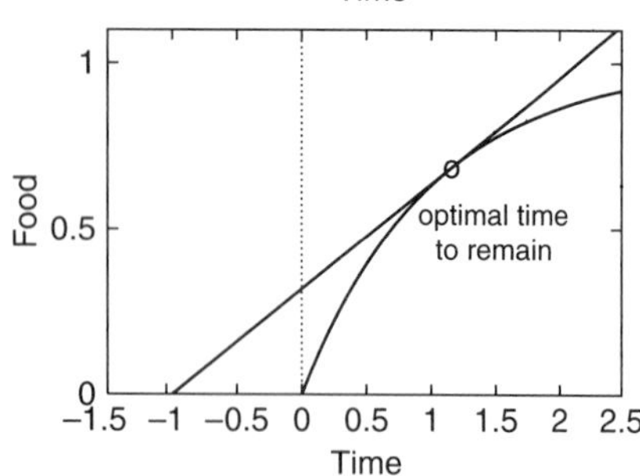

Section 3.5, page 302

1. This is a power function with a negative power, so the limit is 0.

3. This is an exponential function that can be rewritten as $0.8^x = e^{\ln(0.8)x}$, where $\ln(0.8) < 0$. Therefore, this function approaches 0 as x approaches infinity.

5. The power function x^4 has a positive power and thus approaches infinity; adding the constant 1 does not change the limit of ∞.

7. The term in the exponent $-x^2$ approaches $-\infty$, implying that the function approaches 0.

9. e^{2x} approaches infinity faster because an exponential always beats a power function. For x^2, the values are 1, 100, and 10000. For e^{2x}, the values are 7.3, 4.85×10^8, and 7.22×10^{86}, which are always larger.

11. $0.1x^{10}$ approaches infinity faster because the power is larger. For $x^{3.5}$, the values are 1, 3162, and 10^7. For $0.1x^{10}$, the values are 0.1, 10^9, and 10^{19}. By the time $x = 10$, $0.1x^{10}$ is larger.

13. $0.1x^{0.5}$ approaches infinity faster because a power function beats the natural log. For $0.1x^{0.5}$, the values are 0.1, 0.316, and 1.000. For $30\ln(x)$, the values are 0, 69.1, and 138.1. These are much larger. In fact, the two functions don't get into the right order until x is around 3×10^7.

15. e^{-2x} approaches zero faster because exponentials are faster than power functions. For e^{-2x}, the values are 0.135, 2.1×10^{-9}, and 1.38×10^{-87}. For x^{-10}, the values are 1, 10^{-10}, and 10^{-20}. The two functions don't get into the right order until $x = 100$.

17. $x^{-3.5}$ approaches zero faster because the power is more negative. For $1000/x$, the values are 1000, 100, and 10. For $x^{-3.5}$, the values are 1.00, 0.0032, and 1.0×10^{-7}. The two functions are always in the right order.

19. x^{-2} approaches zero faster because power functions are faster than natural logs. For x^{-2}, the values are 1, 0.01, and 0.0001. For $\frac{30}{\ln(x)}$, the values are undefined (divided by 0), 13.03, and 6.51. The two functions are always in the right order.

21. Approaches 0 because the denominator is an exponential function, which increases more quickly than the quadratic in the numerator.

23. Approaches 0 because the denominator is an exponential function with a larger coefficient.

25. The derivative is 5, a constant, consistent with linearity.

27. The derivative is $\alpha'(c) = \frac{10c}{(1+c^2)^2}$, so $\alpha'(0) = 0$ and the limit as c approaches infinity is 0. This is consistent with a graph that starts out flat, then increases and eventually flattens out again.

29. The derivative is $\alpha'(c) = \frac{5(1-c^2)}{(1+c^2)^2}$, so $\alpha'(0) = 5$. The limit as c approaches infinity is 0, but the derivative is negative. This is consistent with a graph that starts out increasing but then decreases to zero.

31. This population increases to infinity because $r > 1$. $b_t = 10^8 1.1^t > 10^{10}$, if

$$\begin{aligned} 1.1^t &> 10^2 \\ e^{\ln(1.1)t} &> 10^2 \\ \ln(1.1)t &> \ln(10^2) \\ t &> \frac{\ln(10^2)}{\ln(1.1)} = 48.3. \end{aligned}$$

The population exceeds the threshold after about 49 generations.

33. This population decreases to zero because $r < 1$. $b_t = 10^8 0.5^t < 10^3$, if

$$\begin{aligned} 0.5^t &< 10^{-5} \\ e^{\ln(0.5)t} &< 10^{-5} \\ \ln(0.5)t &< \ln(10^{-5}) \\ t &> \frac{\ln(10^{-5})}{\ln(0.5)} = 16.60. \end{aligned}$$

It takes 16.60 generations to reach 10^3.

35. $l_t = 2t$ and $r_t = 2^{t+1}$. The ratios are 0.5, 0.156, 0.0098, and 0.000019.

37. Solving $M^* = 0.5M^* + 1$ gives an equilibrium of $M^* = 2.0$.

39. We can solve $M_t = 2.02$, or $0.5^t \times 3.0 = 0.02$, or $0.5^t = 0.0067$. To solve, take natural logs of both sides to find $t\ln(0.5) = \ln(0.0067)$ or $t = \frac{\ln(0.0067)}{\ln(0.5)} = 7.2$.

41. This looks like the saturated model, and the predator is happiest with the largest possible supply of prey.

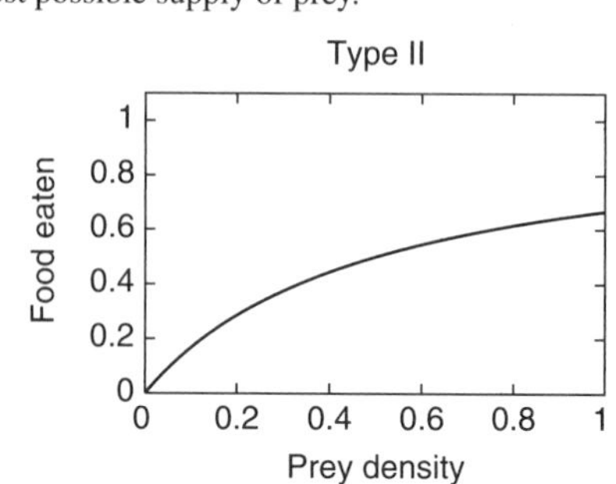

43. We want to maximize $F(p)-cp$. The derivative of this function is $F'(p)-c$ and is equal to 0 when $F'(p)=c$. The function is $F(p)=p-cp=p(1-c)$. If $c>1$, then this is a line with a negative slope and thus a maximum at $p=0$. If $c<1$, then this is a line with a positive slope and no maximum. In this case, we could think of the optimal density as infinity.

45.

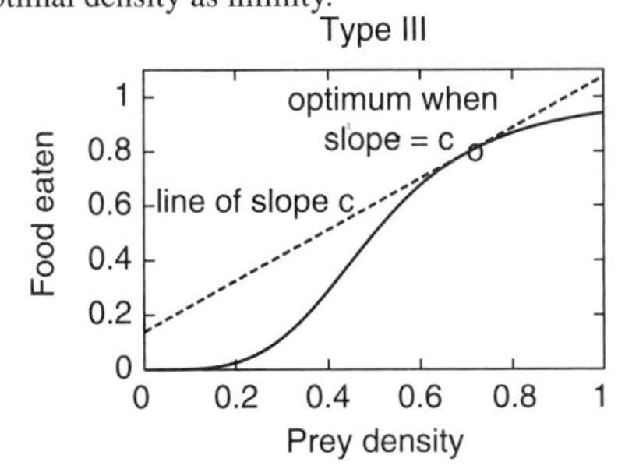

Section 3.6, page 312

1. At 0, the term x decreases to 0 while the constant term 1 does not, so $f_0(x)=1$. At infinity, the term x increases to infinity while the constant term 1 does not, so $f_\infty(x)=x$.

3. At 0, the term z decreases to 0 while the exponential term e^z does not, so $h_0(z)=e^z$. At infinity, the exponential term e^z increases to infinity more quickly than the power function z, so $h_\infty(z)=e^z$.

5. At 0, the term $\frac{1}{a}$ increases to infinity while the terms with positive power do not, so $m_0(a)=\frac{1}{a}$. At infinity, the term $30a^2$ increases to infinity most quickly because it has the largest power, so $m_\infty(a)=30a^2$.

7. The exponential function e^{2x} approaches infinity faster. By L'Hôpital's rule,

$$\lim_{x\to\infty}\frac{e^{2x}}{x^2}=\lim_{x\to\infty}\frac{2e^{2x}}{2x}=\lim_{x\to\infty}\frac{4e^{2x}}{2}=\infty.$$

9. The power function $0.1x^{0.5}$ approaches infinity faster. By L'Hôpital's rule,

$$\lim_{x\to\infty}\frac{0.1x^{0.5}}{30\ln(x)}=\lim_{x\to\infty}\frac{0.05x^{-0.5}}{30x^{-1}}=\lim_{x\to\infty}\frac{1}{600}x^{0.5}=\infty.$$

11. The exponential function e^{-2x} approaches 0 faster. L'Hôpital's rule doesn't really make things simpler directly, but

$$\lim_{x\to\infty}\frac{e^{-2x}}{x^{-2}}=\lim_{x\to\infty}\frac{x^2}{e^{2x}}=\lim_{x\to\infty}\frac{2x}{2e^{2x}}=\lim_{x\to\infty}\frac{2}{4e^{2x}}=0.$$

13. I would guess that the power function approaches infinity faster.

$$\lim_{x\to 0}\frac{x^{-1}}{-\ln(x)}=\lim_{x\to 0}\frac{-x^{-2}}{-x^{-1}}=\lim_{x\to 0}x^{-1}=\infty.$$

Therefore, $\lim_{x\to 0}\frac{1}{x\ln(x)}=\infty$, and $\lim_{x\to 0}x\ln(x)=0$.

15. The power function with the larger power, x^3, approaches 0 more quickly.

$$\lim_{x\to 0}\frac{x^3}{x^2}=\lim_{x\to 0}\frac{3x^2}{2x}=\lim_{x\to 0}\frac{6x}{2}=0.$$

17. The numerator has only one term, so the leading behavior is $2c^2$ at both 0 and ∞. The denominator has leading behavior 1 for c near 0 and c for large c. Therefore,

$$\alpha_0(c)=\frac{2c^2}{1}=2c^2$$
$$\alpha_\infty(c)=\frac{2c^2}{c}=2c$$
$$\lim_{c\to 0}\alpha(c)=0$$
$$\lim_{c\to\infty}\alpha(c)=\infty.$$

L'Hôpital's rule is not appropriate at $c=0$, because the denominator approaches 1. This limit can be found by plugging in. As $c\to\infty$, both the numerator and denominator approach infinity, so we can use L'Hôpital's rule to check

$$\lim_{c\to\infty}\frac{2c^2}{1+c}=\lim_{c\to\infty}\frac{4c}{1}=\infty.$$

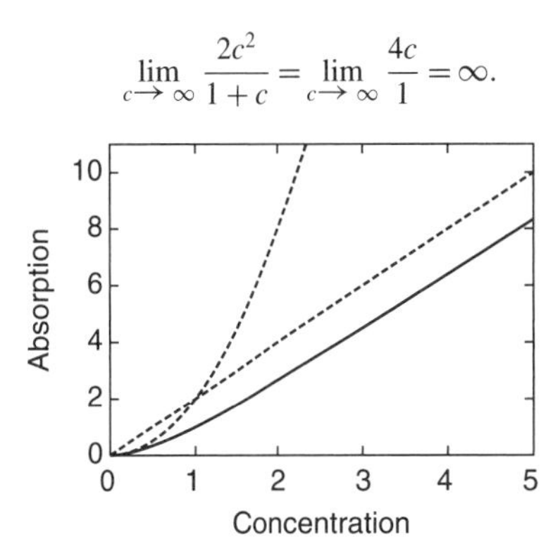

19. The numerator has leading behavior 1 near 0 and c^2 for large c. The denominator has leading behavior 1 for c near 0 and c for large c. Therefore,

$$\alpha_0(c)=\frac{1}{1}=1$$
$$\alpha_\infty(c)=\frac{c^2}{c}=c$$
$$\lim_{c\to 0}\alpha(c)=1$$
$$\lim_{c\to\infty}\alpha(c)=\infty.$$

L'Hôpital's rule is not appropriate at $c=0$, because both the numerator and denominator approach 1. This limit can be found by plugging in. As $c\to\infty$, both the numerator and denominator approach infinity, so we can use L'Hôpital's rule to check

$$\lim_{c\to\infty}\frac{1+c+c^2}{1+c}=\lim_{c\to\infty}\frac{1+2c}{1}=\infty.$$

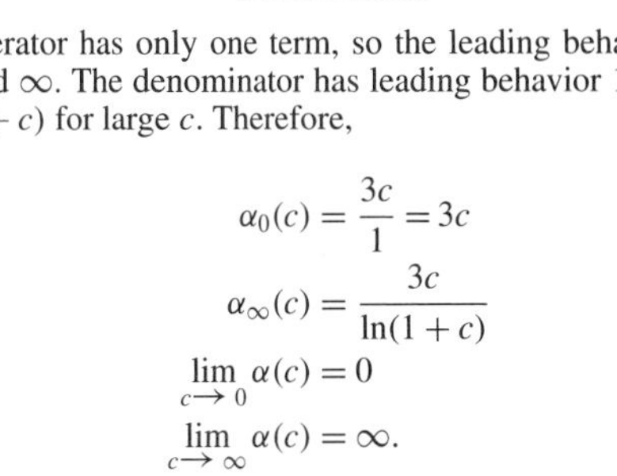

21. The numerator has only one term, so the leading behavior is $3c$ at both 0 and ∞. The denominator has leading behavior 1 for c near 0 and $\ln(1+c)$ for large c. Therefore,

$$\alpha_0(c)=\frac{3c}{1}=3c$$
$$\alpha_\infty(c)=\frac{3c}{\ln(1+c)}$$
$$\lim_{c\to 0}\alpha(c)=0$$
$$\lim_{c\to\infty}\alpha(c)=\infty.$$

L'Hôpital's rule is not appropriate at $c=0$, because the denominator approaches 1. This limit can be found by plugging in. As $c\to\infty$, both the numerator and denominator approach infinity, so we can use L'Hôpital's rule to check

$$\lim_{c\to\infty}\frac{3c}{1+\ln(1+c)}=\lim_{c\to\infty}\frac{3}{\frac{1}{1+c}}=\lim_{c\to\infty}\frac{3(1+c)}{1}=\infty.$$

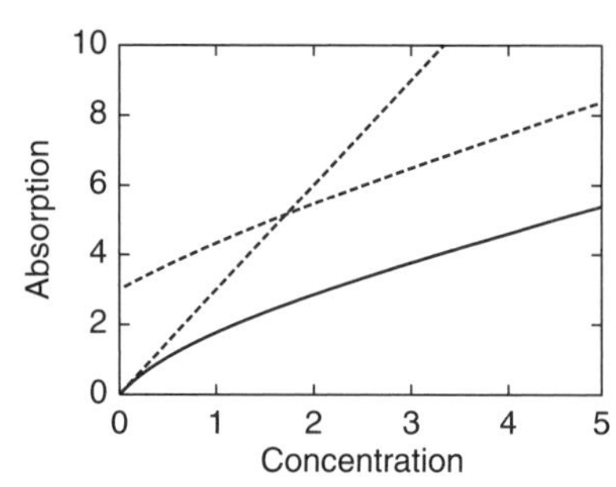

23. The denominator of this function has leading behavior of 1 for c near 0, so $\alpha_0(c) = 5c$. The denominator has leading behavior of c for large c, so $\alpha_\infty(c) = 5$.

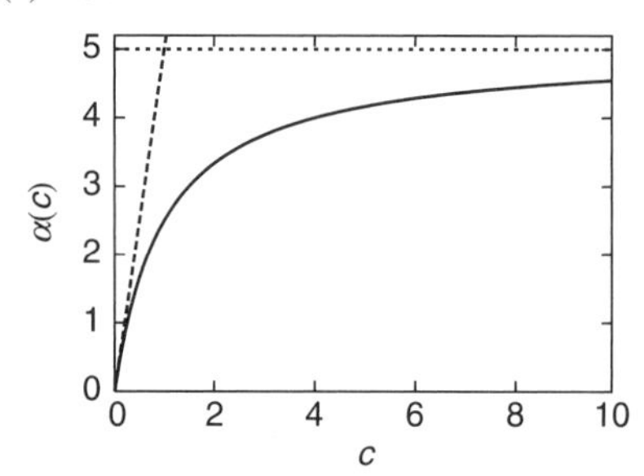

25. The denominator of this function has leading behavior of 1 for c near 0, so $\alpha_0(c) = 5c^2$. The denominator has leading behavior of c^2 for large c, so $\alpha_\infty(c) = 5$.

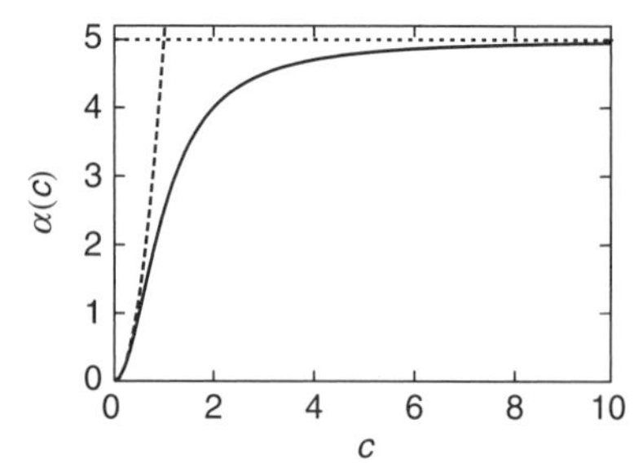

27. The denominator of this function has leading behavior of 1 for c near 0, so $\alpha_0(c) = 5c$. The denominator has leading behavior of c^2 for large c, so $\alpha_\infty(c) = \frac{5}{c}$.

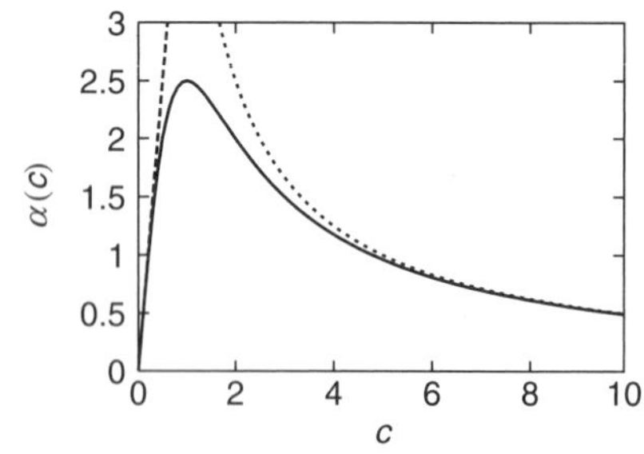

29. The denominator has leading behavior of 1 for x near 0 and of x^3 for large x. Therefore, $h_3(x)_0 = x^3$, $h_3(x)_\infty = 1$.

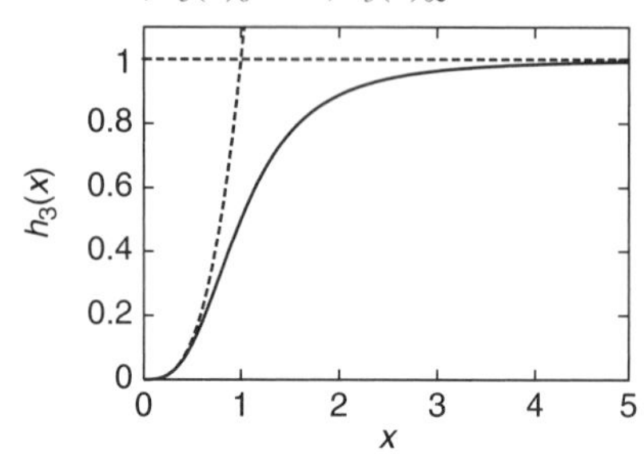

31. The denominator has leading behavior of 1 for x near 0 and of x^{10} for large x. Therefore, $h_{10}(x)_0 = x^{10}$, $h_{10}(x)_\infty = 1$.

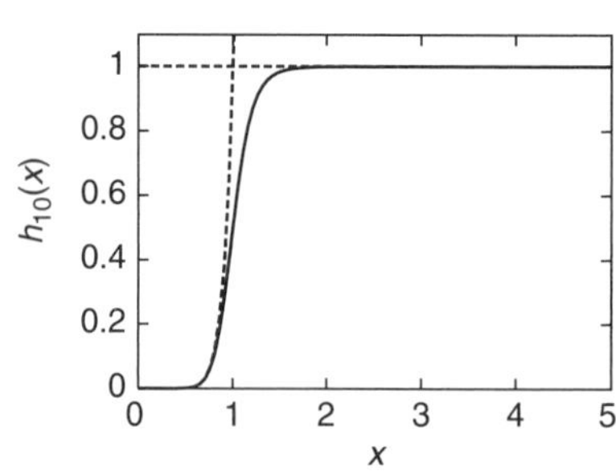

33.
a. Using the general solution of the bacterial discrete-time dynamical system (page 81), we find $a_t = 10^4 \times 2.0^t$, $b_t = 10^6 \times 1.5^t$.
b. $p_t = \frac{10^4 \times 2.0^t}{10^4 \times 2.0^t + 10^6 \times 1.5^t}$.
c. In exponential notation, we can rewrite the denominator as $10^4 e^{\ln(2.0)t} + 10^6 e^{\ln(1.5)t}$, with leading behavior $10^4 e^{\ln(2.0)t}$ because the parameter in the exponent is larger. Therefore, $\lim_{t \to 0} p_t = 1$.
d. $p_0 = 0.01$, $p_{10} = 0.15$, $p_{20} = 0.76$, $p_{50} = 0.9999$. This is mighty close to the limit.

35.
a. Using the general solution of the bacterial discrete-time dynamical system (page 81), we find $a_t = 10^4 \times 0.8^t$, $b_t = 10^5 \times 1.2^t$.
b. $p_t = \frac{10^4 \times 0.8^t}{10^4 \times 0.8^t + 10^5 \times 1.2^t}$.
c. In exponential notation, the denominator is $10^4 e^{\ln(0.8)t} + 10^5 e^{\ln(1.2)t} = 10^4 e^{-0.223t} + 10^5 e^{0.182t}$. The leading behavior is $10^4 e^{0.182t}$ because this term grows and the other term shrinks. Therefore, $\lim_{t \to 0} p_t = 0$.
d. $p_0 = 0.09$, $p_{10} = 0.0017$, $p_{20} = 0.00003$, $p_{50} = 1.5 \times 10^{-10}$. This is extremely close to the limit.

37. $r(c) = c$, the derivative is $\alpha'(c) = \frac{Ak}{(k+c)^2} > 0$. By L'Hôpital's rule,

$$\lim_{c \to \infty} A \frac{c}{k+c} = \lim_{c \to \infty} A \frac{1}{1} = A.$$

Near $c = 0$, the leading behavior of the denominator is k, so $\alpha(c) \approx \frac{c}{k}$. For large c, the leading behavior of the denominator is c, so $\alpha(c) \approx \alpha$.

39. $r(c) = c^n$, the derivative is $\alpha'(c) = \frac{nAkc^{n-1}}{(k+c^n)^2} > 0$. By L'Hôpital's rule,

$$\lim_{c \to \infty} A \frac{c^n}{k+c^2} = \lim_{c \to \infty} A \frac{nc^{n-1}}{nc^{n-1}} = A.$$

Near $c = 0$, the leading behavior of the denominator is k, so $\alpha(c) \approx \frac{c^n}{k}$. For large c, the leading behavior of the denominator is c^n, so $\alpha(c) \approx \alpha$.

Section 3.7, page 324

1. Let $f(x) = x^3$ with base point $a = 2.0$. Then $f'(x) = 3x^2$. To find the tangent line approximation, evaluate $f(2) = 8.0$ and $f'(2) = 12.0$. The tangent line is then $\hat{f}(x) = 8.0 + 12.0(x - 2.0)$ and $\hat{f}(2.02) = 8.0 + 12.0(2.02 - 2.0) = 8.0 + 12.0 \times 0.02 = 8.24$. To find the secant line, we evaluate $f(3) = 27$, so the secant line has slope 19. Therefore, $f_s(t) = 8.0 + 19.0(x - 2.0)$ and $f_s(2.02) = 8.0 + 19.0 \times 0.02 = 8.38$. The exact answer is 8.242408, which is pretty close to the tangent line approximation.

3. Let $f(x) = \sqrt{x}$ with base point 4.0. Then $f(4) = 2$, $f'(x) = \frac{1}{2}x^{\frac{-1}{2}}$, and $f'(4) = 0.25$. So $\hat{f}(x) = 2 + 0.25(x - 4)$ and $\hat{f}(4.01) = 2.0025$. To find the secant line, we evaluate $f(9) = 3$, so the secant line has slope 0.2. Therefore, $f_s(t) = 2.0 + 0.2(x - 4.0)$ and $f_s(4.01) = 2.0 + 0.2 \times 0.01 = 2.002$. The exact answer to six decimal places is 2.002498, which is quite close to the tangent line approximation.

5. Let $f(x) = \sin(x)$. $f(0) = \sin(0) = 0$. $f'(x) = \cos(x)$ and $f'(0) = 1$. So $\hat{f}(x) = 0 + 1(x - 0)$ and $\hat{f}(0.02) = 0.02$. The secant line does not help because there is no easy value of x to evaluate this function. The exact answer to five decimal places is 0.19998.

7. Let $f(x)=x^3$ with base point $a=2.0$. Because $f'(x)=3x^2$, $f''(x)=6x$. Therefore, $f''(2)=12.0$, so $P_2(x)=8.0+12.0(x-2.0)+6.0(x-2.0)^2$ and $P_2(2.02)=8.0+12.0(2.02-2.0)+6.0(2.02-2.0)^2=8.0+12.0\times 0.02+6.0\times 0.0004=8.2424$. The exact answer is 8.242408, which is very close to the quadratic approximation.

9. Let $f(x)=\sqrt{x}$ with base point 4.0. Because $f'(x)=\frac{1}{2}x^{\frac{-1}{2}}$, we can use the power rule again to find $f''(x)=\frac{-1}{4}x^{\frac{-3}{2}}$. Then $f''(4)=\frac{-1}{32}$, so $P_2(x)=2+0.25(x-4)-(x-4)^2/64$ and $P_2(4.01)=2.002498438$. The exact answer to ten decimal places is 2.0024984395.

11. Let $f(x)=\sin(x)$ with base point 0.0. Then $f'(x)=\cos(x)$ and $f''(x)=-\sin(x)$, so $f''(0)=0$. Then $P_2(x)=0+1(x-0)$, which is identical to the tangent line. The exact answer to five decimal places is 0.19998.

13. This is a composition of $f[g(x)]$ where $g(x)=1+3x$ and $f(g)=g^2$. Near the base point $x=1$, we have that $f[g(1)]=16$, and $(f\circ g)'(x)=6(1+3x)$, so $(f\circ g)'(1)=24$. The tangent line approximation is $16+24(x-1)$, which has value 16.24 at $x=1.01$. In steps, the tangent line to g at $x=1$ is $\hat{g}(x)=4+3(x-1)$, so $\hat{g}(1.01)=4.03$. ($\hat{g}=g$ because g is linear). We then evaluate f near 4, finding $f(4)=16$ and $f'(4)=8$. Then $\hat{f}(x)=16+8(x-4)$ and $\hat{f}(4.03)=16.24$, exactly as before.

15. This is a composition of $f[g(x)]$ where $g(x)=\sin(x)$ and $f(g)=e^g$. These have derivatives $g'(x)=\cos(x)$ and $f'(g)=e^g$. Near the base point $x=0$, we have that $f[g(0)]=1$, and $(f\circ g)'(x)=\cos(x)e^{\sin(x)}$, so $(f\circ g)'(0)=1$. The tangent line approximation is $1+x$, which has value 1.02 at $x=0.02$. In steps, the tangent line to g at $x=1$ is $\hat{g}(x)=0+x$ so $\hat{g}(0.02)=0.02$. We then evaluate f near 0, finding $f(0)=1$ and $f'(0)=1$. Then $\hat{f}(x)=1+x$ and $\hat{f}(0.02)=1.02$, exactly as before.

17. $e^{0.1}=1.105>1.1$. $e^{-0.1}=0.905>0.9$. The estimates are low because the graph of e^x is concave up, and lies above the tangent.

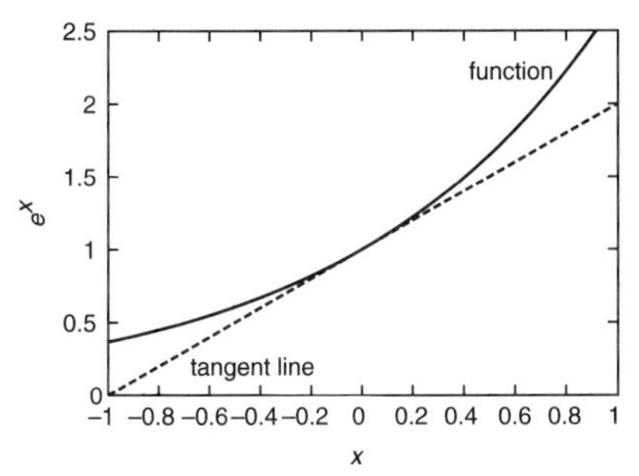

19. $1.1^2=1.21>1.2$. $0.9^2=0.81>0.8$. The estimates are low because the graph of x^2 is concave up and lies above the tangent.

21. We have that $f'(x)=3x^2+8x+3$, $f''(x)=6x+8$, and $f'''(x)=6$. Then $f(0)=1$, $f'(0)=3$, $f''(0)=8$, and $f'''(0)=6$. Therefore,

$$\begin{aligned}P_3(x)&=f(0)+f'(0)x+\frac{f''(0)}{2}x^2+\frac{f'''(0)}{6}x^3\\&=1+3x+4x^2+x^3.\end{aligned}$$

This matches the original polynomial.

23. We have that $f'(x)=8x+3$, $f''(x)=8$, and $f'''(x)=0$. Then $f(1)=8$, $f'(1)=11$, $f''(0)=8$, and $f'''(0)=0$. Therefore,

$$\begin{aligned}P_3(x)&=f(1)+f'(1)(x-1)+\frac{f''(1)}{2}(x-1)^2+\frac{f'''(1)}{6}(x-1)^3\\&=8+11(x_1)+4(x-1)^2.\end{aligned}$$

25. We have that $h'(x)=\frac{1}{x}$, $h''(x)=\frac{-1}{x^2}$, and $h'''(x)=\frac{2}{x^3}$. Then $h(1)=0$, $h'(1)=1$, $h''(1)=-1$, and $h'''(1)=2$. Therefore,

$$\begin{aligned}P_3(x)&=h(1)+h'(1)(x-1)+\frac{h''(1)}{2}(x-1)^2+\frac{h'''(1)}{6}(x-1)^3\\&=0+(x-1)+\frac{1}{2}(x-1)^2+\frac{1}{3}(x-1)^3.\end{aligned}$$

27. This is the Taylor series for $g(x)=e^x$ evaluated at $x=2$, and it should therefore take on the value $e^2=7.389$. The sums of the first 1, 2, 3, 4 and 5 terms are 1.0, 3.0, 5.0, 6.333, 7.0, and 7.267 respectively. This is only 0.122 away from the limit.

29. This is the Taylor series for $f(x)=\frac{1}{1-x}$ evaluated at $x=\frac{1}{3}$, and it should therefore take on the value $\frac{1}{(1-\frac{1}{3})}=1.5$. The sums of the first 1, 2, 3, 4 and 5 terms are 1.0, 1.3333, 1.4444, 1.4815, 1.4938, and 1.4979 respectively. This is 0.0021 from the limit.

31. Using the fact that $\frac{d\cos(x)}{dx}=\sin(x)$ and $\frac{d\sin(x)}{dx}=-\cos(x)$, we find that the Taylor polynomials are

$$\begin{aligned}P_0(x)&=\cos(0)=1\\P_1(x)&=1+\sin(0)x=1\\P_2(x)&=1-\cos(0)\frac{1}{2}x^2=1-\frac{x^2}{2}\\P_3(x)&=1-\frac{x^2}{2}-\sin(0)\frac{x^3}{6}=1-\frac{x^2}{2}\\P_4(x)&=1-\frac{x^2}{2}+\cos(0)\frac{x^4}{24}=1-\frac{x^2}{2}+\frac{x^4}{24}\end{aligned}$$

and the Taylor series is

$$T_h(x)=1-\frac{x^2}{2!}+\frac{x^4}{4!}+\cdots.$$

Because the Taylor polynomials approximate the function h, the sum of the series is $h(1)=\cos(1)=0.5043$.

33. The tangent line to $2x+x^2$ near $x=0$ is $2x$. The tangent line to $3x+2x^2$ near $x=0$ is $3x$. For small x, $f(x)\approx\frac{2x}{3x}=\frac{2}{3}$. With L'Hôpital's rule,

$$\lim_{x\to 0}\frac{2x+x^2}{3x+2x^2}=\lim_{x\to 0}\frac{2+2x}{3+4x}=\frac{2}{3}.$$

35. The tangent line to $\ln(x)$ near $x=1$ is $x-1$. The tangent line to x^2-1 near $x=1$ is $2(x-1)$. For small x, $f(x)\approx\frac{x-1}{2(x-1)}=\frac{1}{2}$. With L'Hôpital's rule,

$$\lim_{x\to 1}\frac{\ln(x)}{x^2-1}=\lim_{x\to 1}\frac{\frac{1}{x}}{2x}=\frac{1}{2}.$$

37. $\alpha(0)=0$, and $\alpha'(c)=\frac{5}{(1+c)^2}$, so $\alpha'(0)=5$. Therefore, the tangent line is $\hat{A}(c)=5c$, matching the result found with leading behavior.

39. $\alpha(0)=0$, and $\alpha'(c)=\frac{10c}{(1+c^2)^2}$, so $\alpha'(0)=0$. Therefore, the tangent line is $\hat{A}(c)=0$, which does not match the result found with leading behavior. We would need the quadratic approximation to match the leading behavior.

41. $\alpha(0)=0$, and $\alpha'(c)=\frac{5}{1+c^2}-\frac{10c^2}{(1+c^2)^2}$, so $\alpha'(0)=5$. Therefore, the tangent line is $\hat{A}(c)=5c$, matching the result found with leading behavior.

43. The derivative is $b'(t)=\frac{-1}{(1+t)^2}$. The tangent line at $t=0$ is $\hat{b}_0(t)=b(0)+b'(0)t=1-t$. The tangent line at $t=1$ is $\hat{b}_1(t)=b(1)+b'(1)t=\frac{1}{2}-\frac{1}{4(t-1)}$. Finally, the secant line connecting them has slope $\frac{-1}{2}$ and passes through the point (0, 1), and thus has equation $b_s(t)=1-\frac{1}{2t}$. Then $\hat{b}_0(0.1)=0.9$, $\hat{b}_1(0.1)=0.725$, and $b_s(0.1)=0.95$. The exact answer is $\frac{10}{11}=0.9091$, closest to the tangent at $t=0$. The secant is in the right ball park, and the tangent at $t=1$ is way off.

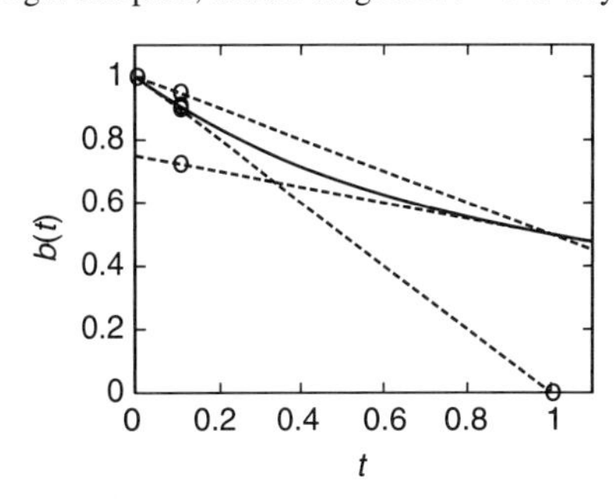

45. The tangent at $t=0$ gives 0.1, the tangent at $t=1$ gives 0.525, and the secant gives 0.55. To three decimal places, the exact answer is 0.526, closest to the tangent at $t=1$. The secant is in the right ball park, and the tangent at $t=0$ is way off.

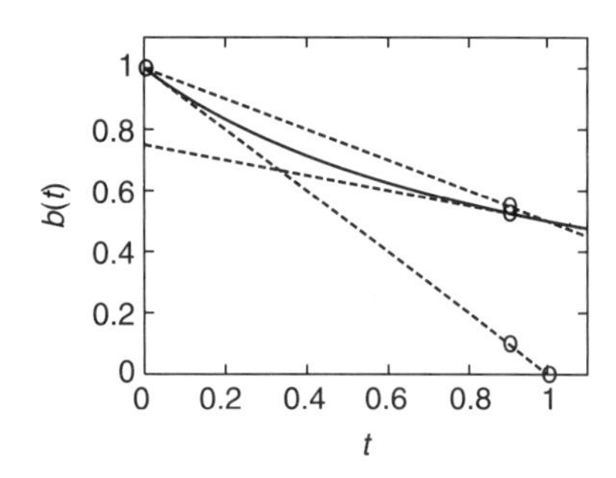

47. Using the base point $a = 1.0$, $M(1) = 1$. $M'(a) = 3a^2$, so $M'(1) = 3$. Then the tangent line is $\hat{M}(a) = 1 + 3(a - 1)$ and $\hat{M}(1.25) = 1 + 3(0.25) = 1.75$. The secant line connecting $a = 1$ and $a = 1.5$ is $M_s(a) = 3.375 + 4.75(a - 1.5)$ and $M_s(1.25) = 3.375 + 4.75(-0.25) = 2.1875$. To three decimal places, the exact value is 1.953. None of the approximations is very close, but only the secant line would be possible if we did not know the formula.

49. First, we can use the equation $T(t) = t + t^2$ to find the tangent line at $t = 0$ as $\hat{T}(t) = t$ and estimate $T(1) = 1$. We can find the tangent line at $t = 2$ as $\hat{T}(t) = 6 + 5(t - 2)$, and again estimate $T(1) = 1$. The secant line between the actual data points has slope $\frac{6.492 - 0.172}{2} = 3.16$, giving the line $T_s(t) = 0.172 + 3.16t$ with the value $T(1) = 3.332$. None of these is very close to the correct answer.

51. Let I_t be the number of people infected in generation t. Then $I_{t+1} = 0.5I_t$ with solution $I_t = 1.0 \times 10^5 \times 0.5^t$. We can use the sum of the geometric series with $r = 0.5$ to find that $I_0 + I_1 + I_2 + \cdots = 1.0 \times 10^5 \times \frac{1}{1-0.5} = 2.0 \times 10^5$. Even though the epidemic is dying out, there will still be as many new people infected as were infected when the treatment was introduced.

Section 3.8, page 334

1.

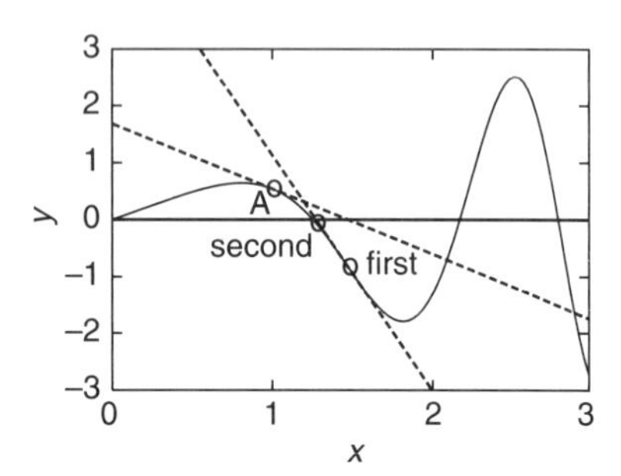

3.

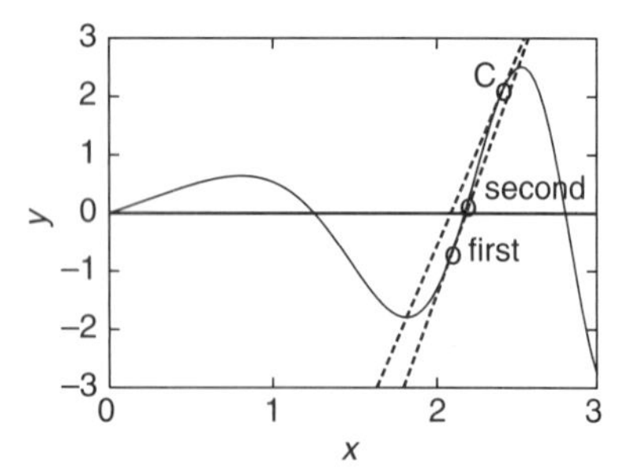

5. Set $f(x) = x^2 - 5x + 1$. Then $f(0) = 1$ and $f(1) = -3$, so there must be a solution in between according to the Intermediate Value Theorem. If we start with a guess of $x = 0$, we find $f'(0) = -5$, so the tangent line is $\hat{f}(x) = 1 - 5x$, which intersects the horizontal axis at $x = 0.2$. The Newton's method discrete-time dynamical system is

$$x_{t+1} = x_t - \frac{f(x_t)}{f'(x_t)} = x_t - \frac{x_t^3 - 5x_t + 1}{2x_t - 5}.$$

If $x_0 = 0$, then $x_1 = 0.2$, $x_2 = 0.2086956522$, and $x_3 = 0.2087121525$. The exact answer to ten decimal places is 0.2087121525.

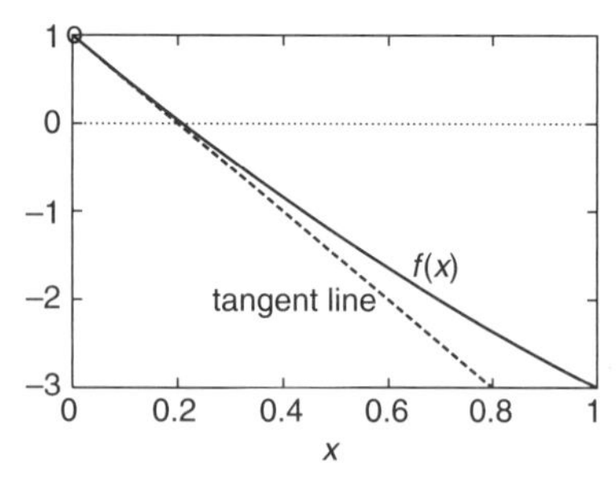

7. Suppose initial guess is $x_0 = 1$. Because $h'(x) = 0.5e^{\frac{x}{2}} - 1$, $h(1) = -0.351$, and $h'(3) = -0.1756$, the tangent line is $\hat{h}(x) = -0.351 - 0.1756(x - 1)$, which intersects the horizontal axis at $x = -1$. The Newton's method discrete-time dynamical system is

$$x_{t+1} = x_t - \frac{h(x_t)}{h'(x_t)} = x_t - \frac{e^{\frac{x_t}{2}} - x_t - 1}{0.5e^{\frac{x_t}{2}} - 1}.$$

If $x_0 = 1$, then $x_1 = -1.00$, $x_2 = -0.1294668027$, $x_3 = -0.0037771286$. This seems to be approaching 0. If we start from $x_0 = 2$, we get $x_1 = 2.7844$, $x_2 = 2.5479$, and $x_3 = 2.51355$. After many steps, the answer converges to 2.512862414.

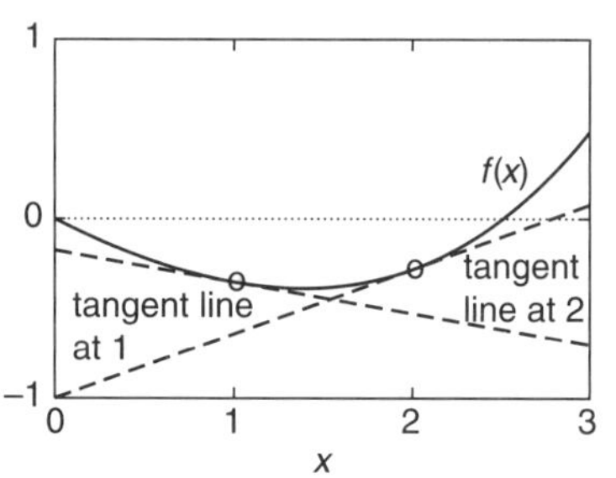

9. It is true that $x = e^x - 2$ if $e^x = x + 2$. But starting from $x_0 = 1$, solutions are $x_1 = 0.718$, $x_2 = 0.051$, and $x_3 = -0.947$. After a while, it seems to converge to another equilibrium at -1.84.

11. It is true that $e^{\frac{x}{2}} = x + 1$ if $h(x) = 0$. Starting from a guess $x_0 = 2$, we find $x_1 = 1.718$, $x_2 = 1.361$, and $x_3 = 0.975$. This seems to be going very slowly to $x = 0$.

13. It should converge most rapidly when the equilibrium is superstable, or the slope is 0. The slope at the equilibrium is $2 - r$, so the most rapid convergence should be when $r = 2$. Starting from $x_0 = 0.75$, we get $x_1 = 0.375$, $x_2 = 0.46875$, $x_3 = 0.4980468750$, and $x_4 = 0.4999923706$.

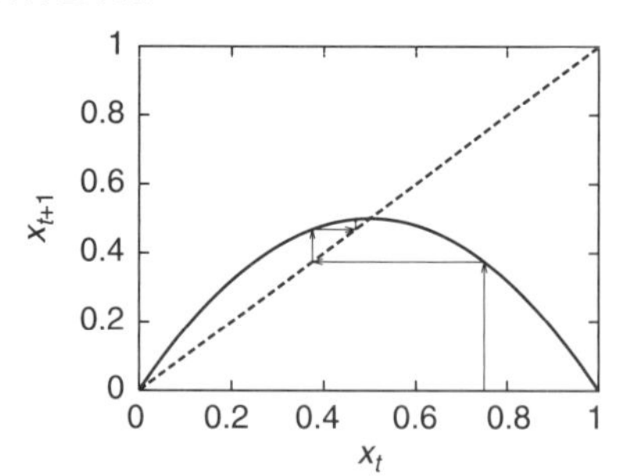

15. The method fails if we start at a critical point of the function, which occur where $x = \pm 0.577$. All values below the lower critical point converge to the negative solution.

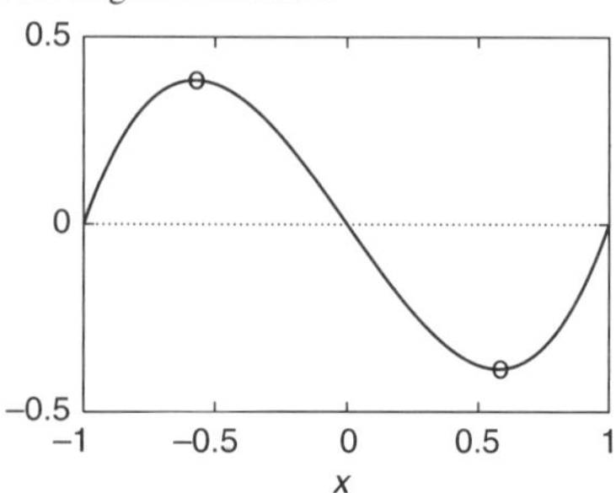

17. The Newton's method discrete-time dynamical system is

$$x_{t+1} = x_t - \frac{x_t^2}{2x_t} = \frac{x_t}{2}.$$

This converges to 0 rather slowly because the derivative of the function x^2 is 0 at the solution.

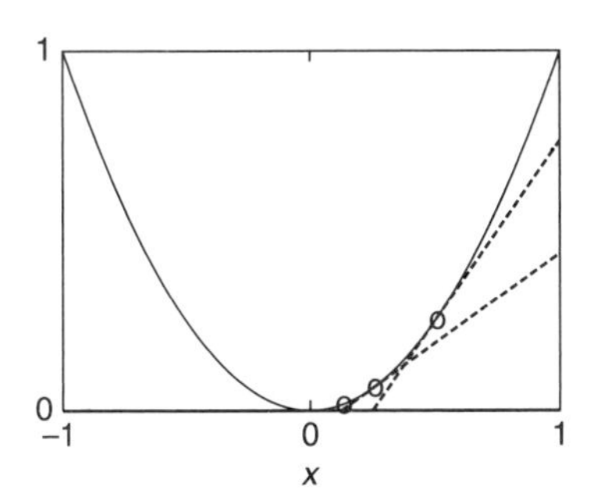

19. $f'(x) \approx e^{x+1} - (x+1) - 2 - (e^x - x - 2) = e^{x+1} - 1 - e^x$. The approximate Newton's method discrete-time dynamical system is

$$x_{t+1} = x_t - \frac{e^{x_t} - x_t - 2}{e^{x_t+1} - 1 - e^{x_t}}.$$

Starting from an initial guess of $x_0 = 1.0$, we find $x_1 = 1.0767$, $x_2 = 1.1118$, $x_3 = 1.1288$, $x_4 = 1.1373$, and $x_5 = 1.1419$. After five steps, it has gotten the first couple of decimal places in the exact answer, to ten decimal places, of 1.146193221.

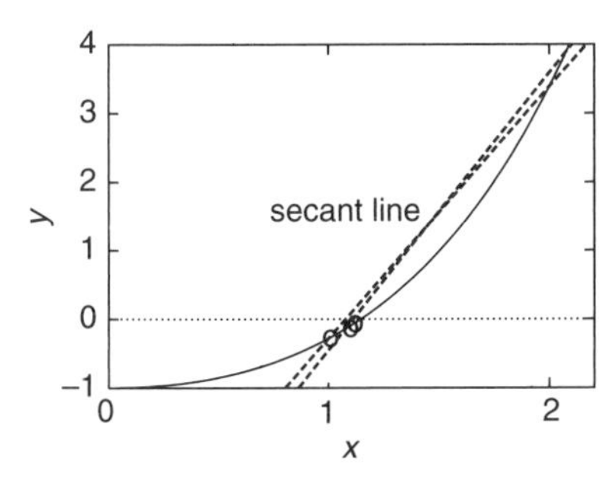

21. The discrete-time dynamical system is

$$x_{t+1} = x_t - \frac{f(x_t)(x_t - x_{t-1})}{f(x_t) - f(x_{t-1})},$$

which depends on both x_t and x_{t-1}, unlike an ordinary discrete-time dynamical system. Starting from $x_0 = 1.0$ and

$$x_1 = x_0 - \frac{e^{x_0} - x_0 - 2}{e^{x_0+1} - 1 - e^{x_0}}$$

from Exercise 19, we find $x_1 = 1.076746253$, $x_2 = 1.154339800$, $x_3 = 1.145768210$, $x_4 = 1.146190691$, and $x_5 = 1.146193221$, which is correct to nine decimal places. This method is better because it uses a more accurate version of the secant line approximation.

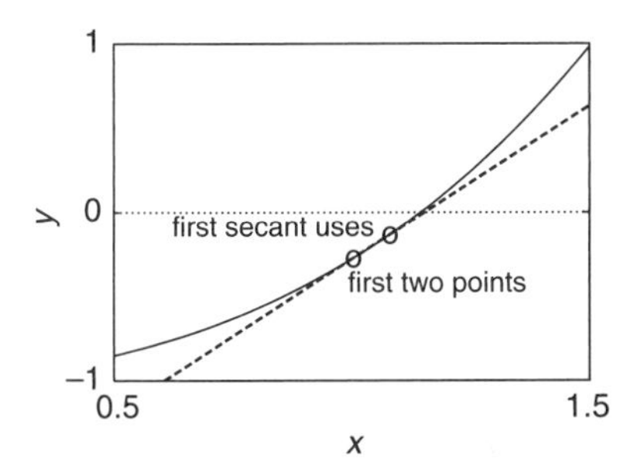

23. This function differs in that it is concave up for small t. The nectar comes out slowly when the bee first arrives. The derivative is $F'(t) = \frac{2t}{(1+t^2)^2}$, so the Marginal Value Theorem equation is

$$\begin{aligned} F'(t) &= \frac{F(t)}{t} \\ \frac{2t}{(1+t^2)^2} &= \frac{t}{1+t^2} \\ \frac{2}{1+t^2} &= 1 \\ 1+t^2 &= 2 \\ t^2 &= 1 \\ t &= 1. \end{aligned}$$

It turned out this could be solved algebraically.

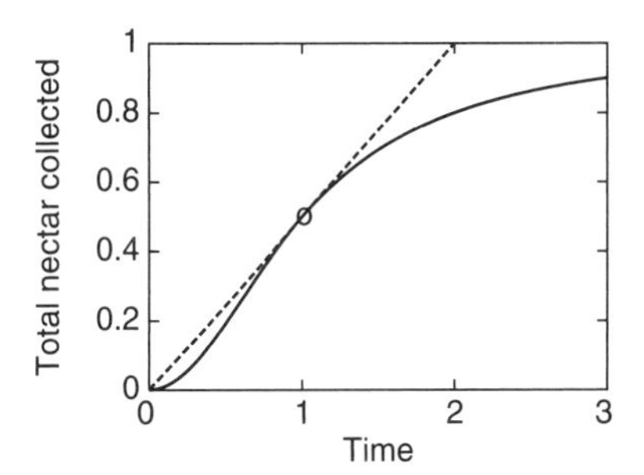

25. $N^* = \ln\left(\frac{2.5}{1+h}\right)$. Therefore, $P(h) = hN^* = h\ln\left(\frac{2.5}{1+h}\right)$. Then

$$P'(h) = \ln\left(\frac{2.5}{1+h}\right) - \frac{h}{1+h} = 0.$$

We can use $P(h)$ and $P'(h)$ in Newton's method, finding the discrete-time dynamical system

$$h_{t+1} = h_t - \frac{P(h_t)}{P'(h_t)}.$$

I used an initial guess of $h_0 = 0.75$ because this is the solution we found when the fish followed the logistic model. After three steps, it had converged to $h = 0.6724$.

27. Let G be the updating function. Then $G(0) = 1$ and $G(5.) = 2.516$. There must be an equilibrium in between. Starting from $M_0 = 2$, we get to 1.726 after about six steps. Solving the equation $0.5e^{-0.1x}x + 1.0 - x = 0$ with Newton's method, it reaches 1.726 after two steps.

29. We need to solve the equation $100e^{0.1t} = 400 + 100t$ or $100e^{0.1t} - 400 - 100t = 0$. The Newton's method discrete-time dynamical system is

$$x_{t+1} = x_t - \frac{100e^{0.1x_t} - 400 - 100x_t}{10e^{0.1x_t} - 100}.$$

If food resources were constant at the initial value of 400, the population would run out of food when $b(t) = 400$, or at time $10\ln(4) = 13.8$. Using an initial guess of $x_0 = 13.8$, the solution shoots off to a negative value. Using an initial guess of $x_0 = 30$ instead, we find the solution as $t = 37.18$ after three steps.

31. We will start with the guess $c_0 = 5.0$ because the equilibrium is between 0 and γ. We need to solve the equation $c^* = 0.75e^{-c^*}c^* + 1.25$ or $f(c) = 0$ where $f(c) = c - 0.75e^{-c^*}c^* - 1.25$. Then $f'(c) = 1 + 0.75e^{-c}c - 0.75e^{-c}$, and the Newton's method discrete-time dynamical system is

$$c_{t+1} = c_t - \frac{c_t - 0.75e^{-c_t}c_t - 1.25}{1 + 0.75e^{-c_t}c_t - 0.75e^{-c_t}}.$$

With $c_0 = 5.0$, $c_1 = 1.349066687$, $c_2 = 1.502145217$, and $c_3 = 1.500942584$.

Section 3.9, page 340

1. The absorption rate is

$$R(T) = \frac{2.5}{1.5 - 0.5T}.$$

The derivative is

$$R'(T) = \frac{1.25}{(1.5 - 0.5T)^2},$$

which is always positive. Therefore, the maximum must be at $T = 1$, where $R(1) = 2.5$.

3. The absorption rate is

$$R(T) = \frac{5.0\alpha}{1.0 + \alpha - \alpha T}.$$

The derivative of $R(T)$ is

$$R'(T) = \frac{5.0\alpha^2}{(1.0 + \alpha - \alpha T)^2},$$

which is always positive. Therefore, the maximum must be at $T = 1$, where

$$R(1) = 5.0\alpha.$$

This is an increasing function of α because α translates breath duration into absorption.

5. Equation 3.9.8 says that

$$c^* = \frac{\gamma r T}{1 - (1 - rT)[1 - \alpha(1 - e^{-kT})]}$$

when the discrete-time dynamical system is

$$c_{t+1} = (1 - rT)[c_t - \alpha c_t(1 - e^{-kT})] + rT\gamma.$$

The equilibrium is

$$\begin{aligned} c^* &= (1 - rT)[c^* - \alpha c^*(1 - e^{-kT})] + r\gamma T && \text{equation for} \\ r\gamma T &= c^* - (1 - rT)[c^* - \alpha c^*(1 - e^{-kT})] && \text{equilibrium move} \\ r\gamma T &= c^*\{1 - (1 - rT)[1 - \alpha(1 - e^{-kT})]\} && \text{all the } c^*\text{'s to one} \\ c^* &= \frac{\gamma r T}{1 - (1 - rT)[1 - \alpha(1 - e^{-kT})]}. && \text{side factor out a } c^* \\ & && \text{solve for } c^* \end{aligned}$$

7. The rate of absorption is

$$R(T) = \frac{1.25\frac{T^2}{0.1+T^2}}{1 - (1 - 0.5T)\left(1 - 0.5\frac{T^2}{0.1+T^2}\right)}.$$

With much patience, it is possible to take the derivative to find the answer.

9. The optimal T does not depend on r at all. I have no idea why.

11. The equilibrium is

$$c^* = \frac{\gamma r T}{1 - (1 - rT)[1 - A(T)]}.$$

13.

$$\begin{aligned} c^* &= \frac{\gamma r T}{1 - (1 - rT)(1 - \alpha T)} \\ \frac{c^* A(T)}{T} &= \frac{\alpha \gamma r T}{1 - (1 - rT)(1 - \alpha T)}. \end{aligned}$$

15.

$$\begin{aligned} c^* &= \frac{\gamma r T}{1 - (1 - rT)\left(1 - \alpha\frac{T^2}{k+T^2}\right)} \\ \frac{c^* A(T)}{T} &= \frac{\alpha \gamma r \frac{T^2}{k+T^2}}{1 - (1 - rT)\left(1 - \alpha\frac{T^2}{k+T^2}\right)}. \end{aligned}$$

17. We cannot plug in $T = 0$ because

$$\frac{c^* A(T)}{T} = \frac{\alpha \gamma r T}{1 - (1 - rT)(1 - \alpha T)}$$

is an indeterminate form. Taking the derivative of the top and bottom with respect to T gives

$$\lim_{T \to 0} \frac{c^* A(T)}{T} = \lim_{T \to 0} \frac{\alpha \gamma r}{r + \alpha - r\alpha T} = \frac{\alpha \gamma r}{r + \alpha}.$$

This is positive because the lung still absorbs some chemical even when breathing is extremely fast.

19. This is an indeterminate form.

$$\begin{aligned} \lim_{T \to 0} \frac{c^* A(T)}{T} &= \lim_{T \to 0} \frac{\alpha \gamma r T^2}{1 - (1 - rT)(1 - \alpha T^2)} \\ &= \lim_{T \to 0} \frac{2\alpha \gamma r T}{2\alpha T(1 - rT) + r(1 - \alpha T^2)} \\ &= \frac{0}{0 + r} = 0. \end{aligned}$$

There is no absorption when the animal pants. This makes sense because this lung absorbs very slowly when T is small.

21. The equilibrium value is

$$c^* = \frac{r\gamma(k + T)}{\alpha + rk + rT - rTa}.$$

The equilibrium absorption rate is then

$$R(T) = \frac{\alpha r \gamma}{\alpha + rk + rT - rT\alpha}.$$

As long as $\alpha < 1$, this is a decreasing function of T, meaning that the organism would do best to pant no matter what the parameter values.

Chapter 4

Section 4.1, page 354

1. This is a pure-time differential equation because the rate of change depends only on t.

3. This is a pure-time differential equation because the rate of change depends only on t.

5.

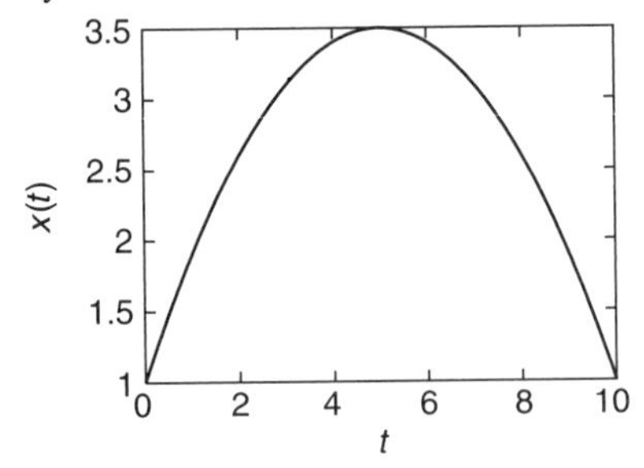

7.

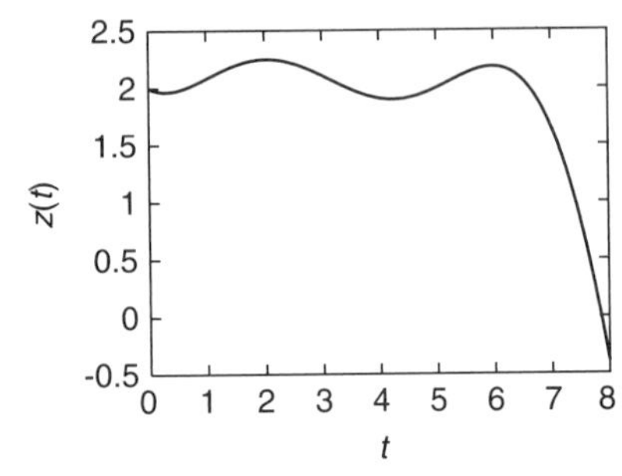

9. Taking the derivative, we find that $\frac{dx}{dt} = t$, as required by the differential equation. The initial condition must have been $x(0) = 1$.

11. With $\Delta t = 1$, $\hat{x}(1) = 1 + x'(0) \cdot 1 = 1 + 0 \cdot 1 = 1$. With $\Delta t = 0.5$, the first step is $\hat{x}(0.5) = 1 + x'(0) \cdot 0.5 = 1 + 0 \cdot 0.5 = 1$. The second step is $\hat{x}(1) = 1 + x'(0.5) \cdot 0.5 = 1 + 0.5 \cdot 0.5 = 1.25$. The exact answer, using the solution $x(t) = 1 + \frac{t^2}{2}$, is $x(1) = 1.5$.

13. The differential equation is $\frac{dV}{dt} = -2$. The function $V(t) = 600 - 2t$ has derivative -2 and satisfies the initial condition $V(0) = 600$. The solution stops making sense after $t = 300$ when the volume becomes negative.

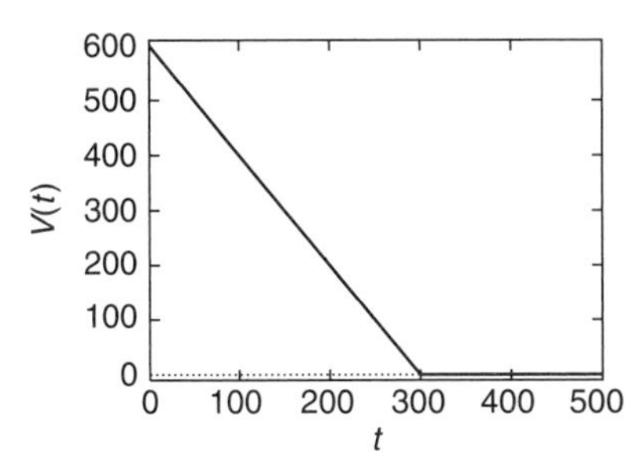

15. The differential equation is $\frac{dV}{dt} = -2t$. The function $V(t) = 900 - t^2$ has derivative $-2t$ and satisfies the initial condition $V(0) = 900$. The solution stops making sense after $t = 30$ when the volume becomes negative.

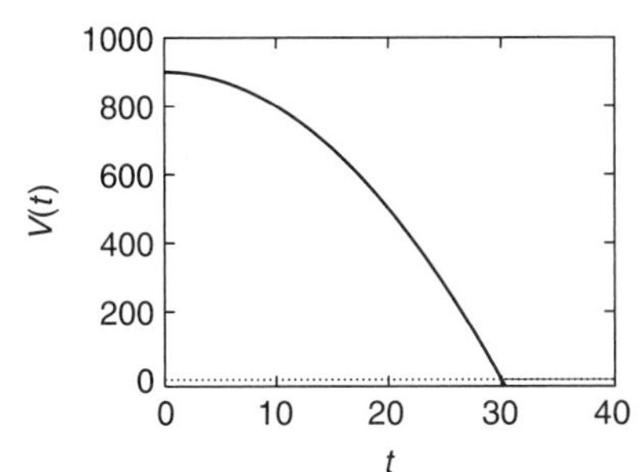

17. a.

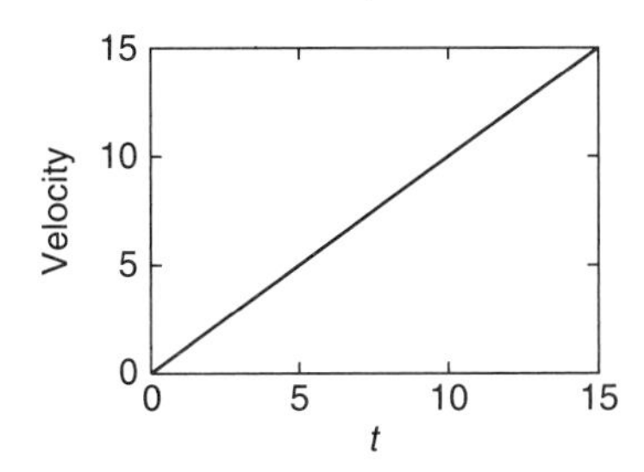

b. $\frac{dp}{dt} = t$.

c. The function $p(t) = t^2$ has derivative $2t$, so we divide by 2 to guess $p(t) = t^2/2$.

d.

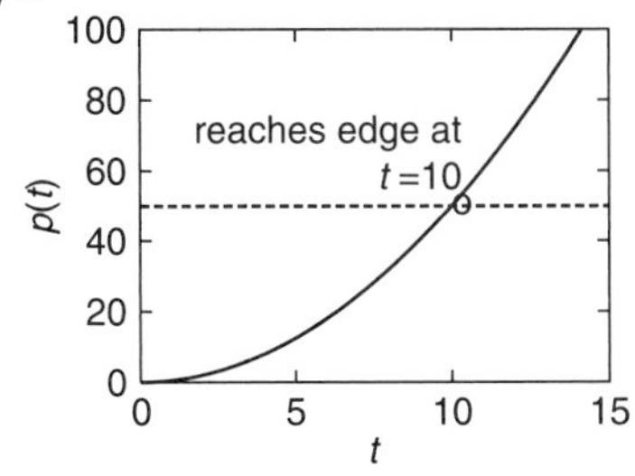

e. It reaches the other side when $p(t) = 50$, or when $t^2/2 = 50$. This has a solution of 10 min.

19. This cell follows the differential equation $\frac{dV}{dt} = -2$ with initial condition $V(0) = 600$. Then

$$\hat{V}(10) = V(0) + V'(0)\Delta t = 600 - 2 \cdot 10 = 580$$
$$\hat{V}(20) = \hat{V}(10) + V'(10)\Delta t = 580 - 2 \cdot 10 = 560$$
$$\hat{V}(30) = \hat{V}(20) + V'(20)\Delta t = 560 - 2 \cdot 10 = 540$$
$$\hat{V}(40) = \hat{V}(30) + V'(30)\Delta t = 540 - 2 \cdot 10 = 520.$$

The solution we guessed was $V(t) = 600 - 2t$, and $V(40) = 520$. In this case, Euler's method gives exactly the right answer.

21. This cell follows the differential equation $\frac{dV}{dt} = -2t$ with initial condition $V(0) = 900$. Then

$$\hat{V}(10) = V(0) + V'(0)\Delta t = 900 - 2 \cdot 0 \cdot 10 = 900$$
$$\hat{V}(20) = \hat{V}(10) + V'(10)\Delta t = 900 - 2 \cdot 10 \cdot 10 = 700$$
$$\hat{V}(30) = \hat{V}(20) + V'(20)\Delta t = 700 - 2 \cdot 20 \cdot 10 = 300.$$

The solution we guessed was $V(t) = 900 - t^2$, and $V(30) = 0$. Euler's method is way off.

23. This snail follows the differential equation $\frac{dp}{dt} = t$ with initial condition $p(0) = 0$. Then

$$\hat{p}(2) = p(0) + p'(0)\Delta t = 0 + 0 \cdot 2 = 0$$
$$\hat{p}(4) = \hat{p}(2) + p'(2)\Delta t = 0 + 2 \cdot 2 = 4$$
$$\hat{p}(6) = \hat{p}(4) + p'(4)\Delta t = 4 + 4 \cdot 2 = 12$$
$$\hat{p}(8) = \hat{p}(6) + p'(6)\Delta t = 12 + 6 \cdot 2 = 24$$
$$\hat{p}(10) = \hat{p}(8) + p'(8)\Delta t = 24 + 8 \cdot 2 = 40.$$

The solution we guessed was $p(t) = \frac{t^2}{2}$, and $p(10) = 50$. Euler's method is in the right ball park.

25. The derivative of e^{-t+1} is $-e^{-t+1}$. We need to multiply by -1, so the derivative of $-e^{-t+1}$ is e^{-t+1}. If $P(t) = -e^{-t+1}$, then $P(0) = -e$, which is e too low. We need to add e. $P(t) = e - e^{-t+1}$. The derivative is $\frac{dP}{dt} = e^{-t+1}$ and $P(0) = e - e = 0$. This checks.

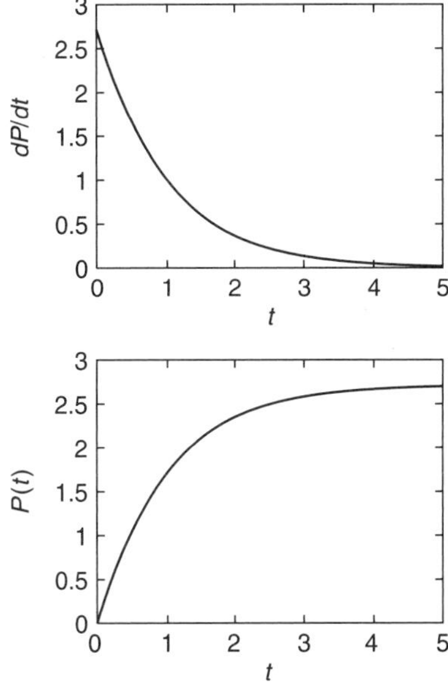

27. a. You could measure position with a map even if your speedometer was broken. b. You could measure speed but not position if your speedometer worked but you were lost.

29. a. You could measure total sodium with a destructive device that separated out sodium. b. You could measure sodium entering a cell but not total sodium if you could track the change in charge.

Section 4.2, page 364

1.

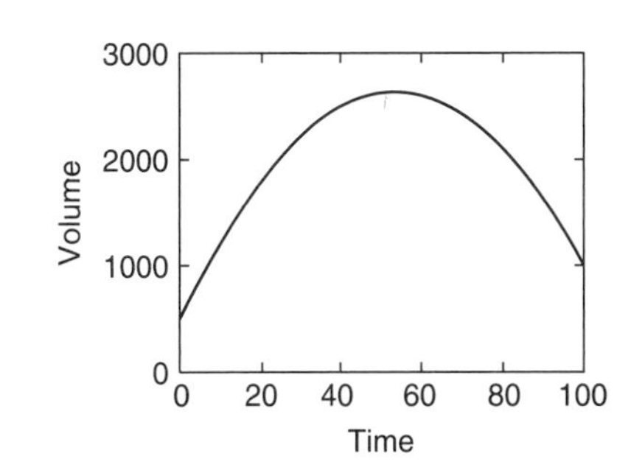

3.

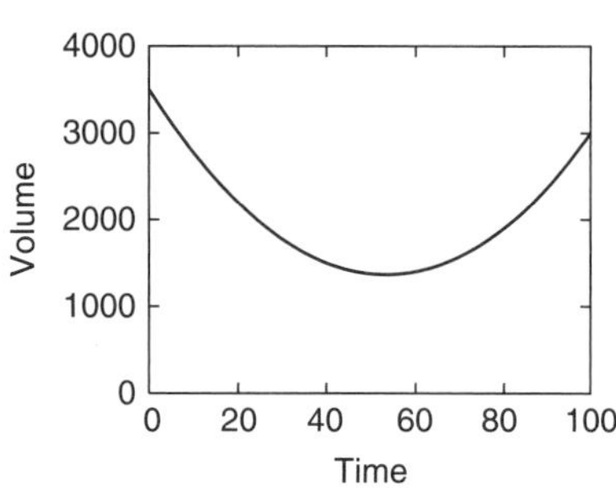

5.

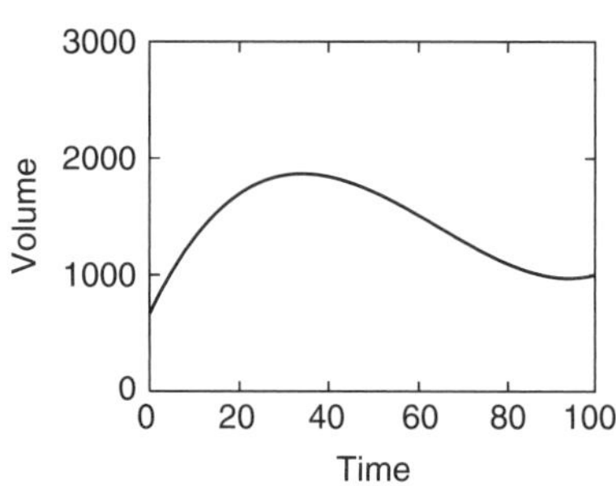

7.

$$\begin{aligned}\int 7x^2dx &= 7\int x^2dx &&\text{constant product rule}\\ &= \frac{7x^3}{3}+c &&\text{power rule}\end{aligned}$$

9.

$$\begin{aligned}\int 72t + 5dt &= 72\int tdt + 5\int dt &&\text{constant product and}\\ &= 36t^2 + 5t + c &&\text{sum rules power rule}\end{aligned}$$

11.

$$\begin{aligned}\int 5x^{-3}dx &= 5\int x^{-3}dx &&\text{constant product rule}\\ &= \frac{-5x^{-2}}{2}+c &&\text{power rule}\end{aligned}$$

13.

$$\begin{aligned}\int \frac{2}{\sqrt[3]{t}} + 3dt &= 2\int t^{-1/3}dt + 3\int dt &&\text{constant product and sum}\\ &= 3t^{2/3} + 3t + c. &&\text{rules power rule}\end{aligned}$$

15. Integrating, we find that $V(t) = 2t^3/3 + 5t + c$. Substituting the initial condition, $V(1) = 2/3 + 5.0 + c = 19.0$ so $c = 40/3$. The solution is $V(t) = 2t^3/3 + 5t + 40/3$.

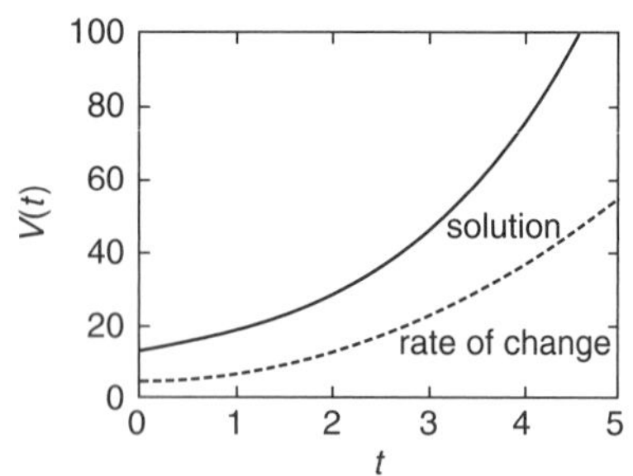

17. Integrating, we find that $f(t) = 1.25t^4 + 2.5t^2 + c$. Substituting the initial condition, $f(0) = c = -12.0$ so $c = -12$. The solution is $f(t) = 1.25t^4 + 2.5t^2 - 12.0$.

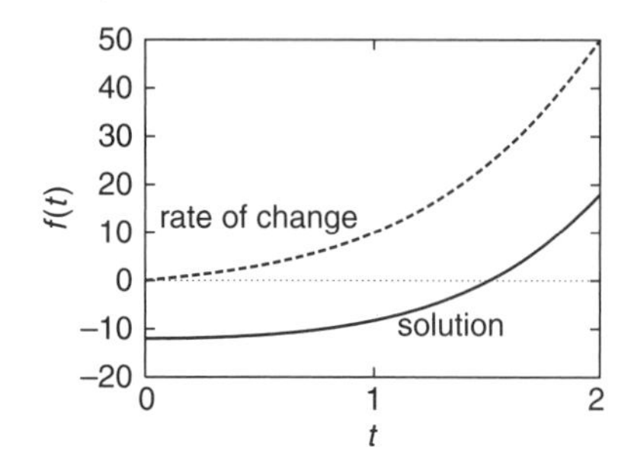

19. Integrating, we find that $M(t) = \frac{t^3}{3} - \frac{1}{t} + c$. Substituting the initial condition, $M(3) = \frac{27}{3} - \frac{1}{3} + c = 10.0$, so $c = 1.33$.

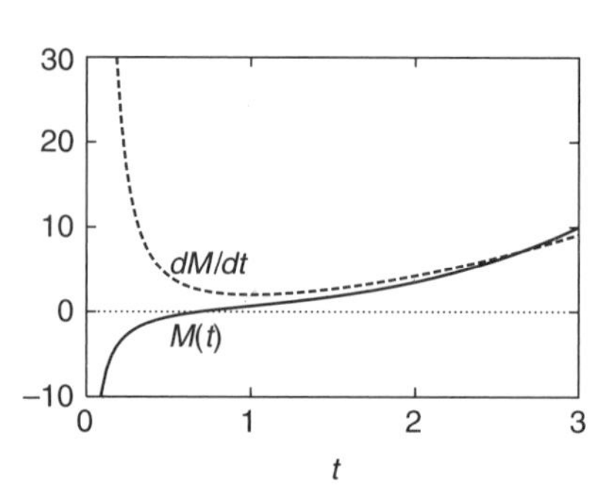

21. Let $F(x) = \int f(x)dx = \frac{x^3}{3} + c$, and $G(x) = \int g(x)dx = \frac{x^4}{4} + c$. The product of the functions is $f(x)g(x) = x^5$, with integral $\int x^5dx = \frac{x^6}{6} + c \neq F(x)G(x)$.

23.
a. $V_1(t) = t^2 + 5.0t + 10.0$, $V_2(t) = 2.5t^2 + 2.0t + 10.0$.
b. $\frac{dV}{dt} = 7.0t + 7.0$ with initial condition $V(0) = V_1(0) + V_2(0) = 20$.
c. $V(t) = 3.5t^2 + 7.0t + 20.0$. This is indeed the sum of $V_1(t)$ and $V_2(t)$.

25.
a. The units of α must be $\frac{\text{grams}}{\text{day}^2}$.
b. $M(t) = t^2 + c$. Substituting the initial conditions, we have $M(0) = c = 5.0$ so $M(t) = t^2 + 5.0$.
c.

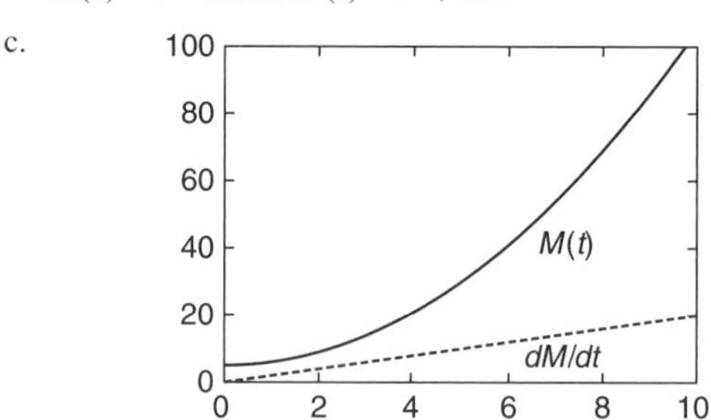

d. The mass increases more and more quickly.

27.
a. Integrating once gives $v(t) = -9.8t + c$. The initial condition $v(0) = 5.0$ implies that $c = 5.0$, so $v(t) = -9.8t + 5.0$. Integrating again gives $p(t) = -4.9t^2 + 5.0t + c$. The initial condition $p(0) = 100$ implies that $c = 100$, so $p(t) = -4.9t^2 + 5.0t + 100$.
b. The maximum height is when $v(t) = 0$, or at $t = \frac{5.0}{9.8} = 0.51$. At this time,

$$p(0.51) = -4.9(0.51)^2 + 5.0 \cdot 0.51 + 100 = 101.27.$$

c. It passes 100 m when $p(t) = 100$. This has solutions

$$\begin{aligned}p(t) = -4.9t^2 + 5.0t + 100 &= 100\\ -4.9t^2 + 5.0t &= 0\\ t(-4.9t + 5.0) &= 0\end{aligned}$$

where $t = 0$ and $t = 1.02$. At this time, the velocity $v(1.02) = -9.8 \cdot 1.02 + 5 = -5.0$ m/s.
d. It hits the ground when $p(t) = 0$, found by solving $p(t) = -4.9t^2 + 5.0t + 100 = 0$. With the quadratic formula, the positive root is 5.06 s to hit the ground, and will be moving $v(5.06) = -9.8 \cdot 5.06 + 5.0 = -44.55$ m/s (which is about 99.65 mph in the downward direction).

e.

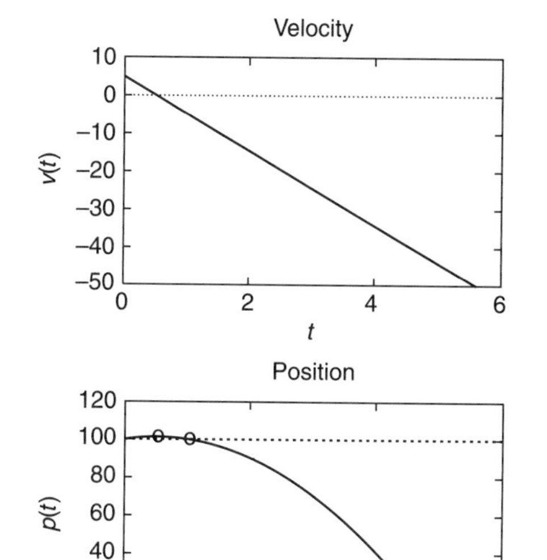

29. a. $v(t) = -22.88t + 5.0$, $p(t) = -11.44t^2 + 5.0t + 100$.
b. The maximum height is when $v(t) = 0$, or at $t = 0.22$. The position is 100.55 m.
c. It passes 100 m when $p(t) = 100$, or at $t = 0$ and $t = 0.44$. The velocity is -5.0 m/s.
d. It take 3.18 s to hit the ground, and will be moving at -67.83 m/s (which is about 151.7 mph in the downward direction).
e.

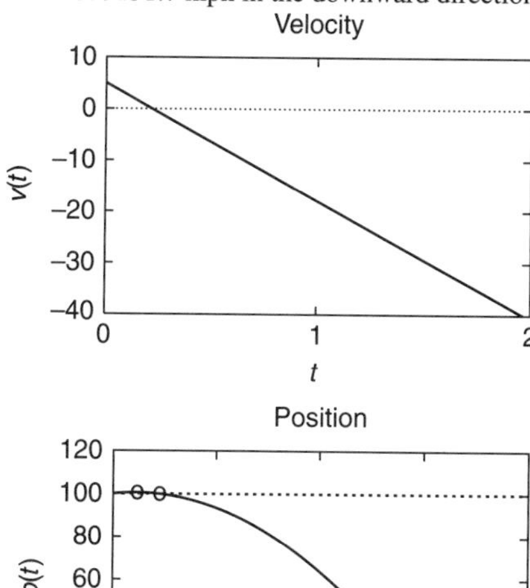

31. With Euler's method, $\hat{p}(1) = 11.0$, $\hat{p}(2) = 14.0$, $\hat{p}(3) = 19.0$, $\hat{p}(4) = 26.0$. The velocities fall on the line $v(t) = 1.0 + 2.0t$. Therefore, position satisfies the differential equation $\frac{dp}{dt} = 1.0 + 2.0t$. This has solution $p(t) = 1.0t + t^2 + 10.0$ when $p(0) = 10$. At $t = 4$, the exact solution is $p(t) = 30.0$.

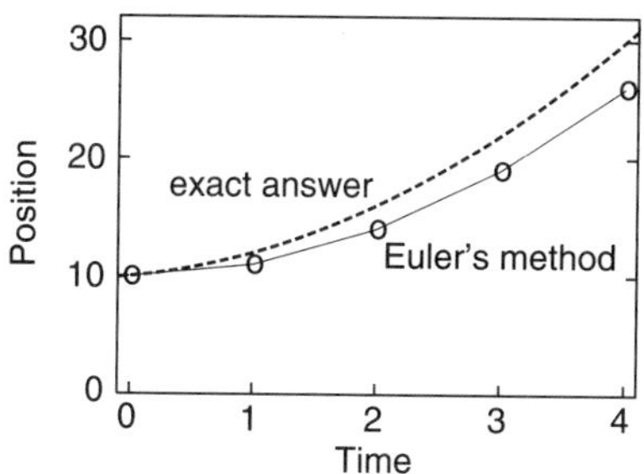

33. With Euler's method, $\hat{p}(1) = 35.0$, $\hat{p}(2) = 51.0$, $\hat{p}(3) = 60.0$, $\hat{p}(4) = 64.0$. The velocities fall on the quadratic $v(t) = (t - 5.0)^2 = t^2 - 10.0t + 25.0$. The position satisfies the differential equation $\frac{dp}{dt} = t^2 - 10t + 25$. This has solution $p(t) = \frac{t^3}{3} - 5.0t^2 + 25.0t + 10.0$ when $p(0) = 10$. At $t = 4$, the exact solution is $p(t) = 51.333$.

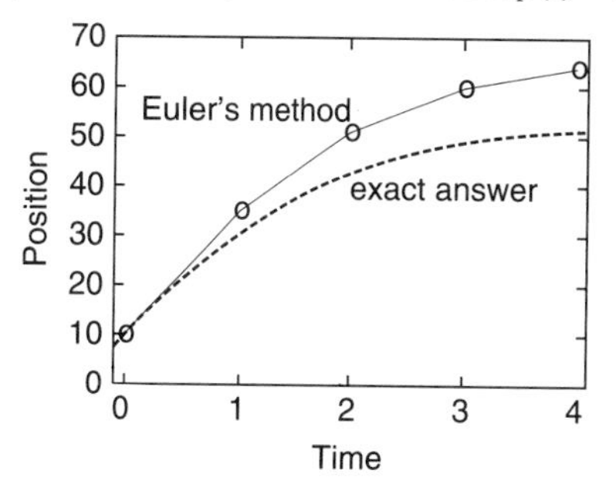

35. We estimate the velocity to be 2.0 during the first minute, 4.0 during the second, 6.0 during the third, and 8.0 during the fourth. Then $\hat{p}(1) = 12.0$, $\hat{p}(2) = 16.0$, $\hat{p}(3) = 22.0$, $\hat{p}(4) = 30.0$. At $t = 4$, this matches the exact solution of $p(t) = 30.0$.

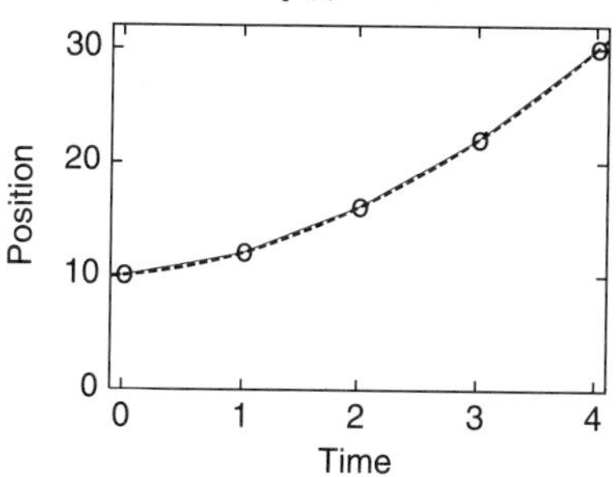

37. We estimate the velocity to be 20.5 during the first minute, 12.5 during the second, 6.5 during the third, and 2.5 during the fourth. Then $\hat{p}(1) = 30.5$, $\hat{p}(2) = 43.0$, $\hat{p}(3) = 49.5$, $\hat{p}(4) = 52.0$. At $t = 4$, the exact solution is $p(t) = 51.333$, so this is much closer.

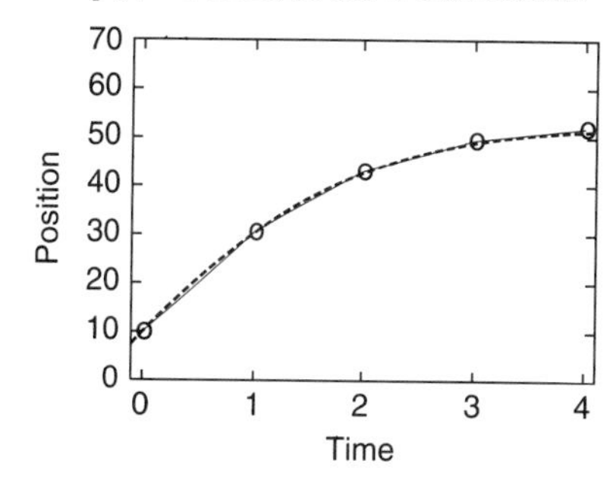

Section 4.3, page 376

1. This can be computed with the sum and power rules as

$$\int \frac{3}{z^2} + \frac{z^2}{3} dz = \int \frac{3}{z^2} dz + \int \frac{z^2}{3} dz = \int \frac{3}{z^2} dz + \int \frac{z^2}{3} dz$$
$$= \frac{-3}{z} + \frac{z^3}{9} + c.$$

3. This can computed with the sum, exponential, and log rules as

$$\int e^x + \frac{1}{x} dx = \int e^x dx + \int \frac{1}{x} dx = e^x + \ln(|x|) + c.$$

5. This can computed with the rules for trigonometric functions as

$$\int 2\sin(x) + 3\cos(x) dx = \int 2\sin(x) dx + \int 3\cos(x) dx$$
$$= -2\cos(x) + 3\sin(x) + c.$$

7. a. Define a new variable $y = \frac{x}{5}$.

b. $\frac{dy}{dx} = \frac{1}{5}$.

c. $dy = \frac{dx}{5}$.

d. To write dx in terms of dy, solve to find $dx = 5dy$ and write

$$\int 3e^{\frac{x}{5}}dx = \int 15e^{y}dy.$$

e. Integrate, finding

$$\int 15e^{y}dy = 15e^{y} + c$$

f. Put everything back in terms of x,

$$\int 3e^{\frac{x}{5}}dx = 15e^{\frac{x}{5}} + c.$$

The derivative of this result is $3e^{\frac{x}{5}}$.

9. Use the substitution $s = 1 + \frac{t}{2}$, to find

$$\int \left(1+\frac{t}{2}\right)^4 dt = 2\frac{\left(1+\frac{t}{2}\right)^5}{5} + c.$$

The derivative of this result is $\left(1+\frac{t}{2}\right)^4$.

11. Use the substitution $y = 4 + t$, and find

$$\int \frac{1}{4+t}dt = \ln(|4+t|) + c.$$

The derivative of this result is $\frac{1}{4+t}$.

13. Substitute $y = 1 + e^x$, finding $dy = e^x\,dx$ so

$$\int \frac{e^x}{1+e^x}dx = \int \frac{1}{1+y}dy$$
$$= \ln(1+y) + c = \ln(1+e^x) + c.$$

15. Using the substitution $z = 1 + y^2$, we find that the integral is $\frac{2}{3}(1+y^2)^{3/2} + c$.

17. Using the substitution $y = \cos(\theta)$, we find that the integral is $-\ln(|\cos(x)|)$.

19. Using the substitution $y = \ln(x)$, we find that the integral is $\ln(|\ln(x)|) + c$.

21. Let $u(x) = x$ and $\frac{dv}{dx} = e^{2x}$, then $\frac{du}{dx} = 1$ and $v(x) = \frac{e^{2x}}{2}$ (using the substitution $y = 2x$ to eliminate the constant 2 in the power). Then,

$$\int xe^{2x}dx = \frac{xe^{2x}}{2} - \int \frac{e^{2x}}{2}dx = \frac{xe^{2x}}{2} - \frac{e^{2x}}{4} + c,$$

where we again used the substitution $y = 2x$ in the last step. Checking,

$$\frac{d\left(\frac{xe^{2x}}{2} - \frac{e^{2x}}{4} + c\right)}{dx} = \frac{e^{2x}}{2} + xe^{2x} - \frac{e^{2x}}{2} = xe^{2x}.$$

23. Pick $u(x) = x^2$ and $\frac{dv}{dx} = e^x$. Then $v(x) = e^x$ and $\frac{du}{dx} = 2x$, giving

$$\int x^2e^xdx = x^2e^x - \int 2xe^xdx = x^2e^x - 2xe^x + 2e^x + c$$

where we used the result from Example 4.3.9. Checking,

$$\frac{d(x^2e^x - 2xe^x + 2e^x + c)}{dx} = x^2e^x + 2xe^x - 2xe^x - 2e^x + 2e^x$$
$$= x^2e^x.$$

25. We can factor the denominator as $1 - x^2 = (1-x)(1+x)$, and then expand $\frac{2}{1-x^2} = \frac{1}{1+x} + \frac{1}{1-x}$. Integrating gives

$$\int \frac{2}{1-x^2}dx = \int \frac{1}{1+x}dx + \int \frac{1}{1-x}dx$$
$$= \ln(|1+x|) - \ln(|1-x|) + c = \ln\left(\frac{1+x}{1-x}\right) + c.$$

27. If $x = \tan(\theta)$, then $\frac{dx}{d\theta} = \sec^2(\theta)$. Also, $1 + x^2 = 1 + \tan^2(\theta) = \sec^2(\theta)$. Then

$$\int \frac{1}{1+x^2}dx = \int \frac{1}{1+\tan^2(\theta)}\sec^2(\theta)d\theta$$
$$= \int \frac{1}{\sec^2(\theta)}\sec^2(\theta)d\theta$$
$$= \int d\theta = \theta + c = \tan^{-1}(x) + c.$$

29. Section 2.10, Exercise 25 gives $\frac{d\tan^{-1}(x)}{dx} = \frac{1}{1+x^2}$. If we break up the integral by setting $u(x) = \tan^{-1}(x)$ and $\frac{dv}{dx} = 1$, we find

$$\int \tan^{-1}(x)dx = x\tan^{-1}(x) - \int \frac{x}{1+x^2}dx$$
$$= x\tan^{-1}(x) - \frac{1}{2}\ln(1+x^2) + c$$

where we used the substitution $y = 1 + x^2$ in the last step.

31. Set $u(x) = x$ and $\frac{dv}{dx} = e^x$. Then $v(x) = \int e^xdx = e^x + c$ and $\frac{du}{dx} = 1$, giving

$$\int xe^x = x(e^x + c) - \int (e^x + c)dx$$
$$= xe^x + cx - e^x - cx + c_1 = xe^x - e^x + c_1.$$

where c_1 is a new arbitrary constant. The original arbitrary constant c canceled, giving the previous answer.

33. Set $u(x) = \sin(x)$ and $\frac{dv}{dx} = e^x$. Then $v(x) = e^x$ and $\frac{du}{dx} = \cos(x)$, giving

$$\int e^x\sin(x)dx = e^x\sin(x) - \int e^x\cos(x)dx.$$

To evaluate this new integral, set $u(x) = \cos(x)$ and $\frac{dv}{dx} = e^x$. Then $v(x) = e^x$ and $\frac{du}{dx} = -\sin(x)$, giving

$$\int e^x\cos(x)dx = e^x\cos(x) + \int e^x\sin(x)dx.$$

Plugging into the original equation, gives

$$\int e^x\sin(x)dx = e^x\sin(x) - \int e^x\cos(x)dx$$
$$= e^x\sin(x) - e^x\cos(x) - \int e^x\sin(x)dx.$$

We can now solve for the original integral, giving

$$2\int e^x\sin(x)dx = e^x\sin(x) - e^x\cos(x) + c.$$

Dividing by 2 gives the result

$$\int e^x\sin(x)dx = \frac{e^x\sin(x) - e^x\cos(x)}{2} + c.$$

35. The Taylor series (from Example 3.7.9) is $e^x = 1 + x + \frac{x^2}{2!} + \frac{x^3}{3!} + \cdots$. Integrating gives

$$\int e^x dx = \int 1 dx + \int x dx + \int \frac{x^2}{2!} dx + \int \frac{x^3}{3!} dx + \cdots$$
$$= x + \frac{x^2}{2!} + \frac{x^3}{3!} + \cdots + c.$$

Choosing the constant $c = 1$ matches the Taylor series for e^x.

37. The Taylor series of e^x (from Example 3.7.9) is $e^x = 1 + x + \frac{x^2}{2!} + \frac{x^3}{3!} + \cdots$. Dividing by x and integrating gives

$$\int \frac{e^x}{x} dx = \int \frac{1}{x} dx + \int 1 dx + \int \frac{x}{2!} dx + \int \frac{x^2}{3!} dx + \cdots$$
$$= \ln(|x|) + x + \frac{x^2}{4} + \frac{x^3}{18} + \cdots + c.$$

This is a novel series that we do not recognize.

39. First we must break up the function with partial fractions as

$$\frac{A_1}{x+1} + \frac{A_2}{x+2} = \frac{A_1(x+2)}{(x+1)(x+2)} + \frac{A_2(x+1)}{(x+1)(x+2)}$$
$$= \frac{(A_1 + A_2)x + 2A_1 + A_2}{(x+1)(x+2)}.$$

We then set this numerator equal to the numerator of the original fraction,

$$x = (A_1 + A_2)x + 2A_1 + A_2.$$

To match each term, we require that $A_1 + A_2 = 1$ and $2A_1 + A_2 = 0$, giving $A_1 = -1$ and $A_2 = 2$. Then

$$\int \frac{x}{(x+1)(x+2)} dx = \int -\frac{1}{x+1} dx + \int \frac{2}{x+2} dx$$
$$= -\ln(|x+1|) + 2\ln(|x+2|) + c$$
$$= \ln\left(\frac{(x+2)^2}{|x+1|}\right) + c.$$

41. Integrating, $P(t) = 2.5\ln(1 + 2.0t) + c$. Substituting the initial condition, we find $c = 0$. The amount of product increases to infinity, but does so rather slowly. In fact, $P(10) = 7.61$ and $P(100) = 13.26$. This increases to infinity because the rate of chemical production decreases to zero rather slowly.

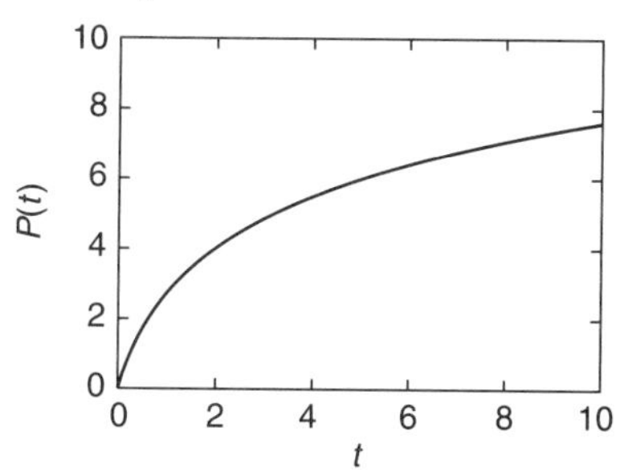

43. Integrating, $P(t) = 2.5\ln(1 + t) - 2.5e^{-t} + c$. Substituting the initial condition, we find $c = 2.5$. The limit is infinity as t approaches infinity. In fact, $P(10) = 8.49$ and $P(100) = 14.04$. The leading behavior of the rate of chemical production decreases to zero rather slowly, like $\frac{1}{1+t}$.

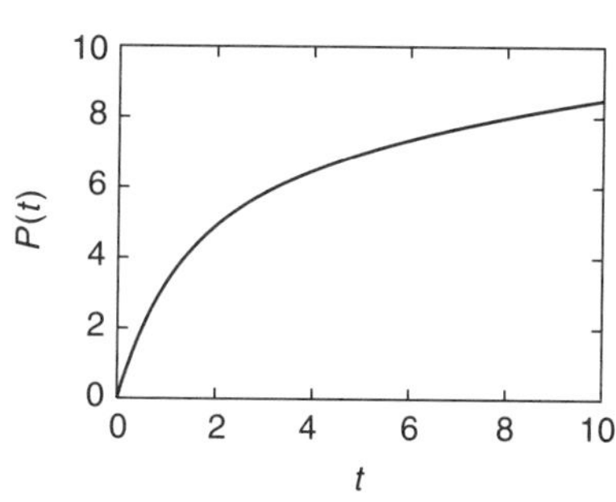

45. Let $g(t) = (t + t^2)e^{-2t}$. To find the time when the toad grows fastest, we compute

$$g'(t) = (1 - 2t^2)e^{-2t}.$$

This is zero when $1 - 2t^2 = 0$ or $t = \frac{\sqrt{2}}{2} = 0.71$. At this point, $g(0.71) = 0.293$. To find the size at $t = 1$, let $u(t) = t + t^2$ and $\frac{dv}{dt} = e^{-2t}$. Then $\frac{du}{dt} = (1 + 2t)$ and $v(t) = -\frac{e^{-2t}}{2}$. Therefore

$$\int_0^1 (t + t^2)e^{-2t} = -(t + t^2)\frac{e^{-2t}}{2}\Big|_0^1 + \int_0^1 (1 + 2t)\frac{e^{-2t}}{2}.$$

To integrate the second piece, let $u(t) = 1 + 2t$ and $\frac{dv}{dt} = \frac{e^{-2t}}{2}$. Then $\frac{du}{dt} = 2$ and $v(t) = -\frac{e^{-2t}}{4}$. Therefore

$$\int_0^1 (1 + 2t)e^{-2t} = -(1 + 2t)\frac{e^{-2t}}{4}\Big|_0^1 + \int_0^1 \frac{e^{-2t}}{2}$$
$$= -(1 + 2t)\frac{e^{-2t}}{4}\Big|_0^1 - \frac{e^{-2t}}{4}\Big|_0^1.$$

Evaluating each term gives $M(1) = 2e^{-2} + \frac{1}{2} = 0.23$. This is about 20% smaller than the toad would be if it grew at its maximum rate the entire time.

47.
a. $L(t) = 54.0(1 - e^{-1.19t})$.
b. The limit is 54.0 cm.
c. These walleye reach maturity when $L(t) = 45$. To solve this,

$$54.0(1 - e^{-1.19t}) = 45.0$$
$$1 - e^{-1.19t} = \frac{45.0}{54.0}$$
$$e^{-1.19t} = 1 - \frac{45.0}{54.0}$$
$$-1.19t = \ln\left(1 - \frac{45.0}{54.0}\right)$$
$$t = \frac{\ln\left(1 - \frac{45.0}{54.0}\right)}{-1.19} = 1.506.$$

d. These walleye reach maturity much more quickly, but then they more or less stop growing.

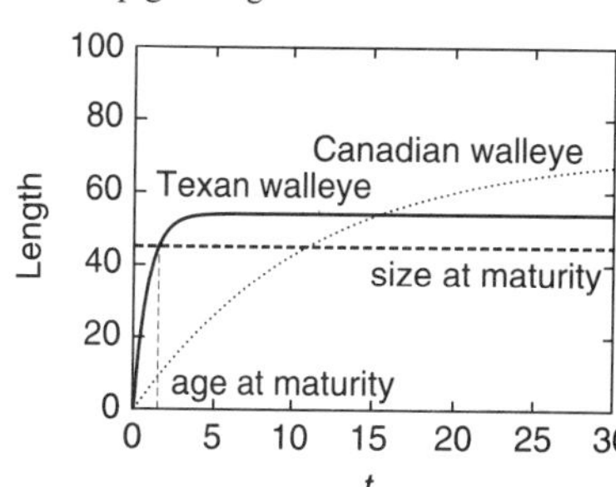

49.
a. $\frac{dB}{dt} = 100e^{0.2t}$.
b. Integrating, $B(t) = 500e^{0.2t} + c$. The initial condition implies $c = -500$, so $B(t) = 500e^{0.2t} - 500$.
c.

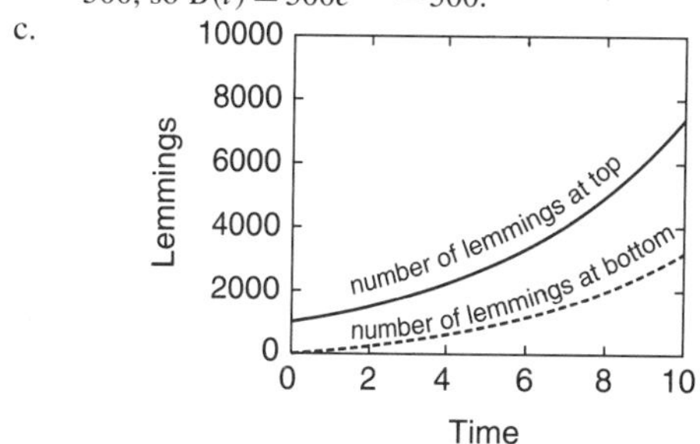

d. The limit is 0.5. There are about half as many lemmings at the bottom as at the top.

51. a.
$$\frac{dB}{dt} = 10t.$$

b. Integrating, $B(t) = 5t^2 + c$. The initial condition implies that $c = 0$, so $B(t) = 5t^2$.

c.

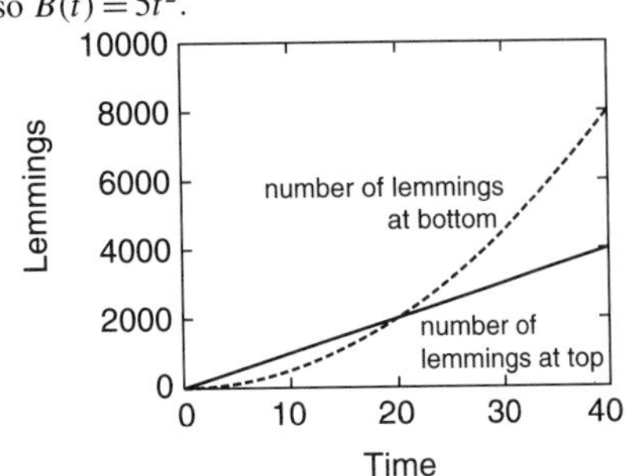

d. The limit of the ratio is infinite. This is probably because the lemmings at the top are reproducing quite slowly.

53. a.

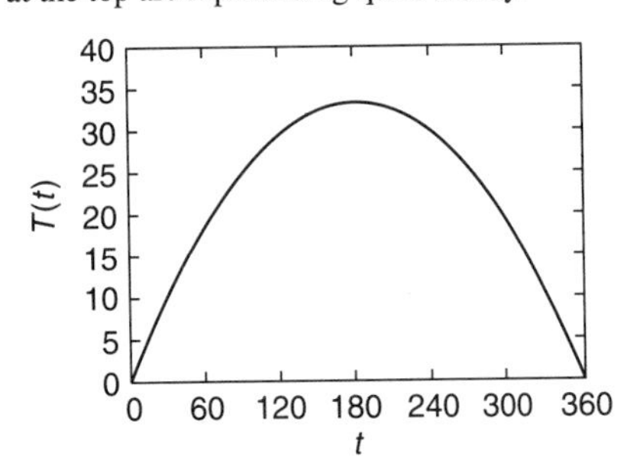

b. The solution is

$$L(t) = \int 0.000001t(365 - t)dt = \int 0.000365t - 0.000001t^2 dt$$
$$= 0.0001825t^2 - 0.00000033t^3 + c$$

Using the initial condition $T(0) = 0.1$, we find $c = 0.1$. The size at $t = 30$ is then 0.25525 cm.

c. Using the initial condition $T(150) = 0.1$, we find that the constant is $c = -2.88125$, and $L(180) = 10.8775$. This bug is a lot bigger because the temperature is warmer.

55. a.

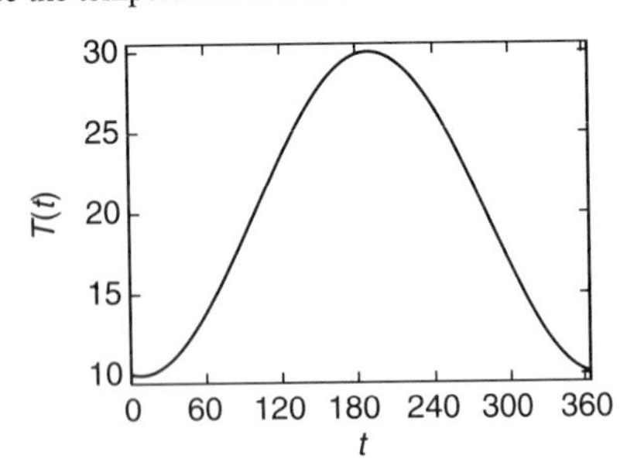

b. The solution is

$$L(t) = \int 0.02 + 0.01 \cos\left[\frac{2\pi(t - 190.0)}{365}\right] dt$$
$$= 0.02t + 0.01 \sin\left[\frac{2\pi(t - 190.0)}{365}\right]\frac{365}{2\pi} + c$$

Using the initial condition $T(0) = 0.1$, we find $c = 0.0252$. At $t = 30$, $L(30.0) = 0.4058$.

c. Using the initial condition $T(150) = 0.1$, we find $c = -2.531$. After 30 days, at $t = 180$, $L(180.0) = 0.969$. This bug is bigger.

Section 4.4, page 386

1. $1 + 1/2 + 1/3 + 1/4 + 1/5 = 2.2833$.

3. $1 + 1/4 + 1/9 + 1/16 + 1/25 = 1.4636$.

5. $\Delta t = 0.4$, $t_0 = 0$, $t_1 = 0.4$, $t_2 = 0.8$, $t_3 = 1.2$, $t_4 = 1.6$, $t_5 = 2.0$.

7. $\Delta t = 0.2$, $t_0 = 2.0$, $t_1 = 2.2$, $t_2 = 2.4$, $t_3 = 2.6$, $t_4 = 2.8$, $t_5 = 3.0$.

9.

	Left-Hand Estimate			Right-Hand Estimate		
Time	Rate	Influx	Net Influx	Rate	Influx	Net Influx
0.2	0.0	0.00	0.00	0.4	0.08	0.08
0.4	0.4	0.08	0.08	0.8	0.16	0.24
0.6	0.8	0.16	0.24	1.2	0.24	0.48
0.8	1.2	0.24	0.48	1.6	0.32	0.80
1.0	1.6	0.32	0.80	2.0	0.40	1.20

11.

	Left-Hand Estimate			Right-Hand Estimate		
Time	Rate	Influx	Net Influx	Rate	Influx	Net Influx
0.4	1.00	0.400	0.400	1.064	0.4256	0.4256
0.8	1.064	0.4256	0.8256	1.512	0.6048	1.0304
1.2	1.512	0.6048	1.4304	2.728	1.0192	2.1216
1.6	2.728	1.0192	2.5216	5.096	2.0384	4.1600
2.0	5.096	2.0384	4.5600	9.000	3.6000	7.7600

13. $I_l = \sum_{i=0}^{4} 2t_i \cdot 0.2$, $I_r = \sum_{i=1}^{5} 2t_i \cdot 0.2$.

15. $I_l = \sum_{i=0}^{4} t_i^2 \cdot 0.4$, $I_r = \sum_{i=1}^{5} 1 + t_i^3 \cdot 0.4$.

17. a.

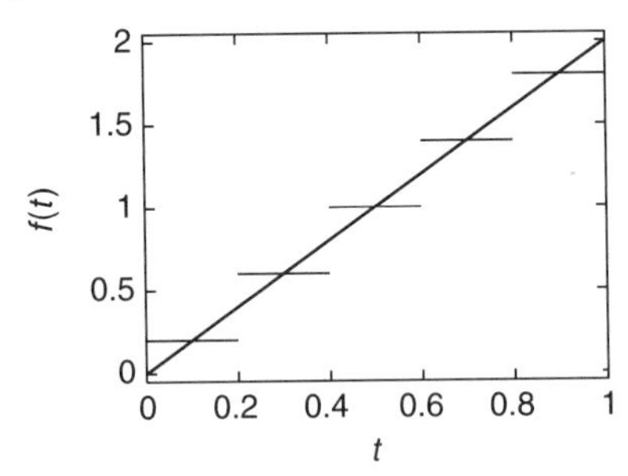

b. $I_a = \sum_{i=0}^{4}(t_i + t_{i+1}) \cdot 0.2$.

c. The sum is 1.0.

19. a.

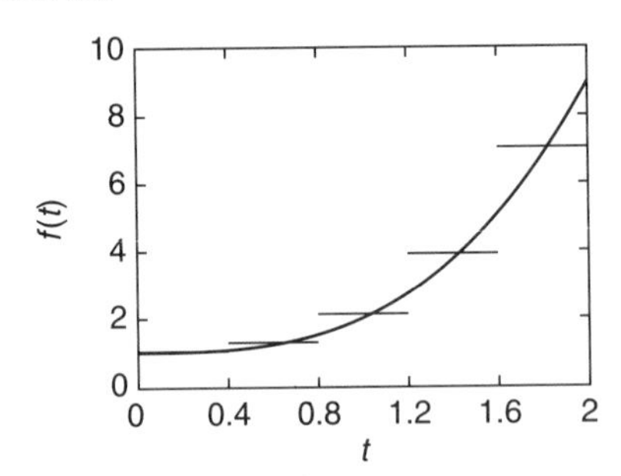

b. $I_a = \sum_{i=0}^{4} \frac{(1 + t_i^3) + (1 + t_{i+1}^3)}{2} \cdot 0.4$.

c. The sum is 6.16.

21. a.

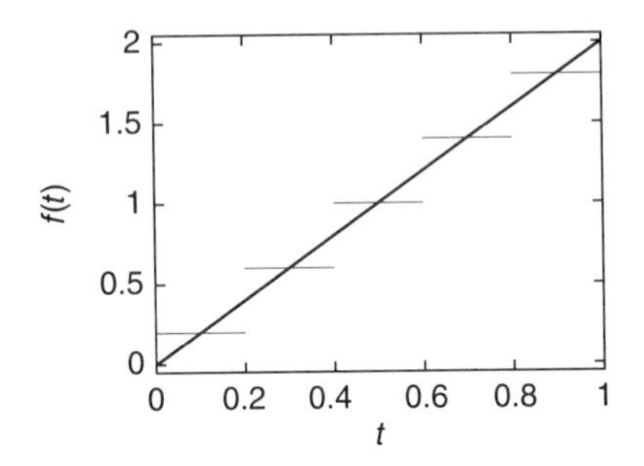

b. $I_a = \sum_{i=0}^{4}(t_i + t_{i+1}) \cdot 0.2$.

c. The sum is 1.0.

23. a.

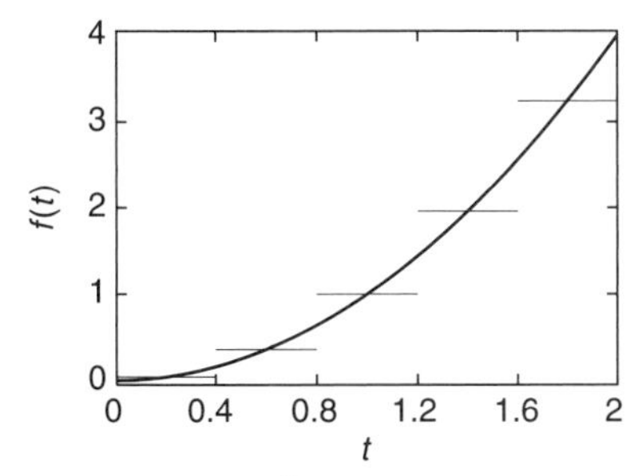

b. $I_a = \sum_{i=0}^{4} 1 + \left(\frac{t_i + t_{i+1}}{2}\right)^3 \cdot 0.4.$

c. The sum is 5.92.

25. Let B_i represent the number of offspring in year i, then the total number of offspring T is $T = \sum_{i=1}^{5} B_i = 2 + 3 + 5 + 4 + 1 = 15$.

27. The total number of offspring T is $T = \sum_{i=0}^{6} B_i = \sum_{i=0}^{6} i(6-i) = 0 + 5 + 8 + 9 + 8 + 5 + 0 = 35$.

29.
$$\begin{aligned}
\hat{V}(0.2) &= V(0) + 2 \cdot 0.0 \cdot 0.2 = 0.0\\
\hat{V}(0.4) &= \hat{V}(0.2) + 2 \cdot 0.2 \cdot 0.2 = 0.08\\
\hat{V}(0.6) &= \hat{V}(0.4) + 2 \cdot 0.4 \cdot 0.2 = 0.24\\
\hat{V}(0.8) &= \hat{V}(0.6) + 2 \cdot 0.6 \cdot 0.2 = 0.48\\
\hat{V}(1.0) &= \hat{V}(0.8) + 2 \cdot 0.8 \cdot 0.2 = 0.8.
\end{aligned}$$

It matches.

31.
$$\begin{aligned}
\hat{V}(0.4) &= V(0) + (1 + 0.0^3) \cdot 0.4 = 0.4\\
\hat{V}(0.8) &= \hat{V}(0.4) + (1 + 0.4^3) \cdot 0.4 = 0.8256\\
\hat{V}(1.2) &= \hat{V}(0.8) + (1 + 0.8^3) \cdot 0.4 = 1.4304\\
\hat{V}(1.6) &= \hat{V}(1.2) + (1 + 1.2^3) \cdot 0.4 = 2.5216\\
\hat{V}(2.0) &= \hat{V}(1.6) + (1 + 1.6^3) \cdot 0.4 = 4.5600.
\end{aligned}$$

Euler's method gives the same answer as the left-hand estimate.

33.
$$\begin{aligned}
I_l &= (127.0 + 118.0 + 113.0 + 112.0 + 116.0) \cdot 2.0 = 1172\\
I_r &= (118.0 + 113.0 + 112.0 + 116.0 + 125.0) \cdot 2.0 = 1168.
\end{aligned}$$

35. The method of Section 4.4, Exercises 17–20 applied to all the measurements gives 1167, as the average of the left-hand and right-hand measurements.

37. The modified left-hand estimate might fill in the last measured value for each of the NA's, giving $12 \cdot 2 + 16 + 17 \cdot 4 + 13 = 121$, multiplying each measurement by the time interval until the next measurement. The total number counted in the four years is 58, so we would estimate 116 with this method. This gives a lower answer because there were three years with NA's following the highest measurement of 17 in 1993.

39. We fill in the NA's using the next measurement, finding $34 \cdot 2 + 40 \cdot 3 + 31 + 37 \cdot 2 = 293$. This is equivalent to multiplying each measurement by the time until the previous measurement. The total number counted in the four years is 142, so we estimate 284 with this method.

Section 4.5, page 397

1. $\int_0^1 2t\,dt = t^2|_0^1 = 1^2 - 0^2 = 1.0$. This is close to the answer to Section 4.4, Exercise 9 and matches the answer to Section 4.4, Exercise 17.

3. $\int_0^2 t^2 dt = \frac{t^3}{3}|_0^2 = \frac{2^3}{3} - \frac{0^3}{3} = 2.667$. This is pretty close to the answer of 2.72 in Section 4.4, Exercise 19.

5. $\int_0^1 7x^2\,dx = \frac{7x^3}{3}|_0^1 = \frac{7}{3}$.

7. $\int_{-1}^{2} 72t + 5\,dt = 36t^2 + 5t|_{-1}^{2} = (36 \cdot 2^2 + 5 \cdot 2) - [36 \cdot (-1)^2 + 5 \cdot (-1)] = 123$.

9. $\int_1^2 \frac{5}{x^3}\,dx = \frac{-5x^{-2}}{2}|_1^2 = \frac{-5 \cdot 2^{-2}}{2} = \frac{-5 \cdot 1^{-2}}{2} = \frac{15}{8}$.

11. $\int_1^8 \frac{2}{\sqrt[3]{t}} + 3\,dt = 3t^{2/3} + 3t|_1^8 = 30$.

13. $\int_2^3 \frac{3}{z^2} + \frac{z^2}{3}\,dz = \frac{-3}{z} + \frac{z^3}{9}|_2^3 = 2.61$.

15. $\int_1^4 e^x + \frac{1}{x}\,dx = e^x + \ln(|x|)|_1^4 = 53.27$.

17. $\int_0^{\pi} 2\sin(x) + 3\cos(x)\,dx = -2\cos(x) + 3\sin(x)|_0^{\pi} = 4$.

19.
$$\int_1^2 t^2\,dt = \frac{t^3}{3}|_1^2 = \frac{7}{3}$$
$$\int_2^3 t^2\,dt = \frac{t^3}{3}|_2^3 = \frac{19}{3}$$
$$\int_1^3 t^2\,dt = \frac{t^3}{3}|_1^3 = \frac{26}{3}.$$

It is true that $\int_1^2 t^2\,dt + \int_2^3 t^2\,dt = \int_1^3 t^2\,dt$ because $\frac{26}{3} = \frac{7}{3} + \frac{19}{3}$.

21. $\int_1^2 L(t)dt = 1.875, \int_2^3 L(t)dt = 0.347, \int_1^3 L(t)dt = 2.222 = 1.875 + 0.347$.

23. $\int_1^2 F(t)dt = 5.36,\ \int_2^3 F(t)dt = 13.10,\ \int_1^3 F(t)dt = 18.46 = 5.36 + 13.10$.

25. $\int_0^x s^2\,ds = \frac{s^3}{3}|_0^x = \frac{x^3}{3}$. But $\frac{d}{dx}\left(\frac{x^3}{3}\right) = x^2 = f(x)$. It worked.

27. Using the substitution $s = \frac{t}{2}$, to find

$$\int_{-1}^{x}\left(1 + \frac{t}{2}\right)^4 dt = 2\frac{\left(1 + \frac{t}{2}\right)^5}{5}|_{-1}^{x} = \frac{2}{5}\left(1 + \frac{x}{2}\right)^5 - \frac{1}{80}.$$

We can take the derivative of this result with the chain rule by setting $\frac{2}{5}\left(1 + \frac{x}{2}\right)^5 = g[h(x)]$ where $h(x) = 1 + \frac{x}{2}$ and $g(h) = \frac{2}{5}h^5 - \frac{1}{80}$ with $h'(x) = 1/2$ and $g'(h) = 2h^4$. Therefore

$$\frac{d}{dx}\left[\frac{2}{5}\left(1 + \frac{t}{2}\right)^5 - \frac{1}{80}\right] = g'(h)h'(x) = \frac{1}{2}2h(x)^4 = \left(1 + \frac{x}{2}\right)^4,$$

matching the original function.

29. Using the indefinite integral, we find that

$$p(t) = \int -9.8t - 5.0dt = -4.9t^2 - 5.0t + c.$$

The initial condition $p(0) = 200$ implies that $c = 200$, so the solution is $p(t) = -4.9t^2 - 5.0t + 200$. The change of position is $p(5) - p(1) = 52.5 - 190.1 = -137.6$. The definite integral is

$$\begin{aligned}
\int_1^5 -9.8t - 5.0\,dt &= -4.9t^2 - 5.0t|_1^5\\
&= -4.9 \cdot 5^2 - 5.0 \cdot 5 - -4.9 \cdot 1^2 - 5.0 \cdot 1\\
&= -137.6.
\end{aligned}$$

31. The solution of the differential equation is $L(t) = 59.0 - 54.0e^{-1.19t}$. The amount grown is $L(1.5) - L(0.5) = 27.72$, which is equal to $\int_{0.5}^{1.5} 64.3e^{-1.19t}\,dt$.

33. The solution of the differential equation is $P(t) = 2.5\ln(1 + 2.0t) + 2.0$. The amount produced is $P(10) - P(5) = 1.616$, which matches $\int_5^{10} \frac{5}{1+2.0t}\,dt$.

35. We found that $p(t) = -4.9t^2 - 5.0t + 200$. The change of position between times 1 and 3 is $p(3) - p(1) = -49.2$, while that between times 3 and 5 is $p(5) - p(3) = -88.4$. We found that $p(5) - p(1) = -137.6$, which is equal to $-49.2 - 88.4$.

37. $L(1.0) - L(0.5) = 13.36$, $L(1.5) - L(1.0) = 7.37$. These add to approximately 20.72.

39. a. The velocity follows $\frac{dv}{dt} = 12.0$, which has solution $v(t) = 12.0t$ if $v(0) = 0$. Then $\frac{dp}{dt} = 12.0t$ is a differential equation for position. This has solution $p(t) = 6.0t^2$ using the initial condition $p(0) = 0$.

b. $v(10) = 120$ and $p(10) = 600$.

c. The velocity follows $\frac{dv}{dt} = -9.8$, which has solution $v(t) = -9.8t + 218$ if $v(10) = 120$. Then $\frac{dp}{dt} = -9.8t + 218$ is a differential equation for position. This has solution $p(t) = -4.9t^2 + 218t - 1090$ using the initial condition $p(10) = 600$.

d. The maximum height is reached when $v = 0$ or when $-9.8t + 218 = 0$. This happens at $t = 22.24$. The height is then 1334 m. It rises more after running out of fuel, because the acceleration of gravity is weaker than the acceleration of the engine.

e. It hits the ground when $p(t) = 0$, or at $t = 38.75$. The velocity is -161.75 m/s.

Section 4.6, page 405

1. After using the substitution $y = \frac{x}{5}$, we find $\int_0^5 3e^{\frac{x}{5}}\,dx = 15e^{\frac{x}{5}}\,|_0^5 = 15e - 15 = 25.77$.

3. Using the substitution $s = \frac{t}{2}$, we find

$$\int_0^4 \left(1 + \frac{t}{2}\right)^4 dt = 2\frac{\left(1+\frac{t}{2}\right)^5}{5}\,|_0^4 = 96.8.$$

5. Using the substitution $y = 4 + t$, we find $\int_{-3}^{0} \frac{1}{4+t}\,dt = \ln(|4+t|)|_{-3}^{0} = \ln(4) = 1.386$.

7. $\int_0^3 3x^3\,dx = \frac{3x^4}{4}\,|_0^3 = 60.75$.

9. Use the substitution $y = x/2$. Then $dx = 2dy$ and the limits of integration are $y = 0$ to $y = \frac{\ln(2)}{2}$.

$$\int_0^{\ln(2)} e^{x/2}dx = \int_0^{\frac{\ln(2)}{2}} 2e^y dy = 2(\sqrt{2} - 1) = 0.828$$

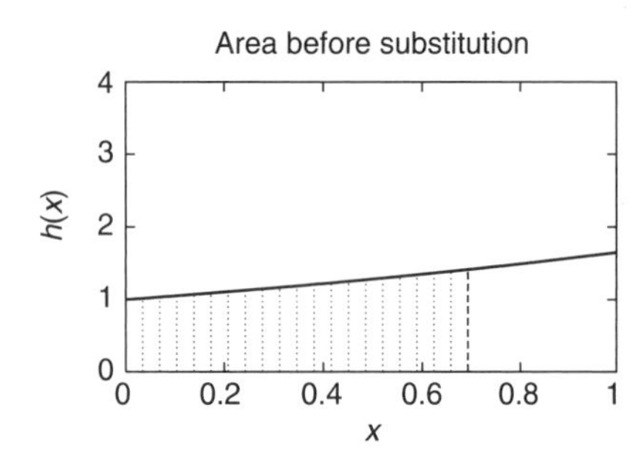

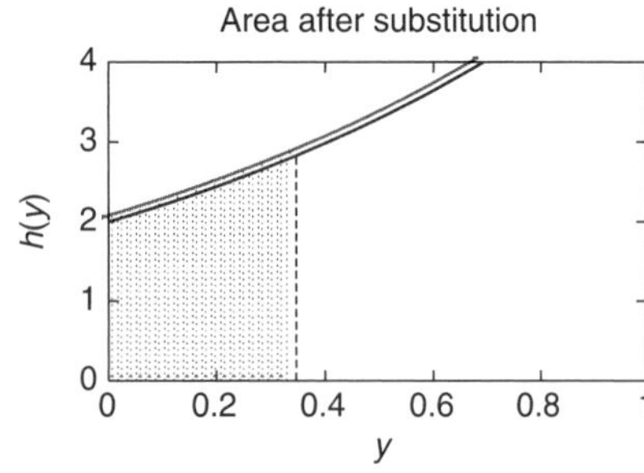

11. Set $z = (3 + 4y)$. Then $dy = dz/4$, and the limits of integration are from $z = 3$ to $z = 11$:

$$\int_0^2 (3 + 4y)^{-2}dy = \int_3^{11} \frac{z^{-2}}{4}\,dz = \frac{-z^{-1}}{4}\,|_3^{11} = 0.061.$$

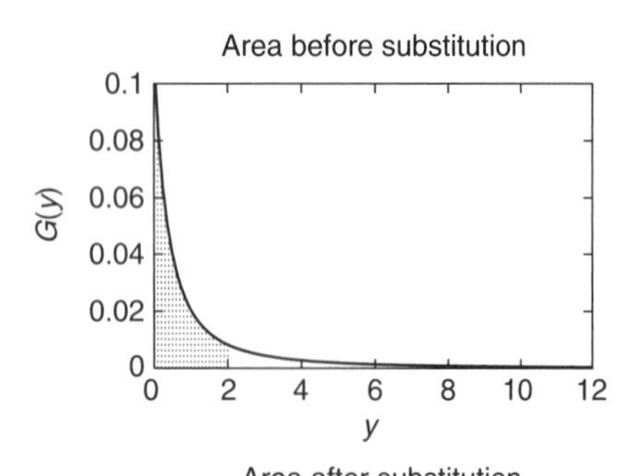

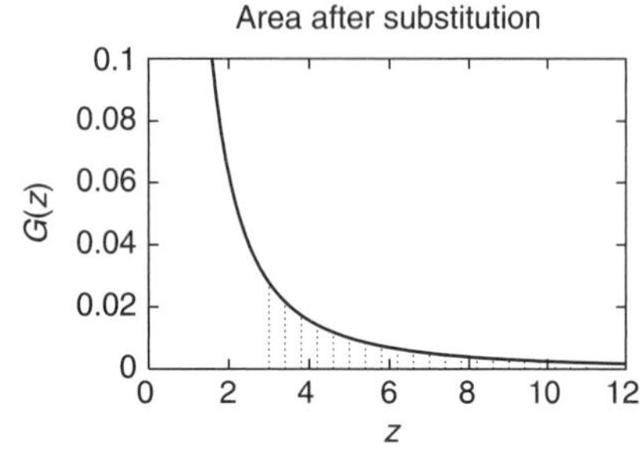

13. a.

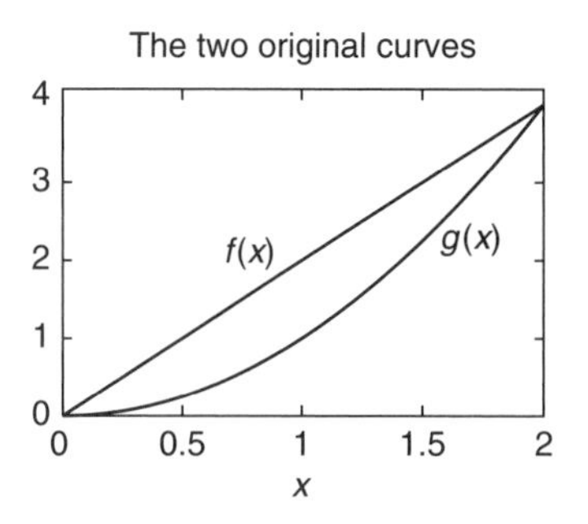

b.

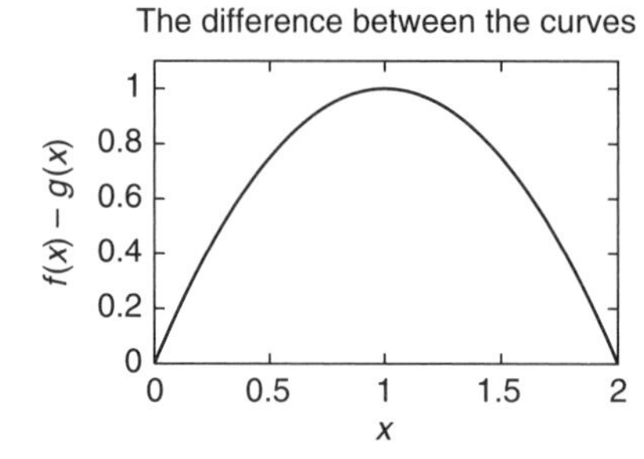

c. $2x - x^2 > 0$ for all $0 \le x \le 2$. Therefore,

$$\int_0^2 2x - x^2\,dx = x^2 - \frac{x^3}{3}\,|_0^2 = \left(4 - \frac{8}{3}\right) = 1.33.$$

15. a.

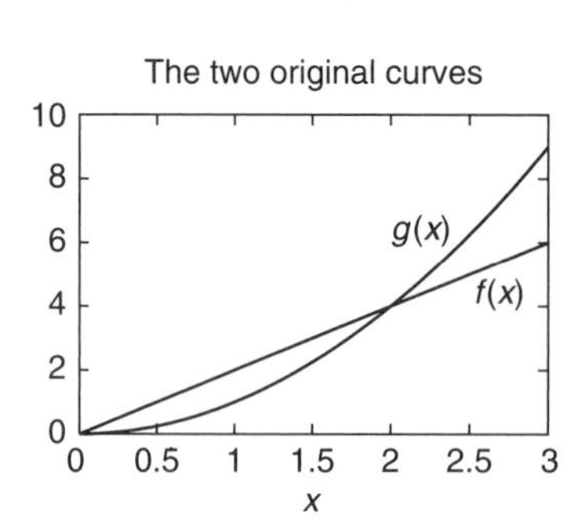

b.

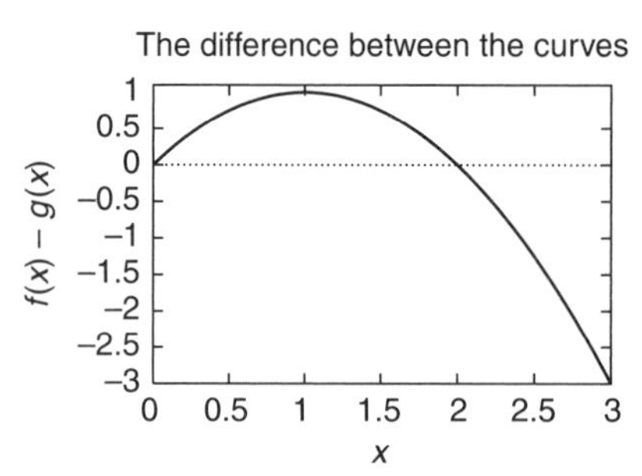

c. $2x - x^2 > 0$ for $0 \le x < 2$, but $2x - x^2 < 0$ for $2 < x \le 4$. Therefore,

$$\int_0^4 |2x - x^2|dx = \int_0^2 (2x - x^2)\,dx + \int_2^4 (x^2 - 2x)\,dx$$

$$= x^2 - \frac{x^3}{3}\,|_0^2 + \left(\frac{x^3}{3} - x^2\right)|_2^4$$

$$= (4 - 8/3) - (0 - 0) + (64/3 - 16)$$

$$- (8/3 - 4) = 8.$$

17. a.

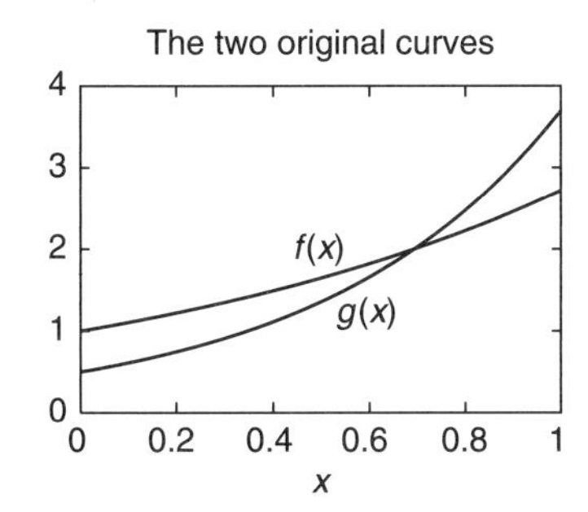

b.

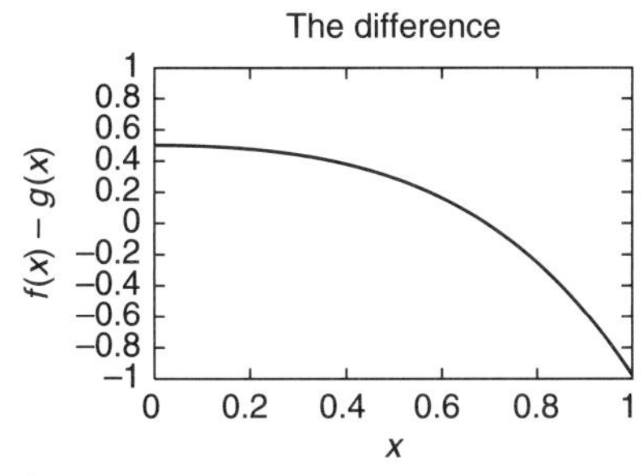

c. $e^x - \frac{e^{2x}}{2} > 0$ if $x < \ln(2)$. Therefore,

$$\int_0^1 |e^x - \frac{e^{2x}}{2}|dx = \int_0^{\ln(2)} \left(e^x - \frac{e^{2x}}{2}\right) dx - \int_{\ln(2)}^1 \left(\frac{e^{2x}}{2} - e^x\right) dx$$

$$= \left(e^x - \frac{e^{2x}}{4}\right) |_0^{\ln(2)} - \left(\frac{e^{2x}}{4} - e^x\right) |_{\ln(2)}^1$$

$$= -e + \frac{e^2}{4} + \frac{5}{4} = 0.378.$$

19. $\int_0^3 x^2 dx = \frac{x^3}{3}|_0^3 = 9$. The average is the integral divided by the width of the interval, or 9/3=3.

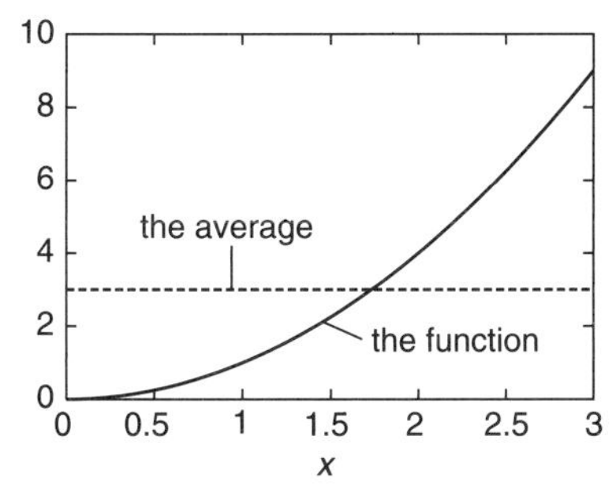

21. $\int_{-1}^1 x - x^3 dx = \frac{x^2}{2} - \frac{x^4}{4}|_{-1}^1 = 0$. The average is the integral divided by the width of the interval, or 0/2=0.

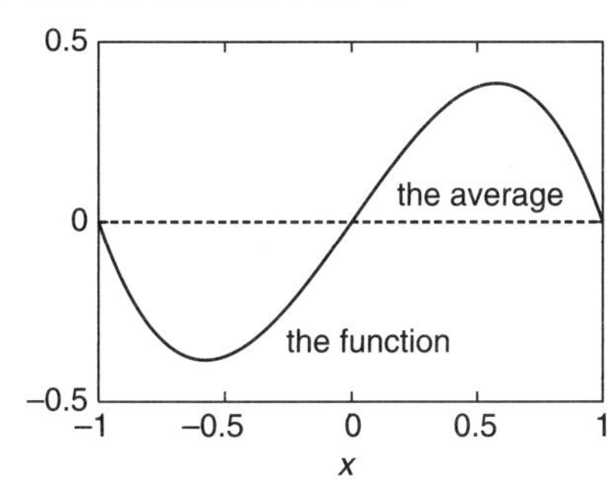

23. We need to compute

$$\text{area} = \int_1^2 x \ln(x) dx.$$

Let $u(x) = \ln(x)$ and $\frac{dv}{dx} = x$, then $\frac{du}{dx} = \frac{1}{x}$ and $v(x) = \frac{x^2}{2}$. Then

$$\int_1^2 x \ln(x) dx = u(x)v(x) - \int v(x)\frac{du}{dx}dx$$

$$= \frac{x^2 \ln(x)}{2}|_1^2 - \int_1^2 \frac{x}{2}dx$$

$$= \frac{x^2 \ln(x)}{2}|_1^2 - \frac{x^2}{4}|_1^2$$

$$= 2\ln(2) - \frac{1}{2}\ln(1) - 1 + \frac{1}{4} = 2\ln(2) - \frac{3}{4} = 0.636.$$

The function itself increases from $g(1) = 0$ to $g(2) = 2\ln(2) = 1.386$.

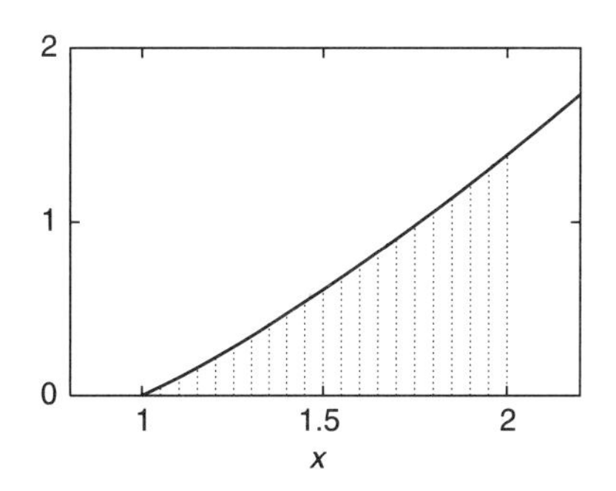

The area is about half the area of a rectangle with width 1 and height 1.386, consistent with the nearly triangular shaped region.

25. a.

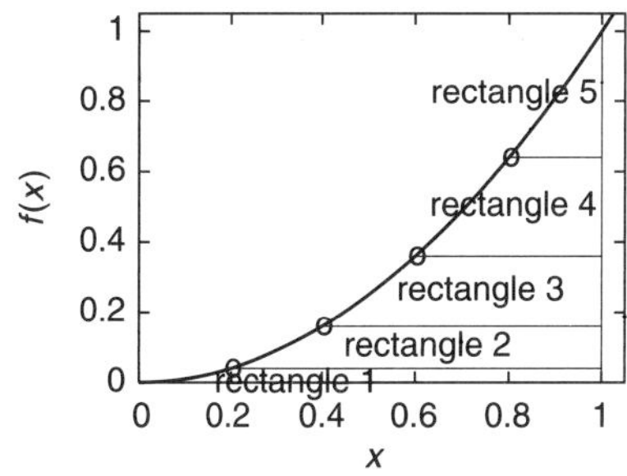

b. Rectangle 1: lower size estimate is $0.8 \cdot f(0.2) = 0.0032$; upper size estimate is $1.0 \cdot f(0.2) = 0.004$. Rectangle 2: lower size estimate is $0.6 \cdot [f(0.4) - f(0.2)] = 0.072$; upper size estimate is $0.8 \cdot [f(0.4) - f(0.2)] = 0.096$. Rectangle 3: lower size estimate is $0.4 \cdot [f(0.6) - f(0.4)] = 0.08$; upper size estimate is $0.6 \cdot [f(0.6) - f(0.4)] = 0.08$. Rectangle 4: lower size estimate is $0.2 \cdot [f(0.8) - f(0.6)] = 0.056$; upper size estimate is $0.4 \cdot [f(0.8) - f(0.6)] = 0.112$. Rectangle 5: lower size estimate is 0.0; upper size estimate is $0.2 \cdot [f(1.0) - f(0.8)] = 0.072$.

c. Lower estimate is 0.211; upper estimate is 0.364.

d. A rectangle at height y goes from the point where $x^2 = y$, or $x = \sqrt{y}$, to $x = 1$. Its length is $1 - \sqrt{y}$.

e. Area=$\int_0^1 (1 - \sqrt{y})\, dy$.

f. $\int_0^1 (1 - \sqrt{y})\, dy = \left(y - \frac{2y^{3/2}}{3}\right)|_0^1 = \frac{1}{3}$. It checks.

27. A little disk at height r will have volume approximately equal to $\pi r^2 \Delta r$. If we pick n disks with thickness Δr (so that $n\Delta r = 1$), the total volume will be approximated by the Riemann sum $\sum_{i=1}^n \pi r^2 \Delta r$. In the limit, this is the definite integral $\int_0^1 \pi r^2\, dr = \pi \frac{r^3}{3}|_0^1 = \frac{\pi}{3}$.

29.

$$l(6) - l(3) = \int_1^6 \frac{1}{x}dx - \int_1^3 \frac{1}{x}dx = \int_3^6 \frac{1}{x}dx = \int_1^2 \frac{1}{y}dy = l(2)$$

where we set $y = \frac{x}{3}$, so $dy = \frac{dx}{3}$ and the limits of integration go from 1 to 2.

31. $\int_1^{10^2} \frac{1}{x}\, dx = l(10^2)$. Substituting $y = \sqrt{x}$, we find that

$$\frac{dy}{dx} = \frac{1}{2}x^{-1/2} = \frac{1}{2}\frac{y}{x}.$$

Then the integrand becomes

$$\frac{1}{x}\,dx = \frac{2}{y}\,dy$$

and the limits of integration go from 1 to 10. So

$$\int_1^{10^2} \frac{1}{x}\,dx = \int_1^{10} \frac{2}{y}\,dy = 2l(10).$$

This matches the law of logs.

33.

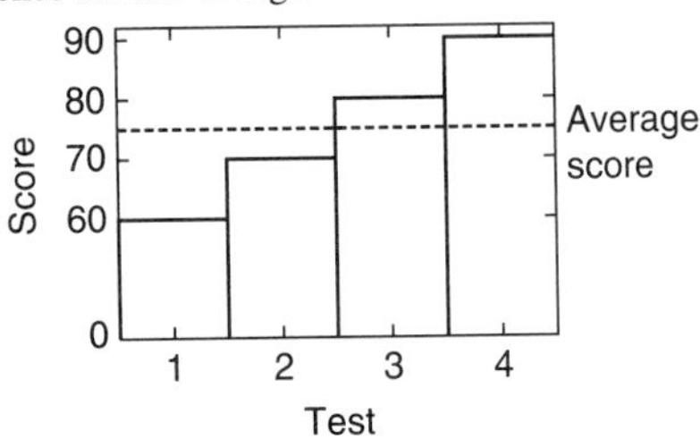

The average is $(60 + 70 + 80 + 90)/4 = 75$. The function is

$$f(x) = \begin{cases} 60 & \text{for } 0 \le x < 1 \\ 70 & \text{for } 1 \le x < 2 \\ 80 & \text{for } 2 \le x < 3 \\ 90 & \text{for } 3 \le x < 4. \end{cases}$$

$$\text{Total score} = \int_0^4 f(x)dx = 60 + 70 + 80 + 90 = 300.$$

$$\text{Average score} = \frac{\text{total score}}{\text{width of interval}} = \frac{300}{4} = 75.$$

35. $$\int_0^{15} 360t - 39t^2 + t^3\,dt = 180t^2 - 13t^3 + \frac{t^4|15}{4} = 9281.25.$$

Dividing this answer by the total time of 15 h gives $\frac{9281.25}{15} = 618.75$ L/h.

37. Using the summation property, $\int_0^{24} g(t)\,dt = \int_0^{15} g(t)\,dt + \int_{15}^{24} g(t)\,dt$ or $9281.25 - 2369.25 = 6912.0$. Over the full 24 h, the average rate is $\frac{6912.0}{24} = 288.0$ L/h.

39. a. The critical points are $x = 0$ and $x = 200$ (both endpoints). Because $\rho(0) = 1.0$ and $\rho(200) = 1.08$, the first is the minimum and the second is the maximum.

b. $\int_0^{200} \rho(x)dx = 208$ g.

c. The average is 1.04 g/cm, which lies right between the minimum and the maximum.

d.

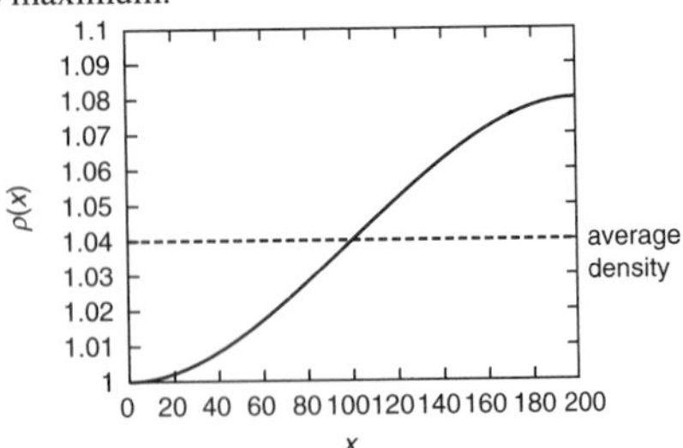

41. a. Let x represent distance along the strand. Then the formula for the line giving the number of A's can be found by finding the slope as

$$\text{slope} = \frac{300 - 150}{4.7 \times 10^3} = 3.19 \times 10^{-2}.$$

Using the point $A(0) = 150$, we find $A(x) = 150 + 3.19 \times 10^{-2}x$. Similarly, $C(x) = 350 - 3.19 \times 10^{-2}x$, $G(x) = 220 + 2.13 \times 10^{-2}x$. Because $A(x) + C(x) + G(x) + T(x) = 1000$,

$$T(x) = 1000 - A(x) - C(x) - G(x) = 280 - 2.13 \times 10^{-2}x.$$

b. The totals can be found by integrating,

$$\int_0^{4.7\times10^3} A(x) = 1.057 \times 10^6.$$

The total number of C's is 1.292×10^6, the total number of G's is 1.269×10^6, and the total number of T's is 1.055×10^6.

c. The mean numbers per thousand are: A, 224; C, 275; G, 270; T, 230.

43. $\int_0^1 t^3\,dt = 0.25$ cm^3. The average rate is 0.25 cm^3/s. The rate at time 0.5 is 0.125, less than the average rate during the first second. This seems to be because the function is concave up (has positive second derivative).

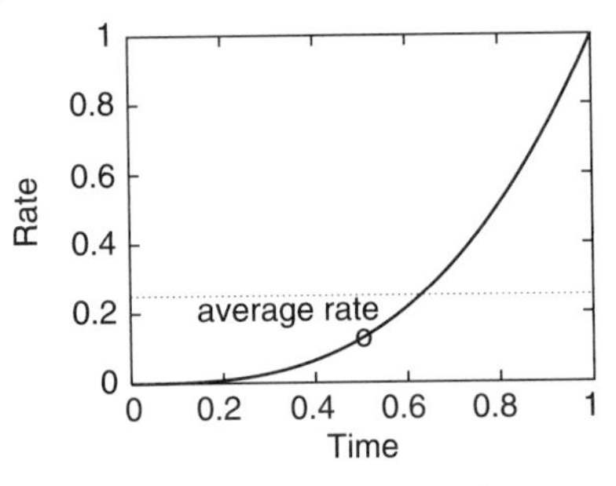

45. $\int_0^1 t\,dt = 0.5$ cm^3. The average rate is 0.5 cm^3/s. The average matches the rate at the average time, probably because this function is linear.

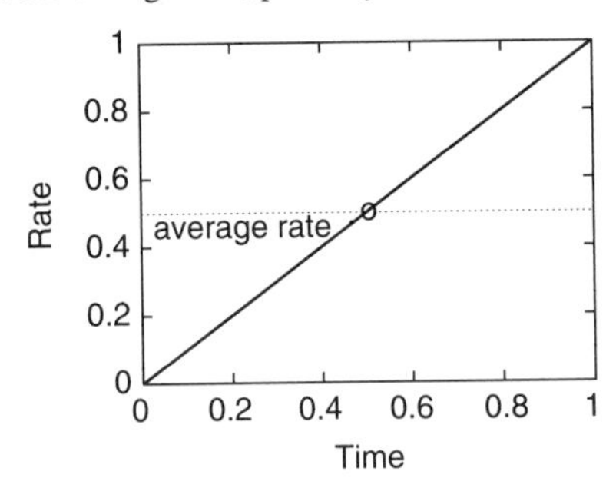

Section 4.7, page 416

1. We know that e^{-x} is faster because exponential functions are faster than power functions. With L'Hôpital's rule,

$$\frac{e^{-x}}{\frac{1}{x}} = \frac{x}{e^x}.$$

Using L'Hôpital's rule, we find that the limit of this quotient (which is an indeterminate form because both the numerator and denominator approach infinity) is

$$\lim_{x\to\infty} \frac{x}{e^x} = \lim_{x\to\infty} \frac{1}{e^x} = 0.$$

e^{-x} does indeed approach 0 faster than $\frac{1}{x}$ as x approaches infinity.

3. The power in the denominator of $\frac{1}{x^2}$ is larger, so it approaches infinity more quickly. Algebraically,

$$\frac{\frac{1}{x^2}}{\frac{1}{x}} = \frac{x}{x^2} = \frac{1}{x},$$

which approaches infinity. Therefore, $\frac{1}{x^2}$ does approach infinity more quickly than $\frac{1}{x}$ as x approaches 0.

5. This integral converges because the integrand decreases to zero exponentially. Using the substitution $u = -3t$, we find

$$\int_0^{\infty} e^{-3t}dt = \frac{1}{3}\int_{-\infty}^{0} e^u\,du = \frac{1}{3}e^u|_{-\infty}^0 = 1/3.$$

7. This does not converge because the integrand approaches zero too slowly as a power function with negative power greater than -1.

9. Using the substitution $u = 1 + 3x$, we find that $dx = du/3$ and the limits of integration are from $u = 1$ to $u = \infty$. Therefore

$$\int_0^\infty \frac{1}{(1+3x)^{3/2}}\,dx = \int_1^\infty \frac{1}{3}u^{-3/2}\,du = \frac{-2}{3}u^{-1/2}|_1^\infty = \frac{2}{3}.$$

11. Diverges because the power is not less than 1.

13. $\int_0^{0.001} \frac{1}{\sqrt[3]{x}}\,dx = 0.015.$

15. For $x < 1$, $\sqrt[3]{x} > x^3$, so

$$\int_0^1 \frac{1}{\sqrt[3]{x}+x^3}dx < \int_0^1 \frac{1}{2\sqrt[3]{x}}dx = 3x^{2/3}|_0^1 = 3.$$

The value is less than 3.

17. For $x > 1$, $x^3 > \sqrt[3]{x}$, so

$$\int_1^\infty \frac{1}{\sqrt[3]{x}+x^3}dx < \int_1^\infty \frac{1}{2x^3}dx = -x^{-2}|_1^\infty = 1.$$

The value is less than 1.

19. Near 0, the leading behavior of the denominator is $\sqrt[3]{x}$. The function then acts like $1/\sqrt[3]{x}$, which converges on the interval from 0 to 1.

21. For large x, the leading behavior of the denominator is x^3. The function then acts like $1/x^3$, which converges on the interval from 1 to infinity. The whole integral converges.

23. The sum corresponds the area under the blocks. Because the blocks lie totally below the curve, and the area under the curve is finite, the sum must also be finite. A bit of experimentation shows that the sum is exactly equal to 2. Integration shows that the area under the curve is 2.885.

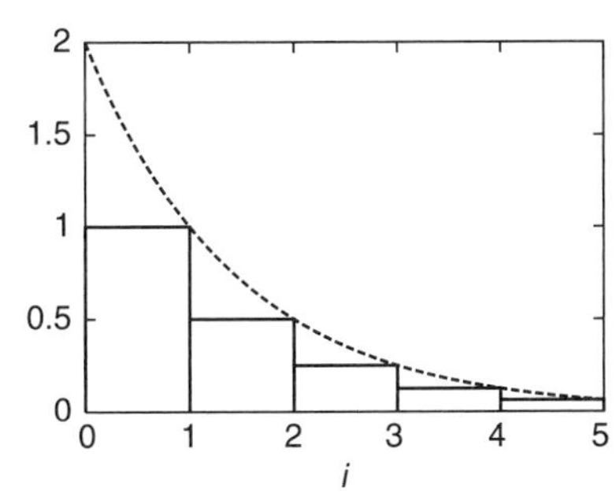

25. The sum corresponds to the area under the blocks. Although the blocks do not lie totally below the curve, if we removed the first block (which has an area of 1) and shifted them all over by one, they would lie below the curve. The area under the curve is finite, so the sum must also be finite. Amazingly, the sum is equal to $\pi^2/6$.

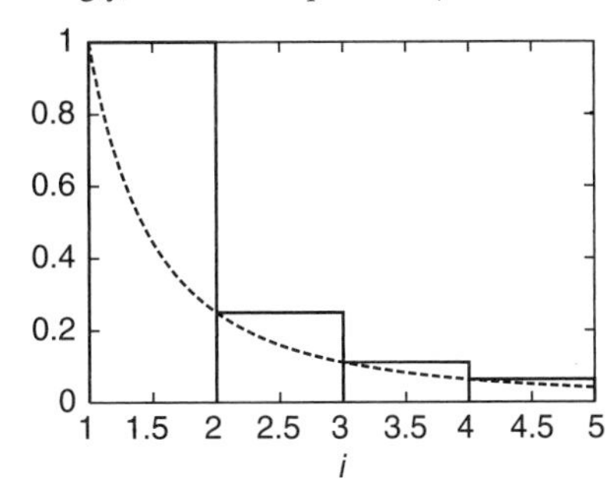

27. The differential equation is

$$\frac{dV}{dt} = \frac{100}{(1+t)^2}.$$

The solution is

$$V(t) = 500 + \int_0^t \frac{100}{(1+s)^2}\,ds = 500 + \frac{-100}{1+s}|_0^t = 600 - \frac{100}{1+t}.$$

This rule could be followed indefinitely, and the limit of the volume is 600.

29. The differential equation is

$$\frac{db}{dt} = \frac{1000}{(2+3t)^{.75}}.$$

The solution is

$$\begin{aligned} b(t) &= 1.0\times10^6 + \int_0^t \frac{1000}{(2+3s)^{.75}}\,ds \\ &= 1.0\times10^6 + 1333(2+3s)^{0.25}|_0^t \\ &= 1.0\times10^6 + 1333(2+3t)^{0.25} - 1333(2^{0.25}) \\ &= 9.98\times10^5 + 1333(2+3t)^{0.25}. \end{aligned}$$

This rule could not be followed indefinitely. The population would reach 2.0×10^6 when $t = 1.06\times10^{11}$. It grows very slowly.

Chapter 5

Section 5.1, page 429

1. This differential equation is nonautonomous because the state variable F and the time t both appear on the right-hand side. The parameter k does not affect the kind of differential equation.

3. This is a pure-time differential equation because the only variable on the right-hand side is the time t (along with the parameter μ).

5. Substituting these values into the right-hand side gives $\frac{dF}{dt} = 1^2 + 1\cdot0 = 1 > 0$. Therefore, F is increasing.

7. Substituting these values into the right-hand side gives $\frac{dy}{dt} = 2e^{-1} - 1 = -0.26 < 0$. Therefore, y is decreasing.

9. $p(t) = \frac{t^2}{2} + 1$. Then $p(1) = 1.5$.

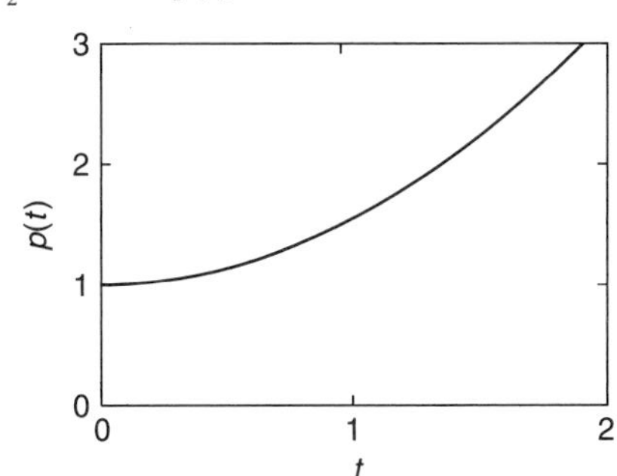

11. The derivative of e^t is e^t, and $e^0 = 1$, so this is the solution that matches the initial conditions. Also, $b(1) = e = 2.718$.

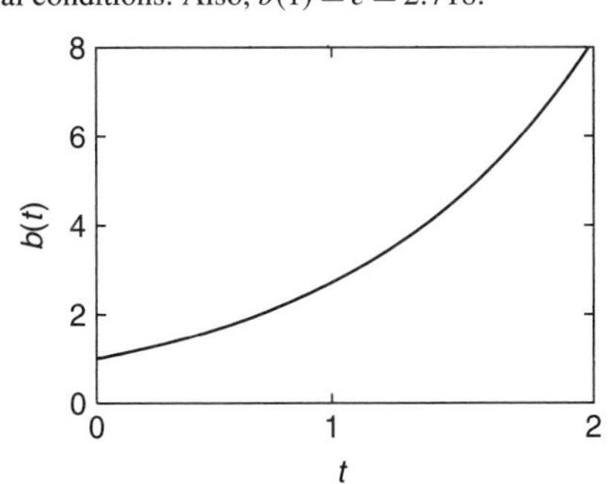

13. They don't match because the rate of change gets larger the longer you wait.

15. First, $x(0) = -\frac{1}{2} + \frac{3}{2}e^0 = 1$, so the initial condition matches. Next,

$$\frac{dx}{dt} = 3e^{2t} = 2\left(-\frac{1}{2} + \frac{3}{2}e^{2t}\right) + 1 = 1 + 2x.$$

This checks.

17. First, $G(0) = 1 + e^0 = 2$, so the initial condition matches. Next, $\frac{dG}{dt} = e^t = (1 + e^t) - 1 = G - 1$. This checks.

19. $\hat{x}(1) = x(0) + x'(0) \cdot 1 = 1 + 3 = 4$. When $x = 4$, the differential equation says that $dx/dt = 1 + 2 \cdot 4 = 9$, so that $\hat{x}(2) = \hat{x}(1) + x'(1) \cdot 1 = 4 + 9 = 13$. The exact answer is $-\frac{1}{2} + \frac{3}{2}e^{2 \cdot 2} = 81.4$. Euler's method is way off.

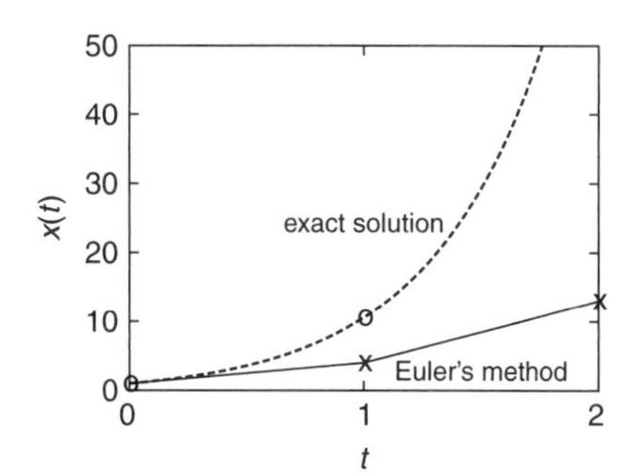

21.

$$\hat{G}(0.2) = G(0) + G'(0) \cdot 0.2 = 2 + 1 \cdot 0.2 = 2.2$$
$$\hat{G}(0.4) = \hat{G}(0.2) + G'(0.2) \cdot 0.2 = 2.2 + 1.2 \cdot 0.2 = 2.44$$
$$\hat{G}(0.6) = \hat{G}(0.4) + G'(0.4) \cdot 0.2 = 2.44 + 1.44 \cdot 0.2 = 2.728$$
$$\hat{G}(0.8) = \hat{G}(0.6) + G'(0.6) \cdot 0.2 = 2.728 + 1.728 \cdot 0.2 = 3.0736$$
$$\hat{G}(1.0) = \hat{G}(0.8) + G'(0.8) \cdot 0.2 = 3.0736 + 2.0736 \cdot 0.2 = 3.48832.$$

The exact answer is $1 + e^1 = 3.7182$. Euler's method is fairly close.

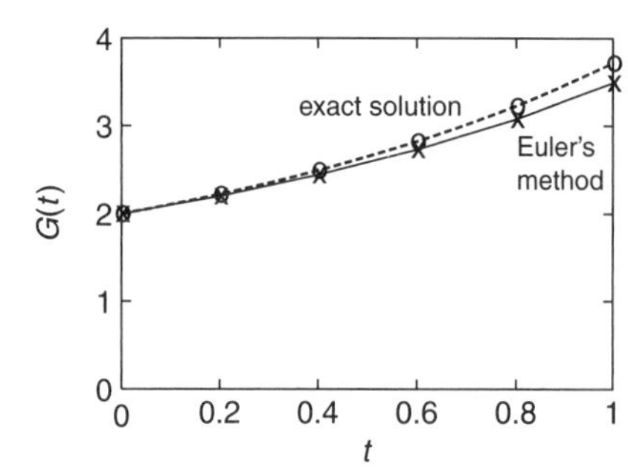

23. The derivative of y can be found with the chain rule, by thinking of $y(t)$ as the composition $F \circ x(t)$ where $F(x) = 2x - 1$. Then

$$\frac{dy}{dt} = \frac{dF}{dx}\frac{dx}{dt} = 2\frac{dx}{dt} = 2(2x - 1).$$

We need to rewrite the right-hand side entirely in terms of y, and get

$$\frac{dy}{dt} = 2y.$$

This is simpler because the -1 term has disappeared.

25. The derivative of y can be found with the chain rule, by thinking of $y(t)$ as the composition $F \circ x(t)$ where $F(x) = 1/x$. Then

$$\frac{dy}{dt} = \frac{dF}{dx}\frac{dx}{dt} = -\frac{1}{x^2}\frac{dx}{dt} = -\frac{1}{x^2}(x + x^2).$$

We need to rewrite the right-hand side entirely in terms of y, and get

$$\frac{dy}{dt} = -y^2\left(\frac{1}{y} + \frac{1}{y^2}\right) = -y + 1.$$

This differential equation is linear.

27. A line with intercept 1 and slope -0.002 has equation $\lambda(b) = 1 - 0.002b$, so that the differential equation is $\frac{db}{dt} = (1 - 0.002b)b$. When $b = 10$, $\frac{db}{dt} = 9.8 > 0$, so this population would increase. When $b = 1000$, $\frac{db}{dt} = -1000 < 0$, so this population would decrease.

29. A line with intercept -2 and slope 0.01 has equation $\lambda(b) = -2 + 0.01b$, so that the differential equation is $\frac{db}{dt} = (-2 + 0.01b)b$. When $b = 100$, $\frac{db}{dt} = -200 < 0$, so this population would decrease. When $b = 300$, $\frac{db}{dt} = 300 > 0$, so this population would increase.

31.

chemical leaves at rate βC
C = internal concentration
Γ = ambient concentration
no chemical enters

The equation is $\frac{dC}{dt} = -\beta C$. This is the same as the differential equation for a shrinking population. The chemical concentration decays exponentially to 0 because there is no source of chemical.

33.

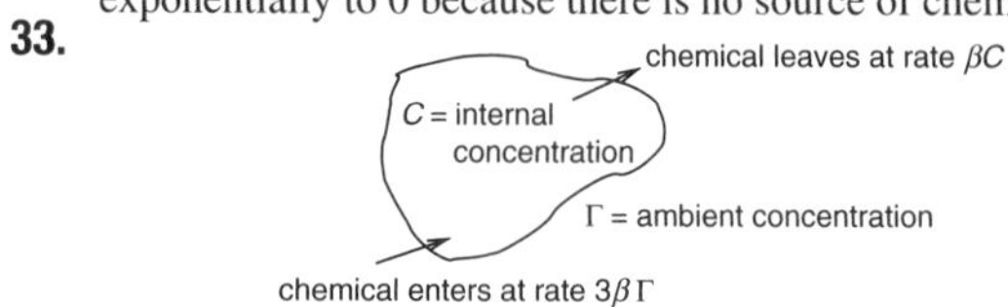

The equation is $\frac{dC}{dt} = -\beta C + 3\beta\Gamma$. If the concentrations were equal, the derivative would be positive, meaning that the internal concentration would increase, and become larger than the external concentration.

35.

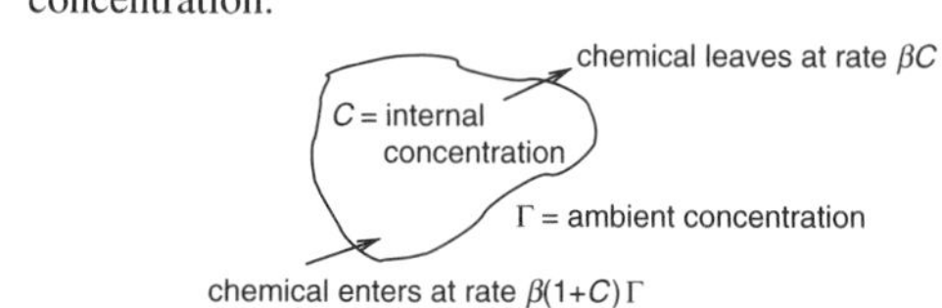

The equation is $\frac{dC}{dt} = -\beta C + \beta(1 + C)\Gamma$. If the concentrations were equal, the derivative would be positive, meaning that the internal concentration would increase, and become larger than the external concentration.

37. The differential equations are

$$\frac{da}{dt} = 2(1 - p)a$$
$$\frac{db}{dt} = 1.5(1 - p)b.$$

Then

$$\begin{aligned}\frac{dp}{dt} &= \frac{b\frac{da}{dt} - a\frac{db}{dt}}{(a+b)^2} \\ &= \frac{2(1-p)ab - 1.5(1-p)ab}{(a+b)^2} \\ &= \frac{(2-1.5)(1-p)ab}{(a+b)^2} \\ &= 0.5p(1-p)^2.\end{aligned}$$

Even though the differential equation is different, the rate of change of p is always positive, and type a still takes over.

39. a. $H(t) = 10 + 30e^{-0.2t}$. $H(0) = 40$, matching the initial condition.

$$\frac{dH}{dt} = -6.0e^{-0.2t}.$$

This is supposed to match $\alpha[A - H(t)]$, which is

$$\begin{aligned}\alpha[A - H(t)] &= 0.2[10 - (10 + 30e^{-0.2t})] = 0.2(-30e^{-0.2t}) \\ &= -6.0e^{-0.2t}.\end{aligned}$$

It checks.

b. $H(1) = 34.56$ and $H(2) = 30.1$.

c. The term $30e^{-0.2t}$ approaches 0, so $\lim_{t \to \infty} H(t) = 10$, the ambient temperature.

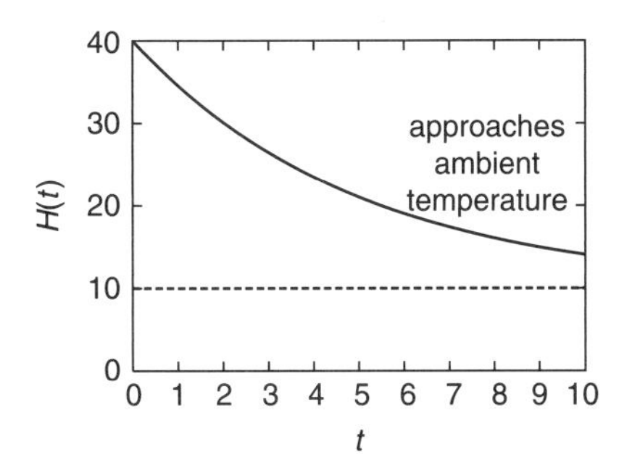

41. $H'(0) = 0.2(10 - 40) = -6.0$. Then $\hat{H}(1.0) = H(0) - 6.0 \cdot 1.0 = 34.0°\text{C}$. $H'(1.0) = 0.2(10 - 34) = -4.8$. Then $\hat{H}(2.0) = \hat{H}(1.0) - 4.8 \cdot 1.0 = 29.2°\text{C}$. This is a bit low.

43.

$$
\begin{aligned}
C(t) &= \Gamma + e^{-\beta t}[C(0) - \Gamma] = 2.0 + e^{-0.01t}(5.0 - 2.0) \\
&= 2.0 + 3.0e^{-0.01t} \\
C(10) &= 2.0 + 3.0e^{-0.1} = 4.714 \\
C(20) &= 2.0 + 3.0e^{-0.2} = 4.456 \\
C(60) &= 2.0 + 3.0e^{-0.6} = 3.646.
\end{aligned}
$$

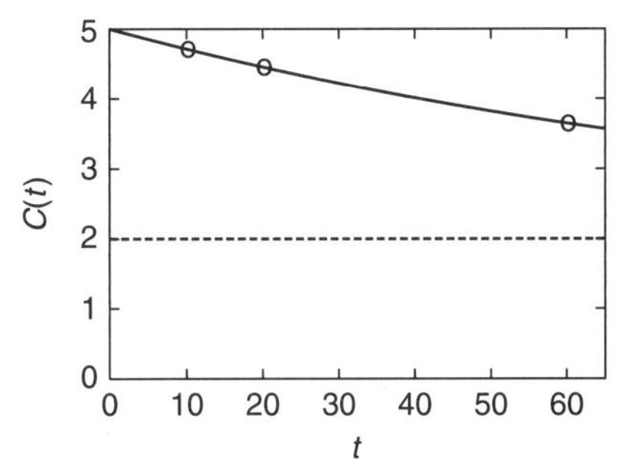

45. We need to rewrite $b_0 r^t$ in terms of the exponential function as $b_0 e^{\ln(r)t}$. This matches when $b_0 = 1.0 \times 10^6$ and $\ln(r) = 2$ or $r = 7.39$.

47. The population described by the differential equation grows when $\lambda > 0$. The population described by the discrete-time dynamical system grows when $r > 1$.

49. We need to take four steps.

$$
\begin{aligned}
\hat{p}(0.5) &= 0.1 + 0.1 \cdot 0.9 \cdot 0.5 = 0.145 \\
\hat{p}(1.0) &= 0.145 + 0.145 \cdot 0.855 \cdot 0.5 = 0.207 \\
\hat{p}(1.5) &= 0.207 + 0.207 \cdot 0.793 \cdot 0.5 = 0.289 \\
\hat{p}(2.0) &= 0.289 + 0.289 \cdot 0.711 \cdot 0.5 = 0.392.
\end{aligned}
$$

The exact solution is

$$
p(t) = \frac{0.1e^{2.0t}}{0.1e^{2.0t} + 0.9e^{1.0t}}
$$

so that $p(2) = 0.45$.

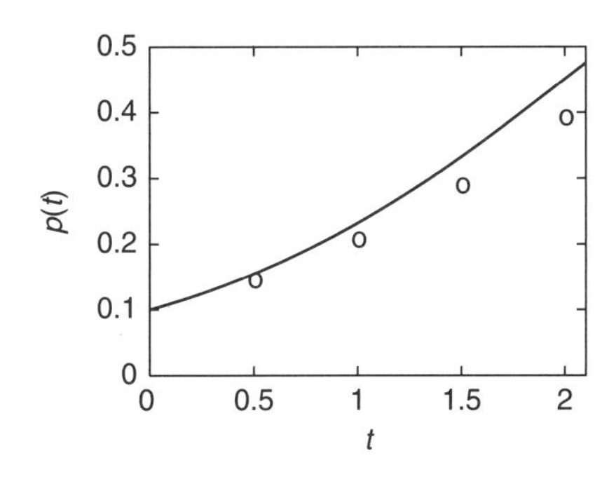

51. At equilibrium, we have that

$$
\begin{aligned}
0 &= c_1(r^3)^{\frac{3}{4}} - c_2 r^2 (T - A) \\
0 &= c_1 r^{\frac{9}{4}} - c_2 r^2 (T - A) \\
0 &= c_1 r^{\frac{1}{4}} - c_2 (T - A) \\
c_2(T - A) &= c_1 r^{\frac{1}{4}} \\
(T - A) &= \frac{c_1}{c_2} r^{\frac{1}{4}} \\
T &= A + \frac{c_1}{c_2} r^{\frac{1}{4}}.
\end{aligned}
$$

The equilibrium temperature is higher when r is larger.

Section 5.2, page 435

1.
1. This equation is autonomous because the only variable on the right-hand side is the state variable x.
2. We must solve $1 - x^2 = 0$.
3. Factoring gives $(1 - x)(1 + x) = 0$.
4. Solving each factor gives $x = 1$ or $x = -1$.

3.
1. This equation is autonomous because the only variable on the right-hand side is the state variable y.
2. We must solve $y\cos(y) = 0$.
3. This is already in factored form.
4. Solving gives $y = 0$ or $y = \frac{\pi}{2} + \pi n$ for any integer value of n.

5.
1. This equation is autonomous because the only variable on the right-hand side is the state variable x.
2. We must solve $1 - ax = 0$.
3. There is only one term, so we don't need to factor.
4. Solving, we find $ax = 1$, which has solution $x = \frac{1}{a}$.

7.
1. This equation is autonomous because the only variable on the right-hand side is the state variable W.
2. We must solve $\alpha e^{\beta W} - 1 = 0$.
3. There is only one term, so we don't need to factor.
4. Solving, we find that $\alpha e^{\beta W} = 1$, which becomes $e^{\beta W} = \frac{1}{\alpha}$. Taking logarithms, $\beta W = -\ln(\alpha)$, which has solution $W = -\frac{\ln(\alpha)}{\beta}$.

9.

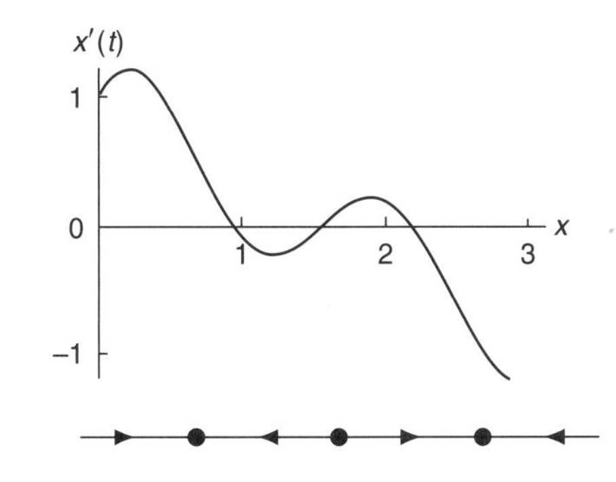

11.

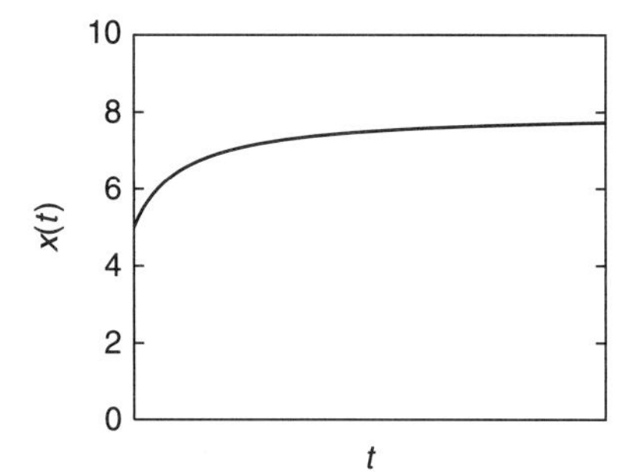

13.

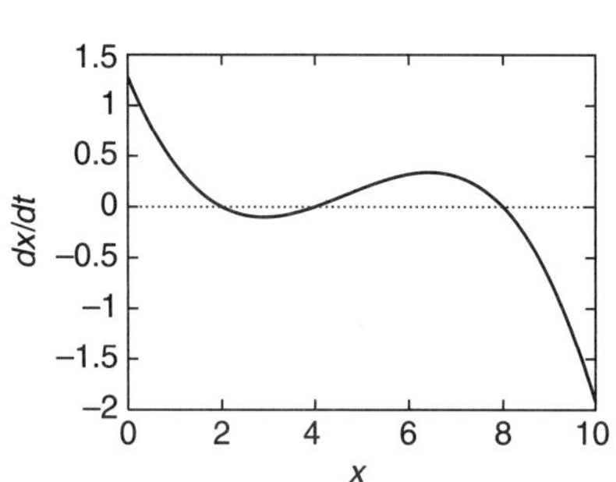

15.

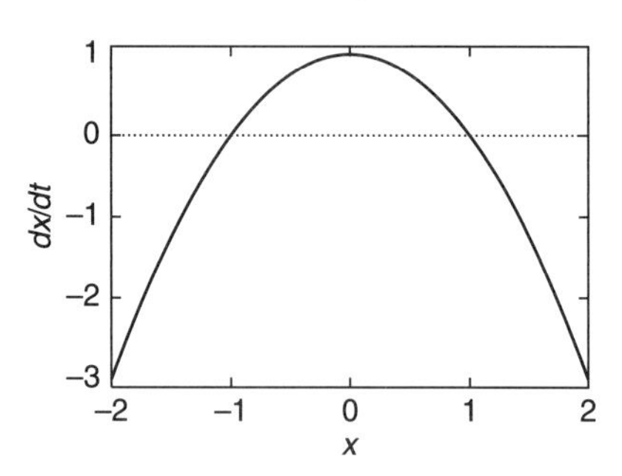

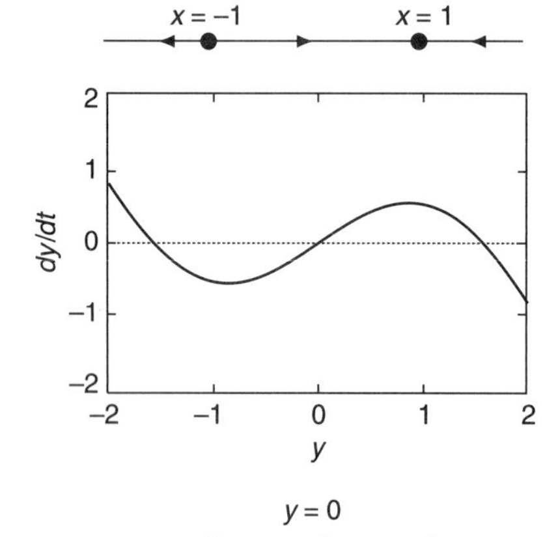

17.

19. The equilibrium solves $\lambda b^* - h = 0$ or $2.0b^* - 1000 = 0$, or $b^* = 500$.

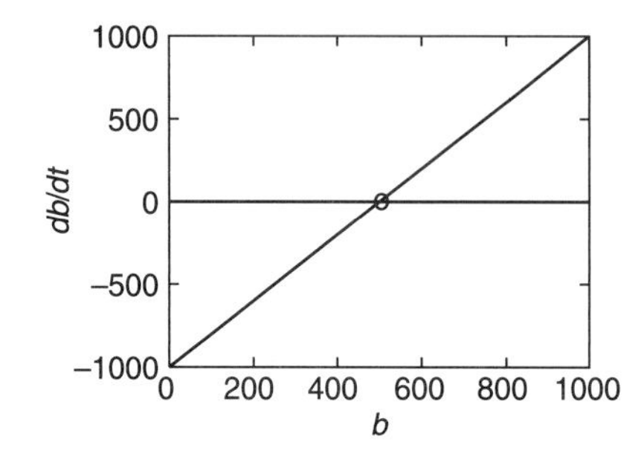

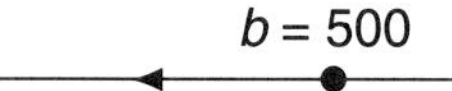

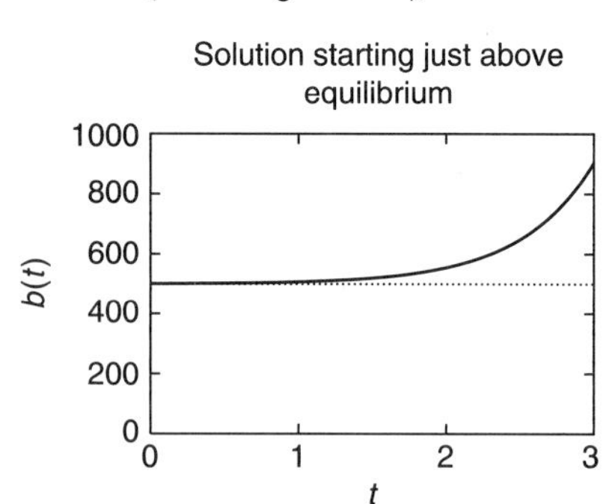

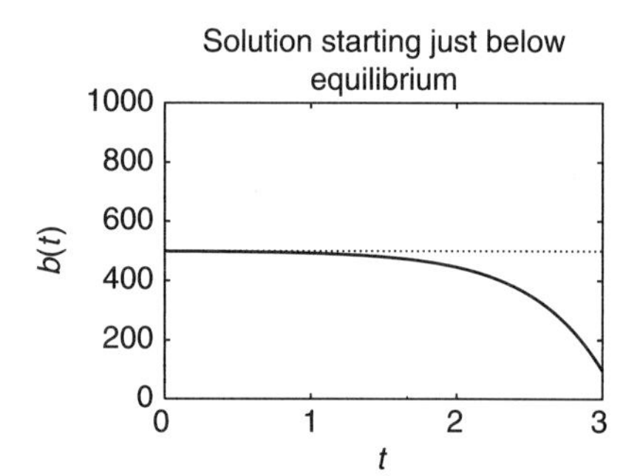

This population can outgrow the harvest if it starts at a large enough value. If it starts too small, the harvest will drive it to extinction.

21. We found that the population obeys the autonomous differential equation $\frac{db}{dt} = (1 - 0.002b)b$. This is in factored form and has equilibria at $b = 500$ and at $b = 0$.

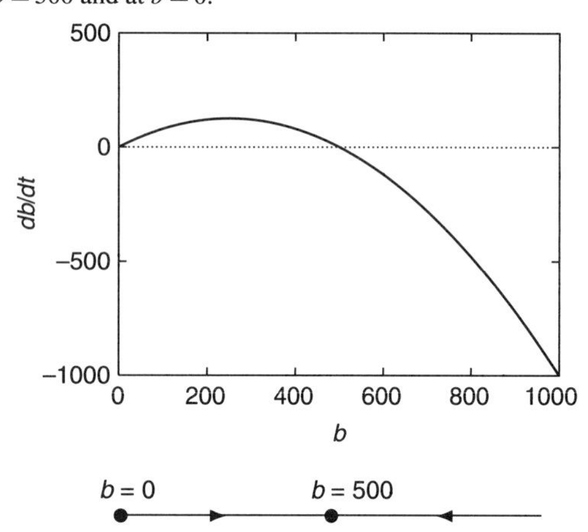

The arrow points up at $b = 10$, consistent with an increasing population, and down at $b = 1000$, consistent with a decreasing population.

23. We found that the population obeys the autonomous differential equation $\frac{db}{dt} = (-2 + 0.01b)b$. This is in factored form, and has equilibria at $b = 200$ and at $b = 0$.

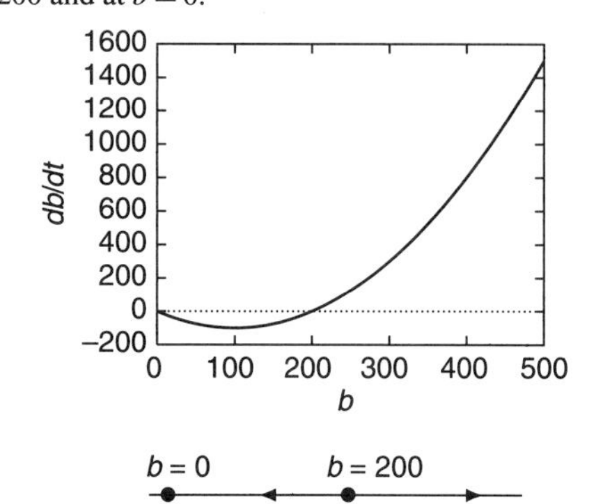

The arrow points down at $b = 100$, consistent with a decreasing population, and up at $b = 300$, consistent with an increasing population.

25. We found that the population obeys the autonomous differential equation $\frac{dC}{dt} = -\beta C + 3\beta\Gamma$. As long as $\beta \neq 0$, this has equilibria at $C = 3\Gamma$.

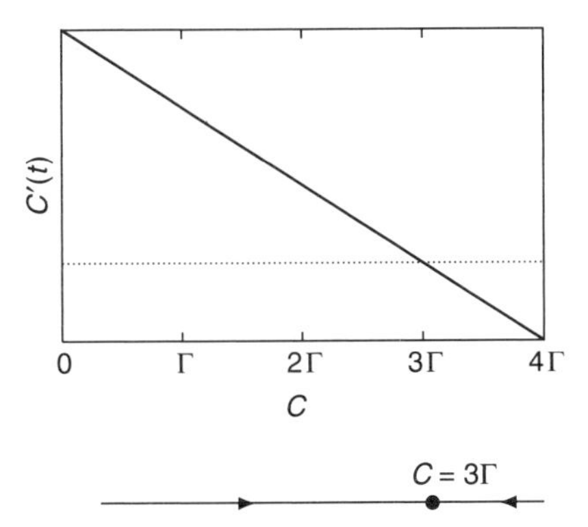

The arrow points up at $C = \Gamma$, consistent with an increasing concentration.

27. We found that the population obeys the autonomous differential equation $\frac{dp}{dt} = 0.5p(1-p)^2$. This has equilibria at $p = 0$ and $p = 1$.

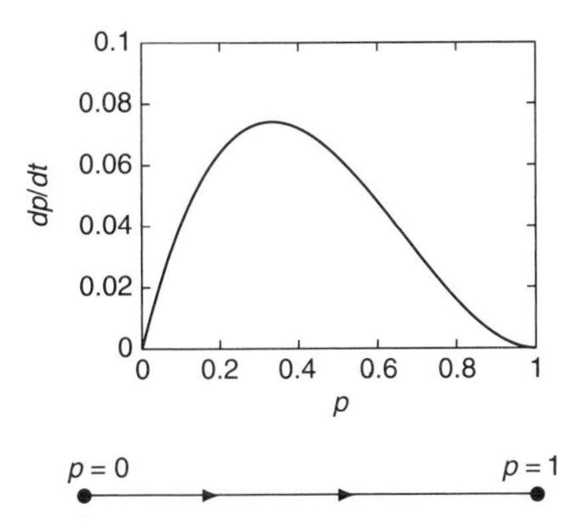

All the arrows point to the right point up, except at the equilibria, so the solution moves up to $p = 1$, meaning that a takes over.

29. The equilibria are $N = 0$, and the solution of $\frac{3N}{2+N^2} - 1 = 0$, which occurs where $N^2 - 3N + 2 = 0$. This factors to have solutions at $N = 1$ and $N = 2$. To see whether N is increasing or decreasing between the equilibria, we need to check whether $\frac{dN}{dt}$ is positive or negative. We find that $f(1/2) = -1/6 < 0$, $f(3/2) = 3/34 > 0$, and $f(3) = -6/11 < 0$. Therefore, the graph of the rate of change and the phase-line diagram must be the following.

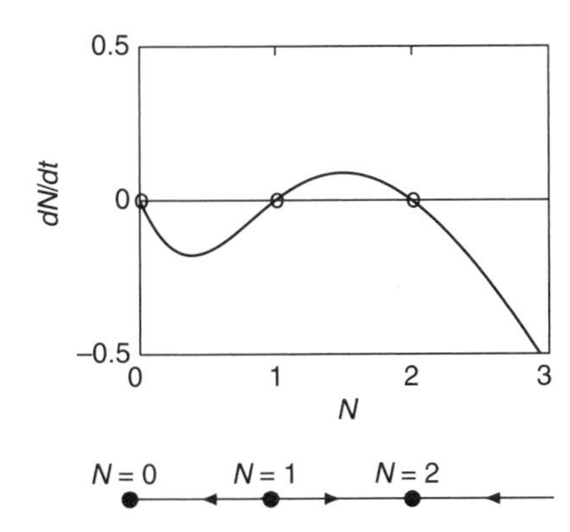

This population dies out if it drops below $N = 1$. Perhaps they cannot find mates when the population gets below 100.

31. Everything has units of meters per second squared. The equilibrium is the solution of $9.8 - 0.0032v^2 = 0$, or $v^* = \sqrt{9.8/0.0032} = 55.3$ m/s. This is the terminal velocity of a sky diver in free fall.

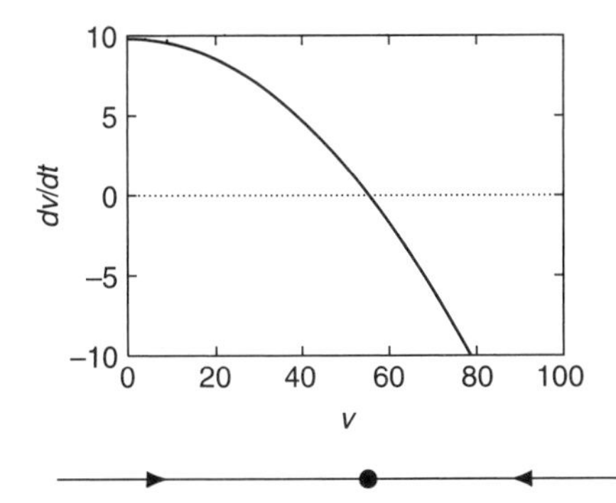

33. Both sides have units of centimeters per second. This checks. The equilibrium is $y^* = 0$, meaning that all water has drained out of the cylinder.

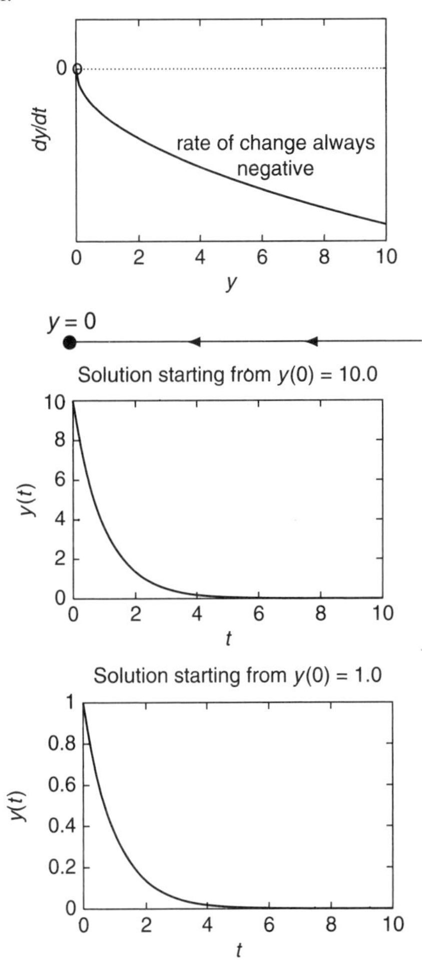

35. The equilibrium is at $S^* = 0$. Eventually, all substrate will be used. In both cases, the rate is always negative. However, the graph of the rate for Torricelli's law of draining is much steeper near a value of 0 for the state variable.

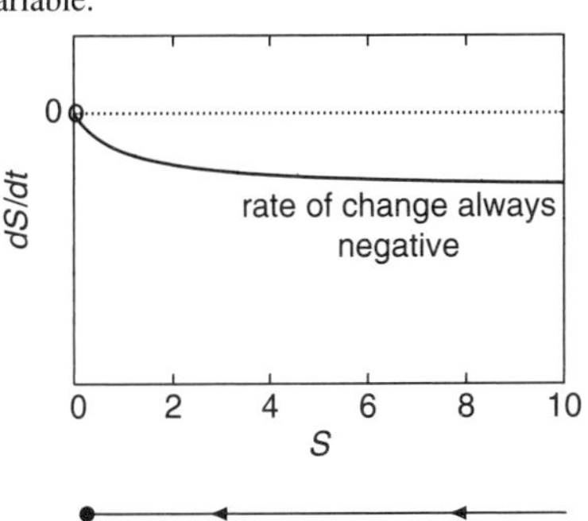

37. Everything has units of cm³/day. The equilibrium is

$$V^* = \left(\frac{a_1}{a_2}\right)^3.$$

The equilibrium gets smaller for smaller values of a_1 because this animal is less effective at collecting food. The equilibrium gets larger when a_2 becomes smaller because this animal is more efficient at using energy.

Section 5.3, page 443

1.

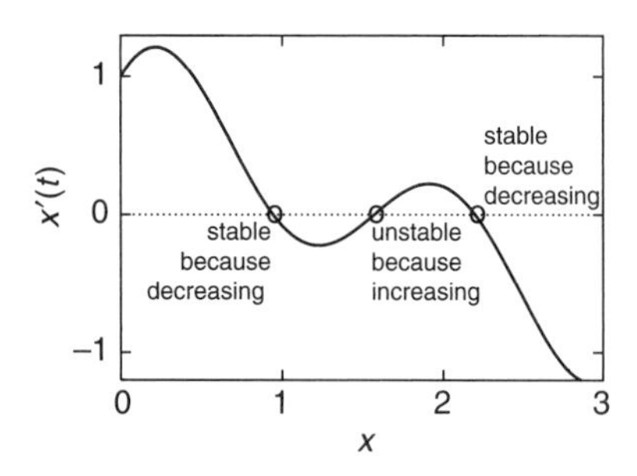

3. We found equilibria at $x = 1$ and $x = -1$. The derivative of the rate-of-change function is $-2x$, which is negative at $x = 1$ and positive at $x = -1$. Therefore, $x = 1$ is stable, consistent with the inward pointing arrows on the phase-line diagram, and $x = -1$ is unstable, consistent with the outward pointing arrows on the phase-line diagram.

5. We found equilibria at $y = 0$ and $y = \frac{\pi}{2} + \pi n$. The derivative of the rate-of-change function is $\cos(y) - y\sin(y)$. This is positive at $y = 0$, negative at $y = \pi/2$, and negative at $y = -\pi/2$. Therefore, $y = 0$ is unstable (inward pointing arrows), $y = \pi/2$ is stable (outward pointing arrows), and $y = -\pi/2$ is stable (outward pointing arrows).

7. The derivative of the rate-of-change function is $-a$, which is negative for all values of x when $a > 0$. Therefore the equilibrium must be stable.

9. The derivative of the rate-of-change function is $\alpha\beta e^{\beta W}$, which is negative for all values of W when $\alpha > 0$ and $\beta < 0$. Therefore, the equilibrium must be stable.

11. The equilibrium occurs where $x^2 = 0$, or at $x = 0$. The derivative of the rate-of-change function is $2x$, which is equal to 0 at the equilibrium. The stability theorem does not apply.

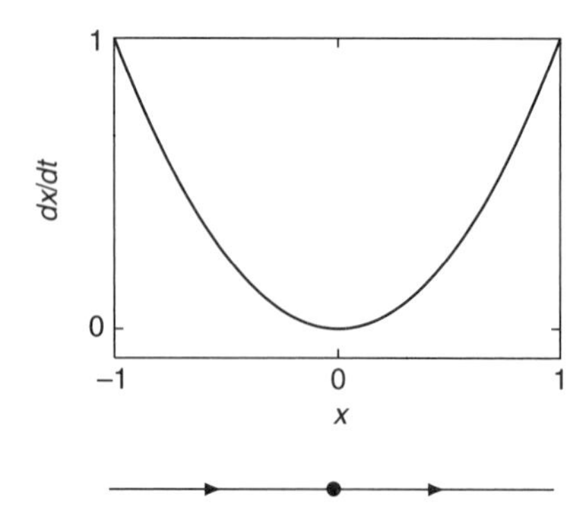

The rate of change is positive except at the equilibrium, so x is increasing except at the equilibrium. This equilibrium is sort of half stable: stable from the left and unstable to the right.

13. The second derivative is 0 at the equilibrium. The third derivative must be negative because the function switches from being concave up for values of x less than the equilibrium to concave down for values of x greater than the equilibrium.

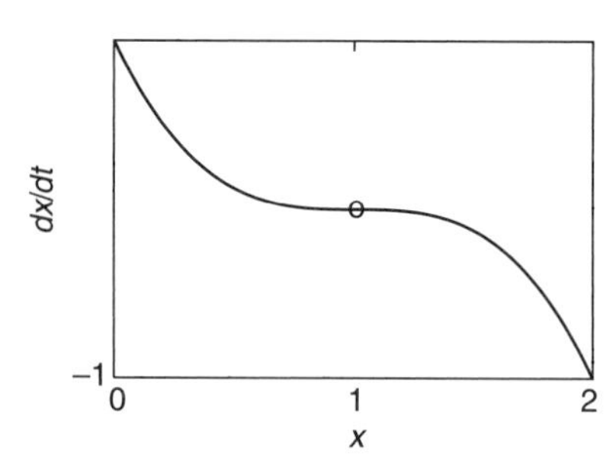

15. Suppose there are two stable equilibria in a row. Then the rate of change must cross from negative to positive at each. In particular, just above the lower one, the value of the rate of change is positive. Just below the upper one, the value of the rate of change is negative. By the Intermediate Value Theorem, there must be another crossing in between.

17. The equilibria are $x = 0$ and $x = a$. The derivative of the rate-of-change function $f(x) = ax - x^2$ is $f'(x) = a - 2x$. Then $f'(0) = a$, so the equilibrium at $x = 0$ is stable if $a < 0$ and unstable if $a > 0$. $f'(a) = -a$, so the equilibrium at $x = a$ is stable if $a > 0$ and unstable if $a < 0$. At $a = 0$, there is an exchange of stability when the two equilibria cross.

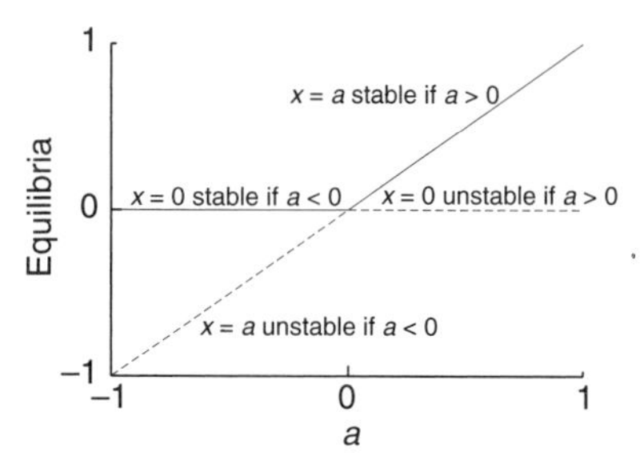

19. The equilibria are $x = 0, x = \pm\sqrt{a}$. The last two equilibria do not exist if $a < 0$. The derivative of the rate-of-change function $f(x) = ax - x^3$ is $f'(x) = a - 3x^2$. $f'(0) = a$, so the equilibrium at $x = 0$ is stable if $a < 0$ and unstable if $a > 0$. $f'(\sqrt{a}) = -2a$, so the equilibrium at $x = \sqrt{a}$ is stable when $a > 0$. $f'(-\sqrt{a}) = -2a$, so the equilibrium at $x = -\sqrt{a}$ is also stable when $a > 0$.

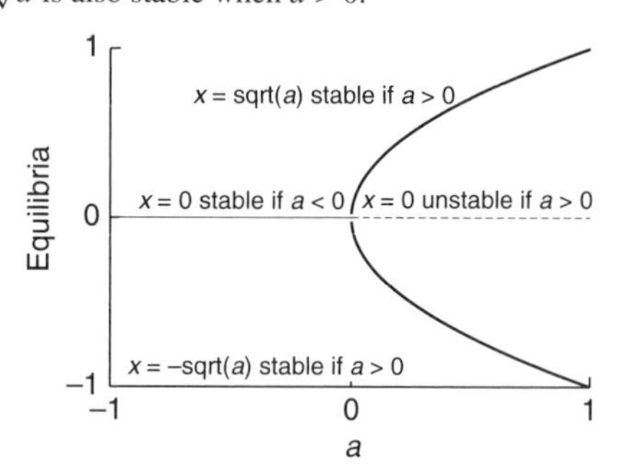

21. This population obeys the autonomous differential equation $\frac{db}{dt} = (1 - 0.002b)b$ and has equilibria at $b = 500$ and at $b = 0$. The derivative of the rate-of-change function $f(b) = (1 - 0.002b)b$ is $f'(b) = 1 - 0.004b$. Then $f'(0) = 1 > 0$, so $b = 0$ is unstable, and $f'(500) = -1 < 0$, so $b = 500$ is stable.

23. This population obeys the differential equation $\frac{db}{dt} = (-2 + 0.01b)b$ and has equilibria at $b = 200$ and at $b = 0$. The derivative of the rate-of-change function $f(b) = (-2 + 0.01b)b$ is $f'(b) = -2 + 0.02b$. Then $f'(0) = -2 < 0$, so $b = 0$ is stable, and $f'(200) = 2 > 0$, so $b = 200$ is unstable.

25. This population obeys the autonomous differential equation $\frac{dp}{dt} = 0.5p(1 - p)^2$ and has equilibria at $p = 0$ and at $p = 1$. The derivative of the rate-of-change function $f(p) = 0.5p(1 - p)^2$ is $f'(p) = 0.5(1 - p)^2 - p(1 - p)$. Then $f'(0) = 0.5 > 0$, so $p = 0$ is unstable, and $f'(1) = 0$, so we can't tell. However, the graph does indicate that this equilibrium is stable.

27. a. This is a case where chemical is used up at a rate proportional to its concentration.
b. Solving $f(C) = (5 - C) - C = 0$ gives $C = 2.5$. The derivative of the rate-of-change function is $f'(C) = -2$, so the equilibrium is stable.
c. Without the reaction, the equilibrium is $C = \Gamma = 5$. Absorption decreases the equilibrium amount.

29. a. This is a case where chemical is created at a rate that gets larger as the concentration gets larger, but reaches a maximum of 0.5.
b. Solving $f(C) = (5 - C) + \frac{C}{2+C} = 0$ gives $C = 2 \pm \sqrt{14}$. Only one of these values is positive, at $C = 2 + \sqrt{14} = 5.74$. The derivative of the rate-of-change function is

$$f'(C) = -1 + \frac{1}{2+C} - \frac{C}{(2+C)^2}$$

and $f'(5.74) = -0.97$, so the equilibrium is stable.
c. Without the reaction, the equilibrium is $C = \Gamma = 5$. Chemical creation increases the equilibrium amount, but only slightly.

31. The derivative of the rate-of-change function is

$$\frac{d}{dN}\left(\frac{3N^2}{2+N^2} - N\right) = \frac{12N}{(2+N^2)^2} - 1.$$

At $N = 0$, this is -1, so the equilibrium at $N = 0$ is stable. At $N = 1$, this is $1/3$, so the equilibrium at $N = 1$ is unstable. At $N = 2$, this is $-1/3$, so the equilibrium at $N = 2$ is stable. This population of 1 acts as a threshold.

33. The derivative of the rate-of-change function is

$$\frac{d}{dv}(9.8 - 0.0032v^2) = -0.0064v,$$

which is negative for any positive speed, including the equilibrium at $v = 55.3$. The equilibrium is stable, consistent with the inward pointing arrows on the phase-line diagram. The falling object will approach its terminal velocity.

35. The derivative of the rate-of-change function is

$$\frac{d}{dy}(-2.0\sqrt{y}) = -\frac{2.0}{2\sqrt{y}}.$$

The equilibrium is at $y = 0$, so this derivative is negative infinity, implying that the equilibrium is very stable, as in our phase-line diagram. The cylinder will drain very quickly.

37. The derivative of the rate-of-change function is

$$\frac{d}{dS}\left(-\frac{S}{1+S}\right) = -\frac{1}{(1+S)^2},$$

which is always negative. Any equilibrium must be stable. In particular, the equilibrium at $S = 0$ is stable, consistent with using up a substance.

39. The derivative of the rate-of-change function is

$$\frac{d}{dV}(a_1 V^{2/3} - a_2 V) = \frac{2}{3}a_1 V^{-1/3} - a_2.$$

At the equilibrium, $V = (\frac{a_2}{a_1})^3$, this is

$$\frac{2}{3}a_1 V^{-1/3} - a_2 = -\frac{1}{3}a_2 < 0.$$

Again, the equilibrium is stable, consistent with the inward pointing arrows on the phase-line diagram. This growing organism will reach a final size when it becomes too big to grow any more.

41. The equilibria are $b = 0$ and $b = 1 - h$. The second equilibrium is positive only if $h < 1$. The rate-of-change function $f(b) = b(1 - b - h)$ has derivative $f'(b) = 1 - h - 2b$. Then $f'(0) = 1 - h$, so $b = 0$ is stable if $h > 1$ and unstable if $h < 1$. Also $f'(1 - h) = h - 1$, so $b = 1 - h$ is stable if $h < 1$ and unstable if $h > 1$ (where it is negative and makes no biological sense).

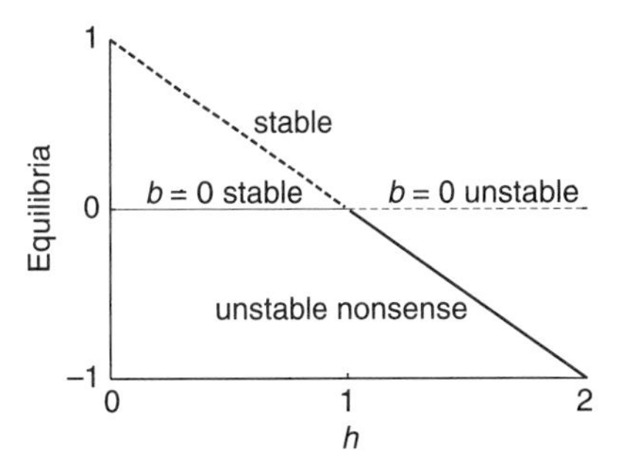

43. The equilibria are $N = 0$, $N = \frac{r+\sqrt{r^2-4}}{2}$, and $N = \frac{r-\sqrt{r^2-4}}{2}$. The last two only exist when $r \geq 2$. The rate-of-change function $f(N) = \frac{rN^2}{1+N^2} - N$ has derivative

$$f'(N) = \frac{2rN}{1+N^2} - \frac{2rN^3}{(1+N^2)^2} - 1.$$

Then $f'(0) = -1$, so $I = 0$ is always stable. At $r = 3$, the equilibria are 2.618 and 0.382. Substituting in gives $f'(2.618) = -0.745$ and $f'(0.382) = 0.745$. The larger equilibrium is stable, and the smaller one is unstable.

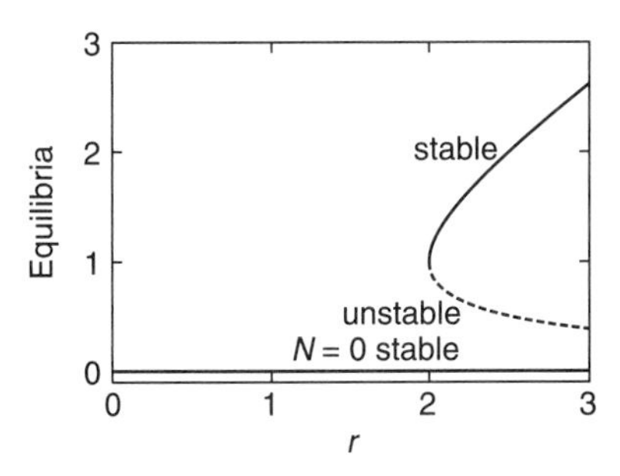

45. With $\alpha = 4$, the differential equation is $\frac{dI}{dt} = 4I^2(1 - I) - I$. The equilibria are $I = 0$ and $I = 1/2$.

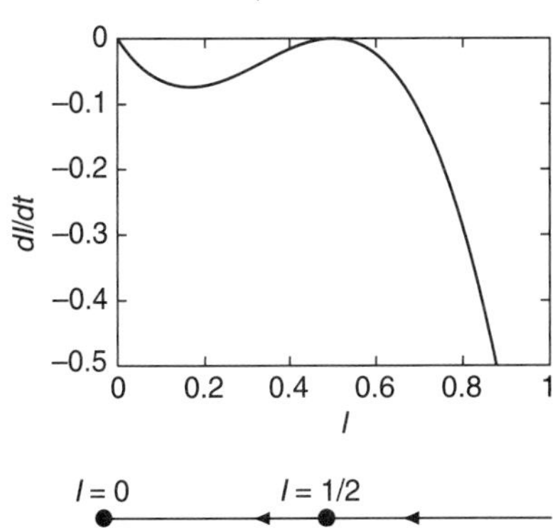

The derivative of the rate-of-change function is $8I(1 - I) - 4I^2 - 1$, which is negative at $I = 0$ and equal to 0 at $I = 1/2$. From our phase-line diagram, however, we see that the rate of change is always negative, implying that the equilibrium at $I = 1/2$ is half stable, stable from the right and unstable to the left.

Section 5.4, page 452

1. $\frac{db}{b} = 0.01dt$, so $\ln(b) = 0.01t + c$ and $b(t) = Ke^{0.01t}$ with $K = 1000$. Checking,

$$\frac{db}{dt} = \frac{d}{dt}(1000e^{0.01t}) = 10e^{0.01t} = 0.01 \cdot 1000e^{0.01t} = 0.01b(t).$$

This checks.

3. $\frac{dN}{1+N} = dt$, so $\ln(1+N) = t + c$, $1 + N(t) = Ke^t$ and $N(t) = Ke^t - 1$. Substituting $t = 0$, we find $K = 2$, so $N(t) = 2e^t - 1$. Checking,

$$\frac{dN}{dt} = \frac{d}{dt}(2e^t - 1) = 2e^t = 1 + (2e^t - 1) = 1 + N(t).$$

This checks.

5. $\frac{db}{1000-b} = dt$, so $-\ln(1000 - b) = t + c$, $1000 - b(t) = Ke^{-t}$, and $b(t) = 1000 - Ke^{-t}$. Substituting $t = 0$, we find $K = 500$, so $b(t) = 1000 - 500e^{-t}$. Checking,

$$\begin{aligned}\frac{db}{dt} &= \frac{d}{dt}(1000 - 500e^{-t}) = 500e^{-t}\\ &= 1000 - (1000 - 500e^{-t}) = 1000 - b(t).\end{aligned}$$

This checks.

7. $dP = \frac{5}{1+2t}dt$. Integrating with the substitution $s = 1 + 2t$, we find $P(t) = 2.5\ln(1 + 2.0t) + c$. The constant is $c = 0$. Checking,

$$\frac{dP}{dt} = \frac{d}{dt}[2.5\ln(1 + 2.0t)] = \frac{5.0}{1 + 2t}.$$

This checks.

9. $dL = 1000e^{0.2t}dt$. Integrating with the substitution $s = 0.2t$, we find $L(t) = 5000e^{0.2t} + c$. The constant is $c = -4000$, giving a solution $L(t) = 5000e^{0.2t} - 4000$. Checking,

$$\frac{dL}{dt} = \frac{d}{dt}(5000e^{0.2t} - 4000) = 1000e^{0.2t}.$$

This checks.

11. $b(t) = \frac{1}{0.01-t}$. It blows up at time $t = 0.01$.

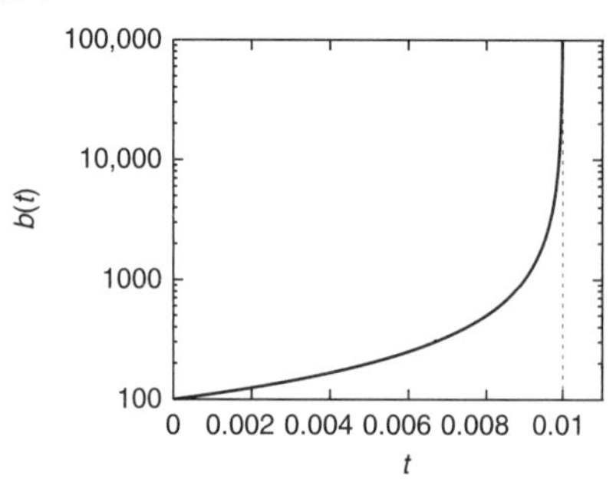

13. Separating variables gives

$$\begin{aligned}\frac{db}{b^{1.1}} &= dt\\ -10b^{-0.1} &= t + c.\end{aligned}$$

Substituting in the initial condition gives $c = -10 \cdot 100^{-0.1} = -6.31$. Solving for $b(t)$ gives $b(t) = \left(\frac{10}{6.31-t}\right)^{10}$. It blows up at time $t = 6.31$.

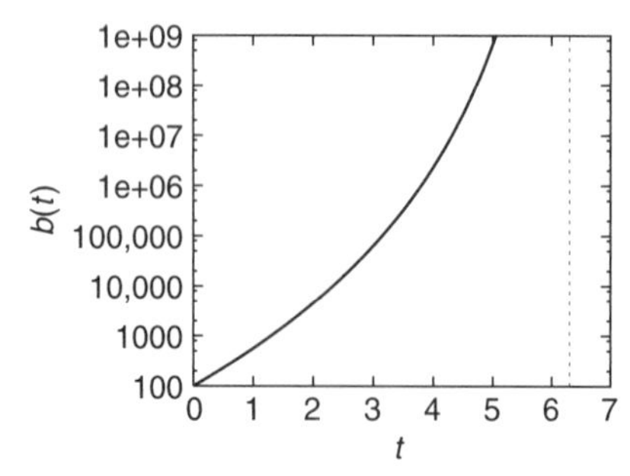

15. a.

$\frac{1+x}{x}dx = dt$	separating variables
$\ln(x) + x + c_1 = t + c_2$	integrating
$\ln(x) + x = t + c$	combining constants

b. Substitute $t = 0$ and $x = 1$, and find $c = 1$.
c. $t = \ln(x) + x - 1$.

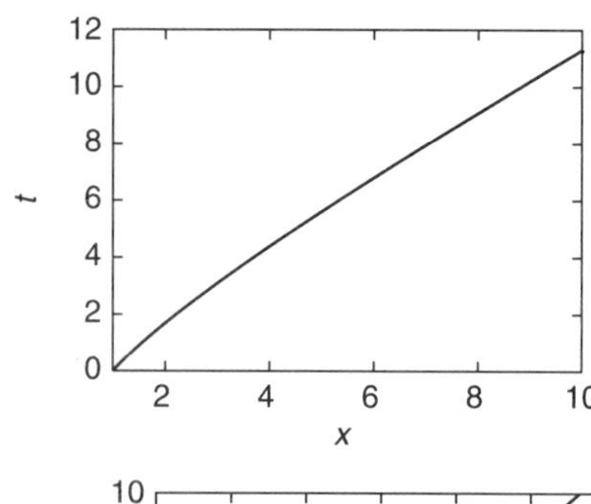

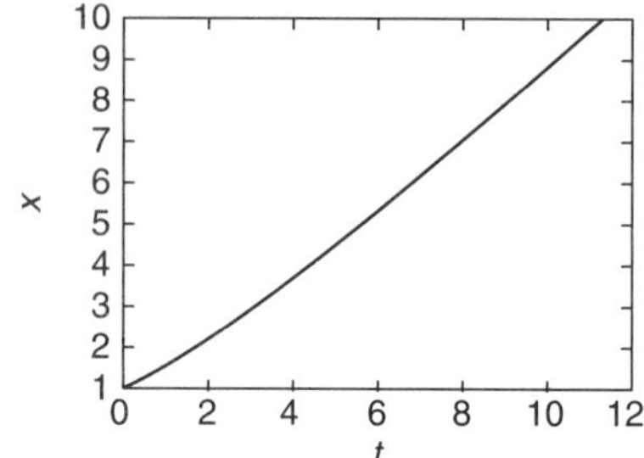

17.

$\frac{dN}{N(1-\frac{N}{K})} = r\,dt$	separating variables
$\frac{dN}{N} + \frac{dN}{K-N} = r\,dt$	breaking into partial
$\ln(N) - \ln(N-K) = rt + c$	fractions integrating and
$\ln\left(\frac{N}{N-K}\right) = rt + c$	combining constants
$\frac{N}{N-K} = Ce^{rt}$	applying a law of logs
$N = Ce^{rt}(N-K)$	exponentiating both sides
$(Ce^{rt} - 1)N = CKe^{rt}$	multiplying out isolating N
$N = \frac{CKe^{rt}}{Ce^{rt}-1}.$	solving for N

This looks different. But solving for C gives

$$\begin{aligned}N_0 &= \frac{CK}{C-1}\\ N_0(C-1) &= CK\\ C(N_0 - K) &= N_0\\ C &= \frac{N_0}{N_0 - K}.\end{aligned}$$

Substituting in and placing over a common denominator gives

$$N = \frac{N_0Ke^{rt}}{K - N_0 + N_0e^{rt}}$$

exactly as before.

19.

$$\begin{aligned}C(t) &= \Gamma + e^{-\beta t}[C(0) - \Gamma]\\ &= 2.0 + e^{-0.01t}(5.0 - 2.0)\\ &= 2.0 + 3.0e^{-0.01t}\\ C(10) &= 2.0 + 3.0e^{-0.1}\\ &= 4.715.\end{aligned}$$

The equilibrium is 2.0, and it is halfway to the equilibrium when $C(t) = 3.5 \times 10^{-5}$. Solving for t,

$$\begin{aligned}3.5 &= 2.0 + 3.0e^{-0.01t}\\ 1.5 &= 3.0e^{-0.01t}\\ 0.5 &= e^{-0.01t}\\ \ln(0.5) &= -0.01t\\ -100 \cdot \ln(0.5) &= t = 69.3.\end{aligned}$$

21. $C(t) = 2.0 + 3.0e^{-0.1t}$, $C(10) = 3.104$. The time to reach 3.5, halfway to the equilibrium, is 6.93 s.

23.

$$\begin{aligned} \frac{dy}{2\sqrt{y}} &= -dt && \text{separating variables} \\ \sqrt{y} + c_1 &= -t + c_2 && \text{integrating} \\ \sqrt{y} &= -t + c && \text{combining constants} \\ y &= (-t + c)^2. && \text{solving for } y \end{aligned}$$

The constant c solves $y(0) = 4 = (-0 + c)^2$, or $c = 2$.

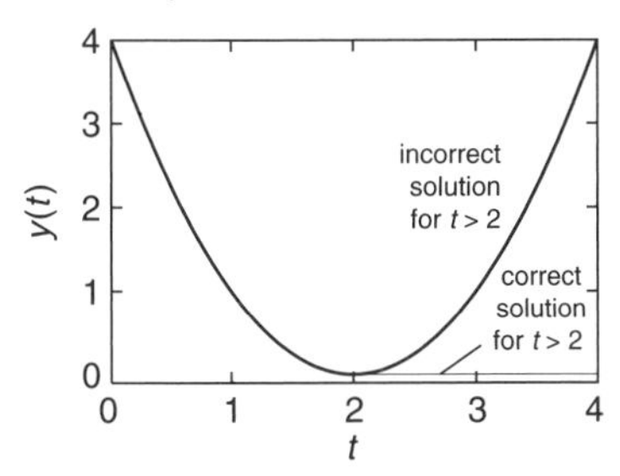

The cylinder is empty at $t = 2$, so the solution must stay at 0. The solution in this case actually reaches 0 at $t = 2$, unlike the exponentially decaying solution of the bacterial death equation, which only approaches 0 as a limit.

25.

$$\begin{aligned} \frac{db}{2.0b - 1000} &= dt && \text{separating variables} \\ \frac{1}{2}\ln(2.0b - 1000) + c_1 &= t + c_2 && \text{integrating combining} \\ \ln(2.0b - 1000) &= 2t + c && \text{constants starting} \\ 2.0b - 1000 &= Ke^{2t} && \text{to solve for } b \text{ solve for } b \\ b &= Ke^{2t} + 500. && \text{(changing the constant } K\text{)} \end{aligned}$$

If the solution starts just above the equilibrium at 500, say with $b(0) = 501$, we find $K = 1$, and a solution $b(t) = 500 + e^{2t}$. This grows exponentially away from the equilibrium. If the solution starts just below the equilibrium at 500, say with $b(0) = 499$, we find $K = -1$, and a solution $b(t) = 500 - e^{2t}$. This declines exponentially away from the equilibrium.

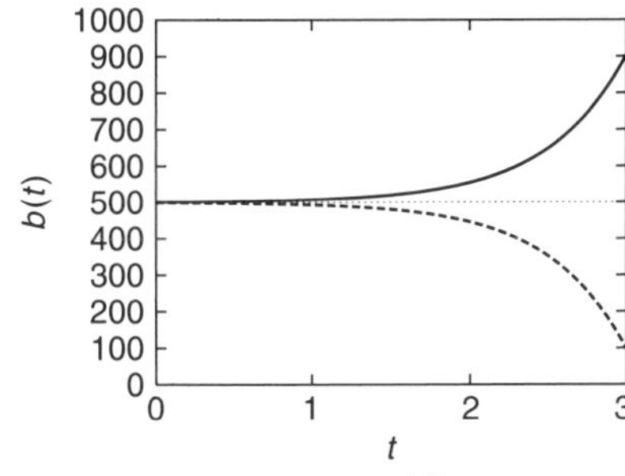

27. We found the differential equation $\frac{dC}{dt} = -\beta C + 3\beta\Gamma$. Separating variables,

$$\begin{aligned} \frac{dC}{C - 3\Gamma} &= -\beta dt && \text{separating variables} \\ \ln(C - 3\Gamma) + c_1 &= -\beta t + c_2 && \text{integrating} \\ C - 3\Gamma &= Ke^{-\beta t} && \text{starting to solve for } C \\ C &= Ke^{-\beta t} + 3\Gamma. && \text{solve for } C \end{aligned}$$

With the initial condition $C(0) = \Gamma$, the constant K is -2Γ. The solution increases from this point up to the equilibrium at $C = 3\Gamma$.

29. The per capita production is an increasing function of time, meaning that things are getting better and better for this population. $\frac{db}{b} = t\,dt$, so $\ln(b) = t^2/2 + c$ and $b = Ke^{t^2/2}$ with $K = 10^6$. This population grows faster than exponentially. The solution is $b(t) = 10^6 e^{t^2/2}$. To check, we must use the chain rule to find the derivative. If we write $b(t) = f[g(t)]$, where $g(t) = t^2/2$ and $f(g) = 10^6 e^g$, then $g'(t) = t$ and $f'(g) = 10^6 e^g$, so

$$b'(t) = f'(g)g'(t) = 10^6 te^g = 10^6 te^{t^2/2} = tb(t).$$

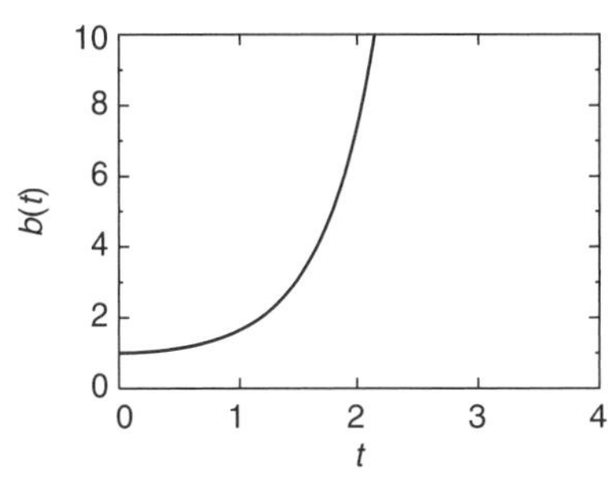

31. The per capita production is an oscillating function of time. $\frac{db}{b} = \cos(t)dt$, so $\ln(b) = \sin(t) + c$ and $b = Ke^{\sin(t)}$ with $K = 10^6$. This population oscillates around its initial value. The solution is $b(t) = 10^6 e^{\sin(t)}$. To check, we must use the chain rule to find the derivative. If we write $b(t) = f[g(t)]$, where $g(t) = \sin(t)$ and $f(g) = 10^6 e^g$, then $g'(t) = \cos(t)$ and $f'(g) = 10^6 e^g$, so

$$b'(t) = f'(g)g'(t) = 10^6 \cos(t)e^g = 10^6 \cos(t)e^{\sin(t)} = \cos(t)b(t).$$

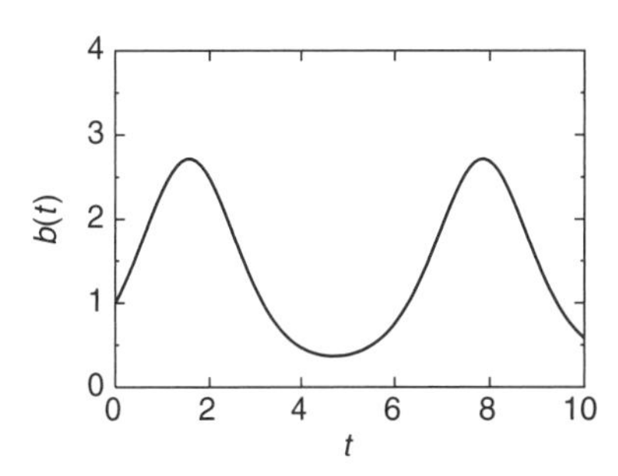

33. The differential equation is $\frac{db}{dt} = (-2 + 0.01b)b$. Separating variables gives

$$\frac{1}{(-2 + 0.01b)b}\,db = dt.$$

We can separate the left-hand side with partial fractions as

$$\begin{aligned} \frac{1}{(-2 + 0.01b)b} &= \frac{A}{-2 + 0.01b} + \frac{C}{b} \\ &= \frac{Ab + C(-2 + 0.01b)}{(-2 + 0.01b)b} \\ &= \frac{-2C + (A + 0.01C)b}{(-2 + 0.01b)b}. \end{aligned}$$

Therefore, $-2C = 1$, $C = -1/2$, and $A + 0.01C = 0$, so that $A = 0.005$.

$$\begin{aligned} \frac{db}{b(-2 + 0.01b)} &= dt && \text{separating variables} \\ -0.5\frac{db}{b} + 0.005\frac{db}{-2 + 0.01b} &= dt && \text{break into partial} \\ -0.5\ln(b) + 0.5\ln(2 - 0.01b) &= t + c && \text{fractions integrating} \\ \ln\left(\frac{-2 + 0.01b}{b}\right) &= 2t + c && \text{and combining} \\ \frac{-2 + 0.01b}{b} &= Ke^{2t} && \text{constants applying a} \\ \frac{-2}{b} + 0.01 &= Ke^{2t} && \text{law of logs expone-} \\ \frac{-2}{b} &= Ke^{2t} - 0.01 && \text{ntiating both sides} \\ b &= \frac{2}{0.01 - Ke^{2t}}. && \text{simplifying fraction} \\ & && \text{move 0.01 to right-} \\ & && \text{hand side solve} \\ & && \text{for } b \end{aligned}$$

Substituting in the initial condition gives $K = -0.01$, and the solution

$$b = \frac{2}{0.01 + 0.01e^{2t}} = \frac{200}{1 + e^{2t}}.$$

This solution approaches 0.

35. The equation

$$\frac{dp}{dt} = (\mu - \lambda)p(1 - p)$$

is a special case of the logistic equation with $r = \mu - \lambda$ and $K = 1$. Using the solution from the example gives

$$p = \frac{p_0 e^{(\mu-\lambda)t}}{1 - p_0 + p_0 e^{(\mu-\lambda)t}}.$$

Multiplying top and bottom by $e^{\lambda t}$ gives

$$p = \frac{p_0 e^{\mu t}}{p_0 e^{\mu t} + (1 - p_0)e^{\lambda t}},$$

which matches the earlier result.

37.

$$\begin{aligned}\frac{dy}{dt} &= e^t H + e^t \frac{dH}{dt}\\ &= e^t H + e^t[-H + A(t)]\\ &= e^t A(t) = e^{(1+\beta)t}.\end{aligned}$$

39. Using $\beta = 0.1$ gives the following graph.

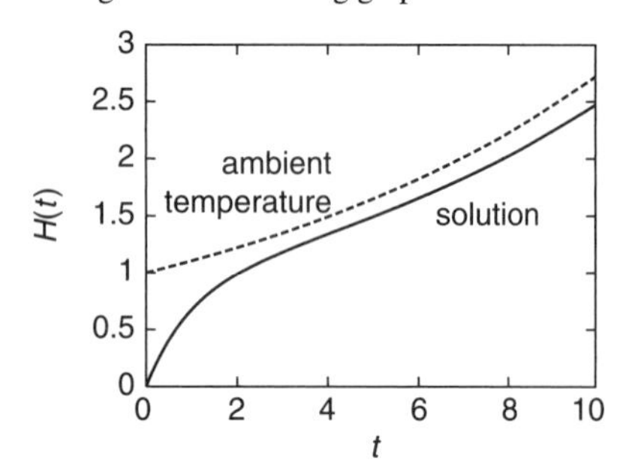

The solution does a pretty good job of tracking the ambient temperature.

Section 5.5, page 462

1. The equations are

$$\begin{aligned}\frac{db}{dt} &= \lambda b\\ \frac{dp}{dt} &= -\delta p.\end{aligned}$$

These two species ignore each other. The prey will grow exponentially, and the predators will die out.

3. The equations are

$$\begin{aligned}\frac{db}{dt} &= (\lambda + \epsilon p)b\\ \frac{dp}{dt} &= -\delta p.\end{aligned}$$

The predators will die out, even though they are eating some of the prey. Apparently, these prey are not at all nutritious.

5. The equation for a is

$$\frac{da}{dt} = 2\left(1 - \frac{a}{1000}\right)a.$$

with equilibria at $a = 0$ and $a = 1000$. The equilibrium at 0 is unstable, and the equilibrium at 1000 is stable.

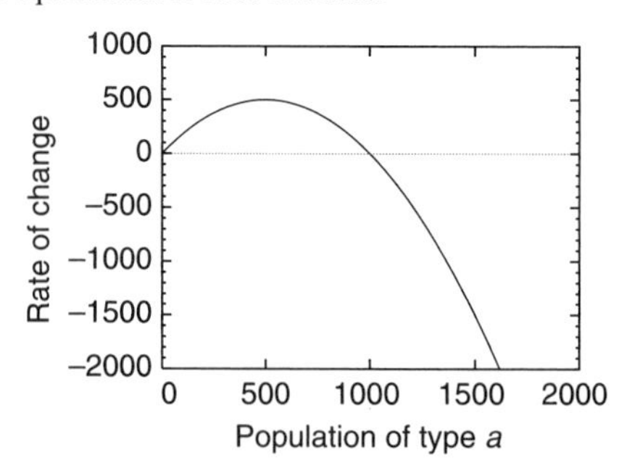

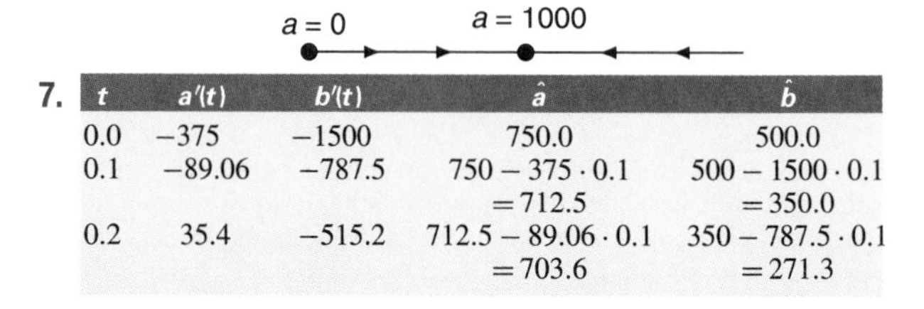

7.

t	$a'(t)$	$b'(t)$	$\hat{a}$	$\hat{b}$
0.0	−375	−1500	750.0	500.0
0.1	−89.06	−787.5	$750 - 375 \cdot 0.1 = 712.5$	$500 - 1500 \cdot 0.1 = 350.0$
0.2	35.4	−515.2	$712.5 - 89.06 \cdot 0.1 = 703.6$	$350 - 787.5 \cdot 0.1 = 271.3$

9.

t	$a'(t)$	$b'(t)$	$\hat{a}$	$\hat{b}$
0.0	−375	−1500	750.0	500.0
0.05	−228.5	−1115.6	$750 - 375 \cdot 0.05 = 731.25$	$500 - 1500 \cdot 0.05 = 425.0$
0.1	−128.2	−869.9	$731.25 - 228.5 \cdot 0.05 = 719.8$	$425 - 1115.6 \cdot 0.05 = 369.2$
0.15	−55.8	−702.4	$719.8 - 128.2 \cdot 0.05 = 713.4$	$369.2 - 869.9 \cdot 0.05 = 325.7$
0.2	−1.73	−582.6	$713.4 - 55.8 \cdot 0.05 = 710.6$	$325.7 - 702.4 \cdot 0.05 = 290.6$

The values are fairly close, but the predators are off by about 10%.

11.

t	$H'(t)$	$A'(t)$	$\hat{H}$	$\hat{A}$
0.0	−12	4	60.0	20.0
0.1	−11.5	3.84	$60 - 12 \cdot 0.1 = 58.8$	$20 + 4 \cdot 0.1 = 20.4$
0.2	−11.1	3.69	$58.8 - 11.5 \cdot 0.1 = 57.6$	$20.4 + 3.84 \cdot 0.1 = 20.8$

13.

t	$H'(t)$	$A'(t)$	$\hat{H}$	$\hat{A}$
0.0	−120	40	60.0	20.0
0.5	0	0	$60 - 120 \cdot 0.25 = 30.0$	$20 + 4 \cdot 0.25 = 30.0$
1.0	0	0	$30.0 + 0 \cdot 0.25 = 30.0$	$30.0 + 0 \cdot 0.25 = 30.0$

This is a bit strange. The solution seems to have gotten stuck after one step. Because the real object would only approach the ambient temperature, this must be a result of using too large a step size.

15. $v = \frac{dx}{dt}$ and $a = \frac{dv}{dt}$. But we know that $a = -x$, so $\frac{dv}{dt} = -x$. As a system,

$$\begin{aligned}\frac{dx}{dt} &= v\\ \frac{dv}{dt} &= -x.\end{aligned}$$

17. $v(t) = -\sin(t) = \frac{dx}{dt}$ and $\frac{dv}{dt} = -\cos(t) = -x(t)$. This works. The initial position is $x(0) = 1$, and the initial velocity is $v(0) = 0$.

19. We find

$$\begin{aligned}\hat{x}(0.2) &= 1.0 & \hat{v}(0.2) &= -0.2\\ \hat{x}(0.4) &= 0.96 & \hat{v}(0.4) &= -0.4\\ \hat{x}(0.6) &= 0.88 & \hat{v}(0.6) &= -0.592\\ \hat{x}(0.8) &= 0.7616 & \hat{v}(0.8) &= -0.768\\ \hat{x}(1.0) &= 0.6080 & \hat{v}(1.0) &= -0.92032\end{aligned}$$

These compare reasonably well with the exact answers,

$$x(0.2) = 0.9801 \qquad v(0.2) = -0.1986$$
$$x(0.4) = 0.9210 \qquad v(0.4) = -0.3894$$
$$x(0.6) = 0.8253 \qquad v(0.6) = -0.5646$$
$$x(0.8) = 0.6967 \qquad v(0.8) = -0.7173$$
$$x(1.0) = 0.5403 \qquad v(1.0) = -0.8414$$

but lag behind by quite a bit.

21. We can write this as

$$\frac{dF}{dt} = F^2 + kt$$
$$\frac{dt}{dt} = 1$$

because the derivative of t with respect to itself is 1.

23.

$$\frac{db}{dt} = (1.0 - 0.05p)b$$
$$\frac{dp}{dt} = (-1.0 + 0.02b)p.$$

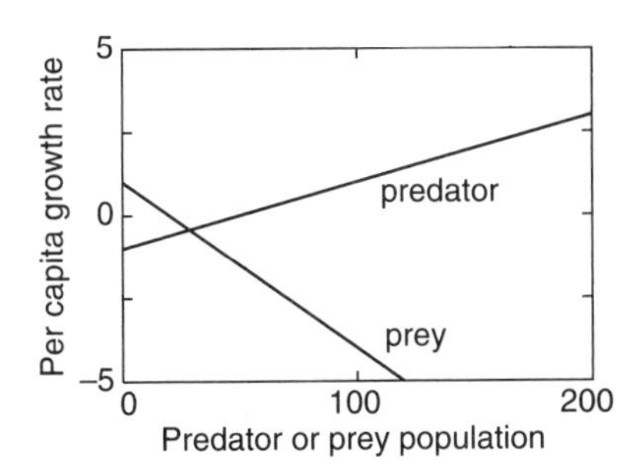

25.

$$\frac{db}{dt} = (2.0 - 0.0001p^2)b$$
$$\frac{dp}{dt} = (-1.0 + 0.01b)p.$$

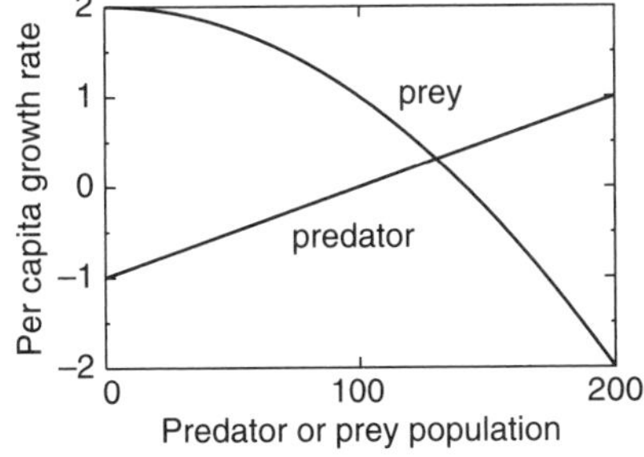

27. Call the predators p_1 and p_2. Then

$$\frac{dp_1}{dt} = (-1.0 + 0.001p_2)p_1$$
$$\frac{dp_2}{dt} = (-1.0 + 0.001p_1)p_2.$$

29.

$$\frac{da}{dt} = \mu\left(1 - \frac{a+b}{K_a}\right)a$$
$$\frac{db}{dt} = \lambda\left(1 - \frac{b}{K_b}\right)b.$$

31. a. $A_1 = C_1V_1 = 2.0C_1$, and $A_2 = C_2V_2 = 5.0C_2$.

b.

$$\frac{dA_1}{dt} = \beta(C_2 - C_1)$$
$$\frac{dA_2}{dt} = \beta(C_1 - C_2).$$

c. Then

$$\frac{dC_1}{dt} = \frac{\beta}{V_1}(C_2 - C_1) = \frac{\beta}{2.0}(C_2 - C_1)$$
$$\frac{dC_2}{dt} = \frac{\beta}{V_2}(C_1 - C_2) = \frac{\beta}{5.0}(C_1 - C_2).$$

d. The concentration changes faster in the smaller cell.

33. The general form is

$$\frac{dH}{dt} = \alpha(A - H)$$
$$\frac{dA}{dt} = \alpha_2(H - A).$$

α_2 will be 10 times smaller due to the size of the room, but 5 times larger due to the specific heat. Therefore, we expect that

$$\frac{dH}{dt} = \alpha(A - H)$$
$$\frac{dA}{dt} = \frac{\alpha}{2}(H - A).$$

35. The equations are

$$\frac{dI}{dt} = \alpha IS - \mu I - kI$$
$$\frac{dS}{dt} = -\alpha IS + \mu I.$$

37. The equations are

$$\frac{dI}{dt} = \alpha IS - \mu I - kI$$
$$\frac{dS}{dt} = -\alpha IS + \mu I - kS.$$

39. The equations are

$$\frac{dI}{dt} = bI + \alpha IS - \mu I$$
$$\frac{dS}{dt} = bS - \alpha IS + \mu I.$$

41. In terms of p, we found differential equations

$$\frac{da}{dt} = 2(1 - p)a$$
$$\frac{db}{dt} = 1.5(1 - p)b.$$

Substituting in that $1 - p = \frac{b}{a+b}$ gives the equations

$$\frac{da}{dt} = 2\frac{b}{a+b}a$$
$$\frac{db}{dt} = 1.5\frac{b}{a+b}b.$$

Section 5.6, page 470

1. Solving each equation for y gives

$$y = 3x - 3$$
$$y = 2x + 1.$$

Setting equal, we find $3x - 3 = 2x + 1$, so $x = 4$. Plugging in gives $y = 9$ in each equation.

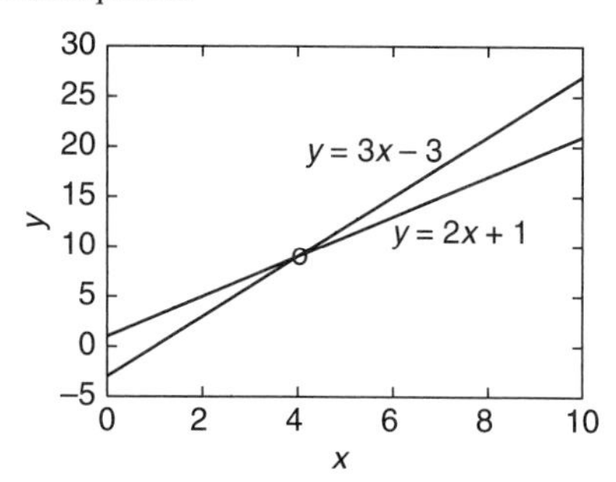

3. Solving each equation for y gives

$$y = x + 1$$
$$y = x + 1.$$

Setting equal, we find $x + 1 = x + 1$. Any value of x is a solution. This happens because the two lines are identical, and any value where $y = x + 1$ is a solution.

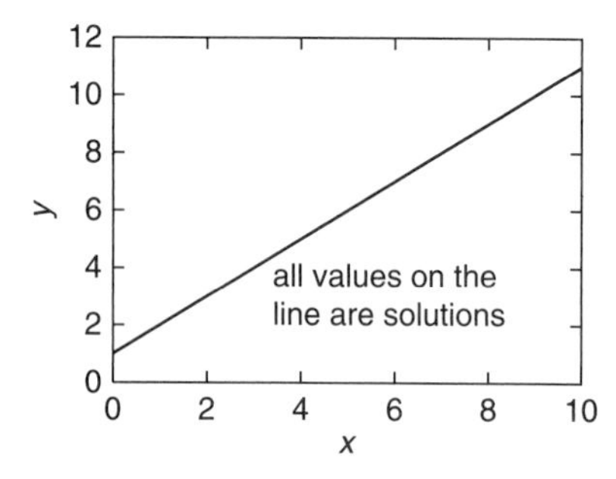

5. Solving each equation for y gives

$$y = 3x^2 + 2x - 5$$
$$y = 2x + 1.$$

Setting equal, we find $3x^2 + 2x - 5 = 2x + 1$, so $3x^2 = 6$, and $x = \pm\sqrt{2}$. With $x = \sqrt{2}$, the first equation gives $y = 1 + 2\sqrt{2}$, as does the second. With $x = -\sqrt{2}$, the first equation gives $y = 1 - 2\sqrt{2}$, as does the second.

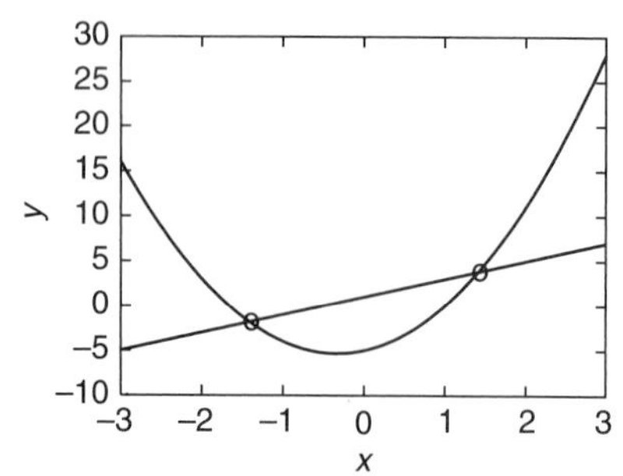

7. Solving the first equation for y gives $y = x$ and $y = -x$. The second gives $y = 2x + 1$. This gives two pairs of equations:

$$y = x$$
$$y = 2x + 1$$

and

$$y = -x$$
$$y = 2x + 1.$$

Setting the first pair equal, we find $x = 2x + 1$, so $x = -1$ and $y = -1$. Setting the second pair equal, we find $-x = 2x + 1$, so $x = -1/3$ and $y = 1/3$.

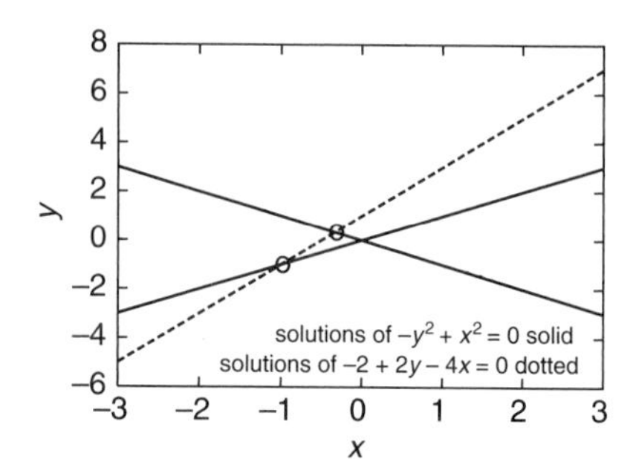

9. Solving the first equation for y gives $y = x$ and $x = 0$. The second gives $y = 2x + 3$. This gives two pairs of equations:

$$y = x$$
$$y = 2x + 3$$

and

$$x = 0$$
$$y = 2x + 3.$$

Setting the first pair equal, we find $x = 2x + 3$, so $x = -3$ and $y = -3$. We can substitute $x = 0$ into the second equation of the second pair to find $y = 3$.

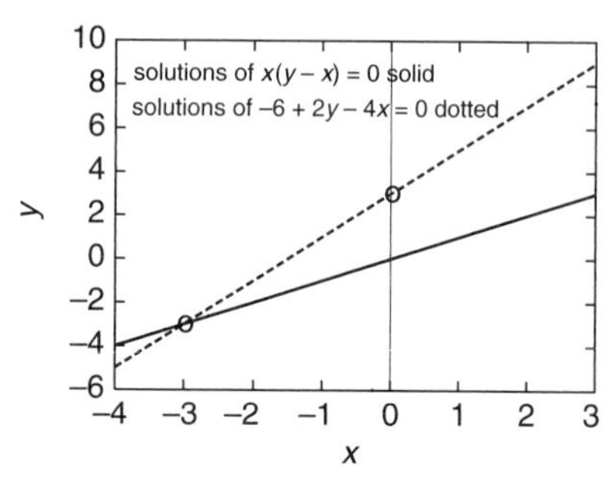

11. The p-nullcline consists of the two pieces $p = 0$ and $b = \delta/\eta = 600$. The b-nullcline consists of the two pieces $b = 0$ and $p = \lambda/\epsilon = 500$. The equilibria are (0, 0) and (600, 500).

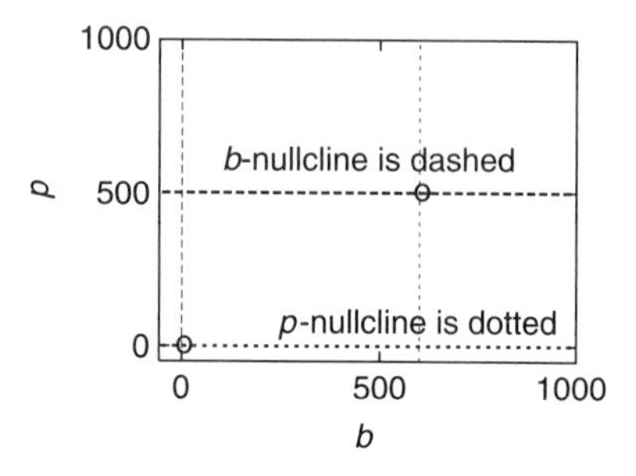

13. Both nullclines are the line $H = A$, which consists entirely of equilibria.

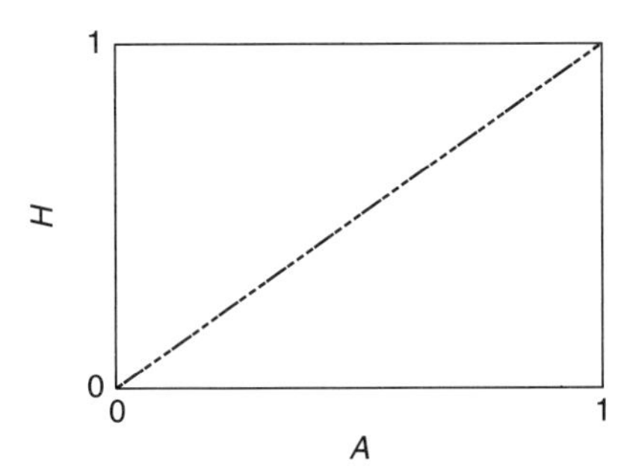

15. Place b on the vertical axis. The a-nullcline is the two pieces $a = 0$ and $b = 10^6 - a$. The b-nullcline is the two pieces $b = 0$ and $b = 10^7 - a$. The equilibria are $(0, 0)$, $(10^6, 0)$, and $(0, 10^7)$.

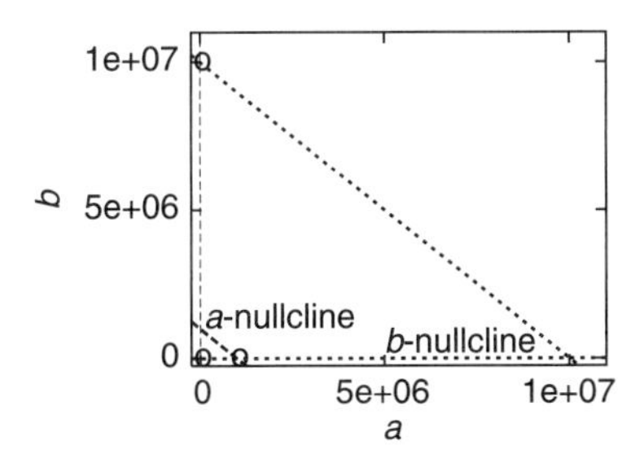

17. Place b on the vertical axis. The p-nullcline consists of the two pieces $p = 0$ and $b = 600$, and the b-nullcline consists of the two pieces $b = 0$ and $p = 500$ as before. The equilibria are $(0, 0)$ and $(500, 600)$.

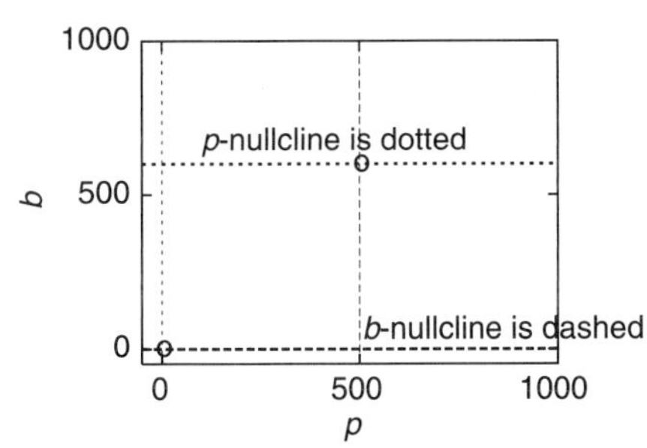

19. Place a on the vertical axis. The a-nullcline is the two pieces $a = 0$ and $a = 10^6 - b$. The b-nullcline is the two pieces $b = 0$ and $a = 10^7 - b$. The equilibria are $(0, 0)$, $(0, 10^6)$, and $(10^7, 0)$.

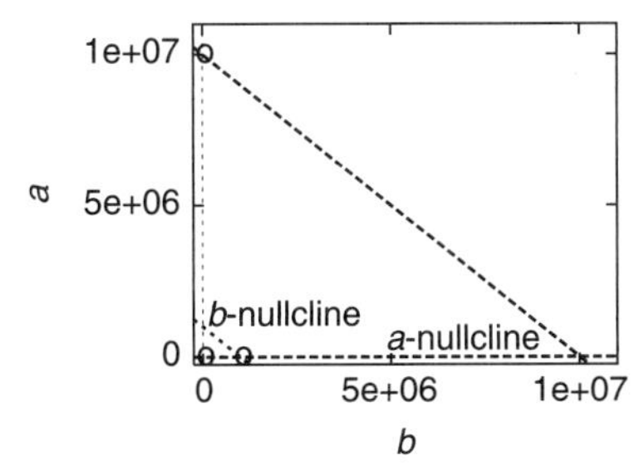

21. The equations are

$$\frac{da}{dt} = 2\left(1 - \frac{a}{1000}\right)a$$

$$\frac{db}{dt} = 3\left(1 - \frac{b}{300}\right)b.$$

The a-nullcline is $a = 0$ and $a = 1000$. The b-nullcline is $b = 0$ and $b = 300$. The equilibria are $(0, 0)$ and $(1000, 300)$.

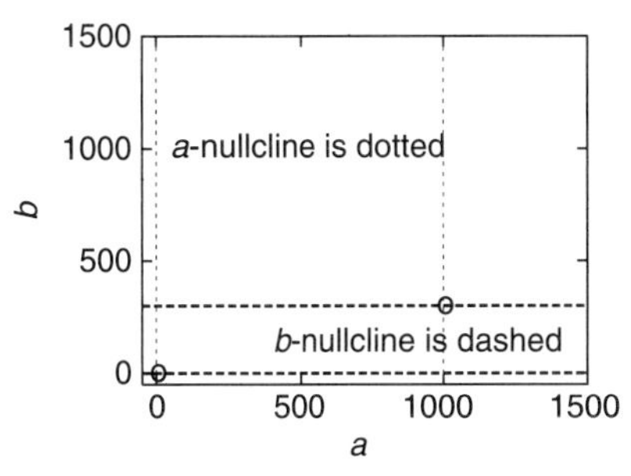

23. With p on the vertical axis, the b-nullcline is $b = 0$ and $p = 20$. The p-nullcline is $p = 0$ or $b = 50$. The only equilibria are $(0, 0)$ and $(50, 20)$.

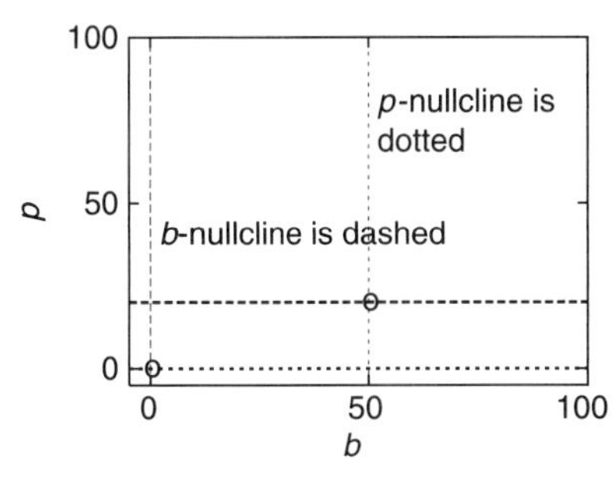

25. With p on the vertical axis, the b-nullcline is $b = 0$ and $p = \sqrt{20,000} = 141$. The p-nullcline is $p = 0$ or $b = 100$. The equilibria are $(0, 0)$ and $(100, 141)$.

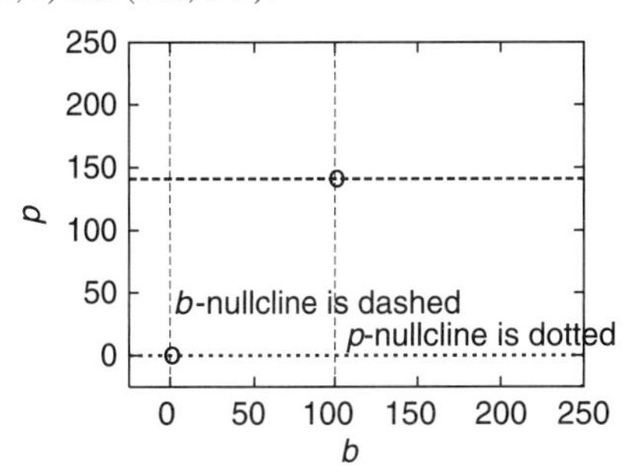

27. We found equations

$$\frac{dp_1}{dt} = (-1.0 + 0.001p_2)p_1$$

$$\frac{dp_2}{dt} = (-1.0 + 0.001p_1)p_2.$$

Putting p_2 on the vertical axis, the p_1-nullcline is $p_1 = 0$ and $p_2 = 1000$. The p_2-nullcline is $p_2 = 0$ or $p_1 = 1000$. The equilibria are $(0, 0)$ and $(1000, 1000)$.

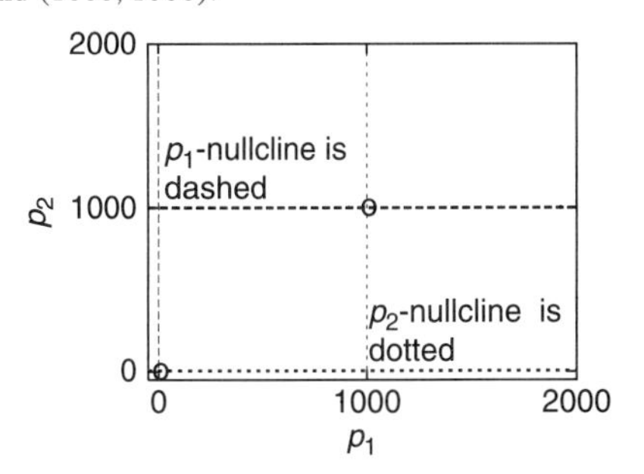

29. We found equations

$$\frac{da}{dt} = \mu\left(1 - \frac{a+b}{K_a}\right)a$$

$$\frac{db}{dt} = \lambda\left(1 - \frac{b}{K_b}\right)b.$$

Putting b on the vertical axis, the a-nullcline is $a = 0$ and $b = K_a - a$. The b-nullcline is $b = 0$ or $a = K_b$. The equilibria are $(0, 0)$, $(K_a, 0)$, and $(K_b, K_a - K_b)$ (if $K_a > K_b$).

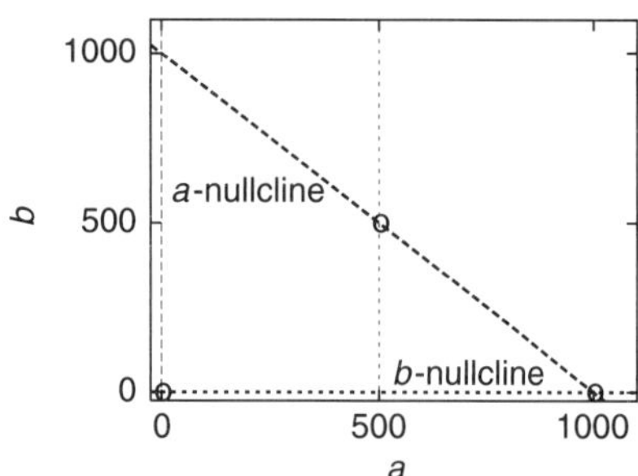

31. a. Let C_1 be the concentration in the first vessel. Then chemical moves from the first to the second at rate βC_1.

b. Let C_2 be the concentration in the second vessel. Then chemical moves from the second to the first at rate $\frac{\beta}{3} C_2$.

c. Let A_1 and A_2 be the total amounts. Then

$$\frac{dA_1}{dt} = \beta \left(\frac{C_2}{3} - C_1 \right)$$

$$\frac{dA_2}{dt} = \beta \left(C_1 - \frac{C_2}{3} \right).$$

d.

$$\frac{dC_1}{dt} = \frac{\beta}{V_1} \left(\frac{C_2}{3} - C_1 \right)$$

$$\frac{dC_2}{dt} = \frac{\beta}{V_2} \left(C_1 - \frac{C_2}{3} \right).$$

e. Place C_2 on the vertical axis; $C_2 = 3C_1$ is the nullcline for both C_1 and C_2.

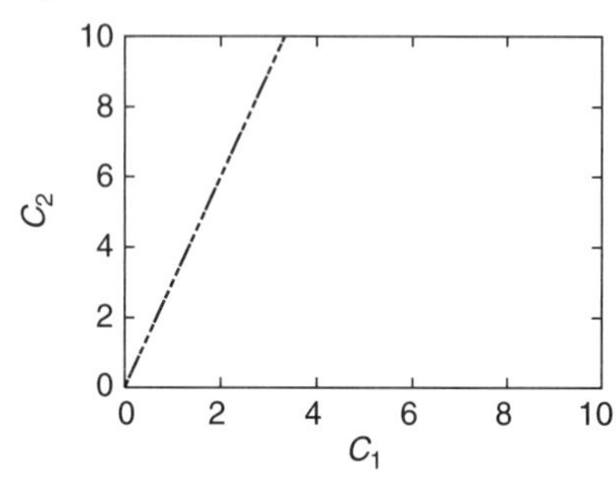

f. All points along the line $C_2 = 3C_1$ are equilibria. The equilibrium concentration in the second vessel is three times that in the first.

33. The equations look right because the per capita growth of a decreases half as quickly as a function of b as it does as a function of a. Therefore, individuals of type b decrease reproduction of individuals of type a only half as much. Similarly, the per capita growth of b decreases twice as quickly as a function of a as it does as a function of b. The a-nullcline is $a = 0$ and $b = 2(1000 - a)$. The b-nullcline is $b = 0$ and $b = 1000 - 2a$.

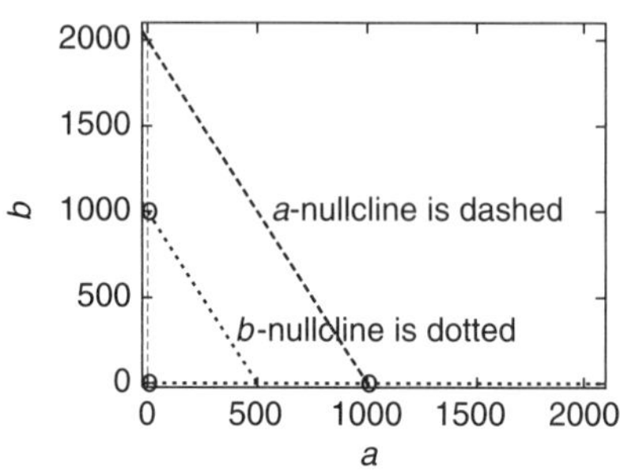

The two nullclines are parallel and do not intersect except at the boundaries.

35. The equations are

$$\frac{dI}{dt} = \alpha IS - \mu I - kI = 2.0IS - 2.0I$$

$$\frac{dS}{dt} = -\alpha IS + \mu I = -2.0IS + 1.0I.$$

The I-nullcline is the two pieces $I = 0$ and $S = 1.0$. The S-nullcline is the two pieces $I = 0$ and $S = 0.5$. There are equilibria wherever $I = 0$.

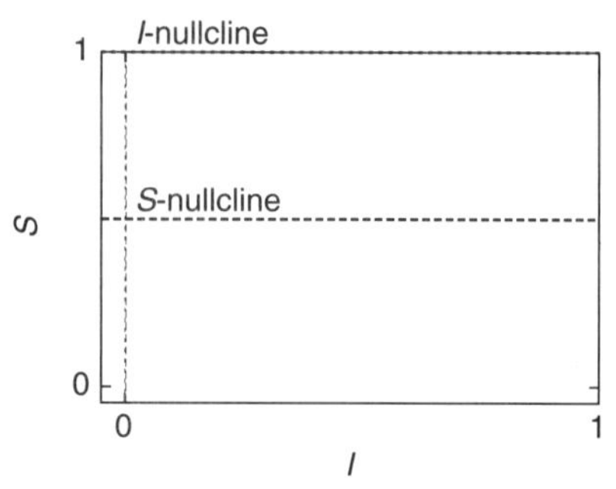

37. The equations are

$$\frac{dI}{dt} = 2IS - 1.5I$$

$$\frac{dS}{dt} = -2IS + I - 0.5S.$$

The I-nullcline is the two pieces $I = 0$ and $S = 3/4$. The S-nullcline is $S = \frac{I}{2I+0.5}$. There is only one equilibrium, at $(0, 0)$, because $\frac{3}{4} = \frac{I}{2I+0.5}$ has no intersection.

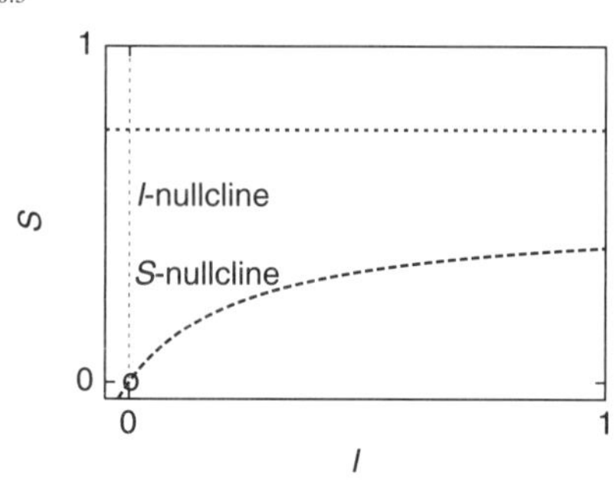

39. The equations are

$$\frac{dI}{dt} = I + 2IS$$

$$\frac{dS}{dt} = 2S - 2IS + I.$$

The I-nullcline has only one reasonable piece at $I = 0$ (along with $S = -1/2$). The S-nullcline is $S = \frac{I}{2(I-1)}$ (which is only defined for $I = 0$ and $I > 1$). There is only one equilibrium at $(0, 0)$.

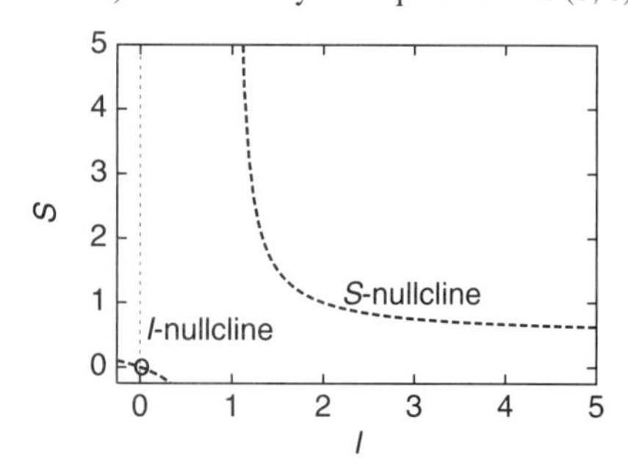

Section 5.7, page 482

1.

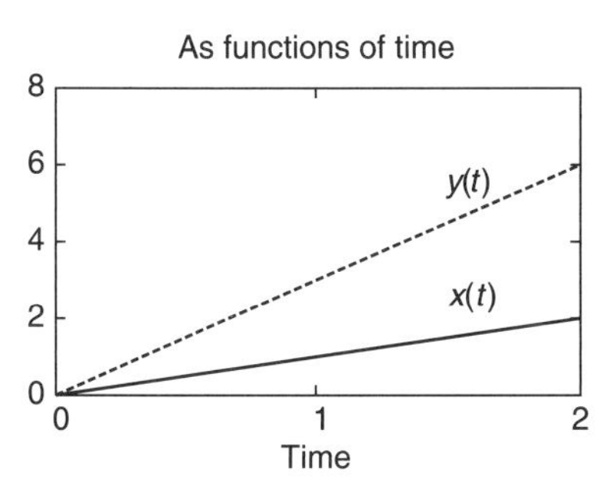

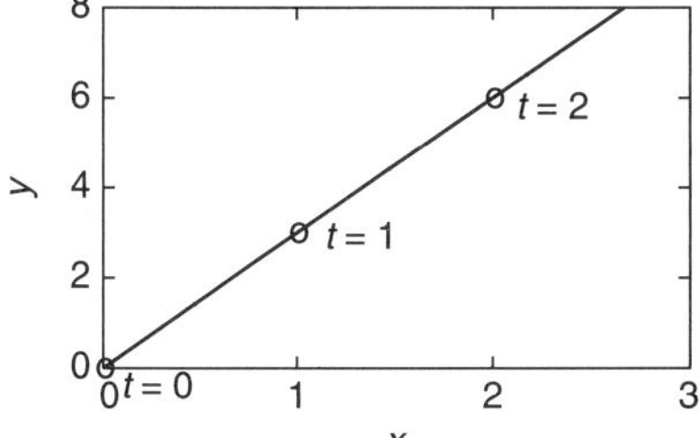

3.

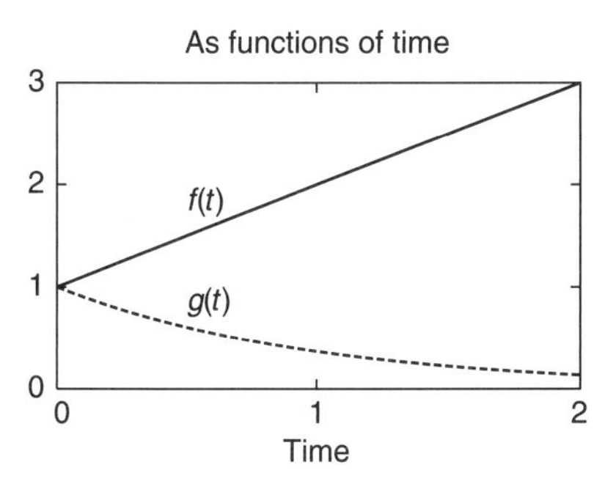

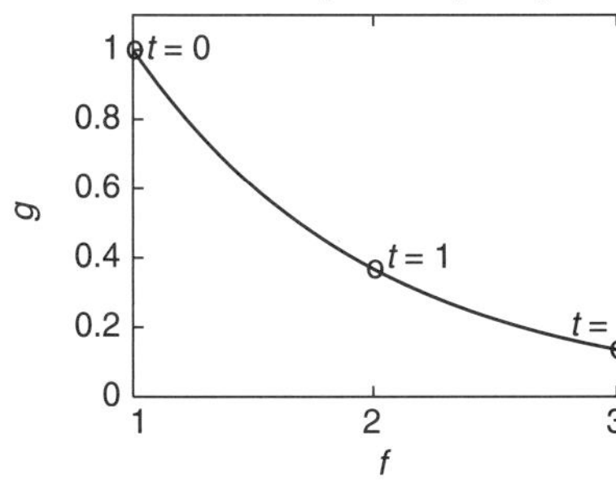

5.

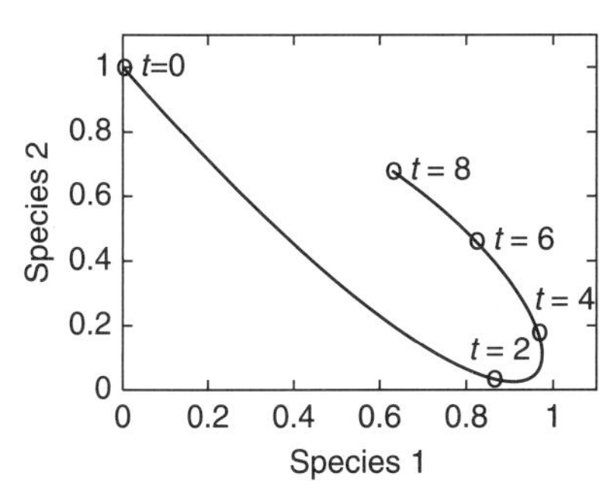

7.

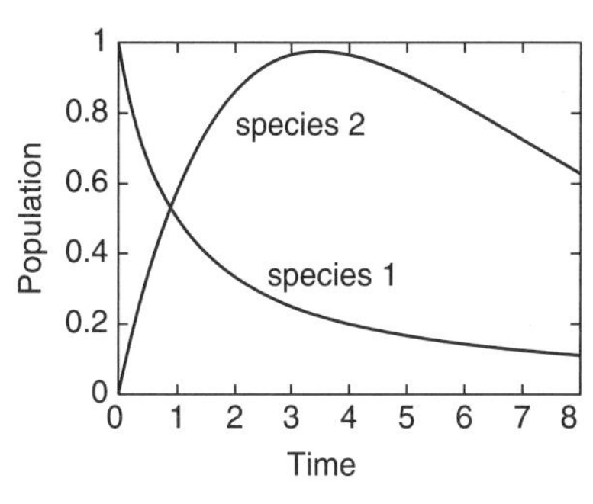

9.

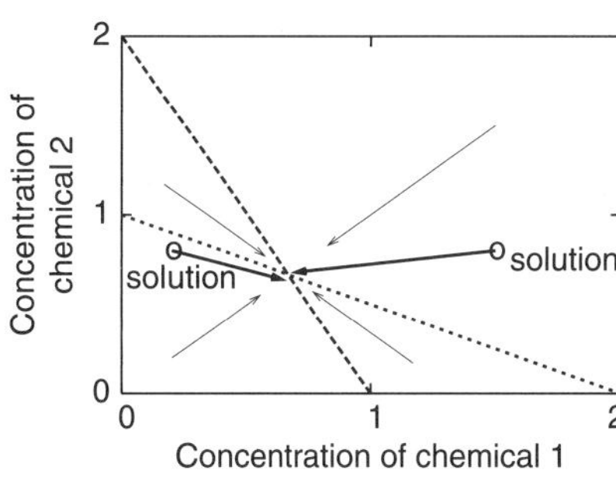

11.

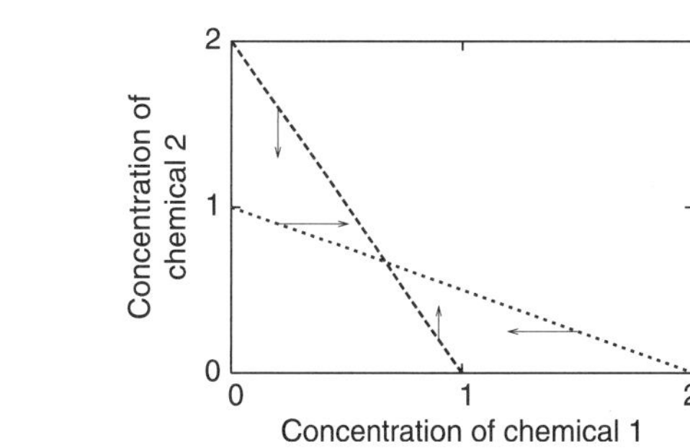

13. We found that $[\hat{a}(0.1), \hat{b}(0.1)] = (712.5, 350)$, with both values having decreased. This is consistent with the fact that our initial conditions lie in region I (because $750 + 500 > 1000$). Similarly, $[\hat{a}(0.2), \hat{b}(0.2)] = (703.6, 271.3)$, with both values having decreased. This is consistent with the fact that our initial conditions lie in region I (because $712.5 + 350 > 1000$). However, the solution has now moved into region II and a will begin to increase.

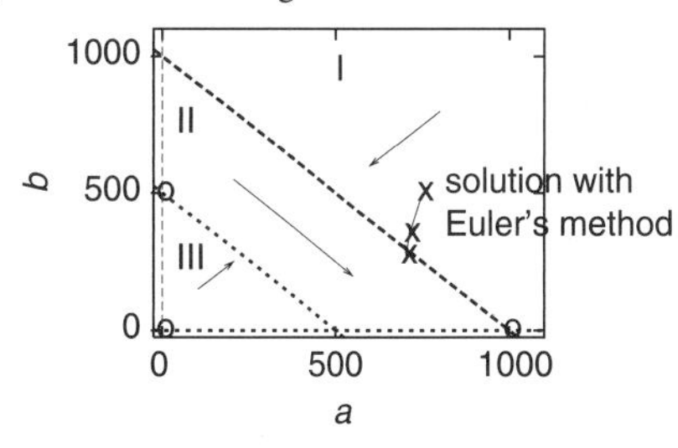

15. We found that $[\hat{H}(0.1), \hat{A}(0.1)] = (58.8, 20.4)$, with H having decreased and A increased. This is consistent with the fact that our initial conditions lie in region II (because $60 > 20$). Similarly, $[\hat{H}(0.2), \hat{A}(0.2)] = (57.6, 20.8)$ because our first step landed in region II.

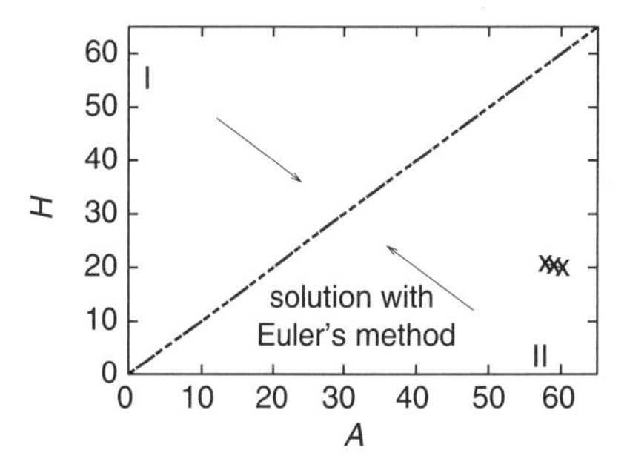

17. We found that $[\hat{H}(0.1), \hat{A}(0.1)] = (30.0, 30.0)$, with H having decreased and A increased. This is consistent with the fact that our initial conditions lie in region II (because $60 > 20$). However, $[\hat{H}(0.2), \hat{A}(0.2)] = (30.0, 30.0)$ because our first step landed right on the equilibrium. This overly long jump is a consequence of taking too big a step. A real solution could not cross the line of equilibria because it would get stuck.

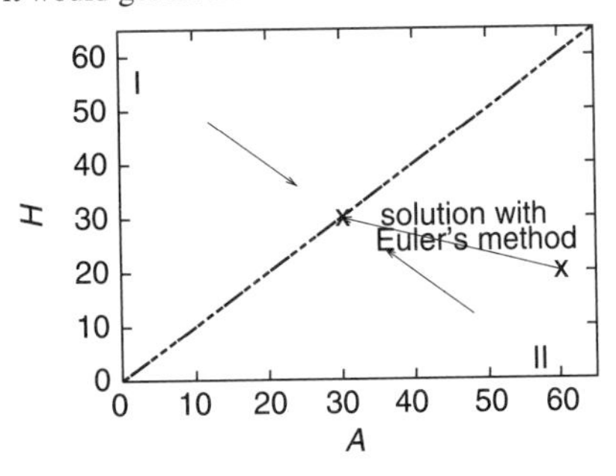

19. The x-nullcline is $v = 0$, and the v-nullcline is $x = 0$. These break the phase plane into four regions. The position is increasing when the velocity is positive, and the velocity is decreasing when the position is positive.

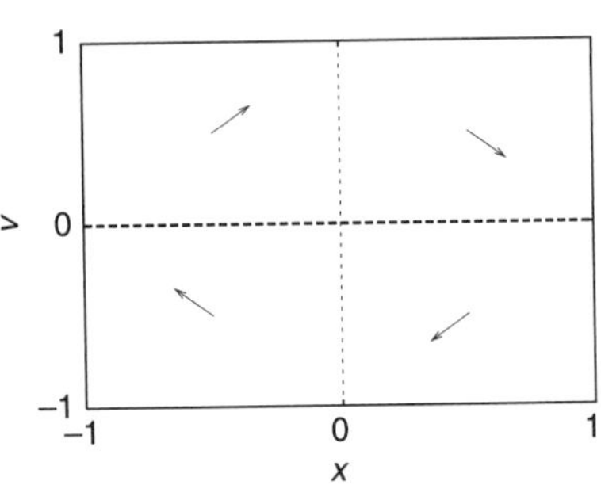

21.

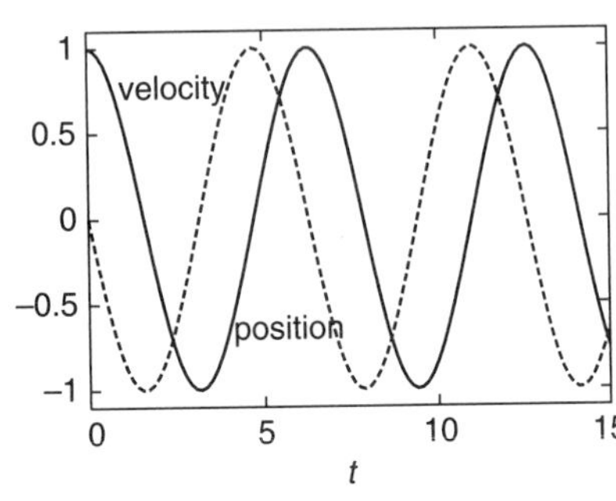

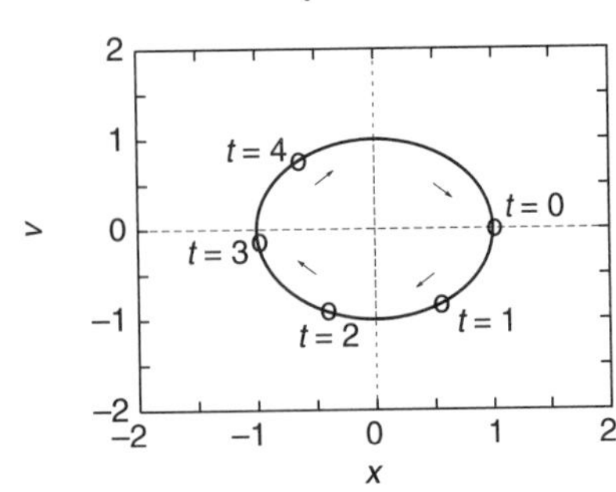

23.

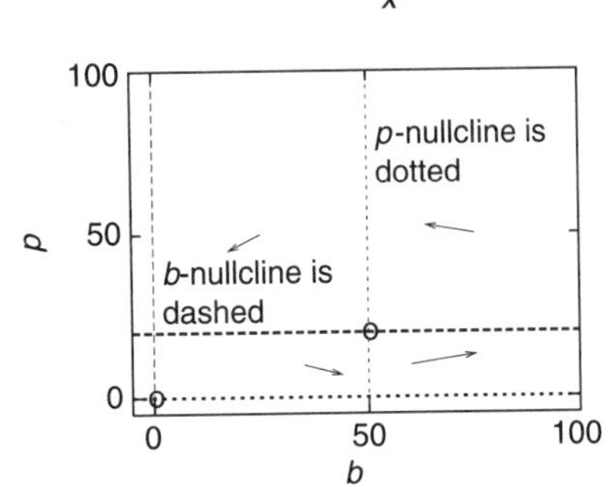

25.

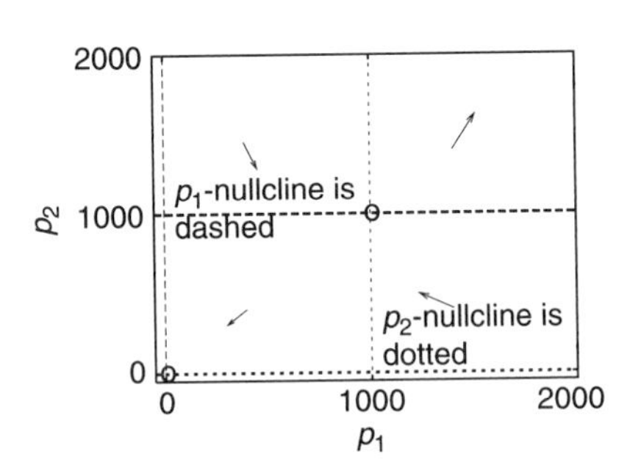

27.

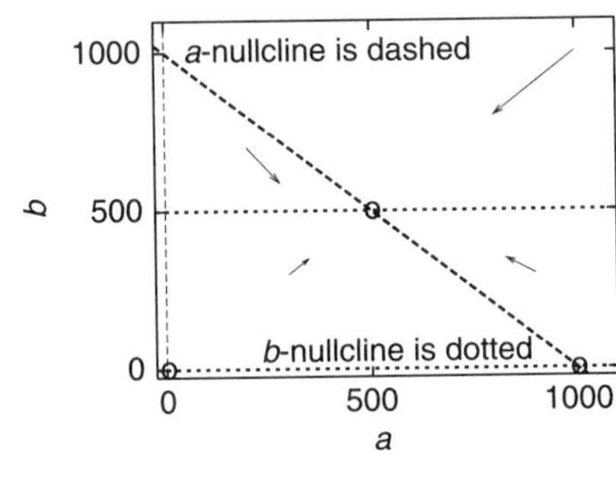

29.

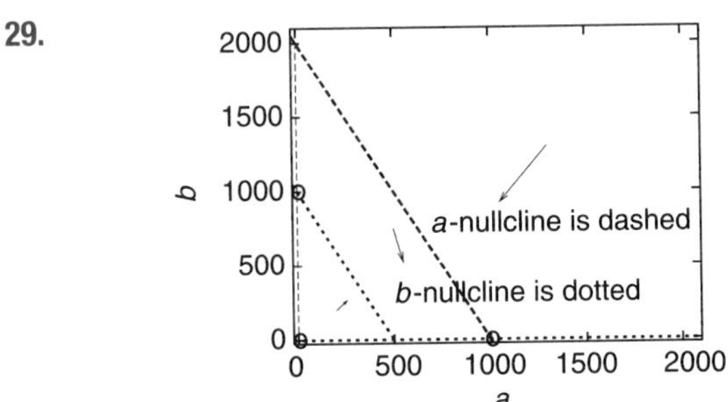

31.

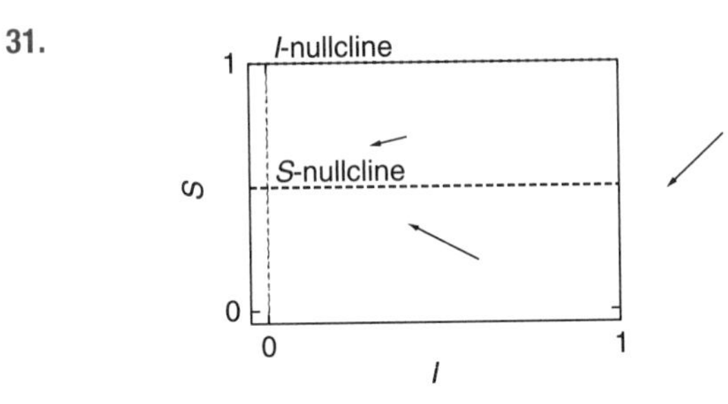

33.

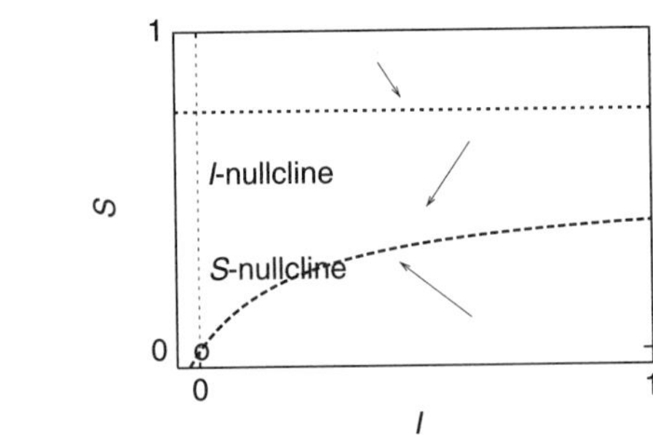

35.

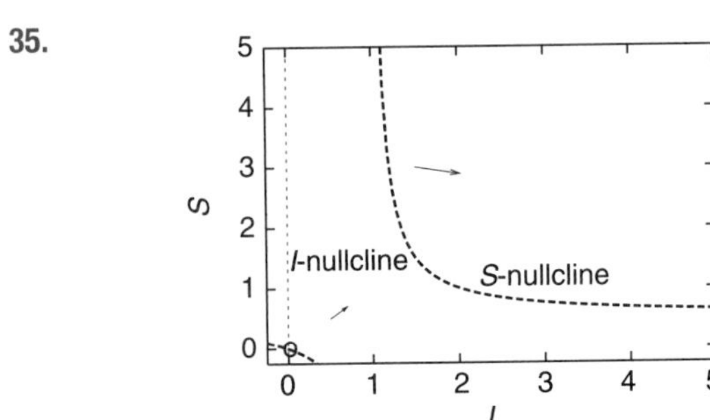

37.

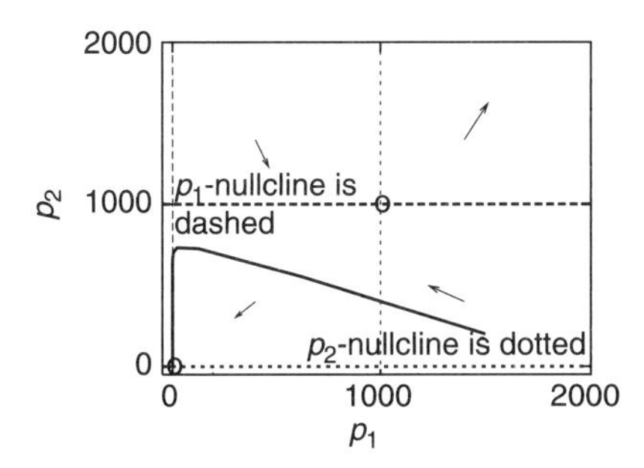

It is also possible that the solution goes to the right of the equilibrium and shoots off the upper right-hand part of the phase plane.

39.

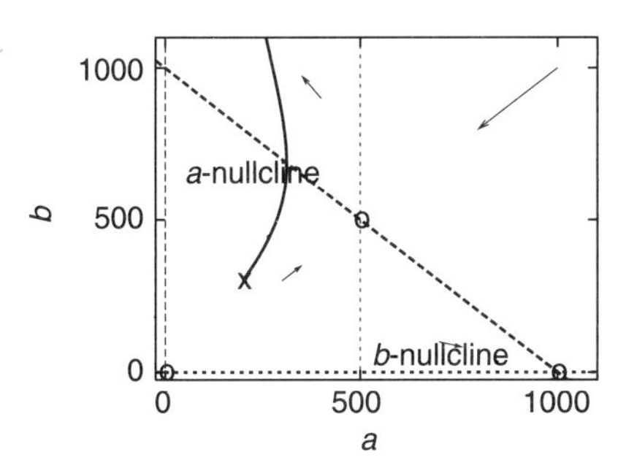

41.

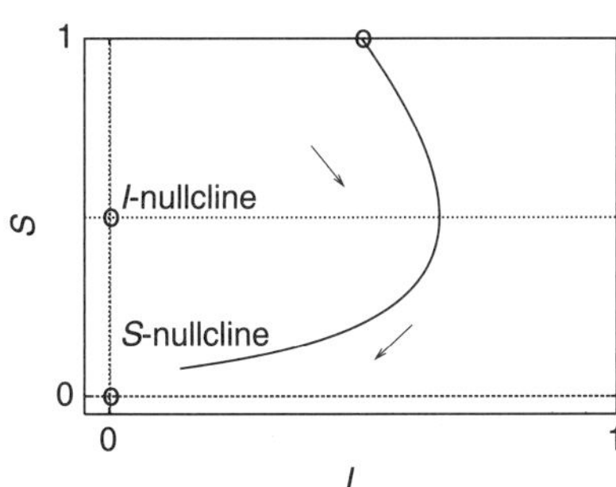

43.

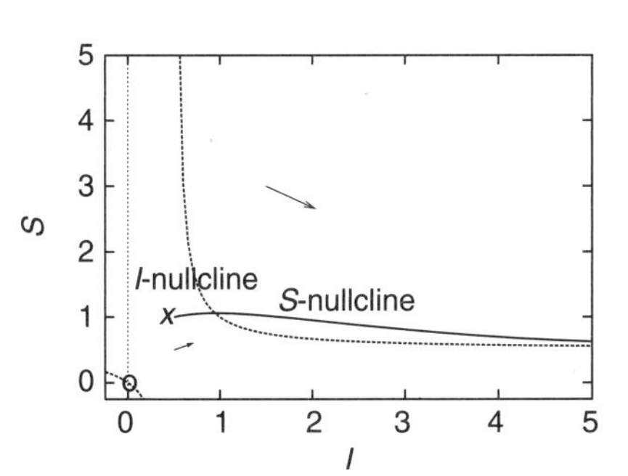

Section 5.8, page 490

1. One such equation is

$$\frac{dv}{dt} = -v(v - a_1)(v - e_1)(v - a_2)(v - e_2)$$

where a_1 and a_2 are the two thresholds and 0, e_1, and e_2 are the three equilibria. The accompanying figure uses $a_1 = 0.3, a_2 = 0.7, e_1 = 0.5$, and $e_2 = 1.0$.

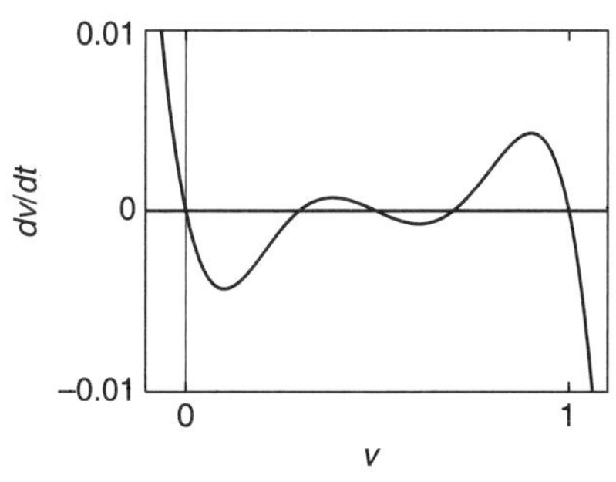

3.

$$\frac{dw}{v - \gamma w} = \epsilon dt$$

$$\frac{-1}{\gamma} \ln|v - \gamma w| = \epsilon t + c.$$

If $w < v/\gamma$,

$$w = \frac{v}{\gamma} - Ke^{-\epsilon\gamma t}.$$

If $w > v/\gamma$,

$$w = \frac{v}{\gamma} + Ke^{-\epsilon\gamma t}.$$

The smaller the value of ϵ, the slower the decay toward the equilibrium at $\frac{v}{\gamma}$.

5.

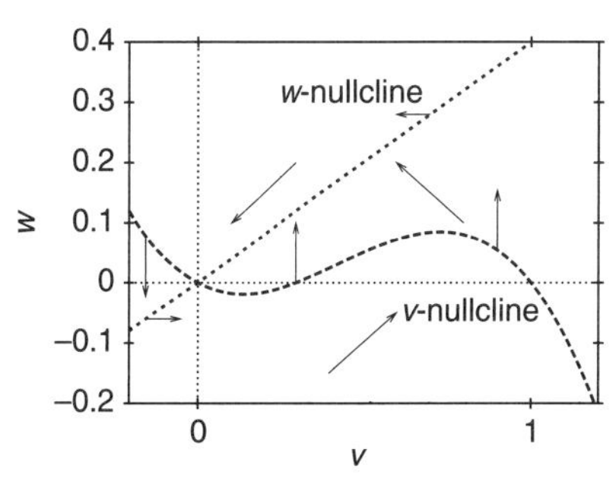

7. a. With ϵ small, the phase-plane trajectory moves very slowly in the vertical direction and follows the v-nullcline. The response is slow but sharp.

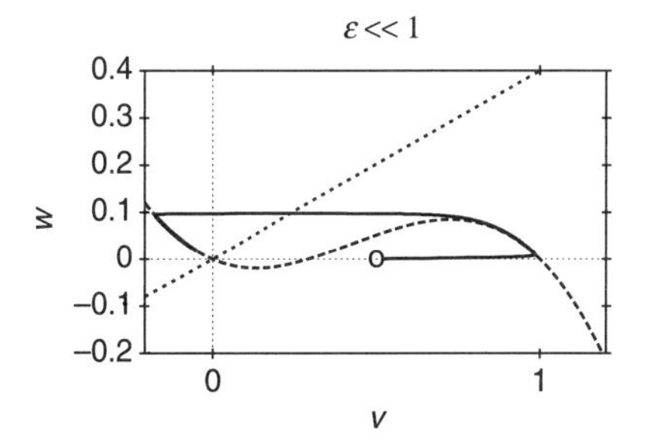

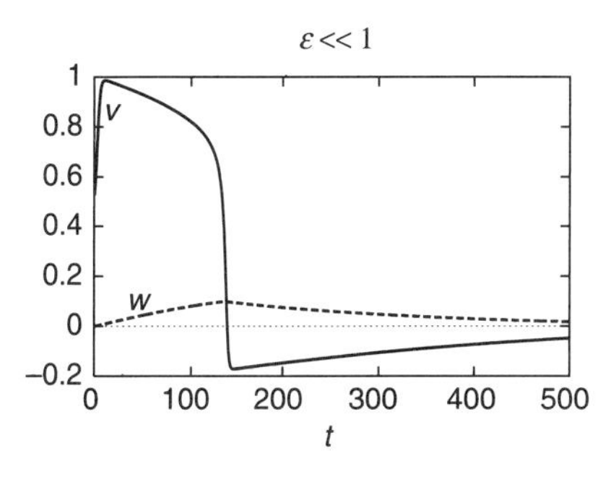

b. With ϵ large, the phase-plane trajectory sort of loops around, producing a poor excuse for an action potential.

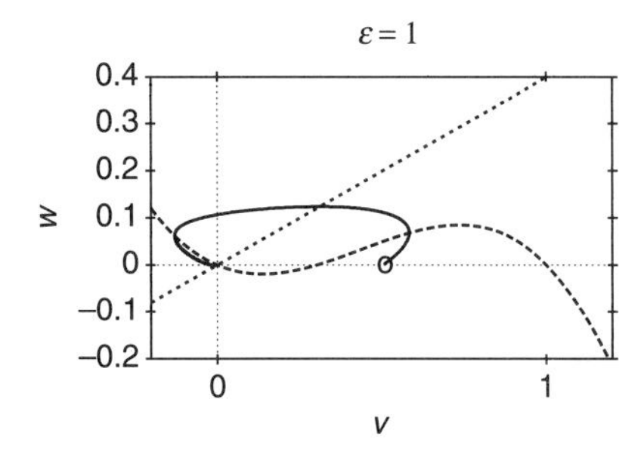

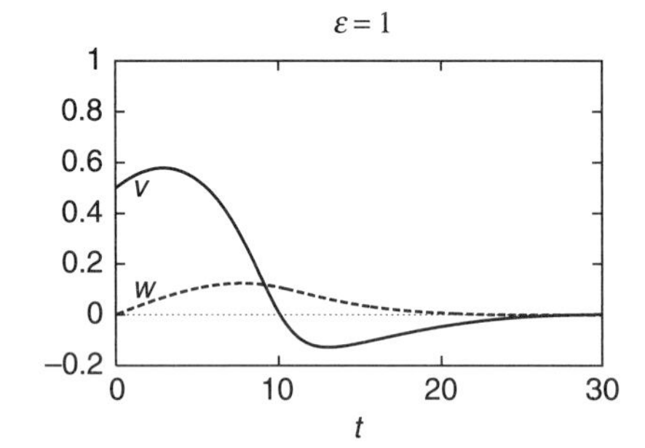

9. If the cell gets pushed far enough below the equilibrium, it creates a downward "action potential" before returning to equilibrium.

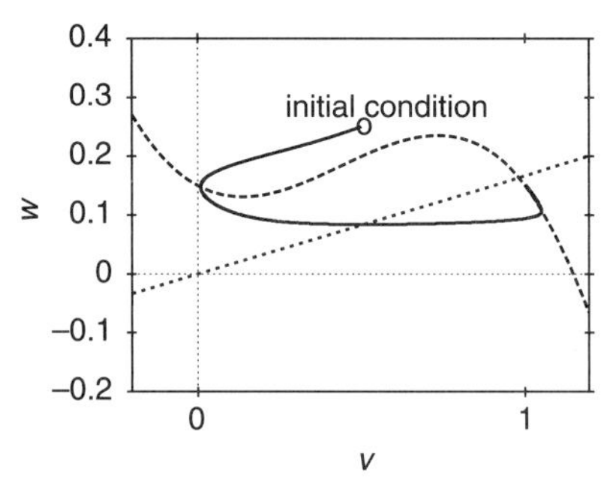

11. Translating level of terror into speed of moving legs might be a useful thing.

Chapter 6

Section 6.1, page 502

1. The population will be $1000 \cdot 0.6 = 600$ after one generation, $600 \cdot 1.5 = 900$ after two, 540 after three, 810 after four, 486 after five, and 729 after six. This population is shrinking. This is surprising because the average per capita production is $(0.6 + 1.5)/2 = 1.05 > 1$.

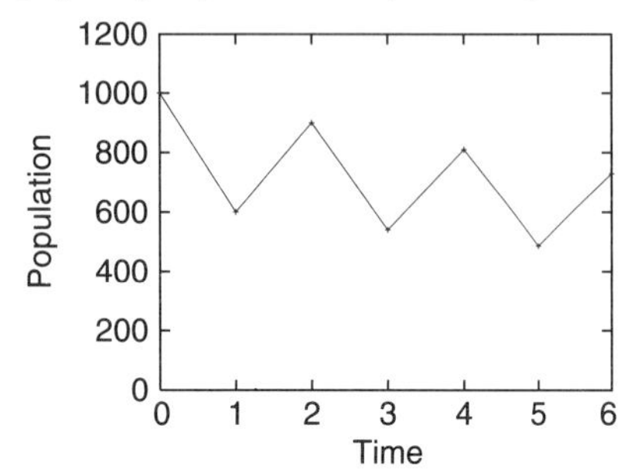

3. The population will be 700 after one generation, 630 after two, 1008 after three, 706 after four, 635 after five, and 1016 after six. This population is growing, but very slowly. The average per capita reproduction is $(0.7 + 0.9 + 1.6)/3 = 1.067 > 1$.

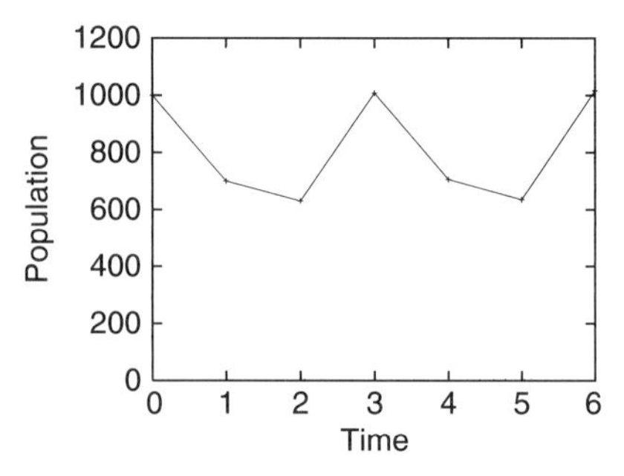

5. The population is 110, 105, 115, 110, 120, and 115 during the first 6 years. It is growing, consistent with the fact that the average number of immigrants is $[10 + (-5)]/2 = 2.5$.

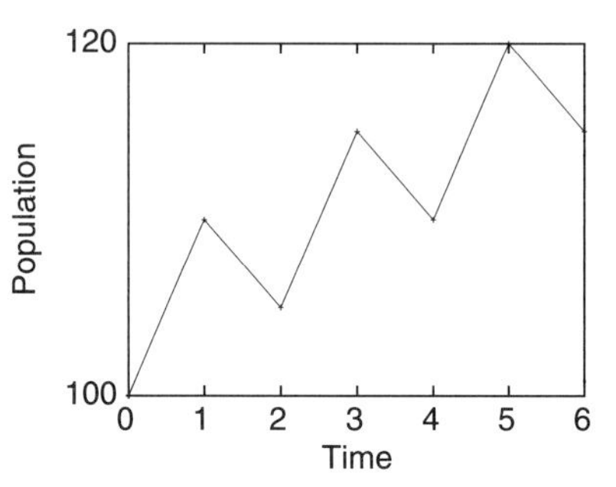

7. The population is 60, 48, 53, 63, 51, and 56 during the first 6 years. It is growing, consistent with the fact that the average number of immigrants is $[10 + (-12) + 5]/3 = 1$.

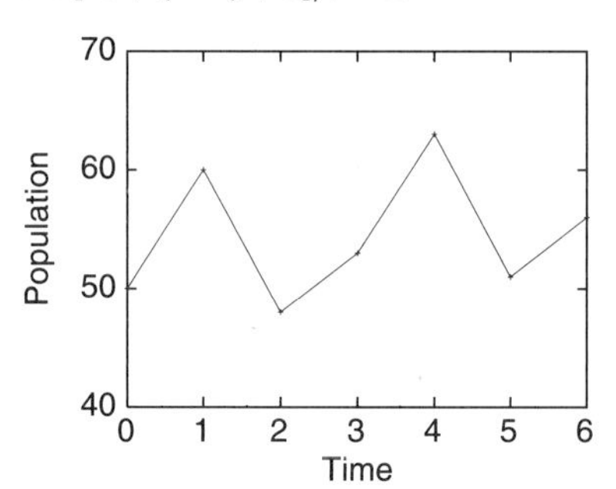

9. The population after one generation is $1.03 \cdot 100 = 103$, and after two generations is $0.886 \cdot 103 = 91.258$, and so forth. After ten generations, the population is 19.262. The average per capita production is 0.891, which is less than 1.0, consistent with the fact that this population shrinks.

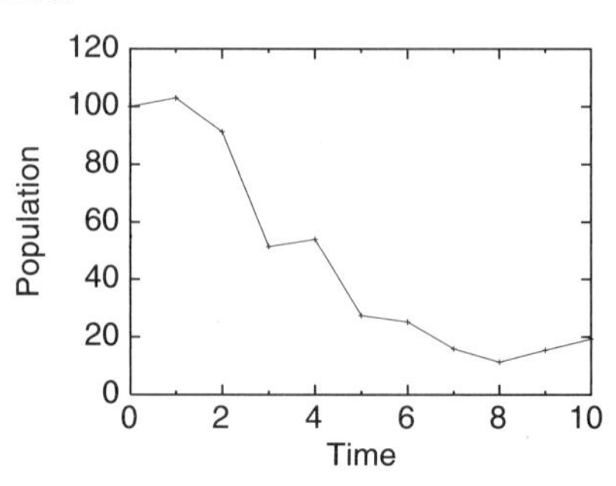

11. After ten generations, the population is 701.055. The average per capita production is 1.055, which is greater than 1.0. This is consistent with the fact that this population grows.

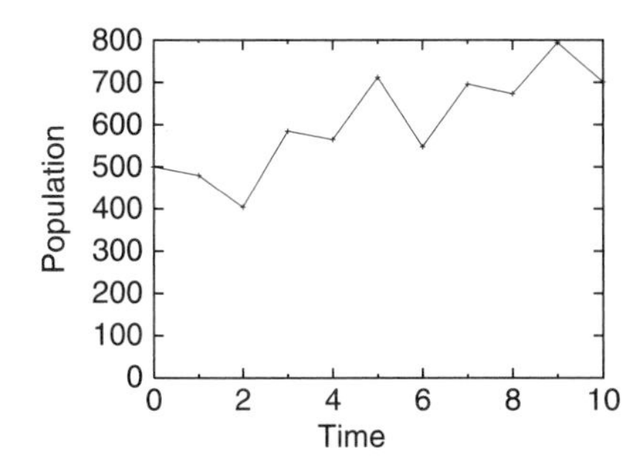

13. After 1 year, the population is $10+2=12$, and after 2 years, it is $12-1=11$, and so forth. Continuing in this way, we find that the population after 10 years is still 10, unchanged. The average immigration is exactly 0, consistent with the lack of change.

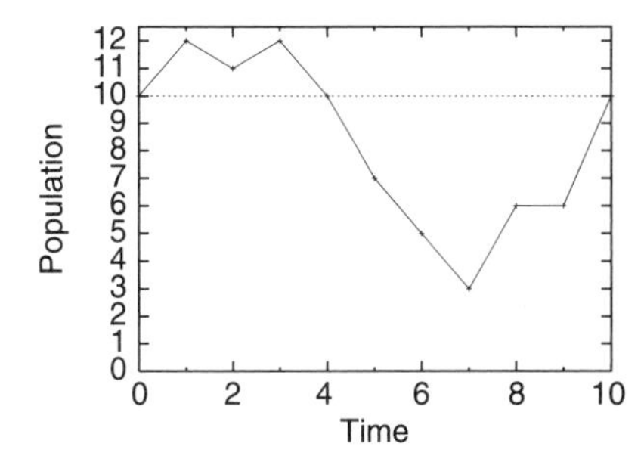

15. The population after 10 years is 8. The average immigration is -0.2, consistent with the decrease in population.

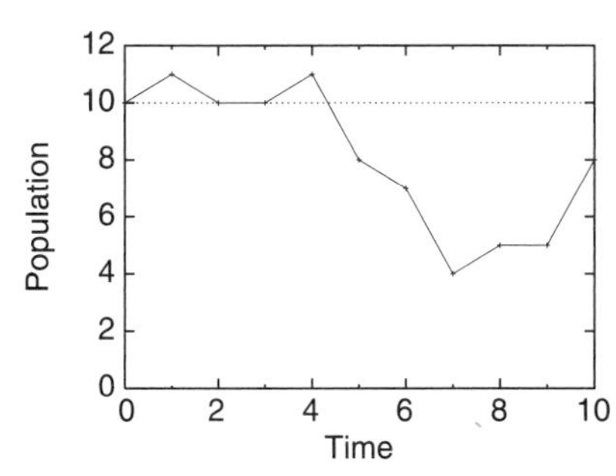

17. Between years 1 and 2, the per capita production is the ratio of the population sizes, or $110/100=1.1$. The same holds during the next interval and so on. This population is growing deterministically with per capita reproduction 1.1. It will reach 500 when $100\cdot 1.1^t=500$, or $1.1^t=5$, or $t=\ln(5)/\ln(1.1)=16.9$, or after 17 years.

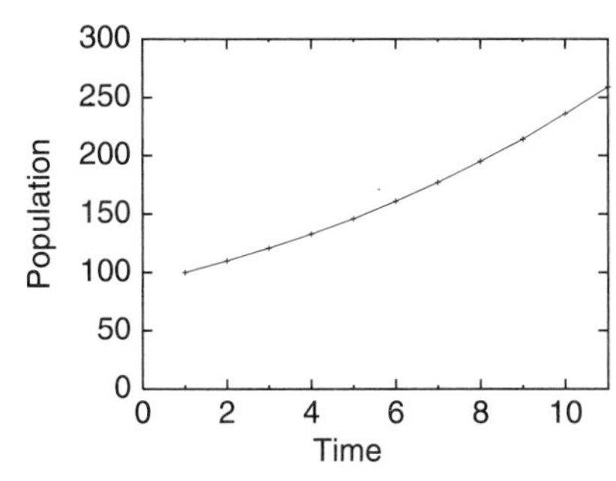

19. During the first 10 years, the per capita production is 0.98, 1.05, 1.15, 1.14, 1.19, 0.95, 1.08, 1.10, 1.26, and 1.02. These values average 1.092, and take on values fairly close to 1.10 every year, but sometimes are less than 1. I would guess it might take longer than 17 years to reach 500, because the bad years slow it down.

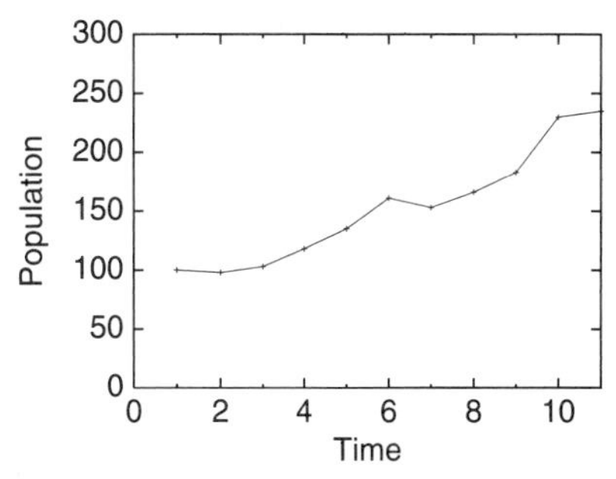

21. During the first 10 years, the change is 2 every year. It would take exactly 25 years to reach 100.

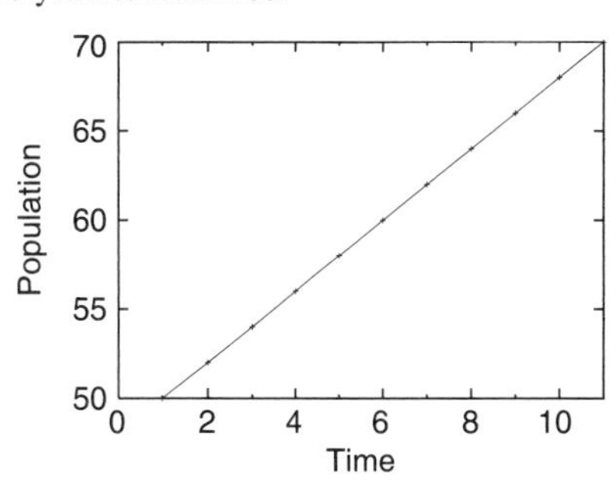

23. During the first 10 years, the change is 4, 2, 4, 2, 2, 4, 5, 2, 2, and 2. These average 2.9 per year, and range from 2 to 5. It would take about $50/2.9=17.2$ years to reach 100.

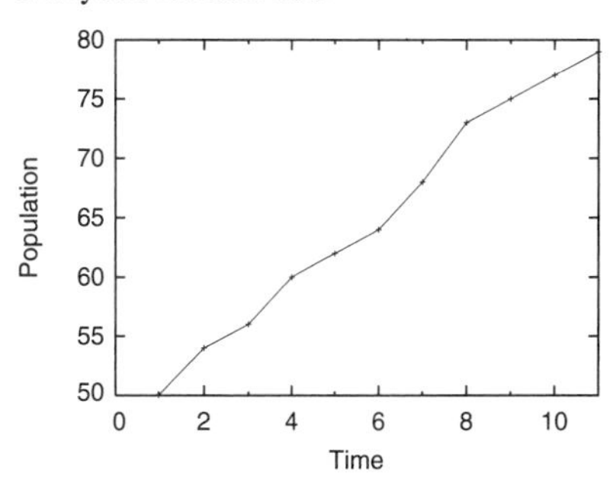

25. The island would jump back and forth between empty and occupied, switching every year.

27.

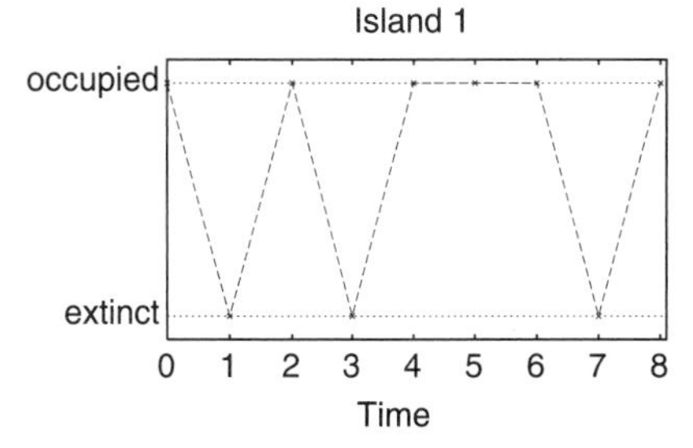

This island jumps back and forth, and seems to be occupied more often than unoccupied.

29.

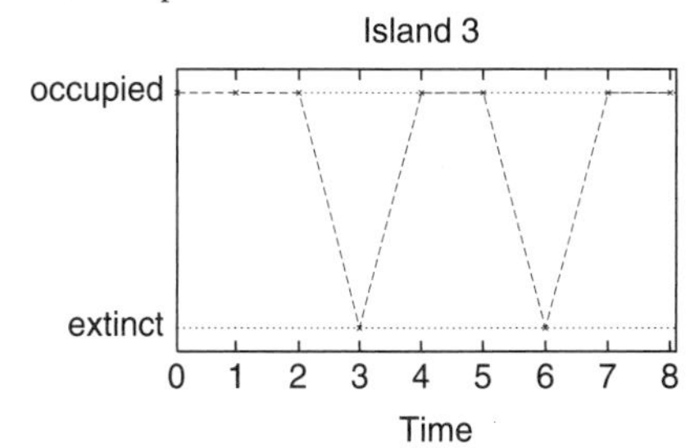

This island tends to be occupied for a few years and then extinct for 1 year.

31. 1. The per capita reproduction does not depend on the population size. 2. There is no immigration or emigration.

33. a. It doesn't take into account the size of the population. b. It doesn't take into account other nearby islands that might also be occupied or unoccupied.

Section 6.2, page 512

1. The discrete-time dynamical system is $p_{t+1} = 0.7p_t$ because the probability it remains is 0.7 each second. The solution is $p_t = 0.7^t$. After 10 seconds, it remains with probability $p_{10} = 0.7^{10} = 0.028$. To find the time when it will have left with probability 0.9, we must solve $p_t = 0.1$ for t, or $0.7^t = 0.1$, $t = \ln(0.1)/\ln(0.7) = 6.45$.

3. The discrete-time dynamical system is $p_{t+1} = 0.7p_t + 0.2(1 - p_t)$. A molecule that starts inside has $p_0 = 1$, so $p_1 = 0.7$ and $p_2 = 0.7 \cdot 0.7 + 0.2 \cdot 0.3 = 0.55$. A molecule that starts outside has $p_0 = 0$, so $p_1 = 0.2$ and $p_2 = 0.7 \cdot 0.2 + 0.2 \cdot 0.8 = 0.30$. The equilibrium is where $p^* = 0.7p^* + 0.2(1 - p^*)$, or $p^* = 0.2 + 0.5p^*$. This has solution $p^* = 0.4$. Out of 100 molecules, about 40 would be inside after a long time.

5.

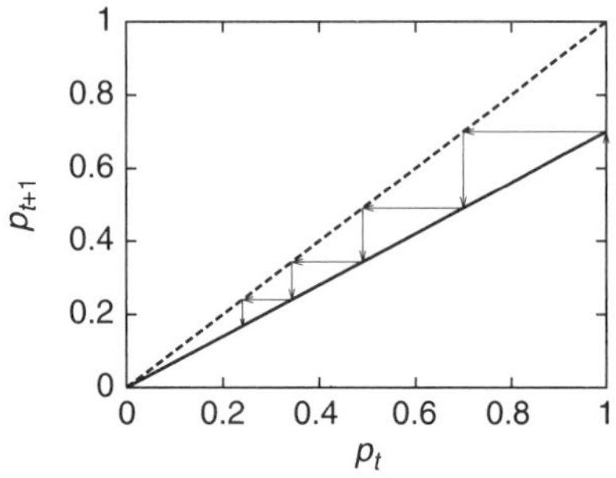

7.

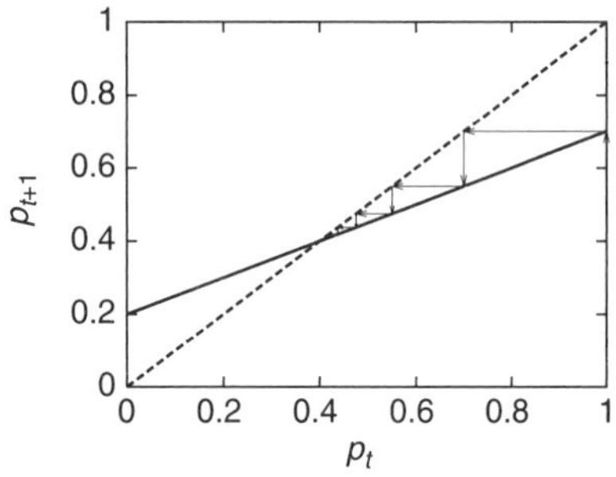

9. The discrete-time dynamical systems for c and d are $c_{t+1} = 0.7c_t + 0.2d_t$ and $d_{t+1} = 0.3c_t + 0.8d_t$. Then

$$\begin{aligned} p_{t+1} &= \frac{c_{t+1}}{c_{t+1} + d_{t+1}} \\ &= \frac{0.7c_t + 0.2d_t}{0.7c_t + 0.2d_t + 0.3c_t + 0.8d_t} \\ &= \frac{0.7c_t + 0.2d_t}{c_t + d_t} = 0.7p_t + 0.2(1 - p_t). \end{aligned}$$

The equilibrium fraction solves $p^* = 0.7p^* + 0.2(1 - p^*)$, with solution $p^* = 0.4$. This describes the same process as in Exercise 3 but is about fluids, which are effectively infinite numbers of molecules. There is nothing random about the results of this discrete-time dynamical system.

11. All the **AA** offspring (0.25)+ one fourth of the heterozygous offspring $(0.25 \cdot 0.5)$ gives $0.25 + 0.125 = 0.375$. Alternatively, we know that $h_2 = 0.5^2 = 0.25$ are heterozygous. The remaining 0.75 are split evenly between **AA** and **aa**, so the fraction is $0.75/2 = 0.375$.

13. All the **AA** great grand-offspring (0.4375) + one fourth of the heterozygous offspring $(0.125 \cdot 0.25)$ gives 0.46875. Alternatively, we know that $h_4 = 0.5^4 = 0.0625$ are heterozygous. The remaining 0.9375 are split evenly between **AA** and **aa**, so the fraction is $0.9375/2 = 0.46875$.

15. There should be an equal fraction, 0.5, of each type produced. However, after mortality, 0.25 of the original offspring are living heterozygotes, and 0.5 of the original offspring are living homozygotes. Out of the total of 0.75 that survive, a fraction $0.25/0.75 = 1/3$ are heterozygotes and $0.5/0.75 = 2/3$ are homozygotes.

17. There are three possible matings, 40 cm with 40 cm, 40 cm with 60 cm, and 60 cm with 60 cm. The first produces offspring with height 40 cm, the second offspring with height 50 cm, and the last offspring with height 60 cm. The probability that both parents have height 40 cm is $0.5 \cdot 0.5 = 0.25$.

19. There are now six possible matings: (40,40), (40,50), (40,60), (50,50), (50,60), and (60,60). They produce offspring of heights 40, 45, 50, 50, 55, and 60, respectively.

21. The probability that any one grandparent has height 40 cm is 0.5, so the probability that all four have height 40 cm is $0.5 \cdot 0.5 \cdot 0.5 \cdot 0.5 = 0.0625$.

23. The discrete-time dynamical system is $p_{t+1} = 0.99p_t$, with solution $p_t = 0.99^t$. After 15 divisions, the probability that a gene has not mutated is $0.99^{15} = 0.860$. The probability that it has mutated is therefore 0.14, or about 14%. Out of 100 genes, about 14 would have mutated in one generation.

25. The discrete-time dynamical system is $p_{t+1} = 0.95p_t$, with solution $p_t = 0.95^t$. After 10 seconds, the probability that a molecule is unbound is $0.95^{10} = 0.599$. The probability that it is bound is therefore 0.401, or about 40%. It has a 95% chance of being bound when it has a 5% or 0.05 chance of remaining unbound. We solve $p_t = 0.95^t = 0.05$ for t as $t = \ln(0.05)/\ln(0.95) = 58.4$, or nearly 1 minute.

27. The discrete-time dynamical system for the fraction of normal genes is $p_{t+1} = 0.99p_t + 0.01(1 - p_t)$. Iterating this function 15 times starting from $p_0 = 1$ gives 0.869. The correction mechanism hardly makes any difference.

29. The discrete-time dynamical system is $p_{t+1} = 0.95p_t + 0.02(1 - p_t)$. The equilibrium solves $p^* = 0.95p^* + 0.02(1 - p^*)$, which has solution $p^* = 0.286$. After 10 seconds, an initially unbound molecule has a 63.1% chance of being bound, or a 36.9% chance of being unbound.

31. The discrete-time dynamical system is $h_{t+1} = 0.6h_t$, with solution $h_t = 0.6^t$. After ten generations, a fraction $h_{10} = 0.006$ will be heterozygous. This is six times as many as for a normal plant.

33. The discrete-time dynamical system is $h_{t+1} = 0.2h_t$, with solution $h_t = 0.2^t$. After ten generations, a fraction $h_{10} = 1.02 \times 10^{-7}$ will be heterozygous. This is vanishingly small.

35. 1/4 of the offspring have genotype **AA** and are tall.

37. Half will be heterozygous, and hence half will be tall.

39. 3/8 of the grand-offspring have genotype **AA** and are tall.

41. After two generations, only 1/4 are heterozygous and tall.

43. The flies are white-eyed and hence have at least one of the white-eyed alleles. Because each has a parent from the red-eyed stock, it must also have one red-eyed allele. Therefore, these flies will also be heterozygous at the eye color locus.

45. In the first generation, all the flies will have genotype **Aa** at the second locus.

47. Half of the **a** alleles will be lost each generation. Let p_t be the probability that a fly has an **a** allele in generation t. Then

$$p_{t+1} = 0.5p_t$$

with initial condition $p_1 = 1$. The solution is $p_t = 0.5^{t-1}$.

49. A heterozygote could have gotten **a** from the ovule and **A** from the pollen or **A** from the ovule and **a** from the pollen. The probability of **a** from ovule is 0.5 and the probability of **A** from pollen is 0.8, giving a probability of 0.4 of a heterozygous offspring in this way. The probability of **A** from ovule is 0.5, and the probability of **a** from the pollen is 0.2, giving a probability of 0.1 of a heterozygous offspring in this way. The total is $0.4 + 0.1 = 0.5$ as in a normal plant.

51. A heterozygote could have gotten **a** from the ovule and **A** from the pollen or **A** from the ovule and **a** from the pollen. The probability of **a** from ovule is 0.8, and the probability of **A** from pollen is 0.8, giving a probability of 0.64 of a heterozygous offspring in this way. The probability of **A** from ovule is 0.2, and the probability of **a** from the pollen is 0.2, giving a probability of 0.04 of a heterozygous offspring in this way. The total is $0.64 + 0.04 = 0.68$.

53. 0.99 (probability it does not mutate).

55. Multiply 0.5 (probability it started out as **A**) times the probability it didn't mutate (0.99) to get 0.495.

Section 6.3, page 521

1. $A \cap B = \{0, 2\}$, $A \cup B = \{0, 1, 2, 4\}$, $A^c = \{3, 4\}$.

3. $A \cap B = \{2\}$, $A \cup B = \{1, 2, 4, 5, 6, 10\}$, $A^c =$ all positive integers except 1, 2, 6, and 10.

5. The probabilities are all positive, and add up to 1. Because we assigned probabilities to the simple events, requirement 3 will take care of itself. $\Pr(A) = 0.2 + 0.3 + 0.4 = 0.9$, $\Pr(A^c) = 1 - \Pr(A) = 1 - 0.9 = 0.1$ and $\Pr(B) = 0.2 + 0.4 + 0.0 = 0.6$. Also, $A \cup B = \{0, 1, 2, 4\}$ so $\Pr(A \cup B) = 0.2 + 0.3 + 0.4 + 0.0 = 0.9 \neq \Pr(A) + \Pr(B)$. This is because A and B are not disjoint.

7. The probabilities are all positive. We are not given the probability of the simple event $A = \{4\}$, so we use the requirement that the probabilities add up to 1.

$$\begin{aligned}&\Pr(\{0\}) + \Pr(\{1\}) + \Pr(\{2\}) + \Pr(\{3\}) + \Pr(\{4\})\\&= 0.2 + 0.1 + 0.4 + 0.1 + \Pr(\{4\}) = 0.8 + \Pr(\{4\}) = 1.\end{aligned}$$

Therefore, $\Pr(\{4\}) = 1 - 0.8 = 0.2$. $\Pr(A^c) = 1 - \Pr(A) = 1 - 0.2 = 0.8$ and $\Pr(B) = 0.1 + 0.2 = 0.3$.

9.

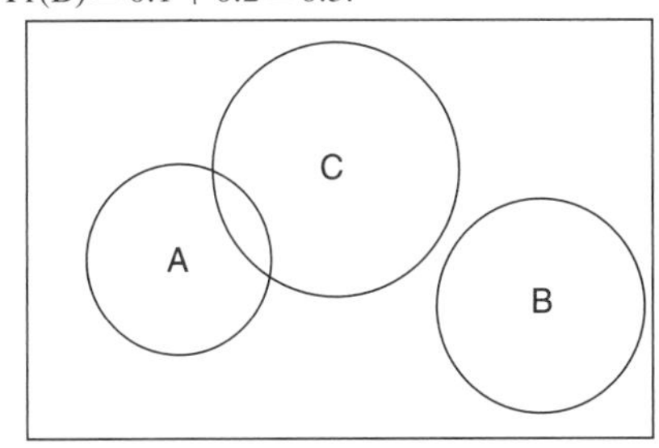

11.

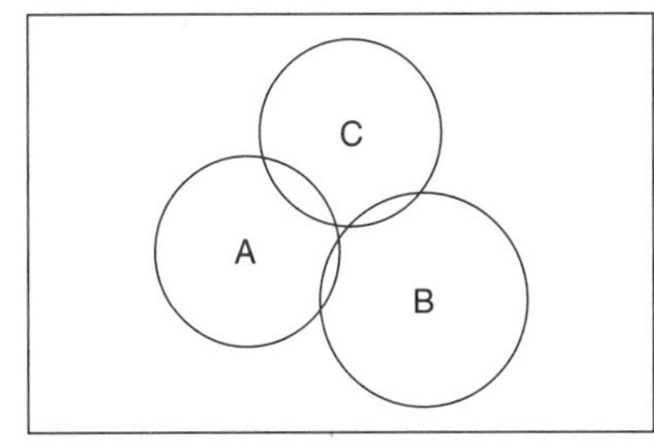

13. We found that $\Pr(A) = 0.9$, $\Pr(B) = 0.6$, and $\Pr(A \cup B) = 0.9$. Also, $\Pr(A \cap B) = \Pr(\{0,2\}) = 0.2 + 0.4 = 0.6$. The formula checks because $\Pr(A \cup B) = 0.9 = \Pr(A) + \Pr(B) - \Pr(A \cap B) = 0.9 + 0.6 - 0.6$.

15. $\Pr(C) = 0.3 + 0.4 + 0.1 = 0.8$, $\Pr(D) = 0.2 + 0.3 + 0.4 = 0.9$, and $\Pr(C \cup D) = 0.2 + 0.3 + 0.4 + 0.1 = 1.0$, Also, $\Pr(C \cap D) = \Pr(\{1,2\}) = 0.3 + 0.4 = 0.7$. The formula checks because $\Pr(C \cup D) = 1.0 = \Pr(C) + \Pr(D) - \Pr(C \cap D) = 0.8 + 0.9 - 0.7$.

17. Three possible genotypes: **bb**, **Bb**, and **BB**.

19. There are four simple events: {In, In}, {In, Out}, {Out, In}, and {Out, Out}.

21. There are nine simple events, described by the number in (which can be 0, 1, or 2) at each time: {0, 0}, {0, 1}, {0, 2}, {1, 0}, {1, 1}, {1, 2}, {2, 0}, {2, 1}, and {2, 2}.

23. There are 17 simple events, the numbers ranging from 0 to 16.

25. $\{N = 1\}$, $\{N = 2\}$, $\{N = 3\}$, $\{N = 4\}$, $\{N = 5\}$.

27. $\{N = 1\}$, $\{N = 3\}$, $\{N = 5\}$, $\{N = 7\}$, $\{N = 9\}$.

29. The union is all numbers between 0 and 100, and the intersection is the values 6, 7, 8, and 9.

31. The union is all numbers greater than 5 and less than or equal to 100, and the intersection is all numbers greater than 10 and less than or equal to 100.

33. This requires that $t_1 < 5$, $t_2 < 5$, $t_3 < 5$, and $t_4 < 5$. Three possible ways this could happen are $\{t_1 = 1, t_2 = 1, t_3 = 1, t_4 = 1\}$, $\{t_1 = 1, t_2 = 2, t_3 = 3, t_4 = 4\}$, and $\{t_1 = 4, t_2 = 3, t_3 = 2, t_4 = 1\}$.

35. The odd-numbered molecules are 1 and 3. Possibilities are $\{t_1 = 1, t_2 = 1, t_3 = 3, t_4 = 1\}$, $\{t_1 = 1, t_2 = 1, t_3 = 5, t_4 = 2\}$, and $\{t_1 = 1, t_2 = 2, t_3 = 5, t_4 = 1\}$.

37. The biologically reasonable assignment is $\Pr(bb) = 0.25$, $\Pr(Bb) = 0.5$, and $\Pr(BB) = 0.25$. Another possibility is $\Pr(bb) = 0.5$, $\Pr(Bb) = 0.25$, and $\Pr(BB) = 0.25$.

39. We could set $\Pr(N = 1) = 0.7$ and all the rest to 0.1. Then $\Pr(N \text{ is odd}) = 0.8$ and $\Pr(N \neq 1) = 0.3$.

41. We could set $\Pr(N = 1) = 0.1$ and all the rest to 0.3. Then $\Pr(N \text{ is odd}) = 0.4$ and $\Pr(N \neq 1) = 0.9$.

Section 6.4, page 530

1. One possibility is $A = \{0, 1\}$, $B = \{2, 3, 4\}$.

3. One possibility is $A = \{1, 3, 5, \ldots\}$, $B = \{2, 4, 6, \ldots\}$.

5. $\Pr(A \cap B) = \Pr(\{0,2\}) = 0.2 + 0.4 = 0.6$. Therefore, $\Pr(A \mid B) = \dfrac{\Pr(A \cap B)}{\Pr(B)} = \frac{0.6}{0.6} = 1$. $\Pr(B \mid A) = \dfrac{\Pr(A \cap B)}{\Pr(A)} = \frac{0.6}{0.9} = 2/3$.

7. $\Pr(A \cap B) = 0$ because the sets are disjoint. Therefore, $\Pr(A \mid B) = 0$, and $\Pr(B \mid A) = 0$.

9. Let S be the event four or more, and let F_1 be the event of a 1 on the first roll, F_2 a 2 on the first role, F_3 a 3 on the first. Then $\Pr(S \mid F_1) = 1/3$ (because the second roll must be a 3), $\Pr(S \mid F_2) = 2/3$ (because the second roll can be a 2 or a 3), $\Pr(S \mid F_3) = 1$ (because the second roll can be anything). Then, by the law of total probability,

$$\begin{aligned}\Pr(S) &= \Pr(S \mid F_1)\Pr(F_1) + \Pr(S \mid F_2)\Pr(F_2) + \Pr(S \mid F_3)\Pr(F_3)\\&= 1/3 \cdot 1/3 + 2/3 \cdot 1/3 + 1 \cdot 1/3 = 2/3.\end{aligned}$$

By direct counting, there are nine possible outcomes, each with probability 1/9. Of these six

$$(1, 3), (2, 2), (2, 3), (3, 1), (3, 2), (3, 3)$$

give a result greater than or equal to 4, for a probability of $6/9 = 2/3$.

11. Let O be the event odd, and let F_1 be the event of a 1 on the first roll, F_2 a 2 on the first role, and F_3 a 3 on the first. Then $\Pr(O \mid F_1) = 1/3$ (because the second roll must be a 2), $\Pr(O \mid F_2) = 2/3$ (because the second roll can be a 1 or a 3), and $\Pr(O \mid F_3) = 1/3$ (because the second roll must be a 2). Then, by the law of total probability,

$$\begin{aligned}\Pr(S) &= \Pr(S \mid F_1)\Pr(F_1) + \Pr(S \mid F_2)\Pr(F_2) + \Pr(S \mid F_3)\Pr(F_3)\\&= 1/3 \cdot 1/3 + 2/3 \cdot 1/3 + 1/3 \cdot 1/3 = 4/9.\end{aligned}$$

By direct counting, there are nine possible outcomes, each with probability 1/9. Of these four

$$(1, 2), (2, 1), (2, 3), (3, 2)$$

give an odd result, for a probability of 4/9.

13. Let S be the event four or more, and F_3 a 3 on the first role. By Bayes' theorem,

$$\Pr(F_3 \mid S) = \frac{\Pr(S \mid F_3)\Pr(F_3)}{\Pr(S)}.$$

From Exercise 9, $\Pr(S \mid F_3) = 1$ and $\Pr(S) = 2/3$. By assumption, $\Pr(F_3) = 1/3$. Therefore, $\Pr(F_3 \mid S) = 1/2$. Out of the six possible ways that the score could be 4 or greater, three start with a roll of 3, for a probability of $3/6 = 1/2$.

15. Let O be the event odd, and F_3 a 3 on the first role. By Bayes' theorem,

$$\Pr(F_3 \mid O) = \frac{\Pr(O \mid F_3)\Pr(F_3)}{\Pr(O)}.$$

From Exercise 11, $\Pr(O \mid F_3) = 1/3$ and $\Pr(O) = 4/9$. By assumption, $\Pr(F_3) = 1/3$. Therefore, $\Pr(F_3 \mid O) = 1/4$. Out of the four possible ways that the score could be odd, one starts with a roll of 3, for a probability of 1/4.

17. Let F be the event "first red," S the event "second red," O the event "one red," and B the event "both red." Then $\Pr(S \mid F) = \dfrac{\Pr(S \cap F)}{\Pr(F)}$. The probability that both are red is 1/6. The probability that the first is red is 1/2. So the requested probability is 1/3.

19. Exactly 1/2, because there are two out of four balls that are red.

21. There are only three events: **bb**, **Bb**, and **BB**. This is the only possible choice.

23. We could break the four elements in the sample space into those with an "Out" at time 1 (two simple events), and the two simple events {In, In} and {In, Out}.

25. Let N denote the number of plants taller than 50 cm. We could use $N = 0$, $0 < N < 16$, and $N = 16$.

27. a.

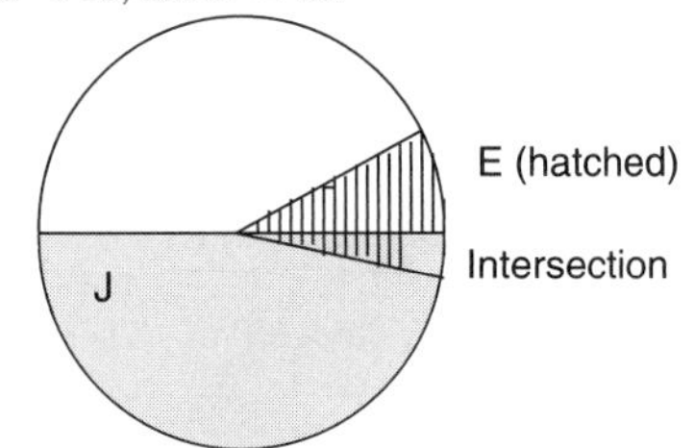

b. Let J be the event of seeing a jack rabbit, and E the event of seeing an eagle. Then

$$\Pr(J \mid E) = \frac{\Pr(J \cap E)}{\Pr(E)} = \frac{0.05}{0.2} = 0.25.$$

It is less likely that she will see a rabbit if she sees an eagle. Perhaps rabbits avoid the eagles.

c.
$$\Pr(E \mid J) = \frac{\Pr(E \cap J)}{\Pr(J)} = \frac{0.05}{0.5} = 0.1.$$

It is less likely that she will see an eagle if she sees a rabbit. Perhaps eagles avoid the rabbits. This seems less likely but cannot be demonstrated from the data.

29. Let S be the event that a cell stains properly, Y that a cell is young, and O that it is old. $\Pr(S) = \Pr(S \mid Y)\Pr(Y) + \Pr(S \mid O)\Pr(O) = 0.9 \cdot 0.3 + 0.7 \cdot 0.7 = 0.76$.

31. Let S be the event "stains," A_1 the event "less than 1 day old," A_2 the event "between 1 and 2 days old," A_3 the event "between 2 and 3 days old," and A_4 the event "more than 3 days old." By the law of total probability,

$$\begin{aligned}\Pr(S) &= \Pr(S \mid A_1)\Pr(A_1) + \Pr(S \mid A_2)\Pr(A_2) + \Pr(S \mid A_3)\Pr(A_3) \\ &\quad + \Pr(S \mid A_4)\Pr(A_4) \\ &= 0.95 \cdot 0.4 + 0.9 \cdot 0.3 + 0.8 \cdot 0.2 + 0.5 \cdot 0.1 = 0.86.\end{aligned}$$

33. We want to find $\Pr(Y \mid S)$. By Bayes' theorem,

$$\Pr(Y \mid S) = \frac{\Pr(S \mid Y)\Pr(Y)}{\Pr(S)} = \frac{0.9 \cdot 0.3}{0.76} = 0.355.$$

A properly staining cell is only slightly more likely to be young.

35.
$$\Pr(A_1 \mid S) = \frac{\Pr(S \mid A_1)\Pr(A_1)}{\Pr(S)} = \frac{0.95 \cdot 0.4}{0.86} = 0.44.$$

A properly staining cell is only slightly more likely to be young.

37. First, find $\Pr(P)$ with the law of total probability,

$$\begin{aligned}\Pr(P) &= \Pr(P \mid D)\Pr(D) + \Pr(P \mid N)\Pr(N) \\ &= (1.0)(0.2) + (0.05)(0.8) = 0.24.\end{aligned}$$

Then, find the conditional probability with Bayes' theorem,

$$\begin{aligned}\Pr(D \mid P) &= \frac{\Pr(P \mid D)\Pr(D)}{\Pr(P)} \\ &= \frac{1.0 \cdot 0.2}{0.24} = 0.833.\end{aligned}$$

This is much better; the test is now identifying mainly people with the disease. This is because the fraction with the disease is significantly greater than the fraction of false positives.

39. First, find $\Pr(P)$ with the law of total probability,

$$\begin{aligned}\Pr(P) &= \Pr(P \mid D)\Pr(D) + \Pr(P \mid N)\Pr(N) \\ &= (0.95)(0.2) + (0.05)(0.8) = 0.23.\end{aligned}$$

Then, find the conditional probability with Bayes' theorem,

$$\Pr(D \mid P) = \frac{\Pr(P \mid D)\Pr(D)}{\Pr(P)} = \frac{0.95 \cdot 0.2}{0.23} = 0.826.$$

Also,

$$\Pr(D \mid P^c) = \frac{\Pr(P^c \mid D)\Pr(D)}{\Pr(P^c)} = \frac{0.05 \cdot 0.2}{0.77} = 0.013.$$

The false negatives do not significantly reduce the fraction of positive tests for those with the disease, but 1.3% of those who test negative are indeed sick.

41. Let T be the event tall and H be the event that the offspring of the **BB** × **Bb** cross is a heterozygote. We want to know $\Pr(Bb \mid T)$. The cross between a **BB** plant and a **Bb** plant has genotype **BB** with probability 1/2 and genotype **Bb** with probability 1/2, $\Pr(H) = 1/2$, and $\Pr(H^c) = 1/2$. The probability that the final plant is tall is

$$\begin{aligned}\Pr(T) &= \Pr(T \mid H)\Pr(H) + \Pr(T|H^c)\Pr(H^c) \\ &= 3/4 \cdot 1/2 + 1 \cdot 1/2 = 7/8\end{aligned}$$

because the cross between two heterozygotes produces a tall plant with probability 3/4. The unconditional probability that the final plant has genotype **Bb** is

$$\begin{aligned}\Pr(Bb) &= \Pr(Bb \mid H)\Pr(H) + \Pr(Bb|H^c)\Pr(H^c) \\ &= 1/2 \cdot 1/2 + 1/2 \cdot 1/2 = 1/2.\end{aligned}$$

We also know that $\Pr(T \mid Bb) = 1$, so we have all the information to use Bayes' theorem to find

$$\Pr(Bb \mid T) = \frac{\Pr(T \mid Bb)\Pr(Bb)}{\Pr(T)} = \frac{1/2}{7/8} = 4/7.$$

43. Let T be the event tall. There are four possible parents, (H,H), (M,H), (H,M), and (M,M), where H represents **Bb**, and M represents **BB**. Then $\Pr(H,H) = \Pr(M,H) = \Pr(H,M) = \Pr(M,M) = 1/4$. The probability that the final plant is tall is

$$\begin{aligned}\Pr(T) &= \Pr[T|(H,H)]\Pr[(H,H)] + \Pr[T|(M,H)]\Pr[(M,H)] + \\ &\quad \Pr[T|(H,M)]\Pr[(H,M)] + \Pr[T|(M,M)]\Pr[(M,M)] \\ &= 3/4 \cdot 1/4 + 1 \cdot 1/4 + 1 \cdot 1/4 + 1 \cdot 1/4 = 15/16.\end{aligned}$$

The unconditional probability that the final plant has genotype **Bb** is

$$\begin{aligned}\Pr(Bb) &= \Pr[Bb|(H,H)]\Pr[(H,H)] + \Pr[Bb|(M,H)]\Pr[(M,H)] + \\ &\quad \Pr[Bb|(H,M)]\Pr[(H,M)] + \Pr[Bb|(M,M)]\Pr[(M,M)] \\ &= 1/2 \cdot 1/4 + 1/2 \cdot 1/4 + 1/2 \cdot 1/4 + 0 \cdot 1/4 = 3/8.\end{aligned}$$

We know that $\Pr(T \mid Bb) = 1$, so we have all the information to use Bayes' theorem to find

$$\Pr(Bb \mid T) = \frac{\Pr(T \mid Bb)\Pr(Bb)}{\Pr(T)} = \frac{3/8}{15/16} = 2/5.$$

45. She ends up with the false eyelashes.

47. Yes. The probability the new car is behind door 1 is 1/3, and the probability that it is behind door 2 or door 3 is 2/3. Because Monty showed that is wasn't behind door 3, the probability it is behind door 2 is 2/3.

49. A contestant who switched would only do so when she was right in the first place. She would never get the new car.

Section 6.5, page 538

1. We have found that $\Pr(A) = 0.9$, $\Pr(B) = 0.6$, $\Pr(A \cap B) = 0.6$, $\Pr(A \mid B) = 1$, and $\Pr(B \mid A) = 2/3$. Therefore, $\Pr(A) = 0.9 \neq \Pr(A \mid B) = 1$, $\Pr(B) = 0.6 \neq \Pr(B \mid A) = 2/3$, and $\Pr(A \cap B) = 0.6 \neq \Pr(A)\Pr(B) = 0.54$. None of the three conditions are satisfied. The events A and B are not independent.

3. $\Pr(A)=0.4$, $\Pr(B)=0.8$, and $\Pr(A\cap B)=0.32$. Therefore, $\Pr(A\mid B)=0.32/0.8=0.4$ and $\Pr(B\mid A)=0.32/0.4=0.8$. In this case, $\Pr(A)=0.4=\Pr(A\mid B)$, $\Pr(B)=0.8=\Pr(B\mid A)$, and $\Pr(A\cap B)=0.32=\Pr(A)\Pr(B)$. All three of the conditions are satisfied, and the events A and B are independent.

5. The probability of rolling a 1 is 1/3, as is the probability of rolling a 3. Because the rolls are independent, the probability of both events is the product of the probabilities, or $1/3\cdot 1/3=1/9$.

7. The probability of rolling an odd value is 2/3, because it could be either a 1 or a 3. The probability of rolling two in a row is the product $2/3\cdot 2/3=4/9$.

9. This probability is the product of the six probabilities 1/3, or $\left(\frac{1}{3}\right)^6=0.00137$.

11. $A\cap B=\{1\}$. Because A and B are independent, $\Pr(A\cap B)=\Pr(A)\Pr(B)=0.4\cdot 0.6=0.24$. Also, $A=\{1\}\cup\{2\}$, so $\Pr(A)=\Pr(\{1\})+\Pr(\{2\})$ because the simple events {1} and {2} are disjoint. Therefore, $0.4=0.24+\Pr(\{2\})$ and $\Pr(\{2\})=0.16$. Similarly, we can find $\Pr(\{3\})=0.36$. Because all the probabilities must add to 1, $\Pr(\{4\})=0.24$.

13. Because the events are disjoint, the intersection is the null set, and its probability is zero. But $\Pr(A\cap B)=0\neq\Pr(A)\Pr(B)$ if $\Pr(A)>0$ and $\Pr(B)>0$.

15. Let M_t denote the event "mutant at generation t" and N_t denote the event "non-mutant at generation t." Then

$$\Pr(M_{t+1}|M_t)=0.99$$
$$\Pr(M_{t+1}|N_t)=0.01$$
$$\Pr(N_{t+1}|M_t)=0.01$$
$$\Pr(N_{t+1}|N_t)=0.99.$$

If $p_t=\Pr(M_t)$, then

$$p_{t+1}=0.99p_t+0.01(1-p_t).$$

The equilibrium is where $p^*=0.99p^*+0.01(1-p^*)$, which has solution $p^*=0.5$.

17. Let B_t denote the event "bound at hour t" and U_t denote the event "unbound at second t." Then

$$\Pr(B_{t+1}|B_t)=0.98$$
$$\Pr(B_{t+1}|U_t)=0.05$$
$$\Pr(U_{t+1}|B_t)=0.02$$
$$\Pr(U_{t+1}|U_t)=0.95.$$

If $p_t=\Pr(B_t)$, then

$$p_{t+1}=0.98p_t+0.05(1-p_t).$$

The equilibrium is where $p^*=0.98p^*+0.05(1-p^*)$, which has solution $p^*=0.714$.

19. $A\cup B=\{1,2,3\}$. From the probabilities found in Exercise 11, $\Pr(\{1,2,3\})=\Pr(\{1\})+\Pr(\{2\})+\Pr(\{3\})=0.24+0.16+0.36=0.76$. By the formula, $\Pr(A\cup B)=0.4+0.6-0.24=0.76$. It worked.

21. The probability of both is the product of the probabilities, or $\Pr(E\cap J)=\Pr(E)\Pr(J)=0.2\cdot 0.5=0.1$.

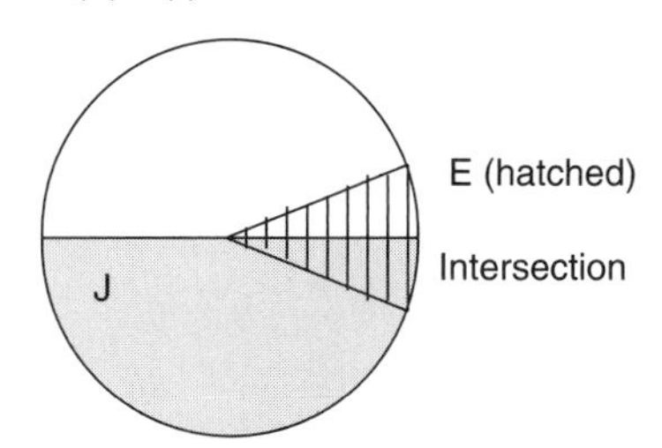

The probability of seeing a jack rabbit conditional on seeing an eagle is 0.5, because $\Pr(J\mid E)=\Pr(J)$ when events are independent. Similarly, the probability of seeing an eagle conditional on seeing a jackrabbit is 0.2, because $\Pr(E\mid J)=\Pr(E)$ when events are independent. It seems that rabbits and eagles ignore each other.

23. a. First, find Pr(P) with the law of total probability,

$$\Pr(P)=\Pr(P\mid D)\Pr(D)+\Pr(P\mid N)\Pr(N)$$
$$=(0.5)(0.01)+(0.5)(0.99)=0.5.$$

Then, find the conditional probability with Bayes' theorem,

$$\Pr(D\mid P)=\frac{\Pr(P\mid D)\Pr(D)}{\Pr(P)}=\frac{0.5\cdot 0.01}{0.5}=0.01.$$

The test gives no new information.

b. We have $\Pr(P)=\Pr(P\mid D)$. Therefore, these events are independent and $\Pr(D)=\Pr(D\mid P)=0.01$.

25. The probability that both molecules remain inside matches the probability that one molecule remains inside for two consecutive seconds, or $0.7\cdot 0.7=0.49$.

27. The probability that both molecules are inside is the product of the probability that the first is inside with the probability that the second is inside. These probabilities can be found by finding p_2 from the discrete-time dynamical system $p_{t+1}=0.7p_t+0.2(1-p_t)$ with $p_0=1$. Then $p_1=0.7$, and $p_2=0.55$. The probability that both molecules are inside is $0.55\cdot 0.55=0.3025$.

29. With meiotic drive, the probability of an **A** from the pollen is independent of the allele that came from the ovule. Let **aA** represent a plant that got an **A** from the ovule and an **a** from the pollen. By the multiplication rule, $\Pr(aA)=\Pr(a)\Pr(A)=0.6\cdot 0.4=0.24$. Similarly, $\Pr(Aa)=\Pr(A)\Pr(a)=0.6\cdot 0.4=0.24$. Therefore, the total probability of a heterozygote is $0.24+0.24=0.48$. With non-independent assortment, $\Pr(aA)=\Pr(a\mid A)\Pr(A)=0.4\cdot 0.5=0.2$ and $\Pr(Aa)=\Pr(A\mid a)\Pr(a)=0.4\cdot 0.5=0.2$. The total is 0.4.

31. Assume she runs into males independently. She will mate with a green bird only if both males are green, which occurs with probability $0.5\cdot 0.5=0.25$.

33. Using the multiplication law, all come with probability 0.729, and none come with probability 0.001.

35. Let S_1 be the event student 1 comes, N_1 that student 1 does not come, etc. By the law of total probability,

$$\Pr(S_2)=\Pr(S_2\mid S_1)\Pr(S_1)+\Pr(S_2\mid N_1)\Pr(N_1)$$
$$0.9=\frac{8}{9}\cdot 0.9+\Pr(S_2\mid N_1)0.1.$$

The only unknown part is $\Pr(S_2\mid N_1)$, which can be solved for as 1.0. The second student therefore always comes when the first does not, meaning that the class will never be empty. Both 1 and 2 come with probability

$$\Pr(\text{both 1 and 2 come})=\Pr(S_2\mid S_1)\Pr(S_1)=0.8.$$

Multiplying by 0.9 (because 3 still ignores them), the probability that all come is 0.72.

37.

$$\Pr(GC_{t+1}|GC_t)=0.999$$
$$\Pr(GC_{t+1}|AT_t)=0.002$$
$$\Pr(AT_{t+1}|GC_t)=0.001$$
$$\Pr(AT_{t+1}|AT_t)=0.998.$$

Let p be the fraction of GC pairs. Then

$$p_{t+1}=0.999p_t+0.002(1-p_t).$$

39. Label the three cells **a**, **b**, and **c**, and let the events a_t, b_t, and c_t mean that it was in cell **a**, **b**, or **c**, respectively, at time t. Suppose it always goes to **a** with probability 0.8, and to **b** or **c** with probability 0.1. Then

$$\Pr(a_{t+1}|a_t)=0.8,\ \Pr(a_{t+1}|b_t)=0.8,\ \Pr(a_{t+1}|c_t)=0.8$$
$$\Pr(b_{t+1}|a_t)=0.1,\ \Pr(b_{t+1}|b_t)=0.1,\ \Pr(b_{t+1}|c_t)=0.1$$
$$\Pr(c_{t+1}|a_t)=0.1,\ \Pr(c_{t+1}|b_t)=0.1,\ \Pr(c_{t+1}|c_t)=0.1.$$

41. Suppose **b** is to the right of **a**, **c** to the right of **b**, and **a** to the right of **c**.

$$\Pr(a_{t+1}|a_t) = 0.9, \Pr(a_{t+1}|b_t) = 0.0, \Pr(a_{t+1}|c_t) = 0.1$$

$$\Pr(b_{t+1}|a_t) = 0.1, \Pr(b_{t+1}|b_t) = 0.9, \Pr(b_{t+1}|c_t) = 0.0$$

$$\Pr(c_{t+1}|a_t) = 0.0, \Pr(c_{t+1}|b_t) = 0.1, \Pr(c_{t+1}|c_t) = 0.9.$$

43. One way to estimate the probability is to note that 3 of the 10 molecules left during the first second, or an estimated probability of leaving of 0.3. The probability of remaining is then 0.7, so the discrete-time dynamical system is $p_{t+1} = 0.7p_t$. The probability of being inside after 3 seconds is $0.7^3 = 0.343$, so we expect 3 or 4 to still be inside at time 3. In fact, 4 out of 10 remain inside, close to what we expected. An alternative, and more accurate, way to estimate the probability is to note that the first molecule, for example, remained in between $t = 0$ and $t = 1$, between $t = 1$ and $t = 2$ for 10 times, and left once. Adding all these up, there are 26 observations of a molecule remaining inside, and 10 of it leaving. We thus estimate that molecules remain with probability $26/36 = 0.72$. Then $0.72^3 = 0.37$, giving pretty much the same result as the simpler estimate.

45. It was "In" 11 times. After being "In," it remained "In" 5 times, for a probability of remaining "In" of about $5/11 = 0.45$. It was "Out" 9 times, but we don't know what happened after time 20. Of the other 8, it jumped "In" 5 times, for a probability of $5/8 = 0.625$. The discrete-time dynamical system is

$$p_{t+1} = 0.45p_t + 0.625(1 - p_t).$$

The equilibrium probability is $p^* = 0.53$. The observed fraction "In" is similar, $11/20 = 0.55$.

Section 6.6, page 553

1.

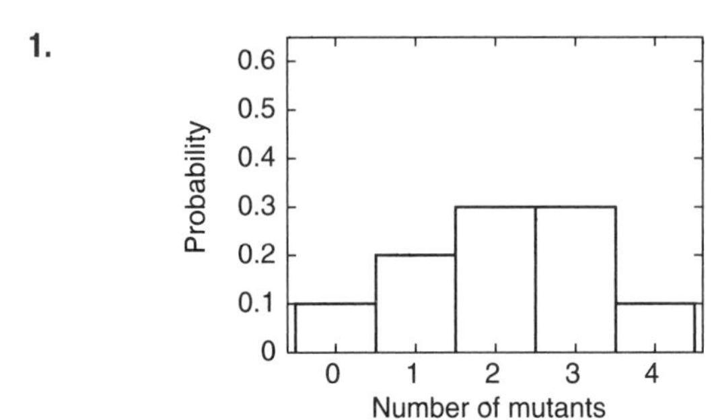

3.

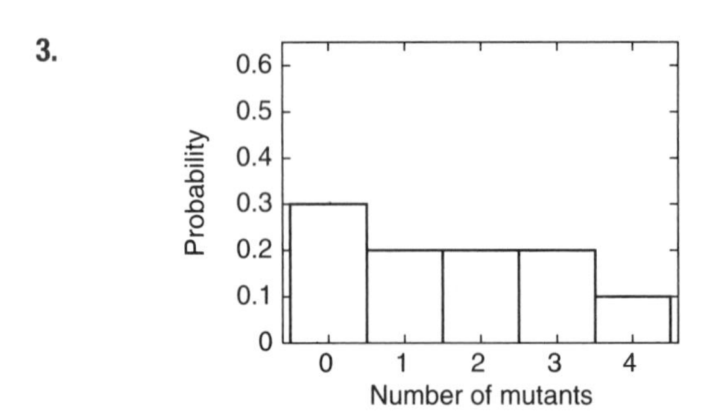

5.

Number of Mutants	Probability Number Is Less Than or Equal
0	0.1
1	0.3
2	0.6
3	0.9
4	1.0

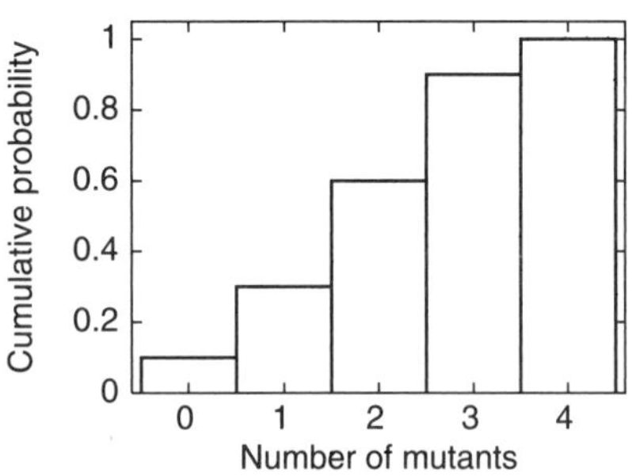

7.

Number of Mutants	Probability Number Is Less Than or Equal
0	0.3
1	0.5
2	0.7
3	0.9
4	1.0

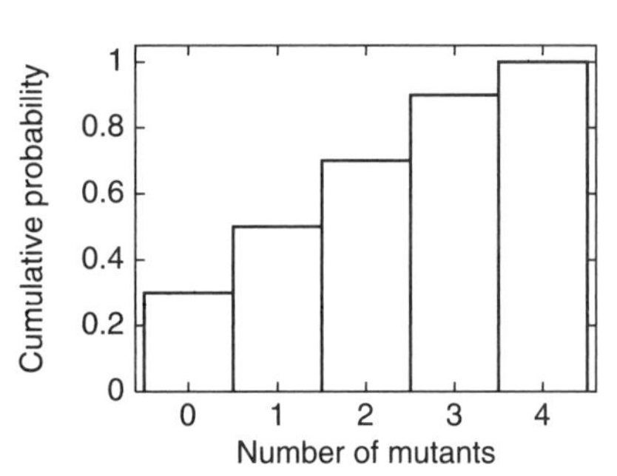

9. The most likely simple event is that the measurement is 5. 1 and 9 tie for least likely. It is symmetric.

11. The most likely simple event is that the measurement is 2. 9 is least likely. It is not symmetric.

13. **a.** The height of the bar above 7 is about 0.11. **b.** The sum of the heights of the bars above 1, 2, 3, and 4 is about 0.38. **c.** This is the complement of the event in **b** with probability of $1 - 0.38 = 0.62$.

15. **a.** The height of the bar above 7 is about 0.08. **b.** The sum of the heights of the bars above 1, 2, 3, and 4 is about 0.64. **c.** This is the complement of the event in **b** with probability of $1 - 0.64 = 0.36$.

17. The area under the curve is

$$\int_0^1 2x\,dx = x^2|_0^1 = 1^2 - 0^2 = 1,$$

as it must be. The maximum is at $x = 1$, where $f(1) = 2 > 1$. However, the probability density can be greater than 1 because probabilities are areas under the curve, which are always less than 1.

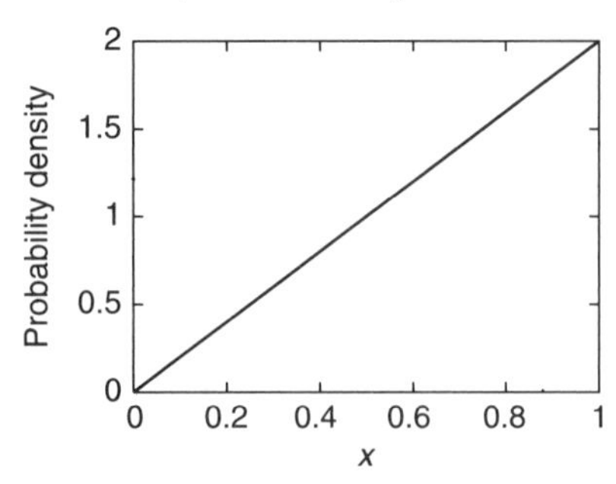

19. The area under the curve is

$$\int_1^e \frac{1}{t}\,dt = \ln(t)|_1^e = \ln(e) - \ln(1) = 1.$$

The maximum is at $t = 1$, where $h(1) = 1$. Because this is a p.d.f., this does not mean that the probability is 1, because probabilities correspond only to areas under the curve.

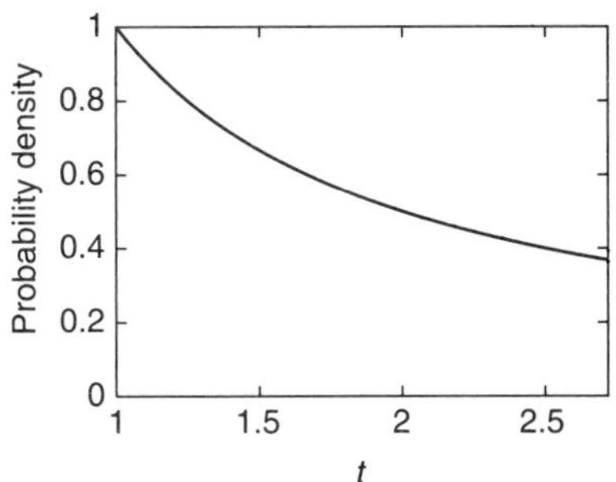

21. The c.d.f. is

$$F(x) = \int_0^x f(y)dy = \int_0^x 2y\,dy = y^2|_0^x = x^2.$$

This function is increasing because $f(x)$ is positive and increases from 0 at $x = 0$ to 1 at $x = 1$.

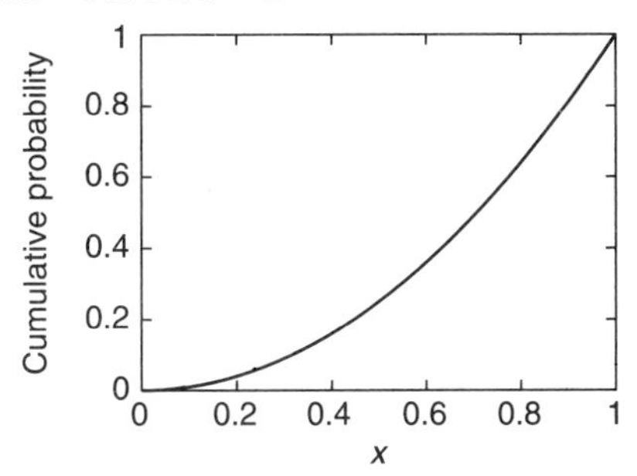

23. The c.d.f. is

$$H(t) = \int_1^t h(s)ds = \int_1^t \frac{1}{s}ds = \ln(s)|_1^t = \ln(t).$$

This function is increasing because $h(t)$ is positive and increases from 0 at $t = 1$ to 1 at $t = e$.

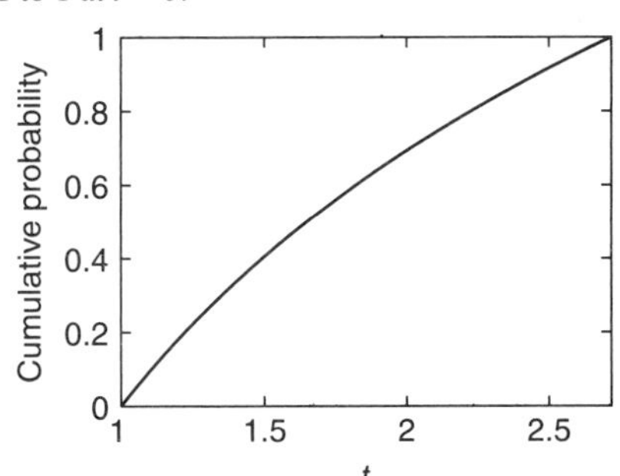

25. Integrating,

$$\int_{0.2}^{0.6} f(x)dx = \int_{0.2}^{0.6} 2x\,dx = x^2|_{0.2}^{0.6} = 0.32.$$

The c.d.f. is $F(x) = x^2$, so

$$\int_{0.2}^{0.6} f(x)dx = F(0.6) - F(0.2) = 0.6^2 - 0.2^2 = 0.32.$$

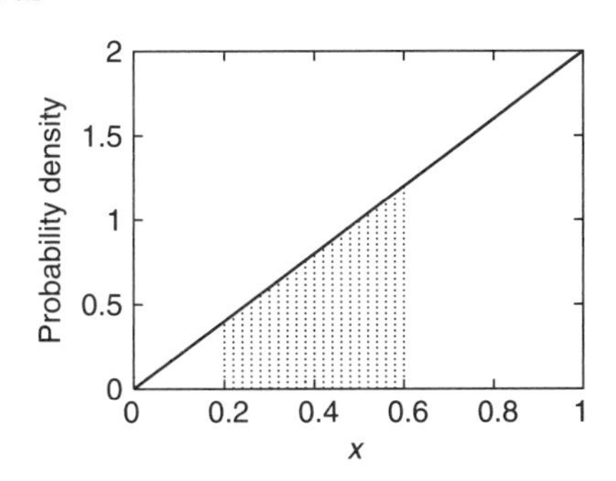

27. Integrating

$$\int_{2.0}^{2.5} \frac{1}{t}dt = \ln(t)|_{2.0}^{2.5} = \ln(2.5) - \ln(2.0) = 0.223.$$

The c.d.f. is $H(t) = \ln(t)$, so

$$\int_{2.0}^{2.5} \frac{1}{t}dt = H(2.5) - H(2.0) = 0.223.$$

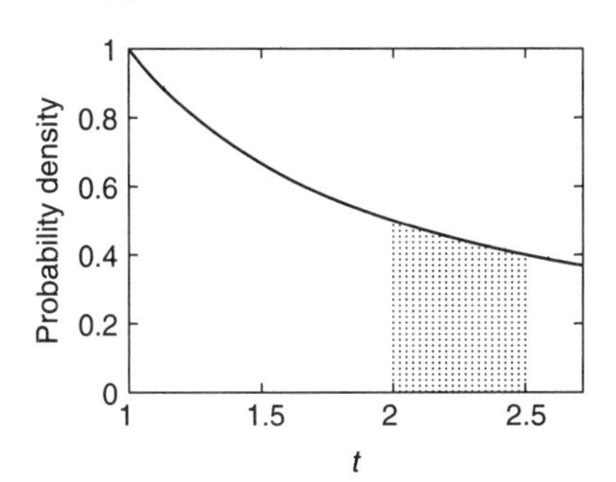

29.

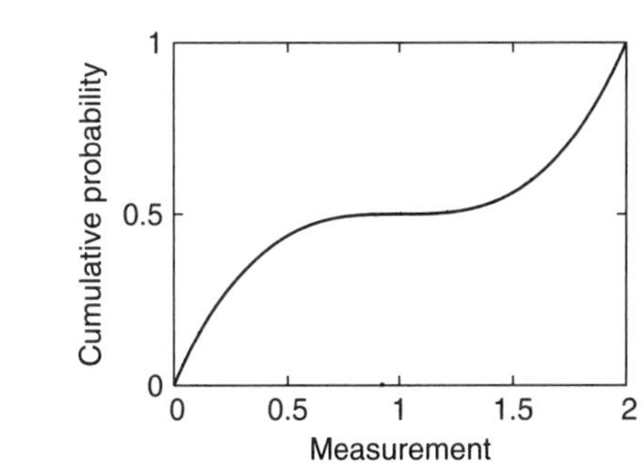

31.

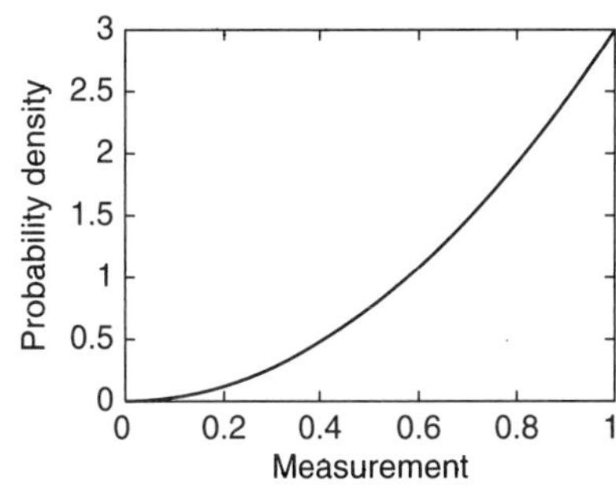

33.

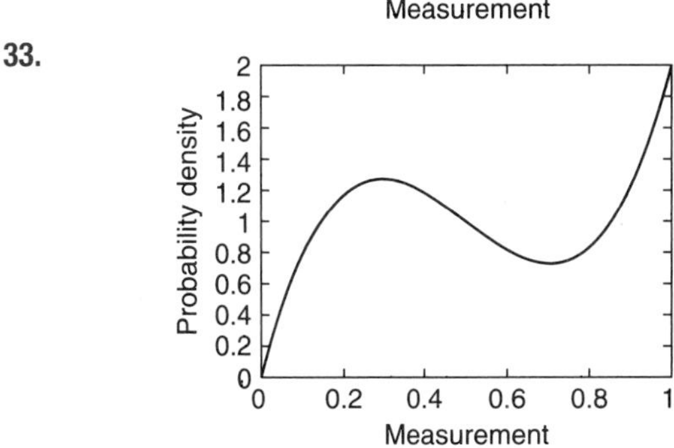

35. The cumulative distribution is

Pr(cell is less than or equal to 0 days old) = 0.4

Pr(cell is less than or equal to 1 day old) = 0.7

Pr(cell is less than or equal to 2 days old) = 0.9

Pr(cell is less than or equal to 3 days old) = 1.0.

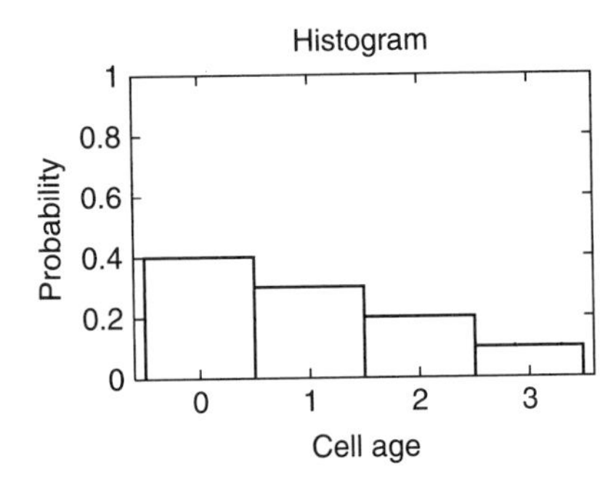

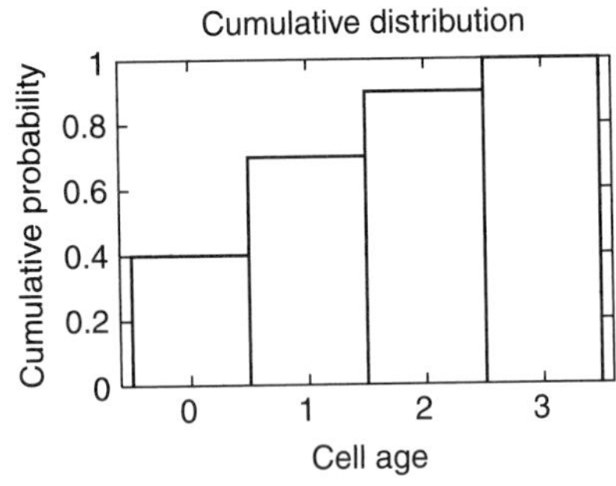

37. The probability that between four and six plants are tall is $0.02 + 0.08 + 0.09 = 0.19$.

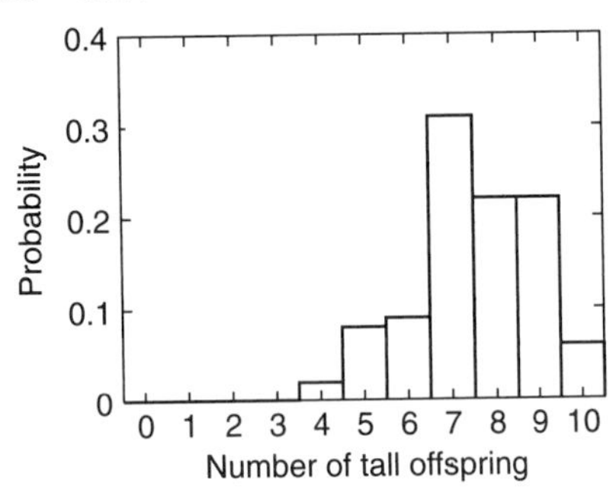

39. The probability that between four and six plants are tall is $0.21 + 0.27 + 0.19 = 0.67$.

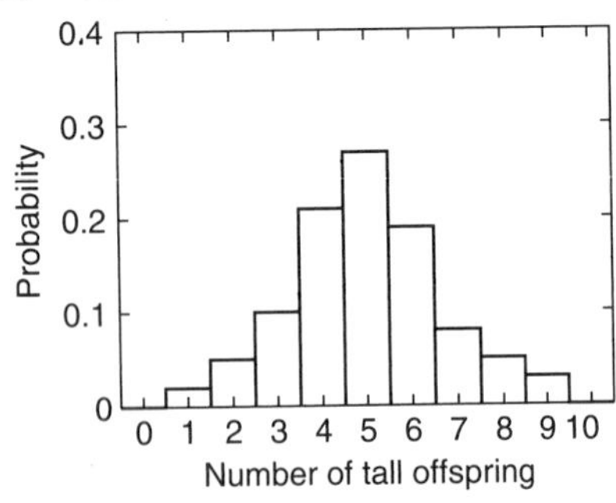

41. In the first experiment, 50 mated with red, 30 with blue, and 20 with green. In the second experiment, 40 mated with red, 35 with blue, and 25 with green. The total is 90 with red, 65 with blue, and 45 with green. Dividing by 200 gives a probability of 0.45 with red, 0.325 with blue, and 0.225 with green.

43. Let R_1 be the event of mating with a red male in the first experiment, R_2 of mating with a red male in the second experiment, and R the event of mating with a red male in the combined experiment (and similarly for B and G). Let p be the probability a female came from the first experiment and $1 - p$ the probability she came from the second. Because the experiments are equally large, $p = 0.5$. Then

$$\begin{aligned}\Pr(R) &= \Pr(R_1) \cdot p + \Pr(R_2) \cdot (1 - p) = 0.5 \cdot 0.5 + 0.4 \cdot 0.5 \\ &= 0.45\end{aligned}$$

$$\begin{aligned}\Pr(B) &= \Pr(B_1) \cdot p + \Pr(B_2) \cdot (1 - p) = 0.3 \cdot 0.5 + 0.35 \cdot 0.5 \\ &= 0.325\end{aligned}$$

$$\begin{aligned}\Pr(G) &= \Pr(G_1) \cdot p + \Pr(G_2) \cdot (1 - p) = 0.2 \cdot 0.5 + 0.25 \cdot 0.5 \\ &= 0.225.\end{aligned}$$

45. a. The c.d.f. is

$$\begin{aligned}G(x) &= \int_0^x g(y)dy = \int_0^x 0.5e^{-0.5y}dy = -e^{-0.5y}|_0^x \\ &= -e^{-0.5x} + 1.\end{aligned}$$

b.

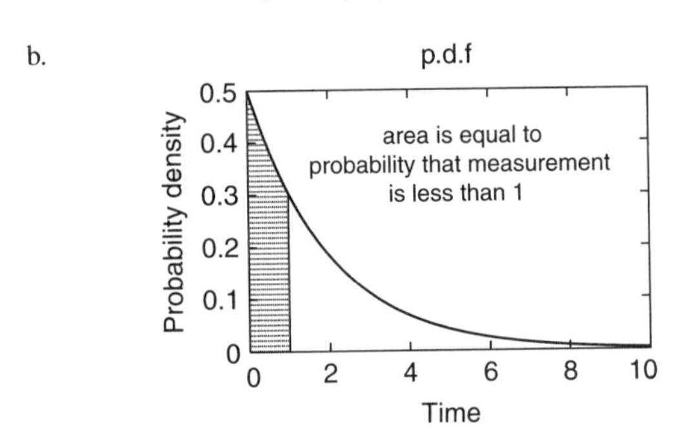

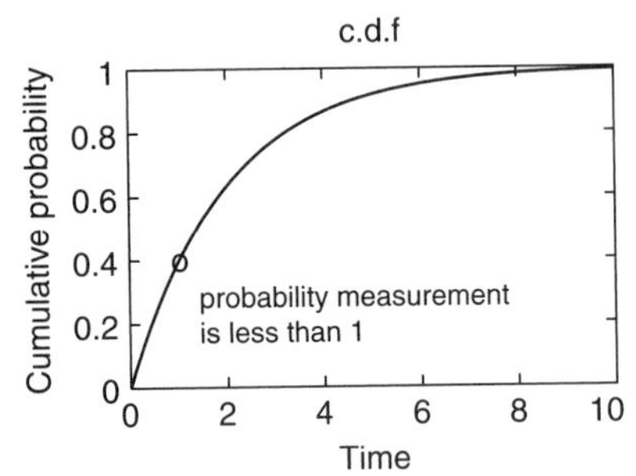

c. $G'(x) = 0 - (-0.5e^{-0.5x}) = 0.5e^{-0.5x}$.
d. This is $G(1.0) = 0.393$.
e. This is $G(3) - G(1) = 0.383$.
f. This is $G(1.01) - G(1) = 0.0030$, quite close to $0.01 \cdot g(1.0) = 0.01 \cdot 0.30 = 0.0030$.

Section 6.7, page 565

1. Denote the number of mutants with the random variable M. Then $\Pr(M = 0) = 0.1$, $\Pr(M = 1) = 0.2$, $\Pr(M = 2) = 0.3$, $\Pr(M = 3) = 0.3$, and $\Pr(M = 4) = 0.1$.

$$E(M) = 0 \cdot 0.1 + 1 \cdot 0.2 + 2 \cdot 0.3 + 3 \cdot 0.3 + 4 \cdot 0.1 = 2.1.$$

All experiments where $M \le 2$ have $M < E(M)$, and $\Pr(M \le 2) = 0.1 + 0.2 + 0.3 = 0.6$. Most of the results are below average.

3. Denote the number of mutants with the random variable M. Then $\Pr(M = 0) = 0.3$, $\Pr(M = 1) = 0.2$, $\Pr(M = 2) = 0.2$, $\Pr(M = 3) = 0.2$, and $\Pr(M = 4) = 0.1$.

$$E(M) = 0 \cdot 0.3 + 1 \cdot 0.2 + 2 \cdot 0.2 + 3 \cdot 0.2 + 4 \cdot 0.1 = 1.6.$$

All experiments where $M \le 1$ have $M < E(M)$, and $\Pr(M \le 1) = 0.3 + 0.2 = 0.5$. Exactly half of the results are below average.

5.

$$\begin{aligned}E(N_4) &= 0 \cdot 0.0 + 1 \cdot 0.0 + 2 \cdot 0.004 + 3 \cdot 0.019 + 4 \cdot 0.064 \\ &\quad + 5 \cdot 0.147 + 6 \cdot 0.234 + 7 \cdot 0.255 + 8 \cdot 0.183 \\ &\quad + 9 \cdot 0.077 + 10 \cdot 0.015 = 6.55.\end{aligned}$$

7.

$$E(X) = \int_0^1 xf(x)dx = \int_0^1 2x^2dx = \frac{2x^3}{3}|_0^1 = \frac{2}{3}.$$

The probability that X is less than 2/3 is

$$\Pr\left(X < \frac{2}{3}\right) = \int_0^{\frac{2}{3}} 2x\,dx = x^2|_0^{\frac{2}{3}} = \frac{4}{9}.$$

Less than half of individuals are below the expectation.

9.

$$E(T) = \int_1^e t h(t)\,dt = \int_1^e 1\,dt = t|_1^e = e - 1.$$

The probability that T is less than $e-1$ is

$$\Pr(T < e-1) = \int_0^{e-1} \frac{1}{t}\,dt = \ln(t)_1^{e-1} = \ln(e-1) = 0.541.$$

Slightly more than half of individuals are below the expectation.

11. Let Y be the imprecise random variable, which takes on just the two values 0.25 and 0.75. Then

$$\Pr(Y = 0.25) = \Pr(X \le 0.5) = \int_0^{0.5} 2x\,dx = x^2|_0^{0.5} = 0.25.$$

Because probabilities must add to 1, $\Pr(Y = 0.75) = 0.75$. Then $E(Y) = 0.25 \cdot 0.25 + 0.75 \cdot 0.75 = 0.625$. This is smaller than the expectation of the continuous random variable.

13. Let Y be the imprecise random variable, which takes on the four values 0.25, 0.5, 0.75, and 1.0. Then

$$\Pr(Y = 0.25) = \Pr(X \le 0.25) = \int_0^{0.25} 2x\,dx = x^2|_0^{0.25} = 0.0625$$

$$\Pr(Y = 0.5) = \Pr(0.25 < X \le 0.5) = \int_{0.25}^{0.5} 2x\,dx = x^2|_{0.25}^{0.5} = 0.1875$$

$$\Pr(Y = 0.75) = \Pr(0.5 < X \le 0.75) = \int_{0.5}^{0.75} 2x\,dx = x^2|_{0.5}^{0.75} = 0.3125$$

$$\Pr(Y = 1.0) = \Pr(0.75 < X \le 1.0) = \int_{0.75}^{1.0} 2x\,dx = x^2|_{0.75}^{1.0} = 0.4375.$$

Then

$$E(Y) = 0.25 \cdot 0.0625 + 0.5 \cdot 0.1875 + 0.75 \cdot 0.3125 + 1.0 \cdot 0.4375 = 0.78125.$$

This is larger than the expectation of the continuous random variable.

15. This p.d.f. approaches infinity at $x = 0$. However,

$$\int_0^1 \frac{1}{2\sqrt{x}}\,dx = \sqrt{x}|_0^1 = 1,$$

so this is a good p.d.f. The expectation is

$$\int_0^1 x\frac{1}{2\sqrt{x}}\,dx = \frac{x^{3/2}}{3}|_0^1 = 1/3.$$

17. Let $M = 1$ represent "in" and $M = 0$ represent "out." Then $\Pr(M = 1) = 0.9$, $\Pr(M = 0) = 0.1$, and $E(M) = 1 \cdot 0.9 + 0 \cdot 0.1 = 0.9$.

19. Let $M = 1$ represent together and $M = 0$ represent separate. The probability that both are in is 0.81, and that both are out is 0.01. Then $\Pr(M = 1) = 0.82$ and $\Pr(M = 0) = 0.18$. Then $E(M) = 1 \cdot 0.82 + 0 \cdot 0.18 = 0.82$.

21. Count the number M of molecules in at time 1. Both are in with probability 0.81, so $\Pr(M = 2) = 0.81$. Both are out with probability 0.01, so $\Pr(M = 0) = 0.01$. Therefore, $\Pr(M = 1) = 0.18$. $E(M) = 0 \cdot 0.01 + 1 \cdot 0.18 + 2 \cdot 0.81 = 1.8$.

23. Let P represent the number out of 10 that are tall. Then $\Pr(P = 4) = 0.02$, $\Pr(P = 5) = 0.08$, $\Pr(P = 6) = 0.09$, $\Pr(P = 7) = 0.31$, $\Pr(P = 8) = 0.22$, $\Pr(P = 9) = 0.22$, and $\Pr(P = 10) = 0.06$. Then

$$E(P) = 4 \cdot 0.02 + 5 \cdot 0.08 + 6 \cdot 0.09 + 7 \cdot 0.31 + 8 \cdot 0.22 + 9 \cdot 0.22 + 10 \cdot 0.06 = 7.53.$$

Exactly half are less than this expectation.

25. Let P represent the number out of 10 that are tall. Then $\Pr(P = 1) = 0.02$, $\Pr(P = 2) = 0.05$, $\Pr(P = 3) = 0.10$, $\Pr(P = 4) = 0.21$, $\Pr(P = 5) = 0.27$, $\Pr(P = 6) = 0.19$, $\Pr(P = 7) = 0.08$, $\Pr(P = 8) = 0.05$, and $\Pr(P = 9) = 0.03$. Then

$$E(P) = 1 \cdot 0.02 + 2 \cdot 0.05 + 3 \cdot 0.10 + 4 \cdot 0.21 + 5 \cdot 0.27 + 6 \cdot 0.19 + 7 \cdot 0.08 + 8 \cdot 0.05 + 9 \cdot 0.03 = 4.98.$$

38% are less than this expectation, but 27% are very close to it.

27. Let the random variable S describe the brightness. Then $\Pr(S = 9) = 0.3$ and $\Pr(S = 7) = 0.7$. $E(S) = 9 \cdot 0.3 + 7 \cdot 0.7 = 7.6$.

29. Let the random variable S describe the brightness. Then $\Pr(S = 9.5) = 0.4$, $\Pr(S = 9.0) = 0.3$, $\Pr(S = 8.0) = 0.2$, and $\Pr(S = 5.0) = 0.1$. $E(S) = 5 \cdot 0.1 + 8 \cdot 0.2 + 9 \cdot 0.3 + 9.5 \cdot 0.4 = 8.6$.

31. Let the random variable be I. Then $\Pr(I = -1) = 0.4$, $\Pr(I = 0) = 0.2$, $\Pr(I = 1) = 0.3$, and $\Pr(I = 2) = 0.1$. $E(I) = -1 \cdot 0.4 + 0 \cdot 0.2 + 1 \cdot 0.3 + 2 \cdot 0.1 = 0.1$. About one individual would arrive over the course of 10 years, and the population should grow slowly.

33. Let the random variable be I. Then $\Pr(I = -1) = 0.4$, $\Pr(I = 0) = 0.2$, $\Pr(I = 1) = 0.3$, and $\Pr(I = 100) = 0.1$. $E(I) = -1 \cdot 0.4 + 0 \cdot 0.2 + 1 \cdot 0.3 + 100 \cdot 0.1 = 9.9$. About 99 individuals would arrive in 10 years, meaning that this population will grow. The expectation seems quite far from the middle of the distribution, having been pulled way up by the rare years with many immigrants.

35.

$$E(X) = \int_0^\infty x g(x)\,dx = \int_0^\infty 0.5xe^{-0.5x}\,dx = \frac{-0.5xe^{-0.5x} - e^{-0.5x}}{0.5}|_0^\infty = 2.0$$

(all the terms except one are zero).

37. The total mass is 100 kg, so the child is 0.2 of the total and the adult is 0.8 of the total. The positions are 3 for the child and -1 for the adult. Therefore, the center of mass is $3 \cdot 0.2 + (-1) \cdot 0.8 = -0.2$. The center of mass is to the left of the center, meaning that the adult will go down and the child will go up.

39. The total mass is 100 kg, so the child is 0.2 of the total and the adult is 0.8 of the total. The positions are 3 for the child and $-x$ for the adult. Therefore, the center of mass is $3 \cdot 0.2 + (-x) \cdot 0.8 = 0.6 - 0.8x$. This is 0 if $x = 0.75$. The adult must sit 75 cm to the left of center for them to balance.

41. The total mass is

$$\int_0^1 x(1 - x^2)\,dx = \frac{x^2}{2} - \frac{x^4}{4}|_0^1 = 1/4.$$

The p.d.f. is then $f(x) = 4x(1 - x^2)$. The expectation of a random variable X with this p.d.f. is

$$E(X) = \int_0^1 x f(x)\,dx = \int_0^1 4x^2(1 - x^2)\,dx = \frac{4x^3}{3} - \frac{4x^5}{5}|_0^1 = \frac{8}{15} = 0.5333.$$

This is the point where the bar would balance, and is to the right of the center. The density has a maximum where $\rho'(x) = 0$. Because $\rho'(x) = 1 - 3x^2$, this occurs at $x = \sqrt{3}/3 = 0.577$. The center of mass is between the actual center of the bar and the point of maximum density.

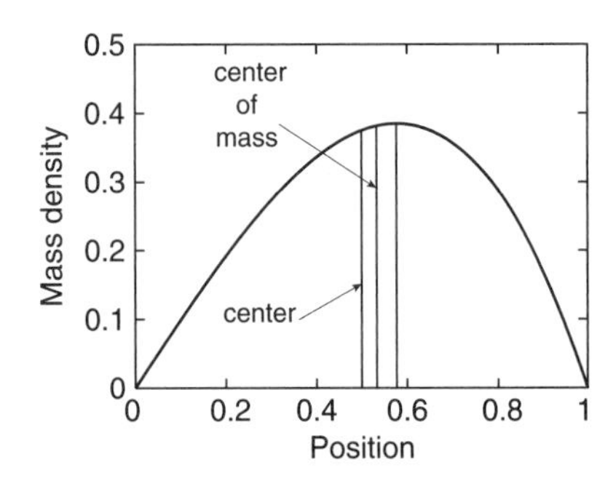

43. The board weighs 20 kg and has center of mass at $x=2$. The total mass is 120 kg, so the child is 1/6 of the total, the board is 1/6 of the total, and the adult is 2/3 of the total. The positions are 3 for the child, 2 for the board, and -1 for the adult. Therefore, the center of mass is $3\cdot 1/6+2\cdot 1/6+(-1)\cdot 2/3=1/6$. The center of mass is to the right of the center, meaning that the adult will go up and the child will go down.

Section 6.8, page 579

1. Denoting the number of mutants with the random variable M, we have that $\Pr(M\le 1)=0.3$ and $\Pr(M\le 2)=0.6$. The median is then between 1 and 2, which is smaller than the expectation of 2.1. The mode is shared by $M=2$ and $M=3$.

3. Denoting the number of mutants with the random variable M, $\Pr(M\le 1)=0.5$, so the median is $M=1$. This is less than the expectation of 1.6. The mode is $M=0$.

5. This histogram is symmetric around a peak at 5, so the mean, median, and mode are all equal to 5.

7. The mode is 2, the median is between 3 and 4, and the mean is probably a bit larger than 4.

9. The mode is at $x=1$. The c.d.f. is $F(x)=x^2$ (Section 6.6, Exercise 21). Therefore, the median solves $F(x)=1/2$, or $x=\sqrt{2}/2=0.707$. This is slightly larger than the mean of 2/3.

11. The mode is at $t=1$. The c.d.f. is $H(t)=\ln(t)$ (Section 6.6, Exercise 23). The median solves $H(t)=1/2$, so $t=e^{1/2}=1.649$, which is slightly less than the expectation of $e-1=1.718$.

13. The c.d.f. is

$$F(x)=\int_0^x \frac{1}{2\sqrt{y}}dy=\sqrt{y}|_0^x=\sqrt{x}.$$

The median solves $F(x)=1/2$, or $x=1/4$, smaller than the mean of 1/3.

15. First, we find

$$\mathrm{E}[\ln(X)]=\ln(1)\cdot 0.3+\ln(2)\cdot 0.7=0.485.$$

Then the geometric mean is $e^{0.485}=1.624$. The arithmetic mean is

$$\mathrm{E}(X)=1\cdot 0.3+2\cdot 0.7=1.7,$$

which is greater than the geometric mean.

17. First, we find

$$\mathrm{E}[\ln(X)]=\ln(1)\cdot 0.3+\ln(2)\cdot 0.3+\ln(3)\cdot 0.4=0.647.$$

Then the geometric mean is $e^{0.647}=1.910$. The arithmetic mean is

$$\mathrm{E}(X)=1\cdot 0.3+2\cdot 0.3+3\cdot 0.4=2.1,$$

which is greater than the geometric mean.

19. First, we find $\mathrm{E}[\ln(X)]$ as

$$\int_{0.75}^{1.25}\ln(x)f(x)dx=\int_{0.75}^{1.25}2\ln(x)dx$$
$$=2x\ln(x)-2x|_{0.75}^{1.25}=-0.0106.$$

The geometric mean is then $e^{-0.0106}=0.9895$. Because this is a symmetric distribution centered at $x=1.0$, the expectation is 1.0, which is greater than the geometric mean.

21. First, we find $\mathrm{E}[\ln(X)]$ as

$$\int_0^1\ln(x)f(x)dx=\int_0^1 2x\ln(x)dx=x^2\ln(x)-\frac{1}{2}x^2|_0^1=-\frac{1}{2}.$$

The term $x^2\ln(x)$ at $x=0$ is equal to 0 because

$$\lim_{x\to 0}x^2\ln(x)=\lim_{x\to 0}\frac{\ln(x)}{\frac{1}{x^2}}=\lim_{x\to 0}\frac{\frac{1}{x}}{\frac{-2}{x^3}}=\lim_{x\to 0}\frac{-x^2}{2}=0$$

where we used L'Hôpital's rule in the second step. The geometric mean is then $e^{-0.5}=0.606$. This is smaller than the expectation of 2/3 found earlier.

23. $\mathrm{E}[\ln(R)]=\ln(1)\cdot 0.5+\ln(2)\cdot 0.5.=\frac{\ln(2)}{2}$. The geometric mean is then $e^{\ln(2)/2}=[e^{\ln(2)}](1/2)=2^{1/2}=\sqrt{2}$. The area of a rectangle with sides of length 1 and 2 is 2. The area of a square with sides of $\sqrt{2}$ is $\sqrt{2}^2=2$.

25. The arithmetic mean is $(1+r_2)/2$, and the geometric mean is $\sqrt{r_2}$. The ratio has maximum where

$$\frac{d}{dr_2}\frac{2\sqrt{r_2}}{1+r_2}=\frac{\frac{1+r_2}{\sqrt{r_2}}-2\sqrt{r_2}}{(1+r_2)^2}=0,$$

which has solution at $r_2=1$.

27. The mean is \$57,000, the median is \$30,000, and the mode is \$20,000. The mode tells what you'll probably get and the median gives a good idea of what happens in the middle. The mean is heavily affected by the top salary.

29. The cumulative distribution is $\Pr(P\le 4)=0.02$, $\Pr(P\le 5)=0.10$, $\Pr(P\le 6)=0.19$, $\Pr(P\le 7)=0.50$, $\Pr(P\le 8)=0.72$, $\Pr(P\le 9)=0.94$, and $\Pr(P\le 10)=1.00$. The median is $P=7$, and the mode is also $P=7$.

31. The cumulative distribution is $\Pr(P\le 1)=0.02$, $\Pr(P\le 2)=0.07$, $\Pr(P\le 3)=0.17$, $\Pr(P\le 4)=0.38$, $\Pr(P\le 5)=0.65$, $\Pr(P\le 6)=0.84$, $\Pr(P\le 7)=0.92$, $\Pr(P\le 8)=0.97$, and $\Pr(P\le 9)=1.00$. The median is between 4 and 5, and the mode is 5.

33. The c.d.f. is

$$G(x)=\int_0^x g(y)dy=\int_0^x 0.5e^{-0.5y}dy=-e^{-0.5y}|_0^x=1-e^{-0.5x}.$$

The median solves $G(x)=1/2$, or

$$1-e^{-0.5x}=1/2$$
$$e^{-0.5x}=1/2$$
$$-0.5x=\ln(1/2)$$
$$x=2\ln(2)=1.386.$$

This is much smaller than the expectation of 2.0.

35. The average city size is just the average of these two, or 550,000. However, the average crowding a person experiences is different. Only 1/9 of people live in the smaller city, so the average crowding encountered is

$$100,000\cdot\frac{1}{9}+1,000,000\cdot\frac{8}{9}=900,000.$$

Nonetheless, 8/9 or 88.9% of people live in a city more crowded than this average.

37. The arithmetic mean is

$$\mathrm{E}(R)=4\cdot 0.5+0.25\cdot 0.5=2.125.$$

To find the geometric mean, we first compute the expectation of $\ln(R)$ as

$$\mathrm{E}[\ln(R)]=\ln(4)\cdot 0.5+\ln(0.25)\cdot 0.5=0.$$

The geometric mean is then $e^{E[\ln(R)]} = e^0 = 1.0$, well less than the arithmetic mean. This population is neither growing nor shrinking, on average.

39. The arithmetic mean is

$$E(R) = 4 \cdot 0.75 + 0.25 \cdot 0.25 = 3.0625.$$

To find the geometric mean, we first compute the expectation of $\ln(R)$ as

$$E[\ln(R)] = \ln(4) \cdot 0.75 + \ln(0.25) \cdot 0.25 = \ln(2).$$

The geometric mean is $e^{\ln(2)} = 2.0$, far less than the arithmetic mean. This population is growing quickly.

41. Because the geometric mean is 1.0, the population will remain at about 100.

43. Because the geometric mean is 2.0, the population is $P(t) = 100 \cdot 2.0^t$. After 50 generations, it would be 1.126×10^{17}, which is hugely large.

45.

a. \$50, \$75, \$37.50, \$56.25.

b. After an even number of weeks, the price p_t is

$$p_t = 0.75^{\frac{t}{2}} \cdot 100.$$

After an odd number of weeks, the price is

$$p_t = 0.5 \cdot 0.75^{\frac{t-1}{2}} \cdot 100.$$

c. The "low price" manager wins out.

d. The price change takes the value 0.5 with probability 0.5 and the value 1.5 with probability 0.5; the change in price has geometric mean of 0.866. The price declines because this is less than 1.

Section 6.9, page 592

1. The range is 0 to 4. We found that the expectation is 2.1. Therefore,

$$\text{MAD} = |0 - 2.1| \cdot 0.1 + |1 - 2.1| \cdot 0.2 + |2 - 2.1| \cdot 0.3 + |3 - 2.1| \cdot 0.3 + |4 - 2.1| \cdot 0.1 = 0.92.$$

With the direct method, the variance is

$$\text{Var}(M) = (0 - 2.1)^2 0.1 + (1 - 2.1)^2 0.2 + (2 - 2.1)^2 0.3 + (3 - 2.1)^2 0.3 + (4 - 2.1)^2 0.1 = 1.29.$$

With the computational formula,

$$\text{Var}(M) = 0^2 0.1 + 1^2 0.2 + 2^2 0.3 + 3^2 0.3 + 4^2 0.1 - 2.1^2 = 1.29.$$

The standard deviation is $\sqrt{1.29} = 1.136$. The coefficient of variation is $1.136/2.1 = 0.541$.

3. The range is 0 to 4. We found that the expectation is 1.6. Therefore,

$$\text{MAD} = |0 - 1.6| \cdot 0.3 + |1 - 1.6| \cdot 0.2 + |2 - 1.6| \cdot 0.2 + |3 - 1.6| \cdot 0.2 + |4 - 1.6| \cdot 0.1 = 1.2.$$

With the direct method, the variance is

$$\text{Var}(M) = (0 - 1.6)^2 0.3 + (1 - 1.6)^2 0.2 + (2 - 1.6)^2 0.2 + (3 - 1.6)^2 0.2 + (4 - 1.6)^2 0.1 = 1.84.$$

With the computational formula,

$$\text{Var}(M) = 0^2 0.3 + 1^2 0.2 + 2^2 0.2 + 3^2 0.2 + 4^2 0.1 - 1.6^2 = 1.84.$$

The standard deviation is $\sqrt{1.84} = 1.356$. The coefficient of variation is $1.357/1.6 = 0.848$.

5. Call the random variable B. Then $\Pr(B = 1) = 1/3$ and $\Pr(B = 0) = 2/3$. Then $E(B) = 1/3$, $E(B^2) = 1/3$, so $\text{Var}(B) = \frac{1}{3} - (\frac{1}{3})^2 = \frac{2}{9}$. The standard deviation is $\sqrt{\frac{2}{9}} = 0.471$, and the coefficient of variation is $0.471/0.333 = 1.414$. $\text{MAD} = 2/3 \cdot 1/3 + 1/3 \cdot 2/3 = 4/9$.

7. Call the random variable B. Then $\Pr(B = 10) = 1/3$ and $\Pr(B = 0) = 2/3$. Then $E(B) = 10/3$, $E(B^2) = 100/3$, so $\text{Var}(B) = \frac{100}{3} - (\frac{10}{3})^2 = \frac{200}{9}$. The standard deviation is $\sqrt{\frac{200}{9}} = 4.71$, and the coefficient of variation is $4.71/3.33 = 1.414$. $\text{MAD} = 10/3 \cdot 2/3 + 20/3 \cdot 1/3 = 40/9$. Compared with Exercise 5, the standard deviation, MAD, and expectation increased by a factor of 10; the variance increased by a factor of 100; and the coefficient of variation did not change.

9. The c.d.f. is $F(x) = x^2$. The lower quartile solves $F(x) = 0.25$ and is equal to $\sqrt{0.25} = 0.5$. The median solves $F(x) = 0.5$ and is equal to $\sqrt{0.5} = 0.707$. The upper quartile solves $F(x) = 0.75$ and is equal to $\sqrt{0.75} = 0.866$.

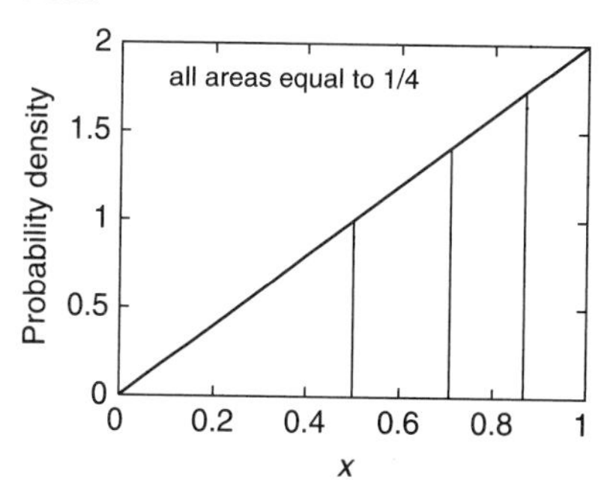

11. The c.d.f. is $H(t) = \ln(t)$. The lower quartile solves $H(t) = 0.25$ and is equal to $e^{0.25} = 1.284$. The median solves $H(t) = 0.5$ and is equal to $e^{0.5} = 1.649$. The upper quartile solves $H(t) = 0.75$ and is equal to $e^{0.75} = 2.117$.

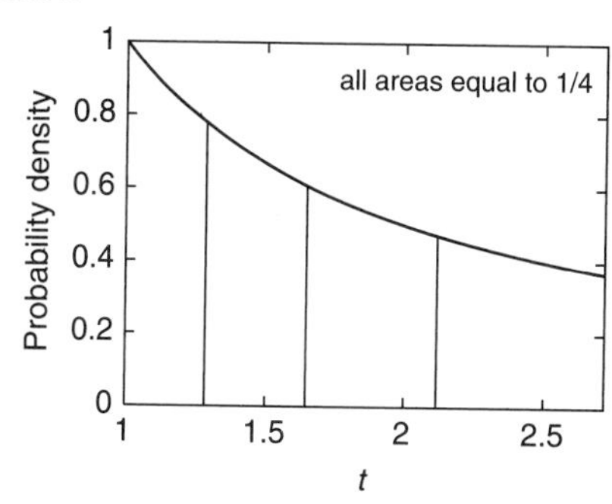

13. We found that $E(X) = 2/3$. Using the computational formula,

$$\text{Var}(X) = \int_0^1 x^2 f(x)dx - E(X)^2 = \int_0^1 2x^3 dx - E(X)^2 = \frac{x^4}{2}\Big|_0^1 - \left(\frac{2}{3}\right)^2 = \frac{1}{2} - \frac{4}{9} = \frac{1}{18}.$$

The standard deviation is $\sqrt{1/18} = 0.236$.

15. We found that $E(T) = e - 1$. Using the computational formula,

$$\text{Var}(T) = \int_1^e t^2 h(t)dt - E(T)^2 = \int_1^e t dt - E(T)^2 = \frac{t^2}{2}\Big|_1^e - (e-1)^2 = \frac{e^2}{2} - \frac{1}{2} - (e-1)^2 = 0.242.$$

The standard deviation is $\sqrt{0.242} = 0.492$.

17. We found that $E(X) = 2/3$. Then

$$\text{MAD}(X) = \int_0^1 |x - 2/3| f(x)dx = \int_0^{\frac{2}{3}} 2x(2/3 - x)dx + \int_{\frac{2}{3}}^1 2x(x - 2/3)dx = \frac{2}{3}x^2 - \frac{2x^3}{3}\Big|_0^{\frac{2}{3}} + \frac{2x^3}{3} - \frac{2}{3}x^2\Big|_{\frac{2}{3}}^1 = \frac{16}{81}.$$

MAD is $16/81 = 0.198$, which is slightly smaller than the standard deviation.

19. We have that $E(X) = 2/3 = 0.667$ and $\sigma = 0.236$. Then one standard deviation below the mean is $0.667 - 0.236 = 0.431$ and two standard deviations below the mean is $0.667 - 2 \cdot 0.236 = 0.195$. Using the c.d.f. of $F(x) = x^2$, we have that $F(0.431) = 0.186$ and $F(0.195) = 0.038$. This p.d.f. has no points of inflection, but the percentiles are pretty close to those expected for a bell-shaped curve, where they would be 0.16 and 0.025.

21. We found that $E(T) = e - 1 = 1.718$ and $\sigma = 0.492$. Then one standard deviation below the mean is $1.718 - 0.492 = 1.226$ and two standard deviations below the mean is $1.718 - 2 \cdot 0.492 = 0.734$. Using the c.d.f. of $H(t) = \ln(t)$, we have that $H(1.226) = 0.204$. Because the value that is two standard deviations below the mean is not in the range, its percentile is not defined. This p.d.f. has no points of inflection, but the percentile for the first point is close to that expected for a bell-shaped curve, while the second is completely different.

23. $(x_i - \overline{X})^2 = x_i^2 - 2x_i\overline{X} + \overline{X}^2$, and

$$\sum_{i=1}^{n}(x_i - \overline{X})^2 p_i = \sum_{i=1}^{n} x_i^2 p_i - \sum_{i=1}^{n} 2x_i\overline{X} p_i + \sum_{i=1}^{n} \overline{X}^2 p_i.$$

25. It says that the probability that the value of the random variable is more than one standard deviation σ away from the mean μ is less than $1/1^2 = 1$. This is useless—all probabilities are less than 1.

27. At least 8/9, or 89%.

29. With the top salary of \$450,000, MAD =\$46,080, $\sigma^2 = 7.441 \times 10^9$ square dollars, $\sigma =$ \$86,264, and CV = 1.51 (no units).

31. Using the computational formula, and the calculation of the expectation as 7.53,

$$\text{Var}(P) = 4^2 \cdot 0.02 + 5^2 \cdot 0.08 + 6^2 \cdot 0.09 + 7^2 \cdot 0.31 + 8^2 \cdot 0.22 + 9^2 \cdot 0.22 + 10^2 \cdot 0.06 - 7.53^2 = 1.949.$$

The standard deviation is $\sqrt{1.949} = 1.396$. Only the two values of 4 lie more than two standard deviations away from the mean, or 2 out of 100.

33. Using the computational formula, and the calculation of the expectation as 4.98,

$$\text{Var}(P) = 1^2 \cdot 0.02 + 2^2 \cdot 0.05 + 3^2 \cdot 0.10 + 4^2 \cdot 0.21 + 5^2 \cdot 0.27 + 6^2 \cdot 0.19 + 7^2 \cdot 0.08 + 8^2 \cdot 0.05 + 9^2 \cdot 0.03 - 4.98^2 = 2.820.$$

The standard deviation is 1.679. The values of 1 and 9 lie more than two standard deviations away from the mean, or 5 out of 100.

35. Mean is about 400, $\sigma = 40$, CV= 0.1, 2.5th percentile at 320, 97.5th percentile at 480.

37. Mean is about 10, $\sigma = 1$, CV = 0.1, 2.5th percentile at 8, 97.5th percentile at 12.

39. The coefficient of variation is 0.02.

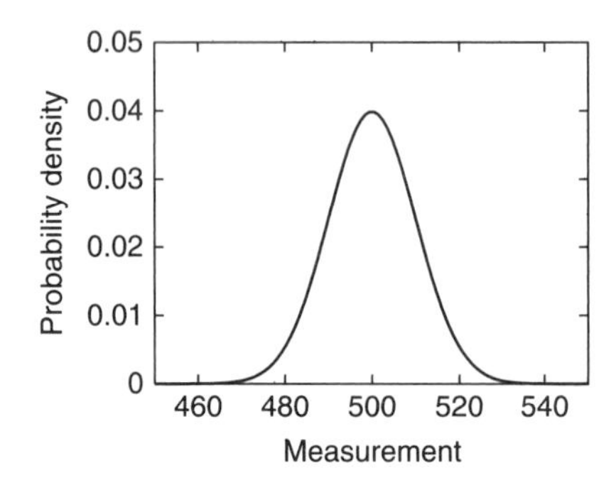

41. The standard deviation is $0.4 \cdot 50 = 20$.

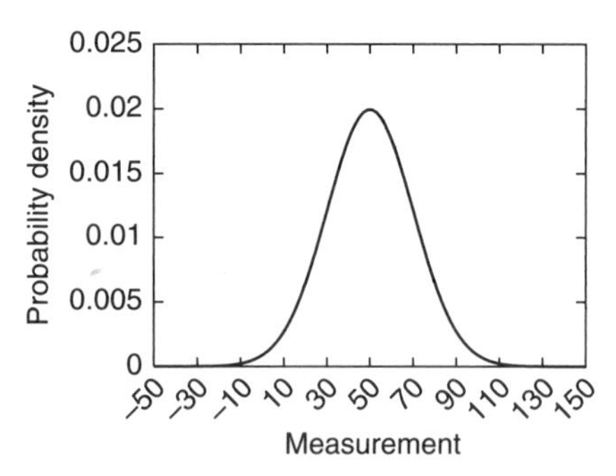

43. N_1 is exactly the same as R_0. The mean is 1.1 and the variance is

$$\text{Var}(N_1) = 1.5^2 \cdot 0.6 + 0.5^2 \cdot 0.4 - 1.1^2 = 0.24.$$

Chapter 7

Section 7.1, page 609

1. The marginal distributions are

	$X=0$	$X=1$		
$Y=0$	0.14	0.26	→	$\Pr(Y=0)=0.4$
$Y=1$	0.21	0.39	→	$\Pr(Y=1)=0.6$
	↓	↓		
	$\Pr(X=0)=0.35$	$\Pr(X=1)=0.65$		

The random variables are independent because the joint distribution is equal to the product of the marginal distributions. For example,

$$\Pr(X=0 \text{ and } Y=0) = 0.14 = \Pr(X=0) \cdot \Pr(Y=0) = 0.35 \cdot 0.4 = 0.14.$$

The distribution of Y conditional on $X = 0$ is

$$\Pr(Y=0 \mid X=0) = \frac{\Pr(Y=0 \text{ and } X=0)}{\Pr(X=0)} = \frac{0.14}{0.35} = 0.4$$

$$\Pr(Y=1 \mid X=0) = \frac{\Pr(Y=1 \text{ and } X=0)}{\Pr(X=0)} = \frac{0.21}{0.35} = 0.6.$$

This matches the marginal distribution for these independent random variables.

3. The marginal distributions are

	$X=0$	$X=1$	$X=2$	
$Y=1$	0.1	0.2	0.3	→ $\Pr(Y=1)=0.6$
$Y=2$	0.05	0.15	0.2	→ $\Pr(Y=2)=0.4$
	↓	↓	↓	
	$\Pr(X=0)=0.15$	$\Pr(X=1)=0.35$	$\Pr(X=2)=0.5$	

The random variables are not independent because

$$\Pr(X=0 \text{ and } Y=1) = 0.1 \neq \Pr(X=0) \cdot \Pr(Y=1) = 0.15 \cdot 0.6 = 0.09.$$

The distribution of Y conditional on $X = 1$ is

$$\Pr(Y=1 \mid X=1) = \frac{\Pr(Y=1 \text{ and } X=1)}{\Pr(X=1)} = \frac{0.2}{0.35} = 0.571$$

$$\Pr(Y=2 \mid X=1) = \frac{\Pr(Y=2 \text{ and } X=1)}{\Pr(X=1)} = \frac{0.15}{0.35} = 0.429.$$

5. $E(X) = 0 \cdot 0.35 + 1 \cdot 0.65 = 0.65$. $E(Y) = 0 \cdot 0.4 + 1 \cdot 0.6 = 0.6$.

7. $E(X) = 0 \cdot 0.15 + 1 \cdot 0.35 + 2 \cdot 0.5 = 1.35$. $E(Y) = 1 \cdot 0.6 + 2 \cdot 0.4 = 1.4$.

9. Because the random variables are independent, we can find the joint distribution by multiplying the marginal distributions.

	$X=0$	$X=1$
$Y=0$	0.21	0.09
$Y=1$	0.49	0.21

Independence also ensures that the conditional distribution of X when $Y=1$ matches the marginal distribution of X, so $\Pr(X=0 \mid Y=1) = \Pr(X=0) = 0.7$ and $\Pr(X=1 \mid Y=1) = \Pr(X=1) = 0.3$.

11. Because the random variables are independent, we can find the joint distribution by multiplying the marginal distributions.

	$X=0$	$X=1$
$Y=1$	0.24	0.06
$Y=2$	0.4	0.1
$Y=3$	0.16	0.04

Independence also ensures that the conditional distribution of Y when $X=0$ matches the marginal distribution of Y, so $\Pr(Y=1 \mid X=0) = \Pr(Y=1) = 0.3$, $\Pr(Y=2 \mid X=0) = \Pr(Y=2) = 0.5$, and $\Pr(Y=3 \mid X=0) = \Pr(Y=3) = 0.2$.

13. First, we use the fact that marginal probabilities for both X and Y must add to 1 to find that $\Pr(X=1) = 0.8$ and $\Pr(Y=1) = 0.6$. We can then include the following information on our joint distribution.

	$X=0$	$X=1$		
$Y=0$	0.1	?	→	$\Pr(Y=0)=0.4$
$Y=1$	?	?	→	$\Pr(Y=1)=0.6$
	↓	↓		
	$\Pr(X=0)=0.2$	$\Pr(X=1)=0.8$		

Because the probabilities in the first row must add to 0.4, we have that $\Pr(X=1 \text{ and } Y=0) = 0.3$. Because the probabilities in the first column must add to 0.2, we have that $\Pr(X=0 \text{ and } Y=1) = 0.1$. For all the probabilities to add to 1, the remaining probability is $\Pr(X=1 \text{ and } Y=1) = 0.5$. The final result is

	$X=0$	$X=1$		
$Y=0$	0.1	0.3	→	$\Pr(Y=0)=0.4$
$Y=1$	0.1	0.5	→	$\Pr(Y=1)=0.6$
	↓	↓		
	$\Pr(X=0)=0.2$	$\Pr(X=1)=0.8$		

15. First, we use the fact that marginal probabilities for both X and Y must add to 1 to find that $\Pr(X=1) = 0.7$ and $\Pr(Y=1) = 0.4$. We find

$$\Pr(X=0 \text{ and } Y=0) = \Pr(X=0|Y=0)\Pr(Y=0) = 0.5 \cdot 0.6 = 0.3.$$

We can then include the following information on our joint distribution.

	$X=0$	$X=1$		
$Y=0$	0.3	?	→	$\Pr(Y=0)=0.6$
$Y=1$	?	?	→	$\Pr(Y=1)=0.4$
	↓	↓		
	$\Pr(X=0)=0.3$	$\Pr(X=1)=0.7$		

Using the row and column sums, we fill in the rest of the information as

	$X=0$	$X=1$		
$Y=0$	0.3	0.3	→	$\Pr(Y=0)=0.6$
$Y=1$	0.0	0.4	→	$\Pr(Y=1)=0.4$
	↓	↓		
	$\Pr(X=0)=0.3$	$\Pr(X=1)=0.7$		

17. The event $T>0$ includes the simple events $T=1$ and $T=2$. Then

$$\begin{aligned}\Pr(T>0 \text{ and } N=0) &= \Pr(T=1 \text{ and } N=0) \\ &\quad + \Pr(T=2 \text{ and } N=0) \\ &= 0.03 + 0.06 = 0.09.\end{aligned}$$

Similarly,

$$\begin{aligned}\Pr(T \le 0 \text{ and } N=0) &= \Pr(T=-1 \text{ and } N=0) \\ &\quad + \Pr(T=0 \text{ and } N=0) \\ &= 0.02 + 0.10 = 0.12.\end{aligned}$$

Finding $\Pr(T \le 0 \text{ and } N>0)$ requires adding up the six probabilities in the lower left-hand corner of the matrix, giving 0.29. Finding $\Pr(T>0 \text{ and } N>0)$ requires adding up the six probabilities in the lower right-hand corner of the matrix, giving 0.50. The joint distribution is then

	$T\le 0$	$T>0$
$N=0$	0.12	0.09
$N>0$	0.29	0.50

19. Using the notation H_1 means the first player hit, M_1 means that the first player missed, and so forth, the maximum is

	H_1	M_1
H_2	0.25	0.1
M_2	0.0	0.65

The conditional distribution for the second player if the first gets a hit is

$$\Pr(H_2 \mid H_1) = \frac{\Pr(H_2 \text{ and } H_1)}{\Pr(H_1)} = \frac{0.25}{0.25} = 1$$
$$\Pr(M_2 \mid H_1) = \frac{\Pr(M_2 \text{ and } H_1)}{\Pr(H_1)} = \frac{0.0}{0.25} = 0.$$

The second player is sure to hit if the first does. The conditional distribution for the first player if the second gets a hit is

$$\Pr(H_1 \mid H_2) = \frac{\Pr(H_1 \text{ and } H_2)}{\Pr(H_2)} = \frac{0.25}{0.35} = 5/7$$
$$\Pr(M_1 \mid H_2) = \frac{\Pr(M_1 \text{ and } H_2)}{\Pr(H_2)} = \frac{0.1}{0.35} = 2/7.$$

The first player is very likely, but not certain, to hit if the second does.

21. Let M_t denote the event "mutant at generation t" and N_t denote the event "nonmutant at generation t." We know that

$$\begin{aligned}\Pr(M_{t+1}|M_t) &= 0.99 \\ \Pr(M_{t+1}|N_t) &= 0.01 \\ \Pr(N_{t+1}|M_t) &= 0.01 \\ \Pr(N_{t+1}|N_t) &= 0.99.\end{aligned}$$

Also, we found that $\Pr(M_t) = 0.5$ after a long time. Therefore,

$$\begin{aligned}\Pr(M_{t+1} \text{ and } M_t) &= \Pr(M_{t+1}|M_t)\Pr(M_t) = 0.99 \cdot 0.5 = 0.495 \\ \Pr(M_{t+1} \text{ and } N_t) &= \Pr(M_{t+1}|N_t)\Pr(N_t) = 0.01 \cdot 0.5 = 0.005 \\ \Pr(N_{t+1} \text{ and } M_t) &= \Pr(N_{t+1}|M_t)\Pr(M_t) = 0.01 \cdot 0.5 = 0.005 \\ \Pr(N_{t+1} \text{ and } N_t) &= \Pr(N_{t+1}|N_t)\Pr(N_t) = 0.99 \cdot 0.5 = 0.495.\end{aligned}$$

In the form of a joint distribution,

	M_t	N_t
M_{t+1}	0.495	0.005
N_{t+1}	0.005	0.495

23. With 0 mites,

$$\Pr(L=0 \mid M=0) = \frac{\Pr(L=0 \text{ and } M=0)}{\Pr(M=0)} = \frac{0.21}{0.5} = 0.42$$
$$\Pr(L=1 \mid M=0) = \frac{\Pr(L=1 \text{ and } M=0)}{\Pr(M=0)} = \frac{0.13}{0.5} = 0.26$$
$$\Pr(L=2 \mid M=0) = \frac{\Pr(L=2 \text{ and } M=0)}{\Pr(M=0)} = \frac{0.16}{0.5} = 0.32.$$

With $M=1$,

$$\Pr(L=0 \mid M=1) = \frac{\Pr(L=0 \text{ and } M=1)}{\Pr(M=1)} = \frac{0.13}{0.3} = 0.433$$
$$\Pr(L=1 \mid M=1) = \frac{\Pr(L=1 \text{ and } M=1)}{\Pr(M=1)} = \frac{0.07}{0.3} = 0.233$$
$$\Pr(L=2 \mid M=1) = \frac{\Pr(L=2 \text{ and } M=1)}{\Pr(M=1)} = \frac{0.10}{0.3} = 0.333.$$

With $M = 2$,

$$\Pr(L = 0 \mid M = 2) = \frac{\Pr(L = 0 \text{ and } M = 2)}{\Pr(M = 2)} = \frac{0.06}{0.2} = 0.3$$

$$\Pr(L = 1 \mid M = 2) = \frac{\Pr(L = 1 \text{ and } M = 2)}{\Pr(M = 2)} = \frac{0.10}{0.2} = 0.5$$

$$\Pr(L = 2 \mid M = 2) = \frac{\Pr(L = 2 \text{ and } M = 2)}{\Pr(M = 2)} = \frac{0.04}{0.2} = 0.2.$$

These distributions differ, showing, for example, that birds with two mites are much more likely to have one louse than are other birds.

25. a.

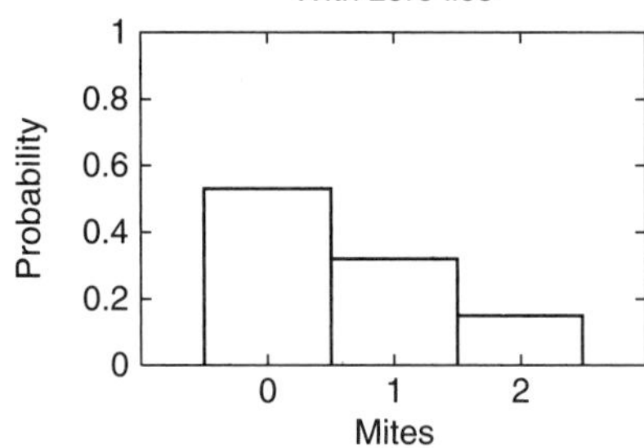

b.

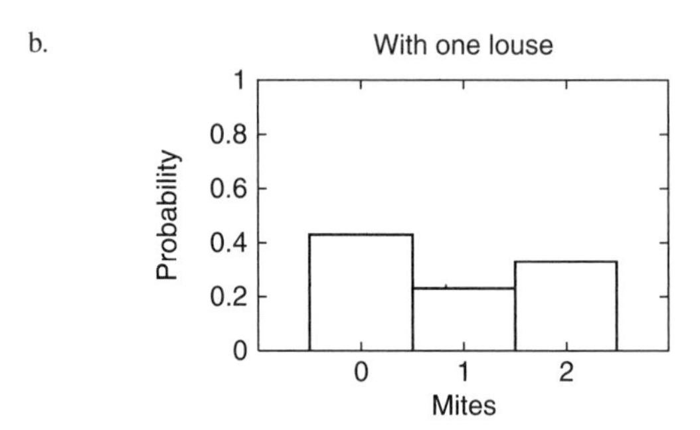

c.

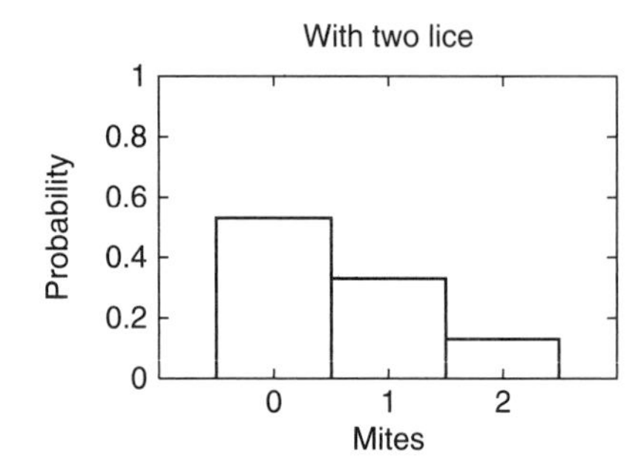

27. The joint distribution is

	$E = 0$	$E = 1$		
$J = 0$	0.35	0.15	→	0.5
$J = 1$	0.45	0.05	→	0.5
	↓	↓		
	0.8	0.2		

The marginal distributions are $\Pr(J = 0) = \Pr(J = 1) = 0.5$, $\Pr(E = 0) = 0.8$, and $\Pr(E = 1) = 0.2$. The conditional distributions for the eagle are

$$\Pr(E = 0 \mid J = 0) = \frac{\Pr(E = 0 \text{ and } J = 0)}{\Pr(J = 0)} = \frac{0.35}{0.5} = 0.7$$

$$\Pr(E = 1 \mid J = 0) = \frac{\Pr(E = 1 \text{ and } J = 0)}{\Pr(J = 0)} = \frac{0.15}{0.5} = 0.3$$

and

$$\Pr(E = 0 \mid J = 1) = \frac{\Pr(E = 0 \text{ and } J = 1)}{\Pr(J = 1)} = \frac{0.45}{0.5} = 0.9$$

$$\Pr(E = 1 \mid J = 1) = \frac{\Pr(E = 1 \text{ and } J = 1)}{\Pr(J = 1)} = \frac{0.05}{0.5} = 0.1.$$

a.

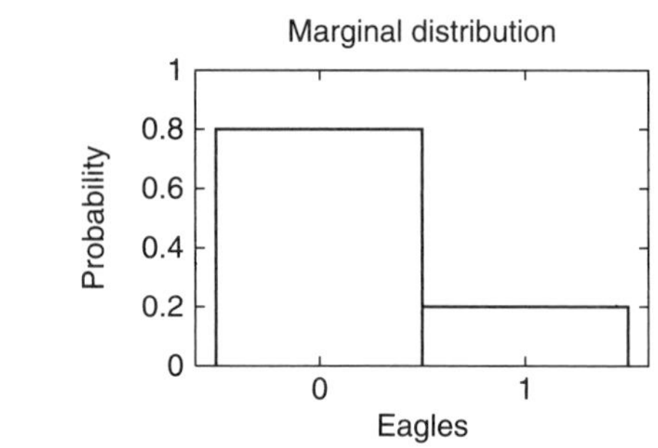

b.

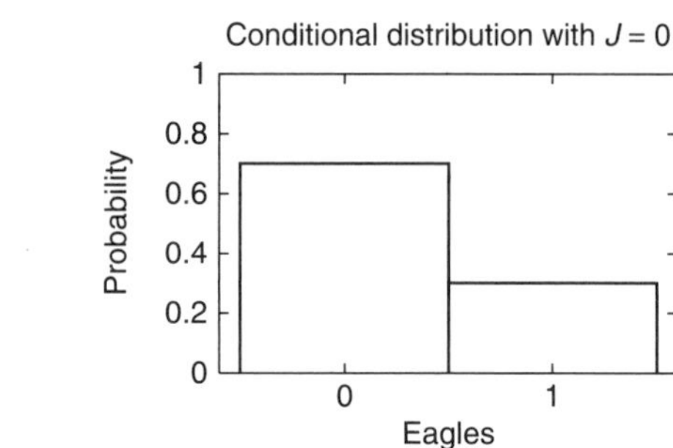

c.

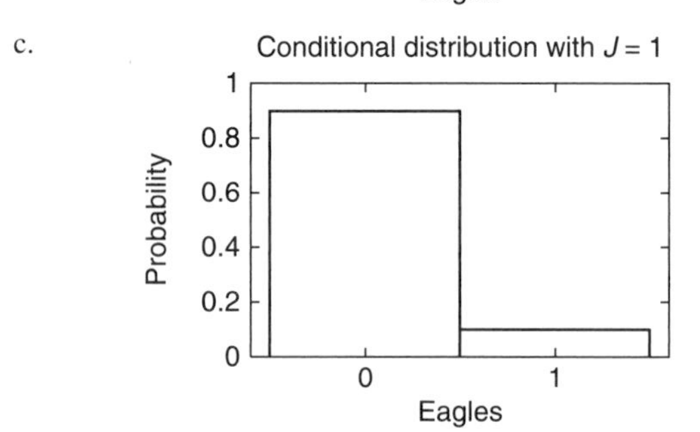

For the rabbit, the conditional distributions are

$$\Pr(J = 0 \mid E = 0) = \frac{\Pr(J = 0 \text{ and } E = 0)}{\Pr(E = 0)} = \frac{0.35}{0.8} = 0.4375$$

$$\Pr(J = 1 \mid E = 0) = \frac{\Pr(J = 1 \text{ and } E = 0)}{\Pr(E = 0)} = \frac{0.45}{0.8} = 0.5625$$

and

$$\Pr(J = 0 \mid E = 1) = \frac{\Pr(J = 0 \text{ and } E = 1)}{\Pr(E = 1)} = \frac{0.15}{0.2} = 0.75$$

$$\Pr(J = 1 \mid E = 1) = \frac{\Pr(J = 1 \text{ and } E = 1)}{\Pr(E = 1)} = \frac{0.05}{0.2} = 0.25.$$

d.

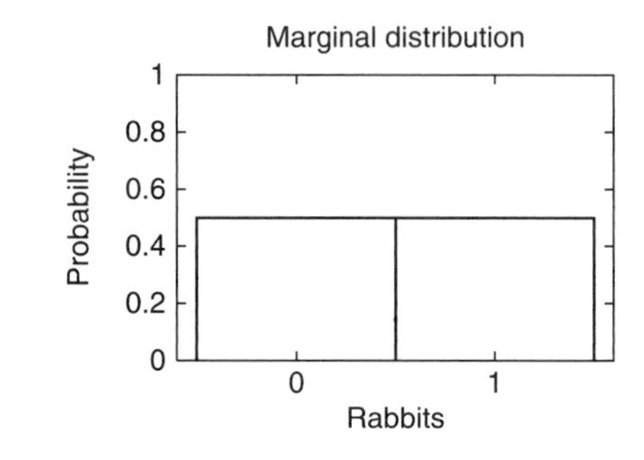

e.

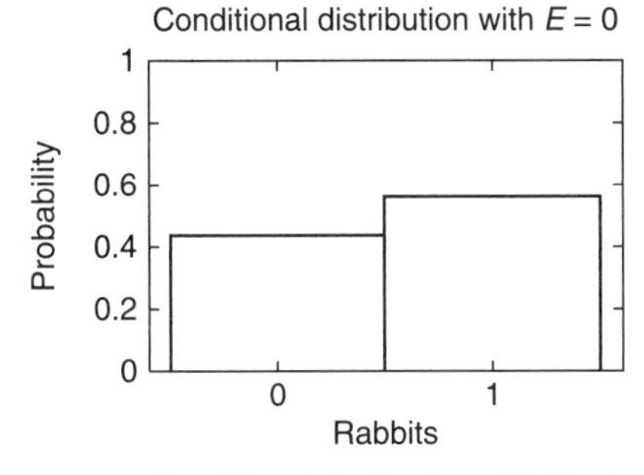

f.

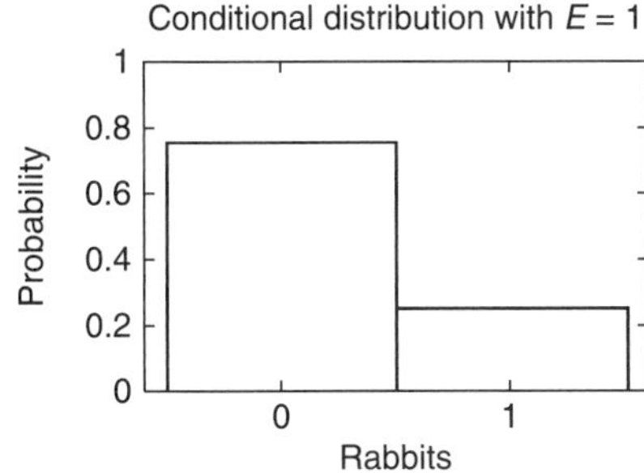

It is more likely that she will see an eagle when no rabbits are seen, and more likely that she will see a rabbit when no eagles are seen. One must be avoiding the other.

29. We used the law of total probability to find $\Pr(\mathrm{P}) = 0.24$. We can then find

$$\Pr(\mathrm{P} \cap \mathrm{D}) = \Pr(\mathrm{P} \mid \mathrm{D})\Pr(\mathrm{D}) = 1.0 \cdot 0.2 = 0.2.$$

We can then fill in the rest of the joint distribution as

	D	N
P	0.2	0.04
P^c	0.0	0.76

Therefore, $\Pr(\mathrm{D} \mid \mathrm{P}) = \frac{\Pr(\mathrm{D} \cap \mathrm{P})}{\Pr(\mathrm{P})} = \frac{0.2}{0.24} = 0.833.$

31. We used the law of total probability to find $\Pr(\mathrm{P}) = 0.23$. We can then find

$$\Pr(\mathrm{P} \cap \mathrm{D}) = \Pr(\mathrm{P} \mid \mathrm{D})\Pr(\mathrm{D}) = 0.95 \cdot 0.2 = 0.19.$$

We can then fill in the rest of the joint distribution as

	D	N
P	0.19	0.04
P^c	0.01	0.76

Therefore, $\Pr(\mathrm{D} \mid \mathrm{P}) = \frac{\Pr(\mathrm{D} \cap \mathrm{P})}{\Pr(\mathrm{P})} = \frac{0.19}{0.23} = 0.826.$

33. The random variables are age A, taking on values 0, 1, 2, and 3, and stain S, taking on value 1 for proper staining and 0 otherwise. The joint distribution is

	$S = 1$	$S = 0$
$A = 0$	0.38	0.02
$A = 1$	0.27	0.03
$A = 2$	0.16	0.04
$A = 3$	0.05	0.05

The marginal distribution for cell age is given in the problem, and the marginal distribution for staining is $\Pr(S = 1) = 0.86$ and $\Pr(S = 0) = 0.14$. The conditional distributions for age are given in the following columns.

	$S = 1$	$S = 0$
$A = 0$	0.44	0.14
$A = 1$	0.31	0.21
$A = 2$	0.19	0.29
$A = 3$	0.06	0.36

a.

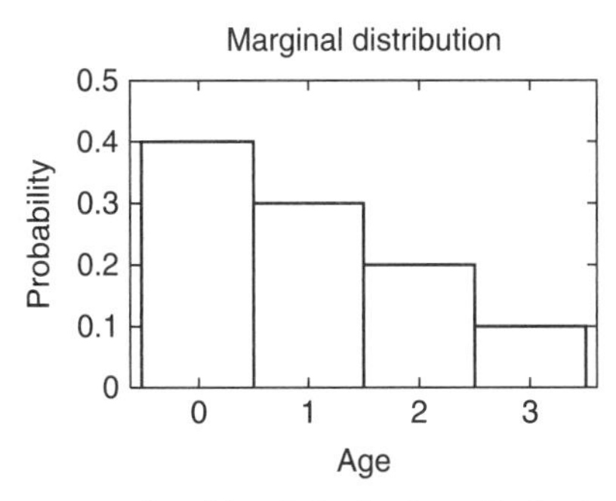

b.

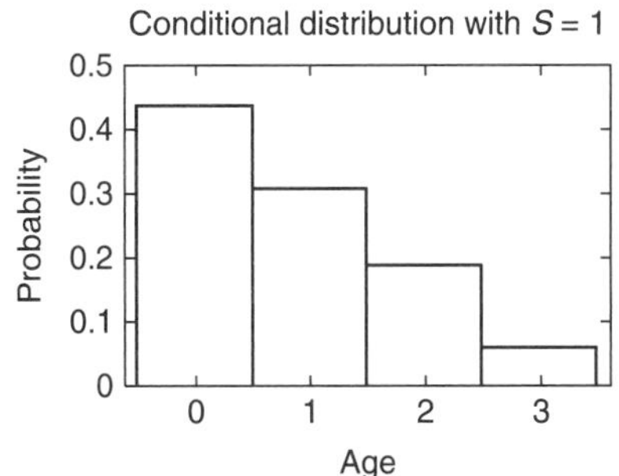

c.

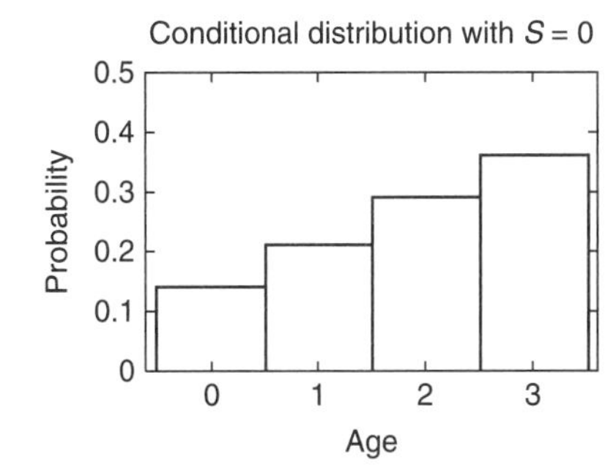

The conditional distribution for age when $S = 1$ looks a lot like the marginal distribution, but with $S = 0$ it is very different. Cells that do not stain properly tend to be much older.

35. Multiplying the marginal probabilities gives

	$I_a = -1$	$I_a = 0$	$I_a = 1$	$I_a = 2$
$I_b = -1$	0.04	0.02	0.03	0.01
$I_b = 0$	0.12	0.06	0.09	0.03
$I_b = 1$	0.08	0.04	0.06	0.02
$I_b = 2$	0.16	0.08	0.12	0.04

37. With meiotic drive, the marginal probabilities of an **A** from either the pollen or ovule is 0.6. With independence, the joint probabilities are the product of the marginal probabilities.

		Ovule	
		A	a
Pollen	A	0.36	0.24
	a	0.24	0.16

The probability of a heterozygote is 0.48. With nonindependent assortment, we know that the ovule provides **A** with probability 0.5. The pollen provides **A** with probability 0.5 also, from the law of total probability. Furthermore, $\Pr(\mathbf{aA}) = \Pr(\mathbf{a} \mid \mathbf{A})\Pr(\mathbf{A}) = 0.4 \cdot 0.5 = 0.2$. The joint distribution, after filling in the remaining values to match the marginal distributions, is

		Ovule	
		A	a
Pollen	A	0.3	0.2
	a	0.2	0.3

The probability of a heterozygote is 0.4.

39. The marginal distribution for females is R with probability 0.5, B with probability 0.3, and G with probability 0.2. For males it is R with probability 0.44, B with probability 0.324, and G with probability 0.236. There seem to be slightly fewer of the red males than expected; perhaps they are vulnerable to attack by hawks. The conditional distribution are: with R females 0.25 R, 0.39 B, 0.36 G; with B females 0.75 R, 0.09 B, 0.16 G; with G females 0.45 R, 0.51 B, 0.04 G. The females seem to prefer to mate with dissimilar males.

Section 7.2, page 621

1. We found that $E(X) = 0.65$ and $E(Y) = 0.6$. Then we can make a table of values.

X	Y	$X-\overline{X}$	$Y-\overline{Y}$	$(X-\overline{X})(Y-\overline{Y})$	Probability
0	0	−0.65	−0.6	0.39	0.14
0	1	−0.65	0.4	−0.26	0.21
1	0	0.35	−0.6	−0.21	0.26
1	1	0.35	0.4	0.14	0.39

Then the covariance is

$$\text{Cov}(X, Y) = 0.39 \cdot 0.14 - 0.26 \cdot 0.21 - 0.21 \cdot 0.26 + 0.14 \cdot 0.39 = 0.$$

The covariance is 0 because these random variables are independent.

3. We found that $E(X) = 1.35$ and $E(Y) = 1.4$. Then we can make a table of values.

X	Y	$X-\overline{X}$	$Y-\overline{Y}$	$(X-\overline{X})(Y-\overline{Y})$	Probability
0	1	−1.35	−0.4	0.54	0.1
0	2	−1.35	0.6	−0.81	0.05
1	1	−0.35	−0.4	0.14	0.2
1	2	−0.35	0.6	−0.21	0.15
2	1	0.65	−0.4	−0.26	0.3
2	2	0.65	0.6	0.39	0.2

Then the covariance is

$$\text{Cov}(X, Y) = 0.54 \cdot 0.1 - 0.81 \cdot 0.05 + 0.14 \cdot 0.2 - 0.21 \cdot 0.15 - 0.26 \cdot 0.3 + 0.39 \cdot 0.2 = 0.01.$$

5. We have that $E(X) = 0.65$ and $E(Y) = 0.6$. We can make a table of values.

X	Y	XY	Probability
0	0	0	0.14
0	1	0	0.21
1	0	0	0.26
1	1	1	0.39

Then the covariance is

$$\text{Cov}(X, Y) = E(XY) - \overline{X}\overline{Y} = 0.39 - 0.65 \cdot 0.6 = 0.$$

7. We have that $E(X) = 1.35$ and $E(Y) = 1.4$. Then we can make a table of values.

X	Y	XY	Probability
0	1	0	0.1
0	2	0	0.05
1	1	1	0.2
1	2	2	0.15
2	1	2	0.3
2	2	4	0.2

Then the covariance is

$$\text{Cov}(X, Y) = 1 \cdot 0.2 + 2 \cdot 0.15 + 2 \cdot 0.3 + 4 \cdot 0.2 - 1.35 \cdot 1.4 = 0.01.$$

9. The covariance is 0, so the correlation is also.

11. We have that $\text{Cov}(X, Y) = 0.01$. To find the correlation of X and Y we need to find the variances of X and Y.

$$\text{Var}(X) = 0^2 \cdot 0.15 + 1^2 \cdot 0.35 + 2^2 \cdot 0.5 - 1.35^2 = 0.5275$$
$$\text{Var}(Y) = 1^2 \cdot 0.6 + 2^2 \cdot 0.4 - 1.4^2 = 0.24.$$

The standard deviations are $\sigma_X = \sqrt{0.5275} = 0.726$ and $\sigma_Y = \sqrt{0.24} = 0.490$. Then

$$\rho = \frac{\text{Cov}(X, Y)}{\sigma_X \sigma_Y} = \frac{0.01}{0.726 \cdot 0.490} = 0.028.$$

13. The joint distribution is

	$P_1 = 1$	$P_1 = 0$
$P_2 = 1$	0.25	0.1
$P_2 = 0$	0.0	0.65

We know from the marginal distributions that $E(P_1) = 0.25$ and $E(P_2) = 0.35$. Also, $E(P_1 P_2) = 0.25$. Then the covariance is $\text{Cov}(P_1, P_2) = 0.25 - 0.25 \cdot 0.35 = 0.1625$. The covariance is positive because the two players tend to do the same thing.

15. Let $M_t = 1$ denote the event "mutant at generation t" and $M_t = 0$ denote the event "non-mutant at generation t." The joint distribution is

	$M_t = 1$	$M_t = 0$
$M_{t+1} = 1$	0.495	0.005
$M_{t+1} = 0$	0.005	0.495

Also, we found that $\Pr(M_t = 1) = 0.5$ after a long time. Therefore, the covariance is $\text{Cov}(M_t, M_{t+1}) = 0.495 - 0.5 \cdot 0.5 = 0.245$. Both M_t and M_{t+1} are Bernoulli random variables with $p = 0.5$, so the variance of each is 0.25, and the standard deviation of each is $\sqrt{0.25} = 0.5$. The correlation is $\rho = \frac{0.245}{0.5 \cdot 0.5} = 0.98$.

17. The joint distribution is

	$X=0$	$X=1$	$X=2$
$Y=1$	0.2	0.0	0.0
$Y=4$	0.0	0.3	0.0
$Y=7$	0.0	0.0	0.5

Using the marginal distributions, we find $\overline{X} = 1.3$, $\overline{Y} = 4.9$. Then

$$\text{Cov}(X, Y) = 0 \cdot 1 \cdot 0.2 + 1 \cdot 4 \cdot 0.3 + 2 \cdot 7 \cdot 0.5 - 1.3 \cdot 4.9 = 1.83.$$

Again using the marginals, we find $\sigma_X = \sqrt{0.61}$, $\sigma_Y = \sqrt{5.49}$, so

$$\rho_{X,Y} = \frac{1.83}{\sqrt{0.61}\sqrt{5.49}} = 1$$

In accord with Theorem 7.3, these linearly related random variables have a correlation of 1.

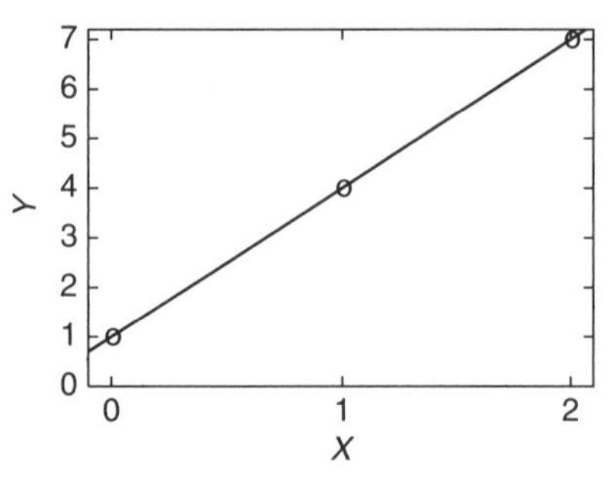

19. The joint distribution is

	$X=0$	$X=1$	$X=2$
$Y=1$	0.2	0.0	0.0
$Y=0$	0.0	0.3	0.0
$Y=1$	0.0	0.0	0.5

Using the marginal distributions, we find $\overline{X} = 1.3$, $\overline{Y} = 0.7$. Then

$$\text{Cov}(X, Y) = 0 \cdot 1 \cdot 0.2 + 1 \cdot 0 \cdot 0.3 + 2 \cdot 1 \cdot 0.5 - 1.3 \cdot 0.7 = 0.09.$$

Again using the marginals, we find $\sigma_X = \sqrt{0.61}$, $\sigma_Y = \sqrt{0.21}$, so

$$\rho_{X,Y} = \frac{0.09}{\sqrt{0.61}\sqrt{0.21}} = 0.251.$$

Even though knowing the value of X means that you know exactly the value of Y, the correlation is far from 1 because the relation because X and Y is not linear. In fact, the relation between them is not even increasing, so the correlation is quite poor.

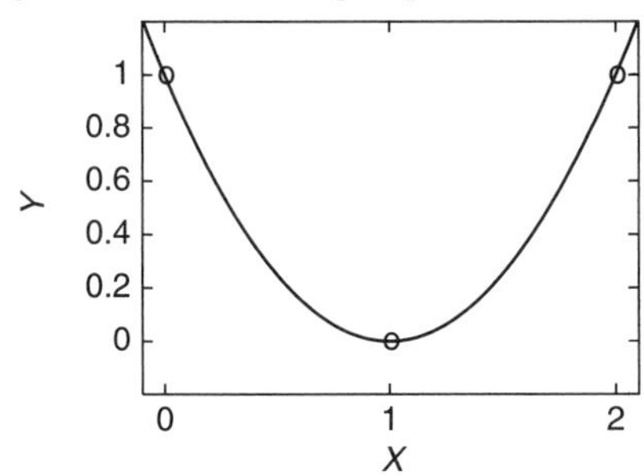

21.

$$E(L) = 0.4 \cdot 0 + 0.3 \cdot 1 + 0.3 \cdot 2 = 0.9$$

$$E(M) = 0.5 \cdot 0 + 0.3 \cdot 1 + 0.2 \cdot 2 = 0.7$$

$$\text{Cov}(L, M) = 0.07 \cdot 1 + 0.10 \cdot 2 + 0.10 \cdot 2 + 0.04 \cdot 4 - 0.9 \cdot 0.7 = 0.0.$$

23. Using the marginal distributions, we find

$$\text{Var}(L) = 0.4 \cdot 0^2 + 0.3 \cdot 1^2 + 0.3 \cdot 2^2 - 0.9^2 = 0.69$$

$$\text{Var}(M) = 0.5 \cdot 0^2 + 0.3 \cdot 1^2 + 0.2 \cdot 2^2 - 0.7^2 = 0.61$$

$$\rho_{L,M} = \frac{\text{Cov}(L, M)}{\sqrt{\text{Var}(L)\text{Var}(M)}} = 0.0.$$

25. The joint distribution is

	$E=0$	$E=1$
$J=0$	0.35	0.15
$J=1$	0.45	0.05

Then $\text{Cov}(E, J) = 0.05 - 0.5 \cdot 0.2 = -0.05$. Also, $\text{Var}(J) = 0.25$ and $\text{Var}(E) = 0.16$, so $\rho_{E,J} = \frac{-0.05}{0.2} = -0.25$. The negative correlation indicates that the two species tend to behave differently.

27. The joint distribution was

	$S=1$	$S=0$
$A=0$	0.38	0.02
$A=1$	0.27	0.03
$A=2$	0.16	0.04
$A=3$	0.05	0.05

Then $E(S) = 0.86$, $E(A) = 1.0$, $E(AS) = 0.74$, and $\text{Cov}(A, S) = 0.74 - 0.86 = -0.12$. The covariance is negative because older cells (with larger values of A) tend to have smaller values of S.

29. The joint distribution of S and P is

		Sleeping		
		10	14	18
Preening	6	0.0	0.0	0.1
	10	0.0	0.2	0.0
	14	0.7	0.0	0.0

$E(S) = 11.6$, $E(P) = 12.4$, $\text{Var}(S) = 7.04$, $\sigma_S = 2.65$, $\text{Var}(P) = 7.04$, and $\sigma_P = 2.65$. $\text{Cov}(P, S) = -7.04$ and $\rho_{P,S} = -1$.

31. The joint distribution of S and P is

		Sleeping		
		10	14	18
Preening	0	0.0	0.0	0.1
	9	0.0	0.2	0.0
	13	0.7	0.0	0.0

$E(S) = 11.6$, $E(P) = 10.9$, $\text{Var}(S) = 7.04$, $\sigma_S = 2.65$, $\text{Var}(P) = 15.69$, $\sigma_P = 3.96$. $\text{Cov}(P, S) = -10.24$ and $\rho_{P,S} = -0.97$. The amount of eating depends on the weather, and destroys the perfect negative correlation of S and P.

33. Using the marginal distributions, we find $E(I_a) = 0.1$ and $E(I_b) = 0.9$. Using the joint distribution, we find $E(I_a I_b) = 0.09$, so $\text{Cov}(I_a, I_b) = 0$. This checks with the fact that the random variables are independent.

35. $E(S) = 33.5$, $\text{Var}(S) = 965.25$, $E(T) = 20.0$, $\text{Var}(T) = 60.0$, $\text{Cov}(S, T) = 150$, and $\rho = 0.623$.

37. $E(N) = 1.07$, $\text{Var}(N) = 0.605$, $E(A) = 2.84$, $\text{Var}(A) = 1.43$, $\text{Cov}(A, N) = 0.07$, and $\rho = 0.076$.

Section 7.3, page 631

1. The random variable $X + Y$ takes on the value 0 with probability 0.14, the value 1 with probability 0.47, and the value 2 with probability 0.39. Therefore, $E(X + Y) = 0 \cdot 0.14 + 1 \cdot 0.47 + 2 \cdot 0.39 = 1.25$. This is indeed equal to $E(X) + E(Y) = 0.65 + 0.6 = 1.25$.

3. The random variable $X + Y$ takes on the value 1 with probability 0.1, the value 2 with probability 0.25, the value 3 with probability 0.45, and the value 4 with probability 0.2. Therefore, $E(X + Y) = 1 \cdot 0.1 + 2 \cdot 0.25 + 3 \cdot 0.45 + 4 \cdot 0.2 = 2.75$. This is indeed equal to $E(X) + E(Y) = 1.35 + 1.4 = 2.75$.

5. The random variable XY takes on the value 0 with probability 0.61 and the value 1 with probability 0.39. Therefore, $E(XY) = 0 \cdot 0.61 + 1 \cdot 0.39 = 0.39$. This is equal to $E(X)E(Y) = 0.65 \cdot 0.6 = 0.39$, consistent with the fact that the covariance is 0 and the random variables are independent.

7. The random variable XY takes on the value 0 with probability 0.15, the value 1 with probability 0.2, the value 2 with probability 0.45, and the value 4 with probability 0.2. Therefore, $E(XY) = 0 \cdot 0.15 + 1 \cdot 0.2 + 2 \cdot 0.45 + 4 \cdot 0.2 = 1.9$. This is not equal to $E(X)E(Y) = 1.35 \cdot 1.4 = 1.89$, consistent with the fact that the covariance is not 0.

9. The random variable $X - Y$ takes on the value 0 with probability 0.53, the value -1 with probability 0.21, and the value 1 with probability 0.26. Therefore, $E(X - Y) = 0 \cdot 0.53 - 1 + 1 = 0.05$, equal to $E(X) - E(Y) = 0.65 - 0.6$. The variance of $X - Y$ can be found as

$$E[(X - Y)^2] - (0.05)^2 = 0 \cdot 0.53 + (-1)^2 \cdot 0.21 + 1^2 \cdot 0.26 = 0.4675.$$

Using the marginal distributions, we can use the fact that X and Y are Bernoulli random variables to find $\text{Var}(X) = 0.35 \cdot 0.65 = 0.2275$ and $\text{Var}(Y) = 0.4 \cdot 0.6 = 0.24$. In this case, it is true that $\text{Var}(X - Y) = \text{Var}(X) + \text{Var}(Y)$, consistent with the fact that $\text{Cov}(X, Y) = 0$.

11. The random variable $X - Y$ takes on the value 0 with probability 0.4, the value -1 with probability 0.25, the value -2 with probability 0.05, and the value 1 with probability 0.3. Therefore, $E(X - Y) = 0 \cdot 0.4 - 1 \cdot 0.25 - 2 \cdot 0.05 + 1 \cdot 0.3 = -0.05$, equal to $E(X) - E(Y) = 1.35 - 1.4$. The variance of $X - Y$ can be found as

$$E[(X - Y)^2] - (-0.05)^2 = 0^2 \cdot 0.4 + (-1)^2 \cdot 0.25 + (-2)^2 \cdot 0.05 + 1^2 \cdot 0.3 - (-0.05)^2 = 0.7475.$$

We found that $\text{Var}(X) = 0.5275$ and $\text{Var}(Y) = 0.24$, so $\text{Var}(X - Y) \neq \text{Var}(X) + \text{Var}(Y)$, consistent with the fact that the covariance is not 0.

13. $\text{Cov}(X, X) = E(X \cdot X) - \overline{X}^2 = E(X^2) - \overline{X}^2 = \text{Var}(X)$. In general, the covariance of a random variable with itself is equal to the variance.

15. For these Bernoulli random variables, $E(X) = E(Y) = 0.5$ and $\text{Var}(X) = \text{Var}(Y) = 0.25$.

a. Z takes the value 0 with probability 0.5 and the value 2 with probability 0.5. Then $E(Z) = 2 \cdot 0.5 = 1$, and $Var(Z) = 2^2 \cdot 0.5 - 1 = 1.0$.

b. S takes the value 0 with probability 0.25, the value 1 with probability 0.5, and the value 2 with probability 0.25. Then $E(S) = 1 \cdot 0.5 + 2 \cdot 0.25 = 1.0$. $Var(S) = 1^2 \cdot 0.5 + 2^2 \cdot 0.25 - 1.0^2 = 0.5$.

c. $E(Z) = E(X) + E(Y) = 0.5 + 0.5 = 1.0$.

$$\begin{aligned} Var(X + X) &= Var(X) + Var(X) + 2Cov(X, X) = Var(X) \\ &+ Var(X) + 2Var(X) = 4Var(X) = 1.0. \end{aligned}$$

d. $E(Z)=E(2X)=2E(X)=2 \cdot 0.5 = 1.0$. By Theorem 7.11, $Var(Z) = Var(2X) = 4Var(X) = 1.0$.

e. $E(S) = E(X) + E(Y) = 0.5 + 0.5 = 1.0$. $Var(S)=Var(X)+Var(Y) = 0.5$. This is smaller than $Var(Z)$ because X and Y are independent, not perfectly correlated. The values in the probability distribution for S are much more likely to be close to the mean of 2.0.

17. The geometric mean of the product XY is $e^{E[\ln(XY)]}$

19. $e^{E[\ln(X)+\ln(Y)]} = e^{E[\ln(X)]+E[\ln(Y)]}$.

21. The random variable $1/X$ takes the value 1 with probability 0.5, and the value 1/2 with probability 0.5. Then $E(1/X) = 0.75$, so $H(X) = 1/0.75 = 1.33$. This is indeed less than the expectation $E(X) = 1.5$.

23. $E(X) = E(Y) = 1.5$. The random variable X/Y takes the value 1 with probability 0.5, the value 1/2 with probability 0.25, and the value 2 with probability 0.25. Then $E(X/Y) = 1.125 \neq \frac{1.5}{1.5} = 1$. We expect that $E(X/Y) = E(X) \cdot E(1/Y)$, but $E(1/Y) > 1/E(Y)$ because the harmonic mean is less than the ordinary mean.

25.

Number	Probability
0	0.21
1	0.26
2	0.29
3	0.20
4	0.04

$$E(P) = 0 \cdot 0.21 + 1 \cdot 0.26 + 2 \cdot 0.29 + 3 \cdot 0.20 + 4 \cdot 0.04 = 1.6$$

The expectations of the sum must be the sum of the expectations, or $0.9 + 0.7$.

27.

$$\begin{aligned} Var(P) &= 0^2 \cdot 0.21 + 1^2 \cdot 0.26 + 2^2 \cdot 0.29 + 3^2 \cdot 0.20 \\ &+ 4^2 \cdot 0.04 - 1.6^2 = 1.3. \end{aligned}$$

Although L and M are not independent for bird C, the covariance is 0, so $Var(P) = Var(L) + Var(M) + 0 = 1.3$.

29.

W	Probability
0.00	0.20
0.02	0.12
0.04	0.08
0.05	0.15
0.07	0.09
0.09	0.06
0.10	0.15
0.12	0.09
0.14	0.06

The expectation is

$$\begin{aligned} E(W) &= 0.00 \cdot 0.20 + 0.02 \cdot 0.12 + 0.04 \cdot 0.08 + 0.05 \cdot 0.15 \\ &+ 0.07 \cdot 0.09 + 0.09 \cdot 0.06 + 0.10 \cdot 0.15 \\ &+ 0.12 \cdot 0.09 + 0.14 \cdot 0.06 \\ &= 0.059. \end{aligned}$$

This checks.

31. The expectations are 0.1 and 9.9, so $E(I_a + I_c) = 10.0$.

33. Let I_1, I_2, and I_3 represent the number of immigrants of each species. Then $E(I_1) = 0.8$, $E(I_2) = 0.6$, and $E(I_1) = 1.5$. The total number of immigrants is $I = I_1 + I_2 + I_3$. Therefore,

$$E(I) = E(I_1) + E(I_2) + E(I_3) = 0.8 + 0.6 + 1.5 = 2.9.$$

35. a. We can find the joint distribution by multiplying the marginal distributions, giving

		Species 1		
		0	1	2
Species 2	0	0.18	0.36	0.06
	1	0.06	0.12	0.02
	2	0.06	0.12	0.02

b.

$$\begin{aligned} Pr(0 \text{ immigrants}) &= 0.18 \\ Pr(1 \text{ immigrant}) &= 0.06 + 0.36 = 0.42 \\ Pr(2 \text{ immigrants}) &= 0.06 + 0.12 + 0.06 = 0.24 \\ Pr(3 \text{ immigrants}) &= 0.12 + 0.02 = 0.14 \\ Pr(4 \text{ immigrants}) &= 0.02. \end{aligned}$$

c. The expected total is 1.4.

d. The variances of the individual species are

$$\begin{aligned} Var(S_1) &= 0^2 \cdot 0.3 + 1^2 \cdot 0.6 + 2^2 \cdot 0.1 - 0.8^2 = 0.36 \\ Var(S_2) &= 0^2 \cdot 0.6 + 1^2 \cdot 0.2 + 2^2 \cdot 0.2 - 0.6^2 = 0.64. \end{aligned}$$

And the variance of the total number is

$$\begin{aligned} Var(S_1 + S_2) &= 0^2 \cdot 0.18 + 1^2 \cdot 0.42 + 2^2 \cdot 0.24 \\ &+ 3^2 \cdot 0.14 + 4^2 \cdot 0.14 - 1.4^2 = 1.0, \end{aligned}$$

matching the sum of the variances.

37. By putting probabilities as large as possible in the top and bottom corners, we can produce a case with maximum covariance and a large variance.

		Species 1		
		0	1	2
	0	0.3	0.3	0.0
	1	0.0	0.2	0.0
Species 2	2	0.0	0.1	0.1

The expectation of the sum is 1.4. The variance of the sum is 1.64, much larger than the value of 1.0 in the independent case.

39. Bird 2, because the joint distribution is the product of the marginal distributions.

41. For bird 1, the total calories C has probability distribution $Pr(C = 0.5) = 0.35$, $Pr(C = 1.0) = 0.05$, $Pr(C = 1.5) = 0.25$, and $Pr(C = 3.0) = 0.35$. The expectation is 1.65. For bird 2, the total calories C has probability distribution $Pr(C = 0.5) = 0.24$, $Pr(C = 1.0) = 0.16$, $Pr(C = 1.5) = 0.36$, and $Pr(C = 3.0) = 0.24$. The expectation is 1.54. For bird 3, the total calories C has probability distribution $Pr(C = 0.5) = 0.1$, $Pr(C = 1.0) = 0.3$, $Pr(C = 1.5) = 0.5$, and $Pr(C = 3.0) = 0.1$. The expectation is 1.4.

43. Keeping only the non-zero terms, we get

$$\begin{aligned} Pr(\text{Cancerous}) &= 0.1 \cdot 0.05 + 0.2 \cdot 0.03 + 0.2 \cdot 0.06 + 0.4 \cdot 0.06 \\ &+ 0.3 \cdot 0.12 + 0.6 \cdot 0.08 + 0.4 \cdot 0.15 + 0.8 \cdot 0.15 \\ &= 0.311 \end{aligned}$$

This is the same as the law of total probability where we take the probability that a cell is cancerous conditional on A and N and add up the products of the conditional probabilities times the joint probabilities.

Section 7.4, page 645

1. $\binom{4}{1} = \frac{4!}{1!3!} = \frac{24}{1 \cdot 6} = 4.$

3. $\binom{5}{2} = \frac{5!}{2!3!} = \frac{120}{2 \cdot 6} = 10.$

5. $\binom{7}{1} = \frac{7!}{1!6!} = \frac{5040}{1 \cdot 720} = 7.$

7. $b(1; 4, 0.3) = \binom{4}{1} 0.3^1 0.7^3 = 4 \cdot 0.3^1 0.7^3 = 0.4116.$

9. $b(2; 5, 0.4) = \binom{5}{2} 0.4^2 0.6^3 = 10 \cdot 0.4^2 0.6^3 = 0.3456.$

11. $b(1; 7, 0.2) = \binom{7}{1} 0.2^1 0.8^6 = 7 \cdot 0.2^1 0.8^6 = 0.3670.$

13. $E(B) = np = 1.2$, $Var(B) = np(1-p) = 0.84$, and the mode is the smallest integer greater than $np - 1 + p = 0.5$ or $B = 1$.

15. $E(B) = np = 4.9$, $Var(B) = np(1-p) = 1.47$, and the mode is the smallest integer greater than $np - 1 + p = 4.6$ or $B = 5$.

17. The probability distribution for B is

$$\Pr(B=0) = b(0; 2, 0.7) = \frac{2!}{0!2!} 0.7^0 0.3^2 = 0.09$$
$$\Pr(B=1) = b(1; 2, 0.7) = \frac{2!}{1!1!} 0.7^1 0.3^1 = 0.42$$
$$\Pr(B=2) = b(2; 2, 0.7) = \frac{2!}{2!0!} 0.7^2 0.3^0 = 0.49.$$

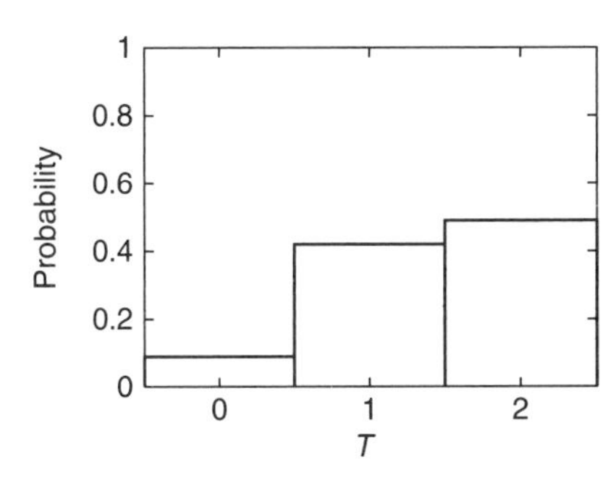

19. The probability distribution for B is

$$\Pr(B=0) = b(0; 3, 0.7) = \frac{3!}{0!3!} 0.7^0 0.3^3 = 0.027$$
$$\Pr(B=1) = b(1; 3, 0.7) = \frac{3!}{1!2!} 0.7^1 0.3^2 = 0.189$$
$$\Pr(B=2) = b(2; 3, 0.7) = \frac{3!}{2!1!} 0.7^2 0.3^1 = 0.441$$
$$\Pr(B=3) = b(3; 3, 0.7) = \frac{3!}{3!0!} 0.7^3 0.3^0 = 0.343.$$

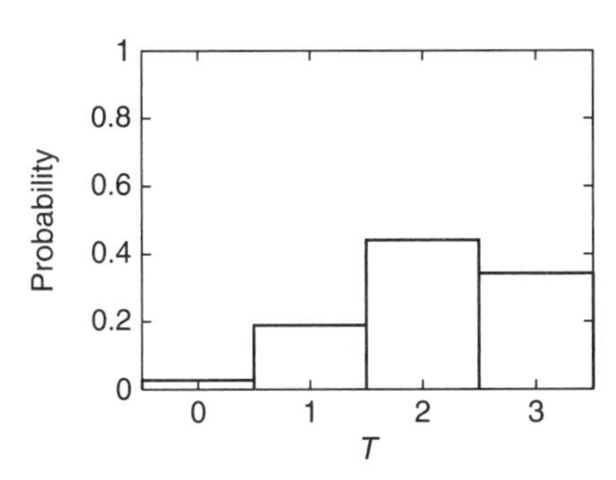

21. Using the probability distribution for B, we find

$$E(B) = 0 \cdot 0.09 + 1 \cdot 0.42 + 2 \cdot 0.49 = 1.4.$$

Checking, we find that $E(B) = np = 2 \cdot 0.7$, as it should. Also,

$$Var(B) = 0^2 \cdot 0.09 + 1^2 \cdot 0.42 + 2^2 \cdot 0.49 - 1.4^2 = 0.42.$$

Checking, we find that $Var(B) = np(1-p) = 2 \cdot 0.7 \cdot 0.3$, as it should.

23. Using the probability distribution for B, we find

$$E(B) = 0 \cdot 0.027 + 1 \cdot 0.189 + 2 \cdot 0.441 + 3 \cdot 0.343 = 2.1.$$

Checking, we find that $E(B) = np = 3 \cdot 0.7$, as it should. Also,

$$Var(B) = 0^2 \cdot 0.027 + 1^2 \cdot 0.189 + 2^2 \cdot 0.441 + 3^2 \cdot 0.343 - 2.1^2 = 0.63.$$

Checking, we find that $Var(B) = 3p(1-p) = 3 \cdot 0.7 \cdot 0.3$, as it should.

25.

Year 1	Year 2	Year 3	Year 4	Probability
1	1	1	0	$p^3(1-p)$
1	1	0	1	$p^3(1-p)$
1	0	1	1	$p^3(1-p)$
0	1	1	1	$p^3(1-p)$

There are four ways to get three successes, each of which has a probability of $p^3(1-p)$, for a total of $b(3; 4, p) = 4p^3(1-p)$. This matches the formula for the binomial.

27. The probability of 2 successes is $0.3 \cdot 0.7 = 0.21$. Similarly, the probability of 0 successes is $0.7 \cdot 0.3 = 0.21$. The probability of 1 success is then 0.58. If the trials were identical with probability 0.5, the probabilities for 0 and 2 success would be 0.25, and for 1 success would be 0.5. In each case, the expected number of successes is 1.0. However, the variance is 0.42 rather than 0.5.

29. Let I_1 be a Bernoulli random variable that takes the value 1 when the first trial is a success, and the value 0 when the first trial is a failure, and similarly for I_2. Then $\Pr(I_2 = 1 | I_1 = 1) = 0.8$ and $\Pr(I_2 = 1 | I_1 = 0) = 0.2$. By the law of total probability,

$$\begin{aligned}\Pr(I_2 = 1) &= \Pr(I_2 = 1 | I_1 = 1) \Pr(I_1 = 1) \\ &\quad + \Pr(I_2 = 1 | I_1 = 0) \Pr(I_1 = 0) \\ &= 0.8 \cdot 0.5 + 0.2 \cdot 0.5 = 0.5.\end{aligned}$$

The probability of 0 successes is

$$\begin{aligned}\Pr(I_2 = 0 \text{ and } I_1 = 0) &= \Pr(I_2 = 0 | I_1 = 0) \Pr(I_1 = 0) \\ &= 0.8 \cdot 0.5 = 0.4.\end{aligned}$$

Similar, the probability of 2 successes is

$$\begin{aligned}\Pr(I_2 = 1 \text{ and } I_1 = 1) &= \Pr(I_2 = 1 | I_1 = 1) \Pr(I_1 = 1) \\ &= 0.8 \cdot 0.5 = 0.4.\end{aligned}$$

The probability of 1 success is then 0.2. If the trials were independent, the probabilities for 0 and 2 successes would be 0.25, and for 1 success would be 0.5. The expected number of successes in each case is 1.0 (thanks to Theorem 7.4), but the variance in the non-independent model is 0.8, greater than the variance of 0.5 in the independent case.

31.

$$\binom{n}{k} = \frac{n!}{(n-k)!k!} = \frac{n!}{k!(n-k)!} = \binom{n}{n-k}$$

because $k = n - (n-k)$. Choosing a subset of with k elements corresponds to leaving out a subset with $n-k$ elements. For every subset with k elements there is a unique subset of with $n-k$ elements (the complement of that set). There must therefore be exactly the same number of groupings of each.

33. $(x+1)^3 = (x+1)(x+1)(x+1) = x^3 + 3x^2 + 3x + 1$. These coefficients exactly match $\binom{3}{k}$. $(x+1)^4 = x^4 + 4x^3 + +6x^2 + 4x + 1$. These coefficients exactly match $\binom{4}{k}$.

35. The probability of seven girls is $b(7; 8, 0.5) = \binom{8}{7} 0.5^7 \times 0.5^1 = 8 \cdot 0.5^8 = 0.03125$. The probability of four girls is $b(4; 8, 0.5) = \binom{8}{4} 0.5^4 \times 0.5^4 = 70 \cdot 0.5^8 = 0.27344$. The second type of family is more probable.

37. There are $5 = \binom{5}{1}$ ways: A, B, C, D, E.

39. There are $10 = \binom{5}{3}$ ways: (A,B,C), (A,B,D), (A,B,E), (A,C,D), (A,C,E), (A,D,E), (B,C,D), (B,C,E), (B,D,E), (C,D,E).

41. There are 3! = 6 ways. Three possibilities are Speedy, Blinky, Sparky; Blinky, Speedy, Sparky; Blinky, Sparky, Speedy.

43. There are 5! = 120 ways. Three possibilities are ace, 2, 5, king, 10; 2, ace, 5, king, 10; 5, king, 2, 10, ace.

45. There are 5! = 120 ways to order the prescriptions, only one of which is right. The probability is 1/120.

47. The conditional probability is 1/4, because there are now four medications to distribute among the remaining four patients. The unconditional probability for the second patient is 1/5, so the events are not independent. This is because getting the first patient right reduces the number of possibilities for the second.

49. The conditional probability is 1/3. The probability that all three get the right medication is 1/60.

51. If the patient with the different medication gets the right stuff (probability 1/5), then all the rest must also get the right stuff, for a total probability of 1/5.

53. Let H be the number. Then

$$\begin{aligned}
\Pr(H=0) &= b(0;5,0.5) = 0.031\\
\Pr(H=1) &= b(1;5,0.5) = 0.156\\
\Pr(H=2) &= b(2;5,0.5) = 0.313\\
\Pr(H=3) &= b(3;5,0.5) = 0.313\\
\Pr(H=4) &= b(4;5,0.5) = 0.156\\
\Pr(H=5) &= b(5;5,0.5) = 0.031.
\end{aligned}$$

The expectation is 2.5, and the variance is 1.25.

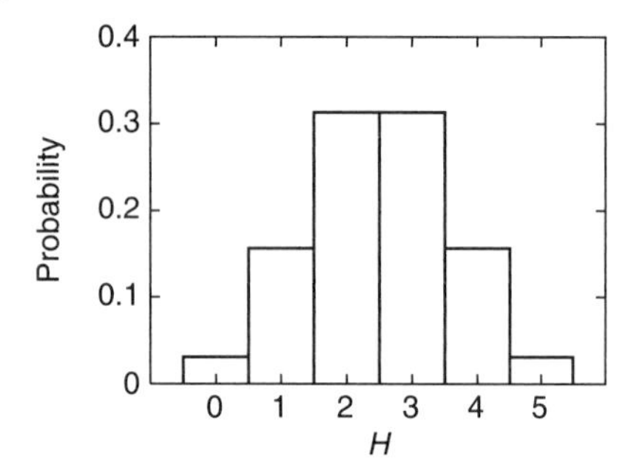

55. Before mortality, a fraction 0.5 will be heterozygotes, and there will be a fraction of 0.25 of each type of homozygote. After mortality, 0.25 of the original offspring are living heterozygotes, and 0.25 of the original offspring are each type of living homozygotes. Out of the total of 0.75 that survive, a fraction 0.25/0.75=1/3 are heterozygotes and 0.25/0.75=1/3 are each type of homozygote. Hence, the probability of no **A** alleles is 1/3, of one **A** allele is 1/3, and of two **A** alleles is 1/3. The expected number of **A** alleles is 1.0, so the binomial would have to have $p = 0.5$. However, the probabilities do not match those with $n = 2$ and $p = 0.5$, so this does not follow a binomial distribution.

57. The probability of no **A** alleles is 0.1, of one **A** allele is 0.5, and of two **A** alleles is 0.4. This does not follow a binomial distribution because the two probabilities do not match.

Section 7.5, page 653

1. $b(3;4,0.75) = \binom{4}{3}0.75^3 \cdot 0.25^1 = 4 \cdot 0.75^3 \cdot 0.25^1 = 0.422.$

3. We must add the probability that one is tall and the probability that zero are tall. $b(1;4,0.75) = \binom{4}{1}0.75^1 \cdot 0.25^3 = 4 \cdot 0.75^1 \cdot 0.25^3 = 0.0469.$ $b(0;4,0.75) = \binom{4}{0}0.75^0 \cdot 0.25^4 = 1 \cdot 0.75^0 \cdot 0.25^4 \approx 0.0039$. The total is 0.0508.

5. Mean $= 100 \cdot 0.5 = 50$, variance $= 100 \cdot 0.5 \cdot 0.5 = 25$, standard deviation $= \sqrt{25} = 5$, and CV = sd/mean = 5/50 = 0.1.

7. Mean $= 25 \cdot 0.9 = 22.5$, variance $= 25 \cdot 0.9 \cdot 0.1 = 2.25$, standard deviation $= \sqrt{2.25} = 1.5$, and CV = sd/mean $= 1.5/22.5 = 0.067$.

9. If we list the possible orders, we find six: TIS, TSI, ITS, IST, STI, SIT. The probability of each of these is $0.25^1 \cdot 0.5^1 \cdot 0.25^1 = 0.0313$, for a total of $6 \cdot 0.0313 = 0.1878$.

11. As with the binomial distribution, there are six possible ways to choose two talls and two intermediates out of four: TTII, TITI, TIIT, ITTI, ITIT, IITT. The probability of each of these is $0.25^2 \cdot 0.5^2 = 0.0156$, for a total of $6 \cdot 0.0156 = 0.0936$.

13. In this case, $n = 3$, $k_1 = 1$, $k_2 = 1$, $k_3 = 1$, $p_1 = 0.25$, $p_2 = 0.5$, and $p_3 = 0.25$. Then the probability is

$$M(1,1,1;3,0.25,0.5,0.25) = \frac{3!}{1!1!1!}0.25^1 0.5^1 0.25^1 = 0.1875.$$

This matches the answer in Exercise 9.

15. In this case, $n = 4$, $k_1 = 2$, $k_2 = 2$, $k_3 = 0$, $p_1 = 0.25$, $p_2 = 0.5$, and $p_3 = 0.25$. Then the probability is

$$M(2,2,0;4,0.25,0.5,0.25) = \frac{4!}{2!2!0!}0.25^2 0.5^2 0.25^0 = 0.0938.$$

This matches the answer in Exercise 11.

17. The expectation is $np = 1.5$. The mode is the smallest integer bigger than $np - 1 + p = 1.5 - 1 + 0.25 = 0.75$, and so is equal to 1.

19. This is $b(0,6,0.25) + b(1,6,0.25)$. $b(1;6,0.25) = \binom{6}{1}0.25^1 \cdot 0.75^5 = 6 \cdot 0.25^1 \cdot 0.75^5 = 0.356.$ $b(0;6,0.25) = \binom{6}{0}0.25^0 \cdot 0.75^6 = 1 \cdot 0.25^0 \cdot 0.75^6 = 0.178$. The total is 0.534.

21. For the first set of islands, this is $b(7;10,2/3) = 0.26$. The second set of islands are either all occupied or all empty, so the probability is 0.

23. For the first set of islands, the mean is $np = 10 \cdot 2/3 = 6.67$. For the second set it is $10 \cdot 2/3 + 0 \cdot \frac{1}{3} = 6.67$.

25. For the first set of islands, $np - 1 + p = 6.67 - 1 + 0.67 = 6.33$. The mode is the next integer, or 7. For the second set of islands, the mode is 10.

27. The probability that a gene is a mutant obeys the discrete-time dynamical system

$$p_{t+1} = 0.99p_t + 0.01(1 - p_t)$$

with $p_0 = 0$. Then $p_1 = 0.01$ and $p_2 = 0.0198$. We found that p_t approaches an equilibrium of 0.5 after a long time. At time 1, the number of mutants follows the binomial with $n = 4$ and $p = 0.01$. The probability of two mutants is 0.00059, of three mutants is 3.96×10^{-6}, and of four mutants is 1.0×10^{-8}, for a total of about 3.97×10^{-6}. At time 2, the number of mutants follows the binomial with $n = 4$ and $p = 0.0198$. The probability of two mutants is 0.00226, of three mutants is 3.0×10^{-5} and of four mutants is 1.5×10^{-7}, for a total of about 0.00229. After a long time, $n = 4$, $p = 0.5$. The probability of two mutants is 0.375, of three mutants is 0.25, and of four mutants is 0.0625, so that the probability of two or more mutants is 0.6875.

29. At time 1, this is a binomial distribution with $n = 5$ and $p = 0.8$. The probabilities for 0 through 5 inside are 0.0003, 0.0064, 0.051, 0.205, 0.410, and 0.328.

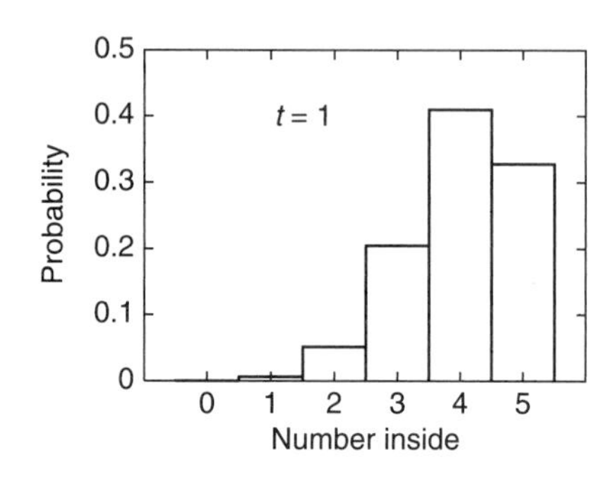

31. At time 5, this is a binomial distribution with $n=5$ and $p=0.8^5=0.32$. The probabilities for 0 through 5 inside are 0.137, 0.335, 0.326, 0.159, 0.0388, and 0.004.

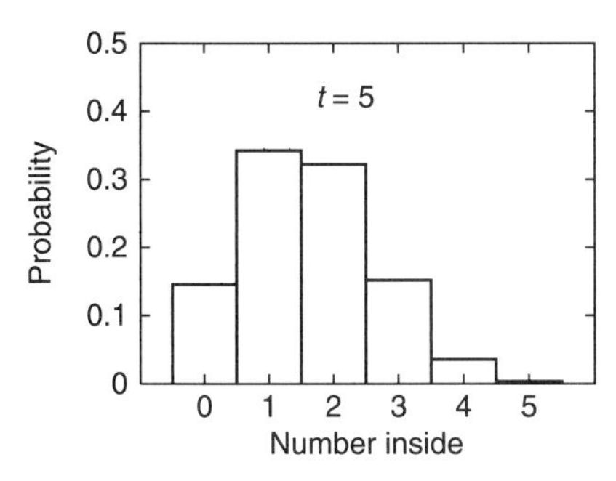

33. As a function of time, the probability that exactly one remains is $b(1; 5, 0.8^t) = 5 \cdot 0.8^t(1-0.8^t)^4$. We want to find the value of t that maximizes this expression. It is easier to replace 0.8^t with x and write the probability as $f(x)=5x(1-x)^4$. Then $f'(x) = 5(1-x)^4 - 20x(1-x)^3$, which is equal to 0 when $x = 1/5 = 0.2$. Then $t = \ln(0.2)/\ln(0.8) = 7.21$.

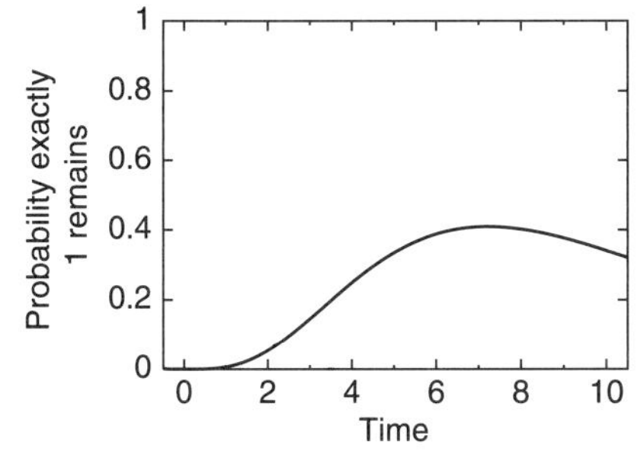

35. The probability is $b(3; 10, 0.410) = 0.205$.
37. $E(N_t) = 40 \cdot 0.8^t$.
39. The variance is $40 \cdot 0.8^t(1-0.8^t)$. This has a maximum at $t = \ln(0.5)/\ln(0.8) = 3.106$.
41. Let P be the random variable describing the number of immigrants. Because they arrive in pairs, $P = 2I$, where I is a binomial random variable with $p = 0.5$ and $n = 100$.
43. $\mathrm{Var}(P) = \mathrm{Var}(2I) = 2^2\mathrm{Var}(I) = 4 \cdot 100 \cdot 0.5 \cdot 0.5 = 100$.
45. There must be two inside (both of the second type). The expectation is 2 and the variance is 0.

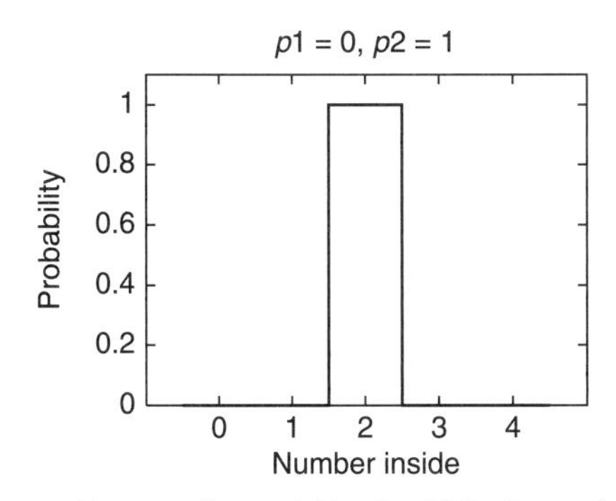

47. We have to add two random variables. Let N_1 be the number of the first type and N_2 be the number of the second type. N_1 has a binomial distribution with $p = 0.25$ and $n = 2$, and N_2 has a binomial distribution with $p = 0.75$ and $n = 2$. Then

$$\Pr(N_1 + N_2 = 0) = \Pr(N_1 = 0) \cdot \Pr(N_2 = 0) = 0.035$$
$$\Pr(N_1 + N_2 = 1) = 0.234$$
$$\Pr(N_1 + N_2 = 2) = 0.461$$
$$\Pr(N_1 + N_2 = 3) = 0.234$$
$$\Pr(N_1 + N_2 = 4) = 0.035.$$

The expectation and variance are

$$\begin{aligned}E(N_1 + N_2) &= E(N_1) + E(N_2) = 2 \cdot 0.25 + 2 \cdot 0.75 = 2\\ \mathrm{Var}(N_1 + N_2) &= \mathrm{Var}(N_1) + \mathrm{Var}(N_2)\\ &= 2 \cdot 0.25 \cdot 0.75 + 2 \cdot 0.75 \cdot 0.25 = 0.75\end{aligned}$$

The expectation is the same as in Exercise 45, but the variance is larger.

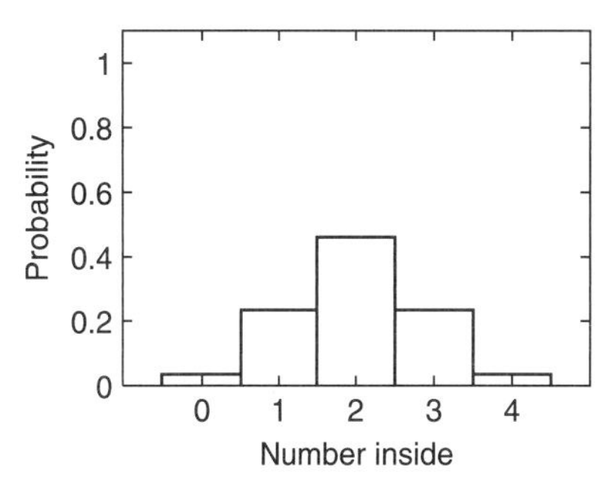

Section 7.6, page 665

1. This follows the geometric distribution with $q = 0.2$. Then $g_3 = (1-q)^2q = 0.8^2 \cdot 0.2 = 0.128$.

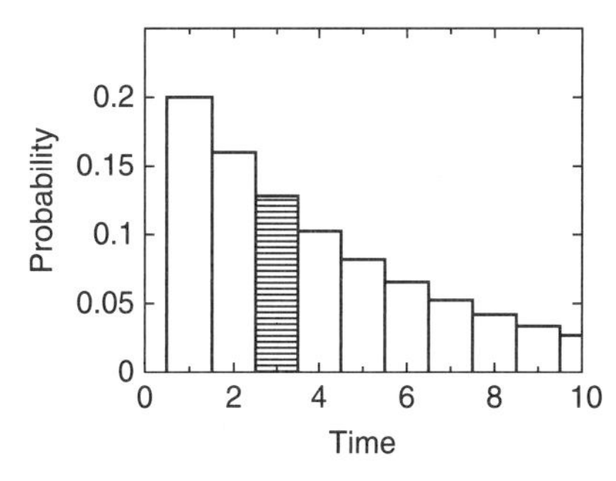

3. This follows the cumulative geometric distribution with $q = 0.2$. Then $G_4 = 1 - (1-q)^4 = 1 - 0.8^4 = 0.5904$.

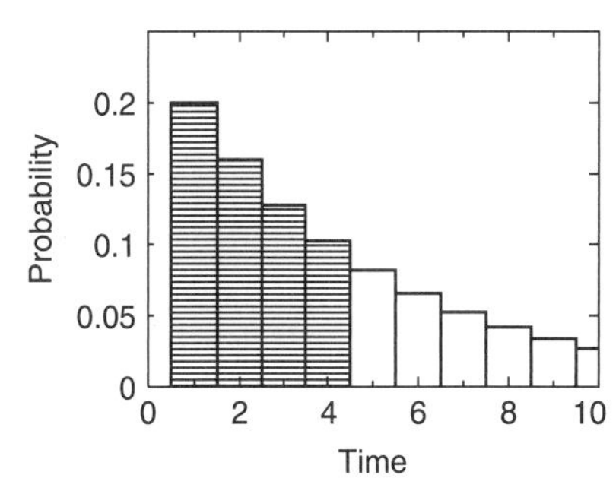

5. This is 1 minus the probability that the first success occurs on or before the second trial, or $1 - G_2 = (1-q)^2 = 0.8^2 = 0.64$.

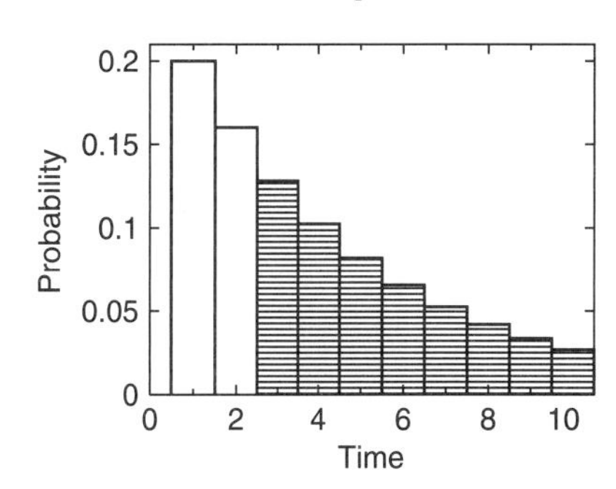

7. This follows the exponential distribution with $\lambda = 0.5$. Then $f(t) = \lambda e^{-\lambda t}$ and $F(t) = 1 - e^{-\lambda t}$. Then the probability that the first event occurs between times 1.0 and 2.0 is

$$\int_1^2 f(t)dt = F(2) - F(1) = e^{-0.5 \cdot 1.0} - e^{-0.5 \cdot 2.0} = 0.239.$$

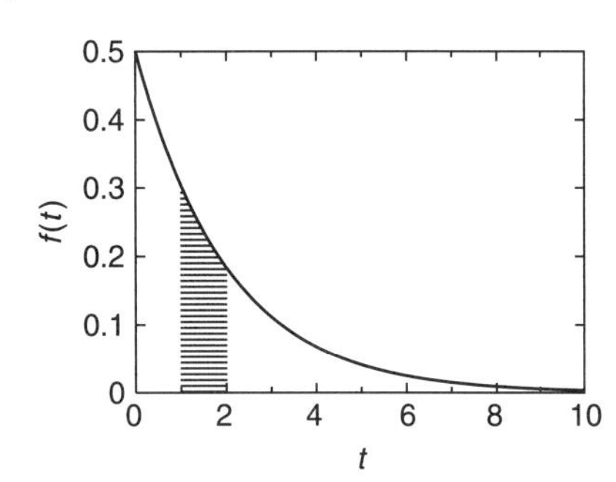

9. This follows the exponential distribution with $\lambda = 0.2$. Then $f(t) = \lambda e^{-\lambda t}$ and $F(t) = 1 - e^{-\lambda t}$. The probability that the first event occurs before time 1.0 is

$$F(1.0) = 1 - e^{-0.2 \cdot 1.0} = 0.181.$$

The probability that the first event occurs after time 3.0 is

$$1 - F(3.0) = 1 - (1 - e^{-0.2 \cdot 3.0}) = 0.549,$$

for a total of about 0.730.

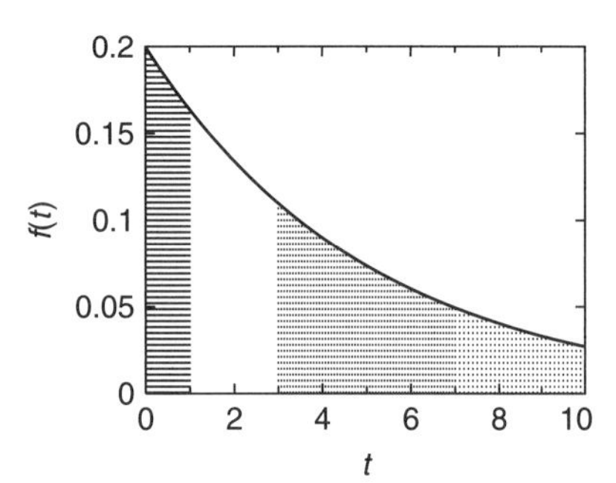

11. The mean is $\mathrm{E}(T) = \frac{1}{q} = 5.0$. The variance is $\mathrm{Var}(T) = \dfrac{1-q}{q^2} = 20$, and the standard deviation is $\sqrt{20} = 4.472$. The coefficient of variation is $4.472/5.0 = 0.894$. The mode is at $T = 1$.

13. The mean is $\mathrm{E}(T) = \frac{1}{q} = 1.429$. The variance is $\mathrm{Var}(T) = \dfrac{1-q}{q^2} = 0.612$, and the standard deviation is $\sqrt{0.612} = 0.782$. The coefficient of variation is $0.782/1.429 = 0.548$. The mode is at $T = 1$.

15. The mean is $\mathrm{E}(T) = \frac{1}{\lambda} = 5.0$. The variance is $\mathrm{Var}(T) = \frac{1}{\lambda^2} = 25.0$, the standard deviation is 5, and the coefficient of variation is 1. The median is $\dfrac{\ln(2.0)}{0.2} = 3.46$. The mode is $T = 0$.

17. The mean is $\mathrm{E}(T) = \frac{1}{\lambda} = 0.667$. The variance is $\mathrm{Var}(T) = \frac{1}{\lambda^2} = 0.444$, the standard deviation is 0.667, and the coefficient of variation is 1. The median is $\dfrac{\ln(2.0)}{1.5} = 0.462$. The mode is $T = 0$.

19. If you multiply $\sum_{i=0}^{n} x^i$ by $1 - x$, all terms cancel except the $1 - x^{n+1}$.

21. Substituting $1 - q$ for x, we get

$$\begin{aligned} T_f(1-q) &= 1 + (1-q) + (1-q)^2 + (1-q)^3 + (1-q)^4 + \cdots \\ &= f(1-q) = \frac{1}{1-(1-q)} = \frac{1}{q}. \end{aligned}$$

Then $\sum_{t=1}^{\infty} q(1-q)^{t-1} = q\frac{1}{q} = 1$.

23. $q\Delta t$

25. This is $(1 - q\Delta t)^{\frac{t}{\Delta t}}$.

27. Probability it leaves immediately is q, and the time is 1.

29. $\mathrm{E}(T) = 1 \cdot q + [1 + \mathrm{E}(T)](1 - q)$.

31. It takes on the value 1 with probability q, just like T. $\Pr(R = 2) = (1-q)\Pr(T = 1) = q(1-q)$, and $\Pr(R = k) = (1-q)\Pr(T = k - 1) = q^{k-1}(1-q)$. The probability distributions for R and T match perfectly.

33. Because R^2 and T^2 have the same distribution.

35. To integrate $\lambda t e^{-\lambda t}$, set $u = t$ and $dv = \lambda e^{-\lambda t}$. Then $du = dt$ and $v = -e^{-\lambda t}$. Therefore,

$$\begin{aligned} \int_0^{\infty} \lambda t e^{-\lambda t} dt &= \int_0^{\infty} u dv = uv|_0^{\infty} - \int_0^{\infty} v du \\ &= -te^{-\lambda t}|_0^{\infty} + \int_0^{\infty} e^{-\lambda t} dt = \frac{1}{\lambda}. \end{aligned}$$

37. The probability of mutating in division 15 is $g_{15} = 0.01 \cdot 0.99^{14} = 0.0087$. The probability of mutating at or before division 15 is $G_{15} = 1 - 0.99^{19} = 0.140$.

39. The probability of leaving at time 10 is $g_{10} = 0.05 \cdot 0.95^9 = 0.0315$. The probability of leaving at or before time 10 is $G_{10} = 1 - 0.95^{10} = 0.401$.

41. The probability of blowing out on day 50 is $g_{50} = 0.01 \cdot 0.99^{49} = 0.0061$. The probability of not blowing out by day 200 is $1 - G_{200} = 0.99^{200} = 0.134$.

43. After n flies have been inspected, the number of purple-eyed flies has a binomial distribution with $p = 0.1$. The expected number is $0.1n$. The expectation is equal to 1 when $n = 10$.

45. This is a case for Bayes' Theorem.

$$\begin{aligned} &\Pr(\text{mutants} \mid \text{none found out of 20}) \\ &= \frac{\Pr(\text{none found out of 20} \mid \text{mutants})\Pr(\text{mutants})}{\Pr(\text{none found out of 20})} \end{aligned}$$

The only piece we do not know is the denominator, which we find with the law of total probability.

$$\begin{aligned} \Pr(\text{none found out of 20}) &= \Pr(\text{none found out of 20} \mid \text{mutant}) \\ &\quad \times \Pr(\text{mutant}) \\ &\quad + \Pr(\text{none found out of 20} \mid \text{not mutant}) \\ &\quad \times \Pr(\text{not mutant}) \\ &= (0.9^{20})(0.5) + 1 \cdot 0.5 \approx 0.5608. \end{aligned}$$

Putting it all together, $\Pr(\text{mutants} \mid \text{none found out of 20}) = \frac{0.9^{20} \cdot 0.5}{0.5608} = 0.108$. The probability that mutant parents are thrown out is $0.9^{20} = 0.12$.

47. The first one had to be OK (probability 0.9), the second one OK (probability 0.8), and the third one bad (probability 0.3), for a total of 0.216.

49. The p.d.f. is $f(t) = 0.3e^{-0.3t}$, the c.d.f. is $F(t) = 1 - e^{-0.3t}$, and the survivorship function is $S(t) = e^{-0.3t}$. The probability it has left by the end of the third second is $F(3.0) = 0.41$. The probability it has left by the end of the first millisecond is $F(0.3 \cdot 0.001) = 1 - e^{-0.0003} = 0.00029996$. According to the definition of the rate, this should be very close to $\lambda t = 0.0003$, as it is. The expectation is 3.333 s and the variance is 11.11 s^2.

51. The p.d.f. is $f(t) = 0.2e^{-0.2t}$, the c.d.f. is $F(t) = 1 - e^{-0.2t}$, and the survivorship function is $S(t) = e^{-0.2t}$. Expectation is 5 h and variance is 25 h^2. The probability of no calls in 10 h is $e^{-2} = 0.135$. 45 s is 45/3600 = 0.0125 h, so the probability of a call is about $0.0125 \cdot 0.2 = 0.0025$.

53. $P(t) = e^{-2.0t}$.

55. The same as the probability it mutates during the first five divisions, or $G_5 = 1 - 0.99^5 = 0.049$.

57. The same as the probability it leaves during the first 3 s, or $F(3) = 1 - e^{-0.3 \cdot 3} = 0.593$.

59. The probability is 0.5; the new a.s.d. is just as likely to leave as the original a.s.d. This would happen if all drivers had the same probability of exiting at every exit.

61. $\dfrac{dP}{dt} = -tP$.

63.

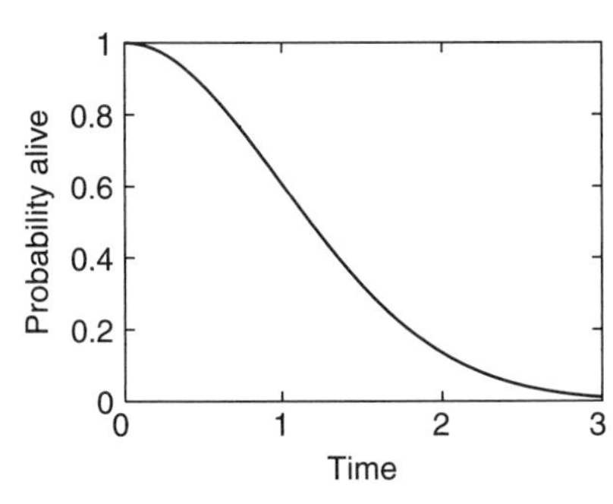

65. The probability it survives to age 1 is $P(1) = 0.606$.

Section 7.7, page 676

1. The numbers of hits in each of the 10 s are 2, 1, 0, 0, 1, 3, 2, 4, 4, and 0. The average is the total of 17 divided by the number of seconds, or 1.7. This is pretty close to what we expect with $\lambda = 1.5$.

3. The numbers of hits in each of the 20 s are 0, 1, 0, 0, 1, 1, 0, 2, 0, 0, 0, 1, 1, 0, 0, 0, 1, 0, 0, and 0. The average is 8/20=0.4, which is close to what we expect with $\lambda = 0.5$.

5. The probabilities in the Poisson distribution with $\Lambda = \lambda t = 1.5 \cdot 1 = 1.5$ are

$$\Pr(N=0) = p(0; 1.5) = \frac{e^{-1.5}(1.5^0)}{0!} = 0.2231$$
$$\Pr(N=1) = p(1; 1.5) = \frac{e^{-1.5}(1.5^1)}{1!} = 0.3347$$
$$\Pr(N=2) = p(2; 1.5) = \frac{e^{-1.5}(1.5^2)}{2!} = 0.2510$$
$$\Pr(N=3) = p(3; 1.5) = \frac{e^{-1.5}(1.5^3)}{3!} = 0.1255$$
$$\Pr(N=4) = p(4; 1.5) = \frac{e^{-1.5}(1.5^4)}{4!} = 0.0471.$$

We got three 0's, two 1's, two 2's, one 3, and two 4's, or observed fractions of 0.3, 0.2, 0.2, 0.1, and 0.2.

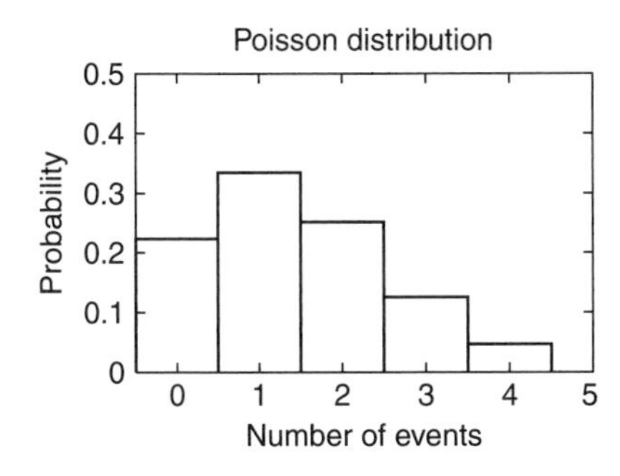

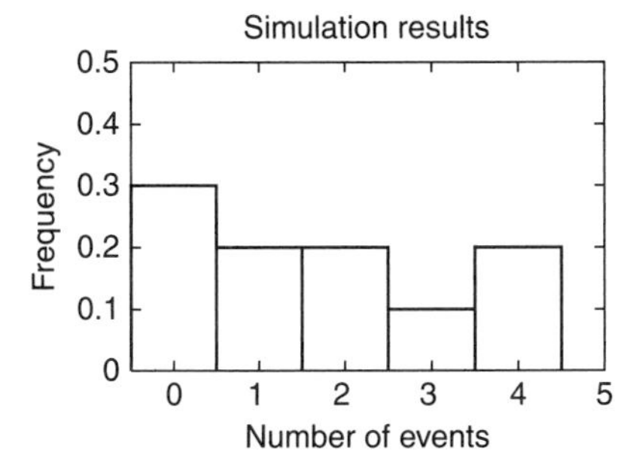

7. The probabilities in the Poisson distribution with $\Lambda = \lambda t = 0.5 \cdot 1 = 0.5$ are

$$\Pr(N=0) = p(0; 0.5) = \frac{e^{-0.5}(0.5^0)}{0!} = 0.6065$$
$$\Pr(N=1) = p(1; 0.5) = \frac{e^{-0.5}(0.5^1)}{1!} = 0.3033$$
$$\Pr(N=2) = p(2; 0.5) = \frac{e^{-0.5}(0.5^2)}{2!} = 0.0758$$
$$\Pr(N=3) = p(3; 0.5) = \frac{e^{-0.5}(0.5^3)}{3!} = 0.0126.$$

We got 13 0's, six 1's, and one 2, or observed fractions of 0.65, 0.3, and 0.05.

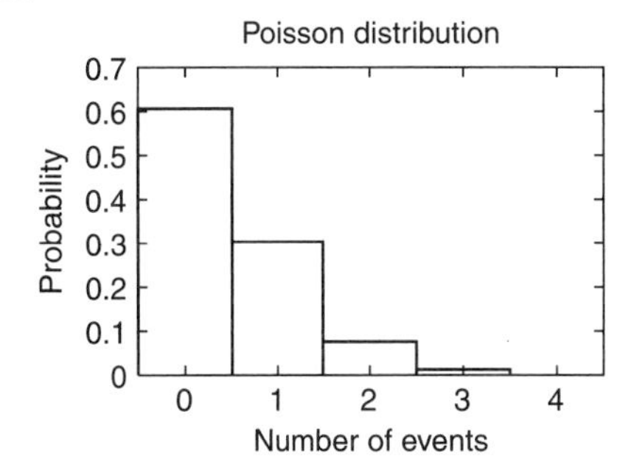

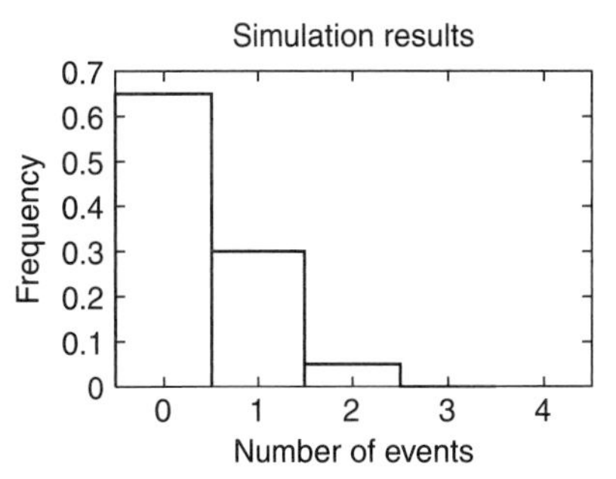

9. The probabilities in the Poisson distribution with $\Lambda = \lambda t = 0.5 \cdot 2 = 1.0$ are

$$\Pr(N=0) = p(0; 1.0) = \frac{e^{-1.0}(1.0^0)}{0!} = 0.3679$$
$$\Pr(N=1) = p(1; 1.0) = \frac{e^{-1.0}(1.0^1)}{1!} = 0.3679$$
$$\Pr(N=2) = p(2; 1.0) = \frac{e^{-1.0}(1.0^2)}{2!} = 0.1839$$
$$\Pr(N=3) = p(3; 1.0) = \frac{e^{-1.0}(1.0^3)}{3!} = 0.0613.$$

Regrouping, the first two seconds have $0 + 1 = 1$ event and so forth, or 1, 0, 2, 2, 0, 1, 1, 0, 1, and 0. We have four 0's, four 1's and two 2's, or observed fractions of 0.4, 0.4, and 0.2.

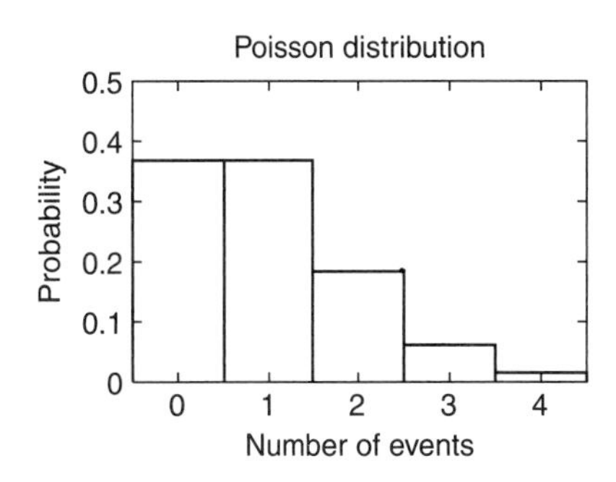

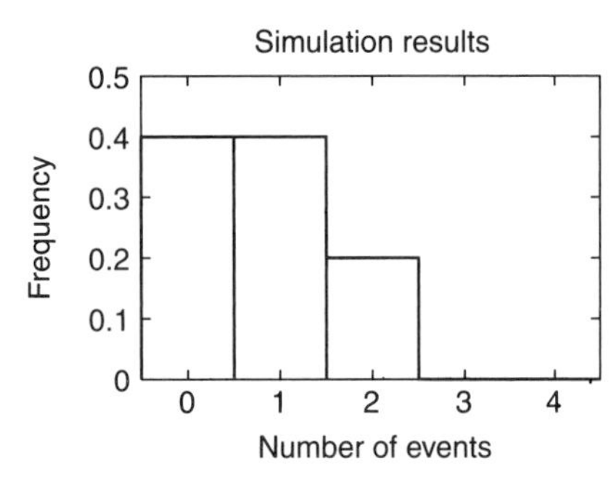

11. The hits are too evenly spaced for a Poisson distribution, with waiting times of about 1 time unit.

13. The hits seem to come faster and faster.

15. $\Lambda = \lambda t = 0.3 \cdot 3 = 0.9$. The expectation and variance are both 0.9, the standard deviation is $\sqrt{0.9} = 0.949$, the coefficient of variation is $0.949/0.9 = 1.054$, and the mode is the greatest integer less than 0.9, or 0.

17. $\Lambda = 1.2 \cdot 7 = 8.4$. The expectation and variance are both 8.4, the standard deviation is $\sqrt{8.4} = 2.898$, the coefficient of variation is $2.898/8.4 = 0.345$, and the mode is the greatest integer less than 8.4, or 8.

19. $\Lambda = \lambda t = 0.3 \cdot 3 = 0.9$. Then $p(2; 0.9) = e^{-0.9}0.9^2/2! = 0.165$.

21. $\Lambda = 1.2 \cdot 7 = 8.4$. $p(10; 8.4) = 0.108$.

23. The first five terms of the Poisson distribution with $\Lambda = 0.9$ are

$$\Pr(N=0) = p(0; 0.9) = \frac{e^{-0.9}(0.9^0)}{0!} = 0.4066$$
$$\Pr(N=1) = p(1; 0.9) = \frac{e^{-0.9}(0.9^1)}{1!} = 0.3659$$
$$\Pr(N=2) = p(2; 0.9) = \frac{e^{-0.9}(0.9^2)}{2!} = 0.1647$$
$$\Pr(N=3) = p(3; 0.9) = \frac{e^{-0.9}(0.9^3)}{3!} = 0.0494$$
$$\Pr(N=4) = p(4; 0.9) = \frac{e^{-0.9}(0.9^4)}{4!} = 0.0111.$$

These add to 0.9977, the probability of four or fewer.

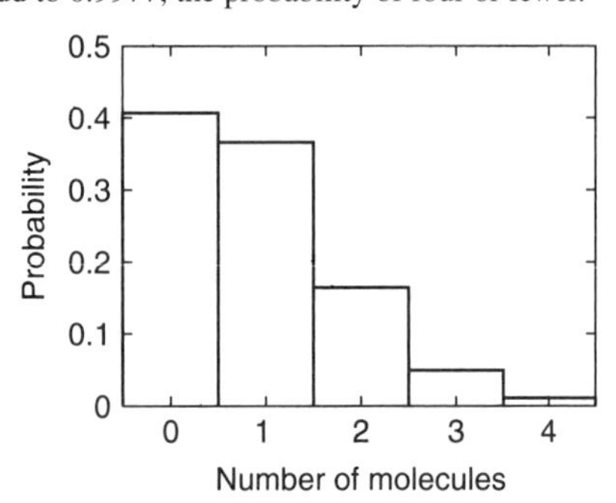

25. The distribution is

$$\Pr(N=0) = p(0; 8.4) = \frac{e^{-8.4}(8.4^0)}{0!} = 0.0002$$
$$\Pr(N=1) = p(1; 8.4) = \frac{e^{-8.4}(8.4^1)}{1!} = 0.0019$$
$$\Pr(N=2) = p(2; 8.4) = \frac{e^{-8.4}(8.4^2)}{2!} = 0.0079$$
$$\Pr(N=3) = p(3; 8.4) = \frac{e^{-8.4}(8.4^3)}{3!} = 0.0222$$
$$\Pr(N=4) = p(4; 8.4) = \frac{e^{-8.4}(8.4^4)}{4!} = 0.0466$$
$$\Pr(N=5) = p(5; 8.4) = \frac{e^{-8.4}(8.4^5)}{5!} = 0.0784$$
$$\Pr(N=6) = p(6; 8.4) = \frac{e^{-8.4}(8.4^6)}{6!} = 0.1097$$
$$\Pr(N=7) = p(7; 8.4) = \frac{e^{-8.4}(8.4^7)}{7!} = 0.1317$$
$$\Pr(N=8) = p(8; 8.4) = \frac{e^{-8.4}(8.4^8)}{8!} = 0.1382$$
$$\Pr(N=9) = p(9; 8.4) = \frac{e^{-8.4}(8.4^9)}{9!} = 0.1290$$
$$\Pr(N=10) = p(10; 8.4) = \frac{e^{-8.4}(8.4^{10})}{10!} = 0.1084.$$

Adding the values associated with 5 through 10 gives 0.695.

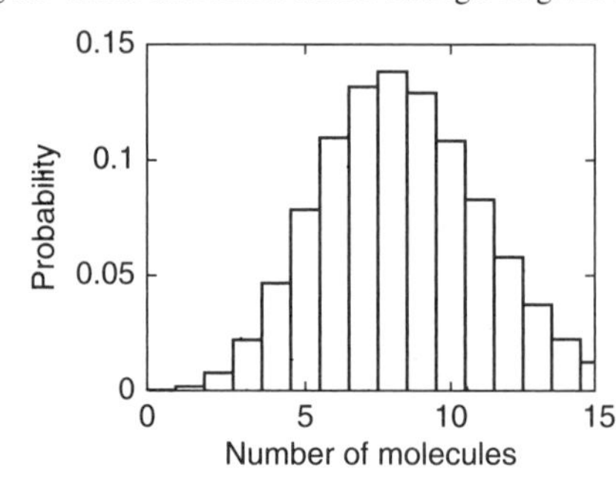

27. $E[N(t)] = 0.3t$.

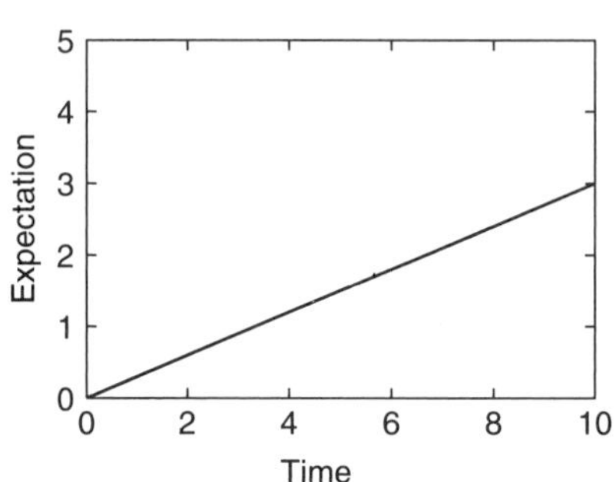

29. $\Pr[N(t) = 1] = 0.3te^{-0.3t}$. The graph increases to a maximum at $t = 3.33$ and decreases to 0 thereafter. It increases because the probability of any events is small near $t = 0$, and then decreases because we expect more than one event for $t > 3.33$.

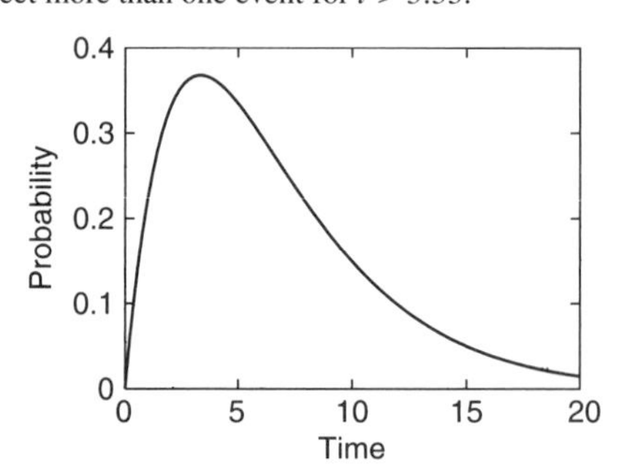

31. M_c will have a Poisson distribution with $\Lambda = 0.3 \cdot 7 = 2.1$ and will have mean and variance equal to 2.1. M_x will have a Poisson distribution with $\Lambda = 0.2 \cdot 7 = 1.4$ and will have mean and variance equal to 1.4. $N = M_c + M_x$ will have a Poisson distribution with $\Lambda = 2.1 + 1.4 = 3.5$ and will have mean and variance equal to 3.5.

33. With the binomial, $p = 0.1$ and $n = 16$, so the probability is $b(1; 16, 0.1) = 16 \cdot 0.9^{15} \cdot 0.1 = 0.329$. The Poisson approximation uses $\Lambda = 0.1 \cdot 16 = 1.6$. We expect that $b(k; 16, 0.1) = p(k; 1.6)$. In this case, the approximation is $p(1, 1.6) = 1.6e^{-1.6} = 0.323$, which is quite close.

35. $P_1(t + \Delta t) \approx (1 - \lambda \Delta t) P_1(t) + \lambda \Delta t P_0(t)$.

37.

$$\frac{d\lambda te^{-\lambda t}}{dt} = \lambda e^{-\lambda t} - \lambda^2 te^{-\lambda t}$$
$$= \lambda P_0(t) - \lambda P_1(t).$$

This is the only solution with the right initial condition of $P_1(0) = 0$.

39. One event during time Δt occurs with probability $\lambda\Delta t$, and zero events occur with probability $1-\lambda\Delta t$. The expected additional number of events is $\lambda\Delta t$, so

$$E(t+\Delta t)=E(t)+\lambda\Delta t.$$

Subtracting $E(t)$ from both sides and dividing by Δt gives

$$\frac{E(t+\Delta t)-E(t)}{\Delta t}=\lambda.$$

Taking the limit as $\Delta t\to 0$ gives

$$\frac{dE}{dt}=\lambda.$$

With the initial condition $E(0)=0$, this has solution $E(t)=\lambda t$.

41. The number of mutations will follow a Poisson distribution with parameter $\Lambda=0.002\cdot 2000=4.0$. The expected number of mutations is 4.0, as is the variance. The probability of zero mutations is $e^{-4.0}=0.018$, and the probability of exactly one mutation is $4.0e^{-4.0}=0.073$.

43. The number of mutations will follow a Poisson distribution with parameter $\Lambda=0.002\cdot t$, which is equal to 1.0 when $t=500$. The probability of zero mutations is $e^{-\Lambda}=0.001$ when $\Lambda=-\ln(0.001)=6.91$. $\Lambda=6.91=0.002\cdot t$ when $t=3454$. It takes 3454 generations for 99.9% of individuals to have a mutation.

45. The number of mutations in each population will follow a Poisson distribution, the first with mean $\Lambda_1=0.002\cdot 10000=20.0$, and the second with mean $\Lambda_2=0.0004\cdot 10000=4.0$. The expected total number of differences is 24.0.

47. We multiply the number of years by the number of sites by the rate to find the expected number of mutations at synonymous sites, or $\Lambda_s=1.0\times 10^6\cdot 100\cdot 6.0\times 10^{-7}=60$. The probability of zero mutations is $e^{-60.0}\approx 8.8\times 10^{-27}$, which is incredibly small. At nonsynonymous sites, $\Lambda_n=1.0\times 10^6\cdot 200\cdot 3.0\times 10^{-9}=0.6$. The probability of zero mutations is $e^{-0.6}\approx 0.549$. Most genes will not have had a change in their protein.

49. The number of mutations in the first year follows a Poisson distribution with $\Lambda=1.3\times 10^{-6}\cdot 4.7\cdot 10^6=6.11$. The number of mutations in the second year follows a Poisson distribution with $\Lambda=2.2\times 10^{-6}\cdot 4.7\cdot 10^6=10.34$. The total number of mutations will also follow a Poisson distribution with $\Lambda=6.11+10.34=16.45$.

51. The line connecting the two values is $\lambda(t)=1.0\times 10^{-6}+1.0\times 10^{-12}t$. Then

$$\Lambda=\int_0^{2.0\cdot 10^6} 1.0\times 10^{-6}+1.0\times 10^{-12}t\,dt$$
$$=1.0\times 10^{-6}t+0.5\times 10^{-13}t^2|_0^{2.0\cdot 10^6}=4.$$

This is the same average number of mutations that would have occurred if the rate had been 2.0×10^{-6} for the entire 2 million years.

53.

a. The random variable has $n=100{,}000$ and $p=0.0067$.

b. It has $\Lambda=np=100,000\cdot 0.0067=670$. The expected number to leave is 670.

c. The variance of the binomial is $np(1-p)=100,000\cdot 0.0067\cdot 0.9933=665.5$. The variance of the Poisson is $\Lambda=670.0$.

55. Probability is $e^{-\lambda t}$.

57. Probability is

$$(1-q)^2p(2;\lambda t)=(1-q)^2\frac{(\lambda t)^2e^{-\lambda t}}{2}.$$

59.

$$p[0,(1-q)\lambda t]=e^{-(1-q)\lambda t}$$
$$p[1,(1-q)\lambda t]=(1-q)\lambda te^{-(1-q)\lambda t}$$
$$p[2,(1-q)\lambda t]=\frac{[(1-q)\lambda t]^2e^{-(1-q)\lambda t}}{2}.$$

These must add up to 1 because it is a probability distribution.

Section 7.8, page 688

1. For these Bernoulli random variables, $E(X_i)=0.5$ and $Var(X_i)=0.25$. Then $E(S_{16})=16\cdot 0.5=8$ and $\sigma^2=Var(S_{16})=16\cdot 0.25=4$. Therefore, $\sigma=2$, and the p.d.f. is

$$f(x)=\frac{1}{2\sqrt{2\pi}}e^{-\frac{(x-8)^2}{8}}.$$

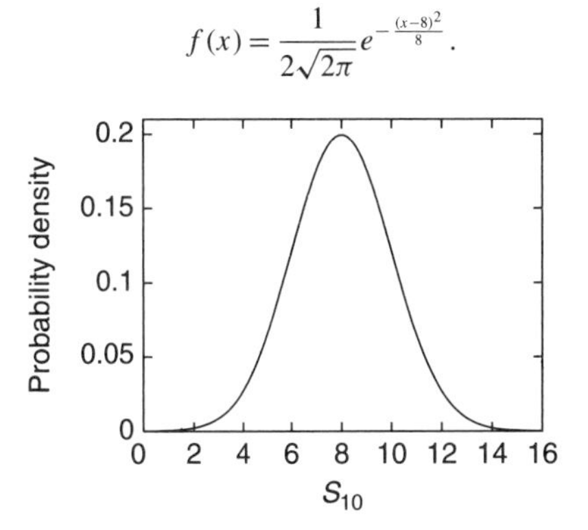

3. We have $E(X_i)=0.5$ and $Var(X_i)=0.25$. Then $E(A_{16})=E(X_i)=0.5$ and $\sigma^2=Var(A_{16})=\frac{0.25}{16}=0.0156$. Therefore, $\sigma=0.125$, and the p.d.f. is

$$f(x)=\frac{1}{0.125\sqrt{2\pi}}e^{-\frac{(x-0.5)^2}{0.03125}}.$$

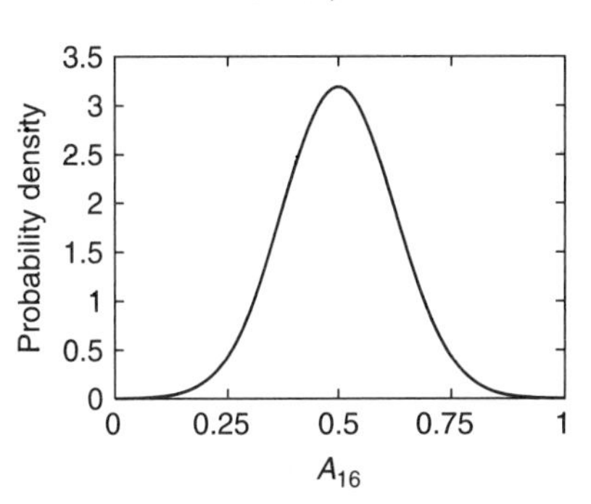

5. Let M_i be a random variable giving the number of mutants in the ith experiment. We found that $E(M_i)=2.1$ and $Var(M_i)=1.29$. Then $S_{20}=\sum_{i=0}^{i=20}M_i$ has mean $E(S_{20})=20\cdot 2.1=42$ and variance $\sigma^2=20\cdot 1.29=25.8$. Therefore $\sigma=5.08$, and the p.d.f. approximating S_{20} is

$$f(x)=\frac{1}{5.08\sqrt{2\pi}}e^{-\frac{(x-42)^2}{51.6}}.$$

The probability that the number is between 30 and 40 is approximately

$$Pr(30\le S_{20}\le 40)=\int_{x=30}^{x=40}f(x)dx.$$

The area looks like about 0.4 of the total.

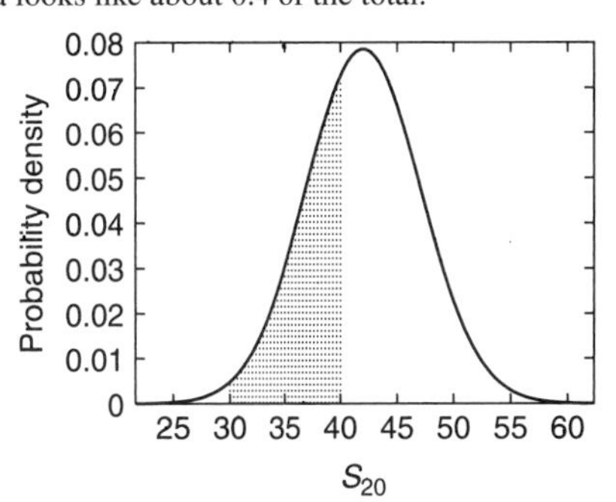

7. Let M_i be a random variable giving the number of mutants in the ith experiment. We found that $E(M_i) = 1.6$ and $Var(M_i) = 1.84$. Therefore $A_{100} = \frac{1}{100} \sum_{i=0}^{i=100} M_i$ has mean $E(A_{100}) = 1.6$ and variance $\sigma^2 = \frac{1.84}{100} = 0.0184$. Then $\sigma = 0.136$, and the p.d.f. approximating A_{100} is

$$f(x) = \frac{1}{0.136\sqrt{2\pi}} e^{-\frac{(x-1.6)^2}{0.0368}}.$$

The probability that the average is less than 1.8 is

$$\Pr(A_{100} \leq 1.8) = \int_{x=-\infty}^{x=1.8} f(x)dx.$$

The area looks like about 0.9 of the total.

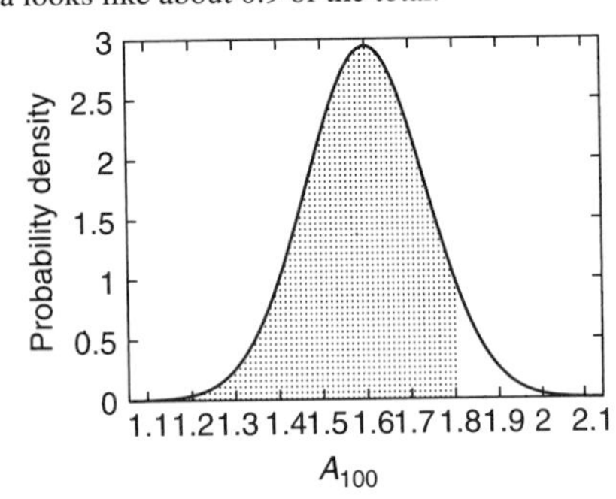

9. $3X \sim N(15.0, 144.0)$. The variance is multiplied by 9.

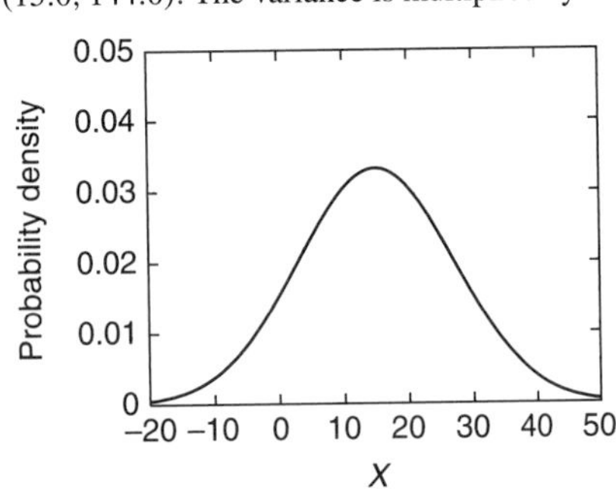

11. Let S_9 be the sum of nine independent samples from X. Then $S_9 \sim N(45.0, 144.0)$.

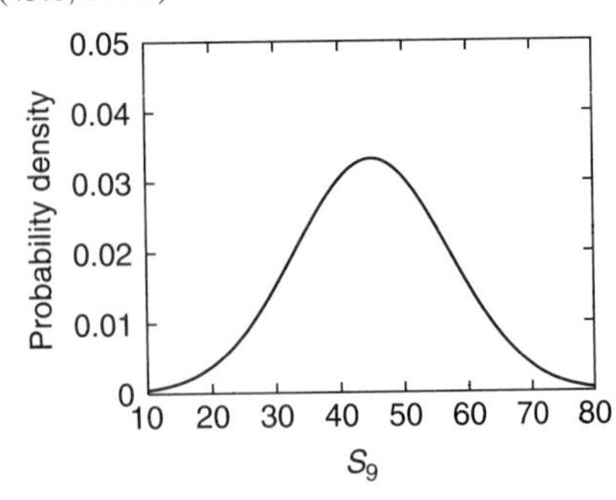

13. The p.d.f. is

$$f(x) = \frac{1}{\sqrt{2\pi}} e^{-\frac{x^2}{2}}.$$

Taking the derivative with respect to x, we find

$$\frac{df}{dx} = -\frac{x}{\sqrt{2\pi}} e^{-\frac{x^2}{2}},$$

which is 0 only at $x = 0$. Furthermore, $f(x)$ approaches 0 at $x = -\infty$ and $x = \infty$ because the power in the exponent approaches negative infinity. Therefore, $x = 0$ must be a maximum.

15. The second derivative is

$$\frac{d^2 f}{dx^2} = \frac{-1}{\sqrt{2\pi}} e^{-\frac{x^2}{2}} + \frac{x^2}{\sqrt{2\pi}} e^{-\frac{x^2}{2}}$$
$$= (x^2 - 1) \frac{1}{\sqrt{2\pi}} e^{-\frac{x^2}{2}}$$

which is 0 where $x^2 = 1$, or $x = -1$ and $x = 1$.

17. If the second event occurs at time t, the first had to occur at some time s before that, and the second wait had to make up the difference of $t - s$ exactly.

19.

$$f_3(t) = \int_{s=0}^{t} f_2(s) f_1(t-s) ds = \int_{s=0}^{t} s e^{-s} e^{-t+s} ds$$
$$= \int_{s=0}^{t} s e^{-t} ds = \frac{s^2}{2} e^{-t} \Big|_0^t = \frac{t^2}{2} e^{-t}.$$

21. For these Bernoulli random variables, $E(X_i) = 0.5$ and $Var(X_i) = 0.25$. But if all the random variables take on the same value (so $X_2 = X_1$, $X_3 = X_1$, and so forth), $S_{16} = 16X_1$. Then $E(S_{16}) = 16 \cdot E(X_1) = 16 \cdot 0.5 = 8$, and $Var(S_{16}) = Var(16X_1) = 256 \cdot 0.25 = 64$. More explicitly, $S_{16} = 0$ with probability 0.5 and $S_{16} = 16$ with probability 0.5. The result has a much higher variance and bears no resemblance to a normal distribution.

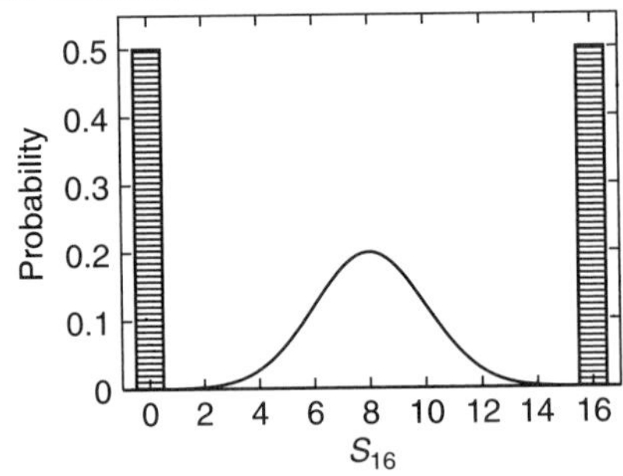

23. Five 1's and five 0's, or two 2.5's and eight 0's.

25. The probability of the five 1's and five 0's is $252 \cdot 0.5^5 \times 0.25^5 = 0.0077$. The probability of two 2.5's and eight 0's is $45 \cdot 0.25^2 \times 0.25^8 = 0.0004$.

27. $N(11.25, 7.97)$. The value of the normal is $f(5.0; 11.25, 7.97) = 0.012$.

29. Each locus gives a mean of 1.875 points, with variance of 1.172. The normal approximation is $IQ \sim N(98.75, 11.72)$.

31. Each factor gives a mean of 0.45 points, with variance of 0.203. The normal approximation is $IQ \sim N(93.5, 6.09)$.

33. Let I_i be a random variable giving the number of immigrants in the ith year. We found that $E(I_i) = 0.1$ and can compute that $Var(I_i) = 1.09$. Then $A_{20} = \frac{1}{20} \sum_{i=0}^{i=20} I_i$ has mean $E(A_{20}) = 0.1$ and variance $\sigma^2 = \frac{1.09}{20} = 0.0545$. Therefore, $\sigma = 0.233$, and the p.d.f. approximating A_{20} is

$$f(x) = \frac{1}{0.233\sqrt{2\pi}} e^{-\frac{(x-0.1)^2}{1.09}}.$$

The maximum possible change is 2. The area looks like about 0.6 of the total.

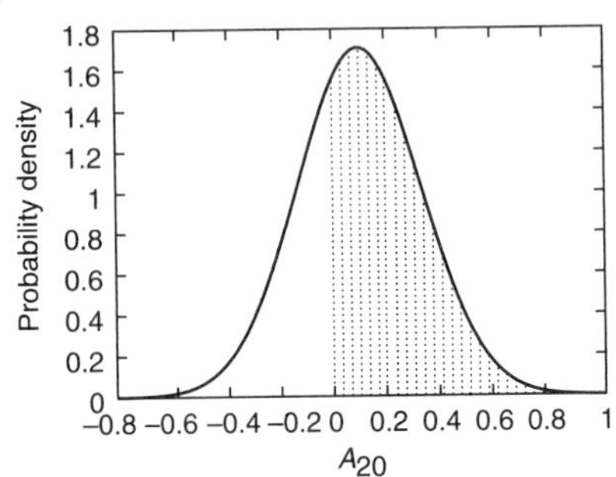

35. Let I_i be a random variable giving the number of immigrants in the ith year. We found that $\mathrm{E}(I_i) = 9.9$ and can compute that $\mathrm{Var}(I_i) = 902.69$. Then $A_{10} = \frac{1}{10}\sum_{i=0}^{i=10} I_i$ has mean $\mathrm{E}(A_{10}) = 9.9$ and variance $\sigma^2 = \frac{902.69}{10} = 90.269$. Therefore, $\sigma = 9.501$, and the p.d.f. approximating A_{10} is

$$f(x) = \frac{1}{9.501\sqrt{2\pi}} e^{-\frac{(x-9.9)^2}{180.538}}.$$

The area looks like about 0.8. The maximum possible average decrease is -1, but this is only about 1 standard deviation below the mean. At least this part of the normal approximation is not very accurate.

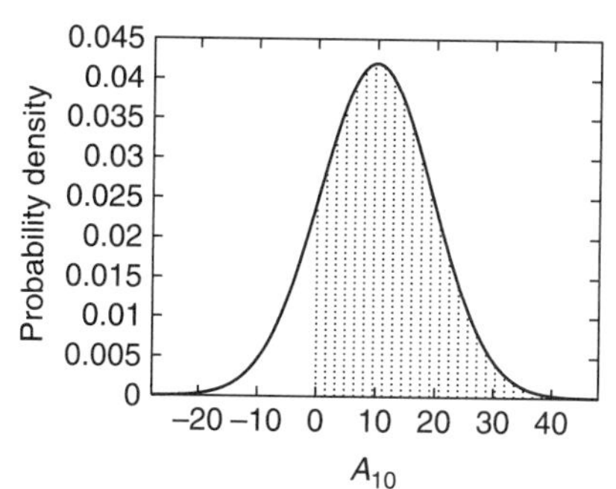

37. We have that $P_{50} = R_1 \cdot R_2 \cdots R_{50} P_0$, so

$$\ln(P_{50}) = \ln(100) + \sum_{i=1}^{50} \ln(R_i).$$

By the Central Limit Theorem for sums, we can find the normal distribution that approximates $\ln(P_{50})$ from the mean and variance of $\ln(R_i)$. But $\mathrm{E}[\ln(R_i)] = \ln(4) \cdot 0.5 + \ln(0.25) \cdot 0.5 = 0$ and $\mathrm{Var}[\ln(R_i)] = \ln(4)^2 \cdot 0.5 + \ln(0.25)^2 \cdot 0.5 = 1.92$. After adding on the $\ln(100) = 4.6$ to the mean, we get that $\ln(P_{50}) \sim N(4.6 + 50 \cdot 0, 1.92 \cdot 50) = N(4.6, 96)$. The standard deviation is 9.8, so $\ln(P_{50})$ will lie between $4.6 + 2 \cdot 9.8 = 24.2$ and $4.6 - 2 \cdot 9.8 = -15.0$ about 95% of the time. Exponentiating, the population itself will lie between $e^{24.2} = 3.24 \times 10^{10}$ and $e^{-15.0} = 3.06 \times 10^{-7}$ about 95% of the time. Even though it doesn't change on average, there is a huge range of possible populations.

39. The log population is approximately

$$\log(P_{50}) \sim N\{50\mathrm{E}[\ln(R)], 50\mathrm{Var}[\ln(R)]\}$$

But

$$\mathrm{E}[\ln(R)] = \int_{1.0}^{1.2} 5.0\ln(x)dx = 5.0[x\ln(x) - x]|_{1.0}^{1.2} = 0.0939$$

$$\mathrm{Var}[\ln(R)] = \int_{1.0}^{1.2} 5.0\ln(x)^2 dx - 0.0939^2$$

$$= 5.0[x\ln(x)^2 - 2x\ln(x) + 2x]|_{1.0}^{1.2} - 0.0939^2 = 0.0028.$$

Therefore,

$$\ln[P_{50}(t)] \sim N(50 \cdot 0.0939, 50 \cdot 0.0028) = N(4.695, 0.140).$$

The standard deviation is $\sqrt{0.140} = 0.37$, so the log population size is likely to lie between $4.695 + 2 \cdot 0.37 = 5.44$ and $4.695 - 2 \cdot 0.37 = 3.96$, and the actual population size is likely to lie between $e^{5.44} = 230$ and $e^{3.96} = 52$. The simulations in Example 6.1.2 end up at about 90 and 190, right in this range.

Section 7.9, page 699

1. The z-score is $\frac{11.0 - 13.0}{1.2} = -1.67$.

3. The standard deviation is $\sqrt{25.0} = 5.0$, so the z-score is $\frac{12.0 - 10.0}{5.0} = 0.4$.

5. Because the normal distribution with mean 0 and variance 1 is the standard normal, this probability is $\Phi(0.7)$ by definition, or 0.7508.

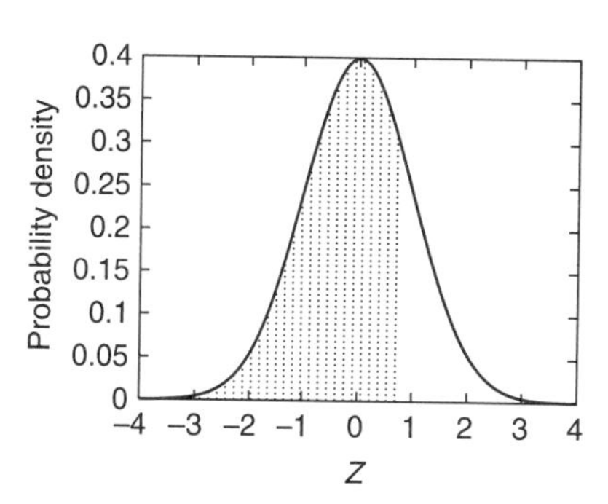

7. We found the z-score to be -1.67. Then $\Phi(-1.67)$ is the probability of a measurement less than 11.0 and $1 - \Phi(-1.67) = 1 - 0.0475 = 0.9525$ is the probability of a value greater than 11.0.

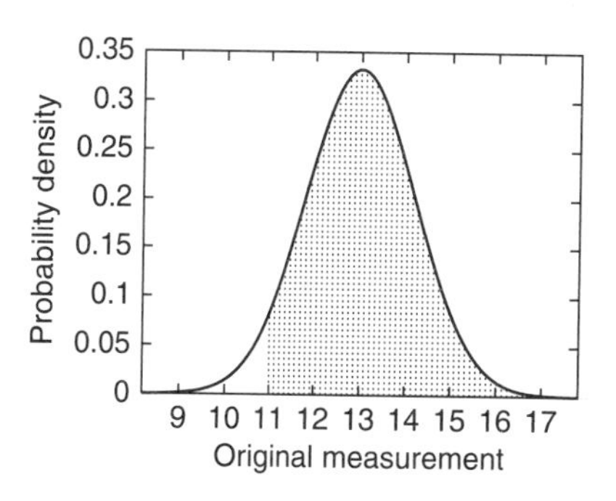

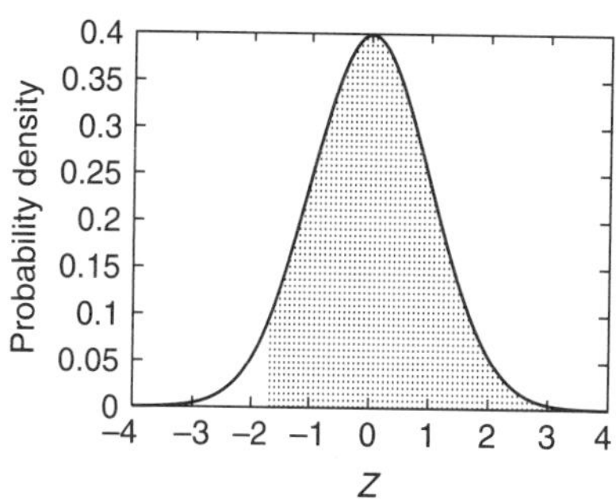

9. We found the z-score associated with 12.0 to be 0.4. The z-score for the value 10.0 is 0 because 10.0 is the mean. The probability is then $\Phi(0.4) - \Phi(0.0) = 0.6554 - 0.5 = 0.1554$.

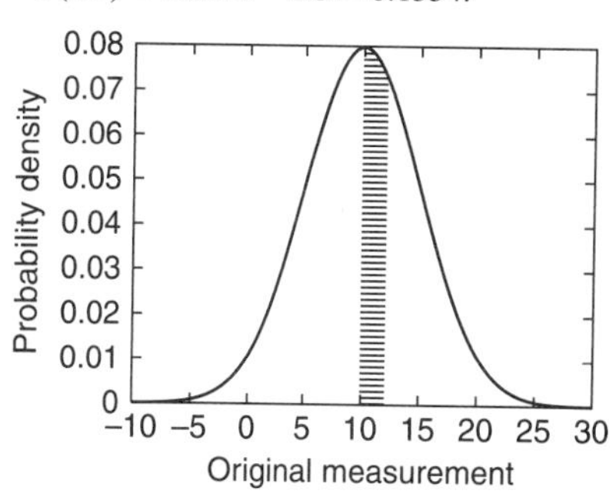

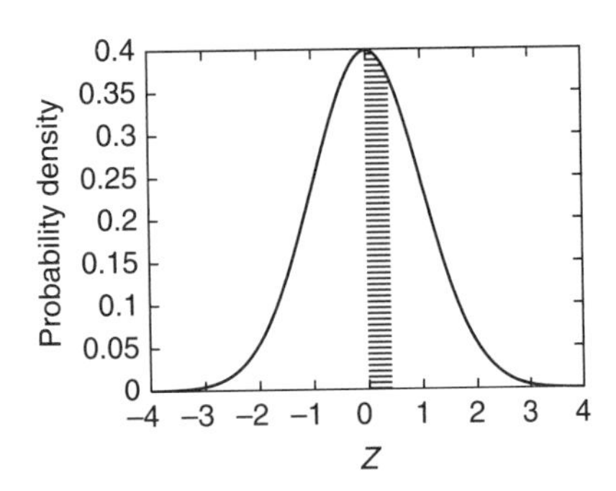

11. Let M represent the mass. The z-score is

$$z = \frac{0.40 - 0.38}{0.09} = 0.222.$$

Then

$$\Pr(M \le 0.4) = \Pr(Z \le 0.222) = \Phi(0.222) = 0.588.$$

13. Let E represent the error. The z-scores are

$$z_1 = \frac{-0.012 - 0}{0.01} = -1.2, \quad z_2 = \frac{0.012 - 0}{0.01} = 1.2.$$

Then

$$\begin{aligned}\Pr(-0.012 \le E \le 0.012) &= \Pr(-1.2 \le Z \le 1.2) = \Phi(1.2) \\ &\quad - \Phi(-1.2) \\ &= 0.8849 - 0.1151 = 0.7698.\end{aligned}$$

15. $n = 40$ and $p = 0.43$. The mean is $np = 17.2$, and the variance is $np(1-p) = 9.804$. The standard deviation is $\sqrt{9.804} = 3.13$. Suppose $X \sim N(17.2, 9.804)$. With the continuity correction,

$$\begin{aligned}\Pr(T \le 10) &= \Pr(X \le 10.5) = \Phi\left(\frac{10.5 - 17.2}{3.13}\right) \\ &= \Phi(-2.14) = 0.0162.\end{aligned}$$

Without the continuity correction,

$$\Pr(T \le 10) = \Pr(X \le 10) = \Phi\left(\frac{10 - 17.2}{3.13}\right) = \Phi(-2.30) = 0.011.$$

Neither is extremely close in this case.

17. $n = 100$ and $p = 0.75$. The mean is $np = 75.0$, and the variance is $np(1-p) = 18.75$. The standard deviation is $\sqrt{18.75} = 4.33$. Suppose $X \sim N(75.0, 18.75)$. With the continuity correction,

$$\begin{aligned}\Pr(70 \le F \le 80) &\approx \Pr(69.5 \le X \le 80.5) \\ &= \Phi\left(\frac{69.5 - 75.0}{4.33}\right) - \Phi\left(\frac{80.5 - 75.0}{4.33}\right) \\ &= \Phi(1.27) - \Phi(-1.27) = 0.7960.\end{aligned}$$

Without the continuity correction,

$$\begin{aligned}\Pr(70 \le F \le 80) &\approx \Pr(70 \le X \le 80) \\ &= \Phi\left(\frac{70 - 75.0}{4.33}\right) - \Phi\left(\frac{80 - 75.0}{4.33}\right) \\ &= \Phi(1.15) - \Phi(-1.15) = 0.7499.\end{aligned}$$

The answer with the continuity correction is very close to the exact answer.

19. Let S represent the number of seeds that fall, following a Poisson distribution with $\Lambda = 20$. The mean is 20, and the standard deviation is $\sqrt{20} = 4.47$. Let X be a normally distributed random variable with matching mean and standard deviation. With the continuity correction,

$$\begin{aligned}\Pr(S \le 15) \approx \Pr(X \le 15.5) &= \Phi\left(\frac{15.5 - 20.0}{4.47}\right) \\ &= \Phi(-1.007) = 0.1570.\end{aligned}$$

Without the continuity correction,

$$\begin{aligned}\Pr(S \le 15) \approx \Pr(X \le 15) &= \Phi\left(\frac{15 - 20.0}{4.47}\right) \\ &= \Phi(-1.119) = 0.1316.\end{aligned}$$

The answer with the continuity correction is much closer.

21. Let I represent the number of insects caught in an hour, with $\Lambda = 0.21 \cdot 60 = 12.6$. The mean is 12.6, and the standard deviation is $\sqrt{12.6} = 3.55$. Let X be a normally distributed random variable with matching mean and standard deviation. With the continuity correction,

$$\begin{aligned}\Pr(10 \le I \le 15) &\approx \Pr(9.5 \le X \le 15.5) \\ &= \Phi\left(\frac{9.5 - 12.6}{3.55}\right) - \Phi\left(\frac{15.5 - 12.6}{3.55}\right) \\ &= \Phi(0.817) - \Phi(-0.873) = 0.6014.\end{aligned}$$

Without the continuity correction,

$$\begin{aligned}\Pr(10 \le I \le 15) &\approx \Pr(10 \le X \le 15) \\ &= \Phi\left(\frac{10 - 12.6}{3.55}\right) - \Phi\left(\frac{15 - 12.6}{3.55}\right) \\ &= \Phi(0.676) - \Phi(-0.732) = 0.5184.\end{aligned}$$

The answer with the continuity correction is much closer.

23. The normal approximation cannot be used because $np = 4 < 5$. The probability of no more than one head is $b(0; 8, 0.5) + b(1; 8, 0.5) = 0.0352$. The mean is $np = 4$, the variance is $np(1-p) = 2.0$, and the standard deviation is $\sqrt{2.0} = 1.414$. Suppose $X \sim N(4.0, 2.0)$. With the continuity correction,

$$\Pr(H \le 1) \approx \Pr(X \le 1.5) = \Phi\left(\frac{1.5 - 4}{1.414}\right) = \Phi(-1.768) = 0.0385.$$

This turns out to be pretty close.

25. In 1 min the expected number is only 0.21, far less than 5. The exact probability of more than 1 is $1 - p(0; 0.21) - p(1; 0.21) = 0.192$. With the normal approximation, $X \sim N(0.21, 0.21)$. The probability of more than 1 is

$$\Pr\left(Z \ge \frac{1.5 - 0.21}{\sqrt{0.21}}\right) = 1 - \Phi(2.815) = 0.0024,$$

which is way off.

27.

$$\begin{aligned}E(Z) = E\left(\frac{X - \mu}{\sigma}\right) &= E\left(\frac{X}{\sigma}\right) - E\left(\frac{\mu}{\sigma}\right) = \frac{E(X)}{\sigma} - \frac{\mu}{\sigma} \\ &= \frac{\mu}{\sigma} - \frac{\mu}{\sigma} = 0.\end{aligned}$$

We used Theorem 7.3.3 and the fact that the expectation of a constant is the constant itself.

29. We found that $S_{20} \sim N(42, 25.8)$. The standard deviation is $\sqrt{25.8} = 5.08$. Then

$$\begin{aligned}\Pr(30 \le S_{20} \le 40) &\approx \Phi\left(\frac{40 - 42}{5.08}\right) - \Phi\left(\frac{30 - 42}{5.08}\right) \\ &= \Phi(-0.394) - \Phi(-2.36) = 0.338.\end{aligned}$$

This is pretty close to the estimate of 0.4 found earlier.

31. We found that $A_{100} \sim N(1.6, 0.0184)$. The standard deviation is $\sqrt{0.0184} = 0.136$. Then

$$\Pr(A_{100} \leq 1.8) \approx \Phi\left(\frac{1.8 - 1.6}{0.136}\right) = \Phi(1.47) = 0.929.$$

This is reasonably close to the estimate of 0.9 found earlier.

33. We found that $A_{20} \sim N(0.1, 0.0545)$, with a standard deviation of $\sigma = 0.233$. Then

$$\Pr(A_{20} \geq 0) \approx 1 - \Phi\left(\frac{0 - 0.1}{0.233}\right) = 1 - \Phi(-0.429) = 0.666.$$

After 40 years, $A_{40} \sim N(0.2, 0.109)$, with a standard deviation of $\sigma = \sqrt{0.109} = 0.330$. Then

$$\Pr(A_{40} \geq 0) \approx 1 - \Phi(\frac{0 - 0.2}{0.330}) = 1 - \Phi(-0.606) = 0.728.$$

This population grows on average, and is thus more likely to grow if given more time.

35.

$$\Pr(\text{IQ} \geq 100) = 1 - \Phi\left(\frac{100.0 - 98.75}{\sqrt{11.72}}\right)$$
$$= 1 - \Phi(0.365) = 1 - 0.6425 = 0.3576.$$

37.

$$\Pr(\text{IQ} \geq 100) = 1 - \Phi\left(\frac{100.0 - 93.5}{\sqrt{6.09}}\right)$$
$$= 1 - \Phi(2.634) = 1 - 0.9958 = 0.0042.$$

39. We found that $\ln(P_{50}) \sim N(4.6, 96)$ with a standard deviation of 9.8. We want to find the probability that $\ln(1) \leq \ln(P_{50}) \leq \ln(10000)$ or $0 \leq \ln(P_{50}) \leq 9.21$.

$$\Pr(0 \leq \ln(P_{50}) \leq 9.21) \approx \Phi\left(\frac{0 - 4.6}{9.8}\right) - \Phi\left(\frac{9.21 - 4.6}{9.8}\right)$$
$$= \Phi(0.470) - \Phi(-0.469) = 0.362.$$

41. We found that $\ln(P_{50}) \sim N(4.695, 0.140)$ with a standard deviation of 0.37. We want to find the probability that $\ln(50) \leq \ln(P_{50}) \leq \ln(150)$ or $3.912 \leq \ln(P_{50}) \leq 5.011$.

$$\Pr(3.912 \leq \ln(P_{50}) \leq 5.011) \approx \Phi\left(\frac{3.912 - 4.695}{0.37}\right) - \Phi\left(\frac{5.011 - 4.695}{0.37}\right)$$
$$= \Phi(0.854) - \Phi(-2.116) = 0.786.$$

43. The binomial has $n = 36$ and $p = 0.25$. The mean is then $np = 9$, the variance is $np(1 - p) = 6.75$, and the standard deviation is 2.598. The number $T \approx X \sim N(9.0, 6.75)$. With the continuity correction,

$$\Pr(T \leq 10) \approx \Pr(X \leq 10.5)$$
$$= \Pr[(X - 9)/2.598 \leq (10.5 - 9)/2.598]$$
$$= \Pr(Z \leq 0.577) = \Phi(0.577) = 0.718.$$

45. The binomial has $n = 50$ and $p = 0.8$. The mean is then $np = 40$, the variance is $np(1 - p) = 8$, and the standard deviation is 2.82. The number $M \approx X \sim N(40.0, 8.0)$. With the continuity correction,

$$\Pr(M \geq 40) \approx \Pr(X \geq 39.5)$$
$$= \Pr[(X - 40)/2.82 \geq (39.5 - 40)/2.82]$$
$$= \Pr(Z \geq -0.177) = 1 - \Phi(-0.177) = 0.570.$$

47. The mean number added per generation is 1.0, and the variance is also 1.0. After 100 generations, the mean is 100.0, the variance is 100.0, and the standard deviation is 10.0. The number $I \approx X \sim N(100.0, 100.0)$.

$$\Pr(96 \leq I \leq 106) = \Pr\left(\frac{96 - 100}{10} \leq \frac{I - 100}{10} \leq \frac{106 - 100}{10}\right)$$
$$= \Pr(-0.4 \leq Z \leq 0.6)$$
$$= \Phi(0.6) - \Phi(-0.4) = 0.7257 - 0.3446 = 0.3811.$$

49. The number of mutations will follow a Poisson distribution with parameter $\Lambda = 0.02 \cdot 2000 = 40.0$. The approximating normal is $X \sim N(40.0, 40.0)$ with a standard deviation of 6.32. Then

$$\Pr(M \geq 50) \approx \Pr(X \geq 49.5)$$
$$= \Pr[(X - 40)/6.32 \geq (49.5 - 40)/6.32]$$
$$= \Pr(Z \geq 1.503) = 1 - \Phi(1.503) = 0.0664.$$

Chapter 8

Section 8.1, page 715

1. This is terrible, because it samples mainly the winter (or the summer in Australia).

3. This is terrible, because March 15 is not representative of the whole year.

5. The true average is $40,000 · 0.2 + $20,000 · 0.8 = $24,000.

7. $N(40000, 2.0 \times 10^5)$ by the Central Limit Theorem for Averages (the variance is divided by $n = 20$). The standard deviation is $447, so the range is from about $39,100 to $40,900. If 0.2 of the people average 40,000 and 0.8 of the people average 20,000, the overall average is $0.2 \cdot 40{,}000 + 0.8 \cdot 20{,}000 = 24{,}000$, which is far outside the range resulting from people with cellular phones.

9. $L(p) = b(2; 4, p) = 6p^2(1 - p)^2$, and $L(0.5) = 0.375$.

11. $L(p) = b(2; 12, p) = 66p^2(1 - p)^{10}$. For a fair coin, $p = 0.5$ and $L(0.5) = 0.016$.

13. The probability distribution for H follows the binomial distribution, with $b(0; 4, 0.5) = 0.0625$, $b(1; 4, 0.5) = 0.25$, $b(2; 4, 0.5) = 0.375$, $b(3; 4, 0.5) = 0.25$, $b(4; 4, 0.5) = 0.0625$. The probability of two heads with a fair coin matches the likelihood of $p = 0.5$ when there are two heads.

15. $L(p) = b(5; 6, p) = 6p^5(1 - p)$. The derivative is

$$L'(p) = 30p^4(1 - p) - 6p^5 = 6p^4[5(1 - p) - p]$$

which is 0 at $p = 0$ and $5(1 - p) + p = 6p - 5 = 0$ or $p = 5/6$. The maximum cannot occur at $p = 0$ because the likelihood there is 0 (or at the other endpoint $p = 1$ for the same reason), so the maximum likelihood estimator of the probability is 5/6. If this is the real probability that team A wins, you win $1 with probability 5/6 and lose $6 with probability 1/6, for expected winnings of $1 \cdot 5/6 - 6 \cdot 1/6 = -1/6$. The bet would lose money on average.

17. $L(\Lambda) = p(20; \Lambda) = \frac{e^{-\Lambda}\Lambda^{20}}{20!}$. Then

$$L'(\Lambda) = \frac{1}{20!}\left(-e^{-\Lambda}\Lambda^{20} + 20e^{-\Lambda}\Lambda^{19}\right)$$
$$= \frac{e^{-\Lambda}\Lambda^{19}}{20!}(-\Lambda + 20)$$

which is equal to 0 when $\Lambda = 0$ and when $\Lambda = 20.0$. Because $L(0) = 0$, $\Lambda = 20.0$ is the maximum likelihood estimator. Then $L(20.0) = 0.089$ and $L(10.0) = 0.0019$. The likelihood of $\Lambda = 20.0$ is higher.

19. The likelihood of all three measurements is the product of the likelihoods of each, so

$$\begin{aligned} L(\Lambda) &= p(20; \Lambda)p(16; \Lambda)p(21; \Lambda) \\ &= \frac{e^{-\Lambda}\Lambda^{20}}{20!}\frac{e^{-\Lambda}\Lambda^{16}}{16!}\frac{e^{-\Lambda}\Lambda^{21}}{21!} \\ &= \frac{e^{-3\Lambda}\Lambda^{57}}{20!16!21!}. \end{aligned}$$

Then

$$\begin{aligned} L'(\Lambda) &= \frac{1}{20!16!21!}\left(-3e^{-3\Lambda}\Lambda^{57} + 57e^{-3\Lambda}\Lambda^{56}\right) \\ &= \frac{e^{-3\Lambda}\Lambda^{56}}{20!16!21!}(-3\Lambda + 57) \end{aligned}$$

which is equal to 0 when $\Lambda = 0$ and when $\Lambda = 19.0$. Because $L(0) = 0$, $\Lambda = 19.0$ is the maximum likelihood estimator. Then $L(19.0) = 0.000524$ and $L(20.0) = 0.000485$. Although both are tiny, the likelihood of $\Lambda = 19.0$ is higher.

21. The likelihood function is $L(q) = q(1-q)^{12}$, with derivative

$$L'(q) = (1-q)^{12} - 12q(1-q)^{11} = (1-q)^{11}(1-q-12q).$$

This is 0 at $q = 1$, where the likelihood is 0, and at $q = 1/13$. $L(1/13) = 0.0294$ and $L(0.1) = 0.0282$, which is slightly smaller.

23. The likelihood function is the product of the likelihoods for each experiment, so

$$L(q) = q(1-q)^{12}q(1-q)^{7}q(1-q)^{11} = q^3(1-q)^{30}$$

with derivative

$$L'(q) = 3q^2(1-q)^{30} - 30q^3(1-q)^{29} = q^2(1-q)^{29}(3-33q).$$

This is 0 at $q = 0$ and $q = 1$, where the likelihood is 0, and at $q = 1/11$. $L(1/11) = 0.0000431$ and $L(0.1) = 0.0000424$, which is slightly smaller.

25. If the true proportion were q, the waiting time would be n with probability $q(1-q)^n$. In this case, we estimate $\hat{q} = \frac{1}{n}$. We would need to show

$$q = \sum_{n=1}^{\infty} \frac{1}{n} q(1-q)^n.$$

27. The likelihood function is $L(p) = 2p(1-p)$.

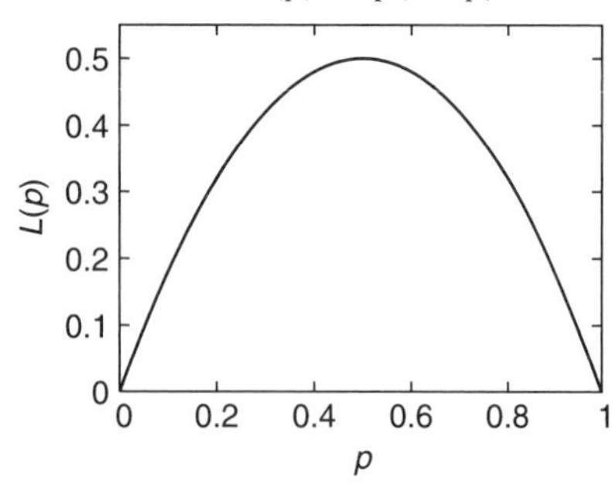

The maximum is at $p = 0.5$. The likelihood is 0 if $p = 0$ or $p = 1$ because it is impossible to get 1 out of 2 if either nobody has the allele ($p = 0$) or everybody has the allele ($p = 1$).

29. We estimate that 40% of women and 20% of men are infected and estimate a proportion of $\hat{p} = 0.4 \cdot 0.5 + 0.2 \cdot 0.5 = 0.3$ in the total population (by the law of total probability). This is what we would get if we ignored the sex of individuals in the sample.

31. We estimate that 100% of women and 12.5% of men are infected and estimate a proportion of $1.0 \cdot 0.5 + 0.125 \cdot 0.5 = 0.5625$ in the total population.

33. The likelihood function is $L(q) = q(1-q)^7$, with maximum at $q = 1/8$. Then $L(1/8) = 0.049$ and $L(0.5) = 0.004$, which looks a lot smaller.

35. The likelihood function is $L(\lambda) = \lambda^{10}e^{-12.307\lambda}$. The maximum is at $\lambda = 0.813$. The likelihood is 5.70×10^{-6}. The likelihood at $\lambda = 1.0$ is 4.52×10^{-6}, which is pretty close. This seems perfectly plausible.

37. $L(\Lambda) = \frac{\Lambda^{14}e^{-\Lambda}}{14!}$. Taking the derivative, we get

$$L'(\Lambda) = \frac{14\Lambda^{13}e^{-\Lambda} - \Lambda^{14}e^{-\Lambda}}{14!} = \frac{\Lambda^{13}e^{-\Lambda}}{14!}(14 - \Lambda).$$

This is 0 when $\Lambda = 0$ or $\Lambda = 14$. The maximum is at $\Lambda = 14$.

39. The likelihood is the product, or

$$\begin{aligned} L(\Lambda) &= \frac{\Lambda^{14}e^{-\Lambda}}{14!}\frac{\Lambda^{17}e^{-\Lambda}}{17!} \\ &= \frac{\Lambda^{31}e^{-2\Lambda}}{14!17!}. \end{aligned}$$

Taking the derivative, we get

$$L'(\Lambda) = \frac{31\Lambda^{30}e^{-2\Lambda} - 2\Lambda^{31}e^{-2\Lambda}}{14!17!} = \frac{\Lambda^{30}e^{-2\Lambda}}{14!17!}(31 - 2\Lambda).$$

This is 0 when $\Lambda = 0$ or $\Lambda = 15.5$. The maximum is at $\Lambda = 15.5$ which is the average of the estimated expected numbers in the first and second pieces.

41. In the set of 100 synonymous sites the number of mutations follows a Poisson distribution with parameter $\Lambda_s = 100\lambda_s$. The likelihood function is

$$L_s(\lambda_s) = \frac{e^{-100\lambda_s}(100\lambda_s)^{15}}{15!} = \frac{100^{15}}{15!}e^{-100\lambda_s}\lambda_s^{15}$$

which has a maximum at $\lambda_s = 0.15$.

43. There are 300 sites, so the number of mutations follows a Poisson distribution with parameter $\Lambda = 300\lambda$. There are a total of 27 mutations, so the likelihood function is

$$L(\lambda) = \frac{e^{-300\lambda}(300\lambda)^{27}}{27!} = \frac{300^{27}}{27!}e^{-300\lambda}\lambda^{27}$$

which has a maximum at $\lambda = 0.09$. This value lies between those for λ_s and λ_n, but closer to λ_n because there are more nonsynonymous sites.

45. The likelihood function for the males is

$$L_m(p) = \binom{1000}{90}p^{90}(1-p)^{910}.$$

The maximum occurs at $p = 90/1000 = 0.09$.

47. The total likelihood is the product, or

$$\begin{aligned} L(p) &= L_m(p)L_f(p) \\ &= \binom{1000}{90}\binom{1000}{13}p^{90}(1-p)^{910}(p^2)^{13}(1-p^2)^{987}. \end{aligned}$$

Section 8.2, page 727

1. The probability of a result as extreme as or more extreme is the probability of 5 heads (a more extreme result is not possible), or $b(5; 5, 0.5) = 0.5^5 = 0.03125$. This is greater than 0.025, so $p = 0.5$ lies within the 95% confidence limits.

3. The probability of a result as extreme as or more extreme is the probability of winning the first or second time, or $g_1 + g_2 = q + q(1-q) = 0.0002$. This is less than 0.005, so lies outside the 99% confidence limits.

5. The probability of a result as extreme as or more extreme is the probability of 0, 1, 2, or 3 cosmic rays. With the rate $\lambda = 10.0$ the number of rays that hits follows a Poisson distribution with $\Lambda = 10.0$. The probability is

$$\begin{aligned} &p(0; 10) + p(1; 10) + p(2; 10) + p(3; 10) \\ &= e^{-10} + 10e^{-10} + \frac{10^2e^{-10}}{2} + \frac{10^3e^{-10}}{3!} = 0.0103. \end{aligned}$$

This is just barely greater than 0.01, so lies just barely within the 98% confidence limits.

7. Because all were successes, the upper confidence limit is $p_h = 1$. With values of $p < 1$, the probability of a result as extreme as or more extreme is the probability of 5 heads (a more extreme result is not possible), or $b(5; 5, p)$ To find the lower confidence limit, we must solve $b(5; 5, p_l) = p_l^5 = 0.025$. This has solution $p_l = 0.478$. The value $p = 0.5$ does lie within the 95% confidence limits.

9. The maximum likelihood estimator of q is 0.5. For values of q less than 0.5, the probability of a result as extreme as or more extreme is the probability of winning the first or second time, or $g_1 + g_2 = q + q(1-q)$. This is equal to 0.005 when $q + q(1-q) = 0.005$ which has solution $q = 0.0025$. For values of q greater than 0.5, the probability of a result as extreme as or more extreme is winning after the first time, or $1 - q$, which is equal to 0.005 when $q = 0.995$. The value $q = 0.001$ is less than 0.0025, and thus lies outside the 99% confidence limits.

11. The likelihood function is $L(p) = p^5$, so the support is $S(p) = 5\ln(p)$. The maximum occurs at $p = 1$, so the maximum support is 0. The support takes on a value 2 less than this when $S(p) = -2$, or $5\ln(p) = -2$, or 0.67. This value is rather larger than the lower confidence limit found in Exercise 7.

13. The maximum likelihood estimator of q is 0.5, with likelihood function $L(q) = q(1-q)$ and support $S(q) = \ln[q(1-q)]$. At $q = 0.5$, the support is -1.386. The support takes on a value 2 less than this when $S(p) = -3.386$, or $q(1-q) = 0.0338$. Solving with the quadratic formula gives confidence limits of 0.035 and 0.965. These are narrower than the 99% confidence limits found in Exercise 9.

15. I would test computer coins with different values of the parameter p to find the smallest value of p that produced five heads in five tries in more than 2.5% of my trials. For example, I could run a series of tests where I repeated flipping five computer coins 1000 times, with $p = 1.0$, $p = 0.95$, $p = 0.9$, and so forth, and waited for the last value of p that produced more than 25 trials out of 1000 with all heads.

17. I would test computer coins with different values of the parameter q to find the smallest value of q that produced a win on the first or second try with probability greater than 0.005 (to find the lower confidence limit) and the largest value of q that produced a win on the second try or after with probability greater than 0.005 (to find the upper confidence limit).

19. The maximum likelihood estimator of p is 0.5. This has to lie within any confidence limits. We could also check that $L(p) = b(2; 4, p) = 6p^2(1-p)^2$, $S(p) = \ln[6p^2(1-p)^2]$, so $S(0.5) = -0.98$, and note that the support for the value $p = 0.5$ is certainly within 2 of this maximum.

21. $L(p) = b(2; 12, p) = 66p^2(1-p)^{10}$, with maximum at $p = 1/6$. The support is $S(p) = \ln[66p^2(1-p)^{10}]$. Then $S(1/6) = -1.217$ and $S(0.5) = -4.128$, which is more than 2 less. Thus $p = 0.5$ lies outside the approximate 95% confidence limits.

23. We found that $L(20.0) = 0.089$ and $L(10.0) = 0.0019$. Then $S(20.0) = \ln(0.089) = -2.419$ and $S(10.0) = \ln(0.0019) = -6.266$. These differ by more than 2, so the value $\Lambda = 10$ lies outside the approximate 95% confidence limits.

25. Experimenting, we plug various values into the equation

$$f(p) = 20\ln(0.2) + 80\ln(0.8) - 2 - 20\ln(p) - 80\ln(1-p),$$

looking for a value p where $f(p) = 0$. We find that $f(0.10) = 2.44$, $f(0.12) = 0.59$, $f(0.13) = -0.1$. The lower confidence limit is close to 0.13. Also $f(0.30) = 0.57$, $f(0.28) = -0.3$, and $f(0.29) = 0.12$. The upper confidence limit is close to 0.29. To use Newton's method, the Newton's method discrete-time dynamical system is

$$p_{t+1} = p_t - \frac{f(p_t)}{f'(p_t)}.$$

Starting from a guess of $p_0 = 0.15$, we get subsequent estimates of 0.122, 0.128, 0.128. It has converged to the lower confidence limit. Starting from a guess of 0.25, we get estimates 0.299, 0.288, 0.287, 0.287, which has converged to the upper confidence limit.

27. $L(p) = 6p^2(1-p)^2$, so $S(p) = \ln(6) + 2\ln(p) + 2\ln(1-p)$ (using various laws of logs). Then $S'(p) = \frac{2}{p} - \frac{2}{1-p}$ which is equal to 0 at $p = 1/2$. Furthermore, $S''(p) = \frac{-2}{p^2} - \frac{2}{(1-p)^2} < 0$, hence this is a maximum. This matches the location of the maximum of the likelihood function itself.

29. $L(\Lambda) = \frac{e^{-3\Lambda}\Lambda^{57}}{20!16!21!}$, so $S(\Lambda) = 57\ln(\Lambda) - 3\Lambda - \ln(20!16!21!)$ (using various laws of logs). Then $S'(\Lambda) = \frac{57}{\Lambda} - 3$ which is equal to 0 at $\Lambda = 19$. Furthermore, $S''(\Lambda) = \frac{-57}{\Lambda^2} < 0$, hence this is a maximum. This matches the location of the maximum of the likelihood function itself.

31. The probability of one or more out of two is equal to $2p(1-p) + p^2$. The lower confidence limit satisfies

$$2p(1-p) + p^2 = 0.025,$$

which has solution $p_l = 0.0126$ (by the quadratic formula). Similarly,

$$\text{the probability of one or fewer out of two } = 2p(1-p) + (1-p)^2.$$

The upper confidence limit satisfies

$$2p(1-p) + (1-p)^2 = 0.025,$$

which has solution $p_h = 0.987$.

33. To find p_l, I would try the experiment 1000 times with small values of p, seeking one so small that a result of one or more out of two occurred only five times. To find p_h, I would try the experiment ten times with large values of p, seeking one so large that a result of one or fewer out of two occurred only five times.

35. The probability of three or more out of three $= p^3$. The lower confidence limit solves $p^3 = 0.025$, which has solution $p_l = 0.292$. Furthermore,

$$\text{the probability of three or fewer out of three } = 1.$$

This is always true, so the upper confidence limit is $p_h = 1$.

37. The maximum likelihood estimator is the largest possible value.

39. The likelihood function is $L(\lambda) = \lambda e^{-20\lambda}$, which has maximum at $\lambda = 0.05$.

41. The lower confidence limit λ_l satisfies

$$\Pr(\text{wait is less than or equal to 20 if } \lambda = \lambda_l) = 0.01$$
$$1 - e^{-20\lambda_l} = 0.01$$
$$\lambda_l = \frac{\ln(0.99)}{-20} = 0.0005.$$

The upper confidence limit λ_h satisfies

$$\Pr(\text{wait is greater than or equal to 20 if } \lambda = \lambda_h) = 0.01$$
$$e^{-20\lambda_h} = 0.01$$
$$\lambda_h = \frac{\ln(0.01)}{-20} = 0.230.$$

43. The confidence limits Λ_l and Λ_h satisfy

$$\Pr(14 \text{ or more}) = \sum_{k=14}^{\infty} p(k; \Lambda_l) = 0.025$$
$$\Pr(14 \text{ or fewer}) = \sum_{k=0}^{14} p(k; \Lambda_h) = 0.025.$$

By substituting various values into the computer, we find $\Lambda_l = 7.65$ and $\Lambda_h = 23.5$.

45. If q is less than $q = 0.125$, the wait for a girl would be greater than 8 on average. The lower confidence limit q_l satisfies

$$\Pr(\text{wait is less than or equal to 8 if } q = q_l) = 0.025$$
$$1 - (1 - q_l)^7 = 0.025$$
$$q_l = 1 - 0.975^{1/7} = 0.0036.$$

The upper confidence limit q_h satisfies

$$\begin{aligned} \Pr(\text{wait is greater than or equal to 8 if } q = q_h) &= 0.025 \\ (1-q_h)^7 &= 0.025 \\ q_h &= 1 - 0.025^{1/7} \\ &= 0.410. \end{aligned}$$

The normal probability $q = 0.5$ seems to lie outside these confidence limits. Perhaps the father does have some genetic abnormality.

Section 8.3, page 741

1. Adding the weights and dividing by 20 gives $\overline{W} = 67.44$. After sorting, we see that $\tilde{W} = 43.8$ because the 10th and 11th values are both 43.8. Trimming off the highest value of 298.8 and the lowest value of 7.3, adding the remaining 18 values and dividing by 18 gives $\overline{W}_{tr(5)} = 57.92$. Similarly, we find that $\overline{W}_{tr(10)} = 51.17$ and $\overline{W}_{tr(20)} = 42.57$.

3. Adding the yields and dividing by 20 gives $\overline{Y} = 0.442$. After sorting, we see that $\tilde{Y} = 0.304$ because the 10th and 11th values are 0.298 and 0.310 which average to 0.304. Trimming off the highest value of 1.745 and the lowest value of 0.045, adding the remaining 18 values and dividing by 18 gives $\overline{Y}_{tr(5)} = 0.392$. Similarly, we find that $\overline{Y}_{tr(10)} = 0.369$ and $\overline{Y}_{tr(20)} = 0.350$.

5. $s^2 = 5726$ and $s = 75.7$. The standard error is the sample standard deviation divided by $\sqrt{n}$, or $75.7/\sqrt{20} = 16.9$

7. $s^2 = 0.152$ and $s = 0.390$. The standard error is $0.39/\sqrt{20} = 0.087$.

9. Seventeen of the values lie in the range from $\overline{W} - s = -8.26$ to $\overline{W} + s = 143.1$. This is more than the 68% (about 13 or 14) we would expect with a normal distribution. Only the last value, 298.8, lies outside the range from $\overline{W} - 2s = -83.96$ to $\overline{W} + 2s = 218.84$. This is what we would expect with the normal distribution, even though these values are far from normal.

11. Seventeen of the values lie in the range from $\overline{Y} - s = 0.052$ to $\overline{Y} + s = 0.832$. This is more than we expect with a normal distribution. Only the last value, 1.745, lies outside the range from $\overline{Y} - 2s = -0.338$ to $\overline{Y} + 2s = 1.222$. This is what we would expect with the normal distribution.

13. The 95% confidence limits are 1.96 standard errors above and below the mean, or $\overline{W} - 1.96 \cdot 16.9 = 34.3$ and $\overline{W} + 1.96 \cdot 16.9 = 100.6$.

15. $\overline{Y} - 1.96 \cdot 0.087 = 0.27$ and $\overline{Y} + 1.96 \cdot 0.087 = 0.61$.

17. There are $n - 1 = 19$ degrees of freedom, so the 95% confidence limits lies 2.093 standard errors above and below the mean, or $\overline{W} - 2.093 \cdot 16.9 = 32.1$ and $\overline{W} + 2.093 \cdot 16.9 = 102.8$.

19. $\overline{Y} - 2.093 \cdot 0.087 = 0.26$ and $\overline{Y} + 2.093 \cdot 0.087 = 0.62$.

21. To find 98% confidence limits, we need to know where the standard normal c.d.f. crosses 0.01 and 0.99. From a table or computer, we find these to be at -2.326 and 2.326. Then the limits are from $\overline{W} - 2.326s/\sqrt{20} = 28.07$ to $\overline{W} + 2.326s/\sqrt{20} = 106.8$.

23. To find 99.8% confidence limits, we need to know where the standard normal c.d.f. crosses 0.001 and 0.999. From a table or computer, we find these to be at -3.090 and 3.090. Then the limits are from $\overline{Y} - 3.09s/\sqrt{20} = 0.173$ to $\overline{Y} + 3.09s/\sqrt{20} = 0.711$.

25. The mean and variance of an exponentially distributed number with $\lambda = 2.0$ are 0.5 and 0.25. The average of 30 numbers will be approximately normal with mean 0.5 and variance $\frac{0.25}{30} = 0.0083$, or $N(0.5, 0.0083)$.

27. Lower limit is $0.4 - 1.96 \cdot \sqrt{0.0028} = 0.596$. Upper limit is $0.4 + 1.96 \cdot \sqrt{0.0028} = 0.804$. These do not include the true mean of 0.5.

29. $\hat{p} = 0.44$ and $s^2 = 0.44 \cdot 0.56 = 0.246$. Then

$$p_l = 0.44 - 1.96\frac{\sqrt{0.246}}{\sqrt{100}} = 0.44 - 0.097 = 0.343$$

$$p_h = 0.44 + 1.96\frac{\sqrt{0.246}}{\sqrt{100}} = 0.44 + 0.097 = 0.540.$$

31. 99% confidence limits with the normal distribution lie 2.576 standard errors above and below the sample mean. This just about matches the width of the 95% confidence limits using the t distribution when there are 5 degrees of freedom, or 6 measurements.

33. The variance of a Bernoulli random variable is $p(1-p)$, so the variance of each measurement is $0.5(1-0.5) = 0.25$.

35. The variance of the first and last is 0. The mean squared deviation from the mean (0.5) of the others is

$$\frac{(0-0.5)^2 + (1-0.5)^2}{2} = 0.25.$$

The average is $0 \cdot 0.25 + 0.25 \cdot 0.25 + 0.25 \cdot 0.25 + 0 \cdot 0.25 = 0.125$.

37. The sample mean is -0.2. Using the computational formula, we find that the sample variance is

$$s^2 = \frac{\sum_{i=1}^{20} P_i^2 - 20 \cdot (-0.2)^2}{19} = 0.905.$$

The sample standard deviation is $\sqrt{0.905} = 0.951$. These are reasonably close to the true mean of 0.1, variance of 1.09, and standard deviation of 1.044.

39. The sample mean is 5.1, the sample variance is 499.8, and the sample standard deviation is 22.4. These are about half of the true mean of 9.9 and variance of 902.7 because we only got 1 yr with 100 and would have expected 2.

41. The sample mean is -0.2 and the true standard deviation is 1.044. These give 95% confidence limits of

$$\mu_l = -0.2 - 1.96 \cdot 1.044/\sqrt{20} = -0.658,$$
$$\mu_h = -0.2 + 1.96 \cdot 1.044/\sqrt{20} = 0.258.$$

The sample standard deviation is 0.951. These give 95% confidence limits of

$$\mu_l = -0.2 - 1.96 \cdot 0.951/\sqrt{20} = -0.617,$$
$$\mu_h = -0.2 + 1.96 \cdot 0.951/\sqrt{20} = 0.217.$$

The true mean lies within the confidence limits in both cases.

43. The sample mean is 5.1 and the true standard deviation is 30.0. These give 95% confidence limits of

$$\mu_l = 5.1 - 1.96 \cdot 30.0/\sqrt{20} = -8.048,$$
$$\mu_h = 5.1 + 1.96 \cdot 30.0/\sqrt{20} = 18.25.$$

The sample standard deviation is 22.4. These give 95% confidence limits of

$$\mu_l = 5.1 - 1.96 \cdot 22.4/\sqrt{20} = -4.72,$$
$$\mu_h = 5.1 + 1.96 \cdot 22.4/\sqrt{20} = 14.92.$$

The true mean lies within the confidence limits in both cases.

45. $\hat{p} = 0.7$ and $s^2 = 0.7 \cdot 0.3 = 0.21$. Then

$$p_l = 0.7 - 2.576\frac{\sqrt{0.21}}{\sqrt{50}} = 0.533$$

$$p_h = 0.7 + 2.576\frac{\sqrt{0.21}}{\sqrt{50}} = 0.867.$$

47. $\hat{p} = 0.52$ and $s^2 = 0.52 \cdot 0.48 = 0.2496$. Then

$$p_l = 0.52 - 2.576\frac{\sqrt{0.2496}}{\sqrt{100}} = 0.391$$

$$p_h = 0.52 + 2.576\frac{\sqrt{0.2496}}{\sqrt{100}} = 0.649.$$

49.

$$\Lambda_l = 14 - 1.96 \cdot \sqrt{14} = 6.7$$
$$\Lambda_h = 14 + 1.96 \cdot \sqrt{14} = 21.3.$$

These are reasonably close to the values 7.65 and 23.5 found earlier.

Section 8.4, page 750

1. The probability of a result as extreme as or more extreme than what was observed is 0.2, even if the null hypothesis is true. If we reject the null hypothesis, there is the chance we do so erroneously and the probability of a type I error.

3. This is the power of the test, meaning that there is a 0.4 chance of failing to reject the false null hypothesis, thereby committing a type II error.

5. The probability of a result as extreme as or more extreme is the probability of 5 heads or 0 heads, $b(5; 5, 0.5) + b(0; 5, 0.5) = 0.5^5 + 0.5^5 = 0.0625$. This is greater than 0.05, so the result is not significantly different from the null hypothesis.

7. The probability of a result as extreme as or more extreme is the probability of 9 or 10 heads (this is a one-tailed test). $b(9; 10, 0.5) + b(10; 10, 0.5) = 0.011$, which is significant.

9. If there were 8 heads, the p-value is

$$b(8; 10, 0.5) + b(9; 10, 0.5) + b(10; 10, 0.5) = 0.055.$$

This is just barely not significant, so the cutoff level is 9 heads.

11. $b(9; 10, 0.6) + b(10; 10, 0.6) = 0.046$. The power is very low.

13. With a one-tailed test, the probability of a result as extreme as or more extreme is the probability of 0 or 1 cosmic rays. With the rate $\lambda = 5.0$/yr the number of rays that hits follows a Poisson distribution with $\Lambda = 5.0$. The probability is $p(0; 5) + p(1; 5) = e^{-5} + 5e^{-5} = 0.04$, which is significant. With a two-tailed test, the probability of a result as extreme as or more extreme is the probability of 0 or 1 cosmic rays, or 9 or more cosmic rays. The only reasonable way to compute this probability is to subtract all other possibilities (2 through 8) from 1, or $1 - [p(2; 5) + p(3; 5) + p(4; 5) + p(5; 5) + p(6; 5) + p(7; 5) + p(8; 5)] = 0.109$. This is not significant.

15. We need the probability of a wait more than 4000 h if the true mean is 1000. The rate λ associated with an exponential distribution is the reciprocal of the mean, or 0.001. Then the probability that the wait is greater than 4000 h is

$$\int_{4000}^{\infty} 0.001e^{-0.001t}\,dt = -e^{-0.001t}\big|_{4000}^{\infty} = e^{-4} = 0.018.$$

The result is significant.

17. We need the probability of a wait more than 25 if the true mean is 10. The probability q is the reciprocal of the mean wait of 10, and is thus 0.1 under the null hypothesis. The wait is greater than or equal to 25 with probability

$$\sum_{t=25}^{\infty} 0.1 \cdot 0.9^t.$$

The easiest way to compute this probability is by using the cumulative distribution for the geometric distribution, which says that the probability of a wait of 25 or more is equal to the probability of 24 failures in a row or $0.9^{24} = 0.08$. This is not quite significant.

19. If any abnormal cell triggers a positive response, the test will catch all cancers but make many type I errors when cells look a bit odd for other reasons. If only grossly abnormal cells triggered a positive response, the test would give very few false positives, but would miss many less obvious cancers. Deciding where to set the threshold would depend on the risk from that particular cancer, the cost and difficulty of follow-up testing, and the abundance of the cancer.

21. The probability of a result as extreme as or more extreme is the probability of 0, 1, 7, or 8 girls out of 8. Using the binomial distribution, the probability is

$$b(0; 8, 0.5) + b(1; 8, 0.5) + b(7; 8, 0.5) + b(8; 8, 0.5)$$

$$= 1 \cdot 0.5^8 + 8 \cdot 0.5^8 + 8 \cdot 0.5^8 + 0.5^8 = 18 \cdot 0.5^8 = 0.070.$$

This is greater than 0.05, so the result is not significantly different from the null hypothesis.

23. a. Null: $\lambda = 3.5$; alternative: $\lambda > 3.5$.
b. $p(7; 3.5) = 0.0385$.
c. $\sum_{k=7}^{\infty} p(k; 3.5) = 0.065$.
d. The result is not quite significant. The Web might have worked, but it is hard to be sure.

25. With 8 calls, the significance level is 0.0267. This is the cutoff value for a significant result. The power is

$$\Pr(8 \text{ or more if } \Lambda = 4.0) = 0.051$$

$$\Pr(8 \text{ or more if } \Lambda = 7.0) = 0.401$$

$$\Pr(8 \text{ or more if } \Lambda = 10.0) = 0.780.$$

Larger values of Λ are further from the original value of $\Lambda = 3.5$.

27. The null hypothesis is that the cell simply got lucky ($\lambda = 0.005$). The alternative is that the cell was improved ($\lambda < 0.005$). Pr(cell lasts more than 800 h) $= e^{-0.005 \cdot 800} = 0.0183$. This is the significance level, and the result looks significant.

29. This occurs when

$$\Pr(\text{cell lasts more than } t \text{ hr}) = e^{-0.005t} = 0.05$$

which has solution $t = 599$.

31. Let m represent the true mean. Then $\lambda = \frac{1}{m}$ and

$$\Pr(\text{cell lasts more than 599 h}) = e^{-599/m}.$$

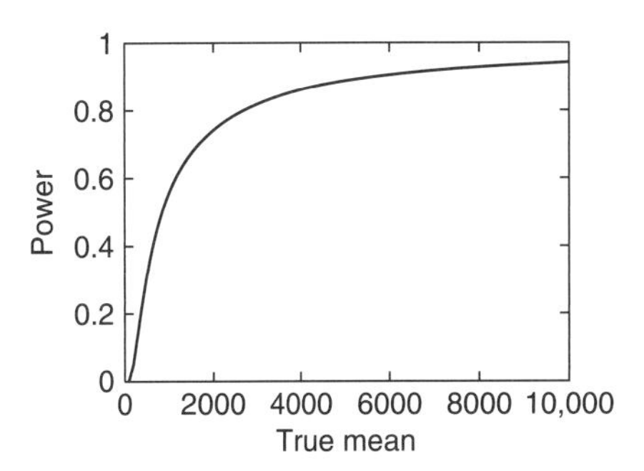

If the true mean is 500 h, $\lambda = 0.002$. Pr(cell lasts more than 599 h) $= e^{-0.002 \cdot 599} = 0.302$. If the true mean is 1000 h, $\lambda = 0.001$. Pr(cell lasts more than 599 h) $= e^{-0.001 \cdot 599} = 0.549$.

33. We need to solve $e^{-599/m} = 0.95$. Taking logs gives

$$\frac{-599}{m} = \log(0.95) = -0.0512$$

$$m = \frac{-599}{-0.0512} = 1.17 \times 10^4.$$

The true mean would have to be very large to be almost sure of detecting an improvement.

35. The 95% confidence limits are 1.96 standard errors above and below the sample mean. Therefore, we require that standard error $= \frac{2.0}{3.92} = 0.51$. But

$$\text{standard error} = \frac{3.2}{\sqrt{n}},$$

so $n = (3.2/0.51)^2 = 39$ and 39 plants are required.

37. Improving by a factor of 4 takes 16 times more plants, or about 624 plants.

Section 8.5, page 760

1. Let ω represent the mean weight of experimental plants. The one-tailed alternative hypothesis is $\omega > 10.0$. The two-tailed alternative hypothesis is $\omega \neq 10.0$. Let W be a random variable representing the mean of ten plants each with mean size 10.0 and variance 9.0 (the null hypothesis). The standard error of W is $\sqrt{9.0}/\sqrt{10} = 0.949$. The sample mean of the numbers in the table is 10.99. For the one-tailed test,

$$\Pr(W > 10.99) = \Pr\left(\frac{W - 10.00}{0.949} > \frac{10.99 - 10.0}{0.949}\right)$$
$$= \Pr(Z > 1.043) = 1 - \Phi(1.043) = 0.148.$$

This is far from significant. For the two-tailed test,

$$\Pr(W > 10.99 \text{ or } W < 9.01) = \Pr\left(\frac{W - 10.00}{0.949} > \frac{10.99 - 10.0}{0.949}\right) + \Pr\left(\frac{W - 10.00}{0.949} < \frac{9.01 - 10.0}{0.949}\right)$$
$$= \Pr(Z < 1.043) + \Pr(Z < -1.043)$$
$$= 1 - \Phi(1.043) + \Phi(-1.043) = 0.297.$$

This is even worse. We have no reason to think our treatment does anything.

3. The sample mean is 10.26. The standard error is $2.5/\sqrt{10} = 0.791$. The difference is $1.26/0.791 = 1.593$ standard errors above the mean. With a one-tailed test, the significance is 0.056 which is not quite significant. With a two-tailed test, it is 0.112.

5. Let ω represent the mean weight of experimental plants. The two-tailed alternative hypothesis is $\omega \neq 10.0$. The sample variance is $s^2 = 10.98$, so the standard error is $\sqrt{10.98/10} = 1.048$. The value of the t statistic is

$$t = \frac{10.99 - 10.00}{1.048} = 0.945.$$

This is nowhere close to the critical value 2.262 of the t distribution with $\nu = 9$ degrees of freedom.

7. The sample mean is 10.26 and the sample variance is 4.268. The standard error is $\sqrt{4.268/10} = 0.653$. The value of the t statistic is

$$t = \frac{10.26 - 9.0}{0.653} = 1.930.$$

This does not exceed the critical value 2.262 of the t distribution with $\nu = 9$ degrees of freedom, so the result is not significant.

9. We are looking for the value $\overline{W}$ such that $\Pr(W > \overline{W}) = 0.01$. This occurs where $\overline{W}$ is 2.326 standard errors above 10.0, or at $10.0 + 2.326 \cdot 0.949 = 12.21$.

11. From a table, we solve where $\Phi(z) = 0.0005$ for $z = -3.29$. The solution is thus 3.29 standard errors above the mean, or 11.60.

13. We need the probability that the mean weight of ten plants exceeds the critical value of 12.21 if the true mean is 13.0. Let X represent the mean weight of ten plants with true mean 10.0. The standard error is still 0.949. So

$$\Pr(X > 12.21) = \Pr\left(\frac{X - 13.0}{0.949} > \frac{12.21 - 13.0}{0.949}\right)$$
$$= \Pr(Z > -0.832)$$
$$= 1 - \Phi(-0.832) = 0.797.$$

We have nearly an 80% chance of finding a difference of 3.0 in the mean height with a one-tailed test at the 0.01 significance level.

15. We need the probability that the mean yield is between 11.60 and 6.4. This is 0.776, for a power of 0.224.

17. The p-value is 0.0017, which is highly significant.

19. $\hat{p} = 0.44$, Under the null hypothesis, $s^2 = 0.5 \cdot 0.5 = 0.25$, $s = 0.5$, and the standard error is $s/\sqrt{n} = 0.5/10 = 0.05$. The value 0.5 is 0.06 away from $\hat{p}$, which is 1.2 standard errors. With a two-tailed test, the p-value is $1 - \Phi(1.2) + \Phi(-1.2) = 0.23$, which is not significant.

21. The mean is 0.936, the sample standard deviation is $s = 1.156$, the standard error is 0.211. The difference from the null hypothesis is 0.064, which is only 0.303 standard errors from the mean. The p-value is $1 - \Phi(0.303) + \Phi(-0.303) = 0.76$. The difference is completely insignificant.

23. The mean is 0.750, the sample standard deviation is $s = 0.555$, the standard error is 0.103. The difference from the null hypothesis is 0.25, which is 2.43 standard errors from the mean. The p-value is $1 - \Phi(2.43) + \Phi(-2.43) = 0.015$. The difference is significant.

25. $\hat{p} = 0.6$, Under the null hypothesis, $s^2 = 0.45 \cdot 0.55 = 0.2475$, $s = 0.497$, and the standard error is $s/\sqrt{n} = 0.497/\sqrt{50} = 0.07$. The value 0.6 is 0.15 away from 45%, which is 2.14 standard errors. With a two-tailed test, the p-value is $1 - \Phi(2.14) + \Phi(-2.14) = 0.032$, which is significant.

27. First, we need to find the minimum number out of 50 that would produce a significant result. This corresponds to a fraction 1.96 standard errors above 0.45, or $0.45 + 1.96 \cdot 0.07 = 0.587$. The power is the probability that a sample of 50 people with true proportion 0.6 produces a proportion of successes greater than 0.587. In this case, $s^2 = 0.24$, $s = 0.49$, the standard error is $0.49/\sqrt{50} = 0.069$. The difference $0.6 - 0.587 = 0.013$ is 0.188 standard errors below the mean. The probability that the result exceeds this value is $1 - \Phi(-0.188) = 0.575$.

29. The null hypothesis has the normal approximation $X \sim N(35.0, 35.0)$. Using the continuity correction and a one-tailed test,

$$\Pr(X \le 27.5) = \Pr\left(Z \le \frac{27.5 - 35}{\sqrt{35}}\right) = \Pr(Z \le -1.268) = 0.102.$$

The difference is not significant.

31. It would have to be at least 1.644 standard deviations below 35, or $35 - 1.644\sqrt{35} = 25.3$. There would have to be 25 or fewer errors.

33. The variance of a single measurement under the null hypothesis is 35.0, so the standard error is $\sqrt{35}/\sqrt{10} = 1.87$. The mean found is 30, which is 5.0 below the mean with the null hypothesis, which is 2.67 standard errors. The associated p-value is $\Phi(-2.67) = 0.0037$, which is highly significant.

35. The null hypothesis is that the log population size is $\ln(P_{50}) \sim N(4.695, 0.135)$. So

$$\Pr(P_{50} \ge 250.0) = \Pr[\ln(P_{50}) \ge \ln(250.0)]$$
$$= \Pr[\ln(P_{50}) \ge 5.521]$$
$$= \Pr\left(Z \ge \frac{5.521 - 4.695}{\sqrt{0.135}}\right) = 1 - \Phi(2.248) = 0.012.$$

The improvements do seem to have helped.

Section 8.6, page 771

1. The distribution of the sample mean from the first sample is $N(\mu_1, 0.5)$ and for the second is $N(\mu_2, 0.5)$, because the standard error is $5/\sqrt{50} = \sqrt{1/2}$ for both populations. The distribution of the difference D is then $D \sim N(\mu_1 - \mu_2, 1.0)$. We want to find the probability that $D \ge 4.0$. Transforming into a standard normal,

$$\Pr(D \ge 4.0) = \Pr\left(\frac{D}{\sqrt{1.0}} \ge \frac{4.0}{\sqrt{1.0}}\right) = \Pr(Z \ge 4.0)$$
$$= 1 - \Phi(4.0) = 0.00003.$$

The difference is very highly significant.

3. The distribution of the sample mean from the first sample is $N(\mu_1, 0.5)$ and for the second is $N(\mu_2, 0.5)$, because the standard error is $\sqrt{250/500} = \sqrt{1/2}$ for both populations. The distribution of the difference D is then $D \sim N(\mu_1 - \mu_2, 1.0)$. We want to find the probability that $D \ge 4.0$. Transforming into a standard normal,

$$\Pr(D \ge 4.0) = \Pr\left(\frac{D}{\sqrt{1.0}} \ge \frac{4.0}{\sqrt{1.0}}\right)$$
$$= \Pr(Z \ge 4.0) = 1 - \Phi(4.0) = 0.00003.$$

The difference is very highly significant, and matches the earlier problem because the samples are ten times larger, which cancels the variances that are ten times larger.

5. Because the two sample sizes are equal, the pooled variance is the ordinary average of the sample variances, or $s_p^2 = 3.0$.

7. Because the two sample sizes are not equal, we must use Equation 8.6.2, finding

$$s_p^2 = \frac{8 \cdot 2.5 + 14 \cdot 3.5}{22} = 3.14.$$

This value is larger than that in Exercise 5 because population 2, with the larger sample variance, also has a larger sample size.

9. The standard error of the difference of the means is

$$s_{\bar{X}_1 - \bar{X}_2} = \sqrt{\frac{3.0}{15} + \frac{3.0}{15}} = 0.632.$$

11. The standard error of the difference of the means is

$$s_{\bar{X}_1 - \bar{X}_2} = \sqrt{\frac{3.14}{9} + \frac{3.14}{15}} = 0.747.$$

13. Using the standard error from Exercise 9,

$$t = \frac{3.2 - 2.1}{0.632} = 1.740.$$

This is less than the critical value 2.048 of the t distribution with $n_1 + n_2 - 2 = 15 + 15 - 2 = 28$ degrees of freedom. The difference is not significant.

15. Using the standard error from Exercise 11,

$$t = \frac{3.8 - 2.1}{0.812} = 2.276.$$

This exceeds the critical value 2.074 of the t distribution with $n_1 + n_2 - 2 = 9 + 15 - 2 = 22$ degrees of freedom, and the difference is significant.

17. The distribution of the sample mean from the experimental plot is $N(\mu_1, 0.9)$ and for the control is $N(\mu_2, 0.9)$ because the variance is $\frac{\sigma^2}{n} = \frac{9.0}{10}$ in both cases. The distribution of the difference D is $D \sim N(\mu_1 - \mu_2, 1.8)$. The difference in sample means is 0.99 (from Section 8.5, Exercise 1), so we must find the probability that $D \geq 0.99$ or $D \leq -0.99$ for a two-tailed test. Transforming into a standard normal distribution, the probability of a result as extreme as or more extreme is

$$\begin{aligned}\Pr(D \geq 0.99 \text{ or } D \leq -0.99) &= \Pr\left(\frac{D}{\sqrt{1.8}} \geq \frac{0.99}{\sqrt{1.8}} \text{ or } \frac{D}{\sqrt{1.8}} \leq \frac{-0.99}{\sqrt{1.8}}\right)\\ &= \Pr(Z \geq 0.738) + \Pr(Z \leq -0.738)\\ &= 1 - \Phi(0.738) + \Phi(-0.738) = 0.46.\end{aligned}$$

The difference is not significant. This p-value is larger than the 0.297 found in Section 8.5, Exercise 1 because of the uncertainty about the true mean of the control population.

19. The distribution of the sample mean from the experimental plot is $N(\mu_1, 0.625)$ and for the control is $N(\mu_2, 0.417)$. The distribution of the difference D is $D \sim N(\mu_1 - \mu_2, 1.042)$. The difference in sample means is 2.06, so we must find the probability that $D \geq 2.06$ or $D \leq -2.06$ for a two-tailed test. Transforming into a standard normal distribution, the probability of a result as extreme as or more extreme is

$$\begin{aligned}&\Pr(D \geq 2.06 \text{ or } D \leq -2.06)\\ &= \Pr\left(\frac{D}{\sqrt{1.042}} \geq \frac{2.06}{\sqrt{1.042}} \text{ or } \frac{D}{\sqrt{1.042}} \leq \frac{-2.06}{\sqrt{1.042}}\right)\\ &= \Pr(Z \geq 2.02) + \Pr(Z \leq -2.02)\\ &= 1 - \Phi(2.02) + \Phi(-2.02) = 0.043.\end{aligned}$$

The difference is significant, because our control population has lower yield than the value used in Section 8.5, Exercise 3.

21. The pooled variance, in this case with equal sample sizes, is the average of the sample variances or

$$s_p^2 = \frac{10.98 + 8.80}{2} = 9.89.$$

Let W_c represent the weights of the control plants. The standard error of the difference in means is

$$s_{\overline{W} - \overline{W}_c} = \sqrt{\frac{9.89}{10} + \frac{9.89}{10}} = 1.41.$$

The observed difference in means is 0.99, so

$$t = \frac{0.99}{1.41} = 0.70.$$

We compare this value with the critical value of the t distribution with $10 + 10 - 2 = 18$ degrees of freedom. This is not even close to the critical value of 2.101 needed for significance, consistent with our earlier result.

23. The sample variance of the experimental plot is 4.268 (Section 8.5, Exercise 7), so the pooled variance is

$$s_p^2 = \frac{9 \cdot 4.268 + 14 \cdot 8.2}{23} = 6.70.$$

Let Y_c represent the yields of the control plants. The standard error of the difference in means is

$$s_{\overline{Y} - \overline{Y}_c} = \sqrt{\frac{6.70}{10} + \frac{6.70}{15}} = 1.06.$$

The observed difference in means is 2.06, so

$$t = \frac{2.06}{1.06} = 1.94.$$

This is less than the critical value 2.069 of the t distribution with $10 + 15 - 2 = 23$ degrees of freedom. The result is not significant, in contrast to our finding in Exercise 19. This is due in part to the uncertainty in the estimate of the sample variance with these small sample sizes, and in part to the large sample variance observed in the control population.

25. The distribution of the sample mean from the first sample is $N(\mu_1, 1.0)$ and for the second is $N(\mu_2, 1.0)$, because the standard error is $\sigma/\sqrt{n} = 1.0$ in both cases. The distribution of the difference D is then $D \sim N(\mu_1 - \mu_2, 2.0)$. We want to find the probability that $D \geq 5.0$ or $D \leq -5.0$. Transforming into a standard normal distribution,

$$\begin{aligned}&\Pr(D \geq 5.0 \text{ or } D \leq -5.0)\\ &= \Pr\left(\frac{D}{\sqrt{2.0}} \geq \frac{5.0}{\sqrt{2.0}} \text{ or } \frac{D}{\sqrt{2.0}} \leq \frac{-5.0}{\sqrt{2.0}}\right)\\ &= \Pr(Z \geq 3.54) + \Pr(Z \leq -3.54)\\ &= 1 - \Phi(3.54) + \Phi(-3.54) = 0.0040.\end{aligned}$$

The difference is very highly significant.

27. The distribution of the sample mean from the first sample is $N(\mu_1, 0.32)$ and for the second is $N(\mu_2, 0.16)$, because the standard error is $4.0/\sqrt{50} = 0.566$ for the first population and $4.0/\sqrt{100} = 0.4$ for the second. The distribution of the difference D is then $D \sim N(\mu_1 - \mu_2, 0.48)$. We want to find the probability that $D \geq 1.0$ or $D \leq -1.0$. Transforming into a standard normal distribution, we get

$$\begin{aligned}&\Pr(D \geq 1.0 \text{ or } D \leq -1.0)\\ &= \Pr\left(\frac{D}{\sqrt{0.48}} \geq \frac{1.0}{\sqrt{0.48}} \text{ or } \frac{D}{\sqrt{0.48}} \leq \frac{-1.0}{\sqrt{0.48}}\right)\\ &= \Pr(Z \geq 1.443) + \Pr(Z \leq -1.443)\\ &= 1 - \Phi(1.443) + \Phi(-1.443) = 0.149.\end{aligned}$$

The difference is not significant.

29. The distribution of the sample proportion from the first sample is $N[p_1, p_1(1-p_1)/n_1]$ and for the second is $N[p_2, p_2(1-p_2)/n_2]$. Using $\hat{p}_1 = 0.7$, $\hat{p}_2 = 0.8$, and $n_1 = n_2 = 50$, we find that the distributions of the sample proportions are $N(p_1, 0.0042)$ and $N(p_2, 0.0032)$. The distribution of the difference D is then $D \sim N(p_1 - p_2, 0.0074)$. We want to find the probability that $D \geq 0.1$ or $D \leq -0.1$. Transforming into a standard normal distribution, we get

$$\begin{aligned}
&\Pr(D \geq 0.1 \text{ or } D \leq -0.1)\\
&= \Pr\left(\frac{D}{\sqrt{0.0074}} \geq \frac{0.1}{\sqrt{0.0074}} \text{ or } \frac{D}{\sqrt{0.0074}} \leq \frac{-0.1}{\sqrt{0.0074}}\right)\\
&= \Pr(Z \geq 1.16) + \Pr(Z \leq -1.16)\\
&= 1 - \Phi(1.16) + \Phi(-1.16) = 0.246.
\end{aligned}$$

The difference is not significant.

31. The distribution of the sample proportion from the first sample is $N[p_1, p_1(1-p_1)/n_1]$ and for the second is $N[p_2, p_2(1-p_2)/n_2]$. Using $\hat{p}_1 = 0.7$, $\hat{p}_2 = 0.8$, $n_1 = 50$, and $n_2 = 500$, we find that the distributions of the sample proportions are $N(p_1, 0.0042)$ and $N(p_2, 0.00032)$. The distribution of the difference D is then $D \sim N(p_1 - p_2, 0.00452)$. We want to find the probability that $D \geq 0.1$ or $D \leq -0.1$. Transforming into a standard normal distribution, we get

$$\begin{aligned}
&\Pr(D \geq 0.1 \text{ or } D \leq -0.1)\\
&= \Pr\left(\frac{D}{\sqrt{0.00452}} \geq \frac{0.1}{\sqrt{0.00452}} \text{ or } \frac{D}{\sqrt{0.00452}} \leq \frac{-0.1}{\sqrt{0.00452}}\right)\\
&= \Pr(Z \geq 1.487) + \Pr(Z \leq -1.487)\\
&= 1 - \Phi(1.487) + \Phi(-1.487) = 0.137.
\end{aligned}$$

The difference is not significant.

33. The null hypothesis is based on the idea that the treatment and control are really the same. Using the pooled proportion treats them together. There is a total of 150 out of 300, so $\hat{p} = 0.5$.

35. The null hypothesis is

$$D \sim N\left(0.0, \frac{\sigma_1^2}{n_1} + \frac{\sigma_2^2}{n_2}\right).$$

As n_1 becomes large, the null hypothesis becomes

$$D \sim N\left(0.0, \frac{\sigma_2^2}{n_2}\right).$$

37. The null hypothesis is that the proteins do not act as transporters, and that the proportion with and without the transporter will be equal. In comparison with the expectation, the null hypothesis is that molecules end up inside with probability 0.5.

39. If the mutation rates are Λ_1 and Λ_2, the numbers M_1 and M_2 are approximately

$$M_1 \sim N(\Lambda_1, \Lambda_1), \quad M_2 \sim N(\Lambda_2, \Lambda_2).$$

The difference D is then $D \sim N(\Lambda_1 - \Lambda_2, \Lambda_1 + \Lambda_2)$. The null hypothesis has $\Lambda_1 = \Lambda_2$. For the sum, we have no better guess than the sum of the estimators $\hat{\Lambda}_1 + \hat{\Lambda}_2 = 8 + 18 = 26$, for a null hypothesis of $D \sim N(0, 26)$. The measured difference is 10, which happens to be exactly 1.96 standard deviations. With a two-tailed test, the significance level is 0.05, right on the borderline of significance.

41. We find that $s_1^2 = 1.473$ and $s_2^2 = 3.101$. Because the two sample sizes are equal, the pooled variance is the ordinary average of these values, so

$$s_p^2 = \frac{s_1^2 + s_2^2}{2} = 2.287.$$

The standard error of the difference is

$$s_{\overline{X}_1 - \overline{X}_2} = \sqrt{\frac{2.287}{10} + \frac{2.287}{10}} = 0.676.$$

The sample means are $\overline{X}_1 = 5.488$ and $\overline{X}_2 = 6.614$. Therefore

$$\begin{aligned}
t &= \frac{\bar{X}_1 - \bar{X}_2}{s_{\bar{X}_1 - \bar{X}_2}}\\
&= \frac{5.488 - 6.614}{0.676} = -1.666.
\end{aligned}$$

The number of degrees of freedom is $\nu = n_1 + n_2 - 2 = 18$. The value of t is a bit less than the critical value 2.101 from Table 8.2. We have no convincing evidence that this medication has an effect.

43. The differences are 0.625, 0.865, 0.342, 0.087, 2.751, 0.312, 0.157, 2.252, 2.357, 1.515. These values have sample variance $s_Y^2 = 1.022$, and thus standard error

$$s_{\overline{Y}} = \sqrt{\frac{s_Y}{10}} = 0.320.$$

Our null hypothesis is that the mean of Y is equal to 0, while the actual sample mean of Y is 1.126. The t statistic is

$$t = \frac{1.126}{0.320} = 3.519.$$

This value exceeds the critical value for a p-value of 0.01 with 9 degrees of freedom. This medication does produce a significant change.

Section 8.7, page 783

1. There are a total of 100 cells, so the expected numbers are 40 with 0, 30 with 1, 20 with 2, and 10 with 3. Then

$$\begin{aligned}
\chi^2 &= \frac{(35-40)^2}{40} + \frac{(25-30)^2}{30} + \frac{(25-20)^2}{20} + \frac{(15-10)^2}{10}\\
&= 5.208.
\end{aligned}$$

We require three measurements to fully specify the data (because the numbers must add to 100), and thus 3 degrees of freedom. The value $\chi^2 = 5.208$ is less than the critical value 7.815 of the χ^2 distribution with 3 degrees of freedom, so there is no reason to suspect that our data deviate from the prior model.

3. There are 60 with 1 or fewer, and 40 with 2 or more. Adding the probabilities from the hypothesis, $\Pr(N \leq 1) = 0.7$ and $\Pr(N \geq 2) = 0.3$. Therefore we expect 70 with 1 or fewer and 30 with 2 or more. Then

$$\chi^2 = \frac{(60-70)^2}{70} + \frac{(40-30)^2}{30} = 4.76.$$

We require only one measurement to fully specify the data and thus 1 degree of freedom. The value $\chi^2 = 4.762$ is greater than the critical value 3.841 of the χ^2 distribution with 1 degree of freedom, so there is some reason to doubt that our data deviate from the prior model with these pooled data. It must give a different result because we pooled the two measurements that are lower than the expected ($N = 0$ and $N = 1$) and the two measurements that are higher ($N = 2$ and $N = 3$).

5. The probabilities with the given binomial are

$$\begin{aligned}
\Pr(N = 0) &= b(0; 3, 0.6) = 0.4^3 = 0.064\\
\Pr(N = 1) &= b(1; 3, 0.6) = 3 \cdot 0.4^2 \cdot 0.6 = 0.288\\
\Pr(N = 2) &= b(2; 3, 0.6) = 3 \cdot 0.4 \cdot 0.6^2 = 0.432\\
\Pr(N = 3) &= b(3; 3, 0.6) = 0.6^3 = 0.216.
\end{aligned}$$

The expected numbers are these values multiplied by 80, or 5.12 for $N = 0$, 23.04 for $N = 1$, 34.56 for $N = 2$, and 17.28 for $N = 3$. Then

$$\begin{aligned}
\chi^2 &= \frac{(10-5.12)^2}{5.12} + \frac{(20-23.04)^2}{23.04} + \frac{(30-34.56)^2}{34.56}\\
&\quad + \frac{(20-17.28)^2}{17.28} = 6.082.
\end{aligned}$$

This does not exceed the critical value of 7.815 with 3 degrees of freedom, so there is no evidence of a significant deviation from the model.

7. With the continuity correction, we reduce the deviations between observed and expected by 0.5, finding

$$\chi^2 = \frac{(9.5-5.12)^2}{5.12} + \frac{(20.5-23.04)^2}{23.04} + \frac{(30.5-34.56)^2}{34.56} + \frac{(19.5-17.28)^2}{17.28} = 4.789.$$

The continuity correction decreases the value by quite a bit, so the result is still not significant.

9. Out of the total of 240 molecules, there are a total of 140 inside, giving an estimate of $\hat{p} = \frac{140}{240} = 0.5833$. Then

$$\Pr(N=0) = b(0; 3, 0.5833) = 0.4167^3 = 0.072$$
$$\Pr(N=1) = b(1; 3, 0.5833) = 3 \cdot 0.4167^2 \cdot 0.5833 = 0.304$$
$$\Pr(N=2) = b(2; 3, 0.5833) = 3 \cdot 0.4167 \cdot 0.5833^2 = 0.425$$
$$\Pr(N=3) = b(3; 3, 0.5833) = 0.5833^3 = 0.198.$$

The expected numbers are these values multiplied by 80, or 5.76 for $N=0$, 24.32 for $N=1$, 34.00 for $N=2$, and 15.84 for $N=3$. Then

$$\chi^2 = \frac{(10-5.76)^2}{5.76} + \frac{(20-24.32)^2}{24.32} + \frac{(30-34.00)^2}{34.00} + \frac{(20-15.84)^2}{15.84} = 5.452.$$

Because we estimated the parameter $\hat{p}$, there are now only 2 degrees of freedom. The value of χ^2 does not exceed the critical value, so there is no reason to believe that these data do not follow a binomial distribution.

11. With $\Lambda = 4.5$, the expected numbers are $200p(k, 4.5)$, or

Number of Events	Number Observed in Experiment 1	Number Expected
0	5	2.222
1	15	9.998
2	32	22.496
3	34	33.744
4	41	37.962
5	26	34.165
6	25	25.624
7	12	16.473
8	7	9.266
9	2	4.633
10	0	2.085
11	1	0.853

We pool the first two and then all measurements greater than or equal to 9 to make sure the expected values are greater than 5.0, giving

Number of Events	Number Observed in Experiment 1	Number Expected
0 or 1	20	12.220
2	32	22.496
3	34	33.744
4	41	37.962
5	26	34.165
6	25	25.624
7	12	16.473
8	7	9.266
≥9	3	8.051

Then

$$\chi^2 = 16.118.$$

Our data set, after pooling, has 9 elements, so there are 8 degrees of freedom. This exceeds the critical value for p = 0.05, so the results are significantly different from the hypothesis $\Lambda = 4.5$.

13. We observed 794 events in 200 trials, or a mean of $\Lambda = 3.97$. Using this estimate, we find the number expected as before. We must pool the two smallest values and all values greater than or equal to 8, giving

Number of Events	Number observed in Experiment 1	Number Expected
0 or 1	20	18.760
2	32	29.746
3	34	39.364
4	41	39.069
5	26	31.021
6	25	20.525
7	12	11.641
≥8	10	9.873

Then

$$\chi^2 = 2.880.$$

There are now 6 degrees of freedom, but this value is far from the threshold. The data provide insufficient reason to reject the Poisson distribution.

15. There are a total of 100 birds, so the probabilities are

	0 Lice	1 Louse	2 Lice
0 Mites	0.23	0.11	0.10
1 Mite	0.11	0.14	0.08
2 Mites	0.07	0.05	0.11

The conditional distribution for lice with 0 mites is

$$\Pr(0 \text{ lice} \mid 0 \text{ mites}) = \frac{0.23}{0.44} = 0.523$$
$$\Pr(1 \text{ louse} \mid 0 \text{ mites}) = \frac{0.11}{0.44} = 0.25$$
$$\Pr(2 \text{ lice} \mid 0 \text{ mites}) = \frac{0.10}{0.44} = 0.227.$$

With 1 mite,

$$\Pr(0 \text{ lice} \mid 1 \text{ mite}) = \frac{0.11}{0.33} = 0.333$$
$$\Pr(1 \text{ louse} \mid 1 \text{ mite}) = \frac{0.14}{0.33} = 0.424$$
$$\Pr(2 \text{ lice} \mid 1 \text{ mite}) = \frac{0.08}{0.33} = 0.242.$$

With 2 mites,

$$\Pr(0 \text{ lice} \mid 2 \text{ mites}) = \frac{0.07}{0.23} = 0.304$$
$$\Pr(1 \text{ louse} \mid 2 \text{ mites}) = \frac{0.05}{0.23} = 0.217$$
$$\Pr(2 \text{ lice} \mid 2 \text{ mites}) = \frac{0.11}{0.23} = 0.478.$$

These look very different, even though the test in Example 8.7.14 showed no evidence of lack of independence.

17. For the expected numbers if birds mate independently, we must find the fraction in each column. There are a total of 98 birds, with 47 red males, 25 blue males, and 26 green males, or fractions of 0.480, 0.255, and 0.265 respectively. Multiplying by the number of females (33 red, 33 blue, and 32 green) gives

	Red Male	Blue Male	Green Male
Red female	15.84	8.415	8.745
Blue female	15.84	8.415	8.745
Green female	15.36	8.160	8.480

Then

$$\chi^2 = 24.59.$$

This table has $R=3$ rows and $C=3$ columns, so there are $(R-1)(C-1)=2\cdot 2=4$ degrees of freedom. This value far exceeds the critical value for p = 0.01 with 4 degrees of freedom, so these matings are not independent. Female birds mate with male birds with matching color.

19. 20% of 50 people is 10, so the result of finding 10 out of 50 people with the particular allele can be written as the following table.

	Number with Allele	Number with Other Allele
Observed	10	40
Expected	6.5	43.5

The chi-squared statistic is

$$\chi^2 = \frac{(10-6.5)^2}{6.5} + \frac{(40-43.5)^2}{43.5} = 2.166.$$

This value is much less than that found with the larger sample size, and is far from significant.

21. 20% of n people is $0.2n$. Our results can be written as the following table.

	Number with Allele	Number with Other Allele
Observed	$0.2n$	$0.8n$
Expected	$0.13n$	$0.87n$

The chi-squared statistic is

$$\chi^2 = \frac{(0.2n-0.13n)^2}{0.13n} + \frac{(0.8n-0.87n)^2}{0.87n} = 0.0433n.$$

The statistic χ^2 is proportional to the sample size.

23. $E(C_1)=E(X_1^2)$. But

$$\text{Var}(X_1)=E(X_1^2)-[E(X_1)]^2=E(X_1^2)=1.$$

because $E(X_1)=0$ and $\text{Var}(X_1)=1$ for a standard normal.

25. We need to solve $\Pr(C_1>c)=0.05$. From the equation,

$$\Pr(C_1>c)=\Pr(X_1^2>c) \quad \text{definition of } C_1$$
$$\Pr(X_1^2>c)=\Pr(|X_1|>\sqrt{c}) \quad \text{carefully taking the square root.}$$

We know that $\Pr(|X_1|>\sqrt{c})=0.05$ has solution $\sqrt{c}=1.96$. Therefore, the critical value is $c=1.96^2=3.8416$, matching the result in Table 8.4.

27. To compute the expected numbers under the null model, we must find the fraction in each column. There are 125 out of 150 worms that produce eggs, or a fraction 0.8333. Multiplying by the number in each row gives

	Produce Eggs	Sterile
Wild type	41.67	8.333
Mutant	83.33	16.67

We compute the statistic χ^2 by summing the squared deviations from the expected divided by the expected, finding

$$\chi^2 = \frac{(45-41.67)^2}{41.67} + \frac{(5-8.33)^2}{8.33} + \frac{(80-83.33)^2}{83.33} + \frac{(20-16.67)^2}{16.67} = 2.396.$$

This table has $R=2$ rows and $C=2$ columns, so there is $(R-1)(C-1)=1\cdot 1=1$ degree of freedom. Our value does not exceed the critical value for p = 0.05 with 1 degree of freedom, and thus is not significant.

29. To compute the expected numbers under the null model, we must find the fraction in each column. There are 143 out of 200 worms that produce eggs and sperm for a fraction of 0.715, 32 out of 200 producing eggs only for a fraction of 0.16, and 25 out of 200 that are sterile for a fraction of 0.125. Multiplying by the number in each row gives

	Produce Eggs and Sperm	Produce Eggs Only	Sterile
Wild type	57.2	12.8	10.0
Mutant	85.8	19.2	15.0

We compute the statistic χ^2 by summing the squared deviations from the expected divided by the expected, finding

$$\chi^2 = 6.960.$$

This table has $R=2$ rows and $C=3$ columns, so there are $(R-1)(C-1)=1\cdot 2=2$ degree of freedom. Our value exceeds the critical value for p = 0.05 with 2 degrees of freedom, and thus is significant.

31. We have 10 plants expressing the recessive trait and 50 that do not. We expect 15 to express the recessive trait and 45 that do not. Then

$$\chi^2 = \frac{(10-15)^2}{15} + \frac{(50-45)^2}{45} = 2.222.$$

This is far from the critical value with 1 degree of freedom, so there is no reason to doubt the Mendelian ratio.

33. We expect 1/4 to be white, 1/2 to be pink, and 1/4 to be red, or 22.5, 45, and 22.5 respectively. Then

$$\chi^2 = \frac{(18-22.5)^2}{22.5} + \frac{(40-45)^2}{45} + \frac{(32-22.5)^2}{22.5} = 5.467.$$

This is slightly less than the critical value with 2 degrees of freedom, so we cannot reject the hypothesis. However, there are more red plants than expected.

35. The probability of having yellow flowers is 1/4, as is the probability of being short. So the probabilities of the four possible phenotypes are

$$\Pr(\text{yellow and short}) = \frac{1}{4}\cdot\frac{1}{4} = \frac{1}{16}$$
$$\Pr(\text{yellow and tall}) = \frac{1}{4}\cdot\frac{3}{4} = \frac{3}{16}$$
$$\Pr(\text{white and short}) = \frac{3}{4}\cdot\frac{1}{4} = \frac{3}{16}$$
$$\Pr(\text{white and tall}) = \frac{3}{4}\cdot\frac{3}{4} = \frac{9}{16}.$$

The expected numbers are then 5 yellow and short, 15 yellow and tall, 15 white and short, and 45 white and tall. Then

$$\chi^2 = \frac{(3-5)^2}{5} + \frac{(12-15)^2}{15} + \frac{(17-15)^2}{15} + \frac{(48-45)^2}{45} = 1.867.$$

This is far from the critical value with 3 degrees of freedom, so there is no reason to suspect that anything strange is going on.

37. No eagles are seen with probability 0.8, and one is seen with probability 0.2. Multiplying by the 50 samples with no jack rabbits and the 50 with one jack rabbit gives an expected table of

	$E=0$	$E=1$
$J=0$	40	10
$J=1$	40	10

Then

$$\chi^2 = 6.25.$$

There is 1 degree of freedom in this table. This value of χ^2 exceeds the critical value for p = 0.05, and is close to the critical value for p = 0.01. This means that the major differences in the conditional distributions found in Section 7.1, Exercise 27 are probably meaningful and worth interpreting.

39. The table of outcomes is

	Male First	Female First
Male second	16	7
Female second	7	14

The fraction of males among the first offspring is $\frac{23}{44} = 0.523$. The expected numbers are then

	Male First	Female First
Male second	12.03	10.98
Female second	10.98	10.02

Then $\chi^2 = 5.776$. This is slightly lower than the value found in Example 8.7.15, but is still significant at the p = 0.05 level.

Section 8.8, page 795

1. The likelihood function is $L(p) = p^5$ and the support is $S(p) = 5\ln(p)$. The maximum likelihood estimate of p is 1.0, so $S(1.0) = 0$. $S(0.5) = -3.47$, which is more than 2 below the maximum likelihood estimate. This indicates that the null hypothesis might not be true.

3. The likelihood function is $L(p) = 10p^9(1-p)$ and the support is $S(p) = \ln(10) + 9\ln(p) + \ln(1-p)$. The maximum likelihood estimate of p is 0.9, so $S(0.9) = -0.948$. $S(0.5) = -4.63$, which is more than 2 below the maximum likelihood estimate. The null hypothesis might not be true.

5. The likelihood function is $L(\lambda) = \lambda e^{-\lambda}$, and the support is $S(\lambda) = \ln(\lambda) - \lambda$. The maximum likelihood estimate of the rate is $\lambda = 1.0$, giving a support of -1.0. With $\lambda = 5.0$, the support is -3.39, slightly more than 2 below the maximum. There is reason to doubt the null hypothesis.

7.

$$S(\lambda) = \ln(\lambda e^{-4000\lambda}) = \ln(\lambda) - 4000\lambda.$$

The maximum likelihood hypothesis is $\lambda = 0.00025$, with a null of $\lambda = 0.001$. Then $S(0.00025) - S(0.001) = 1.61$. Although this difference is less than 2, the result was significant with the p-value approach.

9. Support for a particular q is

$$S(q) = \ln(q) + (25-1)\ln(1-q).$$

At $q = 0.1$, we get $\ln(0.1) + (25-1)\ln(1-0.1) = -4.831$. The maximum occurs where

$$\frac{1}{q} - \frac{24}{1-q} = 0$$

which has solution $q = 0.04$. The support is $\ln(0.04) + (25-1)\ln(1-0.04) = -4.199$. The difference is only 0.633. This difference is less than 2, consistent with the insignificant result using the p-value approach.

11. The difference in support is 1.044 (the number of standard errors separating the two hypotheses). The data provide no reason to prefer one over the other.

13. The difference in support is 1.593, rather less than the threshold of 2.

15. 2 standard errors away. $\Pr(Z < -2 \text{ or } Z > 2) = 1 - \Phi(2) + \Phi(-2) = 0.0455$.

17. 3.5 standard errors away. $\Pr(Z < -3.5 \text{ or } Z > 3.5) = 1 - \Phi(3.5) + \Phi(-3.5) = 0.00047$.

19. First, split up the quadratic,

$$\sum_{i=1}^{n}(X_i - \mu)^2 = \sum_{i=1}^{n} X_i^2 - \sum_{i=1}^{n} 2X_i\mu + \sum_{i=1}^{n}\mu^2.$$

The first term is

$$\sum_{i=1}^{n} X_i^2 = (n-1)s^2 + n\overline{X}^2$$

by solving for $\sum_{i=1}^{n} X_i^2$ in terms of the sample variance. The second term is

$$\sum_{i=1}^{n} 2X_i\mu = 2n\overline{X}\mu$$

because $\sum_{i=1}^{n} X_i = n\overline{X}$, from the definition of sample variance. The last term is

$$\sum_{i=1}^{n}\mu^2 = n\mu^2.$$

Putting it together,

$$\begin{aligned}\sum_{i=1}^{n}(X_i - \mu)^2 &= (n-1)s^2 + n\overline{X}^2 - 2n\overline{X}\mu + n\mu^2 \\ &= (n-1)s^2 + n(\overline{X}^2 - 2\overline{X}\mu + \mu^2) \\ &= (n-1)s^2 + n(\overline{X} - \mu)^2.\end{aligned}$$

21. The likelihood function is $L(\Lambda) = \frac{e^{-\Lambda}\Lambda^7}{7!}$. The support for the null hypothesis is $\ln[L(3.5)] = -3.26$. From the given data, the maximum likelihood estimate is $\Lambda = 7.0$, with support $\ln[L(7.0)] = -1.90$. The difference is only 1.36, meaning that the null hypothesis might well be fine.

23. The average wait for type a is 0.936, so our maximum likelihood estimate of λ is the reciprocal, or 1.069. The likelihood function is

$$L(\lambda) = \lambda^{30}e^{-30\cdot 0.936\lambda} = \lambda^{30}e^{-28.08\lambda}$$

and the support is

$$S(\lambda) = 30\ln(\lambda) - 28.08\lambda.$$

Then $S(1.068) = -27.93$ and $S(1.0) = -28.00$. The difference is tiny, consistent with the fact that there is no reason to reject the null hypothesis.

25. The average is now 0.750, so our maximum likelihood estimate of λ is the reciprocal, or 1.334. The likelihood function is

$$L(\lambda) = \lambda^{29}e^{-29\cdot 0.750\lambda} = \lambda^{29}e^{-21.75\lambda}$$

and the support is

$$S(\lambda) = 29\ln(\lambda) - 21.75\lambda.$$

Then $S(1.334) = -20.66$ and $S(1.0) = -21.75$. The difference is still less than 2. The estimator of the rate became much larger because the single long wait has been excluded. Much of the low support came because that single long wait was quite unlikely.

27. The maximum likelihood estimator for the first player is $\hat{p}_1 = 0.5$, for the second is $\hat{p}_2 = 0.9$, and for the two players pooled is $\hat{p}_1 = 0.7$. The likelihood function is

$$L(p_1, p_2) = \binom{10}{5} p_1^5(1-p_1)^5 \binom{10}{9} p_2^9(1-p_2).$$

Because we are comparing hypotheses, we can ignore the constants in the support function $S(p_1, p_2) = 5\ln(p_1) + 5\ln(1-p_1) + 9\ln(p_2) + \ln(1-p_2)$. Using the individuals separately, we find $S(0.5, 0.9) = -10.18$. The support for the pooled hypothesis is $S(0.7, 0.7) = -12.22$. The difference is slightly greater than 2, giving reason to suspect that the second player is a better shot.

29. The maximum likelihood estimator for Utah is $\hat{\Lambda}_1 = 4.0$, for the North Pole is $\hat{\Lambda}_2 = 10.0$, and for the two sites pooled is $\hat{\Lambda} = 7.0$. The likelihood function is

$$L(\Lambda_1, \Lambda_2) = \frac{e^{-\Lambda_1}\Lambda_1^4}{4!}\frac{e^{-\Lambda_2}\Lambda_2^{10}}{10!}.$$

Because we are comparing hypotheses, we can ignore the factorials in the support function

$$S(\Lambda_1, \Lambda_2) = 4\ln(\Lambda_1) - \Lambda_1 + 10\ln(\Lambda_2) - \Lambda_2.$$

Using the individuals separately, we find $S(4, 10) = 14.57$ (the value can be positive because we left out constants). The support for the pooled hypothesis is $S(7, 7) = 13.24$. The difference is less than 2, giving no reason to believe that the North Pole gets more cosmic rays.

31. The two players together make 14 out of 20, so we expect each to make 7 out of 10. The full table of observed and expected outcomes is then

Observed		
	Player 1	Player 2
Made	5	9
Missed	5	1

Expected		
	Player 1	Player 2
Made	7	7
Missed	3	3

Then

$$G = 2\sum_{i=1}^{m} O_i \ln\left(\frac{O_i}{E_i}\right)$$
$$= 2\left[5\ln\left(\frac{5}{7}\right) + 9\ln\left(\frac{9}{7}\right) + 5\ln\left(\frac{5}{3}\right) + 1\ln\left(\frac{1}{3}\right)\right] = 4.070.$$

This is twice the value found in Exercise 27, and exceeds the value of the χ^2 distribution with 1 degree of freedom. We have reason to think that the second player is better

33. The maximum likelihood estimator for the first organism is $\hat{\Lambda}_1 = 8.0$, for the second organism is $\hat{\Lambda}_2 = 18.0$, and for the two organisms pooled is $\hat{\Lambda} = 13.0$. The likelihood function is

$$L(\Lambda_1, \Lambda_2) = \frac{e^{-\Lambda_1}\Lambda_1^8}{8!}\frac{e^{-\Lambda_2}\Lambda_2^{18}}{18!}.$$

Because we are comparing hypotheses, we can ignore the factorials in the support function

$$S(\Lambda_1, \Lambda_2) = 8\ln(\Lambda_1) - \Lambda_1 + 18\ln(\Lambda_2) - \Lambda_2.$$

Using the individuals separately, we find $S(8.0, 18.0) = 42.66$ (the value can be positive because we left out constants). The support for the pooled hypothesis is $S(13.0, 13.0) = 40.69$. The difference is very close to 2, consistent with the marginally significant difference in Section 8.6, Exercise 39.

35. The null hypothesis is that the two die at the same rate λ, with an alternative that the rates are different. The average mortality rate is $1/17.5 = 0.057$ (see Subsection 7.6.2). The support is

$$S_n(\lambda) = \ln(\lambda e^{-30\lambda}) + \ln(\lambda e^{-5\lambda}).$$

With $\lambda = 0.057$, this is -7.72. The best guess for λ_1 is $1/30 = 0.0333$ and the best guess for λ_2 is $1/5 = 0.2$. The support is

$$S_a(\lambda_1, \lambda_2) = \ln(\lambda_1 e^{-30\lambda_1}) + \ln(\lambda_2 e^{-5\lambda_2}).$$

With the maximum likelihood estimators of λ_1 and λ_2, this is -7.01. The difference is far too small to be convincing.

37. The null hypothesis is that the two die at the same rate λ, with an alternative that the rates are different. The average mortality rate in the bath is $1/45 = 0.022$, the average mortality rate without the bath is $1/4 = 0.25$, and the overall average mortality rate is $1/24.5 = 0.041$. The support for the null hypothesis is

$$S_n(\lambda) = \ln(\lambda e^{-30\lambda}) + \ln(\lambda e^{-5\lambda}) + \ln(\lambda e^{-60\lambda}) + \ln(\lambda e^{-3\lambda}).$$

With $\lambda = 0.022$, this is -16.79. The support for the alternative is

$$S_a(\lambda_1, \lambda_2) = \ln(\lambda_1 e^{-30\lambda_1}) + \ln(\lambda_2 e^{-5\lambda_2}) + \ln(\lambda_1 e^{-60\lambda_1}) + \ln(\lambda_2 e^{-3\lambda_2}).$$

With the maximum likelihood estimators of λ_1 and λ_2, this is -14.39. The difference is greater than 2, and there is reason to think it is significant.

Section 8.9, page 806

1.

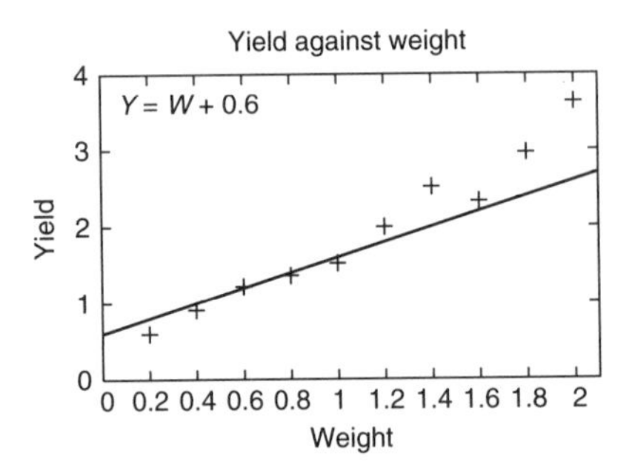

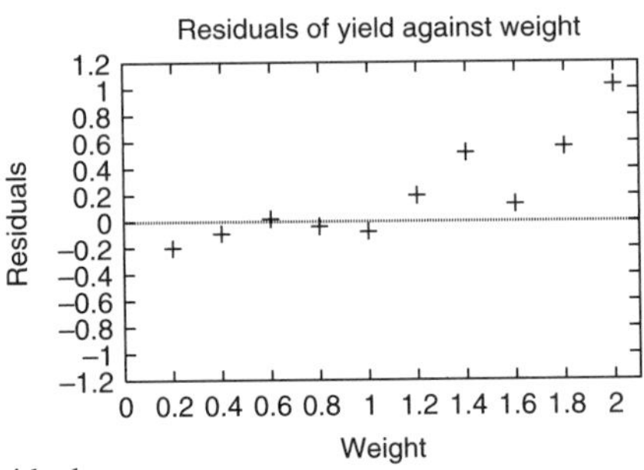

The residuals are

w	Y	$\hat{Y}$	$Y - \hat{Y}$
0.2	0.599	0.8	−0.201
0.4	0.909	1.0	−0.091
0.6	1.220	1.2	0.020
0.8	1.363	1.4	−0.037
1.0	1.523	1.6	−0.077
1.2	1.995	1.8	0.195
1.4	2.518	2.0	0.518
1.6	2.330	2.2	0.130
1.8	2.963	2.4	0.563
2.0	3.628	2.6	1.028

3. Adding up the squares of the residuals gives SSE = 1.753.

5. The null model uses the mean yield, or $Y = \overline{Y} = 1.905$. The sum of the squares of the residuals is SST = 8.259.

7. $r^2 = 1 - \frac{\text{SSE}}{\text{SST}} = 0.788.$

9. The best fitting line is found by computing

$$\widehat{\text{Cov}}(W, Y) = 0.568$$
$$s_W^2 = 0.367$$
$$\overline{W} = 1.10$$
$$\overline{Y} = 1.905.$$

From Theorem 8.6,

$$\hat{a} = \frac{0.568}{0.367} = 1.548$$
$$\hat{b} = 1.905 - 1.548 \cdot 1.10 = 0.202.$$

The line is then $Y = 1.548W + 0.202$.

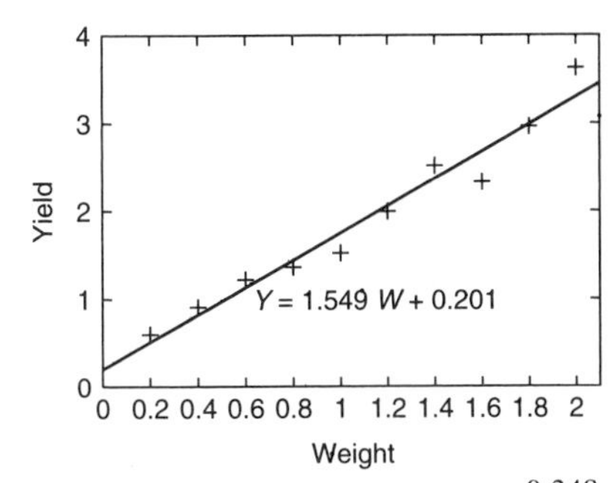

11. Using Theorem 8.7, SSE $=0.348$. $r^2 = 1 - \frac{0.348}{8.26} = 0.96$, much larger than before.

13. The only additional information we need to find the correlation is the variance of Y, which is 0.917. The correlation is then

$$\rho_{Y,W} = \frac{\widehat{\mathrm{Cov}}(W, Y)}{\sqrt{s_W^2}\sqrt{s_Y^2}} = \frac{0.568}{\sqrt{0.367}\sqrt{0.917}} = 0.979.$$

The square is $0.979^2 = 0.958$, closely matching the value found earlier.

15. The best fitting line is $S = 11.85D + 5.81$, and $r^2 = 0.66$. The mean with diet 1 is 17.66 and with diet 2 is 29.51. The variance in each of the means is 25/6, so the variance of the difference between the means is 25/3=8.333. The standard deviation of the difference between the means is then 2.89, so the means differ by 4.1 standard deviations. The associated p-value is tiny, 4.1×10^{-5}.

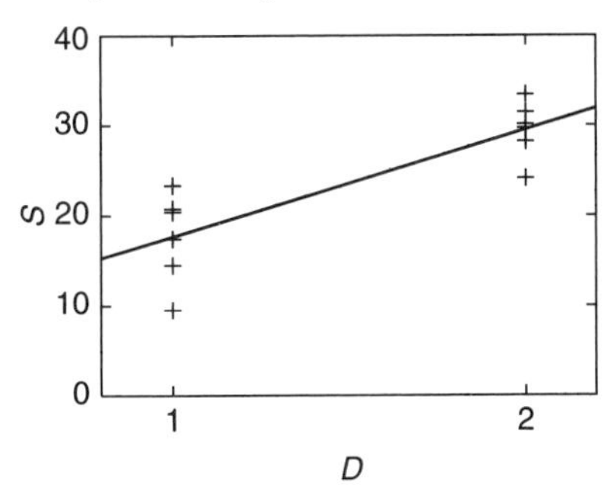

17. The best fitting horizontal line has intercept equal to the mean. Then

$$\sum_{i=1}^{n} (y_i - \overline{Y}) = n\overline{Y} - n\overline{Y} = 0.$$

19. Let the data be $y_1, \ldots, y_n$. Then for a given b,

$$\mathrm{SST} = \sum_{i=1}^{n} (y_i - b)^2.$$

Taking the derivative with respect to b,

$$\begin{aligned}\frac{d(\mathrm{SST})}{db} &= -2\sum_{i=1}^{n}(y_i - b)\\ &= -2(n\overline{Y} - nb)\\ &= -2n(\overline{Y} - b)\end{aligned}$$

which is 0 when $\overline{Y} = b$. Furthermore, the second derivative of SST with respect to b is equal to $2n$, which is positive. Therefore, $b = \overline{Y}$ is a minimum.

21. The best fitting line is found by computing

$$\begin{aligned}\widehat{\mathrm{Cov}}(X, Y) &= 8.517\\ s_X^2 &= 1.667\\ \overline{X} &= 2.50\\ \overline{Y} &= 7.575.\end{aligned}$$

From Theorem 8.6,

$$\begin{aligned}\hat{a} &= \frac{8.517}{1.667} = 5.110\\ \hat{b} &= 7.575 - 5.11 \cdot 2.50 = -5.20.\end{aligned}$$

The line is then

$$Y = 5.11X - 5.20.$$

Using Theorem 8.7, SSE $=6.027$. Furthermore, the sum of the squared differences of Y from the mean 7.575 is 136.6. Therefore, $r^2 = 1 - \frac{6.207}{136.6} = 0.95$, which looks pretty good.

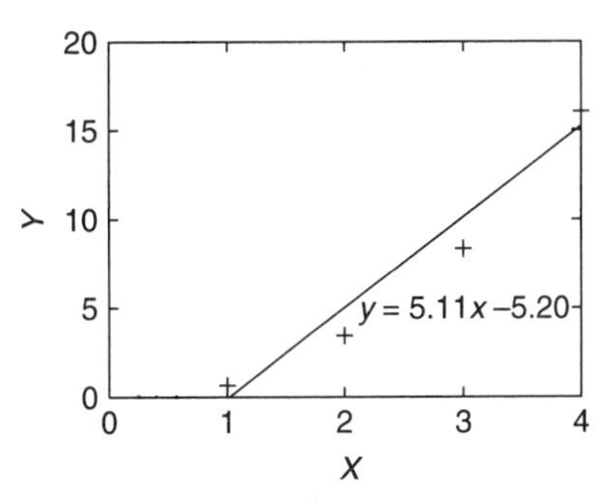

23. $\hat{a}$ has dimensions

$$\begin{aligned}\hat{a} &= \frac{\widehat{\mathrm{Cov}}(X, Y)}{s_X^2}\\ &= \frac{\text{mass} \cdot \text{tolerance}}{\text{mass}^2}\\ &= \frac{\text{tolerance}}{\text{mass}}.\end{aligned}$$

25. The residuals are 0.01, −0.2, 0.04, 0.0, 0.1, and −0.04, giving SSE $= 0.0533$. Using the horizontal line at the mean of b_{t+1}, 3.125, gives residuals of −2.175, 3.275, −1.625, 2.475, −0.025 and −1.925, so SST $= 27.93$. This gives $r^2 = 0.998$.

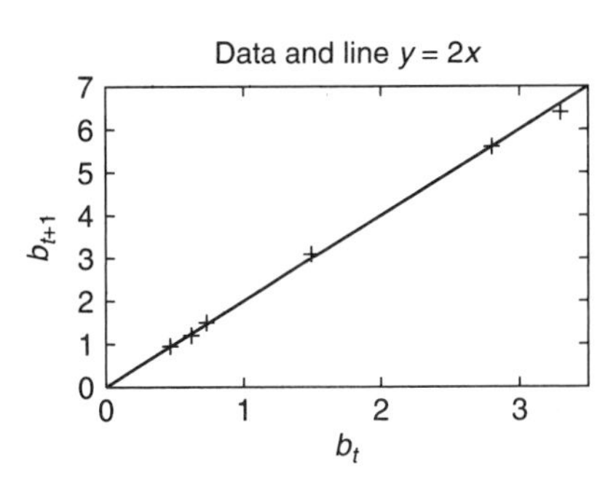

27. The best fitting line has slope 51 and intercept 71.5 with SSE $= 4148$. SST uses the horizontal line 224.5, and gives SST $= 77046$. Then $r^2 = 0.946$.

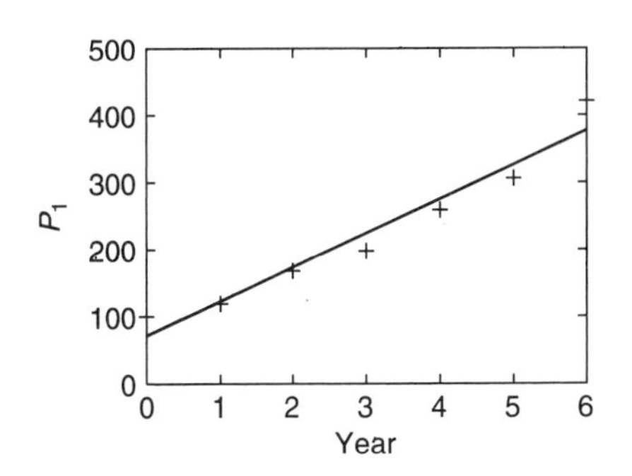

29. Using the logarithms of the population sizes, we find the best fitting line $\ln(P) = 4.59 + 0.237Y$. The r^2 value is 0.99. This fit looks a lot better.

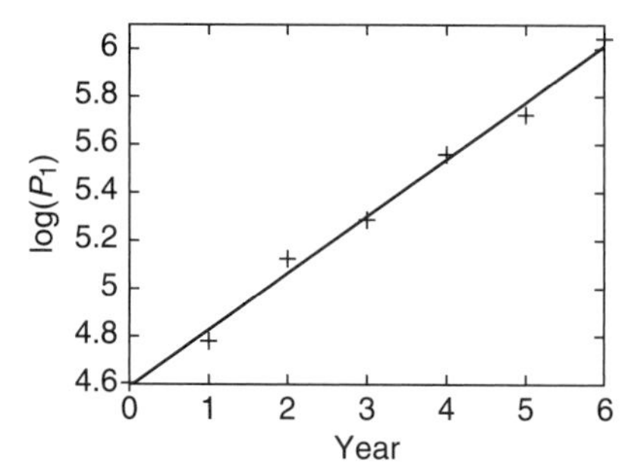

31. The line with all the data is $S = 8.95F + 6.86$, and $r^2 = 0.63$. Removing the fourth point, the best fitting line is $S = 10.1F + 1.16$ with $r^2 = 0.97$. The outlier really messes things up.

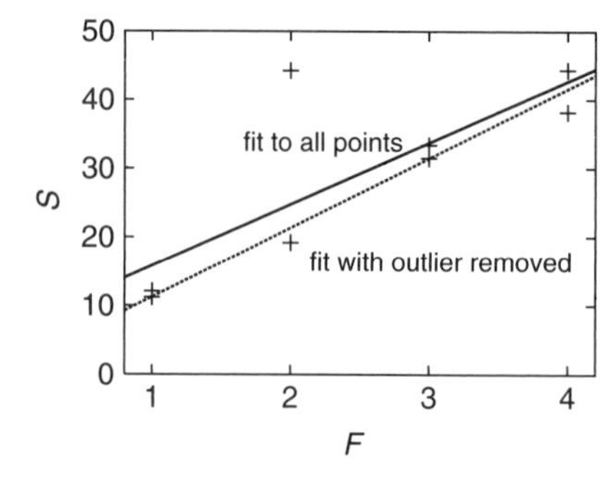

Chapter 1

Anderson, R. M. and R. M. May. *Infectious Diseases of Humans.* Oxford University Press, Oxford, 1992.

Glass, L. and M. C. Mackey. *From Clocks to Chaos: The Rhythms of Life*. Princeton University Press, Princeton, N.J., 1988.

Hartl, D. L. and A. G. Clark. *Principles of Population Genetics*. Sinauer Associates, Sunderland, Mass., 1989.

Hecht, E. *Physics*. Brooks/Cole Publishing Co., Pacific Grove, Calif., 1994.

Hodgkin, A. L. and A. F. Huxley. A quantitative description of membrane current and its application to conduction and excitation in nerve. *Journal of Physiology*, 117:500–544, 1952. (Reprinted in *Bulletin of Mathematical Biology*, 52:25–71, 1990.)

Hoppensteadt, F. C. and C. S. Peskin. *Mathematics in Medicine and the Life Sciences.* Springer-Verlag, New York, 1992.

Keener, J. P. On cardiac arrhythmias: AV conduction block. *Journal of Mathematical Biology*, 12:215–225, 1981.

Paulos, J. A. *Innumeracy*. Hill and Wang, New York, 1988.

Provine, W. B. *The Origins of Theoretical Population Genetics*. University of Chicago Press, Chicago, 1971.

Rinzel, J. Electrical excitability of cells, theory and experiment: review of the Hodgkin-Huxley foundation and an update. *Bulletin of Mathematical Biology*, 52:5–23, 1990.

Ross, R. *The Prevention of Malaria*. Murray, London, 1911.

Stampi, C. *Why We Nap*. Birkhauser, Boston, 1990.

Chapter 3

Charnov, E. L. Optimal foraging: the marginal value theorem. *Theoretical Population Biology*, 9:129–136, 1976.

Clark, C. W. *Mathematical Bioeconomics: The Optimal Management of Renewable Resources* J. Wiley, New York, 1990.

Devaney, R. L. *An Introduction to Chaotic Dynamical Systems*. Benjamin/Cummings, Menlo Park, Calif., 1986.

Edelstein-Keshet, L. *Mathematical Models in Biology*. Random House, New York, 1988.

Hoppensteadt, F. C. and C. S. Peskin. *Mathematics in Medicine and the Life Sciences*. Springer-Verlag, New York, 1992.

May, R. M. Simple mathematical models with very complicated dynamics. *Nature*, 261:459–467, 1976.

Press, W. H., S. A. Teukolsy, W. T. Vetterling, and B. P. Flannery. *Numerical Recipes: The Art of Scientific Computing*. Cambridge University Press, Cambridge, Mass., 1992.

Stephens, D. W. and J. R. Krebs. *Foraging Theory*. Princeton University Press, N.J., 1986.

Chapter 4

Blanchard, P., R. Devaney, and G. Hall. *Differential Equations*. Brooks/Cole, Pacific Grove, Calif., 1998.

Charnov, E. L. *Life History Invariants: Some Explorations of Symmetry in Evolutionary Ecology*. Oxford University Press, New York, 1993.

Hyman, J. M. and E. A. Stanley. Using Mathematical Models to Understand the AIDS Epidemic. *Mathematical Biosciences*, 90:415–473, 1988.

Chapter 6

Falconer, D. S. and T. F. C. Mackay. *Introduction to Quantitative Genetics.* Longman, Essex, England, 1996.

Hartl, D. L. and A. G. Clark. *Principles of Population Genetics.* Sinauer Associates, Sunderland, Mass., 1989.

Huff, D. *How to Lie with Statistics.* Norton, New York, 1982.

Zar, J. H. *Biostatistical Analysis.* Prentice-Hall, Englewood Cliffs, N.J., 1984.

Chapter 8

Devore, J. L. *Statistics: The Exploration and Analysis of Data.* Duxbury Press, Belmont, Calif., 1997.

Edwards, A. W. F. *Likelihood.* Cambridge University Press, Cambridge, England, 1972.

Zar, J. H. *Biostatistical Analysis.* Prentice-Hall, Englewood Cliffs, N.J., 1984.

Letters of the Greek Alphabet Used in the Text

Capital Letter	Small Letter	Pronunciation
	α	alpha
	β	beta
Γ	γ	gamma
Δ	δ	delta
	ϵ	epsilon
	ζ	zeta
Θ	θ	theta
	κ	kappa
Λ	λ	lambda
	μ	mu
	ν	nu
Π	π	pi
	ρ	rho
Σ	σ	sigma
	τ	tau
Φ	ϕ	phi
Ω	ω	omega

Index

Theorems of Probability

Law of Total Probability: Suppose that $E_1, E_2, \ldots, E_n$ form a set of mutually exclusive and collectively exhaustive events. Then for any event A, $\Pr(A) = \sum_{i=1}^{n} \Pr(A \mid E_i)\Pr(E_i)$.

Bayes' Theorem: For any events A and B where $\Pr(A) \neq 0$,

$$\Pr(B \mid A) = \frac{\Pr(A \mid B)\Pr(B)}{\Pr(A)}.$$

Multiplication Rule for Independent Events: Suppose A and B are any two independent events. Then $\Pr(A \cap B) = \Pr(A)\Pr(B)$.

Expectation of the Sum of Random Variables: The expectation of the sum of random variables is equal to the sum of the expectations.

Variance of the Sum of Independent Random Variables: The variance of the sum of independent random variables is equal to the sum of the variances.

Central Limit Theorem: The sum of sufficiently many independent, identically distributed random variables with finite mean and variance has approximately a normal distribution.

Probability Distributions

Distribution	Type of Random Variable	When to Use	Formula	Expectation	Variance
Binomial	Discrete	Number of successes N in n trials each with probability p	$\Pr(N = k) = b(k : n, p) = \binom{n}{k} p^k (1-p)^{n-k}$	np	$np(1-p)$
Poisson	Discrete	Number of events N in time t at rate λ	$\Pr(N = k) = p(k; \lambda t) = \frac{e^{-\lambda t}(\lambda t)^k}{k!}$	λt	λt
Geometric	Discrete	Time T before a success in a series of trials each with probability q	$\Pr(T = t) = g_t = q(1-q)^{t-1}$	$\frac{1}{q}$	$\frac{1-q}{q^2}$
Exponential	Continuous	Waiting time T for an event that occurs at rate λ	p.d.f. $= \lambda e^{-\lambda t}$	$\frac{1}{\lambda}$	$\frac{1}{\lambda^2}$
Normal	Continuous	Value X of many continuous measurements	p.d.f. $= \frac{1}{\sqrt{2\pi}\sigma} e^{-\frac{(x-\mu)^2}{2\sigma^2}}$	μ	σ^2

Exponentials and Logarithms

Laws of Exponents for Any Base $a > 0$		Laws of Logarithms for Base e	
Law 1	$a^x \cdot a^y = a^{x+y}$	Law 1	$\ln(xy) = \ln x + \ln y$
Law 2	$(a^x)^y = a^{xy}$	Law 2	$\ln(x^y) = y \ln x$
Law 3	$a^{-x} = 1/a^x$	Law 3	$\ln(1/x) = -\ln x$
Law 4	$a^y/a^x = a^{y-x}$	Law 4	$\ln(x/y) = \ln x - \ln y$
Law 5	$a^1 = a$	Law 5	$\ln(e) = 1$
Law 6	$a^0 = 1$	Law 6	$\ln(1) = 0$

Rules of Differentiation

Suppose $f(x)$ and $g(x)$ are differentiable functions and c is a constant.

Rule	Function	Derivative
Sum rule	$f(x)+g(x)$	$f'(x)+g'(x)$
Constant sum rule	$f(x)+c$	$f'(x)$
Power rule	x^n	nx^{n-1}
Product rule	$f(x)g(x)$	$f(x)g'(x)+g(x)f'(x)$
Constant product rule	$cf(x)$	$cf'(x)$
Quotient rule	$\frac{f(x)}{g(x)}$	$\frac{f'(x)g(x)-g'(x)f(x)}{(g(x))^2}$
Chain rule	$f(g(x))$	$f'(g(x))g'(x)$
Exponential	e^x	e^x
Natural logarithm	$\ln(x)$	$\frac{1}{x}$
Sine	$\sin(x)$	$\cos(x)$
Cosine	$\cos(x)$	$-\sin(x)$

Basic Indefinite Integrals

$$\int x^n dx = \frac{x^{n+1}}{n+1}+c \quad \text{if } n \neq -1$$

$$\int \frac{1}{x} dx = \ln(|x|)+c$$

$$\int e^{\alpha x} dx = \frac{e^{\alpha x}}{\alpha}+c \quad \text{if } \alpha \neq 0$$

$$\int \sin(\alpha x) dx = -\frac{\cos(\alpha x)}{\alpha}+c \quad \text{if } \alpha \neq 0$$

$$\int \cos(\alpha x) dx = \frac{\sin(\alpha x)}{\alpha}+c \quad \text{if } \alpha \neq 0$$

Definitions in Probability Theory

Conditional Probability: The probability of event A conditional on event B is $\Pr(A \mid B) = \Pr(A \cap B)/(\Pr(B))$ if $\Pr(B) \neq 0$.

Independence: Event A is **independent** of event B if $\Pr(A \mid B) = \Pr(A)$.

Expectation: The mean of a random variable is the sum of the values weighted by their probabilities.

Median: The value of a random variable that is exceeded by 50% of the measurements.

Mode: The most probable value of a random variable.

Variance: A measure of the spread of a random variable equal to the mean squared deviation from the mean.

Standard Deviation: The square root of the variance.